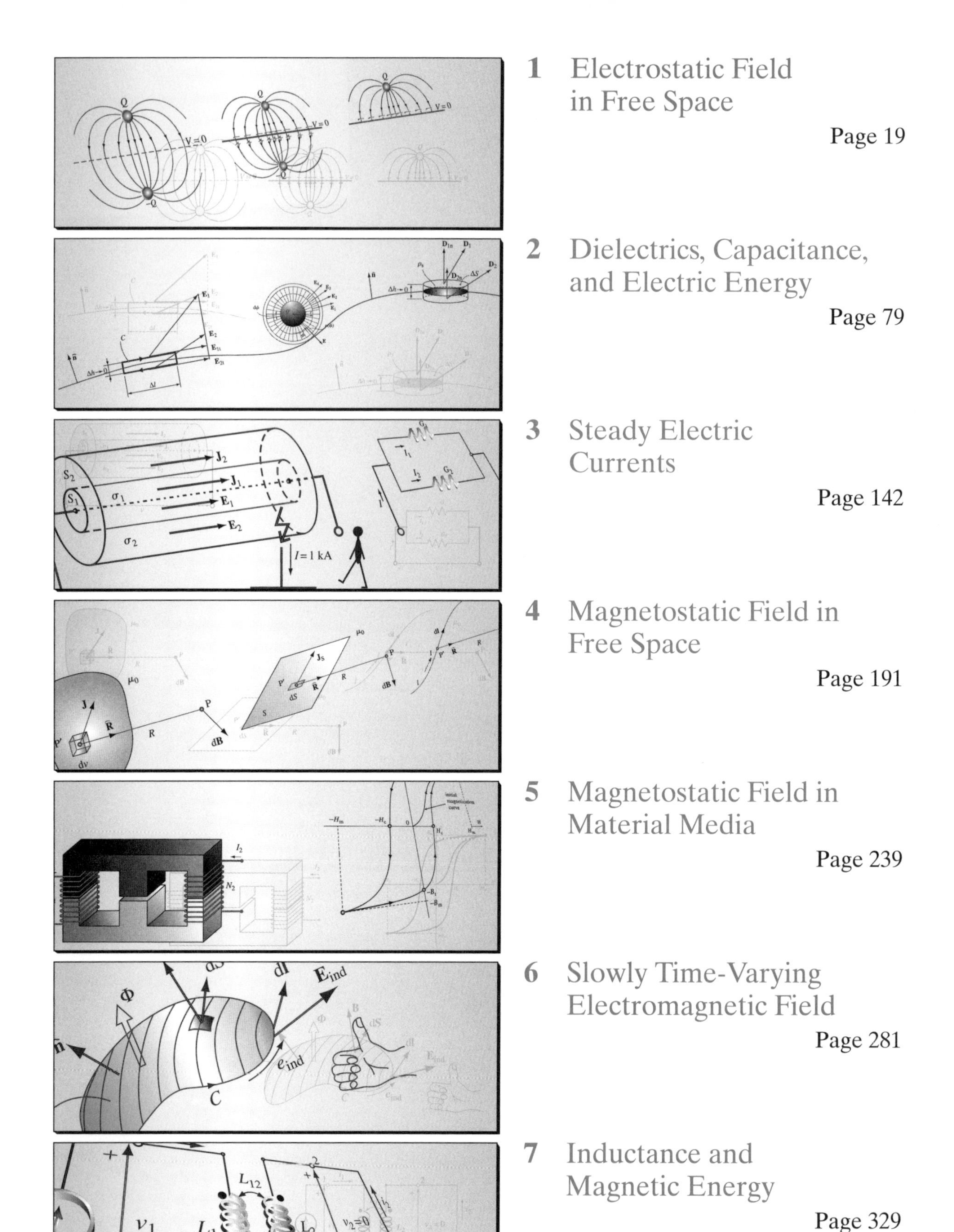

Electromagnetics

Branislav M. Notaroš

Department of Electrical and Computer Engineering
Colorado State University

Boston Columbus Indianapolis New York San Francisco Upper Saddle River
Amsterdam Cape Town Dubai London Madrid Milan Munich Paris Montreal Toronto
Delhi Mexico City Sao Paulo Sydney Hong Kong Seoul Singapore Taipei Tokyo

Vice President and Editorial Director, Engineering and Computer Science: *Marcia J. Horton*
Senior Editor: *Andrew Gilfillan*
Editorial Assistant: *William Opaluch*
Vice President, Production: *Vince O'Brien*
Senior Managing Editor: *Scott Disanno*
Production Editor: *Pavithra Jayapaul, TexTech International*
Senior Operations Supervisor: *Alan Fischer*
Operations Specialist: *Lisa McDowell*
Executive Marketing Manager: *Tim Galligan*
Art Director: *Kenny Beck*
Cover Designer: *Jodi Notowitz*
Art Editor: *Greg Dulles*
Media Editor: *Dan Sandin*
Composition/Full-Service Project Management: *TexTech International*

10 9 8 7 6 5 4 3 2 1

ISBN-13: 978-0-13-247364-4
ISBN-10: 0-13-247364-X

To the pioneering giants of electromagnetics
Michael Faraday, James Clerk Maxwell, and others
for providing the foundation of this book.

To my professors and colleagues
Branko Popović (late), Antonije Djordjević, and others
for making me nearly understand and fully love this stuff.

To all my students in all my classes over all these years
for teaching me to teach.

To Olivera, Jelena, and Milica
for everything else.

Contents

Preface

Electromagnetic theory is a fundamental underpinning of technical education, but, at the same time, one of the most difficult subjects for students to master. In order to help address this difficulty and contribute to overcoming it, here is another textbook on electromagnetic fields and waves for undergraduates, entitled, simply, *Electromagnetics*. This text provides engineering and physics students and other users with a comprehensive knowledge and firm grasp of electromagnetic fundamentals by emphasizing both mathematical rigor and physical understanding of electromagnetic theory, aimed toward practical engineering applications.

The book is designed primarily (but by no means exclusively) for junior-level undergraduate university and college students in electrical and computer engineering, physics, and similar departments, for both two-semester (or two-quarter) course sequences and one-semester (one-quarter) courses. It includes 14 chapters on electrostatic fields, steady electric currents, magnetostatic fields, slowly time-varying (low-frequency) electromagnetic fields, rapidly time-varying (high-frequency) electromagnetic fields, uniform plane electromagnetic waves, transmission lines, waveguides and cavity resonators, and antennas and wireless communication systems.

Apparently, there are an extremely large number of quite different books for undergraduate electromagnetics available (perhaps more than for any other discipline in science and engineering), which are all very good and important. This book, however, aims to combine the best features and advantages of all of them. It also introduces many new pedagogical features not present in any of the existing texts.

This text provides many nonstandard theoretically and practically important sections and chapters, new style and approaches to presenting challenging topics and abstract electromagnetic phenomena, innovative strategies and pedagogical guides for electromagnetic field and wave computation and problem solving, and, most importantly, outstanding (by the judgment of students so far) worked-out examples, homework problems, conceptual questions, and MATLAB exercises. The goal is to significantly improve students' understanding of electromagnetics and their attitude toward it. Overall, the book is meant as an ***"ultimate resource" for undergraduate electromagnetics***.

The distinguishing features of the book are:

- *371 realistic examples with very detailed and instructive solutions, tightly coupled to the theory, including strategies for problem solving*
- *650 realistic end-of-chapter problems, strongly and fully supported by solved examples (there is a demo example for every homework problem)*
- *Clear, rigorous, complete, and logical presentation of material, balance of breadth and depth, balance of static (one third) and dynamic (two thirds) fields, with no missing steps*
- *Flexibility for different options in coverage, emphasis, and ordering the material in a course or courses, including the transmission-lines-first approach*
- *Many nonstandard topics and subtopics and new derivations, explanations, proofs, interpretations, examples, pedagogical style, and visualizations*
- *500 multiple-choice conceptual questions (on the Companion Website), checking conceptual understanding of the book material*
- *400 MATLAB computer exercises and projects (on the Companion Website), many with detailed solutions (tutorials) and MATLAB codes (m files)*

www.pearsonhighered.com/notaros

The following sections explain these and other features in more detail.

WORKED EXAMPLES AND HOMEWORK PROBLEMS

The most important feature of the book is an extremely large number of realistic examples, with detailed and pedagogically instructive solutions, and end-of-chapter (homework) problems, strongly and fully supported by solved examples. There are a total of 371 worked examples, all tightly coupled to the theory, strongly reinforcing theoretical concepts and smoothly and systematically developing the problem-solving skills of students, and a total of 650 end-of-chapter problems, which are essentially offered and meant as end-of-section problems (indications appear at the ends of sections as to which problems correspond to that section).

Most importantly, for each homework problem or set of problems, there is always an example or set of examples in the text whose detailed solution provides the students and other readers with all necessary instruction and guidance to be able to solve the problem on their own, and to complete all homework assignments and practice for tests and exams. The abundance and quality of examples and problems are enormously important for the success of the course and class: students always ask for more and more solved examples, which must be relevant for the many problems that follow (for homework and exam preparation) – and this is exactly what this book attempts to offer.

Examples and problems in the book emphasize physical conceptual reasoning and mathematical synthesis of solutions, and not pure formulaic (plug-and-chug) solving. They also do not carry dry and too complicated pure mathematical formalisms. The primary goal is to teach the readers to reason through different (more or less challenging) situations and to help them gain confidence and really understand and like the material. Many examples and problems have a strong practical engineering context.

Solutions to examples show and explain every step, with ample discussions of approaches, strategies, and alternatives. Very often, solutions are presented in more than one way to aid understanding and development of true electromagnetic problem-solving skills. By acquiring such skills, which are definitely not limited to a skillful browsing through the book pages in a quest for a suitable "black-box" formula or set of formulas nor a skillful use of pocket calculators to plug-and-chug, the reader also acquires true confidence and pride in electromagnetics, and a strong appreciation for both its theoretical fundamentals and its practical applications.

"Physical" nontrivial examples are good also for instructors – for lectures and recitations – as they are much more interesting and suitable for logical presentation and discussion in the class than the "plug-and-chug" or purely "mathematical" examples.

CLARITY, RIGOR, AND COMPLETENESS

Along with the number and type of examples and problems (and questions and exercises), the most characteristic feature of the book is its consistent attention to clarity, completeness, and pedagogical soundness of presentation of the material throughout the entire text, aiming for an optimal balance of breadth and depth. Electromagnetics, as a fundamental science and engineering discipline, provides complete physical explanations for (almost) everything within its scope and rigorous mathematical models for everything it covers. Thus, besides a couple of experimental fundamental laws (like Coulomb's law) that have to be taken for granted for the model to build on, all other steps in building the most impressive and exciting structure called the electromagnetic theory can be readily presented to the reader in a consistent and meaningful manner and with enough detail to be understandable and appreciable. This is exactly what this book attempts to do.

Simply speaking, literally everything is derived, proved, and explained (except for a couple of experimental facts), with many new derivations, explanations, proofs, interpretations, and visualizations. Difficult and important concepts and derivations are regularly presented in more than one way to help students understand and master the subject at hand. Maximum effort has been devoted to a continuous logical flow of topics, concepts, equations, and ideas, with practically no "intentionally skipped" steps and parts. This, however, is done in a structural and modular manner, so that the reader who feels that some steps, derivations, and proofs can be bypassed at the time (with an opportunity of redoing it later) can do so, but this is left to the reader's discretion (or to the discretion and advice of the course instructor), not the author's.

Overall, my approach is to provide all possible (or all necessary) explanations, guidance, and detail in the theoretical parts and examples in the text, whereas students' actual understanding of the material, their thinking "on their own feet," and ability to do independent work are tested and challenged through numerous and relevant end-of-chapter problems and conceptual questions, and not through filling the missing gaps in the text.

On the other hand, I am fully aware that brevity may seem attractive to students at first glance because it typically means fewer pages for reading assignments. However, most students will readily acknowledge that it is indeed much easier and faster to read, grasp, and use several pages of thoroughly explained and presented material as opposed to a single page of condensed material with too many missing parts. During my dealings with students over so many years, I have been constantly told that they in fact prefer having everything derived and explained, and host of sample problems solved, to a lower page count and too many important parts, steps, and explanations missing, and too few detailed solutions, and this was the principal motivation for my writing this book.

This approach, in my opinion, is also good for instructors, as they have a self-contained, ready-to-use continuous "story" for each of their lectures, instead of a set of discrete formulas and sample facts with little or no explanations and detail. On the other hand, the instructor may choose to present only main facts for a given topic in class and rely on students for the rest, as they will be able to quickly and readily understand all reading assignments from the book. Indeed, I expect that every instructor using this text will have different "favorite" topics presented in class with all details and in great depth, including a number of examples, while "giving away" some other topics to students to cover on their own, with more or less depth, including worked examples.

OPTIONS IN COVERAGE OF THE MATERIAL

This book promotes and implements the direct or chronological and not inverse order of topics in teaching/learning electromagnetics, which can briefly be characterized as: first static and then dynamic topics, or first fields (static, quasistatic, and rapidly time-varying) and then waves (uniform plane waves, transmission lines, waveguides, and antennas). In addition, the book features a favorable balance of static (one third) and dynamic (two thirds) fields. Ideally, a course or a sequence of courses using this text would completely cover the book material, with a likelihood that some portions would be given to students as a reading assignment only. However, the book allows a lot of flexibility and many different options in actually covering the material, or parts of it, and ordering the topics in a course (or courses).

One scenario is to quickly go through Chapters 1–7, do just basic concepts and equations, and a couple of examples in each chapter, quickly reach Chapter 8 (general Maxwell's equations, etc.), and then do everything else as applications of general Maxwell's equations, including selected topics from Chapters 1–7 and more or less complete coverage of all other chapters. This scenario would essentially reflect the inverse (nonchronological) order of topics in teaching/learning electromagnetics. In fact, there may be many different scenarios suitable for different areas of emphasis and specialized outcomes of the course and the available time, all of them advancing in chronological order, through Chapters 1–14 of the book, just with different speeds and different levels of coverage of individual chapters.

To help the instructors create a plan for using the book material in their courses and students and other readers prioritize the book contents in accordance with their learning objectives and needs, Tables 1 and 2 provide classifications of all book chapters and sections, respectively, in two levels, indicating which chapters and sections within chapters are suggested as more likely candidates to be skipped or skimmed (covered lightly). This is just a guideline, and I expect that there will be numerous extremely creative, effective, and diverse combinations of book topics and subtopics constituting course outlines and learning/training plans, customized to best meet the preferences, interests, and needs of instructors, students, and other book users.

Most importantly, if chapters and sections are skipped or skimmed in the class, they are not skipped nor skimmed in the book, and the student will always be able to quickly find and apprehend additional information and fill any missing gaps using pieces of the book material from chapters and sections that are not planned to be covered in detail.

Table 1. **Classification of book chapters in two groups**, where "mandatory" chapters are those that would likely be covered in most courses, while some of the "elective" chapters could be skipped (or skimmed) based on specific areas of emphasis and desired outcomes of the course or sequence of courses and the available time. In selecting the material for the course(s), this classification at the chapter level could be combined with the classification at the section level given in Table 2.

"Mandatory" Chapters: 1, 3, 4, 6, 8, 9, 12	"Elective" Chapters: 2, 5, 7, 10, 11, 13, 14
1. Electrostatic Field in Free Space	2. Dielectrics, Capacitance, and Electric Energy
3. Steady Electric Currents	5. Magnetostatic Field in Material Media
4. Magnetostatic Field in Free Space	7. Inductance and Magnetic Energy
6. Slowly Time-Varying Electromagnetic Field	10. Reflection and Transmission of Plane Waves
8. Rapidly Time-Varying Electromagnetic Field	11. Field Analysis of Transmission Lines
9. Uniform Plane Electromagnetic Waves	13. Waveguides and Cavity Resonators
12. Circuit Analysis of Transmission Lines	14. Antennas and Wireless Communication Systems

Table 2. **Classification of book sections in two "tiers"** in terms of the suggested priority for coverage; if one or more sections in any of the chapters are to be skipped (or skimmed) given the areas of emphasis and specialized outcomes of the course or courses and the available time, then it is suggested that they be selected from the "tier two" sections, which certainly does not rule out possible omission (or lighter coverage) of some of the "tier one" sections as well.

Chapter	"Tier One" Sections	"Tier Two" Sections
1. Electrostatic Field in Free Space	1.1–1.4, 1.6, 1.8–1.10, 1.13–1.16	1.5, 1.7, 1.11, 1.12, 1.17–1.21
2. Dielectrics, Capacitance, and Electric Energy	2.1, 2.6, 2.7, 2.9, 2.10, 2.12, 2.13, 2.15, 2.16	2.2–2.5, 2.8, 2.11, 2.14, 2.17
3. Steady Electric Currents	3.1–3.4, 3.8, 3.10, 3.12	3.5–3.7, 3.9, 3.11, 3.13
4. Magnetostatic Field in Free Space	4.1, 4.2, 4.4–4.7, 4.9	4.3, 4.8, 4.10–4.13
5. Magnetostatic Field in Material Media	5.1, 5.5, 5.6, 5.8, 5.11	5.2–5.4, 5.7, 5.9, 5.10
6. Slowly Time-Varying Electromagnetic Field	6.2–6.5	6.1, 6.6–6.8
7. Inductance and Magnetic Energy	7.1, 7.4, 7.5	7.2, 7.3, 7.6
8. Rapidly Time-Varying Electromagnetic Field	8.2, 8.4, 8.6–8.8, 8.11, 8.12	8.1, 8.3, 8.5, 8.9, 8.10
9. Uniform Plane Electromagnetic Waves	9.3–9.7, 9.11, 9.14	9.1, 9.2, 9.8–9.10, 9.12, 9.13
10. Reflection and Transmission of Plane Waves	10.1, 10.2, 10.4–10.7	10.3, 10.8, 10.9
11. Field Analysis of Transmission Lines	11.4–11.6, 11.8	11.1–11.3, 11.7, 11.9, 11.10
12. Circuit Analysis of Transmission Lines	12.1–12.6, 12.11, 12.12, 12.15	12.7–12.10, 12.13, 12.14, 12.16–12.18
13. Waveguides and Cavity Resonators	13.1–13.3, 13.6, 13.8, 13.9, 13.12	13.4, 13.5, 13.7, 13.10, 13.11, 13.13, 13.14
14. Antennas and Wireless Communication Systems	14.1, 14.2, 14.4–14.6, 14.8, 14.14, 14.15	14.3, 14.7, 14.9–14.13

Table 3. Ordering the book material for the transmission-lines-first approach; Chapter 12 (Circuit Analysis of Transmission Lines) is written using only pure circuit-theory concepts (all field-theory aspects of transmission lines are placed in Chapter 11 – Field Analysis of Transmission Lines), so it can be taken at the very beginning of the course (or at any other time in the course). Note that two sections introducing (or reviewing) complex representatives of time-harmonic voltages and currents (Sections 8.6 and 8.7) must be done before Chapter 12.

Section 8.6: Time-Harmonic Electromagnetics
Section 8.7: Complex Representatives of Time-Harmonic Field and Circuit Quantities
Chapter 12: Circuit Analysis of Transmission Lines (or a selection of sections from Chapter 12 – see Table 2)
Chapters 1-11, 13, 14 or a selection of chapters (see Table 1) and sections (see Table 2)

TRANSMISSION-LINES-FIRST APPROACH

One possible exception from the chronological sequence of chapters (topics) in using this text implies a different placement of Chapter 12 (Circuit Analysis of Transmission Lines), which is written in such a manner that it can be taken at any time, even at the very beginning of the course, hence constituting the transmission-lines-first approach to teaching the course and learning the material. Namely, the field and circuit analyses of transmission lines are completely decoupled in the book, so that any field-theory aspects are placed in Chapter 11 (Field Analysis of Transmission Lines) and only pure circuit-theory concepts are used in Chapter 12 with per-unit-length characteristics (distributed parameters) of the lines being taken for granted (are assumed to be known) from the field analysis if the circuit analysis is done first. Table 3 shows the transmission-lines-first scenario using this book.

MULTIPLE-CHOICE CONCEPTUAL QUESTIONS

The book provides, on the *Companion Website*, a total of 500 conceptual questions. These are multiple-choice questions that focus on the core concepts of the material, requiring conceptual reasoning and understanding rather than calculations. They serve as checkpoints for readers following the theoretical parts and worked examples (like homework problems, conceptual questions are referred to at the ends of sections). Generally, conceptual questions may appear simple, but students often find them harder than the standard problems. Pedagogically, they are an invaluable resource. They are also ideal for in-class questions and discussions (so-called active teaching and learning) to be combined with traditional lecturing – if so desired.

In addition, conceptual questions are perfectly suited for class assessments, namely, to assess students' performance and evaluate the effectiveness of instruction, usually as the "gain" between the course "pretest" and "posttest" scores, and especially in light of ABET and similar accreditation criteria (the key word in these criteria is "assessment"). Selected conceptual questions from the large collection provided in the book can readily be used by instructors as partial and final assessment instruments for individual topics at different points in the course and for the entire class.

MATLAB EXERCISES, TUTORIALS, AND PROJECTS

The book provides, on the *Companion Website*, a very large and comprehensive collection of MATLAB computer exercises, strongly coupled to the book material, both the theory and the worked examples, and designed to help students develop a stronger intuition and a deeper understanding of electromagnetics, and find it more attractive and likable. MATLAB is chosen principally because it is a generally accepted standard in science and engineering education worldwide.

There are a total of 400 MATLAB exercises, which are referred to regularly within all book chapters, at the ends of sections, to supplement problems and conceptual questions. Each section of this collection starts with a comparatively very large number of tutorial exercises with detailed completely worked out solutions, as well as MATLAB codes (m files). This resource provides abundant opportunities for instructors for assigning in-class and homework projects – if so desired.

www.pearsonhighered.com/notarosinternational

VECTOR ALGEBRA AND CALCULUS

Elements of vector algebra and vector calculus are presented and used gradually across the book sections with an emphasis on physical insight and immediate links to electromagnetic field theory concepts, instead of having a purely mathematical review in a separate chapter. They are fully integrated with the development of the electromagnetic theory, where they actually belong and really come to life.

The mathematical concepts of gradient, divergence, curl, and Laplacian, as well as line (circulation), surface (flux), and volume integrals, are literally derived from physics (electromagnetics), where they naturally emanate as integral parts of electromagnetic equations and laws and where their physical meaning is almost obvious and can readily be made very visual. Furthermore, the text is written in such a way that even a reader with little background in vector algebra and vector calculus will indeed be able to learn or refresh vector analysis concepts directly through the first several chapters (please see Appendix 3 – Vector Algebra and Calculus Index).

LINKS TO CIRCUIT THEORY

The book provides detailed discussions of the links between electromagnetic theory and circuit theory throughout all of its chapters. It contains physical explanations for all elements of circuit theory, for both dc and ac regimes. All basic circuit-theory equations (circuit laws, element laws, etc.) are derived from electromagnetic theory. The goal is for the reader to develop both an appreciation of electromagnetic theory as a foundation of circuit theory and electrical engineering as a whole, as well as an understanding of limitations of circuit theory as an approximation of field theory.

HISTORICAL ASIDES

Throughout almost all chapters of the book, dozens of Historical Asides appear with quite detailed and fascinating biographies of famous scientists and pioneers in the field of electricity and magnetism. There are a total of 40 biographies, which are, in my view, not only very interesting historically and informative in terms of providing the factual chronological review of the development of one of the most impressive, consistent, and complete theories of the entire scientific and technological world – the electromagnetic theory – but they also often provide additional technical facts and explanations that complement the material in the text. I also feel that some basic knowledge about the discoverers – who made such epochal scientific achievements and far-reaching contributions to humanity – like Faraday, Maxwell, Henry, Hertz, Coulomb, Tesla, Heaviside, Oersted, Ampère, Ohm, Weber, and others should be made an irreplaceable part of a sort of "general education" of our engineering and physics students.

SUPPLEMENTS

Pearson offers many different products around the world to facilitate learning. In countries outside the United States, some products and services related to this textbook may not be available due to copyright and/or permissions restrictions. If you have questions, you can contact your local office by visiting www.pearsonhighered.com/international or you can contact your local Pearson representative.

ACKNOWLEDGMENTS

This text is based on my electromagnetics teaching and research over more than 20 years at the University of Belgrade, Yugoslavia (Serbia), University of Colorado at Boulder, University of Massachusetts Dartmouth, and Colorado State University, in Fort Collins, U.S.A. I gratefully acknowledge my colleagues and/or former Ph.D. students at these institutions whose discussions, advice, ideas, enthusiasm, initiatives, co-teaching, and co-authorships have shaped my knowledge, teaching style, pedagogy, and writing in electromagnetics, including: Prof. Branko Popović (late), Prof. Milan Ilić, Prof. Miroslav Djordjević, Prof.

Antonije Djordjević, Prof. Zoya Popović, Gradimir Božilović, Prof. Momčilo Dragović (late), Prof. Branko Kolundžija, Prof. Vladimir Petrović, and Prof. Jovan Surutka (late). All I know in electromagnetics and about its teaching I learned from them or with them or because of them, and I am enormously thankful for that.

I am grateful to all my students in all my classes over all these years for all the joy I have had in teaching them electromagnetics and for teaching me to teach better.

I especially thank my current Ph.D. students Nada Šekeljić, Ana Manić, and Sanja Manić for their invaluable help in writing MATLAB computer exercises, tutorials, and codes, checking the derivations and examples in the book, and solving selected end-of-chapter problems. I owe a particular debt of gratitude to my colleague and former Ph.D. student Prof. Milan Ilić, for his outstanding work and help with the initial electronic artwork in the book. My colleagues and former students Andjelija Ilić and Prof. Miroslav Djordjević, as well as Olivera Notaroš, also contributed very significantly to the artwork, for which I am sincerely indebted.

I would like to express my gratitude to the reviewers of the manuscript for their extremely detailed, useful, positive, and competent comments that I feel helped me to significantly improve the quality of the book, including: Professors Indira Chatterjee, Robert J. Coleman, Cindy Harnett, Jianming Jin, Leo Kempel, Edward F. Kuester, Yifei Li, Krzysztof A. Michalski, Michael A. Parker, Andrew F. Peterson, Costas D. Sarris, and Fernando L. Teixeira.

Special thanks to all members of the Pearson Prentice Hall team, who all have been excellent, and particularly to my editor Andrew Gilfillan, who has been extremely helpful and supportive, and whose input was essential at many stages in the development of the manuscript and book, my production manager Scott Disanno, for expertly leading the book production, Marcia Horton, Vice President and Editorial Director with Prentice Hall, for great conversations and support in the initial phases of the project, and Tom Robbins, former Publisher at Prentice Hall, for the first encouragements. I hope they enjoyed our dealings and discussions as extensively as I did.

I thank my wife Olivera Notaroš, who also teaches in the ECE Department at Colorado State University, not only for her great and constant support and understanding but also for her direct involvement and absolutely phenomenal ideas, advice, and help in all phases of writing the manuscript and production of the book. Without her, this book would not be possible or would, at least, be very different. I also acknowledge extraordinary support by my wonderful daughters Jelena and Milica, and I hope that I will be able to keep my promise to them that I will now take a long break from writing. I am very sad that the writing of this book took me so long that my beloved parents Smilja and Mile did not live to receive the first dedicated copy of the book from me, as had been the case with my previous books.

Finally, on a very personal note as well, I really love electromagnetics and teaching it, and I hope that this book will convey at least a portion of my admiration and enthusiasm to the readers and help more and more students start liking and appreciating this fascinating discipline with endless impacts. I am proud of being able to do that in my classes, and am now excited and eager to try to spread that message to a much larger audience using this text. Please send me your comments, suggestions, questions, and corrections (I hope there will not be many of these) regarding the book to notaros@colostate.edu.

BRANISLAV M. NOTAROŠ
FORT COLLINS, COLORADO

"I believe but cannot explain that the author's confidence is somehow transferred to the student as a trust that the text they are reading and learning from is worth their time."
—Anonymous reviewer of the book manuscript

1 Electrostatic Field in Free Space

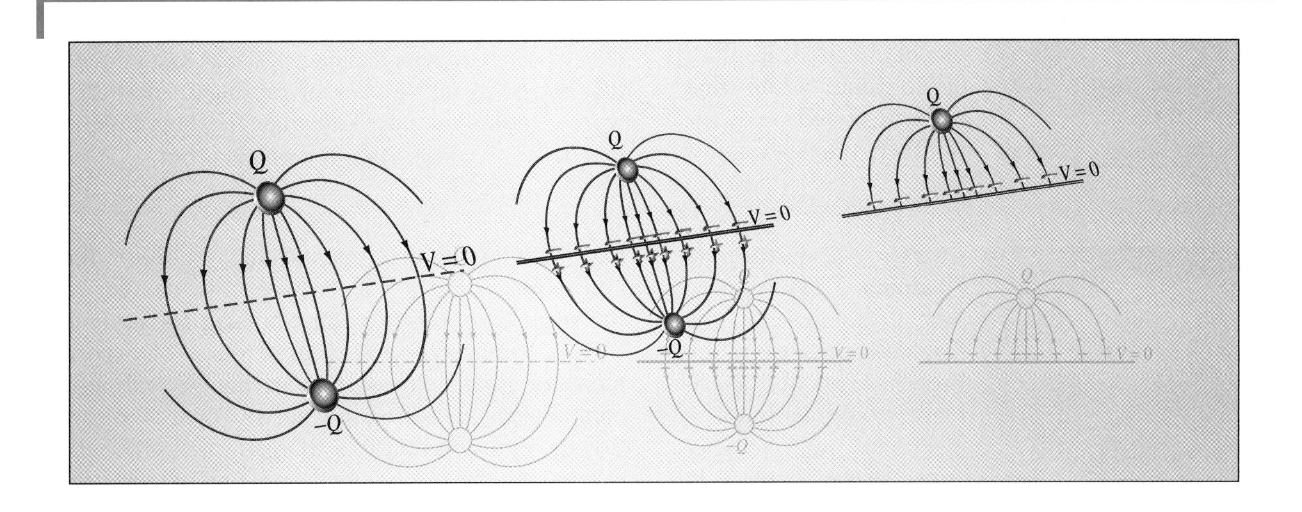

Introduction:

Electrostatics is the branch of electromagnetics that deals with phenomena associated with static electricity, which are essentially the consequence of a simple experimental fact – that charges exert forces on one another. These forces are called electric forces, and the special state in space due to one charge in which the other charge is situated and which causes the force on it is called the electric field. Any charge distribution in space with any time variation is a source of the electric field. The electric field due to time-invariant charges at rest (charges that do not change in time and do not move) is called the static electric field or electrostatic field. This is the simplest form of the general electromagnetic field, and its physics and mathematics represent the foundation of the entire electromagnetic theory. On the other hand, a clear understanding of electrostatics is essential for many practical applications that involve static electric fields, charges, and forces in electrical and electronic devices and systems.

We shall begin our study of electrostatics by investigating the electrostatic field in a vacuum or air (free space), which will then be extended to the analysis of electrostatic structures composed of charged conductors in free space (also in this chapter). In the next chapter, we shall evaluate the electrostatic field in the presence of dielectric materials, and include such materials in our discussion of general electrostatic systems.

HISTORICAL ASIDE

The first record of our experiences with electricity dates back to the sixth century B.C., when **Thales of Miletus** (624 B.C.–546 B.C.), an ancient Greek philosopher and mathematician, one of the greatest minds of all time, wrote that amber rubbed in wool attracts pieces of straw or feathers – which we now know is a manifestation of electrification by friction. In relation to this, the name "electron" for the subatomic particle carrying the smallest amount of (negative) charge comes from the Greek word *ἤλεκτρον* (*elektron*) for amber. Our experiences with magnetism, on the other side, also trace back to ancient times. The origin of the word "magnet" relates to the region in Greece named Magnesia (*Μαγνησία*), in which ancient Greeks first noticed (ca. 800 B.C.) that pieces of the black rock they were standing on, now known as the iron mineral magnetite (Fe_3O_4), attracted one another.

Charles Augustin de Coulomb (1736–1806) was a colonel in the Engineering Corps of the French Army and a brilliant experimentalist in electricity and magnetism. He graduated in 1761 from the School of the Engineering Corps (*École du Génie*), and was in charge of building the Fort Bourbon on Martinique, in the West Indies, where he showed his engineering and organizational skills. In 1772, Coulomb returned to France with impaired health, and began his research in applied mechanics. In 1777, he invented a torsion balance to measure small forces, and as a result of his 1781 memoir on friction, he was elected to the French Academy (*Académie des Sciences*). Between 1785 and 1791, he wrote a series of seven papers on electricity and magnetism, out of which by far the most important and famous is his work on the theory of attraction and repulsion between charged bodies. Namely, Coulomb formulated in 1785 the basic law for the electrostatic force between like and unlike charges (i.e., between two charges of the same or opposite polarity) using his genuine torsion balance apparatus, in a course of experiments originally aimed at improving the mariner's compass. He measured the electric force of attraction or repulsion that two charged small pith balls exerted on one another by the amount of twist produced on the torsion balance, and demonstrated an inverse square law for such forces – the force is proportional to the product of the charges of each of the balls and inversely proportional to the square of the distance between their centers. This result came out to be an underpinning of the whole area of science and engineering now known as electromagnetics, and of all of its applications. Upon the outbreak of the French Revolution in 1789, Coulomb retired to a small estate near Blois, to work in peace on his scientific memoirs. His last post was that of the inspector general of public instruction, under Napoleon, from 1802 to 1806. The law of electric forces on charges now bears his name – Coulomb's law – and his name is further immortalized by the use of coulomb (C) as the unit of charge.

1.1 COULOMB'S LAW

The basis of electrostatics is an experimental result called Coulomb's law. It states that the electric force $\mathbf{F}_{e12}$ on a point charge Q_2 due to a point charge Q_1 in a vacuum (or air) is given by[1] (Fig. 1.1)

$$\boxed{\mathbf{F}_{e12} = \frac{1}{4\pi\varepsilon_0}\frac{Q_1 Q_2}{R^2}\hat{\mathbf{R}}_{12}.} \tag{1.1}$$

Coulomb's law

With $\mathbf{R}_{12}$ denoting the position vector of Q_2 relative to Q_1, $R = |\mathbf{R}_{12}|$ is the distance between the two charges, $\hat{\mathbf{R}}_{12} = \mathbf{R}_{12}/R$ is the unit vector[2] of the vector $\mathbf{R}_{12}$, and ε_0 is the permittivity of a vacuum (free space),

$$\boxed{\varepsilon_0 = 8.8542 \text{ pF/m}} \tag{1.2}$$

permittivity of a vacuum

($\text{p} \equiv 10^{-12}$ and F is farad, the unit for capacitance, which will be studied in the next chapter). By point charges we mean charged bodies of arbitrary shapes whose dimensions are much smaller than the distance between them. The SI (International System of Units[3]) unit for charge is the coulomb (abbreviated C), named in honor of Charles Coulomb. This is a very large unit of charge. The charge of an electron, which is negative, turns out to be

$$\boxed{e = 1.602 \times 10^{-19} \text{ C}} \tag{1.3}$$

charge of electron, magnitude

in magnitude ($Q_{\text{electron}} = -e$). The unit for force ($\mathbf{F}$) is the newton (N). The expression $Q_1 Q_2/(4\pi\varepsilon_0 r^2)$ in Eq. (1.1) represents the algebraic intensity (can be of arbitrary sign) of the vector $\mathbf{F}_{e12}$ with respect to the unit vector $\hat{\mathbf{R}}_{12}$. If Q_1 and Q_2 are of the same sign or polarity (like charges), this intensity is positive, $\mathbf{F}_{e12}$ has the same orientation as $\hat{\mathbf{R}}_{12}$, and the force between charges is repulsive. Conversely, the electric force between unlike charges ($Q_1 Q_2 < 0$) is attractive.

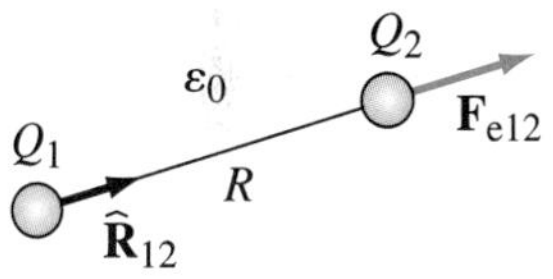

Figure 1.1 Notation in Coulomb's law, given by Eq. (1.1).

By reversing the indices 1 and 2 in Eq. (1.1) and noting that $\hat{\mathbf{R}}_{21} = -\hat{\mathbf{R}}_{12}$, we obtain that $\mathbf{F}_{e21} = -\mathbf{F}_{e12}$; i.e., the force on Q_1 due to Q_2 is equal in magnitude and opposite in direction to the force on Q_2 due to Q_1. This result is essentially an expression of Newton's third law – to every action (force) in nature, there is an opposed equal reaction.

If we have more than two point charges, we can use the principle of superposition, which also is a result of experiments, to determine the resultant force on a particular charge – by adding up vectorially the partial forces exerted on it by each of the remaining charges individually.

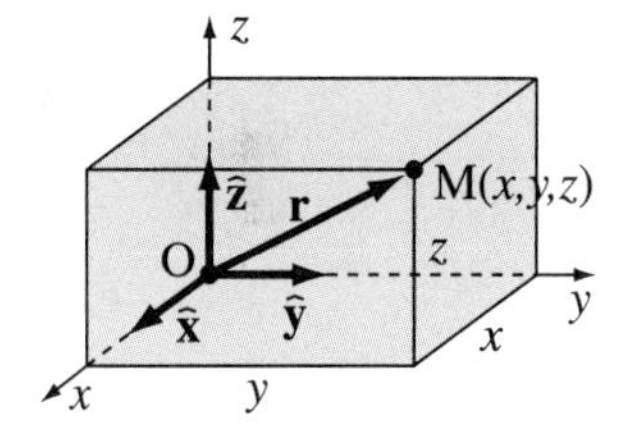

Figure 1.2 Point M(x, y, z) and coordinate unit vectors in the Cartesian coordinate system.

In the general case, vector addition is carried out component by component (for an arbitrary number of vectors), most frequently in the Cartesian coordinate system. Cartesian (or rectangular) coordinates, x, y, and z, and coordinate unit vectors, $\hat{\mathbf{x}}$, $\hat{\mathbf{y}}$, and $\hat{\mathbf{z}}$ (unit vectors along the x, y, and z directions), are shown in Fig. 1.2. The unit vectors are mutually perpendicular, and an arbitrary vector $\mathbf{a}$ in Cartesian coordinates can be represented as

$$\boxed{\mathbf{a} = a_x\,\hat{\mathbf{x}} + a_y\,\hat{\mathbf{y}} + a_z\,\hat{\mathbf{z}}.} \tag{1.4}$$

Cartesian vector components

[1] In typewritten work, vectors are commonly represented by boldface symbols, e.g., **F**, whereas in handwritten work, they are denoted by placing a right-handed arrow over the symbol, as $\vec{F}$.

[2] All unit vectors in this text will be represented using the "hat" notation, so the unit vector in the x direction (in the rectangular coordinate system), for example, is given as $\hat{\mathbf{x}}$ (note that some of the alternative widely used notations for unit vectors would represent this vector as $\mathbf{a}_x$, $\mathbf{i}_x$, and $\mathbf{u}_x$, respectively).

[3] SI is the modernized version of the metric system. The abbreviation is from the French name *Système International d'Unités*.

Here, a_x, a_y, and a_z are the components of vector $\mathbf{a}$ in the Cartesian coordinate system, and its magnitude is

$$a = |\mathbf{a}| = \sqrt{a_x^2 + a_y^2 + a_z^2}. \tag{1.5}$$

The unit vector of $\mathbf{a}$ is $\hat{\mathbf{a}} = \mathbf{a}/a$. Of course, the magnitude of $\hat{\mathbf{a}}$, and of any unit vector, is unity, $|\hat{\mathbf{a}}| = |\mathbf{a}|/a = 1$. The sum of two vectors is given by

$$\mathbf{a} + \mathbf{b} = (a_x + b_x)\,\hat{\mathbf{x}} + \left(a_y + b_y\right)\,\hat{\mathbf{y}} + (a_z + b_z)\,\hat{\mathbf{z}}. \tag{1.6}$$

Shown in Fig. 1.2 is also the position vector $\mathbf{r}$ of an arbitrary point $\mathrm{M}(x, y, z)$ in space, with respect to the coordinate origin (O),

position vector of a point

$$\boxed{\mathbf{r} = x\,\hat{\mathbf{x}} + y\,\hat{\mathbf{y}} + z\,\hat{\mathbf{z}},} \tag{1.7}$$

where, using Eq. (1.5), $r = |\mathbf{r}| = \sqrt{x^2 + y^2 + z^2} = \overline{\mathrm{OM}}$ is the distance[4] between points O and M.

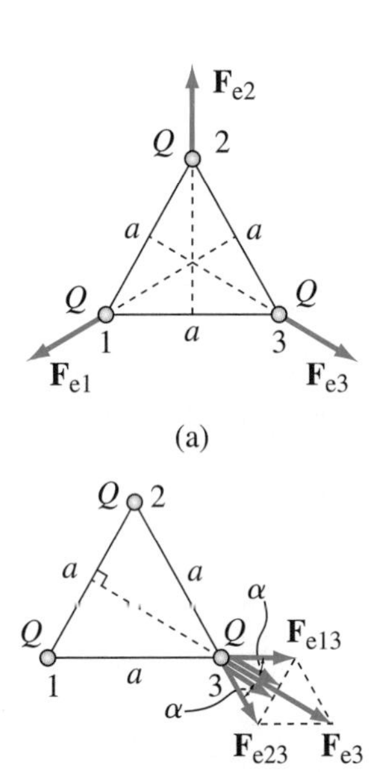

Figure 1.3 (a) Three equal point charges at the vertices of an equilateral triangle and (b) computation of the resultant electric force on one of the charges; for Example 1.1.

Example 1.1 Three Equal Point Charges at Triangle Vertices

Three small charged bodies of charge Q are placed at three vertices of an equilateral triangle with sides a, in air. The bodies can be considered as point charges. Find the direction and magnitude of the electric force on each of the charges.

Solution Even with no computation whatsoever, we can conclude from the symmetry of this problem that the resultant forces on the charges, $\mathbf{F}_{e1}$, $\mathbf{F}_{e2}$, and $\mathbf{F}_{e3}$, all have the same magnitude and are positioned in the plane of the triangle as indicated in Fig. 1.3(a). Let us compute the resultant force on the lower right charge – charge 3. Using the principle of superposition, this force represents the vector sum of partial forces due to charges 1 and 2, respectively, that is [Fig. 1.3(b)],

$$\mathbf{F}_{e3} = \mathbf{F}_{e13} + \mathbf{F}_{e23} \quad \text{(vector superposition)}. \tag{1.8}$$

From Coulomb's law, Eq. (1.1), magnitudes of the individual partial forces are given by

$$F_{e13} = F_{e23} = \frac{Q^2}{4\pi\varepsilon_0 a^2}, \tag{1.9}$$

and both forces are repulsive.

We note that the vector $\mathbf{F}_{e3}$ is positioned along the symmetry line between charges 1 and 2, i.e., between vectors $\mathbf{F}_{e13}$ and $\mathbf{F}_{e23}$, and it makes the angle $\alpha = \pi/6$ with both vectors. The magnitude of the resultant vector is therefore twice the projection of any of the partial vectors on the symmetry line, which yields

$$F_{e3} = 2\,(F_{e13}\cos\alpha) = 2F_{e13}\frac{\sqrt{3}}{2} = F_{e13}\sqrt{3} = \frac{\sqrt{3}Q^2}{4\pi\varepsilon_0 a^2}. \tag{1.10}$$

---

[4] While dealing with a wide variety of vector quantities in electromagnetics, we shall regularly visualize (draw) them as arrows in space, like the force vector $\mathbf{F}_{e12}$ in Fig. 1.1, to aid the analysis and computation. However, we shall always have in mind that only position vectors, like $\mathbf{r}$ in Fig. 1.2, and some other length vectors to be introduced later have this feature of their magnitude being the actual geometrical distance in space. Magnitudes of all other vectors are measured in units different from meter, and the sizes (lengths) of arrows in space that they are associated with can only be indicative of relative magnitudes of quantities of the same nature (with the same unit), like two forces acting on the same body, which is equally useful and will be utilized extensively in this text as well.

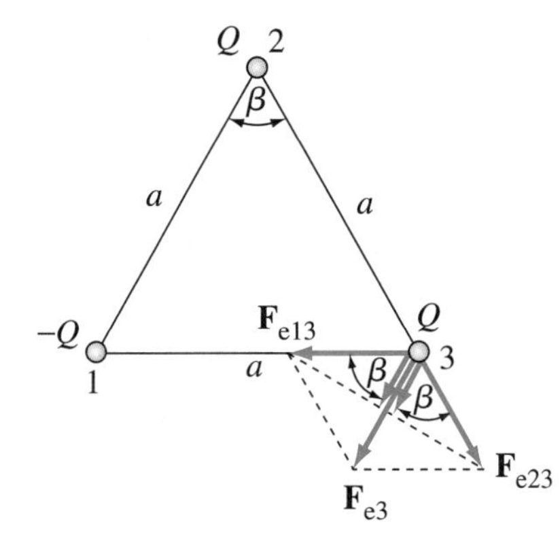

Figure 1.4 Three point charges equal in magnitude but with different polarities at the vertices of an equilateral triangle – computation of the resultant electric force on charge 3; for Example 1.2.

Example 1.2 Three Unequal Point Charges at Triangle Vertices

Determine the resultant force on the lower right charge in the configuration shown in Fig. 1.4. Assume that Q and a are given quantities.

Solution The only difference with respect to the configuration in Fig. 1.3 is that the force $\mathbf{F}_{e13}$ is now attractive, as indicated in Fig. 1.4. The resultant force on charge 3 is parallel to the line connecting charges 2 and 1, and the angle it makes with any of the partial forces, $\mathbf{F}_{e13}$ and $\mathbf{F}_{e23}$, is $\beta = \pi/3$. Its magnitude is hence

$$F_{e3} = 2\,(F_{e13}\cos\beta) = 2F_{e13}\frac{1}{2} = F_{e13} = \frac{Q^2}{4\pi\varepsilon_0 a^2}. \tag{1.11}$$

Example 1.3 Three Point Charges in Cartesian Coordinate System

Point charges $Q_1 = 1\ \mu$C, $Q_2 = -2\ \mu$C, and $Q_3 = 2\ \mu$C are situated in free space at points defined by Cartesian coordinates $(1\text{ m}, 0, 0)$, $(0, 1\text{ m}, 0)$, and $(0, 0, 1\text{ m})$, respectively. Compute the resultant electric force on charge Q_1.

Solution From Coulomb's law and Fig. 1.5(a), the magnitudes of the individual forces on the charge Q_1 are

$$F_{e21} = F_{e31} = \frac{Q_1 Q_3}{4\pi\varepsilon_0 R^2} = 9\text{ mN}, \tag{1.12}$$

where R is the distance of Q_1 from Q_2 (or Q_3). In order to add together vectors $\mathbf{F}_{e21}$ and $\mathbf{F}_{e31}$, we decompose them into convenient components, in this case – into components in the Cartesian coordinate system. Based on Figs. 1.5(b) and (c),

$$\mathbf{F}_{e21} = -F_{e21}\cos\alpha\ \hat{\mathbf{x}} + F_{e21}\sin\alpha\ \hat{\mathbf{y}}, \quad \alpha = \frac{\pi}{4}, \tag{1.13}$$

$$\mathbf{F}_{e31} = F_{e31}\cos\beta\ \hat{\mathbf{x}} - F_{e31}\sin\beta\ \hat{\mathbf{z}}, \quad \beta = \frac{\pi}{4}, \tag{1.14}$$

so the resultant force is

$$\mathbf{F}_{e1} = \mathbf{F}_{e21} + \mathbf{F}_{e31} = F_{e21}\frac{\sqrt{2}}{2}(\hat{\mathbf{y}} - \hat{\mathbf{z}}). \tag{1.15}$$

The Cartesian components of the vector $\mathbf{F}_{e1}$ amount to

$$F_{e1x} = 0, \quad F_{e1y} = -F_{e1z} = F_{e21}\frac{\sqrt{2}}{2} = 6.36\text{ mN}, \tag{1.16}$$

and its magnitude [Eq. (1.5)] comes out to be

$$F_{e1} = \sqrt{F_{e1x}^2 + F_{e1y}^2 + F_{e1z}^2} = F_{e21} = 9\text{ mN}. \tag{1.17}$$

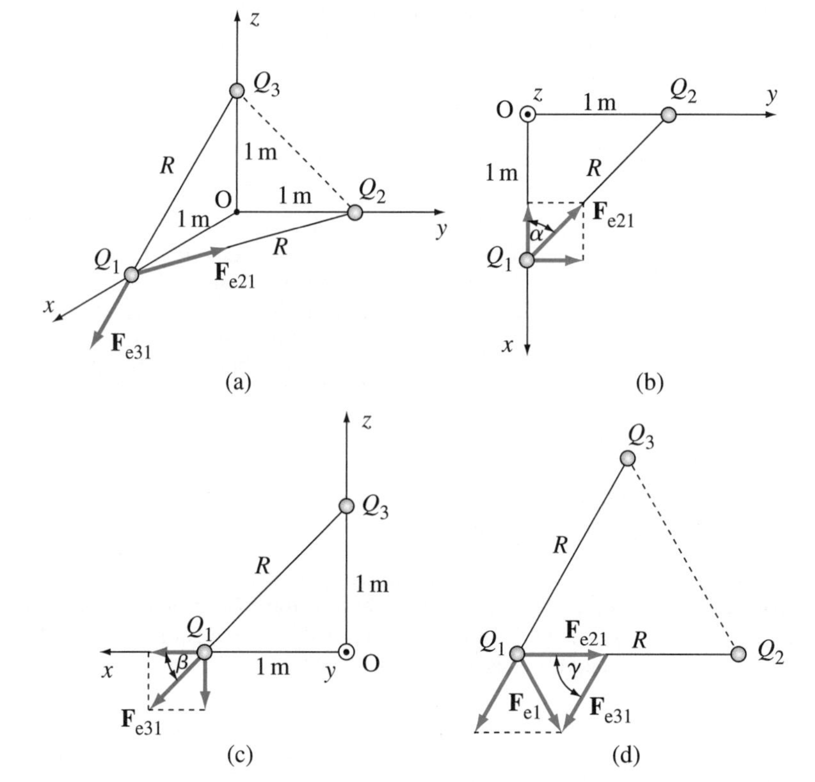

Figure 1.5 Summation of electric forces in the Cartesian coordinate system: (a) three point charges in space, with partial force vectors $\mathbf{F}_{e21}$ and $\mathbf{F}_{e31}$, (b) component decomposition of $\mathbf{F}_{e21}$, (c) decomposition of $\mathbf{F}_{e31}$, and (d) alternative addition of forces using the head-to-tail rule and the cosine formula; for Example 1.3.

Note that F_{e1} can alternatively be obtained using the head-to-tail rule, as portrayed in Fig. 1.5(d),[5] in combination with the cosine formula,[6] which yields

$$F_{e1} = \sqrt{F_{e21}^2 + F_{e31}^2 - 2F_{e21}F_{e31}\cos\gamma} = F_{21} = 9\ \text{mN}, \quad \gamma = \frac{\pi}{3}. \tag{1.18}$$

Note also that the vector $\mathbf{F}_{e1}$ is parallel to the line connecting charges Q_3 and Q_2, and that it is positioned at an angle of $\pi/4$ with respect to the plane xy.

Example 1.4 Four Charges at Tetrahedron Vertices

Four point charges Q are positioned in free space at four vertices of a regular (equilateral) tetrahedron with the side length a. Find the electric force on one of the charges.

---

[5] By the head-to-tail rule for vector addition, to obtain graphically the vector sum $\mathbf{c} = \mathbf{a} + \mathbf{b}$, we first arrange the two vectors (usually translate $\mathbf{b}$ from its original position) in such a way that the tail of $\mathbf{b}$ (second vector) is placed at the head of $\mathbf{a}$ (first vector). In other words, the head of the first vector is "connected" to the tail of the second, and hence the term "head-to-tail" for this arrangement. Then we draw $\mathbf{c}$ (resultant vector) as a vector extending from the tail of $\mathbf{a}$ to the head of $\mathbf{b}$, as in Fig. 1.5(d). An equivalent graphical method to add two vectors together is the parallelogram rule, where $\mathbf{c} = \mathbf{a} + \mathbf{b}$ corresponds to a diagonal of the parallelogram formed by $\mathbf{a}$ and $\mathbf{b}$, which can also be seen in Fig. 1.5(d). To add more than two vectors, e.g., $\mathbf{d} = \mathbf{a} + \mathbf{b} + \mathbf{c}$, we simply apply the head-to-tail rule to add $\mathbf{c}$ to the already found $\mathbf{a} + \mathbf{b}$, and so on – the resultant vector extends from the tail of the first vector to the head of the last vector in the multiple head-to-tail chain, and a polygon is thus obtained, which is why this procedure is often referred to as the polygon rule.

[6] In an arbitrary triangle of side lengths a, b, and c and angles α, β, and γ, the square of the length c of the side opposite to the angle γ equals $c^2 = a^2 + b^2 - 2ab\cos\gamma$ (and analogously for a^2 and b^2 using $\cos\alpha$ and $\cos\beta$, respectively), and this is known as the cosine formula (rule) or law of cosines.

Solution Note that this configuration actually represents a spatial version of the planar configuration of Fig. 1.3. Referring to Fig. 1.6, let us find the force on the charge on the top of the tetrahedron – charge 4. This force is given by

$$\mathbf{F}_{e4} = \mathbf{F}_{e14} + \mathbf{F}_{e24} + \mathbf{F}_{e34}, \tag{1.19}$$

where all the three partial forces are of the same magnitude, equal to $F_{e14} = Q^2/(4\pi\varepsilon_0 a^2)$. The horizontal components of the force vectors all lie in one plane and the angle between each two is 120°, so that they vectorially add up to zero. Thus, the resultant vector $\mathbf{F}_{e4}$ has a vertical component only, whose magnitude amounts to three times that of the vertical component of each partial force,

$$F_{e4} = 3\,(F_{e14}\cos\alpha)\,. \tag{1.20}$$

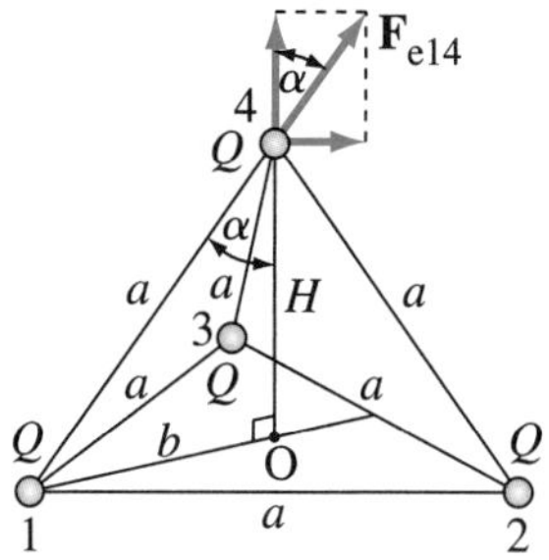

To determine $\cos\alpha$ (as H/a) from the right-angled triangle ΔO14 in Fig. 1.6, we first find the distance b (between charge 1 and point O) from the equilateral triangle Δ123 (the base of the tetrahedron), as 2/3 of the height of this triangle,[7] so we have

$$b = \frac{2}{3}\left(\frac{\sqrt{3}}{2}\,a\right) = \frac{\sqrt{3}}{3}\,a \quad \longrightarrow \quad \cos\alpha = \frac{H}{a} = \frac{\sqrt{a^2 - b^2}}{a} = \sqrt{\frac{2}{3}}, \tag{1.21}$$

which substituted in Eq. (1.20) results in

$$F_{e4} = 3F_{e14}\frac{\sqrt{6}}{3} = \frac{\sqrt{6}Q^2}{4\pi\varepsilon_0 a^2}. \tag{1.22}$$

Figure 1.6 Four point charges at tetrahedron vertices; for Example 1.4.

Problems: 1.1–1.7; *Conceptual Questions* (on Companion Website): 1.1 and 1.2; *MATLAB Exercises* (on Companion Website).

1.2 DEFINITION OF THE ELECTRIC FIELD INTENSITY VECTOR

The electric field is a special physical state existing in a space around charged objects. Its fundamental property is that there is a force (Coulomb force) acting on any stationary charge placed in the space. To quantitatively describe this field, we introduce a vector quantity called the electric field intensity vector, $\mathbf{E}$. By definition, it is equal to the electric force $\mathbf{F}_e$ on a probe (test) point charge Q_p placed in the electric field, divided by Q_p, that is,

$$\mathbf{E} = \frac{\mathbf{F}_e}{Q_p} \quad (Q_p \to 0). \tag{1.23}$$

definition of $\mathbf{E}$ *(unit:* V/m*)*

The probe charge has to be small enough in magnitude in order to practically not affect the distribution of charges which are the sources of $\mathbf{E}$. The unit for the electric field intensity we use is volt per meter (V/m).

From the definition in Eq. (1.23) and Coulomb's law, Eq. (1.1), we obtain the expression for the electric field intensity vector of a point charge Q at a distance R from the charge (Fig. 1.7)

$$\mathbf{E} = \frac{1}{4\pi\varepsilon_0}\frac{Q}{R^2}\,\hat{\mathbf{R}}, \tag{1.24}$$

electric field due to a point charge in free space

[7]Note that the orthocenter (point O in Fig. 1.6) of an equilateral triangle partitions its heights (h) in the ratio 2 : 1, so into segments $2h/3$ and $h/3$ long. Note also that $h = \sqrt{3}a/2$ (in an equilateral triangle), a being the side length of the triangle.

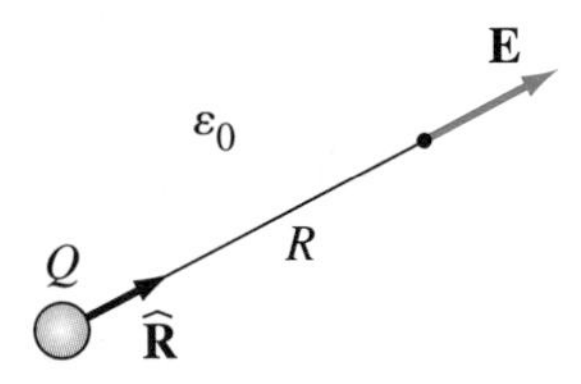

Figure 1.7 Electric field intensity vector due to a point charge in free space.

where $\hat{\mathbf{R}}$ is the unit vector along R directed from the center of the charge (source point) toward the point at which the field is (to be) determined (field or observation point).

By superposition, the electric field intensity vector produced by N point charges $(Q_1, Q_2, \ldots, Q_N)$ at a point that is at distances $R_1, R_2, \ldots, R_N$, respectively, from the charges can be obtained as

$$\mathbf{E} = \mathbf{E}_1 + \mathbf{E}_2 + \cdots + \mathbf{E}_N = \frac{1}{4\pi\varepsilon_0}\sum_{i=1}^{N}\frac{Q_i}{R_i^2}\hat{\mathbf{R}}_i, \quad (1.25)$$

where $\hat{\mathbf{R}}_i$, $i = 1, 2, \ldots, N$, are the corresponding unit vectors.

Problems: 1.8; *MATLAB Exercises* (on Companion Website).

1.3 CONTINUOUS CHARGE DISTRIBUTIONS

A point charge is the simplest case of a charge distribution, which, mathematically, corresponds to the space (three-dimensional) delta function. In the general case, however, charge can be distributed throughout a volume, on a surface, or along a line. Each of these three characteristic continuous charge distributions is described by a suitable charge density function. The volume charge density (in a volume v) is defined as [Fig. 1.8(a)]

volume charge density (unit: C/m^3)

$$\boxed{\rho = \frac{dQ}{dv},} \quad (1.26)$$

the surface charge density (on a surface S) is given by [Fig. 1.8(b)]

surface charge density (unit: C/m^2)

$$\boxed{\rho_s = \frac{dQ}{dS},} \quad (1.27)$$

and the line charge density (along a line l) is [Fig. 1.8(c)]

line charge density (unit: C/m)

$$\boxed{Q' = \frac{dQ}{dl}.} \quad (1.28)$$

Note that the symbol ρ_v is sometimes used instead of ρ, σ instead of ρ_s, and ρ_l instead of Q'. In the above equations, dQ represents the elemental charge in a volume element dv, on a surface element dS, and along a line element dl, respectively, and the corresponding units for charge densities are C/m^3, C/m^2, and C/m.

The total charge Q for the three characteristic charge distributions in Fig. 1.8 is obtained as $\int dQ$ (adding charge elements dQ), which leads to

$$Q_{\text{in } v} = \int_v \rho\, dv, \quad Q_{\text{on } S} = \int_S \rho_s\, dS, \quad \text{and} \quad Q_{\text{along } l} = \int_l Q'\, dl, \quad (1.29)$$

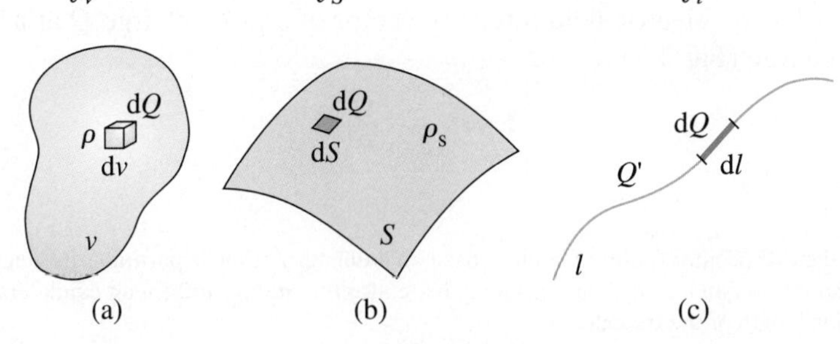

Figure 1.8 Three characteristic continuous charge distributions and charge elements: (a) volume charge, (b) surface charge, and (c) line charge.

respectively. Special, but important, cases of continuous charge distributions are uniform volume, surface, and line charge distributions. A charge distribution is said to be uniform if the associated charge density is constant over the entire region (v, S, or l) with charges. The expressions in Eqs. (1.29) then become much simpler,

$$Q_{\text{in } v} = \rho v \quad (\rho = \text{const}), \quad Q_{\text{on } S} = \rho_{\text{s}} S \quad (\rho_{\text{s}} = \text{const}),$$
$$Q_{\text{along } l} = Q' l \quad (Q' = \text{const}). \tag{1.30}$$

Note that Q' ($Q' = \text{const}$) is also used to represent the so-called charge per unit length (p.u.l.) of a long uniformly charged structure (e.g., thin or thick cylinder), defined as the charge on one meter (unit of length) of the structure divided by 1 m,

$$\boxed{Q' = Q_{\text{p.u.l.}} = \frac{Q_{\text{along } l}}{l} = \frac{Q_{\text{for 1 m length}}}{1\text{ m}},} \tag{1.31}$$

charge per unit length, in C/m

and hence Q' numerically equals the charge on each meter of the structure.

Example 1.5 **Nonuniform Volume Charge Distribution in a Sphere**

A volume charge is distributed in free space inside a sphere of radius a. The charge density is

$$\rho(r) = \rho_0 \frac{r}{a} \quad (0 \le r \le a), \tag{1.32}$$

where r stands for the radial distance from the sphere center, and ρ_0 is a constant. Find the total charge of the sphere.

Solution Since the charge density depends on r only, we need to integrate in the first expression in Eqs. (1.29) only with respect to that coordinate, and we adopt $\mathrm{d}v$ in the form of a thin spherical shell of radius r and thickness $\mathrm{d}r$, as shown in Fig. 1.9. The volume of the shell is

$$\mathrm{d}v = 4\pi r^2 \, \mathrm{d}r, \tag{1.33}$$

which can be visualized as the volume of a thin flat slab of the same thickness ($\mathrm{d}r$) and the same surface area ($S = 4\pi r^2$), $\mathrm{d}v = S\,\mathrm{d}r$. The total charge of the sphere comes out to be

$$Q = \int_v \rho \, \mathrm{d}v = \int_{r=0}^{a} \rho_0 \frac{r}{a} \, 4\pi r^2 \, \mathrm{d}r = \rho_0 \pi a^3. \tag{1.34}$$

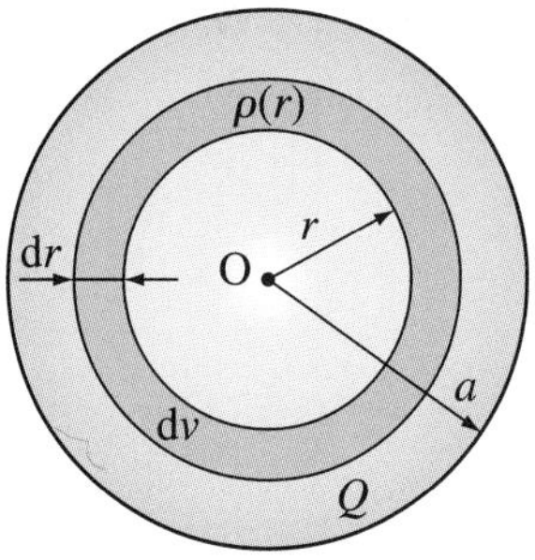

Figure 1.9 Integration of nonuniform volume charge density in a sphere; for Example 1.5.

Problems: 1.9–1.12.

1.4 ON THE VOLUME AND SURFACE INTEGRATION

A few additional comments about the volume integration performed in Eq. (1.34), and similar multiple integrations, may be useful at this point. In general, our strategy for solving volume integrals, $\int_v f \, \mathrm{d}v$, is to adopt as large volume elements $\mathrm{d}v$ as possible, the only restriction being the condition that $f = \text{const}$ in $\mathrm{d}v$. For instance, for the function $\rho = \rho(r)$ in Eq. (1.32) and a sphere as the domain of integration, the largest volume element over which $\rho = \text{const}$ is a thin spherical shell, with $\mathrm{d}v$ given in Eq. (1.33). This adoption enables us to perform the volume integration in Eq. (1.34) by integrating along r only, from 0 to a, whereas the adoption of the standard differential volume element (elementary curvilinear cuboid) in the spherical coordinate system (Fig. 1.10),

$$\mathrm{d}v' = \mathrm{d}r \, (r \, \mathrm{d}\theta) \, (r \sin\theta \, \mathrm{d}\phi) = r^2 \sin\theta \, \mathrm{d}r \, \mathrm{d}\theta \, \mathrm{d}\phi, \tag{1.35}$$

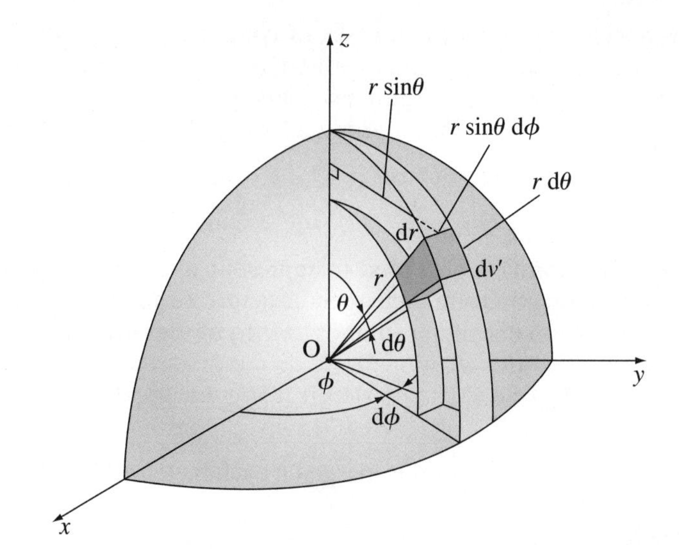

Figure 1.10 Standard differential volume element (dv') in the spherical coordinate system.

would require two additional integrations, i.e., the integration in θ from 0 to π and the integration in ϕ from 0 to 2π. Note, however, that these two integrations are implicitly contained in the expression for dv in Eq. (1.33), as

$$\int_{\theta=0}^{\pi}\int_{\phi=0}^{2\pi} \sin\theta \, d\theta \, d\phi = 4\pi. \tag{1.36}$$

Note also that the elementary cuboid dv' would be necessary for a charge density depending on all three coordinates in the spherical coordinate system, $\rho = \rho(r, \theta, \phi)$. Similar considerations apply to volume integrations associated with the cylindrical and Cartesian coordinate systems.

A trivial example for our integration strategy is the adoption of v in place of dv in cases when $f = \text{const}$ in the entire integration domain v, yielding $\int_v f \, dv = fv$ [e.g., the first expression in Eqs. (1.30)].

The same principle is utilized for adopting surface elements dS for solving surface integrals.

1.5 ELECTRIC FIELD INTENSITY VECTOR DUE TO GIVEN CHARGE DISTRIBUTIONS

Using the superposition principle, the electric field intensity vector due to each of the (uniform or nonuniform) charge distributions ρ, ρ_s, and Q' can be regarded as the vector summation of the field intensities contributed by the numerous equivalent point charges making up the charge distribution. Thus, by replacing Q in Eq. (1.24) with charge element $dQ = \rho \, dv$, $\rho_s \, dS$, or $Q' \, dl$ and integrating, we get

electric field due to volume charge

$$\mathbf{E} = \frac{1}{4\pi\varepsilon_0}\int_v \frac{\rho \, dv}{R^2}\hat{\mathbf{R}}, \tag{1.37}$$

electric field due to surface charge

$$\mathbf{E} = \frac{1}{4\pi\varepsilon_0}\int_S \frac{\rho_s \, dS}{R^2}\hat{\mathbf{R}}, \tag{1.38}$$

electric field due to line charge

$$\mathbf{E} = \frac{1}{4\pi\varepsilon_0}\int_l \frac{Q' \, dl}{R^2}\hat{\mathbf{R}}. \tag{1.39}$$

Note that, in the general case, R and $\hat{\mathbf{R}}$ vary as the integrals in Eqs. (1.37)–(1.39) are evaluated, along with the functions ρ, ρ_s, and Q'. We shall now apply these expressions to some specific charge distributions and field points, for which the integrals can be evaluated analytically. It is very important that we develop analytical skills for solving these (and similar), true three-dimensional (3-D), vector problems. There is not a unique recipe for an optimal solution algorithm. A general advice, however, is to use superposition whenever possible – to break up a complex problem into simpler ones, and then add up (integrate) their solutions to get the solution to the original problem. In doing this, sometimes we use directly the expressions in Eqs. (1.37)–(1.39), which essentially imply breaking up the structure into equivalent point charges. Often, on the other hand, we do not go directly all the way down to point charges. Instead, we decompose the structure and apply superposition "layer by layer," modularly, going down toward simpler structures level by level. In this, we always try to incorporate already known solutions to relevant simpler problems into the solution to the problem under consideration.

Example 1.6 Charged Ring

A line charge of uniform charge density Q' is distributed around the circumference of a ring of radius a in air. Find the electric field intensity vector along the axis of the ring normal to its plane.

Solution We subdivide the ring (contour C) into elemental segments of length $\mathrm{d}l$, as shown in Fig. 1.11, and apply Eq. (1.39). The contribution to the electric field at a point P on the ring axis (z-axis) by each charged segment is

$$\mathrm{d}\mathbf{E} = \frac{Q'\,\mathrm{d}l}{4\pi\varepsilon_0 R^2}\,\hat{\mathbf{R}}, \quad R = \sqrt{z^2 + a^2}, \tag{1.40}$$

with the position of P along the axis being defined by the coordinate z ($-\infty < z < \infty$). The total electric field intensity vector is obtained as

$$\mathbf{E} = \oint_C \mathrm{d}\mathbf{E}. \tag{1.41}$$

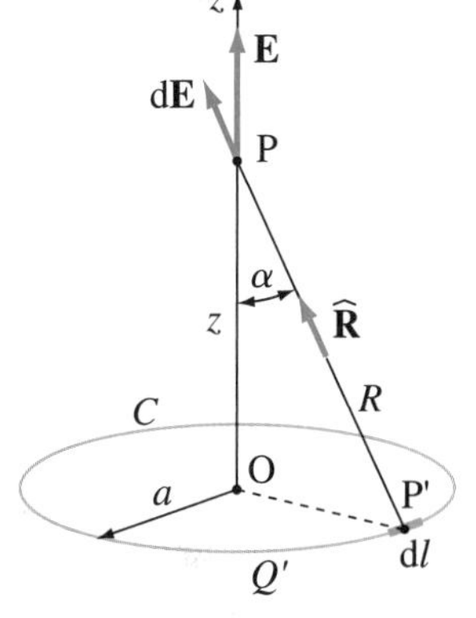

Figure 1.11 Evaluation of the electric field along the axis of a charged ring normal to its plane; for Example 1.6.

Due to symmetry, the horizontal (radial) component of the vector $\mathbf{E}$ is zero (for every element $\mathrm{d}l$ there is a corresponding diametrically opposite element that gives an equal but opposite horizontal electric field component, so that the two contributions cancel each other), and hence $\mathbf{E}$ has a vertical (axial) component only (Fig. 1.11),

$$E = E_z = \oint_C \mathrm{d}E_z, \quad \mathrm{d}E_z = \mathrm{d}E\cos\alpha = \mathrm{d}E\,\frac{z}{R}. \tag{1.42}$$

Substituting the expression for $\mathrm{d}E$ from Eq. (1.40) into Eq. (1.42), we get

$$E_z = \frac{Q'z}{4\pi\varepsilon_0 R^3}\oint_C \mathrm{d}l = \frac{Q'az}{2\varepsilon_0 R^3} \quad \left(\oint_C \mathrm{d}l = l = 2\pi a\right) \tag{1.43}$$

or

$$\boxed{\mathbf{E} = \frac{Qz}{4\pi\varepsilon_0\left(z^2 + a^2\right)^{3/2}}\,\hat{\mathbf{z}},} \tag{1.44}$$

electric field due to a ring of charge

where $Q = Q'2\pi a$ is the total charge of the ring.

Note that for $|z| \gg a$, $z^2 + a^2 \approx z^2$, and Eq. (1.44) yields

$$E \approx \frac{Q}{4\pi\varepsilon_0 z^2} \quad (|z| \gg a). \tag{1.45}$$

This means that far away from the ring, its charge is equivalent to a point charge Q located at its center. In other words, when the distance of the field point from the ring is much larger than the ring dimensions, the ring can be considered as a point charge and the actual shape of the ring (or any other charged body) does not matter. This in fact is the definition of a point charge or a small charged body.

Example 1.7 Semicircular Line Charge

A line charge in the form of a semicircle of radius a is situated in free space, as shown in Fig. 1.12(a). The line charge density is Q'. Compute the electric field intensity vector at an arbitrary point along the z-axis.

Solution The electric field intensity vector along the z-axis due to an elemental charge $Q'\,\mathrm{d}l$ on the semicircle [Fig. 1.12(a)] is given by the expression in Eq. (1.40). To be able to perform the integration in Eq. (1.39), we need to decompose this vector into convenient components. With reference to Fig. 1.12(b), we first represent $\mathbf{dE}$ by its horizontal and vertical components,

$$\mathbf{dE} = \mathbf{dE}_\mathrm{h} + \mathrm{d}E_z\,\hat{\mathbf{z}}, \quad \mathrm{d}E_\mathrm{h} = \mathrm{d}E\sin\alpha, \quad \sin\alpha = \frac{a}{R},$$

$$\mathrm{d}E_z = \mathrm{d}E\cos\alpha, \quad \cos\alpha = \frac{z}{R}. \tag{1.46}$$

Since the direction of the vector $\mathbf{dE}_\mathrm{h}$ varies as the integral is evaluated (i.e., as the point P′ moves along the semicircle), we need to decompose it further [Fig. 1.12(c)],

$$\mathbf{dE}_\mathrm{h} = \mathrm{d}E_x\,\hat{\mathbf{x}} + \mathrm{d}E_y\,\hat{\mathbf{y}}, \quad \mathrm{d}E_x = -\,\mathrm{d}E_\mathrm{h}\cos\phi, \quad \mathrm{d}E_y = -\,\mathrm{d}E_\mathrm{h}\sin\phi \tag{1.47}$$

(the range for the azimuthal angle ϕ is $-\pi/2 \le \phi \le \pi/2$). From the above expressions and the relationship $\mathrm{d}l = a\,\mathrm{d}\phi$ (segment $\mathrm{d}l$ is an arc of radius a defined by the angle $\mathrm{d}\phi$, and its length thus equals the radius times the angle), we obtain

$$E_x = \int_l \mathrm{d}E_x = -\frac{Q'a^2}{4\pi\varepsilon_0 R^3}\int_{-\pi/2}^{\pi/2}\cos\phi\,\mathrm{d}\phi = -\frac{Q'a^2}{2\pi\varepsilon_0 R^3}, \tag{1.48}$$

$$E_y = \int_l \mathrm{d}E_y = -\frac{Q'a^2}{4\pi\varepsilon_0 R^3}\int_{-\pi/2}^{\pi/2}\sin\phi\,\mathrm{d}\phi = 0, \tag{1.49}$$

$$E_z = \int_l \mathrm{d}E_z = \frac{Q'az}{4\pi\varepsilon_0 R^3}\int_{-\pi/2}^{\pi/2}\mathrm{d}\phi = \frac{Q'az}{4\varepsilon_0 R^3}, \tag{1.50}$$

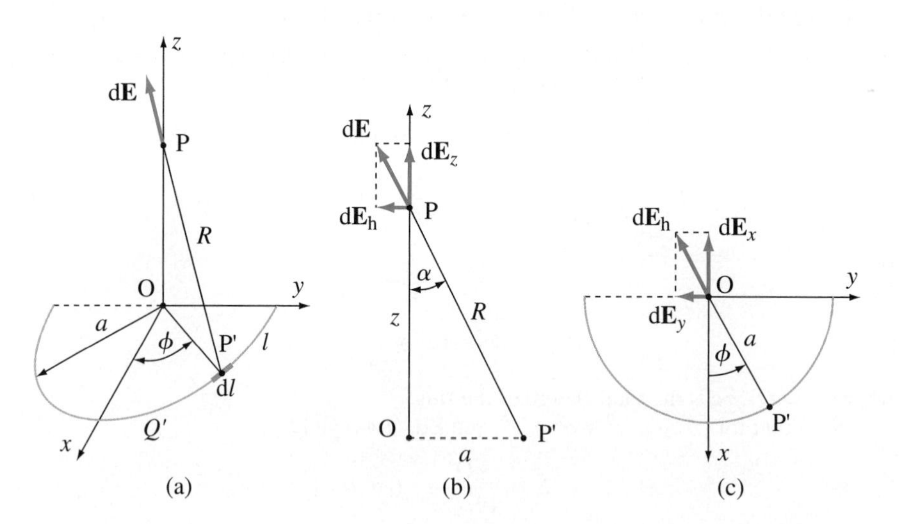

Figure 1.12 Evaluation of the electric field due to a line charge in the form of a semicircle; for Example 1.7.

so that the final expression for the electric field vector is

$$\mathbf{E} = \frac{Q'a}{2\varepsilon_0 R^3}\left(-\frac{a}{\pi}\,\hat{\mathbf{x}} + \frac{z}{2}\,\hat{\mathbf{z}}\right), \quad R = \sqrt{z^2 + a^2}. \tag{1.51}$$

Example 1.8 Straight Line Charge of Finite Length

Find the expression for the **E** field at an arbitrary point in space due to a straight line of length l uniformly charged with a total charge Q. The ambient medium is air.

Solution Let the line charge and the field point (P) be in the plane of drawing (Fig. 1.13). The geometry of this problem can be defined using three parameters: angles θ_1 and θ_2, and the perpendicular distance from the line to the point, d (for the particular position of the point P shown in Fig. 1.13, $\theta_1 < 0$ and $\theta_2 > 0$). In Eq. (1.39), $Q' = Q/l$, $R = \sqrt{z^2 + d^2}$, $dl = dz$, and

$$\hat{\mathbf{R}} = \cos\theta\,\hat{\mathbf{x}} - \sin\theta\,\hat{\mathbf{z}}, \tag{1.52}$$

where z ($z_1 \le z \le z_2$) and θ ($\theta_1 \le \theta \le \theta_2$) are the length and angular coordinates, respectively, each determining the position of the source point, P′, along the line. From Fig. 1.13, the following trigonometric relationship exists between these two coordinates:

$$\tan\theta = \frac{z}{d}, \tag{1.53}$$

and its differential form (by taking the differential of both sides of the equation) reads

$$\frac{\mathrm{d}\theta}{\cos^2\theta} = \frac{\mathrm{d}z}{d} \quad (d = \text{const}), \tag{1.54}$$

which, given that $\cos\theta = d/R$, is equivalent to

$$\frac{\mathrm{d}z}{R^2} = \frac{\mathrm{d}\theta}{d}. \tag{1.55}$$

Finally, substituting the expressions from Eqs. (1.52) and (1.55) into Eq. (1.39), we have

$$\mathbf{E} = \frac{Q}{4\pi\varepsilon_0 l d}\left(\int_{\theta_1}^{\theta_2}\cos\theta\,\mathrm{d}\theta\,\hat{\mathbf{x}} - \int_{\theta_1}^{\theta_2}\sin\theta\,\mathrm{d}\theta\,\hat{\mathbf{z}}\right)$$

$$\longrightarrow \quad \boxed{\mathbf{E} = \frac{Q}{4\pi\varepsilon_0 l d}\left[(\sin\theta_2 - \sin\theta_1)\,\hat{\mathbf{x}} + (\cos\theta_2 - \cos\theta_1)\,\hat{\mathbf{z}}\right].} \tag{1.56}$$

electric field due to a line charge of finite length

Note that the expression in Eq. (1.56) can be combined for computing the electric field due to an arbitrary structure assembled from straight line segments of charge.

Example 1.9 Infinite Line Charge

An infinite line charge of uniform density Q' resides in air. Determine the electric field intensity vector at an arbitrary point in space.

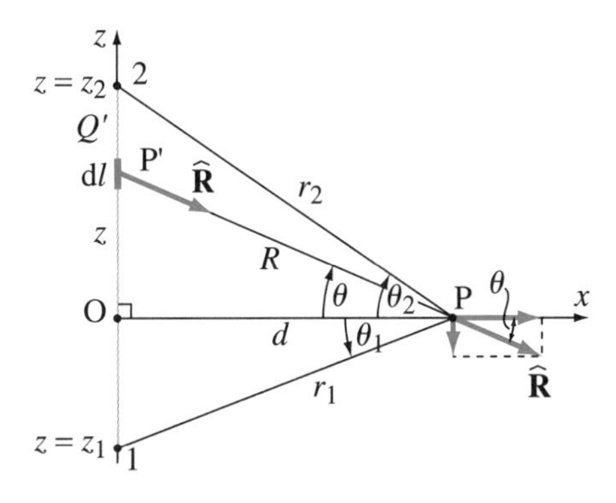

Figure 1.13 Evaluation of the electric field due to a line charge of finite length; for Example 1.8.

Solution Let r be the perpendicular (radial) distance of a field (observation) point, P, from the line charge, and $\hat{\mathbf{r}}$ the unit vector along this distance directed from the line to the point P. We use the expression for the field due to a finite line charge in Eq. (1.56) and let $\theta_1 \to -\pi/2$ and $\theta_2 \to \pi/2$ (the line extends to $z \to -\infty$ and $z \to \infty$, respectively), as well as $Q/l = Q'$, $d = r$, and $\hat{\mathbf{x}} = \hat{\mathbf{r}}$. What we obtain is exactly the field expression for the infinite line charge with density Q':

electric field due to an infinite line charge

$$\boxed{\mathbf{E} = \frac{Q'}{2\pi\varepsilon_0 r}\,\hat{\mathbf{r}}.} \tag{1.57}$$

We see that the vector $\mathbf{E}$ is radial with respect to the line charge axis, and its magnitude varies in space as a function of the radial distance r only, with E being inversely proportional to r.

Example 1.10 Disk of Charge

Consider a very thin charged disk (i.e., a circular sheet of charge), of radius a and a uniform surface charge density ρ_s, in free space. Calculate the electric field intensity vector along the disk axis normal to its surface.

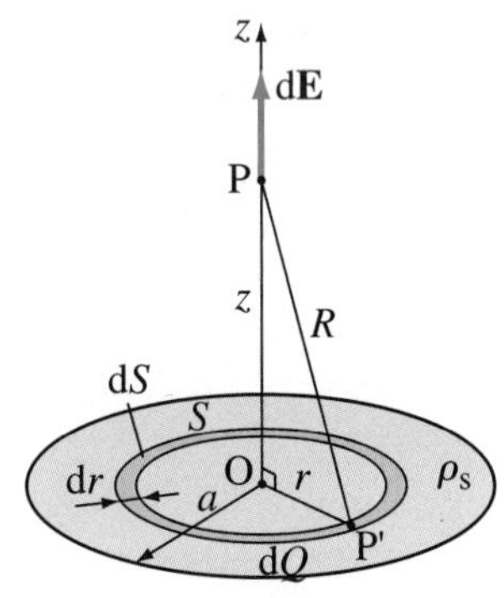

Figure 1.14 Evaluation of the electric field due to a charged disk; for Example 1.10.

Solution Instead of directly applying Eq. (1.38), we subdivide the disk into elemental rings of width $\mathrm{d}r$, as shown in Fig. 1.14. The field due to a ring of radius r $(0 \le r \le a)$ at a point P along the z-axis is [see Eq. (1.44)]

$$\mathrm{d}\mathbf{E} = \frac{\mathrm{d}Qz}{4\pi\varepsilon_0 R^3}\,\hat{\mathbf{z}}, \quad R = \sqrt{r^2 + z^2}, \tag{1.58}$$

with $\mathrm{d}Q$ standing for the charge of the ring, given by

$$\mathrm{d}Q = \rho_s\,\mathrm{d}S. \tag{1.59}$$

The surface area of the ring, $\mathrm{d}S$, can be computed as the area of a thin strip of length equal to the ring circumference, $2\pi r$, and width $\mathrm{d}r$, that is,

$$\mathrm{d}S = 2\pi r\,\mathrm{d}r. \tag{1.60}$$

By superposition,

$$\mathbf{E} = \int_S \mathrm{d}\mathbf{E} = \frac{\rho_s z}{2\varepsilon_0}\,\hat{\mathbf{z}} \int_{r=0}^{a} \frac{r\,\mathrm{d}r}{R^3}. \tag{1.61}$$

Taking the differential of the relationship $R^2 = r^2 + z^2$, we obtain

$$R\,\mathrm{d}R = r\,\mathrm{d}r, \tag{1.62}$$

so that the substitution of $r\,\mathrm{d}r$ by $R\,\mathrm{d}R$ in Eq. (1.61) yields

electric field due to a charged disk

$$\mathbf{E} = \frac{\rho_s z}{2\varepsilon_0} \int_{r=0}^{a} \frac{\mathrm{d}R}{R^2}\,\hat{\mathbf{z}} = \frac{\rho_s z}{2\varepsilon_0}\left(-\frac{1}{R}\right)\Bigg|_{r=0}^{a}\,\hat{\mathbf{z}} \quad \longrightarrow \quad \boxed{\mathbf{E} = \frac{\rho_s}{2\varepsilon_0}\left(\frac{z}{|z|} - \frac{z}{\sqrt{a^2 + z^2}}\right)\hat{\mathbf{z}},} \tag{1.63}$$

where $R|_{r=0} = \sqrt{z^2} = |z|$, to allow a negative z as well $(-\infty < z < \infty)$.

Example 1.11 Infinite Sheet of Charge

Find the electric field intensity vector due to an infinite sheet of charge of density ρ_s in free space.

Solution We note that the infinite sheet of charge can be obtained from the disk in Fig. 1.14 by extending its radius to infinity. Consequently, the field due to the infinite sheet can be obtained from the field due to the disk by letting $a \to \infty$ in Eq. (1.63). With this, the second term in parentheses in the final field expression in Eq. (1.63) vanishes. The first term, $z/|z|$, is either 1 (for $z > 0$) or -1 (for $z < 0$), and we conclude that the field intensity around the infinite sheet of charge is uniform (constant) in both half-spaces cut by the sheet, and given by

$$\boxed{E = \frac{\rho_s}{2\varepsilon_0}} \qquad (1.64)$$

electric field due to an infinite sheet of charge

with respect to the reference directions indicated in Fig. 1.15. Namely, for a positive charge ($\rho_s > 0$) of the sheet, the actual orientations of **E** are those in Fig. 1.15 (field lines are directed outward from the positive charge), while the situation for a negative ρ_s is just opposite (the field vector points toward the negative charge).

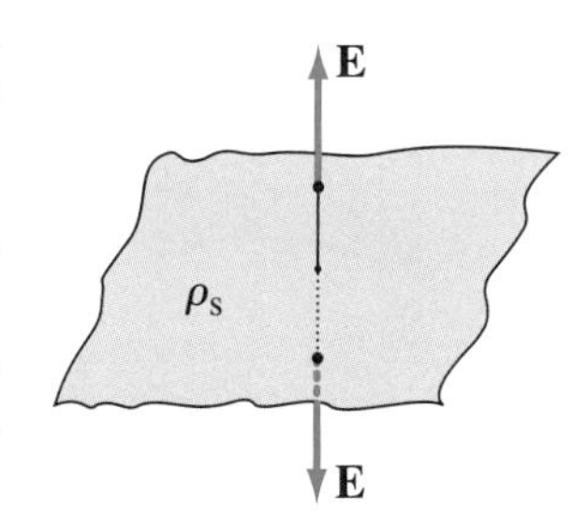

Figure 1.15 Infinite sheet of charge; for Example 1.11.

Example 1.12 Hemispherical Surface Charge

A hemispherical surface of radius a is uniformly charged with a charge density ρ_s. The medium is air. Compute the electric field intensity vector at the center of the hemisphere (center of the corresponding full sphere).

Solution We perform a similar procedure as in Example 1.10 and subdivide the hemisphere into thin rings, as depicted in Fig. 1.16. The radius of a ring whose position on the hemisphere is defined by an angle θ ($0 \le \theta \le \pi/2$) is $a_r = a \sin\theta$. Its charge is

$$\mathrm{d}Q = \rho_s\,\mathrm{d}S = \rho_s \underbrace{2\pi a \sin\theta}_{C_r}\,\underbrace{a\,\mathrm{d}\theta}_{\mathrm{d}l_r}, \qquad (1.65)$$

where C_r and $\mathrm{d}l_r$ denote the ring circumference and width, respectively. Using Eq. (1.44), the electric field intensity $\mathbf{dE}$ at the point O due to the charge $\mathrm{d}Q$ is found as follows:

$$z_r = -a\cos\theta \quad \text{and} \quad R = a \quad \longrightarrow \quad \mathrm{d}\mathbf{E} = \frac{\mathrm{d}Q(-a\cos\theta)}{4\pi\varepsilon_0 a^3}\,\hat{\mathbf{z}}, \qquad (1.66)$$

with z_r being the local z-coordinate of the point O with respect to the ring center (Fig. 1.16). Therefore, the resultant field vector amounts to

$$\mathbf{E} = \int_{\theta=0}^{\pi/2} \mathrm{d}\mathbf{E} = -\frac{\rho_s}{2\varepsilon_0}\int_0^{\pi/2} \underbrace{\sin\theta\cos\theta}_{(\sin 2\theta)/2}\,\mathrm{d}\theta\,\hat{\mathbf{z}} = -\frac{\rho_s}{4\varepsilon_0}\,\hat{\mathbf{z}}. \qquad (1.67)$$

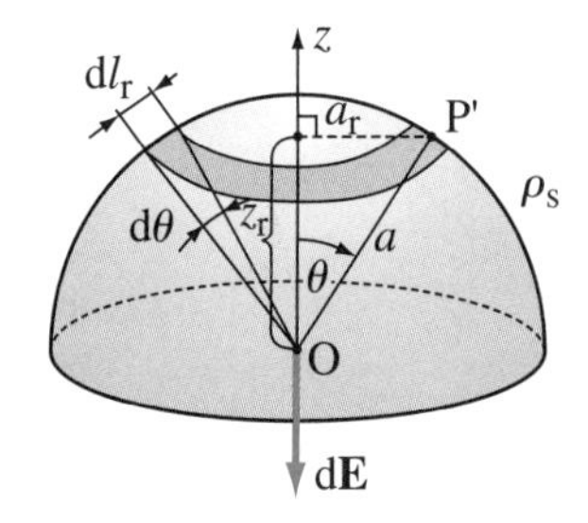

Figure 1.16 Evaluation of the electric field at the center (point O) of a hemispherical surface charge; for Example 1.12.

Example 1.13 Force on a Line Charge due to a Charged Semicylinder

A uniform charge of density ρ_s is distributed in free space over a surface in the form of a half of a very long circular cylinder of radius a, as shown in Fig. 1.17(a). A line charge of uniform density Q' is positioned along the semicylinder axis. Find the electric force on the line charge per unit of its length.

Solution From Eqs. (1.23) and (1.31), the per-unit-length electric force on the line charge, that is, the force on one meter (unit length) of the structure (line charge) divided by 1 m (the unit is N/m), is given by

$$\boxed{\mathbf{F}'_e = Q'\mathbf{E},} \qquad (1.68)$$

per-unit-length Coulomb force on a line charge (unit: N/m)

with **E** standing for the electric field intensity vector due to the charged semicylinder at its axis. We subdivide the semicylinder into elemental strips, of width $\mathrm{d}l$, and find **E** using the superposition principle. The field due to each strip can be approximated by that of a very long line charge with charge density $\mathrm{d}Q' = \rho_s\,\mathrm{d}l$; namely, the charge of the elemental strip of length h, $\rho_s h\,\mathrm{d}l$, must be equal to the charge of the equivalent line charge of the same length, so $\mathrm{d}Q'h$, which yields this expression for $\mathrm{d}Q'$. Having in mind Eq. (1.57), the electric field vector due to the line charge is radial with respect to the line, and its magnitude at the semicylinder axis [Fig. 1.17(b)] comes out to be

$$\mathrm{d}E = \frac{\mathrm{d}Q'}{2\pi\varepsilon_0 a} = \frac{\rho_s\,\mathrm{d}\phi}{2\pi\varepsilon_0}, \qquad (1.69)$$

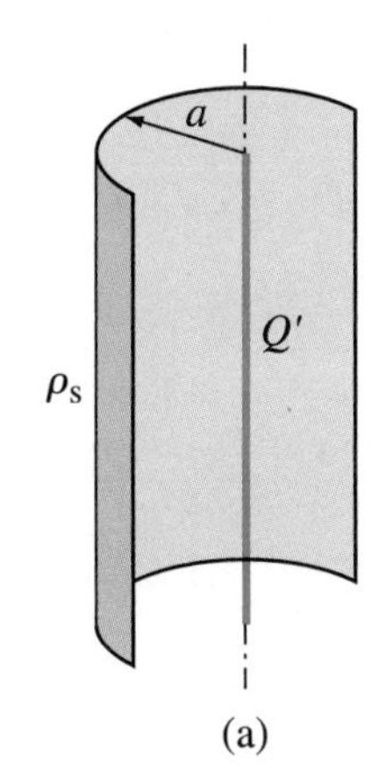

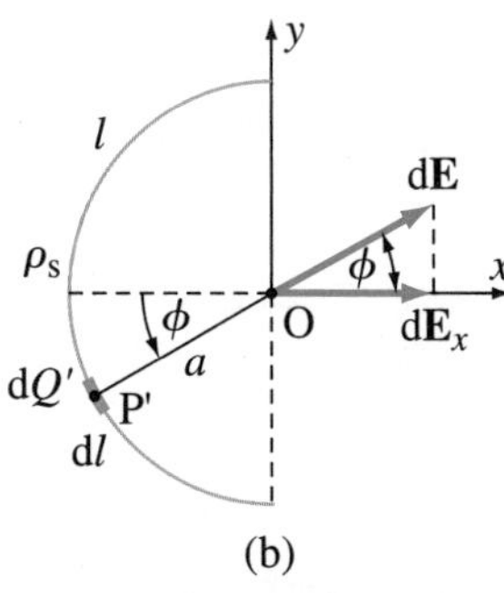

Figure 1.17 Evaluation of the electric force on a line charge positioned along the axis of a charged semicylinder: (a) three-dimensional view showing the structure geometry and (b) cross-sectional view for field computations; for Example 1.13.

since $r = a$ and $dl = a\,d\phi$. By symmetry, the resultant field has an x-component only, computed as

$$E_x = \int_l dE_x = \int_l dE\cos\phi = \frac{\rho_s}{2\pi\varepsilon_0}\int_{\phi=-\pi/2}^{\pi/2}\cos\phi\,d\phi = \frac{\rho_s}{\pi\varepsilon_0}, \tag{1.70}$$

where l denotes the line (semicircle) representing the semicylinder cross section. Finally, Eqs. (1.68) and (1.70) result in

$$\mathbf{F}'_e = \frac{Q'\rho_s}{\pi\varepsilon_0}\,\hat{\mathbf{x}}. \tag{1.71}$$

Problems: 1.13–1.28; *Conceptual Questions* (on Companion Website): 1.3–1.6; *MATLAB Exercises* (on Companion Website).

1.6 DEFINITION OF THE ELECTRIC SCALAR POTENTIAL

The electric scalar potential is a scalar quantity that can be used instead of the electric field intensity vector for the description of the electrostatic field. Dealing with the potential is mathematically simpler than dealing with the field vector. Three different integrations are needed in general for the evaluation of **E**, one integration for each vector component, while a single integration is required for the potential, and **E**, in turn, can easily be found from the potential by differentiation. In addition, using the electric potential we are able to connect the electric field with the voltage, as a fundamental bridge between the field theory and the circuit theory.

The electric scalar potential is defined through the work done by the electric field, that is, by the electric force, $\mathbf{F}_e$, in moving a test point charge, Q_p. The work dW_e done by $\mathbf{F}_e$ while moving Q_p along an elementary vector distance $d\mathbf{l}$ equals the dot product[8] of $\mathbf{F}_e$ and $d\mathbf{l}$,

$$dW_e = \mathbf{F}_e\cdot d\mathbf{l} = F_e\,dl\cos\alpha, \tag{1.72}$$

where α is the angle between the two vectors in the product, as shown in Fig. 1.18. We note that dW_e can be negative (for $\pi/2 < \alpha \le \pi$), meaning that the work $|dW_e|$ is being performed against the electric field by an external agent. Based on Eq. (1.72), the electric scalar potential, V, at a point P in an electric field is defined as the work done by the field in moving a test charge from P to a reference point $\mathcal{R}$ (Fig. 1.18),

$$W_e = \int_P^{\mathcal{R}} \mathbf{F}_e\cdot d\mathbf{l}, \tag{1.73}$$

divided by Q_p. Having in mind Eq. (1.23), this becomes

definition of the electric potential (unit: V)

$$\boxed{V = \frac{W_e}{Q_p} = \int_P^{\mathcal{R}} \frac{\mathbf{F}_e}{Q_p}\cdot d\mathbf{l} = \int_P^{\mathcal{R}} \mathbf{E}\cdot d\mathbf{l}.} \tag{1.74}$$

Thus, the potential V at the point P with respect to the reference point $\mathcal{R}$ turns out to not depend on Q_p and to be equal to the line integral of vector **E** from P to $\mathcal{R}$.[9]

[8]The dot product (also known as the scalar product) of vectors **a** and **b** is a scalar given by $\mathbf{a}\cdot\mathbf{b} = |\mathbf{a}||\mathbf{b}|\cos\alpha$, α being the angle between **a** and **b**.

[9]The line integral of a vector function (field) **a** along a line (curve) l, from a point A to a point B, is defined as $\int_l \mathbf{a}\cdot d\mathbf{l} = \int_A^B \mathbf{a}\cdot d\mathbf{l}$, where $d\mathbf{l}$ is the differential length vector tangential to the curve (as in

Obviously, the potential at the reference point is zero (integral from $P \equiv \mathcal{R}$ to $\mathcal{R}$). The unit for the potential is volt (abbreviated V), and hence the unit V/m for the electric field intensity. Note that Φ is also used to denote the electric potential.

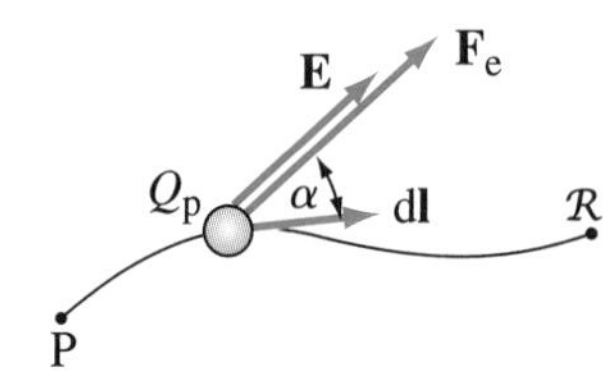

Figure 1.18 Displacement of a test charge in an electrostatic field.

By the principle of conservation of energy, the net work done by the electrostatic field in moving Q_p from a point A to some point B and then moving it back to A along a different path is zero (because after the round trip, the system is the same as at the beginning). This means that the line integral of the electric field intensity vector along an arbitrary closed path (contour) is zero,

conservative nature of the electrostatic field

$$\oint_C \mathbf{E} \cdot d\mathbf{l} = 0, \tag{1.75}$$

which constitutes Maxwell's first equation for the electrostatic field. We see that the circulation (closed line integral) of $\mathbf{E}$ in electrostatics is always zero, and hence the electrostatic field belongs to a class of conservative vector fields.[10]

Eq. (1.75) can alternatively be derived from Coulomb's law, i.e., from the expression for the electric field intensity vector due to a single point charge, Q, given by Eq. (1.24) and Fig. 1.7. To do this, we break the contour C up into elemental segments ($dl \to 0$) parallel and normal to $\mathbf{E}$, as indicated in Fig. 1.19. We realize that contributions to the overall line integral in Eq. (1.75) occur only for the segments parallel to $\mathbf{E}$, while no contribution for the segments normal to $\mathbf{E}$ ($\alpha = 90°$ and $\mathbf{E} \cdot d\mathbf{l} = 0$). In specific, along the segments parallel to $\mathbf{E}$ (segments radial with respect to the charge Q), $\mathbf{E} \cdot d\mathbf{l}$ equals either $E\,dl$ ($\alpha = 0$) or $-E\,dl$ ($\alpha = 180°$), resulting in a zero net line integral (sum) along C. By superposition, this result is

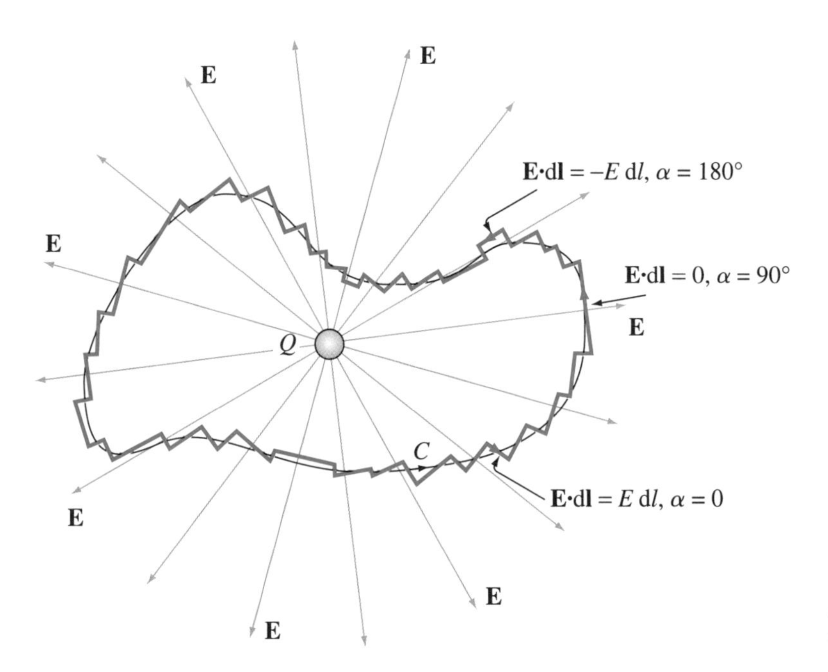

Figure 1.19 Derivation of Maxwell's first equation for the electrostatic field starting with Coulomb's law.

Fig. 1.18) oriented from A toward B. If the line is closed (for example, a circle or a square), we call it contour (and usually mark it C), and the corresponding line integral, $\oint_C \mathbf{a} \cdot d\mathbf{l}$, is termed the circulation of $\mathbf{a}$ along C. The reference direction of $d\mathbf{l}$ coincides with the orientation of the contour.

[10]By definition, a vector field is said to be conservative when its circulation is zero for a closed path of arbitrary shape.

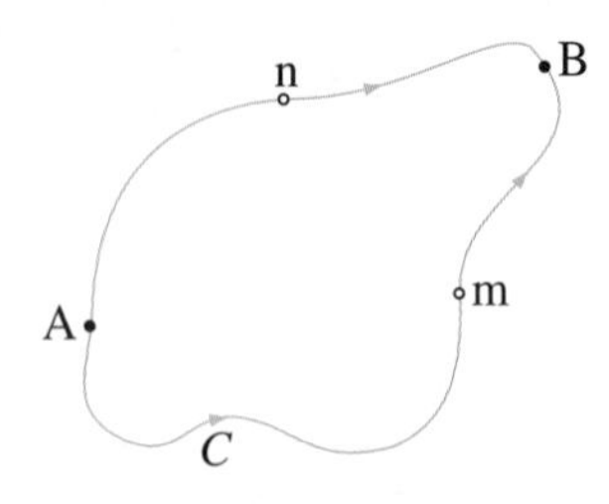

Figure 1.20 For the proof that the line integral of an electrostatic field vector between two points is the same for any path of integration.

also valid for any charge distribution (which, as we know, can be represented as a system of equivalent point charges, and for which the resultant field can be obtained as a vector sum of individual fields due to point charges).

The line integral of **E** between two points in an electrostatic field does not depend on the path of integration. To prove this, let us refer to Fig. 1.20 and write the circulation of **E** along a contour C as

$$\oint_C \mathbf{E} \cdot d\mathbf{l} = \int_{\mathrm{AmBnA}} \mathbf{E} \cdot d\mathbf{l} = \int_{\mathrm{AmB}} \mathbf{E} \cdot d\mathbf{l} + \int_{\mathrm{BnA}} \mathbf{E} \cdot d\mathbf{l}$$
$$= \int_{\mathrm{AmB}} \mathbf{E} \cdot d\mathbf{l} - \int_{\mathrm{AnB}} \mathbf{E} \cdot d\mathbf{l}, \tag{1.76}$$

so that, from Eq. (1.75), the integral from A to B along the path containing the point m is the same as the integral along the path with the point n,

path independence for the line integral of **E**

$$\boxed{\int_{\mathrm{AmB}} \mathbf{E} \cdot d\mathbf{l} = \int_{\mathrm{AnB}} \mathbf{E} \cdot d\mathbf{l},} \tag{1.77}$$

or along any other path. This also means that, for an adopted reference point, the electric potential at any point in the field is a uniquely determined quantity, having the same value for any path of moving the test charge.

Finally, let us see what happens with the potential at an arbitrary point P in the field after a new reference point, $\mathcal{R}'$, is adopted. The new potential is

$$V' = \int_{\mathrm{P}}^{\mathcal{R}'} \mathbf{E} \cdot d\mathbf{l} = \underbrace{\int_{\mathrm{P}}^{\mathcal{R}} \mathbf{E} \cdot d\mathbf{l}}_{V} + \int_{\mathcal{R}}^{\mathcal{R}'} \mathbf{E} \cdot d\mathbf{l}, \tag{1.78}$$

and thus the change in potential amounts to

change in potential due to a new reference point

$$\boxed{\Delta V = V' - V = \int_{\mathcal{R}}^{\mathcal{R}'} \mathbf{E} \cdot d\mathbf{l}.} \tag{1.79}$$

It is very important to note that ΔV does not depend on the position of the point P in the field; if the reference point is changed, the potential at all points changes by the same constant value, i.e., by an additive constant.

Problems: 1.29 and 1.30; *Conceptual Questions* (on Companion Website): 1.7 and 1.8; *MATLAB Exercises* (on Companion Website).

1.7 ELECTRIC POTENTIAL DUE TO GIVEN CHARGE DISTRIBUTIONS

Let us determine the electric scalar potential due to a point charge in free space. In general, it is customary to adopt the reference point, $\mathcal{R}$, at infinity whenever possible, namely, when the charge distribution is finite. As we shall see in an example, this choice is impossible, or at least inadequate, for infinite charge distributions, such as an infinite straight line charge. From Fig. 1.21 and Eqs. (1.74) and (1.24), the potential at a distance R from the charge Q, and with respect to the reference point at infinity, is

electric potential due to a point charge

$$V = \int_{\mathrm{P}}^{\mathcal{R}} \mathbf{E} \cdot d\mathbf{l} = \int_{x=R}^{\infty} E\, dx = \int_{R}^{\infty} \frac{Q}{4\pi\varepsilon_0 x^2}\, dx \quad \longrightarrow \quad \boxed{V = \frac{Q}{4\pi\varepsilon_0 R}.} \tag{1.80}$$

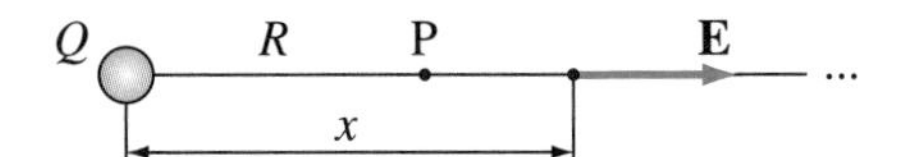

Figure 1.21 Evaluation of the electric potential due to a point charge in free space.

The potential function (of R) obtained for a reference point at an arbitrary (finite) distance from the charge differs from this result by an additive constant, as determined by Eq. (1.79).

Shown in Fig. 1.22 is a sketch of electric-field lines (in general, lines to which vector $\mathbf{E}$ is tangential at all points) and equipotential surfaces (surfaces having the same potential, $V = \text{const}$, at all points) around the point charge in Fig. 1.21. Of course, based on Eqs. (1.24) and (1.80), field lines are radial "beams" starting at the charge and equipotentials are spherical surfaces (graphed as circles – equipotential lines – in the figure) centered at the charge, respectively. Field lines are perpendicular to equipotential surfaces, and this conclusion holds always, for an arbitrary electrostatic field. Namely, since a movement of a probe charge Q_p (in Fig. 1.18) over an equipotential surface results in no work by electric forces (no change in V), $\mathbf{E}$ does not have a component along that movement, meaning that it is perpendicular to the surface. The density of field lines (the number of lines per unit area of the equipotential surface) is representative of the magnitude of the field vector, i.e., it decreases as $1/R^2$ (the area of equipotential surfaces increases as R^2) away from the charge in Fig. 1.22.

Starting with Eq. (1.80) and applying the superposition principle, the expression for the resultant electric potential, similar to the field expression in Eq. (1.25), is obtained for the system of N point charges, which reads

$$V = \frac{1}{4\pi\varepsilon_0}\sum_{i=1}^{N}\frac{Q_i}{R_i}, \tag{1.81}$$

as well as for the three characteristic continuous charge distributions, corresponding to Eqs. (1.37)–(1.39) for the field vector,

$$\boxed{V = \frac{1}{4\pi\varepsilon_0}\int_v \frac{\rho\,\mathrm{d}v}{R},} \tag{1.82}$$

electric potential due to volume charge

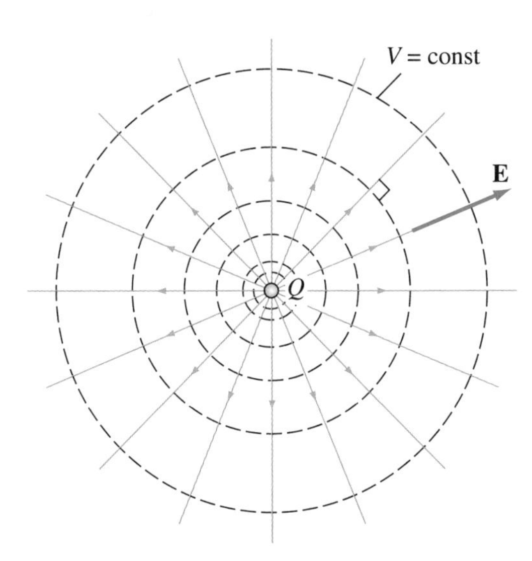

Figure 1.22 Field and equipotential lines for a point charge in Fig. 1.21.

electric potential due to surface charge

$$V = \frac{1}{4\pi\varepsilon_0}\int_S \frac{\rho_s\,dS}{R}, \tag{1.83}$$

electric potential due to line charge

$$V = \frac{1}{4\pi\varepsilon_0}\int_l \frac{Q'\,dl}{R}. \tag{1.84}$$

Obviously, these integrals are substantially simpler than the respective field integrals, and the same holds true for the resulting solutions for the potential due to charge distributions on various characteristic geometries, for which we have already evaluated the electric field vector in Section 1.5.

Example 1.14 Electric Potential due to a Charged Ring

Find the electric scalar potential along the axis of a uniformly charged ring in free space normal to the ring plane. The line charge density of the ring is Q' and its radius is a.

Solution From Eq. (1.84) and Fig. 1.11, the potential at an arbitrary point P on the z-axis (for $-\infty < z < \infty$) is given by

$$V = \frac{1}{4\pi\varepsilon_0}\oint_C \frac{Q'\,dl}{R} = \frac{Q'}{4\pi\varepsilon_0 R}\oint_C dl = \frac{Q'a}{2\varepsilon_0\sqrt{z^2+a^2}}. \tag{1.85}$$

Example 1.15 Potential due to an Infinite Line Charge

Compute the electric potential at an arbitrary point in space due to an infinite uniform line charge of density Q' in air.

Solution The expressions for potential computations given in Eqs. (1.81)–(1.84) all imply that the reference point is at infinity. These expressions cannot therefore be used for the evaluation of the potential due to infinite charge distributions (because the potential due to an infinite charge distribution with respect to the reference point taken at infinity is infinite). In such cases, we invoke Eq. (1.74) instead – to compute the potential, and an infinite straight line charge, the field of which is given by Eq. (1.57), is a typical, and a very important, example. To find the potential due to this charge, at a point whose radial distance from the charge line is r, we apply Eq. (1.74) to a convenient integration path shown in Fig. 1.23, which

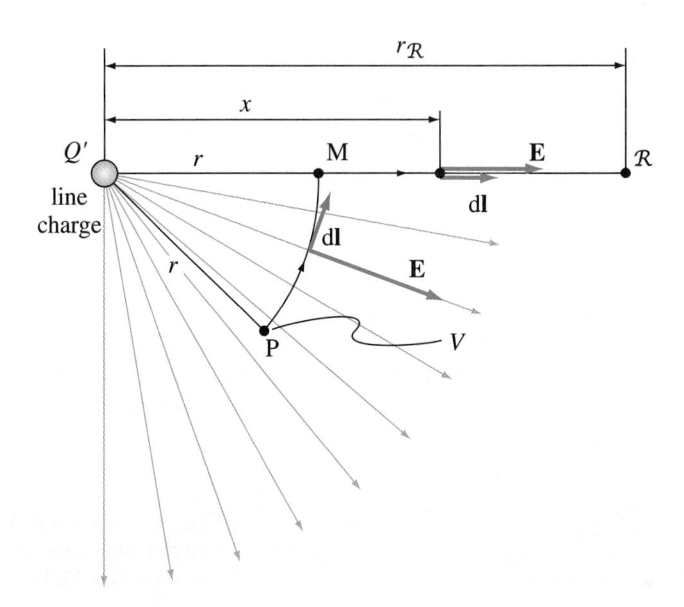

Figure 1.23 Evaluation of the electric potential due to an infinite line charge – cross section of the structure; for Example 1.15.

consists of an arc between points P and M and a straight line segment between points M and $\mathcal{R}$, and obtain

$$V = \int_{\mathrm{P}}^{\mathcal{R}} \mathbf{E} \cdot \mathrm{d}\mathbf{l} = \int_{\mathrm{P}}^{\mathrm{M}} \mathbf{E} \cdot \mathrm{d}\mathbf{l} + \int_{\mathrm{M}}^{\mathcal{R}} \mathbf{E} \cdot \mathrm{d}\mathbf{l} = \int_{x=r}^{r_{\mathcal{R}}} E\,\mathrm{d}x = \int_{r}^{r_{\mathcal{R}}} \frac{Q'}{2\pi\varepsilon_0 x}\,\mathrm{d}x. \tag{1.86}$$

The line integral between points P and M is zero because $\mathbf{E}$ is perpendicular to $\mathrm{d}\mathbf{l}$ along the path. Thus, the potential at a distance r from the line charge with respect to the reference point that is a distance $r_{\mathcal{R}}$ away from it comes out to be

$$\boxed{V = \frac{Q'}{2\pi\varepsilon_0} \ln \frac{r_{\mathcal{R}}}{r}.} \tag{1.87}$$

electric potential due to an infinite line charge

Note that the adoption of the reference point at infinity in this case, which implies that $r_{\mathcal{R}} \to \infty$ in Eq. (1.87), would result in $V \to \infty$. Note, furthermore, that this result, so an infinite potential regardless of the location of the observation point, is also "correct" in a sense that it can be understood as the potential distribution (function of r) given by Eq. (1.87) plus an infinite constant, as a consequence of the change of the reference point from $\mathcal{R}$ in Fig. 1.23 to infinity [see Eqs. (1.78) and (1.79)]. In other words, the potential at all points in space is changed by the same constant value after the new reference point (at infinity) is adopted, which, being infinite ($\Delta V \to \infty$), masks the correct potential distribution. However, this result, although "correct," is useless for the analysis, since the actual function $V(r)$ cannot be extracted from an infinite additive constant, and that is why we say that in potential evaluations due to infinite charge distributions reference point cannot be taken at infinity.

Problems: 1.31–1.36; *MATLAB Exercises* (on Companion Website).

1.8 VOLTAGE

By definition, the voltage between two points is the potential difference between them. The unit is V. The voltage between points A and B is denoted by V_{AB}, so we have

$$\boxed{V_{\mathrm{AB}} = V_{\mathrm{A}} - V_{\mathrm{B}},} \tag{1.88}$$

definition of voltage (unit: V)

where V_{A} and V_{B} are the potentials at point A and point B, respectively, with respect to the same reference point.

Combining Eqs. (1.74) and (1.88), we get

$$V_{\mathrm{AB}} = \int_{\mathrm{A}}^{\mathcal{R}} \mathbf{E} \cdot \mathrm{d}\mathbf{l} - \int_{\mathrm{B}}^{\mathcal{R}} \mathbf{E} \cdot \mathrm{d}\mathbf{l} = \int_{\mathrm{A}}^{\mathcal{R}} \mathbf{E} \cdot \mathrm{d}\mathbf{l} + \int_{\mathcal{R}}^{\mathrm{B}} \mathbf{E} \cdot \mathrm{d}\mathbf{l} \tag{1.89}$$

or

$$\boxed{V_{\mathrm{AB}} = \int_{\mathrm{A}}^{\mathrm{B}} \mathbf{E} \cdot \mathrm{d}\mathbf{l}.} \tag{1.90}$$

voltage via a line integral of **E**

Therefore, the voltage between two points in an electrostatic field equals, in turn, the line integral of the electric field intensity vector along any path between these points. Obviously, $V_{\mathrm{BA}} = -V_{\mathrm{AB}}$.

If a new reference point is adopted for the electric potential in a system, voltages in the system remain unchanged. To prove this statement, we recall that the potential at all points in the system changes by the same constant value (ΔV) after the new reference point is adopted [see Eq. (1.79)]. Since a voltage is the difference of the potential values at the respective points in the system, Eq. (1.88), the new voltage between points A and B is given by

$$V'_{\mathrm{AB}} = V'_{\mathrm{A}} - V'_{\mathrm{B}} = (V_{\mathrm{A}} + \Delta V) - (V_{\mathrm{B}} + \Delta V) = V_{\mathrm{A}} - V_{\mathrm{B}} = V_{\mathrm{AB}}, \tag{1.91}$$

i.e., it equals the old voltage between these points.

Note that, in terms of voltages, Eq. (1.75) can be written as

Kirchhoff's voltage law

$$\sum_{\text{along } C} V_i = 0, \tag{1.92}$$

which, if applied to a closed path in a dc circuit, represents Kirchhoff's circuital law for voltages. This law tells us that the algebraic sum of voltages around any closed path in a circuit is zero.[11] Kirchhoff's voltage law for dc circuits, Eq. (1.92), is therefore just a special form of Maxwell's first equation for the electrostatic field, Eq. (1.75). We shall see in a later chapter that with some restrictions and approximations, Kirchhoff's voltage law in the form in Eq. (1.92) can also be used for the analysis of ac circuits.

Problems: 1.37; *Conceptual Questions* (on Companion Website): 1.9.

HISTORICAL ASIDE

The SI unit for the voltage, electric potential, and electromotive force (to be studied in a later chapter), volt (V), was named in honor of **Alessandro Volta** (1745–1827), an Italian physicist and inventor, a professor of experimental physics at the University of Pavia, who is famous for his invention of the electric battery (voltaic pile) in 1800. In experiments, he used his tongue to sense and measure the electricity and voltage – the first voltmeter. Volta's battery was a series of electrochemical cells made as a pile of alternating silver and zinc plates separated by cardboard disks soaked in salty water (electrolyte). *(Portrait: Edgar Fahs Smith Collection, University of Pennsylvania Libraries)*

Gustav Robert Kirchhoff (1824–1887), a German physicist and chemist, taught at Universities of Berlin, Breslau, and Heidelberg. Kirchhoff received his doctoral degree from the University of Königsberg in 1847, under Neumann (1798–1895). Extending the work of Ohm (1789–1854), he formulated, in 1850, the fundamental relations between currents and voltages in an electric circuit with multiple loops, which we call Kirchhoff's circuital laws for currents and voltages. Kirchhoff also made seminal contributions to spectroscopy and thermal emission. *(Portrait: Edgar Fahs Smith Collection, University of Pennsylvania Libraries)*

1.9 DIFFERENTIAL RELATIONSHIP BETWEEN THE FIELD AND POTENTIAL IN ELECTROSTATICS

Eq. (1.74) represents an integral relationship between the electric field intensity vector and the potential in electrostatics, which enables us to determine V if we know $\mathbf{E}$. In this section, we shall derive an equivalent, differential, relationship between these two quantities, and then use it for evaluating $\mathbf{E}$ from V.

[11] Algebraic sum means that the voltages in the sum can be of arbitrary sign, where the sign with which the voltage is taken depends on the agreement or disagreement of the orientation (polarity) of the voltage with the orientation of the contour.

Consider a point A in an electrostatic field at which the potential is V_A. Let us move from that point for an elementary distance $dl = dx$ along the x-axis, to a point B, as shown in Fig. 1.24. The resulting change in the potential amounts to the potential at B (new potential) minus the potential at A (starting potential), that is,

$$dV = V_B - V_A. \tag{1.93}$$

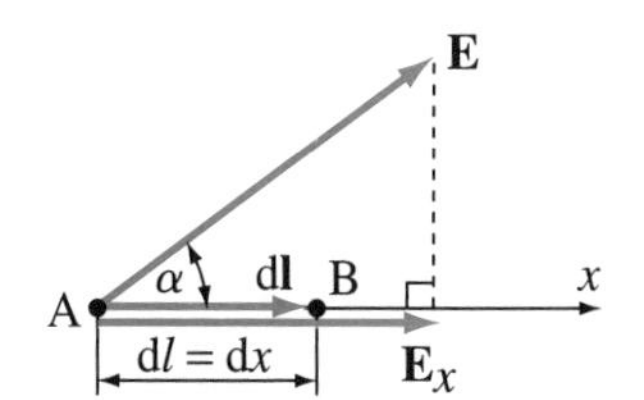

Figure 1.24 Derivation of the differential relationship between **E** and V in electrostatics.

On the other hand, Eq. (1.90) tells us that the potential difference (voltage) between points A and B is related to the electric field intensity vector as

$$V_A - V_B = \mathbf{E} \cdot d\mathbf{l} = E\, dl \cos\alpha = \underbrace{E\cos\alpha}_{E_x}\, dx = E_x\, dx, \tag{1.94}$$

with E_x standing for the x-component of **E**, which equals the projection of **E** on the x-axis, $E\cos\alpha$. Combining Eqs. (1.93) and (1.94), we have

$$\boxed{E_x = -\frac{dV}{dx}.} \tag{1.95}$$

1-D differential relationship between **E** *and* V

Similarly, the projections of the vector **E** on the other two axes of the Cartesian coordinate system are obtained as

$$E_y = -\frac{dV}{dy}, \quad E_z = -\frac{dV}{dz}. \tag{1.96}$$

Hence, the complete vector expression for **E** is given by

$$\mathbf{E} = E_x\,\hat{\mathbf{x}} + E_y\,\hat{\mathbf{y}} + E_z\,\hat{\mathbf{z}} = -\left(\frac{\partial V}{\partial x}\,\hat{\mathbf{x}} + \frac{\partial V}{\partial y}\,\hat{\mathbf{y}} + \frac{\partial V}{\partial z}\,\hat{\mathbf{z}}\right), \tag{1.97}$$

where we use partial derivatives instead of ordinary ones because the potential is a function of all three coordinates (multivariable function), $V = V(x, y, z)$. This differential relationship is an important general means for computing the field **E** from the potential V in electrostatics.

Example 1.16 1-D Problem

In an electrostatic system, the potential is constant in any individual plane normal to the Cartesian x-axis, while it varies along that axis as

$$V(x) = V_0 + V_1\frac{x}{d} + V_2\frac{x^2}{d^2}, \tag{1.98}$$

where V_0, V_1, V_2, and d are constants. Find the electric field intensity vector in the system.

Solution Since $dV/dy = 0$ and $dV/dz = 0$, we are left with an x-component of the vector **E** only, according to Eq. (1.97), which comes out to be

$$E_x = -\frac{dV}{dx} = -\frac{V_1}{d} - \frac{2V_2 x}{d^2}. \tag{1.99}$$

Problems: 1.38; *Conceptual Questions* (on Companion Website): 1.10; *MATLAB Exercises* (on Companion Website).

1.10 GRADIENT

The expression in the parentheses in Eq. (1.97) is called the gradient of the scalar function (V). It is sometimes written as grad V, but much more frequently we write it using the so-called del operator or nabla operator, defined as

del operator

$$\nabla = \frac{\partial}{\partial x}\,\hat{\mathbf{x}} + \frac{\partial}{\partial y}\,\hat{\mathbf{y}} + \frac{\partial}{\partial z}\,\hat{\mathbf{z}}. \tag{1.100}$$

So, we have

$\mathbf{E}$ from V in electrostatics

$$\mathbf{E} = -\operatorname{grad} V = -\nabla V, \tag{1.101}$$

where, in the Cartesian coordinate system (Fig. 1.2),

gradient in Cartesian coordinates

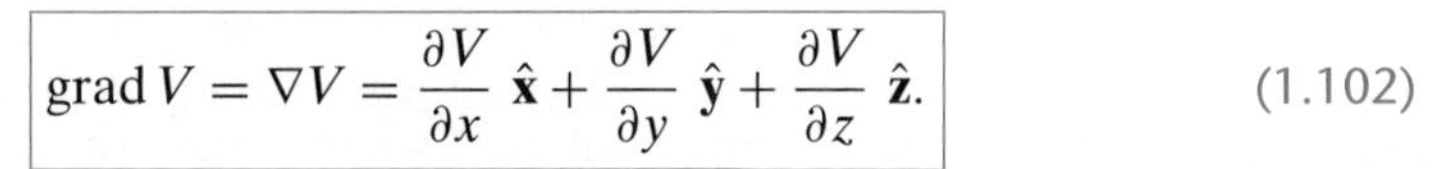

$$\operatorname{grad} V = \nabla V = \frac{\partial V}{\partial x}\,\hat{\mathbf{x}} + \frac{\partial V}{\partial y}\,\hat{\mathbf{y}} + \frac{\partial V}{\partial z}\,\hat{\mathbf{z}}. \tag{1.102}$$

The other two best-known and most commonly used coordinate systems are the cylindrical and the spherical. An arbitrary point (M) in the cylindrical coordinate system is represented as (r, ϕ, z), where r is the radial distance from the z-axis to the point M, ϕ the azimuthal angle measured from the x-axis in the xy-plane, and z the same as in the Cartesian coordinate system, as shown in Fig. 1.25. The ranges of the coordinates are

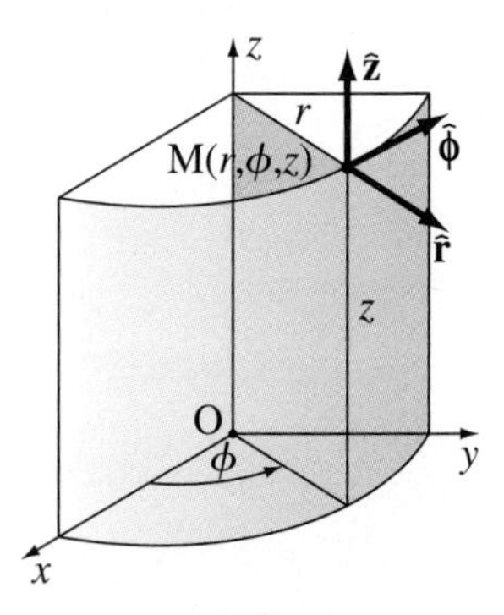

Figure 1.25 Point $\mathrm{M}(r, \phi, z)$ and coordinate unit vectors in the cylindrical coordinate system.

$$0 \le r < \infty, \quad 0 \le \phi < 2\pi \quad (\text{or } -\pi \le \phi < \pi), \quad -\infty < z < \infty. \tag{1.103}$$

The coordinate unit vectors, $\hat{\mathbf{r}}$, $\hat{\boldsymbol{\phi}}$, and $\hat{\mathbf{z}}$, are, by definition, directed in directions of increasing r, ϕ, and z, respectively. They are all mutually perpendicular (the cylindrical coordinate system belongs to the class of orthogonal coordinate systems), and the vector $\mathbf{E}$ in cylindrical coordinates can be expressed as

$$\mathbf{E} = E_r\hat{\mathbf{r}} + E_\phi\hat{\boldsymbol{\phi}} + E_z\,\hat{\mathbf{z}}. \tag{1.104}$$

Since ϕ is not a length coordinate but an angular one, an incremental distance $\mathrm{d}l$ corresponding to an elementary increment in the coordinate, $\mathrm{d}\phi$, equals $\mathrm{d}l = r\,\mathrm{d}\phi$ (the length of an arc of radius r defined by the angle $\mathrm{d}\phi$), and this exactly is the displacement $\mathrm{d}l$ in Fig. 1.24 in computing the change in potential $\mathrm{d}V$ in Eqs. (1.93)–(1.95) now in the ϕ direction. Therefore, the ϕ-component of the electric field vector at the point M in Fig. 1.25 equals $E_\phi = -\,\mathrm{d}V/\mathrm{d}l = -\,\mathrm{d}V/(r\,\mathrm{d}\phi)$ and not just $-\,\mathrm{d}V/\mathrm{d}\phi$. The other two cylindrical coordinates, r and z, are length coordinates, so no modification is needed, $E_r = -\,\mathrm{d}V/\mathrm{d}r$ and $E_z = -\,\mathrm{d}V/\mathrm{d}z$. Consequently, having in mind Eqs. (1.104) and (1.101), the gradient of $V = V(r, \phi, z)$ in the cylindrical coordinate system is, in place of Eq. (1.102), given by

gradient in cylindrical coordinates

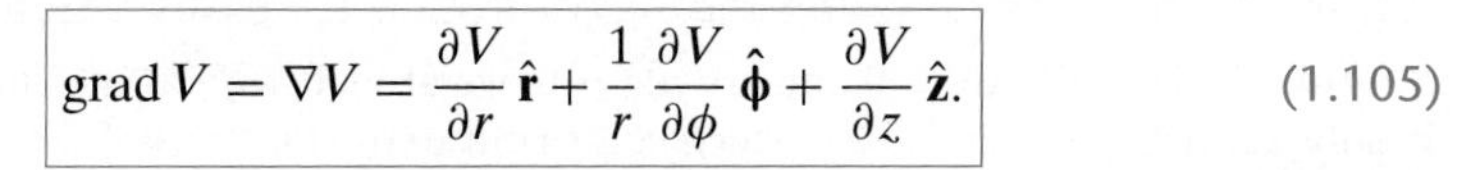

$$\operatorname{grad} V = \nabla V = \frac{\partial V}{\partial r}\,\hat{\mathbf{r}} + \frac{1}{r}\frac{\partial V}{\partial \phi}\,\hat{\boldsymbol{\phi}} + \frac{\partial V}{\partial z}\,\hat{\mathbf{z}}. \tag{1.105}$$

In the spherical coordinate system, a point M is defined by (r, θ, ϕ), with r being the radial distance from the coordinate origin (O) to M, θ the zenith angle between the z-axis and the position vector of M, and ϕ the same as in the cylindrical coordinate system, as illustrated in Fig. 1.26. The ranges of spherical coordinates are

$$0 \le r < \infty, \quad 0 \le \theta \le \pi, \quad 0 \le \phi < 2\pi, \tag{1.106}$$

and the component representation of the vector $\mathbf{E}$ reads

$$\mathbf{E} = E_r\hat{\mathbf{r}} + E_\theta\hat{\boldsymbol{\theta}} + E_\phi\hat{\boldsymbol{\phi}}, \tag{1.107}$$

where $\hat{\mathbf{r}}$, $\hat{\boldsymbol{\theta}}$, and $\hat{\boldsymbol{\phi}}$ are mutually perpendicular coordinate unit vectors (directed in directions of increasing r, θ, and ϕ, respectively). As depicted in Fig. 1.10, the corresponding incremental distances $\mathrm{d}l$ along these vectors equal to $\mathrm{d}r$, $r\,\mathrm{d}\theta$, and $r\sin\theta\,\mathrm{d}\phi$, and hence the following expression for the gradient of $V = V(r, \theta, \phi)$ in the spherical coordinate system:

$$\operatorname{grad} V = \nabla V = \frac{\partial V}{\partial r}\hat{\mathbf{r}} + \frac{1}{r}\frac{\partial V}{\partial \theta}\hat{\boldsymbol{\theta}} + \frac{1}{r\sin\theta}\frac{\partial V}{\partial \phi}\hat{\boldsymbol{\phi}}. \tag{1.108}$$

gradient in spherical coordinates

It is important to note that there is no equivalent in the cylindrical or spherical coordinates to the expression for the del operator, ∇, in Eq. (1.100), essentially because all the unit vectors in Figs. 1.25 and 1.26 except $\hat{\mathbf{z}}$ (of course, their directions and not magnitudes) depend on the location (on coordinates) of the point M. So, ∇ can be formally treated as a vector, given by Eq. (1.100), only in Cartesian coordinates.

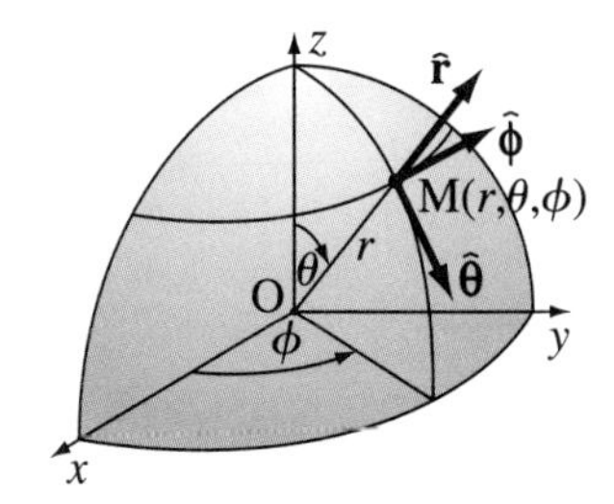

Figure 1.26 Point $\mathrm{M}(r, \theta, \phi)$ and coordinate unit vectors in the spherical coordinate system.

The expression in Eq. (1.102) and the corresponding expressions associated with other coordinate systems enable us to calculate ∇V for any multivariable scalar function V (not necessarily the electric potential). We shall now derive the relationship between the gradient and so-called directional derivative of a scalar field, which will provide us with a new physical interpretation of the gradient operation in general.

Combining Eqs. (1.94), (1.95), and (1.101), we have

$$-\frac{\mathrm{d}V}{\mathrm{d}x} = E_x = \mathbf{E}\cdot\hat{\mathbf{x}} = -\nabla V\cdot\hat{\mathbf{x}} \quad\longrightarrow\quad \frac{\mathrm{d}V}{\mathrm{d}x} = \nabla V\cdot\hat{\mathbf{x}}. \tag{1.109}$$

We can now formally proclaim the Cartesian x-axis to be an arbitrarily positioned linear axis in space, and call it an l-axis, with respect to which (Fig. 1.27)

$$\frac{\mathrm{d}V}{\mathrm{d}l} = \nabla V\cdot\hat{\mathbf{l}} = |\nabla V|\cos\beta, \tag{1.110}$$

directional derivative

where $\hat{\mathbf{l}}$ is the associated unit vector ($|\hat{\mathbf{l}}| = 1$) and β the angle between the vector ∇V and the l-axis. The derivative $\mathrm{d}V/\mathrm{d}l$ is referred to as the directional derivative of V in the l direction. It represents the rate of change of V in this direction, and equals the projection (component) of the vector ∇V on (along) the l-axis. Eq. (1.110) is an equivalent mathematical definition of the gradient of a scalar, and, although obtained specifically for the electrostatic potential, it is valid for any scalar field V (for which the necessary derivatives exist).

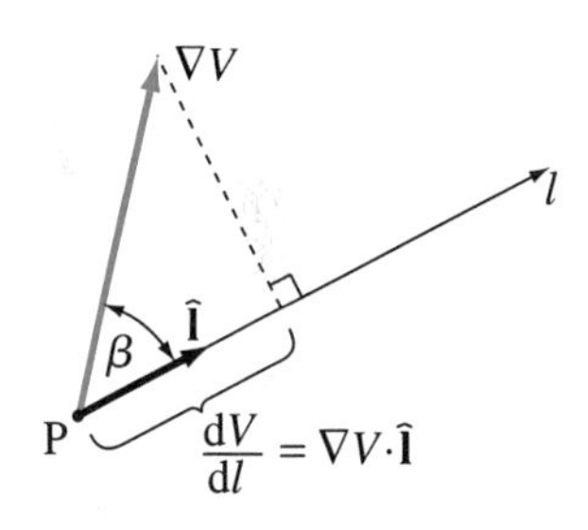

Figure 1.27 Relationship between the gradient and the directional derivative of a scalar field.

In addition, Eq. (1.110) has a very important physical meaning and practical implications. We note that for a given scalar field V, ∇V at a point P in Fig. 1.27 is a fixed vector. By steering the l-axis around P, i.e., by varying the angle β, we obtain different derivatives $\mathrm{d}V/\mathrm{d}l$ at the location P. For $\beta = \pm 90°$, $\mathrm{d}V/\mathrm{d}l = 0$ [from Eq. (1.110)]; for $\beta = 0$, however, we reach the maximum in $\mathrm{d}V/\mathrm{d}l$ ($\cos\beta = 1$), which becomes

$$\left.\frac{\mathrm{d}V}{\mathrm{d}l}\right|_{\max} = |\nabla V| \quad (\beta = 0). \tag{1.111}$$

physical meaning of the gradient

This means that (1) the magnitude of ∇V equals the maximum space rate of change in the function V per unit distance [$|\nabla V| = (\mathrm{d}V/\mathrm{d}l)_{\max}$] and (2) ∇V points in the direction of the maximum space rate of change in V ($\beta = 0$). We conclude that the gradient of a scalar field V is a vector that provides us with both the direction in which V changes most rapidly and the magnitude of the maximum space rate of change. This property of the gradient operation is extensively used in numerous applications, in all areas of science and engineering.

Example 1.17 Field from Potential, Charged Ring

From the electric potential due to a charged ring given by Eq. (1.85), find the electric field intensity vector along the ring axis normal to its plane.

Solution By symmetry, the vector **E** along the z-axis (Fig. 1.11) has a z-component only, so that a combination of Eqs. (1.101), (1.105), and (1.85) gives

$$\mathbf{E} = -\nabla V = -\frac{\mathrm{d}V}{\mathrm{d}z}\,\hat{\mathbf{z}} = -\frac{\mathrm{d}}{\mathrm{d}z}\left(\frac{Q'a}{2\varepsilon_0\sqrt{z^2+a^2}}\right)\hat{\mathbf{z}} = \frac{Q'az}{2\varepsilon_0\left(z^2+a^2\right)^{3/2}}\,\hat{\mathbf{z}}, \tag{1.112}$$

which, of course, is the same result as in Eq. (1.44).

Conversely, according to Eq. (1.74), the expression for the potential due to the ring can be found from the expression for the field, in Eq. (1.44), by integration along the z-axis from the point P (with a coordinate z) to the reference point at infinity ($z \to \infty$), as

$$V = \int_z^\infty E_z\,\mathrm{d}z, \tag{1.113}$$

and it is a simple matter to verify that the solution of this integral is exactly the potential in Eq. (1.85).

Problems: 1.39–1.45; *Conceptual Questions* (on Companion Website): 1.11; *MATLAB Exercises* (on Companion Website).

1.11 3-D AND 2-D ELECTRIC DIPOLES

An electric dipole is a very important, fundamental electrostatic system consisting of two point charges Q of opposite polarities separated by a distance d. We alternatively refer to this system as a three-dimensional (3-D) dipole, to distinguish it from an analogous 2-D system combining line charges, which will be analyzed later in this section. So, let us first derive the expressions for the electrostatic scalar potential and field intensity vector due to a 3-D dipole at large distances compared to d.

Introducing a spherical coordinate system whose origin is at the dipole center as shown in Fig. 1.28 and using Eq. (1.80) for the electric potential due to a single point charge and superposition, the potential due to the dipole at a point P is

$$V = \frac{Q}{4\pi\varepsilon_0}\left(\frac{1}{r_1}-\frac{1}{r_2}\right) = \frac{Q}{4\pi\varepsilon_0}\,\frac{r_2-r_1}{r_1r_2}. \tag{1.114}$$

Since $r \gg d$, $r_2 - r_1 \approx d\cos\theta$ and $r_1r_2 \approx r^2$, and Eq. (1.114) becomes

electric dipole potential

$$\boxed{V \approx \frac{Q}{4\pi\varepsilon_0}\,\frac{d\cos\theta}{r^2} = \frac{p\,\cos\theta}{4\pi\varepsilon_0 r^2} = \frac{\mathbf{p}\cdot\hat{\mathbf{r}}}{4\pi\varepsilon_0 r^2},} \tag{1.115}$$

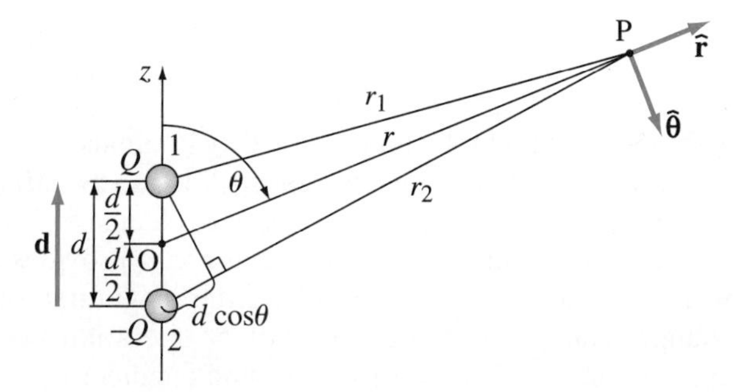

Figure 1.28 Electric dipole.

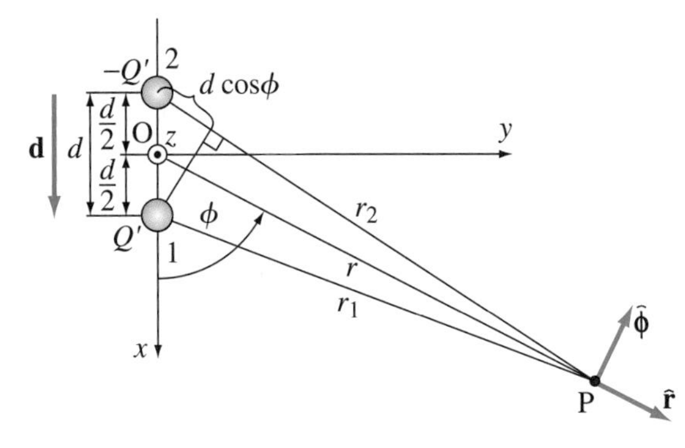

Figure 1.29 Cross section of a line dipole.

where

$$\boxed{\mathbf{p} = Q\mathbf{d}} \tag{1.116}$$

electric dipole moment

is the dipole moment ($p = Qd$). The unit for $\mathbf{p}$ is C · m.

Applying the formula for the gradient in spherical coordinates, Eq. (1.108), to the expression for V in Eq. (1.115) yields

$$\boxed{\mathbf{E} = -\nabla V = -\frac{\partial V}{\partial r}\,\hat{\mathbf{r}} - \frac{1}{r}\frac{\partial V}{\partial \theta}\,\hat{\boldsymbol{\theta}} = \frac{p}{4\pi\varepsilon_0 r^3}\left(2\cos\theta\,\hat{\mathbf{r}} + \sin\theta\,\hat{\boldsymbol{\theta}}\right).} \tag{1.117}$$

electric dipole field

We note that V and $\mathbf{E}$ due to an electric dipole vary with distance as $1/r^2$ and $1/r^3$, respectively.

We shall see in the next chapter that electric dipoles are fundamental for the discussion of dielectric polarization. In addition, a dipole concept is used in electromagnetic interference (EMI) considerations, since the quasistatic (low-frequency) electric field produced by an electrical device can often be approximated by the field due to a single electric dipole, given in Eq. (1.117).

The combination of two equal parallel infinite line charges Q' of opposite polarities separated by a distance d constitutes a line or two-dimensional (2-D) electric dipole. Let us find the potential due to this system.

Fig. 1.29 shows the cross section of a line dipole with the following per-unit-length dipole moment:

$$\mathbf{p}' = Q'\mathbf{d}. \tag{1.118}$$

The cylindrical coordinate system is adopted such that the z-axis is normal to the plane of drawing and coincides with the central line of the dipole. Taking the point O ($r = 0$) for the reference point for potential and applying Eq. (1.87), we have

$$V = \frac{Q'}{2\pi\varepsilon_0}\ln\frac{r_{\mathcal{R}1}}{r_1} + \frac{-Q'}{2\pi\varepsilon_0}\ln\frac{r_{\mathcal{R}2}}{r_2} = \frac{Q'}{2\pi\varepsilon_0}\ln\frac{r_2}{r_1}, \tag{1.119}$$

where $r_{\mathcal{R}1} = d/2$ and $r_{\mathcal{R}2} = d/2$ are the distances of the reference point from charges 1 and 2, respectively. At large distances from the dipole ($r \gg d$),

$$V \approx \frac{Q'}{2\pi\varepsilon_0}\ln\frac{r_1 + d\cos\phi}{r_1} \approx \frac{Q'}{2\pi\varepsilon_0}\ln\left(1 + \frac{d\cos\phi}{r}\right). \tag{1.120}$$

Since $\ln(1+x) \approx x$ for $x \ll 1$, the final expression for the electric potential due to a line dipole turns out to be

$$\boxed{V \approx \frac{p'\cos\phi}{2\pi\varepsilon_0 r},} \tag{1.121}$$

line dipole potential

with $p' = Q'd$.

Note that for quasistatic electric field evaluations, transmission lines (telecommunication lines, power lines, computer buses, etc.) can sometimes be approximately modeled by a single line electric dipole, with the potential computed by Eq. (1.121). Such evaluations are important in studying undesired couplings between transmission lines, as well as associated EMI field levels.

Problems: 1.46–1.50; *MATLAB Exercises* (on Companion Website).

1.12 FORMULATION AND PROOF OF GAUSS' LAW

Gauss' law represents one of the fundamental laws of electromagnetism. It is an alternative statement of Coulomb's law and a direct consequence of the mathematical form of the electric field intensity vector due to a point charge. Gauss' law involves the flux (surface integral) of the vector **E** through a closed mathematical surface,[12] and can equivalently be formulated in a differential form, which is based on a differential operation called divergence on **E**. This important equation, in either form, provides an easy means of calculating the electric field due to highly symmetrical charge distributions, including problems with spherical, cylindrical, and planar symmetry, respectively.

Gauss' law states that the outward flux of the electric field intensity vector through any closed surface in free space is equal to the total charge enclosed by that surface, divided by ε_0. To prove it, let us consider first a single point charge Q in free space and evaluate the flux of vector **E** through an arbitrarily positioned surface element (differentially small patch) of area $\mathrm{d}S$, whose surface area vector is $\mathbf{dS}$, as shown in Fig. 1.30. This elementary flux is given by

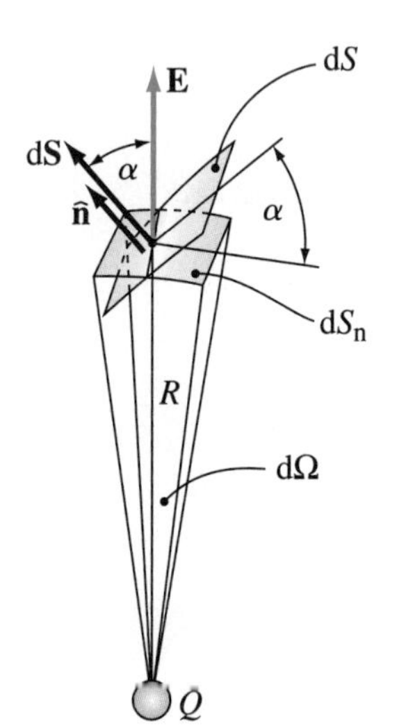

Figure 1.30 Evaluation of the flux of the vector **E** due to a point charge through a surface element $\mathrm{d}S$, whose area vector, $\mathbf{dS}$, is oriented from the charge outward.

$$\mathrm{d}\Psi_E = \mathbf{E} \cdot \mathrm{d}\mathbf{S}, \quad \mathrm{d}\mathbf{S} = \mathrm{d}S\,\hat{\mathbf{n}}, \tag{1.122}$$

where $\hat{\mathbf{n}}$ is the unit vector normal to $\mathrm{d}S$ directed from the charge outward. Using Eq. (1.24), we have

$$\mathrm{d}\Psi_E = E\,\mathrm{d}S\cos\alpha = E\,\mathrm{d}S_\mathrm{n} = \frac{Q}{4\pi\varepsilon_0}\frac{\mathrm{d}S_\mathrm{n}}{R^2}, \tag{1.123}$$

with $\mathrm{d}S_\mathrm{n}$ standing for the projection of $\mathrm{d}S$ on the plane perpendicular to R (Fig. 1.30). This projection can be considered as the base of a cone with the apex at the charge. The solid angle of the cone is, by definition,

elementary solid angle

$$\boxed{\mathrm{d}\Omega = \frac{\mathrm{d}S_\mathrm{n}}{R^2}} \tag{1.124}$$

[a solid angle is measured in steradians (sr or srad)], yielding

$$\mathrm{d}\Psi_E = \frac{Q}{4\pi\varepsilon_0}\,\mathrm{d}\Omega. \tag{1.125}$$

Assume now that the charge Q is enclosed by an arbitrary closed surface S, as depicted in Fig. 1.31(a). From Eq. (1.125), by integration, the outward flux of the

[12]The flux of a vector function **a** through an open or closed surface S is defined as $\int_S \mathbf{a}\cdot\mathrm{d}\mathbf{S}$, where $\mathbf{dS}$ (to be discussed in this section) is the vector element of the surface perpendicular to it, and directed in accordance to the orientation of the surface.

HISTORICAL ASIDE

Johann Karl Friedrich Gauss (1777–1855) was a German mathematician and a director of the Göttingen Observatory. Gauss was born in Braunschweig (Brunswick) as a son of a gardener and a servant girl, and he showed his mathematical genius very early. At the age of seven, to the astonishment of his elementary school teacher, he summed the integers from 1 to 100 within seconds by realizing that the sum equaled 50 pairs of numbers from opposite ends of the list, each pair adding up to 101 ($1 + 100 = 101$, $2 + 99 = 101$, ...), so $50 \times 101 = 5{,}050$. He attended Universities of Braunschweig, Göttingen, and Helmstedt, and received his doctoral degree in 1799. Gauss made numerous seminal contributions to the number theory, algebra, geometry, and calculus, proved several fundamental theorems in different branches of mathematics, and is considered by many to be one of the three greatest mathematicians of all time, others being Archimedes (287 B.C.–212 B.C.) and Newton (1642–1727). He developed in 1813 the divergence theorem, providing the mathematical form for the famous law of electricity that relates the flux of the electric field intensity vector through a closed surface to the enclosed net charge and now carries his name – the law which first came out from experiments with charged concentric metallic spheres by Faraday (1791–1867) and was later translated into the Gaussian mathematical form by Maxwell (1831–1879). Upon Weber's (1804–1891) arrival at Göttingen in 1831, they worked together on a variety of topics in electricity and magnetism (see Weber's biography in Section 4.8).

vector **E** through the entire surface S ($\hat{\mathbf{n}}$ is directed from the surface outward) is

$$\Psi_E = \oint_S \mathrm{d}\Psi_E = \frac{Q}{4\pi\varepsilon_0}\oint_S \mathrm{d}\Omega = \frac{Q}{4\pi\varepsilon_0}\Omega, \tag{1.126}$$

where Ω is the full solid angle, which, in turn, can be interpreted as the angular measure for a spherical surface of arbitrary radius r and area $S_0 = 4\pi r^2$, resulting in (for spherical surfaces, $\mathrm{d}S_\mathrm{n} = \mathrm{d}S$)

$$\boxed{\Omega = \oint_{S_0} \mathrm{d}\Omega = \oint_{S_0} \frac{\mathrm{d}S_\mathrm{n}}{r^2} = \frac{1}{a^2}\oint_{S_0} \mathrm{d}S = \frac{S_0}{r^2} = 4\pi.} \tag{1.127}$$

full solid angle

As the full solid angle turns out to equal 4π (sr), the total flux becomes

$$\Psi_E = \oint_S \mathbf{E}\cdot\mathrm{d}\mathbf{S} = \frac{Q}{\varepsilon_0}, \quad \text{for a point charge } Q \text{ enclosed by } S. \tag{1.128}$$

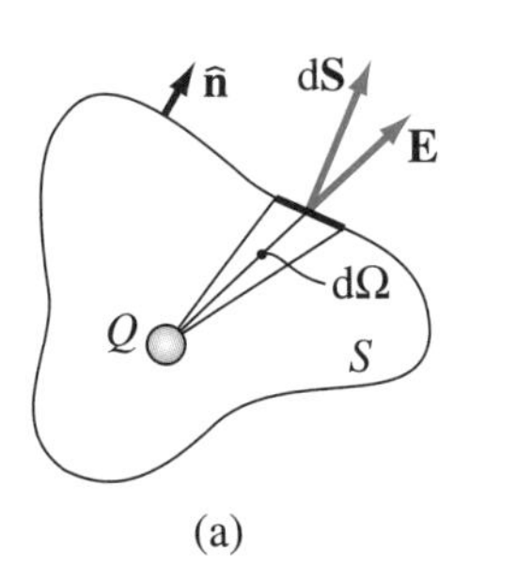

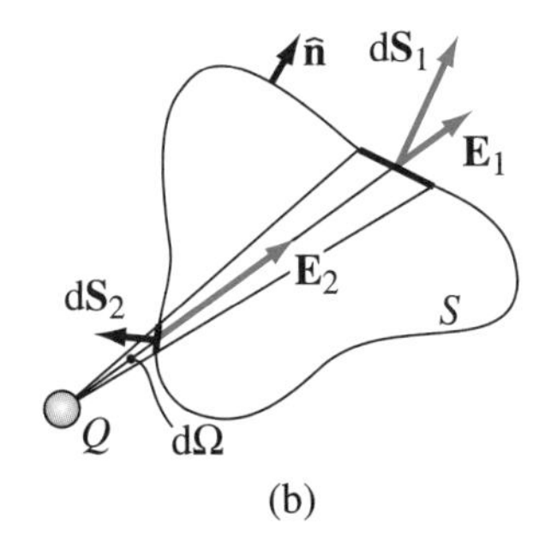

Figure 1.31 Evaluation of the outward flux of **E** due to a point charge through a closed surface S, the charge being enclosed by S (a) or located outside S (b).

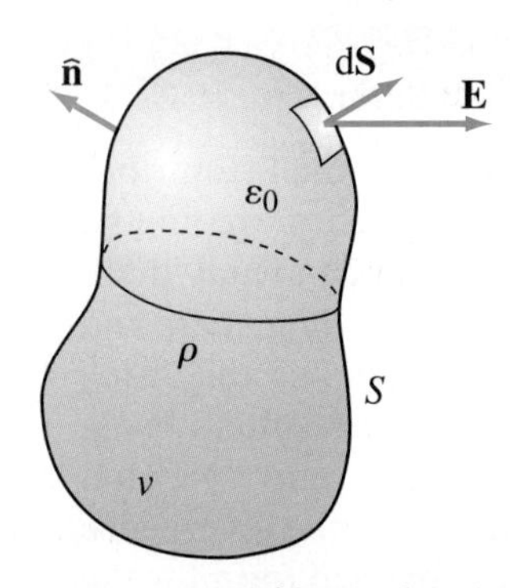

Figure 1.32 Arbitrary closed surface containing a volume charge distribution in free space.

For a point charge Q outside the surface S, we realize that the elementary fluxes through the two surface elements shown in Fig. 1.31(b), which have opposite orientations, are

$$(\mathrm{d}\Psi_E)_{\text{through } \mathrm{d}S_1} = \frac{Q}{4\pi\varepsilon_0}\,\mathrm{d}\Omega \quad \text{and} \quad (\mathrm{d}\Psi_E)_{\text{through } \mathrm{d}S_2} = -\frac{Q}{4\pi\varepsilon_0}\,\mathrm{d}\Omega, \tag{1.129}$$

respectively, so that their contributions to the flux integral cancel each other, and the total flux of $\mathbf{E}$ through S is

$$\Psi_E = \oint_S \mathbf{E}\cdot\mathrm{d}\mathbf{S} = 0, \quad \text{for a point charge outside } S. \tag{1.130}$$

By means of the superposition principle, Eq. (1.128) can then readily be generalized to the case of N point charges enclosed by a (closed) surface S:

$$\oint_S \mathbf{E}\cdot\mathrm{d}\mathbf{S} = \oint_S (\mathbf{E}_1 + \mathbf{E}_2 + \cdots + \mathbf{E}_N)\cdot\mathrm{d}\mathbf{S}$$

$$= \oint_S \mathbf{E}_1\cdot\mathrm{d}\mathbf{S} + \oint_S \mathbf{E}_2\cdot\mathrm{d}\mathbf{S} + \cdots + \oint_S \mathbf{E}_N\cdot\mathrm{d}\mathbf{S} = \frac{Q_1}{\varepsilon_0} + \frac{Q_2}{\varepsilon_0} + \cdots + \frac{Q_N}{\varepsilon_0}, \tag{1.131}$$

where $\mathbf{E}_i$ is the field due to a charge Q_i, $i = 1, 2, \ldots, N$. With the notation

$$Q_S = \sum_{i=1}^{N} Q_i, \tag{1.132}$$

we have

Gauss' law

$$\boxed{\oint_S \mathbf{E}\cdot\mathrm{d}\mathbf{S} = \frac{Q_S}{\varepsilon_0}.} \tag{1.133}$$

There may also be charges located outside the surface S; recall from Eq. (1.130), however, that the contribution to the total flux resulting from such charges is zero. As we know, any (continuous or discontinuous) charge distribution can be represented as a system of equivalent point charges. This means that Eq. (1.133) holds true for any charge distribution (in free space), where

$$Q_S = \text{total charge enclosed by } S \quad \text{(arbitrary charge distribution)}. \tag{1.134}$$

Eq. (1.133) is the mathematical formulation of Gauss' law.

The most general case of continuous charge distributions is the volume charge distribution, Fig. 1.32, in terms of which Gauss' law can be written as

Gauss' law for volume charge

$$\boxed{\oint_S \mathbf{E}\cdot\mathrm{d}\mathbf{S} = \frac{1}{\varepsilon_0}\int_v \rho\,\mathrm{d}v,} \tag{1.135}$$

with v denoting the volume enclosed by the surface S and ρ the volume charge density. This particular form of Gauss' law is usually referred to as Maxwell's third equation for the electrostatic field in free space.

The principle of conservation of energy in the electrostatic field, Eq. (1.75), and Gauss' law, Eq. (1.135), are the two fundamental integral equations (Maxwell's equations) describing the electrostatic field in free space.[13] As shown here and

[13]Gauss' law is the second equation out of a total of two Maxwell's equations used in electrostatics. Nevertheless, we term it Maxwell's third equation because it is customarily positioned as the third equation in the complete set of four general Maxwell's equations for the electromagnetic field, as we shall see in later chapters.

in Fig. 1.19, both equations can be derived from Coulomb's law, i.e., from the expression for the electrostatic field due to a point charge in free space.

Problems: 1.51 and 1.52; *Conceptual Questions* (on Companion Website): 1.12–1.16.

1.13 APPLICATIONS OF GAUSS' LAW

Let us now discuss procedures for using Gauss' law to determine the electric field intensity if the charge distribution is known. Gauss' law is always true, and we can apply it to any charge distribution and any problem. However, it enables us to analytically solve only for the field due to highly symmetrical charge distributions. To understand this, note that Gauss' law is mathematically formulated by an integral equation [Eq. (1.133)] in which the unknown quantity to be determined (**E**) appears inside the integral. We can use the law to obtain a solution, therefore, only in cases in which we are able to bring the field intensity, E, outside the integral sign, and solve for it. These cases involve highly symmetrical charge distributions, for which we are able to choose a closed surface S, known as a Gaussian surface, that satisfies two conditions: (1) **E** is everywhere either normal or tangential to S and (2) $E = \text{const}$ on the portion of S on which **E** is normal. When **E** is tangential to the surface, $\mathbf{E} \cdot \mathrm{d}\mathbf{S}$ in Eq. (1.133) becomes zero. When **E** is normal to the surface, $\mathbf{E} \cdot \mathrm{d}\mathbf{S}$ becomes $E\,\mathrm{d}S$, and since we have also that E is constant, it can be brought outside the integral sign in Eq. (1.133). We shall now apply these basic ideas to problems with spherical, cylindrical, and planar symmetry, respectively, in four characteristic examples.

Example 1.18 Example of a Problem with Spherical Symmetry

The first class of problems for which Gauss' law can be used to solve for the field involve charge distributions that depend only on the radial coordinate in the spherical coordinate system. Due to spherical symmetry, only the radial component of the electric field intensity vector is present, and this component is a function of the radial coordinate only. Based on these facts, the solution procedure is simple to perform. As an example, consider a sphere of radius a with a uniform volume charge density ρ, in free space, and determine (a) the field distribution both inside and outside the sphere and (b) the electric scalar potential at the sphere center.

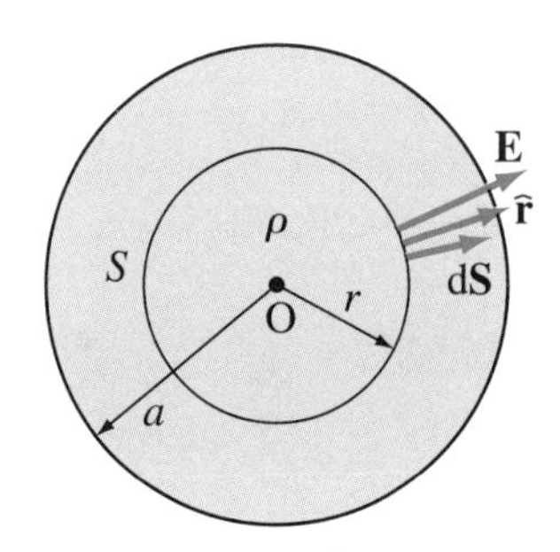

Figure 1.33 Application of Gauss' law to a problem with spherical symmetry – Example 1.18.

Solution

(a) The electric field vector at an arbitrary point in space is of the form

$$\boxed{\mathbf{E} = E(r)\,\hat{\mathbf{r}},} \tag{1.136}$$

$\mathbf{E}$ in a problem with spherical symmetry

where r is the radial coordinate in the spherical coordinate system and $\hat{\mathbf{r}}$ is the radial unit vector, as shown in Fig. 1.33. The Gaussian surface S is a spherical surface of radius r centered at the origin. On S,

$$\mathrm{d}\mathbf{S} = \mathrm{d}S\,\hat{\mathbf{r}} \quad \longrightarrow \quad \mathbf{E} \cdot \mathrm{d}\mathbf{S} = E\hat{\mathbf{r}} \cdot \mathrm{d}S\,\hat{\mathbf{r}} = E\,\mathrm{d}S\,\hat{\mathbf{r}} \cdot \hat{\mathbf{r}} = E\,\mathrm{d}S. \tag{1.137}$$

The outward flux of the vector **E** through S is hence

$$\Psi_E = \oint_S \mathbf{E} \cdot \mathrm{d}\mathbf{S} = \oint_S E(r)\,\mathrm{d}S = E(r) \oint_S \mathrm{d}S = E(r)\,S = E(r)\,4\pi r^2 \quad (0 \le r < \infty). \tag{1.138}$$

Since $\rho = \text{const}$ (uniform charge distribution), the charge enclosed by S is computed, according to Eq. (1.30), as ρ times the corresponding volume, which is that of either the entire domain inside S, for $r < a$, or the charged sphere of radius a (there is no charge outside this sphere), for $r \geq a$, so Q_S amounts to

$$Q_S = \begin{cases} \rho 4\pi r^3/3 & \text{for } r < a \\ \rho 4\pi a^3/3 & \text{for } r \geq a \end{cases}. \tag{1.139}$$

According to Gauss' law, $\Psi_E = Q_S/\varepsilon_0$, yielding

$$E(r) = \begin{cases} \rho r/(3\varepsilon_0) & \text{for } r < a \\ \rho a^3/(3\varepsilon_0 r^2) & \text{for } r \geq a \end{cases}. \tag{1.140}$$

Note that the field for $r \geq a$ is identical to that of a point charge $Q = \rho 4\pi a^3/3$ (the total charge of the sphere) placed at the sphere center. This means that the point charge and the charged sphere are equivalent sources with respect to the region outside the sphere. Concept of equivalent sources is often used in electromagnetics.

(b) To find the electric potential at the sphere center (with respect to the reference point at infinity[14]), we first conclude that the potential outside the charged sphere is identical to the potential of the equivalent point charge, because the fields are identical, so that the potential at the sphere surface ($r = a$) is obtained directly from Eq. (1.80):

$$V(a) = \frac{Q}{4\pi\varepsilon_0 a}. \tag{1.141}$$

Invoking Eq. (1.90), the potential difference (voltage) between the sphere center and surface equals the line integral of $\mathbf{E}$ from the center to the surface, which results in the following for the potential at the center:

$$V(0) = \int_{r=0}^{a} E(r)\,\mathrm{d}r + V(a) = \frac{3Q}{8\pi\varepsilon_0 a} = \frac{\rho a^2}{2\varepsilon_0}. \tag{1.142}$$

Example 1.19 Example of a Problem with Cylindrical Symmetry

The second class of problems we deal with using Gauss' law involve infinitely long charge distributions that depend only on the radial coordinate in the cylindrical coordinate system. As an example, consider an infinitely long charged cylinder of radius a and volume charge density

$$\rho(r) = \rho_0 \frac{r^2}{a^2} \quad (0 \leq r \leq a) \tag{1.143}$$

in free space, where ρ_0 is a constant and r the radial coordinate (i.e., the distance from the cylinder axis), and find the electric field everywhere.

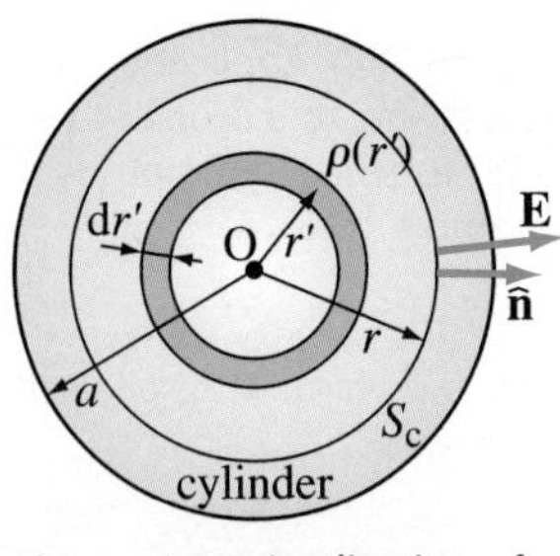

Figure 1.34 Application of Gauss' law to a problem with cylindrical symmetry – Example 1.19.

Solution Shown in Fig. 1.34 is a cross section of the cylinder. Due to cylindrical symmetry of the charge distribution in this case, the field is radial with respect to the cylinder axis and is of the form given in Eq. (1.136), r being here the radial cylindrical coordinate and $\hat{\mathbf{r}}$ the associated unit vector. The Gaussian surface is a cylinder of radius r and height (length) h, positioned coaxially with the charged cylinder.

To determine $\mathbf{E}$ inside the charge distribution, we apply Gauss' law, Eq. (1.135), to a Gaussian cylinder of radius $r \leq a$,

$$E(r) \underbrace{2\pi r h}_{S_c} = \frac{1}{\varepsilon_0} \int_{r'=0}^{r} \underbrace{\rho_0 \frac{r'^2}{a^2}}_{\rho} \underbrace{2\pi r'\,\mathrm{d}r' h}_{\mathrm{d}v}, \tag{1.144}$$

[14]Whenever the reference point for the potential to be determined is not specified, we assume that it is at infinity, except for infinite charge distributions, where such assumption would give us an infinite potential [e.g., Eq. (1.87) for $r_{\mathcal{R}} \to \infty$].

where S_c is the lateral surface area of the cylinder, and dv is the volume of a thin cylindrical shell of radius r' $(0 < r' \leq r)$, thickness dr', and height h (we adopt as large as possible dv, as explained in Section 1.4). This volume is computed as that of a thin flat rectangular slab with edges equal to $2\pi r'$ (circumference of the cylindrical shell), h, and dr' (for the purpose of the volume computation, the shell is flattened into a rectangular slab). The flux of the vector $\mathbf{E}$ through the top and bottom surfaces of the Gaussian cylinder is zero since $\mathbf{E}$ is tangential to those surfaces $(\mathbf{E} \cdot d\mathbf{S} = 0)$. By integration in Eq. (1.144),

$$E(r) = \frac{\rho_0 r^3}{4\varepsilon_0 a^2}, \quad \text{for } r \leq a. \tag{1.145}$$

For a field point outside the charged cylinder $(r > a)$, the upper limit in the integral in Eq. (1.144) becomes a, which results in

$$E(r) = \frac{\rho_0 a^2}{4\varepsilon_0 r}, \quad \text{for } r > a. \tag{1.146}$$

Note that this field is identical to the field due to an infinite line charge positioned along the cylinder axis with the same charge per unit length Q' as the cylinder [see Eqs. (1.31) and (1.57)].

Example 1.20 Example of a Problem with Planar Symmetry

Finally, this and the following example illustrate how problems involving charge distributions that depend on one Cartesian coordinate only – problems with planar symmetry – can be solved using Gauss' law. Consider first a layer of volume charges in free space with the volume charge density being the following even function of the Cartesian coordinate x:

$$\rho(x) = \rho_0 \left(1 - \frac{x^2}{a^2}\right), \quad |x| \leq a, \tag{1.147}$$

where ρ_0 and a $(a > 0)$ are constants and $\rho(x) = 0$ for $|x| > a$. Determine the electric field intensity vector everywhere, i.e., inside and outside the charge layer.

Solution Fig. 1.35 shows the charge distribution. Due to planar symmetry, vector $\mathbf{E}$ at an arbitrary point of space has an x-component only, and E_x depends on the coordinate x only (it is constant in any plane $x = \text{const}$), so we can write

$$\boxed{\mathbf{E} = E_x(x)\,\hat{\mathbf{x}}.} \tag{1.148}$$

$\mathbf{E}$ in a problem with planar symmetry

In addition, since the charge distribution is symmetrical with respect to the plane $x = 0$, vector $\mathbf{E}$ in the plane defined by $x' = x$ $(x > 0)$ in Fig. 1.35 is equal in magnitude and opposite in

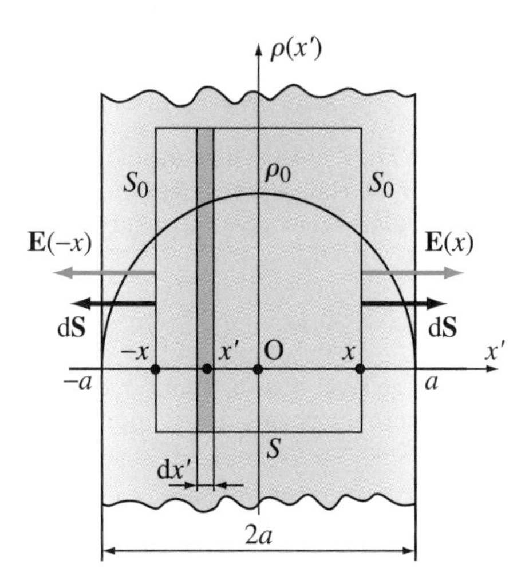

Figure 1.35 Application of Gauss' law to a problem with planar symmetry – Example 1.20.

direction to that in the plane $x' = -x$. The Gaussian surface is a right-angled parallelepiped (rectangular box) that is cut symmetrically by the plane $x = 0$ and has two of its faces, with areas S_0, positioned in the planes $x' = x$ and $x' = -x$, respectively.

Let us first determine the field intensity $E_x(x)$ inside the charge layer ($-a \leq x \leq a$). Noting that $\mathbf{E}$ and $\mathrm{d}\mathbf{S}$ are oriented in the same direction at both faces S_0, as well as that $\mathbf{E}$ and $\mathrm{d}\mathbf{S}$ are mutually perpendicular on the remaining portions of the Gaussian surface, we have

$$2\,E_x(x)\,S_0 = \frac{1}{\varepsilon_0}\int_{x'=-x}^{x} \rho(x')\,\underbrace{S_0\,\mathrm{d}x'}_{\mathrm{d}v}, \tag{1.149}$$

where $\mathrm{d}v$ is the volume of a thin slice of the parallelepiped with thickness $\mathrm{d}x'$. The integration in x' yields

$$E_x(x) = \frac{\rho_0 x}{\varepsilon_0}\left(1 - \frac{x^2}{3a^2}\right), \quad \text{for } |x| \leq a. \tag{1.150}$$

Let us now bring the parallelepiped faces S_0 onto the boundaries of the charge layer. If we then start moving them toward infinity, symmetrically with respect to the plane $x = 0$, nothing will change in the application of Gauss' law, since there is no charge beyond the layer boundaries. This reasoning gives us directly the field intensity outside the charge layer:

$$E_x(x) = E_x(a) = \frac{2\rho_0 a}{3\varepsilon_0}, \quad \text{for } x > a, \tag{1.151}$$

$$E_x(x) = E_x(-a) = -\frac{2\rho_0 a}{3\varepsilon_0}, \quad \text{for } x < -a. \tag{1.152}$$

Example 1.21 Planar Symmetry, Antisymmetrical Charge Distribution

Repeat the previous example but for the following odd function of x as the charge density of the $2a$-thick layer:

$$\rho(x) = \rho_0\,\frac{x}{a}, \quad |x| \leq a\,; \tag{1.153}$$

there is no charge outside the layer.

Solution As this is an antisymmetrical charge distribution with respect to the plane $x = 0$, the total charge of the layer is zero, which implies that the electric field outside the layer is zero; namely, this field is as if due to an equivalent infinite sheet of charge (Fig. 1.15) with zero surface charge density ($\rho_s = 0$). In other words, if we subdivide the charge layer described by Eq. (1.153) into differentially thin layers of thicknesses $\mathrm{d}x'$ and add (integrate) the fields, given by Eq. (1.64), due to all these thin layers (observed outside all of them), the positives and negatives exactly cancel out, and the result (total field for $x < -a$ or $x > a$) is zero. In place of Eqs. (1.151) and (1.152) we thus have

$$\int_{x'=-a}^{a} \rho(x')\,\mathrm{d}x' = 0 \quad \longrightarrow \quad E_x(x) = 0 \quad (|x| > a). \tag{1.154}$$

Positioning the Gaussian rectangular surface in Fig. 1.35 such that one of its parallel faces with area S_0 is in the region $x' < -a$ (on the left of the charge layer) and the other is in the plane $x' = x$ where $-a \leq x \leq a$ (inside the layer), Gauss' law gives [instead of Eqs. (1.149) and (1.150)]

$$E_x(x)\,S_0 = \frac{1}{\varepsilon_0}\int_{x'=-a}^{x} \rho_0\,\frac{x'}{a}\,S_0\,\mathrm{d}x' \quad \longrightarrow \quad E_x(x) = \frac{\rho_0}{2\varepsilon_0 a}\left(x^2 - a^2\right) \quad (|x| \leq a), \tag{1.155}$$

since the flux of $\mathbf{E}$ through the face outside the charge layer is zero, from Eq. (1.154), and the volume integration effectively starts at $x' = -a$, where the charge distribution begins.

Problems: 1.53–1.64; *Conceptual Questions* (on Companion Website): 1.17 and 1.18; *MATLAB Exercises* (on Companion Website).

1.14 DIFFERENTIAL FORM OF GAUSS' LAW

Gauss' law is an integral equation in the spatial domain (the integrations involved are carried out with respect to spatial coordinates), and, in the general case, it represents an integral relationship between the electric field intensity vector, $\mathbf{E}$, and the volume charge density, ρ [Eq. (1.135)]. In this section, we shall derive an equivalent, differential, relationship between $\mathbf{E}$ and ρ, that is, the differential form of Gauss' law.

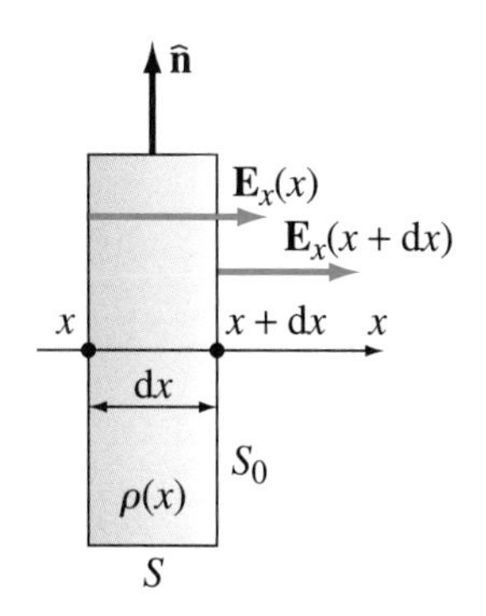

Figure 1.36 For derivation of the one-dimensional Gauss' law in differential form.

Assume first that ρ is a function of the Cartesian coordinate x only, $\rho = \rho(x)$ (1-D charge distribution), so that the only present component of $\mathbf{E}$ is $E_x = E_x(x)$. Let us apply Gauss' law, Eq. (1.135), to a rectangular closed surface S, the dimension of which in the x direction is $\mathrm{d}x$ and the sides normal to that direction are S_0 in area, as indicated in Fig. 1.36. The field being constant on both surfaces S_0, no integration is needed on the left-hand side of Eq. (1.135). In addition, since $\mathrm{d}x$ is differentially small, we can take $\rho(x)$ as constant in the volume enclosed by S, so that no integration is needed on the right-hand side of Eq. (1.135) either. Noting that on the left-hand side of S, $\mathbf{E}$ and $\mathrm{d}\mathbf{S}$ are directed in opposite directions, we have

$$-E_x(x)\,S_0 + E_x(x+\mathrm{d}x)\,S_0 = \frac{1}{\varepsilon_0}\,\rho(x)\,S_0\,\mathrm{d}x. \tag{1.156}$$

The differential of E_x corresponding to the displacement $\mathrm{d}x$,

$$\mathrm{d}E_x = E_x(x+\mathrm{d}x) - E_x(x), \tag{1.157}$$

divided by $\mathrm{d}x$, is, by definition, the derivative of E_x with respect to x. From Eq. (1.156), this derivative comes out to be

$$\boxed{\frac{\mathrm{d}E_x}{\mathrm{d}x} = \frac{\rho}{\varepsilon_0}.} \tag{1.158}$$

1-D differential Gauss' law

This is a differential equation in the spatial domain, and it represents the one-dimensional Gauss' law in differential form. It states that, in cases when the charge distribution changes with a single linear spatial coordinate in free space, the rate of change of the electric field intensity with that coordinate equals the local density of volume charge, divided by ε_0.

The generalization of Eq. (1.158) to the three-dimensional case is straightforward. The charge density is now a function of all three coordinates, $\rho = \rho(x, y, z)$, and the Gaussian surface S has to be differentially small in all three dimensions, as shown in Fig. 1.37. We break the flux integral over S in Eq. (1.135) up into three pairs of integrals, each pair being carried out over two parallel sides of S. All sides being differentially small, the flux over each of them can be approximated by taking a constant value of the field component normal to the side and multiplying it by the side area. This gives us the result for each individual pair of integrals that has exactly the same form as in the one-dimensional case above [Eqs. (1.156)–(1.158)].

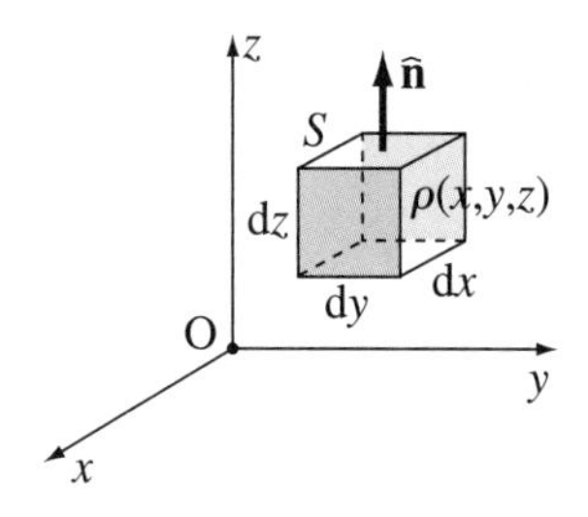

Figure 1.37 For derivation of Gauss' law in differential form for an arbitrary charge distribution.

For the first pair of integrals, we have that the outward flux of $\mathbf{E}$ through sides normal to the x direction (front side and back side in Fig. 1.37) is

$$\int_{\text{front}} \mathbf{E}\cdot\mathrm{d}\mathbf{S} + \int_{\text{back}} \mathbf{E}\cdot\mathrm{d}\mathbf{S} = (\text{change of } E_x \text{ across } \mathrm{d}x) \times (\mathrm{d}y\,\mathrm{d}z)$$

$$= \frac{\partial E_x}{\partial x}\,\mathrm{d}x\,(\mathrm{d}y\,\mathrm{d}z), \tag{1.159}$$

and adding the results for other two pairs of integrals, we obtain the total flux

$$\oint_S \mathbf{E} \cdot d\mathbf{S} = \frac{\partial E_x}{\partial x} dx\,(dy\,dz) + \frac{\partial E_y}{\partial y} dy\,(dx\,dz) + \frac{\partial E_z}{\partial z} dz\,(dx\,dy). \quad (1.160)$$

The volume enclosed by S is $\Delta v = dx\,dy\,dz$, with which the above equation becomes

$$\oint_S \mathbf{E} \cdot d\mathbf{S} = \left(\frac{\partial E_x}{\partial x} + \frac{\partial E_y}{\partial y} + \frac{\partial E_z}{\partial z}\right) \Delta v \quad (\Delta v \to 0). \quad (1.161)$$

Since Δv is very small, the total charge in it can be found with no integration in Eq. (1.135), simply as

$$Q_S = \rho\,\Delta v = \rho\,(dx\,dy\,dz). \quad (1.162)$$

Finally, interconnecting the flux and charge by means of Eq. (1.135), we have

Gauss' law in differential form

$$\boxed{\frac{\partial E_x}{\partial x} + \frac{\partial E_y}{\partial y} + \frac{\partial E_z}{\partial z} = \frac{\rho}{\varepsilon_0}.} \quad (1.163)$$

This is Gauss' law in differential form, i.e., Maxwell's third differential equation for the electrostatic field in free space. It is a partial differential equation or PDE (partial derivatives with respect to individual coordinates enter into the equation) in three unknowns (three components of the vector **E**). This equation provides us with valuable information about the way **E** varies in space. It relates the rate of change of field components with spatial coordinates to the local charge density. We see that changes of individual components along the direction of that component only (change of E_x along x, and not along y and z, etc.) contribute in this relationship.

Problems: 1.65 and 1.66; *Conceptual Questions* (on Companion Website): 1.19; *MATLAB Exercises* (on Companion Website).

1.15 DIVERGENCE

The expression on the left-hand side of Eq. (1.163) is called the divergence of a vector function (**E**), and is written as div **E**. Applying formally the formula for the dot product of two vectors in the Cartesian (rectangular) coordinate system,

$$\mathbf{a} \cdot \mathbf{b} = (a_x\,\hat{\mathbf{x}} + a_y\,\hat{\mathbf{y}} + a_z\,\hat{\mathbf{z}}) \cdot (b_x\,\hat{\mathbf{x}} + b_y\,\hat{\mathbf{y}} + b_z\,\hat{\mathbf{z}}) = a_x b_x + a_y b_y + a_z b_z, \quad (1.164)$$

to the del operator, Eq. (1.100), and vector **E**, we get

$$\nabla \cdot \mathbf{E} = \left(\frac{\partial}{\partial x}\,\hat{\mathbf{x}} + \frac{\partial}{\partial y}\,\hat{\mathbf{y}} + \frac{\partial}{\partial z}\,\hat{\mathbf{z}}\right) \cdot (E_x\,\hat{\mathbf{x}} + E_y\,\hat{\mathbf{y}} + E_z\,\hat{\mathbf{z}})$$
$$= \frac{\partial E_x}{\partial x} + \frac{\partial E_y}{\partial y} + \frac{\partial E_z}{\partial z}, \quad (1.165)$$

and this is exactly div **E**, in Eq. (1.163). Note that the divergence is an operation that is performed on a vector, but the result is a scalar. The differential Gauss' law now can be written in a short form

Gauss' law using divergence notation

$$\boxed{\operatorname{div} \mathbf{E} = \frac{\rho}{\varepsilon_0} \quad \text{or} \quad \nabla \cdot \mathbf{E} = \frac{\rho}{\varepsilon_0},} \quad (1.166)$$

where, in the Cartesian coordinate system,

divergence in Cartesian coordinates

$$\boxed{\operatorname{div} \mathbf{E} = \nabla \cdot \mathbf{E} = \frac{\partial E_x}{\partial x} + \frac{\partial E_y}{\partial y} + \frac{\partial E_z}{\partial z}.} \quad (1.167)$$

In nonrectangular coordinate systems, the corresponding formulas are more complex, due to curvature (or nonrectangularity) of differential volume cells in place of the one in Fig. 1.37. For instance, if, in analogy to the situation in Fig. 1.36, we consider a one-dimensional spherical volume charge distribution with $\rho = \rho(r)$, r being the radial spherical coordinate in Fig. 1.26, the only component of $\mathbf{E}$ is $E_r = E_r(r)$, as in Eq. (1.136), and applying Gauss' law, Eq. (1.135), to the inner (radius r) and outer (radius $r + \mathrm{d}r$) surfaces of a thin spherical shell in Fig. 1.9, Eq. (1.156) becomes

$$-E_r(r)\,4\pi r^2 + E_r(r + \mathrm{d}r)\,4\pi (r + \mathrm{d}r)^2 = \frac{1}{\varepsilon_0}\,\rho(r)\,4\pi r^2\,\mathrm{d}r, \tag{1.168}$$

where the use is made of the expression for $\mathrm{d}v$ in Eq. (1.33). After eliminating 4π on both sides of the equation, the expression on the left-hand side is the differential [see Eq. (1.157)] of the function $r^2 E_r$ corresponding to the coordinate increment $\mathrm{d}r$, which divided by $\mathrm{d}r$ represents the derivative of $r^2 E_r$ with respect to r, and hence

$$\frac{1}{r^2}\frac{\mathrm{d}(r^2 E_r)}{\mathrm{d}r} = \frac{\rho}{\varepsilon_0}, \tag{1.169}$$

analogously to Eq. (1.158). This is a one-dimensional differential Gauss' law in the spherical coordinate system, and its generalization to the 3-D case of $\rho = \rho(r, \theta, \phi)$ mimics the derivation in Eqs. (1.159)–(1.163) accommodated to the elemental spherical cuboid in Fig. 1.10, and similarly for $\rho = \rho(r, \phi, z)$ (Fig. 1.25). The resulting formula for the cylindrical coordinate system is

$$\boxed{\operatorname{div}\mathbf{E} = \nabla\cdot\mathbf{E} = \frac{1}{r}\frac{\partial}{\partial r}(rE_r) + \frac{1}{r}\frac{\partial E_\phi}{\partial \phi} + \frac{\partial E_z}{\partial z},} \tag{1.170}$$

divergence in cylindrical coordinates

and that for spherical coordinates

$$\boxed{\operatorname{div}\mathbf{E} = \nabla\cdot\mathbf{E} = \frac{1}{r^2}\frac{\partial}{\partial r}\left(r^2 E_r\right) + \frac{1}{r\sin\theta}\frac{\partial}{\partial\theta}(\sin\theta\,E_\theta) + \frac{1}{r\sin\theta}\frac{\partial E_\phi}{\partial\phi}.} \tag{1.171}$$

divergence in spherical coordinates

As the vector expression for the operator ∇ in Eq. (1.100) is valid only in the Cartesian coordinate system, the dot product in $\nabla\cdot\mathbf{E}$ actually makes sense only in rectangular coordinates. However, we shall still use extensively the notation $\nabla\cdot\mathbf{E}$ in cylindrical and spherical coordinate systems as well – to simply mean the divergence of $\mathbf{E}$ ($\operatorname{div}\mathbf{E}$) and a symbolic representation of the respective formulas in Eqs. (1.170) and (1.171). In general, we shall often draw conclusions about gradient, divergence, and other operations involving vector field quantities by treating ∇ as a vector, Eq. (1.100), and performing formally vector operations with it, like in Eq. (1.165). Most importantly, these conclusions, properties, and relations, although derived in Cartesian coordinates, hold true (as identities) in all possible (cylindrical, spherical, and other) coordinate systems, because the properties of a physical quantity and relations between two or more quantities are the same regardless of the choice of a coordinate system.

Combining Eqs. (1.161) and (1.167), we obtain the equation

$$\boxed{\operatorname{div}\mathbf{E} = \nabla\cdot\mathbf{E} = \lim_{\Delta v\to 0}\frac{\oint_S \mathbf{E}\cdot\mathrm{d}\mathbf{S}}{\Delta v},} \tag{1.172}$$

physical meaning of the divergence

which tells us that the divergence of $\mathbf{E}$ at a given point is the net outflow of the flux of $\mathbf{E}$ per unit volume as the volume shrinks about the point. Eq. (1.172) is an equivalent mathematical definition of the divergence of a vector. From it, we may regard div $\mathbf{E}$ physically as a measure of how much the field diverges from the point. In general, a positive divergence of any vector field indicates a local source of the field at that point producing radial field components with respect to the point – the outward flow of flux is positive, which means that the vector diverges (spreads out) at the point. Possible field components that are tangential to S do not contribute to the flux, and are not related to the divergence. Similarly, a negative divergence indicates a sink at the point (local negative source) – the outward flux flow is negative, and the vector field converges at the point. Finally, the divergence is zero where there is neither source nor sink of locally radial field components. Gauss' law simply tells us that these positive and negative sources in the case of the electric field, $\mathbf{E}$, are positive and negative charges, $\rho\,\Delta v$. Quantitatively, the divergence represents the volume density of sources, and in our case this density is ρ.

Let us now replace ρ in the integral form of Gauss' law, Eq. (1.135), by its equal, $\varepsilon_0\nabla\cdot\mathbf{E}$ – from the differential form of the law, Eq. (1.166):

divergence theorem

$$\boxed{\oint_S \mathbf{E}\cdot d\mathbf{S} = \int_v \nabla\cdot\mathbf{E}\,dv.} \tag{1.173}$$

This equation is called the divergence theorem (or Gauss-Ostrogradsky theorem). Although we have obtained it specifically for the electrostatic field in free space, the theorem is true for any vector field (for which the appropriate partial derivatives exist) and is one of the basic theorems of vector calculus. It applies to an arbitrary closed surface S and, in words, states that the net outward flux of a vector field through S equals the volume integral of the divergence of that field throughout the volume v bounded by S.

Example 1.22 Problem with Spherical Symmetry Using Differential Gauss' Law

Redo Example 1.18, part (a), but now employing Gauss' law in differential form.

Solution As we already know, from the spherical symmetry of the problem, that the electric field vector both inside the charged sphere and outside it has the form in Eq. (1.136), the formula for the divergence in spherical coordinates, given by Eq. (1.171), retains only the first term. For a point inside the charge distribution in Fig. 1.33, Gauss' law in differential form, Eq. (1.166), thus reduces to the following differential equation in a single coordinate, r:

$$\nabla\cdot\mathbf{E} = \frac{1}{r^2}\frac{\partial}{\partial r}\left[r^2E(r)\right] = \frac{\rho}{\varepsilon_0} \quad (0 \le r < a), \tag{1.174}$$

which can be directly solved by integration. Since $\rho = \text{const}$, we have

$$r^2E(r) = \frac{\rho}{\varepsilon_0}\int r^2\,dr + C_1 = \frac{\rho r^3}{3\varepsilon_0} + C_1 \quad\longrightarrow\quad E(r) = \frac{\rho r}{3\varepsilon_0} + \frac{C_1}{r^2}, \tag{1.175}$$

and the integration constant, C_1, is found from the "initial" condition $E(0) = 0$ at the center of the sphere (for $r = 0$). Namely, given that there is not a point charge (Q_0) at the point O in Fig. 1.33, the field at this point is zero, and hence $C_1 = 0$ [otherwise, the constant would amount to $C_1 = Q_0/(4\pi\varepsilon_0)$, from Eq. (1.24)], which yields the result for E in Eq. (1.140).

Outside the charge distribution in Fig. 1.33, $\rho = 0$, which substituted in Eq. (1.174) results in

$$\frac{\partial}{\partial r}\left[r^2E(r)\right] = 0 \quad\longrightarrow\quad r^2E(r) = C_2 \quad\longrightarrow\quad E(r) = \frac{C_2}{r^2} \quad (a < r < \infty). \tag{1.176}$$

The new constant of integration, C_2, is determined by means of the "boundary" condition at the boundary of the charged sphere, $r = a$; the field intensity on the inner side of the boundary, computed from Eq. (1.175), must be the same (since there exist no surface charges on the boundary) as that on the outer side of it, from Eq. (1.176),

$$E(a^-) = E(a^+) = E(a) \quad \longrightarrow \quad C_2 = \frac{\rho a^3}{3\varepsilon_0}. \tag{1.177}$$

With this, the result for E in Eq. (1.176) agrees with the solution in Eq. (1.140).

Example 1.23 Problem with Planar Symmetry by the Differential Gauss' Law

Redo Example 1.20 but with the use of the differential Gauss' law.

Solution As this is a problem with planar symmetry, the electric field intensity vector everywhere is of the form in Eq.(1.148). Hence, Gauss' law given by the differential equation in Eq. (1.158) applies, which we solve by integration with respect to x (also see the previous example). For the observation point inside the charge layer ($-a \le x \le a$) in Fig. 1.35,

$$E_x(x) = \frac{1}{\varepsilon_0} \int \rho(x)\, dx + C_1 = \frac{\rho_0 x}{\varepsilon_0} \left(1 - \frac{x^2}{3a^2}\right) + C_1, \quad \text{for } |x| \le a. \tag{1.178}$$

Because of symmetry (Fig. 1.35),

$$E_x(a) = -E_x(-a) \quad \longrightarrow \quad \frac{2\rho_0 a}{3\varepsilon_0} + C_1 = \frac{2\rho_0 a}{3\varepsilon_0} - C_1 \quad \longrightarrow \quad C_1 = 0, \tag{1.179}$$

which, substituted in Eq. (1.178), gives the same result as in Eq. (1.150).

For the point outside the layer ($|x| > a$), $\rho = 0$, so that Eq. (1.158) yields $E_x(x) = C_2$. Matching this field value to the one from Eqs. (1.178) and (1.179) in the "boundary" condition at the layer boundary defined by $x = a$, $E_x(a^+) = E_x(a^-) = E_x(a)$, we obtain $C_2 = 2\rho_0 a/(3\varepsilon_0)$, and thus $E_x(x) = 2\rho_0 a/(3\varepsilon_0)$ for $x > a$, the same as in Eq. (1.151). On the other side of the layer, the vector $\mathbf{E}$ has this same magnitude but opposite direction, resulting in the field expression in Eq. (1.152).

Note that the application of Gauss' law in differential form to a problem with planar symmetry and an antisymmetrical charge distribution is presented in Example 2.6.

Problems: 1.67–1.74; *MATLAB Exercises* (on Companion Website).

1.16 CONDUCTORS IN THE ELECTROSTATIC FIELD

So far, we have considered electrostatic fields in free space (a vacuum or air). We shall now extend our theory to electrostatic fields in the presence of materials. As we shall see, most of the formulas derived and solution techniques developed and used for the electrostatic field in free space are directly applicable to the analysis of electrostatic fields in material space, although some require modification. Materials can broadly be classified in terms of their electrical properties as conductors (which conduct electric current) and dielectrics (insulators). In the rest of this chapter, we shall study the interaction of the electrostatic field with conductors, in which case essentially no theoretical modification is needed to the electrostatic equations, whereas the behavior of dielectrics in the electrostatic field will be discussed in the next chapter.

Conductors have a large proportion of freely movable electric charges (free electrons and ions) that make the electric conductivity (ability to conduct electric current) of the material. Best conductors (with highest conductivity) are metals (such as silver, copper, gold, aluminum, etc.). In many applications, we consider

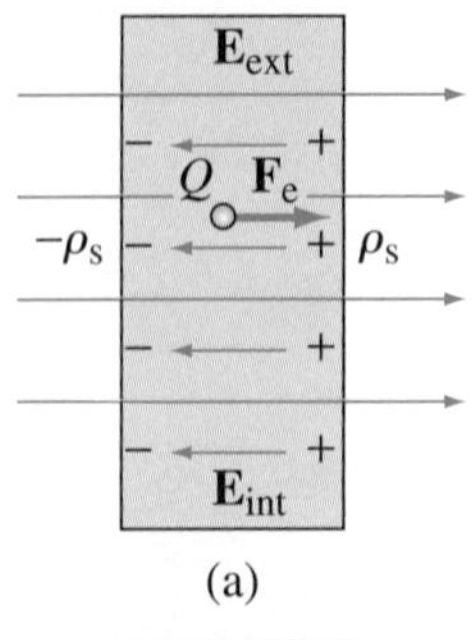

(a)

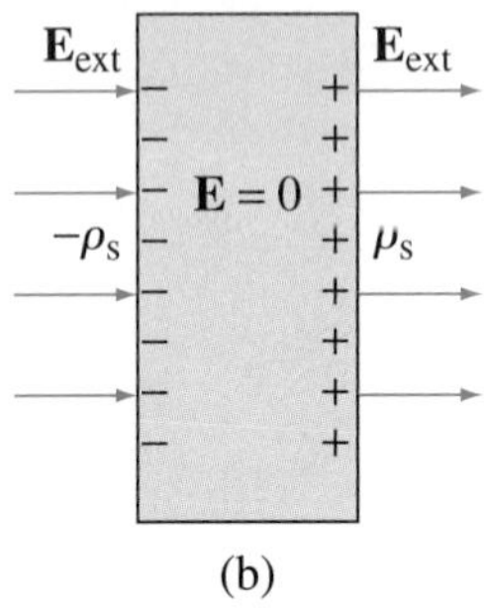

(b)

Figure 1.38 (a) A conductor in an external electrostatic field. (b) After a transitional process, there is no electrostatic field inside the conductor.

metallic conductors as perfect electric conductors (with infinite conductivity). There are also many other, less conductive conductors such as water, earth (ground), so-called semiconductors (e.g., silicon and germanium), etc. Practically all insulators (glass, paper, rubber, etc.) have some (usually extremely low) conductivity, and thus theoretically are (very poor) conductors, although almost always (in electrostatics) they may be considered as perfect dielectrics (with zero conductivity), i.e., nonconductors. In our studies of electrostatic fields, by conductor we normally mean a metallic conductor.

Consider an isolated conductor, shown in Fig. 1.38(a). Assume that it is uncharged or electrically neutral (its net charge is zero). When an external electrostatic field $\mathbf{E}_{ext}$ is applied, the free charges in the conductor are influenced by the electric force [see Eq. (1.23)]

$$\mathbf{F}_e = Q\mathbf{E}_{ext}. \tag{1.180}$$

This force pushes the positive charges along the direction of $\mathbf{E}_{ext}$, that is from left to right, while the negative charges move in the opposite direction. Consequently, the right-hand side of the conductor becomes progressively more positive, and the left-hand side progressively more negative. In fact, for a metallic conductor, what actually happens is movement of free electrons (negative charges) toward the left side leaving an equal number of positive charges, namely, a deficiency of electrons, on the other side of the conductor. Accumulated free charges form two layers of surface charge, of densities $-\rho_s$ ($\rho_s > 0$) and ρ_s, on the conductor sides. Creation of surplus charges in the body caused by an external electrostatic field is called the electrostatic induction. The induced charges, in turn, set up an internal induced electric field in the conductor, $\mathbf{E}_{int}$, which is directed from the positive to the negative layer, i.e., oppositely to $\mathbf{E}_{ext}$. As ρ_s increases, $\mathbf{E}_{int}$ becomes progressively stronger, and it opposes the migration of charges from left to right. In the equilibrium, $\mathbf{E}_{int}$ completely cancels out $\mathbf{E}_{ext}$ in the conductor, so that the total field $\mathbf{E}$ in the conductor is zero, and the motion of charges stops, as illustrated in Fig. 1.38(b). Note that the conductor remains uncharged as a whole. The entire transitional process is extremely fast, and the electrostatic steady state is established practically instantaneously. In fact, based on the length of the time needed for this process of movement of charges to the surface of a material body, i.e., their redistribution in such a way that the total electric field inside the body becomes zero, we determine whether a material is a conductor or dielectric. For example, as we shall see in a later chapter, the time to reach the equilibrium for the most commonly used metallic conductor – copper – is as brief as $\sim 10^{-19}$ s, whereas it takes as long as ~ 50 days for the charge rearrangement across a piece of fused quartz (very good insulator).

In the case of a conductor that had been charged (with a positive or negative excess charge) prior to being situated in the external field, a similar process takes place. All free charges (for a metallic conductor, free electrons of the conductor, which abundantly exist in the material also when it is electrically neutral as a whole, plus excess charge[15]) are exposed to the force $\mathbf{F}_e$ and produce the internal field that cancels out the externally applied field in the electrostatic equilibrium.

[15]Note that excess charge on a metallic body may be produced by bringing electrons to the body (negative excess charge) or by taking some of its free electrons away (positive excess charge), where the number of these extra or missing electrons is always much smaller than the total count of free electrons of the body.

We conclude that under electrostatic conditions, there cannot be electric field in a conductor,

$$\boxed{\mathbf{E} = 0.} \tag{1.181}$$

no electrostatic field inside a conductor

This is the first fundamental property of conductors in electrostatics. Starting from it, we derive all other fundamental conclusions about the behavior of conductors in the electrostatic field.

According to Eqs. (1.181), (1.90), and (1.88), the voltage between any two points in the conductor, including points on its surface, is zero. This means that a conductor is an equipotential body, i.e., the potential is the same everywhere in the conductor and on its surface,

$$\boxed{V = \text{const.}} \tag{1.182}$$

interior and surface of a conductor are equipotential

From Eq. (1.181), $\nabla \cdot \mathbf{E} = 0$ in a conductor, implying that [Eq. (1.166)] there cannot be surplus volume charges inside it,

$$\boxed{\rho = 0.} \tag{1.183}$$

no volume charge inside a conductor

So, any locally surplus charge of a conductor (whether it is neutral as a whole or not) must be located at the surface of the conductor.

Let us now derive so-called boundary conditions that the electric field must satisfy on a conductor surface. The electric field intensity vector $\mathbf{E}$ near the conductor in a vacuum can be decomposed into the normal and tangential components with respect to the boundary surface, as shown in Fig. 1.39(a). The two components are

$$E_\text{n} = E\cos\alpha \quad \text{and} \quad E_\text{t} = E\sin\alpha, \tag{1.184}$$

respectively, where α is the angle that $\mathbf{E}$ makes with the normal to the surface. We apply Eq. (1.75) to the narrow rectangular elementary contour C in Fig. 1.39(a). The field is zero along the lower side of C ($\mathbf{E} = 0$ in conductors), and we let the contour side Δh shrink to zero pressing the sides Δl tightly onto the boundary surface, so that the only contribution to the line integral in Eq. (1.75) is $\mathbf{E} \cdot \Delta\mathbf{l}$ along the upper side of C (no integration is needed, because Δl is small),

$$\oint_C \mathbf{E} \cdot d\mathbf{l} = \mathbf{E} \cdot \Delta\mathbf{l} = E\Delta l \sin\alpha = E_\text{t}\,\Delta l = 0. \tag{1.185}$$

Hence, there is no tangential component of $\mathbf{E}$ over the surface of a conducting body in electrostatics,

$$\boxed{E_\text{t} = 0.} \tag{1.186}$$

zero tangential electric field on a conductor surface

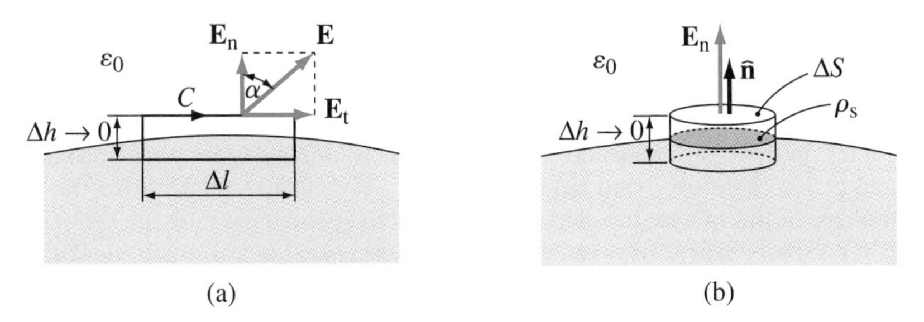

Figure 1.39 Deriving boundary conditions for the electrostatic field ($\mathbf{E}$) near a conductor surface: (a) narrow rectangular elementary contour (used for the boundary condition for the tangential component of $\mathbf{E}$) and (b) pillbox elementary closed surface (for the boundary condition for the normal component of $\mathbf{E}$).

In other words, the electric field intensity vector on the surface of a conductor is always normal to the surface,

$$\mathbf{E} = E_{\mathrm{n}}\hat{\mathbf{n}}, \tag{1.187}$$

where $\hat{\mathbf{n}}$ is the normal unit vector on the surface, directed from the surface outward.

To obtain the boundary condition for the normal (the only existing) component of $\mathbf{E}$, we apply Eq. (1.133) to the pillbox Gaussian surface, with bases ΔS and height Δh (shrinking to zero), shown in Fig. 1.39(b). For similar reasons as in obtaining Eq. (1.185), the flux in Eq. (1.133) reduces to $\mathbf{E} \cdot \Delta\mathbf{S}$ over the upper side of S. Because the charge enclosed by S is $\rho_{\mathrm{s}}\,\Delta S$,

$$\oint_S \mathbf{E} \cdot \mathrm{d}\mathbf{S} = \mathbf{E} \cdot \Delta\mathbf{S} = (E_{\mathrm{n}}\hat{\mathbf{n}}) \cdot (\Delta S\hat{\mathbf{n}}) = E_{\mathrm{n}}\,\Delta S = \frac{1}{\varepsilon_0}\,\rho_{\mathrm{s}}\,\Delta S, \tag{1.188}$$

providing the relationship between the normal component of the electric field intensity vector near a conductor surface and the surface charge density on the surface:

normal electric field component on a conductor surface

$$\boxed{E_{\mathrm{n}} = \frac{\rho_{\mathrm{s}}}{\varepsilon_0}.} \tag{1.189}$$

The lines of the electric field are normal to the surface of a conductor. We should always remember that the normal component E_{n} in Eq. (1.189) is defined with respect to the outward normal $\hat{\mathbf{n}}$. When $\rho_{\mathrm{s}} > 0$, the field lines start from the conductor ($E_{\mathrm{n}} > 0$), whereas they end on it ($E_{\mathrm{n}} < 0$) when $\rho_{\mathrm{s}} < 0$.

In analyzing complex conducting structures, we usually do not know in advance the orientation of the electric field intensity vector at specific portions of conducting surfaces. In such cases, the following expression for ρ_{s} in terms of the field vector, obtained noting that $E_{\mathrm{n}} = \hat{\mathbf{n}} \cdot \mathbf{E}$ from Eq.(1.187), is useful:

$$\rho_{\mathrm{s}} = \varepsilon_0\,\hat{\mathbf{n}} \cdot \mathbf{E}. \tag{1.190}$$

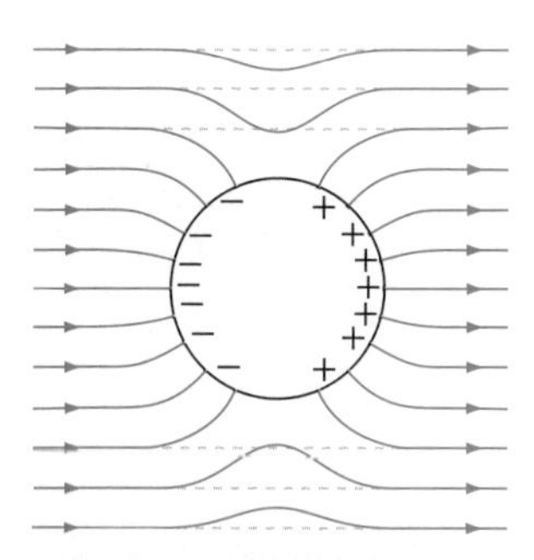

Figure 1.40 Uncharged metallic sphere in a uniform external electrostatic field; for Example 1.24.

Example 1.24 Metallic Sphere in a Uniform Electrostatic Field

An uncharged metallic sphere is brought into a uniform electrostatic field, in air. Sketch the field lines around the sphere after electrostatic equilibrium is reached.

Solution The field lines in the new electrostatic state are sketched in Fig. 1.40. Because the field due to induced charges on the sphere surface (this field exists both inside and outside the sphere) is superimposed to the external field, the field inside the sphere becomes zero, and that outside it is not uniform any more. Negative induced charges are sinks of the field lines on the left-hand side of the sphere, whereas the positive induced charges are sources of the field lines on the right-hand side. The field lines on both sides are normal to the sphere surface, and they therefore bend near the sphere. At points in air close to the left- and right-hand side of the sphere, the electric field is stronger (the field lines are denser) than in the remaining space. This is obvious as well from noting that near the left-hand side of the sphere, in air, the field due to negative induced charges dominates over the field due to positive charges on the opposite side of the sphere, it is directed toward the negative charges, and adds to the external field intensity. The field due to positive induced charges dominates near the right-hand side of the sphere, which results in the same strengthening of the external field at these points in air. The field at distances from the sphere a few times the sphere diameter is practically equal to the external field (the field due to induced charges is negligible).

1.17 EVALUATION OF THE ELECTRIC FIELD AND POTENTIAL DUE TO CHARGED CONDUCTORS

Assume that we know the charge distribution ρ_s over the surface of a conductor situated in free space. The electric field intensity at points close to the conductor surface can be evaluated from Eq. (1.189). How do we obtain the electric field and potential at an arbitrary point in space? The answer is straightforward. Because $\mathbf{E} = 0$ inside the conductor, nothing will change, as far as the field outside the conductor is concerned, if we remove the conductor and fill the space previously occupied by it with a vacuum, keeping the charge distribution ρ_s on the surface unchanged. With this useful equivalence, we are left with the problem of evaluating the field and potential due to a known surface charge distribution in free space, and we can use Eqs. (1.38), (1.83), (1.101), (1.133), and (1.165) to solve the problem.

Example 1.25 Charged Metallic Sphere

A metallic sphere of radius a is situated in air and charged with a charge Q. Find (a) the charge distribution of the sphere, (b) the electric field intensity vector in air, and (c) the potential of the sphere.

Solution

(a) Due to symmetry, the charge distribution over the sphere surface is uniform, and hence the associated surface charge density turns out to be

$$\rho_s = \frac{Q}{S_0} = \frac{Q}{4\pi a^2}, \tag{1.191}$$

where S_0 stands for the surface area of the sphere.

(b) The electric field around the sphere is radial, and has the form given by Eq. (1.136). Applying Gauss' law, Eq. (1.133), to a spherical surface of radius r $(a < r < \infty)$, positioned concentrically with the metallic sphere [see Eqs. (1.137) and (1.138)], we obtain

$$\boxed{E(r) = \frac{Q}{4\pi\varepsilon_0 r^2} \quad (a < r < \infty).} \tag{1.192}$$

electric field due to a charged metallic sphere

Note that

$$E(a^+) = \frac{Q}{4\pi\varepsilon_0 a^2} = \frac{\rho_s}{\varepsilon_0}, \tag{1.193}$$

which is in agreement with Eq. (1.189).[16]

(c) We realize that the field outside the charged metallic sphere, Eq. (1.192), is identical to that due to a point charge Q placed at the sphere center, and so is the potential. The potential at the surface $r = a$ is thus given by Eq. (1.141). This is the potential of the sphere, the same at any point of its interior and surface [see Eq. (1.182)].

Example 1.26 Charged Cylindrical Conductor

Repeat the previous example but for an infinitely long cylindrical conductor of radius a in air, if the charge per unit length of the conductor is Q'.

[16] $E(a^+)$ or $E(a + 0)$ [$E(a + \delta)$, $\delta \to 0$] designates the field in air very close to the conductor surface at $r = a$.

Solution

(a) Using Eq. (1.31), the charge per length h of the conductor is

$$Q_h = Q'h, \tag{1.194}$$

so the surface charge density amounts to

$$\rho_s = \frac{Q_h}{S_0} = \frac{Q_h}{2\pi ah} = \frac{Q'}{2\pi a}, \tag{1.195}$$

where S_0 is the surface area of that part of the conductor.

(b) The electric field is radial (with respect to the conductor axis). From Gauss' law [Eq. (1.133)] applied to the cylindrical surface of radius r ($a < r < \infty$) and height (length) h, coaxial with the conductor [see the left-hand side of Eq. (1.144)],

electric field due to a charged cylindrical conductor

$$\boxed{E(r) = \frac{1}{2\pi rh}\frac{Q_h}{\varepsilon_0} = \frac{Q'}{2\pi\varepsilon_0 r} \quad (a < r < \infty).} \tag{1.196}$$

(c) This is an infinite charge distribution, and the reference point for the potential cannot be adopted at infinity. The field in Eq. (1.196) being identical to the field due to an infinite line charge of density Q', Eq. (1.57), the potential due to the charged cylindrical conductor is given by the expression in Eq. (1.87). In particular, this expression for $r = a$ represents the potential of the conductor:

potential due to a charged cylindrical conductor

$$\boxed{V = \frac{Q'}{2\pi\varepsilon_0}\ln\frac{r_{\mathcal{R}}}{r} \quad (r \geq a)} \quad \longrightarrow \quad V_{\text{conductor}} = \frac{Q'}{2\pi\varepsilon_0}\ln\frac{r_{\mathcal{R}}}{a} \quad (r = a), \tag{1.197}$$

where $r_{\mathcal{R}}$ ($r_{\mathcal{R}} \geq a$) is the distance of the reference point from the conductor axis.

Example 1.27 Charged Sphere Enclosed by an Uncharged Shell

A metallic sphere, of radius a and charge Q ($Q > 0$), is enclosed by an uncharged concentric metallic spherical shell, of inner radius b and outer radius c ($a < b < c$). The medium everywhere is air. Find the potential at the center of the sphere.

Solution As a result of the electrostatic induction, there is induced surface charge on the surfaces of the shell. As we know, there cannot be volume charges in the metal under electrostatic conditions, Eq. (1.183). Let Q_b and Q_c denote the total induced charges on the inner and outer surface of the shell, respectively. On the surface of the sphere, $Q_a = Q$. Since every line of the electric field originating at a positive charge on the sphere terminates at a negative charge on the inner surface of the shell, the relationship between total charges on two surfaces is

$$Q_b = -Q_a. \tag{1.198}$$

This can be obtained also from Gauss' law, Eq. (1.133), applied to a closed surface that is entirely inside the metal of the shell, so that $\Psi_E = 0$ [because of Eq. (1.181)], which implies that $Q_S = 0$, i.e., $Q_a + Q_b = 0$. On the other hand, since the shell is uncharged,

$$Q_b + Q_c = 0, \tag{1.199}$$

and hence $Q_c = -Q_b = Q_a$. Because of symmetry, charges on individual surfaces are distributed uniformly, and the electric field everywhere in air is in the form given by Eq. (1.136). Distributions of the charge and field in the system are sketched in Fig. 1.41.

By means of Gauss' law, the electric field for $a < r < b$ and $c < r < \infty$ is

$$E(r) = \frac{Q_a}{4\pi\varepsilon_0 r^2} \quad \text{(in air)}, \tag{1.200}$$

whereas for $0 \leq r \leq a$ and $b \leq r \leq c$,

$$E(r) = 0 \quad \text{(in metal)}. \tag{1.201}$$

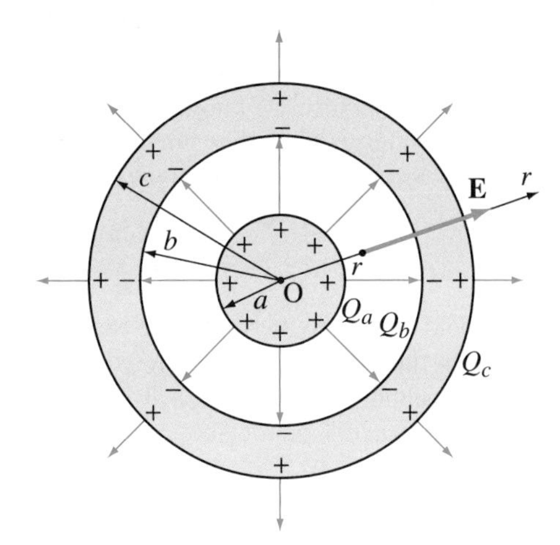

Figure 1.41 Charge and field distributions in the system of Example 1.27.

The potential at the point O in Fig. 1.41 is thus [Eq. (1.74)]

$$V = \int_{r=0}^{\infty} E(r)\,\mathrm{d}r = \frac{Q_a}{4\pi\varepsilon_0}\left(\int_a^b \frac{\mathrm{d}r}{r^2} + \int_c^{\infty} \frac{\mathrm{d}r}{r^2}\right) = \frac{Q}{4\pi\varepsilon_0}\left(\frac{1}{a} - \frac{1}{b} + \frac{1}{c}\right)$$

$$= \frac{Q(bc - ac + ab)}{4\pi\varepsilon_0 abc}. \tag{1.202}$$

Example 1.28 Five Parallel Large Flat Electrodes

Consider five parallel large metallic electrodes situated in air, as shown in Fig. 1.42(a). The thickness of each electrode, as well as the distance between each two adjacent electrodes, is $d = 2$ cm. The surface area of sides of electrodes facing each other is $S = 1\ \mathrm{m}^2$. The first, fourth, and fifth electrodes are grounded, the potential of the second electrode with respect to the ground is $V = 2$ kV, and the charge of the third electrode is $Q = 2\ \mu$C. Find the electric field intensity between the electrodes.

Solution Since the dimensions of the electrodes are much larger than the separation between them, we can neglect the fringing effects, i.e., we can ignore the nonuniformity of the electric field near the electrode edges, as well as the existence ("leakage") of the field lines outside the spaces between electrodes. We assume, thus, that the electric field is localized only in spaces between the electrode sides S, that vector $\mathbf{E}$ is normal to those sides, and that the field in each space is uniform. Charge distributions over electrode sides are also uniform. In

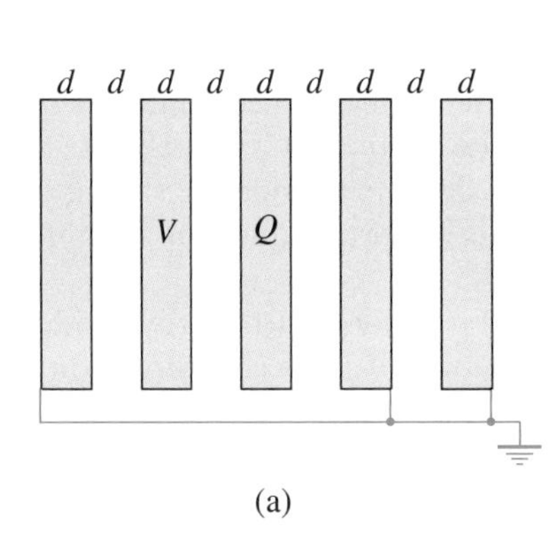

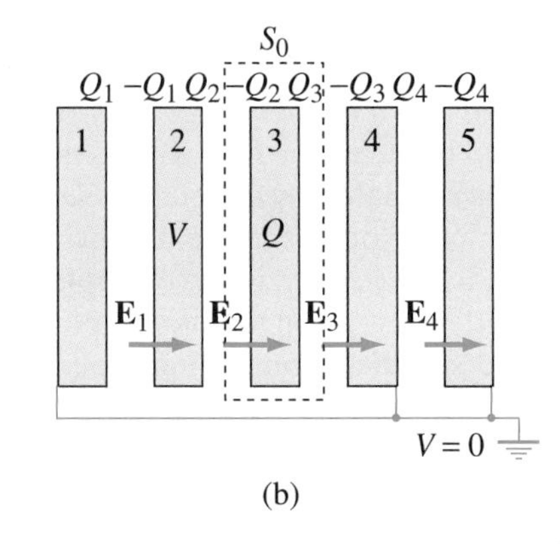

Figure 1.42 Electrostatic analysis of a system of five large electrodes in air: (a) geometry of the system and (b) charges on electrodes and fields between them; for Example 1.28.

each space between electrodes, neighboring sides of electrodes must be charged with charges of equal amounts, but of opposite polarities, as indicated in Fig. 1.42(b).

The potential V (which we know) of the second electrode with respect to the ground can be expressed in terms of the line integral of $\mathbf{E}$ from that electrode, to the left, to the first electrode, which is grounded (its potential is zero). The field is uniform, and the line integral is simply

$$V = \int_{\text{second electrode}}^{\text{first electrode}} \mathbf{E} \cdot d\mathbf{l} = -E_1 d. \tag{1.203}$$

Hence, the field intensity between the first and second electrodes is $E_1 = -V/d = -100$ kV/m. On the other side, the same potential can be expressed as the line integral of $\mathbf{E}$ to the right, to the fourth electrode, which is also grounded, yielding

$$V = E_2 d + E_3 d. \tag{1.204}$$

By a similar token, the field between the fourth and fifth electrodes is zero, $E_4 = 0$, and so are the associated charges on the sides of electrodes facing each other ($Q_4 = 0$), because these electrodes are at the same (zero) potential.

To solve for field intensities E_2 and E_3, we need one more equation with them as unknowns. Applying Gauss' law, Eq. (1.133), to the surface S_0 that encloses the third electrode (the charge of which is known), we obtain

$$-E_2 + E_3 = \frac{Q}{\varepsilon_0 S}. \tag{1.205}$$

The solution of the system of equations composed from Eqs. (1.204) and (1.205) is $E_2 = -62.94$ kV/m and $E_3 = 162.94$ kV/m.

Problems: 1.75–1.82; *Conceptual Questions* (on Companion Website): 1.20–1.22; *MATLAB Exercises* (on Companion Website).

1.18 ELECTROSTATIC SHIELDING

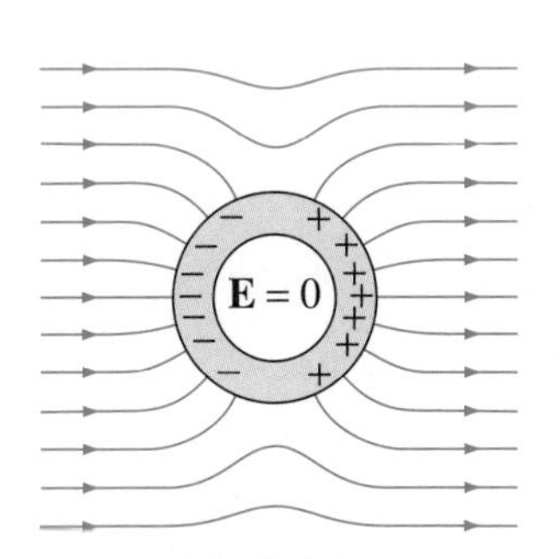

Figure 1.43 Metallic shell in an electrostatic field – Faraday cage.

Let us consider again the metallic sphere in the external electrostatic field shown in Fig. 1.40. Because there is no field throughout the sphere interior, we can remove it, without affecting the field outside the sphere. We thus obtain a domain with no field, bounded by a metallic shell (Fig. 1.43). This means that the space inside the shell cavity is perfectly protected (isolated) from the external electrostatic field. The thickness of the shell can be arbitrary, and its shape does not need to be spherical. Hence, an arbitrary closed conducting shell represents a perfect electrostatic shield or screen for its interior domain. We call such a shield a Faraday cage. If the field outside the cage is changed, the charge on the cage walls will redistribute itself so that the field inside will remain zero.

We see that a Faraday cage provides an absolute protection to its interior from an external electrostatic field. Let us now reverse the problem. Can an electrostatic field be encapsulated by a metallic shell so that the domain outside the shell is protected from the sources inside it? The answer to this important question is twofold. Let us get to it by analyzing two simple examples.

Consider first a single positive point charge Q positioned arbitrarily inside an uncharged spherical conducting shell. The distribution of induced charges on the shell surfaces and the field lines are sketched in Fig. 1.44(a). The total induced charges on the inner and outer shell surfaces are $-Q$ and Q, respectively (see

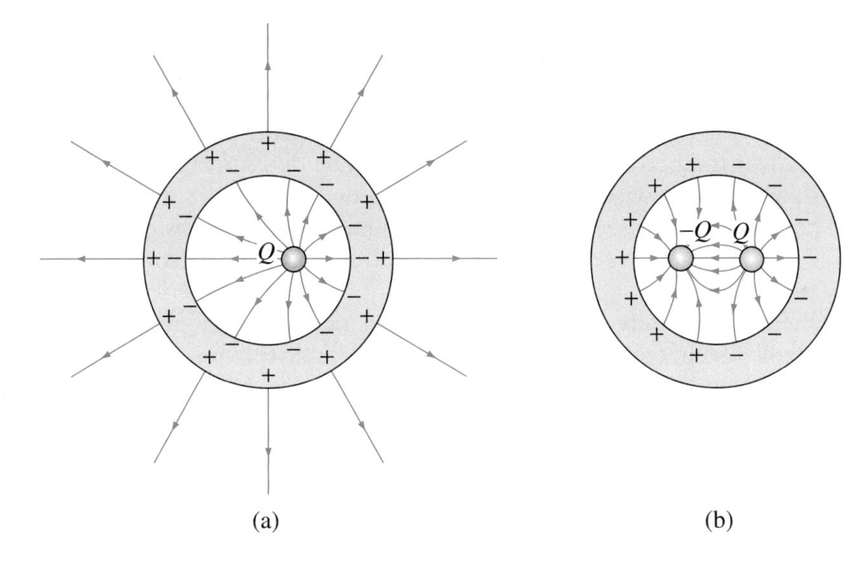

Figure 1.44 Single point charge (a) and two point charges amounting to a zero total charge (b) in a Faraday cage.

Example 1.27). The concentration of induced negative charges is higher on the side of the inner surface closer to the point charge. Because there is no field in the shell wall, the positive charge on the outer shell surface is distributed irrespective of the position of the point charge inside the cavity, which means uniformly in this case (the surface is smooth and symmetrical).

Let us now add another point charge in the cavity, and let it be exactly $-Q$, as shown in Fig. 1.44(b). The total negative and total positive induced charges on the inner surface of the shell are both less than Q in magnitude, because some field lines originating at the positive point charge end at the negative one inside the cavity, but mutually they are equal in magnitude and opposite in polarity. The net induced charge on the inner surface is therefore zero, implying that there is no charges whatsoever on the outer shell surface. This means, in turn, that there is no field outside the shell, which is also in agreement with Gauss' law, applied to a spherical surface enclosing the shell.

We conclude that a Faraday cage can completely encapsulate an interior electrostatic field, with a zero field outside, only if the total charge inside the cage is zero. This is true for any interior charge distribution, provided that the object or the system of objects (devices) inside the cage is electrically neutral (uncharged) as a whole. In cases when the total interior charge is not zero, the exterior domain (and neighboring objects) is in the field of induced charges on the outer surface of the cage. The exterior field, however, is totally independent of the distribution of interior sources. Its relative distribution in space depends only on the shape of the outer cage surface, whereas its absolute values at individual points in space are also proportional to the total amount of interior charge.

It is interesting to note that even very thin metallic shells represent ideal electrostatic shields, either in the mode of operation illustrated in Fig. 1.43 or in the one illustrated in Fig. 1.44(b). However, as we shall see in a later chapter, this is not necessarily the case with time-varying fields, where the effectiveness of a shield of a given thickness depends on the metal conductivity and on the rate with which the field varies in time (that is, frequency in the case of time-harmonic fields).

Conceptual Questions (on Companion Website): 1.23 and 1.24.

1.19 CHARGE DISTRIBUTION ON METALLIC BODIES OF ARBITRARY SHAPES

In the general case of a charged metallic body of an arbitrary shape, the charge distribution over the body surface is not uniform. The determination of this distribution for a given body with nonsymmetrical and/or nonsmooth surface is a rather complex problem. In this section, we shall get some qualitative insight about how the charge is distributed over the surface of an arbitrarily shaped isolated conducting body, and in the next section we shall present a general numerical method for determining approximately the charge density function over conducting objects.

Consider a system composed of two charged metallic spheres of different radii, a and b, whose centers are a distance d apart, in free space. Let the spheres be connected by a very thin conductor, as shown in Fig. 1.45. Assume, for simplicity, that $d \gg a, b$, so that the electric potential of each sphere can be evaluated as if the other one were not present. In addition, we assume that the charge along the connecting conductor is zero, because the conductor is very thin, and ignore its influence on the field between the spheres. Therefore, the potentials of spheres are [see Example 1.25 and Eq. (1.141)]

$$V_a = \frac{Q_a}{4\pi\varepsilon_0 a} \quad \text{and} \quad V_b = \frac{Q_b}{4\pi\varepsilon_0 b}, \tag{1.206}$$

respectively, where Q_a and Q_b are the associated total charges of spheres. The spheres being galvanically connected together, and thus representing a single conducting body, which must be equipotential, Eq. (1.182), these potentials are the same,

$$V_a = V_b. \tag{1.207}$$

So, by equating the expressions in Eq. (1.206), we obtain

$$\frac{Q_a}{Q_b} = \frac{a}{b}. \tag{1.208}$$

Using Eq. (1.191), the sphere charges can be expressed in terms of the corresponding surface charge densities, with which Eq. (1.208) becomes

charge density $\propto$ surface curvature

$$\boxed{\frac{\rho_{sa}}{\rho_{sb}} = \frac{b}{a},} \tag{1.209}$$

and, from Eq. (1.193), the corresponding relationship between the electric field intensities near the surfaces of spheres turns out to be

local field intensity $\propto$ surface curvature

$$\boxed{\frac{E_a}{E_b} = \frac{b}{a}.} \tag{1.210}$$

We see from Eqs. (1.209) and (1.210) that the charge is distributed between the two spheres in Fig. 1.45 in such a way that the surface charge density on and electric

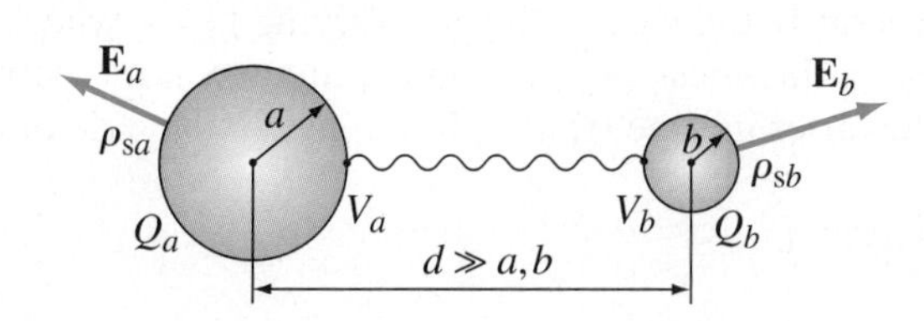

Figure 1.45 Two metallic spheres of different radii at the same potential.

field intensity near the surface of individual spheres is inversely proportional to the sphere radius. The surface charge is denser and the field stronger on the smaller sphere.

The importance of Eqs. (1.209) and (1.210) is much beyond the particular system in Fig. 1.45. They imply a general conclusion that the surface charge density and the nearby field intensity at different parts of the surface of an arbitrarily shaped conducting body are approximately proportional to the local curvature of the surface, as long as it is convex.[17] This means, generally, that the largest concentration of charge and the strongest electric field are around sharp parts of conducting bodies. Note that this phenomenon is essential for the operation of lightning arresters, as we shall see in the next chapter.

Problems: 1.83; *Conceptual Questions* (on Companion Website): 1.25–1.27.

1.20 METHOD OF MOMENTS FOR NUMERICAL ANALYSIS OF CHARGED METALLIC BODIES

Consider a charged metallic body of an arbitrary shape situated in free space. Let the electric potential of the body with respect to the reference point at infinity be V_0. Our goal is to determine the charge distribution of the body. The potential at an arbitrary point on the body surface, S, can be expressed in terms of the surface charge density, ρ_s, over the entire S [Eq. (1.83)]. On the other hand, this potential equals V_0 (the body is equipotential), and hence

$$\frac{1}{4\pi\varepsilon_0}\int_S \frac{\rho_s \, dS}{R} = V_0 \qquad \text{(at an arbitrary point on } S). \tag{1.211}$$

surface integral equation for charge distribution

This is an integral equation with the function ρ_s over S as unknown quantity, to be determined.

Eq. (1.211) cannot be solved analytically – in a closed form, but only numerically, with the aid of a computer. The method of moments (MoM) is a common numerical technique used in solving integral equations such as Eq. (1.211) in electromagnetics and in other disciplines of science and engineering. MoM can be implemented in numerous ways, but the simplest MoM solution in this case implies subdivision of the surface S into small patches ΔS_i, $i = 1, 2, \ldots, N$, with a constant approximation of the unknown function ρ_s on each patch. That is, we assume that each patch is uniformly charged,

$$\rho_s \approx \rho_{si} \quad (\text{on } \Delta S_i), \quad i = 1, 2, \ldots, N. \tag{1.212}$$

piece-wise constant approximation for charge

With this approximation, we reduce Eq. (1.211) to its approximate form:

$$\sum_{i=1}^{N} \rho_{si} \int_{\Delta S_i} \frac{dS}{4\pi\varepsilon_0 R} = V_0, \tag{1.213}$$

in which the unknown quantities are N charge-distribution coefficients, $\rho_{s1}, \rho_{s2}, \ldots,$ ρ_{sN}. Shown in Fig. 1.46 is an example of the application of this method to a metallic

[17] If the surface of a conducting body is concave (curved inward), the effect is just opposite; for a deep incurvature, we actually have a partial effect of a Faraday cage (cavity), Fig.1.43, and a decrease of the local field intensity.

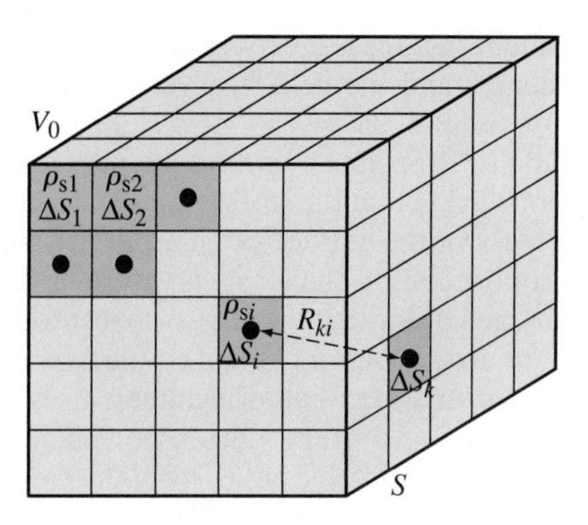

Figure 1.46 Method of moments (MoM) for analysis of charged metallic bodies: approximation of the surface charge distribution on a cube by means of N small square patches with constant charge densities.

cube, where square patches are used [for the particular subdivision shown in the figure, $N = 6 \times (5 \times 5) = 150$, which is a rather coarse model].

By stipulating that Eq. (1.213) be satisfied at centers of every small patch, individually, we obtain[18]

$$\begin{aligned} A_{11}\rho_{s1} + A_{12}\rho_{s2} + \cdots + A_{1N}\rho_{sN} &= V_0 \quad \text{(at the center of } \Delta S_1\text{)}, \\ A_{21}\rho_{s1} + A_{22}\rho_{s2} + \cdots + A_{2N}\rho_{sN} &= V_0 \quad \text{(at the center of } \Delta S_2\text{)}, \\ &\vdots \\ A_{N1}\rho_{s1} + A_{N2}\rho_{s2} + \cdots + A_{NN}\rho_{sN} &= V_0 \quad \text{(at the center of } \Delta S_N\text{)}. \end{aligned} \tag{1.214}$$

This is a system of N linear algebraic equations in N unknowns, $\rho_{s1}, \rho_{s2}, \ldots, \rho_{sN}$. In matrix form,

MoM matrix equation

$$\boxed{[A][\rho_s] = [B],} \tag{1.215}$$

where $[\rho_s]$ is a column matrix whose elements are the unknown charge-distribution coefficients, while elements of the column matrix $[B]$ are known and all equal V_0. Elements of matrix $[A]$, which is a square matrix, are given by

point-matching at centers of patches

$$\boxed{A_{ki} = \int_{\Delta S_i} \frac{\mathrm{d}S}{4\pi\varepsilon_0 R} \quad \text{(at the center of } \Delta S_k\text{)}, \quad k, i = 1, 2, \ldots, N,} \tag{1.216}$$

and they can be computed irrespective of the particular charge distribution. Physically, A_{ki} is the potential at the center of patch ΔS_k due to patch ΔS_i that is uniformly charged with unit (1 C/m^2) surface charge density. In the case of commonly used square or triangular flat patches, this potential can be evaluated analytically (exactly), while it is evaluated numerically (approximately) if some other surface elements are used (for example, curvilinear quadrilateral or triangular patches). Once matrix $[A]$ is filled, i.e., all its elements computed, we can use matrix inversion,

$$[\rho_s] = [A]^{-1}[B], \tag{1.217}$$

or any other standard method for solving systems of linear algebraic equations (e.g., Gaussian elimination method), to obtain the numerical results for the charge-distribution coefficients, which constitute an approximate surface charge distribution of the body, and a numerical solution to the integral equation, Eq. (1.211).

[18]The variant of the method of moments in which the left-hand side and the right-hand side of an integral equation [in our case, Eq. (1.211)] are "matched" to be equal at specific points of the definition domain of the equation (surface S in our case) is called the point-matching method. The idea of point-matching is similar to the concept of taking moments in mechanics, and hence the generic name method of moments.

The larger the number of subdivisions, N, the more accurate (but more computationally demanding in terms of computer resources) solution.

The crudest approximation in computing the elements of matrix $[A]$ is given by

$$A_{ki} = \begin{cases} \Delta S_i/(4\pi\varepsilon_0 R_{ki}) & \text{for } k \neq i \\ \sqrt{\Delta S_i}/(2\sqrt{\pi}\,\varepsilon_0) & \text{for } k = i \end{cases}. \tag{1.218}$$

Here, all nondiagonal elements of $[A]$ $(k \neq i)$ are evaluated by approximating the charged patch ΔS_i by an equivalent point charge, $\Delta Q_i = 1\ (\text{C/m}^2) \times \Delta S_i$, placed at the patch center, and using the expression for the electric potential due to a point charge in free space, Eq. (1.80), with R_{ki} being the distance between centers of patches ΔS_k and ΔS_i (Fig. 1.46). In filling diagonal elements (self terms) of $[A]$ $(k = i)$, the potential due to a (square or triangular) patch ΔS_i at the center of that same patch is evaluated by approximating the patch by the equivalent circular patch of the same surface area and radius $\sqrt{\Delta S_i/\pi}$, and employing the potential expression given in Problem 1.34 with $z = 0$ and $\rho_s = 1\ \text{C/m}^2$.

Starting from the result for $[\rho_s]$, we can now obtain any other quantity of interest (potential and field at any point in space, etc.). For instance, the total charge of the body can be found using the approximate version of the surface integral expression in Eqs. (1.29):

$$Q = \sum_{i=1}^{N} \rho_{si}\Delta S_i. \tag{1.219}$$

Problems: 1.84–1.86; *MATLAB Exercises* (on Companion Website).

1.21 IMAGE THEORY

Often, electrostatic systems include charge configurations in the presence of grounded conducting planes. Examples are charged conductors near grounded metallic plates or large flat bodies, transmission lines in which one of the conductors is a ground plane (such as microstrip transmission lines), various charged objects (power lines, charged clouds, charged airplanes, lightning rods, etc.) above the earth's surface, and so on. There is a very useful theory (theorem) by means of which we can remove the conducting plane from the system, and replace it by the equivalent charge distribution in free space. In this section, we shall derive this theorem, and apply it to problems which otherwise could not be analytically solved.

Consider two point charges of equal magnitudes and opposite polarities, Q and $-Q$, in free space. By symmetry, the total electric field intensity vector due to the charges is normal to the plane of symmetry of the charges (the tangential components due to individual charges are of equal magnitudes and opposite directions, so they cancel each other out), as shown in Fig. 1.47(a). The plane of symmetry is equipotential, and at the potential $V = 0$ (the potentials due to individual charges are of the same magnitudes and opposite signs, and they cancel out). Hence, nothing will change in the entire space if we insert an infinite grounded metallic foil (conducting plane) in the plane of symmetry, as is done in Fig. 1.47(b), because the boundary condition in Eq. (1.186) is automatically satisfied and $V = 0$ over the plane of symmetry (the foil is grounded). In the new system, surface charges will be induced on both sides of the foil, according to Eq. (1.190). We note that the foil actually separates the entire space onto two completely independent half-spaces, i.e., it

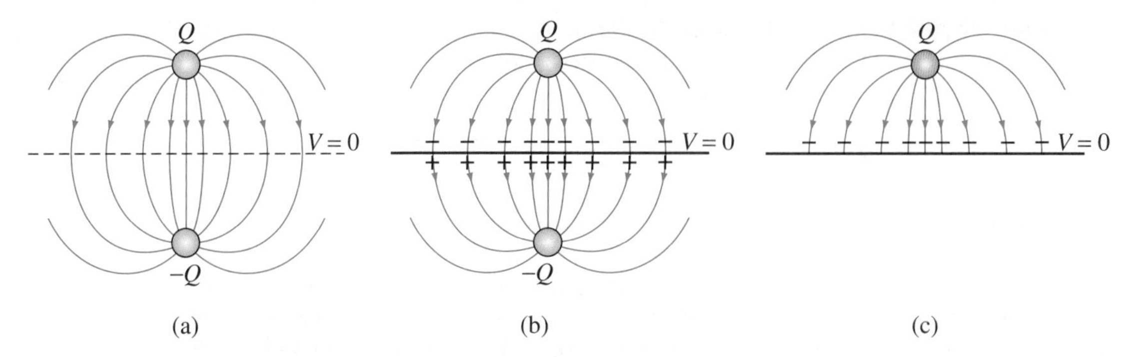

Figure 1.47 Deriving image theory: systems (a), (b), and (c) are equivalent with respect to the electric field in the upper half-space.

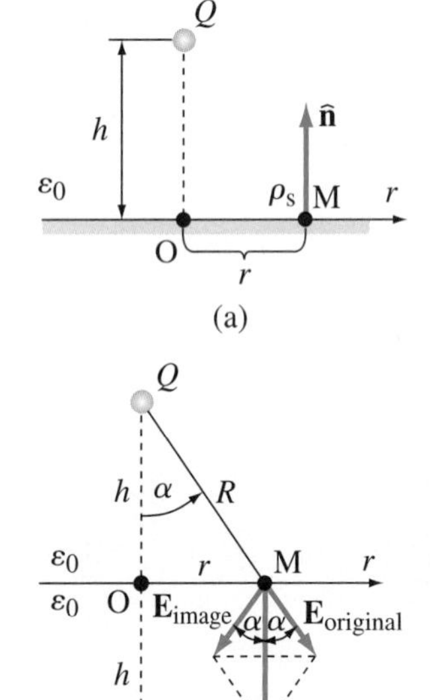

Figure 1.48 Computation of the induced surface charge density, ρ_s, on a conducting plane underneath a point charge Q: (a) original system and (b) equivalent system using image theory; for Example 1.29.

acts as a perfect electrostatic screen (see Section 1.18) between the two half-spaces. Because of that, nothing will change in the upper half-space if we, furthermore, remove the point charge $-Q$ and the induced charge on the lower side of the foil from the system. (Note that we can put whatever we want below the foil, and the field above it will remain the same, in the electrostatic state.) We are thus left with the point charge Q above the foil and the induced charge on its upper side (and nothing in the lower half-space), as depicted in Fig. 1.47(c).

We conclude that, as far as the electrostatic field in the upper half-space is concerned, systems in Figs. 1.47(a) and (c) are equivalent. This is so-called image theory, which, generalized to more than one point charge, i.e., to a (discrete or continuous) charge distribution, states that an arbitrary charge configuration above an infinite grounded conducting plane can be replaced by a new charge configuration in free space consisting of the original charge configuration itself and its negative image in the (former) conducting plane. The equivalence is with respect to the electric field above the conducting plane, whose component due to the induced charge on the plane is equal to the field of the image.

Example 1.29 Induced Charge Distribution on a Conducting Plane

A point charge Q is placed in air at a height h above a grounded conducting plane. (a) Determine the density of induced surface charges at an arbitrary point on the plane. (b) Find the total induced charge on the plane.

Solution

(a) The surface charge density ρ_s at a point M on the conducting plane, in Fig. 1.48(a), is given by Eq. (1.190), with the unit vector $\hat{\mathbf{n}}$ being vertical and directed from the plane upward, and $\mathbf{E}$ representing the electric field intensity vector in air, at a point that is "glued" to the point M from its upper side. This field is produced by the point charge Q and the induced surface charges on the conducting plane. According to the image theory, however, the field due to the induced charges equals that due to the negative image of Q in free space, as shown in Fig. 1.48(b). Let the position of the point M be defined by a radial distance r from the projection of Q on the plane (point O). Vector $\mathbf{E}$ is given by

$$\mathbf{E} = \mathbf{E}_{\text{original}} + \mathbf{E}_{\text{image}} = 2E_Q \cos\alpha\,(-\hat{\mathbf{n}}),$$

$$E_Q = \frac{Q}{4\pi\varepsilon_0 R^2}, \quad R = \sqrt{r^2 + h^2}, \quad \cos\alpha = \frac{h}{R}, \tag{1.220}$$

and the charge density comes out to be

$$\rho_s(r) = \varepsilon_0 \hat{\mathbf{n}} \cdot \mathbf{E} = -\frac{Qh}{2\pi \left(r^2 + h^2\right)^{3/2}}. \tag{1.221}$$

(b) The total induced charge on the conducting plane is

$$Q_{\text{ind}} = \int_S \rho_s(r) \underbrace{2\pi r \, dr}_{dS} = -Qh \int_{r=0}^{\infty} \frac{r \, dr}{R^3} = -Qh \int_{r=0}^{\infty} \frac{dR}{R^2} = Qh \frac{1}{R}\bigg|_{r=0}^{\infty} = -Q, \tag{1.222}$$

where dS is the surface area of an elemental ring of width dr and radius r $(0 \le r < \infty)$ around the point O (see Fig. 1.14), and the use is made of Eq. (1.62) to change variables in integration. The result in Eq. (1.222) is expected, because all field lines terminating on the image, $-Q$, in the equivalent system [(Fig. 1.48(b)] terminate on the surface charges of the conducting plane in the original system [Fig. 1.48(a)].

Example 1.30 Infinite Line Charge above a Conducting Plane

An infinite line charge of uniform density Q' is situated in air and is parallel to a grounded conducting plane at a distance h from it. Compute the electric force on the line charge per unit of its length.

Solution Under the influence of the electric field of the line charge, surface charges are induced on the conducting plane. The electric force on each meter of the line charge is therefore [Eq. (1.68)]

$$\mathbf{F}'_e = Q'\mathbf{E}_2, \tag{1.223}$$

where $\mathbf{E}_2$ represents the electric field vector at points along the line charge due to the induced charges on the plane. By image theory, this field is equal to the field due to a line charge in free space obtained as a negative image of Q', Fig. 1.49, and hence the following for its intensity [see Eq. (1.57)] and the per-unit-length force:

$$E_2 = \frac{Q'}{2\pi\varepsilon_0(2h)} \quad \longrightarrow \quad F'_e = Q'E_2 = \frac{Q'^2}{4\pi\varepsilon_0 h} \tag{1.224}$$

(the distance between the original and the image is $2h$). The force is attractive.

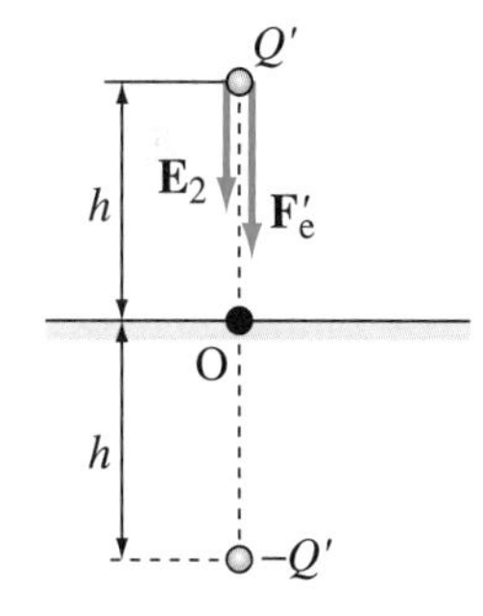

Figure 1.49 Force on a line charge above a conducting plane; for Example 1.30.

Problems: 1.87–1.89; *Conceptual Questions* (on Companion Website): 1.28–1.30; *MATLAB Exercises* (on Companion Website).

Problems

1.1. **Three unequal charges in a triangle.** Repeat Example 1.1 but assuming that one of the three charges in Fig. 1.3(a) amounts to (a) $3Q$ and (b) $-3Q$, respectively.

1.2. **Three charges in equilibrium.** The distance between point charges $Q_1 = 36$ pC and $Q_2 = 9$ pC is $D = 3$ cm. If the third charge, Q_3, is placed at the line connecting Q_1 and Q_2, at a distance d from Q_1, as shown in Fig. 1.50, find Q_3 and d which ensure that all the charges in this system are in the electrostatic equilibrium, i.e., that the resultant Coulomb force on each charge is zero.

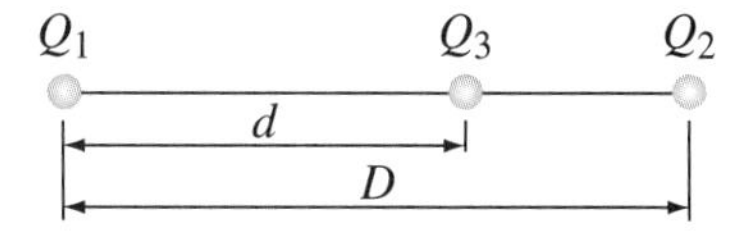

Figure 1.50 Three point charges along a line; for Problem 1.2.

1.3. Four charges at rectangle vertices. Four small charged bodies of equal charges $Q = -1$ nC are placed at four vertices of a rectangle with sides $a = 4$ cm and $b = 2$ cm. Determine the direction and magnitude of the electric force on each of the bodies.

1.4. Five charges in equilibrium. Four small charged balls of equal charges $Q_1 = 5$ pC are positioned at four vertices of a square, whereas the fifth ball of unknown charge Q_2 is at the square center. Find Q_2 such that all the balls are in the electrostatic equilibrium.

1.5. Three point charges in space. (a) For the three charges from Example 1.3, find the resultant electric force on the charge Q_2 ($\mathbf{F}_{e2}$). (b) Determine the force $\mathbf{F}_{e3}$ (on Q_3). (c) What is the sum of all the three forces, $\mathbf{F}_{e1} + \mathbf{F}_{e2} + \mathbf{F}_{e3}$?

1.6. Five charges at pyramid vertices. Four point charges Q are positioned in air at the corners of the square base of a pyramid. A fifth charge $-Q$ is positioned at the top vertex of the pyramid. All sides of the pyramid have the same length, a. Compute the electric force on the top charge.

1.7. Eight charges at cube vertices. Eight small charged bodies of equal charges Q exist at the vertices of a cube with sides of length a, in free space. Find the magnitude and direction of the electric force on one of the charges.

1.8. Electric field due to three point charges in space. For the three charges from Example 1.3, determine the magnitude and direction of the electric field intensity vector at (a) the coordinate origin and (b) the point at the z-axis defined by $z = 100$ m.

1.9. Nonuniform volume charge in a cylinder. An infinitely long cylinder of radius a in free space is charged with a volume charge density $\rho(r) = \rho_0(a - r)/a$ $(0 \le r \le a)$, where ρ_0 is a constant and r the radial distance from the cylinder axis. Find the charge per unit length of the cylinder.

1.10. Nonuniform volume charge in a cube. A cube of edge length a in free space is charged over its volume with a charge density $\rho(x) = \rho_0 \sin(\pi x/a)$, $0 \le x \le a$, where ρ_0 is a constant and x is the normal distance from one of the cube sides. Compute the total charge of the cube.

1.11. Nonuniform surface charge on a disk. A surface charge is distributed in free space over a circular disk of radius a. The charge density is $\rho_s(r) = \rho_{s0} r^2/a^2$ $(0 \le r \le a)$, where r is the radial distance from the disk center, and ρ_{s0} is a constant. Obtain the total charge of the disk.

1.12. Nonuniform line charge along a rod. A rod of length l in air is charged with a line charge of density $Q'(x) = Q'_0[1 - \sin(\pi x/l)]$ $(0 \le x \le l)$, where Q'_0 is a constant and x is the length coordinate along the rod. Calculate the total charge of the rod.

1.13. Field maximum at the axis of a ring. (a) For the charged ring in Fig. 1.11, assume $Q > 0$, and find z for which the electric field intensity along the z-axis is maximum. (b) Plot the function $E_z(z)$, $-\infty < z < \infty$.

1.14. Point charge equivalent to a charged semicircle. Show that far away along the z-axis, the semicircular line charge in Fig. 1.12(a) is equivalent to a point charge with the same amount of charge located at the coordinate origin.

1.15. Charged contour of complex shape. Fig. 1.51 shows a contour consisting of two semicircular parts, of radii a and b $(a < b)$, and two linear parts, each of length $b - a$. The contour is situated in air and carries a charge Q uniformly distributed along its length. Compute the electric field intensity vector at the contour center (point O).

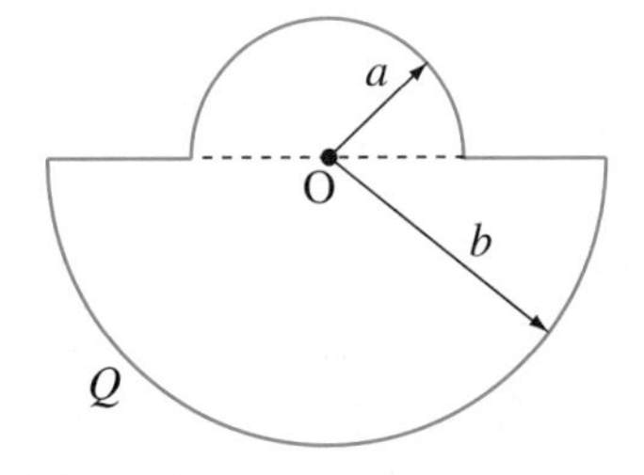

Figure 1.51 Uniformly charged contour with two semicircular and two linear parts; for Problem 1.15.

1.16. Nonuniform line charge along a semicircle. Consider the geometry in Fig. 1.12(a), and assume that the charge along the semicircle is nonuniform, given by $Q'(\phi) = Q'_0 \sin\phi$ $(-\pi/2 \le \phi \le \pi/2)$, where Q'_0 is a constant. (a) Find the total charge of the semicircle.

(b) Prove that the electric field intensity vector along the z-axis equals $\mathbf{E} = -Q_0' a^2 \,\hat{\mathbf{y}}/[8\varepsilon_0(z^2 + a^2)^{3/2}]$.

1.17. Line charge along three-quarters of a circle. A uniform line charge in the form of an arc that is 3/4 of a circle of radius a is situated in air. The total charge of the arc is Q. Calculate the electric field intensity vector at the arc center.

1.18. Line charge along a quarter of a circle. A charge of density Q' in free space is distributed uniformly along an arc representing a quarter of a circle of radius a. Determine the electric field intensity vector at an arbitrary point along the axis that contains the arc center and is normal to the arc plane.

1.19. Semi-infinite line charge. A line charge of uniform charge density Q' is distributed in free space along the negative part of the x-axis in the Cartesian coordinate system $(-\infty < x \leq 0)$. Find the expression for the electric field intensity vector at an arbitrary point in the xy-plane.

1.20. Half-positive, half-negative infinite line charge. A line charge in free space is distributed along the x-axis in the Cartesian coordinate system. The line charge density is Q' $(Q' > 0)$ for $-\infty < x \leq 0$ and $-Q'$ for $0 < x < \infty$. Derive the expression for the electric field intensity vector at a point M with coordinates $(0, d, 0)$, where $d > 0$.

1.21. Charged square contour. A line charge of uniform charge density Q' is distributed along a square contour a on a side. The medium is air. Find the electric field intensity vector at a point that is at a distance a from each of the square vertices.

1.22. Point charge equivalent to a charged disk. Consider the charged disk in Fig. 1.14, and show that for $|z| \gg a$, the $\mathbf{E}$ field in Eq. (1.63) is equivalent to the field of a point charge $Q = \rho_s \pi a^2$ placed at the disk center.

1.23. Field due to a nonuniformly charged disk. Consider the disk with a nonuniform charge distribution from Problem 1.11, and find the electric field intensity vector along the disk axis normal to its plane.

1.24. Nonuniformly charged spherical surface. A sphere of radius a in free space is nonuniformly charged over its surface such that the charge density is given by $\rho_s(\theta) = \rho_{s0} \sin 2\theta$, where ρ_{s0} is a constant and the angle θ $(0 \leq \theta \leq \pi)$ is defined as in Fig. 1.10 or 1.16. Compute (a) the total charge of the sphere and (b) the electric field intensity vector at the sphere center.

1.25. Infinite charged sheet with a circular hole. An infinite sheet of charge with a constant density ρ_s has a hole of radius a in it. The sheet is in the xy-plane of the Cartesian coordinate system and the center of the hole is at the coordinate origin. The ambient medium is air. Under these circumstances, determine the electric field intensity vector at an arbitrary point along the z-axis – in the following two ways, respectively: (a) integrating the fields due to elementary rings as in Fig. 1.14 and (b) combining the results of Examples 1.11 (infinite sheet of charge, with no hole) and 1.10 (charged disk).

1.26. Force on a charged semicylinder due to a line charge. For the structure composed from a line charge and a charged semicylinder shown in Fig. 1.17(a) and described in Example 1.13, find the force per unit length on the semicylinder.

1.27. Charged strip. Consider an infinitely long uniformly charged strip of width a and surface charge density ρ_s in air. Using the geometrical representation of the cross section of the problem as in Fig. 4.11 in Chapter 4 (also see Fig. 1.13) and change of integration variables given by Eqs. (4.43) and (4.44), obtain the expression for the $\mathbf{E}$ field at an arbitrary point in space due to this charge.

1.28. Two parallel oppositely charged strips. Two parallel, very long strips are uniformly charged with charge densities ρ_s and $-\rho_s$, respectively $(\rho_s > 0)$. The cross section of the structure is shown in Fig. 1.52. The width of the strips is the same as the distance between them $(a = d)$, and the medium is air. Find the electric field

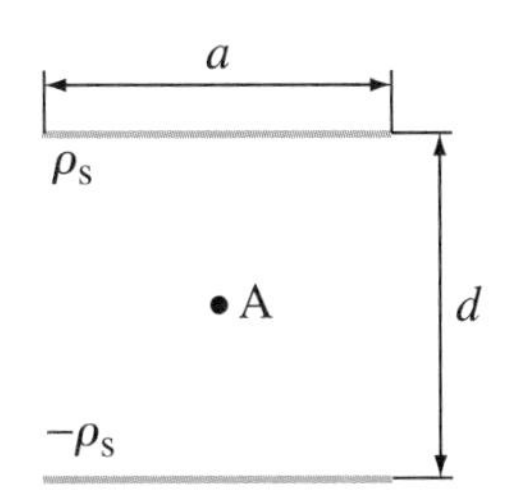

Figure 1.52 Cross section of two parallel, very long charged strips; for Problem 1.28.

intensity vector at the center of the cross section (point A).

1.29. Work in an electrostatic field. What is the work done by electric forces in moving a charge $Q = 1$ nC from the coordinate origin to the point (1 m, 1 m, 1 m) in the electrostatic field given by $\mathbf{E}(x, y, z) = (x\,\hat{\mathbf{x}} + y^2\,\hat{\mathbf{y}} - \hat{\mathbf{z}})$ V/m (x, y, z in m) in the Cartesian coordinate system along the straight line joining the two points?

1.30. Work in the field of a point charge. A point charge $Q_1 = 10$ nC is positioned at the center of a square contour $a = 10$ cm on a side, as shown in Fig. 1.53. Find the work done by electric forces in carrying a charge $Q_2 = -1$ nC from the point M_1 to the point M_2 marked in the figure.

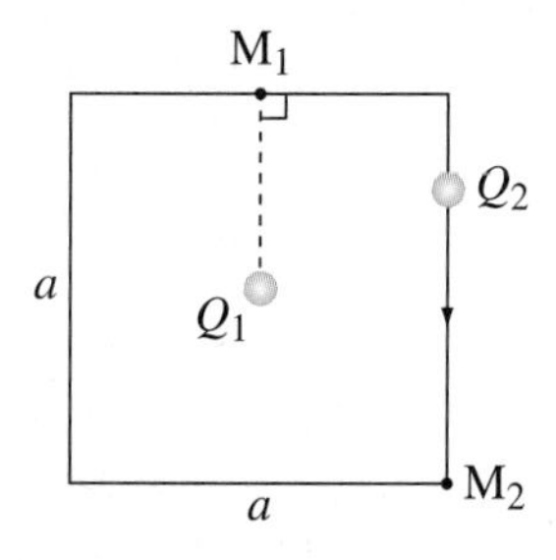

Figure 1.53 Movement of a charge Q_2 in the field of a charge Q_1 positioned at the center of a square contour; for Problem 1.30.

1.31. Electric potential due to three point charges in space. For the three charges from Example 1.3, calculate the electric potential at points defined by (a) (0, 0, 2 m) and (b) (1 m, 1 m, 1 m), respectively.

1.32. Point charge and an arbitrary reference point. Derive the expression for the potential at a distance r from a point charge Q in free space with respect to the reference point which is an arbitrary (finite) distance $r_{\mathcal{R}}$ away from the charge.

1.33. Potential due to a semicircular line charge. Prove that the electric scalar potential at an arbitrary point along the z-axis in the field of the semicircular line charge shown in Fig. 1.12(a) and described in Example 1.7 is $V = Q'a/(4\varepsilon_0\sqrt{z^2 + a^2})$.

1.34. Potential due to a charged disk. For the charged disk from Example 1.10, derive the following expression for the electric scalar potential along the z-axis ($-\infty < z < \infty$): $V = \rho_s(\sqrt{a^2 + z^2} - |z|)/(2\varepsilon_0)$.

1.35. Potential due to a hemispherical surface charge. Consider the hemispherical surface charge from Example 1.12, and find the electric scalar potential at the hemisphere center ($z = 0$).

1.36. Potential due to a nonuniform spherical surface charge. Determine the electric potential at the center of the nonuniformly charged spherical surface from Problem 1.24.

1.37. Voltage due to two point charges. Two point charges, $Q_1 = 7\ \mu$C and $Q_2 = -3\ \mu$C, are located at the two nonadjacent vertices of a square contour $a = 15$ cm on a side. Find the voltage between any of the remaining two vertices of the square and the square center.

1.38. Sketch field from potential. The electrostatic potential V in a region is a function of a single rectangular coordinate x, and $V(x)$ is shown in Fig. 1.54. Sketch the electric field intensity $E_x(x)$ in this region.

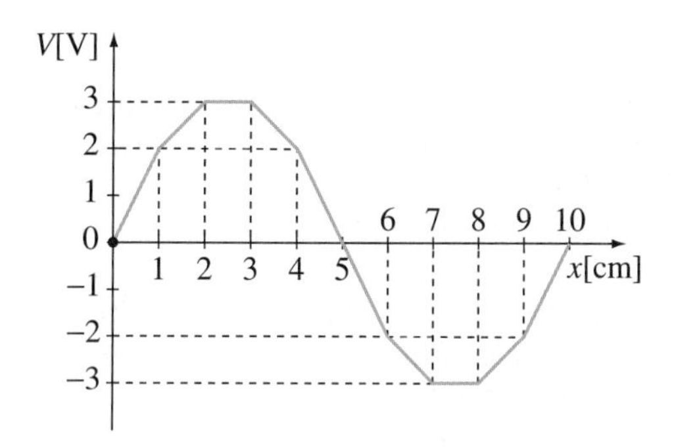

Figure 1.54 1-D potential distribution; for Problem 1.38.

1.39. Field from potential, point charge. For a point charge in free space, obtain the expression for $\mathbf{E}$ in Eq. (1.24) from the expression for V in Eq. (1.80).

1.40. Field from potential, charged semicircle. For the semicircular line charge from Example 1.7, (a) obtain the expression for E_z in Eq. (1.50) from the expression for V given in Problem 1.33 and (b) explain why it is impossible to obtain the expression for E_x in Eq. (1.48) from this same expression for V.

1.41. Field from potential, charged disk. For the charged disk from Example 1.10, obtain the expression for $\mathbf{E}$ in Eq. (1.63) from the expression for V given in Problem 1.34.

1.42. Field from potential, charged hemisphere. For the hemispherical surface charge from Example 1.12, explain why we cannot obtain the expression for **E** at the hemisphere center ($z = 0$), given in Eq. (1.67), from the expression for V computed in Problem 1.35.

1.43. Angle between field lines and equipotential surfaces. Using the concept of gradient, prove that in an arbitrary electrostatic field, field lines are perpendicular to equipotential surfaces (as in Fig. 1.22).

1.44. Direction of the steepest ascent. The terrain elevation in a region is given by a function $h(x, y) = 100x \ln y$ [m] (x, y in km), where x and y are coordinates in the horizontal plane and $1 \text{ km} \leq x, y \leq 10 \text{ km}$. (a) What is the direction of the steepest ascent at (3 km, 3 km)? (b) How steep, in degrees, is the ascent in (a)?

1.45. Maximum increase in electrostatic potential. The electrostatic field intensity vector in a region is given by $\mathbf{E}(x, y, z) = (4\,\hat{\mathbf{x}} - z^2\,\hat{\mathbf{y}} + 2yz\,\hat{\mathbf{z}})$ V/m (x, y, z in m). Find the direction of the maximum increase in the electric scalar potential at a point (1 m, 1 m, −1 m).

1.46. Large and small electric dipole. Two point charges, $Q_1 = 1$ nC and $Q_2 = -1$ nC, are situated in free space at points along the z-axis defined by $z = 1$ m and $z = -1$ m, respectively. Compute the electric potential and field intensity vector at the point defined by Cartesian coordinates (a) $(0, 0, 0)$, (b) $(0, 1 \text{ m}, 0)$, and (c) $(100 \text{ m}, 100 \text{ m}, 100 \text{ m})$, respectively.

1.47. Potential and field due to a small electric dipole. An electric dipole with a moment $\mathbf{p} = 1 \text{ pCm}\,\hat{\mathbf{z}}$ is located at the origin of a spherical coordinate system. The length of the dipole is $d = 1$ cm. Find V and **E** at the following points defined by spherical coordinates: (a) $(1 \text{ m}, 0, 0)$, (b) $(1 \text{ m}, \pi/2, \pi/2)$, (c) $(1 \text{ m}, \pi, 0)$, (d) $(1 \text{ m}, \pi/4, 0)$, (e) $(10 \text{ m}, \pi/4, 0)$, and (f) $(100 \text{ m}, \pi/4, 0)$.

1.48. Dipole equivalent to a nonuniform line charge. Consider the nonuniform line charge distribution along the semicircle from Problem 1.16, and show that far away along the z-axis ($|z| \gg a$) this charge distribution can be replaced by an equivalent electric dipole located at the coordinate origin. Find the moment, **p**, of the equivalent dipole.

1.49. Expression for the electric field due to a line dipole. For the line dipole in Fig. 1.29, obtain the expression for **E** from the expression for V in Eq. (1.121).

1.50. Near and far potential and field due to a line dipole. Two infinite line charges, with densities $Q'_1 = 100$ pC/m and $Q'_2 = -100$ pC/m, are positioned along lines defined by $(1 \text{ m}, 0, z)$ and $(-1 \text{ m}, 0, z)$, $-\infty < z < \infty$, in the Cartesian coordinate system. The medium is air. Calculate V and **E** at the point defined by Cartesian coordinates (a) $(2 \text{ m}, 0, 0)$ and (b) $(100 \text{ m}, 100 \text{ m}, 0)$, respectively.

1.51. Flux of the electric field vector through a cube side. A point charge Q is located at the center of a cube in free space. The cube edges are a long. Find the outward flux of the electric field intensity vector due to this charge through each of the cube sides.

1.52. Flux for a different placement of the point charge. If the point charge Q from the previous problem is placed at the center of a side of the cube, determine the total outward flux of the electric field vector due to the charge through the surface composed of the remaining five sides of the cube.

1.53. Field of a point charge from Gauss' law. Using Gauss' law, derive the expression for the electric field intensity vector of a point charge in free space [Eq. (1.24)].

1.54. Uniformly charged thin spherical shell. An infinitely thin spherical shell of radius a in free space is uniformly charged over its surface with a total charge Q. Determine: (a) the electric field intensity vector inside and outside the shell, (b) the potential of the shell, and (c) the potential at the shell center.

1.55. Sphere with a nonuniform volume charge. Find the distribution of the electric scalar potential inside and outside the sphere with the volume charge density given by Eq. (1.32).

1.56. Field of an infinite line charge from Gauss' law. Using Gauss' law, derive the expression for the electric field intensity vector of an infinite line charge in free space [Eq. (1.57)].

1.57. Uniformly charged thin cylindrical shell. An infinitely long and infinitely thin cylindrical shell of radius a is situated in free space. The shell is charged over its surface with a uniform charge density ρ_s. Find the electric field intensity vector inside and outside the shell.

1.58. Cylinder with uniform volume charge. Compute the voltage between the surface and the axis of a uniformly charged infinite cylinder of radius a in free space, if the volume charge density in the cylinder is ρ.

1.59. Field of an infinite sheet of charge from Gauss' law. Using Gauss' law, derive the expression for the electric field intensity vector of an infinite sheet of charge in free space [Eq. (1.64)].

1.60. Two parallel oppositely charged sheets. Two parallel infinite sheets of charge with densities ρ_s and $-\rho_s$ are situated in free space. (a) Find the electric field intensity vector inside and outside the space between the sheets. (b) What is the voltage between the sheets?

1.61. Equivalent sheet of charge. An infinitely large layer of charge in free space has a uniform volume charge density ρ and thickness d. (a) Compute the electric field vector inside the layer. (b) Show that, as far as the field outside the layer is concerned, the layer can be replaced by an equivalent infinite sheet of charge, and find the surface charge density, ρ_s, of this sheet.

1.62. Layer with a cosine volume charge distribution. The density of a volume charge in free space depends on the Cartesian coordinate x only and is given by $\rho(x) = \rho_0 \cos(\pi x/a)$ ($|x| \leq a$) and $\rho(x) = 0$ ($|x| > a$), where ρ_0 and a ($a > 0$) are constants. (a) Determine the electric field intensity vector in the entire space. (b) Find the voltage between planes $x = -a$ and $x = a$.

1.63. Layer with a sine charge distribution. Repeat the previous problem but for the following charge density function: $\rho(x) = \rho_0 \sin(\pi x/a)$ for $|x| \leq a$ (there is no charge outside the layer).

1.64. Exponential charge distribution in the entire space. A volume charge distribution in free space is described in the rectangular coordinate system as $\rho(x) = \rho_0 \, e^{x/a}$ for $x < 0$, $\rho(0) = 0$, and $\rho(x) = -\rho_0 \, e^{-x/a}$ for $x > 0$, with ρ_0 and a being positive constants. Calculate the electric field intensity vector for $-\infty < x < \infty$.

1.65. Uniform electric field. In a certain region, there is a uniform electric field, $\mathbf{E}_0$. What is the volume charge density in that region?

1.66. Charge distribution from 1-D field distribution. Find the volume charge density $\rho(x)$ in the electrostatic system from Example 1.16, assuming that the permittivity of the medium is ε_0.

1.67. Charge from field, planar symmetry. From the field expressions in Eqs. (1.150)–(1.152), obtain the corresponding charge distribution in free space [Eq. (1.147)].

1.68. Charge from field, cylindrical symmetry. From the field with a radial cylindrical component only given by Eqs. (1.145) and (1.146), obtain the corresponding charge distribution in free space [Eq. (1.143)].

1.69. Charge from field, spherical symmetry. Using Gauss' law in differential form, show that the field with a radial spherical component only given by Eq. (1.140) is produced by a uniformly charged sphere of radius a and charge density ρ in free space.

1.70. Nonuniformly charged sphere using differential Gauss' law. For the nonuniform volume charge distribution in a sphere defined by Eq. (1.32) and analyzed in Problem 1.55 (based on Gauss' law in integral form), compute the electric field intensity vector everywhere using the differential Gauss' law.

1.71. Problem with cylindrical symmetry by differential Gauss' law. Redo Example 1.19 but with the use of Gauss' law in differential form.

1.72. Problem with planar symmetry using differential Gauss' law. Redo Problem 1.61 employing differential Gauss' law.

1.73. Antisymmetrical charge, differential Gauss' law. Redo Example 1.21 applying differential Gauss' law.

1.74. Gauss' law in differential and integral form. In a certain region, the electric field is given by $\mathbf{E} = (4xy\,\hat{\mathbf{x}} + 2x^2\,\hat{\mathbf{y}} + \hat{\mathbf{z}})$ V/m (x, y in m). The medium is air. (a) Calculate the charge density. (b) From the result in (a), find the total charge enclosed in a cube situated in the first coordinate octant ($x, y, z \geq 0$), with one vertex at the coordinate origin, and the edges, of length 1 m, parallel to coordinate axes. (c) Confirm the validity of Gauss' law in integral form and the divergence theorem by evaluating the net outward flux of $\mathbf{E}$ through the surface of the cube defined in (b).

1.75. Excentric charged sphere inside an uncharged shell. Consider the structure from Example 1.27, and assume that the sphere is moved toward the shell wall so that the centers of the sphere and the shell are separated by a distance d. Find the potential of the shell in the new electrostatic state if (a) $d = (b - a)/2$ and (b) $d = b - a$ (the sphere is pressed against the shell wall).

1.76. Point charge inside a charged shell. A point charge $2Q$ is placed at the center of an air-filled spherical metallic shell, charged with Q and situated in air. The inner and outer radii of the shell are a and b ($a < b$). (a) What is the total charge on the inner and on the outer surface of the shell, respectively? (b) Find the potential of the shell.

1.77. Three concentric shells, one uncharged. Three concentric spherical metallic shells are situated in air. The outer radius of the inner shell is $a = 30$ mm, and its charge $Q = 10$ nC. The inner and outer radii of the middle shell are $b = 50$ mm and $c = 60$ mm, and its potential $V = 1$ kV with respect to the reference point at infinity. The inner and outer radii of the outer shell are $d = 90$ mm and $e = 100$ mm, and it is uncharged. Calculate (a) the charge of the middle shell and (b) the voltage between the inner and the outer shell.

1.78. Three concentric shells, two at the same potential. Consider a structure with the same geometry as in the previous problem, and assume that the charges of the inner and outer shells are $Q_1 = 2$ nC and $Q_3 = -2$ nC, respectively, as well as that their potentials are the same ($V_1 = V_3$). Under these circumstances, compute (a) the charge of the middle shell (Q_2) and (b) the potentials of the inner and the middle shells (V_1 and V_2) with respect to the reference point at infinity.

1.79. Four coaxial cylindrical conductors. Four very long conductors, each in the form of a cylindrical shell with thickness $d = 1$ cm, are positioned in air coaxially with respect to each other, as indicated in Fig. 1.55, which shows a detail of the cross section of the system. The first and the fourth conductor are grounded, and the potential of the third conductor with respect to the ground is $V_3 = 1$ kV. The second conductor is uncharged. Find the charges per unit length of the first and the third conductor, Q'_1 and Q'_3.

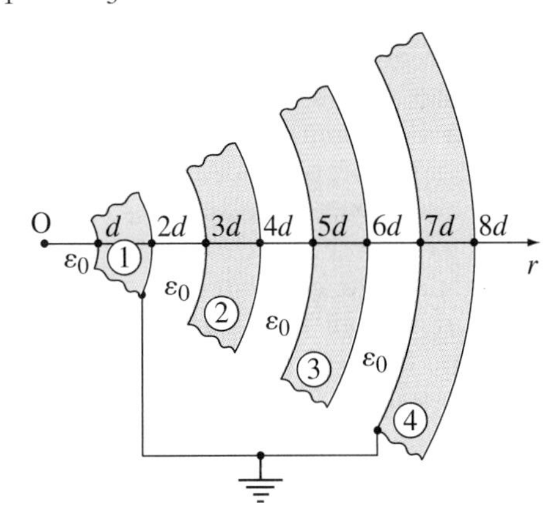

Figure 1.55 Detail of the cross section of a system of four cylindrical conductors; for Problem 1.79.

1.80. Three concentric conductors, one grounded. Shown in Fig. 1.56 is a system consisting of three concentric spherical conductors (the inner conductor is a solid sphere, while the remaining two are spherical shells). The radius of the inner conductor is $a = 2$ mm. The inner radius of the middle conductor is $b = 5$ mm, and outer $c = 6$ mm. The inner radius of the outer conductor is $d = 8$ mm. The space between the conductors is air-filled. The outer conductor is grounded, and the potentials of the inner and middle conductors with respect to the ground are $V_1 = 15$ V and $V_2 = 10$ V, respectively. Determine total charges of the inner and middle conductors, Q_1 and Q_2.

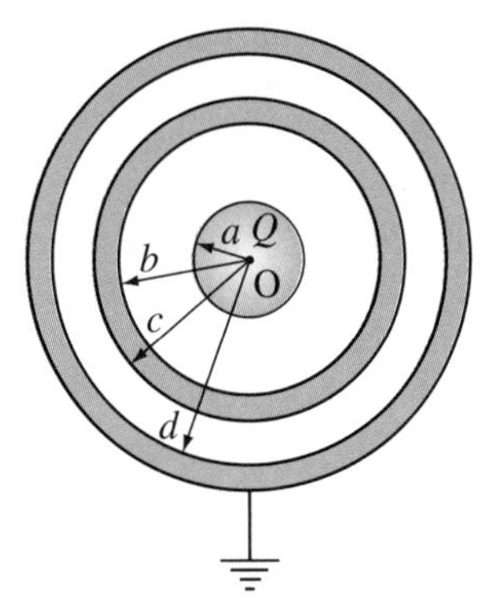

Figure 1.56 System of three concentric spherical conductors; for Problem 1.80.

1.81. Charged metallic foil. An infinitely large flat metallic foil is situated in air and charged uniformly with the surface charge density $\rho_s = 1\ \text{nC/m}^2$. Find the electric field intensity vector everywhere.

1.82. Two metallic slabs. An infinitely large metallic slab of thickness $d = 1$ cm is situated in air and charged such that the surface charge density at each of the slab surfaces is $\rho_s = 1\ \mu\text{C/m}^2$. Another metallic slab of the same thickness, which is uncharged, is then introduced and placed parallel to the charged slab such that the distance between the surfaces of the two slabs facing each other is $D = 3$ cm. In the new electrostatic state, calculate (a) the surface charge densities at all four surfaces of the slabs, (b) electric field intensity vector everywhere, and (c) the voltage between the slabs.

1.83. Two metallic spheres at the same potential. Consider the system in Fig. 1.45, and assume that $a = 5$ cm, $b = 1$ cm, and $d = 1$ m, as well as that the total charge of the two spheres is $Q = 600$ pC. Find (a) the potential of the spheres and (b) the electric field intensities E_a and E_b near the surfaces of the spheres.

1.84. MoM-based computer program for a charged plate. Using the method of moments as presented in Section 1.20, write a computer program to determine the charge distribution on a very thin charged square plate of edge length a at a potential V_0, in free space. Subdivide the plate into N square patches, and assume that $a = 1$ m and $V_0 = 1$ V. (a) Tabulate and plot the results for the surface charge density (ρ_s) of the patches, taking $N = 100$ (ten partitions in each dimension). (b) Compute the total charge of the plate, taking (i) $N = 9$, (ii) $N = 25$, (iii) $N = 49$, and (iv) $N = 100$, respectively.

1.85. MoM computation for a charged cube. Write a computer program for the method-of-moments analysis of a charged metallic cube, Fig. 1.46, with edge length $a = 1$ m, and compute the total charge of the cube for $V_0 = 1$ V and ten, or as many as possible (given available computational resources), subdivisions per cube edge ($N = 600$ if ten subdivisions per edge are adopted).

1.86. Approximate integral expression for the electric field vector. (a) Write the approximate integral expression for the evaluation of the electric field intensity vector at an arbitrary point in space due to a charged body (e.g., the cube in Fig. 1.46), whose charge distribution is approximately described by Eq. (1.212). (b) Using the expression in (a) and the associated computer program, compute the electric field along the axis of the plate from Problem 1.84 perpendicular to its plane at points that are $a/2$, $2a$, and $100a$, respectively, distant from the plate surface (for $N = 100$). (c) Also compute the electric field inside the cube from the previous problem (Problem 1.85), at a quarter of its space diagonal (body diagonal) and at its center.

1.87. Force on a point charge due to its image. Find the electric force on the point charge Q in Fig. 1.48(a).

1.88. Imaging a line charge. For the structure defined in Example 1.30, determine the distribution of induced surface charges on the conducting plane.

1.89. Charged wire parallel to a corner screen. Fig. 1.57 shows a cross section of the structure consisting of a metallic wire of radius a and a 90° corner metallic screen in air. The distance of the wire from both the horizontal and vertical half-planes constituting the screen is h, where $h \gg a$. If the wire is charged with Q' per unit length, calculate the voltage between the wire and the screen.

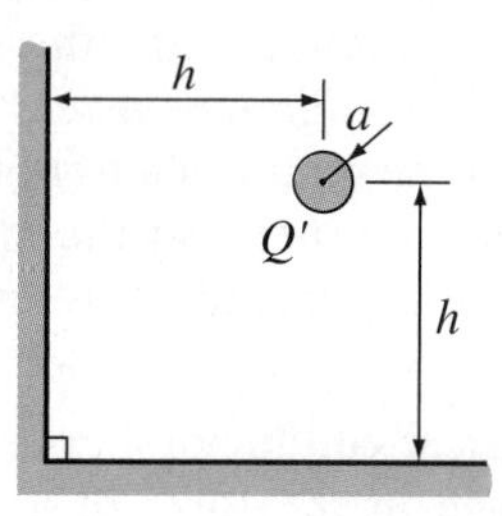

Figure 1.57 Cross section of a charged metallic wire parallel to a metallic corner screen; for Problem 1.89.

2 Dielectrics, Capacitance, and Electric Energy

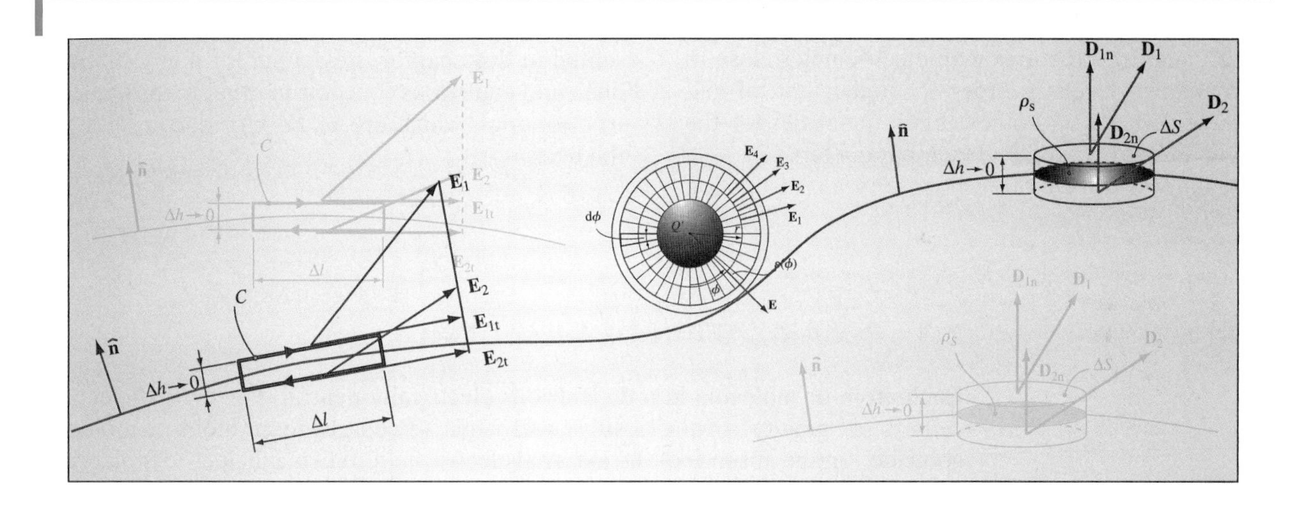

Introduction:

In this chapter, we shall analyze electrostatic fields in the presence of dielectrics and study several important related topics. Dielectrics or insulators are nonconducting materials, having very little free charges inside them (theoretically, perfect dielectrics have no free charges). In addition, redistribution of any free charges (e.g., electrons) deposited inside the material lasts much longer than in metallic conductors, a typical example, as already indicated in Section 1.16, being the charge rearrangement time of ~ 50 days for fused quartz compared to $\sim 10^{-19}$ s for copper. We shall see, however, that another type of charges, called bound or polarization charges, exist in a polarized dielectric. We shall first investigate the mechanisms of the polarization of dielectrics, caused by an external electric field. By introducing the macroscopic quantities such as the polarization vector, and the volume and surface density of bound charges, it is possible to evaluate the electric field intensity vector and the electric potential due to polarized dielectrics using free-space formulas and techniques from the previous chapter. Gauss' law will be generalized for an electrostatic system that includes arbitrary media (conductors and dielectrics), and characterization of dielectric materials in terms of their linearity, homogeneity, and isotropy will be presented. Dielectric-diclectric boundary conditions will be derived and used. The chapter will also introduce Poisson's and Laplace's second-order differential equations for the potential and their solution.

Having both conductors and dielectrics, we shall then put them together to form capacitors and related electrostatic systems. The capacitor is a fundamental element in electrical engineering. Its basic property is its capacitance. We shall analyze capacitors with electrodes of different shapes and with different types of dielectrics, and evaluate capacitance per unit length of various two-conductor transmission lines as well. In addition, we shall evaluate the electric energy contained in charged capacitors, transmission lines, and other systems of conducting bodies, and introduce the electric energy density to help us find and quantify the localization and distribution of the energy in such systems. Dielectric breakdown, occurring for exceedingly strong electric fields in a dielectric material causing it to become conducting, will also be discussed for various electrostatic structures. We shall analyze structures with electric fields close to breakdown levels in order to predict critical values of voltages and other circuit quantities for the structure at breakdown. Such parameters (e.g., the breakdown voltage of a capacitor or a transmission line) define maximal permissible extents of quantities for the safe operation of the structure prior to an eventual breakdown.

The electrostatic analysis of capacitors and transmission lines, to determine their capacitance, energy, and breakdown characteristics, represents a culmination of the theory of the electrostatic field. It represents, on the other hand, a gateway to many practical applications of this theory. Finally, a clear understanding of concepts that will be presented in this chapter is essential for many similar, dual, and analogous concepts in other areas of electromagnetics, which are to be introduced later in the text.

2.1 POLARIZATION OF DIELECTRICS

Each atom or molecule in a dielectric is electrically neutral. For most dielectrics, centers of "gravity" of the positive and negative charges in an atom or molecule coincide – in the absence of the external electric field. When a dielectric is placed in an external field, of intensity $\mathbf{E}_{\text{ext}}$, however, the positive and negative charges shift in opposite directions against their mutual attraction, and produce a small electric dipole (Fig. 1.28), which is aligned with the electric field lines. The moment of this equivalent dipole is given by $\mathbf{p} = Q\mathbf{d}$ [Eq. (1.116)], where Q is the positive charge of the atom or molecule ($-Q$ is the negative charge), and $\mathbf{d}$ is the vector displacement of Q with respect to $-Q$. The charges are displaced from their equilibrium positions by forces

$$\mathbf{F}_{\text{e}1} = Q\mathbf{E}_{\text{ext}} \quad \text{and} \quad \mathbf{F}_{\text{e}2} = -Q\mathbf{E}_{\text{ext}}, \tag{2.1}$$

respectively, and thus Q shifts in the direction of $\mathbf{E}_{\text{ext}}$, while $-Q$ moves in the opposite direction, so that $\mathbf{p}$ and $\mathbf{E}_{\text{ext}}$ are collinear and have the same direction. The displacement $\mathbf{d}$ is very small, smaller than the dimensions of atoms and molecules. The charges Q and $-Q$ are bound in place by atomic and molecular forces and can only shift positions slightly in response to the external field. So, the two charges in an equivalent small dipole cannot separate one from the other and migrate across the material in opposite directions run by the electric field. Hence, these charges are called bound charges (in contrast to free charges).

Some dielectrics, such as water, have molecules with a permanent displacement between the centers of the positive and negative charge, so that they act as small electric dipoles even with no applied electric field. Such molecules are known as polar molecules, and the dielectrics are called polar dielectrics (those with no built-in dipoles are nonpolar dielectrics). In the absence of the electric field, all the dipoles are oriented in a random way. If a polar molecule, which we model by an electric dipole, is brought into an electric field, however, the forces on the two dipole charges, given in Eq. (2.1), act as indicated in Fig. 2.1. The torques (moments) of forces with respect to the center of the dipole (point O) are

$$\mathbf{T}_1 = \mathbf{r}_1 \times \mathbf{F}_{\text{e}1} \quad \text{and} \quad \mathbf{T}_2 = \mathbf{r}_2 \times \mathbf{F}_{\text{e}2}, \tag{2.2}$$

with $\mathbf{r}_1$ and $\mathbf{r}_2$ denoting the position vectors of Q and $-Q$ with respect to the dipole center. We notice that $\mathbf{r}_1 - \mathbf{r}_2 = \mathbf{d}$, and hence the resultant torque on the dipole turns out to be

$$\boxed{\mathbf{T}_{\text{on dipole}} = \mathbf{T}_1 + \mathbf{T}_2 = Q(\mathbf{r}_1 - \mathbf{r}_2) \times \mathbf{E}_{\text{ext}} = Q\mathbf{d} \times \mathbf{E}_{\text{ext}} = \mathbf{p} \times \mathbf{E}_{\text{ext}},} \tag{2.3}$$

torque on an electric dipole in an external electric field

where we assume that $\mathbf{E}_{\text{ext}}$ is practically uniform along the dipole. Vector $\mathbf{T}_{\text{on dipole}}$ is normal to the plane of $\mathbf{p}$ and $\mathbf{E}_{\text{ext}}$ (the plane of drawing in Fig. 2.1), and its magnitude amounts to

$$T_{\text{on dipole}} = |\mathbf{p} \times \mathbf{E}_{\text{ext}}| = p\,E_{\text{ext}} \sin\alpha. \tag{2.4}$$

We see that the torque given by Eq. (2.3) tends to rotate the dipole about the axis passing through the dipole center and being normal to the dipole and the plane of drawing, i.e., about the vector $T_{\text{on dipole}}$. The action of such torques in the dielectric is against random intermolecular thermic forces, and is to align the dipoles, to some extent, in the same direction – toward the field lines. We also see that the stronger the field, the larger $T_{\text{on dipole}}$, and the larger the component of the resultant dipole moment of all the molecules, $\sum \mathbf{p}$, along the direction of $\mathbf{E}_{\text{ext}}$. A sufficiently strong field may even produce an additional displacement between the positive and negative charges in a polar molecule, resulting in a larger p.

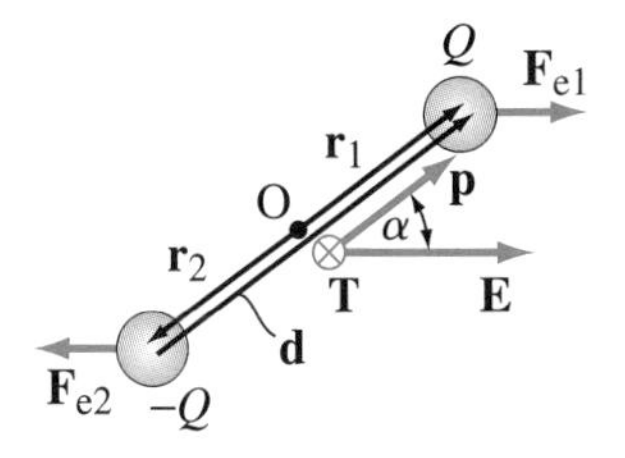

Figure 2.1 Polarization of polar dielectrics: model of a polar molecule in an external electric field.

We conclude that both an unpolar and polar dielectric in an electric field can be viewed as an arrangement of (more or less) oriented microscopic electric dipoles. The process of making atoms and molecules in a dielectric behave as dipoles and orienting the dipoles toward the direction of the external field is termed the polarization of the dielectric, and bound charges are sometimes referred to as polarization charges. This process is extremely fast, practically instantaneous, and the dielectric in the new electrostatic state is said to be polarized or in the polarized state. For almost all materials, the removal of the external electric field results in the return to their normal, unpolarized, state. A very few dielectrics, called electrets, remain permanently polarized in the absence of an applied electric field (an example is a strained piezoelectric crystal).

Conceptual Questions (on Companion Website): 2.1 and 2.2.

2.2 POLARIZATION VECTOR

When polarized (by an external electric field), a dielectric is a source of its own electric field, and the total field at an arbitrary point in space (inside or outside the dielectric) is a sum of the external (primary) field and the field due to the polarized dielectric (secondary field). To determine the secondary field, we replace the dielectric by a collection of equivalent small dipoles, which can be considered to be in a vacuum, as the rest of the material does not produce any field.

Theoretically, we could use the expression for the electric field due to an electric dipole, Eq. (1.117), and obtain the field due to a polarized dielectric by superposition. However, as many atoms or molecules in a dielectric body that many equivalent small dipoles in it, and, with the "microscopic" approach to the evaluation of the field due to the polarized dielectric, we would need to consider every single dipole, which is practically impossible [there is on the order of as many as 10^{30} atoms per unit volume (1 m^3) in solid and liquid dielectrics].

We rather adopt a "macroscopic" approach, and introduce a macroscopic quantity called the polarization vector to describe the polarized state of a dielectric and the resulting field. We first average dipole moments in an elementary volume dv,

average dipole moment in an elementary volume of a polarized dielectric

$$\mathbf{p}_{\text{av}} = \frac{(\sum \mathbf{p})_{\text{in } dv}}{N_{\text{in } dv}}, \tag{2.5}$$

and then multiply this average by the concentration of dipoles (i.e., concentration of atoms or molecules in the dielectric), which equals

$$N_{\text{v}} = \frac{N_{\text{in } dv}}{dv}. \tag{2.6}$$

What we get is, by definition, the polarization vector:

polarization vector (unit: C/m^2)

$$\mathbf{P} = N_{\text{v}}\,\mathbf{p}_{\text{av}} = \frac{(\sum \mathbf{p})_{\text{in } dv}}{dv}. \tag{2.7}$$

Note that $\mathbf{P}$ would represent the resultant dipole moment in a unit volume ($1\ \text{m}^3$) if it were polarized uniformly (equally) throughout the volume. Note also that

$$\mathbf{P}\,dv = \left(\sum \mathbf{p}\right)_{\text{in } dv} \tag{2.8}$$

is the dipole moment of an electric dipole equivalent to an element dv of the polarized dielectric, i.e., to all the dipoles within it.[1] The unit for $\mathbf{P}$ is C/m^2.

In any dielectric material, the polarization vector at a point is a function of the (total) electric field intensity vector at that point,

$$\mathbf{P} = \mathbf{P}(\mathbf{E}). \tag{2.9}$$

For linear (in the electrical sense) materials, this relationship is linear, i.e.,

χ_{e} - electric susceptibility of a linear dielectric

$$\mathbf{P} = \chi_{\text{e}}\varepsilon_0\mathbf{E}, \tag{2.10}$$

where χ_{e} is the electric susceptibility of the dielectric. It is a pure number, i.e., a dimensionless quantity, obtained by measurements on individual materials, and is always nonnegative ($\chi_{\text{e}} \geq 0$). For a vacuum, $\chi_{\text{e}} = 0$, whereas $\chi_{\text{e}} \approx 0$ for air.

2.3 BOUND VOLUME AND SURFACE CHARGE DENSITIES

We shall now derive the expressions for calculating the macroscopic distribution of excess bound charges in a polarized dielectric body from a given distribution of the polarization vector, $\mathbf{P}$, which, in turn, is obtained by averaging the microscopic dipoles in the dielectric material. These expressions will be used in the next section for free-space evaluations of the electric field due to polarized dielectrics.

Let us first find the total bound (polarization) charge $Q_{\text{p}S}$ enclosed by an arbitrary imaginary closed surface S situated (totally or partly) inside a polarized dielectric body, as shown in Fig. 2.2. Knowing that bound charge actually consists of a vast collection of small electric dipoles, each dipole being composed from a

[1] An elementary volume dv, as we use it in macroscopic electromagnetic theory, is small in a physical sense, and cannot be infinitely small in a mathematical sense. Within the definition of the polarization vector, for instance, that means that dv is large enough to contain many small dipoles to be treated "on average," but yet sufficiently small so that $\mathbf{P}$ can be considered constant in dv from the macroscopic point of view. Such dv still contains a vast number (millions) of atoms or molecules.

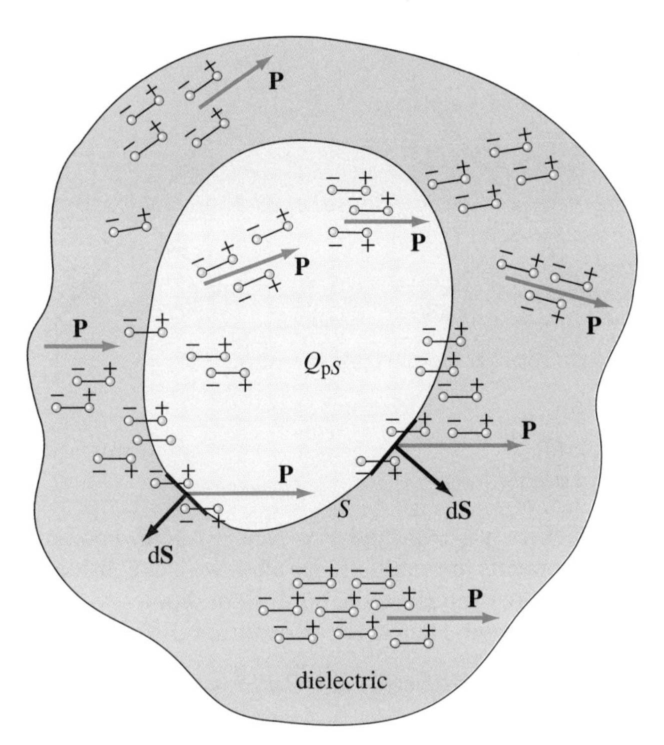

Figure 2.2 Closed surface S in a polarized dielectric body.

positive Q and a negative $-Q$, we realize that all the dipoles that appear inside S with both their ends, Q and $-Q$, as well as dipoles that are totally outside S, contribute with zero net charge to Q_{pS}. Only dipoles whose one end is inside S (and the other end outside S) contribute actually to the total bound charge in S. (We notice right away that $Q_{pS} = 0$ when S encloses the entire dielectric body.) To evaluate Q_{pS} (in the general case), we therefore count the dipoles that cross the surface S. In doing that, we count the contribution of such dipoles as either Q or $-Q$ (note that Q, generally, differ from dipole to dipole), by inspecting which end of the dipole is inside S.

Consider an element $\mathrm{d}S$ of S and the case when the angle α between the vector $\mathbf{P}$ (or vector $\mathbf{p}_{\mathrm{av}}$) and vector $\mathrm{d}\mathbf{S}$, which is oriented from S outward, is less than 90°, as depicted in Fig. 2.3(a). Note that negative ends of dipoles that extend across $\mathrm{d}S$ with one (negative) end inside S are in a cylinder with bases $\mathrm{d}S$ and height

$$h = d\cos\alpha, \tag{2.11}$$

so that the number of these dipoles equals the concentration of dipoles, N_{v}, times the volume of the cylinder, $\mathrm{d}v = \mathrm{d}S\,h$. The dipole ends on the inner side of $\mathrm{d}S$ being all negative, and with an assumption that all dipoles in $\mathrm{d}v$ are with the same moments and charge, the corresponding bound charge is given by

$$\mathrm{d}Q_{\mathrm{p}} = N_{\mathrm{v}}\,\mathrm{d}S\,d\cos\alpha\,(-Q) \qquad (0 \le \alpha < 90^\circ). \tag{2.12}$$

In the case when $\alpha > 90^\circ$, portrayed in Fig. 2.3(b),

$$h = d\,\cos(\pi - \alpha) = d\,(-\cos\alpha), \tag{2.13}$$

and, because the ends of dipoles on the inner side of $\mathrm{d}S$ are all positive,

$$\mathrm{d}Q_{\mathrm{p}} = N_{\mathrm{v}}\,\mathrm{d}S\,d\,(-\cos\alpha)\,Q \qquad (90^\circ < \alpha \le 180^\circ), \tag{2.14}$$

which turns out to be the same result as in Eq. (2.12).

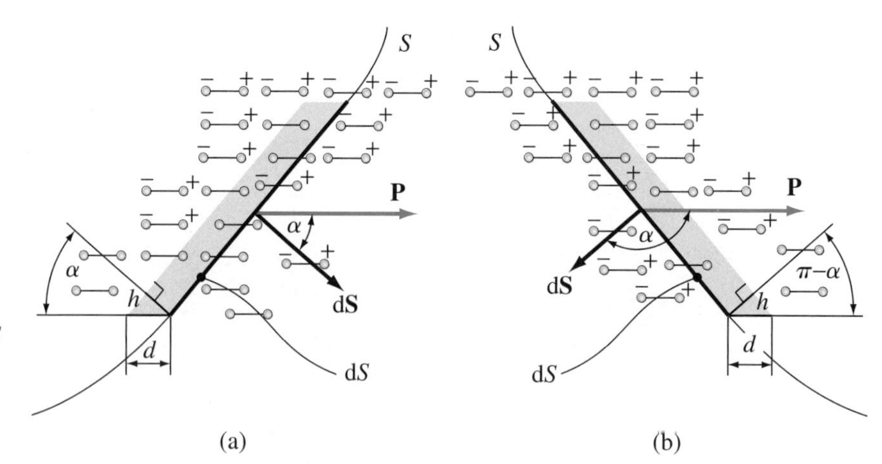

Figure 2.3 Element of surface S in Fig. 2.2, in two cases with regards to the angle α between $\mathbf{P}$ and $\mathrm{d}\mathbf{S}$: (a) $0 \leq \alpha < 90^\circ$ and (b) $90^\circ < \alpha \leq 180^\circ$.

For unpolar dielectrics, $\mathbf{p}_{\mathrm{av}} = \mathbf{p}$, and $\mathbf{P} = N_{\mathrm{v}}\mathbf{p} = N_{\mathrm{v}}Q\mathbf{d}$. For polar dielectrics, where not all dipole moments are mutually parallel, we can consider the dipoles in small cylinders in Figs. 2.3(a) and (b) to be equivalent dipoles with moments $\mathbf{p}_{\mathrm{av}} = Q\mathbf{d}$, so that $\mathbf{P} = N_{\mathrm{v}}\mathbf{p}_{\mathrm{av}} = N_{\mathrm{v}}Q\mathbf{d}$. Hence, for an arbitrary α ($0 \leq \alpha \leq 180^\circ$), we have

$$\mathrm{d}Q_{\mathrm{p}} = -N_{\mathrm{v}}Qd\,\mathrm{d}S\cos\alpha = -P\,\mathrm{d}S\cos\alpha = -\mathbf{P}\cdot\mathrm{d}\mathbf{S} \tag{2.15}$$

(note that the boundary case, $\alpha = 90^\circ$ and $\mathrm{d}Q_{\mathrm{p}} = 0$, is also properly included in this formula). Finally, by integrating the result for $\mathrm{d}Q_{\mathrm{p}}$ over the entire surface S, we obtain

total bound (polarization) charge enclosed by a closed surface S

$$\boxed{Q_{\mathrm{p}S} = -\oint_S \mathbf{P}\cdot\mathrm{d}\mathbf{S}.} \tag{2.16}$$

This is an integral equation similar in form to Gauss' law, Eq. (1.133). It tells us that the outward flux of the polarization vector through an arbitrary closed surface in an electrostatic system that includes dielectric materials is equal to the total bound (polarization) charge enclosed by that surface, multiplied by -1.

Eq. (2.16) is true for any closed surface S. Let us now apply it to the surface S enclosing an elementary volume Δv inside a polarized dielectric:

$$\frac{(Q_{\mathrm{p}})_{\mathrm{in}\,\Delta v}}{\Delta v} = \frac{-\oint_S \mathbf{P}\cdot\mathrm{d}\mathbf{S}}{\Delta v} \quad (\Delta v \to 0), \tag{2.17}$$

with both sides of the equation being also divided by Δv. The expression on the left-hand side of Eq. (2.17) represents the density of excess volume bound charge,

$$\rho_{\mathrm{p}} = \frac{(Q_{\mathrm{p}})_{\mathrm{in}\,\Delta v}}{\Delta v}, \tag{2.18}$$

while the expression on its right-hand side is, by definition [Eq. (1.172)], the negative of the divergence of the polarization vector. Hence,

bound volume charge density

$$\boxed{\rho_{\mathrm{p}} = -\operatorname{div}\mathbf{P} = -\nabla\cdot\mathbf{P}.} \tag{2.19}$$

If $\mathbf{P} = \text{const}$ inside the dielectric (uniformly polarized dielectric), all spatial derivatives of $\mathbf{P}$ are zero, and using Eq. (2.19),

no volume bound charge in a uniformly polarized dielectric

$$\boxed{\mathbf{P} = \text{const} \quad\longrightarrow\quad \rho_{\mathrm{p}} = 0.} \tag{2.20}$$

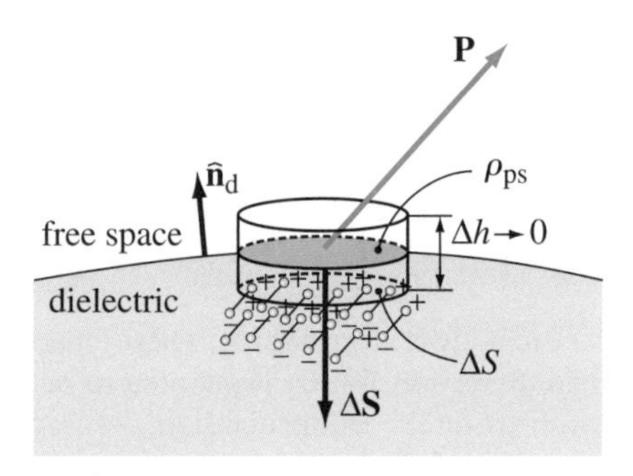

Figure 2.4 Elementary closed surface used for deriving the boundary condition for the vector **P** on a surface dielectric-free space, Eq. (2.23).

If $\mathbf{P} \neq$ const, however, then excess volume bound charge exists only if the polarization vector varies throughout the volume of the dielectric (nonuniformly polarized dielectric) in a way that its divergence is nonzero, otherwise $\rho_\mathrm{p} = 0$.

On the surface of a polarized dielectric, there always exists excess surface bound charge (there are ends of dipoles pressed onto the surface that cannot be compensated by oppositely charged ends of neighboring dipoles), except on parts of the surface where **P** and the dipoles are tangential to the surface. To determine the associated bound (or polarization) surface charge density, ρ_ps, we apply Eq. (2.16) to a small pillbox surface, with bases ΔS and height Δh ($\Delta h \to 0$), shown in Fig. 2.4. There is no polarization in free space (a vacuum or air),

$$\boxed{\mathbf{P} = 0,} \tag{2.21}$$

no polarization in a vacuum or air

so that the flux of vector **P** in Eq. (2.16) is reduced to $\mathbf{P} \cdot \Delta\mathbf{S}$ over the lower side of S, and we have [also see the similar derivation in Eq. (1.188)]

$$\rho_\mathrm{ps}\Delta S = -\mathbf{P} \cdot \Delta\mathbf{S}. \tag{2.22}$$

With $\hat{\mathbf{n}}_\mathrm{d}$ denoting the normal unit vector oriented from the dielectric body outward, $\Delta\mathbf{S} = -\Delta S\hat{\mathbf{n}}_\mathrm{d}$, which yields

$$\boxed{\rho_\mathrm{ps} = \hat{\mathbf{n}}_\mathrm{d} \cdot \mathbf{P}.} \tag{2.23}$$

bound surface charge density; $\hat{\mathbf{n}}_\mathrm{d}$ outward normal on a dielectric surface

This is the boundary condition for the vector **P** on a surface dielectric-free space, connecting the polarization vector in the dielectric near the boundary surface and the bound surface charge density on the surface. Note that only the normal component of **P** contributes to ρ_ps.

Example 2.1 Nonuniformly Polarized Dielectric Cube

The polarization vector in a dielectric cube shown in Fig. 2.5 is given by

$$\mathbf{P}(x, y) = P_0 \frac{xy}{a^2}\,\hat{\mathbf{x}}, \tag{2.24}$$

where P_0 is a constant. The surrounding medium is air. Find the distribution of bound charges of the cube.

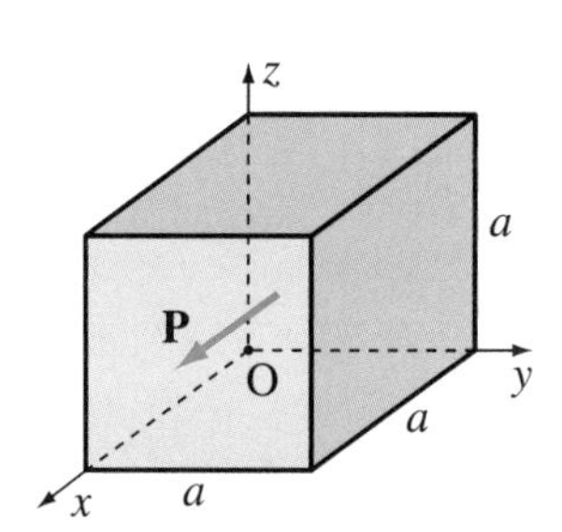

Figure 2.5 Dielectric cube with polarization $\mathbf{P}(x, y)$; for Example 2.1.

Solution Using Eqs. (2.19) and (1.167), the bound volume charge density inside the cube is

$$\rho_\mathrm{p} = -\frac{\partial P_x}{\partial x} = -\frac{P_0 y}{a^2}, \tag{2.25}$$

whereas Eq. (2.23) tells us that bound surface charge exists on the front side of the cube only, and its density amounts to

$$\rho_\mathrm{ps} = \hat{\mathbf{x}} \cdot \mathbf{P}(a^-, y) = \frac{P_0 y}{a}, \tag{2.26}$$

with $\mathbf{P}(a^-, y)$ denoting the polarization vector in the dielectric very close to the boundary surface at $x = a$. On the back side, $\rho_{ps} = 0$ since $P(0^+, y) = 0$, while on the remaining four sides $\rho_{ps} = 0$ because $\hat{\mathbf{n}}_d$ and $\mathbf{P}$ are mutually perpendicular.

Note that, with v designating the volume of the cube and S its boundary surface,

$$Q_p = \int_v \rho_p \, dv + \oint_S \rho_{ps} \, dS = \int_{y=0}^{a} \left(-\frac{P_0 y}{a^2}\right) a^2 \, dy + \int_{y=0}^{a} \frac{P_0 y}{a} a \, dy = 0, \tag{2.27}$$

as expected (the total bound charge of a dielectric body is always zero), where, in accordance with our general integration strategy explained in Section 1.4, dv is adopted to be a slice of the cube of thickness dy, and dS a strip of width dy on the front cube side.

Problems: 2.1.

2.4 EVALUATION OF THE ELECTRIC FIELD AND POTENTIAL DUE TO POLARIZED DIELECTRICS

In this section, we shall evaluate the electric field intensity vector and electric scalar potential due to polarized dielectric bodies in several characteristic cases. We assume that the state of polarization of a dielectric body is described by a given distribution of the polarization vector, $\mathbf{P}$, inside the body. From $\mathbf{P}$, using Eqs. (2.19) and (2.23), we first find the distribution of volume and surface bound charge densities, ρ_p and ρ_{ps}, throughout the body volume and over its surface, respectively. Then, we calculate the field $\mathbf{E}$ and potential V (and any other related quantity of interest) using the appropriate free-space formulas and equations [Eqs. (1.37), (1.38), (1.82), (1.83), (1.133), etc.] and solution techniques suitable to specific geometries and source distributions.

Example 2.2 **Uniformly Polarized Dielectric Disk**

A dielectric disk of radius a and thickness d is situated in free space. The disk is uniformly polarized throughout its volume, the polarization vector being normal to the disk bases and its magnitude being P. Find (a) the distribution of bound charges of the disk and (b) the electric field intensity vector at the disk center.

Solution

(a) Eq. (2.20) tells us that there is no bound volume charge inside the disk. According to Eq. (2.23) and Fig. 2.6, the bound surface charge densities on the upper and lower disk bases are

$$\rho_{ps1} = \hat{\mathbf{n}}_{d1} \cdot \mathbf{P} = P \quad \text{and} \quad \rho_{ps2} = \hat{\mathbf{n}}_{d2} \cdot \mathbf{P} = -P, \tag{2.28}$$

respectively, while on the side disk surface, $\rho_{ps3} = \hat{\mathbf{n}}_{d3} \cdot \mathbf{P} = 0$.

(b) The electric field due to the polarized disk equals the field due to two circular sheets of charge with densities ρ_{ps1} and ρ_{ps2} in free space. We use the expression for the field due to a circular sheet of charge (thin charged disk) in Eq. (1.63) and the superposition principle to add up the fields due to two sheets, $\mathbf{E}_1$ and $\mathbf{E}_2$. The charge densities are $\rho_s = \pm P$ and the distance from each sheet at the disk center (point O) is $d/2$, so the two fields are the same, and the total field comes out to be

$$\mathbf{E} = \mathbf{E}_1 + \mathbf{E}_2 = 2\mathbf{E}_1 = -\frac{P}{\varepsilon_0}\left[1 - \frac{d}{2\sqrt{a^2 + (d/2)^2}}\right]\hat{\mathbf{z}}. \tag{2.29}$$

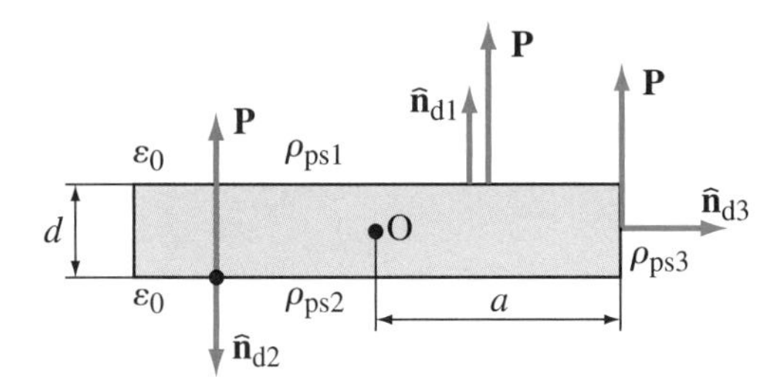

Figure 2.6 Bound surface charge on a uniformly polarized dielectric disk; for Example 2.2.

Example 2.3 Uniformly Polarized Dielectric Sphere

A dielectric sphere of radius a, in free space, is uniformly polarized, and the polarization vector is $\mathbf{P}$. Compute (a) the distribution of bound charge of the sphere, (b) the electric scalar potential at the sphere center, and (c) the electric field intensity vector at the sphere center.

Solution

(a) Let us adopt a spherical coordinate system with the origin at the sphere center and the z-axis parallel to the vector $\mathbf{P}$, as shown in Fig. 2.7. The bound volume charge density is $\rho_p = 0$. The bound surface charge density at a point M on the sphere surface, defined by the angle θ, is

$$\rho_{ps} = \hat{\mathbf{n}}_d \cdot \mathbf{P} = P \cos \angle(\hat{\mathbf{n}}_d, \mathbf{P}) = P \cos\theta, \qquad 0 \le \theta \le \pi. \tag{2.30}$$

We now replace the polarized sphere by a nonuniform spherical sheet of charge in free space, whose charge density is the function of θ given in Eq. (2.30), and compute V and $\mathbf{E}$ at the sphere center (point O) using free-space concepts and equations.

(b) From Eq. (1.83) and Fig. 2.7, the potential at the point O turns out to be

$$V = \frac{1}{4\pi\varepsilon_0} \oint_S \frac{\rho_{ps}\, dS'}{a} = \frac{1}{4\pi\varepsilon_0 a} \oint_S \rho_{ps}\, dS' = \frac{1}{4\pi\varepsilon_0 a} Q_p = 0, \tag{2.31}$$

because the total bound charge of the sphere, Q_p, is zero.

(c) Due to symmetry, vector $\mathbf{E}$ at the point O in Fig. 2.7 has a (negative) z-component only, which is computed essentially in the same way as in Fig. 1.16, subdividing the sphere surface into thin rings and integrating the fields $d\mathbf{E}$ due to individual rings, the only two differences being that the surface charge density is now a function of θ, Eq. (2.30), and

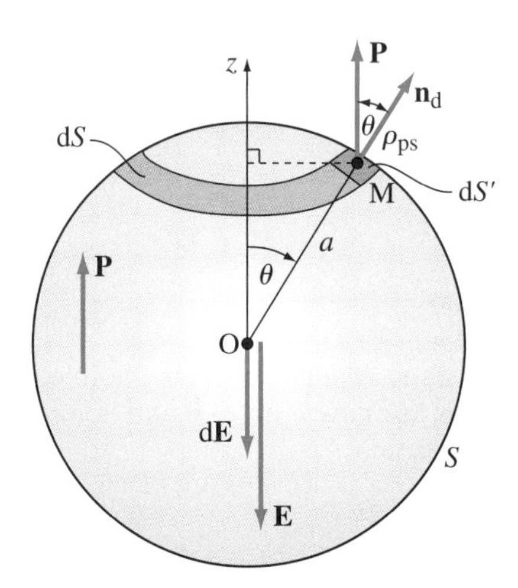

Figure 2.7 Dielectric sphere with a uniform polarization; for Example 2.3.

that the upper limit in the integration is now $\theta = \pi$. With this, Eq. (1.67) becomes

$$\mathbf{E} = \oint_S \mathrm{d}\mathbf{E} = -\frac{1}{2\varepsilon_0}\int_{\theta=0}^{\pi} \rho_{\mathrm{ps}}(\theta)\sin\theta\cos\theta\,\mathrm{d}\theta\,\hat{\mathbf{z}} = -\frac{P}{2\varepsilon_0}\int_0^{\pi} \underbrace{\cos^2\theta}_{u^2}\,\underbrace{\sin\theta\,\mathrm{d}\theta}_{-\mathrm{d}u}\,\hat{\mathbf{z}}$$

electric field inside a uniformly polarized dielectric sphere

$$= -\frac{P}{3\varepsilon_0}\hat{\mathbf{z}} \longrightarrow \boxed{\mathbf{E} = -\frac{\mathbf{P}}{3\varepsilon_0}}, \tag{2.32}$$

where the substitution given by $u = \cos\theta$ is used to solve the integral in θ. It can be shown that **E** has this same (constant) value at any point inside the polarized sphere.

Example 2.4 Nonuniformly Polarized Dielectric Sphere

A nonuniformly polarized dielectric sphere, of radius a, is situated in free space. In a spherical coordinate system whose origin coincides with the sphere center, the polarization vector is given by the expression

$$\mathbf{P}(r) = P_r(r)\,\hat{\mathbf{r}} = P_0\,\frac{r}{a}\,\hat{\mathbf{r}} \tag{2.33}$$

(P_0 is a constant). Determine (a) the bound volume and surface charge densities of the sphere and (b) the electric scalar potential inside and outside the sphere.

Solution

(a) Using the expression for the divergence in spherical coordinates, Eq. (1.171), the bound volume charge density of the sphere amounts to

$$\rho_{\mathrm{p}} = -\nabla\cdot\mathbf{P} = -\frac{1}{r^2}\frac{\partial}{\partial r}\left(r^2 P_r\right) = -\frac{3P_0}{a} \tag{2.34}$$

(it is the same at all points inside the sphere). The bound surface charge density is

$$\rho_{\mathrm{ps}} = \hat{\mathbf{r}}\cdot\mathbf{P}(a^-) = P_0. \tag{2.35}$$

(b) The field outside the sphere (for $r > a$) is zero, because it is identical to the field of the equivalent point charge $Q_{\mathrm{p}} = 0$ (total charge of the sphere) placed at the sphere center. Hence, the potential outside the sphere is also zero,

$$V(r) = 0, \quad a < r < \infty. \tag{2.36}$$

From Eq. (1.140), the electric field inside the sphere is given by

$$E(r) = \frac{\rho_{\mathrm{p}} r}{3\varepsilon_0} = -\frac{P_0 r}{\varepsilon_0 a}, \quad 0 \le r < a. \tag{2.37}$$

The potential inside the sphere is thus

$$V(r) = \int_{r'=r}^{a} E(r')\,\mathrm{d}r' = -\frac{P_0 a}{2\varepsilon_0}\left(1 - \frac{r^2}{a^2}\right), \quad 0 \le r \le a. \tag{2.38}$$

Problems: 2.2–2.6; *Conceptual Questions* (on Companion Website): 2.3 and 2.4; *MATLAB Exercises* (on Companion Website).

2.5 GENERALIZED GAUSS' LAW

We now consider the most general electrostatic system containing both conductors and dielectrics. The equivalent field sources are now both free and bound charges, in free space, and Gauss' law, Eq. (1.133), becomes

Gauss' law for a system with conductors and dielectrics

$$\boxed{\oint_S \mathbf{E}\cdot\mathrm{d}\mathbf{S} = \frac{Q_S + Q_{\mathrm{p}S}}{\varepsilon_0}}, \tag{2.39}$$

where Q_S and Q_{pS} are the total free charge and the total bound charge, respectively, enclosed by an arbitrary closed surface S. Multiplying this equation by ε_0, moving Q_{pS} to the left-hand side of the equation, then substituting it by the negative of the flux of the polarization vector, $\mathbf{P}$, from Eq. (2.16), and finally joining the two integrals over S into a single integral, we obtain the equivalent integral equation:

$$\oint_S (\varepsilon_0 \mathbf{E} + \mathbf{P}) \cdot \mathrm{d}\mathbf{S} = Q_S. \qquad (2.40)$$

To shorten the writing, we define a new vector quantity,

$$\boxed{\mathbf{D} = \varepsilon_0 \mathbf{E} + \mathbf{P},} \qquad (2.41)$$

electric flux density vector (unit: $\mathrm{C/m^2}$)

which is called the electric flux density vector (also known as the electric displacement vector or electric induction vector). Accordingly, the flux of $\mathbf{D}$ is termed the electric flux (symbolized by Ψ),

$$\boxed{\Psi = \int_{S'} \mathbf{D} \cdot \mathrm{d}\mathbf{S},} \qquad (2.42)$$

electric flux (unit: C)

where S' is any designated surface (open or closed). In place of Eq. (2.40),

$$\boxed{\oint_S \mathbf{D} \cdot \mathrm{d}\mathbf{S} = Q_S.} \qquad (2.43)$$

generalized Gauss' law

This is an equivalent form of Gauss' law for electrostatic fields in arbitrary media, which is more convenient than the form in Eq. (2.39) because it has only free charges on the right-hand side of the integral equation, and not the bound charges, and thus is simpler to use. It is referred to as the generalized Gauss' law, and, in words, it states that the outward electric flux through any closed surface in any electrostatic system including conductors and dielectrics equals the total free charge enclosed by the surface. From Eq. (2.43), the unit for the electric flux is C, so that the unit for its density, $\mathbf{D}$, is $\mathrm{C/m^2}$.

In the general case, free charge is represented by means of the volume charge density, ρ, yielding

$$\boxed{\oint_S \mathbf{D} \cdot \mathrm{d}\mathbf{S} = \int_v \rho \, \mathrm{d}v,} \qquad (2.44)$$

generalized Gauss' law in terms of the volume charge density

with v denoting the volume bounded by S. Since this integral relation is true regardless of the choice of v, the divergence theorem, Eq. (1.173), gives the differential form of the generalized Gauss' law:

$$\boxed{\nabla \cdot \mathbf{D} = \rho.} \qquad (2.45)$$

generalized differential Gauss' law

Problems: 2.7–2.11; *Conceptual Questions* (on Companion Website): 2.5.

2.6 CHARACTERIZATION OF DIELECTRIC MATERIALS

The polarization properties of materials can be described by the relationship between the polarization vector, $\mathbf{P}$, and the electric field intensity vector, $\mathbf{E}$, Eq. (2.9). We now employ the electric flux density vector, $\mathbf{D}$, and substituting Eq. (2.9) into Eq. (2.41), obtain the equivalent relationship

$$\boxed{\mathbf{D} = \varepsilon_0 \mathbf{E} + \mathbf{P}(\mathbf{E}) = \mathbf{D}(\mathbf{E}),} \qquad (2.46)$$

constitutive equation of an arbitrary (nonlinear) dielectric which is more often used for characterization of dielectric materials and is termed a constitutive equation of the material. For linear dielectrics, Eq. (2.10) applies, and Eq. (2.46) becomes

constitutive equation of a linear dielectric

$$\boxed{\mathbf{D} = (\chi_e + 1)\varepsilon_0 \mathbf{E} = \varepsilon_r \varepsilon_0 \mathbf{E} \quad \text{or} \quad \mathbf{D} = \varepsilon \mathbf{E},} \tag{2.47}$$

where ε is the permittivity and ε_r the relative permittivity of the medium (ε_r is sometimes referred to as the dielectric constant of the material). The unit for ε is farad per meter (F/m), while ε_r is dimensionless, obtained as

$$\varepsilon_r = \chi_e + 1, \tag{2.48}$$

and hence

$$\varepsilon_r \geq 1. \tag{2.49}$$

The value of ε_r shows how much the permittivity of a dielectric material,

permittivity of a linear dielectric (unit: F/m)

$$\boxed{\varepsilon = \varepsilon_r \varepsilon_0,} \tag{2.50}$$

is higher than the permittivity of free space (vacuum), given in Eq. (1.2). For free space and nondielectric materials (such as metals), $\varepsilon_r = 1$ and

constitutive equation for free space

$$\boxed{\mathbf{D} = \varepsilon_0 \mathbf{E}.} \tag{2.51}$$

Table 2.1 shows values of the relative permittivity of a number of selected materials, for electrostatic or low-frequency time-varying (time-harmonic) applied electric fields,[2] at room temperature (20°C).

For nonlinear dielectrics, the constitutive relation between **D** and **E**, Eq. (2.46), is nonlinear. This also means that the polarization properties of the material depend on the electric field intensity, E (for linear dielectrics, χ_e and ε are constants, independent of E).

In so-called ferroelectric materials, Eq. (2.46) is not only nonlinear, but also shows hysteresis effects. The function $D(E)$ has multiple branches, so that D is not uniquely determined by a value of E, but it depends also on the history of polarization of the material, i.e., on its previous states. A notable example is barium titanate ($BaTiO_3$), used in ceramic capacitors and various microwave devices (e.g., ceramic filters and multiplexers).

Another concept in characterization of materials is homogeneity. A material is said to be homogeneous when its properties do not change from point to point in the region being considered. In a linear homogeneous dielectric, ε is a constant independent of spatial coordinates. Otherwise, the material is inhomogeneous [e.g., $\varepsilon = \varepsilon(x, y, z)$ in the region].

Finally, we introduce the concept of isotropy in classifying dielectric materials. Generally, properties of isotropic media are independent of direction. In a linear isotropic dielectric, ε is a scalar quantity, and hence **D** and **E** are always collinear and in the same direction, regardless of the orientation of **E**. In an anisotropic medium, however, individual components of **D** depend differently on different components of **E**, so that Eq. (2.47) becomes a matrix equation,

[ε] – permittivity tensor of an anisotropic dielectric

$$\boxed{\begin{bmatrix} D_x \\ D_y \\ D_z \end{bmatrix} = \begin{bmatrix} \varepsilon_{xx} & \varepsilon_{xy} & \varepsilon_{xz} \\ \varepsilon_{yx} & \varepsilon_{yy} & \varepsilon_{yz} \\ \varepsilon_{zx} & \varepsilon_{zy} & \varepsilon_{zz} \end{bmatrix} \begin{bmatrix} E_x \\ E_y \\ E_z \end{bmatrix}.} \tag{2.52}$$

[2]At higher frequencies, when viewed over very wide frequency ranges, the permittivity generally (for most materials) is not a constant, but depends on the operating frequency of electromagnetic waves propagating through the material.

Table 2.1. Relative permittivity of selected materials*

Material	ε_r	Material	ε_r
Vacuum	1	Quartz	5
Freon	1	Diamond	5–6
Air	1.0005	Wet soil	5–15
Styrofoam	1.03	Mica (ruby)	5.4
Polyurethane foam	1.1	Steatite	5.8
Paper	1.3–3	Sodium chloride (NaCl)	5.9
Wood	2–5	Porcelain	6
Dry soil	2–6	Neoprene	6.6
Paraffin	2.1	Silicon nitride (Si_3N_4)	7.2
Teflon	2.1	Marble	8
Vaseline	2.16	Alumina (Al_2O_3)	8.8
Polyethylene	2.25	Animal and human muscle	10
Oil	2.3	Silicon (Si)	11.9
Rubber	2.4–3	Gallium arsenide	13
Polystyrene	2.56	Germanium	16
PVC	2.7	Ammonia (liquid)	22
Amber	2.7	Alcohol (ethyl)	25
Plexiglass	3.4	Tantalum pentoxide	25
Nylon	3.6–4.5	Glycerin	50
Fused silica (SiO_2)	3.8	Ice	75
Sulfur	4	Water	81
Glass	4–10	Rutile (TiO_2)	89–173
Bakelite	4.74	Barium titanate ($BaTiO_3$)	1,200

* For static or low-frequency applied electric fields, at room temperature.

Thus, instead of a single scalar ε, we have a tensor $[\varepsilon]$ (permittivity tensor), i.e., nine (generally different) scalars corresponding to different pairs of spatial components of **D** and **E**. Crystalline dielectric materials, in general, are anisotropic; the periodic nature of crystals causes dipole moments to be formed and oriented by means of the applied electric field much more easily along the crystal axes than in other directions. An example is rutile (TiO_2), whose relative permittivity is $\varepsilon_r = 173$ in the direction parallel to a crystal axis and $\varepsilon_r = 89$ at right angles. For many crystals the change in permittivity with direction is small. For example, quartz has $\varepsilon_r = 4.7 - 5.1$, and it is customary to adopt a rounded value $\varepsilon_r = 5$ for its average relative permittivity and treat the material as isotropic.

The theory of dielectrics we have discussed so far assumes normal designed regimes of operation of electrical systems – when the electric field intensity, E, in individual dielectric parts of a system is below a certain "breakdown" level. Namely, the intensity E in a dielectric cannot be increased indefinitely: if a certain value is exceeded, the dielectric becomes conducting. It temporarily or permanently loses its insulating property, and is said to break down. The breaking field value, i.e., the maximum electric field intensity that an individual dielectric material can withstand without breakdown, is termed the dielectric strength of the material. We denote it by E_{cr} (critical field intensity). The values of E_{cr} for different materials are obtained by measurement. For air,

$$\boxed{E_{cr0} = 3\ \mathrm{MV/m}.} \qquad (2.53)$$

dielectric strength of air

In gaseous dielectrics, like air, because of a very strong applied electric field, the free electrons and ions are accelerated, by Coulomb forces [see Eq. (1.23)], to velocities high enough that in collisions with neutral molecules, they are able to knock electrons out of the molecule (so-called impact ionization). The newly created free electrons and positively charged ions are also accelerated by the field, they collide with molecules, liberate more electrons, and the result is an avalanche process of impact ionization and very rapid generation of a vast number of free electrons that constitute a substantial electric current in the gas (usually sparking occurs as well). In other words, the gas, normally a very good insulator, is suddenly transformed into an excellent conductor. Note that many air breakdowns occur at any instant of time in thunderstorms all over the earth. Basically, they are caused by large atmospheric electric fields (fields due to charged clouds), reaching the breakdown value in Eq. (2.53), and their most obvious manifestation is, of course, lightning.

Similar avalanche processes occur at high enough electric field intensities in liquid and solid dielectrics. For solids, these processes are enhanced and the value of the dielectric strength (E_{cr}) of the particular piece of a dielectric is lowered by impurities and structural defects in the material, by certain ways the material is manufactured, and even by microscopic air-filled cracks and voids in the material. In addition, when, under the influence of a strong electric field, the local heat due to leakage currents flowing in lossy (low-loss) dielectrics is generated faster than it can be dissipated in the material, the resulting rise of temperature may cause a change in the material (melting) and lead to a so-called thermal breakdown of the dielectric. Such breakdown processes depend on the duration of the applied strong field and the ambient temperature. Breakdowns in solid dielectrics most often cause a permanent damage to the material (e.g., formation of highly conductive channels of molten material, sometimes including carbonized matter, that irreversibly damage the texture of the dielectric).

The values of E_{cr} for some selected dielectric materials are presented in Table 2.2. Dielectric strengths of dielectrics other than air are larger than the value in Eq. (2.53). Note that, by definition, the dielectric strength of a vacuum is infinite.

Conceptual Questions (on Companion Website): 2.6.

Table 2.2. Dielectric strength of selected materials*

Material	E_{cr} (MV/m)	Material	E_{cr} (MV/m)
Air (atmospheric pressure)	3	Bakelite	25
Barium titanate ($BaTiO_3$)	7.5	Glass (plate)	30
Freon	~8	Paraffin	~30
Germanium	~10	Silicon (Si)	~30
Wood (douglas fir)	~ 10	Alumina	~35
Porcelain	11	Gallium arsenide	~40
Oil (mineral)	15	Polyethylene	47
Paper (impregnated)	15	Mica	200
Polystyrene	20	Fused quartz (SiO_2)	~1000
Teflon	20	Silicon nitride (Si_3N_4)	~1000
Rubber (hard)	25	Vacuum	∞

* At room temperature.

2.7 MAXWELL'S EQUATIONS FOR THE ELECTROSTATIC FIELD

We note that Maxwell's first equation for the electrostatic field, Eq. (1.75), does not depend on the material properties, and is the same in all kinds of dielectrics as it is in free space. Eq. (2.44) is Maxwell's third equation, and we now write down the full set of Maxwell's equations for the electrostatic field in an arbitrary medium, together with the constitutive equation, Eq. (2.46) or (2.47):

$$\begin{cases} \oint_C \mathbf{E} \cdot \mathrm{d}\mathbf{l} = 0 \\ \oint_S \mathbf{D} \cdot \mathrm{d}\mathbf{S} = \int_v \rho \, \mathrm{d}v \\ \mathbf{D} = \mathbf{D}(\mathbf{E}) \; [\mathbf{D} = \varepsilon \mathbf{E}] \end{cases} . \tag{2.54}$$

Maxwell's first equation in electrostatics
Maxwell's third equation
constitutive equation for $\mathbf{D}$

We shall see later in this text that these equations represent a subset of the full set of Maxwell's equations for the electromagnetic field, specialized for the electrostatic case. In the general case, the set contains four Maxwell's equations and three constitutive equations. As we shall see, the third equation (generalized Gauss' law) retains this same form also under nonstatic conditions. Constitutive equations are not Maxwell's equations, but are associated with them and are needed to supply the information about the materials involved.

2.8 ELECTROSTATIC FIELD IN LINEAR, ISOTROPIC, AND HOMOGENEOUS MEDIA

Most often we deal with linear, isotropic, and homogeneous dielectrics, in which Eq. (2.47) applies, and the permittivity ε is independent of the intensity of the applied field, is the same for all directions, and does not change from point to point. For such media, we can bring ε outside the integral sign in the integral form of the generalized Gauss' law, Eq. (2.43),

$$\oint_S \mathbf{E} \cdot \mathrm{d}\mathbf{S} = \frac{Q_S}{\varepsilon}, \tag{2.55}$$

Gauss' law for a homogeneous dielectric

or outside the operator (div) sign in the differential generalized Gauss' law, Eq. (2.45),

$$\nabla \cdot \mathbf{E} = \frac{\rho}{\varepsilon}. \tag{2.56}$$

We notice that Eqs. (2.55) and (2.56) are identical to the corresponding free-space laws, Eqs. (1.133) and (1.165), except for ε_0 being substituted by ε. Recall that the expression for the electric field intensity vector due to a point charge in free space, and with it also Coulomb's law, can be derived from Gauss' law (see Problem 1.53). Based on this, we can now reconsider all charge distributions in free space we have considered so far, and all structures with conductors in free space we have analyzed, and by merely replacing ε_0 with ε in all the equations, obtain the solutions for the same (free) charge distributions and the same conducting structures situated in a homogeneous dielectric of permittivity ε.[3] This is the power of the concept

[3]In what follows (in this entire text), we shall always assume linear and isotropic media, except when we explicitly specify that the medium under consideration is nonlinear and/or anisotropic.

of dielectric permittivity. We emphasize again that, with using the electric flux density vector and the dielectric permittivity, we are left to deal with free charges in the system only, while the contribution of bound charges to the field is properly added through ε. Thus, for example, Eq. (1.82) implies that the potential due to a free volume charge distribution in a homogeneous dielectric with permittivity ε is given by

$$V = \frac{1}{4\pi\varepsilon} \int_v \frac{\rho \, \mathrm{d}v}{R}. \tag{2.57}$$

Also, the free surface charge density on the surface of a conductor surrounded by a dielectric with permittivity ε is [from Eq. (1.190)]

$$\rho_\mathrm{s} = \varepsilon \, \hat{\mathbf{n}} \cdot \mathbf{E} \tag{2.58}$$

(boundary condition for the normal component of $\mathbf{E}$), and so on. Note, however, that the boundary condition for the tangential component of $\mathbf{E}$ near a conductor surface, Eq. (1.186), is always the same, irrespective of the properties (ε) of the surrounding dielectric.

Once we find the electric field in a structure filled with a homogeneous dielectric, we can calculate the polarization vector in the dielectric as [Eqs. (2.41) and (2.47)]

polarization vector in a linear dielectric

$$\boxed{\mathbf{P} = \mathbf{D} - \varepsilon_0 \mathbf{E} = (\varepsilon - \varepsilon_0)\mathbf{E},} \tag{2.59}$$

and then the distribution of volume and surface bound charges of the dielectric – using Eqs. (2.19) and (2.23).

Note that, from Eqs. (2.19), (2.59), (2.56), and (2.50), the bound volume charge density, ρ_p, at a point in the dielectric can be obtained directly from the free volume charge density, ρ, at that point as

$$\rho_\mathrm{p} = -\nabla \cdot \mathbf{P} = -(\varepsilon - \varepsilon_0)\nabla \cdot \mathbf{E} = -\frac{\varepsilon - \varepsilon_0}{\varepsilon}\rho = -\frac{\varepsilon_\mathrm{r} - 1}{\varepsilon_\mathrm{r}}\rho. \tag{2.60}$$

In an analogous manner, we derive the relationship between the bound and free surface charge densities on the surface of a conductor surrounded by a dielectric with relative permittivity ε_r. Shown in Fig. 2.8 is a detail of the surface. Combining Eqs. (2.23), (2.59), (2.58), and (2.50), and noting that $\hat{\mathbf{n}}_\mathrm{d} = -\hat{\mathbf{n}}$ [in Eq. (2.23), $\hat{\mathbf{n}}_\mathrm{d}$ is directed from the dielectric outward; in Eq. (2.58), $\hat{\mathbf{n}}$ is directed from the conductor outward], we obtain

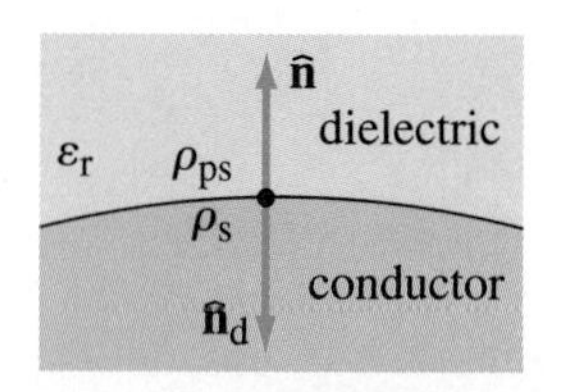

Figure 2.8 Detail of a conductor-linear dielectric surface.

$$\rho_\mathrm{ps} = \hat{\mathbf{n}}_\mathrm{d} \cdot \mathbf{P} = -(\varepsilon - \varepsilon_0)\hat{\mathbf{n}} \cdot \mathbf{E} = -\frac{\varepsilon_\mathrm{r} - 1}{\varepsilon_\mathrm{r}}\rho_\mathrm{s}. \tag{2.61}$$

Although the free surface charge density, ρ_s, is actually localized on the conductor side of the boundary surface and the bound surface charge density, ρ_ps, is localized on the dielectric side of the surface, they can be treated as a single sheet of charge with the total density

$$\rho_\mathrm{s\,tot} = \rho_\mathrm{s} + \rho_\mathrm{ps} = \left(1 - \frac{\varepsilon_\mathrm{r} - 1}{\varepsilon_\mathrm{r}}\right)\rho_\mathrm{s} = \frac{\rho_\mathrm{s}}{\varepsilon_\mathrm{r}}. \tag{2.62}$$

Example 2.5 Dielectric Sphere with Free Volume Charge

A homogeneous dielectric sphere, of radius a and relative permittivity ε_r, is situated in air. There is a free volume charge density $\rho(r) = \rho_0 r/a$ $(0 \le r \le a)$ throughout the sphere volume, where r is the distance from the sphere center (spherical radial coordinate) and ρ_0 is a constant. Determine (a) the electric scalar potential for $0 \le r < \infty$ and (b) the bound charge distribution of the sphere.

Solution

(a) Because of spherical symmetry of the problem, the electric flux density vector, **D**, is purely radial and depends only on r. From the generalized Gauss' law [Eq. (2.44)], applied in a similar fashion to that in Example 1.19 accommodated for spherical symmetry (see also Example 1.18), the magnitude of **D** is found to be

$$D(r) = \begin{cases} \rho_0 r^2/(4a) & \text{for } r \le a \\ \rho_0 a^3/(4r^2) & \text{for } r > a \end{cases} . \tag{2.63}$$

The electric field intensity vector is of the same form, and its magnitude is given by

$$E(r) = \begin{cases} D(r)/(\varepsilon_r \varepsilon_0) & \text{for } r \le a \\ D(r)/\varepsilon_0 & \text{for } r > a \end{cases} . \tag{2.64}$$

The potential at a distance r from the sphere center is hence:

$$V(r) = \frac{1}{\varepsilon_0} \int_{r'=r}^{\infty} D(r')\,\mathrm{d}r' = \frac{\rho_0 a^3}{4\varepsilon_0 r}, \quad \text{for } r \ge a, \tag{2.65}$$

and [also see Eq. (1.142)]

$$V(r) = \frac{1}{\varepsilon_r \varepsilon_0} \int_{r'=r}^{a} D(r')\,\mathrm{d}r' + V(a) = \frac{\rho_0 a^2}{4\varepsilon_0} \left[1 + \frac{1}{3\varepsilon_r}\left(1 - \frac{r^3}{a^3}\right)\right], \quad \text{for } r < a. \tag{2.66}$$

(b) According to Eq. (2.60), the bound volume charge density inside the sphere amounts to

$$\rho_p(r) = -\frac{\varepsilon_r - 1}{\varepsilon_r} \rho(r) = -\frac{\rho_0(\varepsilon_r - 1)r}{\varepsilon_r a}. \tag{2.67}$$

Using Eqs. (2.23), (2.59), (2.47), and (2.63), the bound surface charge density on the sphere surface comes out to be

$$\rho_{ps} = P(a^-) = \frac{\varepsilon_r - 1}{\varepsilon_r} D(a^-) = \frac{\rho_0(\varepsilon_r - 1)a}{4\varepsilon_r}. \tag{2.68}$$

Example 2.6 Model of a *pn* Junction

Sketched in Fig. 2.9(a) is a *pn* junction between two semiconducting half-spaces, doped *p*-type and *n*-type, respectively. The volume charge distribution in the semiconductor can be approximated by the following function:

$$\rho(x) = \begin{cases} -\rho_0\,\mathrm{e}^{x/a} & \text{for } x < 0 \\ 0 & \text{for } x = 0 \\ \rho_0\,\mathrm{e}^{-x/a} & \text{for } x > 0 \end{cases} , \tag{2.69}$$

where ρ_0 and a are positive constants. The permittivity of the semiconductor is ε. Find (a) the electric field intensity vector in the semiconductor, (b) the electric scalar potential in the semiconductor, and (c) the voltage between the ends of the semiconductor, from the end on the *n*-type side to the end on the *p*-type side of the junction.

Solution

(a) This is a problem with planar symmetry (see Examples 1.20, 1.21, and 1.23), and the electric field intensity vector in the semiconductor is given by $\mathbf{E} = E_x(x)\,\hat{\mathbf{x}}$ [Eq. (1.148)]. The differential generalized Gauss' law, Eq. (2.56), becomes

$$\frac{\mathrm{d}E_x(x)}{\mathrm{d}x} = \frac{\rho(x)}{\varepsilon}. \tag{2.70}$$

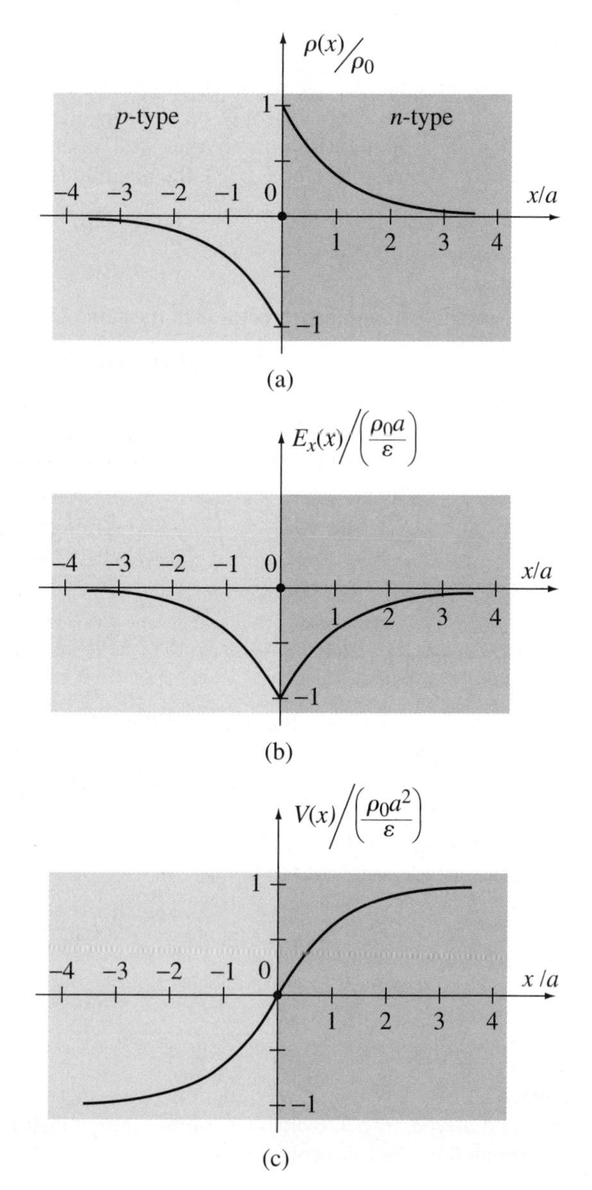

Figure 2.9 Model of a *pn* junction: (a) volume charge density, (b) electric field intensity, and (c) electric scalar potential; for Example 2.6.

This is a first-order differential equation in x, and we solve it by integrating with respect to x [as in Eq. (1.178)]:

$$E_x(x) = \frac{1}{\varepsilon} \int_{x'=-\infty}^{x} \rho(x')\, dx' + C, \tag{2.71}$$

where C is the constant of integration, which represents the field E_x in the plane $x \to -\infty$. We note that

$$\int_{-\infty}^{\infty} \rho(x')\, dx' = 0, \tag{2.72}$$

meaning that the total charge of the semiconductor is zero. This means, in turn, that no field can exist far from the junction [see also Eq. (1.154)],

$$E_x(x \to \mp\infty) = 0, \tag{2.73}$$

and hence $C = 0$. Substituting Eq. (2.69), the integration for the field points in the p-type region of the semiconductor yields

$$E_x(x) = -\frac{\rho_0}{\varepsilon}\int_{-\infty}^{x} e^{x'/a}\,dx' = -\frac{\rho_0 a}{\varepsilon}\,e^{x/a} \quad (-\infty < x \le 0). \tag{2.74}$$

In the n-type region, we have to break the integration up into two parts:

$$E_x(x) = \frac{\rho_0}{\varepsilon}\left(-\int_{-\infty}^{0} e^{x'/a}\,dx' + \int_{0}^{x} e^{-x'/a}\,dx'\right) = -\frac{\rho_0 a}{\varepsilon}\,e^{-x/a} \quad (0 < x < \infty). \tag{2.75}$$

Fig. 2.9(b) shows the electric field intensity $E_x(x)$ in the semiconductor. We see that the field is oriented from the n-type doped region to the p-type doped region. This field is called the built-in field of a pn junction, as it exists in the junction even when an external voltage is not applied (e.g., when the terminals of a pn diode are not connected to an external voltage source). In the equilibrium state established after the pn junction is formed, the built-in field prevents further diffusion of positive charges (holes) to the right and negative charges (electrons) to the left across the junction.[4]

(b) Let us arbitrarily adopt our reference point for potential ($\mathcal{R}$) at the center of the junction, $x = 0$, so that the potential at points in the p-type region is [Eq. (1.74)]

$$V(x) = \int_{\mathrm{P}}^{\mathcal{R}} \mathbf{E}\cdot d\mathbf{l} = \int_{x'=x}^{0} E_x(x')\,dx' = -\frac{\rho_0 a}{\varepsilon}\int_{x}^{0} e^{x'/a}\,dx'$$
$$= \frac{\rho_0 a^2}{\varepsilon}\left(e^{x/a} - 1\right), \quad -\infty < x \le 0. \tag{2.76}$$

In the n-type region, we reverse the direction of integration for convenience,

$$V(x) = \int_{\mathrm{P}}^{\mathcal{R}} \mathbf{E}\cdot d\mathbf{l} = -\int_{\mathcal{R}}^{\mathrm{P}} \mathbf{E}\cdot d\mathbf{l} = -\int_{0}^{x} E_x(x')\,dx'$$
$$= \frac{\rho_0 a}{\varepsilon}\int_{0}^{x} e^{-x'/a}\,dx' = \frac{\rho_0 a^2}{\varepsilon}\left(1 - e^{-x/a}\right), \quad 0 < x < \infty. \tag{2.77}$$

The distribution of the potential, $V(x)$, along the semiconductor is shown in Fig. 2.9(c).

(c) The voltage between the n-type and p-type ends of the semiconductor turns out to be

$$V(x \to \infty) - V(x \to -\infty) = \frac{2\rho_0 a^2}{\varepsilon}. \tag{2.78}$$

This voltage is called the built-in voltage of a pn junction (diode).

Problems: 2.12–2.15; *MATLAB Exercises* (on Companion Website).

2.9 DIELECTRIC-DIELECTRIC BOUNDARY CONDITIONS

So far, we have considered boundary surfaces conductor-free space (Fig. 1.39), dielectric-free space (Fig. 2.4), and conductor-dielectric (Fig. 2.8), and analyzed the fields close to surfaces and surface charge densities on the surfaces [Eqs. (1.186), (1.189), (2.23), and (2.58)]. Let us now consider a dielectric-dielectric boundary

[4] Note that initially there are excess holes to the left of the plane $x = 0$ (p-type semiconductor is positively charged over its entire volume) and excess electrons to the right (n-type semiconductor is negatively charged by doping). In a brief transient, a diffusion of holes occurs from the p-region into the n-region and electrons diffuse in the opposite direction, until an electric field is built up in such a direction that the diffusion current drops to zero.

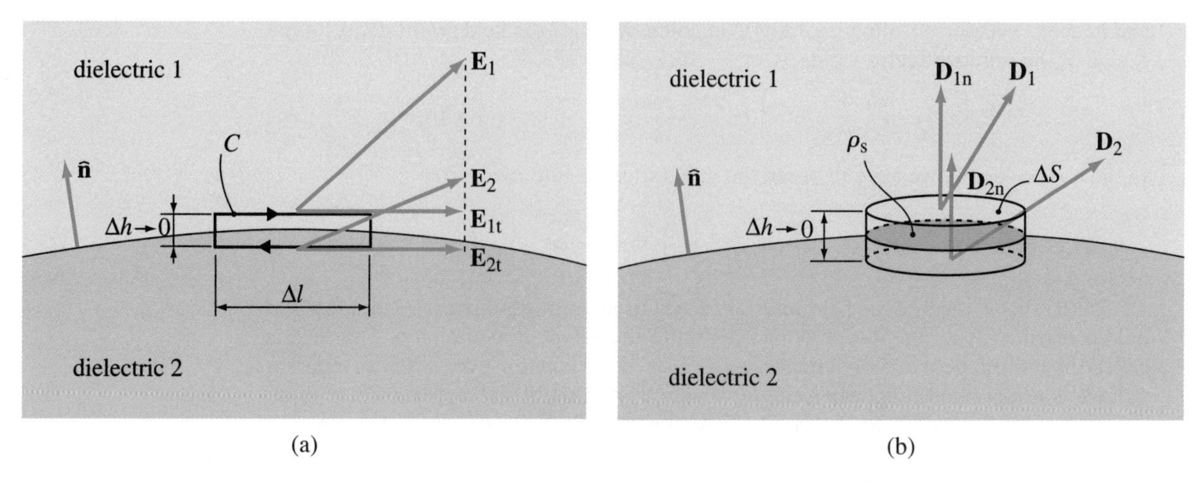

Figure 2.10 Dielectric-dielectric boundary surface: deriving boundary conditions for (a) tangential components of $\mathbf{E}$ and (b) normal components of $\mathbf{D}$.

surface, shown in Fig. 2.10, and derive the boundary conditions for field components near the surface. We apply the same technique as in deriving the corresponding boundary conditions for the conductor-free space case, Eqs. (1.186) and (1.189). The main difference is that now the field exists at both sides of the boundary. Let $\mathbf{E}_1$ and $\mathbf{D}_1$ be, respectively, the electric field intensity vector and electric flux density vector close to the boundary in medium 1, whereas $\mathbf{E}_2$ and $\mathbf{D}_2$ stand for the same quantities in medium 2.

Applying Eq. (1.75) to a narrow rectangular elementary contour C, Fig. 2.10(a), we obtain

continuity of the tangential component of $\mathbf{E}$

$$\oint_C \mathbf{E} \cdot \mathrm{d}\mathbf{l} = E_{1\mathrm{t}}\,\Delta l - E_{2\mathrm{t}}\,\Delta l = 0 \quad \longrightarrow \quad \boxed{E_{1\mathrm{t}} = E_{2\mathrm{t}}.} \tag{2.79}$$

This boundary condition tells us that the tangential components of $\mathbf{E}$ are the same on the two sides of the boundary, i.e., that $\mathbf{E}_\mathrm{t}$ is continuous across the boundary.

On the other hand, an application of Eq. (2.43) to a pillbox Gaussian elementary surface, Fig. 2.10(b), gives

$$\oint_S \mathbf{D} \cdot \mathrm{d}\mathbf{S} = D_{1\mathrm{n}}\,\Delta S - D_{2\mathrm{n}}\,\Delta S = \rho_\mathrm{s}\,\Delta S \quad \longrightarrow \quad D_{1\mathrm{n}} - D_{2\mathrm{n}} = \rho_\mathrm{s} \tag{2.80}$$

(we employ vector $\mathbf{D}$ to avoid dealing with bound charges), where the normal components of $\mathbf{D}$ are defined with respect to the unit normal $\hat{\mathbf{n}}$ directed from region 2 to region 1, and ρ_s is the free surface charge density that may exist on the surface. In the absence of charge,

continuity of the normal component of $\mathbf{D}$, charge-free surface

$$\boxed{D_{1\mathrm{n}} = D_{2\mathrm{n}} \quad (\rho_\mathrm{s} = 0).} \tag{2.81}$$

This boundary condition enforces that the normal components of $\mathbf{D}$ be the same on the two sides of a boundary with no free charge on it. In other words, $\mathbf{D}_\mathrm{n}$ is continuous across the boundary free of charge.

Relationships in Eqs. (2.79) and (2.80) represent two primary boundary conditions for the electrostatic field at the interface between two arbitrary media. When the dielectrics 1 and 2 are linear, $\mathbf{D}_1 = \varepsilon_1\mathbf{E}_1$ and $\mathbf{D}_2 = \varepsilon_2\mathbf{E}_2$, so we obtain an additional pair of (secondary) boundary conditions – for the tangential components of $\mathbf{D}$ and normal components of $\mathbf{E}$. From Eq. (2.79), the boundary condition for $\mathbf{D}_\mathrm{t}$ reads

$$\frac{D_{1\mathrm{t}}}{\varepsilon_1} = \frac{D_{2\mathrm{t}}}{\varepsilon_2}, \tag{2.82}$$

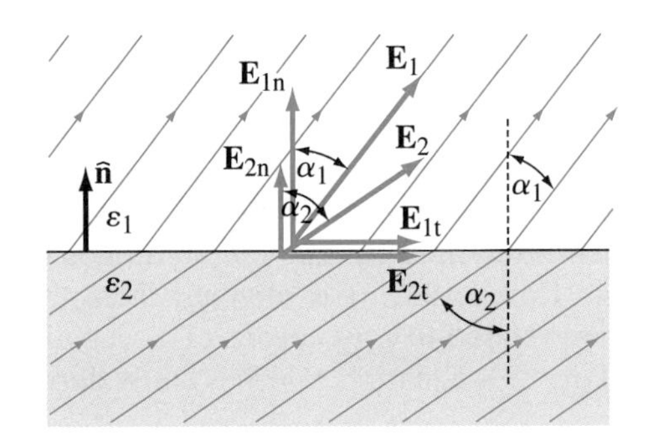

Figure 2.11 Refraction of electric field lines at a dielectric-dielectric interface.

and we see that $\mathbf{D}_t$ is discontinuous across the boundary. Similarly, Eq. (2.81) yields the boundary condition for $\mathbf{E}_n$ (if $\rho_s = 0$ at the interface):

$$\varepsilon_1 E_{1n} = \varepsilon_2 E_{2n}, \tag{2.83}$$

which shows that $\mathbf{E}_n$ is also discontinuous across the boundary.

Note that the boundary conditions for $\mathbf{E}$ and $\mathbf{D}$, that is, for $\mathbf{E}_t$ and $\mathbf{D}_n$, in Eqs. (2.79) and (2.80) can be written in vector form:

$$\hat{\mathbf{n}} \times \mathbf{E}_1 - \hat{\mathbf{n}} \times \mathbf{E}_2 = 0, \quad (\hat{\mathbf{n}} \text{ directed from region 2 to region 1}), \tag{2.84}$$

$$\hat{\mathbf{n}} \cdot \mathbf{D}_1 - \hat{\mathbf{n}} \cdot \mathbf{D}_2 = \rho_s, \tag{2.85}$$

boundary condition for $\mathbf{E}_t$

boundary condition for $\mathbf{D}_n$

which is often more convenient for use in analyzing complex structures.

Let us consider again the interface between two dielectric media and the angles α_1 and α_2 that field lines in region 1 and region 2 make with the normal to the interface, as depicted in Fig. 2.11. The tangents of these angles can be expressed as

$$\tan\alpha_1 = \frac{E_{1t}}{E_{1n}} \quad \text{and} \quad \tan\alpha_2 = \frac{E_{2t}}{E_{2n}}. \tag{2.86}$$

We divide the tangents and use Eqs. (2.79) and (2.83) to get

$$\frac{\tan\alpha_1}{\tan\alpha_2} = \frac{\varepsilon_1}{\varepsilon_2}. \tag{2.87}$$

law of refraction of electric field lines

This is the law of refraction of the electric field lines at a dielectric-dielectric boundary that is free of charge ($\rho_s = 0$). Bending of field lines is, essentially, a result of unequal bound charges on the two sides of the boundary.

Finally, let us find the distribution of bound surface charges on a dielectric-dielectric interface (Fig. 2.10). From Eq. (2.23), the bound charge densities that accumulate on the two sides of the interface are

$$\rho_{ps1} = \hat{\mathbf{n}}_{d1} \cdot \mathbf{P}_1 \quad \text{and} \quad \rho_{ps2} = \hat{\mathbf{n}}_{d2} \cdot \mathbf{P}_2, \tag{2.88}$$

where $\hat{\mathbf{n}}_{d1} = -\hat{\mathbf{n}}$ and $\hat{\mathbf{n}}_{d2} = \hat{\mathbf{n}}$ [$\hat{\mathbf{n}}_d$ in Eq. (2.23) is directed from the dielectric outward]. Hence, by adding ρ_{ps1} and ρ_{ps2} together, the total bound surface charge density at the interface is given by

$$\rho_{ps} = \hat{\mathbf{n}} \cdot \mathbf{P}_2 - \hat{\mathbf{n}} \cdot \mathbf{P}_1. \tag{2.89}$$

boundary condition for $\mathbf{P}_n$

This is the boundary condition for the normal components of vector $\mathbf{P}$ at a dielectric-dielectric boundary.

Problems: 2.16–2.18; *Conceptual Questions* (on Companion Website): 2.7 and 2.8; *MATLAB Exercises* (on Companion Website).

2.10 POISSON'S AND LAPLACE'S EQUATIONS

Consider the electric field intensity vector, $\mathbf{E}$, and scalar potential, V, in a homogeneous dielectric region of permittivity ε. The spatial derivatives of $\mathbf{E}$ at a point in the region are related to the free volume charge density, ρ, that may exist at that point by Eq. (2.56), which is a first-order differential equation. $\mathbf{E}$, in turn, is related to the spatial derivatives of V through Eq. (1.101). It is obvious, then, that the second-order spatial derivatives of V are related to ρ by means of a second-order differential equation. This equation is Poisson's equation, which is easily derived by substituting Eq. (1.101) into Eq. (2.56):

$$\nabla \cdot (\nabla V) = -\frac{\rho}{\varepsilon}. \tag{2.90}$$

The double-∇ operation in Poisson's equation is performed by evaluating first the gradient of V, and then the divergence of the result. In the Cartesian coordinate system, gradient and divergence are calculated using Eqs. (1.102) and (1.167), respectively, and we have

$$\nabla \cdot (\nabla V) = \operatorname{div}(\operatorname{grad} V) = \nabla \cdot \left(\frac{\partial V}{\partial x}\,\hat{\mathbf{x}} + \frac{\partial V}{\partial y}\,\hat{\mathbf{y}} + \frac{\partial V}{\partial z}\,\hat{\mathbf{z}} \right)$$

$$= \frac{\partial}{\partial x}\left(\frac{\partial V}{\partial x}\right) + \frac{\partial}{\partial y}\left(\frac{\partial V}{\partial y}\right) + \frac{\partial}{\partial z}\left(\frac{\partial V}{\partial z}\right) = \frac{\partial^2 V}{\partial x^2} + \frac{\partial^2 V}{\partial y^2} + \frac{\partial^2 V}{\partial z^2}. \tag{2.91}$$

We note that the same result would have been obtained by applying formally the formula for the dot product in the Cartesian coordinate system, Eq. (1.164), to the dot product of two identical vectors (∇) in $(\nabla \cdot \nabla)V$, where ∇ is expressed as in Eq. (1.100). Hence, we can write

$$\nabla \cdot (\nabla V) = (\nabla \cdot \nabla)V. \tag{2.92}$$

The operator $\nabla \cdot \nabla$ is abbreviated ∇^2 ("del squared"), and Poisson's equation becomes

Poisson's equation

$$\boxed{\nabla^2 V = -\frac{\rho}{\varepsilon},} \tag{2.93}$$

where

Laplacian in Cartesian coordinates

$$\boxed{\nabla^2 V = \frac{\partial^2 V}{\partial x^2} + \frac{\partial^2 V}{\partial y^2} + \frac{\partial^2 V}{\partial z^2}.} \tag{2.94}$$

In a charge-free region ($\rho = 0$),

Laplace's equation

$$\boxed{\nabla^2 V = 0,} \tag{2.95}$$

which is known as Laplace's equation. The ∇^2 operator is called the Laplacian. We see that the Laplacian operates on a scalar (e.g., V), and the result is another scalar (e.g., $-\rho/\varepsilon$ or 0).

The expressions for $\nabla^2 V$ in the cylindrical coordinate system (Fig. 1.25) and the spherical coordinate system (Fig. 1.26) can be obtained in the same manner, i.e., by first taking the gradient of V [Eqs. (1.105) and (1.108)], and then the divergence [Eqs. (1.170) and (1.171)] of the result. In cylindrical coordinates, the Laplacian of $V = V(r, \phi, z)$ thus comes out to be

Laplacian in cylindrical coordinates

$$\boxed{\nabla^2 V = \frac{1}{r}\frac{\partial}{\partial r}\left(r\frac{\partial V}{\partial r}\right) + \frac{1}{r^2}\frac{\partial^2 V}{\partial \phi^2} + \frac{\partial^2 V}{\partial z^2},} \tag{2.96}$$

HISTORICAL ASIDE

Siméon Denis Poisson (1781–1840), French mathematician, was a student of Laplace (1749–1827) and Lagrange (1736–1813), and successor of Fourier (1768–1830) as a professor, at École Polytechnique, Paris. He is best known for his work on probability (Poisson's distribution), and his contributions to mathematics as applied to electricity and magnetism. In 1813, he published a generalization of Laplace's differential equation for the potential theory, valid for a nonzero mass density, so inside a solid, in mechanics or for a nonzero charge density, so inside a charge distribution, in electromagnetics.

Pierre Simon de Laplace (1749–1827), French mathematician and astronomer, was President of *Académie des Sciences* (French Academy). His work in mathematical astronomy was summed up in a monumental five-volume book on celestial mechanics (*Mécanique Céleste*), published between 1799 and 1825, in which the second-order partial differential equation for potential that we now name after him, Laplace's equation, appeared. He also wrote a treatise on the theory of probability, and worked on specific heats of substances. Laplace was a minister and senator under Napoleon, and was made a marquis by Louis XVIII.

and that of $V = V(r, \theta, \phi)$ in spherical coordinates

$$\nabla^2 V = \frac{1}{r^2}\frac{\partial}{\partial r}\left(r^2 \frac{\partial V}{\partial r}\right) + \frac{1}{r^2 \sin\theta}\frac{\partial}{\partial \theta}\left(\sin\theta \frac{\partial V}{\partial \theta}\right) + \frac{1}{r^2 \sin^2\theta}\frac{\partial^2 V}{\partial \phi^2}. \quad (2.97)$$

Laplacian in spherical coordinates

Example 2.7 Application of a 1-D Poisson's Equation

A free volume charge of a uniform density ρ exists in a homogeneous dielectric, of permittivity ε, between two flat metallic electrodes, as shown in Fig. 2.12. The electrodes are connected to a voltage V_0, and the distance between them is d. Neglecting the fringing effects, find (a) the electric potential and (b) the electric field intensity vector in the dielectric.

Solution

(a) Neglecting the fringing effects is equivalent to assuming that the electrodes are infinitely large, in which case the potential in the dielectric varies with the distance from the electrodes only. Let x be the normal distance from the left electrode (Fig. 2.12). Poisson's equation, given by Eqs. (2.93) and (2.94), becomes

$$\frac{\mathrm{d}^2 V(x)}{\mathrm{d}x^2} = -\frac{\rho}{\varepsilon} \quad (0 < x < d). \quad (2.98)$$

By integrating it twice, we get

$$V(x) = -\frac{\rho x^2}{2\varepsilon} + C_1 x + C_2, \quad (2.99)$$

where C_1 and C_2 are the constants of integration. The boundary condition $V(0) = V_0$ results in $C_2 = V_0$, and the condition on the other boundary, $V(d) = 0$, gives $C_1 = \rho d/(2\varepsilon) - V_0/d$. Hence,

$$V(x) = \frac{\rho x(d-x)}{2\varepsilon} + V_0\left(1 - \frac{x}{d}\right). \quad (2.100)$$

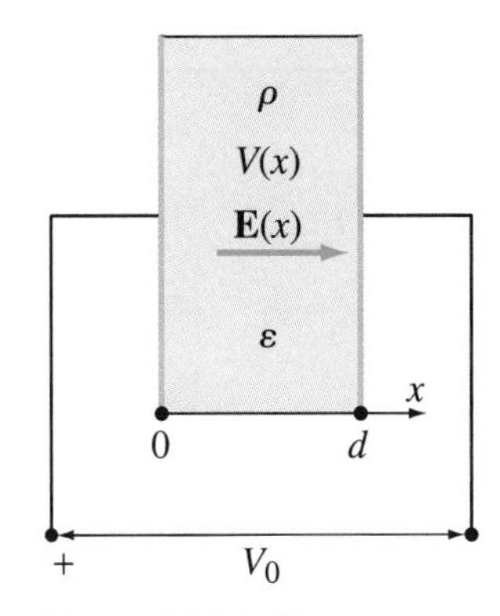

Figure 2.12 Uniform volume charge between metallic plates: application of a 1-D Poisson's equation (fringing neglected); for Example 2.7.

(b) Using Eqs. (1.101) and (1.102), we obtain the electric field vector from the potential:

$$\mathbf{E}(x) = -\nabla V = -\frac{\mathrm{d}V(x)}{\mathrm{d}x}\,\hat{\mathbf{x}} = \left[\frac{\rho}{\varepsilon}\left(x - \frac{d}{2}\right) + \frac{V_0}{d}\right]\hat{\mathbf{x}}. \tag{2.101}$$

Problems: 2.19–2.22; *Conceptual Questions* (on Companion Website): 2.9.

2.11 FINITE-DIFFERENCE METHOD FOR NUMERICAL SOLUTION OF LAPLACE'S EQUATION

In many practical cases, Poisson's or Laplace's equation cannot be solved analytically, but only numerically. The most popular and perhaps the simplest numerical method for solving these equations (and other types of differential equations generally) is the finite-difference (FD) method. It consists of replacing the derivatives in the differential equation by their finite-difference approximations and solving the resulting algebraic equations. To illustrate this, consider an air-filled coaxial cable with conductors of square cross section, shown in Fig. 2.13(a). Let the cross-sectional dimensions of the cable be a and b ($a < b$). The cable is charged by time-invariant charges, and the potentials of the conductors, V_a and V_b, are known. Our goal is to determine the distribution of the potential V in the space between the cable conductors.

This is a two-dimensional electrostatic problem for which, according to Fig. 2.13(b), Laplace's equation, Eqs. (2.95) and (2.94), is given by

$$\nabla^2 V = \frac{\partial^2 V}{\partial x^2} + \frac{\partial^2 V}{\partial y^2} = 0. \tag{2.102}$$

We discretize the region between the conductors by introducing a grid with cells of sides d [Fig. 2.13(b)], and employ the FD method in order to approximately compute the potentials at the grid points (nodes). Obviously, the accuracy of the computation depends on the grid resolution, i.e., the smaller the spacing of the grid, d, the more accurate (but computationally slower) the solution.

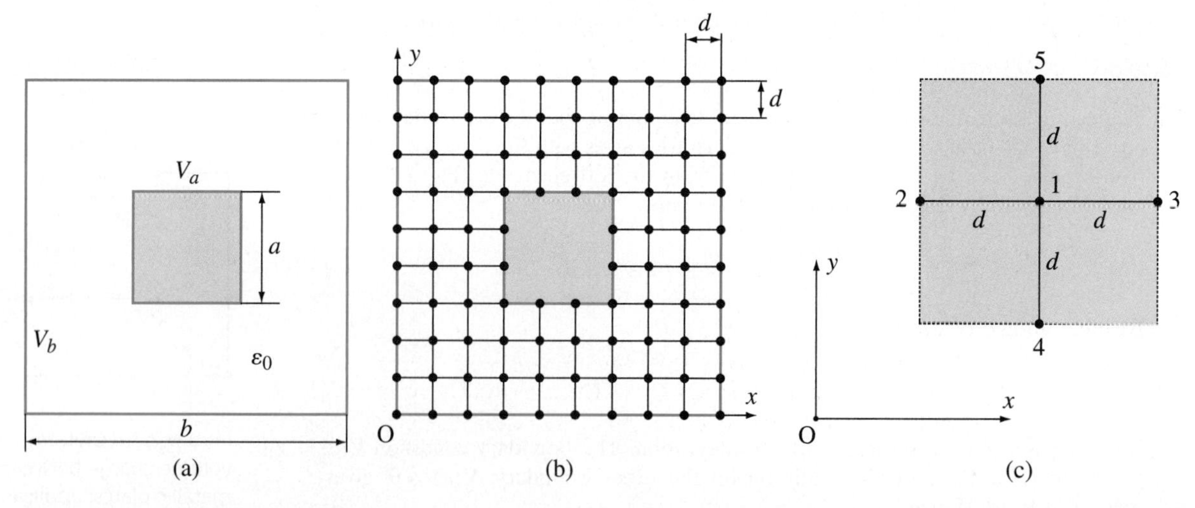

Figure 2.13 Finite-difference analysis of a coaxial cable of square cross section: (a) structure geometry, (b) nodes with discrete potential values as unknowns, and (c) detail of the grid for approximating Laplace's equation in terms of finite differences.

Fig. 2.13(c) shows a detail of the grid in Fig. 2.13(b). At the node 1, the backward-difference approximation for the first partial derivative of V with respect to the x-coordinate turns out to be

$$\left.\frac{\partial V}{\partial x}\right|_1 \approx \frac{V_1 - V_2}{d}, \tag{2.103}$$

finite-difference (FD) approximation of a derivative

which, combined with the forward-difference approximation for the second partial derivative with respect to x, yields

$$\left.\frac{\partial^2 V}{\partial x^2}\right|_1 = \left.\frac{\partial}{\partial x}\left(\frac{\partial V}{\partial x}\right)\right|_1 \approx \frac{1}{d}\left(\left.\frac{\partial V}{\partial x}\right|_3 - \left.\frac{\partial V}{\partial x}\right|_1\right) \approx \frac{1}{d}\left(\frac{V_3 - V_1}{d} - \frac{V_1 - V_2}{d}\right)$$
$$= \frac{V_2 + V_3 - 2V_1}{d^2}. \tag{2.104}$$

Analogously,

$$\left.\frac{\partial^2 V}{\partial y^2}\right|_1 \approx \frac{V_4 + V_5 - 2V_1}{d^2}, \tag{2.105}$$

so that the FD approximation for Laplace's differential equation at point 1 in Fig. 2.13(c) is

$$\left.\nabla^2 V\right|_1 \approx \frac{V_2 + V_3 + V_4 + V_5 - 4V_1}{d^2} = 0 \quad \longrightarrow \quad V_1 = \frac{1}{4}\left(V_2 + V_3 + V_4 + V_5\right). \tag{2.106}$$

FD approximation of Laplace's equation

The simplest technique to solve the above finite-difference equation with the aid of a computer is an iterative technique expressed as

$$V_1^{(k+1)} = \frac{1}{4}\left[V_2^{(k)} + V_3^{(k)} + V_4^{(k)} + V_5^{(k)}\right], \quad k = 0, 1, \ldots \tag{2.107}$$

iterative solution to the FD Laplace's equation

When some of the nodes 2, 3, 4, and 5 belong to one of the surfaces of conductors, the potential at such nodes is in all iteration steps equal to the respective given potential of the conductor (V_a or V_b). For the initial solution, at the zeroth ($k = 0$) iteration step, we can adopt $V^{(0)} = 0$ at all nodes between the conductors. By traversing the grid in a systematic manner, node by node, the average of the four neighboring potentials is computed at the $(k+1)$th step for each node and is used to replace the potential at that node, Eq. (2.107), and thus improve the solution from the kth step. This procedure is repeated until the changes of the solution (residuals) with respect to the previous iteration at all nodes are small enough, i.e., until a final set of values for the unknown potentials consistent with the criterion

$$\left|V_1^{(k+1)} - V_1^{(k)}\right| < \delta_V \tag{2.108}$$

is obtained, where δ_V stands for the specified tolerance of the potential.

Once the approximate solution for the potential distribution is known, numerical results for the electric field intensity vector, $\mathbf{E}$, at the grid nodes can be obtained by approximating the gradient operator involved in Eqs. (1.101) and (1.102) in terms of finite differences. For example, $\mathbf{E}$ at the node 1 in Fig. 2.13(c) is computed approximately as

$$\mathbf{E}_1 = -\left.\nabla V\right|_1 = -\left.\frac{\partial V}{\partial x}\right|_1 \hat{\mathbf{x}} - \left.\frac{\partial V}{\partial y}\right|_1 \hat{\mathbf{y}} \approx \frac{V_2 - V_3}{2d}\,\hat{\mathbf{x}} + \frac{V_4 - V_5}{2d}\,\hat{\mathbf{y}} \tag{2.109}$$

FD approximation of the electric field vector

(central-difference approximation). Additionally, the surface charge density, ρ_s, on the conducting surfaces can be found by means of the boundary condition in Eq. (1.190). For example, assuming that the node 4 in Fig. 2.13(c) belongs to the surface of the inner conductor in Fig. 2.13(a), ρ_s at that point can be approximately evaluated as

$$\rho_{s4} = \varepsilon_0 \, \hat{\mathbf{y}} \cdot \mathbf{E}_4 = \varepsilon_0 \, E_{y4} = -\varepsilon_0 \left. \frac{\partial V}{\partial y} \right|_4 \approx -\varepsilon_0 \frac{V_1 - V_4}{d} = \varepsilon_0 \frac{V_a - V_1}{d}. \tag{2.110}$$

Finally, the total charge per unit length of each of the conductors, Eq. (1.31), can be found by numerically integrating ρ_s, that is, summing ρ_{si}, along the individual conductor contours. Thus, the per-unit-length charge of the inner conductor is given by

$$Q'_a = \oint_{C_a} \rho_s \, dl \approx \sum_{i=1}^{N_a} \rho_{si} d, \tag{2.111}$$

where N_a denotes the total number of nodes along the contour C_a of the conductor, and similarly for the outer conductor.

Problems: 2.23 and 2.24; *MATLAB Exercises* (on Companion Website).

2.12 DEFINITION OF THE CAPACITANCE OF A CAPACITOR

Fig. 2.14 shows a system consisting of two metallic bodies (electrodes) embedded in a dielectric, and charged with equal charges of opposite polarities, Q and $-Q$. This system is referred to as a capacitor. The principal property of a capacitor is its capability to store the charge. The potential difference between the electrode carrying Q and the one with $-Q$ is called the capacitor voltage. The capacitor is linear if its dielectric is linear. The dielectric can be homogeneous or inhomogeneous.

In linear capacitors, the capacitor charge, Q, is linearly proportional to the capacitor voltage, V, i.e., $Q \propto V$. To prove this, assume for the moment that the capacitor charge is changed to $2Q$. The surface charge density, ρ_s, will remain distributed in the same way over the surfaces of electrodes, and its magnitude will

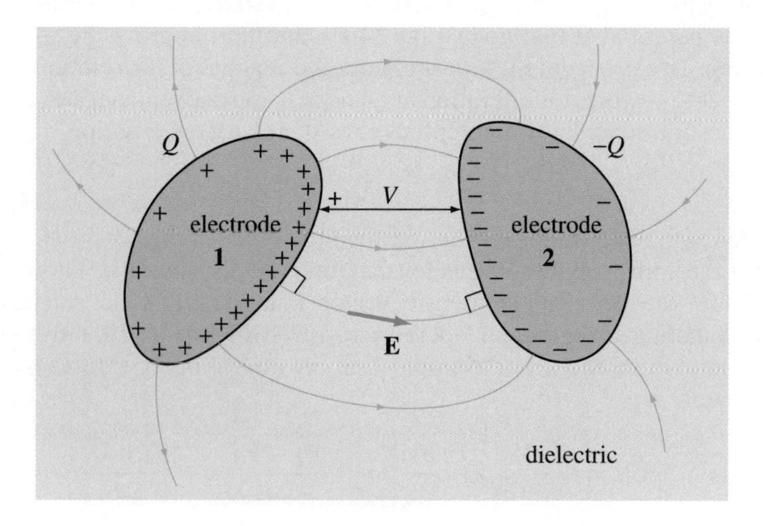

Figure 2.14 Capacitor.

double everywhere. The electric field intensity vector at an arbitrary point in the dielectric also becomes twice its previous value; for homogeneous dielectrics this is obvious from Eq. (1.38) with ε_0 replaced by ε, while for inhomogeneous dielectrics we first note that the vector **D** doubles [e.g., generalized Gauss' law, Eq. (2.43)], and then, since $\mathbf{E} \propto \mathbf{D}$ at any point of the dielectric [Eq. (2.47)], we find the same for the vector **E**. Finally, because the capacitor voltage equals the line integral of **E** (along any path) between the electrodes [Eq. (1.90)], we conclude that the voltage doubles as well. Summarily, V doubles because Q is doubled, meaning that Q and V are linearly proportional to each other. This proportionality is customarily written as

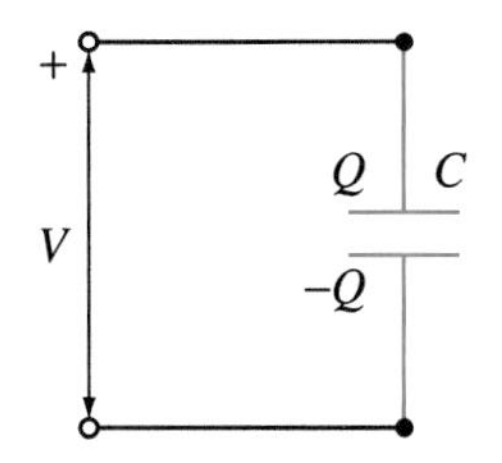

Figure 2.15 Circuit-theory representation of a capacitor.

$$Q = CV, \tag{2.112}$$

from which the constant C, termed the capacitance of the capacitor, is defined as

$$\boxed{C = \frac{Q}{V}.} \tag{2.113}$$

capacitance of a capacitor (unit: F)

The capacitance is a measure of the ability of a capacitor to hold the charge – per a volt of the applied voltage between the electrodes. It depends on the shape, size, and mutual position of the electrodes, and on the properties of the dielectric of the capacitor. For nonlinear capacitors, however, the capacitance depends also on the applied voltage,

$$\boxed{C = C(V),} \tag{2.114}$$

nonlinear capacitor

and a notable example is a varactor diode. The capacitance of a capacitor is always positive ($C > 0$), and the unit is the farad (F). For two-conductor transmission lines (two-body systems with very long conductors of uniform cross section), we define the capacitance per unit length of the line, that is, the capacitance for one meter (unit length) of the structure divided by 1 m,

$$\boxed{C' = C_{\text{p.u.l.}} = \frac{C}{l} = \frac{Q'}{V},} \tag{2.115}$$

capacitance p.u.l. of a transmission line (unit: F/m)

where C, l, and Q' are the total capacitance, length, and charge per unit length of the structure [see Eq. (1.31)]. The unit for C' is F/m.

Shown in Fig. 2.15 is the circuit-theory representation of a capacitor. In circuit theory, it is assumed that the charge is stored only in the capacitors in a circuit (i.e., on the capacitor electrodes), while the connecting conductors are considered as ideal short-circuiting elements with zero capacitance (and also with zero resistance and inductance, as we shall see in later chapters).

Finally, let us introduce another related concept, the capacitance of an isolated metallic body situated in a linear dielectric. It is defined as

$$\boxed{C = \frac{Q}{V_{\text{isolated body}}},} \tag{2.116}$$

capacitance of an isolated metallic body

where Q is the charge of the body and $V_{\text{isolated body}}$ is its potential with respect to the reference point at infinity. Note that this definition can also be regarded as a special case of the definition in Eq. (2.113), for the capacitance of a two-body system, with the assumption that the second body, carrying $-Q$, is at infinity.

2.13 ANALYSIS OF CAPACITORS WITH HOMOGENEOUS DIELECTRICS

We now consider various examples of the analysis of capacitors with homogeneous dielectrics. The examples cover several characteristic types of capacitors and transmission lines of both theoretical and practical importance. The analysis of capacitors and transmission lines with different types of inhomogeneous dielectrics will be presented in the next section.

Example 2.8 Spherical Capacitor

Consider a spherical capacitor, which consists of two concentric spherical conductors, as shown in Fig. 2.16. Let the radius of the inner conductor be a and the inner radius of the outer conductor be b ($b > a$). Find the capacitance of the capacitor if it is filled with a homogeneous dielectric of permittivity ε.

Solution Assume that the capacitor is charged with a charge Q (the inner electrode carries $+Q$). Due to spherical symmetry, the electric field in the dielectric is radial and has the form given by Eq. (1.136). Applying the generalized Gauss' law in Eq. (2.55) to the spherical surface S of radius r ($a < r < b$) positioned concentrically with the capacitor electrodes [see the flux computation in Eq. (1.138)], we obtain

$$E(r) = \frac{Q}{4\pi\varepsilon r^2} \quad (a < r < b). \tag{2.117}$$

The capacitor voltage is [Eq. (1.90)]

$$V = \int_a^b E(r)\,\mathrm{d}r = \frac{Q}{4\pi\varepsilon}\left(\frac{1}{a} - \frac{1}{b}\right), \tag{2.118}$$

so that the capacitance comes out to be

capacitance of a spherical capacitor

$$\boxed{C = \frac{Q}{V} = \frac{4\pi\varepsilon ab}{b-a}.} \tag{2.119}$$

Example 2.9 Capacitance of an Isolated Spherical Conductor

Determine the expression for the capacitance of a metallic sphere of radius a situated in air.

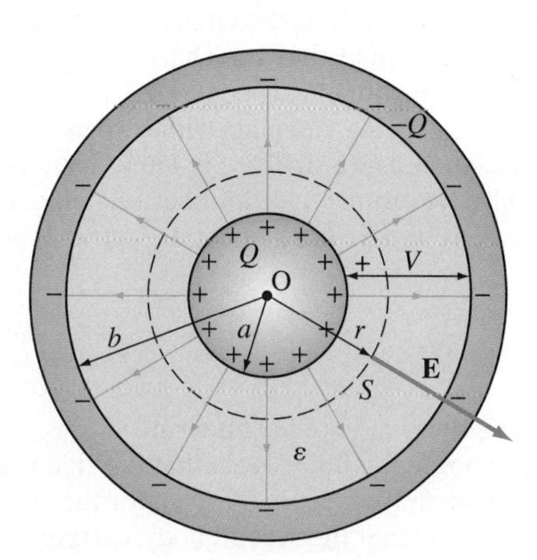

Figure 2.16 Spherical capacitor with a homogeneous dielectric; for Example 2.8.

Solution We use the definition of the capacitance of an isolated body, Eq. (2.116). The potential of the sphere if it carries a charge Q is (see Example 1.25)

$$V_{\text{sphere}} = \frac{Q}{4\pi\varepsilon_0 a}. \tag{2.120}$$

Its capacitance is hence

$$\boxed{C = \frac{Q}{V_{\text{sphere}}} = 4\pi\varepsilon_0 a.} \tag{2.121}$$

capacitance of an isolated metallic sphere in air

Note that an isolated sphere can be regarded as the inner electrode of a spherical capacitor whose outer electrode has an infinite radius, and that the same result for the sphere capacitance is obtained from Eq. (2.119) with $b \to \infty$.

Example 2.10 Coaxial Cable

A coaxial cable is a transmission line consisting of an inner cylindrical conductor (a wire) and an outer hollow cylindrical (tubular) conductor, with the conductors being coaxial with respect to each other, as depicted in Fig. 2.17. Note that this structure is sometimes referred to as a cylindrical capacitor. The inner conductor has a radius a, and the outer conductor has an inner radius b ($b > a$). The cable is filled with a homogeneous dielectric of permittivity ε. Find the capacitance per unit length, C', of the cable.

Solution This is a problem with cylindrical symmetry. The electric field is radial with respect to the cable axis (Fig. 2.17). By means of the generalized Gauss' law applied to a cylindrical surface S positioned in the dielectric coaxially with the cable conductors (see Example 1.26), we obtain the field intensity, and from it the voltage between the conductors,

$$E(r) = \frac{Q'}{2\pi\varepsilon r} \quad (a < r < b) \quad \longrightarrow \quad V = \int_a^b E(r)\,\mathrm{d}r = \frac{Q'}{2\pi\varepsilon}\ln\frac{b}{a}, \tag{2.122}$$

where Q' is the charge per unit length of the cable. Hence, the capacitance per unit length of the cable, Eq. (2.115), equals

$$\boxed{C' = \frac{Q'}{V} = \frac{2\pi\varepsilon}{\ln(b/a)}.} \tag{2.123}$$

capacitance p.u.l. of a coaxial cable

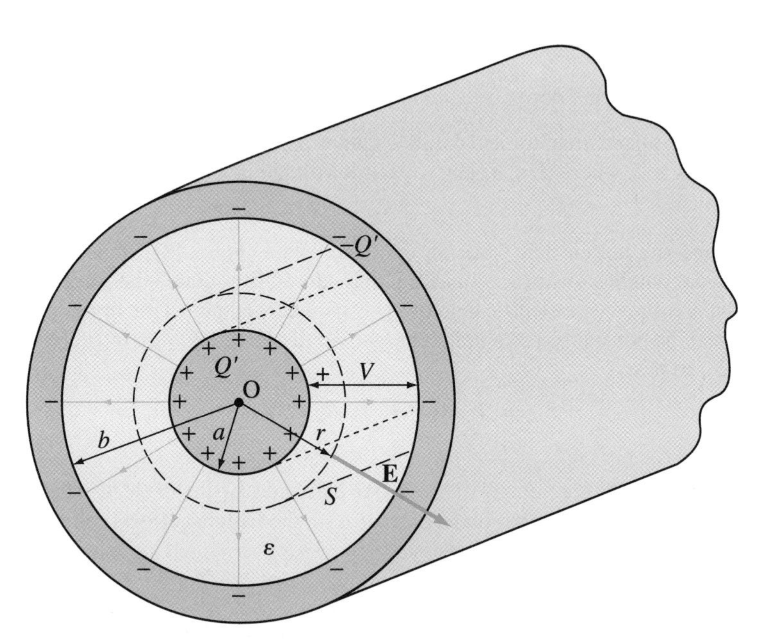

Figure 2.17 Coaxial cable (cylindrical capacitor) with a homogeneous dielectric; for Example 2.10.

We note that, combining together Eqs. (2.122),

electric field in a coaxial cable

$$\boxed{E(r) = \frac{V}{r\ln(b/a)},} \tag{2.124}$$

which is a very useful expression for the field intensity (in terms of the voltage) in a coaxial cable.

Example 2.11 Parallel-Plate Capacitor

Consider a parallel-plate capacitor, which consists of two parallel metallic plates, each of area S, charged with Q and $-Q$. Let the space between the plates be filled with a homogeneous dielectric of permittivity ε, as shown in Fig. 2.18(a). Assume that the plate separation, d, is very small compared to the dimensions of the plates, so that the fringing effects can be neglected. Calculate the capacitance of the capacitor.

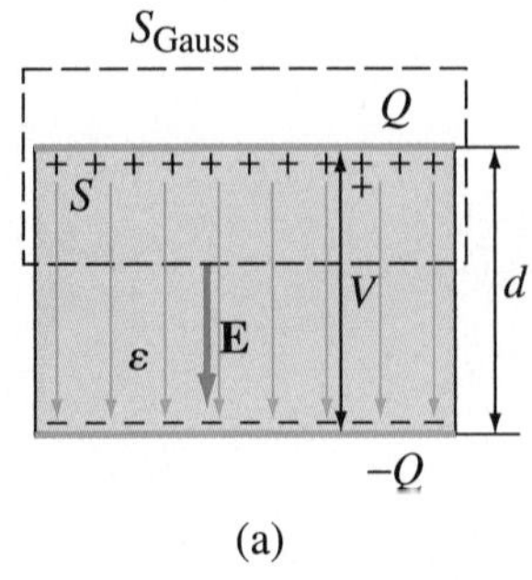

(a)

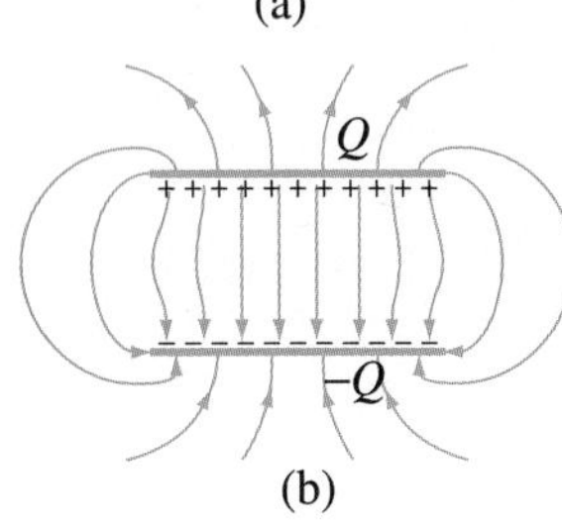

(b)

Figure 2.18 (a) Parallel-plate capacitor with a homogeneous dielectric and (b) fringing field for $\varepsilon_r = 1$; for Example 2.11.

Solution Under given assumptions, the electric field in the dielectric is uniform, and there is no field outside the dielectric. Applying the generalized Gauss' law to a rectangular closed surface that encloses the upper plate [Fig. 2.18(a)], we get [see the left-hand side of Eq. (1.155)]

$$ES = \frac{Q}{\varepsilon}, \tag{2.125}$$

and hence $E = Q/(\varepsilon S)$. The voltage between the plates is

$$V = Ed = \frac{Qd}{\varepsilon S}, \tag{2.126}$$

and the capacitance of the capacitor

capacitance of a parallel-plate capacitor, fringing neglected

$$\boxed{C = \frac{Q}{V} = \varepsilon\frac{S}{d}.} \tag{2.127}$$

For arbitrary capacitor dimensions, there is a considerable fringing field extending far outside the capacitor and the field between the capacitor plates close to plate edges is not uniform (edge effects), as illustrated in Fig. 2.18(b) for $\varepsilon_r = 1$. With this, the actual capacitance is larger than the value obtained from Eq. (2.127).

Example 2.12 Electric Forces on Capacitor Plates

Electrodes of an air-filled capacitor are parallel square plates of edge lengths a. The distance between the plates is d, where $d \ll a$. The capacitor voltage is V. Find the electric forces on electrodes.

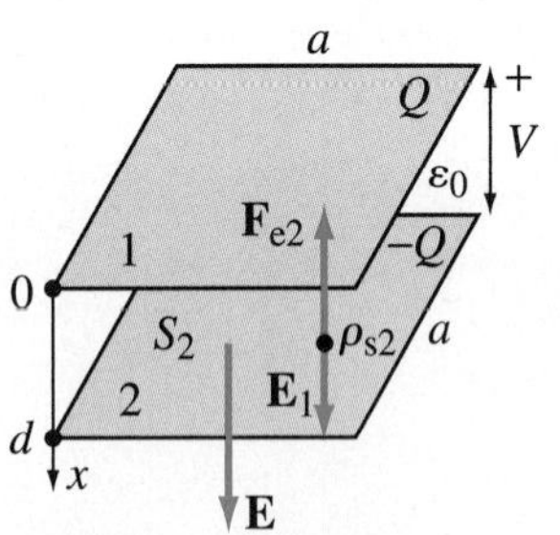

Figure 2.19 Evaluation of the electric force on the lower electrode of an air-filled parallel-plate capacitor; for Example 2.12.

Solution To find the net electric force on the lower plate (Fig. 2.19), we subdivide it into differentially small patches of surface areas dS, and add up (integrate) the forces on individual patches that are due to the electric field of the other charged plate (the upper plate). Each charged patch can be considered as a point charge and the force $d\mathbf{F}_{e2}$ on it can be calculated using Eq. (1.23), so that

$$\mathbf{F}_{e2} = \int_{S_2} d\mathbf{F}_{e2} = \int_{S_2} \underbrace{\rho_{s2}\, dS}_{dQ_2}\, \mathbf{E}_1. \tag{2.128}$$

Here, ρ_{s2} is the surface charge density of the lower plate and $\mathbf{E}_1$ is the electric field intensity vector due to the charge of the upper plate. Since $d \ll a$, the fringing effects can be neglected, and these quantities are

$$\rho_{s2} = -\frac{Q}{S} \quad \text{and} \quad \mathbf{E}_1 = \frac{Q}{2\varepsilon_0 S}\hat{\mathbf{x}} \quad (S = a^2). \tag{2.129}$$

We realize that $\mathbf{E}_1$ equals a half of the total electric field intensity between the plates (the other half is due to the charge of the lower plate) and can also be obtained as the field of an infinite sheet of charge, Eq. (1.64). Q is the charge of the capacitor,

$$Q = CV = \frac{\varepsilon_0 SV}{d}, \tag{2.130}$$

and C is its capacitance. The force on the lower plate is hence

$$\mathbf{F}_{e2} = -\frac{\varepsilon_0 a^2 V^2}{2d^2}\,\hat{\mathbf{x}}. \tag{2.131}$$

The force on the upper electrode is opposite. The forces are attractive.

Note that, from Eq. (2.131),

$$\frac{F_{e2}}{S} = \frac{\varepsilon_0 V^2}{2d^2}. \tag{2.132}$$

By definition, this is the pressure of the force $\mathbf{F}_{e2}$ on the surface of the plate. We term it the electric pressure. The unit for pressure is Pa (pascal), where Pa = N/m^2. By introducing the total electric field intensity of the capacitor, $E = V/d$, the electric pressure can be expressed as

$$\boxed{p_e = \frac{1}{2}\varepsilon_0 E^2.} \tag{2.133}$$

electric pressure (unit: Pa)

This expression is valid for any conducting surface, with E standing for the local electric field intensity in the dielectric (air) near the surface. The pressure acts from the conductor toward the dielectric.

Example 2.13 Microstrip Transmission Line

Consider the transmission line shown in Fig. 2.20. It consists of a conducting strip of width w, resting on a dielectric substrate of permittivity ε and thickness h, and a ground plane beneath the substrate, and is called a microstrip line. Neglecting the fringing effects, determine the capacitance per unit length of this line.

Solution Without taking into account the fringing effects, the electric field of the line is uniform and localized in the dielectric below the strip only. The capacitance of the part of the line with length l is, from Eq. (2.127),

$$C = \varepsilon \frac{wl}{h}, \tag{2.134}$$

and the capacitance per unit length of the line comes out to be

$$\boxed{C' = \frac{C}{l} = \varepsilon \frac{w}{h}.} \tag{2.135}$$

capacitance p.u.l. of a microstrip line, fringing neglected

Note, however, that this expression is accurate only for $h \ll w$. Namely, as pointed out in Example 2.11, the actual electric field distribution in the structure with an arbitrary ratio w/h is quite different from that in Fig. 2.20. In a later chapter (on field analysis of transmission

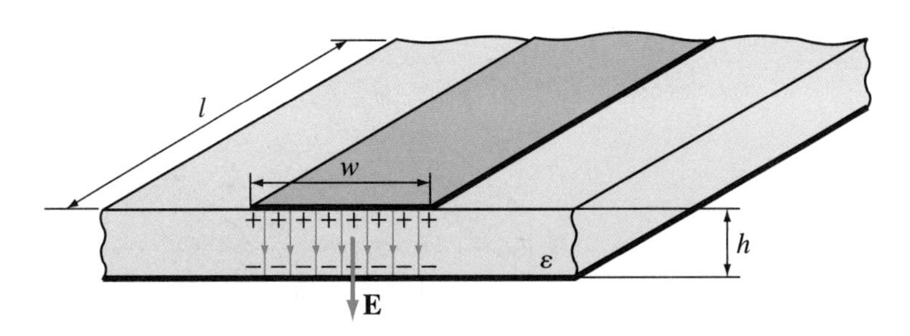

Figure 2.20 Microstrip transmission line; for Example 2.13.

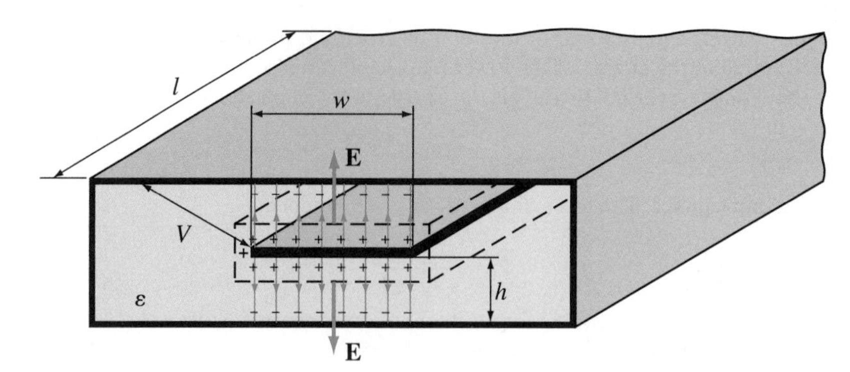

Figure 2.21 Strip transmission line; for Example 2.14.

lines), we shall present accurate empirical formulas for the electrostatic analysis of microstrip lines for all values of w/h.

Example 2.14 Strip Transmission Line

The transmission line consisting of a strip conductor between two conducting planes (large plates) at the same potential, as depicted in Fig. 2.21, is called a strip line. Let the width of the strip be w and its distance from each of the planes be h. The permittivity of the dielectric is ε. Assuming that $h \ll w$, find the capacitance per unit length of the line.

Solution Neglecting the fringing fields (as $h \ll w$) and applying the generalized Gauss' law to a rectangular closed surface of length l (along the line) that encloses the strip (Fig. 2.21), we have [see the left-hand side of Eq. (1.149)]

$$2Ewl = \frac{Q'l}{\varepsilon}, \tag{2.136}$$

where E is the electric field intensity between the line conductors and Q' the charge per unit length of the line. The voltage between the conductors is $V = Eh$, from which the per-unit-length capacitance of the line amounts to

capacitance p.u.l. of a strip line, fringing neglected

$$\boxed{C' = \frac{Q'}{V} = \frac{2\varepsilon w}{h}.} \tag{2.137}$$

Accurate analysis for an arbitrary ratio w/h will be presented in a later chapter.

Example 2.15 Thin Symmetrical Two-Wire Transmission Line

Fig. 2.22 shows a thin symmetrical two-wire transmission line in air. The charge per unit length of the line is Q', and the radii of the wire conductors, a, are much smaller than the distance between the conductor axes, d. Under these circumstances, compute the capacitance per unit length of the line.

Solution By superposition, the total electric field intensity vector in air is given by

$$\mathbf{E} = \mathbf{E}_1 + \mathbf{E}_2, \tag{2.138}$$

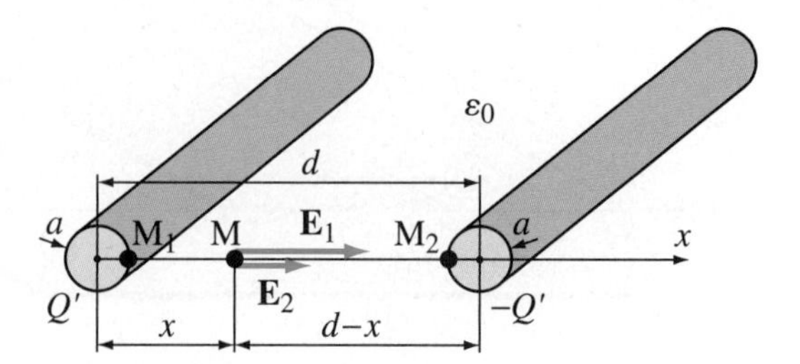

Figure 2.22 Two-wire transmission line with $d \gg a$ in air; for Example 2.15.

where $\mathbf{E}_1$ and $\mathbf{E}_2$ are the fields due to the individual charged conductors. Since $d \gg a$, these fields can be evaluated independently from each other, so we use the expression for the field of an isolated charged wire conductor in air, Eq. (1.196). At a point M in the plane containing the axes of conductors, Fig. 2.22, vectors $\mathbf{E}_1$ and $\mathbf{E}_2$ are collinear, and thus the resultant field intensity turns out to be

$$E = E_1 + E_2 = \frac{Q'}{2\pi\varepsilon_0}\left(\frac{1}{x} + \frac{1}{d-x}\right), \tag{2.139}$$

where x stands for the coordinate defining the position of the point M.

The voltage between the conductors is

$$\begin{aligned} V &= \int_{x=a}^{d-a} E\,\mathrm{d}x = \frac{Q'}{2\pi\varepsilon_0}\left[\int_a^{d-a}\frac{\mathrm{d}x}{x} - \int_a^{d-a}\frac{\mathrm{d}(d-x)}{d-x}\right] \\ &= \frac{Q'}{2\pi\varepsilon_0}\left[\ln x|_a^{d-a} - \ln(d-x)|_a^{d-a}\right] = \frac{Q'}{2\pi\varepsilon_0}\left(\ln\frac{d-a}{a} - \ln\frac{a}{d-a}\right) \\ &= \frac{Q'}{\pi\varepsilon_0}\ln\frac{d-a}{a} \approx \frac{Q'}{\pi\varepsilon_0}\ln\frac{d}{a}, \end{aligned} \tag{2.140}$$

where the use is made of the approximate relation $d - a \approx d$. The capacitance per unit length of the line is

$$\boxed{C' = \frac{Q'}{V} = \frac{\pi\varepsilon_0}{\ln(d/a)}}. \tag{2.141}$$

capacitance p.u.l. of a thin two-wire line

Let us now obtain the same result in a different way, using the expression for the potential due to an isolated charged wire conductor in air, Eq. (1.197) for $r \geq a$, and the superposition principle. We adopt a point M_2 on the surface of the right conductor (Fig. 2.22) to be the reference point for potential ($V_{M_2} = 0$). The potential at a point M_1 on the surface of the other conductor (with respect to the reference point) amounts to

$$V_{M_1} = \frac{Q'}{2\pi\varepsilon_0}\ln\frac{d-a}{a} + \frac{-Q'}{2\pi\varepsilon_0}\ln\frac{a}{d-a}, \tag{2.142}$$

with the two terms representing the potentials [Eq. (1.197)] due to the conductors with per-unit-length charges Q' and $-Q'$, respectively. The capacitance per unit length of the line is

$$C' = \frac{Q'}{V_{M_1} - V_{M_2}} = \frac{Q'}{V_{M_1}}, \tag{2.143}$$

which yields the same expression as in Eq. (2.141).

Example 2.16 Three Parallel Equidistant Wires

Three parallel thin wire conductors are situated in air, as depicted in Fig. 2.23. The wire radii are $a = 1$ mm and the distance between the axes of conductors is $d = 50$ mm. Two wires are galvanically connected to each other. Determine the capacitance per unit length of this system.

Solution This is a two-conductor transmission line, which we analyze as a capacitor of infinite length. The first electrode of the capacitor (the first conductor of the transmission line) is the isolated (free) wire (wire 1), while the other electrode (the other line conductor) consists of the two short-circuited wires (wires 2 and 3), which are at the same potential. Let the electrodes be charged by Q' and $-Q'$ per unit of their length, respectively. Because of symmetry, the charge $-Q'$ of the second electrode is distributed equally between wires 2 and 3, as indicated in Fig. 2.23. We assume that there is no charge on the short-circuiting conductor.

Let us proclaim the second electrode of the capacitor to be at a zero potential. The potential of the first electrode, i.e., the potential at the point M_1 on the surface of wire 1 (Fig. 2.23), with respect to the reference point M_2 taken on the surface of wire 3 ($V_{M_2} = 0$)

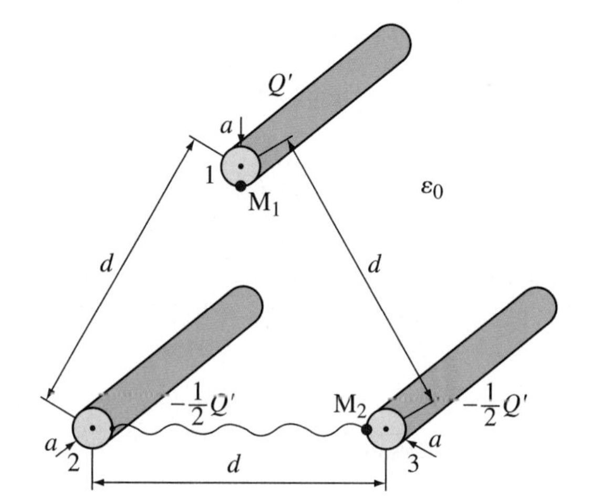

Figure 2.23 Transmission line consisting of an isolated (free) wire and two galvanically interconnected wires in air; for Example 2.16.

is the sum of the corresponding potentials due to wires 1, 2, and 3, respectively. By means of Eq. (1.197) for $r \geq a$,

$$V_{\mathrm{M}_1} = \frac{Q'}{2\pi\varepsilon_0}\ln\frac{d}{a} + \frac{-Q'/2}{2\pi\varepsilon_0}\ln\frac{d}{d} + \frac{-Q'/2}{2\pi\varepsilon_0}\ln\frac{a}{d} = \frac{3Q'}{4\pi\varepsilon_0}\ln\frac{d}{a}. \tag{2.144}$$

The capacitance per unit length of the line (capacitor) is

$$C' = \frac{Q'}{V_{\mathrm{M}_1}} = \frac{4\pi\varepsilon_0}{3\ln(d/a)} = 9.48\ \mathrm{pF/m}. \tag{2.145}$$

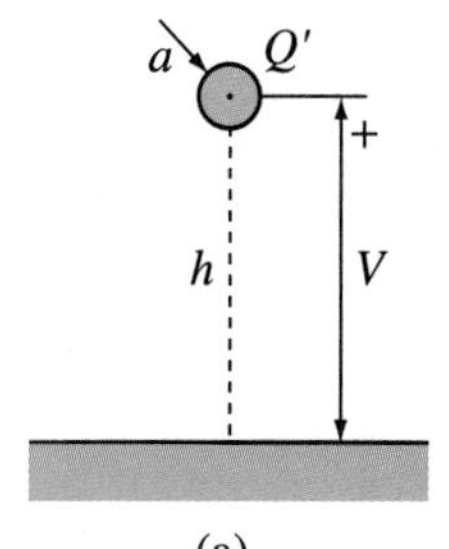

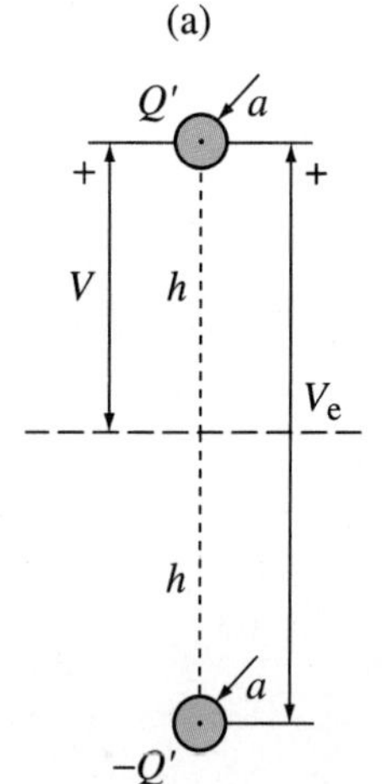

Figure 2.24 (a) Thin wire conductor above a grounded conducting plane in air and (b) equivalent two-wire line; for Example 2.17.

Example 2.17 Thin Wire Conductor above a Ground Plane

Consider a transmission line that consists of a thin wire conductor and a grounded conducting plane, as shown in Fig. 2.24(a). The medium is air, the wire is parallel to the plane, the height of the wire axis with respect to the plane is h, and the wire radius is a ($a \ll h$). Find the capacitance per unit length of this line.

Solution Assume that the charge per unit length of the line is Q', namely, that the wire is charged by Q' and the plane by $-Q'$ per unit length. By image theory (Figs. 1.47 and 1.49), we can replace the conducting plane by a negative image of the charged wire and obtain the equivalent thin two-wire line in Fig. 2.24(b), with the distance between the conductor axes $d = 2h$. Of course, the two systems are equivalent only in the upper half-space. The voltage between the conductors in the original system (wire-plane) equals a half of the voltage in the equivalent system (wire-wire). Hence, the capacitance per unit length of the wire-plane transmission line comes out to be (see Fig. 2.24)

$$C' = \frac{Q'}{V} = \frac{Q'}{V_{\mathrm{e}}/2} = 2\frac{Q'}{V_{\mathrm{e}}} = 2C'_{\mathrm{e}} = \frac{2\pi\varepsilon_0}{\ln(2h/a)}, \tag{2.146}$$

where C'_{e} is the capacitance per unit length of the equivalent two-wire line, which is computed from Eq. (2.141).

Problems: 2.25–2.41; *Conceptual Questions* (on Companion Website): 2.10–2.18; *MATLAB Exercises* (on Companion Website).

2.14 ANALYSIS OF CAPACITORS WITH INHOMOGENEOUS DIELECTRICS

In this section, we deal with systems (capacitors) containing inhomogeneous dielectrics, and, specifically, with two basic classes of systems with dielectric inhomogeneity. The first class includes systems in which the dielectric permittivity varies (abruptly or continuously[5]) in the direction of the electric field lines of the same system if air-filled. In the second class of systems, the dielectric permittivity varies in the direction normal to the electric field lines of the air-filled system. As we shall see, the way in which the electric flux density vector, **D**, varies in the first class of dielectrics is the same as in the corresponding air-filled system, while in the second class of dielectrics this is true for the electric field intensity vector, **E**. Hence, the two classes of systems are referred to as *D*- and *E*-systems, respectively.

To illustrate these concepts, consider the two parallel-plate capacitors with piece-wise homogeneous dielectrics shown in Fig. 2.25. The permittivities of dielectric pieces are ε_1 and ε_2. Assume that the fringing effects are negligible, and that the field in each dielectric piece is uniform.

The capacitor in Fig. 2.25(a) belongs to the first class of systems with dielectric inhomogeneity – the dielectric permittivity (abruptly) changes in the direction of the field lines, perpendicularly to the capacitor plates, so it is a *D*-system. From the generalized Gauss' law, Eq. (2.43), applied to a rectangular closed surface enclosing the plate charged with Q, with the right-hand side positioned in either one of the dielectrics, we have

$$\boxed{D_1 = D_2 = D = \frac{Q}{S},} \tag{2.147}$$

$D = \text{const}$, capacitor in Fig. 2.25(a)

where S is the surface area of the plates. That $D_1 = D_2$ is also obvious from the boundary condition for the normal components of **D**, Eq. (2.81), applied to the interface between the two dielectrics. The electric field intensities in the dielectrics are not the same,

$$E_1 = \frac{D}{\varepsilon_1} = \frac{Q}{\varepsilon_1 S} \quad \text{and} \quad E_2 = \frac{D}{\varepsilon_2} = \frac{Q}{\varepsilon_2 S}. \tag{2.148}$$

The voltage of the capacitor comes out to be

$$V = E_1 d_1 + E_2 d_2 = \frac{Q}{S}\left(\frac{d_1}{\varepsilon_1} + \frac{d_2}{\varepsilon_2}\right), \tag{2.149}$$

with d_1 and d_2 standing for thicknesses of the layers. The capacitance is

$$C = \frac{Q}{V} = \frac{\varepsilon_1 \varepsilon_2 S}{\varepsilon_2 d_1 + \varepsilon_1 d_2}. \tag{2.150}$$

In the capacitor in Fig. 2.25(b), the change of dielectric characteristics is in the direction normal to the field lines – this is an *E*-system (the second class of systems). By means of Eq. (1.90) applied to a straight path between the capacitor plates, positioned in either one of the dielectrics,

$$\boxed{E_1 = E_2 = E = \frac{V}{d},} \tag{2.151}$$

$E = \text{const}$, capacitor in Fig. 2.25(b)

[5]Inhomogeneous dielectrics composed of a number of homogeneous pieces of different permittivities are called piece-wise homogeneous dielectrics; continuously inhomogeneous dielectrics, on the other side, are characterized by continuous spatial variations of permittivity.

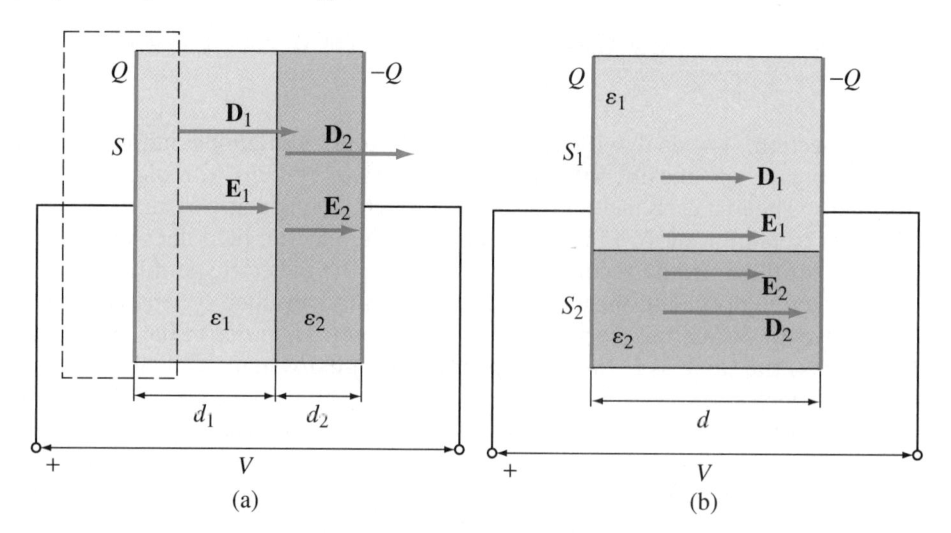

Figure 2.25 Capacitors with two-layer dielectrics representing a D-system (a) and E-system (b).

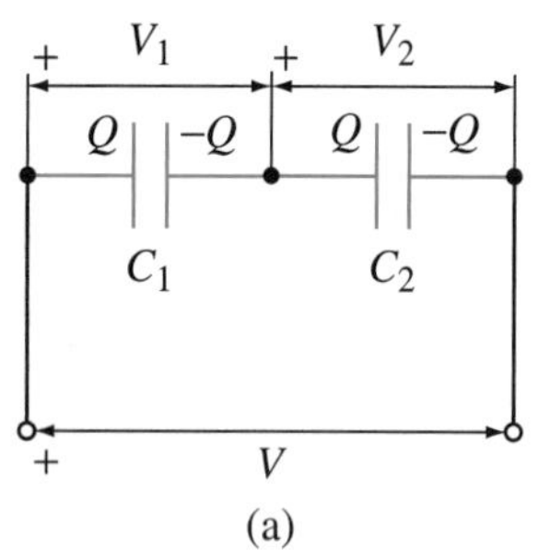

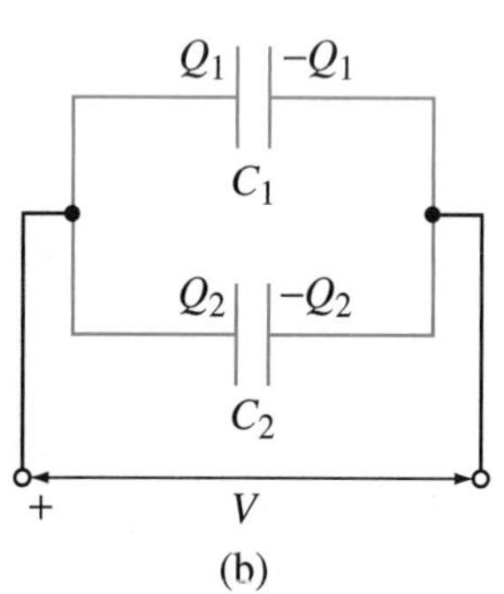

Figure 2.26 Equivalent circuits for two capacitors in series (a) and parallel (b).

where V is the capacitor voltage and d the separation between plates [$E_1 = E_2$ is also obtained from the boundary condition for tangential components of $\mathbf{E}$, Eq. (2.79), applied to the dielectric-dielectric interface]. The electric flux density is discontinuous across the interface,

$$D_1 = \varepsilon_1 E = \frac{\varepsilon_1 V}{d} \quad \text{and} \quad D_2 = \varepsilon_2 E = \frac{\varepsilon_2 V}{d}. \tag{2.152}$$

Applying the generalized Gauss' law to a rectangular surface positioned about the plate with the charge Q yields

$$Q = D_1 S_1 + D_2 S_2 = (\varepsilon_1 S_1 + \varepsilon_2 S_2)\,\frac{V}{d}, \tag{2.153}$$

with S_1 and S_2 being surface areas of the parts of the plate interfacing the individual dielectric layers, and the capacitance is

$$C = \frac{Q}{V} = \frac{\varepsilon_1 S_1 + \varepsilon_2 S_2}{d}. \tag{2.154}$$

The systems in Fig. 2.25 can be explained also from the circuit-theory point of view. Such a view is usually much simpler than the full field-theory view, and is sufficient for some evaluations. Let us compute the capacitances of capacitors in Fig. 2.25 using the associated equivalent circuits.

Note that the interface between the dielectric layers of the capacitor in Fig. 2.25(a) is equipotential, and nothing will change in the system if we metalize this surface, i.e., insert a metallic foil between the layers. We thus obtain two capacitors in series, as indicated in the equivalent circuit in Fig. 2.26(a). The charges of both capacitors are the same, and therefore

$$V = V_1 + V_2 = \frac{Q}{C_1} + \frac{Q}{C_2} = Q\left(\frac{1}{C_1} + \frac{1}{C_2}\right), \tag{2.155}$$

from which the total capacitance of the system amounts to

equivalent capacitance of two capacitors in series

$$\boxed{C = \frac{Q}{V} = \frac{C_1 C_2}{C_1 + C_2}.} \tag{2.156}$$

With the use of Eq. (2.127), the capacitances of individual capacitors are

$$C_1 = \varepsilon_1 \frac{S}{d_1} \quad \text{and} \quad C_2 = \varepsilon_2 \frac{S}{d_2}, \tag{2.157}$$

resulting in the same expression for C as in Eq. (2.150).

On the other hand, the system in Fig. 2.25(b) represents two capacitors in parallel, the equivalent circuit of which is shown in Fig. 2.26(b). The voltages of both capacitors are the same, so that

$$Q = Q_1 + Q_2 = C_1 V + C_2 V = (C_1 + C_2)\, V. \tag{2.158}$$

The total capacitance of the system is

$$\boxed{C = \frac{Q}{V} = C_1 + C_2,} \tag{2.159}$$

equivalent capacitance of two capacitors in parallel

where,

$$C_1 = \varepsilon_1 \frac{S_1}{d} \quad \text{and} \quad C_2 = \varepsilon_2 \frac{S_2}{d}, \tag{2.160}$$

which gives the same result as in Eq. (2.154).

Example 2.18 Spherical Capacitor with Two Concentric Dielectric Layers

A spherical capacitor is filled with two concentric dielectric layers. The relative permittivity of the layer near the inner electrode of the capacitor is $\varepsilon_{r1} = 4$, and that of the other layer $\varepsilon_{r2} = 2$. The radius of the inner electrode is $a = 10$ mm, the radius of the boundary surface between the layers is $b = 25$ mm, and the inner radius of the outer electrode is $c = 35$ mm. If the voltage between the inner and outer electrode is $V = 10$ V, what are (a) the charge of the capacitor and (b) the total bound charge on the interface between the layers?

Solution

(a) This is a D-system, and the electric flux density vector is of the form

$$\mathbf{D} = D(r)\,\hat{\mathbf{r}}, \tag{2.161}$$

where r and $\hat{\mathbf{r}}$ are the radial coordinate and the radial unit vector in the spherical coordinate system with the origin at the capacitor center, as shown in Fig. 2.27. The generalized Gauss' law applied to a spherical surface placed in either the first or the second layer tells us that

$$D(r) = \frac{Q}{4\pi r^2}, \quad a < r < c, \tag{2.162}$$

where Q is the capacitor charge – to be determined.

The electric field intensity vector is $\mathbf{E}_1 = \mathbf{D}/\varepsilon_1$ in the first layer and $\mathbf{E}_2 = \mathbf{D}/\varepsilon_2$ in the second one. The capacitor voltage is given by

$$V = \int_a^b E_1(r)\,\mathrm{d}r + \int_b^c E_2(r)\,\mathrm{d}r = \frac{Q}{4\pi}\left[\frac{1}{\varepsilon_1}\left(\frac{1}{a} - \frac{1}{b}\right) + \frac{1}{\varepsilon_2}\left(\frac{1}{b} - \frac{1}{c}\right)\right], \tag{2.163}$$

from which

$$Q = 4\pi\varepsilon_0 \left(\frac{b-a}{\varepsilon_{r1}ab} + \frac{c-b}{\varepsilon_{r2}bc}\right)^{-1} V = CV = 53.7\ \text{pC}, \tag{2.164}$$

where $C = 5.37$ pF is the capacitance of the capacitor.

Note that the capacitor charge can also be found by computing C as the equivalent total capacitance of a series connection of two spherical capacitors with homogeneous dielectrics, Eq. (2.156), where, using Eq. (2.119),

$$C_1 = \frac{4\pi\varepsilon_{r1}\varepsilon_0 ab}{b-a} = 7.4\ \text{pF} \quad \text{and} \quad C_2 = \frac{4\pi\varepsilon_{r2}\varepsilon_0 bc}{c-b} = 19.5\ \text{pF}. \tag{2.165}$$

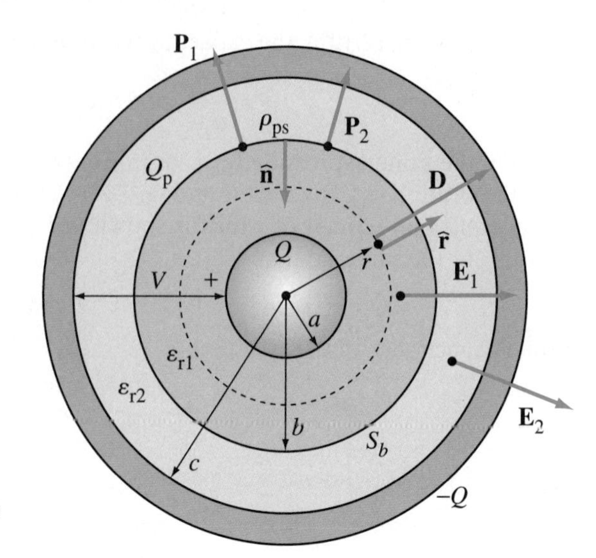

Figure 2.27 Spherical capacitor with two concentric homogeneous dielectric layers; for Example 2.18.

(b) From Eqs. (2.59) and (2.47), the intensities of the polarization vector in the dielectric layers with respect to the positive (outward) radial direction (Fig. 2.27) are

$$P_1(r) = \frac{\varepsilon_{r1} - 1}{\varepsilon_{r1}} D(r) \quad \text{and} \quad P_2(r) = \frac{\varepsilon_{r2} - 1}{\varepsilon_{r2}} D(r). \tag{2.166}$$

By means of Eq. (2.89), the bound surface charge density on the boundary surface S_b between the layers turns out to be

$$\rho_{ps} = \hat{\mathbf{n}} \cdot \mathbf{P}_2 - \hat{\mathbf{n}} \cdot \mathbf{P}_1 = -P_2(b^+) + P_1(b^-) = \left(\frac{\varepsilon_{r1} - 1}{\varepsilon_{r1}} - \frac{\varepsilon_{r2} - 1}{\varepsilon_{r2}} \right) D(b), \tag{2.167}$$

and hence the total bound charge on the surface

$$Q_p = \rho_{ps} S_b = \rho_{ps} 4\pi b^2 = \left(\frac{\varepsilon_{r1} - 1}{\varepsilon_{r1}} - \frac{\varepsilon_{r2} - 1}{\varepsilon_{r2}} \right) Q = 13.43 \text{ pC}. \tag{2.168}$$

Example 2.19 Spherical Capacitor with a Continuously Inhomogeneous Dielectric

A spherical capacitor is filled with a continuously inhomogeneous dielectric, whose permittivity depends on the distance r from the capacitor center and is given by the function $\varepsilon(r) = 3\varepsilon_0 b/r$, $a < r < b$, where a and b are radii of the inner and outer capacitor electrode, respectively. The outer electrode is grounded and the potential of the inner electrode is V. Find (a) the capacitance of the capacitor and (b) the bound charge distribution of the dielectric. (c) By integrating the charge densities in (b), prove that the total bound charge of the dielectric is zero.

Solution

(a) Let us subdivide the dielectric into thin spherical concentric layers of radii r and thicknesses dr, $a < r < b$, as shown in Fig. 2.28. Each thin layer can be considered as being homogeneous, of permittivity $\varepsilon(r)$. It is now obvious that this capacitor represents a generalization of that in Fig. 2.27, which has only two such layers. Therefore, the electric flux density vector in the two capacitors is the same, given by Eqs. (2.161) and (2.162). Here, $\mathbf{E} = \mathbf{D}/\varepsilon(r)$, and the potential of the inner electrode with respect to the ground (outer electrode) and the capacitance of the capacitor can be obtained as

$$V = \int_a^b E(r)\, dr = \frac{Q}{4\pi} \int_a^b \frac{dr}{r^2 \varepsilon(r)} \quad \longrightarrow \quad C = \frac{Q}{V} = 4\pi \left[\int_a^b \frac{dr}{r^2 \varepsilon(r)} \right]^{-1}. \tag{2.169}$$

Specifically, for the given function $\varepsilon(r)$, $C = 12\pi\varepsilon_0 b/\ln(b/a)$.

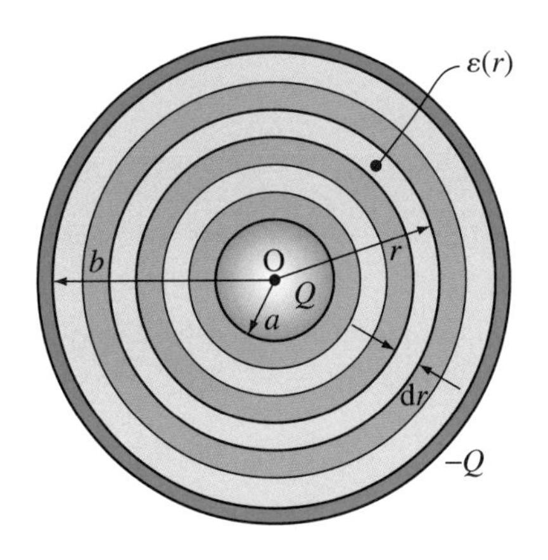

Figure 2.28 Spherical capacitor with a continuously inhomogeneous dielectric of permittivity $\varepsilon(r)$, subdivided into differentially thin homogeneous layers; for Example 2.19.

(b) The polarization vector in the dielectric is radial, with intensity [Eqs. (2.59) and (2.47)]

$$P(r) = \frac{\varepsilon(r) - \varepsilon_0}{\varepsilon(r)} D(r) = \frac{\varepsilon_0 (3b - r) V}{r^2 \ln(b/a)}. \tag{2.170}$$

By means of Eqs. (2.19) and (1.171), the bound volume charge density inside the dielectric amounts to

$$\rho_{\rm p} = -\nabla \cdot \mathbf{P} = -\frac{1}{r^2} \frac{\mathrm{d}}{\mathrm{d}r} \left[r^2 P(r) \right] = \frac{\varepsilon_0 V}{r^2 \ln(b/a)}, \tag{2.171}$$

while, from Eq. (2.23), the bound surface charge densities on the surfaces of the dielectric near the inner and outer capacitor electrodes are

$$\rho_{\mathrm{ps}a} = -P(a^+) = -\frac{\varepsilon_0 (3b - a) V}{a^2 \ln(b/a)} \quad \text{and} \quad \rho_{\mathrm{ps}b} = P(b^-) = \frac{2\varepsilon_0 V}{b \ln(b/a)}, \tag{2.172}$$

respectively.

(c) To verify that the total bound charge of the dielectric is zero, we write

$$Q_{\rm p} = \int_a^b \rho_{\rm p}(r) \underbrace{4\pi r^2 \, \mathrm{d}r}_{\mathrm{d}v} + \rho_{\mathrm{ps}a} \underbrace{4\pi a^2}_{S_a} + \rho_{\mathrm{ps}b} \underbrace{4\pi b^2}_{S_b}$$

$$= \frac{4\pi \varepsilon_0 (b - a) V}{\ln(b/a)} - \frac{4\pi \varepsilon_0 (3b - a) V}{\ln(b/a)} + \frac{8\pi \varepsilon_0 b V}{\ln(b/a)} = 0, \tag{2.173}$$

where $\mathrm{d}v$ is the volume of the thin layer of radius r in Fig. 2.28, while S_a and S_b are the areas of surfaces of inner and outer capacitor electrodes, respectively.

Example 2.20 Spherical Capacitor Half Filled with a Liquid Dielectric

A spherical capacitor has a liquid dielectric occupying a half of the space between the electrodes. The radii of electrodes are a and b ($a < b$), and the permittivity of the dielectric is ε. Determine the capacitance of the capacitor.

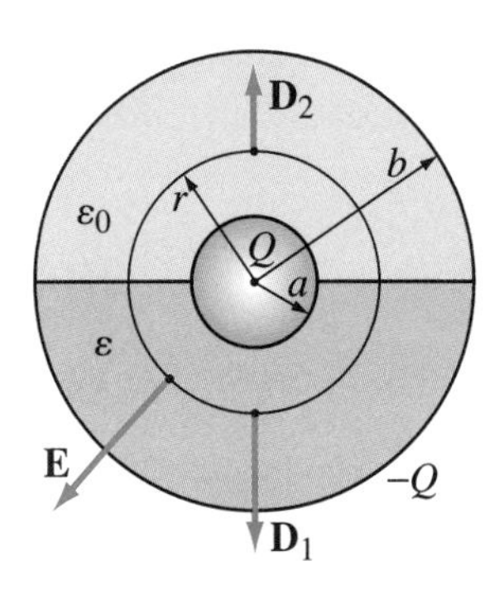

Figure 2.29 Analysis of a spherical capacitor half filled with a liquid dielectric; for Example 2.20.

Solution Assume that the capacitor is charged by Q, as indicated in Fig. 2.29. This is an E-system, and the electric field intensity vector, $\mathbf{E}$, is entirely radial. From the boundary condition for tangential components of the electric field intensity vector, Eq. (2.79), its magnitude depends on the radial spherical coordinate, r, only, i.e., it is the same at all points of a sphere of radius r ($a < r < b$), shown in Fig. 2.29.

The electric flux density vector is $\mathbf{D}_1 = \varepsilon_{\rm r}\varepsilon_0 \mathbf{E}$ in the dielectric and $\mathbf{D}_2 = \varepsilon_0 \mathbf{E}$ in air. Applying the generalized Gauss' law to the sphere of radius r (Fig. 2.29), we obtain

$$D_1 2\pi r^2 + D_2 2\pi r^2 = Q \quad \longrightarrow \quad (\varepsilon_{\rm r} + 1)\varepsilon_0 E(r) 2\pi r^2 = Q, \tag{2.174}$$

and hence

$$E(r) = \frac{Q}{2\pi(\varepsilon_r + 1)\varepsilon_0 r^2}. \tag{2.175}$$

The capacitor voltage and capacitance then come out to be

$$V = \int_a^b E(r)\,dr = \frac{Q}{2\pi(\varepsilon_r + 1)\varepsilon_0}\left(\frac{1}{a} - \frac{1}{b}\right) \quad \longrightarrow \quad C = \frac{Q}{V} = \frac{2\pi(\varepsilon_r + 1)\varepsilon_0 ab}{b - a}. \tag{2.176}$$

We note that this expression for C represents the equivalent total capacitance of a parallel connection of two hemispherical capacitors with homogeneous dielectrics, Eq. (2.159), where $C_1 = 2\pi\varepsilon_r\varepsilon_0 ab/(b - a)$ and $C_2 = 2\pi\varepsilon_0 ab/(b - a)$.

Example 2.21 Coaxial Cable with a Continuously Inhomogeneous Dielectric

Shown in Fig. 2.30(a) is a cross section of a coaxial cable with a continuously inhomogeneous dielectric, the permittivity of which is given by a function $\varepsilon(\phi) = (3 + \sin\phi)\varepsilon_0$, $\phi \in [0, 2\pi]$. The voltage between the inner and outer conductor of the cable is V. Compute (a) the capacitance per unit length of the cable and (b) the distribution of free charges on the cable conductors.

Solution

(a) Let us subdivide the dielectric into thin sectors (slices) defined by elemental azimuthal angles $d\phi$, as depicted in Fig. 2.30(b). Each such sector can be considered as being homogeneous, of permittivity $\varepsilon(\phi)$. This way it becomes obvious that dielectric inhomogeneities in the systems in Figs. 2.29 and 2.30 are of the same type.

Because of symmetry, the electric field intensity vector in the dielectric is radial, and, based on the boundary condition in Eq. (2.79), $\mathbf{E}_1 = \mathbf{E}_2$, $\mathbf{E}_2 = \mathbf{E}_3$, $\mathbf{E}_3 = \mathbf{E}_4$, ... [Fig. 2.30(b)]. This means that E is constant on a cylindrical surface of radius r ($a < r < b$), i.e., it is a function of the radial cylindrical coordinate, r, only. The electric flux density vector is $\mathbf{D}(r, \phi) = \varepsilon(\phi)\mathbf{E}(r)$. Generalized Gauss' law applied to a cylinder of radius r and height (length) h gives

$$\int_{\phi=0}^{2\pi} \underbrace{\varepsilon(\phi)E(r)}_{D}\,\underbrace{r\,d\phi h}_{dS} = \underbrace{Q'h}_{Q_S} \quad \longrightarrow \quad E(r) = \frac{Q'}{r\int_0^{2\pi}\varepsilon(\phi)\,d\phi}, \tag{2.177}$$

where dS is the area of an elemental strip of width $r\,d\phi$ and height h, and Q' is the charge per unit length of the cable. The voltage between the cable conductors and the

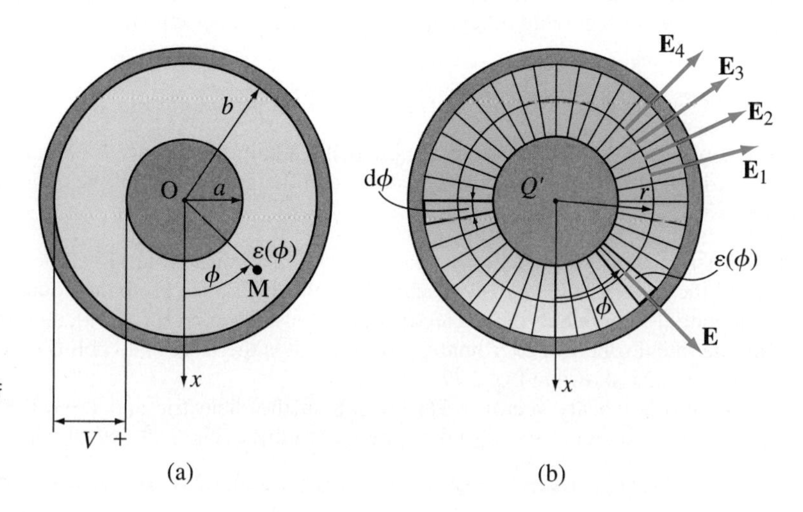

Figure 2.30 Analysis of a coaxial cable with a continuously inhomogeneous dielectric of permittivity $\varepsilon(\phi)$: (a) cross-sectional view of the structure and (b) subdivision of the dielectric into differentially thin homogeneous sectors; for Example 2.21.

capacitance per unit length are then obtained as

$$V = \int_{r=a}^{b} E(r)dr = \frac{Q' \ln(b/a)}{\int_0^{2\pi} \varepsilon(\phi)\, d\phi} \quad \longrightarrow \quad C' = \frac{Q'}{V} = \frac{\int_0^{2\pi} \varepsilon(\phi)\, d\phi}{\ln(b/a)} = \frac{6\pi\varepsilon_0}{\ln(b/a)}. \quad (2.178)$$

(b) From Eqs. (2.177) and (2.178), the electric field intensity, $E(r)$, is the same as in the air-filled coaxial cable, Eq. (2.124), which is expected, because this is an E-system. Using Eq. (2.58), the free surface charge density on the inner conductor of the cable is

$$\rho_{sa}(\phi) = \varepsilon(\phi)E(a^+) = \frac{\varepsilon_0(3+\sin\phi)V}{a\ln(b/a)}, \quad (2.179)$$

while on the outer conductor,

$$\rho_{sb}(\phi) = -\varepsilon(\phi)E(b^-) = -\frac{\varepsilon_0(3+\sin\phi)V}{b\ln(b/a)}. \quad (2.180)$$

Example 2.22 Two-Wire Line with Dielectrically Coated Conductors

Each conductor of a thin symmetrical two-wire transmission line, with the distance between the conductor axes d and conductor radii a ($d \gg a$), is coated by a coaxial dielectric layer of permittivity ε and thickness a. Find the expression for the capacitance per unit length of this line if situated in air.

Solution Let the line be charged by Q' per unit of its length, as shown in Fig. 2.31. Similarly to the analysis of a thin-wire line with bare conductors in Fig. 2.22, the electric fields due to the individual charged conductors can be evaluated independently from each other, because the line is thin. Thus, the electric flux densities due to the conductors 1 and 2 evaluated at the point M in Fig. 2.31 are

$$D_1 = \frac{Q'}{2\pi x} \quad \text{and} \quad D_2 = \frac{Q'}{2\pi(d-x)}, \quad (2.181)$$

respectively. The total electric flux density is $D = D_1 + D_2$. To simplify the computation, however, we take that $D \approx D_1$ in the first dielectric coating (the second conductor is far away) and $D \approx D_2$ in the second coating. Having in mind that the electric field intensity in dielectric coatings is $\mathbf{E} = \mathbf{D}/\varepsilon$, while $\mathbf{E} = \mathbf{D}/\varepsilon_0$ in air, the voltage between the conductors is given by

$$V = \int_a^{d-a} E\, dx = \frac{1}{\varepsilon}\int_a^{2a} D_1\, dx + \frac{1}{\varepsilon_0}\int_{2a}^{d-2a} (D_1 + D_2)\, dx + \frac{1}{\varepsilon}\int_{d-2a}^{d-a} D_2\, dx$$
$$= \frac{Q'}{2\pi}\left[\frac{1}{\varepsilon}\ln 2 + \frac{1}{\varepsilon_0}\left(\ln\frac{d-2a}{2a} - \ln\frac{2a}{d-2a}\right) + \frac{1}{\varepsilon}\ln 2\right] \approx \frac{Q'}{\pi}\left(\frac{1}{\varepsilon}\ln 2 + \frac{1}{\varepsilon_0}\ln\frac{d}{2a}\right) \quad (2.182)$$

[see the integration in Eq. (2.140)]. The capacitance per unit length of the line is

$$C' = \frac{Q'}{V} = \pi\left(\frac{1}{\varepsilon}\ln 2 + \frac{1}{\varepsilon_0}\ln\frac{d}{2a}\right)^{-1}. \quad (2.183)$$

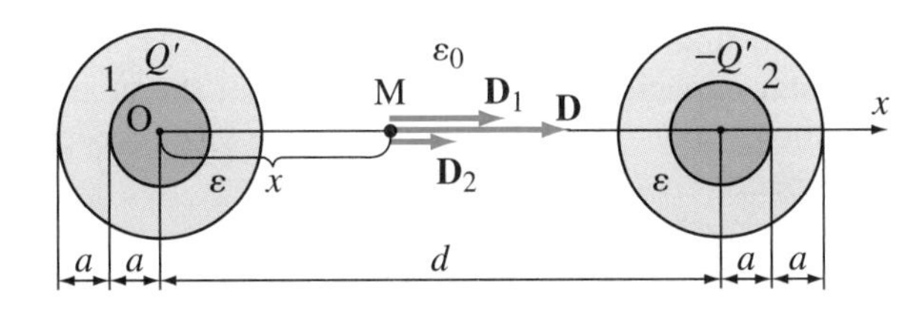

Figure 2.31 Cross section of a thin two-wire line with dielectric coatings over conductors, in air; for Example 2.22.

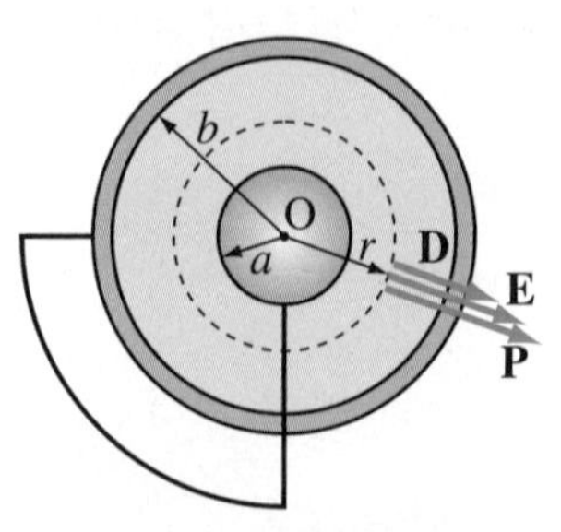

Figure 2.32 Nonlinear spherical capacitor with short-circuited electrodes and remanent polarization; for Example 2.23.

Example 2.23 Spherical Capacitor with a Nonlinear Dielectric

Consider a spherical capacitor with radii of electrodes a and b ($a < b$), filled with a ferroelectric. After being connected to a voltage generator, the capacitor electrodes are short-circuited, and there is a remanent (residual) polarization in the dielectric in the new steady state (Fig. 2.32). The polarization vector is radial, with magnitude $P(r) = P_0 b/r$, where P_0 is a constant and r is the distance from the capacitor center. Calculate the electric field intensity vector in the dielectric.

Solution Since this is a nonlinear capacitor (ferroelectrics are nonlinear materials), relationship in Eq. (2.112) does not hold, and there is a charge, Q and $-Q$, on the capacitor electrodes although the voltage between them is zero in the new state. From spherical symmetry and the given form of the vector $\mathbf{P}$ in the dielectric, the electric flux density vector in the dielectric, $\mathbf{D}$, has that same form (radial vector that depends on r only), as in Eq. (2.161). This means that the capacitor in Fig. 2.32 represents a D-system. Then, using the generalized Gauss' law, Eq. (2.43), which is true for arbitrary (including nonlinear) media, the magnitude of the vector $\mathbf{D}$ is found to be that in Eq. (2.162).

To determine the electric field intensity vector, $\mathbf{E}$, from $\mathbf{D}$, however, we cannot use the relationship in Eq. (2.47), as the dielectric is not linear, but we can use the definition of the vector $\mathbf{D}$ in Eq. (2.41), which yields

$$\mathbf{E} = \frac{\mathbf{D} - \mathbf{P}}{\varepsilon_0}. \tag{2.184}$$

From the condition $V = 0$,

$$\int_{r=a}^{b} E(r)\,dr = \frac{1}{\varepsilon_0}\int_a^b [D(r) - P(r)]\,dr = 0, \tag{2.185}$$

that is,

$$\int_a^b \left(\frac{Q}{4\pi r^2} - \frac{P_0 b}{r}\right) dr = \frac{Q}{4\pi}\left(\frac{1}{a} - \frac{1}{b}\right) - P_0 b \ln\frac{b}{a} = 0. \tag{2.186}$$

Hence, $Q = 4\pi P_0 \ln(b/a)\, ab^2/(b-a)$ and

$$\mathbf{E} = \frac{P_0 b}{\varepsilon_0 r}\left[\frac{ab\ln(b/a)}{(b-a)r} - 1\right]\hat{\mathbf{r}}. \tag{2.187}$$

Problems: 2.42–2.56; *Conceptual Questions* (on Companion Website): 2.19–2.26; *MATLAB Exercises* (on Companion Website).

2.15 ENERGY OF AN ELECTROSTATIC SYSTEM

Every charged capacitor and every system of charged conducting bodies contain a certain amount of energy, which, by the principle of conservation of energy, equals the work done in the process of charging the system. This energy is called the electric energy and is related to the charges and potentials of the conducting bodies in the system. To find this relationship, we perform a numerical experiment in which a system that was initially uncharged is being gradually charged, by bringing elementary charges to the conducting bodies by an external agent. We evaluate the net work done against the electric forces while the charges of the bodies are being changed from zero to their final values.

Consider first a linear capacitor of capacitance C. Let the charges of the electrodes of the capacitor be Q and $-Q$, and their potentials with respect to a reference point be V_1 and V_2, respectively. The capacitor voltage is

$$V = V_1 - V_2 = \frac{Q}{C}. \tag{2.188}$$

If we add an elementary positive charge $\mathrm{d}Q$ to the first electrode and $-\mathrm{d}Q$ to the second one, the change of the energy of the capacitor is equal to the work done against the electric forces in moving $\mathrm{d}Q$ from the reference point (i.e., from the zero potential level) to the first electrode (which is at the potential V_1) and $-\mathrm{d}Q$ to the second electrode (at the potential V_2). By the definition of the electric scalar potential, the potential of a point on either one of the electrodes equals the work done by the electric field in moving a charge from the electrode to the reference point, or, conversely, the work done against the electric field in moving a charge from the reference point to the electrode, divided by the charge [Eq. (1.74)]. This is exactly what we have in our experiment, so the elementary work done while moving $\mathrm{d}Q$ and $-\mathrm{d}Q$ is given by

$$\mathrm{d}W_\mathrm{e} = \mathrm{d}Q\, V_1 + (-\mathrm{d}Q)\, V_2, \tag{2.189}$$

or, equivalently,

$$\mathrm{d}W_\mathrm{e} = \mathrm{d}Q\,(V_1 - V_2) = \mathrm{d}Q\, V = \mathrm{d}Q\,\frac{Q}{C}. \tag{2.190}$$

When uncharged, the capacitor has no energy. The total energy stored in the capacitor when it is charged with Q is therefore equal to the net work done in changing the capacitor charge from $q = 0$ to $q = Q$, and is obtained by adding up all elementary works $\mathrm{d}W_\mathrm{e}$ in Eq. (2.190):

$$W_\mathrm{e} = \int_{q=0}^{q=Q} \mathrm{d}W_\mathrm{e} = \frac{1}{C}\int_{q=0}^{Q} q\,\mathrm{d}q = \frac{1}{C}\frac{Q^2}{2}. \tag{2.191}$$

Hence, the equivalent expressions for the energy of a capacitor are

$$\boxed{W_\mathrm{e} = \frac{Q^2}{2C} = \frac{1}{2}QV = \frac{1}{2}CV^2.} \tag{2.192}$$

energy of a capacitor (unit: J)

The unit for the electric energy is the joule (J). As a special case, the electric energy of an isolated charged metallic body in a linear dielectric is

$$\boxed{W_\mathrm{e} = \frac{1}{2}CV_{\text{isolated body}}^2,} \tag{2.193}$$

energy of an isolated metallic body

where C here is the capacitance of the single body, given by Eq. 2.116, $V_{\text{isolated body}}$ is its potential with respect to the reference point at infinity, and two additional equivalent expressions, like in Eq. 2.192, can be written as well.

Note that the energy of a capacitor can also be expressed in terms of the charges and potentials of individual electrodes as

$$W_\mathrm{e} = \frac{1}{2}QV = \frac{1}{2}Q(V_1 - V_2) = \frac{1}{2}(Q_1V_1 + Q_2V_2). \tag{2.194}$$

Generally, for a linear multibody system with N charged conducting bodies,

$$\boxed{W_\mathrm{e} = \frac{1}{2}\sum_{i=1}^{N} Q_iV_i.} \tag{2.195}$$

energy of a multibody electrostatic system

Finally, if a system also includes a volume charge distributed throughout the dielectric between the conductors, in addition to surface charges over the conductors, the total electric energy of the system is

energy of a multibody system with volume charge in the dielectric

$$W_e = \frac{1}{2}\sum_{i=1}^{N} Q_i V_i + \frac{1}{2}\int_v \rho V \, dv, \tag{2.196}$$

where ρ is the charge density and V the electric scalar potential in the dielectric, and energy $(\rho \, dv) V/2$ of an elemental charge $dQ = \rho \, dV$ is integrated throughout the volume v of the dielectric.

2.16 ELECTRIC ENERGY DENSITY

Expression in Eq. (2.196) enables us to find the total energy associated with the charge distribution of an electrostatic system. As the charges are sources of the electric field, it turns out that the energy of the system can be expressed also in terms of the electric field intensity throughout the system. This leads to an assumption that the electric energy is actually localized in the electric field, and therefore in the dielectric (which can be air and a vacuum) between the conductors of an electrostatic system. Quantitatively, as we shall see, the concentration (density) of energy at specific locations in the dielectric is proportional to the local electric field intensity squared.

Let us consider first a simple case of a homogeneous electric field in the dielectric of a parallel-plate capacitor, with plate areas S, plate separation d (d is small compared to the dimensions of the plates), and dielectric permittivity ε [Fig. 2.18(a)]. From Eqs. (2.192), (2.127), and (2.126), the energy of the capacitor is

$$W_e = \frac{1}{2}CV^2 = \frac{1}{2}\varepsilon\frac{S}{d}(Ed)^2 = \frac{1}{2}\varepsilon E^2 Sd = \frac{1}{2}\varepsilon E^2 v, \tag{2.197}$$

where E is the electric field intensity in the dielectric and $v = Sd$ is the volume of the dielectric, i.e., the volume of the domain where the electric field exists. We define the electric energy density as

$$w_e = \frac{W_e}{v} = \frac{1}{2}\varepsilon E^2, \tag{2.198}$$

or, by employing Eq. (2.47),

electric energy density (unit: J/m^3)

$$w_e = \frac{1}{2}\varepsilon E^2 = \frac{1}{2}ED = \frac{D^2}{2\varepsilon}. \tag{2.199}$$

The unit is J/m^3 (volume density). The energy of the capacitor can now be written in the form

$$W_e = w_e v. \tag{2.200}$$

We shall now generalize this result to obtain a field-based expression for the energy contained in an arbitrary capacitor (Fig. 2.14) and an arbitrary electrostatic system. Let us subdivide the domain between the conducting bodies of a system into elementary flux tubes containing the lines of vector **D**, that start and terminate at surfaces of the bodies. We then cut each tube along its length into small cells of volume dv, such that the interfaces between the neighboring cells are perpendicular to the field lines. We can metalize these interfaces, without changing the field, and get an array of small parallel-plate capacitors along the tube. The entire space between the conducting bodies of the system, i.e., the domain with the electric field, is thus partitioned into volume cells, and each cell represents an elementary parallel-plate capacitor with a homogeneous dielectric and uniform electric field between the plates. The energy contained in each capacitor is

$$dW_e = w_e \, dv, \tag{2.201}$$

where the energy density is given by Eq. (2.199). Summing the energies contained in all capacitors, that is, integrating the energy dW_e throughout the entire dielectric between the conducting bodies of the system (volume v), we obtain the electric energy of the entire system:

$$\boxed{W_e = \int_v w_e \, dv = \frac{1}{2} \int_v ED \, dv.} \tag{2.202}$$

electric energy of an electromagnetic system, via the energy density

The expressions for the electric energy in Eqs. (2.196) and (2.202) are equivalent, namely, they give the same result for the total energy of an electrostatic system. The latter expression, however, implies that the actual localization of electric energy is throughout the electric field, that is, in the dielectric between the conductors of the system (even if the dielectric is a vacuum), and not in the conductors, nor on their surfaces. It also provides us with a means, the energy density in Eq. (2.199), for evaluating and analyzing the exact distribution of energy throughout the dielectric.

On the other hand, reconsidering Eq. (2.196) and reinspecting its terms, we might now come up with an alternative viewpoint on the actual energy localization in electrostatics, which implies that the stored electric energy of a system resides in the system charge, and not the field. Accordingly, $\rho V/2$ might be considered to be the volume energy density at points where $\rho \neq 0$ throughout the volume of the dielectric [from the second term of the energy expression in Eq. (2.196)]. The corresponding surface energy density, equal to $\rho_s V/2$, would then quantify energy localization in the surface charge distribution over the surfaces of the conductors in the system [constituting the first term of the energy expression in Eq. (2.196)]. Both viewpoints have merit and are equally "correct." Nevertheless, the assumption that the energy is actually "contained" in the field, and not in the charge producing it, turns out to be much better suited for energy considerations in the analysis of electromagnetic waves (to be done in a later chapter). Namely, an electromagnetic wave consists of time-varying electric and magnetic fields which travel through space (even in a vacuum) and carry energy independent of the sources (charges and currents) that produced them (the sources might not even exist any more). Therefore, it is much simpler and more natural to describe the energy distribution of a system with traveling waves in terms of the fields, which change in time and travel in space, than to associate it with the stationary distribution of sources at previous instants of time. This is why we have adopted the field-based energy localization approach in the first place and the associated energy density expressions, those in Eq. (2.199), to quantify the energy distribution in an arbitrary (linear) system (with static or time-varying charges and fields).

Example 2.24 Energy of Parallel-Plate Capacitors with Two Dielectric Layers

Find the energy of each of the capacitors in Fig. 2.25, neglecting the fringing effects. Express the energy in terms of the capacitor charge in case (a) and the capacitor voltage in case (b).

Solution Based on Eqs. (2.202), (2.147), (2.192), (2.157), and (2.150), the energy of the capacitor in Fig. 2.25(a) can be found using either (1) the electric flux density in the dielectric, or (2) the capacitances of two capacitors corresponding to individual dielectric layers, or

(3) the equivalent total capacitance of a series connection of capacitors with homogeneous dielectrics:

$$W_e = \underbrace{\frac{D^2}{2\varepsilon_1} S d_1 + \frac{D^2}{2\varepsilon_2} S d_2}_{(1)} = \underbrace{\frac{Q^2}{2C_1} + \frac{Q^2}{2C_2}}_{(2)} = \underbrace{\frac{Q^2}{2C}}_{(3)} = \frac{(\varepsilon_2 d_1 + \varepsilon_1 d_2)Q^2}{2\varepsilon_1\varepsilon_2 S}. \tag{2.203}$$

Similarly, the energy of the capacitor in Fig. 2.25(b) can be computed by employing either (1) the electric field intensity in the dielectric, or (2) the capacitances of the two capacitors with homogeneous dielectrics, or (3) the total capacitance of a parallel connection of capacitors:

$$W_e = \underbrace{\frac{\varepsilon_1 E^2}{2} S_1 d + \frac{\varepsilon_2 E^2}{2} S_2 d}_{(1)} = \underbrace{\frac{C_1 V^2}{2} + \frac{C_2 V^2}{2}}_{(2)} = \underbrace{\frac{CV^2}{2}}_{(3)} = \frac{(\varepsilon_1 S_1 + \varepsilon_2 S_2)V^2}{d} \tag{2.204}$$

[see Eqs. (2.151), (2.160), and (2.154)].

Example 2.25 Energy Per Unit Length of a Coaxial Cable

Determine (a) the energy density and (b) the energy per unit length of a coaxial cable, with conductor radii a and b ($a < b$) and dielectric permittivity ε, if the voltage between the cable conductors is V.

Solution

(a) The electric field intensity in the dielectric of the cable is given by Eq. (2.124), and hence the electric energy density at a distance r from the cable axis is

$$w_e = \frac{1}{2}\varepsilon E(r)^2 = \frac{\varepsilon V^2}{2r^2 \ln^2(b/a)} \quad (a < r < b). \tag{2.205}$$

(b) The electric energy per unit length of the cable can be evaluated as

electric energy per unit length of a transmission line (unit: J/m)

$$\boxed{W'_e = (W_e)_{\text{p.u.l.}} = \frac{W_e}{l} = \frac{1}{l}\int_v w_e \underbrace{\mathrm{d}S\, l}_{\mathrm{d}v} = \int_S w_e \,\mathrm{d}S,} \tag{2.206}$$

where W_e is the energy contained in a part of the cable of length l, S is the cross-sectional area of the dielectric between the cable conductors, and $\mathrm{d}v = \mathrm{d}S\, l$. The unit for W'_e is J/m. Because the energy density is a function of the coordinate r only, we adopt $\mathrm{d}S$ in the form of an elementary ring of radius r and width $\mathrm{d}r$, Eq. (1.60), and obtain

$$W'_e = \int_{r=a}^{b} w_e(r) \underbrace{2\pi r\,\mathrm{d}r}_{\mathrm{d}S} = \frac{\pi\varepsilon V^2}{\ln^2(b/a)} \int_a^b \frac{\mathrm{d}r}{r} = \frac{\pi\varepsilon V^2}{\ln(b/a)}. \tag{2.207}$$

This result is, of course, in agreement with the expression

p.u.l. electric energy via p.u.l. capacitance

$$\boxed{W'_e = \frac{1}{2}C'V^2,} \tag{2.208}$$

where C' is the capacitance per unit length of the cable, given in Eq. (2.123).

Example 2.26 Energy in a Coaxial Cable with a Dielectric Spacer

Shown in Fig. 2.33 is a cross section of a coaxial cable that is partly filled with a dielectric of relative permittivity ε_r. The dielectric is in the form of a spacer between the cable conductors defined by an angle α. The remaining space between the conductors is air-filled. The conductor radii are a and b ($a < b$), and the voltage between the conductors is V. Find

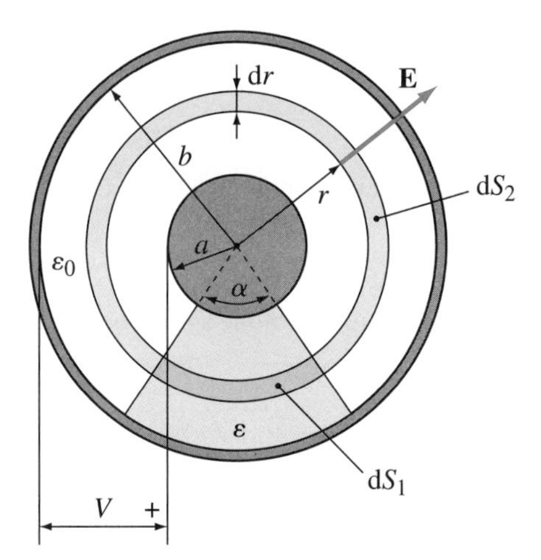

Figure 2.33 Coaxial cable with a dielectric spacer; for Example 2.26.

α for which the electric energy contained in the dielectric equals a half of the total energy in the cable.

Solution Electric energy densities in the dielectric and air part of the cable interior are

$$w_{e1} = \frac{1}{2}\varepsilon_r\varepsilon_0 E^2 \quad \text{and} \quad w_{e2} = \frac{1}{2}\varepsilon_0 E^2, \tag{2.209}$$

respectively, where the electric field intensity E in the cable is continuous across the interface between the dielectric and air. This is an E-system (see Example 2.21), and E is the same as in the air-filled coaxial cable, Eq. (2.124).

By the requirement in the statement of the example, the energy per unit length of the cable contained in the dielectric spacer and that in the air are stipulated to be the same, $W'_{e1} = W'_{e2}$, which means that (Fig. 2.33)

$$\int_{r=a}^{b} \underbrace{\frac{1}{2}\varepsilon_r\varepsilon_0 E(r)^2}_{w_{e1}} \underbrace{\alpha r\,dr}_{dS_1} = \int_{r=a}^{b} \underbrace{\frac{1}{2}\varepsilon_0 E(r)^2}_{w_{e2}} \underbrace{(2\pi - \alpha) r\,dr}_{dS_2}. \tag{2.210}$$

Here, dS_1 and dS_2 are surface areas of the parts of a thin ring of radius r and width dr determined by angles α and $2\pi - \alpha$, respectively. Even without any integration, Eq. (2.210) gives

$$\varepsilon_r\alpha = 2\pi - \alpha \quad \longrightarrow \quad \alpha = \frac{2\pi}{\varepsilon_r + 1}. \tag{2.211}$$

Example 2.27 Energy of a System with Volume Charges

Calculate the electric energy of the system from Example 2.5.

Solution This system does not contain conductors. The electric energy densities in the dielectric sphere and air are

$$w_e = \frac{D(r)^2}{2\varepsilon_r\varepsilon_0} \quad (0 \le r \le a) \quad \text{and} \quad w_{e0} = \frac{D(r)^2}{2\varepsilon_0} \quad (a < r < \infty), \tag{2.212}$$

respectively, where the electric flux density, D, is given in Eq. (2.63). Hence, the electric energy of the system is

$$W_e = \int_0^a w_e(r)\, 4\pi r^2\,dr + \int_a^\infty w_{e0}(r)\, 4\pi r^2\,dr = \frac{\rho_0^2 \pi a^5}{56\varepsilon_0}\left(7 + \frac{1}{\varepsilon_r}\right), \tag{2.213}$$

where dv is adopted in the form of a thin spherical shell of radius r and thickness dr, Eq. (1.33).

The above result can also be obtained employing the expression in Eq. (2.196),

$$W_{\mathrm{e}} = \frac{1}{2}\int_{v} \rho V \, \mathrm{d}v = \frac{1}{2}\int_{0}^{a} \rho(r) V(r) \, 4\pi r^2 \, \mathrm{d}r = \frac{\rho_0^2 \pi a^5}{56\varepsilon_0}\left(7 + \frac{1}{\varepsilon_{\mathrm{r}}}\right), \tag{2.214}$$

with v being the volume of the sphere, ρ the volume charge density given in Example 2.5, and V the electric scalar potential in Eq. (2.66).

Note that the integration in Eq. (2.213) is over both the dielectric sphere and the air, while that in Eq. (2.214) is over the sphere only. This is because the electric field exists in the entire space, whereas the volume charge occupies the volume of the sphere only.

Problems: 2.57–2.68; *Conceptual Questions* (on Companion Website): 2.27–2.31; *MATLAB Exercises* (on Companion Website).

2.17 DIELECTRIC BREAKDOWN IN ELECTROSTATIC SYSTEMS

We shall now analyze electrostatic systems in high-voltage applications, i.e., in situations where the electric field in the dielectric is so strong that there is a danger of dielectric breakdown in the system. As discussed in Section 2.6, dielectric breakdown occurs when the largest field intensity in the dielectric reaches the critical value for that particular material – dielectric strength of the material, E_{cr}. Under the influence of such strong electric fields, a dielectric material is suddenly transformed from an insulator into a very good conductor, causing an intense current to flow. While systems with gaseous and liquid dielectrics can completely recover after a breakdown, those with solid dielectrics are most often permanently damaged by breakdown fields, as insulating properties of the dielectric are irreversibly degraded. In this section, our goal is not to analyze local breakdown processes in materials nor the overall behavior of systems resulting from these processes (these phenomena are largely nonlinear and are not electrostatic), but to evaluate the systems and their performance in linear electrostatic states close to the breakdown occurrences. In fact, our goal here is basically to determine the maximum extent of values of various quantities in a system (or device) that are "permitted" for its safe operation prior to an eventual breakdown.

In systems (devices) with nonuniform electric field distributions, the principal task is to identify the most vulnerable spot for dielectric breakdown, and to relate the corresponding largest electric field intensity to the charges or potentials of conducting bodies in the system. In the case of a capacitor or a two-conductor transmission line, the voltage that corresponds to the critical field in the dielectric is called the breakdown voltage of the capacitor or line. This is the highest possible voltage that can be applied to the system (before it breaks down), and is sometimes also referred to as the voltage rating of the system.[6] In some applications, we optimize individual parameters of a system (e.g., the radius of the inner conductor

[6]Since the dielectric strength of a given material may vary considerably depending on the actual conditions under which the material is used, as well as the way the particular piece of a solid dielectric is manufactured, the voltage rating and critical values of other quantities for a system in practice are always defined with a certain safety factor included. For example, with a safety factor of 10, the voltage rating of a given capacitor equals a tenth of the voltage that would lead to a breakdown under ideal conditions and assumptions.

of a coaxial cable) such that the breakdown voltage is maximum. It is also of interest to evaluate the critical (breakdown) values for the capacitor charge and energy, and the corresponding values for forces on conductors (charge, energy, and forces per unit length for transmission lines), as these are the largest possible charge, energy, and forces for that particular system.

For systems filled with a homogeneous dielectric, the strongest electric field in the dielectric is most frequently right next to one of the conducting bodies of the system. In addition, this is most likely in the vicinity of sharp parts of a conducting surface [see Eq. (1.210)]. In systems with heterogeneous dielectrics, on the other side, the most vulnerable spots can be close to a boundary surface between dielectric parts with different permittivities, where the normal component of the electric field intensity is discontinuous. In such systems, furthermore, the largest field intensity in a given dielectric part is not necessarily the breakdown field, because the dielectric strength vary from material to material, and only an electrostatic analysis (or an experiment) can tell which of the dielectric parts would break down first after a voltage of critical value is applied to the system.

For a system of conducting bodies situated in air (or any other gas), the breakdown region, where the air is ionized by an avalanche breakdown mechanism (see Section 2.6) and behaves like a conducting material, is usually localized only in the immediate vicinity of "hot" spots on conductor surfaces, as the field farther away from the conductor is not sufficiently strong to sustain the breakdown. Due to an avalanche of impact ionization of air molecules, vast numbers of free electrons and positive ions are created, so that the air near the conductor becomes more and more conducting. The ionized air might even glow (at night), appearing as a luminous "crown" around the conductor, and hence such a local discharge close to a conductor surface is referred to as a corona discharge. A corona discharge on a conductor is equivalent to an enlargement of the conductor, since a layer of conducting material (ionized air) is added over the conductor surface. It substantially increases losses on power transmission lines and also emits electromagnetic waves that can interfere with nearby communication devices and systems. On some occasions, the field intensities in the system are so high that breakdowns occur even away from the conductors. A continuously ionized path is formed from a part of a conductor surface with an exceedingly high charge density to the nearest conductor with opposite charge polarity. This path is apparent as a bright luminous arc carrying a current of a very large intensity (typically hundreds to thousands of amperes, sometimes as large as 100 kA). As a result, a very rapid and violent discharge of the conductors occurs, known as arc discharge. The most apparent and spectacular arc discharges certainly are intense cloud-to-ground lightning discharges in the form of giant sparks in thunderstorms. Lightning strikes are, as we know, a frequent cause of loss of human lives and property.

Analysis of electrostatic systems with fields close to breakdown levels and predicting critical values of voltages, charges, and other quantities at breakdown require no new theory, and is just a matter of applying what we already know in an appropriate way and order, having in mind the above general comments. The rest of this section consists therefore of a number of characteristic examples.

Example 2.28 **Breakdown in a Parallel-Plate Capacitor**

Consider the capacitor in Fig. 2.19 for $a = 1$ m and $d = 2$ cm, and determine its breakdown voltage and the corresponding charge and energy, as well as the electric pressure and forces on capacitor electrodes at breakdown.

Solution The electric field between the capacitor plates is approximately uniform, and its breakdown intensity is

$$E = E_{\mathrm{cr0}} = 3\ \mathrm{MV/m} \tag{2.215}$$

[dielectric strength of air, Eq. (2.53)], from which the breakdown voltage of the capacitor, Eq. (2.126), charge of the capacitor, Eq. (2.125), and capacitor energy, Eq. (2.192), come out to be

$$V = E_{\mathrm{cr0}} d = 60\ \mathrm{kV}, \quad Q = \varepsilon_0 a^2 E_{\mathrm{cr0}} = 26.56\ \mu\mathrm{C}, \quad \text{and} \quad W_{\mathrm{e}} = \frac{1}{2} QV = 0.8\ \mathrm{J}, \tag{2.216}$$

respectively. According to Eqs. (2.133) and (2.132), the electric pressure and electric forces on electrodes at breakdown amount to

$$p_{\mathrm{e}} = \frac{1}{2}\varepsilon_0 E_{\mathrm{cr0}}^2 = 39.84\ \mathrm{Pa} \quad \text{and} \quad F_{\mathrm{e}} = p_{\mathrm{e}} a^2 = 39.84\ \mathrm{N}. \tag{2.217}$$

Note that the above values for energy and force are quite small compared to energies and forces of nonelectrical origin, and these are the largest possible values for this (quite large) capacitor.

Example 2.29 Maximum Breakdown Voltage of a Coaxial Cable

Conductor radii of a coaxial cable are a and $b = 7$ mm ($a < b$). The cable is filled with a homogeneous dielectric of relative permittivity $\varepsilon_{\mathrm{r}} = 2.56$ and dielectric strength $E_{\mathrm{cr}} = 20$ MV/m (polystyrene). (a) Find a for which the breakdown voltage of the cable is maximum. (b) What is the maximum breakdown voltage of the cable?

Solution

(a) Assume that the voltage between the cable conductors is V. The electric field intensity in the cable is given by Eq. (2.124). Obviously, the field is the strongest right next to the inner conductor of the cable,

maximum electric field of a coaxial cable

$$\boxed{E(a^+) = \frac{V}{a \ln(b/a)}}, \tag{2.218}$$

meaning that dielectric breakdown occurs when this field intensity reaches the critical value – dielectric strength,

$$E(a^+) = E_{\mathrm{cr}}. \tag{2.219}$$

For a given radius a, the breakdown voltage, i.e., the largest voltage that the cable can withstand, is therefore

$$V_{\mathrm{cr}}(a) = E_{\mathrm{cr}}\, a \ln \frac{b}{a}. \tag{2.220}$$

To find the optimal radius a, for which the voltage V_{cr} is the largest, we impose the condition

$$\frac{\mathrm{d}V_{\mathrm{cr}}}{\mathrm{d}a} = 0 \quad \longrightarrow \quad E_{\mathrm{cr}} \left(\ln \frac{b}{a} - 1 \right) = 0. \tag{2.221}$$

The solution is

coaxial cable radii ratio for a maximum breakdown voltage

$$\boxed{\frac{b}{a} = \mathrm{e} = 2.718} \quad \longrightarrow \quad a_{\mathrm{opt}} = \frac{b}{\mathrm{e}} = 2.57\ \mathrm{mm}. \tag{2.222}$$

Because the second derivative of $V_{\mathrm{cr}}(a)$ for $a = a_{\mathrm{opt}}$ is negative (equals $-E_{\mathrm{cr}}/a_{\mathrm{opt}}$), the function $V_{\mathrm{cr}}(a)$ has indeed a maximum (and not a minimum) at that point.

(b) The maximum breakdown voltage of the cable is

$$(V_{\mathrm{cr}})_{\max} = V_{\mathrm{cr}}(a_{\mathrm{opt}}) = E_{\mathrm{cr}}\, a_{\mathrm{opt}} = 51.5\ \mathrm{kV}. \tag{2.223}$$

HISTORICAL ASIDE

The lightning rod was invented around the middle of the eighteenth century by **Benjamin Franklin** (1706–1790), an American statesman and scientist, a prolific inventor in the area of electricity (e.g., principle of conservation of charge, labeling positive and negative charges, investigation of lightning, etc.), as well as in numerous other areas of scientific inquiry (e.g., bifocal glasses, Franklin stove, improved printing press, Gulf Stream chart, etc.). Franklin speculated that lightning was electricity, and proposed that upright sharp-pointed iron rods be mounted on roofs of buildings and connected by conducting wires to the iron bars buried in the earth (grounding electrodes). One of the first lightning rods with grounding conductors based on his design was that installed in 1752 on the Pennsylvania State House (now Independence Hall), in Philadelphia. With several improvements by Franklin and others in decades to follow, grounded lightning rods soon proved to be an efficient way of protecting buildings from lightning damage. *(Portrait: Library of Congress)*

Example 2.30 Breakdown Voltage of a Thin Two-Wire Line

Consider a symmetrical two-wire line with conductor radii $a = 1$ mm and the distance between conductor axes $d = 0.5$ m. The line is situated in air. Compute (a) the breakdown voltage of the line and (b) the largest possible forces on line conductors per unit length.

Solution

(a) Let $Q' > 0$ and $-Q'$ be the charges per unit length of the line conductors, as in Fig. 2.22. The electric field intensity of the line is the largest very close to the surface of each of the conductors. At these points, the total field (due to both charged conductors), Eq. (2.138), can approximately be evaluated as the field due to a single isolated charged wire conductor, because the other conductor is far away ($d \gg a$). Thus, by means of Eq. (1.196),

$$\boxed{E_{\max} = \frac{Q'}{2\pi\varepsilon_0 a},} \quad (2.224)$$

maximum electric field of a thin two-wire line

and, at breakdown, the critical charge per unit length of the line turns out to be

$$E_{\max} = E_{\text{cr0}} = 3\ \text{MV/m} \quad \longrightarrow \quad Q'_{\text{cr}} = 2\pi\varepsilon_0 a E_{\text{cr0}} = 167\ \text{nC/m}. \quad (2.225)$$

The breakdown voltage of the line amounts to

$$V_{\text{cr}} = \frac{Q'_{\text{cr}}}{C'} = \frac{Q'_{\text{cr}}}{\pi\varepsilon_0} \ln\frac{d}{a} = 2aE_{\text{cr0}} \ln\frac{d}{a} = 37.29\ \text{kV}, \quad (2.226)$$

where C' is the capacitance per unit length of the line [Eq. (2.141)].

(b) Since $a \ll d$, the corresponding intensity of electric forces on line conductors per unit length can be computed as in Eqs. (1.223) and (1.224), and the result is

$$F'_{\text{e}} = \frac{Q'^2_{\text{cr}}}{2\pi\varepsilon_0 d} = 1\ \text{mN/m}, \quad (2.227)$$

which is the largest possible force for this line.

Example 2.31 Grounded Wire Conductor as a Lightning Arrester

A wire conductor, of radius $a = 5$ mm, is positioned horizontally above the surface of the earth, at a height $h = 6$ m, as shown in Fig. 2.34. The conductor is grounded. A uniform

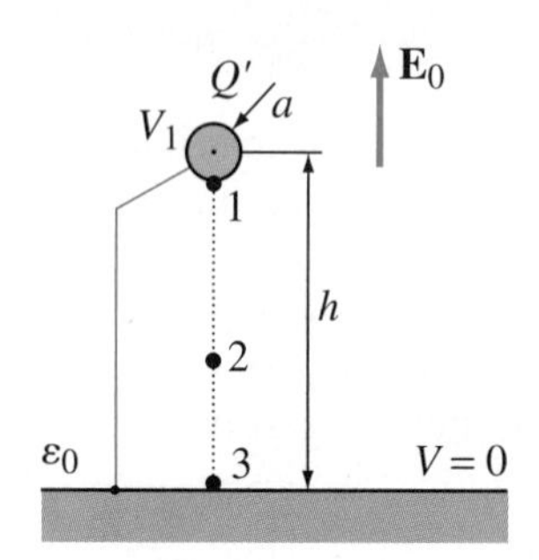

Figure 2.34 Grounded wire conductor in a uniform atmospheric electric field; for Example 2.31.

atmospheric electric field of intensity E_0, due to a large charged cloud, exists above the earth's surface. The vector $\mathbf{E}_0$ is vertical and directed upward.[7] Assuming that E_0 is known ($E_0 > 0$), find (a) the charge induced in the conductor per unit of its length and (b) the electric field intensity on the surface of the conductor. (c) Consider a cloud discharge to the ground and discuss the protection that the wire conductor provides to the space below it.

Solution

(a) The earth's surface represents a conducting plane. Let the potential of the plane be zero. Since the wire conductor is connected to the earth, a charge of opposite polarity to that of the lower part of the cloud, that is, of positive polarity, is induced in the conductor (pulled out from the earth) under the influence of the atmospheric field. Let Q' designate this charge per unit length of the conductor ($Q' > 0$), as indicated in Fig. 2.34. The potential of the conductor, V_1, can be expressed as the sum of the potential due to charge Q' and the potential produced by the external field $\mathbf{E}_0$. This first component can be obtained as the potential at the surface of the wire due to both Q' and its image in the conducting plane, as in Fig. 1.49, and thus using Eq. (1.119) with $r_1 = a$ (distance of the point at which the potential is computed form the axis of the original wire equals the wire radius) and $r_2 = 2h$ (since $2h \gg a$). The other component of the potential can be found from Eq. (1.90), by integrating the field $\mathbf{E}_0$ from the conductor to the earth's surface along the straight, vertical path. As this field is uniform, the voltage is simply $-E_0$ times the length of the path, and the total potential of the conductor is given by

$$V_1 = \frac{Q'}{2\pi\varepsilon_0} \ln \frac{2h}{a} - E_0 h. \tag{2.228}$$

From the condition that the conductor is grounded, we now have

$$V_1 = 0 \quad \longrightarrow \quad Q' = \frac{2\pi\varepsilon_0 E_0 h}{\ln(2h/a)}. \tag{2.229}$$

(b) The electric field intensity on the surface of the wire conductor (point 1 in Fig. 2.34) is approximately [Eq. (2.224)]

$$E_1 = \frac{Q'}{2\pi\varepsilon_0 a} = \frac{E_0 h}{a \ln(2h/a)} = 154 E_0. \tag{2.230}$$

(c) On the other hand, having in mind Eq. (2.139), the electric field intensity below the conductor at a height $H = 2$ m, for instance, with respect to the ground level (point 2 in Fig. 2.34) amounts to

$$E_2 = -\frac{Q'}{2\pi\varepsilon_0}\left(\frac{1}{r} + \frac{1}{2h - r}\right) + E_0 = 0.71 E_0 \tag{2.231}$$

($\mathbf{E}_2$ is directed upward), where $r = h - H = 4$ m. Similarly, the field intensity below the conductor close to the earth's surface (point 3) turns out to be $E_3 = -Q'/(\pi\varepsilon_0 h) + E_0 = 0.743 E_0$ ($r = h$).

As the atmospheric field becomes stronger and stronger, in a thunderstorm, the field intensity E_1 reaches the breakdown value [Eq. (2.225)]

$$E_1 = E_{\text{cr}0}. \tag{2.232}$$

The air ionizes and becomes conductive, so that a portion of the negative charge of the cloud flows through this conducting air channel to the wire conductor and down to the

[7] In cold clouds, that contain water and ice, and cause eventual lightning discharges, there is a positive charge buildup at the top of a cloud and a negative one in its base. The bottom buildup induces another positively charged area at the earth's surface, and these two charge layers, like a giant charged capacitor, generate an electric field above the earth that is directed vertically upward, as in Fig. 2.34.

ground. The field intensities E_2 and E_3, however, are much lower than critical,

$$E_2 = 0.71E_0 = \frac{0.71}{154}E_1 = \frac{E_{\rm cr0}}{217} \quad \text{and} \quad E_3 = 0.743E_0 = \frac{0.743}{154}E_1 = \frac{E_{\rm cr0}}{207}, \qquad (2.233)$$

meaning that a structure or a person that may be underneath the conductor is safe from breakdown and cloud discharge (or lightning strike). The grounded conductor protects therefore the space below it and serves principally as a lightning arrester.[8]

Example 2.32 Breakdown in a Spherical Capacitor with Two Dielectrics

The dielectric of a spherical capacitor consists of two concentric layers. The relative permittivity of the inner layer is $\varepsilon_{\rm r1} = 2.5$ and its dielectric strength $E_{\rm cr1} = 50$ MV/m. For the outer layer, $\varepsilon_{\rm r2} = 5$ and $E_{\rm cr2} = 30$ MV/m. Electrode radii are $a = 3$ cm and $c = 8$ cm, and the radius of the boundary surface between the layers is $b = 5$ cm. Calculate (a) the breakdown voltage and (b) the corresponding energy of the capacitor.

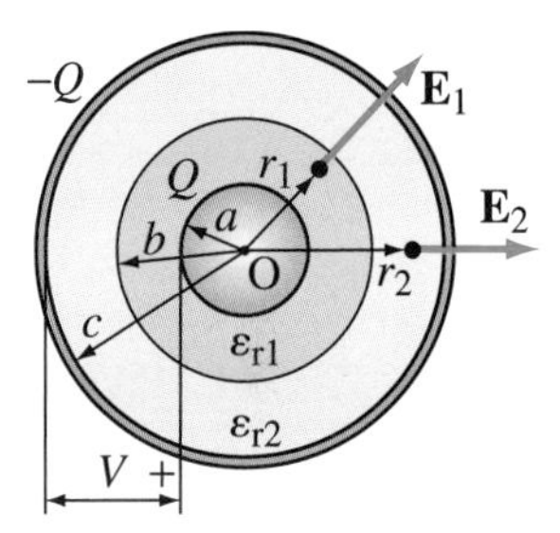

Figure 2.35 Evaluation of the breakdown voltage of a spherical capacitor with a two-layer dielectric; for Example 2.32.

Solution

(a) Let Q be the charge of the capacitor. The electric flux density in the dielectric is given by Eq. (2.162). With reference to Fig. 2.35, the electric field intensities in individual layers are

$$E_1(r) = \frac{Q}{4\pi\varepsilon_{\rm r1}\varepsilon_0 r^2} = E_1(a^+)\frac{a^2}{r^2}, \quad a < r < b, \qquad (2.234)$$

$$E_2(r) = \frac{Q}{4\pi\varepsilon_{\rm r2}\varepsilon_0 r^2} = E_2(b^+)\frac{b^2}{r^2}, \quad b < r < c, \qquad (2.235)$$

where r is the radial spherical coordinate. Note that $E_1(a^+)$ is the largest field intensity in the inner layer, while $E_2(b^+)$ represents the strongest field in the outer layer. Combining Eqs. (2.234) and (2.235), we arrive to the following relationship between these field intensities:

$$\varepsilon_{\rm r1}a^2E_1(a^+) = \varepsilon_{\rm r2}b^2E_2(b^+). \qquad (2.236)$$

We do not know in advance which dielectric layer would break down first after a voltage of critical value is applied across the capacitor electrodes. Let us first check the possibility of a breakdown in the inner layer, which implies that

$$E_1(a^+) = E_{\rm cr1}. \qquad (2.237)$$

At the same time, from Eq. (2.236),

$$E_2(b^+) = E_{\rm cr1}\frac{\varepsilon_{\rm r1}a^2}{\varepsilon_{\rm r2}b^2} = 9 \text{ MV/m}. \qquad (2.238)$$

As $E_2(b^+) < E_{\rm cr2}$, we conclude that the electric field intensity at all points in the outer dielectric layer is lower than the critical value for that dielectric. This means that the assumption of a breakdown occurring in the inner dielectric layer is correct.

[8]Note that this is the principle of operation of lightning arresters, in general. Grounded lightning rods, placed as high as possible, protect exposed objects (buildings or other structures, as well as humans and animals) and their immediate surroundings from violent atmospheric electrical discharges by enforcing that an eventual dielectric breakdown leading to a lightning strike is "induced" near the surface of the rod, and not at some other part of the object. Then, the rod routes the lightning discharge through an insulated wire conductor to a grounding electrode, rather than through the object. In addition to simple vertical metallic rods with pointed ends, designs of lightning arresters include rods with spherical metallic tops or umbrella-like wire extensions, which have capacity for carrying much larger amounts of induced charge.

The other possibility, which assumes that an eventual breakdown occurs in the outer layer, gives

$$E_2(b^+) = E_{\text{cr2}}, \tag{2.239}$$

and $E_1(a^+) = E_{\text{cr2}}\varepsilon_{\text{r2}}b^2/(\varepsilon_{\text{r1}}a^2) = 166.7$ MV/m. This is impossible, as E_1 cannot be larger than E_{cr1} under the assumption that the inner layer is in a normal regime, while the outer layer breaks down.

With the field-intensity values in Eqs. (2.237) and (2.238), the breakdown voltage of the capacitor is

$$V_{\text{cr}} = E_1(a^+)\,a^2 \int_a^b \frac{\mathrm{d}r}{r^2} + E_2(b^+)\,b^2 \int_b^c \frac{\mathrm{d}r}{r^2} = \varepsilon_{\text{r1}}a^2E_{\text{cr1}}\left(\frac{b-a}{\varepsilon_{\text{r1}}ab} + \frac{c-b}{\varepsilon_{\text{r2}}bc}\right) = 769 \text{ kV}. \tag{2.240}$$

This voltage can also be obtained by considering the critical value of the capacitor charge for breakdown. Let us denote by $Q_{\text{cr}}^{(1)}$ and $Q_{\text{cr}}^{(2)}$ the charge in the case of an eventual breakdown in the inner and outer dielectric layer, respectively. Based on Eqs. (2.234), (2.235), (2.237), and (2.239),

$$Q_{\text{cr}}^{(1)} = 4\pi\varepsilon_{\text{r1}}\varepsilon_0 a^2 E_{\text{cr1}} \quad \text{and} \quad Q_{\text{cr}}^{(2)} = 4\pi\varepsilon_{\text{r2}}\varepsilon_0 b^2 E_{\text{cr2}}. \tag{2.241}$$

As Q becomes larger and larger, the breakdown occurs when it reaches the smaller of the two charges in Eqs. (2.241). The critical charge of the capacitor is thus

$$Q_{\text{cr}} = \min\left\{Q_{\text{cr}}^{(1)}, Q_{\text{cr}}^{(2)}\right\}. \tag{2.242}$$

For the given numerical data, $Q_{\text{cr}}^{(1)} < Q_{\text{cr}}^{(2)}$ [$Q_{\text{cr}}^{(1)} = 0.3Q_{\text{cr}}^{(2)}$], meaning that the breakdown occurs in the inner dielectric layer. Hence,

$$Q_{\text{cr}} = Q_{\text{cr}}^{(1)} = 12.52\ \mu\text{C}. \tag{2.243}$$

The corresponding voltage is

$$V_{\text{cr}} = \frac{Q_{\text{cr}}}{C} = 769 \text{ kV}, \tag{2.244}$$

where $C = 16.28$ pF is the capacitance of the capacitor [Eq. (2.164)].

(b) The largest possible energy of the capacitor is

$$W_{\text{e}} = \frac{1}{2}CV_{\text{cr}}^2 = 4.81 \text{ J}. \tag{2.245}$$

Problems: 2.69–2.81; *Conceptual Questions* (on Companion Website): 2.32–2.34; *MATLAB Exercises* (on Companion Website).

Problems

2.1. Nonuniformly polarized dielectric parallelepiped. A rectangular dielectric parallelepiped is situated in air in the first octant of the Cartesian coordinate system $(x, y, z \geq 0)$, with one vertex at the coordinate origin, and the edges, of lengths a, b, and c, parallel to coordinate axes x, y, and z, respectively. The polarization vector in the parallelepiped is given by $\mathbf{P}(x, y, z) = P_0[(x/a)\,\hat{\mathbf{x}} + (y/b)\,\hat{\mathbf{y}} + (z/c)\,\hat{\mathbf{z}}]$, where P_0 is a constant. (a) Find the densities of volume and surface bound charge of the parallelepiped. (b) Show that the total bound charge of the parallelepiped is zero.

2.2. Uniformly polarized disk on a conducting plane. A uniformly polarized dielectric disk surrounded by air is lying at a conducting plane, as shown in Fig. 2.36. The polarization vector in the disk is $\mathbf{P} = P\hat{\mathbf{z}}$, the disk radius is a, and thickness d. Calculate the electric field intensity vector along the disk axis normal to the conducting plane (z-axis).

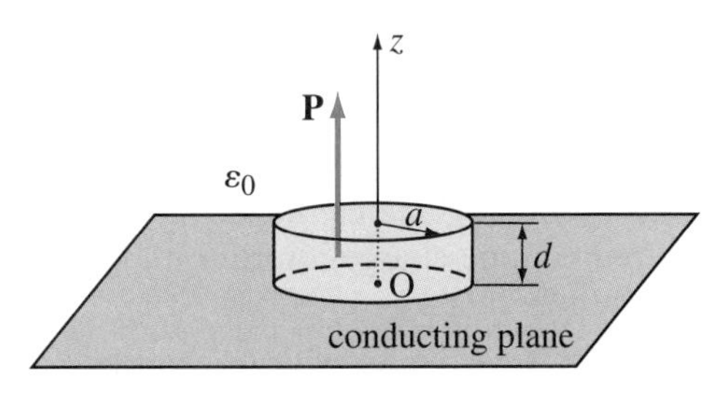

Figure 2.36 Dielectric disk with a uniform polarization lying at a conducting plane; for Problem 2.2.

2.3. Uniformly polarized hollow dielectric cylinder. A hollow dielectric cylinder of radii a and b, and height $2h$, is uniformly polarized and situated in free space. The polarization vector, of magnitude P, is parallel to the cylinder axis, as shown in Fig. 2.37. Find the electric field intensity vector at the center of the cylinder (point O).

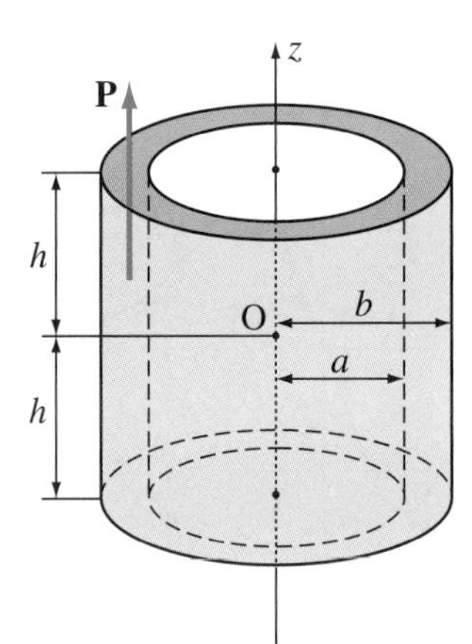

Figure 2.37 Hollow dielectric cylinder with a uniform polarization; for Problem 2.3.

2.4. Nonuniformly polarized thin dielectric disk. A very thin dielectric disk, of radius a and thickness d ($d \ll a$), is polarized throughout its volume. In the cylindrical coordinate system shown in Fig. 2.38, the polarization vector is defined by the expression $\mathbf{P} = P_0 r\,\hat{\mathbf{r}}/a$, where P_0 is a constant. The medium around the disk is air. Determine (a) the distribution of bound charge and (b) the electric field intensity vector along the z-axis.

2.5. Uniformly polarized dielectric hemisphere. A dielectric hemisphere of radius a is situated in

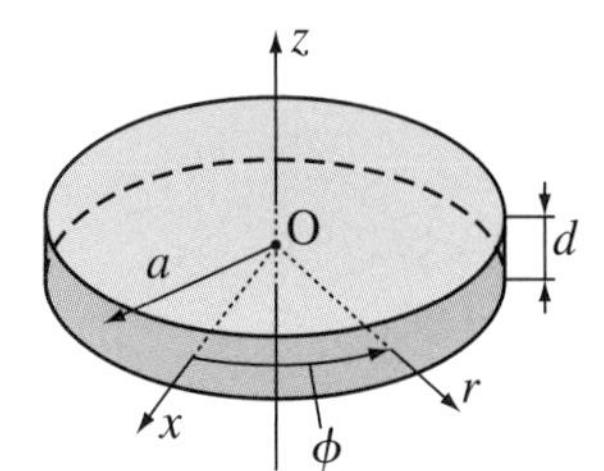

Figure 2.38 Very thin dielectric disk with a nonuniform polarization; for Problem 2.4.

free space and is lying at a conducting plane. The polarization vector is $\mathbf{P}$, and it is normal to the plane, as depicted in Fig. 2.39. Find (a) the bound surface charge density at the flat and spherical surfaces of the hemisphere and (b) the electric field intensity vector at the center of the flat surface (point O in the figure), assuming that it is on the dielectric side of the boundary surface (dielectric-conductor), very close to the surface.

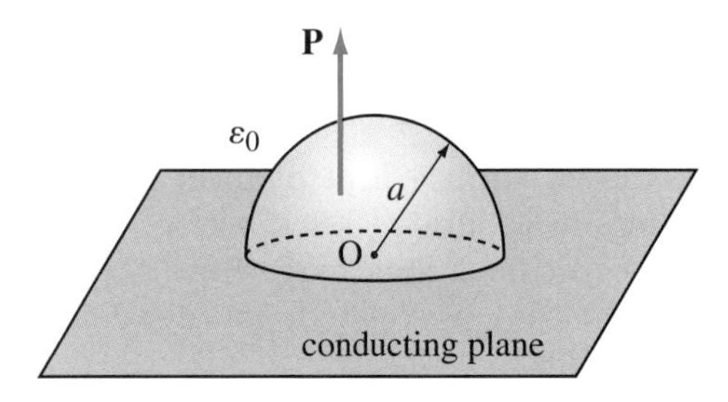

Figure 2.39 Dielectric hemisphere with a uniform polarization lying at a conducting plane; for Problem 2.5.

2.6. Nonuniformly polarized large dielectric slab. An infinitely large dielectric slab of thickness $d = 2a$, shown in Fig. 2.40, is polarized such that the polarization vector is $\mathbf{P} = P_0 x^2\,\hat{\mathbf{x}}/a^2$, where P_0 is a constant. The medium outside the slab is air. Compute (a) the distribution of volume and surface bound charge of the slab, (b) the electric field intensity vector everywhere, and (c) the voltage between the boundary surfaces of the slab.

2.7. Electric flux density vector. Find the electric flux density vector, $\mathbf{D}$, (a) at the center of the polarized dielectric sphere in Fig. 2.7 and (b) along the axis of the polarized dielectric disk (z-axis) in Fig. 2.36.

2.8. Total (free plus bound) volume charge density. The electric field intensity vector in a dielectric, $\mathbf{E}$, is a known function of spatial coordinates. (a) Prove that the total (free plus

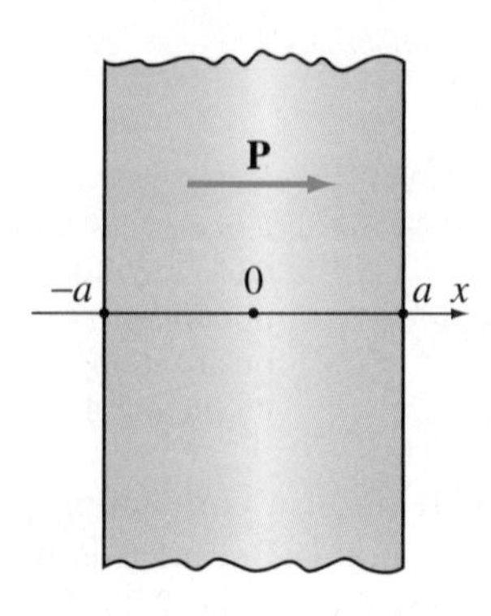

Figure 2.40 Infinitely large dielectric slab with a nonuniform polarization; for Problem 2.6.

bound) volume charge density in the dielectric, $\rho_{\text{tot}} = \rho + \rho_{\text{p}}$, can be obtained as $\rho_{\text{tot}} = \varepsilon_0 \nabla \cdot \mathbf{E}$. (b) Specifically, find ρ_{tot} for $\mathbf{E}$ given as the following function of Cartesian coordinates: $\mathbf{E}(x, y, z) = [4xyz\,\hat{\mathbf{x}} + (2x^2z - y^3)\,\hat{\mathbf{y}} + (2x^2y + z^3)\,\hat{\mathbf{z}}]$ V/m (x, y, z in m).

2.9. Uniform field in a dielectric. There is a uniform electric field in a certain dielectric region. The free volume charge density is ρ. Determine the bound volume charge density, ρ_{p}.

2.10. Closed surface in a uniform field. Considering an arbitrary closed surface in a uniform electric field, in a charge-free ($\rho_{\text{tot}} = 0$) region (Fig. 2.41), prove the following vector identity: $\oint_S \mathbf{dS} = 0$.

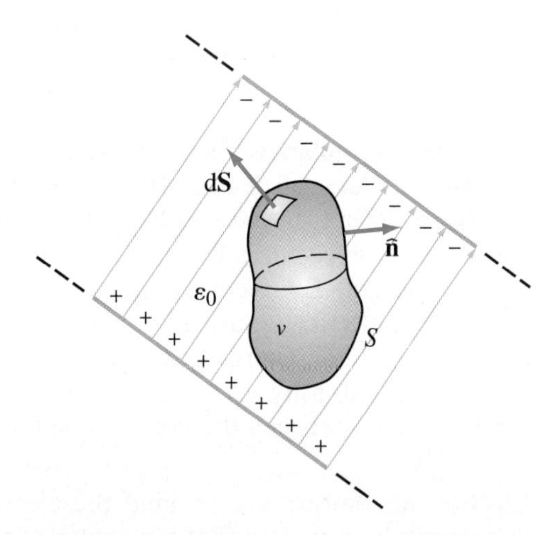

Figure 2.41 Closed surface in a region with a uniform electric field and no volume charge; for Problem 2.10.

2.11. Flux of the electric field intensity vector. The polarization vector, $\mathbf{P}$, and free volume charge density, ρ, are known at every point of a dielectric body. Find the expression for the flux of the electric field intensity vector, Ψ_E, through a closed surface S situated entirely inside the body.

2.12. Total enclosed bound and free charge. Consider an imaginary closed surface S inside a homogeneous dielectric of permittivity ε. The total free charge enclosed by S is Q_S. What is the total bound charge $Q_{\text{p}S}$ enclosed by S?

2.13. Charge-free homogeneous medium. Prove that in a homogeneous linear medium with no free volume charge, there is no bound volume charge either.

2.14. Dielectric cylinder with free volume charge. A very long homogeneous dielectric cylinder, of radius a and relative permittivity ε_{r}, is charged uniformly with free charge density ρ throughout its volume. The cylinder is surrounded by air. (a) Calculate the voltage between the axis and the surface of the cylinder. (b) Find the bound charge distribution of the cylinder.

2.15. Linear-exponential volume charge distribution. Repeat Example 2.6 but for a model of a *pn* junction given by the volume charge density $\rho(x) = \rho_0 (x/a)\, \mathrm{e}^{-|x|/a}$, where ρ_0 and a are positive constants, as shown in Fig. 2.42.

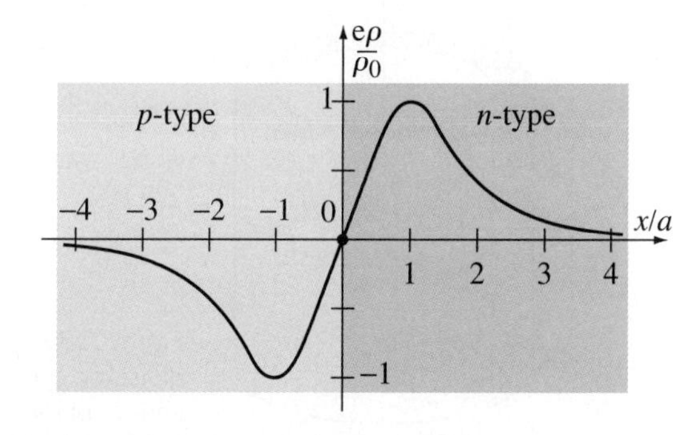

Figure 2.42 Model of a *pn* junction with a linear-exponential charge distribution; for Problem 2.15.

2.16. Dielectric-dielectric boundary conditions. Assume that the plane $z = 0$ separates medium 1 ($z > 0$) and medium 2 ($z < 0$), with relative permittivities $\varepsilon_{\text{r1}} = 4$ and $\varepsilon_{\text{r2}} = 2$, respectively. The electric field intensity vector in medium 1 near the boundary (for $z = 0^+$) is $\mathbf{E}_1 = (4\,\hat{\mathbf{x}} - 2\,\hat{\mathbf{y}} + 5\,\hat{\mathbf{z}})$ V/m. Find the electric field intensity vector in medium 2 near the boundary (for $z = 0^-$), $\mathbf{E}_2$, if (a) no free charge exists on the boundary ($\rho_{\text{s}} = 0$) and (b) there is a surface charge of density $\rho_{\text{s}} = 53.12$ pC/m^2 on the boundary.

2.17. Conductor-dielectric boundary conditions. Obtain conductor-dielectric and conductor-free space boundary conditions, Eqs. (1.186), (2.58), and (1.190), from Eqs. (2.84) and (2.85).

2.18. Water-air boundary. Sketch the field lines emerging from water ($\varepsilon_r = 80$) into air, if the "incident" angle (in water) is $\alpha = 45°$.

2.19. Poisson's equation for inhomogeneous media. (a) Derive Poisson's equation for an inhomogeneous medium. (b) Write Laplace's equation for an inhomogeneous medium.

2.20. Vacuum diode. Fig. 2.43 shows a vacuum diode, which consists of two flat electrodes, the cathode and the anode, and a charge distribution in a vacuum between them. Let the potential of the cathode be zero and the potential of the anode V_0 ($V_0 > 0$). The distribution of the potential in the diode can be described as $V(x) = V_0(x/d)^{4/3}$, $0 \le x \le d$, where d is the distance between the electrodes. Find (a) the volume charge density in the diode, (b) the surface charge density on the cathode, (c) the surface charge density on the anode, and (d) the total charge of the diode.

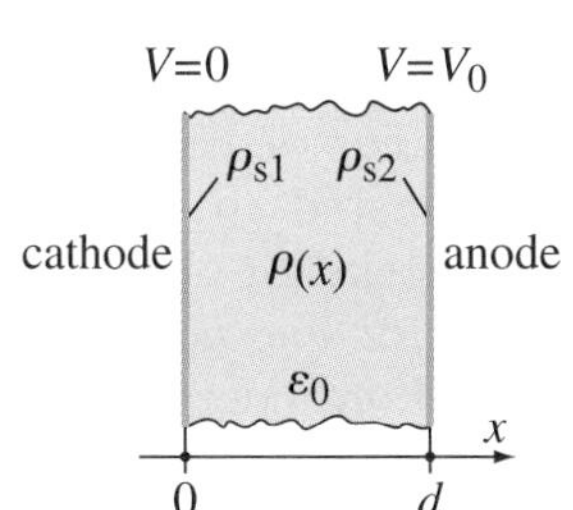

Figure 2.43 Vacuum diode; for Problem 2.20.

2.21. Application of Poisson's equation in spherical coordinates. Consider a system of two spherical concentric metallic electrodes (a solid sphere enclosed by a shell) as the one in Fig. 1.41, with $a = 1$ cm and $b = 5$ cm, and assume that a nonuniform free volume charge whose density is the function $\rho(r)$ given in Eq. (1.32), with $\rho_0 = 3\ \mathrm{C/m^3}$, exists between the electrodes, for $a < r < b$, where the permittivity is ε_0. In addition, let the potential of the inner electrode (sphere) with respect to the outer one (shell) be $V_0 = 10$ V. Under these circumstances, compute (a) the electric potential and (b) the electric field vector at an arbitrary point between the electrodes.

2.22. Application of Laplace's equation in spherical coordinates. Repeat the previous problem but for $\rho(r) = 0$ (no volume charge) between the electrodes (for $a < r < b$).

2.23. FD computer program – iterative solution. Write a computer program for the finite-difference analysis of a coaxial cable of square cross section (Fig. 2.13) based on Eq. (2.107). Assume that $a = 1$ cm, $b = 3$ cm, $V_a = 1$ V, and $V_b = -1$ V. (a) Plot the results for the distribution of the potential and the electric field intensity in the space between the conductors, and the surface charge density on the surfaces of conductors, taking the grid spacing to be $d = a/10$ and the tolerance of the potential $\delta_V = 0.01$ V. (b) Compute the total charge per unit length of the inner and the outer conductor, taking $d = a/N$ and $N = 2, 3, 5, 7, 9, 10,$ and 12, respectively.

2.24. FD computer program – direct solution. As an alternative to the iterative technique based on Eq. (2.107), write a computer program for the FD analysis of a square coaxial cable by directly solving the system of linear algebraic equations with the potentials at interior grid nodes in Fig. 2.13(b) as unknowns [applying Eq. (2.106) to each interior grid node, we get a set of simultaneous equations the number of which equals the number of unknown potentials]. The system of equations, in which known potentials at nodes on the surface of conductors appear on the right-hand side of equations, should be solved by the Gaussian elimination method (or by matrix inversion). Compute and plot the same quantities as in the previous problem, and compare the results obtained by the two programs.

2.25. Capacitance of the earth. Find the capacitance of the earth assuming that it is a conducting sphere of radius $R = 6378$ km (grounds and waters are conducting media).

2.26. Capacitance of a person. It can be shown that the capacitance C of an arbitrary conducting body is in between the capacitance of the sphere inscribed in the body and that of the sphere overscribed about the body, that is [Eq. (2.121)], $4\pi\varepsilon_0 R_{\mathrm{min}} < C < 4\pi\varepsilon_0 R_{\mathrm{max}}$, where R_{min} and R_{max} are the radii of the two spheres. Based on that, estimate the capacitance of an average human body (human tissues are conducting media).

2.27. Capacitance of a metallic cube, computed by the MoM. Find the capacitance of the metallic cube numerically analyzed by the method of moments in Problem 1.85, and compare the result with capacitances of the following metallic spheres, respectively: (a) the sphere inscribed in the cube, (b) the sphere overscribed about the cube, (c) the sphere whose radius is the arithmetic mean of the radii of spheres in (a) and (b), (d) the sphere having the same surface as the cube, and (e) the sphere with the same volume as the cube.

2.28. RG-55/U coaxial cable. A RG-55/U coaxial cable has conductor radii $a = 0.5$ mm and $b = 3$ mm. The dielectric is polyethylene ($\varepsilon_r = 2.25$). Determine the capacitance per unit length of the cable.

2.29. Capacitance p.u.l. of a square coaxial cable, FD analysis. Compute the capacitance per unit length of the coaxial cable of square cross section numerically analyzed by a finite-difference technique in Problems 2.23 and 2.24, and compare the result (using the grid spacing of $d = a/10$) with the per-unit-length capacitance of a (standard) coaxial cable (of circular cross section) having the same ratio of conductor radii ($b/a = 3$) and dielectric (air) as the square cable.

2.30. Parallel-plate capacitor model of a thundercloud. A thundercloud can be approximately represented, as far as its electrical properties are concerned, as a parallel-plate capacitor with horizontal plates of area $S = 15\ \text{km}^2$ and vertical separation $d = 1$ km. Assume that the upper plate has a total charge $Q = 300$ C and the lower plate an equal amount of negative charge. Neglecting the fringing effects, find (a) the capacitance of this capacitor, (b) the voltage between the top and bottom of the cloud, and (c) the electric field intensity in the cloud.

2.31. MoM numerical analysis of a parallel-plate capacitor. Consider the parallel-plate capacitor in Fig. 2.19, and write a computer program based on the method of moments to evaluate its capacitance (C). In specific, subdivide each of the plates into $N = 10 \times 10 = 100$ square patches, and assume constant charge densities on individual patches (as in Fig. 1.46). Additionally, assume that the upper and lower plates in Fig. 2.19 are at potentials $V_1 = 1$ V and $V_2 = -1$ V, respectively, as well as that the charge densities of pairs of corresponding patches that are right above/below each other on the two plates are equal in magnitude and opposite in polarity. With this, the unknowns in the procedure are charge densities ρ_{s1}, ρ_{s2}, $\ldots$, ρ_{sN} on the upper plate only, and matching points, at which the potentials are computed, are centers of the same patches; however, these potentials are due to pairs of patches on both plates, with charge densities ρ_{si} and $-\rho_{si}$ ($i = 1, 2, \ldots, N$). Using the MoM program, find C for $a = 1$ m and the following d/a ratios: (i) 0.1, (ii) 0.5, (iii) 1, (iv) 2, and (v) 10, and compare the results with the corresponding C values obtained from Eq. (2.127), which neglects the fringing effects.

2.32. Nonsymmetrical thin two-wire line. Derive the expression for the capacitance per unit length of a nonsymmetrical thin two-wire transmission line in air. The radii of the conductors are a and b ($a \neq b$), and the distance between the axes of conductors is d ($d \gg a, b$).

2.33. Thick symmetrical two-wire line. Consider a symmetrical two-wire line with conductors of arbitrary thickness (radius of wires, a, is not necessarily small compared to the distance between wire axes, d) in air. Using a version of image theory for line charges in the vicinity of conducting cylinders, it can be shown that the capacitance per unit length of this line is given by $C' = \pi\varepsilon_0 / \ln\{d/(2a) + \sqrt{[d/(2a)]^2 - 1}\}$. Take d/a to be 3, 5, 10, 20, and 100, and calculate C' using this expression and the expression in Eq. (2.141), obtained for thin two-wire lines. Compare the two sets of results and evaluate the error due to thin-wire approximation of the line for individual distance to radius ratios.

2.34. Two small metallic spheres in air. A capacitor consists of two small metallic spheres of equal radii, a, placed in air at a center-to-center distance d ($d \gg a$). Find the capacitance of this capacitor.

2.35. Four parallel wires in air. The transmission line shown in Fig. 2.44 consists of two pairs of galvanically interconnected thin wire conductors, situated in air. The distance between the axes of adjacent wires is $d = 200$ mm and the wire radii are $a = 1$ mm. Compute the capacitance per unit length of the line.

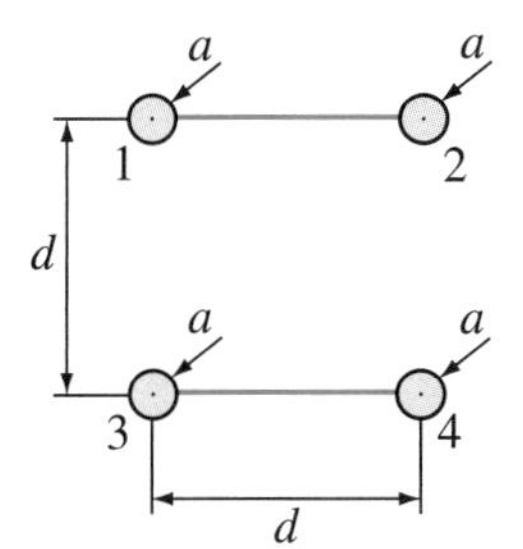

Figure 2.44 Cross section of a transmission line consisting of two pairs of short-circuited wires in air; for Problem 2.35.

2.36. **Two wires at the same potential and a foil.** A short-circuited two-wire line is parallel to a large flat metallic foil, as portrayed in Fig. 2.45. The geometry parameters are $a = 1$ mm, $h = 20$ mm, and $d = 30$ mm, and the medium is air. (a) What is the capacitance per unit length of a two-conductor transmission line whose one conductor is the two-wire line and the other conductor is the foil? (b) If the voltage between the line and the foil is $V = 20$ V, find the induced surface charge density at the central point O on the foil.

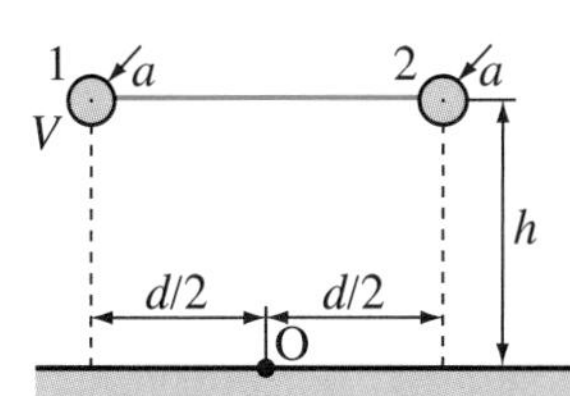

Figure 2.45 Short-circuited horizontal two-wire line over a large metallic foil; for Problem 2.36.

2.37. **Capacitance per unit length of a wire-corner line.** Consider the system composed of a wire parallel to a corner screen in Fig. 1.57, and assume that $a = 2$ mm and $h = 10$ cm. Calculate the capacitance per unit length of this system (transmission line).

2.38. **Equivalent circuit with two spherical capacitors.** (a) Consider the system described in Example 1.27, and show that it can be replaced by the equivalent circuit given in Fig. 2.46. (b) What are the capacitances of the capacitors in this schematic diagram? (c) Obtain the expression for the potential at the center of the structure in Fig. 1.41, Eq. (1.202), that is, repeat Example 1.27, by solving this circuit.

2.39. **Equivalent circuit with three spherical capacitors.** (a) Repeat Problem 1.77 by generating

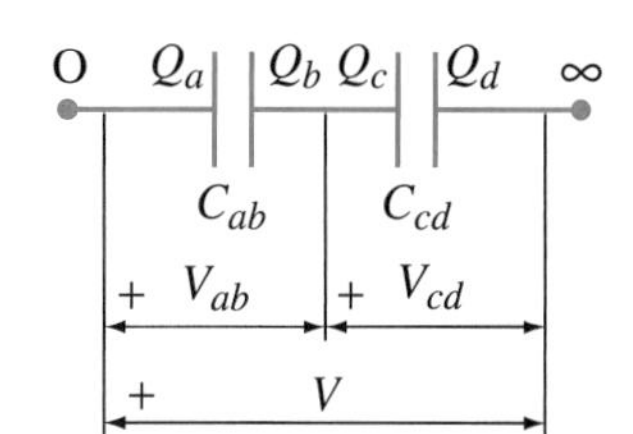

Figure 2.46 Equivalent circuit for the system in Fig. 1.41; for Problem 2.38.

and solving an equivalent circuit with three capacitors. (b) Repeat Problem 1.78 using the same circuit and the corresponding set of conditions.

2.40. **Equivalent circuit with parallel-plate capacitors.** Repeat Example 1.28 by solving the equivalent circuit shown in Fig. 2.47.

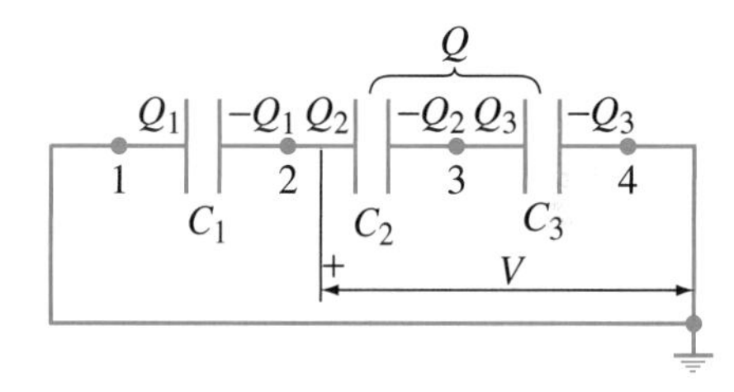

Figure 2.47 Equivalent circuit for the system in Fig. 1.42; for Problem 2.40.

2.41. **Equivalent circuit with cylindrical capacitors.** Repeat Problem 1.79 by generating and solving an equivalent circuit with capacitors.

2.42. **Spherical capacitor with a solid and liquid dielectric.** Consider the spherical capacitor in Fig. 2.27 and assume that the inner dielectric layer is made from mica ($\varepsilon_{r1} = 5.4$), whereas the outer layer is oil ($\varepsilon_{r2} = 2.3$). The geometrical parameters are $a = 2$ cm, $b = 8$ cm, and $c = 16$ cm. The capacitor is connected to a source of voltage $V = 100$ V. The source is then disconnected, and the oil is drained from the capacitor. Find the voltage between the electrodes of the capacitor in the new electrostatic state.

2.43. **Oil drain without disconnecting the source.** If the oil in the capacitor from the previous problem is drained without disconnecting the voltage source, determine the flow of electricity through the source circuit (that is, the difference in the charge of the capacitor) between the two electrostatic states.

2.44. **Metallic sphere with dielectric coating.** A metallic sphere of radius $a = 1$ cm is covered with a concentric dielectric layer of relative permittivity $\varepsilon_r = 4$ and thickness $b - a = 2$ cm and situated in air, as shown in Fig. 2.48. The potential of the sphere with respect to the reference point at infinity is $V = 1$ kV. Compute (a) the capacitance of the sphere, (b) the free surface charge density on the sphere surface, (c) the bound volume charge density in the dielectric, and (d) the bound surface charge densities on dielectric interfaces.

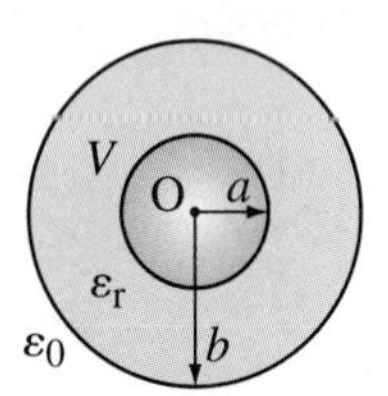

Figure 2.48 Metallic sphere with a dielectric coating in air; for Problem 2.44.

2.45. Charge densities in a half-filled spherical capacitor. Consider the half-filled spherical capacitor from Fig. 2.29, and assume that $a = 2$ cm, $b = 10$ cm, $\varepsilon_r = 3$, and $Q = 10$ nC. Find the distributions of (a) free surface charges on metallic surfaces and (b) bound surface charges on dielectric surfaces.

2.46. Empty and half-filled spherical capacitor. An air-filled spherical capacitor with conductor radii $a = 3$ cm and $b = 15$ cm is connected to a source of voltage $V = 15$ kV. After an electrostatic state is established, the source is disconnected. The capacitor is then half filled with a liquid dielectric of relative permittivity $\varepsilon_r = 2.5$. What is the new voltage between the electrodes of the capacitor?

2.47. Metallic sphere half embedded in a dielectric. A metallic sphere of radius a is pressed into a dielectric half-space of permittivity ε up to a half of its volume, as shown in Fig. 2.49. The medium in the upper half-space is air. (a) Find the capacitance of the sphere. (b) If the sphere is charged with Q, determine what portion of this charge is located on the upper half of the sphere surface.

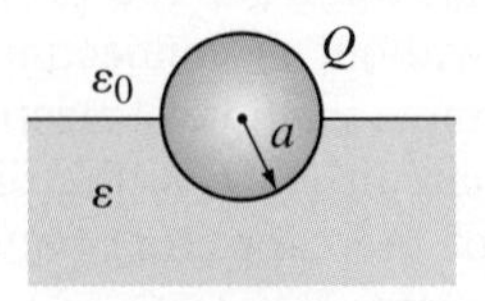

Figure 2.49 Charged metallic sphere half embedded in a dielectric half-space; for Problem 2.47.

2.48. Coaxial cable with two coaxial dielectric layers. Fig. 2.50 shows a cross section of a coaxial cable with two coaxial dielectric layers. The geometry parameters are $a = 1$ mm, $b = 2$ mm, and $c = 4$ mm. The dielectric parameters are $\varepsilon_{r1} = 5$ and $\varepsilon_{r2} = 2$. Calculate the capacitance per unit length of the cable.

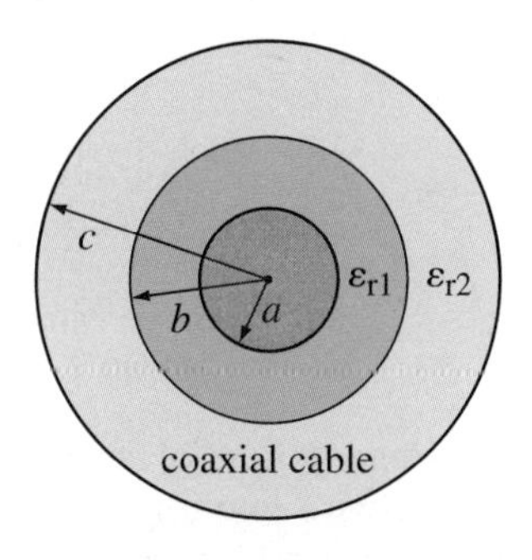

Figure 2.50 Cross section of a coaxial cable with two dielectric layers; for Problem 2.48.

2.49. Coaxial cable with a radial variation of permittivity. Consider a coaxial cable with a continuously inhomogeneous dielectric, whose relative permittivity is given by the following function of the radial distance r from the cable axis: $\varepsilon_r(r) = r/a$ $(a \le r \le b)$, where a and $b = 5a$ are the cable conductor radii. If the potential difference between the cable conductors is V, find (a) the capacitance per unit length of the cable and (b) the bound charge distribution of the dielectric.

2.50. Coaxial cable with four dielectric sectors. A coaxial cable is filled with a piece-wise homogeneous dielectric composed of four 90° sectoral parts with different permittivities, as shown in Fig. 2.51. Let the relative permittivities of the sectors be $\varepsilon_{r1} = 6$, $\varepsilon_{r2} = 2$, $\varepsilon_{r3} = 1$, and $\varepsilon_{r4} = 10$, the radii of the cable conductors $a = 2$ mm and $b = 7$ mm, and the potential of the outer conductor with respect to the inner conductor $V = 25$ V. Compute (a) the capacitance per unit length of the cable and (b) the free charge density at an arbitrary point on the surface of the inner conductor.

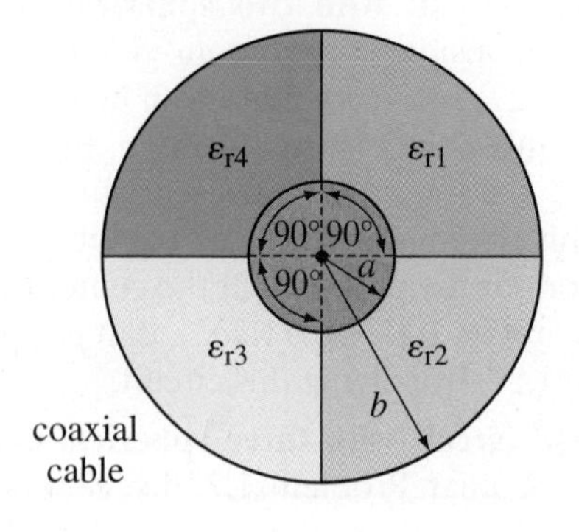

Figure 2.51 Cross section of a coaxial cable with four 90° dielectric sectors; for Problem 2.50.

2.51. Charge distribution for two coated wires. Consider the two-wire line with dielectric coatings in Fig. 2.31, and assume that $a = 1$ mm, $d = 25$ mm, $\varepsilon_r = 4$, and $V = 10$ V. Under these circumstances, find the distribution of (a) free charge and (b) bound charge in the system.

2.52. Two metallic spheres with dielectric coating. If two identical metallic spheres with dielectric coating as the one described in Problem 2.44 are placed in air so that the distance between their centers is $d = 1$ m, determine the capacitance of such a capacitor.

2.53. Two metallic spheres half embedded in a dielectric. Two metallic spheres of radii $a = 5$ mm are half embedded in a dielectric half-space of relative permittivity $\varepsilon_r = 4$, as shown in Fig. 2.52. The distance between the centers of the spheres is $d = 30$ cm. The upper medium is air. The spheres are charged with charges of equal magnitudes and opposite polarities. The potential difference between the spheres is $V = 200$ V. For such a capacitor, find (a) the capacitance, (b) the distribution of free charges over conductors, and (c) distribution of bound charges in the dielectric.

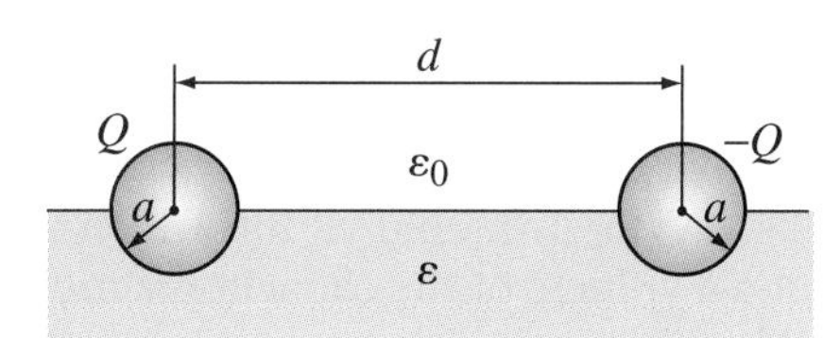

Figure 2.52 Two charged metallic spheres pressed into a dielectric half-space; for Problem 2.53.

2.54. Permittivity gradient normal to capacitor plates. Fig. 2.53 shows a parallel-plate capacitor with rectangular plates of dimensions a and b and continuously inhomogeneous dielectric of thickness d. The permittivity of the dielectric is given by the following function of the x-coordinate: $\varepsilon(x) = 2(1 + 3x/d)\varepsilon_0$ $(0 \le x \le d)$. Neglecting fringing, calculate the capacitance of this capacitor.

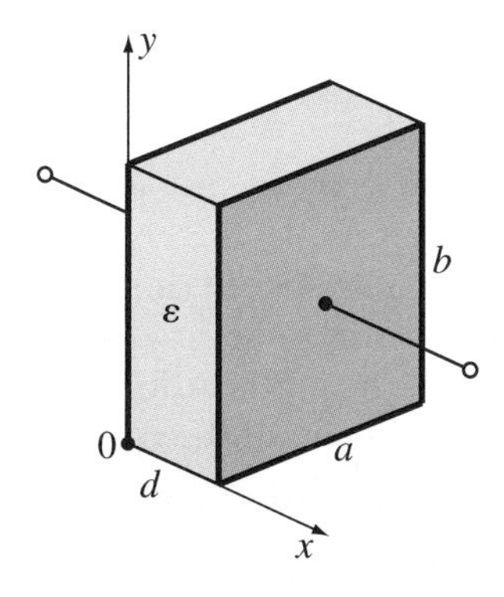

Figure 2.53 Parallel-plate capacitor with a continuously inhomogeneous dielectric; for Problem 2.54.

2.55. Permittivity gradient parallel to capacitor plates. Assume that the permittivity of the dielectric in Fig. 2.53 is a function of the y-coordinate, $\varepsilon(y) = 2[1 + 3\sin(\pi y/b)]\varepsilon_0$ $(0 \le y \le b)$, and find the capacitance of the capacitor. Neglect fringing.

2.56. Capacitor with a nonlinear dielectric layer. Fig. 2.54 shows a parallel-plate capacitor that is half filled with a ferroelectric. The other part of the capacitor is air-filled. The capacitor is charged by being connected to a voltage source. The source is then disconnected, and the capacitor electrodes are short-circuited. In the new electrostatic state, there is a remanent uniform polarization throughout the volume of the dielectric, with the polarization vector being normal to the capacitor plates and its magnitude being P. Determine the electric field intensity vector between the capacitor plates in (a) air and (b) dielectric. Fringing can be neglected.

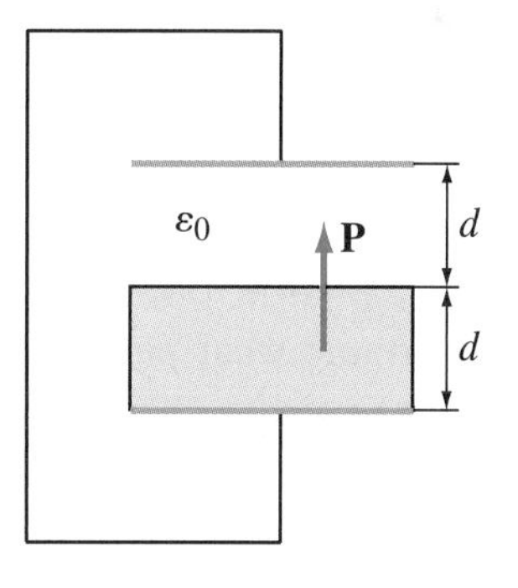

Figure 2.54 Short-circuited parallel-plate capacitor containing a nonlinear dielectric layer with a uniform remanent polarization; for Problem 2.56.

2.57. Energy of a spherical capacitor with two layers. For the spherical capacitor with two concentric dielectric layers from Example 2.18, compute the electric energy stored in each of the layers by (a) integrating the electric energy density over the volume of each layer and (b) representing the capacitor as a series connection of two spherical capacitors with homogeneous dielectrics.

2.58. Change in energy of a spherical capacitor. For the spherical capacitor with solid and liquid dielectric layers from Problem 2.42, find the change in energy of the capacitor between the electrostatic state with the capacitor connected to the voltage source and the final electrostatic state.

2.59. **Energy of a coated metallic sphere.** For the charged metallic sphere with a dielectric coating from Problem 2.44, determine the radius b such that 1/2 of the total energy of the system is stored in the coating.

2.60. **Energy of a coaxial cable with two coaxial layers.** For the coaxial cable with two coaxial dielectric layers from Problem 2.48, assume that the voltage between the cable conductors is $V = 100$ V and find the per-unit-length electric energy contained in each of the layers by (a) integrating the electric energy density over the cross section of each layer and (b) representing the cable as a series connection of two coaxial cylindrical capacitors with homogeneous dielectrics.

2.61. **Energy of a half-filled spherical capacitor.** For the spherical capacitor half filled with a liquid dielectric from Example 2.20, calculate the electric energy stored in the liquid, if the charge of the capacitor is Q.

2.62. **Energy of a coaxial cable with four sectors.** For the coaxial cable with four dielectric sectors from Problem 2.50, find the per-unit-length electric energy contained in each of the sectors.

2.63. **Energy of a capacitor with a variable permittivity.** Consider the parallel-plate capacitor with a permittivity variation normal to plates from Problem 2.54. Determine what percentage of the total electric energy of the capacitor, when charged, is contained in the first half of the dielectric, from $x = 0$ to $x = d/2$.

2.64. **Energy of a capacitor with an inhomogeneous dielectric.** For the parallel-plate capacitor with a permittivity variation parallel to plates from Problem 2.55, find what percentage of the total electric energy of the capacitor, if charged, is stored in the lower half of the dielectric, from $y = 0$ to $y = b/2$.

2.65. **Energy of a system of spherical conductors.** Calculate the energy of the system of three spherical conductors from Problem 1.80.

2.66. **Energy of a system of flat electrodes.** Compute the electric energy stored in the system of five parallel large flat electrodes from Example 1.28 in the following two ways: (a) using the electric field intensities between electrodes (found in Example 1.28) and the corresponding energy densities and (b) using the equivalent circuit in Fig. 2.47 and the involved capacitances, respectively.

2.67. **Energy of a system with free volume charge.** Find the electric energy per unit length of the system with a volume free charge distribution from Problem 2.14.

2.68. **Energy of a *pn* junction.** Calculate the electric energy contained in the *pn* junction from Example 2.6, assuming that the area of the junction cross section perpendicular to the x-axis is S.

2.69. **Breakdown charge and energy of the earth.** Determine the maximum possible charge and electric energy that could be stored on the earth, as described in Problem 2.25, and in the electric field around it, limited by an eventual dielectric breakdown of air near the earth's surface.

2.70. **Maximum breakdown voltage of a spherical capacitor.** A spherical capacitor has electrodes of radii a and $b = 10$ cm ($a < b$). It is filled with a homogeneous dielectric of relative permittivity $\varepsilon_r = 5$ and dielectric strength $E_{cr} = 25$ MV/m. (a) Find a for which the breakdown voltage of the capacitor is maximum. (b) What is the maximum breakdown voltage of the capacitor? (c) What is the energy of the capacitor at breakdown?

2.71. **Breakdown in a wire-plane transmission line.** (a) Compute the breakdown voltage of the wire-plane transmission line in Fig. 2.24(a), assuming that $a = 1$ cm and $h = 1$ m. (b) What is the maximum energy per unit length of this line? (c) Determine the largest possible electric force on the wire conductor per unit of its length.

2.72. **Grounded metallic sphere as a lightning arrester.** Repeat Example 2.31 but for a grounded small metallic sphere (instead of the wire) in a uniform atmospheric electric field above the surface of the earth. Adopt the same radius and the same height of the conductor.

2.73. **Parallel-plate capacitor with two dielectric layers.** Consider the parallel-plate capacitor with two dielectric layers in Fig. 2.25(a), and assume that $d_1 = 2$ mm, $d_2 = 4$ mm, $\varepsilon_{r1} = 3$, and $\varepsilon_{r2} = 5$, as well as that the dielectric strengths for the layers are $E_{cr1} = 20$ MV/m and $E_{cr2} = 11$ MV/m, respectively. Find the breakdown voltage of the capacitor. Neglect fringing.

2.74. Parallel-plate capacitor with two dielectric sectors. Repeat the previous problem but for the parallel-plate capacitor with two dielectric sectors from Fig. 2.25(b) and $d = d_1 + d_2$.

2.75. Breakdown in a spherical capacitor with two layers. Find the breakdown voltage of the spherical capacitor with two concentric dielectric layers from Problem 2.42 in (a) the first state (outer layer is oil) and (b) the second state (outer layer is air). The dielectric strengths for mica and oil are $E_{cr1} = 200$ MV/m and $E_{cr2} = 15$ MV/m, respectively.

2.76. Breakdown potential of a coated metallic sphere. (a) Determine the breakdown potential of the metallic sphere with a dielectric coating from Problem 2.44 if the dielectric strength for the coating is $E_{cr} = 30$ MV/m. (b) What is the maximum energy of this structure?

2.77. Breakdown in a coaxial cable with two layers. (a) Find the breakdown voltage of the coaxial cable with two coaxial dielectric layers from Problem 2.48. The dielectric strengths of the inner and outer layer are $E_{cr1} = 40$ MV/m and $E_{cr2} = 20$ MV/m, respectively. (b) Calculate the maximum energy that this cable can store per unit of its length.

2.78. Simultaneous breakdown in two spherical layers. Consider the spherical capacitor with two concentric dielectric layers in Fig. 2.27. Let the dielectric strengths of the inner and outer layer be E_{cr1} and E_{cr2}, respectively. Find the relationship between the parameters of this capacitor (a, b, c, ε_{r1}, ε_{r2}, E_{cr1}, and E_{cr2}) such that, for sufficiently large capacitor voltage, dielectric breakdown will occur in both dielectric layers simultaneously.

2.79. Simultaneous breakdown in two coaxial layers. Repeat the previous problem but for a coaxial cable with two coaxial dielectric layers (Fig. 2.50).

2.80. Metallic sphere half immersed in a liquid dielectric. A metallic sphere of radius $a = 2$ cm is half immersed in a liquid dielectric, as in Fig. 2.49. The relative permittivity of the dielectric is $\varepsilon_r = 3$ and its dielectric strength is $E_{cr} = 20$ MV/m. The upper medium is air. Calculate the maximum potential of this sphere such that dielectric breakdown will not occur after it is removed from the liquid and raised high above the interface.

2.81. Breakdown in a coaxial cable with a dielectric spacer. Consider the coaxial cable with a dielectric spacer in Fig. 2.33, and assume that $a = 2$ mm, $b = 6$ mm, $\alpha = 60°$, and $\varepsilon_r = 5$, as well as that the dielectric strength for the spacer is $E_{cr} = 200$ MV/m. Find (a) the breakdown voltage of the cable and (b) the electric energy per unit length of the cable at breakdown.

3 Steady Electric Currents

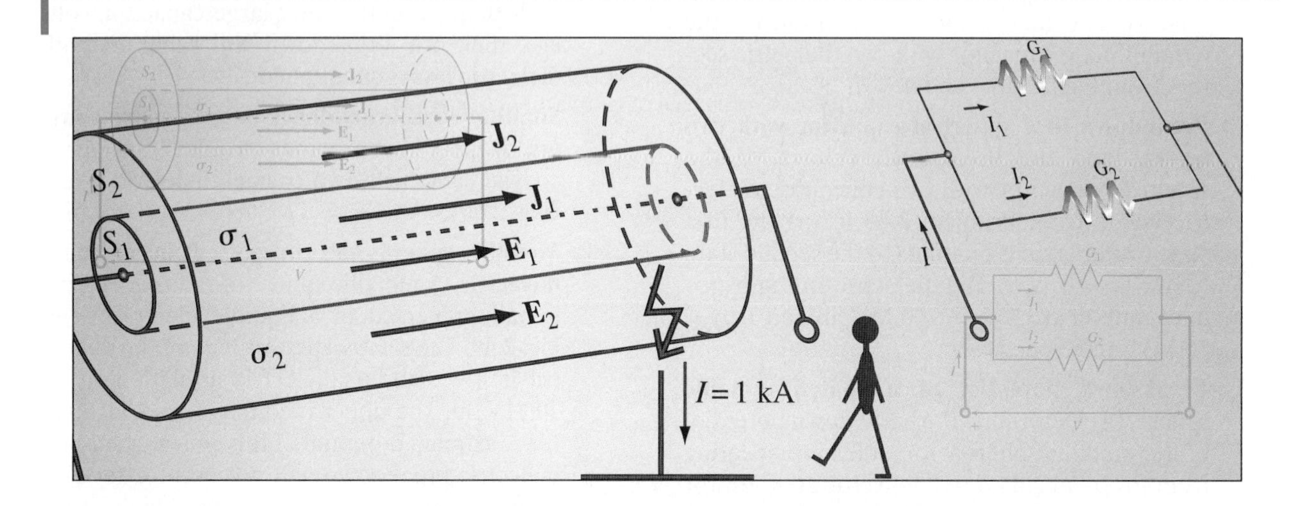

Introduction:

So far, we have dealt with electrostatic fields, associated with time-invariant charges at rest. We now consider the charges in an organized macroscopic motion, which constitute an electric current. Our focus in this chapter is on the steady flow of free charges in conducting materials, i.e., on steady (time-invariant) electric currents, whose macroscopic characteristics (like the amount of current through a wire conductor) do not vary with time. Steady currents are also called direct currents, abbreviated dc. The subject of steady electric currents links field theory to several important concepts of circuit theory, such as Ohm's law, Kirchhoff's current and voltage laws, Joule's law, resistance, conductance, voltage and current generators, and power balance in a circuit. Discussions of steady electric currents will bring us to the field and circuit analysis of transmission lines with losses in a time-invariant (dc) regime, and help us further develop and understand the concept of a transmission line as a circuit with distributed parameters.

We shall derive and discuss integral and differential field equations for steady electric currents and their electric field, along with the corresponding boundary conditions. We shall also study the mechanism of conduction for various materials, introduce models of energy sources, and derive expressions for power and energy calculations. These equations and concepts will enable us to develop and demonstrate techniques for analysis of several general configurations with steady currents, such as resistors of various composition and shapes, capacitors with imperfect inhomogeneous dielectrics, transmission lines with imperfect conductors and imperfect inhomogeneous dielectrics, and grounding electrodes buried in the earth.

3.1 CURRENT DENSITY VECTOR AND CURRENT INTENSITY

In the absence of an externally applied electric field, free charges in a conductor are in a state of random (chaotic) motion due to their thermal energy. This is so-called thermal motion of charges. The corresponding velocity is the thermal velocity, denoted as $\mathbf{v}_t$. Generally, v_t is rather large. In metallic conductors, free electrons move at thermal velocities on the order of $v_t \sim 10^5$ m/s at room temperature, between collisions with the atoms and with one another.[1] Because of the entirely random nature of thermal motion of charges (without any external electric field applied), there is no net macroscopic motion in any given direction, i.e., the macroscopic average vectorial resultant of thermal velocities of individual charges at any point in the conductor is zero. For an electric current, defined as a macroscopic net flow of free charges, to exist, there must be a nonzero macroscopic average velocity of microscopic velocities of charges in some direction in a conductor. This can be achieved, as we shall see, by establishing and maintaining an external (i.e., externally applied) electric field within a conductor.

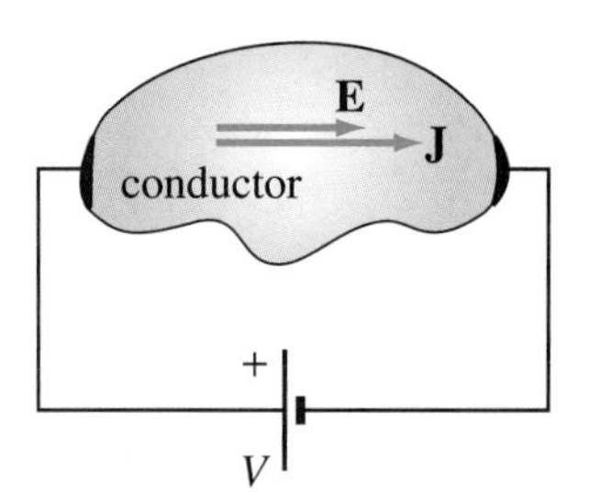

Figure 3.1 Conductor whose two ends are maintained at a potential difference.

Consider a conducting body whose two ends are connected to a generator of voltage V, as shown in Fig. 3.1. Because a potential difference is maintained between the conductor ends, there is an electric field of intensity $\mathbf{E}$ inside the conductor [the line integral of $\mathbf{E}$, Eq. (1.90), through the conductor is nonzero]. Note that this situation is essentially different from the situation in Fig. 1.38, where the transient redistribution of charge occurred and electrostatic equilibrium with zero field inside the conductor was established. Here, the conductor is not isolated but wired to a source of electromotive force (generator), providing a mechanism that forces the free charges to move and prevents them from piling up, which would tend to reduce the field in the conductor. Assume, for simplicity, that the free charge carriers in the conductor are electrons only (as in metallic conductors). The electric force on each charge is thus [Eq. (1.23)]

$$\mathbf{F}_e = -e\mathbf{E}, \tag{3.1}$$

where the charge amount e ($e > 0$) is given in Eq. (1.3). This force compels the electrons to move through the conductor, between its ends. However, since the electrons are not in free space, they cannot accelerate indefinitely under the influence of the electric field. Before they can acquire any appreciable speed, the electrons collide with the atomic lattice and acquire new random velocities $\mathbf{v}_t$. Since the field $\mathbf{E}$ essentially has to start accelerating the electrons all over again every $\Delta t_c \sim 10^{-14}$ s (typical average time interval between collisions at room temperature), it can change the random thermal velocities of the electrons only slightly, but in a systematic manner. This relatively slight systematic drift of the free electrons is the basis of electric conduction (current). After a brief initial transient, the electrons acquire a steady-state average velocity, determined by the balance between the accelerating force of the applied field, $\mathbf{F}_e$, and the scattering effect of the collisions with the lattice. This velocity is termed the drift velocity and symbolized by $\mathbf{v}_d$.

[1]The free electrons in a metallic conductor are so-called conduction-band electrons (or valence electrons), which are very loosely bound to their atoms and are essentially free to move about the crystalline structure of the metal. For example, each atom of copper has 29 electrons, 28 of which are bound electrons (tightly bound in their shells), while the outermost one is a free electron.

In general, $v_d \ll v_t$, since the electric field makes only a slight change in the velocity distribution that existed before the field was applied, and $\mathbf{v}_d$ is a macroscopic resultant of microscopic velocities of free charges in a direction along the electric field lines. In most cases, its magnitude is not larger than $v_d \sim 10^{-4}$ m/s in metals, for reasonable amounts of current carried (as we shall see in an example). The drift velocity is linearly proportional to the electric field intensity vector,

drift velocity

$$\boxed{\mathbf{v}_d = -\mu_e \mathbf{E},} \tag{3.2}$$

where the constant μ_e is the so-called mobility of electrons in the given material. The mobility is measured in the units of $m^2/(Vs)$ and is positive by definition. For electrons, the direction of $\mathbf{v}_d$, as well as the direction of $\mathbf{F}_e$, is opposite to the direction of $\mathbf{E}$. Good conductors have high mobility.

We can now say that charge carriers in the conductor move through its volume with the macroscopic average velocity $\mathbf{v}_d$. This is an organized, directive motion of charges, which constitute an electric current throughout the conductor volume. We introduce then a new field quantity, to describe the current at a point: the current density vector, $\mathbf{J}$. By definition,

current density vector (unit: A/m^2)

$$\boxed{\mathbf{J} = N_v(-e)\mathbf{v}_d,} \tag{3.3}$$

where N_v is the concentration of charge carriers, i.e., their number per unit volume or per 1 m^3 (the unit is m^{-3}).[2]

The current density can alternatively be defined by means of the current intensity, I, which, in turn, is defined as a rate of movement of charge passing through a surface (e.g., cross section of a cylindrical conductor). That is,

current intensity or, simply, current (unit: A)

$$\boxed{I = \frac{dQ}{dt}.} \tag{3.4}$$

In other words, I equals the total amount of charge that flows through the surface during an elementary time dt, divided by dt. The unit for current intensity, which is usually referred to as, simply, current, is ampere or amp (A), equal to C/s. The current density vector is directed along the macroscopic motion of charges, i.e., along the current lines. If we set an elementary surface of area dS perpendicular to the current lines, as in Fig. 3.2(a), the magnitude of the current density vector is given by

current density vs. intensity

$$\boxed{J = \frac{dI}{dS},} \tag{3.5}$$

where dI is the current flowing through dS. We see that the unit for $\mathbf{J}$ is A/m^2, which means that it actually represents a surface density of a volume current.

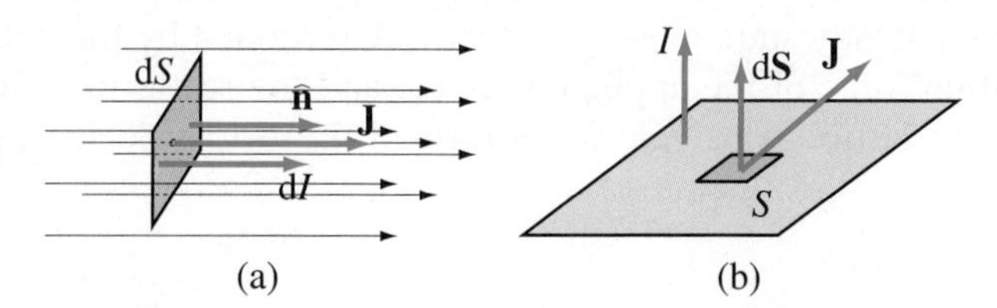

Figure 3.2 (a) Definition of the current density by means of the current intensity. (b) Evaluation of the total current through a surface.

[2]For example, the concentration of conduction electrons in copper is $N_v = 8.45 \times 10^{28}$ m^{-3}. It equals the number of copper atoms per unit volume, since copper has one conduction-band electron per atom. The number of atoms per unit volume is approximately the same for all solids.

To show that the definitions of current density, J, in Eqs. (3.3) and (3.5) are equivalent, we realize that the total amount of charge that crosses $\mathrm{d}S$ during the time interval $\mathrm{d}t$ is

$$\mathrm{d}Q = N_\mathrm{v} \underbrace{\mathrm{d}S v_\mathrm{d}\, \mathrm{d}t}_{\Delta v} (-e) \tag{3.6}$$

(in the time interval $\mathrm{d}t$, a charge moves a distance $v_\mathrm{d}\, \mathrm{d}t$, and all charges in the volume $\Delta v = \mathrm{d}S v_\mathrm{d}\, \mathrm{d}t$, the number of which is the concentration N_v times Δv, pass through $\mathrm{d}S$). Using the definition of J in Eq. (3.3), this becomes

$$\mathrm{d}Q = J\, \mathrm{d}S\, \mathrm{d}t. \tag{3.7}$$

Dividing by $\mathrm{d}t$, we obtain the current intensity $\mathrm{d}I$ through $\mathrm{d}S$ in terms of J:

$$\mathrm{d}I = J\, \mathrm{d}S, \tag{3.8}$$

which indeed is the same as in Eq. (3.5).

In the case where the current density vector is not perpendicular to the surface element, the current through the element is

$$\mathrm{d}I = J\, \mathrm{d}S_\mathrm{n} = J\, \mathrm{d}S \cos\alpha = \mathbf{J} \cdot \mathrm{d}\mathbf{S}, \tag{3.9}$$

where $\mathrm{d}S_\mathrm{n}$ is the projection of $\mathrm{d}S$ on the plane normal to $\mathbf{J}$ and α is the angle between $\mathbf{J}$ and $\mathrm{d}\mathbf{S}$ (see Fig. 1.30). Hence, the total current through an arbitrary surface S, Fig. 3.2(b), equals the flux of the current density vector through the surface,

$$\boxed{I = \int_S \mathbf{J} \cdot \mathrm{d}\mathbf{S}.} \tag{3.10}$$

total current through a surface

If there are several types of free charge carriers in a conductor drifting with different average velocities, the resultant current density vector is a vector sum of current densities in Eq. (3.3),

$$\mathbf{J} = \sum_{i=1}^{M} N_{\mathrm{v}i} q_i \mathbf{v}_{\mathrm{d}i}, \tag{3.11}$$

that correspond to individual types of carriers (e.g., electrons and holes in a semiconductor). Equivalently, the net charge flow is taken in evaluating the resultant current intensity in Eq. (3.4). Positive charges move in the direction of $\mathbf{E}$, negative charges in the opposite direction, but both add to the total current. Eqs. (3.4), (3.5), and (3.10) are therefore valid for any conductor and any combination of charge carriers.

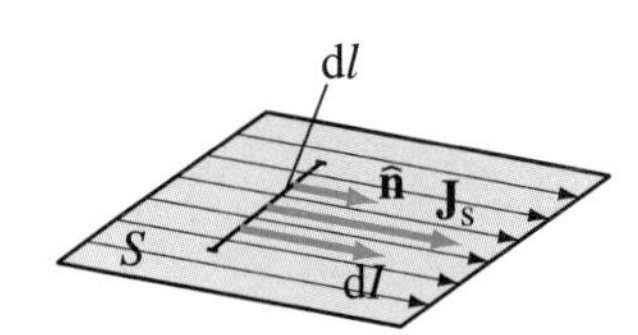

Figure 3.3 Surface current density vector ($\mathbf{J}_\mathrm{s}$).

In many situations, current flow is localized in a very thin (theoretically infinitely thin) film over a surface, as shown in Fig. 3.3. This is so-called surface current, described by the surface current density vector, $\mathbf{J}_\mathrm{s}$, which is defined as

$$\boxed{\mathbf{J}_\mathrm{s} = N_\mathrm{s} q \mathbf{v}_\mathrm{d},} \tag{3.12}$$

surface current density vector (unit: A/m)

where N_s is the surface concentration of charge carriers (number of carriers per unit surface area, in m^{-2}). Note that the surface current density vector is sometimes denoted as $\mathbf{K}$. In terms of the current intensity, the surface current density is given by

$$\boxed{J_\mathrm{s} = \frac{\mathrm{d}I}{\mathrm{d}l},} \tag{3.13}$$

surface current density vs. current intensity

where $\mathrm{d}I$ is the current flowing across a line element $\mathrm{d}l$ set normal to the current flow (Fig. 3.3). The unit for $\mathbf{J}_s$, which represents a line density of a surface current, is A/m. For example, the surface current density of a very thin aluminum strip (foil) with a current I that is uniformly distributed across the strip width, w, equals $J_s = I/w$, and $\mathbf{J}_s$ is directed parallel to the strip axis.

Example 3.1 **Electron Drift along a 1-km Copper Wire**

A copper wire of length $l = 1$ km and radius $a = 3$ mm carries a steady current of intensity $I = 10$ A. The current is uniformly distributed across the wire cross section. Find the time in which the electrons drift along the wire.

Solution As the current is distributed uniformly across the cross section (S) of the wire, the current density in the wire is [Eq. (3.10)]

$$J = \frac{I}{S} = \frac{I}{\pi a^2} = 3.54 \times 10^5 \ \mathrm{A/m^2}. \tag{3.14}$$

From Eq. (3.3), the drift velocity of electrons is

$$v_\mathrm{d} = \frac{J}{N_\mathrm{v} e} = 2.62 \times 10^{-5} \ \mathrm{m/s}, \tag{3.15}$$

where $N_\mathrm{v} = 8.45 \times 10^{28}\ \mathrm{m}^{-3}$ is the concentration of conduction electrons in copper and e is the absolute value of the charge of an electron, Eq. (1.3). The time it takes for an electron to drift along the wire is hence

$$t = \frac{l}{v_\mathrm{d}} = 3.82 \times 10^7 \ \mathrm{s}, \tag{3.16}$$

which is approximately 442 days.[3]

3.2 CONDUCTIVITY AND OHM'S LAW IN LOCAL FORM

Consider again a metallic conductor (the charge is carried by electrons), with the current density $\mathbf{J}$. Substituting Eq. (3.2) into Eq. (3.3), we obtain

current density in a metallic conductor

$$\boxed{\mathbf{J} = N_\mathrm{v} e \mu_\mathrm{e} \mathbf{E}.} \tag{3.17}$$

This equation can be rewritten as

Ohm's law in local form

$$\boxed{\mathbf{J} = \sigma \mathbf{E},} \tag{3.18}$$

where the proportionality constant,

conductivity of metallic conductors (unit: S/m)

$$\boxed{\sigma = N_\mathrm{v} e \mu_\mathrm{e},} \tag{3.19}$$

is a macroscopic parameter of the medium called conductivity. It is always positive, and represents, in general, a measure of the ability of materials to conduct electric current. The unit for conductivity is siemens per meter (S/m). The reciprocal of σ is denoted by the symbol ρ and termed resistivity. The unit is ohm × meter (Ωm).

[3]We know that, of course, we do not need to wait 442 days to receive a communication signal sent via a 1-km long transmission line. We shall see in a later chapter that time-varying signals traveling along transmission lines propagate as electromagnetic waves outside the conductors that constitute the line, and not via the drifting motion of electrons within the conductors. The conductors actually serve as guides for the waves along the line. That is why signals travel at the velocity of electromagnetic waves in the medium surrounding the line conductors. If the medium is air, the velocity is 3×10^8 m/s (speed of light in a vacuum), and the travel time is only 3.33 μs for a 1-km line.

Using resistivity, Eq. (3.18) becomes

$$\boxed{\mathbf{E} = \rho \mathbf{J}, \quad \rho = \frac{1}{\sigma}.} \tag{3.20}$$

ρ – resistivity (unit: Ωm)

Both Eqs. (3.18) and (3.20) are known as Ohm's law in local or point form.

Note that Eq. (3.18) is one of the three general electromagnetic constitutive equations for characterization of materials [another one being Eq. (2.46)]. It can be written in the following form:

$$\boxed{\mathbf{J} = \mathbf{J}(\mathbf{E}),} \tag{3.21}$$

constitutive equation for $\mathbf{J}$, for an arbitrary material

to encompass all possible conducting properties of materials. However, in terms of their conductivity, most conducting materials are linear and isotropic, i.e., $\mathbf{J}(\mathbf{E}) = \sigma \mathbf{E}$, where σ is independent of electric field intensity and current density (the property of linearity), and is the same for all directions (isotropy). In homogeneous conductors, σ does not change from point to point in the region being considered. For inhomogeneous conductors, on the other hand, σ is a function of spatial coordinates [e.g., $\sigma = \sigma(x, y, z)$ in the region].

Almost always, the conductivity is a function of temperature, T. One of the few exceptions is an alloy called constantan (55% copper, 45% nickel), whose conductivity is practically constant in a temperature range 0–100°C. For metallic conductors, the mobility of electrons decreases with an increase in temperature (because the average time interval between collisions with vibrating atoms, Δt_c, decreases). Hence, the conductivity decreases and resistivity increases with a temperature rise. Around a room temperature of $T_0 = 293$ K (20°C), the resistivity varies almost linearly with T, and we can write

$$\rho(T) = \rho_0 \left[1 + \alpha\,(T - T_0)\right], \tag{3.22}$$

where $\rho_0 = \rho(T_0)$. For most metals (copper, aluminum, silver, etc.), the temperature coefficient of resistivity, α, is approximately 0.4% per kelvin.

For some materials the resistivity drops abruptly to zero below a certain temperature:

$$\boxed{\rho(T) = 0 \quad \text{for} \quad T < T_{\text{cr}},} \tag{3.23}$$

superconductors

where T_{cr} is called the critical temperature of the material. This property, discovered by Kamerlingh Onnes in 1911, is called superconductivity, and the materials are said to behave like superconductors. Most superconductors are metallic elements that exhibit transition into superconducting states at a temperature of a few kelvin. Examples are aluminum ($T_{\text{cr}} = 1.2$ K), lead ($T_{\text{cr}} = 7.2$ K), and niobium ($T_{\text{cr}} = 9.2$ K), as well as their alloys and compounds. More recently, new ceramic materials were discovered that become superconducting at considerably higher (and thus less expensive to produce and maintain) temperatures. For example, yttrium-barium-copper oxide ($YBa_2Cu_3O_7$), discovered in 1986, has $T_{\text{cr}} = 80$ K, so its superconductivity can be utilized by cooling with liquid nitrogen. Interestingly, some of the best of the normal conductors, such as silver and copper, cannot become superconducting at any temperature, while the ceramic superconductors are normally good insulators – when they are not at low enough temperatures to be in a superconducting state.

Ohm's law in local form holds also for conductors with more than one type of charge carriers [see Eq. (3.11)]. In plasmas and gases, the charge carriers are electrons and positive ions (electron-deficient atoms or molecules). In liquid conductors, called electrolytes, the charge is carried by positive and negative ions. In all cases, both positively and negatively charged particles (ions and electrons) contribute to the conductivity.

In semiconductors (e.g., silicon and germanium), vacancies in the atomic crystal lattice left by electrons, called holes, can move from atom to atom and behave like positive charge carriers, each hole carrying charge e. The conductivity of a semiconductor is therefore

conductivity of semiconductors

$$\boxed{\sigma = N_{\rm ve} e\mu_{\rm e} + N_{\rm vh} e\mu_{\rm h},} \tag{3.24}$$

where the first term represents the contribution to the conductivity from electrons, moving opposite to the field **E**, while the second term represents the contribution from holes, which move with **E**. The concentrations $N_{\rm ve}$ and $N_{\rm vh}$ rapidly increase with an increase in temperature (temperature rise accelerates generation of free electrons and holes). Consequently, the conductivity of semiconductors increases with increasing the temperature, which is opposite to the temperature behavior of metallic conductors.

By adding very small amounts of impurities to pure (intrinsic) semiconductors, the conductivity may be increased dramatically. Impurities called donors (e.g., phosphorus) provide additional electrons and form n-type semiconductors, while acceptors (e.g., boron) introduce extra holes, forming p-type semiconductors. This procedure is known as doping of semiconductors. Note that the boundary between p-type and n-type parts of a single semiconductor crystal forms a junction region, called *pn* junction (see Fig. 2.9), which is utilized in semiconductor devices (diodes and transistors).[4]

Unlike the relative permittivity ($\varepsilon_{\rm r}$), shown in Table 2.1, the conductivity of materials varies over an extremely wide range of values, as we go from the best insulators to semiconductors, to the finest conductors. In S/m, σ (at room temperature) ranges from around 10^{-17} for fused quartz, 10^{-9} for bakelite, 10^{-2} for fresh water, and 2.2 for germanium to 6.17×10^7 for silver. We see that the range in conductivity from quartz to silver is as large as 25 orders of magnitude (10^{25}), and then it goes to infinity for superconductors. Table 3.1 lists values of the conductivity of selected materials.

Copper (Cu), the most commonly used metallic conductor, has a conductivity of

conductivity of copper, at 20°C

$$\boxed{\sigma_{\rm Cu} = 58 \text{ MS/m}.} \tag{3.25}$$

In many applications, we consider copper and other metallic conductors as perfect electric conductors (PEC), with

perfect electric conductor (PEC)

$$\boxed{\sigma \to \infty.} \tag{3.26}$$

Of course, superconductors also fall under this category. From Eqs. (3.18) and (3.26), we conclude that

no electric field inside a PEC body

$$\boxed{\mathbf{E} = \frac{\mathbf{J}}{\sigma} = 0,} \tag{3.27}$$

[4]Generally, the current density in semiconductor devices is composed of two components: a drift current density, $\mathbf{J} = \sigma\mathbf{E}$, and a diffusion current density, $\mathbf{J}_{\rm dif}$, which depends on the gradient of the concentration of charge carriers in a material and, therefore, does not satisfy Ohm's law in local form. Consequently, the relationship between the total current density vector and the electric field intensity vector, Eq. (3.21), in semiconductor devices is, in general, nonlinear.

Table 3.1. Conductivity of selected materials*

Material	σ (S/m)	Material	σ (S/m)
Quartz (fused)	$\sim 10^{-17}$	Carbon (graphite)	7.14×10^4
Wax	$\sim 10^{-17}$	Bismuth	8.70×10^5
Polystyrene	$\sim 10^{-16}$	Cast iron	$\sim 10^6$
Sulfur	$\sim 10^{-15}$	Nichrome	10^6
Mica	$\sim 10^{-15}$	Mercury (liquid)	1.04×10^6
Paraffin	$\sim 10^{-15}$	Stainless steel	1.1×10^6
Rubber (hard)	$\sim 10^{-15}$	Silicon steel	2×10^6
Porcelain	$\sim 10^{-14}$	Titanium	2.09×10^6
Carbon (diamond)	2×10^{-13}	Constantan (45% Ni)	2.26×10^6
Glass	$\sim 10^{-12}$	German silver	3×10^6
Polyethylene	1.5×10^{-12}	Lead	4.56×10^6
Wood	$10^{-11} - 10^{-8}$	Solder	7×10^6
Bakelite	$\sim 10^{-9}$	Niobium	8.06×10^6
Marble	10^{-8}	Tin	8.7×10^6
Granite	10^{-6}	Platinum	9.52×10^6
Dry soil	10^{-4}	Bronze	10^7
Distilled water	2×10^{-4}	Iron	1.03×10^7
Silicon (intrinsic)	4.4×10^{-4}	Nickel	1.45×10^7
Clay	5×10^{-3}	Brass (30% Zn)	1.5×10^7
Fresh water	10^{-2}	Zinc	1.67×10^7
Wet soil	$\sim 10^{-2}$	Tungsten	1.83×10^7
Animal fat**	4×10^{-2}	Sodium	2.17×10^7
Animal muscle ($\perp$ to fiber)**	8×10^{-2}	Magnesium	2.24×10^7
Animal, body (average)**	0.22	Duralumin	3×10^7
Animal muscle ($\parallel$ to fiber)**	0.4	Aluminum	3.5×10^7
Animal blood**	0.7	Gold	4.1×10^7
Germanium (intrinsic)	2.2	Copper	5.8×10^7
Seawater	3–5	Silver	6.17×10^7
Ferrite	10^2	Mercury (at <4.1 K)	∞
Tellurium	$\sim 5 \times 10^2$	Niobium (at <9.2 K)	∞
Silicon (doped)	1.18×10^3	$YBa_2Cu_3O_7$ (at <80 K)	∞

* For dc or low-frequency currents, at room temperature.
** Also for humans.

i.e., the electric field is always zero in perfect conductors. This, in turn, implies that the voltage between any two points of a perfect conductor [Eq. (1.90)] is zero.

Finally, so-called convection currents, which are the result of the motion of positively or negatively charged particles in a vacuum or rarefied gas (where $\sigma = 0$), are not governed by Ohm's law. Examples are electron beams in a cathode ray tube and a violent motion of charged particles in a thunderstorm. The convection current density is given by

$$\boxed{\mathbf{J} = \rho \mathbf{v},} \tag{3.28}$$

convection current density

where $\mathbf{v}$ is the velocity of particles and ρ is the volume density of charges (charge per unit volume) in the vacuum or rarefied gas. Noting that $\rho = N_v q$, N_v being the concentration of particles and q an elementary charge, we observe the equivalency of the definitions of convection and conduction current densities given by Eqs. (3.28) and (3.3), respectively. However, the velocity $\mathbf{v}$ in Eq. (3.28) is not a drift velocity ($\mathbf{v}_d$) of charges and Eq. (3.2) is not satisfied.

Conceptual Questions (on Companion Website): 3.1 and 3.2.

HISTORICAL ASIDE

Heike Kamerlingh Onnes (1853–1926), Dutch physicist and professor at the University of Leyden, was awarded a Nobel Prize in Physics in 1913 for his investigations of the properties of matter at extremely low temperatures. His experiments led him as close to absolute zero as 0.9 K, which was a fascinating result at that time. He was the first to produce liquid helium (in 1908). Onnes demonstrated in 1911 that the resistivity of mercury absolutely disappears at temperatures below about 4 K, and thus discovered superconductivity.

The SI unit of power, the watt, was named in honor of **James Watt** (1736–1819), a Scottish mechanical engineer and inventor, who is famous for his revolutionary improvements of the steam engine in the 1760s, which led to great advancements in the Industrial Revolution, and is also known for devising the "horsepower" – to measure the power of his steam engines.

3.3 LOSSES IN CONDUCTORS AND JOULE'S LAW IN LOCAL FORM

Let us now consider the current flow in a conductor from the energy point of view. As we know, free charge carriers (e.g., electrons) are accelerated on their paths between collisions with vibrating atoms, and at every collision they lose their acquired kinetic energy. Energy is thus transmitted from the electric field, $\mathbf{E}$, via charge carriers to the atoms, enhancing their thermal vibration and ultimately resulting in a higher temperature of the conductor. This means that in a conductor with electric current, electric energy is constantly converted into heat. We wish to derive the expression for the rate (power) of this energy transformation.

We start with the electric force on a charge dQ given in Eq. (3.7), which equals $dQ\,\mathbf{E}$. The work done by this force in moving dQ a distance dl along the field lines is [Eq. (1.72)]

$$dW_e = dQ\,E\,dl = J\,dS\,dt E\,dl = JE\,dv\,dt \quad (dv = dS\,dl), \tag{3.29}$$

where dv is an elementary volume in the conductor. This work is converted (lost) to heat, known as Joule's heat. The rate of this conversion, dW_e/dt (J/s), is power, called the power of Joule's losses or ohmic losses. Thus, the power of Joule's losses in the volume dv is

$$dP_J = \frac{dW_e}{dt} = JE\,dv. \tag{3.30}$$

The volume density of this power is

Joule's law in local form; p_J – ohmic power density (unit: W/m^3)

$$\boxed{p_J = \frac{dP_J}{dv} = JE = \frac{J^2}{\sigma} = \sigma E^2,} \tag{3.31}$$

and this is known as Joule's law in local (point) form. The unit for power is watt (W), and hence the unit for power density is W/m^3.

The total power of Joule's losses (the electric power that is lost to heat) in a domain of volume v (e.g., in the entire conducting body) is obtained by integration of power $\mathrm{d}P_\mathrm{J}$ throughout v:

$$P_\mathrm{J} = \int_v \underbrace{p_\mathrm{J}\,\mathrm{d}v}_{\mathrm{d}P_\mathrm{J}} = \int_v JE\,\mathrm{d}v. \tag{3.32}$$

power of Joule's (ohmic) losses (unit: W)

Conceptual Questions (on Companion Website): 3.3.

3.4 CONTINUITY EQUATION

We now consider one of the fundamental principles of electromagnetics – the continuity equation, which is the mathematical expression of the principle of conservation of charge. Charge is indestructible, and cannot be lost or created. It can move from place to place, but can never appear from nowhere nor disappear. Let an arbitrary closed surface, S, enclosing volume v, be situated in a region with time-varying currents, as shown in Fig. 3.4(a), and let Q_S and $Q_S + \mathrm{d}Q_S$ denote the net charges in v at instants of time t and $t + \mathrm{d}t$, respectively. The change in charge, $\mathrm{d}Q_S$, cannot be created in v, but only brought in from the domain outside the surface S, and it is brought by the current flowing through S. So, the charge $\mathrm{d}Q_S$ passes the surface S during the time $\mathrm{d}t$, and this is exactly what we have in the definition of current intensity in Eq. (3.4). Therefore,

$$I_\mathrm{in} = \frac{\mathrm{d}Q_S}{\mathrm{d}t} \tag{3.33}$$

is the intensity of the current flowing across the surface S into the region v. The current crossing the surface in the opposite direction, i.e., the current leaving the region, is hence

$$I_\mathrm{out} = -I_\mathrm{in} = -\frac{\mathrm{d}Q_S}{\mathrm{d}t}. \tag{3.34}$$

On the other hand, by virtue of Eq. (3.10), the current leaving v across S equals the total outward flux of the current density vector through S (which is a closed surface), that is,

$$I_\mathrm{out} = \oint_S \mathbf{J}\cdot\mathrm{d}\mathbf{S}. \tag{3.35}$$

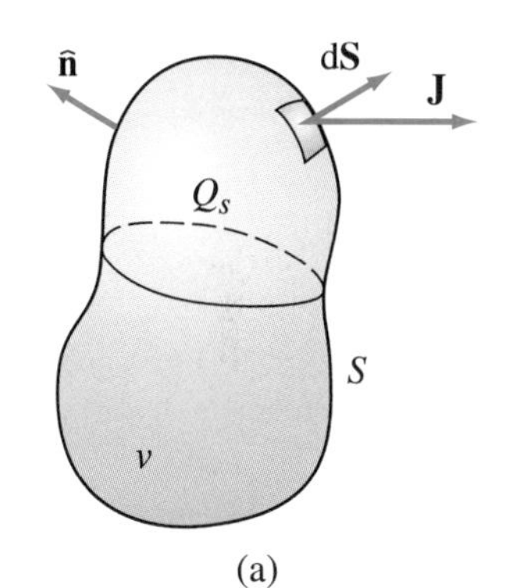

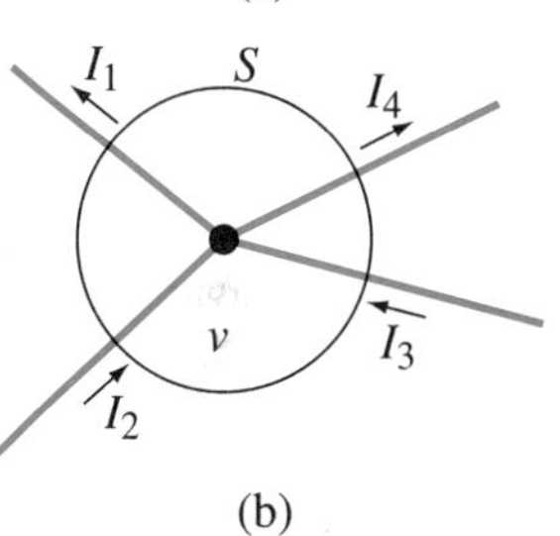

Figure 3.4 Arbitrary closed surface in a region with currents: (a) general case and (b) current flow through wires meeting at a node.

Combining the two preceding equations, we obtain

$$\oint_S \mathbf{J}\cdot\mathrm{d}\mathbf{S} = -\frac{\mathrm{d}Q_S}{\mathrm{d}t}. \tag{3.36}$$

continuity equation, for currents of any time dependence

This is the continuity equation (in integral form). It tells us that the outward flux of the current density vector through any closed surface is equal to the negative of the derivative in time of the total charge enclosed by that surface.

By expressing the charge in terms of the volume charge density, ρ, the continuity equation becomes

$$\oint_S \mathbf{J}\cdot\mathrm{d}\mathbf{S} = -\frac{\mathrm{d}}{\mathrm{d}t}\int_v \rho\,\mathrm{d}v. \tag{3.37}$$

If the surface S does not change in time, the time derivative can be moved inside the volume integral, yielding

continuity equation in terms of the volume charge density

$$\oint_S \mathbf{J} \cdot d\mathbf{S} = -\int_v \frac{\partial \rho}{\partial t}\, dv. \tag{3.38}$$

The ordinary derivative is replaced by a partial derivative because ρ, generally, is a multivariable function of time and space coordinates.

By applying the divergence theorem, Eq. (1.173), to Eq. (3.38) or simply by analogy to the differential form of the generalized Gauss' law, Eq. (2.45), we get the differential form of the continuity equation:

continuity equation in differential form

$$\nabla \cdot \mathbf{J} = -\frac{\partial \rho}{\partial t}. \tag{3.39}$$

It tells us that the divergence of $\mathbf{J}$ at a given point equals the negative of the time rate of variation in charge density at that point, and is also called the continuity equation at a point.

For steady (time-invariant) currents, the charge density is constant in time, $\partial\rho/\partial t = 0$, and the integral form of the continuity equation reduces to

continuity equation for steady currents, integral form

$$\oint_S \mathbf{J} \cdot d\mathbf{S} = 0. \tag{3.40}$$

The differential equation of continuity of steady currents is given by

differential continuity equation for steady currents

$$\nabla \cdot \mathbf{J} = 0. \tag{3.41}$$

Thus, time-invariant current density vector has zero divergence everywhere, and under all circumstances. We say that steady electric currents are divergenceless or solenoidal. The zero divergence of a vector field indicates that there are no sources or sinks in the field for the lines of flux to originate from or terminate on. This means that the streamlines of steady currents close upon themselves (steady currents must flow in closed loops), unlike the streamlines of the electrostatic field intensity, which originate and end on charges.

If the steady current is carried into and out of the volume v by a number (N) of wire conductors meeting at a point, as indicated in Fig. 3.4(b), then Eq. (3.40) implies that the algebraic sum of all the currents leaving the junction is zero,

Kirchhoff's current law

$$\sum_{k=1}^{N} I_k = 0. \tag{3.42}$$

For the situation and notation in Fig. 3.4(b), $I_1 - I_2 - I_3 + I_4 = 0$. Eq. (3.42), if applied to a node in a dc circuit, represents Kirchhoff's circuital law for currents. Like Kirchhoff's voltage law, Eq. (1.92), Kirchhoff's current law in the above form also applies to time-varying situations and ac circuits, with certain restrictions and assumptions, which will be discussed later in this text.

In studying steady current fields, we always have in mind that time-invariant currents in a conductor are produced by a static electric field, which is a conservative field, meaning that the line integral (circulation) of the electric field intensity vector, $\mathbf{E}$, along an arbitrary contour (closed path) is zero,

conservative nature of $\mathbf{E}$ *in steady current field*

$$\oint_C \mathbf{E} \cdot d\mathbf{l} = 0. \tag{3.43}$$

We also have in mind that **J** and **E** are related at any point in the conductor by the constitutive equation for the current density, Eq. (3.21) [or Eq. (3.18) for linear media].

Example 3.2 Element Law for a Capacitor

Prove that the current through a capacitor equals the product of the capacitor capacitance and the time rate of change of the voltage drop across the capacitor.

Solution Fig. 3.5 shows a capacitor of capacitance C connected to a time-varying voltage[5] v. We assume that the capacitor is ideal, i.e., its dielectric is perfect (nonconducting), as well as that there is no excess charge along the connecting conductors in the circuit (charge is localized only on the capacitor electrodes). To relate the current through the capacitor, i, to the capacitor charge, Q, we apply the continuity equation for time-varying currents in integral form, Eq. (3.36), to a surface S enclosing completely only the upper electrode of the capacitor (Fig. 3.5). The total current leaving the enclosed domain (left-hand side of the equation) equals $-i$, whereas the total enclosed charge (appearing on the right-hand side of the equation) equals the charge of the upper electrode, i.e., the capacitor charge: $Q_S = Q$. Thus, Eq. (3.36) becomes

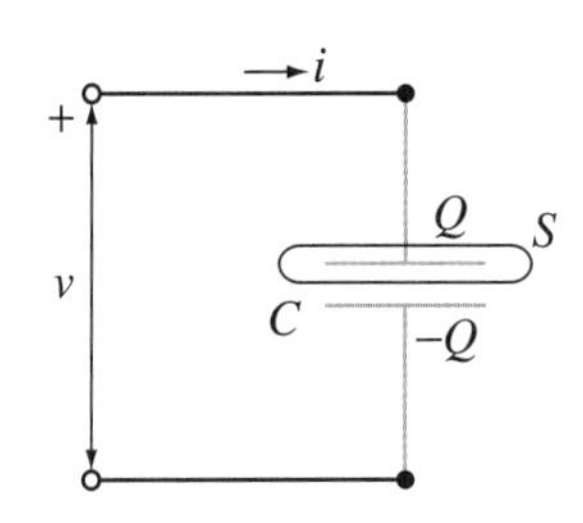

Figure 3.5 Capacitor with a time-varying current; for Example 3.2.

$$-i = -\frac{dQ}{dt}. \tag{3.44}$$

Substituting $Q = Cv$ [Eq. (2.112)] yields

$$\boxed{i = C\frac{dv}{dt}.} \tag{3.45}$$

element law (current-voltage characteristic) for a capacitor with a time-varying current

We see that i is linearly proportional to the rate of change of v in time, with C as the proportionality constant. This is the element law (current-voltage characteristic) for a capacitor, which is widely used in circuit theory, in conjunction with Kirchhoff's laws and other element laws.

Example 3.3 Spherical Capacitor with an Imperfect Dielectric

A spherical capacitor is filled with an imperfect homogeneous dielectric of conductivity σ. The radius of the inner electrode is a and the inner radius of the outer electrode is b ($b > a$). The electrodes have a conductivity that is much larger than σ, so that they can be considered as perfect conductors. The capacitor is connected to a generator of time-invariant voltage V. Find (a) the current density vector in the dielectric and (b) the power of Joule's losses in the capacitor.

Solution

(a) Since $\sigma \neq 0$, there exists a steady current of intensity I in the capacitor circuit, as indicated in Fig. 3.6. To determine the current density vector in the dielectric, we apply the continuity equation for steady currents in integral form, Eq. (3.40). In general, application of the continuity equation is analogous to application of the generalized Gauss' law, Eq. 2.43. This problem is one with spherical symmetry (see Example 1.18), and the current density vector is of the form

$$\boxed{\mathbf{J} = J(r)\,\hat{\mathbf{r}},} \tag{3.46}$$

volume current distribution with spherical symmetry

where r is the radial coordinate in the spherical coordinate system and $\hat{\mathbf{r}}$ is the radial unit vector (Fig. 3.6). The surface S for applying the continuity equation is a spherical surface

[5] We use lowercase (small letter) symbols for the voltage and current here to emphasize that those are time-varying quantities.

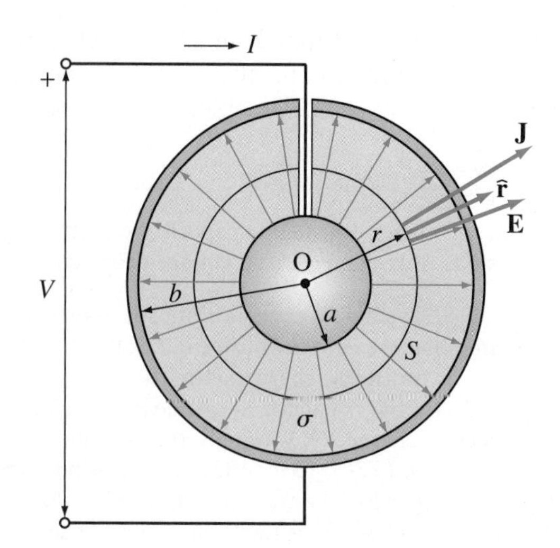

Figure 3.6 Spherical capacitor with an imperfect homogeneous dielectric and time-invariant current; for Example 3.3.

of radius r ($a < r < b$) centered at the origin (the same as the corresponding Gaussian surface in electrostatics, e.g., that in Fig. 2.16). The total current flowing through S in the outward direction is zero, that is,

$$J(r)\,4\pi r^2 - I = 0. \tag{3.47}$$

This means that the outward flux of $\mathbf{J}$ through S equals the current I through the capacitor terminals. Hence,

$$J(r) = \frac{I}{4\pi r^2} \quad (a < r < b). \tag{3.48}$$

From Eq. 3.18, the electric field intensity in the dielectric is

$$E(r) = \frac{J(r)}{\sigma} = \frac{I}{4\pi\sigma r^2}. \tag{3.49}$$

The voltage between the electrodes is given by

$$V = V_a - V_b = \int_{r=a}^{b} E\,\mathrm{d}r = \frac{I}{4\pi\sigma}\left(\frac{1}{a} - \frac{1}{b}\right), \tag{3.50}$$

where V_a and V_b are the potentials of the electrodes (each electrode is equipotential because $\sigma_{\text{electrodes}} \gg \sigma$). Combining Eqs. (3.48) and (3.50), we get

$$J(r) = \frac{\sigma abV}{(b-a)r^2}. \tag{3.51}$$

(b) Using Eq. (3.32), the power of Joule's losses in the dielectric is

$$P_\mathrm{J} = \int_v p_\mathrm{J}\,\mathrm{d}v = \int_{r=a}^{b} \frac{J(r)^2}{\sigma} \underbrace{4\pi r^2\,\mathrm{d}r}_{\mathrm{d}v} = \frac{4\pi\sigma a^2 b^2 V^2}{(b-a)^2} \int_a^b \frac{\mathrm{d}r}{r^2} = \frac{4\pi\sigma abV^2}{b-a}, \tag{3.52}$$

with $\mathrm{d}v$ adopted in the form of a thin spherical shell of radius r and thickness $\mathrm{d}r$ [Eq. (1.33)]. There are no losses in the electrodes (perfect conductors), so this is the overall loss power in the capacitor.

Conceptual Questions (on Companion Website): 3.4–3.6; *MATLAB Exercises* (on Companion Website).

3.5 BOUNDARY CONDITIONS FOR STEADY CURRENTS

Applications of steady current fields involve considerations of interfaces between conducting media of different conductivity. In this section, we shall formulate the boundary conditions that govern the manner in which the current density vector, **J**, and the electric field intensity vector, **E**, behave across such interfaces.

Comparing the integral form of the continuity equation for time-varying currents, Eq. (3.38), to the integral form of the generalized Gauss' law, Eq. (2.44), we conclude that the boundary condition for normal components of the vector **J** is of the same form as the boundary condition for normal components of the vector **D**, Eq. (2.85). The only difference is on the right-hand side of the equation, where ρ_s (the surface charge density that may exist on the surface) is replaced by $-\partial\rho_s/\partial t$. With this,

$$\boxed{\hat{\mathbf{n}} \cdot \mathbf{J}_1 - \hat{\mathbf{n}} \cdot \mathbf{J}_2 = -\frac{\partial \rho_s}{\partial t},} \tag{3.53}$$

boundary condition for $\mathbf{J}_n$, time-varying currents

where $\hat{\mathbf{n}}$ is the normal unit vector on the surface, directed from region 2 to region 1.

For steady currents, $-\partial\rho_s/\partial t = 0$ in Eq. (3.53), and the boundary condition that corresponds to Eq. (3.43) is that in Eq. (2.84) [Eq. (3.43) is the same as for the electrostatic field], so the complete set of boundary conditions for steady currents is given by

$$\hat{\mathbf{n}} \times \mathbf{E}_1 - \hat{\mathbf{n}} \times \mathbf{E}_2 = 0 \quad \text{or} \quad E_{1t} = E_{2t}, \tag{3.54}$$

boundary condition for $\mathbf{E}_t$

$$\hat{\mathbf{n}} \cdot \mathbf{J}_1 - \hat{\mathbf{n}} \cdot \mathbf{J}_2 = 0 \quad \text{or} \quad J_{1n} = J_{2n}. \tag{3.55}$$

boundary condition for $\mathbf{J}_n$, dc regime

We see that in the steady current field, the tangential component of the electric field intensity vector and the normal component of the current density vector, $\mathbf{E}_t$ and $\mathbf{J}_n$, are both continuous across the boundary. As we shall see in a later chapter, the boundary condition for $\mathbf{E}_t$ in Eq. (3.54) has this same form for fields of any time variation.

By analogy to the procedure of deriving Eq. (2.87) for the electrostatic field, we obtain the law of refraction of the current density lines at a boundary between two linear conducting media of conductivities σ_1 and σ_2:

$$\boxed{\frac{\tan\alpha_1}{\tan\alpha_2} = \frac{\sigma_1}{\sigma_2},} \tag{3.56}$$

law of refraction of current streamlines

where α_1 and α_2 are the angles that current lines in region 1 and region 2 make with the normal to the interface, as shown in Fig. 3.7.

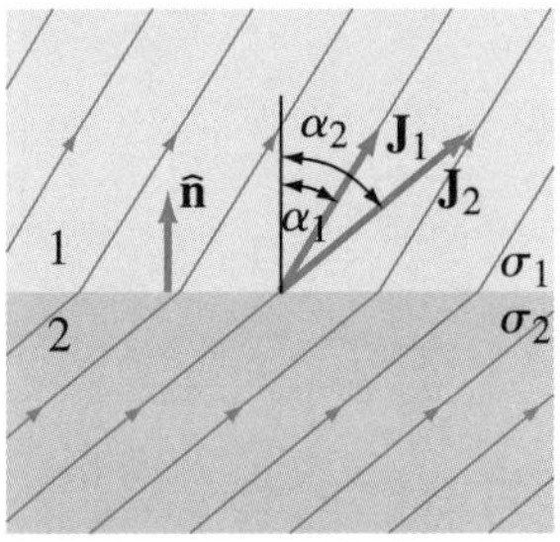

Note that if medium 2 is a good conductor and medium 1 is a low-loss dielectric, then $\sigma_2 \gg \sigma_1$, i.e., $\sigma_1/\sigma_2 \approx 0$, and Eq. (3.56) gives $\tan\alpha_1 \approx 0$ for any α_2. Therefore,

$$\alpha_1 \approx 0 \quad (\sigma_2 \gg \sigma_1), \tag{3.57}$$

meaning that the current lines always leave (or enter) a good conductor at a right angle to the boundary (zero angle to the normal on the boundary).

Note also that if medium 1 is a perfect dielectric (e.g., air), then $\sigma_1 = 0$, and $\tan\alpha_2 \to \infty$, from which

$$\alpha_2 = 90^\circ \quad (\sigma_1 = 0). \tag{3.58}$$

This means that the lines of current flow are always parallel to the surface of a conductor surrounded by a nonconducting medium.

Figure 3.7 Refraction of steady current lines at a conductor-conductor interface.

Conceptual Questions (on Companion Website): 3.7 and 3.8.

3.6 DISTRIBUTION OF CHARGE IN A STEADY CURRENT FIELD

The electric field intensity vector, **E**, in a steady current field is produced by stationary excess charges in the system. In the general case, these charges exist not only on the surfaces of conductors, but also inside their volume. Distribution of charges in the system is conditioned by the distribution of currents, and can be determined only after the current distribution is determined first.

The distribution of the current density vector, **J**, inside conductors can be obtained by solving the basic equations that govern steady current fields, which we summarize here:

Maxwell's first equation for static fields
dc continuity equation
constitutive equation for **J**

$$\begin{cases} \oint_C \mathbf{E} \cdot \mathrm{d}\mathbf{l} = 0 \\ \oint_S \mathbf{J} \cdot \mathrm{d}\mathbf{S} = 0 \\ \mathbf{J} = \mathbf{J}(\mathbf{E}) \ [\mathbf{J} = \sigma \mathbf{E}] \end{cases} . \tag{3.59}$$

Once the solution for **J** is known, the charge distribution can be obtained from the generalized Gauss' law in conjunction with the constitutive equation for the electric flux density vector, **D**:

Maxwell's third equation
constitutive equation for **D**

$$\begin{cases} \oint_S \mathbf{D} \cdot \mathrm{d}\mathbf{S} = Q_S \\ \mathbf{D} = \mathbf{D}(\mathbf{E}) \ [\mathbf{D} = \varepsilon \mathbf{E}] \end{cases} . \tag{3.60}$$

Note that, assuming that the medium is linear, both **D** and **J** are linearly proportional to **E**. As a result, we have a linear relationship between **D** and **J**:

duality of **D** *and* **J**

$$\mathbf{D} = \varepsilon \mathbf{E} = \frac{\varepsilon}{\sigma} \mathbf{J}. \tag{3.61}$$

This duality relationship will be used on many occasions.

Starting from the differential form of the generalized Gauss' law, Eq. (2.45), and using Eqs. (3.61) and (3.41), the volume charge density in the conductor is[6]

volume charge in a steady current field

$$\rho = \nabla \cdot \mathbf{D} = \nabla \cdot \left(\frac{\varepsilon}{\sigma} \mathbf{J} \right) = \left[\nabla \left(\frac{\varepsilon}{\sigma} \right) \right] \cdot \mathbf{J} + \frac{\varepsilon}{\sigma} (\nabla \cdot \mathbf{J}) = \mathbf{J} \cdot \nabla \left(\frac{\varepsilon}{\sigma} \right). \tag{3.62}$$

We see that $\nabla \cdot \mathbf{J} = 0$ does not imply that $\rho = 0$. Volume charge density is nonzero in inhomogeneous conducting media where $\varepsilon/\sigma \neq$ const, and its magnitude is proportional to the gradient of ε/σ. On the other hand, we also conclude that there cannot be volume excess charges ($\rho = 0$) inside homogeneous media (σ and ε do not vary with position) with steady currents.

The corresponding boundary condition for **D**, Eq. (2.85), in combination with Eqs. (3.61) and (3.55), gives the surface charge density on the interface between medium 1 (with parameters σ_1 and ε_1) and medium 2 (with σ_2 and ε_2):

surface charge in a steady current field

$$\rho_s = \hat{\mathbf{n}} \cdot \mathbf{D}_1 - \hat{\mathbf{n}} \cdot \mathbf{D}_2 = \frac{\varepsilon_1}{\sigma_1} \hat{\mathbf{n}} \cdot \mathbf{J}_1 - \frac{\varepsilon_2}{\sigma_2} \hat{\mathbf{n}} \cdot \mathbf{J}_2 = \left(\frac{\varepsilon_1}{\sigma_1} - \frac{\varepsilon_2}{\sigma_2} \right) J_n, \tag{3.63}$$

[6]Having in mind the rule for calculating the derivative of a product of two functions, we apply the divergence operator, which is a differential operator, to a product of a scalar and a vector function, and get $\nabla \cdot (f\mathbf{a}) = (\nabla f) \cdot \mathbf{a} + f(\nabla \cdot \mathbf{a})$.

where $J_n = \hat{\mathbf{n}} \cdot \mathbf{J}_1 = \hat{\mathbf{n}} \cdot \mathbf{J}_2$ is the normal component of the current density vector across the boundary ($\hat{\mathbf{n}}$ is directed as in Figs. 2.10 and 3.7). Note that $\rho_s = 0$ only for the special case of $\varepsilon_1/\sigma_1 = \varepsilon_2/\sigma_2$. If both media are metallic conductors, we have approximately $\varepsilon_1 = \varepsilon_2 = \varepsilon_0$, so that Eq. (3.63) becomes

$$\rho_s = \varepsilon_0 \left(\frac{1}{\sigma_1} - \frac{1}{\sigma_2} \right) J_n. \tag{3.64}$$

Finally, the distribution of bound (polarization) charge in the dielectric can be found by computing first the polarization vector, $\mathbf{P}$, from Eq. (2.59), and then bound volume and surface charge densities, ρ_p and ρ_{ps}, using Eqs. (2.19) and (2.89), respectively.

Problems: 3.1; *Conceptual Questions* (on Companion Website): 3.9 and 3.10; *MATLAB Exercises* (on Companion Website).

3.7 RELAXATION TIME

We know from electrostatics that charge placed in the interior of a conductor will move to the conductor surface and redistribute itself in such a way that $\mathbf{E} = 0$ in the conductor under electrostatic equilibrium conditions. With the continuity equation (for time-varying currents) now in hand, we can quantitatively analyze this nonelectrostatic transitional process and calculate the time it takes to reach an equilibrium.

Consider a homogeneous conducting medium of conductivity σ and permittivity ε. The current density vector, $\mathbf{J}$, and electric flux density vector, $\mathbf{D}$, in the medium are interconnected by the duality relationship in Eq. (3.61). Combining the differential form of the generalized Gauss' law, Eq. (2.45), and that of the continuity equation, Eq. (3.39), we can write

$$\rho = \nabla \cdot \mathbf{D} = \nabla \cdot \left(\frac{\varepsilon}{\sigma} \mathbf{J} \right) = \frac{\varepsilon}{\sigma} \nabla \cdot \mathbf{J} = -\frac{\varepsilon}{\sigma} \frac{\partial \rho}{\partial t}, \tag{3.65}$$

where the ratio ε/σ can be brought outside the divergence sign because it is a constant (the medium is homogeneous). Rewriting, we have that the charge density, ρ, in the medium satisfies the equation

$$\frac{\partial \rho}{\partial t} + \frac{\sigma}{\varepsilon} \rho = 0, \tag{3.66}$$

which is a first-order partial differential equation in time, t. Its solution is given by

$$\boxed{\rho = \rho_0 \, \mathrm{e}^{-(\sigma/\varepsilon)t} = \rho_0 \, \mathrm{e}^{-t/\tau},} \tag{3.67}$$

redistribution of charge

where ρ_0 is the initial value of the charge density at $t = 0$. In general, ρ is a function of the space coordinates as well, e.g., $\rho = \rho(x, y, z, t)$. Eq. (3.67) tells us that the charge density at a given location in the conductor decreases with time exponentially, completely independent of any applied electric field. The time constant

$$\boxed{\tau = \frac{\varepsilon}{\sigma}} \tag{3.68}$$

relaxation time

is referred to as the relaxation time (the unit is s) and it equals the time required for the charge density at any point to decay to 1/e (36.8%) of its initial value. It can easily be shown that ρ decreases to 1% of ρ_0 after approximately 4.6τ, while at $t = 10\tau$, $\rho \approx 4.5 \times 10^{-5} \rho_0$, i.e., $\rho \approx 0$.

For metallic conductors, τ is so short that it can hardly be measured or observed. For example, copper has $\tau_{\rm Cu} \approx 10^{-19}$ s. Even much poorer conductors than metals have very short relaxation times (e.g., $\tau_{\rm H_2O} \approx 10^{-5}$ s for distilled water). On the other hand, the relaxation time for good dielectrics (insulators) is relatively long (e.g., $\tau_{\rm glass} \approx 1$ minute, $\tau_{\rm mica} \approx 15$ hours, and $\tau_{\rm quartz} \approx 50$ days).

While this chapter is devoted to time-invariant currents and fields, we note here that the concept of relaxation time is also used to determine the electrical nature (in terms of conducting and dielectric properties) of materials for time-varying currents and fields. Namely, whether a material of parameters σ and ε is considered a good conductor or a good dielectric is decided on the basis of the relaxation time, as compared to times of interest in a given application. Thus, for a time-harmonic (sinusoidal) field of frequency f, the relaxation time, given by Eq. (3.68), is compared to the time period,[7] $T = 1/f$. If $\tau \ll T$, the medium is classified as a good conductor. In particular, if $\tau = 0$ ($\sigma \to \infty$), the material is said to be a perfect electric conductor (PEC). On the other hand, the material is considered a good dielectric (insulator) if $\tau \gg T$. In a limit, $\tau \to \infty$ for perfect (lossless) dielectrics ($\sigma = 0$). For all other (intermediate) values of τ, the material is classified as a quasi-conductor. We can now understand that some materials that are considered as good conductors at certain frequencies may behave like good dielectrics at sufficiently higher frequencies (i.e., shorter time periods). For example, at frequencies $f_1 = 1$ kHz, $f_2 = 10$ MHz, and $f_3 = 30$ GHz, average rural ground ($\varepsilon_{\rm r} = 14$ and $\sigma = 10^{-2}$ S/m, assuming no change in the parameters as a function of frequency) behaves like (1) a good conductor, (2) a quasi-conductor, and (3) a good dielectric, respectively. Further discussions of conducting and dielectric properties of materials at different frequencies, in a context of general Maxwell's equations and electromagnetic wave propagation, are provided in later chapters.

3.8 RESISTANCE, OHM'S LAW, AND JOULE'S LAW

Consider an arbitrarily shaped conductor made from a linear (generally inhomogeneous) material of conductivity σ, as shown in Fig. 3.8. Let the end surfaces $S_{\rm A}$ and $S_{\rm B}$ of the conductor be coated with perfectly conducting material (or with material of conductivity much greater than σ). If the voltage V is applied between the conductor ends, the current, of density $\mathbf{J}$, in the conductor flows normal to the end surfaces [Eq. (3.57)] and parallel to the sides of the resistor [Eq. (3.58)]. By virtue of the continuity equation, the same total current must pass through every cross section of the conductor. This current is given by

$$I = \int_S \mathbf{J} \cdot \mathrm{d}\mathbf{S} = \int_S \sigma \mathbf{E} \cdot \mathrm{d}\mathbf{S}, \tag{3.69}$$

where $\mathbf{E}$ is the electric field intensity vector in the conductor. On the other hand, the voltage between $S_{\rm A}$ and $S_{\rm B}$ equals the line integral of the same vector, $\mathbf{E}$, along any path in the conductor connecting these surfaces [Eq. (1.90)], that is,

$$V = \int_{\rm A}^{\rm B} \mathbf{E} \cdot \mathrm{d}\mathbf{l}. \tag{3.70}$$

[7]The time period (T) is defined as an interval after which a time-harmonic (or other time-periodic) function repeats itself over time, whereas the frequency (f) is the number of repetitions of the function per unit time (one second), so that f times T equals unity.

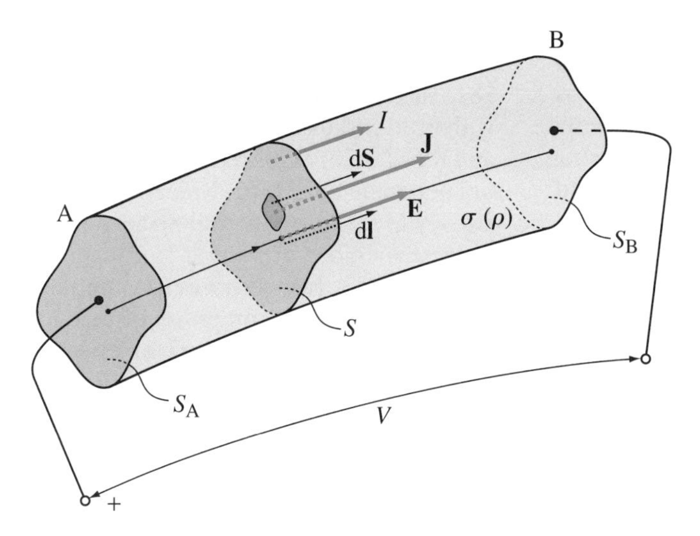

Figure 3.8 Arbitrarily shaped conductor – resistor.

Now, we note that if V is increased (for some reason), the electric field lines do not change shape, but **E** proportionally increases everywhere within the conductor, and, by means of Eq. (3.69), so does the current I. The ratio of V and I is thus a constant,

$$\boxed{R = \frac{V}{I},} \qquad (3.71)$$

resistance (unit: Ω)

called the resistance of the conductor. A conductor with two terminals and a (substantial) resistance R is usually referred to as a resistor. The relation between the voltage, current, and resistance of a resistor is known as Ohm's law:

$$\boxed{V = RI.} \qquad (3.72)$$

Ohm's law

The resistance is always nonnegative ($R \geq 0$), and the unit is the ohm (Ω), equal to V/A. The value of R depends on the shape and size of the conductor (resistor), and on the conductivity σ (or resistivity ρ) of the material.

HISTORICAL ASIDE

Georg Simon Ohm (1789–1854), a German physicist and mathematician, was a professor at the University of Munich. As a child, Ohm received a fine mathematical and scientific education from his father, who was a locksmith and an entirely self-taught man. In 1811, he received a doctorate in mathematics from the University of Erlangen. He was also interested in physics, where he studied analogies between the flow of electricity and flow of heat. In a series of papers in 1825 and 1826, he gave a mathematical description of conduction in electric circuits modeled after Fourier's study of heat conduction. He assumed that, just as the rate at which heat flowed between two points depended on the temperature difference between the points and on the ease with which heat was conducted by the material between the points, the electric current between two points should depend on the difference in electric potential between the points and on the electric conductivity of the material between the points. By experimenting with wires of different lengths and thicknesses, he found that the current intensity through a wire, for a given potential difference between the wire ends, was inversely proportional to the length and directly proportional to the cross-sectional area of the wire. As very knowledgeable of both mathematics and physics, Ohm was able to deduce mathematical relationships based on the experimental evidence that he had tabulated. He defined the resistance of a wire and showed, in 1827, that a simple relation existed among the resistance, potential difference (voltage), and current intensity of a wire. This is now known as Ohm's law. The fully developed presentation of his theory of electric conduction appeared in his famous book "Die Galvanische Kette, Mathematisch Bearbeitet" (The Galvanic Circuit, Analyzed Mathematically), published in 1827 in Berlin. At first, Ohm's work was received with little enthusiasm, and a full acknowledgement and recognition of his results, including Ohm's law, did not come until 1841, when he was awarded the Copley Medal of the Royal Society and soon after became member of several European academies. Only in 1852, two years before his death, Ohm achieved his lifelong ambition of being appointed to the chair of physics at the University of Munich. Ohm's name was further immortalized in 1881, when the International Electrical Congress established the ohm as the unit of resistance. *(Portrait: Edgar Fahs Smith Collection, University of Pennsylvania Libraries)*

The siemens is adopted as the SI unit for conductance in honor of brothers Werner and Wilhelm von Siemens, German engineers and inventors. **Ernst Werner von Siemens** (1816–1892) contributed to then new discipline of electrical engineering with several inventions in telegraphy (needle telegraph), cable transmission (large undersea cables), and energy generation, and was one of the first great entrepreneurs in electrical industry. In 1847, he founded, together with the mechanic Johann Georg Halske (1814–1890), the "Siemens & Halske Telegraph Construction Company" in Berlin, later known as "Siemens." **Karl Wilhelm von Siemens** (1823–1883) was a mechanical engineer by training and successful businessman. His most important inventions are the regenerative furnace and electric pyrometer. He later became British subject, Sir Charles William Siemens.

The reciprocal of resistance is called the conductance and symbolized by G. From Ohm's law,

conductance (unit: S)

$$\boxed{G = \frac{1}{R} = \frac{I}{V}.} \tag{3.73}$$

Its unit is the siemens (S), where $S = \Omega^{-1} = A/V$. Note that sometimes the mho (ohm spelled backwards) is used instead of the siemens.

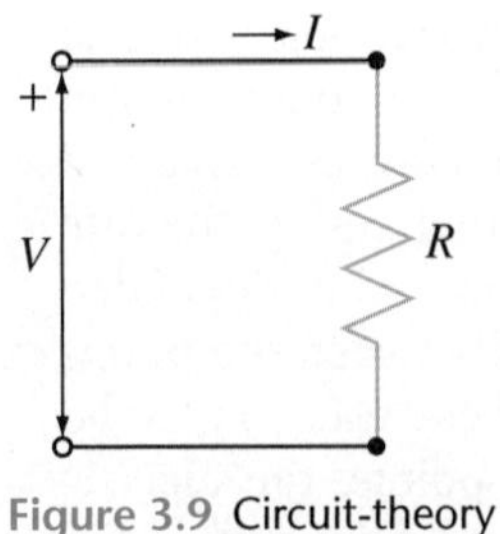

Figure 3.9 Circuit-theory representation of a resistor.

Fig. 3.9 shows the circuit-theory representation of a resistor. In circuit theory, it is assumed that the resistances are located only in the resistors in a circuit, while the interconnecting conductors are considered as perfectly conducting. Each connecting conductor has therefore zero resistance and is equipotential (acts as a short circuit). The only voltage drops in the circuit occur across circuit elements. Note that Ohm's law, Eq. (3.72), is valid also for time-varying voltage (v) and current (i) of a resistor ($v = Ri$). It represents the element law for a resistor, as one of the basic elements in circuit theory [another element law being Eq. (3.45) for a capacitor].

For linear resistors, R remains constant for different voltages and currents, V and I. For nonlinear resistors, however, the conductivity of the material depends on the applied electric field, $\mathbf{E}$, and R therefore depends on the applied voltage (or current),

nonlinear resistor

$$\boxed{R = R(V).} \tag{3.74}$$

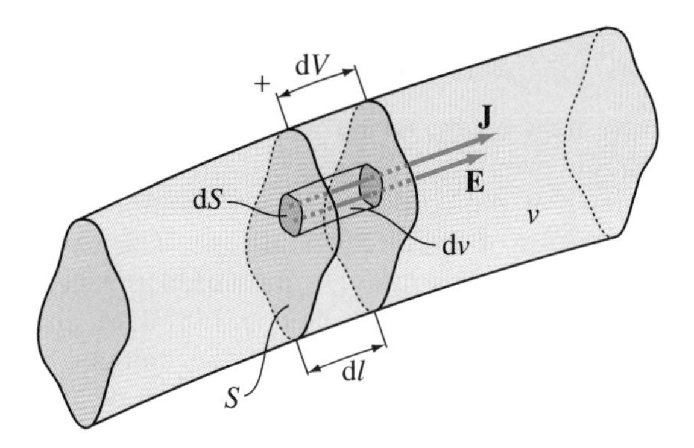

Figure 3.10 Evaluation of the power of Joule's (ohmic) losses in a resistor of general shape.

Examples are semiconductor (pn) diodes, whose current-voltage characteristics, $I = I(V)$, are nonlinear functions.

The total power of Joule's or ohmic losses in the resistor in Fig. 3.8 is given by Eq. (3.32), where v is the volume of the resistor. To carry out the volume integration, we first cut v into thin slices with bases chosen to be equipotential surfaces (perpendicular to current lines). We then subdivide each slice along the current lines into small tubular cells of volume $\mathrm{d}v = \mathrm{d}S\,\mathrm{d}l$, as depicted in Fig. 3.10, and write

$$P_\mathrm{J} = \int_v JE\,\mathrm{d}v = \int_\mathrm{A}^\mathrm{B} \int_S JE\,\underbrace{\mathrm{d}S\,\mathrm{d}l}_{\mathrm{d}v}. \tag{3.75}$$

Note that $E\,\mathrm{d}l$ equals the voltage $\mathrm{d}V$ between the bases of the cell, which is the same for all cells within a slice (because of the equipotentiality of interfaces between adjacent slices). Therefore, $\mathrm{d}V$ is a constant for cross-sectional integration over S, and can be taken out of that integral. We have thus separated the overall integration throughout v into two independent integrals:

$$P_\mathrm{J} = \int_\mathrm{A}^\mathrm{B} \int_S J\,\mathrm{d}S\,\underbrace{E\,\mathrm{d}l}_{\mathrm{d}V} = \left(\int_\mathrm{A}^\mathrm{B} E\,\mathrm{d}l\right)\left(\int_S J\,\mathrm{d}S\right) = VI, \tag{3.76}$$

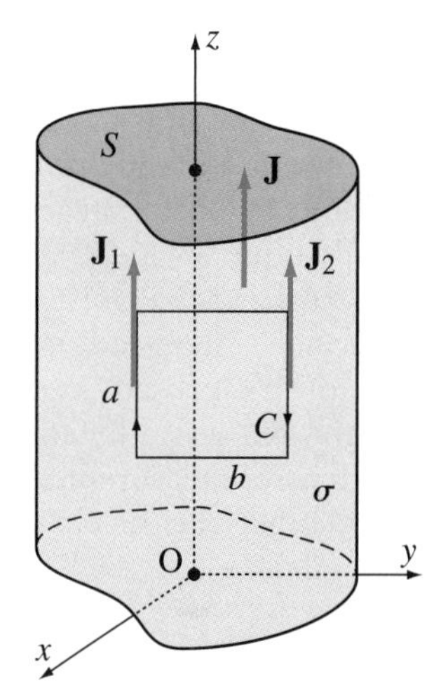

Figure 3.11 Steady current in a homogeneous conductor of a uniform cross section; for Example 3.4.

which equal, respectively, the voltage V [Eq. (3.70)] and the current I [Eq. (3.69)] of the resistor. Employing Eqs. (3.72) and (3.73), the equivalent expressions for the power of Joule's losses in a resistor are

Joule's law

$$\boxed{P_\mathrm{J} = VI = RI^2 = \frac{V^2}{R} = GV^2 = \frac{I^2}{G}.} \tag{3.77}$$

This is known as Joule's law.

Example 3.4 Current Uniformity in Conductors with Uniform Cross Section

A steady current is flowing through a long conductor made of a homogeneous material and having a uniform cross section of an arbitrary shape. Prove that the current density is the same in the entire conductor.

Solution With reference to Fig. 3.11, the current density vector in the conductor has a z-component only, $\mathbf{J} = J_z(x, y, z)\,\hat{\mathbf{z}}$. From the continuity equation for steady currents in differential form, Eq. (3.41), and the expression for divergence in the Cartesian coordinate system, Eq. (1.167), we have

$$\nabla\cdot\mathbf{J} = \frac{\partial J_z}{\partial z} = 0, \tag{3.78}$$

which means that J_z is not a function of z, i.e., it does not vary along the conductor.

HISTORICAL ASIDE

James Prescott Joule (1818–1889), an English scientist, was a member of the Royal Society. As a son of a wealthy Salford brewer, Joule was educated by private tutors, including the famous English chemist and physicist John Dalton (1766–1844). Later, while working in the family brewery, he studied in his spare time the subject of electricity (new at that time) and was especially interested in the efficiency of electric motors. After several attempts to design a superefficient electric motor that would produce infinite power (this possibility had been suggested in previous papers) and thus offer an ideal alternative to steam engines, he realized that this goal was not achievable and became interested in studying the heat generated by electricity. Joule lacked in mathematical rigor, but was a fanatic experimentalist. Based on extensive measurements of heat in electric motors and other electric circuitry, he discovered, in 1840, that the heat produced by a current in an electric circuit over a certain interval of time equals the square of the current intensity multiplied by the resistance of the circuit and time. This came to be called Joule's law. He measured the heat produced by every process he could think of, and studied the relationship between the amount of work entering the system and the amount of heat exiting the system. In his famous "paddle wheel" experiment in 1847, he used a falling weight to spin a paddle wheel in an insulated barrel of water and measured very precisely the increase in the water temperature produced by the friction of the wheel – to determine the mechanical equivalent of the heat dissipated in the water. His general conclusion was that heat was only one of many forms of energy, which can be converted from one form to another but the total energy of a closed system remains constant. With this, he contributed fundamentally to the discovery and recognition of the principle of conservation of energy. We honor Joule also by using joule as the unit for work and energy. *(Portrait: National Bureau of Standards Archives, courtesy AIP Emilio Segrè Visual Archives, E. Scott Barr Collection)*

To prove that $\mathbf{J}$ does not change in a cross section of the conductor either, we apply Eq. (3.43) to a rectangular contour C placed inside the conductor, with sides of length a set parallel to current lines (Fig. 3.11). Using also Eq. (3.18), we can write

$$\oint_C \mathbf{E} \cdot d\mathbf{l} = \oint_C \frac{\mathbf{J}}{\sigma} \cdot d\mathbf{l} = \frac{1}{\sigma} \oint_C \mathbf{J} \cdot d\mathbf{l} = 0, \tag{3.79}$$

where σ is the conductivity of the medium, and $1/\sigma$ can be taken out of the integral because the medium is homogeneous ($\sigma = \text{const}$). Hence,

$$\oint_C \mathbf{J} \cdot d\mathbf{l} = J_1 a - J_2 a = 0, \tag{3.80}$$

with J_1 and J_2 standing for the current densities along the two sides of the contour parallel to $\mathbf{J}$. We conclude that these current densities are the same:

$$J_1 = J_2. \tag{3.81}$$

Note that C can be translated to any position in the conductor, so that J_1 and J_2 can be associated to any pair of points in the conductor cross section. This implies that $\mathbf{J}$ does not depend on x and y, namely, it is the same in the entire cross section of the conductor. Combined with our previous conclusion about the current uniformity in the z-direction, we have that

dc current uniformity in a homogeneous conductor of a uniform cross section

$$\boxed{J = \text{const}} \tag{3.82}$$

everywhere inside the conductor.

Example 3.5 Resistance of a Resistor of a Uniform Cross Section

Determine the resistance of a homogeneous resistor with a uniform cross section of an arbitrary shape and surface area S. The length of the resistor is l and the conductivity of the material is σ.

Solution Eq. (3.82) tells us that the current is distributed uniformly in the resistor. The current density in the resistor is thus

$$J = \frac{I}{S}, \tag{3.83}$$

where I is the current intensity through the resistor, Eq. (3.69). Using Eq. (3.70), the voltage across the resistor is

$$V = El = \frac{J}{\sigma} l = \frac{Il}{\sigma S}. \tag{3.84}$$

Hence, the resistance of the resistor comes out to be

$$\boxed{R = \frac{V}{I} = \frac{l}{\sigma S}}. \tag{3.85}$$

resistance of a homogeneous resistor with a uniform cross section

Example 3.6 Two Resistors in Series

Fig. 3.12(a) shows a cylindrical resistor of radius a consisting of two parts of lengths l_1 and l_2. Conductivities of the parts are σ_1 and σ_2 ($\sigma_1 \neq \sigma_2$). The voltage across the resistor is V. Find the current through the resistor.

Solution The current through the two parts of the resistor is the same, and they can be represented as two homogeneous resistors connected in series, as shown in Fig. 3.12(b). The total resistance of the connection is

$$\boxed{R = R_1 + R_2}, \tag{3.86}$$

equivalent resistance of two resistors in series

where, using Eq. (3.85), the resistances of individual resistors amount to

$$R_1 = \frac{l_1}{\sigma_1 \pi a^2} \quad \text{and} \quad R_2 = \frac{l_2}{\sigma_2 \pi a^2}, \tag{3.87}$$

respectively. The current through the resistor in Fig. 3.12(a) is hence

$$I = \frac{V}{R_1 + R_2} = \frac{\pi \sigma_1 \sigma_2 a^2 V}{\sigma_2 l_1 + \sigma_1 l_2}. \tag{3.88}$$

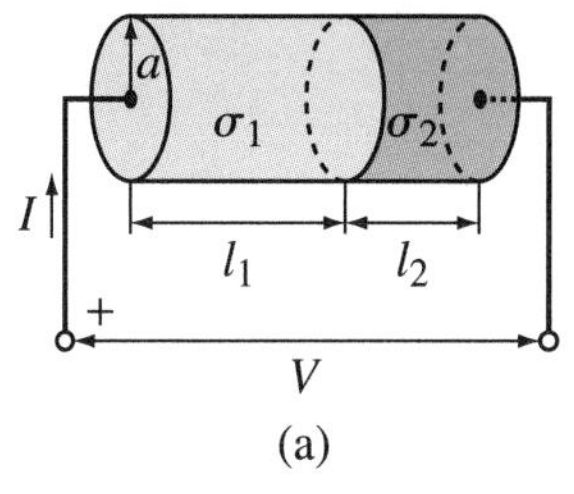

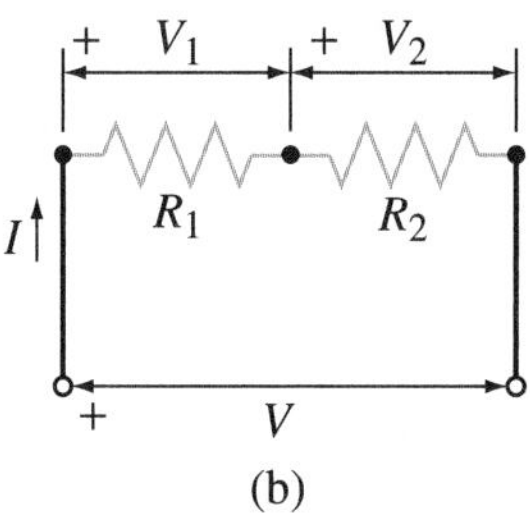

Figure 3.12 Cylindrical resistor with two parts of different conductivities: (a) geometry and (b) network representation; for Example 3.6.

Example 3.7 Two Resistors in Parallel

A cylindrical resistor of length l consists of two coaxial layers of different conductivities, σ_1 and σ_2, as shown in Fig. 3.13(a). The cross-sectional surface areas of the layers are S_1 and S_2. The current through the resistor is I. Compute (a) the current density in the resistor and (b) the voltage across the resistor.

Solution

(a) Eq. (3.54) tells us that the electric field intensities in the two layers are the same:

$$E_1 = E_2 = E. \tag{3.89}$$

Hence, the current densities in the layers are not the same, as $J_1 = \sigma_1 E$ and $J_2 = \sigma_2 E$. The total current through the resistor is given by

$$I = J_1 S_1 + J_2 S_2 = (\sigma_1 S_1 + \sigma_2 S_2)E, \tag{3.90}$$

from which

$$E = \frac{I}{\sigma_1 S_1 + \sigma_2 S_2} \tag{3.91}$$

and

$$J_1 = \frac{\sigma_1 I}{\sigma_1 S_1 + \sigma_2 S_2} \quad \text{and} \quad J_2 = \frac{\sigma_2 I}{\sigma_1 S_1 + \sigma_2 S_2}. \tag{3.92}$$

(b) The voltage across the resistor is

$$V = El = \frac{Il}{\sigma_1 S_1 + \sigma_2 S_2}. \tag{3.93}$$

Note that V can also be found using the equivalent circuit shown in Fig. 3.13(b). Namely, since the voltage across the two layers of the resistor is the same, they can be represented as two homogeneous resistors connected in parallel. The total conductance of the connection turns out to be

equivalent conductance of two resistors in parallel

$$\boxed{G = G_1 + G_2,} \tag{3.94}$$

that is [Eq. (3.85)],

$$G = \frac{\sigma_1 S_1}{l} + \frac{\sigma_2 S_2}{l}. \tag{3.95}$$

The voltage of the resistors is $V = I/G$.

Example 3.8 Conductance of a Spherical Capacitor

Find the conductance of the spherical capacitor with an imperfect (conducting) dielectric (of conductivity σ) from Fig. 3.6.

Solution Using Eqs. (3.73) and (3.50), the conductance of the nonideal spherical capacitor is, by definition,

leakage conductance of a nonideal spherical capacitor

$$\boxed{G = \frac{I}{V} = \frac{4\pi\sigma ab}{b-a}.} \tag{3.96}$$

This, actually, is the so-called leakage conductance of the capacitor. (An ideal capacitor has a perfect dielectric and zero leakage conductance.) Note, however, that the system in Fig. 3.6 may also represent a spherical resistor, with resistance

$$R = \frac{1}{G} = \frac{\rho(b-a)}{4\pi ab}, \tag{3.97}$$

where $\rho = 1/\sigma$ is the resistivity of a (resistive) material between the electrodes.

Note also that, employing the conductance (or resistance) and Joule's law, Eq. (3.77), the power of Joule's losses in the capacitor (resistor) can now easily be found as

$$P_J = GV^2 = \frac{V^2}{R} = \frac{4\pi\sigma abV^2}{b-a}, \tag{3.98}$$

which, of course, is the same result as in Eq. (3.52).

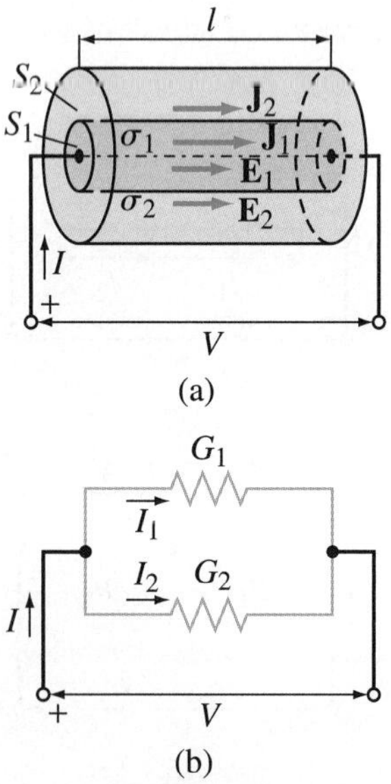

Figure 3.13 Resistor with two coaxial layers (a), which can be represented as two homogeneous resistors connected in parallel (b); for Example 3.7.

Problems: 3.2–3.9; *Conceptual Questions* (on Companion Website): 3.11–3.17; *MATLAB Exercises* (on Companion Website).

3.9 DUALITY BETWEEN CONDUCTANCE AND CAPACITANCE

Consider a pair of metallic bodies (electrodes) placed in a homogeneous conducting medium of conductivity σ and permittivity ε, as shown in Fig. 3.14. The conductivity of electrodes is much larger than σ. Let the voltage between the electrodes be V. From Eq. (3.62), there is no volume charges ($\rho = 0$) in the medium (σ and ε are constants). Having in mind Eq. (3.57), the current (and field) lines are normal to the surfaces of electrodes. We now use the duality relationship between the current

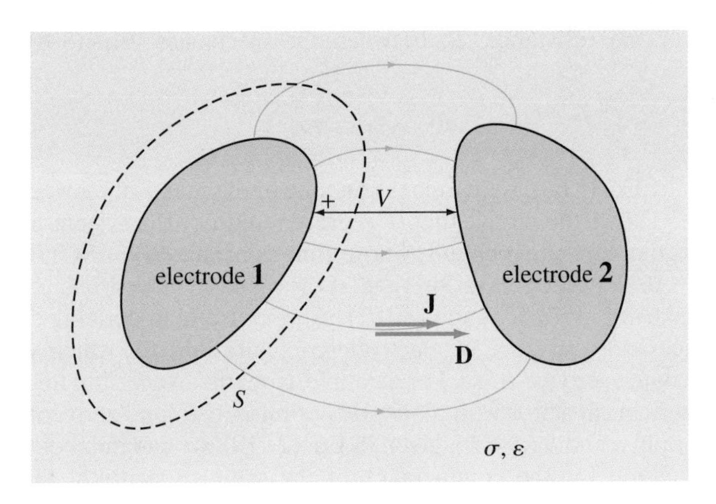

Figure 3.14 Two metallic electrodes in a homogeneous conducting medium.

density vector, **J**, and electric flux density vector, **D**, in the medium, Eq. (3.61), to relate the conductance, G, and capacitance, C, between the electrodes.

Applying the continuity equation for steady currents, Eq. (3.40), and generalized Gauss' law, Eqs. (3.60), to an arbitrary surface S completely enclosing the positive electrode (Fig. 3.14) gives

$$I = \int_S \mathbf{J} \cdot d\mathbf{S} = \frac{\sigma}{\varepsilon} \int_S \mathbf{D} \cdot d\mathbf{S} = \frac{\sigma}{\varepsilon} Q, \tag{3.99}$$

where I is the total current leaving the positive electrode through the conducting medium (and entering the negative electrode), Q is the total charge of the positive electrode (the total charge of the negative electrode is $-Q$), and σ/ε can be brought outside the integral sign because it is a constant, i.e., the medium is homogeneous.[8] By dividing this equation by V, we obtain

$$\frac{I}{V} = \frac{\sigma}{\varepsilon} \frac{Q}{V}, \tag{3.100}$$

which, by means of Eqs. (3.73) and (2.113), yields the following duality relationship between G and C:

$$\boxed{G = \frac{\sigma}{\varepsilon} C.} \tag{3.101}$$

duality of G and C

Note that, since the definition of C, Eq. (2.113), depends on the existence of charges Q and $-Q$ on the electrodes and is independent of whether or not a current also exists in the dielectric medium (of permittivity ε), G and C in Eq. (3.101) do not necessarily represent the conductance and capacitance of a system with an imperfect dielectric of parameters σ and ε. They can also be associated with two independent dual systems, representing the capacitance (C) between a pair of electrodes when placed in a perfect dielectric (of permittivity ε, including the case $\varepsilon = \varepsilon_0$, and zero conductivity) and the conductance (G) between the same electrodes when placed in a conducting medium (of conductivity σ and totally arbitrary permittivity), respectively.

[8]Note, however, that this is true also for inhomogeneous media for which functions σ and ε are such that $\sigma/\varepsilon =$ const.

In terms of the resistance, R, between the electrodes, Eq. (3.101) can be rewritten as

R-C analogy

$$RC = \frac{\varepsilon}{\sigma} = \tau, \tag{3.102}$$

where τ, given by Eq. (3.68), is the relaxation time of the material between the electrodes. Fig. 3.15 shows the circuit-theory representation of the system in Fig. 3.14. This is a first-order capacitive circuit, whose time constant ($\tau_C = RC$) is equal, by virtue of Eq. (3.102), to the relaxation time of the material ($\tau = \varepsilon/\sigma$).

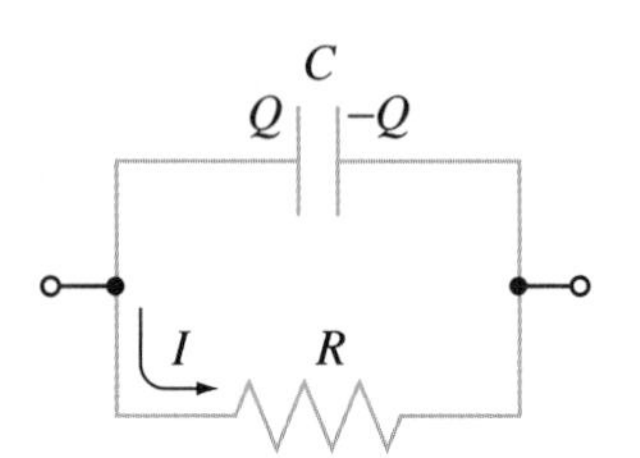

Figure 3.15 Network representation of the system in Fig. 3.14.

The relations in Eqs. (3.101) and (3.102) are very useful in deriving expressions for conductance and resistance of electrode configurations for which we already have the capacitance, or vice versa. For example, from the expression for the capacitance of a spherical capacitor with dielectric permittivity ε (and zero conductivity) and electrode radii a and b ($a < b$), given in Eq. (2.119), we can immediately determine the expression for the conductance of the same capacitor if filled with a conducting medium of conductivity σ:

$$G = \frac{\sigma}{\varepsilon} C = \frac{4\pi\sigma ab}{b - a} \tag{3.103}$$

[the same as in Eq. (3.96)].

Example 3.9 Self-Discharging of a Nonideal Capacitor

A parallel-plate capacitor is filled with an imperfect dielectric of relative permittivity $\varepsilon_r = 6$ and conductivity $\sigma = 10^{-14}$ S/m. The capacitor was connected to an ideal battery and fully charged to a voltage of 20 V between its plates, and then disconnected from the battery. Find the time after which the voltage of the capacitor decays to 1 V.

Solution The capacitor can be represented by the equivalent network in Fig. 3.15, where C and R are the capacitance and resistance between the capacitor plates, respectively. With no battery in the circuit, the current discharging C is also the current across R, which, with help of Eqs. (3.45) and (3.72), gives

$$-C\frac{\mathrm{d}v}{\mathrm{d}t} = \frac{v}{R} \quad \longrightarrow \quad \frac{\mathrm{d}v}{\mathrm{d}t} + \frac{v}{\tau_C} = 0, \tag{3.104}$$

where the time constant $\tau_C = RC$ is found from Eq. (3.102). The solution of this differential equation is

self-discharging of a capacitor with an imperfect dielectric

$$v(t) = v(0)\,\mathrm{e}^{-t/(RC)} = v(0)\,\mathrm{e}^{-(\sigma/\varepsilon)t}. \tag{3.105}$$

Finally, we take the natural logarithms of both sides of the equation, and obtain the time we seek:

$$t = -\frac{\varepsilon_r\varepsilon_0}{\sigma}\ln\frac{v(t)}{v(0)} = 265 \text{ minutes}, \tag{3.106}$$

where $v(0) = 20$ V and $v(t) = 1$ V.

We realize that discharging of a charged capacitor with homogeneous imperfect dielectric that is left to itself (open-circuited) is completely independent of its shape (geometry) and size, and completely analogous to redistribution of charge in a conductor described by Eq. (3.67). The capacitor voltage, $v(t)$, and charge, $Q(t) = Cv(t)$, decrease exponentially with time, at a rate that is set by the relaxation time of the lossy dielectric filling the space between the capacitor electrodes.

Conceptual Questions (on CD): 3.18–3.21; *MATLAB Exercises* (on Companion Website).

3.10 EXTERNAL ELECTRIC ENERGY VOLUME SOURCES AND GENERATORS

We observed in connection with Fig. 3.1 that an external voltage source (e.g., a chemical battery) is required to maintain a steady current in a conductor. This source creates a rise in potential in the circuit and a potential difference between the ends of the conductor. From the energy standpoint, an external source of energy is necessary to maintain steady electric current flow through the circuit by continuously supplying the energy that is then dissipated in the conductor as Joule's heat. In concrete situations, various forms of the external energy (chemical energy, mechanical energy, thermal energy, light energy, etc.) are converted to the energy of the electric field in conductors and ultimately lost to heat. Generally, in analogy to circuit-theory generators, we use two field-theory models of volume-distributed energy sources capable of transmitting energy to electric charges: sources analogous to voltage generators and sources analogous to current generators.

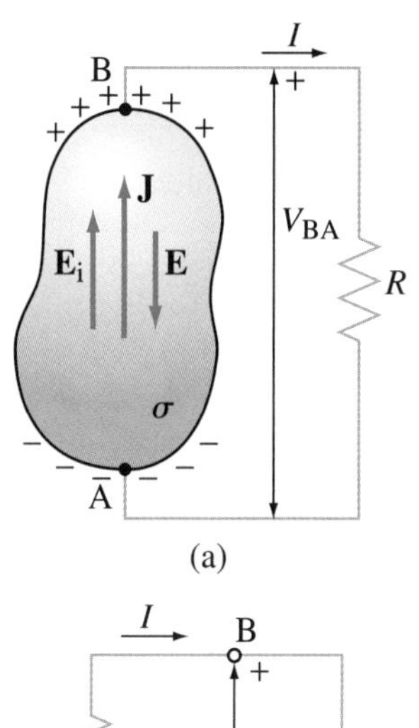

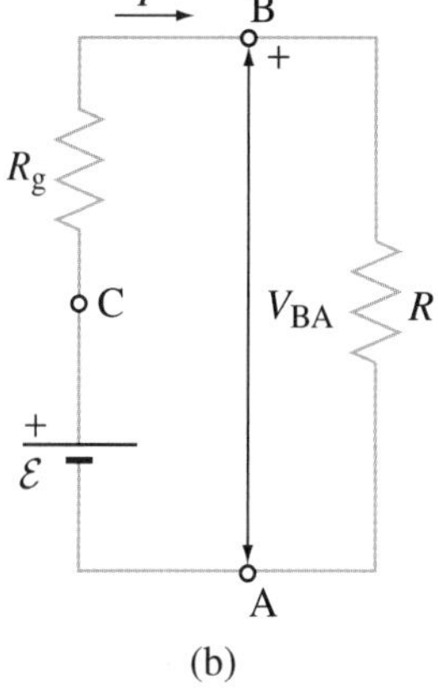

Figure 3.16 (a) External electric energy sources modeled by an impressed electric field, of intensity $\mathbf{E}_i$. (b) Voltage generator in circuit theory.

Consider a source region of volume v shown in Fig. 3.16(a) in which a non-electric external force, $\mathbf{F}_i$, termed the impressed force, acts on charge carriers (e.g., electrons) and separates positive and negative excess charges. An example is the force on electrons in a metallic wire moving in a magnetic field (as we shall see in a later chapter). We can formally divide $\mathbf{F}_i$ by the charge of a carrier, q ($q = -e$ for electrons), and what we get is a quantity expressed in V/m:

$$\mathbf{E}_i = \frac{\mathbf{F}_i}{q}, \tag{3.107}$$

which we call the impressed electric field intensity vector. This field, however, although of the same dimension and unit as $\mathbf{E}$, is not a true electric field (a field due to charges). Since there is also a field due to charges in the domain v, the total force on q is

$$\mathbf{F}_{tot} = \mathbf{F}_i + q\mathbf{E} = q(\mathbf{E}_i + \mathbf{E}), \tag{3.108}$$

and this is the force that compels the charges to move through the region. The current density vector in the region is, therefore, given by

$$\boxed{\mathbf{J} = \sigma(\mathbf{E} + \mathbf{E}_i),} \tag{3.109}$$

$\mathbf{E}_i$ – impressed electric field intensity vector (unit: V/m)

where σ is the conductivity of the material in v. It is very important to note that $\mathbf{E}_i$ does not depend on $\mathbf{J}$.

The model with a volume distribution (throughout v) of energy sources modeled by an impressed electric field, of intensity $\mathbf{E}_i$, can be used in many different electromagnetic situations, as we shall see later in this text. However, we note here that the particular situation in Fig. 3.16(a) actually represents a voltage generator in circuit theory. The generator is connected to a resistor of resistance R. Combining Eqs. (1.90) and (3.109), the voltage between the ends B and A of the source region (generator), can be written as

$$V_{BA} = \int_B^A \mathbf{E} \cdot d\mathbf{l} = \int_B^A \left(-\mathbf{E}_i + \frac{\mathbf{J}}{\sigma}\right) \cdot d\mathbf{l}. \tag{3.110}$$

Reversing the order of integration limits, we have

$$V_{BA} = \int_A^B \mathbf{E}_i \cdot d\mathbf{l} - \int_A^B \frac{\mathbf{J}}{\sigma} \cdot d\mathbf{l}. \tag{3.111}$$

The first term on the right-hand side of this equation is, by definition, the electromotive force (emf) of the generator:

electromotive force – emf (unit: V)

$$\boxed{\mathcal{E} = \int_{\mathrm{A}}^{\mathrm{B}} \mathbf{E}_{\mathrm{i}} \cdot \mathrm{d}\mathbf{l}.} \tag{3.112}$$

Note that $\mathcal{E}$ equals the work done by the external force, $\mathbf{F}_{\mathrm{i}}$, in moving the charge q through the generator, from its negative terminal (A) to its positive terminal (B), divided by q. The unit for emf is V. The second integral in Eq. (3.111) is linearly proportional to the current I in the circuit:

$$\int_{\mathrm{A}}^{\mathrm{B}} \frac{\mathbf{J}}{\sigma} \cdot \mathrm{d}\mathbf{l} = R_{\mathrm{g}} I, \tag{3.113}$$

where the constant of proportionality, R_{g}, is called the internal resistance of the generator. Hence, Eq. (3.111) becomes

voltage generator

$$\boxed{V_{\mathrm{BA}} = \mathcal{E} - R_{\mathrm{g}} I.} \tag{3.114}$$

Fig. 3.16(b) shows the equivalent circuit-theory representation of the generator. Note that if the generator terminals A and B are open-circuited ($R \to \infty$), no current flows through the generator ($I = 0$), and we can write [Fig. 3.16(b)]

$$(V_{\mathrm{BA}})_{\text{open-circuited}} = V_{\mathrm{CA}} = \mathcal{E} \quad (I = 0). \tag{3.115}$$

Thus, the emf of the generator also equals its voltage when open-circuited.

An ideal voltage generator is one with a zero internal resistance ($R_{\mathrm{g}} = 0$), i.e., no internal losses, implying that the voltage of such a generator is always equal to its emf, irrespective of the current flowing through it. That is,

ideal voltage generator

$$\boxed{V_{\mathrm{BA}} = \mathcal{E}} \tag{3.116}$$

for all values of I and all values of R. This, essentially, is a consequence of the impressed electric field intensity, $\mathbf{E}_{\mathrm{i}}$, being completely independent of the current density, $\mathbf{J}$, in the generator.

By Ohm's law, Eq. (3.72), on the other hand, the voltage V_{BA} in Fig. 3.16 can be written as

$$V_{\mathrm{BA}} = RI, \tag{3.117}$$

which, combined with Eq. (3.114), gives

$$\mathcal{E} = R_{\mathrm{g}} I + RI. \tag{3.118}$$

In general, for a closed path in a circuit with many emf sources and resistors (including the internal resistances of the sources), we have

Kirchhoff's voltage law in terms of emf's and voltage drops RI

$$\boxed{\sum_{j=1}^{M} \mathcal{E}_j = \sum_{k=1}^{N} R_k I_k.} \tag{3.119}$$

This equation is an expression of Kirchhoff's voltage law, Eq. (1.92). It states that the algebraic sum of the emf's (voltage rises) around a closed path in a circuit equals the algebraic sum of the voltage drops (RI) across the resistances around the path.

Multiplying both sides of Eq. (3.118) by I, we obtain

$$\mathcal{E} I = R_{\mathrm{g}} I^2 + RI^2. \tag{3.120}$$

By Joule's law, Eq. (3.77), the expressions appearing on the right-hand side of this equation represent the power of Joule's losses in the generator and that in the resistor of resistance R, respectively. Both these powers are lost to heat. Therefore, by the principle of conservation of energy, the expression on the left-hand side of the equation,

$$P_{\rm i} = \mathcal{E} I, \tag{3.121}$$

must be the power generated by the emf of the generator. In local form, the volume density (in $\mathrm{W/m^3}$) of the power of the impressed electric field of intensity $\mathbf{E}_{\rm i}$ is [see Eq. (3.31)]

$$p_{\rm i} = \frac{\mathrm{d}P_{\rm i}}{\mathrm{d}v} = \mathbf{E}_{\rm i} \cdot \mathbf{J}. \tag{3.122}$$

The total power of sources in the domain v is thus [see Eq. (3.32)]

$$\boxed{P_{\rm i} = \int_v \mathbf{E}_{\rm i} \cdot \mathbf{J}\, \mathrm{d}v.} \tag{3.123}$$

power generated by impressed field

Let now the charges in a source region be carried by an impressed nonconduction current of density $\mathbf{J}_{\rm i}$, as depicted in Fig. 3.17(a), independent of the electric field intensity, $\mathbf{E}$. This is the second general model of energy sources we use in electromagnetics.[9] The current density $\mathbf{J}_{\rm i}$ does not enter Ohm's law in local form, and the total current density vector in the region is given by[10]

$$\boxed{\mathbf{J} = \sigma \mathbf{E} + \mathbf{J}_{\rm i},} \tag{3.124}$$

$\mathbf{J}_{\rm i}$ – impressed current density vector (unit: $\mathrm{A/m^2}$)

which, in scalar notation [for the situation in Fig. 3.17(a)], becomes

$$J = -\sigma E + J_{\rm i}. \tag{3.125}$$

This is analogous to a current generator in circuit theory, shown in Fig. 3.17(b) and described by

$$\boxed{I = -G_{\rm g} V + I_{\rm g},} \tag{3.126}$$

current generator

where $I_{\rm g}$ and $G_{\rm g}$ are the current intensity and internal conductance of the generator, respectively. Note that $I_{\rm g}$ equals the current of the generator with its terminals short-circuited ($V = 0$). An ideal current generator has a zero internal conductance ($G_{\rm g} = 0$), or an infinite internal resistance. This means that the current of such a generator is always constant,

$$\boxed{I = I_{\rm g},} \tag{3.127}$$

ideal current generator

i.e., it is independent of the generator terminal voltage (and of the conductance G).

[9] A classical example is so-called Van de Graaff generator, where the impressed current consists of charges transported mechanically on a moving dielectric belt, so that the velocity at which the belt moves actually represents an equivalent drift velocity ($\mathbf{v}_{\rm d}$) of charges. Here, however, the velocity of charges (i.e., the velocity of the belt) obviously does not depend on the electric field intensity vector, $\mathbf{E}$, and Eq. (3.2) does not make sense.

[10] Note that both Eq. (3.109), for a region with energy sources modeled by an impressed electric field, and Eq. (3.124), for a region with an impressed current, can be regarded as versions of the general constitutive equation for the current density, Eq. (3.21). In other words, Eq. (3.21) generally includes models of volume energy sources based on impressed fields and impressed currents, along with conduction properties of materials.

The power density of sources represented by an impressed current of density $\mathbf{J}_\mathrm{i}$ amounts to

$$p_\mathrm{i} = -\mathbf{E} \cdot \mathbf{J}_\mathrm{i}, \tag{3.128}$$

where the minus sign is necessary because vectors $\mathbf{E}$ (the electric field intensity vector, due to positive and negative charges) and $\mathbf{J}_\mathrm{i}$ (the impressed current density vector, which, like any other current density vector, carries positive charges in its direction and negative charges in the opposite direction) are directed oppositely. For the situation in Fig. 3.17(a), $p_\mathrm{i} = EJ_\mathrm{i}$. The total power of sources in v is

power generated by impressed current

$$\boxed{P_\mathrm{i} = -\int_v \mathbf{E} \cdot \mathbf{J}_\mathrm{i} \, \mathrm{d}v.} \tag{3.129}$$

Note that the power generated by I_g in Fig. 3.17(b) is given by

$$P_\mathrm{i} = VI_\mathrm{g}. \tag{3.130}$$

This power is delivered to the rest of the circuit and dissipated to heat in the two resistors, the total power of Joule's losses being $G_\mathrm{g}V^2 + GV^2$.

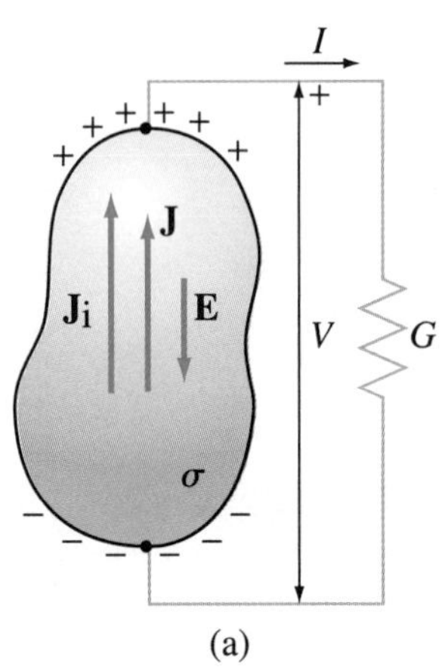

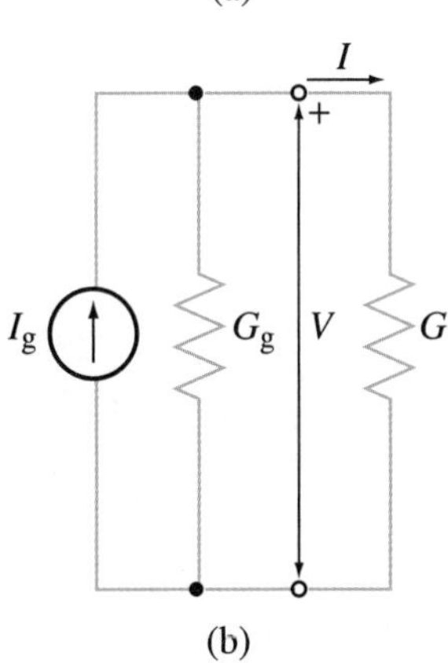

Figure 3.17 (a) External electric energy sources modeled by an impressed electric current, of density $\mathbf{J}_\mathrm{i}$. (b) Current generator in circuit theory.

3.11 ANALYSIS OF CAPACITORS WITH IMPERFECT INHOMOGENEOUS DIELECTRICS

We now deal with steady current fields in capacitors containing piece-wise homogeneous and continuously inhomogeneous imperfect (lossy) dielectrics. Generally, charges in these systems exist not only on the surfaces of capacitor electrodes, but also on the boundary surfaces between homogeneous dielectric layers and over the volume of continuously inhomogeneous dielectrics. The analysis starts with the evaluation of the current distribution, i.e., the computation of the current density vector, $\mathbf{J}$, in the dielectric. The electric field intensity vector, $\mathbf{E}$, is then determined by Ohm's law in local form. By integrating $\mathbf{E}$ through the dielectric, we find the voltage between the electrodes and the conductance (and resistance) of the capacitor. Finally, the distribution of charge in the system can be found from the electric flux density vector, $\mathbf{D}$, using the generalized Gauss' law in differential form and the corresponding boundary condition.

As an illustration, consider a parallel-plate capacitor with two imperfect dielectric layers shown in Fig. 3.18(a). Let the permittivities of the layers be ε_1 and ε_2, their conductivities σ_1 and σ_2, and thicknesses d_1 and d_2, respectively. The surface area of each of the plates is S. The plates can be considered as perfect conductors. Let us analyze this capacitor, assuming that it is connected to an ideal generator of time-invariant voltage V.

The current density vector in the dielectric is normal to the plates [Eq. (3.57)] and uniform in each dielectric layer [Eq. (3.82)]. From the boundary condition for the normal components of $\mathbf{J}$, Eq. (3.55), applied to the interface between the two dielectrics and the continuity equation for steady currents in integral form, Eq. (3.40), applied to a rectangular closed surface enclosing the positive plate [Fig. 3.18(a)], with the right-hand side positioned in either one of the dielectrics, we have

$$J_1 = J_2 = J = \frac{I}{S}, \tag{3.131}$$

where I is the current through the capacitor. The electric field intensities in the dielectrics are

$$E_1 = \frac{J}{\sigma_1} = \frac{I}{\sigma_1 S} \quad \text{and} \quad E_2 = \frac{J}{\sigma_2} = \frac{I}{\sigma_2 S}. \tag{3.132}$$

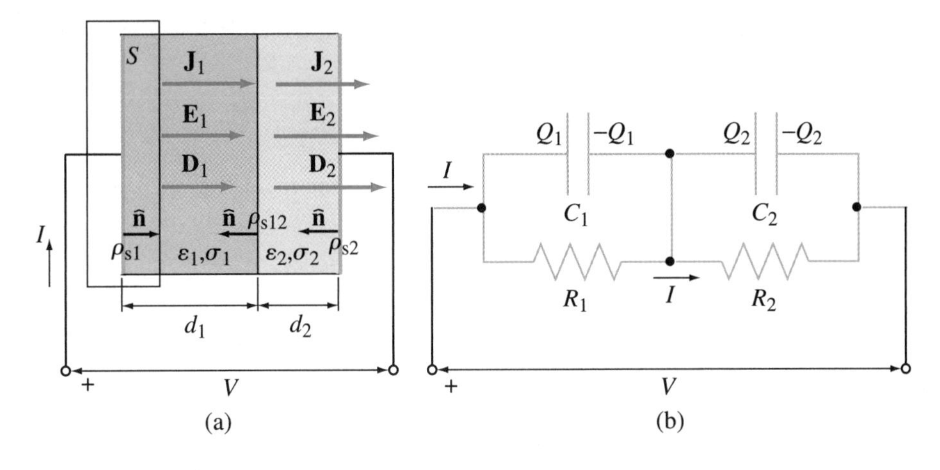

Figure 3.18 Analysis of a capacitor with a two-layer imperfect dielectric: (a) geometry of the structure with vectors **J**, **E**, and **D** in individual layers and (b) equivalent circuit of the system.

The voltage across the capacitor is given by

$$V = E_1 d_1 + E_2 d_2 = \frac{I}{S}\left(\frac{d_1}{\sigma_1} + \frac{d_2}{\sigma_2}\right), \tag{3.133}$$

so that the conductance of the capacitor[11] comes out to be

$$G = \frac{I}{V} = \frac{\sigma_1 \sigma_2 S}{\sigma_2 d_1 + \sigma_1 d_2}. \tag{3.134}$$

The electric flux densities in the layers are[12]

$$D_1 = \varepsilon_1 E_1 = \frac{\varepsilon_1 G V}{\sigma_1 S} \quad \text{and} \quad D_2 = \varepsilon_2 E_2 = \frac{\varepsilon_2 G V}{\sigma_2 S}. \tag{3.135}$$

Eq. (3.62) tells us that, since the dielectric layers are homogeneous, there is no volume charge in them. Using Eq. (3.63), the surface charge on the boundary surface between the layers amounts to

$$\rho_{s12} = D_2 - D_1 = \left(\frac{\varepsilon_2}{\sigma_2} - \frac{\varepsilon_1}{\sigma_1}\right) J = \frac{(\varepsilon_2 \sigma_1 - \varepsilon_1 \sigma_2) V}{\sigma_1 d_2 + \sigma_2 d_1}, \tag{3.136}$$

where we take $J_n = -J$ in Eq. (3.63) because the unit vector $\hat{\mathbf{n}}$ in the corresponding boundary condition is directed from medium 2 to medium 1. Note that if $\varepsilon_1/\sigma_1 = \varepsilon_2/\sigma_2$, $\rho_{s12} = 0$. From Eq. (2.58), the surface charge densities on the plates are

$$\rho_{s1} = D_1 = \frac{\varepsilon_1 \sigma_2 V}{\sigma_1 d_2 + \sigma_2 d_1} \quad \text{and} \quad \rho_{s2} = -D_2 = -\frac{\varepsilon_2 \sigma_1 V}{\sigma_1 d_2 + \sigma_2 d_1}. \tag{3.137}$$

The system in Fig. 3.18(a) can be explained and analyzed also invoking the circuit-theory point of view and concepts. The interface between the dielectric layers is equipotential, and can, therefore, be metalized. We thus get two nonideal capacitors in series, each of them being represented by a parallel connection of an ideal

[11] Note that the current field in the resistor of Fig. 3.12(a) can be analyzed in the same way.

[12] Note that $D_1 \neq D_2$, contrary to Eq. (2.147) for the same capacitor with two perfect dielectric layers, shown in Fig. 2.25(a). The capacitor with a perfect dielectric represents an electrostatic system (with no current), and the analysis starts with the generalized Gauss' law. The capacitor with an imperfect dielectric represents a system with steady current, and the analysis starts with the continuity equation.

capacitor and an ideal resistor, as shown in the equivalent circuit in Fig. 3.18(b). The characteristics of the elements in the circuit are:

$$C_1 = \frac{\varepsilon_1 S}{d_1}, \quad C_2 = \frac{\varepsilon_2 S}{d_2}, \quad R_1 = \frac{d_1}{\sigma_1 S}, \quad R_2 = \frac{d_2}{\sigma_2 S}, \tag{3.138}$$

where the resistances R_1 and R_2 can be obtained either from the corresponding capacitances, C_1 and C_2 [also see Eqs. (2.157)], using the relationship in Eq. (3.102), or from Eq. (3.85).

Using the expression for the equivalent total resistance of a series connection of resistors, Eq. (3.86), the conductance of the system is

$$G = \frac{1}{R_1 + R_2} = \frac{\sigma_1 \sigma_2 S}{\sigma_2 d_1 + \sigma_1 d_2}. \tag{3.139}$$

The charges of the capacitors in Fig. 3.18(b) are computed as

$$Q_1 = C_1 V_1 = C_1 R_1 I = C_1 R_1 G V = \frac{\varepsilon_1}{\sigma_1} G V, \tag{3.140}$$

$$Q_2 = C_2 V_2 = C_2 R_2 I = C_2 R_2 G V = \frac{\varepsilon_2}{\sigma_2} G V, \tag{3.141}$$

with V_1 and V_2 standing for the voltages across individual dielectric layers.[13] Finally, the surface charge densities in Fig. 3.18(a) are given by the expressions

$$\rho_{s1} = \frac{Q_1}{S}, \quad \rho_{s12} = \frac{Q_2 - Q_1}{S}, \quad \text{and} \quad \rho_{s2} = \frac{-Q_2}{S}, \tag{3.142}$$

which give the same results as in Eqs. (3.136) and (3.137).

Example 3.10 Spherical Capacitor with a Continuously Inhomogeneous Imperfect Dielectric

A spherical capacitor is filled with a continuously inhomogeneous imperfect dielectric. The permittivity and conductivity of the dielectric depend on the distance r from the capacitor center and are given by the expressions $\varepsilon(r) = 3\varepsilon_0 b/r$ and $\sigma(r) = \sigma_0 b^2/r^2$ $(a < r < b)$, where a and b are radii of the inner and outer capacitor electrodes, and σ_0 is a (positive) constant. The capacitor is connected to a time-invariant voltage V. Find: (a) the conductance of the capacitor, (b) the free charge distribution of the capacitor, (c) the bound charge distribution of the dielectric, and (d) the total free charge in the capacitor.

Solution

(a) Because of spherical symmetry, the current density vector in the dielectric is of the form given by Eq. (3.46). From the continuity equation, $J(r)$ is the same as in Eq. (3.48), so that the electric field intensity vector in the dielectric is (Fig. 3.19)

$$\mathbf{E} = \frac{\mathbf{J}}{\sigma(r)} = \frac{I}{4\pi\sigma(r)r^2}\hat{\mathbf{r}} = \frac{I}{4\pi\sigma_0 b^2}\hat{\mathbf{r}} = E\hat{\mathbf{r}}, \tag{3.143}$$

with I being the current intensity of the capacitor. We note that the electric field intensity turns out to be the same in the entire dielectric ($E = \text{const}$). The voltage between the

[13]Note that the voltages of the elements in Fig. 3.18(b) are determined by the resistors: $V_1 = VR_1/(R_1 + R_2)$ and $V_2 = VR_2/(R_1 + R_2)$ – resistive voltage divider. In the case of the same system with perfect dielectric layers [Figs. 2.25(a) and 2.26(a)], which is an electrostatic system, the voltages are determined by the capacitors: $V_1 = VC_2/(C_1 + C_2)$ and $V_2 = VC_1/(C_1 + C_2)$ – capacitive voltage divider.

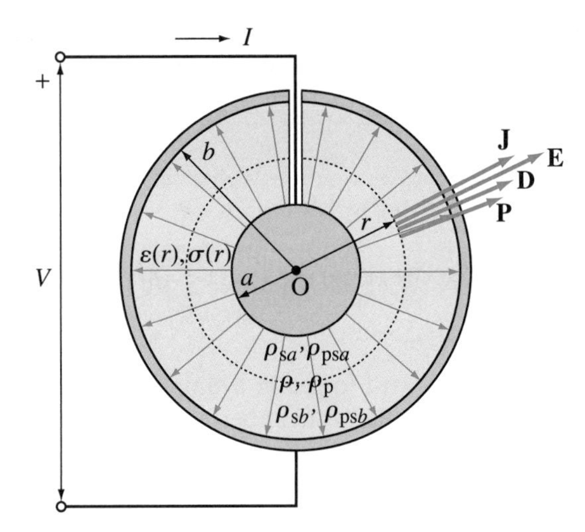

Figure 3.19 Spherical capacitor with a continuously inhomogeneous dielectric with losses; for Example 3.10.

electrodes and the capacitor conductance amount to[14]

$$V = \int_a^b E\,\mathrm{d}r = E(b-a) \quad \longrightarrow \quad G = \frac{I}{V} = \frac{4\pi\sigma_0 b^2}{b-a}. \tag{3.144}$$

(b) The electric flux density vector in the dielectric is given by

$$\mathbf{D} = \varepsilon(r)E\,\hat{\mathbf{r}}, \tag{3.145}$$

where $E = V/(b-a)$. By means of Eqs. (3.62) and (1.171), the free volume charge density inside the dielectric is

$$\rho = \nabla\cdot\mathbf{D} = \frac{1}{r^2}\frac{\mathrm{d}}{\mathrm{d}r}\left[r^2 D(r)\right] = \frac{E}{r^2}\frac{\mathrm{d}}{\mathrm{d}r}\left(r^2\varepsilon\right) = \frac{3\varepsilon_0 bV}{(b-a)r^2}. \tag{3.146}$$

The free surface charge density on the surfaces of the inner and outer capacitor electrodes are

$$\rho_{sa} = D(a^+) = \varepsilon(a^+)E = \frac{3\varepsilon_0 bV}{a(b-a)} \quad \text{and} \quad \rho_{sb} = -D(b^-) = -\varepsilon(b^-)E = -\frac{3\varepsilon_0 V}{b-a}, \tag{3.147}$$

respectively.

(c) From Eq. (2.59), the polarization vector in the dielectric is given by

$$\mathbf{P} = [\varepsilon(r) - \varepsilon_0]E\,\hat{\mathbf{r}}. \tag{3.148}$$

Using Eq. (2.19), the bound volume charge density over the volume of the dielectric is

$$\rho_{\mathrm{p}} = -\nabla\cdot\mathbf{P} = -\frac{1}{r^2}\frac{\mathrm{d}}{\mathrm{d}r}\left[r^2 P(r)\right] = -\frac{\varepsilon_0 V(3b-2r)}{r^2(b-a)}, \tag{3.149}$$

while Eq. (2.23) gives us the bound surface charge density on the surfaces of the dielectric:

$$\rho_{\mathrm{psa}} = -P(a^+) = -\frac{\varepsilon_0(3b-a)V}{a(b-a)} \quad \text{and} \quad \rho_{\mathrm{psb}} = P(b^-) = \frac{2\varepsilon_0 V}{b-a}. \tag{3.150}$$

(d) The total free charge Q in the capacitor is zero. To prove that, we write

$$Q = \int_a^b \rho(r)\,\underbrace{4\pi r^2\,\mathrm{d}r}_{\mathrm{d}v} + \rho_{sa}S_a + \rho_{sb}S_b = 12\pi\varepsilon_0 bV + \frac{12\pi\varepsilon_0 abV}{b-a} - \frac{12\pi\varepsilon_0 b^2 V}{b-a} = 0, \tag{3.151}$$

[14]Note that, since $\varepsilon(r)/\sigma(r) \neq \text{const}$, G is not proportional to C, i.e., Eq. (3.101) is not satisfied. The capacitance C is found in Example 2.19.

where dv is adopted to be the volume of a spherical shell of radius r and thickness dr [see Fig. 1.9 and Eq. (1.33)], $S_a = 4\pi a^2$ (surface area of the inner electrode), and $S_b = 4\pi b^2$ (surface area of the outer electrode).

Problems: 3.10–3.16; *Conceptual Questions* (on Companion Website): 3.22 and 3.23; *MATLAB Exercises* (on Companion Website).

3.12 ANALYSIS OF LOSSY TRANSMISSION LINES WITH STEADY CURRENTS

In this section, we analyze transmission lines with losses in a time-invariant (dc) regime. Consider, as an example, a coaxial cable with imperfect conductors of (finite) conductivity σ_c and an imperfect dielectric of (nonzero) conductivity σ_d. The radius of the inner conductor of the cable is a, the inner and outer radii of the outer conductor are b and c, respectively ($a < b < c$), and the length of the cable is l. The cable is fed by a time-invariant (dc) voltage generator of electromotive force $\mathcal{E}$ and internal resistance R_g, as shown in Fig. 3.20. A load of resistance R_L is connected to the other end of the cable.

Let I designate the current flowing through the inner conductor of the cable. The same current returns through the outer conductor. Since the dielectric in the cable is imperfect, there is a leakage (stray) current through the dielectric, between the cable conductors. Due to cylindrical symmetry of the structure, this current is radial with respect to the cable axis and its density, J, depends only on the radial distance r from the axis. The current leakage, in turn, causes a continual decrease of the current intensity I along the cable, meaning that I is a function of the coordinate z along the cable, $I = I(z)$. Applying the continuity equation, Eq. (3.40), to a cylindrical closed surface S of radius r and length Δz (Fig. 3.20) gives

$$\underbrace{I(z + \Delta z) - I(z)}_{\Delta I} + \underbrace{J2\pi r}_{I'_d}\Delta z = 0 \tag{3.152}$$

[currents $I(z + \Delta z)$ and $I'_d\Delta z$ leave S, while $I(z)$ enters it], where ΔI is the change in current I along a distance Δz, and I'_d is the leakage current per unit length (p.u.l.) of the cable (in A/m). Hence,

current leakage in a transmission line with an imperfect dielectric

$$\boxed{\frac{\Delta I}{\Delta z} = -I'_d = -(I_d)_{\text{p.u.l.}},} \tag{3.153}$$

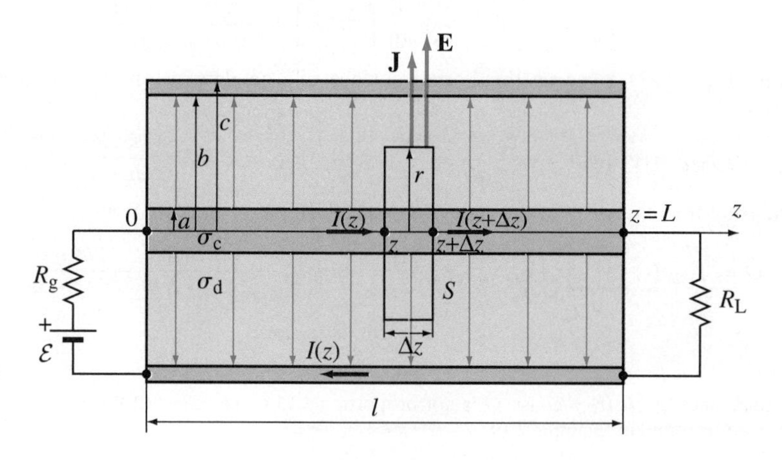

Figure 3.20 Coaxial cable with imperfect conductors and dielectric in a dc regime.

and

$$J = \frac{I'_{\rm d}}{2\pi r}. \tag{3.154}$$

Using Ohm's law in local form, Eq. (3.18), the electric field intensity vector in the dielectric is computed as

$$\mathbf{E} = \frac{\mathbf{J}}{\sigma_{\rm d}} = \frac{I'_{\rm d}}{2\pi\sigma_{\rm d} r}\,\hat{\mathbf{r}}, \tag{3.155}$$

so that the voltage between the inner and outer conductor of the cable amounts to

$$V = \int_a^b E\,{\rm d}r = \frac{I'_{\rm d}}{2\pi\sigma_{\rm d}} \ln\frac{b}{a}. \tag{3.156}$$

By definition, the conductance per unit length of a transmission line, i.e., the conductance for one meter (unit length) of the line divided by 1 m, is

$$\boxed{G' = G_{\rm p.u.l.} = \frac{I'_{\rm d}}{V}.} \tag{3.157}$$

conductance per unit length of a transmission line (unit: S/m)

It is also referred to as the p.u.l. leakage conductance. The unit is S/m. For a coaxial cable, from Eq. (3.156),

$$\boxed{G' = \frac{2\pi\sigma_{\rm d}}{\ln(b/a)}.} \tag{3.158}$$

conductance p.u.l. of a coaxial cable

The drop of the current I along a distance Δz equals $\Delta I = \Delta G V$, where

$$\Delta G = G'\Delta z \tag{3.159}$$

is the conductance through the dielectric at the length Δz.

As Δz approaches zero, the expression on the left-hand side of Eq. (3.153) becomes the derivative of I with respect to z. Therefore, combining Eqs. (3.153) and (3.157), we have

$$\boxed{\frac{{\rm d}I}{{\rm d}z} = -G'V.} \tag{3.160}$$

transmission line with a lossy dielectric

This is a first-order differential equation in z for the current and voltage of a transmission line with a lossy dielectric. It tells us that the rate of change of the current along the line is proportional to the negative of the line voltage, with the conductance per unit length of the line as the proportionality constant. In the case of a transmission line with a perfect dielectric (no losses in the dielectric), $G' = 0$, so that ${\rm d}I/{\rm d}z = 0$, yielding

$$\boxed{I = \text{const.}} \tag{3.161}$$

transmission line with a perfect dielectric

Because of losses in the conductors ($1/\sigma_{\rm c} \neq 0$), the voltage between conductors varies along the cable, i.e., $V = V(z)$. From Eq. (3.85), the total resistance (for time-invariant currents) of the cable turns out to be

$$R = R_1 + R_2 = \frac{l}{\sigma_{\rm c} S_1} + \frac{l}{\sigma_{\rm c} S_2}, \tag{3.162}$$

where R_1 and R_2 are the resistances of the inner and outer conductor, respectively, and S_1 and S_2 are cross-sectional surface areas of the conductors. The resistance per unit length of the line is given by

$$\boxed{R' = R_{\rm p.u.l.} = \frac{R}{l}} \tag{3.163}$$

resistance per unit length of a transmission line (unit: Ω/m) (the unit is Ω/m), so that

dc resistance p.u.l. of a coaxial cable

$$\boxed{R' = \frac{1}{\pi\sigma_c}\left(\frac{1}{a^2} + \frac{1}{c^2 - b^2}\right).} \tag{3.164}$$

The resistance along the length Δz of the line is

$$\Delta R = R'\Delta z, \tag{3.165}$$

and the voltage drop across that resistance

$$V(z) - V(z + \Delta z) = \Delta R I. \tag{3.166}$$

This equation can be rewritten as

$$\underbrace{V(z + \Delta z) - V(z)}_{\Delta V} = -R'\Delta z I, \tag{3.167}$$

which, as Δz approaches zero, becomes

transmission line with lossy conductors

$$\boxed{\frac{dV}{dz} = -R'I.} \tag{3.168}$$

We see that the rate of change of the voltage along a transmission line with lossy conductors is proportional to the negative of the line current, with the resistance per unit length of the line being the proportionality constant. For a transmission line with perfect electric conductors (PEC), $R' = 0$ and $dV/dz = 0$, so that

transmission line with perfect conductors

$$\boxed{V = \text{const.}} \tag{3.169}$$

Eqs. (3.167), (3.153), and (3.157) tell us that each section of a transmission line of length Δz can be represented by a circuit cell consisting of a series resistor of resistance $R'\Delta z$ and a shunt (parallel) resistor of conductance $G'\Delta z$, and the entire transmission line can be replaced by many cascaded equal small cells, as indicated in Fig. 3.21. Thus, a transmission line is said to be a circuit with distributed parameters (parameters per unit length), and Eqs. (3.160) and (3.168) are in fact based on Kirchhoff's laws and Ohm's law for the cells of the circuit.

Eqs. (3.160) and (3.168) are called transmission-line equations or telegrapher's equations for time-invariant currents and voltages on transmission lines. For a transmission line with both $R' \neq 0$ and $G' \neq 0$, these two equations are coupled differential equations in two unknowns: $I(z)$ and $V(z)$. By taking the derivative with respect to z of one equation and substituting its right-hand side by the corresponding expression from the other equation, we can eliminate one unknown (current

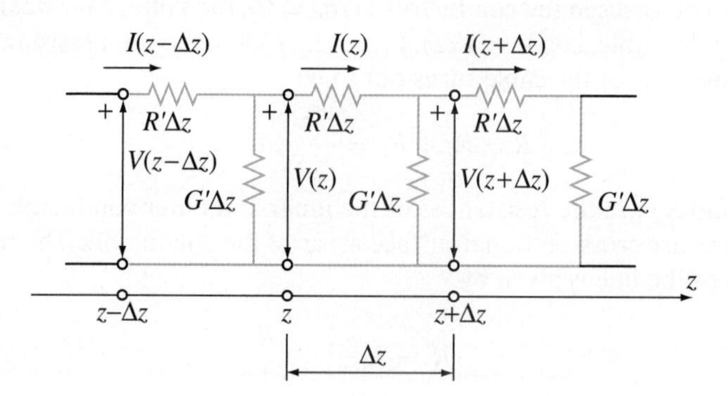

Figure 3.21 Circuit-theory representation of a transmission line with losses in conductors and dielectric in a dc regime.

or voltage) and obtain the following second-order differential equations with only voltage and only current as unknowns:

$$\frac{d^2V}{dz^2} - R'G'V = 0 \quad \text{and} \quad \frac{d^2I}{dz^2} - R'G'I = 0 \quad (R' \neq 0,\ G' \neq 0). \tag{3.170}$$

Their solutions are exponential functions for the voltage and current along the line. If $R' = 0$ and $G' \neq 0$, $V = \text{const}$ along the line [Eq. (3.169)], and the solution to Eq. (3.160) is a linear function for the current along the line. Conversely, if $G' = 0$ and $R' \neq 0$, $I = \text{const}$ along the line [Eq. (3.161)], and the solution to Eq. (3.168) is a linear function for the voltage along the line. Finally, if both $G' = 0$ and $R' = 0$, both current and voltage do not vary along the line.

Note that using the duality relationship in Eq. (3.101), the leakage conductance per unit length of a transmission line with homogeneous imperfect dielectric can be found from the capacitance per unit length of the line as

$$\boxed{G' = \frac{\sigma_d}{\varepsilon} C',} \tag{3.171}$$

duality of G' and C'

where ε is the permittivity of the dielectric [for a coaxial cable, see Eqs. (3.158) and (2.123)]. Note also that $R' \neq 1/G'$ in Eqs. (3.170).

The power of Joule's losses per unit length of a transmission line, P'_J, can be obtained as the power of Joule's losses in one cell in Fig. 3.21, divided by Δz. The unit is W/m. Specifically, in conductors of the line (series resistor in Fig. 3.21),

$$\boxed{(P'_J)_c = \frac{\Delta R I^2}{\Delta z} = R'I^2.} \tag{3.172}$$

Joule's or ohmic losses p.u.l. in line conductors (unit: W/m)

In the dielectric (shunt resistor in Fig. 3.21),

$$\boxed{(P'_J)_d = \frac{\Delta G V^2}{\Delta z} = G'V^2.} \tag{3.173}$$

Joule's (ohmic) losses p.u.l. in line dielectric (unit: W/m)

The total power of Joule's losses per unit length of the line is hence

$$P'_J = R'I^2 + G'V^2. \tag{3.174}$$

Example 3.11 Transmission Line with Perfect Conductors and Imperfect Dielectric

A transmission line of length l and conductance per unit length G' is connected to an ideal voltage generator of time-invariant emf $\mathcal{E}$. The other end of the line is terminated in a load of resistance R_L. The losses in the conductors of the line can be neglected. Find: (a) the distribution of current along the line conductors, (b) the total power of Joule's losses in the line, and (c) the power of the generator.

Solution

(a) With reference to Fig. 3.22,[15] since there are no losses in the line conductors, $R' = 0$, the voltage along the line is [Eq. (3.169)]

$$V(z) = \mathcal{E} \quad (0 \leq z \leq l). \tag{3.175}$$

[15] Although resembles a two-wire line, a pair of parallel horizontal lines in Fig. 3.22 is a symbolic representation of an arbitrary two-conductor transmission line, with conductors of completely arbitrary cross section and generally inhomogeneous dielectric.

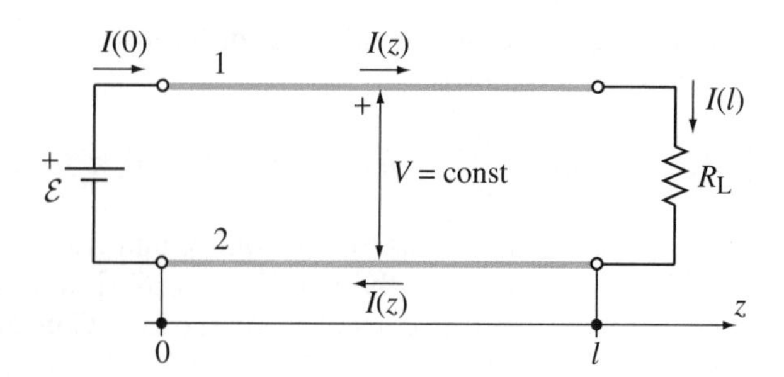

Figure 3.22 Transmission line with lossless conductors and lossy dielectric in a dc regime; for Example 3.11.

Eq. (3.160) becomes

$$\frac{\mathrm{d}I}{\mathrm{d}z} = -G'\mathcal{E}, \tag{3.176}$$

and by integrating it, we obtain

$$I(z) = -G'\mathcal{E}z + I_0 \quad (0 \le z \le l). \tag{3.177}$$

From the condition $I(l) = V/R_\mathrm{L} = \mathcal{E}/R_\mathrm{L}$ (the line current at $z = l$ equals the current through the load), the integration constant is found to be $I_0 = G'\mathcal{E}l + \mathcal{E}/R_\mathrm{L}$ [$I_0 = I(0)$ is the current of the generator], so that

$$I(z) = G'\mathcal{E}(l - z) + \frac{\mathcal{E}}{R_\mathrm{L}} \quad (R' = 0,\ G' \neq 0). \tag{3.178}$$

(b) Based on Eq. (3.173), the power of Joule's losses per unit length of the line is given by

$$P'_\mathrm{J} = (P'_\mathrm{J})_\mathrm{d} = G'V^2 = G'\mathcal{E}^2 = \text{const} \quad (0 \le z \le l). \tag{3.179}$$

As P'_J does not vary along the line, the total power of Joule's losses in the line, including the losses in the load, comes out to be

$$P_\mathrm{J} = \underbrace{P'_\mathrm{J}l}_{\text{line}} + \underbrace{\frac{\mathcal{E}^2}{R_\mathrm{L}}}_{\text{load}} = \left(G'l + \frac{1}{R_\mathrm{L}}\right)\mathcal{E}^2. \tag{3.180}$$

(c) Eq. (3.121) tells us that the power of the generator is

$$P_\mathrm{i} = \mathcal{E}I(0) = \mathcal{E}\left(G'\mathcal{E}l + \frac{\mathcal{E}}{R_\mathrm{L}}\right). \tag{3.181}$$

Note that

$$P_\mathrm{i} = P_\mathrm{J}, \tag{3.182}$$

which, of course, is in agreement with the principle of conservation of energy and power; namely, the power generated by the generator is delivered to the rest of the circuit, and dissipated to heat in the lossy dielectric of the transmission line and in the resistive load.

Example 3.12 Thin Two-Wire Line with Losses

Determine the resistance and conductance per unit length of a thin symmetrical two-wire line with lossy conductors and a lossy dielectric. The conductivities of both conductors and of the dielectric are σ_c and σ_d, respectively. The conductor radii are a and the distance between their axes is d ($d \gg a$).

Solution The resistance per unit length of the line is twice that of a single wire:

dc resistance p.u.l. of a thin two-wire line

$$\boxed{R' = 2\,\frac{1}{\sigma_\mathrm{c}\pi a^2}} \tag{3.183}$$

[see also Eq. (3.164)].

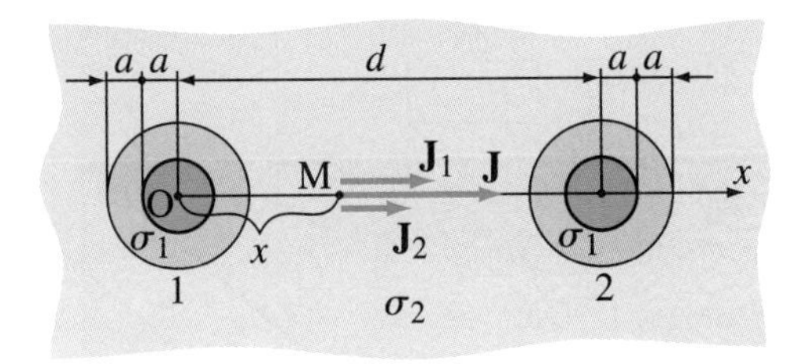

Figure 3.23 Cross section of a thin symmetrical two-wire line with imperfect dielectric coatings immersed in a conducting liquid; for Example 3.13.

From Eqs. (3.171) and (2.141), the conductance per unit length of the line is

$$G' = \frac{\sigma_d}{\varepsilon}C' = \frac{\pi\sigma_d}{\ln(d/a)}. \tag{3.184}$$

conductance p.u.l. of a thin two-wire line

Example 3.13 Two-Wire Line with Imperfect Coatings in a Conducting Liquid

Conductors of a thin two-wire line are coated with thin coaxial layers of imperfect dielectric of conductivity σ_1. The radius of each wire is a, the thickness of coatings is also a, and the distance between wire axes is d ($d \gg a$). The line is immersed in a liquid of conductivity σ_2. What is the conductance per unit length of this line?

Solution The evaluation of the (leakage) conductance per unit length of the line, whose cross section is shown in Fig. 3.23, can be performed completely in parallel to the evaluation of the capacitance per unit length of the line in Fig. 2.31. Here, instead of Q' we have I'_d – the current that leaks from conductor 1 of the line through the coatings and the liquid to conductor 2 of the line, per unit of its length. The leakage current densities due to the individual conductors, i.e., the current density due to the leakage from conductor 1 and that due to the leakage into conductor 2, evaluated at the point M in Fig. 3.23, are

$$J_1 = \frac{I'_d}{2\pi x} \quad \text{and} \quad J_2 = \frac{I'_d}{2\pi(d-x)}, \tag{3.185}$$

respectively. The total current density along the x-axis between the conductors is $J = J_1 + J_2$. The electric field intensity is $E = J/\sigma_1$ in the coatings and $E = J/\sigma_2$ in the liquid. The voltage between the conductors, V, is calculated as in Eq. (2.182), and the conductance per unit length of the line comes out to be

$$G' = \frac{I'_d}{V} = \pi\left(\frac{1}{\sigma_1}\ln 2 + \frac{1}{\sigma_2}\ln\frac{d}{2a}\right)^{-1}. \tag{3.186}$$

Note that this expression for G' can be obtained also by substituting the permittivities in the expression for C' in Eq. (2.183) by the corresponding conductivities. Namely, the inhomogeneity in terms of σ in the system in Fig. 3.23 is of the same form as the inhomogeneity in terms of ε in the system in Fig. 2.31, and a sort of duality between the conductance and capacitance of the two systems can be exploited.

Example 3.14 Planar Line with Two Imperfect Dielectric Layers

Fig. 3.24 shows a planar transmission line consisting of two parallel metallic strips of width w and a two-layer dielectric between the strips. The strips are perfectly conducting, whereas both dielectric layers are imperfect, with conductivities σ_1 and σ_2. The thicknesses of layers are d_1 and d_2, and the length of the line is l. The line is connected at one end to an ideal voltage generator of time-invariant emf $\mathcal{E}$. The other end of the line is open. Calculate (a) the conductance per unit length of the line and (b) the current along the strips.

Solution

(a) An application of the continuity equation to a rectangular closed surface S portrayed in Fig. 3.24, much like that in Eq. (3.152), yields

$$Jw\Delta z = I(z) - I(z+\Delta z) = I'_d\Delta z \tag{3.187}$$

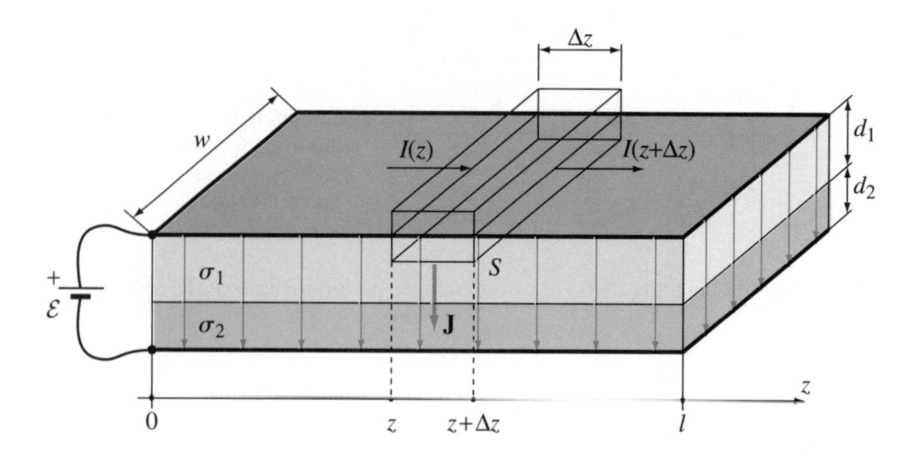

Figure 3.24 Planar transmission line with a two-layer imperfect dielectric; for Example 3.14.

($I'_{\rm d}$ is the leakage current per unit length of the line), from which, the current density in both dielectric layers is

$$J = \frac{I'_{\rm d}}{w}. \tag{3.188}$$

Having in mind Eqs. (3.133) and (3.157), the voltage between the strips and the conductance per unit length of the line are obtained as

$$V = \frac{J}{\sigma_1} d_1 + \frac{J}{\sigma_2} d_2 = \frac{I'_{\rm d}}{w} \left(\frac{d_1}{\sigma_1} + \frac{d_2}{\sigma_2} \right) \quad \longrightarrow \quad G' = \frac{I'_{\rm d}}{V} = w \left(\frac{d_1}{\sigma_1} + \frac{d_2}{\sigma_2} \right)^{-1}. \tag{3.189}$$

(b) The current along the line is given by Eq. (3.178) with $R_{\rm L} \to \infty$ (the line is open):

$$I(z) = G' \mathcal{E}(l - z) = \frac{w\mathcal{E}(l - z)}{d_1/\sigma_1 + d_2/\sigma_2}. \tag{3.190}$$

Problems: 3.17–3.22; *Conceptual Questions* (on Companion Website): 3.24–3.26; *MATLAB Exercises* (on Companion Website).

3.13 GROUNDING ELECTRODES

Consider a metallic body (electrode) of arbitrary shape buried under the flat surface of the earth. This body may represent a grounding electrode for some electrical or electronic device (residing on the earth's surface).[16] Assume that the conductivity of the earth is the same, σ, in the entire lower half-space. Let a time-invariant current of intensity I flow from the electrode into the earth (this current is supplied to the electrode through a thin insulated wire), as shown in Fig. 3.25(a). As the conductivity of the electrode is much larger than σ, Eq. (3.57) tells us that the current lines leaving the electrode are perpendicular to its surface. From Eq. (3.58), on the other side, the current lines in the earth near to its boundary with air are tangential (parallel) to the boundary surface.

[16]Grounding electrodes, in general, are used for draining static electricity induced on parts of a device, for equalizing the electric potential of a device housing with the neighboring conducting objects, for reducing the so-called conduction interference (undesired currents induced in a device, interfering with "regular" currents in the device and with neighboring devices), for routing (together with lightning rods) the charge flow in lightning to the ground, etc.

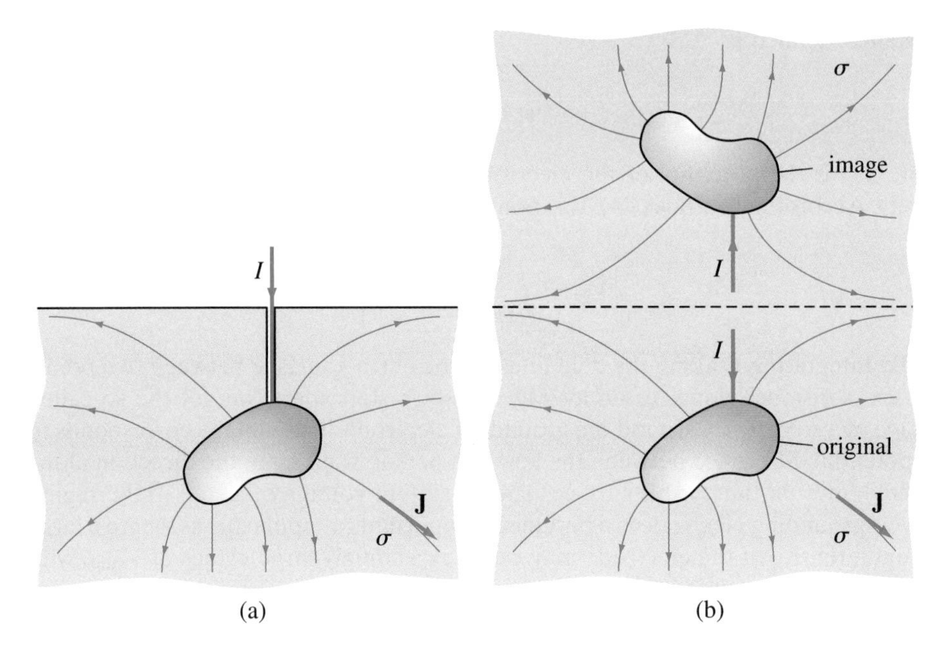

Figure 3.25 Grounding electrode with a steady current (a) and the equivalent system with the earth-air boundary removed (b) – image theory for steady currents.

Consider now the system shown in Fig. 3.25(b), obtained by taking an image of the lower half-space of Fig. 3.25(a) in the earth-air boundary plane. Namely, the grounding electrode in the lower half-space is the same as in the original system and a symmetrical electrode (image electrode) with respect to the boundary plane is introduced in the upper half-space, while the entire space around the electrodes is filled with the medium of conductivity σ. The current of the image electrode is the same as that of the original one (I), and it flows also out of the electrode (positive image). By symmetry, the total current density vector (due to both electrodes) at an arbitrary point of the plane of symmetry in the system in Fig. 3.25(b) has a horizontal component only. This means that the resultant current lines are tangential to the plane of symmetry, and this plane corresponds to the ground-air interface in the system in Fig. 3.25(a). We conclude that the distributions of current in the lower half-space in the systems in Figs. 3.25(a) and (b) are the same. This is the so-called image theory (or theorem) for steady currents, which states that an arbitrary steady-current source (e.g., a grounding electrode with a steady current flowing into the ground) situated in a conducting half-space near a flat boundary surface with a nonconducting medium (e.g., earth-air interface) can be replaced by a new system of sources in an infinite conducting medium. The new system consists of the original source configuration itself and its positive image in the boundary plane. The image electrode (i.e., the charge on its surface) actually represents an equivalent for the surface charge on the boundary between two half-spaces in the original system. There is no direct way to determine this charge and take its contribution to the total electric field into account in the original system, and that is why the equivalent (homogenized) system, with the boundary removed, is generally much simpler for analysis.

After the current density vector, $\mathbf{J}$, in the earth for a given current I of a grounding electrode is determined by analyzing the equivalent system in Fig. 3.25(b), all other quantities of interest for a specific application can be found – in the original system in Fig. 3.25(a). The electric field intensity vector in the ground and on its surface is $\mathbf{E} = \mathbf{J}/\sigma$. An important parameter of a grounding electrode is its grounding resistance, defined as

grounding resistance

$$\boxed{R_{gr} = \frac{V_{gr}}{I},} \tag{3.191}$$

where V_{gr} is the potential of the electrode with respect to the reference point at infinity. According to Eq. (1.74), this potential is given by

potential of a grounding electrode (reference point at infinity)

$$\boxed{V_{gr} = \int_{\text{electrode}}^{\text{infinity}} \mathbf{E} \cdot d\mathbf{l}.} \tag{3.192}$$

By integrating $\mathbf{E}$ along the field lines on the earth's surface between two points that are a distance equal to an average person's step apart, we get the so-called voltage of a step, V_{step}, around the grounding electrode. This voltage corresponds to the potential difference between the feet of a person walking in the direction along the field lines on the earth's surface. The maximum voltage of a step in the region above a grounding electrode is a parameter important for applications where a large current intensity of the electrode may cause exceedingly large values of V_{step}, with even fatal consequences.

Example 3.15 Hemispherical Grounding Electrode

A hemispherical grounding electrode of radius $a = 1$ m is buried in the earth of conductivity $\sigma = 10^{-3}$ S/m, with its base up, as shown in Fig. 3.26(a). The current of the electrode is $I = 1000$ A. Find (a) the grounding resistance of the electrode and (b) the voltage of a step between points on the earth's surface that are $r_1 = 2$ m and $r_2 = r_1 + b$ away from the electrode center, where $b = 0.75$ m (average person's step).

Solution

(a) Using the image theory for steady currents, Fig. 3.25, we get the equivalent system shown in Fig. 3.26(b), which consists of two hemispherical electrodes pressed onto each other, forming thus a spherical electrode with current $2I$ flowing into a homogeneous medium of conductivity σ. Due to spherical symmetry, the current density vector, $\mathbf{J}$, in the medium is radial. Consequently, $\mathbf{J}$ in the original system is also radial and of the form given by Eq. (3.46). Applying the continuity equation to a spherical surface S of radius r

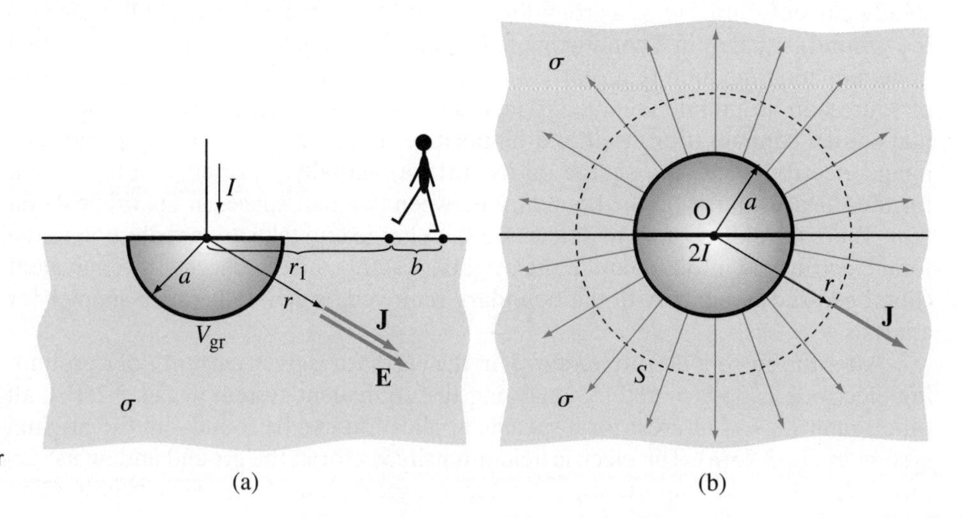

Figure 3.26 (a) Hemispherical grounding electrode and (b) equivalent spherical electrode in a homogeneous medium; for Example 3.15.

positioned around the electrode gives

$$J(r) = \frac{2I}{4\pi r^2} = \frac{I}{2\pi r^2} \quad (a < r < \infty). \tag{3.193}$$

Note that the same result is obtained by applying the continuity equation to a hemispherical surface of radius r in the original system – in Fig. 3.26(a).

Combining Eqs. (3.18) and (3.193), the electric field intensity in the earth is

$$E(r) = \frac{J(r)}{\sigma} = \frac{I}{2\pi\sigma r^2}. \tag{3.194}$$

With the use of Eqs. (3.192) and (3.191), the potential of the electrode with respect to the reference point at infinity and its grounding resistance come out to be

$$V_{gr} = \int_{r=a}^{\infty} E\,dr = \frac{I}{2\pi\sigma a} \longrightarrow \boxed{R_{gr} = \frac{V_{gr}}{I} = \frac{1}{2\pi\sigma a} = 159\ \Omega.} \tag{3.195}$$

grounding resistance of a hemispherical electrode

(b) The voltage of a step is as large as

$$\boxed{V_{step} = \int_{r_1}^{r_2} E\,dr = \frac{I}{2\pi\sigma}\left(\frac{1}{r_1} - \frac{1}{r_2}\right) = \frac{Ib}{2\pi\sigma r_1 r_2} = 21.7\ \text{kV} \quad (b = r_2 - r_1).} \tag{3.196}$$

voltage of a step; b is an average person's step

Example 3.16 Two Hemispherical Grounding Electrodes

The electric circuit shown in Fig. 3.27 consists of an ideal voltage generator of emf $\mathcal{E} = 100$ V, a wire of negligible resistance, and two hemispherical grounding electrodes. The radius of each electrode is $a = 2$ m, the conductivity of the earth is $\sigma = 10$ mS/m, and the distance between the electrode centers is $d = 100$ m. Compute the current in the circuit.

Solution Since $d \gg a$, the potentials with respect to the reference point at infinity of the individual electrodes can be evaluated independently from each other, that is, as for a single isolated grounding electrode. The equivalent resistance seen by the generator equals therefore $2R_{gr}$, where the expression for R_{gr} is given in Eqs. (3.195). As the resistance of the wire is negligible, the current in the circuit is

$$I = \frac{\mathcal{E}}{2R_{gr}} = \pi\sigma a\mathcal{E} = 6.28\ \text{A}. \tag{3.197}$$

Example 3.17 Hemispherical Grounding Electrode in a Two-Layer Earth

Conductivity of a layer of earth around a hemispherical grounding electrode shown in Fig. 3.28 is $\sigma_1 = 5 \times 10^{-3}$ S/m and that of the rest of the earth $\sigma_2 = 10^{-3}$ S/m. The radii of the electrode and the boundary surface between the earth layers are $a = 1$ m and $b = 2$ m, respectively. The current intensity of the electrode is $I = 100$ A. Determine (a) the grounding resistance of the electrode and (b) the power of Joule's losses in the earth.

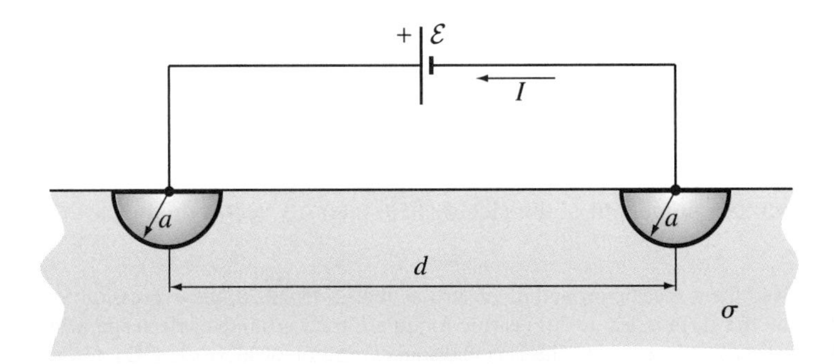

Figure 3.27 Two hemispherical grounding electrodes in a dc circuit; for Example 3.16.

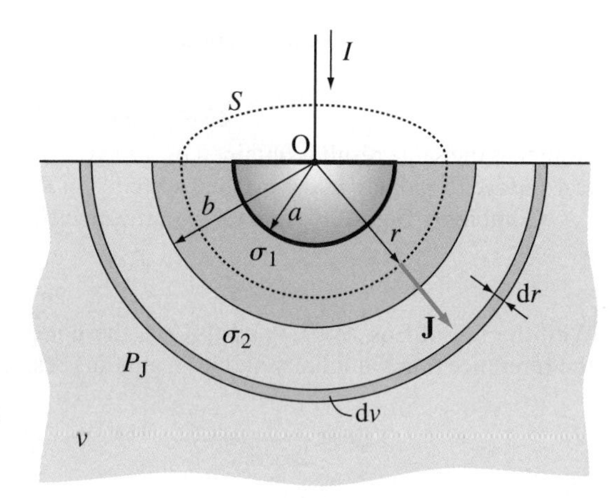

Figure 3.28 Hemispherical grounding electrode in an earth consisting of two concentric layers; for Example 3.17.

Solution

(a) Because of symmetry, the current flow in both layers of earth is radial. Applying the continuity equation to the closed surface S shown in Fig. 3.28, we obtain the same expression for the current density J in the earth as in Eq. (3.193). The electric field intensity vector in the earth is $\mathbf{E}_1 = \mathbf{J}/\sigma_1$ for $a < r < b$ and $\mathbf{E}_2 = \mathbf{J}/\sigma_2$ for $b < r < \infty$. By means of Eq. (3.192), the potential of the electrode with respect to the reference point at infinity is given by

$$V_{\rm gr} = \int_a^\infty E\,{\rm d}r = \int_a^b E_1\,{\rm d}r + \int_b^\infty E_2\,{\rm d}r = \frac{I}{2\pi}\left[\frac{1}{\sigma_1}\left(\frac{1}{a} - \frac{1}{b}\right) + \frac{1}{\sigma_2 b}\right], \tag{3.198}$$

from which the grounding resistance, Eq. (3.191), is

$$R_{\rm gr} = \frac{V_{\rm gr}}{I} = \frac{1}{2\pi b}\left(\frac{b-a}{\sigma_1 a} + \frac{1}{\sigma_2}\right) = 95.5\ \Omega. \tag{3.199}$$

(b) Using Eq. (3.32), and combining it with Eq. (3.193), the power of Joule's losses in the earth turns out to be

$$P_{\rm J} = \int_v JE\,{\rm d}v = \int_a^\infty \frac{I}{2\pi r^2}\,E\,\underbrace{2\pi r^2\,{\rm d}r}_{{\rm d}v} = I\int_a^\infty E\,{\rm d}r, \tag{3.200}$$

where $\mathrm{d}v$ is the volume of a hemispherical shell of radius r and thickness $\mathrm{d}r$ (portrayed in Fig. 3.28). Noting that the last integral in this equation equals $V_{\rm gr}$ [Eq. (3.198)], we have

power of Joule's losses in the earth surrounding a grounding electrode

$$\boxed{P_{\rm J} = V_{\rm gr} I = R_{\rm gr} I^2,} \tag{3.201}$$

and $P_{\rm J} = 955$ kW for given numerical data.[17]

Example 3.18 Spherical Grounding Electrode Deep in the Earth

A spherical grounding electrode of radius $a = 30$ cm is buried in a homogeneous earth of conductivity $\sigma = 10^{-2}$ S/m such that its center is at a depth $d = 6$ m with respect to the ground surface, as shown in Fig. 3.29(a). There is a steady current flowing through the electrode, and its potential with respect to the reference point at infinity is $V_{\rm gr} = 15$ kV. Calculate the maximum tangential component of the electric field intensity vector on the earth's surface.

[17] Although derived for a specific grounding electrode, that in Fig. 3.28, the expression for the power of Joule's losses in the earth in Eq. (3.201) is true for an arbitrary grounding electrode, with grounding resistance $R_{\rm gr}$ and current I. Note that this is actually an expression of Joule's law, Eq. (3.77).

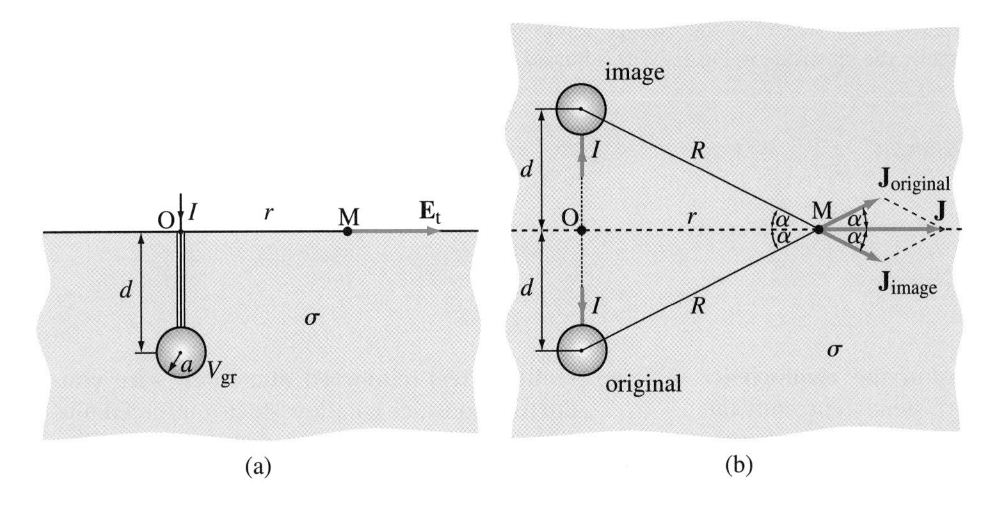

Figure 3.29 (a) Spherical grounding electrode buried deep in the earth and (b) equivalent system – by virtue of the image theory for steady currents; for Example 3.18.

Solution Since $d \gg a$, the influence of the boundary with air on the grounding resistance of the electrode can be neglected, meaning that R_{gr} can be evaluated as for an electrode in the infinite homogeneous medium of conductivity σ. Invoking the duality relationship in Eq. (3.102), we can further relate R_{gr} to the capacitance (C) of the same metallic sphere situated in an infinite dielectric medium of permittivity ε. Eq. (2.121) gives this capacitance for $\varepsilon = \varepsilon_0$, so that

$$\boxed{R_{gr} = \frac{\varepsilon_0}{\sigma C} = \frac{1}{4\pi\sigma a} = 26.5\ \Omega.} \tag{3.202}$$

grounding resistance of a spherical electrode buried deep in the earth

Hence, the current of the electrode is $I = V_{gr}/R_{gr} = 565.5$ A.

Applying the image theory for steady currents (Fig. 3.25) and the superposition principle, the resultant current density vector at a point M in Fig. 3.29(b), the position of which is determined by a radial distance r from the projection of the electrode centers on the plane of symmetry (point O), can be obtained as

$$\mathbf{J} = \mathbf{J}_{original} + \mathbf{J}_{image}, \tag{3.203}$$

where $\mathbf{J}_{original}$ and $\mathbf{J}_{image}$ are the current density vectors due to the current flow from individual electrodes. Making use of the fact that $d \gg a$, these current densities can be evaluated independently from each other, which results in

$$J_{original} = J_{image} = \frac{I}{4\pi R^2} \quad \longrightarrow \quad J = 2J_{original}\cos\alpha = \frac{Ir}{2\pi R^3} \tag{3.204}$$

($\cos\alpha = r/R$ and $R = \sqrt{r^2 + d^2}$).

The tangential component of the electric field intensity vector on the earth's surface in the original system, Fig. 3.29(a), is

$$E_t(r) = \frac{J}{\sigma} = \frac{Ir}{2\pi\sigma(r^2 + d^2)^{3/2}}. \tag{3.205}$$

From the condition

$$\frac{dE_t(r)}{dr} = 0, \tag{3.206}$$

we get that the field intensity E_t is maximum on a circle of radius $r_{max} = d/\sqrt{2} = 4.24$ m, centered at the point O. This maximum field is

$$(E_t)_{max} = E_t(r_{max}) = \frac{I\sqrt{3}}{9\pi\sigma d^2} = 96\ \text{V/m}. \tag{3.207}$$

Note that the maximum voltage of a step, taking $b = 0.75$ m as an average person's step, is approximately

$$\left(V_{step}\right)_{max} \approx (E_t)_{max}\, b = 72\ \text{V}, \tag{3.208}$$

which is much less than the voltage found in Eq. (3.196), for instance. This is expected, because the electrode in Fig. 3.29(a) is buried deep in the earth.

Problems: 3.23–3.29; *Conceptual Questions* (on Companion Website): 3.27–3.30; *MATLAB Exercises* (on Companion Website).

Problems

3.1. Charge density in terms of the conductivity gradient. In a region with steady currents, the conductivity of the material varies with spatial coordinates, $\sigma \neq$ const, whereas permittivity does not, $\varepsilon =$ const. If the electric potential in the region is V, show that the volume charge density is given by $\rho = (\varepsilon/\sigma)\nabla\sigma \cdot \nabla V$.

3.2. Various computations for a copper wire with steady current. For a copper wire with a steady current described in Example 3.1, compute: (a) the number of electrons that pass through a cross section of the wire during one hour, (b) the electric field intensity in the wire, (c) the voltage between the wire ends, (d) the volume density of the power of Joule's losses in the wire material, (e) the total power of ohmic losses in the wire, (f) the resistance of the wire at room temperature, and (g) the wire resistance at the temperature of 100°C ($\alpha_{\text{Cu}} = 0.0039\ \text{K}^{-1}$).

3.3. Parallel-plate capacitor with an imperfect dielectric. A parallel-plate capacitor has a homogeneous imperfect dielectric of permittivity ε and conductivity σ. The separation between plates is d and the plate area is S. If the voltage between the plates is V (time-invariant), find (a) the current distribution in the dielectric, (b) the current through the capacitor terminals, (c) the conductance of the capacitor, (d) the power of Joule's losses in the capacitor, (e) the free charge distribution in the capacitor, and (f) the bound charge distribution in the capacitor.

3.4. Spherical capacitor half filled with a conducting liquid. A spherical capacitor has a poorly conducting liquid dielectric occupying a half of the space between the electrodes. The radii of electrodes are a and b ($a < b$), and the conductivity of the dielectric is σ. Determine the conductance of this capacitor.

3.5. Hollow steel-reinforced aluminum wire conductor. Consider a hollow steel-reinforced aluminum wire conductor of length $l = 100$ m. The radius of the cylindrical hole in the central part of the conductor is $a = 5$ mm, the thickness of the stainless-steel ($\sigma_{\text{steel}} = 1.1$ MS/m) reinforcement is $b - a = 5$ mm, and the rest of the conductor, whose overall radius is $c = 2$ cm, is aluminum ($\sigma_{\text{Al}} = 35$ MS/m). What is the resistance of this wire, and what is the ratio of current intensities $I_{\text{Al}}/I_{\text{steel}}$ in the two materials for a given dc voltage drop between the two ends of the wire?

3.6. Resistor with two cuboidal parts. A resistor is formed from two rectangular cuboids of the same size, with sides $a = 8$ mm, $b = 2$ mm, and $c = 4$ mm. The cuboids are made out from different resistive materials, with conductivities $\sigma_1 = 10^5$ S/m and $\sigma_2 = 4 \times 10^5$ S/m. The voltage between the resistor terminals is $V = 50$ V. Find the electric field intensity, current density, current intensity, and power of Joule's losses in each of the two cuboids if they are connected as in (a) Fig. 3.30(a) and (b) Fig. 3.30(b).

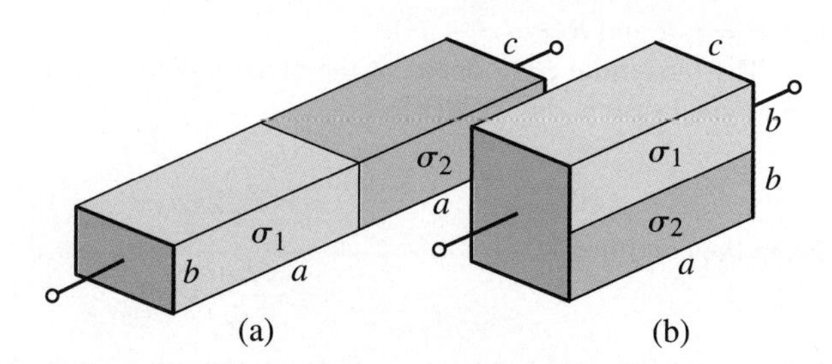

Figure 3.30 Two rectangular cuboids made out from different resistive materials connected in (a) series and (b) parallel; for Problem 3.6.

3.7. Conductivity gradient along the resistor current. Fig. 3.31 shows a right-angled cuboidal resistor of dimensions a, b, and c, made out from a continuously inhomogeneous resistive material. The conductivity of the material is given by the following function of the x-coordinate: $\sigma(x) = \sigma_0/(1 + 9x/a)$ $(0 \le x \le a)$, where σ_0 is a constant. Calculate the resistance of this resistor.

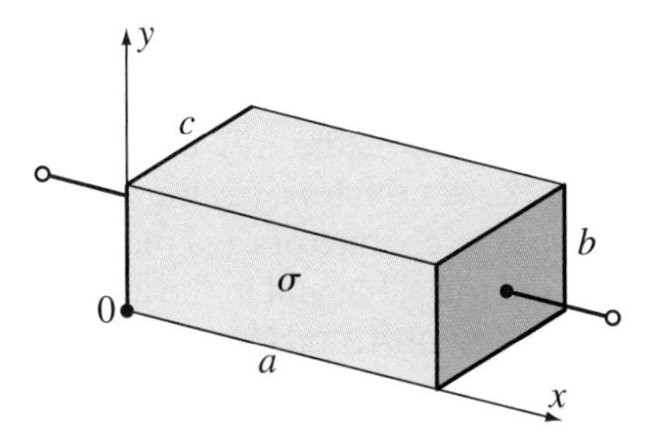

Figure 3.31 Right-angled cuboidal resistor made out from a continuously inhomogeneous resistive material; for Problems 3.7–3.9.

3.8. Conductivity gradient normal to the resistor current. Assume that the conductivity of the material in Fig. 3.31 is a function of the y-coordinate, $\sigma(y) = \sigma_0[1 + 9\sin(\pi y/b)]$ $(0 \le y \le b,\ \sigma_0 = \text{const})$, and find the resistance of the resistor.

3.9. Integrated-circuit resistor with exponential conductivity profile. An integrated-circuit (IC) resistor is built by diffusing a layer of p-type impurity into an n-type background material. As a consequence of the diffusion process, the concentration of impurity is not uniform across the treated region, so that the obtained IC resistor can essentially be represented by the structure in Fig. 3.31 with the conductivity varying as $\sigma = \sigma(y)$. In specific, $\sigma(y)$ exponentially decreases from $\sigma_1 = 100$ S/m at the air-resistor surface to $\sigma_2 = 0.1$ S/m at the interface with the n-type background, which is considered to be nonconducting, and the length, thickness, and width of the p-type region are $a = 4\ \mu$m, $b = 1\ \mu$m, and $c = 2\ \mu$m, respectively. Determine the resistance of the IC resistor.

3.10. Spherical capacitor with two lossy dielectric layers. The dielectric of a spherical capacitor with electrode radii $a = 5$ cm and $c = 15$ cm is composed of two concentric imperfect dielectric layers, where the radius of the boundary surface between the layers is $b = 10$ cm. The relative permittivity and conductivity of the layer near the inner electrode are $\varepsilon_{r1} = 12$ and $\sigma_1 = 4 \times 10^{-4}$ S/m, and those of the other layer $\varepsilon_{r2} = 7$ and $\sigma_2 = 5 \times 10^{-6}$ S/m. The capacitor is connected to a time-invariant voltage $V = 100$ V. (a) Find the current distribution in the dielectric. Using this result and integration (field-theory approach), compute (b) the conductance of the capacitor, (c) the power of Joule's losses in each of the layers, and (d) the free charge on each of the boundaries of radii a, b, and c.

3.11. Solution using equivalent circuit with ideal elements. Repeat parts (b)–(d) of the previous problem but using the equivalent circuit with two ideal capacitors and two ideal resistors in Fig. 3.18(b) (circuit-theory approach).

3.12. Continuously inhomogeneous imperfect dielectric. Fig. 3.32 shows a parallel-plate capacitor with circular plates of radius a and a continuously inhomogeneous imperfect dielectric. The permittivity and conductivity of the dielectric are the following functions of the z-coordinate: $\varepsilon(z) = 2(1 + 3z/d)\varepsilon_0$ and $\sigma(z) = \sigma_0/(1 + 3z/d), 0 \le z \le d$, where σ_0 is a constant and d is the separation between the plates. The capacitor is connected to a time-constant voltage V. Find (a) the current distribution in the dielectric, (b) the conductance of the capacitor, (c) the power of Joule's losses in the capacitor, (d) the free charge distribution in the capacitor, and (e) the bound charge distribution in the capacitor.

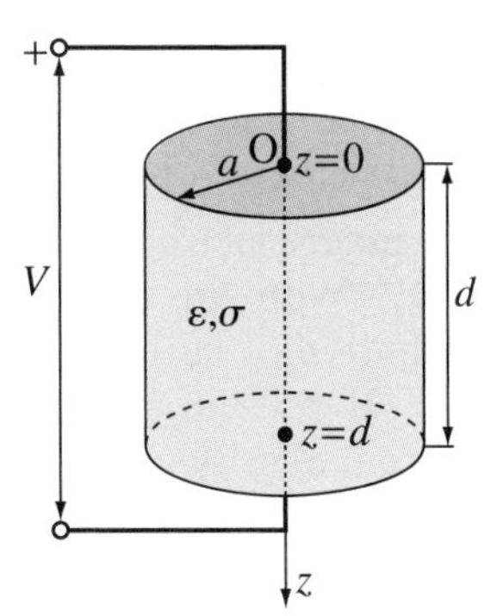

Figure 3.32 Parallel-plate capacitor with a continuously inhomogeneous lossy dielectric; for Problem 3.12.

3.13. Parallel-plate capacitor with two lossy dielectrics. Consider a parallel-plate capacitor with a piece-wise homogeneous dielectric shown in Fig. 3.33. The permittivities of the dielectric parts are ε_1 and ε_2. Both parts are lossy, with conductivities σ_1 and σ_2. The separation between the plates is d and the cross-sectional areas of the dielectric parts are S_1 and S_2. The voltage between the electrodes is V. By employing a field-theory approach (based on determining the current and field distribution in the dielectric), calculate (a) the current through the capacitor terminals, (b) the conductance of the capacitor, (c) the power of Joule's losses in each of the dielectric parts, and (d) the free charge distribution in the capacitor.

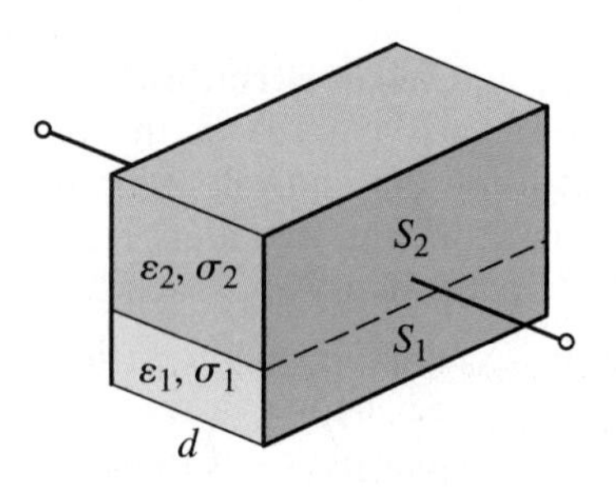

Figure 3.33 Parallel-plate capacitor with two imperfect dielectric parts; for Problem 3.13.

3.14. Equivalent circuit with ideal capacitors and resistors. Repeat the previous problem but employing a circuit-theory approach, i.e., generating and solving an equivalent circuit with ideal capacitors and resistors.

3.15. Current distribution through circular plates. Find the distribution of current through the upper and lower plate of the parallel-plate capacitor with imperfect dielectric from Problem 3.12. Assume that the plates are thin enough so that the current through them can be described using the surface current density vector.

3.16. Current distribution through a thin spherical electrode. Consider the spherical capacitor with imperfect dielectric from Example 3.3, and assume that the inner electrode is hollow, with a very thin wall. With this assumption, determine the distribution of current through the inner electrode. Represent this current using the surface current density vector.

3.17. Coaxial cable with two lossy dielectric layers. Fig. 3.34 shows a longitudinal cross section of a coaxial cable with two coaxial layers of imperfect dielectric. The conductors are perfectly conducting, and their radii are $a = 3$ mm, $c = 10$ mm, and $d = 12$ mm. The radius of the boundary surface between the dielectric layers is $b = 7$ mm and the length of the cable is $l = 100$ m. The relative permittivities of the layers are $\varepsilon_{r1} = 4$ and $\varepsilon_{r2} = 8$, and the conductivities are $\sigma_1 = 10^{-12}$ S/m and $\sigma_2 = 5 \times 10^{-12}$ S/m. The cable is connected at one end to an ideal voltage generator of time-invariant emf $\mathcal{E} = 30$ V, and the other end of the cable is terminated in a load of resistance $R_L = 1$ kΩ. Compute (a) the current density in the dielectric, (b) the conductance per unit length of the cable, (c) the current intensity along the conductors, (d) the current density in each of the conductors, (e) the power of Joule's losses in the cable and in the load, (f) the power of the generator, and (g) the free surface charge density on the boundary between the dielectric layers.

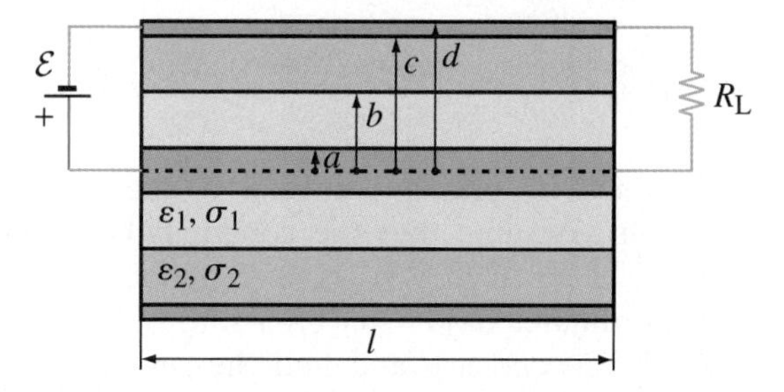

Figure 3.34 Coaxial cable with two coaxial imperfect dielectric layers in a dc regime; for Problem 3.17.

3.18. Coaxial cable with an inhomogeneous imperfect dielectric. A coaxial cable with perfect conductors is filled with an inhomogeneous imperfect dielectric of parameters $\varepsilon = \varepsilon_0(1 + r^2/a^2)$ and $\sigma_d = \sigma_0/(1 + r^2/a^2)$ $(a < r < b)$, where a and b are the radius of the inner cable conductor and the inner radius of the outer conductor, and σ_0 is a constant. The voltage between the cable conductors is V (dc). Calculate (a) the conductance per unit length of the cable and (b) the density of volume free charges in the dielectric.

3.19. Coaxial cable partly filled with a lossy dielectric. A coaxial cable with perfect conductors is partly filled with an imperfect dielectric of conductivity σ, as shown in Fig. 3.35. The conductor radii are a and b, the cable length is l, and the length of the air-filled part of the cable is c. The cable is fed by an ideal dc voltage generator of emf $\mathcal{E}$, and the other end of the cable is open. Find the expression for the current intensity through the cable conductors.

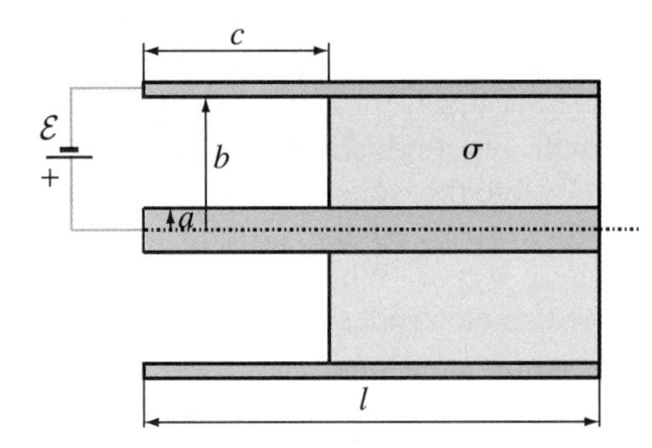

Figure 3.35 Coaxial cable partly filled with a conducting dielectric in a dc regime; for Problem 3.19.

3.20. **Coaxial cable with a poorly conducting spacer.** Shown in Fig. 3.36 is a cross section of an air-filled coaxial cable with a spacer between the conductors. The spacer is made out of an imperfect dielectric of conductivity σ and its cross section is defined by an angle α. The conductor radii are a and b. What is the conductance per unit length of this cable?

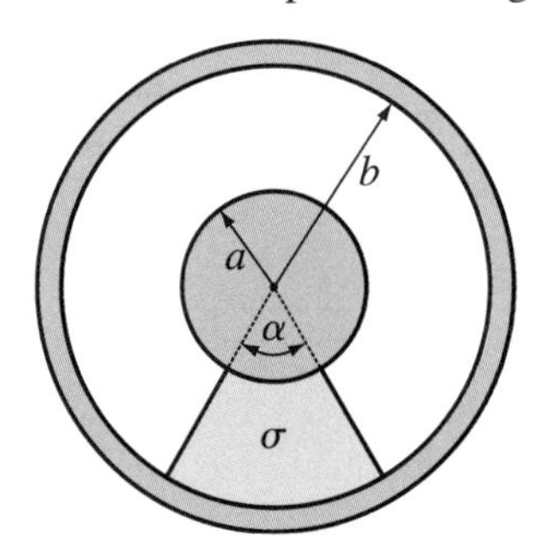

Figure 3.36 Cross section of a coaxial cable with a poorly conducting spacer between conductors; for Problem 3.20.

3.21. **Planar line with imperfect conductors.** If the strips of the planar transmission line in Fig. 3.24 are not perfectly conducting, but have a finite conductivity σ_c, find the dc resistance per unit length of the line. Assume that the thickness of each of the strips is t.

3.22. **Planar line with an inhomogeneous lossy dielectric.** Consider a planar (two-strip) transmission line shown in Fig. 3.37. The strips are perfectly conducting, very thin, w wide, l long, and d apart from each other. The dielectric between the strips is imperfect and continuously inhomogeneous, with parameters $\varepsilon(x) = (4 + 3x/d)\varepsilon_0$ and $\sigma(x) = \sigma_0/(1 + 9x/d)$, $0 \leq x \leq d$, where σ_0 is a constant. The line is connected to an ideal dc voltage generator of emf $\mathcal{E}$, and is open-circuited at the other end. Calculate the distributions of (a) volume current in the dielectric, (b) surface current over the strips, and (c) volume free charge in the dielectric.

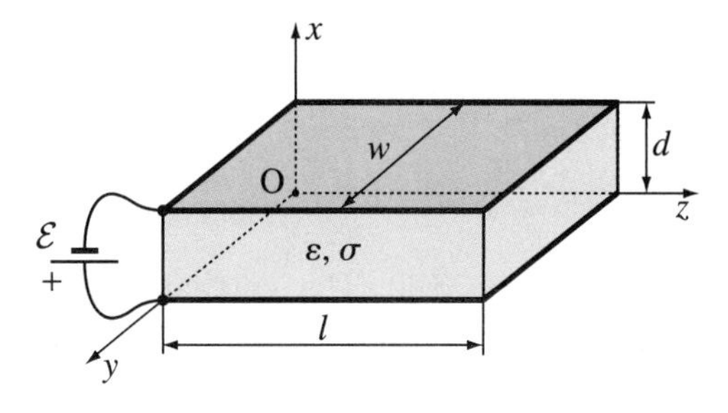

Figure 3.37 Planar line with a continuously inhomogeneous imperfect dielectric in a dc regime; for Problem 3.22.

3.23. **Grounding electrode in an inhomogeneous earth.** Conductivity of the earth around a hemispherical grounding electrode shown in Fig. 3.38 can be described as the following function of the radial coordinate r: $\sigma(r) = \sigma_0\sqrt{a/r}$ $(a < r < \infty)$, where a is the radius of the electrode and σ_0 is a constant. The current intensity of the electrode is I. Find (a) the grounding resistance of the electrode, (b) the total power of Joule's losses in the earth, and (c) the power of Joule's losses in a layer with thickness a around the electrode $(a < r < 2a)$.

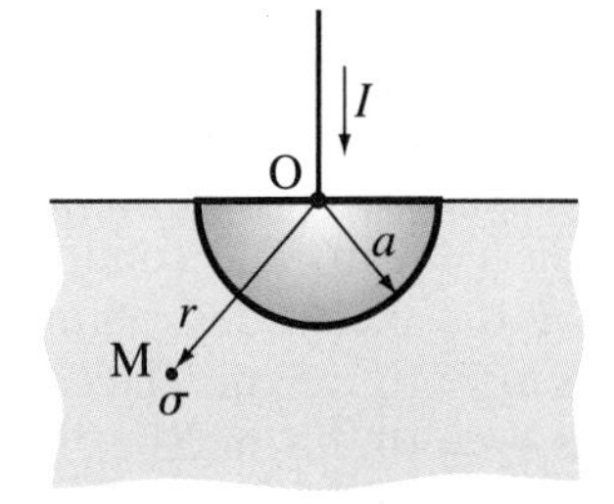

Figure 3.38 Hemispherical grounding electrode in a continuously inhomogeneous earth; for Problem 3.23.

3.24. **Grounding electrode in a two-sector earth.** The earth around a hemispherical grounding electrode of radius $a = 1$ m can be represented as two sectors with conductivities $\sigma_1 = 2 \times 10^{-3}$ S/m and $\sigma_2 = 8 \times 10^{-3}$ S/m, as shown in Fig. 3.39. The current intensity of the electrode amounts to $I = 50$ A. Compute (a) the grounding resistance of the electrode and (b) the power of Joule's losses in each of the two earth sectors.

3.25. **Two short-circuited grounding electrodes.** Two identical hemispherical grounding electrodes of radii a are galvanically connected to each other by a wire of negligible resistance, as

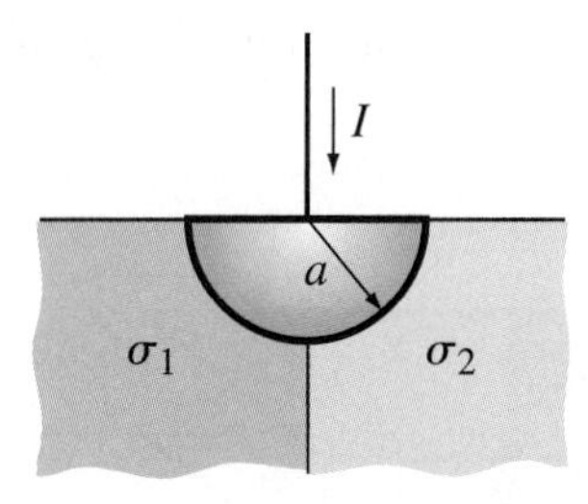

Figure 3.39 Hemispherical grounding electrode in a two-sector earth; for Problem 3.24.

shown in Fig. 3.40. The conductivity of the earth is σ and the distance between the electrode centers is d ($d \gg a$). Determine the grounding resistance of such a system of electrodes.

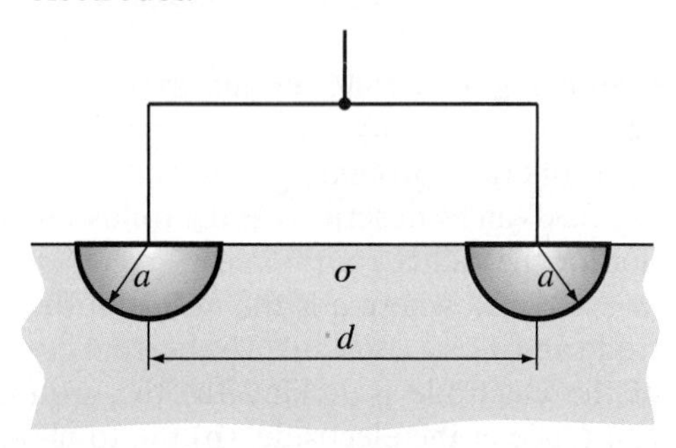

Figure 3.40 Two short-circuited hemispherical grounding electrodes; for Problem 3.25.

3.26. Two grounding electrodes in a layered earth. The electric circuit shown in Fig. 3.41 consists of an ideal current generator of current intensity $I = 100$ A, a wire of negligible resistance, and two identical hemispherical grounding electrodes of radii $a = 2$ m. The conductivity of the earth near the electrodes, within the radius $b = 4$ m, is $\sigma_1 = 10^{-2}$ S/m. The conductivity elsewhere is $\sigma_2 = 10^{-3}$ S/m. The distance between the electrode centers is $d = 80$ m. What is the voltage between the electrodes?

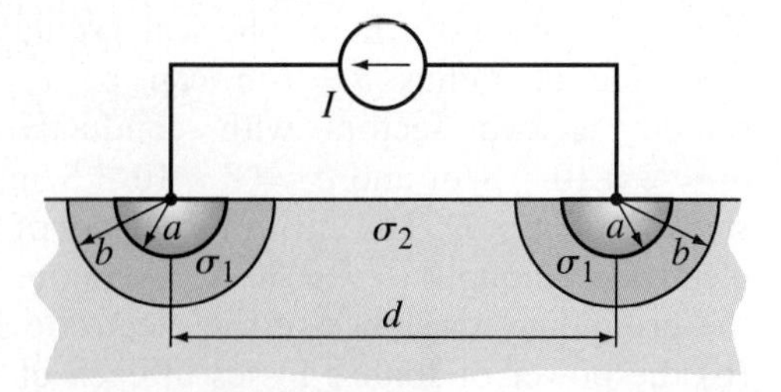

Figure 3.41 Two hemispherical grounding electrodes in a layered earth; for Problem 3.26.

3.27. Two deep spherical grounding electrodes. Consider two identical spherical grounding metallic electrodes of radii a in a homogeneous earth of conductivity σ. Both the distance between the sphere centers and their depth with respect to the ground surface equal d, where $d \gg a$. What is the resistance between the two electrodes?

3.28. Cylindrical grounding electrode deep in the earth. A cylindrical grounding electrode of radius $a = 20$ cm and length $l = 5$ m is buried in a homogeneous earth of conductivity $\sigma = 10^{-2}$ S/m such that its axis is at a depth $d = 4$ m below the ground surface, as depicted in Fig. 3.42. If a steady current of intensity $I = 200$ A flows through the electrode, find the maximum voltage of a step on the earth's surface. Adopt $b = 0.75$ m as an average person's step. Neglect end effects due to the finite length of the cylinder.

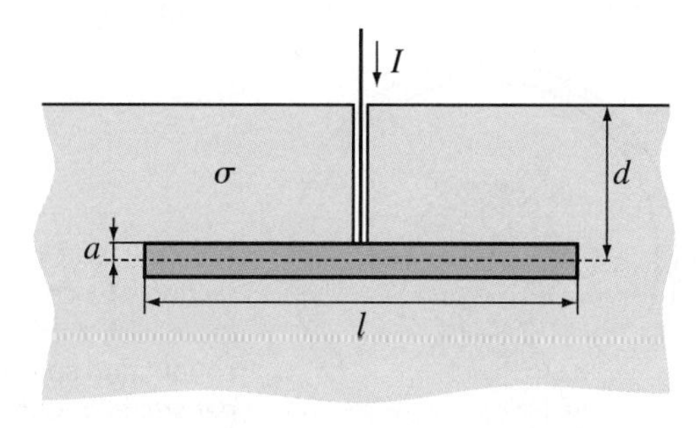

Figure 3.42 Cylindrical grounding electrode buried deep in the earth; for Problem 3.28.

3.29. Two deep parallel cylindrical electrodes. Two identical cylindrical grounding electrodes as the one in Fig. 3.42 are placed parallel to each other (at the same depth with respect to the earth's surface) so that the distance between their axes is $D = 6$ m. If they are connected together in a circuit with an ideal current generator as in Fig. 3.41 (with $I = 200$ A), compute (for the numerical data given in the previous problem) the voltage between the electrodes.

4 Magnetostatic Field in Free Space

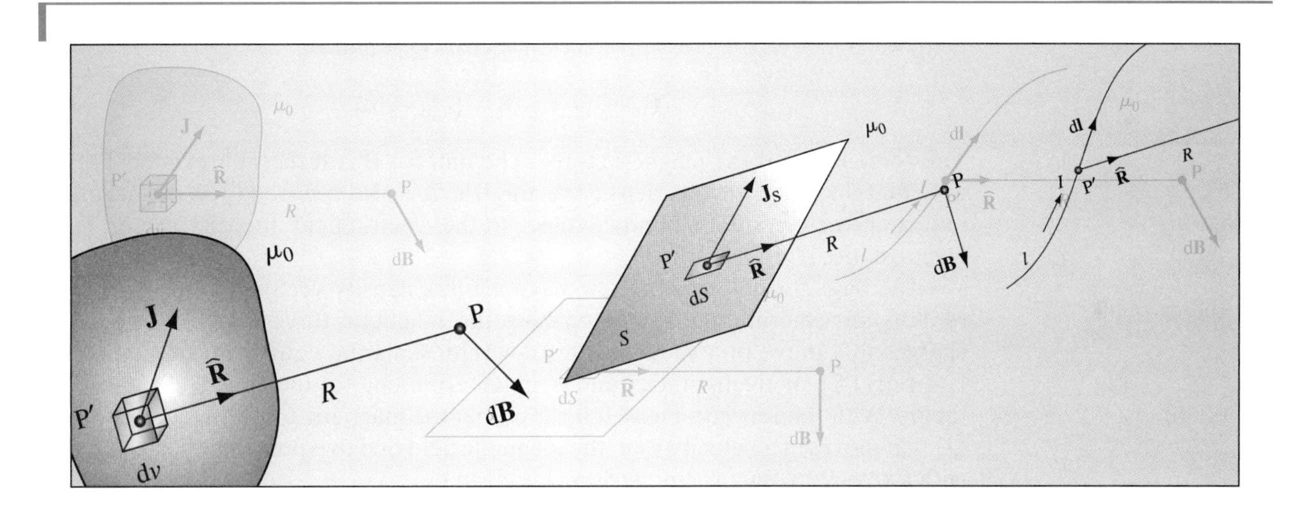

Introduction:

In our studies of steady electric currents in the preceding chapter, we introduced and discussed physical laws and mathematical techniques for determining the distribution of currents for given geometry, material properties, and excitation of various structures. We now introduce a series of new phenomena associated with steady electric currents, which are essentially the consequence of a new simple experimental fact – that conductors with currents exert forces on one another. These forces are called magnetic forces, and the field due to one current conductor in which the other conductor is situated and which causes the force on it is called the magnetic field. Any motion of electric charges and any electric current are followed by the magnetic field. The magnetic field due to steady electric currents is termed the steady (static) magnetic field or magnetostatic field. The theory of the magnetostatic field, the magnetostatics, restricted to a vacuum and nonmagnetic media is the subject of this chapter. The magnetic materials will be discussed in the following chapter.

Starting from the experimental law of the magnetic force between two point charges that move in free space (a vacuum or air) – Coulomb's law for the magnetic field, we shall derive the Biot-Savart law, which, in turn, will serve as a starting point for the derivation of Ampère's law. Both laws represent a means for evaluating the magnetic field due to given steady-current distributions, and we shall apply them to many theoretically and practically important configurations that do not include magnetic materials. The differential form of Ampère's law will introduce a new differential operator, the curl. We shall derive the law of conservation of magnetic flux (Gauss' law for the magnetic field), and complete the full set of Maxwell's equations for the magnetostatic field in nonmagnetic media. The magnetic vector potential will be introduced as a counterpart of the electric scalar potential and the magnetic dipole as the equivalent of the electric dipole. Examples involving evaluation of magnetic forces and torques will also be presented.

4.1 MAGNETIC FORCE AND MAGNETIC FLUX DENSITY VECTOR

The magnetic field is a special physical state existing in a space around moving electric charges and, thus, electric currents in conductors. Its fundamental property is that there is a force acting on any electric charge placed in the space, provided that the charge is moving. That wires carrying electric currents produce magnetic fields was first discovered by Oersted in 1820. To quantitatively describe this field, we introduce a vector quantity called the magnetic flux density vector, **B**. It is defined through the force on a small probe point charge Q_p moving at a velocity **v** in the field, which equals the cross product[1] of vectors $Q_p\mathbf{v}$ and **B**,

definition of **B** *(unit:* T*)*

$$\mathbf{F}_m = Q_p\mathbf{v} \times \mathbf{B} \quad (Q_p \to 0). \tag{4.1}$$

This force is called the magnetic force. The unit for **B** is tesla (abbreviated T). We note that the magnetic flux density vector, which is also referred to as the magnetic induction vector, is defined analogously to the electric field intensity vector, **E**, in electrostatics [Eq. (1.23)].

The basis for determining the magnetic flux density vector due to steady currents is an experimental law describing the magnetic force between two point charges that move in a vacuum. This law represents the equivalent of Coulomb's law, Eq. (1.1), of electrostatics, and is an underpinning for the entire magnetostatic theory. With reference to Fig. 4.1, it states that the magnetic force on a point charge Q_2 that moves at a velocity $\mathbf{v}_2$ in the magnetic field due to a point charge Q_1 moving with a velocity $\mathbf{v}_1$ in a vacuum (or air) is given by

magnetic force

$$\mathbf{F}_{m12} = \frac{\mu_0}{4\pi} \frac{Q_2\mathbf{v}_2 \times \left(Q_1\mathbf{v}_1 \times \hat{\mathbf{R}}_{12}\right)}{R^2}, \tag{4.2}$$

where the notation of vectors is the same as for Coulomb's law in Fig. 1.1, and μ_0 is the permeability of a vacuum (free space),

permeability of a vacuum

$$\mu_0 = 4\pi \times 10^{-7} \text{ H/m} \tag{4.3}$$

(H is henry, the unit for inductance, which will be studied in a later chapter). Eq. (4.2), as mentioned, is the result of experiments, but it can also be derived from Coulomb's law using the special theory of relativity. Note that it is sometimes referred to as Ampère's law of force.

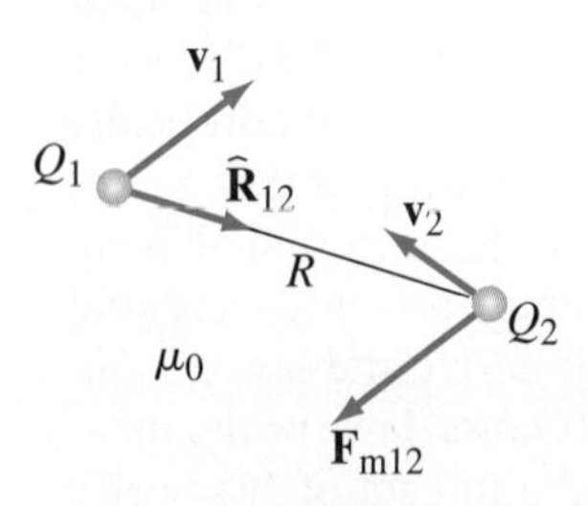

Figure 4.1 Magnetic force between two moving point charges in a vacuum, given by Eq. (4.2).

Although of the similar form, the magnetic force law in Eq. (4.2) is mathematically more complicated than Coulomb's law, because two vector cross products are implied in the equation. We first need to determine the vector resulting from the cross product $Q_1\mathbf{v}_1 \times \hat{\mathbf{R}}_{12}$, and then the cross product of $Q_2\mathbf{v}_2$ and that vector [and, of course, multiply the result by $\mu_0/(4\pi R^2)$]. Let us apply this rule to the situation shown in Fig. 4.2, where vectors $Q_1\mathbf{v}_1$ and $Q_2\mathbf{v}_2$ are parallel to each other and normal to the line joining them. We see that the force between the charges is attractive

[1]The cross product (also referred to as the vector product) of vectors **a** and **b**, $\mathbf{a} \times \mathbf{b}$, is a vector whose magnitude is given by $|\mathbf{a} \times \mathbf{b}| = |\mathbf{a}||\mathbf{b}| \sin\alpha$, where α is the angle between the two vectors in the product. It is perpendicular to the plane defined by the vectors **a** and **b**, and its direction (orientation) is determined by the right-hand rule when the first vector (**a**) is rotated by the shortest route toward the second vector (**b**). In this rule, the direction of rotation is defined by the fingers of the right hand when the thumb points in the direction of the cross product.

HISTORICAL ASIDE

Hans Christian Oersted (1777–1851), a Danish physicist, was a professor of physics and chemistry at the University of Copenhagen. As part of a classroom demonstration, he discovered, apparently by accident, that an electric current flowing through a wire produces a magnetic field around the wire. Namely, while preparing an evening lecture on electricity and magnetism in April of 1820, in which he wanted to demonstrate the heating effects of the electric current in a wire connected between the terminals of a voltaic source (battery), Oersted noticed that the magnetic needle of a compass sitting next to the wire spun off of the north position every time the battery was in use. It was generally believed at that time that electricity and magnetism were not related. Therefore, he was extremely surprised and excited with what he saw. He continued experimenting during that lecture, which eventually led him to the conclusion that an electric current creates a magnetic field – and electromagnetism was born. He also discovered that, not only is a magnetic needle deflected by an electric current, but that a wire with a current is also deflected in a magnetic field (by the magnetic force), thus laying the foundation for the construction of the electric motor. His epoch-making discovery of the magnetic effects of electric currents was announced to the French Academy of Sciences on September 4, 1820. It immediately set off an explosion of research activity by many brilliant minds, including Ampère (1775–1836), Biot (1774–1862), and Savart (1791–1841). Oersted's experiment was the first demonstration of a connection between electricity and magnetism, and is considered the foundation of the modern study of electromagnetism. In 1829, Oersted became the first director of the Polytechnic Institute in Copenhagen, now the Technical University of Denmark. *(Portrait: AIP Emilio Segrè Visual Archives, Brittle Books Collection)*

if the vectors are in the same direction [Fig. 4.2(a)], and repulsive if they are in opposite directions [Fig. 4.2(b)]. This is formally just opposite to Coulomb's law, where like charges repel and unlike attract each other.

Combining Eqs. (4.1) and (4.2), and assuming that the second charge in Fig. 4.1 is a probe charge ($Q_2 = Q_\mathrm{p}$), we can identify the expression for the magnetic flux density vector of a point charge Q moving with a velocity $\mathbf{v}$:

$$\boxed{\mathbf{B} = \frac{\mu_0}{4\pi}\frac{Q\mathbf{v}\times\hat{\mathbf{R}}}{R^2}}, \tag{4.4}$$

B *due to a moving point charge in free space*

where R is the distance from the charge and $\hat{\mathbf{R}}$ is the unit vector along R directed from the source point toward the field point, as shown in Fig. 4.3. Vector $\mathbf{B}$ is perpendicular to R and perpendicular to the plane of the vectors $Q\mathbf{v}$ and $\hat{\mathbf{R}}$. Its

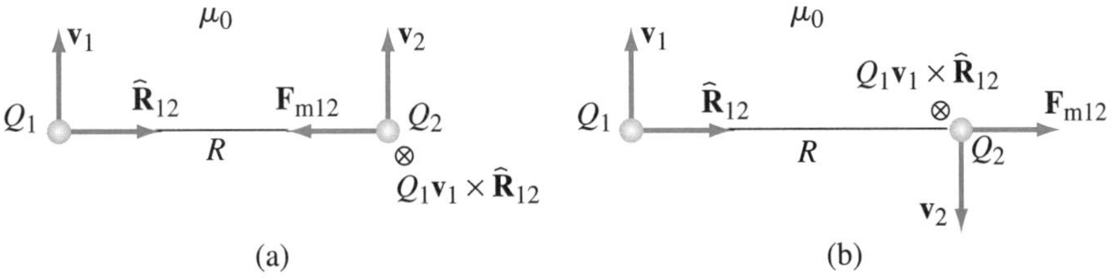

Figure 4.2 Two charges moving parallel to each other in the same direction (a) and in opposite directions (b) in a vacuum.

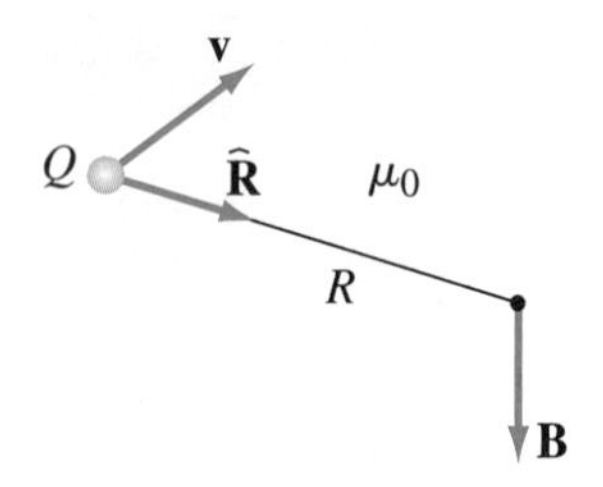

Figure 4.3 Magnetic flux density vector due to a point charge moving in a vacuum.

orientation comes from the definition of the cross product of vectors, i.e., it is determined by the right-hand rule in rotating $Q\mathbf{v}$ by the shortest route to $\hat{\mathbf{R}}$.

Note that the magnetic flux density is proportional to the product $Q\mathbf{v}$, which can be regarded as a measure of sources of the magnetostatic field (the measure of sources of the electrostatic field is a charge Q). Comparing Eq. (4.4) to the expression for the electric field intensity vector due to a point charge, Eq. (1.24), we observe the same dependence on the amount of sources ($Q|\mathbf{v}|$ and Q) and on distance R. The constant μ_0 corresponds to the constant $1/\varepsilon_0$. The only formal difference is the direction of the field vector with respect to the source. The electric field intensity vector, $\mathbf{E}$, due to Q is radial with respect to Q (Fig. 1.7), whereas the magnetic flux density vector, $\mathbf{B}$, due to $Q\mathbf{v}$ is circular with respect to $Q\mathbf{v}$ (Fig. 4.3). In other words, the lines of $\mathbf{E}$ are radials starting at Q, while the lines of $\mathbf{B}$ are circles centered on the line containing the vector $Q\mathbf{v}$.

Conceptual Questions (on Companion Website): 4.1 and 4.2.

HISTORICAL ASIDE

Nikola Tesla (1856–1943), an American inventor, was a brilliant electrical engineer with no complete formal education. He made alternating current (ac) practical and had about 250 patents issued in the U.S. and other countries. Tesla was born in Smiljan near Gospić, Lika, in Croatia (then in the Habsburg Empire of Austria), in a Serbian family. He studied mechanical engineering at the Polytechnic School in Graz from 1875 to 1878 and then natural philosophy at the University of Prague until 1880, but never graduated due to a shortage of funds after his father's death. While working as electrical engineer for companies in Budapest and Paris, he came up with an idea of the principle of the rotating magnetic field and polyphase alternating currents producing it in 1882 and constructed the world's first induction (ac) motor in 1883, both during after-work hours. He emigrated to America in 1884, with four cents in his pocket, and worked for a year for Thomas Edison (1847–1931). He failed to get Edison interested in the induction motor and alternating current, and later established his own "Tesla Electric Company" in New York City. Tesla obtained most of his best known U.S. patents in the area of polyphase alternating currents and polyphase systems, including ac generators, motors, and transformers, as well as principles of power distribution using alternating currents, in 1887 and 1888, and delivered his famous lecture "A New System of Alternate Current Motors and Transformers" before the American Institute of Electrical Engineers (now IEEE) on May 16, 1888. In June of 1888, George Westinghouse (1846–1914), head of the "Westinghouse Electric Company" in Pittsburgh, bought the first seven of Tesla's patents for polyphase systems. This marked the beginning of the five-year "war of currents" to decide whether Edison's existing dc systems or newly proposed Tesla-Westinghouse ac systems would be the chosen technology for the global electric power distribution of the future in America. Tesla invented an air-core resonant transformer known as a Tesla coil (or Tesla transformer) in 1891, which was capable of generating alternating currents of extremely high frequencies for that time (up to a hundred of thousands of cycles per second or 100 kHz in today's notation). The coil could also build up tremendously high voltages and cause spectacular spark discharges. Tesla's new electric lamps based on a high-frequency power supply were the forerunners of our fluorescent tubes and neon signs. In a series of spectacular lectures-demonstrations in America and Europe from 1891 to 1893 aimed at promoting alternating current and high-frequency technology, he was able to power electric lamps without wires, either through air or by allowing high-frequency electricity to flow through his body, produce artificial lightning flashes of different shapes, and even shoot large lightning bolts from his coils to the audience without harm. In early 1893, he proposed a complete radio system with a transmitter and receiver in the form of Tesla coils tuned to resonate at the same frequency as a basis for wireless communications. The "war of currents" ended on the evening of May 1, 1893, with the opening of the Chicago World Exposition, spectacularly illuminated by a hundred thousand lamps with alternating current from Tesla's generators. Only couple of months later, Westinghouse was awarded the contract to build the world's first ac hydroelectric power plant on Niagara Falls, based on Tesla's design, which was put into operation on November 16, 1896. Tesla filed his basic radio patent applications in 1897, and this is, in fact, the invention of radio, although it was Guglielmo Marconi (1874–1937) who put it into practical and commercial use, and was awarded Nobel Prize in Physics in 1909 for the development of wireless telegraphy. In 1898, Tesla demonstrated the world's first wireless remote control (of a boat model) at Madison Square Garden. He continued his investigations of the wireless transmission of power in 1899 and 1900 in Colorado Springs, where he built a huge, 200-kW radio transmitter consisting of a 142-foot metal mast with a spherical top and the world's largest Tesla coil 51 feet in diameter. Tesla also patented a bladeless disk turbine, so-called Tesla turbine, in 1913, as well as many other devices and systems in various areas of science and engineering. The SI unit for the magnetic flux density, the tesla, was named in his honor in 1960. *(Portrait: © Nikola Tesla Museum, Belgrade, Serbia)*

4.2 BIOT-SAVART LAW

If we have many point charges moving in free space, the total magnetic flux density vector is, by the principle of superposition, a vector sum of the magnetic flux density vectors due to individual charges. In a current, a vast number of elementary free charges moves through a conductor in an organized manner with the macroscopic average velocity $\mathbf{v}_d$ (drift velocity). The sum of products $Q\mathbf{v}$ for all charges in an elementary volume dv is given by

$$\left(\sum Q\mathbf{v}\right)_{\text{in } dv} = N_{\text{in } dv}\, Q\mathbf{v}_d = N_v\, dv\, Q\mathbf{v}_d = \mathbf{J}\, dv, \tag{4.5}$$

where $\mathbf{J}$ is the current density vector, defined by Eq. (3.3). We assume that the permeability everywhere[2] is μ_0. Hence, the magnetic flux density vector due to the charges in dv is computed as

$$\boxed{d\mathbf{B} = \frac{\mu_0}{4\pi}\frac{(\mathbf{J}\, dv) \times \hat{\mathbf{R}}}{R^2},} \tag{4.6}$$

Biot-Savart law

which is known as the Biot-Savart law. We see that the product $\mathbf{J}\, dv$ represents a macroscopic volume elemental source of the magnetic field, and we call it accordingly the volume current element. By integrating Eq. (4.6), we obtain the expression

[2]We shall see in the next chapter that most materials we encounter in science and engineering are nonmagnetic materials, that is, their permeability is practically that of a vacuum. Examples are commonly used metallic conductors such as copper, aluminum, silver, gold, etc., most frequently encountered natural environments (air, water, and ground), biological tissues, and practically all insulators and semiconductors. Some materials, such as iron, steel, cobalt, nickel, ferrites, etc., on the other hand, have permeability much greater than μ_0.

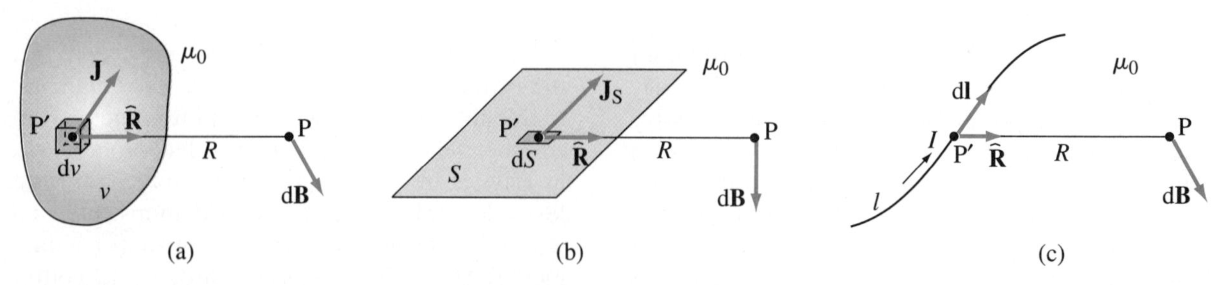

Figure 4.4 Magnetic flux density vector due to three characteristic current distributions – volume current (a), surface current (b), and line current (c).

for the resultant magnetic flux density vector due to a current distribution in an entire volume v [Fig. 4.4(a)]:

Biot-Savart law for volume current

$$\mathbf{B} = \frac{\mu_0}{4\pi} \int_v \frac{(\mathbf{J}\,\mathrm{d}v) \times \hat{\mathbf{R}}}{R^2}. \tag{4.7}$$

HISTORICAL ASIDE

Jean Baptiste Biot (1774–1862), a French physicist, was a student of Lagrange (1736–1813) at the École Polytechnique and a junior professor of mechanics under sponsorship of Laplace (1749–1827) at the Collège de France. Biot's most important contributions to science are in the theory of polarized light and effects on it by organic substances. **Fèlix Savart** (1791–1841), another French physicist, was also a professor at the Collège de France, where he taught experimental physics and acoustics, and a member of the Paris Academy. His most important scientific work was in vibration and acoustics, and particularly in the physics of the violin. Biot and Savart showed experimentally in October of 1820, soon after Oersted's discovery of magnetic effects of electric currents, that the magnetic field produced by a current in a long, straight wire is inversely proportional to the distance from the wire.

In the case of a surface current flowing over a surface S [see Eqs. (3.12) and (3.13)], we have $N_s\,\mathrm{d}S$ instead of $N_v\,\mathrm{d}v$ in Eq. (4.5), yielding $\mathbf{J}_s\,\mathrm{d}S$ as a surface current element, where $\mathbf{J}_s$ is the surface current density vector. The magnetic flux density vector due to a surface current distribution is thus [Fig. 4.4(b)]

Biot-Savart law for surface current

$$\mathbf{B} = \frac{\mu_0}{4\pi} \int_S \frac{(\mathbf{J}_s\,\mathrm{d}S) \times \hat{\mathbf{R}}}{R^2}. \tag{4.8}$$

Finally, for a line current along a line l, e.g., a current of intensity I flowing through a (generally curvilinear) thin wire of length l and cross-sectional area S, $\mathrm{d}v = S\,\mathrm{d}l$, $J = I/S$ [Eq. (3.5)], and $J\,\mathrm{d}v = JS\,\mathrm{d}l = I\,\mathrm{d}l$, where $\mathrm{d}l$ is an elemental segment along l. We conclude that the line current element equals $I\,\mathbf{dl}$, where $\mathbf{dl}$ is oriented in the reference direction of the current flow. For curved lines, $\mathbf{dl}$ is tangential to the line. The magnetic flux density vector due to a line current is hence [Fig. 4.4(c)]

Biot-Savart law for line current

$$\mathbf{B} = \frac{\mu_0}{4\pi} \int_l \frac{I\,\mathbf{dl} \times \hat{\mathbf{R}}}{R^2}. \tag{4.9}$$

In summary, we note that Eqs. (4.7)–(4.9) have the same generic mathematical form: integral of μ_0(current element) $\times\, \hat{\mathbf{R}}/(4\pi R^2)$, with the following three current

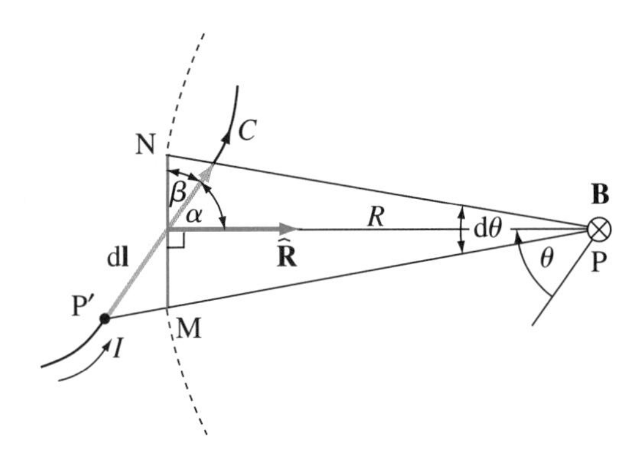

Figure 4.5 Part of a planar current loop C and a field point P in the same plane.

elements as different source functions:

$$\boxed{\mathbf{J}\,\mathrm{d}v \quad\longleftrightarrow\quad \mathbf{J}_\mathrm{s}\,\mathrm{d}S \quad\longleftrightarrow\quad I\,\mathrm{d}\mathbf{l},} \tag{4.10}$$

volume, surface, and line current elements (unit: A · m*)*

and they all represent one physical law – the Biot-Savart law.

In addition, the general form of the Biot-Savart law for line currents, Eq. (4.9), can considerably be simplified in the case of a planar current contour (contour lying in one plane) and a field point P (at which the magnetic flux density vector is calculated) in the same plane, shown in Fig. 4.5. Here, the vector **B** is normal to the plane of the contour (the plane of drawing), and its magnitude is given by

$$B = \frac{\mu_0}{4\pi} \oint_C \frac{I|\mathrm{d}\mathbf{l} \times \hat{\mathbf{R}}|}{R^2}. \tag{4.11}$$

Because (Fig. 4.5)

$$|\mathrm{d}\mathbf{l} \times \hat{\mathbf{R}}| = \mathrm{d}l \sin\alpha = \mathrm{d}l \cos\beta = \overline{\mathrm{MN}} = R\,\mathrm{d}\theta \tag{4.12}$$

($R\,\mathrm{d}\theta$ is the length of an arc representing a differentially small portion of a circle of radius R centered at the point P), we have that

$$\boxed{B = \frac{\mu_0}{4\pi} \oint_C \frac{I\,\mathrm{d}\theta}{R},} \tag{4.13}$$

B for a loop and field point in one plane

where θ is the angle between R and an arbitrarily adopted reference axis in the contour plane. The reference rise of θ corresponds to the reference flow of I, and the reference direction of **B** is related to the direction of I by the right-hand rule (fingers – field; thumb – current).

Note that the integral expression in Eq. (4.13) implies only scalar integration and the function that needs to be integrated is much simpler than that in Eq. (4.9). We shall, therefore, use the simplified form of the Biot-Savart law whenever possible, i.e., when dealing with planar current contours and calculating the magnetic flux density vector in the same plane. Eq. (4.13) can also be used in cases when a planar part of a nonplanar contour and a field point are in one plane.

Conceptual Questions (on Companion Website): 4.3.

4.3 MAGNETIC FLUX DENSITY VECTOR DUE TO GIVEN CURRENT DISTRIBUTIONS

Eqs. (4.7)–(4.9), with Eq. (4.13) added, are general means for evaluating (by superposition and integration) the magnetic flux density vector, **B**, due to given current distributions in free space (or any nonmagnetic medium), just as Eqs. (1.37)–(1.39) are the means for evaluating the electric field intensity vector, **E**, due to given charge distributions in free space. In this section, we shall consider various characteristic examples of the application of the Biot-Savart law. The examples include current distributions that are theoretically and practically important on one hand and for which the integrals can be evaluated analytically on the other. Most of the magnetostatic structures we shall analyze have their electrostatic counterparts in Section 1.5, so that many solution strategies introduced and developed in that section apply also here.

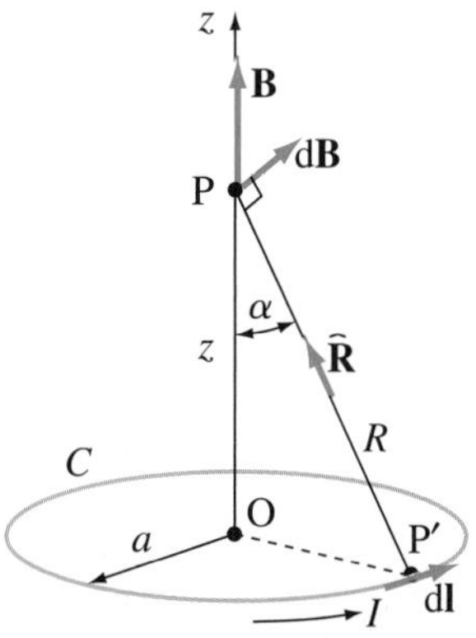

Example 4.1 Magnetic Field of a Circular Current Loop

Consider a circular loop of radius a carrying a steady current of intensity I in air. Calculate the magnetic flux density vector along the axis of the loop normal to its plane.

Figure 4.6 Evaluation of the magnetic flux density vector along the axis of a circular current loop normal to its plane; for Example 4.1.

Solution We subdivide the loop (contour C) into elemental segments and apply Eq. (4.9), analogously to the electric field computation in Example 1.6. With reference to Fig. 4.6, the contribution to the magnetic flux density vector at a point P on the loop axis (z-axis) of a current element $I\,\mathbf{dl}$ at a point P′ on the contour is

$$\mathrm{d}\mathbf{B} = \frac{\mu_0 I\,\mathrm{d}\mathbf{l} \times \hat{\mathbf{R}}}{4\pi R^2}, \tag{4.14}$$

where $R = \sqrt{z^2 + a^2}$ (z is the coordinate of the point P). As $\mathrm{d}\mathbf{l}$ and $\hat{\mathbf{R}}$ are mutually perpendicular and $|\hat{\mathbf{R}}| = 1$ (unit vector), $|\mathrm{d}\mathbf{l} \times \hat{\mathbf{R}}| = \mathrm{d}l$. Hence, the magnitude of the vector $\mathrm{d}\mathbf{B}$ equals

$$\mathrm{d}B = \frac{\mu_0 I\,\mathrm{d}l}{4\pi R^2}. \tag{4.15}$$

The total magnetic flux density vector at the point P is obtained as

$$\mathbf{B} = \oint_C \mathrm{d}\mathbf{B}. \tag{4.16}$$

Because of symmetry, the radial (horizontal) components of the vector $\mathrm{d}\mathbf{B}$ in Fig. 4.6 cancel out in the integral (much like in Fig. 1.11, in the electrical case), and only the axial (vertical) components

$$\mathrm{d}B_z = \mathrm{d}B\cos(90^\circ - \alpha) = \mathrm{d}B\sin\alpha = \mathrm{d}B\frac{a}{R} = \frac{\mu_0 I a\,\mathrm{d}l}{4\pi R^3} \tag{4.17}$$

contribute to the final result. Consequently,

$$\mathbf{B} = \oint_C \mathrm{d}B_z\,\hat{\mathbf{z}} = \frac{\mu_0 I a}{4\pi R^3}\,\hat{\mathbf{z}} \oint_C \mathrm{d}l, \tag{4.18}$$

which yields [see Eq. (1.43)]

B due to a circular current loop

$$\boxed{\mathbf{B} = \frac{\mu_0 I a^2}{2(z^2 + a^2)^{3/2}}\,\hat{\mathbf{z}}.} \tag{4.19}$$

Example 4.2 Magnetic Field of a Finite Straight Wire Conductor

Consider a straight conductor of length l representing a part of a wire contour with a steady current of intensity I in free space. Find the expression for the **B** field at an arbitrary point in space due to this current conductor.

Solution The conductor and an arbitrary field point (P) always determine one plane, which means that we can use the simplified form of the Biot-Savart law in Eq. (4.13). From Fig. 4.7,

$\cos\theta = d/R$, so that the magnetic flux density is given by

$$B = \frac{\mu_0}{4\pi}\int_l \frac{I\,\mathrm{d}\theta}{R} = \frac{\mu_0 I}{4\pi d}\int_{\theta=\theta_1}^{\theta_2} \cos\theta\,\mathrm{d}\theta. \tag{4.20}$$

The solution to this integral is

$$\boxed{B = \frac{\mu_0 I}{4\pi d}(\sin\theta_2 - \sin\theta_1),} \tag{4.21}$$

B due to a straight wire conductor of finite length

where θ_1 and θ_2 are the angles defining the starting and ending point of the conductor, respectively, and d is the perpendicular distance from the conductor to the point P.

The expression in Eq. (4.21) can be combined for computing the magnetic field due to any structure assembled from straight line segments with steady current. In addition, it can be used for the evaluation of the contribution to the total field of straight segments contained in structures that also include curvilinear segments.

By taking $\theta_1 = -\pi/2$ and $\theta_2 = \pi/2$ in Fig. 4.7 and in Eq. (4.21), we obtain the expression for the magnetic flux density due to an infinitely long straight wire conductor carrying a current I in free space. With a notation $d = r$, r standing for the radial coordinate in the cylindrical coordinate system whose z-axis coincides with the axis of the wire conductor, this expression reads

$$\boxed{B = \frac{\mu_0 I}{2\pi r}.} \tag{4.22}$$

B due to an infinite wire conductor

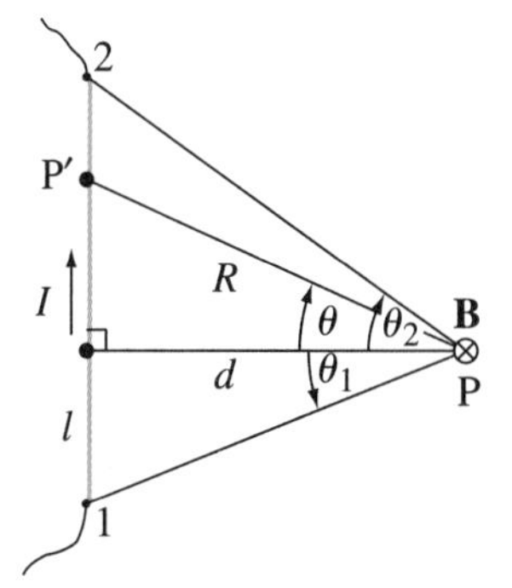

Figure 4.7 Evaluation of the magnetic field of a finite straight wire conductor; for Example 4.2.

Example 4.3 Magnetic Field of a Square Current Loop

A square loop of side length a in free space carries a steady current of intensity I. Obtain the expression for the magnetic flux density vector at the loop center.

Solution The magnetic flux density at the loop center due to the current along one of the square sides, B_1, is given by Eq. (4.21) with $d = a/2$, $\theta_1 = -\pi/4$, and $\theta_2 = \pi/4$, as can be seen in Fig. 4.8. By means of the superposition principle, the magnitude of the total magnetic flux density vector amounts to

$$B = 4B_1 = \frac{2\sqrt{2}\mu_0 I}{\pi a}, \tag{4.23}$$

with respect to the reference direction of $\mathbf{B}$ indicated in Fig. 4.8.

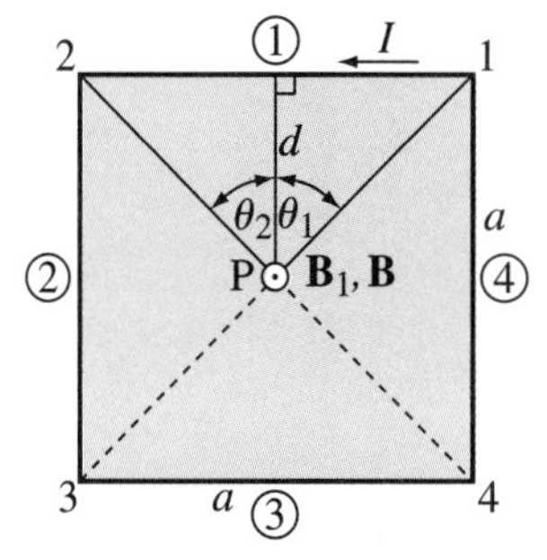

Figure 4.8 Evaluation of the magnetic field at the center of a square current loop; for Example 4.3.

Example 4.4 Magnetic Field of a Loop with a Semicircular Part

Fig. 4.9(a) shows a wire contour consisting of a semicircle (of radius a) and a straight line (of length $2a$). The contour is situated in air and lies in the xy-plane of the Cartesian coordinate system. If the contour carries a steady current of intensity I, calculate the magnetic flux density vector at an arbitrary point along the z-axis.

Solution Let $\mathrm{d}\mathbf{B}'$ denote the magnetic flux density vector at a point P on the z-axis due to a current element $I\,\mathrm{d}\mathbf{l}$ that belongs to the semicircular part of the contour [Fig. 4.9(a)]. This vector is given by Eq. (4.14), and we need to break it up into components suitable for integration, which are x-, y-, and z-components in this case, and this is very similar to the decomposition of the vector $\mathrm{d}\mathbf{E}$ in Fig. 1.12 and Eqs. (1.46) and (1.47). So, we first decompose $\mathrm{d}\mathbf{B}'$ in the plane of the triangle $\triangle$POP′ [Fig. 4.9(b)]:

$$\mathrm{d}\mathbf{B}' = \mathrm{d}\mathbf{B}'_\mathrm{h} + \mathrm{d}B'_z\,\hat{\mathbf{z}}, \quad \mathrm{d}B'_\mathrm{h} = \mathrm{d}B'\cos\alpha, \quad \cos\alpha = \frac{z}{R}, \quad \mathrm{d}B'_z = \mathrm{d}B'\sin\alpha, \quad \sin\alpha = \frac{a}{R}, \tag{4.24}$$

where $\mathrm{d}B'$ is given in Eq. (4.15). We then represent the horizontal vector $\mathrm{d}\mathbf{B}'_\mathrm{h}$ by its x- and y-components [Fig. 4.9(c)]:

$$\mathrm{d}\mathbf{B}'_\mathrm{h} = \mathrm{d}B'_x\,\hat{\mathbf{x}} + \mathrm{d}B'_y\,\hat{\mathbf{y}}, \quad \mathrm{d}B'_x = \mathrm{d}B'_\mathrm{h}\cos\phi, \quad \mathrm{d}B'_y = \mathrm{d}B'_\mathrm{h}\sin\phi. \tag{4.25}$$

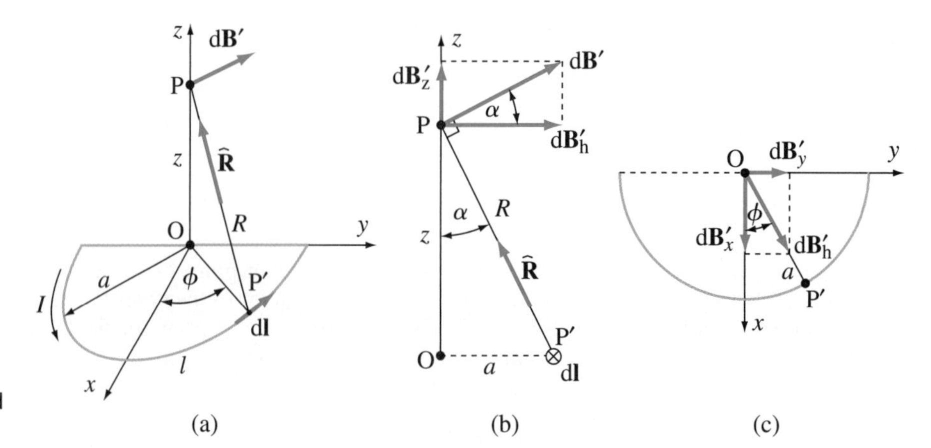

Figure 4.9 Evaluation of the magnetic field due to a current contour with a semicircular and a linear part; for Example 4.4.

Finally, we integrate, as in Eqs. (1.48)–(1.50) in the electrical case, the three Cartesian components of $\mathbf{dB}'$ along the semicircle, with respect to the azimuthal angle ϕ $(-\pi/2 \leq \phi \leq \pi/2)$ as the integration variable [note that $dl = a\,d\phi$ in Eq. (4.15)]:

$$B'_x = \int_l dB'_x = \frac{\mu_0 I a z}{4\pi R^3}\int_{-\pi/2}^{\pi/2} \cos\phi\, d\phi = \frac{\mu_0 I a z}{2\pi R^3}, \quad B'_y = \int_l dB'_y$$
$$= \frac{\mu_0 I a z}{4\pi R^3}\int_{-\pi/2}^{\pi/2} \sin\phi\, d\phi = 0, \quad B'_z = \int_l dB'_z = \frac{\mu_0 I a^2}{4\pi R^3}\int_{-\pi/2}^{\pi/2} d\phi = \frac{\mu_0 I a^2}{4R^3}. \tag{4.26}$$

Hence, the resultant magnetic flux density vector due to the semicircular part of the contour comes out to be

$$\mathbf{B}' = \frac{\mu_0 I a}{2R^3}\left(\frac{z}{\pi}\,\hat{\mathbf{x}} + \frac{a}{2}\,\hat{\mathbf{z}}\right). \tag{4.27}$$

The magnetic flux density vector at the point P due to the linear part of the contour, on the other hand, is determined from Eq. (4.21) with $d = z$, $\theta_1 = -\alpha$, and $\theta_2 = \alpha$:

$$\mathbf{B}'' = \frac{\mu_0 I}{4\pi z} 2\sin\alpha(-\hat{\mathbf{x}}) = -\frac{\mu_0 I a}{2\pi z R}\,\hat{\mathbf{x}}. \tag{4.28}$$

The total $\mathbf{B}$ field along the z-axis is

$$\mathbf{B} = \mathbf{B}' + \mathbf{B}'' = \frac{\mu_0 I a^2}{2R^3}\left(-\frac{a}{\pi z}\,\hat{\mathbf{x}} + \frac{1}{2}\,\hat{\mathbf{z}}\right), \quad R = \sqrt{z^2 + a^2}. \tag{4.29}$$

Example 4.5 Magnetic Field of a Finite Solenoid

Fig. 4.10(a) shows a solenoid (cylindrical coil) consisting of N turns of an insulated thin wire wound uniformly and densely in one layer on a cylindrical nonmagnetic support with a circular cross section of radius a. The length of the solenoid is l and the current through the wire is I. The medium is air. Find the expression for the magnetic flux density vector along the solenoid axis.

Solution Because of the coil being closely wound in a spiral and the wire being thin, the current flowing through the turns of the solenoid can be regarded as a thin cylindrical current sheet with surface current density [Eq. (3.13)]

$$J_s = \frac{NI}{l}. \tag{4.30}$$

The current over an elemental length dz along the solenoid, shown in Fig. 4.10(b),

$$dI = J_s\, dz = \frac{NI\, dz}{l}, \tag{4.31}$$

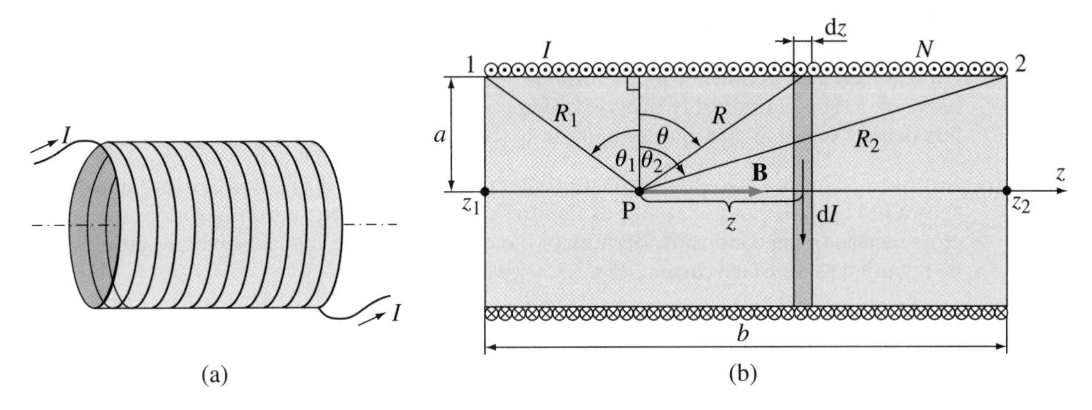

Figure 4.10 (a) Uniformly and densely wound solenoidal coil with a steady current and (b) evaluation of the magnetic flux density vector along the solenoid axis; for Example 4.5.

can be viewed as the current of an equivalent circular current loop of radius a and wire diameter $\mathrm{d}z$. From Eq. (4.19), the magnetic flux density vector of this loop at an arbitrary point P at the solenoid axis [Fig. 4.10(b)] is

$$\mathrm{d}\mathbf{B} = \frac{\mu_0 \, \mathrm{d}I a^2}{2R^3} \, \hat{\mathbf{z}}. \tag{4.32}$$

To find the total field $\mathbf{B}$ at the point P, we integrate $\mathrm{d}\mathbf{B}$ to sum the contributions of all equivalent loops along the solenoid (superposition principle):

$$\mathbf{B} = \int_{z=z_1}^{z_2} \mathrm{d}\mathbf{B} = \frac{\mu_0 N I a^2}{2l} \int_{z_1}^{z_2} \frac{\mathrm{d}z}{R^3} \, \hat{\mathbf{z}}, \qquad R = \sqrt{z^2 + a^2}. \tag{4.33}$$

In order to solve this integral, however, we note that the relationship in Eq. (1.55), with d substituted here by a, exists between the length coordinate z (measured from the point P) and the angular coordinate θ of the position of the equivalent loop along the solenoid axis. Multiplying it by $a/R = \cos\theta$ yields

$$\frac{a^2 \, \mathrm{d}z}{R^3} = \cos\theta \, \mathrm{d}\theta. \tag{4.34}$$

With this, the integral in Eq. (4.33) is reduced to a simple form:

$$\mathbf{B} = \frac{\mu_0 N I}{2l} \int_{\theta=\theta_1}^{\theta_2} \cos\theta \, \mathrm{d}\theta \, \hat{\mathbf{z}}, \tag{4.35}$$

which results in

$$\boxed{\mathbf{B} = \frac{\mu_0 N I}{2l} (\sin\theta_2 - \sin\theta_1) \, \hat{\mathbf{z}}.} \tag{4.36}$$

***B** along the axis of a finite solenoid*

This expression is valid for an arbitrary field point along the solenoid axis, both inside and outside the solenoid, with the position of the point being defined by angles θ_1 and θ_2. Note, however, that the position of the point P can also be defined by coordinates z_1 and z_2, where $z_1 = a\tan\theta_1$ and $z_2 = a\tan\theta_2$.

In the case of an infinitely long solenoid, $\theta_1 = -\pi/2 \, (z_1 \to -\infty)$ and $\theta_2 = \pi/2 \, (z_2 \to \infty)$ in Fig. 4.10(b) and in Eq. (4.36), so that

$$\boxed{B = \mu_0 N' I,} \tag{4.37}$$

B inside an infinite solenoid

where $N' = N/l$ is the number of wire turns per unit length of the solenoid. The magnetic field is uniform, with flux density given in Eq. (4.37), within an infinite solenoid even at points off of its axis, as we shall show in a later example.

Example 4.6 Magnetic Field of an Infinitely Long Strip Conductor

An infinitely long conductor in the form of a thin strip of width a carries a steady current of intensity I. The permeability everywhere is μ_0. Determine the expression for the magnetic flux density vector at an arbitrary point in space.

Solution This is a two-dimensional problem, and we solve it in the cross-sectional plane shown in Fig. 4.11. Eq. (3.82) tells us that the current I must be uniformly distributed in the cross section of the conductor. Because of the conductor being thin, however, its current can be regarded as a surface current, Fig. 3.3, with density

$$J_s = \frac{I}{a}. \tag{4.38}$$

By virtue of the superposition principle, we subdivide the conductor into elemental strips of width $\mathrm{d}l = \mathrm{d}y$, and each such strip can be considered as an infinitely long line current of intensity

$$\mathrm{d}I = J_s\,\mathrm{d}l = \frac{I\,\mathrm{d}y}{a}. \tag{4.39}$$

The magnetic flux density vector due to individual line currents is circular with respect to the line, and, from Eq. (4.22), its magnitude at an arbitrary point P in space (Fig. 4.11) is

$$\mathrm{d}B = \frac{\mu_0\,\mathrm{d}I}{2\pi R}, \qquad R = \sqrt{y^2 + d^2}, \tag{4.40}$$

where d is the perpendicular distance of P from the plane of the strip. We decompose vector $\mathbf{dB}$ into its x- and y-components,

$$\mathrm{d}B_x = \mathrm{d}B\sin\theta, \qquad \mathrm{d}B_y = \mathrm{d}B\cos\theta, \tag{4.41}$$

and integrate them over the width of the strip conductor to sum the contributions of all elemental strips,

$$B_x = \frac{\mu_0 I}{2\pi a}\int_{y=y_1}^{y_2}\frac{\sin\theta\,\mathrm{d}y}{R}, \qquad B_y = \frac{\mu_0 I}{2\pi a}\int_{y_1}^{y_2}\frac{\cos\theta\,\mathrm{d}y}{R}. \tag{4.42}$$

We now use the relationship in Eq. (1.55), where z is y here, to change the integration variable from y to θ. Multiplying its left-hand side by $R\cos\theta$ and right-hand side by d, which is justified by the fact that $\cos\theta = d/R$ in Fig. 4.11, we obtain

$$\frac{\cos\theta\,\mathrm{d}y}{R} = \mathrm{d}\theta, \tag{4.43}$$

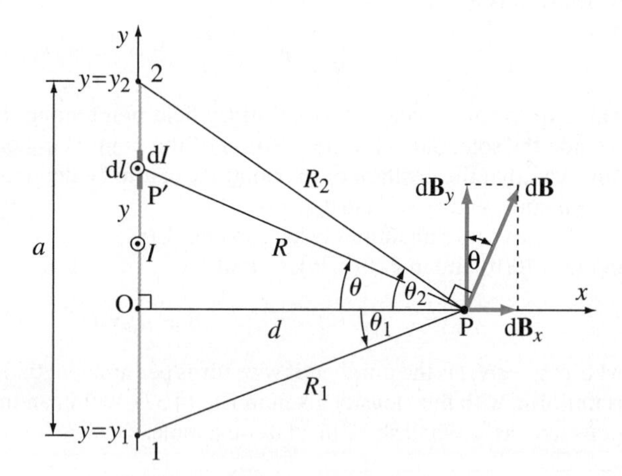

Figure 4.11 Evaluation of the magnetic field due to an infinitely long strip conductor (cross-sectional view); for Example 4.6.

which, multiplied by $\sin\theta/\cos\theta (= \tan\theta)$ gives, in turn,

$$\frac{\sin\theta\,\mathrm{d}y}{R} = \frac{\sin\theta}{\cos\theta}\,\mathrm{d}\theta. \tag{4.44}$$

Based on these two relations, the integrals in Eqs. (4.42) are simple to solve:

$$B_x = \frac{\mu_0 I}{2\pi a}\int_{\theta_1}^{\theta_2}\frac{\sin\theta}{\cos\theta}\,\mathrm{d}\theta = \frac{\mu_0 I}{2\pi a}\ln\frac{\cos\theta_1}{\cos\theta_2}, \quad B_y = \frac{\mu_0 I}{2\pi a}\int_{\theta_1}^{\theta_2}\mathrm{d}\theta = \frac{\mu_0 I}{2\pi a}(\theta_2-\theta_1), \tag{4.45}$$

where the use is made of the substitution $u = \cos\theta$ ($\mathrm{d}u = -\sin\theta\,\mathrm{d}\theta$) in the first integration. Hence, the total magnetic flux density vector due to the strip conductor turns out to be

$$\boxed{\mathbf{B} = \frac{\mu_0 I}{2\pi a}\left[\hat{\mathbf{x}}\ln\frac{R_2}{R_1} + (\theta_2-\theta_1)\,\hat{\mathbf{y}}\right],} \tag{4.46}$$

B due to a strip conductor

with R_1 and R_2 being the distances of the point P from the starting and ending point, respectively, of the line representing the cross section of the strip conductor, in Fig. 4.11.

Note that if we let $a \to \infty$ while keeping $I/a = \text{const}$, we get an infinitely wide and infinitely long planar current sheet with a uniform surface current density $J_s = I/a$. Then, $R_1 = R_2$ (infinite), $\theta_1 = -\pi/2$, and $\theta_2 = \pi/2$, with which Eq. (4.46) becomes

$$\boxed{\mathbf{B} = \frac{\mu_0 J_s}{2}\,\hat{\mathbf{y}}.} \tag{4.47}$$

B due to an infinite current sheet

We conclude that the magnetic field due to an infinite planar sheet of current is uniform at each side of the sheet, with field lines parallel to the sheet.

Problems: 4.1–4.9; *Conceptual Questions* (on Companion Website): 4.4–4.6; *MATLAB Exercises* (on Companion Website).

4.4 FORMULATION OF AMPÈRE'S LAW

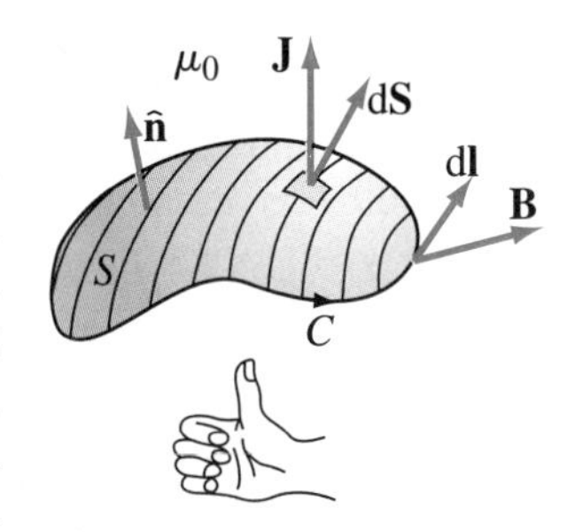

Figure 4.12 Arbitrary contour in a magnetostatic field – for the formulation of Ampère's law.

We saw in Chapter 1 that the electric field due to highly symmetrical charge distributions in free space can be determined much more easily using Gauss' law, Eq. (1.133), than by direct application of Coulomb's law and the superposition principle, i.e., Eqs. (1.37)–(1.39). In magnetostatics, Eqs. (4.7)–(4.9) (the Biot-Savart law) provide general solution procedures analogous to those in Eqs. (1.37)–(1.39), whereas the law that helps us evaluate the magnetic field due to highly symmetrical current distributions in free space more easily is known as Ampère's law (also called Ampère's circuital law or Ampère's work law). It states that the line integral (circulation) of the magnetic flux density vector around any contour (C) in a vacuum (free space) is equal to μ_0 times the total current enclosed by that contour, which we mark as I_C,

$$\boxed{\oint_C \mathbf{B}\cdot\mathrm{d}\mathbf{l} = \mu_0 I_C\,.} \tag{4.48}$$

Ampère's law

The reference direction of the current flow is related to the reference direction of the contour by means of the right-hand rule: the current is in the direction defined by the thumb of the right hand when the other fingers point in the direction of the contour, as shown in Fig. 4.12. This law may be derived from the Biot-Savart law (we recall that Gauss' law is derived from Coulomb's law), and the derivation is carried out in Section 4.10. For the present, we accept Ampère's law temporarily as another law capable of experimental proof. Eq. (4.48) represents Maxwell's second

HISTORICAL ASIDE

André-Marie Ampère (1775–1836), a French mathematician and physicist, was a professor of mathematics, physics, and chemistry in Bourg and Paris, and inspector general of the national university system under Napoleon. Although with little formal education, Ampère acquired the best possible knowledge of mathematics and sciences for that time by reading many books and articles, with a great support and guidance from his father. As a teenager, he was already deriving his own mathematical theories and writing papers on some geometrical problems. However, his life was soon to be shattered when his father was sent to the guillotine by the Jacobins in Lyon in 1793 during the French Revolution. The effect on Ampère of his father's death was devastating, and he did not return to his studies of mathematics for almost two years. After several years of tutoring mathematics in Lyon, he was appointed professor of physics and chemistry at Bourg École Centrale in 1802, and then professor of mathematics at the École Polytechnique in Paris in 1809. His mathematics research included a wide variety of topics in probability, calculus, analytic geometry, and partial differential equations. He contributed also to the theory of light, chemistry, and, most importantly, to electricity and magnetism. Only a few weeks after hearing of Oersted's experimental results, Ampère was ready, before the end of September of 1820, to report to the French Academy of Sciences his discovery of magnetic forces between wires carrying electric currents. He discovered experimentally that two parallel wires carrying currents in the same direction attracted each other, whereas the wires with currents flowing in opposite directions repelled each other. His experiments founded the science of the magnetic field due to electric currents. He described the magnetic force circling around a current-carrying wire in a way which is now known as the right-hand rule. On November 6, 1820, Ampère gave a talk on his circuital law of addition of magnetic forces – a basis of a general equation that we call now Ampère's (circuital) law. He was the first to describe current as the flow of electricity along a wire, analogous to the surge of water through a pipe. He predicted theoretically that a wire helix with current would behave as it were a bar permanent magnet, and called such a helix a solenoid. Ampère also explained permanent magnets as a collection of tiny electric currents circling eternally within them, and in this he was three-quarters of a century ahead of his time. A comprehensive presentation of his findings in electricity and magnetism, including both description of experiments and mathematical derivations, appeared in his most important publication "Memoir on the Mathematical Theory of Electrodynamic Phenomena, Uniquely Deduced from Experience" in 1826. The same year, Ampère was appointed to a chair at the Université de France, which he held until his death. In his honor, the intensity of electric current is measured in amperes. *(Portrait: Edgar Fahs Smith Collection, University of Pennsylvania Libraries)*

equation for static fields in free space (as we know, there is a total of four Maxwell's equations for the general electromagnetic field). By expressing the current in terms of the volume current density, **J**, Ampère's law becomes

Ampère's law for volume current

$$\boxed{\oint_C \mathbf{B} \cdot \mathrm{d}\mathbf{l} = \mu_0 \int_S \mathbf{J} \cdot \mathrm{d}\mathbf{S},} \tag{4.49}$$

where S is a surface of arbitrary shape spanned over (bounded by) the contour C and oriented in accordance to the right-hand rule with respect to the orientation of C (Fig. 4.12).

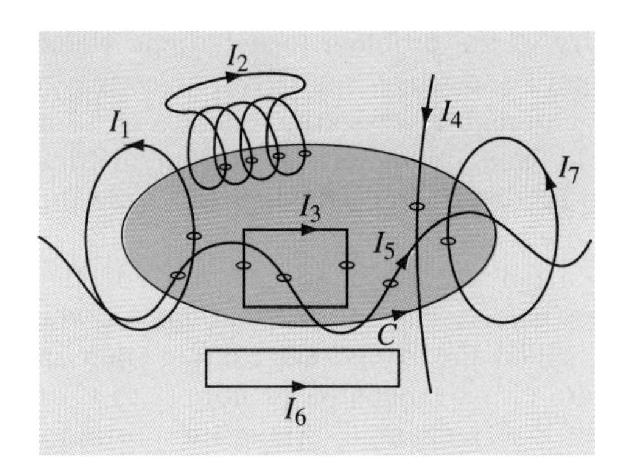

Figure 4.13 Closed path C and seven line currents in air; for Example 4.7.

Example 4.7 Algebraic Total Enclosed Current

What is the circulation of the magnetic flux density vector along the contour C in Fig. 4.13? The medium is air.

Solution According to the right-hand rule given with respect to the reference direction of the contour C indicated in Fig. 4.13, the reference orientation of the surface S spanned over C, i.e., the reference direction of the vector $\mathbf{dS}$ in Eq. (4.49), is from S upward. From Ampère's law, the line integral of the magnetic flux density vector along C equals μ_0 times the algebraic sum of all current intensities passing through S. We note that current I_1 pierces S once in the positive direction (direction in agreement with the orientation of S), I_2 passes it four times, I_4 and I_7 once each but in the negative direction, I_3 once in the positive and once in the negative direction, I_5 twice in the positive and once in the negative direction, while I_6 does not traverse S at all. Hence,

$$\oint_C \mathbf{B} \cdot \mathrm{d}\mathbf{l} = \mu_0 (I_1 + 4I_2 - I_4 + I_5 - I_7). \tag{4.50}$$

What is very important, the same result for the algebraic total current, and thus for the circulation of $\mathbf{B}$, is obtained for any other surface we imagine with the contour C as the perimeter and totally arbitrary shape.

Example 4.8 Contour inside a Conductor

Fig. 4.14 shows the cross section of a very long cylindrical copper conductor that carries a steady current of intensity I ($I > 0$). Is the circulation of the magnetic flux density vector along the contour C less than, equal to, or greater than $\mu_0 I$?

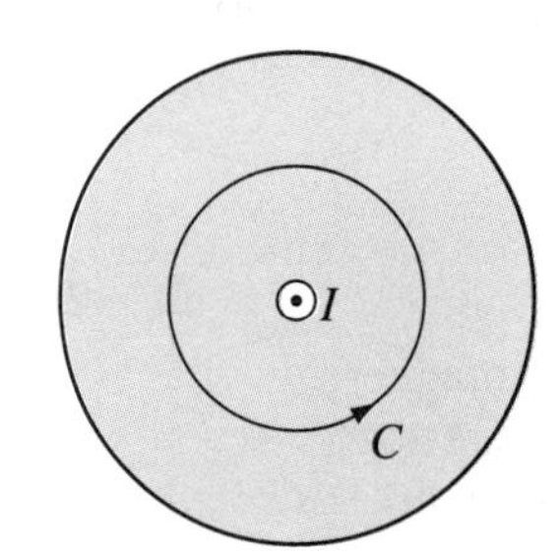

Figure 4.14 Contour inside a conductor with a steady current; for Example 4.8.

Solution From Ampère's law,

$$\oint_C \mathbf{B} \cdot \mathrm{d}\mathbf{l} < \mu_0 I, \tag{4.51}$$

since the total current appearing on the right-hand side of Eq. (4.49) equals exactly that portion of the total current of the conductor (I) enclosed by the contour.

Conceptual Questions (on Companion Website): 4.7–4.10.

4.5 APPLICATIONS OF AMPÈRE'S LAW

This section is devoted to the application of Ampère's law in evaluating the magnetic field due to given steady current distributions in nonmagnetic media. As is the case with Gauss' law, the use of Ampère's law will also require careful consideration of the symmetry of the problem to determine which field components are present in the structure and which spatial variables the present components depend on. Eq. (4.48), although always true, enables us to analytically solve for the field only due to highly symmetrical current distributions. Namely, as the unknown quantity to be determined (**B**) appears inside the integral in Eq. (4.48), we can use Ampère's law to obtain a solution only in cases in which we are able to bring the magnetic flux density, B, outside the integral sign, and solve for it. These cases involve current distributions for which we are able to adopt a closed path C, called the Amperian contour, that satisfies two requirements: (1) **B** is everywhere either tangential or normal to C and (2) $B = \text{const}$ along sections of C where **B** is tangential. Along the portion of the path where **B** is normal to it (if such portion exists), the dot product $\mathbf{B} \cdot \mathbf{dl}$ in Eq. (4.48) becomes zero. Along the remaining part of the contour, $\mathbf{B} \cdot \mathbf{dl}$ becomes $B\,\mathrm{d}l$, and the second requirement (constancy) then permits us to remove B from the integral sign in Eq. (4.48). The integration we are left with is usually trivial and consists of finding the length of that portion of the path to which **B** is tangential.

In this discussion, we notice and exploit the parallelism with the application of Gauss' law, discussed in Section 1.13. All similarities and differences in the mathematical formalism are direct consequences of the mathematical form of the fields **E** and **B** due to the elementary sources Q and $Q\mathbf{v}$, Eqs. (1.24) and (4.4). With Gauss' law, we integrate over a closed surface (Gaussian surface) on the left-hand side of the equation, while Ampère's law implies integration along a closed path (Amperian contour). The field component contributing to the integral is a component normal to the Gaussian surface and tangential to the Amperian contour, respectively. On the right-hand side of the equation, Gauss' law involves finding the total charge enclosed by the surface, whereas the application of Ampère's law involves finding the total current enclosed by the contour. For sources expressed by volume charge and current densities [Eqs. (1.135) and (4.49)], this means volume integration over the volume enclosed by the Gaussian surface and surface integration over the surface bounded by the Amperian contour, respectively.

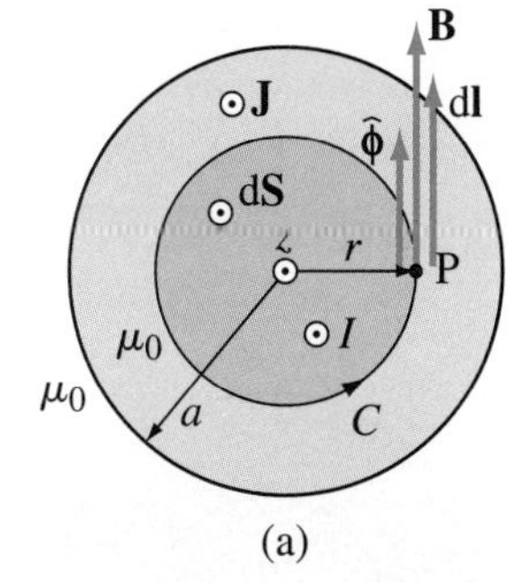

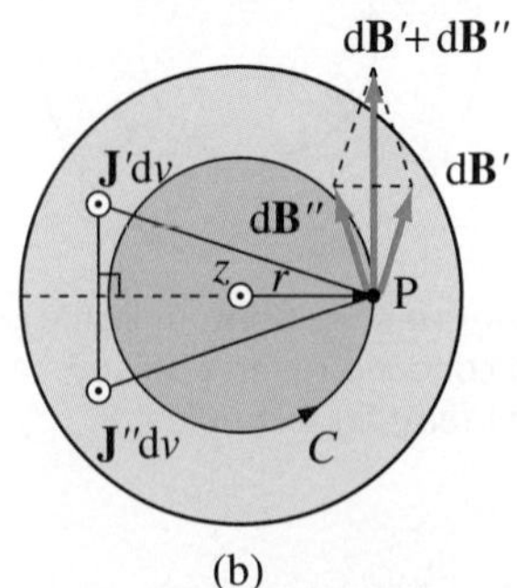

Figure 4.15 Cross section of a cylindrical conductor with a steady current I: (a) application of Ampère's law and (b) proof that the magnetic field lines are circles; for Example 4.9.

Example 4.9 Magnetic Field of a Cylindrical Conductor

An infinitely long cylindrical copper conductor of radius a carries a steady current of intensity I. The conductor is situated in air. Find the magnetic flux density vector inside and outside the conductor.

Solution Fig. 4.15(a) shows a cross section of the conductor. According to Eq. (3.82), the current is uniformly distributed in the cross section, so that its density is

$$J = \frac{I}{\pi a^2}. \tag{4.52}$$

Because of symmetry, the lines of the magnetic field due to the conductor current are circles centered at the conductor axis. To show this, consider the direction of the magnetic flux density vector, **B**, at an arbitrary point P in space, either inside or outside the conductor. Let the distance of the point from the conductor axis be r, and let $\mathbf{dB}'$ and $\mathbf{dB}''$ represent the fields at P due to two symmetrical current elements denoted as $\mathbf{J}'\,\mathrm{d}v$ and $\mathbf{J}''\,\mathrm{d}v$ and shown in Fig. 4.15(b). In accordance to the Biot-Savart law, Eq. (4.6), these two elementary field vectors are such that their sum $\mathbf{dB}' + \mathbf{dB}''$ is tangential to the circular contour C of radius r centered at the conductor axis. The same is true for any other pair of symmetrical current elements, which can also be in a plane that does not contain the point P, and all current elements constituting the current I in the conductor can be grouped in such symmetrical pairs. We conclude that the resultant vector **B** at the point P is tangential to the contour C. In addition, symmetry also implies that $B = \text{const}$ along C, i.e., the magnitude of **B** depends only on the radial coordinate r of the cylindrical coordinate system whose z-axis coincides with the conductor axis. Hence, we can write

$$\mathbf{B} = B(r)\,\hat{\boldsymbol{\phi}}, \qquad (4.53)$$

where $\hat{\boldsymbol{\phi}}$ is the circular unit vector in the system.

Based on the preceding discussion, it is now obvious that the contour C in Fig. 4.15 satisfies both requirements for the Amperian contour for our problem. Along C, $\mathrm{d}\mathbf{l} = \mathrm{d}l\,\hat{\boldsymbol{\phi}}$, so that $\mathbf{B}\cdot\mathrm{d}\mathbf{l} = B\,\mathrm{d}l$ [Fig. 4.15(a)]. The circulation of **B** along C thus turns out to be

$$\oint_C \mathbf{B}\cdot\mathrm{d}\mathbf{l} = \oint_C B(r)\,\mathrm{d}l = B(r)\oint_C \mathrm{d}l = B(r)\,l = B(r)\,2\pi r, \quad 0 \le r < \infty, \qquad (4.54)$$

whereas the current enclosed by C amounts to

$$I_C = \begin{cases} J\pi r^2 & \text{for } r < a \\ I & \text{for } r \ge a \end{cases}. \qquad (4.55)$$

Eqs. (4.54) and (4.55) combined in Ampère's law, Eq. (4.48), finally give

$$\boxed{B = \begin{cases} \mu_0 I r/(2\pi a^2) & \text{for } r < a \\ \mu_0 I/(2\pi r) & \text{for } r \ge a \end{cases}.} \qquad (4.56)$$

B due to a thick cylindrical conductor with a steady current

Note that the magnetic field outside the conductor (for $r \ge a$) is identical to that of a line current of intensity I placed along the conductor axis, Eq. (4.22). This means that the line current (e.g., a thin wire with current I) and the (thick) conductor of Fig. 4.15 are equivalent sources with respect to the region outside the conductor.

Example 4.10 Cylindrical Conductor with an Excentric Cavity

A very long cylindrical nonmagnetic conductor of radius b has an excentric cylindrical cavity of radius a along its entire length, as shown in Fig. 4.16. The axis of the cavity is offset from the axis of the conductor by a vector **d** ($a + d \le b$). The medium in the cavity and outside the conductor is air. Compute the magnetic flux density vector in the cavity, assuming a steady current of density J flowing through the conductor.

Solution The current density in the cavity is zero, and it can be considered to result from two currents of the same density (J) flowing in opposite directions. Accordingly, we can represent the current distribution in the hollow conductor, including the cavity, as a sum of current distributions of a cylindrical conductor of radius b carrying a current of density J without a cavity and another one of radius a with a current of the same density but flowing in the opposite direction, as depicted in Fig. 4.16. By the superposition principle, the magnetic flux density vector of the original (resultant) current distribution, **B**, can be obtained as

$$\mathbf{B} = \mathbf{B}_1 + \mathbf{B}_2, \qquad (4.57)$$

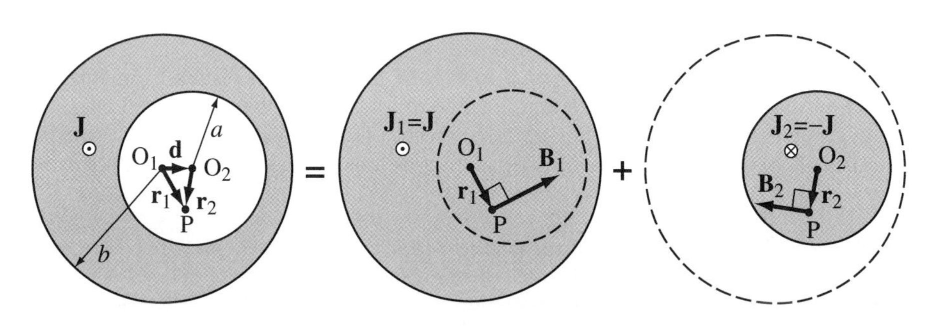

Figure 4.16 Cylindrical conductor with an excentric cylindrical cavity, viewed as a superposition of two solid conductors with currents of the same density but opposite directions; for Example 4.10.

where $\mathbf{B}_1$ and $\mathbf{B}_2$ are the magnetic flux density vectors due to the partial current distributions $\mathbf{J}_1$ and $\mathbf{J}_2$ in Fig. 4.16. These vectors, according to Eq. (4.56), can be written in the form[3]

$$\mathbf{B}_1 = \frac{\mu_0 \mathbf{J} \times \mathbf{r}_1}{2} \quad \text{and} \quad \mathbf{B}_2 = \frac{\mu_0 (-\mathbf{J}) \times \mathbf{r}_2}{2}, \tag{4.58}$$

where $\mathbf{r}_1$ and $\mathbf{r}_2$ are the position vectors of the field point (P) with respect to the conductor axis (O_1) and cavity axis (O_2), respectively. Finally, combining the preceding equations, we get

$$\mathbf{B} = \frac{\mu_0 \mathbf{J} \times (\mathbf{r}_1 - \mathbf{r}_2)}{2} = \frac{\mu_0 \mathbf{J} \times \mathbf{d}}{2}, \tag{4.59}$$

and $B = \mu_0 J d/2$. We conclude that the magnetic field inside the cavity is uniform, as the above expression for the resultant magnetic flux density vector does not depend on the position of the point P. The field lines are parallel to the plane of the conductor cross section and at right angles to the vector $\mathbf{d}$.

Example 4.11 Magnetic Field of a Coaxial Cable

Conductors of a coaxial cable are made from copper, and its dielectric is air. The radius of the inner conductor is a, whereas the inner and outer radii of the outer conductor are b and c, respectively ($a < b < c$). A steady current of intensity I is established in the cable. Find the magnetic flux density vector everywhere.

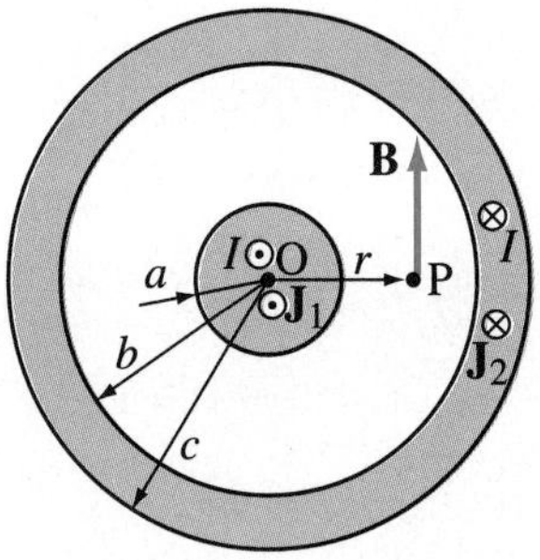

Figure 4.17 Evaluation of the magnetic field of a coaxial cable; for Example 4.11.

Solution Referring to Fig. 4.17, current densities in the cable conductors are

$$J_1 = \frac{I}{\pi a^2} \quad \text{and} \quad J_2 = \frac{I}{\pi (c^2 - b^2)}. \tag{4.60}$$

Due to symmetry, the magnetic flux density vector, $\mathbf{B}$, is circular with respect to the cable axis. Following a similar procedure as in Example 4.9, we apply Ampère's law to the circular contour C of radius r ($0 \le r < \infty$). The circulation of $\mathbf{B}$ along C is the same as in Eq. (4.54), for all values of r, while the expression for the total current enclosed by C is different for four different characteristic positions of the contour. The field inside the inner conductor and between the conductors is identical to that found in Example 4.9 for a single conductor,

B in the dielectric of a coaxial cable

$$B = \frac{\mu_0 J_1 r}{2} = \frac{\mu_0 I r}{2\pi a^2} \quad (0 \le r \le a), \qquad \boxed{B = \frac{\mu_0 I}{2\pi r} \quad (a < r < b).} \tag{4.61}$$

For the contour positioned inside the outer conductor, the enclosed net current equals the entire current of the inner conductor minus that portion of the current of the outer conductor which is enclosed by the contour, so that

$$B = \frac{1}{2\pi r}\,\mu_0 \underbrace{[I - J_2 \pi (r^2 - b^2)]}_{I_C} = \frac{\mu_0 I (c^2 - r^2)}{2\pi (c^2 - b^2) r} \quad (b \le r \le c). \tag{4.62}$$

Finally, if the radius r is larger than the outer radius of the outer conductor,

$$B = \frac{1}{2\pi r}\,\mu_0 (I - I) = 0 \quad (c < r < \infty). \tag{4.63}$$

So, the external magnetic field (for $r > c$) is zero. This, we see, results from currents of equal intensities and opposite directions in cable conductors (making zero total enclosed current in Ampère's law), much as the external electric field of a charged coaxial cable is zero because of charges of equal magnitudes and opposite polarities per unit length of cable

[3]Note that the magnetic flux density inside the solid cylindrical conductor in Fig. 4.15 can be written as $B = \mu_0 J r/2$, and, in vector form, $\mathbf{B} = \mu_0 \mathbf{J} \times \mathbf{r}/2$, where $\mathbf{r}$ is the position vector of the field point with respect to the conductor axis.

conductors (making zero total enclosed charge in Gauss' law). [Such (equal positive and negative) charges and currents are standard in operations of all two-conductor transmission lines.] This is a very important property of a coaxial cable; the electrostatic and magnetostatic fields of the cable are concentrated exclusively inside the cable, and its exterior is perfectly decoupled (shielded) with respect to the interior, and vice versa.

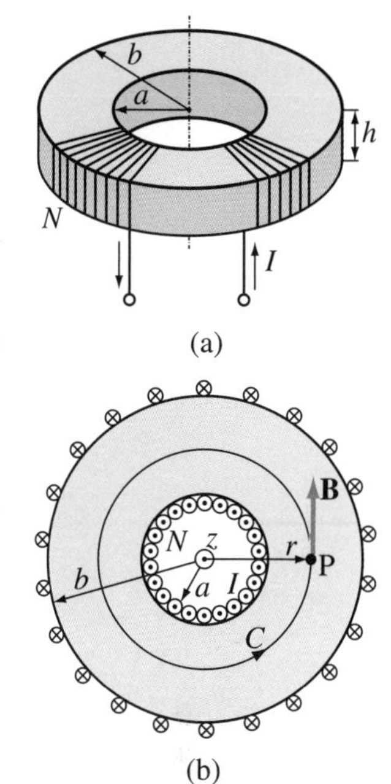

Figure 4.18 Evaluation of the magnetic field of a toroidal coil: (a) three-dimensional view showing the winding and (b) cross-sectional view showing Amperian contours ($0 \le r < \infty$); for Example 4.12.

Example 4.12 Magnetic Field of a Toroidal Coil

Shown in Fig. 4.18(a) is a toroidal ("doughnut"-shaped) coil with a rectangular cross section. The coil consists of N turns of wire that are uniformly and densely wound along the length of the toroid and carry a steady current of intensity I. The inner and outer radii of the toroid are a and b ($a < b$), and its height is h. The medium inside and outside the coil is nonmagnetic. Calculate the magnetic flux density vector inside and outside the coil.

Solution Fig. 4.18(b) portrays a horizontal cut of the toroid in Fig. 4.18(a). Because of symmetry, the lines of the magnetic field density vector, **B**, due to the current in the coil are circles centered at the toroid axis [z-axis in Fig. 4.18(b)], and the magnitude of **B** depends only on the distance r from the axis. This means that the vector **B** is of the form given in Eq. (4.53). Applying Ampère's law to a circular contour of radius r ($0 \le r < \infty$), we obtain the following equations for three characteristic locations of the contour:

$$B(r)\,2\pi r = 0 \quad (0 \le r \le a), \qquad B(r)\,2\pi r = \mu_0 NI \quad (a < r < b),$$

$$B(r)\,2\pi r = \mu_0(NI - NI) = 0 \quad (b \le r < \infty). \tag{4.64}$$

Hence, $B = 0$ outside the toroid, while inside it

$$B(r) = \frac{\mu_0 NI}{2\pi r} \quad \text{(inside a thick toroidal coil).} \tag{4.65}$$

Note that this result is valid for an arbitrary shape of the toroid (vertical) cross section (not only rectangular).

If the toroid is thin, i.e., $b - a \ll a$, we can assume that $r \approx c$ inside it, where $c = (a + b)/2$ is its mean radius. This implies that the magnetic field inside the toroid can be considered to be uniform, with the flux density

$$\boxed{B = \frac{\mu_0 NI}{2\pi c} = \frac{\mu_0 NI}{l} = \mu_0 N' I,} \tag{4.66}$$

B inside a thin toroidal coil

where $l = 2\pi c$ is the length of the toroid and $N' = N/l$ is the number of wire turns per unit length.

We realize that the final expression for B in Eq. (4.66) is the same as that in Eq. (4.37), found for an infinite solenoid. Namely, we can visualize an infinitely long solenoid as a toroid with an infinite radius.

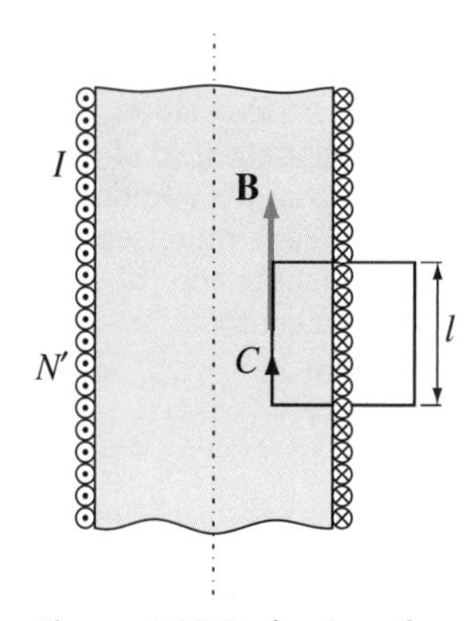

Figure 4.19 Evaluation of the magnetic field inside an infinite solenoid using Ampère's law; for Example 4.13.

Example 4.13 Magnetic Field inside an Infinite Solenoid from Ampère's Law

An infinitely long air-filled solenoid has N' turns of wire per unit of its length. A steady current of intensity I flows through the winding. Using Ampère's law, prove that the magnetic flux density inside the solenoid is given by the expression in Eq. (4.37).

Solution Knowing that the vector **B** inside an infinite solenoid is axial (parallel to the solenoid axis) and that it is zero outside the solenoid, it is now very simple to evaluate the magnitude of **B** in the solenoid from Ampère's law. The Amperian contour is a rectangle positioned partly inside the solenoid as shown in Fig. 4.19. The line integral of **B** along the vertical (axial) edge of the rectangle that is inside the solenoid equals Bl, because **B** is parallel to that edge and does not vary along it (the structure is infinitely long). The line integral of **B** along the remaining three edges of the rectangle is zero, because **B** is perpendicular to portions of horizontal (perpendicular to the axis) edges that are inside the solenoid and is zero outside it. The net current enclosed by the contour, on the other hand, equals the number of turns over the length l, which is $N'l$, times I (the current through each turn). Hence, Ampère's law gives

$$Bl = \mu_0 N' l I, \tag{4.67}$$

i.e., the same as in Eq. (4.37). Note that we did not in any way restrict the location of the left edge of the contour C in Fig. 4.19 to be on the solenoid axis or at a specific distance from it, which means that Eq. (4.37) is valid across the entire cross section of the solenoid, and not only on the solenoid axis (in Example 4.5, the magnetic field is evaluated along the solenoid axis only). In other words, the magnetic field is uniform (the same) throughout the entire volume enclosed by the solenoid winding (while zero outside it). We also note that the cross section of the solenoid in this discussion and calculation is not in any way restricted to be of circular shape, which implies that the magnetic field is the same for arbitrary shape of the solenoid cross section (provided that the solenoid is very long).

Example 4.14 Magnetic Field of a Rotating Charged Cylinder

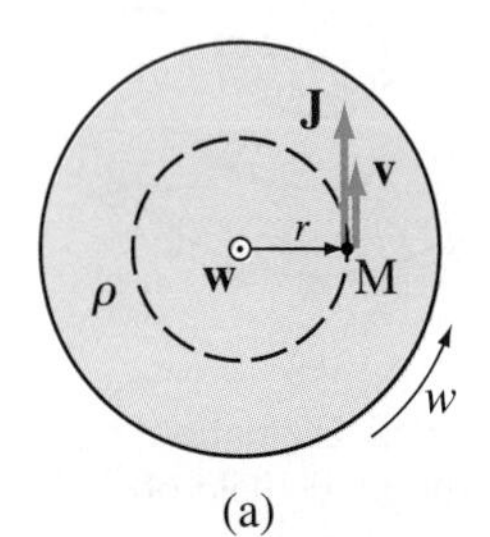

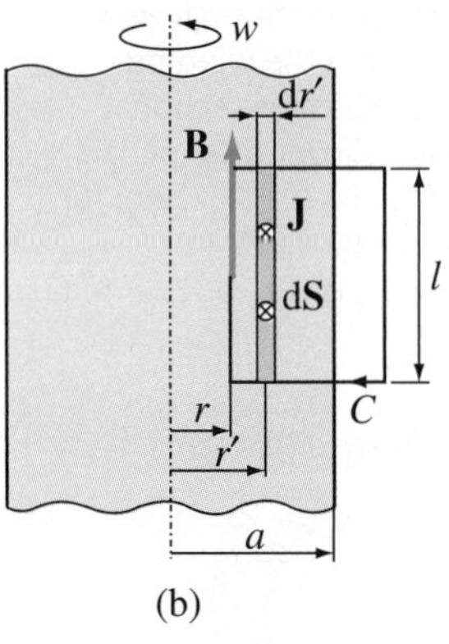

Figure 4.20 Evaluation of the magnetic field of a rotating charged cylinder: (a) equivalent impressed current density and (b) Amperian contour; for Example 4.14.

A very long cylinder of radius a is uniformly charged over its volume by a charge of density ρ. The cylinder uniformly rotates about its axis with an angular velocity w. The permeability is μ_0 everywhere. Find the magnetic flux density vector inside and outside the cylinder, assuming that the charge distribution of the cylinder remains the same during the rotation.

Solution As the charges of the cylinder rotate with it, they form a volume electric current. Referring to Fig. 4.20(a), the velocity at which a point M that is r away from the cylinder axis moves is

$$\mathbf{v} = \mathbf{w} \times \mathbf{r} = wr\,\hat{\boldsymbol{\phi}}, \tag{4.68}$$

and the current density vector at that point can be found from Eq. (3.28). The current lines in the cylinder are circles centered at the cylinder axis [Fig. 4.20(a)], and the current density is a function of r given by

$$J(r) = \rho v(r) = \rho w r \quad (0 \le r \le a). \tag{4.69}$$

We note that this current density, which is independent of the electric field, can be regarded as an impressed electric current density, $\mathbf{J}_\mathrm{i}$ [Fig. 3.17(a)].

Because of the currents in the cylinder being circular, the cylinder can be visualized as a series of many very long coaxial solenoids of radii varying from $r = 0$ to $r = a$. That is why the magnetic flux density vector, $\mathbf{B}$, outside the cylinder (i.e., outside all equivalent solenoids) is zero, whereas $\mathbf{B}$ is axial inside the cylinder. Ampère's law, Eq. (4.49), applied to a rectangular contour C shown in Fig. 4.20(b) results in [left-hand side of the equation is the same as in Eq. (4.67)]

$$B(r)\,l = \mu_0 \int_{r'=r}^{a} J(r')\,\underbrace{l\,\mathrm{d}r'}_{\mathrm{d}S} = \mu_0 \rho w l \int_{r}^{a} r'\,\mathrm{d}r', \tag{4.70}$$

where $\mathrm{d}S$ is the surface area of a thin strip of length l and width $\mathrm{d}r'$ (r' is the integration variable). Hence, the magnitude of $\mathbf{B}$ at points that are r away from the cylinder axis comes out to be

$$B(r) = \begin{cases} \mu_0 \rho w (a^2 - r^2)/2 & \text{for } r < a \\ 0 & \text{for } r \ge a \end{cases}. \tag{4.71}$$

Example 4.15 Magnetic Field of an Infinite Current Sheet from Ampère's Law

Using Ampère's law, prove that the magnetic flux density vector due to an infinite planar current sheet with a uniform surface current density J_s in free space is given by the expression in Eq. (4.47).

Solution Because of symmetry, the magnetic field lines are parallel to the sheet and **B** does not vary in directions parallel to the sheet. We apply Ampère's law to a rectangular contour C portrayed in Fig. 4.21. The line integral of **B** along each of the two edges of the contour that are parallel to the sheet equals Bl and the enclosed current is $J_s l$ [Eq. (3.13)], so that

$$2Bl = \mu_0 J_s l, \tag{4.72}$$

from which the same result as in Eq. (4.47) is obtained.

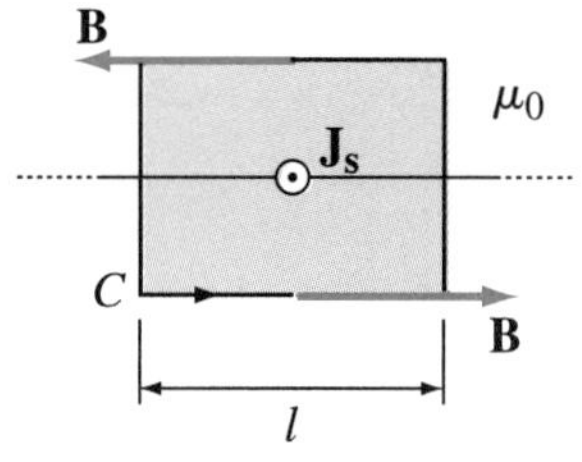

Figure 4.21 Evaluation of the magnetic field of an infinite current sheet using Ampère's law; for Example 4.15.

Problems: 4.10–4.19; *Conceptual Questions* (on Companion Website): 4.11–4.15; *MATLAB Exercises* (on Companion Website).

4.6 DIFFERENTIAL FORM OF AMPÈRE'S LAW

In electrostatics, we derived the differential form of Gauss' law [Eq. (1.163) or (1.166)] starting from its integral form [Eq. (1.135)]. Again, an analogous concept and transformation exist in magnetostatics. In this section, we shall utilize the integral form of Ampère's law, Eq. (4.49), to derive its differential equivalent. What we expect to obtain is a spatial differential relationship between the magnetic flux density vector, **B** (field), and the current density vector, **J** (source), at a point in space.

We consider first the one-dimensional case, and assume that **J** has only a z-component in the Cartesian coordinate system which is a function of the coordinate x only, that is, $\mathbf{J} = J_z(x)\,\hat{\mathbf{z}}$ (1-D current distribution). Then, by symmetry, the only present component of **B** is B_y (like in Example 4.15), i.e., $\mathbf{B} = B_y(x)\,\hat{\mathbf{y}}$. We apply Ampère's law [Eq. (4.49)] to a narrow rectangular contour C lying in the xy-plane, with edges parallel to the x- and y-axes, as shown in Fig. 4.22. The dimension of the contour in the x-direction is $\mathrm{d}x$ and the length of the edges parallel to the y-axis is l. The magnetic field is constant along both edges with length l (B_y does not vary with y), so that essentially no integration is needed on the left-hand side of Eq. (4.49), i.e., the integral along each edge reduces to $\mathbf{B} \cdot \mathbf{l}$ (**l** has the same direction as **dl**). No integration is needed on the right-hand side of Eq. (4.49) either, because $\mathrm{d}x$ is differentially small and we can take $J_z(x)$ as constant over the surface spanned over C. Finally, as **B** is tangential to both edges with length l, **B** and **dl** are directed in opposite directions along the left edge, and in the same direction along the right edge, we can write

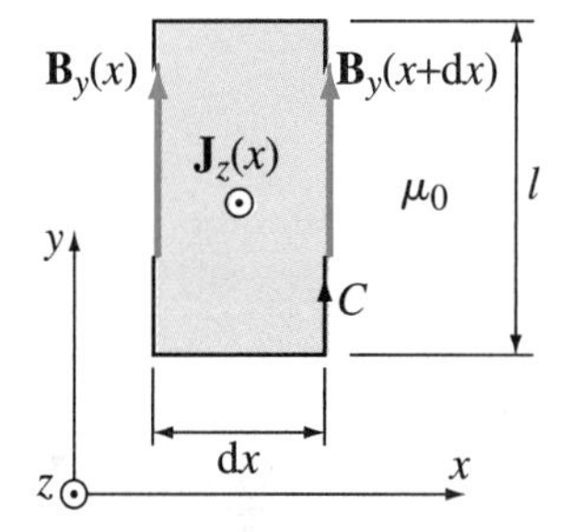

Figure 4.22 For the derivation of the one-dimensional Ampère's law in differential form.

$$-B_y(x)\,l + B_y(x + \mathrm{d}x)\,l = \mu_0 J_z(x)\,l\,\mathrm{d}x. \tag{4.73}$$

Noting that

$$\mathrm{d}B_y = B_y(x + \mathrm{d}x) - B_y(x), \tag{4.74}$$

Eq. (4.73) becomes a differential equation,

$$\boxed{\frac{\mathrm{d}B_y}{\mathrm{d}x} = \mu_0 J_z.} \tag{4.75}$$

1-D differential Ampère's law

This is the one-dimensional Ampère's law in differential form. We observe the analogy with the 1-D Gauss' law in Eq. (1.158).

We now generalize Eq. (4.75) to an arbitrary three-dimensional current distribution. The current density vector has now all three Cartesian components and all of them are functions of all three coordinates,

$$\mathbf{J} = J_x(x, y, z)\,\hat{\mathbf{x}} + J_y(x, y, z)\,\hat{\mathbf{y}} + J_z(x, y, z)\,\hat{\mathbf{z}}. \tag{4.76}$$

The Amperian contour C in Eq. (4.49) must, therefore, be differentially small in both dimensions. Additionally, one contour is not enough; we need three small contours, oriented perpendicularly to each of the three current density components, respectively. All edges of the contours being differentially small, the line integral along each of them can be approximated by taking a constant value of the field component tangential to the edge and multiplying it by plus or minus the edge length. This gives us the result for each pair of integrals along parallel edges within each contour that has exactly the same form as in the 1-D case [Eqs. (4.73)–(4.75)]. For the contour that lies entirely in the plane normal to the x-component of $\mathbf{J}$, shown in Fig. 4.23, we have

$$\oint_C \mathbf{B} \cdot \mathrm{d}\mathbf{l} = -B_z(x, y, z)\,\mathrm{d}z + B_y(x, y, z)\,\mathrm{d}y + B_z(x, y + \mathrm{d}y, z)\,\mathrm{d}z$$
$$-B_y(x, y, z + \mathrm{d}z)\,\mathrm{d}y = \mu_0 J_x(x, y, z)\,\mathrm{d}y\,\mathrm{d}z, \tag{4.77}$$

which, divided by $\mathrm{d}y\,\mathrm{d}z$, results in

$$\frac{\oint_C \mathbf{B} \cdot \mathrm{d}\mathbf{l}}{\mathrm{d}y\,\mathrm{d}z} = \frac{B_z(x, y + \mathrm{d}y, z) - B_z(x, y, z)}{\mathrm{d}y}$$
$$-\frac{B_y(x, y, z + \mathrm{d}z) - B_y(x, y, z)}{\mathrm{d}z} = \frac{\partial B_z}{\partial y} - \frac{\partial B_y}{\partial z} = \mu_0 J_x. \tag{4.78}$$

For contours that are oriented perpendicularly to each of the remaining two coordinate axes, y- and z-axes, analogous procedures lead to equations

$$\frac{\partial B_x}{\partial z} - \frac{\partial B_z}{\partial x} = \mu_0 J_y \quad \text{and} \quad \frac{\partial B_y}{\partial x} - \frac{\partial B_x}{\partial y} = \mu_0 J_z, \tag{4.79}$$

respectively. Multiplying Eq. (4.78) by the unit vector $\hat{\mathbf{x}}$, and Eqs. (4.79) by $\hat{\mathbf{y}}$ and $\hat{\mathbf{z}}$, respectively, and summing the three equations, we obtain the following equation with the vector $\mathbf{J}$ [Eq. (4.76)], multiplied by μ_0, on the right-hand side:

Ampère's law in differential form

$$\boxed{\left(\frac{\partial B_z}{\partial y} - \frac{\partial B_y}{\partial z}\right)\hat{\mathbf{x}} + \left(\frac{\partial B_x}{\partial z} - \frac{\partial B_z}{\partial x}\right)\hat{\mathbf{y}} + \left(\frac{\partial B_y}{\partial x} - \frac{\partial B_x}{\partial y}\right)\hat{\mathbf{z}} = \mu_0 \mathbf{J}.} \tag{4.80}$$

This partial differential equation (PDE) represents Ampère's law in differential form for an arbitrary steady current distribution. It relates the rate of change of $\mathbf{B}$ field components in spatial coordinates to the local current density vector, $\mathbf{J}$. We see that only those variations of individual components that are in directions

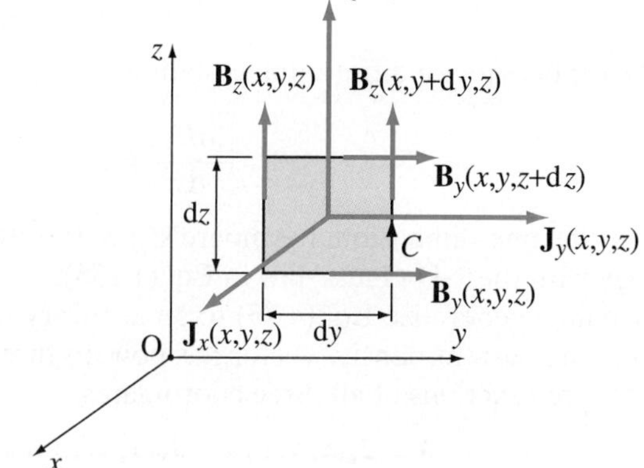

Figure 4.23 For the derivation of Ampère's law in differential form for an arbitrary current distribution.

perpendicular to the direction of the component (change of B_x along y and z, and not along x, etc.) contribute in this relationship, which is just opposite to the dependences in the differential Gauss' law [Eq. (1.163)].

4.7 CURL

The expression on the left-hand side of Eq. (4.80) is the so-called curl of a vector function (**B**), written as curl **B**,[4] analogous to the divergence of a vector field (**E**) as used to express the differential form of Gauss' law [Eq. (1.166)]. In addition, we notice that applying formally the formula for the cross product of two vectors in the Cartesian coordinate system[5] to $\nabla \times \mathbf{B}$, where the del operator is given by Eq. (1.100), we obtain exactly curl **B**. Hence,

$$\operatorname{curl}\mathbf{B} = \nabla \times \mathbf{B} = \left(\frac{\partial B_z}{\partial y} - \frac{\partial B_y}{\partial z}\right)\hat{\mathbf{x}} + \left(\frac{\partial B_x}{\partial z} - \frac{\partial B_z}{\partial x}\right)\hat{\mathbf{y}} + \left(\frac{\partial B_y}{\partial x} - \frac{\partial B_x}{\partial y}\right)\hat{\mathbf{z}}, \tag{4.81}$$

curl in Cartesian coordinates

which can also be written in the form of a determinant,

$$\operatorname{curl}\mathbf{B} = \nabla \times \mathbf{B} = \begin{vmatrix} \hat{\mathbf{x}} & \hat{\mathbf{y}} & \hat{\mathbf{z}} \\ \frac{\partial}{\partial x} & \frac{\partial}{\partial y} & \frac{\partial}{\partial z} \\ B_x & B_y & B_z \end{vmatrix} \tag{4.82}$$

(using the determinant form is a convenient way for memorizing the expression for $\nabla \times \mathbf{B}$ in the Cartesian coordinate system). Note that the curl is an operation that is performed on a vector, and the result is also a vector. Finally, the differential Ampère's law can now be written in a short form as

$$\operatorname{curl}\mathbf{B} = \nabla \times \mathbf{B} = \mu_0 \mathbf{J}. \tag{4.83}$$

Ampère's law using curl notation

In nonrectangular coordinate systems, the differentially small Amperian contour in Fig. 4.23 becomes curvilinear and with different expressions for lengths of its edges for different current density components in each of the systems. For instance, from Fig. 1.10, the contour perpendicular to the radial vector component $J_r\hat{\mathbf{r}}$ in spherical coordinates is a curvilinear quadrilateral of edge lengths $r\,\mathrm{d}\theta$ and $r\sin\theta\,\mathrm{d}\phi$. Carrying out similar derivations as in Eqs. (4.77)–(4.80) for this contour and contours oriented perpendicularly to other unit vectors in Figs. 1.25 and 1.26, the expression for the curl in cylindrical coordinates is obtained to be

$$\operatorname{curl}\mathbf{B} = \nabla \times \mathbf{B} = \left(\frac{1}{r}\frac{\partial B_z}{\partial \phi} - \frac{\partial B_\phi}{\partial z}\right)\hat{\mathbf{r}} + \left(\frac{\partial B_r}{\partial z} - \frac{\partial B_z}{\partial r}\right)\hat{\boldsymbol{\phi}} + \frac{1}{r}\left[\frac{\partial}{\partial r}(rB_\phi) - \frac{\partial B_r}{\partial \phi}\right]\hat{\mathbf{z}}, \tag{4.84}$$

curl in cylindrical coordinates

[4]Note that rot **B** is also used for curl **B**, "rot" being a short for "rotation" (or "rotational").

[5]For vectors given by their Cartesian components,

$$\mathbf{a} \times \mathbf{b} = (a_x\hat{\mathbf{x}} + a_y\hat{\mathbf{y}} + a_z\hat{\mathbf{z}}) \times (b_x\hat{\mathbf{x}} + b_y\hat{\mathbf{y}} + b_z\hat{\mathbf{z}})$$
$$= (a_yb_z - a_zb_y)\,\hat{\mathbf{x}} + (a_zb_x - a_xb_z)\,\hat{\mathbf{y}} + (a_xb_y - a_yb_x)\,\hat{\mathbf{z}}.$$

and that for the spherical coordinate system

curl in spherical coordinates

$$\text{curl}\,\mathbf{B} = \nabla\times\mathbf{B} = \frac{1}{r\sin\theta}\left[\frac{\partial}{\partial\theta}(\sin\theta B_\phi) - \frac{\partial B_\theta}{\partial\phi}\right]\hat{\mathbf{r}} + \frac{1}{r}\left[\frac{1}{\sin\theta}\frac{\partial B_r}{\partial\phi} - \frac{\partial}{\partial r}(rB_\phi)\right]\hat{\boldsymbol{\theta}} + \frac{1}{r}\left[\frac{\partial}{\partial r}(rB_\theta) - \frac{\partial B_r}{\partial\theta}\right]\hat{\boldsymbol{\phi}}. \quad (4.85)$$

In the same way as with $\nabla\cdot\mathbf{E}$ in Section 1.15, we use the notation $\nabla\times\mathbf{B}$ in cylindrical and spherical coordinate systems not to refer to it as the actual cross product of vectors ∇ and $\mathbf{B}$, which is valid only in rectangular coordinates [Eq. (4.81)], but to merely symbolize the curl operation. With this, we also emphasize the fact that relations derived employing such vector formalism in the Cartesian coordinate system hold true (are identities) generally, in all coordinate systems (properties of physical quantities and relations between them are the facts that are independent of the choice of coordinate system).

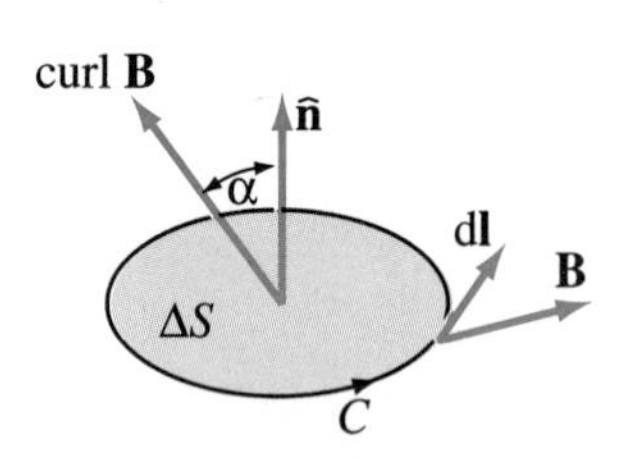

Figure 4.24 For the definition of the curl of a vector field in Eq. (4.87).

Combining Eqs. (4.81) and (4.78), we find that the x-component of curl $\mathbf{B}$, that is, $\hat{\mathbf{x}}\cdot\text{curl}\,\mathbf{B}$, can be expressed as the net circulation (line integral) of $\mathbf{B}$ along the incremental contour C of Fig. 4.23 divided by the area of the surface spanned over the contour,

$$\hat{\mathbf{x}}\cdot\text{curl}\,\mathbf{B} = \frac{\oint_C \mathbf{B}\cdot\text{d}\mathbf{l}}{\text{d}y\,\text{d}z}. \quad (4.86)$$

We can now formally proclaim the Cartesian x-axis to be an arbitrary linear axis (direction) in space and the contour C to be an arbitrarily shaped differentially small contour bounding a surface ΔS, as shown in Fig. 4.24, and write

alternative definition of the curl

$$\hat{\mathbf{n}}\cdot\text{curl}\,\mathbf{B} = \lim_{\Delta S\to 0}\frac{\oint_C \mathbf{B}\cdot\text{d}\mathbf{l}}{\Delta S}, \quad (4.87)$$

where $\hat{\mathbf{n}}$ is the unit vector normal to the surface ΔS ($\Delta\mathbf{S} = \Delta S\hat{\mathbf{n}}$) and determined using the right-hand rule. Eq. (4.87) is an equivalent mathematical definition of the curl of a vector. It enables us to obtain a component of the curl of $\mathbf{B}$ along a desired direction at a given point by computing the circulation of $\mathbf{B}$ along a contour in the plane perpendicular to that direction as the contour shrinks to zero about the point. The circulation of $\mathbf{B}$ about C, per unit area, appearing on the right-hand side of Eq. (4.87) equals $\hat{\mathbf{n}}\cdot\text{curl}\,\mathbf{B} = |\text{curl}\,\mathbf{B}|\cos\alpha$ ($|\hat{\mathbf{n}}| = 1$). Therefore, for the orientation of $\Delta\mathbf{S}$ defined by $\alpha = 0$, we get the maximum in the circulation ($\cos\alpha = 1$), and Eq. (4.87) becomes

physical meaning of the curl

$$|\text{curl}\,\mathbf{B}| = \left(\lim_{\Delta S\to 0}\frac{\oint_C \mathbf{B}\cdot\text{d}\mathbf{l}}{\Delta S}\right)_{\max} \quad (\alpha = 0). \quad (4.88)$$

This means that (1) the magnitude of curl $\mathbf{B}$ equals the maximum (as the direction of the surface element $\Delta\mathbf{S} = \Delta S\hat{\mathbf{n}}$ is varied) net circulation of $\mathbf{B}$ per unit area with the area of the surface element tending to zero and (2) the direction of curl $\mathbf{B}$ is in the direction that gives the maximum value for the magnitude of the net circulation per unit area ($\alpha = 0$). From it, we may regard the curl of a vector field (not only the magnetostatic field) physically as a measure of those local sources of the field at a point in space which produce circular field components with respect to that point. We recall that the divergence is a measure of sources that produce radial field components with respect to the point [Eq. (1.172)]. Ampère's law tells us that the sources producing locally circular field components in the case of the magnetostatic field in free space are elemental currents of intensities $\mathbf{J}\,\Delta S$. Quantitatively, the curl represents the surface density of such sources, and in our case this density is $\mathbf{J}$. The concept of the curl of a vector field is used in numerous applications, in many areas of science and engineering.

Replacing $\mu_0\mathbf{J}$ by $\nabla \times \mathbf{B}$ [from Eq. (4.83)] in the integral form of Ampère's law, Eq. (4.49), leads to

$$\boxed{\oint_C \mathbf{B} \cdot d\mathbf{l} = \int_S (\nabla \times \mathbf{B}) \cdot d\mathbf{S}.} \tag{4.89}$$

Stokes' theorem

Although obtained here specifically for the magnetostatic field in free space, this equation is an identity, holding for any vector field (for which the appropriate partial derivatives exist). It is widely used in electromagnetics and other areas of science and engineering, and is known as Stokes' theorem. In words, it states that the net circulation of a vector field along an arbitrary contour is the same as the net flux of its curl through any surface bounded by the contour, where the reference orientations of the contour and the surface are interconnected by the right-hand rule. We notice the parallelism with the divergence theorem, Eq. (1.173), which applies to a closed surface and relates the flux of a vector field with a volume integral of its divergence.

To prove Stokes' theorem (for an arbitrary vector field $\mathbf{B}$), imagine the surface S subdivided into a large number of differentially small patches ΔS_i ($\Delta S_i \to 0$) which are bounded by infinitesimal contours C_i, as depicted in Fig. 4.25. By applying the definition of the curl in Eq. (4.87) to one of these patches, we can write

$$\oint_{C_i} \mathbf{B} \cdot d\mathbf{l} = \Delta S_i\, \hat{\mathbf{n}} \cdot \operatorname{curl} \mathbf{B} = (\nabla \times \mathbf{B}) \cdot \Delta S_i \hat{\mathbf{n}} = (\nabla \times \mathbf{B}) \cdot \Delta \mathbf{S}_i. \tag{4.90}$$

Now, let us determine this circulation for every ΔS_i comprising S and sum all the results. What we get on the left-hand side of the resulting equation is the line integral of $\mathbf{B}$ along the overall contour C bounding the surface S, since the terms arising from the sides of small contours shared by any two patches cancel out during the summation, as can be seen in Fig. 4.25, and the only boundaries for which the cancelation does not occur are those forming the contour C. Hence,

$$\oint_C \mathbf{B} \cdot d\mathbf{l} = \lim_{\Delta S_i \to 0} \sum_i (\nabla \times \mathbf{B}) \cdot \Delta \mathbf{S}_i. \tag{4.91}$$

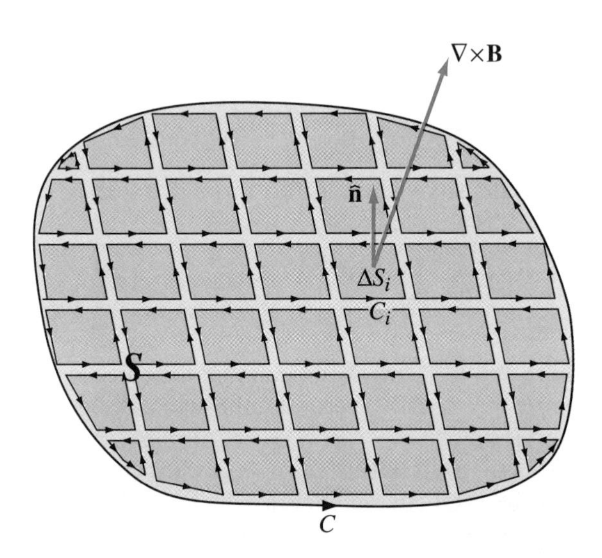

Figure 4.25 Open surface S subdivided into differentially small patches – for the proof of Stokes' theorem, Eq. (4.89).

In the limit, the summation on the right-hand side of this equation becomes an integral, $\Delta \mathbf{S}_i$ becomes $\mathrm{d}\mathbf{S}$, and we have $\int_S (\nabla \times \mathbf{B}) \cdot \mathrm{d}\mathbf{S}$, thus proving the theorem.

By applying Stokes' theorem to Eq. (1.75) or simply by analogy with the differential form of Ampère's law, Eq. (4.83), we arrive to the differential form of Maxwell's first equation for the electrostatic field:

curl of $\mathbf{E}$ *in electrostatics*

$$\boxed{\nabla \times \mathbf{E} = 0.} \tag{4.92}$$

We see that the electrostatic field is a curl-free or irrotational field, and this is a property of any conservative vector field (the field with zero circulation along any closed path).

Example 4.16 Cylindrical Current Conductor Using Differential Ampère's Law

Redo Example 4.9 but now using the differential form of Ampère's law.

Solution This is completely analogous to the application of the differential Gauss' law to solve an electrostatic problem with spherical symmetry in Example 1.22. For the interior of the conductor in Fig. 4.15 (where $\mathbf{J} = J\,\hat{\mathbf{z}}$ and $J = \text{const}$), combining Eqs. (4.53), (4.84), and (4.83), we now have in place of Eqs. (1.174) and (1.175)

$$\nabla \times \mathbf{B} = \frac{1}{r}\frac{\partial}{\partial r}[rB(r)]\,\hat{\mathbf{z}} = \mu_0 J\,\hat{\mathbf{z}} \quad \longrightarrow \quad rB(r) = \mu_0 J \int r\,\mathrm{d}r + C_1 = \frac{\mu_0 J r^2}{2} + C_1$$

$$\longrightarrow \quad B(r) = \frac{\mu_0 J r}{2} + \frac{C_1}{r} \quad (0 \le r < a)\,, \tag{4.93}$$

with $C_1 = 0$, since there exists no line current (of intensity I_0) along the z-axis (for $r = 0$) in Fig. 4.15 [otherwise, this constant would be $C_1 = \mu_0 I_0/(2\pi)$]. Similarly, in the surrounding space (with $\mathbf{J} = 0$),

$$\frac{\partial}{\partial r}[rB(r)] = 0 \quad \longrightarrow \quad rB(r) = C_2 \quad \longrightarrow \quad B(r) = \frac{C_2}{r} \quad (a < r < \infty)\,, \tag{4.94}$$

where $C_2 = \mu_0 J a^2/2$, from $B(a^-) = B(a^+)$. Both Eqs. (4.93) and (4.94) give the same corresponding results as in Eq. (4.56).

Problems: 4.20–4.24; *Conceptual Questions* (on Companion Website): 4.16 and 4.17; *MATLAB Exercises* (on Companion Website).

4.8 LAW OF CONSERVATION OF MAGNETIC FLUX

As vector $\mathbf{B}$ is called the magnetic flux density vector, its flux through a surface S is called, accordingly, the magnetic flux. It is denoted as Φ,

magnetic flux (unit: Wb*)*

$$\boxed{\Phi = \int_S \mathbf{B} \cdot \mathrm{d}\mathbf{S},} \tag{4.95}$$

and measured in webers (Wb), where $\mathrm{Wb} = \mathrm{T} \cdot \mathrm{m}^2$. We recall that the electric flux, Ψ, is defined as the flux of the electric flux density vector, $\mathbf{D}$, through a designated surface [Eq. (2.42)], and that the unit for Ψ is C.

The generalized Gauss' law, Eq. (2.43), tells us that the net outward electric flux through a closed surface is equal to the total charge enclosed by the surface. Our analogy between electric and magnetic fields naturally imposes a question: what

HISTORICAL ASIDE

Sir George Gabriel Stokes (1819–1903), a British mathematician and physicist, was a professor of mathematics at Cambridge University. Stokes was the oldest of the trio of Cambridge professors, James Clerk Maxwell (1831–1879) and Lord Kelvin (1824–1907) being the other two, who especially contributed to the fame of the Cambridge school of mathematical physics in the 19th century. His main contributions were in the areas of viscous fluids, sound, light, spectroscopy, fluorescence, and X rays. During a period of time, he held the post of the Lucasian Chair of Mathematics at Cambridge, was University's Member of Parliament, and President of the Royal Society, the three offices that had only once before been held by one person, Sir Isaac Newton (1642–1727).

Wilhelm Eduard Weber (1804–1891), a German physicist, was a professor at Göttingen University. Weber received his doctoral dissertation from the University of Halle in 1826, with a topic on the acoustic theory of reed organ pipes, and was appointed professor of physics at Göttingen in 1831. In collaboration with Karl Friedrich Gauss (1777–1855), he built the first practically useful telegraph (3 km long) in 1833, to connect his physics laboratory with Gauss' astronomical observatory at Göttingen. They also worked together on investigating terrestrial magnetism (earth's magnetic field). From 1836 to 1841, they organized a network of observation stations around the world to correlate measurements of terrestrial magnetism at different locations. In 1841, he developed the electrodynamometer, which could precisely measure the angular displacement of a coil with a current caused by another coil positioned perpendicularly to it and carrying the same current. Weber used this instrument for a final validation of Ampère's previous conclusions about magnetic forces due to currents in wire loops (in the coils). Furthermore, it was used, in the same or similar realizations, by many researchers and students over decades both to directly measure the magnetic forces and torques due to given currents and to determine (indirectly measure) the (unknown) current from the measured torque (as a sort of ammeter). During his later years at Göttingen, Weber worked on a theoretical generalization and unification of laws describing forces between charges at rest (Coulomb's law) and in motion (Ampère's force law), as well as Faraday's law of electromagnetic induction for time-varying currents (charges in accelerated motion). He also contributed to understanding the connection between light and electromagnetic phenomena and establishing the link between electromagnetism and optics, which was crucial for Maxwell's development of electromagnetic field theory (that includes light). Working again jointly with Gauss, Weber provided a significant impact in the early stage of the development of a new coherent system of units to also include electromagnetic phenomena, which gradually evolved into the present International System of Units (SI). The SI includes the weber, unit of magnetic flux, named in his honor. *(Portrait: AIP Emilio Segrè Visual Archives, Brittle Books Collection)*

is the net magnetic flux through a closed surface equal to? The answer is: zero. To prove this, we consider first the magnetic flux density $d\mathbf{B}$ of a single current element $\mathbf{J}\,dv$ in free space, as depicted in Fig. 4.26. From the Biot-Savart law and the expression for $d\mathbf{B}$ in Eq. (4.6), we know that the lines of this field are circles centered on the straight line containing the current element (Fig. 4.26). We can imagine the entire space surrounding the element divided into thin closed tubes of uniform cross section formed by bunches of field lines, where one such tube is shown in Fig. 4.26. According to Eq. (4.6), again, the magnitude of the magnetic flux density vector is constant along each field line, and thus over the entire volume of each tube (because the tubes are thin). Consequently, the magnetic flux through any tube is the same in magnitude at any cross section of the tube, regardless of whether that cross section is perpendicular to the tube axis or not.

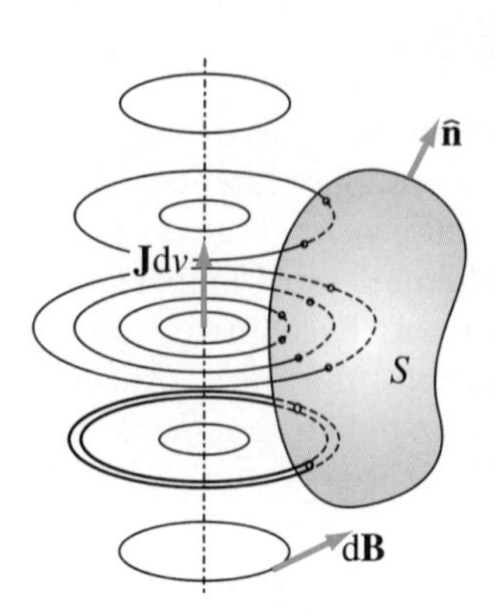

Figure 4.26 For the derivation of the law of conservation of magnetic flux.

Imagine now an arbitrary closed surface S in the field, Fig. 4.26. Some of the elementary tubes pass through S, but always an even number of times. Therefore, the net outward flux of $d\mathbf{B}$ through all small surfaces representing intersections of thin tubes and the surface S equals zero. Since such intersections cover the entire surface S, we have

$$\oint_S d\mathbf{B} \cdot d\mathbf{S} = 0. \tag{4.96}$$

Our proof continues then by invoking the superposition principle, by means of which the actual magnetic flux density $\mathbf{B}$ of an arbitrary current distribution in free space can be decomposed into elementary flux densities $d\mathbf{B}$ due to individual current elements making the current distribution, so that

$$\mathbf{B} = \int_{v_{\text{cur}}} d\mathbf{B}, \tag{4.97}$$

where v_{cur} is the domain with currents (sources of the magnetic field). Applying the integration over v_{cur} as an operator to Eq. (4.96), and interchanging the order of integral signs, we obtain

$$\oint_S \left(\int_{v_{\text{cur}}} d\mathbf{B} \right) \cdot d\mathbf{S} = 0. \tag{4.98}$$

Finally, substituting Eq. (4.97),

law of conservation of magnetic flux

$$\boxed{\oint_S \mathbf{B} \cdot d\mathbf{S} = 0,} \tag{4.99}$$

which completes our proof. This relation is known as the law of conservation of magnetic flux, and also as Maxwell's fourth equation. For obvious reasons, it is sometimes referred to as Gauss' law for the magnetic field. Together with Ampère's law (Maxwell's second equation), it forms a complete set of Maxwell's equations for the magnetostatic field in free space (or any nonmagnetic medium). We notice that the law of conservation of magnetic flux has the identical form as the continuity equation for steady currents, Eq. (3.40).

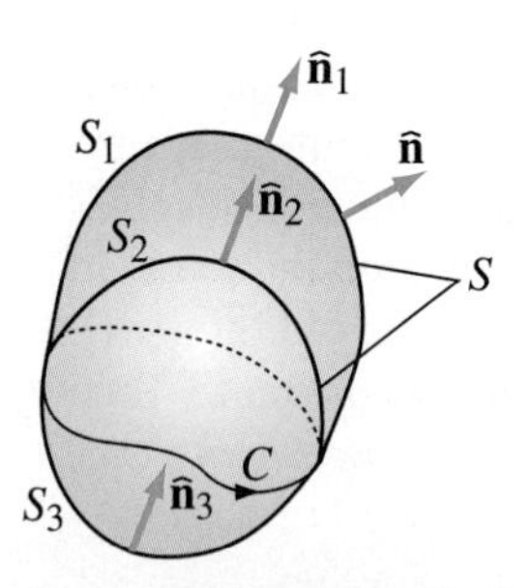

Figure 4.27 For the proof that the magnetic flux through a contour is unique.

Consider next an arbitrary contour C and two open surfaces, S_1 and S_2, with arbitrary shape that are both bounded by the contour and oriented in the same way – according to the right-hand rule with respect to the orientation of the contour, as shown in Fig. 4.27. The magnetic fluxes through the surfaces are

$$\Phi_1 = \int_{S_1} \mathbf{B} \cdot d\mathbf{S} \quad \text{and} \quad \Phi_2 = \int_{S_2} \mathbf{B} \cdot d\mathbf{S}, \tag{4.100}$$

respectively. The total outward flux through the closed surface S formed by S_1 and S_2 ($S = S_1 \cup S_2$) is

$$\oint_S \mathbf{B} \cdot d\mathbf{S} = \Phi_1 - \Phi_2, \tag{4.101}$$

which is zero, from the law of conservation of magnetic flux. Hence,

flux through a contour, Fig. 4.27

$$\boxed{\Phi_1 = \Phi_2,} \tag{4.102}$$

meaning that the magnetic flux through any number of surfaces of arbitrary shape that all have a common contour bounding them is the same, provided that all the surfaces are oriented in the same way. This enables us to link the flux to a contour rather than to a surface bounded by the contour (and there is an infinite number of such surfaces), and to use the term flux through a contour or flux linked by a contour. In other words, the flux (through a contour) is uniquely determined by the shape of a contour and by its orientation, and is the same for any surface spanned over the contour and oriented in accordance to the right-hand rule with respect to the orientation of the contour.

By analogy with the differential form of the continuity equation for steady currents, Eq. (3.41), the law of conservation of magnetic flux in differential form is given by

$$\boxed{\nabla \cdot \mathbf{B} = 0.} \tag{4.103}$$

flux conservation law in differential form

We see that the $\mathbf{B}$ field is another divergenceless (divergence-free) or solenoidal vector field. This means that there cannot be local sources of radial magnetic field components with respect to a point, i.e., that there exist no positive or negative "magnetic charges" and no free north or south magnetic poles, which would correspond to electric charges, with density ρ [Eq. (1.166)]. Equivalently, the magnetic field lines close upon themselves, as there are no "magnetic charges" for the lines to begin and terminate on.

We can now summarize the two Maxwell's differential equations governing the electrostatic field in free space,

$$\boxed{\nabla \times \mathbf{E} = 0 \quad \text{and} \quad \nabla \cdot \mathbf{E} = \frac{\rho}{\varepsilon_0},} \tag{4.104}$$

conservative field

and those for the magnetostatic field in free space,

$$\boxed{\nabla \times \mathbf{B} = \mu_0 \mathbf{J} \quad \text{and} \quad \nabla \cdot \mathbf{B} = 0,} \tag{4.105}$$

solenoidal field

which, in a condensed form, shows all the similarities and fundamental differences between the two fields.

Problems: 4.25 and 4.26; *Conceptual Questions* (on Companion Website): 4.18–4.23; *MATLAB Exercises* (on Companion Website).

4.9 MAGNETIC VECTOR POTENTIAL

In electrostatics, we introduced the electric scalar potential (V) to help us describe electric fields and evaluate the electric field intensity vector ($\mathbf{E}$). The electric scalar potential due to a point charge Q, as an elementary source of the electric field, in free space is [Eq. (1.80)]

$$V = \frac{1}{4\pi\varepsilon_0}\frac{Q}{R}. \tag{4.106}$$

Following the analogy established in connection with Eq. (4.4), which is the basis of the Biot-Savart law, the potential due to an elementary source of the magnetostatic field, $Q\mathbf{v}$, is defined as

$$\boxed{\mathbf{A} = \frac{\mu_0}{4\pi}\frac{Q\mathbf{v}}{R}.} \tag{4.107}$$

magnetic vector potential (unit: $\mathrm{T \cdot m}$)

This quantity is called the magnetic vector potential, and its unit is $\mathrm{T \cdot m}$. It is a vector whose direction is very simple to determine – the same as the direction of $Q\mathbf{v}$, and whose magnitude is proportional to $1/R$ (and not to $1/R^2$, which is present in the expression for **B** due to $Q\mathbf{v}$). By the same reasoning as with obtaining the three versions of the Biot-Savart law for volume, surface, and line currents, Eqs. (4.7)–(4.9), the corresponding integral expressions for the magnetic vector potential are:

A *due to volume current*

$$\mathbf{A} = \frac{\mu_0}{4\pi}\int_v \frac{\mathbf{J}\,\mathrm{d}v}{R}, \tag{4.108}$$

A *due to surface current*

$$\mathbf{A} = \frac{\mu_0}{4\pi}\int_S \frac{\mathbf{J}_\mathrm{s}\,\mathrm{d}S}{R}, \tag{4.109}$$

A *due to line current*

$$\mathbf{A} = \frac{\mu_0}{4\pi}\int_l \frac{I\,\mathrm{d}\mathbf{l}}{R}. \tag{4.110}$$

In general, the solutions for the magnetic vector potential due to given current distributions are substantially simpler than the corresponding solutions for the magnetic flux density vector.

Let us find the curl of **A**. By representing the expression in Eq. (4.107) in the spherical coordinate system shown in Fig. 4.28, in which Q is at the coordinate origin, so $R = r$ (r being the radial spherical coordinate), and **v** is z-directed,

$$\mathbf{A} = A\,\hat{\mathbf{z}} = A\cos\theta\,\hat{\mathbf{r}} - A\sin\theta\,\hat{\boldsymbol{\theta}} = A_r\hat{\mathbf{r}} + A_\theta\hat{\boldsymbol{\theta}}, \quad A = \frac{\mu_0 Qv}{4\pi r}, \tag{4.111}$$

where

$$A_r = A\cos\theta = \frac{\mu_0 Qv\cos\theta}{4\pi r}, \qquad A_\theta = -A\sin\theta = -\frac{\mu_0 Qv\sin\theta}{4\pi r} \tag{4.112}$$

($A_\phi = 0$). Using the expression for the curl in spherical coordinates, Eq. (4.85), we have

$$\begin{aligned}\nabla\times\mathbf{A} &= \frac{1}{r}\left[\frac{\partial}{\partial r}(rA_\theta) - \frac{\partial A_r}{\partial\theta}\right]\hat{\boldsymbol{\phi}} \\ &= \frac{\mu_0 Qv}{4\pi r}\left[\frac{\partial}{\partial r}(-\sin\theta) - \frac{\partial}{\partial\theta}\left(\frac{\cos\theta}{r}\right)\right]\hat{\boldsymbol{\phi}} = \frac{\mu_0 Qv\sin\theta}{4\pi r^2}\,\hat{\boldsymbol{\phi}}.\end{aligned} \tag{4.113}$$

Noting that (Fig. 4.28)

$$Q\mathbf{v}\times\hat{\mathbf{r}} = Qv\,\hat{\mathbf{z}}\times\hat{\mathbf{r}} = Qv\,|\hat{\mathbf{z}}\times\hat{\mathbf{r}}|\,\hat{\boldsymbol{\phi}} = Qv\sin\theta\,\hat{\boldsymbol{\phi}}, \tag{4.114}$$

we obtain

$$\nabla\times\mathbf{A} = \frac{\mu_0}{4\pi}\frac{Q\mathbf{v}\times\hat{\mathbf{r}}}{r^2}, \tag{4.115}$$

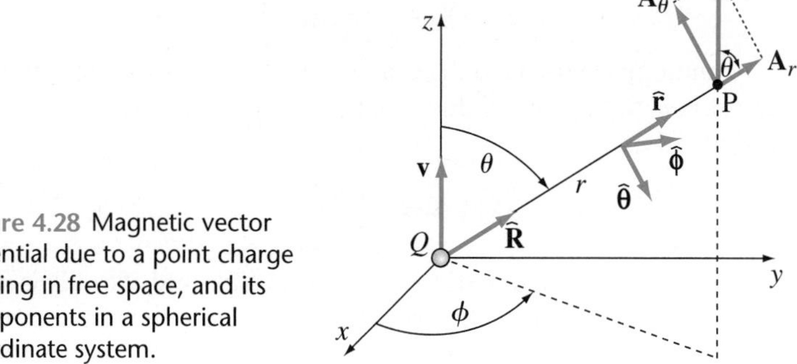

Figure 4.28 Magnetic vector potential due to a point charge moving in free space, and its components in a spherical coordinate system.

which is exactly the magnetic flux density vector due to $Q\mathbf{v}$, Eq. (4.4) with $R = r$ and $\hat{\mathbf{R}} = \hat{\mathbf{r}}$. Hence,

$$\boxed{\mathbf{B} = \nabla \times \mathbf{A},} \tag{4.116}$$

magnetic flux density from potential

and the same is true for the magnetic vector potential $\mathrm{d}\mathbf{A}$ due to an arbitrary volume, surface, or line current element, Eq. (4.10). Integrating the expression $\nabla \times \mathrm{d}\mathbf{A}$ over a domain v with volume currents, we have

$$\int_v (\nabla \times \mathrm{d}\mathbf{A}) = \nabla \times \left(\int_v \mathrm{d}\mathbf{A} \right) = \nabla \times \mathbf{A}, \tag{4.117}$$

where the integration and differentiation (del) operators can readily interchange places because they are completely independent – the integration is performed with respect to the coordinates of the source point (point at which the current density vector is $\mathbf{J}$), while the differentiation is carried out with respect to the coordinates of the field point (point at which the magnetic vector potential is $\mathbf{A}$). Consequently, the magnetic flux density vector ($\mathbf{B}$) and the magnetic vector potential ($\mathbf{A}$) due to an arbitrary volume current distribution, Eq. (4.108), are related at an arbitrary point in space as in Eq. (4.116), and similar proofs can be carried out for the magnetic potential due to surface currents, Eq. (4.109), and line currents, Eq. (4.110). This relationship, in conjunction with Eqs. (4.108)–(4.110), provides an alternative general method for evaluating the $\mathbf{B}$ field produced by steady currents, where $\mathbf{A}$ is evaluated first by integration, and then $\mathbf{B}$ is found from the potential by differentiation. Potentials, generally, are auxiliary quantities that are used to determine fields indirectly. Finally, we note that, contrary to V in electrostatics, $\mathbf{A}$ does not have any simple physical interpretation in magnetostatics.

On the other side, the divergence of the magnetostatic potential $\mathbf{A}$ is always zero, which is, essentially, a consequence of the steady current density $\mathbf{J}$ being a divergenceless vector, i.e., of the continuity equation in Eq. (3.41) or (3.40). Namely, in analogy to Eq. (4.117), we can apply the divergence operator to the integral expression in Eq. (4.108) and write

$$\nabla \cdot \mathbf{A} = \frac{\mu_0}{4\pi} \nabla \cdot \left(\int_v \frac{\mathbf{J}\,\mathrm{d}v}{R} \right) = \frac{\mu_0}{4\pi} \int_v \nabla \cdot \left(\frac{\mathbf{J}}{R} \right) \mathrm{d}v = \frac{\mu_0}{4\pi} \oint_S \frac{\mathbf{J}}{R} \cdot \mathrm{d}\mathbf{S}, \tag{4.118}$$

where S is the closed surface bounding v, and the use is made of the divergence theorem, Eq. (1.173), to convert the volume integral to a surface (flux) one. For the special case when v is a spherical domain centered at the field point (where the potential is being computed), $1/R$ can be brought out of the flux integral, as S in that case is a spherical surface of radius R, and the integral, in turn, becomes zero, $\oint_S \mathbf{J} \cdot \mathrm{d}\mathbf{S} = 0$, by virtue of the continuity equation, Eq. (3.40). In the general case, the arbitrary domain v can be subdivided into a stack of concentric spherical layers with each of them being a part of a full spherical shell of thickness $\mathrm{d}R$ centered at the field point, with the current of density $\mathbf{J}$ flowing through that part and no current in the rest of the shell, outside v. The same conclusion about a zero flux integral based on the continuity equation can then be derived for the surface of each shell, and thus

$$\boxed{\nabla \cdot \mathbf{A} = 0,} \tag{4.119}$$

divergence of $\mathbf{A}$ *due to steady currents*

which holds true for the magnetic potential due to surface and line steady currents as well.

The magnetic flux through an arbitrary contour, Eq. (4.95), can now be expressed in terms of the magnetic vector potential,

$$\Phi = \int_S \mathbf{B} \cdot d\mathbf{S} = \int_S (\nabla \times \mathbf{A}) \cdot d\mathbf{S}, \tag{4.120}$$

which, by using Stokes' theorem, Eq. (4.89), becomes

magnetic flux from the potential

$$\boxed{\Phi = \oint_C \mathbf{A} \cdot d\mathbf{l}.} \tag{4.121}$$

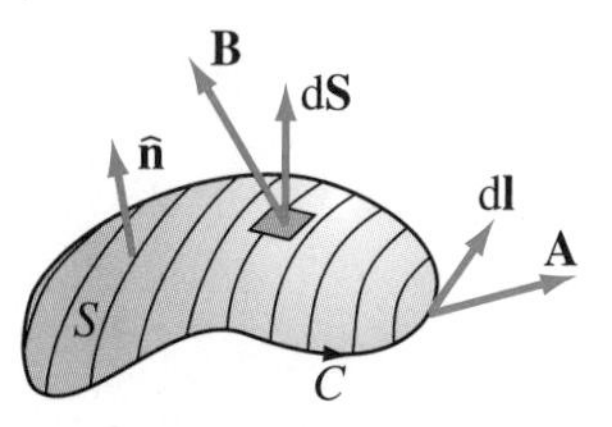

Figure 4.29 For the evaluation of the magnetic flux through a surface by integrating the magnetic vector potential along its boundary.

The orientation of the contour C and the surface S are in accordance to the right-hand rule, as shown in Fig. 4.29. Eq. (4.121) represents a means for determining the magnetic flux by evaluating a contour integral (of $\mathbf{A}$) rather than a surface integral (of $\mathbf{B}$), which is very convenient in some computations and derivations. Note that this is yet another proof that the magnetic flux through a contour is unique (the same for all surfaces bounded by C).

Problems: 4.27; *Conceptual Questions* (on Companion Website): 4.24 and 4.25; *MATLAB Exercises* (on Companion Website).

4.10 PROOF OF AMPÈRE'S LAW

We are now ready to prove Ampère's law by utilizing the magnetic vector potential, $\mathbf{A}$. On the left-hand side of the differential form of the law, Eq. (4.83), we have, by means of Eq. (4.116),

$$\nabla \times \mathbf{B} = \nabla \times (\nabla \times \mathbf{A}). \tag{4.122}$$

Applying formally (symbolically) the vector identity for expanding the vector triple product,[6]

$$\mathbf{a} \times (\mathbf{b} \times \mathbf{c}) = \mathbf{b}(\mathbf{a} \cdot \mathbf{c}) - \mathbf{c}(\mathbf{a} \cdot \mathbf{b}), \tag{4.123}$$

to the product $\nabla \times (\nabla \times \mathbf{A})$, we get the following identity for expanding the curl of the curl of an arbitrary vector field ($\mathbf{A}$):

$$\nabla \times (\nabla \times \mathbf{A}) = \nabla(\nabla \cdot \mathbf{A}) - (\nabla \cdot \nabla)\mathbf{A} = \nabla(\nabla \cdot \mathbf{A}) - \nabla^2 \mathbf{A}. \tag{4.124}$$

The first term in this expansion is grad(div $\mathbf{A}$), and in our case it is zero, because div $\mathbf{A} = \nabla \cdot \mathbf{A} = 0$ [Eq. (4.119)]. With this, Eq. (4.122) becomes

$$\nabla \times \mathbf{B} = -\nabla^2 \mathbf{A}. \tag{4.125}$$

Vector $\nabla^2 \mathbf{A}$ is called the Laplacian of the vector function $\mathbf{A}$, which, having in mind Eq. (1.100), can be written in Cartesian coordinates as

$$\nabla^2 \mathbf{A} = (\nabla \cdot \nabla)\mathbf{A} = \left(\frac{\partial^2}{\partial x^2} + \frac{\partial^2}{\partial y^2} + \frac{\partial^2}{\partial z^2}\right)(A_x \hat{\mathbf{x}} + A_y \hat{\mathbf{y}} + A_z \hat{\mathbf{z}})$$

[6] A cross (vector) product of a vector (**a**) with a cross product of two (other) vectors (**b** and **c**) is called the vector triple product. Note also that the vector triple product identity in Eq. (4.123) is known as the "bac-cab" or "back-cab" identity, as the word "bac-cab" can be read on the right-hand side of the equation.

$$= \left(\frac{\partial^2 A_x}{\partial x^2} + \frac{\partial^2 A_x}{\partial y^2} + \frac{\partial^2 A_x}{\partial z^2}\right)\hat{\mathbf{x}} + \left(\frac{\partial^2 A_y}{\partial x^2} + \frac{\partial^2 A_y}{\partial y^2} + \frac{\partial^2 A_y}{\partial z^2}\right)\hat{\mathbf{y}}$$

$$+ \left(\frac{\partial^2 A_z}{\partial x^2} + \frac{\partial^2 A_z}{\partial y^2} + \frac{\partial^2 A_z}{\partial z^2}\right)\hat{\mathbf{z}}. \quad (4.126)$$

We see that Cartesian components of the Laplacian of **A** equal the Laplacian of the corresponding components of **A**, where the latter operator is the Laplacian of a scalar field, Eq. (2.94), and hence[7]

$$\boxed{\nabla^2 \mathbf{A} = \nabla^2 A_x \, \hat{\mathbf{x}} + \nabla^2 A_y \, \hat{\mathbf{y}} + \nabla^2 A_z \, \hat{\mathbf{z}}.} \quad (4.127)$$

Laplacian of a vector in Cartesian coordinates

From Eq. (4.108),

$$\mathbf{A} = A_x \, \hat{\mathbf{x}} + A_y \, \hat{\mathbf{y}} + A_z \, \hat{\mathbf{z}} = \frac{\mu_0}{4\pi} \int_v \frac{(J_x \, \hat{\mathbf{x}} + J_y \, \hat{\mathbf{y}} + J_z \, \hat{\mathbf{z}}) \, dv}{R}$$

$$= \frac{\mu_0}{4\pi} \left(\hat{\mathbf{x}} \int_v \frac{J_x \, dv}{R} + \hat{\mathbf{y}} \int_v \frac{J_y \, dv}{R} + \hat{\mathbf{z}} \int_v \frac{J_z \, dv}{R} \right), \quad (4.128)$$

and we conclude that each Cartesian component of **A** is actually produced by the same component of the current density vector, namely, A_x is produced by J_x, and so on. Recalling the expression for the electric scalar potential due to a volume charge in free space, Eq. (1.82), we identify the duality between V due to ρ and A_x due to J_x:

$$V = \frac{1}{4\pi\varepsilon_0} \int_v \frac{\rho \, dv}{R} \quad \longleftrightarrow \quad A_x = \frac{\mu_0}{4\pi} \int_v \frac{J_x \, dv}{R}. \quad (4.129)$$

By the same duality principle, in turn, there must be a differential equation for A_x which has the same form as the differential equation for V, that is, Poisson's equation [Eq. (2.93)]. Therefore, by changing the variables, we get

$$\nabla^2 V = -\frac{\rho}{\varepsilon_0} \quad \longleftrightarrow \quad \nabla^2 A_x = -\mu_0 J_x. \quad (4.130)$$

Similarly,

$$\nabla^2 A_y = -\mu_0 J_y \quad \text{and} \quad \nabla^2 A_z = -\mu_0 J_z, \quad (4.131)$$

and substituting all these back in Eq. (4.127),

$$\boxed{\nabla^2 \mathbf{A} = -\mu_0 \mathbf{J}.} \quad (4.132)$$

vector Poisson's equation

This is a second-order vector partial differential equation (PDE) for the magnetic vector potential of a volume current in free space (or any nonmagnetic medium), usually referred to as the vector Poisson's equation. It can be used as a starting point in solving (analytically or numerically) for **A** due to a given current distribution **J**. We use it here for deriving Ampère's law. Namely, returning to Eq. (4.125), we can now substitute for the Laplacian of **A** and obtain

$$\nabla \times \mathbf{B} = -\nabla^2 \mathbf{A} = \mu_0 \mathbf{J}, \quad (4.133)$$

[7]Note that the expansion in Eq. (4.127) does not have equally simple counterparts in cylindrical and spherical coordinate systems (Figs. 1.25 and 1.26). In these systems, the Laplacian of a vector function (**A**), $\nabla^2\mathbf{A}$, is computed, from Eq. (4.124), as $\nabla^2\mathbf{A} = \nabla(\nabla \cdot \mathbf{A}) - \nabla \times (\nabla \times \mathbf{A}) = \text{grad}(\text{div}\,\mathbf{A}) - \text{curl}\,(\text{curl}\,\mathbf{A})$, namely, as the gradient of the divergence of **A** minus the curl of the curl of **A**, using the expressions for the gradient, divergence, and curl in Eqs. (1.105), (1.108), (1.170), (1.171), (4.84), and (4.85).

which is Eq. (4.83), and this completes our proof of the Ampère's law in differential form. Stokes' theorem, Eq. (4.89), which is derived from the mathematical definition of the curl of a vector, gives its integral equivalent, Eq. (4.49).

Note that the proof carried out in this section is essentially based on the concept of the magnetic vector potential, which is defined by Eq. (4.107) and related to the magnetic flux density vector by Eq. (4.116). This latter relation is obtained from Eq. (4.4), which is the rudimentary version of the Biot-Savart law. This means that we have, in fact, derived Ampère's law from the Biot-Savart law.

4.11 MAGNETIC DIPOLE

A small loop with a steady current constitutes the magnetic equivalent of the electric dipole of Fig. 1.28, and is referred to as a magnetic dipole. The reason for this and what we mean by "small" will soon be evident. A magnetic dipole is characterized by its magnetic moment, defined as

magnetic dipole moment

$$\boxed{\mathbf{m} = I\mathbf{S},} \tag{4.134}$$

where I is the current intensity of the loop and $\mathbf{S} = S\hat{\mathbf{n}}$ is the loop surface area vector, oriented in accordance to the right-hand rule with respect to the reference direction of the current. Note that $\mathbf{m}$ is analogous to the electric dipole moment, $\mathbf{p}$, defined by Eq. (1.116). The unit for $\mathbf{m}$ is $\mathrm{A \cdot m^2}$.

We would like to find the expressions for the magnetic vector potential and the magnetic flux density vector due to a magnetic dipole at large distances compared with the loop dimensions. To this end, we consider a rectangular loop with sides a and b, shown in Fig. 4.30. Let the axis of the loop coincide with the z-axis of a spherical coordinate system and the center of the loop be located at the origin of the system. Far away from the loop ($r \gg a, b$), the loop is observed as being small and the magnetic vector potential of the loop can be evaluated as that due to four line current elements, given by $I\mathbf{a}$, $I\mathbf{b}$, $I(-\mathbf{a})$, and $I(-\mathbf{b})$, using Eq. (4.110). Let us first calculate the potential at a point P due to the pair of parallel elements of length a, Fig. 4.30,

$$\mathbf{A}_{aa} = \mathbf{A}_1 + \mathbf{A}_2 = \frac{\mu_0}{4\pi}\frac{I\mathbf{a}}{r_1} + \frac{\mu_0}{4\pi}\frac{I(-\mathbf{a})}{r_2} = \frac{\mu_0 I\mathbf{a}}{4\pi}\left(\frac{1}{r_1} - \frac{1}{r_2}\right), \tag{4.135}$$

where r_1 and r_2 are the distances of the point P from the centers of the first and second element, respectively. Noting that the position vector of the first element with respect to the second element is $\mathbf{d} = -\mathbf{b}$, we can now use the same approximations

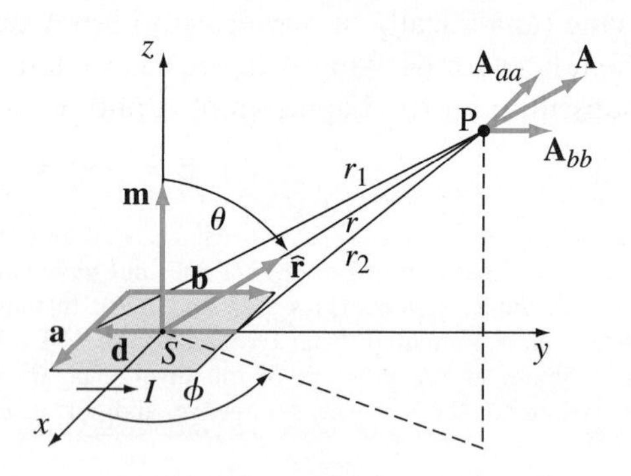

Figure 4.30 Magnetic dipole.

as in deriving Eq. (1.115) for an electric dipole, with which Eq. (4.135) becomes

$$\mathbf{A}_{aa} \approx \frac{\mu_0 I \mathbf{a}}{4\pi} \frac{\mathbf{d} \cdot \hat{\mathbf{r}}}{r^2} = -\frac{\mu_0 I \mathbf{a}(\mathbf{b} \cdot \hat{\mathbf{r}})}{4\pi r^2}. \tag{4.136}$$

Similarly, the potential due to the other pair of elements, those of length b, is given by

$$\mathbf{A}_{bb} \approx \frac{\mu_0 I \mathbf{b}(\mathbf{a} \cdot \hat{\mathbf{r}})}{4\pi r^2}. \tag{4.137}$$

By superposition and using the formula for the vector triple product in Eq. (4.123), the total magnetic vector potential of the dipole turns out to be

$$\mathbf{A} = \mathbf{A}_{aa} + \mathbf{A}_{bb} = \frac{\mu_0 I}{4\pi r^2} [\mathbf{b}(\mathbf{a} \cdot \hat{\mathbf{r}}) - \mathbf{a}(\mathbf{b} \cdot \hat{\mathbf{r}})] = \frac{\mu_0 I}{4\pi r^2} (\mathbf{a} \times \mathbf{b}) \times \hat{\mathbf{r}}. \tag{4.138}$$

As the magnetic moment of the loop is (Fig. 4.30)

$$\mathbf{m} = I\mathbf{S} = Iab\,\hat{\mathbf{z}} = I\mathbf{a} \times \mathbf{b}, \tag{4.139}$$

$\mathbf{A}$ can be expressed in terms of $\mathbf{m}$ as

$$\mathbf{A} = \frac{\mu_0}{4\pi} \frac{\mathbf{m} \times \hat{\mathbf{r}}}{r^2}. \tag{4.140}$$

We see that the magnetic potential depends on the dipole magnetic moment and the position of the field point with respect to the dipole, and not on the shape of the loop, which can be arbitrary (not only rectangular). Finally, since $\mathbf{m} \times \hat{\mathbf{r}} = m \sin\theta\, \hat{\boldsymbol{\phi}}$, which directly comes from Eq. (4.114), Eq. (4.140) can be rewritten as

$$\boxed{\mathbf{A} = A_\phi \hat{\boldsymbol{\phi}} = \frac{\mu_0 m \sin\theta}{4\pi r^2} \hat{\boldsymbol{\phi}}.} \tag{4.141}$$

magnetic dipole potential

We conclude that the magnetic vector potential due to the magnetic dipole has only a ϕ-component in the spherical coordinate system, which is a function of coordinates r and θ.

The magnetic flux density vector of the dipole is now determined applying the formula for the curl in spherical coordinates, Eq. (4.85), to the expression for $\mathbf{A}$ in Eq. (4.141),

$$\boxed{\begin{aligned} \mathbf{B} = \nabla \times \mathbf{A} &= \frac{1}{r\sin\theta} \frac{\partial}{\partial\theta} (\sin\theta A_\phi)\,\hat{\mathbf{r}} - \frac{1}{r} \frac{\partial}{\partial r} (rA_\phi)\,\hat{\boldsymbol{\theta}} \\ &= \frac{\mu_0 m}{4\pi r^3} \left(2\cos\theta\,\hat{\mathbf{r}} + \sin\theta\,\hat{\boldsymbol{\theta}}\right). \end{aligned}} \tag{4.142}$$

magnetic dipole field

Comparing Eqs. (1.117) and (4.142), we observe that the $\mathbf{E}$ and $\mathbf{B}$ field of the electric and magnetic dipoles, respectively, are identical in form, so that the corresponding field lines have identical shape. However, this is true only at large distances from the dipoles (relative to their dimensions), which is illustrated in Fig. 4.31. We see that, although identical far from the sources, the normalized field lines due to the two dipoles are fundamentally different close to them; the electric field lines terminate on the two charges forming the electric dipole, whereas the magnetic field lines close upon themselves through the current loop.

We shall see in the next chapter that the concept of a magnetic dipole is fundamental for understanding the behavior of magnetic materials, much like the electric dipole was used in Chapter 2 in studying the electric field in the presence of dielectric materials. The field of a magnetic dipole, Eq. (4.142), is also used, again in

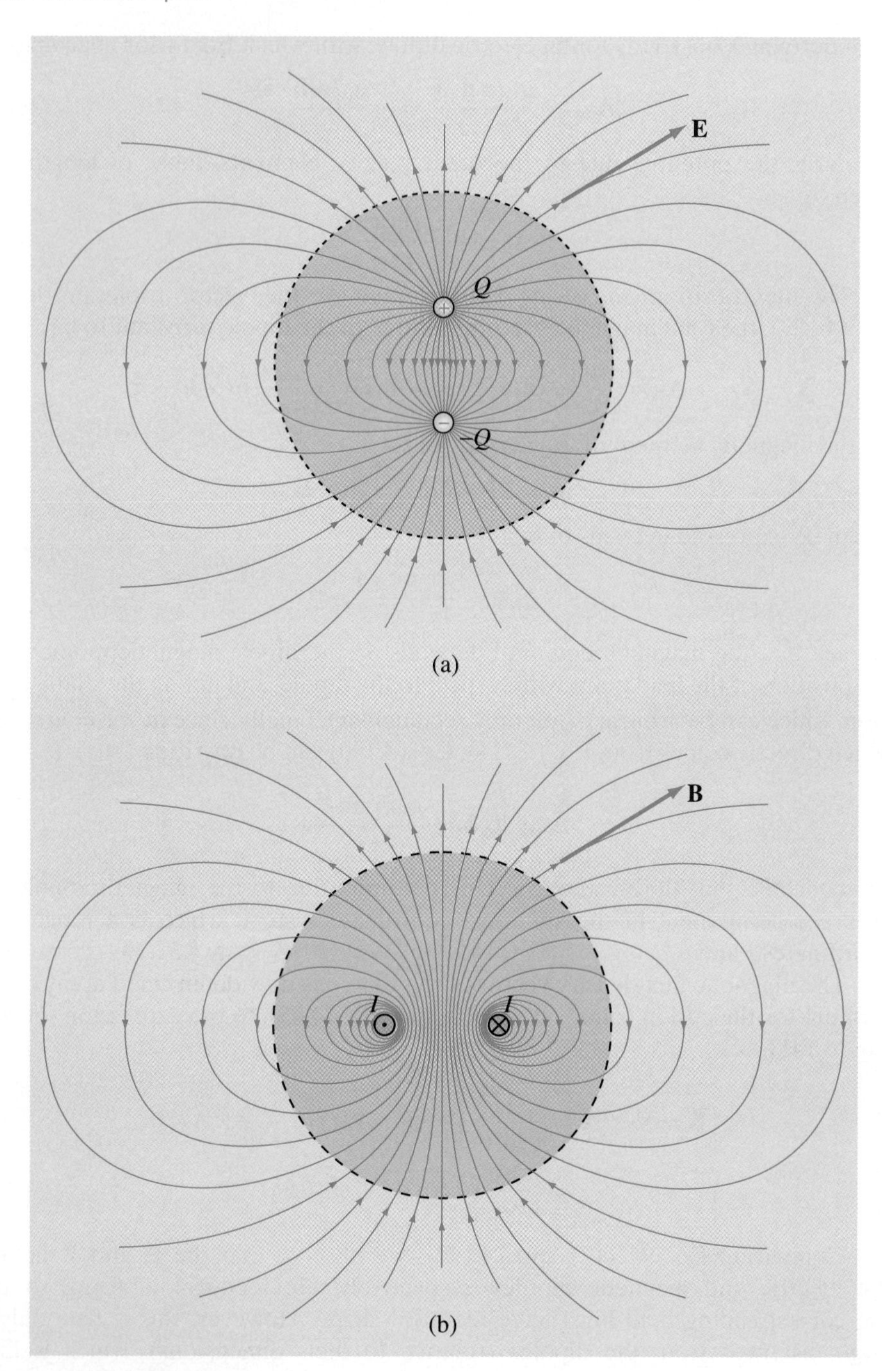

Figure 4.31 Normalized electric field lines of an electric dipole (a) and magnetic field lines of a magnetic dipole (b).

parallel to the electric field of an electric dipole, as an approximation for the static and quasistatic (low-frequency) magnetic field produced by an electrical device, which is important in EMI considerations.

Problems: 4.28–4.31; *Conceptual Questions* (on Companion Website): 4.26; *MATLAB Exercises* (on Companion Website).

4.12 THE LORENTZ FORCE AND HALL EFFECT

From the definition of the electric field intensity vector, Eq. (1.23), we have that the electric force on a point charge Q situated in an electric field of intensity $\mathbf{E}$ is

$$\boxed{\mathbf{F}_e = Q\mathbf{E}.} \tag{4.143}$$

electric force on a particle

Similarly, Eq. (4.1) tells us that the magnetic force on a point charge Q moving at a velocity $\mathbf{v}$ in a magnetic field equals

$$\boxed{\mathbf{F}_m = Q\mathbf{v} \times \mathbf{B},} \tag{4.144}$$

magnetic force on a particle

where $\mathbf{B}$ is the flux density vector of the field. Finally, the force on a moving charge due to both an electric and a magnetic field is obtained by superposition,

$$\boxed{\mathbf{F} = \mathbf{F}_e + \mathbf{F}_m = Q\mathbf{E} + Q\mathbf{v} \times \mathbf{B}.} \tag{4.145}$$

Lorentz force

This equation is known as the Lorentz force equation or law. The total (electric + magnetic = electromagnetic) force on the particle is called the Lorentz force.

An interesting and important manifestation of the motion of free charges in a material under the influence of the Lorentz force is the Hall effect, which we describe briefly here. We consider a conducting strip of width a situated in a uniform steady magnetic field of flux density vector $\mathbf{B}$ that is perpendicular to the strip, as depicted in Fig. 4.32. Let a steady current of density $\mathbf{J}$ flow through the strip. The free charges constituting the current can be positive or negative (e.g., holes and electrons in a semiconductor), and Fig. 4.32 shows both cases. Due to the magnetic force given by Eq. (4.144), the charges move (deflect) across the strip in the direction perpendicular to both $\mathbf{B}$ and $\mathbf{J}$, which results in a charge separation on the two sides of the strip. In Fig. 4.32(a), the free charges are positive ($Q > 0$), $\mathbf{v} = \mathbf{v}_d$ (drift velocity) is in the same direction as $\mathbf{J}$ [Eq. (3.11)], $\mathbf{v}_d \times \mathbf{B}$ is directed to the right, and so is $\mathbf{F}_m$; thus, the positive charges move to the right. In Fig. 4.32(b), on the other hand, the free charges are negative ($Q < 0$), $\mathbf{v}_d$ is in the direction opposite to $\mathbf{J}$, $\mathbf{v}_d \times \mathbf{B}$ is directed to the left, and $\mathbf{F}_m$ is again directed to the right (because $Q < 0$); hence, the negative charges end up at the right edge of the strip. Accumulated charges produce an electric field (of intensity $\mathbf{E}$) across the strip. This field, in turn, acts on the free charges with an electric force, Eq. (4.143), which is in the opposite direction to the magnetic force. In the equilibrium, the two forces are equal in magnitude, i.e., the Lorentz (total) force on the charges is zero,

$$\mathbf{F} = Q(\mathbf{E} + \mathbf{v}_d \times \mathbf{B}) = 0, \tag{4.146}$$

from which,

$$E = v_d B. \tag{4.147}$$

The voltage between the strip edges amounts to

$$\boxed{V_H = Ea = v_d Ba.} \tag{4.148}$$

Hall voltage

This voltage is known as the Hall voltage, and the effect itself is called the Hall effect. The direction of the Hall voltage drop is different for positive and negative charges [from right to left in Fig. 4.32(a), and from left to right in Fig. 4.32(b)]. So, the polarity of the Hall voltage tells us the sign of free charge carriers in a material. Note that this represents a method for determining whether a given semiconductor is p-type or n-type.

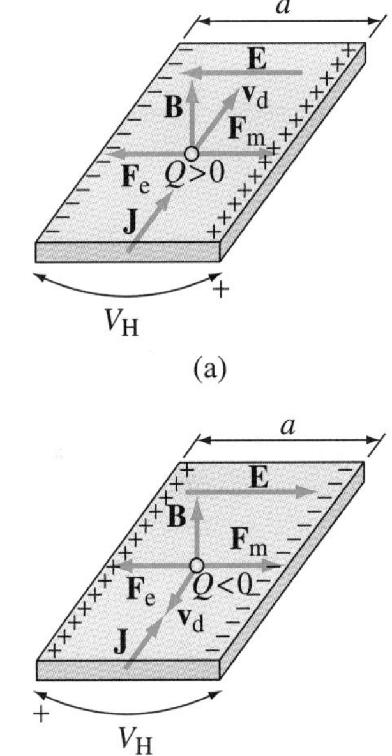

Figure 4.32 Hall effect in a material with (a) positive free charges and (b) negative free charges.

HISTORICAL ASIDE

Hendrik Antoon Lorentz (1853–1928), a Dutch mathematician and physicist, a professor of mathematical physics at the University of Leiden, was the winner of the 1902 Nobel Prize in Physics. He further developed Maxwell's electromagnetic theory of light, and proposed the electron theory according to which oscillating electrons within atoms constitute miniature equivalent Hertzian radiators that emit light. In 1904, he introduced his famous Lorentz transformations (of space and time coordinates), which describe time dilation and length contraction for a body moving at velocities close to the speed of light and represent the foundation of Einstein's (1879–1955) special theory of relativity. He formulated the Lorentz force law for a moving charged particle in the presence of electric and magnetic fields. Lorentz is also the author of the so-called Lorentz-Lorenz formula, jointly with Danish physicist Ludwig Lorenz (1829–1891), who discovered it independently. The formula provides a mathematical relationship between the index of refraction (of light) and density of a medium. *(Portrait: AIP Emilio Segrè Visual Archives, Lande Collection)*

Edwin Herbert Hall (1855–1938), an American physicist, was a professor at Harvard University. He received his doctorate in physics from Johns Hopkins University under Professor Henry Augustus Rowland (1848–1901), who was one of the world's most brilliant physicist of the last quarter of the 19th century. As a part of his dissertation work, Hall pursued the question as to whether the resistance of a current-carrying conductor was affected by the presence of an external magnetic field. In experiments with guidance from Professor Rowland in 1879, he used a conductor in the form of a strip of a gold leaf mounted on a glass plate and placed it between the poles of an electromagnet such that the magnetic field lines were perpendicular to the current flow in the strip. What he observed was the development of a significant transverse electric field in the conductor and the associated voltage across the strip as the result of the applied magnetic field. Hall published his findings in the famous article "On a New Action of the Magnet on Electric Currents" in American Journal of Mathematics in November of 1879, and this phenomenon soon came to be known as the "Hall effect." He was appointed professor of physics at Harvard in 1895. *(Portrait: "Voltiana," Como, Italy - Sept. 10, 1927 issue, courtesy AIP Emilio Segrè Visual Archives)*

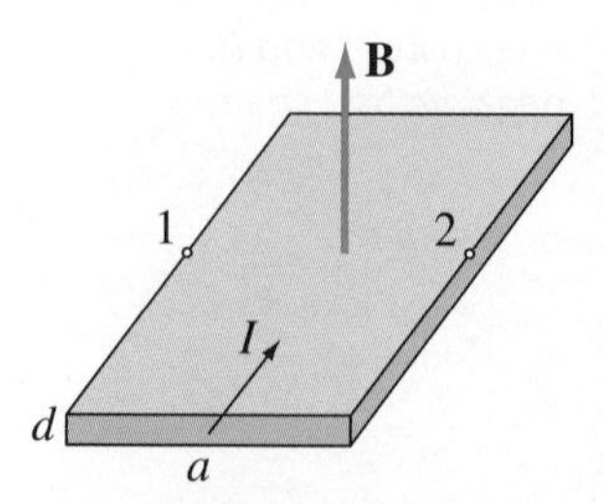

Figure 4.33 Hall element for measuring the magnetic flux density; for Example 4.17.

Example 4.17 Hall Element for Measuring the Magnetic Flux Density

A Hall element for measuring the magnetic flux density is in the form of a strip with width a and thickness d, shown in Fig. 4.33. The concentration of free charge carriers in the strip is N_v and the charge of each carrier is Q. The strip carries a steady current of intensity I, and the magnetic flux density vector is perpendicular to the strip. A voltmeter shows a voltage V_{12} between the strip edges. What is the algebraic intensity of the magnetic flux density vector (B) with respect to the reference direction indicated in Fig. 4.33?

Solution From Eqs. (3.11) and (3.5), the current density in the element can be written as

$$J = N_v Q v_d = \frac{I}{ad}, \tag{4.149}$$

which yields $v_d = I/(N_v a d Q)$. The Hall voltage is given in Eq. (4.148), and $V_H = V_{21}$ in Fig. 4.33, namely, it is opposite to the measured voltage V_{12}. Hence, the magnetic flux density turns out to be

$$B = \frac{V_H}{v_d a} = -\frac{V_{12}}{v_d a} = -\frac{N_v Q d V_H}{I}. \tag{4.150}$$

Example 4.18 Lorentz Force due to a Rotating Charged Contour

A uniformly charged circular contour of radius a and total charge Q rotates in free space about its axis with a uniform angular velocity $\mathbf{w} = w\,\hat{\mathbf{z}}$, as shown in Fig. 4.34. A charged particle q moves with a uniform velocity $\mathbf{v} = v\,\hat{\mathbf{y}}$ along a path that belongs to the plane $z = a$ and is parallel to the y-axis. Find the Lorentz force on the particle at an instant when it is at the point P(0, 0, a), above the center of the contour.

Solution The electric field intensity vector at the point P in Fig. 4.34 is given by Eq. (1.44) with $z = a$,

$$\mathbf{E} = \frac{Q\sqrt{2}}{16\pi\varepsilon_0 a^2}\,\hat{\mathbf{z}} = E\,\hat{\mathbf{z}}. \tag{4.151}$$

The time period for one rotation of the contour is

$$T = \frac{2\pi}{w} \tag{4.152}$$

(full angle divided by the angular velocity). Noting that the total charge of the contour, Q, passes any given reference point on the contour during the time T, we conclude that the rotating charged contour is equivalent to a line current along the contour of intensity [Eq. (3.4)]

$$I = \frac{Q}{T} = \frac{Qw}{2\pi}. \tag{4.153}$$

From Eq. (4.19), the magnetic flux density vector due to this current at the point P ($z = a$) is

$$\mathbf{B} = \frac{\mu_0 I\sqrt{2}}{8a}\,\hat{\mathbf{z}} = \frac{\mu_0 Q w\sqrt{2}}{16\pi a}\,\hat{\mathbf{z}} = B\,\hat{\mathbf{z}}. \tag{4.154}$$

Using Eq. (4.145), the Lorentz force on the charge q comes out to be (Fig. 4.34)

$$\mathbf{F} = q(\mathbf{E} + v\,\hat{\mathbf{y}} \times \mathbf{B}) = \underbrace{qE\,\hat{\mathbf{z}}}_{\mathbf{F}_e} + \underbrace{qvB\,\hat{\mathbf{x}}}_{\mathbf{F}_m} = \frac{qQ\sqrt{2}}{16\pi\varepsilon_0 a^2}\left(\varepsilon_0\mu_0 v w a\,\hat{\mathbf{x}} + \hat{\mathbf{z}}\right). \tag{4.155}$$

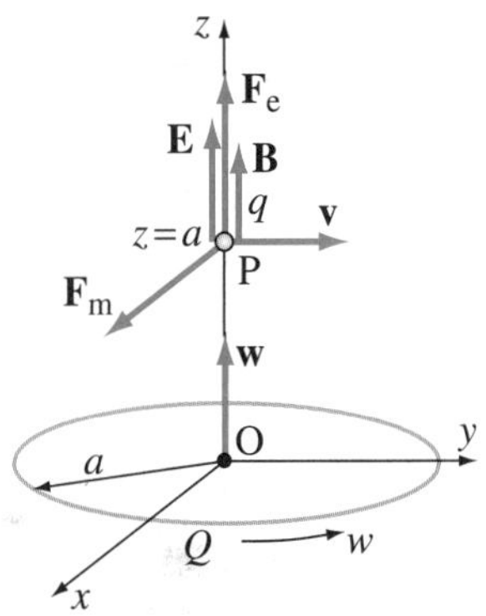

Figure 4.34 Lorentz force on a charged particle moving above a rotating charged contour; for Example 4.18.

Problems: 4.32.

4.13 EVALUATION OF MAGNETIC FORCES

Eq. (4.144) gives the Lorentz magnetic force (i.e., the magnetic component of the total Lorentz force) on a single point charge moving in a magnetic field. If we have many charges constituting a current in some domain, we utilize the superposition principle as in Eq. (4.5) and conclude that the magnetic force on a volume current element $\mathbf{J}\,\mathrm{d}v$ is given by

$$\mathrm{d}\mathbf{F}_m = (\mathbf{J}\,\mathrm{d}v) \times \mathbf{B}, \tag{4.156}$$

where $\mathbf{B}$ is the flux density vector of the external magnetic field. Then, from Eq. (4.10), the magnetic force on a surface current element is

$$\mathrm{d}\mathbf{F}_m = (\mathbf{J}_s\,\mathrm{d}S) \times \mathbf{B}, \tag{4.157}$$

and that on a line current element

$$d\mathbf{F}_m = I\, d\mathbf{l} \times \mathbf{B}. \tag{4.158}$$

Integration of Eqs. (4.156)–(4.158) leads to the following integral formulations for the total magnetic force for volume, surface, and line current distributions:

magnetic force on volume current

$$\mathbf{F}_m = \int_v \mathbf{J} \times \mathbf{B}\, dv, \tag{4.159}$$

magnetic force on surface current

$$\mathbf{F}_m = \int_S \mathbf{J}_s \times \mathbf{B}\, dS, \tag{4.160}$$

magnetic force on line current

$$\mathbf{F}_m = \int_l I\, d\mathbf{l} \times \mathbf{B} = I \int_l d\mathbf{l} \times \mathbf{B}. \tag{4.161}$$

In the last integral, I can be taken out of the integral sign because it is always constant along the line (continuity equation for steady currents).

In the case of a straight homogeneous conductor of arbitrary cross section with a steady current placed in a uniform magnetic field, we have $\mathbf{J} = \text{const}$ [Eq. (3.82)] and $\mathbf{B} = \text{const}$, so that $\mathbf{J} \times \mathbf{B} = \text{const}$ and Eq. (4.159) becomes

$$\mathbf{F}_m = (\mathbf{J} \times \mathbf{B}) \int_v dv = (\mathbf{J}v) \times \mathbf{B} = (\mathbf{J}Sl) \times \mathbf{B} = (JS\mathbf{l}) \times \mathbf{B}, \tag{4.162}$$

where v and l are the volume and length of the conductor, respectively, S is the surface area of its cross section, the direction of the vector $\mathbf{l}$ is the same as the direction of the current flow along the conductor, and $|\mathbf{l}| = l$. Introducing the current intensity of the conductor, $I = JS$, we obtain

$\mathbf{F}_m$ on a straight conductor in a uniform magnetic field

$$\mathbf{F}_m = I\mathbf{l} \times \mathbf{B}. \tag{4.163}$$

Example 4.19 Force Between Two Long Parallel Wires with Current

Two parallel, very long and thin wires in air carry currents of intensities I_1 and I_2, both flowing in the same direction. The distance between the wire axes is d. Find the magnetic forces on wires per unit of their length.

Solution Fig. 4.35 shows the cross section of the two wires. From Eq. (4.22), the magnetic flux density vector due to the current in the first wire (I_1), assuming that it is infinitely long, at the axis of the second wire is

$$\mathbf{B}_1 = \frac{\mu_0 I_1}{2\pi d}\, \hat{\mathbf{y}}. \tag{4.164}$$

Since the wires are thin (as compared to the distance between their axes), we can assume that the magnetic field across the entire cross section of the second wire is uniform and given

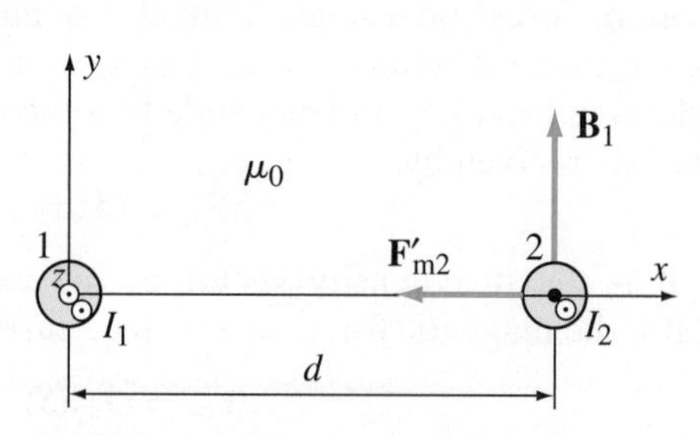

Figure 4.35 Evaluation of the force between two long parallel current-carrying wires; for Example 4.19.

by Eq. (4.164), so that Eq. (4.163) can be used. Hence, the force on the part of the second conductor that is l long equals

$$\mathbf{F}_{m2} = I_2(l\,\hat{\mathbf{z}}) \times \mathbf{B}_1 = \frac{\mu_0 I_1 I_2 l}{2\pi d}\,\hat{\mathbf{z}} \times \hat{\mathbf{y}} = -\frac{\mu_0 I_1 I_2 l}{2\pi d}\,\hat{\mathbf{x}}, \tag{4.165}$$

and the force per unit of its length

$$\mathbf{F}'_{m2} = \frac{\mathbf{F}_{m2}}{l} = -\frac{\mu_0 I_1 I_2}{2\pi d}\,\hat{\mathbf{x}}. \tag{4.166}$$

The per-unit-length force on the first conductor is $\mathbf{F}'_{m1} = -\mathbf{F}'_{m2}$.

We see that the magnetic force between the wires is attractive if the currents are in the same direction, i.e., if both I_1 and I_2 are either positive or negative, and repulsive if they are in opposite directions ($I_1 I_2 < 0$). So, just in contrary to charges Q_1 and Q_2 and Coulomb's law, "like" currents (in parallel wires) attract and "unlike" currents repel each other (also see Fig. 4.2).

Example 4.20 Force on a Loop near a Long Wire

A long straight wire carries a steady current of intensity I. A rectangular conducting loop lies in the same plane as the wire, with two sides (of length b) parallel to the wire and two sides (of length a) perpendicular. The distance between the wire and the closer parallel side of the loop is c. The loop carries a steady current of the same intensity, and the directions of currents are shown in Fig. 4.36. Determine the net magnetic force on the loop.

Solution The magnetic flux density vector due to the current in the long wire at any point in the plane of the loop is normal to the plane, and at a distance x from the wire its magnitude is

$$B(x) = \frac{\mu_0 I}{2\pi x}. \tag{4.167}$$

The forces on each side of the loop are obtained from Eq. (4.163), and their directions are given in Fig. 4.36. From symmetry, it is obvious that the forces on the sides 2 and 4 are equal in magnitude and with opposite directions,

$$\mathbf{F}_{m2} = -\mathbf{F}_{m4}. \tag{4.168}$$

The forces on the sides 1 and 3 are also in opposing directions, but their magnitudes are different because of their different distances from the long wire ($x = c$ and $x = c + a$),

$$F_{m1} = IbB(c) = \frac{\mu_0 I^2 b}{2\pi c} \quad \text{(repulsive)}, \tag{4.169}$$

$$F_{m3} = IbB(c+a) = \frac{\mu_0 I^2 b}{2\pi(c+a)} \quad \text{(attractive)}. \tag{4.170}$$

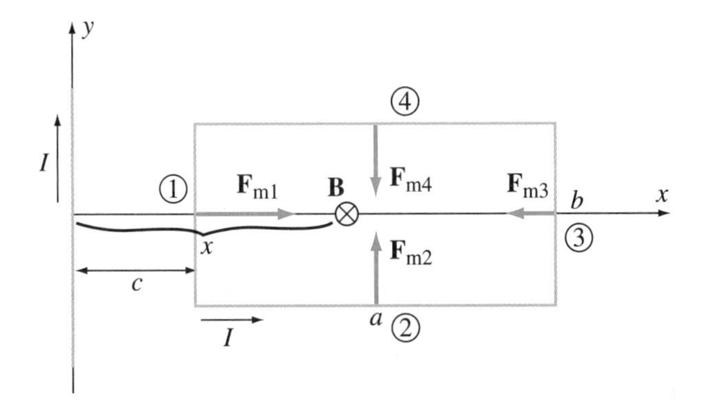

Figure 4.36 Evaluation of the magnetic force on a rectangular current loop near a long wire with current; for Example 4.20.

The total force on the loop, assuming that it is rigid (i.e., the loop maintains its shape even under the influence of the magnetic forces on its sides), is hence given by

$$\mathbf{F}_\text{m} = \mathbf{F}_\text{m1} + \mathbf{F}_\text{m2} + \mathbf{F}_\text{m3} + \mathbf{F}_\text{m4} = \mathbf{F}_\text{m1} + \mathbf{F}_\text{m3}$$
$$= \frac{\mu_0 I^2 b}{2\pi}\left(\frac{1}{c} - \frac{1}{c+a}\right)\hat{\mathbf{x}} = \frac{\mu_0 I^2 ab}{2\pi c(c+a)}\,\hat{\mathbf{x}}. \tag{4.171}$$

Thus, the loop is pushed away from the long wire. Note that if the polarity of either (but not both) of the two currents (in the wire and the loop) were reversed, the loop would be pulled toward the long wire.

Example 4.21 Force on a Loop in a Uniform Magnetic Field

Prove that the net magnetic force on a contour of arbitrary shape with a steady current in a uniform magnetic field is zero.

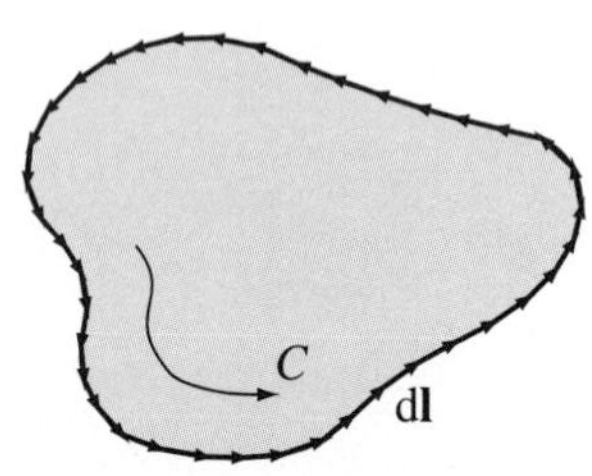

Figure 4.37 In view of the head-to-tail rule for vector addition, it is obvious that the integral of dl along a closed path is always zero; for Example 4.21.

Solution For a uniform magnetic field, $\mathbf{B} = \text{const}$, so that Eq. (4.161) becomes

$$\mathbf{F}_\text{m} = I\oint_C \text{d}\mathbf{l} \times \mathbf{B} = I\left(\oint_C \text{d}\mathbf{l}\right) \times \mathbf{B}. \tag{4.172}$$

For any closed path,

$$\oint_C \text{d}\mathbf{l} = 0, \tag{4.173}$$

which is evident from Fig. 4.37, and hence $\mathbf{F}_\text{m} = 0$.

Example 4.22 Torque on a Current Loop in a Uniform Magnetic Field

A rigid square loop of side length a is situated in a uniform steady magnetic field of flux density $\mathbf{B}$, as shown in Fig. 4.38(a). There is a steady current of intensity I in the loop. The angle between the plane of the loop and the plane normal to the vector $\mathbf{B}$ is θ. The loop is mounted such that it is free to rotate about the axis O-O′ which is perpendicular to the plane of the drawing. Find (a) the net force and (b) the net torque on the loop.

Solution

(a) From Eq. (4.172), there is no net force on the loop, i.e., the vector sum of individual magnetic forces on the loop sides is zero.

(b) The forces on the sides of the loop that are normal to the axis O-O′ tend to stretch the loop, but do not produce torques on it. The torques (moments) of the forces on the sides 1 and 2 of the loop (sides parallel to the axis O-O′) calculated with respect to the center of the loop are

$$\mathbf{T}_1 = \mathbf{r}_1 \times \mathbf{F}_\text{m1} \quad \text{and} \quad \mathbf{T}_2 = \mathbf{r}_2 \times \mathbf{F}_\text{m2}, \tag{4.174}$$

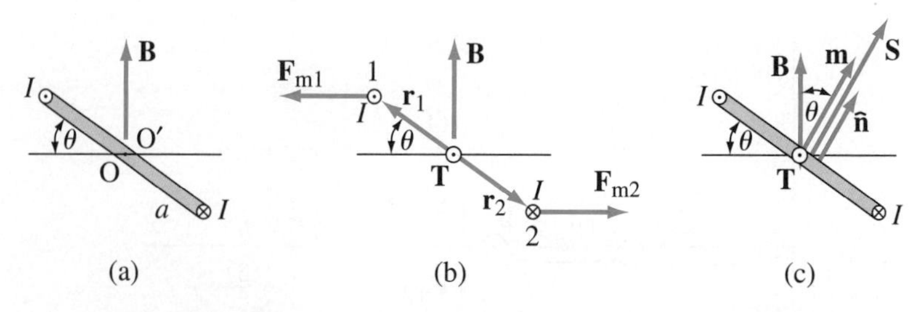

Figure 4.38 Evaluation of the torque **T** of magnetic forces on a current loop in a uniform magnetic field: (a) position of the loop relative to the magnetic field, (b) magnetic forces on loop sides producing a torque, and (c) the relationship with the magnetic moment **m** of the loop; for Example 4.22.

respectively, where $\mathbf{F}_{m1}$ and $\mathbf{F}_{m2}$ are the magnetic forces on the sides, and $\mathbf{r}_1$ and $\mathbf{r}_2$ are the position vectors of the centers of sides with respect to the loop center [Fig. 4.38(b)]. Since

$$\mathbf{F}_{m1} + \mathbf{F}_{m2} = 0, \tag{4.175}$$

the resultant torque on the loop is given by

$$\mathbf{T} = \mathbf{T}_1 + \mathbf{T}_2 = \mathbf{r}_1 \times \mathbf{F}_{m1} + \mathbf{r}_2 \times \mathbf{F}_{m2} = (\mathbf{r}_1 - \mathbf{r}_2) \times \mathbf{F}_{m1} = \mathbf{r}_{12} \times \mathbf{F}_{m1}. \tag{4.176}$$

The vector $\mathbf{r}_{12} = \mathbf{r}_1 - \mathbf{r}_2$ joins the point of application of $\mathbf{F}_{m2}$ to that of $\mathbf{F}_{m1}$ and is independent of the choice of origin of the two vectors $\mathbf{r}_1$ and $\mathbf{r}_2$. Therefore, the torque is also independent of the choice of origin, i.e., it is the same when calculated about any reference point, provided that the total force on the loop is zero.

Using Eq. (4.163), the magnitudes of the forces $\mathbf{F}_{m1}$ and $\mathbf{F}_{m2}$ are

$$F_{m1} = F_{m2} = IaB, \tag{4.177}$$

which, substituted in Eqs. (4.174), gives the magnitudes of the corresponding torque vectors:

$$T_1 = T_2 = |\mathbf{r}_1 \times \mathbf{F}_{m1}| = \frac{a}{2} F_{m1} \sin\theta = \frac{Ia^2 B \sin\theta}{2} \tag{4.178}$$

($|\mathbf{r}_1| = a/2$). Hence, the magnitude of the resultant torque vector $\mathbf{T}$ is

$$T = 2T_1 = Ia^2 B \sin\theta, \tag{4.179}$$

and its direction is shown in Fig. 4.38(b). Of course, the same is obtained from Eq. (4.176).

Noting that the angle between the unit normal $\hat{\mathbf{n}}$ on the loop surface oriented in accordance to the right-hand rule with respect to the reference direction of the loop current and the vector $\mathbf{B}$ is also θ [Fig. 4.38(c)], we conclude that the resultant torque of magnetic forces on the loop can be compactly expressed as

$$\boxed{\mathbf{T} = \mathbf{m} \times \mathbf{B},} \tag{4.180}$$

torque on a current loop in a uniform magnetic field

where $\mathbf{m}$ is the magnetic moment of the loop, given by Eq. 4.134, so that

$$T = |\mathbf{m} \times \mathbf{B}| = mB \sin\theta = Ia^2 B \sin\theta. \tag{4.181}$$

Eq. (4.180) is a general expression for the torque on a current loop of arbitrary shape and size in a uniform magnetic field. We see that the torque on the loop always tend to turn the loop so as to align the vectors $\mathbf{m}$ and $\mathbf{B}$. In other words, it tends to align the magnetic field produced by the loop current (which coincides with the direction of $\mathbf{m}$ – see Figs. 4.6 and 4.8) with the applied (external) magnetic field that is causing the torque. Finally, the magnetic field across a small current loop, i.e., a magnetic dipole (Fig. 4.30), can always be considered as locally uniform, which means that Eq. (4.180) gives us the torque on a magnetic dipole (with any shape and the magnetic moment $\mathbf{m}$) in any (generally nonuniform) magnetic field (with the local flux density $\mathbf{B}$). We observe the analogy with the torque on an electric dipole, $\mathbf{T} = \mathbf{p} \times \mathbf{E}$, in Eq. (2.3).

Problems: 4.33–4.39; *Conceptual Questions* (on Companion Website): 4.27–4.30; *MATLAB Exercises* (on Companion Website).

Problems

4.1. **Rectangular current loop.** Consider a rectangular loop with sides a and b in air. If the loop carries a steady current of intensity I, find the magnetic flux density vector at an arbitrary point along the axis of the loop perpendicular to its plane.

4.2. Triangular current loop. A loop in the form of a triangle representing a half of a square of side a carries a steady current of intensity I, as shown in Fig. 4.39. The medium is air. Calculate the magnetic flux density vector at a point P located at the fourth vertex of the square.

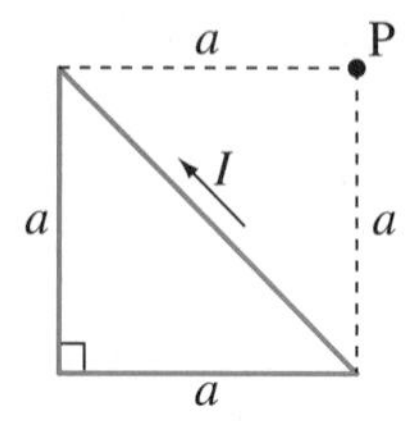

Figure 4.39 Triangular current loop; for Problem 4.2.

4.3. Current loop with circular and linear parts. Fig. 4.40 shows a wire contour composed from a half of a circle of radius a and a half of a square of side $2a$. The contour is situated in air and carries a steady current of intensity I. Find the magnetic flux density vector (a) at the central point O and (b) at a point $2a$ apart from the point O along the line perpendicular to the plane of the contour.

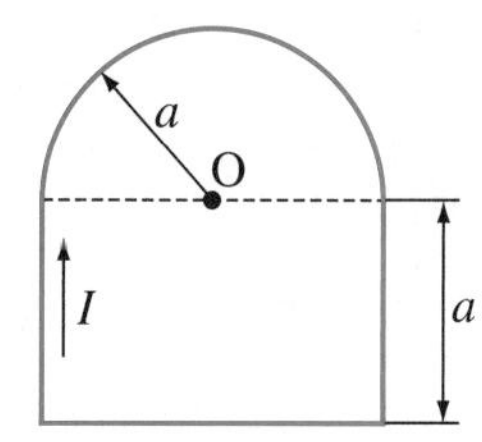

Figure 4.40 Current contour composed from a semicircle and three linear parts; for Problem 4.3.

4.4. Circular surface current distribution. There is a surface current over a circular surface of radius b with a hole of radius a ($a < b$) in free space. The surface lies in the plane $z = 0$ of a cylindrical coordinate system, with the coordinate origin coinciding with the surface center. The surface current density vector is $\mathbf{J}_s = J_{s0}(a/r)\hat{\boldsymbol{\phi}}$ ($a \leq r \leq b$), where J_{s0} is a constant. Compute the magnetic flux density vector along the z-axis.

4.5. Magnetic field of a rotating charged disk. A circular disk of radius a is uniformly charged over its surface by a charge of density ρ_s. The disk uniformly rotates in air about its axis (perpendicular to the disk) with an angular velocity w. Find the magnetic flux density vector at an arbitrary point along the axis of rotation. Assume that the charge distribution over the disk remains the same during the rotation.

4.6. Solenoids with different length-to-diameter ratios. Consider a solenoid with a nonmagnetic core and $N = 1000$ tightly wound turns of wire carrying a steady current $I = 1$ A. The length of the solenoid is $l = 50$ cm. Calculate the **B** field at the solenoid center, at the center of the 250th (or 750th) wire turn, and at the center of the first (or last) wire turn, and sketch the function $B(z)$ along the solenoid axis ($-\infty < z < \infty$) for the solenoid radius equal to (a) $a = 25$ cm and (b) $a = 2$ cm, respectively.

4.7. Helmholtz coils. Fig. 4.41 shows two identical very short coils, each with N circular turns of wire, in air. The distance between the centers of the coils is d, and the wire turn radii are a. The coils carry steady currents of equal intensities, I, and are oriented in the same way. When $d = a$, the magnetic field near the center of the structure (midway between the two coils) is approximately uniform, and the structure is referred to as Helmholtz coils. (a) For an arbitrary distance d ($d \neq a$), find the expression for the magnetic flux density $B(z)$ at an arbitrary point along the axis of the coils (z-axis), assuming that B due to each of the coils equals N times the flux density of a single turn of wire (circular current loop, Fig. 4.6) at the respective coil location. (b) Show that $\mathrm{d}B/\mathrm{d}z = 0$ at the center of the structure (point C), so for $z = d/2$ (and any d, relative to a). (c) Then verify that $\mathrm{d}^2B/\mathrm{d}z^2 = 0$ at the point C when $d = a$ (note that even $\mathrm{d}^3B/\mathrm{d}z^3 = 0$ at C in this case).

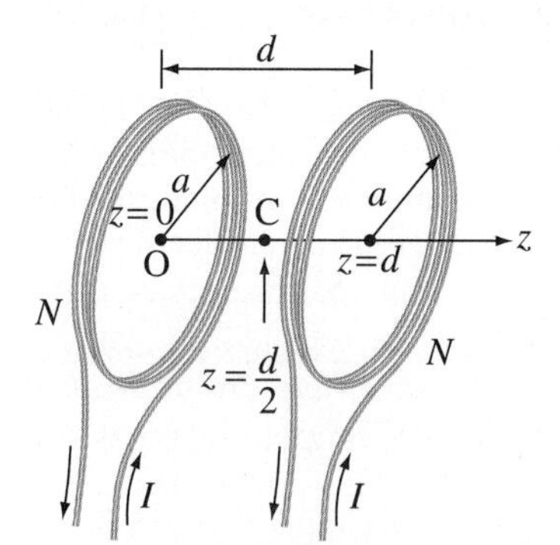

Figure 4.41 Helmholtz coils; for Problem 4.7.

4.8. Spherical coil. Consider a coil consisting of N turns of an insulated thin wire wound uniformly and densely in one layer on a nonmagnetic sphere of radius a. The current through the wire is I. The surrounding medium is air.

Find the magnetic flux density vector at the sphere center.

4.9. Two parallel strips with opposite currents. Two parallel, very long identical strip conductors of width $2a$ carry currents of the same intensity I and opposite directions. The distance between the strips is $2a$ as well. The cross section of the structure is shown in Fig. 4.42. The permeability everywhere is μ_0. Compute the **B** field at the center of the cross section (point O).

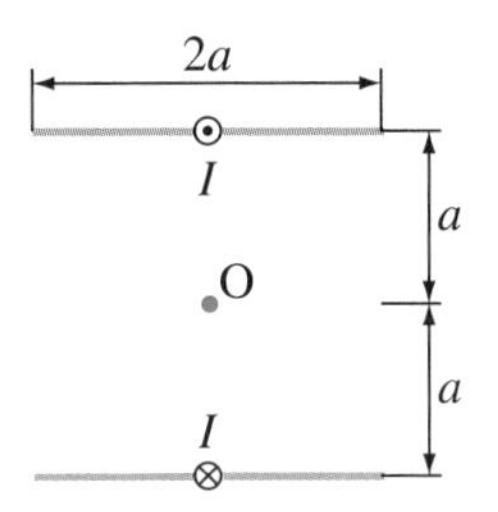

Figure 4.42 Cross section of two parallel strip conductors with currents of the same magnitude and opposite directions; for Problem 4.9.

4.10. Magnetic field of a hollow cylindrical conductor. A steady current of intensity I flows through an infinitely long hollow cylindrical copper conductor of radii a and b $(a < b)$. The cross section of the conductor is shown in Fig. 4.43. The medium in the hole and outside the conductor is nonmagnetic. Find the magnetic flux density vector everywhere.

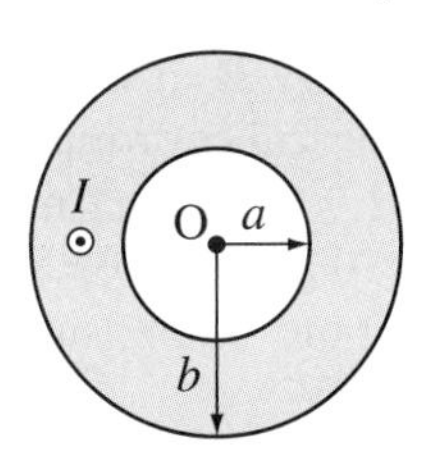

Figure 4.43 Cross section of a hollow cylindrical conductor with a steady current; for Problem 4.10.

4.11. Magnetic field of a triaxial cable. The cross section of a triaxial cable, having three coaxial cylindrical conductors, looks the same as the cross cut of the system of three concentric spherical conductors in Fig. 1.56. The radius of the inner conductor of the cable is $a = 1$ mm, the inner and outer radii of the middle conductor are $b = 2$ mm and $c = 2.5$ mm, and those of the outer conductor $d = 5$ mm and $e = 5.5$ mm. The cable conductors and dielectric, as well as the surrounding medium, are all nonmagnetic. Assuming that steady currents of intensities $I_1 = 2$ A, $I_2 = -1$ A, and $I_3 = -1$ A flow through the inner, middle, and outer conductor, respectively, all given with respect to the same reference direction, calculate the magnetic flux density vector everywhere $(0 \leq r < \infty)$.

4.12. Rotating cylinder with a surface charge. An infinitely long conducting nonmagnetic cylinder of radius a is uniformly charged over its surface with a charge density ρ_s. The cylinder rotates in air about its axis with a uniform angular velocity w. Find the **B** field inside the cylinder and outside it.

4.13. Rotating nonuniformly charged hollow cylinder. An infinitely long hollow cylinder of inner radius a and outer radius b $(a < b)$ in air is charged with a volume charge density $\rho(r) = \rho_0 r/a$ $(a \leq r \leq b)$, where ρ_0 is a constant and r the radial distance from the cylinder axis. The cylinder uniformly rotates about its axis with an angular velocity w. Assuming that the charge distribution of the cylinder does not change during the rotation, determine the magnetic flux density vector everywhere.

4.14. Two parallel infinite planar current sheets. Two parallel infinite planar current sheets in air have the same uniform surface current density J_s. The distance between the sheets is d. Find the magnetic flux density vector everywhere if the currents of the sheets run in (a) the same direction and (b) opposite directions.

4.15. Magnetic field inside a thin plate with current. A thin copper plate of length b, width a, and thickness d $(d \ll a)$ carries a steady current of intensity I, as shown in Fig. 4.44. The plate is situated in air. Neglecting the end effects, use Ampère's law to find the distribution of the magnetic flux density vector inside the plate [note that this is mathematically completely analogous to the application of a similar integral equation in Eq. (6.146) and Fig. 6.25(b)].

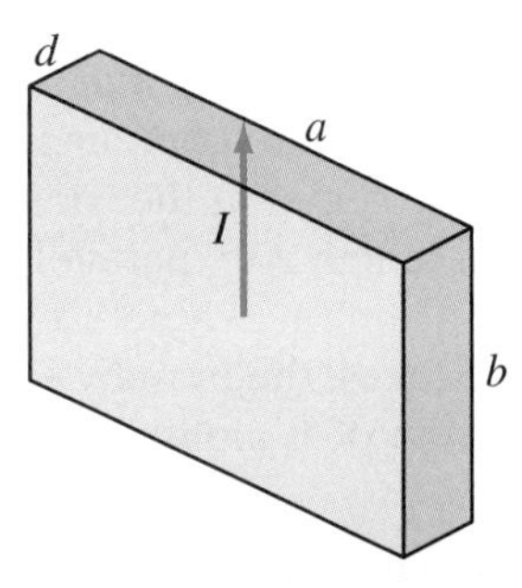

Figure 4.44 Thin conducting plate with a steady current; for Problem 4.15.

4.16. Magnetic field in a leaky parallel-plate capacitor. Calculate the magnetic flux density vector in the imperfect dielectric of the parallel-plate capacitor from Problem 3.12 (Fig. 3.32).

4.17. **Magnetic field in a leaky spherical capacitor.** Find the magnetic flux density vector in the imperfect dielectric of the spherical capacitor from Example 3.3 (Fig. 3.6).

4.18. **Magnetic field in a leaky coaxial cable.** Determine the magnetic flux density vector in the imperfect dielectric of the coaxial cable from Problem 3.17 (Fig. 3.34).

4.19. **Magnetic field around a grounding electrode.** Find the magnetic flux density vector on the surface of the ground for the grounding electrode from Example 3.15 [Fig. 3.26(a)].

4.20. **Current distribution from field distribution.** Using the differential Ampère's law, show that the magnetic field given by Eqs. (4.53) and (4.56) is produced by a uniform volume current of density given by Eq. (4.52) along an infinitely long cylinder of radius a in free space.

4.21. **Coaxial cable using differential Ampère's law.** Redo Example 4.11 but employing Ampère's law in differential form.

4.22. **Rotating charged cylinder by differential Ampère's law.** Redo Example 4.14 but with the use of the differential form of Ampère's law.

4.23. **Thin plate with current, differential Ampère's law.** Redo Problem 4.15 but applying the differential Ampère's law [adopt the vector-component notation as in Fig. 4.22 and Eq. (4.75), and use the analogy with the application of the differential Gauss' law in Example 1.23].

4.24. **Ampère's law in differential and integral form.** In a certain region, the magnetic field is given by $\mathbf{B} = [4(z-1)^2\,\hat{\mathbf{x}} + 2x^3\,\hat{\mathbf{y}} + xy\,\hat{\mathbf{z}}]$ mT (x, y, z in m). The medium is air. (a) Find the current density. (b) From the result in (a), find the total current enclosed by a square contour lying in the xy-plane, with the center at the coordinate origin and sides, of length 2 m, parallel to the x- and y-axes. (c) Confirm Ampère's law in integral form and Stokes' theorem by evaluating the net circulation of $\mathbf{B}$ along the contour defined in (b).

4.25. **Magnetic flux through a cylindrical surface.** Calculate the outward magnetic flux through the lateral surface of a cylinder of radius $a = 10$ cm and height $h = 20$ cm in a uniform magnetic field of flux density $B = 1$ T. The vector $\mathbf{B}$ makes an angle of 60° with the cylinder axis.

4.26. **Law of conservation of magnetic flux.** For the magnetic field defined in Problem 4.24, confirm the law of conservation of magnetic flux in differential and integral forms by evaluating (a) the divergence of $\mathbf{B}$ and (b) the outward flux of $\mathbf{B}$ through the surface of a cube, respectively. Let the cube be centered at the coordinate origin, with edges parallel to coordinate axes and 2 m long.

4.27. **Magnetic flux from vector potential.** In a certain region, the magnetic vector potential is given as the following function in a cylindrical coordinate system: $\mathbf{A} = 2r^2\hat{\boldsymbol{\phi}}$ T · m (r in m). (a) Find the magnetic flux density vector in this region. (b) Obtain the magnetic flux through a circular contour 1 m in radius that lies in the plane $z = 0$ and is centered at the coordinate origin. (c) Check the results by evaluating the circulation of $\mathbf{A}$ along the contour.

4.28. **Potential and field due to a magnetic dipole.** A magnetic dipole with a moment $\mathbf{m} = 400\ \mu\text{Am}^2\,\hat{\mathbf{z}}$ is located at the origin of a spherical coordinate system. Calculate the magnetic potential $\mathbf{A}$ and flux density $\mathbf{B}$ at the following points defined by spherical coordinates: (a) (1 m, 0, 0), (b) (1 m, $\pi/2$, $\pi/2$), (c) (1 m, π, 0), (d) (1 m, $\pi/4$, 0), (e) (10 m, $\pi/4$, 0), and (f) (100 m, $\pi/4$, 0). The dipole dimensions are much smaller than 1 m.

4.29. **Circular current loop as a magnetic dipole.** Refer to the circular current loop in Fig. 4.6, and show that for $|z| \gg a$, Eqs. (4.19) and (4.142) become the same, the latter equation giving the magnetic flux density vector of a magnetic dipole far away from it.

4.30. **Rectangular current loop as a magnetic dipole.** Check the result for the $\mathbf{B}$ field due to a rectangular loop obtained in Problem 4.1 by comparing it with the corresponding dipole-field expression, from Eq. (4.142), at points far away from the loop.

4.31. **Dipole equivalent to a surface current distribution.** Consider the circular nonuniform surface current distribution from Problem 4.4, and show that far away along the z-axis ($|z| \gg a$), this current distribution can be replaced by an equivalent magnetic dipole located at the coordinate origin. Find the moment, **m**, of the equivalent dipole.

4.32. **Lorentz force due to a rotating charged disk.** Refer to the rotating charged disk from Problem 4.5, and assume that a charged particle Q moves with a uniform velocity v along a path parallel to the plane of the disk. Find the Lorentz force on the particle at an instant when it is at the point that belongs to the axis of disk rotation and is at a distance a from the disk center.

4.33. **Forces between three parallel wires with current.** Three parallel very long and thin wires in air carry currents of intensities $I_1 = 1$ A, $I_2 = -1$ A, and $I_3 = 2$ A, all given with respect to the same reference direction. The distance between any two of the wires is $d = 1$ m, so their cross section constitutes an equilateral triangle (with sides d). (a) Determine the direction and magnitude of the per-unit-length magnetic force on the wire with current I_3. (b) Redo the problem if $I_2 = 1$ A.

4.34. **Force on a wire due to a semicylindrical conductor.** A very long aluminum conductor in the form of a half of a thin cylindrical shell of radius $b = 40$ mm and thickness $d = 0.5$ mm is situated in air. Another very long aluminum conductor, in the form of a very long wire of radius $a = 1$ mm, is positioned along the axis of the semicylinder. The two conductors carry steady currents of the same magnitude $I = 100$ A and opposite directions, as shown in Fig. 4.45. Find the magnetic force on the wire conductor per unit length.

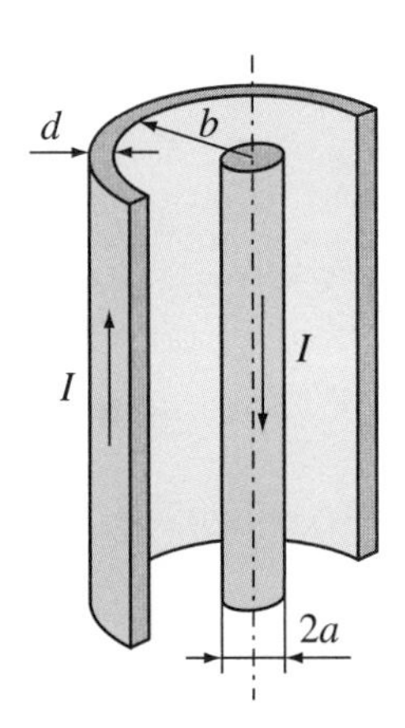

Figure 4.45 System composed of a wire conductor along the axis of a thin semicylindrical conductor; for Problem 4.34.

4.35. **Wire-strip transmission line.** Fig. 4.46 shows a cross section of a two-conductor transmission line consisting of a wire and a strip. The width of the strip is $2a$ and the separation between the conductors is a. The dielectric is air. If the current I runs along the line, calculate the magnetic force on the strip conductor per unit of its length.

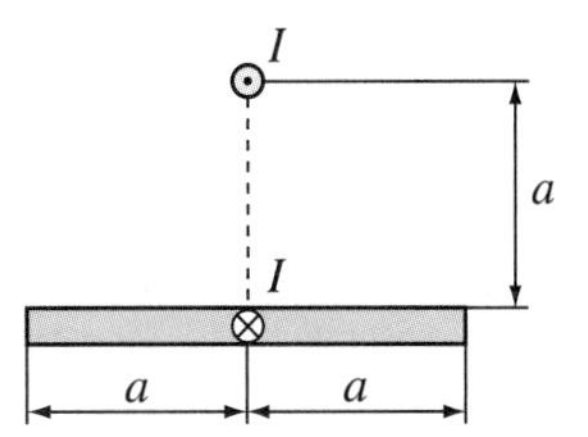

Figure 4.46 Cross section of a transmission line consisting of a wire and a strip; for Problem 4.35.

4.36. **Line composed of a wire and a corner strip conductor.** A wire and a 90° corner conductor, which is made from two identical strips of width a, carry steady currents of intensities I and $-I$ given with respect to the same reference direction. The distance of the wire from both ends of the corner conductor is a. A cross section of such a transmission line is portrayed in Fig. 4.47. The materials and the ambient medium are nonmagnetic. Find the per-unit-length magnetic force on the wire.

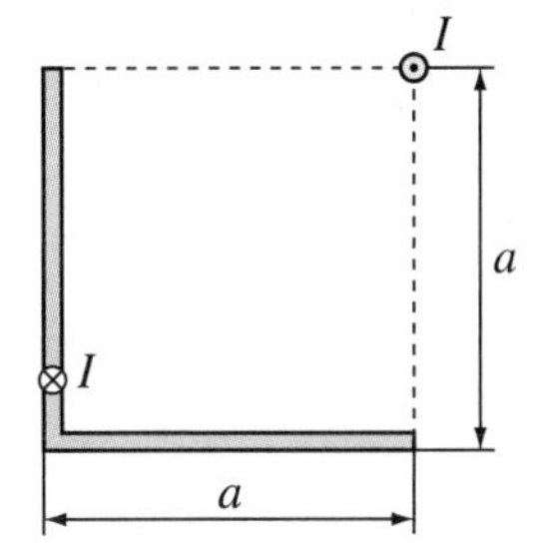

Figure 4.47 Cross section of a transmission line consisting of a wire and a 90° corner strip conductor; for Problem 4.36.

4.37. **Coaxial cable with off-centered cavity.** Shown in Fig. 4.48 is a cross section of a coaxial cable in which the cylindrical cavity of radius b representing the inner surface of the outer conductor is off-centered by a vector **d** with respect to the common axis of the inner conductor and the outer surface of the outer conductor. The other two radii are a and c, and the relationship $b + d \leq c$ is satisfied. The permeability everywhere is μ_0. If a steady current of intensity I is established in the cable, determine the magnetic force on the inner conductor per unit of its length.

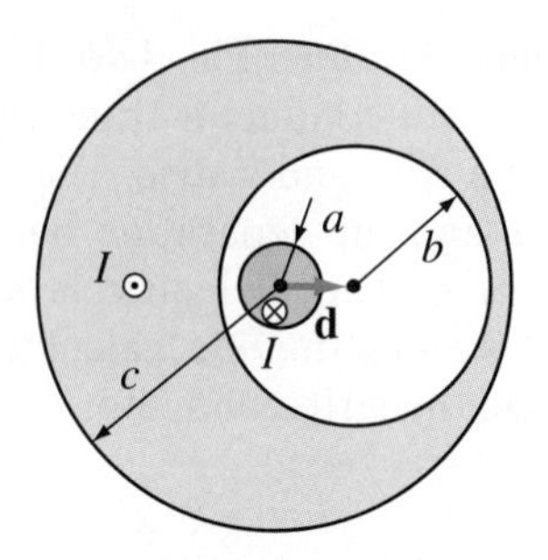

Figure 4.48 Coaxial cable with off-centered inner surface of the outer conductor; for Problem 4.37.

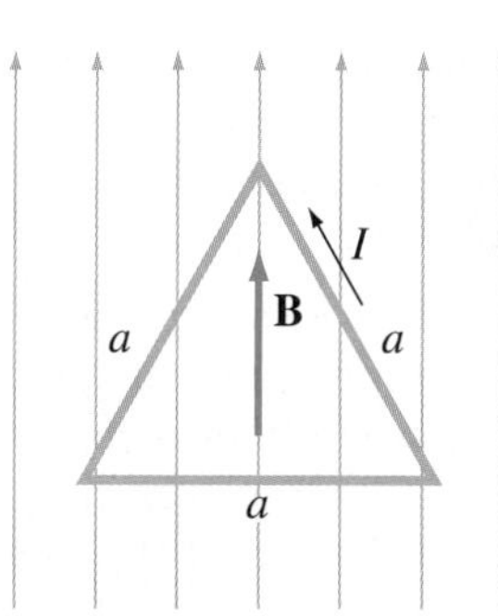

Figure 4.49 Triangular current loop in a uniform magnetic field; for Problem 4.38.

4.38. **Force and torque on a triangular current loop.** A rigid loop in the form of an equilateral triangle of side length a is situated in a uniform steady magnetic field of flux density $\mathbf{B}$. The magnetic field lines are parallel to the plane of the loop and are perpendicular to one of its sides, as shown in Fig. 4.49. If a steady current of intensity I is established in the loop, find (a) the force and (b) the torque on each of the loop sides, as well as (c) the net force and (d) the net torque on the loop.

4.39. **Torque between two magnetic dipoles.** Two small current loops are positioned in free space. The first loop has a magnetic moment $\mathbf{m}_1 = m\,\hat{\mathbf{z}}$ and is centered at the origin (O) of the Cartesian coordinate system. Obtain the torque on the second loop, for the following locations of its center P and directions of its magnetic moment $\mathbf{m}_2$: (a) P$(0, a, 0)$ and $\mathbf{m}_2 = m\,\hat{\mathbf{z}}$, (b) P$(0, a, 0)$ and $\mathbf{m}_2 = m\,\hat{\mathbf{y}}$, (c) P$(0, 0, a)$ and $\mathbf{m}_2 = m\,\hat{\mathbf{z}}$, (d) P$(0, 0, a)$ and $\mathbf{m}_2 = m\,\hat{\mathbf{x}}$, and (e) P$(0, a, a)$ and $\mathbf{m}_2 = m\,\hat{\mathbf{z}}$. Take $m = 0.1$ Am^2 and $a = 10$ m.

5 Magnetostatic Field in Material Media

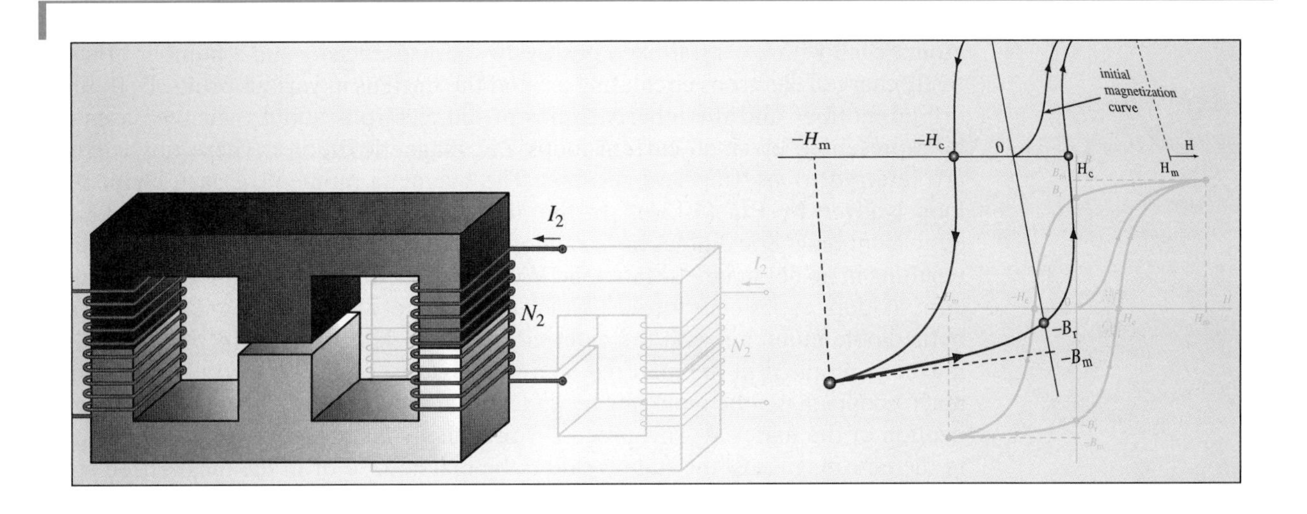

Introduction:

Our study of magnetostatics has so far been restricted to the magnetic field due to steady electric currents in a vacuum and other nonmagnetic media. In this chapter, we shall introduce and discuss phenomena associated with the magnetostatic field in the presence of magnetic materials. Many of the basic concepts, physical laws, and mathematical techniques constituting the analysis of materials in the magnetic field are entirely analogous to the corresponding concepts, laws, and techniques in electrostatics, which makes our discussions in this chapter much easier. The most important difference, however, with respect to the analysis of dielectric materials is the inherent nonlinear behavior of the most important class of magnetic materials, called ferromagnetics. This is a class of materials with striking magnetic properties (many orders of magnitude stronger than in other materials), with iron as a typical example.

We shall start with a qualitative characterization of microscopic magnetic phenomena in substances and describe the behavior of different types of magnetic materials based on the classical atomic model. By analogy with the polarization vector in electrostatics, the magnetization vector will be used to describe the magnetized state of a material on a macroscopic scale. Magnetization volume and surface current density vectors will be defined as macroscopic equivalents to a vast collection of tiny electric currents that are microscopic sources of the magnetization of a magnetic material. These current densities will enable us to evaluate the magnetic field due to magnetized materials using free-space formulas and techniques from the preceding chapter. We shall derive and discuss Maxwell's equations and boundary conditions for magnetostatic systems that include arbitrary media. The concept of permeability of a material will allow for additional macroscopic characterization of magnetic materials. Finally, a section on magnetic circuits (consisting of ferromagnetic cores of different shapes with current-carrying windings) will represent a culmination of the theory of the magnetostatic field in the presence of magnetic materials. Most of the work of the chapter will be applied to perform the analysis of such circuits, which essentially resembles the dc analysis of nonlinear electric circuits.

5.1 MAGNETIZATION VECTOR

According to the elementary atomic model of matter, all materials are composed of atoms, each with a central fixed positively charged nucleus and a number of negatively charged electrons circulating around the nucleus in various orbits. Both these orbital motions and the inherent spins of the electrons about their own axes can be represented by small current loops, i.e., magnetic dipoles. These tiny currents are referred to as Ampère's currents. The magnetic moment of each elementary loop is given by Eq. (4.134).[1] In the absence of an external magnetic field, the equivalent magnetic dipoles have random orientations with respect to one another, resulting in no net magnetic moment. With an applied field, however, the equivalent current loops experience torques, which lead to a net alignment of microscopic magnetic dipole moments with the external magnetic field [see Eq. (4.180)] and a net magnetic moment in the material on a macroscopic scale. The process of inducing macroscopic magnetic moments by an external magnetic field is called the magnetization of the material. This process is practically instantaneous, and the material in the new magnetostatic state is said to be magnetized or in the magnetized state. When magnetized (by an external field), a material is a source of its own magnetic field, and the total field at an arbitrary point in space (inside or outside the material) is a sum of the external (primary) field and the field due to the magnetized material (secondary field). In the analysis, we can replace the material by a collection of microscopic Ampère's current loops (magnetic dipoles) residing in a vacuum, as the rest of the material does not produce any field. Then, the secondary field can, in principle, be determined using the expression for the magnetic field due to a magnetic dipole in free space, Eq. (4.142), and superposition.

Instead of analyzing every single atom and all microscopic magnetic dipole moments, however, we rather introduce a macroscopic quantity termed the magnetization vector to describe the magnetized state of a material and the resulting field. Analogously to the definition of the polarization vector (**P**) for electrically polarized dielectric materials, Eq. (2.7), the magnetization vector, **M**, is defined as the density of the equivalent elementary magnetic moments in a magnetic material at a given point:

magnetization vector (unit: A/m)

$$\mathbf{M} = \frac{(\sum \mathbf{m})_{\text{in } dv}}{dv}. \tag{5.1}$$

We note that $\mathbf{M}\, dv$ represents the dipole moment of a magnetic dipole that is equivalent to an elementary volume dv in the material, i.e., to all the dipoles (microscopic Ampère's currents) within it. The unit for the magnetization vector is A/m.

[1] A third source of equivalent microscopic magnetic dipole moments in atoms is nuclear spin, but it provides a negligible contribution to the macroscopic magnetic properties of materials. Note, however, that nuclear spin represents the basis of magnetic resonance imaging (MRI), used in medicine, and also in many areas of science and engineering.

In any magnetic material, the magnetization vector at a point is a function of the magnetic flux density vector at that point,

$$\mathbf{M} = \mathbf{M}(\mathbf{B}), \tag{5.2}$$

and this relationship is a characteristic of individual materials. It is entirely analogous to the relationship between the polarization vector and the electric field intensity vector, Eq. (2.9), in electrostatics.

5.2 BEHAVIOR AND CLASSIFICATION OF MAGNETIC MATERIALS

A thorough understanding and precise quantitative characterization of microscopic magnetic phenomena in materials require a full quantum mechanical treatment. Here, however, we describe qualitatively the behavior of different types of magnetic materials based on the classical atomic model. Generally, materials can be classified according to their magnetic behavior into diamagnetic, paramagnetic, ferromagnetic, antiferromagnetic, ferrimagnetic, and superparamagnetic materials.

In diamagnetic materials, the magnetic moments of electrons orbiting about their nuclei are dominant compared to the magnetic moments attributed to electron spin. In order to describe diamagnetic behavior, which is present to a greater or lesser extent with all magnetic materials, we consider first a model of an atom with a single electron that circulates about the nucleus along an orbit of radius a with a uniform angular velocity $\mathbf{w}_0 = w_0\,\hat{\mathbf{z}}$, as shown in Fig. 5.1(a). In the absence of an external magnetic field, the outward centrifugal force on the electron, given by

$$F_{\mathrm{cf}} = \frac{m_{\mathrm{e}} v_0^2}{a} = m_{\mathrm{e}} w_0^2 a, \tag{5.3}$$

where m_{e} and v_0 are the mass and velocity of the electron ($v_0 = w_0 a$), respectively, is balanced by the centripetal (attractive) electric (Coulomb) force, F_{e}, between the nucleus and electron. Thus, the balance equation reads

$$F_{\mathrm{e}} = m_{\mathrm{e}} w_0^2 a. \tag{5.4}$$

The orbiting electron is equivalent to a small current loop (magnetic dipole), where the current I of the loop is given by Eq. (4.153) with $Q = -e$ and is

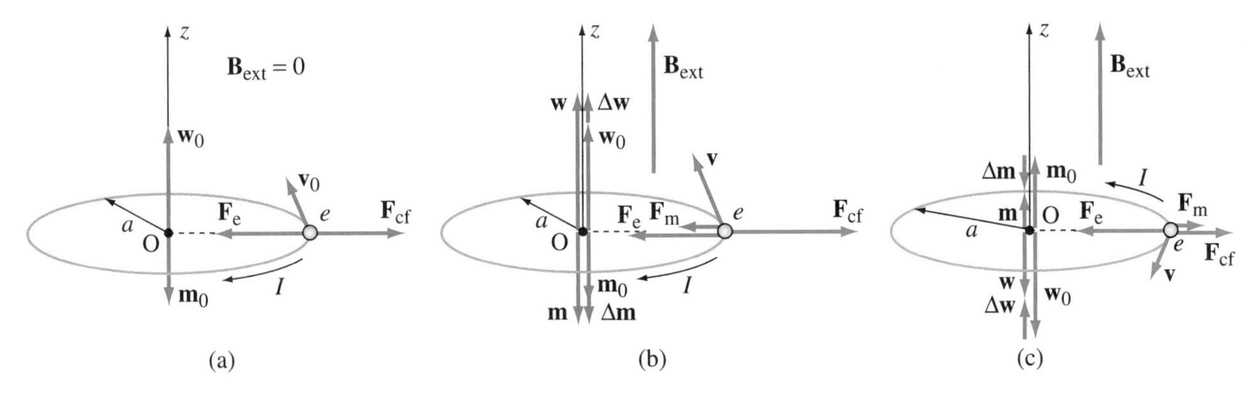

Figure 5.1 Atom with a single electron orbiting about the nucleus, in the absence of an external magnetic field (a), and with an applied magnetic field, whose direction coincides with the direction of the electron angular velocity vector (b) or is opposite to it (c).

directed oppositely to the direction of electron travel (because the electron charge is negative). From Eq. (4.134), the magnetic moment of the dipole is

$$\mathbf{m}_0 = I\pi a^2(-\hat{\mathbf{z}}) = -\frac{ew_0a^2}{2}\,\hat{\mathbf{z}}. \tag{5.5}$$

In the presence of an external magnetic field, there is an additional force on the electron – the magnetic force, Eq. (4.144),

$$\mathbf{F}_\mathrm{m} = Q\mathbf{v} \times \mathbf{B}_\mathrm{ext} = -e\mathbf{v} \times \mathbf{B}_\mathrm{ext}, \tag{5.6}$$

with $\mathbf{B}_\mathrm{ext}$ denoting the magnetic flux density vector of the applied field. For the situation in Fig. 5.1(b), where the electron angular velocity vector is in the same direction as the applied field $\mathbf{B}_\mathrm{ext} = B_\mathrm{ext}\,\hat{\mathbf{z}}$, the magnetic force is centripetal (inward), so that the new balance equation is

$$F_\mathrm{e} + ewaB_\mathrm{ext} = m_\mathrm{e}w^2a, \quad w = w_0 + \Delta w. \tag{5.7}$$

Here, the force unbalance created by the magnetic force is compensated for by an increase Δw of the orbital angular velocity ($\Delta w > 0$). Combining Eqs. (5.4) and (5.7) gives

$$m_\mathrm{e}(w^2 - w_0^2) = ewB_\mathrm{ext}. \tag{5.8}$$

Since the perturbation of the electron velocity is small, i.e., $\Delta w \ll w_0$, even for the strongest applied magnetic fields, we can write

$$w^2 - w_0^2 = (w + w_0)(w - w_0) \approx 2w\Delta w. \tag{5.9}$$

Substituting this into Eq. (5.8), the increase of the angular velocity comes out to be

$$\Delta w = \frac{eB_\mathrm{ext}}{2m_\mathrm{e}}. \tag{5.10}$$

In place of Eq. (5.5), the new equivalent magnetic dipole moment is

$$\mathbf{m} = -\frac{ewa^2}{2}\,\hat{\mathbf{z}} = -\frac{ea^2}{2}\,\mathbf{w}. \tag{5.11}$$

Introducing the angular momentum of the electron,

$$\mathbf{L} = m_\mathrm{e}\mathbf{r} \times \mathbf{v} = \mathcal{I}\mathbf{w}, \tag{5.12}$$

where $\mathbf{r}$ is the instantaneous position vector of the electron with respect to the origin O ($|\mathbf{r}| = a$) and $\mathcal{I} = m_\mathrm{e}a^2$ is the rotational inertia of the electron, Eq. (5.11) becomes

$$\mathbf{m} = -\frac{e}{2m_\mathrm{e}}\mathbf{L} = -\frac{e}{2m_\mathrm{e}}\mathcal{I}\mathbf{w}. \tag{5.13}$$

The increase of the dipole moment due to the increase of the orbiting velocity of the electron, in Eq. (5.10), is given by

$$\Delta\mathbf{m} = -\frac{e}{2m_\mathrm{e}}\mathcal{I}\Delta\mathbf{w} = -\left(\frac{e}{2m_\mathrm{e}}\right)^2\mathcal{I}\,\mathbf{B}_\mathrm{ext}, \tag{5.14}$$

and is called the induced magnetic moment of the electron. We note that $\Delta\mathbf{m}$ is antiparallel (i.e., in the opposite direction) to the applied magnetic field $\mathbf{B}_\mathrm{ext}$.

For an electron whose orbital angular velocity vector and the field $\mathbf{B}_\mathrm{ext}$ are opposed, as in Fig. 5.1(c), the force $\mathbf{F}_\mathrm{m}$ on the electron is in the outward direction, and the force unbalance is compensated for by a reduced velocity, i.e., $\Delta w < 0$. The new magnetic dipole moment is smaller in magnitude than $\mathbf{m}_0$, so that the induced moment $\Delta\mathbf{m}$ is again in the $-z$ direction, that is, still antiparallel to the applied field, and given by the same expression in Eq. (5.14). This same expression for $\Delta\mathbf{m}$ is obtained also in the case of an arbitrary mutual position of the electron orbit and the applied magnetic field, as well as for the model of an atom with N electrons, which travel along orbits that are arbitrarily oriented with respect to the applied magnetic field, where $\mathcal{I}$ should be replaced by the average rotational inertia of all electrons in the atom, $\mathcal{I} = \mathcal{I}_{\rm av}$.

With no external magnetic field applied, the magnetic dipoles of the atoms have random orientations, so that the net magnetic moment and the magnetization vector in the material are zero. With an applied field, however, each atom acquires a differential induced moment given by Eq. (5.14), with $\mathcal{I} = \mathcal{I}_{\rm av}$, and all these moments are antiparallel to the magnetic flux density vector, $\mathbf{B}$, in the material. As a result, the net magnetic moment of the atoms opposes the applied field. Thus, a rod of a diamagnetic material placed in a magnetic field will orient itself perpendicular to the applied field, i.e., across the field lines. The term diamagnetism originates from the Greek word "dia" meaning "across." If a diamagnetic specimen is brought near either pole of a strong bar magnet, it will be repelled by the magnet. From Eqs. (5.1) and (5.14), the magnetization vector is

$$\mathbf{M} = N_{\rm v}\,\Delta\mathbf{m} = -\frac{N_{\rm v}e^2\mathcal{I}_{\rm av}}{4m_{\rm e}^2}\,\mathbf{B}, \tag{5.15}$$

$N_{\rm v}$ being the concentration (number per unit volume) of atoms in the material. We note that this equation represents a special form of Eq. (5.2). In addition, it is customary to write

$$\mathbf{M} = \frac{\chi_{\rm m}}{\mu_0}\,\mathbf{B}, \tag{5.16}$$

where $\chi_{\rm m}$ is a dimensionless proportionality constant given by

$$\boxed{\chi_{\rm m} = -\frac{\mu_0 N_{\rm v}e^2\mathcal{I}_{\rm av}}{4m_{\rm e}^2}.} \tag{5.17}$$

magnetic susceptibility of diamagnetic materials

It is called the magnetic susceptibility of the material. (For a vacuum, which is the only truly nonmagnetic medium, $\chi_{\rm m} = 0$.) We see that the diamagnetic materials are linear magnetic media, because the relationship $\mathbf{M}(\mathbf{B})$ in Eq. (5.16) is a linear one. Typical diamagnetic materials are bismuth, silver, lead, copper, gold, silicon, germanium, graphite, sulfur, hydrogen, helium, sodium chloride, and water. Substituting typical values for the quantities involved in the expression in Eq. (5.17) indicates that $\chi_{\rm m}$ of diamagnetic materials is of order -10^{-6} (water) to -10^{-4} (bismuth). We see that the magnetic susceptibility of diamagnetic materials is negative and very small. The macroscopic magnetic effect in diamagnetic substances is very weak and negligible in most practical situations. In general, the diamagnetic effect is present in all materials when placed in a magnetic field, because it arises from an interaction of orbiting electrons in atoms with the external field. However, in other types of magnetic materials, the diamagnetic reaction is completely masked by other effects.

In paramagnetic materials, the atoms have a small permanent net magnetic dipole moment associated with them, due almost entirely to spin magnetic dipole moments of electrons. In the absence of an external magnetic field, the random orientation of the atoms produces an average magnetic moment of a zero in a finite volume. When an external field is applied, however, there is a small torque given by Eq. (4.180) on each atomic moment, which tends to align the moment in the direction of the applied field. This alignment acts to increase the value of the magnetic flux density vector, **B**, within the material over the external value. However, the external field also causes a diamagnetic effect of the orbiting electrons (all magnetic materials exhibit diamagnetic behavior), which counteracts the increase in **B**. The alignment process is also impeded by the forces of random thermal vibrations, and the resulting increase in **B** is quite small. The overall macroscopic effect is equivalent to that of a small positive magnetization. The magnetization vector, **M**, is in the same direction as the vector **B**. Eq. (5.16), then, tells us that the magnetic susceptibility of paramagnetic materials is positive and very small. When a rod of a paramagnetic material is placed in a magnetic field, it orients itself along the field lines, and hence the term paramagnetism (the word "para" in Greek means "along"). Additionally, if a paramagnetic substance is brought near a pole of a strong bar magnet, it will be attracted to it. Typical paramagnetic materials are palladium, aluminum, and oxygen, and typical values of $\chi_{\rm m}$ are of orders 10^{-6} (oxygen) to 10^{-3} (palladium). Note that air is a paramagnetic medium as well. Obviously, the paramagnetic effect is also very weak and paramagnetic materials can be treated as nonmagnetic media ($\chi_{\rm m} = 0$) in most practical applications.

The remaining four classes of magnetic materials (ferromagnetic, antiferromagnetic, ferrimagnetic, and superparamagnetic) all have strong atomic magnetic dipole moments caused primarily by uncompensated electron spin moments. The interaction of adjacent atoms leads to an alignment of the atomic moments in either an aiding (parallel) or opposing (antiparallel) manner.

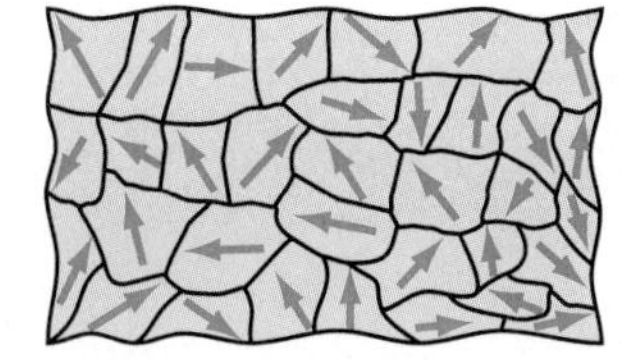

Figure 5.2 Magnetic domains in a polycrystalline ferromagnetic specimen with no external magnetic field applied.

The macroscopic magnetization of ferromagnetic materials can be many orders of magnitude larger than that of paramagnetic materials. Interatomic forces in a ferromagnetic sample cause the atomic moments to line up in a parallel fashion over regions containing large number of atoms (e.g., 10^{15} atoms). These regions, called the magnetic domains (or Weiss' domains), range in size from a few microns to several centimeters in each dimension. With no external field applied, the domain moments vary in direction from domain to domain, as indicated in Fig. 5.2 for a polycrystalline ferromagnetic specimen. Due to overall vector cancelations, the material as a whole has no net magnetization. Upon the application of an external magnetic field, however, the volumes of the domains that have moments aligned or nearly aligned with the applied field grow at the expense of their neighbors, and the magnetic flux density vector of the secondary field (that due to the material) increases greatly over that of the external field alone. For weak applied fields, the movements of magnetic domain walls are reversible, i.e., the domains go back to their initial states after the field is turned off. Above a certain (not large) field magnitude, however, this process becomes irreversible. Additionally, the domains start rotating toward the direction of the applied field, so that completely random domain orientations are no longer attained after the external field has been removed. In other words, there is a residual or remanent net magnetization of the material that remains after the complete removal of the primary field. This means that the magnetization of the material lags behind the field producing it, and also that the magnetization state of the material at an instant of time is a function not only of the magnetic field, i.e., its flux density vector, at that instant, but also of the magnetic history of the material. This phenomenon is called hysteresis (which is derived from a Greek word meaning "to lag"). As the applied field becomes even much stronger, a total alignment of all the domain moments with the applied field occurs. At this point, the ferromagnetic material is said to be saturated. In the state of saturation, further increase of the external magnetic flux density vector no longer causes an increase of the magnetization vector.

Typical ferromagnetic materials are iron, nickel, and cobalt, and their alloys. The name ferromagnetic comes from the Latin word for iron, "ferrum." Ferromagnetics are the most important magnetic materials in engineering. They are widely used in cores of inductors and transformers, and in electric motors, generators, electromagnets, relays, and other devices that use magnetic forces and torques, as well as in magnetic heads and tracks of computer hard disks and other magnetic storage (recording) devices. Note that above a certain (very high) temperature, called the Curie temperature, the thermal vibrations of atoms completely prevent the coupling (parallel alignment) of atomic magnetic moments, so that ferromagnetic materials lose all their ferromagnetic characteristics and revert to paramagnetic materials. The Curie temperature for iron is 770°C.

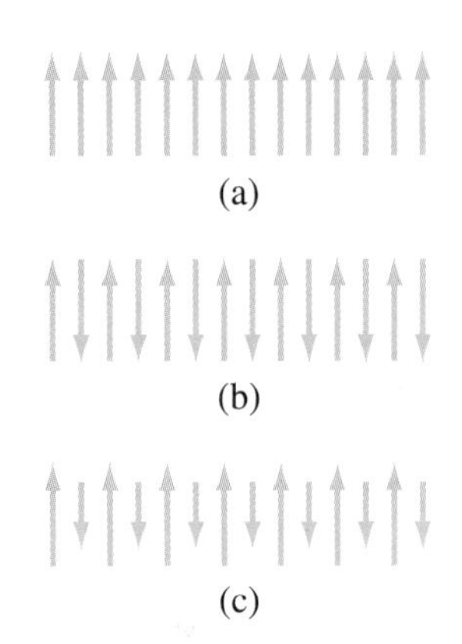

Figure 5.3 Schematic structures of atomic magnetic dipole moments for (a) ferromagnetic, (b) antiferromagnetic, and (c) ferrimagnetic materials.

In antiferromagnetic materials, the forces between adjacent atoms cause the atomic magnetic moments to line up in opposite directions so that the net magnetic moment of a specimen is zero, even in the presence of an applied magnetic field. Chromium and manganese, as well as many oxides, sulfides, and chlorides, belong to this class of materials. Examples are manganese oxide (MnO_2), ferrous sulfide (FeS), and cobalt chloride ($CoCl_2$). Antiferromagnetism is not of practical importance.

In ferrimagnetic materials, the magnetic moments of adjacent atoms are also aligned antiparallel, but the moments are not equal, so there is a net magnetic moment of a specimen. A large response to an externally applied magnetic field therefore occurs, although it is substantially less than in ferromagnetic substances. Fig. 5.3 depicts schematically a comparison of the atomic magnetic dipole moment structure for ferromagnetic, antiferromagnetic, and ferrimagnetic materials. The most important subclass of ferrimagnetics are the ferrites, which have much lower electric conductivity (σ) than the ferromagnetics (for instance, 10^{-4} to 10^2 S/m, compared with 10^7 S/m for iron). Low conductivity limits induced currents in the material when alternating (ac) fields are applied (so-called eddy currents), and this is why the ferrites are used in many applications at high frequencies, such as cores for high-frequency transformers, AM, short-wave, and FM antennas, and phase shifters, in spite of their weaker magnetic effects (compared to ferromagnetics). The reduced eddy currents lead to lower Joule's (ohmic) losses in the material (core). Typical examples of ferrite substances are iron ferrite (Fe_3O_4), nickel ferrite ($NiFe_2O_4$), and cobalt ferrite ($CoFe_2O_4$). Ferrimagnetism also disappears at temperatures above a critical value – the Curie temperature.

Finally, superparamagnetic materials are composed of ferromagnetic particles suspended in a nonmagnetic (dielectric) matrix. Each particle contains many magnetic domains, but the domain walls cannot penetrate the matrix material and there is no coupling with adjacent particles. Thin dielectric (plastic) tapes with suspended ferromagnetic particles can store large amounts of information in magnetic form because the particles are independent from each other and it is possible to change the state of magnetization along the tape abruptly in very small distances. Such superparamagnetic tapes are widely used as audio, video, and data recording tapes.

5.3 MAGNETIZATION VOLUME AND SURFACE CURRENT DENSITIES

In this section, we shall obtain the expressions for evaluation of the macroscopic distribution of volume and surface current densities equivalent to microscopic Ampère's currents in a magnetized body – from a given distribution of the magnetization vector, **M**. The vector **M**, in turn, is obtained by averaging the microscopic magnetic dipoles (magnetic moments of microscopic Ampère's current loops) in the magnetic material. The equivalent macroscopic current is called the magnetization current, and the corresponding volume and surface current densities are denoted by $\mathbf{J}_{\text{m}}$ and $\mathbf{J}_{\text{ms}}$, respectively. The expressions for these current densities are similar to the expressions for calculating the volume and surface bound (polarization) charge densities (ρ_{p} and ρ_{ps}) from the polarization vector (**P**) in a polarized dielectric body. They will be used later for free-space evaluations of the magnetic flux density vector (**B**) due to magnetized bodies.

Let us first find the intensity of the total magnetization current $I_{\text{m}C}$ enclosed by an arbitrary imaginary contour C situated (totally or partly) inside a magnetized magnetic body, as depicted in Fig. 5.4. We note that $I_{\text{m}C}$ is actually the current that passes through any surface S bounded by C. Let the magnetic moments of small Ampère's current loops in a vacuum that constitute the magnetization current be expressed as

$$\mathbf{m} = I\mathbf{S}_{\text{m}}, \tag{5.18}$$

where $S_{\text{m}} = |\mathbf{S}_{\text{m}}|$ is the surface area of the loop. It is obvious in Fig. 5.4 that all the loops that pass through S twice, as well as those that do not pass through S at all, contribute with zero net current intensity to $I_{\text{m}C}$. Only loops that pierce S only once, that is, the loops that encircle C, contribute actually to the total current intensity flowing through S. To evaluate $I_{\text{m}C}$ in the general case, we therefore count the loops that are strung along C (like pearls on a string). In doing that, we count the contribution of such loops as either I or $-I$ (note that I, generally, differs from loop to loop), by inspecting whether the direction of the loop current traversing S is in agreement or disagreement with the reference orientation of S.

Consider an element $\text{d}l$ of C and the case when the angle β between the vector **M** (or the average of elementary dipole moment vectors **m** near $\text{d}l$) and vector $\text{d}\mathbf{l}$, which is oriented in accordance to the orientation of C, is less than 90°, as shown in Fig. 5.5(a). Note that centers of loops that are positioned close to $\text{d}l$ and encircle it are inside an oblique cylinder with bases S_{m} and height

$$\text{d}h = \text{d}l \cos\beta, \tag{5.19}$$

so that the total number of these loops equals the concentration of loops (magnetic dipoles), N_{v} [see Eq. (2.6)], times the volume of the cylinder, $\text{d}v = S_{\text{m}}\,\text{d}h$. The reference orientation of the surface S is related to that of the contour C by means of the right-hand rule, so the reference direction of the unit normal vector $\hat{\mathbf{n}}$ on S in

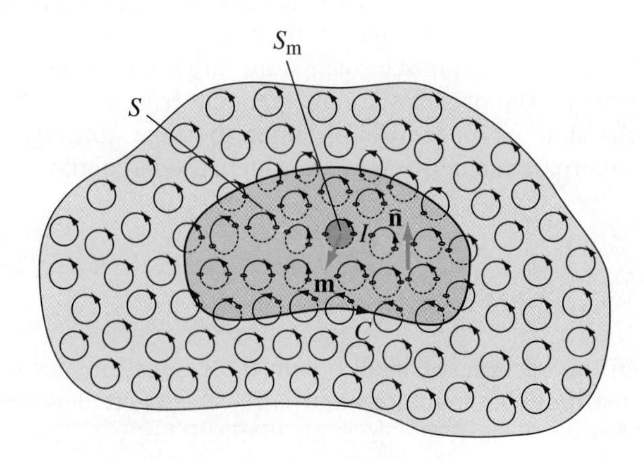

Figure 5.4 Contour C in a magnetized magnetic body.

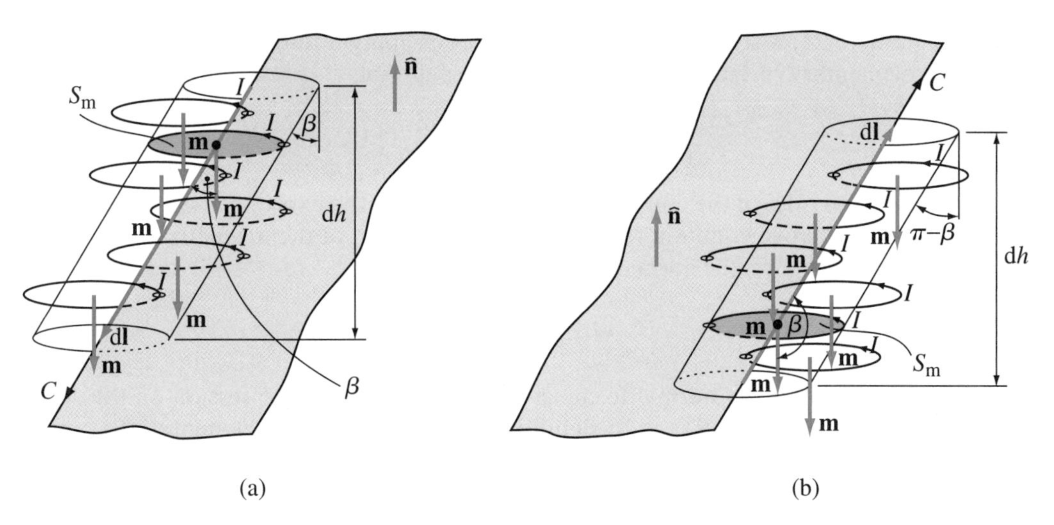

Figure 5.5 Element of the contour C in Fig. 5.4, in two cases with regards to the angle β between $\mathbf{M}$ and $\mathrm{d}\mathbf{l}$: (a) $0 \le \beta < 90°$ and (b) $90° < \beta \le 180°$.

Fig. 5.5(a) is from S upward. We then realize that all the loop currents encircling $\mathrm{d}l$ pass through S in the positive direction (direction in agreement with the orientation of S) and contribute to the total current with the intensity I (we assume that all the loops near $\mathrm{d}l$, which is differentially small, have the same moments and currents). Hence, the corresponding contribution to the magnetization current intensity through S is given by

$$\mathrm{d}I_{\mathrm{m}} = N_{\mathrm{v}} S_{\mathrm{m}}\, dl \cos\beta\, I \quad (0 \le \beta < 90°). \tag{5.20}$$

In the case when $\beta > 90°$, Fig. 5.5(b),

$$\mathrm{d}h = \mathrm{d}l\, \cos(\pi - \beta) = \mathrm{d}l\,(-\cos\beta), \tag{5.21}$$

and, because all the loop currents encircling $\mathrm{d}l$ pierce S in the negative direction and are taken as $-I$,

$$\mathrm{d}I_{\mathrm{m}} = N_{\mathrm{v}} S_{\mathrm{m}}\, \mathrm{d}l\,(-\cos\beta)\,(-I) \quad (90° < \beta \le 180°), \tag{5.22}$$

which is the same result as in Eq. (5.20).

Eq. (5.1) can be interpreted here as $\mathbf{M} = N_{\mathrm{v}}\mathbf{m}$, where $\mathbf{m}$ is given by Eq. (5.18), and hence, for an arbitrary β $(0 \le \beta \le 180°)$, we have

$$\mathrm{d}I_{\mathrm{m}} = N_{\mathrm{v}} m\, \mathrm{d}l \cos\beta = N_{\mathrm{v}}\mathbf{m} \cdot \mathrm{d}\mathbf{l} = \mathbf{M} \cdot \mathrm{d}\mathbf{l} \tag{5.23}$$

[note that the boundary case, $\beta = 90°$ (the contour element $\mathrm{d}\mathbf{l}$ is tangential to the surfaces S_{m} of the loops) and $\mathrm{d}I_{\mathrm{m}} = 0$, is also properly included in this formula]. Finally, by integrating the result for $\mathrm{d}I_{\mathrm{m}}$ along the entire contour C, we get

$$\boxed{I_{\mathrm{m}C} = \oint_C \mathbf{M} \cdot \mathrm{d}\mathbf{l}.} \tag{5.24}$$

total magnetization current enclosed by a contour C

This is an integral equation similar in form to Ampère's law, Eq. (4.48). It tells us that the circulation (line integral) of the magnetization vector along an arbitrary contour in a magnetostatic system that includes magnetic materials is equal to the total magnetization current enclosed by that contour. The reference direction of the current flow is related to the reference orientation of the contour by the right-hand rule.

Eq. (5.24) is true for any contour C. Let us apply it to the contour C enclosing an elementary surface ΔS inside a magnetic material:

$$\frac{(I_{\text{m}})_{\text{through }\Delta S}}{\Delta S} = \frac{\oint_C \mathbf{M}\cdot d\mathbf{l}}{\Delta S} \quad (\Delta S \to 0), \tag{5.25}$$

where both sides of the equation are divided by ΔS. The expression on the left-hand side of the above equation represents the component of the magnetization volume current density vector normal to ΔS,

$$\hat{\mathbf{n}}\cdot\mathbf{J}_{\text{m}} = \frac{(I_{\text{m}})_{\text{through }\Delta S}}{\Delta S} \tag{5.26}$$

($\hat{\mathbf{n}}$ is the unit vector normal to the surface ΔS), while the expression on the right-hand side of the equation is, by definition [Eq. (4.87)], the component of the curl of $\mathbf{M}$ along $\hat{\mathbf{n}}$, that is, $\hat{\mathbf{n}}\cdot\text{curl}\,\mathbf{M}$. Hence,

$$\hat{\mathbf{n}}\cdot\mathbf{J}_{\text{m}} = \hat{\mathbf{n}}\cdot\text{curl}\,\mathbf{M}, \tag{5.27}$$

and, since this is true for any $\hat{\mathbf{n}}$ and thus for all components of the vector $\mathbf{J}_{\text{m}}$, it implies that

magnetization volume current density vector

$$\boxed{\mathbf{J}_{\text{m}} = \text{curl}\,\mathbf{M} = \nabla\times\mathbf{M}.} \tag{5.28}$$

This is a differential form of the integral relationship in Eq. (5.24). It tells us that the magnetization volume current density vector, in A/m^2, at an arbitrary point in a material is equal to the curl of the magnetization vector at that point.

If $\mathbf{M} = \text{const}$ inside the magnetic material (uniformly magnetized material), all spatial derivatives of $\mathbf{M}$ are zero, and from Eq. (5.28),

no volume magnetization current in a uniformly magnetized material

$$\boxed{\mathbf{M} = \text{const} \quad \longrightarrow \quad \mathbf{J}_{\text{m}} = 0.} \tag{5.29}$$

Physically, the currents of adjacent Ampère's current loops that flow in opposite directions cancel everywhere in the interior of a uniformly magnetized material, and there is no net volume current. If $\mathbf{M} \neq \text{const}$, however, then volume macroscopic magnetization current exists only if the magnetization vector varies throughout the volume of the material (nonuniformly magnetized material) in a way that its curl is nonzero, otherwise $\mathbf{J}_{\text{m}} = 0$.

On the surface of a magnetized magnetic body, there always exists surface macroscopic magnetization current (there are parts of small Ampère's current loops pressed onto the surface that cannot be compensated by oppositely flowing currents of neighboring loops). The only exception are parts of the surface where $\mathbf{M}$ and the microscopic magnetic dipoles are normal to the surface, that is, the surfaces S_{m} of the Ampère's current loops are laying in the surface of the body. To determine the magnetization macroscopic surface current density vector, $\mathbf{J}_{\text{ms}}$ (in A/m), equivalent to the microscopic currents, we apply Eq. (5.24) to a narrow rectangular elementary contour, with length Δl and height Δh ($\Delta h \to 0$), shown in Fig. 5.6. The contour being differentially small, the magnetic moments $\mathbf{m}$ of Ampère's current loops near the contour are all parallel to each other. The resultant of all the corresponding Ampère's currents, and thus the vector $\mathbf{J}_{\text{ms}}$, is perpendicular to the local magnetization vector, $\mathbf{M}$. The contour C in Fig. 5.6 is positioned such that $\mathbf{M}$ lies in the plane of the contour and $\mathbf{J}_{\text{ms}}$ is perpendicular to that plane. There is no magnetization ($\mathbf{M} = 0$) in free space (a vacuum, air, or any other nonmagnetic medium) surrounding the body, so that the circulation of vector $\mathbf{M}$ in Eq. (5.24) is reduced to $\mathbf{M}\cdot\Delta\mathbf{l}$ along the lower side of C. By the definition of the surface current density vector

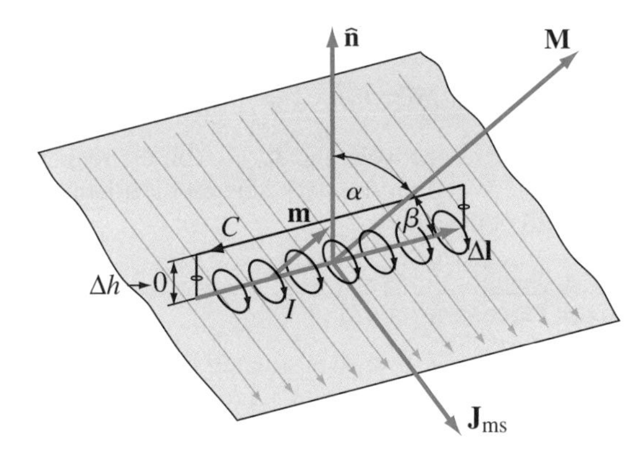

Figure 5.6 Elementary contour used for deriving the boundary condition for the vector **M** on the surface of a magnetic body.

[see Eq. (3.13)], the total current enclosed by C, appearing on the left-hand side of Eq. (5.24), equals $J_{\rm ms}\Delta l$, and we have

$$J_{\rm ms}\Delta l = \mathbf{M}\cdot\Delta\mathbf{l} = M\Delta l\cos\beta = M\Delta l\sin\alpha, \tag{5.30}$$

where α is the angle that **M** makes with the normal on the surface directed from the magnetic body outward ($\alpha = 90° - \beta$). Hence,

$$J_{\rm ms} = M\sin\alpha, \tag{5.31}$$

or, in a vector form,

$$\boxed{\mathbf{J}_{\rm ms} = \mathbf{M}\times\hat{\mathbf{n}},} \tag{5.32}$$

magnetization surface current density vector; $\hat{\mathbf{n}}$ outward normal on a magnetic body surface

with $\hat{\mathbf{n}}$ standing for the outward normal unit vector on the surface. This is the boundary condition for the vector **M** on the surface of a magnetic body, connecting the magnetization vector in the body near the surface and the magnetization surface current density vector on the surface. Note that only the tangential component of **M** contributes to $\mathbf{J}_{\rm ms}$.

The expressions in Eqs. (5.28) and (5.32) can be used for determining the distribution of volume and surface magnetization current densities, $\mathbf{J}_{\rm m}$ and $\mathbf{J}_{\rm ms}$, of a magnetized body, assuming that the state of magnetization of the body is described by a given distribution of the magnetization vector, **M**, inside the body. We can then, considering these macroscopic currents to reside in a vacuum, calculate the magnetic flux density vector, **B**, due to the magnetized body (and any other related quantity of interest) using the appropriate free-space equations (e.g., various forms of the Biot-Savart law and Ampère's law) and solution techniques suitable to specific geometries and current distributions.

Example 5.1 Nonuniformly Magnetized Ferromagnetic Cube

The magnetization vector in a ferromagnetic cube shown in Fig. 5.7 is given by

$$\mathbf{M}(x,y) = M_0\,\frac{xy}{a^2}\,\hat{\mathbf{z}}, \tag{5.33}$$

where M_0 is a constant. The surrounding medium is air. Find the distribution of magnetization currents of the cube.

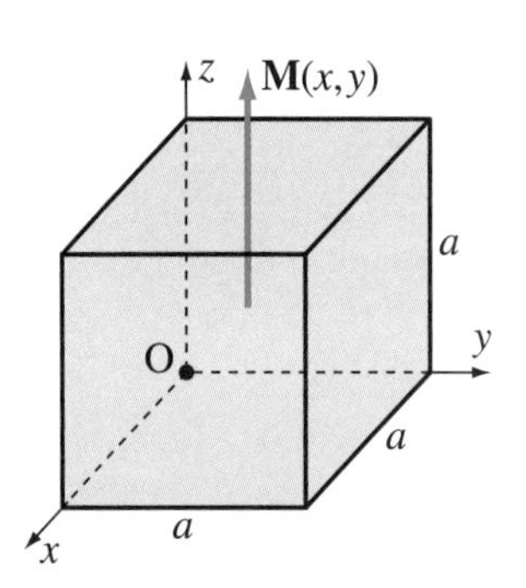

Figure 5.7 Ferromagnetic cube with magnetization $\mathbf{M}(x, y)$; for Example 5.1.

Solution From Eqs. (5.28) and (4.81), the magnetization volume current density vector inside the cube is

$$\mathbf{J}_{\rm m} = \frac{\partial M_z}{\partial y}\,\hat{\mathbf{x}} - \frac{\partial M_z}{\partial x}\,\hat{\mathbf{y}} = \frac{M_0}{a^2}\,(x\,\hat{\mathbf{x}} - y\,\hat{\mathbf{y}}). \tag{5.34}$$

The magnetization surface current density vector is, by means of Eq. (5.32),

$$\mathbf{J}_{\text{ms1}} = \mathbf{M}(a^-, y) \times \hat{\mathbf{x}} = \frac{M_0 y}{a}\,\hat{\mathbf{y}} \quad \text{and} \quad \mathbf{J}_{\text{ms2}} = \mathbf{M}(x, a^-) \times \hat{\mathbf{y}} = -\frac{M_0 x}{a}\,\hat{\mathbf{x}} \tag{5.35}$$

on the front side and right-hand side of the cube, respectively, and $\mathbf{J}_{\text{ms}} = 0$ on the remaining four sides of the cube [in terms of Eq. (5.31), $M = 0$ on the back side and left-hand side, whereas $\sin\alpha = 0$ on the top and bottom sides of the cube].

Example 5.2 Uniformly Magnetized Ferromagnetic Disk

A thin ferromagnetic disk of radius a and thickness d ($d \ll a$) is situated in air. The disk is uniformly magnetized throughout its volume. The magnetization vector is normal to disk bases and its magnitude is M. Determine (a) the distribution of magnetization currents of the disk and (b) the magnetic flux density vector at an arbitrary point along the disk axis normal to the bases.

Solution

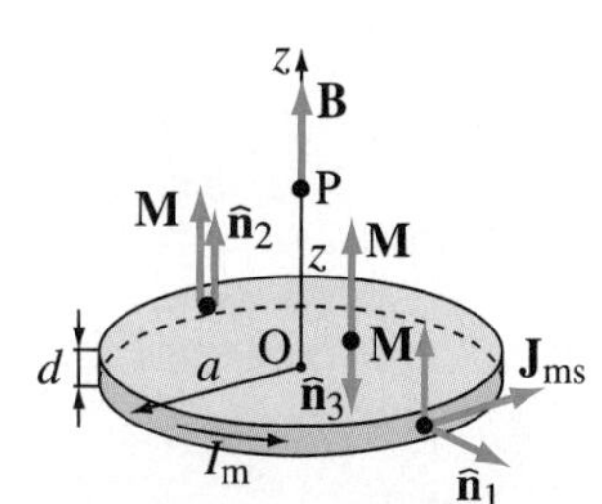

Figure 5.8 Magnetization surface current on a uniformly magnetized ferromagnetic disk; for Example 5.2.

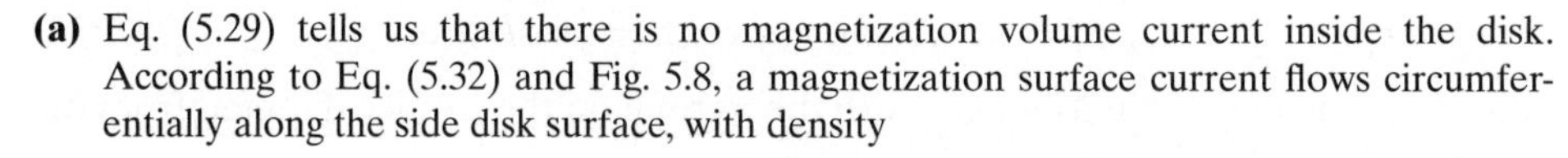

(a) Eq. (5.29) tells us that there is no magnetization volume current inside the disk. According to Eq. (5.32) and Fig. 5.8, a magnetization surface current flows circumferentially along the side disk surface, with density

$$\mathbf{J}_{\text{ms}} = \mathbf{M} \times \hat{\mathbf{n}}_1 = M\,\hat{\mathbf{z}} \times \hat{\mathbf{r}} = M\,\hat{\boldsymbol{\phi}}, \tag{5.36}$$

while on the upper and lower disk bases, $\mathbf{J}_{\text{ms}} = 0$, because $\mathbf{M}$ is collinear with both $\hat{\mathbf{n}}_2$ and $\hat{\mathbf{n}}_3$.

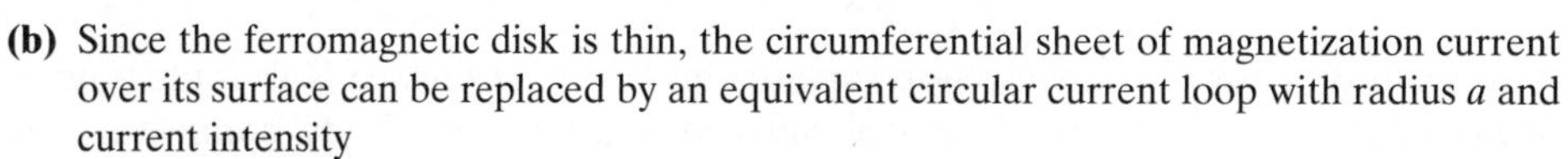

(b) Since the ferromagnetic disk is thin, the circumferential sheet of magnetization current over its surface can be replaced by an equivalent circular current loop with radius a and current intensity

$$I_{\text{m}} = J_{\text{ms}} d = Md. \tag{5.37}$$

Assuming that the loop is in a vacuum, the magnetic flux density vector along the z-axis is given by Eq. (4.19), that is,

$$\mathbf{B} = \frac{\mu_0 M d a^2}{2\left(z^2 + a^2\right)^{3/2}}\,\hat{\mathbf{z}}. \tag{5.38}$$

Example 5.3 Cylindrical Bar Magnet

A cylindrical bar magnet of radius a and length l is permanently magnetized with a uniform magnetization. The magnetization vector, $\mathbf{M}$, is parallel to the bar axis. The medium around the magnet is air. Compute the magnetic flux density vector at the center of the magnet.

Solution The magnetization surface current density over the magnet is the same as that over the magnetized disk in Fig. 5.8, and the vector $\mathbf{B}$ both inside and outside the magnet can be found as that due to a cylindrical current sheet of radius a and length l in a vacuum with the surface current density given in Eq. (5.36). This sheet, on the other hand, is equivalent to the solenoid in Fig. 4.10, so that the flux density along the magnet axis is given by Eq. (4.36) with NI/l substituted by $J_{\text{ms}} = M$ [see Eq. (4.30)]. Specifically, $\mathbf{B}$ at the center of the magnet equals

$$\mathbf{B} = \frac{\mu_0 l}{\sqrt{l^2 + 4a^2}}\,\mathbf{M}. \tag{5.39}$$

Example 5.4 Nonuniformly Magnetized Infinitely Long Cylinder

Fig. 5.9(a) shows an infinitely long cylinder of radius a in air, having a nonuniform magnetization given by $\mathbf{M} = M_0(1 - r^2/a^2)\,\hat{\mathbf{z}}$, where M_0 is a constant. Find (a) the distribution of magnetization currents of the cylinder and (b) the magnetic flux density vector inside and outside the cylinder.

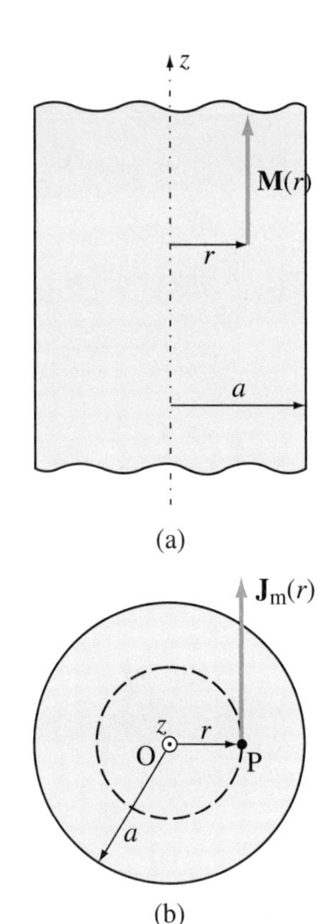

Figure 5.9 Magnetization vector (a) and magnetization volume currents (b) inside a nonuniformly magnetized infinitely long ferromagnetic cylinder; for Example 5.4.

Solution

(a) Applying the formula for the curl in cylindrical coordinates, Eq. (4.84), we obtain that the streamlines of the magnetization volume current inside the cylinder are circles centered at the cylinder axis, as indicated in Fig. 5.9(b), with the following current density vector:

$$\mathbf{J}_{\mathrm{m}} = \nabla \times \mathbf{M} = -\frac{\partial M_z}{\partial r}\,\hat{\boldsymbol{\phi}} = \frac{2M_0}{a^2}\,r\,\hat{\boldsymbol{\phi}}. \tag{5.40}$$

There is no magnetization surface current over the surface of the cylinder, because $M(a^-) = 0$ in the cylinder close to the surface.

(b) We note that the vector $\mathbf{J}_{\mathrm{m}}$ in Eq. (5.40) is of exactly the same form as the current density vector $\mathbf{J}$ inside the rotating charged cylinder in Fig. 4.20 computed in Eq. (4.69). The only difference are the multiplicative constants, ρw in Eq. (4.69) vs. $2M_0/a^2$ in Eq. (5.40). Consequently, the magnetic flux density vectors in the two systems are also of the same form, the only difference being those two multiplicative constants. [The magnetic field due to the rotating charged cylinder in Fig. 4.20 was obtained by visualizing the current distribution $\mathbf{J}$ as a series of coaxial infinitely long solenoids and applying Ampère's law, Eq. (4.49).] By substituting, therefore, ρw in Eq. (4.71) by $2M_0/a^2$, the flux density due to the magnetized cylinder in Fig. 5.9 turns out to be

$$\mathbf{B} = \mu_0 M_0\left(1 - \frac{r^2}{a^2}\right)\hat{\mathbf{z}} = \mu_0 \mathbf{M} \tag{5.41}$$

inside the cylinder ($0 \le r \le a$) and $\mathbf{B} = 0$ outside it.

Example 5.5 Uniformly Magnetized Ferromagnetic Sphere

A ferromagnetic sphere of radius a is uniformly magnetized, and the magnetization vector is $\mathbf{M}$. The sphere is surrounded by air. Calculate (a) the magnetization surface current density vector at an arbitrary point on the sphere surface and (b) the magnetic flux density vector at the sphere center.

Solution

(a) For a spherical coordinate system with the origin at the sphere center and the z-axis parallel to the magnetization vector, as shown in Fig. 5.10, $\mathbf{M} = M\,\hat{\mathbf{z}}$ and the magnetization surface current density vector over the sphere surface at a point P defined by an angle θ is

$$J_{\mathrm{ms}} = \mathbf{M} \times \hat{\mathbf{n}} = M \sin\angle(\mathbf{M}, \hat{\mathbf{n}}) = M\sin\theta, \quad 0 \le \theta \le \pi. \tag{5.42}$$

(b) After the ferromagnetic material has been removed, we apply the superposition principle to find the $\mathbf{B}$ field due to the spherical sheet of current described by the function in Eq. (5.42), which is analogous to the computation of the electric field due to a uniformly polarized dielectric sphere in Fig. 2.7. We subdivide the sphere surface into thin rings of width $\mathrm{d}l_{\mathrm{r}} = a\,\mathrm{d}\theta$, which is depicted in Fig. 5.10. Each such ring can be viewed as an equivalent circular wire loop with the same current. The current of the ring containing the point P equals

$$\mathrm{d}I_{\mathrm{m}} = J_{\mathrm{ms}}\,\mathrm{d}l_{\mathrm{r}} = Ma\sin\theta\,\mathrm{d}\theta. \tag{5.43}$$

Using Eq. (4.19), the magnetic flux density vector of this ring at the sphere center is

$$\mathrm{d}\mathbf{B} = \frac{\mu_0\,\mathrm{d}I_{\mathrm{m}}a_{\mathrm{r}}^2}{2a^3}\,\hat{\mathbf{z}}, \tag{5.44}$$

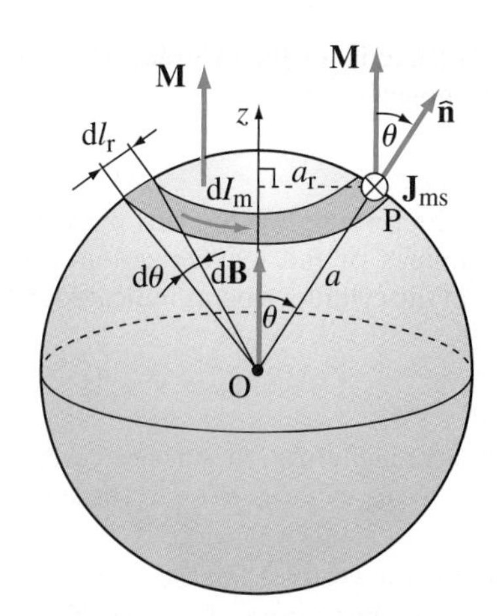

Figure 5.10 Ferromagnetic sphere with a uniform magnetization; for Example 5.5.

with $a_{\rm r} = a \sin\theta$ standing for the ring radius. Hence, the resultant flux density comes out to be

$$\mathbf{B} = \frac{\mu_0 M}{2}\,\hat{\mathbf{z}} \int_{\theta=0}^{\pi} \sin^3\theta \,\mathrm{d}\theta = \frac{2}{3}\,\mu_0 \mathbf{M}, \tag{5.45}$$

where the integral in θ is evaluated as [see also Eq. (2.32)]

$$\int_0^{\pi} \sin^3\theta \,\mathrm{d}\theta = \int_0^{\pi} \left(1 - \cos^2\theta\right) \sin\theta \,\mathrm{d}\theta = \left(-\cos\theta + \frac{\cos^3\theta}{3}\right)\Bigg|_0^{\pi} = \frac{4}{3}. \tag{5.46}$$

Problems: 5.1–5.6; *Conceptual Questions* (on Companion Website): 5.1–5.3; *MATLAB Exercises* (on Companion Website).

5.4 GENERALIZED AMPÈRE'S LAW

We now consider a general magnetostatic system where, in addition to equivalent magnetization currents (bound currents) inside magnetic bodies and over their surfaces, we have conduction currents (free currents) flowing through conductors (including conducting magnetic materials). As sources of the magnetic flux density vector, **B**, all these currents act as if they were in a vacuum, and Ampère's law, Eq. (4.48), becomes

Ampère's law for a system with conductors and magnetic materials

$$\oint_C \mathbf{B} \cdot \mathrm{d}\mathbf{l} = \mu_0 \left(I_C + I_{\mathrm{m}C}\right), \tag{5.47}$$

with I_C and $I_{\mathrm{m}C}$ standing for the total conduction current and the total magnetization current, respectively, enclosed by an arbitrary contour C. Dividing this equation by μ_0, moving $I_{\mathrm{m}C}$ to the left-hand side of the equation, then substituting it by the circulation of the magnetization vector, **M**, from Eq. (5.24), and finally joining the two integrals along C into a single integral, we get the equivalent integral equation:

$$\oint_C \left(\frac{\mathbf{B}}{\mu_0} - \mathbf{M}\right) \cdot \mathrm{d}\mathbf{l} = I_C, \tag{5.48}$$

which is conveniently written as

generalized Ampère's law

$$\oint_C \mathbf{H} \cdot \mathrm{d}\mathbf{l} = I_C. \tag{5.49}$$

This equation is referred to as the generalized Ampère's law. The new quantity on the left-hand side of the equation,

$$\boxed{\mathbf{H} = \frac{\mathbf{B}}{\mu_0} - \mathbf{M},} \tag{5.50}$$

magnetic field intensity vector (unit: A/m)

is called the magnetic field intensity vector and is measured in A/m. It is analogous to the electric flux density vector, **D**, in electrostatics, which is defined by Eq. (2.41). The generalized Ampère's law is valid for magnetostatic fields in arbitrary media, and is easier to use than the form in Eq. (5.47) because it has only free (true) currents on the right-hand side of the integral equation.

The most general representation of conduction current is that by means of the volume current density, **J**, which yields

$$\boxed{\oint_C \mathbf{H} \cdot \mathrm{d}\mathbf{l} = \int_S \mathbf{J} \cdot \mathrm{d}\mathbf{S},} \tag{5.51}$$

generalized Ampère's law in terms of the volume current density

where S is a surface of arbitrary shape bounded by the contour C (orientations of C and S are in accordance to the right-hand rule). Since this integral relation holds for any choice of C, applying Stokes' theorem, Eq. (4.89), results in the following differential relation:

$$\boxed{\nabla \times \mathbf{H} = \mathbf{J},} \tag{5.52}$$

generalized differential Ampère's law

namely, the differential form of the generalized Ampère's law. Eqs. (5.51) and (5.52) represent, respectively, the integral and differential Maxwell's second equation for the magnetostatic field in an arbitrary medium.

Example 5.6 Toroidal Coil with a Ferromagnetic Core

A uniform and dense winding with N turns of wire is placed over a thin toroidal ferromagnetic core of length l. If there is a steady current of intensity I through the winding, find the magnetic field intensity vector in the core.

Solution Due to symmetry, magnetic field lines in the core are circular as in the air-filled toroid in Fig. 4.18. Moreover, the field in the core is uniform, because the toroid is thin [see Eq. (4.66)]. Applying generalized Ampère's law, Eq. (5.49), to a circular contour C of length l along the toroid axis, as shown in Fig. 5.11, gives $Hl = NI$, from which,

$$H = \frac{NI}{l} \quad \text{(thin toroid with an arbitrary core).} \tag{5.53}$$

Note that this result holds true for a core made from an arbitrary magnetic material (including nonlinear and inhomogeneous media).

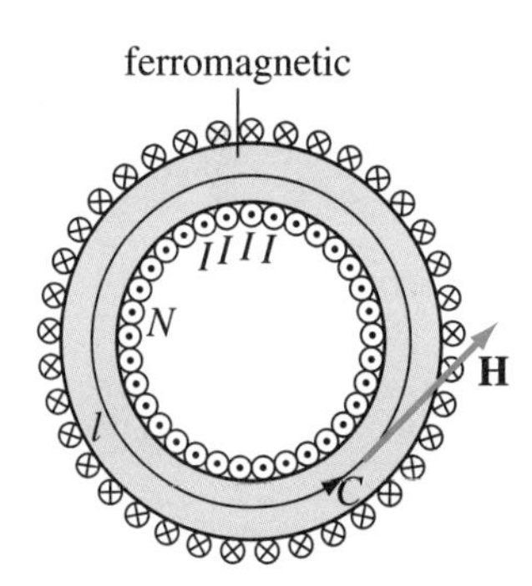

Figure 5.11 Evaluation of the magnetic field intensity vector in a toroidal coil with a ferromagnetic core; for Example 5.6.

Example 5.7 Closed Surface in a Uniformly Magnetized Material

Prove that the flux of the magnetic field intensity vector through a closed surface situated inside a uniformly magnetized ferromagnetic material equals zero.

Solution According to our experience so far with vector calculus and different forms of Maxwell's equations, the flux of a vector through a closed surface in integral notation corresponds, in differential notation, to the divergence of that same vector (with similar correspondence between the circulation along a contour and the curl). Let us therefore consider the divergence of the magnetic field intensity vector, **H**, in the material. Since the material is uniformly magnetized, the divergence of the magnetization vector, **M**, at an arbitrary point in the material is zero ($\mathbf{M} = \text{const}$). The divergence of the magnetic flux density vector, $\mathbf{B}$, on the other hand, is always zero [Eq. (4.103)]. Consequently, the definition of the magnetic field intensity vector, Eq. (5.50), gives

$$\nabla \cdot \mathbf{H} = \frac{1}{\mu_0} \nabla \cdot \mathbf{B} - \nabla \cdot \mathbf{M} = 0, \tag{5.54}$$

i.e., the divergence of $\mathbf{H}$ is also zero in the material, which in integral notation [see Eq. (1.173)] reads

$$\oint_S \mathbf{H} \cdot \mathrm{d}\mathbf{S} = 0, \tag{5.55}$$

for any closed surface S situated inside the material.

Problems: 5.7–5.11; *Conceptual Questions* (on Companion Website): 5.4–5.7; *MATLAB Exercises* (on Companion Website).

5.5 PERMEABILITY OF MAGNETIC MATERIALS

We now introduce the concept of permeability for macroscopic characterization of magnetic materials, which is analogous to the permittivity concept in electrostatics. Substituting Eq. (5.2) into Eq. (5.50), we obtain that the magnetic field intensity vector ($\mathbf{H}$) at a point in any magnetic material is a function of the magnetic flux density vector ($\mathbf{B}$) at that point, or vice versa,

constitutive equation of an arbitrary (nonlinear) magnetic material

$$\boxed{\mathbf{B} = \mathbf{B}(\mathbf{H}).} \tag{5.56}$$

This is the general constitutive equation for characterization of magnetic materials, parallel to Eq. (2.46) in electrostatics. For linear magnetic materials, the magnetization vector, $\mathbf{M}$, is linearly proportional to $\mathbf{B}$, and hence also to $\mathbf{H}$ [see Eq. (5.50)], which is customarily written as

$$\mathbf{M} = \chi_\mathrm{m}\mathbf{H}. \tag{5.57}$$

Here, χ_m is the magnetic susceptibility of the material, that is, the same dimensionless quantity defined in Eq. (5.16). Note, however, that Eq. (5.16) holds only for diamagnetic and paramagnetic materials, where, as we shall see later in this section, $\mathbf{B} \approx \mu_0\mathbf{H}$, while Eq. (5.57) holds for all linear magnetic media. Thus we have

$$\mathbf{B} = \mu_0(\mathbf{H} + \mathbf{M}) = \mu_0(1 + \chi_\mathrm{m})\mathbf{H} = \mu_0\mu_\mathrm{r}\mathbf{H}, \tag{5.58}$$

where $\mu_\mathrm{r} = 1 + \chi_\mathrm{m}$ is a dimensionless proportionality constant called the relative permeability of the material, which is entirely analogous to the relative permittivity ε_r in electrostatics. In practice, μ_r (or χ_m) for a given material can be determined experimentally. In analogy to Eq. (2.50) in electrostatics, we also introduce the permeability (or absolute permeability) of the material,

permeability (unit: H/m)

$$\boxed{\mu = \mu_\mathrm{r}\mu_0,} \tag{5.59}$$

where the value of the permeability of a vacuum, μ_0, is given in Eq. (4.3), with which,

constitutive equation of a linear magnetic material

$$\boxed{\mathbf{B} = \mu\mathbf{H}.} \tag{5.60}$$

The unit for μ is henry per meter (H/m). For free space and other nonmagnetic media, $\mu_\mathrm{r} = 1$ and

constitutive equation for free space

$$\boxed{\mathbf{B} = \mu_0\mathbf{H}.} \tag{5.61}$$

For diamagnetic materials (see Section 5.2), the value of μ is slightly smaller than the permeability of a vacuum (e.g., $\mu_r = 0.999833$ for bismuth, a substance which shows diamagnetism more strongly than most materials), whereas μ of paramagnetic materials is slightly greater than μ_0 (e.g., $\mu_r = 1.0008$ for palladium, one of the strongest paramagnetic materials). Because μ differs only insignificantly from μ_0, it is very common to assume $\mu = \mu_0$ for diamagnetic and paramagnetic materials, as well as for antiferromagnetic substances, in most practical applications. Thus, these three classes of materials are commonly said to be nonmagnetic.

For ferromagnetic, ferrimagnetic, and superparamagnetic materials, on the other hand, μ is much larger than μ_0. These materials, especially ferromagnetic ones, often exhibit permanent magnetization, highly nonlinear behavior, and hysteresis effects, as we shall discuss in more detail in a later section. In ferromagnetics, the function $B(H)$ in Eq. (5.56) is in general nonlinear and has multiple branches. The magnetization properties of the material depend on the applied magnetic field intensity, H, and also on the history of magnetization of the material, i.e., on its previous states. In other words, the value of μ for a ferromagnetic material generally is not unique, but is a function of H and the previous history of the material. Typically, the maximum value of μ_r is around 250 for cobalt, 600 for nickel, and 5000 for iron (with 0.4% impurity), whereas it is as high as about 200,000 for purified iron (0.04% impurity) and 1,000,000 for supermalloy (79.5% Ni, 15% Fe, 5% Mo, 0.5% Mn). In many applications involving ferromagnetics, we assume

$$\boxed{\mu_r \to \infty,} \tag{5.62}$$

perfect magnetic conductor (PMC)

and such media are customarily referred to as perfect magnetic conductors (PMC). As we shall see in the next section, the magnetic flux density vector, **B**, in a nonmagnetic medium near the surface of a ferromagnetic (or PMC) body is always normal to the surface, the same as for the electric field intensity vector, **E**, near the surface of a perfect electric conductor (PEC), with $\sigma \to \infty$. Shown in Table 5.1 are values of the relative permeability of an illustrative set of selected materials,

Table 5.1. Relative permeability of selected materials

Material	μ_r	Material	μ_r
Bismuth	0.999833	Titanium	1.00018
Gold	0.99996	Platinum	1.0003
Mercury	0.999968	Palladium	1.0008
Silver	0.9999736	Manganese	1.001
Lead	0.9999831	Cast iron	150
Copper	0.9999906	Cobalt	250
Water	0.9999912	Nickel	600
Paraffin	0.99999942	Nickel-zinc ferrite (Ni-Zn-Fe_2O_3)	650
Wood	0.9999995	Manganese-zinc ferrite (Mn-Zn-Fe_2O_3)	1200
Vacuum	1	Steel	2000
Air	1.00000037	Iron (0.4% impurity)	5000
Beryllium	1.0000007	Silicon iron (4% Si)	7000
Oxygen	1.000002	Permalloy (78.5% Ni, 21.5% Fe)	7×10^4
Magnesium	1.000012	Mu-metal (75% Ni, 14% Fe, 5% Cu, 4% Mo, 2% Cr)	10^5
Aluminum	1.00002	Iron (purified – 0.04% impurity)	2×10^5
Tungsten	1.00008	Supermalloy (79.5% Ni, 15% Fe, 5% Mo, 0.5% Mn)	10^6

of considerable theoretical and/or practical interest – in terms of their magnetic properties.

A magnetic material is said to be homogeneous when its magnetic properties do not change from point to point in the region being considered. Otherwise, the material is inhomogeneous [e.g., $\mu = \mu(x, y, z)$ in the region]. Finally, some magnetic materials, such as ferrites, are anisotropic. Namely, an applied magnetic field (vector **B**) in one direction can produce magnetization (i.e., vector **M**) in another direction in a material. Accordingly, Eq. (5.60) becomes a matrix equation,

$[\mu]$ – permeability tensor of an anisotropic magnetic material

$$\begin{bmatrix} B_x \\ B_y \\ B_z \end{bmatrix} = \begin{bmatrix} \mu_{xx} & \mu_{xy} & \mu_{xz} \\ \mu_{yx} & \mu_{yy} & \mu_{yz} \\ \mu_{zx} & \mu_{zy} & \mu_{zz} \end{bmatrix} \begin{bmatrix} H_x \\ H_y \\ H_z \end{bmatrix}, \tag{5.63}$$

where $[\mu]$ is the permeability tensor. It is analogous to the permittivity tensor, $[\varepsilon]$, defined by Eq. (2.52). However, Eq. (5.50) remains valid (it represents the definition of the **H** field), although **B**, **H**, and **M** are no longer parallel at a point. The anisotropy of ferrites is used in a number of microwave devices, including some types of gyrators, directional couplers, and isolators.

Note that in a linear, isotropic, and homogeneous magnetic material, we can substitute **H** by $\mathbf{B}/\mu$ in the generalized Ampère's law and bring $1/\mu$ (because it is constant) outside the integral sign in the integral form of the law, Eq. (5.49), or outside the operator (curl) sign in its differential form, Eq. (5.52), yielding

$$\oint_C \mathbf{B} \cdot \mathrm{d}\mathbf{l} = \mu I_C \quad \text{and} \quad \nabla \times \mathbf{B} = \mu \mathbf{J}, \tag{5.64}$$

respectively. We realize that these equations are identical to the corresponding free-space laws, Eqs. (4.48) and (4.83), except for μ_0 being substituted by μ. In the same way, replacing μ_0 with μ in Eq. (4.7) gives the version of the Biot-Savart law for a volume conduction current in a homogeneous magnetic medium of permeability μ:

$$\mathbf{B} = \frac{\mu}{4\pi} \int_v \frac{(\mathbf{J}\,\mathrm{d}v) \times \hat{\mathbf{R}}}{R^2}, \tag{5.65}$$

with $\mathbf{H} = \mathbf{B}/\mu$. Additionally, Eq. (4.132) and the concept of magnetic permeability imply that the Laplacian of the magnetic vector potential of a volume current in a homogeneous magnetic medium satisfies the following differential equation:

$$\nabla^2 \mathbf{A} = -\mu \mathbf{J}. \tag{5.66}$$

Finally, Eq. (4.108) becomes

$$\mathbf{A} = \frac{\mu}{4\pi} \int_v \frac{\mathbf{J}\,\mathrm{d}v}{R}. \tag{5.67}$$

From Eq. (5.50), we find that the magnetization vector in a linear and isotropic magnetic material can be expressed in terms of the magnetic field intensity vector as

$$\mathbf{M} = \frac{\mathbf{B}}{\mu_0} - \mathbf{H} = \left(\frac{\mu}{\mu_0} - 1\right)\mathbf{H} = (\mu_\mathrm{r} - 1)\mathbf{H}. \tag{5.68}$$

If the material is also homogeneous, then the magnetization volume current density vector, $\mathbf{J}_\mathrm{m}$, at a point in the material can be obtained directly from the conduction volume current density vector, **J**, at that point:

$$\mathbf{J}_\mathrm{m} = \nabla \times \mathbf{M} = \nabla \times [(\mu_\mathrm{r} - 1)\mathbf{H}] = (\mu_\mathrm{r} - 1)\nabla \times \mathbf{H} = (\mu_\mathrm{r} - 1)\mathbf{J}. \tag{5.69}$$

We see that there cannot be magnetization volume current ($\mathbf{J}_\mathrm{m} = 0$) in a homogeneous linear magnetic medium with no free current ($\mathbf{J} = 0$).

Problems: 5.12–5.15; *Conceptual Questions* (on Companion Website): 5.8.

5.6 MAXWELL'S EQUATIONS AND BOUNDARY CONDITIONS FOR THE MAGNETOSTATIC FIELD

From the Biot-Savart law for the magnetic flux density $\mathrm{d}\mathbf{B}$ of a single conduction current element $\mathbf{J}\,\mathrm{d}v$ in free space and the superposition principle, we proved in Section 4.8 the law of conservation of magnetic flux, Eq. (4.99), for the magnetostatic field in free space. We can repeat here that same proof for a magnetization current element, $\mathbf{J}_\mathrm{m}\,\mathrm{d}v$ (or $\mathbf{J}_\mathrm{ms}\,\mathrm{d}S$), and get the same result, which means that the net magnetic flux (the flux of the vector $\mathbf{B}$) due to any distribution of conduction and magnetization currents (which, equivalently, reside in a vacuum) through a closed surface always equals zero. In other words, we conclude that the law of conservation of magnetic flux (Maxwell's fourth equation) holds for structures that include arbitrary magnetic materials.

We now write down the full set of Maxwell's equations governing the magnetostatic field in an arbitrary medium, together with the associated constitutive equation:

$$\begin{cases} \oint_C \mathbf{H}\cdot\mathrm{d}\mathbf{l} = \int_S \mathbf{J}\cdot\mathrm{d}\mathbf{S} \\ \oint_S \mathbf{B}\cdot\mathrm{d}\mathbf{S} = 0 \\ \mathbf{B} = \mathbf{B}(\mathbf{H})\ [\mathbf{B} = \mu\mathbf{H}] \end{cases}. \tag{5.70}$$

Maxwell's second equation, static field
Maxwell's fourth equation
constitutive equation for **B**

The corresponding differential Maxwell's equations are:

$$\nabla\times\mathbf{H} = \mathbf{J} \quad \text{and} \quad \nabla\cdot\mathbf{B} = 0. \tag{5.71}$$

Maxwell's differential equations, magnetostatic field

In addition to integral and differential Maxwell's equations for any field, boundary conditions always represent the third form of field equations, and they are derived from the respective integral equations (differential equations apply only at a point). Let us derive here the boundary conditions for the magnetostatic field on the boundary surface between two arbitrary media. Let $\mathbf{J}_\mathrm{s}$ be the density vector of a surface conduction (free) current that may exist on the boundary. We apply the generalized Ampère's law in integral form, Eq. (5.49), to a narrow rectangular elementary contour C positioned such that $\mathbf{J}_\mathrm{s}$ is normal to the plane of the contour, as depicted in Fig. 5.12. Having in mind Eqs. (2.79) and (5.30), we obtain

$$\oint_C \mathbf{H}\cdot\mathrm{d}\mathbf{l} = H_{1\mathrm{t}}\,\Delta l - H_{2\mathrm{t}}\,\Delta l = I_C = J_\mathrm{s}\Delta l, \tag{5.72}$$

which yields

$$H_{1\mathrm{t}} - H_{2\mathrm{t}} = J_\mathrm{s}. \tag{5.73}$$

In vector form [also see Eq. (2.84)],

$$\hat{\mathbf{n}}\times\mathbf{H}_1 - \hat{\mathbf{n}}\times\mathbf{H}_2 = \mathbf{J}_\mathrm{s}, \tag{5.74}$$

boundary condition for $\mathbf{H}_\mathrm{t}$; $\hat{\mathbf{n}}$ directed from region 2 to region 1

where $\hat{\mathbf{n}}$ is the normal unit vector on the surface, directed from region 2 to region 1. If no surface conduction current exists on the boundary, Eq. (5.73) becomes

$$H_{1\mathrm{t}} = H_{2\mathrm{t}} \quad (\mathbf{J}_\mathrm{s} = 0), \tag{5.75}$$

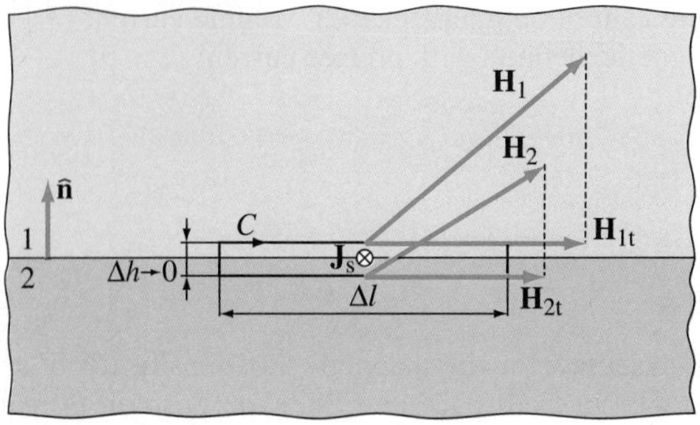

Figure 5.12 Deriving the boundary condition for tangential components of vector **H** on the boundary surface between two arbitrary magnetic media.

that is, the tangential component of **H** is continuous across the boundary free of conduction current.

Noting, on the other hand, that the law of conservation of magnetic flux, Eq. (4.99), and the continuity equation for steady currents, Eq. (3.40), have exactly the same form, we conclude that the corresponding boundary conditions must have exactly the same form as well. Therefore, from the boundary condition for normal components of the vector **J** in the steady current field, Eq. (3.55), we directly write the boundary condition for normal components of the vector **B** in the magnetostatic field:

boundary condition for $\mathbf{B}_\mathrm{n}$

$$\boxed{\hat{\mathbf{n}} \cdot \mathbf{B}_1 - \hat{\mathbf{n}} \cdot \mathbf{B}_2 = 0 \quad \text{or} \quad B_{1\mathrm{n}} = B_{2\mathrm{n}}.} \tag{5.76}$$

It tells us that $\mathbf{B}_\mathrm{n}$ is always continuous across a boundary.

At an interface between two linear magnetic media of permeabilities μ_1 and μ_2 with no surface conduction current ($\mathbf{J}_\mathrm{s} = 0$), the law of refraction of the magnetic field lines holds that is entirely analogous to the corresponding laws in electrostatics, Eq. (2.87), and steady current field, Eq. (3.56). With α_1 and α_2 denoting the angles that field lines in region 1 and region 2 make with the normal to the interface, as shown in Fig. 5.13, we have

law of refraction of magnetic field lines

$$\boxed{\frac{\tan\alpha_1}{\tan\alpha_2} = \frac{\mu_1}{\mu_2}.} \tag{5.77}$$

This relationship indicates that magnetic field lines are bent farther away from the normal in the medium with the higher permeability. Bending of field lines is basically due to unequal magnetization surface currents on the two sides of the interface.

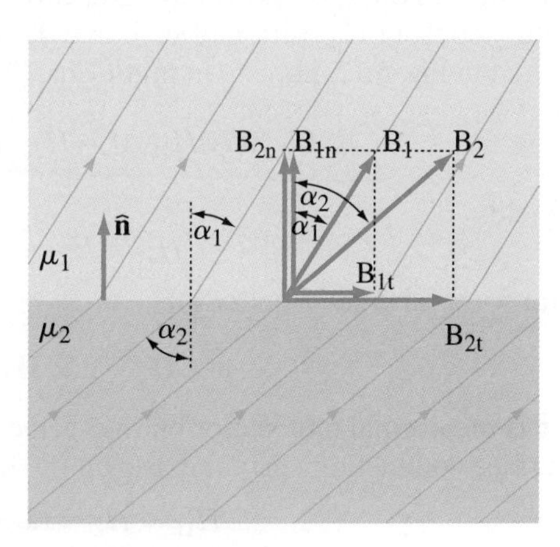

Figure 5.13 Refraction of magnetic field lines at a magnetic-magnetic interface.

Note that if region 2 is filled with a ferromagnetic material and region 1 with a nonmagnetic material, then $\mu_2 \gg \mu_1$ and $\mu_1/\mu_2 \approx 0$, and Eq. (5.77) results in

$$\alpha_1 \approx 0 \quad (\mu_2 \gg \mu_1) \tag{5.78}$$

for any α_2 (except for $\alpha_2 = 90°$, i.e., for field lines in region 2 parallel to the interface). This means that magnetic field lines in a nonmagnetic medium near the interface with a ferromagnetic medium (or PMC) are always normal to the interface.

Finally, using Eq. (5.32) and adding up the magnetization surface current density vectors that accumulate on the two sides of a magnetic-magnetic interface, analogously to deriving Eq. (2.89) in electrostatics, we arrive to the following boundary condition for the tangential components of the magnetization vector:

$$\boxed{\hat{\mathbf{n}} \times \mathbf{M}_1 - \hat{\mathbf{n}} \times \mathbf{M}_2 = \mathbf{J}_{\text{ms}},} \tag{5.79}$$

boundary condition for $\mathbf{M}_t$

where $\hat{\mathbf{n}}$ is directed from medium 2 to medium 1 (Figs. 5.12 and 5.13) and $\mathbf{J}_{\text{ms}}$ is the total magnetization surface current density vector at the interface.

Problems: 5.16; *Conceptual Questions* (on Companion Website): 5.9–5.12; *MATLAB Exercises* (on Companion Website).

5.7 IMAGE THEORY FOR THE MAGNETIC FIELD

Magnetostatic systems often include current conductors in the presence of large flat ferromagnetic bodies. By utilizing image theory, similarly to the procedure described in Section 1.21, we can remove the ferromagnetic body from the system and replace it by an image of the original current distribution. The equivalent problem is then much simpler to solve because it consists of a known current distribution (original plus image) in free space.

Consider a straight current conductor in a nonmagnetic half-space in the vicinity of an infinite planar interface with a ferromagnetic (or PMC) half-space. Let the conductor be parallel to the interface. Eq. (5.78) tells us that the ferromagnetic material in the lower half-space, i.e., the induced magnetization current in the material, influences the resultant magnetic flux density vector, **B**, such that it has no tangential component at the upper side of the interface, as shown in Fig. 5.14(a). This condition remains unaltered, however, if we replace the ferromagnetic block by another conductor parallel to the interface that is positioned symmetrically with respect to the original conductor and carries a current of the same intensity and the same direction, Fig. 5.14(b). We conclude thus that, as far as the magnetic field in the upper half-space is concerned, systems in Fig. 5.14(a) and Fig. 5.14(b) are equivalent. This is an example of the image theory (theorem) for the magnetic field.

The theory is not restricted to conductors parallel to the material interface only. It states that an arbitrary current configuration above an infinite ferromagnetic (or PMC) plane can be replaced by a new current configuration in free space consisting from the original current configuration itself and its positive image in the ferromagnetic plane. The equivalence is with respect to the magnetic field above the ferromagnetic plane, the component of that field due to the induced magnetization current in the ferromagnetic material being equal to the field of the image. As another example, Fig. 5.15 shows the image of a current conductor consisting of three segments with different orientations with respect to the ferromagnetic interface in Fig. 5.15(a), where it is a simple matter to conclude that the vector **B** in the plane of symmetry in Fig. 5.15(b) has no component tangential to the plane.

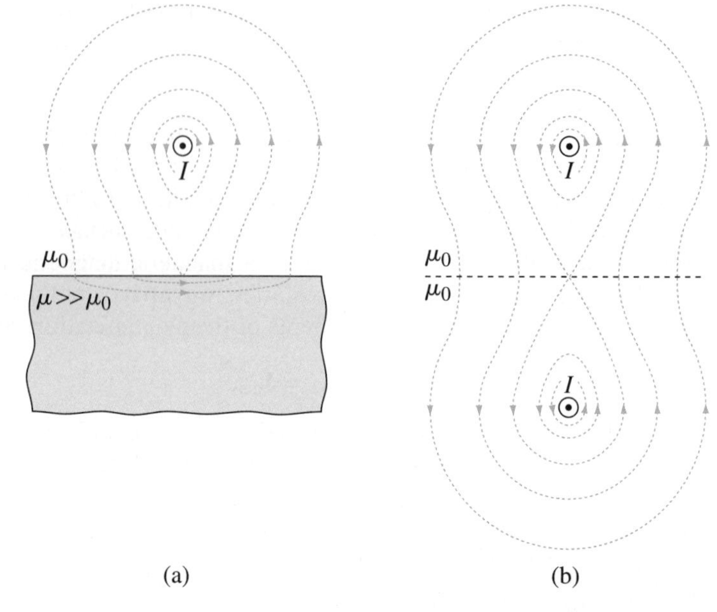

Figure 5.14 (a) Straight current conductor parallel to the interface of a ferromagnetic (or PMC) half-space. (b) By image theory, the influence of the ferromagnetic material on the magnetic field in the upper half-space can be represented by a positive image of the original current.

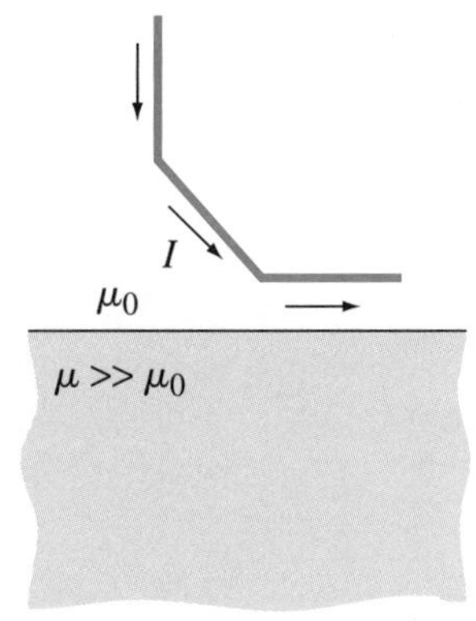

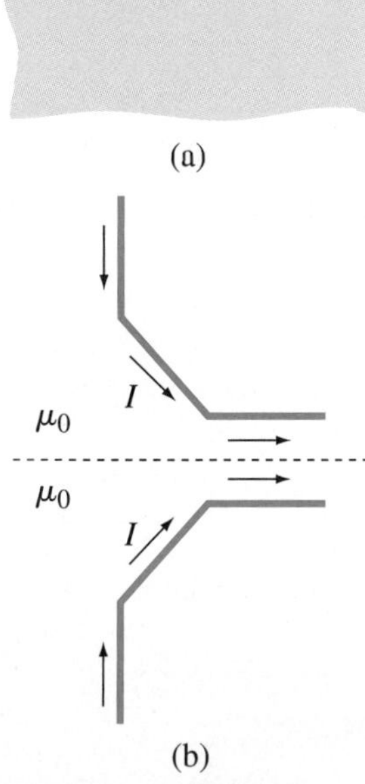

Figure 5.15 Image theory for the magnetic field due to a current conductor with arbitrarily oriented segments above a ferromagnetic (or PMC) plane: (a) original system with the material interface and (b) equivalent free-space system.

Example 5.8 Force on a Conductor above a Ferromagnetic Plane

A very long and thin wire is situated in air at a height h above a ferromagnetic half-space, parallel to its surface, and carries a current of intensity I. Determine the magnetic force on the wire per unit of its length.

Solution By image theory, the system wire-ferromagnetic is equivalent to a symmetric two-wire system in air, as shown in Fig. 5.16. The force on the upper (original) conductor per unit of its length is then obtained using Eq. (4.166) with $I_1 = I_2 = I$ and $d = 2h$, that is,

$$\mathbf{F}'_{\mathrm{m}} = IB = \frac{\mu_0 I^2}{4\pi h}. \tag{5.80}$$

Here, B is the magnetic flux density along the upper conductor due to the lower conductor, i.e., due to the magnetization current in the ferromagnetic material. We see that a large ferromagnetic (or PMC) block always attracts a current conductor running parallel to it.

Example 5.9 Strip Conductor between Two PMC Planes

An infinitely long thin strip conductor of width a carries a steady current of intensity I. The strip is placed between two parallel PMC planes, as shown in Fig. 5.17(a). Find the magnetic flux density vector in the space between the PMC planes, assuming that it is air-filled.

Solution By multiple applications of the image theory for the magnetic field, we obtain the equivalent system in Fig. 5.17(b), which represents an infinite planar current sheet with surface current density $J_s = I/a$ in free space. The magnetic field between the PMC planes is thus uniform, with flux density $B = \mu_0 J_s/2 = \mu_0 I/(2a)$ [see Fig. 4.21 and Eq. (4.47)].

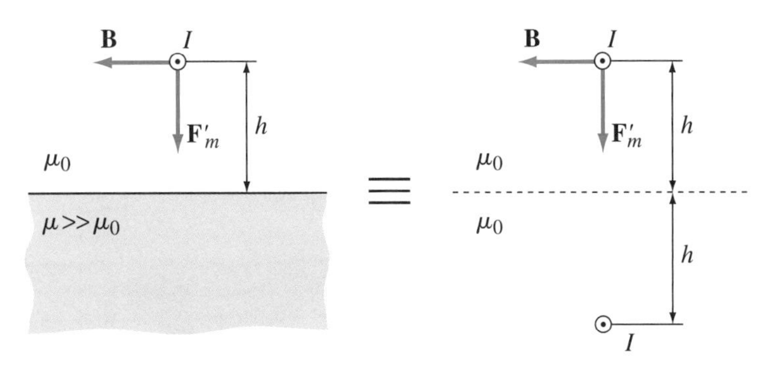

Figure 5.16 Evaluation of the force on a horizontal straight wire with a current above a ferromagnetic plane, using image theory; for Example 5.8.

Example 5.10 **Uniformly Magnetized Hemisphere on a PMC Plane**

A uniformly magnetized ferromagnetic hemisphere of radius a in air is lying on a PMC ($\mu_r \to \infty$) plane. The magnetization vector is $\mathbf{M}$, and it is perpendicular to the plane, as shown in Fig. 5.18. Calculate the magnetic flux density and field intensity vectors at the center of the bottom surface of the hemisphere (point O in the figure).

Solution We first evaluate the distribution of equivalent magnetization currents of the hemisphere. The magnetization volume current density vector inside the material and the magnetization surface current density vector on the bottom surface of the hemisphere are both zero. There is, however, a hemispherical sheet of magnetization current over the upper surface of the hemisphere given by Eq. (5.42), where now $0 \leq \theta \leq \pi/2$. Using the image theory for the magnetic field, we then supplement this sheet, considered to be in a vacuum, with another hemispherical sheet below the plane of symmetry, which is also described by the function in Eq. (5.42), but with $\pi/2 \leq \theta \leq \pi$ (note that $\sin\theta$ is symmetric with respect to $\theta = \pi/2$, which corresponds to a positive image of the current, exactly as required by the image theory). We thus obtain the full spherical sheet of magnetization current in Fig. 5.10, and conclude that the magnetized hemisphere on the PMC plane is equivalent to the magnetized sphere of Fig. 5.10, in free space. Hence, the magnetic flux density vector at the sphere center, as well as that at the point O in Fig. 5.18, is given by Eq. (5.45). Finally, by means of Eq. (5.50), the magnetic field intensity vector at the point O is

$$\mathbf{H} = \frac{\mathbf{B}}{\mu_0} - \mathbf{M} = \frac{2}{3}\mathbf{M} - \mathbf{M} = -\frac{\mathbf{M}}{3}. \tag{5.81}$$

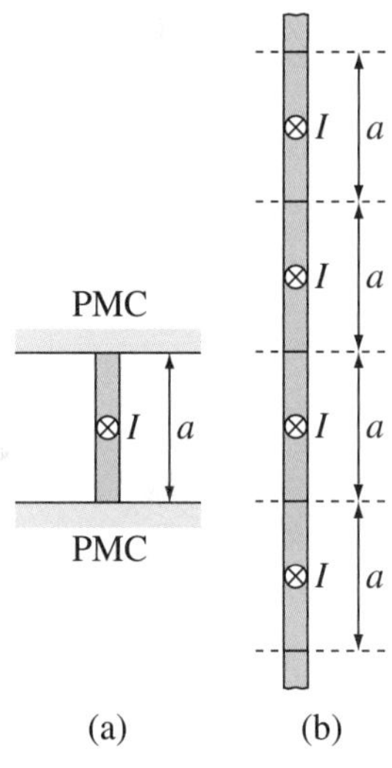

Figure 5.17 (a) Strip conductor between two parallel PMC planes and (b) equivalent infinite planar current sheet in free space; for Example 5.9.

Problems: 5.17–5.19; *Conceptual Questions* (on Companion Website): 5.13; *MATLAB Exercises* (on Companion Website).

5.8 MAGNETIZATION CURVES AND HYSTERESIS

In this section, we consider in more detail the B-H relationship, Eq. (5.56), for ferromagnetic materials. This relationship, being nonlinear in general, is usually given as a graph showing B (ordinate) as a function of H (abscissa). A curve representing

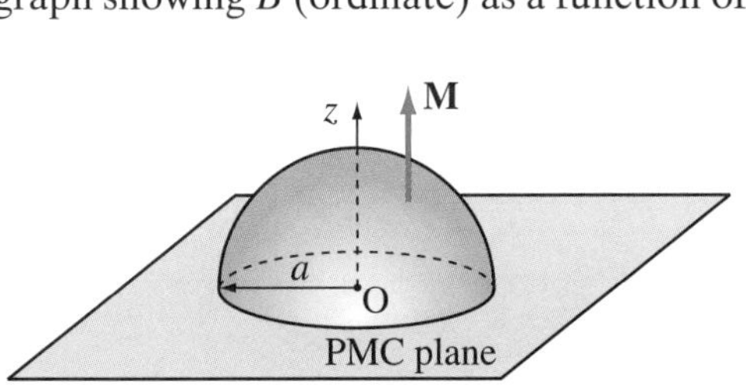

Figure 5.18 Ferromagnetic hemisphere with a uniform magnetization lying on a PMC plane; for Example 5.10.

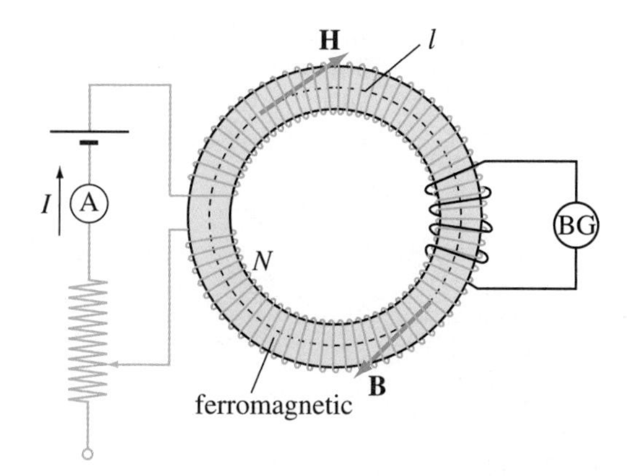

Figure 5.19 Simple apparatus for measurement of magnetization curves.

the function $B(H)$ on such a diagram is called a magnetization curve and is obtained by measurement on a given material specimen.

Fig. 5.19 shows the simplest apparatus for measurement of magnetization curves, where a uniform winding is placed over a thin toroidal core (ring) cut from a ferromagnetic sample we want to measure. If a current I is established in the toroid (primary coil), the magnetic field intensity H in the core is given by Eq. (5.53), where N is the number of wire turns of the coil and l is the mean length of the toroid. By varying I, therefore, we directly vary H (H is proportional to I). The magnetic flux density B in the core is then measured, as a response to H, by a ballistic galvanometer (BG) connected to another winding (secondary coil) that is placed over the core. Namely, the galvanometer measures the charge that passes through the secondary circuit, as a consequence of the change of magnetic flux through the secondary coil. This charge, as we shall see in the next chapter, is proportional to the flux change, so that the galvanometer actually serves as a fluxmeter. Thus, by changing, step by step, the current I in the primary coil and field H in the core, and measuring the corresponding values of the flux density B, we obtain, point by point, the magnetization curve of the material.

Shown in Fig. 5.20 is a typical initial magnetization curve for a ferromagnetic sample, where the material is completely demagnetized and both B and H are zero

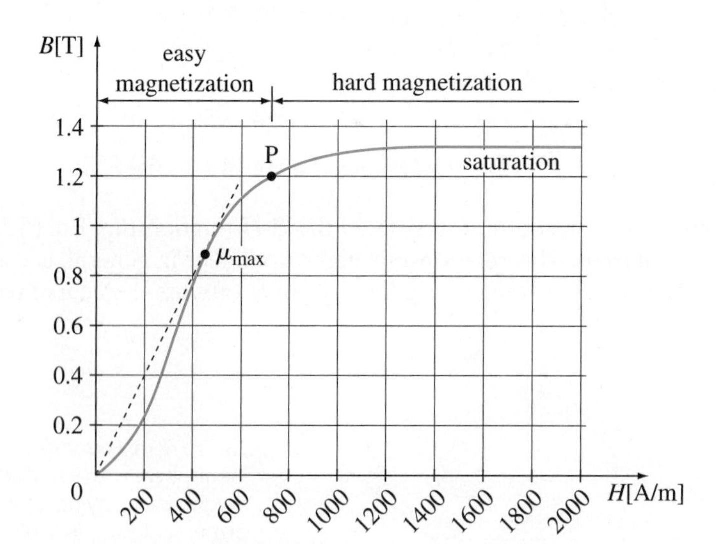

Figure 5.20 Typical initial magnetization curve for a ferromagnetic material.

before a field is applied. As we begin to apply a current in the primary circuit in Fig. 5.19, the magnetic flux density also rises, but not linearly. Moreover, the value of B rises rapidly at first and then more slowly. The first part of the initial magnetization curve (roughly up to the point P in Fig. 5.20) represents the region of easy (steep) magnetization. In the upper section of the curve, the increase in magnetization due to gradual rotations of magnetic domains in the material not already parallel to **H** is more difficult, and the curve tends to become flat. This is the region of hard (flat) magnetization. Very strong magnetic fields are usually required to reach the state of saturation, where all the moments of magnetic domains in the material are parallel to **H** and the magnetization curve flattens off completely.

The permeability at any point on the magnetization curve is given by

$$\mu = \frac{B}{H}, \tag{5.82}$$

where B is the ordinate of the point (in T) and H is the abscissa (in A/m). The relative permeability is then $\mu_r = \mu/\mu_0$. The maximum permeability is at the point on the curve with the largest ratio of B to H, i.e., at the point of tangency with the straight line of steepest slope that passes through the origin and intersects the magnetization curve (Fig. 5.20). Note that the maximum μ does not correspond to the steepest slope of the magnetization curve, because μ is not proportional to the slope of the curve ($\mathrm{d}B/\mathrm{d}H$), but equals the ratio B/H.

Having reached saturation, let us now turn to Fig. 5.21, where we continue our experiment – by reducing I and H. As we do so, the effects of hysteresis begin to show, and we do not retrace the initial-magnetization curve. Hysteresis means that B lags behind H, so that the magnetization curves for increasing and decreasing the applied field are not the same. Even after H becomes zero, B does not go to zero, but to a value $B = B_r$, termed the remanent (residual) magnetic flux density. Note that the existence of a remanent flux density in a ferromagnetic material makes permanent magnets possible. We then reverse H (by reversing the polarity of the battery in Fig. 5.19), and increase it in the negative direction, so that B comes to zero at $H = -H_c$, where H_c is the so-called coercive force. As H is increased in the negative direction further, the material becomes magnetized even more, with negative polarity. It passes through the stages of easy and hard magnetization, the magnetization curve flattens off, and negative saturation is reached. The end of the curve on the left-hand side of the diagram is for the field $H = -H_m$, after which

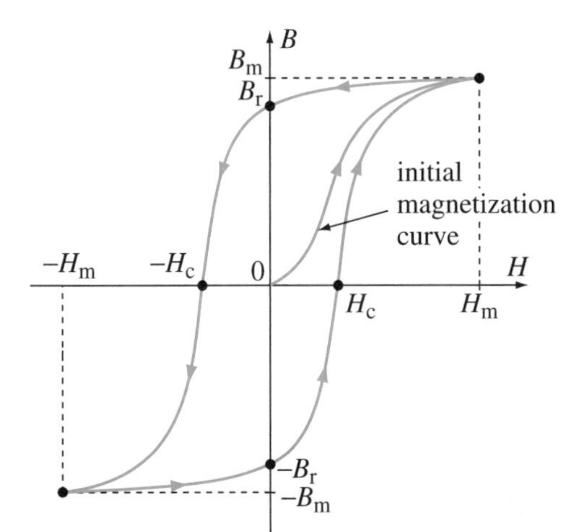

Figure 5.21 Typical hysteresis loop for a ferromagnetic material.

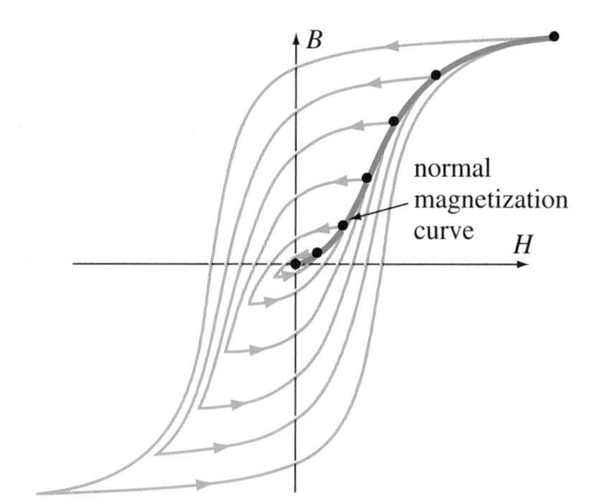

Figure 5.22 Demagnetization of a ferromagnetic sample by reversals of the applied field **H**; definition of the normal magnetization curve.

we start reducing H. The next intercept of the curve with the B axis (for $H = 0$) gives $B = -B_\text{r}$ (negative remanent flux density). At this point, we reverse the battery polarity again and continue – by increasing H in the positive direction. This makes the flux density zero at a positive field $H = H_\text{c}$ (coercive force). With a further increase in H, the material reaches positive saturation (at $H = H_\text{m}$), and a full cycle in the BH diagram is completed. The loop traced out by the magnetization curve during this cycle is referred to as the hysteresis loop.

Having carried our ferromagnetic specimen to saturation at both ends of the magnetization curve in Fig. 5.21, we now move on to Fig. 5.22, to continue to cycle the applied field H, but over successively smaller ranges ($\pm H$ is brought to smaller and smaller amplitudes on each reversal). We obtain thus a series of hysteresis loops that decrease in size, and the residual B for $H = 0$ eventually becomes zero, that is, the material is left in a demagnetized state. Such a process is used for demagnetization of objects that have a residual magnetization (i.e., remanent flux density) under conditions of zero applied magnetic field. In practice, the process can be carried out by inserting the object to be demagnetized inside a coil with a low-frequency ac current, and then gradually reducing the current amplitude in the coil or slowly removing the object from the coil with the current amplitude constant. The curve connecting the tips of the hysteresis loops in Fig. 5.22 is another characteristic of ferromagnetic materials known as the normal magnetization curve. For a particular ferromagnetic material, the normal and initial magnetization curves are very similar.

Ferromagnetic materials having small coercive forces H_c and therefore narrow (thin) hysteresis loops (with small loop areas), as illustrated in Fig. 5.23, are referred to as soft ferromagnetics. As we shall see in a later chapter, the area enclosed by the hysteresis loop is proportional to energy loss per unit volume of the material in one cycle of field variation. This is so-called hysteresis loss, which corresponds to the energy lost in the form of heat in overcoming the friction encountered during the movements of magnetic domain walls and rotations of the domains in a ferromagnetic material. Soft ferromagnetic materials have a large magnetization for a very small applied field and exhibit low hysteresis losses. Additionally, they have very large values of initial permeability [μ in Eq. (5.82) for a very small H]. For example, supermalloy, a typical soft ferromagnetic, has an initial relative permeability on the order of 10^5, $B_\text{r} = 0.6$ T, and $H_\text{c} = 0.4$ A/m. Soft ferromagnetic materials are used for building transformers and ac machines

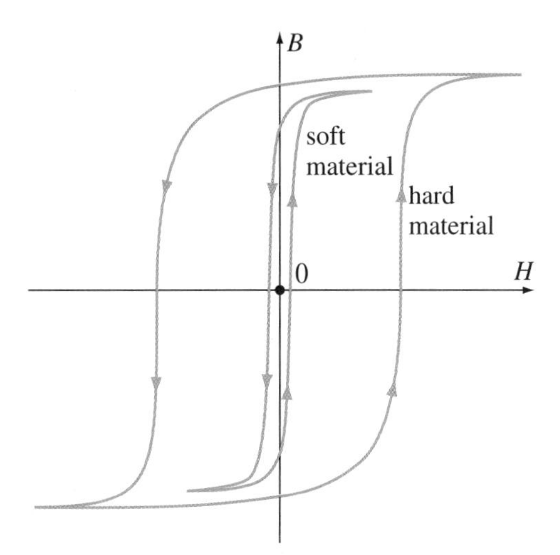

Figure 5.23 Hysteresis loops for soft and hard ferromagnetic materials.

(motors and generators), where the material is permanently exposed to alternating magnetization.

Hard ferromagnetic materials, on the other hand, have large coercive forces H_c and hence broad (fat) hysteresis loops (Fig. 5.23). They have small initial permeabilities, and are used for building permanent magnets and dc machines. In these applications, fields do not change frequently, and hysteresis losses, therefore, are not significant, in spite of large hysteresis loop areas. The essential characteristic of permanent-magnet materials is a high remanent flux density B_r (they are permanently magnetized even with no magnetic field applied), but it is also important that their coercive force be large, so that the material may not be easily demagnetized. Alnico, an aluminum-nickel-cobalt alloy with a small amount of copper, is a typical hard ferromagnetic, having an initial μ_r around 4, B_r around 1 T, and H_c on the order of 50,000 A/m.

Conceptual Questions (on Companion Website): 5.14–5.16.

5.9 MAGNETIC CIRCUITS – BASIC ASSUMPTIONS FOR THE ANALYSIS

Magnetic circuit in general is a collection of bodies and media that form a way along which the magnetic field lines close upon themselves, i.e., it is a circuit of the magnetic flux flow. The name arises from the similarity to electric circuits. In practical applications, including transformers, generators, motors, relays, magnetic recording devices, etc., magnetic circuits are formed from ferromagnetic cores of various shapes, that may or may not have air gaps, with current-carrying windings wound about parts of the cores. Fig. 5.24 shows a typical magnetic circuit.

The analysis of magnetic circuits with steady currents in the windings (dc magnetic circuits) is based, of course, on Maxwell's equations for the magnetostatic field in arbitrary media, Eqs. (5.70). Given, however, great complexity of the rigorous analysis of practical magnetic circuits, we introduce here a set of approximations that will, in conjunction with Maxwell's equations, make the analysis much simpler and yet accurate enough for engineering applications.

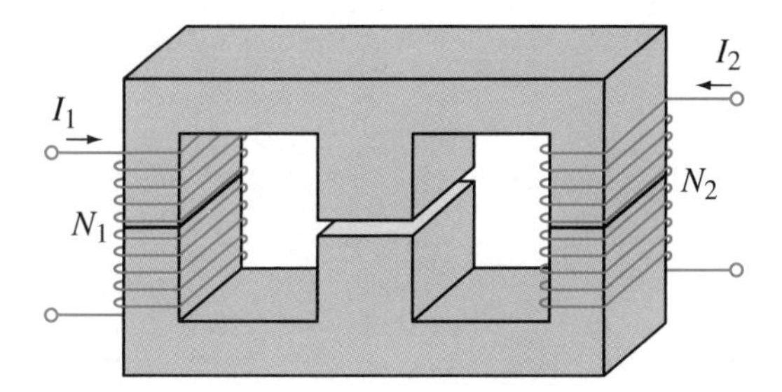

Figure 5.24 Typical magnetic circuit.

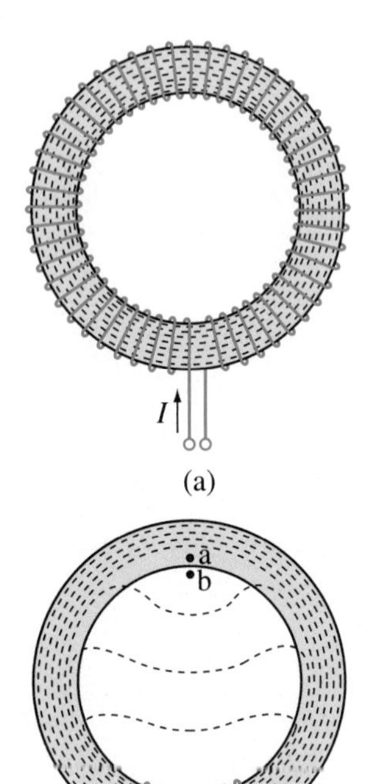

Figure 5.25 Toroidal ferromagnetic core with (a) a uniform winding and (b) concentrated winding.

Firstly, we assume that the magnetic flux is concentrated exclusively inside the magnetic circuit, i.e., in the branches of the ferromagnetic core and air gaps. This is never exactly true. For a toroidal ferromagnetic core with a uniform winding shown in Fig. 5.25(a), however, the flux is restricted to the interior of the toroid as a consequence of the geometrical symmetry of the structure, provided that the wire turns are very tightly wound over the core. In reality, there is always some leakage flux between the wire turns. In addition, in circuits containing coils placed only over parts of the core, some flux lines bridge the space between the core sections through air, as indicated in Fig. 5.25(b). Boundary conditions for the magnetostatic field applied to the interface between the ferromagnetic and air in Fig. 5.25(b) tell us that magnitudes of the magnetic flux density vector at points a and b in the figure are related as $B_{\rm a} \approx \mu_{\rm r} B_{\rm b}$. As $\mu_{\rm r}$ is very large for ferromagnetic materials, $B_{\rm a} \gg B_{\rm b}$, so we conclude that practically the entire magnetic flux is restricted to the ferromagnetic core, and the larger the $\mu_{\rm r}$ the more accurate this assumption. Note that in electric circuits this is always true for the current flow because the conductivity of air is zero, whereas the permeability of air is not.

Secondly, we assume that air gaps in the magnetic circuit are narrow enough so that the fringing flux near the gap edges can be neglected. For more precise analysis, formulas for an effective length and cross-sectional area of the gap may be used to incorporate the fringing effects into the basic equations for the circuit.

Finally, we assume that the magnetic field is uniform throughout the volume of each branch of the circuit. Then, in applying Maxwell's equations, every cross section of a branch is assumed to have the same area, and the path of every flux line along the branch is assumed to be of the same length, equal to the mean length of that part of the circuit. This is true only for thin magnetic circuits. In many applications involving thick cores, however, the error due to thin magnetic circuit approximations is acceptable, which is illustrated in the following examples.

Example 5.11 Thick Toroid with a Linear Ferromagnetic Core

Assume that the toroidal coil in Fig. 4.18 is wound about a core made of a linear ferromagnetic material of permeability μ. (a) Determine the magnetic flux through the core. (b) Find the error made in the flux computation if the magnetic field in the core is assumed to be uniform, specifically for $b - a = 0.1a$ (thin toroid) and $b - a = a$ (thick toroid).

Solution

(a) The magnetic field intensity vector, **H**, is the same as in the air-filled toroid in Fig. 4.18. From Eq. (4.65),

H inside a thick toroidal coil

$$H(r) = \frac{NI}{2\pi r}, \quad a \le r \le b. \tag{5.83}$$

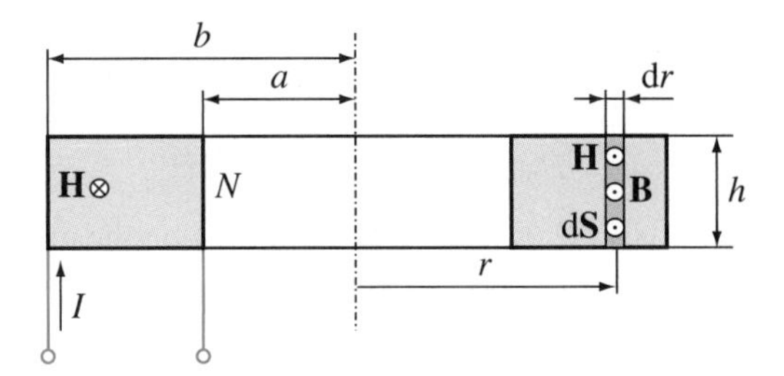

Figure 5.26 Evaluation of the magnetic flux through a cross section of a thick toroidal coil with a linear ferromagnetic core; for Example 5.11.

The magnetic flux through the core is then obtained by integrating the flux density $B(r) = \mu H(r)$ over a cross section of the toroid, as shown in Fig. 5.26,

$$\Phi = \int_{r=a}^{b} B(r)\,\mathrm{d}S = \frac{\mu NIh}{2\pi}\int_{a}^{b}\frac{\mathrm{d}r}{r} = \frac{\mu NIh}{2\pi}\ln\frac{b}{a}, \tag{5.84}$$

where $\mathrm{d}S$ is the surface area of a thin strip of length h and width $\mathrm{d}r$.

(b) Under the assumption of a uniform field distribution in the core, the path of every flux line is assumed to be of the same length, equal to the mean length of the toroid, $l = \pi(a+b)$, and the magnetic field intensity is given by Eq. (5.53). The approximate flux is then obtained by multiplying the constant flux density $B = \mu H$ by the surface area S of the core cross section,

$$\Phi_{\text{approx}} = BS = \mu H(b-a)h = \frac{\mu NIh}{\pi}\frac{b-a}{a+b}. \tag{5.85}$$

The associated relative error in the flux computation is computed as

$$\delta_\Phi = \frac{|\Phi_{\text{approx}} - \Phi|}{\Phi} = \left|\frac{2(b/a-1)}{(1+b/a)\ln(b/a)} - 1\right|. \tag{5.86}$$

For $b - a = 0.1a$, that is, $b/a = 1.1$, the error turns out to be $\delta_\Phi = 0.075\%$, while for $b - a = a$ $(b/a = 2)$, $\delta_\Phi = 3.8\%$. We see that even for a quite thick toroid (the latter case), the approximate expression in Eq. (5.85) is reasonably accurate.

Example 5.12 Thick Toroid with a Nonlinear Ferromagnetic Core

Let the toroidal coil of Fig. 4.18 be wound around a core made from a nonlinear ferromagnetic material. Fig. 5.27(a) shows the cross section of such a structure, with $a = 2$ cm, $b = 4$ cm, $h = 1$ cm, $N = 200$, and $I = 1$ A. Assume that the initial magnetization curve of the material can be approximated by the piece-wise linear curve shown in Fig. 5.27(b). Calculate the magnetic flux through the core (a) rigorously and (b) assuming that the field in the core is uniform.

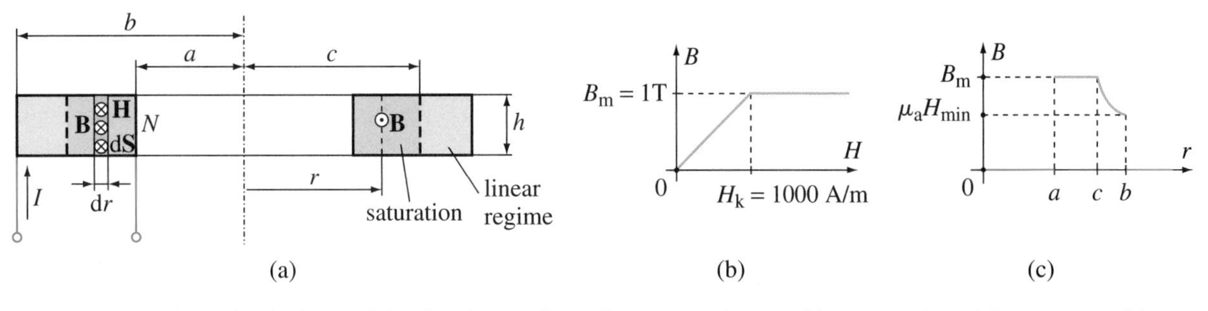

Figure 5.27 Analysis of a thick toroidal coil with a nonlinear ferromagnetic core: (a) cross section of the structure, (b) idealized initial magnetization curve of the material, and (c) distribution of the magnetic flux density along the radial axis; for Example 5.12.

Solution

(a) From Eq. (5.83), the minimum and maximum magnetic field intensities in the core are

$$H_{\text{min}} = H(b) = 796 \text{ A/m} \quad \text{and} \quad H_{\text{max}} = H(a) = 1592 \text{ A/m}, \tag{5.87}$$

respectively. Since the field value $H_{\text{k}} = 1000$ A/m, representing the limit on the magnetization curve in Fig. 5.27(b) up to which the material is in the linear regime and above which is in saturation ("knee" value), satisfies the condition

$$H_{\text{min}} < H_{\text{k}} < H_{\text{max}}, \tag{5.88}$$

we conclude that the part of the core for which $H(r) > H_{\text{k}}$ is in saturation $[B(r) = B_{\text{m}}]$, whereas the remaining part of the core is in the linear regime. In this latter case,

$$B(r) = \mu_{\text{a}} H(r), \quad \mu_{\text{a}} = \frac{B_{\text{m}}}{H_0} = 0.001 \text{ H/m}, \tag{5.89}$$

where μ_{a} is the initial permeability of the material. From the condition

$$H(c) = H_{\text{k}}, \tag{5.90}$$

we obtain the radial distance c that represents the boundary between the two parts of the core:

$$c = \frac{NI}{2\pi H_{\text{k}}} = 3.2 \text{ cm} \tag{5.91}$$

[see Fig. 5.27(a)]. Sketched in Fig. 5.27(c) is the distribution of the magnetic flux density, $B(r)$, along the radial axis.

The magnetic flux through the core is (see Fig. 5.26)

$$\Phi = \int_a^b B(r)\,\text{d}S = \underbrace{B_{\text{m}} h(c-a)}_{\text{saturation}} + \underbrace{\frac{\mu_{\text{a}} NIh}{2\pi} \int_c^b \frac{\text{d}r}{r}}_{\text{linear regime}}$$

$$= B_{\text{m}} h \left[(c-a) + \frac{NI}{2\pi H_{\text{k}}} \ln \frac{b}{c} \right] = 191.4\ \mu\text{Wb}. \tag{5.92}$$

(b) If we assume a uniform field distribution in the core, then the field intensity everywhere in the core equals that for $r = r_{\text{mean}} = (a+b)/2 = 3$ cm. That is, $H = H(r_{\text{mean}}) = 1061$ A/m, which, being greater than H_{k}, implies that the entire core is in saturation. The magnetic flux through the core is hence

$$\Phi_{\text{approx}} = B_{\text{m}}(b-a)h = 200\ \mu\text{Wb}. \tag{5.93}$$

The error relative to the result in Eq. (5.92) is $\delta_\Phi = 4.5\%$, and this is very reasonable given that the core is both thick and nonlinear.

Problems: 5.20; *Conceptual Questions* (on Companion Website): 5.17; *MATLAB Exercises* (on Companion Website).

5.10 KIRCHHOFF'S LAWS FOR MAGNETIC CIRCUITS

With the assumptions made (in the previous section) that the field is restricted to the branches of the magnetic circuit (flux leakage and fringing are negligible) and is uniform in every branch, we now specialize general Maxwell's equations for the magnetostatic field to obtain the laws analogous to Kirchhoff's laws in the electric circuit theory. Thus, applying the law of conservation of magnetic flux, Eq. (4.99), to

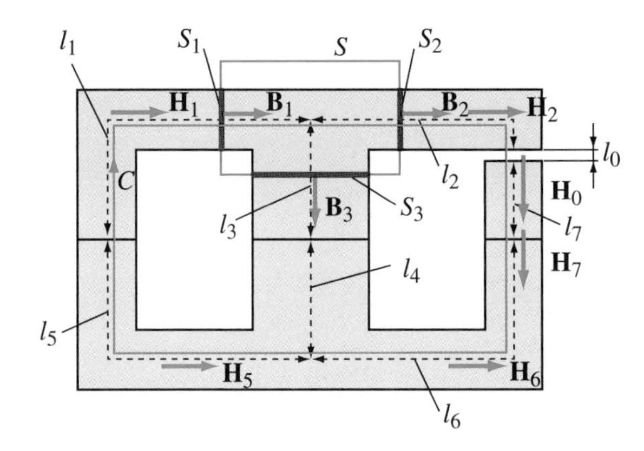

Figure 5.28 A closed surface S about a node and a closed path C along the axes of branches of a magnetic circuit – for the formulation of Kirchhoff's laws for magnetic circuits.

a closed surface S placed about a node (junction of branches) in a magnetic circuit, as shown in Fig. 5.28, yields

$$\oint_S \mathbf{B} \cdot d\mathbf{S} = 0 \quad \longrightarrow \quad \boxed{\sum_{i=1}^{M} B_i S_i = 0,} \tag{5.94}$$

Kirchhoff's "current" law for magnetic circuits

where S_i $(i = 1, 2, \ldots, M)$ are cross-sectional areas of the branches in the junction and B_i $(i = 1, 2, \ldots, M)$ are the magnetic flux densities in these branches.

Applying, on the other hand, the generalized Ampère's law, Eq. (5.49), to a contour C placed along a closed path of flux lines in the circuit (Fig. 5.28) gives

$$\oint_C \mathbf{H} \cdot d\mathbf{l} = I_C \quad \longrightarrow \quad \boxed{\sum_{j=1}^{P} H_j l_j = \sum_{k=1}^{Q} N_k I_k,} \tag{5.95}$$

Kirchhoff's "voltage" law for magnetic circuits

with l_j $(j = 1, 2, \ldots, P)$ standing for the lengths of the branches along the path and H_j $(j = 1, 2, \ldots, P)$ for the magnetic field intensities in the branches, whereas N_k and I_k $(k = 1, 2, \ldots, Q)$ are the numbers of wire turns and current intensities, respectively, in the coils that exist along the path. The product $N_k I_k$, expressed in ampere-turns, is termed a magnetomotive force (mmf), in analogy to an electromotive force (emf) in electric circuits.

Eqs. (5.94) and (5.95) are referred to as Kirchhoff's laws for magnetic circuits. In addition to these circuital laws, we need the "element laws" describing individual parts of the circuit, analogous to current-voltage characteristics for elements (e.g., resistors) in the analysis of electric circuits. These laws are the relationships $B = B(H)$, i.e., the magnetization curves, for the branches of the circuit, including air gaps (where $B = \mu_0 H$). Because of the nonlinear nature of the ferromagnetic portions of the magnetic circuit, and nonlinearity of magnetization curves, the analysis of magnetic circuits often resembles the analysis of nonlinear electric circuits which contain diodes and other elements with nonlinear current-voltage characteristics. This is, at the same time, the most important difference between the analysis of magnetic circuits and the electric circuit theory, which primarily deals with linear electric circuits.

In cases where the ferromagnetic materials in the circuit can be considered as linear, however, an equivalent electric circuit with linear resistors and time-invariant voltage generators can be introduced. By solving the equivalent circuit using some of the standard circuit-theory techniques, we obtain thus all required quantities for the original magnetic circuit. As an illustration, note that the magnetic flux through a simple magnetic circuit in Fig. 5.29(a) can be expressed, having in mind Eq. (5.53), as

$$\Phi = BS = \mu HS = \frac{\mu NIS}{l} = \frac{NI}{\mathcal{R}}, \tag{5.96}$$

where

reluctance (unit: H^{-1})

$$\boxed{\mathcal{R} = \frac{l}{\mu S}} \tag{5.97}$$

is the so-called reluctance of the core. It is defined generally as the ratio of the magnetomotive force (ampere-turns) to the flux. The SI unit for reluctance is ampere per weber (A/Wb) or inverse henry (H^{-1}). We note that the final expression for Φ in Eq. (5.96) is analogous to the expression for the current in a simple electric circuit in which an ideal voltage source of emf NI is connected with a resistor of resistance $\mathcal{R}$, as indicated in Fig. 5.29(b). Eq. (5.97), moreover, has the same form as the expression for the resistance of a resistor with uniform cross section and conductivity σ in Eq. (3.85). The concept of reluctance can be used for the analysis of arbitrary linear magnetic circuits, where the equivalent electric circuit is obtained by replacing the individual parts of the core and the air gaps by resistors with resistances calculated from Eq. (5.97) and representing the coils by voltage generators with electromotive forces equal to the magnetomotive forces NI.

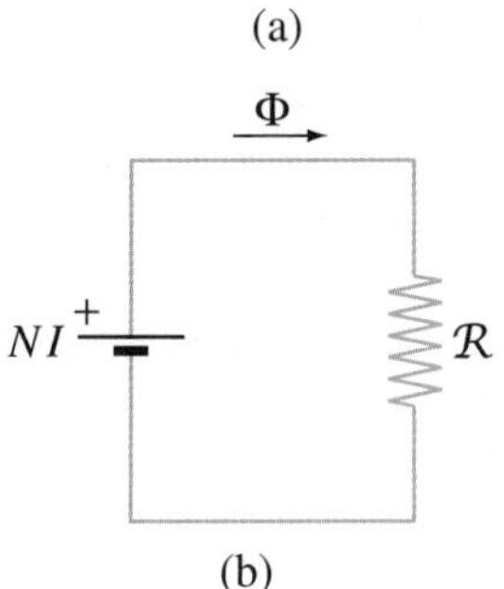

Figure 5.29 Reluctance concept: a simple linear magnetic circuit (a) and the equivalent electric circuit (b).

Example 5.13 Simple Nonlinear Magnetic Circuit with an Air Gap

Consider a magnetic circuit consisting of a thin toroidal ferromagnetic core with a coil and an air gap shown in Fig. 5.30(a). The coil has $N = 1000$ turns of wire with the total resistance $R = 50\ \Omega$. The length of the ferromagnetic portion of the circuit is $l = 1$ m, the thickness (width) of the gap is $l_0 = 4$ mm, the cross-sectional area of the toroid is $S = S_0 = 5\ \mathrm{cm}^2$, and the emf of the generator in the coil circuit is $\mathcal{E} = 200$ V. The idealized initial magnetization curve of the material is given in Fig. 5.30(b). Find the magnetic field intensities in the core and in the air gap.

Solution We adopt standard approximations for the analysis of magnetic circuits and neglect the field nonuniformity in the ferromagnetic core, as well as the flux leakage from the core and fringing around the air gap edges (Section 5.9). Let (B, H) and (B_0, H_0) designate the magnetic flux density and field intensity in the core and the gap, respectively, as

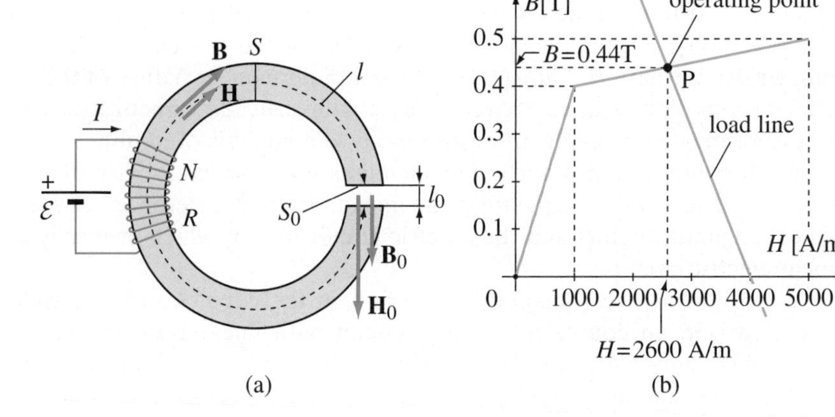

Figure 5.30 Analysis of a simple nonlinear magnetic circuit with an air gap: (a) circuit geometry, with flux density and field intensity vectors in the circuit, and (b) idealized initial magnetization curve of the material, with the load line and operating point for the circuit; for Example 5.13.

indicated in Fig. 5.30(a). From Kirchhoff's laws for magnetic circuits, Eqs. (5.94) and (5.95), we have

$$BS = B_0 S_0 \quad \longrightarrow \quad B = B_0 \tag{5.98}$$

and

$$Hl + H_0 l_0 = NI, \tag{5.99}$$

where $I = \mathcal{E}/R = 4$ A is the current intensity of the coil. The above equations, combined with the constitutive equation for the air gap,

$$B_0 = \mu_0 H_0, \tag{5.100}$$

result in the following relationship between the flux density and field intensity in the core:

$$Hl + B\frac{l_0}{\mu_0} = NI. \tag{5.101}$$

With the numerical data substituted,

$$H + 3183B = 4000 \quad (H \text{ in A/m}; \quad B \text{ in T}), \tag{5.102}$$

and this represents the equation of the load line for the magnetic circuit (the locus of all possible combinations of values B and H for this given circuit configuration and excitation, not yet taking into account the characteristics of the core material). Upon plotting this line in the BH chart [Fig. 5.30(b)], we conclude that its intersection with the magnetization curve of the core material belongs to the second part of the curve (hard magnetization section – also see Fig. 5.20). This intersection (point P) represents the operating point for the circuit, i.e., it determines the actual position of the ferromagnetic core on the magnetization curve for the given circuit configuration and excitation (NI). To find numerically the abscissa and ordinate of the point P, we solve the system of equations composed of the load line equation, Eq. (5.102), and the equation of the line describing the hard magnetization segment of the magnetization curve, read from Fig. 5.30(b),

$$B = 0.4 + 2.5 \times 10^{-5}(H - 1000), \tag{5.103}$$

and what we get is $B = 0.44$ T and $H = 2600$ A/m. By means of Eq. (5.98), $B_0 = 0.44$ T as well, and Eq. (5.100) then gives $H_0 = 350$ kA/m (field intensity in the air gap). We see that $H_0 \gg H$. This can also be concluded without actually solving the circuit, by introducing the permeability of the material at the point P, $\mu = B/H = 1.7 \times 10^{-4}$ H/m, into Eq. (5.98) and expressing the magnetic field intensity in the gap as

$$H_0 = \frac{\mu}{\mu_0} H = 135.3H. \tag{5.104}$$

(Note that μ here is a function of H, not a constant.) It is typical for magnetic circuits with air gaps, in general, that the magnetic field intensities in the gaps are much larger than those in the ferromagnetic material.

Example 5.14 Simple Linear Magnetic Circuit with an Air Gap

Assume that the core of the circuit of Fig. 5.30(a) is made from a linear ferromagnetic material of relative permeability $\mu_r = 1000$, and compute the magnetic flux through the core.

Solution Combining Eq. (5.101) with the relationship $B = \mu H$, we can write now

$$\Phi = BS = NI\left(\frac{l}{\mu S} + \frac{l_0}{\mu_0 S_0}\right)^{-1} = \frac{NI}{\mathcal{R} + \mathcal{R}_0}, \tag{5.105}$$

where the reluctances [Eq. (5.97)] of the core and the gap are

$$\mathcal{R} = \frac{l}{\mu_r \mu_0 S} = 1.6 \times 10^6 \text{ H}^{-1} \quad \text{and} \quad \mathcal{R}_0 = \frac{l_0}{\mu_0 S_0} = 6.37 \times 10^6 \text{ H}^{-1}, \tag{5.106}$$

respectively, and the equivalent electric circuit is shown in Fig. 5.31. Hence, the flux through the core amounts to $\Phi = 5 \times 10^{-4}$ Wb.

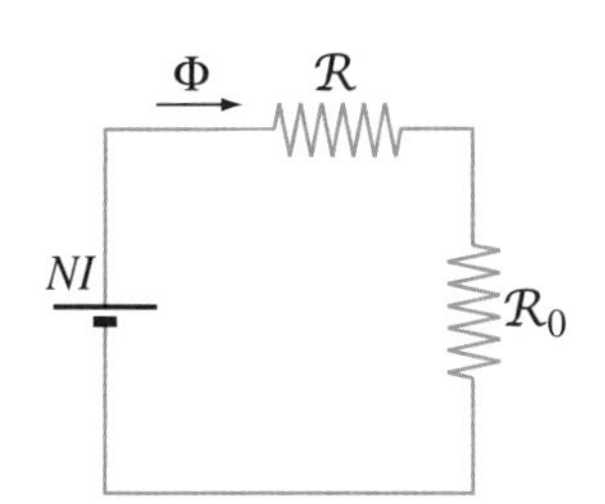

Figure 5.31 Equivalent electric circuit for the magnetic circuit of Fig. 5.30(a), assuming that the ferromagnetic material is linear; for Example 5.14.

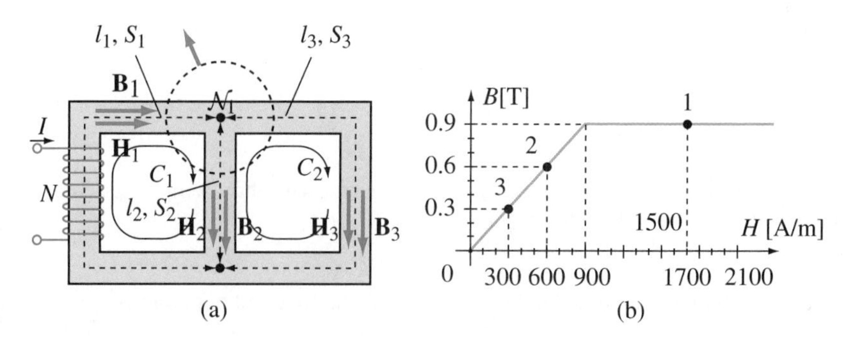

Figure 5.32 Analysis of a nonlinear magnetic circuit with three branches: (a) circuit geometry, with the adopted independent node and closed paths in the circuit, and (b) idealized initial magnetization curve of the material, with computed operating points for the branches; for Example 5.15.

Example 5.15 Nonlinear Magnetic Circuit with Three Branches

Dimensions of the magnetic circuit shown in Fig. 5.32(a) are $l_1 = l_3 = 2l_2 = 20$ cm and $S_1 = S_2 = S_3 = 1$ cm^2. The mmf in the first branch of the circuit is $NI = 400$ ampere-turns. The initial magnetization curve of the core can be linearized as in Fig. 5.32(b). Calculate the magnetic flux densities and field intensities in the three branches of the circuit.

Solution Let us orient the branches of the circuit as in Fig. 5.32(a). There is a total of two nodes and three closed paths in the circuit. Thus, in complete analogy with the analysis of electric circuits, Kirchhoff's "current" and "voltage" laws for magnetic circuits, Eqs. (5.94) and (5.95), are to be applied to one node (independent node) and two closed paths (independent closed paths), respectively. We choose the node $\mathcal{N}_1$ and closed paths C_1 and C_2 in Fig. 5.32(a). The corresponding equations are:

$$-B_1S_1 + B_2S_2 + B_3S_3 = 0 \quad \longrightarrow \quad B_1 = B_2 + B_3, \tag{5.107}$$

$$H_1l_1 + H_2l_2 = NI, \tag{5.108}$$

$$-H_2l_2 + H_3l_3 = 0. \tag{5.109}$$

Depending on whether the operating points for the individual branches of the circuit belong to the linear part or to the saturation part of the magnetization curve in Fig. 5.32(b), we may have

$$B_i = \begin{cases} \mu_a H_i & \text{for } H_i \leq 900 \text{ A/m} \\ 0.9 \text{ T} & \text{for } H_i > 900 \text{ A/m} \end{cases}, \quad i = 1, 2, 3, \tag{5.110}$$

where the initial permeability of the material is $\mu_a = 0.9/900$ H/m $= 0.001$ H/m. So, there exist a total of eight combinations for the magnetization stages of the branches and only one of them is true.

Suppose, first, that none of the magnetizations in the branches is in saturation ($B = \mu_a H$ in all three branches). Eq. (5.107) then becomes

$$H_1 = H_2 + H_3, \tag{5.111}$$

and the solution of the system with Eqs. (5.111), (5.108), and (5.109) is $H_1 = 1500$ A/m, $H_2 = 1000$ A/m, and $H_3 = 500$ A/m. This is contradictory to the assumption of linearity in all three branches, since both H_1 and H_2 appear to be larger than 900 A/m. We conclude that our initial guess is not correct, and the circuit has therefore to be solved again, with another assumption, i.e., that some of the branches are in saturation.

To reduce the number of additional trials (out of remaining seven combinations) to a minimum, we note first that the adopted reference directions of the magnetic flux density vectors in Fig. 5.32(a) likely reflect the actual flux flow in the branches, meaning that B_1, B_2, and B_3 are all positive. Eq. (5.107) then tells us that B_1 must be larger than both B_2 and B_3 individually, so it is logical to expect that the branch with the coil would first reach saturation ($B_1 = 0.9$ T). The other two branches, however, must remain in the linear regime ($B_2 = \mu_a H_2$ and $B_3 = \mu_a H_3$), because $B_2 = 0.9$ T would imply $B_3 = 0$ and vice versa, which is impossible. With this, Eq. (5.107) becomes

$$H_2 + H_3 = 900\ \text{A/m}, \tag{5.112}$$

and the solution of the new system of three equations is $H_1 = 1700$ A/m, $H_2 = 600$ A/m, and $H_3 = 300$ A/m. Obviously, these values are consistent with the new assumption made about the magnetization stages in the branches, and this is the true solution for the circuit. The actual positions of the operating points for the branches on the magnetization curve are indicated in Fig. 5.32(b). The remaining two flux densities in the circuit are $B_2 = 0.6$ T and $B_3 = 0.3$ T.

Example 5.16 Reverse Problem in the Magnetic Circuit Analysis

For the magnetic circuit shown in Fig. 5.33, $l_1 = l_3 = 20$ cm, $l_2 = l_2' + l_2'' = 10$ cm, $l_0 = 1$ mm, $S_1 = S_3 = 1$ cm^2, $S_2 = 2$ cm^2, and $N_2 I_2 = 1000$ A turns. The core is made from the material whose initial magnetization curve can be approximated analytically by the function

$$B = \frac{1.5H}{500 + H}, \quad H \geq 0 \quad (H \text{ in A/m}; \quad B \text{ in T}). \tag{5.113}$$

If the magnetic flux in the central branch of the circuit is $\Phi_2 = 200\ \mu$Wb with respect to the downward reference direction, find $N_1 I_1$.

Solution This is a reverse problem in the magnetic circuit analysis: for a given response (magnetic flux in one or more branches of the circuit), find the unknown excitation (one or more magnetomotive forces) that produces it. In the analysis of magnetic circuits with nonlinear magnetization curves (given either graphically or analytically), reverse problems are generally much simpler to solve than direct problems, because in the latter case we have to simultaneously solve the full set of equations written in accordance to Kirchhoff's circuital laws together with nonlinear material characteristics for individual parts of the circuit. In reverse problems, on the other hand, we start with the given field quantities for one or more branches and then applying the circuital and material equations one at a time we solve for other field quantities, one by one, and for the required magnetomotive forces.

With reference to the notation in Fig. 5.33, the flux densities in the ferromagnetic section of the second (central) branch of the circuit (B_2) and in the air gap (B_0) are

$$B_2 = B_0 = \frac{\Phi_2}{S_2} = 1\ \text{T}. \tag{5.114}$$

From the magnetization curve, i.e., by solving the magnetization equation, Eq. (5.113), for the corresponding field intensity in the material,

$$H_2 = \frac{500B_2}{1.5 - B_2} = 1\ \text{kA/m}, \tag{5.115}$$

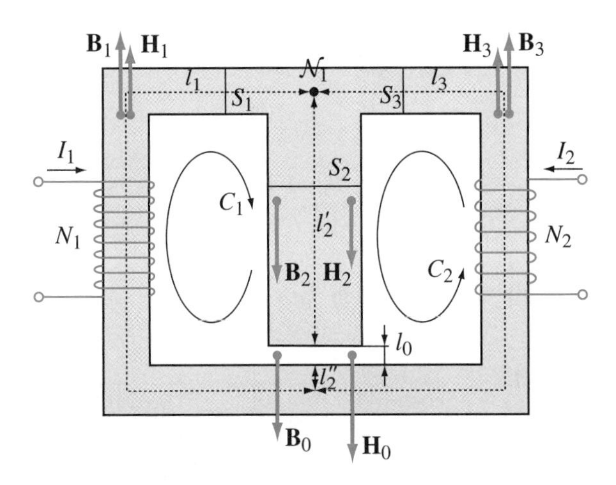

Figure 5.33 Magnetic circuit with a nonlinear magnetization curve given analytically; for Example 5.16.

whereas in the gap,

$$H_0 = \frac{B_0}{\mu_0} = 795.8 \text{ kA/m}. \tag{5.116}$$

Kirchhoff's "voltage" law for the closed path C_2 in Fig. 5.33 now yields

$$H_3 = \frac{N_2 I_2 - H_2 l_2 - H_0 l_0}{l_3} = 521 \text{ A/m}, \tag{5.117}$$

with the corresponding flux density in the third branch given by

$$B_3 = \frac{1.5 H_3}{500 + H_3} = 0.765 \text{ T}. \tag{5.118}$$

The flux density in the first branch is next obtained using Kirchhoff's "current" law for the node $\mathcal{N}_1$ in Fig. 5.33, and from it the associated H value,

$$B_1 = \frac{B_2 S_2 - B_3 S_3}{S_1} = 1.235 \text{ T} \quad \longrightarrow \quad H_1 = \frac{500 B_1}{1.5 - B_1} = 2.33 \text{ kA/m}. \tag{5.119}$$

Finally, we apply Kirchhoff's "voltage" law to the closed path C_1 and obtain the mmf we seek:

$$N_1 I_1 = H_1 l_1 + H_2 l_2 + H_0 l_0 = 1362 \text{ A turns}. \tag{5.120}$$

Example 5.17 Demagnetization in a Magnetic Circuit

The ferromagnetic core of the circuit shown in Fig. 5.34(a) has the cross-sectional area $S = 1 \text{ cm}^2$, mean length $l = 20$ cm, and air-gap thickness $l_0 = 1$ mm. The number of wire turns of the coil is $N = 1000$, the resistance of the winding is $R = 20\ \Omega$, and the emf of the generator is $\mathcal{E} = 20$ V. The core does not have residual magnetization and the switch K is in the off position (open). By turning the switch on, the mmf is applied to the circuit and the magnetic flux through it rises following the initial magnetization curve of the material. This curve can be considered as linear, as shown in Fig. 5.34(b), where the initial permeability is $\mu_a = 0.001$ H/m. The magnetic flux density in the core becomes $B = B_m$ in the stationary state. The switch is then turned off (opened) and a new stationary state established in the

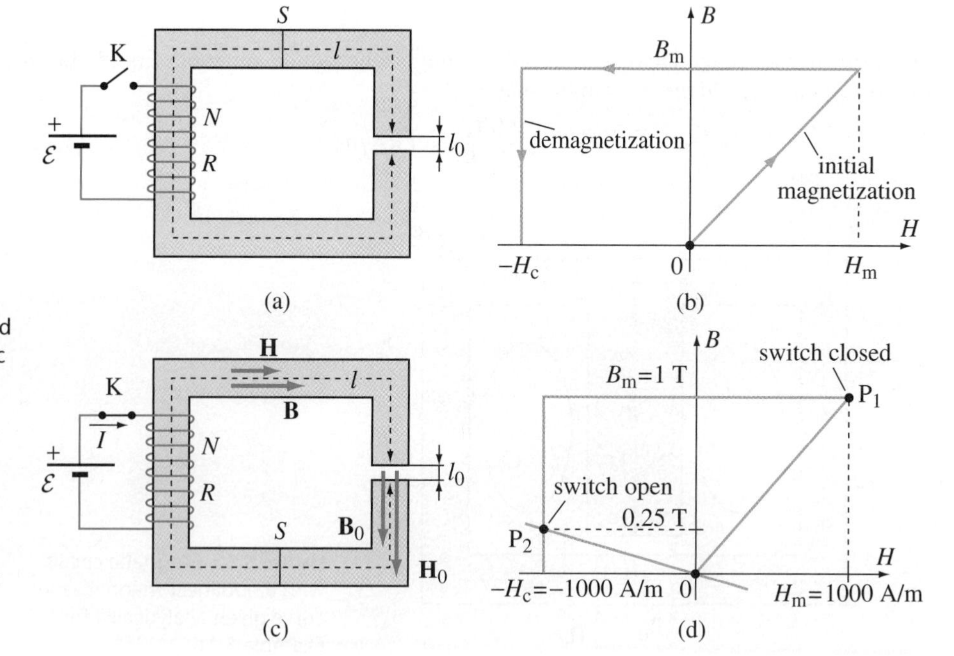

Figure 5.34 Magnetization and demagnetization in a magnetic circuit with an air gap: (a) circuit geometry, (b) idealized initial magnetization and demagnetization curves of the material, (c) magnetic flux density and field intensity vectors in the circuit, and (d) operating points for the circuit in two stationary states; for Example 5.17.

circuit. The demagnetization curve for the material can be approximated by two straight-line segments [Fig. 5.34(b)], where the coercive force is given by $H_c = H_m$. What is the magnetic field intensity in the gap in the new state?

Solution After the switch K is turned on (closed), the current in the winding is [Fig. 5.34(c)] $I = \mathcal{E}/R = 1$ A, and the relationship between the flux density B and field intensity H in the core is the same as in Eq. (5.101). Moreover, as

$$B = \mu_a H, \tag{5.121}$$

we obtain [see also Eq. (5.105)]

$$B = NI\left(\frac{l}{\mu_a} + \frac{l_0}{\mu_0}\right)^{-1} = 1 \text{ T}, \tag{5.122}$$

and this is the maximum value for the flux density, $B = B_m = 1$ T. Hence, $H_m = B_m/\mu_a = 1$ kA/m.

After the switch K is opened, Eq. (5.101) becomes

$$Hl + B\frac{l_0}{\mu_0} = 0, \tag{5.123}$$

and this represents the equation of the load line for the magnetic circuit in this state. We suppose first that the intersection of this line with the demagnetization curve in Fig. 5.34(b) belongs to the horizontal segment of the curve, i.e., that $B = B_m = 1$ T. From Eq. (5.123), $H = -4$ kA/m, which is impossible because $H < -H_c$ ($H_c = H_m = 1$ kA/m). We conclude thus that the operating point for the circuit belongs to the vertical segment of the demagnetization curve, as indicated in Fig. 5.34(d). The magnetic field intensity in the core is therefore $H = -H_c = -1$ kA/m. Eq. (5.123) then gives the corresponding flux density in the core, $B = 0.25$ T. The magnetic field intensity in the gap is $H_0 = B/\mu_0 = 200$ kA/m.

Example 5.18 Complex Linear Magnetic Circuit

For the magnetic circuit shown in Fig. 5.35(a), $l_1 = l_2 = 10$ cm, $l_3 = 5$ cm, $l_0 = 1$ mm, $N_1 = 1000$, $N_2 = 500$, and $I_1 = I_2 = 1$ A. The ferromagnetic material from which the core is made can be considered as linear, with relative permeability $\mu_r = 1000$. The cross-sectional area of all parts of the core is $S = 1$ cm^2. Find the magnetic field intensity in the air gap.

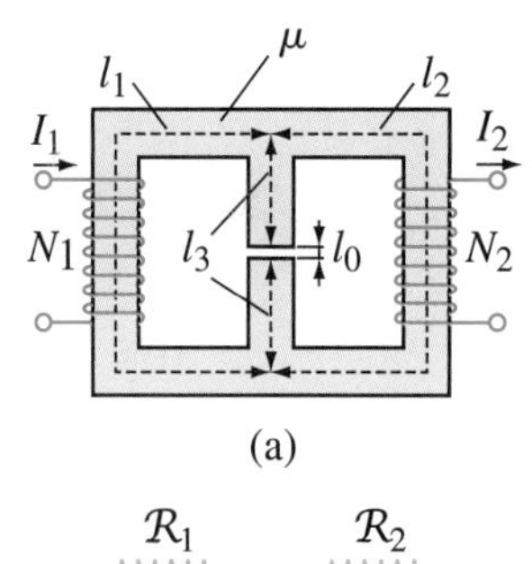

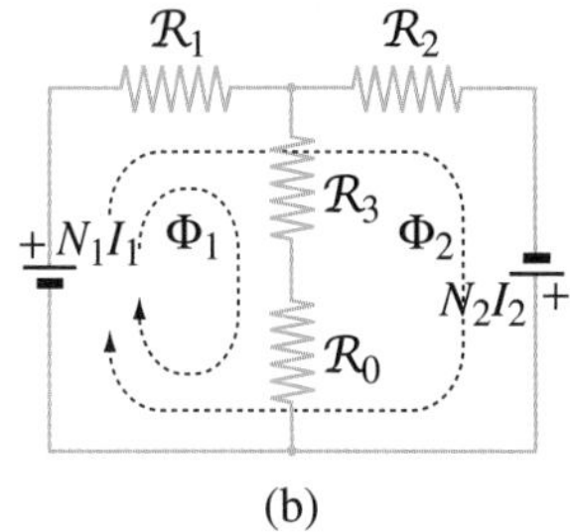

Figure 5.35 Analysis of a linear magnetic circuit with three branches: (a) circuit geometry and (b) equivalent electric circuit for loop analysis; for Example 5.18.

Solution Fig. 5.35(b) shows the equivalent electric circuit for the problem, with reluctances, Eq. (5.97),

$$\mathcal{R}_1 = \mathcal{R}_2 = 2\mathcal{R}_3 = \frac{l_1}{\mu_r\mu_0 S} = 8 \times 10^5 \text{ H}^{-1}, \quad \mathcal{R}_0 = \frac{l_0}{\mu_0 S} = 8 \times 10^6 \text{ H}^{-1}. \tag{5.124}$$

The figure also shows the adopted loops for the loop analysis of the (electric) circuit. The corresponding loop equations are

$$(\mathcal{R}_1 + \mathcal{R}_3 + \mathcal{R}_0)\Phi_1 + \mathcal{R}_1\Phi_2 = N_1I_1, \tag{5.125}$$

$$\mathcal{R}_1\Phi_1 + (\mathcal{R}_1 + \mathcal{R}_2)\Phi_2 = N_1I_1 + N_2I_2, \tag{5.126}$$

and their solution for the flux through the central branch of the circuit

$$\Phi_1 = \frac{\mathcal{R}_2N_1I_1 - \mathcal{R}_1N_2I_2}{\mathcal{R}_0\mathcal{R}_1 + \mathcal{R}_0\mathcal{R}_2 + \mathcal{R}_1\mathcal{R}_2 + \mathcal{R}_1\mathcal{R}_3 + \mathcal{R}_2\mathcal{R}_3} = 2.84 \times 10^{-5} \text{ Wb}. \tag{5.127}$$

The magnetic field intensity in the gap comes out to be

$$H_0 = \frac{\Phi_1}{\mu_0 S} = 226 \text{ kA/m}. \tag{5.128}$$

Problems: 5.21–5.27; *Conceptual Questions* (on Companion Website): 5.18–5.21; *MATLAB Exercises* (on Companion Website).

5.11 MAXWELL'S EQUATIONS FOR THE TIME-INVARIANT ELECTROMAGNETIC FIELD

Maxwell's equations for the magnetostatic field, Eqs. (5.70), represent a general mathematical model for determining the magnetic field ($\mathbf{H}$) from a distribution of steady electric currents ($\mathbf{J}$), which is considered to be known. The distribution of currents, however, for the geometry, material properties, and excitation of a given structure can be obtained from Eqs. (3.59), which also yield the solution for the electric field ($\mathbf{E}$) in the system. In addition, Eqs. (3.60) provide a means for evaluating the associated distribution of charges (ρ) in the system from the electric field distribution. All these equations together represent a full set of Maxwell's equations that govern both the electric and the magnetic field due to steady electric currents. These two fields, moreover, can be considered as components of a more complex field – the electromagnetic field – produced by steady currents. The combined field is called the time-invariant electromagnetic field, and we summarize here the four Maxwell's equations in differential form for this field:

Maxwell's first equation, static field
Maxwell's second equation, static field
Maxwell's third equation
Maxwell's fourth equation

$$\begin{cases} \nabla \times \mathbf{E} = 0 \\ \nabla \times \mathbf{H} = \mathbf{J} \\ \nabla \cdot \mathbf{D} = \rho \\ \nabla \cdot \mathbf{B} = 0 \end{cases} . \qquad (5.129)$$

What is very important, the electric and magnetic fields constituting the time-invariant electromagnetic field are entirely unrelated and independent from each other, and can be analyzed separately, as we have done in Chapters 3 (electric part) and 5 (magnetic part). This is not the case, however, with the time-varying electromagnetic field, as we shall see in the following chapters. Under nonstatic conditions, the third and fourth Maxwell's equations retain the same form as in Eqs. (5.129), whereas the first two equations have additional terms on the right-hand side of equations, which are responsible for the coupling between the electric and magnetic fields that change in time. In chapters that follow, the electric and magnetic fields will always be treated together, as related integral parts of the time-varying electromagnetic field.

Example 5.19 Continuity Equation from Ampère's Law

Starting from Maxwell's second differential equation for the time-invariant electromagnetic field derive the corresponding form of the continuity equation.

Solution Let us take the divergence of both sides of the differential form of Maxwell's second equation (generalized Ampère's law) for static fields, in Eqs. (5.129):

$$\nabla \cdot (\nabla \times \mathbf{H}) = \nabla \cdot \mathbf{J}. \qquad (5.130)$$

The second-order vector derivative on the left-hand side of this equation is equal to zero for any vector field,[2] so the right-hand side of the equation must be zero as well, $\nabla \cdot \mathbf{J} = 0$, and

[2]The divergence of the curl of an arbitrary vector function ($\mathbf{A}$) is always zero, or, in terms of the del operator,

$$\nabla \cdot (\nabla \times \mathbf{A}) = 0.$$

This is obvious if $\mathbf{A}$ is the magnetic vector potential, because then $\mathbf{B} = \nabla \times \mathbf{A}$ [Eq. (4.116)] and $\nabla \cdot \mathbf{B} = 0$ [Eq. (4.103)]. To prove it for $\mathbf{A}$ representing any vector field, however, we apply formally the identity

$$\mathbf{a} \cdot (\mathbf{b} \times \mathbf{c}) = (\mathbf{a} \times \mathbf{b}) \cdot \mathbf{c}$$

this is exactly the differential form of the continuity equation under the static assumption, Eq. (3.41). We see now that Eq. (3.41) is actually included implicitly in the set of Maxwell's equations in Eqs. (5.129).

Problems: 5.28; *MATLAB Exercises* (on Companion Website).

Problems

5.1. **Nonuniformly magnetized parallelepiped.** A rectangular ferromagnetic parallelepiped is situated in air in the first octant of the Cartesian coordinate system ($x, y, z \geq 0$), with one vertex at the coordinate origin, and the edges, of lengths a, b, and c, parallel to coordinate axes x, y, and z, respectively. The magnetization vector in the parallelepiped is given by $\mathbf{M}(x, y, z) = M_0[\sin(\pi y/b)\,\hat{\mathbf{x}} + \sin(\pi z/c)\,\hat{\mathbf{y}} + \sin(\pi x/a)\,\hat{\mathbf{z}}]$, where M_0 is a constant. Compute (a) the volume magnetization current density vector in the parallelepiped and (b) the surface magnetization current density vector over its sides.

5.2. **Hollow cylindrical bar magnet.** A hollow cylindrical bar magnet of radii a and b ($a < b$), and length l, is permanently magnetized with a uniform magnetization and situated in air. The magnetization vector, of magnitude M, is parallel to the bar axis. Find (a) the distribution of magnetization currents of the magnet and (b) the magnetic flux density vector along the axis.

5.3. **Uniformly magnetized square ferromagnetic plate.** A uniformly magnetized square ferromagnetic plate of side length a and thickness d ($d \ll a$) is situated in air. With reference to the coordinate system in Fig. 5.36, the magnetization vector in the plate is given by $\mathbf{M} = M_0\,\hat{\mathbf{z}}$, where M_0 is a constant. Determine the magnetic flux density vector at an arbitrary point along the z-axis.

5.4. **Magnetization parallel to plate faces.** Consider the square ferromagnetic plate in Fig. 5.36, and assume that the magnetization vector is $\mathbf{M} = M_0\,\hat{\mathbf{y}}$ ($M_0 =$ const). Neglecting end effects, find the magnetic flux density vector at the plate center (point O).

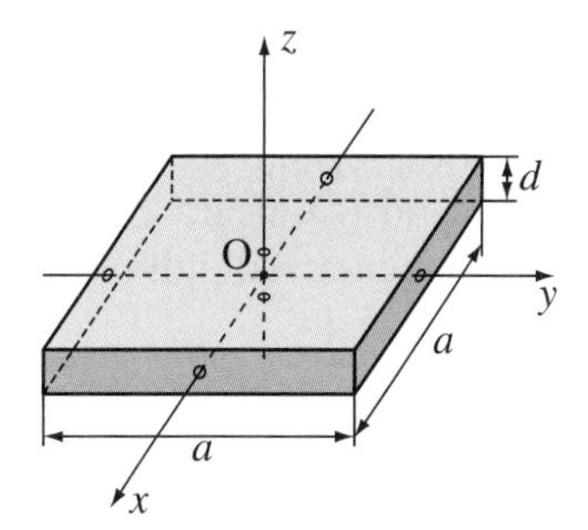

Figure 5.36 Uniformly magnetized square ferromagnetic plate; for Problem 5.3.

5.5. **Nonuniformly magnetized ferromagnetic disk.** A thin ferromagnetic disk of radius a and thickness d ($d \ll a$) in air has a nonuniform magnetization, given by $\mathbf{M} = M_0(r/a)^2\,\hat{\mathbf{z}}$ (Fig. 5.37), where M_0 is a constant. Calculate (a) the distribution of magnetization currents of the disk and (b) the magnetic flux density vector along the z-axis.

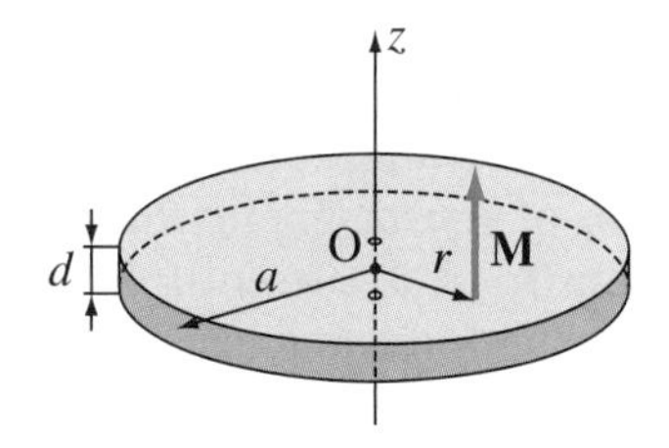

Figure 5.37 Nonuniformly magnetized thin ferromagnetic disk; for Problem 5.5.

5.6. **Infinite cylinder with circular magnetization.** An infinitely long ferromagnetic cylinder of radius a in air has a nonuniform magnetization.

(in a scalar triple product, the cyclic permutation of the order of the three vectors does not change the result) to the scalar triple product $\nabla \cdot (\nabla \times \mathbf{A})$, and obtain

$$\nabla \cdot (\nabla \times \mathbf{A}) = (\nabla \times \nabla) \cdot \mathbf{A} = 0$$

(cross product of a vector with itself is always zero). Alternatively, the divergence, using Eq. (1.167), of the expression for the curl (of any vector) in the Cartesian coordinate system in Eq. (4.81) turns out to be zero.

In a cylindrical coordinate system whose z-axis coincides with the cylinder axis, $\mathbf{M} = M_0(r/a)\hat{\boldsymbol{\phi}}$ $(0 \le r \le a)$, where M_0 is a constant. Find (a) the volume magnetization current density vector in the cylinder, (b) the surface magnetization current density vector on the cylinder surface, (c) the magnetic flux density vector in the cylinder, and (d) the magnetic flux density vector outside the cylinder.

5.7. **Magnetic field intensity vector.** Find the magnetic field intensity vector, $\mathbf{H}$, (a) along the axis of the magnetized disk from Example 5.2, (b) at the center of the cylindrical bar magnet from Example 5.3, and (c) inside and outside the nonuniformly magnetized infinitely long cylinder from Example 5.4.

5.8. **Total (conduction plus magnetization) current density.** The magnetic flux density vector, $\mathbf{B}$, in a ferromagnetic material is a known function of spatial coordinates. (a) Prove that the total (conduction plus magnetization) volume current density in the material, $\mathbf{J}_{\mathrm{tot}} = \mathbf{J} + \mathbf{J}_{\mathrm{m}}$, can be obtained as $\mathbf{J}_{\mathrm{tot}} = \nabla \times \mathbf{B}/\mu_0$. (b) Specifically, compute $\mathbf{J}_{\mathrm{tot}}$ for $\mathbf{B}$ given as the following function of Cartesian coordinates: $\mathbf{B}(x, y, z) = \{[(2x + z)/y^2]\,\hat{\mathbf{x}} + (2/y)\,\hat{\mathbf{y}} + (x + 2y)\,\hat{\mathbf{z}}\}$ T (x, y, z in m).

5.9. **Constant flux density vector in a magnetic region.** In a certain magnetic region, the magnetic flux density vector does not vary with spatial coordinates. The conduction volume current density vector is $\mathbf{J}$. Find the magnetization volume current density vector, $\mathbf{J}_{\mathrm{m}}$.

5.10. **Closed path in a uniform field.** Considering an arbitrary contour in a magnetic field with no spatial variation of the magnetic flux density vector ($\mathbf{B} = \text{const}$), in a current-free ($\mathbf{J}_{\mathrm{tot}} = 0$) region, prove the following vector identity: $\oint_C \mathrm{d}\mathbf{l} = 0$.

5.11. **Circulation of the magnetic flux density vector.** The magnetization vector, $\mathbf{M}$, and conduction current density vector, $\mathbf{J}$, are known at every point of a magnetic body. Find the expression for the circulation of the magnetic flux density vector along a closed path C situated entirely inside the body.

5.12. **Total magnetization and conduction current.** Consider an imaginary closed path C inside a homogeneous magnetic material of relative permeability μ_{r}. The total conduction current enclosed by C is I_C. What is the total magnetization current $I_{\mathrm{m}C}$ enclosed by C?

5.13. **Thin toroidal coil with a linear ferromagnetic core.** A coil with N turns of wire is wound uniformly and densely over a thin toroidal ferromagnetic core of permeability μ. If a steady current of intensity I is established in the coil, find the circulation of (a) the magnetic field intensity vector, (b) the magnetic flux density vector, and (c) the magnetization vector through the core, along its mean length.

5.14. **Solenoidal coil with a linear ferromagnetic core.** There is a uniform and dense solenoidal winding wound over a very long cylindrical ferromagnetic core of permeability μ. The number of wire turns per unit length of the core is N'. A steady current of intensity I flows through the winding. Determine (a) the magnetic field intensity vector, (b) the magnetic flux density vector, (c) the magnetization vector, and (d) the volume magnetization current density vector in the core, as well as (e) the surface magnetization current density vector over the surface of the core.

5.15. **Ferromagnetic cylinder with a conduction current.** A very long ferromagnetic cylinder of radius a and relative permeability μ_{r} is situated in air. There is a uniform conduction current of density J flowing along the cylinder. Find the distribution of (a) the magnetic field intensity vector, (b) the magnetic flux density vector, and (c) the magnetization vector inside and outside the cylinder, as well as (d) the distribution of magnetization currents of the cylinder.

5.16. **Magnetic-magnetic boundary conditions.** Assume that the plane $z = 0$ separates medium 1 ($z > 0$) and medium 2 ($z < 0$), with relative permeabilities $\mu_{\mathrm{r}1} = 600$ and $\mu_{\mathrm{r}2} = 250$, respectively. The magnetic field intensity vector in medium 1 near the boundary (for $z = 0^+$) is $\mathbf{H}_1 = (5\,\hat{\mathbf{x}} - 3\,\hat{\mathbf{y}} + 2\,\hat{\mathbf{z}})$ A/m. Calculate the magnetic field intensity vector in medium 2 near the boundary (for $z = 0^-$), $\mathbf{H}_2$, if (a) no conduction current exists on the boundary ($\mathbf{J}_{\mathrm{s}} = 0$) and (b) there is a surface current of density $\mathbf{J}_{\mathrm{s}} = 3\,\hat{\mathbf{y}}$ A/m on the boundary.

5.17. **Force on a conductor above an infinite PMC corner.** Two PMC ($\mu_{\mathrm{r}} \to \infty$) half-planes are connected together at an angle of 90° with respect to each other, as shown in Fig. 5.38. A very long and thin wire in air runs parallel to the half-planes, at a distance h from each of them. If a steady current of intensity I is established in the wire, find the per-unit-length magnetic force on it.

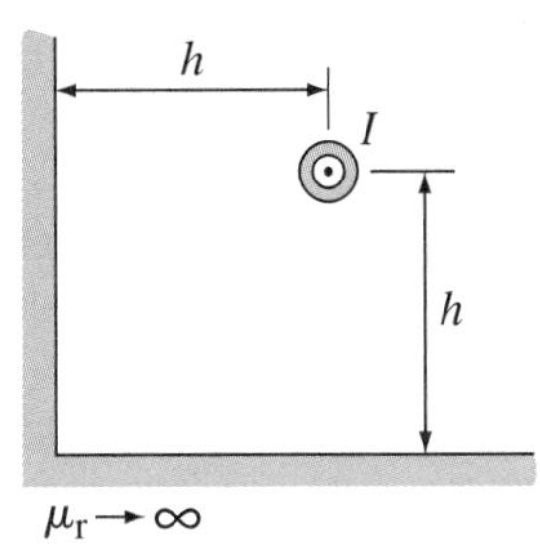

Figure 5.38 Cross section of a system consisting of a current-carrying wire and an infinitely large 90° PMC corner; for Problem 5.17.

5.18. **Uniformly magnetized hollow disk on a PMC plane.** A uniformly magnetized hollow ferromagnetic disk surrounded by air is lying on a PMC plane. The magnetization vector is $\mathbf{M}$, and it is normal to the plane, as shown in Fig. 5.39. The disk radii are a and b, and thickness is d ($d \ll a, b$). Obtain the magnetic flux density vector along the z-axis for $z > 0$.

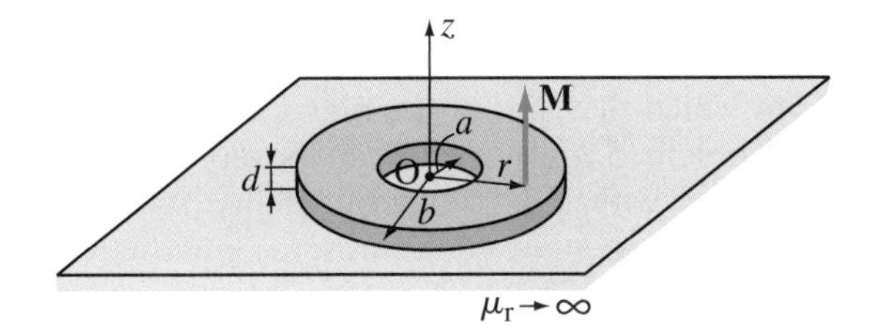

Figure 5.39 Hollow ferromagnetic disk with a uniform magnetization lying on a PMC plane; for Problem 5.18.

5.19. **Magnetized cylinder between two PMC planes.** A ferromagnetic cylinder of radius a and height h is placed between two parallel PMC planes, as portrayed in Fig. 5.40. The medium around the cylinder is air. The cylinder is uniformly magnetized throughout its volume, with the magnetization vector given as $\mathbf{M} = M_0\,\hat{\mathbf{z}}$, where M_0 is a constant. Find the magnetic flux density vector in the space between the PMC planes, both inside and outside the cylinder.

5.20. **Magnetic flux through a thick toroid.** Repeat Example 5.12 but for the piece-wise linear approximation of the initial magnetization curve of the core material given in Fig. 5.30(b).

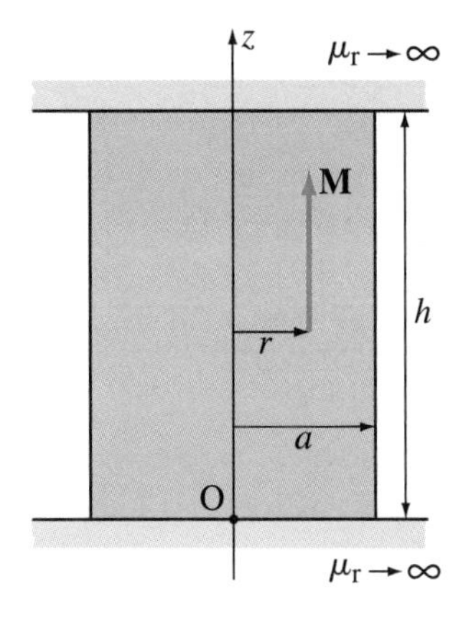

Figure 5.40 Uniformly magnetized cylinder between two parallel PMC planes; for Problem 5.19.

5.21. **Simple nonlinear magnetic circuit.** Fig. 5.41 shows a magnetic circuit consisting of a thin magnetic core of length $l = 40$ cm and cross-sectional area $S = 2.25$ cm^2 and an air gap of thickness $l_0 = 0.25$ mm. The winding has $N = 800$ turns of wire with a steady current of intensity $I = 1$ A. The core is made from a nonlinear ferromagnetic material whose initial magnetization curve can be linearized in parts as in Fig. 5.27(b). Find the magnetic field intensities in the core and in the air gap.

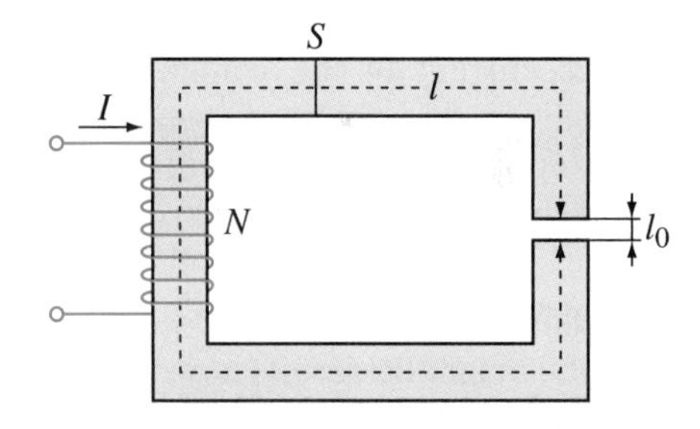

Figure 5.41 Simple nonlinear magnetic circuit with an air gap; for Problem 5.21.

5.22. **Complex nonlinear magnetic circuit.** Dimensions of the magnetic circuit shown in Fig. 5.42(a) are $l_1 = l_3 = 2l_2 = 20$ cm and $S_1 = S_2 = S_3 = 2$ cm^2. The magnetomotive forces in the circuit are $N_1I_1 = 100$ ampere-turns and $N_2I_2 = 300$ ampere-turns. The idealized initial magnetization curve of the core is shown in Fig. 5.42(b). Compute the magnetic flux densities and field intensities in each of the three branches of the circuit.

5.23. **Magnetic circuit with a zero flux in one branch.** Referring to the magnetic circuit in Fig. 5.42(a), let $l_1 = l_3 = 30$ cm, $l_2 = 10$ cm,

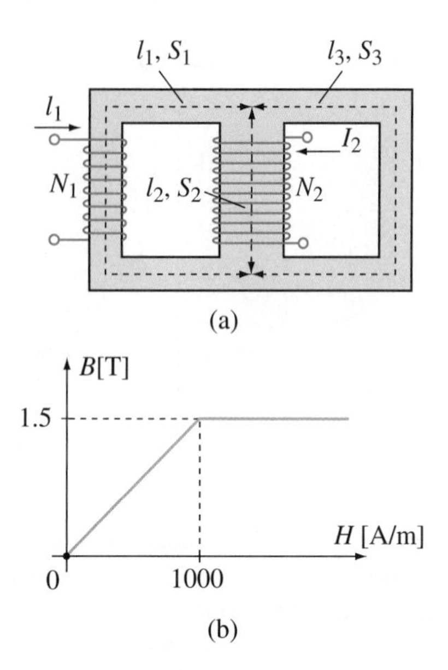

Figure 5.42 Nonlinear magnetic circuit with three branches and two mmf's: (a) circuit geometry and (b) idealized initial magnetization curve of the material; for Problem 5.22.

$S_1 = S_3 = 10\ \text{cm}^2$, $S_2 = 20\ \text{cm}^2$, $N_1 = 1500$, and $N_2 = 1000$. Let the idealized initial magnetization curve of the core be that in Fig. 5.27(b). If the current of the winding in the second branch of the circuit is $I_2 = 0.5$ A, find the current I_1 of the winding in the first branch such that the magnetic flux in that branch is zero.

5.24. Nonlinear magnetic circuit with two air gaps. For the magnetic circuit shown in Fig. 5.43, $l_1' = l_1'' = l_3' = l_3'' = 12$ cm, $l_2 = 5$ cm, $l_0 = 0.6$ mm, and $N_1 I_1 = 1100$ A turns. The cross-sectional area of each of the branches is $S = 2.5\ \text{cm}^2$. The initial magnetization curve of the core can be approximately represented as in Fig. 5.27(b). Determine the mmf $N_2 I_2$ of the second coil such that the magnetic flux through that coil is zero.

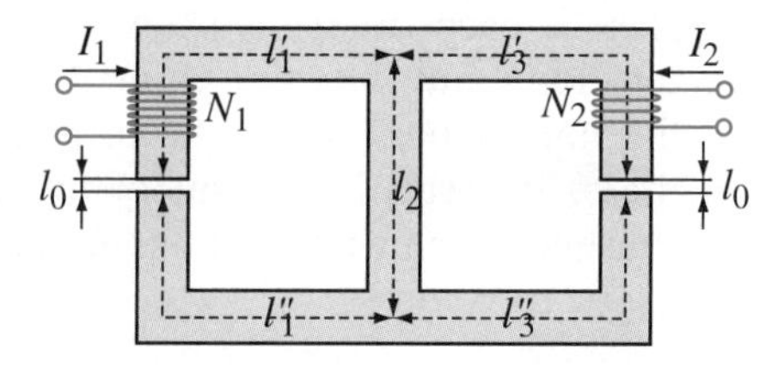

Figure 5.43 Nonlinear magnetic circuit with three branches, two air gaps, and two mmf's; for Problem 5.24.

5.25. Reverse problem with a nonlinear magnetic circuit. Refer to the magnetic circuit shown in Fig. 5.43 and described in the previous problem, and assume that the core is made out from the material whose initial magnetization curve can be approximated analytically by the function $B = \arctan(H/250)$ T (H in A/m), $H \geq 0$. Under these circumstances, find the mmf $N_2 I_2$ in the third branch such that the magnetic flux through the first branch (branch with the given mmf $N_1 I_1$) is $\Phi_1 = 125\ \mu$Wb with respect to the upward reference direction.

5.26. Remanent flux in a circuit with zero mmf. Repeat Example 5.17 but for $l = 50$ cm and $l_0 = 0.5$ mm.

5.27. Linear magnetic circuit with three branches. Assuming that the ferromagnetic material out of which the core of the magnetic circuit from Problem 5.24 is made can be considered as linear, with relative permeability $\mu_r = 1000$, and that $N_2 I_2 = 500$ A turns, (a) find the reluctances of the individual parts of the core and the air gaps and generate an equivalent electric circuit for the problem. (b) By solving the electric circuit in (a), find the magnetic field intensities in each of the air gaps.

5.28. Continuity equation from Ampère's law in integral form. Starting from Maxwell's second equation in integral form for the time-invariant electromagnetic field, derive the time-invariant integral form of the continuity equation.

6 Slowly Time-Varying Electromagnetic Field

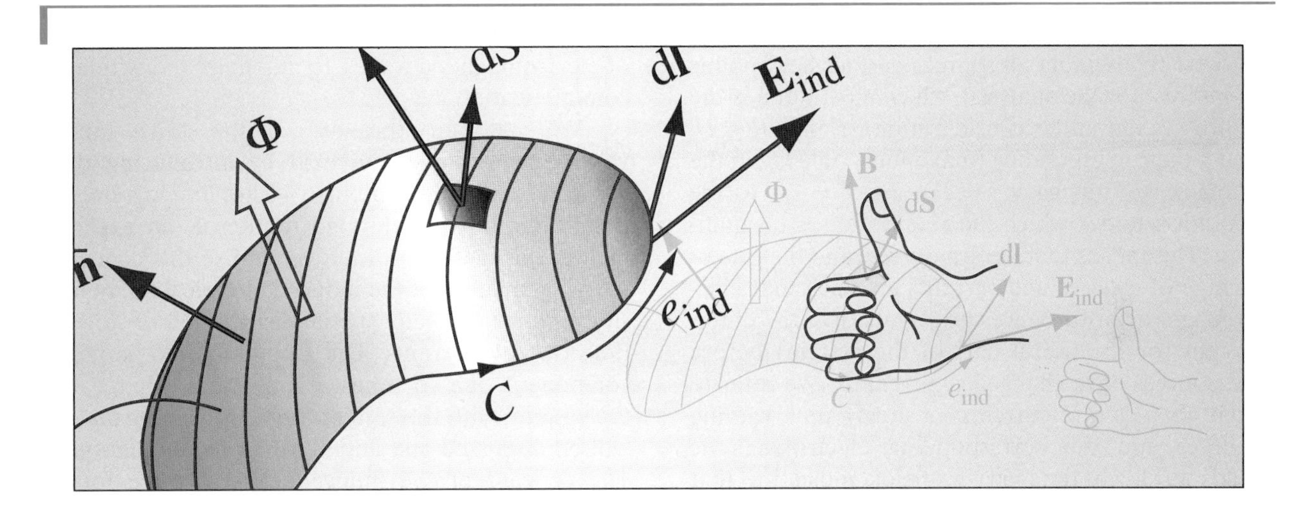

Introduction:

We now introduce time variation of electric and magnetic fields into our electromagnetic model. The new field is the time-varying electromagnetic field, which is caused by time-varying charges and currents. In the new model, all the quantities, in general, change in both space and time. As opposed to static fields, the electric and magnetic fields constituting the time-varying electromagnetic field are coupled to each other and cannot be analyzed separately. Namely, a magnetic field that changes in time produces (induces) an electric field and thus an electric current, and this phenomenon is known as electromagnetic induction. Additionally, a changing electric field induces a magnetic field. As we shall see in a later chapter, this mutual induction of time-varying electric and magnetic fields is the basis of time retardation in electromagnetic systems (lagging in time of time-varying electric and magnetic fields behind their sources), and as such of propagation of electromagnetic waves and of electromagnetic radiation.

The time-retardation concept is one of the most important phenomena in electromagnetics. It basically tells us that there is a time lag between a change of the field sources, i.e., of time-varying charges and currents, and the associated change of the fields, so that the values of field intensities at a distance from the sources depend on the values of charge and current densities at an earlier time. In other words, it takes some time for the effect of a change of charges and currents to be "felt" at distant field points. The time lag equals the time needed for electromagnetic disturbances to propagate over the corresponding distance. We shall see in a later chapter that the velocity of propagation of electromagnetic disturbances in a vacuum or air (free space) equals $c_0 = 3 \times 10^8$ m/s (speed of light and other electromagnetic waves in free space). Hence, the time lag in free space is $\tau = R/c_0$, where

R is the distance between the source and field (observation) points.

If the time τ for all combinations of source and field points in a domain of interest is much shorter than the time of change of the sources [e.g., the period of change of time-harmonic (steady-state sinusoidal) charges and currents, $T = 1/f$, where f is the frequency of the sources], the retardation effect in the system can be neglected. With D designating the maximum dimension of the domain of interest (containing all sources and all field points of interest for the analysis), which most often is the entire system under consideration, so that $R \leq D$, and τ the corresponding (maximum) time lag in the domain, we thus have $\tau = D/c_0 \ll T = 1/f$ as the condition under which the retardation is insignificant. This means that the system size and the rate of change of charges and currents are such that electromagnetic disturbances propagate over the entire system (or the useful part of the system) before the sources have changed significantly. We refer to such charges and currents as slowly time-varying sources and the corresponding electromagnetic fields as slowly time-varying fields, indicating that the rate of their change in time is slow when compared to the velocity of travel of electromagnetic disturbances (waves). In time-harmonic electromagnetics, a slow time variation corresponds to the low frequency, and the field model in which the time retardation is negligible is called accordingly the low-frequency electromagnetic field. On the other side, the rapidly time-varying (e.g., high-frequency time-harmonic) field cannot be analyzed without taking into account the travel time of electromagnetic disturbances from one point in the system to another. In this chapter (and the following one), we restrict our attention to slow time variations and low frequencies of sources and fields.

The slowly time-varying (low-frequency) electromagnetic field has many formal similarities with the time-invariant (static) electromagnetic field. This is why it is also called the quasistatic electromagnetic field. In addition to all the quantities being now time-dependent, the only essentially new feature of the quasistatic electromagnetic field that is not present under the static assumption is electromagnetic induction, and the fundamental governing law of electromagnetics describing this new phenomenon is Faraday's law of electromagnetic induction. However, this law is probably the most important basic experimental fact of electromagnetics. It is the underpinning of all of the dynamic-field practical applications, from electric motors and generators, through propagation of electromagnetic waves, to antennas and wireless communication.

We shall start the study of the slowly time-varying electromagnetic field by introducing the induced electric field intensity vector due to a single point charge in accelerated motion as an experimental postulate, and then generalize this concept to the evaluation of the induced electric field intensity vector due to any spatial distribution of slowly time-varying currents. The Coulomb electric field component due to time-varying excess charge in the system, with the same form as in electrostatics, will be discussed and added to the field equations. The concept of the induced electromotive force will be introduced and Faraday's law of electromagnetic induction derived using the magnetic vector potential. The full set of Maxwell's equations for the slowly time-varying electromagnetic field will be completed, along with the associated version of the continuity equation. In parallel, the induced electric field due to motion of conductors in static magnetic fields will be introduced as an impressed electric field and discussed in the context of Faraday's law of electromagnetic induction. All new concepts and equations will be applied to the analysis of a whole variety of quasistatic electromagnetic systems. Examples will include systems based on transformer induction (stationary conductors in changing magnetic fields) and those involving motional induction (moving conductors in static magnetic fields), as well as structures in which both types of electromagnetic induction are present at the same time (moving conductors in changing magnetic fields).

6.1 INDUCED ELECTRIC FIELD INTENSITY VECTOR

We know from Chapter 1 that a point charge Q in free space is a source of an electric field, predicted by Coulomb's law and described by Eq. (1.24). On the other side, the Biot-Savart law (Chapter 4) tells us that there will also be a magnetic field, given by Eq. (4.4), if this charge moves with some velocity $\mathbf{v}$ in space. We now introduce a third field, which will exist in the space around the charge whenever the velocity $\mathbf{v}$ changes in time, i.e., whenever the acceleration (or deceleration) $\mathbf{a} = \mathrm{d}\mathbf{v}/\mathrm{d}t$ of the charge is not zero. This new field is an electric field in its nature. It is called the induced electric field and its intensity vector is given by

$$\mathbf{E}_{\text{ind}}(t) = -\frac{\mu_0}{4\pi}\frac{Q\,\frac{\mathrm{d}\mathbf{v}}{\mathrm{d}t}}{R}, \tag{6.1}$$

point charge in an accelerated motion

which is an experimental result as well. Of course, the unit for $\mathbf{E}_{\text{ind}}$ is V/m. This is a time-varying field, that is, $\mathbf{E}_{\text{ind}}$ is a function of time.[1] Comparing Eqs. (6.1) and (4.107), we conclude that the induced electric field intensity vector at any instant of time and any point of space is actually equal to the negative of the time rate of change of the magnetic vector potential, $\mathbf{A}$, at that point,[2]

$$\mathbf{E}_{\text{ind}}(t) = -\frac{\partial \mathbf{A}}{\partial t}. \tag{6.2}$$

induced electric field intensity vector (unit: V/m)

Combining then this temporal differential relationship between $\mathbf{A}$ and $\mathbf{E}_{\text{ind}}$ with the spatial integral relationship between the current density vector, $\mathbf{J}$, and $\mathbf{A}$ in Eq. (4.108), we obtain the following integral (more precisely, integro-differential) expression for the induced electric field intensity vector due to an arbitrary volume current distribution:

$$\mathbf{E}_{\text{ind}}(t) = -\frac{\mu_0}{4\pi}\frac{\partial}{\partial t}\int_v \frac{\mathbf{J}\,\mathrm{d}v}{R} = -\frac{\mu_0}{4\pi}\int_v \frac{(\partial \mathbf{J}/\partial t)\,\mathrm{d}v}{R}. \tag{6.3}$$

$\mathbf{E}_{\text{ind}}$ due to volume current

Here, the time derivative operator can be moved inside the integral sign and applied directly to the current density vector because the differentiation with respect to time and spatial integration over the volume v are entirely independent operations and can be performed in an arbitrary order. Similarly, Eqs. (4.109) and (4.110) yield the corresponding expressions for a surface current of density $\mathbf{J}_s$ and line current of intensity i:[3]

$$\mathbf{E}_{\text{ind}}(t) = -\frac{\mu_0}{4\pi}\int_S \frac{(\partial \mathbf{J}_s/\partial t)\,\mathrm{d}S}{R}, \tag{6.4}$$

$\mathbf{E}_{\text{ind}}$ due to surface current

$$\mathbf{E}_{\text{ind}}(t) = -\frac{\mu_0}{4\pi}\int_l \frac{(\mathrm{d}i/\mathrm{d}t)\,\mathrm{d}\mathbf{l}}{R}. \tag{6.5}$$

$\mathbf{E}_{\text{ind}}$ due to line current

We conclude from Eqs. (6.3)–(6.5) that an induced electric field will exist in a system whenever time-varying electric currents (representing accelerated or decelerated motion of electric charges) exist in the conductors. Such currents are said to induce

[1] While using the notation $\mathbf{E}_{\text{ind}}(t)$ for the induced electric field vector to emphasize its time dependence, and similarly designating other time-varying field quantities (field intensity vectors, flux density vectors, potentials, charge and current densities, etc.) that will be introduced in this chapter, we always keep in mind that all these quantities, in general, are functions of spatial coordinates as well, e.g., $\mathbf{E}_{\text{ind}} = \mathbf{E}_{\text{ind}}(x, y, z, t)$.

[2] When dealing with time-varying spatially distributed quantities, we use the partial derivative with respect to time (e.g., $\partial \mathbf{A}/\partial t$) rather than the ordinary (total) derivative ($\mathrm{d}\mathbf{A}/\mathrm{d}t$) to emphasize their multivariable character.

[3] As mentioned in Chapter 3, for circuit-theory quantities (e.g., current intensity and voltage) that vary in time, we use lowercase notation (e.g., i and v) to distinguish it from the same quantities in the time-constant (dc) regime, which are capitalized (I and V).

the field. On the other side, it will be zero at all instants of time when $\mathbf{J}$, $\mathbf{J}_s$, and I are constants with respect to time (time-invariant currents) at all points of the system.

Eqs. (6.3)–(6.5) represent a general means for evaluating (analytically or numerically) the electric field due to any spatial distribution of slowly time-varying currents in free space. In a later chapter, we shall introduce certain corrections in these expressions to extend their validity to rapidly time-varying current distributions. Essentially, these corrections introduce wave propagation effects in the computation of fields due to rapidly time-varying currents.

Example 6.1 Induced Electric Field of a Straight Wire Conductor

Find the expression for the electric field intensity vector at an arbitrary point in space induced by a slowly time-varying current $i(t)$ in a straight segment of length l representing a part of a wire structure in air.

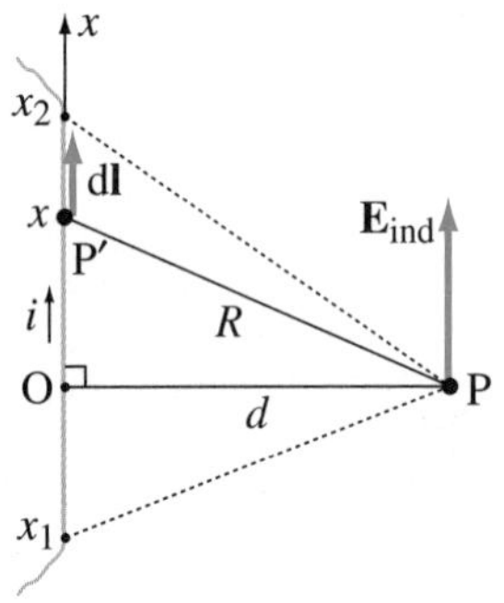

Figure 6.1 Evaluation of the induced electric field due to a finite straight wire conductor with a slowly time-varying current; for Example 6.1.

Solution We use the integral expression in Eq. (6.5) and refer to Fig. 6.1. Let P′ and P denote, respectively, a source point along the segment and a field point positioned arbitrarily in space. The coordinate defining the position of the point P′ is x, $x_1 \le x \le x_2$, where x_1 and x_2 $(x_2 - x_1 = l)$ are the coordinates of the starting and ending point of the segment, respectively. Note that both x_1 and x_2 can be either positive or negative, as well as zero, depending on the actual position of the point P with respect to the segment. Finally, let d mark the perpendicular distance of the point P from the segment. As $R = \sqrt{x^2 + d^2}$ and $\mathbf{dl} = \mathrm{d}x\,\hat{\mathbf{x}}$ (Fig. 6.1), the induced electric field intensity vector at the point P is given by

$$\mathbf{E}_{\mathrm{ind}} = -\frac{\mu_0}{4\pi}\frac{\mathrm{d}i}{\mathrm{d}t}\int_l \frac{\mathbf{dl}}{R} = -\frac{\mu_0}{4\pi}\frac{\mathrm{d}i}{\mathrm{d}t}\,\hat{\mathbf{x}}\int_{x_1}^{x_2}\frac{\mathrm{d}x}{\sqrt{x^2 + d^2}}, \tag{6.6}$$

where $\mathrm{d}i/\mathrm{d}t$ can be brought outside the integral sign because the current intensity $i(t)$ does not change along the wire (given that the current is slowly time-varying).[4] The solution of this integral[5] is

$\mathbf{E}_{\mathrm{ind}}$ – *finite straight wire conductor*

$$\boxed{\mathbf{E}_{\mathrm{ind}} = -\frac{\mu_0}{4\pi}\frac{\mathrm{d}i}{\mathrm{d}t}\ln\frac{x_2 + \sqrt{x_2^2 + d^2}}{x_1 + \sqrt{x_1^2 + d^2}}\,\hat{\mathbf{x}}.} \tag{6.7}$$

We note that the expression in Eq. (6.7) can be combined for computing the induced electric field due to structures containing any number of straight wire segments with a slowly time-varying current.

Example 6.2 Induced Electric Field of an EMI Source (Square Contour)

A source of electromagnetic interference (EMI) can be approximated by a square current contour of side length $a = 5$ cm in free space, as shown in Fig. 6.2(a). The contour carries a current whose intensity, $i(t)$, is a pulse function shown in Fig. 6.2(b). Compute the induced electric field intensity vector at the point M.

Solution We note that the time needed for electromagnetic disturbances to propagate over the domain of interest (from source points at the contour to the field point M) in Fig. 6.2(a) is on the order of $\tau = a/c_0 = 0.167$ ns $(c_0 = 3\times 10^8$ m/s). Since this time is much shorter

[4] As we shall see in a later chapter, the intensity of a rapidly time-varying current in a wire conductor, in general, changes along the conductor.

[5] Note that $\int \mathrm{d}x/\sqrt{x^2 + d^2} = \ln\left(x + \sqrt{x^2 + d^2}\right) + C$ (C being the integration constant), which can be easily verified by differentiation.

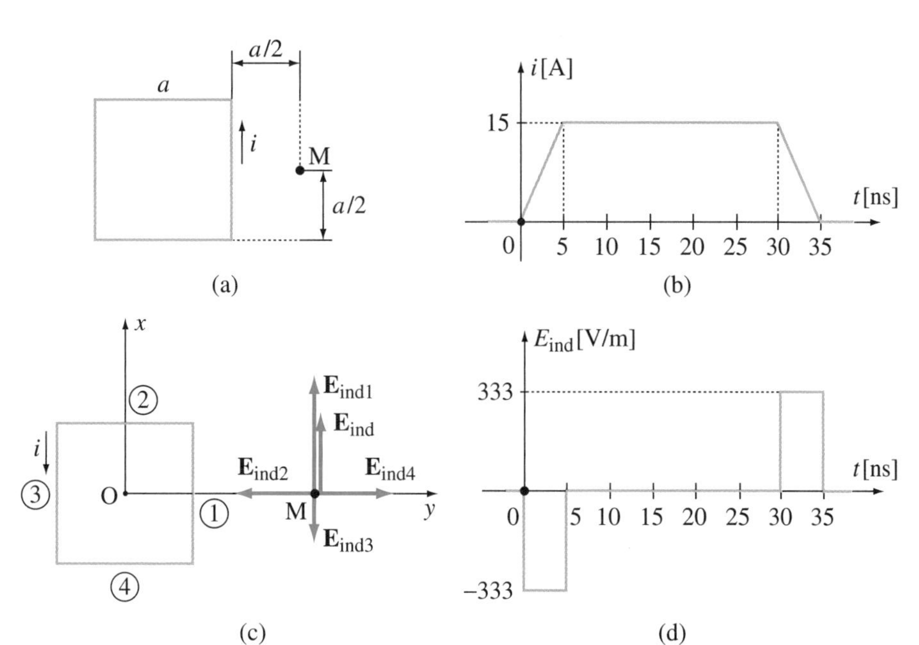

Figure 6.2 Evaluation of the induced electric field near a square current contour: (a) geometry of the problem, (b) contour current intensity as a function of time, (c) field intensity vectors due to individual sides of the contour, and (d) resultant field intensity as a function of time; for Example 6.2.

than the time of change (rise or fall) of the current intensity $i(t)$ in Fig. 6.2(b), $\Delta t = 5$ ns, the current can be considered as slowly time-varying and the system as quasistatic.

By means of the superposition principle, the total induced electric field intensity vector is given by [Fig. 6.2(c)]

$$\mathbf{E}_{\text{ind}} = \mathbf{E}_{\text{ind1}} + \mathbf{E}_{\text{ind2}} + \mathbf{E}_{\text{ind3}} + \mathbf{E}_{\text{ind4}}, \tag{6.8}$$

where the field intensity vectors due to individual sides of the contour, $\mathbf{E}_{\text{ind1}}, \ldots, \mathbf{E}_{\text{ind4}}$, are obtained from Eq. (6.7). Due to mutual antisymmetry of the second and fourth current segments with respect to the point M, $\mathbf{E}_{\text{ind2}} = -\mathbf{E}_{\text{ind4}}$. The distances d in the expressions for computing $\mathbf{E}_{\text{ind1}}$ and $\mathbf{E}_{\text{ind3}}$ equal $a/2$ and $3a/2$, respectively, whereas $x_1 = -a/2$ and $x_2 = a/2$ are the same in both expressions. Hence, the resultant $\mathbf{E}_{\text{ind}}$ at the point M comes out to be

$$\mathbf{E}_{\text{ind}} = \mathbf{E}_{\text{ind1}} + \mathbf{E}_{\text{ind3}} = -\frac{\mu_0}{4\pi}\frac{\mathrm{d}i}{\mathrm{d}t}\left(\ln\frac{1+\sqrt{2}}{-1+\sqrt{2}} - \ln\frac{1+\sqrt{10}}{-1+\sqrt{10}}\right)\hat{\mathbf{x}}$$

$$= -1.11 \times 10^{-7}\frac{\mathrm{d}i}{\mathrm{d}t}\,\hat{\mathbf{x}}\ \text{V/m} \quad (\mathrm{d}i/\mathrm{d}t \text{ in A/s}). \tag{6.9}$$

From Fig. 6.2(b), $\mathrm{d}i/\mathrm{d}t$ is nonzero only during the rise and fall intervals of the current pulse, when $\mathrm{d}i/\mathrm{d}t = \Delta i/\Delta t = \pm 15$ A/(5 ns) $= \pm 3 \times 10^9$ A/s, which yields $E_{\text{ind}} = \mp 333$ V/m. The function $E_{\text{ind}}(t)$ is plotted in Fig. 6.2(d). We see that very strong pulses of the induced electric field in the form of "spikes" are generated in the vicinity of the contour. This field thus may cause a very strong undesirable interference (EMI) into the operation of neighboring circuits in the system.

Example 6.3 Induced Electric Field of a Circular Wire Segment (Arc)

Consider a wire conductor in the form of an arc representing a part of a wire contour with a slowly time-varying current of intensity $i(t)$ in free space. The arc is defined by its radius a and angle α, as shown in Fig. 6.3(a). Find the expression for the induced electric field intensity vector at the arc center (point O) due to this current conductor.

Solution Since the distance of the field point from the source point in this case is always the same, $R = a$ [see Fig. 6.3(a)], we can bring it outside the integral sign in Eq. (6.5), which

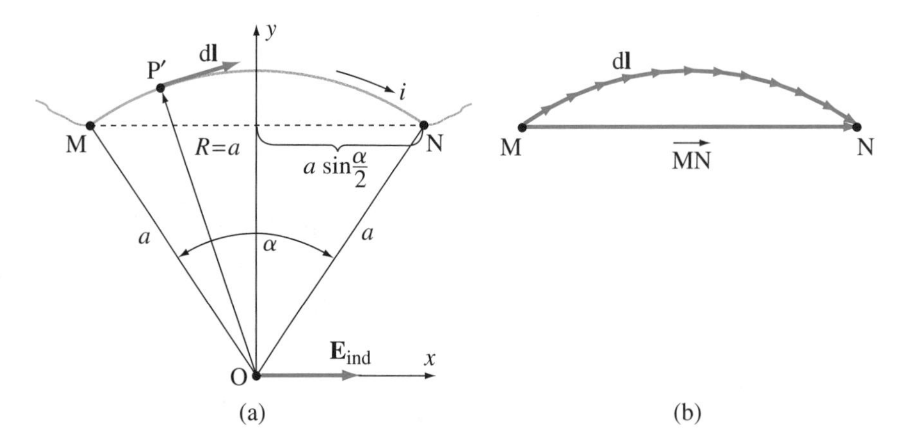

Figure 6.3 (a) Electric field induced by a slowly time-varying current along a circular wire segment and (b) application of the head-to-tail rule for vector addition to solve the integral of $\mathbf{dl}$ along the segment; for Example 6.3.

leads to

$$\mathbf{E}_{\rm ind} = -\frac{\mu_0}{4\pi a}\frac{{\rm d}i}{{\rm d}t}\int_{\rm M}^{\rm N}\mathbf{dl}, \tag{6.10}$$

where M and N are the starting and ending points of the conductor, respectively. Using the head-to-tail rule for vector addition, we observe from Fig. 6.3(b) that

$$\int_{\rm M}^{\rm N}\mathbf{dl} = \overrightarrow{\rm MN} \tag{6.11}$$

(note that this result holds true not only for a path in the form of an arc but for an arbitrarily shaped path, planar or nonplanar, between points M and N). Finally, as the distance between points M and N equals $\overline{\rm MN} = 2a\sin(\alpha/2)$,

$\mathbf{E}_{\rm ind}$ – *circular wire segment*

$$\boxed{\mathbf{E}_{\rm ind} = -\frac{\mu_0}{4\pi a}\frac{{\rm d}i}{{\rm d}t}\,\overrightarrow{\rm MN} = -\frac{\mu_0}{2\pi}\sin\frac{\alpha}{2}\frac{{\rm d}i}{{\rm d}t}\,\hat{\mathbf{x}}.} \tag{6.12}$$

Example 6.4 Current Contour of Complex Shape

Consider the contour consisting of two semicircular and two linear parts in Fig. 1.51, and assume that it carries a time-harmonic current of intensity $i(t) = \cos(9\times10^6 t)$ A (t in s) with respect to the clockwise reference direction, as well as that $a = 10$ cm and $b = 20$ cm. Under these circumstances, determine the induced electric field intensity vector at the contour center (point O).

Solution Since the period of change of the current intensity $i(t)$, $T = 2\pi/\omega = 0.7\ \mu$s (where $\omega = 9\times10^6$ rad/s is the angular frequency), is much longer than the time $\tau = b/c_0 = 0.667$ ns needed for electromagnetic disturbances to propagate from source points at the larger semicircle to the field point O in Fig. 1.51, the time-harmonic current in the contour can be considered as a low-frequency (slowly time-varying) current.

Referring to Fig. 6.4 and employing the superposition principle, the resultant induced electric field intensity vector is given by Eq. (6.8). From Eq. (6.12),

$$\mathbf{E}_{\rm ind1} = -\frac{\mu_0}{4\pi a}\frac{{\rm d}i}{{\rm d}t}\,\overrightarrow{\rm QR} = -\frac{\mu_0}{2\pi}\frac{{\rm d}i}{{\rm d}t}\,\hat{\mathbf{x}}, \quad \mathbf{E}_{\rm ind3} = -\frac{\mu_0}{4\pi b}\frac{{\rm d}i}{{\rm d}t}\,\overrightarrow{\rm SP} = -\frac{\mu_0}{2\pi}\frac{{\rm d}i}{{\rm d}t}(-\hat{\mathbf{x}}), \tag{6.13}$$

that is, $\mathbf{E}_{\rm ind1} + \mathbf{E}_{\rm ind3} = 0$. From Eq. (6.7), on the other hand,

$$\mathbf{E}_{\rm ind2} = -\frac{\mu_0}{4\pi}\frac{{\rm d}i}{{\rm d}t}\ln\frac{b+\sqrt{b^2+0}}{a+\sqrt{a^2+0}}\,\hat{\mathbf{x}} = -\frac{\mu_0}{4\pi}\frac{{\rm d}i}{{\rm d}t}\ln\frac{b}{a}\,\hat{\mathbf{x}}. \tag{6.14}$$

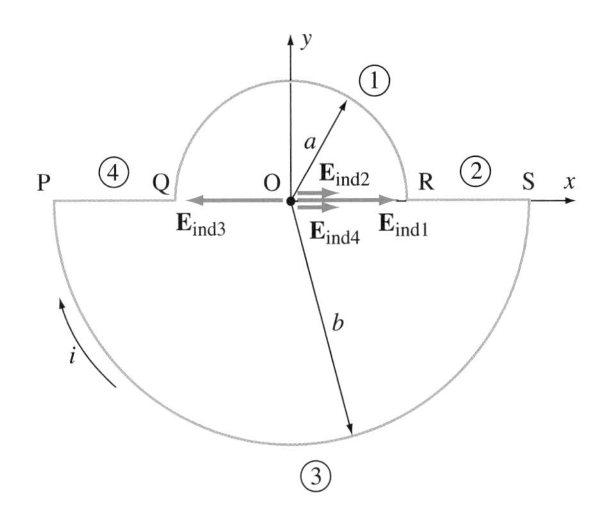

Figure 6.4 Evaluation of the induced electric field due to a current contour with two semicircular and two linear parts; for Example 6.4.

Due to symmetry, $\mathbf{E}_{\text{ind4}} = \mathbf{E}_{\text{ind2}}$, so that the total $\mathbf{E}_{\text{ind}}$ equals

$$\mathbf{E}_{\text{ind}} = 2\mathbf{E}_{\text{ind2}} = -\frac{\mu_0}{2\pi}\frac{\mathrm{d}i}{\mathrm{d}t}\ln\frac{b}{a}\,\hat{\mathbf{x}} = 1.25\sin(9\times10^6 t)\,\hat{\mathbf{x}}\ \text{V/m} \quad (t \text{ in s}). \tag{6.15}$$

Problems: 6.1–6.9; *Conceptual Questions* (on Companion Website): 6.1–6.8; *MATLAB Exercises* (on Companion Website).

6.2 SLOWLY TIME-VARYING ELECTRIC AND MAGNETIC FIELDS

The slowly time-varying electric field, in general, is composed of two components: the induced electric field, given by Eqs. (6.3)–(6.5), and the Coulomb electric field – the field due to excess charge, which we denote here as $\mathbf{E}_q$. The total electric field intensity vector is thus

$$\boxed{\mathbf{E}(t) = \mathbf{E}_{\text{ind}}(t) + \mathbf{E}_q(t).} \tag{6.16}$$

total electric field = induced plus Coulomb fields

The Coulomb field component has the same form as in electrostatics, given by Eqs. (1.37)–(1.39), however, the charge densities are now time-dependent. For instance, the field due to excess volume charge in free space is obtained as

$$\mathbf{E}_q(t) = \frac{1}{4\pi\varepsilon_0}\int_v \frac{\rho(t)\,\mathrm{d}v}{R^2}\,\hat{\mathbf{R}}. \tag{6.17}$$

Equivalently,

$$\boxed{\mathbf{E}_q(t) = -\nabla V(t),} \tag{6.18}$$

electric field due to excess charge

where the electric scalar potential is expressed as

$$\boxed{V(t) = \frac{1}{4\pi\varepsilon_0}\int_v \frac{\rho(t)\,\mathrm{d}v}{R}} \tag{6.19}$$

slowly time-varying electric potential

[see Eqs. (1.101) and (1.82)].

As the spatial distribution of the field $\mathbf{E}_q(t)$ has all the properties of the electrostatic field, we can write

$$\oint_C \mathbf{E}_q(t) \cdot \mathrm{d}\mathbf{l} = 0 \tag{6.20}$$

[Eq. (1.75)] or

$$\nabla \times \mathbf{E}_q(t) = 0 \tag{6.21}$$

[Eq. (4.92)], which are the mathematical expressions of the conservative character of the field $\mathbf{E}_q(t)$. In accordance to Eqs. (1.88) and (1.90), the time-varying potential difference (voltage) between points M and N in space is given by

$$v_{\mathrm{MN}}(t) = V_{\mathrm{M}}(t) - V_{\mathrm{N}}(t) = \int_{\mathrm{M}}^{\mathrm{N}} \mathbf{E}_q(t) \cdot \mathrm{d}\mathbf{l}. \tag{6.22}$$

The field due to excess time-varying charge also obeys Gauss' law. In differential notation [Eq. (1.166)],

$$\nabla \cdot \mathbf{E}_q(t) = \frac{\rho(t)}{\varepsilon_0}. \tag{6.23}$$

However, a combination of Eqs. (6.2) and (4.119) leads to

$$\nabla \cdot \mathbf{E}_{\mathrm{ind}} = -\nabla \cdot \left(\frac{\partial \mathbf{A}}{\partial t}\right) = -\frac{\partial}{\partial t}(\nabla \cdot \mathbf{A}) = 0, \tag{6.24}$$

where the time derivative operator can be brought outside the divergence operator (which implies spatial differentiation) because these two operations are entirely independent from one another and can be performed in an arbitrary order. Hence, the divergence of the vector $\mathbf{E}(t)$ in Eq. (6.16) equals

$$\nabla \cdot \mathbf{E}(t) = \nabla \cdot \mathbf{E}_{\mathrm{ind}}(t) + \nabla \cdot \mathbf{E}_q(t) = \frac{\rho(t)}{\varepsilon_0}, \tag{6.25}$$

which means that Gauss' law holds for the total electric field as well.

In the case of dielectric materials in the time-varying electric field, the polarization vector [Eq. (2.7)] and bound charge densities [Eqs. (2.19) and (2.23)] in the material are time-dependent. The generalized Gauss' law in integral form, Eq. (2.44), is written as

$$\oint_S \mathbf{D}(t) \cdot \mathrm{d}\mathbf{S} = \int_v \rho(t)\, \mathrm{d}v, \tag{6.26}$$

where $\mathbf{D}(t)$ is the total electric flux density vector in the material, given by Eqs. (2.41) and (2.47) with $\mathbf{E}$ representing the total electric field intensity vector.

On the other side, the slowly time-varying magnetic field has the same form as the magnetostatic field. The magnetic flux density vector in free space can thus be obtained using the time-varying version of Eqs. (4.7)–(4.9). For instance, the field due to a volume current distribution in free space is given by

$$\mathbf{B}(t) = \frac{\mu_0}{4\pi} \int_v \frac{[\mathbf{J}(t)\, \mathrm{d}v] \times \hat{\mathbf{R}}}{R^2}. \tag{6.27}$$

The field $\mathbf{B}(t)$ can also be obtained indirectly, via the magnetic vector potential, as [Eq. (4.116)]

slowly time-varying magnetic field

$$\boxed{\mathbf{B}(t) = \nabla \times \mathbf{A}(t),} \tag{6.28}$$

where

slowly time-varying magnetic potential

$$\boxed{\mathbf{A}(t) = \frac{\mu_0}{4\pi} \int_v \frac{\mathbf{J}(t)\, \mathrm{d}v}{R},} \tag{6.29}$$

with analogous expressions for surface and line currents.

In magnetic materials, the magnetization vector [Eq. (5.1)] and magnetization current densities [Eqs. (5.28) and (5.32)] are now time-dependent. The integral form of the generalized Ampère's law, Eq. (5.51), can be rewritten as

$$\oint_C \mathbf{H}(t) \cdot d\mathbf{l} = \int_S \mathbf{J}(t) \cdot d\mathbf{S}, \tag{6.30}$$

and that of the law of conservation of magnetic flux, Eq. (4.99), as

$$\oint_S \mathbf{B}(t) \cdot d\mathbf{S} = 0. \tag{6.31}$$

Problems: 6.10.

6.3 FARADAY'S LAW OF ELECTROMAGNETIC INDUCTION

We now introduce Faraday's law of electromagnetic induction, as the most important governing law of the slowly time-varying electromagnetic field and the explicit relation between the electric and magnetic fields that change in time. Following Oersted's discovery in 1820 that electric currents produced magnetic fields, Michael Faraday was convinced that the reverse was also possible – that a magnetic field could produce an electric current. In 1831, Faraday set up an apparatus consisting of an iron toroidal core (ring), like the one in Fig. 5.19, with two coils wound on it. The primary coil was connected through a switch to a battery (voltaic cell) and the secondary coil was short-circuited by a wire running above a compass, as sketched in Fig. 6.5. Thus, any electric current in the secondary coil would, by means of its magnetic field, deflect the compass needle. Upon closing the switch, Faraday observed a momentary deflection of the needle, indicating a brief surge of current induced in the secondary coil. The same happened when the switch was opened, terminating the current in the primary coil, but the needle deflection was opposite in polarity with respect to the previous one. In steady states, however, i.e., once the current in the primary coil reached its final value (equal to the battery voltage divided by the resistance of the primary circuit or to zero), there was no current in the secondary coil and the compass needle was at its zero position. Faraday realized that a current was produced (induced) in the secondary coil by a changing magnetic field in the iron core (the field was changed from zero to a final steady value, corresponding to a steady current intensity established in the primary circuit, and then back to zero when the current was terminated). This extraordinary discovery led to the formulation of the law of electromagnetic induction, named after Faraday.

The mathematical statement of Faraday's law of electromagnetic induction describes the time variation of the magnetic flux through an arbitrary contour as

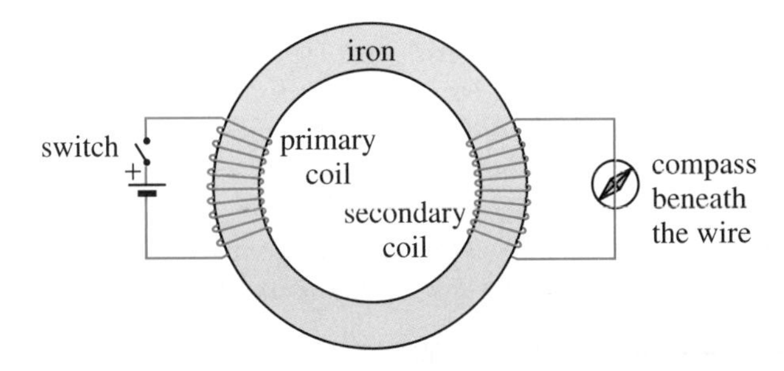

Figure 6.5 Sketch of the apparatus used in Faraday's 1831 experiment that led to the discovery of electromagnetic induction.

HISTORICAL ASIDE

Michael Faraday (1791–1867), an English physicist and chemist, self-educated from books he was binding to earn a living, became a member of the Royal Society at age 34. He was mostly unfamiliar with mathematics, but at the same time he was an enormously gifted experimentalist and imaginative thinker, and certainly one of the greatest scientists ever. Upon a recommendation by Sir Humphry Davy (1778–1829), the discoverer of six chemical elements, Faraday was appointed Chemical Assistant at the Royal Institution on March 1, 1813. After repeating Oersted's (1777–1851) 1820 experiment, he demonstrated in September of 1821 that the magnetic field around a straight wire with an electric current was circular, and, in the same set of experiments, went a large step further – by making a current-carrying wire suspended above a permanent magnet circle around the magnet, he invented the first electric motor. Generalizing from the patterns formed by iron filings around magnets, he introduced the concept of electric and magnetic field lines (he called them lines of force) as a new approach to studying electricity and magnetism. Faraday was elected to the Royal Society in 1824 and was made Director of the Laboratory at the Royal Institution in 1825. In 1826, he started Christmas Lectures for children at the Institution, which not only continue, but are now televised to giant audiences all over the world. In his famous August 29, 1831 experiment, he wound two coils of wire on the same iron ring (see Fig. 6.5) and discovered that a current in one coil if changed in time induced a current in the other coil. He concluded that an electric current could be produced by a time-varying magnetic field and thus discovered electromagnetic induction. Faraday's induction ring was the world's first electric transformer. In subsequent experiments in the fall of 1831, he attempted to create a current using a permanent magnet. He discovered that when a permanent magnet was moved in and out of a wire coil, a current was induced in the coil. He then demonstrated that a continuous current could be generated by rotating a copper disk between the poles of a large permanent magnet and taking leads off the rim and the center of the disk. This invention, referred to as Faraday's wheel (see Fig. 6.35), was the first dynamo (electric generator). Faraday used his concept of lines of force to explain the principle of electromagnetic induction observed in his experiments. He explained that an electric current was induced in a conductor only when magnetic lines of force cut across it, and this could be either because the lines of a changing magnetic field expanded and collapsed in space cutting thus across the conductor (transformer induction) or because the conductor moved across the static field lines (motional induction). He realized that the magnitude of the induced current was dependent on the number of lines of force cut by the conductor in unit time, which is a true equivalent to a more mathematical formulation of what is now known as Faraday's law of electromagnetic induction. It was the discovery of electromagnetic induction in 1831, more than any other, that allowed electricity to be turned, during the remainder of the 19th century, from a scientific curiosity into a powerful technology. In the 1830s, Faraday also studied the relationships between the amount of material deposited on electrodes of an electrolytic cell, the amount of electricity passed through the cell, and chemical properties of different elements, and formulated fundamental principles of electrochemistry (Faraday's laws of electrolysis). He also discovered that light could be affected by a magnetic force – he demonstrated in 1845 that a strong magnetic field could rotate the plane of polarization of polarized light – which later became known as the Faraday magneto-optical effect. In his famous lecture "Thoughts on Ray-vibrations" at the Royal Institution in April of 1846, he suggested that the propagation of light through space consisted of vibrations of lines of force, which, intuitively, was not far from Maxwell's (1831–1879) explanation, given much later (in 1865) and based on rigorous mathematical derivations, that light was an oscillatory electromagnetic disturbance – electromagnetic wave. In a series of studies from 1846 to 1850, Faraday evolved his global theoretical electromagnetic model based on the "force field" – the field of lines of force in tension filling the space around charged bodies, current-carrying conductors, and permanent magnets, and thus established the field theory of electromagnetism. These concepts of electric and magnetic force fields were put into a mathematical form a generation later by Maxwell, who himself made it very clear in his texts that the basic ideas for his classical electromagnetic field equations came directly from Faraday. Despite his epochal scientific achievements and far-reaching contributions to humanity, Faraday remained a modest and humble person throughout his life. We honor Faraday also by using farad (F) as the unit for capacitance. *(Portrait: Edgar Fahs Smith Collection, University of Pennsylvania Libraries)*

the cause of the induced electromotive force along the contour. In developing the general electromagnetic model, it is usually taken as an experimentally based postulate. However, because we started our study of electromagnetic induction by taking the mathematical expression for the induced electric field intensity vector due to a point charge in accelerated motion as an experimental postulate, we are now able to actually derive Faraday's law from the facts that we already know about the induced electric field.

In a region with free charge carriers (e.g., in a conducting wire), the induced electric field intensity vector, $\mathbf{E}_{\text{ind}}$, acts on the carriers by the force $Q\mathbf{E}_{\text{ind}}$ [Eq. (4.143)], where Q is the charge of a carrier (e.g., a free electron). Therefore, the line integral of $\mathbf{E}_{\text{ind}}$ along a line joining any two points M and N in space represents the electromotive force (emf) induced in the line:

$$\boxed{e_{\text{ind}} = \int_{\text{M}}^{\text{N}} \mathbf{E}_{\text{ind}} \cdot \text{d}\mathbf{l}.} \tag{6.32}$$

induced electromotive force (emf), in volts

The induced emf is measured in volts and defined in the same way as the emf of a voltage generator in Eq. (3.112). In fact, the line can be replaced by an equivalent voltage generator whose emf is e_{ind}, as shown in Fig. 6.6.

For a closed line (contour) that does not change or move in time, Fig. 6.7, we use Eq. (6.2) and write

$$e_{\text{ind}} = \oint_C \mathbf{E}_{\text{ind}} \cdot \text{d}\mathbf{l} = -\oint_C \frac{\partial \mathbf{A}}{\partial t} \cdot \text{d}\mathbf{l} = -\frac{\text{d}}{\text{d}t} \oint_C \mathbf{A} \cdot \text{d}\mathbf{l}. \tag{6.33}$$

We now recall that the circulation of the magnetic vector potential along a contour equals the magnetic flux through the contour, Eq. (4.121), so that

$$\boxed{e_{\text{ind}} = -\frac{\text{d}\Phi}{\text{d}t}.} \tag{6.34}$$

Faraday's law of electromagnetic induction

This equation is known as Faraday's law of electromagnetic induction. It is one of the most important experimental pillars of electromagnetics. It shows, first of all, that the electric and magnetic fields are related to one another under nonstatic conditions. More specifically, it states that a magnetic field that changes with time produces (induces) an electric field and an electromotive force, as well as an electric current in conducting media [by virtue of Ohm's law in local form, Eq. (3.18)].

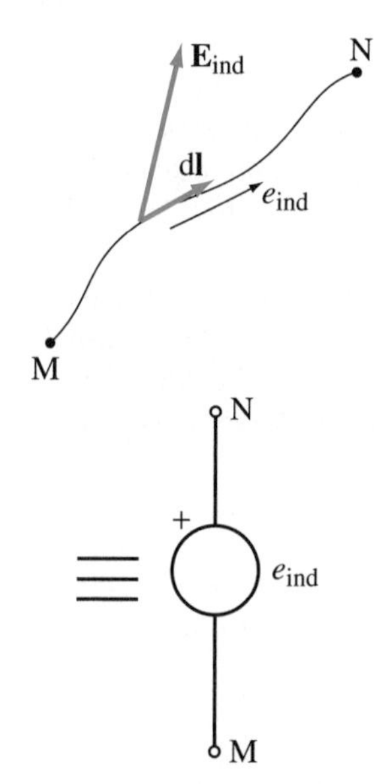

Figure 6.6 Induced emf in a line joining two points in space.

Finally, it quantifies the induced emf in an arbitrary contour as being equal to the negative of the time rate of change of the magnetic flux through the contour, i.e., through a surface of arbitrary shape spanned over the contour and oriented in accordance to the right-hand rule with respect to the orientation of the contour. This rule tells us that the flux is in the direction defined by the thumb of the right hand when the other fingers point in the direction of the emf, as indicated in Fig. 6.7.

Expressing the magnetic flux using the flux density vector [Eq. (4.95)] leads to

$$\oint_C \mathbf{E}_{ind} \cdot \mathrm{d}\mathbf{l} = -\frac{\mathrm{d}}{\mathrm{d}t} \int_S \mathbf{B} \cdot \mathrm{d}\mathbf{S}, \tag{6.35}$$

where the reference directions of $\mathrm{d}\mathbf{l}$ and $\mathrm{d}\mathbf{S}$ are interconnected by the right-hand rule: fingers – $\mathrm{d}\mathbf{l}$, thumb – $\mathrm{d}\mathbf{S}$ (see Fig. 6.7). Since the circulation of the field intensity vector due to excess charge is zero [Eq. (6.20)], we have that

$$\oint_C \mathbf{E} \cdot \mathrm{d}\mathbf{l} = \oint_C \mathbf{E}_{ind} \cdot \mathrm{d}\mathbf{l}, \tag{6.36}$$

which means that Faraday's law can be expressed in terms of the total electric field intensity vector as well. In addition, the time derivative on the right-hand side of Eq. (6.35) can be moved inside the surface integral, provided that the surface S does not change or move in time. Combining these two conclusions, we obtain the following version of Faraday's law of electromagnetic induction in integral form:

Faraday's law in integral form

$$\boxed{\oint_C \mathbf{E} \cdot \mathrm{d}\mathbf{l} = -\int_S \frac{\partial \mathbf{B}}{\partial t} \cdot \mathrm{d}\mathbf{S}.} \tag{6.37}$$

We note the formal similarity between this equation and the generalized Ampère's law in integral form, Eq. (6.30), where $-\partial \mathbf{B}/\partial t$ in Eq. (6.37) stands for $\mathbf{J}$ in Eq. (6.30) in the flux integrals, while the electric and magnetic field intensity vectors, $\mathbf{E}$ and $\mathbf{H}$, appear in the corresponding line integrals on the left-hand side of equations.

The contour C in Fig. 6.7 can be an imaginary (nonmaterial) contour, i.e., it does not need to be a conducting wire loop for Faraday's law of electromagnetic induction to be true. The electric field and emf are induced by a magnetic field that changes with time regardless of whether or not conducting wires are present. However, in the case when C does represent a conducting wire contour, there is a current of intensity

induced current

$$\boxed{i_{ind} = \frac{e_{ind}}{R}} \tag{6.38}$$

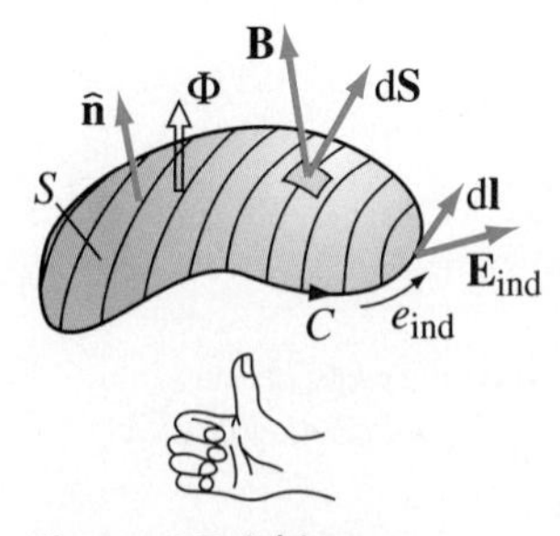

Figure 6.7 Arbitrary contour in a time-varying magnetic field – for the formulation of Faraday's law of electromagnetic induction.

in the wire, as shown in Fig. 6.8, where R is the total resistance of the contour and of the equivalent closed circuit including the ideal voltage generator with emf e_{ind} [this, basically, comes from the version of Kirchhoff's voltage law in Eq. (3.118)]. This current is called the induced current. We can say thus that, in general, time-varying magnetic fields, $\mathbf{B}(t)$, induce electric currents in conducting media, which also change with time.

On the other hand, if the conducting wire loop is not closed (e.g., there is a small air gap in the loop), there is no current flowing through it,[6] and the loop behaves like an open-circuited generator with emf e_{ind}, as in Fig. 6.6. The voltage across the gap equals the induced emf, i.e., $v_{NM}(t) = e_{ind}(t)$ [see Eq. (3.115)].

[6]As opposed to dc and slowly time-varying cases, as we shall see in a later chapter, a rapidly time-varying current can exist even in open-ended wire conductors that do not form closed current circuits.

HISTORICAL ASIDE

Heinrich Friedrich Emil Lenz (1804–1865), a Russian physicist, was a professor of physics at the University of St. Petersburg. He was one of the three great scientists, together with Faraday (1791–1867) and Henry (1797–1878), who independently from each other investigated electromagnetic induction at about the same time at three remote places on the globe. Lenz was born and educated in Dorpat (now Tartu), Estonia, then a part of Russian Empire. As geophysical scientist, he traveled around the world in the 1820s and made extremely accurate measurements of the salinity, temperature, and specific gravity of sea waters. He also studied electricity and magnetism, and discovered, in 1833, that the resistivity of metallic conductors increases with a rise in temperature [see Eq. (3.22)]. In 1834, he discovered that an induced current always produces effects that oppose its cause. This became to be known as Lenz's law. From 1840 to 1863, Lenz was the Dean of Mathematics and Physics at the University of St. Petersburg.

The minus sign in Eq. (6.34) indicates that the induced emf in the contour is in a direction that opposes the change in the magnetic flux through the contour that caused the emf in the first place. This fact, contained in Faraday's law, is an experimental result referred to as Lenz's law. To illustrate it, assume that at an instant t the flux in Fig. 6.8 increases in time, i.e., $\mathrm{d}\Phi/\mathrm{d}t > 0$. From Eq. (6.34), the emf $e_{\mathrm{ind}}(t)$ at that instant is negative, and so is the induced current intensity $i_{\mathrm{ind}}(t)$ in Eq. 6.38. The induced current produces a secondary magnetic field, whose reference direction is determined by another application of the right-hand rule: fingers – current, thumb – field [for example, see Fig. 4.31(b)]. Hence, given the reference direction of i_{ind} in Fig. 6.8, the reference direction of the secondary magnetic field, and its flux, will be the same as that of the primary (original) magnetic field $\mathbf{B}(t)$ and flux $\Phi(t)$. So, with respect to that reference direction, the secondary flux is negative, because $i_{\mathrm{ind}}(t)$ is negative at the time instant considered. We conclude thus that the magnetic field due to the induced current opposes the change (increase in this case) in the primary magnetic field, which caused the induced emf and current in the first place, and this is the statement of Lenz's law. Generally, Lenz's law represents a rule for (quickly) determining the actual direction of an induced current in a loop (circuit), without fully applying Faraday's law. This direction is always such that the magnetic field due to the induced current opposes (tends to cancel) the change in the magnetic flux that induces the current.[7]

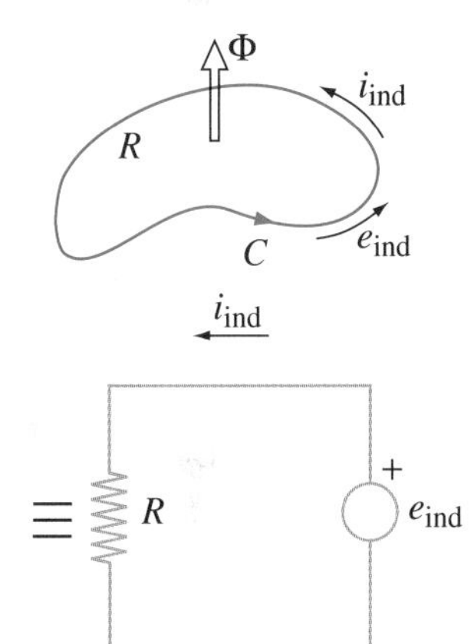

Figure 6.8 Induced current in a conducting wire loop situated in a time-varying magnetic field.

By virtue of Stokes' theorem, Eq. (4.89), or simply by analogy with the differential form of the generalized Ampère's law, Eq. (5.52), we obtain the differential equivalent of Eq. (6.37), namely, Faraday's law of electromagnetic induction in differential form:

$$\boxed{\nabla \times \mathbf{E} = -\frac{\partial \mathbf{B}}{\partial t}.} \tag{6.39}$$

Faraday's law in differential form

[7]The magnetic field due to the induced slowly time-varying current in typical thin-wire circuits is much weaker than the primary magnetic field inducing the current. Therefore, it is usually negligible with respect to the primary field and falls far short of fully canceling the flux change causing the electromagnetic induction.

In words, the curl of the time-varying electric field intensity vector existing at any point of space and any instant of time equals the negative of the time rate of change of the magnetic flux density vector at that point.

Note that the differential form of Faraday's law can also be derived by taking the curl of both sides of Eq. (6.2) and using Eq. (6.28). This results in

$$\nabla \times \mathbf{E}_{\text{ind}} = -\nabla \times \left(\frac{\partial \mathbf{A}}{\partial t}\right) = -\frac{\partial}{\partial t}(\nabla \times \mathbf{A}) = -\frac{\partial \mathbf{B}}{\partial t}, \tag{6.40}$$

which is the version of the law with the induced electric field intensity vector. Since the curl of the field intensity vector due to excess charge is zero [Eq. (6.21)], we conclude that

$$\nabla \times \mathbf{E} = \nabla \times \mathbf{E}_{\text{ind}} + \nabla \times \mathbf{E}_q = \nabla \times \mathbf{E}_{\text{ind}}, \tag{6.41}$$

which gives the version of the law with the total electric field intensity vector, Eq. (6.39).

6.4 MAXWELL'S EQUATIONS FOR THE SLOWLY TIME-VARYING ELECTROMAGNETIC FIELD

Faraday's law of electromagnetic induction, given by Eq. (6.37) or Eq. (6.39), represents Maxwell's first equation for the time-varying electromagnetic field. It is essentially different from Maxwell's first equation for the time-invariant electromagnetic field in Eqs. (5.129). It tells us that a magnetic field changing with time gives rise to an electric field. The remaining three Maxwell's equations are given by Eqs. (6.30), (6.26), and (6.31), and they retain the same form as in the time-invariant case. We now summarize the full set of Maxwell's equations in differential form along with the constitutive equations for the slowly time-varying electromagnetic field in a linear isotropic medium:

Maxwell's first equation

Maxwell's second equation, quasistatic field

Maxwell's third equation

Maxwell's fourth equation

constitutive equation for **D**

constitutive equation for **B**

constitutive equation for **J**

$$\begin{cases} \nabla \times \mathbf{E}(t) = -\frac{\partial \mathbf{B}}{\partial t} \\ \nabla \times \mathbf{H}(t) = \mathbf{J}(t) \\ \nabla \cdot \mathbf{D}(t) = \rho(t) \\ \nabla \cdot \mathbf{B}(t) = 0 \\ \mathbf{D}(t) = \varepsilon \mathbf{E}(t) \\ \mathbf{B}(t) = \mu \mathbf{H}(t) \\ \mathbf{J}(t) = \sigma \mathbf{E}(t) \end{cases} . \tag{6.42}$$

Because of the new term on the right-hand side of the first equation, the time-varying electric and magnetic fields are coupled together and cannot be analyzed separately, as opposed to time-invariant electric and magnetic fields, which are entirely independent from each other. On the other hand, as we shall see in a later chapter, Maxwell's equations for the rapidly time-varying electromagnetic field differ from the corresponding equations for the slowly time-varying field only in the generalized Ampère's law (second equation). Namely, an additional term exists on the right-hand side of this equation in the general case, expressing the fact that an electric field rapidly changing with time gives rise to a magnetic field. Therefore, the second equation in the system of Eqs. (6.42) represents the quasistatic version of

Maxwell's second equation and is true for slowly time-varying (e.g., low-frequency time-harmonic) fields.

Eqs. (6.19) and (6.29) give the expressions for slowly-time varying electromagnetic potentials in free space. Substituting Eqs. (6.2) and (6.18) in Eq. (6.16) leads to the following expression for the electric field intensity vector, **E**, in terms of the potentials:

$$\boxed{\mathbf{E}(t) = -\frac{\partial \mathbf{A}}{\partial t} - \nabla V(t).} \tag{6.43}$$

electric field via potentials

We see that both potentials are needed for **E**, whereas **A** alone suffices for the magnetic flux density vector, **B**, in Eq. (6.28). Eqs. (6.43) and (6.28) represent a means for evaluating (by differentiation) the electromagnetic field (**E**, **B**) from potentials (V, **A**), the evaluation of potentials being, in general, considerably simpler than the direct evaluation of field vectors.

In addition to Maxwell's equations for a given class of electromagnetic fields, we always have in mind the associated version of the continuity equation, which represents one of the fundamental principles of electromagnetics – the principle of conservation of charge, but can also be derived from Maxwell's equations. Thus, by taking the divergence of both sides of the quasistatic version of the generalized Ampère's law in Eqs. (6.42), as in Eq. (5.130) for the static case, we obtain that

$$\nabla \cdot \mathbf{J}(t) = 0, \tag{6.44}$$

which is the differential form of the continuity equation for slowly time-varying currents. We note that it has the same form as the continuity equation for time-invariant (steady) currents, Eq. (3.41). We also note that it can be regarded as the special case of the general continuity equation for time-varying currents, Eq. (3.39), with $\partial\rho/\partial t \approx 0$. Namely, in the slowly time-varying field, the rate of the time-variation in excess charge is slow enough to be neglected while evaluating the current continuity balance. In integral notation,

$$\boxed{\oint_S \mathbf{J}(t) \cdot d\mathbf{S} = 0,} \tag{6.45}$$

continuity equation for slowly time-varying currents

which is the same as Eq. (3.40) for steady currents.

Eq. (6.45) implies that the slowly time-varying current intensity $i(t)$ along a wire conductor, just as the steady current intensity I along a wire, is the same in every cross section of the conductor – the fact that we have already used throughout this chapter. It also means that Kirchhoff's circuital law for slowly time-varying currents is given by

$$\sum_{k=1}^{N} i_k(t) = 0, \tag{6.46}$$

where N is the number of conductors (branches in a circuit) meeting at a node. We see that Kirchhoff's current law for low-frequency ac circuits has the same form as that for dc circuits [Eq. (3.42)].

6.5 COMPUTATION OF TRANSFORMER INDUCTION

This section is devoted to the application of Faraday's law of electromagnetic induction in evaluating the induced emf, e_{ind}, and electric field intensity vector, $\mathbf{E}_{\text{ind}}$, in stationary contours due to given slowly time-varying current distributions and their magnetic fields. This kind of electromagnetic induction is called transformer induction, because it is the basis of current and voltage transformation by magnetic coupling between circuits. Namely, it enables time-varying currents and voltages in one circuit (primary circuit) to be transformed, by induction, to time-varying currents and voltages in another circuit (secondary circuit), where the transfer of energy between the circuits is actually performed by the magnetic field due to the currents in the primary circuit causing the induced electric field in the secondary circuit. The electromagnetic induction due to motion of conductors in magnetic fields will be introduced and studied in the next section.

In some applications, we are interested only in the total emf induced in a contour (e.g., a wire loop). In such cases, we apply the version of Faraday's law of electromagnetic induction in Eq. (6.34), where we only need to evaluate the magnetic flux through the contour and compute its time derivative, which is usually a simple task to do. However, to find the actual distribution of the emf along a contour or to find the distribution of the induced electric field intensity vector in space, which is necessary in many applications, we have to employ the version of the law in Eq. (6.35). This equation, although always true, enables us to analytically solve for the field $\mathbf{E}_{\text{ind}}$ only due to highly symmetrical primary current distributions. These are the cases where the vector $\mathbf{E}_{\text{ind}}$ is tangential to some (or all) sections of the contour and has a constant magnitude along such sections, while being perpendicular to the remaining sections of the contour (if such sections exist). In other words, these are the cases in which we are able to bring the induced electric field intensity, E_{ind}, outside the integral sign on the left-hand side of Eq. (6.35), and solve for it. Note that Eqs. (6.3)–(6.5), on the other hand, provide general solution procedures for computing $\mathbf{E}_{\text{ind}}$.

Because Faraday's law and Ampère's law have the same mathematical form, there is a complete formal analogy in their application. In solving for $\mathbf{E}_{\text{ind}}$ due to $\mathbf{B}$ in highly symmetrical situations, therefore, we shall exploit the parallelism with the application of Ampère's law in solving for $\mathbf{B}$ (or $\mathbf{H}$) due to $\mathbf{J}$, discussed in Section 4.5, whenever possible.

Example 6.5 Induced Electric Field of an Infinite Solenoid

An infinitely long solenoid with a circular cross section of radius a has N' turns of wire per unit of its length. The solenoid is wound about a ferromagnetic core of permeability μ. A slowly time-varying current of intensity $i(t)$ flows through the winding. The medium outside the solenoid is air. The magnetic field due to currents induced in the core can be neglected. Find the induced electric field intensity vector inside and outside the solenoid.

Solution Because of symmetry, the lines of the induced electric field due to the current in the solenoid winding are circles centered at the solenoid axis. To show this, consider an arbitrary point P inside the solenoid, at a distance r from the solenoid axis (Fig. 6.9). Let $\mathrm{d}\mathbf{E}'_{\text{ind}}$ and $\mathrm{d}\mathbf{E}''_{\text{ind}}$ be the field intensity vectors at this point due to two symmetrical current elements denoted as $i\,\mathrm{d}\mathbf{l}'$ and $i\,\mathrm{d}\mathbf{l}''$ and shown in Fig. 6.9. In accordance to Eq. (6.5), these two vectors are such that their sum $\mathrm{d}\mathbf{E}'_{\text{ind}} + \mathrm{d}\mathbf{E}''_{\text{ind}}$ is tangential to the circular contour C of radius r centered at the solenoid axis. The same is true for any other pair of symmetrical current elements, which can also be in a plane that does not contain the point P. As all current elements constituting the current $i(t)$ in the winding can be grouped in such symmetrical pairs, we conclude that the resultant vector $\mathbf{E}_{\text{ind}}$ at the point P is tangential to the contour C. This same conclusion holds for a point outside the solenoid. In addition, $E_{\text{ind}} = \text{const}$ along C, i.e., the magnitude of $\mathbf{E}_{\text{ind}}$ depends only on the radial coordinate r of the cylindrical coordinate system whose z-axis is the solenoid axis. Hence,

$$\mathbf{E}_{\text{ind}} = E_{\text{ind}}(r, t)\,\hat{\boldsymbol{\phi}}, \tag{6.47}$$

where $\hat{\boldsymbol{\phi}}$ is the circular unit vector in the system.

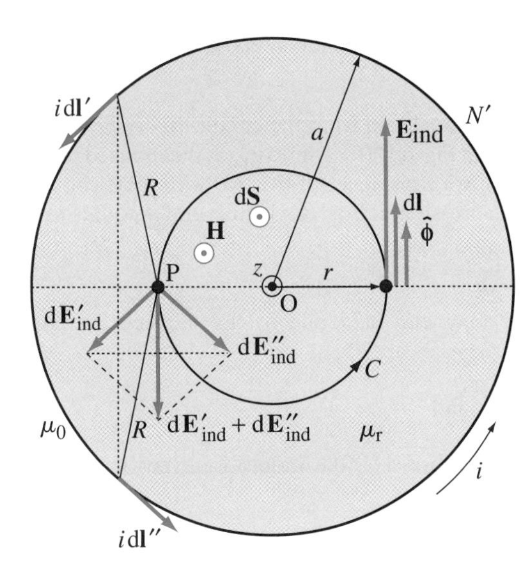

Figure 6.9 Evaluation of the induced electric field due to a slowly time-varying current in an infinite solenoid (cross-sectional view); for Example 6.5.

The magnetic field due to any spatial distribution of slowly time-varying currents has the same form as that due to the same spatial distribution of steady currents. Therefore, we can use the result of the analysis of a solenoid with a steady current of intensity I in the winding, performed in Example 4.13. This means that the magnetic field inside the solenoid is axial (field lines are parallel to the solenoid axis) and uniform, given by

$$\mathbf{H} = H(t)\ \hat{\mathbf{z}} = N'i(t)\ \hat{\mathbf{z}}, \tag{6.48}$$

while $\mathbf{H} = 0$ outside the solenoid. Note that this result does not take into account the magnetic field due to currents induced in the ferromagnetic core, which can be neglected.

We now use Faraday's law of electromagnetic induction, Eq. (6.35), and apply it to the contour C in Fig. 6.9 in a manner completely analogous to the application of Ampère's law in Eqs. (4.54)–(4.56). The circulation of $\mathbf{E}_{\text{ind}}$ along C equals

$$E_{\text{ind}}\, 2\pi r = -\frac{\mathrm{d}\Phi}{\mathrm{d}t}. \tag{6.49}$$

The magnetic flux through the contour is

$$\Phi = \begin{cases} B\pi r^2 & \text{for } r \le a \\ B\pi a^2 & \text{for } r > a \end{cases} \tag{6.50}$$

(there is no flux outside the solenoid), where $B = \mu H$ and H is given in Eq. (6.48). The solutions for the induced electric field intensity inside and outside the solenoid come out to be

$$\boxed{E_{\text{ind}} = -\frac{\mu N' r}{2}\frac{\mathrm{d}i}{\mathrm{d}t} \quad (r \le a) \quad \text{and} \quad E_{\text{ind}} = -\frac{\mu N' a^2}{2r}\frac{\mathrm{d}i}{\mathrm{d}t} \quad (r > a),} \tag{6.51}$$

E_{ind} of an infinite solenoid

respectively.

We note that the magnetic field inside the solenoid varies synchronously with the current in the winding [$H(t) \propto i(t)$], whereas the induced electric field varies synchronously with the time derivative of the current [$E_{\text{ind}}(t) \propto \mathrm{d}i(t)/\mathrm{d}t$].

Example 6.6 Inhomogeneous Conducting Loop around a Solenoid

Consider the solenoid from the previous example, and assume that a wire loop of radius b ($b > a$) is placed coaxially around it, as shown in Fig. 6.10(a). The two halves of the loop are made from different materials, with conductivities σ_1 and σ_2. The cross-sectional area of both wire parts is S. The magnetic field due to currents induced in the core and in the wire loop can be neglected. Calculate the voltage between the junctions of two wire parts (points M and N).

Solution The system in Fig. 6.10(a) can be analyzed from the circuit-theory point of view, using the equivalent circuit diagram shown in Fig. 6.10(b), where e_{ind} is the induced emf in the contour C in Fig. 6.10(a), while R_1 and R_2 are resistances of the two wire parts constituting the contour. Having in mind the second expression in Eqs. (6.51), this emf amounts to

$$e_{\text{ind}} = -\frac{\mathrm{d}\Phi}{\mathrm{d}t} = -\pi\mu N' a^2 \frac{\mathrm{d}i}{\mathrm{d}t}. \tag{6.52}$$

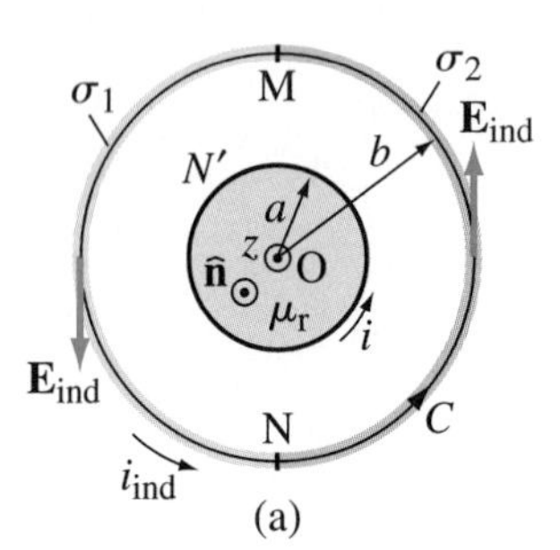

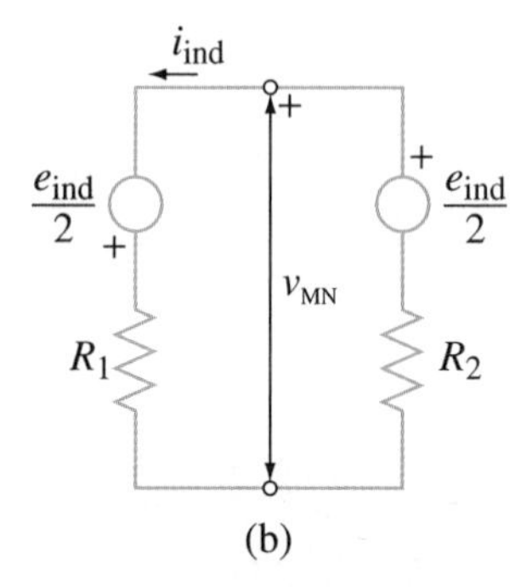

Figure 6.10 (a) Wire loop composed of two parts with different conductivities placed around an infinite solenoid carrying a slowly time-varying current and (b) equivalent circuit diagram; for Example 6.6.

Electromotive forces generated by the magnetic field due to the induced currents are neglected. By means of Eq. (3.85), the resistances are

$$R_1 = \frac{\pi b}{\sigma_1 S} \quad \text{and} \quad R_2 = \frac{\pi b}{\sigma_2 S}. \tag{6.53}$$

Employing Eq. (6.38), the current in the circuit, i.e., the induced current in the loop, is given by

$$i_{\text{ind}} = \frac{e_{\text{ind}}}{R_1 + R_2}. \tag{6.54}$$

Finally, the voltage between points M and N is obtained from Fig. 6.10(b) as

$$v_{\text{MN}} = R_1 i_{\text{ind}} - \frac{e_{\text{ind}}}{2} = \frac{R_1 - R_2}{2(R_1 + R_2)} e_{\text{ind}} = -\frac{\pi\mu(\sigma_2 - \sigma_1)N' a^2}{2(\sigma_1 + \sigma_2)} \frac{\mathrm{d}i}{\mathrm{d}t}. \tag{6.55}$$

Note that in the case of a homogeneous wire loop ($\sigma_1 = \sigma_2$), there is no voltage between different points along the loop.

Example 6.7 Magnetic Field due to Induced Current in a Loop

Refer to the system in Fig. 6.10(a), and assume that $i(t) = 2\cos 1000t$ A (t in s), $N' = 1000$ turns/m, $a = 10$ cm, $b = 20$ cm, $S = 1\ \text{mm}^2$, $\sigma_1 = 57$ MS/m, $\sigma_2 = 15$ MS/m, and $\mu = \mu_0$ (air-filled solenoid). Under these circumstances, find the magnetic flux density vector at the loop center due to the induced current in the loop.

Solution Let us first check whether the current in the solenoid can be considered as slowly time-varying. Since the solenoid is infinitely long, this problem is actually a two-dimensional one, and the check of whether a low-frequency analysis is sufficiently accurate here should be performed in the cross section of the structure containing the wire loop. The maximum dimension of the structure relevant to the field computation in this cross section is $2b$, and the corresponding time of propagation of electromagnetic disturbances is $\tau = 2b/c_0 = 1.33$ ns. This time is much shorter than the period of change of $i(t)$, which equals $T = 2\pi/\omega = 6.28$ ms ($\omega = 1000$ rad/s). We conclude that this structure can indeed be analyzed as a low-frequency (quasistatic) problem. This means that we can use the results of the analysis of the structure performed in the previous example.

From Eq. (6.54), the induced current intensity in the loop is

$$i_{\text{ind}}(t) = -\frac{\mu_0 \sigma_1 \sigma_2 N' a^2 S}{(\sigma_1 + \sigma_2) b} \frac{\mathrm{d}i}{\mathrm{d}t} = 1.5 \sin 1000t \text{ A}. \tag{6.56}$$

The magnetic flux density vector due to this current at the center of the loop is obtained using Eq. (4.19) for the field point defined by $z = 0$ and the contour radius b:

$$\mathbf{B}_{\text{ind}}(t) = \frac{\mu_0 i_{\text{ind}}(t)}{2b} \hat{\mathbf{z}} = 4.7 \sin 1000t\, \hat{\mathbf{z}}\ \mu\text{T}. \tag{6.57}$$

Let us finally compare the field $\mathbf{B}_{\text{ind}}(t)$ to the primary magnetic field – that due to the current $i(t)$ in the solenoid winding. By means of Eq. (6.48), the flux density vector of the primary field inside the solenoid is

$$\mathbf{B}(t) = \mu_0 N' i(t)\, \hat{\mathbf{z}} = 2.5 \cos 1000t\, \hat{\mathbf{z}}\ \text{mT}. \tag{6.58}$$

We see that $|\mathbf{B}_{\text{ind}}(t)|/|\mathbf{B}(t)| = 1.9 \times 10^{-3}$, i.e.,

$$|\mathbf{B}_{\text{ind}}(t)| \ll |\mathbf{B}(t)|. \tag{6.59}$$

In other words, the magnetic field due to the induced current in the loop is absolutely negligible with respect to the magnetic field of the solenoid, which caused the induced current in the first place. We also note that the waveforms of these two fields ($\sin 1000t$ and $\cos 1000t$) are such that $\mathbf{B}_{\text{ind}}(t)$ tends to compensate for the change in $\mathbf{B}(t)$, i.e., it opposes the action that actually created it. Of course, this is in accordance with Lenz's law.

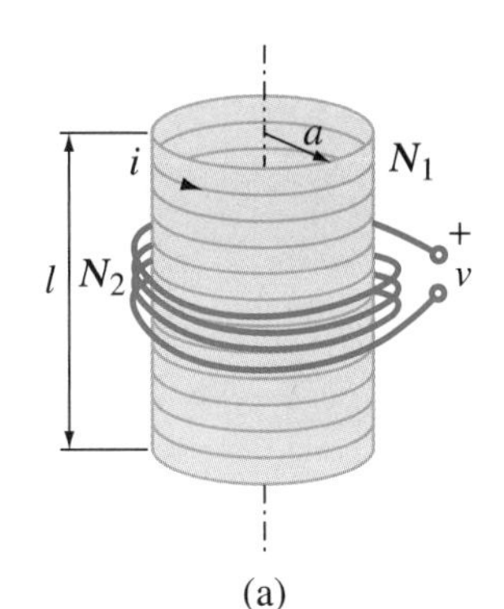

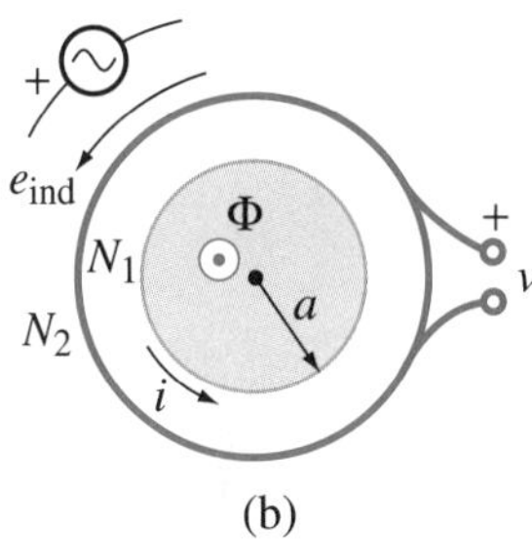

Figure 6.11 An open-circuited short coil placed around a very long solenoid carrying a low-frequency current: (a) three-dimensional view showing the windings and (b) cross-sectional view showing reference directions for the emf and voltage; for Example 6.8.

Example 6.8 Open-Circuited Coil around a Solenoid

An air-filled solenoid of length $l = 2$ m and circular cross section of radius $a = 10$ cm has $N_1 = 1750$ turns of wire. There is a low-frequency time-harmonic current of intensity $i(t) = I_0 \sin \omega t$ flowing through the winding, where $I_0 = 10$ A and $\omega = 10^6$ rad/s. An open-circuited short coil with $N_2 = 10$ turns of wire is placed around the solenoid, as shown in Fig. 6.11(a). Compute the voltage between the terminals of the coil.

Solution The solenoid is very long ($l \gg a$), so that the end effects can be neglected while computing the magnetic field about its center. This means that the solenoid can be considered as infinitely long while computing the magnetic flux through the short coil in Fig. 6.11(a). As the coil consists of N_2 wire turns, this flux is given by

$$\Phi = N_2 \Phi_{\text{single turn}}, \tag{6.60}$$

where $\Phi_{\text{single turn}}$ is the flux through a surface spanned over any of the turns. In other words, the electromotive forces induced in individual turns all add in series, and hence the total emf in the coil is N_2 times that in Eq. (6.52) with $\mu = \mu_0$ and $N' = N_1/l$, which results in

$$e_{\text{ind}} = -N_2 \frac{\mathrm{d}\Phi_{\text{single turn}}}{\mathrm{d}t} = -\frac{\pi \mu_0 N_1 N_2 a^2}{l} \frac{\mathrm{d}i}{\mathrm{d}t} = -\frac{\pi \omega \mu_0 N_1 N_2 a^2 I_0}{l} \cos \omega t. \tag{6.61}$$

There is no current in the coil, because it is open-circuited, so that the voltage across the coil terminals is [Fig. 6.11(b)]

$$v(t) = -e_{\text{ind}}(t) = \frac{\pi \omega \mu_0 N_1 N_2 a^2 I_0}{l} \cos \omega t = 3.45 \cos 10^6 t \text{ kV} \quad (t \text{ in s}). \tag{6.62}$$

Example 6.9 Rectangular Contour near an Infinite Line Current

An infinitely long straight wire carries a slowly time-varying current of intensity $i(t)$. A rectangular contour of side lengths a and b lies in the same plane with the wire, with two sides parallel to it, as shown in Fig. 6.12. The distance between the wire and the closer parallel side of the contour is c. Determine the emf induced in the contour.

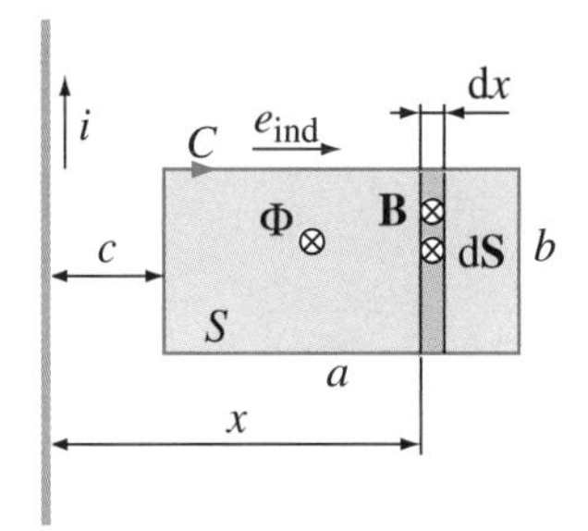

Figure 6.12 Evaluation of the emf in a rectangular contour in the vicinity of an infinitely long wire with a slowly time-varying current; for Example 6.9.

Solution The magnetic flux density vector produced by the current in the wire at any point in the plane of the contour is perpendicular to the plane and at a distance x from the wire (Fig. 6.12) and an instant t, its magnitude is [see Eq. (4.22)]

$$B(x, t) = \frac{\mu_0 i(t)}{2\pi x}. \tag{6.63}$$

By integrating this density across the flat surface spanned over the contour, we obtain the magnetic flux through the contour:

$$\Phi(t) = \int_{x=c}^{c+a} B(x, t) \underbrace{b\,\mathrm{d}x}_{\mathrm{d}S} = \frac{\mu_0 i(t) b}{2\pi} \int_c^{c+a} \frac{\mathrm{d}x}{x} = \frac{\mu_0 i(t) b}{2\pi} \ln \frac{c+a}{c}, \tag{6.64}$$

where $\mathrm{d}S$ is the surface area of a thin strip of length b and width $\mathrm{d}x$ in Fig. 6.12, and the flux is determined with respect to the reference direction into the plane of drawing (note that this integration is similar to that in Fig. 5.26).

For the adopted direction of the flux, the right-hand rule gives the clockwise reference direction for the induced emf in the contour, with respect to which,

$$e_{\text{ind}}(t) = -\frac{\mathrm{d}\Phi}{\mathrm{d}t} = -\frac{\mu_0 b}{2\pi} \ln \frac{c+a}{c} \frac{\mathrm{d}i}{\mathrm{d}t}. \tag{6.65}$$

Example 6.10 Induced Emf in a Coil with a Nonlinear Core

A coil with $N = 400$ turns of wire is wound uniformly and densely about a thin toroidal core of length $l = 40$ cm and cross-sectional area $S = 1\ \text{cm}^2$, as shown in Fig. 6.13(a). A slowly time-varying current whose intensity, $i(t)$, is a periodic alternating triangular-pulse function sketched in Fig. 6.13(b) is established in the coil, where $I_0 = 0.1$ A and $T = 1$ ms. The core is made of a nonlinear ferromagnetic material that exhibits hysteresis effects. In steady state, the operating point (B, H) periodically circumscribes a hysteresis loop that can approximately be represented as in Fig. 6.13(c), where $B_{\text{m}}/H_{\text{m}} = \mu_{\text{h}} = 0.001$ H/m. The resistance of the coil can be neglected. Find the voltage across the coil terminals in the time interval $0 \leq t \leq T$.

Solution The magnetic field intensity in the core varies synchronously with the current in the winding $[H(t) \propto i(t)]$ and is given by

$$H(t) = \frac{Ni(t)}{l} \tag{6.66}$$

[see Eq. (5.53)]. The function $H(t)$ is plotted in Fig. 6.13(d), where $H_{\text{m}} = NI_0/l = 100$ A/m.

From the hysteresis loop in Fig. 6.13(c), we conclude that the magnetic flux density in the core, $B(t)$, first varies as a linear function of time from $B = -B_{\text{m}}$ (for $H = 0$) to $B = B_{\text{m}}$ (for $H = H_{\text{m}}$), then becomes time-invariant while H is reduced from $H = H_{\text{m}}$ to $H = 0$, afterwards it is again a linear (now decaying) function of time when $H(t)$ is reversed and increased in the negative direction from $H = 0$ to $H = -H_{\text{m}}$, and so on. The amplitude of

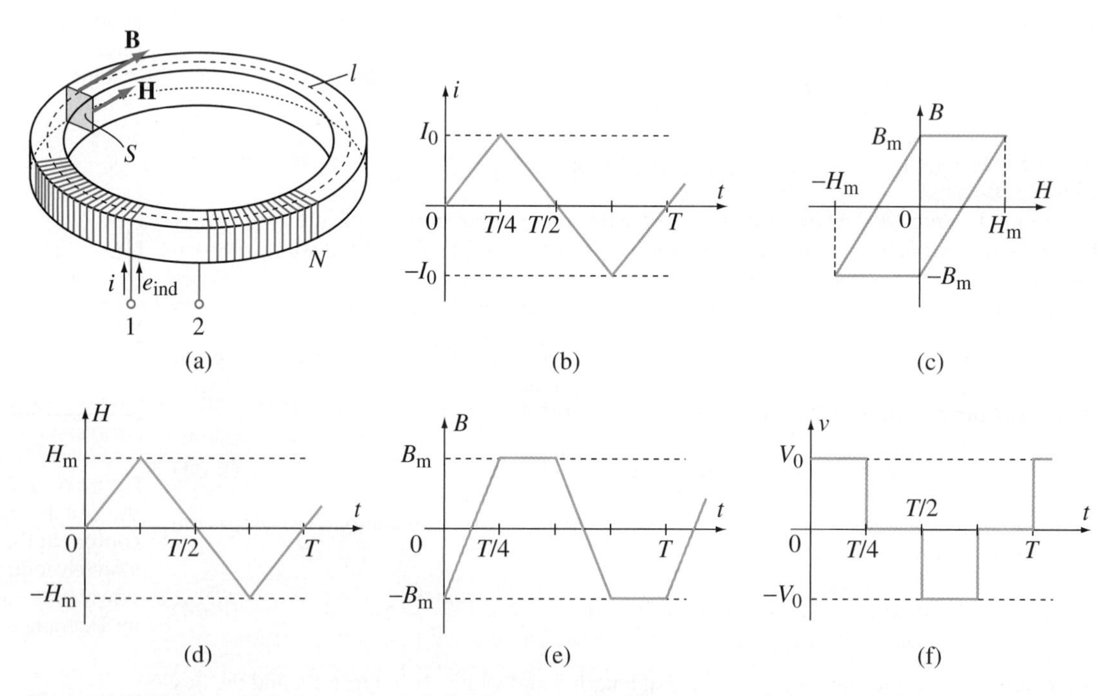

Figure 6.13 Analysis of electromagnetic induction in a coil with a core made from a nonlinear ferromagnetic material (note that $v = v_{12}$); for Example 6.10.

the periodic trapezoidal-pulse waveform of B is $B_{\rm m} = \mu_{\rm h} H_{\rm m} = 0.1$ T, and $B(t)$ is plotted in Fig. 6.13(e).

The magnetic flux through the coil is $\Phi(t) = NB(t)S$ and the induced emf in the coil, $e_{\rm ind}(t)$, is given by Faraday's law of electromagnetic induction, Eq. (6.34). As the resistance of the coil is negligible, the voltage across the coil terminals is

$$v(t) = v_{12}(t) = -e_{\rm ind}(t) = \frac{{\rm d}\Phi(t)}{{\rm d}t} = NS\,\frac{{\rm d}B(t)}{{\rm d}t}. \tag{6.67}$$

This function is proportional to the slope of the function $B(t)$ and is plotted in Fig. 6.13(f). It is a periodic alternating rectangular-pulse function with the same period (T). The pulse amplitude is proportional to the slope of $B(t)$ in the first quarter of the period, that is, $V_0 = NS\,(2B_{\rm m})/(T/4) = 8NSB_{\rm m}/T = 32$ V.

We see that the induced emf and voltage of the coil do not vary synchronously with the time derivative of its current, which is a consequence of the nonlinearity and hysteresis behavior of the core material.

Problems: 6.11–6.17; *Conceptual Questions* (on Companion Website): 6.9–6.16; *MATLAB Exercises* (on Companion Website).

6.6 ELECTROMAGNETIC INDUCTION DUE TO MOTION

Consider a conductor moving with a velocity $\mathbf{v}$ in a static (time-invariant) magnetic field of the flux density $\mathbf{B}$. The field exerts the magnetic Lorentz force, $\mathbf{F}_{\rm m}$, given by Eq. (4.144), on each of the charge carriers in the conductor. This force "pushes" the carriers to move, and separates positive and negative excess charges in the conductor. We can formally divide $\mathbf{F}_{\rm m}$ by the charge of a carrier and obtain $\mathbf{F}_{\rm m}/Q = \mathbf{v} \times \mathbf{B}$. This new quantity, although expressed in V/m, is not a true electric field intensity vector, because it is not produced by an excess charge [Eq. (6.18)] or by a time-varying current [Eq. (6.2)]. By definition given by Eq. (3.107), it represents an impressed electric field intensity vector, which we term here the induced electric field intensity vector due to motion, and write

$$\boxed{\mathbf{E}_{\rm ind} = \mathbf{v} \times \mathbf{B}.} \tag{6.68}$$

induced electric field intensity vector due to motion (unit: V/m)

This field generates an induced electromotive force, as given by Eq. (6.32). Hence, the emf along a line through a conductor between points M and N (Fig. 6.6) due to motion in a time-invariant magnetic field is

$$e_{\rm ind} = \int_{\rm M}^{\rm N} \mathbf{E}_{\rm ind} \cdot {\rm d}\mathbf{l} = \int_{\rm M}^{\rm N} (\mathbf{v} \times \mathbf{B}) \cdot {\rm d}\mathbf{l}. \tag{6.69}$$

This emf is referred to as the emf due to motional induction or simply motional emf. For a moving contour (closed line),

$$\boxed{e_{\rm ind} = \oint_C (\mathbf{v} \times \mathbf{B}) \cdot {\rm d}\mathbf{l}.} \tag{6.70}$$

motional emf (unit: V)

Note that the velocity of different parts of the contour need not be the same, including cases when some parts are stationary while other move in arbitrary directions. In other words, the motion of the contour may include translation, rotation, and deformation (changing shape and size) of the contour in an arbitrary manner.

When a contour moves and/or changes in a static magnetic field, the magnetic flux through the contour changes with time. It is possible to relate the total

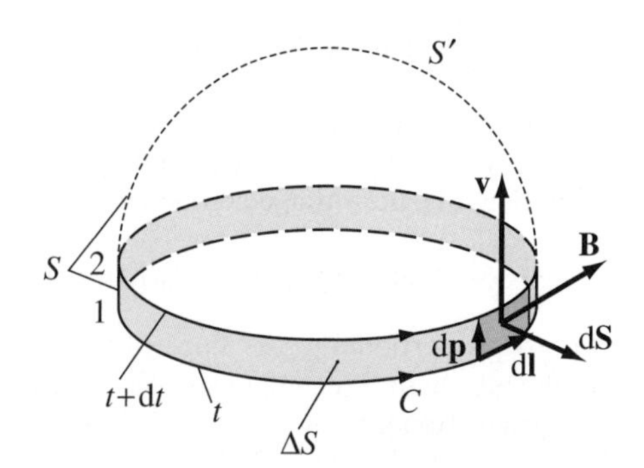

Figure 6.14 A contour moving in a time-invariant magnetic field.

emf induced in the contour to the rate of change of the flux, i.e., to express the motional emf in terms of Faraday's concept of changing flux through the contour, as in Eq. (6.34). To see this, consider the moving contour C in Fig. 6.14. Let $\mathbf{v}$ be the velocity of an element $\mathrm{d}\mathbf{l}$ of the contour. In a time interval $\mathrm{d}t$, this element moves a distance $\mathrm{d}\mathbf{p} = \mathbf{v}\,\mathrm{d}t$ (see Fig. 6.14). Eq. (6.70) thus becomes

$$e_{\mathrm{ind}} = \oint_C (\mathbf{v} \times \mathbf{B}) \cdot \mathrm{d}\mathbf{l} = \frac{1}{\mathrm{d}t} \oint_C (\mathrm{d}\mathbf{p} \times \mathbf{B}) \cdot \mathrm{d}\mathbf{l}. \tag{6.71}$$

Applying the identity $(\mathbf{a} \times \mathbf{b}) \cdot \mathbf{c} = (\mathbf{c} \times \mathbf{a}) \cdot \mathbf{b}$ (scalar triple product is unaffected by cyclic permutation of the order of vectors) to the scalar triple product $(\mathrm{d}\mathbf{p} \times \mathbf{B}) \cdot \mathrm{d}\mathbf{l}$ and noting that $\mathrm{d}\mathbf{l} \times \mathrm{d}\mathbf{p}$ equals the vector surface element $\mathrm{d}\mathbf{S}$ shown in Fig. 6.14, we have

$$e_{\mathrm{ind}} = \frac{1}{\mathrm{d}t} \oint_C \mathbf{B} \cdot (\mathrm{d}\mathbf{l} \times \mathrm{d}\mathbf{p}) = \frac{1}{\mathrm{d}t} \oint_C \mathbf{B} \cdot \mathrm{d}\mathbf{S}. \tag{6.72}$$

This last integral represents the magnetic flux through the strip ΔS swept out by the contour C during the interval $\mathrm{d}t$ (tinted strip between the positions 1 and 2 of the contour in Fig. 6.14). Let us mark this flux by $\mathrm{d}\Phi_{\mathrm{through}\ \Delta S}$, so that

$$e_{\mathrm{ind}} = \frac{\mathrm{d}\Phi_{\mathrm{through}\ \Delta S}}{\mathrm{d}t}. \tag{6.73}$$

The strip ΔS represents the difference in area between the surface S bounded by the contour in position 1 (at instant t) and the surface S' bounded by the contour in position 2 (at instant $t + \mathrm{d}t$), as

$$S = S' \cup \Delta S. \tag{6.74}$$

Designating the magnetic flux through S and S' by Φ and Φ', respectively, the increment in flux from t to $t + \mathrm{d}t$ equals

$$\mathrm{d}\Phi = \Phi' - \Phi, \tag{6.75}$$

and this is exactly the negative of the flux through the strip ΔS, i.e., $\mathrm{d}\Phi = -\mathrm{d}\Phi_{\mathrm{through}\ \Delta S}$. Finally, substituting this equation in Eq. (6.73) leads to $e_{\mathrm{ind}} = -\mathrm{d}\Phi/\mathrm{d}t$, which is the same as in Eq. (6.34). We conclude that the same form of Faraday's law of electromagnetic induction holds for both transformer and motional emf in a contour.

Example 6.11 Moving Metallic Bar in a Static Magnetic Field

A metallic bar of length $a = 2$ m slides without friction at a constant velocity over parallel metallic rails, as shown in Fig. 6.15. The bar is perpendicular to the rails and the mechanical force acting on the bar is $F_{mech} = 4$ N. The whole system is situated in a uniform time-invariant magnetic field of flux density $B = 1$ T. The field lines are perpendicular to the plane of the rails and directed out of the page. A load of resistance $R = 5\ \Omega$ is connected between the rails. The losses in the bar and in the rails, as well as the magnetic field due to induced currents in the system, can be neglected. (a) Find the velocity of the bar. (b) Evaluate the power of Joule's losses in the load and discuss the energy balance in this system.

Solution

(a) The bar moves in a static magnetic field and the emf is induced in it due to motional induction. Since the bar is straight and the field is uniform (i.e, the same everywhere), the expression for the motional emf in Eq. (6.69) becomes

$$\boxed{e_{ind} = (\mathbf{v} \times \mathbf{B}) \cdot \mathbf{l},} \tag{6.76}$$

emf in a straight conductor moving in a uniform magnetic field

where the emf is directed from point M to point N along the bar (Fig. 6.15), $\mathbf{l} = \overrightarrow{MN}$, and $\mathbf{v}$ is the velocity of the bar (which is to be determined). Moreover, since the vectors $\mathbf{v}$, $\mathbf{B}$, and $\mathbf{l}$ are all orthogonal with respect to each other and $|\mathbf{l}| = a$, we can write

$$e_{ind} = vBa. \tag{6.77}$$

The bar and the rails constitute a conducting loop, that is, an electric circuit of the form shown in Fig. 6.8. Hence, there will be a time-invariant induced current in the loop of intensity

$$I_{ind} = \frac{e_{ind}}{R} = \frac{vBa}{R} \tag{6.78}$$

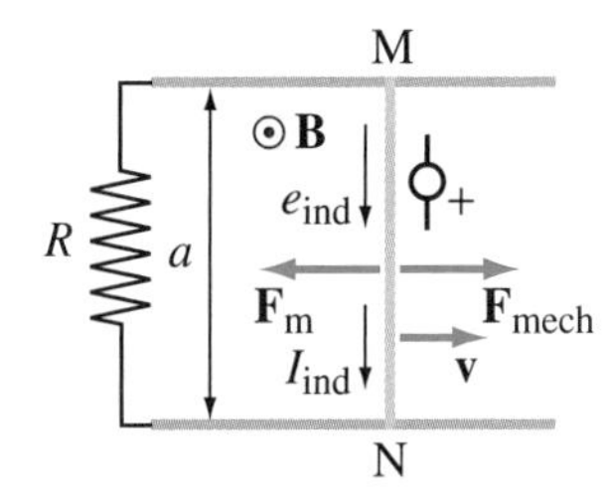

Figure 6.15 A metallic bar moving in a uniform time-invariant magnetic field (elementary electric generator); for Example 6.11.

[see Eq. (6.38)], given with respect to the same reference direction as the emf, where we neglect the resistance of the bar and the rails and the corresponding Joule's losses. The current I_{ind}, on the other hand, produces a secondary magnetic field, which is neglected in evaluating the emf in Eq. (6.76). Note that this field opposes the increasing flux of the primary field $\mathbf{B}$ through the loop, the area of which expands as the bar moves to the right, as yet another example of Lenz's law.

We know that, in the presence of a magnetic field, a current-carrying conductor experiences a magnetic force. From Eq. (4.163), the magnetic force on the metallic bar in Fig. 6.15 comes out to be

$$\mathbf{F}_m = I_{ind}\mathbf{l} \times \mathbf{B} \quad \longrightarrow \quad F_m = I_{ind}aB. \tag{6.79}$$

This force opposes the motion of the bar and the generation of the emf in it, which again is in accordance with Lenz's law. In other words, $\mathbf{F}_m$ is directed oppositely to the mechanical force $\mathbf{F}_{mech}$ on the bar. Furthermore, as the velocity of the bar is constant, these two forces must be exactly equal in magnitude (Newton's second law), that is,

$$F_{mech} = F_m = \frac{vB^2a^2}{R} \quad (v = \text{const}). \tag{6.80}$$

Hence, the velocity we seek is

$$v = \frac{F_{mech}R}{a^2B^2} = 5 \text{ m/s}. \tag{6.81}$$

Eqs. (6.77) and (6.78) now give the values for the induced emf and current in the loop: $e_{ind} = 10$ V and $I_{ind} = 2$ A.

(b) By Joule's law, Eq. (3.77), the power of Joule's losses in the load resistor is

$$P_J = RI_{ind}^2 = 20 \text{ W}. \tag{6.82}$$

Note that, on the other side, the power generated by the induced emf in the bar, i.e., the power of the equivalent ideal voltage generator of emf e_{ind} in Fig. 6.8, equals

$$P_{ind} = e_{ind}I_{ind} = 20 \text{ W} \tag{6.83}$$

[see Eq. (3.121)]. Finally, the mechanical power used to move the bar at the velocity v is obtained as

$$P_{\rm mech} = F_{\rm mech} v = 20\ {\rm W}. \tag{6.84}$$

We see that, as expected,

$$P_{\rm mech} = P_{\rm ind} = P_{\rm J}, \tag{6.85}$$

which is in compliance with the principle of conservation of energy. The system in Fig. 6.15 is a simple example of an electric generator based on electromagnetic induction, where the applied mechanical energy is converted into the electric energy and delivered to the load. The agent by which the energy transfer is carried out is the induced emf in the moving bar, and the energy of this emf is ultimately dissipated to heat in the resistor.

Example 6.12 Rotating Wire Loop in a Static Magnetic Field

A rectangular wire loop of edge lengths a and b rotates with a constant angular velocity ω about its axis in a uniform time-invariant magnetic field of flux density B, as depicted in Fig. 6.16(a). The vector $\mathbf{B}$ is perpendicular to the plane of drawing and is directed out of the page. At an instant $t = 0$, the loop lies in the plane of drawing. The resistance of the loop is R. The magnetic field due to induced currents can be neglected. Calculate (a) the induced emf in the loop and (b) the instantaneous and time-average mechanical power of loop rotation.

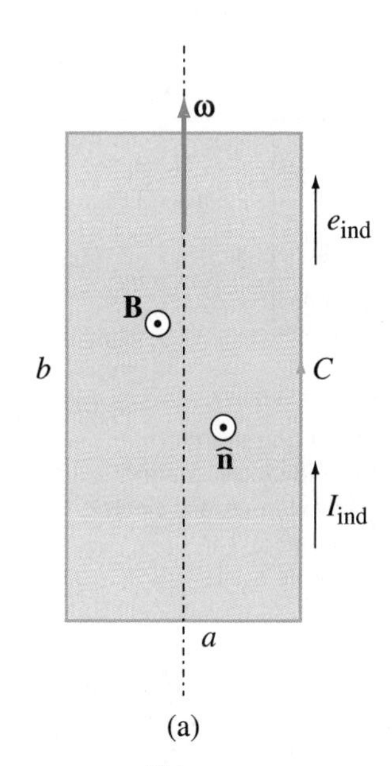

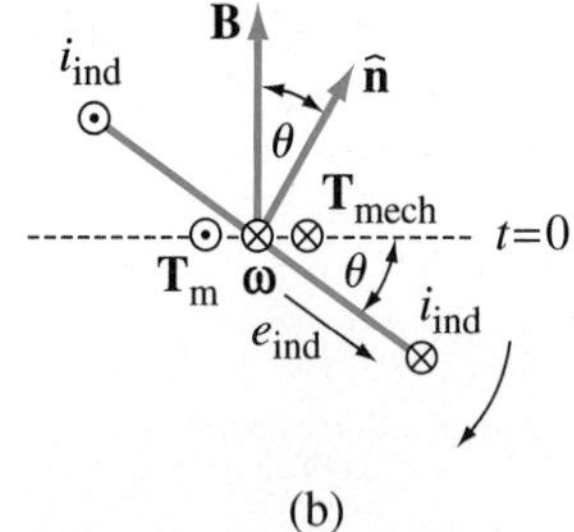

Figure 6.16 A wire loop rotating in a uniform time-invariant magnetic field (elementary ac generator): (a) top view at an instant $t = 0$ and (b) cross-sectional view at an arbitrary instant t; for Example 6.12.

Solution

(a) Referring to Fig. 6.16(b), the magnetic flux through the loop is

$$\Phi = \mathbf{B} \cdot \mathbf{S} = \mathbf{B} \cdot ab\hat{\mathbf{n}} = abB\cos\theta, \tag{6.86}$$

where θ is the angle between the plane of the loop at an instant t and the plane perpendicular to the vector $\mathbf{B}$. From the definition of angular velocity,

$$\frac{{\rm d}\theta}{{\rm d}t} = \omega. \tag{6.87}$$

Since $\omega = {\rm const}$, the solution for θ is

$$\theta(t) = \theta_0 + \omega t, \tag{6.88}$$

where $\theta_0 = 0$ ($\theta = 0$ for $t = 0$). Thus,

$$\Phi(t) = abB\cos\omega t, \tag{6.89}$$

which, substituted in Eq. (6.34), leads to the following expression for the induced motional emf in the loop:

$$e_{\rm ind}(t) = -\frac{{\rm d}\Phi}{{\rm d}t} = \omega abB\sin\omega t. \tag{6.90}$$

(b) The current intensity in the loop amounts to

$$i_{\rm ind}(t) = \frac{e_{\rm ind}}{R} = \frac{\omega abB}{R}\sin\omega t. \tag{6.91}$$

The magnetic field due to this current is neglected in the computation of the emf in Eq. (6.90).

From Fig. 4.38 and Eq. (4.181), the instantaneous torque of magnetic forces acting on the loop is given by

$$T_{\rm m}(t) = i_{\rm ind}abB\sin\theta = \frac{\omega a^2b^2B^2}{R}\sin^2\omega t. \tag{6.92}$$

Eq. (4.180) tells us that the direction of this torque is such that it opposes the rotation of the loop (Lenz's law), i.e., it is opposite to the direction of an externally applied mechanical torque, $\mathbf{T}_{\rm mech}$, that rotates the loop. In order to sustain the rotation at a constant rate,

$\mathbf{T}_{\text{mech}}$ must be exactly equal in magnitude to $\mathbf{T}_{\text{m}}$ (Newton's second law in angular form). Hence, the instantaneous mechanical power used to rotate the loop is

$$P_{\text{mech}}(t) = T_{\text{mech}}\omega = T_{\text{m}}\omega = \frac{\omega^2 a^2 b^2 B^2}{R} \sin^2 \omega t. \qquad (6.93)$$

The same result can also be obtained from energy conservation, as the mechanical power of loop rotation equals the electric power of the circuit, i.e., the power of Joule's losses dissipated in the loop [as in Eq. (6.85)]. This yields

$$P_{\text{mech}}(t) = P_{\text{J}}(t) = Ri_{\text{ind}}^2 = \frac{\omega^2 a^2 b^2 B^2}{R} \sin^2 \omega t. \qquad (6.94)$$

Given that the time-average value of the function $\sin^2 \omega t$ is

$$\frac{1}{T}\int_0^T \sin^2 \omega t \, \mathrm{d}t = \frac{1}{T}\int_0^T \frac{1-\cos 2\omega t}{2} \, \mathrm{d}t = \frac{1}{2T}\left(\int_0^T \mathrm{d}t - \int_0^T \cos 2\omega t \, \mathrm{d}t\right) = \frac{1}{2}, \qquad (6.95)$$

where $T = 2\pi/\omega$ is the period of time-harmonic variation of the emf and current in the loop, the time-average mechanical power of loop rotation equals

$$(P_{\text{mech}})_{\text{ave}} = \frac{\omega^2 a^2 b^2 B^2}{2R}. \qquad (6.96)$$

Note that the system in Fig. 6.16 represents a rotational version of the generator based on translational motion in a static magnetic field in Fig. 6.15. It illustrates the basic principle of an alternating current (ac) generator, where the loop is mechanically rotated in a static magnetic field at a constant rate and the emf and current are induced in the loop of a sinusoidal (time-harmonic) waveform. The angular frequency of this waveform equals the angular velocity ω of the loop rotation.

Example 6.13 Moving Contour near an Infinite dc Line Current

Assume that the current in the straight wire conductor from Fig. 6.12 is time-invariant, with intensity I, and that the contour moves away from the wire at a constant velocity v, as shown in Fig. 6.17. At $t = 0$, the distance of the closer parallel side of the contour from the wire is $x = c$. Determine the emf induced in the contour.

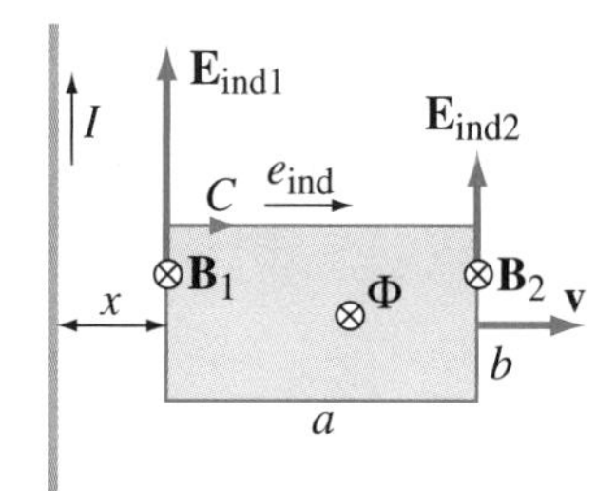

Figure 6.17 Evaluation of the emf in a rectangular contour moving in the magnetic field due to an infinitely long wire with a steady current; for Example 6.13.

Solution The magnetic field produced by the current in the wire is time-invariant, and the system in Fig. 6.17 represents a motional induction version of the same geometry with transformer induction in Fig. 6.12. The magnetic field is nonuniform (magnetic flux density changes in space), and that is why the magnetic flux through the contour is time-varying. As the contour moves uniformly away from the wire, the distance of its closer parallel side from the wire increases linearly in the course of time, and is given by

$$x(t) = c + vt. \qquad (6.97)$$

The magnetic flux density vector around the wire has the same spatial distribution as in Fig. 6.12, so that the magnetic flux through the contour has the same form as in Eq. (6.64), with $i(t)$ substituted by I and c by $x(t)$:

$$\Phi(t) = \frac{\mu_0 I b}{2\pi} \ln \frac{x+a}{x} = \frac{\mu_0 I b}{2\pi} \ln \frac{c+a+vt}{c+vt}. \qquad (6.98)$$

From Eq. (6.34), the motional emf induced in the contour, given with respect to the clockwise reference direction (Fig. 6.17), is

$$e_{\text{ind}}(t) = -\frac{\mathrm{d}\Phi}{\mathrm{d}t} = -\frac{\mathrm{d}\Phi}{\mathrm{d}x}\frac{\mathrm{d}x}{\mathrm{d}t} = -\frac{\mathrm{d}\Phi}{\mathrm{d}x}v = \frac{\mu_0 I a b v}{2\pi}\frac{1}{x(x+a)} = \frac{\mu_0 I a b v}{2\pi(c+vt)(c+a+vt)}. \qquad (6.99)$$

The emf in the contour can also be computed using Eq. (6.70). Note that the induced electric field intensity vector $\mathbf{E}_{\text{ind}}$ due to motion, given by Eq. (6.68), is perpendicular to the pair of contour edges of length a, and hence there is no emf in these edges. On the other hand, $\mathbf{E}_{\text{ind}}$ does not change along each of the remaining two contour edges, which are parallel to it. Its magnitude along the left and right edges in Fig. 6.17 equals

$$E_{\text{ind1}} = vB_1 = \frac{\mu_0 I v}{2\pi x} \quad \text{and} \quad E_{\text{ind2}} = vB_2 = \frac{\mu_0 I v}{2\pi (x+a)}, \tag{6.100}$$

respectively, where B_1 and B_2 are the corresponding magnetic flux densities. Finally, the total emf in the contour is obtained as

$$e_{\text{ind}} = \oint_C \mathbf{E}_{\text{ind}} \cdot d\mathbf{l} = E_{\text{ind1}} b - E_{\text{ind2}} b = \frac{\mu_0 I b v}{2\pi} \left(\frac{1}{x} - \frac{1}{x+a} \right), \tag{6.101}$$

which is the same result as in Eq. (6.99).

Example 6.14 Conducting Fluid Flow in a Static Magnetic Field

Fig. 6.18 shows a system for measurement of fluid velocity that consists of a parallel-plate capacitor situated in a uniform time-invariant magnetic field. A liquid flows between the capacitor plates and the voltage between the plates is measured by a voltmeter. The velocity of the fluid can be considered to be uniform. The magnetic field lines are parallel to the plates. The conductivity and permeability of the liquid are σ and μ_0, respectively. The capacitor plate area is S and the separation between plates is d. Fringing effects in the capacitor can be neglected. The flux density of the applied magnetic field is B. The internal resistance of the voltmeter is R_V, while the resistance of the interconnecting conductors is negligible. Show that the velocity of the fluid is linearly proportional to the voltage V indicated by the voltmeter and find the proportionality constant.

Solution We have a motion (flow) of the conducting liquid (with an unknown velocity $\mathbf{v}$) in the static magnetic field, and therefore an electric field due to motion is induced, given by Eq. (6.68). Vectors $\mathbf{v}$ and $\mathbf{B}$ are mutually orthogonal (Fig. 6.18), and we can write

$$E_{\text{ind}} = vB. \tag{6.102}$$

This field forces the charge carriers in the liquid to move perpendicularly to the direction of the liquid flow, so that positive and negative excess charges are accumulated on the lower and upper capacitor plates, respectively. These charges produce a Coulomb field $\mathbf{E}_q$ (field due to excess charge), which is perpendicular to the plates and practically uniform in the space between them.

Since the voltmeter is not ideal, i.e., its internal resistance is not infinite, there is a steady current flowing through its terminals. By the continuity equation for steady currents, this current continues with the current of the same intensity through the conducting liquid, that is,

$$I_V = I_{\text{liquid}} = JS, \tag{6.103}$$

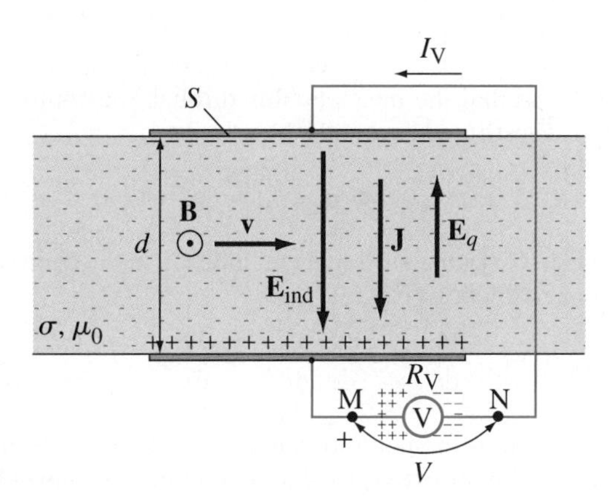

Figure 6.18 Measurement of fluid velocity based on motional electromagnetic induction; for Example 6.14.

with J standing for the current density in the liquid. This current density, in turn, can be expressed in terms of the total electric field in the liquid, given by Eq. (6.16), as follows:

$$\mathbf{J} = \sigma\mathbf{E} = \sigma(\mathbf{E}_q + \mathbf{E}_{\text{ind}}) \quad \longrightarrow \quad J = \sigma(E_{\text{ind}} - E_q), \tag{6.104}$$

where the scalar form of the equation is obtained for the reference directions of vectors $\mathbf{J}$, $\mathbf{E}_{\text{ind}}$, and $\mathbf{E}_q$ adopted in Fig. 6.18.

The voltage that the voltmeter indicates equals the potential difference between its terminals (M and N):

$$V = V_{\text{M}} - V_{\text{N}}, \tag{6.105}$$

which, if evaluated using Eq. (6.22) along a (straight) path through the liquid, can also be expressed as

$$V_{\text{M}} - V_{\text{N}} = E_q d \tag{6.106}$$

(the interconnecting conductors in the voltmeter circuit are ideal, and thus equipotential). Finally, Ohm's law gives

$$V = R_{\text{V}} I_{\text{V}} = R_{\text{V}} JS = R_{\text{V}} S\sigma(E_{\text{ind}} - E_q) = R_{\text{V}} S\sigma\left(vB - \frac{V}{d}\right), \tag{6.107}$$

from which the solution for the velocity of liquid flow turns out to be

$$v = \frac{\sigma S R_{\text{V}} + d}{\sigma S d R_{\text{V}} B} V. \tag{6.108}$$

We see that, indeed, v is linearly proportional to V, where the proportionality constant depends on the parameters of the system in Fig. 6.18 and the conductivity of the fluid.

Problems: 6.18–6.28; *Conceptual Questions* (on Companion Website): 6.17–6.22; *MATLAB Exercises* (on Companion Website).

6.7 TOTAL ELECTROMAGNETIC INDUCTION

Consider now the most general case of electromagnetic induction – that of a moving conductor in a time-varying magnetic field. This is the case where both sources of the induced electric field, namely, the magnetic field change and conductor motion, act simultaneously. Hence, the induced electromotive force in a contour that is moved and/or deformed in a magnetic field that itself varies with time is the sum of the transformer emf, Eq. (6.37), and the motional emf, Eq. (6.70). We thus write

$$e_{\text{ind}} = \underbrace{-\int_S \frac{\partial \mathbf{B}}{\partial t} \cdot \text{d}\mathbf{S}}_{\text{transformer emf}} + \underbrace{\oint_C (\mathbf{v} \times \mathbf{B}) \cdot \text{d}\mathbf{l}}_{\text{motional emf}}, \tag{6.109}$$

and call e_{ind} here the total (transformer plus motional) or complex (combined) emf in the contour.

The motional emf term of Eq. (6.109) can be transformed as described by Eqs. (6.71)–(6.75) and Fig. 6.14, so that the total emf in the contour can be expressed as $e_{\text{ind}} = -\,\text{d}\Phi/\text{d}t$ [same as in Eq. (6.34)], or

$$\boxed{e_{\text{ind}} = \oint_C \mathbf{E}_{\text{ind}} \cdot \text{d}\mathbf{l} = -\frac{\text{d}}{\text{d}t}\int_S \mathbf{B} \cdot \text{d}\mathbf{S}.} \tag{6.110}$$

total induction

This last expression on the right-hand side of the equation represents the total derivative of the magnetic flux through the contour with respect to time, where the change in flux with time is partly due to a change in the magnetic field and partly due to a change in the shape, orientation, and/or position of the contour. These two parts of the flux change correspond to the transformer and motional emf terms of Eq. (6.109).[8] It appears, however, that Eq. (6.34) is the most general form of Faraday's law of electromagnetic induction, which includes both mechanisms by which the magnetic flux through a contour could change. These two mechanisms are the magnetic field variation and contour motion, and except for them, there are no other possibilities that may result in an induced emf in the contour.

Example 6.15 Moving Contour near a Time-Varying Line Current

Refer to Fig. 6.17 and assume that the current in the infinite wire conductor is slowly time-varying, with intensity $i(t)$. Find the emf induced in the moving contour.

Solution We now have a motion of the contour in a time-varying magnetic field, produced by the current in the wire, i.e., a combination of systems in Figs. 6.12 and 6.17. Therefore, the emf is induced in the contour due to combined (transformer plus motional) induction. Combining Eqs. (6.64) and (6.98), the magnetic flux through the contour is

$$\Phi(t) = \frac{\mu_0 i(t) b}{2\pi} \ln \frac{c+a+vt}{c+vt}. \tag{6.111}$$

From Eq. (6.110) [or Eq. (6.34)], the total (combined) emf in the contour is

$$e_{\text{ind}}(t) = -\frac{\mathrm{d}\Phi}{\mathrm{d}t} = \underbrace{-\frac{\mu_0 b}{2\pi} \ln \frac{c+a+vt}{c+vt} \frac{\mathrm{d}i}{\mathrm{d}t}}_{\text{transformer emf}} + \underbrace{\frac{\mu_0 i(t) abv}{2\pi(c+vt)(c+a+vt)}}_{\text{motional emf}}. \tag{6.112}$$

We note that the first term in this expression represents the transformer part of the total emf; it becomes zero in the case of a steady current in the wire and becomes the same as in Eq. (6.65) in the case of a stationary contour. The second term represents the motional part of the total emf; it becomes zero in the case of a stationary contour and becomes the same as in Eq. (6.99) in the case of a steady current in the wire.

Example 6.16 Rotating Loop in a Time-Harmonic Magnetic Field

Assume that the applied field in Fig. 6.16 is a low-frequency time-harmonic magnetic field with flux density $B(t) = B_0 \sin \omega t$, and obtain the emf induced in the rotating loop. Identify the parts of the emf corresponding to the transformer and motional induction.

Solution Obviously, we are now adding a transformer induction component to the system in Fig. 6.16. The magnetic flux in Eq. (6.86) becomes

$$\Phi(t) = abB(t)\cos\theta = abB(t)\cos\omega t. \tag{6.113}$$

Hence, the emf induced in the contour is

$$e_{\text{ind}}(t) = -\frac{\mathrm{d}\Phi}{\mathrm{d}t} = \underbrace{-ab\frac{\mathrm{d}B}{\mathrm{d}t}\cos\omega t}_{\text{transformer emf}} + \underbrace{\omega abB(t)\sin\omega t}_{\text{motional emf}}. \tag{6.114}$$

[8]Note that the division of the induced emf between the transformer and motional parts depends on the chosen frame of reference. The particular division in Eq. (6.109) is given for the stationary frame of reference, attached to the field **B**, with respect to which the contour moves at the velocity **v**. In other words, it is given as measured by a stationary observer (so-called laboratory observer). The total emf, however, is unique and the same for any chosen frame of reference and any observer, including the one moving with the contour.

For the given time-variation $B(t)$, the terms corresponding to the transformer and motional induction appear to be

$$e_{\text{ind(transformer)}} = -\omega ab B_0 \cos^2 \omega t \quad \text{and} \quad e_{\text{ind(motional)}} = \omega ab B_0 \sin^2 \omega t, \tag{6.115}$$

respectively, and the total emf

$$e_{\text{ind}}(t) = -\omega ab B_0 \left(\cos^2 \omega t - \sin^2 \omega t\right) = -\omega ab B_0 \cos 2\omega t = \omega ab B_0 \cos(2\omega t + \pi). \tag{6.116}$$

We see that the frequency of the induced emf (and current) in the contour is twice the frequency of the applied magnetic field.

Example 6.17 Stationary Loop in a Rotating Magnetic Field

A rectangular loop of resistance R is situated in the magnetic field produced by two mutually perpendicular large coils with low-frequency time-harmonic currents. The field due to each coil can be considered to be uniform. The currents in the coils are of equal amplitudes and 90° out of phase, so that the magnetic flux densities they produce are given as $B_1(t) = B_0 \cos \omega t$ and $B_2(t) = B_0 \sin \omega t$, respectively, and shown in Fig. 6.19(a). The sides of the loop are a and b long. Neglecting the magnetic field due to induced current in the loop, find the time-average torque of magnetic forces on the loop.

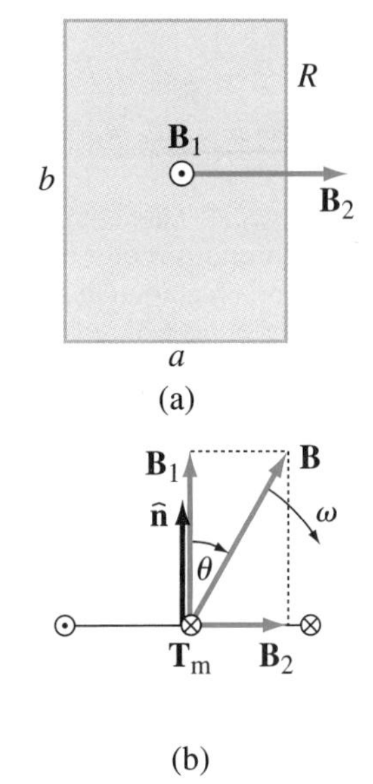

Figure 6.19 A rectangular wire loop exposed to two time-harmonic magnetic fields of equal amplitudes and 90° out of phase (a), which superposed to each other represent a rotating magnetic field (b); for Example 6.17.

Solution From Fig. 6.19(b), the magnitude of the resultant magnetic flux density vector,

$$\mathbf{B}(t) = \mathbf{B}_1(t) + \mathbf{B}_2(t), \tag{6.117}$$

at an arbitrary instant of time equals

$$|\mathbf{B}(t)| = \sqrt{B_1^2(t) + B_2^2(t)} = \sqrt{B_0^2(\cos^2 \omega t + \sin^2 \omega t)} = B_0, \tag{6.118}$$

i.e., it is constant with respect to time. This means that the tip of the vector $\mathbf{B}(t)$ traces a circle of radius B_0 in the course of time. Such a vector belongs to a class of so-called circularly polarized time-harmonic vectors. Let θ mark the angle between vectors $\mathbf{B}(t)$ and $\mathbf{B}_1(t)$ at time t, Fig. 6.19(b). The tangent of this angle is

$$\tan \theta(t) = \frac{B_2(t)}{B_1(t)} = \frac{B_0 \sin \omega t}{B_0 \cos \omega t} = \tan \omega t, \tag{6.119}$$

and hence the rate at which it changes in time is given by Eq. (6.87). This means that $\mathbf{B}$ rotates at a constant angular velocity, equal to the angular frequency ω of the individual magnetic flux densities and currents in the coils. At $t = 0$, $\mathbf{B}_2 = 0$ and $\mathbf{B} = \mathbf{B}_1$, which implies that $\theta(0) = 0$ and $\theta(t) = \omega t$ [see Eq. (6.88)].

As the contour is stationary and vector $\mathbf{B}$ changes in time (its magnitude is constant, but its direction changes), this is a system based on transformer induction. On the other hand, for the generation of emf it is irrelevant whether $\mathbf{B}$ rotates around a stationary loop or a loop rotates (at the same rate) in a static $\mathbf{B}$. This latter case is exactly the system based on motional induction in Fig. 6.16. Exploiting this equivalency,[9] the flux through the loop in Fig. 6.19, emf, current, and instantaneous torque of magnetic forces on the loop are given by Eqs. (6.89), (6.90), (6.91), and (6.92), respectively, with B substituted by B_0. Using Eq. (6.95), the time-average torque is

$$(T_{\text{m}})_{\text{ave}} = \frac{\omega a^2 b^2 B_0^2}{2R}. \tag{6.120}$$

The direction of this torque is the same as the direction of the field rotation [Fig. 6.19(b)], which is in accordance with Lenz's law. Namely, the induced current in the loop and the associated magnetic moment of the loop produce a torque ($\mathbf{T}_{\text{m}}$) that tends to rotate the loop

[9]The possibility to approach the problem of a stationary contour in a rotating magnetic field as a motional induction case is the reason for which we analyze it in the section devoted to total induction.

along with the applied rotating field [i.e., to rotate $\hat{\mathbf{n}}$ closer to $\mathbf{B}$ in Fig. 6.19(b) and thus decrease the angle θ], which is in opposition to the change in the magnetic flux through the contour (caused by the increase in θ) that generated the emf in the first place.

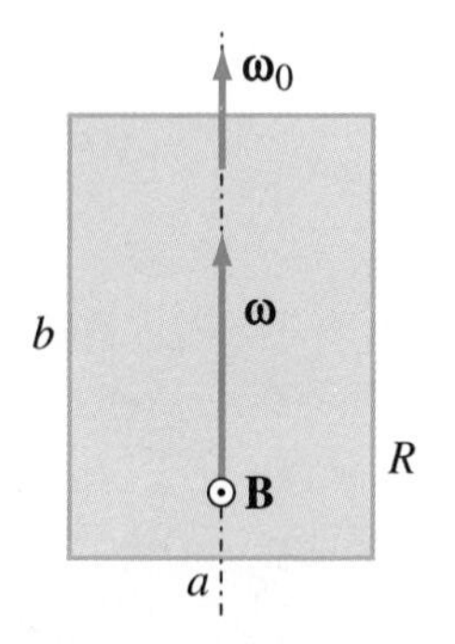

Figure 6.20 Elementary asynchronous motor in the form of a rectangular wire loop rotating with an angular velocity ω_0 in a rotating magnetic field of angular frequency ω, where $\omega > \omega_0$ (top view at $t = 0$); for Example 6.18.

Example 6.18 Rotating Loop in a Rotating Field – Asynchronous Motor

Assume that the loop exposed to the rotating magnetic field from Fig. 6.19 also rotates in the same direction with an angular velocity ω_0 ($\omega_0 < \omega$), as indicated in Fig. 6.20. This device represents an elementary asynchronous motor. Calculate (a) the time-average power of Joule's losses dissipated in the loop, (b) the time-average torque of magnetic forces on the loop, and (c) the efficiency of the motor.

Solution

(a) This is a system based on total (mixed) induction – the magnetic field changes (rotates) in time and the loop moves (rotates). It is called the asynchronous motor because the loop does not rotate in synchronism with the field. The relative rate of rotation of the field with respect to the rotating part of the motor (the loop in our case), called the rotor, equals

$$\Delta\omega = \omega - \omega_0, \tag{6.121}$$

which is referred to as the slipping angular velocity of the asynchronous motor. Consequently, this system can be replaced by either an equivalent system with a stationary loop and a magnetic field rotating with a velocity $\Delta\omega$ (transformer induction case, as in Fig. 6.19) or an equivalent system with a loop rotating with a velocity $\Delta\omega$ in a static magnetic field (motional induction case, as in Fig. 6.16). From Eqs. (6.89), (6.118), and (6.121), the magnetic flux through the loop in Fig. 6.20 is

$$\Phi(t) = ab|\mathbf{B}|\cos\Delta\omega t = abB_0\cos(\omega - \omega_0)t. \tag{6.122}$$

Eq. (6.94) then tells us that the time-average power of Joule's losses in the loop can be written as

$$(P_\mathrm{J})_\mathrm{ave} = k(\omega - \omega_0)^2, \quad \text{where} \quad k = \frac{a^2b^2B_0^2}{2R}. \tag{6.123}$$

(b) By means of Eq. (6.92), the time-average torque of magnetic forces on the loop is

$$(T_\mathrm{m})_\mathrm{ave} = k(\omega - \omega_0), \tag{6.124}$$

where the coefficient k is that in Eq. (6.123). This torque has the same direction as the slipping velocity of the motor.

(c) The rate of the loop rotation in Fig. 6.20 is ω_0, so that Eq. (6.93) gives the following expression for the time-average mechanical power used to rotate the loop:

$$(P_\mathrm{mech})_\mathrm{ave} = (T_\mathrm{m})_\mathrm{ave}\,\omega_0 = k\,\omega_0(\omega - \omega_0). \tag{6.125}$$

The efficiency of the motor is given by

$$\eta = \frac{(P_\mathrm{mech})_\mathrm{ave}}{(P_\mathrm{mech})_\mathrm{ave} + (P_\mathrm{J})_\mathrm{ave}} = \frac{\omega_0}{\omega}, \tag{6.126}$$

where we neglect the losses in the stationary part of the motor (the stator).

Example 6.19 Charge Flow due to a Magnetic Flux Change

Consider a wire contour of resistance R situated in a magnetic field, as shown in Fig. 6.21. If this field is changed and/or the contour is moved in the field during an arbitrary interval of time so that the corresponding net change of the magnetic flux through the contour is $\Delta\Phi$, find the total charge flow Q in the contour during this process.

Solution Since the magnetic flux of the contour, Φ, varies in time during the considered process, an emf e_{ind} is induced in the contour (due to the total induction in the general case), for which we can write

$$e_{ind} = -\frac{d\Phi}{dt} \quad \text{and} \quad e_{ind} = Ri, \tag{6.127}$$

where i is the intensity of current in the contour (Fig. 6.21). Combining these two equations gives

$$d\Phi = -Ri\,dt. \tag{6.128}$$

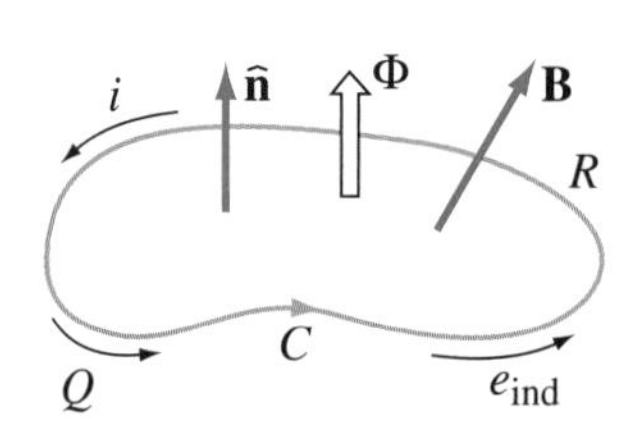

Figure 6.21 Evaluation of the charge flow in a wire contour as a consequence of a change of the magnetic flux through the contour; for Example 6.19.

From Eq. (3.4), the charge that flows through the contour during an elementary time dt is $dQ = i\,dt$, resulting in

$$d\Phi = -R\,dQ. \tag{6.129}$$

We then integrate both sides of the above equation. On the left-hand side of the equation, we thus obtain the total change of flux, $\Delta\Phi$, from its starting value (Φ_1) to the ending value (Φ_2) in the process,

$$\int_{\Phi_1}^{\Phi_2} d\Phi = \Phi_2 - \Phi_1 = \Delta\Phi. \tag{6.130}$$

On the right-hand side of the equation, the integral of dQ equals the total charge flow, Q, and hence

$$\boxed{Q = -\frac{\Delta\Phi}{R},} \tag{6.131}$$

charge flow in a wire contour due to a magnetic flux change

where the reference directions of the charge flow and the magnetic flux are interconnected by the right-hand rule, as indicated in Fig. 6.21.

Example 6.20 Fluxmeter Based on a Charge-Flow Measurement

A fluxmeter consists of a small coil (magnetic sonde) connected to a ballistic galvanometer, as in Fig. 6.22. The cross-sectional area of the coil is S and the number of wire turns is N. The total resistance of the coil and the galvanometer is R. The coil is placed in a uniform time-invariant magnetic field such that the magnetic field lines are perpendicular to the flat surface spanned over the coil cross section (Fig. 6.22). The coil is then removed from the field, and the charge flow indicated by the galvanometer is Q. What is the flux density of the magnetic field?

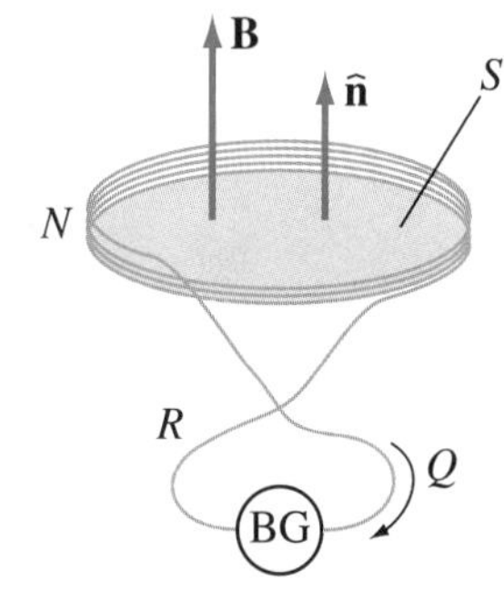

Figure 6.22 Fluxmeter consisting of a small coil and a ballistic galvanometer (BG); for Example 6.20.

Solution The magnetic flux of the coil while it is in the field is

$$\Phi_1 = NBS, \tag{6.132}$$

whereas $\Phi_2 = 0$ after the coil is removed from the field. Using Eq. (6.131), the charge flow through the galvanometer is given by

$$Q = -\frac{\Delta\Phi}{R} = -\frac{\Phi_2 - \Phi_1}{R} = \frac{\Phi_1}{R} = \frac{NBS}{R}, \tag{6.133}$$

which yields the following expression for the magnetic flux density of the field:

$$B = \frac{R}{NS}Q. \tag{6.134}$$

We see that B is linearly proportional to Q, where the proportionality constant is determined by the parameters (R, N, and S) of the fluxmeter in Fig. 6.22.

Thus, by directly measuring the charge flow through its terminals, a ballistic galvanometer connected to a magnetic sonde can be used to indirectly measure the flux density of an unknown magnetic field. In some cases, like in this example, the measurement is based on motional electromagnetic induction (field is time-constant and a coil is either removed from the field or brought into it). In other applications, like in the apparatus for measurement of magnetization curves in Fig. 5.19, a coil is stationary in a field that is either established from zero to its final value B (to be measured) or reduced from some value (B) to zero, so that the measurement is based on transformer electromagnetic induction.

Example 6.21 **Magnetic Flux through a Superconducting Contour**

Prove that the magnetic flux through a superconducting wire contour cannot be changed.

Solution A superconducting wire has zero resistivity [see Eq. (3.23)] and zero total resistance, $R = 0$, so that Eqs. (6.127) give

$\Phi = \text{const}$ through a superconducting contour

$$\boxed{\frac{d\Phi}{dt} = 0 \quad (R = 0),} \tag{6.135}$$

that is, $\Phi = \text{const}$, which completes our proof.

Having in mind that Φ denotes the total existing flux through the contour, this result can be explained as follows. If the magnetic field in which a superconducting contour resides (external or primary magnetic field) is changed and/or the contour is moved in the field, a current is induced in the contour whose magnetic field (secondary field) completely cancels the change of the magnetic flux through the contour (Lenz's law in its extreme form), such that the total flux Φ through the contour remains constant.

Problems: 6.29–6.33; *Conceptual Questions* (on Companion Website): 6.23–6.25; *MATLAB Exercises* (on Companion Website).

6.8 EDDY CURRENTS

Whenever electric field is induced in a conducting medium, electric current is also established, with the same time-dependence as the field (we assume that the medium is linear in terms of its conductivity). An example is a conducting wire loop in a time-varying magnetic field (Fig. 6.8), where the intensity of the induced current in the wire is given by Eq. (6.38). This section is devoted to studying volume induced currents in solid conducting bodies, where many current contours are established throughout the volume of the body as a result of the induced electric field. These currents are perpendicular to the magnetic flux in the body and, since they flow like "eddies" (in water), we call them eddy currents. The eddy current density vector, $\mathbf{J}_{\text{eddy}}$, is related to the electric field intensity vector, $\mathbf{E}$, through Ohm's law in local form:

density of eddy currents

$$\boxed{\mathbf{J}_{\text{eddy}} = \sigma \mathbf{E},} \tag{6.136}$$

where σ is the conductivity of the material. The vector $\mathbf{E}$ represents the actual (measurable) electric field in the material, which, in general, is composed of the induced electric field, $\mathbf{E}_{\text{ind}}$, and the field due to excess charge, $\mathbf{E}_q$ [see Eq. (6.16)]. The vector $\mathbf{J}_{\text{eddy}}$ in Eq. (6.136), therefore, is the actual (total) current density vector in the material, the corresponding components of which are $\sigma \mathbf{E}_{\text{ind}}$ and $\sigma \mathbf{E}_q$. Note that the component $\sigma \mathbf{E}_{\text{ind}}$ alone is usually identified as the induced current density vector in the material. However, as the accumulation of excess charge in the majority of practical systems with an induced electric field is also a result of electromagnetic induction, both components of the eddy current density vector can be said to be induced by the same cause that induces the field $\mathbf{E}_{\text{ind}}$ and electromotive force in the system, i.e., to represent the induced current in the material. This cause, on the other hand, can be related to either transformer or motional induction, as well as to total (combined) induction.

As a consequence of eddy currents, electric power is lost to heat in the material, according to Joule's law, and this is the principle of induction heating. In so-called induction furnaces, for instance, eddy currents are created on purpose to produce local heating in metal pieces and high enough temperatures to melt the metal. In ac machines and transformers, however, the power loss due to eddy currents induced in ferromagnetic[10] cores is undesirable. From Eq. (3.31), the volume density of the power of Joule's losses at a point in the material is proportional to the square of the density of eddy currents, $J_{\rm eddy}$, at that point. Using Eq. (3.32), the total instantaneous power of Joule's (ohmic) losses in the entire body is obtained as

$$\boxed{P_{\rm J} = \int_v \frac{J_{\rm eddy}^2}{\sigma}\, {\rm d}v,} \tag{6.137}$$

Joule's (ohmic) losses due to eddy currents

where v denotes the volume of the body.

Another important consequence of eddy currents is the magnetic field that they produce. By Lenz's law, this field (secondary magnetic field) opposes the change in the primary magnetic flux inside the body, which caused the eddy currents in the first place. While the secondary magnetic field due to induced currents in thin-wire circuits is practically always negligible with respect to the primary magnetic field, this often is not the case with volume eddy currents in solid bodies. The larger the volume of the body and areas of eddy current contours (eddies) in the material the larger the induced emf along the contours and the current intensities, as well as the magnetic field they produce. This effect is also usually not desirable. For instance, the magnetic field due to eddy currents in ferromagnetic cores of ac machines and transformers tends to cancel the primary magnetic flux in the core and thus considerably reduces the efficiency of the device. As an illustration, consider a large core of a rectangular parallelepipedal shape in a uniform time-varying primary magnetic field $\mathbf{B}$, as depicted in Fig. 6.23(a). The secondary (induced) magnetic field, $\mathbf{B}_{\rm ind}$, is the strongest at the center of the cross section of the core, i.e., at the center of all eddy current contours, where all the fields due to these contours add up. Hence, the resultant (primary plus secondary) magnetic flux density is not uniformly distributed over the core cross section; it is the smallest at the center and the largest near the core surface. In other words, practically only the "skin" region below the surface of the core carries the magnetic flux and is effectively used for the operation of the device. This phenomenon is referred to as the skin effect in ferromagnetic cores.

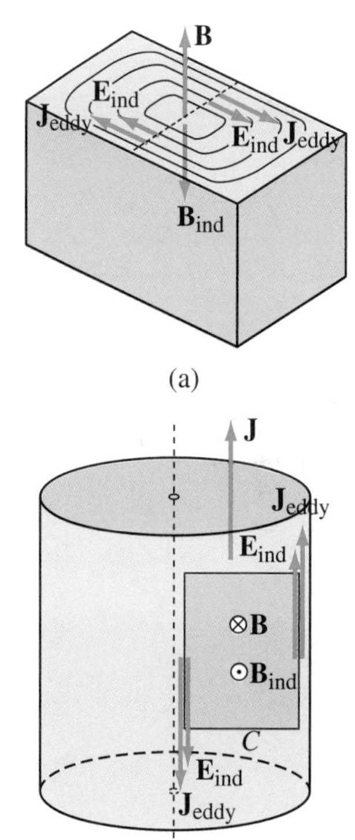

Figure 6.23 Illustration of the skin effect in a parallelepipedal ferromagnetic core with a time-varying magnetic field (a) and in a cylindrical conductor with a time-harmonic current (b).

Note that the skin effect in current conductors at high frequencies is also a consequence of induced (eddy) currents and their magnetic field. To show this, consider a cylindrical conductor carrying a time-harmonic (ac) current of density $\mathbf{J}$, Fig. 6.23(b). Using the analogy with the dc case in Fig. 4.15, the lines of the magnetic field $\mathbf{B}$ due to this current are circles centered at the conductor axis. This field induces an electric field $\mathbf{E}_{\rm ind}$, which is axial in the conductor (field lines are parallel to the conductor axis) and can be qualitatively analyzed by applying Faraday's law of electromagnetic induction, Eq. (6.35), to the rectangular contour C shown in Fig. 6.23(b). The direction of vectors $\mathbf{E}_{\rm ind}$ and $\mathbf{J}_{\rm eddy}$ is determined by Lenz's law (i.e., by the minus sign in Faraday's law). It is such that the secondary magnetic field, $\mathbf{B}_{\rm ind}$, opposes the primary field $\mathbf{B}$. Hence, the eddy current density vector tends to cancel

[10]Ferromagnetic materials are electrically conducting, with large conductivities (e.g., $\sigma_{\rm Fe} = 10$ MS/m for iron).

the current density **J** inside the conductor volume, while adding to its magnitude near the conductor surface. Therefore, the magnitude of the total current density vector is small at the conductor axis and increases towards the conductor surface. The higher the frequency (i.e., the faster the time rate of change $\mathrm{d}/\mathrm{d}t$ in Faraday's law) the larger the induced emf in the contour C and the more pronounced the skin effect. At very high frequencies, the current is restricted to a very thin layer ("skin") near the conductor surface,[11] practically on the surface itself, and can be considered therefore as a surface current and described using the surface current density vector, $\mathbf{J}_\mathrm{s}$ [see Eqs. (3.12) and (3.13)].

Examples in this section include evaluation of eddy current distributions and Joule's losses in several characteristic systems based on each of the three types of electromagnetic induction (transformer, motional, and total induction). In all cases, we shall neglect the magnetic field due to eddy currents and the associated skin effect in conducting bodies. Taking this field into account in the evaluation of the distribution of eddy currents in the body would require a much more complex analysis based on numerical field-computation techniques.

Example 6.22 Eddy Currents in a Thin Conducting Disk

A thin conducting disk of radius a, thickness δ ($\delta \ll a$), conductivity σ, and permeability μ_0 is positioned inside an infinitely long air-filled solenoid, as shown in Fig. 6.24(a). A low-frequency time-harmonic current of intensity $i(t) = I_0 \cos \omega t$ flows through the winding. The number of wire turns per unit of the solenoid length is N'. (a) Determine the distribution of eddy currents induced in the disk, neglecting the magnetic field that they produce. (b) Find the time-average power of Joule's losses dissipated in the disk. (c) Evaluate the magnetic field due to eddy currents at the disk center.

Solution

(a) Eddy currents are induced in the disk due to transformer induction in this structure. The induced electric field inside the solenoid is given by the first expression in Eqs. (6.51). Using Eq. (6.136) and referring to Fig. 6.24(b), the distribution of eddy currents in the

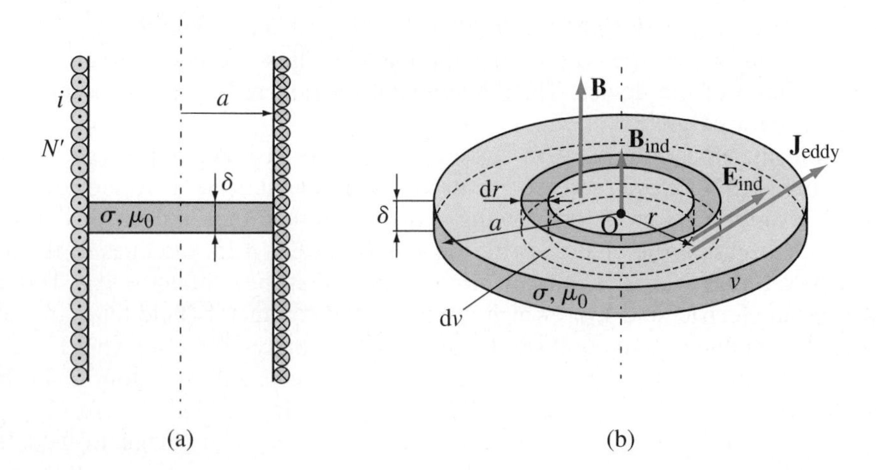

Figure 6.24 (a) Thin conducting disk inside an infinitely long solenoid with a low-frequency time-harmonic current and (b) evaluation of eddy currents in the disk and their magnetic field at the disk center; for Example 6.22.

[11] For example, we shall see in a later chapter that the thickness of the layer that carries most of the current in copper conductors at frequencies higher than about 1 MHz is less than a fraction of a millimeter.

disk is described by the following expression for the induced current density:

$$J_{\text{eddy}}(r,t) = \sigma E_{\text{ind}}(r,t) = -\frac{\mu_0 \sigma N' r}{2}\frac{\text{d}i}{\text{d}t} = \frac{\omega \mu_0 \sigma N' I_0 r}{2}\sin \omega t \quad (0 \le r \le a), \tag{6.138}$$

where r is the radial distance from the solenoid axis and we assume that the electric field due to excess charge is practically zero.

(b) From Eq. (6.137), the instantaneous power of Joule's losses in the disk is

$$P_{\text{J}}(t) = \int_{r=0}^{a} \frac{J_{\text{eddy}}^2(r,t)}{\sigma} \underbrace{2\pi r\,\text{d}r\,\delta}_{\text{d}v} = \frac{\pi\sigma(\omega\mu_0 N' I_0)^2\delta}{2}\sin^2\omega t \int_0^a r^3\,\text{d}r$$

$$= \frac{\pi\sigma(\omega\mu_0 N' I_0)^2 a^4 \delta}{8}\sin^2\omega t, \tag{6.139}$$

with $\text{d}v$ being the volume of an elementary hollow disk of radius r, width $\text{d}r$, and thickness (height) δ [Fig. 6.24(b)]. Having in mind Eq. (6.95), the time-average of this power amounts to

$$(P_{\text{J}})_{\text{ave}} = \frac{\pi\sigma(\omega\mu_0 N' I_0)^2 a^4 \delta}{16}. \tag{6.140}$$

(c) As $\delta \ll a$, the elementary hollow disk of volume $\text{d}v$ in Fig. 6.24(b) can be replaced by an equivalent circular current contour (wire) of radius r and current intensity

$$\text{d}I_{\text{eddy}}(r,t) = J_{\text{eddy}}(r,t)\,\delta\,\text{d}r \tag{6.141}$$

(cross section of the hollow disk through which the current of density $\mathbf{J}_{\text{eddy}}$ flows is a small rectangle of side lengths δ and $\text{d}r$, and surface area $\delta\,\text{d}r$). The magnetic flux density due to this current at the center of the disk is obtained using Eq. (4.19) for $z = 0$ and r substituting a:

$$\text{d}B_{\text{ind}}(r,t) = \frac{\mu_0\,\text{d}I_{\text{eddy}}(r,t)}{2r}. \tag{6.142}$$

By virtue of the superposition principle, the resultant magnetic field due to eddy currents is given by

$$B_{\text{ind}}(t) = \int_{r=0}^{a} \text{d}B_{\text{ind}}(r,t) = \frac{\omega\mu_0^2\sigma N' I_0 \delta}{4}\sin\omega t \int_0^a \text{d}r = \frac{\omega\mu_0^2\sigma N' I_0 a \delta}{4}\sin\omega t. \tag{6.143}$$

Comparing the amplitude B_{ind0} of the field $B_{\text{ind}}(t)$ to the amplitude B_0 of the primary field $B(t) = \mu_0 N' I_0 \cos\omega t$ [Eq. (6.48)] inside the solenoid, we see that

$$\frac{B_{\text{ind0}}}{B_0} = \frac{\omega\mu_0\sigma a\delta}{4}. \tag{6.144}$$

Hence, the magnetic field due to eddy currents in the disk is negligible with respect to the magnetic field due to primary currents in the solenoid only if

$$\frac{\pi f \mu_0 \sigma a \delta}{2} \ll 1, \tag{6.145}$$

where $f = \omega/(2\pi)$ is the frequency of the currents, and whether or not this condition is satisfied depends on the numerical values of the parameters of the structure in Fig. 6.24(a). As an example, for $f = 60$ Hz (power frequency), $\sigma = 58$ MS/m (copper), and $\delta = a/20$, we obtain that only for disks with quite small radii ($a \ll 5$ cm), $B_{\text{ind0}} \ll B_0$, whereas eddy currents in larger disks produce magnetic fields that cannot be neglected with respect to primary fields (at least at the disk center, where the secondary magnetic field is maximum).

Note that the magnetic field $\mathbf{B}_{\text{ind}}$ in Eq. (6.143) is evaluated based on the distribution of eddy currents given by Eq. (6.138). However, in the case when the condition in Eq. (6.145) is not satisfied, this evaluation is not accurate enough and provides only qualitative results, because the starting expression for the induced electric field in Eqs. (6.51) is obtained taking into account only the primary magnetic field on the right-hand side of Faraday's law of electromagnetic induction.

Example 6.23 Eddy Currents in a Thin Ferromagnetic Plate

A thin conducting ferromagnetic plate of length b, width a, and thickness d ($d \ll a$) is situated in a uniform low-frequency time-harmonic magnetic field of flux density $B(t) = B_0 \cos \omega t$. The field lines are perpendicular to the plate cross section, as shown in Fig. 6.25(a). The conductivity of the plate is σ. Neglecting the end effects and the magnetic field produced by eddy currents, find (a) the distribution of these currents throughout the plate and (b) the total time-average power of Joule's losses associated with them.

Solution

(a) This is another example of a structure with eddy currents due to transformer induction. By neglecting the end effects (since $d \ll a$), we assume that the current streamlines in the plate are straight and parallel to the plate surfaces, as indicated in Fig. 6.25(b). From Eq. (6.136), the same is true for the lines of the total electric field intensity vector, $\mathbf{E}$, in the plate. By the same token, the magnitude E of this vector does not depend on the coordinate y in Fig. 6.25(b). Applying Faraday's law of electromagnetic induction, Eq. (6.37), to the rectangular contour C shown in Fig. 6.25(b), we get

$$2E(x,t)\,l = -\frac{\mathrm{d}B}{\mathrm{d}t}\,2xl, \qquad -\frac{d}{2} < x < \frac{d}{2}, \tag{6.146}$$

which yields

$$E(x,t) = -x\,\frac{\mathrm{d}B}{\mathrm{d}t} = \omega B_0 x \sin \omega t, \tag{6.147}$$

where we neglect the magnetic field due to eddy currents in the plate. The density of these currents is given by

$$J_{\text{eddy}}(x,t) = \sigma E(x,t) = \omega\sigma B_0 x \sin \omega t. \tag{6.148}$$

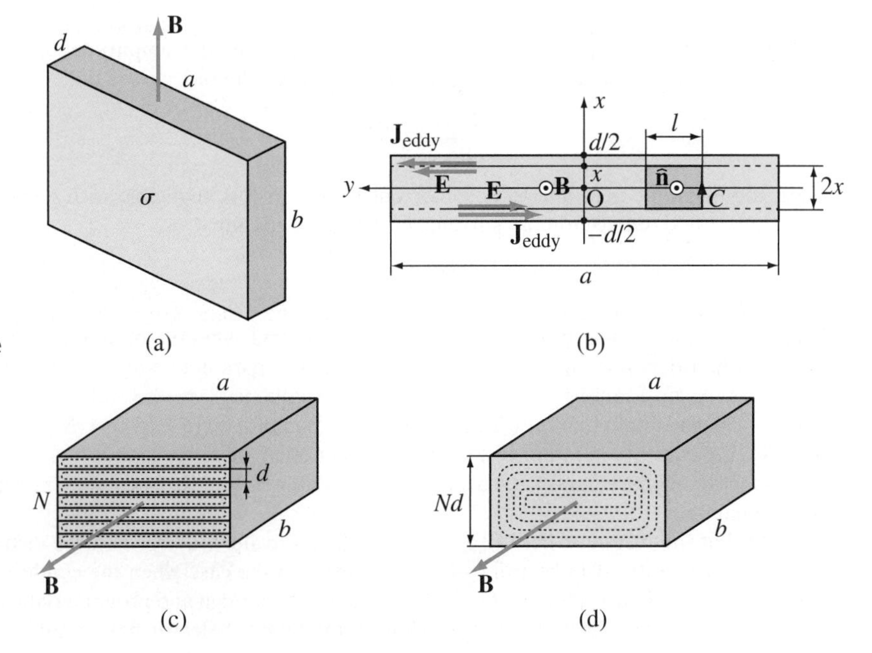

Figure 6.25 (a) Thin conducting ferromagnetic plate in a uniform low-frequency time-harmonic magnetic field, (b) distribution of eddy currents in the plate, (c) N insulated thin plates forming a core of an ac machine or transformer, and (d) a core with the same dimensions made of a single piece of material; for Example 6.23.

(b) The total instantaneous power of Joule's losses dissipated throughout the volume v of the plate is

$$P_J(t) = \int_v \frac{J_{\text{eddy}}^2(x,t)}{\sigma} \underbrace{ab\,dx}_{dv} = \omega^2 \sigma ab B_0^2 \sin^2 \omega t \int_{x=-d/2}^{d/2} x^2\,dx = \frac{\omega^2 \sigma abd^3 B_0^2}{12} \sin^2 \omega t, \tag{6.149}$$

where dv represents the volume of a differentially thin slab of thickness dx used in the volume integration. Finally, averaging in time [Eq. (6.95)] results in

$$(P_J)_{\text{ave}} = \frac{\omega^2 \sigma abd^3 B_0^2}{24}. \tag{6.150}$$

Note that ferromagnetic cores of ac machines and transformers are made of mutually insulated stacked thin plates, as portrayed in Fig. 6.25(c), rather than of a single piece of material, shown in Fig. 6.25(d). With this, the areas of eddy current contours in the core are considerably reduced, and so are the electromotive forces and current intensities along the contours. Consequently, both undesirable effects of eddy currents (Joule's losses and secondary magnetic field) are reduced significantly as well. In specific, the time-average power of Joule's losses in the laminated core in Fig. 6.25(c) can be obtained as N times the power given in Eq. (6.150) for a single thin plate, where N is the number of insulated thin plates. On the other hand, the time-average power of Joule's losses in the homogeneous core in Fig. 6.25(d) can roughly be estimated using the same thin-plate expression in Eq. (6.150) for the plate thickness Nd. Hence, we can write

$$\boxed{(P_J)_{\text{ave1}} \propto Nd^3 \quad \text{and} \quad (P_J)_{\text{ave2}} \propto (Nd)^3,} \tag{6.151}$$

ohmic losses in a laminated core vs. homogeneous core – of an ac machine or transformer

respectively, for these two powers. We see that the reduction of Joule's losses in the laminated core is estimated to be as large as by roughly a factor of N^2 as compared to the homogeneous core with the same dimensions. Also, the skin effect caused by the magnetic field due to eddy currents is much more pronounced in the core in Fig. 6.25(d), where the resultant magnetic flux is "pushed" to the "skin" region near the surface of the core only, so that the entire interior of the core is practically flux-free. In the core in Fig. 6.25(c), on the other side, the "skin" regions are formed in each of the thin insulated plates, so that, although not entirely uniform over the cross section of the core, the resultant flux is much more densely distributed throughout the volume of the core and the ferromagnetic material is much more effectively used for the machine or transformer operation.

Note also that, from Eq. (6.150),

$$\boxed{(P_J)_{\text{ave}} \propto f^2 \sigma,} \tag{6.152}$$

dependence of ohmic losses due to eddy currents on frequency and conductivity

which means that the power loss due to eddy currents in the core increases very rapidly with an increase in the operating frequency of the device (f) and also that it can be reduced by using core materials that have low conductivity (σ). That is why ferrites (which have high permeability but low conductivity) are used instead of ferromagnetics in some applications at high frequencies (e.g., for the cores of high-frequency transformers or multiturn loop antennas).

Example 6.24 Eddy Currents in a Rotating Strip

A very long, thin conducting strip of length l, width a, and thickness δ ($\delta \ll a \ll l$) rotates about its axis at a constant angular velocity ω in a uniform time-invariant magnetic field of flux density B, as shown in Fig. 6.26(a). At an instant $t = 0$, the strip is perpendicular to the vector $\mathbf{B}$. The conductivity of the strip is σ, and permeability μ_0. Determine (a) the distribution of eddy currents and (b) the instantaneous power of Joule's losses in the strip. Neglect the end effects and the magnetic field produced by eddy currents.

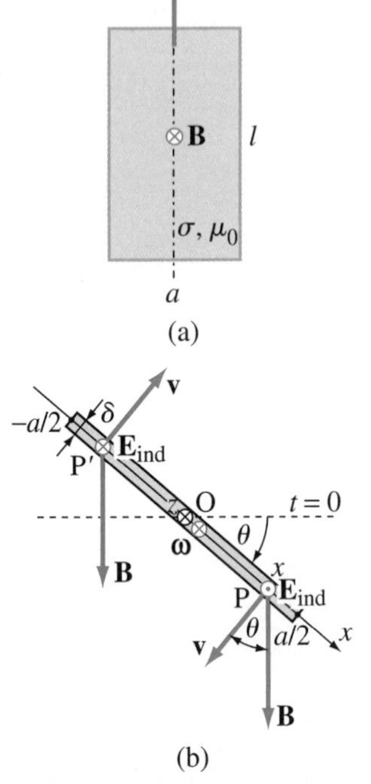

Figure 6.26 Evaluation of eddy currents in a strip rotating in a uniform time-invariant magnetic field: (a) top view at $t = 0$ and (b) cross-sectional view at an arbitrary time t; for Example 6.24.

Solution

(a) This is a structure with eddy currents generated because of motional induction. The induced electric field in the strip is given by Eq. (6.68), where we neglect the magnetic field due to currents in the strip. The vector $\mathbf{E}_{\text{ind}}$ is perpendicular to the cross section of the strip, shown in Fig. 6.26(b). At a point P, the velocity of which is $v = \omega x$,

$$\mathbf{E}_{\text{ind}} = \mathbf{v} \times \mathbf{B} = vB\sin\theta(-\hat{\mathbf{z}}) = -\omega x B \sin\omega t\,\hat{\mathbf{z}}, \quad -\frac{a}{2} < x < \frac{a}{2}, \tag{6.153}$$

where $\theta = \omega t$ is the angle between the strip and the reference horizontal plane at time t [see Eq. (6.88)]. Note that, for the position of the strip in Fig. 6.26(b), the direction of $\mathbf{E}_{\text{ind}}$ is out of the page for $x > 0$ and into the page for $x < 0$ (point P′). The density of eddy currents in the strip, given with respect to the reference direction out of the page in Fig. 6.26(b), is

$$J_{\text{eddy}} = \sigma E_{\text{ind}} = \omega\sigma x B \sin\omega t, \tag{6.154}$$

where we neglect the end effects, i.e., the electric field due to excess charge that accumulates near the ends of the strip and causes the current streamlines to bend and close into themselves near the ends.

(b) The total instantaneous power of Joule's losses in the strip comes out to be

$$P_{\text{J}}(t) = \int_{x=-a/2}^{a/2} \frac{J_{\text{eddy}}^2}{\sigma}\,\mathrm{d}v = \omega^2\sigma l\delta B^2 \sin^2\omega t \int_{-a/2}^{a/2} x^2\,\mathrm{d}x = \frac{\omega^2\sigma a^3 l\delta B^2}{12}\sin^2\omega t, \tag{6.155}$$

where $\mathrm{d}v = l\delta\,\mathrm{d}x$ is the volume of an elementary strip of length l, thickness δ, and width $\mathrm{d}x$.

Example 6.25 Rotating Cylinder in a Rotating Magnetic Field

A very long conducting cylinder of length l, radius a, and conductivity σ rotates at a constant angular velocity ω_0 about its axis, while being exposed to a rotating magnetic field of flux density B and angular frequency ω ($\omega > \omega_0$), as shown in Fig. 6.27(a). Neglect the end effects and the magnetic field due to eddy currents, and calculate the total instantaneous power of Joule's losses in the cylinder.

Solution This is a system based on total (transformer plus motional) induction. However, we can use the concept of slipping velocity given in Eq. (6.121) and replace this system by a cylinder rotating in a static magnetic field with the rate $\Delta\omega = \omega - \omega_0$ in the opposite

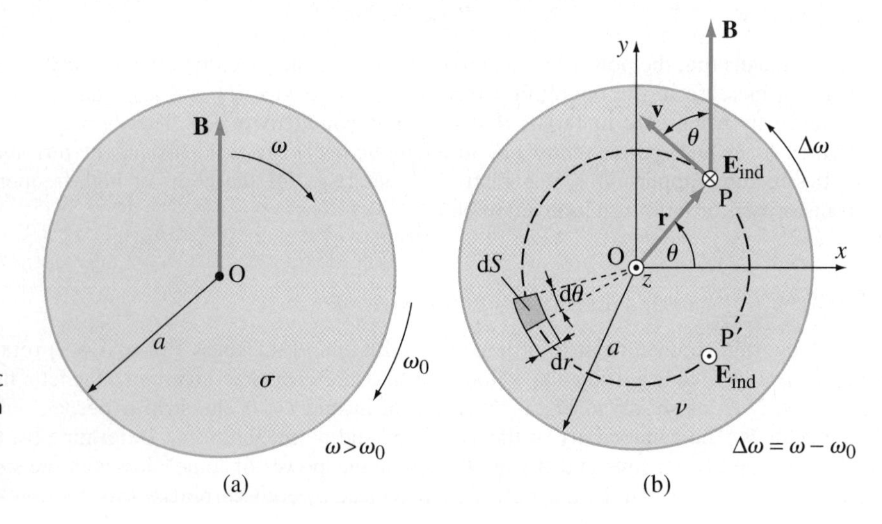

Figure 6.27 (a) Rotating cylinder in a rotating magnetic field and (b) equivalent system with the cylinder rotating in a static magnetic field; for Example 6.25.

direction, as depicted in Fig. 6.27(b). In the equivalent system, which is based on motional induction only, the induced electric field is found in a similar fashion as in Eq. (6.153). At a point P in Fig. 6.27(b),

$$\mathbf{E}_{\text{ind}} = \mathbf{v} \times (B\,\hat{\mathbf{y}}) = vB\sin\theta(-\hat{\mathbf{z}}) = -\Delta\omega r B\sin\theta\,\hat{\mathbf{z}}$$
$$= -\Delta\omega r B\sin(\Delta\omega t + \theta_0)\,\hat{\mathbf{z}}, \quad 0 \le r < a, \quad -\pi \le \theta < \pi, \tag{6.156}$$

where r is the radial distance from the cylinder axis ($r = |\mathbf{r}|$), $v = \Delta\omega r$ is the velocity of the point, and θ is the angle between the vector $\mathbf{r}$ and the x-axis, which equals θ_0 at $t = 0$ and is given by Eq. (6.88) with $\Delta\omega$ as rotation rate. Note the opposite directions of $\mathbf{E}_{\text{ind}}$ at points P and P′ in the cylinder cross section, i.e., for θ positive and negative, respectively, which tells us how the streamlines of eddy currents ($\mathbf{J}_{\text{eddy}}$) close throughout the cylinder.

The total instantaneous power of Joule's losses dissipated in the cylinder is obtained by integration:

$$P_{\text{J}}(t) = \int_v \sigma E_{\text{ind}}^2\,\mathrm{d}v = \int_{r=0}^{a}\int_{\theta=-\pi}^{\pi} \sigma E_{\text{ind}}^2(r,\theta)\,l\,\underbrace{r\,\mathrm{d}\theta\,\mathrm{d}r}_{\mathrm{d}S}$$
$$= (\Delta\omega)^2\sigma B^2 l\int_0^a r^3\,\mathrm{d}r\int_{-\pi}^{\pi}\sin^2\theta\,\mathrm{d}\theta = \frac{\pi(\omega-\omega_0)^2\sigma a^4 l B^2}{4}, \tag{6.157}$$

where $\mathrm{d}S$ is an elementary patch in the cylinder cross section [Fig. 6.27(b)], the sides of which are $\mathrm{d}r$ and $r\,\mathrm{d}\theta$ long, and $\mathrm{d}v = l\,\mathrm{d}S$. We see that the dissipated power is constant with respect to time, which is a consequence of the cylindrical (rotational) symmetry of this problem.

Problems: 6.34–6.45; *Conceptual Questions* (on Companion Website): 6.26–6.30; *MATLAB Exercises* (on Companion Website).

Problems

6.1. Induced electric field of a circular current loop. Assuming that the current in the circular loop in Fig. 4.6 is not steady but slowly time-varying, with intensity $i(t)$, find the induced electric field intensity vector at an arbitrary point along the z-axis (point P).

6.2. Induced electric field of a triangular current loop. For the triangular current loop in Fig. 4.39, calculate the electric field intensity vector at the point P induced by a slowly time-varying current of intensity $i(t)$ in the loop.

6.3. Magnetic field of an EMI source (square contour). Consider the square current contour described in Example 6.2, and find the magnetic field intensity vector at the point M in Fig. 6.2(a). Compare the result with that for the induced electric field intensity vector in Fig. 6.2(d).

6.4. Induced electric field above a square current contour. Determine the electric field intensity vector at a point N placed at a height a above the vertex representing the junction of sides 1 and 2 of the square contour in Fig. 6.2(c), the coordinates of the point thus being $x = y = a/2$ and $z = -a$, induced by the pulse current $i(t)$ in Fig. 6.2(b).

6.5. Magnetic field of a current contour of complex shape. For the wire contour with semicircular and linear parts carrying a low-frequency time-harmonic current from Example 6.4, compute the magnetic field intensity vector at the point O in Fig. 6.4.

6.6. Induced electric field of a semicircular-rectangular loop. Find the induced electric field intensity vector at the point O in Fig. 4.40, assuming that a slowly time-varying current of intensity $i(t)$ flows along the loop.

6.7. Current contour with circular and straight segments. A current of intensity $i(t) = \sin(10^8 t)$ A (t in s) flows along a wire contour with two circular (quarter-circle and 3/4-circle) and two linear parts, shown in Fig. 6.28, where $a = 3$ cm and $b = 9$ cm. (a) Verify that this is a low-frequency current, and compute (b) the induced electric field intensity vector and (c) the magnetic field intensity vector at the point O.

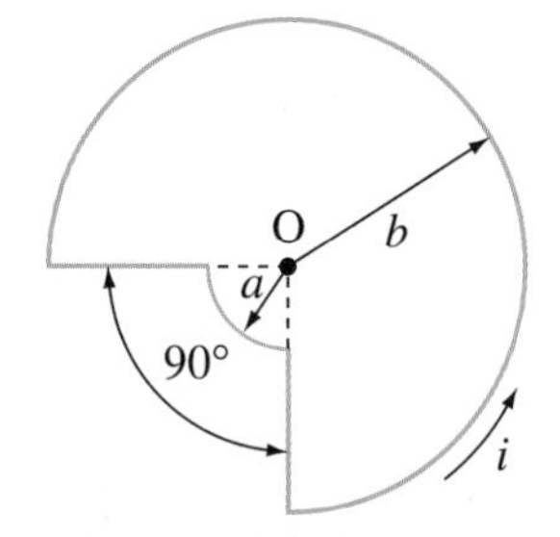

Figure 6.28 Wire contour with a quarter-circle, 3/4-circle, and two linear parts carrying a low-frequency time-harmonic current; for Problem 6.7.

6.8. Induced electric field at the axis of a circular segment. Repeat Example 6.3 but for the field point at an arbitrary location (defined by the coordinate z) along the z-axis (normal to the plane of drawing) in Fig. 6.3(a).

6.9. Induced electric field above or below a semicircular loop. Consider the wire contour made of a semicircle and a straight line from Example 4.4, and assume that it carries a slowly time-varying current of intensity $i(t)$. Determine the induced electric field intensity vector at an arbitrary point along the z-axis in Fig. 4.9(a).

6.10. Voltage from current distribution and total electric field. In a domain v in free space, there is a slowly time-varying distribution of volume currents and charges. We know the current density vector, $\mathbf{J}$, at every point of v, as well as the (total) electric field intensity vector (due to currents and charges in v), $\mathbf{E}$, at every point outside it. Find the voltage between any two points, M and N, outside v.

6.11. Voltage along a straight wire in a quasistatic field. A straight metallic wire is placed in a quasistatic electromagnetic field, for which the magnetic vector potential, $\mathbf{A}$, is known at every point of space. What are (a) the total electric field inside the wire and (b) the voltage between the ends, M and N, of the wire?

6.12. Rectangular wire loop around a solenoid. Consider the solenoid described in Example 6.5, and assume that a rectangular wire loop of edge lengths b and c is placed coaxially around it ($b, c > 2a$), as shown in Fig. 6.29. The magnetic field due to induced currents can be neglected. (a) What is the total induced emf in the loop? Find the emf in the edge MN of the loop, e_{indMN}, in the following two ways, respectively: (b) by integrating along the edge the induced electric field intensity vector due to the current in the solenoid winding [use the relationship in Eq. (4.43) to solve the integral] and (c) by showing that e_{indMN} equals the induced emf in the triangle ΔOMN in Fig. 6.29, and then computing this latter emf as a part of the total emf from (a).

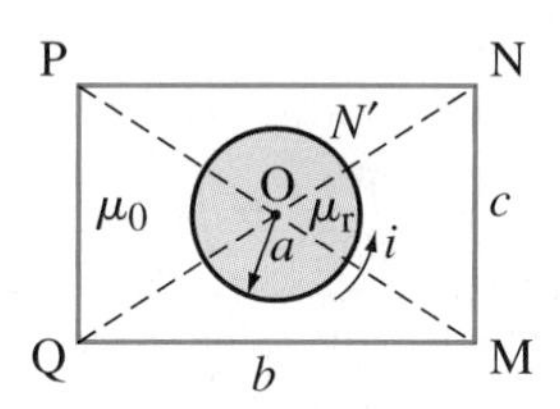

Figure 6.29 Rectangular wire loop placed coaxially around the solenoid in Fig. 6.9; for Problem 6.12.

6.13. Solenoid and a loop of wire with nonuniform cross section. Repeat Example 6.6 but assuming that the wire loop around the solenoid is composed of two parts with the same conductivity, σ, but with different cross-sectional areas, S_1 and S_2 ($S_1 \neq S_2$), and different lengths, determined by angles α and $2\pi - \alpha$, respectively, as shown in Fig. 6.30.

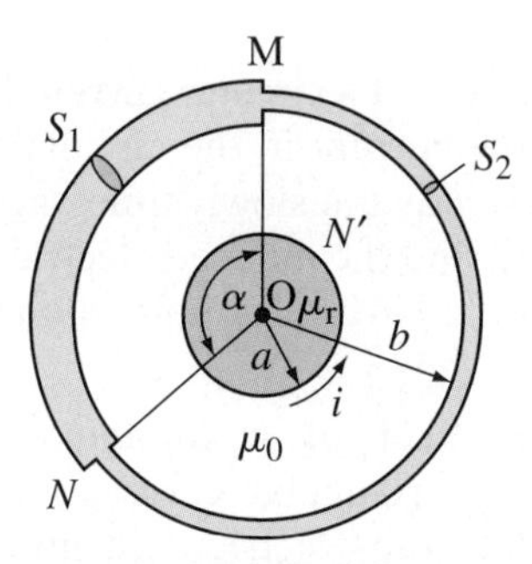

Figure 6.30 Structure in Fig. 6.10(a) but with a loop of wire of nonuniform cross section around the solenoid; for Problem 6.13.

6.14. Complex wire assembly inside a solenoid. Let the wire loop composed of two semicircular parts with conductivities σ_1 and σ_2 from Example 6.6 have radius $a/2$ and be placed coaxially inside an air-filled solenoid, and let two additional linear pieces of wire, with conductivities σ_3 and σ_4, be attached to it, at points M and N, as depicted in Fig. 6.31. Linear wire segments are positioned radially with respect to the solenoid axis, and their ends (points

P and Q) are very close to one another (the gap between them is much smaller than a). Compute the voltage between points P and Q.

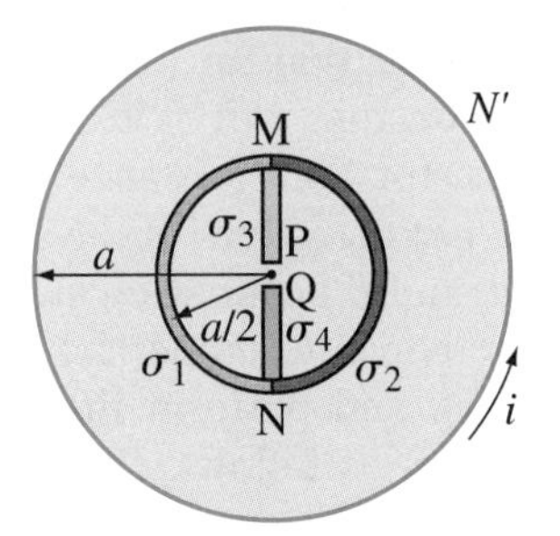

Figure 6.31 Wire assembly of four parts with different conductivities placed inside an air-filled solenoid (the gap between points P and Q is very small); for Problem 6.14.

6.15. **Emf in a rectangular loop due to a two-wire line current.** A very long lossless thin two-wire line in air, with distance between axes of conductors equal to $4a$, is fed at one end by an ideal current generator of low-frequency time-harmonic current intensity $i_g(t) = I_{g0} \cos \omega t$, while the other end of the line is short-circuited. A rectangular wire loop of side lengths a and b is placed in the plane of the line, such that its two sides are parallel to the line and the distance of one of the sides from one of the line conductors is a. Neglecting end and propagation effects, i.e., computing the magnetic field of the line as if it were infinitely long and assuming that the line current is the same in every cross section, as well as the magnetic field due to induced current, find the emf induced in the loop for situations in (a) Fig. 6.32(a) and (b) Fig. 6.32(b), respectively.

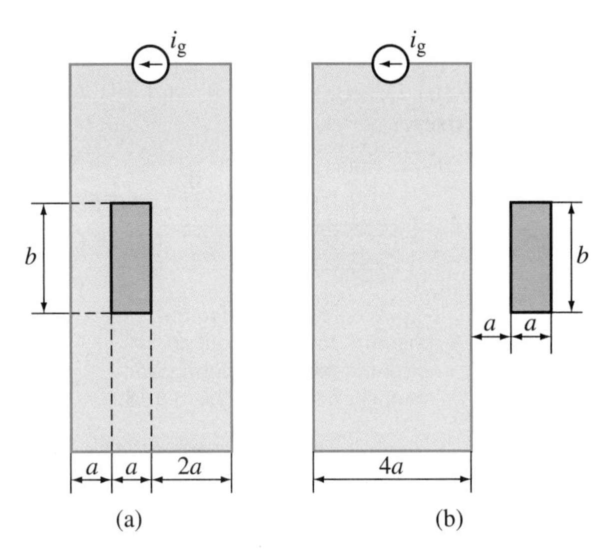

Figure 6.32 Electromagnetic induction in a rectangular loop due to a time-varying current of a thin two-wire line: (a) loop between line conductors and (b) loop on a side of the line; for Problem 6.15.

6.16. **Large square and small circular concentric coplanar loops.** Fig. 6.33 shows two concentric wire loops lying in the same plane, in free space, a large square loop of side length a and a small circular one of radius b ($b \ll a$). The loops are oriented in the same, counterclockwise, direction. The square loop carries a low-frequency time-harmonic current of intensity $i(t) = I_0 \sin \omega t$, and the resistance of the circular loop is R. Determine the induced current in the circular loop, neglecting its own magnetic field.

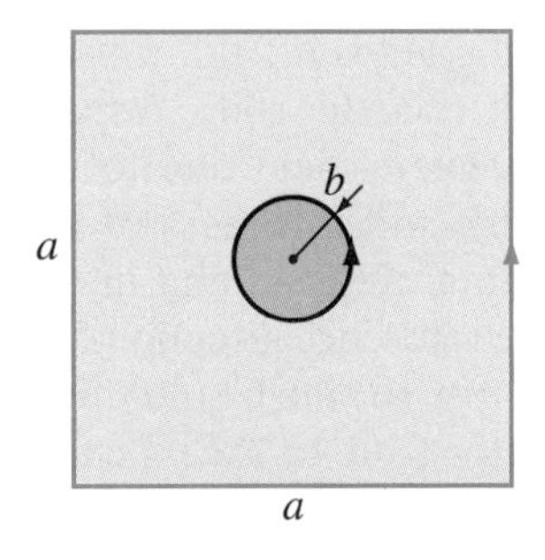

Figure 6.33 Magnetically coupled large square and small circular concentric coplanar loops in free space; for Problem 6.16.

6.17. **Electromagnetic induction in a nonlinear magnetic circuit.** Consider a magnetic circuit in the form of a thin toroidal ferromagnetic core with two windings, like the one in Fig. 5.19. The length and cross-sectional area of the core are $l = 50$ cm and $S = 1$ cm^2, respectively. The primary coil, with $N_1 = 850$ turns of wire wound uniformly and densely along the entire core, is fed by a low-frequency time-harmonic current of intensity $i(t) = I_0 \sin \omega t$, where $I_0 = 0.1$ A and $\omega = 10^6$ rad/s. The secondary coil has only $N_2 = 4$ wire turns encircling the primary winding, and is open-circuited. The idealized hysteresis loop of the core material is that in Fig. 6.13(c), with $B_m = 0.5$ T and $H_m = 170$ A/m. Sketch roughly the voltage waveform across the secondary coil terminals within one period of time-harmonic variation of the current in the primary circuit ($T = 2\pi/\omega$), that is, for $0 \leq \omega t \leq 2\pi$.

6.18. **Rotating rod in a uniform magnetic field.** A conducting rod rotates uniformly with angular velocity w about an axis that splits it onto two unequal parts of lengths l_1 and l_2 ($l_1 \neq l_2$) in a uniform time-invariant magnetic field of flux density B, as shown in Fig. 6.34. The axis of rotation is perpendicular to the rod, and vector **B** is parallel to the axis. (a) Find the total induced emf in the rod. What is the total emf if (b) $l_1 = l_2$, (c) $l_2 = 0$, and (d) $l_1 = 0$, respectively?

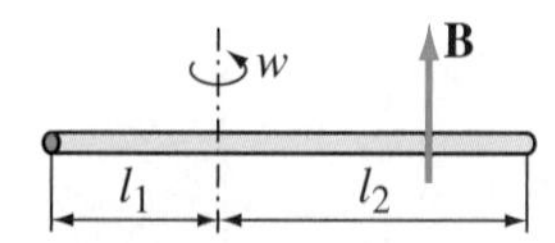

Figure 6.34 Conducting rod uniformly rotating about an excentric axis in a uniform static magnetic field; for Problem 6.18.

6.19. Faraday's wheel. Fig. 6.35 shows Faraday's wheel, consisting of a copper disk of radius a that rotates at a constant angular velocity w about its axis together with an attached axial copper rod, between the poles of a large permanent magnet producing a uniform magnetic field of flux density B, with **B** perpendicular to the disk surface. A pair of terminals, 1 and 2, is defined by taking leads off the rim and the center of the disk (via the rod), respectively. Neglecting the magnetic field due to induced currents in the disk and thickness of the rod, compute (a) the total induced emf in the disk with respect to the reference direction from the disk center to the rim, (b) the Coulomb electric field intensity vector (due to excess charge) at an arbitrary point of the disk, and (c) the voltage ($V = V_{12}$) across the open terminals of the wheel.

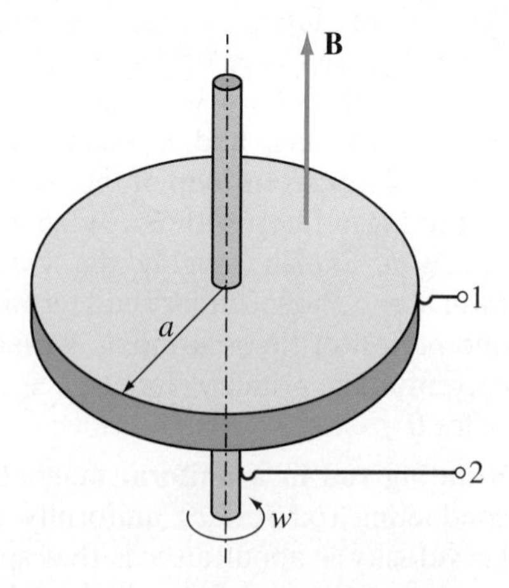

Figure 6.35 Faraday's wheel (a copper disk uniformly rotating between the poles of a large permanent magnet), with open terminals; for Problem 6.19.

6.20. Electric motor – with a linearly sliding bar in a magnetic field. Consider the system with a metallic bar moving in a uniform static magnetic field described in Example 6.11 (Fig. 6.15), and assume that an ideal voltage generator of time-constant emf $\mathcal{E}$ is added in series with the resistor (of resistance R), as shown in Fig. 6.36. Let the values of the system parameters $\mathcal{E}$, R, a, and B be all given and positive (disregard the concrete numerical values from Example 6.11), and perform the analysis of this new system as follows. (a) For the bar at rest ($v = 0$), find the current in the circuit (I) and mechanical force acting on the bar. (b) For the bar sliding (uniformly), sketch the dependence of I on the velocity of the bar (v), for v both positive and negative (movement away from and toward the voltage generator), respectively. (c) Determine the mechanical power of the bar movement (P_{mech}) in terms of the algebraic intensity of the mechanical force on the bar (F_{mech}), for $-\infty < F_{\text{mech}} < \infty$, and sketch this dependence (note that v is a function of F_{mech}). (d) What are the ranges of values of F_{mech} in which the system in Fig. 6.36 operates as a motor ($P_{\text{mech}} < 0$ – the main mode of operation of the system) and as a generator ($P_{\text{mech}} > 0$), respectively? (e) Compute F_{mech} for which the mechanical power of the motor ($|P_{\text{mech}}|$) is maximum. (f) If for the safe operation of the motor, its current has to be smaller in magnitude than I_{max} (so that the motor does not burn out), what is the corresponding range of velocity v? (g) Establish and discuss the power balance for the system, in both motor and generator mode of operation.

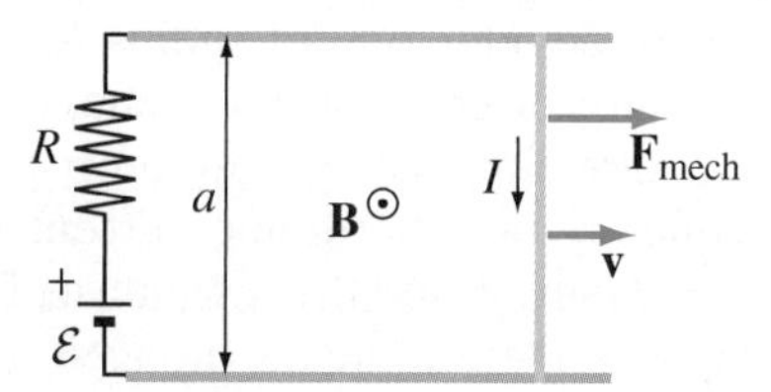

Figure 6.36 Electric motor – with a linearly sliding metallic bar in a static magnetic field (as in Fig. 6.15), and a dc voltage generator added in the circuit; for Problem 6.20.

6.21. Computation for a sliding-bar electric motor. For the electric motor from the previous problem, let $\mathcal{E} = 5$ V, $R = 2\ \Omega$, $a = 1$ m,

$F_{\text{mech}} = 2.5$ N, and $v = 10$ m/s. Find B and the power that the voltage generator $(\mathcal{E}, R)$ delivers to the rest of the structure.

6.22. Electric motor – with a rotating bar in a magnetic field. A metallic bar of length a is attached at its one end to a vertical metallic rod, about which, as an axis, it can rotate so that its other end slides without friction along a circular horizontal metallic rail (of radius a), as portrayed in Fig. 6.37. A voltage generator of time-invariant emf $\mathcal{E}$ and internal resistance R is connected between the rail and the rod (axis), and the whole system is situated in a uniform static magnetic field, whose field lines are perpendicular to the plane of the rail and flux density is B. The algebraic intensity of an externally applied mechanical torque on the rail is T_{mech}, for the reference direction of vector $\mathbf{T}_{\text{mech}}$ in Fig. 6.37. The losses in the bar, rod, and rail, and the magnetic field due to the current in the circuit (I) can be neglected. (a) What are I and T_{mech} for the bar at rest? (b) If the bar rotates uniformly, find its angular velocity (w). (c) Sketch the dependence of the mechanical power of the bar rotation (P_{mech}) on T_{mech}, for $-\infty < T_{\text{mech}} < \infty$, and mark the ranges of system operation as a motor and as a generator, respectively. (d) Compute all relevant powers in the system, and discuss the overall power balance.

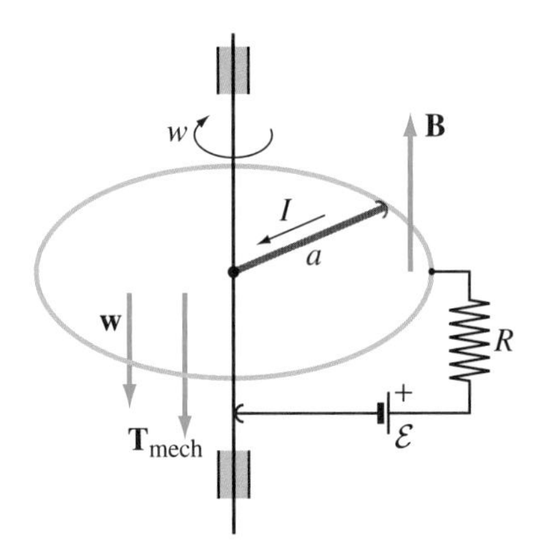

Figure 6.37 Electric motor – with a uniformly rotating metallic bar in a static magnetic field, forming an electric circuit with a dc voltage generator; for Problem 6.22.

6.23. Rotating loop near an infinite dc line current. An infinitely long straight wire conductor situated in air carries a time-invariant current of intensity I. A rectangular wire loop of edge lengths a and b and resistance R rotates with a constant angular velocity w about its axis that is parallel to the conductor and at a distance c from it, as shown in Fig. 6.38(a). At an instant $t = 0$, the loop and conductor lie in the same plane (plane of drawing). Neglecting the magnetic field due to induced currents in the loop, determine (a) the induced emf in the loop and (b) the instantaneous mechanical power of loop rotation. [To compute the magnetic flux through the loop, adopt the integration surface consisting of a cylindrical part of radius r_1 and a flat part of width $r_2 - r_1$, as indicated in Fig. 6.38(b), where r_1 and r_2 can be found using the cosine rule. See also the flux computation in Fig. 7.9(c).]

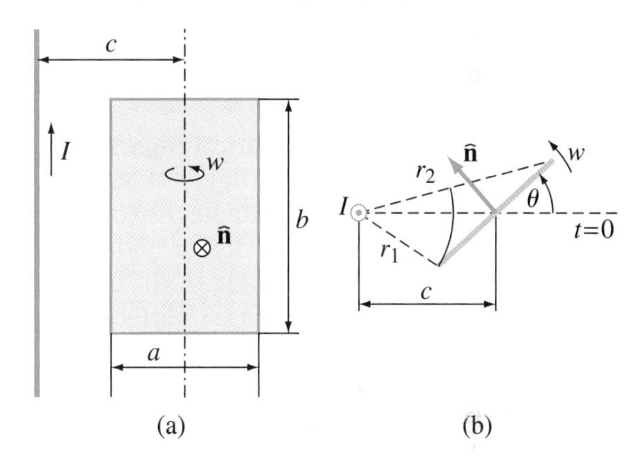

Figure 6.38 Rectangular loop rotating in the magnetic field due to an infinitely long wire with a steady current: (a) side view at an instant $t = 0$ and (b) top (cross-sectional) view at an arbitrary instant t, with a suggested integration surface (in two parts) bounded by the loop; for Problem 6.23.

6.24. System for measurement of fluid velocity with an ideal voltmeter. If in the system for measurement of fluid velocity based on motional electromagnetic induction described in Example 6.14 the voltmeter is ideal, i.e., the intensity of the current flowing through its terminals is so low that it can be assumed to be zero, express (a) the electric field due to excess charge in the region between the capacitor plates and (b) the voltage indicated by the voltmeter in terms of the fluid velocity, v, and other parameters of the system.

6.25. Thévenin generator for fluid flow in a magnetic field. Consider the Thévenin equivalent generator for the system with motional electromagnetic induction from Example 6.14 – that replaces the rest of the system with respect to the voltmeter in Fig. 6.18, as indicated in Fig. 6.39. (a) Show that the emf and internal resistance of this generator, computed as the open-circuit voltage of the circuit (in Fig. 6.18) it represents and input resistance of the circuit (with all the excitations shut down), equal $\mathcal{E}_T = vBd$ and $R_T = d/(\sigma S)$, respectively. (b) Obtain the expression for the velocity of the fluid in Eq. (6.108) using the generator from (a).

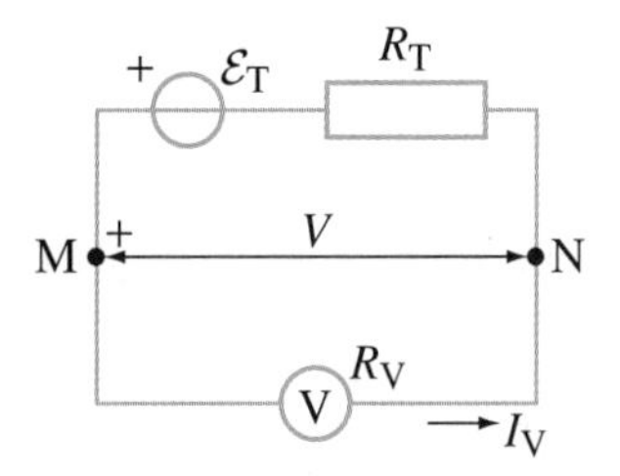

Figure 6.39 Thévenin equivalent generator at terminals (M and N) of the voltmeter in Fig. 6.18; for Problem 6.25.

6.26. Fluid flow through a cylindrical capacitor with dc current. A nonmagnetic liquid of conductivity σ flows between the conductors of a very long coaxial cable of conductor radii a and b ($a < b$), as depicted in Fig. 6.40. Both the inner conductor and the inner surface of the outer conductor of the cable are insulated by a thin layer of perfect dielectric. At one end, the cable is fed by an ideal current generator of time-constant current intensity I_g, while the other end is short-circuited. A cylindrical capacitor of length l is placed inside the cable such that its electrodes (thin cylindrical plates), with radii a and b, are tightly pressed against the respective (insulated) cable conductors. A voltmeter, whose internal resistance, including the resistance of the interconnecting conductors, is R_V, is connected to the capacitor. The velocity of the fluid, considered to be uniform, is v. Calculate (a) the magnetic flux density vector, (b) the induced electric field intensity vector, (c) the current density vector, and (d) the Coulomb electric field intensity vector (due to excess charge) – in the region between the electrodes of the cylindrical capacitor, as well as (e) the voltage indicated by the voltmeter.

6.27. Nonuniform fluid flow and motional induction. Consider the structure from Example 6.14, and assume that the velocity of the fluid is nonuniform, given by $v(x) = v_0[1 - (2x/d)^2]$, $-d/2 \leq x \leq d/2$, where v_0 is a constant, as well as that a variable resistor (rheostat) is connected to the capacitor plates in place of the voltmeter in Fig. 6.18, as shown in Fig. 6.41. (a) Find the parameters of the Thévenin equivalent generator – to replace the rest of the structure with respect to the rheostat, as in Fig. 6.39. (b) If the resistance of the rheostat is R, what is the voltage across it?

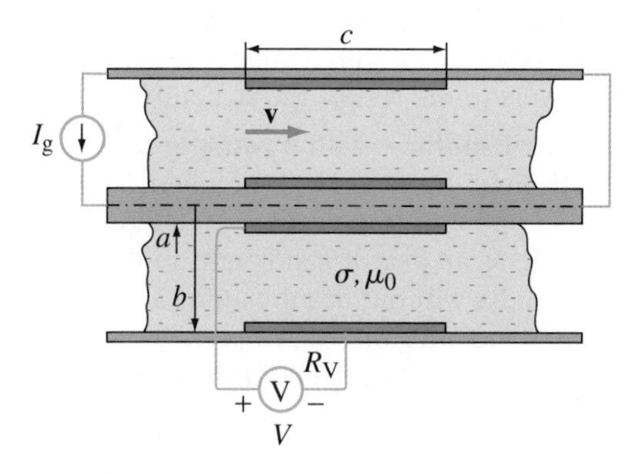

Figure 6.40 System for measurement of conducting fluid flow velocity using motional electromagnetic induction inside a cylindrical capacitor with steady current; for Problem 6.26.

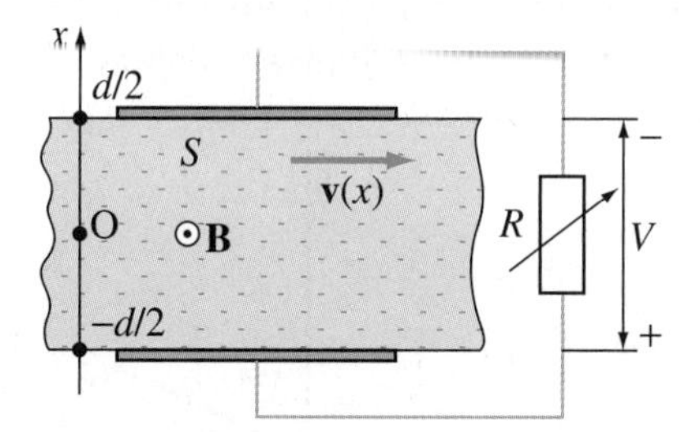

Figure 6.41 Structure in Fig. 6.18 but with a nonuniform fluid velocity, $v(x)$, and a variable resistor connected to the capacitor plates; for Problem 6.27.

6.28. Measurement of fluid velocity and conductivity. For the structure described in the previous problem, let $S = 0.5\ \mathrm{m}^2$, $d = 10$ cm, and $B = 0.1$ T. When the resistance of the rheostat is set to $R = R_0 = 40\ \mathrm{m}\Omega$, its voltage is $V = V_0 = 30$ mV, whereas $V = 2V_0$ if $R = 4R_0$. Based on these data, compute the central velocity (for $x = 0$), v_0, and conductivity, σ, of the fluid, using the Thévenin equivalent generator (from the previous problem).

6.29. Moving contour in a nonuniform dynamic magnetic field. A rectangular contour of side lengths a and b moves along the x-axis with a constant velocity v in a time-harmonic magnetic field of angular frequency ω, which can be considered to be low, and flux density $B(x, t) = B_0 \cos kx \cos \omega t$, where B_0 and k are constants, and vector $\mathbf{B}$ is normal to the plane of the contour, as shown in Fig. 6.42. At $t = 0$, the center of the contour coincides with the coordinate origin ($x = 0$). Find the emf induced in the contour.

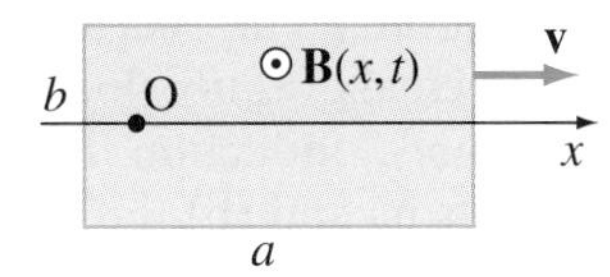

Figure 6.42 Rectangular contour moving in a nonuniform low-frequency time-harmonic magnetic field; for Problem 6.29.

6.30. Rotating loop near an infinite ac line current. Repeat Problem 6.23 but for a low-frequency time-harmonic current of intensity $i(t) = I_0 \cos \omega t$ flowing through the infinitely long wire conductor in Fig. 6.38(a), where $\omega = w$.

6.31. Small loop in the magnetic field of a rotating large loop. Assume that the current in the large square loop from Problem 6.16 is time-constant, of intensity I, and that this loop uniformly rotates with an angular velocity ω about its axis of symmetry that is parallel to a pair of its sides. At an instant $t = 0$, it is in the same plane with the small circular loop (as in Fig. 6.33). The magnetic field due to induced currents in the circular loop can be neglected. Find (a) the induced emf in the circular loop and (b) the instantaneous and time-average torque of magnetic forces acting on it.

6.32. Two rotating loops. If in the previous problem the small loop situated in the magnetic field of the rotating large loop also rotates in the same direction with an angular velocity ω_0 ($\omega_0 < \omega$), determine (a) the time-average power of Joule's (ohmic) losses in the small loop and (b) the time-average mechanical power of its rotation.

6.33. Charge flow through the secondary coil on a magnetic core. Fig. 6.43 shows a thin linear ferromagnetic core with cross-sectional area $S = 1\ \text{cm}^2$, mean length $l = 20$ cm, and two windings each having $N = 100$ wire turns and resistance $R = 10\ \Omega$. The first winding is connected via a switch K to a dc voltage generator of emf $\mathcal{E} = 9$ V and internal resistance $R_1 = 20\ \Omega$. The second winding is terminated in a ballistic galvanometer, whose resistance is $R_2 = 5\ \Omega$. The switch K is first open, and there is no residual magnetization in the core. The switch is then closed, and the charge flow indicated by the galvanometer is $Q = 600\ \mu$C. What is the relative permeability of the core?

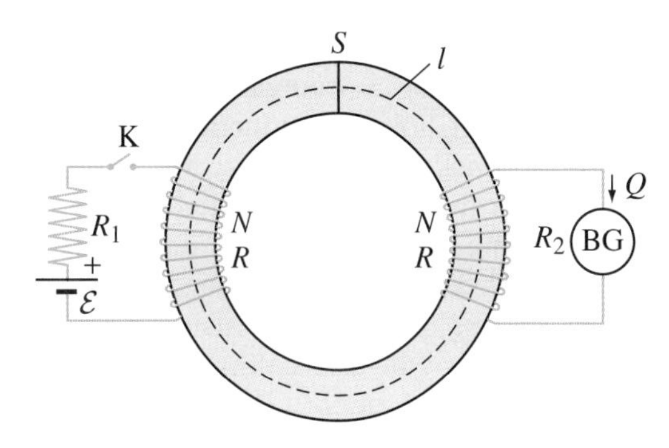

Figure 6.43 Linear ferromagnetic core with two coils, and a voltage generator and ballistic galvanometer (BG) in the primary and secondary circuits, respectively; for Problem 6.33.

6.34. Thin conducting disk in the gap of a magnetic circuit. For the simple linear magnetic circuit from Example 5.14, assume that the cross section of the core in Fig. 5.30(a) is a circle of radius a ($S = \pi a^2$), that the air gap is completely filled with a thin conducting disk of radius a, thickness l_0, conductivity σ, and permeability μ_0, like the one in Fig. 6.24, and that the current in the coil (with N turns of wire) is low-frequency time-harmonic, with intensity $i(t) = I_0 \cos \omega t$. The length and relative permeability of the ferromagnetic portion of the circuit are l and μ_r, respectively. Neglecting the magnetic field due to eddy currents in the disk and core, find l_0 such that the time-average power of Joule's losses dissipated in the disk is maximum, and find that maximum power.

6.35. Eddy currents in Faraday's wheel. (a) Compute the time-average power of Joule's losses due to induced (eddy) currents in Faraday's wheel described in Problem 6.19. (b) What is the magnetic field that these currents produce at the center of the wheel?

6.36. Eddy currents in an infinite conducting cylinder. An infinitely long solenoid with N' turns of wire per unit of its length is wound about a conducting ferromagnetic (infinitely long) cylinder of radius a, permeability μ, and conductivity σ. The winding carries a low-frequency time-harmonic current of intensity $i(t) = I_0 \sin \omega t$, and the medium outside the solenoid is air. (a) Determine the time-average per-unit-length power of ohmic losses due to eddy currents induced in the cylinder, neglecting the magnetic field they produce, and then (b) find this magnetic field at the axis of the cylinder (use the procedure from Example 4.14).

6.37. Hollow disk in a triangular-pulse magnetic field. A thin hollow conducting disk of radii a and b, thickness δ ($\delta \ll a < b$), conductivity σ, and permeability μ_0 is situated in a uniform slowly time-varying magnetic field, such that the field lines are perpendicular to the disk, as shown in Fig. 6.44. The intensity of this field, $H(t)$, is a periodic alternating triangular-pulse time function of amplitude H_m and period T, sketched in Fig. 6.13(d). Neglecting the magnetic field produced by eddy currents, find (a) the distribution of these currents throughout the disk and (b) the total instantaneous and time-average powers of Joule's losses associated with them, and then (c) compute the magnetic field due to eddy currents at the disk center (point O).

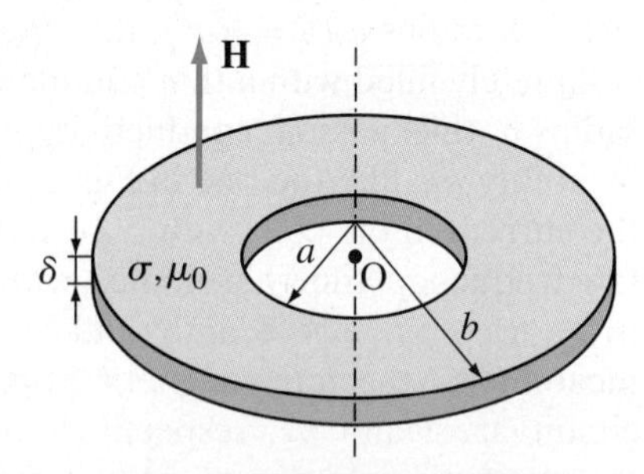

Figure 6.44 Thin hollow conducting disk in a uniform slowly time-varying magnetic field, whose intensity, $H(t)$, is sketched in Fig. 6.13(d); for Problem 6.37.

6.38. Induction furnace. Fig. 6.45 shows an induction furnace consisting of a very long air-filled solenoid, with circular cross section of radius a and length l ($l \gg a$), and a toroidal channel (carrying a metal piece that is heated), with a rectangular cross section, placed coaxially around the solenoid and centrally with respect to its length. The inner and outer radii of the toroid are b and c ($a < b < c$), and its height is h. The solenoid has N turns of wire with a low-frequency time-harmonic current of intensity $i(t) = I_0 \cos \omega t$. The channel is completely filled with a metal of conductivity σ and permeability μ_0. Neglecting the end effects (i.e., computing the induced electric field of a solenoid as if it were infinitely long) and the magnetic field due to eddy currents, find (a) the distribution of eddy currents in the channel (these currents produce local heating in the metal, namely, induction heating, that ultimately melts it) and (b) the total time-average power of Joule's losses dissipated to heat in the metal.

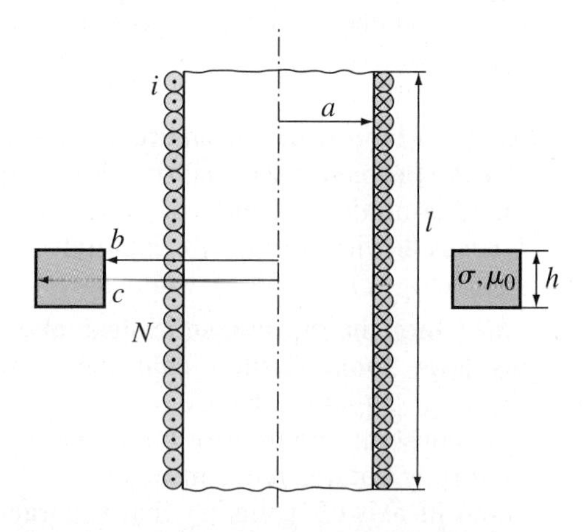

Figure 6.45 Induction furnace: eddy currents accompanying the induced electric field of a very long solenoid with a low-frequency time-harmonic current in the winding heat and melt metal in a toroidal channel of rectangular cross section placed around the solenoid; for Problem 6.38.

6.39. Eddy currents in a thin conducting spherical shell. Consider the solenoid described in Example 6.22, and assume that a thin conducting spherical shell of radius b ($b < a$), thickness δ ($\delta \ll b$), conductivity σ, and permeability μ_0 is placed inside it, such that the sphere center lies on the solenoid axis. Determine the time-average power of Joule's losses in the shell, due to eddy currents, neglecting the magnetic field that they produce.

6.40. Loss power in a laminated ferromagnetic core. A laminated conducting ferromagnetic core in the form of a packet of N insulated stacked thin plates of length b, width a, thickness d ($d \ll a$), and conductivity σ is placed in a uniform slowly time-varying magnetic field, such that the field lines are perpendicular to the packet cross section, as shown in Fig. 6.25(c). The flux density of this field, $B(t)$, is a periodic alternating triangular-pulse time function of amplitude B_m and period T, like the function sketched in Fig. 6.13(d). End effects and magnetic field due to eddy currents can be neglected. (a) Find the total time-average power of Joule's losses in the core. (b) How large has to be N for a given total thickness of the packet, c, so that the total loss power does not exceed a given value P?

6.41. **Torque on a rotating strip in a magnetic field.** For the conducting strip rotating in a uniform time-constant magnetic field from Example 6.24, find the instantaneous torque of magnetic forces on the strip – (a) by integrating torques of magnetic forces on elementary strips (with eddy currents) of width dx in Fig. 6.26(b) (see Example 4.22) and (b) from energy conservation (see Example 6.12), respectively. (c) What is the time-average torque on the strip?

6.42. **Inhomogeneous strip in two orthogonal magnetic fields.** Fig. 6.46 shows a very long, thin nonmagnetic conducting strip of length l, width a, and thickness δ ($\delta \ll a \ll l$) lying in the xy-plane of a Cartesian coordinate system. The conductivity of the strip varies with the x coordinate, and is given by $\sigma(x) = 4\sigma_0 x^2/a^2$, $-a/2 \le x \le a/2$, where σ_0 is a positive constant. The strip is exposed to two uniform low-frequency time-harmonic magnetic fields of flux density vectors $\mathbf{B}_1(t) = B_0 \cos\omega t\,\hat{\mathbf{x}}$ and $\mathbf{B}_2(t) = B_0 \sin\omega t\,\hat{\mathbf{z}}$, respectively. Neglecting the end effects and the magnetic field due to eddy currents, compute (a) the time-average power of Joule's losses dissipated in the strip and (b) the time-average torque of magnetic forces acting on the strip.

6.43. **Eddy currents in two crossed rotating strips.** Fig. 6.47 shows a cross section of a very long conductor, of length l, conductivity σ, and permeability μ_0, consisting of two crossed (at a right angle) thin strips, each of width a and thickness δ ($\delta \ll a \ll l$). The conductor uniformly rotates in air about its axis, with an angular velocity ω, in a uniform magnetostatic field, of flux density B (Fig. 6.47). At $t = 0$, one of the strips is parallel to the field lines. Neglecting the end effects and the magnetic field due to eddy currents in the conductor, find (a) the induced electric field vector at an arbitrary point of each of the two strips, (b) the instantaneous power of Joule's losses in each of the strips, (c) the total instantaneous torque of magnetic forces on the conductor, and (d) the time-average torque on the conductor.

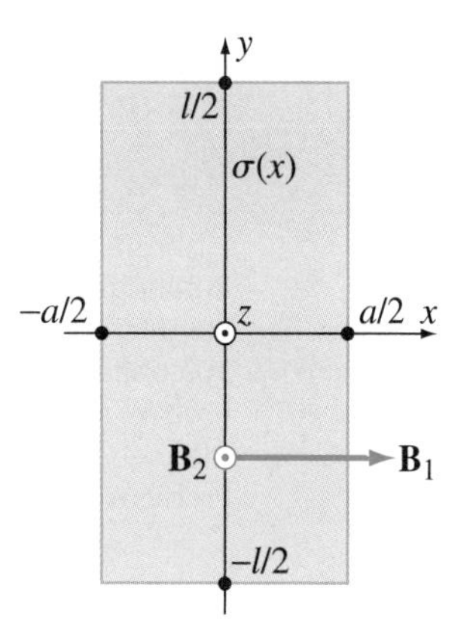

Figure 6.46 Continuously inhomogeneous conducting strip situated in two mutually orthogonal time-harmonic magnetic fields of equal amplitudes and 90° out of phase; for Problem 6.42.

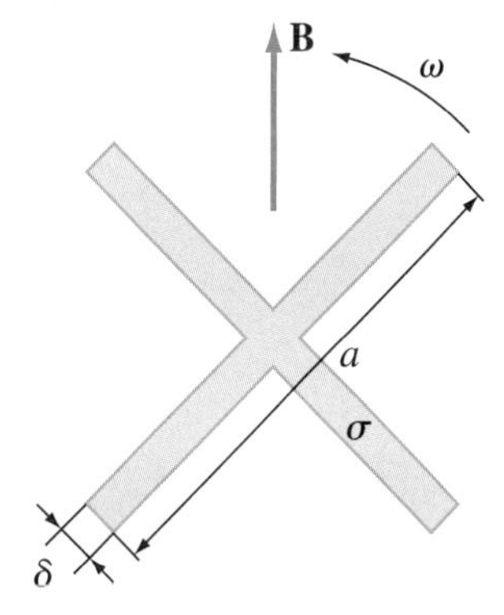

Figure 6.47 Cross section of a conductor composed of two crossed strips rotating in a uniform time-invariant magnetic field; for Problem 6.43.

6.44. **Continuously inhomogeneous rotating cylinder.** Repeat Example 6.25 but for a continuously inhomogeneous cylinder whose conductivity is given by the following function of the radial coordinate r [Fig. 6.27(b)]: $\sigma(r) = \sigma_0 r/a$, $0 \le r \le a$, where σ_0 is a positive constant.

6.45. **Eddy currents in a rotating cylindrical shell.** An infinitely long thin conducting cylindrical shell of radius a, thickness δ ($\delta \ll a$), conductivity σ, and permeability μ_0 is rotated in air uniformly about its axis by an externally applied mechanical torque, $\mathbf{T}'_{\text{mech}}$, per unit length of the shell and against the torque of forces of a uniform time-invariant magnetic field of flux

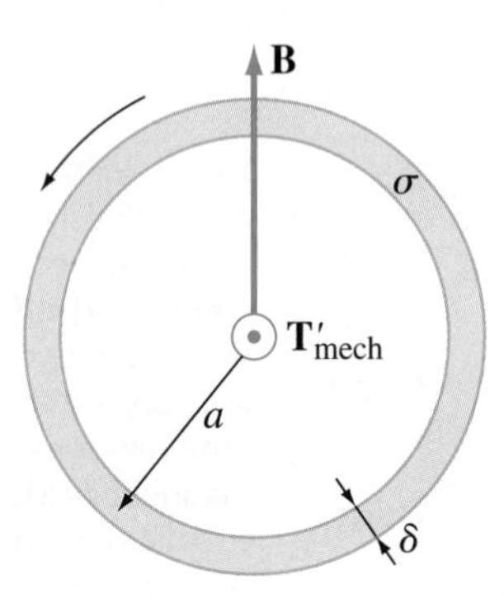

Figure 6.48
Rotating cylindrical conducting shell in a uniform time-invariant magnetic field; for Problem 6.45.

density B, in which the shell resides, as shown in Fig. 6.48. Neglecting the magnetic field due to eddy currents in the shell, find (a) the angular velocity of rotation of the shell (ω) and (b) the per-unit-length mechanical power used to rotate the shell (P'_{mech}).

7 Inductance and Magnetic Energy

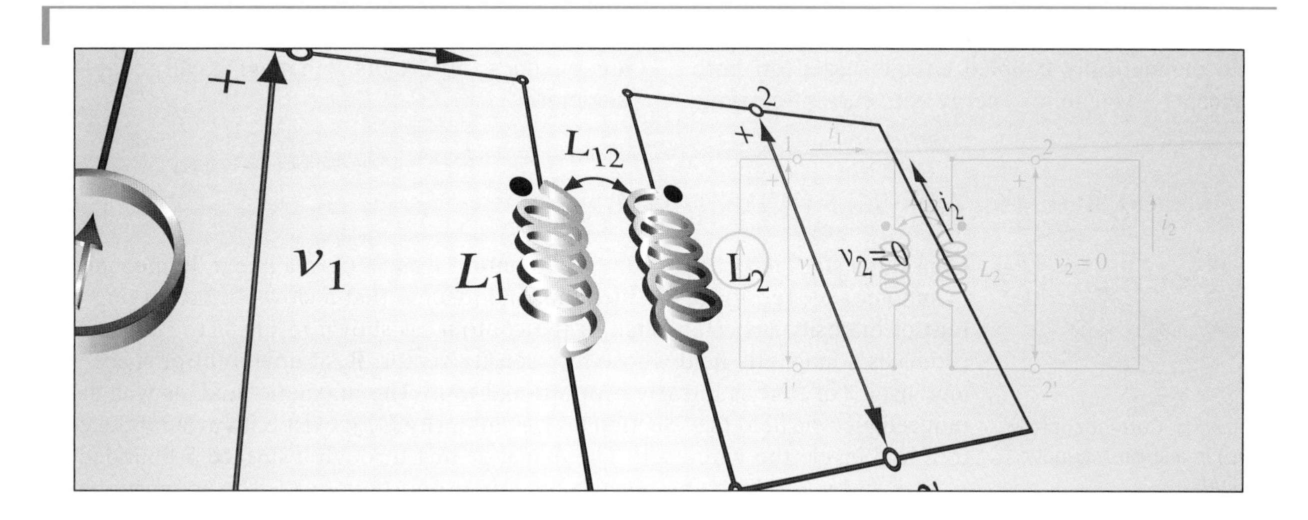

Introduction:

In this chapter, we introduce and study the concepts of self- and mutual inductance. In general, inductance can be interpreted as a measure of transformer electromagnetic induction in a system of conducting contours (circuits) with slowly time-varying currents in a linear magnetic medium. Briefly, self-inductance is a measure of the magnetic flux and induced emf in a single isolated contour (or in one of the contours in a system) due to its own current. Similarly, a current in one contour causes magnetic flux through another contour and induced emf in it, and mutual inductance is used to characterize this coupling between the contours. Some conductor configurations, called inductors, are specially designed to have a desired (large) inductance (they can have many turns of wire and can be loaded with magnetic cores). Along with the resistor and capacitor, the inductor represents another fundamental element in circuit theory and a basic building block for ac electric circuits. The inductance of an inductor is dual to the capacitance of a capacitor.

An equally important concept of energy in magnetic fields is also discussed. We shall see that, just as configurations of charged bodies store electric energy, configurations of current-carrying conductors store magnetic energy. Using circuit-theory terminology, inductors (magnetically coupled or uncoupled) in a circuit contain magnetic energy, in a manner analogous to capacitors as electric energy "containers." We shall introduce magnetic energy density as well. In the case of conductors in the presence of linear magnetic materials, magnetic energy will be related to self- and mutual inductances of the conductors. For systems that contain ferromagnetic materials with pronounced nonlinearity and hysteresis behavior, on the other side, special attention will be paid to establishing a clear physical understanding and precise mathematical characterization of the energy balance in the system, including so-called hysteresis losses in the material.

The material of this chapter represents a culmination of our investigations of steady and slowly time-varying magnetic fields and electromagnetic induction (Chapters 4–6). On the other hand, it parallels, to a large extent, the electrostatic analysis of capacitors and other systems of charged conducting bodies, including electric energy considerations (Chapter 2). Therefore, a substantial portion of the previous work will be referenced and used here, in both the theoretical narrative and examples.

We shall first study self- and mutual inductance in two separate sections, and then analyze magnetically coupled circuits based on both concepts. Magnetic energy of current-carrying conductors and magnetic energy density in the field will bediscussed next. Finally, the concept of self-inductance will be revisited from the energy standpoint. Examples will include inductance and energy computation for a large variety of theoretically and practically important electromagnetic structures with slowly time-varying and time-invariant currents and fields, ranging from various contours and coils with cores of different shapes and material compositions to several types of transmission lines and circuits with magnetically coupled inductors.

7.1 SELF-INDUCTANCE

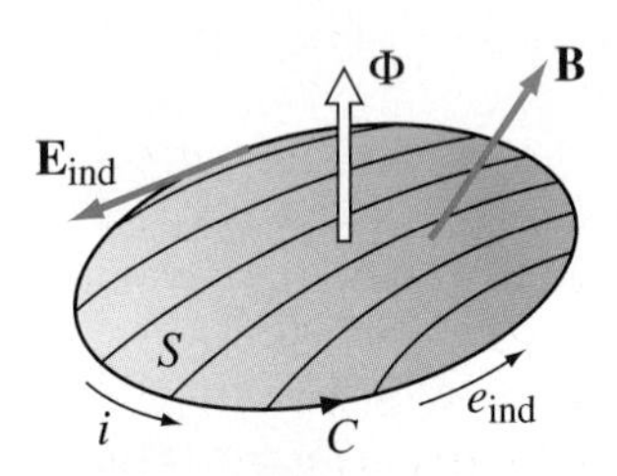

Figure 7.1 Current contour (loop) in a linear magnetic medium.

Consider a stationary conducting wire contour (loop), C, in a linear, homogeneous or inhomogeneous, magnetic medium, and assume that a slowly time-varying current of intensity i is established in the contour, as shown in Fig. 7.1. This current produces a magnetic field whose flux density vector, $\mathbf{B}$, at any point of space and any instant of time is linearly proportional to i.[1] The magnetic field, as well as an induced electric field (see Section 6.1), of intensity $\mathbf{E}_{\text{ind}}$, exist both around the contour and inside the wire itself. The magnetic flux through a surface S bounded by C is given by Eq. (4.95) and is also linearly proportional to i. The proportionality constant,

self-inductance (unit: H)

$$\boxed{L = \frac{\Phi}{i},} \tag{7.1}$$

is termed the self-inductance or just inductance of the contour. More precisely, the inductance defined by Eq. (7.1) is the so-called external inductance, since it takes into account only the flux Φ of the magnetic field that exists outside the conductor of the loop. There is also an internal inductance of the loop, due to the flux inside the conductor. We shall introduce the concept of internal inductance in terms of the magnetic energy stored inside the conductor in a later section. Because the surrounding medium is magnetically linear, L depends only on the medium permeability and on the shape and dimensions of the contour, and not on the current intensity i. It is always positive, provided, of course, that the reference orientations of C and S (i.e., the reference directions of i and Φ) are interconnected by the right-hand rule, as in Fig. 7.1.

The induced emf along the loop C is given by Eq. (6.33) and is linearly proportional to $\mathrm{d}i/\mathrm{d}t$. It is interrelated with the flux Φ by means of Faraday's law of electromagnetic induction, and we can write

emf due to self-induction

$$\boxed{e_{\text{ind}} = -\frac{\mathrm{d}\Phi}{\mathrm{d}t} = -L\frac{\mathrm{d}i}{\mathrm{d}t}.} \tag{7.2}$$

This emf is referred to as the emf due to self-induction (or self-induced emf) in the circuit (it is caused by the magnetic field due to the current in the circuit

[1]Since the current is slowly time-varying, its intensity (i) is only a function of time and does not change along the contour.

itself, and not due to currents in other circuits). The unit for the self-inductance is henry (H). From Eqs. (7.1) and (7.2), which represent two equivalent definitions of L, $\mathrm{H} = \mathrm{Wb/A} = \mathrm{V \cdot s/A}$. One henry is a very large unit. Typical values of self-inductances in practice are on the order of mH, μH, and nH.

HISTORICAL ASIDE

Joseph Henry (1797–1878), an American physicist and one of the greatest scientists and inventors in the area of electricity and magnetism ever, was a professor of natural philosophy at Princeton and the first secretary of the Smithsonian Institution. From 1819 to 1822, he attended the Albany Academy, where he was appointed professor of mathematics and natural philosophy in 1826. Although his teaching duties were extremely heavy, Henry soon became the foremost American scientist of his time. He discovered electromagnetic induction independently of Faraday (1791–1867). In fact, Henry had performed the key experiments that led to the discovery of induction ahead of Faraday, in 1830, but Faraday published his discovery first, in 1831. On the other side, Henry is fully credited for the discovery of self-induction. It was his idea to wind many layers of insulated wire on an iron core and thus obtain electromagnets of unmatched power. In a demonstration at Yale University in 1831, his electromagnet lifted more than a ton of iron (previous electromagnets were capable of lifting only a couple of kilograms). Experimenting with electromagnets, he observed a large spark that was generated whenever the circuit was broken, and thus discovered self-induction (in 1831). He realized that, in general, a time-varying current in a circuit not only induces electromotive force in another circuit (mutual induction), but also in itself (self-induction). He also defined the associated property of a circuit, its self-inductance, as a measure of its ability to "self-induce" electromotive force. He found that "coiling" of the wire greatly enhances the self-inductance of the circuit. The same year, he demonstrated his electric motor based on a continuous oscillating motion of a straight electromagnet with two coils at its ends. The electromagnet rocked back and forth on a horizontal axis with its ends being alternately attracted and repelled by two vertical permanent magnets and its polarity being reversed automatically in the same rhythm by alternately connecting the coils to electrochemical cells (sources). Although still an experimental laboratory device, Henry's motor was much closer to a mechanically useful practical machine than Faraday's 1821 motor. It was just one year later, in 1832, that William Sturgeon (1783–1850) invented the commutator and the first motor with continuous rotary motion, which was a rotary analogue of Henry's oscillating motor and essentially a "prototype" of our modern dc motors. In another stunning demonstration to his students at the Albany Academy, in 1831, Henry strung a mile of wire all around the inside of the lecture hall to connect an electromagnet to a battery. The magnet was placed close to one end of a pivot mounted steel bar, whose other end, in turn, was next to a bell. After the circuit was closed and a current "sent" from the battery to the coil of the electromagnet, a steel bar swung toward the magnet, striking the bell on the other end. Breaking the connection, next, made the electromagnet lose its force and release the bar, which was then free to strike again. This was the world's first electrical relay (electromechanical switch). In addition, by connecting and disconnecting the battery to the circuit in a particular pattern, the steel bar, a mile away, could be made to ring the same series of signals on the bell, and this is nothing else but telegraphy. Although, obviously, Henry invented the telegraph in 1831, it is Samuel Morse (1791–1872) who is credited for this invention [Henry did not patent any of his devices, and actually helped Morse to put his telegraph model to practical use and transmit the first telegraph message using the Morse code (invented in 1838) from

Baltimore to Washington, D.C. on May 24, 1844]. In 1832, Henry became professor of natural philosophy (physics) at Princeton University (then College of New Jersey). After the Smithsonian Institution was established in 1846, Henry was named its first secretary (director). He was also the second president (elected in 1868) of the National Academy of Sciences, and he held both positions until his death. In his honor, we use henry (H) as the unit for inductance. *(Portrait: Library of Congress, Brady-Handy Photograph Collection)*

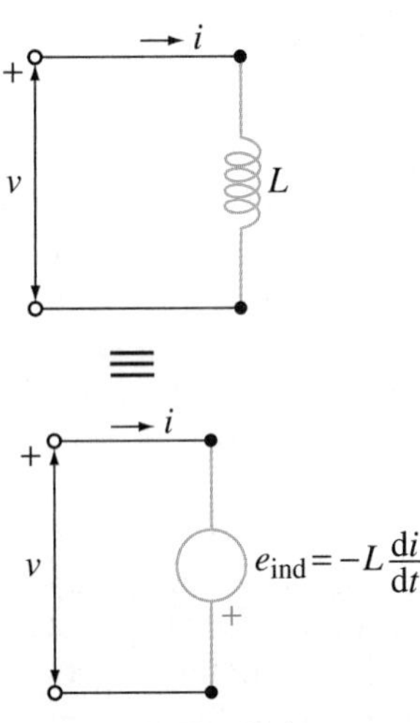

Figure 7.2 Circuit-theory representation of an inductor and equivalent controlled voltage generator.

Note that, while the emf definition of self-inductance in Eq. (7.2) does not make any sense for steady currents, the flux definition in Eq. (7.1) can be used in practically the same way under both dynamic and static conditions. Namely, a dc current I that is assumed to flow in the contour produces the flux Φ through the contour in the same way a (slowly) time-varying current does, and the inductance obtained as $L = \Phi/I$ is equal to that obtained, for the same contour, using either Eq. (7.1) or (7.2) and time-varying current. On the other hand, the flux Φ due to the current I is time-invariant, and thus no emf is generated in the contour.

Every conducting loop or circuit has some inductance (self-inductance), usually as an undesirable side effect, which can often be neglected. In practical applications, however, we frequently design and use conductors that are specially arranged and shaped (such as a conducting wire wound as a coil), and sometimes loaded with magnetic cores, to supply a certain (large) amount of inductance. Such a device, with its inductance L as its basic property, is called an inductor. Just as a capacitor can store electric energy, an inductor can store magnetic energy, as we shall see in a later section. Resistors, capacitors, and inductors are basic circuit elements that are combined, together with voltage and current generators, to form arbitrary RLC circuits.

Fig. 7.2 shows the circuit-theory representation of an inductor. When a time-varying current of intensity i flows through the inductor terminals, the emf e_{ind} is induced in the inductor, given by Eq. (7.2), where L is the inductance of the inductor. With respect to its terminals, the inductor can now be replaced by an equivalent ideal voltage generator whose emf equals e_{ind} (this is a voltage generator controlled by the time derivative of a current), as indicated in Fig. 7.2. The reference direction of this emf is the same as the reference direction of the current i (see Fig. 7.1). The voltage v across the inductor terminals in Fig. 7.2 is hence

element law for an inductor

$$\boxed{v = -e_{ind} = L\frac{di}{dt}.} \tag{7.3}$$

This is the element law (current-voltage characteristic) for an inductor. It tells us that v is linearly proportional to the rate of change of i in time, with L as the proportionality constant. Its form is just opposite to the element law for a capacitor, Eq. (3.45), and we say that an inductor and a capacitor are dual elements. These two laws, the element law for a resistor (Ohm's law), Eq. (3.72), Kirchhoff's current and voltage laws, Eqs. (3.42) and (1.92), and element laws for an ideal voltage generator, Eq. (3.116), and an ideal current generator, Eq. (3.127), represent a full set of basic equations of circuit theory for analysis of linear ac circuits.

Note that the model in Fig. 7.2 actually represents an ideal inductor, which does not include any parasitic effects, such as parasitic capacitances between the adjacent turns in a coil and Joule's losses (in wires and magnetic cores). These effects are present to a greater or lesser extent in all real inductors, but can be neglected in many practical situations. The only effect modeled by an ideal inductor is the emf due to self-induction, e_{ind}. On the other side, it is assumed in circuit theory that the magnetic flux, induced emf, and magnetic energy are concentrated only in the inductors in a circuit, while the magnetic field and the associated flux, emf, and energy due to the connecting conductors are assumed to be negligible. In practice, there is always an induced emf distributed along conductors in the circuit (in the ac regime), and it depends on the shape and dimensions of the conductors. However, this emf, i.e., the inductance of the conductors, can, again, be neglected in many practical applications. We recall that similar assumptions of zero capacitance and zero resistance of the conductors have been made while introducing circuit-theory representations of a capacitor and a resistor in Figs. 2.15 and 3.9, respectively. The circuit-theory model, therefore, deals with connecting conductors (lines between the elements in circuit layouts) as if they were all ideal short-circuits, where the shape, dimensions, and material properties of the interconnects are assumed to be completely irrelevant for the operation (and analysis and design) of the circuit.

Finally, inductors filled with magnetically nonlinear materials (e.g., the coil of Fig. 6.13) are nonlinear circuit elements. In such cases, the magnetic properties (permeability) of the material depend on the applied magnetic field intensity, H, whereas H is always proportional to i. Consequently, the inductance of the inductor depends on the current intensity,

$$\boxed{L = L(i),} \tag{7.4}$$

nonlinear inductor

which is analogous to the relations in Eqs. (2.114) and (3.74) for a nonlinear capacitor and a nonlinear resistor, respectively.

Example 7.1 Inductance of a Very Long Air-Filled Solenoid

An air-filled solenoidal coil has $N = 200$ turns of wire. The length of the solenoid is $l = 20$ cm and the surface area of its cross section is $S = 4\ \text{cm}^2$. The solenoid can be considered as very long, so that the end effects can be neglected. Under these circumstances, find the inductance of the coil.

Solution Let us assume a slowly time-varying current of intensity i in the coil, find the magnetic flux through all of its turns, and use Eq. (7.1) for the inductance. By neglecting the end effects, we also assume that the magnetic field produced by the current i is uniform inside the entire solenoid (as for an infinitely long solenoid). This field is given by Eq. (6.48), which leads to the following expressions for the magnetic flux through a surface spanned over a single turn of the coil and the total flux through the coil:

$$\Phi_{\text{single turn}} = \mu_0 HS \quad \longrightarrow \quad \Phi = N\Phi_{\text{single turn}} = \frac{\mu_0 N^2 Si}{l}, \tag{7.5}$$

where the reference direction for Φ is adopted in accordance to the right-hand rule with respect to the reference direction for i. The inductance of the coil is hence

$$\boxed{L = \frac{\Phi}{i} = \frac{\mu_0 N^2 S}{l},} \tag{7.6}$$

L – very long solenoid

that is, $L = 100\ \mu$H for the given numerical data. It is the same for any shape of the solenoid cross section (provided that the solenoid is very long).

Example 7.2 Coil on an Inhomogeneous Thick Toroidal Core

A coil with N turns of wire is wound over a thick toroidal core of rectangular cross section made of two ferromagnetic layers of permeabilities μ_1 and μ_2, as portrayed in Fig. 7.3. The

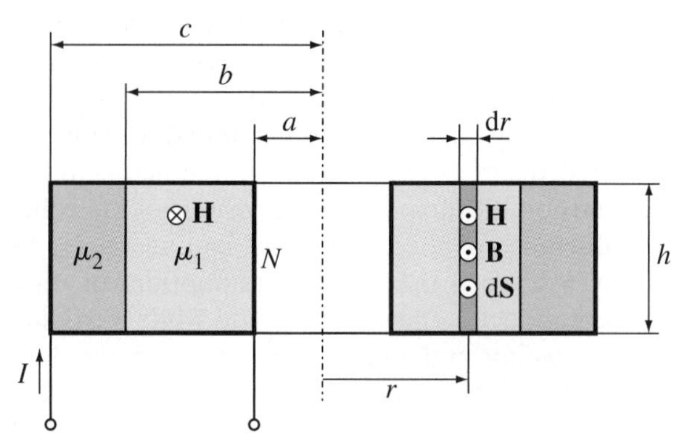

Figure 7.3 Evaluation of the inductance of a coil wound on a thick toroidal core with two linear ferromagnetic layers; for Example 7.2.

inner and outer radii of the toroid are a and c, its height is h, and the radius of the boundary surface between the layers is b ($a < b < c$). Calculate the inductance of the coil.

Solution If a steady current of intensity I is assumed to flow in the coil (Fig. 7.3), the magnetic field intensity $H(r)$ inside the core (in both ferromagnetic layers) is given by Eq. (5.83). The magnetic flux through the coil is then [see Fig. 7.3 and Eqs. (5.84) and (5.92)]

$$\Phi = N\int_{r=a}^{c} B(r)\,\mathrm{d}S = N\left[\int_a^b \mu_1 H(r)\,\mathrm{d}S + \int_b^c \mu_2 H(r)\,\mathrm{d}S\right] = \frac{N^2 I h}{2\pi}\left(\mu_1 \ln\frac{b}{a} + \mu_2 \ln\frac{c}{b}\right), \tag{7.7}$$

and the inductance of the coil

$$L = \frac{\Phi}{I} = \frac{N^2 h}{2\pi}\left(\mu_1 \ln\frac{b}{a} + \mu_2 \ln\frac{c}{b}\right). \tag{7.8}$$

Example 7.3 External Inductance p.u.l. of a Thin Two-Wire Line

Find the external inductance per unit length of a thin symmetrical two-wire transmission line in air. The conductor radii are a and the distance between their axes is d ($d \gg a$).

Solution The two-wire line can be considered as an infinitely long wire loop that closes upon itself at both ends of the line at infinity. Fig. 7.4 shows the cross section of the line. The procedure of evaluation of the line inductance (external self-inductance) per unit length is analogous (dual) to the procedure of evaluation of the capacitance per unit length of a thin two-wire line in Example 2.15. We assume that a slowly time-varying current of intensity i (or a steady current of intensity I) is established in the line (Fig. 7.4). At a point M in the plane containing the axes of conductors, the magnetic flux density vectors $\mathbf{B}_1$ and $\mathbf{B}_2$ due to currents (of the same magnitude and opposite directions) in individual conductors of the line are collinear, so that the resultant magnetic flux density is

$$B = B_1 + B_2 = \frac{\mu_0 i}{2\pi}\left(\frac{1}{x} + \frac{1}{d-x}\right), \tag{7.9}$$

where x is the distance from the axis of the first conductor.

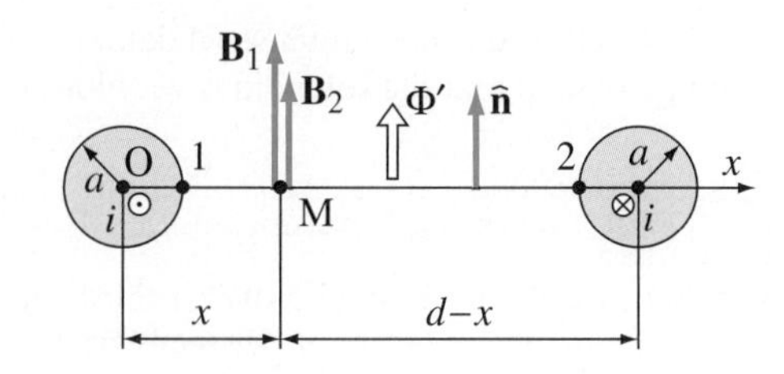

Figure 7.4 Evaluation of the external inductance per unit length of a thin two-wire transmission line in air (cross section of the structure); for Example 7.3.

As the structure is infinitely long, we consider only a part of it that is l long and compute the magnetic flux through the flat surface of length l that is spanned between the line conductors (such integration surface is the right choice because **B** is perpendicular to the surface) and oriented as required by the right-hand rule with respect to the adopted reference direction of the line current. This flux is given by

$$\Phi = \int_{x=a}^{d-a} B \underbrace{l\,\mathrm{d}x}_{\mathrm{d}S} = \frac{\mu_0 i l}{2\pi} \int_a^{d-a} \left(\frac{1}{x} + \frac{1}{d-x} \right) \mathrm{d}x = \frac{\mu_0 i l}{\pi} \ln \frac{d-a}{a} \approx \frac{\mu_0 i l}{\pi} \ln \frac{d}{a} \quad (7.10)$$

[see the integration in Eq. (2.140)]. Hence, the external inductance per unit length (per each meter) of the line comes out to be

$$L' = L_{\mathrm{p.u.l.}} = \frac{\Phi'}{i} = \frac{\Phi}{il} = \frac{\mu_0}{\pi} \ln \frac{d}{a}, \quad (7.11)$$

L' – thin two-wire line (unit: H/m)

where Φ' is the flux per unit length of the line.

As a numerical example, for $d/a = 30$, $L' = 1.36\ \mu\mathrm{H/m}$. If the line is 100 m long, its total external inductance amounts to $L = 136\ \mu\mathrm{H}$.

Example 7.4 External Inductance p.u.l. of a Coaxial Cable

Consider an air-filled coaxial cable. The radius of the inner conductor of the cable is a and the inner radius of the outer conductor is b ($b > a$). Obtain the expression for the external inductance per unit length of the cable.

Solution Let us assume a dc current of intensity I flowing through the line conductors, as shown in Fig. 7.5. The associated magnetic flux density vector, **B**, is circular with respect to the cable axis and its magnitude, $B(r)$, in the dielectric is given by the second expression in Eq. (4.61), where r is the distance from the axis and $a < r < b$. Computing the magnetic flux through a flat surface of length l spanned between the conductors (Fig. 7.5) then yields the following expression for the external inductance per unit length of the cable:

$$L' = \frac{\Phi}{Il} = \frac{1}{Il} \int_{r=a}^{b} B(r)\,\mathrm{d}S = \frac{1}{Il} \int_a^b \frac{\mu_0 I}{2\pi r} l\,\mathrm{d}r = \frac{\mu_0}{2\pi} \ln \frac{b}{a}, \quad (7.12)$$

L' – coaxial cable

where the dielectric can be any nonmagnetic material ($\mu = \mu_0$). The procedure is dual to the voltage and capacitance computation in Example 2.10.

As a numerical example, the inductance per unit length of an RG-11 coaxial cable, having $a = 0.6$ mm and $b = 3.62$ mm (dielectric is polyethylene), is computed to be $L' = 359$ nH/m.

Comparing the expression for L' in Eq. (7.12) and the expression for the capacitance per unit length of the same cable, $C' = 2\pi\varepsilon_0 / \ln(b/a)$ [see Eq. (2.123)], we note that the following

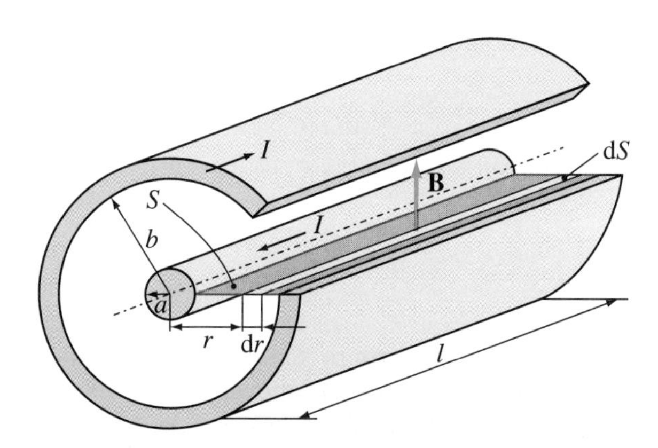

Figure 7.5 Evaluation of the external inductance per unit length of a coaxial cable; for Example 7.4.

relationship exists between the two parameters:

duality of L' and C' for air dielectric

$$\boxed{L'C' = \varepsilon_0\mu_0.} \tag{7.13}$$

Comparing then the expressions in Eqs. (7.11) and (2.141), we also note that the same is true for a thin two-wire line in air. As a matter of fact, as we shall prove in a later chapter, the same relationship exists between L' and C' for any two-conductor transmission line with air dielectric. For a transmission line having homogeneous linear dielectric of arbitrary permittivity and permeability, $L'C' = \varepsilon\mu$. Using these relations, we can now very easily obtain the inductance per unit length of transmission lines (with homogeneous linear dielectrics) for which we already have the capacitance per unit length (found in Chapter 2). For example, from the expression in Eq. (2.146) for the capacitance per unit length of the transmission line consisting of a thin wire conductor and a grounded conducting plane in Fig. 2.24(a), we can directly write the expression for the external inductance per unit length of this line:

$$L' = \frac{\varepsilon_0\mu_0}{C'} = \frac{\mu_0}{2\pi}\ln\frac{2h}{a}. \tag{7.14}$$

Example 7.5 Superconducting Contour in a Magnetic Field

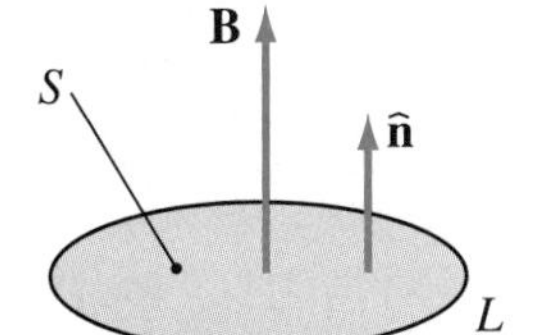

Figure 7.6
Superconducting contour of inductance L in a uniform magnetostatic field; for Example 7.5.

A superconducting planar contour of area S and inductance L is situated in a uniform magnetostatic field of flux density B. The contour is first positioned such that the vector $\mathbf{B}$ is perpendicular to the plane of the contour, as in Fig. 7.6. In this state, a steady current of intensity I_1 flows along the contour. The contour is then turned such that its plane becomes parallel to $\mathbf{B}$. Find the current I_2 in the contour in the new steady state.

Solution The total magnetic flux through the contour in the first state is

$$\Phi_1 = LI_1 + BS, \tag{7.15}$$

where the first term is the flux due to the current (I_1) in the contour (self-flux) and the second term is the flux of the external magnetic field ($\mathbf{B}$), i.e., the flux due to other currents (or permanent magnets) in the system (mutual flux).

In the second state,

$$\Phi_2 = LI_2, \tag{7.16}$$

since the mutual flux is zero ($\mathbf{B}\cdot\mathbf{S} = 0$).

The total magnetic flux through a superconducting contour cannot be changed [see Eq. (6.135)], which means that

$$\Phi_1 = \Phi_2. \tag{7.17}$$

Combining Eqs. (7.15)–(7.17), the current in the contour in the second state is hence

$$I_2 = I_1 + \frac{BS}{L}. \tag{7.18}$$

Note that the increment in current between the two states, $\Delta I = BS/L$, is induced in the contour exactly at the amount that completely compensates, via the associated magnetic field, the change of the mutual flux (from BS to zero), maintaining thus a constant total flux through the contour.

Problems: 7.1–7.9; *Conceptual Questions* (on Companion Website): 7.1–7.10; *MATLAB Exercises* (on Companion Website).

7.2 MUTUAL INDUCTANCE

Consider now two stationary conducting wire contours, C_1 and C_2, in a linear (homogeneous or inhomogeneous) magnetic medium, as shown in Fig. 7.7. Let the first contour carry a slowly time-varying current of intensity i_1. As a result, a magnetic field, of flux density $\mathbf{B}_1$, is produced everywhere, and $\mathbf{B}_1$, which is a function of both the spatial coordinates and time, is linearly proportional to i_1. Some of the lines of $\mathbf{B}_1$ pass through the second contour, i.e., through a surface S_2 bounded by C_2. These lines constitute the magnetic flux through the second contour due to the current i_1, which can be expressed as

$$\Phi_2 = \int_{S_2} \mathbf{B}_1 \cdot \mathrm{d}\mathbf{S}_2. \tag{7.19}$$

The flux Φ_2 is linearly proportional to i_1 as well:

$$\Phi_2 = L_{21} i_1, \tag{7.20}$$

where the proportionality constant L_{21} is called the mutual inductance between the two contours and is obtained as

$$\boxed{L_{21} = \frac{\Phi_2}{i_1}.} \tag{7.21}$$

mutual inductance (unit: H)

Note that the symbol M is also used to denote mutual inductance. It is a measure of magnetic coupling between the contours. Its magnitude depends on the shape, size, and mutual position of the contours, and on the magnetic properties (permeability) of the medium. The mutual inductance can be both positive and negative, depending on the adopted reference orientation of each of the contours for their given mutual position. Namely, if a positive current i_1 in the contour C_1 gives rise to a positive magnetic flux Φ_2 for the orientation of the surface S_2 that is in accordance to the right-hand rule with respect to the orientation of the contour C_2, the mutual inductance is positive. Otherwise, it is negative.[2] The mutual inductance is also expressed in henrys.

By applying Faraday's law of electromagnetic induction, Eq. (6.34), to the flux in Eq. (7.20), we obtain the induced emf (due to transformer induction) in the contour C_2:

$$\boxed{e_{\text{ind2}} = -\frac{\mathrm{d}\Phi_2}{\mathrm{d}t} = -L_{21}\frac{\mathrm{d}i_1}{\mathrm{d}t}.} \tag{7.22}$$

emf due to mutual induction

Eqs. (7.21) and (7.22) should be regarded as equivalent definitions of mutual inductance. They indicate that the mutual inductance between two magnetically coupled contours (circuits) can be evaluated by assuming a current (i_1) to flow in one contour (primary circuit) and computing or measuring the magnetic flux (Φ_2) or the induced emf (e_{ind2}) in the other contour (secondary circuit). Again, note that, while the flux definition of L_{21} in Eq. (7.21) can be used for both time-varying

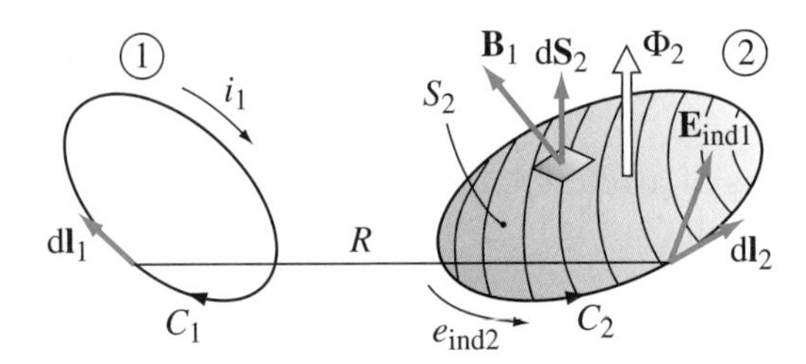

Figure 7.7 Two magnetically coupled conducting contours.

[2] In some texts, the information about the sign of the mutual inductance for specific reference orientations of contours is not included in its definition, i.e., mutual inductance is defined as always being nonnegative.

and steady currents, the emf definition in Eq. (7.22) makes sense for time-varying currents only.

Conversely, if we assume that a current i_2 flows along the contour C_2 and determine the associated magnetic flux Φ_1 through the contour C_1, the mutual inductance between the contours is given by

$$L_{12} = \frac{\Phi_1}{i_2}, \tag{7.23}$$

with an analogous expression using the induced emf $e_{\text{ind}1}$ along the contour C_1. Because of reciprocity (in a linear system, transfer functions remain the same if the source location and the location at which the response to the source is observed are interchanged),

reciprocity of mutual induction

$$\boxed{L_{12} = L_{21}.} \tag{7.24}$$

If the medium around the contours is air (or any nonmagnetic medium), the induced electric field intensity vector due to the current in the first contour is given by [see Eq. (6.5) and Fig. 7.7]

$$\mathbf{E}_{\text{ind}1} = -\frac{\mu_0}{4\pi}\oint_{C_1}\frac{\mathrm{d}i_1}{\mathrm{d}t}\frac{\mathrm{d}\mathbf{l}_1}{R}, \tag{7.25}$$

so that the induced emf in the second contour can be written as [see Eq. (6.33)]

$$e_{\text{ind}2} = \oint_{C_2}\mathbf{E}_{\text{ind}1}\cdot\mathrm{d}\mathbf{l}_2 = -\frac{\mu_0}{4\pi}\oint_{C_2}\oint_{C_1}\frac{\mathrm{d}\mathbf{l}_1\cdot\mathrm{d}\mathbf{l}_2}{R}\frac{\mathrm{d}i_1}{\mathrm{d}t} = -L_{21}\frac{\mathrm{d}i_1}{\mathrm{d}t}. \tag{7.26}$$

This yields the following expression for the mutual inductance between the contours C_1 and C_2:

Neumann formula for mutual inductance

$$\boxed{L_{21} = \frac{\mu_0}{4\pi}\oint_{C_2}\oint_{C_1}\frac{\mathrm{d}\mathbf{l}_1\cdot\mathrm{d}\mathbf{l}_2}{R},} \tag{7.27}$$

which is known as the Neumann[3] formula for mutual inductance. It implies the evaluation of a double line integral along the contours and underscores the fact that the mutual inductance is only a property of the geometrical shape and the physical arrangement of coupled contours, as well as of the permeability of the medium [if the contours are situated in a homogeneous linear magnetic medium of permeability μ, the constant μ_0 needs to be replaced by μ in Eq. (7.27)]. It is obvious from Eq. (7.27) that the change of the reference orientation of one of the contours in Fig. 7.7 (but not both of them), which means the change of the reference direction of one of the vectors $\mathrm{d}\mathbf{l}_1$ and $\mathrm{d}\mathbf{l}_2$, changes the polarity (sign) of the mutual inductance L_{21}. It is also obvious that interchanging the subscripts 1 and 2 in Eq. (7.27) does not change the algebraic value (which includes both the magnitude and the sign) of the double integral (since the dot product is commutative and the two line integrations can be performed in an arbitrary order). This proves the identity in Eq. (7.24). The Neumann formula represents a good basis for numerical evaluation of the mutual inductance of contours of arbitrary shapes, where the involved line integrals along the contours are computed using numerical integration methods.

[3]Franz Ernst Neumann (1798–1895), a German mineralogist, physicist, and mathematician, a professor of mineralogy and physics at the University of Königsberg. Neumann contributed to the theory of electromagnetic induction, and derived, in 1845, the formula for the mutual inductance between two equal parallel coaxial polygons of wire, the generalization of which is the Neumann formula for mutual inductance of arbitrary wire contours (loops), as we use it today.

Example 7.6 Mutual Inductance between a Loop and an Infinite Wire

Find the mutual inductance between the rectangular loop and the infinitely long straight wire in Fig. 6.12.

Solution The infinitely long wire can be considered as a loop that closes upon itself at infinity. Designating it as C_1 and the rectangular loop as C_2, the current i_1 and magnetic flux Φ_2 in Eq. (7.21) are actually the current i carried by the wire in Fig. 6.12 and the flux Φ through the rectangular contour given by Eq. (6.64), respectively. The mutual inductance between the contours for their reference orientations given in Fig. 6.12 (upward orientation for the infinitely long wire and clockwise direction for the rectangular loop) is hence

$$L_{21} = \frac{\Phi_2}{i_1} = \frac{\mu_0 b}{2\pi} \ln \frac{c+a}{c}. \tag{7.28}$$

Note that the same result is obtained by applying the emf definition of mutual inductance in Eq. (7.22) to the expression for the induced emf in the rectangular contour in Eq. (6.65). Note also that computing the mutual inductance as L_{12} in this case would be prohibitively complicated. It would require either finding the induced electric field intensity vector $\mathbf{E}_{\text{ind}2}$ at an arbitrary point of the infinite wire due to an assumed current i_2 in the rectangular contour, and then integrating it along the wire (which, in fact, is the Neumann formula for L_{12}) or finding and integrating the magnetic flux density vector $\mathbf{B}_2$ due to i_2 across a half-plane bounded by the loop C_1.

Example 7.7 Mutual Inductance of Two Coils on a Thin Toroidal Core

Two coils are wound uniformly and densely in two layers, one on top of the other, about a thin toroidal core (such as the one shown in Fig. 5.11). The core is made from a linear ferromagnetic material of relative permeability $\mu_r = 500$. Its length is $l = 50$ cm and cross-sectional area $S = 1\ \text{cm}^2$. The number of wire turns is $N_1 = 400$ for the first coil and $N_2 = 600$ for the second one. Compute the mutual inductance of the coils.

Solution Let us adopt the same reference orientations for the two coils (this will give us a positive mutual inductance of the coils) and assume that the first coil carries a current of intensity i_1. This current produces a magnetic field of the same intensity (H_1) everywhere inside the core. From Eq. (5.53), $H_1 = N_1 i_1/l$, so that the total magnetic flux through the second coil (Φ_2) is given by [see also Eq. (5.96)]

$$\Phi_{\text{single turn}} = \mu H_1 S \quad (\mu = \mu_r \mu_0) \quad \longrightarrow \quad \Phi_2 = N_2 \Phi_{\text{single turn}} = \frac{\mu N_1 N_2 S i_1}{l}, \tag{7.29}$$

and the mutual inductance between the coils comes out to be

$$\boxed{L_{21} = \frac{\Phi_2}{i_1} = \frac{\mu N_1 N_2 S}{l}.} \tag{7.30}$$

L_{21} – two coupled coils on a thin toroidal core

Finally, substituting the numerical data gives $L_{21} = 30$ mH.

Note that the same result is obtained by assuming a current i_2 to flow in the second coil and computing the mutual inductance L_{12} from the total magnetic flux through the first coil (Φ_1).

Example 7.8 Magnetic Coupling between a Toroidal Coil and a Loop

An open-circuited toroidal coil with a rectangular cross section has $N = 500$ turns of wire that are uniformly and densely wound about a ferromagnetic core. The relative permeability of the material is $\mu_r = 1000$. The inner and outer radii of the toroid are $a = 2$ cm and $b = 4$ cm (thick toroid), and its height is $h = 1$ cm. A wire loop is placed around the toroid, as shown in Fig. 7.8. There is a low-frequency time-harmonic current of intensity $i = I_0 \cos \omega t$

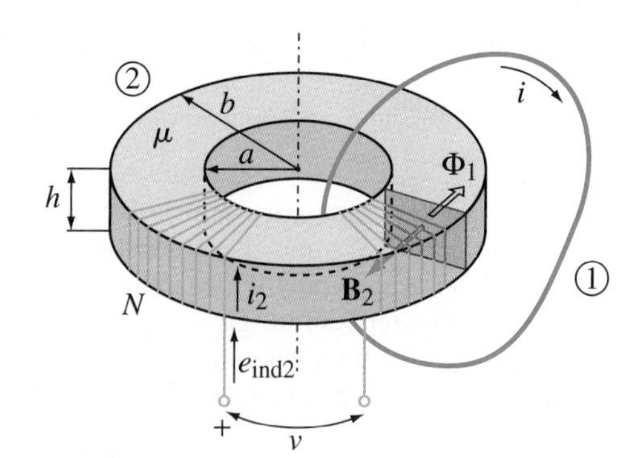

Figure 7.8 Magnetic coupling between a coil wound over a thick toroidal core and a wire loop encircling the toroid; for Example 7.8.

flowing along the loop, where $I_0 = 2$ A and $\omega = 7 \times 10^3$ rad/s. Find the voltage between the terminals of the coil.

Solution Let us denote the wire loop as circuit 1 and the toroidal coil as circuit 2. There is no current in the coil (it is open-circuited) and the voltage across the coil terminals is [see Eqs. (6.62) and (7.22)]

$$v = -e_{\text{ind2}} = L_{21}\frac{\mathrm{d}i_1}{\mathrm{d}t} \quad (i_1 \equiv i). \tag{7.31}$$

However, the mutual inductance L_{21} is extremely difficult to find using Eq. (7.21). Namely, the field $\mathbf{B}_1$ due to i_1, which would have to be integrated through each of the turns of the toroidal coil in order to compute the flux Φ_2, is not only highly nonuniform but also impossible to find analytically for a loop of arbitrary, irregular shape.

On the other side, we can use the identity in Eq. (7.24) and find the inductance L_{12} instead. To this end, we assume a current of intensity i_2 in the toroidal coil, and no current in the loop ($i_1 = 0$), find the flux Φ_1 through a surface bounded by the loop, and use Eq. (7.23). Under these circumstances, there is no field outside the coil [see Example 4.12], so that the flux Φ_1 equals the negative of the flux Φ through a cross section of the toroid given by Eq. (5.84), where the minus sign (negative) comes from different reference directions of the fluxes in Figs. 7.8 and 5.26. Hence,

$$L_{12} = \frac{\Phi_1}{i_2} = -\frac{\Phi}{I} = -\frac{\mu_r\mu_0 Nh}{2\pi}\ln\frac{b}{a} = -693\ \mu\text{H}. \tag{7.32}$$

Finally, combining Eqs. (7.31), (7.24), and (7.32), the time-harmonic voltage between the coil terminals in Fig. 7.8 comes out to be

$$v = L_{12}\frac{\mathrm{d}i_1}{\mathrm{d}t} = -\omega L_{12} I_0 \sin\omega t = 9.7\sin(7 \times 10^3 t)\ \text{V} \quad (t \text{ in s}). \tag{7.33}$$

Note that the amplitude of the voltage v can be written as

Z_{21} – transfer impedance between two circuits (unit: Ω)

$$\boxed{V_0 = |Z_{21}|I_0,} \quad \text{where} \quad Z_{21} = \omega L_{21} = \omega L_{12} = -\frac{\omega\mu_r\mu_0 Nh}{2\pi}\ln\frac{b}{a} = -4.85\ \Omega. \tag{7.34}$$

This amplitude (or the corresponding rms value[4]) can be measured by a voltmeter connected to the terminals of the coil. Thus, by directly measuring V_0, we can indirectly measure the amplitude (or rms value) of the current (I_0) in an arbitrary conductor, without inserting an ammeter in its circuit. In such cases, the coil in Fig. 7.8 is used as a test transformer, where the current to be determined is transformed to the voltage that is indicated by the voltmeter by means of magnetic coupling and transformer electromagnetic induction. The constant Z_{21} in Eq. (7.34), expressed in ohms, is said to be the mutual impedance or transfer impedance

[4]Note that most instruments show rms (root-mean-square) values of measured quantities instead of their amplitudes (maximum values).

between the two circuits ($Z_{12} = Z_{21}$). It is negative here, as the mutual inductance between the circuits is, because of the particular reference orientations of the contours and reference directions of i and v in Fig. 7.8.

Example 7.9 Mutual Inductance p.u.l. of Two Two-Wire Lines

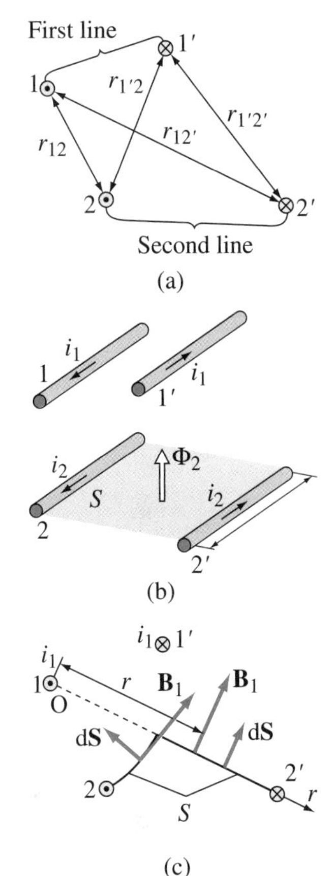

Figure 7.9 Evaluation of the mutual inductance per unit length of two parallel infinitely long thin two-wire lines in air: (a) cross section of the system with given geometrical data, (b) three-dimensional view showing a part of the system of length l that is only considered, and (c) integration of the magnetic field due to the conductor 1 over a conveniently chosen surface bounded by the conductors of the line 2–2′; for Example 7.9.

Fig. 7.9(a) shows a cross section of a system composed of two infinitely long thin two-wire lines running parallel to each other in air. The first line has conductors marked as 1 and 1′, and the second one as 2 and 2′. The distances between axes of the first and second conductor of the first line and each of the conductors of the second line are r_{12}, $r_{12'}$, $r_{1'2}$, and $r_{1'2'}$, respectively. Find the mutual inductance per unit length between the two lines.

Solution We assume that the first line (line 1–1′) carries a current of intensity i_1, as indicated in Fig. 7.9(b). Our goal is to find the magnetic flux Φ_2 due to this current through a surface S of length l bounded by the conductors of the second line (line 2–2′), and to apply Eq. (7.21) for the mutual inductance between the lines.

Let us first find the flux Φ_a through S due to the conductor 1 alone. The corresponding magnetic flux density vector, $\mathbf{B}_1$, is circular with respect to the axis of the conductor and its magnitude is [Eq. (4.22)]

$$B_1 = \frac{\mu_0 i_1}{2\pi r}, \tag{7.35}$$

where r is the distance from the axis. To integrate $\mathbf{B}_1$, however, it is not convenient to adopt S simply as the flat surface spanned across the conductors 2 and 2′. Instead, a more complicated surface, the cross section of which is shown in Fig. 7.9(c), is adopted. It consists of a cylindrical part of radius r_{12} and a flat part (strip) whose width is equal to $r_{12'} - r_{12}$. The flux through the cylindrical part of the surface is zero, because $\mathbf{B}_1$ is tangential to the surface ($\mathbf{B}_1 \perp \mathrm{d}\mathbf{S}$) in this part of integration (note the analogy with the electric potential computation in Fig. 1.23). The integration over the rest of S (flat part) is quite simple to perform, because $\mathbf{B}_1$ is now perpendicular to the integration surface ($\mathbf{B}_1 \parallel \mathrm{d}\mathbf{S}$). Hence, the flux computation practically reduces to the one in Eq. (6.64), and we have

$$\Phi_a = \int_{r=r_{12}}^{r_{12'}} B_1\,\mathrm{d}S = \int_{r_{12}}^{r_{12'}} \frac{\mu_0 i_1}{2\pi r}\,l\,\mathrm{d}r = \frac{\mu_0 i_1 l}{2\pi}\ln\frac{r_{12'}}{r_{12}}. \tag{7.36}$$

Similarly, the flux Φ_b due to the current i_1 in the conductor 1′ alone can be found by integrating the corresponding magnetic field through a surface conveniently placed with respect to this conductor. However, instead of repeating the above computation for a new position of the conductor (now conductor 1′), we can use the final result in Eq. (7.36) with just changing the notation:

$$\Phi_b = -\frac{\mu_0 i_1 l}{2\pi}\ln\frac{r_{1'2'}}{r_{1'2}}, \tag{7.37}$$

where the minus sign comes from the reference direction of the current i_1 in the conductor 1′ being opposite to the direction of the current in the conductor 1.

By superposition, the total magnetic flux through the line 2–2′ at the length l of the system amounts to

$$\Phi_2 = \Phi_a + \Phi_b = \frac{\mu_0 i_1 l}{2\pi}\ln\frac{r_{12'}r_{1'2}}{r_{12}r_{1'2'}}. \tag{7.38}$$

The corresponding flux per unit length (for one meter) of the system is obtained as

$$\Phi_2' = (\Phi_2)_{\text{p.u.l.}} = \frac{\Phi_2}{l}, \tag{7.39}$$

so that the mutual inductance per unit length of the two lines for the orientations of the lines given in Fig. 7.9 is

$$\boxed{L_{21}' = (L_{21})_{\text{p.u.l.}} = \frac{L_{21}}{l} = \frac{\Phi_2'}{i_1} = \frac{\mu_0}{2\pi}\ln\frac{r_{12'}r_{1'2}}{r_{12}r_{1'2'}}.} \tag{7.40}$$

L_{21}' – two parallel thin two-wire lines (unit: H/m)

Depending on the actual mutual position of the lines, L'_{21} can be both positive (when $r_{12'}r_{1'2} > r_{12}r_{1'2'}$) and negative (when $r_{12'}r_{1'2} < r_{12}r_{1'2'}$). The unit for L'_{21} is H/m.

This formula is of great importance in assessing (usually undesired) magnetic coupling and mutual electromagnetic induction between different combinations of parallel two-wire transmission lines in practical applications (e.g., coupling between a phone line and nearby power line or between pairs of wire conductors in an electronic device). It can be used also for other types of transmission lines that can be approximated by wire lines in some considerations (e.g., rough computation of mutual inductances within multiconductor microstrip and strip transmission lines that give rise to the so-called inductive "cross talk" between conductors in printed circuit boards in computers).

Problems: 7.10–7.15; *Conceptual Questions* (on Companion Website): 7.11–7.21; *MATLAB Exercises* (on Companion Website).

7.3 ANALYSIS OF MAGNETICALLY COUPLED CIRCUITS

Having now in hand the concepts of both self- and mutual inductance, let us assume that slowly time-varying currents exist in both contours (circuits) of Fig. 7.7 at the same time, as shown in Fig. 7.10. The total magnetic flux Φ_1 through the first circuit is now caused by both i_1 and i_2. Using Eqs. (7.1) and (7.23) and the superposition principle, we have

$$\Phi_1 = L_1 i_1 + L_{12} i_2, \tag{7.41}$$

with L_1 being the self-inductance of the first circuit and L_{12} the mutual inductance between the circuits. Similarly, the total magnetic flux through the second circuit is

$$\Phi_2 = L_{21} i_1 + L_2 i_2, \tag{7.42}$$

where $L_{21} = L_{12}$ and L_2 is the self-inductance of the second circuit. By the same token, the induced emf in each of the circuits is composed of both self-induction and mutual-induction terms, which can be written as

$$e_{\text{ind1}} = -\frac{d\Phi_1}{dt} = -L_1\frac{di_1}{dt} - L_{12}\frac{di_2}{dt}, \tag{7.43}$$

$$e_{\text{ind2}} = -\frac{d\Phi_2}{dt} = -L_{21}\frac{di_1}{dt} - L_2\frac{di_2}{dt}. \tag{7.44}$$

Finally, from Fig. 7.10 [see also Figs. 6.6 and 6.11(b)], voltages across the terminals of the circuits are given by

current-voltage characteristic of two magnetically coupled circuits

$$v_1 = -e_{\text{ind1}} = L_1\frac{di_1}{dt} + L_{12}\frac{di_2}{dt}, \tag{7.45}$$

$$v_2 = -e_{\text{ind2}} = L_{21}\frac{di_1}{dt} + L_2\frac{di_2}{dt}. \tag{7.46}$$

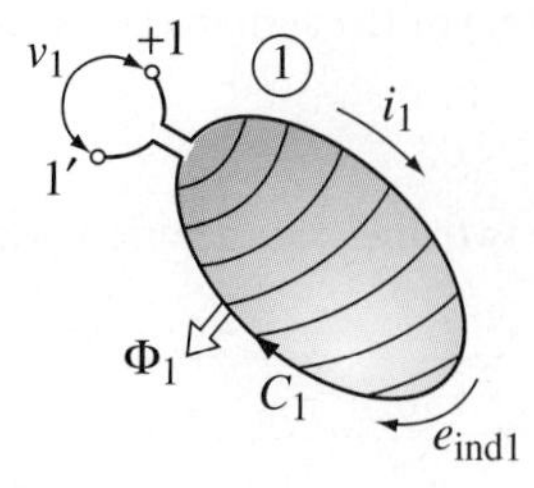

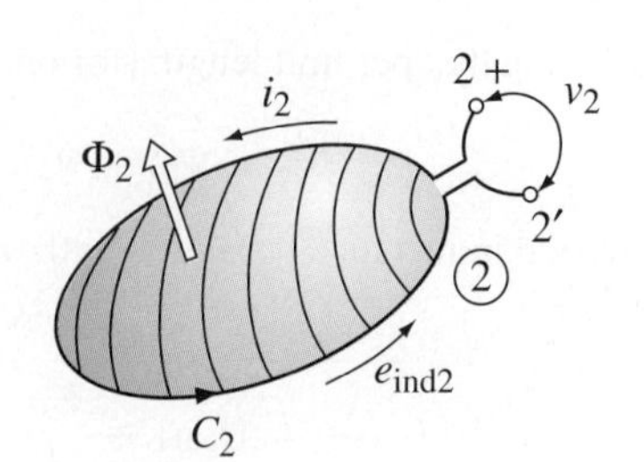

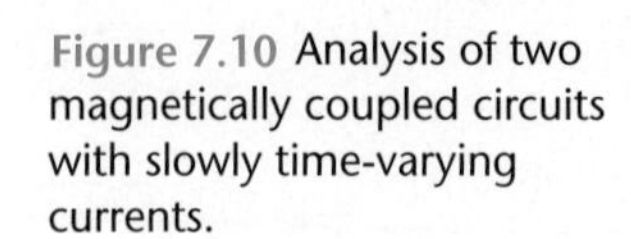
Figure 7.10 Analysis of two magnetically coupled circuits with slowly time-varying currents.

In words, the voltage across the terminals of each of the circuits is a linear combination of time derivatives of currents in both circuits, with L_1, L_2, and $L_{12} = L_{21}$ as linearity (proportionality) constants.

Shown in Fig. 7.11 is the circuit-theory representation of the two magnetically coupled circuits of Fig. 7.10. This is a two-port network whose current-voltage characteristic is given by Eqs. (7.45) and (7.46). It consists of two coupled ideal inductors, where, in addition to modeling the emf due to self-induction in each inductor, the effect of the emf due to mutual induction between the inductors is also modeled. The mutual inductance between the inductors is customarily written as

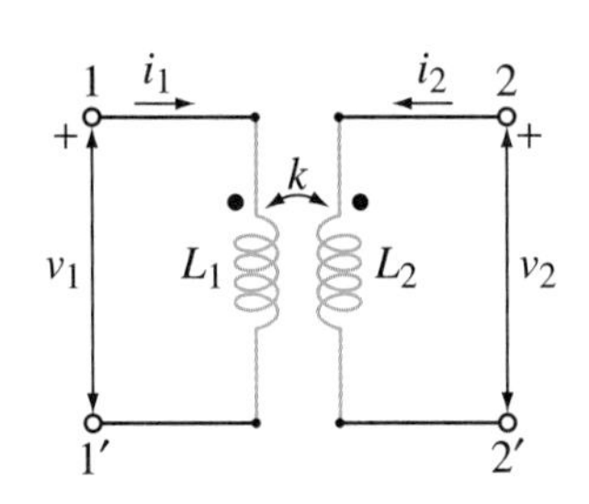

Figure 7.11 Circuit-theory representation of two coupled inductors.

$$L_{12} = \pm k\sqrt{L_1 L_2}, \tag{7.47}$$

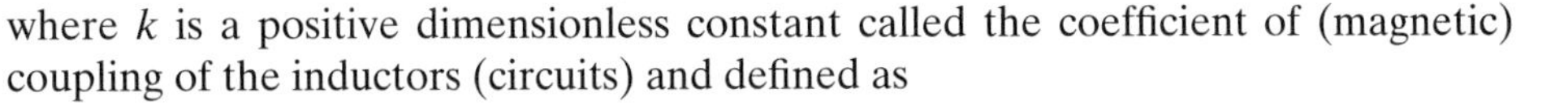

where k is a positive dimensionless constant called the coefficient of (magnetic) coupling of the inductors (circuits) and defined as

$$\boxed{k = \frac{|L_{12}|}{\sqrt{L_1 L_2}}, \quad 0 \le k \le 1.} \tag{7.48}$$

coefficient of magnetic coupling between two circuits (dimensionless)

We shall see in the next section that k is always less than or eventually equal to unity. The coefficient k in Fig. 7.11 thus provides the information about the magnitude of the mutual inductance between the circuits of Fig. 7.10. The sign of L_{12}, however, depends on the adopted reference directions of currents i_1 and i_2, and therefore cannot be given as a single piece of information (positive or negative) along with k, independently from the current directions. That is why we use a so-called two-dot notation to include the information about the sign of L_{12} in the representation in Fig. 7.11, by placing two big dots near the particular ends of the two inductors. According to this notation (convention), if both currents (i_1 and i_2) enter the inductors at ends marked by a big dot (as in Fig. 7.11), the mutual inductance, for that particular combination of reference directions of currents, is positive (note that $L_{12} > 0$ in Fig. 7.10). The same is true if both currents leave the inductors at marked (dotted) ends. Otherwise, if one current enters and the other leaves the inductor at marked ends, the mutual inductance is negative (note that a change of the reference direction of one of the currents in Figs. 7.10 and 7.11 would result in L_{12} becoming negative). Finally, if $k = 0$, the two inductors in Fig. 7.11 become decoupled and independently described by the current-voltage characteristic in Eq. (7.3) for a single inductor [Eqs. (7.45) and (7.46) with $L_{12} = 0$].

Note that Eqs. (7.41) and (7.42) can be represented in matrix form:

$$[\Phi] = [L][i], \tag{7.49}$$

where $[\Phi]$ and $[i]$ are column matrices whose elements are the fluxes and currents of the coupled circuits, respectively, and $[L]$ is a symmetrical square matrix of inductances,

$$\boxed{[L] = \begin{bmatrix} L_{11} & L_{12} \\ L_{21} & L_{22} \end{bmatrix} = \begin{bmatrix} L_1 & L_{12} \\ L_{12} & L_2 \end{bmatrix}} \tag{7.50}$$

inductance matrix

($L_{11} \equiv L_1$ and $L_{22} \equiv L_2$). Note also that these equations, as well as Eqs. (7.43)–(7.46) for the electromotive forces and voltages, can be generalized to the system with an arbitrary number (N) of coupled contours (circuits), in which case $[L]$ is an $N \times N$ matrix.

Example 7.10 Two Coupled Inductors Connected in Series and Parallel

Find the equivalent inductance of two coupled inductors of inductances L_1 and L_2 and coupling coefficient k if they are connected in (a) series, as in Fig. 7.12(a), and (b) parallel, as in Fig. 7.12(b).

Solution

(a) Voltages v_1 and v_2 of individual inductors are given by Eqs. (7.45) and (7.46). The current through the inductors in Fig. 7.12(a) is the same, $i_1 = i_2 = i$, so that the voltage across their connection is

$$v = v_1 + v_2 = \underbrace{L_1 \frac{\mathrm{d}i}{\mathrm{d}t} + L_{12} \frac{\mathrm{d}i}{\mathrm{d}t}}_{\text{inductor 1}} + \underbrace{L_{12} \frac{\mathrm{d}i}{\mathrm{d}t} + L_2 \frac{\mathrm{d}i}{\mathrm{d}t}}_{\text{inductor 2}} = (L_1 + L_2 + 2L_{12}) \frac{\mathrm{d}i}{\mathrm{d}t}. \quad (7.51)$$

By comparison with the current-voltage characteristic for a single inductor, Eq. (7.3), we conclude that the total (equivalent) inductance of the connection in series equals

equivalent inductance of two coupled inductors in series

$$\boxed{L = L_1 + L_2 + 2L_{12},} \quad (7.52)$$

where the mutual inductance between the inductors, for the situation given in Fig. 7.12(a), is positive (the current enters both inductors at their ends marked by a big dot) and, using Eq. (7.47), amounts to

$$L_{12} = k\sqrt{L_1 L_2}. \quad (7.53)$$

(b) The voltage across the inductors in Fig. 7.12(b) is the same, $v_1 = v_2 = v$. Assuming it to be known, Eqs. (7.45) and (7.46) provide the following system of equations with $\mathrm{d}i_1/\mathrm{d}t$ and $\mathrm{d}i_2/\mathrm{d}t$ as unknowns:

$$L_1 \frac{\mathrm{d}i_1}{\mathrm{d}t} + L_{12} \frac{\mathrm{d}i_2}{\mathrm{d}t} = v \quad \text{and} \quad L_{12} \frac{\mathrm{d}i_1}{\mathrm{d}t} + L_2 \frac{\mathrm{d}i_2}{\mathrm{d}t} = v. \quad (7.54)$$

Its solution is

$$\frac{\mathrm{d}i_1}{\mathrm{d}t} = \frac{L_2 - L_{12}}{L_1 L_2 - L_{12}^2} v \quad \text{and} \quad \frac{\mathrm{d}i_2}{\mathrm{d}t} = \frac{L_1 - L_{12}}{L_1 L_2 - L_{12}^2} v. \quad (7.55)$$

As the current through the terminals of the parallel connection of inductors is $i = i_1 + i_2$, we have

$$\frac{\mathrm{d}i}{\mathrm{d}t} = \frac{\mathrm{d}i_1}{\mathrm{d}t} + \frac{\mathrm{d}i_2}{\mathrm{d}t} = \frac{L_1 + L_2 - 2L_{12}}{L_1 L_2 - L_{12}^2} v. \quad (7.56)$$

By expressing v in terms of $\mathrm{d}i/\mathrm{d}t$,

$$v = \frac{L_1 L_2 - L_{12}^2}{L_1 + L_2 - 2L_{12}} \frac{\mathrm{d}i}{\mathrm{d}t}, \quad (7.57)$$

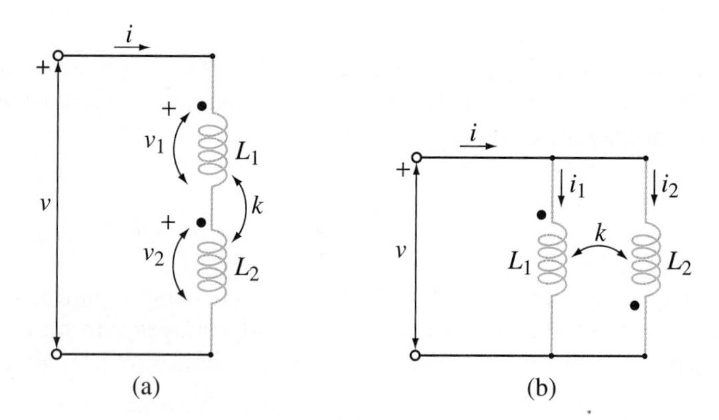

Figure 7.12 Evaluation of the equivalent inductance of two coupled inductors connected in (a) series and (b) parallel; for Example 7.10.

we conclude that the two coupled inductors of Fig. 7.12(b) can be replaced by a single equivalent inductor of inductance

$$L = \frac{L_1 L_2 - L_{12}^2}{L_1 + L_2 - 2L_{12}}. \quad (7.58)$$

equivalent inductance of two coupled inductors in parallel

The particular (given) placement of big dots in Fig. 7.12(b) tells us that the mutual inductance L_{12} is negative, and hence Eq. (7.47) yields

$$L_{12} = -k\sqrt{L_1 L_2}. \quad (7.59)$$

Note that for $k = 0$ and $L_{12} = 0$, Eqs. (7.52) and (7.58) reduce to expressions for equivalent inductances of two ordinary inductors of inductances L_1 and L_2 (that are not magnetically coupled together) connected in series and parallel, respectively. Note also that these expressions (with $L_{12} = 0$) have the same form as the corresponding expressions for two connected resistors of resistances R_1 and R_2 [see Eqs. (3.86) and (3.94)].

Example 7.11 Coupling Coefficient of Two Coils on a Toroidal Core

Compute the coefficient of coupling between the coils on the thin toroidal core from Example 7.7.

Solution Using the expression in Eq. (7.29) for the magnetic flux through the cross section of the core due to the current i_1 in the first coil, the self-inductance of the first coil is

$$L_1 = \frac{\Phi_1}{i_1} = \frac{N_1 \Phi_{\text{single turn}}}{i_1} = \frac{\mu N_1^2 S}{l}. \quad (7.60)$$

L – coil on a thin toroidal core

Similarly, the self-inductance of the second coil is $L_2 = \mu N_2^2 S/l$. The mutual inductance between the coils is given by Eq. (7.30). Hence, their coupling coefficient turns out to be

$$k = \frac{|L_{12}|}{\sqrt{L_1 L_2}} = \frac{N_1 N_2}{\sqrt{N_1^2 N_2^2}} = 1. \quad (7.61)$$

maximum coupling – coils on a toroidal core

This maximum coupling is a consequence of the magnetic flux due to the current of each of the coils being concentrated entirely inside the core (flux leakage from the core is negligible), so that the entire flux passes through every turn of both coils.

Example 7.12 Voltage Transformation by Two Coupled Coils

Consider a magnetic circuit in the form of a thin toroidal ferromagnetic core with two windings shown in Fig. 7.13(a). The windings have $N_1 = 1000$ and $N_2 = 500$ turns of wire, respectively. The ferromagnetic material can be considered to be linear. Losses in the

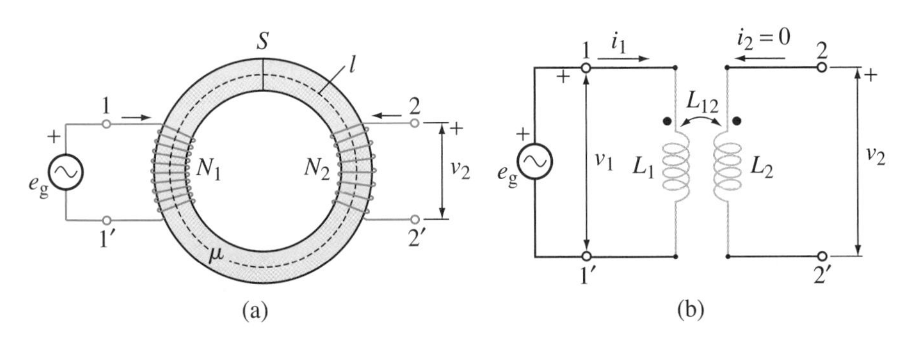

Figure 7.13 (a) Magnetic circuit with two windings and (b) equivalent schematic diagram with two coupled inductors; for Example 7.12.

windings and the core can be neglected. The primary winding is connected to an ideal ac voltage generator of emf $e_g = 100\cos 377t$ V (t in s). The secondary winding is open-circuited. Find the voltage across the secondary winding.

Solution Fig. 7.13(b) shows the equivalent schematic diagram with two coupled inductors, where L_{12} is positive for the reference orientations of windings in Fig. 7.13(a). As the secondary winding is open-circuited, Eqs. (7.45) and (7.46) give

$$v_1 = L_1 \frac{di_1}{dt} \quad \text{and} \quad v_2 = L_{21}\frac{di_1}{dt} \quad (i_2 = 0). \tag{7.62}$$

Using Eqs. (7.60) and (7.30), we can write

voltage transformation by coupled coils

$$\boxed{\frac{v_1}{v_2} = \frac{L_1}{L_{12}} = \frac{N_1^2}{N_1N_2} = \frac{N_1}{N_2},} \tag{7.63}$$

that is, the ratio of the voltages across the primary and secondary windings is equal to the ratio of the numbers of wire turns. The voltage across the secondary winding is hence

$$v_2 = \frac{N_2}{N_1}v_1 = \frac{N_2}{N_1}e_g = 50\cos 377t \text{ V} \quad (t \text{ in s}). \tag{7.64}$$

We can say that the circuit in Fig. 7.13 operates as a transformer of voltage from its primary port to its secondary port, with the transformation (multiplication) factor being equal to the wire turns ratio N_2/N_1.

Example 7.13 Current Transformation by Two Coupled Coils

Assume that the primary winding in the magnetic circuit in Fig. 7.13(a) is connected to an ideal ac current generator of current intensity $i_g = 15\sin 377t$ mA (t in s), while the secondary winding is short-circuited, and determine the current of the secondary winding.

Solution From the equivalent schematic diagram shown in Fig. 7.14 and Eq. (7.46),

$$L_{12}\frac{di_1}{dt} + L_2\frac{di_2}{dt} = 0 \quad (v_2 = 0). \tag{7.65}$$

This means that

current transformation by coupled coils

$$\boxed{\frac{i_1}{i_2} = -\frac{L_2}{L_{12}} = -\frac{N_2^2}{N_1N_2} = -\frac{N_2}{N_1},} \tag{7.66}$$

i.e., the ratio of the currents in the primary and secondary windings is equal to the negative inverse ratio of the numbers of wire turns. Solving for the current in the secondary winding, we have

$$i_2 = -\frac{N_1}{N_2}i_1 = -\frac{N_1}{N_2}i_g = -30\sin 377t \text{ mA} \quad (t \text{ in s}). \tag{7.67}$$

We see that the circuit in Fig. 7.14 performs a transformation of current between its ports, with the magnitude transformation factor being the reciprocal of that for the voltage transformation in Eq. (7.64).

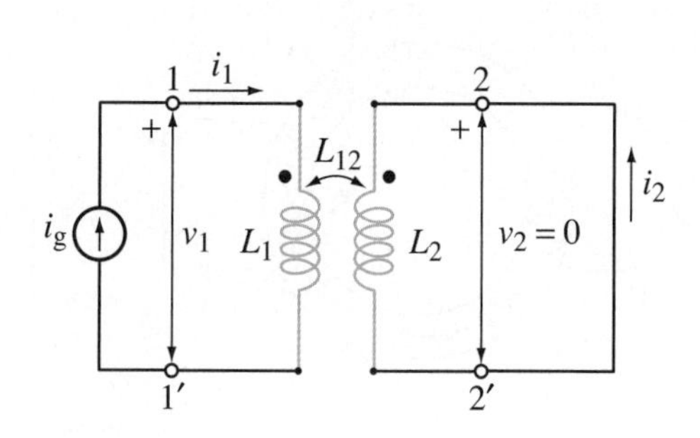

Figure 7.14 Current transformation by two magnetically coupled coils; for Example 7.13.

Example 7.14 Coupling Coefficient of Two Coaxial Solenoids

Fig. 7.15 shows a cross section of two very long solenoidal coils positioned coaxially with respect to each other. Radii of solenoids are a_1 and a_2, their length is the same, l, and numbers of wire turns are N_1 and N_2, respectively. The inner coil is wound on a core made of a linear ferromagnetic material of relative permeability μ_r, while the space between the coils is air-filled. Neglecting the end effects, calculate the coefficient of coupling between the two coils.

Solution Let us adopt counter-clockwise reference directions of currents in both coils (Fig. 7.15). Assume first that a current, of intensity I_1 (steady current), exists only in the first (inner) coil, while $I_2 = 0$. Under this assumption, the magnetic field is nonzero only inside the first coil and is given by [see Eq. (6.48)]

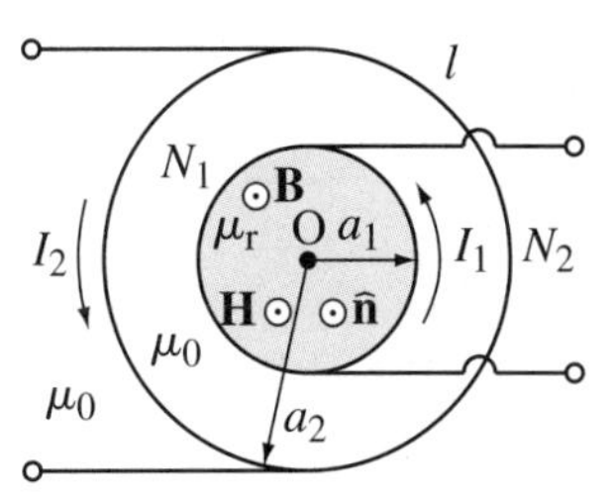

Figure 7.15 Cross section of two coupled coaxial solenoids; for Example 7.14.

$$H_1 = \frac{N_1 I_1}{l}. \tag{7.68}$$

Therefore, the magnetic fluxes through a single turn of both coils are the same,

$$\Phi_{\text{single turn}} = \mu_r \mu_0 H_1 \pi a_1^2 = \frac{\mu_r \mu_0 N_1 I_1 \pi a_1^2}{l}. \tag{7.69}$$

Using this expression, the self-inductance of the first coil is found to be

$$L_1 = \frac{\Phi_1}{I_1} = \frac{N_1 \Phi_{\text{single turn}}}{I_1} = \frac{\mu_r \mu_0 N_1^2 \pi a_1^2}{l} \tag{7.70}$$

[note that this can also be obtained by multiplying the inductance in Eq. (7.6) by μ_r]. Using the same expression (for $\Phi_{\text{single turn}}$), the mutual inductance between the coils is obtained as

$$L_{21} = \frac{\Phi_2}{I_1} = \frac{N_2 \Phi_{\text{single turn}}}{I_1} = \frac{\mu_r \mu_0 N_1 N_2 \pi a_1^2}{l}. \tag{7.71}$$

On the other hand, the assumption of a current, of intensity I_2, existing only in the second (outer) coil (while $I_1 = 0$) in Fig. 7.15 gives a nonzero magnetic field everywhere inside the second coil, of intensity

$$H_2 = \frac{N_2 I_2}{l}. \tag{7.72}$$

The magnetic flux density in the core is $B_2 = \mu_r \mu_0 H_2$, while $B_{20} = \mu_0 H_2$ in the air-filled space between the coils, which leads to the following for the total magnetic flux through the second coil:

$$\Phi_2 = N_2[\underbrace{B_2 \pi a_1^2}_{\text{core}} + \underbrace{B_{20} \pi (a_2^2 - a_1^2)}_{\text{air}}] = \frac{\mu_0 N_2^2 \pi I_2 (\mu_r a_1^2 + a_2^2 - a_1^2)}{l}. \tag{7.73}$$

The self-inductance of the second coil is hence

$$L_2 = \frac{\Phi_2}{I_2} = \frac{\mu_0 N_2^2 \pi [(\mu_r - 1) a_1^2 + a_2^2]}{l}, \tag{7.74}$$

whereas computing Φ_1 through the first coil and $L_{12} = \Phi_1 / I_2$ would, of course, give the same result for the mutual inductance as in Eq. (7.71).

Finally, the coefficient of coupling between the coils, Eq. (7.48), comes out to be

$$k = \frac{|L_{12}|}{\sqrt{L_1 L_2}} = \sqrt{\frac{\mu_r}{\mu_r - 1 + (a_2/a_1)^2}}. \tag{7.75}$$

Note that for very large relative permeabilities μ_r (e.g., $\mu_r = 1000$), k is very close to unity (coupling is very strong), which is a consequence of the magnetic flux Φ_2 due to I_2 in the second coil being practically entirely concentrated in the core of the first coil for $\mu_r \gg 1$ [$\mu_r a_1^2 \gg a_2^2 - a_1^2$ in Eq. (7.73)].

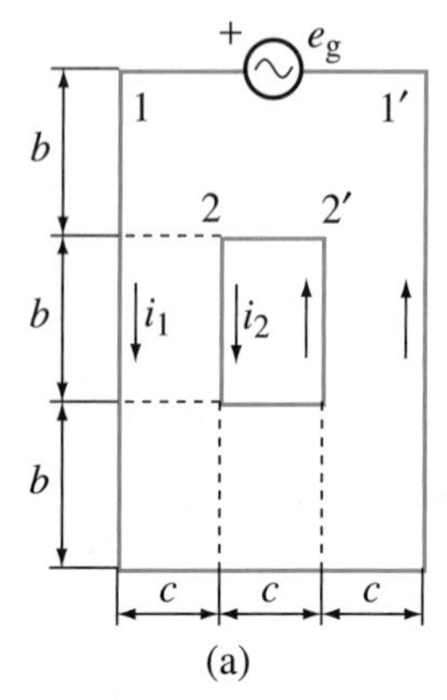

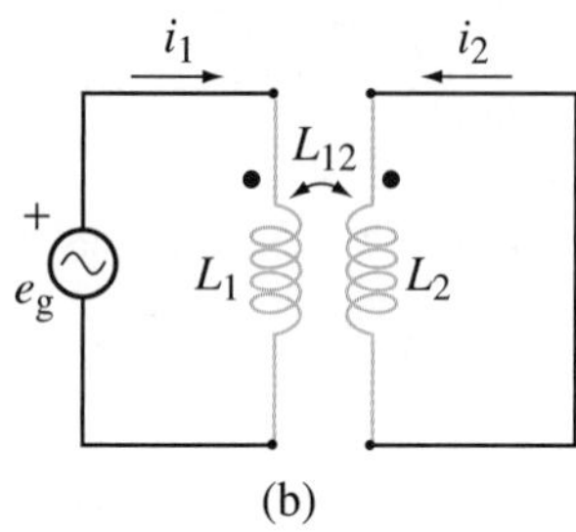

Figure 7.16 Analysis of magnetically coupled circuits containing two thin two-wire lines: (a) structure geometry and (b) equivalent schematic diagram with two coupled inductors; for Example 7.15.

Example 7.15 Magnetically Coupled Circuits Containing Two-Wire Lines

Fig. 7.16(a) depicts two parallel thin two-wire lines (1–1′ and 2–2′) in air, where $b = 20$ cm and $c = 5$ mm. The radii of all wires are $a = 0.3$ mm. The first line is connected at one end to an ideal voltage generator of time-harmonic emf $e_g(t) = \mathcal{E}_{g0} \cos \omega t$, where $\mathcal{E}_{g0} = 3$ V and $\omega = 10^6$ rad/s, and the other end of the line is short-circuited. The second line is short-circuited at both ends. Neglecting internal inductances, losses in the wires, capacitive coupling between the lines, end effects, and propagation effects, find the amplitudes of currents in the lines.

Solution Using Eq. (7.11), self-inductances per unit length of the lines in Fig. 7.16(a) are

$$L_1' = \frac{\mu_0}{\pi} \ln \frac{3c}{a} = 1.565\ \mu\text{H/m} \quad \text{and} \quad L_2' = \frac{\mu_0}{\pi} \ln \frac{c}{a} = 1.125\ \mu\text{H/m}, \tag{7.76}$$

respectively, where we neglect internal inductances of the lines. Eq. (7.40) gives the mutual inductance per unit length of the lines ($r_{12'} = r_{1'2} = 2c$ and $r_{12} = r_{1'2'} = c$):

$$L_{12}' = \frac{\mu_0}{2\pi} \ln \frac{(2c)^2}{c^2} = \frac{\mu_0}{\pi} \ln 2 = 277\ \text{nH/m}. \tag{7.77}$$

We neglect the end effects and the propagation effects, so that the total self-inductances and mutual inductance of the magnetically coupled circuits formed by the lines are obtained by multiplying the per-unit-length inductances by the corresponding lengths of the lines, $L_1 = L_1' 3b = 939$ nH, $L_2 = L_2' b = 225$ nH, and $L_{12} = L_{12}' b = 55.4$ nH (the length of the shorter of the two lines, b, is relevant for the mutual inductance between the lines).

Fig. 7.16(b) shows the equivalent schematic diagram of the two coupled circuits (we neglect the losses in wires and capacitive coupling between the circuits). According to this diagram,

$$L_1 \frac{di_1}{dt} + L_{12} \frac{di_2}{dt} = e_g, \quad L_{12} \frac{di_1}{dt} + L_2 \frac{di_2}{dt} = 0. \tag{7.78}$$

The solution of these two equations for the time derivatives of currents in the circuits is

$$\frac{di_1}{dt} = \frac{L_2}{L_1 L_2 - L_{12}^2} \mathcal{E}_{g0} \cos \omega t \quad \text{and} \quad \frac{di_2}{dt} = -\frac{L_{12}}{L_1 L_2 - L_{12}^2} \mathcal{E}_{g0} \cos \omega t, \tag{7.79}$$

which is then integrated with respect to time to obtain the solution for the currents. Integration in time of time-harmonic quantities results in an additional factor $1/\omega$ in the expressions for amplitudes, so that the amplitudes (peak-values) of the currents in the circuits (and the lines) are, respectively,

$$I_{01} = \frac{L_2 \mathcal{E}_{g0}}{\omega (L_1 L_2 - L_{12}^2)} = 3.24\ \text{A} \quad \text{and} \quad I_{02} = \frac{L_{12} \mathcal{E}_{g0}}{\omega (L_1 L_2 - L_{12}^2)} = 0.8\ \text{A}. \tag{7.80}$$

Example 7.16 Inductance p.u.l. of a Two-Wire Line above a PMC Plane

A symmetrical thin two-wire line is positioned in air over a ferromagnetic ($\mu \gg \mu_0$) plane, as shown in Fig. 7.17(a). The wire radii are a, the height of the axes of both wires with respect to the plane is h ($h \gg a$), and the distance between the wire axes is d ($d \gg a$). Compute the inductance per unit length of the line (in the presence of the plane).

Solution Assume that a slowly time-varying current of intensity i is established in the line [Fig. 7.17(a)]. By image theory for the magnetic field, Fig. 5.14, we can remove the ferromagnetic plane by introducing positive images of the original current conductors. The equivalent system in Fig. 7.17(b) is thus obtained, which consists of two (magnetically coupled) parallel thin two-wire lines with the same current (i) in air. The total magnetic flux per unit length of the upper (original) line is therefore [see Eq. (7.41)]

$$\Phi' = L_1' i + L_{12}' i, \tag{7.81}$$

where L_1' is the self-inductance p.u.l. of the upper line when isolated in free space and L_{12}' is the mutual inductance p.u.l. of the upper and lower lines in the equivalent system in air. These inductances are computed using Eqs. (7.11) and (7.40), respectively, so that the inductance p.u.l. of the original line in Fig. 7.17(a) comes out to be

$$L' = \frac{\Phi'}{i} = L_1' + L_{12}' = \frac{\mu_0}{\pi}\left(\ln\frac{d}{a} + \ln\frac{\sqrt{d^2+4h^2}}{2h}\right) = \frac{\mu_0}{\pi}\ln\frac{d\sqrt{d^2+4h^2}}{2ah}. \qquad (7.82)$$

Obviously, L_{12}' represents the influence of the ferromagnetic material in the lower half-space on the line inductance.

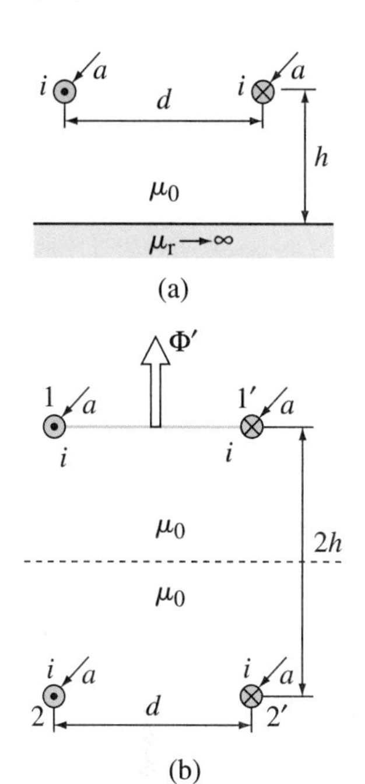

Figure 7.17 Evaluation of the inductance per unit length of a two-wire line above a ferromagnetic (or PMC) plane: (a) original system and (b) equivalent free-space system with two magnetically coupled two-wire lines; for Example 7.16.

Problems: 7.16–7.24; *Conceptual Questions* (on Companion Website): 7.22–7.28; *MATLAB Exercises* (on Companion Website).

7.4 MAGNETIC ENERGY OF CURRENT-CARRYING CONDUCTORS

Every system of conducting loops with currents contains a certain amount of energy, called magnetic energy, in a manner analogous to a system of charged conducting bodies storing electric energy. In other words, current-carrying (coupled or uncoupled) inductors in an ac or dc circuit store magnetic energy, much like charged capacitors store electric energy. By the principle of conservation of energy, the magnetic energy of a system of loops (or inductors) with currents equals the work done to the system in the process of establishing these currents from zero to their final values. Thus, by simulating this process of "loading" the loops by currents, we can find general expressions for computing the magnetic energy in terms of the current intensities and associated magnetic fluxes (final values) of the loops.

Let us consider first a single loop with a slowly time-varying current of intensity i. The current flow is maintained in the circuit by an ideal voltage generator of emf e_g, as shown in Fig. 7.18. As the magnetic flux through the loop, Φ, changes in time, an emf e_{ind} is induced in the loop, given by Faraday's law of electromagnetic induction, Eq. (6.34). Using Kirchhoff's voltage law, Eq. (3.119), we can write

$$e_g = Ri - e_{ind}, \qquad (7.83)$$

where R is the resistance of the loop. The emf e_{ind} (emf due to self-induction) opposes the current change in the loop (Lenz's law), that is, it acts against the emf e_g. Therefore, an amount of work must be done in establishing the current in the loop to overcome this induced emf. This work is done by en external agent, with the energy transfer to the circuit being modeled by the generator (of emf e_g) in Fig. 7.18.

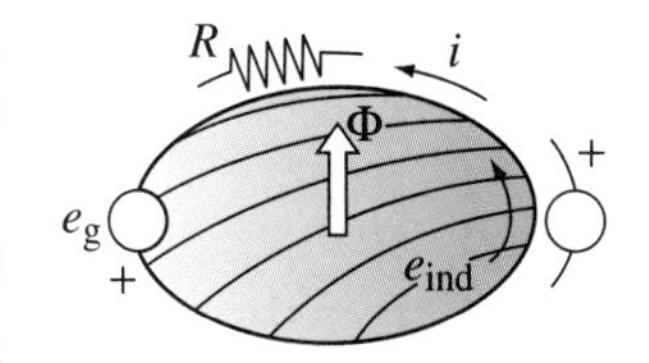

Figure 7.18 Wire loop of inductance L and resistance R with a slowly time-varying current of intensity i, maintained by an ideal voltage generator of emf e_g.

To investigate the energy balance in the circuit during a differentially short time interval dt, we multiply both sides of Eq. (7.83) by $i\,dt$,

$$e_g i\,dt = Ri^2\,dt + (-e_{ind} i\,dt). \qquad (7.84)$$

By means of Eq. (3.121), the term on the left-hand side of this equation is the work done by the emf of the generator during the time interval dt. It equals the energy of external sources delivered to the circuit during that fraction of time. From Joule's law, Eq. (3.77), the first term on the right-hand side of the equation represents Joule's (ohmic) losses in the loop during dt. Finally, the last term in the equation (including the minus sign) is the work done against the emf e_{ind} in dt. Eq. (7.84) thus expresses the principle of conservation of energy for the circuit in Fig. 7.18, telling us that the work done by the generator during dt is partly converted into Joule's losses and partly used to overcome the induced emf in the loop. While the first part is lost to heat, the second part is given to the emf e_{ind}, i.e., to the flux Φ of the loop. Ultimately, it represents the energy delivered to the magnetic field of the current i in the loop during the time dt. Using Eq. (6.34), $-e_{\text{ind}}\,dt = d\Phi$, so that the elementary work done by external sources to the magnetic field of the loop can be written as

$$dW_{\text{m}} = -e_{\text{ind}} i\,dt = i\,d\Phi. \tag{7.85}$$

If the medium surrounding the loop is magnetically linear, the use of Eq. (7.1) yields

$$dW_{\text{m}} = i\,d\Phi = i\,d(Li) = Li\,di, \tag{7.86}$$

where L is the inductance of the loop. The magnetic field, on the other hand, is capable of storing the received energy, and this stored energy is the magnetic energy of the loop, expressed in joules (J). Consequently, the work dW_{m} in Eq. (7.86) represents an increment of the magnetic energy of the circuit in Fig. 7.18.

When the current in the loop is zero, the loop has no energy. The total energy stored in the magnetic field of the loop when its current is i equals therefore the net work done to the magnetic field in changing the loop current from $i = 0$ to its final value i, and is obtained by adding up all elementary works dW_{m}:

$$W_{\text{m}} = \int_{i=0}^{i} dW_{\text{m}} = L\int_0^i i\,di = L\left.\frac{i^2}{2}\right|_0^i = \frac{1}{2}Li^2. \tag{7.87}$$

From the circuit-theory standpoint, this is also the energy of a linear inductor of inductance L in an arbitrary electric circuit. Employing Eq. (7.1), the equivalent expressions for the energy of an inductor are

energy of inductor (unit: J)

$$\boxed{W_{\text{m}} = \frac{1}{2}Li^2 = \frac{1}{2}\Phi i = \frac{\Phi^2}{2L}.} \tag{7.88}$$

We note the complete analogy (duality) with the corresponding expressions for the electric energy (W_{e}) of a capacitor in Eq. (2.192).

Generally, for a system with N ($N \geq 1$) magnetically coupled loops (inductors),

energy of N magnetically coupled loops

$$\boxed{W_{\text{m}} = \frac{1}{2}\sum_{k=1}^{N}\Phi_k i_k,} \tag{7.89}$$

where the magnetic fluxes through the loops are given by linear relationships in Eqs. (7.41)–(7.42) or (7.49). Using these relationships, the energy of the system can be expressed only in terms of the currents and self- and mutual inductances of the loops. For example, for $N = 2$, the magnetic energy of two coupled loops can be written in the form

magnetic energy of two coupled inductors

$$\boxed{W_{\text{m}} = \frac{1}{2}[\underbrace{(L_1 i_1 + L_{12} i_2)}_{\Phi_1} i_1 + \underbrace{(L_{12} i_1 + L_2 i_2)}_{\Phi_2} i_2] = \frac{1}{2}L_1 i_1^2 + \frac{1}{2}L_2 i_2^2 + L_{12} i_1 i_2.} \tag{7.90}$$

This energy can be both larger and smaller than the sum of energies of the two loops when magnetically isolated from each other (when $L_{12} = 0$), because L_{12} can be both positive and negative. Note that the expression in Eq. (7.89) is entirely analogous to the expression for the electric energy of a linear multibody system in Eq. (2.195).

The magnetic energy can also be expressed in terms of distributions of the current density vector, **J**, and the magnetic vector potential, **A**, throughout the volume of current-carrying loops. Namely, the magnetic flux Φ_k through the loop C_k of the system with N loops equals the circulation of **A** along C_k [see Eq. (4.121)], so that Eq. (7.89) becomes

$$W_{\mathrm{m}} = \frac{1}{2}\sum_{k=1}^{N} i_k \underbrace{\oint_{C_k} \mathbf{A}\cdot \mathrm{d}\mathbf{l}_k}_{\Phi_k} = \frac{1}{2}\sum_{k=1}^{N} \oint_{C_k} \mathbf{A}\cdot i_k\, \mathrm{d}\mathbf{l}_k, \tag{7.91}$$

where the current i_k can be brought inside the integral sign in the line integral because it does not change along the wire (currents are slowly time-varying). As $i_k\,\mathrm{d}\mathbf{l}_k = \mathbf{J}\,\mathrm{d}v_k$, Eq. (4.10), we then have

$$W_{\mathrm{m}} = \frac{1}{2}\sum_{k=1}^{N} \int_{v_k} \mathbf{A}\cdot \mathbf{J}_k\, \mathrm{d}v_k, \tag{7.92}$$

with v_k standing for the volume of the kth loop. Finally, the sum of N integrals over volumes of individual loops can be joined together into a single volume integral:

$$\boxed{W_{\mathrm{m}} = \frac{1}{2}\int_{v_{\mathrm{current}}} \mathbf{J}\cdot \mathbf{A}\, \mathrm{d}v.} \tag{7.93}$$

magnetic energy in terms of volume currents

Since any slowly time-varying volume current distribution can be considered as consisting of an infinite number of filamentary current loops, this is a general expression for evaluation of the magnetic energy of a system of current conductors of arbitrary shapes in a linear medium. In general, the volume integration is performed over all parts of the system populated by current (v_{current}). We again note the duality with the corresponding expression for the electric energy, namely, with the volume integral expression in Eq. (2.196) for W_{e} in terms of the charge density (ρ) and the electric scalar potential (V) for an electrostatic system with a volume charge.

Example 7.17 Magnetic Energy of Two Coupled Two-Wire Lines

Find the time-average magnetic energy of the system of two coupled two-wire lines shown in Fig. 7.16 and described in Example 7.15.

Solution From Eqs. (7.79), instantaneous intensities of currents in the lines have the following forms: $i_1(t) = I_{01}\sin\omega t$ and $i_2(t) = -I_{02}\sin\omega t$, where the current amplitudes I_{01} and I_{02} are given in Eqs. (7.80). Using Eq. (7.90), the instantaneous magnetic energy contained in the magnetic field of the two lines is

$$\begin{aligned} W_{\mathrm{m}}(t) &= \frac{1}{2}L_1 i_1^2(t) + \frac{1}{2}L_2 i_2^2(t) + L_{12} i_1(t) i_2(t) \\ &= \frac{1}{2}L_1 I_{01}^2 \sin^2\omega t + \frac{1}{2}L_2 I_{02}^2 \sin^2\omega t - L_{12} I_{01} I_{02}\sin^2\omega t, \end{aligned} \tag{7.94}$$

where the self-inductances and mutual inductance of the coupled circuits in Fig. 7.16(b) are also found in Example 7.15. Eq. (6.95) tells us that the time-average value of the function $\sin^2\omega t$ is 1/2, so that the time-average magnetic energy of the system is

$$(W_{\mathrm{m}})_{\mathrm{ave}} = \frac{1}{2}\left(\frac{1}{2}L_1 I_{01}^2 + \frac{1}{2}L_2 I_{02}^2 - L_{12} I_{01} I_{02}\right) = 2.43\ \mu\mathrm{J}. \tag{7.95}$$

This energy can also be obtained using the equivalent inductance, L, seen by the generator in Fig. 7.16. To find L, we rewrite the first equation of Eqs. (7.79) in the following form:

$$e_g = \left(L_1 - \frac{L_{12}^2}{L_2}\right)\frac{di_1}{dt} = L\frac{di_1}{dt}. \tag{7.96}$$

Hence,

$$L = L_1 - \frac{L_{12}^2}{L_2} = 925.4 \text{ nH}. \tag{7.97}$$

In other words, this is the inductance of an inductor that, as far as the generator is concerned, can replace the two coupled inductors in Fig. 7.16(b). Consequently, the energy of the two coupled two-wire lines is the same as the energy of the equivalent inductor, and Eq. (7.88) can be used. Since the current of the equivalent inductor is i_1,

$$W_m(t) = \frac{1}{2}Li_1^2(t), \tag{7.98}$$

which, averaged in time, gives the same result as in Eq. (7.95).

Example 7.18 Proof that the Maximum Coupling Coefficient is Unity

Prove that the largest possible value for the magnitude of the mutual inductance of two magnetically coupled circuits equals the geometric mean of the self-inductances of circuits.

Solution The inequality that we have to prove, $|L_{12}| \le \sqrt{L_1L_2}$, is a consequence of the simple fact that the stored magnetic energy is always positive (or eventually zero). To show this, we start with the expression for the energy stored in the magnetic field of the two circuits in Eq. (7.90) and write it in the following form:

$$W_m = \frac{1}{2}L_1i_1^2\left[1 + \frac{L_2}{L_1}\left(\frac{i_2}{i_1}\right)^2 + 2\frac{L_{12}}{L_1}\frac{i_2}{i_1}\right] = \frac{1}{2}L_1i_1^2 f(x), \tag{7.99}$$

where

$$f(x) = ax^2 + bx + 1, \quad x = \frac{i_2}{i_1}, \quad a = \frac{L_2}{L_1}, \quad \text{and} \quad b = 2\frac{L_{12}}{L_1}. \tag{7.100}$$

From the fact that $W_m \ge 0$, we conclude that the quadratic function $f(x)$ must be nonnegative for all values of x. This, in turn, is satisfied only if the minimum of f, given by

$$f_{\min} = f(-b/2a) = f(-L_{12}/L_2) = \frac{L_2}{L_1}\frac{L_{12}^2}{L_2^2} - 2\frac{L_{12}}{L_1}\frac{L_{12}}{L_2} + 1 = -\frac{L_{12}^2}{L_1L_2} + 1, \tag{7.101}$$

is nonnegative. Hence,

$$f_{\min} \ge 0 \quad \longrightarrow \quad L_{12}^2 \le L_1L_2 \quad \text{or} \quad |L_{12}| \le \sqrt{L_1L_2} \quad \longrightarrow \quad k \le 1, \tag{7.102}$$

which concludes our proof. We realize that these inequalities can also be written in terms of the coefficient of coupling of the two circuits, defined by Eq. (7.48), as $k \le 1$ (that the coupling coefficient cannot be larger than unity we noted earlier, in Section 7.3, but without a proof).

Problems: 7.25 and 7.26; *Conceptual Questions* (on Companion Website): 7.29 and 7.30; *MATLAB Exercises* (on Companion Website).

7.5 MAGNETIC ENERGY DENSITY

In analogy with the concept of electric energy density (Section 2.16), we shall now define and use the magnetic energy density to describe the actual localization and distribution of the magnetic energy of a system of current-carrying conductors, the total amount of which is given by Eqs. (7.89) and (7.93). To this end, let us consider first a simple case of a magnetic field of uniform intensity in a thin toroidal ferromagnetic core with a winding that carries a slowly time-varying current of intensity i. Let the number of wire turns of the winding be N, the length of the core l, and its cross section S in area. From Eq. (5.53), the current in the winding can be expressed in terms of the magnetic field intensity H in the core as $i = Hl/N$. On the other hand, the magnetic flux through the winding (all of its turns) can be written as $\Phi = N\Phi_{\text{single turn}} = NBS$, where B is the magnetic flux density in the core. Substituting these two expressions in Eq. (7.85) leads to the following expression for the work of external sources needed for a change $\mathrm{d}\Phi$ in the magnetic flux of the winding (not counting Joule's losses in the winding):

$$\mathrm{d}W_{\mathrm{m}} = i\,\mathrm{d}\Phi = \frac{Hl}{N}\,\mathrm{d}(NBS) = H\,\mathrm{d}B\,Sl = H\,\mathrm{d}B\,v, \qquad (7.103)$$

where $v = Sl$ is the volume of the core, i.e., the volume of the domain where the magnetic field exists. The net work in changing the flux from zero to its final value Φ or, equivalently, the flux density from zero to the final value B is hence

$$W_{\mathrm{m}} = \int_{B=0}^{B} \mathrm{d}W_{\mathrm{m}} = v \int_{0}^{B} H\,\mathrm{d}B. \qquad (7.104)$$

Dividing it by v, we get the magnetic energy density (energy per unit volume) of the core (in $\mathrm{J/m^3}$),

$$\boxed{w_{\mathrm{m}} = \frac{W_{\mathrm{m}}}{v} = \int_{0}^{B} H\,\mathrm{d}B.} \qquad (7.105)$$

magnetic energy density, arbitrary medium (unit: $\mathrm{J/m^3}$)

More precisely, for an arbitrary magnetic material of the core, this is the energy per unit volume of the material spent by external sources in establishing the field, and not the energy stored in the field per unit volume of the material. Namely, as we shall see later in this section, in the case of materials that exhibit hysteresis effects (see Fig. 5.21), this energy can only partially be returned by the field to the sources in the reverse process of reducing the field intensity H to zero, because of losses occurring during the magnetization (while establishing H) and demagnetization (while reducing H) of the material. These losses, appearing as heat, are a consequence of microscopic frictions encountered as elementary magnetic domains (see Fig. 5.2) change their size and rotate in the magnetization-demagnetization process of the material and are known as hysteresis losses. For materials with no hysteresis losses, on the other side, Eq. (7.105) gives the energy that is contained in the field per unit volume of the core and can be obtained from it at any time in its entirety by reducing to zero the current in the coil. While using the term magnetic energy density for the integral expression in Eq. (7.105), we shall always keep in mind this distinction in its actual meaning between materials with pronounced hysteresis behavior and those for which hysteresis effects are not present or can be neglected.

Although derived for a special case of a coil on a thin toroidal core, the result in Eq. (7.105) can be generalized to an arbitrary magnetic field, which can be visualized as a collection of elementary flux tubes (toroids) formed by the lines of vector $\mathbf{B}$ [see Fig. 4.26 and the proof of the law of conservation of magnetic flux, Eq. (4.99)]. In general, the magnetic energy of a differentially small cell of volume $\mathrm{d}v$ in an arbitrary (nonuniform) magnetic field in an arbitrary (nonlinear) material is

$$\mathrm{d}W_{\mathrm{m}} = w_{\mathrm{m}}\,\mathrm{d}v, \qquad (7.106)$$

where the energy density is given by Eq. (7.105). By summing up the energies of all of the cells, that is, by integrating the energy $\mathrm{d}W_{\mathrm{m}}$ over the entire domain with the magnetic field (volume v), we obtain the total magnetic energy of the system:

$$W_m = \int_v w_m \, dv = \int_v \left(\int_0^B H \, dB \right) dv. \tag{7.107}$$

In the case of a linear magnetic material of permeability μ, $\mathbf{B} = \mu \mathbf{H}$, so that the integral with respect to B in the expression for the magnetic energy density in Eq. (7.105) can easily be solved,

magnetic energy density, linear medium

$$w_m = \int_0^B H \, dB = \frac{1}{\mu} \int_0^B B \, dB = \frac{B^2}{2\mu} = \frac{1}{2} BH = \frac{1}{2} \mu H^2. \tag{7.108}$$

Note that these expressions are entirely analogous to the corresponding expressions in Eq. (2.199) for the electric energy density w_e in a linear dielectric of permittivity ε.[5] The expression for the total magnetic energy becomes

$$W_m = \frac{1}{2} \int_v \mathbf{B} \cdot \mathbf{H} \, dv. \tag{7.109}$$

For nonlinear magnetic materials, the integral in Eq. (7.105), in general, cannot be solved analytically (in a closed form). Having in mind a typical initial magnetization curve of a nonlinear ferromagnetic material (e.g., that in Fig. 5.20), we note that $H \, dB$ is proportional to the area of a thin strip of "length" H (length measured in A/m) and "width" dB (width measured in T) positioned between the curve and the B-axis at the "height" B with respect to the H-axis, as indicated in Fig. 7.19(a). This means that the integral in Eq. (7.105) represents the sum of areas of all such strips as the point P′ with abscissa H and ordinate B moves in the integration process from the coordinate origin to its final position (P). We conclude, thus, that the magnetic energy density in the material is proportional to the area between the magnetization curve and the B-axis, that is, to the area of the curvilinear triangle OPQ in Fig. 7.19(a). The proportionality constant (w_m/area) can be expressed in terms of the ratio of the magnetic field intensity H in A/m that corresponds to a certain physical length along the H-axis and the length (e.g., a 1-cm division along the axis may represent a 100-A/m field intensity) and the similar ratio (B/length) for the B-axis. This conclusion, of course, applies also to linear materials, which can always be considered as a special case of nonlinear ones. In a linear case, the hypotenuse of the triangle OPQ becomes straight, and the area of the triangle is computed as $BH/2$, Fig. 7.19(b), which is the same result as in Eq. (7.108).

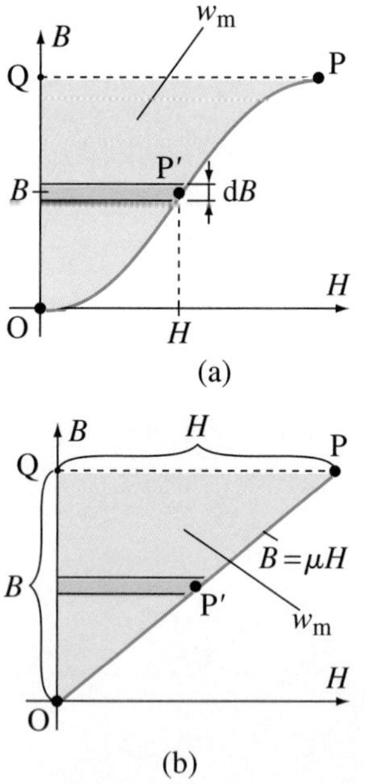

In ferromagnetic materials that exhibit hysteresis effects, the function $B(H)$ is not only nonlinear, but also has multiple branches. Thus, if H is reduced from its value at the point P in Fig. 7.19(a) to zero, B does not go to zero, but to the remanent

Figure 7.19 Correspondence between the magnetic energy density given by the integral in Eq. (7.105) and the area between the magnetization curve and the B-axis for (a) nonlinear and (b) linear magnetic materials.

[5] As in the electrostatic case, the magnetic energy of a system of current-carrying conductors might alternatively be viewed to reside in the system current, and not the magnetic field [see the corresponding discussion on the two energy localization viewpoints (field-based and charge-based) for an electrostatic system, in Section 2.16]. With this approach, the magnetic energy density would be evaluated as being equal to $\mathbf{J} \cdot \mathbf{A}/2$ at points where the current exists in the system [from Eq. (7.93)] and would be zero elsewhere. Although this viewpoint is also "correct" and has its merit, we choose to describe the magnetic energy localization of a system in terms of the magnetic field distribution (energy exists wherever and whenever the field exists) and use the associated energy density expressions in Eq. (7.108). As noted in the discussion of the electric energy localization, the field-based approach is much better suited to modeling electromagnetic wave propagation and the associated energy flow, where both electric and magnetic fields of a wave exist (and carry energy) independent of the charge and current that produced them at previous instants of time.

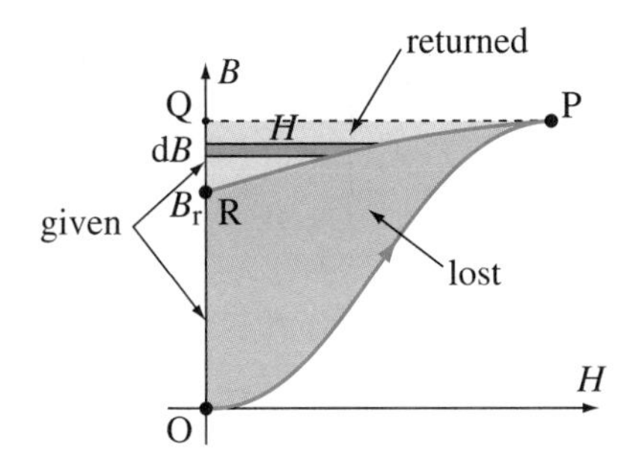

Figure 7.20 Evaluation of hysteresis losses in a ferromagnetic material as the difference between the energy given to the field and the energy returned to the sources in the process of magnetization and demagnetization of the material.

magnetic flux density $B = B_r$ (see Fig. 5.21). As B also decreases, dB is negative. This means that the energy $H\,dB$ per unit volume of the material given to the magnetic field in this process is negative, that is, the field is returning its energy to the external sources (e.g., to the generator in Fig. 7.18). However, this returned energy, which is proportional to the area of the curvilinear triangle RPQ shown in Fig. 7.20, is smaller than the energy spent by the sources in increasing the magnetic field in the material (curvilinear triangle OPQ). The difference in energy is lost to heat in the material in the process of its magnetization and demagnetization (hysteresis losses), and the difference in area in Fig. 7.20 is the area of the curvilinear triangle OPR. Consequently, the energy of hysteresis losses per unit volume of the material is proportional to the area of the triangle OPR.

Finally, let us consider a full hysteresis cycle in the material. The density of energy spent on changing the magnetic field in this cycle is given by the integral in Eq. (7.105) with the integration being carried out all around the hysteresis loop. Dividing the loop (contour C_h) into four characteristic segments, as indicated in Fig. 7.21, we note that $H\,dB > 0$ along the first line segment ($H > 0$ and $dB > 0$) and the third segment ($H < 0$ and $dB < 0$), while $H\,dB < 0$ along the second line segment ($H > 0$ and $dB < 0$) and the fourth one ($H < 0$ and $dB > 0$). The areas between line segments 1 and 3 and the B-axis are therefore a measure of the energy density given to the field at a point in the material, while the areas between line segments 2 and 4 and the B-axis correspond to the energy density returned to the sources at the same point. The difference, the area enclosed by the hysteresis loop, S_h, represents hysteresis losses in the material. In other words, the energy density

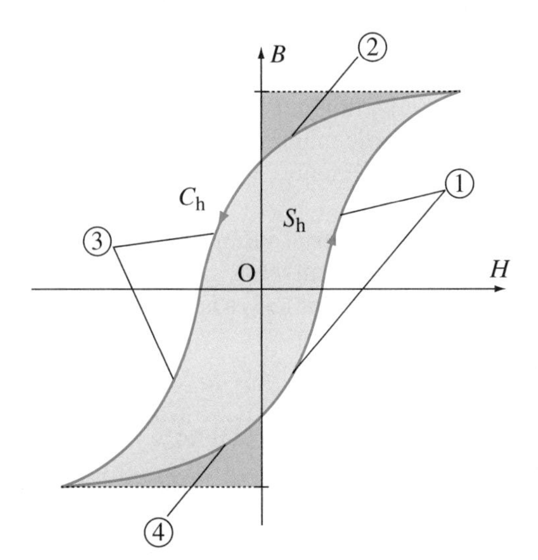

Figure 7.21 Evaluation of hysteresis losses in a complete magnetization-demagnetization hysteresis cycle of the material.

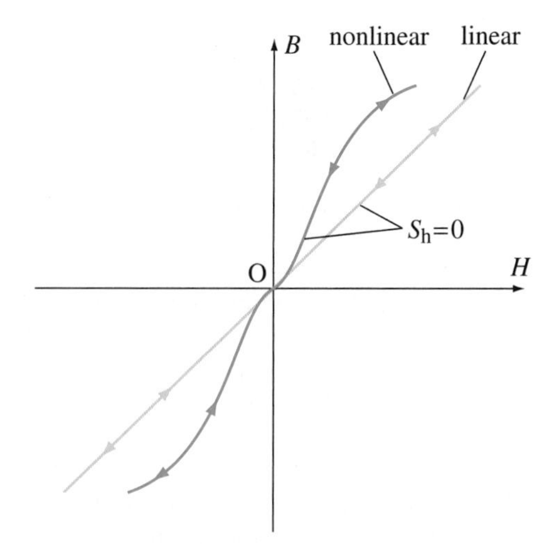

Figure 7.22 Cases with no hysteresis losses: nonlinear material with retracing of the initial magnetization curve in the periodic magnetization-demagnetization of the material, which results in zero loop area, and linear material (or material whose magnetization curve can be approximated by a linear function).

of hysteresis losses, w_h, in one complete magnetization-demagnetization hysteresis cycle is proportional to the area S_h,

energy density of hysteresis losses (see Fig. 7.21)

$$w_h = w_m = \oint_{C_h} H\,dB \propto S_h. \tag{7.110}$$

If the field is time-harmonic (ac field), with a frequency f, the time of one cycle of the periodic magnetization-demagnetization of the material, that is, the time during which the point P′ circumscribes once the hysteresis loop, equals the period of the field variation, $T = 1/f$. The time-average power of hysteresis losses, $(P_h)_{ave}$, is obtained by dividing the energy lost within a cycle by T, i.e., by multiplying it by f. Thus, in the case of a uniform field intensity in the material, we can write

time-average power of hysteresis losses for an ac field

$$(P_h)_{ave} \propto f\,S_h v, \tag{7.111}$$

where v is the volume of the material with losses.

We now recall that hard ferromagnetic materials, having large S_h (see Fig. 5.23), are used primarily for dc applications, so that $f = 0$ in Eq. (7.111) and hysteresis losses do not represent any problem. Soft ferromagnetic materials have small S_h, and that is why they are very suitable for ac applications. In the limiting case, if no hysteresis is present and the initial magnetization curve is retraced, as depicted in Fig. 7.22, $S_h = 0$ in Eq. (7.111) and the periodic magnetization-demagnetization process is accomplished with no hysteresis losses. Finally, $S_h = 0$ for linear magnetic materials as well (Fig. 7.22).

In general, total losses in a ferromagnetic material in an ac field are the sum of hysteresis losses and Joule's losses due to eddy currents, given by Eq. (6.137). We recall that the time-average power of eddy-current losses is proportional to the frequency squared [Eq. (6.152)].

Example 7.19 Energy Distribution in a Thick Linear Toroidal Core

Consider the toroidal coil from Example 5.11, and assume that a slowly time-varying current of intensity i is established in the coil. Under these circumstances, find (a) the distribution of magnetic energy in the core and (b) the total energy of the coil.

Solution

(a) The distribution of energy in the core is described by the magnetic energy density, w_m. As the material of the core is linear, we use Eq. (7.108). The magnetic field intensity in the core is given by Eq. (5.83), and hence

$$w_\mathrm{m}(r) = \frac{1}{2}\mu H^2(r) = \frac{\mu N^2 i^2}{8\pi^2 r^2} \quad (a < r < b), \tag{7.112}$$

where r is the distance from the toroid axis.

(b) The energy of the coil, that is, the total magnetic energy stored in the core, can be obtained by integrating the energy density w_m over the volume v of the core [Eq. (7.109)]. Because w_m is a function of the coordinate r only, we adopt $\mathrm{d}v$ in the form of a differentially thin toroid of radius r and thickness $\mathrm{d}r$, the cross section of which is the thin strip of length h and width $\mathrm{d}r$ shown in Fig. 5.26. The volume of this elementary toroid is thus $\mathrm{d}v = l\,\mathrm{d}S$, where $l = 2\pi r$ (length of the toroid) and $\mathrm{d}S = h\,\mathrm{d}r$, so that the magnetic energy comes out to be

$$W_\mathrm{m} = \int_{r=a}^{b} w_\mathrm{m}(r)\,\underbrace{2\pi r\,h\,\mathrm{d}r}_{\mathrm{d}v} = \frac{\mu N^2 i^2 h}{4\pi}\int_a^b \frac{\mathrm{d}r}{r} = \frac{\mu N^2 i^2 h}{4\pi}\ln\frac{b}{a}. \tag{7.113}$$

The above result can also be obtained from Eq. (7.88), as $W_\mathrm{m} = \Phi_\mathrm{total} i/2$, where Φ_total is N times the flux through the cross section of the core, found in Eq. (5.84).

Example 7.20 Energy of a Simple Nonlinear Magnetic Circuit

Calculate the energy spent for establishing the field in the magnetic circuit shown in Fig. 5.30 and described in Example 5.13.

Solution Final (established) values of the magnetic flux density and field intensity in the ferromagnetic core and the air gap, (B, H) and (B_0, H_0), are found in Example 5.13. As the material of the core is nonlinear and the operating point P of the circuit, in Fig. 5.30(b), does not belong to the first segment of the idealized initial magnetization curve (which could be described by an initial permeability), we must use Eq. (7.105) for the density of energy spent to change the field in the core from zero to (B, H). Having in mind Fig. 7.19(a), this density is proportional to the area of a polygon formed by the curve and the B-axis, from the coordinate origin to the point P, as shown in Fig. 7.23. This polygon, in turn, is a sum of a triangle and a trapezoid, which gives the following solution for the integral in Eq. (7.105):

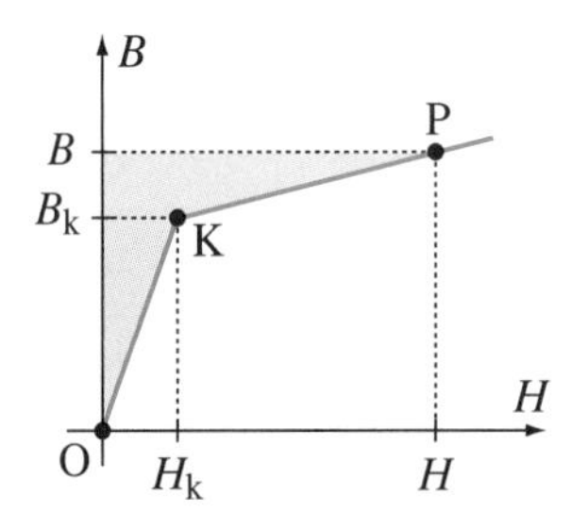

Figure 7.23 Evaluation of the density of energy spent for establishing the field in the core in Fig. 5.30; for Example 7.20.

$$w_\mathrm{m} = \int_0^B H\,\mathrm{d}B = \underbrace{\frac{1}{2}H_\mathrm{k}B_\mathrm{k}}_{\text{triangle}} + \underbrace{\frac{1}{2}(H_\mathrm{k} + H)(B - B_\mathrm{k})}_{\text{trapezoid}} = 272\ \mathrm{J/m^3}, \tag{7.114}$$

where $B_\mathrm{k} = 0.4$ T and $H_\mathrm{k} = 1000$ A/m are the magnetic flux density and field intensity at the point K ("knee" point between the two segments of the curve) in Figs. 7.23 and 5.30(b).

Air is a linear medium, so that the magnetic energy density in the gap can be obtained using Eq. (7.108),

$$w_\mathrm{m0} = \frac{1}{2}B_0 H_0 = 77\ \mathrm{kJ/m^3}. \tag{7.115}$$

Finally, the total magnetic energy of the circuit amounts to

$$W_\mathrm{m} = w_\mathrm{m} S l + w_\mathrm{m0} S_0 l_0 = 290\ \mathrm{mJ}. \tag{7.116}$$

This energy can be completely returned by the field only if the initial magnetization curve is retraced in the process of reducing the field intensities in the circuit, as in Fig. 7.22.

Example 7.21 Energy Lost in Magnetization and Demagnetization

In a thin toroidal core of cross-sectional area $S = 1\ \mathrm{cm^2}$ and length $l = 20$ cm, a magnetic field is established of intensity $H_\mathrm{m} = 1$ kA/m, and then reduced to $H = 0$. In this process,

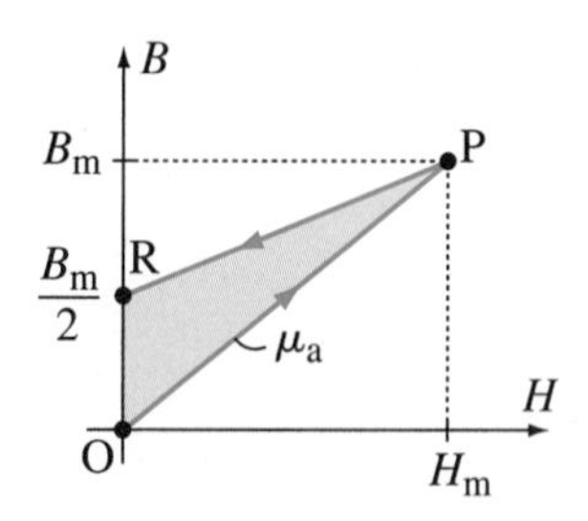

Figure 7.24 Evaluation of the net magnetic energy spent in the magnetization-demagnetization of a ferromagnetic core; for Example 7.21.

the operating point described the path shown in Fig. 7.24, where the initial permeability is $\mu_a = 0.001$ H/m. Find the net magnetic energy spent in the magnetization-demagnetization of the core.

Solution The magnetic energy spent to establish $H = H_m$ is positive, while it is negative in the process of reducing H to zero. The net magnetic energy, W_m, spent in the entire magnetization-demagnetization process represents hysteresis losses in the core (see Fig. 7.20). Its density is proportional to the area of the shaded triangle OPR in Fig. 7.24 and is given by

$$w_m = \underbrace{\int_O^P H\,dB}_{\text{magnetization}} + \underbrace{\int_P^R H\,dB}_{\text{demagnetization}} = \underbrace{\frac{1}{2}\frac{B_m}{2}H_m}_{\Delta\text{OPR}} = \frac{1}{4}B_m H_m = 250\ \text{J/m}^3, \tag{7.117}$$

where $B_m = \mu_a H_m = 1$ T, so that $W_m = w_m Sl = 5$ mJ.

Example 7.22 Time-Average Power of Hysteresis Losses in a Core

Consider the core depicted in Fig. 6.13(a) and (c), and assume that a low-frequency time-harmonic current of intensity $i = I_0 \sin(2\pi ft)$ is established in the coil, where $I_0 = 0.1$ A and $f = 1$ kHz. Compute the time-average power of hysteresis losses in the core.

Solution The energy density of hysteresis losses in one magnetization-demagnetization hysteresis cycle of the material is given by Eq. (7.110), with S_h now being the area of the parallelogram in Fig. 6.13(c). The current amplitude is the same as in Example 6.10, so that the peak-values of $H(t)$ and $B(t)$ are also the same, $H_m = 100$ A/m and $B_m = 0.1$ T [note that $B(t)$ is not a time-harmonic function, due to the nonlinearity and hysteresis behavior of the core material]. Thus, computing the parallelogram area, we obtain

$$w_h = \oint_{C_h} H\,dB = 2B_m H_m = 20\ \text{J/m}^3. \tag{7.118}$$

Based on Eq. (7.111), the time-average power of hysteresis losses in the volume $v = Sl$ of the core is

$$(P_h)_{ave} = f w_h Sl = 0.8\ \text{W}. \tag{7.119}$$

Example 7.23 Evaluation of Force from Energy for an Electromagnet

An electromagnet consisting of an iron core in the shape of a horseshoe and a coil is sketched in Fig. 7.25. The cross-sectional area of the core is S. With a steady current established in the coil, the electromagnet is capable of lifting a weight W (made also of iron). Assuming that there are two tiny air gaps between the core and the weight and that the magnetic flux density in the core and the gaps is B, find the lifting force of the electromagnet.

Solution The lifting force of the electromagnet is a consequence of the magnetic field in the magnetic circuit in Fig. 7.25, namely, it is the magnetic force, $\mathbf{F}_m$, on the lower part of the circuit that is lifting its weight. This force can be determined from the magnetic energy of the system using the principle of conservation of energy. Let us suppose that $\mathbf{F}_m$ moves the weight by an elementary distance dx upward. We recall the energy considerations in connection with Fig. 7.18. Here, however, the elementary work of external sources $i\,d\Phi$ from Eq. (7.85) is split to the change in the magnetic energy of the system dW_m and the work dW_F of the force $\mathbf{F}_m$ along the displacement dx. As $dW_F = F_m\,dx$, we can write

$$i\,d\Phi = dW_m + dW_F = dW_m + F_m\,dx. \tag{7.120}$$

If the magnetic flux is maintained at a constant value (Φ) in this experiment, which means that the current i in the coil is varied appropriately, then $d\Phi = 0$ and the above equation yields

magnetic force from energy

$$\boxed{F_m = -\left.\frac{dW_m}{dx}\right|_{\Phi=\text{const}}}. \tag{7.121}$$

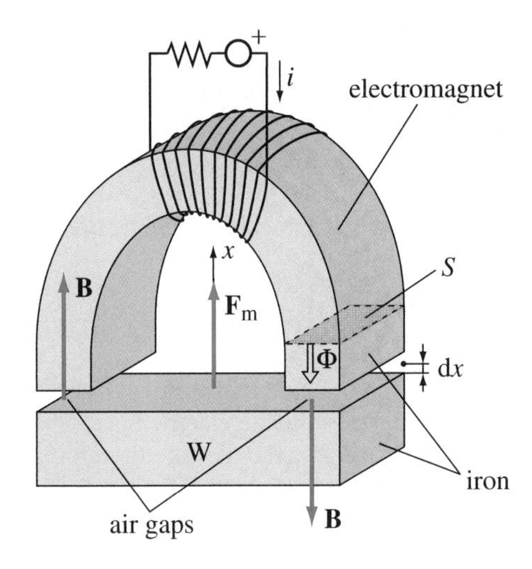

Figure 7.25 Evaluation of the lifting force of an electromagnet from the magnetic energy contained in the air gaps; for Example 7.23.

When the weight is moved upward by $\mathrm{d}x$, the only change in magnetic energy of the system is the reduction in energy contained in the two air gaps due to their decreased length. The energy change $\mathrm{d}W_\mathrm{m}$ in Eq. (7.121) is thus negative and corresponds to the energy contained in the parts of the air gaps that are $\mathrm{d}x$ long and that vanish in our experiment. In other words, it equals the negative of the magnetic energy density w_m in the air gaps, given by Eq. (7.108) with $\mu = \mu_0$, times the change in the volume of the gaps:

$$\mathrm{d}W_\mathrm{m} = -\underbrace{\frac{B^2}{2\mu_0}}_{w_\mathrm{m}} \underbrace{2S\,\mathrm{d}x}_{\mathrm{d}v} = -\frac{B^2 S\,\mathrm{d}x}{\mu_0} \tag{7.122}$$

($2S$ is the total cross-sectional area of the gaps). From Eq. (7.121), the lifting force of the electromagnet is

$$F_\mathrm{m} = -\frac{\mathrm{d}W_\mathrm{m}}{\mathrm{d}x} = \frac{B^2 S}{\mu_0}. \tag{7.123}$$

As a numerical example, for $B = 1$ T and $S = 0.125$ m^2, $F_\mathrm{m} \approx 100$ kN, which means that this electromagnet can lift a weight of $m = F_\mathrm{m}/g \approx 10$ tons ($g = 9.81$ m/s^2 – standard acceleration of free fall). Such powerful electromagnets are used in cranes for lifting large pieces of iron.

Example 7.24 Magnetic Pressure

(a) For the electromagnet of Fig. 7.25, determine the magnetic pressure, that is, the pressure of the lifting force $\mathbf{F}_\mathrm{m}$, on the surface of the iron piece that is being lifted. (b) Based on the result in (a), compare the maximal values of the magnetic and electric pressures attained in practical situations.

Solution

(a) The force acts on the parts of the surface of the iron piece that form the air gaps with the core in Fig. 7.25, that is, on the surface of area $2S$ in total. The corresponding pressure is therefore given by

$$\boxed{p_\mathrm{m} = \frac{F_\mathrm{m}}{2S} = \frac{B^2}{2\mu_0},} \tag{7.124}$$

magnetic pressure (unit: Pa)

and is called the magnetic pressure. This expression is valid for any boundary surface between a ferromagnetic material (with $\mu \gg \mu_0$) and air (or any other nonmagnetic medium), with B being the local magnetic flux density in air near the surface. We see that the pressure $p_{\rm m}$ actually equals the local magnetic energy density on the nonmagnetic side of the surface. It acts from the ferromagnetic material toward the medium with $\mu = \mu_0$.

(b) Note that the expression in Eq. (7.124) is entirely analogous to the expression in Eq. (2.133) for the electric pressure on a metallic surface in air. These two expressions provide us now with an opportunity to compare the electric and magnetic pressures and forces. Combining them, we get

$$\frac{p_{\rm m}}{p_{\rm e}} = \frac{1}{\varepsilon_0 \mu_0} \left(\frac{B}{E}\right)^2. \tag{7.125}$$

To estimate the ratio of the maximal magnetic and maximal electric pressure attained in practical situations, we recall that the maximal permissible electric field intensity, E, is determined by the dielectric strength of air, Eq. (2.53), and we take therefore $E_{\rm max} = 3$ MV/m in this estimation. On the other side, there is no such limit for B. However, maximal magnetic flux densities that are normally attained in typical magnetic circuits are on the order of one tesla and it is reasonable to assume that $B_{\rm max} = 1$ T for the comparison. Hence,

ratio of maximal magnetic and electric pressures attained in practice

$$\frac{(p_{\rm m})_{\rm max}}{(p_{\rm e})_{\rm max}} = \frac{1}{\varepsilon_0 \mu_0} \left(\frac{B_{\rm max}}{E_{\rm max}}\right)^2 \approx 10{,}000. \tag{7.126}$$

We conclude that practically attainable magnetic pressures are several orders of magnitude stronger than electric ones. This is why magnetic forces, and not electric ones, are the actual workhorses of our industrial world. Practically all devices for electromechanical energy conversion, such as different types of electric motors and generators, are based on magnetic forces and their work and power.

Problems: 7.27–7.35; *Conceptual Questions* (on Companion Website): 7.31–7.33; *MATLAB Exercises* (on Companion Website)

7.6 INTERNAL AND EXTERNAL INDUCTANCE IN TERMS OF MAGNETIC ENERGY

In this section, we revisit the concept of self-inductance and techniques to compute it, now from the energy standpoint. We consider, thus, an arbitrary conductor in a linear magnetic medium. From Eq. (7.88), if we assume a slowly time-varying (or steady) current of intensity i to flow in the conductor, the inductance (self-inductance) of the conductor, L, can be expressed in terms of the energy contained in the magnetic field due to i, $W_{\rm m}$, as

inductance from magnetic energy

$$L = \frac{2W_{\rm m}}{i^2}. \tag{7.127}$$

This expression can be viewed as the third equivalent definition of self-inductance, the other two being the flux definition in Eq. (7.1) and emf definition in Eq. (7.2). It can hence be used, as an alternative general means for computing the inductance of different structures, where the magnetic energy of the structure is computed using the integral expression in Eq. (7.109).

Since the energy $W_{\rm m}$ can be written as the sum of energies localized inside and outside the conductor,

$$W_{\rm m} = W_{\rm mi} + W_{\rm me}, \tag{7.128}$$

the inductance L can be decomposed accordingly into the internal inductance, L_i, and external inductance, L_e, of the conductor:

$$L = L_\mathrm{i} + L_\mathrm{e}. \tag{7.129}$$

In other words,

$$\boxed{L_\mathrm{i} = \frac{2W_\mathrm{mi}}{i^2} \quad \text{and} \quad L_\mathrm{e} = \frac{2W_\mathrm{me}}{i^2},} \tag{7.130}$$

internal and external inductances

where the internal and external magnetic energies of the conductor are obtained by integrating the energy density over the conductor interior (volume v_i) and exterior (volume v_e), respectively, that is,

$$W_\mathrm{mi} = \frac{1}{2}\int_{v_\mathrm{i}} \mathbf{B}\cdot\mathbf{H}\,\mathrm{d}v \quad \text{and} \quad W_\mathrm{me} = \frac{1}{2}\int_{v_\mathrm{e}} \mathbf{B}\cdot\mathbf{H}\,\mathrm{d}v. \tag{7.131}$$

More precisely, v_e is only that part of the conductor exterior which is occupied by the magnetic field (there is no need to integrate zero energy density).

In all of the examples of Section 7.1, we only evaluated the external inductance of conductors, by employing the flux (or emf) definition of self-inductance with only the external magnetic flux taken into account. Evaluation of L_i from the internal magnetic flux is also possible, but is physically less clear and mathematically more complicated than from the internal stored magnetic energy.

At high frequencies, due to skin effect (see Fig. 6.23), the current and magnetic field in a conductor are confined to a very thin region on the surface of the conductor, which considerably reduces the high-frequency value of the internal inductance of the conductor from its low-frequency (or dc) value. Therefore, in most high-frequency applications, L_i is negligible as compared to L_e, and assumption that $L \approx L_\mathrm{e}$ yields very accurate results.

Example 7.25 Internal Inductance p.u.l. of a Cylindrical Conductor

Find the internal inductance per unit length of an infinitely long cylindrical conductor of radius a. The permeability of the conductor is μ. Skin effect is not pronounced.

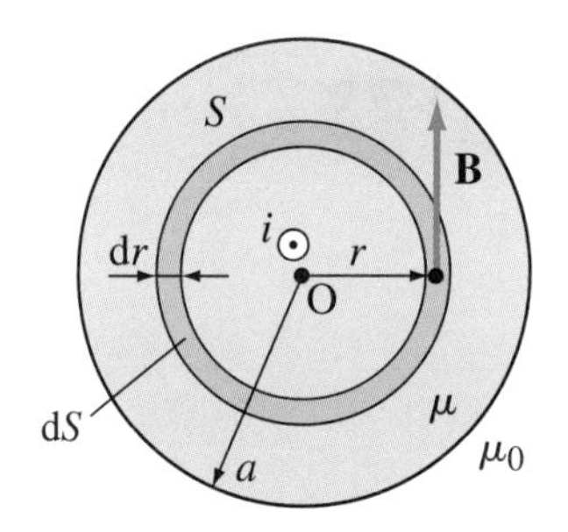

Figure 7.26 Evaluation of the low-frequency internal inductance per unit length of an infinitely long cylindrical conductor of permeability μ; for Example 7.25.

Solution Assume that the conductor carries a current of intensity i that is uniformly distributed over its cross section (skin effect not pronounced). The magnetic flux density in the conductor is given by the expression for $r < a$ in Eq. (4.56) with μ_0 substituted by μ and I by i, that is,

$$B(r) = \frac{\mu i r}{2\pi a^2}, \tag{7.132}$$

with r being the distance from the conductor axis, as shown in Fig. 7.26. From the first expression in Eqs. (7.131), the internal magnetic energy per unit length of the conductor is obtained by integrating the magnetic energy density over the cross section S of the conductor [see Eqs. (2.206) and (2.207) for the similar integration of the electric energy density],

$$W'_\mathrm{mi} = (W_\mathrm{mi})_\mathrm{p.u.l.} = \int_S w_\mathrm{m}\,\mathrm{d}S = \int_0^a \frac{B^2(r)}{2\mu}\underbrace{2\pi r\,\mathrm{d}r}_{\mathrm{d}S} = \frac{\mu i^2}{4\pi a^4}\int_0^a r^3\,\mathrm{d}r = \frac{\mu i^2}{16\pi}, \tag{7.133}$$

where $\mathrm{d}S$ is the surface area of an elementary ring of radius r and width $\mathrm{d}r$ (Fig. 7.26).

The first equation of Eqs. (7.130) then gives the following expression for the low-frequency internal inductance per unit length of the conductor:

$$\boxed{L'_\mathrm{i} = \frac{2W'_\mathrm{mi}}{i^2} = \frac{\mu}{8\pi}.} \tag{7.134}$$

low-frequency internal inductance p.u.l. of a cylindrical conductor

We note that this inductance is independent of the conductor radius. For nonmagnetic conductors, $L'_{\rm i} = \mu_0/(8\pi) = 50$ nH/m.

Example 7.26 Internal Inductance p.u.l. of a Thin Two-Wire Line

Determine the total (internal plus external) low-frequency inductance per unit length of the two-wire line in Fig. 7.4, assuming that the conductors are nonmagnetic.

Solution Since the distance between the conductors of the line is much larger than the conductor radii, the internal magnetic energies of individual conductors (for an assumed low-frequency current in the line) can be evaluated independently from each other. Therefore, the low-frequency internal inductance per unit length of the line can be found as twice that of a single isolated conductor, Eq. (7.134) with $\mu = \mu_0$ (conductors are nonmagnetic), which yields

$$L'_{\rm i} = L'_{\rm i1} + L'_{\rm i2} = 2L'_{\rm i1} = \frac{\mu_0}{4\pi} = 100 \text{ nH/m}. \tag{7.135}$$

Using the expression for the external inductance p.u.l. of the line, Eq. (7.11), its total low-frequency per-unit-length inductance is

low-frequency total inductance p.u.l. of a thin two-wire line

$$\boxed{L' = L'_{\rm i} + L'_{\rm e} = \frac{\mu_0}{\pi}\left(\frac{1}{4} + \ln\frac{d}{a}\right).} \tag{7.136}$$

Example 7.27 Inductance from Energy for a Coaxial Cable

Find the total low-frequency inductance per unit length of the coaxial cable from Example 4.11.

Solution If we assume that a dc current of intensity I is established in the cable conductors, as in Fig. 4.17, the magnetic flux density at a distance r from the cable axis, $B(r)$, is given by Eqs. (4.61)–(4.63). We use the energy definition of inductance, Eq. (7.127). The total (internal plus external) magnetic energy per unit length of the cable can be obtained by integrating the magnetic energy density over the cross section of the cable (the magnetic field outside the cable, for $r > c$, is zero) in the same manner as in Eq. (7.133),

$$W'_{\rm m} = \int_{r=0}^{c} \underbrace{\frac{B^2(r)}{2\mu_0}}_{w_{\rm m}} \underbrace{2\pi r\,{\rm d}r}_{{\rm d}S}. \tag{7.137}$$

As the function $B(r)$ for $0 \le r \le c$ is given by three different expressions in Eqs. (4.61) and (4.62), we break the integration with respect to r up into three parts: from 0 to a, from a to b, and from b to c. This yields

$$W'_{\rm m} = \underbrace{\frac{\mu_0 I^2}{16\pi}}_{W'_{\rm mi1}} + \underbrace{\frac{\mu_0 I^2}{4\pi}\ln\frac{b}{a}}_{W'_{\rm me}} + \underbrace{\frac{\mu_0 I^2}{4\pi(c^2-b^2)}\left(\frac{c^4}{c^2-b^2}\ln\frac{c}{b} - \frac{3c^2-b^2}{4}\right)}_{W'_{\rm mi2}}, \tag{7.138}$$

where $W'_{\rm mi1}$, $W'_{\rm me}$, and $W'_{\rm mi2}$ are the magnetic energies residing in the inner conductor, dielectric, and outer conductor of the cable per unit of its length, respectively. The total dc (or low-frequency) inductance per unit length of the cable is hence

low-frequency total inductance p.u.l. of a coaxial cable

$$\boxed{L' = \frac{2W'_{\rm m}}{I^2} = \underbrace{\frac{\mu_0}{2\pi}\left[\frac{1}{4} + \frac{1}{c^2-b^2}\left(\frac{c^4}{c^2-b^2}\ln\frac{c}{b} - \frac{3c^2-b^2}{4}\right)\right]}_{L'_{\rm i}} + \underbrace{\frac{\mu_0}{2\pi}\ln\frac{b}{a}}_{L'_{\rm e}},} \tag{7.139}$$

where $L'_{\rm i}$ and $L'_{\rm e}$ are the internal and external inductance, respectively, per unit length of the cable.

The above result for L'_e is, of course, the same as that in Eq. (7.12), which was obtained using the flux definition of inductance. Note that the first component of L'_i, that corresponding to the magnetic energy in the inner conductor, equals the internal inductance per unit length of an isolated cylindrical conductor in air, Eq. (7.134) with $\mu = \mu_0$.

As a numerical example, the internal and external inductances per unit length of a coaxial cable with $a = 1$ mm, $b = 4$ mm, and $c = 5$ mm amount to $L'_i = 66.6$ nH/m and $L'_e = 277.3$ nH/m.

Problems: 7.36–7.38; *Conceptual Questions* (on Companion Website): 7.34; *MATLAB Exercises* (on Companion Website).

Problems

7.1. Induced emf and voltage of an inductor. A current of intensity $i(t) = I_0 \, e^{-t/\tau}$, where I_0 and $\tau > 0$ are constants, flows through an inductor of inductance L. Calculate (a) the magnetic flux, (b) the induced emf, and (c) the voltage of the inductor. Specify the reference directions/orientations for all quantities.

7.2. Inductance of a solenoid with a two-layer core. If the solenoidal coil described in Example 7.1 is wound over a ferromagnetic core with two coaxial layers of relative permeabilities $\mu_{r1} = 500$ and $\mu_{r2} = 1000$, as shown in Fig. 7.27, where the cross-sectional area of the inner layer is $S_1 = 1 \text{ cm}^2$, find the inductance of the coil.

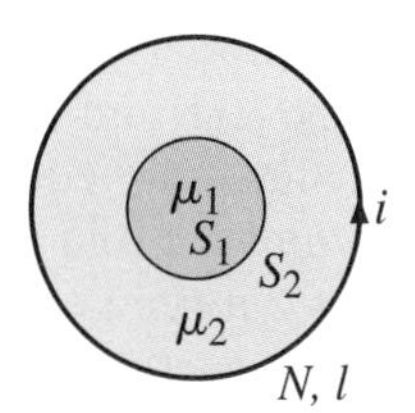

Figure 7.27 Solenoid with a core composed of two coaxial linear ferromagnetic layers; for Problem 7.2.

7.3. Toroidal coil with a two-layer core. Repeat Example 7.2 but for the two-layer core shown in Fig. 7.28.

7.4. Inductance of a coil in a simple linear magnetic circuit. Find the inductance of the coil in the simple linear magnetic circuit with an air gap from Example 5.14.

7.5. Two-wire line with ferromagnetic coatings over conductors. Let the conductors of the thin symmetrical two-wire transmission line in Fig. 7.4 be coated by ferromagnetic layers of permeability μ and thickness a, analogously to the line with dielectrically coated conductors in Fig. 2.31. Determine the external inductance per unit length of the new line.

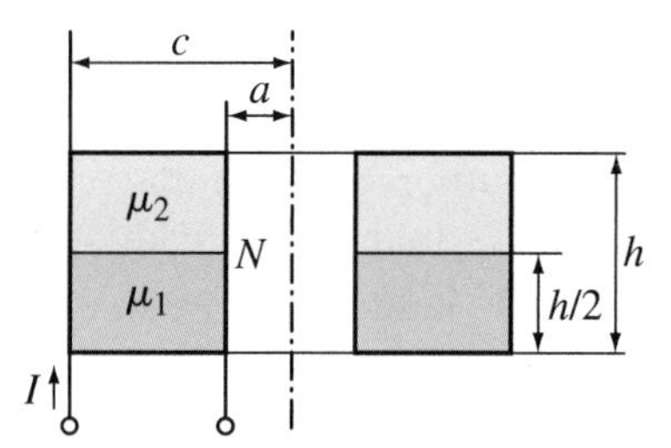

Figure 7.28 The same toroidal coil as in Fig. 7.3 but with ferromagnetic layers stacked on top of one another; for Problem 7.3.

7.6. Coaxial cable filled with an inhomogeneous magnetic material. Compute the external inductance per unit length of a coaxial cable filled with a piece-wise homogeneous ferromagnetic material composed of two coaxial layers of relative permeabilities μ_{r1} and μ_{r2} and dimensions as in Fig. 2.50.

7.7. Planar line with two magnetic layers. For the planar transmission line in Fig. 3.24, assume that the two material layers between metallic strips are lossless ($\sigma_1 = \sigma_2 = 0$), but magnetic, with permeabilities μ_1 and μ_2. Let $w \gg d_1 + d_2$, so that fringing effects can be neglected, i.e., the magnetic field of the line can be considered to be uniform and localized only in the two layers, and not in the air around (also see the magnetic field computation in

Example 11.10). Under these circumstances, find the p.u.l. external inductance of the line.

7.8. External inductance p.u.l. of a three-wire line. What is the external inductance per unit length of the transmission line consisting of an isolated wire and two short-circuited wires in air described in Example 2.16?

7.9. Bringing a superconducting contour in a magnetic field. A superconducting square contour of edge length a and inductance L is first situated outside any magnetic field, and there is no current in it. The contour is then brought in a uniform time-invariant magnetic field of flux density B and positioned so that the vector $\mathbf{B}$ is perpendicular to the plane of the contour. Orienting the contour in accordance to the right-hand rule with respect to the direction of $\mathbf{B}$, find (a) the magnetic flux through the contour and (b) the current intensity along it – in the new steady state.

7.10. Mutual inductance between a loop and two-wire line. Find the mutual inductance between the rectangular loop and the two-wire line in both cases in Fig. 6.32, in the following two ways: (a) from the definition of mutual inductance in Eq. (7.21) or (7.23) and (b) using Eq. (7.40), respectively.

7.11. Mutual inductance of two coils on a thick toroidal core. Repeat Example 7.7 but for a thick toroidal core of a rectangular cross section, as the one shown in Fig. 5.26, with the inner and outer radii and height of the toroid being $a = 3$ cm, $b = 6$ cm, and $h = 2$ cm, respectively.

7.12. Mutual inductance between a toroidal coil and axial wire. Consider the thick toroidal coil with a rectangular cross section in Fig. 4.18, and assume that an infinitely long straight wire runs along the axis of the toroid (z-axis). Find the mutual inductance between the coil and the wire, computing it both as (a) L_{12} and (b) L_{21}.

7.13. Mutual inductance between a solenoid and rectangular loop. Assume that the length of the solenoid in Fig. 6.29 is l ($l \gg a$), so that the total number of wire turns amounts to $N = N'l$, as well as that its terminals are open. If the rectangular loop positioned around the solenoid at the middle of its length (Fig. 6.29) carries a low-frequency time-harmonic current of intensity $i = I_0 \cos \omega t$, determine the voltage between the terminals of the solenoid.

7.14. Mutual inductance between large and small concentric loops. If in the two-loop system in Fig. 6.33 a slowly time-varying current of intensity $i(t)$ flows along the small circular loop, while the large square loop is open-circuited, compute the induced emf in the large loop.

7.15. Emf in a phone line due to a nearby power line. A power conductor of a cable car and track form a transmission line that can be approximated by a two-wire line, in a vertical plane, with distance between wire axes $h = 5$ m. In the same horizontal plane containing the power conductor, running in parallel to it, there is a two-wire telephone line, and the distances of axes of wires of this line from the power-conductor axis are $d_1 = 5.5$ m and $d_2 = 5.9$ m, respectively. Both lines can be considered to be thin. If a time-harmonic current of amplitude (peak-value) $I_0 = 150$ A and frequency $f = 60$ Hz is established in the power line, find the amplitude of the induced emf per $l = 1$ km length of the phone line, neglecting the influence of the earth, i.e., assuming that the two lines are in free space.

7.16. Equivalent input inductance of structures with coupled coils. Two magnetically coupled coils, wound on a cardboard core, have inductances $L_1 = L_2 = 50\ \mu$H and coupling coefficient $k = 0.1$. Calculate the equivalent inductance between terminals 1 and 2 of the structure for connections between coils as in Fig. 7.29(a), (b), and (c), respectively.

7.17. Equivalent inductance for a unity coupling coefficient. Repeat the previous problem assuming that the cardboard core is replaced by a linear ferromagnetic one, with which $L_1 = L_2 = 60$ mH and $k = 1$.

7.18. Coupling coefficient of two circuits from charge flow. Fig. 7.30 shows the circuit-theory representation of two magnetically coupled circuits, where the inductances of the circuits are $L_1 = 10$ mH and $L_2 = 3$ mH, and their resistances $R_1 = 10\ \Omega$ and $R_2 = 8\ \Omega$, respectively. When the switch K in

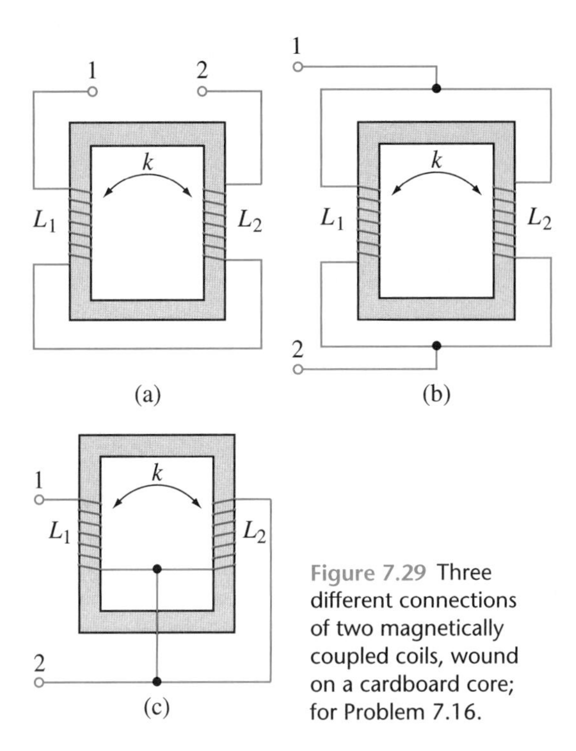

Figure 7.29 Three different connections of two magnetically coupled coils, wound on a cardboard core; for Problem 7.16.

the primary circuit is closed, connecting it to an ideal dc voltage generator of emf $\mathcal{E} = 20$ V, the charge flow indicated by the ballistic galvanometer (whose resistance is included in R_2) in the secondary circuit is $Q = 1$ mC. What is the coefficient of coupling (k) between the circuits?

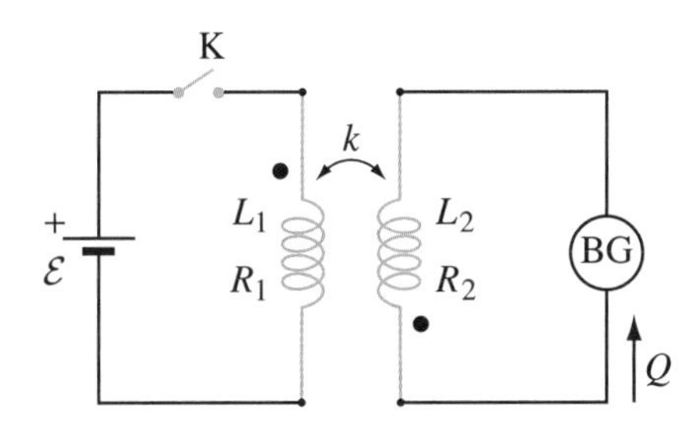

Figure 7.30 Magnetically coupled circuits with a dc voltage generator and a ballistic galvanometer (BG) on the primary and secondary sides, respectively; for Problem 7.18.

7.19. Thick toroidal coil and three-turn loop around it. An air-filled thick toroidal coil of a rectangular cross section, defined by the inner and outer radii $a = 5$ cm and $b = 15$ cm and height $h = 7$ cm of the toroid, has $N = 1800$ wire turns of negligible resistance. The toroid is encircled by three turns of wire that are connected to an ideal low-frequency time-harmonic current generator, as shown in Fig. 7.31. When the switch K between the terminals P and Q of the coil is open, the voltage across it is $v_{PQ} = 100 \sin 377t$ mV (t in s). (a) What is the current intensity $i_g(t)$ of the current generator? (b) If the switch K is then closed, find the current through it in the new steady state.

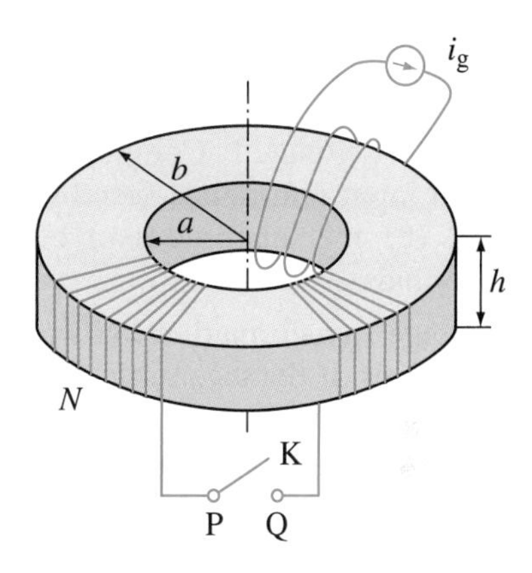

Figure 7.31 Magnetically coupled circuits with a thick toroidal coil and a three-turn loop around it; for Problem 7.19.

7.20. Coupling coefficient of two coaxial toroidal coils. Shown in Fig. 7.32 is a system of two air-filled thick toroidal coils, wound uniformly and densely on cardboard cores with rectangular cross sections, that are positioned coaxially one inside the other. The inner and outer radii of the inner toroid are a_1 and b_1, while those of the outer toroid are a_2 and b_2 ($a_2 < a_1 < b_1 < b_2$), and the heights of the

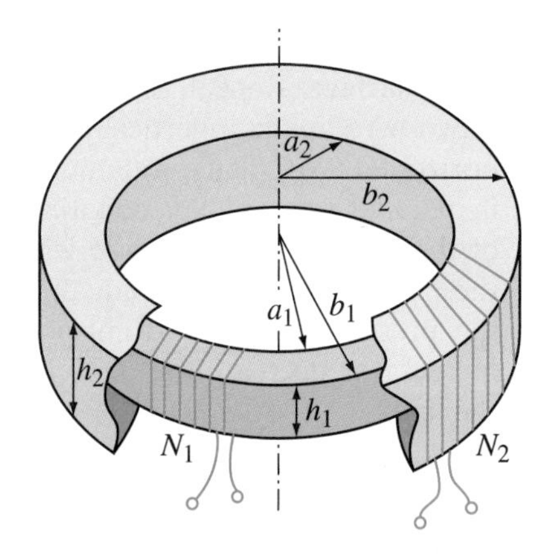

Figure 7.32 Two coupled coaxial air-filled thick toroidal coils; for Problem 7.20.

toroids are h_1 and h_2 ($h_1 < h_2$), respectively. The numbers of wire turns are N_1 for the inner and N_2 for the outer coil. Determine the coefficient of coupling between the two coils.

7.21. Magnetically coupled circuits with coaxial solenoids. For the system of two coupled coaxial solenoids in Fig. 7.15, let $a_1 = 2$ cm, $a_2 = 6$ cm, $l = 1$ m, $N_1 = 2000$, $N_2 = 4000$, and $\mu_r = 200$. In addition, let the outer solenoid be connected to an ideal time-harmonic voltage generator of emf $e_g = 50 \cos 377t$ V (t in s), and the inner solenoid be short-circuited. Compute (a) the equivalent input inductance seen by the generator and (b) the amplitudes of currents in the solenoids.

7.22. Voltage transformation by two coupled two-wire lines. If the second (inner) two-wire line in Fig. 7.16 is short-circuited at one end but open at the other, compute the amplitude of the current in the first line and voltage across the open terminals of the second circuit.

7.23. Current transformation by two coupled two-wire lines. Considering the system in Fig. 6.32 as two magnetically coupled circuits containing two two-wire lines, assume that $a = 8$ mm, $b = 25$ cm, $I_{g0} = 1$ A, and $\omega = 10^5$ rad/s, as well as that the length of the longer line is $l = 90$ cm and the radii of all wires $r_w = 0.4$ mm, and compute the amplitude of the current in the shorter line (of length b) for cases in Fig. 6.32(a) and Fig. 6.32(b), respectively. Neglect internal inductances, losses in the wires, capacitive coupling between the lines, end effects, and propagation effects.

7.24. Two-wire line in a vertical plane above a PMC surface. Repeat Example 7.16 but for a two-wire line in a vertical plane above the horizontal PMC plane, with the wires running in parallel to it, and h denoting the height of the axis of the lower wire with respect to the material surface and equaling $d/2$. Does the presence of the ferromagnetic material increase or decrease the p.u.l. inductance of the line?

7.25. Energy of magnetically coupled coil and loop. For magnetically coupled circuits containing an air-filled thick toroidal coil of a rectangular cross section and three-turn wire loop around it shown in Fig. 7.31 and described in Problem 7.19, find the total instantaneous and time-average magnetic energies of the circuits in both steady states, namely, with the switch K open and closed, respectively.

7.26. Energy of magnetically coupled two-wire lines – three cases. Compute the time-average magnetic energy of systems with two coupled two-wire lines from (a) Problem 7.22, (b) Problem 7.23 for the case in Fig. 6.32(a), and (c) Problem 7.23 for the case in Fig. 6.32(b), respectively.

7.27. Magnetic energy of two coupled coaxial solenoids. Consider the system of two coupled coaxial solenoids described in Problem 7.21, and find the time-average magnetic energy of the system – in the following three ways, respectively: (a) using the equivalent input inductance seen by the voltage generator, (b) by means of self- and mutual inductances of the coupled circuits and Eq. (7.90), and (c) from the magnetic field intensity and energy density throughout the system. (d) What portion of the total energy is stored in the ferromagnetic core, inside the inner coil, and what in the air-filled region between the two coils?

7.28. Energy distribution in thick linear two-layer toroidal cores. Find the magnetic energy density in the material, energy stored in each linear ferromagnetic layer, and total energy of the coil for thick toroidal coils with two-layer cores in (a) Fig. 7.3 and (b) Fig. 7.28, respectively, assuming that a slowly time-varying current of intensity $i(t)$ is established in the coil. Compute the energies by integrating the energy density throughout the volumes of individual layers, as well as using the expressions for inductances where appropriate.

7.29. Energy distribution in a planar line with two magnetic layers. Assuming that the planar transmission line with two magnetic layers between metallic strips described in Problem 7.7 carries a dc current of intensity I, compute the magnetic energy density and total energy per unit length of each of the layers, as well as the p.u.l. energy of the entire line.

7.30. Energy of a nonlinear magnetic circuit with three branches. Compute the magnetic energy spent for establishing the field in each of the three branches, as well as the total energy, of the nonlinear magnetic circuit shown in Fig. 5.32 and described in Example 5.15.

7.31. Energy of another nonlinear magnetic circuit. Repeat the previous problem but for the magnetic circuit from Problem 5.22 (Fig. 5.42).

7.32. Magnetization-demagnetization energy of a thick toroidal core. Consider the thick toroidal coil with a rectangular cross section and nonlinear ferromagnetic core described in Example 5.12, and assume that its current is first established at an intensity of $I = 1$ A, so that the distribution of the magnetic flux density, $B(r)$, in the core is exactly that in Fig. 5.27(c), and then reduced to $I = 0$. In this process, the operating point at different locations in the core first moves up the idealized initial magnetization curve, and then back to the B-axis, where the magnetic flux densities drop to a half of the previously reached value, so either to $B_m/2$ in the part of the core that is in saturation ($r \leq c$) or to $B(r)/2$ [$B(r) < B_m$] in the part of the core in the linear regime ($r > c$), as indicated in Fig. 7.33 (similarly to the process in Fig. 7.24). Find (a) the density of the net magnetic energy spent in the magnetization-demagnetization process at every point of the core (for every r, $a \leq r \leq b$) and (b) the energy spent in the entire core.

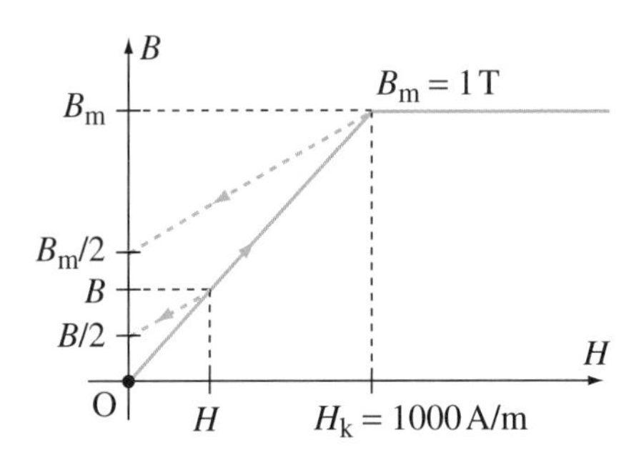

Figure 7.33 Magnetization and demagnetization of the thick toroidal nonlinear ferromagnetic core in Fig. 5.27; for Problem 7.32.

7.33. Power of hysteresis losses in the core of a solenoid. An infinitely long solenoid with a circular cross section of radius a and N' turns of wire per unit of its length is wound about a nonlinear ferromagnetic core whose idealized hysteresis loop is shown in Fig. 7.34, where $B_m/H_m = \mu_h =$ const. There is a low-frequency time-harmonic current of amplitude I_0 and angular frequency ω flowing through the winding. Saturation is not reached in the core. Determine the time-average power of hysteresis losses per unit length of the core.

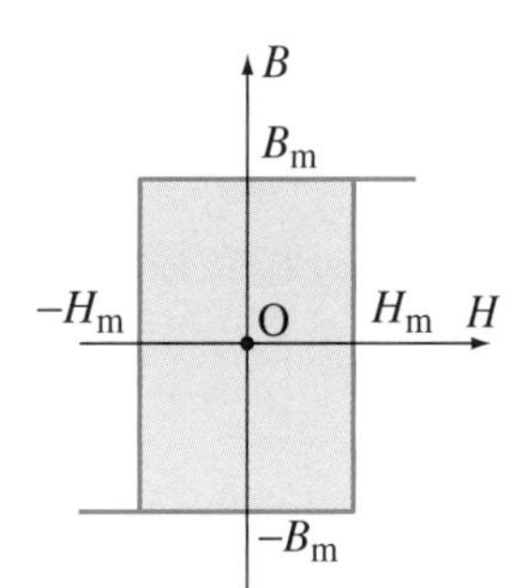

Figure 7.34 Idealized hysteresis loop of a ferromagnetic core filling an infinite solenoid; for Problem 7.33.

7.34. Power of hysteresis losses in a thick toroidal core. Assume that a low-frequency time-harmonic current of intensity $i = I_0 \cos(2\pi f t)$ is established in the thick toroidal coil with a nonlinear ferromagnetic core in Fig. 5.27(a), where $I_0 = 1$ A and $f = 10$ kHz (a, b, h, and N are specified in Example 5.12), that the idealized hysteresis loop of the core material is that in Fig. 6.13(c), where $B_m/H_m = \mu_h = 0.001$ H/m for all locations in the core, and that saturation is not reached at any location. Note that, from Eq. (5.83), $H_m = NI_0/(2\pi r)$ ($a \leq r \leq b$). Under these circumstances, find (a) the energy density of hysteresis losses in one magnetization-demagnetization hysteresis cycle at every point of the material (for every r) and (b) the total time-average power of hysteresis losses in the core.

7.35. Frequency dependence of hysteresis and eddy-current losses. A conducting ferromagnetic body is placed in a uniform low-frequency time-harmonic magnetic field, and the total time-average power of losses in the body is measured to be P_1 and P_2 at two different frequencies, f_1 and f_2, respectively, for the same amplitude of the applied magnetic flux density. What are the time-average powers of (a) hysteresis losses and (b) Joule's losses due to eddy currents in the body, at each of the frequencies?

7.36. **Magnetic energy p.u.l. of a triaxial cable.** Find the internal, external, and total magnetic energy per unit length of the triaxial cable from Problem 4.11.

7.37. **Internal inductance p.u.l. of a hollow cylindrical conductor.** Determine the dc or low-frequency internal inductance per unit length of the infinitely long hollow cylindrical copper conductor in Fig. 4.43.

7.38. **Total inductance of a coaxial cable with magnetic inhomogeneity.** From energy, find the total low-frequency inductance per unit length of a coaxial cable filled with two coaxial magnetic layers, described in Problem 7.6. Assume that the conductors of the cable are nonmagnetic and that the inductance of the outer conductor (which is very thin) is negligible.

8 Rapidly Time-Varying Electromagnetic Field

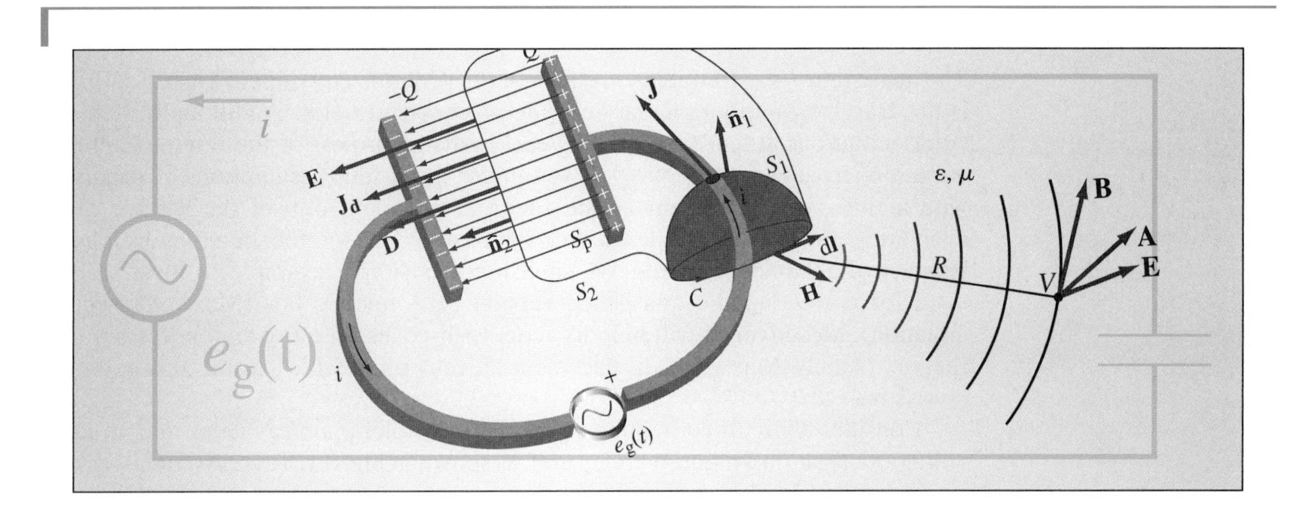

Introduction:

This chapter is devoted to the rapidly time-varying (e.g., high-frequency time-harmonic) electromagnetic field, which cannot be analyzed without taking into account the electromagnetic retardation effect (lagging in time of fields and potentials behind their sources). We shall first correct the quasistatic version of the generalized Ampère's law (Maxwell's second equation) by adding a new type of current, so-called displacement current, in parallel to the conduction current. The introduction of the notion of a displacement current was, in fact, a crucial step in Maxwell's development of the electromagnetic theory. The addition of this new term in Maxwell's equations corresponds to the inclusion of the time retardation in the expressions for the potentials and field vectors. It enables modeling of electromagnetic wave propagation and radiation. We shall then summarize and discuss the full set of Maxwell's equations for the most general field – rapidly time-varying electromagnetic field – in integral and differential notation, as well as in the form of boundary conditions at an interface between two arbitrary electromagnetic media. The existence of electromagnetic waves, as predicted by general Maxwell's equations, will be viewed as a process of successive mutual induction of electric and magnetic fields in space and time. Different forms of the continuity equation for rapidly time-varying currents will also be studied. Next, we shall focus on time-harmonic (steady-state sinusoidal) electromagnetic fields in linear media and introduce Maxwell's equations in the complex domain for such fields. As these equations contain no time and no time derivatives (nor integrals), the complex-domain (or frequency-domain) analysis of linear electromagnetic systems (including linear electric circuits) with time-harmonic excitations is considerably simpler than the time-domain analysis of the same structures. The expressions for Lorenz (retarded) electromagnetic potentials will be derived from Maxwell's equations and the associated wave equations for potentials. The potentials due to volume, surface, and line distributions of rapidly time-varying (high-frequency) currents and charges will be evaluated in both time and complex domains. The electric and magnetic field vectors will be computed from the potentials. Finally, Poynting's theorem, which expresses the principle of conservation of energy for electromagnetic phenomena, will be derived from Maxwell's equations and applied to computations of generated and stored energies, losses, and power transfers, as well as overall power balances, in dynamic and static electromagnetic systems.

8.1 DISPLACEMENT CURRENT

The quasistatic (low-frequency) version of the generalized Ampère's law, Eq. (6.30), is not true for rapidly time-varying (high-frequency) fields. Additionally, in some (rare) situations, it leads to meaningless results regardless of the rate of the time-variation (frequency) of fields, so even under quasistatic assumptions. A capacitor with a time-varying current is one such example, but only if the surface S for evaluating the total current on the right-hand side of the equation is placed between the capacitor plates. We shall use this simple example to illustrate the need for correcting the quasistatic version of Ampère's law (Maxwell's second equation). Moreover, it will help us actually discover on our own a necessary correction, namely, Maxwell's displacement-current term, which makes this equation always true.

Consider a circuit containing an air-filled parallel-plate capacitor and an ideal voltage generator of a time-varying emf, as shown in Fig. 8.1. There is a time-varying current in the circuit, equal to the product of the capacitor capacitance and the time rate of change of the voltage across the capacitor, Eq. (3.45), where the voltage equals the emf of the generator. This current, in turn, is accompanied by a time-varying magnetic field, of intensity $\mathbf{H}$, in the space surrounding the circuit. Let us apply the version of Ampère's law in Eq. (6.30) to a contour C encircling the wire conductor of the circuit (Fig. 8.1) and two different characteristic surfaces, S_1 and S_2, bounded by the contour (S_1 intersects the conductor, while S_2 is completely in air, with a part of it being placed between the capacitor plates). This yields

$$\oint_C \mathbf{H} \cdot d\mathbf{l} = \begin{cases} i & \text{for surface } S_1 \\ 0 & \text{for surface } S_2 \end{cases}, \tag{8.1}$$

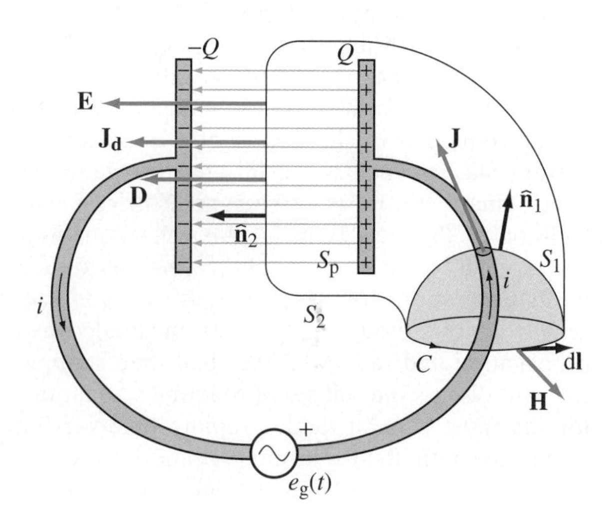

Figure 8.1 Application of Ampère's law to a circuit with an air-filled capacitor and time-varying current.

which, of course, is a contradictory result, as $i \neq 0$. We conclude that something is indeed wrong with Ampère's law, at least if this example is concerned.

It seems that there must be a sort of nonconduction current in the air-filled space between the capacitor plates as a continuation of the conduction current flow i through the wire conductors. This new type of current has to be included on the right-hand side of Eq. (6.30) in a manner that ensures that the choice of either S_1 or S_2 gives the same result in Eq. (8.1). To find its density, we note that the current i actually does not vanish at the capacitor terminals. Instead, it is terminated by time-varying charges Q and $-Q$ that accumulate on the capacitor plates. This termination is governed by the continuity equation for time-varying currents, Eq. (3.36), which gives [see Fig. 3.5 and Eq. (3.44)]

$$i = \frac{\mathrm{d}Q}{\mathrm{d}t}. \tag{8.2}$$

The charges on the plates, on the other hand, produce an electric field, of intensity $\mathbf{E}$. Neglecting the fringing effects, this field is localized in the capacitor only and is uniform. From Gauss' law [see Fig. 2.18(a) and Eq. (2.125)],

$$Q = DS_{\mathrm{p}}, \tag{8.3}$$

where D is the electric flux density in the capacitor ($D = \varepsilon_0 E$) and S_{p} is the plate area. Combining Eqs. (8.2) and (8.3), we can write

$$i = \frac{\partial D}{\partial t} S_{\mathrm{p}}. \tag{8.4}$$

The expression $\partial D/\partial t$ has the dimension of a current density (it equals i/S_{p}, and is thus expressed in $\mathrm{A/m^2}$), and this is the density of the new type of current that exists in air between the capacitor plates. It is called the displacement current density and is denoted by J_{d}. In vector form,

$$\boxed{\mathbf{J}_{\mathrm{d}} = \frac{\partial \mathbf{D}}{\partial t}.} \tag{8.5}$$

displacement current density vector (unit: $\mathrm{A/m^2}$)

The conduction current (of density $\mathbf{J}$) is the only current in the wire conductors in Fig. 8.1, whereas the displacement current is the only current in the capacitor. In the general case, however, both types of current may exist in the same material [e.g., if we fill the capacitor in Fig. 8.1 with an imperfect (lossy) dielectric, a conduction current will flow through the dielectric along with the displacement current]. The total current density vector is

$$\boxed{\mathbf{J}_{\mathrm{tot}} = \mathbf{J} + \mathbf{J}_{\mathrm{d}} = \mathbf{J} + \frac{\partial \mathbf{D}}{\partial t}.} \tag{8.6}$$

total (conduction plus displacement) current density vector

If we now correct Eq. (6.30) by including both types of current as sources of the magnetic field, we obtain the following form of Ampère's law:

$$\boxed{\oint_C \mathbf{H} \cdot \mathrm{d}\mathbf{l} = \int_S \left(\mathbf{J} + \frac{\partial \mathbf{D}}{\partial t} \right) \cdot \mathrm{d}\mathbf{S},} \tag{8.7}$$

corrected generalized Ampère's law

which, reapplied to the contour C and the two surfaces in Fig. 8.1, results in

$$\oint_C \mathbf{H} \cdot \mathrm{d}\mathbf{l} = \begin{cases} i & \text{for surface } S_1 \\ J_{\mathrm{d}} S_{\mathrm{p}} & \text{for surface } S_2 \end{cases} \tag{8.8}$$

(note that $\int_{S_2} \mathbf{J}_d \cdot d\mathbf{S} = J_d S_p$ because the fringing effects are neglected and $\mathbf{J}_d$ is uniform in the capacitor while zero outside it). There is no inconsistency any more between the applications of Eq. (8.7) to S_1 and S_2 because $i = J_d S_p$ [Eqs. (8.4) and (8.5)].

Note that the same meaningless result $i = 0$ obtained from the application of the quasistatic version of the generalized Ampère's law in Eq. (8.1) would have been obtained also from the quasistatic version of the continuity equation, Eq. (6.45), if applied to the structure in Fig. 8.1. Namely, if we consider a surface S enclosing completely one of the capacitor plates (like in Fig. 3.5), the total conduction current leaving the domain enclosed equals $\pm i$ (the sign depends on which plate is enclosed), so that Eq. (6.45) gives $i = 0$. This shows that essentially the same problems exist with quasistatic versions of the two equations. In other words, the addition of the displacement current in Ampère's law, Eq. (8.7), is equivalent to the addition of the time derivative of the enclosed charge in the continuity equation, Eq. (3.36).

Let us now introduce the same correction into the differential form of the generalized Ampère's law in Eqs. (6.42), which becomes

corrected generalized differential Ampère's law

$$\boxed{\nabla \times \mathbf{H} = \mathbf{J} + \frac{\partial \mathbf{D}}{\partial t}}. \tag{8.9}$$

This equation tells us that the sources producing locally circular components of the time-varying magnetic field intensity vector at a point in space (see Fig. 4.24) are described by the densities of both conduction and displacement currents at that point. In other words, the net curl of **H** exists at a point whenever a time-varying electric field (of flux density **D**) is present, even in the absence of **J**. This relationship between **H** and **D** is analogous to that between **E** and **B** in Faraday's law of electromagnetic induction in differential form, Eq. (6.39), which states that curl **E** exists at a point whenever a time-varying magnetic field (of flux density **B**) is present at that point.

Eqs. (8.8), (8.5), (8.3), and (8.2) indicate that the full (high-frequency) version of the generalized Ampère's law is consistent with the corresponding version of the continuity equation, at least for the structure in Fig. 8.1. To show that the same is true in general, we consider the two equations in their differential form and take the divergence of both sides of Eq. (8.9), as in Eq. (5.130) for the static case. This leads to

$$0 = \nabla \cdot \left(\mathbf{J} + \frac{\partial \mathbf{D}}{\partial t}\right) = \nabla \cdot \mathbf{J} + \nabla \cdot \left(\frac{\partial \mathbf{D}}{\partial t}\right) = \nabla \cdot \mathbf{J} + \frac{\partial}{\partial t}(\nabla \cdot \mathbf{D}), \tag{8.10}$$

where the time derivative operator in the expression for the displacement current density can be brought outside the divergence operator because these two operations are entirely independent from each other and can be performed in an arbitrary order. Combined with the generalized Gauss' law in differential form, Eq. (2.45),

$$0 = \nabla \cdot \mathbf{J} + \frac{\partial \rho}{\partial t}, \tag{8.11}$$

and this is the same as Eq. (3.39), that is, the full (high-frequency) version of the continuity equation in differential form.

In the slowly time-varying field, the rate of the time-variation in the electric flux density vector, $\partial \mathbf{D}/\partial t$, at a point is slow enough to be neglected with respect to the conduction current density vector, $\mathbf{J}$, so that Eq. (8.6) becomes $\mathbf{J}_{\text{tot}} \approx \mathbf{J}$ and Ampère's law in Eq. (8.7) can be approximated by its quasistatic version in Eq. (6.30). Exceptions are slowly time-varying fields in nonconducting media (such as the air dielectric of the capacitor in Fig. 8.1), where there cannot be any conduction current, and the displacement current density, $\mathbf{J}_\mathrm{d} = \partial \mathbf{D}/\partial t$, as the only current term cannot be omitted in Eq. (8.7). In addition, in poorly conducting media (e.g., in an imperfect dielectric of a capacitor), $\mathbf{J}$ is so small that $\mathbf{J}_\mathrm{d}$ cannot be neglected even in the slowly time-varying case. Similarly, assumption $\partial \rho/\partial t \approx 0$ for slow time-variations in the charge density leads to the quasistatic version of the continuity equation, Eq. (6.45). Exceptions here are large reservoirs of time-varying charges, such as the capacitor plates in Fig. 8.1, that serve as terminations of conduction currents at the boundaries with nonconducting media [e.g., $\mathrm{d}Q/\mathrm{d}t$ cannot be neglected in Eq. (8.2) for any rate of the time-variation in charge].

From the definition of the electric flux density vector, Eq. (2.41), the displacement current density vector can be written as the sum of the following two components:

$$\mathbf{J}_\mathrm{d} = \frac{\partial \mathbf{D}}{\partial t} = \varepsilon_0 \frac{\partial \mathbf{E}}{\partial t} + \frac{\partial \mathbf{P}}{\partial t}. \tag{8.12}$$

We know that the polarization vector, $\mathbf{P}$, is proportional to the average of moments of small electric dipoles representing a polarized material [see Eq. (2.7) and Fig. 2.2], where each dipole moment, in turn, is proportional to a displacement between the positive and negative charges in an atom or molecule [see Eq. (1.116) and Fig. 1.28]. If the electric field changes with time, these microscopic displacements also change with time, such that the second component in Eq. (8.12), $\partial \mathbf{P}/\partial t$, characterizes the average motion of bound (polarization) charges in the dielectric (note that a time variation of small dipoles in Fig. 2.3 would result in a motion of bound charges through the surface $\mathrm{d}S$). This motion of charges as a result of time-varying displacements of microscopic electric dipoles in a dielectric material thus constitutes a macroscopic current, so-called polarization current, of density

$$\boxed{\mathbf{J}_\mathrm{p} = \frac{\partial \mathbf{P}}{\partial t}.} \tag{8.13}$$

polarization current density vector

The other component of the displacement current density in Eq. (8.12),

$$\boxed{\mathbf{J}_{\mathrm{d}0} = \varepsilon_0 \frac{\partial \mathbf{E}}{\partial t},} \tag{8.14}$$

displacement current density in a vacuum

does not represent any motion of charges and, although expressed in $\mathrm{A/m^2}$, is not a real current density. It exists also in air, where $\mathbf{P} = 0$, and even in a vacuum, in the complete absence of material media, provided, of course, that a time-varying electric field is present. The component $\mathbf{J}_{\mathrm{d}0}$ is therefore termed the displacement current density in a vacuum. In fact, the most brilliant achievement of Maxwell was to imagine that a displacement current could also occur in a vacuum. Most importantly, the displacement current in a vacuum produces a magnetic field in the same fashion as the conduction and polarization currents. However, we note that the term "displacement current density," used for both components of $\partial \mathbf{D}/\partial t$ in Eq. (8.12), has an associated physical meaning only for the second component (polarization current density), while the component $\varepsilon_0\, \partial \mathbf{E}/\partial t$ is not related whatsoever to the displacement process and polarization charge, whether being considered in a vacuum or in an arbitrary dielectric medium.

Example 8.1 Displacement Current in an Ideal Capacitor

A parallel-plate capacitor of plate area S is filled with a perfect, homogeneous dielectric of permittivity ε. The distance between the plates, d, is much lesser than the dimensions of the plates, so the fringing effects can be neglected. The capacitor is connected to a slowly time-varying voltage $v(t) = V_0 \cos \omega t$. Find the displacement current density in the dielectric.

Solution As the voltage $v(t)$ is slowly time-varying and the fringing effects are negligible, the electric field in the capacitor can be considered as quasistatic and uniform throughout the dielectric. From Eq. (2.126), the electric field intensity in the dielectric is $E(t) = v(t)/d$ (the same everywhere). Eq. (8.5) then gives us the following expression for the displacement current density between the capacitor plates:

$$J_\mathrm{d}(t) = \frac{\mathrm{d}D}{\mathrm{d}t} = \varepsilon \frac{\mathrm{d}E}{\mathrm{d}t} = \frac{\varepsilon}{d}\frac{\mathrm{d}v}{\mathrm{d}t} = -\frac{\omega\varepsilon V_0}{d} \sin \omega t. \tag{8.15}$$

This result can also be obtained using the capacitance of the capacitor, $C = \varepsilon S/d$ [see Eq. (2.127)]. Namely, the current intensity through the capacitor terminals (leads) is [Eq. (3.45)]

$$i(t) = C\frac{\mathrm{d}v}{\mathrm{d}t} = -\frac{\varepsilon S \omega V_0}{d} \sin \omega t. \tag{8.16}$$

Hence,

$$J_\mathrm{d}(t) = \frac{i(t)}{S} = -\frac{\omega\varepsilon V_0}{d} \sin \omega t. \tag{8.17}$$

Example 8.2 Evaluation of the Time-Varying Magnetic Field in a Capacitor

Assume that the electrodes of the capacitor from the previous example are parallel circular plates of radius a, as shown in Fig. 8.2, and that the dielectric between the plates is imperfect, with parameters ε, σ, and μ_0. Under these circumstances, obtain the magnetic field intensity vector in the dielectric.

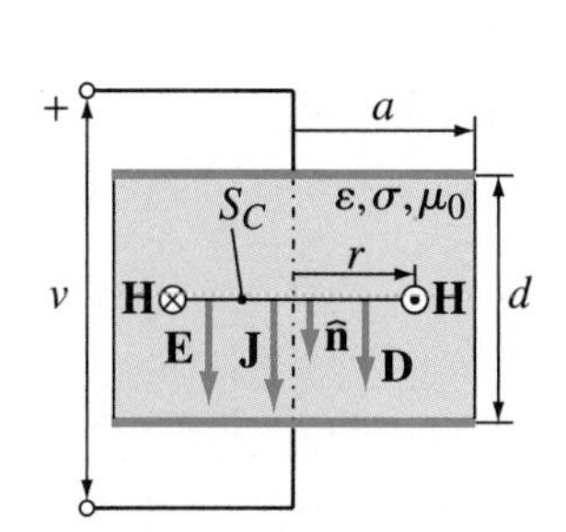

Figure 8.2 Evaluation of the magnetic field in a nonideal capacitor connected to a time-varying voltage; for Example 8.2.

Solution Since the dielectric is now imperfect (conducting), there is also a conduction current flowing (leaking) between the capacitor plates. Its density is determined by Ohm's law in local form, Eq. (3.18),

$$J(t) = \sigma E(t) = \frac{\sigma v(t)}{d} = \frac{\sigma V_0}{d} \cos \omega t. \tag{8.18}$$

Both this current and the displacement current with density given in Eq. (8.15) are sources of the magnetic field. Due to symmetry, the lines of the vector **H** in the dielectric are circles centered at the capacitor axis perpendicular to the plates [see Fig. 4.15(b)]. Applying the corrected generalized Ampère's law in integral form, Eq. (8.7), to a circular contour C of radius r ($r < a$) centered at the capacitor axis and the flat surface (S_C) spanned over the contour (Fig. 8.2), in exactly the same way as in Fig. 4.15(a) and Eqs. (4.54)–(4.56), yields

$$H\, 2\pi r = \left(J + \frac{\mathrm{d}D}{\mathrm{d}t}\right)\pi r^2, \tag{8.19}$$

and the following solution for H:

$$H = H(r,t) = \frac{[J(t) + J_\mathrm{d}(t)]\, r}{2} = \frac{V_0 r}{2d}(\sigma \cos \omega t - \omega\varepsilon \sin \omega t). \tag{8.20}$$

We note that the conduction current is in phase with the electric field in the dielectric (and capacitor voltage), whereas the displacement current is 90° out of phase. We also note that the amplitude (peak value) of the conduction current density in practical capacitors, which have excellent (almost nonconducting) dielectrics, is much smaller than the amplitude of the displacement current density ($\sigma \ll \omega\varepsilon$) even at very low frequencies.

Example 8.3 Conduction to Displacement Current Ratio for Rural Ground

For a sample of rural ground with $\varepsilon_\mathrm{r} = 14$ and $\sigma = 10^{-2}$ S/m that is occupied by a time-varying electric field of intensity $E(t) = E_0 \cos \omega t$, find the frequency $f = \omega/2\pi$ at which the amplitude of the density of the conduction current in the ground is n times that of the displacement current, where n takes values 0.001, 1, and 1000, respectively. Assume no changes in the material parameters as a function of frequency.

Solution From Eq. (8.20), we conclude that the ratio of the amplitudes of the conduction and displacement current densities is given by

$$\boxed{\frac{|J|_{\max}}{|J_{\mathrm{d}}|_{\max}} = \frac{\sigma}{\omega\varepsilon},} \tag{8.21}$$

conduction to displacement current ratio

which, for the given parameters of the ground sample, becomes

$$n = \frac{\sigma}{2\pi f \varepsilon_{\mathrm{r}}\varepsilon_0} = \frac{12.84 \times 10^6}{f} \quad (f \text{ in Hz}). \tag{8.22}$$

Hence, frequencies at which $n = 0.001$, 1, and 1000 are $f_1 = 12.84$ GHz, $f_2 = 12.84$ MHz, and $f_3 = 12.84$ kHz, respectively.

Problems: 8.1–8.6; *Conceptual Questions* (on Companion Website): 8.1–8.10; *MATLAB Exercises* (on Companion Website).

8.2 MAXWELL'S EQUATIONS FOR THE RAPIDLY TIME-VARYING ELECTROMAGNETIC FIELD

Having now in place the corrected version of the generalized Ampère's law that includes the displacement-current term and is always true, we are ready to summarize the full set of Maxwell's equations for the most general field – rapidly time-varying electromagnetic field. Ampère's law is Maxwell's second equation. The remaining three equations are the same as in the slowly time-varying case, described by Eqs. (6.42). Maxwell's equations for the slowly time-varying field, in turn, have the same form as the corresponding equations governing the time-invariant field, Eqs. (5.129), except for the first one (Faraday's law of electromagnetic induction). Hence, the four equations in integral notation for the rapidly time-varying electromagnetic field in arbitrary electromagnetic media are Eqs. (6.37), (8.7), (2.44), and (4.99). We list them here, together with the three constitutive equations describing material properties of electromagnetic media, Eqs. (2.46), (5.56), and (3.21):

$$\boxed{\begin{cases} \oint_C \mathbf{E}\cdot \mathrm{d}\mathbf{l} = -\int_S \frac{\partial \mathbf{B}}{\partial t}\cdot \mathrm{d}\mathbf{S} \\ \oint_C \mathbf{H}\cdot \mathrm{d}\mathbf{l} = \int_S \left(\mathbf{J} + \frac{\partial \mathbf{D}}{\partial t}\right)\cdot \mathrm{d}\mathbf{S} \\ \oint_S \mathbf{D}\cdot \mathrm{d}\mathbf{S} = \int_v \rho\, \mathrm{d}v \\ \oint_S \mathbf{B}\cdot \mathrm{d}\mathbf{S} = 0 \\ \mathbf{D} = \mathbf{D}(\mathbf{E}) \ [\mathbf{D} = \varepsilon\mathbf{E}] \\ \mathbf{B} = \mathbf{B}(\mathbf{H}) \ [\mathbf{B} = \mu\mathbf{H}] \\ \mathbf{J} = \mathbf{J}(\mathbf{E}) \ [\mathbf{J} = \sigma\mathbf{E}] \end{cases}} \tag{8.23}$$

Maxwell's first equation, integral
Maxwell's second equation, integral
Maxwell's third equation, integral
Maxwell's fourth equation, integral
constitutive equation for **D**
constitutive equation for **B**
constitutive equation for **J**

[ε, μ, and σ are the permittivity, permeability, and conductivity, respectively, of linear materials, defined by Eqs. (2.47), (5.60), and (3.18)]. The first equation tells us that, briefly, a magnetic field that changes in time produces an electric field. In parallel, the second equation states that, in general, both conduction currents and a time-varying electric field are the sources of a magnetic field. According to the third equation (generalized Gauss' law), the sources of an electric field are also electric charges, whereas the fourth equation (law of conservation of magnetic flux) expresses the fact that there are no analogous "magnetic charges." Finally, constitutive equations describe polarization, magnetization, and conduction properties, respectively, of materials and include such concepts in characterization of materials as linearity (and nonlinearity), homogeneity (and inhomogeneity), and isotropy (and anisotropy), as well as hysteresis effects.

We recall that external electric energy volume sources, analogous to ideal voltage and current generators in circuit theory, are modeled by impressed electric fields and currents (see Figs. 3.16 and 3.17). These sources are incorporated in Eqs. (8.23) through the constitutive equation for vector $\mathbf{J}$. Specifically, for linear conducting materials, the impressed electric field intensity vector, $\mathbf{E}_\mathrm{i}$, is included as in Eq. (3.109), while Eq. (3.124) takes into account the impressed current density vector, $\mathbf{J}_\mathrm{i}$, in the source region.

HISTORICAL ASIDE

James Clerk Maxwell (1831–1879), a Scottish physicist and the greatest name in electromagnetic theory, was the first Cavendish Professor of Physics at Cambridge. He possessed exceptional mathematical skills and talent but also a profound understanding and appreciation of physical reality. His early education took place at the Edinburgh Academy, and then at the University of Edinburgh. In 1850, he went to Cambridge University, where he graduated in mathematics from Trinity College in 1854. His interest in electricity and magnetism began immediately after graduation. In 1856, he became professor of natural philosophy (physics) at Marischal College in Aberdeen. Upon reading Faraday's three volumes of "Experimental Researches in Electricity" (published from 1839 to 1855), Maxwell was fascinated with Faraday's experimental results and theoretical speculations, and especially with his field approach to electromagnetism. In his papers "On Faraday's Lines of Force" (1856) and "On Physical Lines of Force" (1861), Maxwell translated Faraday's concept of lines of force into a mathematical form and established it as a general analytical tool for describing electric and magnetic phenomena. In 1860, he moved to London to teach natural philosophy and astronomy at King's College. In a series of brilliant papers in the 1860s culminating in his famous book "A Treatise on Electricity and Magnetism" (1873), he formulated the complete classical electromagnetic theory. He provided a unified mathematical framework for all fundamental laws of electricity and magnetism discovered experimentally by his predecessors and compiled and completed the four fundamental equations of electromagnetics that bear his name. As stated by Albert Einstein (1879–1955), "the formulation of these equations is the most important event in physics since Newton's (1642–1727) time." Within his mathematical derivations, Maxwell concluded that a changing electric field must always be accompanied by a changing magnetic field, even in situations where a conduction current (flowing through conductors) is not present. This brought him to his famous idea to introduce a hypothetical entity, a displacement current, as an equivalent source of the magnetic field. The notion of a displacement current enabled him to theoretically explain the propagation of electromagnetic energy in space, i.e., to mathematically predict the existence of electromagnetic waves. As a matter of fact, he calculated the exact speed of light from his equations. He showed that oscillations of electric charges were sources of electromagnetic radiation and, since the charges could oscillate at any rate, he believed that a whole family of radiated electromagnetic waves (that is, the electromagnetic spectrum, as we call it today) was possible. Finally, he identified visible light as an electromagnetic radiation and a part of the family of electromagnetic waves, which is brilliantly summarized in the following sentence from his classical work "A Dynamical Theory of the Electromagnetic Field" (1864): "We have strong reason to conclude that light itself – including radiant heat and other radiation, if any – is an electromagnetic disturbance in the form of waves propagated through the electromagnetic field according to electromagnetic laws." In 1871, he accepted a position as the first professor of experimental physics at Cambridge, where he founded the world famous Cavendish Laboratory in 1874. His theoretical predictions were experimentally verified after his death, in laboratory experiments by Heinrich Hertz (1857–1894), whose first transmission and reception of radio waves in 1887 came as a glorious confirmation of the entire Maxwell's work in electromagnetics. It was not later than by the early 1900s, that Guglielmo Marconi (1874–1937) turned Hertz's laboratory demonstration and subsequent inventions in radio engineering by Nikola Tesla (1856–1943) into a practical means of wireless communication over long distances. Maxwell also provided great contributions to thermodynamics. In his paper "On the Dynamical Theory of Gases" (1867), he proposed a new mathematical basis for the kinetic theory of gases. He developed a formula that determines the distribution of molecular speeds for a system of gaseous molecules of a given molecular weight at a given temperature, which, in a combination with the work of Ludwig Boltzmann (1844–1906), is now known as the Maxwell-Boltzmann distribution of molecular speeds. In addition, as his major contribution to astronomy, Maxwell showed mathematically in 1857 that Saturn's rings must consist of a vast number of small solid particles in order to be dynamically stable (this was confirmed more than a century later in the spacecraft Voyager explorations of Saturn in 1980–1981). However, it is his equations for the electromagnetic field of an arbitrary time variation, above all, that place Maxwell as one of the top few greatest contributors to the prosperity and progress of humanity ever, which cannot be emphasized better than by the following quote of Richard Feynman (1918–1988) from "The Feynman Lectures on Physics" (1963–1965): "From a long view of the history of mankind, seen from, say, ten thousand years from now, there can be little doubt that the most significant event of the nineteenth century will be judged as Maxwell's discovery of the laws of electrodynamics." *(Portrait: AIP Emilio Segré Visual Archives)*

Maxwell's equations in differential form relate the curl and divergence of basic field vectors (**E**, **H**, **D**, and **B**) at a point to the corresponding field sources at the same point. They are referred to also as Maxwell's equations at a point and are derived from their integral counterparts. Note, however, that the original notation that James Clerk Maxwell used in "A Treatise on Electricity and Magnetism" in 1873, where his fully developed system of electromagnetic equations first appeared, was in the form of partial differential equations. For the rapidly time-varying electromagnetic field, Maxwell's equations at a point in an arbitrary electromagnetic medium are given by Eqs. (6.39), (8.9), (2.45), and (4.103), which completes the following general list of four partial differential equations of space coordinates and time:

$$\boxed{\begin{cases} \nabla \times \mathbf{E} = -\frac{\partial \mathbf{B}}{\partial t} \\ \nabla \times \mathbf{H} = \mathbf{J} + \frac{\partial \mathbf{D}}{\partial t} \\ \nabla \cdot \mathbf{D} = \rho \\ \nabla \cdot \mathbf{B} = 0 \end{cases}} . \tag{8.24}$$

Maxwell's first equation, differential
Maxwell's second equation, differential
Maxwell's third equation, differential
Maxwell's fourth equation, differential

The above differential equations are valid always and everywhere, except at points where material properties of electromagnetic media change abruptly from one value to another, that is, at boundary surfaces between electromagnetically different media. At such points, basic field vector functions change abruptly as well across the boundary surface, which means that their spatial derivatives in the direction normal to the surface are not defined. Namely, a jump in a field function between two close points on the two sides of the boundary surface corresponds to an infinite partial derivative (rate of change) of that function with respect to a local coordinate normal to the boundary, and thus to undefined differential operators in Eqs. (8.24).

Example 8.4 Continuity Equation from Integral Maxwell's Equations

Starting from Maxwell's equations in integral notation for the rapidly time-varying electromagnetic field, derive the corresponding form of the continuity equation.

Solution Consider an arbitrary closed surface S in the field and a contour C that splits S into two parts, the upper part S_1 and the lower part S_2 (similarly to Fig. 4.27 for S_1 and S_3). Let the surfaces S_1 and S_2, which are both enclosed by C, be oriented in the same way, that in accordance with the right-hand rule with respect to the orientation of the contour. Applying Maxwell's second equation (corrected generalized Ampère's law) in integral form, Eqs. (8.23), to the contour C and either S_1 or S_2 must give the same result, i.e., the fluxes through S_1 and S_2 of the total current density vector, $\mathbf{J} + \partial\mathbf{D}/\partial t$, are the same. This means in turn that [note the similarity with Eq. (4.101)]

$$\oint_S \left(\mathbf{J} + \frac{\partial \mathbf{D}}{\partial t}\right) = 0, \tag{8.25}$$

which itself is an interesting general conclusion. Hence,

$$\oint_S \mathbf{J} \cdot \mathrm{d}\mathbf{S} = -\oint_S \frac{\partial \mathbf{D}}{\partial t} \cdot \mathrm{d}\mathbf{S} = -\frac{\mathrm{d}}{\mathrm{d}t} \oint_S \mathbf{D} \cdot \mathrm{d}\mathbf{S}, \tag{8.26}$$

where the time derivative in the expression for the displacement current density can be brought outside the surface integral because these two operations are entirely independent from each other (provided that the surface S is stationary). Combining with Maxwell's third equation (generalized Gauss' law) in integral form [Eqs. (8.23)] then gives

$$\oint_S \mathbf{J} \cdot \mathrm{d}\mathbf{S} = -\frac{\mathrm{d}}{\mathrm{d}t} \int_v \rho \, \mathrm{d}v = -\int_v \frac{\partial \rho}{\partial t} \, \mathrm{d}v, \tag{8.27}$$

that is, the general (high-frequency) continuity equation in integral form, Eq. (3.38).

We note that this derivation parallels in its entirety the one given by Eqs. (8.10) and (8.11) in differential notation.

Example 8.5 Maxwell's Fourth Differential Equation from the First One

Consider general Maxwell's equations in differential form, and derive the fourth equation from the first one.

Solution By taking the divergence of both sides of the differential form of Maxwell's first equation (Faraday's law of electromagnetic induction), Eqs. (8.24), in the same way as in Eq. (5.130) or Eq. (8.10), we get

$$0 = -\nabla \cdot \left(\frac{\partial \mathbf{B}}{\partial t}\right) = -\frac{\partial}{\partial t} (\nabla \cdot \mathbf{B}), \tag{8.28}$$

and hence

$$\nabla \cdot \mathbf{B} = \text{const.} \tag{8.29}$$

This equation tells us that the magnetic flux density vector at the point considered has had the same value always. However, if we go back far enough in time, we have that $\mathbf{B} = 0$ – before the field was created, which means that the constant in Eq. (8.29) is equal to zero, as in Maxwell's fourth equation (law of conservation of magnetic flux) in differential form [Eqs. (8.24)], and thus completes our derivation.

Example 8.6 No Dynamic Fields in Perfect Conductors

Prove that in a perfect electric conductor (PEC) there can be no time-varying electromagnetic field nor its sources (time-varying currents and charges).

Solution For perfect electric conductors, $\sigma \to \infty$, and Eq. (3.27) proves that there can be no (time-varying or time-constant) electric field ($\mathbf{E} = 0$) inside a PEC body. From Maxwell's first equation in differential form [Eqs. (8.24)], we then get $\partial\mathbf{B}/\partial t = -\nabla \times \mathbf{E} = 0$, i.e., $\mathbf{B} = \text{const}$. This means that no time-varying magnetic field is possible in a perfect conductor, but only a time-constant magnetic field. Furthermore, $\mathbf{E} = 0$ and $\mathbf{B} = 0$ imply, through the constitutive equations in Eqs. (8.23), that $\mathbf{D} = 0$ and $\mathbf{H} = 0$, respectively, which, substituted in Maxwell's second equation in differential form, yield $\mathbf{J} = \nabla \times \mathbf{H} - \partial\mathbf{D}/\partial t = 0$. More precisely, this is true only for time-varying currents, while $\mathbf{H} = \text{const}$ gives $\mathbf{J} = \nabla \times \mathbf{H} = \text{const}$ in the static case. This tells us that a time-constant volume current, as a source of a time-constant magnetic field, can flow through a PEC material. Finally, using Maxwell's third equation in differential form, we obtain $\rho = \nabla \cdot \mathbf{D} = 0$.

In summary, we have proved using general Maxwell's equations in differential form that there can be no time-varying electromagnetic field ($\mathbf{E} = 0$, $\mathbf{D} = 0$, $\mathbf{B} = 0$, $\mathbf{H} = 0$) nor its sources ($\mathbf{J} = 0$, $\rho = 0$) inside perfectly conducting materials, whereas magnetostatic fields ($\mathbf{B} = \text{const}$) and steady currents ($\mathbf{J} = \text{const}$) can exist in PEC bodies.

Problems: 8.7–8.9; *Conceptual Questions* (on Companion Website): 8.11–8.15; *MATLAB Exercises* (on Companion Website).

8.3 ELECTROMAGNETIC WAVES

The most important implication of Maxwell's equations for the rapidly time-varying electromagnetic field is the concept of electromagnetic waves that can exist in free space and in material media and represent a means of transmitting energy and information over a distance. Electromagnetic waves consist of electric and magnetic fields which, once created by rapidly time-varying sources (currents and charges), travel through space independent of the sources that produced them. To qualitatively illustrate how electromagnetic waves expand based on Maxwell's equations,[1] consider, for example, Eqs. (8.23) for a linear, homogeneous, and lossless ($\sigma = 0$) medium, with permittivity ε and permeability μ. Away from the source region, the first two equations (circulation equations) can be written as

$$\oint_C \mathbf{E} \cdot \mathrm{d}\mathbf{l} = -\mu \int_S \frac{\partial \mathbf{H}}{\partial t} \cdot \mathrm{d}\mathbf{S}, \tag{8.30}$$

$$\oint_C \mathbf{H} \cdot \mathrm{d}\mathbf{l} = \varepsilon \int_S \frac{\partial \mathbf{E}}{\partial t} \cdot \mathrm{d}\mathbf{S}. \tag{8.31}$$

[1]The complete analysis and quantitative characterization of electromagnetic wave propagation in different homogeneous media based on Maxwell's equations for the rapidly time-varying electromagnetic field will be provided in the next chapter.

Their cyclic combination gives rise to the phenomenon of electromagnetic wave propagation: in an electromagnetic wave, a time-varying magnetic field produces a time-varying electric field, which in turn generates a magnetic field, and so on, with a resulting propagation of the electric and magnetic (electromagnetic) energy. An electromagnetic wave is initiated whenever a rapid time change of either the electric or magnetic field takes place at a point in space.

In specific, suppose that a time-varying electric field $\mathbf{E}_1$ is detected at a point in space as a result of a change in the current through a distant vertical wire conductor (transmitting antenna), as shown in Fig. 8.3. According to Eq. (8.31), this field produces a time-varying magnetic field, $\mathbf{H}_1$, along a small horizontal contour C_1 surrounding it. On the other hand, Eq. (8.30) tells us that the field $\mathbf{H}_1$ itself generates a time-varying electric field, $\mathbf{E}_2$, along a small contour C_2 lying in a vertical plane. Invoking Eq. (8.31) again, we have that $\mathbf{E}_2$ is then accompanied by $\mathbf{H}_2$, and the electric and magnetic fields thus continue to generate one another indefinitely, as indicated in Fig. 8.3. However, we note that $\mathbf{E}_2$ must be different from $\mathbf{E}_1$ and $\mathbf{H}_2$ cannot be the same as $\mathbf{H}_1$ for this process to be carried on, because the net line integrals on the left-hand sides of Eqs. (8.30) and (8.31) must be nonzero. Therefore, Fig. 8.3 actually shows that the time variations of $\mathbf{E}$ (e.g., $\partial\mathbf{E}_2/\partial t$) produce the space variations of $\mathbf{H}$ (e.g., $\Delta\mathbf{H} = \mathbf{H}_2 - \mathbf{H}_1$), and the same for time variations of $\mathbf{H}$ causing space variations of $\mathbf{E}$, so that both $\mathbf{E}$ and $\mathbf{H}$ travel during the course of time away from the sources (antenna currents). We also note that the wave, once created, continues to propagate in space with no connection whatsoever with its sources, which might not even exist any more (e.g., the current in the antenna in Fig. 8.3 might be shut down while the wave still propagates away from the antenna[2]).

The theoretical and practical importance of the concept of electromagnetic waves as traveling electromagnetic fields can hardly be overemphasized. Most of the material in the chapters to follow will be devoted to the analysis of various forms of electromagnetic wave propagation and its application in high-frequency electromagnetic systems.

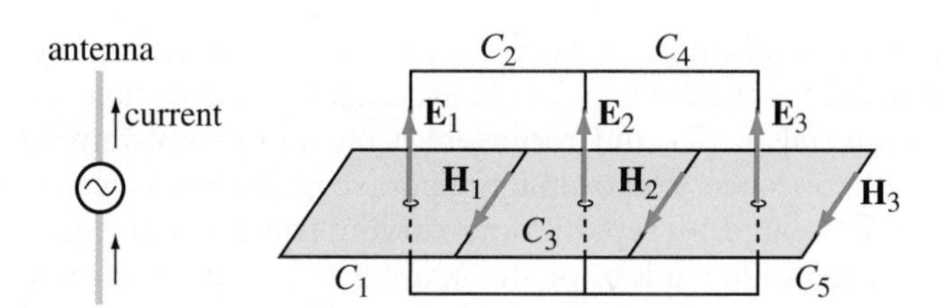

Figure 8.3 Electromagnetic wave propagation viewed as a process of successive mutual induction of electric and magnetic fields in space and time, as dictated by Maxwell's equations for the rapidly time-varying electromagnetic field. (Note that the imaginary contours, presented here drastically enlarged for the clarity of the figure and a better visualization of the process, are supposed to be very small and are far away from the antenna.)

[2]As another example, a star that we see in a clear night sky might have been "dead" for a long time as the electromagnetic waves constituting visible light once launched from the star continue to propagate indefinitely through space away from it.

8.4 BOUNDARY CONDITIONS FOR THE RAPIDLY TIME-VARYING ELECTROMAGNETIC FIELD

Boundary conditions are essentially the third form of Maxwell's equations, formulated at boundary surfaces between different media. They represent the relations between the tangential or normal components of basic field vectors at two close points on the two sides of the boundary surface and, like Maxwell's equations in differential form, are also derived from the respective equations in integral form, out of Eqs. (8.23). If we apply Faraday's law of electromagnetic induction in integral form to a narrow rectangular elementary contour of Fig. 2.10(a), we get the same boundary condition as in Eq. (2.84) for the static case, because the flux of the vector $\partial B/\partial t$ appearing on the right-hand side of the equation reduces to zero when the contour side Δh shrinks to zero. Similarly, if we reconsider the application of the generalized Ampère's law in integral form to the elementary contour of Fig. 5.12, but now with the general version of the law that includes the displacement-current term, what we get is the same boundary condition as in Eq. (5.74) for the static case, because the displacement current enclosed by the contour (flux of $\partial D/\partial t$ through the contour) is zero, while the conduction current may be nonzero if a surface conduction current, of density $\mathbf{J}_\mathrm{s}$, exists on the boundary. The remaining two Maxwell's equations have the same form for both static and dynamic fields, including the high-frequency case, which means that the corresponding boundary conditions for the rapidly time-varying electromagnetic field are the same as in Eqs. (2.85) and (5.76) for the time-invariant field. In summary, the four general electromagnetic boundary conditions, for the tangential components of vectors **E** and **H** and normal components of vectors **D** and **B**, respectively, at a boundary surface between two electromagnetic media (regions 1 and 2) are given by

$$\begin{cases} \hat{\mathbf{n}} \times \mathbf{E}_1 - \hat{\mathbf{n}} \times \mathbf{E}_2 = 0 \\ \hat{\mathbf{n}} \times \mathbf{H}_1 - \hat{\mathbf{n}} \times \mathbf{H}_2 = \mathbf{J}_\mathrm{s} \\ \hat{\mathbf{n}} \cdot \mathbf{D}_1 - \hat{\mathbf{n}} \cdot \mathbf{D}_2 = \rho_\mathrm{s} \\ \hat{\mathbf{n}} \cdot \mathbf{B}_1 - \hat{\mathbf{n}} \cdot \mathbf{B}_2 = 0 \end{cases} \quad (\hat{\mathbf{n}} \text{ directed from region 2 to region 1}), \tag{8.32}$$

boundary condition for $\mathbf{E}_\mathrm{t}$

boundary condition for $\mathbf{H}_\mathrm{t}$

boundary condition for $\mathbf{D}_\mathrm{n}$

boundary condition for $\mathbf{B}_\mathrm{n}$

where $\hat{\mathbf{n}}$ is the normal unit vector on the surface, directed from region 2 to region 1.

As we shall see in later chapters, boundary conditions for the rapidly time-varying electric and magnetic field vectors are essential for the analysis of different types of electromagnetic waves in the presence of interfaces between different material media, including PEC surfaces.

Example 8.7 Dynamic Boundary Conditions on a PEC Surface

Write down the full set of boundary conditions for a surface of a perfect electric conductor in a dynamic electromagnetic field.

Solution Let us mark the PEC as region 2. The electromagnetic field in a perfect electric conductor is always zero under dynamic conditions (see Example 8.6), so that $\mathbf{E}_2 = 0$, $\mathbf{H}_2 = 0$, $\mathbf{D}_2 = 0$, and $\mathbf{B}_2 = 0$ in Eqs. (8.32). Designating by $\mathbf{E}_1 = \mathbf{E}$, $\mathbf{H}_1 = \mathbf{H}$, $\mathbf{D}_1 = \mathbf{D}$, and $\mathbf{B}_1 = \mathbf{B}$ the field vectors close to the boundary in the surrounding medium (region 1), we can write

$$\begin{array}{llll} \hat{\mathbf{n}} \times \mathbf{E} = 0 & (E_\mathrm{t} = 0), & \hat{\mathbf{n}} \times \mathbf{H} = \mathbf{J}_\mathrm{s} & (H_\mathrm{t} = J_\mathrm{s}), \\ \hat{\mathbf{n}} \cdot \mathbf{D} = \rho_\mathrm{s} & (D_\mathrm{n} = \rho_\mathrm{s}), & \hat{\mathbf{n}} \cdot \mathbf{B} = 0 & (B_\mathrm{n} = 0), \end{array} \tag{8.33}$$

boundary conditions on a PEC surface; $\hat{\mathbf{n}}$ outward normal with $\hat{\mathbf{n}}$ standing for the normal unit vector on the conductor surface, directed from the conductor outward. In words, the tangential component of the electric field intensity vector and normal component of the magnetic flux density vector at the PEC surface are both zero, whereas the tangential component of the magnetic field intensity vector and normal component of the electric flux density vector equal the local surface current and charge densities, respectively, over the surface. We also note that the electric field lines near the PEC surface are normal to the surface, while the magnetic field lines are tangential to it. Finally, we recall from Example 8.6 that $\mathbf{J} = 0$ and $\rho = 0$ inside the PEC volume, so that $\mathbf{J}_s$ and ρ_s are the only local sources of the field near the PEC surface.

Conceptual Questions (on Companion Website): 8.16 and 8.17; *MATLAB Exercises* (on Companion Website).

8.5 DIFFERENT FORMS OF THE CONTINUITY EQUATION FOR RAPIDLY TIME-VARYING CURRENTS

The continuity equation expresses the principle of conservation of electric charges and can be derived from Maxwell's equations. In analysis of electromagnetic systems, we often use it as the fifth equation added to the set of four Maxwell's equations. In this section, we discuss different forms of the continuity equation, for rapid time variations of currents and charges. Eqs. (3.38) and (3.39) represent the integral and differential forms, respectively, of the equation, for volume spatial distributions of currents and charges. We write them down here again, together with the corresponding boundary condition, Eq. (3.53), for the normal components of the current density vector at a conductor-conductor boundary:

continuity equation, integral

continuity equation, differential

boundary condition for $\mathbf{J}_n$

$$\left\{\begin{array}{l} \oint_S \mathbf{J} \cdot d\mathbf{S} = -\int_v \frac{\partial \rho}{\partial t}\, dv \\ \nabla \cdot \mathbf{J} = -\frac{\partial \rho}{\partial t} \\ \hat{\mathbf{n}} \cdot \mathbf{J}_1 - \hat{\mathbf{n}} \cdot \mathbf{J}_2 = -\frac{\partial \rho_s}{\partial t} \end{array}\right., \tag{8.34}$$

with $\hat{\mathbf{n}}$ being directed again from region 2 to region 1. This last equation can be added as the fifth general electromagnetic boundary condition to the set of four standard conditions in Eqs. (8.32). We note that only this boundary condition, having the time derivative of the surface charge density (on the right-hand side of the equation), differs from its static version, Eq. (3.55). We also note that it is valid only if there is no surface current flowing over the boundary ($\mathbf{J}_s = 0$). Otherwise, this current must also be included in the current continuity balance.

The continuity equation in differential form in Eqs. (8.34) relates the spatial derivatives of the volume current density vector, $\mathbf{J}$, at a point to the time derivative of the volume charge density, ρ, at that point. Similar relationships can be derived also for line and surface currents and charges. Consider first a line current of intensity i along a line l of arbitrary shape (e.g., current through a thin metallic wire in a nonconducting medium). In the general case (rapidly time-varying current), i is a function of both the position (length coordinate) along the line (wire) and time, $i = i(l, t)$. We apply the continuity equation in integral form to a closed surface S enclosing a differentially small segment Δl along the line, as shown in Fig. 8.4(a). Let i_1 and i_2 designate the intensities of current entering and leaving S, at the beginning and end of the segment Δl, respectively. The net outflow of current through S is hence

$$i_{\text{out}} = i_2 - i_1 = \Delta i. \tag{8.35}$$

The change Δi in current intensity over the distance Δl along the line is enabled by an excess charge that accumulates along the line. The line density of this charge, Q', is also dependent on both l and t. However, as Δl is very small, the total charge along it, that is, the total charge enclosed by the surface S, can be found essentially with no integration, simply as

$$Q_S = Q'\,\Delta l, \tag{8.36}$$

so that the continuity equation, Eq. (3.34), becomes

$$\Delta i = -\frac{\mathrm{d}}{\mathrm{d}t}\left(Q'\,\Delta l\right). \tag{8.37}$$

Dividing both sides of this relationship by Δl and letting $\Delta l \to 0$, we get

$$\boxed{\frac{\partial i}{\partial l} = -\frac{\partial Q'}{\partial t}.} \tag{8.38}$$

continuity equation for wires (for line currents)

This is the continuity equation in differential form for line currents, also known as the continuity equation for wires. It is very important for analysis of transmission lines and wire antennas. It tells us that, indeed, the intensity of a rapidly time-varying current in a wire conductor can change along the conductor. Consequently, a rapidly time-varying current can exist even in wires with open ends (such as an open-ended two-wire line or a wire dipole antenna), where the current intensity drops to zero at wire ends. More precisely, Eq. (8.38) quantifies the rate of change of the current intensity with the length coordinate (partial derivative of i with respect to l) at a point along the wire as being equal to the negative of the rate of change of the line charge density with time (negative partial derivative of Q' with respect to t) at the same point. In the slowly time-varying (low-frequency) case, on the other hand, $\partial Q'/\partial t \approx 0$ in Eq. (8.38), which means that $\partial i/\partial l \approx 0$ as well, i.e., that a slowly time-varying current practically does not change along the wire. Of course, this is also true (exactly) for a steady (time-invariant) current along the wire.

For a surface current distribution described by the surface current density vector, $\mathbf{J}_\mathrm{s}$, over a surface S [e.g., current through a thin metallic plate or through the "skin" region of a solid metallic body with the skin effect pronounced – see Fig. 6.23(b)], we evaluate the local net current outflow through the contour C enclosing a differentially small patch ΔS on S, as depicted in Fig. 8.4(b). From the definition of the surface current density vector in Eq. (3.13), the total current leaving ΔS across the contour C is given by

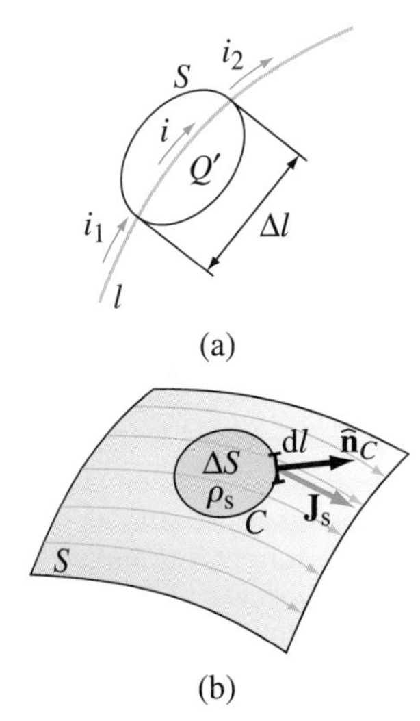

Figure 8.4 Derivation of the continuity equation in differential form for rapidly time-varying line currents (a) and surface currents (b).

$$i_{\text{out}} = \oint_C \mathbf{J}_\mathrm{s} \cdot \hat{\mathbf{n}}_C \,\mathrm{d}l, \tag{8.39}$$

where $\hat{\mathbf{n}}_C$ is the unit vector normal to an element $\mathrm{d}l$ along C. This vector is locally tangential to S and directed from the patch outward. In other words, the elementary vector $\hat{\mathbf{n}}_C\,\mathrm{d}l$ in Eq. (8.39) replaces the corresponding vector $\mathrm{d}\mathbf{S} = \hat{\mathbf{n}}\,\mathrm{d}S$ in the analogous equation for volume currents, Eq. (3.35). The total charge on the patch amounts to

$$Q_S = \rho_\mathrm{s}\,\Delta S, \tag{8.40}$$

with ρ_s being the surface charge density of the patch, so that Eq. (3.34) with $\Delta S \to 0$ yields

$$\lim_{\Delta S \to 0} \frac{\oint_C \mathbf{J}_\mathrm{s} \cdot \hat{\mathbf{n}}_C \,\mathrm{d}l}{\Delta S} = -\frac{\partial \rho_\mathrm{s}}{\partial t}. \tag{8.41}$$

The expression on the left-hand side of this equation represents, by definition, the so-called surface divergence of a vector ($\mathbf{J}_s$), namely, the surface version of the (volume) divergence operator in Eq. (1.172). It is denoted as $\text{div}_s\mathbf{J}_s$ or $\nabla_s \cdot \mathbf{J}_s$, where ∇_s is the surface del (nabla) operator, i.e., the surface version of the standard (volume) del operator. Hence, the differential form of the continuity equation for rapidly time-varying surface currents can be written as

continuity equation for plates (for surface currents)

$$\boxed{\nabla_s \cdot \mathbf{J}_s = -\frac{\partial \rho_s}{\partial t}.} \tag{8.42}$$

This equation is used in analysis of antennas and scatterers composed of metallic surfaces (plates), and is also referred to as the continuity equation for plates.

For example, if the surface current density vector is given by its x- and y-components in the xy-plane in the Cartesian coordinate system, the surface del operator is obtained by omitting the z-component in Eq. (1.100),

$$\nabla_s = \frac{\partial}{\partial x}\,\hat{\mathbf{x}} + \frac{\partial}{\partial y}\,\hat{\mathbf{y}}, \tag{8.43}$$

and the continuity equation for this case becomes [see Eq. (1.165)]

$$\frac{\partial J_{sx}}{\partial x} + \frac{\partial J_{sy}}{\partial y} = -\frac{\partial \rho_s}{\partial t}. \tag{8.44}$$

Example 8.8 Application of the Continuity Equation for Surface Currents

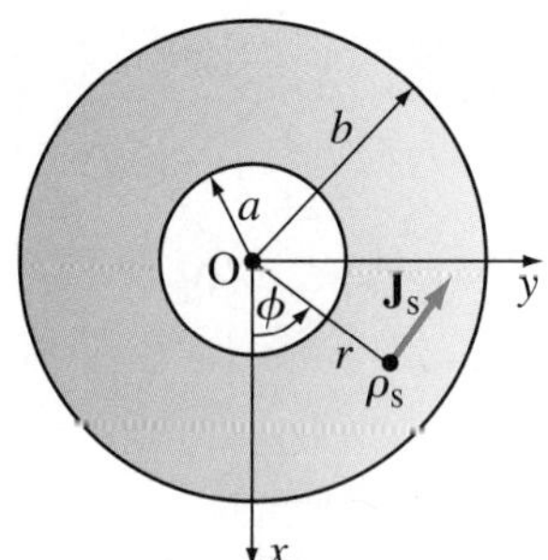

Figure 8.5 Evaluation of the surface charge distribution from a current distribution over a hollow circular plate; for Example 8.8.

A time-harmonic surface current of angular frequency ω flows circularly over a hollow circular plate of radii a and b, as shown in Fig. 8.5. The surface current density vector is given by $\mathbf{J}_s(r,\phi,t) = J_{s0}\cos(\phi/2)\cos(\omega t + kr)\,\hat{\boldsymbol{\phi}}$ ($a \le r \le b$, $-\pi < \phi \le \pi$), where J_{s0} and k are constants. Find the associated charge distribution over the plate.

Solution We use the expression for the divergence in cylindrical coordinates, Eq. (1.170), with only the second term left, as $\mathbf{J}_s$ in Fig. 8.5 has only a ϕ-component ($J_{s\phi}$). The continuity equation for plates, Eq. (8.42), and differentiation with respect to ϕ then give

$$\nabla_s \cdot \mathbf{J}_s = \frac{1}{r}\frac{\partial J_{s\phi}}{\partial \phi} = -\frac{J_{s0}}{2r}\sin\frac{\phi}{2}\cos(\omega t + kr) = -\frac{\partial \rho_s}{\partial t}. \tag{8.45}$$

Hence, integrating with respect to time, we obtain the following expression for the surface charge density over the plate:

$$\rho_s(r,\phi,t) = \frac{J_{s0}}{2r}\sin\frac{\phi}{2}\int \cos(\omega t + kr)\,dt = \frac{J_{s0}}{2\omega r}\sin\frac{\phi}{2}\sin(\omega t + kr). \tag{8.46}$$

Problems: 8.10; *Conceptual Questions* (on Companion Website): 8.18–8.20; *MATLAB Exercises* (on Companion Website).

8.6 TIME-HARMONIC ELECTROMAGNETICS

Maxwell's equations hold true for electromagnetic quantities with an arbitrary time dependence. The actual type of time functions that the fields (**E**, **H**, **D**, and **B**) assume depends on the current and charge density functions, **J** and ρ, in the system, and ultimately on the time variation of external sources, i.e., on the impressed electric field intensity and current density functions, $\mathbf{E}_i$ and $\mathbf{J}_i$, in source regions [see Eqs. (3.109) and (3.124)]. Very often, the time variation of external sources in the system is sinusoidal. If the system is also linear, meaning that all electromagnetic materials in the system are linear, then all fields in the system vary in time sinusoidally as well. Namely, since Maxwell's equations for linear systems are linear integral or differential equations, sinusoidal time variations of source functions (excitations) of a given frequency produce, in the steady state, sinusoidal variations of the fields (responses to excitations) at all points of the system, with the same frequency. By the same token, in a linear electric circuit, as a special case of linear electromagnetic systems, with all excitations (voltage and current generators) being sinusoidal time functions of the same frequency, all responses (voltages across circuit elements and currents in circuit branches) in the steady state are also sinusoids with the same frequency. Here, we focus on steady-state sinusoidal, also known as time-harmonic, field and circuit quantities, discuss their properties, and then (in the next section) introduce their complex equivalents. Maxwell's equations in the complex domain will be introduced in the section to follow. These equations are much simpler to use than the equivalent equations in the time domain, thus providing an extremely convenient basis for solving problems in time-harmonic electromagnetics.

Consider a time-harmonic voltage of frequency (repetition rate) f and amplitude (peak-value) V_0. Its instantaneous value, that is, the value at an instant t, can be written as either a cosine or a sine function, as they both give the same general shape for the waveform over time. In this text, we choose the cosine function (so-called cosine reference) and write

$$\boxed{v(t) = V_0 \cos(\omega t + \theta),} \tag{8.47}$$

time-harmonic voltage

where ω is the angular frequency (or radian frequency) of time-harmonic oscillation, given by

$$\boxed{\omega = 2\pi f.} \tag{8.48}$$

ω – angular or radian frequency (unit rad/s); f – frequency (unit: Hz)

Units are the hertz (Hz) for f ($\text{Hz} = 1/\text{s}$) and radian per second (rad/s) for ω. The function $\phi(t) = \omega t + \theta$ is the instantaneous phase (in radians) of the voltage $v(t)$, while θ is the initial phase (phase at an instant $t = 0$) for the cosine reference.[3] Since the period of change of a cosine function, $\cos\phi$, is 2π (for ϕ as independent variable), the time period of change (for t as independent variable) of $v(t)$ is defined by the relation $\omega T = 2\pi$, which yields

$$\boxed{T = \frac{2\pi}{\omega} = \frac{1}{f}.} \tag{8.49}$$

time period

In other words, after each T, a time-harmonic function repeats itself over time.

We note that the time-average value of $v(t)$ is zero (over one complete cycle, the average value of $\cos\phi$ is zero). On the other hand, the root-mean-square (rms) value of $v(t)$, which, by definition, is found as the square root of the time-average of the voltage squared, amounts to

$$\boxed{V_{\text{rms}} = \sqrt{\frac{1}{T}\int_0^T v^2(t)\,\mathrm{d}t} = \sqrt{\frac{1}{T}\int_0^T V_0^2 \cos^2(\omega t + \theta)\,\mathrm{d}t} = \frac{V_0}{\sqrt{2}}} \tag{8.50}$$

rms (root-mean-square) value

[3]Note that, as $\cos\phi$ and $\sin\phi$ are shifted in phase by $90°$ ($\pi/2$) with respect to each other, but otherwise the same waveforms, the same voltage $v(t)$ in Eq. (8.47) written as a sine function is $v(t) = V_0 \sin(\omega t + \theta')$, with $\theta' = \theta + \pi/2$ being the initial phase for the sine reference.

[the average value of $\cos^2\phi$ or $\sin^2\phi$ is 1/2, as shown in Eq. (6.95)]. We can now rewrite Eq. (8.47) as[4]

V – rms value (used much more frequently than the peak-value)

$$\boxed{v(t) = V\sqrt{2}\cos(\omega t + \theta) \quad (V = V_{\rm rms}).} \tag{8.51}$$

In fact, rms values of time-harmonic quantities are used much more frequently than their maximum values (amplitudes). Most instruments are calibrated to read rms values of measured quantities. For instance, if an ac voltmeter plugged into a household electric outlet reads 110 V, that is an rms voltage, so that the maximum value of the voltage at the outlet is $\sqrt{2} \times 110\ \text{V} \approx 155\ \text{V}$. In addition, it is very convenient to use rms values of field and circuit quantities in the expressions for time-average power and energy in the time-harmonic operation of electromagnetic systems. To illustrate this, note that the instantaneous power of Joule's (ohmic) losses in a resistor with a time-varying current can be expressed using Joule's law as

$$P_{\rm J}(t) = Ri^2(t), \tag{8.52}$$

where R is the resistance of the resistor and $i(t)$ is the instantaneous intensity of the current. If the current is time-harmonic, with amplitude I_0 and rms value $I = I_0/\sqrt{2}$, the time-average power of Joule's losses in the resistor turns out to be [see Eq. (8.50)]

time-average powers and energies in terms of rms quantities

$$\boxed{(P_{\rm J})_{\rm ave} = \frac{1}{T}\int_0^T P_{\rm J}(t)\,{\rm d}t = R\frac{1}{T}\int_0^T i^2(t)\,{\rm d}t = \frac{1}{2}RI_0^2 = RI^2.} \tag{8.53}$$

We see that the expression for $(P_{\rm J})_{\rm ave}$ via the rms current has the same form as the expression for the instantaneous power in Eq. (8.52) and looks the same as the expression for the power of Joule's losses in a resistor with a time-invariant (dc) current, Eq. (3.77), with I standing there for the time-constant current intensity. Generally, all expressions for time-average powers, energies, and power and energy densities in the time-harmonic operation in circuit theory and electromagnetics can be computed just as for the time-invariant operation, if rms values of currents, voltages, field intensities, and other circuit and field quantities are used. For example, the time-average power density of Joule's losses at a point in a conducting material can be obtained using Joule's law in local form, Eq. (3.31), for the static case with J and E now being the rms current density and electric field intensity, respectively, at the point. In the same manner, time-average electric and magnetic energy densities for linear media and time-harmonic field variations are given by the corresponding time-constant expressions in Eqs. (2.199) and (7.108) with E, D, H, and B now representing rms quantities. On the other hand, the expression for $(P_{\rm J})_{\rm ave}$ using the current amplitude (I_0) in Eq. (8.53) contains an additional factor 1/2, and this extra factor, in general, inconveniently appears in all kinds of time-average power and energy calculations under the steady-state sinusoidal assumption if peak (and not rms) values are used.

Time-harmonic expressions for electromagnetic quantities that vary also in space are written in a completely analogous way to that in Eq. (8.51), while keeping in mind that both the rms value and initial phase are, in general, functions of spatial coordinates. In addition, for a vector, separate expressions are written for each of its components. For example, if the Cartesian x-component of the electric

[4]As rms quantities will be used regularly throughout the rest of this text, we drop the subscripts ("rms") identifying them. With such convention, $V_{\rm rms}$ will be denoted simply as V, $H_{\rm rms}$ as H, and so on.

field intensity vector is a sinusoidal function of time and an arbitrary function of coordinates x, y, and z, we have

$$E_x(x, y, z, t) = E_x(x, y, z)\sqrt{2}\cos[\omega t + \theta_x(x, y, z)], \tag{8.54}$$

and similarly for E_y and E_z.

Conceptual Questions (on Companion Website): 8.21–8.23.

8.7 COMPLEX REPRESENTATIVES OF TIME-HARMONIC FIELD AND CIRCUIT QUANTITIES

Time-harmonic quantities can be graphically represented as uniformly rotating vectors. To show this, consider a vector of magnitude V_0 rotating in the Cartesian xy-plane about the coordinate origin with a constant angular velocity ω in the counter-clockwise (mathematically positive) direction, as in Fig. 8.6(a). If the angle between the vector and the x-axis at an instant $t = 0$ is θ, then this angle at an arbitrary instant t equals $\phi(t) = \omega t + \theta$ [see Eq. (6.88)]. This means that the projection of this vector on the x-axis equals $V_0 \cos(\omega t + \theta)$, that is, $v(t)$ in Eq. (8.47). Thus, all

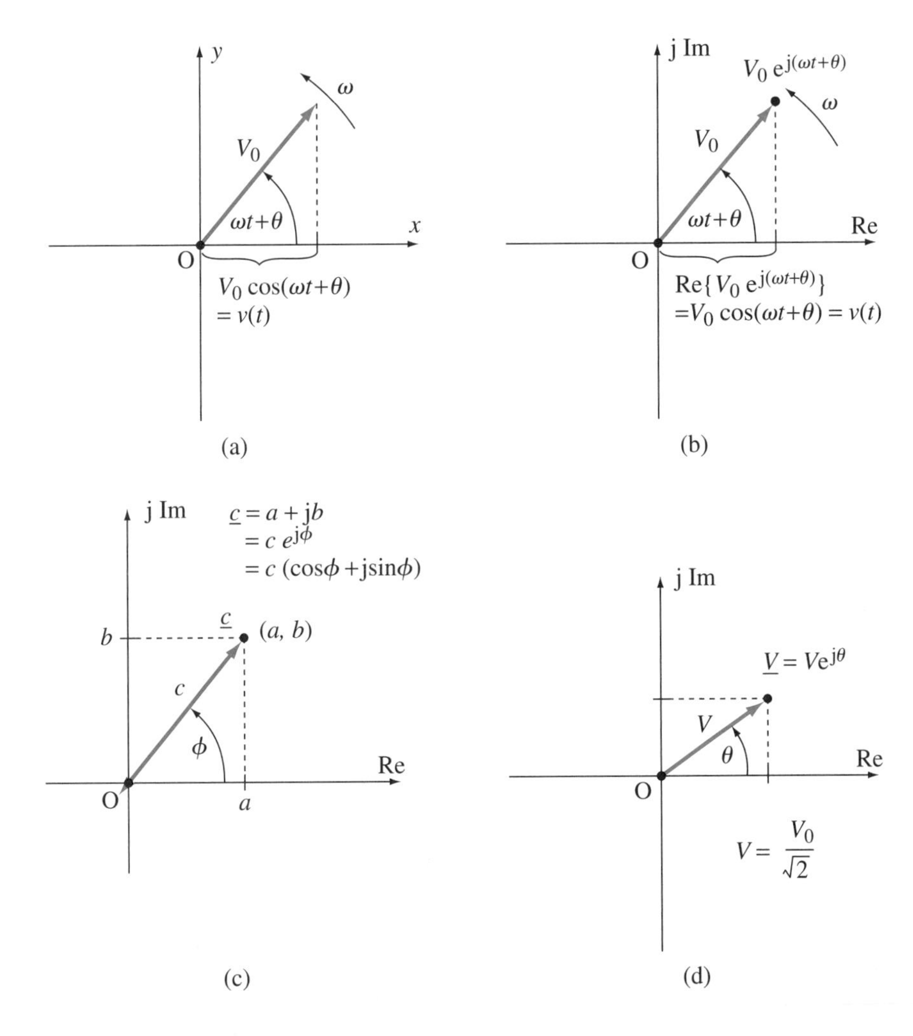

Figure 8.6 Representing time-harmonic quantities by phasors and complex numbers: (a) a phasor (rotating vector) whose magnitude and angular velocity equal the amplitude (peak-value) and angular frequency, respectively, of the instantaneous quantity [see Eq. (8.47)], (b) a complex number with magnitude (modulus) and phase angle (argument) equal to the amplitude and instantaneous phase of the instantaneous quantity, (c) different forms of a complex number, in general, and (d) the final adopted complex root-mean-square (rms) representative of a time-harmonic quantity with the time factor $e^{j\omega t}$ suppressed.

the quantities in a linear system with time-harmonic excitations can be visualized as projections of vectors with magnitudes equal to the amplitudes of the respective sinusoids that rotate in the same plane with a constant angular velocity, equal to the angular frequency of the sinusoids. These rotating vectors are called phasors. We can manipulate with phasors in a phasor diagram [drawing like the one in Fig. 8.6(a), but with all relevant quantities for the system under consideration represented] just as with any other geometrical vectors. For example, we can sum time-harmonic quantities with different amplitudes and different initial phases by vectorially adding their phasors and taking the projection of the resultant vector.

However, instead of drawing complicated phasor diagrams for realistic problems and geometrically or analytically manipulating with phasors (vectors), we can translate the problem from phasor to complex domain and employ complex algebra to analyze circuits and fields under the time-harmonic assumption. Namely, we can formally proclaim the x- and y-axes of Fig. 8.6(a) to be the real (Re) and imaginary (Im) axes of the complex plane, as indicated in Fig. 8.6(b), and use complex numbers to represent time-harmonic quantities.

A complex number $\underline{c}$ is a number composed of two real numbers, a and b.[5] Its rectangular (algebraic) form reads

complex number ($\underline{c}$); j – imaginary unit

$$\boxed{\underline{c} = a + \mathrm{j}b, \quad \text{where} \quad a = \mathrm{Re}\{\underline{c}\}, \quad b = \mathrm{Im}\{\underline{c}\}, \quad \text{and} \quad \mathrm{j} = \sqrt{-1},} \tag{8.55}$$

so a and b represent the real and imaginary parts, respectively, of $\underline{c}$, and j stands for the imaginary unit ($\mathrm{j}^2 = -1$). Every complex number corresponds to a point, (a, b), in the complex plane, as illustrated in Fig. 8.6(c), or, equivalently, to a vector representing the position vector of the point (a, b) with respect to the coordinate origin. Alternatively, $\underline{c}$ can be written in polar (exponential) form as

$\underline{c}$ in polar form (c – magnitude, ϕ – argument)

$$\boxed{\underline{c} = c\,\mathrm{e}^{\mathrm{j}\phi},} \tag{8.56}$$

where $c = |\underline{c}|$ is the magnitude (or modulus) of $\underline{c}$, that is, the magnitude of the vector $\underline{c}$ in Fig. 8.6(c), and ϕ is the phase angle (argument) of $\underline{c}$, which is the angle between the vector $\underline{c}$ and the real axis. From the right-angled triangle with arms $|a|$ and $|b|$ in Fig. 8.6(c), we obtain the following formulas for transforming a complex number from its rectangular to polar form:

$$c = \sqrt{a^2 + b^2} \quad \text{and} \quad \phi = \arg(a, b), \tag{8.57}$$

where the argument (arg) function equals the inverse tangent of b/a if the point (a, b) is in the first or fourth quadrant, while some modifications are necessary for the point in other quadrants,

$$\arg(a, b) = \begin{cases} \arctan(b/a) & \text{for } a > 0 \\ \pi/2 & \text{for } a = 0 \text{ and } b > 0 \\ -\pi/2 & \text{for } a = 0 \text{ and } b < 0 \\ \arctan(b/a) + \pi & \text{for } a < 0 \text{ and } b \geq 0 \\ \arctan(b/a) - \pi & \text{for } a < 0 \text{ and } b < 0 \\ \text{not defined} & \text{for } a = b = 0 \end{cases} \tag{8.58}$$

($\arctan \equiv \tan^{-1}$). From the same triangle,

$$a = c\cos\phi \quad \text{and} \quad b = c\sin\phi. \tag{8.59}$$

---

[5]In this text, letters denoting complex numbers and complex variables are underlined, which is in compliance with the recommendation of the International Electrotechnical Commission (IEC).

Combining Eqs. (8.55) and (8.59) leads to the following expression for $\underline{c}$:

$$\underline{c} = c(\cos\phi + \mathrm{j}\sin\phi), \tag{8.60}$$

which is referred to as the trigonometric form of a complex number. Omitting the magnitude of $\underline{c}$, we obtain the relation

$$\boxed{\mathrm{e}^{\mathrm{j}\phi} = \cos\phi + \mathrm{j}\sin\phi,} \tag{8.61}$$

Euler's identity

known as Euler's identity and frequently used in time-harmonic electromagnetics.

With the above review of complex numbers, it is now clear that the rotating vector in Fig. 8.6(a) can be identified as a complex number equal to $V_0\,\mathrm{e}^{\mathrm{j}\phi} = V_0\,\mathrm{e}^{\mathrm{j}(\omega t+\theta)}$. The projection of the vector on the real axis in Fig. 8.6(b) equals the real part of the complex number,

$$\mathrm{Re}\left\{V_0\,\mathrm{e}^{\mathrm{j}\phi}\right\} = V_0\cos\phi = V_0\cos(\omega t+\theta) = v(t), \tag{8.62}$$

which means that representations in Fig. 8.6(a) and Fig. 8.6(b) are indeed formally equivalent. As all the phasors representing time-harmonic quantities in a system rotate with the same angular velocity (all the quantities have the same frequency), they always have the same relative positions with respect to each other. Hence, the picture with all the vectors frozen at instant $t = 0$ contains all relevant data for the quantities in the system: their amplitudes, initial phases, and phase differences between individual quantities. We can, therefore, disregard rotation of phasors in Fig. 8.6(a) or increase in phase angles with time in Fig. 8.6(b), and use the complex representation shown in Fig. 8.6(d). In other words, as

$$\boxed{V_0\,\mathrm{e}^{\mathrm{j}(\omega t+\theta)} = V_0\,\mathrm{e}^{\mathrm{j}\theta}\,\mathrm{e}^{\mathrm{j}\omega t},} \tag{8.63}$$

extraction of the common time factor

all the quantities in the system contain the same factor $\mathrm{e}^{\mathrm{j}\omega t}$, which then appears on both sides of all of the field/circuit equations governing the system. Ignoring the rotation in time of vectors in Fig. 8.6(d) is thus equivalent to dropping the time factor $\mathrm{e}^{\mathrm{j}\omega t}$ from the associated equations. The resulting complex term $V_0\,\mathrm{e}^{\mathrm{j}\theta}$, which does not contain time, is called the complex magnitude of the instantaneous voltage $v(t)$. Dividing it by $\sqrt{2}$, we obtain the complex root-mean-square value of $v(t)$,

$$\boxed{\underline{V} = \frac{V_0}{\sqrt{2}}\,\mathrm{e}^{\mathrm{j}\theta} = V\,\mathrm{e}^{\mathrm{j}\theta}.} \tag{8.64}$$

complex rms value

Inversely, the instantaneous voltage, Eq. (8.51), can be obtained from the complex rms voltage as

$$\boxed{v(t) = \mathrm{Re}\left\{\underline{V}\sqrt{2}\,\mathrm{e}^{\mathrm{j}\omega t}\right\}.} \tag{8.65}$$

instantaneous from complex rms voltage

We shall use complex rms values as complex representatives of time-harmonic quantities, Fig. 8.6(d). The above equation pair, Eqs. (8.64) and (8.65), can be regarded as a direct/inverse transform pair for switching between time and complex domains. The time-complex correspondence can also be summarized as

$$\boxed{\underbrace{V\sqrt{2}\cos(\omega t+\theta)}_{\text{instantaneous } v(t)} \quad\longleftrightarrow\quad \underbrace{V\,\mathrm{e}^{\mathrm{j}\theta}}_{\text{complex } \underline{V}}.} \tag{8.66}$$

time-complex conversion

In words, the magnitude and phase angle (argument) of the complex representative equal the rms value and initial phase of the time-harmonic quantity, respectively.

The magnitude of the complex quantity is represented with the rms value rather than with the amplitude of the corresponding instantaneous quantity because engineers and scientists normally assume that all time-harmonic quantities are reported as rms values and, as already mentioned and illustrated on two examples (reading of instruments and computation of time-average power) in the previous section, it is more convenient to deal with rms quantities.

From Eq. (8.65),

$$\frac{\mathrm{d}v}{\mathrm{d}t} = \frac{\mathrm{d}}{\mathrm{d}t}\mathrm{Re}\left\{\underline{V}\sqrt{2}\,\mathrm{e}^{\mathrm{j}\omega t}\right\} = \mathrm{Re}\left\{\underline{V}\sqrt{2}\,\frac{\mathrm{d}}{\mathrm{d}t}\,\mathrm{e}^{\mathrm{j}\omega t}\right\} = \mathrm{Re}\left\{\mathrm{j}\omega\underline{V}\sqrt{2}\,\mathrm{e}^{\mathrm{j}\omega t}\right\}, \tag{8.67}$$

which means that taking the time derivative of the instantaneous quantity in the time domain is equivalent to multiplying its representative by $\mathrm{j}\omega$ in the complex domain, that is,

replacement of time derivatives

$$\boxed{\frac{\mathrm{d}v}{\mathrm{d}t} \quad\longleftrightarrow\quad \mathrm{j}\omega\underline{V}.} \tag{8.68}$$

This feature of the time-complex conversion allows us to replace all time derivatives in field/circuit equations by the factor $\mathrm{j}\omega$ (second time derivatives are replaced by $\mathrm{j}\omega\times\mathrm{j}\omega = -\omega^2$), which enormously simplifies the analysis. For instance, the complex-domain equivalent of the time-domain element law for an inductor, Eq. (7.3), reads

complex current-voltage characteristic of an inductor

$$\boxed{\underline{V} = \mathrm{j}\omega L\underline{I},} \tag{8.69}$$

i.e., its form is as simple as that of Ohm's law for a resistor. Namely, Eq. (8.69) tells us that the voltage of the inductor equals a constant times the current, the same as in Eq. (3.72), but with all the quantities now being complex. Similarly,

$$\int v\,\mathrm{d}t \quad\longleftrightarrow\quad \frac{\underline{V}}{\mathrm{j}\omega}, \tag{8.70}$$

showing that integration in time is equivalent to division by $\mathrm{j}\omega$ in the complex domain.

For time-harmonic electromagnetic quantities that change also in space, the time-complex conversion is performed in a usual way, as the spatial dependences of the rms value and initial phase are, of course, not affected by the conversion. In the case of time-harmonic spatially distributed vectors, the conversion is done for each vector component separately (three components in general). For example, the complex representative (complex rms value) of the Cartesian vector component $E_x(x, y, z, t)$ in Eq. (8.54) is

$$\underline{E}_x(x, y, z) = E_x(x, y, z)\,\mathrm{e}^{\mathrm{j}\theta_x(x, y, z)}. \tag{8.71}$$

Analogous expressions hold for the y- and z-components of $\mathbf{E}$, and the complex electric field intensity vector is given by

$$\underline{\mathbf{E}}(x, y, z) = \underline{E}_x(x, y, z)\,\hat{\mathbf{x}} + \underline{E}_y(x, y, z)\,\hat{\mathbf{y}} + \underline{E}_z(x, y, z)\,\hat{\mathbf{z}}. \tag{8.72}$$

Of course,

$$E_x(x, y, z, t) = \mathrm{Re}\left\{\underline{E}_x(x, y, z)\sqrt{2}\,\mathrm{e}^{\mathrm{j}\omega t}\right\}, \tag{8.73}$$

and similarly for the other two components of $\mathbf{E}$. Hence, the following relationship can be established between an instantaneous time-harmonic vector and its complex rms representative:

$\underline{\mathbf{E}}$ – complex electric field vector

$$\boxed{\mathbf{E}(x, y, z, t) = \mathrm{Re}\left\{\underline{\mathbf{E}}(x, y, z)\sqrt{2}\,\mathrm{e}^{\mathrm{j}\omega t}\right\}.} \tag{8.74}$$

Note that a complex vector, in general, is a set of six numbers, three real and three imaginary parts of its components. This is why a complex vector, unlike its instantaneous counterpart, cannot be drawn as an arrow in space, except in some special cases (as we shall see in the next chapter).

Example 8.9 Time-to-Complex and Complex-to-Time Transformations

Find (a) the complex rms equivalent of a time-harmonic voltage given by $v(t) = 2\sqrt{2}\sin(10^8 t + \pi/3)$ V (t in s) and (b) the instantaneous current $i(t)$ if $\underline{I} = (-1 + \mathrm{j})$ A in the complex domain and the angular frequency is $\omega = 5 \times 10^6$ rad/s.

Solution

(a) We use the cosine reference for representing time-harmonic quantities, Eq. (8.51), and, therefore, the instantaneous voltage $v(t)$ needs first to be written as a cosine function:

$$v(t) = 2\sqrt{2}\sin\left(10^8 t + \frac{\pi}{3}\right) \text{ V} = 2\sqrt{2}\cos\left(10^8 t - \frac{\pi}{6}\right) \text{ V} \quad (t \text{ in s}), \tag{8.75}$$

where the trigonometric identity $\sin\alpha = \cos(\alpha - \pi/2)$ is employed. According to Eq. (8.66), the complex rms voltage is then found to be

$$\underline{V} = 2\,\mathrm{e}^{-\mathrm{j}\pi/6} \text{ V} = 2\left(\cos\frac{\pi}{6} - \mathrm{j}\sin\frac{\pi}{6}\right) \text{ V} = (\sqrt{3} - \mathrm{j}) \text{ V}. \tag{8.76}$$

(b) The magnitude of the complex current $\underline{I}$ is

$$I = |\underline{I}| = \sqrt{(-1)^2 + 1^2} \text{ A} = \sqrt{2} \text{ A}. \tag{8.77}$$

As the point $(-1, 1)$ lies in the second quadrant of the complex plane, we refer to Eq. (8.58) to obtain the phase angle (argument) of $\underline{I}$ as follows:

$$\psi = \arg(-1, 1) = \arctan(-1) + \pi = -\frac{\pi}{4} + \pi = \frac{3\pi}{4} \tag{8.78}$$

(this can also be read directly from the position of the complex number $-1 + \mathrm{j}$ in the complex plane). Hence, the instantaneous current is [see Eq. (8.66)]

$$i(t) = 2\cos\left(5 \times 10^6 t + \frac{3\pi}{4}\right) \text{ A} \quad (t \text{ in s}). \tag{8.79}$$

Problems: 8.11 and 8.12; *Conceptual Questions* (on Companion Website): 8.24; *MATLAB Exercises* (on Companion Website).

8.8 MAXWELL'S EQUATIONS IN COMPLEX DOMAIN

We now assume the steady-state sinusoidal regime for the rapidly time-varying electromagnetic field, which gives rise to the high-frequency time-harmonic electromagnetic field, and convert the associated time-domain Maxwell's equations into their complex equivalents. We use the time-complex correspondence rule compactly expressed by Eq. (8.66), and elaborated in Eqs. (8.64)–(8.74). Of course, complex equations make sense only for linear electromagnetic media (of parameters ε, μ, and σ). From Eqs. (8.23), we thus obtain the following complete set of integral Maxwell's equations in the complex domain (with the constitutive equations added):

Maxwell's equations in complex domain, integral form

$$\begin{cases} \oint_C \underline{\mathbf{E}} \cdot \mathrm{d}\mathbf{l} = -\mathrm{j}\omega \int_S \underline{\mathbf{B}} \cdot \mathrm{d}\mathbf{S} \\ \oint_C \underline{\mathbf{H}} \cdot \mathrm{d}\mathbf{l} = \int_S (\underline{\mathbf{J}} + \mathrm{j}\omega \underline{\mathbf{D}}) \cdot \mathrm{d}\mathbf{S} \\ \oint_S \underline{\mathbf{D}} \cdot \mathrm{d}\mathbf{S} = \int_v \underline{\rho}\, \mathrm{d}v \\ \oint_S \underline{\mathbf{B}} \cdot \mathrm{d}\mathbf{S} = 0 \\ \underline{\mathbf{D}} = \varepsilon \underline{\mathbf{E}},\ \underline{\mathbf{B}} = \mu \underline{\mathbf{H}},\ \underline{\mathbf{J}} = \sigma(\underline{\mathbf{E}} + \underline{\mathbf{E}}_\mathrm{i}) \end{cases}, \tag{8.80}$$

where the external energy sources are included as an impressed electric field, of complex rms intensity $\underline{\mathbf{E}}_\mathrm{i}$ [Eq. (3.109)]. Note that the material parameters, in general, are functions of frequency, $\varepsilon = \varepsilon(\omega)$, $\mu = \mu(\omega)$, and $\sigma = \sigma(\omega)$, which will be discussed in the following chapter. In the low-frequency case, the displacement-current term, $\mathrm{j}\omega\underline{\mathbf{D}}$, in the second equation can be neglected with respect to the conduction-current term, that is, we can assume that $\omega \approx 0$ on the right-hand side of the equation. The complete set of differential Maxwell's equations for the high-frequency time-harmonic electromagnetic field in the complex domain, equivalent to Eqs. (8.24), reads

Maxwell's equations in complex domain, differential form

$$\begin{cases} \nabla \times \underline{\mathbf{E}} = -\mathrm{j}\omega \underline{\mathbf{B}} \\ \nabla \times \underline{\mathbf{H}} = \underline{\mathbf{J}} + \mathrm{j}\omega \underline{\mathbf{D}} \\ \nabla \cdot \underline{\mathbf{D}} = \underline{\rho} \\ \nabla \cdot \underline{\mathbf{B}} = 0 \end{cases}. \tag{8.81}$$

The conversion of the boundary conditions in Eqs. (8.32) and different forms of the continuity equation, Eqs. (8.34), (8.38), and (8.42), from time to complex domain is also straightforward. For instance, the complex-domain continuity equation in differential form for high-frequency time-harmonic volume currents is given by

$$\nabla \cdot \underline{\mathbf{J}} = -\mathrm{j}\omega\underline{\rho}. \tag{8.82}$$

All these equations contain no time and, most importantly, no time derivatives (nor integrals), and are, therefore, considerably simpler to work with than the corresponding equations in the time domain. Note that, as an alternative to the analysis in the time domain, the electromagnetic analysis using complex representatives is often referred to as the analysis in the frequency domain. In the rest of this text, we shall deal predominantly with time-harmonic fields and signals and use frequency-domain field and circuit equations and solution techniques.

Phasors and complex representatives can also be used for analyzing linear electromagnetic systems and electric circuits driven by arbitrary (nonsinusoidal) periodic time functions (such as a sequence of rectangular pulses). By expanding these functions (excitations) into a Fourier series of sinusoidal components of different frequencies, we can solve for desired field and circuit quantities (responses) using the complex-domain analysis for each Fourier component separately. By virtue of the superposition principle (which holds for linear systems), the total solution for any of the quantities is obtained as the sum of the partial (single-frequency) solutions due to all of the excitation components. Moreover, transient nonperiodic functions (such as a single pulse in time) can be expressed as Fourier integrals, and a similar application of the principle of superposition can be used as well. Finally, in many practical applications with more than one frequency present in the spectrum of time-varying signals (fields) in a system, the bandwidth of the signals is in fact very small, with a carrier frequency accompanied by some form of modulation giving a narrow spread of frequencies around the carrier. In such cases, the analysis in the complex domain at a single frequency, that of the carrier, is usually sufficient for the characterization of the system and evaluation of the fields and other quantities of interest, provided that the material parameters in the system do not vary significantly with frequency over the bandwidth.

Example 8.10 No Time-Harmonic Charge in Homogeneous Media

Prove that the interior of a homogeneous lossy medium with a time-harmonic electromagnetic field and no external energy volume sources (impressed electric fields or currents) is always charge-free.

Solution Transforming the charge-relaxation differential equation for a homogeneous medium of conductivity σ and permittivity ε, Eq. (3.66), to the complex domain, we have

$$\mathrm{j}\omega\underline{\rho} + \frac{\sigma}{\varepsilon}\underline{\rho} = \underline{\rho}\,\frac{\sigma + \mathrm{j}\omega\varepsilon}{\varepsilon} = 0 \qquad (8.83)$$

[note that this equation can also be derived directly from differential Maxwell's equations and the continuity equation in the complex domain, Eqs. (8.81) and (8.82)]. As σ, ε, and ω are all nonnegative real quantities, this condition can be satisfied only if $\underline{\rho} = 0$, which proves that there cannot be volume excess time-harmonic charges in homogeneous lossy media.

In addition, the differential form of the generalized Gauss' law in the complex domain can now be written as

$$\nabla \cdot \underline{\mathbf{D}} = \varepsilon \nabla \cdot \underline{\mathbf{E}} = 0, \qquad (8.84)$$

which means that the time-harmonic electric field in a homogeneous conducting medium is always divergenceless (div $\mathbf{E} = 0$).

Note that this is true for any frequency, that is, for both high-frequency and low-frequency time-harmonic fields, including the static case [see Eq. (3.62)].

Example 8.11 Complex Surface Charge Density on a Material Interface

Find the complex surface charge density on the boundary surface between medium 1 (with conductivity σ_1 and permittivity ε_1) and medium 2 (with σ_2 and ε_2), if the normal component of the complex current density close to the boundary in medium 1 is $\underline{J}_{1n}$ (defined with respect to the normal directed from medium 2 to medium 1) and the angular frequency of the field is ω.

Solution The solution procedure is similar to that in Eq. (3.63), given for the static case, and we write

$$\underline{\rho}_s = \hat{\mathbf{n}} \cdot \underline{\mathbf{D}}_1 - \hat{\mathbf{n}} \cdot \underline{\mathbf{D}}_2 = \frac{\varepsilon_1}{\sigma_1}\hat{\mathbf{n}} \cdot \underline{\mathbf{J}}_1 - \frac{\varepsilon_2}{\sigma_2}\hat{\mathbf{n}} \cdot \underline{\mathbf{J}}_2 = \frac{\varepsilon_1}{\sigma_1}\underline{J}_{1n} - \frac{\varepsilon_2}{\sigma_2}\underline{J}_{2n}. \qquad (8.85)$$

However, $\mathbf{J}_n$, in general, is not continuous across material interfaces in the dynamic case, so we use here the high-frequency form of the boundary condition for normal components of the current density vector [the third equation of Eqs. (8.34)], whose complex-domain version is

$$\hat{\mathbf{n}} \cdot \underline{\mathbf{J}}_1 - \hat{\mathbf{n}} \cdot \underline{\mathbf{J}}_2 = -\mathrm{j}\omega\underline{\rho}_s \quad \text{or} \quad \underline{J}_{1n} - \underline{J}_{2n} = -\mathrm{j}\omega\underline{\rho}_s. \qquad (8.86)$$

Solving for $\underline{J}_{2n}$,

$$\underline{J}_{2n} = \underline{J}_{1n} + \mathrm{j}\omega\underline{\rho}_s, \qquad (8.87)$$

and substituting this expression in Eq. (8.85), we obtain

$$\underline{\rho}_s = \frac{\varepsilon_1/\sigma_1 - \varepsilon_2/\sigma_2}{1 + j\omega\varepsilon_2/\sigma_2} \underline{J}_{1n}. \tag{8.88}$$

Problems: 8.13–8.16.

8.9 LORENZ ELECTROMAGNETIC POTENTIALS

Consider an arbitrary distribution of rapidly time-varying volume currents and charges in a source domain of volume v, as shown in Fig. 8.7. Let the current and charge densities, $\mathbf{J}$ and ρ, in v be known functions of spatial coordinates and time. With $\mathbf{r}'$ denoting the position vector of a source point P′ with respect to the coordinate origin (O), we can write $\mathbf{J} = \mathbf{J}(\mathbf{r}', t)$ and $\rho = \rho(\mathbf{r}', t)$. Of course, the source distributions $\mathbf{J}$ and ρ cannot be specified independently, but must be related to one another through the continuity equation. Our goal is to derive the expressions for the electric and magnetic field intensity vectors due to these sources, at an arbitrary point, P, in space (field or observation point), assuming that the medium in which the sources reside is linear, homogeneous, and lossless, with permittivity ε and permeability μ ($\sigma = 0$). Instead of solving the original form of Maxwell's equations directly for the fields, we shall first evaluate the electromagnetic potentials, namely, the electric scalar potential, V,[6] and magnetic vector potential, $\mathbf{A}$, at the point P, as an intermediate step in finding fields. Once the potentials, as functions of spatial coordinates and time, so $V = V(\mathbf{r}, t)$ and $\mathbf{A} = \mathbf{A}(\mathbf{r}, t)$, $\mathbf{r}$ being the position vector of the point P with respect to O, are known, the field vectors $\mathbf{E}$ and $\mathbf{B}$ can be evaluated (by differentiation in space and time) using Eqs. (6.43) and (6.28).

In the slowly time varying case, the potentials (for $\varepsilon = \varepsilon_0$ and $\mu = \mu_0$) are given by Eqs. (6.19) and (6.29). These are quasistatic potentials (evaluated in the same way as static ones), with the time retardation (lagging in time of time-varying potentials and fields behind their sources) being neglected. In the rapidly time-varying case, however, the retardation effect must be taken into account [time variations in the system are so fast that even slight time delays between the sources and fields (potentials) cannot be ignored]. Mathematically, the time retardation in the expressions for the potentials and field vectors is a consequence of the addition of the displacement-current term, $\partial\mathbf{D}/\partial t$, in Maxwell's equations for the rapidly time-varying electromagnetic field. In this section, we derive and discuss the time-domain

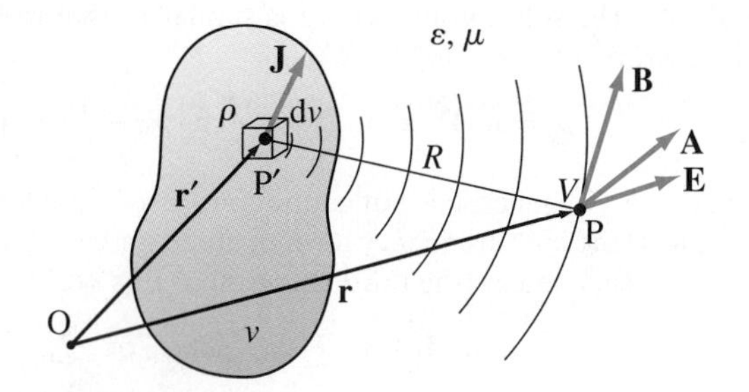

Figure 8.7 Evaluation of electromagnetic potentials and field vectors due to rapidly time-varying volume currents and charges in a linear, homogeneous, and lossless medium.

[6] As mentioned in Section 1.6, the electric potential is symbolized by Φ as well.

expressions for V and $\mathbf{A}$ in the most general case (for arbitrary time variations of sources), while the complex expressions for time-harmonic potentials and fields at high frequencies will be derived and used in the next section.

Let us substitute Eq. (6.43) into the generalized Gauss' law in differential form [the third equation of Eqs. (8.24)] with $\mathbf{D}$ previously replaced by $\varepsilon\mathbf{E}$. The result is

$$\nabla \cdot (\nabla V) = \nabla^2 V = -\frac{\rho}{\varepsilon} - \frac{\partial}{\partial t}(\nabla \cdot \mathbf{A}), \tag{8.89}$$

where ε and $\partial/\partial t$ can be brought outside the divergence sign because the permittivity is a constant (the medium is homogeneous) and the time derivative operator is entirely independent from the divergence operator (which implies spatial differentiation), respectively, and the identity in Eq. (2.92) defining the Laplacian operator (∇^2) is used [the expression for $\nabla^2 V$ in the Cartesian coordinate system is given in Eq. (2.94)]. In a similar manner, substituting Eqs. (6.28) and (6.43) into the corrected generalized Ampère's law in differential form [the second equation of Eqs. (8.24)] with $\mathbf{H}$ represented as $\mathbf{B}/\mu$ leads to the following equation:

$$\nabla \times (\nabla \times \mathbf{A}) = \nabla(\nabla \cdot \mathbf{A}) - \nabla^2 \mathbf{A} = \mu\mathbf{J} - \varepsilon\mu\frac{\partial^2 \mathbf{A}}{\partial t^2} - \varepsilon\mu\nabla\frac{\partial V}{\partial t}, \tag{8.90}$$

where the use is made of the identity in Eq. (4.124) defining the Laplacian of a vector function [Eq. (4.126) gives the expression for $\nabla^2\mathbf{A}$ in Cartesian coordinates].

In order for the vector potential $\mathbf{A}$ to be uniquely defined, we must specify both its curl and its divergence. While the curl of $\mathbf{A}$ is already specified as $\nabla \times \mathbf{A} = \mathbf{B}$, we are at liberty to choose $\operatorname{div}\mathbf{A}$ in an arbitrary way. The choice that leads to the solution of Eqs. (8.89) and (8.90) that is mathematically the simplest and physically the most meaningful is given by

$$\boxed{\nabla \cdot \mathbf{A} = -\varepsilon\mu\frac{\partial V}{\partial t}.} \tag{8.91}$$

Lorenz condition (gauge) for potentials

This differential relation between $\mathbf{A}$ and V is termed the Lorenz condition (or Lorenz gauge) for electromagnetic potentials. It is consistent with the analogous relation between the source distributions $\mathbf{J}$ and ρ given by the continuity equation in differential form in Eqs. (8.34) and is also called the continuity equation for potentials. With $\nabla \cdot \mathbf{A}$ replaced by the expression from Eq. (8.91), Eqs. (8.89) and (8.90) become

$$\nabla^2 V - \varepsilon\mu\frac{\partial^2 V}{\partial t^2} = -\frac{\rho}{\varepsilon}, \tag{8.92}$$

wave equation for V

$$\nabla^2 \mathbf{A} - \varepsilon\mu\frac{\partial^2 \mathbf{A}}{\partial t^2} = -\mu\mathbf{J}. \tag{8.93}$$

wave equation for $\mathbf{A}$

These are second-order partial differential equations with spatial coordinates and time as independent variables. They are called the wave equations for the potentials, because, as we shall see, their solutions represent waves traveling from source points toward field points in the system. What is very important, the equations are uncoupled, that is, each wave equation has only one of the potentials as unknown. Moreover, although one of them is a scalar equation and the other a vector one, they have practically identical forms, so that the scalar solution for V in terms of ρ (more precisely, ρ/ε) will have the same form as the vector solution for $\mathbf{A}$ in terms of $\mathbf{J}$ (or $\mu\mathbf{J}$). For instance, we can decompose Eq. (8.93) into three scalar differential equations in the Cartesian coordinate system, with the equation for the x-component of the magnetic vector potential being

$$\nabla^2 A_x - \varepsilon\mu \frac{\partial^2 A_x}{\partial t^2} = -\mu J_x \tag{8.94}$$

and analogous equations for the y- and z-components. Eqs. (8.94) and (8.92) have the same form, and we can use the solution for V due to ρ to directly write the analogous solutions for A_x, A_y, and A_z due to J_x, J_y, and J_z, respectively. In what follows, therefore, we solve only Eq. (8.92) for the potential V.

HISTORICAL ASIDE

Ludwig Valentine Lorenz (1829–1891), a Danish mathematician and physicist, was a professor at the Royal Military Academy in Copenhagen. In his studies of light based on Maxwell's electromagnetic field equations, Lorenz proposed in 1867 vector and scalar potentials that included the time of propagation from the sources – the retarded potentials. He showed that such potentials were related to each other by the condition given in Eq. (8.91), which became known as the Lorenz condition or Lorenz gauge. This was published in his paper "On the Identity of the Vibrations of Light With Electrical Currents" (*Annalen der Physik und Chemie*, June 1867). The retarded potentials themselves are also referred to as the Lorenz potentials. Note, however, that it is a common mistake in just about all textbooks and research articles dealing with retarded potentials that Dutch physicist and Nobelist Hendrik A. Lorentz (1853–1928), whose historical aside appears in Section 4.12 of this text, is erroneously credited for the invention of the Lorenz gauge and Lorenz potentials. Lorenz, now jointly with Lorentz, is also known for the so-called Lorentz-Lorenz formula for the index of refraction as a function of the density of a medium, devised independently by the two scientists.

In our solution procedure, we subdivide the domain v into differentially small cells of volume dv and evaluate the electric scalar potential due to a charge $\rho\, dv$ in a cell whose center is at a point P′ (source point) in Fig. 8.7. By the superposition principle, the total potential equals the sum (integral) of the elementary potentials due to all of the cells over v. While considering the potential due to the charged cell at P′ alone, $\rho = 0$ everywhere outside the cell, so that the potential at the field point P outside the cell satisfies the version of Eq. (8.92) with the source term on the right-hand side of the equation annulled. In addition, as the charged cell can be considered as a point charge at the point P′, the function V must be spherically symmetrical with respect to that point, i.e., it depends only on the distance R between P′ and P. Therefore, we use the expression for the Laplacian of V in the spherical coordinate system, Eq. (2.97), with the coordinate origin at P′, so $r = R$, and the terms containing partial derivatives with respect to θ and ϕ dropped, which simplifies the source-free version of Eq. (8.92) to

$$\frac{1}{R^2}\frac{\partial}{\partial R}\left(R^2\frac{\partial V}{\partial R}\right) - \varepsilon\mu\frac{\partial^2 V}{\partial t^2} = 0 \quad (\rho = 0). \tag{8.95}$$

With a new variable defined by

$$V = \frac{U}{R}, \tag{8.96}$$

it further becomes

1-D source-free wave equation; c – velocity of electromagnetic waves

$$\boxed{\frac{\partial^2 U}{\partial R^2} - \frac{1}{c^2}\frac{\partial^2 U}{\partial t^2} = 0, \quad \text{where} \quad c = \frac{1}{\sqrt{\varepsilon\mu}}.} \tag{8.97}$$

The constant c is the velocity of propagation of electromagnetic disturbances in the medium (of parameters ε and μ). If the medium is air (vacuum), substituting the values for ε_0 and μ_0 from Eqs. (1.2) and (4.3) yields $c_0 = 1/\sqrt{\varepsilon_0\mu_0} \approx 3 \times 10^8$ m/s, which equals the speed of light (and other electromagnetic waves) in free space. Eq. (8.97) is the well-known one-dimensional source-free wave equation (it is one-dimensional because it has only one spatial dimension or "degree of freedom," that in terms of the variable R). In various notations, it is often encountered in electromagnetics, as well as in other areas of science and engineering.

Any twice-differentiable function f of the variable $t' = t - R/c$ is a solution of Eq. (8.97), which can be verified by direct substitution. Namely, writing U as

$$U(R,t) = f(t') = f\left(t - \frac{R}{c}\right), \tag{8.98}$$

and using the chain rule for taking derivatives, the partial derivatives of U with respect to t and R can be expressed as

$$\frac{\partial U}{\partial t} = \frac{\mathrm{d}f}{\mathrm{d}t'}\frac{\partial t'}{\partial t} = \frac{\mathrm{d}f}{\mathrm{d}t'}, \quad \frac{\partial U}{\partial R} = \frac{\mathrm{d}f}{\mathrm{d}t'}\frac{\partial t'}{\partial R} = -\frac{1}{c}\frac{\mathrm{d}f}{\mathrm{d}t'}. \tag{8.99}$$

In an analogous fashion, the expressions for the second partial derivatives are

$$\frac{\partial^2 U}{\partial t^2} = \frac{\partial}{\partial t}\frac{\mathrm{d}f}{\mathrm{d}t'} = \frac{\mathrm{d}^2 f}{\mathrm{d}t'^2}\frac{\partial t'}{\partial t} = \frac{\mathrm{d}^2 f}{\mathrm{d}t'^2}, \quad \frac{\partial^2 U}{\partial R^2} = \frac{\partial}{\partial R}\left(-\frac{1}{c}\frac{\mathrm{d}f}{\mathrm{d}t'}\right) = -\frac{1}{c}\frac{\mathrm{d}^2 f}{\mathrm{d}t'^2}\frac{\partial t'}{\partial R} = \frac{1}{c^2}\frac{\mathrm{d}^2 f}{\mathrm{d}t'^2}. \tag{8.100}$$

Substituting these expressions into Eq. (8.97), we see that the wave equation is indeed satisfied for the function f in Eq. (8.98).[7] In addition, noting that

$$U(R+\Delta R, t+\Delta t) = f\left(t + \Delta t - \frac{R+\Delta R}{c}\right) = f\left(t - \frac{R}{c}\right) \quad \text{if} \quad \Delta R = c\Delta t, \tag{8.101}$$

we conclude that after a time Δt, the function f retains the same value at a point that is $\Delta R = c\Delta t$ away from the previous position in space (defined by R), as illustrated in Fig. 8.8. This means that an arbitrary function of the form $f(t - R/c)$ represents a wave traveling with a velocity c ($c = \Delta R/\Delta t = 1/\sqrt{\varepsilon\mu}$) in the positive R direction as the time t advances. From Eq. (8.96), we now have

$$V = \frac{U}{R} = \frac{f(t - R/c)}{R}. \tag{8.102}$$

The particular solution for the function $f(t')$ is determined from the condition that its value in the vicinity of the source point must be the same as the quasistatic

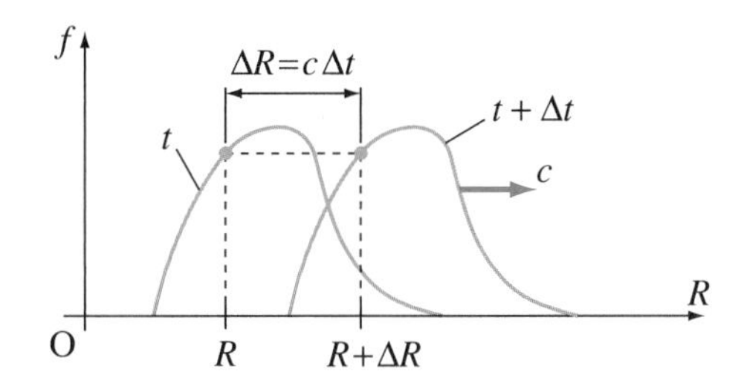

Figure 8.8 Sketch of a function $f(t - R/c)$ at two instants of time, according to Eq. (8.101), which illustrates that any function of this form represents a wave traveling with velocity c along the R-axis.

[7] It can be verified in a similar way that any twice-differentiable function of the form $f(t + R/c)$ is also a solution of Eq. (8.97). However, as we shall see later in this section, such a function does not correspond to a physically meaningful solution for the potentials, and that is why we do not consider it in the analysis.

solution for the potential V due to $\rho\,\mathrm{d}v$ as $R \to 0$. By means of Eq. (6.19), the quasistatic potential is given by

$$\frac{f(t)}{R} = \frac{1}{4\pi\varepsilon}\frac{\rho(t)\,\mathrm{d}v}{R} \quad (R \to 0). \tag{8.103}$$

Hence, for an arbitrary R,

$$f(t') = \frac{1}{4\pi\varepsilon}\rho(t')\,\mathrm{d}v \quad \longrightarrow \quad V = \frac{1}{4\pi\varepsilon}\frac{\rho(t - R/c)\,\mathrm{d}v}{R}. \tag{8.104}$$

Finally, the total electric scalar potential is

retarded electric scalar potential

$$\boxed{V(t) = \frac{1}{4\pi\varepsilon}\int_v \frac{\rho(t - R/c)\,\mathrm{d}v}{R},} \tag{8.105}$$

where, to shorten the writing, by $V(t)$ we actually mean $V(\mathbf{r}, t)$, and $\rho(t - R/c)$ stands for $\rho(\mathbf{r}', t - R/c)$ in Fig. 8.7. Note also that the source-to-field distance R in the integral can be expressed as $R = |\mathbf{r} - \mathbf{r}'|$.

By analogy (duality), the solution of Eq. (8.94) for the x-component of the magnetic vector potential is

$$A_x(t) = \frac{\mu}{4\pi}\int_v \frac{J_x(t - R/c)\,\mathrm{d}v}{R}, \tag{8.106}$$

which gives rise to the following expression for the total magnetic vector potential at the point P:

retarded magnetic vector potential

$$\boxed{\mathbf{A}(t) = \frac{\mu}{4\pi}\int_v \frac{\mathbf{J}(t - R/c)\,\mathrm{d}v}{R}.} \tag{8.107}$$

The potentials given by Eqs. (8.105) and (8.107) are called the Lorenz or retarded electromagnetic potentials. Since the time t in the equations is the time at the field (observation) point (P), while $t' = t - R/c$ is the time at the source point (P$'$), we conclude that there is a time delay (retardation) between the sources and the potentials (and therefore the fields) equal to

time lag between source and observation points

$$\boxed{\tau = \frac{R}{c}.} \tag{8.108}$$

In other words, the electromagnetic disturbances caused by a time variation of elementary sources $\rho\,\mathrm{d}v$ and $\mathbf{J}\,\mathrm{d}v$ at the point P$'$ propagate over the distance R in the form of electromagnetic waves of velocity c and are conveyed to the potentials (and fields) at the point P after the propagation time τ in Eq. (8.108).[8] The velocity of waves, given in Eq. (8.97), depends only on the parameters ε and μ of the medium in which the sources are embedded. As the waves originated by the elementary sources at the point P$'$ expand at the same rate in all radial directions with respect to P$'$ and the corresponding disturbances have uniform properties at all points of a sphere of radius R (Fig. 8.7), they belong to a class of so-called spherical electromagnetic waves. In addition, we note that the magnitudes of potentials due to individual elementary sources are inversely proportional to the distance R from the sources.

Problems: 8.17 and 8.18; *Conceptual Questions* (on Companion Website): 8.25.

[8] It is now clear that a function of the argument $t' = t + R/c$ cannot be a physically useful solution, since it would imply a negative time delay, i.e., that the effects of ρ and $\mathbf{J}$ are felt at the field point before they occur at the source point.

8.10 COMPUTATION OF HIGH-FREQUENCY POTENTIALS AND FIELDS IN COMPLEX DOMAIN

We now assume a steady-state sinusoidal regime in the system in Fig. 8.7 at high frequencies and convert the time-domain expressions for the Lorenz potentials in Eqs. (8.105) and (8.107) into their complex equivalents. To see how the time retardation is translated into the complex domain, consider a time-harmonic current of intensity $i(t) = I\sqrt{2}\cos(\omega t + \psi)$ and note that

$$i(t-\tau) = I\sqrt{2}\cos[\omega(t-\tau)+\psi] = \mathrm{Re}\left\{\underline{I}\sqrt{2}\,\mathrm{e}^{\mathrm{j}\omega(t-\tau)}\right\} = \mathrm{Re}\left\{\underline{I}\sqrt{2}\,\mathrm{e}^{-\mathrm{j}\omega\tau}\,\mathrm{e}^{\mathrm{j}\omega t}\right\}, \tag{8.109}$$

where $\underline{I}$ is the complex rms value of the current [see Eqs. (8.64) and (8.65)]. This means that shifting (retardation) in time by τ is equivalent to multiplication by $\mathrm{e}^{-\mathrm{j}\omega\tau}$ in the complex domain. This complex factor contributes only to the phase angle (argument) of the complex representative of the retarded time-harmonic quantity, while its magnitude is unity. For τ in Eq. (8.108),

$$\mathrm{e}^{-\mathrm{j}\omega\tau} = \mathrm{e}^{-\mathrm{j}\omega R/c} = \mathrm{e}^{-\mathrm{j}\beta R}. \tag{8.110}$$

The constant β, in units of rad/m, is called the phase coefficient or wavenumber[9] and is given by

$$\boxed{\beta = \frac{\omega}{c} = \omega\sqrt{\varepsilon\mu} = \frac{2\pi f}{c} = \frac{2\pi}{\lambda},} \tag{8.111}$$

phase coefficient or wavenumber (unit: rad/m)

where λ is the wavelength of the waves, measured in meters and defined as the distance traveled by a wave during one time period T of time-harmonic variations [see Eq. (8.49)],

$$\boxed{\lambda = cT = \frac{c}{f}.} \tag{8.112}$$

wavelength (unit: m)

Hence, the complex rms Lorenz scalar potential, $\underline{V}$, and vector potential, $\underline{\mathbf{A}}$, can be written as

$$\underline{V} = \frac{1}{4\pi\varepsilon}\int_v \frac{\underline{\rho}\,\mathrm{e}^{-\mathrm{j}\beta R}\,\mathrm{d}v}{R}, \tag{8.113}$$

complex Lorenz electric potential

$$\underline{\mathbf{A}} = \frac{\mu}{4\pi}\int_v \frac{\underline{\mathbf{J}}\,\mathrm{e}^{-\mathrm{j}\beta R}\,\mathrm{d}v}{R}. \tag{8.114}$$

complex Lorenz magnetic potential

The complex-domain Lorenz condition comes out to be

$$\nabla\cdot\underline{\mathbf{A}} = -\mathrm{j}\omega\varepsilon\mu\underline{V}, \tag{8.115}$$

and the associated version of the continuity equation is that in Eq. (8.82).

In the case of time-harmonic surface currents and charges of complex densities $\underline{\mathbf{J}}_{\mathrm{s}}$ and $\underline{\rho}_{\mathrm{s}}$ over a surface S, the expressions for complex potentials in Eqs. (8.113) and (8.114) become [see Eqs. (1.83) and (4.109)]

$$\boxed{\underline{V} = \frac{1}{4\pi\varepsilon}\int_S \frac{\underline{\rho}_{\mathrm{s}}\,\mathrm{e}^{-\mathrm{j}\beta R}\,\mathrm{d}S}{R} \quad \text{and} \quad \underline{\mathbf{A}} = \frac{\mu}{4\pi}\int_S \frac{\underline{\mathbf{J}}_{\mathrm{s}}\,\mathrm{e}^{-\mathrm{j}\beta R}\,\mathrm{d}S}{R},} \tag{8.116}$$

complex Lorenz potentials due to surface sources

where $\underline{\mathbf{J}}_{\mathrm{s}}$ and $\underline{\rho}_{\mathrm{s}}$ are related to each other through the complex-domain version of the continuity equation for surface currents, Eq. (8.42).

[9]Note that the symbol k is also used to denote the wavenumber.

Finally, for time-harmonic line currents of complex intensity $\underline{I}$ and charges of complex density $\underline{Q}'$ along a line (e.g., a metallic wire) l, the complex potentials are obtained as [see Eqs. (1.84) and (4.110)]

complex Lorenz potentials for wire structures

$$\underline{V} = \frac{1}{4\pi\varepsilon}\int_l \frac{\underline{Q}'\,\mathrm{e}^{-\mathrm{j}\beta R}\,\mathrm{d}l}{R} \quad \text{and} \quad \underline{\mathbf{A}} = \frac{\mu}{4\pi}\int_l \frac{\underline{I}\,\mathrm{d}\mathbf{l}\,\mathrm{e}^{-\mathrm{j}\beta R}}{R}, \tag{8.117}$$

and the continuity equation for wires, Eq. (8.38), applies.

The electric field intensity vector, **E**, and magnetic flux density vector, **B**, at the point P in Fig. 8.7 can now be evaluated using Eqs. (6.43) and (6.28), respectively. In the time-harmonic case,

high-frequency field vectors from potentials

$$\underline{\mathbf{E}} = -\mathrm{j}\omega\underline{\mathbf{A}} - \nabla\underline{V} = -\mathrm{j}\omega\left[\underline{\mathbf{A}} + \frac{1}{\beta^2}\nabla(\nabla\cdot\underline{\mathbf{A}})\right], \tag{8.118}$$

$$\underline{\mathbf{B}} = \nabla\times\underline{\mathbf{A}}, \tag{8.119}$$

where the complex form of the Lorenz condition, Eq. (8.115), is used in Eq. (8.118) to express the complex electric field intensity vector in terms of the complex magnetic vector potential only.

Eqs. (8.118) and (8.119) together with Eqs. (8.113), (8.114), (8.116), and (8.117) represent a general means for computing the electromagnetic field ($\underline{\mathbf{E}}$, $\underline{\mathbf{B}}$) due to an arbitrary distribution of high-frequency time-harmonic sources in a homogeneous, linear, and lossless medium. The computation involves integration over the source domain (to find the potentials), and then differentiation at the field point (to find the fields from potentials). However, differential operators in Eqs. (8.118) and (8.119) can be moved inside the integrals and applied directly to the functions of the distance R appearing in the integrands, which gives rise to an alternative way for field computation, with the reversed order of operations – first differentiation (with respect to R) and then integration (over the source domain). The two operations can be performed in an arbitrary order, independently from each other, because the gradient and curl in Eqs. (8.118) and (8.119) are taken with the coordinates of the field point (**r**) as independent variables, while the integrals in Eqs. (8.113), (8.114), (8.116), and (8.117) are evaluated with respect to the coordinates of the source point (**r**′). Within this alternative approach, the gradient of V in the case of volume charges can be expressed as

$$\nabla\underline{V} = \frac{1}{4\pi\varepsilon}\nabla\int_v \frac{\underline{\rho}\,\mathrm{e}^{-\mathrm{j}\beta R}\,\mathrm{d}v}{R} = \frac{1}{4\pi\varepsilon}\int_v \nabla\frac{\underline{\rho}\,\mathrm{e}^{-\mathrm{j}\beta R}\,\mathrm{d}v}{R} = \frac{1}{4\pi\varepsilon}\int_v \underline{\rho}\,\mathrm{d}v\,\nabla\frac{\mathrm{e}^{-\mathrm{j}\beta R}}{R}, \tag{8.120}$$

where the elementary charge $\underline{\rho}\,\mathrm{d}v$ can be brought outside the gradient sign because it is a function of the coordinates of the source point (**r**′) and thus represents a constant for taking derivatives at the field point. Using the chain rule,

$$\nabla\frac{\mathrm{e}^{-\mathrm{j}\beta R}}{R} = \frac{\mathrm{d}}{\mathrm{d}R}\left(\frac{\mathrm{e}^{-\mathrm{j}\beta R}}{R}\right)\nabla R = -\frac{(1+\mathrm{j}\beta R)\,\mathrm{e}^{-\mathrm{j}\beta R}}{R^2}\nabla R. \tag{8.121}$$

For a spherical coordinate system centered at the point P′ in Fig. 8.7, the formula in Eq. (1.108) for the gradient in spherical coordinates with the radial coordinate r denoted as R ($r = R$) gives

gradient of source-to-field distance

$$\nabla R = \frac{\mathrm{d}R}{\mathrm{d}R}\hat{\mathbf{R}} = \hat{\mathbf{R}}, \tag{8.122}$$

where, according to our usual notation, $\hat{\mathbf{R}}$ stands for the unit vector along R directed from the source point (P′) to the field point (P). Note that the gradient of the source-to-field distance R can be computed also by expressing the position vectors of the source and field points in Fig. 8.7, $\mathbf{r}'$ and $\mathbf{r}$, in terms of their Cartesian (rectangular) coordinates (x', y', z') and (x, y, z) [see Eq. (1.7)],

$$\mathbf{r}' = x'\,\hat{\mathbf{x}} + y'\,\hat{\mathbf{y}} + z'\,\hat{\mathbf{z}} \quad \text{and} \quad \mathbf{r} = x\,\hat{\mathbf{x}} + y\,\hat{\mathbf{y}} + z\,\hat{\mathbf{z}}, \tag{8.123}$$

respectively, from which

$$R = |\mathbf{r} - \mathbf{r}'| = \sqrt{(x - x')^2 + (y - y')^2 + (z - z')^2}. \tag{8.124}$$

The operator ∇ acts on the coordinates of the field point, and a straightforward application of the formula for gradient in the Cartesian coordinate system, Eq. (1.102), with x, y, and z as independent variables (x', y', and z' are fixed in this operation) leads to the result in Eq. (8.122).[10] Hence, the complex electric field intensity vector due to high-frequency volume sources becomes

$$\boxed{\underline{\mathbf{E}} = -\mathrm{j}\omega\frac{\mu}{4\pi}\int_v \frac{\underline{\mathbf{J}}\,\mathrm{e}^{-\mathrm{j}\beta R}\,\mathrm{d}v}{R} + \frac{1}{4\pi\varepsilon}\int_v \underline{\rho}\,\mathrm{d}v\,\frac{(1 + \mathrm{j}\beta R)\,\mathrm{e}^{-\mathrm{j}\beta R}}{R^2}\,\hat{\mathbf{R}},} \tag{8.125}$$

complex high-frequency electric field

and analogous expressions can be written for surface and line sources.

In a similar fashion, the complex magnetic flux density vector due to high-frequency volume currents can be expressed as

$$\underline{\mathbf{B}} = \nabla \times \underline{\mathbf{A}} = \frac{\mu}{4\pi}\,\nabla \times \int_v \frac{\underline{\mathbf{J}}\,\mathrm{e}^{-\mathrm{j}\beta R}\,\mathrm{d}v}{R} = \frac{\mu}{4\pi}\int_v \nabla \times \frac{\underline{\mathbf{J}}\,\mathrm{e}^{-\mathrm{j}\beta R}\,\mathrm{d}v}{R}. \tag{8.126}$$

The curl of the product of the scalar function $\mathrm{e}^{-\mathrm{j}\beta R}/R$ and the current element $\underline{\mathbf{J}}\,\mathrm{d}v$, which is a vector function of the coordinates of the source point, can be expanded as[11]

$$\nabla \times \left(\frac{\mathrm{e}^{-\mathrm{j}\beta R}}{R}\,\underline{\mathbf{J}}\,\mathrm{d}v\right) = \left(\nabla\frac{\mathrm{e}^{-\mathrm{j}\beta R}}{R}\right) \times (\underline{\mathbf{J}}\,\mathrm{d}v) + \frac{\mathrm{e}^{-\mathrm{j}\beta R}}{R}\,[\nabla \times (\underline{\mathbf{J}}\,\mathrm{d}v)]$$
$$= \left(\nabla\frac{\mathrm{e}^{-\mathrm{j}\beta R}}{R}\right) \times (\underline{\mathbf{J}}\,\mathrm{d}v) = (\underline{\mathbf{J}}\,\mathrm{d}v) \times \hat{\mathbf{R}}\,\frac{(1 + \mathrm{j}\beta R)\,\mathrm{e}^{-\mathrm{j}\beta R}}{R^2}, \tag{8.127}$$

where the gradient of $\mathrm{e}^{-\mathrm{j}\beta R}/R$ is substituted by the expression given by Eqs. (8.121) and (8.122) and the curl of $\underline{\mathbf{J}}\,\mathrm{d}v$ is zero because $\underline{\mathbf{J}}\,\mathrm{d}v$ is a constant for differentiation

[10] A third way to obtain this important result (identity), in Eq. (8.122), in a less formal, and more physical, fashion, not associated with any particular coordinate system, invokes Eq. (1.111), which expresses the physical meaning of the gradient operator in relation to the direction in which the function on which it operates changes most rapidly in spatial coordinates and the magnitude of the maximum spatial rate of change. Namely, if we perform a virtual experiment of moving the field point P for an elementary displacement $\mathrm{d}l$ in various directions around its current location, the function R, that is, the distance between points P′ and P, increases the most if P is moved along the existing (before the movement) R line in Fig. 8.7, which means, according to Eq. (1.111), that the gradient of R is in that direction, or that the unit vector of ∇R is $\hat{\mathbf{R}}$ ($\nabla R = |\nabla R|\,\hat{\mathbf{R}}$). Eq. (1.111) then also tells us that the magnitude of ∇R equals this maximum change of R, and this exactly is $\mathrm{d}l$ (R increases exactly by $\mathrm{d}l$ if P moves for $\mathrm{d}l$ along R). Consequently, $|\nabla R| = \mathrm{d}l/\mathrm{d}l = 1$, i.e., ∇R is a unit vector, and $\nabla R = 1 \times \hat{\mathbf{R}} = \hat{\mathbf{R}}$.

[11] Using the rule for calculating the derivative of the product of two functions, the curl of the product of a scalar function, f, and a vector function, $\mathbf{a}$, can be written as the following expansion in terms of the gradient of f and the curl of $\mathbf{a}$: $\nabla \times (f\mathbf{a}) = (\nabla f) \times \mathbf{a} + f(\nabla \times \mathbf{a})$.

at the field point. This yields the following integral expression for $\underline{\mathbf{B}}$:

complex high-frequency magnetic field

$$\underline{\mathbf{B}} = \frac{\mu}{4\pi} \int_v (\underline{\mathbf{J}}\,\mathrm{d}v) \times \hat{\mathbf{R}} \frac{(1 + \mathrm{j}\beta R)\,\mathrm{e}^{-\mathrm{j}\beta R}}{R^2}, \tag{8.128}$$

with analogous expressions for $\underline{\mathbf{B}}$ in terms of $\underline{\mathbf{J}}_s$ and $\underline{I}$.

If the time lag τ in Eq. (8.108) for all combinations of source and field points in a domain of interest is much shorter than the time period T in Eq. (8.49) for time-harmonic sources, the retardation effect in the system can be neglected and the system can be considered as quasistatic. The definition of quasistatic systems can be expressed also in terms of the electrical size of the system (or the useful part of the system), that is, by comparing the maximum dimension D of the domain that contains all sources and all field points of interest, Fig. 8.9, to the wavelength λ given in Eqs. (8.112) and (8.111). Namely, if

quasistatic condition (Fig. 8.9)

$$D \ll \lambda, \tag{8.129}$$

in which case the domain is said to be electrically small, we can write

$$\beta R = 2\pi \frac{R}{\lambda} \ll 1 \tag{8.130}$$

for all distances R in the domain ($R \leq D$). Since $\mathrm{e}^x \approx 1$ for $|x| \ll 1$, we conclude that $\mathrm{e}^{-\mathrm{j}\beta R}$ can be approximated by 1 in Eqs. (8.113) and (8.114), which then simplify to

quasistatic potentials

$$\underline{V} = \frac{1}{4\pi\varepsilon} \int_v \frac{\underline{\rho}\,\mathrm{d}v}{R} \quad \text{and} \quad \underline{\mathbf{A}} = \frac{\mu}{4\pi} \int_v \frac{\underline{\mathbf{J}}\,\mathrm{d}v}{R}. \tag{8.131}$$

These are nonretarded, quasistatic complex electromagnetic potentials, the same as in Eqs. (6.19) and (6.29) in the time domain (and free space). They can also be obtained by solving the quasistatic versions of Eqs. (8.92) and (8.93), which have the same form as the corresponding static equations – scalar and vector Poisson's equations, Eqs. (2.93) and (4.132). The quasistatic forms of the Lorenz condition and continuity equation for volume currents are those in Eqs. (4.119) and (6.44), respectively.

With $\beta R \approx 0$ in Eqs. (8.125) and (8.128), we obtain the following expressions for the quasistatic (low-frequency) complex field vectors:

quasistatic fields

$$\underline{\mathbf{E}} = -\mathrm{j}\omega \frac{\mu}{4\pi} \int_v \frac{\underline{\mathbf{J}}\,\mathrm{d}v}{R} + \frac{1}{4\pi\varepsilon} \int_v \frac{\underline{\rho}\,\mathrm{d}v}{R^2} \hat{\mathbf{R}} \quad \text{and} \quad \underline{\mathbf{B}} = \frac{\mu}{4\pi} \int_v \frac{(\underline{\mathbf{J}}\,\mathrm{d}v) \times \hat{\mathbf{R}}}{R^2}. \tag{8.132}$$

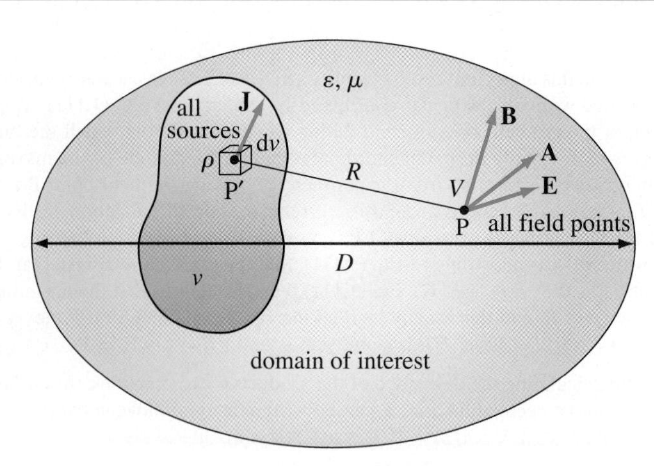

Figure 8.9 For the definition of a quasistatic system in Eq. (8.129); D is the maximum dimension of the domain that contains all sources and all field points of interest.

The first term on the right-hand side of the first equation is the complex-domain expression for the induced electric field intensity vector ($\mathbf{E}_{\rm ind}$) due to slowly time-varying volume currents, Eq. (6.3), and the second term corresponds to the Coulomb electric field ($\mathbf{E}_q$) due to excess volume charges, Eq. (6.17), for arbitrary medium parameters ε and μ. The second equation (for $\underline{\mathbf{B}}$) represents the Biot-Savart law for low-frequency volume currents in the complex domain, the time-domain free-space version of which is given in Eq. (6.27). Of course, expressions analogous to those in Eqs. (8.131) and (8.132) can also be written for low-frequency complex potentials and field vectors due to surface and line sources.[12]

In practice, a useful rule of thumb for quantifying the quasistatic condition in Eq. (8.129) is that a given electromagnetic system (domain) can be considered to be electrically small and quasistatic analysis used when $D < 0.1\lambda$ or, only in very stringent applications, when $D < 0.01\lambda$. The electrical size of the system depends on its physical size and material properties, as well as on the frequency of time-harmonic sources. For instance, at $f = 60$ Hz in free space ($c = c_0 = 3 \times 10^8$ m/s), the wavelength is as large as $\lambda = c_0/f = 5000$ km, so that the retardation even across distances on the order of 100 km can be disregarded and such large electromagnetic systems that extend over hundreds of kilometers can be treated as electrically small domains. At $f = 10$ GHz, on the other hand, even very small electromagnetic systems whose physical dimensions are comparable to $\lambda = c_0/f = 3$ cm (assuming propagation at $c = c_0$) cannot be analyzed without taking into account the travel time of electromagnetic disturbances (waves) from one point in the system to another.

Example 8.12 Loop with a High-Frequency Current of Constant Amplitude

A circular wire loop of radius a in free space carries a high-frequency time-harmonic current of intensity $i(t) = I\sqrt{2}\cos\omega t$. Find (a) the line charge density along the loop, and then compute (b) the electric scalar potential, (c) the magnetic vector potential, (d) the electric field intensity vector, and (e) the magnetic flux density vector at an arbitrary point along the axis of the loop perpendicular to its plane.

Solution

(a) We shall perform the analysis in the complex domain. By means of Eq. (8.66), the complex representative (rms value) of the current intensity $i(t)$ is $\underline{I} = I\,\mathrm{e}^{\mathrm{j}0} = I$. Noting that the complex-domain version of the continuity equation for wires, Eq. (8.38), reads

$$\frac{\mathrm{d}\underline{I}}{\mathrm{d}l} = -\mathrm{j}\omega\underline{Q}' \tag{8.133}$$

continuity equation for wires in complex form

and that $\underline{I}$ does not change, i.e., $\mathrm{d}\underline{I}/\mathrm{d}l = 0$, along the loop, we conclude that there is no excess charge along the loop.

(b) As $\underline{Q}' = 0$ at all points of the loop, the first expression in Eqs. (8.117) tells us that the electric scalar potential is zero everywhere in space.

[12]It is important to have in mind that the low-frequency expressions for electromagnetic potentials and fields are included in the corresponding high-frequency expressions. In other words, we can use high-frequency expressions also at low frequencies. However, although the result will be (approximately) the same, it is mathematically much simpler and physically more clear to use low-frequency solutions whenever the quasistatic criterion in Eq. (8.129) is satisfied.

(c) From the second expression in Eqs. (8.117), the magnetic vector potential along the axis of the loop normal to the loop plane (see Fig. 4.6) appears to be

$$\underline{\mathbf{A}} = \frac{\mu_0}{4\pi} \oint_C \frac{\underline{I}\,\mathrm{d}\mathbf{l}\,\mathrm{e}^{-\mathrm{j}\beta R}}{R} = \frac{\mu_0 \underline{I}\,\mathrm{e}^{-\mathrm{j}\beta R}}{4\pi R} \oint_C \mathrm{d}\mathbf{l} = 0, \tag{8.134}$$

because both $\underline{I}$ and R are constants for the integration process and the integral of $\mathrm{d}\mathbf{l}$ along any closed path is zero [see Eq. (4.173) and Fig. 4.37]. However, no such symmetry exists and hence $\underline{\mathbf{A}} \neq 0$ for observation points that do not belong to the axis ($R \neq$ const as the source point is moved around the loop in the integration process).

(d) Since $\underline{V} = 0$ everywhere, the same is true for $\nabla \underline{V}$, so that Eq. (8.118) yields $\underline{\mathbf{E}} = 0$ at the loop axis (note that $\underline{\mathbf{E}} = -\mathrm{j}\omega\underline{\mathbf{A}} \neq 0$ elsewhere).

(e) Finally, applying Eq. (8.128) to the geometry in Fig. 4.6 and carrying out the same integration (that includes decomposition of $\mathrm{d}\underline{\mathbf{B}}$ into radial and axial components) as in Eqs. (4.14)–(4.19), we obtain the following expression for the complex magnetic flux density vector at the loop axis:

$$\underline{\mathbf{B}} = \frac{\mu_0}{4\pi} \oint_C \underline{I}\,\mathrm{d}\mathbf{l} \times \hat{\mathbf{R}}\, \frac{(1 + \mathrm{j}\beta R)\,\mathrm{e}^{-\mathrm{j}\beta R}}{R^2} = \frac{\mu_0 \underline{I} a^2 (1 + \mathrm{j}\beta R)\,\mathrm{e}^{-\mathrm{j}\beta R}}{2R^3}\,\hat{\mathbf{z}} \quad \left(R = \sqrt{a^2 + z^2}\right), \tag{8.135}$$

where $\beta = \omega\sqrt{\varepsilon_0 \mu_0}$. Using Eq. (8.65), its instantaneous value is

$$\mathbf{B}(t) = \mathrm{Re}\left\{\underline{\mathbf{B}}\sqrt{2}\,\mathrm{e}^{\mathrm{j}\omega t}\right\} = \frac{\mu_0 I \sqrt{2} a^2 \sqrt{1 + \beta^2 R^2}}{2R^3} \cos(\omega t - \beta R + \arctan \beta R)\,\hat{\mathbf{z}}. \tag{8.136}$$

Note that $\mathbf{B}$ is nonzero at the loop axis in spite of $\mathbf{A}$ being zero at the axis. These two results are not contradictory with respect to each other, given that $\mathbf{A} \neq 0$ at points off of the axis. Namely, the curl of a vector function at a point is determined not only by the value of the function at that point or along a specific line (e.g., loop axis) but also by the values and dynamics of the function in a small volume region around the point (which, in our example, includes points off of the loop axis as well). In other words, translated to a one-dimensional situation, we cannot judge on the derivative (slope) of a $f(x)$ curve at $x = x_0$ just based on the $f(x_0)$ value.

In the case of a low-frequency current in the loop, $\beta R \ll 1$ [see Eq. (8.130)] in Eqs. (8.135) and (8.136), which leads to the following quasistatic expressions for the magnetic flux density vector at the loop axis: $\underline{\mathbf{B}} = \mu_0 \underline{I} a^2\,\hat{\mathbf{z}}/(2R^3)$ (complex) and $\mathbf{B}(t) = \mu_0 I \sqrt{2} a^2 \cos \omega t\,\hat{\mathbf{z}}/(2R^3)$ (instantaneous). We observe that these expressions have the same form as their static (dc) counterpart in Eq. (4.19).

Example 8.13 Loop with a High-Frequency Current Changing Along the Wire

Repeat the previous example but for a loop with a high-frequency current of intensity $i(t,\phi) = I_0\sqrt{2}\cos(\phi/2)\cos\omega t$ $(-\pi < \phi \leq \pi)$, shown in Fig. 8.10(a).

Solution

(a) The complex rms current intensity along the loop is $\underline{I}(\phi) = I_0\cos(\phi/2)$. From Eq. (8.133) and the relationship $\mathrm{d}l = a\,\mathrm{d}\phi$, the corresponding complex rms line charge density amounts to

$$\underline{Q}' = \frac{\mathrm{j}}{\omega}\frac{\mathrm{d}\underline{I}}{\mathrm{d}l} = \frac{\mathrm{j}}{\omega a}\frac{\mathrm{d}\underline{I}}{\mathrm{d}\phi} = -\frac{\mathrm{j}I_0}{2\omega a}\sin\frac{\phi}{2}. \tag{8.137}$$

(b) Using the first expression in Eqs. (8.117), the electric scalar potential at the loop axis [z-axis in Fig. 8.10(a)], where $R = \sqrt{a^2 + z^2}$, turns out to be

$$\underline{V} = \frac{1}{4\pi\varepsilon_0} \oint_C \frac{\underline{Q}'(\phi)\,\mathrm{e}^{-\mathrm{j}\beta R}\,\mathrm{d}l}{R} = -\frac{\mathrm{j}I_0\,\mathrm{e}^{-\mathrm{j}\beta R}}{8\pi\varepsilon_0\omega R} \int_{\phi=-\pi}^{\pi} \sin\frac{\phi}{2}\,\mathrm{d}\phi = 0. \tag{8.138}$$

(c) Referring to Fig. 8.10(b), the elementary line vector **dl** along the loop can be represented as

$$\mathbf{dl} = -\,dl \sin\phi\,\hat{\mathbf{x}} + dl\cos\phi\,\hat{\mathbf{y}}. \quad (8.139)$$

With this, the vector integral in the second expression in Eqs. (8.117) can be decomposed into two scalar ones, for the x- and y-components of the magnetic vector potential ($\underline{A}_z = 0$). For an observation point at the z-axis,

$$\underline{A}_x = -\frac{\mu_0}{4\pi}\oint_C \frac{\underline{I}(\phi)\,dl \sin\phi\,\mathrm{e}^{-\mathrm{j}\beta R}}{R} = -\frac{\mu_0 \underline{I}_0 a\,\mathrm{e}^{-\mathrm{j}\beta R}}{4\pi R}\int_{-\pi}^{\pi}\cos\frac{\phi}{2}\sin\phi\,\mathrm{d}\phi = 0 \quad (8.140)$$

(integral of an odd function of ϕ within symmetrical limits with respect to $\phi = 0$ is zero) and

$$\underline{A}_y = \frac{\mu_0}{4\pi}\oint_C \frac{\underline{I}(\phi)\,dl\cos\phi\,\mathrm{e}^{-\mathrm{j}\beta R}}{R} = \frac{\mu_0 \underline{I}_0 a\,\mathrm{e}^{-\mathrm{j}\beta R}}{4\pi R}\int_{-\pi}^{\pi}\cos\frac{\phi}{2}\cos\phi\,\mathrm{d}\phi$$

$$= \frac{\mu_0 \underline{I}_0 a\,\mathrm{e}^{-\mathrm{j}\beta R}}{8\pi R}\left(\int_{-\pi}^{\pi}\cos\frac{3\phi}{2}\,\mathrm{d}\phi + \int_{-\pi}^{\pi}\cos\frac{\phi}{2}\,\mathrm{d}\phi\right) = \frac{\mu_0 \underline{I}_0 a\,\mathrm{e}^{-\mathrm{j}\beta R}}{3\pi R} \quad (8.141)$$

$\{\cos A\cos B = [\cos(A+B) + \cos(A-B)]/2\}$, where $\beta = \omega\sqrt{\varepsilon_0\mu_0}$, so that $\underline{\mathbf{A}} = \mu_0 \underline{I}_0 a\,\mathrm{e}^{-\mathrm{j}\beta R}\,\hat{\mathbf{y}}/(3\pi R)$.

(d) At the point P in Fig. 8.10(a), the first part of the expression in Eq. (8.118) for the complex electric field intensity vector equals $-\mathrm{j}\omega\underline{A}_y\,\hat{\mathbf{y}}$, where $\underline{A}_y$ is given in Eq. (8.141). The second part, $-\nabla\underline{V}$, is not zero, although $\underline{V} = 0$ [Eq. (8.138)] at the same point. Moreover, it is impossible to find grad V from V in this case, because we know V, i.e., that $V = 0$, only at points along the z-axis, which is not enough, as pointed out in the discussion following Eqs. (8.135) and (8.136) in the previous example. Therefore, we shall find grad V by directly computing the field due to the charge along the loop, as in Eq. (8.125).

The x-, y-, and z-components of the unit vector $\hat{\mathbf{R}}$ along the line connecting the source and field points in Fig. 8.10(a) are obtained in the same way as for the elementary electric field intensity vector $\mathrm{d}\mathbf{E}$ in Fig. 1.12 [see Eqs. (1.46) and (1.47)],

$$\hat{\mathbf{R}} = -\frac{a}{R}\cos\phi\,\hat{\mathbf{x}} - \frac{a}{R}\sin\phi\,\hat{\mathbf{y}} + \frac{z}{R}\,\hat{\mathbf{z}}. \quad (8.142)$$

Substituting this expression in Eq. (8.125), the y-component of $\underline{\mathbf{E}}$ at the z-axis is computed as

$$\underline{E}_y = -\mathrm{j}\omega\underline{A}_y + \frac{1}{4\pi\varepsilon_0}\oint_C \underline{Q}'(\phi)\,dl\,\frac{(1+\mathrm{j}\beta R)\,\mathrm{e}^{-\mathrm{j}\beta R}}{R^2}\left(-\frac{a}{R}\sin\phi\right) = -\mathrm{j}\omega\underline{A}_y$$

$$+\frac{\mathrm{j}\underline{I}_0 a(1+\mathrm{j}\beta R)\,\mathrm{e}^{-\mathrm{j}\beta R}}{8\pi\omega\varepsilon_0 R^3}\int_{-\pi}^{\pi}\sin\frac{\phi}{2}\sin\phi\,\mathrm{d}\phi = -\frac{\mathrm{j}\omega\mu_0 \underline{I}_0 a\,\mathrm{e}^{-\mathrm{j}\beta R}}{3\pi R}\left(1 - \frac{1+\mathrm{j}\beta R}{\beta^2 R^2}\right). \quad (8.143)$$

$\{\sin A\sin B = [\cos(A-B) - \cos(A+B)]/2\}$. The other two Cartesian componets of $\underline{\mathbf{E}}$ are zero, because

$$\underline{E}_x \propto \int_{-\pi}^{\pi}\sin\frac{\phi}{2}\cos\phi\,\mathrm{d}\phi = 0 \quad \text{and} \quad \underline{E}_z \propto \int_{-\pi}^{\pi}\sin\frac{\phi}{2}\,\mathrm{d}\phi = 0, \quad (8.144)$$

so $\underline{\mathbf{E}} = \underline{E}_y\,\hat{\mathbf{y}}$.

(e) Finally, the complex magnetic flux density vector at the loop axis is found from the integral expression in Eq. (8.135) with $\underline{I}$ (constant) substituted by $\underline{I}(\phi)$. To represent the elementary vector $\mathbf{dl}\times\hat{\mathbf{R}}$ by its Cartesian components, suitable for integration, we can either use the decomposition of the elementary magnetic flux density vector $\mathrm{d}\mathbf{B}'$ in Fig. 4.9, given by Eqs. (4.24) and (4.25), or take the cross product of the expressions in Eqs. (8.139) and (8.142). The result is

$$\mathbf{dl}\times\hat{\mathbf{R}} = \mathrm{d}l\,\frac{z}{R}\cos\phi\,\hat{\mathbf{x}} + \mathrm{d}l\,\frac{z}{R}\sin\phi\,\hat{\mathbf{y}} + \mathrm{d}l\,\frac{a}{R}\,\hat{\mathbf{z}}. \quad (8.145)$$

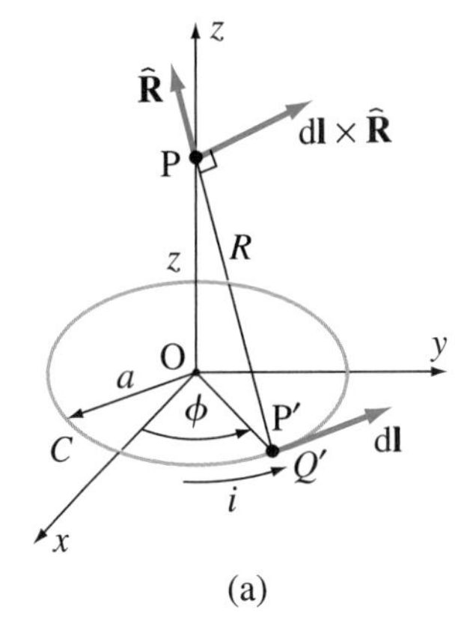

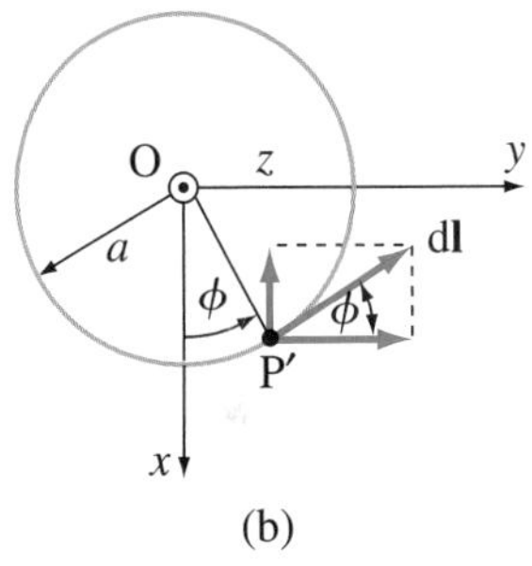

Figure 8.10 Evaluation of potentials and fields at an axis of a wire loop with a high-frequency current whose intensity changes along the wire; for Example 8.13.

Hence,

$$\underline{B}_x = \frac{\mu_0 I_0 a z (1 + \mathrm{j}\beta R)\,\mathrm{e}^{-\mathrm{j}\beta R}}{4\pi R^3} \int_{-\pi}^{\pi} \cos\frac{\phi}{2}\cos\phi\,\mathrm{d}\phi = \frac{\mu_0 I_0 a z (1 + \mathrm{j}\beta R)\,\mathrm{e}^{-\mathrm{j}\beta R}}{3\pi R^3}, \tag{8.146}$$

$$\underline{B}_y = \frac{\mu_0 I_0 a z (1 + \mathrm{j}\beta R)\,\mathrm{e}^{-\mathrm{j}\beta R}}{4\pi R^3} \int_{-\pi}^{\pi} \cos\frac{\phi}{2}\sin\phi\,\mathrm{d}\phi = 0, \tag{8.147}$$

$$\underline{B}_z = \frac{\mu_0 I_0 a^2 (1 + \mathrm{j}\beta R)\,\mathrm{e}^{-\mathrm{j}\beta R}}{4\pi R^3} \int_{-\pi}^{\pi} \cos\frac{\phi}{2}\,\mathrm{d}\phi = \frac{\mu_0 I_0 a^2 (1 + \mathrm{j}\beta R)\,\mathrm{e}^{-\mathrm{j}\beta R}}{\pi R^3}, \tag{8.148}$$

and the final expression for the complex B-vector:

$$\underline{\mathbf{B}} = \frac{\mu_0 I_0 a (1 + \mathrm{j}\beta R)\,\mathrm{e}^{-\mathrm{j}\beta R}}{3\pi R^3}\,(z\,\hat{\mathbf{x}} + 3a\,\hat{\mathbf{z}}). \tag{8.149}$$

Note that $\underline{\mathbf{B}}$ at the point P cannot be obtained from the expression for $\underline{\mathbf{A}}$ in Eq. (8.141) using Eq. (8.119). To be able to find $\underline{\mathbf{B}}$ as curl $\underline{\mathbf{A}}$, it is necessary to know $\underline{\mathbf{A}}$ also at neighboring points off of the loop axis.

Example 8.14 High-Frequency Volume Current in a Sphere

A high-frequency time-harmonic volume current is uniformly distributed inside a sphere of radius a, as shown in Fig. 8.11. The complex rms density of the current is $\underline{\mathbf{J}} = \underline{J}\,\hat{\mathbf{z}}$, and its radian frequency is ω. The medium parameters are ε_0 and μ_0 everywhere. Find (a) the distributions of volume and surface charges inside the sphere and over its surface, respectively, (b) the Lorenz potentials at the sphere center, and (c) the electric field intensity vector at the sphere center.

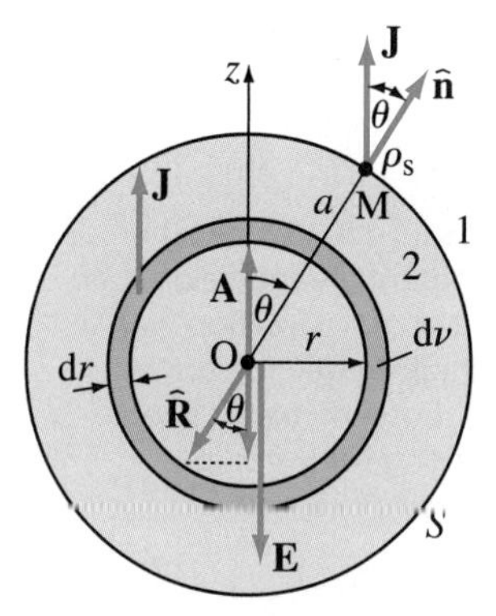

Figure 8.11 Evaluation of the Lorenz potentials and electric field at the center of a sphere with a uniformly distributed high-frequency time-harmonic volume current; for Example 8.14.

Solution

(a) As $\underline{\mathbf{J}} = \text{const}$ at all points inside the sphere, the divergence of $\underline{\mathbf{J}}$ is zero. The continuity equation in differential form and complex domain for high-frequency time-harmonic volume currents, Eq. (8.82), then tells us that $\underline{\rho} = 0$, i.e., there is no excess volume charge in the sphere.

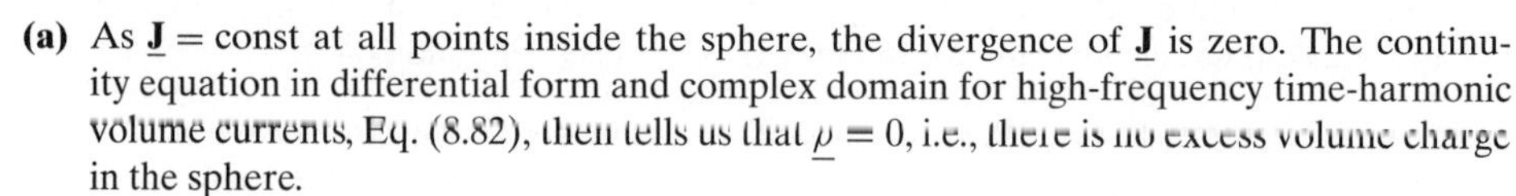

To find the surface charge distribution on the sphere surface, we consider the boundary condition for normal components of the high-frequency current density vector in the complex domain, Eq. (8.86). Marking the external space (with no current) as region 1 ($\underline{\mathbf{J}}_1 = 0$) and sphere interior as region 2 ($\underline{\mathbf{J}}_2 = \underline{\mathbf{J}}$), the complex surface charge density at the point M in Fig. 8.11 is obtained in practically the same way as the bound surface charge density ρ_{ps} in Fig. 2.7 and Eq. (2.30):

$$\underline{\rho}_{\mathrm{s}} = \frac{\mathrm{j}}{\omega}\left(\hat{\mathbf{n}}\cdot\underline{\mathbf{J}}_1 - \hat{\mathbf{n}}\cdot\underline{\mathbf{J}}_2\right) = -\frac{\mathrm{j}}{\omega}\,\hat{\mathbf{n}}\cdot\underline{\mathbf{J}} = -\frac{\mathrm{j}}{\omega}\,\underline{J}\cos\theta, \quad 0 \le \theta \le \pi. \tag{8.150}$$

(b) To find the complex electric scalar potential at the sphere center (point O in Fig. 8.11), we use the corresponding integral expression for surface charges, in Eqs. (8.116). Noting that $R = a$ in this case, we have

$$\underline{V} = \frac{1}{4\pi\varepsilon_0}\oint_S \frac{\underline{\rho}_{\mathrm{s}}\,\mathrm{e}^{-\mathrm{j}\beta R}\,\mathrm{d}S}{R} = \frac{\mathrm{e}^{-\mathrm{j}\beta a}}{4\pi\varepsilon_0 a}\oint_S \underline{\rho}_{\mathrm{s}}(\theta)\,\mathrm{d}S = 0 \quad (R = a), \tag{8.151}$$

as it is obvious, because the charge distribution given by Eq. (8.150) is antisymmetrical with respect to the plane $z = 0$, that the total charge over S is zero.

For the magnetic vector potential, we use the integral expression for $\underline{\mathbf{A}}$ due to volume currents, Eq. (8.114). Since $\underline{\mathbf{J}}$, as a constant, can be brought outside the integral sign, we need to integrate only with respect to R. Moreover, since the observation point is the sphere center, R equals the radial coordinate r in Fig. 8.11, and we adopt $\mathrm{d}v$ in the form of a thin spherical shell of radius r and thickness $\mathrm{d}r$. The volume of the shell is given in

Eq. (1.33). Hence,

$$\underline{\mathbf{A}} = \frac{\mu_0}{4\pi}\int_v \frac{\underline{\mathbf{J}}\,\mathrm{e}^{-\mathrm{j}\beta R}\,\mathrm{d}v}{R} = \frac{\mu_0\underline{\mathbf{J}}}{4\pi}\int_{r=0}^{a}\frac{\mathrm{e}^{-\mathrm{j}\beta r}}{r}\underbrace{4\pi r^2\,\mathrm{d}r}_{\mathrm{d}v} = \mu_0\underline{\mathbf{J}}\int_0^a r\,\mathrm{e}^{-\mathrm{j}\beta r}\,\mathrm{d}r$$

$$= \frac{\mu_0\underline{\mathbf{J}}}{\beta^2}\left[(1+\mathrm{j}\beta a)\,\mathrm{e}^{-\mathrm{j}\beta a} - 1\right], \quad \beta = \omega\sqrt{\varepsilon_0\mu_0} \quad (R = r), \tag{8.152}$$

where the last integral is solved by integration by parts $[\int x\,\mathrm{e}^{-x}\,\mathrm{d}x = -(1+x)\,\mathrm{e}^{-x} + C]$.

(c) Finally, to find the complex electric field intensity vector at the point O, we start with Eq. (8.125) and carry out essentially the same surface integration procedure as in Eq. (2.32) for the uniformly polarized dielectric sphere under static conditions in Fig. 2.7:

$$\underline{\mathbf{E}} = -\mathrm{j}\omega\underline{\mathbf{A}} + \frac{1}{4\pi\varepsilon_0}\oint_S \rho_{\mathrm{s}}(\theta)\,\mathrm{d}S'\,\frac{(1+\mathrm{j}\beta R)\,\mathrm{e}^{-\mathrm{j}\beta R}}{R^2}\,\hat{\mathbf{R}} = -\mathrm{j}\omega\underline{\mathbf{A}}$$

$$+\frac{\mathrm{j}\underline{J}(1+\mathrm{j}\beta a)\,\mathrm{e}^{-\mathrm{j}\beta a}}{2\varepsilon_0\omega}\,\hat{\mathbf{z}}\int_{\theta=0}^{\pi}\cos^2\theta\sin\theta\,\mathrm{d}\theta = \frac{\mathrm{j}\underline{\mathbf{J}}}{\omega\varepsilon_0}\left[1 - \frac{2}{3}(1+\mathrm{j}\beta a)\,\mathrm{e}^{-\mathrm{j}\beta a}\right] \quad (R = a), \tag{8.153}$$

where $\hat{\mathbf{R}}$ in the integral is replaced by its z-component, $\cos\theta\,\hat{\mathbf{z}}$, as it is obvious (before the integration) that $\mathbf{E}$ has a z-component only, due to symmetry. In addition, since the function in the integrand depends on the angle θ only, the surface $\mathrm{d}S'$ in Fig. 2.7 is extended to the surface $\mathrm{d}S$ in the form of a thin ring of radius $a\sin\theta$ and width $a\,\mathrm{d}\theta$ [see the expression for $\mathrm{d}S$ in Eq. (1.65)] – our surface-integration strategy is always to adopt as large as possible $\mathrm{d}S$ (Section 1.4).

Note that the spherical domain with a uniform volume current in Fig. 8.11 may represent a uniformly polarized dielectric sphere with a high-frequency time-harmonic polarization vector $\underline{\mathbf{P}} = \underline{P}\,\hat{\mathbf{z}}$ of angular frequency ω. In such a model, $\underline{\mathbf{J}}$ would stand for the polarization current density vector in the dielectric, $\underline{\mathbf{J}} \mapsto \underline{\mathbf{J}}_{\mathrm{p}} = \mathrm{j}\omega\underline{\mathbf{P}}$ [see Eq. (8.13)] and $\underline{\rho}_{\mathrm{s}}$ in Eq. (8.150) for the associated polarization (bound) surface charge density on the dielectric surface, $\underline{\rho}_{\mathrm{s}} \mapsto \underline{\rho}_{\mathrm{ps}} = \hat{\mathbf{n}}_{\mathrm{d}}\cdot\underline{\mathbf{P}}$ [see Eq. (2.30)]. While computing the potentials $\underline{V}$ and $\underline{\mathbf{A}}$ and the field $\underline{\mathbf{E}}$ due to the polarized dielectric, current and charge distributions $\underline{\mathbf{J}}_{\mathrm{p}}$ and $\underline{\rho}_{\mathrm{ps}}$ could be considered to be in a vacuum [medium parameters in Eqs. (8.151)–(8.153) are ε_0 and μ_0].

If the dielectric sphere is electrically small, that is, if $a \ll \lambda$ or $\beta a \ll 1$, then the result in Eq. (8.153) reduces to

$$\underline{\mathbf{E}} = \frac{\mathrm{j}\underline{\mathbf{J}}}{3\omega\varepsilon_0} = -\frac{1}{3\varepsilon_0}\frac{\underline{\mathbf{J}}_{\mathrm{p}}}{\mathrm{j}\omega} = -\frac{\underline{\mathbf{P}}}{3\varepsilon_0} \quad (a \ll \lambda). \tag{8.154}$$

This is the quasistatic (low-frequency) result of the problem of a uniformly polarized dielectric sphere in the time-harmonic regime. We realize that it has the same form as the expression in Eq. (2.32) for the electric field intensity vector at the center of a dielectric sphere with static (time-invariant) uniform polarization.

Problems: 8.19–8.27; *Conceptual Questions* (on Companion Website): 8.26–8.29; *MATLAB Exercises* (on Companion Website).

8.11 POYNTING'S THEOREM

The rest of this chapter is devoted to energy and power considerations associated with the general electromagnetic field and its sources. The focal point of the material is Poynting's theorem, which represents the mathematical expression of the principle of conservation of energy as applied to electromagnetic fields. It is derived directly from Maxwell's equations in either time or complex domain, and can be used also in static situations. The theorem details the power balance for the fields and sources in a region of interest including the transfer of electromagnetic power into and out of the region. In this section, we assume an arbitrary time dependence of sources and discuss Poynting's theorem in the time domain; formulation of the theorem in the complex (frequency) domain, for time-harmonic sources, is presented in the next section.

Consider an arbitrary spatial domain of volume v with rapidly time-varying currents of density $\mathbf{J}$, charges of density ρ, and electromagnetic field vectors $\mathbf{E}$, $\mathbf{H}$, $\mathbf{D}$, and $\mathbf{B}$. We assume that the domain is filled with a linear, isotropic, generally inhomogeneous, and lossy material of permittivity ε, permeability μ, and conductivity σ. Let the external electric energy volume sources (excitations) in the domain be represented by an impressed electric field (analogous to an ideal voltage generator in circuit theory) of intensity $\mathbf{E}_\mathrm{i}$.

We start with Maxwell's equations for the rapidly time-varying electromagnetic field in differential form and time domain, Eqs. (8.24), and take the dot product of both sides of the first equation with $\mathbf{H}$ and the second equation with $-\mathbf{E}$, and then add up the two equations thus obtained. What we get is

$$\mathbf{H}\cdot(\nabla\times\mathbf{E})-\mathbf{E}\cdot(\nabla\times\mathbf{H})=-\mathbf{H}\cdot\frac{\partial\mathbf{B}}{\partial t}-\mathbf{E}\cdot\mathbf{J}-\mathbf{E}\cdot\frac{\partial\mathbf{D}}{\partial t}. \tag{8.155}$$

The expression on the left-hand side of this equation equals the divergence of the cross product of $\mathbf{E}$ and $\mathbf{H}$.[13] In addition, using the constitutive equations for $\mathbf{B}$ and $\mathbf{D}$ in linear media [see Eqs. (8.23)], the two terms with the time derivatives can be written as

$$\mathbf{H}\cdot\frac{\partial\mathbf{B}}{\partial t}=\mu H\frac{\partial H}{\partial t}=\frac{\partial}{\partial t}\left(\frac{1}{2}\mu H^2\right) \tag{8.156}$$

and similarly for $\mathbf{E}\cdot\partial\mathbf{D}/\partial t$. Finally, from the constitutive equation for $\mathbf{J}$ (in a linear conducting medium) that incorporates the impressed electric field intensity vector, Eq. (3.109),

$$\mathbf{E}\cdot\mathbf{J}=\left(\frac{\mathbf{J}}{\sigma}-\mathbf{E}_\mathrm{i}\right)\cdot\mathbf{J}=\frac{J^2}{\sigma}-\mathbf{E}_\mathrm{i}\cdot\mathbf{J}. \tag{8.157}$$

Hence, Eq. (8.155) can be recast to read

$$\mathbf{E}_\mathrm{i}\cdot\mathbf{J}=\frac{J^2}{\sigma}+\frac{\partial}{\partial t}\left(\frac{1}{2}\varepsilon E^2+\frac{1}{2}\mu H^2\right)+\nabla\cdot(\mathbf{E}\times\mathbf{H}). \tag{8.158}$$

[13]The divergence of the cross product of vectors $\mathbf{E}$ and $\mathbf{H}$ (or any two vectors), $\nabla\cdot(\mathbf{E}\times\mathbf{H})$, can be expressed in terms of the curl of $\mathbf{E}$ and curl of $\mathbf{H}$ using the formula for the derivative of the product of two functions, in which the derivative (del operator) first acts on one function ($\mathbf{E}$) and then on the other ($\mathbf{H}$). This combined with the fact that cyclic permutation of the order of vectors does not change the scalar triple product of three vectors, $\mathbf{a}\cdot(\mathbf{b}\times\mathbf{c})=\mathbf{c}\cdot(\mathbf{a}\times\mathbf{b})=-\mathbf{b}\cdot(\mathbf{a}\times\mathbf{c})$, with $\mathbf{a}$, $\mathbf{b}$, and $\mathbf{c}$ substituted formally by vectors ∇, $\mathbf{E}$, and $\mathbf{H}$, respectively, gives

$$\nabla\cdot(\mathbf{E}\times\mathbf{H})=\mathbf{H}\cdot(\nabla\times\mathbf{E})-\mathbf{E}\cdot(\nabla\times\mathbf{H}).$$

Of course, this identity, like all other similar identities of vector calculus, can be verified representing $\mathbf{E}$ and $\mathbf{H}$ by their components in the Cartesian coordinate system and applying formulas for the divergence and curl in Eqs. (1.167) and (4.81).

Multiplying both sides of this new equation by an elementary volume dv and integrating them over the total volume v yield

$$\int_v \mathbf{E}_i \cdot \mathbf{J}\, dv = \int_v \frac{J^2}{\sigma}\, dv + \frac{d}{dt}\int_v \left(\frac{1}{2}\varepsilon E^2 + \frac{1}{2}\mu H^2\right) dv + \oint_S (\mathbf{E} \times \mathbf{H}) \cdot d\mathbf{S}, \quad (8.159)$$

Poynting's theorem

where S is the closed surface bounding v and the divergence theorem, Eq. (1.173), is applied to convert the volume integral of the divergence of the vector $\mathbf{E} \times \mathbf{H}$ over v to the net outward flux (surface integral) of the same vector through S.

HISTORICAL ASIDE

John Henry Poynting (1852–1914), an English physicist, was a professor of physics at Mason College (later the University of Birmingham). Poynting studied at Owens College (now the University of Manchester) from 1867 to 1872, and then entered Cambridge University's Trinity College and graduated there in 1876. In 1880, he was appointed as first professor of physics at Mason College, subsequently the University of Birmingham, the position he held until his death. Based on Faraday's (1791–1867) and Maxwell's (1831–1879) field and energy concepts [in "A Dynamical Theory of the Electromagnetic Field" (1865), Maxwell stated that "the energy in electromagnetic phenomena resided in the electromagnetic field, in the space surrounding the electrified and magnetic bodies, as well as in those bodies themselves"], Poynting developed the theory of electromagnetic energy transfer, known today as Poynting's theorem. In his famous paper "On the Transfer of Energy in the Electromagnetic Field" (1884), he showed that the flow of electromagnetic power (rate of energy) at a point can be expressed by a simple formula in terms of the electric and magnetic forces (or field vectors) at that point, and the vector describing this flow bears his name as well (Poynting vector). He also made accurate measurements of the density of the earth (in 1891) and the gravitational constant (in 1893) using the torsion balance method pioneered by Henry Cavendish (1731–1810). In 1903, Poynting suggested the existence of a combined effect of the sun's gravity and radiation that causes small particles orbiting the sun to slowly spiral into it, the idea later (in 1937) developed by Howard Percy Robertson (1903–1961), and now known as the Poynting-Robertson effect. *(Portrait: Library of Congress)*

Eq. (8.159) is known as Poynting's theorem. Having in mind Eq. (3.123), we recognize that the integral on the left-hand side of Eq. (8.159) equals the total instantaneous generated power P_i of the sources (represented by the impressed field) in the domain v. The terms on the right-hand side of the equation show how this power is used. The first term equals the total instantaneous power of Joule's (ohmic) losses, P_J, in v [see Eq. (3.32)], i.e., the part of P_i that is lost to heat. To interpret the second term, we recall that the total stored instantaneous electric and magnetic energies in v can be obtained from the integrals over v given by Eqs. (2.202) and (7.109), respectively. Summing up these energies, we arrive to the total stored instantaneous electromagnetic energy W_{em} in v,

$$W_{em} = W_e + W_m = \int_v w_e\, dv + \int_v w_m\, dv = \int_v \frac{1}{2}\varepsilon E^2\, dv + \int_v \frac{1}{2}\mu H^2\, dv. \quad (8.160)$$

stored electromagnetic energy in a domain v

This means that the second term on the right-hand side of Eq. (8.159) represents the time-rate of change of the total energy localized in the electromagnetic field in v. Finally, from conservation of energy, the last term in Eq. (8.159) must equal the total instantaneous net power (rate of energy) leaving the domain through the surface S enclosing it. The vector $\mathbf{E} \times \mathbf{H}$ has the dimension of a surface power density (power per unit area), and is thus expressed in W/m^2 (the unit for E, V/m, times the unit for H, A/m). It is called the Poynting vector and is designated by $\mathcal{P}$ (calligraphic **P**),

Poynting vector (unit: W/m^2)

$$\boxed{\mathcal{P} = \mathbf{E} \times \mathbf{H}.} \tag{8.161}$$

Note that **S** is also widely used to denote the Poynting vector. We see that $\mathcal{P}$ is perpendicular to both **E** and **H**. Its magnitude depends on the magnitudes of the electric and magnetic field intensity vectors and on the angle between them: $\mathcal{P} = EH \sin \angle(\mathbf{E}, \mathbf{H})$.

The power transferred through S (power flow), P_f, can now be written as

power flow through a closed surface S (unit: W)

$$\boxed{P_\mathrm{f} = \oint_S \mathcal{P} \cdot \mathrm{d}\mathbf{S}.} \tag{8.162}$$

This assertion alone, namely, that the outward net flux of the Poynting vector through any closed surface (S) is equal to the instantaneous power leaving the enclosed domain (v), is also frequently considered as Poynting's theorem. Locally, the Poynting vector is a measure of the rate of energy flow per unit area at any point on the surface S. More precisely, as the power flow through a surface element $\mathrm{d}S$ on S is

$$\mathrm{d}P_\mathrm{f} = \mathcal{P} \cdot \mathrm{d}\mathbf{S} = \mathcal{P}\,\mathrm{d}S \cos\alpha = \underbrace{\mathcal{P}\cos\alpha}_{\mathcal{P}_\mathrm{n}}\,\mathrm{d}S = \mathcal{P}_\mathrm{n}\,\mathrm{d}S, \tag{8.163}$$

where α is the angle between $\mathcal{P}$ and $\mathrm{d}\mathbf{S}$ (i.e., the normal to $\mathrm{d}S$), the local surface power density at a point on S, $\mathrm{d}P_\mathrm{f}/\mathrm{d}S$, equals the normal component of $\mathcal{P}$ with respect to the patch $\mathrm{d}S$, $\mathcal{P}_\mathrm{n}$, while the tangential component of $\mathcal{P}$, $\mathcal{P}_\mathrm{t} = \mathcal{P}\sin\alpha$, does not contribute to the power transfer. We notice that $\mathrm{d}P_\mathrm{f}$ can be positive (for $0 \le \alpha < \pi/2$), meaning that this amount of power indeed leaves the domain v through $\mathrm{d}S$, negative (for $\pi/2 < \alpha \le \pi$), in which case the actual direction of power flow is from the surrounding space into the domain and the power $|\mathrm{d}P_\mathrm{f}|$ enters v through $\mathrm{d}S$, or zero (for $\alpha = \pi/2$), implying no power flow through that particular patch. In addition, since $\mathrm{d}P_\mathrm{f}$ can also be represented as

$$\mathrm{d}P_\mathrm{f} = \mathcal{P}\,\underbrace{\mathrm{d}S\cos\alpha}_{\mathrm{d}S_\mathrm{n}} = \mathcal{P}\,\mathrm{d}S_\mathrm{n}, \tag{8.164}$$

where $\mathrm{d}S_\mathrm{n}$ is the projection of $\mathrm{d}S$ on the plane normal to $\mathcal{P}$ (see Fig. 1.30), the full magnitude of $\mathcal{P}$ equals the surface power density if computed through an elementary surface positioned normal to $\mathcal{P}$. In other words, the power flow through a surface element of a given area is maximum when it is set normal to the lines of $\mathcal{P}$. We conclude that (1) the direction of the Poynting vector at a point coincides with the direction of the power flow at that point and (2) the magnitude of the Poynting vector equals the maximum transferred surface power density at a point.

If the excitations in v are represented by an impressed electric volume current (analogous to an ideal current generator in circuit theory) of density $\mathbf{J}_\mathrm{i}$, Eq. (8.157) becomes

$$\mathbf{E} \cdot \mathbf{J} = \mathbf{E} \cdot (\sigma\mathbf{E} + \mathbf{J}_\mathrm{i}) = \sigma E^2 + \mathbf{E} \cdot \mathbf{J}_\mathrm{i}, \tag{8.165}$$

where the version of the linear constitutive equation for **J** that takes into account the impressed current density vector, Eq. (3.124), is used. This leads to the following formulation of Poynting's theorem:

$$-\int_v \mathbf{E}\cdot\mathbf{J}_\mathrm{i}\,\mathrm{d}v = \int_v \sigma E^2\,\mathrm{d}v + \frac{\mathrm{d}}{\mathrm{d}t}\int_v \left(\frac{1}{2}\varepsilon E^2 + \frac{1}{2}\mu H^2\right)\mathrm{d}v + \oint_S \boldsymbol{\mathcal{P}}\cdot\mathrm{d}\mathbf{S}, \qquad (8.166)$$

with the term on the left-hand side of the equation (including the minus sign) being the total instantaneous source power P_i generated by the impressed current in v [see Eq. (3.129)].

Poynting's theorem, given by Eqs. (8.159) or (8.166), can concisely be written as

$$\boxed{P_\mathrm{i}(t) = P_\mathrm{J}(t) + \frac{\mathrm{d}W_\mathrm{em}}{\mathrm{d}t} + P_\mathrm{f}(t).} \qquad (8.167)$$

Poynting's theorem in condensed form

At an instant t, all the terms in this equation can be both positive and negative except $P_\mathrm{J}(t)$, which is always nonnegative (there cannot be negative Joule's losses). A negative $P_\mathrm{i}(t)$ means that the impressed sources (generators) receive energy from the electromagnetic field in v and/or its surroundings. When $\mathrm{d}W_\mathrm{em}/\mathrm{d}t < 0$, the electromagnetic energy in v decreases, i.e., the field in v delivers some of its stored energy to the sources in v and/or to the region outside S, in addition to the power conversion to heat throughout v in the form of Joule's losses. Finally, a negative $P_\mathrm{f}(t)$ implies that the net instantaneous power flow is directed from the space exterior to v inwards, where the energy is partly received by the sources and/or field and partly dissipated as heat in v.

Let us list several special cases of the application of the general Poynting's theorem in Eq. (8.167). (1) If the region of concern (v) is occupied by the time-invariant (static) electromagnetic field, the term $\mathrm{d}W_\mathrm{em}/\mathrm{d}t$ vanishes and the remaining terms are time-constant. (2) When the region v does not contain impressed sources, $P_\mathrm{i}(t) = 0$ at all times. (3) In the case of a lossless material throughout v, $P_\mathrm{J}(t) = 0$ always. (4) Finally, for a domain enclosed by a PEC surface (S), $P_\mathrm{f}(t) = 0$ for any t and any field distribution in v, because the electromagnetic field in a PEC is always zero under nonstatic conditions (see Example 8.6), and consequently there cannot be energy exchange through S with a PEC environment.

Poynting's theorem, in its various forms, is of great theoretical and practical importance in electromagnetics, and will be used on many occasions throughout the remaining chapters of this text. The examples in this and the next section are aimed at illustrating its application and usefulness in several characteristic configurations with different geometries and different regimes of operation.

Example 8.15 Poynting Vector in a Leaky Coaxial Cable in a dc Regime

A coaxial cable with perfect conductors and an imperfect dielectric of conductivity σ_d and permeability μ_0 is connected at one end to an ideal voltage generator of time-invariant emf $\mathcal{E}$, while the other end of the cable is open, as shown in Fig. 8.12(a). The radius of the inner conductor and inner radius of the outer conductor are a and b, respectively, and the length of the cable is l. Find the flux of the Poynting vector through an arbitrary cross section of this cable.

Solution There is a leakage volume steady (dc) current through the dielectric of the cable, as shown in Fig. 3.20. From Eqs. (3.175) and (3.178) with $R_\mathrm{L} \to \infty$ (the line is open), and employing the expression for the conductance per unit length of the cable (G'), Eq. (3.158), the voltage and current intensity along the cable ($0 \le z \le l$) are given by

$$V(z) = V = \mathcal{E} \quad \text{and} \quad I(z) = G'\mathcal{E}(l - z), \quad \text{where} \quad G' = \frac{2\pi\sigma_\mathrm{d}}{\ln(b/a)}. \qquad (8.168)$$

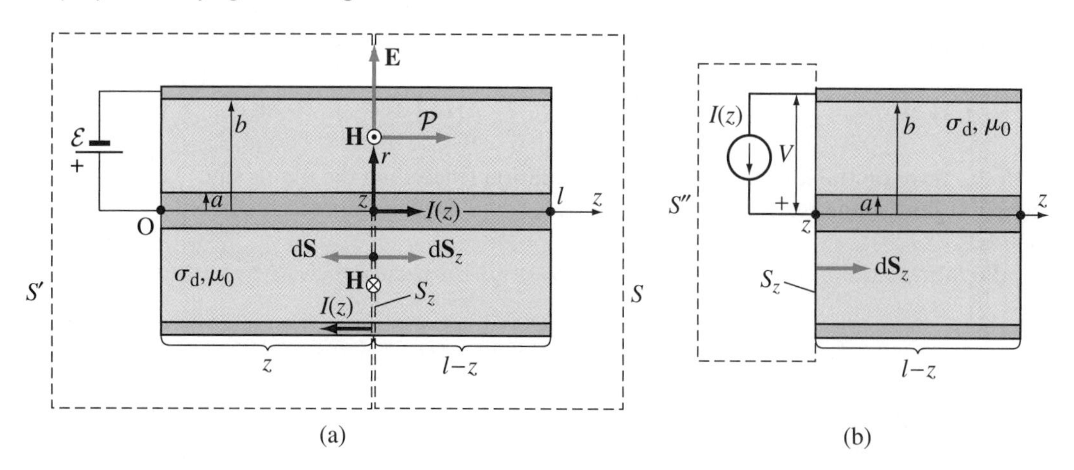

Figure 8.12 Applications of Poynting's theorem to a leaky coaxial cable in a dc regime; for Example 8.15.

With the use of Eqs. (3.155) and (3.157), the electric field intensity vector in the dielectric, Fig. 8.12(a), equals

$$\mathbf{E}(r) = \frac{G'V}{2\pi\sigma_\mathrm{d} r}\,\hat{\mathbf{r}} = \frac{V}{r\ln(b/a)}\,\hat{\mathbf{r}}. \tag{8.169}$$

Due to symmetry of the current distribution in the cable (see Fig. 3.20), the magnetic field in the dielectric is circular with respect to the cable axis (z-axis), as in Fig. 4.17, so that the generalized Ampère's law, Eq. (5.49), gives the following expression for the magnetic field intensity vector between the cable conductors [see also Eq. (4.61)]:

$$\mathbf{H}(r,z) = \frac{I(z)}{2\pi r}\,\hat{\boldsymbol{\phi}}. \tag{8.170}$$

By means of Eq. (8.161), the Poynting vector in the cable [Fig. 8.12(a)] is then

$$\boldsymbol{\mathcal{P}}(r,z) = \mathbf{E}(r)\times\mathbf{H}(r,z) = \frac{VI(z)}{2\pi r^2\ln(b/a)}\,\hat{\mathbf{r}}\times\hat{\boldsymbol{\phi}} = \frac{\sigma_\mathrm{d}\mathcal{E}^2(l-z)}{r^2\ln^2(b/a)}\,\hat{\mathbf{z}} \tag{8.171}$$

[note that $\angle(\mathbf{E},\mathbf{H}) = 90^\circ$]. The direction of $\boldsymbol{\mathcal{P}}$, which coincides with the direction of the power flow through the dielectric between the conductors of the cable, is, of course, from the generator toward the rest of the cable. The flux of the Poynting vector through a cross section of the cable defined by the coordinate z for the reference direction toward the load comes out to be

first way: computing the flux of $\mathcal{P}$ through the cross section S_z in Fig. 8.12(a)

$$\boxed{\Psi_\mathcal{P} = \int_{S_z}\boldsymbol{\mathcal{P}}\cdot\mathrm{d}\mathbf{S}_z = \int_{r=a}^{b}\mathcal{P}(r,z)\,2\pi r\,\mathrm{d}r = \frac{VI(z)}{\ln(b/a)}\int_a^b\frac{\mathrm{d}r}{r} = VI(z) = \frac{2\pi\sigma_\mathrm{d}\mathcal{E}^2(l-z)}{\ln(b/a)},} \tag{8.172}$$

where $\mathrm{d}S_z = 2\pi r\,\mathrm{d}r$ is the surface area of an elementary ring of radius r and width $\mathrm{d}r$ centered at the z-axis [see Fig. 7.26].

Let us now obtain this same result in a different way – applying Poynting's theorem, Eq. (8.167), to the closed surface S shown in Fig. 8.12(a). Since there are no generators in the domain enclosed by S and the electromagnetic energy in the domain is time-invariant (as are the electric and magnetic fields, voltage, and currents in the cable), $P_\mathrm{i} = 0$ and $\mathrm{d}W_\mathrm{em}/\mathrm{d}t = 0$ in Eq. (8.167), and the theorem reduces to

$$0 = P_\mathrm{J} + P_\mathrm{f}. \tag{8.173}$$

Invoking the per-unit-length power of Joule's losses in Eq. (3.179), the total loss power P_J in the part of the cable enclosed by S is

$$P_\mathrm{J} = P'_\mathrm{J}(l-z) = G'\mathcal{E}^2(l-z). \tag{8.174}$$

Considering the term P_f in Eq. (8.173), we note that there is no field outside the cable, which means that the flux of $\mathcal{P}$ through the parts of S outside the cable is zero. In addition, $\mathcal{P}$ is tangential to the parts of S that "cut" the conductors of the cable, because the vector $\mathbf{E}$ in the conductors is axial (parallel to the z-axis) like the current streamlines in the conductors, so that the flux of $\mathcal{P}$ through these parts of S is also zero ($\mathcal{P} \cdot \mathrm{d}\mathbf{S} = 0$). Consequently, the net flux of $\mathcal{P}$ through the entire surface S appears to be equal to the flux just through the part of the surface that is in the dielectric of the cable, that is,

$$P_\mathrm{f} = \oint_S \mathcal{P} \cdot \mathrm{d}\mathbf{S} = -\Psi_\mathcal{P}, \tag{8.175}$$

where the minus sign comes from $\mathrm{d}\mathbf{S} = -\,\mathrm{d}\mathbf{S}_z$ (the reference direction of the flux of $\mathcal{P}$ in Poynting's theorem is outward with respect to the enclosed domain). Hence,

$$\boxed{\Psi_\mathcal{P} = -P_\mathrm{f} = P_\mathrm{J} = G'\mathcal{E}^2(l - z).} \tag{8.176}$$

second way: applying Poynting's theorem to the closed surface S in Fig. 8.12(a)

The third way to compute $\Psi_\mathcal{P}$ is to apply Poynting's theorem to another closed surface containing the cross section S_z as its part, that (S') enclosing the generator in Fig. 8.12(a). This yields

$$P_\mathrm{i} = P_\mathrm{J1} + \Psi_\mathcal{P}, \tag{8.177}$$

where P_J1 is now the power of Joule's losses in the part of the cable between the generator and the cross section S_z and P_i is the power of the generator (ideal voltage generator), given by Eq. (3.181). The result is

$$\boxed{\Psi_\mathcal{P} = P_\mathrm{i} - P_\mathrm{J1} = \mathcal{E}I(0) - P'_\mathrm{J}z = \mathcal{E}G'calEl - G'\mathcal{E}^2 z = G'\mathcal{E}^2(l - z).} \tag{8.178}$$

third way: applying Poynting's theorem to the closed surface S′ in Fig. 8.12(a)

Finally, as the fourth way to solve this problem, we can substitute (compensate) the part of the cable to the left of S_z by an ideal current generator of current intensity $I_\mathrm{g} = I(z)$, as shown in Fig. 8.12(b). The flux $\Psi_\mathcal{P}$ can now be found as being equal to the power generated by this compensating current generator (this power is transferred through the cross section S_z and delivered to the rest of the cable), which is actually yet another application of Poynting's theorem – to the closed surface S'' in Fig. 8.12(b). Having in mind the expression for the power of an ideal current generator, Eq. (3.130), we obtain

$$\boxed{\Psi_\mathcal{P} = VI(z),} \tag{8.179}$$

fourth way: applying Poynting's theorem to the closed surface S″ in Fig. 8.12(b)

which again is the same result as in Eq. (8.172).

Example 8.16 Poynting Vector on the Earth's Surface around a Grounding Electrode

Consider the hemispherical grounding electrode in a two-layer earth from Example 3.17, and determine the flux of the Poynting vector through the surface of the earth.

Solution The expressions for the electric field intensity vector, $\mathbf{E}$, in each of the layers of the earth (expressions for $\mathbf{E}_1$ and $\mathbf{E}_2$) are found in Example 3.17. Because of symmetry, lines of the magnetic field intensity vector, $\mathbf{H}$, on the earth's surface are circles centered at the point O in Fig. 3.28, as indicated in Fig. 8.13, and the generalized Ampère's law yields

$$\mathbf{H} = \frac{I}{2\pi r}\,\hat{\boldsymbol{\phi}} \quad (a < r < \infty). \tag{8.180}$$

The Poynting vector on the earth's surface (S_gr) is $\mathcal{P}_1 = E_1 H\,\hat{\mathbf{z}}$ for $a < r < b$ and $\mathcal{P}_2 = E_2 H\,\hat{\mathbf{z}}$ for $b < r < \infty$, so that its flux through S_gr for the reference direction downward (into the earth) is

$$\Psi_\mathcal{P} = \int_{S_\mathrm{gr}} \mathcal{P} \cdot \mathrm{d}\mathbf{S}_\mathrm{gr} = \int_{r=a}^{b} \mathcal{P}_1\, 2\pi r\,\mathrm{d}r + \int_b^\infty \mathcal{P}_2\, 2\pi r\,\mathrm{d}r = \frac{I^2}{2\pi b}\left(\frac{b-a}{\sigma_1 a} + \frac{1}{\sigma_2}\right) = 955\ \mathrm{kW}. \tag{8.181}$$

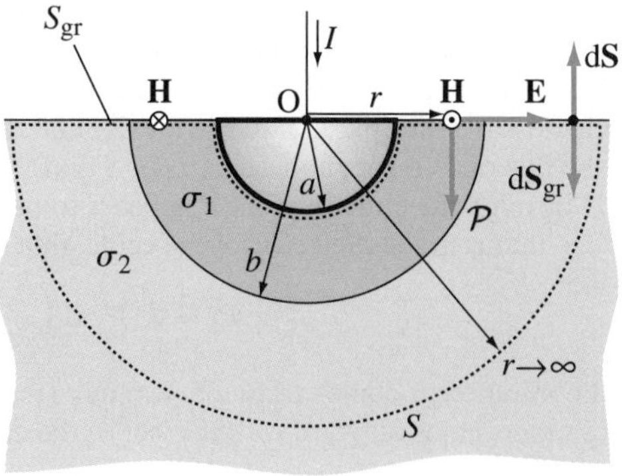

Figure 8.13 Evaluation of the flux of the Poynting vector through the surface of a two-layer earth around a hemispherical grounding electrode with a steady current; for Example 8.16.

The flux $\Psi_{\mathcal{P}}$ can be evaluated also from Poynting's theorem. Let us apply it to the closed surface S shown in Fig. 8.13, which consists of a hemispherical part of radius a (right on the hemispherical surface of the grounding electrode), another hemispherical surface of infinite radius, and the flat surface of the earth, $S_{\rm gr}$. The flux of $\mathcal{P}$ through the first hemispherical part of S (for $r = a$) is zero, because the vector $\mathbf{E}$ is normal to the electrode surface, which makes $\mathcal{P} = \mathbf{E} \times \mathbf{H}$ tangential to the surface no matter how the vector $\mathbf{H}$ at that point is directed. The flux of $\mathcal{P}$ through the other hemispherical surface is also zero, because the electromagnetic field is zero for $r \to \infty$. Eq. (8.167) then results in

$$\Psi_{\mathcal{P}} = -\oint_S \mathcal{P} \cdot \mathrm{d}\mathbf{S} = P_{\rm J} = R_{\rm gr} I^2 = 955\ \mathrm{kW} \tag{8.182}$$

(note that $\mathrm{d}\mathbf{S} = -\,\mathrm{d}\mathbf{S}_{\rm gr}$), where $P_{\rm J}$ is the power of Joule's losses in the earth, given in Eq. (3.201), and $R_{\rm gr}$ is the grounding resistance of the electrode in Fig. 8.13, given in Eq. (3.199).

Example 8.17 Poynting Vector inside a Solenoid with a Low-Frequency Current

A very long air-filled solenoid of length l and circular cross section of radius a ($l \gg a$) carries a time-harmonic current of intensity $i(t) = I_0 \cos \omega t$ in its winding. The frequency is low, such that $\omega\sqrt{\varepsilon_0 \mu_0}\, a \ll 1$ [quasistatic condition, Eq. (8.129), for this case]. The number of wire turns per unit of the solenoid length is N'. Calculate (a) the Poynting vector inside and outside the solenoid and (b) the time-average electromagnetic energy stored inside the solenoid.

Solution

(a) The induced electric field intensity vector, $\mathbf{E}_{\rm ind}$, inside the solenoid is given by Eq. (6.47) and the first expression in Eqs. (6.51), and the magnetic field intensity vector, $\mathbf{H}$, by Eq. (6.48), so that the instantaneous Poynting vector inside the solenoid comes out to be

$$\mathcal{P}(r,t) = \mathbf{E}_{\rm ind} \times \mathbf{H} = E_{\rm ind} H\, \hat{\boldsymbol{\phi}} \times \hat{\mathbf{z}} = -\frac{\mu_0 N'^2}{2}\, r\, i \frac{\mathrm{d}i}{\mathrm{d}t}\, \hat{\mathbf{r}} = \frac{\mu_0 N'^2 \omega I_0^2}{4}\, r \sin 2\omega t\, \hat{\mathbf{r}} \quad (r < a). \tag{8.183}$$

Outside the solenoid, $\mathbf{H} = 0$, and hence $\mathcal{P} = 0$.

(b) The instantaneous electric energy inside the solenoid is [see Eq. (8.160)]

$$W_{\rm e} = \int_v w_{\rm e}\, \mathrm{d}v = \int_v \frac{1}{2} \varepsilon_0 E_{\rm ind}^2\, 2\pi r\, \mathrm{d}r l = \frac{\pi \varepsilon_0 \mu_0^2 N'^2 l}{4} \left(\frac{\mathrm{d}i}{\mathrm{d}t}\right)^2 \int_0^a r^3\, \mathrm{d}r$$
$$= \frac{\pi \omega^2 \varepsilon_0 \mu_0^2 N'^2 a^4 l I_0^2}{16} \sin^2 \omega t, \tag{8.184}$$

where v stands for the volume of the solenoid interior and $\mathrm{d}v$ for an element of that volume in the form of a thin cylindrical shell of radius r, thickness $\mathrm{d}r$, and length l [as in

Eq. (1.144) and Fig. 1.34]. The instantaneous magnetic energy amounts to

$$W_{\mathrm{m}} = w_{\mathrm{m}} v = \frac{1}{2}\mu_0 H^2 \pi a^2 l = \frac{\pi \mu_0 N'^2 a^2 l}{2} i^2 = \frac{\pi \mu_0 N'^2 a^2 l I_0^2}{2} \cos^2 \omega t. \tag{8.185}$$

As the time-average of both $\sin^2 \omega t$ and $\cos^2 \omega t$ is 1/2 [Eq. (6.95)], the time-average total electromagnetic energy inside the solenoid is

$$(W_{\mathrm{em}})_{\mathrm{ave}} = (W_{\mathrm{e}})_{\mathrm{ave}} + (W_{\mathrm{m}})_{\mathrm{ave}} = \frac{\pi \omega^2 \varepsilon_0 \mu_0^2 N'^2 a^4 l I_0^2}{32} + \frac{\pi \mu_0 N'^2 a^2 l I_0^2}{4}$$

$$= \frac{\pi \mu_0 N'^2 a^2 l I_0^2}{4}\left[\frac{(\beta a)^2}{8} + 1\right] \approx \frac{\pi \mu_0 N'^2 a^2 l I_0^2}{4} = (W_{\mathrm{m}})_{\mathrm{ave}}, \tag{8.186}$$

where $\beta = \omega\sqrt{\varepsilon_0 \mu_0}$ is the phase coefficient, Eq. (8.111), in air (we are given that $\beta a \ll 1$).

Let us now check the power balance for this system by using the above results and applying Poynting's theorem to a closed cylindrical surface S laid on the solenoid winding from its interior side (the radius of the surface is $r = a^-$). From Eqs. (8.162) and (8.183), the outward net flux of the Poynting vector through S is

$$P_{\mathrm{f}} = \oint_S \mathcal{P} \cdot \mathrm{d}\mathbf{S} = \mathcal{P}(a^-, t)\, 2\pi a l = \frac{\pi \omega \mu_0 N'^2 a^2 l I_0^2}{2} \sin 2\omega t \tag{8.187}$$

(the flux of $\mathcal{P}$ through the cylinder bases is zero since $\mathcal{P}$ is tangential to S at those points). Combining Eqs. (8.184) and (8.185), on the other hand, we have

$$\frac{\mathrm{d}W_{\mathrm{em}}}{\mathrm{d}t} = \frac{\mathrm{d}W_{\mathrm{e}}}{\mathrm{d}t} + \frac{\mathrm{d}W_{\mathrm{m}}}{\mathrm{d}t} = \frac{\pi \omega \mu_0 N'^2 a^2 l I_0^2}{2}\left[\frac{(\beta a)^2}{8} - 1\right] \sin 2\omega t$$

$$\approx -\frac{\pi \omega \mu_0 N'^2 a^2 l I_0^2}{2} \sin 2\omega t = \frac{\mathrm{d}W_{\mathrm{m}}}{\mathrm{d}t} \tag{8.188}$$

($2 \sin \alpha \cos \alpha = \sin 2\alpha$). We see that

$$\frac{\mathrm{d}W_{\mathrm{em}}}{\mathrm{d}t} + P_{\mathrm{f}} = 0, \tag{8.189}$$

which is the same as Eq. (8.167) for this case ($P_{\mathrm{i}} = 0$ and $P_{\mathrm{J}} = 0$, as there are no generators nor losses inside the solenoid), i.e., the result for the Poynting vector in Eq. (8.183) and energy expressions in Eqs. (8.184) and (8.185) comply with Poynting's theorem applied to S.

Problems: 8.28–8.33; *Conceptual Questions* (on Companion Website): 8.30–8.32; *MATLAB Exercises* (on Companion Website).

8.12 COMPLEX POYNTING VECTOR

In the case of time-harmonic sources and fields in the region v, Poynting's theorem can be formulated also in the complex domain, as the mathematical expression of the balance of complex powers for the region and its surroundings. It is derived from complex Maxwell's equations for the high-frequency electromagnetic field in differential form and complex domain, Eqs. (8.81), and is based on a complex equivalent of the instantaneous Poynting vector in Eq. (8.161). As an introduction to dealing with the complex Poynting vector and complex electromagnetic powers in the general case, let us consider first a simple example of an arbitrary lumped impedance load with a time-harmonic voltage $v(t) = V\sqrt{2}\cos(\omega t + \theta)$ and current $i(t) = I\sqrt{2}\cos(\omega t + \psi)$. The instantaneous power of the load (power that the load receives) is obtained as the product of the instantaneous voltage and current:

instantaneous power

$$P(t) = v(t)i(t) = 2VI\cos(\omega t + \theta)\cos(\omega t + \theta - \phi) = VI\{\cos\phi + \cos[2(\omega t + \theta) - \phi]\}, \quad (8.190)$$

where $\phi = \theta - \psi$ (the phase difference between the voltage and current of the load) and the trigonometric identity $2\cos\alpha\cos\beta = \cos(\alpha+\beta) + \cos(\alpha-\beta)$ is used. We see that $P(t)$ periodically oscillates in time about the constant value $VI\cos\phi$, which represents the time-average power of the load, P_{ave} [also see Eq. (6.95)]. On the other hand, the complex power of the load ($\underline{S}$) is defined as the product of the complex rms voltage and conjugate[14] rms current:

complex power

$$\underline{S} = \underline{P}_{\text{complex}} = \underline{V}\,\underline{I}^* = V\,\text{e}^{\text{j}\theta}\left(I\,\text{e}^{\text{j}\psi}\right)^* = VI\,\text{e}^{\text{j}(\theta-\psi)} = VI\,\text{e}^{\text{j}\phi} = VI\cos\phi + \text{j}\,VI\sin\phi, \quad (8.191)$$

so that its real part equals the time-average power, also referred to as the active power, of the load,

time-average power

$$P_{\text{ave}} = P_{\text{active}} = \text{Re}\{\underline{S}\} = VI\cos\phi. \quad (8.192)$$

The imaginary part of the complex power is equal to the so-called reactive power (Q) of the load,

reactive power

$$Q = P_{\text{reactive}} = \text{Im}\{\underline{S}\} = VI\sin\phi, \quad (8.193)$$

which describes the periodic oscillation of energy between the load and the rest of the circuit.[15]

Analogously, the complex Poynting vector is defined as the cross product of the complex rms electric field intensity vector, $\underline{\mathbf{E}}$, and complex conjugate rms magnetic field intensity vector, $\underline{\mathbf{H}}^*$:

complex Poynting vector

$$\underline{\mathcal{P}} = \underline{\mathbf{E}} \times \underline{\mathbf{H}}^*, \quad (8.194)$$

so that the time average of the instantaneous Poynting vector, in Eq. (8.161), equals the real part of the complex Poynting vector,

time-average from complex Poynting vector

$$\mathcal{P}_{\text{ave}} = \frac{1}{T}\int_0^T \mathcal{P}(t)\,\text{d}t = \frac{1}{T}\int_0^T [\mathbf{E}(t) \times \mathbf{H}(t)]\,\text{d}t = \text{Re}\{\underline{\mathcal{P}}\} = \text{Re}\{\underline{\mathbf{E}} \times \underline{\mathbf{H}}^*\}, \quad (8.195)$$

while the imaginary part of $\underline{\mathcal{P}}$ represents the reactive power flow per unit area.

The derivation of the complex Poynting's theorem is similar to that given by Eqs. (8.155)–(8.159) for the theorem in the time domain. The principal difference is that, in taking dot products of both sides of the two curl equations with the respective field intensity vectors, the first equation is multiplied by $\underline{\mathbf{H}}^*$ (and not $\underline{\mathbf{H}}$) and the second equation is first transformed to the complex conjugate form and then multiplied by $-\underline{\mathbf{E}}$. The result (for the sources represented by an impressed field) is

Poynting's theorem in the complex domain

$$\int_v \underline{\mathbf{E}}_{\text{i}} \cdot \underline{\mathbf{J}}^*\,\text{d}v = \int_v \frac{J^2}{\sigma}\,\text{d}v + \text{j}\omega\int_v (\mu H^2 - \varepsilon E^2)\,\text{d}v + \oint_S (\underline{\mathbf{E}} \times \underline{\mathbf{H}}^*) \cdot \text{d}\mathbf{S}, \quad (8.196)$$

[14] The complex conjugate of $\underline{c} = a + \text{j}b = c\,\text{e}^{\text{j}\alpha}$ (denoted by asterisk) is $\underline{c}^* = a - \text{j}b = c\,\text{e}^{-\text{j}\alpha}$.

[15] Note that only the phase difference ϕ between the voltage and current, and not the individual initial phases θ and ψ, is relevant for both the active and reactive powers of the load, and that is why the current in the definition of complex power in Eq. (8.191) is set in the complex conjugate form, which enables θ and ψ to be replaced by $\phi = \theta - \psi$ in the final expression for $\underline{S}$.

with J, H, and E standing here for the rms current density, magnetic field intensity, and electric field intensity, respectively, in v, which are obtained as $J = J_{\rm rms} = |\underline{\mathbf{J}}| = \sqrt{\underline{\mathbf{J}} \cdot \underline{\mathbf{J}}^*}$ and similarly for H and E.[16] Eq. (8.196) expresses the general principle of conservation of complex power in time-harmonic electromagnetic systems. Analogous complex equation can be written for a region with an impressed current of complex rms density $\underline{\mathbf{J}}_{\rm i}$ [see Eq. (8.166)]. The integral on the left-hand side of Eq. (8.196) is the complex power generated by the impressed field in v. The first integral on the right-hand side of the equation is the time-average power of Joule's or ohmic losses in v. The second integral equals the difference of maximum (peak) magnetic and electric energies in v (H and E are rms values). Multiplied by ω, it represents the reactive power oscillating back and forth between the electromagnetic field in v on one side and the impressed sources in v and/or the fields and sources outside S on the other side. Finally, the last integral in Eq. (8.196) equals the complex power flow through S out of the domain v.

Example 8.18 Poynting Vector in a Lossless Coaxial Cable in an ac Regime

A lossless coaxial cable with conductor radii a and b ($a < b$) is connected at one end to an ideal voltage generator of low-frequency time-harmonic emf $e_{\rm g}(t) = \mathcal{E}\sqrt{2}\cos\omega t$, and the other end of the cable is terminated in a capacitor of capacitance C. Find the flux of the Poynting vector through a cross section of the cable.

Solution As the frequency is low, we can neglect the propagation effects along the cable and assume the same voltage, $v(t) = e_{\rm g}(t)$, between the conductors and the same current intensity, $i(t) = C\,{\rm d}v/{\rm d}t$ [Eq. (3.45)], in the conductors in every cross section of the cable. In the complex domain,

$$\underline{V} = \mathcal{E}\,{\rm e}^{{\rm j}0} = \mathcal{E} \quad \text{and} \quad \underline{I} = {\rm j}\omega C\underline{V} = {\rm j}\omega C\mathcal{E}. \tag{8.197}$$

The electromagnetic field in the dielectric of the cable is quasistatic and has the same spatial distribution as in Fig. 2.17 and Eq. (2.124) for the electric field and in Fig. 4.17 and Eq. (4.61) for the magnetic field. The electric field intensity vector is radial with respect to the cable axis, the magnetic field intensity vector is circular, and their complex rms values are

$$\underline{E} = \frac{\underline{V}}{r\ln(b/a)} \quad \text{and} \quad \underline{H} = \frac{\underline{I}}{2\pi r} \quad (a < r < b), \tag{8.198}$$

respectively, where r is the radial distance from the axis.

The complex Poynting vector, $\underline{\mathcal{P}}$, in the dielectric, computed by Eq. (8.194), has an axial component only, which equals

$$\boxed{\underline{\mathcal{P}} = \underline{E}\,\underline{H}^* = \frac{\underline{V}\,\underline{I}^*}{2\pi r^2\ln(b/a)}}. \tag{8.199}$$

complex Poynting vector (axial) in a coaxial cable

Noting that $\underline{\mathcal{P}}$ has the same dependence on r as in Eq. (8.171) and that it does not depend on z (axial coordinate), its flux through any cross section of the cable (S_z) is obtained in exactly the same way as in Eq. (8.172) for the leaky coaxial cable with a steady current, and the result is

$$\boxed{\underline{S} = \int_{S_z} \underline{\mathcal{P}} \cdot {\rm d}\mathbf{S}_z = \underline{V}\,\underline{I}^*,} \tag{8.200}$$

complex power flow along a transmission line

[16]The validity of this formula for time-harmonic vectors having just one (say x) Cartesian component [see Eq. (8.54)], $\mathbf{J}(t) = J_{\rm rms}\sqrt{2}\cos(\omega t + \psi)\,\hat{\mathbf{x}}$, is obvious, since $\underline{\mathbf{J}} \cdot \underline{\mathbf{J}}^* = (J_{\rm rms}\,{\rm e}^{{\rm j}\psi}\,\hat{\mathbf{x}}) \cdot (J_{\rm rms}\,{\rm e}^{-{\rm j}\psi}\,\hat{\mathbf{x}}) = J_{\rm rms}^2$. That it is also true for time-harmonic vectors represented by two or three mutually orthogonal components with arbitrary peak-values and initial phases will be proved in the next chapter (when studying polarization of electromagnetic waves and their field vectors).

which, given the expressions for $\underline{V}$ and $\underline{I}$ in Eqs. (8.197), becomes $\underline{S} = -\mathrm{j}\omega C\mathcal{E}^2$ in our case. We know from Poynting's theorem in complex form, Eq. (8.196), that this is the complex power flow along the cable through the dielectric from the generator toward the load.[17] In our case, it is purely reactive (negative imaginary), which is to be expected because there are no losses in the cable and the load is purely reactive (capacitive).

Example 8.19 Poynting's Theorem for a Simple RLC Circuit

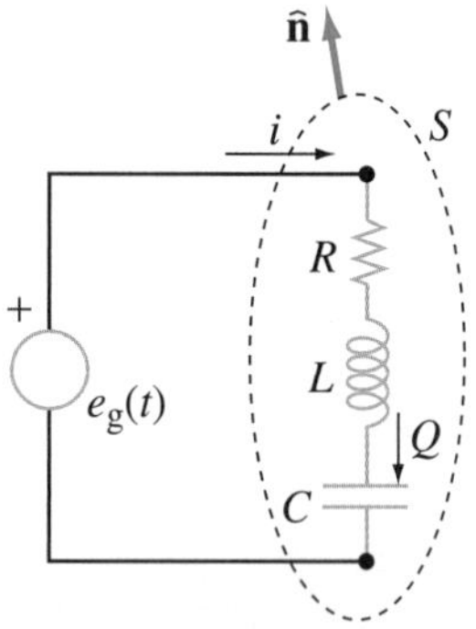

Consider a series RLC circuit with an ideal voltage generator of slowly time-varying emf $e_\mathrm{g}(t)$, as in Fig. 8.14. Apply Poynting's theorem to a surface enclosing the RLC load and discuss the individual terms in the power-balance equation.

Figure 8.14 Application of Poynting's theorem to a series RLC circuit with a slowly time-varying current; for Example 8.19.

Solution Assuming a time dependence of arbitrary form (but slowly varying) for $e_\mathrm{g}(t)$, we first carry out the analysis in the time domain. Then, we also discuss the complex-power terms for the circuit (and a time-harmonic emf).

In Eq. (8.167), as applied to the surface S in Fig. 8.14, $P_\mathrm{i} = 0$ (the generator is outside S), P_J is the instantaneous power of Joule's losses in the resistor, and $W_\mathrm{em} = W_\mathrm{e} + W_\mathrm{m}$, with W_e and W_m being the instantaneous electric and magnetic energies stored in the capacitor and inductor, respectively. From Eqs. (8.52), (2.192), and (7.88),

$$P_\mathrm{J} = Ri^2, \quad W_\mathrm{e} = \frac{Q^2}{2C}, \quad \text{and} \quad W_\mathrm{m} = \frac{1}{2}Li^2. \tag{8.201}$$

The net outward flux of the Poynting vector through S (P_f) multiplied by -1 equals the power delivered through S from the generator to the load in the circuit. Hence, using Eq. (8.200), we have

$$-\oint_S \boldsymbol{\mathcal{P}} \cdot \mathrm{d}\mathbf{S} = vi = e_\mathrm{g}i, \tag{8.202}$$

where the last expression is the instantaneous power generated by the emf e_g in Fig. 8.14 [see Eq. (3.121)], so that Eq. (8.167) becomes

conservation of instantaneous power in a series RLC circuit

$$\boxed{e_\mathrm{g}i = Ri^2 + \frac{\mathrm{d}}{\mathrm{d}t}\left(\frac{Q^2}{2C} + \frac{1}{2}Li^2\right).} \tag{8.203}$$

This equation expresses the theorem of conservation of instantaneous power in ac circuits applied to the simple RLC circuit in Fig. 8.14.[18] It tells us how the instantaneous power of the generator in Fig. 8.14 is used in the RLC load and quantifies the portions of this power that are transformed into heat in the resistor and spent to change (increase if positive, decrease if negative) the electric energy stored in the capacitor and magnetic energy stored in the inductor.

Note that Eq. (8.203) can be rewritten as

$$e_\mathrm{g}i = Ri^2 + Li\frac{\mathrm{d}i}{\mathrm{d}t} + \frac{Q}{C}\frac{\mathrm{d}Q}{\mathrm{d}t}, \tag{8.204}$$

and then both sides of the equation divided by $i = \mathrm{d}Q/\mathrm{d}t$ [see Eq. (8.2)] to yield

$$e_\mathrm{g} = Ri + L\frac{\mathrm{d}i}{\mathrm{d}t} + \frac{Q}{C}. \tag{8.205}$$

[17] Although derived for a specific geometry, that of a coaxial cable, the expression for the complex power flow in Eq. (8.200) is true for an arbitrary transmission line with complex rms voltage $\underline{V}$ and current $\underline{I}$ in a cross section of the line. Note that this expression holds also in the high-frequency case, where $\underline{V}$ and $\underline{I}$ are functions of the z-coordinate along the line.

[18] In general, the theorem on conservation of (instantaneous or complex) power in circuit theory is a special case of Poynting's theorem (in time- or complex-domain) in electromagnetics.

This is the loop equation for the RLC circuit in Fig. 8.14, and the fact that it has been derived from Eq. (8.167) can be considered as a proof of Poynting's theorem for this circuit.

Let us now assume a time-harmonic variation for $e_{\mathrm{g}}(t)$ in Fig. 8.14 and check the complex-power balance for the circuit in light of the general Poynting's theorem in complex form, in Eq. (8.196). To this end, we start with the complex equivalent of Eq. (8.205),

$$\underline{\mathcal{E}}_{\mathrm{g}} = R\underline{I} + \mathrm{j}\omega L\underline{I} - \frac{\mathrm{j}}{\omega C}\underline{I}, \tag{8.206}$$

and multiply both sides of this equation by $\underline{I}^*$. What we obtain is

$$\boxed{\underline{\mathcal{E}}_{\mathrm{g}}\underline{I}^* = RI^2 + \mathrm{j}\omega LI^2 - \mathrm{j}\omega C\left(\frac{I}{\omega C}\right)^2 = RI^2 + \mathrm{j}\omega\left(LI^2 - CV_C^2\right),} \tag{8.207}$$

conservation of complex power in a series RLC circuit

where I is the rms current intensity through the circuit ($I^2 = \underline{I}\underline{I}^*$) and $V_C = I/(\omega C)$ is the rms voltage across the capacitor. Noting that the maximum energy in the inductor equals $LI_0^2/2 = LI^2$, with $I_0 = I\sqrt{2}$ being the amplitude of the current i [see Eq. (8.50)], and, similarly, that the maximum energy in the capacitor equals CV_C^2, we conclude that the two-term expression in parentheses in Eq. (8.207) equals the difference of maximum magnetic and electric energies in the domain enclosed by the surface S in Fig. 8.14. With this, it is obvious that Eq. (8.207) represents a version of Eq. (8.196) for the circuit in Fig. 8.14, i.e., that it expresses the complex Poynting's theorem for this case.

Example 8.20 General Impedance Load

A general impedance load in the form of an arbitrary three-dimensional material object with a pair of terminals and no impressed sources inside it is positioned in air and fed at the terminals by a time-harmonic current of an arbitrary frequency. Given are the complex rms intensity of the input current, $\underline{I}$, and complex rms electric and magnetic field intensity vectors, $\underline{\mathbf{E}}$ and $\underline{\mathbf{H}}$, at every point of the surface S of the object. Find (a) the time-average power of Joule's losses and reactive power inside the object, and (b) resistance and internal reactance of the object with respect to its terminals.

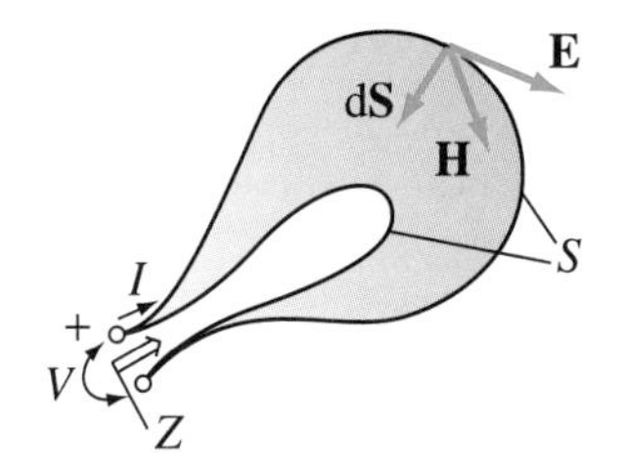

Figure 8.15 General impedance load; for Example 8.20.

Solution

(a) Using Poynting's theorem in complex form, Eq. (8.196), the time-average power of Joule's losses, $(P_{\mathrm{J}})_{\mathrm{ave}}$, and reactive power, Q, inside the object are

$$(P_{\mathrm{J}})_{\mathrm{ave}} = \mathrm{Re}\left\{\oint_S (\underline{\mathbf{E}} \times \underline{\mathbf{H}}^*) \cdot \mathrm{d}\mathbf{S}\right\} \quad \text{and} \quad Q = \mathrm{Im}\left\{\oint_S (\underline{\mathbf{E}} \times \underline{\mathbf{H}}^*) \cdot \mathrm{d}\mathbf{S}\right\}, \tag{8.208}$$

where the vector surface element $\mathrm{d}\mathbf{S}$ is oriented into the object (inward flux), as shown in Fig. 8.15.

(b) Denoting by $\underline{Z}$ the complex impedance of the object (load) with respect to its terminals (see Fig. 8.15), the complex rms voltage across the terminals is $\underline{V} = \underline{Z}\underline{I}$, and

$$\underline{Z} = R + \mathrm{j}X_{\mathrm{i}}, \tag{8.209}$$

where R is the resistance and X_{i} internal reactance of the load. From Eq. (8.191), the complex power of the load, whose real and imaginary parts are given in Eqs. (8.208), can now also be written as

$$\underline{S} = \underline{V}\underline{I}^* = \underline{Z}\underline{I}\underline{I}^* = (R + \mathrm{j}X_{\mathrm{i}})|\underline{I}|^2. \tag{8.210}$$

Hence, R and X_{i} are obtained as

$$\boxed{R = \frac{(P_{\mathrm{J}})_{\mathrm{ave}}}{|\underline{I}|^2} \quad \text{and} \quad X_{\mathrm{i}} = \frac{Q}{|\underline{I}|^2}.} \tag{8.211}$$

resistance and internal reactance of a general impedance load

Note that X_{i} is referred to as the internal reactance because it takes into account the reactive power inside the load (Q) only. In general, the electromagnetic field exists also outside the load, and the associated external reactive power (in the vicinity of the object in Fig. 8.15) is not included in the power balance in Eqs. (8.208). In circuit theory, on the other hand, it is assumed that the field and the power are concentrated only in the circuit elements, so that the imaginary part of $\underline{Z}$ in Eq. (8.209) is simply called reactance and marked as X ($X \equiv X_i$) in the circuit-theory model of the load in Fig. 8.15.

Example 8.21 Complex Poynting Vector from Magnetic Potential

The high-frequency complex magnetic vector potential in a region in the form of a cube of edge length a is given by $\underline{\mathbf{A}} = A_0(x^2\,\hat{\mathbf{x}} + \mathrm{j}y^2\,\hat{\mathbf{z}})/a^2$, where A_0 is a real constant and the region (cube) is defined as $0 \leq x, y, z \leq a$. The medium is air, the angular frequency is ω, and there are no impressed sources in the region. Compute (a) the complex Poynting vector in this region and (b) the complex power flow into the region.

Solution

(a) Using Eqs. (8.194), (8.118), and (8.119), the complex Poynting vector can be expressed in terms of $\underline{\mathbf{A}}$ as

$$\underline{\mathcal{P}} = \underline{\mathbf{E}} \times \underline{\mathbf{H}}^* = -\frac{\mathrm{j}\omega}{\mu_0}\left[\underline{\mathbf{A}} + \frac{1}{\beta^2}\nabla(\nabla\cdot\underline{\mathbf{A}})\right] \times (\nabla \times \underline{\mathbf{A}}^*). \tag{8.212}$$

The expressions for divergence, gradient, and curl in Cartesian coordinates, Eqs. (1.167), (1.102), and (4.81), then yield

$$\underline{\mathcal{P}} = -\frac{\mathrm{j}\omega}{\mu_0}\left[\frac{A_0}{a^2}\left(x^2\,\hat{\mathbf{x}} + \mathrm{j}y^2\,\hat{\mathbf{z}}\right) + \frac{2A_0}{\beta^2 a^2}\,\hat{\mathbf{x}}\right] \times \left(-\frac{2\mathrm{j}A_0 y}{a^2}\,\hat{\mathbf{x}}\right) = -\frac{2\mathrm{j}\omega A_0^2 y^3}{\mu_0 a^4}\,\hat{\mathbf{y}}. \tag{8.213}$$

(b) The complex power flow into the region equals the inward net flux of $\underline{\mathcal{P}}$ through the boundary surface of the region. This net flux reduces to the flux through the cube side defined by $y = a$ only ($\underline{\mathcal{P}}$ is zero for $y = 0$ and is tangential to the remaining four cube sides), so that the inward power flow is

$$\underline{S}_{\mathrm{in}} = \underline{\mathcal{P}}\big|_{y=a}\cdot(-a^2\,\hat{\mathbf{y}}) = \frac{2\mathrm{j}\omega a A_0^2}{\mu_0}. \tag{8.214}$$

We note that $\underline{S}_{\mathrm{in}}$ is purely imaginary, which means that the power inside the region is entirely reactive. This can also be concluded from the Poynting's theorem in complex form, Eq. (8.196), since there are no losses (air) nor impressed sources in the region.

Problems: 8.34–8.41; *Conceptual Questions* (on Companion Website): 8.33–8.35; *MATLAB Exercises* (on Companion Website).

Problems

8.1. **Displacement current in a capacitor with two dielectric layers.** A parallel-plate capacitor of plate area S is connected to a time-harmonic generator operating at a low frequency f. The capacitor is filled with a two-layer perfect dielectric. The thickness of the first layer is d_1 and its permittivity ε_1, while these parameters are d_2 and ε_2 for the second layer, as in Fig. 2.25(a). The amplitude (peak-value) of the conduction current intensity in the capacitor terminals is I_0. Neglecting the fringing effects, find (a) the amplitude of the displacement current density vector in each of the dielectric layers, (b) the amplitude of the electric field intensity vector in each of the layers, and (c) the amplitude of the voltage across the capacitor.

8.2. **Magnetic field due to the displacement current.** If the plates of the capacitor from the previous problem are circular, a in radius, and so is the cross section of each of the two dielectric layers, and the voltage between the plates is given by $v(t) = V_0 \sin \omega t$, compute the magnetic field intensity vector at an arbitrary point in the dielectric. In particular, what is the magnetic field at the dielectric-air interface?

8.3. **Displacement current in an ideal spherical capacitor.** A spherical capacitor with inner electrode radius a and inner radius of the outer electrode b ($b > a$), filled with a perfect, homogeneous dielectric of permittivity ε, is connected to a low-frequency time-harmonic voltage $v(t) = V_0 \cos \omega t$. Find the displacement current density vector at an arbitrary point in the dielectric.

8.4. **Displacement current in a nonideal spherical capacitor.** Repeat the previous problem but for an imperfect dielectric, with parameters ε and σ, filling the capacitor. Also compute the total current density vector in the dielectric, and the conduction current intensity in the capacitor terminals.

8.5. **Displacement current in a coaxial cable.** A coaxial cable with conductor radii a and b ($a < b$) and dielectric permittivity ε is connected to a slowly time-varying voltage $v(t)$. Find the displacement and total current density vectors at an arbitrary point in the cable dielectric if it is (a) perfect and (b) lossy, with conductivity σ, respectively. Cable conductors are perfect.

8.6. **Conduction to displacement current ratio for water.** Repeat Example 8.3 but for samples of (a) fresh water with $\varepsilon_r = 80$ and $\sigma = 10^{-3}$ S/m and (b) seawater with $\varepsilon_r = 80$ and $\sigma = 4$ S/m, respectively.

8.7. **Divergence Maxwell's equations from curl equations.** Starting from the two curl Maxwell's equations for the rapidly time-varying electromagnetic field and the continuity equation, derive the two divergence Maxwell's equations.

8.8. **Flux Maxwell's equations from circulation equations.** Repeat the previous problem but for the integral form of equations, namely, obtain the two flux general Maxwell's equations combining the two circulation equations and the continuity equation.

8.9. **Magnetic from electric field of an antenna using Maxwell's equations.** The electric field intensity vector radiated by an antenna placed at the coordinate origin of a spherical coordinate system is given, far away from the antenna, by $\mathbf{E}(r, \theta, t) = E_0 \sin\theta \cos(\omega t - \beta r)\,\hat{\boldsymbol{\theta}}/r$, where E_0 is a constant and $\beta = \omega\sqrt{\varepsilon_0 \mu_0}$. Using Maxwell's equations in differential form, find the magnetic field intensity vector of the antenna, at the same far point.

8.10. **Charge accompanying volume current with pulse time dependence.** A volume current flowing in a region has the density vector given by $\mathbf{J}(x, t) = J_0 x\, \Pi(t)\, \hat{\mathbf{x}}$, where J_0 is a constant and $\Pi(t)$ is the unit rectangular pulse time function of duration t_0, so $\Pi(t) = 1$ for $0 < t < t_0$ and $\Pi(t) = 0$ for $t < 0$ and $t > t_0$. Determine the associated volume charge density in this region.

8.11. **Transferring time-harmonic field vectors to complex domain.** Determine the complex rms equivalents of the following time-harmonic electric and magnetic field vectors: (a) $\mathbf{E} = 10\,\mathrm{e}^{-0.02x} \cos(3 \times 10^{10}\, t - 250x + 30°)\,\hat{\mathbf{y}}$ V/m, (b) $\mathbf{H} = \left[\cos(10^8 t - z)\,\hat{\mathbf{x}} + \sin(10^8 t - z)\,\hat{\mathbf{y}}\right]$ A/m, and (c) $\mathbf{E} = -0.5 \sin 0.01y \sin(3 \times 10^6 t)\,\hat{\mathbf{z}}$ V/m (t in s; x, y, z in m).

8.12. **Converting complex vectors to instantaneous expressions.** Obtain the instantaneous counterparts of the following complex rms field intensity vectors, assuming that the operating angular frequency is ω: (a) $\underline{\mathbf{E}} = \mathrm{j}\underline{E}_0 \sin\beta z\, \mathrm{e}^{-\mathrm{j}\beta x}\,\hat{\mathbf{x}} + \underline{E}_0 \cos\beta z\, \mathrm{e}^{-\mathrm{j}\beta x}\,\hat{\mathbf{z}}$ ($\underline{E}_0 = E_0\,\mathrm{e}^{\mathrm{j}\theta_0}$), (b) $\underline{\mathbf{H}} = \mathrm{j}h\underline{H}_0 \sin(\pi x/a)\,\mathrm{e}^{-\mathrm{j}\beta z}\,\hat{\mathbf{x}} + \underline{H}_0 \cos(\pi x/a)\,\mathrm{e}^{-\mathrm{j}\beta z}\,\hat{\mathbf{z}}$ ($\underline{H}_0 = H_0\,\mathrm{e}^{\mathrm{j}\psi_0}$), and (c) $\underline{\mathbf{E}} = b\underline{I}\,\mathrm{e}^{-\mathrm{j}\beta r}\{2[1/(\mathrm{j}\beta r)^2 + 1/(\mathrm{j}\beta r)^3]\,\hat{\mathbf{r}} + [1/(\mathrm{j}\beta r) + 1/(\mathrm{j}\beta r)^2 + 1/(\mathrm{j}\beta r)^3]\,\hat{\boldsymbol{\theta}}\}$ ($\underline{I} = I\,\mathrm{e}^{\mathrm{j}\psi}$).

8.13. **Divergence-free equivalent electric displacement vector.** Consider a high-frequency time-harmonic electromagnetic field of angular frequency ω in an inhomogeneous lossy medium of parameters ε, μ, and σ, and a vector defined as $\underline{\varepsilon}_\mathrm{e}\underline{\mathbf{E}}$, where $\underline{\varepsilon}_\mathrm{e} = \varepsilon - \mathrm{j}\sigma/\omega$ (this is the so-called equivalent complex permittivity of the medium, which will be introduced and discussed in the next chapter) and $\underline{\mathbf{E}}$ is the complex electric field vector. (a) Using complex Maxwell's equations in differential form, show that $\underline{\varepsilon}_\mathrm{e}\underline{\mathbf{E}}$ (which may be termed the equivalent electric displacement vector) is a divergence-free vector, namely, that $\nabla \cdot (\underline{\varepsilon}_\mathrm{e}\underline{\mathbf{E}}) = 0$, at any point of the medium.

(b) From the corresponding integral Maxwell's equations, also show that the flux of $\underline{\varepsilon}_{\rm e}\underline{\mathbf{E}}$ through any closed surface is zero.

8.14. **Boundary condition for the equivalent displacement vector.** Consider a boundary surface between two media, whose electromagnetic properties are described by ε_1, μ_1, and σ_1 for medium 1 and ε_2, μ_2, and σ_2 for medium 2, and which, in general, is not free of surface charges ($\rho_{\rm s} \neq 0$). For a time-harmonic variation of electromagnetic fields with angular frequency ω, use the appropriate boundary conditions to show that the normal component of the vector $\underline{\varepsilon}_{\rm e}\underline{\mathbf{E}}$ ($\underline{\varepsilon}_{\rm e} = \varepsilon - {\rm j}\sigma/\omega$) is continuous across the boundary, that is, $\hat{\mathbf{n}} \cdot (\underline{\varepsilon}_{\rm e1}\underline{\mathbf{E}}_1) - \hat{\mathbf{n}} \cdot (\underline{\varepsilon}_{\rm e2}\underline{\mathbf{E}}_2) = 0$, where $\underline{\varepsilon}_{{\rm e}k} = \varepsilon_k - {\rm j}\sigma_k/\omega$ ($k = 1, 2$).

8.15. **Analysis of a nonideal capacitor in the complex domain.** Redo Example 8.2 but in the complex domain, assuming that the angular frequency of the applied voltage v in Fig. 8.2 equals $\omega = \sigma/\varepsilon$.

8.16. **Antenna magnetic field from electric – in the complex domain.** Redo Problem 8.9 but in the complex domain: (a) find the complex representative of the antenna instantaneous electric field $\mathbf{E}(r, \theta, t)$, and (b) use complex Maxwell's equations in differential form to obtain the complex magnetic field vector ($\underline{\mathbf{H}}$) of the antenna.

8.17. **Wave equations for Lorenz potentials in complex form.** (a) Transfer wave equations for Lorenz potentials, Eqs. (8.92) and (8.93), to the complex domain. (b) Derive complex wave equations in (a) from complex Maxwell's equations in differential form, paralleling the time-domain derivation in Eqs. (8.89)–(8.93).

8.18. **One-dimensional source-free wave equation in complex form.** Write the complex-domain equivalent of the one-dimensional wave equation (for U) in Eq. (8.97). Show that $\underline{U} = {\rm e}^{-{\rm j}\beta R}$, with β (phase coefficient) given in Eq. (8.111), is a solution of this equation. Why is then $\underline{V} = {\rm e}^{-{\rm j}\beta R}/R$ a solution of the three-dimensional wave equation in complex form for the electric potential (from the previous problem)?

8.19. **Gradient of the source-to-field distance.** Obtain the gradient of the source-to-field distance (R) in Fig. 8.7 using Eq. (8.124) and the formula for gradient in Cartesian coordinates, and verify that the result matches that in Eq. (8.122).

8.20. **High-frequency current in a semicircular wire conductor.** A wire conductor in the form of a semicircle of radius a, representing a part of a more complex wire contour, carries a time-harmonic current of a high frequency, f, and complex rms intensity $\underline{I}(\phi) = I_0 \cos\phi$, where I_0 is a constant and $0 \leq \phi \leq \pi$, as shown in Fig. 8.16, and the surrounding medium is air. Find (a) the complex rms line charge density accompanying $\underline{I}(\phi)$, as well as the complex expressions for (b) the electric scalar potential, (c) the magnetic vector potential, (d) the electric field intensity vector, and (e) the magnetic field intensity vector at an arbitrary point along the z-axis due to the current and charge of the semicircular wire. (f) Write down the instantaneous (time-domain) expressions of the results from (c) and (d).

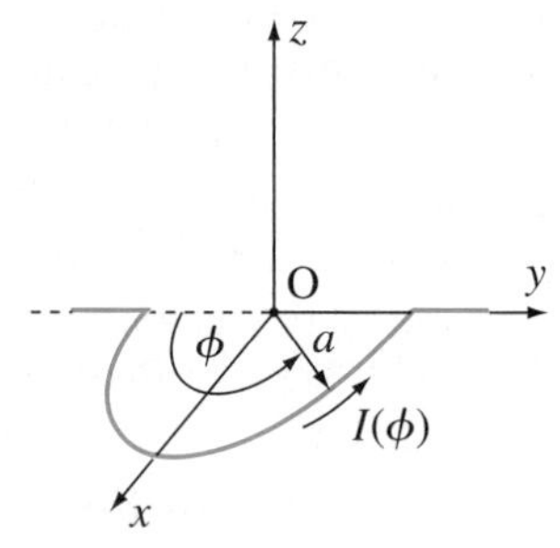

Figure 8.16 Semicircular wire conductor with a high-frequency current whose intensity varies along the wire; for Problem 8.20.

8.21. **High-frequency line current along 3/4 of a circle.** Repeat the previous problem but for a wire conductor along three quarters of the circle, Fig. 8.17, and current intensity $\underline{I}(\phi) = I_0 \sin\phi$ ($0 \leq \phi \leq 3\pi/2$).

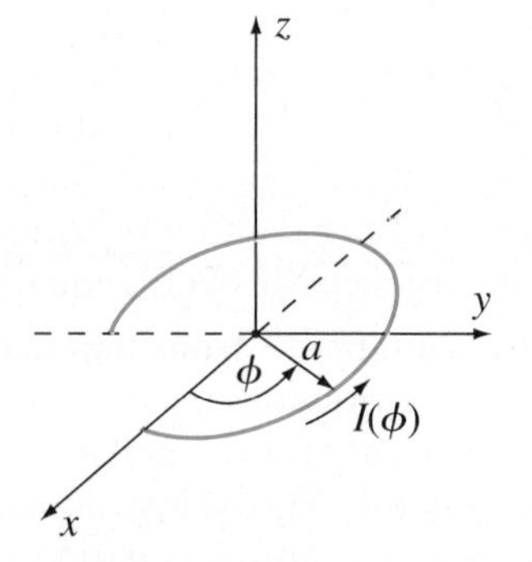

Figure 8.17 High-frequency current with a varying magnitude along three quarters of a circle; for Problem 8.21.

8.22. **Abrupt change of current intensity in a circular loop.** A current of intensity $i(t, \phi) = \cos(3\phi/2)\sin(10^{10}t)$ A ($0 < \phi < 2\pi$; t in s) flows in the positive ϕ direction along a circular wire loop of radius $a = 10$ cm in free space. (a) Verify that this is a high-frequency current. (b) From the complex rms current intensity along the loop, $\underline{I}(\phi)$, determine the charge distribution of the wire, namely, the complex rms line charge density, $\underline{Q}'(\phi)$, for $0 < \phi < 2\pi$ and the point charge, $\underline{Q}_0$, at the point defined by $\phi = 0$ [apply the continuity equation in integral form to a small closed surface enclosing this point, as in Fig. 8.4(a), and identify the intensities of current entering the surface, at $\phi = 2\pi^-$, and leaving it, at $\phi = 0^+$]. (c) Compute both electromagnetic potentials at an arbitrary point (P) along the axis of the loop perpendicular to its plane (z-axis). (d) Show that the electric field intensity vector at the point P does not have a z-component.

8.23. **High-frequency circular surface current over a hollow plate.** Assume that the high-frequency time-harmonic surface current in Fig. 8.5 is given by $\mathbf{J}_s(r, t) = J_{s0}\sqrt{2}(a/r)\cos\omega t\,\hat{\boldsymbol{\phi}}$ ($a \le r \le b$), where J_{s0} is a constant. Using the result for the magnetic field due to a loop with a current of constant magnitude in Eq. (8.135), along with Eq. (1.62) to change variables in integration (see Fig. 1.14) and the expression for the derivative in R of $\mathrm{e}^{-\mathrm{j}\beta R}/R$ in Eq. (8.121) to carry out the integration (instead of differentiation), find the magnetic flux density vector along the z-axis (in Fig. 8.5).

8.24. **Uniform high-frequency plate current.** A high-frequency time-harmonic surface current flows uniformly over a circular plate of radius a in free space. The complex rms density of the current is $\underline{\mathbf{J}}_s$, the same at every point of the plate, and its angular frequency is ω. (a) Determine the magnetic vector potential at an arbitrary point along the plate axis normal to its plane. (b) Starting with Eq. (8.128) and performing a similar integration procedure as in the previous problem, compute the magnetic flux density vector at the same point. (c) Obtain the result in (b) from that in (a), using Eq. (8.119).

8.25. **High-frequency surface currents over a sphere, θ-directed.** Consider the distribution of high-frequency time-harmonic surface currents over the surface of a sphere of radius a in free space given by the following expression for the instantaneous current density vector in the spherical coordinate system with origin at the sphere center: $\mathbf{J}_s(\theta, t) = J_{s0}\sqrt{2}\sin\theta\sin\omega t\,\hat{\boldsymbol{\theta}}$ ($0 \le \theta \le \pi$), where J_{s0} is a constant. Carrying out a similar surface integration procedure as that in Eqs. (2.32) and (8.153), find the magnetic vector potential at the sphere center.

8.26. **High-frequency surface currents over a sphere, ϕ-directed.** If the current distribution over the sphere from the previous problem is given by $\mathbf{J}_s(t) = J_{s0}\sqrt{2}\cos\omega t\,\hat{\boldsymbol{\phi}}$, calculate (a) the magnetic vector potential and (b) the magnetic flux density vector at the sphere center.

8.27. **High-frequency volume current in a hollow hemisphere.** A high-frequency time-harmonic current is uniformly distributed throughout the volume of a hollow hemisphere of inner and outer radii a and b ($b > a$), respectively, as shown in Fig. 8.18. The complex rms current density vector, given by $\underline{\mathbf{J}} = \underline{J}\,\hat{\mathbf{z}}$, is perpendicular to the flat surface of the hemisphere. The angular frequency of the current is ω, and the medium parameters are ε_0 and μ_0 everywhere. Under these circumstances, find (a) the distribution of volume and surface charges of the hemisphere and (b) the electric scalar potential at its center (point O).

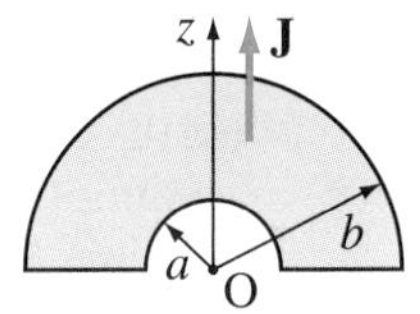

Figure 8.18 Hollow hemisphere with a uniformly distributed high-frequency time-harmonic volume current; for Problem 8.27.

8.28. **Poynting's theorem for a leaky coaxial cable.** Consider the coaxial cable with perfect conductors and a continuously inhomogeneous imperfect dielectric described in Problem 3.18, and assume that it is driven at one end by an ideal voltage generator of time-constant emf $\mathcal{E}$ and terminated at the other end in a load of resistance R_L. The permeability of the dielectric is μ_0 and the length of the cable is l. Under these circumstances, find the flux of the Poynting vector through an arbitrary cross section of the cable – in the following four ways, respectively: (a) integrating $\mathcal{P}$, as in Eq. (8.172), (b) applying Poynting's theorem to a closed surface S enclosing the load as shown in Fig. 8.12(a), (c) applying the theorem to a closed surface S' enclosing the generator, also shown in Fig. 8.12(a), and finally (d) to a closed surface S'' enclosing a compensating ideal current generator of current intensity $I_g = I(z)$, as in Fig. 8.12(b).

8.29. **Poynting's theorem for another coaxial cable.** Repeat the previous problem but for the coaxial cable with two coaxial homogeneous layers of imperfect dielectric shown in Fig. 3.34 and described in Problem 3.17.

8.30. **Poynting's theorem for a leaky planar transmission line.** Repeat Problem 8.28 but for the leaky planar transmission line with a two-layer dielectric (of permeability μ_0) from Fig. 3.24 and Example 3.14. Assume that $w \gg d_1 + d_2$, so that fringing effects can be neglected, and compute the magnetic field between the metallic strips as suggested in Problem 7.7.

8.31. **Poynting's theorem for a grounding electrode.** Find the flux of the Poynting vector through the surface of the continuously inhomogeneous earth around the hemispherical grounding electrode shown in Fig. 3.38 and described in Problem 3.23, in the following two ways: (a) integrating $\mathcal{P}$, as in Eq. (8.181), and (b) applying Poynting's theorem to the closed surface S in Fig. 8.13, respectively.

8.32. **Flux of the Poynting vector for a deeply buried electrode.** For the spherical grounding electrode in a homogeneous earth shown in Fig. 3.29(a) and described in Example 3.18, compute the flux of the Poynting vector through the surface of the earth.

8.33. **Poynting vector due to slowly time-varying line currents.** Consider (a) the square wire contour carrying an EMI pulse current from Example 6.2 and Problem 6.3 and (b) the contour composed of semicircular and linear parts with a low-frequency time-harmonic current from Example 6.4 and Problem 6.5, and find the Poynting vector at the point M in Fig. 6.2(a) and point O in Fig. 6.4, respectively, neglecting the electric field due to excess charge in the contours.

8.34. **Deriving complex Poynting's theorem.** Derive Poynting's theorem in complex form, Eq. (8.196), in the way suggested in the text preceding this equation.

8.35. **Poynting vector in a lossless planar line in an ac regime.** Consider an air-filled lossless planar transmission line, with the width of both conducting strips being w, and the separation between them d. The line is driven, at one end, by a low-frequency time-harmonic generator, and is terminated at the other end in a complex impedance load. The complex rms voltage and current of the line are $\underline{V}$ and $\underline{I}$, respectively, in every cross section. Assuming that $w \gg d$, and thus neglecting the fringing effects, determine the complex Poynting vector between the strips of the line, and its flux through the line cross section.

8.36. **Poynting vector of an electromagnetic wave in a coaxial cable.** An electromagnetic wave propagates along a lossless coaxial cable with conductor radii a and b ($a < b$) and homogeneous dielectric of parameters ε and μ. In the cylindrical coordinate system whose z-axis coincides with the axis of the cable, the electric field intensity vector of the wave in the cable dielectric is given by $\mathbf{E}(r, z, t) = E_0 \cos(\omega t - \beta z)\,\hat{\mathbf{r}}/r$, with E_0 being a constant and $\beta = \omega\sqrt{\varepsilon\mu}$. Under these circumstances, find: (a) the magnetic field intensity vector in the dielectric, (b) the complex Poynting vector in the dielectric, (c) the complex power flow along the cable (through the dielectric), and (d) the time-average power flow of the cable.

8.37. **Poynting's theorem for a parallel RLC circuit.** Repeat Example 8.19 but for a simple parallel *RLC* or *GLC* circuit driven by an ideal current generator of slowly time-varying current intensity $i_g(t)$. Write equivalents of all equations in the analysis and discussion in both the time and the complex domains, Eqs. (8.201)–(8.207), for this circuit.

8.38. **Poynting vector in a nonideal capacitor in an ac regime.** For the nonideal capacitor with slowly time-varying displacement and conduction currents in the dielectric from Example 8.2 and Problem 8.15, find both the instantaneous and complex Poynting vectors at an arbitrary point of the dielectric (take $\omega = \sigma/\varepsilon$). Show that the two vectors are related as in Eq. (8.195).

8.39. **Poynting's theorem in complex form for a nonideal capacitor.** Consider the capacitor from the previous problem, and assume that its dielectric is low-loss (and nonmagnetic) so that $\mu_0\sigma^2a^2 \ll \varepsilon$. Then apply Poynting's theorem in complex form, Eq. (8.196), to a closed cylindrical surface coinciding with the surface of the dielectric (in Fig. 8.2). Compute all individual terms in the power-balance equation, and show that the theorem holds true.

8.40. **Poynting vector due to a high-frequency line current.** For the loop with a high-frequency cosine current distribution along the wire described in Example 8.13, calculate the instantaneous and complex Poynting vectors at the point P (along the z-axis) in Fig. 8.10(a), and show that they satisfy the relationship in Eq. (8.195).

8.41. **Poynting vector due to antenna radiation.** Repeat the previous problem but for the antenna field at a far point from Problems 8.9 and 8.16.

9 Uniform Plane Electromagnetic Waves

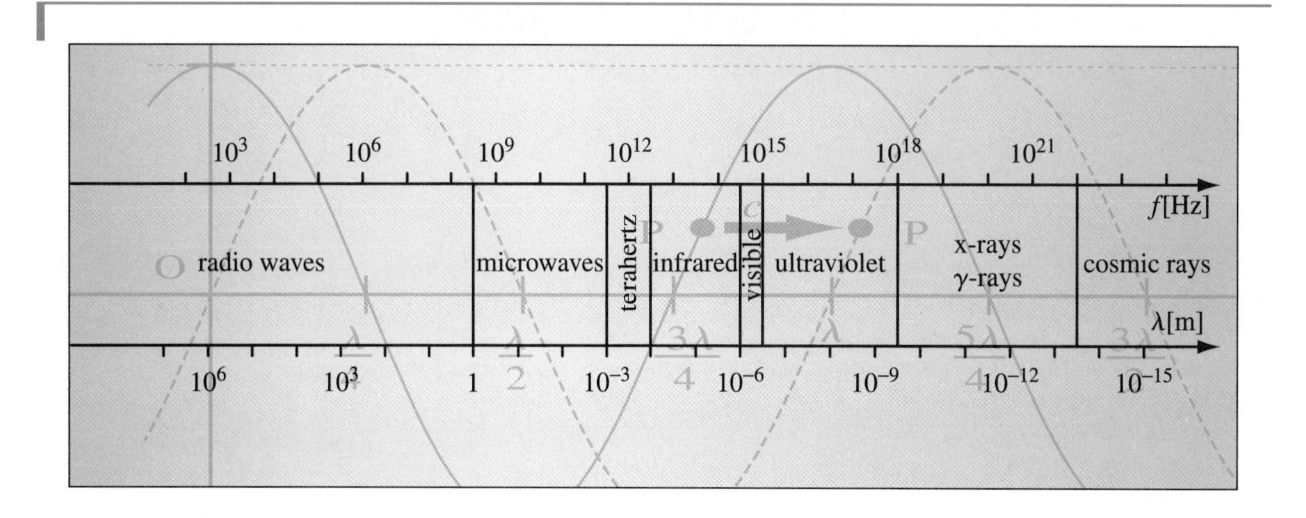

Introduction:

Electromagnetic waves, i.e., traveling electric and magnetic fields, are the most important consequence of general Maxwell's equations, discussed in the preceding chapter. In Section 8.3, we studied qualitatively the propagation of electromagnetic waves as a process of successive mutual induction of electric and magnetic fields in space and time based on the first two integral Maxwell's equations for the rapidly time-varying electromagnetic field. We now proceed with a formal (quantitative) analysis of electromagnetic wave propagation starting with general Maxwell's equations in differential form. Our goal here is not to investigate how the electromagnetic radiation is originated (undoubtedly by rapidly time-varying currents and charges in a source region, such as the one in Fig. 8.7, which is simply a transmitting antenna), but to describe the properties of waves as they propagate away from their sources. In this chapter, we study propagation of electromagnetic waves in unbounded media; analysis of wave interaction with interfaces between material regions with different electromagnetic properties, namely, wave reflection and transmission at such interfaces, follows in the next chapter.

We shall first derive three-dimensional wave equations for the electric and magnetic field intensity vectors, as a starting point for the characterization of electromagnetic waves in free space and in material media. Uniform-plane-wave approximation of nonuniform spherical waves will then be introduced, as a means for much simpler analysis of waves radiated by remote sources (antennas). We shall next perform detailed discussions of uniform plane electromagnetic waves in lossless media, based on computations in both time and complex domains. A general theory of time-harmonic waves in media that exhibit ohmic losses, derived from the concept of equivalent complex permittivity, will be followed by several special cases of wave interaction with electromagnetic materials. These include investigations of wave propagation in good dielectrics, good conductors, and ionized gases (plasmas). We shall also discuss frequency behavior of electromagnetic materials and associated wave propagation effects, as well as dispersion (signal distortion due to a frequency-dependent wave velocity in the medium). Finally, polarization of time-harmonic electromagnetic waves will be studied, based on the analysis of the curve traced in the course of time by the tip of the electric field intensity vector of the wave.

Theory of uniform plane electromagnetic waves has direct application in radio communication systems, wireless propagation, radar engineering, optics, etc. However, the theoretical and practical importance and usefulness of the topics to be covered in this chapter are beyond the model of uniform plane waves and their interaction with various electromagnetic media, as this material is crucial for understanding all other wave topics that will follow in later chapters, including guided electromagnetic waves (e.g., in a coaxial cable) and electromagnetic radiation (antennas).

9.1 WAVE EQUATIONS

We consider an electromagnetic wave whose electric and magnetic field intensity vectors are **E** and **H**, respectively, in an unbounded region filled with a linear, homogeneous, and lossless ($\sigma = 0$) material of permittivity ε and permeability μ. The region is assumed to be completely free of sources, whether that be impressed generators ($\mathbf{E}_\mathrm{i} = 0$ and $\mathbf{J}_\mathrm{i} = 0$) or induced currents and charges ($\mathbf{J} = 0$ and $\rho = 0$). For such a source-free region, general Maxwell's equations in differential form, Eqs. (8.24), can be written as (note that $\mathbf{J} = 0$ also follows from $\sigma = 0$)

$$\nabla \times \mathbf{E} = -\mu \frac{\partial \mathbf{H}}{\partial t}, \tag{9.1}$$

$$\nabla \times \mathbf{H} = \varepsilon \frac{\partial \mathbf{E}}{\partial t}, \tag{9.2}$$

$$\nabla \cdot \mathbf{E} = 0, \tag{9.3}$$

$$\nabla \cdot \mathbf{H} = 0. \tag{9.4}$$

source-free Maxwell's equations in time domain

These are first-order partial differential equations with spatial coordinates and time as independent variables and **E** and **H** as unknowns (unknown functions, to be determined, of space and time). They can be combined to give second-order partial differential equations in terms of **E** or **H** alone. Namely, taking the curl of Eq. (9.1) and substituting $\nabla \times \mathbf{H}$ on the right-hand side of thus obtained equation by the expression on the right-hand side of Eq. (9.2), along with transformations similar to those in Eq. (8.90), yield

$$\nabla \times (\nabla \times \mathbf{E}) = \nabla(\nabla \cdot \mathbf{E}) - \nabla^2 \mathbf{E} = -\varepsilon\mu \frac{\partial^2 \mathbf{E}}{\partial t^2}. \tag{9.5}$$

The divergence of **E** is zero [Eq. (9.3)], and hence

$$\nabla^2 \mathbf{E} - \varepsilon\mu \frac{\partial^2 \mathbf{E}}{\partial t^2} = 0. \tag{9.6}$$

wave equation for **E**

In an entirely analogous fashion, starting with the curl of Eq. (9.2), we can obtain a second-order partial differential equation in **H**:

$$\nabla^2 \mathbf{H} - \varepsilon\mu \frac{\partial^2 \mathbf{H}}{\partial t^2} = 0. \tag{9.7}$$

wave equation for **H**

Eqs. (9.6) and (9.7) are three-dimensional source-free wave equations for the electric and magnetic field intensity vectors, respectively. Note that these equations, which are commonly referred to simply as wave equations, have the same form as the source-free version (with $\mathbf{J} = 0$) of Eq. (8.93). They have natural solutions in the form of traveling waves in lossless homogeneous material media (of parameters ε and μ) with no sources. The velocity of the waves is given in Eq. (8.97).

The advantage of using wave equations in the analysis of wave propagation is in that each of them is an equation with one unknown (**E** or **H**), whereas Maxwell's equations represent a system of simultaneous equations with two unknowns (**E** and **H**). However, Eqs. (9.6) and (9.7) are not independent from each other, because they are both obtained from the same two curl Maxwell's equations, Eqs. (9.1) and (9.2). Therefore, the system of two wave equations is not sufficient for obtaining both **E** and **H**. In other words, while Eqs. (9.6) and (9.7) are a consequence of Eqs. (9.1)–(9.4), so that any solution to the full set of Maxwell's equations automatically satisfies both wave equations, an equivalent reasoning in the opposite direction is not valid [since Eqs. (9.1)–(9.4) cannot be obtained from Eqs. (9.6) and (9.7)]. Consequently, a possible approach to solving wave propagation problems makes use of one of the two wave equations and two (one curl and one divergence) of the four Maxwell's differential equations. In specific, we can find from Eq. (9.6) a solution for **E** that satisfies Eq. (9.3), and then solve Eq. (9.1) for **H** (by "back substitution"). An analogous solution procedure starting with Eq. (9.7) is, of course, also possible.

In the case of time-harmonic electromagnetic fields (waves) of angular frequency ω, Eqs. (9.6) and (9.7) can be converted into their complex equivalents [or, alternatively, derived from Maxwell's differential equations in the complex domain, Eqs. (8.81)]. We recall that taking the time derivative of an instantaneous quantity is equivalent to multiplying its complex representative by $\mathrm{j}\omega$ [see Eq. (8.68)], so that the second time derivatives in Eqs. (9.6) and (9.7) ought to be replaced by $\mathrm{j}\omega \times \mathrm{j}\omega = -\omega^2$ in the complex domain, which results in

Helmholtz equation for **E**

$$\boxed{\nabla^2 \underline{\mathbf{E}} + \beta^2 \underline{\mathbf{E}} = 0,} \quad (9.8)$$

Helmholtz equation for **H**

$$\boxed{\nabla^2 \underline{\mathbf{H}} + \beta^2 \underline{\mathbf{H}} = 0,} \quad (9.9)$$

where β is the phase coefficient (wavenumber) given in Eq. (8.111). Wave equations in complex form are known as the Helmholtz equations. Eqs. (9.8) and (9.9) are also mutually dependent, and the same comments given above with regards to the possible solution procedures in the time domain using the corresponding wave and Maxwell's equations together apply also in complex-domain analysis of wave propagation based on Helmholtz equations.

The solutions of wave (or Helmholtz) equations describe the characteristics of electromagnetic waves as dictated by Maxwell's equations, and Eqs. (9.6)–(9.9) thus represent a starting point in many branches of theoretical, computational, and applied electromagnetics concerned with wave propagation. In addition, second-order partial differential equations of similar types occur in many other disciplines of science and engineering, making a wave equation (in different forms and notations) one of the mathematical concepts with the most important and comprehensive implications in the modeling of physical processes.

Example 9.1 Scalar Helmholtz Equations for Cartesian Field Components

For an electromagnetic wave whose electric and magnetic complex field vectors, $\underline{\mathbf{E}}$ and $\underline{\mathbf{H}}$, are expressed as functions of Cartesian coordinates and lossless propagation medium, show that each of the two vector Helmholtz equations reduces to three scalar partial differential equations with individual Cartesian components of $\underline{\mathbf{E}}$ or $\underline{\mathbf{H}}$ as unknowns.

Solution Since, in general, the Cartesian components of the Laplacian (∇^2) of a vector ($\underline{\mathbf{E}}$ or $\underline{\mathbf{H}}$ in this case) equal the Laplacian of the corresponding vector components (scalars), Eq. (4.127), the vector Helmholtz equation for the electric field, Eq. (9.8), can be written as

$$\nabla^2 \underline{E}_x \hat{\mathbf{x}} + \nabla^2 \underline{E}_y \hat{\mathbf{y}} + \nabla^2 \underline{E}_z \hat{\mathbf{z}} + \beta^2 \left(\underline{E}_x \hat{\mathbf{x}} + \underline{E}_y \hat{\mathbf{y}} + \underline{E}_z \hat{\mathbf{z}} \right) = 0$$

$$\longrightarrow \quad \nabla^2 \underline{E}_x + \beta^2 \underline{E}_x = 0, \quad \nabla^2 \underline{E}_y + \beta^2 \underline{E}_y = 0, \quad \text{and} \quad \nabla^2 \underline{E}_z + \beta^2 \underline{E}_z = 0, \qquad (9.10)$$

i.e., it decouples onto three scalar Helmholtz equations, for $\underline{E}_x$, $\underline{E}_y$, and $\underline{E}_z$, respectively [also see Eqs. (8.93) and (8.94), for a similar decomposition]. The H-field vector Helmholtz equation, in Eq. (9.9), is decoupled onto scalar equations in an analogous manner.

Problems: 9.1; *Conceptual Questions* (on Companion Website): 9.1–9.3.

9.2 UNIFORM-PLANE-WAVE APPROXIMATION

Consider an arbitrary distribution of rapidly time-varying sources (currents and charges) in an unbounded linear, homogeneous, and lossless medium (Fig. 8.7). Far away from the source domain v, the elementary spherical electromagnetic waves originated by the elementary sources $\rho\,\mathrm{d}v$ and $\mathbf{J}\,\mathrm{d}v$ over v form a unified global spherical wavefront[1] with respect to the center of v, as illustrated in Fig. 9.1(a). In other words, the superposition of elementary waves due to a faraway distribution of sources can be replaced by an equivalent spherical wave emanating from a single point source (note that this discussion is not restricted to volume sources, shown in Fig. 8.7, but applies also to surface and line source distributions). To a distant observer (e.g., at the receiving end of a wireless link), the wavefront of a spherical wave appears to be approximately planar, as if it were a part of a plane (or planar) wave – the wave whose wavefront is a plane (sphere with infinite radius). In addition, the electromagnetic energy (transmitted signal) is actually received at the

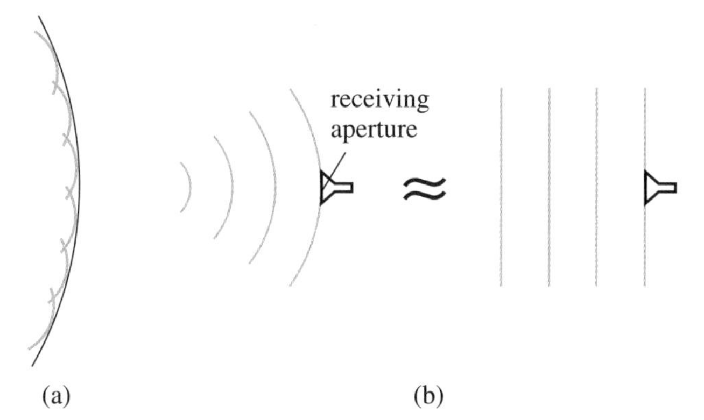

Figure 9.1 (a) A unified global spherical wavefront formed from many elementary spherical waves due to a faraway distribution of sources and (b) an approximation, over a receiving aperture, of the radiated global spherical wave by a uniform plane wave propagating through the entire space.

[1]The wavefront of a wave, in general, is the geometrical locus of points in space reached by the wave in all directions at a given time.

receiving end only over a finite surface area representing a small portion of the entire spherical wavefront, called a receiving aperture, as indicated in Fig. 9.1(b). Therefore, although the global spherical wave is generally nonuniform (fields at the same distance from the center of v are stronger in some directions than in others), its part over a receiving aperture can be considered to be uniform (fields are the same at every point). Overall, the nonuniform spherical global electromagnetic wave produced (radiated) by the sources in Fig. 8.7, if considered only over a receiving aperture far away from the sources, can be treated as if it were a part of a uniform plane wave. Such a wave has planar wavefronts and uniform (constant) distributions of fields over every plane perpendicular to the direction of wave propagation. Most importantly, we can completely remove the spherical wave from the analysis and assume that a uniform plane wave illuminating the aperture exists in the entire space [Fig. 9.1(b)].

The uniform-plane-wave approximation of nonuniform spherical waves enables considerably simpler analysis of waves radiated by remote rapidly time-varying sources. Once this model is established, we then deal with uniform plane waves only, and study their propagation not only in unbounded media with and without losses (this chapter) but also in the presence of planar interfaces between material regions with different electromagnetic properties (next chapter). However, the importance and usefulness of the concept of uniform plane waves is beyond the approximation of spherical waves in Fig. 9.1. For instance, it can be shown that an arbitrary, nonuniform and/or nonplanar (not necessarily spherical), wave may be expressed as a superposition of uniform-plane-wave components, and some analysis techniques use such wave decomposition to reduce complex wave propagation and interaction problems to essentially the analysis of uniform plane waves.

9.3 TIME-DOMAIN ANALYSIS OF UNIFORM PLANE WAVES

We start with an arbitrary time dependence of uniform plane electromagnetic waves propagating in unbounded media without losses and perform time-domain analysis of such waves based on wave equations. Complex-domain analysis of time-harmonic uniform plane waves will be introduced in the next section.

A plane wave can be described using rectangular (Cartesian) coordinates. If the coordinate axes are oriented such that the direction of wave propagation is along the z-axis, then the electric field intensity vector of the wave at any instant of time is constant in every plane perpendicular to the z-axis (uniform wave), i.e., $\mathbf{E}$ depends only on z and time. Furthermore, $\mathbf{E}$ cannot have a z-component ($E_z = 0$), because Eq. (9.3) must be satisfied (source-free region). Namely, as $\partial E_x/\partial x = 0$ and $\partial E_y/\partial y = 0$ (since $\mathbf{E}$ does not depend on x and y), Eq. (9.3) reduces to $\partial E_z/\partial z = 0$ [see Eq. (1.167)], which in turn implies that $E_z = 0$ [the possibility of a constant solution (with respect to z) for E_z is not of interest, because the electric field intensity must change along the direction of the wave propagation (see Fig. 8.3)]. Consequently, $\mathbf{E}$ lies entirely in a plane perpendicular to the z-axis, and we can position the other two axes of the Cartesian coordinate system such that $\mathbf{E}$ be directed along any one of them. For instance, if we choose the x-axis to represent the direction of $\mathbf{E}$, as shown in Fig. 9.2, and eliminate the coordinates on which the field is not dependent, we are left with

$$\mathbf{E} = E_x(z, t)\,\hat{\mathbf{x}}. \tag{9.11}$$

With this, Eq. (9.6) simplifies to [see Eqs. (4.126) and (9.10), and note that $\partial^2 E_x/\partial x^2 = 0$ and $\partial^2 E_x/\partial y^2 = 0$]

$$\boxed{\frac{\partial^2 E_x}{\partial z^2} - \varepsilon\mu\frac{\partial^2 E_x}{\partial t^2} = 0.} \tag{9.12}$$

1-D wave equation for E

This is a one-dimensional (1-D) scalar wave equation of the same form as Eq. (8.97), so that its solution must have the same form as that in Eq. (8.98):

$$\boxed{E_x = f\left(t - \frac{z}{c}\right),} \tag{9.13}$$

E-field of a uniform plane wave

where $f(\cdot)$ is an arbitrary twice-differentiable function.

To find the solution for the magnetic field intensity vector of the wave, we invoke Eq. (9.1) and note that curl **E** on the left-hand side of the equation becomes $\partial E_x/\partial z\,\hat{\mathbf{y}}$ [see Eqs. (9.11) and (4.81)]. This means that **H** (on the right-hand side of the equation) must be in the following form:

$$\mathbf{H} = H_y(z, t)\,\hat{\mathbf{y}}, \tag{9.14}$$

as indicated in Fig. 9.2, with which Eq. (9.1) becomes

$$\frac{\partial E_x}{\partial z} = -\mu\frac{\partial H_y}{\partial t}. \tag{9.15}$$

From Eqs. (8.99), $\partial f/\partial z = -(1/c)\partial f/\partial t$, so that

$$\frac{\partial H_y}{\partial t} = -\frac{1}{\mu}\frac{\partial f}{\partial z} = \frac{1}{\mu c}\frac{\partial f}{\partial t}, \tag{9.16}$$

and finally

$$\boxed{H_y = \frac{1}{\mu c} f = \sqrt{\frac{\varepsilon}{\mu}}\, f\left(t - \frac{z}{c}\right).} \tag{9.17}$$

H-field of a uniform plane wave

Having the solutions for both **E** and **H** in Fig. 9.2, we now summarize the basic properties of uniform plane electromagnetic waves in general, regardless of any particular coordinate system. Based on Eqs. (9.11), (9.13), (9.14), and (9.17), a uniform plane wave consists of electric and magnetic fields that are uniform in planes perpendicular to the direction of wave propagation (uniform wave), belong to these planes (plane wave), and are perpendicular to each other and to the direction of propagation. Such a wave also belongs to a class of so-called TEM (transverse electromagnetic) waves, since both **E** and **H** are transverse to the direction of propagation (planes z = const are called transversal planes of the wave). We see that the variations in space and time of the electric and magnetic fields of the wave in Fig. 9.2 are identical, that is, both field intensities are proportional to the same function (f) of z and t. This means that both **E** and **H** propagate in unison along z, having their maxima and minima at the same points of space and at the same instants of time. The orientation of the field vectors is such that their cross product, $\mathbf{E} \times \mathbf{H}$, is in the positive z direction in Fig. 9.2 ($\hat{\mathbf{x}} \times \hat{\mathbf{y}} = \hat{\mathbf{z}}$), that is, in the direction of the wave propagation.

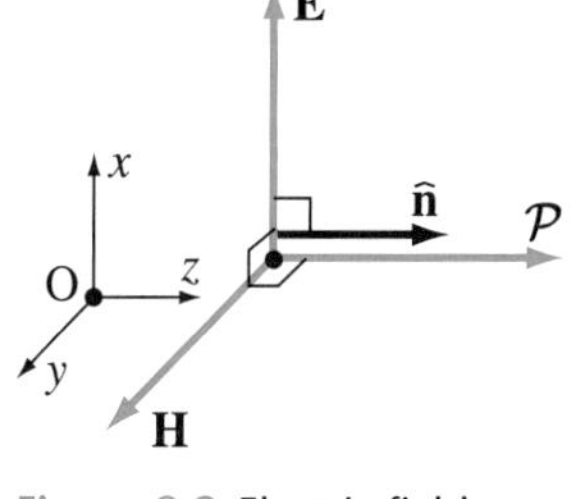

Figure 9.2 Electric field intensity vector (**E**), magnetic field intensity vector (**H**), propagation unit vector ($\hat{\mathbf{n}}$), and Poynting vector ($\mathcal{P}$) of a uniform plane electromagnetic wave propagating in an unbounded medium.

That the electric and magnetic fields described by Eqs. (9.13) and (9.17) move with the velocity c in the positive z direction is evident from the analogy with Eq. (8.101) and Fig. 8.8. This velocity is given by

$$\boxed{c = \frac{1}{\sqrt{\varepsilon\mu}},} \tag{9.18}$$

velocity of uniform plane electromagnetic waves (unit: m/s) i.e., it is determined solely by the properties (ε and μ) of the medium. For a vacuum or air (free space), Eqs. (1.2) and (4.3) give

velocity of electromagnetic waves in free space

$$c_0 = \frac{1}{\sqrt{\varepsilon_0 \mu_0}} \approx 3 \times 10^8 \text{ m/s}. \tag{9.19}$$

This constant, first calculated by Maxwell, is commonly referred to as the speed of light.[2] However, we see that light travels with this velocity only in free space and, more importantly, that this is the speed of not only (visible and invisible) light but electromagnetic waves in general, in free space. This very fact led Maxwell to suggest that visible light is nothing but an electromagnetic radiation.

Eqs. (9.11), (9.13), (9.14), and (9.17) also tell us that the ratio of the electric and magnetic field intensities at any instant of time and any point of space equals a constant, which is denoted by η,

proportionality of E and H of a plane wave

$$\frac{E}{H} = \eta, \tag{9.20}$$

and evaluated from the medium parameters as

intrinsic impedance of a medium (unit: Ω)

$$\eta = \sqrt{\frac{\mu}{\varepsilon}}. \tag{9.21}$$

This new quantity has the unit of impedance, Ω [the units for E and H are V/m and A/m, respectively, and $(\text{V/m})/(\text{A/m}) = \text{V/A} = \Omega$], and is called the intrinsic impedance of the medium (of parameters ε and μ). Using η, the following vector relations between the electric and magnetic field vectors of the wave can be written independently of any given coordinate system:

vector relations between $\mathbf{E}$ and $\mathbf{H}$ of a TEM wave

$$\mathbf{H} = \frac{1}{\eta}\,\hat{\mathbf{n}} \times \mathbf{E} \quad \text{and} \quad \mathbf{E} = \eta \mathbf{H} \times \hat{\mathbf{n}}, \tag{9.22}$$

where $\hat{\mathbf{n}}$ is the unit vector defining the direction of the wave propagation (in Fig. 9.2, $\hat{\mathbf{n}} = \hat{\mathbf{z}}$), which we refer to as the propagation unit vector. If the medium is air (vacuum),

intrinsic impedance of free space

$$\eta_0 = \sqrt{\frac{\mu_0}{\varepsilon_0}} \approx 120\pi\ \Omega \approx 377\ \Omega \tag{9.23}$$

$[\varepsilon_0 = 1/(\mu_0 c_0^2) = 10^{-9}/(36\pi)$ F/m, so that $\mu_0/\varepsilon_0 = 144\pi^2 \times 10^2$ H/F$]$.

Using Eqs. (2.199), (7.108), (9.20), and (9.21), the electric and magnetic energy densities of the wave are given, respectively, by

instantaneous electric and magnetic energy densities of a plane wave

$$w_\text{e} = \frac{1}{2}\,\varepsilon E^2 \quad \text{and} \quad w_\text{m} = \frac{1}{2}\,\mu H^2 = \frac{1}{2}\,\mu\left(\frac{E}{\eta}\right)^2 = \frac{1}{2}\,\varepsilon E^2. \tag{9.24}$$

It comes out that they are the same at all points and at all instants. In other words, the energy of the wave is equally distributed between the electric and magnetic fields. The total electromagnetic energy density of the wave is hence

$$w_\text{em} = w_\text{e} + w_\text{m} = 2w_\text{e} = \varepsilon E^2 = \varepsilon E_x^2(z,t) = \varepsilon f^2\left(t - \frac{z}{c}\right). \tag{9.25}$$

[2]A more accurate value of the speed of light in a vacuum, adopted (in 1983) as the "exact" constant, is $c_0 = 299{,}792{,}458$ m/s. Note that meter as a unit of length is defined relative to this value – as equaling the length of the path traveled by light in a vacuum during a time interval of 1/299,792,458 of a second.

The time-rate with which the electromagnetic energy flows through space carried by the wave is described by the associated Poynting vector. From Eq. (8.161), the Poynting vector of the wave is

$$\boxed{\mathcal{P} = \mathbf{E} \times \mathbf{H} = EH\hat{\mathbf{n}} = \frac{E^2}{\eta}\,\hat{\mathbf{n}} = \eta H^2\hat{\mathbf{n}} = \sqrt{\frac{\varepsilon}{\mu}}\,f^2\left(t - \frac{z}{c}\right)\hat{\mathbf{n}} = \mathcal{P}(z,t)\hat{\mathbf{n}}.} \tag{9.26}$$

instantaneous Poynting vector of a plane wave

Of course, the direction of $\mathcal{P}$ (Fig. 9.2) coincides with the direction of the wave propagation ($\hat{\mathbf{n}}$) and its instantaneous magnitude, $\mathcal{P}$, equals the surface power density transported by the wave, i.e., the power per unit area of the wavefront (plane perpendicular to $\hat{\mathbf{n}}$), at a given point of space and instant of time. Comparing Eqs. (9.26) and (9.25), we note that $\mathcal{P}$ is proportional to the electromagnetic energy density of the wave. The constant of proportionality is the wave velocity, c, which is evident from

$$\mathcal{P} = \sqrt{\frac{\varepsilon}{\mu}}\,E^2 = \frac{1}{\sqrt{\varepsilon\mu}}\,\varepsilon E^2 = c w_{\text{em}}. \tag{9.27}$$

Finally, let us write the field expressions for a uniform plane wave having an x-oriented electric field intensity vector, as in Eq. (9.11), but propagating in the negative z direction, oppositely to the wave in Fig. 9.2. Noting that travel with velocity c in the negative (backward) z direction can be interpreted as travel with velocity $-c$ in the positive (forward) z direction[3] and knowing that the orientation of the magnetic field intensity vector of the wave must be such that the Poynting vector is in the propagation direction ($\hat{\mathbf{n}}_{\text{backward}} = -\hat{\mathbf{z}}$), we have

$$\mathbf{E}_{\text{backward}} = f\left(t + \frac{z}{c}\right)\hat{\mathbf{x}} \quad \text{and} \quad \mathbf{H}_{\text{backward}} = -\frac{1}{\eta}f\left(t + \frac{z}{c}\right)\hat{\mathbf{y}} \tag{9.28}$$

$[\hat{\mathbf{x}} \times (-\hat{\mathbf{y}}) = -\hat{\mathbf{z}}]$, which is also in agreement with Eqs. (9.22).

Example 9.2 1-D Wave Equations from 1-D Maxwell's Equations

Starting from the one-dimensional curl Maxwell's equations in scalar form specialized for a uniform plane electromagnetic wave with field components $E_x(z,t)$ and $H_y(z,t)$ propagating in a lossless homogeneous medium of permittivity ε and permeability μ, Fig. 9.2, derive the associated 1-D wave equations, with only E_x and only H_y as unknowns, respectively.

Solution The specialized 1-D Maxwell's first equation in this case is that in Eq. (9.15), and the 1-D version of the general Maxwell's second differential equation, Eq. (9.2), simplified for **E** and **H** in Eqs. (9.11) and (9.14) using the formula for the curl in Cartesian coordinates, Eq. (4.81), reads

$$\frac{\partial H_y}{\partial z} = -\varepsilon\frac{\partial E_x}{\partial t}. \tag{9.29}$$

We now take the derivative with respect to z of both sides of Eq. (9.15), and combine the result with Eq. (9.29) as follows:

$$\frac{\partial^2 E_x}{\partial z^2} = -\mu\frac{\partial}{\partial z}\left(\frac{\partial H_y}{\partial t}\right) = -\mu\frac{\partial}{\partial t}\left(\frac{\partial H_y}{\partial z}\right) = \varepsilon\mu\frac{\partial^2 E_x}{\partial t^2}, \tag{9.30}$$

and this exactly is the 1-D wave equation for the electric field, Eq. (9.12). The H-field 1-D wave equation, which has the identical form, in terms of H_y alone, as Eq. (9.12), is obtained

[3] Alternatively, we can introduce a new axis, z', in the direction of the backward wave travel, and (since $z' = -z$) write $f(t - z'/c) = f(t + z/c)$ for this wave.

in a similar way – by taking the derivative with respect to z of Eq. (9.29) and substituting its right-hand side by the corresponding expression from Eq. (9.15).

Problems: 9.2; *Conceptual Questions* (on Companion Website): 9.4–9.6; *MATLAB Exercises* (on Companion Website).

9.4 TIME-HARMONIC UNIFORM PLANE WAVES AND COMPLEX-DOMAIN ANALYSIS

In the case of harmonic (steady-state sinusoidal) time variations of uniform plane electromagnetic waves, the function $f(t')$ in Eqs. (9.13) and (9.17), where $t' = t - z/c$, acquires the form given by Eq. (8.51), so that the expression for the electric field becomes

$$E_x = E_{\rm m}\cos(\omega t' + \theta_0) = E_0\sqrt{2}\cos\left[\omega\left(t - \frac{z}{c}\right) + \theta_0\right], \tag{9.31}$$

where $E_{\rm m}$ is the amplitude,[4] E_0 the rms value ($E_0 = E_{\rm m}/\sqrt{2}$), and θ_0 the initial (for $t = 0$) phase in the plane $z = 0$ of the electric field intensity of the wave, ω is its angular frequency, and $H_y = E_x/\eta$. Using the phase coefficient (wavenumber), β, defined in Eq. (8.111), we have

instantaneous field intensities of a time-harmonic uniform plane wave

$$\boxed{E_x = E_0\sqrt{2}\cos(\omega t - \beta z + \theta_0) \quad \text{and} \quad H_y = \frac{E_0}{\eta}\sqrt{2}\cos(\omega t - \beta z + \theta_0).} \tag{9.32}$$

From Eqs. (9.32), we see that β represents the rate with which the instantaneous phase of the wave,

instantaneous phase of a uniform plane wave

$$\boxed{\phi(z, t) = \omega t - \beta z + \theta_0,} \tag{9.33}$$

changes (decreases) with z, and that is why it is called the phase coefficient [the decrease with z of the phase is a consequence of a time retardation (lagging) of the fields – see Eq. (8.110)]. The initial phase of the wave, $\theta(z) = -\beta z + \theta_0$, is constant for $z = \text{const}$, and therefore each transversal plane of a uniform plane time-harmonic wave is said to be an equiphase plane of the wave. This phase is the same for the electric and magnetic fields of the wave, i.e., **E** and **H** are in phase at every point of space. Note also that the field intensities are periodic in both time (t) and space (along z). The time period, T, is given in Eq. (8.49). The space period is obtained from the relation $\beta z = 2\pi$, namely, it comes out to be the wavelength of the wave, λ, defined in Eq. (8.112). The wavelength equals the distance between two adjacent (closest) transversal planes in which the field intensities are in phase with respect to each other. Finally, from $\beta = 2\pi/\lambda$ [Eq. (8.111)], we see that β can be interpreted as a measure of the number of wavelengths in a complete cycle of a cosine function, 2π, and hence its other name, the wavenumber. The wave behavior in time and space is illustrated in Fig. 9.3. In Fig. 9.3(a), we observe that the same sinusoid in time appears at $z = 0$ (solid line) and then at $z = \lambda/4$ (dashed line) delayed by $\Delta t = \Delta z/c = (\lambda/4)/c = T/4$. In Fig. 9.3(b), we notice that the curve

---

[4]In the analysis of wave propagation, it is convenient to use symbols with a subscript "0" (zero), e.g., E_0, to denote rms wave quantities at $z = 0$ (or some other reference position). For this reason, the corresponding amplitudes (maximum values) of the wave quantities will regularly be denoted using a subscript "m" (e.g., $E_{\rm m}$) throughout the rest of this text.

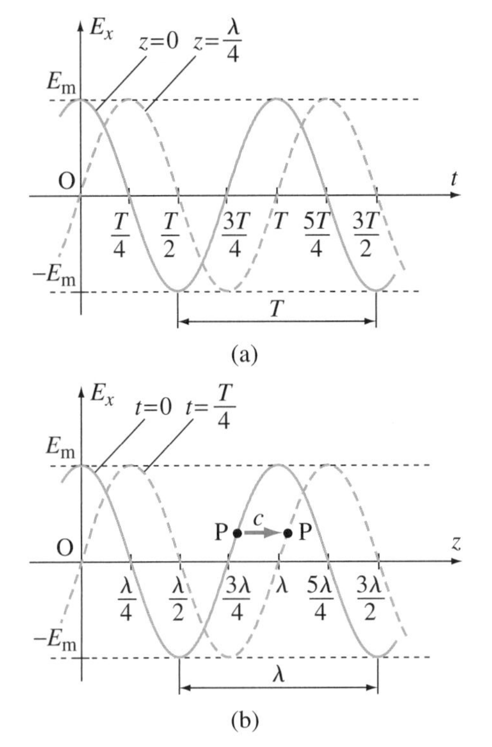

Figure 9.3 Behavior in time at two positions (a) and in space at two instants (b) of a time-harmonic uniform plane wave [$E_x(z, t)$ is given in Eqs. (9.32), with $E_m = E_0\sqrt{2}$ (amplitude) and $\theta_0 = 0$].

for $t = T/4$ (dashed line) is the same sinusoid in space as the one for $t = 0$ (solid line), but displaced to the right by $\Delta z = c\Delta t = c(T/4) = \lambda/4$. In other words, any constant-phase point (e.g., the point P in the figure), for which $\omega t - \beta z = \text{const}$, moves to the right with a velocity c, equal to ω/β. Analytically,

$$\omega t - \beta z = \text{const} \quad \xrightarrow{\text{d}/\text{d}t} \quad \omega - \beta\frac{\text{d}z}{\text{d}t} = 0 \quad \longrightarrow \quad \frac{\text{d}z}{\text{d}t} = \frac{\omega}{\beta}. \tag{9.34}$$

This velocity, equal to $\text{d}z/\,\text{d}t$, is hence also called the phase velocity of the wave and designated by v_p. In the general case,

$$\boxed{v_\text{p} = \frac{\omega}{\beta},} \tag{9.35}$$

phase velocity (unit: m/s)

with $v_\text{p} = c = 1/\sqrt{\varepsilon\mu}$ for wave propagation in a lossless unbounded medium of permittivity ε and permeability μ. However, as we shall see in later sections and chapters, for waves in some other electromagnetic media and structures (e.g., lossy materials, plasmas, and metallic waveguides), the phase velocity is not constant, but depends on frequency.

Applying the time-complex conversion in Eq. (8.66) to the expressions for the instantaneous field intensities in Eqs. (9.32), we obtain the following expressions for complex rms field intensities of a time-harmonic uniform plane wave:

$$\boxed{\underline{E}_x = \underline{E}_0\,\text{e}^{-\text{j}\beta z} \quad \text{and} \quad \underline{H}_y = \frac{\underline{E}_0}{\eta}\,\text{e}^{-\text{j}\beta z}, \quad \text{where} \quad \underline{E}_0 = E_0\,\text{e}^{\text{j}\theta_0}} \tag{9.36}$$

complex rms field intensities of a uniform plane wave

($\underline{E}_0$ is the complex rms electric field intensity of the wave in the plane $z = 0$). Note that these same expressions could have alternatively been derived from Helmholtz equations, Eqs. (9.8) and (9.9), combining them with the source-free version of complex Maxwell's equations in differential form, Eqs. (8.81), e.g., as in Eqs. (9.11)–(9.17) in the time domain. In specific, the assumption that the field vectors of a wave are of the form $\underline{\mathbf{E}} = \underline{E}_x(z)\,\hat{\mathbf{x}}$ and $\underline{\mathbf{H}} = \underline{H}_y(z)\,\hat{\mathbf{y}}$, as in Fig. 9.2 (uniform-plane-wave approximation), leads to the following simplified form of Eqs. (9.8)–(9.10):

1-D Helmholtz equations

$$\frac{\mathrm{d}^2\underline{E}_x}{\mathrm{d}z^2} + \beta^2\underline{E}_x = 0, \qquad \frac{\mathrm{d}^2\underline{H}_y}{\mathrm{d}z^2} + \beta^2\underline{H}_y = 0, \tag{9.37}$$

and it is now a very simple task to verify by direct substitution that the expressions in Eqs. (9.36) are solutions of these equations (to be done for the electric field in an example), as well as that the corresponding Maxwell's equations are also satisfied. In addition, note that the phase propagation factor $\mathrm{e}^{-\mathrm{j}\beta z}$ in Eqs. (9.36), indicating the travel of the corresponding fields along the z-axis with the velocity $c = \omega/\beta$ (in the positive z direction), has the identical form as the factor $\mathrm{e}^{-\mathrm{j}\beta R}$ present in the expressions for the complex Lorenz potentials in Eqs. (8.113), (8.114), (8.116), and (8.117). Note finally that a factor $\mathrm{e}^{\mathrm{j}\beta z}$, on the other hand, would indicate plane wave propagation in the negative z direction [see Eqs. (9.28)].

We know that, in general, time-average powers, energies, and power and energy densities for linear media and time-harmonic field variations can be obtained by means of the corresponding time-constant expressions if rms values of the quantities involved in the expressions are used [see Eq. (8.53)]. The time-average electric and magnetic energy densities of the wave defined by Eqs. (9.36) are thus

time-average electric and magnetic energy densities of a time-harmonic plane wave

$$(w_\mathrm{e})_\mathrm{ave} = \frac{1}{2}\,\varepsilon E_0^2 \quad \text{and} \quad (w_\mathrm{m})_\mathrm{ave} = \frac{1}{2}\,\mu H_0^2, \tag{9.38}$$

respectively, where $H_0 = E_0/\eta$ is the rms magnetic field intensity of the wave. Of course, $(w_\mathrm{e})_\mathrm{ave} = (w_\mathrm{m})_\mathrm{ave}$ and the total time-average electromagnetic energy density of the wave is

$$(w_\mathrm{em})_\mathrm{ave} = 2(w_\mathrm{e})_\mathrm{ave} = 2(w_\mathrm{m})_\mathrm{ave} = \varepsilon E_0^2 = \mu H_0^2. \tag{9.39}$$

Note that, although the instantaneous energy densities of the wave, obtained from Eqs. (9.24), (9.25), and (9.32), depend on the spatial coordinate z, their time averages are constants over the entire space.

Finally, using Eq. (8.194), the complex Poynting vector of the wave is

complex Poynting vector of a plane wave

$$\underline{\mathcal{P}} = \underline{\mathbf{E}} \times \underline{\mathbf{H}}^* = \underline{E}_x\underline{H}_y^*\,\hat{\mathbf{z}} = \underline{E}_0\,\mathrm{e}^{-\mathrm{j}\beta z}\,\frac{\underline{E}_0^*}{\eta}\,\mathrm{e}^{\mathrm{j}\beta z}\,\hat{\mathbf{z}} = \frac{E_0^2}{\eta}\,\hat{\mathbf{n}} \tag{9.40}$$

($\hat{\mathbf{n}} = \hat{\mathbf{z}}$). We see that $\underline{\mathcal{P}}$ comes out to be purely real, and thus it equals the time average of the instantaneous Poynting vector of the wave [see Eq. (8.195)]. Namely, $\mathcal{P}_\mathrm{ave} = (E_0^2/\eta)\hat{\mathbf{n}}$, which can also be obtained by taking the time average of $\mathcal{P}(z, t)$ expressed in terms of the instantaneous field intensities in Eqs. (9.32). We also note that, as in the corresponding energy density expressions, $\mathcal{P}_\mathrm{ave}$ is not a function of z.

In the rest of this chapter and in the next one, we shall deal almost exclusively with time-harmonic uniform plane waves. The analysis will regularly be carried out in the complex (frequency) domain. However, while working with complex representatives of the field vectors and other time-harmonic quantities associated with the waves, we shall always have in mind (and sometimes write) the corresponding instantaneous (time-domain) expressions, in order to fully observe and understand the physical behavior and characteristics of time-harmonic waves in different situations and problems.

Example 9.3 Verification of Solutions of Wave (Helmholtz) Equations

Verify (directly) that the expressions for the electric field intensity of a time-harmonic uniform plane electromagnetic wave propagating in a lossless medium of parameters ε and μ in Eqs. (9.32) and (9.36) are solutions of the corresponding wave (or Helmholtz) equations.

Solution In the time domain, the second partial derivatives in z and t of the electric field intensity $E_x(z, t)$ given in Eqs. (9.32) come out to be

$$\frac{\partial^2 E_x}{\partial z^2} = E_0\sqrt{2}\,\frac{\partial^2}{\partial z^2}\cos(\omega t - \beta z + \theta_0) = -\beta^2 E_0\sqrt{2}\cos(\omega t - \beta z + \theta_0) \quad \text{and}$$

$$\frac{\partial^2 E_x}{\partial t^2} = E_0\sqrt{2}\,\frac{\partial^2}{\partial t^2}\cos(\omega t - \beta z + \theta_0) = -\omega^2 E_0\sqrt{2}\cos(\omega t - \beta z + \theta_0), \qquad (9.41)$$

respectively. Since $\beta^2 = \varepsilon\mu\omega^2$, from Eq. (8.111), the first result equals $\varepsilon\mu$ times the second, hence verifying that the wave equation for the electric field in Eq. (9.12) is indeed satisfied.

In the complex domain, we similarly take the second derivative of $\underline{E}_x(z)$ in Eqs. (9.36),

$$\frac{\mathrm{d}^2 \underline{E}_x}{\mathrm{d}z^2} = \underline{E}_0\,\frac{\mathrm{d}^2}{\mathrm{d}z^2}\,\mathrm{e}^{-\mathrm{j}\beta z} = (-\mathrm{j}\beta)^2 \underline{E}_0\,\mathrm{e}^{-\mathrm{j}\beta z} = -\beta^2 \underline{E}_x, \qquad (9.42)$$

and what we obtain is simply the E-field equation in Eqs. (9.37), which proves that this expression for $\underline{E}_x(z)$ is a solution of the respective governing Helmholtz equation.

Of course, that the expressions for $H_y(z, t)$ and $\underline{H}_y(z)$ in Eqs. (9.32) and (9.36) also satisfy the respective wave (Helmholtz) equations can be shown in an analogous manner.

Example 9.4 Magnetic Field from Electric Field in Complex Domain

Using Maxwell's equations, obtain $\underline{H}_y$ from the solution for $\underline{E}_x$ in Eqs. (9.36).

Solution We use Maxwell's first equation, written as the complex-domain version of Eq. (9.15), or, equivalently, as the one-dimensional version (for uniform plane waves) of the first equation in Eqs. (8.81), and substitute in it the expression for $\underline{E}_x$, Eqs. (9.36), to obtain

$$\frac{\mathrm{d}\underline{E}_x}{\mathrm{d}z} = -\mathrm{j}\omega\mu\underline{H}_y \quad \longrightarrow \quad \underline{H}_y = \frac{\mathrm{j}\underline{E}_0}{\omega\mu}\frac{\mathrm{d}}{\mathrm{d}z}\,\mathrm{e}^{-\mathrm{j}\beta z} = \frac{\beta}{\omega\mu}\,\underline{E}_0\,\mathrm{e}^{-\mathrm{j}\beta z} = \sqrt{\frac{\varepsilon}{\mu}}\,\underline{E}_x = \frac{\underline{E}_x}{\eta}, \qquad (9.43)$$

where β and η are the phase coefficient and intrinsic impedance of the propagation medium, given by Eqs. (8.111) and (9.21), respectively. Of course, this is the same result for $\underline{H}_y$ as in Eqs. (9.36). Note that Eq. (9.43) represents the complex equivalent of the manipulations performed in computing H_y from E_x in Eqs. (9.15)–(9.17) in the time domain.

Example 9.5 Oscillation in Time of the Poynting Vector

The electric and magnetic field vectors of a timc-harmonic uniform traveling plane wave oscillate in time at an angular frequency ω. At what angular frequency does the Poynting vector of the wave oscillate?

Solution From Eqs. (9.26) and (9.32), the expression for the instantaneous Poynting vector of the wave in Fig. 9.2, if time-harmonic, can be written as

$$\boldsymbol{\mathcal{P}} = E_x H_y\,\hat{\mathbf{z}} = \frac{2E_0^2}{\eta}\cos^2(\omega t - \beta z + \theta_0)\,\hat{\mathbf{z}} = \frac{E_0^2}{\eta}\left[1 + \cos(2\omega t - 2\beta z + 2\theta_0)\right]\hat{\mathbf{z}}, \qquad (9.44)$$

where the use is made of the trigonometric identity $\cos^2\alpha = (1 + \cos 2\alpha)/2$. We see that $\boldsymbol{\mathcal{P}}$ oscillates in time at twice the angular frequency of the field oscillation, so at 2ω. Of course, this oscillation is not time-harmonic, and its time average is not zero, but that in Eq. (9.40).

We also note that the periodicity of the Poynting vector in Eq. (9.44) in space (along z), being determined by 2β, is as well at twice the respective repetition rate for fields.

Example 9.6 Computation of Various Parameters of a Plane Wave

The electric field of an electromagnetic wave propagating through a lossless nonmagnetic material is given by

$$\mathbf{E} = 10\sqrt{2}\cos(10^8 t + y)\,\hat{\mathbf{z}}\ \mathrm{V/m} \quad (t \text{ in s};\ y \text{ in m}). \tag{9.45}$$

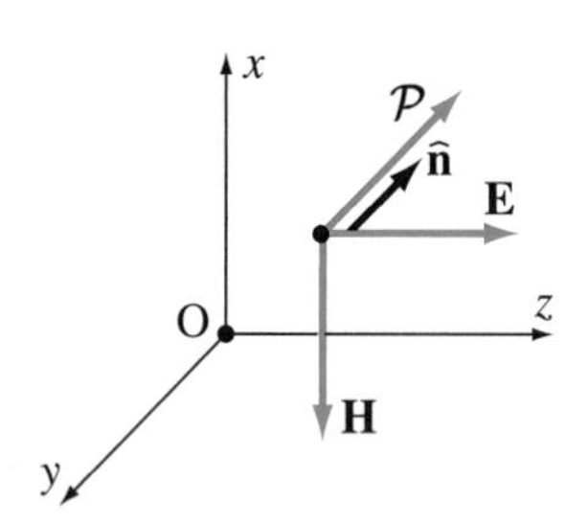

Figure 9.4 Electric and magnetic field intensity vectors and Poynting vector of a uniform plane time-harmonic wave traveling in the negative y direction; for Example 9.6.

Find: (a) the direction of propagation, time period, wavelength, and phase velocity of the wave, the relative permittivity and intrinsic impedance of the material, (b) the instantaneous magnetic field intensity vector, and complex electric and magnetic field vectors of the wave, and (c) the instantaneous and time-average electromagnetic energy densities and time-average Poynting vector of the wave.

Solution

(a) The wave propagates in the negative y direction, as shown in Fig. 9.4. Having in mind Eqs. (9.32), the angular frequency in Eq. (9.45) is $\omega = 10^8$ rad/s and the phase coefficient $\beta = 1$ rad/m, so that, using Eqs. (8.49) and (8.111), the time period and wavelength of the wave come out to be

$$T = \frac{2\pi}{\omega} = 62.8\ \mathrm{ns} \quad \text{and} \quad \lambda = \frac{2\pi}{\beta} = 6.28\ \mathrm{m}, \tag{9.46}$$

respectively. From Eq. (9.35), the phase velocity of the wave amounts to $v_\mathrm{p} = c = \omega/\beta = 10^8$ m/s. As the propagation medium is nonmagnetic, Eq. (9.18) reduces to

c – nonmagnetic medium

$$\boxed{c = \frac{1}{\sqrt{\varepsilon_\mathrm{r}\varepsilon_0\mu_0}} = \frac{c_0}{\sqrt{\varepsilon_\mathrm{r}}} \quad \left(\mu = \mu_0;\quad c_0 = 3\times10^8\ \mathrm{m/s}\right),} \tag{9.47}$$

with c_0 standing for the speed of light in free space, Eq. (9.19), and hence the relative permittivity of the medium (dielectric) $\varepsilon_\mathrm{r} = (c_0/c)^2 = 3^2 = 9$. Similarly, employing Eqs. (9.21) and (9.23), the intrinsic impedance of the dielectric is given by

η – nonmagnetic medium

$$\boxed{\eta = \sqrt{\frac{\mu_0}{\varepsilon_\mathrm{r}\varepsilon_0}} = \frac{\eta_0}{\sqrt{\varepsilon_\mathrm{r}}} \quad (\eta_0 = 120\pi\ \Omega),} \tag{9.48}$$

that is, $\eta = \eta_0/3 = 40\pi\ \Omega = 125.7\ \Omega$.

(b) Eq. (9.20) tells us that the rms magnetic field intensity of the wave is $H_0 = E_0/\eta = 0.08$ A/m ($E_0 = 10$ V/m), and we see in Fig. 9.4 that the vector $\mathbf{H}$ is oriented in the negative x direction – such that the cross product, $\mathbf{E}\times\mathbf{H}$, is in the direction of the wave propagation. Therefore, the instantaneous value of $\mathbf{H}$ is

$$\mathbf{H} = 0.08\sqrt{2}\cos(10^8 t + y)\,(-\hat{\mathbf{x}})\ \mathrm{A/m} \quad (t \text{ in s};\ y \text{ in m}). \tag{9.49}$$

On the other side, complex rms field intensity vectors of the wave, Eqs. (9.36), are

$$\underline{\mathbf{E}} = 10\,\mathrm{e}^{\mathrm{j}y}\,\hat{\mathbf{z}}\ \mathrm{V/m} \quad \text{and} \quad \underline{\mathbf{H}} = 0.08\,\mathrm{e}^{\mathrm{j}y}\,(-\hat{\mathbf{x}})\ \mathrm{A/m} \quad (y \text{ in m}). \tag{9.50}$$

(c) Finally, combining Eqs. (9.25) and (9.45), the instantaneous electromagnetic energy density of the wave is equal to

$$w_\mathrm{em} = \varepsilon_\mathrm{r}\varepsilon_0 E^2 = 16\cos^2(10^8 t + y)\ \mathrm{nJ/m^3} \quad (t \text{ in s};\ y \text{ in m}), \tag{9.51}$$

and, by means of Eqs. (9.39) and (9.40), its time average and the time-average Poynting vector (Fig. 9.4), respectively, are

$$(w_\mathrm{em})_\mathrm{ave} = \varepsilon_\mathrm{r}\varepsilon_0 E_0^2 = 8\ \mathrm{nJ/m^3} \quad \text{and} \quad \mathcal{P}_\mathrm{ave} = \frac{E_0^2}{\eta}\,(-\hat{\mathbf{y}}) = 0.8\,(-\hat{\mathbf{y}})\ \mathrm{W/m^2}. \tag{9.52}$$

Note that $\mathcal{P}_\mathrm{ave} = c\,(w_\mathrm{em})_\mathrm{ave}$, which is the same proportionality as between the corresponding instantaneous quantities in Eq. (9.27).

Example 9.7 Finding Material Parameters from Wave Properties

A uniform plane wave travels at a velocity $c = 2 \times 10^8$ m/s through a lossless medium, and its fields are expressed as

$$\mathbf{E} = 754 \sin(10^7 t + \beta z)\,\hat{\mathbf{x}}\ \text{mV/m} \quad \text{and} \quad \mathbf{H} = -3 \sin(10^7 t + \beta z)\,\hat{\mathbf{y}}\ \text{mA/m}, \tag{9.53}$$

where t is measured in s, and z in m. Under these circumstances, compute (a) the phase coefficient and wavelength of the wave, and intrinsic impedance of the medium, (b) the relative permittivity and permeability of the medium, and (c) the initial phase of the electric field intensity of the wave in the plane $z = 0$.

Solution

(a) We see in Eqs. (9.53) that the angular frequency of the wave is $\omega = 10^7$ rad/s, and hence the phase coefficient, Eq. (8.111), of $\beta = \omega/c = 0.05$ rad/m and wavelength of $\lambda = 2\pi/\beta = 125.66$ m. From Eq. (9.20), on the other side, the intrinsic impedance of the propagation medium amounts to $\eta = E/H = 251.33\ \Omega$.

(b) Combining Eqs. (9.21) and (9.18), the product of η and c and their ratio turn out to be

$$\eta c = \frac{1}{\varepsilon} \quad \text{and} \quad \frac{\eta}{c} = \mu, \tag{9.54}$$

from which the relative permittivity and permeability of the medium are $\varepsilon_r = 1/(\eta c \varepsilon_0) = 2.25$ and $\mu_r = \eta/(c\mu_0) = 1$, respectively (note that these are parameters of a widely used dielectric polyethylene).

(c) With the help of the trigonometric identity $\sin\alpha = \cos(\alpha - \pi/2)$, we rewrite the expression for the electric field of the wave in Eqs. (9.53) as $\mathbf{E} = 754\cos(10^7 t + \beta z - \pi/2)\,\hat{\mathbf{x}}$ V/m, and, with reference to Eqs. (9.32), realize that the initial phase of the field for $z = 0$ equals $\theta_0 = -90°$.

Example 9.8 Induced Emf in a Large Contour due to a Plane Wave

A rectangular contour of side lengths a and b is placed in the field of a uniform plane time-harmonic electromagnetic wave of angular frequency ω and rms electric field intensity E_0 propagating in free space. The magnetic field vector of the wave is perpendicular to the plane of the contour, and the electric field vector is parallel to the pair of contour edges that are b long. The electrical dimensions of the contour are arbitrary (i.e., the contour cannot be considered to be electrically small). Find the emf induced in the contour.

Solution This situation is shown in Fig. 9.5. For the adopted Cartesian coordinate system, the field vectors of the wave are those in Eqs. (9.32). As we know, the induced emf in a contour, given by Eq. (6.34), can, in fact, be computed from either the left- or right-hand side of Faraday's law of electromagnetic induction in integral form, in Eq. (6.37), i.e., using either the electric or magnetic field in which the contour is situated. In the case in Fig. 9.5, however, the former approach involves simpler computation, so let us pursue it first.

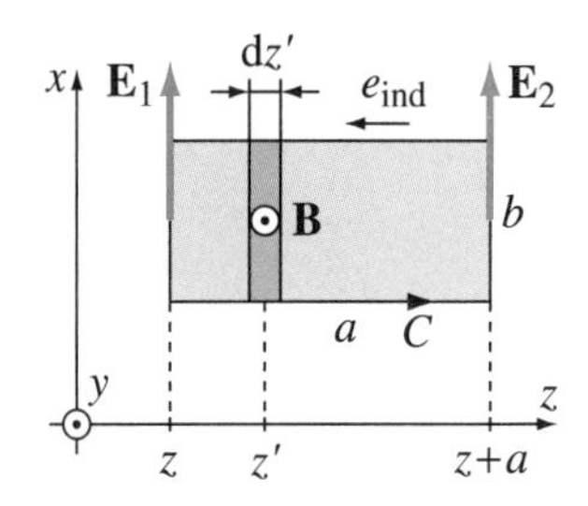

Figure 9.5 Computation of the induced emf in a rectangular contour, of an arbitrary electrical size, situated in the field of a uniform plane electromagnetic wave; for Example 9.8.

The x-directed electric field vector of the wave being perpendicular to the pair of contour edges of length a, while parallel to and constant along each of the remaining two edges (fields $\mathbf{E}_1$ and $\mathbf{E}_2$ in Fig. 9.5), much like the field $\mathbf{E}_{\text{ind}}$ in Fig. 6.17, the emf in the contour for the adopted counterclockwise reference orientation of the contour equals [see Eq. (6.101)]

$$e_{\text{ind}} = \oint_C \mathbf{E}\cdot d\mathbf{l} = -E_1 b + E_2 b = -E_x(z,t)\,b + E_x(z+a,t)\,b$$
$$= E_0 b\sqrt{2}\,[\cos(\omega t - \beta z - \beta a) - \cos(\omega t - \beta z)], \tag{9.55}$$

where the phase coefficient, Eq. (8.111), is $\beta = \omega\sqrt{\varepsilon_0\mu_0}$, and the initial phase of the field for $z = 0$ in Eqs. (9.32) is adopted to be $\theta_0 = 0$.

Let us also evaluate the emf based on the right-hand side of Faraday's law. The y-directed magnetic field vector of the wave is perpendicular to the plane of the contour, and its spatial variation is only in terms of the coordinate z in Fig. 9.5. Therefore, the computation of the magnetic flux through the contour is very similar to that performed in Fig. 6.12 and Eq. (6.64). For the positive y reference direction of the flux, Fig. 9.5, which is interconnected by the right-hand rule, Fig. 6.7, with the direction of the induced emf in the contour, we have

$$\Phi = \int_S \mathbf{B} \cdot d\mathbf{S} = \int_{z'=z}^{z+a} \mu_0 H_y(z', t)\, b\, dz' = \frac{\mu_0 E_0 b\sqrt{2}}{\eta_0} \int_z^{z+a} \cos(\omega t - \beta z')\, dz'$$
$$= \frac{\mu_0 E_0 b\sqrt{2}}{\eta_0 \beta} [\sin(\omega t - \beta z) - \sin(\omega t - \beta z - \beta a)], \tag{9.56}$$

with η_0 standing for the free-space intrinsic impedance, Eq. (9.23). Taking the negative of the time derivative of the flux, like in Eq. (6.65), gives

$$e_{\text{ind}} = -\frac{d\Phi}{dt} = \frac{\omega \mu_0 E_0 b\sqrt{2}}{\eta_0 \beta} [\cos(\omega t - \beta z - \beta a) - \cos(\omega t - \beta z)], \tag{9.57}$$

which is the same result as in Eq. (9.55), since $\omega\mu_0/(\eta_0\beta)$ is identically equal to unity.

In essence, the equality of the results for e_{ind} obtained from the left- and right-hand sides of the integral Faraday's law comes from the proportionality of the electric and magnetic fields of the plane wave, Eq. (9.20). Conversely, this proportionality of fields is derived from Eq. (9.15), that is, from a 1-D version of Faraday's law of electromagnetic induction in differential form. So, as the two ways of computing the emf in Eqs. (9.55)–(9.57) represent one physical phenomenon, the electromagnetic induction, the two fields in Fig. 9.5 are also just the two faces of the same one traveling electromagnetic wave.

Example 9.9 Induced Emf in a Small Contour

Repeat the previous example but for an electrically small rectangular contour.

Solution If the contour in Fig. 9.5 is electrically small, meaning that both a and b are small relative to the free-space wavelength at the frequency of the wave, λ_0, then the application of the right-hand side of Faraday's law in Eq. (6.37) provides a very simple solution for the induced emf in the contour. Namely, the magnetic field of the wave can be assumed to be uniform all over the surface S bounded by the contour in Fig. 9.5, so that the integral in Eq. (9.56) reduces to

$$\Phi \approx \mathbf{B} \cdot \mathbf{S} = \mu_0 H_y(z, t)\, ab = \frac{\mu_0 E_0 ab\sqrt{2}}{\eta_0} \cos(\omega t - \beta z) \quad (a, b \ll \lambda_0), \tag{9.58}$$

and the emf, Eq. (9.57), becomes

$$e_{\text{ind}} = -\frac{d\Phi}{dt} \approx \frac{\omega \mu_0 E_0 ab\sqrt{2}}{\eta_0} \sin(\omega t - \beta z) = \beta E_0 ab\sqrt{2} \sin(\omega t - \beta z). \tag{9.59}$$

On the other side, to obtain this same emf from the electric field of the wave, we cannot assume that $E_1 \approx E_2$ in Fig. 9.5, as this would result in $e_{\text{ind}} = 0$, but have to take into account the difference between E_1 and E_2 in Eq. (9.55), which, however, is now very small. Hence, we first use the trigonometric identity $\cos(A - B) = \cos A \cos B + \sin A \sin B$ (with $A = \omega t - \beta z$ and $B = \beta a$) to expand the first cosine term in Eq. (9.55), and then the fact that βa is much smaller than unity [see Eq. (8.130)] to obtain the same result as in Eq. (9.59),

$$\beta a \ll 1 \quad \longrightarrow \quad e_{\text{ind}} = E_0 b\sqrt{2} [\cos(\omega t - \beta z) \cos\beta a + \sin(\omega t - \beta z) \sin\beta a - \cos(\omega t - \beta z)]$$
$$\approx \beta E_0 ab\sqrt{2} \sin(\omega t - \beta z) \quad (\cos\beta a \approx 1; \ \sin\beta a \approx \beta a). \tag{9.60}$$

Example 9.10 Absorbed Energy in a Screen Illuminated by a Plane Wave

A uniform plane time-harmonic electromagnetic wave of frequency f and rms electric field intensity E_0 propagates in air, and is incident obliquely, at an angle θ, on a flat, perfectly

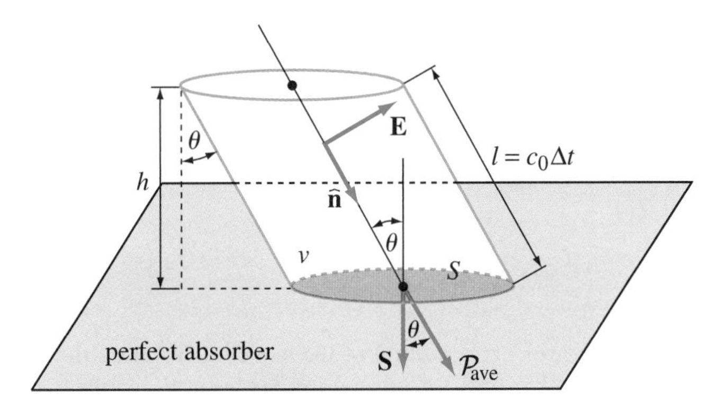

Figure 9.6 Evaluation of the absorbed energy during time Δt in a part of area S of a perfectly absorbing screen illuminated by an obliquely incident uniform plane electromagnetic wave; for Example 9.10.

absorbing screen, as shown in Fig. 9.6. Find the absorbed energy in a part of the screen of surface area S during a time interval Δt, if $\Delta t \gg 1/f$.

Solution Since this is a perfect absorber (also referred to as a black body), it absorbs all of the incident energy carried by the incoming wave. This also means that there is only one traveling wave (the incident wave), and no reflected wave, in the region above the screen in Fig. 9.6. The local surface power density delivered by the wave through the surface S to the absorbing material equals the normal component of the Poynting vector, $\boldsymbol{\mathcal{P}}$, of the wave on S. Since the time-average of this vector, $\boldsymbol{\mathcal{P}}_{\rm ave}$, given in Eq. (9.40), is not a function of spatial coordinates, the time-average power absorbed by the part of area S of the screen, using Eq. (8.162), can be found as

$$(P_{\rm abs})_{\rm ave} = \int_S \boldsymbol{\mathcal{P}}_{\rm ave} \cdot \mathrm{d}\mathbf{S} = \boldsymbol{\mathcal{P}}_{\rm ave} \cdot \mathbf{S} = \mathcal{P}_{\rm ave} S \cos\theta \quad \left(\mathcal{P}_{\rm ave} = \frac{E_0^2}{\eta_0}\right), \tag{9.61}$$

with $\mathbf{S}$ standing for the surface area vector of this part of the screen, which makes an angle θ with $\boldsymbol{\mathcal{P}}$ (Fig. 9.6), and η_0 for the free-space intrinsic impedance, Eq. (9.23). Furthermore, since the time we are considering, Δt, is much longer than the period T of time-harmonic variation of the incident wave, in Eq. (8.49), $\Delta t \gg T = 1/f$, the absorbed energy on S, $W_{\rm abs}$, equals the corresponding time-average power multiplied by Δt. To explain this, let us write the time interval as $\Delta t = t_2 - t_1 = NT + \delta t$, where $0 \le \delta t < T$ and $\delta t \ll \Delta t$, yielding

$$W_{\rm abs} = \int_{t=t_1}^{t_2} P_{\rm abs}(t)\,\mathrm{d}t = NT \underbrace{\frac{1}{NT}\int_{t_1}^{t_1+NT} P_{\rm abs}(t)\,\mathrm{d}t}_{(P_{\rm abs})_{\rm ave}} + \underbrace{\int_{t_2-\delta t}^{t_2} P_{\rm abs}(t)\,\mathrm{d}t}_{\text{negligibly small}}$$

$$\approx (P_{\rm abs})_{\rm ave} NT \approx (P_{\rm abs})_{\rm ave}\Delta t \quad (\delta t \ll NT). \tag{9.62}$$

Note that for f on the order of GHz ($T \sim$ ns) and Δt on the order of s, for example, N is as large as $N \sim 10^9$. So, indeed, we can readily compute $W_{\rm abs}$ as

$$W_{\rm abs} = (P_{\rm abs})_{\rm ave}\Delta t = \frac{E_0^2}{\eta_0} S\cos\theta\,\Delta t \quad (\Delta t \gg T). \tag{9.63}$$

The absorbed energy can alternatively be determined via the electromagnetic energy stored in the electric and magnetic fields of the wave. Namely, we realize that the energy that the wave delivers to the part of the screen with area S from an instant t_1 to $t_2 = t_1 + \Delta t$, while progressing toward the screen at velocity c_0, Eq. (9.19), in fact equals the total electromagnetic energy contained at time t_1 in an oblique cylinder with the basis S and length $l = c_0\Delta t$, as indicated in Fig. 9.6. This energy can be found by integrating the electromagnetic energy density of the wave, $w_{\rm em}$, throughout the volume v of the cylinder, like in Eq. (8.160). However, because of the condition $\Delta t \gg T$, or, equivalently, because l is much larger than the wavelength (space period) of the wave, λ_0, Eq. (8.112), $l = c_0\Delta t \gg c_0 T = \lambda_0$, we can conveniently use the time-average energy density of the wave, $(w_{\rm em})_{\rm ave} = \varepsilon_0 E_0^2$, Eq. (9.39), in the integral. This density, in turn, being constant in space, the total time-average energy stored in the cylinder comes out to be simply $(w_{em})_{ave}$ times $v = Sh$, where $h = l\cos\theta$ is the height of the cylinder. Hence,

$$W_{abs} = \int_v (w_{em})_{ave}\, dv = (w_{em})_{ave} v = \varepsilon_0 E_0^2 S c_0 \Delta t \cos\theta, \tag{9.64}$$

which, of course, is the same result as in Eq. (9.63) [note that, from the first relationship in Eqs. (9.54), $\varepsilon_0 c_0 = 1/\eta_0$].

Example 9.11 Solar Power Density on the Earth's Surface

The time-average surface power density of all of the sun's radiation on the earth's surface is measured to be about 1.35 kW/m^2, for normal incidence on the surface. (a) Assuming, for simplicity, that this radiation can be represented by a single plane wave (at a single frequency), find the rms electric field intensity of this equivalent wave, as well as the absorbed energy per meter squared of a large black (perfectly absorbing) plate positioned, on the earth's surface, perpendicularly to the direction of light beam – in one hour. (b) Assuming also that the sun radiates isotropically (equally in all directions), compute its total time-average radiated power (the radius of the earth's orbit around the sun is approximately 1.5×10^8 km). (c) With an assumption that the entire energy of the sun's radiation reaching the earth is absorbed by it, determine the energy "received" by the earth in one day (the average radius of the earth is about 6378 km). (d) Calculate the time-average surface power density of the solar illumination at the surface of Mercury (Mercury to sun distance is approximately 6×10^7 km). (e) How long does it take the sunlight to reach Mercury and the earth, respectively?

Solution

(a) The time-average Poynting vector magnitude of the equivalent wave is $\mathcal{P}_{ave} \approx$ 1.35 kW/m^2. This corresponds, by way of Eq. (9.61), to the rms electric field intensity of $E_0 = \sqrt{\eta_0 \mathcal{P}_{ave}} = 713.4$ V/m. From Eq. (9.63) with $\theta = 0$ (normal incidence), the absorbed energy in the black plate for $S = 1$ m^2 and $\Delta t_1 = 1$ h $= 3600$ s amounts to $W_{abs} = \mathcal{P}_{ave} S \Delta t_1 = 4.86$ MJ.

(b) The total time-average power radiated by the sun (P_{rad}) can be computed using Poynting's theorem in complex form, Eq. (8.196), as the flux of the real part of the complex Poynting vector, that is, the time-average Poynting vector, $\boldsymbol{\mathcal{P}}_{ave}$, of the equivalent wave in (a) through a spherical surface, S_0, with the sun at the center and the earth on its perimeter, as shown in Fig. 9.7. The sphere radius is thus $r_{sun-earth} = 1.5 \times 10^8$ km. The vector $\boldsymbol{\mathcal{P}}_{ave}$ being radial with respect to the center of the sphere, and its magnitude being the same at every point of the surface (isotropic radiation), the integration over S_0 is carried out in exactly the same way as in the application of Gauss' law in Fig. 1.33 and Eq. (1.138), yielding

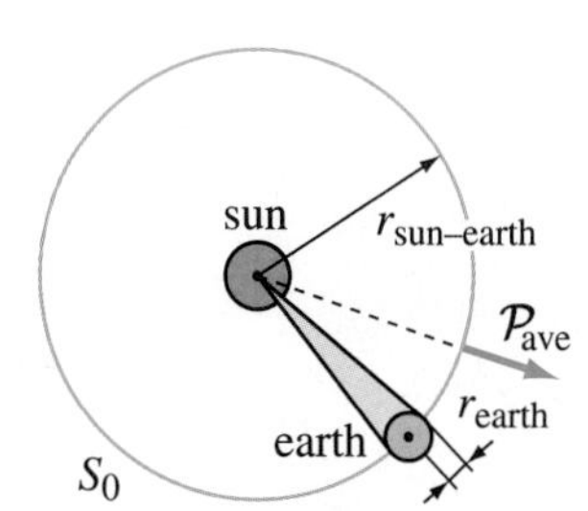

Figure 9.7 Computation of the total power radiated by the sun and the portion of it absorbed by the earth – via the flux of the associated equivalent Poynting vector measured at the earth's surface; for Example 9.11.

$$P_{rad} = \oint_{S_0} \boldsymbol{\mathcal{P}}_{ave} \cdot d\mathbf{S} = \mathcal{P}_{ave} S_0 = \mathcal{P}_{ave} 4\pi r_{sun-earth}^2 = 3.82 \times 10^{26}\ \mathrm{W}. \tag{9.65}$$

(c) The time-average power absorbed by the earth (P_{earth}) equals the portion of the flux of the Poynting vector in Eq. (9.65) within a cone with the sun at its apex and the earth defining its opening (Fig. 9.7). Because the earth's diameter is much smaller than the circumference of its orbit around the sun, the time-average power radiated in this cone can be found as $\mathcal{P}_{ave}$ times the area of the earth's cross section, so that the absorbed energy, Eq. (9.62), for the time duration $\Delta t_2 = 24$ h $= 86{,}400$ s comes out to be

$$P_{earth} = \mathcal{P}_{ave} \pi r_{earth}^2 = 1.72 \times 10^{17}\ \mathrm{W} \longrightarrow W_{earth} = P_{earth} \Delta t_2 = 1.5 \times 10^{22}\ \mathrm{J} \tag{9.66}$$

($r_{earth} = 6378$ km), which, of course, is an enormously large energy.

(d) Since the total radiated power of the sun is the same whether its portions are "received," in terms of the illumination surface power density, on earth or Mercury, we have from Eq. (9.65) and $r_{sun-Mercury} = 6 \times 10^7$ km, $(\mathcal{P}_{ave})_{Mercury} = P_{rad}/(4\pi r_{sun-Mercury}^2) =$

8.44 kW/m^2. Of course, the illumination is considerably stronger (power density higher) on Mercury than on the earth, as Mercury is much closer to the sun.

(e) Given the speed of light in free space (c_0) in Eq. (9.19), the sunlight travels times $t_1 = r_{\text{sun-Mercury}}/c_0 = 200\ \text{s} = 3.33\ \text{min}$ and $t_2 = r_{\text{sun-earth}}/c_0 = 500\ \text{s} = 8.33\ \text{min}$ from the sun to Mercury and the earth, respectively.

Problems: 9.3–9.18; *Conceptual Questions* (on Companion Website): 9.7–9.12; *MATLAB Exercises* (on Companion Website).

9.5 THE ELECTROMAGNETIC SPECTRUM

As already discussed on many occasions throughout this text, Maxwell's equations, and thus the electromagnetic wave theory they govern, are valid and are nowadays effectively used at frequencies spanning dc to optics to astrophysics, for system sizes ranging from subatomic to intergalactic, and for an extremely broad range of application areas. Examples are antennas, RF/microwave circuits, electronics, wireless communication systems, radar, remote sensing, electromagnetic compatibility, signal integrity, materials, nanoelectromagnetics, bioelectromagnetics, and radio astronomy. In specific, the equations we have derived so far, based on Maxwell's mathematical model, for describing and analyzing uniform plane electromagnetic waves and properties of these waves, are identical over the entire electromagnetic spectrum. Most notably, regardless of their frequency, all waves propagate through free space (a vacuum or air) with the same velocity, c_0, given in Eq. (9.19). Of course, the corresponding free-space wavelengths, λ_0, are obtained from the operating frequency, f, of the wave using Eq. (8.112), namely,

$$\lambda_0 = \frac{c_0}{f} = \frac{3 \times 10^8\ \text{m/s}}{f}. \qquad (9.67)$$

free-space wavelength

The frequencies of the electromagnetic waves that have been investigated experimentally range from a fraction of a hertz to as high as about 10^{30} Hz, and the range of wavelengths in the known electromagnetic spectrum is equally impressive. For example, the wavelength in free space corresponding to $f = 1$ Hz is as large as 300,000 km (more than 20 earth's diameters), and that at $f = 10^{24}$ Hz is as small as 0.3 fm (a millionth of the diameter of a typical atom; f $\equiv 10^{-15}$). Having covered in the previous section the fundamentals of time-harmonic computation of plane waves, we now reemphasize the huge span of frequencies and wavelengths encompassed by our derived theory and used in practical applications, and introduce some classification and subdivision of the electromagnetic spectrum that will help us in further studies. This is shown in Table 9.1, where a detailed designation of frequency bands, with their full names and acronyms, as they are used in engineering and scientific practice is given with their ranges expressed in terms of both f and λ_0, along with some selected applications for individual bands.

The frequencies within the ultra high frequency (UHF), super high frequency (SHF), and extremely high frequency (EHF) bands combined, so those from 300 MHz to 300 GHz, or wavelengths (in free space) ranging from 1 m down to 1 mm, constitute the microwave region of the electromagnetic spectrum. The further subdivision within UHF, SHF, and EHF bands in the form of bands with alphabetical designations (e.g., L-band, S-band, etc.) originate from radar work,

Table 9.1. The electromagnetic spectrum

Frequency	Free-space wavelength	Band	Selected applications
< 3 Hz	> 100 Mm		Geophysical sensing
3–30 Hz	10–100 Mm	Extremely low frequency (ELF)	Detection of buried metallic objects
30–300 Hz	1–10 Mm	Super low frequency (SLF)	Electric power distribution (50 or 60 Hz), submarine communications, ionospheric sensing
0.3–3 kHz	0.1–1 Mm	Ultra low frequency (ULF)	Telephone, audio systems, geomagnetic sensing
3–30 kHz	10–100 km	Very low frequency (VLF)	Navigation, positioning, ship/submarine comm.
30–300 kHz	1–10 km	Low frequency (LF)	Long-wave broadcasting, radio beacons, navigation
0.3–3 MHz	0.1–1 km	Medium frequency (MF)	AM radio broadcasting (0.535–1.605 MHz)
3–30 MHz	10–100 m	High frequency (HF)	Short-wave broadcasting, amateur radio
30–300 MHz	1–10 m	Very high frequency (VHF) TV channels 2–4 (54–72 MHz) TV channels 5–6 (76–88 MHz) FM radio (88–108 MHz) TV chann. 7–13 (174–216 MHz)	TV broadcasting (all TV channels have a 6-MHz bandwidth), FM radio broadcasting, mobile radio communication, air traffic control, navigation
0.3–3 GHz	0.1–1 m	Ultra high frequency (UHF) TV chann. 14–69 (470–806 MHz) Cellular (824–894 MHz) PCS (1850–1990 MHz) L-band (1–2 GHz) S-band (2–4 GHz)	Radar, TV broadcasting, cellular telephone, personal communication service (PCS), global positioning system – GPS (1.23 and 1.58 GHz), microwave cooking (2.45 GHz), satellite radio
3–30 GHz	1–10 cm	Super high frequency (SHF) C-band (4–8 GHz) X-band (8–12 GHz) K_u-band (12–18 GHz) K-band (18–27 GHz)	Radar, satellite communications, direct TV, wireless communication systems, wireless networks
30–300 GHz	1–10 mm	Extremely high frequency (EHF) K_a-band (27–40 GHz) V-band (40–75 GHz) W-band (75–110 GHz) Millimeter-wave (110–300 GHz)	Radar, remote sensing, radio astronomy, satellite communications
0.3–3 THz	0.1–1 mm	Submillimeter wave or terahertz	Meteorology, sensors, imaging, astronomy
3–400 THz	0.75–100 μm	Infrared (IR)	IR heating, night vision, optical communications
400–789 THz	380–750 nm	Visible light Red (620–750 nm) Orange (590–620 nm) Yellow (570–590 nm) Green (495–570 nm) Blue (450–495 nm) Violet (380–450 nm)	Vision, optical devices and systems, lasers
10^{15}–10^{18} Hz	0.3–300 nm	Ultraviolet (UV)	UV sterilization, lasers, semiconductor processing
10^{17}–10^{21} Hz	0.3 pm–3 nm	X-rays	Medical diagnostics
10^{19}–10^{22} Hz	0.03–30 pm	γ-rays	Radiation medical therapy, astrophysics
> 10^{22} Hz	< 0.03 pm	Cosmic rays	Astrophysics

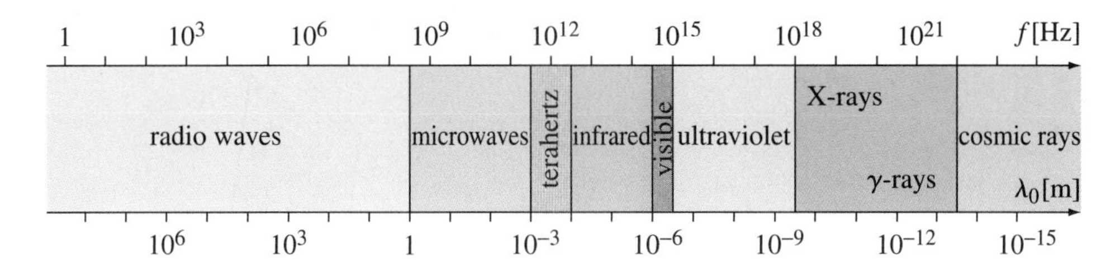

Figure 9.8 Graphical representation of major regions of the electromagnetic spectrum, from Table 9.1.

but are used in other applications as well. Electromagnetic waves with frequencies lower than 300 MHz and wavelengths longer than 1 m are called radio waves or radio-frequency (RF) waves, as indicated in the graphical representation of the electromagnetic spectrum in Fig. 9.8. On the other side of the microwave range, the parts of the spectrum pertaining to light, i.e., infrared, visible, and ultraviolet light, from $\lambda_0 = 100\ \mu$m to $\lambda_0 = 0.3$ nm, are referred to as optical wavelengths or frequencies, whereas submillimeter or terahertz waves occupy the "gap" between microwave and optical frequencies (Fig. 9.8). Above light, we have X-, γ-, and cosmic rays. Note that, while, of course, there is an one-to-one correspondence between f and λ_0, Eq. (9.67), frequency is more often used than wavelength in RF applications, waves in the optical region are almost exclusively characterized by their wavelength and not frequency, and both are used in the microwave range. Note, however, that the terms radio waves and RF are often used to mean all waves up to the optical region. Note, finally, that the general applications listed in Table 9.1 are only meant to be illustrative, and by no means exhaustive, of a host of practical devices, processes, and systems using the electromagnetic waves in various frequency bands.

MATLAB Exercises (on Companion Website).

9.6 ARBITRARILY DIRECTED UNIFORM TEM WAVES

In cases with more than one uniform plane wave propagating in different directions, it is impossible to adopt a global rectangular coordinate system such that the propagation direction of each of the waves coincides with an axis (x, y, or z) of the system, except in some special situations (e.g., two waves traveling in the forward and backward directions along the same line or traveling in mutually orthogonal directions). This is why we need expressions describing a uniform TEM wave whose propagation direction is completely arbitrary with respect to a given coordinate system, as is the wave shown in Fig. 9.9. The propagation unit vector of the wave is $\hat{\mathbf{n}}$ and its complex electric field rms intensity vector at a reference point (coordinate origin) O is $\underline{\mathbf{E}}_0$. Of course, $\mathbf{E}_0$ must be perpendicular to $\hat{\mathbf{n}}$, and thus it also is not positioned along any of the coordinate axes. To find the expressions for the field vectors at an arbitrary point P in space, the position vector of which (with respect to O) is $\mathbf{r}$, we realize that the normal distance of the transversal (equiphase) plane of the wave that contains this point from the origin, i.e., from the reference transversal plane (that contains the origin), amounts to

$$l = \mathbf{r} \cdot \hat{\mathbf{n}} \tag{9.68}$$

(see the right-angled triangle $\triangle$OP′P). As this distance obviously plays the role of z in Eqs. (9.36), the complex electric field intensity vector at the point P (and in the

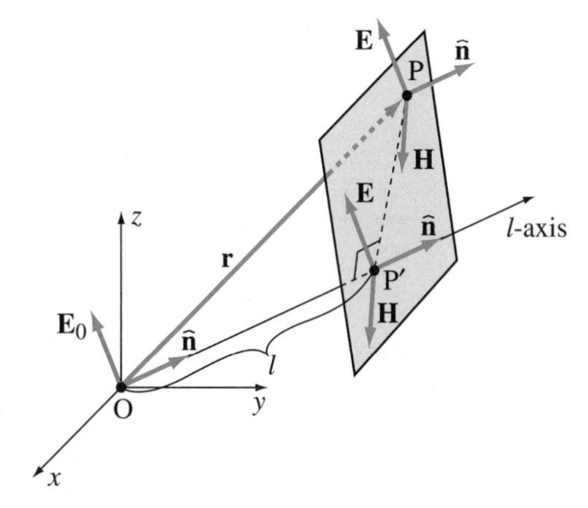

Figure 9.9 Uniform plane wave whose propagation does not coincide with any of the axes of the adopted global Cartesian coordinate system.

entire plane $l = \text{const}$) is given by

electric field vector of an arbitrarily directed TEM wave

$$\boxed{\underline{\mathbf{E}} = \underline{\mathbf{E}}_0\, \mathrm{e}^{-\mathrm{j}\beta l} = \underline{\mathbf{E}}_0\, \mathrm{e}^{-\mathrm{j}\beta \mathbf{r}\cdot\hat{\mathbf{n}}}.} \tag{9.69}$$

From Eq. (9.22), the associated complex magnetic field intensity vector of the wave is

magnetic field vector of an arbitrarily directed wave

$$\boxed{\underline{\mathbf{H}} = \frac{1}{\eta}\,\hat{\mathbf{n}} \times \underline{\mathbf{E}} = \frac{1}{\eta}\,\hat{\mathbf{n}} \times \underline{\mathbf{E}}_0\, \mathrm{e}^{-\mathrm{j}\beta \mathbf{r}\cdot\hat{\mathbf{n}}}.} \tag{9.70}$$

For an explicit dependence of the expressions in Eqs. (9.69) and (9.70) on the coordinates x, y, and z of the point P at which the fields are computed, we write [Eqs. (1.7) and (1.4)]

$$\mathbf{r} = x\,\hat{\mathbf{x}} + y\,\hat{\mathbf{y}} + z\,\hat{\mathbf{z}} \quad \text{and} \quad \hat{\mathbf{n}} = n_x\,\hat{\mathbf{x}} + n_y\,\hat{\mathbf{y}} + n_z\,\hat{\mathbf{z}}, \tag{9.71}$$

where n_x, n_y, and n_z are constants (Cartesian components of $\hat{\mathbf{n}}$) defining the wave propagation direction ($\hat{\mathbf{n}}$ generally does not coincide with any of the coordinate unit vectors or their opposites) and $n_x^2 + n_y^2 + n_z^2 = 1$ (unit vector). With this, the phase propagation factor in Eqs. (9.69) and (9.70) can be written as the following function of x, y, and z:

phase factor for an arbitrary propagation direction

$$\boxed{\mathrm{e}^{-\mathrm{j}\beta \mathbf{r}\cdot\hat{\mathbf{n}}} = \mathrm{e}^{-\mathrm{j}\beta(x n_x + y n_y + z n_z)}.} \tag{9.72}$$

Equivalently, the instantaneous phase of the wave in Eq. (9.33) becomes

$$\phi(x, y, z, t) = \omega t - \beta \mathbf{r}\cdot\hat{\mathbf{n}} + \theta_0 = \omega t - \beta(x n_x + y n_y + z n_z) + \theta_0. \tag{9.73}$$

Note that $x n_x + y n_y + z n_z = \text{const}$ represents the equation of an equiphase plane of the wave (that is l distant from the origin O). Note also that the vector cross product in (9.70) is calculated from the expression for $\hat{\mathbf{n}}$ in Eqs. (9.71) and the corresponding expression in terms of the Cartesian components for $\underline{\mathbf{E}}_0$. Note finally that a wave propagating along any of the coordinate axis is, of course, just a special case of the general representation in Fig. 9.9. For example, Eqs. (9.69) and (9.70) with $\hat{\mathbf{n}} = \hat{\mathbf{z}}$ ($n_x = n_y = 0$ and $n_z = 1$) and $\underline{\mathbf{E}}_0 = \underline{E}_0\,\hat{\mathbf{x}}$ simplify to Eqs. (9.36) for the wave in Fig. 9.2 (traveling in the positive z direction).

Example 9.12 Plane Wave Propagation in an Arbitrary Direction

A time-harmonic uniform plane wave of frequency $f = 300$ MHz propagates in the direction defined by the vector $\hat{\mathbf{x}} + \hat{\mathbf{y}}$ in a rectangular coordinate system. The complex electric field intensity vector of the wave at the coordinate origin is $\underline{\mathbf{E}}_0 = (1 + \mathrm{j})\,\hat{\mathbf{z}}$ V/m, and the medium is air. Find the expressions for electric and magnetic field vectors and Poynting vector of the wave at an arbitrary point in space, P(x, y, z), and specifically for $x = 10$ m, $y = 1$ m, and $z = 5$ m.

Solution From Eq. (8.111), the phase coefficient of the wave amounts to $\beta = 2\pi f/c_0 = 2\pi$ rad/m, where $c_0 = 3 \times 10^8$ m/s, Eq. (9.19), is the free-space wave velocity. The propagation unit vector of the wave is $\hat{\mathbf{n}} = \sqrt{2}(\hat{\mathbf{x}} + \hat{\mathbf{y}})/2$ (to ensure that $|\hat{\mathbf{n}}| = 1$), and its Cartesian components are $n_x = n_y = \sqrt{2}/2$ and $n_z = 0$. Eqs. (9.69)–(9.72) then lead to the following expressions for the field vectors for an arbitrary (x, y, z):

$$\underline{\mathbf{E}} = \underline{\mathbf{E}}_0\, \mathrm{e}^{-\mathrm{j}\beta \mathbf{r}\cdot\hat{\mathbf{n}}} = (1 + \mathrm{j})\, \mathrm{e}^{-\mathrm{j}\pi\sqrt{2}(x+y)}\, \hat{\mathbf{z}} \text{ V/m} = \sqrt{2}\, \mathrm{e}^{\mathrm{j}\pi[-\sqrt{2}(x+y)+1/4]}\, \hat{\mathbf{z}} \text{ V/m},$$

$$\underline{\mathbf{H}} = \frac{1}{\eta_0}\hat{\mathbf{n}} \times \underline{\mathbf{E}} = \frac{\sqrt{2}}{377}\, \mathrm{e}^{\mathrm{j}\pi[-\sqrt{2}(x+y)+1/4]}\frac{\sqrt{2}}{2}(\hat{\mathbf{x}} + \hat{\mathbf{y}}) \times \hat{\mathbf{z}} \text{ A/m}$$

$$= 2.65\, \mathrm{e}^{\mathrm{j}\pi[-\sqrt{2}(x+y)+1/4]}(\hat{\mathbf{x}} - \hat{\mathbf{y}}) \text{ mA/m} \quad (x, y \text{ in m}). \tag{9.74}$$

Fig. 9.10 shows vectors **E**, **H**, and $\hat{\mathbf{n}}$ (at the coordinate origin). At the particular given point,

$$x = 10 \text{ m}, \quad y = 1 \text{ m}, \quad z = 5 \text{ m} \quad \longrightarrow \quad \underline{\mathbf{E}} = 1.41\, \mathrm{e}^{-\mathrm{j}15.31\pi}\, \hat{\mathbf{z}} \text{ V/m}$$

$$= 1.41\, \mathrm{e}^{\mathrm{j}0.69\pi}\, \hat{\mathbf{z}} \text{ V/m} = 1.41\, \mathrm{e}^{\mathrm{j}125^\circ}\, \hat{\mathbf{z}} \text{ V/m}, \quad \underline{\mathbf{H}} = 2.65\, \mathrm{e}^{\mathrm{j}125^\circ}(\hat{\mathbf{x}} - \hat{\mathbf{y}}) \text{ mA/m}. \tag{9.75}$$

Note that the phase angle of $8 \times 2\pi = 16\pi$ is added to the initial phase of the wave computed from Eqs. (9.74), to make it, as is customary, fall in the range $-\pi < \theta \le \pi$.

Combining Eqs. (8.194), (9.69), and (9.70), the complex Poynting vector of the wave comes out to be (Fig. 9.10)

$$\underline{\mathcal{P}} = \underline{\mathbf{E}} \times \underline{\mathbf{H}}^* = \frac{\underline{E}_0 \underline{E}_0^*}{\eta}\, \mathrm{e}^{-\mathrm{j}\beta\mathbf{r}\cdot\hat{\mathbf{n}}}\, \mathrm{e}^{\mathrm{j}\beta\mathbf{r}\cdot\hat{\mathbf{n}}}\, \hat{\mathbf{n}} = \frac{|\underline{E}_0|^2}{\eta}\, \hat{\mathbf{n}} = 3.75(\hat{\mathbf{x}} + \hat{\mathbf{y}}) \text{ mW/m}^2 \tag{9.76}$$

(it does not depend on the coordinates x, y, and z, and is purely real, and thus equal to the time-average Poynting vector, $\mathcal{P}_{\text{ave}}$). Of course, this result can as well be obtained from the results for $\underline{\mathbf{E}}$ and $\underline{\mathbf{H}}$ in Eqs. (9.74).

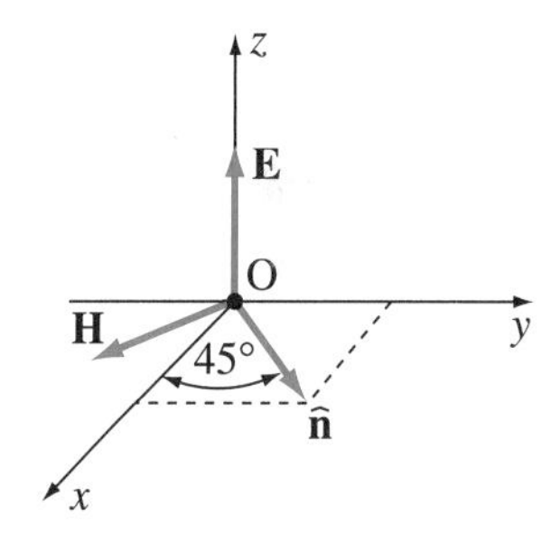

Figure 9.10 Uniform plane wave whose direction of travel is determined by the vector $\hat{\mathbf{x}} + \hat{\mathbf{y}}$; for Example 9.12.

Problems: 9.19; *Conceptual Questions* (on Companion Website): 9.13; *MATLAB Exercises* (on Companion Website).

9.7 THEORY OF TIME-HARMONIC WAVES IN LOSSY MEDIA

We now consider uniform plane time-harmonic electromagnetic waves in a linear and homogeneous medium that exhibits losses ($\sigma \neq 0$). The conduction volume current in the medium is now nonzero, with a complex density vector $\underline{\mathbf{J}} = \sigma\underline{\mathbf{E}}$, where $\underline{\mathbf{E}}$ is the complex electric field intensity vector of the wave. However, from Eq. (8.83), we know that this medium must be charge-free ($\underline{\rho} = 0$), provided that there are no external energy volume sources (impressed electric fields or currents) in the medium. Consequently, looking at complex Maxwell's differential equations, Eqs. (8.81), and comparing their versions for lossy and lossless homogeneous media (both with $\underline{\rho} = 0$), we realize that the only difference between the two sets of equations is in the second equation (corrected generalized Ampère's law). Its version for a lossy medium of parameters ε, μ, and σ reads

$$\nabla \times \underline{\mathbf{H}} = \underline{\mathbf{J}} + \mathrm{j}\omega\underline{\mathbf{D}} = \sigma\underline{\mathbf{E}} + \mathrm{j}\omega\varepsilon\underline{\mathbf{E}} = \mathrm{j}\omega\left(\varepsilon - \mathrm{j}\frac{\sigma}{\omega}\right)\underline{\mathbf{E}}. \tag{9.77}$$

However, introducing a new quantity,

equivalent complex permittivity (unit: F/m*)*

$$\underline{\varepsilon}_{\mathrm{e}} = \varepsilon - \mathrm{j}\,\frac{\sigma}{\omega}, \tag{9.78}$$

called the equivalent complex permittivity of the medium, Eq. (9.77) can be written in the form

Ampère's law for a lossy medium

$$\nabla \times \underline{\mathbf{H}} = \mathrm{j}\omega\underline{\varepsilon}_{\mathrm{e}}\underline{\mathbf{E}}. \tag{9.79}$$

This equation is formally identical to its lossless ($\sigma = 0$) version,

Ampère's law for a lossless medium

$$\nabla \times \underline{\mathbf{H}} = \mathrm{j}\omega\varepsilon\underline{\mathbf{E}}. \tag{9.80}$$

Based on this, we can now analyze the propagation of uniform plane TEM time-harmonic waves in lossy media in the same way as in the lossless case (Sections 9.4 and 9.6) by merely replacing ε with $\underline{\varepsilon}_{\mathrm{e}}$ in all the existing expressions containing this parameter. We note that the equivalent complex permittivity of a medium depends on the permittivity (ε) and conductivity (σ) of the medium, as well as on the angular frequency (ω) of the wave propagating through it.[5] We also note that, while the real part of $\underline{\varepsilon}_{\mathrm{e}}$ is always positive ($\varepsilon \geq \varepsilon_0$), its imaginary part ($-\sigma/\omega$) must be either negative or eventually zero (for $\sigma = 0$).

From Eqs. (9.36), (8.111), and (9.78), the complex electric field intensity of a TEM wave in a lossy medium is

complex E-field of a wave in a lossy medium

$$\underline{E}_x = \underline{E}_0\,\mathrm{e}^{-\mathrm{j}\omega\sqrt{\underline{\varepsilon}_{\mathrm{e}}\mu}\,z} = \underline{E}_0\,\mathrm{e}^{-\underline{\gamma}z} = \underline{E}_0\,\mathrm{e}^{-\alpha z}\,\mathrm{e}^{-\mathrm{j}\beta z}, \tag{9.81}$$

where $\underline{\gamma}$ is termed the complex propagation coefficient and is evaluated as

complex propagation coefficient (unit: m^{-1}*)*

$$\underline{\gamma} = \mathrm{j}\omega\sqrt{\underline{\varepsilon}_{\mathrm{e}}\mu} = \alpha + \mathrm{j}\beta. \tag{9.82}$$

Its real and imaginary parts,

attenuation and phase coefficients (units: Np/m *and* rad/m*)*

$$\alpha = \mathrm{Re}\{\underline{\gamma}\} \quad \text{and} \quad \beta = \mathrm{Im}\{\underline{\gamma}\}, \tag{9.83}$$

are called, respectively, the attenuation coefficient and phase coefficient (this same term is used also for waves in lossless media) of the wave. In the time domain, using Eq. (8.66),

instantaneous E-field

$$E_x(z,t) = E_0\sqrt{2}\,\mathrm{e}^{-\alpha z}\cos(\omega t - \beta z + \theta_0). \tag{9.84}$$

We see that, as in the lossless case, β is the spatial rate of decrease of the instantaneous phase of the wave, and thus it retains its name – the phase coefficient. The term $\mathrm{e}^{-\alpha z}$ obviously represents an exponential spatial decrease (attenuation) of the wave amplitude,

$$E_{\mathrm{m}}(z) = E_{\mathrm{m}}(0)\,\mathrm{e}^{-\alpha z} \quad [E_{\mathrm{m}}(0) = E_0\sqrt{2}], \tag{9.85}$$

and hence its name – the attenuation coefficient of the wave. In other words, the wave propagating in the positive z direction has a decaying amplitude (of its electric and magnetic fields) with increasing the distance z at any fixed instant of time, and the rate of this attenuation is determined by the coefficient α. Fig. 9.11 shows a time

[5] At higher frequencies, so-called polarization losses in dielectric materials may result in an effective conductivity that is larger than σ, and that addition to the conductivity is also involved in the imaginary part of $\underline{\varepsilon}_{\mathrm{e}}$, whose frequency behavior becomes rather complicated, including resonance effects. Moreover, when viewed over very wide frequency ranges, the real part of the complex permittivity generally is not a constant, but depends on frequency as well.

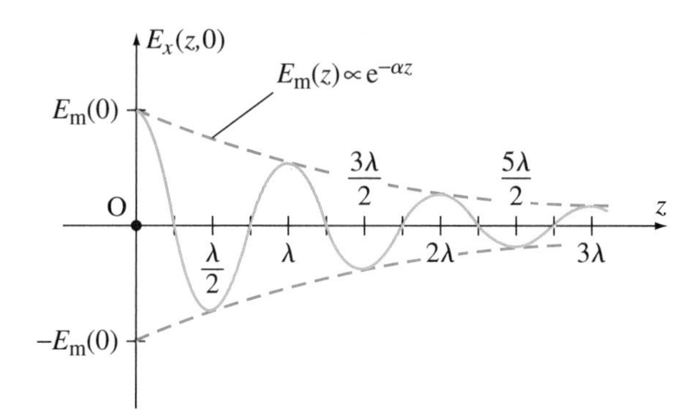

Figure 9.11 Plot of the electric field intensity of a uniform plane time-harmonic wave in a lossy medium, Eq. (9.84), as a function of z at $t = 0$ for $\theta_0 = 0$ ($\lambda = 2\pi/\beta$).

snapshot (at $t = 0$) of the electric field intensity in Eq. (9.84) as a function of z. Note that the curves $\pm E_m(z)$ form an envelope of the oscillating wave pattern along z.

Substituting ε by $\underline{\varepsilon}_e$ in Eq. (8.111) and then in Eqs. (9.8) and (9.9), we obtain Helmholtz equations for a lossy medium,

Helmholtz equations – lossy medium

$$\boxed{\nabla^2 \underline{\mathbf{E}} - \underline{\gamma}^2 \underline{\mathbf{E}} = 0, \quad \nabla^2 \underline{\mathbf{H}} - \underline{\gamma}^2 \underline{\mathbf{H}} = 0,} \tag{9.86}$$

with $\underline{\gamma}$ being that in Eq. (9.82). The electric field intensity of a uniform plane wave propagating through the medium can now be found as a solution of the one-dimensional version (with the uniform-plane-wave approximation) of the first of these two equations, and, of course, the result is that in Eq. (9.81).

Although $\underline{\gamma}$, α, and β are the quantities of the same nature, we use different units, m^{-1} for $\underline{\gamma}$, Np/m (neper per meter) for α, and rad/m for β, to distinguish one from another. Note that the ratio of the amplitudes of E in two different transversal planes separated by a distance d along the z-axis is constant, i.e., it does not depend on z. It is given by

$$\frac{E_m(z)}{E_m(z+d)} = e^{\alpha d} \quad \text{or} \quad \alpha d = \ln \frac{E_m(z)}{E_m(z+d)}. \tag{9.87}$$

Since αd equals the natural (Napierian) logarithm of the ratio of field intensities, it is commonly expressed in nepers (although it is a dimensionless number), and hence Np/m as the unit for α. Alternatively, the attenuation can be expressed in decibels,[6] as

$$A_{\text{dB}} = 20 \log \frac{E_m(z)}{E_m(z+d)} = (20 \log e)\, \alpha d = 8.686\, \alpha d \tag{9.88}$$

[$\log x \equiv \log_{10} x$ (common or decadic logarithm)[7]], so the dB attenuation comes out to be 8.686 times the attenuation expressed in nepers. Dividing both attenuations by d (or assuming that $d = 1$ m), we have

conversion from Np/m to dB/m

$$\boxed{\alpha \text{ in dB/m} = 8.686 \times (\alpha \text{ in Np/m}),} \tag{9.89}$$

which means that in order to express the attenuation coefficient in dB/m, we simply multiply the value obtained in Eq. (9.83) by 8.686. Note that A_{dB} is always positive or eventually 0 dB (for lossless case). For instance, for the ratio of field intensities

[6]The decibel (dB) is a unit for attenuation expressed using the decadic logarithm (logarithm with base 10) of the ratio of relevant field or circuit quantities (e.g., electric field intensities at different positions). It equals 1/10 of the bel (B), which was used in early work on telephone systems, but is now superseded by dB.

[7]Whereas "log" in our mathematical notation means "$\log_{10}$," note that many computer programming languages, including C and MATLAB, use "log" in place of "ln," to denote the natural logarithm.

amounting to 2 in natural numbers, $A_{\text{dB}} = 6$ dB. Logarithmic units, such as decibels or nepers, appear to be extremely convenient in a multitude of applications in both field and circuit theory. For example, the total attenuation (or gain) of several cascaded components of a system (e.g., a medium composed of several layers of lossy material with different conductivities or a series of amplifiers in a circuit) can simply be found as an algebraic sum of the attenuations (or gains) of individual components expressed in dB or Np.

HISTORICAL ASIDE

Hermann von Helmholtz (1821–1894), a German physicist and physiologist, taught at the Universities of Königsberg, Bonn, Heidelberg, and Berlin. He is considered the last great scholar whose work traversed practically all the natural sciences, philosophy, and fine arts.

The neper (Np), used to express the values of quantities based on the Napierian or natural logarithm, was named after **John Napier** (also spelled Neper) (1550–1617), a Scottish mathematician who invented logarithms.

The bel was named in honor of **Alexander Graham Bell** (1847–1922), a Scottish-American inventor, who invented in 1876 the first practical telephone, and introduced it to the world at the Centennial Exhibition in Philadelphia the same year. To exploit the invention, Bell cofounded in 1877 the "Bell Telephone Company," which would eventually become the "American Telephone & Telegraph Corporation" (AT&T).

Combining Eqs. (9.36), (9.81), and (9.21) with ε substituted by $\underline{\varepsilon}_{\text{e}}$, we obtain the following expression for the complex magnetic field intensity of the wave:

$$\underline{H}_y = \sqrt{\frac{\underline{\varepsilon}_{\text{e}}}{\mu}}\,\underline{E}_x = \frac{E_0}{\underline{\eta}}\,\text{e}^{-\underline{\gamma} z}, \tag{9.90}$$

where $\underline{\eta}$ is the complex constant – the complex intrinsic impedance of the medium – defined as the ratio of the electric and magnetic complex rms field intensities at any point of space. It can be written as

complex intrinsic impedance (unit: Ω)

$$\underline{\eta} = \sqrt{\frac{\mu}{\underline{\varepsilon}_{\text{e}}}} = \sqrt{\frac{\mu}{\varepsilon - \text{j}\frac{\sigma}{\omega}}} = |\underline{\eta}|\,\text{e}^{\text{j}\phi}, \tag{9.91}$$

which yields

complex H-field of a wave in a lossy medium

$$\underline{H}_y = \frac{E_0}{|\underline{\eta}|}\,\text{e}^{-\alpha z}\,\text{e}^{-\text{j}\beta z}\,\text{e}^{-\text{j}\phi}. \tag{9.92}$$

The instantaneous magnetic field intensity of the wave is hence

instantaneous H-field

$$H_y(z,t) = \frac{E_0}{|\underline{\eta}|}\sqrt{2}\,\text{e}^{-\alpha z}\cos(\omega t - \beta z + \theta_0 - \phi). \tag{9.93}$$

We see that the magnitude of the complex intrinsic impedance of the medium, $|\underline{\eta}|$, determines the ratio of the amplitudes of the electric and magnetic field intensities of the wave, whereas its phase angle (argument), ϕ, equals the phase difference between the field intensities. Namely, in lossy media, the electric and magnetic fields are not in phase – the magnetic field lags the electric field by a phase lag ϕ. In lossless media, $\phi = 0$.

The complex magnitude of the Poynting vector of the wave, $\underline{\mathcal{P}} = \underline{\mathcal{P}}_z\,\hat{\mathbf{z}}$, is given by

$$\underline{\mathcal{P}}_z = \underline{E}_x \underline{H}_y^* = \underline{E}_x \frac{\underline{E}_x^*}{\underline{\eta}^*} = \frac{|\underline{E}_x|^2}{\underline{\eta}^*} = \underline{\eta}\underline{H}_y\underline{H}_y^* = \underline{\eta}|\underline{H}_y|^2 \tag{9.94}$$

($|\underline{E}_x|$ and $|\underline{H}_y|$ are, respectively, electric and magnetic rms field intensities of the wave, for an arbitrary z), and we can use either one of these equivalent formulas to further develop the expression for $\underline{\mathcal{P}}_z$. Thus, using the first one,

$$\underline{\mathcal{P}}_z = \underline{E}_x \underline{H}_y^* = \underline{E}_0\,\mathrm{e}^{-\alpha z}\,\mathrm{e}^{-\mathrm{j}\beta z}\,\frac{\underline{E}_0^*}{|\underline{\eta}|}\,\mathrm{e}^{-\alpha z}\,\mathrm{e}^{\mathrm{j}\beta z}\,\mathrm{e}^{\mathrm{j}\phi} = \frac{E_0^2}{|\underline{\eta}|}\,\mathrm{e}^{-2\alpha z}\,\mathrm{e}^{\mathrm{j}\phi}, \tag{9.95}$$

and hence the time average of the corresponding instantaneous vector magnitude [Eq. (8.195)]

$$(\mathcal{P}_z)_{\mathrm{ave}} = \mathrm{Re}\{\underline{\mathcal{P}}_z\} = \frac{E_0^2}{|\underline{\eta}|}\,\mathrm{e}^{-2\alpha z}\cos\phi, \tag{9.96}$$

time-average Poynting vector magnitude of a wave in a lossy medium

representing real power flow (by the electromagnetic wave) in the positive z direction. We see that the rate of the attenuation of $\mathcal{P}_{\mathrm{ave}}$, i.e., of the time-average surface power density of the wave in the direction of its propagation, is determined by twice the attenuation coefficient, α,

$$\mathcal{P}_{\mathrm{ave}}(z) = \mathcal{P}_{\mathrm{ave}}(0)\,\mathrm{e}^{-2\alpha z}. \tag{9.97}$$

However, the attenuation in decibels is the same whether decibels are defined for power ratios or the corresponding field intensity ratios,

$$A_{\mathrm{dB}} = 10\log\frac{\mathcal{P}_1}{\mathcal{P}_2} = 20\log\frac{E_1}{E_2} \tag{9.98}$$

decibel attenuation

(note different scale factors, 10 vs. 20 in the definitions), and equals $8.686\alpha d$ (dB) for a distance d between the two reference transversal planes along the z-axis [Eq. (9.88)].

Finally, the time-average power density of Joule's (or ohmic) losses at a point in a lossy medium can be obtained from Eq. (3.31) using the rms electric field intensity of the wave at that point, which in turn can be found either from Eq. (9.81), as the magnitude of the complex rms electric field intensity, $|\underline{E}_x|$, or from Eq. (9.84), simply identifying the rms value of the time-harmonic field intensity,

$$(p_{\mathrm{J}})_{\mathrm{ave}} = \sigma|\underline{E}_x|^2 = \sigma E_0^2\,\mathrm{e}^{-2\alpha z}. \tag{9.99}$$

Note that this loss-power volume density has the same dependence on the spatial coordinate z as the time-average surface power density carried by the wave, Eq. (9.96).

Problems: 9.20–9.23; *Conceptual Questions* (on Companion Website): 9.14–9.16.

9.8 EXPLICIT EXPRESSIONS FOR BASIC PROPAGATION PARAMETERS

To obtain the explicit expressions for the attenuation and phase coefficients, α and β, for a given (lossy) medium at a given frequency, we need to solve for the real and imaginary parts of the complex expression for $\underline{\gamma}$ in Eq. (9.82). Squaring the left- and right-hand sides of this equation and using Eq. (9.78), we have

$$\underline{\gamma}^2 = (\alpha + \mathrm{j}\beta)^2 = \alpha^2 - \beta^2 + \mathrm{j}2\alpha\beta = -\omega^2\underline{\varepsilon}_\mathrm{e}\mu = -\omega^2\varepsilon\mu + \mathrm{j}\omega\mu\sigma, \tag{9.100}$$

so that equating the real parts gives

$$\alpha^2 - \beta^2 = -\omega^2\varepsilon\mu. \tag{9.101}$$

On the other side,

$$|\underline{\gamma}|^2 = |\alpha + \mathrm{j}\beta|^2 = \alpha^2 + \beta^2, \tag{9.102}$$

and, from Eq. (9.100),

$$|\underline{\gamma}^2| = |-\omega^2\varepsilon\mu + \mathrm{j}\omega\mu\sigma| = \sqrt{\omega^4\varepsilon^2\mu^2 + \omega^2\mu^2\sigma^2}. \tag{9.103}$$

The two above results combined, since $|\underline{\gamma}^2| = |\underline{\gamma}|^2$, lead to

$$\alpha^2 + \beta^2 = \omega^2\varepsilon\mu\sqrt{1 + \left(\frac{\sigma}{\omega\varepsilon}\right)^2}. \tag{9.104}$$

Now, adding and subtracting Eqs. (9.104) and (9.101), we obtain the expressions for $2\alpha^2$ and $2\beta^2$, respectively, and hence

α – arbitrary medium

$$\alpha = \omega\sqrt{\frac{\varepsilon\mu}{2}}\left[\sqrt{1 + \left(\frac{\sigma}{\omega\varepsilon}\right)^2} - 1\right]^{1/2}, \tag{9.105}$$

β – arbitrary medium

$$\beta = \omega\sqrt{\frac{\varepsilon\mu}{2}}\left[\sqrt{1 + \left(\frac{\sigma}{\omega\varepsilon}\right)^2} + 1\right]^{1/2}. \tag{9.106}$$

Using these expressions, we can compute the values for α (in Np/m) and β (in rad/m) for an arbitrary medium based on its electromagnetic parameters, ε, μ, and σ, as well as the angular frequency, ω.

To find the explicit expressions for the magnitude and phase angle, $|\underline{\eta}|$ and ϕ, of the complex intrinsic impedance in Eq. (9.91), we first look at those of the equivalent complex permittivity in Eq. (9.78),

$$\underline{\varepsilon}_\mathrm{e} = \varepsilon\left(1 - \mathrm{j}\frac{\sigma}{\omega\varepsilon}\right) = \varepsilon\sqrt{1 + \left(\frac{\sigma}{\omega\varepsilon}\right)^2}\,\mathrm{e}^{-\mathrm{j}\arctan[\sigma/(\omega\varepsilon)]}, \tag{9.107}$$

and then of its square root,

$$\sqrt{\underline{\varepsilon}_\mathrm{e}} = \sqrt{\varepsilon}\left[1 + \left(\frac{\sigma}{\omega\varepsilon}\right)^2\right]^{1/4}\mathrm{e}^{-\mathrm{j}\frac{1}{2}\arctan[\sigma/(\omega\varepsilon)]}. \tag{9.108}$$

This, substituted in Eq. (9.91), results in

$|\underline{\eta}|$, ϕ – arbitrary medium

$$|\underline{\eta}| = \frac{\sqrt{\frac{\mu}{\varepsilon}}}{\left[1 + \left(\frac{\sigma}{\omega\varepsilon}\right)^2\right]^{1/4}}, \quad \phi = \frac{1}{2}\arctan\frac{\sigma}{\omega\varepsilon}, \tag{9.109}$$

which completes the expressions for direct computation of basic propagation parameters for a uniform plane time-harmonic wave in an arbitrary linear and homogeneous medium.

Example 9.13 Computation of Wave Parameters for a Lossy Medium

The amplitude of a uniform plane time-harmonic electromagnetic wave traveling through a lossy nonmagnetic medium at a frequency of 5.2 MHz reduces by 25% every meter. The electric field of the wave leads the magnetic field by 30°. Under these circumstances, find (a) the complex propagation coefficient of the wave and (b) the relative permittivity and conductivity of the medium.

Solution

(a) From Eq. (9.87) and the given per-meter reduction of the wave amplitude, we solve for the attenuation coefficient of the wave,

$$-\alpha d = \ln \frac{E_{\mathrm{m}}(z+d)}{E_{\mathrm{m}}(z)} = \ln(\underbrace{1 - 25\%}_{0.75}) \quad (d = 1 \text{ m}) \quad \longrightarrow \quad \alpha = 0.288 \text{ Np/m}. \quad (9.110)$$

Using Eqs. (9.109), we next realize that the term $\sigma/(\omega\varepsilon)$ in expressions for both α and the phase coefficient, β, of the wave, Eqs. (9.105) and (9.106), equals the tangent of twice the phase angle (ϕ) of the complex intrinsic impedance of the propagation medium, which we know, as it equals the phase lag of the magnetic field behind the electric field of the wave, so $\phi = 30°$. Therefore, we can replace this term by a known value in Eqs. (9.105) and (9.106). Furthermore, we note that the ratio of α and β, which is needed for finding β from the already computed α, does not depend on the permittivity of the medium (ε), as $\sqrt{\varepsilon}$ at the beginning of both expressions cancels out in division. Combining these facts, the ratio α/β and then the coefficient β are evaluated as follows:

$$\frac{\sigma}{\omega\varepsilon} = \tan 2\phi = \tan 60° = \sqrt{3} \quad \longrightarrow \quad u = \sqrt{1 + \tan^2 2\phi} = 2$$

$$\longrightarrow \quad \frac{\alpha}{\beta} = \sqrt{\frac{u-1}{u+1}} = 0.577 \quad \longrightarrow \quad \beta = \frac{\alpha}{0.577} = 0.5 \text{ rad/m}, \quad (9.111)$$

where the temporary parameter u is introduced just to shorten the writing. The complex propagation coefficient of the wave, Eq. (9.82), is now

$$\underline{\gamma} = \alpha + \mathrm{j}\beta = (0.288 + \mathrm{j}0.5) \text{ m}^{-1}. \quad (9.112)$$

(b) Expressing α alone, Eq. (9.105), in terms of the parameter u, Eqs. (9.111), and taking into account that $\mu = \mu_0$ (nonmagnetic medium), the relative permittivity of the medium can be solved from α,

$$\alpha = \omega\sqrt{\frac{\varepsilon_{\mathrm{r}}\varepsilon_0\mu_0}{2}}\sqrt{u-1} \quad (\omega = 2\pi f) \quad \longrightarrow \quad \varepsilon_{\mathrm{r}} = \frac{\alpha^2 c_0^2}{2\pi^2 f^2 (u-1)} = 14, \quad (9.113)$$

with c_0 standing for the free-space wave velocity, Eq. (9.19). Upon the substitution of this result into the first relationship in Eqs. (9.111), the material conductivity comes out to be

$$\sigma = 2\pi f \varepsilon_{\mathrm{r}} \varepsilon_0 \tan 2\phi = 7 \text{ mS/m}. \quad (9.114)$$

Note that the obtained values for ε_{r} and σ indicate that the propagation medium might be a block of rural ground with certain characteristics (including the water content and salinity).

Example 9.14 Fields and Poynting Vector of a Wave in a Lossy Case

The instantaneous magnetic field intensity vector of a wave propagating in a lossy nonmagnetic medium of relative permittivity $\varepsilon_{\mathrm{r}} = 10$ is given by

$$\mathbf{H} = 5\,\mathrm{e}^{-\alpha x}\cos(2.4 \times 10^8 t - 2.83x)\,\hat{\mathbf{z}} \text{ A/m} \quad (t \text{ in s};\ x \text{ in m}). \quad (9.115)$$

Determine (a) the attenuation coefficient, (b) the instantaneous electric field intensity vector, and (c) the time-average Poynting vector of the wave.

Solution

(a) Comparing Eq. (9.115) to the corresponding general H-field expression in Eq. (9.93), we read an angular frequency of $\omega = 2.4 \times 10^8$ rad/s and phase coefficient of

$\beta = 2.83$ rad/m. Having in mind Eq. (9.106) and expressing β in the same way the attenuation coefficient, α, is expressed in Eq. (9.113), we solve for the parameter u, which is defined in Eqs. (9.111),

$$\beta = \omega\sqrt{\frac{\varepsilon_r\varepsilon_0\mu_0}{2}}\sqrt{u+1} \quad \longrightarrow \quad u = \frac{2\beta^2 c_0^2}{\omega^2\varepsilon_r} - 1 = 1.5, \tag{9.116}$$

and then use the expression in terms of u of the ratio α/β from Eqs. (9.111) to obtain α,

$$\alpha = \beta\sqrt{\frac{u-1}{u+1}} = 1.26 \text{ Np/m}. \tag{9.117}$$

(b) Combining Eqs. (9.109), (9.111), and (9.48), the magnitude and phase angle of the complex intrinsic impedance of the medium are computed as

$$|\underline{\eta}| = \frac{\sqrt{\mu/\varepsilon}}{\sqrt{u}} = \frac{\eta_0}{\sqrt{\varepsilon_r u}} = 97.34\ \Omega, \quad \phi = \frac{1}{2}\arctan\sqrt{u^2-1} = 24.1^\circ. \tag{9.118}$$

From Eq. (9.115), the amplitude of the electric field of the wave for $x = 0$ [see Eq. (9.85)] is now $E_m(0) = |\underline{\eta}|H_m(0) = 97.34\ \Omega \times 5$ A/m $= 486.7$ V/m. In addition, as the vector $\mathbf{H}$ is oriented in the positive z direction, $\mathbf{E}$ must be in the positive y direction to ensure that the Poynting vector of the wave is in the direction (positive x) of the wave travel ($\hat{\mathbf{y}} \times \hat{\mathbf{z}} = \hat{\mathbf{x}}$), and hence the instantaneous E-field vector

$$\mathbf{E} = 486.7\,\mathrm{e}^{-1.26x}\cos(2.4\times10^8 t - 2x + 24.1^\circ)\,\hat{\mathbf{y}}\ \text{V/m} \quad (t \text{ in s};\ x \text{ in m}). \tag{9.119}$$

(c) By means of Eq. (9.96), the time-average Poynting vector of the wave equals

$$\mathcal{P}_{\text{ave}} = \frac{E_0^2}{|\underline{\eta}|}\mathrm{e}^{-2\alpha x}\cos\phi\,\hat{\mathbf{x}} = 1.11\,\mathrm{e}^{-2.52x}\,\hat{\mathbf{x}}\ \text{kW/m}^2 \quad (x \text{ in m}), \tag{9.120}$$

where $E_0 = E_m(0)/\sqrt{2} = 344.1$ V/m is the rms electric field intensity of the wave in the plane $x = 0$.

Problems: 9.24 and 9.25; *Conceptual Questions* (on Companion Website): 9.17 and 9.18; *MATLAB Exercises* (on Companion Website).

9.9 WAVE PROPAGATION IN GOOD DIELECTRICS

In many applications, we do not need to use the exact expressions for α, β, $|\underline{\eta}|$, and ϕ in Eqs. (9.105), (9.106), and (9.109), but much simpler approximate ones, specialized for specific ranges of values of material parameters and frequencies. One such important special case of lossy materials are good dielectrics, whose permittivity (ε) and conductivity (σ) at a given frequency (f) satisfy the following condition:

criterion for good dielectrics

$$\boxed{\sigma \ll \omega\varepsilon,} \tag{9.121}$$

where $\omega = 2\pi f$, and this is the case for usual dielectrics (insulators) in practical situations (note that all real dielectrics have some, however small, losses). Of course, perfect dielectrics, where losses can be completely neglected ($\sigma = 0$), also fall under the category of good dielectrics. Although setting a more specific numerical expression of the condition in Eq. (9.121) is rather arbitrary, a useful rule of thumb is that a given material should be classified as good dielectric if $\sigma/(\omega\varepsilon) < 1/100$. For these materials, the amplitude of the conduction current density is much smaller than the amplitude of the displacement current density (see Example 8.2). By the same token, the relaxation time, Eq. (3.68), is much longer than the time period, Eq. (8.49). With the condition in Eq. (9.121), the complex propagation coefficient can approximately be evaluated [using Eqs. (9.82) and (9.78)] as

$$\underline{\gamma} = \mathrm{j}\omega\sqrt{\varepsilon\mu}\left(1 - \mathrm{j}\frac{\sigma}{\omega\varepsilon}\right)^{1/2} \approx \mathrm{j}\omega\sqrt{\varepsilon\mu}\left(1 - \mathrm{j}\frac{\sigma}{2\omega\varepsilon}\right) = \frac{\sigma}{2}\sqrt{\frac{\mu}{\varepsilon}} + \mathrm{j}\omega\sqrt{\varepsilon\mu}, \quad (9.122)$$

since[8] $(1 + \underline{a})^{1/2} \approx 1 + \underline{a}/2$ for $|\underline{a}| \ll 1$. Eq. (9.83) then gives us the following simplified expressions for the attenuation and phase coefficients for good dielectrics:

$$\alpha \approx \frac{\sigma}{2}\sqrt{\frac{\mu}{\varepsilon}}, \quad \beta \approx \omega\sqrt{\varepsilon\mu}. \quad (9.123)$$

α, β – good dielectrics

Note that these same expressions can be obtained also from Eqs. (9.105), (9.106), and (9.121).

Although the intrinsic impedance of good dielectrics is a complex quantity, given by Eq. (9.91) or (9.109), its imaginary part is much smaller than the real part, yielding

$$\underline{\eta} \approx \sqrt{\frac{\mu}{\varepsilon}} \quad (9.124)$$

$\underline{\eta}$ – good dielectrics

or $|\underline{\eta}| \approx \sqrt{\mu/\varepsilon}$ and $\phi \approx 0$. We note that expressions for both the phase coefficient and intrinsic impedance for good dielectrics are practically the same as those for perfect dielectrics.

In practice, the losses in dielectrics are identified by specifying the so-called loss tangent of the material (rather than conductivity),

$$\tan\delta_{\mathrm{d}} = \frac{\sigma}{\omega\varepsilon}, \quad (9.125)$$

loss tangent

which actually equals the ratio of the amplitudes of the conduction and displacement current densities, Eq. (8.21). We recall that the conduction current density is in phase with the electric field intensity in the dielectric, while the displacement current density leads the electric field intensity by 90° (see Example 8.2), and this phase relationship is shown in Fig. 9.12. The angle δ_{d} in Eq. (9.125) is now identified in the figure as the angle by which the displacement current density leads the total (conduction plus displacement) current density, Eq. (8.6). Using this new parameter, the low-loss criterion in Eq. (9.121) can alternatively be written as $\tan\delta_{\mathrm{d}} \ll 1$. Typical values for $\tan\delta_{\mathrm{d}}$ of dielectric materials used in most RF and microwave applications are on the order of $10^{-4} - 10^{-3}$. Combining Eqs. (9.123) and (9.125), the attenuation coefficient for good dielectrics can now be expressed also as

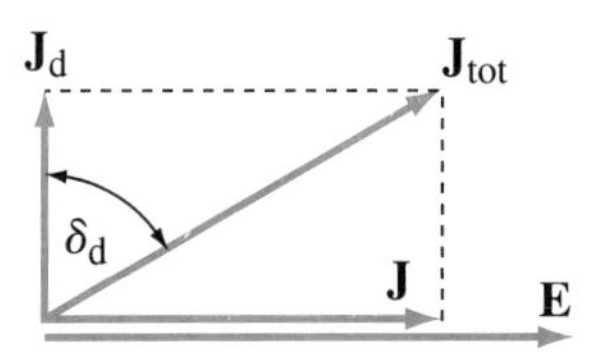

Figure 9.12 Phasor diagram for the electric field and current in a lossy dielectric material, showing the relationship between the conduction, displacement, and total current densities; the tangent of the angle δ_{d}, by which the displacement current density leads the total current density, is referred to as the loss tangent of the material at a given frequency.

$$\alpha \approx \frac{\sigma}{2}\sqrt{\frac{\mu}{\varepsilon}} = \frac{1}{2}\frac{\sigma}{\omega\varepsilon}\omega\varepsilon\sqrt{\frac{\mu}{\varepsilon}} \approx \frac{\beta}{2}\tan\delta_{\mathrm{d}}. \quad (9.126)$$

Of course, $\tan\delta_{\mathrm{d}} = 0$ and $\alpha = 0$ for perfect dielectrics.

In addition to the conduction of free electrons and holes, modeled by the conductivity (σ) of the medium, losses in dielectric materials, and the associated attenuation of electromagnetic waves, may also occur from the polarization vector, **P**, failing to keep in phase with the electric field intensity vector of the wave at

[8]From the binomial series identity,

$$(1 + x)^k = 1 + kx + \frac{k(k-1)}{2}x^2 + \frac{k(k-1)(k-2)}{6}x^3 + \ldots$$

high frequencies. This, in turn, results in a component of the polarization current, Eq. (8.13), that is in phase with the electric field intensity, much like the conduction current density, $\underline{\mathbf{J}} = \sigma\underline{\mathbf{E}}$. Namely, at low enough frequencies, bound charges, whose displacements give rise to the vector $\mathbf{P}$ in a polarized dielectric [see Fig. 2.2 and Eq. (2.7)], can oscillate in phase with the applied time-harmonic field; however, at higher frequencies, their inertia tends to prevent them from following the rapid oscillations. More specifically, the charges tend to resist acceleration by the rapidly changing field due to their masses and frictional forces that keep them attached to the molecules, so that charge displacements, and hence $\mathbf{P}$, are out of phase with $\mathbf{E}$. As a result, a portion of the electromagnetic energy of the wave is lost to heat in overcoming the friction in molecules (the work must be done by the electric field against the frictional damping forces). The phase mismatch between $\mathbf{P}$ and $\mathbf{E}$ and the associated losses (due to frictional damping), referred to as polarization losses, can be characterized by the complex electric susceptibility of the dielectric, $\underline{\chi}_e$, in Eq. (2.10), with its phase angle being equal to the phase difference between the two vectors. This implies that the permittivity of the dielectric [Eqs. (2.50) and (2.48)] is complex as well, and such permittivity that includes polarization losses is customarily written in the following form:

high-frequency permittivity

$$\boxed{\underline{\varepsilon} = \varepsilon' - \mathrm{j}\varepsilon'',} \tag{9.127}$$

where the imaginary part is always negative or zero ($\varepsilon'' \geq 0$). Of course, the relative complex permittivity is $\underline{\varepsilon}_r = \underline{\varepsilon}/\varepsilon_0 = \varepsilon_r' - \mathrm{j}\varepsilon_r''$. At low frequencies, $\varepsilon' \approx \varepsilon$ (ε is the electrostatic or low-frequency permittivity) and $\varepsilon'' \approx 0$. At high frequencies, ε'' is involved in a current density component in the dielectric that is in phase with $\underline{\mathbf{E}}$, so it effectively acts like the conduction current density, $\underline{\mathbf{J}}$. This is apparent from Ampère's law in Eq. (9.77), which now becomes

$$\nabla \times \underline{\mathbf{H}} = \sigma\underline{\mathbf{E}} + \mathrm{j}\omega(\varepsilon' - \mathrm{j}\varepsilon'')\underline{\mathbf{E}} = (\sigma + \omega\varepsilon'')\underline{\mathbf{E}} + \mathrm{j}\omega\varepsilon'\underline{\mathbf{E}}. \tag{9.128}$$

Alternatively, we can look at the polarization current density vector in the dielectric, $\underline{\mathbf{J}}_p = \mathrm{j}\omega\underline{\mathbf{P}}$ [see Eq. (8.13)], and realize that an out-of-phase $\underline{\mathbf{P}}$ results in an in-phase component of $\underline{\mathbf{J}}_p$ (both with respect to $\underline{\mathbf{E}}$). We see that the term $\omega\varepsilon''$ has the dimension of conductivity, and that the power losses in the dielectric are the same as if the dielectric had effective conductivity amounting to $\sigma_{\mathrm{eff}} = \sigma + \omega\varepsilon''$. The loss tangent, Eq. (9.125), becomes $\tan\delta_d = (\sigma + \omega\varepsilon'')/(\omega\varepsilon')$, and the criterion for judging whether the medium behaves like a good dielectric reads $\sigma + \omega\varepsilon'' \ll \omega\varepsilon'$. Ampère's law can also be written as

$$\nabla \times \underline{\mathbf{H}} = \mathrm{j}\omega\left(\varepsilon' - \mathrm{j}\varepsilon'' - \mathrm{j}\frac{\sigma}{\omega}\right)\underline{\mathbf{E}}, \tag{9.129}$$

and hence, in comparison with Eq. (9.79), the equivalent complex permittivity of the medium, Eq. (9.78), is now identified to be $\underline{\varepsilon}_e = \varepsilon' - \mathrm{j}\varepsilon'' - \mathrm{j}\sigma/\omega$. In practice, for dielectrics that exhibit both (nonnegligible) ohmic losses (due to nonzero σ) and polarization losses due to appreciable $\omega\varepsilon''$, it is seldom necessary, and even possible, to distinguish between the relative contributions of the two loss mechanisms to the total power dissipation. The term σ/ω is thus usually incorporated into ε'' in such cases, and the values of the effective conductivity, given by $\sigma_{\mathrm{eff}} = \omega\varepsilon''$ (with $\sigma = 0$), or loss tangent, given by

loss tangent via high-frequency permittivity

$$\boxed{\tan\delta_d = \frac{\varepsilon''}{\varepsilon'} = \frac{\varepsilon_r''}{\varepsilon_r'},} \tag{9.130}$$

are most frequently determined by measurement. In general, at higher frequencies, both the real and imaginary parts of $\underline{\varepsilon}$ in Eq. (9.127) for dielectric materials are not constants, but functions of frequency. In fact, when observed over very wide frequency (or wavelength) ranges across the electromagnetic spectrum (Table 9.1), real dielectrics may exhibit several resonances and associated permittivity changes.

Example 9.15 Propagation through an Alumina Substrate

A 5-GHz uniform plane time-harmonic electromagnetic wave travels through the ceramic substrate of a printed circuit board made of alumina (aluminum oxide, Al_2O_3) that is characterized by a relative permittivity and loss tangent of 9 and 5×10^{-4}, respectively. Find the attenuation and phase coefficients and the phase velocity of the wave, as well as the complex intrinsic impedance of the material.

Solution The alumina sample is a good dielectric, since

$$\tan \delta_{\rm d} = 5 \times 10^{-4} \ll 1 \quad \longrightarrow \quad \text{good dielectric.} \tag{9.131}$$

So, the phase coefficient of the propagating wave is practically the same as if no losses at all, Eqs. (9.123), and the attenuation coefficient can be computed from Eq. (9.126),

$$\beta \approx 2\pi f \sqrt{\varepsilon_{\rm r}\varepsilon_0\mu_0} = \frac{2\pi f \sqrt{\varepsilon_{\rm r}}}{c_0} = 314 \text{ rad/m}, \quad \alpha \approx \frac{\beta}{2} \tan \delta_{\rm d} = 0.0785 \text{ Np/m}. \tag{9.132}$$

The phase velocity of the wave and intrinsic impedance of the dielectric are as well approximately given by their no-loss expressions, Eqs. (9.35) and (9.124), which in the nonmagnetic ($\mu = \mu_0$) case reduce to those in Eqs. (9.47) and (9.48), respectively, and hence $v_{\rm p} \approx c_0/\sqrt{\varepsilon_{\rm r}} = 10^8$ m/s and $\eta \approx \eta_0/\sqrt{\varepsilon_{\rm r}} = 125.7\ \Omega$. We see that both $v_{\rm p}$ and η are three times smaller than the corresponding free-space values, Eqs. (9.19) and (9.23).

Problems: 9.26 and 9.27; *Conceptual Questions* (on Companion Website): 9.19; *MATLAB Exercises* (on Companion Website).

9.10 WAVE PROPAGATION IN GOOD CONDUCTORS

Another important special case of lossy materials are good conductors, with

$$\boxed{\sigma \gg \omega\varepsilon.} \tag{9.133}$$

criterion for good conductors

In contrast to good dielectrics, here the conduction current dominates over the displacement current, and the relaxation time is very short when compared to the time period of time-harmonic variation of electric and magnetic fields. Perfect electric conductors (PEC), with $\sigma \to \infty$, can be said to belong to good conductors, as an extreme case. Analogously to the numerical expression of the condition in Eq. (9.121) for good dielectrics, the one in Eq. (9.133) for good conductors may be written as $\sigma/(\omega\varepsilon) > 100$ for use in practice. Note that under conditions midway between these two groups of materials – when the conduction current is of the same or similar order of magnitude as the displacement current, i.e., when $1/100 \le \sigma/(\omega\varepsilon) \le 100$, the material is said to be a quasi-conductor.[9] With

[9]Of course, we always have in mind that, in addition to the material parameters (conductivity and permittivity), frequency of the electromagnetic wave is an important factor in determining whether a medium acts like a good dielectric, good conductor, or quasi-conductor.

Eq. (9.133), the equivalent complex permittivity, Eq. (9.78), becomes $\underline{\varepsilon}_{\mathrm{e}} \approx -\mathrm{j}\sigma/\omega$ for good conductors. Hence, the complex propagation coefficient can be approximated as

$$\underline{\gamma} \approx \mathrm{j}\omega\sqrt{-\mathrm{j}\,\frac{\sigma}{\omega}\,\mu} = \sqrt{\mathrm{j}\omega\mu\sigma} = \sqrt{\frac{\omega\mu\sigma}{2}}\,(1+\mathrm{j}), \tag{9.134}$$

since[10] $\sqrt{\mathrm{j}} = (1+\mathrm{j})/\sqrt{2}$. This means that the attenuation and phase coefficients for good conductors are approximately the same and equal to

α, β – good conductors

$$\alpha \approx \beta \approx \sqrt{\frac{\omega\mu\sigma}{2}} = \sqrt{\pi\mu f\sigma}. \tag{9.135}$$

On the other side, the complex intrinsic impedance of a good conductor can, using the condition in Eq. (9.133), be simplified to the following expression:

η – good conductors

$$\underline{\eta} \approx \sqrt{\frac{\mu}{-\mathrm{j}\frac{\sigma}{\omega}}} = \sqrt{\frac{\mathrm{j}\omega\mu}{\sigma}} = \sqrt{\frac{\omega\mu}{\sigma}}\,\mathrm{e}^{\mathrm{j}\pi/4} = \sqrt{\frac{\pi\mu f}{\sigma}}\,(1+\mathrm{j}). \tag{9.136}$$

We note that $\underline{\eta}$ (like $\underline{\gamma}$) also has the real and imaginary parts approximately equal to each other. Its magnitude and phase angle (argument) are

|η|, φ – good conductors

$$|\underline{\eta}| \approx \sqrt{\frac{\omega\mu}{\sigma}}, \quad \phi \approx 45^\circ. \tag{9.137}$$

Since $|\underline{\eta}|$ is inversely proportional to the square root of the conductivity, which is large, it is quite small when compared to the intrinsic impedance of free space, namely, $|\underline{\eta}| \ll \eta_0$ ($\eta_0 \approx 377\ \Omega$), for all frequencies up to the visible light (note that a PEC has zero intrinsic impedance, which is analogous to a short circuit in circuit theory). This means that the magnetic field intensity of a wave penetrating into a good conductor from air undergoes a dramatic increase in amplitude (or rms value) upon crossing the boundary between the two media assuming the same (or nearly same) amplitudes (and rms values) of the electric field intensity of the wave on the two sides of the boundary. For example, copper[11] ($\sigma = 58$ MS/m, $\varepsilon_{\mathrm{r}} = 1$, $\mu_{\mathrm{r}} = 1$) at $f = 100$ MHz has $|\underline{\eta}| \approx 3.7$ mΩ, so that the magnetic field in copper is about 10^5 times stronger than that in air, for the same electric field strengths on the two sides of the boundary. On the other hand, we see that the magnetic field lags the electric field by very nearly 45° in all good conductors.

Example 9.16 Finding if a Material Is a Good Dielectric or Conductor

For each of the following combinations of material parameters and frequency, determine if the material is a good dielectric, good conductor, or quasi-conductor: (a) glass ($\varepsilon_{\mathrm{r}} = 5$ and $\sigma = 10^{-12}$ S/m), fresh water ($\varepsilon_{\mathrm{r}} = 80$ and $\sigma = 10^{-3}$ S/m), and copper ($\varepsilon_{\mathrm{r}} = 1$ and $\sigma = 58$ MS/m) at a frequency of $f = 100$ kHz, and (b) rural ground from Example 8.3 at frequencies of $f_1 = 12.84$ GHz, $f_2 = 12.84$ MHz, and $f_3 = 12.84$ kHz, respectively.

[10]For the square root of the imaginary unit, we can write

$$\mathrm{j} = \mathrm{e}^{\mathrm{j}\pi/2} \quad\longrightarrow\quad \sqrt{\mathrm{j}} = \left(\mathrm{e}^{\mathrm{j}\pi/2}\right)^{1/2} = \mathrm{e}^{\mathrm{j}\pi/4} = \cos\frac{\pi}{4} + \mathrm{j}\sin\frac{\pi}{4} = \frac{1+\mathrm{j}}{\sqrt{2}}.$$

[11]Note that, at all frequencies up to and including the visible-light region, the permittivity of metallic conductors (such as copper) can be assumed to be that of a vacuum, $\varepsilon = \varepsilon_0$.

Solution

(a) For glass, fresh water, and copper at 100 kHz, $\sigma/(\omega\varepsilon)$ turns out to be 3.6×10^{-8}, 2.25, and 10^{13}, respectively, so glass behaves as a good dielectric, water as a quasi-conductor, and copper as a good conductor. This is an example of three materials exhibiting extremely different conduction properties at the same frequency.

(b) For the sample of rural ground at frequencies f_1, f_2, and f_3, $\sigma/(\omega\varepsilon)$ amounts, respectively, to (parameter n in Example 8.3) 0.001 (good dielectric), 1 (quasi-conductor), and 1000 (good conductor). This is an example of a single piece of matter acting effectively as three completely different conducting/dielectric media at different operating frequencies.

Problems: 9.28; *Conceptual Questions* (on Companion Website): 9.20–9.23; *MATLAB Exercises* (on Companion Website).

9.11 SKIN EFFECT

The losses in good conductors are considerable, and an electromagnetic wave incident on the surface of a good conductor attenuates rapidly with distance from the surface, quickly reaching rather negligible field intensities. To quantitatively express the degree to which the wave can penetrate into the conductor (with still substantial intensity), we introduce a simple parameter defined as the depth into the conductor (distance from the conductor surface) at which the amplitude of the electric field of the wave is attenuated to 1/e (or about 36.8%) of its initial value, i.e., value at the surface. The same applies to the amplitude of the current density, $\mathbf{J} = \sigma\mathbf{E}$, in the conductor. From Eq. (9.85), it is a simple matter to conclude that this parameter, denoted by δ, equals the reciprocal of the attenuation coefficient,

$$\delta = \frac{1}{\alpha} \quad \left[E_{\mathrm{m}}(\delta) = E_{\mathrm{m}}(0)\,\mathrm{e}^{-\alpha\delta} = \frac{E_{\mathrm{m}}(0)}{\mathrm{e}}\right]. \tag{9.138}$$

definition of skin depth, δ (unit: m)

It is termed the skin depth, to emphasize that the substantial wave penetration is confined to a very thin layer near the conductor surface ("skin" of the conductor), as illustrated in Fig. 9.13. We see from the figure that the electric field (or current density) amplitude is multiplied by a factor of 0.368 for every new distance equal to δ farther into the conductor. Remembering that the time-average surface power density (real part of the complex Poynting vector intensity) of the wave, $\mathcal{P}_{\mathrm{ave}}$, carries an exponential term $\mathrm{e}^{-2\alpha z}$ [Eq. (9.97)], we also see that the power density at $z = \delta$

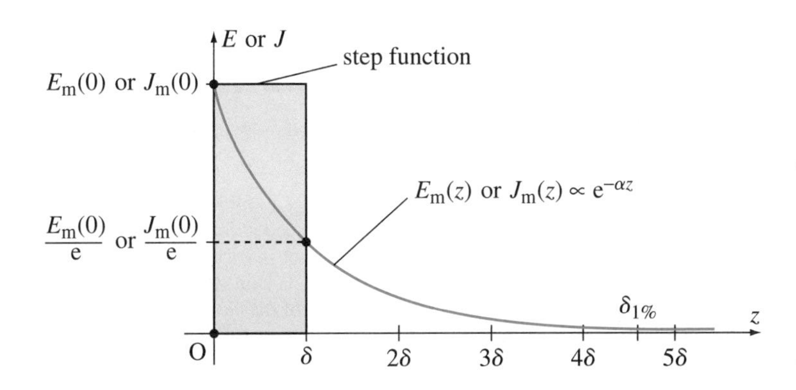

Figure 9.13 Illustration of the definition of the skin depth, δ, for a conducting medium – the plot represents the amplitude of the electric field of the wave, Eq. (9.85), or the associated current density ($\mathbf{J} = \sigma\mathbf{E}$); note that the areas under the exponential curve and the step function are equal.

is damped to $0.368^2 = 13.5\%$ of its initial value (at $z = 0$), and is further attenuated by the same factor (i.e., multiplied by 0.135) at every following integer multiple of δ. The phenomenon of predominant localization of fields, currents, and power in the skin of a conducting body is referred to as the skin effect. By the same token, ac currents of higher frequencies flow practically only through the skin region below the surface of metallic cylindrical conductors (with circular or any other cross-sectional shape), and the thickness of the layer that carries most of the current is on the order of the skin depth, δ. Note that the skin effect in current conductors is also discussed in Section 6.8, as a consequence of induced (eddy) currents in the conductor and their magnetic field, using Faraday's law of electromagnetic induction [see Fig. 6.23(b)]. A similar discussion is provided in the same section explaining the skin effect for the magnetic flux in ferromagnetic cores of ac machines and transformers [see Fig. 6.23(a)].

Combining Eqs. (9.138) and (9.135) now gives a very useful expression for the skin depth in good conductors:

skin depth for good conductors

$$\delta \approx \frac{1}{\sqrt{\pi \mu f \sigma}}, \tag{9.139}$$

from which we see that δ is inversely proportional to the square root of both the frequency and conductivity.[12] In a limit, $\delta = 0$ for a PEC, at any frequency, that is, an electromagnetic wave cannot penetrate at all into a PEC [see also Example 8.6, proving, based on general Maxwell's equations, that there can be no time-varying (including time-harmonic) electromagnetic field inside perfectly conducting materials]. For copper (Cu), $\sigma = 58$ MS/m and $\mu = \mu_0$, so that

skin depth for copper

$$\delta_{\text{Cu}} \approx \frac{66}{\sqrt{f}} \text{ mm} \quad (f \text{ in Hz}), \tag{9.140}$$

which is another useful formula (since copper is so widely used in engineering practice). The computed values of δ_{Cu} at various frequencies are shown in Table 9.2. We see that the skin depth at 1 GHz is as small as about 2 μm. Even at much lower frequencies, e.g., those of 100 kHz and 60 Hz, δ_{Cu} is quite small,

Table 9.2. Skin depth for some materials, δ (m), at different frequencies, f

Material	$f = 60$ Hz	1 kHz	100 kHz	1 MHz	100 MHz	1 GHz
Copper	8.61×10^{-3}	2.11×10^{-3}	2.11×10^{-4}	6.67×10^{-5}	6.67×10^{-6}	2.11×10^{-6}
Iron	6.5×10^{-4}	1.6×10^{-4}	1.6×10^{-5}	5.03×10^{-6}	5.03×10^{-7}	1.6×10^{-7}
Seawater	32.5	7.96	0.796	0.252	2.66×10^{-2}	1.28×10^{-2}
Rural ground	649.7	159.2	15.92	5.233	1.99	1.986

[12]Note that the definition of the skin depth in Eq. (9.138) applies to all lossy media, not only to good conductors. In the general lossy case, the attenuation coefficient needed for Eq. (9.138) is obtained from Eq. (9.105). Note also that even in situations where $\delta = 1/\alpha$ is not particularly small (compared to the relevant dimensions of the conductor and wavelength), the terms skin depth and skin effect are still used, in a wider sense, to indicate an exponential spatial decrease of the electric-field and current-density amplitudes into the conducting medium, not necessarily a rapid decrease to negligible levels within the skin of the conductor only.

namely, about 0.2 mm and 8.5 mm, respectively. This means that at the power frequency (60 Hz), skin effect becomes pronounced in copper conductors whose cross-sectional dimensions are on the order of 10 mm. So strong skin effect even at rather low frequencies is obviously a consequence of a very high conductivity for copper. It is important, however, to have in mind that the skin effect and skin depth also depend on the permeability of the conductor, not only on its conductivity. A typical example is iron, having about six times lower conductivity than copper ($\sigma_{\rm Fe} = 10$ MS/m), but much smaller skin depth because of its very high permeability ($\mu_{\rm Fe} = 1000\mu_0$), as can be seen from Table 9.2 as well. For instance, $\delta_{\rm Fe} \approx 0.65$ mm at the power frequency, so that even a couple of millimeters thick iron plates in laminated cores of 60-Hz ac machines and transformers exhibit a substantial magnetic-flux nonuniformity due to skin effect. Shown in Table 9.2 are also the corresponding values of δ for two nonmetallic conductors, seawater with $\sigma = 4$ S/m and $\varepsilon_{\rm r} = 81$ and rural ground with $\sigma = 10^{-2}$ S/m and $\varepsilon_{\rm r} = 14$. Since these materials do not behave like good conductors at some of the higher frequencies in the table [i.e., when $\sigma/(\omega\varepsilon) \leq 100$], the skin depth in such cases is computed using Eqs. (9.138) and (9.105), rather than Eq. (9.139). Note that the values for skin depth for seawater in the table clearly indicate that it is virtually impossible to use higher radio frequencies (greater than 100 kHz) for undersea radio communications (with submerged submarines). Consequently, submarines typically use the very low frequency (VLF) band, 3–30 kHz (see Table 9.1), if close to the water surface, and even lower frequencies, down to about 10 Hz in the extremely low frequency (ELF) range, are required for communication with deeply submerged submarines.

The rapid spatial decrease of the wave amplitude in a good conductor is, of course, caused by local Joule's losses (electric power is lost to heat) throughout the volume of the material, Eq. (9.99). The total power of Joule's losses in the conductor (for a given incident time-harmonic electromagnetic field), as we shall see in the next chapter, can be assessed using another important parameter related to the skin effect, the so-called surface resistance of the conductor (at a given frequency), which actually equals the real part of the complex intrinsic impedance, $\underline{\eta}$, in Eq. (9.136). The losses in a conductor, if the skin effect is pronounced, are directly proportional to the square root of the frequency and permeability, and inversely proportional to the square root of the conductivity. For a PEC, surface resistance is zero at any frequency. In other words, there are no losses in a PEC object, since there cannot be wave penetration into it, and both $\mathbf{E} = 0$ and $\mathbf{J} = 0$ throughout its volume (also see Example 8.6).

Skin effect in good conductors at higher frequencies is an extremely important phenomenon in the theory and practice of electromagnetic waves. It is essential for understanding the operation of many high-frequency electromagnetic devices and systems, as well as some low-frequency ones (e.g., ac machines and transformers), and is used to advantage in many applications. For example, there is practically no difference in the performance between a silver-plated brass waveguide component and a pure silver (much more expensive) component. For the same reason, hollow metallic (e.g., copper or aluminum) conductors are used instead of solid conductors for various types of antennas, transmission lines, and other radio-frequency and microwave structures. In addition, effective electromagnetic shielding, i.e., protecting a domain from external electromagnetic fields (waves) or, conversely, encapsulating the fields produced by internal sources so that the domain outside the shield is protected from their influence, can be provided by metallic enclosures only several skin depths thick. On the other hand, the attenuation in transmission lines and other wave-guiding structures is determined by the degree to which the skin effect is pronounced and is proportional to the surface resistance of the conductors of the line, which will be discussed in later chapters.[13]

Example 9.17 **One-Percent Depth of Penetration**

If the skin depth for a wave penetrating into a conducting material is δ, find the one-percent depth of penetration.

Solution In addition to the skin depth, i.e., 1/e depth of penetration, we can define other depths into a conductor – at which the electric field intensity (and current density) decreases to an arbitrarily specified fraction of its initial value. For the 1% depth of penetration ($\delta_{1\%}$), in particular, the wave attenuation is to 1% of the initial value. This condition can be written as [see Eqs. (9.85) and (9.138)]

$$E_{\rm m}(\delta_{1\%}) = E_{\rm m}(0)\,{\rm e}^{-\alpha\delta_{1\%}} = E_{\rm m}(0)\,{\rm e}^{-\delta_{1\%}/\delta} = 0.01E_{\rm m}(0), \tag{9.141}$$

and hence the 1% depth of penetration for a conductor whose skin depth is δ at a given frequency equals

one-percent depth of penetration

$$\boxed{\delta_{1\%} = \delta\,\ln 100 \approx 4.6\,\delta} \tag{9.142}$$

(see Fig. 9.13). We see that at locations more than about 5 skin depths (δ) away from the surface of a conductor, the penetrating wave retains less than one percent of its intensity at the surface.

Example 9.18 **Phase Velocity and Wavelength for Good Conductors**

A time-harmonic uniform plane wave of frequency f travels in a good conductor of conductivity σ and permeability μ. (a) Determine the phase velocity ($v_{\rm p}$) and wavelength (λ) of the wave, and compare them to the corresponding free-space quantities. (b) Show that λ is proportional to the skin depth (δ) in the conductor. (c) Compute the values for $v_{\rm p}$, λ, δ, and $\delta_{1\%}$ (one-percent depth of penetration) for copper ($\sigma = 58$ MS/m and $\mu = \mu_0$) at three different frequencies, $f_1 = 60$ Hz, $f_2 = 300$ MHz, and $f_3 = 300$ GHz.

Solution

(a) The phase velocity and wavelength in good conductors can be obtained from the phase coefficient, Eq. (9.135), using Eqs. (9.35) and (8.111), as

$v_{\rm p}$, λ – good conductors

$$\boxed{v_{\rm p} = \frac{\omega}{\beta} \approx 2\sqrt{\frac{\pi f}{\mu\sigma}} \quad \text{and} \quad \lambda = \frac{2\pi}{\beta} = \frac{v_{\rm p}}{f} \approx 2\sqrt{\frac{\pi}{\mu f\sigma}},} \tag{9.143}$$

respectively. We see that both quantities have different dependences on frequency (f) than in free space (in free space, $v_{\rm p} = $ const and $\lambda = $ const/f). We also realize that, since both quantities are proportional to $\sigma^{-1/2}$, they both are significantly smaller than their free-space values (in a PEC, both $v_{\rm p} = 0$ and $\lambda = 0$ – no propagation at all). In general, the ratio of the phase velocity of a wave in free space ($c_0 = 3 \times 10^8$ m/s) and that in a material medium (good conductor in this case),

index of refraction (dimensionless)

$$\boxed{n = \frac{c_0}{v_{\rm p}},} \tag{9.144}$$

[13]In a coaxial cable (Fig. 2.17), for instance, a guided electromagnetic wave propagates along the cable through its dielectric and penetrates into the conductors only to the degree that the conductors are imperfect (the wave is merely guided by the cable conductors). The larger the penetration into the conductors, i.e., the skin depth, the larger the losses and attenuation along the cable. On the other side, an ideal (largely hypothetical) cable, with perfect conductors (PEC), so absolutely no penetration into the conductors, and perfect (lossless) dielectric, would enable unattenuated wave propagation through the cable along arbitrary distances.

Table 9.3. Various wave parameters for copper at three different frequencies*

Frequency	v_p (m/s)	λ (m)	λ_0 (m)	n	δ (m)	$\delta_{1\%}$ (m)
60 Hz	3.22	5.36×10^{-2}	5×10^6	9.33×10^7	8.5×10^{-3}	3.92×10^{-2}
300 MHz	7.19×10^3	2.4×10^{-5}	1	4.17×10^4	3.82×10^{-6}	1.76×10^{-5}
300 GHz	2.27×10^5	7.58×10^{-7}	10^{-3}	1.32×10^3	1.21×10^{-7}	5.55×10^{-7}

* Values for the free-space wavelength (λ_0) are also given.

where the same ratio holds for the corresponding wavelengths, is called the index of refraction (or refractive index) of the (conducting) medium (a dimensionless quantity). So, $n \gg 1$ for good conductors.

(b) Since the phase and attenuation coefficients for good conductors are approximately the same, Eq. (9.135), there is a very simple approximate relation between the wavelength and skin depth for the material,

$$\lambda = \frac{2\pi}{\beta} \approx \frac{2\pi}{\alpha} = 2\pi\delta, \tag{9.145}$$

and hence $\delta \approx \lambda/(2\pi) \approx 0.16\lambda$. Thus, in a good conductor, the skin depth is always much smaller than the wavelength. This means that the wave amplitude in the conductor decays to small fractions of its value at the conductor surface even within the first space period of the wave cosine function, Eq. (9.84). In specific, the electric field of the wave is damped to 1% of its initial amplitude in about $3\lambda/4$ in the conductor [from Eq. (9.142), $\delta_{1\%} \approx 0.73\lambda$]. This also means that, while evaluating the skin effect in good conductors, it is enough to compare the skin depth to the relevant physical dimensions (usually thickness) of the conductor (e.g., for a solid metallic wire carrying an ac current, the skin effect is said to be pronounced and must be taken into account for analysis if δ is smaller than the wire diameter), and there is no need to evaluate δ in terms of its electrical size – it is always electrically small, i.e., much smaller than λ.

(c) Table 9.3 shows the computed values for v_p, λ, n, δ, and $\delta_{1\%}$ for copper at the three given frequencies; values for the free-space wavelength, λ_0, computed from Eq. (9.67), are also given for comparison. We see that the reduction in v_p and λ for copper with respect to free-space values is especially dramatic at lower frequencies. At the power frequency (60 Hz), the index of refraction of copper is as large as nearly 10^8. Even at a quite high radio frequency of 300 MHz, the phase velocity and wavelength in copper, $v_p \approx 7.2$ km/s and $\lambda \approx 24\ \mu$m, are very small when compared to the velocity of 3×10^8 m/s and wavelength of 1 m, respectively, in free space.[14]

Example 9.19 VLF Submarine Communications

A 10-kHz VLF (very low frequency) plane wave is launched into the ocean, as a ship is trying to communicate a message to a submerged submarine. The position of the ship is approximately at the vertical line relative to the submarine, and the rms electric field intensity of the transmitted wave immediately below the water surface is 4 kV/m. (a) Compute the attenuation and phase coefficients, the skin depth, phase velocity, wavelength, index of refraction, and complex intrinsic impedance of the wave in seawater, taking $\varepsilon_r = 81$

[14]In a transmission line (e.g., coaxial cable) with copper conductors, the travel of a 300-MHz signal is not described by the computed propagation parameters for copper, but those for the dielectric between the conductors. Namely, since the signal (energy) propagates along the transmission line as an electromagnetic wave outside the conductors, its velocity equals $c = c_0/\sqrt{\varepsilon_r}$ for the line filled with a good dielectric of relative permittivity ε_r (and $\mu = \mu_0$), and the wavelength of the wave is found from Eq. (8.112).

and $\sigma = 4$ S/m for its relative permittivity and conductivity, respectively. (b) Determine the distribution of complex electric and magnetic fields in the ocean. (c) If the submarine's receiving antenna requires a minimum rms electric field intensity of 0.01 μV/m, find the maximum depth of the antenna with respect to the water surface to which successful communication is still possible, and the time-average surface power density of the incoming wave at that depth.

Solution

(a) The operating angular frequency of the wave being $\omega = 2\pi f = 6.283 \times 10^4$ rad/s, Eq. (9.133) gives

$$\frac{\sigma}{\omega\varepsilon} = 88{,}766 \gg 1 \quad \longrightarrow \quad \text{good conductor,} \tag{9.146}$$

i.e., the seawater with given material parameters behaves like a good conductor at 10 kHz. Therefore, we can use Eq. (9.135) to find the attenuation and phase coefficients for the medium,

$$\alpha = \beta = \sqrt{\pi\mu_0 f\sigma} = 0.4 \text{ Np/m or rad/m.} \tag{9.147}$$

Using Eqs. (9.138), (9.143), and (9.144), the skin depth, phase velocity, wavelength, and refractive index of the wave/material amount, respectively, to $\delta = 2.5$ m [note that the corresponding one-percent depth of penetration, Eq. (9.142), is $\delta_{1\%} = 11.5$ m], $v_p = 1.58 \times 10^5$ m/s, $\lambda = 15.8$ m (free-space wavelength is 30 km at 10 kHz), and $n = 1900$.[15] Apparently, even in nonmetallic good conductors, the free space to conductor ratio for v_p and λ (index of refraction) is very large. The complex intrinsic impedance of the water, Eq. (9.136), is $\underline{\eta} = 0.1\,(1 + \mathrm{j})\ \Omega = 0.14\,\mathrm{e}^{\mathrm{j}\pi/4}\ \Omega$, so $|\underline{\eta}|$ is about 2693 times smaller than the free-space intrinsic impedance, Eq. (9.23), and, as for all good conductors, the phase angle of $\underline{\eta}$ is (approximately) $\phi = 45°$.

(b) By means of Eqs. (9.81) and (9.92), with a z-axis placed perpendicularly to the ocean surface and oriented downward, and the rms electric and magnetic field intensities for $z = 0^+$ (just below the surface) being $E_{\text{surface}} = 4$ kV/m and $H_{\text{surface}} = E_{\text{surface}}/|\underline{\eta}| = 28.6$ kA/m, respectively, the distribution of the complex fields in the water is given by

$$\underline{E} = 4\,\mathrm{e}^{-0.4z}\,\mathrm{e}^{-\mathrm{j}0.4z}\ \text{kV/m}, \quad \underline{H} = 28.6\,\mathrm{e}^{-0.4z}\,\mathrm{e}^{-\mathrm{j}0.4z}\,\mathrm{e}^{-\mathrm{j}\pi/4}\ \text{kA/m} \quad (z \text{ in m}). \tag{9.148}$$

(c) From Eq. (9.87), the maximum depth of the submarine's antenna that would enable the required rms E-field for reception ($E_{\text{submarine}} = 0.01\ \mu$V/m) for the available rms field at the sea surface turns out to be

$$d_{\max} = \frac{1}{\alpha} \ln \frac{E_{\text{surface}}}{E_{\text{submarine}}} = 66.8 \text{ m.} \tag{9.149}$$

[15] As already discussed with reference to the data for skin depth of seawater in Table 9.2, we are forced to use rather low frequencies for submarine communications, because of the large attenuation in seawater. The lower the frequency the larger the wavelength, as well as the physical size of the antennas used for transmission or reception of signals (as we shall see in a later chapter), which is one of the drawbacks, in general, of using lower frequencies for radio communications. Another drawback are small bandwidths for communication implied by low frequencies, meaning in turn slow signal data rates – namely, a much smaller number of words per minute can be transmitted as compared to radio communications at higher frequencies (for example, signal data rates at ELF band are so slow that a single word can take several minutes to transmit). However, the large phase coefficient in seawater (as compared to the corresponding phase coefficient in free space) is now very helpful for feasibility and practicality of low-frequency undersea submarine communications. Namely, since the travel of electromagnetic waves transmitted or received by an antenna that is submerged in seawater is described by the propagation parameters for seawater (at a given frequency), the electrical size of the antenna is expressed in terms of the wavelength in seawater, not in free space. The importance of this fact is best illustrated by a simple example: a half-wave dipole antenna (wire antenna that is a half of the wavelength long) operating in seawater at 10 kHz ($\lambda = 15.8$ m) within a VLF submarine communication system needs to be only about 7.9 m long, whereas the physical length of a 10-kHz VLF half-wave dipole antenna operating in air ($\lambda = 30$ km) would be as large as 15 km or $n = 1900$ times the undersea dimension.

At this depth, the associated time-average surface power density, equal to the Poynting vector magnitude of the incident wave in Eq. (9.96), is

$$\mathcal{P}_{\text{ave}} = \frac{E_{\text{submarine}}^2}{|\underline{\eta}|} \cos\phi = 5 \times 10^{-16}\ \text{W/m}^2. \tag{9.150}$$

We see that, due to an extremely large attenuation in the salty water, only a tiny portion of the power density entering the ocean, which is as high as $80\ \text{MW/m}^2$, reaches the submarine. Obviously, it is practically unfeasible to use 10-kHz electromagnetic signals to communicate with a submarine that is about 60 m deep.

Example 9.20 Decibel Attenuation of an Aluminum Foil

Find the wave attenuation in decibels of a 5-mil thick (1 mil = 0.001 inch = 25.4 μm) aluminum ($\sigma_{\text{Al}} = 38$ MS/m, $\varepsilon_{\text{r}} = \mu_{\text{r}} = 1$) foil at a frequency of $f = 100$ MHz, and evaluate the shielding effectiveness of the foil.

Solution At the given frequency, the loss tangent, Eq. (9.125), of aluminum is as large as

$$\tan\delta_{\text{d}} = \frac{\sigma_{\text{Al}}}{2\pi f \varepsilon_0} = 6.8 \times 10^9 \gg 1, \tag{9.151}$$

so the condition in Eq. (9.133) is definitely satisfied, and aluminum acts like a good conductor, as expected. This also means that the wave attenuation coefficient, α, in the material can be computed using Eq. (9.135). Hence, Eq. (9.88) results in the following dB attenuation of the foil, with thickness $d = 5$ mil $= 0.127$ mm:

$$A_{\text{dB}} = 8.686\alpha d = 8.686\sqrt{\pi\mu_0 f \sigma_{\text{Al}}}\, d = 135\ \text{dB}. \tag{9.152}$$

This is a very large attenuation, implying that, as given by Eq. (9.98), only a tiny (negligible) portion of the electric (or magnetic) field intensities on one (incident) side of the shield is passed to the other (protected) side ($E_2/E_1 = 1.8 \times 10^{-7}$). We conclude that even a rather thin aluminum foil provides a very effective electromagnetic shielding at this, relatively low, frequency. Moreover, as can be seen in Eq. (9.152), $A_{\text{dB}} \propto \sqrt{f}$, so the shielding effectiveness is even greater at higher frequencies.[16]

Problems: 9.29–9.31; *Conceptual Questions* (on Companion Website): 9.24–9.27; *MATLAB Exercises* (on Companion Website).

9.12 WAVE PROPAGATION IN PLASMAS

Plasmas are ionized gases which, in addition to neutral atoms and molecules, include a large enough number of ionized atoms and molecules and free electrons that macroscopic electromagnetic effects caused by Coulomb forces between charged particles are notable. The plasma medium is often considered to be the fourth state of matter (along with solid, liquid, and gaseous states). It, of course, has many properties common with the gaseous state; however, its basic distinction with

[16]Note that such aluminum foils find broad application in electromagnetically shielded rooms used for electromagnetic and other types of measurements and testing, where the foils are attached (glued) to walls, ceiling, and floor of the room, to prevent the electric and magnetic fields and waves from entering or leaving it. The two examples, out of many, are shielded anechoic chambers for measurements of radiation properties of antennas or for electromagnetic compatibility (EMC) testing, and shielded rooms housing equipment for medical examinations based on magnetic resonance imaging (MRI) technology in hospitals.

respect to normal gases is a high level of collective electromagnetic interaction effects between charged particles, enabled by a sufficiently high degree of ionization of the gas.[17] The plasma thus resembles a fluid having higher mass density, and hence higher concentration of charge carriers, than a normal gas. In this section, we investigate the propagation of electromagnetic waves through plasmas, starting with the electromagnetic force on a moving charged particle (Lorentz force) in an ionized gas.

The most important example of a plasma medium, as far as the electromagnetic wave propagation and radio communications are concerned, is the upper region of the earth's atmosphere called ionosphere. This part of the atmosphere, from about 50 to 500 km altitude above the earth's surface, consists of a highly rarefied gas that is ionized by the sun's radiation. It plays an essential role in a number of radio-wave applications, as it selectively reflects back or passes through waves of certain frequencies and incident angles. On the other hand, the most apparent manifestation of plasmas in nature is probably that taking place temporarily in atmosphere at every lightning strike. In general, plasma physics is a very intense and quite broad research area in physics and engineering dealing with various aspects of plasma science and technology, including research in controlled thermonuclear fusion to develop a new source of energy for the world. Finally, in the most global view, note that more than 99% of the matter in the universe is in plasma state.

Consider a uniform plane time-harmonic electromagnetic wave of frequency f propagating through a plasma medium. The Lorentz force on a charged particle (free electron or ion) moving at a velocity $\mathbf{v}$ in the medium can be obtained using Eq. (4.145), so that Newton's second law then gives the following equation of motion of the particle:

$$m\frac{\mathrm{d}\mathbf{v}}{\mathrm{d}t} = q\mathbf{E} + \mu_0 q\mathbf{v} \times \mathbf{H}, \tag{9.153}$$

where m is the mass of the particle, q is its charge, and $\mathbf{E}$ and $\mathbf{H}$ are the instantaneous electric and magnetic field intensities of the wave, respectively. These intensities are related through the intrinsic impedance of the medium, which is larger (as we shall see) than that of free space, that is, $E/H > \eta_0$. This, in turn, means that $\mu_0|\mathbf{v} \times \mathbf{H}|/E \leq \mu_0 vH/E < \mu_0 v/\eta_0 = v/c_0 \ll 1$ [see Eqs. (9.23) and (9.19)], since the velocities v that the charged particles can acquire in the field $(\mathbf{E}, \mathbf{H})$ are much smaller than the speed of light in free space (c_0). We see, therefore, that the second term on the right-hand side of Eq. (9.153), i.e., the magnetic force on the particle, can be neglected. Since the only remaining term on the right-hand side of thus obtained equation is a time-harmonic quantity, the time derivative of the particle velocity and the velocity itself are also time-harmonic functions. Consequently, we can introduce the complex equivalent $\underline{\mathbf{v}}$ of the velocity (velocity phasor) and convert the equation of motion from time to complex domain,

particle motion equation, in the complex domain

$$\boxed{\mathrm{j}\omega m\underline{\mathbf{v}} = q\underline{\mathbf{E}},} \tag{9.154}$$

where $\omega = 2\pi f$ is the angular frequency of the wave. The current constituted by this motion of charges in the gas is an example of convection current, since $\sigma = 0$ in the medium and Ohm's law in local form, Eq. (3.18), does not apply. The current

[17]The degree of ionization of a gas can be expressed as the ratio of the concentration of ionized atoms and molecules to the total concentration (ionized plus neutral) of atoms and molecules. Note that even very low degrees of ionization, on the order of 10^{-5} or so, are sufficient for a gas to exhibit notable macroscopic electromagnetic properties and behave as a plasma.

density is thus given by Eq. (3.28) with the charge density $\rho = Nq$ and N being the concentration of charged particles. Solving Eq. (9.154) for $\underline{\mathbf{v}}$ and substituting the solution in Eq. (3.28), we obtain the complex convection current density vector in the medium,

$$\underline{\mathbf{J}} = Nq\underline{\mathbf{v}} = -\frac{\mathrm{j}}{\omega m} Nq^2\underline{\mathbf{E}}. \tag{9.155}$$

Ampère's law in Eqs. (8.81) now becomes

$$\nabla \times \underline{\mathbf{H}} = \underline{\mathbf{J}} + \mathrm{j}\omega\varepsilon_0\underline{\mathbf{E}} = \mathrm{j}\omega \underbrace{\varepsilon_0 \left(1 - \frac{Nq^2}{\omega^2\varepsilon_0 m}\right)}_{\varepsilon_\mathrm{p}} \underline{\mathbf{E}} = \mathrm{j}\omega\varepsilon_\mathrm{p}\underline{\mathbf{E}}. \tag{9.156}$$

We see that, similarly to the concept of the equivalent complex permittivity in Eq. (9.79), the presence of the charged particles in the gas and their influence on the electromagnetic wave propagation can be taken into account by means of a new permittivity ε_p, which, here, is smaller than ε_0. The effective reduction in permittivity is inversely proportional to the mass m of the particle, which tells us that the contribution of ions to this reduction is negligible in comparison with the contribution of electrons (ions are much heavier and relatively immobile). So m, q, and N in Eq. (9.156) can be assumed to be the mass, charge, and concentration of electrons exclusively. Since the ionized gas can electromagnetically be completely represented by the permittivity ε_p, it is termed the effective or apparent permittivity of the plasma. It can be regarded as the permittivity of a hypothetical dielectric (e.g., a solid one) that would be described by formally identical expression for Ampère's law (and other Maxwell's equations), thus providing the same interaction with electromagnetic waves. The plasma effective permittivity can be written as

$$\boxed{\varepsilon_\mathrm{p} = \varepsilon_0 \left(1 - \frac{f_\mathrm{p}^2}{f^2}\right),} \tag{9.157}$$

plasma effective permittivity (unit: F/m)

where f_p is called the plasma frequency and is computed as

$$f_\mathrm{p} = \frac{1}{2\pi}\sqrt{\frac{Nq^2}{\varepsilon_0 m}}. \tag{9.158}$$

This important parameter of a plasma medium is a direct measure of the concentration of free electrons (N) in the gas. Namely, substituting the values of constants $q = -e$, Eq. (1.3), $m = m_\mathrm{e} = 9.1094 \times 10^{-31}$ kg (electron mass at rest), and ε_0, Eq. (1.2), the expression becomes

$$\boxed{f_\mathrm{p} = 9\sqrt{N} \quad (N \text{ in m}^{-3};\ \ f_\mathrm{p} \text{ in Hz}).} \tag{9.159}$$

plasma frequency

We can now use all previous equations for the propagation of uniform plane time-harmonic waves in a perfect dielectric of permittivity ε, and by replacing ε with ε_p obtain the corresponding solutions for wave propagation in plasmas. For instance, using Eqs. (8.111) and (9.157), the phase coefficient in a plasma medium is given by

$$\boxed{\beta = \omega\sqrt{\varepsilon_\mathrm{p}\mu_0} = \frac{\omega}{c_0}\sqrt{1 - \frac{f_\mathrm{p}^2}{f^2}}.} \tag{9.160}$$

phase coefficient in a plasma medium

For $f > f_\mathrm{p}$, β is purely real, and the wave propagates through the plasma like through an ordinary lossless dielectric. For $f < f_\mathrm{p}$, on the other hand, β becomes purely imaginary and thus effectively acts like an attenuation coefficient α in Eq. (9.82), so that the wave does not propagate. Such strongly attenuated (nonpropagating) waves in a medium that does not exhibit any losses (or the losses are assumed to be negligible and are not taken into account in the model), and hence the attenuation does not represent any absorbtion of electromagnetic energy and conversion into heat but is brought about by the physical properties of the medium or wave-guiding structure, are referred to as evanescent waves.[18] We conclude that only electromagnetic waves whose frequency is higher than the plasma frequency can propagate through the medium, which apparently behaves like a high-pass filter. The frequency f_p is thus also called the cutoff or critical frequency of the plasma. At $f = f_\mathrm{p}$, $\beta = 0$, meaning that the cutoff frequency itself also belongs to the nonpropagating frequency range. From Eq. (9.21), the intrinsic impedance of the plasma medium is

intrinsic impedance of a plasma medium

$$\boxed{\eta = \sqrt{\frac{\mu_0}{\varepsilon_\mathrm{p}}} = \frac{\eta_0}{\sqrt{1 - f_\mathrm{p}^2/f^2}}.} \tag{9.161}$$

We see that for all propagating frequencies, that is, for $f > f_\mathrm{p}$, η is purely real (like in a perfect dielectric), and is indeed larger than η_0 [fact already mentioned and used in obtaining Eq. (9.154)]. However, it is frequency dependent: very large at frequencies close to the cutoff frequency (at $f = f_\mathrm{p}$, $\eta \to \infty$, and the plasma behaves analogously to an open circuit in circuit theory) and approaching η_0 at frequencies much farther away from the cutoff (for $f \gg f_\mathrm{p}$, the plasma acts much like free space).

The concentration of electrons, N, in the ionosphere is a function of the altitude h above the earth's surface, but also greatly depends on the hour of the day and season of the year, as well as the geographical location on the earth (latitude) and the activity of the sun. The typical maximum of the function $N(h)$ ranges roughly from 10^{11} to 10^{12} m^{-3}, so that Eq. (9.159) gives the corresponding maximum plasma cutoff frequency roughly in the range $(f_\mathrm{p})_\mathrm{max} = 3$–$9$ MHz. This means that, if $(f_\mathrm{p})_\mathrm{max} = 7$ MHz, for instance, at a given time and site, all waves of frequencies f below 7 MHz cannot pass through the ionosphere in either upward or downward direction (and this will be illustrated in an example), but are bounced off it, and the ionosphere represents a perfect shield for such waves. Waves incident obliquely (not vertically) to the ionosphere are reflected back at frequencies even higher than $(f_\mathrm{p})_\mathrm{max}$.

Example 9.21 Wave Propagation in a Parabolic Ionospheric Slab

The ionosphere can approximately be represented by a parabolic profile $N(h)$, of the concentration of free electrons (N) as a function of the altitude (h) above the earth's surface, within a plasma slab of thickness $2d$ as follows:

$$N(h) = N_\mathrm{m}\left[1 - \frac{(h - h_\mathrm{m})^2}{d^2}\right], \quad h_\mathrm{m} - d \le h \le h_\mathrm{m} + d, \tag{9.162}$$

where h_m is the altitude at the middle of the slab and N_m is the maximum electron concentration – at this altitude, which is illustrated in Fig. 9.14. A uniform plane time-harmonic

[18] In addition to the evanescent attenuation for $f < f_\mathrm{p}$, the plasma always (at all frequencies) exhibits some real attenuation due to losses caused by collisions of moving charged particles (primarily electrons) with neutral atoms and molecules, ions, and (other) electrons. In these collisions, a portion of the electromagnetic energy of the wave is lost to heat, which results in a nonzero attenuation coefficient (α) of the wave in the medium.

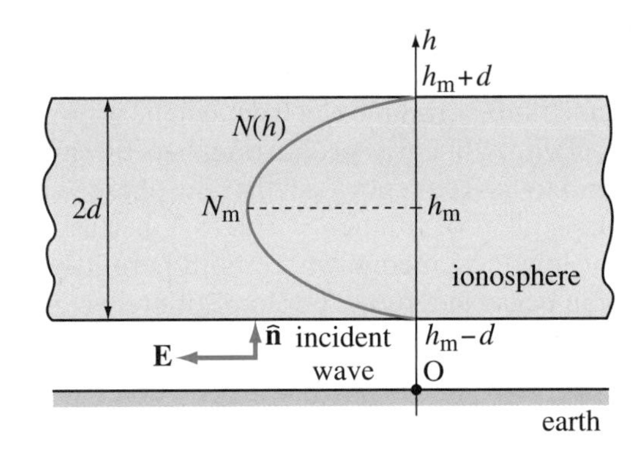

Figure 9.14 Normal incidence of a uniform plane wave on the lower boundary of a parabolic ionospheric slab with the concentration of free electrons (N) given by Eq. (9.162), at two different frequencies; for Example 9.21.

electromagnetic wave of frequency f, launched from the earth's surface, is incident vertically (normally) on the ionosphere (on its lower boundary). For $h_m = 300$ km, $d = 200$ km, and $N_m = 6 \times 10^{11}$ m^{-3}, find whether the wave can pass through the ionosphere, and if not, determine the altitude h at which it bounces off, when (a) $f = 8$ MHz and (b) $f = 6$ MHz, respectively.

Solution

(a) From Eq. (9.159), maximum plasma cutoff frequency of the ionosphere is

$$(f_p)_{max} = 9\sqrt{N_m}\ \text{Hz} \approx 7\ \text{MHz} \quad (N \text{ in m}^{-3}). \tag{9.163}$$

Since the first considered operating frequency of the incident wave, $f = 8$ MHz, is higher than $(f_p)_{max}$, the propagating condition for a plasma [$f > f_p$, resulting in a purely real phase coefficient (β) in the plasma in Eq. (9.160)] is satisfied at every layer of the ionosphere (100 km $\leq h \leq$ 500 km),

$$f > (f_p)_{max} \quad \longrightarrow \quad f > f_p(h) \quad \text{for every } h \quad \longrightarrow \quad \text{wave goes through ionosphere.} \tag{9.164}$$

Therefore, the wave will pass through the entire slab in Fig. 9.14, and leave the ionosphere through its upper boundary.

(b) Now, $f = 6$ MHz is lower than $(f_p)_{max}$, so there is an altitude in the first half of the ionospheric slab, between $h_m - d$ and h_m, in Fig. 9.14 at which $f = f_p$. This means that the wave will propagate through the ionosphere to the layer at which the propagating condition in the plasma is no longer met (for $f = f_p$, β is zero, and would become purely imaginary, effectively a large attenuation coefficient, if $f < f_p$ for even higher altitudes), and bounce back off that layer. Combining Eqs. (9.159) and (9.162), the bounce-off altitude (h_b) is thus determined as

$$f < (f_p)_{max} \quad \longrightarrow \quad f = f_p(h_b) = 9\sqrt{N(h_b)} = 9\sqrt{N_m\left[1 - \frac{(h_b - h_m)^2}{d^2}\right]}$$

$$\longrightarrow \quad \text{wave bounces at } h_b = h_m \pm d\sqrt{1 - \frac{f^2}{81 N_m}} = 198\ \text{km} \quad (N \text{ in m}^{-3};\ f \text{ in Hz}), \tag{9.165}$$

where the solution $h_b = 402$ km $> h_m$ (corresponding to the plus sign in the expression for h_b) is eliminated, as the wave cannot reach that altitude if incident from the lower side of the ionosphere.

Problems: 9.32 and 9.33; *Conceptual Questions* (on Companion Website): 9.28 and 9.29; *MATLAB Exercises* (on Companion Website).

9.13 DISPERSION AND GROUP VELOCITY

The phase coefficient of a uniform plane time-harmonic electromagnetic wave propagating through free space, $\beta = \omega\sqrt{\varepsilon_0\mu_0}$, is linearly proportional to the angular (radian) frequency (ω) of the wave, and to its frequency f, so that the phase velocity in the medium, v_p, is a constant, independent of frequency, Eq. (9.35). This is also true for other simple and lossless (or low-loss) media, where both permittivity, ε, and permeability, μ, of the medium can be considered to be simple, purely real constants and Joule's (ohmic) losses, expressed by conductivity, σ, of the medium, are either zero or small. Typical examples are perfect dielectrics, Eq. (8.111), and low-loss dielectrics, Eq. (9.123), at low frequencies. If an electromagnetic signal that can be decomposed onto multiple time-harmonic waves of different frequencies (and generally different amplitudes and phases) – Fourier components of the signal – is transmitted through such a medium, the different frequency components propagate at equal phase velocities, $v_\mathrm{p} = \text{const}$, and acquire equal phase delays as the signal travels through the medium. Hence, the relative phase distribution of the individual components at signal reception is preserved, and their superposition is an exact replica of the original signal.

On the other hand, the phase velocity in some lossy and/or complex media depends on frequency, since the phase coefficient is a nonlinear function of ω. Examples are media with arbitrary ohmic losses, Eq. (9.106), including good (imperfect) conductors, Eq. (9.143), dielectrics at higher frequencies, where both the real and imaginary parts of the complex permittivity, $\underline{\varepsilon}$, in Eq. (9.127) are functions of frequency, and plasmas, Eq. (9.160). Here, different frequency components of a signal propagate at different phase velocities, $v_\mathrm{p}(\omega)$, and thus arrive with different phase delays to the receiving point. This means that the relative phases of Fourier components are changed, so that the signal shape is changed as well (signal is distorted). For instance, a signal in the form of a rectangular pulse in time loses its sharp edges and broadens as it propagates through the medium.[19] In other words, the medium, as a result of a frequency-dependent phase velocity, causes the distortion of the signal by dispersing its frequency components. This phenomenon is generally known as dispersion, and media (or wave-guiding structures) with

phase velocity in dispersive media

$$\boxed{v_\mathrm{p} = \frac{\omega}{\beta(\omega)} = v_\mathrm{p}(\omega)} \tag{9.166}$$

are referred to as dispersive media (or structures).[20] The graphical representation of a nonlinear β-ω relationship for the medium is called the dispersion diagram, Fig. 9.15.

For the propagation of an information-carrying electromagnetic wave (signal) in a dispersive medium, in addition to the phase velocity in Eq. (9.166), it is necessary to also define a new, special velocity that would serve as a measure of the speed of propagation of the whole wave packet containing many frequency components in a certain frequency band. To this end, let us consider, for simplicity, a

[19]Note that broadening of time pulses in a sequence as they travel in a digital system with a frequency-dependent phase velocity can result in an overlap of adjacent pulses (bits) at reception ("tail" of one pulse spreads into the leading edge of another), causing ambiguities and errors in communication.

[20]As we shall see in a later chapter, metallic waveguides are a typical example of a structure exhibiting wave dispersion that is not due to the nature and parameters of an electromagnetic material through which the wave propagates but is rather a consequence of the presence and configuration of metallic boundaries between which the wave progresses by bouncing back and forth off them.

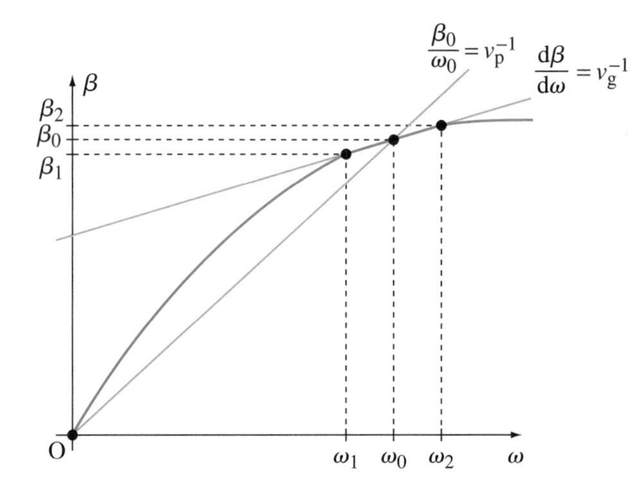

Figure 9.15 Sketch of the β-ω relationship of a dispersive medium, so-called dispersion diagram, with graphical interpretations of the definitions of phase and group velocities for the medium.

wave packet composed of only two components. In specific, consider a superposition of two time-harmonic waves with equal field amplitudes and slightly different angular frequencies, ω_1 and ω_2, propagating through a dispersive medium whose dispersion diagram is that sketched in Fig. 9.15. The two frequencies are labeled on the diagram, along with the frequency ω_0 midway between them. The corresponding phase coefficients, β_1, β_2, and β_0, are also marked. The electric field vectors of the two waves are directed in the same way, and their instantaneous intensities [see Eqs. (9.31) and (9.32)] are given by

$$E_1 = E_m \cos(\omega_1 t - \beta_1 z) \quad \text{and} \quad E_2 = E_m \cos(\omega_2 t - \beta_2 z), \tag{9.167}$$

respectively. The total electric field intensity in the medium is

$$E_{tot} = E_1 + E_2 = \underbrace{2E_m \cos(\Delta\omega t - \Delta\beta z)}_{\text{modulation envelope}} \underbrace{\cos(\omega_0 t - \beta_0 z)}_{\text{carrier wave}}, \tag{9.168}$$

where

$$\Delta\omega = \omega_0 - \omega_1 = \omega_2 - \omega_0 \quad \text{and} \quad \Delta\beta = \beta_0 - \beta_1 = \beta_2 - \beta_0, \tag{9.169}$$

and the use is made of the trigonometric identity $\cos\alpha_1 + \cos\alpha_2 = 2\cos[(\alpha_2 - \alpha_1)/2]\cos[(\alpha_1 + \alpha_2)/2]$. The above equation for $\Delta\beta$, indicating that β_0 obtained as $\beta(\omega_0)$, Fig. 9.15, falls midway between β_1 and β_2, is approximately true since $\Delta\omega$ is assumed to be small. We see that the amplitude of the total field in Eq. (9.168) is not a constant, but varies slowly with both time (t), at a frequency $\Delta\omega$ ($\Delta\omega \ll \omega$), and position in space (z), at a rate defined by $\Delta\beta$ ($\Delta\beta \ll \beta$), between $-2E_m$ and $2E_m$. The resultant wave can thus be interpreted as a rapidly varying carrier wave [of frequency $\omega_0 = (\omega_1 + \omega_2)/2$] sinusoidally modulated by a slowly varying wave [of frequency $\Delta\omega = (\omega_2 - \omega_1)/2$], with the modulating wave representing the amplitude envelope for the modulated one. In other words, the two original waves in Eq. (9.167), with equal amplitudes E_m and slightly different frequencies ω_1 and ω_2, are "beating" each other such that a slow amplitude modulation (AM) of the carrier wave is formed[21] (the phenomenon of beats, in general, refers to a slow variation superimposed on a more rapid one). Fig. 9.16 shows the shape of the resultant signal

[21]Note that, mathematically obvious but physically quite abstract, effect of two slightly different frequencies beating each other to essentially create a new, much lower, frequency has an acoustical manifestation that might be closer to our everyday experience. Namely, if the same tone is played by two slightly out-of-tune string instruments, what we hear is a new, modulated, tone at the beat frequency.

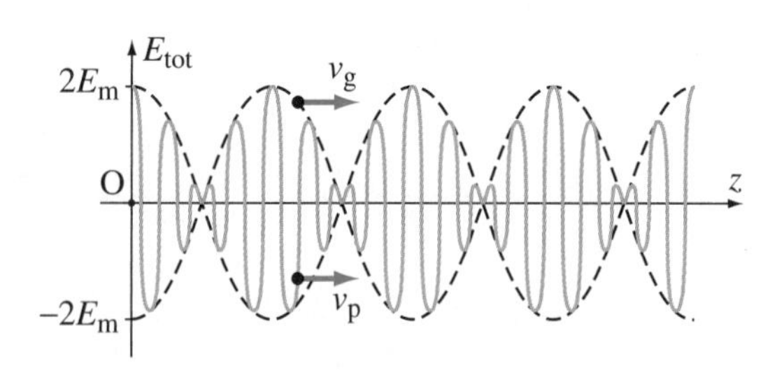

Figure 9.16 Plot along z for $t = 0$ of the electric field intensity in Eq. (9.168) obtained as a superposition of two time-harmonic waves with equal field amplitudes and slightly different angular frequencies, Eq. (9.167); the amplitude envelope of the rapidly oscillating carrier wave pattern travels with the group velocity ($v_{\rm g}$) of the medium, while the carrier wave itself travels with the phase velocity ($v_{\rm p}$).

in Eq. (9.168) frozen in time (at instant $t = 0$), which clearly reveals a beat pattern of the signal transfer through the medium.

We can now compute two phase velocities for the signal in Fig. 9.16: the phase velocity of the carrier wave and that of the modulation envelope. From Eq. (9.35), these velocities are

$$(v_{\rm p})_{\rm carrier} = \frac{\omega_0}{\beta_0} \quad \text{and} \quad (v_{\rm p})_{\rm envelope} = \frac{\Delta\omega}{\Delta\beta}, \tag{9.170}$$

respectively. With reference to Fig. 9.15, we note that the reciprocal of the carrier phase velocity, $(v_{\rm p})^{-1}_{\rm carrier}$, geometrically represents the slope of the straight line connecting the coordinate origin and the operating point (ω_0, β_0) on the β-ω curve. The reciprocal of the envelope velocity, $(v_{\rm p})^{-1}_{\rm envelope}$, approximates the slope of the curve at the operating point.[22] In the limit of $\Delta\omega \to 0$, $(v_{\rm p})^{-1}_{\rm envelope}$ becomes exactly the slope of the curve, i.e., the line tangential to the curve, for $\omega = \omega_0$,

$$\lim_{\Delta\omega \to 0} \frac{\Delta\beta}{\Delta\omega} = \left.\frac{{\rm d}\beta}{{\rm d}\omega}\right|_{\omega=\omega_0} = v_{\rm g}^{-1}(\omega_0). \tag{9.171}$$

Since, in an AM communication system, the modulating wave (with slow variations) contains information that is transmitted, the envelope velocity in Fig. 9.16 is actually the velocity with which the information-carrying signal is conveyed in the system. It also represents a measure of the speed of propagation of a group of frequencies (frequencies ω_1 and ω_2 of the two original waves in our case) that constitute a wave packet, and is hence called the group velocity and denoted as $v_{\rm g}$. In general, this is the velocity of travel of electromagnetic energy and information carried by an electromagnetic wave through a given medium, and is also often called the energy velocity or signal velocity.[23] Since the slope of the β-ω curve for a dispersive medium changes with frequency (ω), as in Fig. 9.15, the group velocity is obviously a function of frequency,

group velocity (unit: m/s*)*

$$\boxed{v_{\rm g}(\omega) = \frac{1}{{\rm d}\beta(\omega)/{\rm d}\omega}}. \tag{9.172}$$

When evaluated at a specified frequency $\omega = \omega_0$, it represents the velocity of a group of frequencies within a wave packet centered at ω_0. Although derived for the

[22]Note that, instead of inverting the velocities, we could have "inverted" the dispersion diagram in Fig. 9.15 and alternatively shown the definitions of the two velocities in thus obtained ω-β diagram for the medium. However, wave dispersion characteristics of media in practice are almost exclusively represented using β-ω diagrams, as in Fig. 9.15.

[23]Note that the general difference between the phase and group velocities, shown in Fig. 9.16 as the carrier and envelope velocities, respectively, is perhaps best illustrated by a simple example from nature – that of a crawling caterpillar. Namely, as the caterpillar crawls, the humps on its back ("ripples") move forward with a "phase velocity," while the caterpillar as a whole progresses with a "group velocity."

simplest possible group of waves, that in Eq. (9.167), the concept of group velocity and the definition in Eq. (9.172) apply to any number of frequencies (with generally different amplitudes) in a group, including waves that have a continuous frequency spectrum, provided that all frequency components are confined to a narrow band around a carrier frequency.

For nondispersive media, such as perfect and low-loss dielectrics with $\beta = \omega\sqrt{\varepsilon\mu}$, both $\mathrm{d}\beta/\mathrm{d}\omega = \sqrt{\varepsilon\mu}$ and $\beta/\omega = \sqrt{\varepsilon\mu}$, so that the group velocity is equal to the phase velocity. Therefore, there is no real need to distinguish between these two velocities in electromagnetic wave applications associated with nondispersive media (and such applications are predominant in antennas and wireless communications), so we simply use c, Eq. (9.18), as the velocity of electromagnetic waves in a nondispersive medium of parameters ε and μ. In summary,

$$\boxed{v_\mathrm{g} = v_\mathrm{p} = c = \frac{1}{\sqrt{\varepsilon\mu}}.} \tag{9.173}$$

nondispersive media

For dispersive media, however, where β-ω relationship is nonlinear, the phase and group velocities are different. Using Eqs. (9.35) and (9.172), they can be related to each other as

$$v_\mathrm{g} = \left(\frac{\mathrm{d}\beta}{\mathrm{d}\omega}\right)^{-1} = \left[\frac{\mathrm{d}}{\mathrm{d}\omega}\left(\frac{\omega}{v_\mathrm{p}}\right)\right]^{-1} = \frac{v_\mathrm{p}}{1 - (\omega/v_\mathrm{p})\,\mathrm{d}v_\mathrm{p}/\mathrm{d}\omega}. \tag{9.174}$$

We see from this equation that $v_\mathrm{g} = v_\mathrm{p}$ only if $\mathrm{d}v_\mathrm{p}/\mathrm{d}\omega = 0$ (nondispersive media). Media for which the phase velocity decreases with increasing frequency ($\mathrm{d}v_\mathrm{p}/\mathrm{d}\omega < 0$) or, equivalently, for which $v_\mathrm{g} < v_\mathrm{p}$ from Eq. (9.174) are called normally dispersive media. The remaining dispersive media, i.e., those with a phase velocity increasing as frequency is increased ($\mathrm{d}v_\mathrm{p}/\mathrm{d}\omega > 0$) and with $v_\mathrm{g} > v_\mathrm{p}$, which is the case in Fig. 9.15, are thus anomalously (or abnormally) dispersive media. Typical examples of normally and anomalously dispersive media are plasmas [as we shall see in an example, based on Eq. (9.160)] and good conductors [see Eq. (9.143)], respectively. The terms normal and anomalous are arbitrary, and are brought about for purely historical reasons.

Example 9.22 Wave Dispersion in Good Conductors

Consider a uniform plane time-harmonic electromagnetic wave propagating in a good conductor. (a) Show that the group velocity of the wave is twice the phase velocity. (b) For the VLF undersea ship-to-submarine communication system from Example 9.19, find the relative phase delay between two different frequency components, at $f_1 = 7.77$ kHz and $f_2 = 10$ kHz, respectively, of a signal that is launched at the ocean surface and received at the depth in Eq. (9.149).

Solution

(a) From Eqs. (9.135) and (9.143), the phase coefficient and velocity of waves in good conductors can, respectively, be written as

$$\beta(\omega) = k\sqrt{\omega} \quad \text{and} \quad v_\mathrm{p}(\omega) = \frac{\sqrt{\omega}}{k}, \quad \text{where} \quad k = \sqrt{\frac{\mu\sigma}{2}}, \tag{9.175}$$

and the dispersion diagram of a good conductor looks like that in Fig. 9.15. Using Eq. (9.172), the group velocity comes out to be

$$\boxed{v_\mathrm{g}(\omega) = \frac{1}{\mathrm{d}\beta(\omega)/\mathrm{d}\omega} = \frac{2\sqrt{\omega}}{k} = 2v_\mathrm{p}(\omega),} \tag{9.176}$$

v_g – good conductors

namely, twice the phase velocity. Of course, the same result is obtained by means of the expression for v_g in terms of v_p in Eq. (9.174). Also, as mentioned in discussions of this expression, good conductors are anomalously dispersive media, since their phase velocity increases with increasing frequency, Eqs. (9.175), or, equivalently, v_g is larger (in fact, twice as large) than v_p, Eq. (9.176).

(b) With the conductivity of seawater specified in Example 9.19 and $\mu = \mu_0$, the constant k in Eqs. (9.175) becomes $k = 1.58 \times 10^{-3}\ \sqrt{\text{s}}/\text{m}$. So, having as well in mind Eq. (9.33), the relative phase delay between the two signals for the path traveled by the wave equal to the depth (with respect to the water surface) of the submarine's antenna of $d = d_{\text{max}} = 66.8$ m, Eq. (9.149), amounts to

phase shift between different frequency components due to dispersion

$$\Delta\phi = -\beta(\omega_1)d - [-\beta(\omega_2)d] = k\sqrt{2\pi}\left(\sqrt{f_2} - \sqrt{f_1}\right)d = 180°. \tag{9.177}$$

In general, as already explained earlier, such large phase shifts between different frequency (Fourier) components of a signal, due to their propagation at different phase velocities, $v_p(\omega)$, may cause significant distortions of signals at the receiving point (the superposition of the individual components is not an exact replica of the original signal). In particular, the phase difference between the two components in Eq. (9.177), appearing to be exactly 180°, is very illustrative of signal distortion due to dispersion of its frequency components. Namely, it makes the respective electric and magnetic fields, at the two frequencies, be in counter-phase and actually cancel each other at signal reception, instead of adding in phase if propagated in a nondispersive medium, which, obviously, are diametrally different results.

Example 9.23 Wave Dispersion in Plasmas

A uniform plane time-harmonic electromagnetic wave of frequency f travels through a plasma medium whose cutoff frequency is f_p. (a) Sketch the dispersion diagram of the medium and (b) calculate the phase and group velocities of the wave.

Solution

(a) Based on the β-ω relationship for a plasma in Eq. (9.160), the dispersion diagram is sketched in Fig. 9.17, where $\omega = 2\pi f$ and $\omega_p = 2\pi f_p$ are, respectively, the operating radian frequency of the wave and cutoff radian frequency of the plasma. Of course, the phase coefficient is zero at cutoff ($f = f_p$), and purely imaginary (not shown) below it ($f < f_p$). In the propagating region ($f > f_p$), β increases with increasing frequency, asymptotically approaching the free-space value of $\beta_{\text{free-space}} = \omega/c_0$ for $f \to \infty$.

(b) Combining Eqs. (9.35), (9.172), and (9.160), the phase and group velocities of a propagating wave in the plasma, at frequency f ($f > f_p$), are

v_p, v_g – plasmas

$$v_p = \frac{\omega}{\beta} = \frac{c_0}{\sqrt{1 - f_p^2/f^2}} \quad \text{and} \quad v_g = \frac{1}{d\beta/d\omega} = c_0\sqrt{1 - \frac{f_p^2}{f^2}}, \tag{9.178}$$

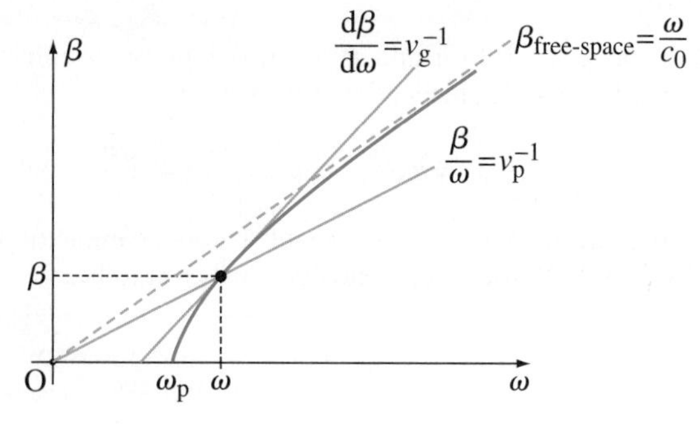

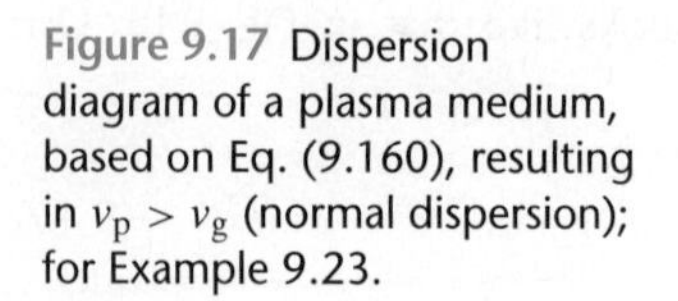
Figure 9.17 Dispersion diagram of a plasma medium, based on Eq. (9.160), resulting in $v_p > v_g$ (normal dispersion); for Example 9.23.

respectively. We see that both $v_p \to c_0$ and $v_g \to c_0$ for $f \gg f_p$, which is another confirmation that plasmas behave much like free space far above cutoff. In addition, it is obvious from Eqs. (9.178) that $v_p > c_0$ and $v_g < c_0$, in the entire propagating region. Hence, $v_p > v_g$, which is also illustrated graphically in Fig. 9.17, based on the definitions of phase and group velocities, in Eqs. (9.35) and (9.172), and means that the plasma is a normally dispersive propagation medium. Finally, note that the fact that the phase velocity in plasmas is larger (and even much larger near cutoff) than the speed of light in free space, Eq. (9.19), although it might seem alarming at the first glance, indeed does not violate the theory of special relativity, which states that energy and matter cannot travel faster than c_0. Namely, v_p, as explained in this section, is just a velocity with which the constant-phase plane moves in the given direction. Rather, it is v_g the velocity at which the electromagnetic energy and information are transported by the wave, and $v_g \leq c_0$ always.

Example 9.24 Dispersion of Pulsar Radio Waves in the Interstellar Plasma

A distant rotating neutron star, called pulsar, emits radio waves (pulses) that can be detected on the earth. The interstellar medium can roughly be considered to be a plasma with an average concentration of free electrons of $N = 3 \times 10^4\ \text{m}^{-3}$. Because of the dispersion in such propagation medium, a difference in arrival time of $\Delta t = 1$ s is detected of waves emitted simultaneously at frequencies $f_1 = 200$ MHz and $f_2 = 500$ MHz by the pulsar. What is the distance to the pulsar?

Solution Eq. (9.159) tells us that the cutoff frequency of the interstellar plasma amounts to $f_p = 1.56$ kHz, and both measured radio frequencies are, of course, well above it (so in the propagating region). The associated group velocities are thus given by Eq. (9.178), at the two frequencies, and the time lag between the two signals can be expressed as

$$\boxed{\Delta t = \frac{d}{v_g(f_1)} - \frac{d}{v_g(f_2)},} \tag{9.179}$$

time lag between signals due to dispersions

with d designating the unknown distance they travel from the pulsar to the earth (that we seek). In fact, frequencies f_1 and f_2 are so high relative to f_p that these velocities are practically the same, and equal to c_0, $v_g(f_1) \approx v_g(f_2) \approx c_0$, which would result in a zero time difference. However, to still be able to capture and use the difference of $v_g(f_1)$ and $v_g(f_2)$, namely, of their reciprocals, we turn to the following approximate expression for $1/v_g$, obtained using the binomial series identity like in Eq. (9.122):

$$f \gg f_p \quad \longrightarrow \quad \frac{1}{v_g(f)} = \frac{1}{c_0}\left(1 - \frac{f_p^2}{f^2}\right)^{-1/2} \approx \frac{1}{c_0}\left(1 + \frac{f_p^2}{2f^2}\right). \tag{9.180}$$

With this in Eq. (9.179), d is computed as follows:

$$\Delta t = \frac{d f_p^2}{2c_0}\left(\frac{1}{f_1^2} - \frac{1}{f_2^2}\right) \quad \longrightarrow \quad d = \frac{2c_0 \Delta t}{f_p^2\,(1/f_1^2 - 1/f_2^2)} = 1.173 \times 10^{19}\ \text{m}. \tag{9.181}$$

As we know, such vast cosmic distances are customarily measured in light-years,

$$\boxed{\frac{d}{c_0 \times 365 \times 24 \times 60 \times 60\ \text{s}} = 1241\ \text{light-years}.} \tag{9.182}$$

expressing cosmic distances in light-years

So, the pulsar radio waves received today were emitted more than a millennium ago!

Problems: 9.34–9.38; *Conceptual Questions* (on Companion Website): 9.30–9.35; *MATLAB Exercises* (on Companion Website).

9.14 POLARIZATION OF ELECTROMAGNETIC WAVES

All time-harmonic electromagnetic waves considered so far in this chapter exhibit so-called linear polarization (LP), which means that the tip of the electric field intensity vector of the wave (and the same is true for the magnetic one) at a given point in space traces a straight line in the course of time.[24] For instance, we see that the tip of the vector $\mathbf{E}$ of a uniform plane wave propagating in a lossless medium given by

linear polarization (LP)

$$\boxed{\mathbf{E}(z,t) = E_{\mathrm{m}} \cos(\omega t - \beta z + \theta_0)\, \hat{\mathbf{x}}} \tag{9.183}$$

[Eq. (9.32)] for a fixed z oscillates, in the course of time, along the x-axis of the Cartesian coordinate system. The same is true for the vector $\mathbf{E}$ in Eq. (9.84), for a wave propagating in a lossy medium. As shown in Fig. 9.18 for $z = 0$ and $\theta_0 = 0$, at instants that are multiples of $T/8$, where T is the time period defined in Eq. (8.49), $\mathbf{E}(0,t)$ in Eq. (9.183) periodically assumes values $E_{\mathrm{m}}\, \hat{\mathbf{x}}$, $(E_{\mathrm{m}}/\sqrt{2})\, \hat{\mathbf{x}}$, 0, $-(E_{\mathrm{m}}/\sqrt{2})\, \hat{\mathbf{x}}$, $-E_{\mathrm{m}}\, \hat{\mathbf{x}}$, $-(E_{\mathrm{m}}/\sqrt{2})\, \hat{\mathbf{x}}$, ... as the time progresses from $t = 0$ on. We say that the vector $\mathbf{E}$ is linearly polarized in the x direction. By the same token, the magnetic field intensity vector of the wave, $\mathbf{H}(z,t)$, given in Eqs. (9.32) or (9.93), is linearly polarized in the y direction. However, the overall polarization of a time-harmonic wave is determined by the polarization of its electric field vector, so for the wave considered here we say that it is an x-polarized LP wave.

However, if two time-harmonic uniform plane waves with mutually orthogonal linear polarizations at the same frequency copropagate in the same direction, the polarization of the resultant time-harmonic wave depends on the relative amplitudes and phases of its individual LP components. For instance, if the total electric field intensity vector of a wave is represented as

$$\mathbf{E}(z,t) = E_x(z,t)\, \hat{\mathbf{x}} + E_y(z,t)\, \hat{\mathbf{y}}, \tag{9.184}$$

and the two transverse components have the same amplitudes but are out of phase by 90° (i.e., they are in time-phase quadrature),

circular polarization (CP)

$$\boxed{E_x = E_{\mathrm{m}} \cos(\omega t - \beta z), \quad E_y = E_{\mathrm{m}} \cos(\omega t - \beta z \mp 90^\circ) = \pm E_{\mathrm{m}} \sin(\omega t - \beta z),} \tag{9.185}$$

the wave is circularly polarized (CP). Namely, the resultant vector $\mathbf{E}$ for $z =$ const rotates and its tip describes a circle as a function of time in a transversal plane defined by the fixed coordinate z, as discussed in an example of a rotating magnetic field in Fig. 6.19(b). Referring to Eqs. (6.118) and (6.119), we realize that the radius of the circle traced by the tip of $\mathbf{E}$ (so-called polarization circle) equals the amplitude of each of the wave components, E_{m}, and that the rotation of $\mathbf{E}$ is uniform, with an angular velocity equal to the angular frequency ω of the individual wave components. With δ denoting the relative phase of $\mathbf{E}_y$ with respect to $\mathbf{E}_x$,

$$\delta = \theta_{y0} - \theta_{x0}, \tag{9.186}$$

where θ_{x0} and θ_{y0} are the absolute initial phases in the plane $z = 0$ of $\mathbf{E}_x$ and $\mathbf{E}_y$, respectively, with θ_{x0} adopted to be zero in Eq. (9.185) for simplicity, Fig. 9.19(a)

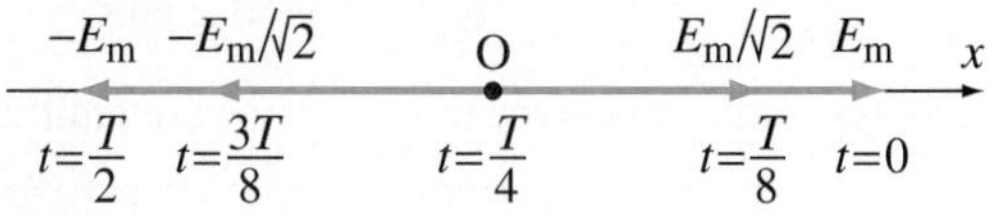

Figure 9.18 Linearly polarized time-harmonic uniform plane wave, Eq. (9.183), for $z = 0$ and $\theta_0 = 0$.

[24] Polarization, in general, is defined for arbitrary time-harmonic vectors; not necessarily electric and magnetic field vectors of a plane wave; not necessarily related to electromagnetic fields at all.

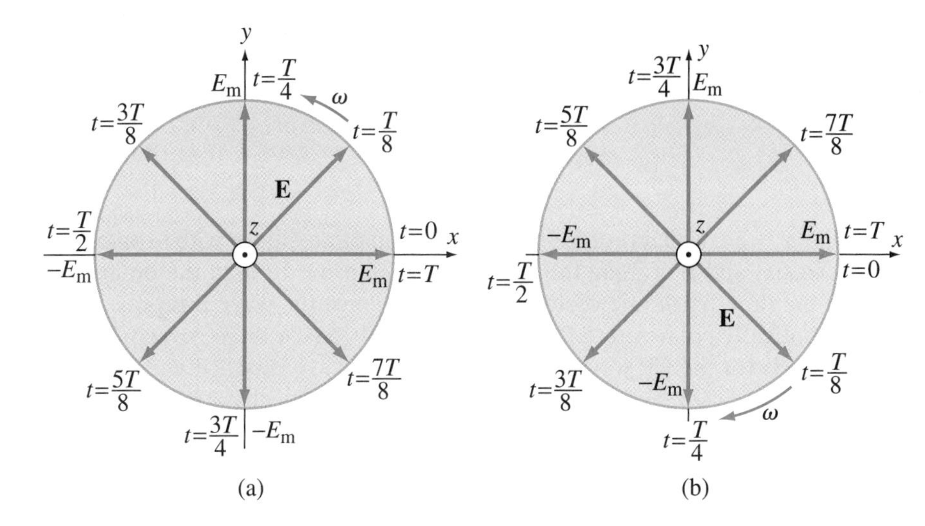

Figure 9.19 Right-hand (a) and left-hand (b) circularly polarized time-harmonic uniform plane wave (propagating in the positive z direction), Eq. (9.185) with $\delta = -90°$ (RHCP) or $\delta = 90°$ (LHCP); in determining the polarization handedness, the field rotation must be viewed as the wave travels away from the observer (out of the page).

shows the rotation of the total field vector for the case $\delta = -90°$ [$E_y = E_m \sin(\omega t - \beta z)$]. At instants $kT/8$, where $k = 0, 1, 2, \ldots$, $\mathbf{E}(0, t)$ cyclically acquires values $E_m \hat{\mathbf{x}}$, $E_m(\hat{\mathbf{x}} + \hat{\mathbf{y}})/\sqrt{2}$, $E_m \hat{\mathbf{y}}$, $E_m(-\hat{\mathbf{x}} + \hat{\mathbf{y}})/\sqrt{2}, \ldots$ We see that in this case, when the wave is observed from the rear (receding), i.e., toward the positive z direction, the vector **E** rotates in the clockwise (CW) direction as a function of time. Such CP wave is said to be right-hand circularly polarized (RHCP), given that, when the thumb of the right hand points into the direction of the wave travel, the other fingers curl in the direction of rotation of **E**. Conversely, the case with $\delta = 90°$ in Eq. (9.185) [$E_y = -E_m \sin(\omega t - \beta z)$] gives rise to **E** rotating in the counterclockwise (CCW) direction when the wave is viewed receding, Fig. 9.19(b). Such rotation is in the direction defined by the fingers of the left hand when the thumb points along the direction of the wave propagation, which classifies this wave as a left-hand circularly polarized (LHCP) one. In other words, if the sense of the electric field vector rotation and the direction of wave propagation comply with the right-hand rule (thumb – wave propagation; fingers – field rotation), the wave is RHCP; otherwise it is LHCP (it agrees with the corresponding left-hand rule).[25]

If we change one of the amplitudes of the transverse components of the wave in Eq. (9.185) so that they are no longer the same,

elliptical polarization (EP)

$$\boxed{E_x = E_1 \cos(\omega t - \beta z), \quad E_y = \pm E_2 \sin(\omega t - \beta z) \quad (E_1 \neq E_2, \ \delta = \mp 90°),} \tag{9.187}$$

[25]The (right- vs. left-hand) rule for determining whether a CP wave is right- or left-handed adopted in this text is referred to as the IEEE (Institute of Electrical and Electronics Engineers) convention for polarization handedness. It is largely arbitrary, as we could have adopted to look at the sense of field rotation by viewing the wave approaching (instead of receding) and pointing the thumb (of the right or left hand) to where the wave is coming from (and not to where it is heading). This latter definition, exactly opposite to the IEEE one, is the classical optics convention for polarization handedness, which is a preferred choice for most physicists and optical engineers. Note, however, that the IEEE definition is in perfect agreement with the radiation of so-called axial-mode helical antennas, and this is perhaps the basic motivation for its adoption as an IEEE standard. Namely, a helical antenna with right-hand sensed winding radiates a right-handed CP wave, while the polarization of the wave radiated by a left-hand wound helix is LH circular. Most importantly, although it is certainly unfortunate that we have two widely adopted and used definitions, we should understand that this is just a matter of convention, and simply exercise caution in interpreting what is meant when polarization handedness is stated (in different texts and communications).

the tip of the resultant vector $\mathbf{E}$, as it rotates in time with angular velocity equal to ω, traces an ellipse in the plane $z = \text{const}$. This is evident from rewriting Eq. (9.187) in the form

polarization ellipse

$$\boxed{\frac{E_x^2}{E_1^2} + \frac{E_y^2}{E_2^2} = 1} \tag{9.188}$$

($\cos^2\alpha + \sin^2\alpha = 1$), which represents the equation of an ellipse with semi-axes E_1 (along the x-axis) and E_2 (along the y-axis). This ellipse is termed the polarization ellipse, and the time-harmonic vector $\mathbf{E}(z, t)$, and hence the wave it represents, are said to be elliptically polarized (EP). The same definition for the polarization handedness applies as for the CP waves: for $\delta = -90°$, the wave is right-hand elliptically polarized (RHEP), whereas $\delta = 90°$ yields a left-hand elliptically polarized (LHEP) wave, as shown in Fig. 9.20.

The most general expressions for the two orthogonal transverse components of the electric field vector $\mathbf{E}(z, t)$ in Eq. (9.184) of a uniform plane wave are

the most general EP plane wave

$$\boxed{E_x = E_1 \cos(\omega t - \beta z), \quad E_y = E_2 \cos(\omega t - \beta z + \delta).} \tag{9.189}$$

Namely, in the most general case, the phase difference between the components, Eq. (9.186), is arbitrary ($-180° < \delta \leq 180°$), in addition to an arbitrary ratio between their amplitudes. This wave is also elliptically polarized, with $\delta > 0$ implying a left-handed polarization and $\delta < 0$ the right-handed one, and the polarization ellipse being tilted with respect to the x- and y-axes. So, elliptical polarization is the most general polarization of time-harmonic vectors. Circular polarization is a special case; when the semi-axes of the polarization ellipse are the same, the ellipse degenerates into a circle. Linear polarization is a collapsed ellipse with infinite major to minor axis ratio, and the other special case of Eq. (9.189). It occurs when $E_1 = 0$ (wave is linearly polarized in the y direction) or $E_2 = 0$ (x-polarized wave). The wave with both orthogonal components nonzero but in phase ($\delta = 0$) or counter-phase ($\delta = 180°$) with respect to each other, so that

$$\mathbf{E} = E_1 \cos(\omega t - \beta z)\,\hat{\mathbf{x}} \pm E_2 \cos(\omega t - \beta z)\,\hat{\mathbf{y}} = (E_1\,\hat{\mathbf{x}} \pm E_2\,\hat{\mathbf{y}}) \cos(\omega t - \beta z), \tag{9.190}$$

is also linearly polarized – the tip of $\mathbf{E}$ oscillates, in the course of time, along the line defined by the constant vector $E_1\,\hat{\mathbf{x}} \pm E_2\,\hat{\mathbf{y}}$ (if $E_1 = E_2$, for example, this line makes an angle of 45° or 135° with the x-axis).

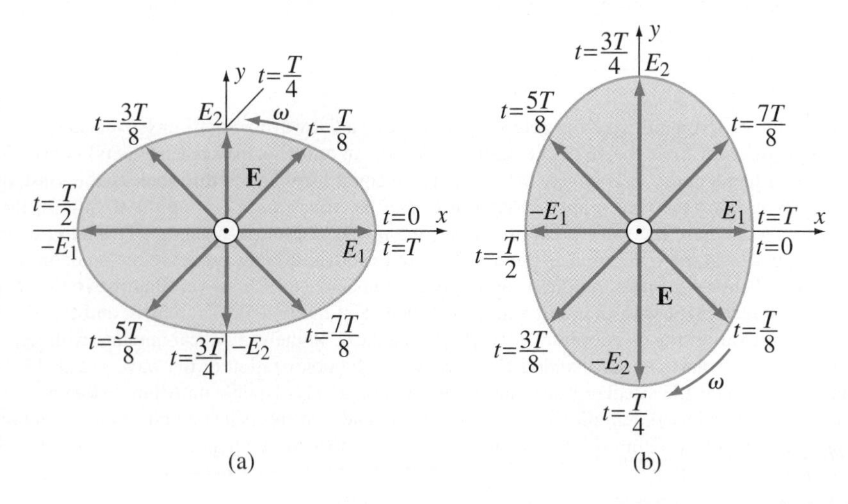

Figure 9.20 Elliptical polarization of a time-harmonic uniform plane wave (propagating out of the page), Eq. (9.187): (a) RHEP ($\delta = -90°$) with the major axis of the polarization ellipse, Eq. (9.188), adopted to be in the x direction ($E_1 > E_2$) and (b) LHEP ($\delta = 90°$) with a y-directed major axis ($E_2 > E_1$); the polarization handedness is defined in the same way as for the CP wave in Fig. 9.19.

In radio communication systems, in general, the generators and receivers of electromagnetic waves, which carry the energy or information to a distance, are antennas (to be studied in a later chapter). The polarization of an antenna, which is one of its most important properties, is defined as the polarization of the electromagnetic wave radiated in a given direction (most notably in the direction of maximum radiation) by the antenna when transmitting. In the receiving mode of operation, an antenna receives best the electromagnetic wave whose polarization is the same as the polarization of the antenna (if it were transmitting) in the direction of the wave approaching [the antenna is (perfectly) polarization-matched to the wave]. This is achieved, for example, if both the antenna and the wave are linearly polarized along the same direction.

Typical examples of linearly polarized antennas are wire dipole antennas (see Fig. 8.3). Linear polarization is, in communication systems, often referred to as vertical or horizontal when being perpendicular or parallel, respectively, to the earth's surface. AM radio broadcast stations launch vertically polarized LP waves. Most TV broadcast transmitting antennas are horizontally polarized, and that is why typical rooftop receiving antennas are composed of parallel horizontal wires (antenna arrays). Helical antennas are classical representatives of circularly polarized radiators, with the sense of rotation of the radiated wave being the same as that of the helix winding. With reference to Eq. (9.185), it is a simple matter to realize that two crossed (at 90°) wire dipole antennas fed with equal input powers but in time-phase quadrature constitute a CP antenna system as well. For example, FM radio broadcast stations, as well as recent TV transmitters, typically emit CP waves, which can be received by both horizontally and vertically oriented LP antennas (note that most receiving antennas attached to vehicles are vertically polarized).

Finally, if the amplitudes E_1 and E_2 of the transverse wave components E_x and E_y in Eq. (9.189) and the phase difference δ between them are not constants but undergo random variations, such electromagnetic wave is said to be randomly polarized or, simply, unpolarized. Note that if we choose any direction in a transversal plane of an unpolarized wave, exactly half of the time-average power carried by the wave will be contained in a wave component polarized along that direction, and the other half of the power will be polarized along the transverse direction orthogonal to the first polarization direction. Hence, unpolarized waves can be received equally well using an LP antenna of any transverse orientation, with a constant half-power polarization mismatch obtained at the reception. Most of natural light is unpolarized, as well as most cosmic radio sources and their emission.

Example 9.25 Magnetic Field of an Elliptically Polarized Plane Wave

An elliptically polarized time-harmonic uniform plane electromagnetic wave propagates in a lossless medium whose intrinsic impedance is η. The electric field of the wave is described in Eqs. (9.187). Determine the magnetic field intensity vector of the wave.

Solution Referring to Eq. (9.22), the magnetic field vector of a uniform plane wave whose electric field vector consists of an x and y Cartesian components can be obtained as

$$\mathbf{H} = \frac{1}{\eta}\,\hat{\mathbf{n}} \times \mathbf{E} = \frac{1}{\eta}\,\hat{\mathbf{z}} \times (E_x\,\hat{\mathbf{x}} + E_y\,\hat{\mathbf{y}}) = \frac{E_x}{\eta}\,\hat{\mathbf{y}} - \frac{E_y}{\eta}\,\hat{\mathbf{x}}, \tag{9.191}$$

so that H_y is associated with E_x and H_x with $-E_x$. Hence, the components of $\mathbf{H}$ for the elliptically polarized time-harmonic wave with electric field in Eqs. (9.187) are

$$\boxed{H_x = -\frac{E_y}{\eta} = \mp\frac{E_2}{\eta}\sin(\omega t - \beta z), \quad H_y = \frac{E_x}{\eta} = \frac{E_1}{\eta}\cos(\omega t - \beta z).} \tag{9.192}$$

magnetic field of an EP plane wave

Note that this is equivalent to applying the concept of the intrinsic impedance of the medium and the rule by which the directions of **E**, **H**, and wave propagation are interrelated in Fig. 9.2 to each of the two linearly-polarized waves (x- and y-polarized waves) constituting the total EP wave separately, and then adding together the such obtained LP magnetic field vectors. The total magnetic field vector rotates in synchronism with the total electric field vector (with the same angular velocity and same sense of rotation). From Eqs. (9.192), their instantaneous intensities [see Eq. (6.118)] are related as

E-H proportionality for an EP wave

$$|\mathbf{H}(z,t)| = \sqrt{H_x^2 + H_y^2} = \sqrt{\left(-\frac{E_y}{\eta}\right)^2 + \left(\frac{E_x}{\eta}\right)^2} = \frac{1}{\eta}\sqrt{E_x^2 + E_y^2} = \frac{|\mathbf{E}(z,t)|}{\eta}, \tag{9.193}$$

that is, $H = E/\eta$ at every instant of time and every point of space, like for an ordinary (linearly polarized) TEM wave, Eq. (9.20).

Example 9.26 Instantaneous Poynting Vector of a Circularly Polarized Wave

Find the instantaneous Poynting vector of a circularly polarized uniform plane wave of frequency f and electric field amplitude E_{m} traveling through a lossless electromagnetic material of permittivity ε and permeability μ.

Solution Using Eqs. (8.161), (9.187), and (9.192), with $E_1 = E_2 = E_{\mathrm{m}}$ (circular polarization) and $\eta = \sqrt{\mu/\varepsilon}$ [Eq. (9.21)], the instantaneous Poynting vector of the wave is given by

$$\mathcal{P} = \mathbf{E}\times\mathbf{H} = [E_{\mathrm{m}}\cos(\omega t - \beta z)\,\hat{\mathbf{x}} \pm E_{\mathrm{m}}\sin(\omega t - \beta z)\,\hat{\mathbf{y}}] \times \left[\mp\frac{E_{\mathrm{m}}}{\eta}\sin(\omega t - \beta z)\,\hat{\mathbf{x}}\right.$$

$$\left. + \frac{E_{\mathrm{m}}}{\eta}\cos(\omega t - \beta z)\,\hat{\mathbf{y}}\right] = \frac{E_{\mathrm{m}}^2}{\eta}\left[\cos^2(\omega t - \beta z) + \sin^2(\omega t - \beta z)\right]\hat{\mathbf{z}} = \frac{E_{\mathrm{m}}^2}{\eta}\,\hat{\mathbf{z}}. \tag{9.194}$$

We see that $\mathcal{P}$ is constant in time. Note that, therefore, its time average is this same value, $\mathcal{P}_{\mathrm{ave}} = (E_{\mathrm{m}}^2/\eta)\,\hat{\mathbf{z}}$.

Example 9.27 EP Wave as a Sum of Two Counter-Rotating CP Waves

Show that an elliptically polarized time-harmonic uniform plane wave can be represented as a superposition of two oppositely rotating circularly polarized waves propagating in the same direction.

Solution As the amplitudes of transverse components of an elliptically polarized electric field intensity vector given by Eqs. (9.187) can always be represented as $E_1 = E' + E''$ and $E_2 = E' - E''$, the vector can be written as

$$\mathbf{E} = \underbrace{(E' + E'')}_{E_1}\cos(\omega t - \beta z)\,\hat{\mathbf{x}} \pm \underbrace{(E' - E'')}_{E_2}\sin(\omega t - \beta z)\,\hat{\mathbf{y}}$$

$$= \underbrace{E'\cos(\omega t - \beta z)\,\hat{\mathbf{x}} \pm E'\sin(\omega t - \beta z)\,\hat{\mathbf{y}}}_{\text{CP wave 1}} + \underbrace{E''\cos(\omega t - \beta z)\,\hat{\mathbf{x}} \mp E''\sin(\omega t - \beta z)\,\hat{\mathbf{y}}}_{\text{CP wave 2}} \tag{9.195}$$

[and a similar decomposition can be performed for the magnetic field vector of the wave, starting with Eqs. (9.192)]. This indeed is a sum of two circularly polarized electric field vectors, Eqs. (9.185). The two waves have the same angular frequency (ω) and propagate in the same direction (positive z direction) as the original (total) electromagnetic wave, but have unequal amplitudes ($E' \neq E''$) and opposite senses of rotation (note the sign "$\mp$" in place of "$\pm$" in the second term) with respect to each other.

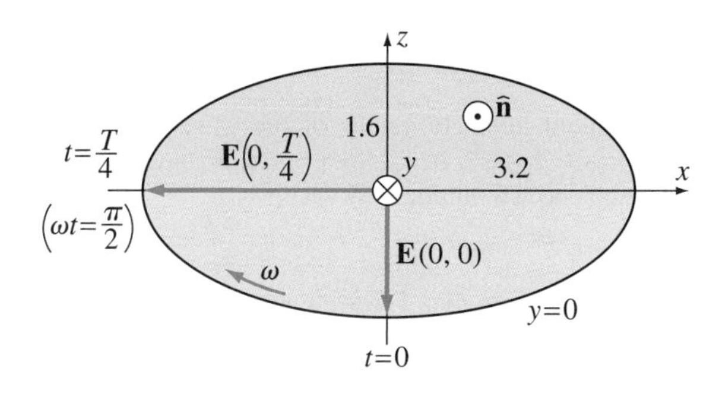

Figure 9.21 Polarization ellipse of the electric field vector given by Eq. (9.198); for Example 9.28.

Example 9.28 Determination of Polarization Handedness

Two linearly polarized electromagnetic waves propagate in unison in free space, and their magnetic field vectors are given by

$$\underline{\mathbf{H}}_1 = 3\,\mathrm{e}^{\mathrm{j}y}\,\hat{\mathbf{x}}\ \mathrm{mA/m} \quad \text{and} \quad \underline{\mathbf{H}}_2 = \mathrm{j}6\,\mathrm{e}^{\mathrm{j}y}\,\hat{\mathbf{z}}\ \mathrm{mA/m} \quad (y \text{ in m}), \tag{9.196}$$

respectively. Determine the instantaneous electric field intensity vector of the resultant wave, and the type (linear, circular, or elliptical) and handedness (right- or left-handed) of the polarization of this wave.

Solution Using Eq. (8.66), the resultant magnetic field intensity vector in the time domain is

$$\mathbf{H}(y,t) = \mathbf{H}_1(y,t) + \mathbf{H}_2(y,t) = 3\sqrt{2}\,[\cos(\omega t + \beta y)\,\hat{\mathbf{x}} - 2\sin(\omega t + \beta y)\,\hat{\mathbf{z}}]\ \ \mathrm{mA/m}, \tag{9.197}$$

and hence, having in mind Eq. (9.22) and that the wave travel is in the negative y direction ($\hat{\mathbf{n}} = -\,\hat{\mathbf{y}}$), the associated instantaneous electric field vector comes out to be

$$\mathbf{E}(y,t) = \eta_0 \mathbf{H}(y,t) \times \hat{\mathbf{n}} = \eta_0 \mathbf{H}(y,t) \times (-\,\hat{\mathbf{y}}) = -1.6\,[2\sin(\omega t + \beta y)\,\hat{\mathbf{x}} + \cos(\omega t + \beta y)\,\hat{\mathbf{z}}]\ \ \mathrm{V/m} \quad (t \text{ in s};\ y \text{ in m}), \tag{9.198}$$

where $\eta_0 = 377\ \Omega$, Eq. (9.23). This is an elliptically polarized wave, as its two transverse components are of unequal amplitudes and out of phase with respect to each other, which is obvious from Eq. (9.196) as well. The semi-axes of the polarization ellipse are 3.2 V/m and 1.6 V/m along the x- and z-axes, respectively, as shown in Fig. 9.21. To determine, however, the handedness of the ellipse (sense of rotation of the total field), we sketch (as is done in Fig. 9.20) the vector $\mathbf{E}(y,t)$, Eq. (9.198), at several characteristic instants of time, that are multiples of $T/4$, T being the time period in Eq. (8.49), for $y = 0$. From this sketch, in Fig. 9.21, we realize that this is a left-handed polarization (the total wave is an LHEP one). Simply, from the position at $t = 0$, at which $\mathbf{E} = -1.6\,\hat{\mathbf{z}}$ V/m, to that at $t = T/4$ ($\omega t = \pi/2$), where $\mathbf{E} = -3.2\,\hat{\mathbf{x}}$ V/m, the vector (arrow) $\mathbf{E}$ rotates in the counterclockwise direction when the wave is observed receding (see the orientation of the propagation unit vector $\hat{\mathbf{n}}$), and this corresponds to an LH polarization (see the explanation given with Figs. 9.19 and 9.20).

Note that the same type and handedness of the wave polarization would have been obtained by considering the magnetic field vector of the total wave, Eq. (9.197), in the first place. As already pointed out in Example 9.25, $\mathbf{H}$ always (for EP and CP waves) rotates in synchronism with $\mathbf{E}$. However, the two field vectors being orthogonal to each other in space, when $\mathbf{E}$ is maximum, $\mathbf{H}$ is minimum (for EP waves), and vice versa, so that the polarization ellipse for the magnetic field vector of the wave is orthogonal to that for the electric field vector (major semi-axis of the H-field ellipse is along the minor semi-axis of the E-field ellipse, and vice versa).

Example 9.29 Complex Poynting Vector of an Elliptically Polarized Wave

Derive the expression for the complex Poynting vector of an elliptically polarized uniform plane electromagnetic wave in a lossless medium of intrinsic impedance η, if the rms values of the two mutually orthogonal components of the electric field vector of the wave are E_{x0} and E_{y0}. What is the associated time-average Poynting vector?

Solution Let the instantaneous electric field intensity vector of the wave be given by Eq. (9.189), and hence $E_{x0} = E_1/\sqrt{2}$ and $E_{y0} = E_2/\sqrt{2}$. In complex notation, having in mind Eq. (8.66), the corresponding complex rms electric field intensity vector is

$$\underline{\mathbf{E}} = \underline{E}_x\,\hat{\mathbf{x}} + \underline{E}_y\,\hat{\mathbf{y}}, \quad \text{where} \quad \underline{E}_x = E_{x0}\,\mathrm{e}^{-\mathrm{j}\beta z} \quad \text{and} \quad \underline{E}_y = E_{y0}\,\mathrm{e}^{-\mathrm{j}\beta z}\,\mathrm{e}^{\mathrm{j}\delta}. \tag{9.199}$$

Combining Eqs. (8.194), (9.192), and (9.199), the complex Poynting vector of an arbitrarily (elliptically) polarized wave, which includes linear and circular polarizations as special cases, comes out to be

complex and time-average Poynting vector of an EP wave

$$\underline{\mathcal{P}} = \underline{\mathbf{E}} \times \underline{\mathbf{H}}^* = (\underline{E}_x\,\hat{\mathbf{x}} + \underline{E}_y\,\hat{\mathbf{y}}) \times (\underline{H}_x\,\hat{\mathbf{x}} + \underline{H}_y\,\hat{\mathbf{y}})^* = \underline{E}_x\underline{H}_y^*\,\hat{\mathbf{z}} - \underline{E}_y\underline{H}_x^*\,\hat{\mathbf{z}}$$
$$= \frac{1}{\eta}\,(\underline{E}_x\underline{E}_x^* + \underline{E}_y\underline{E}_y^*)\,\hat{\mathbf{z}} = \frac{|\underline{E}_x|^2 + |\underline{E}_y|^2}{\eta}\,\hat{\mathbf{z}} = \frac{E_{x0}^2 + E_{y0}^2}{\eta}\,\hat{\mathbf{z}} = \frac{|\underline{\mathbf{E}}|^2}{\eta}\,\hat{\mathbf{z}}, \tag{9.200}$$

and the time average of the associated instantaneous Poynting vector ($\mathcal{P}_{\text{ave}}$), Eq. (8.195), is the same ($\underline{\mathcal{P}}$ is purely real). As $|\underline{\mathbf{E}}|$, the magnitude of the complex vector $\underline{\mathbf{E}}$, equals the rms value of its instantaneous time-harmonic counterpart $\mathbf{E}(t)$, we realize that the complex Poynting vector of a wave with any polarization is computed in the same way as that of a linearly polarized wave, Eq. (9.40), that is, as the square of the rms value of $\mathbf{E}(t)$ divided by η (and multiplied by the propagation unit vector). Note also that for a circularly polarized wave [Eq. (9.185)], $E_{x0} = E_{y0} = E_{\text{m}}/\sqrt{2}$ in Eq. (9.199), and hence $E_{x0}^2 + E_{y0}^2 = E_{\text{m}}^2$ in Eq. (9.200), yielding the same result for $\mathcal{P}_{\text{ave}}$ as in Eq. (9.194).

Example 9.30 Change of Wave Polarization due to Material Anisotropy

A linearly polarized time-harmonic uniform plane electromagnetic wave propagating, with the free-space wavelength of $\lambda_0 = 1\ \mu\text{m}$, in the positive z direction enters an anisotropic crystalline dielectric material. The electric field intensity vector of the wave, $\mathbf{E}$, is at 45° to two mutually orthogonal transverse directions in the crystal, namely, x and y directions, along which the crystal relative permittivity has different values [Eq. (2.52)], $\varepsilon_{\text{r}x} = 2.25$ and $\varepsilon_{\text{r}y} = 2.12$, respectively. (a) Find the shortest length of the crystal (along the z-axis) for which the wave emerging on the other side of it is circularly polarized, and whether this polarization is right- or left-handed. (b) Determine the polarization state of the output wave if the crystal length in (a) is doubled.

Solution

(a) Using Eq. (8.111), the phase coefficients in x and y directions in the crystal are, respectively,

$$\beta_x = \omega\sqrt{\varepsilon_{\text{r}x}\varepsilon_0\mu_0} = \frac{2\pi}{\lambda_0}\sqrt{\varepsilon_{\text{r}x}} \quad \text{and} \quad \beta_y = \omega\sqrt{\varepsilon_{\text{r}y}\varepsilon_0\mu_0} = \frac{2\pi}{\lambda_0}\sqrt{\varepsilon_{\text{r}y}} \tag{9.201}$$

($\mu_\text{r} = 1$). The vector $\mathbf{E}$ can be decomposed onto E_x and E_y components that have equal amplitudes (equal to the amplitude of $\mathbf{E}$ times $\cos 45°$), and travel along the z-axis with different phase velocities, corresponding, by means of Eq. (9.35), to the phase coefficients in Eqs. (9.201). Therefore, a phase difference between the components is accumulated as the wave progresses into the crystal. Having in mind Eq. (9.33), the relative phase δ of E_y with respect to E_x, Eq. (9.186), for a length d of the crystal amounts to

$$\delta = \phi_y - \phi_x = -\beta_y d - (-\beta_x d) = (\beta_x - \beta_y)\,d = \frac{2\pi d}{\lambda_0}\left(\sqrt{\varepsilon_{\text{r}x}} - \sqrt{\varepsilon_{\text{r}y}}\right) > 0 \quad (\varepsilon_{\text{r}x} > \varepsilon_{\text{r}y}), \tag{9.202}$$

and it turns out to be positive for the given permittivities of the material. The two field components will be exactly 90° out of phase, which is circular polarization, Eq. (9.185), for the following (smallest possible) value of d:

$$\delta_1 = \frac{\pi}{2} \quad \text{(LHCP output wave)} \quad \longrightarrow \quad d_1 = \frac{1}{\sqrt{\varepsilon_{rx}} - \sqrt{\varepsilon_{ry}}} \frac{\lambda_0}{4} = 5.68\ \mu\text{m}. \quad (9.203)$$

As $\delta_1 = +90°$, this is an LHCP wave, Fig. 9.19(b). Note that crystal pieces cut at such length are used in optics to generate circularly polarized light at the output.

(b) If the crystal is made to be twice as long ($d_2 = 2d_1 = 11.36\ \mu$m), the phase shift between components, Eq. (9.202), doubles as well, so $\delta_2 = 2\delta_1 = 180°$. Eq. (9.190) then tells us that the output wave is now linearly polarized, like the input wave. However, since the field transformation across the crystal can in this case be represented as

$$\mathbf{E}_{\text{in}} = E'(\hat{\mathbf{x}} + \hat{\mathbf{y}}) \quad \longrightarrow \quad \mathbf{E}_{\text{out}} = E''(\hat{\mathbf{x}} - \hat{\mathbf{y}}) \quad (\delta_2 = 180°), \quad (9.204)$$

we realize that the output electric field vector is polarized (directed) along a different line – that is perpendicular to the polarization axis of the LP wave entering the crystal. Obviously, anisotropic crystals of this length can be used to change the electric field direction, i.e., to rotate the field (by 90°).

Problems: 9.39–9.43; *Conceptual Questions* (on Companion Website): 9.36–9.40; *MATLAB Exercises* (on Companion Website).

Problems

9.1. 3-D Helmholtz equations from Maxwell's equations. Derive both 3-D Helmholtz equations for a lossless electromagnetic medium, Eqs. (9.8) and (9.9), from the source-free version of Maxwell's complex differential equations, Eqs. (8.81).

9.2. Field expressions for a different propagation direction. Consider a uniform plane (not necessarily time-harmonic) electromagnetic wave propagating in free space in the negative x direction, with the electric field having only a y-component, E_y. (a) Write down the expression for E_y, and show that it satisfies the corresponding 1-D wave equation. (b) Use Maxwell's equations to obtain the accompanying magnetic field vector component from E_y, and verify that the relationships in Eqs. (9.22) hold true for the result. (c) What is the total electromagnetic energy density of the wave, and what its Poynting vector?

9.3. Complex expressions for an x-directed wave travel. Repeat the previous problem but for a time-harmonic plane wave, traveling in the negative x direction. Parts (a) and (b) should be done in the complex domain. In part (c), use time-average energy and power density expressions.

9.4. Plane wave in a lossless nonmagnetic medium. The complex rms magnetic field intensity vector of an electromagnetic wave propagating through a lossless nonmagnetic medium is given by $\underline{\mathbf{H}} = \mathrm{e}^{-\mathrm{j}\pi x}\,\hat{\mathbf{z}}$ A/m (x in m) at a frequency of $f = 75$ MHz. Find: (a) the direction of propagation, time period, wavelength, and phase velocity of the wave, the relative permittivity and intrinsic impedance of the medium, (b) the complex electric field intensity vector, and instantaneous electric and magnetic field vectors of the wave, and (c) the instantaneous and time-average electromagnetic energy densities and complex Poynting vector of the wave.

9.5. Plane wave computation in free space. A uniform plane wave travels through free space, and its instantaneous electric field intensity vector is expressed as $\mathbf{E} = 15\cos(\omega t + 10\pi z + \theta_0)\,\hat{\mathbf{y}}$ V/m (t in s; z in m). The magnetic field intensity of the wave amounts to $H = 0.02$ A/m at $t = 0$ and $z = 1.15$ m. Determine (a) the operating frequency and initial (for $t = 0$) phase in the plane $z = 0$ of the electric field intensity (θ_0), as well as (b) the complex magnetic field intensity vector and (c) the time-average Poynting vector, of the wave.

9.6. Finding material parameters of a propagation medium. Field vectors of an electromagnetic wave traveling through a lossless medium are given by $\mathbf{E} = 1.333\cos(\omega t - \beta y)\,\hat{\mathbf{x}}$ V/m and $\mathbf{H} = -4.243\cos(\omega t - \beta y)\,\hat{\mathbf{z}}$ mA/m. At a point of space and instant of time, the surface power density transported by the wave and electromagnetic energy density of the wave amount to $\mathcal{P} = 1.7$ mW/m^2 and $w_{\rm em} = 16.97$ pJ/m^3, respectively. What are the relative permittivity and permeability of the medium?

9.7. Complex emf in a large contour due to a plane wave. Redo Example 9.8 but in the complex domain.

9.8. Complex emf in a small contour. Redo Example 9.9 but in the complex domain.

9.9. Large contour positioned obliquely w.r.t. the wave travel. A square contour of edge length $a = 1$ m is placed in free space in the field of a uniform plane time-harmonic electromagnetic wave of frequency $f = 500$ MHz and rms electric field intensity $E_0 = 5$ V/m so that its plane makes an angle of $\alpha = 30°$ with the direction of wave propagation, as shown in Fig. 9.22. Find the emf induced in the contour, from (a) the left- and (b) the right-hand side of Faraday's law of electromagnetic induction in integral form, i.e., using the electric and magnetic field of the wave, respectively.

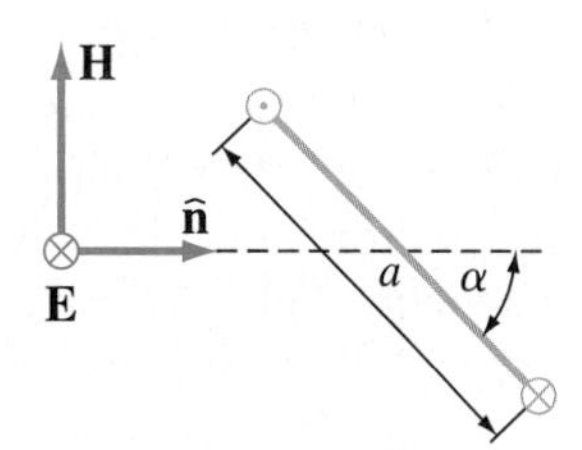

Figure 9.22 Large square contour positioned obliquely with respect to the direction of propagation of a uniform plane wave; for Problem 9.9.

9.10. Electrically small oblique contour. Repeat the previous problem but for the contour edge length of $a = 2$ cm. In part (b), assume that the magnetic field of the wave is uniform over a surface bounded by the contour (in Fig. 9.22) to directly compute the induced emf. In part (a), specialize the result from the previous problem to obtain the same result as in (b).

9.11. Induced current in a small wire loop. A uniform plane time-harmonic electromagnetic wave propagates in free space carrying a time-average power of $\mathcal{P}_{\rm ave} = 1$ W/m^2 per unit area of the wavefront, at a frequency of $f = 40$ MHz. A circular wire loop of radius $a = 10$ cm is positioned in the field of the wave such that the rms emf induced in it is maximum. The inductance and resistance of the loop are $L = 0.6$ μH and $R = 0.8$ Ω, respectively. Compute the rms intensity of the current induced in the loop, neglecting the magnetic field it produces (use the impedance of the loop).

9.12. Circulation of the magnetic field vector of a plane wave. Consider the electrically large rectangular contour in the field of a traveling plane wave in Fig. 9.5. Find (a) the circulation (line integral) of the magnetic field intensity vector of the wave along the contour and (b) the flux of the electric field vector through a surface spanned over the contour. (c) Can the result in (b) be obtained directly from that in (a), and vice versa? (d) For what position of the contour in Fig. 9.5 would the integral in (a) be maximum?

9.13. Displacement current bounded by a large contour. A uniform plane time-harmonic electromagnetic wave of angular frequency ω and rms magnetic field intensity H_0 travels through a lossless medium of permittivity ε and permeability μ. (a) Compute the line integral of the complex magnetic field intensity vector of the wave along a square contour of edge length a (arbitrary electrical size) positioned such that its plane is perpendicular to the electric field vector of the wave, and that two of its edges are parallel to the magnetic field vector. (b) What is the total rms displacement current intensity through a surface spanned over the contour?

9.14. Displacement current bounded by a small contour. Repeat the previous problem but for an electrically small circular contour of surface area S (the contour plane is still perpendicular to the electric field vector of the wave).

9.15. Power flow through a large rectangular aperture. A uniform plane time-harmonic electromagnetic wave propagating in free space has an amplitude of the electric field intensity $E_{\rm m} = 1$ V/m and frequency $f = 30$ GHz. The wave is incident normally onto an infinitely large flat screen (the screen is perpendicular to the direction of wave propagation) with an electrically large rectangular aperture (opening) of edge lengths $a = 20$ cm and $b = 10$ cm. (a) Neglecting the scattered electromagnetic field, due to induced electric surface currents and charges on the screen, calculate the energy delivered by the wave to the other side of the screen in one hour. (b) Repeat (a) but for the wave propagation direction making an angle of $\alpha = 60°$ with the normal to the screen (or the aperture).

9.16. **Safety limits for human exposure to electromagnetic radiation.** There exist various effects of electromagnetic fields on humans and many quite different recommendations and standards for the safety limits of human exposure to electromagnetic radiation. One of them is the IEEE (Institute of Electrical and Electronics Engineers) standard, whose recommendations for the maximum permissible time-average intensity of the Poynting vector in uncontrolled environments (where individuals generally have no knowledge or control of their exposure to electromagnetic fields) at frequencies between 100 MHz and 300 GHz are as follows: (i) $(\mathcal{P}_{\text{ave}})_{\text{max}} = 2\ \text{W/m}^2$ for $100\ \text{MHz} \leq f < 300\ \text{MHz}$, (ii) $(\mathcal{P}_{\text{ave}})_{\text{max}} = (f/150)\ \text{W/m}^2$ (f in MHz) for $300\ \text{MHz} \leq f < 15\ \text{GHz}$, and (iii) $(\mathcal{P}_{\text{ave}})_{\text{max}} = 100\ \text{W/m}^2$ for $15\ \text{GHz} \leq f \leq 300\ \text{GHz}$. Based on this standard, compute the maximum permissible levels for the rms intensities of the electric and magnetic fields in air at frequencies of $f_1 = 150$ MHz, $f_2 = 1.5$ GHz, and $f_3 = 15$ GHz, respectively.

9.17. **Unsafe zone around a radar antenna.** A radar antenna radiates a 15-GHz electromagnetic wave with the amplitude of the electric field intensity approximately given by the following function of the distance r from the radar: $E_{\text{m}}(r) = (5/r)$ kV/m (r in m). What is the radius of the unsafe zone around the antenna, according to the IEEE standard presented in the previous problem?

9.18. **FCC limit for EMI radiation.** All electrical and electronic equipment must be tested against stringent requirements for controlling electromagnetic interference (EMI), i.e., unintentional electromagnetic radiation from the equipment, which can interfere with other devices and systems, and degrade and jeopardize their performance and operation. A requirement of the FCC (U.S. Federal Communications Commission) is that the EMI radiation at a distance of 3 m from the equipment be less than $E_{\text{EMI}}(r = 3\ \text{m}) = 100\ \mu\text{V/m}$ (peak-value). Calculate the corresponding time-average power density, $\mathcal{P}_{\text{ave}}$. How does this FCC "interference-free" safe power level for equipment compare to the IEEE health-safety level for humans?

9.19. **Propagation along the main diagonal of the first octant.** Repeat Example 9.12 but for the direction of wave propagation defined by the vector $\hat{\mathbf{x}} + \hat{\mathbf{y}} + \hat{\mathbf{z}}$ and the complex electric field vector of the wave at the coordinate origin expressed as $\underline{\mathbf{E}}_0 = (1 - \text{j}\sqrt{3})(-\hat{\mathbf{x}} - \hat{\mathbf{y}} + 2\,\hat{\mathbf{z}})$ V/m.

9.20. **More on Helmholtz equations for a lossy medium.** For a lossy homogeneous medium and time-harmonic variation of the electromagnetic field, (a) derive the Helmholtz equation for the electric field, in Eqs. (9.86), from the governing Maxwell's equations and (b) show that the field in (9.81) is its solution, and (c) then use this solution and Maxwell's equations to obtain the accompanying magnetic field.

9.21. **Different propagation direction in a lossy medium.** A uniform plane time-harmonic electromagnetic wave propagates in the negative y direction through a lossy medium of parameters ε, μ, and σ. The electric field of the wave has only a z-component, $\underline{E}_z$, whose complex rms value at the coordinate origin is $\underline{E}_0$. (a) Write down the expression for $\underline{E}_z$, and show that it satisfies the corresponding 1-D Helmholtz equation. (b) Use Maxwell's equations to obtain the accompanying complex magnetic field vector component from $\underline{E}_z$, and verify that the relationships in Eqs. (9.22), employing the complex intrinsic impedance of the medium, Eq. (9.91), hold true for the result. (c) What is the time-average Poynting vector of the wave?

9.22. **3-D Helmholtz equations for lossy medium.** (a) Derive 3-D Helmholtz equations for a lossy medium, Eqs. (9.86), from the corresponding Maxwell's equations. (b) Comparing Helmholtz equations in (a) with their versions for a lossless medium (with $\sigma = 0$), Eqs. (9.8) and (9.9), obtain the equivalent complex permittivity of the lossy medium, as is done with Ampère's law in Eqs. (9.77)–(9.80).

9.23. Power ratio of two waves in decibels. The rms electric field intensities of two time-harmonic uniform plane waves propagating in a lossless medium are E_1 and E_2, respectively, and the corresponding time-average Poynting vector magnitudes are $\mathcal{P}_1$ and $\mathcal{P}_2$. (a) Compute the power ratio for the two waves in decibels ($A_{\rm dB}$), Eq. (9.98), for the following values of E_1/E_2: 100, 10, 2, 1.41, 1, 0.707, 0.5, 0.1, and 0.01. (b) Find $\mathcal{P}_1/\mathcal{P}_2$ and E_1/E_2 for the following values of $A_{\rm dB}$: 60 dB, 14 dB, 6 dB, 1 dB, 0 dB, −3 dB, −14 dB, and −100 dB.

9.24. Finding parameters of a lossy medium from wave travel. The instantaneous electric field intensity vector of a wave traveling in a lossy nonmagnetic medium is given by $\mathbf{E} = \mathrm{e}^{\alpha y}\cos(6.28\times10^9 t + 204y)\,\hat{\mathbf{x}}$ V/m (t in s; y in m). The magnetic field of the wave lags the electric field by 21°. Under these circumstances, find (a) the relative permittivity and conductivity of the medium, as well as (b) the complex propagation coefficient, (c) the instantaneous magnetic field intensity vector, and (d) the time-average Poynting vector of the wave.

9.25. Finding parameters of a biological tissue. A uniform plane time-harmonic electromagnetic wave of frequency $f = 1.9$ GHz propagates in the positive z direction through a biological tissue of unknown parameters. The magnetic field of the wave has only an x-component, and its rms intensity at the coordinate origin is $H_0 = 25$ mA/m. The wave amplitude is reduced by 3.25 dB for every centimeter traveled, and the phase coefficient of the wave amounts to $\beta = 260$ rad/m. Determine (a) the permittivity and conductivity of the tissue and (b) the time-average Poynting vector of the wave.

9.26. Finding parameters of glacier ice. The phase velocity and phase lag of the magnetic field behind the electric field of a uniform plane time-harmonic electromagnetic wave of frequency $f = 10$ GHz and rms electric field intensity $E_0 = 15$ V/m propagating through glacier ice are found to be $v_{\rm p} = 1.73\times10^8$ m/s and $\phi = 5\times10^{-4}$ rad, respectively. Compute: (a) the relative permittivity and loss tangent of ice, as well as (b) the complex propagation coefficient and (c) the time-average Poynting vector of the wave.

9.27. Wave absorption in a phantom head. To investigate the electromagnetic coupling of cellular telephone antennas and a human head, a phantom head – a plastic container filled with a solution that approximately resembles the dielectric and conductive properties of a human head – is used for measurements. In particular, solutions are made that have the relative permittivity and loss tangent equal to the corresponding average head tissue parameters at two frequency bands allocated for wireless communications in North America: (i) $\varepsilon_{\rm r} = 44.8$ and $\tan\delta_{\rm d} = 0.408$ at $f = 835$ MHz and (ii) $\varepsilon_{\rm r} = 41.9$ and $\tan\delta_{\rm d} = 0.293$ at $f = 1.9$ GHz. (a) Find the attenuation coefficient of a uniform plane wave propagating through the phantom solution and complex intrinsic impedance of the material, at each of the two wireless communication frequencies. (b) If the rms electric field intensity of the wave at its entry into the solution is $E_0 = 50$ V/m, use Poynting's theorem in complex form, Eq. (8.196), to determine the time-average power absorbed (lost to heat) in the first 1 cm of depth into the material per 1 cm^2 of cross-sectional area, that is, in the first 1 cm × 1 cm × 1 cm of the material past the interface, at each of the frequencies.

9.28. Various combinations of material parameters and frequency. For each of the combinations of material parameters and frequency from Example 9.16, calculate the attenuation and phase coefficients, dB attenuation per meter traveled, wavelength, and phase velocity of a uniform plane wave propagating in the material, as well as the complex intrinsic impedance of the material. Use approximate (simpler) expressions for good dielectrics or good conductors whenever appropriate.

9.29. 1/1000 depth of penetration in seawater. Find the ocean depth at which the electric field amplitude of a radio wave decreases to 1/1000 of its value at the ocean surface, as well as the phase velocity, wavelength, and index of refraction in the medium, at each of the frequencies $f_1 = 1$ kHz, $f_2 = 10$ kHz, $f_3 = 100$ kHz, $f_4 = 1$ MHz, and $f_5 = 10$ MHz, taking $\varepsilon_r = 81$ and $\sigma = 4$ S/m for seawater.

9.30. Radio communication with a submerged submarine. Repeat Example 9.19 but for (a) a 1-kHz ULF (ultra low frequency) wave launched into the ocean from the ship and (b) a less salty water with $\sigma = 0.4$ S/m (near a river delta), respectively; all other parameters of the system are the same.

9.31. Decibel attenuation of a microwave oven wall. (a) What is the one-percent depth of penetration into the stainless steel ($\sigma = 1.2$ MS/m, $\mu_r = 500$, and $\varepsilon_r = 1$) wall of a microwave oven at the standard frequency for microwave cooking, $f = 2.45$ GHz? (b) Calculate the dB attenuation of a 1-mm thick wall, and evaluate its shielding effectiveness.

9.32. Wave incidence from space onto a parabolic ionospheric slab. Consider the parabolic ionospheric slab in Fig. 9.14, and assume that $h_m = 250$ km, $d = 100$ km, and $N_m = 10^{12}$ m^{-3}. For a uniform plane wave vertically incident from space onto the upper boundary of the slab, find if and where (at what altitude h) it bounces back to space at each of the frequencies separated by steps of $\Delta f = 0.5$ MHz within a frequency range 8 MHz $\leq f \leq$ 12 MHz.

9.33. Wave propagation in a linear ionospheric layer. A uniform plane electromagnetic wave of frequency $f = 5$ MHz emitted by an antenna at the earth's surface is incident normally on a layer of the ionosphere whose concentration of free electrons can be approximated by the following linear function of the altitude above the earth's surface: $N(h) = N_0(h - h_1)/(h_2 - h_1)$ for $h_1 \leq h \leq h_2$, where $h_1 = 100$ km, $h_2 = 200$ km, and $N_0 = 5 \times 10^{11}$ m^{-3}, as shown in Fig. 9.23. Find the highest altitude h reached by the wave, at which it bounces off and returns to the antenna.

9.34. Wave dispersion in a lossy nonmagnetic medium. For the plane wave in a lossy nonmagnetic medium (rural ground) from Example 9.13, find the phase and group velocities at the operating frequency of the wave, and whether the medium is nondispersive, normally dispersive, or anomalously dispersive at that frequency.

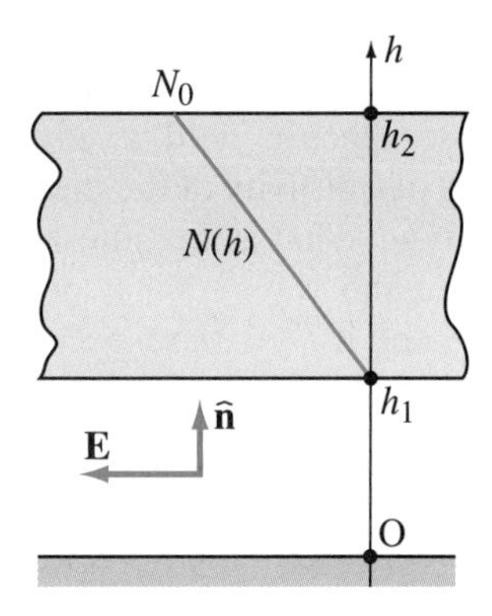

Figure 9.23 Normal incidence of a uniform plane wave launched from the earth's surface on an ionospheric layer with a linear profile $N(h)$; for Problem 9.33.

9.35. Phase velocity as a function of wavelength. The phase velocity of a uniform plane wave propagating in a lossless medium is given by $v_p(\lambda) = k\sqrt{\lambda}$, where λ is the wavelength in the medium and k is a constant. If $v_p = 10^8$ m/s at a given frequency, what is the group velocity of the wave at that frequency? Is the dispersion of the medium normal or anomalous?

9.36. Frequency-dependent index of refraction. The index of refraction of a lossless nonmagnetic medium is expressed as $n(f) = n_1 + (n_2 - n_1)(f - f_1)/(f_2 - f_1)$ between frequencies f_1 and f_2 ($f_1 < f_2$), where n_1 and n_2 are positive constants. In this frequency range, find the phase and group velocities of a propagating plane wave, and whether the medium is nondispersive, normally dispersive, or anomalously dispersive.

9.37. Time lag between signals due to dispersion. The carrier frequencies of two narrow-band signals launched at the same instant of time to travel as free (unguided) waves along the same path in a marshy soil of parameters $\varepsilon_r = 20$, $\sigma = 10^{-2}$ S/m, and $\mu_r = 1$ are $f_1 = 40$ kHz and $f_2 = 50$ kHz, respectively. What is the time lag between the signals as they are received on the other end of the path, if its length is $l = 50$ m?

9.38. Time lag between signals in a plasma medium. (a) Repeat the previous problem but for the signals propagating in a plasma medium with concentration of free electrons of $N = 10^7$ m^{-3}. (b) Then repeat (a) but for the carrier frequencies of signals changed to $f_1 = 40$ MHz and $f_2 = 50$ MHz, respectively.

9.39. Instantaneous Poynting vector of an elliptically polarized wave. Find the instantaneous

Poynting vector of an elliptically polarized time-harmonic uniform plane electromagnetic wave whose electric field is given by Eqs. (9.187), if the permittivity of the medium is ε and permeability μ. What is the time-average Poynting vector of the wave? What are both results if $E_1 = E_2$ and if $E_2 = 0$, respectively?

9.40. Superposition of two counter-rotating CP waves. Two oppositely rotating, right- and left-handed, circularly polarized time-harmonic uniform plane electromagnetic waves of the same frequency travel in the same direction, and the amplitudes of their electric field intensity vectors are E_1 and E_2, respectively. Determine the type (linear, circular, or elliptical) and handedness (right- or left-handed) of the polarization of the resultant wave (obtained by superposition of the two CP waves) for different combinations of values of E_1 and E_2.

9.41. Determination of polarization state in different cases. The electric field intensity vectors of two linearly polarized electromagnetic waves traveling in unison through free space are given by $\underline{\mathbf{E}}_1 = \mathrm{e}^{-\mathrm{j}\pi(z-0.25)}\,\hat{\mathbf{x}}$ V/m and $\underline{\mathbf{E}}_2 = a\,\mathrm{e}^{-\mathrm{j}\pi(z+0.25b)}\,\hat{\mathbf{y}}$ V/m (z in m), respectively, where a and b are constants. Determine the complex magnetic field intensity vector, instantaneous electric and magnetic field vectors, and polarization state (type and handedness) of the resultant wave in the following cases: (a) $a = 1$ and $b = 1$; (b) $a = -1$ and $b = 1$; (c) $a = -1$ and $b = 3$; (d) $a = 3$ and $b = 1$; (e) $a = -3$ and $b = 1$; and (f) $a = 3$ and $b = -1$.

9.42. More different polarization states. Consider the electromagnetic wave whose electric field vector is the sum of vectors $\underline{\mathbf{E}}_1$ and $\underline{\mathbf{E}}_2$ defined in the previous problem. Determine the polarization state of this wave for the following values of constants a and b in the expression for $\underline{\mathbf{E}}_2$: (a) $a = 0.5$ and $b = 1$; (b) $a = 0.5$ and $b = -1$; (c) $a = -2$ and $b = 2$; (d) $a = 0$ and $b = 0$; (e) $a = 5$ and $b = 5$; (f) $a = 1$ and $b = -5$; (g) $a = 1$ and $b = 7$; (h) $a = 1$ and $b = 0$; and (i) $a = -1$ and $b = 0$.

9.43. Poynting vector of waves in various polarization states. Compute the time-average Poynting vector of all resultant waves, in cases (a)–(i), from the previous problem.

10 Reflection and Transmission of Plane Waves

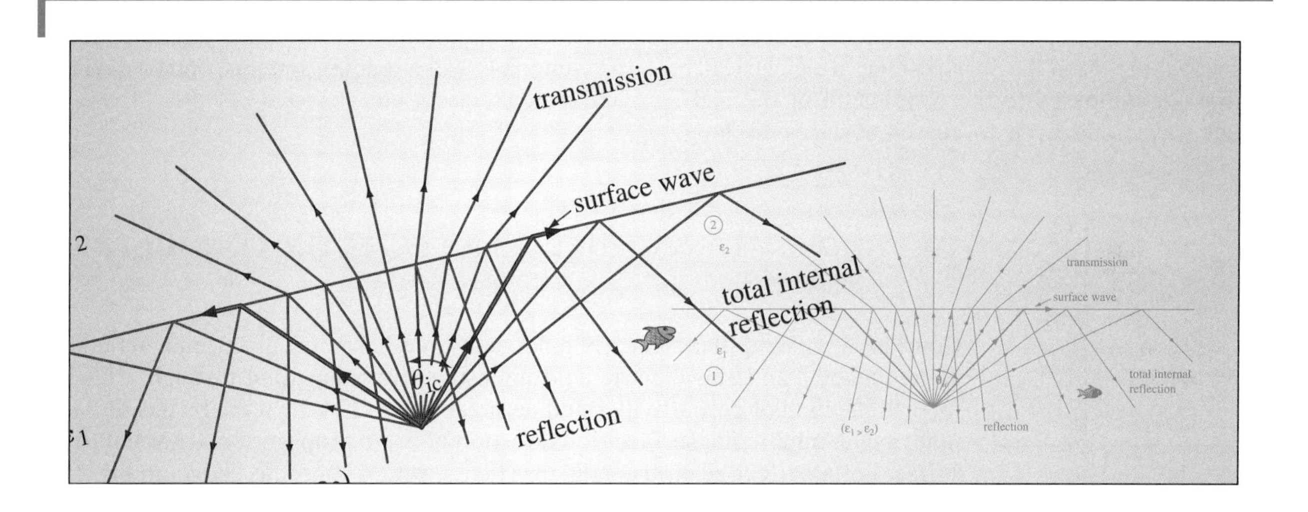

Introduction:

This chapter is a continuation of our studies of the theory and applications of uniform plane electromagnetic waves, in a two-chapter sequence. Capitalizing on the concepts and techniques of the analysis of wave propagation in homogeneous and unbounded media of various electromagnetic properties from the previous chapter, we now proceed to develop the concepts and techniques for the analysis of wave interaction with planar boundaries between material regions. In general, as a wave encounters an interface separating two different media, it is partly reflected back to the incident medium (wave reflection) and partly transmitted to the medium on the other side of the interface (wave transmission), and hence the title of this chapter. Our evaluations of reflection and transmission at material interfaces in various situations will, in addition to the underlying physical processes and mathematical relations between the local fields and waves at the interfaces, also include all implications to the distributions of total fields and waves in each of the material regions in the system, as well as the related energy and power considerations. In all problems, however, the core of the solution will be the use of appropriate general electromagnetic boundary conditions, as a "connection" between the fields on different sides of the interfaces.

The material will be presented as several separate cases of reflection and transmission (also referred to as refraction) of plane waves, in order of increasing complexity, from normal incidence (wave propagation direction is normal to the interface) on a perfectly conducting plane and normal incidence on a penetrable interface (between two arbitrary media), to oblique incidence (at an arbitrary angle) on these two types of interfaces, to wave propagation in multilayer media (with multiple interfaces). Along this way, we shall also discuss a number of related phenomena and problems, including the surface resistance of good conductors (associated with the skin effect) and the resulting perturbation method for evaluation of (small) losses in good conductors, total internal reflection (no transmission) at dielectric interfaces, total transmission (no reflection) at the Brewster incident angle, etc.

Although our discussions are confined, theoretically, to wave reflection and transmission in the presence of flat boundaries of infinite extent, they are practically applicable, with sufficient accuracy, also to finite-sized flat surfaces that are electrically large (as compared to the wavelength of the incident wave) and even to curved surfaces, as long as the radii of curvature are large in comparison to the wavelength. Even if these conditions, for a given material object impinged by a wave and a given frequency of the wave, are not met, the equations and results derived for planar boundaries can still be used, in a more or less approximative fashion, in solutions to much more complex problems than those involving only planar or almost planar surfaces. Overall, the topics of this chapter have countless immediate applications in such areas as indoor and outdoor wireless propagation, radio and microwave communication systems, radar engineering, antennas, wave-guiding systems, optical devices and systems, lasers, etc.

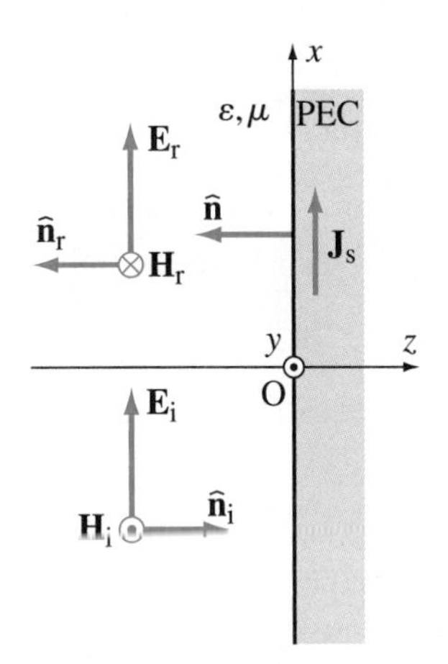

Figure 10.1 Normal incidence of a uniform plane time-harmonic electromagnetic wave on a planar interface between a perfect dielectric and a perfect conductor.

10.1 NORMAL INCIDENCE ON A PERFECTLY CONDUCTING PLANE

Consider a uniform plane linearly polarized[1] time-harmonic electromagnetic wave of frequency f and rms electric field intensity $E_{\rm i0}$ propagating through a lossless ($\sigma = 0$) medium of permittivity ε and permeability μ. Let the wave be incident normally on an infinite flat surface (the direction of wave propagation is normal to the surface) of a perfect electric conductor (PEC), with $\sigma \to \infty$, as shown in Fig. 10.1. Due to the presence of the conductor, in the way of the wave propagation, the electromagnetic field in the medium in front of it is changed. Namely, since the wave cannot penetrate into the conductor [the skin depth, Eq. (9.139), for perfect conductors is infinitely small, $\delta = 0$; see also Example 8.6], and the electromagnetic energy it carries cannot disappear, it must be bounced back. Our goal is to determine the total electromagnetic field in the entire half-space on the left-hand side of the PEC screen.

Let the plane $z = 0$ of the adopted Cartesian coordinate system in Fig. 10.1 be at the interface between the two media, and let complex electric and magnetic field intensity vectors of the wave, which we refer to as the incident (or forward) wave, be written as [see Eq. (9.36)]

incident wave

$$\underline{\mathbf{E}}_{\rm i} = \underline{E}_{\rm i0}\,{\rm e}^{-{\rm j}\beta z}\,\hat{\mathbf{x}} \quad \text{and} \quad \underline{\mathbf{H}}_{\rm i} = \frac{\underline{E}_{\rm i0}}{\eta}\,{\rm e}^{-{\rm j}\beta z}\,\hat{\mathbf{y}} \tag{10.1}$$

(the propagation unit vector is $\hat{\mathbf{n}}_{\rm i} = \hat{\mathbf{z}}$), where $\underline{E}_{\rm i0} = E_{\rm i0}\,{\rm e}^{{\rm j}\xi}$ (ξ is the initial phase of the electric field intensity of the wave for $z = 0$), $\beta = \omega\sqrt{\varepsilon\mu}$ [Eq. (8.111)] and $\eta = \sqrt{\mu/\varepsilon}$ [Eq. (9.21)] are, respectively, the phase coefficient of the wave and intrinsic impedance of the incident medium, and $\omega = 2\pi f$ is the angular or radian frequency of the wave. The wave excites currents to flow on the PEC surface.

[1] In what follows, in our statements of problems to be analyzed, the polarization state of time-harmonic uniform plane electromagnetic waves (see Section 9.14) will be assumed to be linear whenever it is not explicitly specified that the wave is circularly or elliptically polarized (which might as well be obvious from the given equations).

Namely, the incident electric field ($\underline{\mathbf{E}}_{\rm i}$), via the electric force in Eq. (3.1), compels free charge carriers in the conductor (electrons) to move (oscillate) along the electric field lines (much like in Fig. 3.1 for a static situation). This is equivalent to an x-directed current flow, as indicated in Fig. 10.1. However, since the conductor is assumed to be perfect, the current is confined to practically only the surface of the material (there can be no time-varying volume current inside perfectly conducting materials – see Example 8.6, and this is also evident from the zero skin depth), so we have surface currents of density $\underline{\mathbf{J}}_{\rm s} = \underline{J}_{\rm s}\,\hat{\mathbf{x}}$ flowing in the plane $z = 0$, over the PEC surface. These currents, in turn, are sources of a so-called scattered electromagnetic field, which can be represented as a reflected (or backward) wave, propagating in the negative z direction, opposite to the incident wave ($\hat{\mathbf{n}}_{\rm r} = -\hat{\mathbf{n}}_{\rm i}$).[2] Of course, there is no wave propagating to the right of the current sheet (for $z > 0$), within the PEC.

Let us adopt the reference direction for the electric field intensity vector of the reflected wave ($\underline{\mathbf{E}}_{\rm r}$) to be the same as that of the incident wave. Then, the reference direction for the reflected magnetic field intensity vector ($\underline{\mathbf{H}}_{\rm r}$) must be opposite to that of the incident one, as indicated in Fig. 10.1, because the reflected Poynting vector must be oriented backward (along $\hat{\mathbf{n}}_{\rm r}$). We now translate this picture of the reflected wave and its field vectors into equations:

$$\underline{\mathbf{E}}_{\rm r} = \underline{E}_{\rm r0}\,{\rm e}^{{\rm j}\beta z}\,\hat{\mathbf{x}} \quad \text{and} \quad \underline{\mathbf{H}}_{\rm r} = \frac{\underline{E}_{\rm r0}}{\eta}\,{\rm e}^{{\rm j}\beta z}(-\,\hat{\mathbf{y}}) \tag{10.2}$$

reflected wave

(note the minus sign in the expression for $\underline{\mathbf{H}}_{\rm r}$, coming, as explained, from the orientation of the associated Poynting vector), where $\underline{E}_{\rm r0}$ is the complex rms electric field intensity of the reflected wave for $z = 0$ (with respect to the same reference direction as for the incident wave). To determine this complex constant, we apply, in the plane $z = 0$, the boundary condition for the vector $\mathbf{E}$ (more precisely, for its tangential component, $\mathbf{E}_{\rm t}$) in Eqs. (8.33), for a surface of a perfect electric conductor in a dynamic electromagnetic field. Here, it tells us that the tangential component of the total (incident plus scattered) electric field intensity vector over the material interface, for $z = 0$, must be zero. Note that the incident wave alone would not be able to satisfy this boundary condition, which confirms that there must be another (reflected) wave in front of the PEC surface, radiated by the current sheet of density $\underline{\mathbf{J}}_{\rm s}$. Since both $\underline{\mathbf{E}}_{\rm i}$ and $\underline{\mathbf{E}}_{\rm r}$ are entirely tangential to the boundary, the condition is simplified to

$$\hat{\mathbf{n}} \times \underline{\mathbf{E}}_{\rm tot} = 0 \quad \longrightarrow \quad \left(\underline{\mathbf{E}}_{\rm i} + \underline{\mathbf{E}}_{\rm r}\right)\big|_{z=0} = 0, \tag{10.3}$$

where $\hat{\mathbf{n}}$ is the normal unit vector on the surface, directed from the PEC to the other medium. This yields

$$\underline{E}_{\rm i0} + \underline{E}_{\rm r0} = 0, \tag{10.4}$$

boundary condition

[2]Note that, in this chapter, like in the previous one, the time-harmonic plane electromagnetic waves are analyzed in the steady state, that is, after all the initial transient processes have already occurred, and the resultant steady-state sinusoidal fields have been established in the entire domain considered. Therefore, in the problem in Fig. 10.1, we are not looking at the transient processes of incident wave first approaching the PEC screen [of course, with the velocity of electromagnetic wave propagation in the given material, Eq. (9.18)], then the currents being induced in the screen, and finally the reflected wave traveling back – as a sequence of events in time. Instead, we are looking at the steady state with the sinusoidal incident field, surface currents, and reflected field already established in the entire incident half-space (fields) or PEC plane (currents) and coexisting together at all times as different components of the overall time-harmonic solution to the problem.

or $\underline{E}_{r0} = -\underline{E}_{i0}$ (the scattered electric field at $z = 0$ is exactly the same in amplitude as the incident field, but is 180° out of phase).[3] Hence, the total electric and magnetic field intensity vectors in the medium in front of the PEC (for $z \leq 0$), i.e., the field vectors of the resultant (incident plus reflected) plane wave, are

$$\underline{\mathbf{E}}_{tot} = \underline{\mathbf{E}}_i + \underline{\mathbf{E}}_r = \underline{E}_{i0}\left(e^{-j\beta z} - e^{j\beta z}\right)\hat{\mathbf{x}}, \tag{10.5}$$

$$\underline{\mathbf{H}}_{tot} = \underline{\mathbf{H}}_i + \underline{\mathbf{H}}_r = \frac{\underline{E}_{i0}}{\eta}\left(e^{-j\beta z} + e^{j\beta z}\right)\hat{\mathbf{y}}. \tag{10.6}$$

The resulting two-term complex expressions in parentheses are proportional to either sine or cosine of βz. Namely, using Euler's identity, Eq. (8.61), it is easily shown that

$$e^{j\phi} - e^{-j\phi} = 2j\sin\phi \quad \text{and} \quad e^{j\phi} + e^{-j\phi} = 2\cos\phi, \tag{10.7}$$

with which,

complex fields of a standing wave

$$\boxed{\underline{\mathbf{E}}_{tot} = -2j\underline{E}_{i0}\sin\beta z\,\hat{\mathbf{x}} \quad \text{and} \quad \underline{\mathbf{H}}_{tot} = 2\frac{\underline{E}_{i0}}{\eta}\cos\beta z\,\hat{\mathbf{y}}.} \tag{10.8}$$

Using Eq. (8.66), and recalling that $j = e^{j\pi/2}$ and $\cos(\alpha + \pi/2) = -\sin\alpha$, we convert the complex field expressions in Eqs. (10.8) to their time-domain[4] counterparts:

instantaneous fields of a standing wave

$$\boxed{\mathbf{E}_{tot}(t) = 2\sqrt{2}E_{i0}\sin\beta z\sin\omega t\,\hat{\mathbf{x}} \quad \text{and} \quad \mathbf{H}_{tot}(t) = 2\sqrt{2}\,\frac{E_{i0}}{\eta}\cos\beta z\cos\omega t\,\hat{\mathbf{y}},} \tag{10.9}$$

where a zero initial phase of the incident electric field intensity in the $z = 0$ reference plane, $\xi = 0$, is assumed for simplicity, so that $\underline{E}_{i0}$ is a purely real constant ($\underline{E}_{i0} = E_{i0}$).[5] Fig. 10.2 shows snapshots at different time instants of the resultant electric and magnetic field intensities in Eqs. (10.9) as a function of z. We note that there are planes in which $E_{tot}(t)$ is zero at all times. These planes are defined by

planes of zero electric field

$$\boxed{\sin\beta z = 0 \quad \longrightarrow \quad \beta z = -m\pi \quad \longrightarrow \quad z = -m\frac{\lambda}{2} \quad (m = 0, 1, 2, \ldots),} \tag{10.10}$$

[3] Apart from the solution using the boundary condition in Eq. (10.3), the existence of a backward propagating wave in Fig. 10.1, as well as the mathematical expressions in Eqs. (10.2) and (10.4) describing this wave, can alternatively be deduced by considering an equivalent structure with the infinite current sheet (of density $\underline{\mathbf{J}}_s$) radiating in both directions away from the sheet. Namely, since there is no electromagnetic field in the half-space $z > 0$ in Fig. 10.1, we can substitute the PEC by the material occupying the other half-space, and consider the surface currents flowing in the plane $z = 0$ to exist in an unbounded homogeneous medium (of parameters ε, μ, and $\sigma = 0$). In this equivalent model (more precisely, the model is equivalent to the original structure only for $z < 0$), the currents must be such that, for $z > 0$, the field they produce exactly cancels the incident field in Eq. (10.1) (zero total field requirement). This means that the currents radiate to the right a uniform plane wave (propagating along with the incident wave) that has the electric field vector equal in amplitude but opposite in polarity (180° out of phase) to that of the incident wave, and the same is true for the magnetic field vectors of the two waves. However, because of symmetry, the currents radiate another wave back into the $z < 0$ half-space – the reflected wave in Eq. (10.2). Its electric field intensity vector near the $z = 0$ symmetry plane, that is, for $z = 0^-$, must be, again due to symmetry, the same as that of the wave radiated by the currents to the right for $z = 0^+$, which gives the relationship in Eq. (10.4).

[4] As has already been done at several places throughout the previous four chapters, we use here the notation $\mathbf{E}_{tot}(t)$, $\mathbf{H}_{tot}(t)$, etc., to emphasize the time-domain format of these field expressions, which, of course, does not mean that time is the only independent variable [e.g., $\mathbf{E}_{tot} = \mathbf{E}_{tot}(z, t)$].

[5] Otherwise, the time-dependent terms in Eqs. (10.9) would be $\sin(\omega t + \xi)$ and $\cos(\omega t + \xi)$ instead of $\sin\omega t$ and $\cos\omega t$, respectively, which essentially amounts to a simple shift of time reference ($t = 0$).

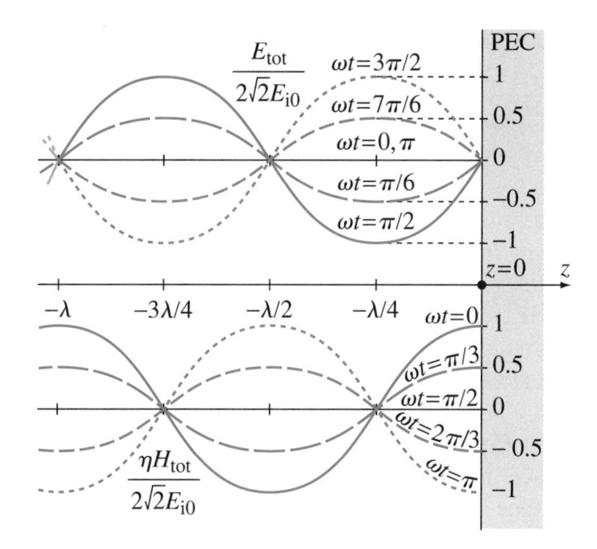

Figure 10.2 Plots of normalized total electric and magnetic field intensities, $E_{tot}/(2\sqrt{2}E_{i0})$ and $H_{tot}\eta/(2\sqrt{2}E_{i0})$, in Eqs. (10.9) against z at different instants of time.

where $\lambda = 2\pi/\beta$ is the wavelength [Eq. (8.112)] in the incident medium (with parameters ε and μ). Note that E_{tot} can be visualized as the one sine function of coordinate z, $\sin\beta z$, with a varying amplitude, $2\sqrt{2}E_{i0}\sin\omega t$. Similarly, $H_{tot}(t)$ is always zero in planes where $\cos\beta z = 0$, that is, in planes defined by $z = -(2m+1)\lambda/4$ $(m = 0, 1, 2, \ldots)$. This means that the fields in Eqs. (10.9) do not travel as the time advances, but stay where they are, only oscillating in time between the stationary zeros. In other words, they do not represent a traveling wave in either direction. The resultant wave, which is a superposition of two traveling waves with opposite directions of travel, is thus termed a standing wave.[6]

Note that, in general, standing waves are easily recognized by the absence of a propagation (retardation) argument of the form $t - l/c$ [Eq. (9.13)] or $\omega t - \beta l$ [Eq. (9.32)] in the time domain or the corresponding factor $e^{-j\beta l}$ in the complex domain, where l is an arbitrary length coordinate [see Eq. (9.69)]. Note also that, unlike for traveling waves, where the phases of both electric and magnetic fields depend on the spatial coordinate z, but as the same (linear) function of z, so that the two fields are in phase for every z, the electric field of a standing wave has a constant phase (same at all points), and this is also true for the magnetic field. However, these constant phases differ by 90° [$\sin\omega t$ vs. $\cos\omega t$ terms in time-domain field expressions in Eqs. (10.9) or an extra "j" in the expression for $\underline{\mathbf{E}}_{tot}$ when compared to that for $\underline{\mathbf{H}}_{tot}$, Eqs. (10.8)], so the two fields are in time-phase quadrature at every point of space.

Knowing the instantaneous and complex electric and magnetic field intensities of the resultant standing wave in the material half-space in front of the PEC screen in Fig. 10.1, we are now able to determine and discuss, in the rest of this section (including examples), its various properties, like we did for a traveling wave in an unbounded medium in Sections 9.3 and 9.4. Of course, some concepts and quantities to be analyzed, like the distribution of surface currents and charges in the PEC plane associated with the standing wave, do not have their counterparts in unbounded media.

[6]The wave in Fig. 10.2 is also referred to as a pure standing wave, since, as we shall see in later sections, there are also "impure" standing waves, with combined standing-wave and traveling-wave properties.

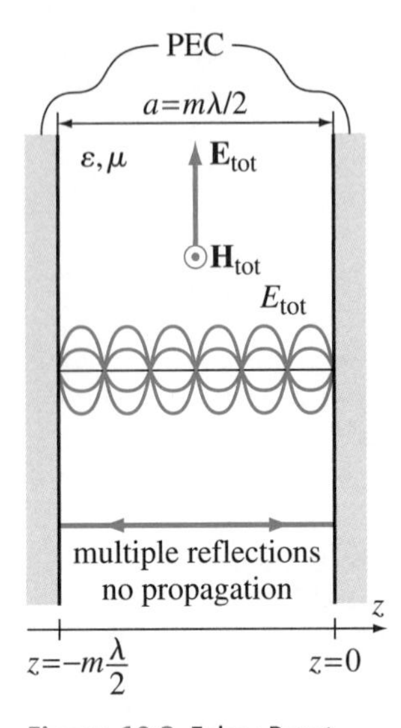

Figure 10.3 Fabry-Perot resonator, obtained by metalizing one of the planes defined by Eq. (10.10), for $m \neq 0$, in the field of a standing uniform plane wave in Figs. 10.1 and 10.2.

First, it is important to always have in mind that, unlike for traveling waves, the ratio of the electric and magnetic field intensities in either time or complex domain for a standing wave, e.g., that in Fig. 10.2, does not equal η,

$$\frac{E_{\text{tot}}(t)}{H_{\text{tot}}(t)} \neq \eta \quad \text{and} \quad \frac{\underline{E}_{\text{tot}}}{\underline{H}_{\text{tot}}} \neq \eta, \tag{10.11}$$

or any other constant [see Eqs. (10.9) and (10.8)], where $\underline{E}_{\text{tot}}$ and $\underline{H}_{\text{tot}}$ stand for complex rms field intensities, including the phase terms [in Eqs. (10.8), $\underline{\mathbf{E}}_{\text{tot}} = \underline{E}_{\text{tot}}\,\hat{\mathbf{x}}$ and $\underline{\mathbf{H}}_{\text{tot}} = \underline{H}_{\text{tot}}\,\hat{\mathbf{y}}$]. That is why, in Eqs. (10.8), we were not able to obtain $\underline{\mathbf{H}}_{\text{tot}}$ from $\underline{\mathbf{E}}_{\text{tot}}$, by dividing $\underline{E}_{\text{tot}}$ by η. Instead, $\underline{\mathbf{H}}_{\text{tot}}$ was found by adding together the incident and reflected magnetic field vectors, which, representing traveling waves, were both first computed by dividing by η the associated electric field intensities, Eqs. (10.1) and (10.2).

Let us now have another look at Eq. (10.10) specifying the planes of zero net electric field in the material half-space in front of the PEC plane, where we realize that we can insert another PEC surface (foil) in any of these planes, i.e., for any m (of course, except the trivial case of $m = 0$), and nothing will change in the entire half-space. This is because the electric field in Eqs. (10.9) is zero at all times for $z = -m\lambda/2$, so that the boundary condition for the vector $\mathbf{E}$ in Eqs. (8.33) is automatically satisfied. Note that the boundary condition for the vector $\mathbf{B}$ in Eqs. (8.33), stipulating that the normal component of $\mathbf{B} = \mu\mathbf{H}$ must be zero at a PEC surface, is also automatically satisfied, as the vector $\mathbf{H}_{\text{tot}}$ in Eqs. (10.9) is entirely tangential to the planes in Eq. (10.10). The PEC foil (at $z = -m\lambda/2$) subdivides the half-space into two mutually isolated regions, shown in Fig. 10.3. We can now remove the field from the region on the left-hand side of the foil, and so obtain a self-contained structure, on the right-hand side of the foil, with a standing electromagnetic plane wave trapped between the two parallel PEC planes (like two mirrors) that are a multiple of half-wavelengths apart. This structure behaves like an electromagnetic resonator, and is known as the Fabry-Perot resonator. It is used extensively in optics (e.g., in lasers) and at higher microwave frequencies. Note that the field in the structure can also be thought of as an infinite sequence of reflections of a traveling wave, back and forth (at normal incidence) between the mirrors, as indicated in Fig. 10.3. The resonant frequency, f_{res}, of the Fabry-Perot resonator, for its given length (separation between the PEC planes), a, is the frequency at which the wavelength in the medium (of parameters ε and μ), λ, satisfies the zero-field condition (at the resonator's newly added wall) $a = m\lambda/2$ (or $\beta a = m\pi$). Since $\lambda = c/f$, where $c = 1/\sqrt{\varepsilon\mu}$ [Eq. (9.18)] is the velocity of (traveling) electromagnetic waves in the medium, we have

resonant frequencies of a Fabry-Perot resonator

$$\boxed{f_{\text{res}} = m\frac{c}{2a} \quad (m = 1, 2, \ldots).} \tag{10.12}$$

Example 10.1 Current and Charge Distributions in an Illuminated PEC Screen

A uniform plane time-harmonic electromagnetic wave of angular frequency ω and rms electric field intensity E_{i0} is normally incident on a PEC plane from a lossless medium of intrinsic impedance η. Find the distribution of surface currents and charges in the plane.

Solution Knowing the total magnetic field intensity vector near the PEC surface, we compute the surface current density vector $\mathbf{J}_{\text{s}}$ in Fig. 10.1 as follows. From the boundary condition for the vector $\mathbf{H}$ (its tangential component) in Eqs. (8.33) and complex magnetic field expression in Eqs. (10.8),

finding surface currents on a PEC boundary

$$\boxed{\underline{\mathbf{J}}_{\text{s}} = \hat{\mathbf{n}} \times \underline{\mathbf{H}} = (-\hat{\mathbf{z}}) \times \underline{\mathbf{H}}_{\text{tot}}\big|_{z=0} = 2\,\frac{E_{\text{i0}}}{\eta}\,\hat{\mathbf{x}},} \tag{10.13}$$

where we assume that $\underline{E}_{i0} = E_{i0}$. The instantaneous surface current density vector is hence $\mathbf{J}_s(t) = (2\sqrt{2}E_{i0}/\eta)\cos\omega t\,\hat{\mathbf{x}}$, which, of course, can also be obtained directly by applying the same boundary condition to the instantaneous total magnetic field intensity vector in Eqs. (10.9). We see that the normally incident uniform plane wave induces a uniform current sheet in the PEC plane ($\mathbf{J}_s$ does not depend on x or y). Therefore, this current is not associated with any excess surface charge, which is obvious from the continuity equation for high-frequency surface currents (continuity equation for plates), Eq. (8.42), whose complex-domain version reads

$$\boxed{\nabla_s \cdot \underline{\mathbf{J}}_s = -j\omega\underline{\rho}_s.} \tag{10.14}$$

continuity equation for plates in complex form

Namely, the surface divergence of a constant vector is zero, and hence $\underline{\rho}_s = 0$. This is also obvious from the boundary condition for the vector $\mathbf{D}$ (namely, for its normal component, $\mathbf{D}_n$) in Eqs. (8.33),

$$\boxed{\underline{\rho}_s = \hat{\mathbf{n}} \cdot \underline{\mathbf{D}} = \varepsilon(-\hat{\mathbf{z}}) \cdot \hat{\mathbf{x}}\,\underline{E}_{\text{tot}}\big|_{z=0} = 0.} \tag{10.15}$$

finding surface charges on a PEC boundary

Example 10.2 Electric and Magnetic Energy Densities of a Standing Wave

Consider the standing electromagnetic wave in front of a PEC screen in Fig. 10.1. (a) Determine the expressions for instantaneous electric and magnetic energy densities of the wave, and plot them in space for three time instants that are an eighth of the time period apart. (b) Show that the total time-average electromagnetic energy density of the wave is constant in space.

Solution

(a) As in Eq. (9.24), the electric and magnetic instantaneous stored energy densities at a point (defined by the coordinate z in front of the conductor) are obtained from the instantaneous electric and magnetic field intensities at that point, which, here, are the field intensities of the resultant wave in Eqs. (10.9). Thus,

$$w_e(t) = \frac{1}{2}\varepsilon E_{\text{tot}}(t)^2 = 4\varepsilon E_{i0}^2 \sin^2\beta z \sin^2\omega t, \tag{10.16}$$

$$w_m(t) = \frac{1}{2}\mu H_{\text{tot}}(t)^2 = 4\varepsilon E_{i0}^2 \cos^2\beta z \cos^2\omega t. \tag{10.17}$$

instantaneous electric and magnetic energy densities of a standing wave

where, in the second equation, the use is made of the fact that $\mu/\eta^2 = \varepsilon$, like in Eq. (9.24), and the total electromagnetic energy density of the resultant wave is $w_{em} = w_e + w_m$. We note that, since $\sin^2\alpha = (1 - \cos 2\alpha)/2$ and $\cos^2\alpha = (1 + \cos 2\alpha)/2$, where α stands for either ωt or βz, the periodicity of both w_e and w_m in time and space (along z) is determined by 2ω and 2β, respectively, that is, at twice the corresponding repetition rates (in time and space) for fields [see also Eq. (9.44)]. However, the oscillations of energy densities in time are not time-harmonic, because their time averages are not zero (the 1/2 constant terms in the expansions of $\sin^2\omega t$ and $\cos^2\omega t$).

In Fig. 10.4, functions w_e and w_m in Eqs. (10.16) and (10.17) are plotted against z for time instants $t = 0$, $t = T/8$, and $t = T/4$, respectively, T being the time period of the incident wave in Eq. (8.49). These plots are another confirmation that there is no net transfer of electromagnetic energy in a standing wave – the energy does not travel, but oscillates back and forth along the z-axis between the peak locations of the electric and magnetic fields, respectively (like water slops in a pail). At one instant (e.g., $t = 0$), the magnetic field is maximum and the energy is all magnetic ($\cos\omega t = 1$), while the electric field and its energy density are zero everywhere ($\sin\omega t = 0$). One quarter of a period later (at $t = T/4$), the energy is all electric (the magnetic field is zero everywhere), with energy density maxima shifted in space by $\lambda/4$. After one more quarter-period, the energy is all back in the magnetic field, and so on. At intermediate times, the energy is moving from the electric to the magnetic field, and conversely (note that at $t = T/8$, the

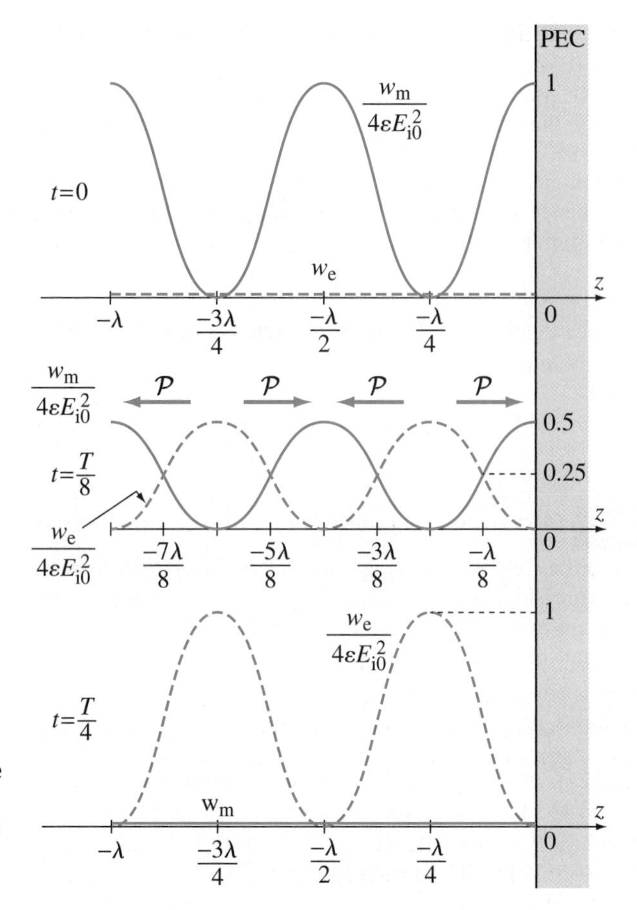

Figure 10.4 Sketch of instantaneous electric and magnetic energy densities, Eqs. (10.16) and (10.17), of the resultant wave in Fig. 10.1 (divided by $4\varepsilon E_{i0}^2$) as a function of z at three characteristic time instants; for Example 10.2.

energy is exactly half electric and half magnetic). So, this complete periodic exchange of stored energy between the electric and magnetic field of the standing electromagnetic wave is exactly what occurs in the Fabry-Perot resonator, Fig. 10.3, at its resonant frequency, Eq. (10.12), and such energy fluctuation is characteristic for all electromagnetic resonators, including resonant electric circuits (e.g., a simple LC circuit).[7]

(b) The time-average electric and magnetic energy densities of the resultant (standing) wave at a point with coordinate z $(-\infty < z < 0)$ are obtained, as in Eqs. (9.38), using the rms electric and magnetic field intensities, which in turn are found as magnitudes of complex field expressions in Eqs. (10.8). Hence, the total time-average electromagnetic energy density of the resultant wave is

time-average electromagnetic energy density of a standing wave

$$\boxed{(w_{\text{em}})_{\text{ave}} = (w_{\text{e}})_{\text{ave}} + (w_{\text{m}})_{\text{ave}} = \frac{1}{2}\,\varepsilon|\underline{\mathbf{E}}_{\text{tot}}|^2 + \frac{1}{2}\,\mu|\underline{\mathbf{H}}_{\text{tot}}|^2 = 2\varepsilon E_{i0}^2} \qquad (10.18)$$

$(\sin^2\alpha + \cos^2\alpha = 1)$, so it indeed is the same for every z. Of course, this same result can be obtained by averaging in time [see Eq. (6.95)] the instantaneous electromagnetic

[7]Note that in a simple resonant LC circuit, the current and magnetic field of an inductor are 90° out of phase (difference in "j" in complex expressions) with respect to the voltage and electric field of a capacitor [see Eq. (8.69) or (3.45)]. At the resonant angular frequency of the circuit, $\omega = 1/\sqrt{LC}$, the energy oscillates between the inductor (all magnetic energy) and capacitor (all electric energy), as the inductor current and capacitor voltage cyclically assume maximum/zero values [see Eqs. (7.88) and (2.192), as well as Eq. (8.207)].

energy density of the resultant wave, given by Eqs. (10.16) and (10.17), and shown in Fig. 10.4.

Example 10.3 Poynting Vector of a Standing Wave

For the same situation as in the previous example, find the expressions for the instantaneous, complex, and time-average Poynting vector, respectively, of the resultant wave, and discuss their standing-wave features. At what frequency does the instantaneous vector oscillate?

Solution From Eqs. (8.161) and (10.9), the resultant instantaneous Poynting vector in Fig. 10.1 is

$$\mathcal{P}(t) = \mathbf{E}_{\text{tot}}(t) \times \mathbf{H}_{\text{tot}}(t) = 8\,\frac{E_{\text{i0}}^2}{\eta}\sin\beta z\cos\beta z\sin\omega t\cos\omega t\,\hat{\mathbf{z}} = 2\,\frac{E_{\text{i0}}^2}{\eta}\sin 2\beta z\sin 2\omega t\,\hat{\mathbf{z}} \tag{10.19}$$

($2\sin\alpha\cos\alpha = \sin 2\alpha$). Obviously, much like the energy densities in Fig. 10.4, $\mathcal{P}(t)$ oscillates in time at twice the frequency of fields ($2f = \omega/\pi$), and similarly for spatial periodicity (2β instead of β). Of course, zeros of the Poynting vector encompass both electric and magnetic field intensity zeros in Fig. 10.2. Knowing that the direction of the vector $\mathcal{P}$ coincides with the direction of the wave propagation and that its instantaneous magnitude equals the time-rate with which the electromagnetic energy is carried by the wave per unit area of the wavefront, Eq. (10.19) provides one more proof that the resultant wave in Fig. 10.1 does not propagate, but is standing. It shows again that the energy is not transported by the wave in either direction, but just moved back and forth in an oscillatory fashion, within a distance of $\lambda/4$, between locations of electric and magnetic energy maxima. For instance, the expression for the Poynting vector at $t = T/8$ reads $\mathcal{P} = (2E_{\text{i0}}^2/\eta)\sin 2\beta z\,\hat{\mathbf{z}}$ ($\sin 2\omega t = 1$), and thus $\mathcal{P} = (2E_{\text{i0}}^2/\eta)\,\hat{\mathbf{z}}$ (forward direction) for $z = -(4m+1)\lambda/8$ and $\mathcal{P} = -(2E_{\text{i0}}^2/\eta)\,\hat{\mathbf{z}}$ (backward direction) for $z = -(4m+3)\lambda/8$, where $m = 0, 1, 2, \ldots$, as indicated in Fig. 10.4. This is an intermediate stage at which the energy is being redistributed, via the Poynting vector, from the all-magnetic energy stage at $t = 0$ to the all-electric stage at $t = t/4$ in Fig. 10.4. At these two latter stages, however, $\mathcal{P} = 0$ everywhere ($\sin 2\omega t = 0$) and the fluctuation of energy stops for a moment, as $\mathcal{P}$ changes the direction at all points, leading to the next all-electric or all-magnetic energy stage.

Using Eqs. (8.194) and (10.8), the complex Poynting vector of the standing wave is

$$\underline{\mathcal{P}} = \underline{\mathbf{E}}_{\text{tot}} \times \underline{\mathbf{H}}_{\text{tot}}^* = -\text{j}2\,\frac{E_{\text{i0}}^2}{\eta}\sin 2\beta z\,\hat{\mathbf{z}} \tag{10.20}$$

complex Poynting vector of a standing wave

($\underline{E}_{\text{i0}}\underline{E}_{\text{i0}}^* = E_{\text{i0}}^2$). We see that $\underline{\mathcal{P}}$ is purely imaginary, which, of course, is again in agreement with the fact that there is no net real power flow (by the resultant electromagnetic wave) in Fig. 10.1. It represents the reactive power fluctuation determining the rate of energy exchange in time and space between the electric and magnetic fields, like in every electromagnetic resonator. The time average of the instantaneous total Poynting vector [see Eq. (8.195)] is

$$\mathcal{P}_{\text{ave}} = \text{Re}\{\underline{\mathcal{P}}\} = 0, \tag{10.21}$$

which is evident also from Eq. (10.19).

Example 10.4 Energy in an Imaginary Cylinder and Poynting's Theorem

A time-harmonic uniform plane wave with the rms intensity of the electric field E_{i0} and wavelength λ travels in a dielectric with permittivity ε and is incident normally on a perfectly conducting plane. Compute the instantaneous resultant electromagnetic energy stored in an imaginary cylinder with the basis area S and length $l = \lambda/4$, placed in the dielectric along the wave travel such that one of its bases lies in the PEC plane.

Solution With the use of the expressions for electric and magnetic instantaneous stored energy densities (w_e and w_m) of the standing wave, Eqs. (10.16) and (10.17), the total instantaneous electromagnetic energy contained in the cylinder is computed as

$$W_{em}(t) = W_e(t) + W_m(t) = \int_{z=-l}^{0} w_e(z,t) \underbrace{S\,dz}_{dv} + \int_{-l}^{0} w_m(z,t)\,S\,dz$$

$$= 4\varepsilon E_{i0}^2 S \sin^2\omega t \underbrace{\int_{-\lambda/4}^{0} \sin^2\beta z\,dz}_{\lambda/8} + 4\varepsilon E_{i0}^2 S \cos^2\omega t \underbrace{\int_{-\lambda/4}^{0} \cos^2\beta z\,dz}_{\lambda/8}$$

$$= \frac{\varepsilon\lambda E_{i0}^2 S}{2}\left(\sin^2\omega t + \cos^2\omega t\right) = \frac{\varepsilon\lambda E_{i0}^2 S}{2}, \tag{10.22}$$

with the two integrals in z, evaluated in practically the same way as the integral in time (t) in Eq. (6.95), coming out to be the same, equal to $l/2 = \lambda/8$, which eliminates the time dependence in the result, so the total energy is constant in time.

Note that this fact, that $W_{em} = \text{const}$, can be obtained without actually computing the energy, namely, by applying instead Poynting's theorem, Eq. (8.159) or (8.167), to the cylinder. As there are no external energy volume sources (impressed electric fields or currents) nor losses inside the cylinder, the theorem yields

$$\frac{dW_{em}}{dt} = -\oint_{S_{cyl}} \mathcal{P} \cdot d\mathbf{S}, \tag{10.23}$$

where $\mathcal{P}$ is the Poynting vector of the standing wave in Fig. 10.1, given by Eq. (10.19). In addition, since $\mathcal{P}$ is tangential to the lateral surface of the cylinder and is zero at both cylinder bases, at $z = 0$ and $z = -\lambda/4$ ($2\beta\lambda/4 = \pi$), respectively, at all times, its outward net flux through the entire (closed) surface of the cylinder (S_{cyl}) is zero, and so is, from Eq. (10.23), the time rate of change of the energy in the cylinder – hence $W_{em} = \text{const}$. However, Poynting's theorem cannot give us the actual value of W_{em}, in Eq. (10.22).

Example 10.5 Standing Wave Reception by Small Loop and Short Dipole Antennas

A uniform plane electromagnetic wave is incident normally from air on the earth's surface, which can be assumed to be perfectly flat, of infinite extent, and perfectly conducting, so a PEC ground plane. The rms electric field intensity of the wave is $E_0 = 1$ V/m and its wavelength is $\lambda_0 = 20$ m. To receive the wave, that is, the signal it carries, a small wire loop antenna of surface area $S = 100$ cm^2 is placed at a height h with respect to the plane. (a) For what orientation and position (h) of the antenna is the rms emf induced in it maximum, and what is this maximum emf (neglecting the magnetic field due to the induced current in the loop)? (b) If, instead, the wave is to be received by a short wire dipole antenna (two straight wire arms of total length much smaller than λ_0 with a small gap between the antenna terminals), what are the orientation and position of this antenna for maximum reception?

Solution

(a) Of course, we cannot extract energy (signal) from the incident wave only, but from the total existing wave in air, and that is the standing electromagnetic wave in Fig. 10.1. In addition, although the associated electric and magnetic fields are just the two faces of the electromagnetic wave, and a loop antenna receives (if oriented and positioned properly) both fields equally, as discussed in Example 9.8, if the loop is small it is much simpler to evaluate the reception of the magnetic field, as shown in Example 9.9. Therefore, we consider the resultant magnetic field vector, $\underline{\mathbf{H}}_{tot}$, given in Eqs. (10.8). Having in mind Eq. (9.58), we then realize that for the maximum reception of the wave, the loop has to be oriented such that its surface is perpendicular to $\underline{\mathbf{H}}_{tot}$, as depicted in Fig. 10.5,

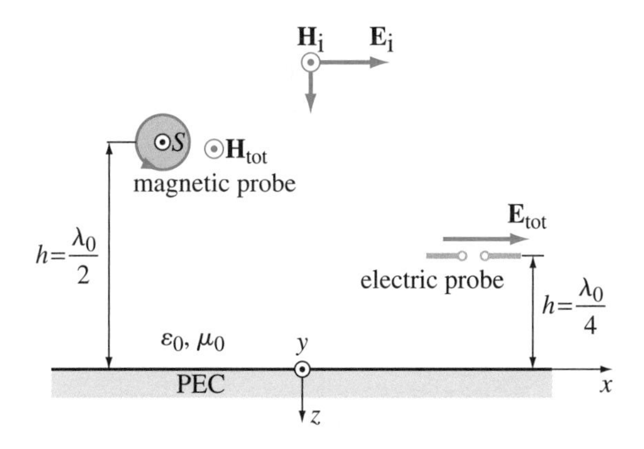

Figure 10.5 Reception of a uniform plane wave normally incident on the earth's surface (or a PEC ground plane) by a small loop antenna and short dipole antenna, respectively, with optimal orientations and positions; for Example 10.5.

with which, and the complex form of the expression for Faraday's law of electromagnetic induction from Eq. (9.59), the complex rms emf induced in the antenna comes out to be

$$\underline{\mathcal{E}}_{\text{ind}} = -\text{j}\omega\underline{\Phi} = -\text{j}\omega\underline{\mathbf{B}}\cdot\mathbf{S} = -\text{j}\omega\mu_0\underline{\mathbf{H}}_{\text{tot}}\cdot\mathbf{S} = \mp\text{j}\omega\mu_0\underline{H}_{\text{tot}}S = \frac{\mp 2\text{j}\omega\mu_0\underline{E}_{\text{i}0}S}{\eta_0}\cos\beta z$$

$$\text{for} \quad \mathbf{S} = \pm S\,\hat{\mathbf{y}}. \tag{10.24}$$

Furthermore, the magnitude of this emf is maximum when the antenna, also called a magnetic probe, is positioned at the magnetic field maxima of the standing wave, i.e., when $\cos\beta z = \pm 1$, and this, in turn, is exactly at electric field zeros, in Eqs. (10.10). The lowest corresponding height of the antenna above the earth's surface ($h = -z$) is thus a half-wavelength (Fig. 10.5), and, combining Eqs. (10.24), (9.23), and (8.111), the maximum rms emf amounts to

$$\left|\underline{\mathcal{E}}_{\text{ind}}\right|_{\max} = \frac{2\omega\mu_0|\underline{E}_{\text{i}0}|S}{\eta_0} = 2\beta E_{\text{i}0}S = \frac{4\pi E_{\text{i}0}S}{\lambda_0} = 6.28\ \text{mV} \quad \text{for} \quad h = \frac{\lambda_0}{2} = 10\ \text{m}. \tag{10.25}$$

The same result is, of course, obtained for h equal to any multiple of $\lambda_0/2$.

(b) For the receiving antenna in the form of a short wire dipole, referred to as an electric probe, the induced emf in the antenna can be found from Eq. (6.32), using the electric field vector of the total (standing) wave, $\underline{\mathbf{E}}_{\text{tot}}$, in Eqs. (10.8). It is then obvious that the wave reception is maximum when the dipole is aligned with the vector $\underline{\mathbf{E}}_{\text{tot}}$, and placed at electric field maxima, in one of the planes where $\sin\beta z = \pm 1$. This gives $z = -(2m+1)\lambda_0/4$ ($m = 0, 1, 2, \ldots$), so the lowest height above the PEC plane is now $h = \lambda_0/4 = 5$ m, Fig. 10.5.

Example 10.6 Moving Contour in the Field of a Standing Wave

A uniform plane time-harmonic electromagnetic wave of angular frequency ω and rms magnetic field intensity $H_{\text{i}0}$ propagates in air and impinges a PEC plane at normal incidence. Determine the instantaneous emf induced in an electrically small contour of surface area S that travels with velocity v in the direction of propagation of the reflected wave. The plane of the contour is perpendicular to the magnetic field lines. Identify what portions of the emf are due to the transformer and motional induction, respectively.

Solution We note first that, as opposed to the situation in the previous example, we now cannot apply Faraday's law of electromagnetic induction in the complex domain, Eq. (10.24), but in the time domain, to be able to take into account the motion of the contour. We then realize that the moving contour in the total magnetic field of the standing wave in Fig. 10.1 represents, as far as the induced emf in the contour is concerned, a system based on total (mixed) electromagnetic induction, that is, a combination of the transformer induction (due to the time variation of the magnetic field) and motional induction (due to the motion of the contour). As such, this system is similar to that in Example 6.15. Having in mind Eqs. (9.58)

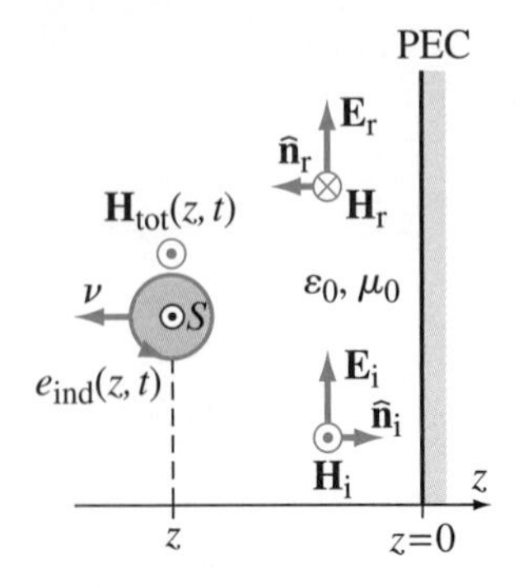

Figure 10.6 Moving contour ($S \ll \lambda_0^2$) in the magnetic field of a standing uniform plane time-harmonic wave in front of a PEC screen; for Example 10.6.

and (10.9), as well as that $E_{i0}/\eta_0 = H_{i0}$, the magnetic flux through the contour when it is at a location defined by coordinate z ($-\infty < z < 0$) at instant t, as shown in Fig. 10.6, equals

$$\Phi(z,t) = \mu_0 H_{tot}(z,t) S = 2\sqrt{2}\mu_0 H_{i0} S \cos\beta z \cos\omega t. \tag{10.26}$$

Since $dz/dt = -v$ (the contour travels in the negative z direction), the total (combined) emf in the contour is, much like in Eq. (6.112), given by

$$e_{ind}(z,t) = -\frac{d\Phi}{dt} = -\frac{\partial\Phi(z,t)}{\partial t} - \frac{\partial\Phi(z,t)}{\partial z}\frac{dz}{dt} = \underbrace{-\frac{\partial\Phi(z,t)}{\partial t}}_{\text{transformer emf}} + \underbrace{v\frac{\partial\Phi(z,t)}{\partial z}}_{\text{motional emf}}$$

$$= 2\sqrt{2}\mu_0 H_{i0} S\left[\omega\cos\beta z \sin\omega t - \beta v \sin\beta z \cos\omega t\right], \tag{10.27}$$

with the first term (negative partial derivative of Φ with respect to t) amounting to the transformer part of the total emf (it would be zero if $\omega = 0$ – no time variation of H_{tot}) and the second term (negative partial derivative of Φ with respect to z times dz/dt) to the motional emf (it becomes zero if $v = 0$ – stationary contour or if $\beta = 0$ – no spatial variation of H_{tot}).

Assuming that the reference position of the contour for $t = 0$ is $z = 0$ in Fig. 10.6, we have that $z = -vt$ for $t \geq 0$, with which we can express the emf in Eq. (10.27) as a function of time solely,

$$e_{ind}(t) = 2\sqrt{2}\mu_0 \beta H_{i0} S\left[c_0 \cos(\beta v t)\sin\omega t + v \sin(\beta v t)\cos\omega t\right]. \tag{10.28}$$

Here, the use is made also of the relationship $\omega/\beta = c_0$, Eq. (8.111), c_0 standing for the velocity of the incident and reflected waves (in air), Eq. (9.19), to make it obvious in the result that the partial emf due to transformer induction is, overall, dominant over that due to motional induction – as long as $v \ll c_0$ (and this is most likely the case).

Example 10.7 Reflection of a Circularly Polarized Wave

A right-hand circularly polarized uniform plane electromagnetic wave of frequency f and electric field amplitude E_m propagates in air and is incident normally on a PEC plane. The instantaneous electric field vector of the wave is given by the two field expressions in Eqs. (9.185), with $\delta = -90°$. Find: (a) complex electric and magnetic field vectors of the reflected wave, (b) polarization state of the reflected wave, (c) complex electric and magnetic field vectors of the resultant wave in air, and (d) complex surface current and charge densities in the PEC plane.

Solution

(a) From Eqs. (9.185) and (8.66), the complex rms electric field intensity vectors of the two linearly polarized waves constituting the incident RHCP wave are expressed as

$$\underline{\mathbf{E}}_1 = E' \, e^{-j\beta z}\,\hat{\mathbf{x}}, \quad \underline{\mathbf{E}}_2 = -jE' \, e^{-j\beta z}\,\hat{\mathbf{y}} \quad \left(E' = \frac{E_m}{\sqrt{2}}\right). \tag{10.29}$$

Using Eqs. (9.192), namely, $\underline{H}_y = \underline{E}_x/\eta_0$ and $\underline{H}_x = -\underline{E}_y/\eta_0$, the accompanying magnetic field vectors are

$$\underline{\mathbf{H}}_1 = H' \, e^{-j\beta z}\,\hat{\mathbf{y}}, \quad \underline{\mathbf{H}}_2 = jH' \, e^{-j\beta z}\,\hat{\mathbf{x}} \quad \left(H' = \frac{E'}{\eta_0}\right), \tag{10.30}$$

with η_0 being the free-space intrinsic impedance, Eq. (9.23). Consequently, vectors $\underline{\mathbf{E}}_i$ and $\underline{\mathbf{H}}_i$ of the incident wave can be written as

incident CP wave

$$\boxed{\underline{\mathbf{E}}_i = E' \, e^{-j\beta z}(\hat{\mathbf{x}} - j\,\hat{\mathbf{y}}), \quad \underline{\mathbf{H}}_i = H' \, e^{-j\beta z}(\hat{\mathbf{y}} + j\,\hat{\mathbf{x}}).} \tag{10.31}$$

With reference to Fig. 10.1, and by means of the boundary condition in Eq. (10.3), we have, for the reflected wave, $\underline{\mathbf{E}}_r = -\underline{\mathbf{E}}_i$ for $z = 0$, and hence

$$\underline{\mathbf{E}}_r = -E' \, e^{j\beta z}(\hat{\mathbf{x}} - j\,\hat{\mathbf{y}}), \quad \underline{\mathbf{H}}_r = H' \, e^{j\beta z}(\hat{\mathbf{y}} + j\,\hat{\mathbf{x}}), \tag{10.32}$$

where $\underline{\mathbf{H}}_{\rm r}$ is computed from $\underline{\mathbf{E}}_{\rm r}$ either by the vector relationship in Eq. (9.22) with the propagation unit vector in the backward direction, $\hat{\mathbf{n}}_{\rm r} = -\hat{\mathbf{z}}$, or by determining each of the transverse components of $\underline{\mathbf{H}}_{\rm r}$ from the corresponding (orthogonal) component of $\underline{\mathbf{E}}_{\rm r}$ using the intrinsic impedance η_0 and making sure that the associated Poynting vector is properly directed (in the negative z direction).

(b) Obviously, the reflected wave is also circularly polarized. Since there is exactly the same phase relationship between the x- and y-components of $\underline{\mathbf{E}}_{\rm r}$, in Eqs. (10.32), as for $\underline{\mathbf{E}}_{\rm i}$, in Eqs. (10.31), the vector $\mathbf{E}_{\rm r}(t)$ for $z =$ const rotates in the same direction as $\mathbf{E}_{\rm i}(t)$, and the same is true for magnetic field vectors. Of course, the sense of rotation of $\mathbf{E}_{\rm r}(t)$ can as well be found by sketching this vector at several characteristic instants of time, as in Fig. 9.21. However, as the two waves travel in opposite directions ($\hat{\mathbf{n}}_{\rm r} = -\hat{\mathbf{n}}_{\rm i}$), the handedness of the reflected wave is just opposite to that of the incident wave (see the definition of RH and LH polarizations given with Fig. 9.19), so it is a LHCP wave. In general, a reflecting surface (mirror) changes the sense of polarization of an impinging wave.

(c) The total electric and magnetic field vectors are obtained as in Eqs. (10.5) and (10.6),

standing CP wave

$$\boxed{\underline{\mathbf{E}}_{\rm tot} = \underline{\mathbf{E}}_{\rm i} + \underline{\mathbf{E}}_{\rm r} = -2{\rm j}E' \sin\beta z(\hat{\mathbf{x}} - {\rm j}\,\hat{\mathbf{y}}), \quad \underline{\mathbf{H}}_{\rm tot} = \underline{\mathbf{H}}_{\rm i} + \underline{\mathbf{H}}_{\rm r} = 2H' \cos\beta z(\hat{\mathbf{y}} + {\rm j}\,\hat{\mathbf{x}})} \tag{10.33}$$

[also see Eqs. (10.8)], so the resultant wave is a CP standing wave. Our definition of polarization handedness (in Fig. 9.19), however, does not make sense for standing waves, as they do not propagate.

(d) The surface current density vector in the PEC plane is obtained from the boundary condition in Eq. (10.13),

$$\underline{\mathbf{J}}_{\rm s} = (-\hat{\mathbf{z}}) \times \underline{\mathbf{H}}_{\rm tot}\big|_{z=0} = 2H'(\hat{\mathbf{x}} - {\rm j}\,\hat{\mathbf{y}}). \tag{10.34}$$

As expected, $\mathbf{J}_{\rm s}(t)$ is also a CP vector. On the other side, boundary condition in Eq. (10.15) tells us that there are no excess surface charges in the plane, $\underline{\rho}_{\rm s} = 0$.

Problems: 10.1–10.8; *Conceptual Questions* (on Companion Website): 10.1–10.10; *MATLAB Exercises* (on Companion Website).

10.2 NORMAL INCIDENCE ON A PENETRABLE PLANAR INTERFACE

We now consider a more general case with the medium on the right-hand side of the interface in Fig. 10.1 being penetrable for the (normally) incident wave. Moreover, let us add the possibility of having arbitrary losses in both media, and describe their electromagnetic properties by ε_1, μ_1, and σ_1, for medium 1, and ε_2, μ_2, and σ_2, for medium 2, as indicated in Fig. 10.7. We assume, of course, that at least one of the three material parameters has different values for the two regions, and because of this discontinuity in the propagation medium (abrupt change of some of the material properties) in the way of the incident wave travel, the incident wave will be partially reflected back from the interface, and partially transmitted through it (medium 2 is penetrable). Hence, we have three traveling waves in Fig. 10.7: two in medium 1 (incident and reflected), like in the case with the PEC boundary in Fig. 10.1, and one in medium 2 (which, of course, does not exist in Fig. 10.1). This new wave is called the transmitted wave; it propagates away from the material interface, with wave propagation parameters based on the material properties of medium 2. Adopting the same reference direction of the electric field intensity vector for all three waves, their graphical representations in Fig. 10.7 are translated into the

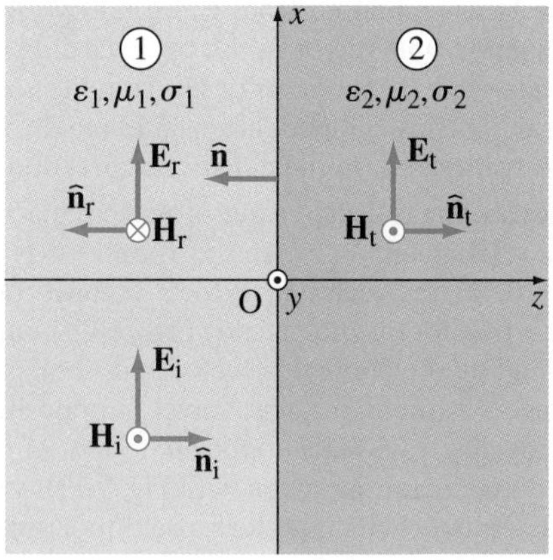

Figure 10.7 Normal incidence of a uniform plane time-harmonic electromagnetic wave on a planar interface between two linear homogeneous media with arbitrary electromagnetic parameters.

following equations:

incident wave

$$\underline{\mathbf{E}}_{\mathrm{i}} = \underline{E}_{\mathrm{i}0}\,\mathrm{e}^{-\underline{\gamma}_1 z}\,\hat{\mathbf{x}}, \quad \underline{\mathbf{H}}_{\mathrm{i}} = \frac{\underline{E}_{\mathrm{i}0}}{\underline{\eta}_1}\,\mathrm{e}^{-\underline{\gamma}_1 z}\,\hat{\mathbf{y}}, \tag{10.35}$$

reflected wave

$$\underline{\mathbf{E}}_{\mathrm{r}} = \underline{E}_{\mathrm{r}0}\,\mathrm{e}^{\underline{\gamma}_1 z}\,\hat{\mathbf{x}}, \quad \underline{\mathbf{H}}_{\mathrm{r}} = \frac{\underline{E}_{\mathrm{r}0}}{\underline{\eta}_1}\,\mathrm{e}^{\underline{\gamma}_1 z}(-\,\hat{\mathbf{y}}), \tag{10.36}$$

transmitted wave

$$\underline{\mathbf{E}}_{\mathrm{t}} = \underline{E}_{\mathrm{t}0}\,\mathrm{e}^{-\underline{\gamma}_2 z}\,\hat{\mathbf{x}}, \quad \underline{\mathbf{H}}_{\mathrm{t}} = \frac{\underline{E}_{\mathrm{t}0}}{\underline{\eta}_2}\,\mathrm{e}^{-\underline{\gamma}_2 z}\,\hat{\mathbf{y}}, \tag{10.37}$$

where $\underline{E}_{\mathrm{i}0}$, $\underline{E}_{\mathrm{r}0}$, and $\underline{E}_{\mathrm{t}0}$ are, respectively, complex rms electric field intensities of the incident, reflected, and transmitted wave in the plane $z = 0$. The complex propagation coefficients and intrinsic impedances for the two media are given [see Eqs. (9.82), (9.91), and (9.78)] by $\underline{\gamma}_k = \mathrm{j}\omega\sqrt{\underline{\varepsilon}_{\mathrm{e}k}\mu_k} = \alpha_k + \mathrm{j}\beta_k$ and $\underline{\eta}_k = \sqrt{\mu_k/\underline{\varepsilon}_{\mathrm{e}k}}$, respectively, where $k = 1$ for medium 1 and $k = 2$ for medium 2, $\underline{\varepsilon}_{\mathrm{e}k} = \varepsilon_k - \mathrm{j}\sigma_k/\omega$ $(k = 1, 2)$ are the equivalent complex permittivities, α_k attenuation coefficients, and β_k phase coefficients of the two media.

In Eqs. (10.36) and (10.37), we have two unknown field intensities at $z = 0$, $\underline{E}_{\mathrm{r}0}$ and $\underline{E}_{\mathrm{t}0}$, so we need two linear algebraic equations with these two complex constants as unknowns to solve, for a given $\underline{E}_{\mathrm{i}0}$, whereas for the case in Fig. 10.1, one equation was sufficient. So, in addition to the boundary condition for the vector $\mathbf{E}$, we invoke here the boundary condition for $\mathbf{H}$ as well. Of course, both are for the tangential vector components and for the general case of a boundary surface between two arbitrary electromagnetic media, Eqs. (8.32). Since all vectors in Fig. 10.7 are entirely tangential to the boundary, the conditions apply to the entire vectors, and we have

$$\hat{\mathbf{n}} \times \underline{\mathbf{E}}_1 - \hat{\mathbf{n}} \times \underline{\mathbf{E}}_2 = 0 \quad \longrightarrow \quad (\underline{\mathbf{E}}_{\mathrm{i}} + \underline{\mathbf{E}}_{\mathrm{r}})\big|_{z=0} = \underline{\mathbf{E}}_{\mathrm{t}}\big|_{z=0}, \tag{10.38}$$

$$\hat{\mathbf{n}} \times \underline{\mathbf{H}}_1 - \hat{\mathbf{n}} \times \underline{\mathbf{H}}_2 = \underline{\mathbf{J}}_{\mathrm{s}} \quad \xrightarrow{(\underline{\mathbf{J}}_{\mathrm{s}}=0)} \quad (\underline{\mathbf{H}}_{\mathrm{i}} + \underline{\mathbf{H}}_{\mathrm{r}})\big|_{z=0} = \underline{\mathbf{H}}_{\mathrm{t}}\big|_{z=0}, \tag{10.39}$$

with $\hat{\mathbf{n}} = -\,\hat{\mathbf{z}}$ being the normal unit vector on the interface (directed from medium 2 to medium 1). The vector $\underline{\mathbf{J}}_{\mathrm{s}}$ in the second boundary condition is taken to be zero since surface currents in the plane $z = 0$ can only exist if one of the two media is a perfect conductor, like in Fig. 10.1. Substituting the field expressions in Eqs. (10.35)–(10.37) into Eqs. (10.38) and (10.39) then yields the following two equations with unknowns $\underline{E}_{r0}$ and $\underline{E}_{t0}$:

$$\underline{E}_{i0} + \underline{E}_{r0} = \underline{E}_{t0}, \quad \frac{\underline{E}_{i0}}{\underline{\eta}_1} - \frac{\underline{E}_{r0}}{\underline{\eta}_1} = \frac{\underline{E}_{t0}}{\underline{\eta}_2}, \tag{10.40}$$

boundary conditions

and their solution is

$$\underline{E}_{r0} = \frac{\underline{\eta}_2 - \underline{\eta}_1}{\underline{\eta}_1 + \underline{\eta}_2}\underline{E}_{i0}, \quad \underline{E}_{t0} = \frac{2\underline{\eta}_2}{\underline{\eta}_1 + \underline{\eta}_2}\underline{E}_{i0}. \tag{10.41}$$

This completes the computation of the reflected and transmitted fields in Fig. 10.7.

We now define the reflection and transmission coefficients, $\underline{\Gamma}$ and $\underline{\tau}$, as the ratios of the corresponding complex rms electric field intensities, reflected over incident and transmitted over incident, respectively, at the material interface ($z = 0$). So, for the situation in Fig. 10.7 (normal incidence from a medium with complex intrinsic impedance $\underline{\eta}_1$ to a medium with impedance $\underline{\eta}_2$), these coefficients come out to be [from Eqs. (10.41)]

$$\underline{\Gamma} = \frac{\underline{E}_{r0}}{\underline{E}_{i0}} = \frac{\underline{\eta}_2 - \underline{\eta}_1}{\underline{\eta}_1 + \underline{\eta}_2}, \tag{10.42}$$

reflection coefficient (dimensionless)

$$\underline{\tau} = \frac{\underline{E}_{t0}}{\underline{E}_{i0}} = \frac{2\underline{\eta}_2}{\underline{\eta}_1 + \underline{\eta}_2}, \tag{10.43}$$

transmission coefficient (dimensionless)

where it is obvious from the first boundary condition in Eqs. (10.40), dividing both sides of the equation by $\underline{E}_{i0}$, that $1 + \underline{\Gamma} = \underline{\tau}$.[8]

The reflection coefficient, being generally complex, can be represented as

$$\underline{\Gamma} = |\underline{\Gamma}|\,e^{j\psi} \quad (0 \leq |\underline{\Gamma}| \leq 1;\ -180^\circ < \psi \leq 180^\circ), \tag{10.44}$$

reflection coefficient in polar form

where the phase angle ψ amounts to the phase shift between the reflected and incident electric field intensities at the interface, in addition to their amplitude ratio, determined by the magnitude of the reflection coefficient, $|\underline{\Gamma}|$. That $|\underline{\Gamma}| \leq 1$ will be obvious from the expression for the Poynting vector in the incident medium, which will be derived in an example, and from the fact that the time-average power carried back by the reflected wave cannot be (by the principle of conservation of power) larger than the time-average power of the incident wave. In decibels,

$$\Gamma_{\mathrm{dB}} = 20\log|\underline{\Gamma}| \quad (-\infty < \Gamma_{\mathrm{dB}} \leq 0\ \mathrm{dB}). \tag{10.45}$$

dB reflection coefficient

A similar representation, in terms of the magnitude and phase, can be given for the complex transmission coefficient, $\underline{\tau}$.

If both media in Fig. 10.7 are lossless ($\sigma_1 = \sigma_2 = 0$), their intrinsic impedances are purely real, $\underline{\eta}_1 = \eta_1 = \sqrt{\mu_1/\varepsilon_1}$ and $\underline{\eta}_2 = \eta_2 = \sqrt{\mu_2/\varepsilon_2}$, which results in both coefficients in Eqs. (10.42) and (10.43) being purely real ($\underline{\Gamma} = \Gamma$ and $\underline{\tau} = \tau$) as well. For $\eta_1 < \eta_2$, Γ is positive ($\psi = 0$), whereas $\eta_1 > \eta_2$ yields a negative Γ ($\psi = 180^\circ$).

[8]Let us point out again that the expressions for $\underline{\Gamma}$ and $\underline{\tau}$ in terms of $\underline{\eta}_1$ and $\underline{\eta}_2$ in Eqs. (10.42) and (10.43) are defined for the same reference direction of vectors $\underline{\mathbf{E}}_{i0}$, $\underline{\mathbf{E}}_{r0}$, and $\underline{\mathbf{E}}_{t0}$ (Fig. 10.7). Such adoption of vector orientations is arbitrary; however, once we have made this choice, we must use it consistently and always have it in mind in interpreting numerical values for the coefficients. Note also that the coefficients are defined for the electric field intensities of the three waves and not magnetic ones, which, as well, is a matter of convention.

On the other side, τ is nonnegative for any combination of purely real impedances (which themselves must, of course, be nonnegative). It is interesting to note that $\tau > 1$, meaning that the amplitude of the transmitted electric field is larger than that of the incident field, when $\eta_1 < \eta_2$. Although this gain in the electric field intensity across the boundary might seem at the first glance like an alarming violation of the principle of conservation of energy in the system in Fig. 10.7, we instantly recall that both electric and magnetic fields, namely, the Poynting vector, must be considered in assessing the energy carried by a wave and in any power-balance evaluations. Indeed, the amplitude of the magnetic field in this case is smaller for the transmitted than for the incident wave [the amplitude ratio, transmitted over incident, for magnetic fields is $2\eta_1/(\eta_1 + \eta_2) < 1$ (since $\eta_1 < \eta_2$)]. Moreover, it is small enough to make the magnitude of the time-average Poynting vector of the transmitted wave smaller than that of the incident wave, so the power conservation, of course, holds true [the transmitted to incident power flow ratio, as we shall see in an example, amounts to $\tau^2\eta_1/\eta_2$, and it is a simple matter to show, substituting the expression for τ from Eq. (10.43), that this ratio is always smaller than unity]. Shown in Fig. 10.8 is a time snapshot (at $t = 0$) of the instantaneous electric field intensities of incident, reflected, and transmitted waves as a function of z, for two combinations of permittivities, $\varepsilon_1 > \varepsilon_2$ and $\varepsilon_1 < \varepsilon_2$, of the two media in Fig. 10.7, assuming that both media are nonmagnetic ($\mu_1 = \mu_2 = \mu_0$) and lossless. We observe matching of field intensities at the interface according to the boundary condition for E-fields in Eqs. (10.40) for the two situations, as well as different wavelengths ($\lambda_1 \neq \lambda_2$) for the waveforms in the two media.

If medium 2 is a good conductor, so that its intrinsic impedance is given by Eq. (9.137), and medium 1 is a perfect dielectric (e.g., air), $|\underline{\eta}_2| \ll \eta_1$ (at frequencies

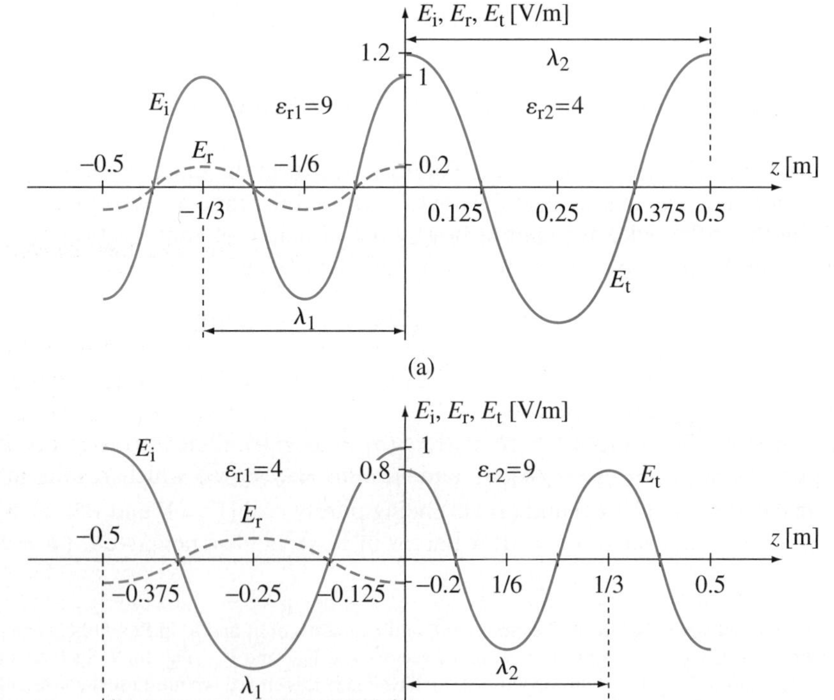

Figure 10.8 Plots of instantaneous electric field intensities of incident, reflected, and transmitted waves in Fig. 10.7 against z at $t = 0$ for $\underline{E}_{i0} = E_{i0} = (\sqrt{2}/2)$ V/m (incident electric field has a zero initial phase at $z = 0$), $f = 300$ MHz, $\mu_1 = \mu_2 = \mu_0$, $\sigma_1 = \sigma_2 = 0$, and two combinations of relative permittivities of the two media: (a) $\varepsilon_{r1} = 9$ and $\varepsilon_{r2} = 4$ and (b) $\varepsilon_{r1} = 4$ and $\varepsilon_{r2} = 9$.

up to the visible-light region), and Eqs. (10.42) and (10.43) become

$$\boxed{\underline{\Gamma} \approx \frac{-\eta_1}{\eta_1} = -1, \quad \underline{\tau} \approx \frac{2\underline{\eta}_2}{\eta_1}.} \tag{10.46}$$

reflecting medium good conductor

If, furthermore, the conductor can be considered to be perfect, $\underline{\eta}_2 = 0$, and hence

$$\boxed{\underline{\Gamma} = -1, \quad \underline{\tau} = 0,} \tag{10.47}$$

reflecting medium PEC

where $\underline{\Gamma} = -1$ means, of course, $|\Gamma| = 1$ and $\psi = 180°$, and is equivalent to the relationship in Eq. (10.4).

The electromagnetic field in medium 2 in Fig. 10.7 consists of electric and magnetic fields of the transmitted wave, $\underline{\mathbf{E}}_2 = \underline{\mathbf{E}}_\mathrm{t}$ and $\underline{\mathbf{H}}_2 = \underline{\mathbf{H}}_\mathrm{t}$, where $\underline{\mathbf{E}}_\mathrm{t}$ and $\underline{\mathbf{H}}_\mathrm{t}$ are given by Eqs. (10.37) and (10.43), since this is the only wave past the boundary at $z = 0$. This wave, of course, is a pure traveling wave. The electromagnetic field in medium 1, on the other hand, results from the superposition of two waves (incident and reflected) traveling in opposite directions, like in Eqs. (10.5) and (10.6). However, as $\underline{E}_{\mathrm{r}0} \neq -\underline{E}_{\mathrm{i}0}$ in Fig. 10.7, the resultant wave is not a pure standing wave, which is also obvious from the fact that at least a portion of its energy must be carried over to the transmitted wave. To quantitatively investigate the standing-wave content in the total field in medium 1, let us assume, for simplicity, that it exhibits no losses ($\sigma_1 = 0$), that is, no wave attenuation ($\alpha_1 = 0$), so that $\gamma_1 = \mathrm{j}\beta_1$, where $\beta_1 = \omega\sqrt{\varepsilon_1\mu_1}$.[9] By adding and subtracting the term $\underline{\Gamma}\underline{E}_{\mathrm{i}0}\,\mathrm{e}^{-\mathrm{j}\beta_1 z}\,\hat{\mathbf{x}}$ in the expression for the incident electric field intensity vector in Eqs. (10.35), we obtain

$$\underline{\mathbf{E}}_\mathrm{i} = \underline{E}_{\mathrm{i}0}[\underbrace{(1 + \underline{\Gamma})}_{\underline{\tau}}\,\mathrm{e}^{-\mathrm{j}\beta_1 z} - \underline{\Gamma}\,\mathrm{e}^{-\mathrm{j}\beta_1 z}]\,\hat{\mathbf{x}}. \tag{10.48}$$

The reflected field is given by Eqs. (10.36) and (10.42), so the total electric field intensity vector in the incident medium can now be written as

$$\underline{\mathbf{E}}_1 = \underline{\mathbf{E}}_\mathrm{i} + \underline{\mathbf{E}}_\mathrm{r} = \underline{E}_{\mathrm{i}0}[\underline{\tau}\,\mathrm{e}^{-\mathrm{j}\beta_1 z} + \underline{\Gamma}(\underbrace{-\,\mathrm{e}^{-\mathrm{j}\beta_1 z} + \mathrm{e}^{\mathrm{j}\beta_1 z}}_{2\mathrm{j}\sin\beta_1 z})]\,\hat{\mathbf{x}}$$

$$= \underbrace{\underline{\tau}\underline{E}_{\mathrm{i}0}\,\mathrm{e}^{-\mathrm{j}\beta_1 z}\,\hat{\mathbf{x}}}_{\text{traveling wave}} + \underbrace{2\mathrm{j}\underline{\Gamma}\underline{E}_{\mathrm{i}0}\sin\beta_1 z\,\hat{\mathbf{x}}}_{\text{standing wave}}, \tag{10.49}$$

and the magnetic field of the resultant wave, $\underline{\mathbf{H}}_1$, can be decomposed in a similar fashion. This tells us that the resultant wave can be visualized as a superposition of a traveling wave, with rms electric field intensity $|\underline{\tau}||\underline{E}_{\mathrm{i}0}|$, and a standing wave, whose maximum rms electric field intensity is $2|\underline{\Gamma}||\underline{E}_{\mathrm{i}0}|$. In the time domain,

$$\mathbf{E}_1(t) = |\underline{\tau}||\underline{E}_{\mathrm{i}0}|\sqrt{2}\cos(\omega t - \beta_1 z + \psi_\tau + \xi)\,\hat{\mathbf{x}} - 2\sqrt{2}|\underline{\Gamma}||\underline{E}_{\mathrm{i}0}|\sin\beta_1 z\sin(\omega t + \psi + \xi)\,\hat{\mathbf{x}}, \tag{10.50}$$

where ψ_τ and ξ are phase angles of $\underline{\tau}$ and $\underline{E}_{\mathrm{i}0}$, respectively. Furthermore, we can say, based on Eq. (10.49) or (10.50), that the reflected wave interferes with a portion of the incident wave having the electric field vector equal in amplitude but in counterphase [the second term in Eq. (10.48)] to form a standing wave, with twice larger

[9]Although we obviously lose some generality with the assumption that the incident medium is lossless, we note that cases with (nonnegligible) losses in both media in Fig. 10.7 occur relatively rarely in practical situations. In addition, in cases where one of the media is lossy, it most frequently is the reflecting medium, as in our assumption, with the incident medium being air or some other perfect dielectric.

amplitude in its maxima. The rest of the incident wave [the first term in Eq. (10.48)], which does not interfere with the reflected wave, travels toward the interface and sustains the transmitted wave in medium 2.

Example 10.8 Field Maxima and Minima, and Standing Wave Pattern

A uniform plane time-harmonic electromagnetic wave of frequency f and rms electric field intensity E_{i0} propagates in a lossless medium of parameters ε_1 and μ_1 and impinges normally the planar surface of a lossy medium of parameters ε_2, μ_2, and σ_2. (a) Determine the expressions for maxima and minima of the rms electric field intensity of the resultant wave in the incident medium, and their locations. (b) Sketch the spatial distribution of the total rms electric field intensity in the incident medium. (c) What are locations of maxima and minima of the accompanying resultant magnetic field intensity vector?

Solution

(a) In addition to the representation in Eq. (10.49), and with the definition of all involved parameters as in that representation, the total electric field intensity vector in the incident medium (medium 1) can also be written as

total electric field in front of a penetrable interface

$$\underline{\mathbf{E}}_1 = \underline{E}_{i0}\,\hat{\mathbf{x}}\left(e^{-j\beta_1 z} + \underline{\Gamma}\,e^{j\beta_1 z}\right) = \underbrace{\underline{E}_{i0}\,e^{-j\beta_1 z}\,\hat{\mathbf{x}}}_{\text{incident wave}}\underbrace{(1 + \underline{\Gamma}\,e^{j2\beta_1 z})}_{\text{array factor}}, \tag{10.51}$$

where the field intensity of the incident wave is multiplied by a factor that takes into account the reflected wave and its interaction with the incident one, resulting in standing wave behavior. This factor, ultimately, is due to the existence of the second medium in Fig. 10.7. Using Eq. (10.44), it is expressed as

array factor for waves (dimensionless)

$$\underline{F}_a = 1 + |\underline{\Gamma}|\,e^{j(2\beta_1 z + \psi)} \quad (-\infty < z \le 0). \tag{10.52}$$

We refer to it as the array factor and symbolize it by $\underline{F}_a$, in analogy with the analysis of antenna arrays, where the total far electric field (radiation pattern) of a spatial array of identical antenna elements is obtained as a product of the field (wave) radiated by a single element (reference element) and the array factor.[10] In our case, we have an "array" of two waves, with the incident wave playing the role of a reference element in the array. In Fig. 10.9(a), Eq. (10.52) is represented in the complex plane [see also Fig. 8.6(d)], with $\underline{F}_a$ corresponding to a point on a circle of radius $|\underline{\Gamma}|$, centered at $(1, 0)$. As the coordinate z varies from $z = 0$ to $z \to -\infty$, the point $\underline{F}_a$, whose position on the circle is determined by the angle $\phi_a = 2\beta_1 z + \psi$, rotates in the mathematically negative (clockwise) direction. The magnitude of the array factor, for any z, equals, in the complex plane, the magnitude of the position vector of the point $\underline{F}_a$ with respect to the coordinate origin. It is thus obvious from Fig. 10.9(a) that the maxima and minima of $|\underline{F}_a|$ are, respectively, $1 + |\underline{\Gamma}|$ for $\phi_a = -2m\pi$ and $1 - |\underline{\Gamma}|$ for $\phi_a = -(2m + 1)\pi$ $(m = 0, 1, 2, \ldots)$. Hence, the maxima

[10] As we shall see in a later chapter, the array factor of an antenna array is a sum of terms specifying the relative amplitudes and phases of radiated fields (waves) of individual antenna elements in the array (due to their location and feed) with respect to the reference element, whose term equals unity (if it is adopted to be one of the array elements). Thus, the second term in the array factor in Eq. (10.52) represents the phase difference of the reflected wave (its electric field) at the point defined by an arbitrary coordinate z in the incident medium $(-\infty < z \le 0)$ with respect to the incident wave at the same point, along with the amplitude ratio $|\underline{\Gamma}|$, introduced at the bounce off the interface $(z = 0)$. The phase difference results from the round trip of the wave traveling a distance equal to $-2z$ (z is negative), from the point with coordinate z to the interface and back, and reflection phase shift ψ added at the interface.

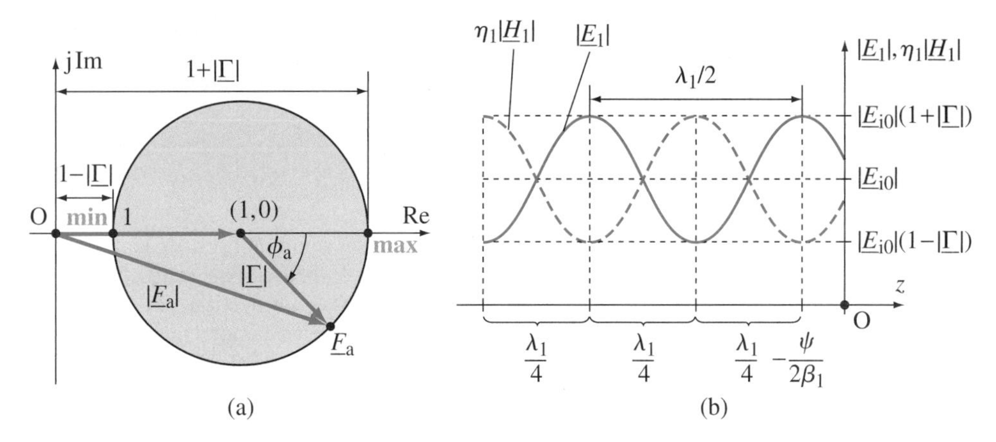

Figure 10.9 (a) Representation in the complex plane of the array factor ($\underline{F}_a$) in Eq. (10.52) for the total electric field intensity vector in the incident medium in Fig. 10.7 and (b) sketch of the magnitudes of the total electric and magnetic field intensity vectors in Eqs. (10.51) and (10.55) as functions of z (standing wave patterns), for an arbitrary phase (ψ) of the reflection coefficient in Eq. (10.44); for Example 10.8.

of the total rms electric field intensity in Eq. (10.51), $|\underline{\mathbf{E}}_1| = |\underline{E}_{i0}||\underline{F}_a|$, are

$$|\underline{\mathbf{E}}_1|_{\max} = |\underline{E}_{i0}|\left(1+|\underline{\Gamma}|\right) \quad \text{for} \quad z = -m\frac{\lambda_1}{2} - \frac{\psi}{2\beta_1}, \quad \begin{cases} m = 0, 1, 2, \ldots \text{ if } \psi \geq 0 \\ m = 1, 2, \ldots \text{ if } \psi < 0 \end{cases} \tag{10.53}$$

electric-field maxima

(the case $m = 0$ for $\psi < 0$ is excluded since it gives a positive z), and minima

$$|\underline{\mathbf{E}}_1|_{\min} = |\underline{E}_{i0}|\left(1-|\underline{\Gamma}|\right) \quad \text{(shifted by } \lambda_1/4 \text{ with respect to maxima)}, \tag{10.54}$$

electric-field minima

where $\lambda_1 = 2\pi/\beta_1$.

(b) Fig. 10.9(b) shows the plot, so-called electric-field standing wave pattern, of the field magnitude $|\underline{\mathbf{E}}_1|$ as a function of z, for an arbitrary ψ. Specifically, when $\psi = 0$ (Γ real and positive), the location of the first field maximum [$m = 0$ in Eq. (10.53)] is right at the boundary plane ($z = 0$). On the other hand, the boundary plane contains the first field minimum [Eq. (10.54)] when $\psi = 180°$ (Γ real but negative), since $\psi/(2\beta_1) = \lambda_1/4$ for this reflection coefficient phase. Note that the situation in Fig. 10.1 (the second medium PEC) also falls under this latter case [see Eq. (10.47)], with minima in Eq. (10.54) now being field zeros ($|\underline{\mathbf{E}}_1|_{\min} = 0$), as in Eq. (10.10), and the first one of them lying in the PEC plane.

(c) Similarly, the total magnetic field intensity vector in the incident medium can be expressed as

$$\underline{\mathbf{H}}_1 = \underline{\mathbf{H}}_i + \underline{\mathbf{H}}_r = \frac{\underline{E}_{i0}}{\eta_1}\,\hat{\mathbf{y}}\left(e^{-j\beta_1 z} - \underline{\Gamma}\,e^{j\beta_1 z}\right) = \frac{\underline{E}_{i0}}{\eta_1}\,e^{-j\beta_1 z}\,\hat{\mathbf{y}}\left(1 - \underline{\Gamma}\,e^{j2\beta_1 z}\right). \tag{10.55}$$

total magnetic field in front of a penetrable interface

Because of the extra minus sign (180° phase difference) in the reflected-field term, when compared to Eq. (10.51), the magnetic field maxima occur at the locations of the electric field minima, and vice versa (like in Fig. 10.2 for the PEC boundary). The magnetic-field standing wave pattern, $\eta_1|\underline{\mathbf{H}}_1(z)|$, is also shown in Fig. 10.9(b).

Example 10.9 Standing Wave Ratio

For the situation described in the previous example, find the ratios of maxima to minima for the total electric and magnetic fields, respectively, in the incident medium. What is the range of their possible values?

Solution Using Eqs. (10.53)–(10.55), we realize that these ratios, for $|\underline{\mathbf{E}}_1|$ and $|\underline{\mathbf{H}}_1|$, are the same. In fact, the ratio of (electric or magnetic) field maxima to minima is a key numerical parameter describing a standing wave pattern, called the standing wave ratio (SWR), and for the patterns in Fig. 10.9(b) it equals

standing wave ratio (dimensionless)

$$s = \frac{|\underline{\mathbf{E}}_1|_{\max}}{|\underline{\mathbf{E}}_1|_{\min}} = \frac{|\underline{\mathbf{H}}_1|_{\max}}{|\underline{\mathbf{H}}_1|_{\min}} = \frac{1+|\underline{\Gamma}|}{1-|\underline{\Gamma}|} \quad (1 \le s < \infty). \tag{10.56}$$

The range of possible numerical values for s (for different combinations of materials in Fig. 10.7) is given by $1 \le s < \infty$, which comes from the corresponding range for $|\underline{\Gamma}|$ in Eq. (10.44). Note that a unity SWR ($s = 1$) occurs for $\underline{\Gamma} = 0$ (no reflection), i.e., when the material parameters of the second medium are such that its intrinsic impedance is the same as (matches) the intrinsic impedance of the incident medium. We say that the second medium is "matched" (by impedance) to the incident one, so that the incident wave is totally transmitted through the boundary ($\underline{\tau} = 1$). As we shall see in later sections and chapters, there are many different practical situations where the no-reflection (or almost no-reflection) feature for the (unbounded or guided) propagation medium discontinuity is desirable (or critical), and SWR as close to unity as possible is often the most important design requirement. The other extreme case is a total reflection situation, with an infinitely large SWR ($s \to \infty$), which occurs for $\underline{\Gamma} = -1$ (PEC boundary).

Example 10.10 Power Computations in the Two-Media Problem

(a) For the two-media problem from Example 10.8, determine the time-average Poynting vector in the incident medium, and from that expression show that the magnitude of the associated reflection coefficient for the material interface cannot be larger than unity. (b) Also find the complex Poynting vector in the second medium, and show that if this medium is lossless, the reflection and transmission coefficients for the interface are related as $1 - \Gamma^2 = \eta_1\tau^2/\eta_2$, η_1 and η_2 being the intrinsic impedances of the two media.

Solution

(a) Combining Eqs. (8.194), (10.51), (10.55), (10.44), and (10.7), the complex Poynting vector in the incident medium in Fig. 10.7 is

$$\underline{\mathcal{P}}_1 = \underline{\mathbf{E}}_1 \times \underline{\mathbf{H}}_1^* = \frac{E_{\mathrm{i0}}^2}{\eta_1}\left(1+\underline{\Gamma}\,\mathrm{e}^{\mathrm{j}2\beta_1 z}\right)\left(1-\underline{\Gamma}^*\,\mathrm{e}^{-\mathrm{j}2\beta_1 z}\right)\hat{\mathbf{z}} = \frac{E_{\mathrm{i0}}^2}{\eta_1}\Big\{1-\underline{\Gamma}\underline{\Gamma}^*$$

$$+|\underline{\Gamma}|\left[\mathrm{e}^{\mathrm{j}(2\beta_1 z+\psi)} - \mathrm{e}^{-\mathrm{j}(2\beta_1 z+\psi)}\right]\Big\}\hat{\mathbf{z}} = \frac{E_{\mathrm{i0}}^2}{\eta_1}\left[1-|\underline{\Gamma}|^2+2\mathrm{j}|\underline{\Gamma}|\sin(2\beta_1 z+\psi)\right]\hat{\mathbf{z}}, \tag{10.57}$$

where $E_{\mathrm{i0}}^2 = |\underline{E}_{\mathrm{i0}}|^2 = \underline{E}_{i0}\underline{E}_{i0}^*$. The imaginary part of $\underline{\mathcal{P}}_1$, given by the same type of expression as in Eq. (10.20) for the incidence on a PEC interface, represents the reactive power fluctuation in the medium characteristic for standing waves. Its real part, equal to the time-average Poynting vector for $-\infty < z < 0$, i.e., to the net time-average (real) power flow in the positive z direction (toward the media interface) by the resultant electromagnetic wave per unit area of the wavefront, turns out to be

$$(\mathcal{P}_1)_{\mathrm{ave}} = \mathrm{Re}\{\underline{\mathcal{P}}_1\} = \frac{E_{\mathrm{i0}}^2}{\eta_1}\left(1-|\underline{\Gamma}|^2\right)\hat{\mathbf{z}} = (\mathcal{P}_{\mathrm{i}})_{\mathrm{ave}} + (\mathcal{P}_{\mathrm{r}})_{\mathrm{ave}}. \tag{10.58}$$

This power is delivered to the electromagnetic field in the second medium.

We note that $(\mathcal{P}_1)_{\mathrm{ave}}$ can be written as a sum of the time-average Poynting vector of the incident wave, $(\mathcal{P}_{\mathrm{i}})_{\mathrm{ave}}$, and that of the reflected wave, $(\mathcal{P}_{\mathrm{r}})_{\mathrm{ave}}$, which are expressed as

$$(\mathcal{P}_{\mathrm{i}})_{\mathrm{ave}} = \frac{E_{\mathrm{i0}}^2}{\eta_1}\hat{\mathbf{z}}, \quad (\mathcal{P}_{\mathrm{r}})_{\mathrm{ave}} = |\underline{\Gamma}|^2\frac{E_{\mathrm{i0}}^2}{\eta_1}(-\hat{\mathbf{z}}) = -|\underline{\Gamma}|^2(\mathcal{P}_{\mathrm{i}})_{\mathrm{ave}}. \tag{10.59}$$

We now see that, indeed, $|\underline{\Gamma}| \le 1$, since $|(\mathcal{P}_{\mathrm{r}})_{\mathrm{ave}}| \le |(\mathcal{P}_{\mathrm{i}})_{\mathrm{ave}}|$ (conservation of power).

(b) The complex Poynting vector in the second medium in Fig. 10.7 is found from the electric and magnetic field vectors of the transmitted wave, Eqs. (10.37). In analogy with Eq. (9.95),

$$\underline{\mathcal{P}}_2 = \underline{\mathcal{P}}_\mathrm{t} = \frac{E_{\mathrm{t0}}^2}{|\underline{\eta}_2|}\,\mathrm{e}^{-2\alpha_2 z}\,\mathrm{e}^{\mathrm{j}\phi_2}\,\hat{\mathbf{z}}, \tag{10.60}$$

where ϕ_2 is the phase angle of $\underline{\eta}_2$.

If there are no losses in medium 2 ($\alpha_2 = 0$, $\phi_2 = 0$), with the use of Eqs. (10.60), (10.43), and (10.59), the time-average Poynting vector of the transmitted wave is

$$(\mathcal{P}_\mathrm{t})_\mathrm{ave} = \frac{E_{\mathrm{t0}}^2}{\eta_2}\,\hat{\mathbf{z}} = \tau^2\,\frac{E_{\mathrm{i0}}^2}{\eta_2}\,\hat{\mathbf{z}} = \frac{\eta_1}{\eta_2}\,\tau^2(\mathcal{P}_\mathrm{i})_\mathrm{ave}. \tag{10.61}$$

Since both media are now lossless, the time-average power flow in either one of them does not depend on the z-coordinate. By the conservation of power principle, or, more formally, from Poynting's theorem in complex form, Eq. (8.196), applied to a pillbox closed surface, like the one in Fig. 2.10(b), with one of the bases lying in the plane $z = 0^-$ and the other in the plane $z = 0^+$, this power flow must be continuous across the boundary surface at $z = 0$. Hence, having in mind Eqs. (10.58), (10.59), and (10.61), we obtain the following relationship between the reflection and transmission coefficients:

$$(\mathcal{P}_1)_\mathrm{ave} = (\mathcal{P}_2)_\mathrm{ave} \quad\longrightarrow\quad 1 - \Gamma^2 = \frac{\eta_1}{\eta_2}\,\tau^2 \quad \text{(both media lossless).} \tag{10.62}$$

Note that this relationship can be derived also using the expressions for the coefficients in terms of the media intrinsic impedances, Eqs. (10.42) and (10.43).

Example 10.11 Wave Impedance

This example introduces a new quantity, equal to the ratio of the total electric to magnetic complex field intensities at a point in space. It is called the wave impedance and is denoted by $\underline{\eta}_\mathrm{w}$. With this definition, find $\underline{\eta}_\mathrm{w}$ in the incident medium in (a) Fig. 10.7 and (b) Fig. 10.1, respectively.

Solution

(a) As pointed out in the previous section, the electric over magnetic complex field intensity ratio for standing waves is not constant, Eq. (10.11), but depends on the spatial coordinate z (along the direction of wave propagation). It, of course, has the dimension of impedance, like the intrinsic impedance of the medium ($\underline{\eta}$).[11] From Eqs. (10.51) and (10.55), taken with $\mathrm{j}\beta_1$ replaced by $\underline{\gamma}_1$, the wave impedance in (possibly lossy) medium 1 in Fig. 10.7 is

$$\boxed{\underline{\eta}_{\mathrm{w}1} = \frac{\underline{E}_1}{\underline{H}_1} = \underline{\eta}_1\,\frac{\mathrm{e}^{-\underline{\gamma}_1 z} + \underline{\Gamma}\,\mathrm{e}^{\underline{\gamma}_1 z}}{\mathrm{e}^{-\underline{\gamma}_1 z} - \underline{\Gamma}\,\mathrm{e}^{\underline{\gamma}_1 z}}}. \tag{10.63}$$

wave impedance in the incident medium, general two-media case (unit: Ω)

We note that the wave impedance in medium 2, with only one traveling wave, is $\underline{\eta}_{\mathrm{w}2} = \underline{\eta}_2$ [from Eqs. (10.37)].

The concept of wave impedance and the result in Eq. (10.63) are especially useful in the analysis of plane-wave propagation in multilayer media, that involve more than two different material regions (separated by parallel planar interfaces). In such problems,

[11] Note that, in general, for non-TEM waves (to be studied in later chapters), that also have a longitudinal field component (along the wave propagation direction) of either electric or magnetic field, the wave impedance is defined as the ratio of transverse only field intensities (electric to magnetic) at a given point.

as we shall see in a later section, $\underline{\eta}_{\rm w}$ computed at a certain plane (e.g., at one of the multiple material interfaces) can be used as an equivalent "input" impedance, in analogy to the input impedance of a part of an electric circuit, to replace the rest of the structure (beyond that plane) by a homogeneous medium described by an equivalent intrinsic impedance equal to $\underline{\eta}_{\rm w}$.

(b) For the situation in Fig. 10.1, Eqs. (10.8) give the following expression for the wave impedance in the incident medium:

$$\underline{\eta}_{\rm w} = \frac{\underline{E}_{\rm tot}}{\underline{H}_{\rm tot}} = -{\rm j}\,\eta\tan\beta z \quad \text{(incidence from a lossless medium on a PEC)}, \tag{10.64}$$

and this can also be obtained from Eq. (10.63) as a special case – for a perfect dielectric and a perfect conductor as media 1 and 2, respectively, namely, with $\underline{\gamma}_1 = {\rm j}\beta$, $\underline{\eta}_1 = \eta$, and $\underline{\Gamma} = -1$ [Eq. (10.47)].

Problems: 10.9–10.15; *Conceptual Questions* (on Companion Website): 10.11–10.26; *MATLAB Exercises* (on Companion Website).

10.3 SURFACE RESISTANCE OF GOOD CONDUCTORS

Having now the tools for the analysis of reflection and transmission of uniform plane time-harmonic electromagnetic waves at the interface between two arbitrary media (for normal incidence), we revisit our study of skin effect in good conductors at higher frequencies from Section 9.11, and extend it to evaluate the losses in a conducting half-space illuminated by a normally incident plane wave. These losses are an essential attribute that distinguishes wave propagation in the presence of real conductors with respect to the corresponding situations when the conductors can be considered to be perfect (with no losses).

The approximate values for the reflection and transmission coefficients for a good conductor as the reflecting medium are given in Eqs. (10.46). We use the same notation for fields as in Fig. 10.7, but, to simplify writing, let the parameters of the conducting medium be marked here as ε, μ, and σ, and let the medium in front of it be air, as indicated in Fig. 10.10. Nevertheless, all the results derived in this section will be valid also for any other perfect dielectric as medium 1. To express the current density in the conductor, we first note that the complex propagation coefficient ($\underline{\gamma}$) of the transmitted wave and intrinsic impedance ($\underline{\eta}$) of the medium (good conductor), given in Eqs. (9.134) and (9.136), respectively, can be represented in terms of the skin depth (δ) in the conductor, Eq. (9.139), as

$\underline{\gamma}$, $\underline{\eta}$ – good conductors

$$\boxed{\underline{\gamma} = \frac{1}{\delta}(1+{\rm j}), \quad \underline{\eta} = \frac{1}{\sigma\delta}(1+{\rm j}).} \tag{10.65}$$

With this expression for $\underline{\gamma}$, the expression for the complex rms electric field intensity vector for $0 \le z < \infty$ [see Eqs. (10.37), (10.43), and (10.46)] becomes

$$\underline{\mathbf{E}}_2 = \underline{\mathbf{E}}_{\rm t} = \underline{\tau}\,\underline{E}_{\rm i0}\,{\rm e}^{-\underline{\gamma}z}\,\hat{\mathbf{x}} \approx \frac{2\underline{\eta}}{\eta_0}\,\underline{E}_{\rm i0}\,{\rm e}^{-z/\delta}\,{\rm e}^{-{\rm j}z/\delta}\,\hat{\mathbf{x}}, \tag{10.66}$$

where η_0 is the intrinsic impedance of free space [Eq. (9.23)]. Using then the expression for $\underline{\eta}$ in Eqs. (10.65), the associated complex rms current density vector in the conductor is

$$\underline{\mathbf{J}}_2 = \sigma\underline{\mathbf{E}}_2 \approx \frac{2\underline{E}_{\rm i0}}{\delta\eta_0}(1+{\rm j})\,{\rm e}^{-z/\delta}\,{\rm e}^{-{\rm j}z/\delta}\,\hat{\mathbf{x}}. \tag{10.67}$$

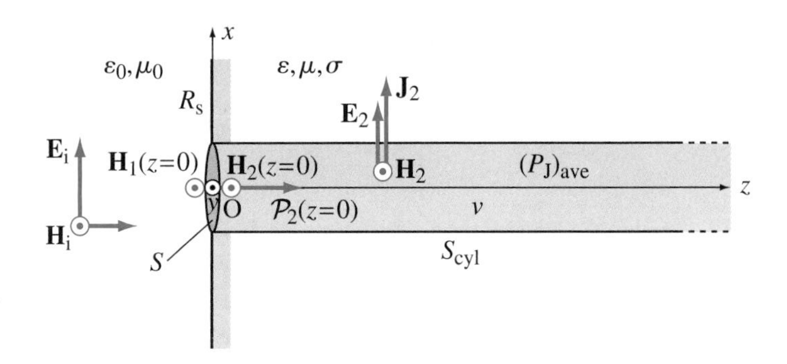

Figure 10.10 Evaluation of losses in a half-space occupied by a good conductor; the conductor is illuminated from air by a normally incident uniform plane time-harmonic electromagnetic wave.

This current, which theoretically exists in the entire conducting half-space, but is practically negligible at locations more than several skin depths away from the conductor surface (see Fig. 9.13), causes the local Joule's losses (to heat) of electromagnetic energy (carried by the transmitted wave) inside the conductor.

As the overall quantitative measure of these losses, it is of interest to find their total time-average power, $(P_J)_{ave}$, for a given (finite) area, S, of the conductor surface. In other words, our goal is to compute the time-average power lost in the volume of a semi-infinite cylinder whose one basis (of surface area S) is placed in the plane $z = 0$ such that the cylinder extends through the entire depth of the conductor ($0 \le z < \infty$), as sketched in Fig. 10.10. The most straightforward way for this is to integrate the local loss power density throughout the cylinder volume, which will be done in the example. Alternatively, we can invoke Poynting's theorem in complex form, Eq. (8.196), and apply it to the same cylinder. Since there are no external energy volume sources (impressed electric fields or currents) inside the cylinder (domain v), the theorem reads

$$0 = \underbrace{\int_v \frac{|\underline{\mathbf{J}}_2|^2}{\sigma}\, dv}_{(P_J)_{ave}} + \underbrace{j\omega \int_v \left(\mu|\underline{\mathbf{H}}_2|^2 - \varepsilon|\underline{\mathbf{E}}_2|^2\right) dv}_{P_{reactive}} + \underbrace{\oint_{S_{cyl}} \underline{\mathcal{P}}_2 \cdot d\mathbf{S}}_{(\underline{P}_{complex})_{out}} . \tag{10.68}$$

Of course, the time-average power of Joule's losses we seek equals the first term on the right-hand side of the equation. The net outward flux of the complex Poynting vector of the transmitted wave (the last term in the equation) represents the complex power flow through the (closed) surface of the cylinder (S_{cyl}) out of the domain v. The complex power flow into the cylinder is the negative of that, and thus Eq. (10.68) gives

$$(\underline{P}_{complex})_{in} = -(\underline{P}_{complex})_{out} = (P_J)_{ave} + jP_{reactive}, \tag{10.69}$$

where $P_{reactive}$ is the reactive power in v [see also Eqs. (8.191)–(8.193)]. Hence, both active (loss) and reactive powers in the cylinder are obtained as

$$(P_J)_{ave} = P_{active} = \text{Re}\{(\underline{P}_{complex})_{in}\}, \quad P_{reactive} = \text{Im}\{(\underline{P}_{complex})_{in}\}. \tag{10.70}$$

We note that $\underline{\mathcal{P}}_2 = \underline{\mathcal{P}}_2\,\hat{\mathbf{z}}$ is tangential to the lateral surface of the cylinder and is zero at the faraway ($z \to \infty$) cylinder basis [due to the exponential attenuation with z in Eq. (10.60)]. Therefore, the inward net flux of $\underline{\mathcal{P}}_2$ through the entire surface of the cylinder reduces to the flux just through the basis (S) lying in the material interface ($z = 0$), counted with respect to the positive z (forward) direction,

$$(\underline{P}_{complex})_{in} = \underline{\mathcal{P}}_2(z=0) \cdot S\,\hat{\mathbf{z}} = \underline{\mathcal{P}}_2(z=0)S. \tag{10.71}$$

We now wish to express the input power in terms of the total magnetic field intensity in the incident medium, right at the interface, which, from Eqs. (10.55) and (10.46), is approximately given by

$$\underline{\mathbf{H}}_1(z=0) \approx \frac{2\underline{E}_{i0}}{\eta_0}\,\hat{\mathbf{y}}. \tag{10.72}$$

Namely, while the magnetic fields of the incident and reflected waves are added together constructively (in phase) at the interface, the corresponding electric fields practically cancel each other, resulting in a very small total electric field for $z = 0$. That is why it appears convenient to use this large quantity (total magnetic field intensity) rather than the vanishingly small one (total electric field intensity) for further power expressions and calculations. Indeed, Eq. (9.94) tells us that the complex magnitude of the Poynting vector in the second medium can be expressed in terms of the constituting magnetic field ($\underline{\mathbf{H}}_2$) only:

$$\underline{\mathcal{P}}_2 = \underline{\eta}|\underline{\mathbf{H}}_2|^2. \tag{10.73}$$

In addition, the boundary condition in Eq. (10.39) gives

$$\underline{\mathbf{H}}_1(z=0) = \underline{\mathbf{H}}_2(z=0), \tag{10.74}$$

and hence

$$\underline{\mathcal{P}}_2(z=0) = \underline{\eta}|\underline{\mathbf{H}}_1(z=0)|^2. \tag{10.75}$$

Combining Eqs. (10.70), (10.71), and (10.75), the time-average loss power in the cylinder in Fig. 10.10 becomes

$$(P_\mathrm{J})_\mathrm{ave} = \mathrm{Re}\{\underline{\mathcal{P}}_2(z=0)\}S = \mathrm{Re}\{\underline{\eta}\}|\underline{\mathbf{H}}_1(z=0)|^2 S, \tag{10.76}$$

and a similar expression, with $\mathrm{Im}\{\underline{\eta}\}$ instead of $\mathrm{Re}\{\underline{\eta}\}$, is obtained for the reactive power in the cylinder, P_reactive. Dividing $(P_\mathrm{J})_\mathrm{ave}$ by S, we obtain the loss or ohmic power in the conductor per unit area of its surface,

time-average loss power per unit area of good-conductor surface (unit: $\mathrm{W/m^2}$)

$$\boxed{\frac{(P_\mathrm{J})_\mathrm{ave}}{S} = R_\mathrm{s}|\underline{\mathbf{H}}_1(z=0)|^2 \approx \frac{4R_\mathrm{s}|\underline{E}_{i0}|^2}{\eta_0^2},} \tag{10.77}$$

with R_s standing for the real part of the complex intrinsic impedance of the conductor. This new quantity is called the surface resistance of the conductor, and its unit appears to be Ω (the same unit as for $\underline{\eta}$). From Eq. (9.136),

surface resistance of a good conductor (unit: Ω/square)

$$\boxed{R_\mathrm{s} = \mathrm{Re}\{\underline{\eta}\} \approx \sqrt{\frac{\pi\mu f}{\sigma}},} \tag{10.78}$$

where, most frequently, $\mu = \mu_0$ (nonmagnetic conductors). The surface resistance of a conductor (at a given frequency) and its skin depth (δ) are the most important parameters related to the skin effect in good conductors. We see that R_s is inversely proportional to the square root of conductivity – the better the conductor (larger σ) the lower the losses (in the limit, $R_\mathrm{s} = 0$ for a perfect conductor). We also see that, unlike δ in Eq. (9.139), R_s is directly proportional to the square root of frequency – the losses grow with an increase of frequency. Because of this general trend, as will be discussed in later chapters, there is always an upper frequency limit for practical usability of a given transmission line or metallic waveguide aimed to transport electromagnetic signals (energy or information) to a certain distance (except for structures made of superconductors).

Since $R_s = 1/(\sigma\delta)$ [see Eqs. (10.78) and (10.65)], it can be visualized as the standard (dc) resistance of a square-shaped conducting layer (under the surface of the conductor) with thickness δ and conductivity σ, as depicted in Fig. 10.11. Namely, using the expression for the dc resistance of a resistor of uniform cross section in Eq. (3.85), the layer in Fig. 10.11, considered as a resistor whose length is l and cross section (perpendicular to the current flow) $S_0 = l\delta$ in area, exhibits

$$R_{\text{on } l^2} = \frac{l}{\sigma S_0} = \frac{l}{\sigma l\delta} = \frac{1}{\sigma\delta} = R_s. \tag{10.79}$$

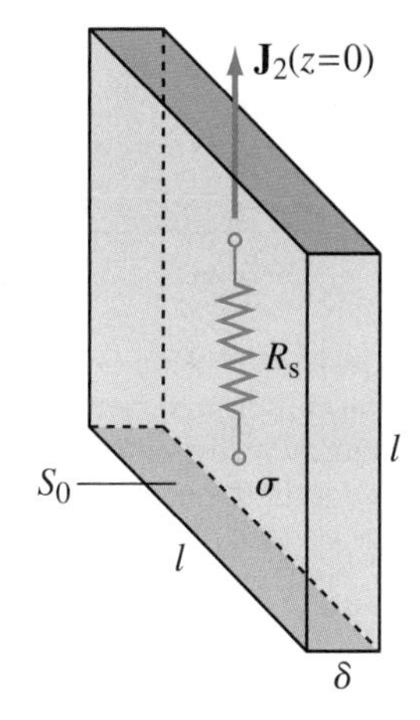

Figure 10.11 Visualization of the surface resistance R_s of a conductor with the skin effect pronounced (Fig. 10.10) as the dc resistance of a resistor formed by a square-shaped conducting layer (under the surface of the conductor), according to Eq. (10.79).

We see that this resistance does not depend on the size of the face of the layer, as long as it is square in shape. This is why R_s is often stated in "ohms per square" (Ω/square), rather than just ohms, where, again, the actual size of the "square" is immaterial. Using ohms-per-square for surface resistances and impedances in general, including the surface resistance (or impedance) of thin material (resistive/reactive) films, while ohms stay reserved for usual resistances (impedances), is also helpful in avoiding any confusion between the two different physical concepts. In addition, we note that this illustration of the concept of the surface resistance (in Fig. 10.11) is in line with the illustration of the definition of the skin depth in Fig. 9.13. Indeed, Fig. 9.13 points out that the conductor current with exponentially decaying amplitude can be replaced by an equivalent uniform current, with an amplitude equal to that at the conductor surface in the original case, in a layer of thickness δ, and no current in the rest of the conductor. So this uniform current, represented by a step function in Fig. 9.13, can be thought of as flowing through the resistor of resistance R_s in Fig. 10.11.

The surface resistance of good conductors is generally a very small quantity. For instance, we know that $R_s \ll \eta_0$ since $|\eta| \ll \eta_0$. In addition, a formula, from Eq. (10.78), for R_s for copper ($\sigma = 58$ MS/m, $\mu = \mu_0$) as a function of frequency, analogous to the formula for δ_{Cu} in Eq. (9.140), reads

$$\boxed{(R_s)_{\text{Cu}} \approx 261\sqrt{f}\ \text{n}\Omega/\text{square} \quad (f \text{ in Hz}).} \tag{10.80}$$

surface resistance of copper

It comes out that $(R_s)_{\text{Cu}}$ at 60 Hz is as small as about 2 $\mu\Omega$/square. Even at the frequency of 1 THz, it amounts to only about 261 mΩ/square.

Because the imaginary part of $\underline{\eta}$ in Eq. (9.136) is the same as the real part, the reactive power [Eq. (10.70)] per unit area of the conductor surface is equal to the corresponding active power,

$$\boxed{\frac{P_{\text{reactive}}}{S} = X_s|\underline{\mathbf{H}}_1(z=0)|^2 = \frac{(P_J)_{\text{ave}}}{S} \quad (X_s = \text{Im}\{\underline{\eta}\} = R_s),} \tag{10.81}$$

surface density of reactive power

where X_s is termed the surface reactance of the conductor. We can also define the complex surface impedance of the conductor,

$$\boxed{\underline{Z}_s = R_s + \text{j}X_s \quad (\underline{Z}_s = \underline{\eta}),} \tag{10.82}$$

complex surface impedance

expressed in Ω (or Ω-per-square).

Next, let us compute the total complex current intensity across a section of the conductor in Fig. 10.10 that is l wide in the y direction. This is done by integrating the current density $\underline{\mathbf{J}}_2$ in Eq. (10.67) over the area S_l of width l in the plane $x = 0$ (perpendicular to $\underline{\mathbf{J}}_2$), from the surface of the conductor to infinity, as shown in

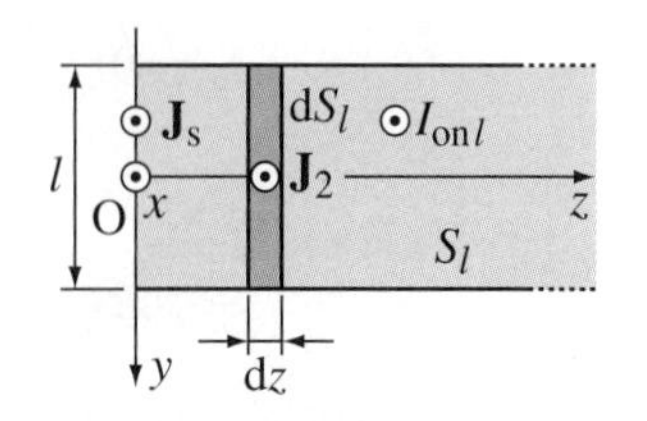

Figure 10.12 Finding the total complex current intensity across an l-wide section of the conductor in Fig. 10.10 by integration in Eq. (10.83).

Fig. 10.12. The result is

$$\underline{I}_{\mathrm{on}\,l} = \int_{S_l} \underbrace{\underline{J}_2\, l\, \mathrm{d}z}_{\mathrm{d}S_l} \approx \frac{2\underline{E}_{\mathrm{i}0} l}{\delta \eta_0} (1+\mathrm{j}) \int_{z=0}^{\infty} \mathrm{e}^{-z(1+\mathrm{j})/\delta}\, \mathrm{d}z = \frac{2\underline{E}_{\mathrm{i}0} l}{\eta_0}, \tag{10.83}$$

with $\mathrm{d}S_l$ being the area of an elemental surface for integration in the form of a thin strip perpendicular to the z-axis. Dividing $\underline{I}_{\mathrm{on}\,l}$ by l, we obtain the current intensity per unit width (in the y direction), in A/m, of the infinitely wide conductor in Figs. 10.10 and 10.12. Having in mind Eq. (3.13), this current-intensity to conductor-width ratio can also be thought of as the line density of a surface current, $\underline{J}_s$, localized (Fig. 10.12) in an infinitely thin film over the conductor surface (in the plane $z = 0$),

$$\underline{J}_s = \frac{\underline{I}_{\mathrm{on}\,l}}{l} \approx \frac{2\underline{E}_{\mathrm{i}0}}{\eta_0} \quad (\underline{\mathbf{J}}_s = \underline{J}_s\, \hat{\mathbf{x}}). \tag{10.84}$$

We see that this surface current density is approximately the same as that in Eq. (10.13) for a perfect conductor in Fig. 10.1 (where $\delta = 0$). We also see that, although $\underline{\mathbf{J}}_2$, of course, depends on σ, both $\underline{I}_{\mathrm{on}\,l}$ and $\underline{\mathbf{J}}_s$ are independent of it. This interesting result tells us that, as the conductivity of a real conductor increases, the same amount of current is being redistributed more and more towards the surface of the material, and in the limit of $\sigma \to \infty$ (PEC), a surface current of density $\underline{\mathbf{J}}_s$, with the same current intensity per unit width of the conductor, is obtained.

Comparing Eqs. (10.84) and (10.72), we realize that $|\underline{\mathbf{J}}_s| = |\underline{\mathbf{H}}_1(z = 0)|$, so that the loss power relationship in Eq. (10.77) can now be written as

"surface" form of Joule's law

$$\boxed{\frac{(P_\mathrm{J})_\mathrm{ave}}{S} = R_s |\underline{\mathbf{J}}_s|^2,} \tag{10.85}$$

which may be viewed as a "surface" form of Joule's law [see Eqs. (3.77) and (8.53)] – surface density of time-average power in a conductor (with the skin effect pronounced) equals the surface resistance times the rms surface current density squared. Similarly, from Eqs. (10.38), (10.66), (10.82), and (10.84), the total complex electric field intensity vector at the surface of a conductor, which is very small, can be expressed as the product of the complex surface impedance and complex surface current density vector,

"surface" Ohm's law in local form

$$\underline{\mathbf{E}}_1(z = 0) = \underline{\mathbf{E}}_2(z = 0) \approx \frac{2\underline{\eta}\,\underline{E}_{\mathrm{i}0}}{\eta_0}\,\hat{\mathbf{x}} \quad \longrightarrow \quad \boxed{\underline{\mathbf{E}}_1(z = 0) = \underline{Z}_s \underline{\mathbf{J}}_s,} \tag{10.86}$$

and this may be identified with Ohm's law in local form, Eq. (3.20), as its "surface" counterpart. In other words, it can be regarded as the constitutive equation characterizing the intrinsic material properties of a conducting surface. Note that the approximation for Γ in Eqs. (10.46) leads to $\underline{\mathbf{E}}_1(z = 0) \approx 0$ [see Eq. (10.51)]; obviously, Eq. (10.86) provides a better approximation for this small quantity.

Example 10.12 Direct Computation of the Loss Power in a Good Conductor

(a) Consider Eq. (10.68) and obtain the time-average power of Joule's losses in the cylinder in Fig. 10.10 directly, by performing the indicated volume integration. (b) Find the reactive power in the cylinder from the respective volume integral in Eq. (10.68).

Solution

(a) Finding $|\underline{\mathbf{J}}_2|$ from Eq. (10.67), given that $|1+\mathrm{j}| = \sqrt{2}$, we have

$$(P_\mathrm{J})_\mathrm{ave} = \int_v \frac{|\underline{\mathbf{J}}_2|^2}{\sigma} \underbrace{S\,\mathrm{d}z}_{\mathrm{d}v} \approx \frac{8|\underline{E}_{\mathrm{i}0}|^2 S}{\sigma\delta^2\eta_0^2} \int_{z=0}^{\infty} \mathrm{e}^{-2z/\delta}\,\mathrm{d}z = \frac{4|\underline{E}_{\mathrm{i}0}|^2 S}{\sigma\delta\eta_0^2}, \tag{10.87}$$

where $\mathrm{d}v$ is the volume of a thin slice of the cylinder (of thickness $\mathrm{d}z$), and this, of course, is the same result as in Eqs. (10.77) and (10.79).

(b) The reactive power term in Eq. (10.68) can be integrated in a similar fashion. However, having in mind Eqs. (9.90), (9.137), and (9.133), we do not need to carry out the actual integration to realize that

$$P_\mathrm{reactive} = \omega \int_v \left(\mu|\underline{\mathbf{H}}_2|^2 - \varepsilon|\underline{\mathbf{E}}_2|^2\right) \mathrm{d}v = \omega \int_v \left(\frac{\mu}{|\underline{\eta}|^2} - \varepsilon\right) |\underline{\mathbf{E}}_2|^2\,\mathrm{d}v$$

$$\approx \omega \int_v \underbrace{\left(\frac{\sigma}{\omega} - \varepsilon\right)}_{\approx\sigma/\omega} |\underline{\mathbf{E}}_2|^2\,\mathrm{d}v \approx \int_v \sigma|\underline{\mathbf{E}}_2|^2\,\mathrm{d}v = (P_\mathrm{J})_\mathrm{ave}, \tag{10.88}$$

which is the same conclusion $[P_\mathrm{reactive} = (P_\mathrm{J})_\mathrm{ave}]$ as in Eq. (10.81), where, of course, the result of integration for $(P_\mathrm{J})_\mathrm{ave}$ is given in Eq. (10.87). We see that the reactive power in the conductor is predominantly concentrated in the magnetic field of the transmitted electromagnetic wave. Namely, since $\sigma/\omega \gg \varepsilon$ (for good conductors), it appears that $\mu|\underline{\mathbf{H}}_2|^2 \gg \varepsilon|\underline{\mathbf{E}}_2|^2$ in Eq. (10.88).

Conceptual Questions (on Companion Website): 10.27 and 10.28; *MATLAB Exercises* (on Companion Website).

10.4 PERTURBATION METHOD FOR EVALUATION OF SMALL LOSSES

This section uses the concept of surface resistance developed in the previous section and presents an approximate method for evaluation of (small) losses in good conductors using the fields in front of the conductor surface obtained as if the conductor were perfect (Fig. 10.1). In addition, this method is generalized to conductors of practically arbitrary shapes in arbitrary high-frequency electromagnetic fields (not necessarily uniform plane waves).

We emphasize the fact that – in order to evaluate the loss (ohmic) and reactive powers in a good conductor (in Fig. 10.10) based on Poynting's theorem, that is, using Eqs. (10.77) and (10.81), we do not need to know the distribution of either electric or magnetic field in the conductor; we only need to know the magnitude of the resultant magnetic field $\underline{\mathbf{H}}_1(z=0)$ on the conductor surface, Eq. (10.72). However, this field is practically the same whether the conductor is good or perfect [see the expression for $\underline{\mathbf{H}}_\mathrm{tot}(z=0)$ in Eqs. (10.8), for a PEC case], and in both cases its magnitude equals the magnitude of the surface current density vector $\underline{\mathbf{J}}_\mathrm{s}$ in Eq. (10.84). This gives rise to the so-called perturbation method for evaluation of losses (and reactive power) in good conductors, according to which the magnetic field at the conductor surface is first computed assuming that the conductor is perfect (which greatly simplifies the analysis), and then the loss and reactive powers in the conductor are determined using Eqs. (10.77) and (10.81). So, we first carry out the analysis of waves and fields in front of the conductor surface completely neglecting the losses in the material, and then perturb that ideal picture by allowing the (small) losses to occur inside the conductor; however, the losses (and reactive power) are evaluated using the no-loss (ideal) field picture outside the conductor.

Moreover, this method applies not only to reflections of plane waves from (infinite) flat conducting surfaces, but to conductors of arbitrary shapes in arbitrary electromagnetic fields. The only restriction is that the skin effect in the conductor must be pronounced, i.e., the skin depth (δ) must be small when compared to the relevant dimensions of the object. Since, in the most general case, the field around the conductor is nonuniform and the conductor surface is nonplanar, the loss power in Eq. (10.77) must be computed on a differentially small surface $\mathrm{d}S$ (rather than the arbitrarily large surface S),

surface density of ohmic power in an arbitrary conductor with the skin effect pronounced

$$\frac{\mathrm{d}(P_\mathrm{J})_\mathrm{ave}}{\mathrm{d}S} = R_\mathrm{s}|\underline{\mathbf{H}}_\mathrm{tang}|^2, \tag{10.89}$$

where the surface resistance, R_s, is computed using Eq. (10.78), and $\underline{\mathbf{H}}_\mathrm{tang}$ is the tangential component of the complex rms magnetic field intensity vector on the conductor surface, that is, on the patch $\mathrm{d}S$ [the normal component of $\underline{\mathbf{H}}$ is zero or negligibly small, which is apparent from the boundary condition for the vector $\mathbf{B}$ in Eqs. (8.33)]. Most importantly, this field is computed as if the object were nonpenetrable, i.e., made of a PEC. The total time-average power of Joule's losses in the conductor amounts to

total ohmic losses and reactive power in an arbitrary conductor

$$(P_\mathrm{J})_\mathrm{ave} = \int_S \mathrm{d}(P_\mathrm{J})_\mathrm{ave} = \int_S R_\mathrm{s}|\underline{\mathbf{H}}_\mathrm{tang}|^2\,\mathrm{d}S = P_\mathrm{reactive}, \tag{10.90}$$

with S now standing for the surface of the conductor or the part of the surface that is exposed to the electromagnetic field. Of course, the total reactive power in the conductor is the same. In addition, the tangential component of the local complex rms electric field intensity vector on the conductor surface, which, although rather small, is sometimes of interest for the analysis, can also be found from $\underline{\mathbf{H}}_\mathrm{tang}$, via the associated rms surface current density vector [see Eqs. (10.13) and (10.86)],

tangential electric field on a conducting surface

$$\underline{\mathbf{J}}_\mathrm{s} = \hat{\mathbf{n}} \times \underline{\mathbf{H}}_\mathrm{tang} \quad \longrightarrow \quad \underline{\mathbf{E}}_\mathrm{tang} = \underline{Z}_\mathrm{s}\underline{\mathbf{J}}_\mathrm{s}, \tag{10.91}$$

where $\hat{\mathbf{n}}$ is the outward local normal unit vector on the surface, and the complex surface impedance of the conductor, $\underline{Z}_\mathrm{s}$, is given in Eq. (10.82). Of course, $\underline{\mathbf{E}}_\mathrm{tang} = 0$ for a PEC object.

The perturbation method for evaluation of losses in good conductors at higher frequencies, at which the skin effect is pronounced, will be used on many occasions throughout the rest of this text. For instance, in the next chapter, we shall use Eq. (10.90) to determine the per-unit-length power of Joule's losses and the associated wave attenuation along transmission lines due to the (small) penetration of guided electromagnetic waves, more precisely – their electric field, into the conductors of the line. Based on this power, the high-frequency resistance per unit length of different transmission lines will be computed. The attenuation of guided waves due to Joule's losses in conductors for very long lines may be so large that the structure becomes impractical to convey signals at a given frequency or in a given frequency range. Similarly, the per-unit-length reactive power inside the conductors, which is due to the (small) penetration of the magnetic field of the guided wave into the conductors and will be obtained from Eq. (10.90) as well, will give us the high-frequency internal inductance p.u.l. of the line. In all cases, $\underline{\mathbf{H}}_\mathrm{tang}$ on the surface of conductors will be found assuming that they are perfect.

Example 10.13 **Loss Power in a Copper Plane for Normal Wave Incidence**

A uniform plane time-harmonic electromagnetic wave of frequency $f = 1$ GHz and rms electric field intensity $E_{i0} = 10$ V/m is normally incident from air on the planar surface of a large copper conductor. Assuming that the skin effect is pronounced, find the time-average power of Joule's losses in the conductor per unit area of its surface.

Solution At the given frequency, $\sigma/(2\pi f\varepsilon_0) \approx 10^9$ (for copper, $\sigma = 58$ MS/m), so the condition in Eq. (9.133) is definitely satisfied, and, as expected, copper can be treated as a good (low-loss) conductor. With the skin effect pronounced, we determine the loss power in the conductor using Eq. (10.89). In addition, because the losses are small, the perturbation method for their evaluation can be applied, with the tangential component of the complex rms magnetic field intensity vector near the conductor surface, $\underline{\mathbf{H}}_{\text{tang}}$, found as if the conductor were perfect (PEC), and this exactly is the situation in Fig. 10.1. Therefore, $\underline{\mathbf{H}} = \underline{\mathbf{H}}_{\text{tang}}$ is the total magnetic field vector (of the standing wave) in front of the PEC screen given in Eqs. (10.8) for $z = 0$ ($\underline{\mathbf{H}}_{\text{tot}}$ is entirely tangential to the plane). Finally, we can equivalently employ Eq. (10.85) to evaluate losses, and, by the perturbation method, the actual complex rms surface current density vector, $\underline{\mathbf{J}}_{\text{s}}$, over the conductor surface can approximately be replaced by $\underline{\mathbf{J}}_{\text{s}}$ for the no-loss case – from Eq. (10.13). With all this in mind, the time-average loss power per unit area of the conductor surface comes out to be

$$\frac{\mathrm{d}(P_{\mathrm{J}})_{\text{ave}}}{\mathrm{d}S} = R_{\text{s}}|\underline{\mathbf{H}}_{\text{tang}}|^2 = R_{\text{s}}|\underline{\mathbf{J}}_{\text{s}}|^2 = R_{\text{s}}\left|\underline{\mathbf{H}}_{\text{tot}}\right|^2_{z=0} = \frac{4R_{\text{s}}E_{\text{i0}}^2}{\eta_0^2} = 23.22\ \mu\text{W/m}^2, \qquad (10.92)$$

where $\eta_0 = 377\ \Omega$, Eq. (9.23), and the surface resistance of copper, $R_{\text{s}} = (R_{\text{s}})_{\text{Cu}} = 8.25\ \text{m}\Omega/\text{square}$, is computed from Eq. (10.80).

Problems: 10.16; *Conceptual Questions* (on Companion Website): 10.29; *MATLAB Exercises* (on Companion Website).

10.5 OBLIQUE INCIDENCE ON A PERFECT CONDUCTOR

In this section, we generalize the analysis of plane-wave reflections upon perfectly conducting surfaces for the normal incidence on the surface (Fig. 10.1) to the case of an arbitrary, oblique, incidence. Namely, we now let an incident uniform plane time-harmonic electromagnetic wave approach the PEC boundary at an arbitrary angle, so-called incident angle, θ_{i} ($0 \le \theta_{\text{i}} \le 90°$) with respect to the normal on the boundary (for the normal incidence, $\theta_{\text{i}} = 0$). Furthermore, let the incident electric field intensity vector, $\mathbf{E}_{\text{i}}$, be normal to the plane of incidence, defined by the direction of incident-wave propagation (incident ray) and normal on the boundary, as shown in Fig. 10.13(a). An obliquely incident wave with such orientation of $\mathbf{E}_{\text{i}}$ is said to be normally (or perpendicularly) polarized. The other characteristic case, with $\mathbf{E}_{\text{i}}$ lying in the plane of incidence (and, of course, being perpendicular to the direction of wave travel), is depicted in Fig. 10.13(b). It is referred to as the parallel polarization of the incident wave, and will be studied in an example.[12] These two cases need to

[12]Note that the normal (perpendicular) and parallel polarizations of the incident wave in Fig. 10.13(a) and Fig. 10.13(b) are sometimes referred to as the horizontal and vertical polarizations, respectively. This comes from a frequent situation where the reflection plane is the earth's surface, with the case in Fig. 10.13(a) corresponding to a horizontal $\mathbf{E}_{\text{i}}$ and the case in Fig. 10.13(b) to $\mathbf{E}_{\text{i}}$ in a vertical plane. In addition, these two cases are sometimes also labeled, respectively, as s- and p-polarization ("s" being an abbreviation for the German *senkrecht*, meaning perpendicular, and "p" for the German word for parallel, which is, also, *parallel*). Finally, some texts use TE versus TM polarization terminology, which refers to the electric field, in case (a), versus magnetic field, in case (b), being transverse to the observation direction normal to the PEC interface in Fig. 10.13.

be considered separately, since the structures of the electric and magnetic fields for these two mutually orthogonal polarizations are distinctly different. For instance, as can be seen in Fig. 10.13, the electric field in case (a) has only a component parallel to the PEC interface and the magnetic field has both parallel and normal components (with respect to the interface), whereas the combination of field components in case (b) is just opposite.

In general, any uniform plane wave with an oblique incidence on a PEC surface and arbitrary direction of the vector $\mathbf{E}_{\rm i}$ in the plane perpendicular to the incident ray can be decomposed onto a uniform plane wave with normal polarization and another one with parallel polarization. Simply, $\mathbf{E}_{\rm i}$ in the general case can be represented as a superposition of two mutually orthogonal vectors, one normal to the plane of incidence (normal polarization) and the other parallel to it (parallel polarization).[13] Therefore, once the analysis of the two basic polarization cases in Fig. 10.13 is complete, the resultant electric and magnetic fields in the incident medium in the case of an arbitrary orientation of $\mathbf{E}_{\rm i}$ can be obtained by superposition of the corresponding resultant field expressions for individual basic polarizations.

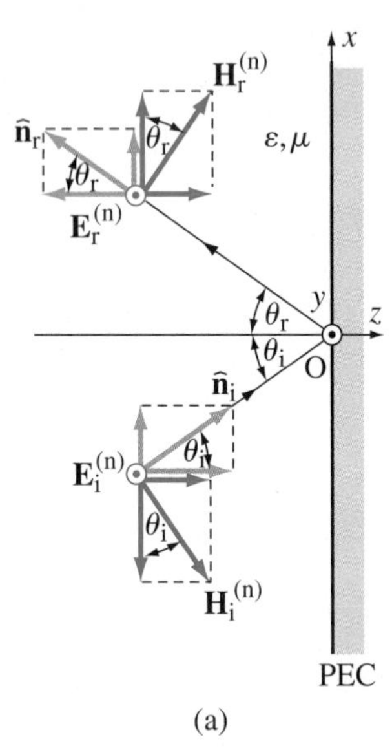

(a)

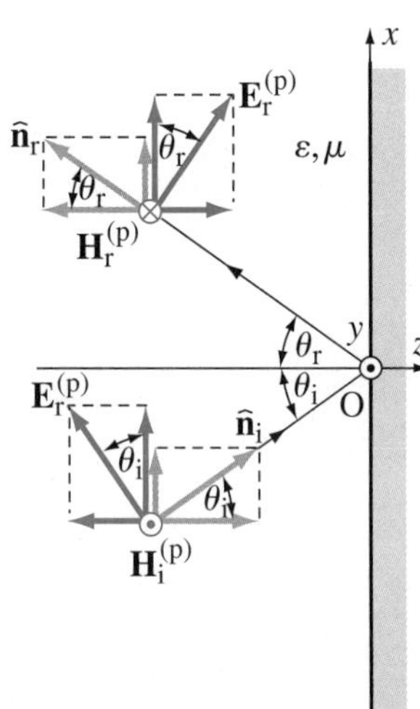

(b)

Figure 10.13 Oblique incidence, on a planar perfect electric conductor, of a uniform plane time-harmonic electromagnetic wave with (a) normal (perpendicular) polarization and (b) parallel polarization; The incident medium is a perfect dielectric, and the plane of drawing is also the plane of incidence.

Because of symmetry, the reflected ray in Fig. 10.13 is also in the plane of incidence. The angle it forms with the normal on the interface is $\theta_{\rm r}$ – the reflected angle. Since (unlike the situation in Fig. 10.1 for the normal incidence) it is impossible to have both incident and reflected rays along one or two axes of a single Cartesian coordinate system, the global coordinate system in Fig. 10.13 is adopted irrespective of the direction of any of them and such (as in Fig. 10.1) that the plane $z = 0$ coincides with the PEC interface. Hence, we must use the field expressions for an arbitrarily directed TEM wave derived in association with Fig. 9.9 – for both the incident and reflected waves in Fig. 10.13. The Cartesian x- and z-components of propagation unit vectors of the waves (of course, their y-components are zero) are easily identified in terms of the incident and reflected angles in Fig. 10.13, yielding

$$\hat{\mathbf{n}}_{\rm i} = \sin\theta_{\rm i}\,\hat{\mathbf{x}} + \cos\theta_{\rm i}\,\hat{\mathbf{z}}, \quad \hat{\mathbf{n}}_{\rm r} = \sin\theta_{\rm r}\,\hat{\mathbf{x}} - \cos\theta_{\rm r}\,\hat{\mathbf{z}}, \tag{10.93}$$

for both wave polarizations.

In the normal polarization case, Fig. 10.13(a), the electric field vectors of the incident and reflected waves, $\underline{\mathbf{E}}_{\rm i}^{(n)}$ and $\underline{\mathbf{E}}_{\rm r}^{(n)}$, have only y-components. Using Eqs. (9.69), (9.71), and (9.72), the expressions for these vectors at an arbitrary point in the incident medium, defined by the position vector $\mathbf{r}$ (with respect to the coordinate origin) or by coordinates x, y, and z ($z \le 0$), are given by

$$\underline{\mathbf{E}}_{\rm i}^{(n)} = \underline{E}_{\rm i0}\,{\rm e}^{-{\rm j}\beta\mathbf{r}\cdot\hat{\mathbf{n}}_{\rm i}}\,\hat{\mathbf{y}} = \underline{E}_{\rm i0}\,{\rm e}^{-{\rm j}\beta(x\sin\theta_{\rm i}+z\cos\theta_{\rm i})}\,\hat{\mathbf{y}}, \tag{10.94}$$

$$\underline{\mathbf{E}}_{\rm r}^{(n)} = \underline{E}_{\rm r0}\,{\rm e}^{-{\rm j}\beta\mathbf{r}\cdot\hat{\mathbf{n}}_{\rm r}}\,\hat{\mathbf{y}} = \underline{E}_{\rm r0}\,{\rm e}^{-{\rm j}\beta(x\sin\theta_{\rm r}-z\cos\theta_{\rm r})}\,\hat{\mathbf{y}}. \tag{10.95}$$

[13] Note that the decomposition onto two (linearly polarized) waves with normal and parallel polarizations, respectively, applies directly even to an elliptically polarized obliquely incident wave. Namely, an arbitrary elliptically polarized time-harmonic vector can be represented as a superposition of two linearly polarized vectors oscillating, in the course of time, along mutually orthogonal straight lines (local axes), where the direction of one of the two axes can be adopted at will in the plane of the polarization ellipse. So if the axis for one of the two linearly polarized components of the elliptically polarized incident wave is adopted to be normal to the plane of incidence (normal polarization), the other one will lie in the plane of incidence (parallel polarization). Of course, any particular polarization state of the incident wave implies a particular magnitude ratio and phase shift between the vectors $\mathbf{E}_{\rm i}$ for the two linearly polarized components.

The boundary condition in Eq. (10.3) now gives the following relationship between complex constants $\underline{E}_{i0}$ and $\underline{E}_{r0}$:

$$\left[\underline{\mathbf{E}}_{i}^{(n)} + \underline{\mathbf{E}}_{r}^{(n)}\right]\Big|_{z=0} = 0 \quad \longrightarrow \quad \underline{E}_{i0}\, e^{-j\beta x \sin\theta_i} + \underline{E}_{r0}\, e^{-j\beta x \sin\theta_r} = 0, \tag{10.96}$$

which must be satisfied for any x at the interface $(-\infty < x < \infty)$. This is possible only if the two exponential functions of x in Eq. (10.96) are the same, that is, only if the reflected angle is the same as incident:

$$\boxed{\theta_r = \theta_i.} \tag{10.97}$$

Snell's law of reflection

The fact that the reflected wave propagates away from the interface along the path that is the mirror image (with respect to the normal on the interface) of the path of the incident wave is known as Snell's law of reflection. Once the exponential terms in Eq. (10.96) are thus eliminated, we are left with the condition $\underline{E}_{r0} = -\underline{E}_{i0}$.[14] Substituting it, along with Eq. (10.97), back in Eq. (10.95), the total electric field vector in the incident medium (for $z \le 0$) in Fig. 10.13(a) is

$$\underline{\mathbf{E}}_{tot}^{(n)} = \underline{\mathbf{E}}_{i}^{(n)} + \underline{\mathbf{E}}_{r}^{(n)} = \underline{E}_{i0}\left(e^{-j\beta z\cos\theta_i} - e^{j\beta z\cos\theta_i}\right) e^{-j\beta x\sin\theta_i}\,\hat{\mathbf{y}} = \underline{E}_{tot\,y}^{(n)}\,\hat{\mathbf{y}}, \tag{10.98}$$

and the expression for its complex magnitude (y-component) can be written [see Eqs. (10.7)] as

$$\boxed{\underline{E}_{tot\,y}^{(n)} = -2j\underline{E}_{i0}\sin(\beta z\cos\theta_i)\, e^{-j\beta x\sin\theta_i}.} \tag{10.99}$$

$\underline{E}_y$ – normal polarization

Magnetic field vectors in Fig. 10.13(a), $\underline{\mathbf{H}}_{i}^{(n)}$ and $\underline{\mathbf{H}}_{r}^{(n)}$, have both x- and z-components, and so do the unit vectors in their respective directions. The components of the unit vectors can be either identified directly from Fig. 10.13(a), like in Eq. (10.93) for propagation unit vectors, or computed as $\hat{\mathbf{n}}_i \times \hat{\mathbf{y}}$ and $\hat{\mathbf{n}}_r \times \hat{\mathbf{y}}$, respectively, from Eq. (9.70), since the unit vector in the direction of both $\underline{\mathbf{E}}_{i}^{(n)}$ and $\underline{\mathbf{E}}_{r}^{(n)}$ is $\hat{\mathbf{y}}$. We thus have

$$\underline{\mathbf{H}}_{i}^{(n)} = \frac{\underline{E}_{i0}}{\eta}\, e^{-j\beta(x\sin\theta_i + z\cos\theta_i)} \underbrace{(-\cos\theta_i\,\hat{\mathbf{x}} + \sin\theta_i\,\hat{\mathbf{z}})}_{\text{unit vector for } H_i^{(n)}}, \tag{10.100}$$

$$\underline{\mathbf{H}}_{r}^{(n)} = \frac{\underline{E}_{r0}}{\eta}\, e^{-j\beta(x\sin\theta_r - z\cos\theta_r)} \underbrace{(\cos\theta_r\,\hat{\mathbf{x}} + \sin\theta_r\,\hat{\mathbf{z}})}_{\text{unit vector for } H_r^{(n)}}, \tag{10.101}$$

with $\theta_r = \theta_i$ and $\underline{E}_{r0} = -\underline{E}_{i0}$. Hence, the x- and z-components of the total magnetic field vector,

$$\underline{\mathbf{H}}_{tot}^{(n)} = \underline{\mathbf{H}}_{i}^{(n)} + \underline{\mathbf{H}}_{r}^{(n)} = \underline{H}_{tot\,x}^{(n)}\,\hat{\mathbf{x}} + \underline{H}_{tot\,z}^{(n)}\,\hat{\mathbf{z}}, \tag{10.102}$$

---

[14] Note that both obtained relationships $\theta_r = \theta_i$ and $\underline{E}_{r0} = -\underline{E}_{i0}$ can alternatively be deduced by looking at the scattered electromagnetic field due to surface currents (of density $\underline{\mathbf{J}}_s$) induced in the PEC plane, similarly to the corresponding discussion associated with Eqs. (10.2) and (10.4) in the normal incidence case in Fig. 10.1. Namely, we can remove the PEC medium in Fig. 10.13(a), and consider the induced currents flowing in the plane $z = 0$ to exist in an unbounded homogeneous medium (of parameters ε, μ, and $\sigma = 0$). These currents, in the equivalent model, produce two waves, one of which propagates to the right of the plane $z = 0$ in the direction of the unit vector $\hat{\mathbf{n}}_i$ and cancels the incident wave (zero total electromagnetic field in the PEC). The other wave radiated by the currents, the reflected wave with the electric field given in Eq. (10.95), propagates symmetrically back into the $z < 0$ half-space, and hence $\theta_r = \theta_i$. In addition, because of the cancelation of fields for $z > 0$, we have $\underline{E}_{r0} = -\underline{E}_{i0}$ in the incident medium.

can be written as

$\underline{H}_x$ – *normal polarization*

$\underline{H}_z$ – *normal polarization*

$$\underline{H}^{(\mathrm{n})}_{\mathrm{tot}\,x} = -\frac{2\underline{E}_{\mathrm{i}0}}{\eta}\cos\theta_{\mathrm{i}}\cos(\beta z\cos\theta_{\mathrm{i}})\,\mathrm{e}^{-\mathrm{j}\beta x\sin\theta_{\mathrm{i}}}, \tag{10.103}$$

$$\underline{H}^{(\mathrm{n})}_{\mathrm{tot}\,z} = -\frac{2\mathrm{j}\underline{E}_{\mathrm{i}0}}{\eta}\sin\theta_{\mathrm{i}}\sin(\beta z\cos\theta_{\mathrm{i}})\,\mathrm{e}^{-\mathrm{j}\beta x\sin\theta_{\mathrm{i}}}. \tag{10.104}$$

Note that, in addition to the fact that $\underline{E}^{(\mathrm{n})}_{\mathrm{tot}\,y}(z=0)=0$, from Eq. (10.99), we also have that $\underline{H}^{(\mathrm{n})}_{\mathrm{tot}\,z}(z=0)=0$, which is in accordance with the boundary condition for the vector $\mathbf{B}=\mu\mathbf{H}$ in Eqs. (8.33) – normal component of the magnetic field must vanish on a PEC surface. While the former condition is directly imposed by Eq. (10.96), the latter one comes out as a part of the overall solution for the field in the incident medium in Fig. 10.13(a) and, ultimately, as a consequence of Maxwell's equations. We can thus regard it as a check of the correctness of our solution.

Example 10.14 Various Computations for Oblique Incidence, Normal Polarization

A normally polarized uniform plane time-harmonic electromagnetic wave of angular frequency ω and rms electric field intensity $E_{\mathrm{i}0}$ is incident from air at an angle θ_{i} on a flat horizontal screen made from a good nonmagnetic conductor, of conductivity σ. Assuming that the skin effect in the screen is pronounced, find: (a) the distribution of surface currents and charges induced in the screen, (b) the time-average electromagnetic energy density above the screen, (c) the instantaneous emf induced in a small contour of surface area S moving with velocity v in parallel to the screen, at a height h, as shown in Fig. 10.14, and (d) the time-average power of Joule's losses per unit area of the screen.

Solution As the losses in the screen in Fig. 10.14 are small, we can utilize the expressions for the electric and magnetic field distribution in air for the no-loss case, that is, assuming that the screen is perfectly conducting, and the small losses in the screen will then be evaluated, in (d), invoking the perturbation method for evaluation of losses in good conductors.

(a) Substituting the complex rms field intensities in Eqs. (10.102)–(10.104), (10.98), and (10.99) into the boundary conditions in Eqs. (10.13) and (10.15), the complex rms surface current and charge densities in the conducting screen are

$$\underline{\mathbf{J}}_{\mathrm{s}} = \hat{\mathbf{n}}\times\underline{\mathbf{H}}^{(\mathrm{n})}_{\mathrm{tot}}\Big|_{z=0} = (-\hat{\mathbf{z}})\times\left[\underline{H}^{(\mathrm{n})}_{\mathrm{tot}\,x}\,\hat{\mathbf{x}}+\underline{H}^{(\mathrm{n})}_{\mathrm{tot}\,z}\,\hat{\mathbf{z}}\right]_{z=0} = \underline{H}^{(\mathrm{n})}_{\mathrm{tot}\,x}\Big|_{z=0}(-\hat{\mathbf{y}})$$

$$= \frac{2E_{\mathrm{i}0}}{\eta_0}\cos\theta_{\mathrm{i}}\,\mathrm{e}^{-\mathrm{j}\beta x\sin\theta_{\mathrm{i}}}\,\hat{\mathbf{y}},\quad \underline{\rho}_{\mathrm{s}} = \hat{\mathbf{n}}\cdot\varepsilon_0\,\underline{\mathbf{E}}^{(\mathrm{n})}_{\mathrm{tot}}\Big|_{z=0} = (-\hat{\mathbf{z}})\cdot\varepsilon_0\,\underline{E}^{(\mathrm{n})}_{\mathrm{tot}\,y}\Big|_{z=0}\hat{\mathbf{y}} = 0, \tag{10.105}$$

where it is assumed that $\underline{E}_{\mathrm{i}0}=E_{\mathrm{i}0}$ (incident electric field has a zero initial phase at $x=z=0$), and η_0 is the intrinsic impedance, Eq. (9.23), and $\beta=\omega/c_0$ the phase coefficient of free space, with $c_0=3\times10^8$ m/s standing for the free-space wave velocity.

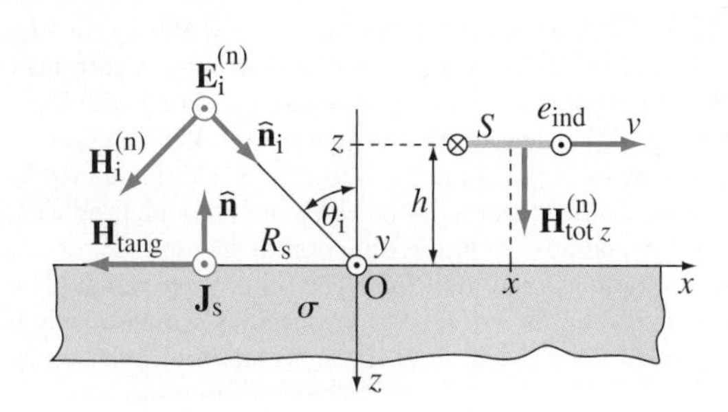

Figure 10.14 Oblique incidence of a normally polarized uniform plane electromagnetic wave on a low-loss conducting screen, and a small moving contour parallel to the screen; for Example 10.14.

(b) Having in mind Eqs. (10.18), (9.200), and (9.24), the total time-average electromagnetic energy density in air equals

$$(w_{\rm em})_{\rm ave} = \frac{1}{2}\varepsilon_0\left|\underline{\mathbf{E}}_{\rm tot}^{(\rm n)}\right|^2 + \frac{1}{2}\mu_0\left|\underline{\mathbf{H}}_{\rm tot}^{(\rm n)}\right|^2 = \frac{1}{2}\varepsilon_0\left|\underline{E}_{{\rm tot}\,y}^{(\rm n)}\right|^2 + \frac{1}{2}\mu_0\left[\left|\underline{H}_{{\rm tot}\,x}^{(\rm n)}\right|^2 + \left|\underline{H}_{{\rm tot}\,z}^{(\rm n)}\right|^2\right]$$
$$= 2\varepsilon_0 E_{\rm i0}^2\left[(1+\sin^2\theta_{\rm i})\sin^2(\beta z\cos\theta_{\rm i}) + \cos^2\theta_{\rm i}\cos^2(\beta z\cos\theta_{\rm i})\right]. \quad (10.106)$$

(c) Given the orientation of the small moving contour in Fig. 10.14, we realize that only the z-component of the resultant magnetic field in air contributes to the magnetic flux through the contour. From Eqs. (10.104) and (8.66), the instantaneous intensity of this field component is [see also Eqs. (10.9)]

$$H_{{\rm tot}\,z}^{(\rm n)}(x,z,t) = \frac{2\sqrt{2}E_{\rm i0}}{\eta_0}\sin\theta_{\rm i}\sin(\beta z\cos\theta_{\rm i})\sin(\omega t - \beta x\sin\theta_{\rm i}). \quad (10.107)$$

Similarly to the situation in Fig. 10.6, as the contour moves, the coordinates of its center are $x = vt$ (adopting that $x = 0$ for $t = 0$), y, and $z = -h$ (Fig. 10.14), so that the magnetic flux through it varies in time as

$$\Phi(t) = \mu_0 H_{{\rm tot}\,z}^{(\rm n)}(x,z,t)S = -\frac{2\sqrt{2}\mu_0 E_{\rm i0}S}{\eta_0}\sin\theta_{\rm i}\sin(\beta h\cos\theta_{\rm i})\sin[(\omega - \beta v\sin\theta_{\rm i})t], \quad (10.108)$$

and hence, like in Eqs. (10.27) and (10.28), the emf induced in the contour

$$e_{\rm ind}(t) = -\frac{{\rm d}\Phi}{{\rm d}t} = -\frac{2\sqrt{2}\omega E_{\rm i0}S}{c_0^2}(c_0 - v\sin\theta_{\rm i})\sin\theta_{\rm i}\sin(\beta h\cos\theta_{\rm i})\cos[(\omega - \beta v\sin\theta_{\rm i})t]. \quad (10.109)$$

Here, it is also obvious that the portion of the total emf that is due to transformer induction dominates over the portion corresponding to motional induction, provided that $v \ll c_0$.

(d) As the tangential component of the total magnetic field vector near the screen surface is its x-component, the per-unit-area time-average loss power in the screen, Eq. (10.92), now becomes

$$\frac{{\rm d}(P_{\rm J})_{\rm ave}}{{\rm d}S} = R_{\rm s}|\underline{\mathbf{H}}_{\rm tang}|^2 = R_{\rm s}|\underline{\mathbf{J}}_{\rm s}|^2 = R_{\rm s}\left|\underline{H}_{{\rm tot}\,x}^{(\rm n)}\right|_{z=0}^2 = \frac{4R_{\rm s}E_{\rm i0}^2\cos^2\theta_{\rm i}}{\eta_0^2}, \quad (10.110)$$

with $R_{\rm s}$ being the surface resistance of the conductor, Eq. (10.78).

Example 10.15 Parallel Polarization Case

For a uniform plane time-harmonic electromagnetic wave impinging a PEC plane at an oblique incidence with parallel polarization, as depicted in Fig. 10.13(b), calculate (a) the total electromagnetic field in the incident medium and (b) the distribution of induced sources in the PEC plane.

Solution

(a) In the parallel polarization case, in Fig. 10.13(b), the electric field intensity vectors of the incident and reflected waves have two Cartesian components (x and z). Noting that the unit vector in the direction of $\underline{\mathbf{E}}_{\rm i}^{(\rm p)}$ is opposite to that along $\underline{\mathbf{H}}_{\rm i}^{(\rm n)}$ in Eq. (10.100) for the normal polarization, Fig. 10.13(a), as well as that the unit vectors for $\underline{\mathbf{E}}_{\rm r}^{(\rm p)}$ and $\underline{\mathbf{H}}_{\rm r}^{(\rm n)}$ [Eq. (10.101)] are the same, we can write

$$\underline{\mathbf{E}}_{\rm i}^{(\rm p)} = \underline{E}_{\rm i0}\,{\rm e}^{-{\rm j}\beta\mathbf{r}\cdot\hat{\mathbf{n}}_{\rm i}}\underbrace{(\cos\theta_{\rm i}\,\hat{\mathbf{x}} - \sin\theta_{\rm i}\,\hat{\mathbf{z}})}_{\text{unit vector for } E_{\rm i}^{(\rm p)}}, \quad \underline{\mathbf{E}}_{\rm r}^{(\rm p)} = \underline{E}_{\rm r0}\,{\rm e}^{-{\rm j}\beta\mathbf{r}\cdot\hat{\mathbf{n}}_{\rm r}}\underbrace{(\cos\theta_{\rm r}\,\hat{\mathbf{x}} + \sin\theta_{\rm r}\,\hat{\mathbf{z}})}_{\text{unit vector for } E_{\rm r}^{(\rm p)}}. \quad (10.111)$$

On the other hand, the magnetic field vectors for the parallel polarization are simpler, having only one component (y),

$$\underline{\mathbf{H}}_{\rm i}^{\rm (p)} = \frac{\underline{E}_{\rm i0}}{\eta}\,{\rm e}^{-{\rm j}\beta\mathbf{r}\cdot\hat{\mathbf{n}}_{\rm i}}\,\hat{\mathbf{y}}, \quad \underline{\mathbf{H}}_{\rm r}^{\rm (p)} = \frac{\underline{E}_{\rm r0}}{\eta}\,{\rm e}^{-{\rm j}\beta\mathbf{r}\cdot\hat{\mathbf{n}}_{\rm r}}\,(-\,\hat{\mathbf{y}}) \qquad (10.112)$$

[note that the minus sign in the expression for $\underline{\mathbf{H}}_{\rm r}^{\rm (p)}$ due to its orientation in Fig. 10.13(b) is as in Eqs. (10.2) for the normal incidence].

Since only the x-components of $\underline{\mathbf{E}}_{\rm i}^{\rm (p)}$ and $\underline{\mathbf{E}}_{\rm r}^{\rm (p)}$ are tangential to the PEC interface, the boundary condition in Eq. (10.3) becomes

$$\left[\underline{E}_{{\rm i}x}^{\rm (p)} + \underline{E}_{{\rm r}x}^{\rm (p)}\right]\Big|_{z=0} = 0 \quad \longrightarrow \quad \underline{E}_{\rm i0}\cos\theta_{\rm i}\,{\rm e}^{-{\rm j}\beta x\sin\theta_{\rm i}} + \underline{E}_{\rm r0}\cos\theta_{\rm r}\,{\rm e}^{-{\rm j}\beta x\sin\theta_{\rm r}} = 0 \qquad (10.113)$$

($-\infty < x < \infty$), which tells us that Snell's law of reflection, Eq. (10.97), must be satisfied for the parallel polarization as well. With this, Eq. (10.113) reduces to the same final relationship between $\underline{E}_{\rm i0}$ and $\underline{E}_{\rm r0}$ as in the normal polarization case, $\underline{E}_{\rm r0} = -\underline{E}_{\rm i0}$.[15] In a way similar to obtaining expressions in Eqs. (10.99), (10.103), and (10.104), the expressions for the nonzero components of the total electric and magnetic field intensity vectors in the incident medium for the parallel polarization are then finalized in the following form:

$\underline{E}_x$ – parallel polarization

$$\underline{E}_{{\rm tot}\,x}^{\rm (p)} = -2{\rm j}\underline{E}_{\rm i0}\cos\theta_{\rm i}\sin(\beta z\cos\theta_{\rm i})\,{\rm e}^{-{\rm j}\beta x\sin\theta_{\rm i}}, \qquad (10.114)$$

$\underline{E}_z$ – parallel polarization

$$\underline{E}_{{\rm tot}\,z}^{\rm (p)} = -2\underline{E}_{\rm i0}\sin\theta_{\rm i}\cos(\beta z\cos\theta_{\rm i})\,{\rm e}^{-{\rm j}\beta x\sin\theta_{\rm i}}, \qquad (10.115)$$

$\underline{H}_y$ – parallel polarization

$$\underline{H}_{{\rm tot}\,y}^{\rm (p)} = \frac{2\underline{E}_{\rm i0}}{\eta}\cos(\beta z\cos\theta_{\rm i})\,{\rm e}^{-{\rm j}\beta x\sin\theta_{\rm i}}. \qquad (10.116)$$

(b) Similarly to the computation in Eqs. (10.105), surface current and charge densities induced in the PEC plane in Fig. 10.13(b) come out to be

$$\underline{\mathbf{J}}_{\rm s} = (-\,\hat{\mathbf{z}}) \times \underline{H}_{{\rm tot}\,y}^{\rm (p)}\Big|_{z=0}\hat{\mathbf{y}} = \frac{2\underline{E}_{\rm i0}}{\eta}\,{\rm e}^{-{\rm j}\beta x\sin\theta_{\rm i}}\,\hat{\mathbf{x}}, \quad \underline{\rho}_{\rm s} = (-\,\hat{\mathbf{z}})\cdot\varepsilon\left[\underline{E}_{{\rm tot}\,x}^{\rm (p)}\,\hat{\mathbf{x}} + \underline{E}_{{\rm tot}\,z}^{\rm (p)}\,\hat{\mathbf{z}}\right]_{z=0}$$

$$= -\varepsilon\,\underline{E}_{{\rm tot}\,z}^{\rm (p)}\Big|_{z=0} = 2\varepsilon\underline{E}_{\rm i0}\sin\theta_{\rm i}\,{\rm e}^{-{\rm j}\beta x\sin\theta_{\rm i}}. \qquad (10.117)$$

Example 10.16 Normal Incidence as a Special Case of Oblique Incidence

Show that the total-field expressions for the normal incidence on a PEC plane, in Fig. 10.1, can be obtained as a special case of the corresponding field expressions for an oblique incidence with both normal and parallel polarizations, in Fig. 10.13(a) and Fig. 10.13(b).

Solution Upon substituting $\theta_{\rm i} = 0$ (normal incidence) in Eqs. (10.99), (10.103), (10.104), and (10.114)–(10.116), the corresponding field expressions for the normal and parallel polarizations of the incident wave become the same (with x- and y-axes swapped), and equal to the fields in Eqs. (10.8), obtained assuming a normal incidence on a PEC plane. Of course, this result was to be expected, and it also confirms that the classification of incident waves into those with normal and parallel polarizations, respectively, does not apply to normally incident waves.

Problems: 10.17–10.22; *Conceptual Questions* (on Companion Website): 10.30–10.32; *MATLAB Exercises* (on Companion Website).

[15] Of course, the same deliberations as in the normal polarization case leading to finding the unknowns $\theta_{\rm r}$ and $\underline{E}_{\rm r0}$ (in terms of $\theta_{\rm i}$ and $\underline{E}_{\rm i0}$) alternatively, based on the radiation of induced surface currents in the two half-spaces, are here in place as well.

10.6 CONCEPT OF A RECTANGULAR WAVEGUIDE

We now continue our discussion of a uniform plane wave at an oblique incidence on a PEC boundary, Fig. 10.13, and investigate a possibility of guiding this wave, to long distances, parallel to the boundary. Within the study, we shall also introduce and evaluate a number of important propagation parameters of the resultant wave.

We consider the normal polarization of the incident wave, Fig. 10.13(a), first. By inspecting the exponential and sinusoidal functions of spatial coordinates in the field expressions in Eqs. (10.99), (10.103), and (10.104), we realize that the total electromagnetic wave in the incident medium exhibits a traveling-wave character (term $e^{-j\beta x \sin\theta_i}$) along the x-axis, while behaving as a standing wave [see Eqs. (10.8)] along the z-axis. The equivalent phase coefficient (determined as the entire factor multiplying $-jx$ in the exponents), wavelength, and phase velocity [Eq. (9.35)] of the wave travel in the positive x direction are

$$\boxed{\beta_x = \beta \sin\theta_i, \quad \lambda_x = \frac{2\pi}{\beta_x} = \frac{\lambda}{\sin\theta_i}, \quad v_{px} = \frac{\omega}{\beta_x} = \frac{c}{\sin\theta_i},} \tag{10.118}$$

β_x, λ_x, v_{px} – along the waveguide

where $\lambda = 2\pi/\beta$ and $c = 1/\sqrt{\varepsilon\mu}$ are the wavelength and velocity of the incident (and reflected) wave. Note that if the incident medium is air (or a vacuum), we have that $v_{px} = c_0/\sin\theta_i$, where c_0 is the speed of light in free space, Eq. (9.19), implying that $v_{px} > c_0$ for any incident angle θ_i. Although this might seem at the first glance like a paradox, we recall (see Section 9.13) that v_{px} is not a "real" velocity of travel of electromagnetic energy or information (carried by the wave) near the PEC boundary, but just a velocity with which the constant-phase plane (wavefront) moves in this (given) direction (along the x-axis). In other words, this is the velocity with which an imaginary observer would need to move (in the direction of the x-axis) in order to always see the same phase of the total field.

On the other hand, from the equivalent phase coefficient β_z of the standing wave along the z-axis [which represents the entire factor of z in the sinusoids in Eqs. (10.99), (10.103), and (10.104)], the corresponding wavelength is

$$\lambda_z = \frac{2\pi}{\beta_z} = \frac{2\pi}{\beta\cos\theta_i} = \frac{\lambda}{\cos\theta_i} \tag{10.119}$$

(of course, the phase velocity is not defined for standing waves). In analogy with the construction of the Fabry-Perot resonator in Fig. 10.3 [see Eqs. (10.10) and (10.12)], we can insert a PEC sheet in any one of the planes defined by

$$\boxed{z = -m\frac{\lambda_z}{2} = -\frac{m\lambda}{2\cos\theta_i} \quad (m = 1, 2, \ldots),} \tag{10.120}$$

position of the second reflecting plane, forming a waveguide

as $\underline{E}_{\rm tang} = \underline{E}_{{\rm tot}\,y}^{(\rm n)} = 0$ and $\underline{H}_{\rm norm} = \underline{H}_{{\rm tot}\,z}^{(\rm n)} = 0$ at these planes, and remove the field from the region to the left of the sheet. However, thus obtained structure to the right of the sheet is not a resonator in this case, given that, due to an oblique (and not normal) wave incidence in Fig. 10.13(a), it supports a traveling resultant wave along the x-axis. Namely, the two parallel PEC planes (mirrors) constitute a simple wave-guiding system, known as a parallel-plate waveguide, in which a uniform plane wave bounces back and forth between the waveguide walls (plates), with all of the multiple reflections occurring (alternately from the two walls) at the same oblique incident angle (θ_i). As a result, the total electromagnetic field progresses in the positive x direction, with phase velocity v_{px} in Eq. (10.118). Since $\underline{\mathbf{E}}_{\rm tot}^{(\rm n)}$ in

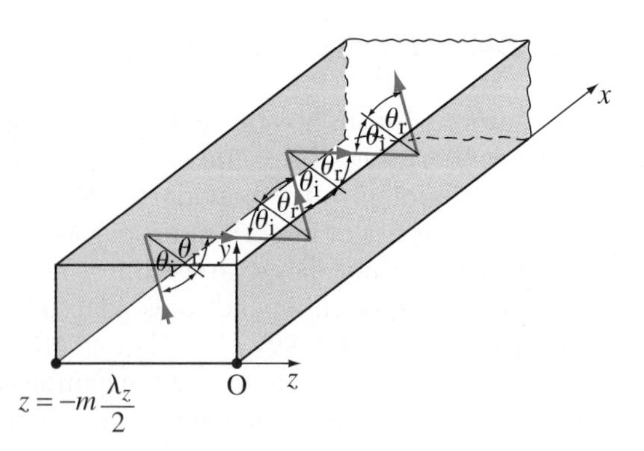

Figure 10.15 Rectangular metallic (PEC) waveguide, formed by four PEC planes at which the tangential component of the electric field vector and normal component of the magnetic field vector of the resultant wave in Fig. 10.13(a) are zero.

the waveguide has only a y-component and $\underline{\mathbf{H}}_{\text{tot}}^{(\text{n})}$ only an x- and z-components, we can insert two more PEC sheets perpendicularly to the y-axis, and the boundary conditions for vectors $\mathbf{E}$ and $\mathbf{B} = \mu\mathbf{H}$ in Eqs. (8.33) will also be automatically satisfied ($\mathbf{E}_{\text{tang}} = 0$ and $\mathbf{H}_{\text{norm}} = 0$ at a PEC surface). The four PEC planes now form an infinitely long PEC tube of a rectangular cross section, as shown in Fig. 10.15. This turns out to be a rectangular metallic (in this case, PEC) waveguide, which is used extensively in the microwave area for transporting large electromagnetic powers (e.g., in radar systems). The electromagnetic field inside the tube, given by Eqs. (10.99), (10.103), and (10.104), represents a nonuniform, non-TEM plane wave propagating along the waveguide (in the positive x direction). The wave is nonuniform since the field components in Eqs. (10.99), (10.103), and (10.104) depend on a coordinate (z) in transversal (equiphase) planes of the wave. It belongs to a class of so-called transverse electric (TE) waves, since the $\mathbf{E}$ field is transverse to the direction of wave propagation, whereas the $\mathbf{H}$ field has a longitudinal (x) component (along the wave travel) as well. Based on Eq. (10.63), the ratio of the transverse components of the electric and magnetic fields defines the wave impedance of this particular TE wave in the waveguide,

TE wave impedance in a rectangular waveguide

$$\boxed{\underline{\eta}_{\text{w}} = \frac{\underline{E}_{\text{tot}\,y}^{(\text{n})}}{\underline{H}_{\text{tot}\,z}^{(\text{n})}} = \frac{\eta}{\sin\theta_{\text{i}}},} \tag{10.121}$$

which comes out to be constant (the same at all points of the field inside the waveguide) and purely real (since the transverse field components are in phase). As we shall see in a later chapter, there are an infinite number of different wave modes (with different electric and magnetic fields) that can propagate through a rectangular waveguide.

For the parallel polarization case, Fig. 10.13(b), the field expressions in Eqs. (10.114)–(10.116) tell us that the total wave is again a combination of a traveling wave along the x-axis and a standing wave along the z-axis. The equivalent phase coefficients in the two directions and the associated wavelengths, as well as the phase velocity of the traveling wave, are the same as for the normal polarization, given by Eqs. (10.118) and (10.119). Since the component of the total electric field $\underline{\mathbf{E}}_{\text{tot}}^{(\text{p})}$ parallel to the PEC boundary in Fig. 10.13(b), $\underline{E}_{\text{tot}\,x}^{(\text{p})}$, is identically zero in the same planes defined by Eq. (10.120), we can again insert a PEC sheet in one of these planes (there will be no tangential electric field on the sheet). With this, a parallel-plate waveguide is formed for guiding the field in Eqs. (10.114)–(10.116) in the positive x direction, like in the normal polarization case. However, as $\underline{\mathbf{E}}_{\text{tot}}^{(\text{p})}$ always lies in one of the planes $y = \text{const}$, and the magnitudes of both of its components in Eqs. (10.114) and (10.115) are functions of only z, it is impossible to specify a y for which the tangential electric field would be identically zero in the plane $y = \text{const.}$ Therefore, as opposed to the normal polarization case (see Fig. 10.15), the field in Eqs. (10.114)–(10.116) cannot be enclosed by the four PEC surfaces to form a rectangular metallic waveguide.

Example 10.17 Poynting Vector in a Rectangular Metallic Waveguide

(a) Find the complex and time-average Poynting vectors in the rectangular metallic waveguide in Fig. 10.15. (b) What is the time-average Poynting vector in the parallel polarization case in Fig. 10.13(b)?

Solution

(a) Like in Eq. (10.20), the complex Poynting vector in the incident medium in Fig. 10.13(a), and thus in the waveguide in Fig. 10.15, is computed as follows, using Eqs. (10.99), (10.103), and (10.104):

$$\underline{\mathcal{P}}^{(\mathrm{n})} = \underline{\mathbf{E}}_{\mathrm{tot}}^{(\mathrm{n})} \times \left[\underline{\mathbf{H}}_{\mathrm{tot}}^{(\mathrm{n})}\right]^* = -\underline{E}_{\mathrm{tot}\,y}^{(\mathrm{n})}\left[\underline{H}_{\mathrm{tot}\,x}^{(\mathrm{n})}\right]^* \hat{\mathbf{z}} + \underline{E}_{\mathrm{tot}\,y}^{(\mathrm{n})}\left[\underline{H}_{\mathrm{tot}\,z}^{(\mathrm{n})}\right]^* \hat{\mathbf{x}}$$

$$= -\mathrm{j}\,\frac{2|\underline{E}_{\mathrm{i}0}|^2}{\eta}\cos\theta_{\mathrm{i}}\sin(2\beta z\cos\theta_{\mathrm{i}})\,\hat{\mathbf{z}} + \frac{4|\underline{E}_{\mathrm{i}0}|^2}{\eta}\sin\theta_{\mathrm{i}}\sin^2(\beta z\cos\theta_{\mathrm{i}})\,\hat{\mathbf{x}}. \quad (10.122)$$

This result shows again that the total electromagnetic field in Fig. 10.13(a) is a standing wave in the z direction, and a traveling wave in the x direction. Namely, the z-component of $\underline{\mathcal{P}}^{(\mathrm{n})}$ is purely imaginary, which is characteristic for standing waves [see Eq. (10.20)], and is, of course, in agreement with the fact that no power is delivered to the PEC medium. The x-component of $\underline{\mathcal{P}}^{(\mathrm{n})}$, on the other hand, is purely real and positive for all values of the coordinate z, which indicates a net real power flow in the positive x direction, parallel to the boundary and along the waveguide, by the resultant electromagnetic wave. This component therefore exactly comprises the resultant time-average Poynting vector in the structure in Fig. 10.15,

$$\mathcal{P}_{\mathrm{ave}}^{(\mathrm{n})} = \mathrm{Re}\{\underline{\mathcal{P}}^{(\mathrm{n})}\} = \frac{4|\underline{E}_{\mathrm{i}0}|^2}{\eta}\sin\theta_{\mathrm{i}}\sin^2(\beta z\cos\theta_{\mathrm{i}})\,\hat{\mathbf{x}}. \quad (10.123)$$

Note that the flux of $\mathcal{P}_{\mathrm{ave}}^{(\mathrm{n})}$ through an arbitrary cross section (for an arbitrary x) of the waveguide, obtained by integration in terms of z between the upper and lower side of the waveguide, that is, from one of the values specified in Eq. (10.120), e.g., $z = -\lambda_z/2$ (for $m = 1$), to $z = 0$, gives the total time-average power transmitted along the structure.

(b) For the situation in Fig. 10.13(b), similarly to obtaining Eq. (10.123), the time-average Poynting vector associated with the total electromagnetic field in Eqs. (10.114)–(10.116) comes out to be

$$\mathcal{P}_{\mathrm{ave}}^{(\mathrm{p})} = \mathrm{Re}\{\underline{\mathcal{P}}^{(\mathrm{p})}\} = -\underline{E}_{\mathrm{tot}\,z}^{(\mathrm{p})}\left[\underline{H}_{\mathrm{tot}\,y}^{(\mathrm{p})}\right]^* \hat{\mathbf{x}} = \frac{4|\underline{E}_{\mathrm{i}0}|^2}{\eta}\sin\theta_{\mathrm{i}}\cos^2(\beta z\cos\theta_{\mathrm{i}})\,\hat{\mathbf{x}}, \quad (10.124)$$

which is another confirmation that, in the parallel polarization case as well, the real power flow by the resultant electromagnetic wave is parallel to the PEC boundary, in the positive x direction.

10.7 OBLIQUE INCIDENCE ON A DIELECTRIC BOUNDARY

If the medium on the right-hand side of the interface in Fig. 10.13 is another lossless dielectric,[16] a part of the incident energy will be transmitted through the interface, like in Fig. 10.7 for the normal incidence case. However, it is intuitively obvious,

[16] As on many occasions so far, while using the term "dielectric" for a medium, we allow here that both ε and μ may have non-free-space values.

that, for a given oblique incident angle, θ_i ($\theta_\text{i} \neq 0$), the propagation direction of the transmitted wave cannot be the same for all combinations of material parameters of the two half-spaces. Therefore, unlike the situation in Fig. 10.7, the transmitted wave must be deflected with respect to the normal on the interface, and the angle between the transmitted ray and the normal is called the transmitted angle and designated as θ_t ($\theta_\text{t} \neq 0$). For the same reason, i.e., because θ_t cannot be constant but must depend on the material parameters on the two sides of the interface, the transmitted wave is deflected also with respect to the incident wave direction ($\theta_\text{t} \neq \theta_\text{i}$), as shown in Fig. 10.16. This phenomenon, that the incident ray breaks upon transmission, either toward the interface ($\theta_\text{t} > \theta_\text{i}$) or away from it ($\theta_\text{t} < \theta_\text{i}$), is known as wave refraction, and the transmitted wave is also referred to as the refracted wave.

To find θ_t, as well as the reflected angle, θ_r, in Fig. 10.16, let us perform a simple geometrical analysis of the propagation of constant-phase wavefronts of the incident, reflected, and transmitted waves that is independent of their polarization (normal or parallel). In specific, let us consider two equiphase points, A and B, of the incident wave, where one of them (B) belongs to the material interface (Fig. 10.16). After some time, Δt, these points become, respectively, points C and D at a wavefront of the reflected wave, on one side of the interface, as well as the equiphase points C and E of the transmitted wave, on the other side, with the common point C belonging to the interface. During this time the individual waves travel distances $\overline{\text{AC}}$ (incident), $\overline{\text{BD}}$ (reflected), and $\overline{\text{BE}}$ (transmitted), the first two with velocity $c_1 = 1/\sqrt{\varepsilon_1\mu_1}$, and the third with $c_2 = 1/\sqrt{\varepsilon_2\mu_2}$. We thus have

$$\Delta t = \frac{\overline{\text{AC}}}{c_1} = \frac{\overline{\text{BD}}}{c_1} = \frac{\overline{\text{BE}}}{c_2}. \tag{10.125}$$

From the trigonometry of the right-angled triangles $\triangle$ACB, $\triangle$BDC, and $\triangle$BEC (with the common hypotenuse BC) in Fig. 10.16,

$$\overline{\text{AC}} = \overline{\text{BC}} \sin\theta_\text{i}, \quad \overline{\text{BD}} = \overline{\text{BC}} \sin\theta_\text{r}, \quad \overline{\text{BE}} = \overline{\text{BC}} \sin\theta_\text{t}. \tag{10.126}$$

Substituting the first two expressions, for $\overline{\text{AC}}$ and $\overline{\text{BD}}$, into Eq. (10.125) leads to Eq. (10.97), namely, Snell's law of reflection. The first and third expressions give

Snell's law of refraction

$$\boxed{\frac{\sin\theta_\text{i}}{\sin\theta_\text{t}} = \frac{c_1}{c_2} = \sqrt{\frac{\varepsilon_2\mu_2}{\varepsilon_1\mu_1}}.} \tag{10.127}$$

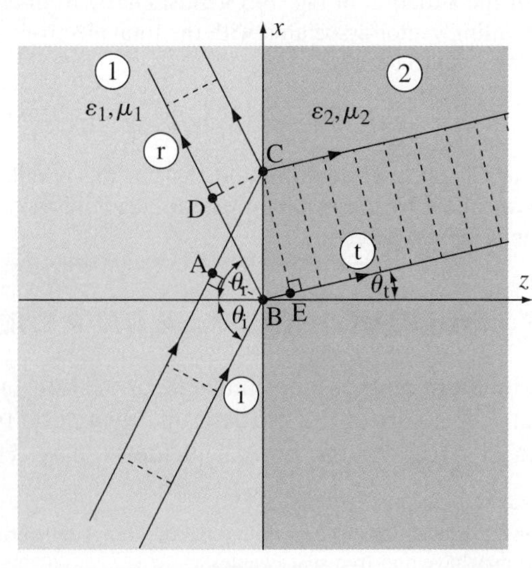

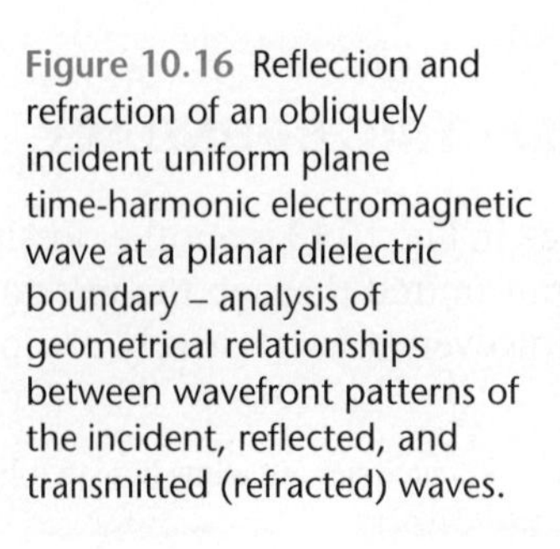

Figure 10.16 Reflection and refraction of an obliquely incident uniform plane time-harmonic electromagnetic wave at a planar dielectric boundary – analysis of geometrical relationships between wavefront patterns of the incident, reflected, and transmitted (refracted) waves.

This relationship enables us to find the transmitted (refracted) angle in terms of an incident angle, for given characteristics of the two media in Fig. 10.16, and is called Snell's law of refraction. It is often expressed in terms of indices of refraction (or refractive indices), Eq. (9.144), of the two media, $n_1 = c_0/c_1 = \sqrt{\varepsilon_{1r}\mu_{1r}}$ and $n_2 = c_0/c_2 = \sqrt{\varepsilon_{2r}\mu_{2r}}$,

$$\boxed{\frac{\sin\theta_i}{\sin\theta_t} = \frac{n_2}{n_1}.} \qquad (10.128)$$

refraction law via refractive indices

Note that, although the definition of the index of refraction in Eq. (9.144) applies to arbitrary materials, it is most frequently used in the case of nonmagnetic ($\mu_r = 1$) lossless media (dielectrics), where it becomes

$$\boxed{n = \sqrt{\varepsilon_r}} \qquad (10.129)$$

index of refraction of a nonmagnetic lossless material

(of course, $n = 1$ for free space). Hence, there is an one-to-one correspondence between the index of refraction and (relative) permittivity if $\mu_r = 1$. In fact, while the index of refraction is relatively rarely used at radio and microwave frequencies and practically never in static (dc) applications, dielectric materials at optical frequencies are almost exclusively characterized using n (instead of ε_r). Eq. (10.128) is therefore the most frequently used version of Snell's law of refraction in optics.

We note that Fig. 10.16 and the associated relationships in Eqs. (10.125) and (10.126) essentially imply that the intersections of the wavefronts of the incident, reflected, and transmitted waves with the interface coincide with each other, and move along the interface together, at the same speed, which is further illustrated in Fig. 10.17. The common speed is $c_1/\sin\theta_i = c_2/\sin\theta_t$, Eq. (10.127). We can think of this velocity match along the interface also as being imposed by the boundary conditions for the electric and magnetic fields of the three waves at the interface, which, assumed to be satisfied at one instant of time, must remain satisfied at all times. This is possible only if the picture of the wavefronts moving in the positive x direction in Fig. 10.17 remains unchanged during the course of time, that is, if the phase velocities of the three waves as measured along the x-axis are the same.

To find the unknown complex rms electric field intensities of the reflected and transmitted waves in the plane $z = 0$, $\underline{E}_{r0}$ and $\underline{E}_{t0}$, for a given intensity $\underline{E}_{i0}$ at $z = 0$ of the incident wave, we apply the boundary conditions for tangential components

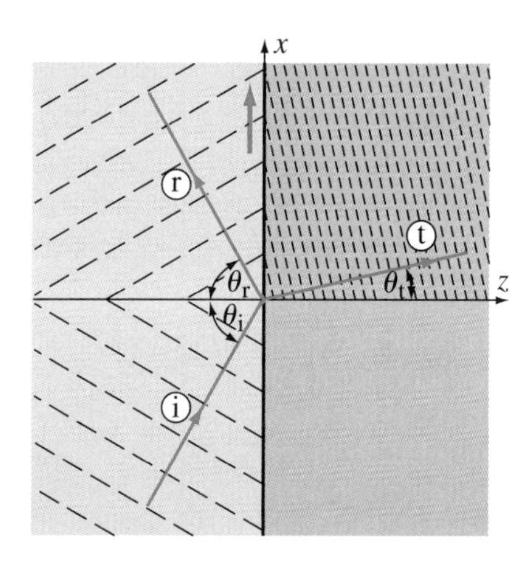

Figure 10.17 More complete picture of wavefront patterns in Fig. 10.16, illustrating movement in unison of wavefront intersections with the interface of the three waves in the positive x direction.

of both electric and magnetic field vectors, as in Eqs. (10.38) and (10.39) for the normal incidence on a dielectric interface. Here, however, we need to distinguish between the normal and parallel polarizations of the incident wave, which are depicted, respectively, in Fig. 10.18(a) and Fig. 10.18(b). The expressions for fields (or their components) entering the boundary conditions are similar to those in Eqs. (10.96) and (10.113) for the oblique incidence on a PEC interface. So, for the dielectric interface and normal polarization, Fig. 10.18(a), the boundary condition in Eq. (10.96) becomes

$$\left[\underline{\mathbf{E}}_{\mathrm{i}}^{(\mathrm{n})} + \underline{\mathbf{E}}_{\mathrm{r}}^{(\mathrm{n})}\right]\Big|_{z=0} = \underline{\mathbf{E}}_{\mathrm{t}}^{(\mathrm{n})}\Big|_{z=0} \longrightarrow \underline{E}_{\mathrm{i}0}\,\mathrm{e}^{-\mathrm{j}\beta_1 x \sin\theta_{\mathrm{i}}} + \underline{E}_{\mathrm{r}0}\,\mathrm{e}^{-\mathrm{j}\beta_1 x \sin\theta_{\mathrm{r}}} = \underline{E}_{\mathrm{t}0}\,\mathrm{e}^{-\mathrm{j}\beta_2 x \sin\theta_{\mathrm{t}}}. \tag{10.130}$$

Because of the requirement that it must be satisfied for any x, we have

phase-matching condition at a material interface

$$\boxed{\beta_1 \sin\theta_{\mathrm{i}} = \beta_1 \sin\theta_{\mathrm{r}} = \beta_2 \sin\theta_{\mathrm{t}}.} \tag{10.131}$$

This equation comprises both Snell's laws, with the first equality giving the law of reflection, Eq. (10.97), and the equality between the first and third term representing, given Eq. (8.111), the law of refraction, Eq. (10.127). We note that matching the phases of the three waves at the interface (for $z = 0$) is equivalent to matching the corresponding phase velocities along the interface. With the elimination of exponential terms, Eq. (10.130) yields

$$\underline{E}_{\mathrm{i}0} + \underline{E}_{\mathrm{r}0} = \underline{E}_{\mathrm{t}0}. \tag{10.132}$$

On the other side, since only the x-components of the magnetic field vectors in Fig. 10.18(a) are tangential to the boundary surface, the corresponding boundary condition for $\mathbf{H}$ yields [note the similarity with Eq. (10.113)]

$$\left[\underline{H}_{\mathrm{i}x}^{(\mathrm{n})} + \underline{H}_{\mathrm{r}x}^{(\mathrm{n})}\right]\Big|_{z=0} = \underline{H}_{\mathrm{t}x}^{(\mathrm{n})}\Big|_{z=0} \longrightarrow -\frac{\underline{E}_{\mathrm{i}0}}{\eta_1}\cos\theta_{\mathrm{i}} + \frac{\underline{E}_{\mathrm{r}0}}{\eta_1}\cos\theta_{\mathrm{r}} = -\frac{\underline{E}_{\mathrm{t}0}}{\eta_2}\cos\theta_{\mathrm{t}}, \tag{10.133}$$

where $\eta_1 = \sqrt{\mu_1/\varepsilon_1}$ and $\eta_2 = \sqrt{\mu_2/\varepsilon_2}$ are the intrinsic impedances of the two media in Fig. 10.18 (they are purely real, because of the no-loss assumption for

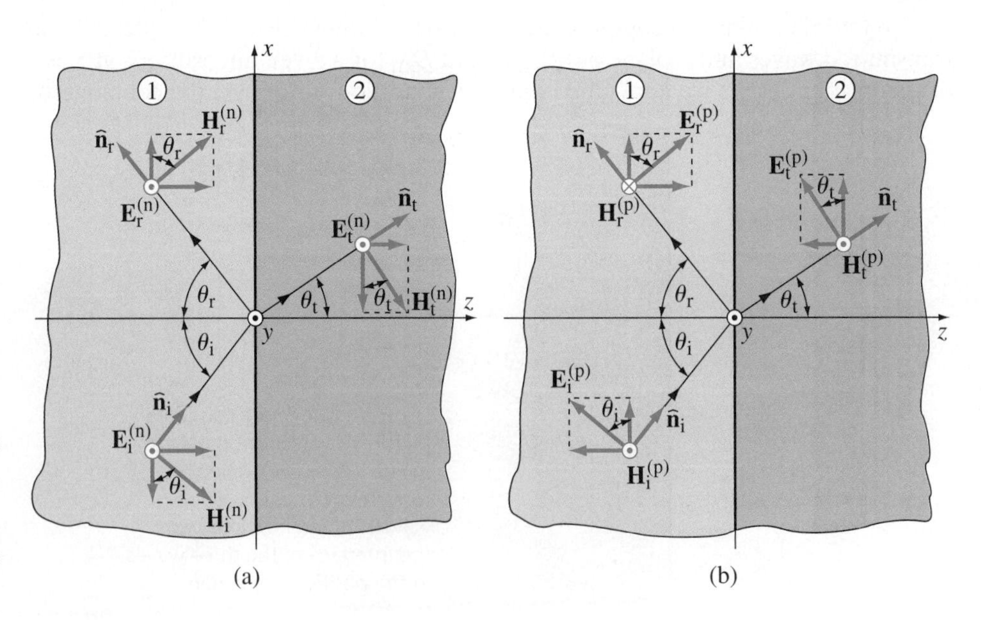

Figure 10.18 Electric and magnetic field vectors of the incident, reflected, and transmitted waves in Fig. 10.16 for normal (a) and parallel (b) polarizations of waves.

both media). Much like Eqs. (10.40) for the corresponding normal-incidence case, Eqs. (10.132) and (10.133) are two linear algebraic equations with unknowns $\underline{E}_{r0}$ and $\underline{E}_{t0}$ (for a given $\underline{E}_{i0}$), and their solution can be represented in the form of reflection and transmission coefficients – for normal polarization,

Fresnel's coefficients for normal polarization

$$\boxed{\Gamma_n = \left(\frac{\underline{E}_{r0}}{\underline{E}_{i0}}\right)_n = \frac{\eta_2 \cos\theta_i - \eta_1 \cos\theta_t}{\eta_1 \cos\theta_t + \eta_2 \cos\theta_i}, \quad \tau_n = \left(\frac{\underline{E}_{t0}}{\underline{E}_{i0}}\right)_n = \frac{2\eta_2 \cos\theta_i}{\eta_1 \cos\theta_t + \eta_2 \cos\theta_i},} \tag{10.134}$$

with $1 + \Gamma_n = \tau_n$ (as in the normal incidence case). These coefficients are known as Fresnel's coefficients for normal polarization.

For parallel polarization, Fig. 10.18(b), the boundary conditions, at $z = 0$, for x-components of the electric field vectors and for entire magnetic field vectors, respectively, give the following two equations:

$$\underline{E}_{i0} \cos\theta_i + \underline{E}_{r0} \cos\theta_r = \underline{E}_{t0} \cos\theta_t, \quad \frac{\underline{E}_{i0}}{\eta_1} - \frac{\underline{E}_{r0}}{\eta_1} = \frac{\underline{E}_{t0}}{\eta_2}, \tag{10.135}$$

and hence Fresnel's (reflection and transmission) coefficients for parallel polarization

Fresnel's coefficients for parallel polarization

$$\boxed{\Gamma_p = \left(\frac{\underline{E}_{r0}}{\underline{E}_{i0}}\right)_p = \frac{\eta_2 \cos\theta_t - \eta_1 \cos\theta_i}{\eta_1 \cos\theta_i + \eta_2 \cos\theta_t}, \quad \tau_p = \left(\frac{\underline{E}_{t0}}{\underline{E}_{i0}}\right)_p = \frac{2\eta_2 \cos\theta_i}{\eta_1 \cos\theta_i + \eta_2 \cos\theta_t}.} \tag{10.136}$$

Note that here $1 + \Gamma_p$ is not equal to τ_p, but instead $1 + \Gamma_p = \tau_p \cos\theta_t / \cos\theta_i$, which is a direct consequence of the form of the boundary condition for the vector **E** (first condition) in Eqs. (10.135).

Although derived assuming that both media in Fig. 10.18 are lossless, the expressions for Fresnel's coefficients in Eqs. (10.134) and (10.136) are valid for lossy media as well, provided that complex intrinsic impedances $\underline{\eta}_1$ and $\underline{\eta}_2$ are used, as in Eqs. (10.42) and (10.43) for the normal incidence case. Of course, if either one of the two media is lossy, all the coefficients are complex ($\underline{\Gamma}_n$, $\underline{\Gamma}_p$, $\underline{\tau}_n$, $\underline{\tau}_p$). The only other difference with respect to the lossless case is in Snell's law of refraction. Namely, since the exponential terms in Eq. (10.130) now have $\underline{\gamma}_1$ and $\underline{\gamma}_2$ (assuming that both media are lossy) instead of $j\beta_1$ and $j\beta_2$, respectively, as in Eqs. (10.35)–(10.37) for the normal incidence, the refraction-law relationship in Eq. (10.131) becomes

Snell's law of refraction in complex domain

$$\boxed{\underline{\gamma}_1 \sin\theta_i = \underline{\gamma}_2 \sin\underline{\theta}_t,} \tag{10.137}$$

while the reflection law remains the same, Eq. (10.97). This complex equation is referred to as Snell's law of refraction in the complex domain, with the "ordinary" Snell's law of refraction in Eq. (10.127) being its special case. Note that, from Eq. (10.137), $\sin\underline{\theta}_t$ in general is a complex quantity, since $\underline{\gamma}_1/\underline{\gamma}_2$ is in general complex ($\sin\theta_i$ is real). Consequently, $\cos\underline{\theta}_t$, needed for Fresnel's coefficients $\underline{\Gamma}_n$, $\underline{\Gamma}_p$, $\underline{\tau}_n$, and $\underline{\tau}_p$, is a complex quantity as well, and so is $\underline{\theta}_t$ itself. It can be shown that the direction of propagation of the refracted wave in the second medium is determined by an angle (measured from the normal on the interface) equal to the real part of $\underline{\theta}_t$.

By inspecting the expressions in Eqs. (10.134) and (10.136), modified so that both intrinsic impedances are complex (the most general case), we note that, on one side, $\underline{\Gamma}_{\rm n} = \underline{\Gamma}_{\rm p} = -1$ and $\underline{\tau}_{\rm n} = \underline{\tau}_{\rm p} = 0$ for the second medium PEC ($\underline{\eta}_2 = 0$). Here, $\underline{\Gamma} = -1$ corresponds to the relationship $\underline{E}_{\rm r0} = -\underline{E}_{\rm i0}$, obtained from either Eq. (10.96) for normal polarization or Eq. (10.113) for parallel polarization in the analysis of obliquely incident waves on a PEC boundary (in Section 10.5). On the other side, if $\theta_{\rm i} = 0$ (normal incidence), $\underline{\Gamma}_{\rm n} = \underline{\Gamma}_{\rm p} = \underline{\Gamma}$ and $\underline{\tau}_{\rm n} = \underline{\tau}_{\rm p} = \underline{\tau}$, where $\underline{\Gamma}$ and $\underline{\tau}$ stand for the reflection and transmission coefficients in Eqs. (10.42) and (10.43), obtained in the analysis of normally incident waves on a dielectric boundary (in Section 10.2). Of course, all these results were to be expected, since both the oblique incidence on a PEC and normal incidence on a dielectric boundary can be regarded as special cases of the general solution for an oblique incidence on an interface between two arbitrary media.

Example 10.18 Exit Angle for a Light Beam Passing through a Glass Slab

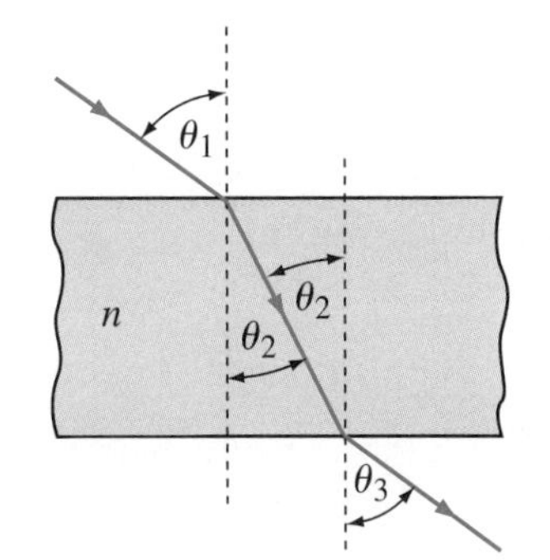

Figure 10.19 Oblique incidence of light on a glass slab in air – proof that the incoming and outgoing beams are parallel to each other; for Example 10.18.

A light beam is incident obliquely on a glass slab surrounded by air. Prove that the emerging beam on the other side of the slab is always parallel to the incident beam.

Solution With reference to Fig. 10.19, Snell's law of refraction in Eq. (10.128) applied successively to each of the surfaces of the slab (assuming that $\mu_{\rm r} = 1$ for glass) gives

$$\frac{\sin\theta_1}{\sin\theta_2} = n = \sqrt{\varepsilon_{\rm r}} \quad \text{and} \quad \frac{\sin\theta_2}{\sin\theta_3} = \frac{1}{n} \quad \longrightarrow \quad \theta_3 = \theta_1, \tag{10.138}$$

where n is the index of refraction, Eq. (10.129), of the glass ($n = 1$ for air). So, the exit angle for the light beam on the other side of the slab (θ_3) turns out to be the same as the incident angle from air (θ_1), and, indeed, the incoming and outgoing beams are parallel to each other.

Example 10.19 Apparent Depth of a Coin at the Bottom of a Fountain

A coin lies at the bottom of a water (the refractive index is $n = 1.33$) fountain at a depth of $d = 50$ cm. What is the apparent depth of the coin below the water surface when viewed from above the water at an angle of 60° with respect to the normal on the surface?

Solution This is shown in Fig. 10.20. On their way to the viewer's eyes, sunlight rays reflected from the coin refract in air, upon passing through the water-air interface, farther away from the normal on the interface – according to Snell's law of refraction, as air is optically less dense than water ($n > 1$). Having in mind Eqs. (10.138), the incident angle (in water) equals

$$\theta_1 = \arcsin\frac{\sin\theta_2}{n} = 40.63^\circ \quad (\theta_2 = 60^\circ) \tag{10.139}$$

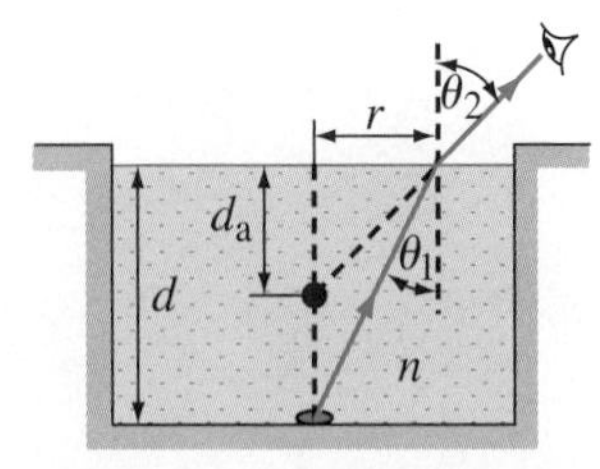

Figure 10.20 Determining the apparent depth of a coin at the bottom of a water fountain as viewed from above the water surface; for Example 10.19.

($\arcsin \equiv \sin^{-1}$). However, the coin appears to the viewer higher than it is, because the brain perceives the information about the received light beam from the eyes without, of course, taking into account Snell's law, so as if there were a direct (straight) line of sight between the coin and the eyes, and this determines the apparent location of the coin below the water surface. To find this location, we then express the distance r in Fig. 10.20 via the true depth of the coin (d) and angle θ_1, on one side, and via its apparent depth ($d_{\rm a}$) and angle θ_2, on the other side, which gives

$$d_{\rm a} = \frac{\cot\theta_2}{\cot\theta_1}\,d \approx 25 \text{ cm}, \tag{10.140}$$

so $d_{\rm a}$ is approximately a half of the actual depth. In other words, the fountain is twice as deep as it looks like (and this is a very common misperception in situations like the one in Fig. 10.20).

Example 10.20 Transmitted Fields for an Oblique Incidence and Parallel Polarization

A uniform plane time-harmonic electromagnetic wave of frequency f and rms electric field intensity E_{i0} propagates in a lossless medium of parameters ε_1 and μ_1, and is incident at an angle θ_i on the interface of another lossless medium, with parameters ε_2 and μ_2. Assuming that the polarization of the wave is parallel, determine the expressions for complex electric and magnetic field intensity vectors of the transmitted wave at an arbitrary point in the second medium. In specific, calculate these fields for $\varepsilon_{r1} = 4$, $\varepsilon_{r2} = 2$, $\mu_{r1} = \mu_{r2} = 1$, $f = 1$ GHz, $E_{i0} = 1$ V/m, and $\theta_i = 30°$ at the point defined by $x = y = z = 1$ m, if the coordinate system is adopted as in Fig. 10.18.

Solution The expressions for the electric and magnetic field vectors of the transmitted (refracted) wave, shown in Fig. 10.18(b), are similar to those derived in Example 10.15 for the corresponding field vectors of the incident wave for the oblique incidence on a PEC plane in the parallel polarization case. So, either from Fig. 10.18(b) or Eqs. (10.93), the propagation unit vector of the transmitted wave is $\hat{\mathbf{n}}_t = \sin\theta_t\,\hat{\mathbf{x}} + \cos\theta_t\,\hat{\mathbf{z}}$, where θ_t is the transmitted angle, obtained by Snell's law of refraction, Eq. (10.127). Having then in mind Eqs. (10.111), (10.112), and (10.94), the transmitted fields at an arbitrary point in the second medium, defined by the position vector $\mathbf{r}$ with respect to the coordinate origin, Eqs. (9.71), with $z \geq 0$, can be written as

$$\underline{\mathbf{E}}_t^{(p)} = \underline{E}_{t0}\,e^{-j\beta_2\mathbf{r}\cdot\hat{\mathbf{n}}_t}\,\hat{\mathbf{e}} = \tau_p\underline{E}_{i0}\,e^{-j\beta_2(x\sin\theta_t + z\cos\theta_t)}(\cos\theta_t\,\hat{\mathbf{x}} - \sin\theta_t\,\hat{\mathbf{z}}),$$
$$\underline{\mathbf{H}}_t^{(p)} = \frac{\underline{E}_{t0}}{\eta_2}\,e^{-j\beta\mathbf{r}\cdot\hat{\mathbf{n}}_t}\,\hat{\mathbf{y}} = \frac{\tau_p\underline{E}_{i0}}{\eta_2}\,e^{-j\beta_2(x\sin\theta_t + z\cos\theta_t)}\,\hat{\mathbf{y}}, \qquad (10.141)$$

where $\hat{\mathbf{e}}$ designates the unit vector for $\underline{\mathbf{E}}_t^{(p)}$ in Fig. 10.18(b), $\beta_2 = 2\pi f\sqrt{\varepsilon_2\mu_2}$ [Eq. (8.111)] and $\eta_2 = \sqrt{\mu_2/\varepsilon_2}$ [Eq. (9.21)] are, respectively, the phase coefficient of the transmitted wave and intrinsic impedance of the second medium, and τ_p is Fresnel's transmission coefficient for parallel polarization, in Eqs. (10.136). For the given numerical data, $\theta_t = 45°$, $\underline{\mathbf{E}}_t^{(p)} = 0.928\,e^{j118.3°}(\hat{\mathbf{x}} - \hat{\mathbf{z}})$ V/m, and $\underline{\mathbf{H}}_t^{(p)} = 4.95\,e^{j118.3°}\,\hat{\mathbf{y}}$ mA/m.

Note that the expressions for the field vectors of the reflected wave in Fig. 10.18(b) are practically the same as those for $\underline{\mathbf{E}}_r^{(p)}$ and $\underline{\mathbf{H}}_r^{(p)}$ for the incidence on a PEC plane in Eqs. (10.111) and (10.112), with $\underline{E}_{r0} = \Gamma_p\underline{E}_{i0}$, Γ_p being the corresponding Fresnel's reflection coefficient, in Eqs. (10.136). The field expressions for both media in the normal polarization case, Fig. 10.18(a), can be written in an analogous manner.

Problems: 10.23–10.28; *Conceptual Questions* (on Companion Website): 10.33–10.36; *MATLAB Exercises* (on Companion Website).

10.8 TOTAL INTERNAL REFLECTION AND BREWSTER ANGLE

Let us now restrict our attention to a frequent practical situation when the two media in Fig. 10.18 are magnetically identical ($\mu_1 = \mu_2$), which, most importantly, includes the case with both media nonmagnetic ($\mu_1 = \mu_2 = \mu_0$). Assuming that both media are also lossless, Snell's law of refraction, Eq. (10.127), can now be written as

$$\boxed{\frac{\sin\theta_i}{\sin\theta_t} = \sqrt{\frac{\varepsilon_2}{\varepsilon_1}} \quad (\mu_1 = \mu_2).} \qquad (10.142)$$

Snell's law of refraction for magnetically equal media

It appears here that, if $\varepsilon_1 < \varepsilon_2$, in which case we say that the first medium is electromagnetically less dense than the second, $\theta_t < \theta_i$ (since $\sin\theta_t < \sin\theta_i$ and both angles are in the range $0 \leq \theta_i, \theta_t \leq 90°$). This means that for an electromagnetic wave incident onto a denser medium (from a less dense one), the transmitted wave is bent toward the normal on the interface. The transmission of incident energy occurs for any incident angle in Fig. 10.16, except for $\theta_i = 90°$ (propagation of the incident wave is parallel to the interface – so-called grazing incidence).

On the other hand, $\varepsilon_1 > \varepsilon_2$ (incidence onto a less dense medium) yields $\theta_t > \theta_i$, meaning that the wave in medium 2 is refracted away from the normal. So, if we start increasing θ_i to larger and larger values, θ_t will always be even larger, and at some point become 90° (where it cannot increase further). In this border-line case, the refracted wave flows along the interface (so-called surface wave), as shown in Fig. 10.21, and no energy is transmitted into medium 2. In other words, the incident wave is totally reflected. Since this phenomenon requires that a wave must be incident from a medium of higher index of refraction than that of the medium beyond the boundary, most frequently it is for an incidence from the interior of a dielectric object onto its boundary surface with air as the surrounding medium. That is why this type of total reflection is known as total internal reflection. From Eq. (10.142), the incident angle corresponding to the maximum value of the transmitted angle is determined by

θ_{ic} – critical angle

$$\boxed{\theta_t = 90° \quad \longrightarrow \quad \sin\theta_{ic} = \sqrt{\frac{\varepsilon_2}{\varepsilon_1}}, \quad \text{for} \quad \varepsilon_1 > \varepsilon_2,} \tag{10.143}$$

and is termed the critical angle. For any incident angle exceeding this critical value, there is no transmission into medium 2 as well, Fig. 10.21, so that the condition for total internal reflection can be written as

total internal reflection

$$\boxed{\theta_i \geq \theta_{ic}.} \tag{10.144}$$

Also, for $\theta_i = \theta_{ic}$ the refracted wave becomes a surface wave traveling along the boundary (Fig. 10.21). The refracted wave must exist so that the boundary conditions in Eqs. (10.130), (10.133), and (10.135) can be satisfied, and thus the traveling surface wave serves simply as the matching field at the interface from its $z = 0^+$ side.[17] As portrayed in Fig. 10.21, only the incident rays inside the cone defined by

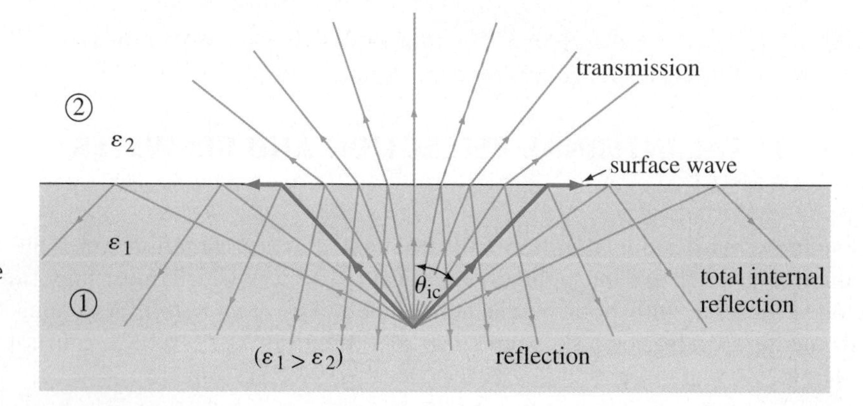

Figure 10.21 Illustration of the critical angle, Eq. (10.143), and total internal reflection, Eq. (10.144), accompanied by a refracted surface wave.

[17]More precisely, it can be shown from the expressions for the electric and magnetic fields of the refracted wave in medium 2, in the case of total internal reflection, that this is a quite complicated electromagnetic wave. The planes of constant wave amplitude (i.e., E- and H-field amplitudes) are parallel

half-angle θ_{ic} (or total angle $2\theta_{ic}$) can pass through the interface and propagate in the second medium. Of course, they are not transmitted totally (except for a single special case, as we shall see later in this section) but as determined by the corresponding transmission coefficient τ_n [Eqs. (10.134)] or τ_p [Eqs. (10.136)], depending on the polarization of the incident wave.

As will be shown in an example, if the condition in Eq. (10.144) is satisfied, reflection coefficients for both normal and parallel polarizations of the incident wave, $\underline{\Gamma}_n$ and $\underline{\Gamma}_p$, in Eqs. (10.134) and (10.136), are unity in magnitude ($|\underline{\Gamma}_n| = |\underline{\Gamma}_p| = 1$). This is another confirmation that, if $\theta_i \geq \theta_{ic}$, the entire energy of the incident wave is reflected back into medium 1 for both polarization cases. The phase angles of $\underline{\Gamma}_n$ and $\underline{\Gamma}_p$ in total internal reflection, however, are functions of the incident angle, so that an incident wave acquires a phase shift (different from 180°) upon total reflection, that depends on a given θ_i. This phase shift also depends on the polarization (normal or parallel) of the incident wave, which can be used to generate a circularly or elliptically polarized reflected wave from a linearly polarized incident wave possessing components with both normal and parallel polarizations (by properly choosing θ_i, ε_1, and ε_2).

In general, total internal reflection finds many applications in optics, in various optical devices, such as beam-steering glass prisms (where light is totally internally reflected from glass-air interfaces), and in optical dielectric waveguides, such as optical fibers. In a typical optical fiber, light is confined to traveling inside a cylindrical dielectric (glass or transparent-plastic) rod by means of multiple total internal reflections from the interface with a surrounding coaxial layer made of a different dielectric material, as shown in Fig. 10.22. With n_{core} and $n_{cladding}$, respectively, denoting the refraction indices of the fiber core and outer layer (called cladding), it is necessary that $n_{core} > n_{cladding}$ for total reflection to be possible (usually, n_{core} is just slightly higher than $n_{cladding}$). In addition, from Eq. (10.144), the incident angle θ_i in the core must be equal or greater than the critical angle (θ_{ic}) for the core-cladding interface, defined by $\sin\theta_{ic} = n_{cladding}/n_{core}$.

Having now completed the discussion of the concept of total reflection of obliquely incident waves on a dielectric boundary, a natural question arises whether an opposite phenomenon, that of a total transmission (no reflection), may occur for certain incident angles and/or certain combination of dielectric parameters (ε_1 and ε_2) – for a given polarization (normal or parallel) of the incident wave. To answer this question, we start with the expressions for Γ_n and Γ_p in Eqs. (10.134) and (10.136) and use the fact that $\eta_1/\eta_2 = \sqrt{\varepsilon_2/\varepsilon_1}$ (for $\mu_1 = \mu_2$) and Eq. (10.142) to obtain alternative expressions – in terms of only the incident angle and the ratio of

---

to the boundary surface (planes defined by $z =$ const in Fig. 10.18), and the planes of constant phase perpendicular to it (planes defined by $x =$ const, Fig. 10.18). The field amplitudes exhibit a rapid exponential decay in the positive z direction (away from the surface), and these attenuated fields travel along the surface (surface wave), in the positive x direction, with velocity $c_2/\sin\theta_t$. Since this is a wave whose amplitude varies as a function of position (function of z) in the constant-phase planes (wavefronts), it is a nonuniform plane wave. It can also be shown that the complex Poynting vector in the second medium has a purely imaginary z-component and a purely real and positive x-component, much like the vector $\underline{\mathcal{P}}^{(n)}$ in Eq. (10.122). This reinforces the facts that no time-average power is transmitted into the second medium and that the real power flows (unattenuated) in the positive x direction, parallel to the interface, carried by the refracted surface wave. However, this x-directed power-flow density (that is, x-component of the time-average Poynting vector) is attenuated in the transversal direction (along the z-axis), with the rate of attenuation determined by twice the attenuation coefficient for the field vectors [like in Eq. (9.97)].

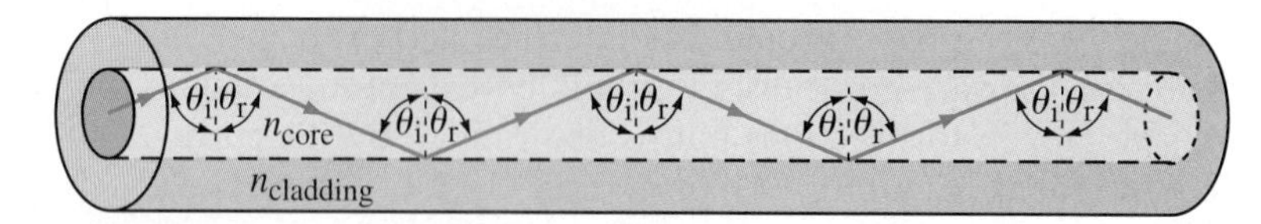

Figure 10.22 Sketch of wave propagation inside the core of an optical fiber, by means of multiple total internal reflections from the outer layer (cladding), made of optically less dense dielectric material ($n_{cladding} < n_{core}$).

two permittivities:

Fresnel's reflection coefficients for $\mu_1 = \mu_2$

$$\Gamma_n = \frac{\cos\theta_i - \sqrt{\varepsilon_2/\varepsilon_1 - \sin^2\theta_i}}{\cos\theta_i + \sqrt{\varepsilon_2/\varepsilon_1 - \sin^2\theta_i}}, \quad \Gamma_p = \frac{-\varepsilon_2\cos\theta_i/\varepsilon_1 + \sqrt{\varepsilon_2/\varepsilon_1 - \sin^2\theta_i}}{\varepsilon_2\cos\theta_i/\varepsilon_1 + \sqrt{\varepsilon_2/\varepsilon_1 - \sin^2\theta_i}}. \tag{10.145}$$

We then try to satisfy the no-reflection condition in the normal polarization case, by setting the numerator in the expression for Γ_n to zero. This yields

$$\Gamma_n = 0 \quad \longrightarrow \quad \cos^2\theta_i = \frac{\varepsilon_2}{\varepsilon_1} - \sin^2\theta_i \quad \longrightarrow \quad \varepsilon_1 = \varepsilon_2, \tag{10.146}$$

which is impossible, as $\varepsilon_1 \neq \varepsilon_2$ (we have two different media), or can be regarded as a trivial solution for total transmission (of course, there is no reflection if there is no discontinuity in the propagation medium). So, total transmission cannot occur for normal polarization of the incident wave, any value of θ_i, and any two (different) dielectric media. On the other hand, note that in the case of magnetically different materials ($\mu_1 \neq \mu_2$), there exists an incident angle for which $\Gamma_n = 0$ (and this angle will be found as an example); however, this is of less practical interest.

For parallel polarization,

$$\Gamma_p = 0 \quad \longrightarrow \quad \frac{\varepsilon_2^2}{\varepsilon_1^2}\cos^2\theta_i = \frac{\varepsilon_2}{\varepsilon_1} - \sin^2\theta_i \quad \longrightarrow \quad \sin\theta_i = \sqrt{\frac{\varepsilon_2}{\varepsilon_1 + \varepsilon_2}}, \tag{10.147}$$

where $\cos\theta_i$ is eliminated using the identity $\sin^2\theta_i + \cos^2\theta_i = 1$, from which then $\cos\theta_i = \sqrt{\varepsilon_1/(\varepsilon_1 + \varepsilon_2)}$. Dividing $\sin\theta_i$ by $\cos\theta_i$, we finally obtain

Brewster angle – total transmission, parallel polarization

$$\tan\theta_{iB} = \sqrt{\frac{\varepsilon_2}{\varepsilon_1}}. \tag{10.148}$$

We call this special incident angle, at which total transmission occurs for parallel polarization of the incident wave, the Brewster angle, and symbolize it by θ_{iB}. For incidences at angles close to θ_{iB}, the amount of reflected energy (for parallel polarization) will not be zero, but still very small and often negligible. Again, there is no counterpart of the Brewster angle for normally polarized incident waves and nonmagnetic materials.

If an electromagnetic wave having components with both normal and parallel polarizations is incident at $\theta_i = \theta_{iB}$ on a dielectric interface, the component with parallel polarization will be totally transmitted into the second medium and the other component partially transmitted and partially reflected. The result is that the reflected wave in the first medium is entirely normally polarized. This means that a piece of dielectric material illuminated from air at the Brewster angle may act as a polarizer, producing, for instance, linearly polarized light from unpolarized (randomly polarized) light, and hence θ_{iB} is often referred to as the polarizing angle. The polarizing capacity associated with the Brewster condition is responsible for many notable effects in nature and has numerous applications in optical devices and systems. For example, essentially because of the Brewster phenomenon most reflected sunlight that we see outside is predominantly horizontally polarized, that is, linearly polarized in a direction parallel to the surface of the earth. This is enabled also by the fact that large interfaces between different media outside are most frequently horizontal – these are air-ground, air-water, and other interfaces constituting the surface of the earth, the most notable example being the surface of a sea or ocean. Namely, we know that most of the sunlight is randomly polarized, and can be represented as a superposition of light with normal polarization and that with parallel polarization, for the plane of incidence containing the sun beam and the normal on the horizontal surface. The two components are with approximately equal energies. With respect to the earth's surface, these polarizations are horizontal and vertical, respectively. Therefore, the component with parallel (vertical) polarization incident on the surface at the Brewster angle and angles around it is largely absorbed into the water, soil, or other material by virtue of total or nearly total transmission, and that is why dominant polarization of light that we see is horizontal.[18]

HISTORICAL ASIDE

Willebrord van Royen Snell (1580–1626), a Dutch mathematician and astronomer, a professor of mathematics at the University of Leiden, discovered in 1621 the law of refraction of light, which now bears his name.

Augustin Jean Fresnel (1788–1827), a French physicist, played an essential role in the establishment of wave optics. Extending the work of Thomas Young (1773–1829), Fresnel explained that light was a transverse wave, oscillating in planes perpendicular to the direction of wave propagation, and suggested that it could be decomposed onto two components with mutually orthogonal oscillations (polarizations).

Sir David Brewster (1781–1868), a Scottish physicist and writer, found that light could be polarized by reflection and refraction, and established, in 1815, the condition under which the light beam with parallel polarization was totally transmitted into a medium – the Brewster or polarizing incident angle. However, it was his invention of the kaleidoscope in 1816 that instantly earned him attention and fame among the general public worldwide.

As a graphical summary of angular and material variations of Fresnel's reflection coefficients for nonmagnetic (or magnetically identical) lossless media, Eqs. (10.145), Fig. 10.23 shows plots of the magnitude and phase angle of both $\underline{\Gamma}_{\rm n}$ and $\underline{\Gamma}_{\rm p}$ versus the incident angle ($0 \le \theta_{\rm i} \le 90°$) for different combinations of permittivities of media 1 and 2 in Fig. 10.18. Specifically, in Fig. 10.23(a), $\varepsilon_1 < \varepsilon_2$,

[18]Note that a high content of horizontally polarized light, taking place upon reflections of sunlight from horizontal surfaces, often creates a high concentration of glare, which essentially is optical noise that prevents human eyes to clearly see colors and contrasts. In relation to this, note also that Polaroid sunglasses, known to enable (almost) glare-free vision, are designed as polarizing lenses that selectively filter out horizontal light, which is why they are so efficient at reducing glare.

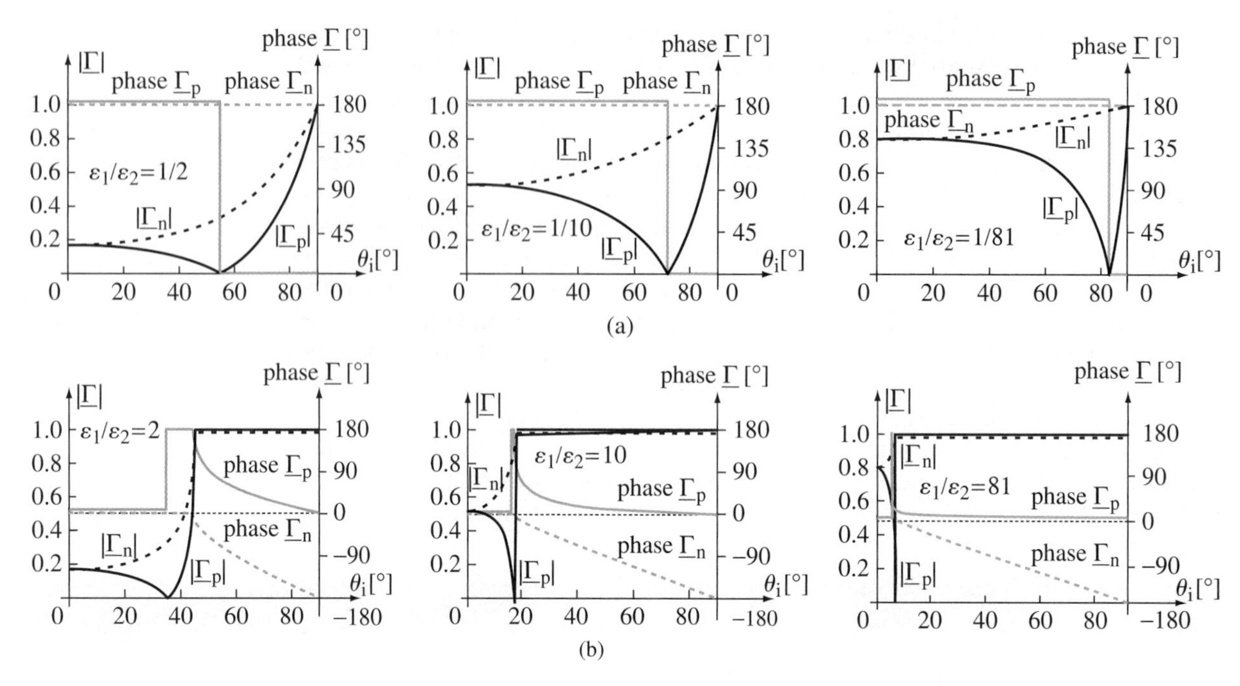

Figure 10.23 Plots of the magnitude and phase angle of Fresnel's reflection coefficients in Eqs. (10.145) against the incident angle (θ_i) for (a) incidence from air onto three perfect-dielectric materials, with relative permittivities $\varepsilon_r = 2$ (paraffin), 10 (flint glass), and 81 (distilled water), respectively, and (b) incidence from each of the materials from (a) to air.

and the incidence is from air ($\varepsilon_1 = \varepsilon_0$) onto three characteristic dielectric materials (assumed to exhibit no losses); in Fig. 10.23(b), the media from (a) are interchanged, so that $\varepsilon_1 > \varepsilon_2$ for all three materials ($\varepsilon_2 = \varepsilon_0$). We observe the Brewster condition [Eq. (10.148)] in the parallel polarization case, and note that, as θ_i is varied from 0 to 90°, Γ_p changes polarity at $\theta_i = \theta_{iB}$ (from positive to negative for $\varepsilon_1 > \varepsilon_2$ and vice versa for $\varepsilon_1 < \varepsilon_2$), whereas Γ_n retains the same polarity for all incident angles (it is positive for $\varepsilon_1 > \varepsilon_2$ and negative for $\varepsilon_1 < \varepsilon_2$). Of course, θ_{iB} is larger for higher values of ε_2 in Fig. 10.23(a), whereas it is shifted toward smaller incident angles as ε_1 increases in Fig. 10.23(b). For $\varepsilon_1 > \varepsilon_2$, the total internal reflection phenomenon [Eq. (10.144)] is apparent (for both polarizations), and the Brewster condition still occurs for parallel polarization, at an incident angle smaller than the corresponding critical angle [note that it can be shown from Eqs. (10.148) and (10.143) that $\theta_{iB} < \theta_{ic}$ in general]. We also see that the phase angles in total reflection are not constant (180°), and are not the same for the two polarizations, as already discussed earlier.

Example 10.21 Magnitude of Reflection Coefficients in Total Internal Reflection

Show that in the case of total internal reflection, Fig. 10.21, the magnitude of the reflection coefficient for both normal and parallel polarizations of the incident wave is equal to unity.

Solution From Eqs. (10.142)–(10.144), $\sin\theta_t > 1$ for total internal reflection, and hence $\cos\theta_t = \sqrt{1 - \sin^2\theta_t}$, as the square root of a negative real number, is a purely imaginary complex number. With this, expressions for both reflection coefficients in Eqs. (10.134) and (10.136) acquire the form $\underline{\Gamma} = \pm(a + jb)/(a - jb)$, where a and b are real numbers, different for different (normal vs. parallel) polarizations. Since $|\underline{\Gamma}| = \sqrt{a^2 + b^2}/\sqrt{a^2 + b^2} = 1$, we have indeed that $|\underline{\Gamma}_n| = |\underline{\Gamma}_p| = 1$.

Example 10.22 Hiding Gold Fish under a Floating Leaf

A large leaf of approximately circular shape with a diameter of 50 cm floats on the surface of the water ($n = 1.33$) in a fountain. What is the maximum depth in the water directly under the center of the leaf for a small gold fish to be totally invisible from above the water?

Solution This situation is illustrated in Fig. 10.24, and it is essentially the limiting (critical) case of that in Fig. 10.20. We realize that the fish will be totally invisible for a viewer anywhere above the water surface if the condition of total internal reflection, Eq. (10.144), for the light ray emanating from the fish and impinging the water-air interface is satisfied. However, the depth of the fish is maximum when the incident angle θ_1 in water exactly equals the critical angle, $\theta_{\rm ic}$, in Eqs. (10.143). Recalling Eqs. (10.139) and (10.140) as well, this maximum depth, in Fig. 10.24, is computed as

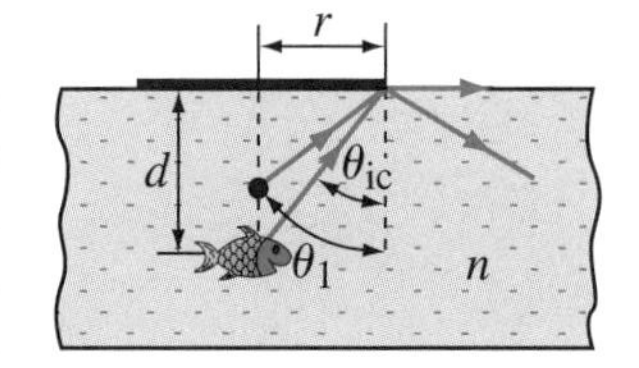

Figure 10.24 Determining the maximum depth of a gold fish underneath a large floating leaf such that it is invisible from above the water; for Example 10.22.

$$\theta_1 = \theta_{\rm ic} = \arcsin\frac{1}{n} = 48.75^\circ \quad \longrightarrow \quad d = r\cot\theta_1 \approx 22 \text{ cm} \quad (r = 25 \text{ cm}) . \tag{10.149}$$

Of course, any location of the fish above the critical one gives θ_1 that is greater than $\theta_{\rm ic}$, as in Eq. (10.144), and hence is also in the invisible zone.

Example 10.23 Total Transmission for Magnetically Different Materials

Show that in the case of magnetically different materials ($\mu_1 \neq \mu_2$) in Fig. 10.13, there is an incident angle ($\theta_{\rm i}$) for which total transmission for normal polarization of the incident wave ($\Gamma_{\rm n} = 0$) occurs, and find that angle.

Solution Combining Eqs. (10.127) and (10.134) for arbitrary parameters ε_1, μ_1, ε_2, and μ_2 of the two (lossless) media, we obtain that the condition $\Gamma_{\rm n} = 0$, i.e., $\eta_2\cos\theta_{\rm i} = \eta_1\cos\theta_{\rm t}$, is satisfied for the angle $\theta_{\rm i}$ given by

$$\sin^2\theta_{\rm i} = \frac{\varepsilon_2\mu_2}{\varepsilon_1\mu_1}\sin^2\theta_{\rm t} = \frac{\varepsilon_2\mu_2}{\varepsilon_1\mu_1}\left(1-\cos^2\theta_{\rm t}\right) = \frac{\varepsilon_2\mu_2}{\varepsilon_1\mu_1}\left(1-\frac{\varepsilon_1\mu_2}{\varepsilon_2\mu_1}\cos^2\theta_{\rm i}\right) = \frac{\varepsilon_2\mu_2}{\varepsilon_1\mu_1}$$
$$-\frac{\mu_2^2}{\mu_1^2}\left(1-\sin^2\theta_{\rm i}\right) \quad \longrightarrow \quad \sin\theta_{\rm i} = \sqrt{\frac{1-\varepsilon_2\mu_1/(\varepsilon_1\mu_2)}{1-(\mu_1/\mu_2)^2}} \quad (\mu_1 \neq \mu_2), \tag{10.150}$$

with a restriction that ε_1, μ_1, ε_2, and μ_2 must be such that the expression under the square root sign is nonnegative. With $\mu_1 = \mu_2$, however, this expression becomes infinite, confirming that there is no $\theta_{\rm i}$ yielding $\Gamma_{\rm n} = 0$ for magnetically identical (e.g., nonmagnetic) media [see Eq. (10.146)].

Example 10.24 Brewster Condition at Both Sides of an Illuminated Glass Slab

A beam of polarized light is incident onto a glass slab in air at the Brewster angle (polarization is parallel). Prove that the other slab interface is also at the Brewster condition for the beam transmitted into glass, so that there is no reflected light in the incident region.

Solution We again refer to Fig. 10.19, and express the Brewster condition at the first (upper) interface of the slab through the sine of the incident angle (θ_1), as in Eqs. (10.147). Combining this with the Snell's law of refraction in Eq. (10.142), applied to the same interface, we obtain the following expression for the sine of the associated transmitted angle (θ_2):

$$\sin\theta_1 = \sqrt{\frac{\varepsilon}{\varepsilon_0+\varepsilon}} \quad \longrightarrow \quad \sin\theta_2 = \sqrt{\frac{\varepsilon_0}{\varepsilon}}\sin\theta_1 = \sqrt{\frac{\varepsilon_0}{\varepsilon_0+\varepsilon}} . \tag{10.151}$$

Since θ_2 is also the incident angle at the second interface of the slab, this expression is exactly the Brewster condition in Eqs. (10.147) for the total transmission from glass to air in Fig. 10.19. So, if the dielectric (glass) slab is illuminated by an electromagnetic wave (light) with parallel polarization at the Brewster angle, total transmission occurs through both surfaces of the slab, and there is no reflected waves in any of the regions.

Example 10.25 **Reflection and Refraction of a CP Wave Incident at Brewster Angle**

A right-hand circularly polarized uniform plane wave of electric field amplitude $E_{\mathrm{m}} = 2$ V/m and frequency $f = 3$ GHz propagates in air and is incident at the Brewster angle on a dielectric half-space with parameters $\varepsilon_{\mathrm{r}} = 4$, $\mu_{\mathrm{r}} = 1$, and $\sigma = 0$. Decompose the electric field vector of the wave onto two linearly polarized components, with normal and parallel polarizations, respectively, and find the corresponding E-field components for the reflected and refracted waves as well. What are the polarization states of these waves?

Solution We represent this CP incident wave as a superposition of two linearly polarized waves, with the electric field vector of the first wave normal to the plane of incidence (normal polarization) and that of the second wave parallel to it (parallel polarization), as in Fig. 10.13(a) and Fig. 10.13(b), respectively. The two vectors have the same amplitudes but are out of phase by 90°. In particular, for the RH circular polarization, the complex rms electric field intensities of the two LP waves can be written as

right-hand circular polarization at oblique incidence

$$\boxed{\underline{E}_{\mathrm{i0}}^{(\mathrm{n})} = -\mathrm{j}E' \quad \text{and} \quad \underline{E}_{\mathrm{i0}}^{(\mathrm{p})} = E',} \tag{10.152}$$

where $E' = E_{\mathrm{m}}/\sqrt{2} = 1.414$ V/m [see Eqs. (10.29)]. Note that with $\theta_{\mathrm{i}} = 0$ in Fig. 10.13, the vector $\underline{\mathbf{E}}_{\mathrm{i}}$ becomes that of the normally incident RHCP wave in Eqs. (10.31). Namely, the component of the total wave with parallel polarization becomes x directed, while the normally polarized component is y directed, and the wave propagates along the z-axis. From Eq. (10.148), the Brewster incident angle for the air-dielectric interface is $\theta_{\mathrm{i}} = \arctan\sqrt{\varepsilon_{\mathrm{r}}} = 63.46°$, and Snell's law of refraction, Eq. (10.142), then gives the refracted (transmitted) angle of $\theta_{\mathrm{t}} = 26.58°$, so that Fresnel's coefficients, Eqs. (10.134) and (10.136), amount to $\Gamma_{\mathrm{n}} = -0.6$, $\tau_{\mathrm{n}} = 0.4$, $\Gamma_{\mathrm{p}} = 0$ (Brewster condition), and $\tau_{\mathrm{p}} = 0.5$ in this situation. Consequently, the components with normal and parallel polarizations of the reflected complex rms electric field intensity vector at the coordinate origin, so not taking into account the propagation factors in Eqs. (10.95) and (10.111), are

parallel polarization filtered out by reflection (Brewster condition)

$$\boxed{\underline{E}_{\mathrm{r0}}^{(\mathrm{n})} = \Gamma_{\mathrm{n}}\underline{E}_{\mathrm{i0}}^{(\mathrm{n})} = -\mathrm{j}\Gamma_{\mathrm{n}}E' = \mathrm{j}0.848 \text{ V/m} \quad \text{and} \quad \underline{E}_{\mathrm{r0}}^{(\mathrm{p})} = \Gamma_{\mathrm{p}}\underline{E}_{\mathrm{i0}}^{(\mathrm{p})} = 0,} \tag{10.153}$$

respectively. As expected, this is a linearly polarized wave with the electric field vector normal to the plane of incidence (normal polarization); the component of the incident wave with parallel polarization is filtered out by reflection. Similarly, the transmitted electric field is given by

$$\underline{E}_{\mathrm{t0}}^{(\mathrm{n})} = \tau_{\mathrm{n}}\underline{E}_{\mathrm{i0}}^{(\mathrm{n})} = -\mathrm{j}\tau_{\mathrm{n}}E' = -\mathrm{j}0.565 \text{ V/m} \quad \text{and} \quad \underline{E}_{\mathrm{t0}}^{(\mathrm{p})} = \tau_{\mathrm{p}}\underline{E}_{\mathrm{i0}}^{(\mathrm{p})} = \tau_{\mathrm{p}}E' = 0.707 \text{ V/m}, \tag{10.154}$$

and this is a right-hand elliptically polarized (RHEP) wave. It consists of both normal and parallel linear polarizations, with the larger magnitude of the electric field for the parallel polarization, which is again to be expected (the component of the incident wave with parallel polarization is entirely transmitted through the interface).

Problems: 10.29–10.35; *Conceptual Questions* (on Companion Website): 10.37–10.39; *MATLAB Exercises* (on Companion Website).

10.9 WAVE PROPAGATION IN MULTILAYER MEDIA

All cases of reflections and transmissions (refractions) of uniform plane waves studied so far in this chapter have included only one planar material interface (between two half-spaces with different electromagnetic properties). However, in many practical situations, we have more than one interface, that is, more than two different planar material regions (layers) separated by multiple parallel interfaces. Such inhomogeneous regions are called planar multilayer media. In this section, we study plane-wave propagation in a three-layer medium, where a layer of thickness d is placed between two semi-infinite regions, as shown in Fig. 10.25; generalization to an arbitrary number of layers is straightforward. In practice, the situation in Fig. 10.25 comes about whenever a coating (medium 2) is placed on a material object (medium 3) that is illuminated by a plane wave from air (medium 1), or when a wave impinges a sheet or wall (slab) of material (medium 2) in air (media 1 and 3). We allow that all three media in the figure be lossy, in general, with material parameters ε_k, μ_k, and σ_k ($k = 1, 2, 3$), and let a uniform plane time-harmonic electromagnetic wave be incident from medium 1 normally on interface 1-2 (at $z = -d$). The complex propagation coefficients ($\underline{\gamma}_k$) and intrinsic impedances ($\underline{\eta}_k$) of the three media ($k = 1, 2, 3$) are computed as in Section 10.2 (for the two-media problem).

In steady state, the resultant wave in the first region can be expressed in terms of two traveling waves, an incident (forward) wave (propagating to the right) and a reflected (backward) wave (propagating to the left), and the same holds true for the second region, whereas only an incident wave exists in the third region, as indicated in Fig. 10.25. In the figure, $\underline{E}_{1i}$ and $\underline{E}_{1r}$ denote, respectively, complex rms electric field intensities of the incident and reflected waves in medium 1 at interface 1-2 (at $z = -d^-$), while $\underline{E}_{2i}$ and $\underline{E}_{2r}$ are the corresponding field intensities in medium 2 also at the same interface, but on its other side (at $z = -d^+$), and $\underline{E}_3$ designates the complex intensity of the forward (and total) electric field in medium 3 at the second interface (at $z = 0^+$). Similarly to Eqs. (10.40) for the two-media situation, the boundary conditions for electric and magnetic fields on interfaces 1-2 and 2-3, respectively, yield a system of four simultaneous linear algebraic equations with four complex unknowns, $\underline{E}_{1r}$, $\underline{E}_{2i}$, $\underline{E}_{2r}$, and $\underline{E}_3$, assuming that $\underline{E}_{1i}$ is given. Once these unknowns are determined, the spatial distributions of the incident, reflected, and total electric and magnetic fields in any of the three regions in Fig. 10.25 in terms of the coordinate z can be obtained, if needed, using these constants, along with the propagation coefficients ($\underline{\gamma}_k$) and intrinsic impedances ($\underline{\eta}_k$) of the media. However, we solve here for $\underline{E}_{1r}$ alone, since the solution for this unknown (determining reflection back into the incident medium) in fact suffices for most practical applications. To this end, let us introduce the equivalent intrinsic impedance of the combination of media 2 and 3, $\underline{\eta}_e$, which can be used, much like the equivalent input impedance

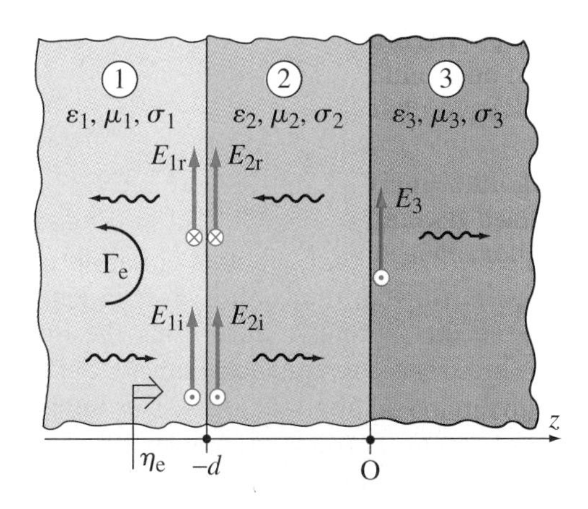

Figure 10.25 Normal incidence, from medium 1, of a uniform plane time-harmonic electromagnetic wave on two parallel interfaces separating three different homogeneous material regions with arbitrary electromagnetic parameters.

of a part of an electric circuit, to replace these two media by a semi-infinite homogeneous material region with intrinsic impedance equal to $\underline{\eta}_e$. By means of Eq. (10.42), the corresponding equivalent reflection coefficient at the first interface, which takes into account the presence of both material regions on its right, is

equivalent reflection coefficient at the first interface

$$\underline{\Gamma}_e = \frac{\underline{E}_{1r}}{\underline{E}_{1i}} = \frac{\underline{\eta}_e - \underline{\eta}_1}{\underline{\eta}_1 + \underline{\eta}_e}. \tag{10.155}$$

Since the ratio of the total electric to magnetic complex field intensities at a point in space constitutes the wave impedance at that point, $\underline{\eta}_e$ can be obtained directly from Eq. (10.63), as it equals $\underline{E}_2/\underline{H}_2$, with $\underline{E}_2$ and $\underline{H}_2$ standing, respectively, for total (incident plus reflected) complex rms electric and magnetic field intensities in medium 2 at the first interface. So, substituting $z = -d$ (or, more precisely, $z = -d^+$) and using the appropriate notation (Fig. 10.25) for the involved wave parameters in Eq. (10.63), we have

$$\underline{\eta}_e = \frac{\underline{E}_2}{\underline{H}_2} = \underline{\eta}_{w2}(z = -d) = \underline{\eta}_2 \frac{e^{\underline{\gamma}_2 d} + \underline{\Gamma}_{23}\, e^{-\underline{\gamma}_2 d}}{e^{\underline{\gamma}_2 d} - \underline{\Gamma}_{23}\, e^{-\underline{\gamma}_2 d}} \quad \left(\underline{\Gamma}_{23} = \frac{\underline{\eta}_3 - \underline{\eta}_2}{\underline{\eta}_2 + \underline{\eta}_3}\right), \tag{10.156}$$

where $\underline{\Gamma}_{23}$ represents the reflection coefficient for the interface between media 2 and 3 as if they both were semi-infinite, i.e., as if there were no medium 1 and interface 1-2 on the left. Once the equivalent reflection coefficient $\underline{\Gamma}_e$ is computed from Eqs. (10.155) and (10.156), the solution for the reflected electric field intensity in medium 1 at the first interface is $\underline{E}_{1r} = \underline{\Gamma}_e \underline{E}_{1i}$. Consequently, as in Eq. (10.59) for the two-media situation, the time-average Poynting vector of the reflected wave in the incident medium, that is, the real power (per unit area of the interface) bounced back off the interface, is proportional to $|\underline{\Gamma}_e|^2$.

Given the expression for $\underline{\Gamma}_{23}$, the expression for $\underline{\eta}_e$ in Eq. (10.156) can easily be manipulated to read

$$\begin{aligned}\underline{\eta}_e &= \underline{\eta}_2 \frac{(\underline{\eta}_2 + \underline{\eta}_3)\, e^{\underline{\gamma}_2 d} + (\underline{\eta}_3 - \underline{\eta}_2)\, e^{-\underline{\gamma}_2 d}}{(\underline{\eta}_2 + \underline{\eta}_3)\, e^{\underline{\gamma}_2 d} - (\underline{\eta}_3 - \underline{\eta}_2)\, e^{-\underline{\gamma}_2 d}} \\ &= \underline{\eta}_2 \frac{\underline{\eta}_2 (e^{\underline{\gamma}_2 d} - e^{-\underline{\gamma}_2 d}) + \underline{\eta}_3 (e^{\underline{\gamma}_2 d} + e^{-\underline{\gamma}_2 d})}{\underline{\eta}_2 (e^{\underline{\gamma}_2 d} + e^{-\underline{\gamma}_2 d}) + \underline{\eta}_3 (e^{\underline{\gamma}_2 d} - e^{-\underline{\gamma}_2 d})}.\end{aligned} \tag{10.157}$$

From the definitions of hyperbolic sine and cosine functions,

$$\sinh x = \frac{e^x - e^{-x}}{2}, \quad \cosh x = \frac{e^x + e^{-x}}{2}, \tag{10.158}$$

we finally have

equivalent intrinsic impedance of media 2 and 3

$$\underline{\eta}_e = \underline{\eta}_2 \frac{\underline{\eta}_2 \sinh \underline{\gamma}_2 d + \underline{\eta}_3 \cosh \underline{\gamma}_2 d}{\underline{\eta}_2 \cosh \underline{\gamma}_2 d + \underline{\eta}_3 \sinh \underline{\gamma}_2 d}. \tag{10.159}$$

For given parameters of the central layer (d, $\underline{\gamma}_2$, and $\underline{\eta}_2$), this equation can be regarded as a function $\underline{\eta}_e = \underline{\eta}_e(\underline{\eta}_3)$. This means that the central layer in Fig. 10.25 essentially acts as an impedance transformer, in that it transforms the intrinsic impedance of the third medium, $\underline{\eta}_3$, to a different impedance, namely, the equivalent input impedance $\underline{\eta}_e$ seen at interface 1-2 looking to the right. This impedance transformation, using a specially designed material (dielectric) layer between two given media, finds many applications in optics, antennas, radars, and other areas of applied electromagnetics. In addition, an analogous impedance transformation (to be discussed in a later chapter) is extensively used in the analysis and design of transmission lines. Note also that Eq. (10.159) can readily be applied to problems involving more than three material regions, where the overall equivalent input impedance at the first interface is found by successive impedance transformations applied layer by layer in the backward direction.

In a frequent practical situation when all media in Fig. 10.25 are lossless, their individual intrinsic impedances are purely real, and $\underline{\gamma}_2 = \mathrm{j}\beta_2$ $(\alpha_2 = 0)$ in Eq. (10.159). Combining Eqs. (10.158) and (10.7), it is easily shown that $\sinh \mathrm{j}x = \mathrm{j}\sin x$ and $\cosh \mathrm{j}x = \cos x$, so that the equivalent impedance of media 2 and 3 in Eq. (10.159) becomes

$$\boxed{\underline{\eta}_\mathrm{e} = \eta_2 \frac{\eta_3 \cos\beta_2 d + \mathrm{j}\eta_2 \sin\beta_2 d}{\eta_2 \cos\beta_2 d + \mathrm{j}\eta_3 \sin\beta_2 d}.} \tag{10.160}$$

equivalent impedance, lossless case

It is interesting that, even in the lossless case, $\underline{\eta}_\mathrm{e}$ is in general (for an arbitrary thickness of the central layer, d, and arbitrary permittivities and permeabilities of the media) complex. The same thus holds true for the equivalent reflection coefficient $\underline{\Gamma}_\mathrm{e}$, unlike the case with "ordinary" reflection coefficients, Eq. (10.42), which are always purely real for lossless media. On the other hand, as will be shown in examples, there are special cases where $\underline{\eta}_\mathrm{e}$ and $\underline{\Gamma}_\mathrm{e}$ are purely real, and, moreover, the central layer can be designed so that $\Gamma_\mathrm{e} = 0$ (total transmission from medium 1 into media 2 and 3).

Example 10.26 Quarter-Wave Transformer for Total Transmission

(a) Assuming that the three electromagnetic media in Fig. 10.25 are lossless, as well as that media 1 and 3 have different parameters, design the central layer such that there is no reflection of the incident wave back to medium 1. (b) What are the parameters of the central layer in the case of all nonmagnetic media?

Solution

(a) To satisfy the condition of total transmission from medium 1 into the combination of media 2 and 3 in Fig. 10.25, Eq. (10.155) results in

$$\boxed{\underline{\Gamma}_\mathrm{e} = 0 \quad \longrightarrow \quad \underline{\eta}_\mathrm{e} = \eta_1,} \tag{10.161}$$

no reflection at the first interface

so the equivalent impedance $\underline{\eta}_\mathrm{e}$ has to be purely real and equal to the intrinsic impedance of the incident medium. With this, Eq. (10.160) is then reduced to

$$\eta_2 \left(\eta_3 \cos\beta_2 d + \mathrm{j}\eta_2 \sin\beta_2 d\right) = \eta_1 \left(\eta_2 \cos\beta_2 d + \mathrm{j}\eta_3 \sin\beta_2 d\right). \tag{10.162}$$

Equating the real and imaginary parts, respectively, of the right- and left-hand side of this complex equation, it decouples onto the following two real equations:

$$(\eta_1 - \eta_3)\cos\beta_2 d = 0 \quad \text{and} \quad \left(\eta_2^2 - \eta_1\eta_3\right)\sin\beta_2 d = 0. \tag{10.163}$$

Since $\eta_1 \neq \eta_3$, from the first equation we have

$$\cos\beta_2 d = 0 \quad \longrightarrow \quad \beta_2 d = (2m+1)\frac{\pi}{2} \quad \longrightarrow \quad d = (2m+1)\frac{\lambda_2}{4} \quad (m = 0, 1, 2, \ldots), \tag{10.164}$$

and hence the smallest value of the thickness of the central layer is $d = \lambda_2/4$, λ_2 being the wavelength in medium 2 at the operating frequency of the incident wave. Furthermore, as it now turns out that $\sin\beta_2 d = \pm 1$, the second equation in Eqs. (10.163) gives that the required intrinsic impedance of the central layer equals the geometric mean of the intrinsic impedances of the two semi-infinite regions on its sides,

quarter-wave matching, $\eta_1 \neq \eta_3$

$$\eta_2 = \sqrt{\eta_1 \eta_3}, \quad d = \frac{\lambda_2}{4}. \tag{10.165}$$

We note that the designed layer essentially matches the intrinsic impedance of a medium (region 3), η_3, to that, η_1, of another medium (region 1) in order to eliminate the reflection due to the material discontinuity between the regions ($\eta_1 \neq \eta_3$). In other words, it transforms the intrinsic impedance η_3 to the equivalent input impedance at interface 1-2 (η_e) equal to η_1. Being a quarter of a wavelength (or an odd integer multiple of that) thick, the layer is referred to as a quarter-wave transformer. As we shall see in a later chapter, a completely analogous transformer in the form of a quarter-wavelength long section of a transmission line (e.g., a coaxial cable, Fig. 2.17) is utilized for impedance matching of transmission lines and their loads.

(b) If all media in Fig. 10.25 are nonmagnetic (or if all have the same permeability), combining Eqs. (10.165), (9.21), and (8.112) leads to the relative permittivity and minimum thickness of the central layer amounting to

$$\varepsilon_{r2} = \sqrt{\varepsilon_{r1}\varepsilon_{r3}} \quad \text{and} \quad d = \frac{\lambda_2}{4} = \frac{\lambda_0}{4\sqrt{\varepsilon_{r2}}} = \frac{c_0}{4\sqrt{\varepsilon_{r2}}\, f} \quad (\mu_{r1} = \mu_{r2} = \mu_{r3}), \tag{10.166}$$

respectively, where c_0 is the free-space wave velocity, Eq. (9.19).

Finally, let us emphasize that the principal drawback of quarter-wave transformers when used for impedance matching is their resonant (or narrow-band) operation. Namely, the condition in Eqs. (10.161) is satisfied only at a single frequency f, in Eq. (10.166), at which the thickness of the middle layer in Fig. 10.25 is exactly $d = \lambda_2/4$. Around this frequency, the equivalent reflection coefficient $\underline{\Gamma}_e$ is not zero, but might still be within an acceptable value, as imposed by the impedance-matching or low-reflection requirements in the concrete application, over a narrow band of frequencies. On the other hand, the transformer design can be made more broadband, with $|\underline{\Gamma}_e|$ small enough within a broader frequency range, by cascading multiple quarter-wave (at the central frequency of the range) layers (or transmission-line sections), whose intrinsic impedances are gradually varied (increased or decreased) from η_1 to η_3.

Example 10.27 Design of Anti-Reflective Coatings for Optical Devices

We wish to design an anti-reflective coating for a glass surface in an optical device at a free-space wavelength of $\lambda_0 = 600$ nm. The refractive index of the glass is $n_{\text{glass}} = 1.9$. (a) Determine the refractive index and minimum thickness for the coating. (b) What percentage of the incident power is reflected from the coated glass if the wavelength is changed to 500 nm?

Solution

(a) We adopt a quarter-wave matching transformer design. Adopting also that the coating be nonmagnetic, like the other two media (glass and air), Eqs. (10.166) and (10.129) give the following for its index of refraction and minimum thickness:

optical anti-reflective coating

$$n_{\text{coating}} = \sqrt{n_{\text{air}} n_{\text{glass}}} = \sqrt{n_{\text{glass}}} = 1.38, \quad d = \frac{\lambda_0}{4 n_{\text{coating}}} = 108.7 \text{ nm}. \tag{10.167}$$

So, if a dielectric film with these obtained parameters is deposited on the glass surface (e.g., that of a camera lens), no normally incident visible light is reflected at $\lambda_0 = 600$ nm. Of course, this is practically a single-frequency (single-wavelength) or narrow-band anti-reflective structure, as discussed in the previous example. Note that the refractive index of $n = 1.38$ happens to be exactly that of magnesium fluoride (MgF_2), which is commonly used as anti-reflective film material in optical applications.

(b) If the free-space wavelength of the incident wave is changed to $\lambda_0 = 500$ nm, the phase coefficient in the film material changes as well, to $\beta_2 = 2\pi/\lambda_2 = 2\pi n_{\text{coating}}/\lambda_0 = 17.34 \times 10^6$ rad/m, and hence the equivalent intrinsic impedance of the combination of media 2 and 3 in Fig. 10.25 is no longer matched to that of free space ($\eta_0 = 377\ \Omega$). Using Eq. (10.160), it equals $\underline{\eta}_e = 352.77\,\mathrm{e}^{-\mathrm{j}10.82^\circ}\ \Omega$ [note that for both the glass and coating, $\eta = \eta_0/\sqrt{\varepsilon_r} = \eta_0/n$, from Eq. (9.48)], and Eq. (10.155) then gives the equivalent reflection coefficient of $\underline{\Gamma}_e = 0.1\,\mathrm{e}^{-\mathrm{j}109.5^\circ}$. Finally, having in mind Eqs. (10.59), the percentage of the time-average incident power that is bounced back off the coating comes out to be

$$\frac{|(\mathcal{P}_r)_{\text{ave}}|}{|(\mathcal{P}_i)_{\text{ave}}|} = |\underline{\Gamma}_e|^2 = 1\%. \tag{10.168}$$

percentage of incident power bouncing back

So, this is still a rather low reflection.

Example 10.28 Half-Wave Matching Plate

Determine the thickness of a dielectric slab in free space such that no normally incident uniform plane electromagnetic waves of frequency f are reflected from it.

Solution With reference to Fig. 10.25, we now have that $\eta_1 = \eta_3 = \eta_0$, η_0 being the free-space intrinsic impedance, Eq. (9.23), and hence the first equation in Eqs. (10.163) is satisfied for any thickness d of the slab. In the second equation, however, $\eta_2^2 \neq \eta_1\eta_3$, since $\eta_2 \neq \eta_0$ (dielectric slab), which means that [see Eqs. (10.10)]

$$d = m\frac{\lambda_2}{2} \quad (m = 1, 2, \ldots), \tag{10.169}$$

half-wave matching, $\eta_1 = \eta_3$

namely, that the slab has to be an integer multiple of a half-wavelength (computed in the slab material) thick, the minimum thickness being $d = \lambda_2/2$, where $\lambda_2 = 2\pi/\beta_2 = 1/(f\sqrt{\varepsilon_2\mu_2})$. This structure is thus called a half-wave matching slab (or plate), and it is extensively used for designs of protective dielectric housings (enclosures) of antennas that, built (in air) around an antenna, should be as transparent as possible for electromagnetic waves traveling through it in both directions. The name of such an enclosure, a radome, comes from their frequent applications as domes housing large radar antennas – radar dome.

Example 10.29 Design of Airplane Antenna Radome

Find the minimum thickness of a fiberglass ($\varepsilon_r = 4.9$, $\sigma \approx 0$) radome for an antenna on an airplane (the antenna needs to be protected from weather) such that (1) the radome appears transparent to the antenna radiation at $f = 10$ GHz and (2) the mechanical requirement that the radome be at least 2 cm thick is met. Assume that the electromagnetic waves radiated (or received) by the antenna are normally incident and planar at the radome surface.

Solution The no-reflection requirement, at the given frequency, is expressed in Eq. (10.169). Having in mind Eqs. (10.166), the half-wave thickness of the radome amounts to $\lambda_2/2 = c_0/(2\sqrt{\varepsilon_r}f) = 6.78$ mm. Hence, $m = 3$ in Eq. (10.169) determines the minimum thickness that also satisfies the mechanical requirement ($d \geq 2$ cm) – it is equal to $d_{\min} = 3\lambda_2/2 = 2.03$ cm.

Example 10.30 Reflection from a Copper Plate with Teflon Coating

A large copper plate is coated by a 1-cm thick teflon ($\varepsilon_r = 2.1$) layer. The surrounding medium is air. Assuming that both the conductor and the dielectric are perfect, find the magnitude and phase of the equivalent reflection coefficient ($\underline{\Gamma}_e$) at the air-teflon interface for an 8-GHz uniform plane wave at normal incidence.

Solution Treating copper as a perfect conductor, we know from Eq. (9.137) that its intrinsic impedance (for $\sigma \to \infty$) is zero, and using $\eta_3 = 0$ we then specialize Eq. (10.160) for the equivalent impedance of the teflon-coated plate and the associated reflection coefficient, Eq. (10.155), at the air-teflon interface,

$$\underline{\eta}_e = j\eta_2 \tan\beta_2 d \quad \longrightarrow \quad \underline{\Gamma}_e = \frac{\underline{\eta}_e - \eta_0}{\eta_0 + \underline{\eta}_e} = -\frac{\eta_0 - j\eta_2\tan\beta_2 d}{\eta_0 + j\eta_2 \tan\beta_2 d} \tag{10.170}$$

($\eta_2 = \eta_0/\sqrt{\varepsilon_r}$; $\beta_2 = 2\pi f\sqrt{\varepsilon_r}/c_0$), where this expression for $\underline{\eta}_e$ is in agreement with that in Eq. (10.64). As shown in Example 10.21, complex fractions of the form above are unity in magnitude, so $|\underline{\Gamma}_e| = 1$. Substituting the given numerical data into Eqs. (10.170), we compute the phase of the reflection coefficient as well, namely, $\underline{\Gamma}_e = e^{-j118.5^\circ}$.

Note that the fact that $|\underline{\Gamma}_e| = 1$, which means that the entire power carried by the incident wave is reflected back into the incident medium (air), can also be obtained from the conservation of power principle, since no power is transmitted into the copper plate (perfect conductor) and no power is absorbed (lost to heat) in the teflon layer (perfect dielectric).

Example 10.31 No Reflection in a Four-Media Problem

A slab of thickness d made of a good conductor of parameters ε, μ, and σ is placed in air parallel to a PEC plane. A time-harmonic uniform plane electromagnetic wave with free-space wavelength λ_0 propagates in air and is normally incident on the slab. The distance of the slab from the plane is $3\lambda_0/4$, as shown in Fig. 10.26. Assuming that $|\underline{\gamma}_2 d| \ll 1$ (the slab is electrically thin), $\underline{\gamma}_2$ being the complex propagation coefficient in the slab, find d so that there is no reflected wave in the incident region.

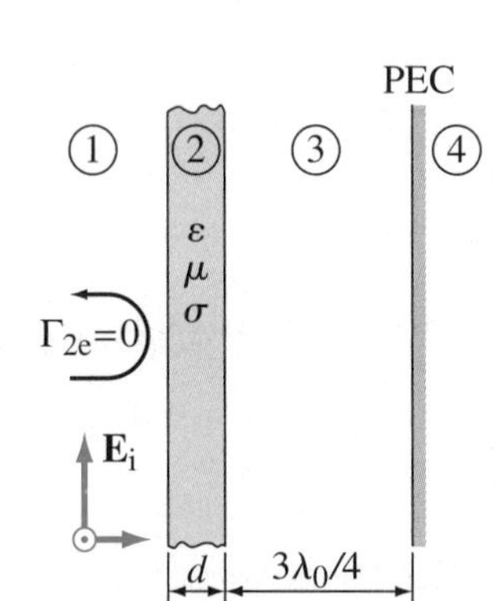

Figure 10.26 Normal incidence of a uniform plane wave on an electrically thin slab made of a good conductor, placed in parallel to a PEC plane, at a distance of three quarter-wavelengths from the plane; for Example 10.31.

Solution The analysis can be performed using the final formula in Eq. (10.159) for the equivalent intrinsic impedance for the three-media situation in Fig. 10.25, by successive impedance transformations applied across the two layers in Fig. 10.26 in the backward direction, layer 3 and then layer 2. Namely, medium 4 being a PEC implies that $\eta_4 = 0$ (see the previous example), so that the no-loss version of the transformation formula in Eq. (10.160) written for media 2-3-4, where the thickness of the central layer (region 3) is given by $d_3 = 3\lambda_3/4$ and $\lambda_3 = \lambda_0$ (air), yields an infinite equivalent impedance at interface 2-3,

$$\eta_4 = 0 \quad (\text{PEC}) \quad \text{and} \quad \beta_3\frac{3\lambda_3}{4} = \frac{3\pi}{2} \quad \longrightarrow \quad \underline{\eta}_{3e} = \eta_3\frac{\eta_4\cos\frac{3\pi}{2} + j\eta_3\sin\frac{3\pi}{2}}{\eta_3\cos\frac{3\pi}{2} + j\eta_4\sin\frac{3\pi}{2}} \to \infty \tag{10.171}$$

($\eta_3 = \eta_0$). This impedance, $\underline{\eta}_{3e} \to \infty$, is then transformed further (backward), across layer 2, invoking the transformation formula for the general (lossy) case, Eq. (10.159), for media 1-2-3 in Fig. 10.26, resulting in the following for the equivalent wave impedance $\underline{\eta}_{2e}$ seen at interface 1-2, looking to the right:

$$\underline{\eta}_{2e} = \underline{\eta}_2\frac{\underline{\eta}_2\sinh\underline{\gamma}_2 d + \underline{\eta}_{3e}\cosh\underline{\gamma}_2 d}{\underline{\eta}_2\cosh\underline{\gamma}_2 d + \underline{\eta}_{3e}\sinh\underline{\gamma}_2 d} = \underline{\eta}_2\frac{\cosh\underline{\gamma}_2 d}{\sinh\underline{\gamma}_2 d} = \underline{\eta}_2\frac{e^{\underline{\gamma}_2 d} + e^{-\underline{\gamma}_2 d}}{e^{\underline{\gamma}_2 d} - e^{-\underline{\gamma}_2 d}} \approx \frac{\underline{\eta}_2}{\underline{\gamma}_2 d}, \tag{10.172}$$

where the use is also made of the fact that $|\underline{\gamma}_2 d| \ll 1$ and that $e^{\underline{a}} \approx 1 + \underline{a}$ for $|\underline{a}| \ll 1$. Combining the final expression for $\underline{\eta}_{2e}$ with the no-reflection condition at this interface, as in Eqs. (10.161), and using the approximate expressions for the complex intrinsic impedance ($\underline{\eta}_2$) and propagation coefficient ($\underline{\gamma}_2$) of good conductors, in Eqs. (9.136) and (9.134), respectively, we obtain the required thickness of the slab,

$$\underline{\eta}_{2e} = \eta_1 = \eta_0 \quad \longrightarrow \quad d = \frac{\underline{\eta}_2}{\underline{\gamma}_2\eta_0} \approx \frac{1}{\sigma\eta_0} = \frac{1}{\sigma}\sqrt{\frac{\varepsilon_0}{\mu_0}}. \tag{10.173}$$

Problems: 10.36–10.39; *Conceptual Questions* (on Companion Website): 10.40; *MATLAB Exercises* (on Companion Website).

Problems

10.1. Normal incidence on a PEC – derivation in time domain. A uniform plane time-harmonic electromagnetic wave with the electric field given by $\mathbf{E}_{\mathrm{i}} = E_{\mathrm{i}0}\sqrt{2}\cos(\omega t - \beta z)\,\hat{\mathbf{x}}$ travels for $z < 0$ in a medium with intrinsic impedance η and is incident on a perfectly conducting plane at $z = 0$. Derive the expressions for the instantaneous total electric and magnetic field intensity vectors in the incident medium, Eqs. (10.9), carrying out the analysis entirely in the time domain, in place of the complex-domain derivation in Eqs. (10.1)–(10.8). In particular, assume that the electric field vector of the reflected wave has the form $\mathbf{E}_{\mathrm{r}} = E_{\mathrm{r}0}\sqrt{2}\cos(\omega t + \beta z + \xi_{\mathrm{r}})\,\hat{\mathbf{x}}$, and find $E_{\mathrm{r}0}$ and ξ_{r} from a boundary condition on the PEC surface, then obtain the incident and reflected instantaneous magnetic field vectors from the corresponding electric field vectors, and finally the total instantaneous vectors using superposition and the appropriate trigonometric identities.

10.2. Various properties of a standing wave. A uniform plane time-harmonic electromagnetic wave of frequency $f = 1$ GHz and rms magnetic field intensity $H_{\mathrm{i}0} = 1$ A/m propagates in a lossless dielectric medium of relative permittivity $\varepsilon_{\mathrm{r}} = 2.25$ ($\mu_{\mathrm{r}} = 1$) and impinges a PEC screen at normal incidence. Assuming a zero initial phase of $\mathbf{H}_{\mathrm{i}}$ at the surface of the screen, find: (a) locations of zeros of the total electric field in the dielectric, (b) times at which zeros of the total electric field at specific locations occur, (c) locations of magnetic field zeros in the dielectric, (d) maximum amplitude of the total magnetic field, (e) rms surface current and charge densities in the screen, (f) electric and magnetic energy densities and Poynting vector in the dielectric at a distance $d = 11.25$ cm from the screen and instant $t = 2.15$ ns, and (g) time-average electromagnetic energy density and Poynting vector at the same location.

10.3. Energy in an imaginary cylinder of arbitrary length. Repeat Example 10.4 but for an arbitrary length l of the imaginary cylinder placed in the field of a standing plane electromagnetic wave (assume that $\mu = \mu_0$).

10.4. Time-average energy of a cylinder. Compute the time-average resultant electromagnetic energy stored in the cylinder from the previous problem (arbitrary l), in the following two ways: (a) averaging in time the instantaneous energy of the cylinder (result of the previous problem) and (b) integrating in space the time-average electromagnetic energy density of the standing wave.

10.5. Induced emf in a large contour due to a standing wave. Consider the standing electromagnetic wave in front of a PEC screen in Fig. 10.1 (rms electric field intensity of the incident wave is $E_{\mathrm{i}0}$ and frequency f) with free space as the propagation medium, and assume that an electrically large (arbitrarily sized) rectangular contour of edge lengths a and b is placed in the plane $y = 0$. The contour edges that are b long are parallel to the screen, one being at the coordinate z and the other at $z + a$ ($z < -a$). Find the emf induced in the contour, from (a) the left- and (b) the right-hand side of Faraday's law of electromagnetic induction in integral form (in the complex domain), respectively. (c) Show that the result in (a) or (b) becomes that in Eq. (10.24) when the contour is electrically small.

10.6. Standing-wave measurements with a magnetic probe. To determine the frequency, f, and electric-field rms intensity, $E_{\mathrm{i}0}$, of a uniform plane wave traveling in air, a perfectly conducting plate is introduced normally to the wave propagation and emf induced in a small square wire loop 2.5 cm on a side is measured. By varying the orientation and location of the loop, it is found that the rms emf in it has a maximum, of 5 mV, at a distance of 80 cm from the conductor (with the plane of the loop being perpendicular to the magnetic field vector of the wave). It is also found that the first adjacent minimum (zero) of the rms emf is at 60 cm from the conductor (for the same orientation of the loop). What are f and $E_{\mathrm{i}0}$?

10.7. Rotating contour in the field of a standing wave. A uniform plane time-harmonic electromagnetic wave of rms electric field intensity E_{i0} and angular frequency ω is incident normally from air on a PEC plane. An electrically small square contour of edge length a rotates, in the incident region, with a constant angular velocity w_0 about its axis. The center of the contour is at an arbitrary distance d from the conductor, and the axis of rotation is parallel to the direction of wave propagation. At a reference instant $t = 0$, the plane of the contour is perpendicular to the magnetic field lines. Find the instantaneous emf induced in the contour, and identify the portions of it corresponding to the transformer and motional induction, respectively.

10.8. Reflection of an elliptically polarized wave. The magnetic field of a plane electromagnetic wave impinging a PEC plane at $z = 0$ from a nonmagnetic medium is given by $\mathbf{H}_\text{i} = [3\cos(\omega t - \beta z)\,\hat{\mathbf{x}} - \sin(\omega t - \beta z)\,\hat{\mathbf{y}}]$ A/m $(z < 0)$, where $\omega = 6\pi \times 10^8$ rad/s and $\beta = 4\pi$ rad/m. Determine (a) complex and instantaneous electric and magnetic field intensity vectors of the reflected wave, (b) the polarization state (type and handedness) of the reflected wave, (c) complex and instantaneous electric and magnetic field vectors of the resultant wave in the incident medium, (d) the polarization state of the resultant wave, (e) the total time-average Poynting vector in the incident medium, and (f) rms surface current and charge densities in the PEC plane.

10.9. Air-glass interface. The magnetic field intensity vector of a TEM wave propagating in air for $x > 0$ is given by $\mathbf{H}_\text{i} = 5\sqrt{2}\sin(\omega t + 10\pi x)\,\hat{\mathbf{z}}$ A/m (x in m). The wave is incident at $x = 0$ on the planar interface of a glass medium with relative permittivity $\varepsilon_\text{r} = 4$ $(\mu_\text{r} = 1)$ and negligible losses, occupying the $x < 0$ half-space. Compute (a) the frequency of the wave, (b) the wavelength in the second (glass) medium, (c) the instantaneous and complex total electric field intensity vector in air, (d) the maxima and minima of the total rms magnetic field intensity in air and glass, respectively, (e) the time-average Poynting vector for $-\infty < x < \infty$, and (f) the wave impedance in a plane defined by $x = 10$ cm.

10.10. Glass-air interface. Repeat the previous problem but for glass as the incident medium $(x > 0)$, with the given field $\mathbf{H}_\text{i}$, and air as the second medium.

10.11. Air-concrete interface. A uniform plane time-harmonic electromagnetic wave of frequency $f = 1$ GHz and rms electric field intensity $E_{i0} = 1$ V/m propagates in air and impinges normally the planar surface of a large concrete block with material parameters $\varepsilon_\text{r} = 6$, $\sigma = 2.5 \times 10^{-3}$ S/m, and $\mu_\text{r} = 1$. Determine if the concrete is a good dielectric, good conductor, or quasi-conductor, and find (a) the distributions of electric and magnetic fields in both media, (b) the standing wave ratio in air, (c) the time-average Poynting vector in the second medium, and (d) the percentages of the time-average incident power that are reflected from the interface and transmitted into the material block, respectively.

10.12. Air-seawater interface. Repeat the previous problem but for seawater with $\varepsilon_\text{r} = 81$ and $\sigma = 4$ S/m as the reflecting medium (flat ocean surface).

10.13. Air-seawater interface at a different frequency. Repeat the previous problem (ocean surface) but for the frequency of $f = 1$ MHz (assuming that the parameters of seawater do not change with frequency).

10.14. Aircraft-to-submarine radio communication. A 15-kHz VLF aircraft antenna transmits a wave that approaches the ocean surface at normal incidence in the form of a uniform plane wave with a rms electric field intensity of 1 kV/m. Find the maximum depth below the surface to which successful communication with a submarine is possible, if the receiving antenna on the submarine requires a minimum rms magnetic field intensity of 0.05 μA/m, the aircraft and submarine are approximately at the same vertical line, and parameters of seawater are $\varepsilon_\text{r} = 81$ and $\sigma = 4$ S/m.

10.15. SWR measurement to determine unknown permittivity. To determine the permittivity of a nonmagnetic lossless dielectric, a time-harmonic plane wave is launched to propagate through air and impinge normally upon and partially reflects from the surface of this material. The relative rms electric field intensities of the resultant wave are measured by an electric probe in the region in front of the material, along the incident wave propagation. By such measurement, it is found that the field maximum amounts to three times the minimum, and that the distance between successive maxima is 1 m. What is the relative permittivity of the dielectric and what the frequency of the wave?

10.16. Loss power in copper walls of a Fabry-Perot resonator. Consider a Fabry-Perot resonator, in Fig. 10.3, with a standing plane electromagnetic wave given by Eqs. (10.9), where $E_{i0} = 100$ V/m, $\beta = \pi \times 10^2$ rad/m, and $\omega = \pi \times 10^{10}$ rad/s. In addition, let the length of the resonator be $a = 3$ cm, and let its walls be made of copper ($\sigma = 58$ MS/m and $\mu = \mu_0$). Under these circumstances, compute the time-average power of Joule's losses in the resonator per $S = 1\ \text{cm}^2$ area of its surface (viewed across both walls).

10.17. Total fields for oblique incidence, normal polarization. A TEM wave propagating in air for $y > 0$ is incident obliquely on a PEC screen occupying the plane $y = 0$. The complex rms electric field intensity vector of the wave is given by $\underline{\mathbf{E}}_{\mathrm{i}} = \mathrm{e}^{\mathrm{j}\pi(y+\sqrt{3}z)}\,\hat{\mathbf{x}}$ V/m (y, z in m). Calculate the electric and magnetic field vectors of the resultant wave at an arbitrary point in air.

10.18. Energy computation for oblique incidence, normal polarization. (a) Consider the oblique incidence on a PEC plane of a normally polarized uniform plane time-harmonic wave in Fig. 10.13(a), and find the expression for the instantaneous electromagnetic energy density of the resultant wave in the incident medium. (b) Then obtain the expression for the time-average total energy density in Eq. (10.106) by averaging in time the result in (a).

10.19. Energy in an imaginary cylinder for oblique incidence. A uniform plane time-harmonic electromagnetic wave of frequency $f = 300$ MHz and rms electric field intensity $E_{i0} = 1$ V/m propagates in air and impinges obliquely at an angle $\theta_{\mathrm{i}} = 60°$ and normal polarization a horizontal PEC screen. Compute the time-average resultant electromagnetic energy stored in an imaginary cylinder of height $h = 50$ cm and basis area $S = 100\ \text{cm}^2$ positioned vertically in air with one of its bases lying in the screen.

10.20. Vertically moving contour for oblique incidence. Assume that the small contour in Fig. 10.14 lies in the plane defined by $x = a =$ const, where it moves (with velocity v) in the vertical direction away from the screen, and find the instantaneous emf induced in it.

10.21. Total fields for oblique incidence, parallel polarization. The complex rms magnetic field intensity vector of a TEM wave traveling in air for $x > 0$ and impinging obliquely a PEC boundary located at the $x = 0$ plane is expressed as $\underline{\mathbf{H}}_{\mathrm{i}} = \mathrm{e}^{\mathrm{j}\sqrt{2}\pi(x-z)}\,\hat{\mathbf{y}}$ A/m (x, z in m). Determine the distributions of the electric and magnetic fields in air.

10.22. Various computations for oblique incidence, parallel polarization. A uniform plane time-harmonic electromagnetic wave of frequency $f = 75$ MHz and rms magnetic field intensity $H_{i0} = 1$ A/m is incident from air at an angle $\theta_{\mathrm{i}} = 30°$ and parallel polarization upon a flat horizontal surface of a large copper block. Find (a) the instantaneous and time-average electromagnetic energy densities of the resultant wave above the block, (b) the time-average power of Joule's losses per unit area of the copper surface, (c) the orientation and position (height with respect to the conducting surface) of a small receiving wire loop antenna that results in the maximum reception of the wave, and (d) the maximum rms emf induced in the loop if its edge length is $a = 8$ cm and the magnetic field due to the induced current in the loop is negligible.

10.23. Angular dispersion of white light by a glass prism. At optical frequencies, glass is a weakly dispersive medium, as its index of refraction, n, slightly varies with the free-space wavelength, λ_0. In particular, for the visible light spectrum and a type of flint glass, n decreases from approximately $n_{\text{violet}} = 1.66$ for violet light ($\lambda_0 = 400$ nm) to $n_{\text{red}} = 1.62$ for red light ($\lambda_0 = 700$ nm) according to the following equation: $n(\lambda_0) = 1.6 + 9.5 \times 10^{-15}/\lambda_0^2$ (λ_0 in m). Taking advantage of this property, an optical prism made of the flint glass and with apex angle $\alpha = 60°$ (equiangular prism) is used to disperse white light, i.e., to separate in space the colors that constitute it, as shown in Fig. 10.27. The white light beam is incident at an angle $\theta_i = 65°$ on one surface of the prism, and, upon double refraction, emerges on the other side of the prism with different exit angles for different light colors – due to the wave dispersion in glass. Find the deviation angle δ, measured with respect to the incident beam, of the outgoing beam for each of the colors in the visible spectrum: violet, blue ($\lambda_0 = 470$ nm), green ($\lambda_0 = 540$ nm), yellow ($\lambda_0 = 590$ nm), orange ($\lambda_0 = 610$ nm), and red. What is the total angular dispersion of the prism, defined as $\gamma = \delta_{violet} - \delta_{red}$?

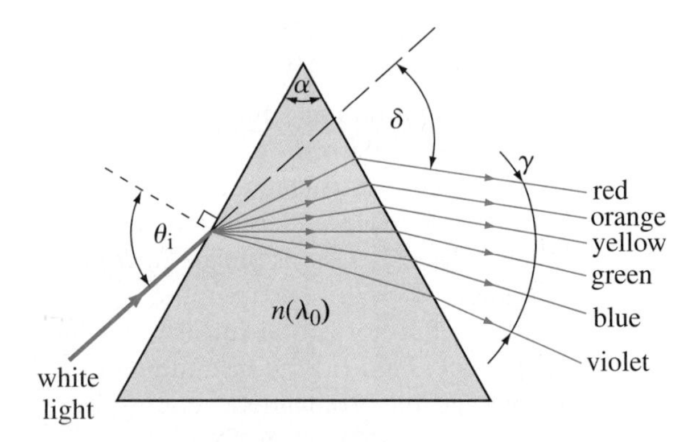

Figure 10.27 Angular dispersion of a beam of white light into its constituent colors in the visible spectrum using a glass prism; for Problem 10.23.

10.24. Apparent length of a vertical stick immersed in water. What is the actual length of a vertical stick that is completely immersed in a freshwater lake, with a 1.33 index of refraction, right below the water surface, if it appears to be 1 m long to a person in a boat viewing it at an angle of 45° with respect to the normal on the surface?

10.25. Oblique incidence on a dielectric boundary, parallel polarization. (a) For the oblique incidence at parallel polarization upon the interface between two lossless dielectric media described in Example 10.20 (and for the specified numerical values of the parameters of the media and the wave), compute the electric and magnetic field vectors of the resultant wave at an arbitrary point in the incident medium, and specifically at the point defined by $x = y = z = -1$ m. Also find (b) the time-average total energy density and Poynting vector for $-\infty < x, y, z < \infty$ and (c) the rms emf induced in an electrically small circular contour of radius $a = 1$ cm that is positioned and oriented for maximum wave reception in the incident medium.

10.26. Oblique incidence on a dielectric boundary, normal polarization. (a) Repeat Example 10.20 but assuming that the polarization of the incident wave is normal. Also calculate (b) the electric and magnetic field vectors of the reflected and resultant waves, respectively, at the point $x = y = z = -1$ m in the incident medium and (c) the time-average surface power densities transported by the incident, reflected, and transmitted waves.

10.27. Refracted wave computation for normal polarization. Assuming that the reflecting medium (for $y < 0$) in Problem 10.17 is penetrable, with parameters $\varepsilon_r = 9$, $\mu_r = 1$, and $\sigma = 0$, find the electric and magnetic transmitted field vectors at an arbitrary point in this medium.

10.28. Refracted wave computation for parallel polarization. If the second medium (for $x < 0$) in Problem 10.21 is a nonmagnetic lossless dielectric of relative permittivity $\varepsilon_r = 5$, compute the complex Poynting vector of the refracted wave.

10.29. Beam-steering glass prisms in a periscope. In a periscope, two beam-steering 45°-90°-45° glass prisms in air are used as shown in Fig. 10.28. Each prism turns a beam of light by 90°, by means of total internal reflection.

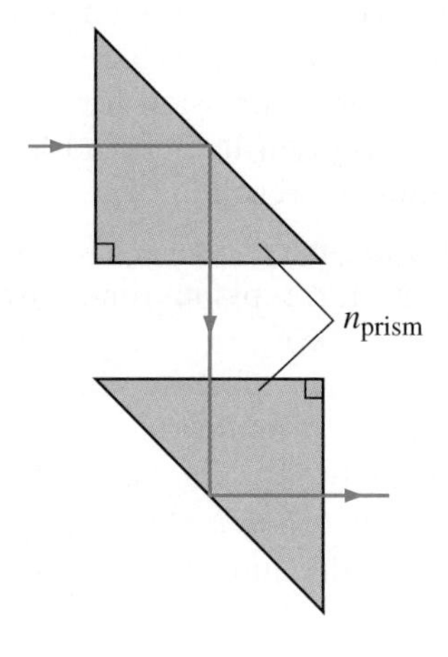

Figure 10.28 Light path through beam-steering glass prisms in a periscope, based on total internal reflection; for Problem 10.29.

(a) Determine the minimum required index of refraction of the prisms. (b) Repeat (a) but for water ($n_{water} = 1.33$) as the surrounding medium. (c) What is the percentage of the time-average incident power that passes through the periscope, neglecting multiple internal reflections, for cases (a) and (b) and the obtained respective minimum values of n_{prism}?

10.30. Total internal reflection for oil-covered water surface. There is a layer of oil floating on the surface of fresh water, and the corresponding indices of refraction are $n_{oil} = 1.54$ and $n_{water} = 1.33$. A beam of light is obliquely incident from water, at an angle θ_i, upon the water-oil interface. (a) Find θ_i for which total internal reflection occurs at the oil-air interface, so there is no transmission into air. (b) Repeat (a) but assuming that the beam is incident (at an angle θ_i) from air on the air-oil interface, and determine θ_i that results in no transmission into water, that is, in total reflection at the oil-water interface.

10.31. Acceptance angle of an optical fiber. Consider the optical fiber of Fig. 10.22, and assume that incident light enters it from a surrounding dielectric medium of refractive index n_0 (most frequently, air) through the front surface (face) of the fiber at oblique incidence defined by the angle θ_1 measured with respect to the normal to the surface, that is, to the axis of the fiber, as shown in Fig. 10.29. Find the maximum value of θ_1, in terms of n_0, n_{core}, and $n_{cladding}$ ($n_{core} > n_{cladding}$), such that the total internal reflection condition is satisfied at the core-cladding interface. This maximum incident angle is called the acceptance angle of the optical fiber, $\theta_{accept} = \theta_{1\,max}$, as all light beams incident upon the face of the fiber that fall into the cone determined by this angle ($\theta_1 \leq \theta_{accept}$) can propagate (are "accepted") by means of multiple total internal reflections along the fiber, confined inside the core. In specific, compute θ_{accept} for a fiber with $n_{core} = 1.49$ and $n_{cladding} = 1.46$, and air ($n_0 = 1$) and water ($n_0 = 1.33$), respectively, as the incident medium.

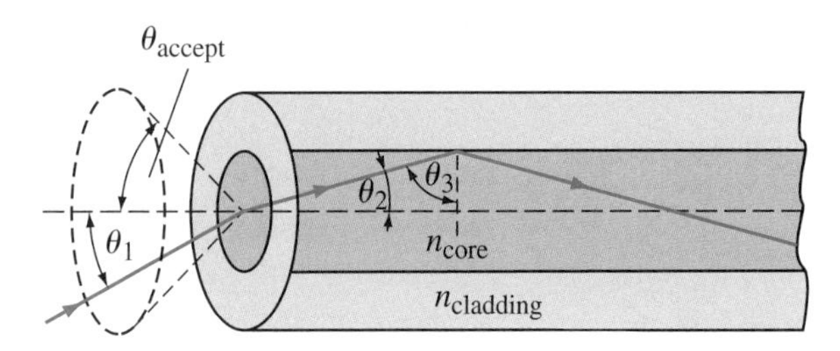

Figure 10.29 Finding the maximum incident angle (the acceptance angle) of a light beam impinging the face of an optical fiber such that the total internal reflection condition (for the angle θ_3) at the core-cladding boundary is still met; for Problem 10.31.

10.32. Refracted angle for Brewster incident angle. A plane wave with parallel polarization is incident upon an interface between two arbitrary lossless nonmagnetic media at the Brewster angle, θ_{iB}. The refracted angle is θ_t. Prove that $\sin\theta_t = \cos\theta_{iB}$, i.e., that the sum of the incident and refracted angles at the Brewster condition is always 90°.

10.33. Polarizing light by parallel glass slabs at Brewster condition. Light is incident at the Brewster angle from air on a system of six parallel glass slabs, with air separation (equal to the slab thickness) between all adjacent slabs. Light is unpolarized, and it can be represented as a superposition of the beam with normal polarization and that with parallel polarization, the two components being with approximately equal energies. (a) What is the polarization of light reflected from the system of slabs? (b) Describe the polarization of light transmitted through the system of slabs. (c) What would change if another six slabs were added?

10.34. Elimination of ground-bounced wave in a communication link. In a wireless communication link at a frequency of $f = 1.5$ GHz, the height of the transmitting antenna with respect to the earth's surface is $h_t = 1$ km, and the horizontal distance between the transmitting and receiving antennas is $d = 10$ km. The soil below the antennas is very dry, with parameters $\varepsilon_r = 3.2$ and $\sigma = 0$. The transmitting antenna emits waves with parallel polarization, and, in general, they travel to the receiving antenna either as a direct ray through air or as a reflected ray, bouncing off the earth's surface. Find the height of the receiving antenna, h_r, for which the reflected wave is eliminated, so that only the direct wave is received.

10.35. Reflection and refraction of an EP wave at Brewster condition. Write the expressions for complex rms electric field intensities of two uniform plane time-harmonic waves with normal and parallel polarizations whose superposition represents a left-hand elliptically polarized wave incident from air obliquely on the planar surface of a nonmagnetic lossless dielectric of relative permittivity $\varepsilon_r = 3$ (adopt all necessary parameters to define the waves). If the incident angle satisfies the Brewster condition, determine (a) the polarization states of the reflected and refracted waves and (b) the percentages of the time-average incident power that are reflected from the air-dielectric interface and transmitted into the dielectric medium, respectively.

10.36. Anti-reflective coatings for periscope prisms. Propose anti-reflective coatings to increase the power throughput of the periscope in Fig. 10.28, for each of the cases (a) and (b) from Problem 10.29, and the free-space wavelengths of $\lambda_0 = 400$ nm and 700 nm, respectively.

10.37. X-band antenna radome. If the fiberglass radome from Example 10.29 is used to weather-protect an X-band (8–12 GHz) antenna, compute the percentage of the incident power (radiated by the antenna) that is reflected back from the radome (assuming normal incidence) at the central frequency and at each of the end frequencies of the band.

10.38. Reflection from a lossy slab backed by an aluminum foil. A lossy dielectric slab with thickness $d = 20$ cm and relative complex permittivity $\underline{\varepsilon}_r = 3 - \mathrm{j}2$ ($\mu_r = 1$) is backed by an aluminum foil, which can be regarded as nonpenetrable. Find the equivalent reflection coefficient in decibels at the air-slab interface for a uniform plane wave of frequency $f =$ 1 GHz at normal incidence.

10.39. Wave computations in a four-media problem. Consider the four-media structure in Fig. 10.26 and assume that the thickness of the slab (d) is that in Eq. (10.173). (a) If the rms electric field intensity of the plane wave incident on the slab is E_0, find the expression for the total time-average power of Joule's losses in the slab per unit area of its surface. (b) If the distance of the slab from the PEC plane is changed to $\lambda_0/2$, obtain the expression for the equivalent wave impedance at interface 1-2 looking to the right ($\underline{\eta}_{2e}$).

11 Field Analysis of Transmission Lines

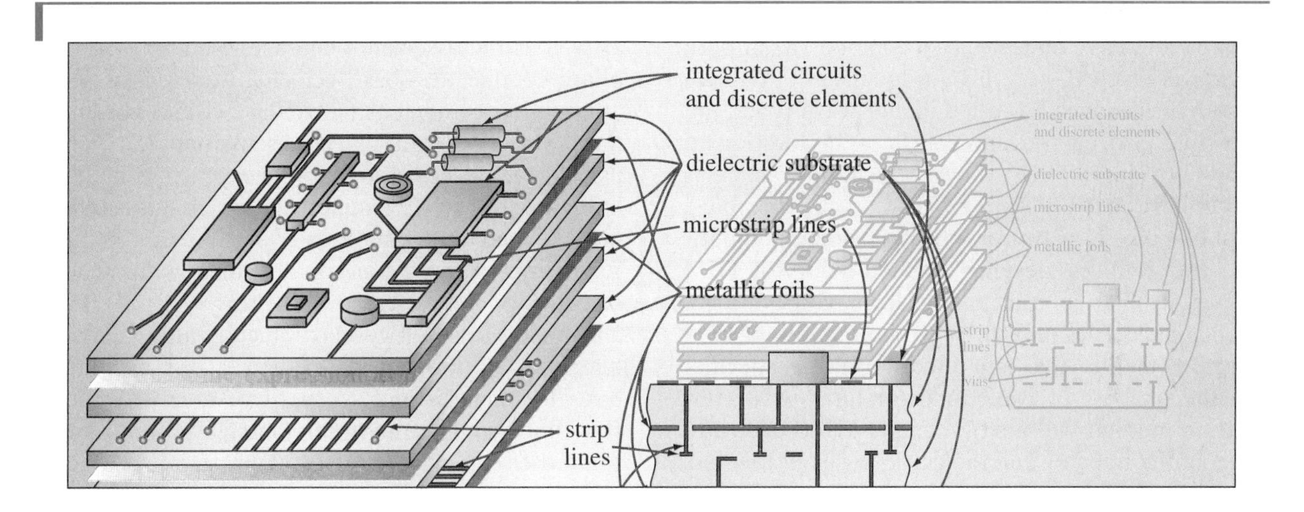

Introduction:

In addition to wireless links, which use free (unbounded) electromagnetic waves propagating in free space or material media (previous two chapters), electromagnetic signals and energy can be transported to a distance also using guided electromagnetic waves. Such waves are channeled through a guiding system composed of conductors and dielectrics. Guiding systems normally have a uniform cross section, and are classified into transmission lines and waveguides, according to the number of conductors in the system. Transmission lines have two or more separate conductors (e.g., a coaxial cable, Fig. 2.17), whereas waveguides consist of a single conductor (e.g., a rectangular metallic waveguide, Fig. 10.15) or only dielectrics (e.g., an optical fiber, Fig. 10.22). In this chapter, we present a field analysis of transmission lines, with a focus on two-conductor lines. Field theory of transmission lines is important for understanding physical processes that constitute the propagation and attenuation along a line of a given geometry and material composition. The principal result of the analysis are the parameters of a circuit model of an arbitrary two-conductor transmission line, in the form of a network of many cascaded equal small cells with lumped elements. This network is then solved, in the next chapter, using circuit-theory concepts and equations, as the starting point of the frequency-domain (complex-domain) and transient (time-domain) analysis of transmission lines as circuits with distributed parameters (circuit analysis of two-conductor transmission lines).

Waves propagating along transmission lines are of either transverse electromagnetic (TEM) type (for lossless lines with homogeneous dielectrics) or quasi-TEM type (for lines with small losses and/or inhomogeneous dielectrics). This means that the components along the direction of wave propagation (i.e., along the axis of the guiding structure) of both electric and magnetic field vectors of the wave are either zero or very small when compared to the corresponding transverse field components (perpendicular to the axis). On the other side, waveguides carry non-TEM waves, which include transverse electric (TE), transverse magnetic (TM), and hybrid waves. TE waves have a zero electric and nonzero magnetic field component along the waveguide axis, whereas the situation for TM waves is just opposite (magnetic field vector is in a transversal plane, perpendicular to the axis). Hybrid waves are combinations of TE and TM waves, and have nonzero axial components of both field vectors. At very high frequencies, TE, TM, and hybrid waves, so-called higher wave types, are possible also on transmission lines, in addition to TEM (or quasi-TEM) waves, which, in practice, is an undesirable situation (what we want is to have just one type of waves propagating along the line).[1]

We shall first develop the field theory of TEM waves in lossless transmission lines with homogeneous dielectrics regardless of the number of conductors in the line, and then study the specifics of the analysis of two-conductor lines. A perturbation method will next be employed to take into account conductor and dielectric losses in low-loss lines. Modifications of the theory to approximately analyze transmission lines with inhomogeneous dielectrics (which carry quasi-TEM waves) will also be introduced. Based on the developed concepts and analysis procedures, we shall evaluate and discuss circuit parameters of a variety of practically important classes of two-conductor transmission lines, with homogeneous and inhomogeneous dielectrics. The parameters include the capacitance, inductance, resistance, and conductance per unit length of the line (primary circuit parameters), as well as the associated characteristic impedance, phase coefficient, phase velocity, wavelength, and attenuation coefficient (secondary circuit parameters). These evaluations and discussions lean heavily on the electrostatic analysis of transmission lines in Chapter 2, magnetostatic (or quasistatic) analysis and inductance computation in Chapter 7, analysis of lossy transmission lines with steady currents, Section 3.12, and analysis of skin effect and Joule's losses in good conductors at higher frequencies, Sections 9.11, 10.3, and 10.4. Finally, the basis for analysis and synthesis of microstrip and strip lines constituting interconnects in multilayer printed circuit boards, with taking fringing effects into account, will be provided.

11.1 TEM WAVES IN LOSSLESS TRANSMISSION LINES WITH HOMOGENEOUS DIELECTRICS

Consider a transmission line consisting of M ($M \geq 2$) conductors of arbitrary cross section, shown in Fig. 11.1, in a homogeneous dielectric of permittivity ε and permeability μ. We assume that the line is uniform, i.e., that the cross section in Fig. 11.1 is the same along the entire line (theoretically, for every coordinate z, $-\infty < z < \infty$) and lossless (both the conductors and dielectric are perfect). We also assume a time-harmonic variation of the electromagnetic field in the line, of frequency f, and perform the analysis in the complex domain (see Sections 8.6–8.8). This field is governed by the corresponding set of Maxwell's equations for the dielectric region in Fig. 11.1 and boundary conditions for the dielectric-conductor boundary surfaces. We note that there are no induced volume currents and charges ($\underline{\mathbf{J}} = 0$ and $\underline{\rho} = 0$) in the dielectric, since it is lossless [see Eqs. (3.18) and (8.82)]. Assuming finally that the dielectric region is completely source-free, i.e., free of any impressed sources [$\underline{\mathbf{E}}_{\mathrm{i}} = 0$ and $\underline{\mathbf{J}}_{\mathrm{i}} = 0$ – see Eqs. (3.109) and (3.124)] as well, the underpinning of our

[1]For a coaxial cable with a homogeneous dielectric (Fig. 2.17), for instance, the frequency limit below which no higher wave types can propagate is determined by $\lambda \approx \pi(a + b)$, where a and b are the radii of the cable dielectric, and λ is the wavelength in the dielectric, Eq. (8.112).

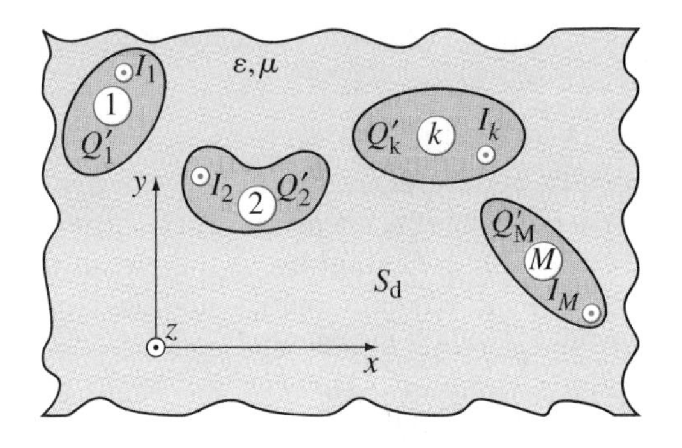

Figure 11.1 Cross section of a transmission line with M conductors and a homogeneous dielectric.

analysis is the complex-domain version of Eqs. (9.1)–(9.4),

$$\nabla \times \underline{\mathbf{E}} = -\mathrm{j}\omega\mu\underline{\mathbf{H}}, \tag{11.1}$$

$$\nabla \times \underline{\mathbf{H}} = \mathrm{j}\omega\varepsilon\underline{\mathbf{E}}, \tag{11.2}$$

$$\nabla \cdot \underline{\mathbf{E}} = 0, \tag{11.3}$$

$$\nabla \cdot \underline{\mathbf{H}} = 0, \tag{11.4}$$

source-free Maxwell's equations in complex domain

along with the boundary conditions in Eqs. (8.33), for a surface of a perfect electric conductor (PEC) in a dynamic electromagnetic field. In these equations, $\underline{\mathbf{E}}$ and $\underline{\mathbf{H}}$ are complex rms electric and magnetic field intensity vectors in the dielectric, and $\omega = 2\pi f$ is the angular (radian) frequency of the field.

We would like to find a solution for an electromagnetic wave that propagates along the line (along the z-axis), through the dielectric – guided by the conductors. Choosing the propagation in the positive z direction, we know from the analysis of plane waves in unbounded media that the field dependence on the z-coordinate for such a wave is given by the propagation factor $\mathrm{e}^{-\underline{\gamma} z}$ [see Eqs. (9.81) and (9.90)], and hence

$$\underline{\mathbf{E}}(x, y, z) = \underline{\mathbf{E}}(x, y, 0)\,\mathrm{e}^{-\underline{\gamma} z}, \quad \underline{\mathbf{H}}(x, y, z) = \underline{\mathbf{H}}(x, y, 0)\,\mathrm{e}^{-\underline{\gamma} z}, \tag{11.5}$$

wave propagation along a transmission line

where $\underline{\gamma}$ is the complex propagation coefficient – to be determined. The dependences on the Cartesian coordinates x and y of field vectors $\underline{\mathbf{E}}$ and $\underline{\mathbf{H}}$, in the line cross section in Fig. 11.1, are also to be determined. Note that these dependences can alternatively be expressed in terms of cylindrical (polar) coordinates r and ϕ (see Fig. 1.25), which is especially convenient for transmission lines exhibiting cylindrical symmetry (e.g., a coaxial cable with a homogeneous dielectric in Fig. 2.17). With this, taking the partial derivative with respect to z in the expression for the del (nabla) operator, ∇, in Cartesian coordinates, Eq. (1.100), is equivalent to multiplication by $-\underline{\gamma}$ [similarly to multiplication by jω in Eq. (8.67)], so that the longitudinal (z-) component of ∇ becomes $\nabla_z = -\underline{\gamma}\,\hat{\mathbf{z}}$. The same is true for the expression for ∇ in cylindrical coordinates [see Eq. (1.105)]. Moreover, we can combine the transverse components (x- and y-components or r- and ϕ-components) of ∇ into a transverse del operator, ∇_t, similar to the surface del operator (∇_s) in Eq. (8.43), used in the continuity equation for plates, Eq. (8.42) or (10.14), and write

$$\nabla = \underbrace{\frac{\partial}{\partial x}\hat{\mathbf{x}} + \frac{\partial}{\partial y}\hat{\mathbf{y}}}_{\nabla_\mathrm{t}} + \underbrace{\frac{\partial}{\partial z}\hat{\mathbf{z}}}_{\nabla_z} = \nabla_\mathrm{t} - \underline{\gamma}\,\hat{\mathbf{z}}. \tag{11.6}$$

HISTORICAL ASIDE

The mathematical foundation of the theory of guided electromagnetic waves was laid out by **Oliver Heaviside** (1850–1925), an English electrical engineer, mathematician, and physicist. Heaviside suffered from increasing deafness, was almost completely self-educated, and his only paid job was that of a telegrapher over a short period of time. Although unappreciated by the scientific establishment for the most of his life, he changed the electromagnetic theory and electrical engineering forever. In a series of excellent papers in the 1880s and in his 1893 three-volume work "Electromagnetic Theory," Heaviside introduced the vector notation in Maxwell's equations, reformulated the mathematical description of wave propagation, provided several important contributions to the circuit theory and the involved mathematics, and developed the transmission-line model and the associated telegrapher's equations. He was forced to publish his papers at his own expense, primarily due to the unorthodoxy of his work in terms of both his ideas and his vector notation. He was elected to the Royal Society in 1891, and was the first recipient, in 1922, of the Faraday Medal awarded by the Institution of Electrical Engineers (IEE). However, he spent his last years poor and alone. In many aspects, the significance of Heaviside's achievements to science and engineering has never been given the proper recognition and praise. *(Portrait: AIP Emilio Segrè Visual Archives, Brittle Books Collection)*

By analogous decomposition of the field vectors onto transverse[2] and longitudinal (axial) components,

transverse and longitudinal fields

$$\boxed{\underline{\mathbf{E}} = \underline{\mathbf{E}}_{\mathrm{t}} + \underline{\mathbf{E}}_z, \quad \underline{\mathbf{H}} = \underline{\mathbf{H}}_{\mathrm{t}} + \underline{\mathbf{H}}_z,} \tag{11.7}$$

Eqs. (11.1)–(11.4) can now be rewritten into a new set of differential equations with $\underline{\mathbf{E}}_{\mathrm{t}}$, $\underline{\mathbf{H}}_{\mathrm{t}}$, $\underline{\mathbf{E}}_z$, and $\underline{\mathbf{H}}_z$ as unknowns, using the operators ∇_{t} and ∇_z. These equations can be solved for different types of guided electromagnetic waves, i.e., for TE, TM, and TEM waves.

Here, we seek the solution in the form of a TEM (transverse electromagnetic) wave, where both **E** and **H** are transverse to the direction of wave propagation (i.e., to the line axis), that is,

guided TEM wave

$$\boxed{\underline{\mathbf{E}} = \underline{\mathbf{E}}_{\mathrm{t}} \quad (\underline{\mathbf{E}}_z = 0) \quad \text{and} \quad \underline{\mathbf{H}} = \underline{\mathbf{H}}_{\mathrm{t}} \quad (\underline{\mathbf{H}}_z = 0),} \tag{11.8}$$

with which Eq. (11.1) becomes

$$\nabla \times \underline{\mathbf{E}}_{\mathrm{t}} = (\nabla_{\mathrm{t}} - \underline{\gamma}\,\hat{\mathbf{z}}) \times \underline{\mathbf{E}}_{\mathrm{t}} = \nabla_{\mathrm{t}} \times \underline{\mathbf{E}}_{\mathrm{t}} - \underline{\gamma}\,\hat{\mathbf{z}} \times \underline{\mathbf{E}}_{\mathrm{t}} = -\mathrm{j}\omega\mu\underline{\mathbf{H}}_{\mathrm{t}}. \tag{11.9}$$

We note that $\nabla_{\mathrm{t}} \times \underline{\mathbf{E}}_{\mathrm{t}}$ is a z-directed vector (∇_{t} is in a transversal plane), whereas $\hat{\mathbf{z}} \times \underline{\mathbf{E}}_{\mathrm{t}}$ has only a transverse component, so that equating transverse and longitudinal components, respectively, on the two sides of this equation results in

$$\underline{\gamma}\,\hat{\mathbf{z}} \times \underline{\mathbf{E}}_{\mathrm{t}} = \mathrm{j}\omega\mu\underline{\mathbf{H}}_{\mathrm{t}}, \quad \nabla_{\mathrm{t}} \times \underline{\mathbf{E}}_{\mathrm{t}} = 0. \tag{11.10}$$

Similar separation of transverse and longitudinal components in Eq. (11.2) yields

$$\underline{\gamma}\,\hat{\mathbf{z}} \times \underline{\mathbf{H}}_{\mathrm{t}} = -\mathrm{j}\omega\varepsilon\underline{\mathbf{E}}_{\mathrm{t}}, \quad \nabla_{\mathrm{t}} \times \underline{\mathbf{H}}_{\mathrm{t}} = 0. \tag{11.11}$$

[2]The notation used here for the field components transverse to the direction of wave propagation, $\mathbf{E}_{\mathrm{t}}$ and $\mathbf{H}_{\mathrm{t}}$, should not be confused with the same notation used in previous chapters in various forms of boundary conditions for the field components tangential to boundary surfaces between electromagnetically different media, e.g., in Eqs. (8.33).

Finally, since $\hat{\mathbf{z}} \cdot \underline{\mathbf{E}}_t = \hat{\mathbf{z}} \cdot \underline{\mathbf{H}}_t = 0$, Eqs. (11.3)–(11.4) reduce to

$$\nabla_t \cdot \underline{\mathbf{E}}_t = 0, \quad \nabla_t \cdot \underline{\mathbf{H}}_t = 0. \tag{11.12}$$

From Eqs. (11.10) and (11.11),

$$\underline{\mathbf{H}}_t = \frac{\underline{\gamma}}{j\omega\mu}\,\hat{\mathbf{z}} \times \underline{\mathbf{E}}_t, \quad \underline{\mathbf{E}}_t = -\frac{\underline{\gamma}}{j\omega\varepsilon}\,\hat{\mathbf{z}} \times \underline{\mathbf{H}}_t, \tag{11.13}$$

and each of these equations tells us that the electric and magnetic field vectors of this guided wave are perpendicular to each other,

$$\boxed{\underline{\mathbf{E}}_t \perp \underline{\mathbf{H}}_t,} \tag{11.14}$$

field orthogonality in a transmission line

as indicated in Fig. 11.2(a). Combining the two equations, we obtain

$$\underline{\mathbf{E}}_t = -\frac{\underline{\gamma}^2}{(j\omega\varepsilon)(j\omega\mu)}\,\hat{\mathbf{z}} \times (\hat{\mathbf{z}} \times \underline{\mathbf{E}}_t) = -\frac{\underline{\gamma}^2}{\omega^2\varepsilon\mu}\,\underline{\mathbf{E}}_t, \tag{11.15}$$

since $\hat{\mathbf{z}} \times (\hat{\mathbf{z}} \times \underline{\mathbf{E}}_t) = -\underline{\mathbf{E}}_t$, which is shown in Fig. 11.2(a). Obviously, the factor multiplying $\underline{\mathbf{E}}_t$ in Eq. (11.15) must be unity, and hence the following solution for the propagation coefficient of the transmission line:

$$\boxed{\underline{\gamma} = j\omega\sqrt{\varepsilon\mu} = j\beta \quad \longrightarrow \quad \beta = \omega\sqrt{\varepsilon\mu} \quad (\alpha = 0).} \tag{11.16}$$

β – TEM wave

We see that $\underline{\gamma}$ and the resulting phase coefficient, β, are the same [see Eq. (8.111)] as for a uniform plane wave propagating in an unbounded medium having the same parameters, ε and μ ($\sigma = 0$), as the dielectric of the transmission line in Fig. 11.1 (as expected, $\underline{\gamma}$ is purely imaginary – no losses in the system). The phase velocity, v_p, and wavelength, λ_z, along the line (along the z-axis) are then also the same as for the uniform plane wave in the unbounded medium of parameters ε and μ [see Eqs. (9.35), (9.18), and (8.111)], namely,

$$\boxed{v_p = \frac{\omega}{\beta} = \frac{1}{\sqrt{\varepsilon\mu}} = c, \quad \lambda_z = \frac{2\pi}{\beta} = \frac{v_p}{f} = \frac{c}{f}.} \tag{11.17}$$

v_p, λ_z – TEM wave

In addition, substituting the solution for $\underline{\gamma}$ in Eqs. (11.16) back into Eqs. (11.13), the vector relations between the electric and magnetic field vectors of the wave become

$$\boxed{\underline{\mathbf{H}}_t = \sqrt{\frac{\varepsilon}{\mu}}\,\hat{\mathbf{z}} \times \underline{\mathbf{E}}_t, \quad \underline{\mathbf{E}}_t = \sqrt{\frac{\mu}{\varepsilon}}\,\underline{\mathbf{H}}_t \times \hat{\mathbf{z}},} \tag{11.18}$$

vector relations between **E** and **H** in a transmission line

i.e., they acquire the same form as those in Eqs. (9.22) for a uniform plane wave in the unbounded medium. As in Eq. (9.20) for the plane wave, the ratio of the electric and magnetic complex rms field intensities of a guided TEM wave at any point in the cross section of the dielectric in Fig. 11.1 and for any z along the line (and the same is true for instantaneous field intensities) equals a real constant, an impedance. It is denoted by Z_{TEM} and called the wave impedance of TEM waves. It has the same value as the intrinsic impedance (η) in Eq. (9.21) of the medium of parameters ε and μ,

$$\boxed{Z_{TEM} = \frac{\underline{E}_t}{\underline{H}_t} = \sqrt{\frac{\mu}{\varepsilon}}.} \tag{11.19}$$

wave impedance of a TEM wave

As we shall see in a later section, a similar relation exists between surface charges and currents of the conductors in the line [Fig. 11.2(b)].

Note that β, c, and λ of a uniform plane wave are also, like η, referred to as the intrinsic phase coefficient, intrinsic phase velocity, and intrinsic wavelength,

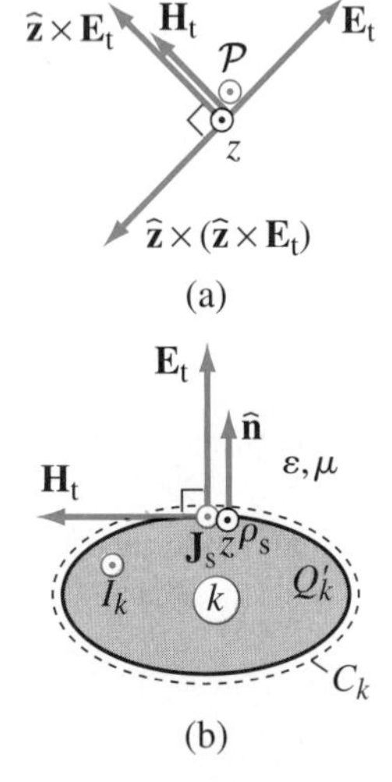

Figure 11.2 Details of the transmission line in Fig. 11.1, with a TEM wave propagating in the positive z direction: (a) electric and magnetic field vectors in the dielectric and (b) charges and currents on the surface of a conductor.

respectively, of a given unbounded medium (of parameters ε and μ). Therefore, in summary, we can say that all four parameters in Eqs. (11.16), (11.17), and (11.19) of a transmission line (β, v_p, λ_z, and Z_{TEM}) equal the corresponding intrinsic values of the line dielectric.

As for plane waves in unbounded lossless media, Eq. (9.40), the complex Poynting vector of a TEM wave traveling along the transmission line in Fig. 11.1 is purely real, given by

$$\underline{\mathcal{P}} = \underline{\mathbf{E}} \times \underline{\mathbf{H}}^* = \underline{E}_t \underline{H}_t^* \, \hat{\mathbf{z}} = \underline{E}_t \frac{\underline{E}_t^*}{Z_{TEM}} \, \hat{\mathbf{z}} = \frac{|\underline{E}_t|^2}{Z_{TEM}} \, \hat{\mathbf{z}} \tag{11.20}$$

[Fig. 11.2(a)]. From Eq. (8.195), it equals the time average of the instantaneous Poynting vector of the wave. As the conductors of the line are considered to be perfect, the wave does not penetrate at all into them [skin depth for a PEC is zero – see Eqs. (9.139) and (3.26)], so that the energy flows only through the line dielectric (guided by the conductors). Quantitatively, noting that the vector $\underline{\mathcal{P}}$ has a longitudinal component only, Poynting's theorem in complex form, Eq. (8.196), tells us that the complex power transported by the wave along the line is [see also Eq. (8.200)]

$$\underline{S} = \int_{S_d} \underline{\mathcal{P}} \cdot d\mathbf{S} = \frac{1}{Z_{TEM}} \int_{S_d} |\underline{E}_t|^2 \, dS, \tag{11.21}$$

where S_d stands for the cross section of the dielectric in Fig. 11.1 and $d\mathbf{S} = dS\,\hat{\mathbf{z}}$. Of course, $\underline{S}$ is purely real as well, representing the time-average power flow along the line, $\underline{S} = P$. Using Eqs. (11.5) and (11.16), we see that $|\underline{\mathbf{E}}|$ does not depend on z ($|\mathrm{e}^{-\mathrm{j}\beta z}| = 1$). This means that $P = \text{const}$ along the entire line, which is to be expected as there are no losses in the line.

Conceptual Questions (on Companion Website): 11.1–11.4.

11.2 ELECTROSTATIC AND MAGNETOSTATIC FIELD DISTRIBUTIONS IN TRANSVERSAL PLANES

The distribution of the electric field intensity vector, $\underline{\mathbf{E}}_t$, in a cross section of the line in Fig. 11.1 is determined, from Eqs. (11.10) and (11.12), by its transverse curl and divergence in the dielectric, along with the boundary condition for its tangential component, in Eqs. (8.33), on the boundaries of conductors,

as in a 2-D electrostatic system

$$\boxed{\nabla_t \times \underline{\mathbf{E}}_t = 0, \quad \nabla_t \cdot \underline{\mathbf{E}}_t = 0, \quad \hat{\mathbf{n}} \times \underline{\mathbf{E}}_t = 0,} \tag{11.22}$$

where $\hat{\mathbf{n}}$ is the normal unit vector on the boundary surface, directed from the conductor toward the dielectric, as shown in Fig. 11.2(b). We realize that these equations have the same form as the corresponding equations for the electrostatic field [see Eqs. (4.92), (2.56), and (1.186), and recall that $\rho = 0$ in the dielectric in Fig. 11.1] in the same two-dimensional system, representing one cross section of the transmission line.[3] The only difference is that $\underline{\mathbf{E}}_t$ in Eqs. (11.22) is a complex

[3]Note that in an infinitely long electrostatic system with the same cross section as in Fig. 11.1, $\mathbf{E} = \mathbf{E}_t$ ($E_z = 0$) and $\nabla = \nabla_t$ ($\nabla_z = 0$), because of its two-dimensional nature (z-coordinate is irrelevant for the analysis). This can also be explained by the superposition principle and 2-D nature of the electrostatic field due to an infinite line charge with uniform charge density, Eq. (1.57). Namely, the surfaces of all conductors in the electrostatic equivalent of the system in Fig. 11.1 can be subdivided into very thin infinitely long uniformly charged strips, and the field $d\mathbf{E}$ of each of the strips can be computed using Eq. (1.57). Obviously, $d\mathbf{E}$ has only a transverse component and does not depend on z. The same is true for the total field ($\mathbf{E}$) at any point in the dielectric, since it is a superposition of the elementary fields $d\mathbf{E}$.

vector, whereas the corresponding field vector in the electrostatic system is real. Consequently, we can find the electric field of a TEM wave in a transmission line performing a standard electrostatic analysis in its cross section. In doing so, we treat the field and related quantities (electric potential, charge, etc.) formally as complex quantities,[4] and have in mind that all of them, for the TEM wave, depend also on the longitudinal coordinate, Eqs. (11.5).

Similarly, based on Eqs. (11.11), (11.12), and (8.33), the magnetic field intensity vector, $\underline{\mathbf{H}}_t$, of a TEM wave can be found from the magnetostatic analysis in a cross section of the transmission line (recall that $\mathbf{J} = 0$ in the dielectric). However, once the electric field is known, $\underline{\mathbf{H}}_t$ can alternatively be found simply from Eq. (11.18).

Conceptual Questions (on Companion Website): 11.5 and 11.6.

11.3 CURRENTS AND CHARGES OF LINE CONDUCTORS

From the proportionality of the electric and magnetic fields of the TEM wave and the fact that the electrostatic field, in general, is due to charges and magnetostatic field due to currents, we conclude that there must be a similar proportionality between the charges and currents of the conductors in the line. Note that there are no volume charges and currents in the conductors, as they are assumed to be perfect (see Example 8.6), so this proportionality is to be explored for surface charges and currents of the conductors. Indeed, Eqs. (11.18) and the boundary conditions in Eqs. (8.33) tell us that the complex rms surface current density vector ($\underline{\mathbf{J}}_s$) at a point on the surface of a conductor is related to the complex rms surface charge density ($\underline{\rho}_s$) at the same point as

$$\underline{\mathbf{J}}_s = \hat{\mathbf{n}} \times \underline{\mathbf{H}}_t = \sqrt{\frac{\varepsilon}{\mu}}\, \hat{\mathbf{n}} \times (\hat{\mathbf{z}} \times \underline{\mathbf{E}}_t) = \sqrt{\frac{\varepsilon}{\mu}}\, \hat{\mathbf{n}} \times \left(\hat{\mathbf{z}} \times \frac{\underline{\rho}_s}{\varepsilon}\hat{\mathbf{n}}\right) = \frac{\underline{\rho}_s}{\sqrt{\varepsilon\mu}}\hat{\mathbf{z}}, \tag{11.23}$$

since $\underline{\mathbf{E}}_t = \underline{\rho}_s \hat{\mathbf{n}}/\varepsilon$ [$\underline{\mathbf{E}}_t$ is normal to the surface; see also Eq. (2.58)] and $\hat{\mathbf{n}} \times (\hat{\mathbf{z}} \times \hat{\mathbf{n}}) = \hat{\mathbf{z}}$, as can be seen in Fig. 11.2(b). Therefore, having in mind Eq. (11.17), the constant of proportionality between surface current and charge densities (note that $\underline{\mathbf{J}}_s$ has only a z-component) is the wave velocity (c),

$$\boxed{\underline{J}_s = c\underline{\rho}_s.} \tag{11.24}$$

current-charge proportionality at a point of a line conductor surface

Let us denote by $\underline{I}_k$ the total complex current intensity of the kth conductor in the line ($k = 1, 2, \ldots, M$), with respect to the positive z reference direction, and let $\underline{Q}'_k$ stand for the total complex rms charge per unit length of that conductor (Fig. 11.1). Using the definitions of the surface current and charge densities in Eqs. (3.13) and (1.27), respectively, $\underline{I}_k$ and $\underline{Q}'_k$ can be evaluated by integrating $\underline{J}_s$ and $\underline{\rho}_s$ along the contour C_k of the conductor [Fig. 11.2(b)]. Hence, the same proportionality as in Eq. (11.24) is established for the total current and per-unit-length charge of each of the conductors in the line,

$$\boxed{\underline{I}_k = \oint_{C_k} \underline{J}_s \,\mathrm{d}l = c\oint_{C_k} \underline{\rho}_s \,\mathrm{d}l = c\underline{Q}'_k \quad (k = 1, 2, \ldots, M).} \tag{11.25}$$

overall current-charge proportionality for line conductors

[4]We can also understand $\underline{\mathbf{E}}_t$ as the complex representative of the quasistatic time-varying (time-harmonic) electric field due to excess charge, $\mathbf{E}_q(t)$, given in Eq. (6.18), and analogously for the related quantities.

Considering the transmission line in Fig. 11.1 as the associated 2-D electrostatic system, governed by Eqs. (11.22), let us denote by Q'_{tot} the sum of all charges of the line conductors per unit length of the line. The electric field far away from all conductors in the line cross section can be approximated by that of an infinitely long uniform filamentary charge, Eq. (1.57), positioned along the z-axis in Fig. 11.1 with the same charge per unit length as the line:

$$E_{\text{t}} \approx \frac{Q'_{\text{tot}}}{2\pi\varepsilon r} \quad \text{(far away from all conductors)}, \tag{11.26}$$

where r is the (large) radial distance from the z-axis. Therefore, if $Q'_{\text{tot}} \neq 0$, the electric energy density, w_{e}, at distant points is inversely proportional to r^2 [see Eq. (2.199)], and the electric energy per unit length of the line comes out to be infinite, $W'_{\text{e}} \to \infty$, which is evident, for example, from a similar integration as in Eq. (2.207) with an infinite upper limit. Of course, W'_{e} must be finite, which implies $Q'_{\text{tot}} = 0$, and hence

from conservation of energy, total charge of the line is zero

$$\boxed{\sum_{k=1}^{M} \underline{Q}'_k = 0.} \tag{11.27}$$

From Eq. (11.25), the same must hold true for currents in Fig. 11.1,

sum of currents of line conductors is also zero

$$\boxed{\sum_{k=1}^{M} \underline{I}_k = 0,} \tag{11.28}$$

i.e., the algebraic sum of all current intensities of line conductors, with respect to the same reference direction (positive z direction), is zero. These relations remain the same in the time domain as well.

Finally, note that conditions in Eqs. (11.27) and (11.28) can be satisfied only if $M \geq 2$. Since electromagnetic waveguides have either a single conductor, as in rectangular metallic waveguides (Fig. 10.15), or no conductors, as in optical fibers (Fig. 10.22), we conclude that TEM waves cannot propagate along waveguides.

Conceptual Questions (on Companion Website): 11.7–11.11.

11.4 ANALYSIS OF TWO-CONDUCTOR TRANSMISSION LINES

For two-conductor transmission lines, $M = 2$ in Eqs. (11.27), (11.28), and (11.25), so that

charge and current of a two-conductor transmission line

$$\boxed{\underline{Q}'_1 = -\underline{Q}'_2 = \underline{Q}', \quad \underline{I}_1 = -\underline{I}_2 = \underline{I}, \quad \underline{I} = c\underline{Q}',} \tag{11.29}$$

i.e., the per-unit-length charges/currents of line conductors are equal in magnitude but with opposite polarities/directions. The electric and magnetic field lines in a cross section of the system are as in electrostatics and magnetostatics, respectively (see Section 11.2). Since the electric and magnetic field vectors are mutually orthogonal, Eq. (11.14), the magnetic field lines correspond to the equipotential lines of the electrostatic field, as shown in Fig. 11.3.

From Eqs. (11.5), (11.16), (11.22), and (1.101), the complete distribution of the complex electric field intensity vector in the line dielectric is given by

electric field distribution

$$\boxed{\underline{\mathbf{E}}(x, y, z) = \underline{\mathbf{E}}_{\text{t}}(x, y, z) = \underline{\mathbf{E}}_{\text{t}}(x, y, 0)\, \text{e}^{-\text{j}\beta z} = -\nabla_{\text{t}}\underline{V}(x, y)\, \text{e}^{-\text{j}\beta z},} \tag{11.30}$$

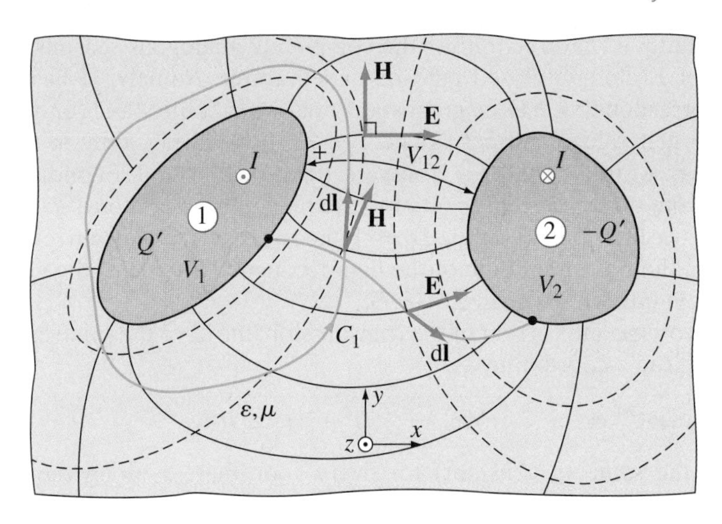

Figure 11.3 Cross section of a two-conductor ($M = 2$) transmission line.

where $\underline{V}$ is the complex potential in the line cross section with the same dependence on the transverse coordinates, x and y (of course, cylindrical coordinates r and ϕ could have been used as well), as the associated electrostatic potential (V). Eqs. (11.30), (1.88), and (1.90) tell us that the voltage $\underline{V}_{12}$ between the line conductors, that is, the difference between their potentials, $\underline{V}_1$ and $\underline{V}_2$, can be uniquely determined for an arbitrary cross section of the line (for any z) as a line integral of $\underline{\mathbf{E}}$,

$$\underline{V}_{12} = \underline{V}_1 - \underline{V}_2 = \int_1^2 \underline{\mathbf{E}} \cdot \mathrm{d}\mathbf{l}, \qquad (11.31)$$

line voltage

along any path lying entirely in that cross section, as indicated in Fig. 11.3. Namely, with an integration path not belonging to one transversal plane, the field $\underline{\mathbf{E}}$ along the path would not have the same form as the electrostatic field, i.e., the integral would take into account the propagation factor $\mathrm{e}^{-\mathrm{j}\beta z}$ in Eq. (11.30), which, of course, is a nonelectrostatic dependence. As a result, the integral, and the voltage $\underline{V}_{12}$, would depend on the shape of the path between the conductors, and thus would not have been uniquely determined. Equivalently, the line integral of $\underline{\mathbf{E}}$ along a closed path (contour) of arbitrary shape is zero [as in Eq. (1.75)], as long as the contour is entirely in a transversal plane, which comes from the electrostatic form of the electric field of the TEM wave in any cross section of the line. This can also be explained by the fact that the magnetic field vector, $\underline{\mathbf{H}} = \underline{\mathbf{H}}_\mathrm{t}$, of the wave is tangential to the plane of the contour, and thus does not produce flux on the right-hand side of Faraday's law of electromagnetic induction in integral form, Eq. (6.37); on the other hand, the magnetic flux would be nonzero through a nontransversal contour of arbitrary shape.

As a counterpart of the integral relationship between $\underline{\mathbf{E}}$ and $\underline{V}_{12}$ in Eq. (11.31), the generalized Ampère's law in integral form in Eq. (5.51) gives the following relationship between $\underline{\mathbf{H}}$ and $\underline{I}$:

$$\underline{I} = \oint_{C_1} \underline{\mathbf{H}} \cdot \mathrm{d}\mathbf{l}, \qquad (11.32)$$

line current

where C_1 is a contour of arbitrary shape lying entirely in a transversal plane (defined by a coordinate z) and enclosing conductor 1 (Fig. 11.3), and similar equation can be written for a contour C_2 around conductor 2. The contours must not extend into the longitudinal (z) direction for the completely analogous reasons as for the line integral of $\underline{\mathbf{E}}$ along a closed path discussed above. Namely, $\underline{\mathbf{H}}$ has the same form as the corresponding 2-D magnetostatic field, so that the static (or quasistatic) version of the generalized Ampère's law, Eq. (5.51), applies – only in transversal planes. In other words, only for a transversal contour around a conductor, as in Fig. 11.3, $\underline{\mathbf{E}}$ in the dielectric is tangential to the plane of the contour, so there is no flux of the electric flux density vector, $\underline{\mathbf{D}} = \varepsilon\underline{\mathbf{E}}$, i.e., no displacement current, through the contour on the right-hand side of the high-frequency version of the generalized Ampère's law in integral form, in Eq. (8.7).

Since the voltage and current of the transmission line have the same exponential dependence on the z-coordinate,

voltage and current waves

$$\boxed{\underline{V}_{12} = \underline{V}_{12}(z) = \underline{V}_{12}(0)\,\mathrm{e}^{-\mathrm{j}\beta z}, \quad \underline{I} = \underline{I}(z) = \underline{I}(0)\,\mathrm{e}^{-\mathrm{j}\beta z},} \tag{11.33}$$

their ratio is the same (a constant) for every coordinate z along the line. This can also be concluded from Eqs. (11.31) and (11.32), given the proportionality between electric and magnetic fields of the TEM wave, in Eqs. (11.18), at every point of the line dielectric. Moreover, as the wave impedance Z_{TEM} in Eq. (11.19) is real, the voltage to current ratio (for lossless lines) must be real as well. It is termed the characteristic impedance of the line[5] and is designated by Z_0,

characteristic impedance of a transmission line (unit: Ω)

$$\boxed{Z_0 = \frac{\underline{V}_{12}}{\underline{I}}.} \tag{11.34}$$

As we shall see in this and the following chapter, the characteristic impedance is one of the most important parameters of a transmission line. Its reciprocal is the characteristic admittance of the line,

characteristic admittance (unit: S)

$$\boxed{Y_0 = \frac{1}{Z_0}.} \tag{11.35}$$

Of course, the units for Z_0 and Y_0 are Ω and S, respectively.

From the electrostatic analysis,

$$\underline{V}_{12} = \frac{\underline{Q}'}{C'}, \tag{11.36}$$

where C' is the capacitance per unit length of the line, Eq. (2.115). Eqs. (11.34) and (11.29) then give the following expression for Z_0 via C':

characteristic impedance via C'

$$\boxed{Z_0 = \frac{\underline{V}_{12}}{c\underline{Q}'} = \frac{1}{cC'} = \frac{\sqrt{\varepsilon\mu}}{C'}.} \tag{11.37}$$

So, using this simple relation, we can now obtain the characteristic impedance of every transmission line with homogeneous dielectric for which we already have the capacitance per unit length found in Chapter 2.

[5] Note that the proportionality between the voltage and current of a transmission line, in Eq. (11.34), takes place only for a single traveling TEM wave, given by Eqs. (11.33), in the line. As we shall see in the next chapter, if there is also a backward (reflected) wave, propagating in the negative z direction, the ratio of the voltage and current (in either complex or time domain) for the resultant (forward plus backward) TEM wave does not equal Z_0, or any other constant, but is a function of z. This is analogous to Eq. (10.11) in the analysis of free (unguided) uniform plane waves. The same is true for the ratio of the electric and magnetic field intensities in the line dielectric in Eq. (11.19), as well as for the proportionality between the currents and charges of the line conductors in Eqs. (11.24) and (11.29), i.e., these relations too hold only for a single traveling wave.

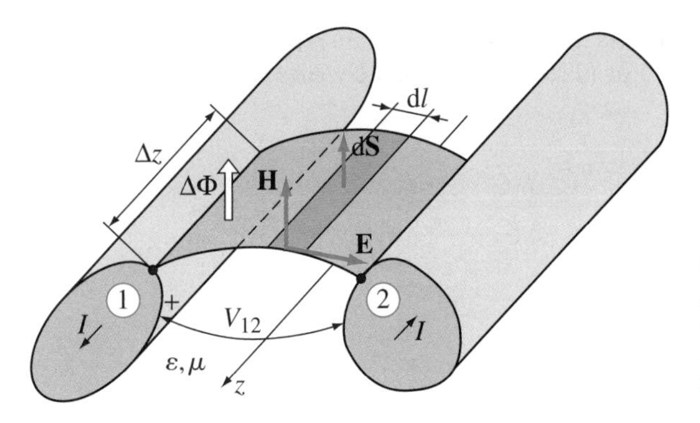

Figure 11.4 Derivation of the duality relationship between the inductance and capacitance per unit length of the transmission line in Fig. 11.3.

Again, the proportionality between $\underline{E}$ and $\underline{H}$ indicates that Z_0 can be expressed in terms of the inductance per unit length of the line, L', as well [L' of different transmission lines (with homogeneous dielectrics) is evaluated in Chapter 7]. To show this, we need the magnetic flux per unit length of the line, $\underline{\Phi}'$, and to find this flux, we adopt an integration surface that follows the lines of vector $\underline{\mathbf{E}}$ between the conductors, as depicted in Fig. 11.4. Note that, from Eq. (11.14), the magnetic flux density vector, $\underline{\mathbf{B}} = \mu\underline{\mathbf{H}}$, in the dielectric is perpendicular to this surface at every point in the flux integral, Eq. (4.95). Hence, assuming that the surface is Δz long (along the line), and using Eqs. (11.19) and (11.31), the flux of $\underline{\mathbf{B}}$ through it is computed as [see also Eqs. (7.10) and (7.12)]

$$\Delta\underline{\Phi} = \int_{\Delta S} \underline{\mathbf{B}} \cdot \mathrm{d}\mathbf{S} = \mu \int_1^2 \underline{H} \underbrace{\Delta z\, \mathrm{d}l}_{\mathrm{d}S} = \frac{\mu \Delta z}{Z_{\mathrm{TEM}}} \int_1^2 \underline{E}\, \mathrm{d}l = \frac{\Delta z}{c} \underline{V}_{12}, \tag{11.38}$$

where $\mathrm{d}S$ is the area of an elemental surface in the form of a thin strip of width $\mathrm{d}l$ and length Δz, and the flux is determined with respect to the upward reference direction in Fig. 11.4 (as required by the right-hand rule and the reference direction of the line current). The external inductance per unit length of the line is [see Eq. (7.11)]

$$L' = \frac{\underline{\Phi}'}{\underline{I}} = \frac{\Delta\underline{\Phi}}{\underline{I}\Delta z} = \frac{\underline{V}_{12}}{c\underline{I}} = \frac{Z_0}{c} = \frac{\varepsilon\mu}{C'}, \tag{11.39}$$

and hence

$$\boxed{Z_0 = cL' = \frac{L'}{\sqrt{\varepsilon\mu}}.} \tag{11.40}$$

characteristic impedance via L'

Note that the assumption that the line conductors in Fig. 11.3 are perfect means no penetration of the magnetic field of the TEM wave into the conductors, and hence zero internal inductance of the line. Consequently, L' in Eq. (11.39) represents also the total inductance, Eq. (7.129), per unit length of the line.

Combining Eqs. (11.37) and (11.40), we obtain the following general relationship between the inductance and capacitance per unit length of an arbitrary two-conductor transmission line with a homogeneous dielectric:

$$\boxed{L'C' = \varepsilon\mu.} \tag{11.41}$$

duality of L' and C' for a line with a homogeneous dielectric

Note that this duality relationship, which provides L' for a known C' and vice versa, is identified in Chapter 7, Eq. (7.13), comparing the expressions for L' and C' for two specific geometries of transmission lines (coaxial cable and thin two-wire line) with air dielectric. We can now substitute the product of ε and μ by the product of

L' and C' in all pertinent expressions in the TEM-wave analysis. The characteristic impedance of the line, Eq. (11.37) or (11.40), can thus be written as

Z_0 – arbitrary dielectric

$$Z_0 = \sqrt{\frac{L'}{C'}}. \tag{11.42}$$

Similarly, the phase coefficient, phase velocity, and wavelength along the line, Eqs. (11.16) and (11.17), become

β, v_p, λ_z – arbitrary dielectric

$$\beta = \omega\sqrt{L'C'}, \quad v_p = \frac{1}{\sqrt{L'C'}}, \quad \lambda_z = \frac{1}{f\sqrt{L'C'}}. \tag{11.43}$$

As we shall see in a later section, the expressions in Eqs. (11.42) and (11.43) can be used also for an approximate analysis of two-conductor transmission lines with inhomogeneous dielectrics.

From Eqs. (8.200), (11.34), and (11.35), the power carried by a traveling TEM wave along the two-conductor transmission line in Fig. 11.3 (through its dielectric) can be obtained as

power carried by a traveling TEM wave

$$\underline{S} = \underline{V}_{12}\underline{I}^* = \underline{V}_{12}\frac{\underline{V}_{12}^*}{Z_0} = Y_0|\underline{V}_{12}|^2 = Z_0|\underline{I}|^2 = P \tag{11.44}$$

[see also Eq. (9.40)]. It is real (P is the time-average power of the wave) because Z_0 and Y_0 are real, i.e., because the line is lossless. Of course, this power can be found also using Eq. (11.21).

Example 11.1 TEM Wave on a Lossless Coaxial Cable with a Homogeneous Dielectric

Fig. 11.5 shows a cross section of an infinitely long lossless coaxial cable carrying a TEM wave of angular frequency ω. The radius of the inner conductor of the cable is a and the inner radius of the outer conductor is b ($b > a$). The dielectric of the cable is homogeneous, with permittivity ε and permeability μ. With the z-axis of a cylindrical coordinate system adopted along the cable axis, the complex rms voltage in the cross section of the cable defined by $z = 0$ is $\underline{V}_0$. For this transmission line, find (a) the voltage and current along the line, (b) the

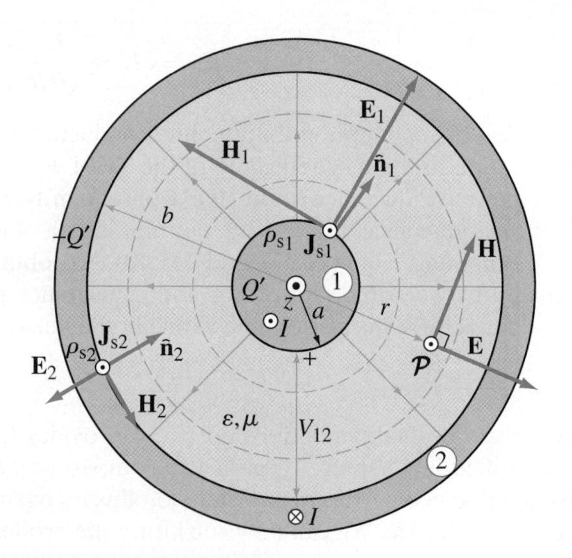

Figure 11.5 Cross section of a lossless coaxial cable with a homogeneous dielectric and TEM wave propagating in the positive z direction; for Example 11.1.

electric and magnetic field intensities in the line dielectric, (c) the charge per unit length of the line, (d) the surface charge and current on line conductors, and (e) the Poynting vector in the dielectric and total power transported by the wave.

Solution

(a) From Eqs. (11.33) and (11.16), the line voltage in an arbitrary cross section is

$$\underline{V}_{12}(z) = \underline{V}_0\, e^{-j\beta z} \quad (-\infty < z < \infty), \quad \text{where} \quad \beta = \omega\sqrt{\varepsilon\mu}. \tag{11.45}$$

As the capacitance and inductance per unit length of the cable in Fig. 11.5 are, Eqs. (2.123) and (7.12),

$$C' = \frac{2\pi\varepsilon}{\ln(b/a)} \quad \text{and} \quad L' = \frac{\mu}{2\pi}\ln\frac{b}{a}, \tag{11.46}$$

any one of Eqs. (11.42), (11.37), and (11.40) gives the following expression for the line characteristic impedance:

$$\boxed{Z_0 = \sqrt{\frac{L'}{C'}} = \frac{\sqrt{\varepsilon\mu}}{C'} = \frac{L'}{\sqrt{\varepsilon\mu}} = \frac{1}{2\pi}\sqrt{\frac{\mu}{\varepsilon}}\ln\frac{b}{a}.} \tag{11.47}$$

Z_0 – coaxial cable

Note that if the dielectric in Fig. 11.5 is nonmagnetic (which is true in most practical situations), Z_0 can conveniently be expressed in terms of the dielectric relative permittivity, ε_r, and cable conductor radii ratio, as

$$\boxed{Z_0 = \frac{60\ \Omega}{\sqrt{\varepsilon_r}}\ln\frac{b}{a},} \tag{11.48}$$

Z_0 – coaxial cable if $\mu = \mu_0$

where the value of the intrinsic impedance of free space ($120\pi\ \Omega$), Eq. (9.23), is incorporated. Using either one of these expressions and Eq. (11.34), the line current is now given by

$$\underline{I}(z) = \frac{\underline{V}_{12}(z)}{Z_0} = \frac{\underline{V}_0}{Z_0}\, e^{-j\beta z} \quad (-\infty < z < \infty). \tag{11.49}$$

(b) The distribution of the electric field in a cross section of the cable is as in electrostatics (Section 11.2), so as in Fig. 2.17 and Eq. (2.124). It also is the same as for the quasistatic (low-frequency time-harmonic) case in Eq. (8.198). Combining this with the line voltage expression in Eq. (11.45), the complex electric field intensity at an arbitrary point in the dielectric in Fig. 11.5 is

$$\boxed{\underline{E}(r,z) = \frac{\underline{V}_{12}(z)}{r\ln(b/a)} = \frac{\underline{V}_0}{r\ln(b/a)}\, e^{-j\beta z} \quad (a < r < b, \ -\infty < z < \infty),} \tag{11.50}$$

E-field distribution of a coaxial cable

where r is the radial coordinate in the cylindrical system, and $\underline{\mathbf{E}}$ is a radial vector. Similarly, based on the magnetostatic or quasistatic field distribution in a cable cross section in Fig. 4.17 and Eq. (4.61) or (8.198), and line current expression in Eq. (11.49), the vector $\underline{\mathbf{H}}$ of the TEM wave in the dielectric is circular with respect to the z-axis, and its intensity is

$$\boxed{\underline{H}(r,z) = \frac{\underline{I}(z)}{2\pi r} = \frac{\underline{V}_0}{2\pi r Z_0}\, e^{-j\beta z} \quad (a < r < b, \ -\infty < z < \infty),} \tag{11.51}$$

H-field distribution of a coaxial cable

with the expression for Z_0 in Eq. (11.47). The electric and magnetic field vectors at any point in the dielectric are mutually orthogonal, as can be seen in Fig. 11.5, and the ratio of their complex rms intensities equals the wave impedance of the TEM wave along the line,

$$\frac{\underline{E}(r,z)}{\underline{H}(r,z)} = \frac{2\pi Z_0}{\ln(b/a)} = \sqrt{\frac{\mu}{\varepsilon}} = Z_{\mathrm{TEM}}, \tag{11.52}$$

which, of course, is in agreement with Eqs. (11.14) and (11.19).

(c) Combining Eqs. (11.36) and (11.45), the charge per unit length of the cable at an arbitrary coordinate z is

$$\underline{Q}'(z) = C'\underline{V}_{12}(z) = C'\underline{V}_0\,\mathrm{e}^{-\mathrm{j}\beta z}. \tag{11.53}$$

Note that the same result for the electric field in Eq. (11.50) can now be obtained from Eqs. (2.122), (11.53), and (11.46) as well,

$$\underline{E}(r,z) = \frac{\underline{Q}'(z)}{2\pi\varepsilon r} = \frac{C'\underline{V}_0}{2\pi\varepsilon r}\,\mathrm{e}^{-\mathrm{j}\beta z} = \frac{\underline{V}_0}{r\ln(b/a)}\,\mathrm{e}^{-\mathrm{j}\beta z}. \tag{11.54}$$

Note also that Eqs. (11.53), (11.49), and (11.47) yield

$$\frac{\underline{Q}'(z)}{\underline{I}(z)} = C'Z_0 = \sqrt{\varepsilon\mu} = \frac{1}{c}, \tag{11.55}$$

namely, the charge to current ratio in Eq. (11.25) or (11.29), with c standing for the velocity of the TEM wave in Fig. 11.5.

(d) Using the boundary condition in Eq. (2.58) for the normal (and the only existing) component of the electric field vector at the surfaces of conductors in Fig. 11.5, as in Eqs. (2.179) and (2.180), the surface charge densities on the surfaces ($\underline{\rho}_{s1}$ on the inner conductor and $\underline{\rho}_{s2}$ on the outer) are

$$\underline{\rho}_{s1}(z) = \varepsilon\underline{E}(a^+,z) = \frac{\varepsilon\underline{V}_0}{a\ln(b/a)}\,\mathrm{e}^{-\mathrm{j}\beta z}, \quad \underline{\rho}_{s2}(z) = -\varepsilon\underline{E}(b^-,z) = -\frac{\varepsilon\underline{V}_0}{b\ln(b/a)}\,\mathrm{e}^{-\mathrm{j}\beta z}. \tag{11.56}$$

Analogously, the boundary condition for the tangential component of the magnetic field vector at a PEC interface in Eqs. (8.33) applied to the conducting surfaces in Fig. 11.5 results in the following surface current densities, defined with respect to the positive z reference direction for the vector $\underline{\mathbf{J}}_s$ on both surfaces:

$$\underline{J}_{s1}(z) = \underline{H}(a^+,z) = \frac{\underline{V}_0}{2\pi aZ_0}\,\mathrm{e}^{-\mathrm{j}\beta z}, \quad \underline{J}_{s2}(z) = -\underline{H}(b^-,z) = -\frac{\underline{V}_0}{2\pi bZ_0}\,\mathrm{e}^{-\mathrm{j}\beta z}, \tag{11.57}$$

where $\hat{\mathbf{n}}\times\underline{\mathbf{H}}$ in the boundary condition turns out to be in the positive z direction on the surface of the inner conductor, and oppositely on the surface of the outer conductor of the cable (this is why $\underline{J}_s$ is a negative of $\underline{H}$ in the second equation). Comparing the corresponding current and charge expressions in Eqs. (11.57) and (11.56), while having in mind the characteristic impedance expression in Eq. (11.47), we identify the current-charge proportionality in Eq. (11.24) for each of the conductors,

$$\underline{J}_{s1}(z) = \frac{\ln(b/a)}{2\pi\varepsilon Z_0}\,\underline{\rho}_{s1}(z) = c\underline{\rho}_{s1}(z), \quad \underline{J}_{s2}(z) = c\underline{\rho}_{s2}(z). \tag{11.58}$$

(e) From Eqs. (11.20) and (11.50), the time-average Poynting vector (the complex Poynting vector is purely real) at an arbitrary location in the dielectric of the cable in Fig. 11.5 is

$$\mathcal{P}(r) = \frac{|\underline{E}(r,z)|^2}{Z_{\mathrm{TEM}}} = \frac{|\underline{V}_0|^2}{r^2\ln^2(b/a)Z_{\mathrm{TEM}}}. \tag{11.59}$$

Using Eq. (11.44), the total time-average power flow along the line is

$$P = Y_0|\underline{V}_{12}(z)|^2 = \frac{|\underline{V}_0|^2}{Z_0}, \tag{11.60}$$

where Y_0 is the characteristic admittance of the cable, Eq. (11.35), and this can alternatively be obtained by integrating $\mathcal{P}(r)$ over a cross section of the cable, as in Eq. (8.200).

Conceptual Questions (on Companion Website): 11.12–11.16; *MATLAB Exercises* (on Companion Website).

11.5 TRANSMISSION LINES WITH SMALL LOSSES

All real transmission lines have some losses (Joule's or ohmic losses), which, in general, consist of losses in conductors and losses in the dielectric of the line. However, for lines used in engineering practice, these losses, evaluated per unit length of the line, are small. Simply, the conductors and dielectrics in practical transmission lines, if not perfect, are good – by design. Denoting the conductivity of the line conductors by σ_c, and that of the line dielectric by σ_d, as indicated in Fig. 11.6, σ_c is very large and σ_d is very small. More precisely, the conditions in Eqs. (9.133) and (9.121) are met, at the operating frequency, f, of the line. Since in practically all designs and applications of transmission lines, the permittivity of line conductors can be assumed to be that of a vacuum ($\varepsilon_c = \varepsilon_0$), these conditions read

$$\sigma_c \gg \omega\varepsilon_0 \quad \text{and} \quad \sigma_d \ll \omega\varepsilon, \tag{11.61}$$

transmission line with small losses

where $\omega = 2\pi f$ (angular frequency) and ε stands for the permittivity of the line dielectric. Of course, $\sigma_c \to \infty$ for a PEC, and $\sigma_d = 0$ for a perfect dielectric. In many situations, the losses in either conductors or dielectrics, or both, can be completely neglected (i.e., the conductors and/or dielectrics can be treated as perfect) in the analysis of transmission lines with TEM waves. In other cases, the (small) losses per unit length of the line are taken into account based on an approximate analysis, described in this and the following section. However, as we shall see, the losses in both conductors and dielectrics increase with frequency, and thus even the lines that are treated and analyzed as having small losses (locally and per-unit-length) may, if very long, become impractical to convey electromagnetic signals (energy or information) above certain (high) frequency limits. Hence, it is very important to always have in mind that the term "small losses" in the analysis of lossy transmission lines does not necessarily and automatically mean that the losses are negligible. They can, overall, be very significant (even prohibitively large) in a given application, but the structure can still be treated by the theory of lines with "small losses" or "low-loss" lines presented here.[6]

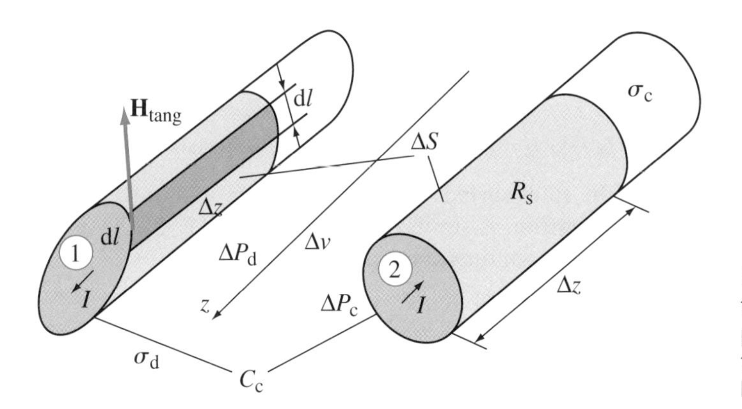

Figure 11.6 Evaluation of the time-average ohmic loss power per unit length of a transmission line with small losses.

[6]Note that, in general, transmission lines radiate (like antennas), so that losses due to their radiation also contribute to the overall losses of the line. Simply, some part of the power carried by the TEM wave along the line is radiated into the external space instead of being delivered to the load (receiver) at the end of the line. Radiation losses grow with frequency. However, in most applications of transmission lines, they can be neglected. Otherwise, radiation of a line must be analyzed using the concepts and techniques of antenna theory, which will be presented in a later chapter.

The (ohmic) losses in a transmission line result in the attenuation of TEM waves along the line, as in Fig. 9.11 in the analysis of plane waves in unbounded lossy media. For the wave traveling in the positive z direction in Fig. 11.6, the attenuation factor is $\mathrm{e}^{-\alpha z}$ [as in Eq. (9.81)], where α is the attenuation coefficient of the line (for the given geometry of its cross section and given material parameters), so that the propagation coefficient in Eq. (11.16) becomes

$$\underline{\gamma} = \alpha + \mathrm{j}\beta. \tag{11.62}$$

Accordingly, the complex current intensity along the line in Eqs. (11.33), for instance, is now given by

attenuated (current) wave along a transmission line

$$\boxed{\underline{I}(z) = \underline{I}(0)\,\mathrm{e}^{-\underline{\gamma} z} = \underline{I}(0)\,\mathrm{e}^{-\alpha z}\,\mathrm{e}^{-\mathrm{j}\beta z},} \tag{11.63}$$

and similarly for the voltage, field intensity vectors, and other z-dependent quantities in the analysis. The losses in the line also result in different distributions of electric and magnetic fields in a cross section of the line with respect to the corresponding distributions for the same line with no losses, in Fig. 11.4. In addition, due to losses, the field vectors, $\underline{\mathbf{E}}$ and $\underline{\mathbf{H}}$, in Fig. 11.6 have longitudinal (z-) components, Eqs. (11.7), as well, i.e., the wave is not TEM any more (it is a hybrid wave).[7] However, since the losses are small, the differences in transversal distributions of fields (with respect to the lossless case), as well as the longitudinal (axial) field components in Eqs. (11.7), are small (so-called quasi-TEM wave), and can be neglected. This constitutes the approximate technique, a perturbation method, for the analysis of transmission lines with small conductor and dielectric losses – based on an assumption that the fields in every transversal plane of the line are practically the same as if there were no losses, and just attenuate along the line as $\mathrm{e}^{-\alpha z}$. Moreover, the loss power per unit length of the line and coefficient α are determined using the field distributions for the lossless case (perturbation method).

To evaluate the loss power in the line conductors, we use the perturbation method for approximate computation of losses in good conductors at higher frequencies (with the skin effect pronounced) from Section 10.4. These losses, and the associated wave attenuation along the line, are due to the (small) penetration of the guided TEM wave, i.e., its electric field, into the conductors [skin depth, Eq. (9.139), is small, but nonzero]. Using Eq. (10.90), the time-average power of Joule's or ohmic losses ΔP_{c} in the conductors in Fig. 11.6 for a part of the line Δz long is obtained as

$$\Delta P_{\mathrm{c}} = [\Delta(P_{\mathrm{J}})_{\mathrm{ave}}]_{\text{in conductors}} = \int_{\Delta S} R_{\mathrm{s}} |\underline{\mathbf{H}}_{\mathrm{tang}}|^2 \Delta z \,\mathrm{d}l, \tag{11.64}$$

where ΔS stands for the total surface of the conductors within the part of the line considered and the integration is similar to that in Eq. (11.38). From Eq. (10.78), the surface resistance of the conductors, R_{s}, is given by

skin-effect surface resistance of line conductors

$$\boxed{R_{\mathrm{s}} = \sqrt{\frac{\pi \mu_{\mathrm{c}} f}{\sigma_{\mathrm{c}}}},} \tag{11.65}$$

where μ_{c} is the permeability of the conductors, which is practically always equal to μ_0 (conductors are nonmagnetic). Most importantly, the tangential component

[7]Note, for instance, that since there is a volume current of density $\underline{\mathbf{J}}$ in lossy conductors of the transmission line, whose direction is axial, the same as the direction of current intensities $\underline{I}_1$ and $\underline{I}_2$ along the conductors, there is also an axial (longitudinal) electric field vector in conductors, given by $\underline{\mathbf{E}}_{\mathrm{c}} = \underline{\mathbf{J}}/\sigma_{\mathrm{c}}$. By means of the boundary condition for the tangential electric field, in Eqs. (8.32), this then results in a nonzero axial component of the vector $\underline{\mathbf{E}}$ on the surface of conductors, so in the line dielectric, as well.

of the complex rms magnetic field intensity vector on the conductor surface, $\underline{\mathbf{H}}_{\rm tang}$, is computed as if the conductors in Fig. 11.6 were perfect (perturbation method), and ultimately from the magnetostatic analysis in a cross section of the transmission line [see Eqs. (11.11), (11.12), and (11.32)]. Dividing the power $\Delta P_{\rm c}$ by Δz, the surface integral in Eq. (11.64) reduces to a line integral (circulation) along the contour $C_{\rm c}$ of both conductors in the line cross section ($\mathrm{d}l$ is an elemental segment along $C_{\rm c}$), and the result is the time-average ohmic loss power in the conductors per unit length of the line, $P'_{\rm c}$. From the first equation of Eqs. (8.211), dividing then $P'_{\rm c}$ by the square of the magnitude of the complex rms current intensity of the line, $\underline{I}$, gives the high-frequency resistance per unit length of the line [see also the same relationship between the loss power and resistance p.u.l. for transmission lines in a dc regime in Eq. (3.172)],

$$R' = \frac{P'_{\rm c}}{|\underline{I}|^2} = \frac{\Delta P_{\rm c}}{|\underline{I}|^2 \Delta z} = \frac{1}{|\underline{I}|^2} \oint_{C_{\rm c}} R_{\rm s} |\underline{\mathbf{H}}_{\rm tang}|^2 \,\mathrm{d}l. \quad (11.66)$$

high-frequency resistance p.u.l. of a line

Since the magnetic field in the line is proportional to the line current, we have that $|\underline{\mathbf{H}}_{\rm tang}| \propto |\underline{I}|$ in Eq. (11.66), so that R', in addition to being proportional to $R_{\rm s}$, depends on the geometry of the line cross section (i.e., on the shape, size, and mutual position of the conductors), but not on the line current. Eq. (11.66) represents the general recipe for computing the resistance per unit length of two-conductor transmission lines with TEM waves, assuming that the skin effect is pronounced [note that the low-frequency (or dc) resistance per unit length of transmission lines is evaluated in Section 3.12]. For transmission lines with perfect conductors, $R' = 0$ ($\sigma_{\rm c} \to \infty$, so that $R_{\rm s} = 0$).

From Eq. (8.196), considering again a line segment of length Δz, the time-average power of Joule's losses $\Delta P_{\rm d}$ in the line dielectric is obtained by integrating the corresponding power density [see also Eq. (9.99)] over the volume Δv of the dielectric region in Fig. 11.6 (that is Δz long),

$$\Delta P_{\rm d} = [\Delta (P_{\rm J})_{\rm ave}]_{\rm in\ dielectric} = \int_{\Delta v} \sigma_{\rm d} |\underline{\mathbf{E}}|^2 \underbrace{\mathrm{d}S \Delta z}_{\mathrm{d}v}, \quad (11.67)$$

where $\mathrm{d}v$ is a volume element for the integration, and $\mathrm{d}S$ is the corresponding surface element in the cross section of the dielectric ($S_{\rm d}$). The distribution of the electric field intensity vector, $\underline{\mathbf{E}}$, over $S_{\rm d}$ is assumed to be the no-loss one, so this is also a perturbation method – for evaluation of dielectric losses. Divided by Δz, the integral in Eq. (11.67) amounts to the time-average ohmic loss power in the dielectric per unit length of the line, $P'_{\rm d}$. As in Eq. (3.173) in the analysis of lossy transmission lines in a dc regime, $P'_{\rm d} = G' |\underline{V}_{12}|^2$, which can also be obtained from the first relationship in Eqs. (8.211), so that the conductance per unit length of the line is

$$G' = \frac{P'_{\rm d}}{|\underline{V}_{12}|^2} = \frac{\Delta P_{\rm d}}{|\underline{V}_{12}|^2 \Delta z} = \frac{1}{|\underline{V}_{12}|^2} \int_{S_{\rm d}} \sigma_{\rm d} |\underline{\mathbf{E}}|^2 \,\mathrm{d}S, \quad (11.68)$$

conductance p.u.l. of a line

where the complex rms voltage of the line, $\underline{V}_{12}$, is given in Eq. (11.31). Since $|\underline{\mathbf{E}}| \propto |\underline{V}_{12}|$, G' is a constant, depending on the geometry of the line cross section and the conductivity of the dielectric ($\sigma_{\rm d}$). Moreover, as $\underline{\mathbf{E}}$ can be determined from a 2-D electrostatic analysis (in a transversal plane), G' equals the leakage conductance per unit length of the line in Eq. (3.157), determined under static conditions (see examples of evaluation of G' of different transmission lines in Section 3.12).

Finally, provided that the line dielectric is homogeneous, it can be computed from the capacitance per unit length of the line (C'), in Eq. (11.36), using the duality relationship in Eq. (3.171).

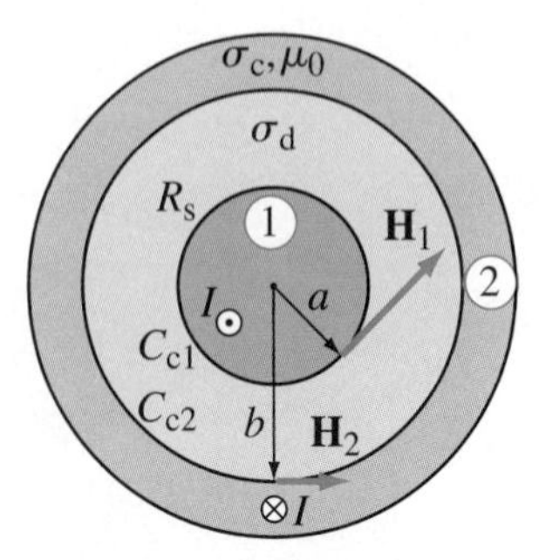

Figure 11.7 Computation of the high-frequency resistance per unit length of a coaxial cable, Eq. (11.70); for Example 11.2.

Example 11.2 **Low-Loss Coaxial Cable at High Frequencies**

Assuming that the coaxial cable in Fig. 11.5 has small losses, with the conductivity and permeability of the conductors σ_c and μ_0, respectively, and conductivity of the dielectric σ_d, find (a) the resistance (with the skin effect pronounced) and (b) the conductance per unit length of the cable.

Solution

(a) From Eq. (11.51), the complex rms magnetic field intensity on the surface of the inner conductor of the cable (where $r = a^+$) and that on the inner surface of the outer conductor (where $r = b^-$) are, respectively,

$$\underline{H}_1 = \frac{\underline{I}}{2\pi a} \quad \text{and} \quad \underline{H}_2 = \frac{\underline{I}}{2\pi b}. \tag{11.69}$$

The vector $\underline{\mathbf{H}}$ is entirely tangential to both surfaces, as shown in Fig. 11.7, so that $\underline{\mathbf{H}}_{\text{tang}} = \underline{\mathbf{H}}$ in Eq. (11.66), which results in the following expression for the high-frequency resistance per unit length of the cable:

R′ – coaxial cable, skin effect pronounced

$$R' = \frac{1}{|\underline{I}|^2}\left(\oint_{C_{c1}} R_s|\underline{H}_1|^2\,dl + \oint_{C_{c2}} R_s|\underline{H}_2|^2\,dl\right) = \frac{R_s}{|\underline{I}|^2}\left[\left(\frac{|\underline{I}|}{2\pi a}\right)^2 \underbrace{\oint_{C_{c1}} dl}_{2\pi a} + \left(\frac{|\underline{I}|}{2\pi b}\right)^2 \underbrace{\oint_{C_{c2}} dl}_{2\pi b}\right] = \frac{R_s}{2\pi}\left(\frac{1}{a} + \frac{1}{b}\right), \quad \text{where} \quad R_s = \sqrt{\frac{\pi\mu_0 f}{\sigma_c}}. \tag{11.70}$$

In this computation, C_{c1} and C_{c2} denote contours of the cable conductors (circles of radii a and b, respectively), and R_s stands for the surface resistance of the conductors, given in Eq. (11.65), with $\mu_c = \mu_0$. Note that the portion of R' corresponding to losses in the inner conductor of the cable, being proportional to $1/a$, is considerably larger than that for losses in the outer conductor, which depends on $1/b$. Note also that the expression for the low-frequency (or dc) resistance p.u.l. of the cable appears in Eq. (3.164).

(b) The per-unit-length conductance, G', of the cable is that in Eq. (3.158).

Conceptual Questions (on Companion Website): 11.17–11.21.

11.6 ATTENUATION COEFFICIENTS FOR LINE CONDUCTORS AND DIELECTRIC

Having now R' and G' of a transmission line, in this section we determine the attenuation coefficient (α) of the line. We first realize that, because of the losses, the time-average power transported by the TEM wave through the line cross section [see Eq. (11.21)] varies along the z-axis, $P \neq$ const. However, as both the wave impedance of the TEM wave along the line, Z_{TEM}, and characteristic impedance of the line, Z_0, for small losses are approximately real and given respectively by Eqs. (11.19) and (11.42) for the lossless line, the complex power ($\underline{S}$) along the line, Eq. (11.21) or (11.44), is also approximately real and equal to P. Combining

Eqs. (11.44) and (11.63), the dependence of P on the z-coordinate reads

$$P(z) = P(0)\,\mathrm{e}^{-2\alpha z}, \quad \text{where} \quad P(0) = Z_0|\underline{I}(0)|^2, \tag{11.71}$$

i.e., its rate of attenuation in the direction of wave propagation is determined by twice α. Note that this z-dependence is the same as that of the time-average Poynting vector of a uniform plane wave in a lossy medium, Eq. (9.97).

We next realize that, by the conservation of power principle, the time-average Joule's power ΔP_{J} dissipated (to heat) in the line conductors, Eq. (11.64), and dielectric, Eq. (11.67), along Δz is equal, in magnitude, to the corresponding change ΔP of the power $P(z)$ in Eq. (11.71). Since ΔP_{J} is positive [power of Joule's losses is always positive (or zero)] and ΔP is negative [P decreases with z in Eq. (11.71), due to losses],

$$\boxed{\Delta P_{\mathrm{J}} = \Delta P_{\mathrm{c}} + \Delta P_{\mathrm{d}} = -\Delta P.} \tag{11.72}$$

by conservation of power principle

Dividing this equation by Δz, letting Δz approach zero, and taking the derivative of P with respect to z from Eq. (11.71), we get

$$P'_{\mathrm{J}}(z) = \frac{\mathrm{d}P_{\mathrm{J}}(z)}{\mathrm{d}z} = -\frac{\mathrm{d}P(z)}{\mathrm{d}z} = 2\alpha P(0)\,\mathrm{e}^{-2\alpha z} = 2\alpha P(z), \tag{11.73}$$

and hence the following expression for the attenuation coefficient along the line (that may have any number of conductors):

$$\boxed{\alpha = \frac{P'_{\mathrm{J}}}{2P}.} \tag{11.74}$$

attenuation coefficient of a wave-guiding structure

This expression provides a general means for evaluating the attenuation along structures for guiding electromagnetic waves, not necessarily transmission lines with TEM waves. For instance, we shall use it, in a later chapter, to find α of rectangular metallic (non-PEC) waveguides (filled with lossy dielectrics) with TE (transverse electric) and TM (transverse magnetic) waves. For a general structure, we can write, from Eqs. (11.74) and (11.72),

$$\alpha = \frac{P'_{\mathrm{c}} + P'_{\mathrm{d}}}{2P} = \alpha_{\mathrm{c}} + \alpha_{\mathrm{d}}, \tag{11.75}$$

where α_{c} and α_{d} are the attenuation coefficients (namely, portions of α) corresponding to the losses in the conductors and dielectric in the structure, given by

$$\boxed{\alpha_{\mathrm{c}} = \frac{P'_{\mathrm{c}}}{2P} \quad \text{and} \quad \alpha_{\mathrm{d}} = \frac{P'_{\mathrm{d}}}{2P},} \tag{11.76}$$

attenuation coefficients for conductor and dielectric losses

respectively. Of course, the unit for these coefficients, as they are obtained in the above equations, is Np/m. However, the attenuation along a transmission line (or a waveguide) is often expressed in dB/m, for which we use the conversion in Eq. (9.89).

Finally, combining Eqs. (11.76), (11.66), and (11.44), the attenuation coefficient representing the losses in conductors of a two-conductor transmission line (Fig. 11.6) is found as

$$\boxed{\alpha_{\mathrm{c}} = \frac{R'}{2Z_0}.} \tag{11.77}$$

α for transmission-line conductors

Eq. (11.66) tells us that the high-frequency resistance per unit length of the line, R', depends on frequency in the same way as the surface resistance of the line conductors, R_{s}, in Eq. (11.65), that is, it is proportional to the square root of frequency. Since the characteristic impedance of the line (with small losses), Z_0, is not a function of frequency [see Eq. (11.37)], frequency dependence of α_c is as well the same as that of R_s,

frequency dependence of α_c

$$\alpha_c \propto \sqrt{f}. \tag{11.78}$$

Similarly, from Eqs. (11.76), (11.68), and (11.44), the attenuation coefficient for the losses in the dielectric of the line in Fig. 11.6 becomes

α for line dielectric

$$\alpha_d = \frac{G'}{2Y_0}. \tag{11.79}$$

The phase coefficient of the line, β, is, for small losses, given by its no-loss expression in Eq. (11.16), so that, using Eqs. (3.171), (11.35), (11.37), and (11.19), α_d can alternatively be expressed as

α_d – transmission lines and unbounded waves

$$\alpha_d = \frac{\sigma_d C'}{2\varepsilon Y_0} = \frac{\sigma_d\sqrt{\varepsilon\mu}}{2\varepsilon} = \frac{\sigma_d}{2} Z_{TEM} = \frac{\beta}{2}\tan\delta_d, \tag{11.80}$$

where $\tan\delta_d$ is the loss tangent of the dielectric, Eq. (9.125). We see that α_d is the same as the attenuation coefficient for free uniform plane waves in the same dielectric (good dielectric), Eq. (9.123) or (9.126). In other words, it does not depend on the shape and size of the line conductors, and is the same for all transmission lines with small losses – and the same dielectric, at the same frequency. Note that this can be obtained also directly comparing Eqs. (11.68) and (11.21), and identifying the following relationship between P'_d and P:

$$P'_d = \sigma_d Z_{TEM} P, \tag{11.81}$$

which, substituted in Eq. (11.79), leads to Eq. (11.80). As explained in Section 9.9, for dielectrics that exhibit (at high frequencies) both ohmic and polarization losses, they both are usually specified through a single parameter – the imaginary part of the high-frequency complex permittivity, Eq. (9.127). The associated loss tangent, given in Eq. (9.130), is frequency dependent. For good dielectrics at microwave frequencies, it is, roughly, linearly proportional to frequency. In addition, ε in the expression for β in Eq. (11.16) can roughly be taken to be the electrostatic permittivity, even in the microwave region, and hence α_d, for applications at microwave frequencies, can be said to be proportional to frequency squared,

frequency dependence of α_d

$$\alpha_d \propto f^2. \tag{11.82}$$

Comparing frequency dependencies in Eqs. (11.78) and (11.82), we conclude (e.g., visualizing the graphs of the two functions) that for every real transmission line there exists a frequency at which $\alpha_c = \alpha_d$. For instance, this frequency for standard coaxial cables (with homogeneous dielectrics) is on the order of 10 GHz. Below it, the losses in conductors are dominant; above it, the losses in the dielectric prevail. However, because of losses, transmission lines are, in general, mostly used at radio and lower microwave frequencies, up to several GHz. Therefore, in typical applications of transmission lines in engineering practice, α_c is considerably larger than α_d, and, most often, α_d can be neglected in Eq. (11.75), that is, $\alpha \approx \alpha_c$. Losses in the conductors and α_c are thus a typical limiting factor for practical usability of a given transmission line in terms of a combination of frequency and length of the line (except for lines made of superconductors). Namely, for certain (high) frequencies and (large) distances along the line, the attenuation in the conductors becomes so large that the energy/information transfer using the line is considered inefficient (if not impossible), and alternative means to convey the signal are explored (e.g., using waveguides or waves in free space, radiated by an antenna).

Example 11.3 Attenuation Coefficient of a Coaxial Cable with Small Losses

For the coaxial cable with small losses in Fig. 11.7, let $a = 1$ mm, $b = 3.5$ mm, $\sigma_c = 58$ MS/m (copper), $\varepsilon_r = 2.25$ (polyethylene), $\tan\delta_d = 10^{-4}$, $\mu = \mu_0$, and $f = 1$ GHz. Compute the attenuation coefficient of the cable.

Solution Note that $\sigma_c/(2\pi f \varepsilon_0) = 1.043 \times 10^9 \gg 1$ and $\tan\delta_d \ll 1$, which means [see also Eq. (9.125)] that low-loss conditions in Eqs. (11.61) are indeed satisfied for this set of data. Using Eq. (11.77) and the expressions for the per-unit-length high-frequency resistance (R') and characteristic impedance (Z_0) of a low-loss coaxial cable, in Eqs. (11.70) and (11.47), the attenuation coefficient for the losses in conductors of the cable is evaluated as

$$\boxed{\alpha_c = \frac{R'}{2Z_0} = \frac{R_s}{2Z_{TEM}} \frac{1/a + 1/b}{\ln(b/a)},} \qquad (11.83)$$

α_c – coaxial cable

with R_s standing for the surface resistance of the conductors (copper), Eq. (10.80), and Z_{TEM} for the wave impedance of the TEM wave along the cable, Eq. (11.19). As $\varepsilon = \varepsilon_r\varepsilon_0$ and $\mu = \mu_0$, we have $Z_{TEM} = \sqrt{\mu_0/(\varepsilon_r\varepsilon_0)}$. For the given numerical data, $\alpha_c = 1.685 \times 10^{-2}$ Np/m $=$ 0.146 dB/m, where the use is made of the relationship in Eq. (9.89) to convert Np/m to dB/m.

From Eq. (11.80), the attenuation coefficient for the losses in the cable dielectric amounts to

$$\alpha_d = \frac{\beta}{2} \tan\delta_d = \pi f \sqrt{\varepsilon_r\varepsilon_0\mu_0} \tan\delta_d = 1.57 \times 10^{-3} \text{ Np/m} = 0.01365 \text{ dB/m}, \qquad (11.84)$$

where β is the phase coefficient of the cable, Eq. (11.16).

Combining Eqs. (11.75), (11.83), and (11.84), the total attenuation coefficient of the cable is $\alpha = \alpha_c + \alpha_d = 0.0184$ Np/m $= 0.16$ dB/m.

Example 11.4 Coaxial Cable Design for Minimum Attenuation Coefficient

Consider the coaxial cable in Fig. 11.7. (a) For given materials in the structure, and assuming a fixed outer radius b and variable inner radius a of the cable, design the cable (find a) so that its attenuation coefficient is minimum. (b) For a in (a) and $\mu = \mu_0$ (in the dielectric), compute the cable characteristic impedance.

Solution

(a) Eq. (11.84) tells us that the attenuation coefficient for the losses in the cable dielectric does not depend on the cable radii (in fact, it is the same for all transmission lines with the same low-loss dielectric, at the same frequency). Therefore, our task is reduced to the optimization (minimization) of the attenuation coefficient for the cable conductors, in Eq. (11.83). Denoting by x the outer to inner radii ratio, we can write

$$\alpha_c(x) = \frac{R_s}{2bZ_{TEM}} \frac{1+x}{\ln x} \quad \left(x = \frac{b}{a}\right). \qquad (11.85)$$

To find the optimal x, for which α_c is minimum,[8] we perform a standard procedure of equating to zero the derivative of α_c with respect to x, which yields a transcendental

[8]Note that the attenuation coefficient α_c can also be reduced by increasing the radius b, that is, the cross-sectional size of the cable. This, however, would raise the weight and cost of the cable, which is most often undesirable or unfeasible for a given application. In addition, it would reduce the flexibility (bendability) of the cable, which may be a limiting factor too. Finally, the larger the transversal dimensions of the cable the lower the frequency limit above which the cable is unusable due to emergence of higher wave types that may propagate along the cable. Accordingly, it makes a lot of practical sense to fix the outer radius b of the cable, and, for a fixed b and given operating frequency and materials (most frequently, copper and polyethylene) of the cable, minimize the attenuation coefficient by optimizing the inner radius a of the cable.

equation,

coaxial cable optimization for minimum attenuation

$$\frac{d\alpha_c}{dx} = 0 \quad \longrightarrow \quad \ln x = \frac{1}{x} + 1 \quad \longrightarrow \quad x = x_{opt} \approx 3.59 \quad [\alpha_c = (\alpha_c)_{min}], \tag{11.86}$$

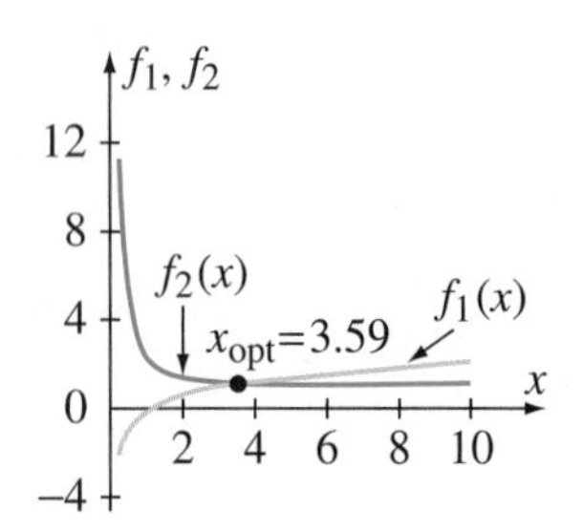

Figure 11.8 Graphical solution ($x = x_{opt}$) of the transcendental equation $\ln x = 1/x + 1$, in Eqs. (11.86), corresponding to the intersection point of curves $f_1(x)$ and $f_2(x)$; for Example 11.4.

whose approximate solution ($x \approx 3.59$) is obtained graphically, at the intersection point of curves $f_1(x) = \ln x$ and $f_2(x) = 1/x + 1$ in Fig. 11.8 (note that this transcendental equation can be solved numerically as well). It is a simple matter to verify that the second derivative of $\alpha_c(x)$ for $x = x_{opt}$ is positive, meaning that the optimization result is indeed a minimum (and not a maximum) of the function. So, the optimal inner radius of the cable is $a_{opt} = b/3.59$, and the minimum attenuation coefficient (for conductors) is $(\alpha_c)_{min} = \alpha_c(x_{opt}) = 1.8R_s/(bZ_{TEM})$.

(b) From Eqs. (11.48) and (11.86), the characteristic impedance of the cable is

$$Z_0 = \frac{60\ \Omega}{\sqrt{\varepsilon_r}} \ln x_{opt} \approx \frac{76.7\ \Omega}{\sqrt{\varepsilon_r}}, \tag{11.87}$$

where ε_r is the relative permittivity of the cable dielectric. For polyethylene (with $\varepsilon_r = 2.25$) and $a = a_{opt}$, we thus obtain Z_0 of about 50 Ω. Given that polyethylene is the most frequently used dielectric in coaxial cables, as well as that reducing the attenuation of electromagnetic signals (information or energy) along the cable is crucial in many applications, this is exactly why the value of $Z_0 = 50\ \Omega$ is the standard characteristic impedance of professional coaxial cables.[9] Note that the other two dielectrics used often in coaxial cables, teflon ($\varepsilon_r = 2.1$) and polystyrene ($\varepsilon_r = 2.56$), also result in Z_0 close to 50 Ω in Eq. (11.87). Moreover, apart from coaxial cables, 50 Ω is established as the standard general reference impedance in radio-frequency (RF) and microwave engineering.

Example 11.5 Coaxial Cable Design for Maximum Breakdown Rms Voltage

In Fig. 11.7, the dielectric strength of the cable dielectric is E_{cr}. (a) For a fixed b, optimize a such that the cable can withstand the maximum possible applied rms voltage (before its dielectric breaks down). (b) What is this maximum voltage? (c) What is the attenuation coefficient for the cable conductors for the optimized cable?

Solution

(a) Let $\underline{V}$ denote the applied complex rms voltage at the beginning (at generator terminals) of the cable. Because of losses, the cable voltage decreases in magnitude (attenuates) away from this point, so $\underline{V}$ and the associated electric field represent the strongest signal of the cable, that is relevant for breakdown. On the other side, the distribution of the electric field in any cross section of the cable is the same as in electrostatics, which means that the optimization of the radius a for the maximum breakdown voltage is that already carried out in Example 2.29. Practically the only difference is an additional factor $\sqrt{2}$ in the breakdown condition in Eq. (2.219). Namely, in the dynamic (TEM-wave) case, dielectric breakdown occurs when the peak-value (amplitude) of the electric field intensity on the surface of the inner conductor of the cable at the generator terminals reaches the critical field value (dielectric strength), E_{cr}, for the dielectric. This peak-value equals the corresponding rms field intensity, $|\underline{E}(a^+)|$, times $\sqrt{2}$, and therefore Eqs. (2.218) and

[9]On the other side, $Z_0 = 75\ \Omega$ is the standard characteristic impedance of commercial coaxial cables for TV and radio antennas, and this value comes from the fact that in antenna applications it is often necessary to make transitions between coaxial cables and two-wire transmission lines that feed into symmetrical antennas (e.g., symmetrical wire dipole antennas). The standard characteristic impedance of commercial two-wire lines for TV and radio antennas is $Z_0 = 300\ \Omega$, and the transition between the lines is usually made using symmetrizing circuits that transform impedance in the ratio 1 : 4, and hence the choice of 75 Ω, equal to a quarter of 300 Ω, for the standard for commercial coaxial cables.

(2.219) combined now read [see also Eq. (11.50)]

$$E_{\rm cr} = \underbrace{|\underline{E}(a^+)|\sqrt{2}}_{\text{peak-value}} = \frac{|\underline{V}|\sqrt{2}}{a\ln(b/a)}. \tag{11.88}$$

The maximization of the breakdown voltage $|\underline{V}|_{\rm cr}$ is given in Eqs. (2.220)–(2.222). Using the notation from Eq. (11.85), that is, $x = b/a$, we have

$$\boxed{|\underline{V}(x)|_{\rm cr} = \frac{E_{\rm cr}b}{\sqrt{2}}\frac{\ln x}{x} \quad \longrightarrow \quad x'_{\rm opt} = {\rm e} = 2.718 \quad \left[|\underline{V}|_{\rm cr} = \left(|\underline{V}|_{\rm cr}\right)_{\max}\right],} \tag{11.89}$$

coaxial cable optimization for maximum permissible voltage

with the optimal inner radius of the cable being $a'_{\rm opt} = b/{\rm e}$.[10]

(b) From Eqs. (11.89), the maximum permissible rms voltage at the beginning of the cable amounts to

$$\left(|\underline{V}|_{\rm cr}\right)_{\max} = |\underline{V}(x'_{\rm opt})|_{\rm cr} = \frac{E_{\rm cr}b}{\sqrt{2}\,{\rm e}} = 0.26E_{\rm cr}b. \tag{11.90}$$

(c) The attenuation coefficient $\alpha_{\rm c}$, Eq. (11.85), for the cable optimized for the maximum breakdown voltage is

$$\alpha_{\rm c}(x'_{\rm opt}) = \frac{R_{\rm s}}{2bZ_{\rm TEM}}(1+{\rm e}) = \frac{1.86R_{\rm s}}{bZ_{\rm TEM}}. \tag{11.91}$$

Note that if a high-voltage cable is intended to be used in low-frequency (power) applications, then the attenuation along the cable should be evaluated based on the low-frequency (or dc) analysis of conductor losses and the p.u.l. resistance of the cable in Eq. (3.164).

Example 11.6 Coaxial Cable Design for Maximum Permissible Power Flow

Assume that the cable defined in the previous example is l long, that the losses in the cable dielectric can be neglected, and that there is only a single traveling TEM wave on the line. (a) Redo the optimization of the inner radius of the cable such that the time-average power flow along the cable is maximal permissible for the safe operation of the cable prior to an eventual dielectric breakdown, and find that power flow. (b) For the optimized radius in (a), compute the characteristic impedance and attenuation coefficient of the cable. (c) What is the time-average power at the end of the cable?

Solution

(a) The time-average power P carried by the TEM wave along the cable is given by Eqs. (11.71) and (11.60). Since the cable characteristic impedance, Eq. (11.47), is a function of the inner radius a of the cable, or of the outer to inner conductor radii ratio, $x = b/a$,

$$Z_0(x) = \frac{Z_{\rm TEM}}{2\pi}\ln x, \tag{11.92}$$

where $Z_{\rm TEM}$ is the wave impedance of the traveling TEM wave, Eq. (11.19), maximization of P (with x as an optimization parameter) is different from that of $|\underline{V}|^2$ or $|\underline{V}|$, in Eq. (11.89), and the result is different from $x = x'_{\rm opt}$. Namely, the power corresponding

[10] Note that this optimization of coaxial cables – for the maximum permissible applied voltage, that the cable can carry with no breakdown of its insulation (dielectric), is important in power applications, so at very low frequencies, an example being high-voltage coaxial cables in underground power distribution systems. Some electromagnetic systems, however, use high-voltage coaxial cables at higher frequencies as well. On the other side, for communication coaxial cables the principal concern is the high-frequency attenuation of signals, and hence the optimization in Eq. (11.86), for the minimum attenuation factor (for conductors) of the line, constitutes the main criterion used in design and construction of cables for communication systems and related electronic devices.

to the breakdown rms voltage of the cable, at the beginning of the line, for a conductor radii ratio x, $|\underline{V}(x)|_{\text{cr}}$, in Eq. (11.89), is given by the following expression:

$$P_{\text{cr}}(x) = \frac{|\underline{V}(x)|_{\text{cr}}^2}{Z_0(x)} = \frac{\pi E_{\text{cr}}^2 b^2}{Z_{\text{TEM}}} \frac{\ln x}{x^2}. \tag{11.93}$$

Its maximization results in

coaxial cable optimization for maximum permissible power

$$\boxed{\frac{\mathrm{d}P_{\text{cr}}}{\mathrm{d}x} = 0 \quad \longrightarrow \quad \ln x = \frac{1}{2} \quad \longrightarrow \quad x = x''_{\text{opt}} = \sqrt{\mathrm{e}} = 1.649 \quad [P = (P_{\text{cr}})_{\text{max}}],} \tag{11.94}$$

i.e., in the optimal inner radius $a''_{\text{opt}} = b/\sqrt{\mathrm{e}}$. The maximum power, limited by the breakdown, is

$$(P_{\text{cr}})_{\text{max}} = P_{\text{cr}}(x''_{\text{opt}}) = \frac{\pi E_{\text{cr}}^2 b^2}{2\,\mathrm{e} Z_{\text{TEM}}} = \frac{0.578 E_{\text{cr}}^2 b^2}{Z_{\text{TEM}}}. \tag{11.95}$$

Note that the power corresponding to $x = x'_{\text{opt}}$ in Eq. (11.89) is, of course, smaller, amounting to $P_{\text{cr}}(x'_{\text{opt}}) = 0.425 E_{\text{cr}}^2 b^2 / Z_{\text{TEM}} < P_{\text{cr}}(x''_{\text{opt}})$.

(b) From Eqs. (11.92), (11.85), and (11.94), the characteristic impedance of the cable and attenuation coefficient for cable conductors for the cable optimized for the maximum permissible power flow are $Z_0(x''_{\text{opt}}) = Z_{\text{TEM}}/(4\pi)$ and $\alpha_{\text{c}}(x''_{\text{opt}}) = R_{\text{s}}(1 + \sqrt{\mathrm{e}})/(bZ_{\text{TEM}}) = 2.65 R_{\text{s}}/(bZ_{\text{TEM}})$, respectively.

(c) Since there is not a reflected wave on the line and $\alpha = \alpha_{\text{c}}$ ($\alpha_{\text{d}} = 0$), Eq. (11.71) tells us that the time-average power at the end of the line is

$$P_l = (P_{\text{cr}})_{\text{max}}\, \mathrm{e}^{-2\alpha_{\text{c}}(x''_{\text{opt}})\,l}, \tag{11.96}$$

where $(P_{\text{cr}})_{\text{max}}$ and $\alpha_{\text{c}}(x''_{\text{opt}})$ are found in (a) and (b), respectively.

Problems: 11.1 and 11.2; *Conceptual Questions* (on Companion Website): 11.22–11.27; *MATLAB Exercises* (on Companion Website).

11.7 HIGH-FREQUENCY INTERNAL INDUCTANCE OF TRANSMISSION LINES

Another consequence of the (small) penetration of the guided TEM wave, now its magnetic field, into the imperfect conductors of a transmission line (Fig. 11.6) is a nonzero reactive power (P_{reactive}) inside the conductors, and the associated high-frequency (with the skin effect pronounced) internal inductance of the line. From Eq. (10.90), this power is the same as the time-average power of Joule's losses in the line conductors. Per unit length of the line,

$$P'_{\text{reactive}} = P'_{\text{c}}, \tag{11.97}$$

where P'_{c} is computed in Eq. (11.66). Using the second relationship in Eqs. (8.211), P'_{reactive} divided by the magnitude of $\underline{I}$ squared equals the high-frequency internal reactance per unit length of the line, X'_{i}. Combining then Eqs. (11.97) and (11.66), we get

skin-effect internal reactance p.u.l. of a transmission line

$$\boxed{X'_{\text{i}} = \frac{P'_{\text{reactive}}}{|\underline{I}|^2} = \frac{P'_{\text{c}}}{|\underline{I}|^2} = R',} \tag{11.98}$$

i.e., X'_{i} equals the high-frequency resistance per unit length of the line, R'. Having in mind Eqs. (8.209) and (8.69),

$$X'_{\text{i}} = \omega L'_{\text{i}}, \tag{11.99}$$

so that the high-frequency internal inductance per unit length of the line can be obtained from R' simply as

$$L'_{\rm i} = \frac{R'}{\omega}. \qquad (11.100)$$

high-frequency internal inductance p.u.l. of a line

In general, due to the skin effect (internal magnetic field is confined to a very thin region below the surface of the conductors), $L'_{\rm i}$ in Eq. (11.100) is considerably smaller than its low-frequency (or dc) value, evaluated in Section 7.6. Moreover, as $R' \propto \sqrt{f}$, Eq. (11.100) tells us that $L'_{\rm i} \propto 1/\sqrt{f}$, i.e., the high-frequency internal inductance of the line decreases with an increase of frequency. Consequently, in most applications of transmission lines with TEM waves, $L'_{\rm i}$ can be neglected with respect to the external inductance per unit length of the line (L') in Eq. (11.39). In what follows, we shall always assume that $L'_{\rm i} = 0$ in the analysis of transmission lines, except when we explicitly specify otherwise.

Example 11.7 High-Frequency Internal Inductance p.u.l. of a Coaxial Cable

Find the high-frequency internal inductance per unit length of the coaxial cable described in Example 11.3.

Solution Combining Eqs. (11.100) and (11.70), the high-frequency internal p.u.l. inductance of the cable is given by

$$L'_{\rm i} = \frac{R'}{\omega} = \frac{R_{\rm s}}{2\pi\omega}\left(\frac{1}{a} + \frac{1}{b}\right) = \frac{1}{4\pi}\sqrt{\frac{\mu_0}{\pi\sigma_{\rm c} f}}\left(\frac{1}{a} + \frac{1}{b}\right), \qquad (11.101)$$

$L'_{\rm i}$ – coaxial cable

where R' and $R_{\rm s}$ stand for the high-frequency p.u.l. resistance of the cable and surface resistance of its conductors, respectively. Substituting the numerical data, $L'_{\rm i} = 0.269$ nH/m. Note that, from Eq. (11.46), the external inductance per unit length of the cable amounts to $L' = \mu_0 \ln(b/a)/(2\pi) = 250.55$ nH/m, i.e., it indeed is much larger than $L'_{\rm i}$. Note also that Eq. (7.139), adopting $c = 4.5$ mm for the outer radius of the outer cable conductor, yields $L'_{\rm i} = 68.92$ nH/m for the low-frequency (or dc) internal inductance p.u.l. of this cable, which, again, is a substantially larger value than the high-frequency $L'_{\rm i}$ (with the skin effect pronounced).

11.8 EVALUATION OF PRIMARY AND SECONDARY CIRCUIT PARAMETERS OF TRANSMISSION LINES

As we shall see in the next chapter, an arbitrary two-conductor transmission line with TEM waves can be analyzed as an electric circuit with distributed parameters, based on a representation of the line by a network of cascaded equal small cells, of length Δz, with lumped elements. These elements are characterized by per-unit-length parameters C', L', R', and G' of the line (discussed in Sections 11.4–11.7), multiplied by Δz. Such a model is a generalization of the circuit-theory representation of a lossy transmission line in a dc regime in Fig. 3.21. As C', L', R', and G' are a basis for the circuit analysis of transmission lines (to be presented in the next chapter), they are referred to as primary circuit parameters of a line. The other parameters that will be used in the circuit analysis are the characteristic impedance, Z_0, phase coefficient, β, phase velocity, $v_{\rm p}$, wavelength, λ_z, and attenuation coefficient, α, of the line (studied in Sections 11.1, 11.4, 11.5, and 11.6). As these parameters can be derived from the primary parameters, they are called secondary circuit parameters of transmission lines. Moreover, once the secondary parameters are known for a given line, they suffice for the analysis (i.e., primary parameters are not needed).

We now consider various examples of evaluation and discussion of circuit parameters of two-conductor transmission lines with small losses and homogeneous dielectrics (note that the parameters of a coaxial cable are already computed and discussed in Examples 11.1–11.7). Analysis of transmission lines with inhomogeneous dielectrics and evaluation of their circuit parameters will be presented in the next section.

Example 11.8 Circuit Parameters of a Thin Two-Wire Line with a TEM Wave

A thin symmetrical two-wire line has lossy conductors, of radii a and conductivity σ_c, and a lossy homogeneous dielectric, of relative permittivity ε_r and conductivity σ_d. The distance between the conductor axes is d ($d \gg a$), and the permeability everywhere is μ_0. The line carries a TEM wave of frequency f. The losses can be considered to be small. (a) Find the primary per-unit-length circuit parameters of the line. (b) What are the characteristic impedance, phase coefficient, phase velocity, wavelength, and attenuation coefficient (secondary circuit parameters) of the line?

Solution

(a) From Eqs. (2.141), (7.11), and (3.184), the capacitance, inductance, and conductance per unit length of the line are, respectively, given by

$$C' = \frac{\pi \varepsilon_r \varepsilon_0}{\ln(d/a)}, \quad L' = \frac{\mu_0}{\pi} \ln \frac{d}{a}, \quad G' = \frac{\pi \sigma_d}{\ln(d/a)}. \tag{11.102}$$

To find the high-frequency p.u.l. line resistance using Eq. (11.66), we assume a time-harmonic current of complex rms intensity $\underline{I}$ on the line, which, in any cross section of the line, gives the same distribution of the magnetic field as in Fig. 7.4. Since $d \gg a$, the field $\underline{\mathbf{H}}$ on the surface of each of the line conductors (wires) can be computed as if the other conductor is not present, and hence is the same as the field intensity $\underline{\mathbf{H}}_1$ in Fig. 11.7 and Eq. (11.69), on the surface of the inner conductor (with radius a) of a coaxial cable. Accordingly, each of the two integrals in Eq. (11.66) along contours of the two wires in Fig. 7.4 equals the integral along the contour C_{c1} in Eq. (11.70), and R' of the two-wire line is exactly twice the portion of the expression for R' of the coaxial cable that results from the integration along C_{c1},

R′ – thin two-wire line, at high frequencies

$$R' = 2\frac{R_s}{2\pi a} = \frac{R_s}{\pi a}, \tag{11.103}$$

where the surface resistance of the line conductors, R_s, is also given in Eq. (11.70). Note that the expression for the low-frequency or dc resistance p.u.l. of the line is the one in Eq. (3.183).

(b) Having in mind Eqs. (11.37), (11.102), (11.19), and (9.23), the characteristic impedance of the two-wire line is

Z_0 – thin two-wire line

$$Z_0 = \frac{\sqrt{\varepsilon_r \varepsilon_0 \mu_0}}{C'} = \frac{Z_{TEM}}{\pi} \ln \frac{d}{a} = \frac{120\ \Omega}{\sqrt{\varepsilon_r}} \ln \frac{d}{a}, \tag{11.104}$$

with $Z_{TEM} = \sqrt{\mu_0/(\varepsilon_r \varepsilon_0)}$ being the wave impedance of the traveling TEM wave along the line. Most frequently, $\varepsilon_r = 1$ (air two-wire lines). Note that Z_0 of two-wire lines used in telephony is on the order of 100 Ω, whereas $Z_0 = 240$ Ω and 300 Ω are standard characteristic impedances of two-wire lines for antenna applications.

Eqs. (11.16) and (11.17) tell us that the phase coefficient, phase velocity, and wavelength of the line are $\beta = 2\pi f\sqrt{\varepsilon_r\varepsilon_0\mu_0}$, $v_p = 1/\sqrt{\varepsilon_r\varepsilon_0\mu_0}$, and $\lambda_z = 1/(\sqrt{\varepsilon_r\varepsilon_0\mu_0}f)$, respectively.

Combining Eqs. (11.77), (11.103), and (11.104), the attenuation coefficient for the losses in the line conductors is

$$\boxed{\alpha_c = \frac{R'}{2Z_0} = \frac{R_s}{2Z_{\text{TEM}}\, a\ln(d/a)},} \tag{11.105}$$

α_c – thin two-wire line

and that for the losses in the dielectric, equals, with the use of Eq. (11.80), $\alpha_d = \sigma_d Z_{\text{TEM}}/2 = \sigma_d\sqrt{\mu_0/(\varepsilon_r\varepsilon_0)}/2$. The total attenuation coefficient of the line is $\alpha = \alpha_c + \alpha_d$. Of course, $\alpha_d = 0$ ($\sigma_d = 0$) and $\alpha = \alpha_c$ for an air two-wire line.

Note that standard two-wire lines have lower attenuation than standard coaxial cables. On the other hand, the principal disadvantage of two-wire lines is that they radiate (like antennas), especially at higher frequencies, and are susceptible to interference with external signals as well, whereas coaxial cables are immune to both radiation losses and pickup of electromagnetic noise from the environment. Namely, in the latter case, the cable dielectric, in which the electromagnetic field resides, is shielded by the outer conductor. Note also that the impedance Z_0 in Eq. (11.104) can be increased and coefficient α_c in Eq. (11.105) reduced by increasing the distance d between the conductor axes of a two-wire line. This, however, may considerably enhance the radiation of the line (the radiated fields due to the two wires, which carry currents and charges that are equal in magnitude but with opposite directions/polarities, do not cancel each other in substantially large regions of space). By the same token, the line whose conductors are too far apart is a better "receiver" of external interference. So, a compromise has to be made in the design of the line between the low attenuation and low radiation/interference requirements, for a given application. Note finally that an increase of the wire radii a would, on the other side, also result in a reduction of α_c, through a reduction of the resistance R' in Eq. (11.103), but a cannot be arbitrarily enlarged, due to cost (more copper or other metal for wires) and mechanical requirements.

Example 11.9 Wire-Plane Transmission Line

For the wire-plane transmission line in Fig. 2.24(a), assume that the skin-effect surface resistance of the wire is R_{s1} and that of the ground plane R_{s2}, as well as that the material parameters of the line dielectric are ε, μ_0, and σ_d. The line can be treated as a low-loss one. Under these circumstances, find the primary circuit parameters of the line at high frequencies (for TEM waves).

Solution By electrostatic analysis, using image theory, the capacitance per unit length of the line if air-filled is found to be that in Eq. (2.146), and we now substitute ε_0 by ε in the result to obtain C' for this present case. The inductance and conductance p.u.l. of the line are then computed from Eqs. (11.41) and (3.171), respectively, and we can write

$$C' = \frac{2\pi\varepsilon}{\ln(2h/a)}, \quad L' = \frac{\mu_0}{2\pi}\ln\frac{2h}{a}, \quad G' = \frac{2\pi\sigma_d}{\ln(2h/a)}. \tag{11.106}$$

Since the upper conductor (wire) of the line is thin ($a \ll h$), the complex magnetic field intensity on its surface is (approximately) $\underline{H} = \underline{I}/(2\pi a)$, $\underline{I}$ being the complex current intensity of the line, i.e., the current of the wire and that flowing in the opposite direction over the conducting plane, as shown in Fig. 11.9(a). Accordingly, the part of the high-frequency p.u.l. line resistance taking into account losses in this wire (R'_1) equals the part of the resistance in Eq. (11.103) corresponding to one wire of a thin two-wire line, and hence $R'_1 = R_{s1}/(2\pi a)$.

To find the remaining part of R' of the wire-plane line, that (R'_2) quantifying losses in the conducting plane, let us again apply image theory to the p.u.l. complex charge $\underline{Q}'$ of the wire in Fig. 11.9(a), but now considered together with the associated complex current $\underline{I}$

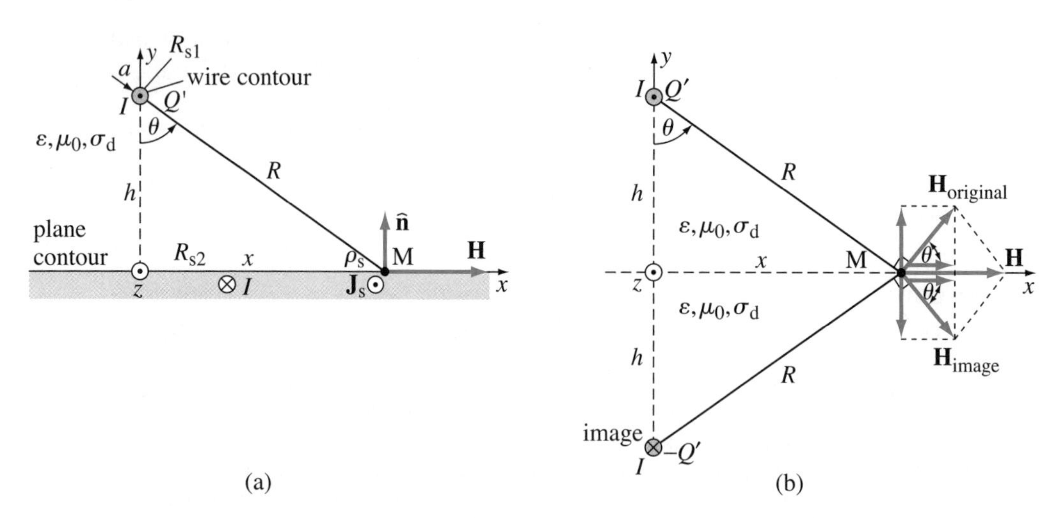

Figure 11.9 Evaluation of the high-frequency resistance per unit length of a wire-plane transmission line: (a) original system, with the conductor contours for integration in Eq. (11.66), and (b) equivalent system, to find the magnetic field of the line near the ground plane, using image theory; for Example 11.9.

along the wire. This charge and current, for the original wire, are related as in Eq. (11.29), and the same relationship holds for the charge $\underline{Q}'_{\text{image}}$ and current $\underline{I}_{\text{image}}$ of the image wire in Fig. 11.9(b). Namely, $\underline{Q}'_{\text{image}}$ and $\underline{I}_{\text{image}}$ substitute, respectively, the surface charge and current of densities $\underline{\rho}_s$ and $\underline{\mathbf{J}}_s$ on the conducting plane in the original system, and the relationship $\underline{J}_s = c\underline{\rho}_s$ between these densities, Eq. (11.24), results in $\underline{I}_{\text{image}} = c\underline{Q}'_{\text{image}}$ in the equivalent system, where c is the intrinsic phase velocity of the line dielectric. Additionally, the image of the charge is negative (with the opposite polarity to the original charge), so we have

image theory for a current above a conducting plane

$$\boxed{\underline{I} = c\underline{Q}', \quad \underline{I}_{\text{image}} = c\underline{Q}'_{\text{image}}, \quad \text{and} \quad \underline{Q}'_{\text{image}} = -\underline{Q}' \quad \longrightarrow \quad \underline{I}_{\text{image}} = -\underline{I},} \tag{11.107}$$

that is, the image of the current $\underline{I}$ in the equivalent system with the ground plane removed is the current of the same complex intensity flowing in the opposite direction relative to the original current (electrically negative image). It flows along a virtual wire that is the mirror image of the original wire in the symmetry plane (former conducting plane). Note that, although derived for the current of a conductor of a transmission line with a TEM wave whose other conductor is a conducting ground plane, this image theory is valid and applicable for an arbitrary current (at low or high frequencies) flowing in parallel to a conducting plane. More precisely, the theory is exact only for a perfectly conducting (PEC) plane ($\sigma_c \to \infty$). In a later chapter, when analyzing antennas above ground planes, we shall present the image theory for arbitrary current distributions in the presence of a PEC plane, which are not necessarily associated with TEM waves and transmission lines, nor are parallel to the plane.

In the equivalent system in Fig. 11.9(b), which is a symmetrical thin two-wire line with conductor axes distance $2h$, the total magnetic field, $\underline{\mathbf{H}}$, is the superposition of the fields due to the original and image currents, $\underline{I}$ and $-\underline{I}$, respectively, along the two wires, with the latter field representing the contribution of the surface current ($\underline{\mathbf{J}}_s$) on the original conducting plane. At a point M defined by the coordinate x in Fig. 11.9(b), $\underline{\mathbf{H}}$ is entirely tangential to the plane and given by

$$\underline{\mathbf{H}} = \underline{\mathbf{H}}_{\text{original}} + \underline{\mathbf{H}}_{\text{image}} = 2\underline{H}_{\text{wire}} \cos\theta\, \hat{\mathbf{x}} = \frac{\underline{I}\cos\theta}{\pi R}\, \hat{\mathbf{x}} = \underline{\mathbf{H}}_{\text{tang}} \quad \left(\underline{H}_{\text{wire}} = \frac{\underline{I}}{2\pi R}\right). \tag{11.108}$$

Hence, back in the original system, the integration, Eq. (11.66), along the contour of the conducting plane in Fig. 11.9(a) yields

$$R_2' = \frac{1}{|\underline{I}|^2} \int_{\text{plane contour}} R_{s2} |\underline{\mathbf{H}}_{\text{tang}}|^2 \, dx = \frac{R_{s2}}{\pi^2} \int_{x=-\infty}^{\infty} \cos^2\theta \underbrace{\frac{dx}{R^2}}_{d\theta/h}$$

$$= \frac{R_{s2}}{\pi^2 h} \underbrace{\int_{\theta=-\pi/2}^{\pi/2} \cos^2\theta \, d\theta}_{\pi/2} = \frac{R_{s2}}{2\pi h}, \tag{11.109}$$

where the use is made of Eq. (1.55), reading $dx/R^2 = d\theta/h$ for the notation in Fig. 11.9, to transform the integral in x into one in θ, which is much simpler to compute [see Eq. (6.95) for the actual integration].

Finally, the total resistance per unit length of the wire-plane line is

$$R' = R_1' + R_2' = \frac{1}{2\pi}\left(\frac{R_{s1}}{a} + \frac{R_{s2}}{h}\right). \tag{11.110}$$

We note that $R_2' \ll R_1'$ (because $h \gg a$), and therefore the losses in the conducting ground plane can usually be neglected in the evaluations of the attenuation of this and similar types of transmission lines.

Example 11.10 Circuit Parameters of a Microstrip Line Neglecting Fringing Effects

A TEM wave of frequency f propagates along a microstrip transmission line, Fig. 2.20, with small losses. The width of the conducting strip and thickness of the dielectric substrate are w and h, respectively, where $h \ll w$, so that fringing effects can be neglected. The conductivity and permeability of the strip and the ground plane are σ_c and μ_0, and the relative permittivity, permeability, and conductivity of the substrate are ε_r, μ_0, and σ_d, respectively. Under these circumstances, determine the primary and secondary circuit parameters of the line.

Solution The capacitance per unit length of the line, with the fringing effects neglected, is that in Eq. (2.135), from which the line p.u.l. inductance and conductance are then computed using Eqs. (11.41) and (3.171), respectively, so all three parameters are given by

$$\boxed{C' = \varepsilon_r \varepsilon_0 \frac{w}{h}, \quad L' = \mu_0 \frac{h}{w}, \quad G' = \sigma_d \frac{w}{h}.} \tag{11.111}$$

C', L', G' – microstrip line, fringing neglected

Note that G' can alternatively be obtained directly as the leakage conductance per unit length of the line from the analysis of the steady current field in the imperfect dielectric substrate, as is done in Example 3.14 for a similar transmission line (with a two-layer dielectric). In fact, the expression for G' in Eqs. (11.111) is a special case of the result for the two-layer structure in Eq. (3.189).

With no fringing effects taken into account, the magnetic field of the microstrip line is assumed to be uniform and localized in the dielectric below the strip only. To find this field, we apply the generalized Ampère's law in integral form, Eq. (5.49), to a rectangular contour C completely enclosing the strip, as shown in Fig. 11.10. The line integral of $\underline{\mathbf{H}}$ along the contour equals $\underline{H}w$ (under the no-fringing assumption, the field is nonzero only in the $w \times h$ large region of the substrate cross section) and the enclosed current is $\underline{I}$, and hence

$$\underline{H} = \frac{\underline{I}}{w}, \tag{11.112}$$

i.e., $\underline{\mathbf{H}}$ in the substrate is as if due to two infinitely wide planar current sheets with uniform surface current densities $\underline{J}_s = \underline{I}/w$ [see Eqs. (4.47) and (4.72)] flowing in opposite directions. Of course, $\underline{\mathbf{H}} = \underline{\mathbf{H}}_{\text{tang}}$ (entirely tangential) on both the lower surface of the strip and part of the upper surface of the ground plane where the field $\underline{\mathbf{H}}$ and current $\underline{\mathbf{J}}_s$ exist. Denoting the contour lines of these surfaces by l_{c1} and l_{c2}, where they are both w long and the complex

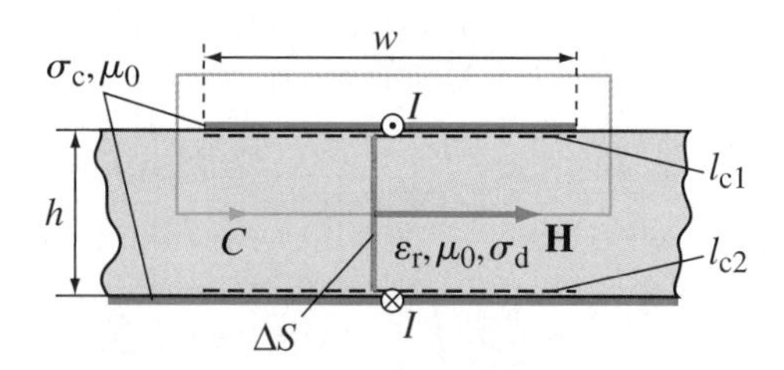

Figure 11.10 Evaluation of circuit parameters for TEM waves of a microstrip line, neglecting fringing effects; for Example 11.10.

magnetic field intensities along them are both given by Eq. (11.112), Eq. (11.66) tells us that the high-frequency resistance per unit length of the line is

R′ – microstrip line, fringing neglected

$$R' = \frac{1}{|\underline{I}|^2}\left(\int_{l_{c1}} R_s|\underline{\mathbf{H}}_{\text{tang1}}|^2\, dl + \int_{l_{c2}} R_s|\underline{\mathbf{H}}_{\text{tang2}}|^2\, dl\right) = \frac{R_s}{|\underline{I}|^2}\left(\frac{|\underline{I}|}{w}\right)^2 2\underbrace{\int_{l_{c1}} dl}_{w} = \frac{2R_s}{w}, \tag{11.113}$$

where the surface resistance R_s is that in Eq. (11.70).

Note that, using the magnetic field of the line in Eq. (11.112), we can now alternatively find the line inductance L' by its definition, as the magnetic flux per unit length of the line, $\underline{\Phi}'$, divided by the line current. The flux is computed through the vertical flat surface $\Delta S = h\Delta z$ spanned between the strip and the ground plane (Fig. 11.10), where Δz is the length of the surface (along the line), and since $\underline{\mathbf{H}}$ is uniform and perpendicular to that surface, we have [see also Eqs. (11.38) and (11.39)]

$$L' = \frac{\underline{\Phi}'}{\underline{I}} = \frac{\Delta\underline{\Phi}}{\underline{I}\Delta z} = \frac{\mu_0 \underline{H} h\Delta z}{\underline{I}\Delta z} = \frac{\mu_0(\underline{I}/w)h}{\underline{I}} = \mu_0\frac{h}{w}, \tag{11.114}$$

which, of course, is the same result as in Eqs. (11.111).

Combining Eqs. (11.37) and (11.111), the characteristic impedance of the line amounts to

$$Z_0 = \frac{\sqrt{\varepsilon_r\varepsilon_0\mu_0}}{C'} = \frac{\eta_0}{\sqrt{\varepsilon_r}}\frac{h}{w}, \tag{11.115}$$

where $\eta_0 = \sqrt{\mu_0/\varepsilon_0}$ is the intrinsic impedance of free space, Eq. (9.23). With this expression for Z_0 and that for R' in Eq. (11.113) substituted into Eq. (11.77), the attenuation coefficient for the losses in the line conductors comes out to be

$$\alpha_c = \frac{R'}{2Z_0} = \frac{R_s}{wZ_0} = \frac{R_s\sqrt{\varepsilon_r}}{\eta_0 h}. \tag{11.116}$$

The line phase coefficient (β), phase velocity (v_p), wavelength (λ_z), and attenuation coefficient for the dielectric (α_d) are given in Eqs. (11.16), (11.17), and (11.80).

Finally, note that the evaluation of the circuit parameters of a microstrip line with an arbitrary ratio w/h, so with taking into account the fringing effects, will be presented in a later section. For such a line, the electric and magnetic fields extend to substantial amounts also into the air region above the dielectric substrate, so that the dielectric of the line is actually inhomogeneous, and, contrary to the analysis in this example, the microstrip line cannot be treated as a transmission line with a homogeneous dielectric.

Example 11.11 Strip Line Neglecting Fringing

Repeat the previous example but for a strip line, Fig. 2.21.

Solution Without taking into account the fringing effects ($h \ll w$), the p.u.l. capacitance of the strip line is given in Eq. (2.137), so $C' = 2\varepsilon_r\varepsilon_0 w/h$, and Eqs. (11.41) and (3.171) then result, respectively, in the following expressions for the inductance and conductance p.u.l. of the line: $L' = \mu_0 h/(2w)$ and $G' = 2\sigma_d w/h$.

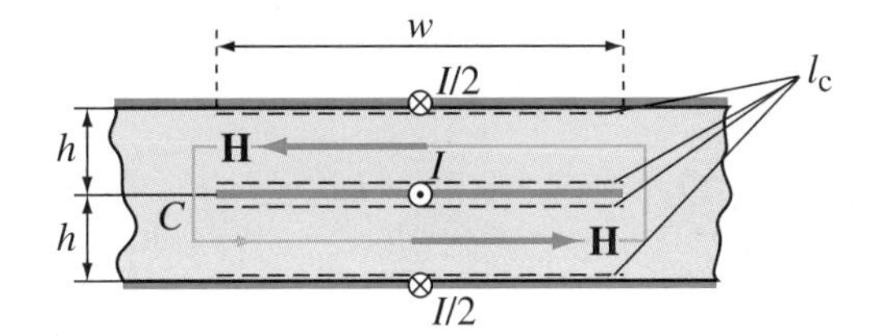

Figure 11.11 Evaluation of the high-frequency resistance per unit length of a strip line, neglecting fringing effects; for Example 11.11.

Applying the generalized Ampère's law to a rectangular contour C around the strip, shown in Fig. 11.11, we get, neglecting fringing, $\underline{H} = \underline{I}/(2w)$ for the magnetic field of the line in the two regions between the strip and the two conducting planes. Integrating the square of this field times the surface resistance R_s of the line conductors, as in Eq. (11.113), along the contour lines of the upper and lower surfaces of the strip and the corresponding two parts, w long, of the inner surfaces of the two conducting planes (Fig. 11.11) gives the high-frequency resistance per unit length of the line as follows:

$$R' = \frac{1}{|\underline{I}|^2} \int_{l_c} R_s |\underline{H}|^2 \, dl = \frac{1}{|\underline{I}|^2} 4 \int_w R_s \left(\frac{|\underline{I}|}{2w} \right)^2 dl = \frac{R_s}{w}. \tag{11.117}$$

R′ – strip line, fringing neglected

With the similar computations as in Eqs. (11.115) and (11.116), the characteristic impedance of the line and attenuation coefficient for its conductors are

$$Z_0 = \frac{\eta_0}{\sqrt{\varepsilon_r}} \frac{h}{2w}, \quad \alpha_c = \frac{R_s}{2wZ_0} = \frac{R_s \sqrt{\varepsilon_r}}{\eta_0 h}, \tag{11.118}$$

so α_c turns out to be the same as for the microstrip line, in Fig. 11.10. The remaining circuit parameters of the line (β, v_p, λ_z, and α_d) are determined as in the previous example.

A strip line with an arbitrary w/h will be discussed in a later section, taking into account the fringing effects.

Problems: 11.3–11.11; *Conceptual Questions* (on Companion Website): 11.28–11.30; *MATLAB Exercises* (on Companion Website).

11.9 TRANSMISSION LINES WITH INHOMOGENEOUS DIELECTRICS

Consider a two-conductor transmission line with an inhomogeneous dielectric, like the one the cross section of which is shown in Fig. 11.12(a). The inhomogeneity of the dielectric is arbitrary – in a transversal plane; however, the dielectric properties do not change in the longitudinal (axial) direction (the line is uniform). For such a line, the wave propagating through the dielectric is a hybrid wave, with both $\underline{\mathbf{E}}_z \neq 0$ and $\underline{\mathbf{H}}_z \neq 0$ in Eqs. (11.7). However, as in the case of transmission lines with small losses and homogeneous dielectrics (Fig. 11.6), these components are rather small in comparison with the corresponding transverse field components, and we refer to the wave as a quasi-TEM wave. Therefore, an approximate analysis treating quasi-TEM waves as pure TEM waves, and applying the theory of TEM waves from Sections 11.1–11.8, with certain modifications that will be described here, is sufficiently accurate for most practical applications.

In particular, primary circuit parameters of the line (C', L', R', and G') are found as follows. The capacitance C' in Eq. (11.36) is determined from a 2-D electrostatic analysis in the cross section of the line, in Fig. 11.12(a), taking into account

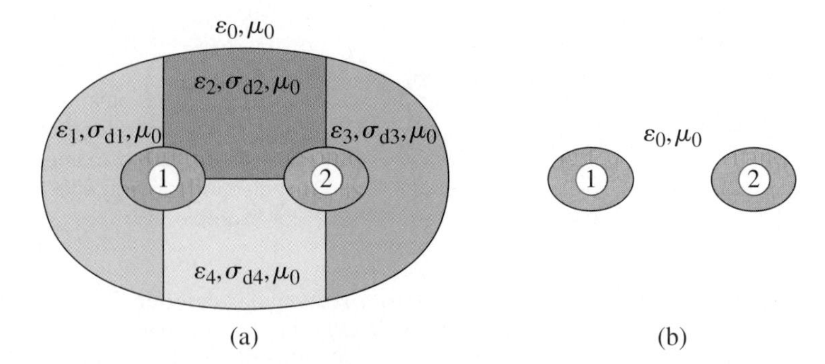

Figure 11.12 Cross section of (a) an arbitrary two-conductor transmission line with an inhomogeneous dielectric, and (b) the same line if air-filled.

the inhomogeneity of the dielectric (C' of transmission lines with different types of dielectric inhomogeneity is evaluated in Section 2.14). The electric field ($\underline{\mathbf{E}}$) from the same electrostatic analysis can then be used in Eq. (11.68) to find the conductance G', provided that the dielectric in Fig. 11.12(a) exhibits the same form of inhomogeneity in terms of its conductivity (σ_d) and permittivity (ε). Alternatively, G' can be obtained directly as the leakage conductance per unit length of the line from the analysis of the steady current field in Fig. 11.12(a) (G' of transmission lines with inhomogeneous imperfect dielectrics is evaluated in Section 3.12).

As the distribution of the magnetic field ($\underline{\mathbf{H}}$) in the line cross section is the same as under static (or quasistatic) conditions, the electric field distribution does not influence $\underline{\mathbf{H}}$ [the displacement current term is zero in Eq. (8.7)]. This is why the permittivity ε of the dielectric in Fig. 11.12(a) does not influence the inductance L' per unit length of the line. Provided furthermore that $\mu = \mu_0$ in Fig. 11.12(a), i.e., that the dielectric is nonmagnetic (which is true in most practical situations), this means that L' is the same as if the dielectric in Fig. 11.12(a) were homogeneous ($\varepsilon =$ const in the entire cross section of the dielectric). Ultimately, it is the same as the inductance per unit length, L_0', of the same line if air-filled ($\varepsilon = \varepsilon_0$). Denoting by C_0' the capacitance per unit length of this new line, in Fig. 11.12(b), Eq. (11.41) tells us that L_0' and C_0' are related as $L_0'C_0' = \varepsilon_0\mu_0$. Hence, the inductance L' we seek can simply be obtained as

inductance p.u.l. of a transmission line with an inhomogeneous nonmagnetic dielectric

$$\boxed{L' = L_0' = \frac{\varepsilon_0\mu_0}{C_0'},} \tag{11.119}$$

where C_0' is calculated from a 2-D electrostatic analysis in Fig. 11.12(b).

By the same token, the magnetic field $\underline{\mathbf{H}}_{\text{tang}}$ used in Eq. (11.66) to compute the resistance R' per unit length of the line in Fig. 11.12(a) does not depend on the inhomogeneity of the dielectric (if it is nonmagnetic). Consequently, R' equals the resistance per unit length of the air-filled line in Fig. 11.12(b), $R' = R_0'$.

Secondary circuit parameters of the line (Z_0, β, v_p, λ_z, and α) are then found from Eqs. (11.42), (11.43), (11.75), (11.77), and (11.79).

Another useful parameter of transmission lines with inhomogeneous dielectrics, Fig. 11.12(a), is the so-called effective relative permittivity of the line, $\varepsilon_{\text{reff}}$, defined as

effective relative permittivity of a line with an inhomogeneous dielectric

$$\boxed{\varepsilon_{\text{reff}} = \frac{C'}{C_0'}.} \tag{11.120}$$

This is a dimensionless quantity that can be interpreted as the relative permittivity of an equivalent homogeneous dielectric material which if occupying the space between the conductors in Fig. 11.12(b) would give the same capacitance per unit length, C', as the inhomogeneous dielectric of the actual line, in Fig. 11.12(a). Note that for lines with homogeneous dielectrics (ε = const), as in Fig. 11.3, $\varepsilon_{\rm reff} = \varepsilon_{\rm r}$, $\varepsilon_{\rm r} = \varepsilon/\varepsilon_0$ being the relative permittivity of the (actual) dielectric. With this interpretation of $\varepsilon_{\rm reff}$, or from

$$L'C' = L'_0 C'_0 \varepsilon_{\rm reff} = \varepsilon_{\rm reff}\varepsilon_0\mu_0, \tag{11.121}$$

β, $v_{\rm p}$, and λ_z of the line in Fig. 11.12(a) can alternatively be expressed as

$$\boxed{\beta = \frac{\omega}{c_0}\sqrt{\varepsilon_{\rm reff}}, \quad v_{\rm p} = \frac{c_0}{\sqrt{\varepsilon_{\rm reff}}}, \quad \lambda_z = \frac{\lambda_0}{\sqrt{\varepsilon_{\rm reff}}},} \tag{11.122}$$

β, $v_{\rm p}$, λ_z – via the effective relative permittivity

where c_0 is the wave velocity (speed of light) in free space, Eq. (9.19), and $\lambda_0 = c_0/f$, Eq. (9.67), the free-space wavelength at the frequency f of the line ($\omega = 2\pi f$).

Example 11.12 Microstrip Line with a Two-Layer Dielectric Substrate

Fig. 11.13 shows a cross section of a microstrip transmission line whose substrate consists of two dielectric layers. The thicknesses of the layers are $h_1 = 0.5$ mm and $h_2 = 0.25$ mm, and the strip width is $w = 8$ mm. The relative permittivities of the layers are $\varepsilon_{\rm r1} = 4$ and $\varepsilon_{\rm r2} = 8$, and the conductivities are $\sigma_{\rm d1} = 10^{-12}$ S/m and $\sigma_{\rm d2} = 5 \times 10^{-12}$ S/m, while both layers are nonmagnetic. The strip and ground plane are made of copper. If the frequency of the propagating quasi-TEM wave on the line is $f = 1$ GHz, compute the primary and secondary circuit parameters of the line.

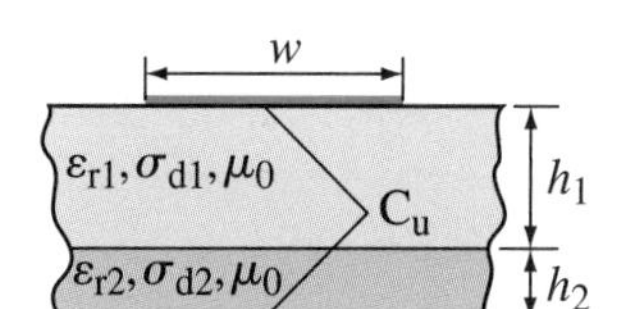

Figure 11.13 Cross section of a microstrip line with a two-layer dielectric substrate and small losses; for Example 11.12.

Solution We first note that $\sigma_{\rm c}/(\omega\varepsilon_0) = 1.043 \times 10^9 \gg 1$ ($\sigma_{\rm c} = 58$ MS/m – for copper), $\sigma_{\rm d1}/(\omega\varepsilon_{\rm r1}\varepsilon_0) = 4.5 \times 10^{-12} \ll 1$, and $\sigma_{\rm d2}/(\omega\varepsilon_{\rm r2}\varepsilon_0) = 1.12 \times 10^{-11} \ll 1$ ($\omega = 2\pi f$), namely, that conditions in Eqs. (11.61) are satisfied, which means that the line in Fig. 11.13 can be analyzed as having small losses. In addition, since $h_1 + h_2 \ll w$, fringing effects in the structure can be neglected. The capacitance of the part of the line with length l is given in Eq. (2.150) where $S = wl$ [see also Eq. (2.134)], and dividing this expression for C by l results in the capacitance per unit length (C') of the line. The p.u.l. line conductance (G') is that in Eq. (3.189). The dielectric layers in Fig. 11.13 being nonmagnetic, both the inductance and high-frequency resistance per unit length of the line (L' and R') are the same as those of the same line if air-filled (L'_0 and R'_0) or if with a homogenous nonmagnetic substrate, Fig. 11.10, so that the respective expressions in Eqs. (11.111) and (11.113) apply. Hence, all primary per-unit-length circuit parameters of the line come out to be

$$C' = \frac{\varepsilon_{\rm r1}\varepsilon_{\rm r2}\varepsilon_0 w}{\varepsilon_{\rm r2}h_1 + \varepsilon_{\rm r1}h_2} = 453~\text{pF/m}, \quad G' = \frac{\sigma_{\rm d1}\sigma_{\rm d2}w}{\sigma_{\rm d2}h_1 + \sigma_{\rm d1}h_2} = 14.5~\text{pS/m},$$
$$L' = L'_0 = \mu_0\frac{h_1 + h_2}{w} = 118~\text{nH/m}, \quad R' = R'_0 = \frac{2R_{\rm s}}{w} = 2.062~\Omega/\text{m}, \tag{11.123}$$

where the surface resistance $R_{\rm s}$ is computed using Eq. (10.80).

For the secondary circuit parameters of the line, we use Eqs. (11.42) and (11.43) to determine the characteristic impedance and phase coefficient of the line,

$$Z_0 = \sqrt{\frac{L'}{C'}} = 16.14~\Omega, \quad \beta = \omega\sqrt{L'C'} = 45.94~\text{rad/m}, \tag{11.124}$$

and then Eqs. (11.17), (11.77), (11.79), and (9.89) yield the following for the phase velocity, wavelength, and attenuation coefficients for line conductors and dielectric: $v_{\rm p} = \omega/\beta = 1.37 \times 10^8$ m/s, $\lambda_z = v_{\rm p}/f = 13.7$ cm, $\alpha_{\rm c} = R'/(2Z_0) = 0.064$ Np/m $= 0.56$ dB/m, and $\alpha_{\rm d} = G'Z_0/2 = 1.17 \times 10^{-10}$ Np/m $= 10^{-9}$ dB/m, respectively.

Example 11.13 Effective Permittivity of a Coaxial Cable with Inhomogeneous Dielectric

Consider the coaxial cable with a continuously inhomogeneous dielectric of permittivity $\varepsilon(\phi)$ from Example 2.21, and assume that the conductivity of the cable conductors, which exhibit small losses, is σ_c, that there are no losses in the dielectric, and that all materials are nonmagnetic. A quasi-TEM wave of frequency f propagates along the cable. (a) Find the cable characteristic impedance and attenuation coefficient. (b) What are the effective relative permittivity and phase velocity of the cable?

Solution

(a) The capacitance per unit length of the cable is given in Eq. (2.178), inductance in Eq. (7.12), and resistance in Eq. (11.70), while the conductance is zero (perfect dielectric). Using Eqs. (11.42) and (11.77), the cable characteristic impedance and attenuation coefficient are then found to be

$$Z_0 = \sqrt{\frac{L'}{C'}} = \frac{\eta_0}{2\sqrt{3}\pi} \ln\frac{b}{a}, \quad \alpha = \alpha_c = \frac{R'}{2Z_0} = \frac{\sqrt{3}R_s}{2\eta_0}\frac{1/a + 1/b}{\ln(b/a)} \quad (\alpha_d = 0), \quad (11.125)$$

with R_s and η_0 standing for the surface resistance of the cable conductors, Eq. (11.70), and intrinsic impedance of free space, Eq. (9.23), respectively.

(b) Since the p.u.l. capacitance of the cable in Fig. 2.30(a) if air-filled is $C_0' = 2\pi\varepsilon_0/\ln(b/a)$ [Eq. (2.123) with $\varepsilon = \varepsilon_0$], the effective relative permittivity of the cable and phase velocity of the propagating wave, from Eqs. (11.120) and (11.122), are

$$\varepsilon_{reff} = \frac{C'}{C_0'} = 3, \quad v_p = \frac{c_0}{\sqrt{\varepsilon_{reff}}} = \frac{c_0}{\sqrt{3}}, \quad (11.126)$$

where c_0 is the free-space wave velocity, Eq. (9.19). Note that ε_{reff} turns out to equal the average value of the relative permittivity of the cable dielectric, $\varepsilon(\phi)/\varepsilon_0 = 3 + \sin\phi$, over the full range of the angle ϕ, $0 \le \phi \le 2\pi$, in Fig. 2.30(a). Note also that the characteristic impedance in Eqs. (11.125) amounts to $1/\sqrt{\varepsilon_{reff}} = 1/\sqrt{3}$ of Z_0 for the air-filled cable [Eq. (11.48) with $\varepsilon_r = 1$].

Example 11.14 Effective Permittivity and Attenuation of a Coated Two-Wire Line

For the thin two-wire line with dielectrically coated conductors in air in Fig. 2.31, assume that $d = 40$ mm, $a = 1$ mm, and $\varepsilon_r = 4$ (relative permittivity of dielectric coatings), as well as that the coatings are nonmagnetic and wires made of copper. Under these circumstances, calculate the effective relative permittivity of the line and attenuation coefficient for a TEM wave of frequency $f = 100$ MHz.

Solution Using Eq. (2.183), the capacitance per unit length of the line in Fig. 2.31 is $C' = 8.78$ pF/m, and that of the same line with no dielectric coatings is $C_0' = 7.54$ pF/m, from Eq. (2.141), so that the effective relative permittivity of the former line equals $\varepsilon_{reff} = C'/C_0' = 1.164$. We note that ε_{reff} is quite close to unity, as expected, since the coatings of the wires are thin, relative to the distance between wire axes. The p.u.l. line inductance and high-frequency resistance do not depend on the inhomogeneity of the dielectric (because it is nonmagnetic), and hence Eqs. (7.11) and (11.103) can be employed, giving $L' = 1.48$ μH/m and $R' = 0.83$ Ω/m, respectively. Finally, as the coated two-wire line is situated in air (perfect insulator), there cannot be leakage current between the wires even if the coatings are made of an imperfect dielectric,[11] and the attenuation along the line is thus solely due to conductor losses, with Eqs. (11.77), (11.42), and (9.89) telling us that the line attenuation coefficient is $\alpha = R'\sqrt{C'/L'}/2 = 0.001$ Np/m $= 0.0088$ dB/m.

Problems: 11.12–11.17; *Conceptual Questions* (on Companion Website): 11.31 and 11.32; *MATLAB Exercises* (on Companion Website).

[11] Note that for the same structure but with lossy dielectric coatings and lossy ambient medium, the conductance per unit length of the line is given in Eq. (3.186).

11.10 MULTILAYER PRINTED CIRCUIT BOARD

Fig. 11.14 shows a typical multilayer printed circuit board, which is widely used in digital electronics (e.g., in computers). Its top surface consists of active and passive electronic components, such as integrated circuits (chips) and discrete circuit elements, interconnected by strip conductors, called traces. Traces are printed on a layer of dielectric (dielectric substrate), beneath which is a metallic foil extending throughout the entire board, Fig. 11.14(a). A similar configuration may exist on the bottom of the board. There are then one or more layers of conductors (traces) sandwiched in dielectric layers between pairs of metallic foils (electronic components can also be placed in these layers). The foils serve as ground planes for high-speed signals along traces. They can also be used to distribute the power supply to integrated circuits. Finally, the components, traces, and foils at different levels in the board are (selectively) connected together by metalized holes through the board, called vias, as indicated in Fig. 11.14(b). Note that a typical relative permittivity of the dielectric in Fig. 11.14 is $\varepsilon_\mathrm{r} = 3.5 - 4.5$ (most frequently, various fiberglass materials). Traces, foils, and vias are normally made out of copper ($\sigma_\mathrm{c} = 58$ MS/m). For high-speed signals, the interconnects formed by traces and foils in individual layers of the board have to be considered as transmission lines, and not just as short-circuiting conductors. In this section, we provide the basis for circuit analysis of transmission lines in a typical circuit board.

Observing the interconnects in different layers in Fig. 11.14, we identify two types of two-conductor lines making up the structure. Namely, each trace on the top of the board, above which is air, and the highest foil represent a microstrip line, Fig. 2.20, and similarly for traces on the bottom of the board (if they exist). Note that electric connections within the integrated circuits themselves in Fig. 11.14 (so-called on-chip interconnects) are also realized as microstrip lines.[12] On the other side,

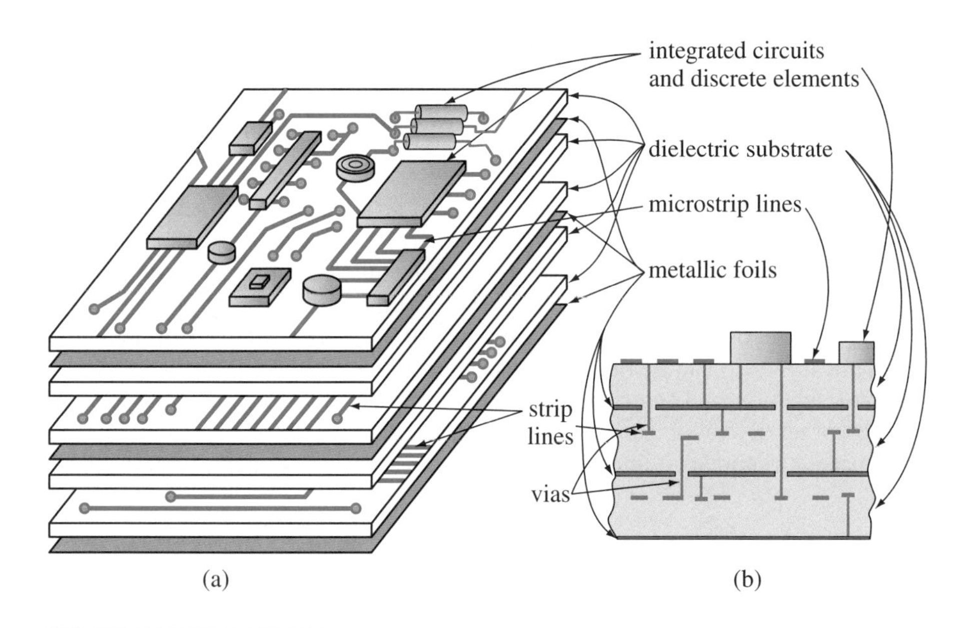

Figure 11.14 Sketch of a typical multilayer printed circuit board: (a) three-dimensional view of the structure and (b) detail of its cross section.

[12]Substrates in integrated circuits are commonly made from silicon (Si). However, silicon exhibits large losses at high frequencies, resulting in a large attenuation coefficient in the dielectric (α_d), Eq. (11.79), for microstrip lines in the circuit. This is why gallium arsenide (GaAs), which is much less lossy, is frequently used as a substitute for Si for very fast integrated circuits.

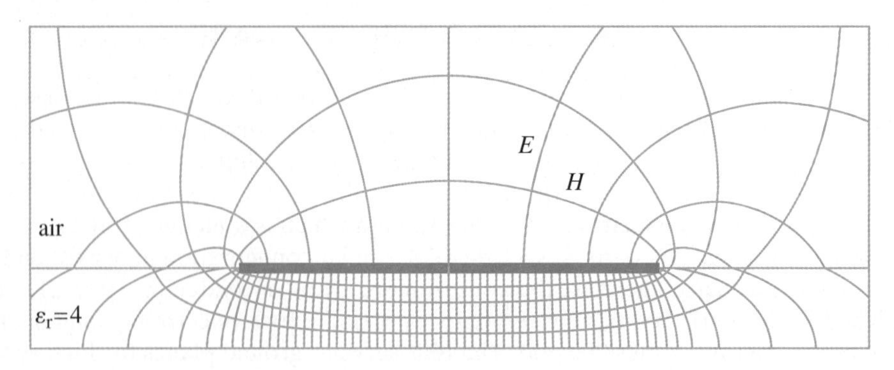

Figure 11.15 Electric (E) and magnetic (H) field lines in a cross section of a microstrip line with the strip width to substrate height ratio $w/h = 5.4$ and substrate relative permittivity $\varepsilon_r = 4$; field pattern plots are obtained by numerical analysis based on a method of moments (see Section 1.20).

each trace between adjacent foils, which is completely surrounded by the dielectric, constitutes, together with the foils, a strip line, Fig. 2.21.[13]

In Examples 11.10 and 11.11, which, in turn, refer to Examples 2.13 and 2.14, the primary and secondary circuit parameters of a microstrip and strip line, respectively, are found – neglecting the fringing effects. These values are, thus, accurate only for $h \ll w$, h being the distance of the strip conductor (trace) from the foil (ground plane) in Fig. 2.20 or from each of the planes in Fig. 2.21, and w the width of the strip in both cases. On the other side, the lines with $w \ll h$ can be analyzed replacing the strip by a thin wire of equivalent radius (see Example 11.9). In practice, however, w and h are of the same order of magnitude, a typical value for their ratio being $w/h = 1 - 3$ for microstrip lines and $w/h \approx 1$ for strip lines. For such values of w/h, and, generally speaking, for an arbitrary ratio w/h, the actual electrostatic field distributions in the lines are quite different from those in Figs. 2.20 and 2.21. In particular, there is a considerable fringing field outside the region below the strip in Fig. 2.20, and the field in this region close to the strip edges is not uniform (edge effects), as illustrated in Fig. 11.15, and similarly for the line in Fig. 2.21. The magnetic fields in the two lines are related to the corresponding electric fields through Eqs. (11.18), and hence the same discrepancy of the actual magnetic field distributions from those in Figs. 11.10 and 11.11 exists (see Fig. 11.15). Consequently, the results for the capacitance per unit length, C', of the two lines, in Eqs. (2.135) and (2.137), as well as for their other parameters, in Eqs. (11.111)–(11.118), are not accurate enough for practical microstrip and strip lines, and for an arbitrary ratio w/h.

Accurate electrostatic analysis of arbitrary transmission lines can be performed by use of numerical electrostatic techniques. Based on such analysis, results for all primary and secondary circuit parameters[14] of the lines in Figs. 2.20 and 2.21 (and of arbitrary lines) can be obtained taking into account the exact geometry and material composition and realistic field distributions of the line. Alternatively, we present here a set of available empirical closed-form formulas for the circuit parameters of microstrip and strip lines. These useful formulas (obtained as curve-fitting

[13]Note that, in general, strip lines, being shielded between metallic foils, are much less liable to radiate and are much less susceptible to interference with external signals than microstrip lines, which are "open" to the upper half-space.

[14]Note that even the resistance R' in Eq. (11.66) of a transmission line with an inhomogeneous dielectric can be obtained based on electrostatics, starting from the electrostatic field distribution of the line when air-filled, as in Fig. 11.12(b), and using the relation between electric and magnetic fields in Eqs. (11.18) with $\varepsilon = \varepsilon_0$ and $\mu = \mu_0$.

approximations of experimental and/or analytical data) agree very well with the numerical solution (or measurement) for all practical values of the ratio w/h.

Since the electric field in a microstrip line, Fig. 11.15, is only partly in the dielectric substrate, of relative permittivity ε_r, and the rest is in air above it, the dielectric of the line is actually inhomogeneous, and must be treated as such for an accurate analysis. Note that in the approximate analysis for $h \ll w$ in Fig. 2.20, the dielectric can be treated as homogeneous as the electric field, neglecting the fringing effects, is assumed to be localized in the dielectric below the strip only. Therefore, the effective relative permittivity $\varepsilon_{\text{reff}}$ of the line with an arbitrary w/h, Eq. (11.120), is between unity (for air) and ε_r. It can be computed from the empirical formula given by

$$\varepsilon_{\text{reff}} = \frac{\varepsilon_r + 1}{2} + \frac{\varepsilon_r - 1}{2}\left[\left(1 + 12\frac{h}{w}\right)^{-1/2} + p\right], \tag{11.127}$$

$\varepsilon_{\text{reff}}$ – *microstrip line*

where $p = 0.04(1 - w/h)^2$ if $w/h < 1$, and $p = 0$ otherwise. The characteristic impedance of the line, Z_0, is then found as

$$Z_0 = \frac{\eta_0}{2\pi\sqrt{\varepsilon_{\text{reff}}}} \ln\left(\frac{8h}{w} + \frac{w}{4h}\right) \quad \text{for} \quad \frac{w}{h} \leq 1,$$

$$Z_0 = \frac{\eta_0}{\sqrt{\varepsilon_{\text{reff}}}}\left[\frac{w}{h} + 1.393 + 0.667\ln\left(\frac{w}{h} + 1.444\right)\right]^{-1} \quad \text{for} \quad \frac{w}{h} > 1, \tag{11.128}$$

Z_0 – *microstrip line – analysis*

where η_0 is the intrinsic impedance of free space, Eq. (9.23). For design (synthesis) purposes, namely, to find w/h for a desired Z_0 of the line and given ε_r of the substrate dielectric, the following formulas are used:

$$\frac{w}{h} = 8\left(\mathrm{e}^{A} - 2\,\mathrm{e}^{-A}\right)^{-1} \quad \text{for} \quad \frac{w}{h} \leq 2, \quad \frac{w}{h} = \frac{\varepsilon_r - 1}{\pi\varepsilon_r}$$

$$\times\left[\ln(B-1) + 0.39 - \frac{0.61}{\varepsilon_r}\right] + \frac{2}{\pi}\left[B - 1 - \ln(2B-1)\right] \quad \text{for} \quad \frac{w}{h} > 2,$$

$$A = \pi\sqrt{2(\varepsilon_r + 1)}\,\frac{Z_0}{\eta_0} + \frac{\varepsilon_r - 1}{\varepsilon_r + 1}\left(0.23 + \frac{0.11}{\varepsilon_r}\right), \quad B = \frac{\pi}{2\sqrt{\varepsilon_r}}\frac{\eta_0}{Z_0}. \tag{11.129}$$

microstrip line – synthesis

The phase coefficient, β, phase velocity, v_p, and wavelength, λ_z, along the microstrip line are found from Eqs. (11.122), using $\varepsilon_{\text{reff}}$ of the line, Eq. (11.127), and the corresponding free-space values of these propagation parameters. Note that v_p (and analogously for β and λ_z) is between $c_0/\sqrt{\varepsilon_r}$ (wave velocity in a homogeneous dielectric of relative permittivity ε_r) and c_0 (wave velocity in free space), and depends on the ratio w/h of the line. The larger w/h the less field in air above the substrate in Fig. 11.15 and the smaller v_p (closer to $c_0/\sqrt{\varepsilon_r}$), which can also be seen from Eq. (11.127).

The attenuation coefficient representing the losses in conductors, α_c, of the line in Fig. 11.15 is computed approximately using the expression in Eq. (11.116) for the same line with the fringing effects neglected,

$$\alpha_c = \frac{R_s}{wZ_0} \tag{11.130}$$

α_c – *microstrip line*

[R_s is the surface resistance of the conductors, Eq. (11.65)], but with the values for the characteristic impedance Z_0 that take into account fringing, from Eqs. (11.128). The attenuation coefficient for the losses in the dielectric, α_d, of the line is found from Eq. (11.80),

α_d – microstrip line

$$\alpha_d = q\frac{\beta}{2}\tan\delta_d = \frac{\pi q f \tan\delta_d\sqrt{\varepsilon_{\text{reff}}}}{c_0}, \quad q = \frac{\varepsilon_{\text{reff}}-1}{\varepsilon_{\text{reff}}}\frac{\varepsilon_r}{\varepsilon_r-1}, \tag{11.131}$$

where q is the empirical effective dielectric filling factor of the line, taking approximately into account the inhomogeneity of its dielectric, $\tan\delta_d$ is the loss tangent of the substrate, and f is the frequency of the wave.

The dielectric of a strip line, Fig. 2.21, is homogeneous, and hence $\varepsilon_{\text{reff}} = \varepsilon_r$. The empirical formulas for the characteristic impedance of the line read

Z_0 – strip line – analysis

$$Z_0 = \frac{\eta_0}{4\sqrt{\varepsilon_r}\,[w/(2h)+0.441-s]}, \quad s = \left(0.35 - \frac{w}{2h}\right)^2 \quad \text{for} \quad \frac{w}{h} < 0.7,$$
$$s = 0 \quad \text{for} \quad \frac{w}{h} \geq 0.7. \tag{11.132}$$

To design a line with a certain Z_0 for a given ε_r, these equations can, unlike Eqs. (11.127) and (11.128) for the microstrip line, be easily solved for w/h (assuming that Z_0 and ε_r are known), as will be shown in one of the examples. Note, however, that the formulas in Eq. (11.132), as well as those in Eqs. (11.127) and (11.128), can be modified to also take into account a nonzero thickness of the strip in Figs. 2.20 and 2.21.

For the strip line, standard expressions for β, v_p, λ_z, and α_d for lines with homogeneous dielectrics apply, given in Eqs. (11.16), (11.17), and (11.80).[15] Of course, the same values can also be obtained from Eqs. (11.122) and (11.131) – with $\varepsilon_{\text{reff}} = \varepsilon_r$ and $q = 1$. Similarly to Eq. (11.130), $\alpha_c = R_s/(2wZ_0)$, from Eq. (11.118).

In printed boards for high-speed digital circuits (Fig. 11.14), a large number of traces usually run parallel to each other (e.g., in computer data buses). The distance between adjacent traces is on the order of the strip width, w, so that the coupling between the traces is very strong and, in general, cannot be neglected. This coupling, in turn, is a cause of crosstalk between the traces and other undesirable effects in the circuit. Therefore, for a proper design and reliable operation of the circuit, the analysis of isolated microstrip and strip lines, whose circuit parameters are discussed in this section, needs to be generalized to include the coupling effects. In other words, each group of interconnects in a layer of the board consisting of several parallel microstrip or strip lines must be analyzed (and designed) as a network of coupled two-conductor transmission lines or, equivalently, as a multiconductor transmission line. The same coupling effects take place between interconnects in integrated circuits.

Example 11.15 Analysis of a Microstrip Line Including Fringing Effects

The width of the conducting strip of a microstrip line is $w = 2$ mm, the thickness of the dielectric substrate is $h = 1$ mm, and both the strip and ground plane are made out of copper. The relative permittivity, loss tangent, and permeability of the dielectric are $\varepsilon_r = 4$, $\tan\delta_d = 10^{-4}$, and $\mu = \mu_0$, respectively. Calculate (a) the effective relative permittivity, (b) the characteristic impedance, (c) the phase coefficient and velocity, and (d) the attenuation coefficient of this line, for a quasi-TEM wave of frequency $f = 3$ GHz on the line.

[15] Note that, in general, microstrip lines are "faster" than strip lines, i.e., for the same dielectric material, signals travel faster along a microstrip line than along a strip line ($c_0/\sqrt{\varepsilon_{\text{reff}}} > c_0/\sqrt{\varepsilon_r}$).

Solution Since $\sigma_c/(2\pi f\varepsilon_0) = 3.5 \times 10^8 \gg 1$ (for copper, $\sigma_c = 58$ MS/m) and $\tan\delta_d \ll 1$, low-loss conditions in Eqs. (11.61) and (9.125) are satisfied in this case. On the other hand, the condition $h \ll w$ is not met, and hence fringing effects cannot be neglected.

(a) In particular, Eq. (11.127) with $w/h = 2$ and $p = 0$ gives the following for the effective relative permittivity of the line: $\varepsilon_{\rm reff} = 3.07$.

(b) With the use of the second expression (for $w/h > 1$) in Eqs. (11.128) and the fact that $\eta_0 = 377\ \Omega$, the line characteristic impedance is $Z_0 = 51\ \Omega$. We note that this result differs very considerably from $(Z_0)_{\rm approx} = \eta_0 h/(\sqrt{\varepsilon_r}w) = 94.25\ \Omega$, Eq. (11.115), obtained neglecting fringing effects (the relative error in computation is 85%).

(c) From Eqs. (11.122), the phase coefficient and velocity of the line are $\beta = 110.09$ rad/m and $v_p = 1.71 \times 10^8$ m/s, respectively. Note that $(v_p)_{\rm approx} = c_0/\sqrt{\varepsilon_r} = 1.5 \times 10^8$ m/s ($c_0 = 3 \times 10^8$ m/s), with an assumption that the electromagnetic field of the line resides entirely in the dielectric substrate.

(d) The surface resistance of the line conductors (R_s) is that in Eq. (10.80), with which Eq. (11.130) yields the attenuation coefficient for the conductors of $\alpha_c = 0.14$ Np/m. Using Eqs. (11.131), the effective dielectric filling factor of the line (q) comes out to be $q = 0.9$, and the attenuation coefficient for the dielectric $\alpha_d = 0.005$ Np/m. These results for α_c and α_d are quite different from the respective results according to Eqs. (11.116) and (11.80), $(\alpha_c)_{\rm approx} = R_s\sqrt{\varepsilon_r}/(\eta_0 h) = 0.076$ Np/m and $(\alpha_d)_{\rm approx} = \pi f\sqrt{\varepsilon_r}\tan\delta_d/c_0 = 0.0063$ Np/m, which completely neglect fringing effects. Finally, having in mind Eqs. (11.75) and (9.89), the total attenuation coefficient of the line is $\alpha = \alpha_c + \alpha_d = 0.145$ Np/m $= 1.26$ dB/m.

Example 11.16 Primary Circuit Parameters of a Microstrip Line with Fringing

Find the primary per-unit-length circuit parameters of the microstrip line from the previous example.

Solution From Eq. (11.121), we can write for the p.u.l. inductance of the line

$$L' = \frac{\varepsilon_{\rm reff}\varepsilon_0\mu_0}{C'}. \tag{11.133}$$

Substituting this in Eq. (11.42), we obtain the p.u.l. capacitance via the effective relative permittivity and characteristic impedance of the line, which are given in the previous example, as follows:

$$Z_0 = \sqrt{\frac{L'}{C'}} = \sqrt{\frac{\varepsilon_{\rm reff}\varepsilon_0\mu_0/C'}{C'}} = \frac{\sqrt{\varepsilon_{\rm reff}}}{c_0 C'} \quad \longrightarrow \quad C' = \frac{\sqrt{\varepsilon_{\rm reff}}}{c_0 Z_0} = 114.5\ \text{pF/m}. \tag{11.134}$$

Knowing C', L' is now found using Eq. (11.133), and it turns out to be $L' = 297$ nH/m.

On the other side, solving for the high-frequency resistance and conductance p.u.l. of the line in Eqs. (11.77) and (11.79), respectively, and using the results for the attenuation coefficients for the line conductors and dielectric obtained in the previous example, we have

$$R' = 2Z_0\alpha_c = 14.28\ \Omega/\text{m}, \quad G' = 2Y_0\alpha_d = 0.2\ \text{mS/m}, \tag{11.135}$$

where Y_0 stands for the characteristic admittance of the line, Eq. (11.35).

Note that, apart from microstrip lines, the above equations can be used to find C', L', R', and G' from the known $\varepsilon_{\rm reff}$, Z_0, α_c, and α_d, so primary from secondary circuit parameters, of an arbitrary transmission line (with an inhomogeneous dielectric).

Example 11.17 Microstrip Line Design

Design a microstrip line that has a characteristic impedance of (a) $Z_0 = 75\ \Omega$ and (b) $Z_0 = 50\ \Omega$, respectively, for a given relative permittivity of the substrate dielectric, $\varepsilon_r = 4$.

In both cases, find the corresponding effective relative permittivity, and phase coefficient and velocity for the line if the operating frequency is $f = 3$ GHz.

Solution

(a) To find the strip width to substrate height ratio, w/h, that results in $Z_0 = 75\ \Omega$ for $\varepsilon_r = 4$, we use Eqs. (11.129), from which $A = 2.131$ and $B = 3.948$. The value for A gives, in turn, $w/h = 0.977$ from the first expression in Eqs. (11.129), the one for $w/h \leq 2$. Substituting B in the second expression, leads to $w/h = 0.962$, which is an impossible result, since this expression is valid for $w/h > 2$ only. So, the required w/h ratio is $w/h = 0.977$.

The corresponding effective relative permittivity of the line, Eq. (11.127), is $\varepsilon_{\text{reff}} = 2.91$. From Eqs. (11.122), the phase coefficient and velocity of a quasi-TEM wave on the line are $\beta = 107.18$ rad/m and $v_p = 1.76 \times 10^8$ m/s, respectively.

Finally, as a check of our designed w/h ratio, we compute the line characteristic impedance, using Eqs. (11.128), for $w/h = 0.977$ and $\varepsilon_{\text{reff}} = 2.91$, and what we get is indeed $Z_0 = 74.95\ \Omega \approx 75\ \Omega$, the desired impedance.

(b) To design a line with $Z_0 = 50\ \Omega$, a reuse of Eqs. (11.129) gives $A = 1.472$ and $B = 5.922$, and then $w/h = 2.0516$ from the expression for $w/h \leq 2$, which is contradictory, whereas $w/h = 2.0531$ from the other expression, so this latter result is the required w/h ratio in this case. Eqs. (11.127) and (11.122) then yield $\varepsilon_{\text{reff}} = 3.07$, $\beta = 110.03$ rad/m, and $v_p = 1.71 \times 10^8$ m/s, and a check in Eqs. (11.128) confirms that $Z_0 = 50.23\ \Omega \approx 50\ \Omega$, as desired.

Example 11.18 Analysis of a Strip Line Including Fringing

Repeat Example 11.15 but for a strip line, in Fig. 2.21.

Solution (a)–(d) The effective relative permittivity of the line is $\varepsilon_{\text{reff}} = \varepsilon_r = 4$ (the line dielectric is homogeneous). Eq. (11.132) gives the line characteristic impedance of $Z_0 = 32.7\ \Omega$ [note the difference relative to the result obtained neglecting fringing effects, from Eq. (11.118), $(Z_0)_{\text{approx}} = \eta_0 h/(2\sqrt{\varepsilon_r}w) = 47.1\ \Omega$]. From Eqs. (11.16), (11.17), and (11.80), the phase coefficient, phase velocity, and attenuation coefficients for the line dielectric are $\beta = 125.6$ rad/m, $v_p = 1.5 \times 10^8$ m/s, and $\alpha_d = 62.82 \times 10^{-4}$ Np/m, respectively. Having in mind Eq. (11.118), the attenuation coefficient for line conductors is computed as $\alpha_c = R_s/(2wZ_0) = 0.109$ Np/m, and the total attenuation coefficient of the line amounts to $\alpha = \alpha_c + \alpha_d = 0.116$ Np/m $= 1$ dB/m.

Example 11.19 Deriving Synthesis Formulas for a Strip Line

Consider a strip line (Fig. 2.21) whose fringing effects are not negligible. (a) Derive the synthesis formulas, analogous to those in Eqs. (11.129), for finding the geometrical ratio w/h for a desired characteristic impedance, Z_0, of the line and given relative permittivity, ε_r, of the line dielectric. (b) Using the formulas in (a), design strip lines with $Z_0 = 50\ \Omega$ and $Z_0 = 75\ \Omega$, respectively, if $\varepsilon_r = 4$ in both cases.

Solution

(a) If $w/h \geq 0.7$, $s = 0$ in Eqs. (11.132), from which

strip line – synthesis

$$\boxed{\frac{w}{h} = \frac{\eta_0}{2\sqrt{\varepsilon_r}Z_0} - 0.882, \quad \sqrt{\varepsilon_r}Z_0 \leq 0.316\eta_0,} \tag{11.136}$$

where the last inequality (condition) is obtained by requiring that this solution for w/h, in terms of Z_0 and ε_r, be greater than or equal to 0.7. If $w/h < 0.7$, on the other side, Eqs. (11.132) give the following quadratic equation in $x = w/h$, which we solve in a standard fashion,

$$(0.7-x)^2-2x=1.764-\frac{\eta_0}{\sqrt{\varepsilon_r}Z_0} \quad \longrightarrow \quad x^2-3.4x-1.274+\frac{\eta_0}{\sqrt{\varepsilon_r}Z_0}=0$$

$$\longrightarrow \quad x=\frac{w}{h}=1.7-\sqrt{4.164-\frac{\eta_0}{\sqrt{\varepsilon_r}Z_0}}, \quad \sqrt{\varepsilon_r}Z_0>0.316\eta_0, \tag{11.137}$$

strip line – synthesis (continued)

where the other solution, $x=1.7+\sqrt{4.164-\eta_0/(\sqrt{\varepsilon_r}Z_0)}$, is eliminated, because of the condition $x<0.7$. So, substituting the known Z_0 and ε_r in Eq. (11.136) or (11.137), depending whether $\sqrt{\varepsilon_r}Z_0$ is smaller or larger than $0.316\eta_0$, we find w/h.

(b) The required w/h ratio for $Z_0=50\ \Omega$ and $\varepsilon_r=4$, in which case $\sqrt{\varepsilon_r}Z_0=100\ \Omega<0.316\eta_0\approx 120\ \Omega$ and Eq. (11.136) applies, turns out to be $w/h=1$, whereas $w/h=0.415$ for $Z_0=75\ \Omega$ ($\sqrt{\varepsilon_r}Z_0=150\ \Omega>120\ \Omega$) – from Eq. (11.137).

Problems: 11.18–11.20; *Conceptual Questions* (on Companion Website): 11.33 and 11.34; *MATLAB Exercises* (on Companion Website).

Problems

11.1. Circuit/field quantities in the time domain for a coaxial cable. For the low-loss coaxial cable described in Example 11.3, assume that the rms value and initial phase of the voltage of the traveling TEM wave in the cross section of the cable defined by $z=0$ are $V_0=1$ V and zero, respectively. Determine the time-domain (instantaneous) expressions for (a) the voltage and current along the cable, (b) electric and magnetic field intensities in the cable dielectric, (c) surface charge and current densities on the conductors, (d) Poynting vector in the dielectric, and (e) total power transported by the TEM wave.

11.2. Three different optimizations of a coaxial cable. Consider a coaxial cable with a polyethylene ($\varepsilon_r=2.25$) dielectric and copper ($\sigma_c=58$ MS/m) conductors at a frequency of $f=100$ MHz. The outer radius of the cable is $b=8.6$ mm, the dielectric strength of the dielectric is $E_{cr}=47$ MV/m, and the losses in the dielectric can be neglected. Compute the attenuation coefficient ($\alpha=\alpha_c$), breakdown rms voltage ($|\underline{V}|_{cr}$), and maximum permissible (breakdown) time-average transferred power (P_{cr}) of the cable for the following values of the inner radius of the cable: (a) $a=b/3.59$ (for which α_c is minimum), (b) $a=b/\mathrm{e}$ (for which $|\underline{V}|_{cr}$ is maximum), and (c) $a=b/\sqrt{\mathrm{e}}$ (for which P_{cr} is maximum).

11.3. Circuit parameters of a nonsymmetrical two-wire line. Consider a nonsymmetrical thin two-wire transmission line with conductor radii a and b ($a\neq b$), the distance between conductor axes d ($d\gg a,b$), and small losses in both the conductors and the dielectric. Let the conductivity of wires be σ_c and permeability μ_0, and let the dielectric around them be homogeneous and nonmagnetic of relative permittivity ε_r and conductivity σ_d. If the frequency of the propagating TEM wave on the line is f, find (a) the primary and (b) the secondary circuit parameters of the line.

11.4. Maximum power transfer along a two-wire line. For the nonsymmetrical two-wire line from the previous problem, let $a=6$ mm, $b=3$ mm, $d=90$ mm, $\sigma_c=30$ MS/m, $\varepsilon_r=3$, $\tan\delta_d=10^{-4}$, and $f=75$ MHz. In addition, let the length of the line be $l=50$ m, and dielectric strength of its dielectric $E_{cr}=20$ MV/m. Under these circumstances, compute the maximum time-average power that the line can receive from a generator at one of its ends for the safe operation of the structure, i.e., prior to an eventual dielectric breakdown, as well as the corresponding maximum time-average power delivered to a load at the other end of the line.

11.5. Charge and current distributions on the ground plane. Assuming that the complex rms current intensity of the wire-plane transmission line in Fig. 11.9(a) is $\underline{I}(z)$ and that its ground plane is perfectly conducting, find the distributions of surface charge and current on the plane.

11.6. Satisfaction of the continuity equation on the ground plane. (a) Consider the computed surface charge and current densities on the PEC plane in the wire-plane transmission line from the previous problem, and show that they satisfy the continuity equation for surface currents (for plates). (b) Also show, by integrating the results from the previous problem, that the surface charge and current on the plane total $-\underline{I}(z)\sqrt{\varepsilon\mu_0}$ per unit length of the line and $-\underline{I}(z)$, respectively.

11.7. Three-wire transmission line. Consider the system of three parallel thin wires in air shown in Fig. 2.23 and described in Example 2.16, which constitutes a two-conductor transmission line with wire 1 being one conductor and galvanically connected wires 2 and 3 the other conductor of the line, and assume that a TEM wave of frequency $f = 300$ MHz propagates along the line. The wires are made out of copper. Compute (a) the primary circuit parameters of the line and (b) the line characteristic impedance and attenuation coefficient.

11.8. Maximum permissible power delivered to a load. If the transmission line from the previous problem is $l = 2$ m long, what is the maximum permissible time-average power, limited by the dielectric breakdown of the line, that can be delivered to a load terminating the line?

11.9. Four-wire transmission line. Assume that a TEM wave of frequency $f = 200$ MHz is established on the two-conductor transmission line consisting of two pairs of galvanically interconnected thin wires shown in Fig. 2.44 (and described in Problem 2.35), as well as that the four wires are made from copper. Compute (a) the primary circuit parameters of the line and (b) the line attenuation coefficient.

11.10. Two-wire line and a foil. For the two-conductor transmission line whose one conductor is a two-wire line with galvanically connected wires and the other conductor is a metallic foil in Fig. 2.45, assume that the skin-effect surface resistance of the two wires is $R_s = 8.25$ mΩ/square, and that the losses in the foil can be neglected. In addition, let the rms current intensity of the traveling TEM wave on the line be $I_0 = 1$ A at the beginning of the line (for $z = 0$). Determine (a) the primary circuit parameters of the line and (b) the induced rms surface charge and current densities at the central point O on the foil in an arbitrary cross section of the line (for arbitrary z).

11.11. Wire-corner transmission line. Take the two-conductor transmission line with an isolated wire as one conductor and a 90° corner metallic screen as the other in Fig. 1.57 with $a = 0.5$ mm and $h = 4$ cm, and also assume that both the wire and the screen are made out of aluminum ($\sigma_c = 35$ MS/m). (a) Find the attenuation coefficient of the line at a frequency of $f = 375$ MHz, neglecting the contribution of the losses in the screen. (b) If the complex rms current intensity of the line is $\underline{I}(z)$, what is the complex rms surface current density vector at each of the two points on the screen that are closest to the wire?

11.12. Planar TEM line with a continuously inhomogeneous dielectric. Assume that the conductors of the planar transmission line with continuously inhomogeneous imperfect dielectric in Fig. 3.37 are also imperfect but homogeneous, of conductivity σ_c, as well as that a quasi-TEM wave of frequency f is established on the line. The losses in the strips and dielectric can be considered to be small, the permeability everywhere is μ_0, and the width of the strips and their separation are such that $w \gg h$, so that fringing effects can be neglected. Find the primary and secondary circuit parameters of this line.

11.13. Quasi-TEM wave on a coaxial cable with two dielectric layers. For the coaxial cable with two coaxial layers of imperfect dielectric shown in Fig. 3.34 or 2.50 and described in Problem 3.17, let its conductors be made out of copper and the dielectric be nonmagnetic. Calculate (a) the primary circuit parameters of the cable for quasi-TEM waves at a frequency of $f = 2$ GHz, (b) the effective relative permittivity of the cable, and (c) the phase velocity and attenuation coefficient.

11.14. **Power capacity of a coaxial cable with a two-layer dielectric.** Consider the coaxial cable with two coaxial dielectric layers from Problem 2.77, and assume that the conductivity of cable conductors is $\sigma_c = 30$ MS/m, the length of the cable is $l = 10$ m, and there are no losses in the dielectric. At a frequency of $f = 5$ GHz, determine (a) the maximum time-average power, limited by an eventual dielectric breakdown in the structure, that the cable can receive from a generator and (b) the corresponding power delivered to a load at the other end of the cable.

11.15. **Quasi-TEM analysis of coaxial cables with dielectric sectors.** Find the effective relative permittivity and attenuation coefficient for quasi-TEM waves at a frequency f of the coaxial cable with (a) two and (b) four dielectric sectors shown in Figs. 2.33 and 2.51, respectively, assuming that all conductors are made of copper and that losses in the dielectrics are negligible.

11.16. **Coated wire-ground plane quasi-TEM transmission line.** Fig. 11.16 shows a transmission line whose one conductor is a wire of radius a with a coaxial dielectric coating of thickness b and the other conductor is a ground plane. The height of the wire axis with respect to the plane is h ($h \gg a, b$). The skin-effect surface resistances of the wire and plane are R_{s1} and R_{s2}, respectively. The relative permittivity and conductivity of the wire coating are ε_{r1} and σ_{d1} and those of the rest of the dielectric are ε_{r2} and σ_{d2}, whereas both materials are nonmagnetic, and the line can be treated as a low-loss one. The complex rms current intensity of a quasi-TEM wave traveling along the line is $\underline{I}(z)$. Determine (a) the attenuation coefficient of this line and (b) the distributions of surface charge and current on the ground plane.

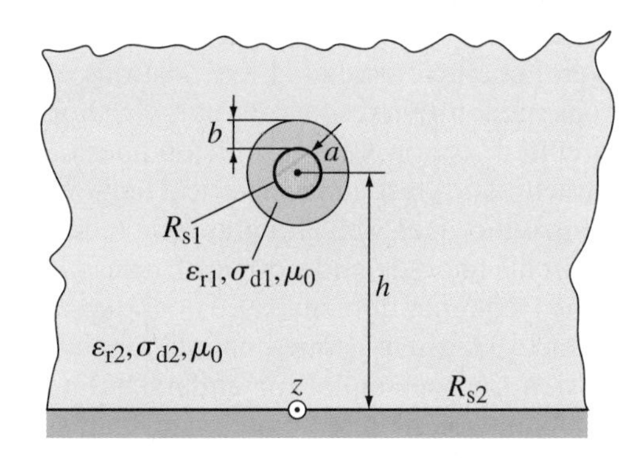

Figure 11.16 Cross section of a low-loss quasi-TEM transmission line consisting of a coated wire and ground plane; for Problem 11.16.

11.17. **High-frequency internal inductance of three different lines.** Calculate the high-frequency internal inductance per unit length of the two-wire line from Problem 11.4, coaxial cable from Problem 11.13, and microstrip line from Example 11.12, and compare the results with the corresponding values of their per-unit-length external inductances.

11.18. **Microstrip lines with different strip width to height ratios.** Consider a microstrip line with a copper strip and ground plane, dielectric substrate parameters $\varepsilon_r = 4$ and $\tan\delta_d = 10^{-4}$ ($\mu_r = 1$), strip width w, and substrate thickness $h = 2$ mm. Compute the primary and secondary circuit parameters of the line, taking into account the fringing effects, for the following w/h ratios: (a) 0.05, (b) 0.1, (c) 0.5, (d) 1, (e) 2, (f) 10, and (g) 20. (h) Compare the results in cases (d)–(g) with the corresponding values of circuit parameters of the line obtained neglecting the fringing effects (see Example 11.10). (i) For cases (a)–(d), compare the results to those obtained for a wire-plane transmission line (see Example 11.9) with the conducting strip in Fig. 2.20 replaced by a thin wire of an equivalent radius equal to $a = w/4$.

11.19. **Primary circuit parameters of a strip line with fringing.** Find the primary circuit parameters of the strip line from Example 11.18.

11.20. **Design of microstrip and strip lines.** Design (a) a microstrip line and (b) a strip line that have the same characteristic impedance for the same relative permittivity of the dielectric as the coaxial cable from Example 11.3 and two-wire line from Problem 11.4, respectively.

12 Circuit Analysis of Transmission Lines

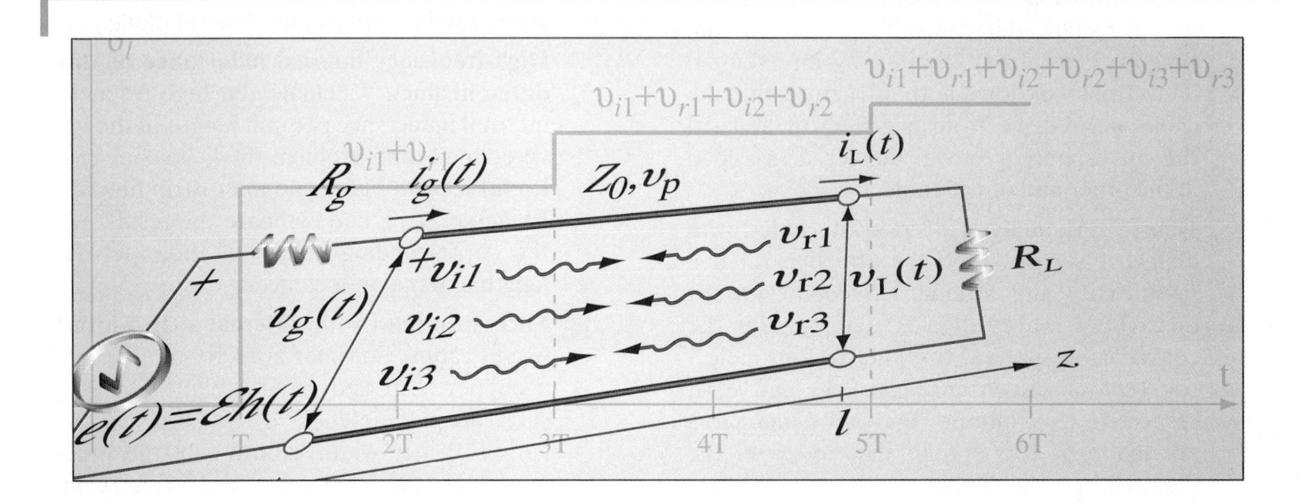

Introduction:

This chapter takes over the primary and secondary circuit parameters of transmission lines computed in the field analysis of lines in the previous chapter, and uses them to solve for the voltage, current, and power along lossless and lossy lines, with various excitations and load terminations. Most importantly, this is a circuit analysis of transmission lines, using only pure circuit-theory concepts to develop the complete frequency-domain and transient analysis of lines as circuits with distributed parameters whose per-unit-length characteristics are already known.

The chapter starts with a circuit model of an arbitrary two-conductor transmission line in the form of a ladder network of elementary circuit cells with lumped elements. Transmission-line equations, termed telegrapher's equations, will be derived for voltages and currents on this network and solved in the complex domain (using complex representatives of time-harmonic voltages and currents introduced in Section 8.7). The analysis will then be specialized to practically important cases of lossless and low-loss transmission lines, respectively, and focused on important concepts and details including the reflection coefficient for the line, power flow in the structure, transmission-line impedance, and several characteristic load terminations of the line. Transmission-line resonators, namely, short- or open-circuited sections of transmission lines of certain characteristic (resonant) electrical lengths, will also be introduced, as well as a graphical technique for the circuit analysis and design of transmission lines in the frequency domain based on the so-called Smith chart. Transient (time-domain) analysis of transmission lines, essential for understanding the transient behavior of interconnects in high-speed digital circuits, will cover step and pulse excitations of lines and a variety of line terminations, including reactive and nonlinear loads, and both matched and unmatched conditions at either end of the line.

12.1 TELEGRAPHER'S EQUATIONS AND THEIR SOLUTION IN COMPLEX DOMAIN

Fig. 12.1 (upper part) shows a circuit-theory representation of an arbitrary two-conductor lossy transmission line, where (as in Fig. 3.22) a pair of parallel horizontal thick lines in the schematic diagram, although resembling a two-wire transmission line, symbolizes a structure with conductors of completely arbitrary cross sections and a generally inhomogeneous dielectric. We assume a time-harmonic (steady-state sinusoidal) regime in the structure (see Section 8.6), with f denoting the operating frequency of the line. To develop the equations for general analysis of transmission lines with time-harmonic waves based on circuit theory, we first subdivide the line under consideration into short sections, of length Δz (Fig. 12.1). In specific, let Δz be much shorter than the wavelength along the line, so that the changes of the voltage and current along Δz are small. Using the resistance R' and leakage conductance G' per unit length of the line, losses in each such section can then be represented by a circuit cell consisting of a series resistor of resistance $\Delta R = R'\Delta z$ and a shunt (parallel) resistor of conductance $\Delta G = G'\Delta z$, as in the circuit model of the line in Fig. 3.21 – for a dc (time-constant) regime. In an ac (including time-harmonic) regime, however, capacitive and inductive effects in the section can be modeled by an additional parallel capacitor, of capacitance $\Delta C = C'\Delta z$, and series inductor, of inductance $\Delta L = L'\Delta z$, as indicated in Fig. 12.1 (lower part), where C' and L' are the capacitance and inductance per unit length of the line. This capacitor and inductor are irrelevant for the dc analysis, since they are an open and short circuit [see Eqs. (3.45) and (7.3)], respectively, for time-constant voltages and currents. Of course, C', L', R', and G' are primary circuit parameters of the line studied in the previous chapter based on the field analysis of the line, at high frequencies, and computed, in Sections 11.8 and 11.9, for a variety of line geometries and material compositions. In Fig. 12.1, R' thus denotes the high-frequency per-unit-length resistance (assuming that the skin effect is pronounced) of the line, Eq. (11.66), rather than the low-frequency (or dc) one (studied in Section 3.12). We also recall that L' generally includes the high-frequency per-unit-length internal inductance, $L'_{\rm i}$, of the line, Eq. (11.100), along with the external one, that in Eq. (11.39), although $L'_{\rm i}$ can be neglected in most high-frequency applications of transmission lines. Cascading equal small cells with lumped elements of parameters

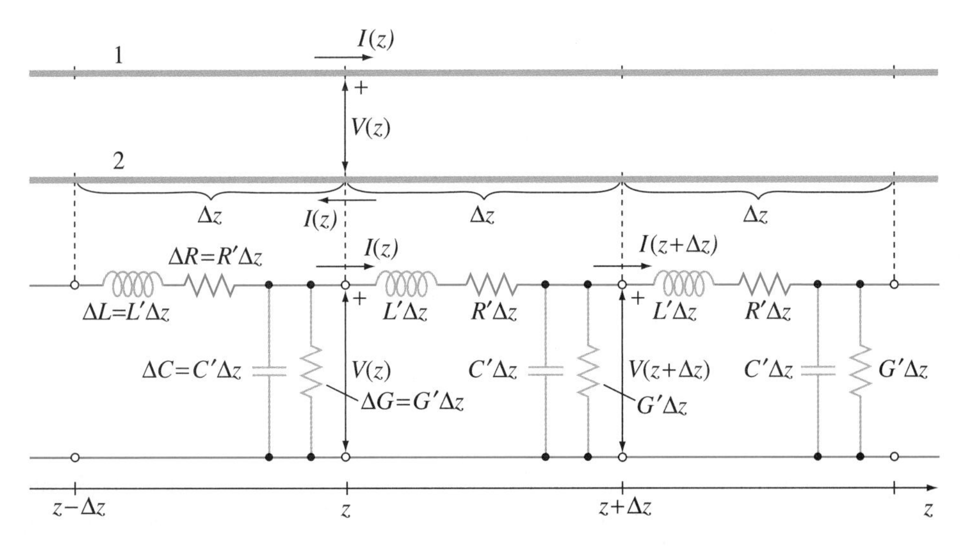

Figure 12.1 Circuit model of a two-conductor lossy transmission line in an ac regime.

ΔC, ΔL, ΔR, and ΔG in Fig. 12.1 for every Δz long section of the line, we finally obtain a high-frequency circuit model of the line, which is a basis for the circuit analysis of the structure. Like the model in Fig. 3.21, this ladder network (of elementary circuit cells) is said to be a circuit with distributed parameters.

For the frequency-domain analysis of the network in Fig. 12.1, we employ complex representatives of the voltage and current along the line (see Section 8.7), namely, the complex rms voltage, $\underline{V}(z)$, and current, $\underline{I}(z)$, which are unknown functions (to be determined) of the coordinate z. Adding to the right-hand side of Eq. (3.166) the expression for the voltage drop across the series inductor [Eq. (8.69)], $\Delta\underline{V}_L = \mathrm{j}\omega\Delta L\underline{I}$, we have

$$\underline{V}(z) - \underline{V}(z + \Delta z) = \Delta R\underline{I} + \mathrm{j}\omega\Delta L\underline{I}, \tag{12.1}$$

where $\omega = 2\pi f$ is the angular (radian) frequency of the voltage and current. In other words, the voltages in the line cross sections at the coordinates z and $z + \Delta z$ in Fig. 12.1 differ for the emf induced in the enclosed portion of the line (term $\mathrm{j}\omega\Delta L\underline{I}$), in addition to the voltage drop (loss) along Δz due to a finite (nonzero) resistance of the line conductors (term $\Delta R\underline{I}$). Similarly, combining the expression for the current drop through the parallel capacitor [Eq. (3.45)], $\Delta\underline{I}_C = \mathrm{j}\omega\Delta C\underline{V}$, with Eqs. (3.152) and (3.157) results in

$$\underline{I}(z) - \underline{I}(z + \Delta z) = \Delta G\underline{V} + \mathrm{j}\omega\Delta C\underline{V}. \tag{12.2}$$

Here, the difference in current intensities between cross sections at z and $z + \Delta z$ equals the sum of intensities, within the portion of the line considered, of the leakage current (term $\Delta G\underline{V}$) through the imperfect dielectric of the line and drainage current (term $\mathrm{j}\omega\Delta C\underline{V}$) that exists even if the dielectric is a vacuum. This latter term is actually the displacement current in the dielectric, with density given in Eq. (8.5), necessary to drain the enclosed charge on line conductors, $\Delta\underline{Q} = \underline{Q}'\Delta z$, where $\underline{Q}'$ is the complex rms charge per unit length of the line and $\Delta\underline{I}_C = \mathrm{j}\omega\Delta\underline{Q}$ [see Eq. (8.2)]. As with Eqs. (3.168) and (3.160) in the dc case, in the limit of $\Delta z \to 0$ Eqs. (12.1) and (12.2) become

telegrapher's equations

$$\boxed{\frac{\mathrm{d}\underline{V}}{\mathrm{d}z} = -\underline{Z}'\underline{I}, \quad \frac{\mathrm{d}\underline{I}}{\mathrm{d}z} = -\underline{Y}'\underline{V},} \tag{12.3}$$

where $\underline{Z}'$ and $\underline{Y}'$, termed the per-unit-length complex impedance and admittance, respectively, of the line, are given by

p.u.l. complex impedance and admittance of a transmission line

$$\boxed{\underline{Z}' = R' + \mathrm{j}\omega L', \quad \underline{Y}' = G' + \mathrm{j}\omega C'.} \tag{12.4}$$

Of course, $\underline{Z}' \neq 1/\underline{Y}'$. Being a generalization of Eqs. (3.168) and (3.160), Eqs. (12.3) are the transmission-line equations or telegrapher's equations for complex voltages and currents (representing the time-harmonic ones) on two-conductor transmission lines. They constitute a system of two coupled first-order differential equations in z with $\underline{V}(z)$ and $\underline{I}(z)$ as unknowns.

In the same way Eqs. (3.170) are obtained from Eqs. (3.168) and (3.160) in the dc analysis, Eqs. (12.3) can be combined together into second-order differential equations in terms of $\underline{V}$ and $\underline{I}$ alone,

wave equations for $\underline{V}$ and $\underline{I}$ along a line

$$\boxed{\frac{\mathrm{d}^2\underline{V}}{\mathrm{d}z^2} - \underline{\gamma}^2\underline{V} = 0, \quad \frac{\mathrm{d}^2\underline{I}}{\mathrm{d}z^2} - \underline{\gamma}^2\underline{I} = 0.} \tag{12.5}$$

These are the wave equations for the complex voltage and current, respectively, on the line. The complex propagation coefficient, $\underline{\gamma}$, of the line, Eq. (11.62), comes out

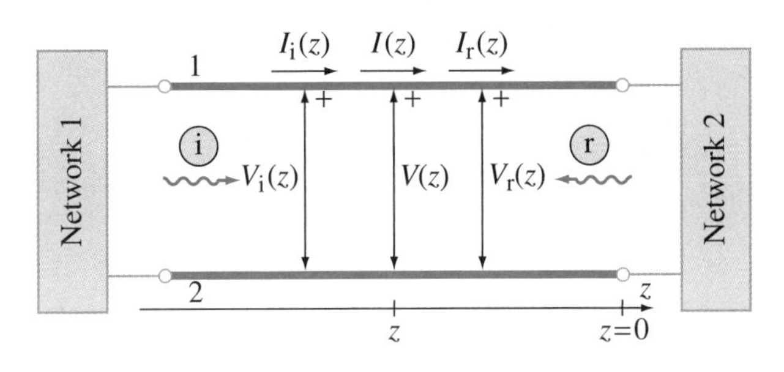

Figure 12.2 Incident, reflected, and total voltages and currents on the transmission line in Fig. 12.1.

to be

$$\boxed{\underline{\gamma} = \sqrt{\underline{Z}'\underline{Y}'} = \alpha + j\beta,} \tag{12.6}$$

complex propagation coefficient of a transmission line

with α and β standing for the line attenuation and phase coefficients, respectively.

General solutions of Eqs. (12.5) are complex exponential functions in z. For the line voltage, the solution is

$$\boxed{\underline{V}(z) = \underbrace{\underline{V}_{i0}\,e^{-\underline{\gamma}z}}_{\text{incident wave}} + \underbrace{\underline{V}_{r0}\,e^{\underline{\gamma}z}}_{\text{reflected wave}} = \underline{V}_i(z) + \underline{V}_r(z) = \underline{V}_{tot}(z),} \tag{12.7}$$

solution for line voltage

which can be verified by direct substitution in the voltage wave equation, like in Eq. (9.42). As expected, the total voltage wave along the line, $\underline{V}_{tot}$, is, in general, a sum of two oppositely directed traveling waves, an incident (forward) wave, $\underline{V}_i$, propagating in the positive z direction, and a reflected (backward) wave, $\underline{V}_r$, progressing in the negative z direction [analogously to Eqs. (10.35) and (10.36)], as shown in Fig. 12.2. The complex constants $\underline{V}_{i0}$ and $\underline{V}_{r0}$ equal, respectively, the incident and reflected complex rms voltages in the line cross section defined by $z = 0$. They depend on the terminal networks connected at the ends of the line (Fig. 12.2), including the operating frequency of the wave – imposed by one or more (voltage and/or current) generators within the networks, as well as on the length and other characteristics of the line itself. As in Eq. (9.81) or (11.63), the expression for $\underline{V}_i$ can be written as

$$\boxed{\underline{V}_i(z) = \underline{V}_{i0}\,e^{-\underline{\gamma}z} = \underline{V}_{i0}\,e^{-\alpha z}\,e^{-j\beta z},} \tag{12.8}$$

incident voltage

and similarly for $\underline{V}_r$. The instantaneous incident voltage along the line is [in parallel to Eq. (9.84)],

$$v_i(z, t) = V_{i0}\sqrt{2}\,e^{-\alpha z}\cos(\omega t - \beta z + \theta_{i0}), \tag{12.9}$$

where V_{i0} and θ_{i0} are the rms value and initial phase, respectively, of this voltage for $z = 0$ ($\underline{V}_{i0} = V_{i0}\,e^{j\theta_{i0}}$). The term $e^{-\alpha z}$ represents an exponential spatial decrease of the wave amplitude, as given in Eq. (9.85) and illustrated in Fig. 9.11. While the unit for β is rad/m, α is measured in Np/m [it can also be expressed in dB/m, for which the conversion in Eq. (9.89) is used]. From Eqs. (9.35) and (8.111), the phase velocity, v_p, and wavelength, λ_z, along the line for each of the traveling waves (incident and reflected) are computed as

$$v_p = \frac{\omega}{\beta}, \quad \lambda_z = \frac{2\pi}{\beta} = \frac{v_p}{f}. \tag{12.10}$$

Taking the derivative with respect to z of the solution for the total voltage of the line in Eq. (12.7), and substituting it in telegrapher's first equation, i.e., the first relationship in Eqs. (12.3) – yield the following solution for the total complex rms current of the line:

solution for line current

$$\underline{I}(z) = -\frac{1}{R' + \mathrm{j}\omega L'}\frac{\mathrm{d}\underline{V}}{\mathrm{d}z} = \underbrace{\frac{\underline{V}_{\mathrm{i}0}}{\underline{Z}_0}\,\mathrm{e}^{-\underline{\gamma} z}}_{\text{incident wave}} + \underbrace{\left(-\frac{\underline{V}_{\mathrm{r}0}}{\underline{Z}_0}\,\mathrm{e}^{\underline{\gamma} z}\right)}_{\text{reflected wave}} = \underline{I}_{\mathrm{i}}(z) + \underline{I}_{\mathrm{r}}(z), \tag{12.11}$$

where all three current intensities, namely, that of the incident wave ($\underline{I}_{\mathrm{i}}$), reflected wave ($\underline{I}_{\mathrm{r}}$), and resultant wave ($\underline{I} = \underline{I}_{\mathrm{tot}}$), are given with respect to the same reference direction (Fig. 12.2). The complex parameter $\underline{Z}_0$ is the characteristic impedance of the line, defined by Eq. (11.34), which in Eq. (12.11) appears to be

complex characteristic impedance

$$\underline{Z}_0 = \sqrt{\frac{\underline{Z}'}{\underline{Y}'}} = |\underline{Z}_0|\,\mathrm{e}^{\mathrm{j}\phi}, \tag{12.12}$$

with ϕ designating the phase angle (argument) of the impedance. The complex incident current can be written as [also see Eq. (9.92)]

incident current

$$\underline{I}_{\mathrm{i}}(z) = \frac{\underline{V}_{\mathrm{i}0}}{\underline{Z}_0}\,\mathrm{e}^{-\underline{\gamma} z} = \frac{\underline{V}_{\mathrm{i}0}}{|\underline{Z}_0|}\,\mathrm{e}^{-\alpha z}\,\mathrm{e}^{-\mathrm{j}\beta z}\,\mathrm{e}^{-\mathrm{j}\phi}. \tag{12.13}$$

In the time domain,

$$i_{\mathrm{i}}(z,t) = \frac{V_{\mathrm{i}0}}{|\underline{Z}_0|}\sqrt{2}\,\mathrm{e}^{-\alpha z}\cos(\omega t - \beta z + \theta_{\mathrm{i}0} - \phi), \tag{12.14}$$

and analogous expressions hold for the reflected current.

Note that, combining Eqs. (12.7) and (12.11), we have

voltage to current ratio for incident and reflected waves

$$\frac{\underline{V}_{\mathrm{i}}(z)}{\underline{I}_{\mathrm{i}}(z)} = -\frac{\underline{V}_{\mathrm{r}}(z)}{\underline{I}_{\mathrm{r}}(z)} = \underline{Z}_0, \tag{12.15}$$

where the minus sign in the second voltage to current ratio comes from the same adopted reference orientations of voltage-current pairs and opposite propagation directions for the incident and reflected waves – in Fig. 12.2. In other words, when both waves are viewed in the same way relative to their propagation, namely, either receding from or approaching an observer, the mutual orientations of their voltage and current look the same for voltage-current pairs ($\underline{V}_{\mathrm{i}}, \underline{I}_{\mathrm{i}}$) and ($\underline{V}_{\mathrm{r}}, -\underline{I}_{\mathrm{r}}$) [rather than ($\underline{V}_{\mathrm{r}}, \underline{I}_{\mathrm{r}}$)].[1] That is why the ratio between $\underline{V}_{\mathrm{r}}$ and $-\underline{I}_{\mathrm{r}}$ (and not $+\underline{I}_{\mathrm{r}}$) is the same (equal to $\underline{Z}_0$) as the ratio between $\underline{V}_{\mathrm{i}}$ and $\underline{I}_{\mathrm{i}}$. We also see from Eqs. (12.7) and (12.11) that

$$\frac{\underline{V}_{\mathrm{tot}}(z)}{\underline{I}_{\mathrm{tot}}(z)} \neq \underline{Z}_0, \tag{12.16}$$

i.e., the ratio of the total voltage and current along the line in general does not equal $\underline{Z}_0$ (or any other constant), but is a function of the coordinate z.

Conceptual Questions (on Companion Website): 12.1–12.5.

[1] From the point of view of the power transfer along a transmission line based on its voltages and currents (which will be discussed in a later section), note that the voltage-current orientations in Fig. 12.2 all give, in Eq. (11.44), the complex power flow in the forward (positive z) reference direction. Hence, the pair ($\underline{V}_{\mathrm{r}}, -\underline{I}_{\mathrm{r}}$) [and not ($\underline{V}_{\mathrm{r}}, \underline{I}_{\mathrm{r}}$)] must be used in Eq. (11.44) for the actual reflected power, since the actual power transfer by the reflected wave is in the opposite (backward) direction.

12.2 CIRCUIT ANALYSIS OF LOSSLESS TRANSMISSION LINES

In many practical situations, losses in both conductors and dielectric of a transmission line in an ac regime can be completely neglected, that is, the line conductors and dielectric can be treated as perfect. For such a lossless line, both the series resistance R' and leakage (shunt) conductance G' per unit length of the line are zero, so that both per-unit-length series impedance $\underline{Z}'$ and shunt admittance $\underline{Y}'$ of the line, in Eqs. (12.4), are purely imaginary (reactive),

$$\boxed{R', G' = 0 \quad \longrightarrow \quad \underline{Z}' = \mathrm{j}\omega L' \quad \text{and} \quad \underline{Y}' = \mathrm{j}\omega C'.} \tag{12.17}$$

lossless transmission line

With this, the expressions for secondary circuit parameters ($\underline{Z}_0$, α, β, v_p, and λ_z) of lossy transmission lines from the previous section become those obtained in Sections 11.1–11.4 based on the field analysis of lossless transmission lines.

In specific, the characteristic impedance of the line ($\underline{Z}_0$), Eq. (12.12), becomes purely real, and given by

$$\boxed{\underline{Z}_0 = \sqrt{\frac{L'}{C'}} = Z_0 \qquad (\phi = 0),} \tag{12.18}$$

Z_0 – lossless line

which is the same expression as in Eq. (11.42). In addition, the complex propagation coefficient of the line ($\underline{\gamma}$), Eq. (12.6), is now purely imaginary, yielding a zero attenuation coefficient (α) and the expression for the line phase coefficient (β) in Eqs. (11.43),

$$\boxed{\underline{\gamma} = \mathrm{j}\omega\sqrt{L'C'} \quad \longrightarrow \quad \alpha = 0 \quad \text{and} \quad \beta = \omega\sqrt{L'C'}.} \tag{12.19}$$

β – lossless line

The corresponding lossless expressions for the phase velocity (v_p) and wavelength (λ_z) along the line are those also given in Eqs. (11.43).

Finally, for a lossless transmission line with a homogeneous dielectric, of permittivity ε and permeability μ, the duality relationship between the per-unit-length inductance (L') and capacitance (C') of the line in Eq. (11.41) leads to the simplified expressions, using ε and μ, for Z_0, β, v_p, and λ_z in Eqs. (11.37), (11.40), (11.16), and (11.17).

12.3 CIRCUIT ANALYSIS OF LOW-LOSS TRANSMISSION LINES

As explained in Section 11.5, conductor and dielectric losses in transmission lines used in practical applications can be considered to be small, and sometimes even zero (previous section). For lines with small (but nonzero) losses, the primary circuit parameters of the line, namely, the per-unit-length capacitance, inductance, resistance, and conductance of the line (C', L', R', and G'), and the operating angular frequency, ω, of the time-harmonic wave on the line satisfy the following condition:

$$\boxed{R' \ll \omega L' \quad \text{and} \quad G' \ll \omega C'.} \tag{12.20}$$

low-loss condition

Based on this, i.e., assuming that, by design, transmission lines are fabricated from good conductors and good dielectrics [see Eqs. (11.61)], we are able to transform the general expressions for secondary circuit parameters of lossy transmission lines from Section 12.1 ($\underline{\gamma}$, $\underline{Z}_0$, ...) into much simpler approximate ones, that are much easier to use. Of course, the lossless case, Eq. (12.17), is included in the low-loss condition.

Using Eq. (12.20) and binomial expansion, the expression for the complex propagation coefficient of a transmission line in Eqs. (12.6) and (12.4) can be simplified, under the low-loss assumption, to an approximate form:

$$\underline{\gamma} = \sqrt{(R' + \mathrm{j}\omega L')(G' + \mathrm{j}\omega C')} = \mathrm{j}\omega\sqrt{L'C'}\sqrt{\left(1 - \mathrm{j}\frac{R'}{\omega L'}\right)\left(1 - \mathrm{j}\frac{G'}{\omega C'}\right)}$$
$$\approx \mathrm{j}\omega\sqrt{L'C'}\left[1 - \mathrm{j}\left(\frac{R'}{\omega L'} + \frac{G'}{\omega C'}\right)\right]^{1/2} \approx \mathrm{j}\omega\sqrt{L'C'}\left[1 - \mathrm{j}\left(\frac{R'}{2\omega L'} + \frac{G'}{2\omega C'}\right)\right]$$
$$= \frac{R'}{2}\sqrt{\frac{C'}{L'}} + \frac{G'}{2}\sqrt{\frac{L'}{C'}} + \mathrm{j}\omega\sqrt{L'C'}, \tag{12.21}$$

similar to the procedure in Eq. (9.122). In particular, the use is made here of the facts that $(1 + \underline{a})(1 + \underline{b}) \approx 1 + \underline{a} + \underline{b}$ ($|\underline{a}||\underline{b}| \ll |\underline{a}|, |\underline{b}|$) and $(1 + \underline{a} + \underline{b})^{1/2} \approx 1 + (\underline{a} + \underline{b})/2$ (from the binomial series identity) for $|\underline{a}| \ll 1$ and $|\underline{b}| \ll 1$. Eq. (9.83) then gives the following expressions for the attenuation and phase coefficients of low-loss transmission lines:

α, β – line with small losses

$$\alpha \approx \underbrace{\frac{R'}{2}\sqrt{\frac{C'}{L'}}}_{\alpha_c} + \underbrace{\frac{G'}{2}\sqrt{\frac{L'}{C'}}}_{\alpha_d}, \qquad \beta \approx \omega\sqrt{L'C'}. \tag{12.22}$$

The first term in the expression for α is the attenuation coefficient representing the losses in conductors (α_c) of the line and the second term that for the losses in the line dielectric (α_d), like in Eq. (11.75). On the other side, having in mind Eqs. (12.22), (12.19), and (12.10), we note that the phase coefficient, as well as the phase velocity and wavelength, along low-loss lines are practically the same as those for the corresponding lossless lines (at the same frequency).

In a similar fashion, starting with Eqs. (12.12) and (12.4), an approximate expression for the characteristic impedance of the line with small losses is derived, whose real part is given by

Z_0 – line with small losses

$$Z_0 \approx \sqrt{\frac{L'}{C'}}, \tag{12.23}$$

while the imaginary part can be neglected. This is analogous to the evaluation of the intrinsic impedance of good dielectrics, in Eq. (9.124). So, Z_0 is computed as in the lossless case, Eq. (12.18).

Combined with Eq. (12.23), the attenuation coefficients for line conductors and dielectric in Eqs. (12.22) become

$$\alpha_c \approx \frac{R'}{2Z_0}, \qquad \alpha_d \approx \frac{G'}{2Y_0}, \tag{12.24}$$

where Y_0 is the characteristic admittance of the line, defined by Eq. (11.35). We see that these are exactly the expressions for α_c in Eq. (11.77) and α_d in Eq. (11.79), obtained from the field analysis of transmission lines with small losses, in Sections 11.5 and 11.6.

However, if, for some reason, the condition in Eq. (12.20) is not met for a given lossy transmission line, at a given frequency of its operation, then the "full" expressions for secondary circuit parameters of the line from Section 12.1 are employed.

12.4 REFLECTION COEFFICIENT FOR TRANSMISSION LINES

In this and the following section, we define and discuss the reflection coefficient and standing wave patterns and compute power flow on a transmission line in Fig. 12.2. For this analysis, let the terminal network at the beginning of the line be a voltage generator of complex rms electromotive force (open-circuit voltage) $\underline{\mathcal{E}}$ and complex internal (series) impedance $\underline{Z}_g$, as shown in Fig. 12.3. In general, such a generator represents the Thévenin equivalent generator (circuit), with respect to the line input terminals, of an arbitrary input network in Fig. 12.2. In addition, let the other end of the line be terminated in a load of complex impedance $\underline{Z}_L$, which, in general, is an equivalent (input) impedance of an arbitrary passive (with no generators) output network in Fig. 12.2. Finally, we adopt the origin of the z-axis to be, as in Fig. 12.2, at the output terminals of the line (i.e., at the load), so that, denoting the length of the line by l, the location of the line input terminals (generator) is defined by $z = -l$ (Fig. 12.3).

Boundary conditions at the load terminals for the total (incident plus reflected) voltage and current waves, $\underline{V}(z)$ and $\underline{I}(z)$, of the line simply stipulate that this voltage and current, given by Eqs. (12.7) and (12.11), for $z = 0$ equal the load voltage, $\underline{V}_L$, and current, $\underline{I}_L$, respectively. In addition, $\underline{V}_L$ and $\underline{I}_L$ are related via the load impedance. We thus write

$$\underline{V}(0) = \underline{V}_L, \quad \underline{I}(0) = \underline{I}_L, \quad \underline{V}_L = \underline{Z}_L \underline{I}_L, \tag{12.25}$$

which, combined with Eqs. (12.7) and (12.11), gives

$$\boxed{\underline{V}_{i0} + \underline{V}_{r0} = \underline{V}_L, \quad \frac{\underline{V}_{i0}}{\underline{Z}_0} - \frac{\underline{V}_{r0}}{\underline{Z}_0} = \frac{\underline{V}_L}{\underline{Z}_L}.} \tag{12.26}$$

boundary conditions at the load

In what follows, we assume that $\underline{V}_{i0}$ is a given (known) constant, and solve for $\underline{V}_{r0}$ and $\underline{V}_L$. However, we have in mind that for the complete solution for the voltage and current along the line, we also need to employ the boundary conditions at the generator terminals ($z = -l$) in Fig. 12.3, to express $\underline{V}_{i0}$, and then $\underline{V}_{r0}$ and $\underline{V}_L$, in terms of the emf $\underline{\mathcal{E}}$ and impedance $\underline{Z}_g$ of the generator (in addition to other parameters of the structure). This will be done in a later section.

Eqs. (12.26) with $\underline{V}_{r0}$ and $\underline{V}_L$ as unknowns have the identical form as Eqs. (10.40), and the same thus holds true for their respective solutions, so, analogously to Eq. (10.42), the ratio of $\underline{V}_{r0}$ and $\underline{V}_{i0}$ turns out to be

$$\boxed{\underline{\Gamma}_L = \frac{\underline{V}_{r0}}{\underline{V}_{i0}} = \frac{\underline{Z}_L - \underline{Z}_0}{\underline{Z}_L + \underline{Z}_0}.} \tag{12.27}$$

load voltage reflection coefficient

Having in mind that this is the reflected to incident voltage ratio of the line at the load terminals ($z = 0$) in Fig. 12.3, namely, that $\underline{V}_{i0} = \underline{V}_i(0)$ and $\underline{V}_{r0} = \underline{V}_r(0)$, we

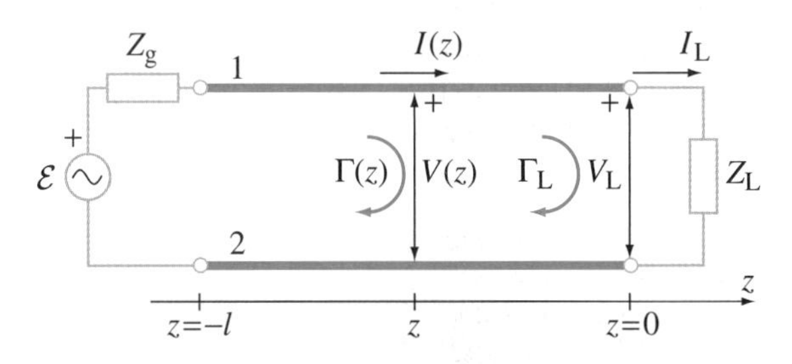

Figure 12.3 Transmission line of Fig. 12.2 with a voltage generator (at $z = -l$) and complex impedance load (at $z = 0$) as terminal networks.

term it the load voltage reflection coefficient of the line. Note that the obtained expression for $\underline{\Gamma}_\mathrm{L}$ in terms of $\underline{Z}_0$ and $\underline{Z}_\mathrm{L}$ assumes the same reference orientation of $\underline{V}_{\mathrm{i}0}$ and $\underline{V}_{\mathrm{r}0}$ (Fig. 12.2). Note also that $\underline{\Gamma}_\mathrm{L}$ is defined for the line voltages, and not currents, in the first place. Both these choices (conventions) are, of course, important for the use of this parameter and interpretation of its numerical values. For instance, from Eqs. (12.11) and (12.27), the load reflection coefficient of the line for currents, defined as the ratio of the reflected and incident complex rms current intensities, instead of voltages, of the line at the load (for $z = 0$),

$$\underline{\Gamma}_{\text{for currents}} = \frac{\underline{I}_\mathrm{r}(0)}{\underline{I}_\mathrm{i}(0)} = -\frac{\underline{V}_{\mathrm{r}0}}{\underline{V}_{\mathrm{i}0}} = -\underline{\Gamma}_\mathrm{L}, \tag{12.28}$$

comes out to be just opposite to the voltage coefficient. As in Eq. (10.44), the complex $\underline{\Gamma}_\mathrm{L}$ can be written in the exponential or polar form:

$\underline{\Gamma}_\mathrm{L}$ in polar form

$$\boxed{\underline{\Gamma}_\mathrm{L} = |\underline{\Gamma}_\mathrm{L}|\,\mathrm{e}^{\mathrm{j}\psi_\mathrm{L}} \quad (0 \le |\underline{\Gamma}_\mathrm{L}| \le 1;\ -180^\circ < \psi_\mathrm{L} \le 180^\circ),} \tag{12.29}$$

where ψ_L denotes its phase angle. The representation in decibels for its magnitude, $(\Gamma_\mathrm{L})_\mathrm{dB}$, is computed by Eq. (10.45). The negative of $(\Gamma_\mathrm{L})_\mathrm{dB}$,

return loss, in dB

$$\boxed{\mathrm{RL} = -(\Gamma_\mathrm{L})_\mathrm{dB} = -20\log|\underline{\Gamma}_\mathrm{L}| \quad (0\ \mathrm{dB} \le \mathrm{RL} < \infty),} \tag{12.30}$$

referred to as the return loss (RL) of the line (for its termination load impedance $\underline{Z}_\mathrm{L}$), is also frequently used.

If both $\underline{Z}_0$ and $\underline{Z}_\mathrm{L}$ in Eq. (12.27) are purely real, that is, if the transmission line in Fig. 12.3 is lossless or with small losses, Eq. (12.18) or (12.23), and terminated in a purely resistive load ($\underline{Z}_\mathrm{L} = R_\mathrm{L} + \mathrm{j}0$), the coefficient $\underline{\Gamma}_\mathrm{L}$ becomes purely real as well,

lossless line and purely resistive load

$$\boxed{\Gamma_\mathrm{L} = \frac{R_\mathrm{L} - Z_0}{R_\mathrm{L} + Z_0}.} \tag{12.31}$$

Moreover, if $R_\mathrm{L} > Z_0$, Γ_L is positive, and this corresponds to the situation in Fig. 10.8(a). The load with $R_\mathrm{L} < Z_0$ gives a negative Γ_L, as in Fig. 10.8(b). Finally, $\Gamma_\mathrm{L} = 0$ for $R_\mathrm{L} = Z_0$, in which case we say that the load is matched (impedance-matched) to the line.

We can also define, and compute based on Eqs. (12.26), the load voltage transmission coefficient of the line,

load voltage transmission coefficient

$$\boxed{\underline{\tau}_\mathrm{L} = \frac{\underline{V}_\mathrm{L}}{\underline{V}_{\mathrm{i}0}} = \frac{2\underline{Z}_\mathrm{L}}{\underline{Z}_0 + \underline{Z}_\mathrm{L}} = 1 + \underline{\Gamma}_\mathrm{L}.} \tag{12.32}$$

It determines the portion of the incident voltage (and the associated power) delivered to the load, or to an arbitrary passive output network in Fig. 12.2, whose equivalent impedance equals $\underline{Z}_\mathrm{L}$. Most importantly, this network does not need to be a classical *RLC* electric circuit, composed of lumped elements. Rather, $\underline{Z}_\mathrm{L}$ may, in general, be the complex impedance of an arbitrary three-dimensional electromagnetic material object with a pair of terminals (and no impressed sources), in Fig. 8.15. For example, it may stand for the input impedance of another transmission line (terminated in another load) or an antenna radiating in free space.

Assuming, for simplicity, that the line in Fig. 12.3 is lossless, so that $\alpha = 0$, Eq. (12.19), and $\gamma = \mathrm{j}\beta$ in Eq. (12.7), we now generalize the concept of the line voltage reflection coefficient at the load, Eq. (12.27), to that at an arbitrary position, defined by the coordinate z, along the line,

$$\underline{\Gamma}(z) = \frac{\underline{V}_{\mathrm{r}}(z)}{\underline{V}_{\mathrm{i}}(z)} = \frac{\underline{V}_{\mathrm{r0}}\,\mathrm{e}^{\mathrm{j}\beta z}}{\underline{V}_{\mathrm{i0}}\,\mathrm{e}^{-\mathrm{j}\beta z}} = \underline{\Gamma}_{\mathrm{L}}\,\mathrm{e}^{\mathrm{j}2\beta z} = |\underline{\Gamma}_{\mathrm{L}}|\,\mathrm{e}^{\mathrm{j}(2\beta z+\psi_{\mathrm{L}})}. \tag{12.33}$$

generalized voltage reflection coefficient

Representing then the generalized coefficient in terms of the magnitude and phase,

$$\underline{\Gamma}(z) = |\underline{\Gamma}(z)|\,\mathrm{e}^{\mathrm{j}\psi(z)} \quad (-l \le z \le 0), \tag{12.34}$$

we have that

$$|\underline{\Gamma}(z)| = |\underline{\Gamma}_{\mathrm{L}}| = \text{const}, \quad \psi(z) = 2\beta z + \psi_{\mathrm{L}}. \tag{12.35}$$

We see that the reflection coefficient does not change in magnitude along the line (if lossless), whereas its phase is a linear function of z (note that z is negative in Fig. 12.3), with the proportionality constant, multiplying z, equaling twice the phase coefficient of the line, β. The generalized voltage transmission coefficient, for an arbitrary z, $\underline{\tau}(z)$, can be defined in a similar way. With the use of the coefficient $\underline{\Gamma}(z)$, the total voltage and current along the line, Eqs. (12.7) and (12.11), can be written as

$$\underline{V}(z) = \underline{V}_{\mathrm{i}}(z)\,[1 + \underline{\Gamma}(z)], \quad \underline{I}(z) = \underline{I}_{\mathrm{i}}(z)\,[1 - \underline{\Gamma}(z)]. \tag{12.36}$$

total voltage and current along a line

Example 12.1 Total Line Voltage as a Sum of Traveling and Standing Waves

Write both the complex and instantaneous total voltages along a lossless transmission line as a sum of traveling and standing waves.

Solution Being the superposition of two traveling waves with generally different amplitudes (and opposite propagation directions), the transmission-line voltage $\underline{V}(z)$ in Eq. (12.7) [or current $\underline{I}(z)$ in Eq. (12.11)] is not a pure standing wave. For a lossless line (in Fig. 12.3), $\underline{V}(z)$ can be written as in Eq. (10.49) for $\underline{\mathbf{E}}_1(z)$ in Fig. 10.7,

$$\underline{V}(z) = \underbrace{\underline{\tau}_{\mathrm{L}}\,\underline{V}_{\mathrm{i0}}\,\mathrm{e}^{-\mathrm{j}\beta z}}_{\text{traveling wave}} + \underbrace{2\mathrm{j}\underline{\Gamma}_{\mathrm{L}}\underline{V}_{\mathrm{i0}}\sin\beta z}_{\text{standing wave}} \tag{12.37}$$

[note that an analogous decomposition holds for $\underline{I}(z)$]. In the time domain [see Eqs. (10.50) and (12.9)],

$$v(z,t) = |\underline{\tau}_{\mathrm{L}}|V_{\mathrm{i0}}\sqrt{2}\cos(\omega t - \beta z + \xi_{\mathrm{L}} + \theta_{\mathrm{i0}}) - 2\sqrt{2}|\underline{\Gamma}_{\mathrm{L}}|V_{\mathrm{i0}}\sin\beta z\sin(\omega t + \psi_{\mathrm{L}} + \theta_{\mathrm{i0}}), \tag{12.38}$$

where ξ_{L} is the phase angle of $\underline{\tau}_{\mathrm{L}}$. We see that the resultant voltage wave along the line consists of a traveling wave, with amplitude $|\underline{\tau}_{\mathrm{L}}|V_{\mathrm{i0}}\sqrt{2}$, and a standing wave, whose maximum amplitude (for $\sin\beta z = \pm 1$) is $2\sqrt{2}|\underline{\Gamma}_{\mathrm{L}}|V_{\mathrm{i0}}$.

Example 12.2 Voltage and Current Maxima and Minima, and SWR

A time-harmonic wave of rms voltage V_{i0} propagates with phase coefficient β along a lossless transmission line of characteristic impedance Z_0. The line is terminated in an impedance load whose reflection coefficient has magnitude $|\underline{\Gamma}_{\mathrm{L}}|$ and phase angle ψ_{L}. (a) Find the maxima and minima of the line total voltage and current, respectively, and their locations. (b) What is the standing wave ratio of the line?

Solution

(a) From the analogy with Eq. (10.53), or from the analysis of the expression for $\underline{V}(z)$ in Eqs. (12.36), with the help of Fig. 10.9(a), the voltage maxima on the line are

$$|\underline{V}|_{\max} = V_{\mathrm{i0}}\left(1 + |\underline{\Gamma}_{\mathrm{L}}|\right) \quad \text{at} \quad z_{\max} = -m\frac{\lambda_z}{2} - \frac{\psi_{\mathrm{L}}}{2\beta}, \quad \begin{cases} m = 0, 1, 2, \dots \text{ if } \psi_{\mathrm{L}} \ge 0 \\ m = 1, 2, \dots \text{ if } \psi_{\mathrm{L}} < 0 \end{cases}, \tag{12.39}$$

voltage maxima on a transmission line

where $m \geq 1$ for $\psi_L < 0$ to ensure that $z_{max} \leq 0$, and $\lambda_z = 2\pi/\beta$ is the wavelength along the line, Eqs. (12.10). The minima are

voltage minima

$$|\underline{V}|_{min} = |\underline{V}_{i0}|\left(1 - |\underline{\Gamma}_L|\right) \quad \text{at} \quad z_{min} = -(2m+1)\frac{\lambda_z}{4} - \frac{\psi_L}{2\beta}, \qquad m = 0, 1, 2, \ldots,$$
$$\text{and} \quad z_{min} = 0 \quad \text{if} \quad \psi_L = 180°, \tag{12.40}$$

i.e., they are shifted by $\lambda_z/4$ with respect to the adjacent maxima [see Eq. (10.54)]. Because of the extra minus sign (180° phase difference) in the reflected-wave term in the expression for $\underline{I}(z)$ in Eq. (12.11) or Eqs. (12.36), when compared to the corresponding expressions for $\underline{V}(z)$, the current maxima occur at the locations of the voltage minima, and vice versa,

locations of current maxima and minima

$$|\underline{V}|_{min} \quad \longleftrightarrow \quad |\underline{I}|_{max}, \quad |\underline{V}|_{max} \quad \longleftrightarrow \quad |\underline{I}|_{min}. \tag{12.41}$$

The maximum and minimum rms current intensities amount to

$$|\underline{I}|_{max} = \frac{V_{i0}}{Z_0}\left(1 + |\underline{\Gamma}_L|\right), \quad |\underline{I}|_{min} = \frac{V_{i0}}{Z_0}\left(1 - |\underline{\Gamma}_L|\right). \tag{12.42}$$

Of course, successive voltage or current maxima, as well as successive minima, are separated by $\lambda_z/2$.

(b) The standing wave ratio (SWR) of a transmission line, for a given load impedance, is defined as the ratio of voltage or current maxima to minima on the line. Using Eqs. (12.39)–(12.42), we thus have

SWR of a line

$$s = \frac{|\underline{V}|_{max}}{|\underline{V}|_{min}} = \frac{|\underline{I}|_{max}}{|\underline{I}|_{min}} = \frac{1 + |\underline{\Gamma}_L|}{1 - |\underline{\Gamma}_L|} \qquad (1 \leq s < \infty). \tag{12.43}$$

Example 12.3 Voltage and Current Standing Wave Patterns

Consider the transmission line from the previous example and sketch normalized voltage and current standing wave patterns, given by $|\underline{V}(z)|/V_{i0}$ and $|\underline{I}(z)|Z_0/V_{i0}$, for the line if the load impedance is (a) $\underline{Z}_L = 2Z_0 + \text{j}0$, (b) $\underline{Z}_L = Z_0/4 + \text{j}0$, (c) $\underline{Z}_L = Z_0(1 + \text{j})$, and (d) $\underline{Z}_L = Z_0(1 - \text{j})$, respectively.

Solution (a)–(d) Combining Eqs. (12.36) and (12.33), the line voltage can alternatively be written as

$$\underline{V}(z) = \underline{V}_{i0}\left[\text{e}^{-\text{j}\beta z} + |\underline{\Gamma}_L|\,\text{e}^{\text{j}(\beta z + \psi_L)}\right]. \tag{12.44}$$

This way, it is very easy to identify the real and imaginary parts of $\underline{V}(z)$, which, in turn, give the following expression for the rms voltage magnitude [Eq. (8.57)] along the (lossless) line (as a function of z):

$$|\underline{V}(z)| = V_{i0}\sqrt{[\cos\beta z + |\underline{\Gamma}_L|\cos(\beta z + \psi_L)]^2 + [|\underline{\Gamma}_L|\sin(\beta z + \psi_L) - \sin\beta z]^2}. \tag{12.45}$$

In a similar fashion, the magnitude of the line current is expressed as

$$|\underline{I}(z)| = \frac{V_{i0}}{Z_0}\sqrt{[\cos\beta z - |\underline{\Gamma}_L|\cos(\beta z + \psi_L)]^2 + [\sin\beta z + |\underline{\Gamma}_L|\sin(\beta z + \psi_L)]^2}. \tag{12.46}$$

These expressions for $|\underline{V}(z)|$ and $|\underline{I}(z)|$ are used, in conjunction with expressions in Eqs. (12.27) and (12.29) for the load reflection coefficient ($\underline{\Gamma}_L$), to plot the voltage and current standing wave patterns in Fig. 12.4 for the four given different load impedances, $\underline{Z}_L = R_L + \text{j}X_L$. We see from the plots that a purely resistive load ($X_L = 0$) with $R_L > Z_0$, Fig. 12.4(a), makes the location of the first voltage maximum (current minimum) be exactly at the load terminals ($z = 0$). If, however, $R_L < Z_0$, Fig. 12.4(b), then the first voltage minimum (current maximum) falls right at the load location. For an inductive complex load ($X_L > 0$), Fig. 12.4(c), the first voltage maximum on the line is closer to the load than the first minimum. On the other hand, the first voltage minimum comes first, looking from the load toward the

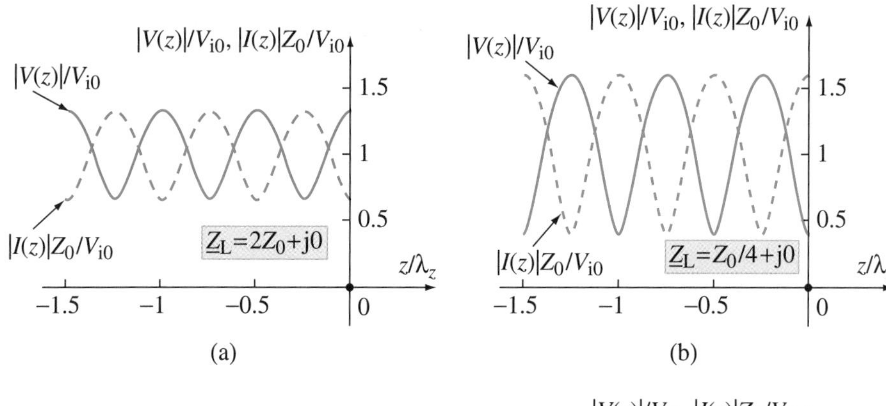

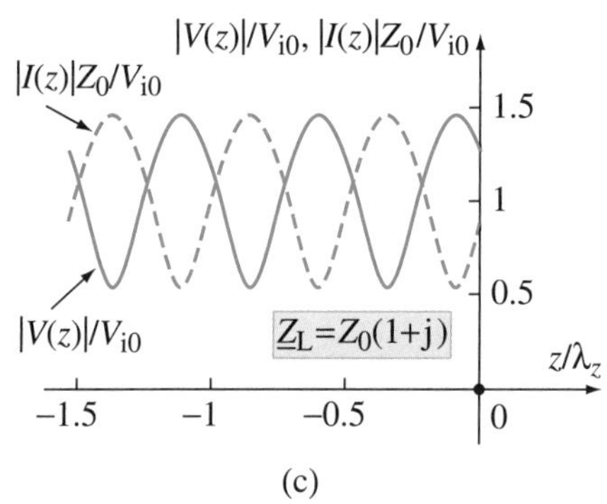

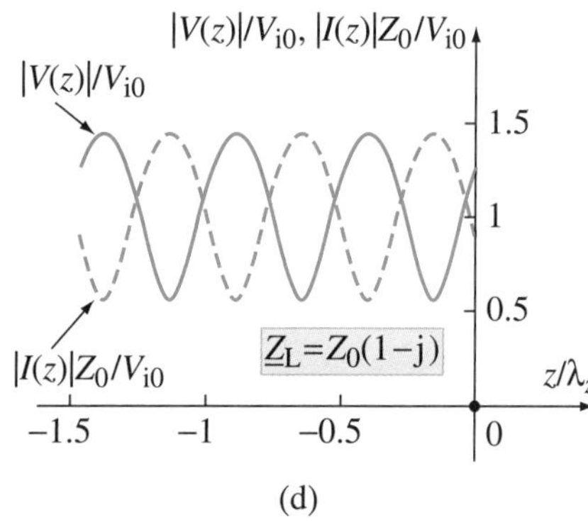

Figure 12.4 Normalized voltage and current standing wave patterns [$|\underline{V}(z)|/V_{i0}$ and $|\underline{I}(z)|Z_0/V_{i0}$, Eqs. (12.45) and (12.46), as functions of z/λ_z] for a lossless transmission line (Fig. 12.3) with (purely real) characteristic impedance Z_0 and load impedance of (a) $\underline{Z}_L = 2Z_0 + j0$, (b) $\underline{Z}_L = Z_0/4 + j0$, (c) $\underline{Z}_L = Z_0(1 + j)$, and (d) $\underline{Z}_L = Z_0(1 - j)$; for Example 12.3.

generator, for a capacitive complex load ($X_L < 0$), Fig. 12.4(d). Note that these conclusions can also be drawn from Eqs. (12.39) and (12.40), with the help of Eqs. (12.31) and (12.27). Namely, in cases (a) and (b) they are obvious [via Eqs. (12.39) and (12.40)] if we have in mind [from Eq. (12.31)] that $R_L > Z_0$ and $X_L = 0$ give $\psi_L = 0$ (positive Γ_L), whereas $\psi_L = 180°$ (negative Γ_L) results from $R_L < Z_0$ and $X_L = 0$. For complex load impedances, we first write $\underline{\Gamma}_L$ [Eq. (12.27)] in the following form, in terms of R_L, X_L, and Z_0:

$$\underline{\Gamma}_L = \frac{R_L - Z_0 + jX_L}{R_L + Z_0 + jX_L} = \frac{R_L^2 + X_L^2 - Z_0^2 + j2X_L Z_0}{(R_L + Z_0)^2 + X_L^2} \qquad (\underline{Z}_L = R_L + jX_L), \tag{12.47}$$

from which we realize that, as Z_0 is always positive, the sign of X_L determines the sign of the imaginary part of $\underline{\Gamma}_L$. Hence, $X_L > 0$ results in $\psi_L > 0$, while $\psi_L < 0$ for $X_L < 0$, and this [in Eqs. (12.39) and (12.40)] makes the conclusions in cases (c) and (d) obvious as well. This discussion shows how just a brief (qualitative) inspection of a measured voltage (or current) standing wave diagram of a transmission line can be used to determine the nature of an unknown load connected to the line (at a certain frequency), e.g., the existence (zero or nonzero) and polarity (inductive or capacitive) of the reactive component of the load impedance. For a quantitative determination of the unknown load, as will be shown in an example, based on Eqs. (12.43) and (12.39) or (12.40) and the measured SWR and locations of voltage (or current) maxima (or minima) on the line we can easily find the complete complex $\underline{Z}_L$, even if the operating frequency of the line is not known. The special cases with $\underline{Z}_L = 0$, $|\underline{Z}_L| \to \infty$, and $\underline{Z}_L = Z_0$ will be discussed in a separate section.

Example 12.4 Measurement of Load Impedance Using a Slotted Line

An instrument consisting of an air-filled section of a rigid coaxial line (cable) with a narrow longitudinal slot in the outer conductor through which a movable (sliding) electric probe (short wire antenna) is inserted to sample the electric field and hence voltage between the line conductors, at different locations along the line, is used to measure load impedances at high frequencies. By measurements on such a slotted line terminated in an unknown load, it is determined that the standing wave ratio of the line is $s = 3$, the distance between successive voltage minima $\Delta l = 40$ cm, and the distance of the first voltage minimum from the load $l_{\text{min}} = 12$ cm. The characteristic impedance of the line is $Z_0 = 50\ \Omega$, and losses in the line conductors can be neglected. What is the complex impedance of the load?

Solution From Eq. (12.43), the magnitude of the load reflection coefficient amounts to

$$|\underline{\Gamma}_{\text{L}}| = \frac{s-1}{s+1} = 0.5. \tag{12.48}$$

We see in Fig. 12.4 that adjacent minima (or maxima) on the line are $\Delta l = \lambda_z/2$ apart, which tells us that the wavelength along the line is $\lambda_z = 2\Delta l = 80$ cm. With the use of the expression for the location of the first minimum ($m = 0$) in Eq. (12.40) and relationship $\beta = 2\pi/\lambda_z$, the phase angle of $\underline{\Gamma}_{\text{L}}$ comes out to be

$$\psi_{\text{L}} = -2\beta\left(z_{\text{min}} + \frac{\lambda_z}{4}\right) = \frac{4\pi l_{\text{min}}}{\lambda_z} - \pi = -0.4\pi = -72^\circ \qquad (m = 0;\ \ z_{\text{min}} = -l_{\text{min}}). \tag{12.49}$$

The complex coefficient, Eq. (12.29), is then

$$\underline{\Gamma}_{\text{L}} = |\underline{\Gamma}_{\text{L}}|\,\text{e}^{\text{j}\psi_{\text{L}}} = 0.5\,\text{e}^{-\text{j}72^\circ} = 0.154 - \text{j}0.475, \tag{12.50}$$

and solving Eq. (12.27) for the load complex impedance we obtain

$$\underline{Z}_{\text{L}} = Z_0\,\frac{1+\underline{\Gamma}_{\text{L}}}{1-\underline{\Gamma}_{\text{L}}} = (39.87 - \text{j}50.46)\ \Omega. \tag{12.51}$$

Note that $\underline{Z}_{\text{L}}$ is found even not knowing in advance the operating frequency of the slotted line. The frequency, in fact, can as well be identified from this measurement – as the line dielectric is air, Eqs. (11.17) and (9.19) give $f = c_0/\lambda_z = 375$ MHz.

Example 12.5 Standing Wave Patterns for a Lossy Transmission Line

Sketch voltage and current standing wave patterns for a lossy transmission line with attenuation coefficient and phase angle of the complex characteristic impedance ($\underline{Z}_0$) given by (a) $\alpha = 0.3$ Np/λ_z and $\phi = 0$ and (b) $\alpha = 1.2$ Np/λ_z and $\phi = 20^\circ$, respectively, and a purely resistive load with $R_{\text{L}} = |\underline{Z}_0|/4$.

Solution (a)–(b) If the line in Fig. 12.3 is lossy, the generalized voltage reflection coefficient in Eq. (12.33) becomes (with $\text{j}\beta$ replaced by $\underline{\gamma} = \alpha + \text{j}\beta$)

generalized voltage reflection coefficient for a lossy line

$$\boxed{\underline{\Gamma}(z) = \underline{\Gamma}_{\text{L}}\,\text{e}^{2\underline{\gamma}z} = |\underline{\Gamma}_{\text{L}}|\,\text{e}^{2\alpha z}\,\text{e}^{\text{j}(2\beta z+\psi_{\text{L}})}.} \tag{12.52}$$

With this expression for $\underline{\Gamma}(z)$ and expressions for the incident voltage $\underline{V}_{\text{i}}(z)$ and current $\underline{I}_{\text{i}}(z)$ from Eqs. (12.8) and (12.13) now substituted in Eqs. (12.36), we are able to compute the total rms voltage and current magnitudes, $|\underline{V}(z)|$ and $|\underline{I}(z)|$, as functions of z, along lossy transmission lines, terminated in arbitrary loads. Of course, we also need Eqs. (12.27) and (12.29), to find the magnitude and phase angle of the load reflection coefficient, $\underline{\Gamma}_{\text{L}}$, from $\underline{Z}_{\text{L}}$ and $\underline{Z}_0$. Fig. 12.5 shows the voltage and current standing wave patterns for the two given attenuation coefficients, α, corresponding to a relatively low-loss (practically meaningful) line and a line with large losses (which would seldom be used in practice), respectively, and the load with $\underline{Z}_{\text{L}} = |\underline{Z}_0|/4 + \text{j}0$. We see that, as expected, the values of $|\underline{V}|_{\text{max}}$, $|\underline{V}|_{\text{min}}$, $|\underline{I}|_{\text{max}}$, and $|\underline{I}|_{\text{min}}$ are no longer constant, but increase with distance from the load (or decrease with distance from the generator, toward the load). Hence, for lossy lines, these (different) values, except the ones closest to the generator or load, represent local (and not global) maxima and minima (for a distance of $\lambda_z/2$) of the functions $|\underline{V}(z)|$ and $|\underline{I}(z)|$. However, while the patterns in the low-loss case, Fig. 12.5(a), are not much different from the corresponding no-loss patterns in Fig. 12.4(b), the increase of local maximum and minimum voltage and current values with moving away from the load is quite significant for the high-loss line, Fig. 12.5(b). Moreover, the differences between adjacent maxima and minima for both $|\underline{V}|$ and $|\underline{I}|$ also

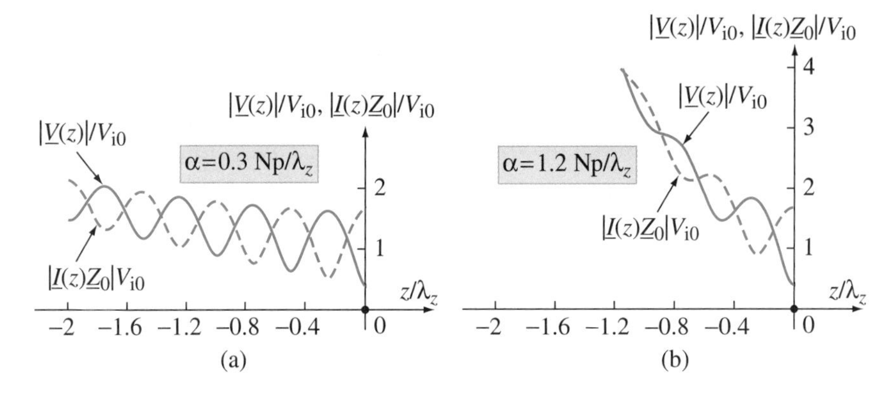

Figure 12.5 Normalized standing wave patterns [computed from Eqs. (12.36), (12.52), (12.8), (12.13), (12.27), and (12.29)] for a lossy transmission line with (a) $\alpha = 0.3\ \text{Np}/\lambda_z$ and $\phi = 0$ and (b) $\alpha = 1.2\ \text{Np}/\lambda_z$ and $\phi = 20°$ (ϕ is the phase angle of $\underline{Z}_0$), and a purely resistive load with $R_L = |\underline{Z}_0|/4$; for Example 12.5.

become smaller and smaller as the coordinate z becomes more and more negative, which, again, is especially evident in the high-loss case. Far away from the load, these differences virtually vanish, and the SWR, Eq. (12.43), becomes nearly unity, as if the line were matched ($\underline{\Gamma}_L = 0$) or infinitely long. This is very simple to explain based on the expression for $\underline{\Gamma}(z)$ in Eq. (12.52), where $|\underline{\Gamma}(z)| \propto e^{2\alpha z}$, and at some point toward the generator – for a sufficiently large $|z|$ (z is negative), $|\underline{\Gamma}(z)| \approx 0$, i.e., $|\underline{V}_r(z)| \approx 0$. In other words, this far away from the load, the reflected wave is so strongly attenuated that the effects of the reflection occurring at the load are unnoticeable. Viewed from that point (z) to the right, the line appears like an infinitely long or matched one. Finally, note that for the line with large losses [case (b)], whose characteristic impedance cannot be assumed to be purely real, a nonzero phase angle ϕ of $\underline{Z}_0$, Eq. (12.12), leads to a phase difference between the total voltage and current on the line. This results in, as opposed to Eq. (12.41), the locations of voltage maxima do not exactly coinciding with the locations of current minima, and vice versa, in Fig. 12.5(b).

Problems: 12.1–12.5; *Conceptual Questions* (on Companion Website): 12.6–12.14; *MATLAB Exercises* (on Companion Website).

12.5 POWER COMPUTATIONS OF TRANSMISSION LINES

We now discuss power flow along a transmission line, in Fig. 12.3, based on its voltage and current. First, let us assume that the line is lossless. Its (purely real) characteristic impedance is Z_0, its complex termination load impedance $\underline{Z}_L = R_L + jX_L$, and the corresponding load reflection coefficient $\underline{\Gamma}_L$, Eq. (12.27). Let the complex rms voltage and current on the line, $\underline{V}(z)$ and $\underline{I}(z)$, be given by Eqs. (12.36), (12.33), (12.8), and (12.13). Using Eq. (11.44) and carrying out the same simple algebraic manipulations as in Eq. (10.57), the net complex power flow in the positive z direction (toward the load) through the cross section of the line defined by the coordinate z ($-l \le z \le 0$) is

$$\begin{aligned}\underline{S}(z) &= \underline{V}(z)[\underline{I}(z)]^* = \underline{V}_i(z)\,[1+\underline{\Gamma}(z)]\,\frac{[\underline{V}_i(z)]^*}{Z_0}\,[1-\underline{\Gamma}(z)]^* \\ &= \frac{|\underline{V}_i(z)|^2}{Z_0}\left(1+\underline{\Gamma}_L\,e^{j2\beta z}\right)\left(1-\underline{\Gamma}_L^*\,e^{-j2\beta z}\right) \\ &= \frac{V_{i0}^2}{Z_0}\left[1-|\underline{\Gamma}_L|^2+2j|\underline{\Gamma}_L|\sin(2\beta z+\psi_L)\right],\end{aligned} \tag{12.53}$$

where $|\underline{V}_i(z)| = |\underline{V}_{i0}| = V_{i0} = \text{const}$ since $|e^{-j\beta z}| = 1$. The imaginary part of $\underline{S}$ represents the reactive power on the line characteristic for standing waves. It determines the rate of energy exchange in time and space between the electric and magnetic fields in the line dielectric, that is, between the line voltage and current, like in an electromagnetic resonator (e.g., a simple resonant LC circuit). On the other side, the real part of $\underline{S}$ equals the net time-average power transfer in the positive z direction by the resultant voltage and current waves along the line,

$$P = P_{\text{ave}} = \text{Re}\{\underline{S}\} = \frac{V_{\text{i0}}^2}{Z_0}\left(1 - |\underline{\Gamma}_{\text{L}}|^2\right). \tag{12.54}$$

This power can be written as

net time-average power flow on a transmission line

$$\boxed{P = P_{\text{i}} - P_{\text{r}},} \tag{12.55}$$

in terms of the time-average powers carried by the incident and reflected waves, P_{i} and P_{r}, respectively. Namely, from Eq. (11.44),

incident and reflected powers, no losses

$$\boxed{P_{\text{i}} = (P_{\text{i}})_{\text{ave}} = \frac{V_{\text{i0}}^2}{Z_0}, \quad P_{\text{r}} = (P_{\text{r}})_{\text{ave}} = \frac{V_{\text{r0}}^2}{Z_0} = |\underline{\Gamma}_{\text{L}}|^2\,\frac{V_{\text{i0}}^2}{Z_0} = |\underline{\Gamma}_{\text{L}}|^2 P_{\text{i}}.} \tag{12.56}$$

Note that all three time-average powers are nonnegative ($P,\ P_{\text{i}},\ P_{\text{r}} \geq 0$), which confirms that $|\underline{\Gamma}_{\text{L}}|$ cannot be greater than unity, Eq. (12.29). Note also that P_{r} appears with a minus sign in Eq. (12.55) because the actual power flow by the reflected wave is in the backward (negative z) direction, opposite to the reference direction (positive z) of the power flow P (as well as P_{i}) in this equation. In other words, given the same reference orientations of all voltages ($\underline{V}$, $\underline{V}_{\text{i}}$, and $\underline{V}_{\text{r}}$) and currents ($\underline{I}$, $\underline{I}_{\text{i}}$, and $\underline{I}_{\text{r}}$) in Fig. 12.2, all resulting in the reference power flow in the forward direction, the reflected current $\underline{I}_{\text{r}}$, Eq. (12.11), must be introduced with a minus sign in Eq. (11.44) to obtain the proper expression for P_{r}. Therefore, with the same minus sign as in Eq. (12.15), the reflected complex power (flowing in the backward direction in Fig. 12.2) is computed as

$$\underline{S}_{\text{r}} = \underline{V}_{\text{r}}(z)[-\underline{I}_{\text{r}}(z)]^* = \underline{V}_{\text{r}}(z)\left[\frac{\underline{V}_{\text{r}}(z)}{Z_0}\right]^* = \frac{|\underline{V}_{\text{r}}(z)|^2}{Z_0} = \frac{V_{\text{r0}}^2}{Z_0} = P_{\text{r}}, \tag{12.57}$$

and this is, of course, the same result as in Eq. (12.56).

We see that the power P in Eq. (12.54) does not depend on z, i.e., $P = \text{const}$ along the entire line, as expected – since the line is assumed to be lossless (no power is lost to heat in series and shunt resistors in Fig. 12.1). This power is, in its entirety, delivered to the load, at $z = 0$, i.e., it is equal to the time-average power dissipated in the load, P_{L}, which, in turn, can be expressed as [see Eq. (8.210)]

time-average load power

$$\boxed{P = P_{\text{L}} = \text{Re}\{\underline{V}_{\text{L}}\underline{I}_{\text{L}}^*\} = \text{Re}\{\underline{Z}_{\text{L}}\}|\underline{I}_{\text{L}}|^2 = R_{\text{L}}|\underline{I}(0)|^2,} \tag{12.58}$$

where $\underline{V}_{\text{L}}$ and $\underline{I}_{\text{L}}$ are the voltage and current of the load, Fig. 12.3, and R_{L} is the load resistance. Moreover, for a given incident power on the line (P_{i}), the load power P_{L} is maximum when $R_{\text{L}} = Z_0$ and $X_{\text{L}} = 0$ (impedance-matched load), that is, when $\underline{\Gamma}_{\text{L}} = 0$ and $P_{\text{r}} = 0$ [Eqs. (12.47) and (12.56)], and hence $P = P_{\text{i}} = P_{\text{L}}$ in Eqs. (12.55) and (12.58). This means that all of the power P_{i} is delivered to the load, or to a network or device (e.g., a transmitting antenna) represented by the resistance R_{L}. In general, the power efficiency of the load impedance match can be expressed through the following ratio:

$$\eta_{\text{load match}} = \frac{P_{\text{L}}}{P_{\text{i}}} = 1 - |\underline{\Gamma}_{\text{L}}|^2 \quad (0 \leq \eta_{\text{load match}} \leq 1), \tag{12.59}$$

with the efficiency coefficient $\eta_{\text{load match}}$ being maximum (unity) for a perfect match of the load impedance to the characteristic impedance of the line ($\underline{\Gamma}_{\text{L}} = 0$), and

$\eta_{\text{load match}} = 0$ resulting from a complete mismatch of impedances at $z = 0$ ($|\underline{\Gamma}_L| = 1$). However, note that the maximum power-handling capability of the system in Fig. 12.3 requires impedance matching of the generator to the line, at the line input terminals ($z = -l$), as well.

Example 12.6 Power Transfer along a Lossy Line with a Matched Load

Find the net time-average power flow in the forward direction on a lossy transmission line with a time-harmonic wave traveling with attenuation and phase coefficients α and β, respectively, along the line. The rms voltage of the wave for $z = 0$ is V_{i0}. Assume that there is no reflected wave on the line, as well as that the line complex characteristic impedance is $\underline{Z}_0$.

Solution For only a single traveling (forward) wave on a lossy line (Fig. 12.3), which means that the line is either terminated in a matched load, or infinitely long, or so lossy that the forward wave practically does not reach the load, and thus there is no reflected wave on the line, $\underline{V}(z) = \underline{V}_i(z)$ and $\underline{I}(z) = \underline{I}_i(z)$ in Eq. (12.53). Using Eqs. (12.8) and (12.13), we then have

$$\underline{S}(z) = \underline{V}_i(z)[\underline{I}_i(z)]^* = \underline{V}_{i0}\, e^{-\alpha z}\, e^{-j\beta z}\, \frac{\underline{V}_{i0}^*}{|\underline{Z}_0|}\, e^{-\alpha z}\, e^{j\beta z}\, e^{j\phi} = \frac{V_{i0}^2}{|\underline{Z}_0|}\, e^{-2\alpha z}\, e^{j\phi}. \tag{12.60}$$

Hence, the time-average power along the line is [also see Eq. (9.96)]

$$\boxed{P(z) = P_i(z) = \text{Re}\{\underline{S}(z)\} = \frac{V_{i0}^2}{|\underline{Z}_0|}\, e^{-2\alpha z} \cos\phi,} \tag{12.61}$$

time-average incident power flow on a lossy line

which is the same dependence on the coordinate z, i.e., an exponential decrease of P with z (due to uniform z-distributed losses along the line), as in Eq. (11.71). As expected, the effective attenuation coefficient for the power is twice that, α, for the voltage and current, as well as for the electric and magnetic fields, of the lossy line. Note that, assuming that the losses on the line are small, the characteristic impedance of the line can approximately be taken to be purely real, Eq. (12.23), and the phase angle ϕ approximately zero ($\cos\phi \approx 1$).

Example 12.7 Power Flow on a Low-Loss Line with an Unmatched Load

Repeat the previous example but for a transmission line with small losses that is terminated in a load whose reflection coefficient with respect to the line is $\underline{\Gamma}_L$.

Solution For a line that is both lossy and with an arbitrary (unmatched) load, both the incident and reflected waves are exponentially attenuated with distance, and the generalized voltage reflection coefficient, $\underline{\Gamma}(z)$, is given by Eq. (12.52). The complex power on the line in Eq. (12.53), with $\underline{Z}_0 \approx Z_0$ (low-loss assumption), now becomes

$$\begin{aligned}\underline{S}(z) &= \frac{|\underline{V}_i(z)|^2}{Z_0}\left(1 + \underline{\Gamma}_L\, e^{2\alpha z}\, e^{j2\beta z}\right)\left(1 - \underline{\Gamma}_L^*\, e^{2\alpha z}\, e^{-j2\beta z}\right) \\ &= \frac{V_{i0}^2}{Z_0}\, e^{-2\alpha z}\left[1 - |\underline{\Gamma}_L|^2\, e^{4\alpha z} + 2j|\underline{\Gamma}_L|\, e^{2\alpha z} \sin(2\beta z + \psi_L)\right].\end{aligned} \tag{12.62}$$

Its real part is

$$P(z) = \frac{V_{i0}^2}{Z_0}\, e^{-2\alpha z}\left(1 - |\underline{\Gamma}_L|^2\, e^{4\alpha z}\right) = P_i(z) - P_r(z), \tag{12.63}$$

and the incident- and reflected-wave power components,

$$\boxed{P_i(z) = \frac{V_{i0}^2}{Z_0}\, e^{-2\alpha z}, \quad P_r(z) = |\underline{\Gamma}_L|^2\, \frac{V_{i0}^2}{Z_0}\, e^{2\alpha z} = |\underline{\Gamma}(z)|^2 P_i(z).} \tag{12.64}$$

incident and reflected powers on a low-loss line

Note that both $\alpha \neq 0$ and $\underline{\Gamma}_L \neq 0$ cause respective reductions in $P(z)$, when compared to the ideal case (with no Joule's losses and no reflection at the load). Obviously, the expression for $P(z)$ in Eq. (12.63) resumes its no-loss form in Eq. (12.54) when $\alpha = 0$, and reduces to its perfect-match, low-loss equivalent in Eq. (12.61) with $\phi = 0$ when $\underline{\Gamma}_{\rm L} = 0$. Note also that far away from the load (theoretically, for $z \to -\infty$), $P_{\rm r}(z) \approx 0$ (see Fig. 12.5).

Problems: 12.6; *Conceptual Questions* (on Companion Website): 12.15 and 12.16; *MATLAB Exercises* (on Companion Website).

12.6 TRANSMISSION-LINE IMPEDANCE

As pointed out on several occasions so far, the proportionality between the voltage and current of a transmission line (Fig. 12.3) via its characteristic impedance, $\underline{Z}_0$, takes place only for a single traveling (forward or backward) wave, and not for a general solution for $\underline{V}(z)$ and $\underline{I}(z)$, on the line, Eq. (12.16). Combining Eqs. (12.36), (12.15), and (12.52), the total voltage to current ratio expressed in terms of $\underline{Z}_0$ and either the generalized voltage reflection coefficient, $\underline{\Gamma}(z)$, or the load voltage reflection coefficient, $\underline{\Gamma}_{\rm L}$, and the complex propagation coefficient, $\underline{\gamma}$, of the line amounts to

transmission-line impedance (unit: Ω)

$$\underline{Z}(z) = \frac{\underline{V}(z)}{\underline{I}(z)} = \underline{Z}_0 \frac{1+\underline{\Gamma}(z)}{1-\underline{\Gamma}(z)} = \underline{Z}_0 \frac{1+\underline{\Gamma}_{\rm L}\,{\rm e}^{2\underline{\gamma} z}}{1-\underline{\Gamma}_{\rm L}\,{\rm e}^{2\underline{\gamma} z}} \quad (-l \le z \le 0), \tag{12.65}$$

with $\underline{\Gamma}_{\rm L}$ given by Eq. (12.27). This ratio, denoted simply by $\underline{Z}(z)$, represents the so-called transmission-line impedance, seen at a line cross section defined by the coordinate z looking toward the load. In other words, $\underline{Z}(z)$ equals the complex impedance of an equivalent load that can be used to completely replace the portion of the line beyond this cross section, including the (original) load of impedance $\underline{Z}_{\rm L}$, with respect to the rest of the line (and the generator), as illustrated in Fig. 12.6. By such equivalency, the voltage $\underline{V}(z')$ and current $\underline{I}(z')$ on the first portion of the line, for $-l \le z' \le z$, are identical in the original structure, Fig. 12.6(a), and the structure with the equivalent load of impedance $\underline{Z}(z)$, Fig. 12.6(b). From Eq. (12.65),

generalized reflection coefficient via line impedance

$$\underline{\Gamma}(z) = \frac{\underline{Z}(z) - \underline{Z}_0}{\underline{Z}(z) + \underline{Z}_0}, \tag{12.66}$$

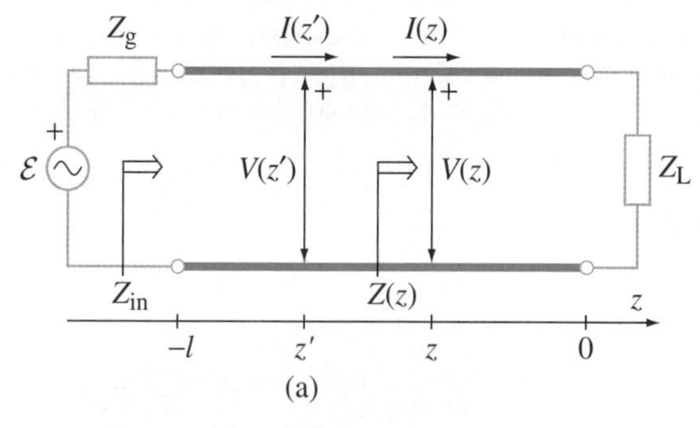

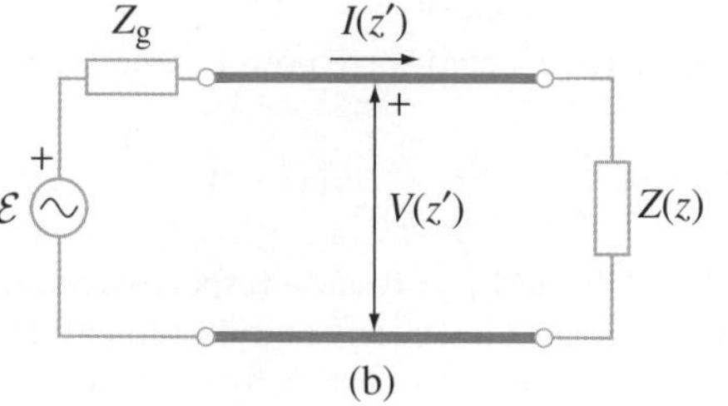

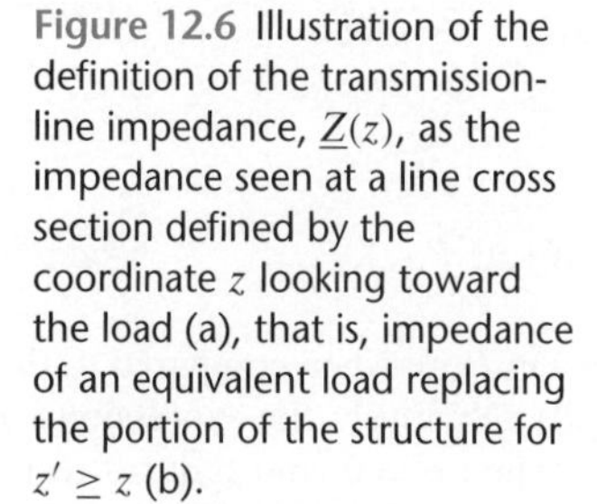

Figure 12.6 Illustration of the definition of the transmission-line impedance, $\underline{Z}(z)$, as the impedance seen at a line cross section defined by the coordinate z looking toward the load (a), that is, impedance of an equivalent load replacing the portion of the structure for $z' \ge z$ (b).

and this can also be obtained from Fig. 12.6(b), in the context of which $\underline{\Gamma}(z)$ stands for the load voltage reflection coefficient, for the load impedance $\underline{Z}(z)$, and thus Eq. (12.27) applies. Like any (load) impedance [Eq. (8.209)], $\underline{Z}(z)$ can be written as

$$\underline{Z}(z) = R(z) + \mathrm{j}X(z), \tag{12.67}$$

where $R(z)$ and $X(z)$ are termed, respectively, the line resistance and reactance (at a given coordinate z along the line). The line admittance (at any z) is

$$\underline{Y}(z) = \frac{1}{\underline{Z}(z)} = \underline{Y}_0 \frac{1 - \underline{\Gamma}_{\mathrm{L}} \, \mathrm{e}^{2\underline{\gamma} z}}{1 + \underline{\Gamma}_{\mathrm{L}} \, \mathrm{e}^{2\underline{\gamma} z}} = G(z) + \mathrm{j}B(z), \tag{12.68}$$

line admittance (unit: S)

with the representation in terms of the line conductance (G) and susceptance (B), and $\underline{Y}_0 = 1/\underline{Z}_0$ being the (complex) characteristic admittance of the line [see Eq. (11.35)].

In a limit of $z = 0$, the line impedance at the load terminals in Fig. 12.6(a) becomes, using Eqs. (12.65) and (12.51),

$$\underline{Z}(0) = \underline{Z}_0 \frac{1 + \underline{\Gamma}_{\mathrm{L}}}{1 - \underline{\Gamma}_{\mathrm{L}}} = \underline{Z}_{\mathrm{L}}, \tag{12.69}$$

i.e., it turns out to be equal to the load impedance, $\underline{Z}_{\mathrm{L}}$, as expected. In the other limit, for $z = -l$, $\underline{Z}$ at the generator terminals represents the equivalent input impedance of the entire line in Fig. 12.6(a), replacing the combination of the line and load with respect to the generator, and we mark it as $\underline{Z}_{\mathrm{in}}$. From Eq. (12.65), it is computed as

$$\underline{Z}_{\mathrm{in}} = \underline{Z}(-l) = \underline{Z}_0 \frac{1 + \underline{\Gamma}_{\mathrm{L}} \, \mathrm{e}^{-2\underline{\gamma} l}}{1 - \underline{\Gamma}_{\mathrm{L}} \, \mathrm{e}^{-2\underline{\gamma} l}}. \tag{12.70}$$

We note that this expression for the impedance $\underline{Z}_{\mathrm{in}}$ has the identical form as the one given by Eqs. (10.156), and hence the alternative expression analogous to that in Eq. (10.159):

$$\underline{Z}_{\mathrm{in}} = \underline{Z}_0 \frac{\underline{Z}_{\mathrm{L}} \cosh \underline{\gamma} l + \underline{Z}_0 \sinh \underline{\gamma} l}{\underline{Z}_0 \cosh \underline{\gamma} l + \underline{Z}_{\mathrm{L}} \sinh \underline{\gamma} l}. \tag{12.71}$$

input impedance – lossy line

Similarly to the structure in Fig. 10.25, this equation defines the impedance transformation property of transmission lines. Namely, a transmission line [Fig. 12.6(a)] of given parameters $\underline{Z}_0$, $\underline{\gamma}$, and l transforms the termination impedance $\underline{Z}_{\mathrm{L}}$ to the equivalent impedance $\underline{Z}_{\mathrm{in}}$ at its input terminals. Here, $\underline{Z}_{\mathrm{L}}$ may be the input impedance of another passive network (Fig. 12.2), such as an arbitrary RLC electric circuit, another transmission line, a transmitting antenna, etc. As we shall see throughout the rest of this chapter, there are numerous applications of transmission lines as impedance transformers.

If the line in Fig. 12.6(a) is (or can approximately be treated as) lossless, $\underline{\gamma}$ is purely imaginary ($\underline{\gamma} = \mathrm{j}\beta$), Eq. (12.19), and $\underline{Z}_0$ purely real ($\underline{Z}_0 = Z_0$), Eq. (12.18), so that, in analogy with Eq. (10.160), the input impedance of the line becomes

$$\underline{Z}_{\mathrm{in}} = Z_0 \frac{\underline{Z}_{\mathrm{L}} \cos \beta l + \mathrm{j} Z_0 \sin \beta l}{Z_0 \cos \beta l + \mathrm{j} \underline{Z}_{\mathrm{L}} \sin \beta l}. \tag{12.72}$$

input impedance – lossless line

Since, from Eqs. (12.10),

$$\beta l = 2\pi \frac{l}{\lambda_z} \quad \longrightarrow \quad l_{\mathrm{e}} = \frac{l}{\lambda_z}, \tag{12.73}$$

electrical length of a line

we realize that $\underline{Z}_{\text{in}}$ actually depends on the electrical length of the line, l_e, defined as the ratio of the physical length of the line, l, to the wavelength along the line, λ_z. Note that the line impedance in Eq. (12.65), for an arbitrary z along the line, can now be written as

no-loss impedance at an arbitrary z

$$\underline{Z}(z) = Z_0 \frac{\underline{Z}_L \cos\beta z - jZ_0 \sin\beta z}{Z_0 \cos\beta z - j\underline{Z}_L \sin\beta z}. \quad (12.74)$$

In the analysis and design of transmission lines, it is often convenient to deal with all impedances using their normalized values, with the characteristic impedance of the line (Z_0) serving as the normalization constant. From Eq. (12.67), the normalized transmission-line impedance, denoted by $\underline{z}_n$, is given by

normalized impedance (dimensionless)

$$\underline{z}_n = \frac{\underline{Z}}{Z_0} = \frac{R}{Z_0} + j\frac{X}{Z_0} = r + jx, \quad (12.75)$$

where we keep in mind that all the quantities except Z_0 are functions of the coordinate z. The real and imaginary parts of $\underline{z}_n$, equal to

$$r = \frac{R}{Z_0}, \quad x = \frac{X}{Z_0}, \quad (12.76)$$

are termed the normalized line resistance and reactance, respectively. Similarly, using Eq. (12.68), we introduce the normalized line admittance,

normalized admittance (dimensionless)

$$\underline{y}_n = \frac{\underline{Y}}{Y_0} = \frac{Z_0}{\underline{Z}} = \frac{1}{\underline{z}_n} = g + jb, \quad (12.77)$$

with the normalized line conductance and susceptance given, respectively, by

$$g = \frac{G}{Y_0}, \quad b = \frac{B}{Y_0}. \quad (12.78)$$

Of course, all lowercase (normalized) quantities in Eqs. (12.75) and (12.77) are dimensionless. In the case of a transmission line with substantial losses such that imaginary parts of complex $\underline{Z}_0$ and $\underline{Y}_0$ cannot be neglected, $\underline{Z}$ and $\underline{Y}$ are normalized to $|\underline{Z}_0|$ and $|\underline{Y}_0|$, respectively.

Example 12.8 Quarter-Wave Transformer Matching

At a frequency of $f_1 = 1$ GHz, the input impedance of an antenna is purely real and equal to $R_A = 200\ \Omega$. This antenna needs to be impedance-matched to a transmission line having a characteristic impedance of $Z_{01} = 50\ \Omega$. To this end, (a) design a transmission-line section (find its characteristic impedance and length) such that it transforms R_A to Z_{01}, namely, that the input impedance of the line section with the antenna as load equals Z_{01}. (b) If the frequency is changed to $f_2 = 1.4$ GHz, determine the percentage of the incident power that is bounced back off the connection of the matching line in (a) and the feeding line (of characteristic impedance Z_{01}), assuming that the antenna impedance is unchanged.

Solution

(a) We realize that this impedance-matching (no reflection) requirement is completely analogous to that in Eqs. (10.161), rewritten here, of course, in terms of the input impedance of the line section, $\underline{Z}_{\text{in}}$, given by Eq. (12.72). Therefore, we need a line acting like a quarter-wave transformer, as in Eqs. (10.165), which become

quarter-wave transformer

$$\underline{Z}_{\text{in}} = Z_{01} \longrightarrow Z_{0t} = \sqrt{Z_{01} R_L} \quad \text{and} \quad l = \frac{\lambda_t}{4}, \quad (12.79)$$

where $R_L = R_A$ and λ_t denotes the wavelength of the transformer. For the given numerical data, the characteristic impedance of the transformer amounts to $Z_{0t} = 100\ \Omega$. Its length, combining Eqs. (8.112) and (9.47) and assuming that the line dielectric is polyethylene ($\varepsilon_r = 2.25$), for instance, comes out to be $l = \lambda_0/(4\sqrt{\varepsilon_r}) = c_0/(4\sqrt{\varepsilon_r} f_1) = 5$ cm [or any odd integer multiple of this – see Eqs. (10.164)], with c_0 and λ_0 standing for the free-space wave velocity and wavelength at the operating frequency (f_1) of the structure.

(b) As discussed in Example 10.26, the principal drawback of quarter-wave transformer matching in general is its inherent resonant (single-frequency) operation (which can be made more broadband by cascading multiple quarter-wave sections). In particular, if the operating frequency of the transformer in (a) is changed to $f_2 = 1.4$ GHz, βl ($l = 5$ cm) in Eqs. (10.164) is no longer $\pi/2$, but $\beta l = 2\pi f_2 \sqrt{\varepsilon_r}\, l/c_0 = 2.2$ (rad). With this, the input impedance of the line, Eq. (12.72), the associated generalized voltage reflection coefficient, Eq. (12.66), and the percentage of the time-average incident power that is reflected from the transformer with the antenna as load, Eqs. (12.56), are computed as follows:

$$\underline{Z}_{in} = Z_{0t} \frac{R_A \cos\beta l + jZ_{0t} \sin\beta l}{Z_{0t} \cos\beta l + jR_A \sin\beta l} = 83\, e^{j35.5^\circ}\ \Omega \longrightarrow \underline{\Gamma}_{in} = \frac{\underline{Z}_{in} - Z_{01}}{\underline{Z}_{in} + Z_{01}}$$

$$= 0.404\, e^{j47.7^\circ} \longrightarrow \frac{P_r}{P_i} = |\underline{\Gamma}_{in}|^2 = 16.31\% \qquad (12.80)$$

(see the similar computation in Example 10.27).

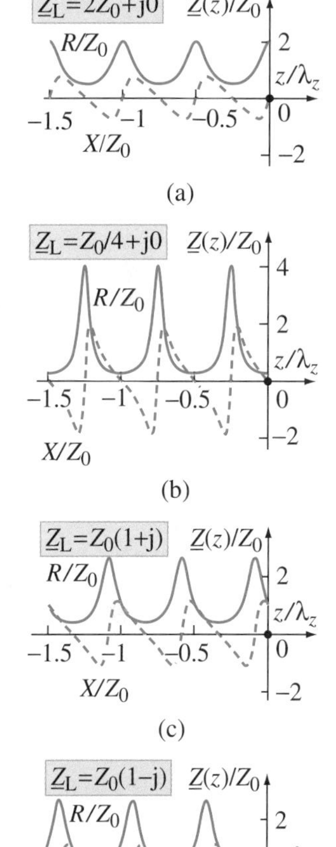

Figure 12.7 Plots of the real (R) and imaginary (X) parts of the impedance $\underline{Z}$ in Eq. (12.74), normalized according to Eq. (12.75), against z/λ_z for the lossless transmission line and four different load impedances ($\underline{Z}_L$) in Fig. 12.4; for Example 12.10.

Example 12.9 Impedance Inverter

Show that the normalized input impedance of a quarter-wave transformer (with no losses) is exactly the reciprocal of its normalized complex load impedance.

Solution Substituting $l = \lambda_t/4$ or $\beta l = \pi/2$ in the impedance expression in Eq. (12.72), we obtain

$$\underline{Z}_{in} = \frac{Z_0^2}{\underline{Z}_L} \longrightarrow \frac{\underline{Z}_{in}}{Z_0} = \frac{1}{\underline{Z}_L/Z_0} \longrightarrow \underline{z}_{in} = \frac{1}{\underline{z}_L}, \qquad (12.81)$$

where $\underline{z}_{in}$ and $\underline{z}_L$ are normalized (to Z_0) values ($\underline{z}_n$), according to Eq. (12.75), of the input and load impedances, $\underline{Z}_{in}$ and $\underline{Z}_L$, respectively, of the transformer. So, indeed, a quarter-wave long transmission line transforms the normalized termination complex impedance $\underline{z}_L$ to its reciprocal (also normalized) at the line input terminals. This is why the quarter-wave transformer is also referred to as the impedance inverter.

Note that Eqs. (12.81) can as well be derived from Eqs. (12.79) in the case of a purely resistive load. However, we also note that the quarter-wave matching requirements, in Eqs. (12.79), are more stringent than the impedance inverting property. In other words, every $l = \lambda_t/4$ long section of a transmission line inverts normalized impedances, but does not necessarily provide impedance matching.

Example 12.10 Maximum and Minimum Resistances of a Lossless Line

Consider the lossless transmission line and four different load impedances, $\underline{Z}_L$, from Example 12.3, and sketch the dependence of the real and imaginary parts of the transmission-line impedance, $\underline{Z}$, on the coordinate z along the line. Also show that both the maximum and minimum value of the real part of $\underline{Z}(z)$ (line resistance) can be expressed in terms of the characteristic impedance and standing wave ratio of the line.

Solution Fig. 12.7 shows the plots of the real and imaginary parts of the impedance $\underline{Z}(z)$, Eqs. (12.74) and (12.67), corresponding to the standing wave patterns in Fig. 12.4. We see from the plots that at the locations of voltage maxima on the line, $z = z_{max}$, which coincide with the locations of current minima, $\underline{Z}$ is purely resistive ($X = 0$) and R attains its maximum value, R_{max}, whereas at the locations of voltage minima (current maxima), $z = z_{min}$, $\underline{Z}$ is purely resistive and equal to R_{min}. Combining Eqs. (12.36), (12.8), (12.33), (12.39), (12.9), (12.13), and (12.42), we then realize that

$$\underline{V}(z_{\text{max}}) = \underline{V}_{\text{i}}(z_{\text{max}})\,[1+\underline{\Gamma}(z_{\text{max}})] = \underline{V}_{\text{i0}}\,\text{e}^{-\text{j}\beta z_{\text{max}}}\,(1+|\underline{\Gamma}_{\text{L}}|) = |\underline{V}|_{\text{max}}\,\text{e}^{\text{j}(\theta_{\text{i0}}-\beta z_{\text{max}})},$$

$$\underline{I}(z_{\text{max}}) = \frac{\underline{V}_{\text{i0}}}{Z_0}\,\text{e}^{-\text{j}\beta z_{\text{max}}}\,(1-|\underline{\Gamma}_{\text{L}}|) = |\underline{I}|_{\text{min}}\,\text{e}^{\text{j}(\theta_{\text{i0}}-\beta z_{\text{max}})} \qquad \left[\text{e}^{\text{j}(2\beta z_{\text{max}}+\psi_{\text{L}})}=1\right], \quad (12.82)$$

namely, that $\underline{V}(z_{\text{max}})$ and $\underline{I}(z_{\text{max}})$ are in phase. With this, and the help of Eqs. (12.65), (12.67), and (12.43), the maximum of the line resistance is given by

maximum resistance of a transmission line

$$\boxed{R_{\text{max}} = R(z_{\text{max}}) = \frac{\underline{V}(z_{\text{max}})}{\underline{I}(z_{\text{max}})} = \frac{|\underline{V}|_{\text{max}}}{|\underline{I}|_{\text{min}}} = Z_0\,\frac{1+|\underline{\Gamma}_{\text{L}}|}{1-|\underline{\Gamma}_{\text{L}}|} = sZ_0 \quad [X(z_{\text{max}})=0],} \quad (12.83)$$

where s is the standing wave ratio of the line. In a similar fashion, its minimum is

minimum resistance

$$\boxed{R_{\text{min}} = R(z_{\text{min}}) = \frac{\underline{V}(z_{\text{min}})}{\underline{I}(z_{\text{min}})} = \frac{|\underline{V}|_{\text{min}}}{|\underline{I}|_{\text{max}}} = Z_0\,\frac{1-|\underline{\Gamma}_{\text{L}}|}{1+|\underline{\Gamma}_{\text{L}}|} = \frac{Z_0}{s} \quad [X(z_{\text{min}})=0].} \quad (12.84)$$

So, $R(z)$ varies between Z_0/s and sZ_0 along the line (note that $s \geq 1$). On the other side, the line reactance $X(z)$ changes its sign at $z = z_{\text{max}}$ and z_{min}, i.e., the complex $\underline{Z}(z)$ alternates from an inductive impedance $[X(z) > 0]$ to a capacitive one $[X(z) < 0]$ or vice versa at intervals of $\lambda_z/4$. For purely resistive loads ($X_{\text{L}} = 0$), Fig. 12.7(a) and Fig. 12.7(b), either the first voltage maximum or minimum is at the load terminals ($z = 0$), and we have that either $R_{\text{max}} = R_{\text{L}}$ (when $R_{\text{L}} > Z_0$) or $R_{\text{min}} = R_{\text{L}}$ (when $R_{\text{L}} < Z_0$). Hence, R_{L} equals either sZ_0 or Z_0/s, which, for instance, can be used to determine the resistance of an unknown (purely resistive) load based on the known Z_0 for the line and measured SWR. (The case $R_{\text{L}} = Z_0$ will be discussed in a later section.) For complex loads ($X_{\text{L}} \neq 0$), Fig. 12.7(c) and Fig. 12.7(d), the locations of $R = R_{\text{max}}$ and $R = R_{\text{min}}$ ($X = 0$) are shifted with respect to the load position, with the resistance maximum being closer to the load than the minimum if $X_{\text{L}} > 0$ (inductive load), and vice versa if $X_{\text{L}} < 0$ (capacitive load). Note that each of the graphs in Fig. 12.7 can as well be considered as representing the line input impedance $\underline{Z}_{\text{in}}$, Eq. (12.72), for different lengths l of the line ($l = |z|$) and a particular load impedance. As such, they show a wide range of possible impedance transformations from $\underline{Z}_{\text{L}}$ to $\underline{Z}_{\text{in}}$ that can be obtained with different choices of the electrical length of the line, Eq. (12.73).

Example 12.11 Quarter-Wave Transformer Circuit for a Complex Load

Design a matching circuit using a quarter-wave transformer for the antenna from Example 12.8 if its input impedance equals $\underline{Z}_{\text{L}} = (15 + \text{j}35)\ \Omega$ at the frequency f_1.

Solution Since the load (antenna) impedance now is not purely real, we first need to compensate (annul) its imaginary part, which can be done by moving along the feeding transmission line (of characteristic impedance $Z_{01} = 50\ \Omega$) away from the load (toward the generator) to a location where the transmission-line impedance, $\underline{Z}$, is purely resistive ($X = 0$). In other words, we insert another line section between the load and a quarter-wave transformer, as shown in Fig. 12.8. In general, all possible locations at which $X = 0$ are those in Eqs. (12.83) and (12.84). However, to make the length l_{c} of the compensating section the smallest possible, we recall from the discussion of voltage and current standing wave patterns

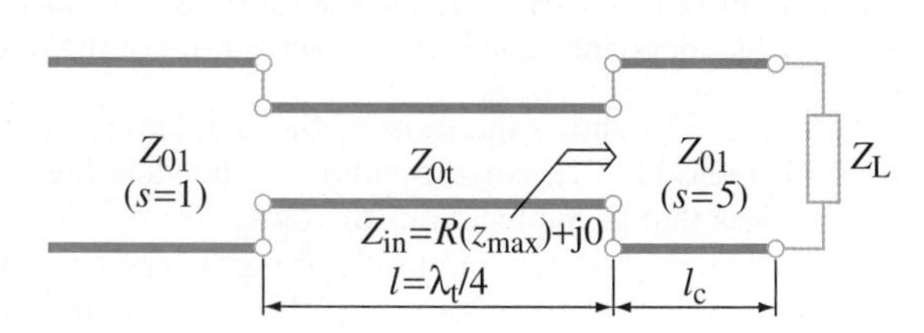

Figure 12.8 Inserting a quarter-wave transformer ($l = \lambda_{\text{t}}/4$) in the feeding transmission line of an antenna with an inductive complex input impedance, at a location on the line where its impedance ($\underline{Z}$) is purely resistive; for Example 12.11.

in Fig. 12.4 that in the case of an inductive complex load ($X_L > 0$), such as the one in Fig. 12.8 ($X_L = 35\ \Omega$), the first voltage maximum (and not minimum) comes first, moving away from the load, and that is where we choose to connect the transformer. Using Eqs. (12.27) and (12.29), on the other side, the magnitude and phase angle of the reflection coefficient at the antenna terminals are $|\underline{\Gamma}_L| = 0.67$ and $\psi_L = 106.7° = 1.86$ rad, respectively, and the associated SWR, Eq. (12.43), amounts to $s = 5$. Eq. (12.39) with $m = 0$ then tells us that the location of the first maximum is given by

$$z_{max} = -\frac{\psi_L}{2\beta} = -\frac{1.86\lambda_z}{4\pi} = -0.148\lambda_z \qquad (m = 0). \tag{12.85}$$

Hence, the length of the compensating line section is $l_c = 0.148\lambda_z \approx 3$ cm, where we assume that the dielectric of this section is polyethylene as well, so the wavelength along it is the same as along the transformer ($\lambda_z = \lambda_t = 20$ cm). From Eq. (12.83), the input impedance of the section is

$$\underline{Z}_{in} = R(z_{max}) + j0 = sZ_{01} = 250\ \Omega \qquad [X(z_{max}) = 0]. \tag{12.86}$$

Finally, to transform this purely resistive impedance to Z_{01} (for the main feeding line), and fulfill the overall no-reflection condition ($s = 1$), the characteristic impedance of the quarter-wave transformer in Fig. 12.8 is computed as in Eqs. (12.79), and it turns out to be

$$Z_{0t} = \sqrt{Z_{01}R(z_{max})} = 112\ \Omega \qquad \left(l = \frac{\lambda_t}{4} = 5\ \text{cm}\right). \tag{12.87}$$

Example 12.12 Impedance Plots for Lossy Lines

Sketch the behavior of the real and imaginary parts of the transmission-line impedance along the line for the two cases of lossy transmission lines from Example 12.5.

Solution Using Eqs. (12.65) and (12.75), this impedance behavior, namely, the dependence of the impedance $\underline{Z}(z)$ [or $\underline{Z}_{in}(l)$] on z (or l), is illustrated in Fig. 12.9. We see that the periodicity in repetitions of the same values in R and X with successive movements by $\lambda_z/2$ away from the load (observed for the lossless line in Fig. 12.7) does not take place any more, which is especially evident in the high-loss case. The values of R_{max} and R_{min} are not constant, but decrease with distance from the load, due to losses in the line. Far away from the load, $R_{max} \approx R_{min}$, and $\underline{Z}(z) \approx \underline{Z}_0$, as if the line were infinitely long or terminated in a matched load [$\underline{\Gamma}_L = 0$ in Eq. (12.65)]. Note that this phenomenon is explained in the discussion of patterns in Fig. 12.5. In addition, because of a nonzero (nonnegligible) phase angle ϕ of $\underline{Z}_0$ for the high-loss line, the locations where $R = R_{max}$ and $R = R_{min}$ do not exactly coincide with the zeros of X in this case.

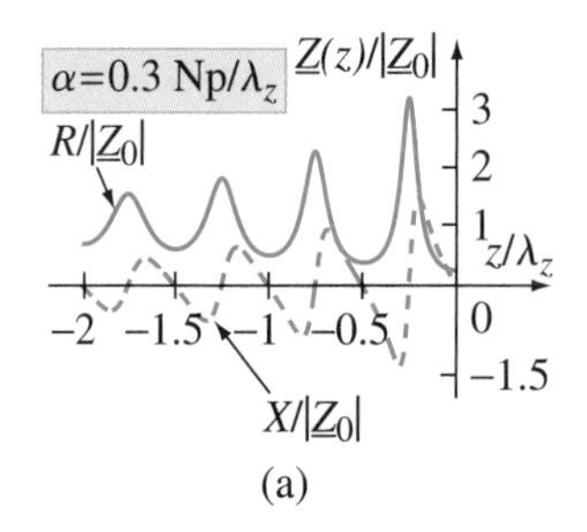

(a)

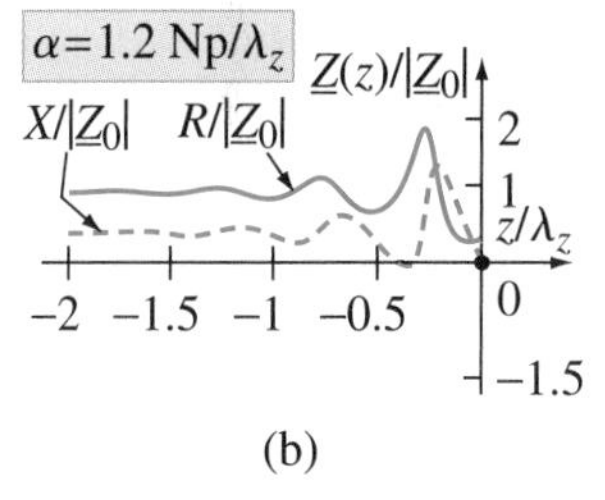

(b)

Figure 12.9 Normalized impedance (resistance and reactance) plots [computed from Eqs. (12.65) and (12.75), with $|\underline{Z}_0|$ as the normalization constant] for two lossy transmission lines in Fig. 12.5; for Example 12.12.

Problems: 12.7–12.11; *Conceptual Questions* (on Companion Website): 12.17–12.20; *MATLAB Exercises* (on Companion Website).

12.7 COMPLETE SOLUTION FOR LINE VOLTAGE AND CURRENT

With the concept of the input impedance of a transmission line ($\underline{Z}_{in}$) in hand, and the expressions in Eqs. (12.70)–(12.72) for its computation, it is now a very simple matter to express the constant $\underline{V}_{i0}$ in Eqs. (12.26) using the parameters of the voltage generator at the beginning of the line in Fig. 12.3, namely, its complex rms emf $\underline{\mathcal{E}}$ and internal impedance $\underline{Z}_g$. Fig. 12.10 shows a version of the equivalent circuit in Fig. 12.6(b) – for $z = -l$, so with the entire line and the load in Fig. 12.6(a) replaced by $\underline{Z}_{in}$. In this circuit, by which we are essentially taking into account the boundary

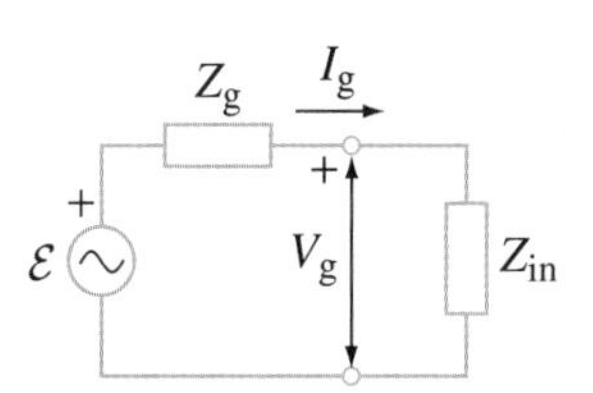

Figure 12.10 Equivalent circuit of a transmission line in Fig. 12.6(a) as seen from the generator terminals.

conditions at the generator terminals ($z = -l$) in Fig. 12.3, $\underline{Z}_{\rm in}$ and $\underline{Z}_{\rm g}$ form a voltage divider, that gives the following expression for the voltage $\underline{V}_{\rm g}$ of the generator:

$$\underline{V}_{\rm g} = \frac{\underline{Z}_{\rm in}}{\underline{Z}_{\rm in} + \underline{Z}_{\rm g}}\,\underline{\mathcal{E}}. \tag{12.88}$$

From Eqs. (12.36), (12.8), and (12.52), on the other side, this same voltage is

$$\underline{V}_{\rm g} = \underline{V}(-l) = \underline{V}_{\rm i0}\,{\rm e}^{\underline{\gamma} l}\left(1 + \underline{\Gamma}_{\rm L}\,{\rm e}^{-2\underline{\gamma} l}\right), \tag{12.89}$$

and hence the solution for $\underline{V}_{\rm i0}$,

solution for $\underline{V}_{\rm i0}$ in Eq. (12.8)

$$\boxed{\underline{V}_{\rm i0} = \frac{\underline{Z}_{\rm in}\underline{\mathcal{E}}\,{\rm e}^{-\underline{\gamma} l}}{(\underline{Z}_{\rm in} + \underline{Z}_{\rm g})(1 + \underline{\Gamma}_{\rm L}\,{\rm e}^{-2\underline{\gamma} l})}.} \tag{12.90}$$

With it, and the solution for $\underline{V}_{\rm r0}$ [$\underline{V}_{\rm r0} = \underline{\Gamma}_{\rm L}\underline{V}_{\rm i0}$, Eq. (12.27)], we are then able to express the voltage $\underline{V}(z)$ and current $\underline{I}(z)$ along the line, Eqs. (12.7) and (12.11), in terms of the parameters of the line terminal networks and operating frequency (f) of the structure, in addition to the length (l) and other characteristics of the line itself. Examples in this section illustrate obtaining such complete solutions for line voltage and current. Of course, we can also use the expression for $\underline{V}_{\rm i0}$ in Eq. (12.90) to rewrite the solutions for all other associated quantities of interest, like the time-average power $P(z)$ along the line, Eq. (12.63).

Example 12.13 Complete Circuit Analysis of a Lossless Transmission Line

A lossless transmission line of length $l = 4.25$ m and characteristic impedance $Z_0 = 50\ \Omega$ is driven by a time-harmonic voltage generator of frequency $f = 75$ MHz. The emf of the generator has rms value of $\mathcal{E} = 20$ V and zero initial phase; its internal impedance is purely real and equal to $R_{\rm g} = 20\ \Omega$. At the other end, the line is terminated in a load whose complex impedance is $\underline{Z}_{\rm L} = (100 + {\rm j}50)\ \Omega$. The relative permittivity of the line dielectric is $\varepsilon_{\rm r} = 4$ ($\mu_{\rm r} = 1$) Find (a) complex and instantaneous total voltages and currents along the line, and (b) time-average loss powers in the load and generator, respectively.

Solution

(a) Using Eqs. (11.17) and (9.47) to compute the phase coefficient of the line, we can write

$$\beta = \frac{\omega}{v_{\rm p}} = \frac{2\pi f\sqrt{\varepsilon_{\rm r}}}{c_0} = \pi\ {\rm rad/m} \quad\longrightarrow\quad \tan\beta l = \tan 4.25\pi = \tan\frac{\pi}{4} = 1, \tag{12.91}$$

so that the input impedance of the line, Eq. (12.72), amounts to

$$\underline{Z}_{\rm in} = Z_0\frac{\underline{Z}_{\rm L} + {\rm j}Z_0\tan\beta l}{Z_0 + {\rm j}\underline{Z}_{\rm L}\tan\beta l} = (50 - {\rm j}50)\ \Omega. \tag{12.92}$$

From Eq. (12.27), on the other side, the load reflection coefficient is $\underline{\Gamma}_{\rm L} = 0.4 + {\rm j}0.2$, and hence its magnitude and phase angle, Eq. (12.29), come out to be $|\underline{\Gamma}_{\rm L}| = 0.447$ and $\psi_{\rm L} = 26.6^\circ$, respectively. Having in mind that $\underline{\mathcal{E}} = \mathcal{E}\,{\rm e}^{{\rm j}0}$ and $\underline{Z}_{\rm g} = R_{\rm g} + {\rm j}0$, the version of Eq. (12.90) for lossless lines then yields

$$\underline{V}_{\rm i0} = \frac{\underline{Z}_{\rm in}\mathcal{E}\,{\rm e}^{-{\rm j}\beta l}}{(\underline{Z}_{\rm in} + R_{\rm g})\left[1 + |\underline{\Gamma}_{\rm L}|\,{\rm e}^{{\rm j}(-2\beta l + \psi_{\rm L})}\right]} = 13\,{\rm e}^{-{\rm j}36^\circ}\ {\rm V}. \tag{12.93}$$

With this, the expression in Eq. (12.44) for the total complex rms voltage of the line (in Fig. 12.3) becomes

$$\underline{V}(z) = \underline{V}_{\rm i0}\left[{\rm e}^{-{\rm j}\beta z} + |\underline{\Gamma}_{\rm L}|\,{\rm e}^{{\rm j}(\beta z + \psi_{\rm L})}\right] = 13\,{\rm e}^{-{\rm j}36^\circ}\left[{\rm e}^{-{\rm j}\pi z} + 0.447\,{\rm e}^{{\rm j}(\pi z + 26.6^\circ)}\right]{\rm V}$$

$$(-l \le z \le 0;\quad z \text{ in m}). \tag{12.94}$$

Similarly, the complex current distribution along the line is given by

$$\underline{I}(z) = \frac{\underline{V}_{i0}}{Z_0}\left[e^{-j\beta z} - |\underline{\Gamma}_L|\, e^{j(\beta z+\psi_L)}\right] = 260\, e^{-j36°}\left[e^{-j\pi z} - 0.447\, e^{j(\pi z+26.6°)}\right] \text{mA}. \tag{12.95}$$

Finally, by means of Eq. (8.66), we convert these complex expressions to their time-domain (instantaneous) counterparts:

$$v(z,t) = [18.37\cos(4.71\times10^8 t - \pi z - 36°) + 8.21\cos(4.71\times10^8 t + \pi z - 9.4°)]\text{V},$$

$$i(z,t) = [367.6\cos(4.71\times10^8 t - \pi z - 36°) - 164.3\cos(4.71\times10^8 t + \pi z - 9.4°)]\,\text{mA}$$

$$(t \text{ in s};\ z \text{ in m}). \tag{12.96}$$

(b) Specifying, respectively, $z = 0$ and $z = -l = -4.25$ m in Eq. (12.95), we obtain the complex current intensities of the load and generator: $\underline{I}_L = \underline{I}(0) = 164.4\, e^{-j54.5°}$ mA and $\underline{I}_g = \underline{I}(-l) = 233\, e^{j35.5°}$ mA. Invoking Eq. (12.58), the time-average power dissipated in the load and generator (its internal resistance) are

$$P_L = R_L|\underline{I}_L|^2 = 2.7 \text{ W} \quad (R_L = \text{Re}\{\underline{Z}_L\} = 100\ \Omega) \quad \text{and} \quad P_{R_g} = R_g|\underline{I}_g|^2 = 1.08 \text{ W}, \tag{12.97}$$

respectively. Since the transmission line itself is lossless, the sum of the above two loss powers equals, by the conservation of power principle, the time-average power delivered by the emf of the generator to the rest of the circuit. This input power (P_{input}) can, on the other hand, be evaluated in terms of the complex emf and generator current as in Eq. (8.207), so indeed we have

$$P_{\text{input}} = \text{Re}\{\underline{\mathcal{E}}\underline{I}_g^*\} = P_{R_g} + P_L = 3.78 \text{ W}. \tag{12.98}$$

Example 12.14 Complete Solution for a Low-Loss Line

Consider the transmission line and its excitation and load from the previous example, and assume that the line conductors have small losses described by the high-frequency resistance per unit length of the line equal to $R' = 1.2\ \Omega/\text{m}$, whereas the line dielectric is the same (lossless). Under these circumstances, compute (a) the complex resultant current intensity along the line and (b) the total time-average power of Joule's losses in the conductors.

Solution

(a) Since the losses along the line can be considered to be small, the line phase coefficient is (approximately) the same as in the no-loss case, so that in Eqs. (12.91). The low-loss attenuation coefficient (α) of the line is found from Eqs. (11.75) and (12.24), where $\alpha_d = 0$ (perfect dielectric), which results in the following complex propagation coefficient ($\underline{\gamma}$):

$$\alpha = \alpha_c = \frac{R'}{2Z_0} = 0.012 \text{ Np/m} \quad \longrightarrow \quad \underline{\gamma} = \alpha + j\beta = (0.012 + j3.14)\ \text{m}^{-1}. \tag{12.99}$$

With the use of Eq. (12.70), the input impedance of the line is now

$$\underline{Z}_{in} = Z_0\,\frac{1 + |\underline{\Gamma}_L|\, e^{-2\alpha l}\, e^{j(-2\beta l+\psi_L)}}{1 - |\underline{\Gamma}_L|\, e^{-2\alpha l}\, e^{j(-2\beta l+\psi_L)}} = (52.18 - j45.05)\ \Omega \tag{12.100}$$

($\underline{\Gamma}_L$ is the same as in the previous example), in place of Eq. (12.92). Similarly, Eq. (12.90) gives

$$\underline{V}_{i0} = \frac{\underline{Z}_{in}\underline{\mathcal{E}}\, e^{-\alpha l}\, e^{-j\beta l}}{(\underline{Z}_{in} + R_g)\left[1 + |\underline{\Gamma}_L|\, e^{-2\alpha l}\, e^{j(-2\beta l+\psi_L)}\right]} = 12.47\, e^{-j36.8°}\ \text{V}, \tag{12.101}$$

and the expression for the total complex rms current intensity along the line in Eq. (12.95) becomes

$$\underline{I}(z) = \frac{\underline{V}_{i0}}{Z_0}\left[e^{-\alpha z}\,e^{-j\beta z} - |\underline{\Gamma}_L|\,e^{\alpha z}\,e^{j(\beta z+\psi_L)}\right] = 249\,e^{-j36.8^\circ}\left[e^{-0.012z}\,e^{-j\pi z}\right.$$
$$\left.-0.447\,e^{0.012z}\,e^{j(\pi z+26.6^\circ)}\right]\ \text{mA} \quad (-l \le z \le 0;\ \ z \text{ in m}). \tag{12.102}$$

(b) The time-average loss power in the load is computed as in Eqs. (12.97), and, using Eq. (12.102), it amounts to $P_L = R_L|\underline{I}(0)|^2 = 2.5$ W. The time-average power that the generator (including its internal resistance) delivers to the line (and load), P_g, is found as the time-average (real) power dissipated in the real part of the impedance $\underline{Z}_{in}$ in the equivalent circuit in Fig. 12.10. In Fig. 12.3, this power is split between P_L and the power dissipated along the line, so we have

$$P_g = \text{Re}\{\underline{Z}_{in}\}|\underline{I}(-l)|^2 = 2.9\ \text{W} \quad \longrightarrow \quad P_{\text{line losses}} = P_g - P_L = 0.4\ \text{W}. \tag{12.103}$$

Note that $P_{\text{line losses}}$ can also be obtained directly, by integrating the per-unit-length loss power $P'_c(z) = R'|\underline{I}(z)|^2$, Eq. (11.66), along the line, that is, by adding up the time-average dissipated power in series resistors of resistances $\Delta R = R'\Delta z$ in the circuit model of the line in Fig. 12.1.

Example 12.15 Two Cascaded Transmission Lines

Shown in Fig. 12.11 are two cascaded lossless transmission lines thorough which a time-harmonic voltage generator drives a complex impedance load. For the parameters of the circuit given in the figure, find the time-average power delivered to the load.

Solution By means of Eq. (9.67), the free-space wavelength at the operating frequency of the circuit is $\lambda_0 = c_0/f = 1$ m, and hence the wavelengths along the two cascaded transmission lines $\lambda_{z1} = \lambda_0/\sqrt{\varepsilon_r} = 50$ cm and $\lambda_{z2} = \lambda_0 = 1$ m, respectively. From Eqs. (12.10), the corresponding phase coefficients are $\beta_1 = 2\pi/\lambda_{z1} = 4\pi$ rad/m and $\beta_2 = 2\pi/\lambda_{z2} = 2\pi$ rad/m. Since $\tan\beta_2 l_2 = \tan 1.55\pi = \tan 0.55\pi = -6.31$, Eq. (12.92) gives the following for the input impedance of the second line in Fig. 12.11:

$$\underline{Z}_{in2} = Z_{02}\frac{\underline{Z}_L + jZ_{02}\tan\beta_2 l_2}{Z_{02} + j\underline{Z}_L\tan\beta_2 l_2} = (18.62 - j7.9)\ \Omega, \tag{12.104}$$

which, in turn, represents a load impedance for the first line. With $\tan\beta_1 l_1 = \tan 3.1\pi = \tan 0.1\pi = 0.325$, the input impedance of the first line, and thus of the series connection of the two lines (plus, of course, the load $\underline{Z}_L$), amounts to

cascaded transmission lines

$$\boxed{\underline{Z}_{in} = \underline{Z}_{in1} = Z_{01}\frac{\underline{Z}_{in2} + jZ_{01}\tan\beta_1 l_1}{Z_{01} + j\underline{Z}_{in2}\tan\beta_1 l_1} = (19.13 + j14.45)\ \Omega.} \tag{12.105}$$

Eq. (12.88) then tells us that the voltage of the generator in Fig. 12.11, including its internal resistance, is

$$\underline{V}_g = \frac{\underline{Z}_{in}}{\underline{Z}_{in} + R_g}\,\underline{\mathcal{E}} = 17\,e^{j25.3^\circ}\ \text{V}. \tag{12.106}$$

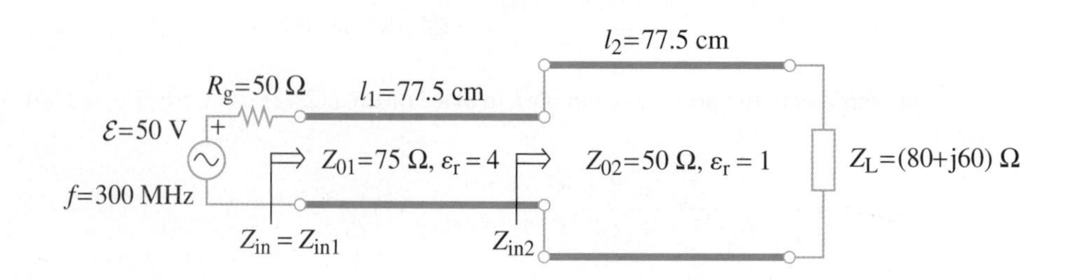

Figure 12.11 Two cascaded lossless transmission lines of different characteristic impedances and different electrical lengths in a time-harmonic regime; for Example 12.15.

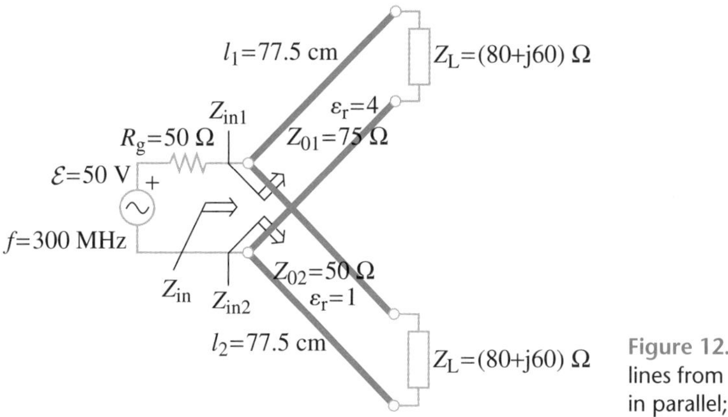

Figure 12.12 Transmission lines from Fig. 12.11 connected in parallel; for Example 12.16.

Like in Eq. (12.103), the time-average power delivered by the generator to the rest of the circuit (P_g) equals the real power dissipated in the impedance $\underline{Z}_{in}$ in the equivalent circuit as seen by the generator (Fig. 12.10). Because both transmission lines are lossless, this power equals the time-average power of the load in Fig. 12.11 (P_L), which we seek,

$$P_L = P_g = \text{Re}\{\underline{Z}_{in}\}|\underline{I}_g|^2 = \text{Re}\{\underline{Z}_{in}\}\left|\frac{\underline{V}_g}{\underline{Z}_{in}}\right|^2 = 9.6 \text{ W}. \quad (12.107)$$

Example 12.16 Transmission Lines Connected in Parallel

Repeat the previous example but for the two transmission lines connected in parallel, as depicted in Fig. 12.12 (find the time-average power delivered to each of the loads).

Solution The input impedance of the second line in Fig. 12.12, $\underline{Z}_{in2}$, is the same as in the previous example, so given in Eq. (12.104); the input impedance of the first line is computed in a similar fashion, and it amounts to $\underline{Z}_{in1} = (132.45 + j52)\ \Omega$. The equivalent impedance in Fig. 12.10 is now the parallel combination of the two input impedances,

$$\boxed{\underline{Z}_{in} = \frac{\underline{Z}_{in1}\underline{Z}_{in2}}{\underline{Z}_{in1} + \underline{Z}_{in2}} = (17.4 - j5.6)\ \Omega,} \quad (12.108)$$

transmission lines in parallel

and the voltage of the generator, Eq. (12.106), amounts to $\underline{V}_g = 13.51\,e^{-j13^\circ}$ V. Finally, as this voltage is the same for the two parallel lines, we can apply the power computation from Eq. (12.107) to each of the lines to obtain the time-average powers delivered to the two loads in Fig. 12.12 as follows:

$$P_{L1} = P_{g1} = \text{Re}\{\underline{Z}_{in1}\}\left|\frac{\underline{V}_g}{\underline{Z}_{in1}}\right|^2 = 1.19 \text{ W}, \quad P_{L2} = P_{g2} = \text{Re}\{\underline{Z}_{in2}\}\left|\frac{\underline{V}_g}{\underline{Z}_{in2}}\right|^2 = 8.32 \text{ W}. \quad (12.109)$$

Problems: 12.12–12.15; *MATLAB Exercises* (on Companion Website).

12.8 SHORT-CIRCUITED, OPEN-CIRCUITED, AND MATCHED TRANSMISSION LINES

This section discusses three important special cases of load terminations of a transmission line, in Fig. 12.3: a short-circuited, open-circuited, and matched line, with the load impedance given by $\underline{Z}_L = 0$, $|\underline{Z}_L| \to \infty$, and $\underline{Z}_L = \underline{Z}_0$, respectively. We first

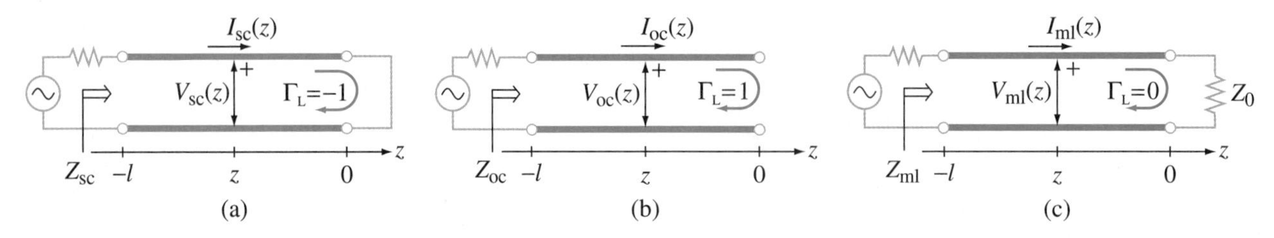

Figure 12.13 Three important special cases of the impedance load termination of a transmission line: (a) short circuit, (b) open circuit, and (c) impedance-matched load.

assume that the line is lossless. Let its phase coefficient be β, Eq. (12.19), and (purely real) characteristic impedance Z_0, Eq. (12.18). The length of the line is l.

If the line is terminated in a short circuit (sc), i.e., if its output terminals (at $z = 0$) are galvanically connected together (are at the same potential), as shown in Fig. 12.13(a), the load voltage, $\underline{V}_L$, is zero in Eqs. (12.25), and so is the load impedance. Therefore, the load voltage reflection coefficient ($\underline{\Gamma}_L$), Eq. (12.27), equals $(0 - Z_0)/(0 + Z_0) = -1$,

short-circuited transmission line

$$\boxed{\underline{Z}_L = 0 \quad \longrightarrow \quad \underline{\Gamma}_L = -1.} \tag{12.110}$$

Referring to Eq. (12.29), we then have that the magnitude and phase angle of $\underline{\Gamma}_L$ are $|\underline{\Gamma}_L| = 1$ and $\psi_L = 180°$, respectively. In addition, the return loss of the line, Eq. (12.30), comes out to be $RL = 0$ dB, and the standing wave ratio, Eq. (12.43), $s \to \infty$. From Eq. (12.27), the reflected complex rms voltage at the load terminals is related to the incident one as $\underline{V}_{r0} = -\underline{V}_{i0}$. Also, Eq. (12.32) tells us that the load voltage transmission coefficient of the line is $\underline{\tau}_L = 0$. With this, and having in mind Eqs. (12.37), (12.11), and (12.38), the total complex rms voltage and current on a short-circuited line are given by

voltage, current – shorted line

$$\boxed{\underline{V}_{sc}(z) = -2j\underline{V}_{i0} \sin\beta z, \quad \underline{I}_{sc}(z) = 2\frac{\underline{V}_{i0}}{Z_0}\cos\beta z,} \tag{12.111}$$

and their instantaneous counterparts by

$$v_{sc}(z, t) = 2\sqrt{2}V_{i0} \sin\beta z \sin\omega t, \quad i_{sc}(z, t) = 2\sqrt{2}\frac{V_{i0}}{Z_0}\cos\beta z \cos\omega t. \tag{12.112}$$

Here, a zero initial phase, $\theta_{i0} = 0$, of the incident voltage in the $z = 0$ cross section of the line (reference plane) is assumed for simplicity ($\underline{V}_{i0} = V_{i0}$). Of course, this can be done without any loss of generality, since a $\theta_{i0} \neq 0$ only takes into account a shift of the time reference ($t = 0$).

The resultant voltage and current waves along the shorted line are pure standing waves. The instantaneous $v_{sc}(z, t)$ and $i_{sc}(z, t)$, Eqs. (12.112), are in time-phase quadrature (90° out of phase with respect to each other) in every cross section of the line (for every z), which corresponds to the difference in "j" in the complex expressions – for $\underline{V}_{sc}(z)$ and $\underline{I}_{sc}(z)$, Eqs. (12.111). The line cross sections in which $v_{sc}(z, t)$ is zero at all times are defined by Eq. (10.10) with λ substituted by λ_z (wavelength along the line), Eqs. (11.43), and these same locations are obtained from Eq. (12.40) with $\psi_L = 180°$. The locations at which $i_{sc}(z, t)$ is always zero are shifted by $\lambda_z/4$ with respect to the voltage zeros. Note that the very same snapshots at different times of normalized electric and magnetic field intensities against z in Fig. 10.2 represent as well the snapshots of the normalized voltage and current $v_{sc}/(2\sqrt{2}V_{i0})$ and $i_{sc}Z_0/(2\sqrt{2}V_{i0})$ in Fig. 12.13(a). In addition, Fig. 12.14(a) shows

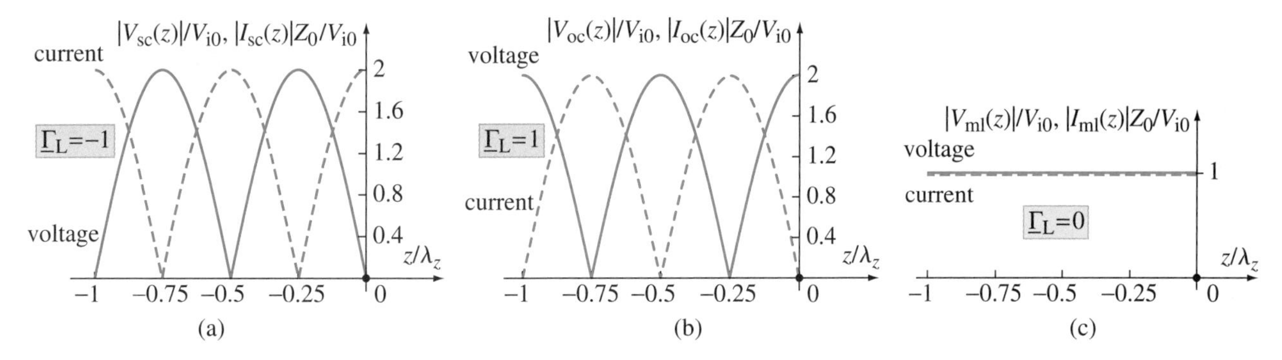

Figure 12.14 Normalized voltage and current standing wave patterns, $|\underline{V}|/V_{i0}$ and $|\underline{I}|Z_0/V_{i0}$ against z/λ_z, for a lossless transmission line and three special cases of the impedance load termination in Fig. 12.13, computed from Eqs. (12.111), (12.114), and (12.116).

the dependence of the magnitudes of $\underline{V}_{sc}(z)$ and $\underline{I}_{sc}(z)$ on z along the line (standing wave patterns). Compared with the patterns in Fig. 12.4(b), where the load is given by $0 < R_L < Z_0$ ($X_L = 0$) and the first voltage minimum and current maximum are also at the load location, the principal difference is that the voltage and current minima are now zero, $|\underline{V}|_{min}, |\underline{I}|_{min} = 0$ [see Eqs. (12.40) and (12.42)]. The maxima amount to $|\underline{V}|_{max} = 2V_{i0}$ and $|\underline{I}|_{max} = 2V_{i0}/Z_0$ [Eqs. (12.39) and (12.42)], and these are the largest possible values for these quantities (for given V_{i0} and Z_0, and any load), as the largest possible value for $|\underline{\Gamma}_L|$ is unity, Eq. (12.29).

On the other side, if the line is terminated in an open circuit (oc), that is, if the output terminals ($z = 0$) in Fig. 12.3 are left open, as in Fig. 12.13(b), then the load current, $\underline{I}_L$, is forced to be zero, and hence an infinite load impedance [see Eqs. (12.25)]. This, in turn, gives [in Eq. (12.27)] a unity load voltage reflection coefficient of the line [it equals $(\underline{Z}_L - 0)/(\underline{Z}_L - 0) = 1$, since Z_0 can be treated as a zero value in comparison with the infinitely large $|\underline{Z}_L|$],

$$\boxed{|\underline{Z}_L| \to \infty \quad \longrightarrow \quad \underline{\Gamma}_L = 1.} \qquad (12.113)$$

open-circuited transmission line

Using the same equations as for the shorted line, we now obtain that, for the situation in Fig. 12.13(b), $|\underline{\Gamma}_L| = 1$, $\psi_L = 0$, $\mathrm{RL} = 0$ dB, $s \to \infty$, $\underline{V}_{r0} = \underline{V}_{i0}$, and $\underline{\tau}_L = 2$. Thus, with the help of Eqs. (12.7) and (12.11), the complex voltage and current on an open-circuited line are

$$\boxed{\underline{V}_{oc}(z) = 2\underline{V}_{i0}\cos\beta z, \quad \underline{I}_{oc}(z) = -2\mathrm{j}\frac{\underline{V}_{i0}}{Z_0}\sin\beta z.} \qquad (12.114)$$

voltage, current – open line

These are also pure standing waves, as in Eqs. (12.111). Moreover, note that the expression for the current on the open line has the same form (the only difference is a constant multiplier $1/Z_0$) as the voltage expression for the shorted one, and vice versa, so that $\underline{V}_{oc} = Z_0\underline{I}_{sc}$ and $\underline{I}_{oc} = \underline{V}_{sc}/Z_0$. The voltage and current standing wave patterns based on Eqs. (12.114) are shown in Fig. 12.14(b), where it is obvious that they can be obtained by shifting the patterns in Fig. 12.14(a), for the short-circuited line, by $\lambda_z/4$ along the z-axis.

As the last special case, if the line is terminated in a load whose impedance is equal to the line characteristic impedance (Z_0), Fig. 12.13(c), the load voltage reflection coefficient is zero,

$$\boxed{\underline{Z}_L = Z_0 \quad \longrightarrow \quad \underline{\Gamma}_L = 0.} \qquad (12.115)$$

matched transmission line

We say that the load is matched (by its impedance) to the line – a matched load (ml), and also refer to the line itself as a matched line. Obviously, for lossless and low-loss lines, whose characteristic impedance is, or can be considered as, purely real, Eqs. (12.18) and (12.23), the matched load must be purely real as well. In Fig. 12.13(c), RL $\to \infty$ and $s = 1$. Since $\underline{V}_{r0} = 0$, i.e., there is no reflected wave on the line, the total voltage and current on the line equal the incident ones, Eqs. (12.8) and (12.13),

voltage, current – matched load

$$\boxed{\underline{V}_{ml}(z) = \underline{V}_{i0}\, e^{-j\beta z}, \quad \underline{I}_{ml}(z) = \frac{\underline{V}_{i0}}{Z_0}\, e^{-j\beta z}.} \tag{12.116}$$

So, there is only one traveling wave (in the forward direction) in Fig. 12.13(c). Given that $|\underline{V}_{ml}(z)| = Z_0|\underline{I}_{ml}(z)| = V_{i0} = \text{const}$, the standing wave patterns degenerate into a constant (normalized patterns amount to unity) – along the entire line, as shown in Fig. 12.14(c). Overall, a matched line appears as if it were infinitely long, i.e., the load impedance in Fig. 12.13(c) may be considered as an equivalent of an infinite extension of the line to the right (for $0 \le z < \infty$).

The voltages and currents in Fig. 12.14(a) and Fig. 12.14(b), being pure standing waves, do not carry any net real power (in any direction) along the line. This is also obvious from Eq. (12.54), where the resultant time-average power is $P = 0$ for $|\underline{\Gamma}_L| = 1$, so for both shorted and open lines. On the other side, as discussed in Section 12.5, the transfer of real (time-average) power (P) along the line and its delivery to the load is maximized for a matched load, in which case the power efficiency coefficient $\eta_{\text{load match}}$ in Eq. (12.59) is unity. For shorted or open lines, $\eta_{\text{load match}} = 0$.

Using Eqs. (12.72) and (12.110), the input impedance of a short-circuited line, Fig. 12.13(a), is

input impedance – shorted line

$$\boxed{\underline{Z}_{sc} = \underline{Z}_{in}\big|_{\underline{Z}_L=0} = jZ_0 \tan\beta l.} \tag{12.117}$$

This impedance is purely imaginary (reactive), $\underline{Z}_{sc} = jX_{sc}$ ($R_{sc} = 0$). Fig. 12.15(a) shows the plot of the line reactance,

$$X_{sc}(z) = -Z_0 \tan\beta z \quad (-l \le z \le 0), \tag{12.118}$$

as a function of the z-coordinate in Fig. 12.13(a) [l substituted by $-z$ in Eq. (12.117)]. We conclude that any input reactance X_{sc} ($-\infty < X_{sc} < \infty$) can be realized by simply varying the length l of a short-circuited line, for a given operating frequency (f) of the structure (given λ_z). If $0 < l < \lambda_z/4$, the input impedance is inductive ($X_{sc} > 0$); in the next interval of $\lambda_z/4$ ($\lambda_z/4 < l < \lambda_z/2$), it is capacitive ($X_{sc} < 0$), and so on. Note that X_{sc} can be varied by varying λ_z (or f) for a fixed l as well; in general, a desired X_{sc} can be obtained by adjusting both f and l, i.e., by adopting a proper electrical length of the line, Eq. (12.73). Owing to these features, short-circuited transmission line segments (called stubs), connected usually in parallel (shunt stubs) to the existing circuit or device, are extensively used as tuning and compensating elements in impedance-matching applications at higher (microwave) frequencies.

We note that for $l = \lambda_z/4$, $X_{sc} \to \pm\infty$, where the sign, plus or minus, depends on the side from which this vertical asymptote, at $z = -\lambda_z/4$, in Fig. 12.15(a) is approached (for line lengths slightly shorter than $\lambda_z/4$, X_{sc} is very large and positive, while being very large and negative for lengths slightly longer than $\lambda_z/4$). This means that a quarter-wave (quarter-wavelength long) short-circuited line appears at its input terminals ($z = -l$) as an open circuit. We may say that a quarter-wave line transforms a short to an open, which, of course, is in agreement with the

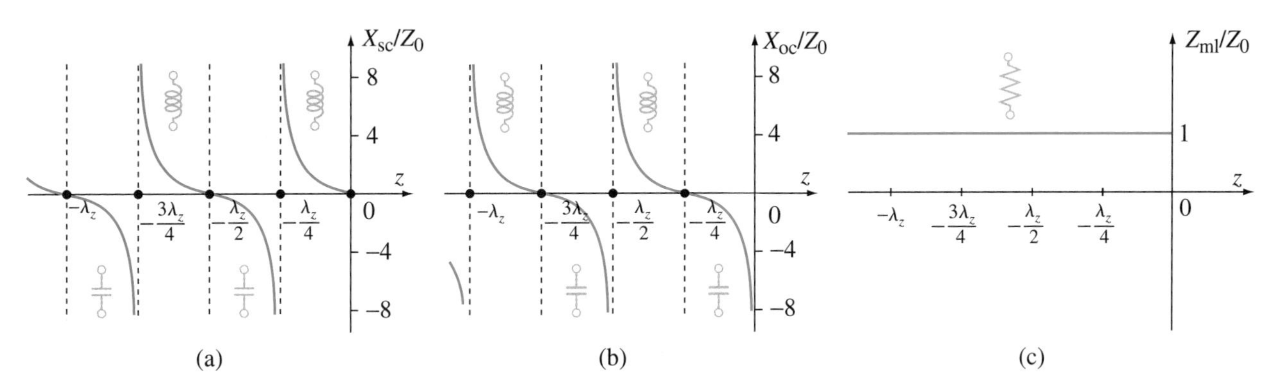

Figure 12.15 Line impedance plots, Eqs. (12.118), (12.120), and (12.122), corresponding to standing wave patterns in Fig. 12.14 [normalization using Eqs. (12.76)].

impedance inverting property of such line sections, in Eqs. (12.81). In general, this is true for a line with length $l = (2m+1)\lambda_z/4$ $(m = 0, 1, 2, \ldots)$. Similarly, a half-wave $(l = \lambda_z/2)$ line in Fig. 12.13(a) behaves, looking from the cross section at $z = -l$, as a short circuit $(\underline{Z}_{sc} = 0)$. Again, this can be any of the lengths given by $l = m\lambda_z/2$ $(m = 1, 2, \ldots)$. Short-circuited quarter- and half-wave transmission lines (as well as their extensions by any integer multiple of $\lambda_z/2$) are commonly used as electromagnetic resonators, which will be discussed in the next two sections.

In a similar fashion, the input impedance of an open-circuited line, Fig. 12.13(b), is

$$\boxed{\underline{Z}_{oc} = \underline{Z}_{in}\big|_{|\underline{Z}_L|\to\infty} = -\mathrm{j}Z_0 \cot \beta l} \tag{12.119}$$

input impedance – open line

$(\underline{Z}_{oc} = \mathrm{j}X_{oc})$, and the line reactance, whose plot against z along the line is presented in Fig. 12.15(b),

$$X_{oc}(z) = Z_0 \cot \beta z \quad (-l \le z \le 0). \tag{12.120}$$

As for the corresponding standing wave patterns in Fig. 12.14, we see that the X_{oc} diagram can be obtained from that in Fig. 12.15(a), for X_{sc}, by a simple translation by $\lambda_z/4$ along the z-axis. Shunt open-circuited stubs, providing a desired (capacitive or inductive) reactance X_{oc} by adjusting the electrical length of the line segment [Eq. (12.73)], are as well routinely utilized for impedance matching. Open stubs are preferred over shorted ones when realized as sections of microstrip or strip lines (studied in Section 11.10), due to a more difficult fabrication of shorted sections in this case. If $l = \lambda_z/4$, $X_{oc} = 0$ (line appears as a short circuit at $z = -l$); for $l = \lambda_z/2$, $X_{oc} \to \pm\infty$ (open circuit). Such sections (of lengths $\lambda_z/4$, $\lambda_z/2, \ldots$), like their short-circuited counterparts, find application as transmission-line resonators.

Finally, from either Eq. (12.70) or (12.72), the input impedance of a matched line, Fig. 12.13(c), is

$$\boxed{\underline{Z}_{ml} = \underline{Z}_{in}\big|_{\underline{Z}_L = Z_0} = Z_0.} \tag{12.121}$$

input impedance – matched load

We see that it does not depend on the length of the line. Equivalently, the line impedance in Eq. (12.74) is constant along the entire line,

$$\underline{Z}(z) = Z_0 = \text{const} \quad (-l \le z \le 0), \tag{12.122}$$

and this is illustrated in Fig. 12.15(c).

For lossy lines, all the characteristic effects attributed to losses in voltage and current standing wave patterns and impedance plots in Fig. 12.5 and 12.9, given

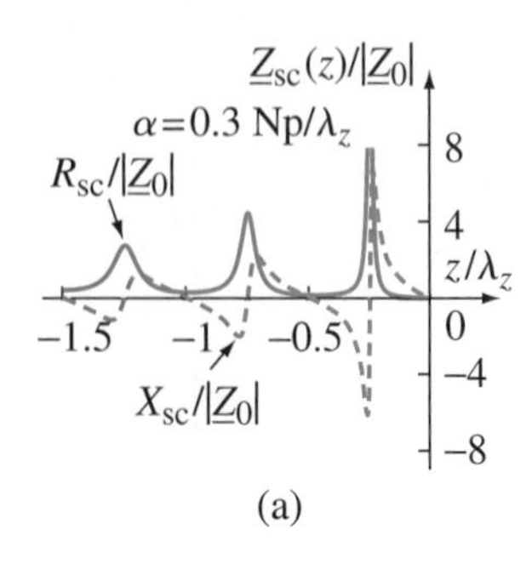

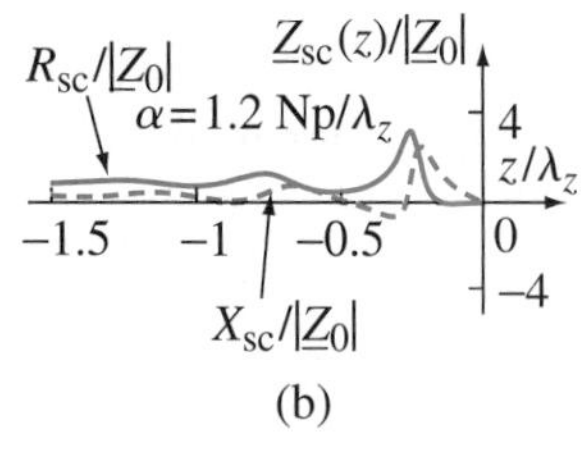

Figure 12.16 Normalized resistance and reactance plots for two lossy transmission lines from Fig. 12.5 and a short-circuit termination [Eqs. (12.123) and (12.75)].

for a purely resistive load with $R_L = |\underline{Z}_0|/4$, are present also for such lines when terminated in a short or open circuit, Fig. 12.13(a) and Fig. 12.13(b). On the other hand, when the termination is a matched load, Fig. 12.13(c), the voltage and current magnitudes along the line are, from Eqs. (12.8) and (12.13), $|\underline{V}_{ml}(z)| = |\underline{V}_i(z)| = V_{i0}\,e^{-\alpha z} \neq \text{const}$ and $|\underline{I}_{ml}(z)| = (V_{i0}/|\underline{Z}_0|)\,e^{-\alpha z} \neq \text{const}$, where α is the attenuation coefficient of the line [Eq. (12.6)]. Hence, the wave patterns in Fig. 12.14(c) are no longer uniform along the line, but exponentially decaying functions of the coordinate z. However, as $\underline{Z}(z) = \underline{Z}_0 = \text{const}$ [from Eq. (12.65) or (12.71)], the associated line impedance (resistance and reactance) diagrams are uniform. For low-loss lines, the imaginary part of $\underline{Z}_0$, and hence the line reactance, are approximately zero. When the termination is a short or open, Eq. (12.71) gives

$$\underline{Z}_{sc} = \underline{Z}_0 \tanh \underline{\gamma} l, \quad \underline{Z}_{oc} = \underline{Z}_0 \coth \underline{\gamma} l, \tag{12.123}$$

with $\underline{\gamma}$ standing for the complex propagation coefficient of the line.[2] As an illustration, Fig. 12.16 shows the impedance diagrams for the two lossy lines in Fig. 12.5 and a short-circuit termination. The input impedance of a quarter-wave (or nearly so long) line (as well as of lines with $l \approx 3\lambda_z/4$, etc.) is not infinite as in Fig. 12.15(a) any more, but is a large finite value. Similarly, the input impedance of a (nearly) half-wave line is not zero, but a small nonzero value. This can also be explained based on the circuit model of the lossy line in Fig. 12.1, where the shunt resistors of conductances ΔG and series resistors of resistances ΔR make infinite and zero input impedances from Fig. 12.15 finite (large) and nonzero (small), respectively, in Fig. 12.16. The larger the losses [case (b)] the more pronounced this effect.

Example 12.17 Measuring the Line Characteristic Impedance

Show that it is possible to obtain the (unknown) characteristic impedance of any lossless transmission line by measuring the input impedance of a section of the line when it is short-circuited, and when it is open-circuited.

Solution We note that, although both $\underline{Z}_{sc}$ and $\underline{Z}_{oc}$ undergo quite vigorous variations along a transmission line, Fig. 12.15(a) and Fig. 12.15(b), their product comes out to be independent of l (or z) in Fig. 12.13. Namely, combining Eqs. (12.117), (12.119), and (12.121), we have

$$\underline{Z}_{sc}\underline{Z}_{oc} = \underline{Z}_{ml}^2 = Z_0^2 \quad \longrightarrow \quad Z_0 = \sqrt{\underline{Z}_{sc}\underline{Z}_{oc}}. \tag{12.124}$$

So, the line characteristic impedance equals the geometric mean of the input impedance of the line when shorted and when open, respectively, and by measuring the latter two impedances, we can indeed determine the unknown Z_0.

Example 12.18 Transmission Line as a Lumped Capacitor or Inductor

Consider a lossless coaxial cable of characteristic impedance $Z_0 = 50\ \Omega$, for which the phase velocity of propagating waves equals $v_p = 2 \times 10^8$ m/s. (a) Find the smallest possible length of an open-circuited section of this cable such that it is equivalent to a lumped capacitor of capacitance $C_{eq} = 10$ pF at a frequency of $f_1 = 500$ MHz. (b) What lumped element is the section in (a) equivalent to if the frequency is changed to $f_2 = 1$ GHz?

[2]Note that "tanh" and "coth" stand, respectively, for the hyperbolic tangent and cotangent functions, which are defined via the hyperbolic sine and cosine, Eqs. (10.158), as $\tanh x = \sinh x/\cosh x$ and $\coth x = \cosh x/\sinh x$.

Solution

(a) The operating angular frequency being $\omega_1 = 2\pi f_1 = 3.14 \times 10^9$ rad/s, the associated phase coefficient of the line, using Eq. (12.10), amounts to $\beta_1 = \omega_1/v_p = 15.7$ rad/m. Equating the expression for the input impedance of an open-circuited transmission line in Eq. (12.119) to the impedance $\underline{Z}_C$ of the equivalent lumped capacitor, we obtain the required length (l) of the line section:

$$\underline{Z}_{\text{oc1}} = -\text{j}Z_0 \cot \beta_1 l = \underline{Z}_C = -\frac{\text{j}}{\omega_1 C_{\text{eq}}} \quad \longrightarrow \quad l = \frac{1}{\beta_1} \arctan\left(\omega_1 C_{\text{eq}} Z_0\right) = 6.4 \text{ cm}, \tag{12.125}$$

and this is the shortest possible section.

(b) As now $\omega_2 = 6.28 \times 10^9$ rad/s and $\beta_2 = 31.4$ rad/m, $\underline{Z}_{\text{oc}}$ is changed as well. In particular, it turns out to be purely inductive, so the line becomes equivalent to a lumped inductor, whose inductance, L_{eq}, is computed as follows:

$$\underline{Z}_{\text{oc2}} = -\text{j}Z_0 \cot \beta_2 l = \text{j}23.31\ \Omega = \underline{Z}_L = \text{j}\omega_2 L_{\text{eq}} \quad \longrightarrow \quad L_{\text{eq}} = \frac{|\underline{Z}_{\text{oc2}}|}{\omega_2} = 3.71 \text{ nH}. \tag{12.126}$$

Example 12.19 Admittance-Matching by a Shunt Short-Circuited Stub

For a lossless transmission line, the magnitude and phase angle of the load reflection coefficient are $|\underline{\Gamma}_L| = 0.38$ and $\psi_L = 138°$, respectively, and the wavelength along the line is $\lambda_z = 60$ mm. At a distance of $l = 20.8$ mm from the load, the real part of the complex line admittance appears to be equal to the characteristic admittance of the line, $Y_0 = 10$ mS. Design an admittance-matching shunt short-circuited stub that is to be connected at this location.

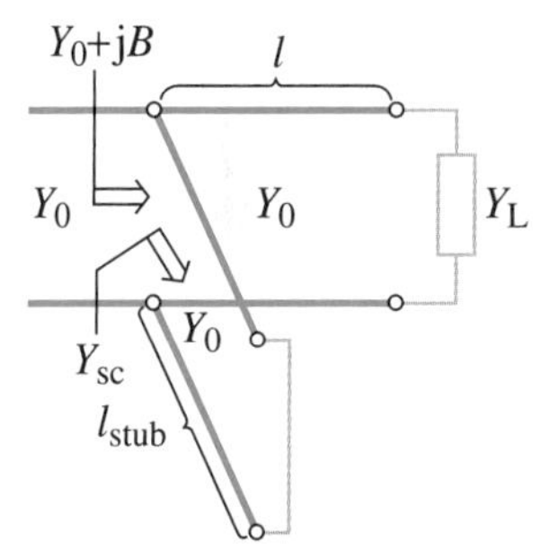

Figure 12.17 Matching by a shunt short-circuited stub connected at a location on a transmission line where the real part of the complex line admittance equals $Y_0 = 1/Z_0$; for Example 12.19.

Solution From Eqs. (12.68), (12.29), and (12.10), the line admittance at the given location amounts to

$$\underline{Y}(-l) = Y_0 \frac{1 - |\underline{\Gamma}_L|\, \text{e}^{\text{j}(-2\beta l + \psi_L)}}{1 + |\underline{\Gamma}_L|\, \text{e}^{\text{j}(-2\beta l + \psi_L)}} = (10 + \text{j}8.2) \text{ mS} = Y_0 + \text{j}B, \tag{12.127}$$

so, indeed, its real part is exactly Y_0. The imaginary part, $B = 8.2$ mS, needs to be compensated (canceled) by a short-circuited transmission line segment (stub) connected in parallel – as shown in Fig. 12.17. In other words, the sum of the input admittances of the two parallel lines at their junction (looking toward their loads) has to match Y_0 (or $Y_0 + \text{j}0$). Adopting that the stub be cut of the same transmission line (the same $Y_0 = 1/Z_0$ and β) as the main section and using the expression for the input impedance of a short-circuited line in Eq. (12.117), we thus write

$$\underline{Y}(-l) + \frac{1}{\underline{Z}_{\text{sc}}} = Y_0 \quad \longrightarrow \quad Y_0 + \text{j}B - \text{j}\frac{Y_0}{\tan \beta l_{\text{stub}}} = Y_0 \quad \longrightarrow \quad \tan \beta l_{\text{stub}} = \frac{Y_0}{B} = \frac{1}{b}, \tag{12.128}$$

where $b = B/Y_0 = 0.82$ is the normalized line susceptance, Eqs. (12.78). Hence the required (minimal) length of the stub

$$l_{\text{stub}} = \frac{1}{\beta} \arctan \frac{1}{b} = \frac{\lambda_z}{2\pi} 0.884 = 0.141\lambda_z = 8.4 \text{ mm}. \tag{12.129}$$

Example 12.20 Impedance-Matching Using a Series Stub

Repeat the previous example but for an impedance-matching series short-circuited stub that is to be connected at a location $l = 17$ mm away from the load.

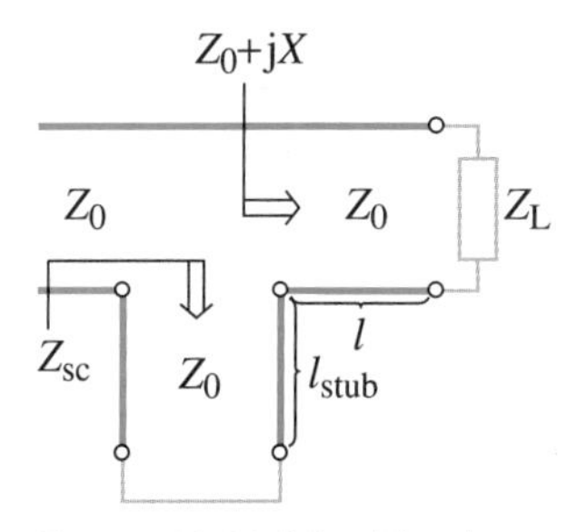

Figure 12.18 Matching by a series short-circuited stub, at a distance from the load where the real part of the transmission-line impedance equals Z_0; for Example 12.20.

Solution This situation is depicted in Fig. 12.18. The line impedance at the new given location is obtained from Eq. (12.70), and, in an analogy with Eqs. (12.127)–(12.129), the stub length is determined as follows:

$$\underline{Z}(-l) = (100 - \mathrm{j}82.06)\ \Omega = Z_0 + \mathrm{j}X \quad \longrightarrow \quad \underline{Z}(-l) + \underline{Z}_{\mathrm{sc}} = Z_0 \quad \longrightarrow \quad \tan\beta l_{\mathrm{stub}}$$
$$= -\frac{X}{Z_0} = -x \quad \longrightarrow \quad l_{\mathrm{stub}} = \frac{\lambda_z}{2\pi}\arctan(-x) = 0.109\lambda_z = 6.56\ \mathrm{mm}, \qquad (12.130)$$

with $x = X/Z_0 = -0.821$ being the normalized line reactance, Eqs. (12.76).

Problems: 12.16–12.20; *Conceptual Questions* (on Companion Website): 12.21–12.26; *MATLAB Exercises* (on Companion Website).

12.9 TRANSMISSION-LINE RESONATORS

As mentioned in the previous section, short- or open-circuited transmission lines of certain characteristic electrical lengths behave like electromagnetic resonators, and as such find application in engineering practice. Examples include their use as elements of microwave circuits (e.g., filters and tuned amplifiers), devices for measuring the frequency or wavelength (frequency-meters or wavemeters), and components in impedance-matching networks in antenna systems. Overall, transmission-line resonators are typically used at frequencies between about 300 MHz and 3 GHz. Below that range, lumped resonant *RLC* circuits suffice; above it, waveguide resonators are a preferable choice. So, in continuation of studies of shorted and open transmission lines from the previous section, we take here, and in the next section, a closer look at transmission-line resonators, and define and discuss a number of their parameters and properties with theoretical and practical importance.

Considering the short-circuited transmission line in Fig. 12.13(a), let us first assume that there are no losses along the line, i.e., that both the conductors and dielectric of the line are perfect. As already discussed in the previous section, at locations defined by $z = -m\lambda_z/2$ $(m = 1, 2, \ldots)$, with λ_z being the wavelength along the line, Eqs. (11.43), the total instantaneous voltage of the line, $v(z,t) = v_{\mathrm{sc}}(z,t)$, given in Eqs. (12.112), is zero at all times. Consequently, we can short-circuit (galvanically connect together) the line conductors in any of these cross sections, and nothing will change in the entire structure; simply, the newly added short circuit forces the condition $v = 0$ to be always satisfied. This can also be explained from the impedance point of view. Namely, we know that, looking from the above specified locations in Fig. 12.13(a) to the right, the line behaves as a short circuit, because its input impedance ($\underline{Z}_{\mathrm{in}} = \underline{Z}_{\mathrm{sc}}$), Eq. (12.117), is zero there, and actually creating a short in these transversal planes will not affect the voltage and current on the line. We can now remove the part of the transmission line on the left-hand side of the new short circuit, to obtain a self-contained structure (shorted at both ends), with a standing wave trapped between the two short circuits (multiple of half-wavelengths apart). With the no-loss assumption, this wave, once generated, exists in the structure (theoretically) indefinitely. In analogy to Eqs. (10.10) and (10.12), the resonant frequency of the transmission-line resonator, f_{res}, is given by

transmission-line resonance

$$\boxed{\sin\beta l = 0 \quad \longrightarrow \quad l = m\,\frac{\lambda_z}{2} \quad \longrightarrow \quad f = f_{\mathrm{res}} = m\,\frac{v_{\mathrm{p}}}{2l} \quad (m = 1, 2, \ldots),} \qquad (12.131)$$

where l is the length of the resonator, and v_{p} the phase velocity along the line, Eqs. (11.43). Note that m (arbitrary positive integer) equals the number of half-wavelengths along the z-axis in Fig. 12.13(a) that fit into l. Each value of m

determines a possible resonant voltage/current distribution along the line, at a different frequency f_{res} (there are an infinite number of resonant frequencies). Most often, we use a half-wave ($l = \lambda_z/2$) resonator, shown in Fig. 12.19. Its resonant frequency and the associated phase coefficient of the line are

$$f_{res} = \frac{v_p}{2l} = \frac{1}{2l\sqrt{L'C'}}, \quad \beta_{res} = \frac{\pi}{l}, \tag{12.132}$$

half-wave resonator, $m = 1$

where C' and L' are the capacitance and inductance per unit length of the line.

The principal property of an electromagnetic resonator is its capability to store the energy. The total stored energy periodically oscillates between the electric and magnetic energies in the structure, as they cyclically assume maximum and zero values. In what follows, we compute the total instantaneous electromagnetic energy, $W_{em}(t)$, of the transmission-line resonator in Fig. 12.19. The instantaneous electric energy of every short section of the transmission line, with length Δz, equals the energy stored in the associated capacitor of capacitance ΔC in the circuit model of the line in Fig. 12.1, which, using Eq. (2.192), is given by

$$\Delta W_e(z,t) = \frac{1}{2}\Delta C v^2(z,t) = \frac{1}{2}C'\Delta z\, v^2(z,t), \tag{12.133}$$

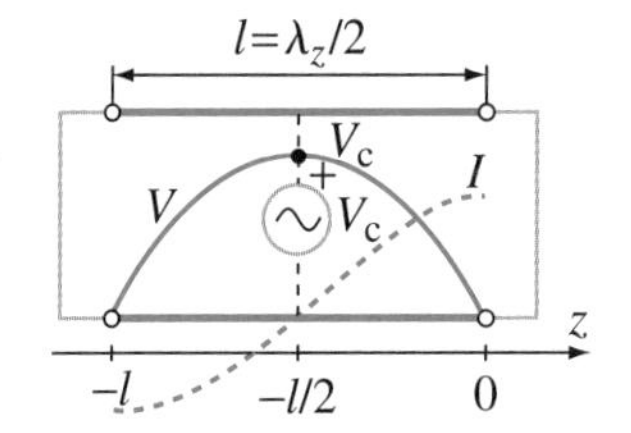

Figure 12.19 Half-wave transmission-line resonator, short-circuited at both ends; once the structure is charged with voltage and current, the generator at its center can be removed, and a standing electromagnetic wave oscillates trapped between the two short circuits.

with the expression for the voltage $v(z,t)$ being that in Eqs. (12.112). Hence, the electric energy per unit length of the line is [see also Eq. (2.208)]

$$W'_e(z,t) = \frac{\Delta W_e(z,t)}{\Delta z} = \frac{1}{2}C'v^2(z,t) = 4C'V_{i0}^2 \sin^2\beta z \sin^2\omega t, \tag{12.134}$$

V_{i0} being the rms value of the incident voltage on the line. Similarly, given the instantaneous current distribution along the line, $i(z,t)$, in Eqs. (12.112), the per-unit-length instantaneous magnetic energy of the line [see Eq. (7.88)] is

$$W'_m(z,t) = \frac{1}{2}L'i^2(z,t) = 4L'\frac{V_{i0}^2}{Z_0^2}\cos^2\beta z\cos^2\omega t = 4C'V_{i0}^2\cos^2\beta z\cos^2\omega t, \tag{12.135}$$

where the use is made of the relation between L' and C' via the characteristic impedance of the line, Z_0, in Eq. (12.18). With $\omega = \omega_{res} = 2\pi f_{res}$ (resonant angular frequency of the resonator) and the resonant value for β in Eqs. (12.132), the total energy in the resonator in Fig. 12.19 is now obtained performing essentially the same integrations as in Eq. (10.22), and the result is

$$W_{em} = W_{em}(t) = \int_{z=-l}^{0} W'_e(z,t)\,dz + \int_{z=-l}^{0} W'_m(z,t)\,dz = 2C'V_{i0}^2 l. \tag{12.136}$$

stored energy in a transmission-line resonator

As expected, it comes out to be constant in time ($W_{em} = \text{const}$). Note that it is customary to express the energy in terms of the magnitude (rms value) of the total voltage, Eqs. (12.111) and (12.112), at the center of the resonator in Fig. 12.19, V_c, i.e., for $z = -l/2 = -\lambda_z/4$,

$$W_{em} = \frac{1}{2}C'V_c^2 l, \quad \text{where} \quad V_c = 2V_{i0}. \tag{12.137}$$

In the example, we shall obtain this same result by computing W_{em} at instants of time when it is all electric, $W_{em} = (W_e)_{max}$ ($W_m = 0$), and also as the maximum magnetic energy of the resonator, $W_{em} = (W_m)_{max}$ ($W_e = 0$).

In order to excite the electromagnetic field in the resonator in Fig. 12.19, we can, for instance, use an electric probe (short wire antenna), like the one in Fig. 10.5, positioned at the electric field (and voltage) maximum of the structure, so in the middle of the line (at $z = -l/2$), and directed in parallel to the electric field lines. This probe can be modeled by an ideal voltage generator, of the rms emf equal to V_c, Eqs. (12.137), inserted between the line conductors at the same location, as indicated in Fig. 12.19. Moreover, the same probe can be used to extract (receive) the energy (signal) from the resonator, and deliver it to an external device or system.

Finally, note that similar resonators consisting of open-circuited transmission lines are also possible (as mentioned in the previous section, open ends are preferable to shorted ones for microstrip or strip lines), as well as lines with a short circuit at one end and open at the other. Namely, both a short and open circuit as line terminations lead to the voltage reflection coefficients with a unity magnitude, $|\Gamma_L| = 1$, Eqs. (12.110) and (12.113), so we have a total reflection of the incoming waves at both ends of the line, resulting in everlasting standing waves, in all cases.

Example 12.21 Maximum Electric and Magnetic Energies of a Resonator

Obtain the expression for the electromagnetic energy stored in a transmission-line resonator (Fig. 12.19) by computing the maximum electric (a) and magnetic (b) energies of the structure, respectively.

Solution

(a) At instants $t = T/4 + kT/2$, T being the time period of the time-harmonic wave in Eq. (8.49) and k any integer, we have that $\sin\omega t = \pm 1$ and $\cos\omega t = 0$, which means that the voltage and per-unit-length electric energy of a transmission-line resonator, in Eqs. (12.112) and (12.134), respectively, both attain their maximum values for any z $(-l \leq z \leq 0)$. At the same times, the current intensity and magnetic energy per unit length of the resonator, Eqs. (12.112) and (12.135), are zero at every location. Therefore, the stored electromagnetic energy of the resonator (W_{em}) is all electric, and we can compute it as

$$W_{em} = (W_e)_{max} = \int_{z=-l}^{0} (W'_e)_{max}(z)\,dz = \int_{-l}^{0} \frac{1}{2}C' v_{max}^2(z)\,dz = 4C'V_{i0}^2 \underbrace{\int_{-l}^{0} \sin^2\left(\frac{\pi}{l}z\right) dz}_{l/2}$$

$$= 2C'V_{i0}^2 l \quad (W_m = 0), \tag{12.138}$$

where the use is also made of the expression for $\beta = \beta_{res}$ in Eqs. (12.132). Of course, this is the same result as in Eq. (12.136).

(b) Similarly, W_{em} can be evaluated at instants $t = kT/2$ ($\sin\omega t = 0$, $\cos\omega t = \pm 1$), when it is all magnetic (both i and W'_m are maximum for every z along the resonator, and $W_e = 0$). Having in mind Eq. (12.135), this integration gives

$$W_{em} = (W_m)_{max} = \int_{-l}^{0} (W'_m)_{max}(z)\,dz = 4C'V_{i0}^2 \int_{-l}^{0} \cos^2\left(\frac{\pi}{l}z\right) dz = 2C'V_{i0}^2 l. \tag{12.139}$$

12.10 QUALITY FACTOR OF RESONATORS WITH SMALL LOSSES

In an ideal (lossless) transmission-line resonator, that is once charged with voltage and current, and the accompanying fields (e.g., using the generator in Fig. 12.19), and then left to itself (the generator is turned off or removed), its electromagnetic energy, $W_{\rm em}(t)$, stored in distributed capacitors and inductors in Fig. 12.1, stays constant forever (to $t \to \infty$). In a real resonator, however, $W_{\rm em}(t)$ decays with time, due to Joule's (ohmic) losses in the structure. If these losses are small, which is always the case in practical applications, we can assume that the line voltage and current and the resonator energy are given by their no-loss expressions, e.g., Eqs. (12.112) and (12.136) for the resonator in Fig. 12.19, and just decrease exponentially with time, as the capacitors and inductors (slowly) discharge through resistors in Fig. 12.1. Denoting by τ the time constant, also known as the relaxation time, of the resonator, this exponential decay for v and i can be written as

$$v(z,t) = v(z,0)\,{\rm e}^{-t/\tau}, \quad i(z,t) = i(z,0)\,{\rm e}^{-t/\tau} \quad (0 < t < \infty). \tag{12.140}$$

The reciprocal of τ is termed the damping factor of the resonator, and marked by $\delta_{\rm r}$. From Eqs. (12.134) and (12.135), the effective damping factor for $W_{\rm em}$ is twice that for the voltage and current,

$$\boxed{W_{\rm em}(t) = W_{\rm em}(0)\,{\rm e}^{-2\delta_{\rm r}t}, \quad \text{where} \quad \delta_{\rm r} = \frac{1}{\tau}.} \tag{12.141}$$

damping of a resonator

By the conservation of power principle, the time-average Joule's power $P_{\rm J}$ lost (to heat) in the resistors in Fig. 12.1 is equal to the negative of the time rate of change of the stored energy $W_{\rm em}$ (the loss power $P_{\rm J}$ is positive and the change of energy ${\rm d}W_{\rm em}$ over an elemental time ${\rm d}t$ is negative). Taking the derivative of $W_{\rm em}$ with respect to t from Eq. (12.141), we get

$$P_{\rm J}(t) = -\frac{{\rm d}W_{\rm em}(t)}{{\rm d}t} = 2\delta_{\rm r}W_{\rm em}(0)\,{\rm e}^{-2\delta_{\rm r}t} = 2\delta_{\rm r}W_{\rm em}(t), \tag{12.142}$$

and hence the following expression for the time constant of the resonator:

$$\boxed{\tau = \frac{1}{\delta_{\rm r}} = \frac{2W_{\rm em}}{P_{\rm J}}.} \tag{12.143}$$

time constant of a resonator

The larger the ratio $W_{\rm em}/P_{\rm J}$ the slower the damping (discharge) of the resonator and, simply, the better the resonator, in terms of its ability to store and keep the energy. Accordingly, this ratio determines the so-called quality factor (or Q factor) of the resonator.[3] More precisely, we define the Q factor as 2π times the ratio of $W_{\rm em}$ and the energy lost in one cycle of time-harmonic variation of the voltage and current on the line, $W_{\rm lost/cycle}$,

$$Q = 2\pi\,\frac{W_{\rm em}}{W_{\rm lost/cycle}}. \tag{12.144}$$

Symbolizing the cycle duration (time period), Eq. (8.49), at resonance by $T_{\rm res}$ and having in mind Eq. (8.53), this lost energy can be expressed as

$$W_{\rm lost/cycle} = P_{\rm J}T_{\rm res}, \quad \text{where} \quad T_{\rm res} = \frac{1}{f_{\rm res}} = \frac{2\pi}{\omega_{\rm res}}, \tag{12.145}$$

[3]In general, the quality factor of electromagnetic resonators, including those with lumped elements, also defines the bandwidth (BW) of the resonator around the resonant frequency, and is thus a measure of the sharpness of the resonance and frequency selectivity of the device. BW is inversely proportional to the Q factor, and the higher the Q of a resonant structure (circuit) the narrower its bandwidth and sharper the resonance. This means, in turn, that a high-Q structure (device) is more frequency selective, that is, more tuned to a single (resonant) frequency, than a similar structure with a lower Q.

and finally

quality factor of a resonator (dimensionless)

$$Q = \omega_{\rm res} \frac{W_{\rm em}}{P_{\rm J}} . \tag{12.146}$$

Note that $Q \to \infty$ for an ideal resonator ($P_{\rm J} = 0$). Using Eqs. (12.143) and (12.146), the damping factor of the resonator is now

$$\delta_{\rm r} = \frac{\omega_{\rm res}}{2Q} , \tag{12.147}$$

which yields, with the help of Eqs. (12.145), an explicit proportionality relation between Q and τ:

Q factor via the time constant and period

$$Q = \pi \frac{\tau}{T_{\rm res}} . \tag{12.148}$$

Specifying $t = QT_{\rm res}$ in Eq. (12.141),

$$W_{\rm em}(t = QT_{\rm res}) = W_{\rm em}(0)\,{\rm e}^{-2\pi} = 0.00187 W_{\rm em}(0) , \tag{12.149}$$

so we see that after Q time periods at the resonant frequency, the stored energy of the resonator is damped to less than 0.2% of its initial value (at $t = 0$), which is another physical interpretation of the quality factor. The energy is further reduced by the same factor (i.e., multiplied by 0.00187) at every following $QT_{\rm res}$ long interval of time. Overall, the Q factor of a resonator is a quantitative measure of its ability to keep (store) the energy during the course of time in spite of losses.

With the resonant angular frequency of the resonator in Fig. 12.19 being given by $\omega_{\rm res} = 2\pi f_{\rm res}$ and Eq. (12.132), and the energy in the resonator, $W_{\rm em}$, by Eq. (12.136), we yet need the expression for the time-average power of Joule's losses in the structure, $P_{\rm J}$, to be able to find its quality factor using Eq. (12.146). Under the low-loss assumption, we compute $P_{\rm J}$, and thus Q, based on the current and voltage distributions for the lossless case (perturbation method). Let us consider first the conductor losses. From Eq. (11.66), the time-average dissipated power in the line conductors in Fig. 12.19, or, equivalently, in series resistors of resistances ΔR in the circuit model of the line in Fig. 12.1, per unit length of the structure is obtained as $P'_{\rm c}(z) = R'|\underline{I}(z)|^2$, with $\underline{I}(z)$ in Eqs. (12.111) and R' designating the high-frequency resistance per unit length of the line. Hence, neglecting the losses in the short circuits at both ends of the line, the total time-average conductor loss power in the resonator is

$$P_{\rm c} = \int_{z=-l}^{0} P'_{\rm c}(z)\,{\rm d}z = \int_{-l}^{0} R'|\underline{I}(z)|^2\,{\rm d}z = \frac{4R'V_{\rm i0}^2}{Z_0^2} \int_{-l}^{0} \cos^2\left(\frac{\pi}{l}\,z\right){\rm d}z = \frac{R'V_{\rm c}^2 l}{2Z_0^2} , \tag{12.150}$$

where the integral in z is already computed in Eq. (12.136) or (12.139), and $V_{\rm c}$ is the rms voltage at the center of the resonator, Eqs. (12.137). Combining Eqs. (12.146), (12.136), and (12.150), the quality factor $Q_{\rm c}$ associated with the losses in the resonator conductors alone (as if there were no losses in the dielectric) is

Q factor for line conductors

$$Q_{\rm c} = \omega_{\rm res} \frac{W_{\rm em}}{P_{\rm c}} = \omega_{\rm res} \frac{C'Z_0^2}{R'} = \omega_{\rm res} \frac{L'}{R'} . \tag{12.151}$$

Note that $Q_{\rm c}$ can be written as $Q_{\rm c} = \omega_{\rm res}\Delta L/\Delta R$, and viewed as the Q factor of a series resonant RLC circuit whose resistance, inductance, and capacitance are $\Delta R = R'\Delta z$, $\Delta L = L'\Delta z$, and $\Delta C = C'\Delta z$, respectively. This can as well be identified in Fig. 12.1, where each circuit cell in the transmission-line model with the shunt resistor of conductance ΔG removed ($\Delta G = 0$) represents such a resonant circuit.

Similarly, based on Eq. (11.68), the total time-average loss power in the dielectric of the resonator in Fig. 12.19 (i.e., in the shunt resistors of conductances ΔG in Fig. 12.1) is

$$P_\mathrm{d} = \int_{-l}^{0} P'_\mathrm{d}(z)\,\mathrm{d}z = \int_{-l}^{0} G'|\underline{V}(z)|^2\,\mathrm{d}z = 4G'V_{\mathrm{i}0}^2 \int_{-l}^{0} \sin^2\left(\frac{\pi}{l}z\right)\mathrm{d}z = \frac{G'V_\mathrm{c}^2 l}{2}, \tag{12.152}$$

with G' being the leakage conductance per unit length of the line. Combined with Eqs. (12.146) and (12.136), this gives the following expression for the quality factor Q_d for the resonator dielectric:

$$\boxed{Q_\mathrm{d} = \omega_\mathrm{res}\frac{W_\mathrm{em}}{P_\mathrm{d}} = \omega_\mathrm{res}\frac{C'}{G'}.} \tag{12.153}$$

Q factor for line dielectric

Again, this Q factor can be recognized as that of a parallel resonant RLC or GLC circuit with parameters $\Delta G = G'\Delta z$, ΔL, and ΔC, which is in agreement with the model in Fig. 12.1 with $\Delta R = 0$. For transmission-line resonators with a homogeneous (lossy) dielectric, of permittivity ε and conductivity σ_d, using the duality relationship between G' and C' in Eq. (3.171), Q_d can alternatively be expressed as

$$\boxed{Q_\mathrm{d} = \omega_\mathrm{res}\frac{\varepsilon}{\sigma_\mathrm{d}} = \frac{1}{\tan\delta_\mathrm{d}},} \tag{12.154}$$

Q_d using dielectric parameters

where $\tan\delta_\mathrm{d}$ is the loss tangent of the dielectric, in Eq. (9.125) or (9.130).

Once Q_c and Q_d are known, the overall Q factor of the resonator is obtained from the sum of conductor and dielectric losses making up the total loss power in the structure,

$$P_\mathrm{J} = P_\mathrm{c} + P_\mathrm{d} = \omega_\mathrm{res}W_\mathrm{em}\left(\frac{1}{Q_\mathrm{c}} + \frac{1}{Q_\mathrm{d}}\right) \quad\longrightarrow\quad \frac{1}{Q} = \frac{1}{Q_\mathrm{c}} + \frac{1}{Q_\mathrm{d}}, \tag{12.155}$$

so that

$$\boxed{Q = \frac{Q_\mathrm{c}Q_\mathrm{d}}{Q_\mathrm{c} + Q_\mathrm{d}}.} \tag{12.156}$$

total Q factor in terms of Q_c and Q_d

This is a general expression for the quality factor of any structure that has both imperfectly conducting and lossy dielectric parts. In particular, substituting the results from Eqs. (12.151) and (12.153), we have, for the resonator in Fig. 12.19,

$$Q = \frac{\omega_\mathrm{res}L'C'}{R'C' + G'L'}. \tag{12.157}$$

Having in mind Eqs. (12.22), Q_c and Q_d can alternatively be evaluated in terms of β_res, Eqs. (12.132), and the respective attenuation coefficients, for the line conductors and dielectric, α_c and α_d,

$$Q_\mathrm{c} = \frac{\omega_\mathrm{res}L'}{R'} = \frac{1}{2}\frac{\omega_\mathrm{res}\sqrt{L'C'}}{R'\sqrt{C'/L'}/2} = \frac{\beta_\mathrm{res}}{2\alpha_\mathrm{c}}, \tag{12.158}$$

$$Q_\mathrm{d} = \frac{\omega_\mathrm{res}C'}{G'} = \frac{1}{2}\frac{\omega_\mathrm{res}\sqrt{L'C'}}{G'\sqrt{L'/C'}/2} = \frac{\beta_\mathrm{res}}{2\alpha_\mathrm{d}}. \tag{12.159}$$

Adding up the partial attenuation coefficients, according to Eq. (11.75), the total α for the line is

$$\alpha = \alpha_\mathrm{c} + \alpha_\mathrm{d} = \frac{\beta_\mathrm{res}}{2}\left(\frac{1}{Q_\mathrm{c}} + \frac{1}{Q_\mathrm{d}}\right) = \frac{\beta_\mathrm{res}}{2Q}, \tag{12.160}$$

which results in yet another useful simple expression to compute the Q factor of transmission-line resonators:

Q factor via attenuation and phase coefficients

$$\boxed{Q = \frac{\beta_{\text{res}}}{2\alpha}.} \tag{12.161}$$

Of course, this can as well be derived directly from Eq. (12.157). Q factors of practical transmission-line resonators can be very high, up to several thousand, which is by more than an order of magnitude higher than what can be achieved (up to about 100) with lumped resonant circuits (e.g., a series or parallel resonant RLC circuit).

Although the presented theory of transmission-line resonators applies to a line of an arbitrary cross section, in Fig. 11.3, the quality factor in Eq. (12.146) may not adequately represent the behavior of open structures that significantly radiate (like antennas) into the external space. For instance, a resonator made out of a section of a two-wire transmission line with air (or any other) dielectric, Fig. 2.22, may exhibit nonnegligible radiation losses at high frequencies, because it is entirely open in transversal planes. On the other side, a half-wave coaxial cable, Fig. 2.17, completely closed (using metallic plates) at both ends is immune to radiation, since its dielectric, in which the electromagnetic field resides, is shielded by the outer conductor. Such closed metallic structures, of different shapes (not necessarily transmission-line segments), are called cavity resonators. A notable example is a rectangular cavity resonator, to be studied in the next chapter, which is a metallic box made of a resonant section of a rectangular waveguide in Fig. 10.15.

Example 12.22 Quality Factor of a Coaxial-Cable Resonator

Find the Q factor, as well as its portions associated with conductor and dielectric losses, of a resonator made by short-circuiting both ends of a half-wavelength long section of the coaxial cable described in Example 11.3 (neglect the losses in the short circuits).

Solution Combining Eqs. (11.17) and (9.47), the length of the cable resonator comes out to be $l = \lambda_z/2 = v_{\text{p}}/(2f) = c_0/(2\sqrt{\varepsilon_{\text{r}}}f) = 10$ cm, where $f = f_{\text{res}} = 1$ GHz is the resonant frequency of the structure, so that the associated phase coefficient of the cable, from Eqs. (12.132), equals $\beta_{\text{res}} = \pi/l = 31.4$ rad/m. With this, and the result for the attenuation coefficient for the cable conductors and the given loss tangent of the cable dielectric from Example 11.3, Eqs. (12.160), (12.161), and (12.154) yield the following for the respective partial Q factors of the resonator:

$$Q_{\text{c}} = \frac{\beta_{\text{res}}}{2\alpha_{\text{c}}} = 931.7 \quad \text{and} \quad Q_{\text{d}} = \frac{1}{\tan\delta_{\text{d}}} = 10{,}000. \tag{12.162}$$

By means of Eq. (12.156), the total quality factor amounts to $Q = 852.3$, which, of course, can also be obtained directly from Eq. (12.161), as well as from Eq. (12.157) using the values of primary circuit parameters (C', L', R', and G') of the cable.

Problems: 12.21 and 12.22; *Conceptual Questions* (on Companion Website): 12.27–12.29; *MATLAB Exercises* (on Companion Website).

12.11 THE SMITH CHART – CONSTRUCTION AND BASIC PROPERTIES

As the last topic in the circuit theory of two-conductor transmission lines in a time-harmonic regime, in this and the following section we present an alternative, graphical, technique for the circuit analysis and design of transmission lines in the frequency domain. The technique is based on the so-called Smith[4] chart. It enables approximate determination, based on graphical manipulations on the chart, of the reflection coefficients, line impedances, voltages, currents, and other quantities of interest for a given transmission-line problem – without actually performing any complex algebra. Of course, it is possible (and advisable) to combine computations in the complex domain using concepts and equations developed in previous sections of this chapter with visualizations and graphical evaluations on the Smith chart. In addition, electronic forms of the chart are often used as a presentation medium for displays of antenna and microwave laboratory test equipment, as well as output interfaces of computational electromagnetic software. While the graphical analysis of lossy transmission lines employing the Smith chart is also possible, we confine our discussions to the lossless case. In fact, this graphical tool is seldom used for transmission-line calculations where the losses are not negligible. This section develops understanding of the construction and basic properties of the Smith chart; its use to solve transmission-line problems is presented in the next section, with a number of application examples.

The Smith chart is, essentially, a polar plot of the generalized voltage reflection coefficient, $\underline{\Gamma}(z)$, given by Eq. (12.33), along a transmission line (in Fig. 12.3). Equivalently, representing $\underline{\Gamma}$ via its real and imaginary parts,

$$\boxed{\underline{\Gamma} = \Gamma_{\rm r} + {\rm j}\Gamma_{\rm i},} \tag{12.163}$$

real and imaginary parts of the reflection coefficient

the chart lies in the complex plane of $\underline{\Gamma}$, i.e., the $\Gamma_{\rm r}$–$\Gamma_{\rm i}$ plane, as shown in Fig. 12.20. The magnitude of $\underline{\Gamma}$ is constant along the line (with no losses), which corresponds to a circle of radius $|\underline{\Gamma}|$, centered at the coordinate origin in the $\Gamma_{\rm r}$–$\Gamma_{\rm i}$ plane.[5] The position of a particular point $\underline{\Gamma}$ on the circle is determined by the angle $\psi = 2\beta z + \psi_{\rm L}$, Eq. (12.35), where β is the phase coefficient of the line, Eq. (12.19), and $\psi_{\rm L}$ is the phase angle of the load (for $z = 0$) voltage reflection coefficient, Eq. (12.29). Employing the standing wave ratio, s, of the line, Eq. (12.43), we have

$$\boxed{|\underline{\Gamma}| = \frac{s-1}{s+1} = \text{const} \qquad (-l \le z \le 0),} \tag{12.164}$$

constant-$|\underline{\Gamma}|$ or s circle

where l is the length of the line, and hence the constant-$|\underline{\Gamma}|$ circle in Fig. 12.20 is also referred to as the s circle (or SWR circle). Of course, the values of both $|\underline{\Gamma}|$ and s are set by the given (complex) impedance of the load ($\underline{Z}_{\rm L} = R_{\rm L} + {\rm j}X_{\rm L}$) and (purely real) characteristic impedance of the line (Z_0), Eq. (12.18). Since $0 \le |\underline{\Gamma}| \le 1$, Eq. (12.29), the Smith chart is bounded by the circle defined by $|\underline{\Gamma}| = 1$, called the unit circle.

As there is an one-to-one correspondence between $\underline{\Gamma}(z)$ and the transmission-line impedance, $\underline{Z}(z)$, at the same location (defined by the coordinate z) on the line, Eqs. (12.65) and (12.66), the Smith chart provides graphical representation of $\underline{Z}$ as

[4]Philip H. Smith (1905–1987), an American electrical engineer, graduated from Tufts College in 1928. While working at Bell Laboratories, Smith devised in 1939 a chart for graphical calculations on transmission lines – the famous Smith chart.

[5]For a lossy transmission line, Eq. (12.52) gives $|\underline{\Gamma}(z)| \propto {\rm e}^{2\alpha z}$, and as the coordinate z becomes more and more negative going away from the load in Fig. 12.3, $|\underline{\Gamma}|$ becomes smaller and smaller, i.e., the point $\underline{\Gamma}$ in the $\Gamma_{\rm r}$–$\Gamma_{\rm i}$ plane in Fig. 12.20 becomes closer and closer to the coordinate origin. In other words, the circle of radius $|\underline{\Gamma}|$ becomes a spiral in the lossy case. However, if the losses per unit length of the line are small, then the deviation of the spiral (for any z on the line) from the circle for the load position ($z = 0$) is small, and in many applications, especially if the line is not too long, can be neglected.

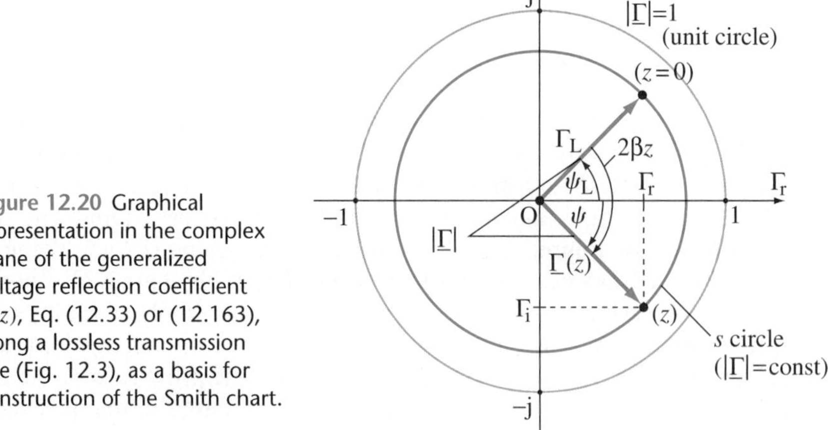

Figure 12.20 Graphical representation in the complex plane of the generalized voltage reflection coefficient $\underline{\Gamma}(z)$, Eq. (12.33) or (12.163), along a lossless transmission line (Fig. 12.3), as a basis for construction of the Smith chart.

well. However, impedances are displayed on the chart using their normalized (to Z_0) values, $\underline{z}_n$, according to Eq. (12.75). For instance, the normalized impedance of the load in Fig. 12.3, $\underline{z}_L$, is given by

normalized load impedance

$$\underline{z}_L = \frac{\underline{Z}_L}{Z_0} = \frac{R_L + jX_L}{Z_0} = r_L + jx_L, \tag{12.165}$$

where r_L and x_L are the normalized load resistance and reactance, respectively. At any location along the line, we have, from Eqs. (12.75), (12.65), and (12.66),

mapping between complex $\underline{\Gamma}$ and $\underline{z}_n$

$$\underline{z}_n = \frac{1+\underline{\Gamma}}{1-\underline{\Gamma}}, \quad \underline{\Gamma} = \frac{\underline{z}_n - 1}{\underline{z}_n + 1}. \tag{12.166}$$

These simple relationships, defining in the complex plane a mapping of $\underline{\Gamma}$ to $\underline{z}_n$, and vice versa, are the underpinning of the utility of the Smith chart.

With the use of the representation of $\underline{\Gamma}$ in Eq. (12.163), the first relationship in Eqs. (12.166) can be written in the following form [note the similarity with Eq. (12.47)]:

$$\underline{z}_n = r + jx = \frac{1 + \Gamma_r + j\Gamma_i}{1 - \Gamma_r - j\Gamma_i} = \frac{1 - \Gamma_r^2 - \Gamma_i^2 + j2\Gamma_i}{(1-\Gamma_r)^2 + \Gamma_i^2}. \tag{12.167}$$

Equating r to the real part of the last expression in this equation, we obtain

$$\Gamma_r^2 - 2\Gamma_r \frac{r}{1+r} + \Gamma_i^2 = \frac{1-r}{1+r}, \tag{12.168}$$

which is further rearranged to read

equation of an r circle

$$\left(\Gamma_r - \frac{r}{1+r}\right)^2 + (\Gamma_i - 0)^2 = \left(\frac{1}{1+r}\right)^2. \tag{12.169}$$

We realize that this is the equation of a circle in the Γ_r–Γ_i plane, whose center and radius are

$$r \text{ circle:} \quad \text{center at } (\Gamma_r, \Gamma_i) = \left(\frac{r}{1+r}, 0\right); \quad \text{radius} = \frac{1}{1+r}. \tag{12.170}$$

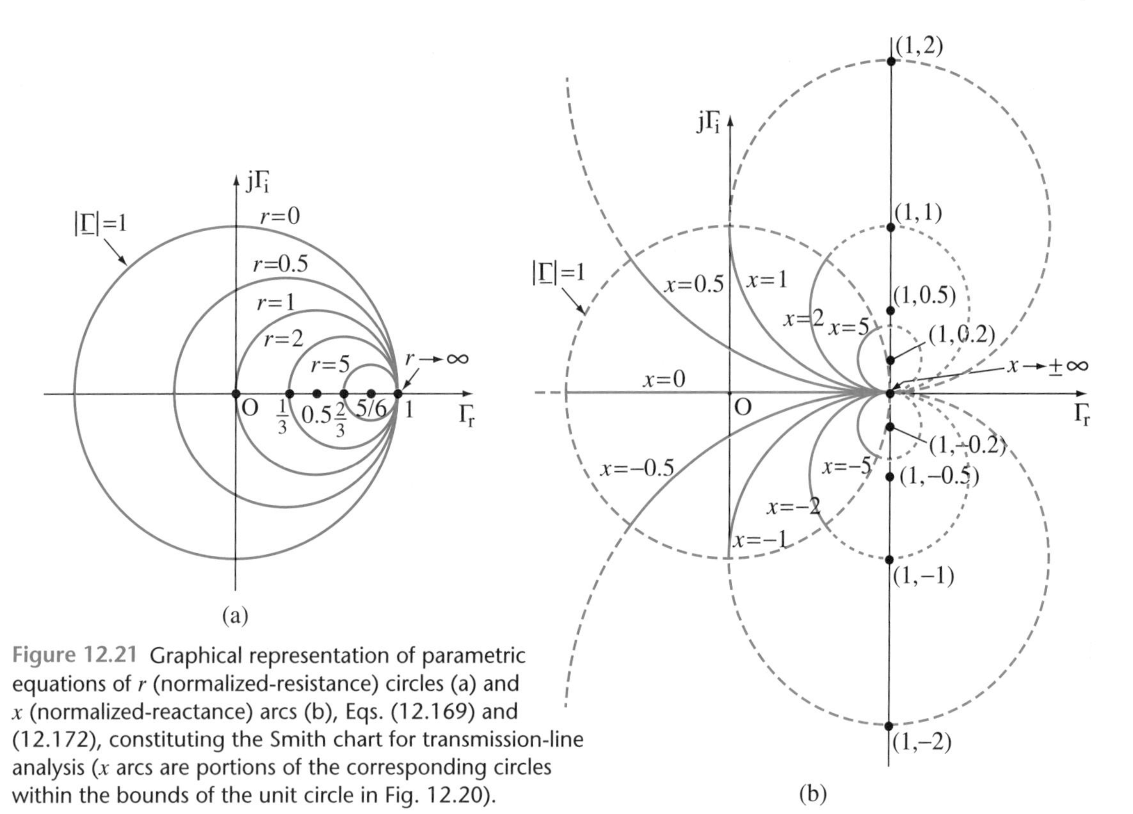

Figure 12.21 Graphical representation of parametric equations of r (normalized-resistance) circles (a) and x (normalized-reactance) arcs (b), Eqs. (12.169) and (12.172), constituting the Smith chart for transmission-line analysis (x arcs are portions of the corresponding circles within the bounds of the unit circle in Fig. 12.20).

Moreover, this is a parametric equation corresponding to a given r (as a parameter), and hence the name of the resulting family of circles (for the values of the parameter r in the range of $0 \leq r < \infty$) in the complex plane – r circles (normalized-resistance circles). Fig. 12.21(a) shows several characteristic r circles. We see that the circle centers are all on the positive Γ_r-axis ($0 \leq \Gamma_\mathrm{r} \leq 1$, $\Gamma_\mathrm{i} = 0$), and they move toward the center of the chart, $(\Gamma_\mathrm{r}, \Gamma_\mathrm{i}) = (0, 0)$, as r decreases. The smaller the r the larger the circle, with the largest one being that for $r = 0$ – it coincides with the unit circle ($|\underline{\Gamma}| = 1$) in Fig. 12.20. For $r \to \infty$, on the other side, the circle degenerates into a point, defined by $(\Gamma_\mathrm{r}, \Gamma_\mathrm{i}) = (1, 0)$. Note that the r circle passing through the chart center corresponds to $r = 1$ ($R = Z_0$); it is centered at the point $(0.5, 0)$, and has a radius of 0.5.

Similarly, equating the imaginary parts in Eq. (12.167) leads to

$$(\Gamma_\mathrm{r} - 1)^2 + \Gamma_\mathrm{i}^2 - 2\Gamma_\mathrm{i}\frac{1}{x} = 0, \tag{12.171}$$

which can be written as the parametric equation describing a circle, an x circle (normalized-reactance circle), for a given value of x, in the plane Γ_r–Γ_i,

$$\boxed{(\Gamma_\mathrm{r} - 1)^2 + \left(\Gamma_\mathrm{i} - \frac{1}{x}\right)^2 = \left(\frac{1}{x}\right)^2.} \tag{12.172}$$

equation of an x circle

Its center and radius are

$$x \text{ circle}: \quad \text{center at } (\Gamma_\mathrm{r}, \Gamma_\mathrm{i}) = \left(1, \frac{1}{x}\right); \quad \text{radius} = \frac{1}{|x|}. \tag{12.173}$$

Several typical representatives of the family of x circles are shown in Fig. 12.21(b). The circle centers now lie on the vertical line defined by $\Gamma_r = 1$, and they move away from the Γ_r-axis as $|x|$ decreases. For $x > 0$ (inductive line impedance), the centers are above the point $(1, 0)$, while below it when $x < 0$ (capacitive line impedance). Again, like in Fig. 12.21(a), the smaller the $|x|$ the larger the circle. The largest one, that for $x = 0$, degenerates into a line, coinciding with the Γ_r-axis ($\Gamma_i = 0$). The circles defined by $x \to \pm\infty$ both degenerate into a point, the same as for $r \to \infty$ in Fig. 12.21(a). Note that the x circles tangential to the Γ_i-axis (at $\Gamma_i = \pm 1$) correspond to $x = \pm 1$ ($X = \pm Z_0$); they have a unity radius, and are centered at points $(1, \pm 1)$ in the Γ_r–Γ_i plane. Of course, only portions of x circles fall within the domain determined by $|\underline{\Gamma}| \leq 1$ (Fig. 12.20). Therefore, while using the term "x circles," what we mean are actually only the contours within the bounds of the unit circle ($|\underline{\Gamma}| = 1$), in Fig. 12.21(b). Alternatively, we shall refer to these contours as x arcs or circular segments.

Superposing the r circles and x arcs, from Fig. 12.21(a) and Fig. 12.21(b), we obtain the Smith chart – shown in Fig. 12.22. On the chart, a normalized line impedance given by $\underline{z}_{n0} = r_0 + jx_0$ corresponds to the point of intersection of the $r = r_0$ circle and $x = x_0$ arc. Note that Fig. 12.22 provides a plain chart, on top of which, as we shall see in the next section (in examples), the user of the chart draws the s circle and other lines and points pertinent for the particular application or problem. A rather fine grid of contours of constant r and x is plotted and labeled within the chart, so that $\underline{z}_n$ at any position along the s circle can be easily and accurately read. Overall, the Smith chart simultaneously displays values of $\underline{z}_n$ and $\underline{\Gamma}$, according to the relationships in Eqs. (12.166), in a convenient format – for graphical calculations on transmission lines and/or visualization of measured or simulated data.

12.12 CIRCUIT ANALYSIS OF TRANSMISSION LINES USING THE SMITH CHART

In continuation of discussions of the Smith chart from the previous section, to help us efficiently use the chart in analysis and design of transmission lines, which is the subject of this section, Fig. 12.23 highlights several key features of the chart in Fig. 12.22. For instance, the three important special cases of load terminations of a transmission line in Fig. 12.13, a short circuit, open circuit, and matched load, are marked in Fig. 12.23 – points P_{sc}, P_{oc}, and P_{ml}, respectively, at the leftmost and rightmost positions, and the center of the chart. The figure also reemphasizes that the upper (lower) half of the chart corresponds to inductive (capacitive) line impedances. By inspection of Figs. 12.22 and 12.23, we can see how movements of the point P (with reflection coefficient $\underline{\Gamma}$ and normalized line impedance $\underline{z}_n$) around the s circle (i.e., along the transmission line in Fig. 12.3), for different values of s or $|\underline{\Gamma}|$ (different circle radii), translate into changes of line impedances. These changes, which, analytically, are given by Eq. (12.74), constitute impedance transformations on transmission lines discussed in Sections 12.6 and 12.8. Note, however, that for a given s circle on the chart, the numerical value for $|\underline{\Gamma}|$ is obtained by measuring the length (e.g., in cm) of the radius of that circle, and comparing (normalizing) it to the length of the unit ($|\underline{\Gamma}| = 1$) circle,

reflection coefficient magnitude on the Smith chart

$$|\underline{\Gamma}| = \frac{\overline{OP}}{\overline{OQ}}, \tag{12.174}$$

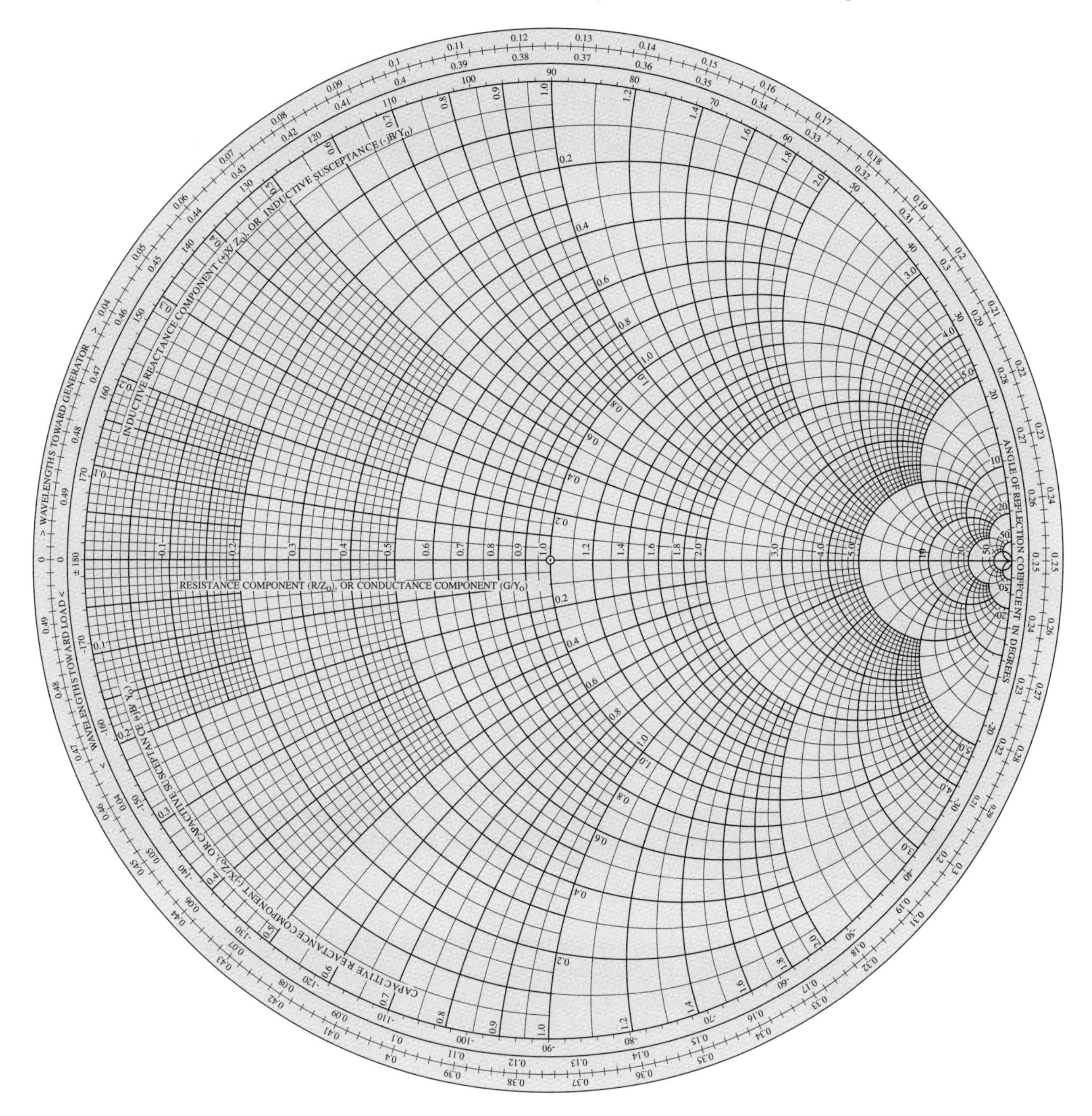

Figure 12.22 The Smith chart.

where O is the chart center and Q an arbitrary point on the unit circle (which is also the $r = 0$ circle) in Fig. 12.23.

From Eq. (12.35), the change in the coordinate z due to a movement along the line, in Fig. 12.3, results in the following change of the phase angle ψ of the reflection coefficient $\underline{\Gamma}(z)$:

$$\Delta\psi = 2\beta\Delta z \qquad \left(\beta = \frac{2\pi}{\lambda_z}\right). \tag{12.175}$$

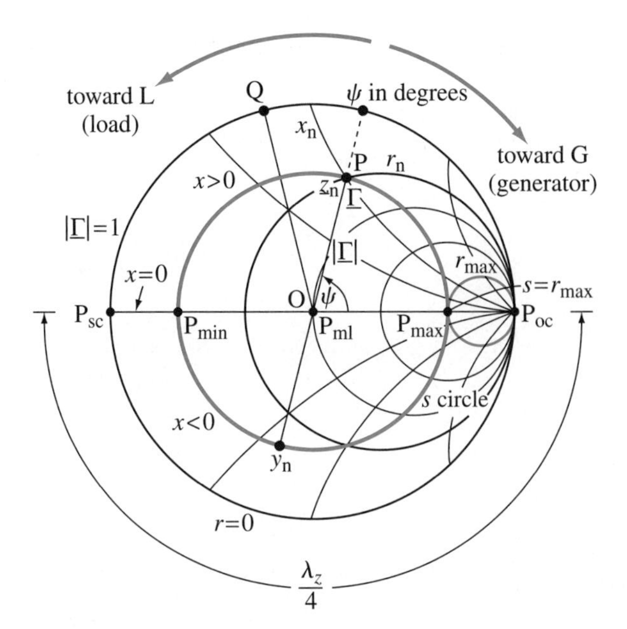

Figure 12.23 Highlighting several key features of the Smith chart in Fig. 12.22 and its use.

Expressing Δz in terms of the wavelength along the line (λ_z), Eqs. (11.43), $\Delta\psi$ in degrees is evaluated according to the following formula:

motion around the Smith chart in degrees

$$\boxed{\Delta\psi = 720^\circ\,\frac{\Delta z}{\lambda_z}.} \tag{12.176}$$

The orientation of the rotation on the chart is determined by the direction of motion along the line. Namely, if z varies from $z = 0$ (load) to $z = -l$ (generator), Δz is negative, and so is $\Delta\psi$, which means that the point P on the chart rotates in the negative (clockwise) direction. So, a clockwise rotation on the chart in Fig. 12.23 corresponds to a movement toward the generator (G) along the line in Fig. 12.3,

clockwise rotation on the chart

$$\boxed{\Delta z < 0 \;\; \text{(move toward G)} \quad \longleftrightarrow \quad \Delta\psi < 0.} \tag{12.177}$$

By the same token, moving toward the load (L) in Fig. 12.3 gives the counterclockwise direction of rotation on the chart,

counterclockwise rotation

$$\boxed{\Delta z > 0 \;\; \text{(move toward L)} \quad \longleftrightarrow \quad \Delta\psi > 0.} \tag{12.178}$$

In movements in either direction, one complete rotation around the chart corresponds, using Eq. (12.176), to a half-wave shift along the line,

one full rotation

$$\boxed{360^\circ \text{ around the Smith chart} \quad \longleftrightarrow \quad \frac{\lambda_z}{2} \text{ along the transmission line.}} \tag{12.179}$$

To help us perform various manipulations on the Smith chart, there are several concentric scales around its perimeter, Fig. 12.22. The scale labeled "angle of reflection coefficient in degrees" measures (as a protractor) the angle ψ (in degrees). The two outermost scales, as indicated by their names, serve for determining a new position on the chart based on the old position and a given movement along the line expressed in either "wavelengths toward generator" or "wavelengths toward load." According to Eq. (12.179), these latter scales range from zero to $(0.5\lambda_z)/\lambda_z = 0.5$

for one full circle on the chart, in their respective directions [see Eqs. (12.177) and (12.178)].

With the use of Eqs. (12.36) and (12.39), we realize that voltage maxima on the line in Fig. 12.3 occur at locations where the generalized voltage reflection coefficient is purely real and positive, $\underline{\Gamma}(z_{\max}) = |\underline{\Gamma}_{\rm L}|$, and this corresponds to the point $\rm P_{max}$ on the Smith chart (Fig. 12.23) where the s circle intersects the positive $\Gamma_{\rm r}$-axis (i.e., the line $\overline{\rm OP_{oc}}$),

$$\boxed{|\underline{V}| = |\underline{V}|_{\max} \quad \longleftrightarrow \quad (\Gamma_{\rm r}, \Gamma_{\rm i}) = (|\underline{\Gamma}_{\rm L}|, 0).} \tag{12.180}$$

point $\rm P_{max}$ on the chart

On the other side, Eq. (12.40) tells us that $\underline{\Gamma}$ at locations of voltage minima on the line is purely real and negative, $\underline{\Gamma}(z_{\max}) = -|\underline{\Gamma}_{\rm L}|$, so that the corresponding point $\rm P_{min}$ on the chart is at the intersection of the s circle and the negative $\Gamma_{\rm r}$-axis ($\overline{\rm OP_{sc}}$),

$$\boxed{|\underline{V}| = |\underline{V}|_{\min} \quad \longleftrightarrow \quad (\Gamma_{\rm r}, \Gamma_{\rm i}) = (-|\underline{\Gamma}_{\rm L}|, 0).} \tag{12.181}$$

point $\rm P_{min}$ on the chart

So, as the point P moves around the s circle in Fig. 12.20, every crossing of the positive (negative) $\Gamma_{\rm r}$-axis means a maximum (minimum) in the voltage standing wave patterns, and thus minimum (maximum) in the current patterns, in Figs. 12.4 and 12.14. From Fig. 12.23, distances between successive maxima (or minima) are $\lambda_z/2$ (a full rotation), whereas the adjacent maxima and minima are separated by $\lambda_z/4$ (a half rotation), which, of course, is in an agreement with Eqs. (12.39)–(12.41) and Figs. 12.4 and 12.14. Furthermore, from Eq. (12.83), the line SWR equals the normalized maximum resistance of the line,

$$\boxed{s = r_{\max} = \frac{R_{\max}}{Z_0} \quad [x(z_{\max}) = 0] \quad \longrightarrow \quad s = r|_{\text{at } \rm P_{max}},} \tag{12.182}$$

SWR on the chart

and this resistance is attained at locations of voltage maxima on the line. Hence, the constant-$|\underline{\Gamma}|$ circle of the line and the r circle for $r = s$ pass through the same point ($\rm P_{max}$) on the positive $\Gamma_{\rm r}$-axis, that is, simply, s numerically equals the value of r at $\rm P_{max}$. This can be used to, in place of Eq. (12.43), graphically determine s from a given $|\underline{\Gamma}|$, or vice versa. In fact, a multitude of interpretations and discussions of voltage, current, and impedance patterns based on various analytical expressions and relationships in Sections 12.4, 12.6, and 12.8 now become directly (visually) understandable from the Smith chart. As an example, let us look back at the discussion in Example 12.3 of voltage (and current) standing wave patterns in Fig. 12.4. With the Smith chart now available, note that, as the point L (load point) in the chart for a purely resistive load ($x_{\rm L} = 0$) must be on the horizontal diameter of the unit circle, it coincides with the location of either the first voltage maximum (point $\rm P_{max}$ in Fig. 12.23), which occurs if $r_{\rm L} > 1$, or the first voltage minimum (point $\rm P_{min}$), when $r_{\rm L} < 1$. On the other hand, L for an inductive complex load ($x_{\rm L} > 0$) is somewhere in the upper half of the chart, and if we move from it toward G (generator point), so in the clockwise direction, we would sooner encounter the first voltage maximum than the first voltage minimum on the line. Finally, $\rm P_{min}$ would come first if the starting point (L) is in the lower half of the chart, signifying a capacitive complex load ($x_{\rm L} < 0$).

We note that the relationship expressing the impedance inverting property (for normalized impedances) of a quarter-wave transformer ($\lambda_z/4$ long section of a transmission line) in Eqs. (12.81) can be written as

$$\underline{z}_{\rm n}(-\lambda_z/4) = \frac{1}{\underline{z}_{\rm n}(0)}, \tag{12.183}$$

where $\underline{z}_n(0) = \underline{z}_L$ (normalized load impedance). Using the normalized line admittance ($\underline{y}_n$), Eq. (12.77), we have

impedance-admittance conversion

$$\underline{z}_n(-\lambda_z/4) = \underline{y}_n(0) \quad \text{or} \quad \underline{y}_n(-\lambda_z/4) = \underline{z}_n(0), \tag{12.184}$$

with the second relationship being obtained from the first one by inverting each of the sides of the equation. This means that a rotation by $\lambda_z/4$ around the Smith chart, in either direction, transforms $\underline{z}_n$ into $\underline{y}_n$, and vice versa. Therefore, as shown in Fig. 12.23, the points representing $\underline{z}_n$ and $\underline{y}_n$ are diametrally opposite to each other on the s circle (inverting a normalized line impedance in the complex domain is equivalent to "jumping" to the opposite side of the s circle in the Smith chart). Note that this can also be realized from the expressions for $\underline{z}_n$ and $\underline{y}_n$ in terms of $\underline{\Gamma}$, the first one being that in Eqs. (12.166) and the other one, from Eqs. (12.77) and (12.68), given by

$$\underline{y}_n = \frac{1-\underline{\Gamma}}{1+\underline{\Gamma}}. \tag{12.185}$$

Namely, these two expressions can be obtained from one another by replacing $\underline{\Gamma}$ with $-\underline{\Gamma}$, that is, by adding a minus sign to $\underline{\Gamma}$. Since $-1 = \mathrm{e}^{\mathrm{j}(\pm\pi)}$, this is equivalent to a change of $\Delta\psi = \pm 180°$ in the phase angle of $\underline{\Gamma}$, so a half rotation about the s circle (or a "jump" to its opposite side) in Fig. 12.23. This complex-inverting transformation on the chart can be used to determine any normalized admittance (impedance) from the corresponding impedance (admittance). Moreover, the Smith chart in Fig. 12.22 can readily be used as an admittance chart (sometimes it is more convenient to work with admittances than with impedances), with r circles and x arcs being treated as g circles and b arcs, respectively.

Combining Eqs. (12.36), (12.8), and (12.13), normalized magnitudes of the voltage and current along a lossless transmission line can be written as

$$\frac{|\underline{V}(z)|}{V_{i0}} = \left|1+\underline{\Gamma}(z)\right|, \quad \frac{|\underline{I}(z)|Z_0}{V_{i0}} = \left|1-\underline{\Gamma}(z)\right|, \tag{12.186}$$

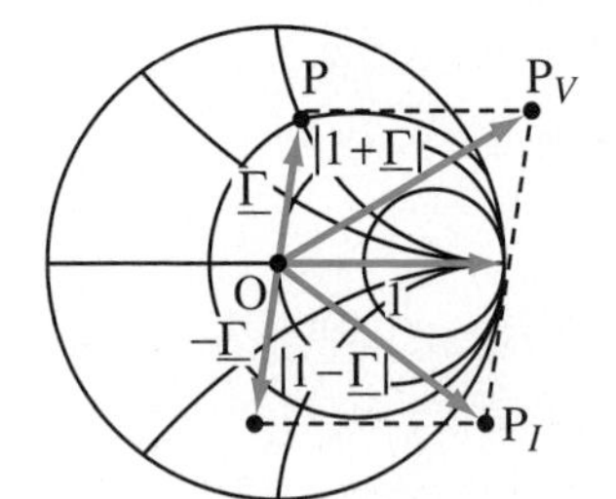

Figure 12.24 Graphical representation in the Smith chart of normalized line voltage and current magnitudes given by Eqs. (12.186).

where V_{i0} is the magnitude (rms value) of the incident voltage on the line. Their graphical representation in the Smith chart is shown in Fig. 12.24 [note the similarity with Fig. 10.9(a)]. We see that $|\underline{V}|/V_{i0}$ for any z, being equal, in the Γ_r–Γ_i complex plane, to the magnitude of the complex number $1+\underline{\Gamma}$ (i.e., the sum of the real number 1 and complex $\underline{\Gamma}$), can be determined graphically by measuring the length of the position vector of the point P_V in Fig. 12.24 with respect to the coordinate origin. Likewise, the relative current magnitudes for different positions along the line are obtained by measuring the length of the vector $1-\underline{\Gamma}$, the tip of which is at the point P_I, in Fig. 12.24. As the coordinate z varies along the transmission line in Fig. 12.3, and the point P rotates about the s circle in Fig. 12.23, it is obvious from Fig. 12.24 that the maxima and minima of $|1+\underline{\Gamma}|$ are $1+|\underline{\Gamma}|$ and $1-|\underline{\Gamma}|$, respectively, for the positions P_{max} and P_{min} in Fig. 12.23 [see also Fig. 10.9(a)]. These values give the expressions for $|\underline{V}|_{max}$ and $|\underline{V}|_{min}$ in Eqs. (12.39) and (12.40). The locations and normalized values of $|\underline{I}|_{max}$ and $|\underline{I}|_{min}$ (maxima and minima of $|1-\underline{\Gamma}|$), from Fig. 12.24, are reversed, which is consistent with Eq. (12.41). Note that the normalized line voltage and current standing wave patterns in Figs. 12.4 and 12.14, computed and plotted using Eqs. (12.45), (12.46), (12.111), (12.114), and (12.116), can as well be obtained using the Smith chart and the explained graphical procedure.

Example 12.23 Calculations on a Transmission Line Using the Smith Chart

A lossless transmission line of length $l = 2.8$ m and characteristic impedance $Z_0 = 50\ \Omega$, fed by a time-harmonic generator of frequency $f = 150$ MHz, is terminated at the other end in a load with impedance $\underline{Z}_L = (30 + j60)\ \Omega$. The parameters of the line dielectric are $\varepsilon_r = 4$ and $\mu_r = 1$. Using the Smith chart, find: (a) the load reflection coefficient, (b) the input impedance of the line, (c) locations of the first voltage maximum and minimum on the line, (d) total numbers of voltage maxima and minima, (e) the standing wave ratio of the line, and (f) the line input admittance.

Solution

(a) The normalized load impedance, Eq. (12.165), is $\underline{z}_L = \underline{Z}_L/Z_0 = 0.6 + j1.2$, so we mark the corresponding point L (load point) at the intersection of the $r = 0.6$ circle and $x = 1.2$ arc in the Smith chart – in Fig. 12.25. Then we draw (using a ruler) a radial line

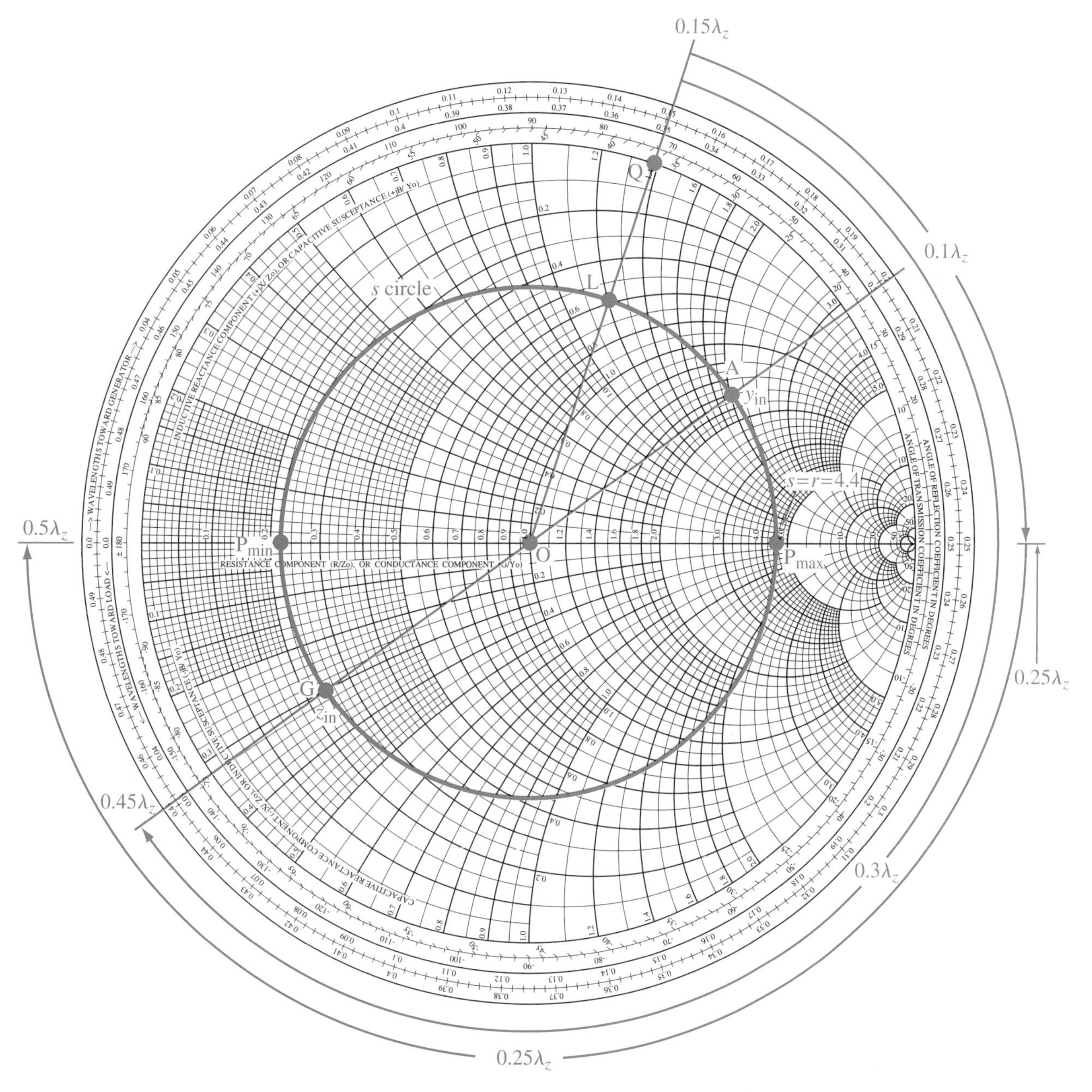

Figure 12.25 Graphical calculations on a lossless transmission line in a time-harmonic regime using the Smith chart; for Example 12.23.

and measure distances from the chart center (point O) to points L and Q, respectively, the latter point lying on the unit circle, to compute, as in Eq. (12.174) and Fig. 12.23, the magnitude of the load reflection coefficient, $|\underline{\Gamma}_{\rm L}| = \overline{\rm OL}/\overline{\rm OQ} = 0.62$. We also read at the scale labeled "angle of reflection coefficient in degrees" that the phase angle of the coefficient is $\psi_{\rm L} = 72°$, and hence $\underline{\Gamma}_{\rm L} = 0.62\,{\rm e}^{{\rm j}72°}$.

(b) Next, we draw (using a compass) in Fig. 12.25 the s circle for the given load and transmission line (the circle centered at point O and passing through L). The wavelength along the line being $\lambda_z = c_0/(\sqrt{\varepsilon_{\rm r}}f) = 1$ m [Eqs. (11.17) and (9.47)], the line length spans $l = 2.8\lambda_z$. Because every half-wave ($\lambda_z/2$) shift along the structure (in Fig. 12.3) translates, according to Eq. (12.179), to a full circle on the Smith chart, the input impedance (measured at the generator) of this line ($\underline{Z}_{\rm in}$) is identical to that of a line with length equal to $l_1 = 2.8\lambda_z - 5 \times 0.5\lambda_z = 0.3\lambda_z$ ($0 < l_1 \le 0.5\lambda_z$). So, we move around the s circle in the direction toward the generator (clockwise direction) by $0.3\lambda_z$, that is, from $0.15\lambda_z$ mark to $0.45\lambda_z$ mark on the "wavelengths toward generator" scale in Fig. 12.25, to obtain point G (generator point), where we read the normalized input impedance of $\underline{z}_{\rm in} = 0.24 - {\rm j}0.3$, and thus $\underline{Z}_{\rm in} = \underline{z}_{\rm in} Z_0 = (12 - {\rm j}15)\ \Omega$.

(c)–(e) As in Fig. 12.23 and Eqs. (12.180) and (12.181), locations of voltage maxima and minima on the transmission line under consideration correspond to points $\rm P_{max}$ and $\rm P_{min}$ on the Smith chart in Fig. 12.25, where the s circle intersects the positive and negative $\Gamma_{\rm r}$-axis, respectively. From point L ($0.15\lambda_z$ mark on the "wavelengths toward generator" scale) to $\rm P_{max}$ ($0.25\lambda_z$ mark), the distance from the load of the first voltage maximum on the line is $l_{\rm max} = 0.25\lambda_z - 0.15\lambda_z = 0.1\lambda_z = 10$ cm (going toward the generator), where, with reference to the notation in Eqs. (12.39), $l_{\rm max} = -z_{\rm max}$. Similarly, since $\rm P_{min}$ is at $0.5\lambda_z$ mark on the "wavelengths toward generator" scale, it takes $l_{\rm min} = 0.5\lambda_z - 0.15\lambda_z = 0.35\lambda_z = 35$ cm (or $l_{\rm min} = l_{\rm max} + 0.25\lambda_z$) from the load to the first voltage minimum. A slide along the full length of the line resulting in 5 full (clockwise) rotations about the s circle on the chart plus $0.3\lambda_z$, there is a total of $5 + 1 = 6$ voltage maxima and 5 voltage minima on the line. From Eq. (12.182), the line SWR equals the value of the normalized line resistance at locations of voltage maxima, $s = r|_{\rm at\ P_{max}} = 4.4$.

(f) Finally, having in mind Eq. (12.184) and Fig. 12.23, the normalized input admittance of the line is read at a location (point A) diametrally opposite (rotation by $\lambda_z/4$) to point G on the s circle as $\underline{y}_{\rm in} = 1.6 + {\rm j}2$. Therefore, Eq. (12.77) and the fact that the line characteristic admittance, Eq. (11.35), is $\underline{Y}_0 = 1/\underline{Z}_0 = 20$ mS tell us that the input admittance of the line amounts to $\underline{Y}_{\rm in} = \underline{y}_{\rm in} Y_0 = (32 + {\rm j}40)$ mS.

Example 12.24 Finding Load Impedance Using the Smith Chart

Redo Example 12.4 but now with the use of the Smith chart.

Solution From the given SWR of the transmission line and Eq. (12.182), we have $r|_{\rm at\ P_{max}} = s = 3$, so the point $\rm P_{max}$, denoting locations of voltage maxima on the line, is at the intersection of the $r = 3$ circle and the horizontal diameter of the unit circle in the Smith chart, as shown in Fig. 12.26. We are now able to draw the constant-$|\underline{\Gamma}|$ circle ($s = 3$ circle) of the line, as it also passes through this same point. On the opposite side of the s circle, we mark the point $\rm P_{min}$, and move from it, i.e., from the location of the first voltage minimum on the line, around the circle toward the load (counterclockwise). To actually reach the load, we need to move a distance of $l_{\rm min} = 12$ cm $= 0.15\lambda_z$ ($\lambda_z = 2\Delta l = 80$ cm), from 0 mark to $0.15\lambda_z$ mark on the "wavelengths toward load" scale. At such obtained point L in Fig. 12.26, the normalized impedance is read to be $\underline{z}_{\rm L} = 0.8 - {\rm j}1$, yielding, by means of Eq. (12.165), the load impedance of $\underline{Z}_{\rm L} = \underline{z}_{\rm L} Z_0 = (40 - {\rm j}50)\ \Omega$, which is approximately the same result as in Eq. (12.51).

Example 12.25 Quarter-Wave Matching Design Using the Smith Chart

Repeat Example 12.11 using the Smith chart.

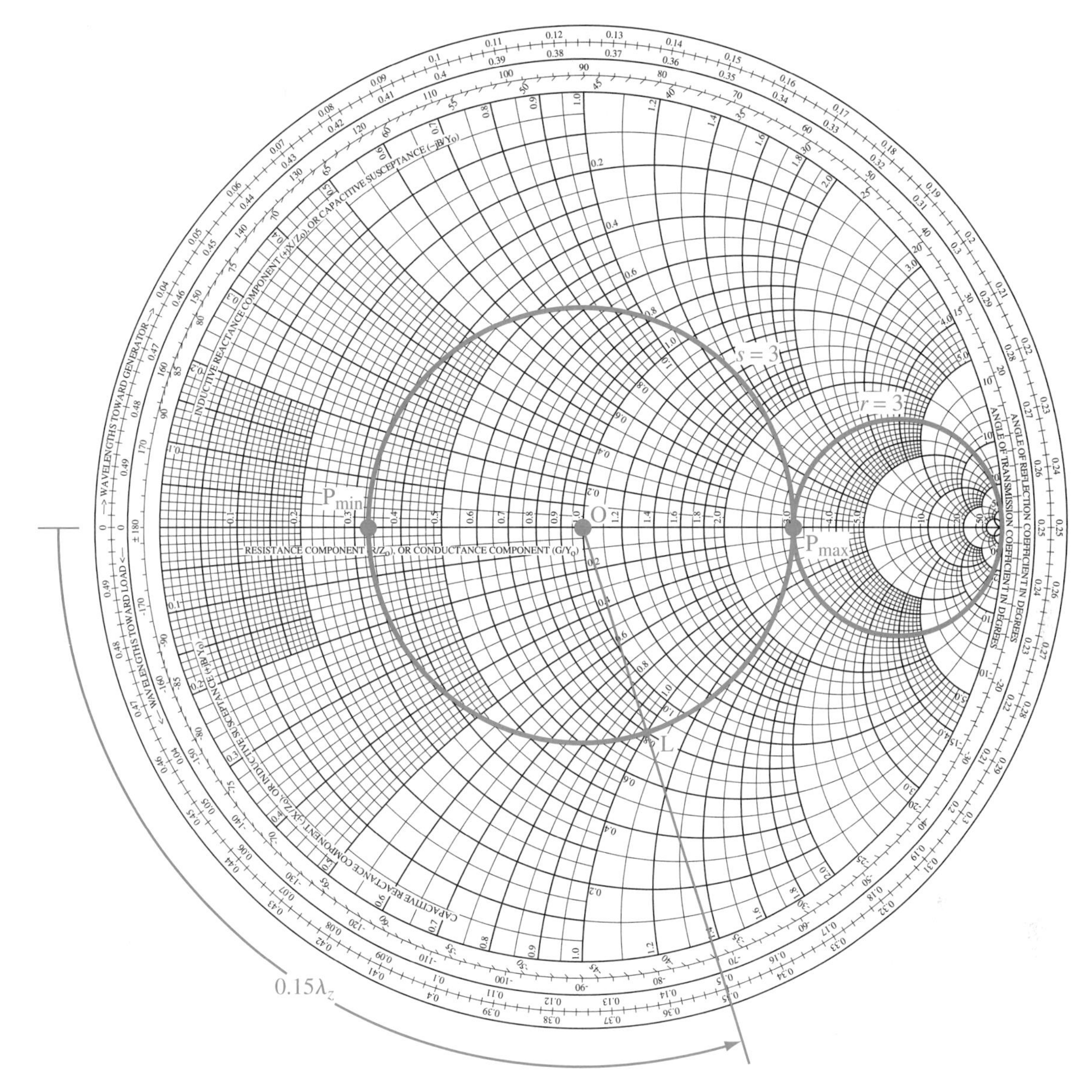

Figure 12.26 Finding an unknown load impedance by measurements on a slotted line (see Example 12.4) and graphical calculations on the Smith chart; for Example 12.24.

Solution This graphical solution is shown in Fig. 12.27. From point A on the Smith chart, with the normalized load impedance $\underline{z}_A = \underline{Z}_L/Z_{01} = 0.3 + j0.7$ ($Z_{01} = 50\ \Omega$), we move around the s circle toward the generator (clockwise) to arrive to point B, where $\underline{z}_B = s = 5$ $(5 + j0)$ and $\underline{Z}_B = sZ_{01} = 250\ \Omega$ (purely resistive impedance). The length of this compensating line section in Fig. 12.8 is $l_c = 0.25\lambda_z - 0.102\lambda_z = 0.148\lambda_z$ (read on the "wavelengths toward generator" scale in the chart). Then, continuing our movement toward the generator along the structure (Fig. 12.8), we enter the transformer line section, and while on this section we need to normalize the associated impedances to the characteristic impedance of the transformer, $Z_{0t} = \sqrt{Z_{01}(sZ_{01})} = 112\ \Omega$ [as in Eq. (12.87)]. Consequently, at the beginning of the transformer, looking from the load, this normalization yields $\underline{z}_C = \underline{Z}_B/Z_{0t} = 2.23$ $(2.23 + j0)$, which means a horizontal shift in the Smith chart to point C in Fig. 12.27. The length of the

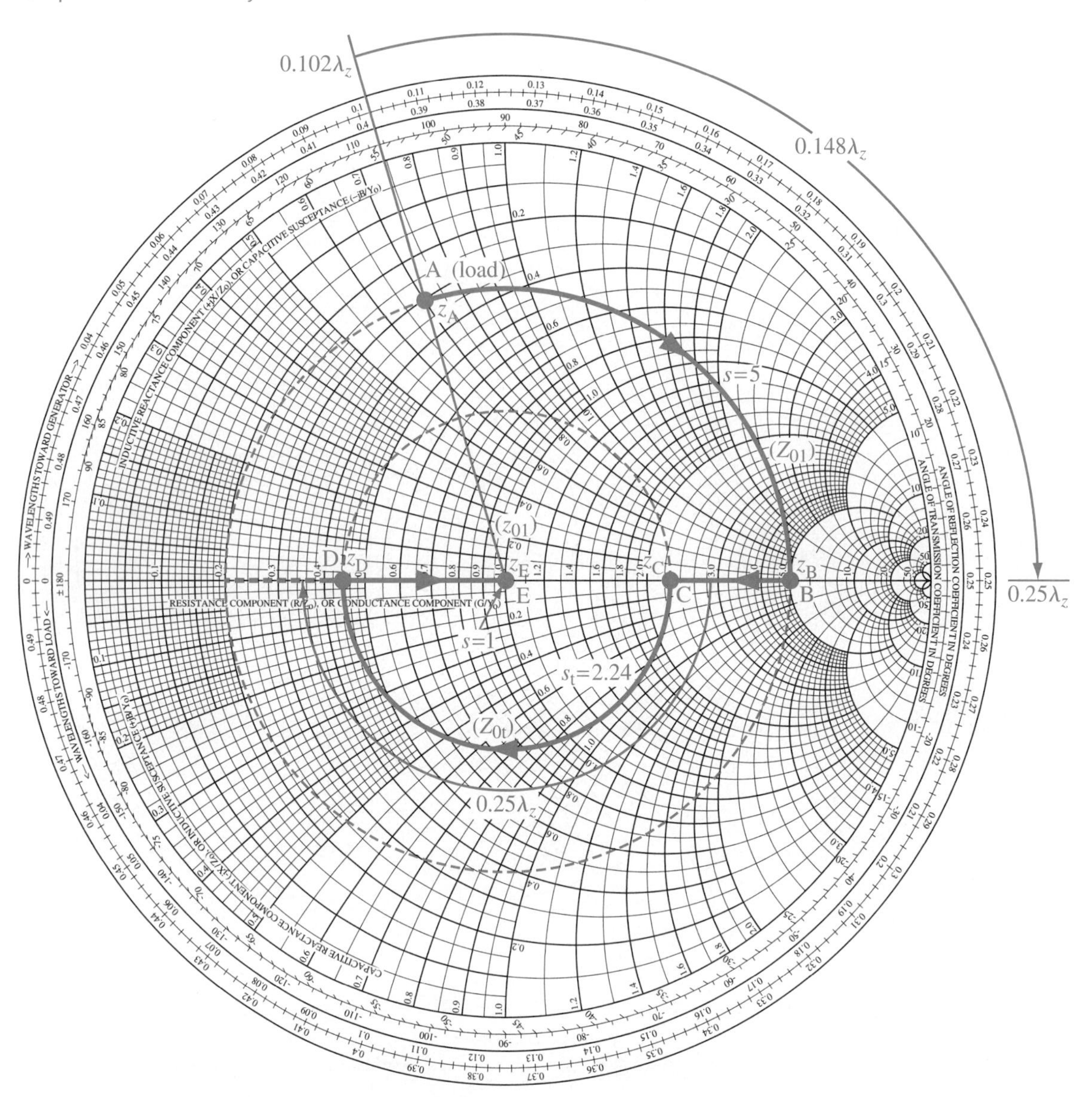

Figure 12.27 Design of the impedance-matching transmission-line circuit with a quarter-wave transformer in Fig. 12.8 using the Smith chart; for Example 12.25.

transformer being $l = \lambda_z/4$ (assuming that the wavelength is the same along the entire structure in Fig. 12.8), the complete slide along it translates to a half-circle rotation about the s circle of the transformer ($s_t = 2.23$ within the transformer) on the chart, to point D, where $\underline{z}_D = 0.446$ and $\underline{Z}_D = \underline{z}_D Z_{0t} = 50\ \Omega$. At this location, we leave the transformer section and return to the main transmission line, so the impedance normalization should again be to Z_{01}, resulting in $\underline{z}_E = \underline{Z}_D/Z_{01} = 1$, which implies another horizontal move in the chart, ending at its center (point E), and hence the matched load.

Example 12.26 Design of a Shunt Short-Circuited Stub Using the Smith Chart

Consider the circuit with a shunt short-circuited stub in Fig. 12.17, and assume that the load impedance is $\underline{Z}_L = (40 + j30)\ \Omega$, the wavelength along the transmission line is $\lambda_z = 1$ m, and

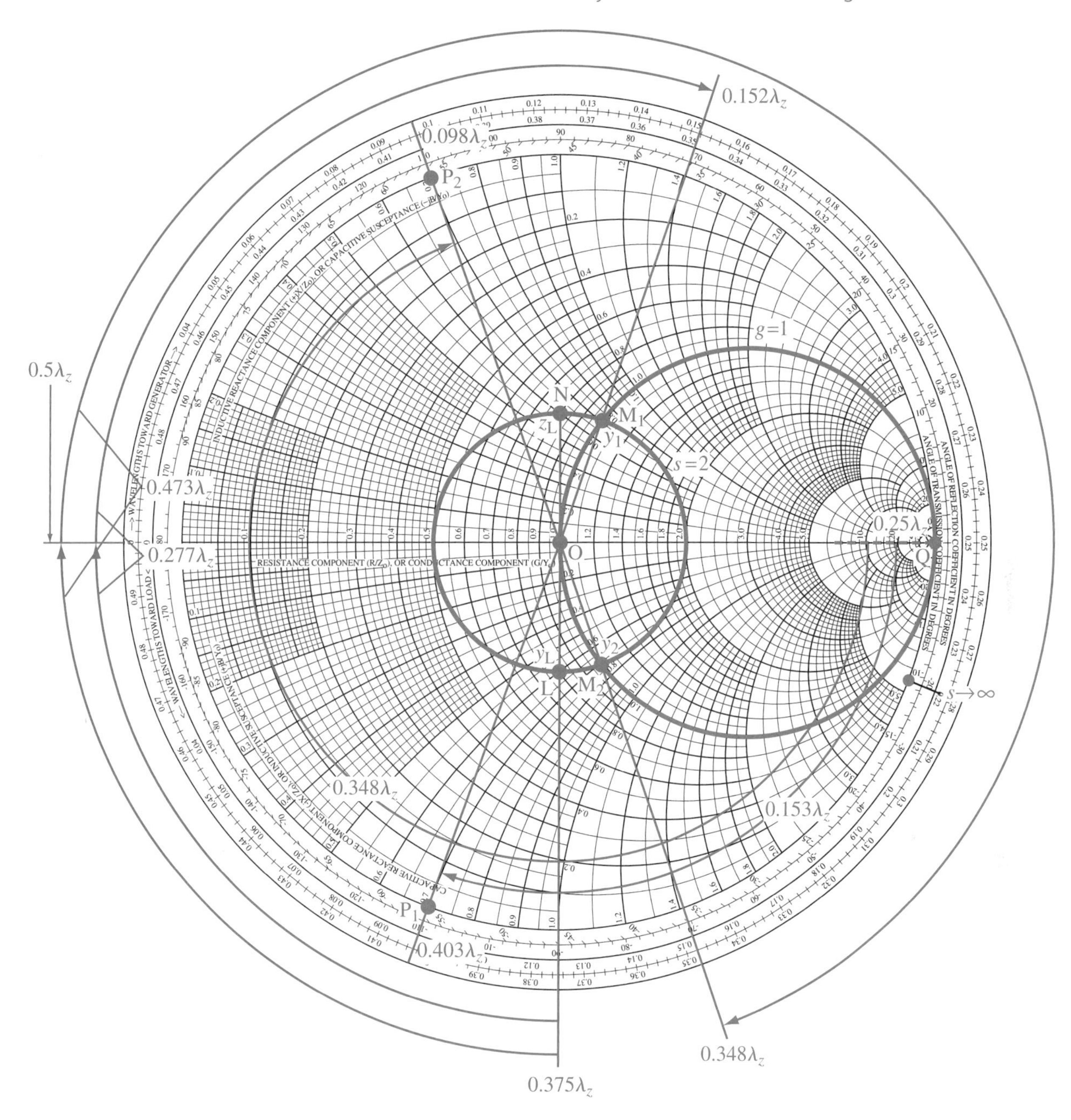

Figure 12.28 Design of an admittance-matching transmission-line circuit with a shunt short-circuited stub using the Smith chart; for Example 12.26.

the characteristic impedance of the line is $Z_0 = 50\ \Omega$. Under such circumstances, find both the distance of the stub junction from the load and the length of the stub – to match the load to the feeding line.

Solution We start by entering the Smith chart, in Fig. 12.28, at the load impedance point (point N), representing the normalized impedance $\underline{z}_{\rm L} = \underline{Z}_{\rm L}/Z_0 = 0.8 + {\rm j}0.6$, and draw the associated s circle (note that the SWR on the line is $s = 2$). Since the matching stub is connected to the main transmission line in parallel, it is more convenient to work with admittances than with impedances. We therefore find the normalized load admittance, $\underline{y}_{\mathrm{L}} = 0.8 - \mathrm{j}0.6$, at the point L across from N on the circle (also see Fig. 12.25). In graphical calculations to follow, the Smith chart in Fig. 12.28 is essentially used as an admittance chart, with r and x interpreted as g and b, respectively.

We then recall that the transmission-line admittance at the location where the stub is attached has to be of the form $\underline{Y} = Y_0 + \mathrm{j}B$, Eq. (12.127), namely, its real part must equal the characteristic admittance of the line, $Y_0 = 1/Z_0$, so that the overall matching (with the stub) is achieved. Normalized, $\underline{y} = 1 + \mathrm{j}b$, which means, in turn, that the distance l of the stub junction from the load should be at the intersection of the $g = 1$ and s circles in the Smith chart. This gives rise to two solutions to the problem: matching points M_1 and M_2 in Fig. 12.28, with $\underline{y}_1 = 1 + \mathrm{j}0.7$ and $\underline{y}_2 = 1 - \mathrm{j}0.7$, respectively. The corresponding values for the normalized input admittance of the shunt stub are $\underline{y}_{\mathrm{stub1}} = -\mathrm{j}0.7$ and $\underline{y}_{\mathrm{stub2}} = \mathrm{j}0.7$, such that $\underline{y} + \underline{y}_{\mathrm{stub}} = 1$ (Y_0 when unnormalized) in both cases. From L ($0.375\lambda_z$ mark on the "wavelengths toward generator" scale) to M_1 ($0.152\lambda_z$ mark), the required distance of the stub from the load (moving clockwise) is $l_1 = 0.5\lambda_z - 0.375\lambda_z + 0.152\lambda_z = 0.277\lambda_z = 27.7$ cm, for the first solution. Similarly, point M_2 (second solution) is found to be $l_2 = 0.5\lambda_z - 0.375\lambda_z + 0.348\lambda_z = 0.473\lambda_z = 47.3$ cm away from the load.[6]

The shunt stub in Fig. 12.17 is a separate transmission line, and assuming that its characteristic admittance and wavelength are the same as for the main line, we perform a separate graphical calculation in Fig. 12.28 to determine its length (l_{stub}). Its load termination is a short circuit, so $\underline{y} \to \infty$, and this admittance appears at the rightmost position of the chart (marked here as Q), which in fact is the open-circuit position when the chart is used for impedances. We start from the point Q and move about the unit circle, which is the s circle for the stub (note that $s \to \infty$ along the stub) and coincides with the $g = 0$ circle in an admittance chart, toward the generator until, first, the $b = -0.7$ arc is reached; at this location (point P_1), $\underline{y}_{\mathrm{stub1}} = 0 - \mathrm{j}0.7$, as desired. We read from the "wavelengths toward generator" scale that this movement, i.e., the stub length, equals $l_{\mathrm{stub1}} = 0.403\lambda_z - 0.25\lambda_z = 0.153\lambda_z = 15.3$ cm. For the other solution (stub junction at distance l_2 from the load along the main line), we trace the rotation from the point Q to the intersection (point P_2) of the unit circle and $b = 0.7$ arc ($\underline{y}_{\mathrm{stub2}} = 0 + \mathrm{j}0.7$) on the "wavelengths toward generator" scale, and obtain $l_{\mathrm{stub2}} = 0.5\lambda_z - 0.25\lambda_z + 0.098\lambda_z = 0.348\lambda_z = 34.8$ cm.

Problems: 12.23–12.30; *Conceptual Questions* (on Companion Website): 12.30–12.36; *MATLAB Exercises* (on Companion Website).

12.13 TRANSIENT ANALYSIS OF TRANSMISSION LINES

In our studies of high-frequency transmission lines so far, the time-harmonic waves on the lines are analyzed in the steady state, that is, after all the initial transitional processes have already occurred, and the resultant steady-state sinusoidal voltage and current have been established along the entire line under consideration. However, it is sometimes important to analyze the waves on the line during these transitional periods of time. Temporary variations of voltages and currents on transmission lines, or, in general, of various field and circuit quantities in arbitrary electromagnetic systems (including electric circuits), in establishing the

[6]Note that this part of the stub matching design (determination of the location on the main line where the stub should be attached) is skipped in Example 12.19, where the length l (one solution) is given in advance.

steady state of time-harmonic or any other forms of signals in the structure are called transients. A notable example are transients produced by step-like (on or off) abrupt changes of the input voltage or current at the beginning of a transmission line, which corresponds to establishing a time-constant (dc) voltage and current along the line. In integrated electronic circuits, step excitations of lines occur, for instance, when the output voltage of a source logic gate driving another logic gate via an interconnect (transmission line) switches from a "low" to "high" state, or vice versa, at a certain time (change of status of the driver logic gate). Moreover, digital signals consisting of a sequence of rectangular pulses in time can be represented as a combination of step time functions. Therefore, studies of the step response of transmission lines terminated in arbitrary loads are a basis for understanding the transient behavior of interconnects in high-speed digital circuits in many applications of digital electronics, communications, and computer engineering.

The material in this and the following sections is devoted to discussions of step and pulse responses of two-conductor transmission lines with different terminations. The analysis is performed directly, in the time domain. Alternatively, transmission lines with linear terminal networks driven by time-pulse excitations (generators) can be analyzed in the frequency (complex) domain, at a set of frequencies, which is followed by the use of the inverse Fourier transform to obtain the time-domain (transient) response. We recall that complex representatives of voltages and currents do not make sense for nonlinear circuits, so that lines with nonlinear terminal networks, on the other hand, are best analyzed directly in the time domain. In any case, in our studies of transients on transmission lines, we assume that the lines are lossless. Note that a direct transient analysis of lossy lines, i.e., lines where losses cannot be neglected, is a quite difficult problem, and the inverse Fourier transform of the frequency-domain solution is a preferable approach for such structures.

We start with telegrapher's equations for the instantaneous voltage and current, $v(z,t)$ and $i(z,t)$, on the line in Fig. 12.3, obtained as the time-domain equivalent of Eqs. (12.3). Namely, with a no-loss assumption for the line, its per-unit-length series resistance and shunt (leakage) conductance are zero, R', $G' = 0$, so that, using Eqs. (12.17) and (8.68), telegrapher's equations in the time domain read

$$\boxed{\frac{\partial v(z,t)}{\partial z} = -L'\frac{\partial i(z,t)}{\partial t}, \quad \frac{\partial i(z,t)}{\partial z} = -C'\frac{\partial v(z,t)}{\partial t},} \tag{12.187}$$

temporal telegrapher's equations, no losses

where L' and C' are the inductance and capacitance per unit length of the line. Either combining together these equations or converting Eqs. (12.5) from complex to time domain with the help of Eq. (12.19), we then obtain the temporal form of transmission-line wave equations. For the line voltage,

$$\frac{\partial^2 v(z,t)}{\partial z^2} - \frac{1}{v_\mathrm{p}^2}\frac{\partial^2 v(z,t)}{\partial t^2} = 0, \quad \text{where} \quad v_\mathrm{p} = \frac{1}{\sqrt{L'C'}} \tag{12.188}$$

[v_p is the phase velocity along the line, Eqs. (11.43)], and analogously for $i(z,t)$. The general solution for v along the line has the same form as that in Eq. (8.98), or Eqs. (9.13) and (9.28),

$$\boxed{v(z,t) = \underbrace{f_\mathrm{i}\left(t - \frac{z}{v_\mathrm{p}}\right)}_{v_\mathrm{i}(z,t)} + \underbrace{f_\mathrm{r}\left(t + \frac{z}{v_\mathrm{p}}\right)}_{v_\mathrm{r}(z,t)},} \tag{12.189}$$

instantaneous total line voltage

where $f_i(\cdot)$ and $f_r(\cdot)$ are arbitrary twice-differentiable functions.[7] The two terms, like in the complex voltage expression in Eq. (12.7), represent, respectively, a solution for the incident (forward) voltage wave, $v_i(z, t)$, propagating in the positive z direction, and one for the reflected (backward) wave, $v_r(z, t)$, with the opposite (negative z) direction of travel.

Having in mind Eqs. (9.17) and (9.28), the associated general solution for the instantaneous current, $i(z, t)$, along the line, is given by

instantaneous current

$$i(z, t) = \underbrace{\frac{1}{Z_0} f_i\left(t - \frac{z}{v_p}\right)}_{i_i(z,t)} + \underbrace{\left[-\frac{1}{Z_0} f_r\left(t + \frac{z}{v_p}\right)\right]}_{i_r(z,t)}, \tag{12.190}$$

with Z_0 being the characteristic impedance of the line, Eq. (12.18). Of course, the same result is obtained by substituting the solution for $v(z, t)$ from Eq. (12.189) into telegrapher's first equation in the time domain, i.e., the first relationship in Eqs. (12.187), and performing the same manipulations as in Eqs. (9.15)–(9.17). Note that particular functions f_i and f_r in Eqs. (12.189) and (12.190) are determined by the terminal networks at the two ends of the line (Fig. 12.2), along with the characteristics of the line itself.

From Eqs. (12.189) and (12.190), we have, analogously to Eq. (12.15) in the complex domain,

$$\frac{v_i(z, t)}{i_i(z, t)} = -\frac{v_r(z, t)}{i_r(z, t)} = Z_0 \qquad \left(Z_0 = \sqrt{\frac{L'}{C'}}\right), \tag{12.191}$$

meaning that if there is only one traveling wave on the line, the ratio of the voltage and current intensity in an arbitrary cross section of the line and at any instant of time equals a constant ($\pm Z_0$). In general, when both forward and backward waves are present, we define the time-domain counterpart of the complex transmission-line impedance, $\underline{Z}(z)$, in Eq. (12.65), as the total instantaneous voltage to current ratio at the coordinate z in Fig. 12.3,

dynamic transmission-line impedance

$$Z_{\text{dynamic}}(z, t) = \frac{v(z, t)}{i(z, t)}. \tag{12.192}$$

This impedance is referred to as the dynamic impedance (resistance), or impedance for transients, of the line.

12.14 THÉVENIN EQUIVALENT GENERATOR PAIR AND REFLECTION COEFFICIENTS FOR LINE TRANSIENTS

To further discuss the transients on lossless transmission lines, let us now assume that the voltage generator at the beginning of the line in Fig. 12.3 is given by its time-varying emf $e(t)$ and internal resistance R_g, as well as that $e(t)$ is zero for $t < 0$ and nonzero (with an arbitrary time variation) for $t \geq 0$ (the generator is switched

[7] Note that in the case of a time-harmonic (steady-state sinusoidal) regime on a lossless transmission line, both functions f_i and f_r acquire the form $f(t') = V_m \cos(\omega t' + \theta_0)$, where $t' = t \mp z/v_p$, and hence the expression for v_i in Eq. (12.9) with $\alpha = 0$ (no losses): $v_i(z, t) = f_i(t - z/v_p) = V_{i0}\sqrt{2}\cos[\omega(t - z/v_p) + \theta_{i0}] = V_{i0}\sqrt{2}\cos(\omega t - \beta z + \theta_{i0})$ [also see Eqs. (9.31) and (9.32)], and similarly for $f_r(t + z/v_p)$.

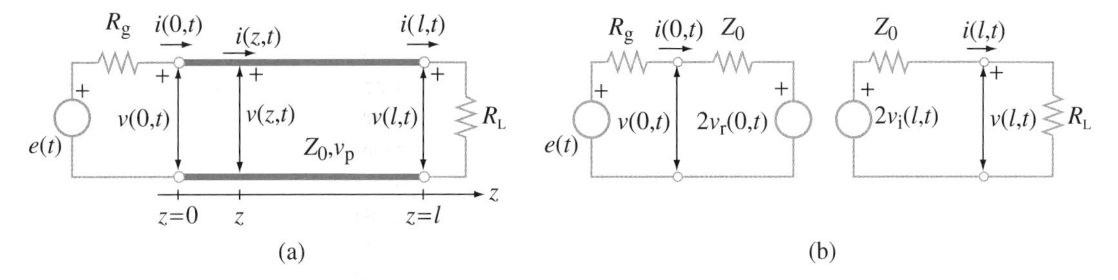

Figure 12.29 (a) Transmission line fed by a time-varying (nonsinusoidal) voltage generator with a purely resistive internal impedance and terminated in a purely resistive load and (b) its Thévenin equivalent representation for transients.

on at $t = 0$). In addition, let us, for the convenience of the discussion in this and the sections to follow, adopt the origin of the z-axis ($z = 0$) to be at the generator terminals, as shown in Fig. 12.29(a). The load terminals are then at the position defined by $z = l$, with l being the length of the line. Finally, let (for now) the load be purely resistive, of resistance R_L.

Prior to the time $t = 0$, the voltage and current are identically zero at every position along the line. At $t = 0$, the generator launches a voltage v_i, and the accompanying current i_i, to propagate toward the load. The incident wavefront reaches the load at instant $t = T$, where T designates the full one-way transit time along the line,

$$\boxed{T = \frac{l}{v_p}.} \tag{12.193}$$

one-way delay time of a transmission line

Since this time, in general, represents a delay of a signal because of the presence of a transmission line between the two points in an electric circuit (or any other electromagnetic structure), it is often called the delay time of the line [see the analogy with the delay time in Eq. (8.108)]. In the general case for Fig. 12.29(a), the load is not matched to the line, so that a reflected wave, of voltage v_r and current i_r, is generated at the load. With the use of Eqs. (12.189) and (12.190), the total voltage and current at the load terminals (for $z = l$) and any time after the arrival of the incident wave (for $t > T$) are given by

$$v(l,t) = v_i(l,t) + v_r(l,t), \quad i(l,t) = \frac{1}{Z_0}[v_i(l,t) - v_r(l,t)]. \tag{12.194}$$

Combining these two equations in a way to eliminate v_r, we get the following relationship between v and i (assuming that v_i is known):

$$\boxed{v(l,t) = 2v_i(l,t) - Z_0 i(l,t) \qquad (T < t < \infty),} \tag{12.195}$$

Thévenin generator at the right end of a line

which tells us that the line, with respect to the load, behaves like a real (nonideal) voltage generator – Thévenin equivalent generator, whose emf amounts to twice the incident voltage at the load location, $2v_i(l,t) = 2f_i(t - l/v_p)$, and internal resistance to the characteristic impedance of the line, Z_0, as in Fig. 12.29(b). Note that the parameters of the equivalent generator can be found also from its general definition, namely, by Thévenin's theorem. First, the emf of the Thévenin generator equals, in general, the open-circuit voltage of the circuit it represents. So, if the transmission line in Fig. 12.29(a) is open ($R_L \to \infty$), $i = 0$, the voltages of the incident and reflected waves across the open terminals are the same, and the total voltage is twice v_i. Second, the internal resistance of the generator equals the input dynamic impedance [see Eq. (12.192)] of the circuit (with all the generators shut down), and for the circuit in Fig. 12.29(a) that is Z_0. Regardless of the way we identify its parameters, the Thévenin generator on the right-hand side of Fig. 12.29(b) and the relationship in Eq. (12.195) are dictated by the transmission line in Fig. 12.29(a), and are independent of the load. The load, however, imposes another relationship between $v(l,t)$ and $i(l,t)$, constituting the load boundary condition for the line, so that a system of two equations with these two unknowns is obtained. For the purely resistive load in Fig. 12.29(a),

boundary condition imposed by the load

$$\boxed{v(l,t) = R_\mathrm{L} i(l,t),} \tag{12.196}$$

which, together with Eq. (12.195), yields

$$v(l,t) = \frac{2R_\mathrm{L}}{R_\mathrm{L}+Z_0}\, v_\mathrm{i}(l,t). \tag{12.197}$$

This is also obvious from Fig. 12.29(b), where R_L and Z_0 form a voltage divider. The first equation in Eqs. (12.194) then leads to the solution for v_r, for a given v_i,

$$v_\mathrm{r}(l,t) = v(l,t) - v_\mathrm{i}(l,t) = \frac{R_\mathrm{L}-Z_0}{R_\mathrm{L}+Z_0}\, v_\mathrm{i}(l,t). \tag{12.198}$$

Hence, as a major conclusion from this analysis, we realize that the reflected to incident instantaneous-voltage ratio of the line at the load terminals,

load reflection coefficient for transients

$$\boxed{\frac{v_\mathrm{r}(l,t)}{v_\mathrm{i}(l,t)} = \frac{R_\mathrm{L}-Z_0}{R_\mathrm{L}+Z_0} = \Gamma_\mathrm{L},} \tag{12.199}$$

namely, the load reflection coefficient of the line (Γ_L) for instantaneous voltages, is the same as for complex rms voltages (and a lossless or low-loss line and purely resistive load), Eq. (12.31).

As the reflected wave travels to the left, its wavefront arrives to the generator at $t = 2T$. Rewriting Eqs. (12.194) for the generator position ($z = 0$) and time $t > 2T$ (i.e., any time after the arrival of the reflected wave), and eliminating v_i from the equations, we obtain the relationship between the total voltage and current at $z = 0$, analogous to that in Eq. (12.195),

Thévenin generator at the left end

$$\boxed{v(0,t) = 2v_\mathrm{r}(0,t) + Z_0 i(0,t) \qquad (2T < t < \infty).} \tag{12.200}$$

This relationship is, again, dictated by the line, and is the same as the current-voltage characteristic of a real voltage generator, now of emf $2v_\mathrm{r}(0,t) = 2f_\mathrm{r}(t)$ and internal resistance Z_0. The Thévenin generator replacing the line looking to the right at the position $z = 0$ in Fig. 12.29(a) is thus obtained, forming the equivalent circuit on the left-hand side of Fig. 12.29(b). In addition, the relationship between v and i dictated by the generator in Fig. 12.29(a), i.e., the generator boundary condition for the line, is

boundary condition by the generator

$$\boxed{v(0,t) = e(t) - R_\mathrm{g} i(0,t),} \tag{12.201}$$

and the solution for v of the two equations comes out to be

$$v(0,t) = \frac{Z_0}{R_\mathrm{g}+Z_0}\, e(t) + \frac{2R_\mathrm{g}}{R_\mathrm{g}+Z_0}\, v_\mathrm{r}(0,t). \tag{12.202}$$

Of course, the same result can be obtained also solving the simple circuit with two voltage generators in Fig. 12.29(b). This gives the following expression for $v_{\rm i}(0, t)$ (for $t > 2T$):

$$v_{\rm i}(0,t) = v(0,t) - v_{\rm r}(0,t) = \frac{Z_0}{R_{\rm g} + Z_0}\, e(t) + \frac{R_{\rm g} - Z_0}{R_{\rm g} + Z_0}\, v_{\rm r}(0,t). \qquad (12.203)$$

The first term in the solution is the voltage $v_{\rm i}$ that is directly due to the emf e; if, for instance, $e = 0$ for $t > 2T$, then this voltage is zero. The second term is the component of $v_{\rm i}$ existing because of a difference (mismatch) of the generator internal resistance and line characteristic impedance ($R_{\rm g} \neq Z_0$) in the general case. This wave can be interpreted as due to the reflection (namely, re-reflection) of the reflected voltage $v_{\rm r}$ from the generator. Hence another major result of our analysis: the expression for the generator reflection coefficient of the line ($\Gamma_{\rm g}$) for instantaneous voltages, defined as the re-reflected to reflected transient voltage ratio at the generator terminals,

$$\Gamma_{\rm g} = \left.\frac{v_{\rm i}(0,t)}{v_{\rm r}(0,t)}\right|_{e(t)=0} = \frac{R_{\rm g} - Z_0}{R_{\rm g} + Z_0}. \qquad (12.204)$$

generator reflection coefficient

Namely, it is the same as for the load coefficient in Eq. (12.199) with $R_{\rm g}$ in place of the load resistance. If $R_{\rm g} = Z_0$, then $\Gamma_{\rm g} = 0$ (we say that the generator is matched to the line), and this component of $v_{\rm i}$ is zero.

Most importantly, the same Thévenin equivalent generator pair in Fig. 12.29(b), being dependent only on the transmission line in Fig. 12.29(a), can be used for any two networks connected at the left ($z = 0$) and right ($z = l$) ends of the line, as in Fig. 12.2. Thus, the equivalent schematic diagram in Fig. 12.30 is obtained, in which each of the terminal networks in the two Thévenin circuits can include both resistive and reactive lumped elements, as well as other transmission lines, and one or more (voltage and/or current) generators. Moreover, the networks can be both linear and nonlinear. In general, the time-domain analysis of the line can be performed by solving (analytically or numerically) Eqs. (12.195), (12.200), (12.189), and (12.190) simultaneously with the (linear or nonlinear) equations dictated by the terminal networks. In some simpler, but theoretically and practically important, cases, on the other hand, the analysis can be carried out by a simple reflection tracking method for transients on the line. The method is founded on the discussions in this section, and consists, as we shall see in the following sections, of tracking (analytically or graphically) the transient processes as they occur on the line. It uses the Thévenin equivalent representation in Fig. 12.29(b) or 12.30 and reflection coefficients in Eqs. (12.199) and (12.204). However, instead of solving all coupled space-time equations for the line and its terminal networks simultaneously, it solves them sequentially, by tracing the travel of partial transient waves in space (along the line) and multiple reflections at the line terminals as a sequence of events in time.

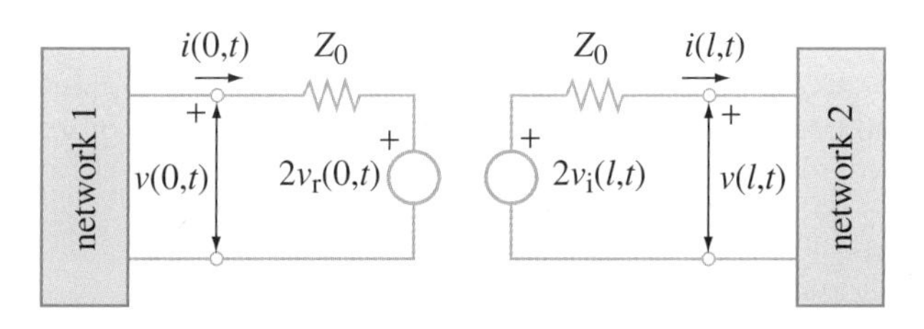

Figure 12.30 The same Thévenin equivalent generator pair from Fig. 12.29 – holds for arbitrary (resistive or reactive, linear or nonlinear) terminal networks with any number of generators.

12.15 STEP RESPONSE OF TRANSMISSION LINES WITH PURELY RESISTIVE TERMINATIONS

In this and the next section, we employ the reflection tracking method to study step and pulse responses of lossless transmission lines terminated in purely resistive loads, which also includes open- and short-circuited lines as special cases; the section to follow will then introduce a graphical tool, based on so-called bounce diagrams, to aid the analysis. Starting with a step excitation, let the emf in Fig. 12.29(a) be a step time function, as indicated in Fig. 12.31(a). The emf is defined as

step excitation

$$\boxed{e(t) = \mathcal{E}h(t), \quad h(t) = \begin{cases} 0 & \text{for } t < 0 \\ 1 & \text{for } t > 0 \end{cases},} \tag{12.205}$$

thus representing a voltage generator switched on at an instant $t = 0$, from a zero emf to a time-constant (for $t > 0$) value, $\mathcal{E}$. The unit step function, $h(t)$, is known as the Heaviside[8] function.

Prior to the return of any backward propagating wave reflected from the load, the only wave on the line is the incident one, traveling in the positive z direction in Fig. 12.31(a). Denoting its voltage and current intensity by v_{i1} and i_{i1}, respectively, we can write $v = v_{i1}$ and $i = i_{i1}$ – for the total instantaneous voltage and current of the line. Using Eq. (12.192), we then realize that the dynamic line impedance at the generator terminals, that is, the input dynamic impedance of the line, amounts to

input dynamic impedance of a transmission line

$$\boxed{(Z_{\text{in}})_{\text{dynamic}} = \frac{v_{i1}}{i_{i1}} = Z_0.} \tag{12.206}$$

In other words, the generator sees, looking into the line, a purely resistive impedance equal to the characteristic impedance of the line, Z_0. Hence, the line can be replaced, with respect to the generator, by an equivalent purely resistive load of resistance Z_0, as illustrated in Fig. 12.31(b). From this equivalent circuit, i.e., the voltage divider formed by Z_0 and R_g, the incident voltage in this initial period of time (before any reflections have occurred) is given by

initial incident voltage of a line

$$\boxed{v_{i1} = \frac{Z_0}{Z_0 + R_g}\mathcal{E}} \tag{12.207}$$

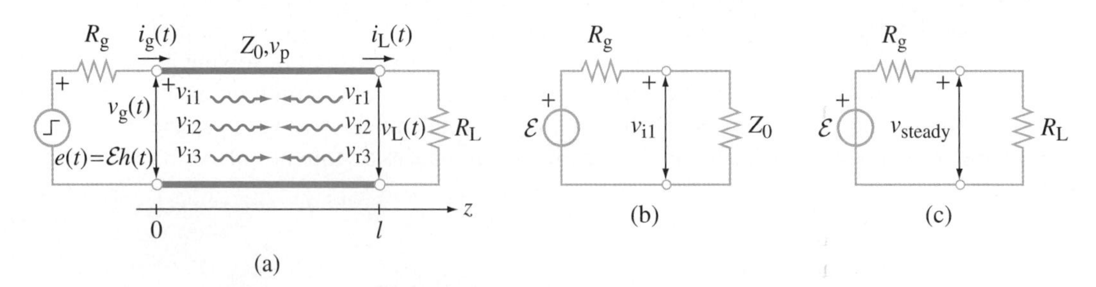

Figure 12.31 Determining the step response of a lossless transmission line with a purely resistive load by the multiple reflection tracking method: (a) components of the incident and reflected voltage waves on the line, (b) initial equivalent circuit as seen from the generator, and (c) steady-state equivalent circuit of the line.

[8] After Oliver Heaviside (see the historical aside in Section 11.1).

[note that this equation is a time-domain (transient) counterpart of Eq. (12.88) in the complex domain].

The incident wave progresses down the line at the velocity v_p, Eq. (12.188), so that at the time $t = T/2$, for instance, with T denoting the (one-way) delay period of the line, Eq. (12.193), the voltage picture along the line is: $v = v_{i1}$ for the first half of the line and $v = 0$ for the rest of it, as depicted in Fig. 12.32(a). Assuming that $R_L \neq Z_0$ (unmatched load), the incident wave (partially) bounces back off the load at $t = T$. The reflected voltage, Fig. 12.31(a), is

$$v_{r1} = \Gamma_L v_{i1}, \tag{12.208}$$

where the load reflection coefficient Γ_L is given in Eq. (12.199). Note that, for a positive v_{i1}, the sign of v_{r1} equals that of Γ_L, so $v_{r1} > 0$ for $R_L > Z_0$ and $v_{r1} < 0$ when $R_L < Z_0$. By the time $t = 3T/2$, this backward traveling wave covers a half of the return path from the load to the generator, along which the total voltage equals $v = v_{i1} + v_{r1}$, while it still amounts to $v = v_{i1}$ on the other half of the line (near the generator end), as shown in Fig. 12.32(b) – for the case $R_L > Z_0$. At $t = 2T$, the voltage v_{r1} reaches the generator and reflects from it, provided that $R_g \neq Z_0$ (unmatched generator). With the generator reflection coefficient Γ_g in Eq. (12.204), the voltage of this new incident (forward) wave in Fig. 12.31(a) is

$$v_{i2} = \Gamma_g v_{r1} = \Gamma_L \Gamma_g v_{i1}. \tag{12.209}$$

Again, the outgoing voltage (v_{i2}) changes the sign with respect to the incoming one (v_{r1}) if the reflection coefficient (Γ_g) is negative (i.e., $R_g < Z_0$); otherwise (for $R_g > Z_0$), v_{i2} keeps the same sign as v_{r1}. The total voltage distribution along the line at $t = 5T/2$, when the wavefront of the voltage step increment v_{i2} is halfway down the line, is given by

$$v(z, 5T/2) = \begin{cases} v_{i1} + v_{r1} + v_{i2} = \left(1 + \Gamma_L + \Gamma_L \Gamma_g\right) v_{i1} & \text{for } 0 \le z < l/2 \\ v_{i1} + v_{r1} = (1 + \Gamma_L)\, v_{i1} & \text{for } l/2 \le z \le l \end{cases} \tag{12.210}$$

[sketch in Fig. 12.32(c), assuming that $R_g > Z_0$]. Of course, these are algebraic sums of voltage increments, as they, in general, can be both positive and negative. Next, v_{i2} bounces off the load, with the reflection coefficient Γ_L, at $t = 3T$, so that the new reflected voltage [Fig. 12.31(a)] is

$$v_{r2} = \Gamma_L v_{i2} = \Gamma_L^2 \Gamma_g v_{i1}, \tag{12.211}$$

and so on.

While tracing the voltage distribution along the line, it is very important to always have in mind that the voltages $v_{i1}, v_{i2}, \ldots$ in Fig. 12.31(a) are the simultaneously existing components at a given instant of time and given position on the line of the voltage v_i in Eq. (12.189) of the actual forward propagating wave. By the same token, $v_{r1}, v_{r2}, \ldots$ in Fig. 12.31(a) constitute the actual reflected voltage v_r in Eq. (12.189). Accordingly, we refer to v_{in} and v_{rn} ($n = 1, 2, \ldots$) as component or partial line voltages.

Obviously, for each round trip of the wave along the line, the component voltage gets multiplied by a factor of $x = \Gamma_L \Gamma_g$. None of the reflection coefficients can be greater than unity [see Example 10.10 and Eqs. (12.55) and (12.56)], and aside from the special cases of loads and generators resulting in $|\Gamma_L| = |\Gamma_g| = 1$, which are to be addressed in an example in this section, we have that $|x| < 1$. Therefore, as the multiple reflection process on the line continues (indefinitely), the contribution of new added reflected (re-reflected) components, propagating in either direction in Fig. 12.31(a), to the total voltage at any position along the line diminishes with every

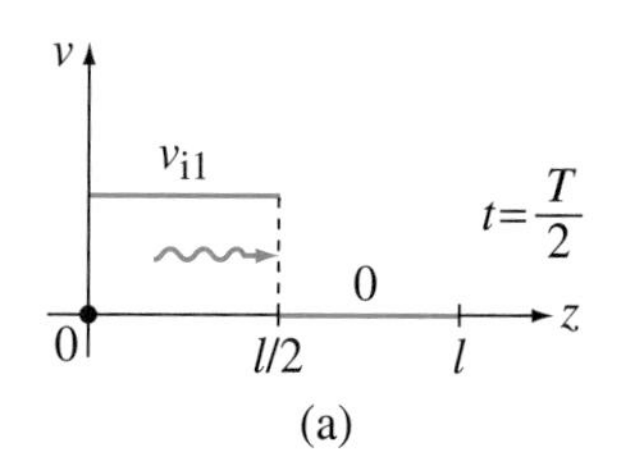

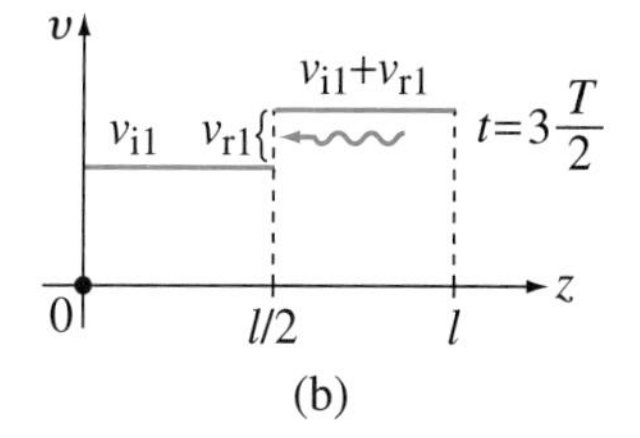

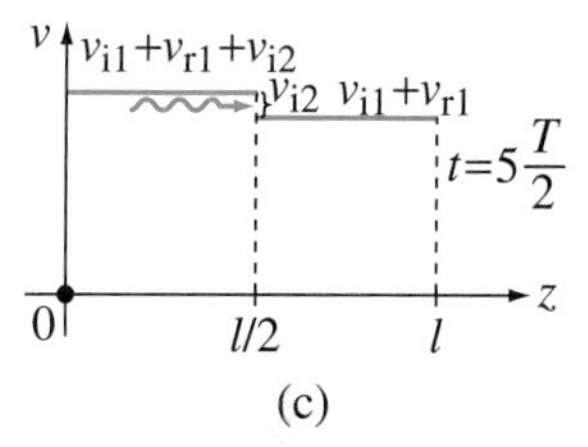

Figure 12.32 Total voltage distribution, $v(z, t)$, $0 \le z \le l$, on the transmission line in Fig. 12.31(a) for R_L, $R_g > Z_0$ at (a) $t = T/2$, (b) $t = 3T/2$, and (c) $t = 5T/2$.

new bounce from the load or generator. At $t \to \infty$, in the steady state, the total voltage for any z $(0 \le z \le l)$ is given by

$$\begin{aligned} v_{\text{steady}} &= v(z, t \to \infty) = v_{\text{i1}} + v_{\text{r1}} + v_{\text{i2}} + v_{\text{r2}} + v_{\text{i3}} + v_{\text{r3}} + \dots \\ &= \left(1 + \Gamma_{\text{L}} + \Gamma_{\text{L}}\Gamma_{\text{g}} + \Gamma_{\text{L}}^2\Gamma_{\text{g}} + \Gamma_{\text{L}}^2\Gamma_{\text{g}}^2 + \Gamma_{\text{L}}^3\Gamma_{\text{g}}^2 + \dots\right) v_{\text{i1}} \\ &= (1 + \Gamma_{\text{L}})\left[1 + \Gamma_{\text{L}}\Gamma_{\text{g}} + (\Gamma_{\text{L}}\Gamma_{\text{g}})^2 + \dots\right] v_{\text{i1}} = \frac{1 + \Gamma_{\text{L}}}{1 - \Gamma_{\text{L}}\Gamma_{\text{g}}}\, v_{\text{i1}}, \end{aligned} \quad (12.212)$$

where the use is made of the well-known formula for the sum of a converging infinite geometric series, $1 + x + x^2 + \dots = 1/(1 - x)$ for $|x| < 1$ $(x = \Gamma_{\text{L}}\Gamma_{\text{g}})$. Substituting the expressions for Γ_{L}, Γ_{g}, and v_{i1} from Eqs. (12.199), (12.204), and (12.207) into the result for the steady-state voltage in Eq. (12.212), and simplifying thus obtained expression [multiplying both the numerator and denominator by $(R_{\text{L}} + Z_0)(R_{\text{g}} + Z_0)$], we have

steady-state voltage on a line

$$\boxed{v_{\text{steady}} = \frac{R_{\text{L}}}{R_{\text{L}} + R_{\text{g}}}\,\mathcal{E}.} \quad (12.213)$$

This simple result is expected, since the steady state for a structure excited by a voltage step generator, Eq. (12.205), is actually a dc regime, in which a lossless transmission line can be considered simply as a pair of ideal short-circuiting conductors. Namely, for time-constant voltages and currents, all the distributed series inductors and shunt capacitors in the circuit model of the line in Fig. 12.1 behave like short and open circuits [see Eqs. (7.3) and (3.45), where $\mathrm{d}/\mathrm{d}t \equiv 0$], respectively. In addition, if losses are zero or negligible, there are no distributed resistors in the model, i.e., the series and shunt resistors in Fig. 12.1 are shorted (bypassed) and open (disconnected), respectively (ΔR, $\Delta G = 0$), which is also equivalent to the dc line representation in Fig. 3.21 with R', $G' = 0$. So, at $t \to \infty$, the line appears as if nonexistent, that is, as if the load is directly connected to the generator. This gives rise to the equivalent circuit in Fig. 12.31(c), from which Eq. (12.213) is directly obtained. However, in establishing and using such an equivalency, we always have in mind that the line, of course, is present in the circuit all the time, including the steady state $(t \to \infty)$, in which it is fully charged (for a given step emf $\mathcal{E}$), with all the distributed capacitors in Fig. 12.1 charged to a voltage v_{steady} and all inductors carrying a current of intensity $v_{\text{steady}}/R_{\text{L}}$.

Fig. 12.33 shows the waveforms during the time of several transit periods T of the voltages at the generator and load terminals, respectively, $v_{\text{g}}(t) = v(0, t)$ and $v_{\text{L}}(t) = v(l, t)$, for the line in Fig. 12.31(a), sketched based on Fig. 12.32 and Eqs. (12.207)–(12.213). Note that the signal $v_{\text{g}}(t)$ for $0 \le t < 2T$ is the same for any load at the other end of the line; the generator does not know what is connected as a load until the voltage v_{r1} returns at $t = 2T$ to it with the information (Γ_{L}) about the load. Note also that, although the transitional process, i.e., establishing a steady (dc) state, in the general case ($R_{\text{L}} \ne Z_0$ and $R_{\text{g}} \ne Z_0$) theoretically lasts indefinitely, the changes of v_{g} and v_{L}, as well as of the total voltage and current at any position along the line, become practically negligible after a finite time, namely, a finite number of transit periods T. The smaller the magnitudes of the load and generator reflection coefficients, $|\Gamma_{\text{L}}|$ and $|\Gamma_{\text{g}}|$, the smaller this number of periods T by which the signal (voltage or current) on the entire line practically reaches its steady-state value. Equivalently, the smaller the magnitude of $x = \Gamma_{\text{L}}\Gamma_{\text{g}}$ the faster the convergence of the geometric series in Eq. (12.212).

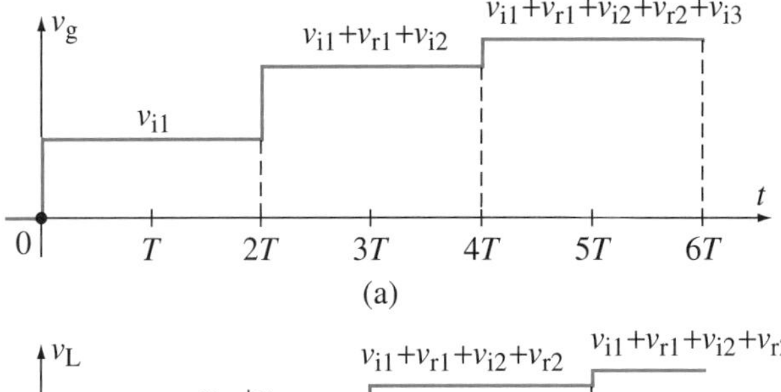

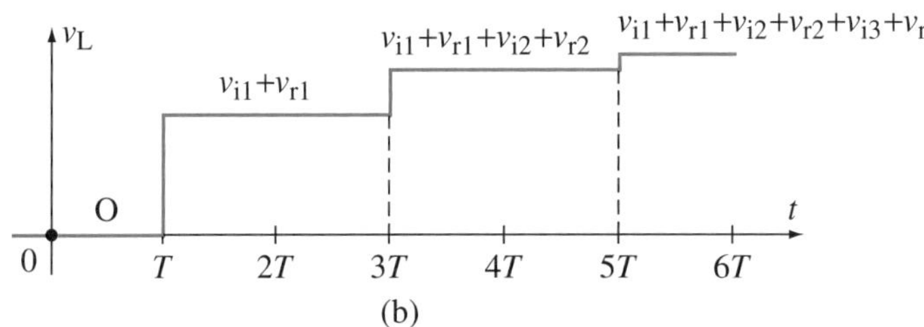

Figure 12.33 Voltage waveforms at the generator (a) and load (b) for the transmission line in Fig. 12.31(a) corresponding to the line voltage snapshots in Fig. 12.32.

If the load in Fig. 12.31(a) is matched to the line, so $R_{\rm L} = Z_0$ and $\Gamma_{\rm L} = 0$ [see Fig. 12.13(c) and Eq. (12.115)], we have that $v_{\rm r1} = 0$ in Eq. (12.208), and the total voltage along the entire line equals $v = v_{\rm i1}$ for $t > T$, meaning that the transitional regime lasts only one delay time T, whether the generator is matched or not. On the other hand, a combination of an unmatched load ($R_{\rm L} \neq Z_0$) and matched generator, with $R_{\rm g} = Z_0$ and $\Gamma_{\rm g} = 0$, leads to $v_{\rm i2} = 0$ in Eq. (12.209), and a steady state is established after only $2T$ for the whole structure. From Eq. (12.207),

$$v_{\rm i1} = \frac{\mathcal{E}}{2} \quad (R_{\rm g} = Z_0), \tag{12.214}$$

and hence Eq. (12.212) becomes

$$\boxed{v_{\rm steady} = (1 + \Gamma_{\rm L})\,\frac{\mathcal{E}}{2}.} \tag{12.215}$$

matched generator, arbitrary resistive load

At the load, this final voltage is established even earlier, at instant $t = T$. Fig. 12.34 shows the input and output waveforms, $v_{\rm g}(t)$ and $v_{\rm L}(t)$, for the cases of $R_{\rm L} > Z_0$ and $R_{\rm L} < Z_0$, respectively. Note that the load voltage for $t > T$ can also be written as $v_{\rm L} = \tau_{\rm L} v_{\rm i1}$, where $\tau_{\rm L}$ stands for the load voltage transmission coefficient of the line [see Eq. (12.32)].

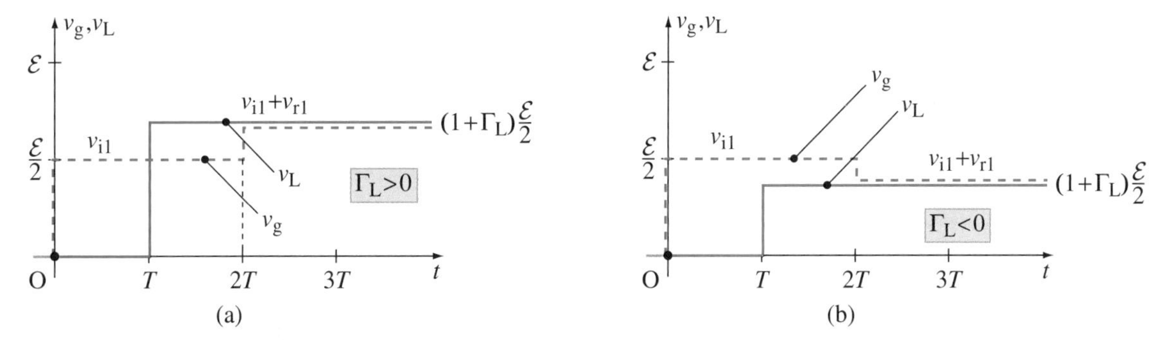

Figure 12.34 Input and output voltage waveforms, $v_{\rm g}(t)$ and $v_{\rm L}(t)$, for the line in Fig. 12.31(a) and matched generator ($R_{\rm g} = Z_0$) and unmatched load, with (a) $R_{\rm L} > Z_0$ and (b) $R_{\rm L} < Z_0$.

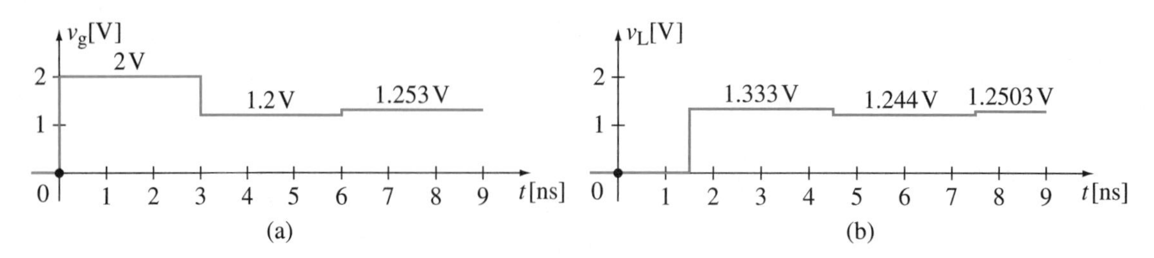

Figure 12.35 Voltage waveforms at the generator (a) and load (b) for the transmission line in Fig. 12.31(a) for $R_L < Z_0$ ($\Gamma_L < 0$) and $R_g > Z_0$ ($\Gamma_g > 0$); for Example 12.27.

Example 12.27 **Coaxial Cable with Both Load and Generator Unmatched**

A lossless coaxial cable of length $l = 30$ cm has a homogeneous dielectric of parameters $\varepsilon_r = 2.25$ and $\mu_r = 1$ (polyethylene). The ratio of the outer to inner conductor radii is such that $\ln b/a = 1.25$. The line is fed by a voltage generator of step emf $\mathcal{E} = 5$ V applied at $t = 0$ and internal resistance $R_g = 75\ \Omega$. The other end of the line is terminated in a purely resistive load of resistance $R_L = 25\ \Omega$. Sketch the voltage waveforms at both ends of the line within a time interval $0 \le t \le 9$ ns.

Solution From Eqs. (11.17), (9.47), and (11.48), $v_p = 2 \times 10^8$ m/s and $Z_0 = 50\ \Omega$, so that Eqs. (12.193), (12.207), (12.199), and (12.204) then give $T = 1.5$ ns, $v_{i1} = 2$ V, $\Gamma_L = -1/3$ (note that we now have a negative load reflection coefficient, as opposed to the situation in Fig. 12.33), and $\Gamma_g = 1/5$, respectively. Using Eqs. (12.208), (12.209), and (12.211), $v_{r1} = -0.667$ V (note that this is a negative increment to the previous total voltage on the line), $v_{i2} = -0.133$ V, $v_{r2} = 0.0444$ V, $v_{i3} = 8.89$ mV, and $v_{r3} = -2.96$ mV, with which the waveforms in Fig. 12.35 are sketched. In the steady state ($t \to \infty$), $v_{\text{steady}} = 1.25$ V, by means of Eq. (12.213).

Example 12.28 **Time-Domain Reflectometry**

A time-domain reflectometer (TDR) consisting of a step voltage generator and an oscilloscope is used to determine the resistance of an unknown load resistor connected to a lossless transmission line. The characteristic impedance of the line is $Z_0 = 50\ \Omega$, and so is the internal resistance of the generator. If the line is fed by this generator, with emf $\mathcal{E} = 4$ V applied at $t = 0$, the oscilloscope displays voltage $v_g(t)$ in Fig. 12.36 at the generator end of the line. What is the load resistance?

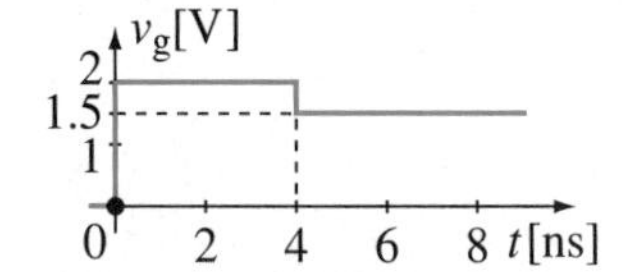

Figure 12.36 Measuring unknown load termination of a transmission line by time-domain reflectometry: oscilloscope display of the voltage step response of the line at its generator end ($R_g = Z_0$), from which Eqs. (12.215) and (12.199) give the load resistance, R_L; for Example 12.28.

Solution The generator being matched to the transmission line, the total voltage at the generator end of the line in the steady state, that is, after two one-way wave travel time periods in Eq. (12.193), is given by Eq. (12.215). From Fig. 12.36, $v_{\text{steady}} = 1.5$ V, which, combined with Eqs. (12.215) and (12.199), yields the load reflection coefficient and resistance of $\Gamma_L = -1/4$ and $R_L = 30\ \Omega$, respectively.

Note that time-domain reflectometry is widely used for measurement of characteristics of unknown (resistive and/or reactive) load terminations of transmission lines, as well as line discontinuities (between the generator and load), such as a break on a buried coaxial cable (or invisible internal partial break on any other coaxial cable), an undesired parasitic capacitance on a microstrip line, an inductance of a bonding wire connecting two lines, etc. Note also that, by means of Eqs. (12.193) and (9.47) and Fig. 12.36, TDR can provide information on the (unknown) length of a transmission line (with a known dielectric) or location of the discontinuity on the line.

Example 12.29 Step Response of Open- and Short-Circuited Lines

A lossless transmission line of length l is fed by a step voltage generator with emf magnitude $\mathcal{E}$, Eq. (12.205), that is matched to the line. The phase velocity of electromagnetic waves on the line is v_p. Sketch the voltage waveforms at both the generator and load if the other end of the line is (a) open and (b) shorted, respectively.

Solution

(a) Voltage waveforms for an open circuit as load in Fig. 12.31(a) are sketched in Fig. 12.37(a), as the limit of the case in Fig. 12.34(a) for $R_L \to \infty$, which results in a unity load reflection coefficient, $\Gamma_L = 1$, as in Eq. (12.113) in the frequency domain [see Fig. 12.13(b)]. Note that in practice this situation also occurs when the input dynamic impedance of a network or device connected to the farther end of the line is very high compared to Z_0. We then have $v_{r1} = v_{i1}$ in Eq. (12.208), so that the total voltage across the open line terminals (at $z = l$) is $v_{steady} = 2v_{i1} = \mathcal{E}$ for $t > T$, i.e., after one delay period of the line, Eq. (12.193), which is obtained from Eq. (12.215) as well. We see that

$$\boxed{v_L(t) = \mathcal{E}h(t - T),} \qquad (12.216)$$

matched generator, open-circuited line

i.e., that the output voltage of the line is exactly the same as the emf of the generator in Eq. (12.205), just delayed by the period T. In other words, an open-circuited lossless transmission line acts as an ideal delay line for a matched generator.

(b) On the other side, as the extreme case of the situation in Fig. 12.34(b), Fig. 12.37(b) shows the voltage step response of a short-circuited line [as in Fig. 12.13(a)], with $R_L = 0$ and thus $\Gamma_L = -1$ [also see Eq. (12.110)]. A short circuit can as well approximate a very low input dynamic impedance seen into a network/device connected to the right. Eq. (12.208) now gives $v_{r1} = -v_{i1}$, namely, that the reflected voltage completely cancels the incident one at $z = l$ and $t = T$. The total voltage is zero along the entire line for $t > 2T$ [$v_{steady} = 0$ in Eq. (12.215)]. Of course, $v_L = 0$ at all times, as dictated by the short circuit (or a very low input dynamic impedance seen to the right) at the load terminals.

Note that in both cases in Fig. 12.37, we have a total reflection of the incident wave at the load end of the transmission line. In other words, line terminations given by both $R_L \to \infty$ and $R_L = 0$ lead to $|\Gamma_L| = 1$, and the entire energy of the incident wave is reflected back to the line.

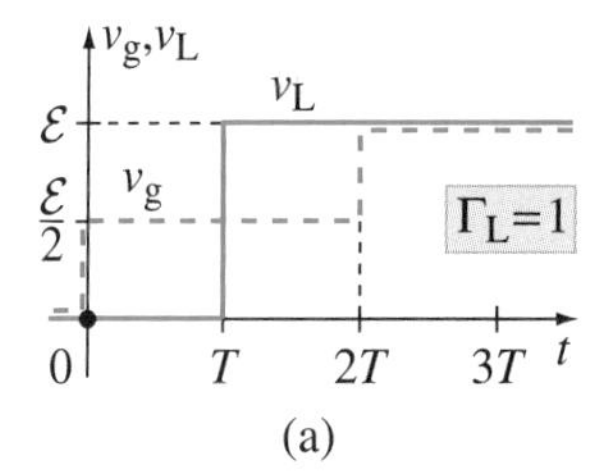

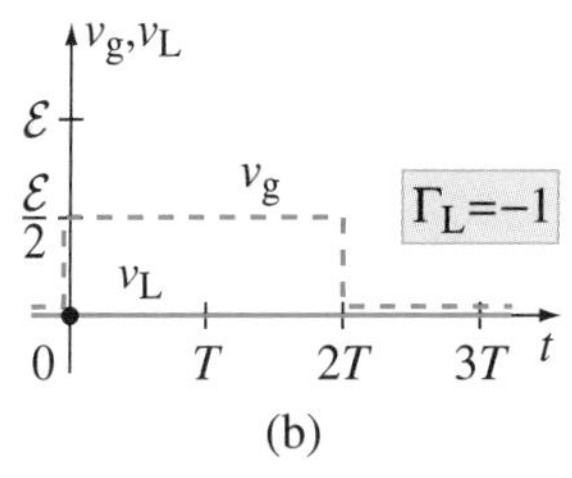

Figure 12.37 Step response of the line in Fig. 12.31(a) with $R_g = Z_0$ (matched generator) and (a) $R_L \to \infty$ (open-circuited line) and (b) $R_L = 0$ (short-circuited line); for Example 12.29.

Example 12.30 Open- or Short-Circuited Line with an Ideal Generator

Repeat the previous example but for an ideal step voltage generator feeding the line. For case (b), also sketch the voltage snapshots along the line at times $t = 2.5T$ and $t = 5.75T$, respectively, T being the one-way time delay of the line.

Solution

(a) An ideal voltage generator has a zero internal resistance, and with $R_g = 0$, Eqs. (12.207) and (12.204) yield $v_{i1} = \mathcal{E}$ and $\Gamma_g = -1$, respectively. In addition, since $\Gamma_L = 1$, as in Fig. 12.37(a), $v_{r1} = v_{i1}$, and hence the total voltage at the load for $T < t < 3T$ equals $v_{tot1} = v_{i1} + v_{r1} = 2\mathcal{E}$. From Eq. (12.209), $v_{i2} = -v_{r1} = -\mathcal{E}$, so that $v_{tot2} = v_{tot1} + v_{i2} + v_{r2} = 0$ ($v_{r2} = v_{i2}$) after two more delay times T. Starting at $t = 5T$, we again have $v_{tot3} = 2\mathcal{E}$, and so on. So, the voltage at the generator terminals is constant ($\mathcal{E}$), while the voltage at the open end of the line consists of a periodic sequence of rectangular pulses of amplitude $2\mathcal{E}$ in time, as shown in Fig. 12.38(a).

(b) For the short-circuited line and ideal voltage generator, $v_g(t) = \mathcal{E}$ and $v_L(t) = 0$ for $0 < t < \infty$, which can as well be obtained from a similar analysis to that in case (a), with $\Gamma_L = \Gamma_g = -1$.

All signals on the line periodically repeat themselves in time after each interval $2T$, which means that the voltage scans on the line, $v(z, t)$, $0 \le z \le l$, at times $t = 2.5T$ and

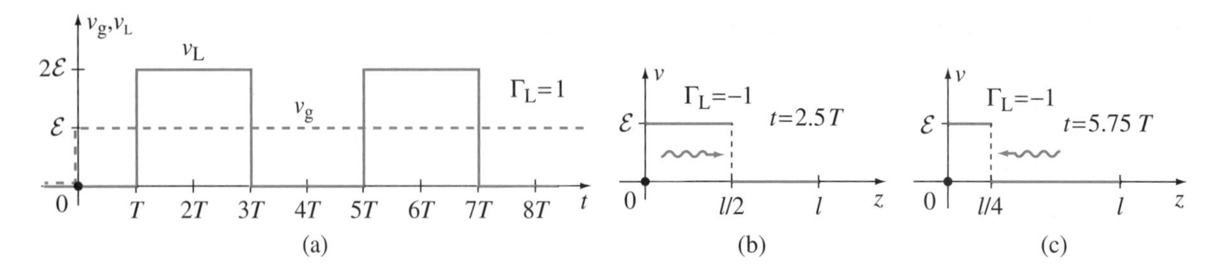

Figure 12.38 Input and output voltage waveforms for an open-circuited transmission line fed by an ideal step voltage generator (a) and voltage snapshots at times $t = 2.5T$ (or $t = 0.5T$) and $t = 5.75T$ (or $t = 1.75T$), respectively, along the line when short-circuited (b)–(c); for Example 12.30.

$t = 5.75T$ are the same as those at $t = 0.5T$ and $t = 1.75T$, respectively. Having this, as well as Fig. 12.32(a) and Fig. 12.32(b), in mind, the two scans are sketched in Fig. 12.38(b) and Fig. 12.38(c).

Example 12.31 **Current-Intensity Transient Response of a Line**

Sketch transient diagrams for the transmission-line current corresponding to those in Figs. 12.32 and 12.33 for the line voltage.

Solution Of course, the transient response of a transmission line can be traced for the line current as well. The solution for $i(z, t)$ of the line can be obtained starting with the current intensity of the initial incident wave launched by the generator in Fig. 12.31(a) at $t = 0$, which equals, from Fig. 12.31(b), $i_{i1} = \mathcal{E}/(Z_0 + R_g)$. Multiple reflections are then tracked at the load and generator as for the voltages in Eqs. (12.208)–(12.211), having in mind that the reflection coefficients for currents are the negative of the corresponding voltage coefficients in Eqs. (12.199) and (12.204) [see Eq. (12.28)]. Hence, the first reflected component current, associated with the component voltage in Eq. (12.208), is $i_{r1} = -\Gamma_L i_{i1}$, the current accompanying the voltage in Eq. (12.209) comes out to be $i_{i2} = -\Gamma_g i_{r1} = \Gamma_L \Gamma_g i_{i1}$, and so on. Alternatively, the transient current can be found directly from the voltage waveforms given that the voltage to current ratio for a traveling wave on the line equals $\pm Z_0$, Eq. (12.191), with the plus (minus) sign corresponding to the wave propagating in the positive (negative) z direction in Fig. 12.31(a). The components of the total forward propagating current are thus obtained from the corresponding voltage components in Eqs. (12.207)–(12.211) as $i_{i1} = v_{i1}/Z_0$, $i_{r1} = -v_{r1}/Z_0$, $i_{i2} = v_{i2}/Z_0$, $i_{r2} = -v_{r2}/Z_0$, etc. So, using either one approach we sketch in Fig. 12.39 the current-intensity snapshots (scans), $i(z, t)$, along the line at different characteristic instants of time and input/output current waveforms, $i_g(t)$ and $i_L(t)$, that correspond to voltage snapshots and waveforms in Figs. 12.32 and 12.33.

Problems: 12.31–12.37; *Conceptual Questions* (on Companion Website): 12.37–12.43; *MATLAB Exercises* (on Companion Website).

12.16 ANALYSIS OF TRANSMISSION LINES WITH PULSE EXCITATIONS

Before considering pulse excitations of lossless transmission lines in this section, let us point out that, from the point of view of the general system theory, an arbitrary line made from linear electromagnetic materials represents a linear time-invariant system. This comes from the linearity of the governing telegrapher's equations, Eqs. (12.187), and the coefficients in these equations, namely, the line

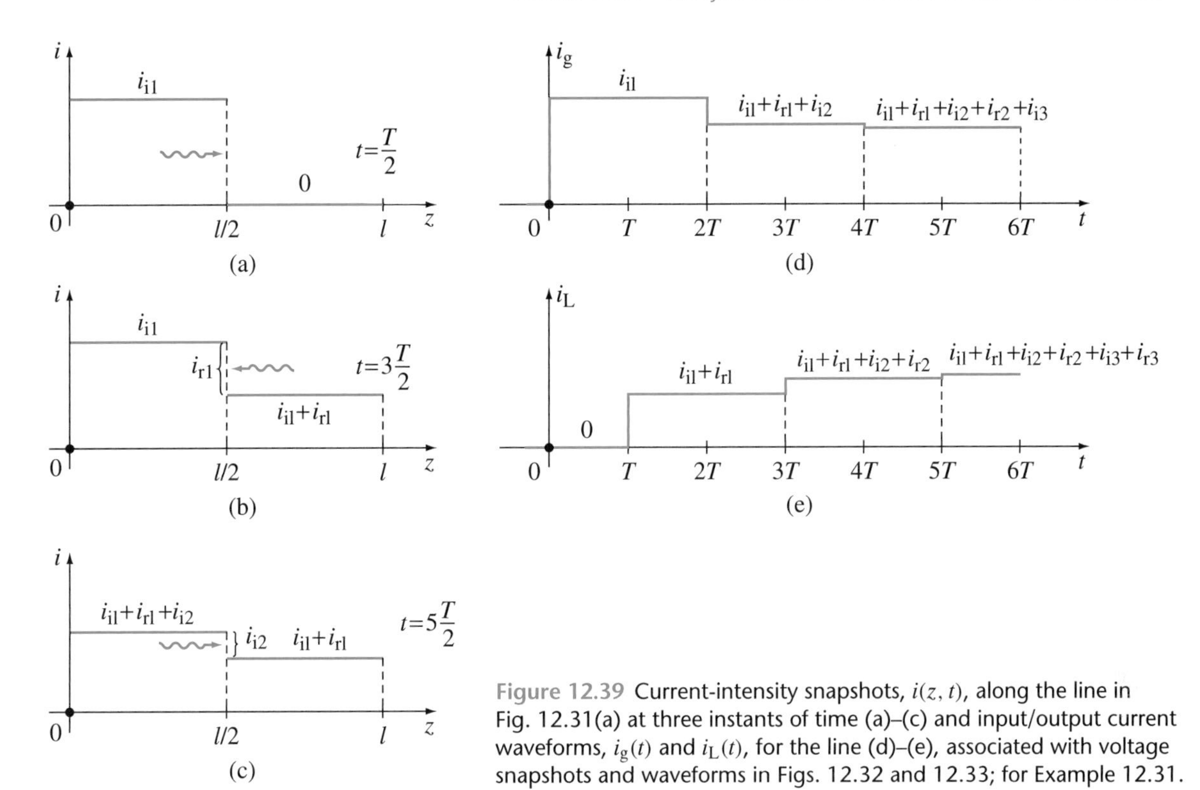

Figure 12.39 Current-intensity snapshots, $i(z, t)$, along the line in Fig. 12.31(a) at three instants of time (a)–(c) and input/output current waveforms, $i_g(t)$ and $i_L(t)$, for the line (d)–(e), associated with voltage snapshots and waveforms in Figs. 12.32 and 12.33; for Example 12.31.

per-unit-length inductance and capacitance, L' and C', being time-invariant (not changing with time). In other words, all the distributed elements (inductors and capacitors for the lossless case) in the equivalent schematic diagram in Fig. 12.1 are linear (because of the linearity of the line dielectric in Fig. 11.3) and do not change with time. By the superposition principle, which holds because of the system linearity,[9] if the input (excitation) to the system, $x(t)$, can be represented as a linear combination of two component signals, $x_1(t)$ and $x_2(t)$, then the output (response) of the system, $y(t)$, can be obtained as the same combination of the individual responses to the component inputs if applied alone,

$$\boxed{x(t) = ax_1(t) + bx_2(t) \quad \longrightarrow \quad y(t) = ay_1(t) + by_2(t),} \tag{12.217}$$

linear system

where a and b are arbitrary constants. On the other side, time invariance means that whether we apply an excitation to the system at a reference instant $t = 0$ or after some time t_0, the outputs will be identical, just shifted in time (delayed) relative to each other by t_0. Simply, if the output due to an input $x(t)$ is $y(t)$, then the following equally time-shifted input/output pair of signals takes place:

$$\boxed{x(t - t_0) \quad \longrightarrow \quad y(t - t_0).} \tag{12.218}$$

time-invariant system

Both properties of transmission lines, in Eqs. (12.217) and (12.218), are useful to have in mind, and can often help in time-domain analysis of transmission-line

[9]Note that in a general linear electromagnetic system, the superposition principle holds for the electric and magnetic fields, excited by impressed electric fields and currents, Eqs. (3.109) and (3.124), in linear media, and is a consequence of the linear character of the governing Maxwell's equations in that case. Of course, a transmission line with linear materials is such a system.

systems, in conjunction with other equations and techniques. In particular, we use them here to find the pulse response of transmission lines based on what we already know about the step analysis of lines from the preceding section.

Let us assume that the emf in Fig. 12.31(a) is a rectangular pulse function of time, of magnitude $\mathcal{E}$ and duration t_0, triggered at $t = 0$, as shown in Fig. 12.40(a). This function is analytically given by

pulse excitation

$$e(t) = \mathcal{E}\Pi(t), \quad \Pi(t) = \begin{cases} 1 & \text{for } 0 < t < t_0 \\ 0 & \text{for } t < 0 \text{ and } t > t_0 \end{cases}, \tag{12.219}$$

with $\Pi(t)$ standing for the corresponding unit rectangular pulse time function, which is also called the gate or window function.

It is obvious from Fig. 12.40(b) that $e(t)$ can be viewed as a superposition of two step functions, Eq. (12.205), with opposite polarities and a time shift t_0 between them,

$$e(t) = \mathcal{E}h(t) - \mathcal{E}h(t - t_0). \tag{12.220}$$

Due to the linearity and time invariance of the transmission line in Fig. 12.31(a), including its terminal networks (generator and load), as expressed by Eqs. (12.217) and (12.218), the response of the line to $e(t)$ can be computed combining the individual responses to the two step inputs if applied alone. Namely, marking by $v_{L1}(t)$ the line output response [load voltage in Fig. 12.31(a)] to the input $e_1(t) = \mathcal{E}h(t)$ alone, the resultant output response to the combined excitation in Eq. (12.220) is obtained as

pulse response from step analysis

$$v_L(t) = v_{L1}(t) - v_{L1}(t - t_0), \tag{12.221}$$

i.e., the load voltage is the same superposition of $v_{L1}(t)$ and its flipped-over (multiplied by -1) and delayed (by t_0) version as in Eq. (12.220) for the excitation.

Consequently, transient analysis of a transmission line with a rectangular pulse excitation can essentially be reduced to the computation on the same line if excited by a step generator, and this latter task is generally simpler to perform. In particular, we can use Eq. (12.221) and the output voltage waveforms in Figs. 12.33(b)–12.38(a) to obtain the corresponding waveforms for the pulse excitation in Eq. (12.219). Analogous transformations from a step response to the pulse one, as in Eq. (12.221), can as well be applied for the generator voltage, $v_g(t)$, and for the total voltage and current, $v(z, t)$ and $i(z, t)$, at any position (z) along the line. In addition, the generalization to an excitation in the form of a sequence of rectangular pulses in time is straightforward. Note, however, that pulse responses can also be found directly, by applying the multiple reflection tracking technique to the original pulse function, i.e., by simultaneously tracing the travel and load/generator reflections of its two (rise and fall) edges. Finally, exploiting the linearity and time-invariance of transmission lines under consideration, we can obtain transient response to pulses of nonrectangular shapes, such as triangular pulses in time.

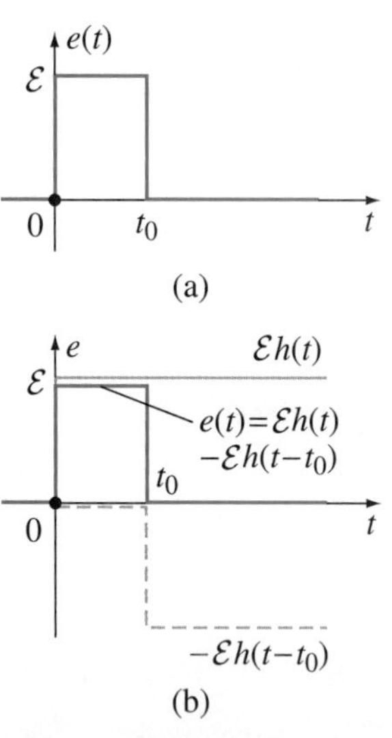

Figure 12.40 Pulse excitation of a lossless transmission line [Fig. 12.31(a)]: (a) rectangular pulse emf function in time and (b) its representation using two step functions, Eq. (12.205).

Example 12.32 Pulse Response of a Line with Unmatched Load and Generator

A lossless transmission line of length $l = 45$ cm and characteristic impedance $Z_0 = 50\ \Omega$ is driven by an ideal voltage generator of rectangular pulse emf in Fig. 12.40(a) with magnitude $\mathcal{E} = 10$ V and duration $t_0 = 2$ ns. At its other end, the line is terminated in a purely resistive load of resistance $R_L = 200\ \Omega$. The line dielectric has relative permittivity $\varepsilon_r = 4$ and is nonmagnetic. Sketch the voltage waveform at the load within a time interval $0 \leq t \leq 20$ ns.

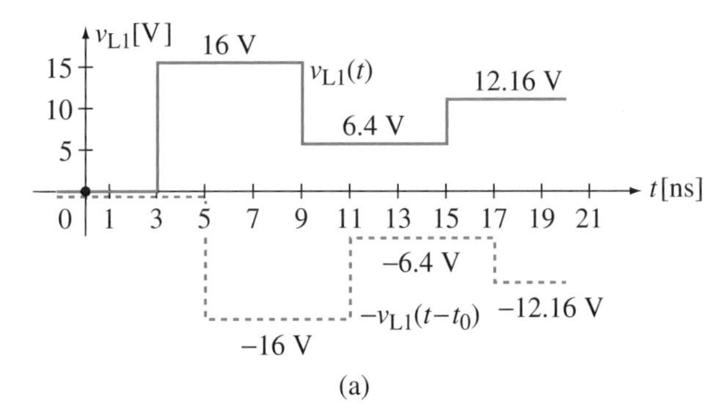

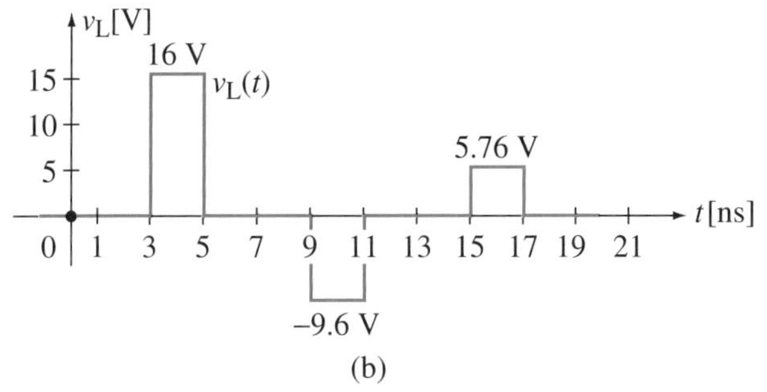

Figure 12.41 Evaluation of the output pulse response of a lossless transmission line: (a) the line step response and its delayed (by the pulse width) negative, and (b) the resultant output waveform, obtained by superposition in Eq. (12.221); for Example 12.32.

Solution Using Eqs. (9.47) and (12.193), the one-way delay period of the line comes out to be $T = 3$ ns. Given, furthermore, that $R_g = 0$, $\Gamma_g = -1$, and $\Gamma_L = 3/5$, the step response of the line at the load terminals, $v_{L1}(t)$, obtained as that in Fig. 12.33, is sketched in Fig. 12.41(a). Shown in the same figure is also the flipped-over and delayed by t_0 version of this signal, namely, $-v_{L1}(t - t_0)$. Having in mind the decomposition of the pulse excitation onto two step ones in Eq. (12.220) and Fig. 12.40(b), the output pulse response of the line, $v_L(t)$, is obtained, according to Eq. (12.221), as a sum of the two waveforms in Fig. 12.41(a), and the result is shown, for $0 \le t \le 20$ ns, in Fig. 12.41(b). We see that the output response of the line to a pulse excitation in Fig. 12.40(a) consists of multiple pulses (theoretically, an infinite series of pulses) of decaying magnitudes with alternating polarity, separated in time by $2T - t_0 = 4$ ns. Since $t_0 < 2T$, the adjacent pulses in the sequence do not overlap (or touch each other).

Note that if the generator were matched to the line, $R_g = Z_0$, the voltage $v_{L1}(t)$ would have the form as in Fig. 12.34, meaning that only one pulse (the first one) would be observed at the load, in Fig. 12.41(b), and the same would hold true if $R_L = Z_0$ (matched load). So, in the case considered in this example (unmatched terminations at both ends of the line), multiple reflections from the load and generator result in the reception of additional pulses at the load, which is normally unintended and undesirable.

Example 12.33 Overlapping Pulses at the Load Terminals

Repeat the previous example but for $t_0 = 12$ ns and $R_g = 200\ \Omega$. Show the output voltage during time $0 \le t \le 30$ ns.

Solution By the same procedure as in Fig. 12.41, the voltage at the load is sketched in Fig. 12.42. This is an illustration of a case with $t_0 > 2T$ (specifically, $t_0 = 4T$), in which the adjacent pulse responses due to multiple reflections on the line in the output sequence in Fig. 12.41(b) now overlap, resulting in a distorted pulse that is being received by the load. Namely, the load and then generator reflection of the rise edge of the incident pulse comes back, after one round trip, to the load terminals before the arrival of the fall edge of the pulse, distorting it. The principal part of this pulse lasts the same as the generator pulse (from $t = T$ to $t = T + t_0$), but the overall signal lasts theoretically indefinitely (to $t \to \infty$).

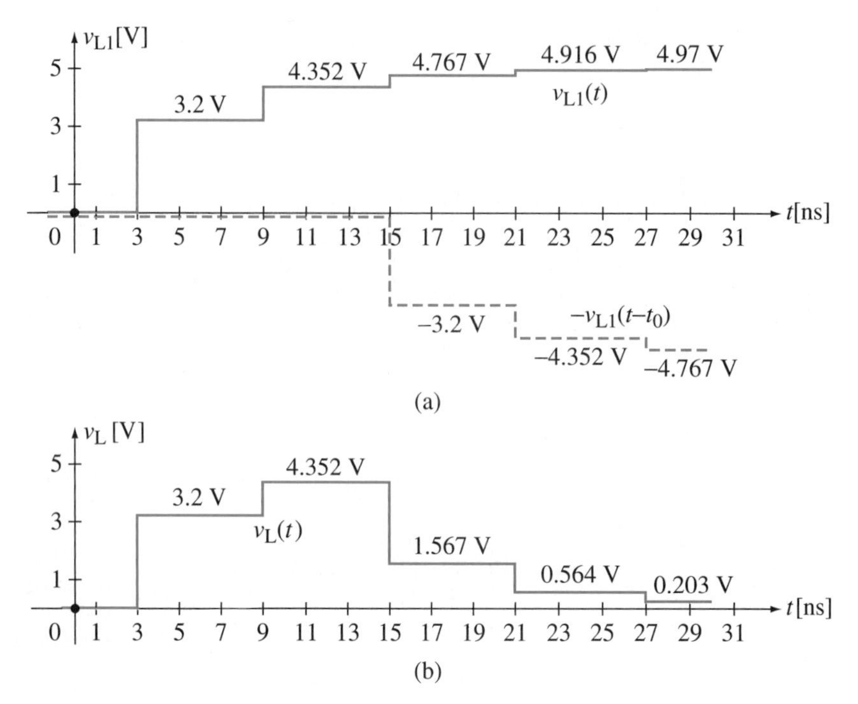

Figure 12.42 The same as in Fig. 12.41 but for $t_0 > 2T$ ($t_0 = 4T$) and $R_g > Z_0$ ($\Gamma_g > 0$); for Example 12.33.

Example 12.34 Bipolar Triangular Pulse Response of a Transmission Line

An air-filled lossless transmission line of length $l = 60$ cm and characteristic impedance $Z_0 = 100\ \Omega$ is connected at its one end to a voltage generator with emf in the form of a bipolar triangular pulse of magnitude $\mathcal{E} = 4$ V and duration $t_0 = 3$ ns, applied at $t = 0$, as shown in Fig. 12.43(a). The internal resistance of the generator is $R_g = 300\ \Omega$. The load, at the other end of the line, is purely resistive, of resistance $R_L = 400\ \Omega$. Sketch the load voltage waveform for $0 \leq t \leq 10$ ns.

Solution From Eqs. (12.193) and (9.19), the one-way time delay of the (air-filled) line is $T = l/c_0 = 2$ ns. Because the bipolar pulse in Fig. 12.43(a) lasts shorter than twice the delay, $t_0 < 2T$, there is no overlapping of responses due to multiple reflections on the line, which means that the load voltage consists of a sequence of pulses of the same (undistorted) shape, so bipolar triangular, as the excitation pulse, just with decaying magnitudes, similarly to the

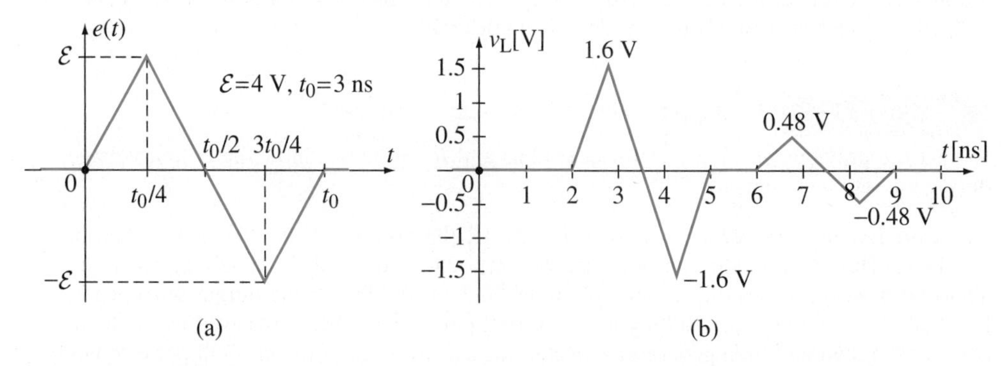

Figure 12.43 Evaluation of the transient response of a transmission line to a triangular pulse excitation: (a) bipolar triangular emf function in time and (b) resulting voltage waveform at the load terminals; for Example 12.34.

situation in Fig. 12.41(b). Therefore, we obtain the overall output response of the line to the given excitation function by essentially finding the corresponding step response, and then replacing each (delayed) step increment in the output sequence in time by a bipolar triangular pulse of the same magnitude.

Reflection coefficients in Eqs. (12.199) and (12.204) for the line come out to be both positive, $\Gamma_L = 3/5$ and $\Gamma_g = 1/2$. Using Eqs. (12.207) and (12.208), we have $v_{i1} = 1$ V and $v_{r1} = 0.6$ V, and hence the total magnitude of the first bipolar triangular pulse at the load, lasting from $t = T = 2$ ns to $t = T + t_0 = 5$ ns, equals $v_{triang1} = v_{i1} + v_{r1} = 1.6$ V. For the second pulse in the output sequence, starting at $t = 3T = 6$ ns and ending at $t = 3T + t_0 = 9$ ns, Eqs. (12.209) and (12.211) give $v_{i2} = 0.3$ V and $v_{r2} = 0.18$ V, respectively, from which $v_{triang2} = v_{i2} + v_{r2} = 0.48$ V. Only these two pulses fall into the time interval considered ($0 \le t \le 10$ ns), and they are sketched in Fig. 12.43(b).

Example 12.35 Effects of a Finite Rise Time of a Step Voltage Excitation

A lossless transmission line with characteristic impedance $Z_0 = 100\ \Omega$ and one-way wave travel time $T = 0.2$ ns is excited at one end by a voltage generator of step emf with magnitude $\mathcal{E} = 3$ V and finite linear-rise time, t_r, as shown in Fig. 12.44(a). This emf is applied at $t = 0$. At its other end, the line is terminated in a purely resistive load of resistance $R_L = 400\ \Omega$. The internal resistance of the generator is $R_g = 20\ \Omega$. Sketch the waveform during time $0 \le t \le 2.2$ ns of the voltage across the load for (a) $t_r = 0.4$ ns and (b) $t_r = 0.8$ ns, respectively.

Solution

(a)–(b) Similarly to the procedure carried out in the previous example, we consider the signal buildup on the line as if it were excited by an ideal step emf, with a zero rise time ($t_r = 0$), and merely "multiply" the ideal-step increments as they appear (delayed) at the load by a step signal form with a finite rise time ($t_r \neq 0$) and unit magnitude. So, each increment in the output voltage sequence in time linearly rises during time t_r from zero to the same voltage magnitude as in the ideal-step sequence. These voltage magnitudes are computed using Eqs. (12.207), (12.199), (12.208), (12.204), (12.209), and (12.211), and they amount to (note that $\Gamma_L = 3/5$ and $\Gamma_g = -2/3$) $v_{i1} + v_{r1} = 4$ V, $v_{i2} + v_{r2} = -1.6$ V, $v_{i3} + v_{r3} = 0.64$ V, $v_{i4} + v_{r4} = -0.256$ V, $v_{i5} + v_{r5} = 0.1$ V, and so on. The corresponding finite-rise-time voltage increments for $t_r = 0.4$ ns $= 2T$ are sketched in Fig. 12.44(b). The total voltage waveform at the load terminals, in Fig. 12.44(c), is then obtained as a superposition of all partial waveforms in Fig. 12.44(b). In Fig. 12.44(d) and Fig. 12.44(e), the same procedure is repeated for $t_r = 0.8$ ns $= 4T$. Shown in both Fig. 12.44(c) and Fig. 12.44(e) are also the voltage waveforms that would be observed at the load if the transmission-line effects in the circuit in Fig. 12.44(a) were neglected, that is, if the load were directly connected to the generator, as in Fig. 12.31(c); these voltages are obtained combining Eq. (12.213) and the emf time function in Fig. 12.44(a). Of course, similar analysis can be performed for pulse excitations with both rise and fall times nonzero, $t_r \neq 0$ and $t_f \neq 0$.

We see that, while the actual voltage waveform at the load in Fig. 12.44(c), for $t_r/T = 2$, is very different from the same voltage with no transmission-line effects, such difference is quite small in Fig. 12.44(e). This implies that for a larger t_r over T ratio, namely, if $t_r/T = 4$, the line appears almost as if nonexistent. In general, a rule of thumb in transient analysis of digital circuits is that the one-way signal delay time T along an interconnect in the circuit can be neglected ($T \approx 0$) if the rise and fall times of pulse signals in the circuit are such that $t_r/T > 5$ and similarly for t_f; otherwise, the interconnect must be treated as a transmission line.

Problems: 12.38–12.42; *MATLAB Exercises* (on Companion Website).

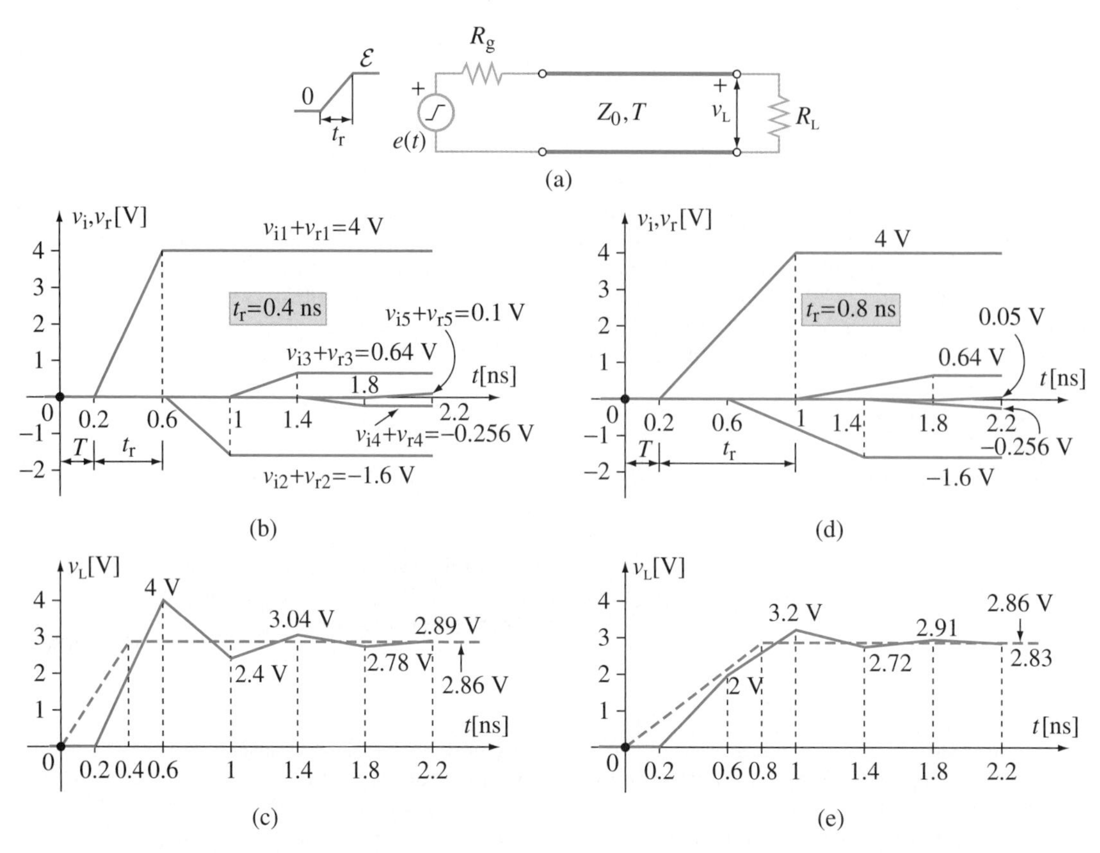

Figure 12.44 Evaluation of effects of a finite rise time of a step voltage excitation of a transmission line: (a) circuit schematic diagram and source emf function in time, (b)–(c) time sequence of voltage increments and total voltage waveform, $v_L(t)$, at the load terminals for $t_r = 2T$, and (d)–(e) the same as (b)–(c) but for $t_r = 4T$ [shown in dashed line in (c) and (e) are also the corresponding output voltages for the same circuit but with the transmission-line effects neglected, as in Fig. 12.31(c)]; for Example 12.35.

12.17 BOUNCE DIAGRAMS

We now present a graphical tool for recording multiple reflection transient processes on lossless transmission lines and computing the total voltage and current intensity at an arbitrary location on the line and any instant of time, called a bounce diagram (also known as a lattice diagram) of the line. This is a space-time (i.e., distance-time) plot of the voltage (or current) state of a transmission line, with the distance (z) from the generator end measured on the horizontal axis and time ($t > 0$) on the vertical axis oriented downward, as shown in Fig. 12.45. We assume a step excitation and purely resistive load termination of the line, as in Fig. 12.31(a). The zigzag line in Fig. 12.45 indicates the progress of the voltage wave along the line, where each line segment (sloping downward from left to right or right to left) represents a component traveling voltage (forward or backward) and is labeled with the magnitude of the voltage step increment. Starting from the top of the diagram, these voltages are successively obtained by multiplying the magnitude of the preceding component by one of the line reflection coefficients Γ_L and Γ_g, given in Eqs. (12.199) and (12.204), depending on the position (load or generator) where the reflection occurs. Of course, the expressions for component voltages in the labels of segments in Fig. 12.45 are the same as those in Eqs. (12.207)–(12.211).

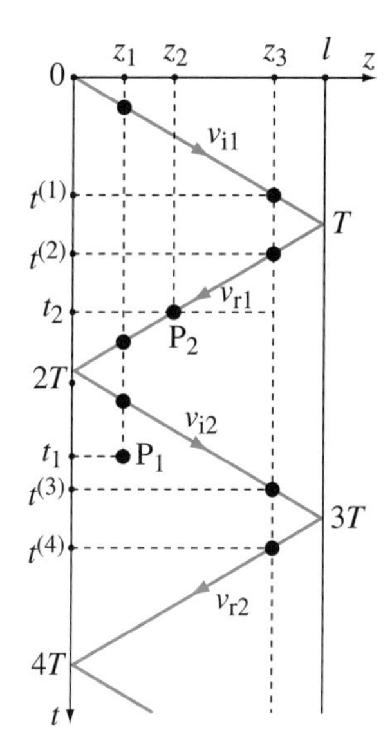

Figure 12.45 Voltage bounce diagram of a lossless transmission line with a step excitation, in Fig. 12.31(a).

Once the bounce diagram is constructed, the total voltage at a location $z = z_1$ and time $t = t_1$, $v(z_1, t_1)$, can be easily found from the position of the point (z_1, t_1) in Fig. 12.45, point P_1, relative to the zigzag voltage line. Namely, $v(z_1, t_1)$ equals the sum of the component voltages of all sloping line segments intersected by the vertical line $z = z_1$ between points $t = 0$ and $t = t_1$, so above P_1, on that line (these points are also included in the count if on the zigzag line). Hence, for the choice of z_1 and t_1 in Fig. 12.45, we see that $v(z_1, t_1) = v_{i1} + v_{r1} + v_{i2} = \left(1 + \Gamma_L + \Gamma_L\Gamma_g\right) v_{i1}$, since the vertical (dashed) line starting at the point marked z_1 on the z-axis intersects the first three sloping line segments along its part above P_1. Similarly, the total voltage scan $v(z, t_2)$ along the transmission line at any given time (t_2), as well as the voltage waveform $v(z_3, t)$ at any fixed location (z_3) on the line, can also be obtained from the diagram, in a straightforward fashion. In the first case, we draw a horizontal (dashed) line from t_2 on the t-axis to the right, and identify its intersection with the zigzag voltage line, point P_2, and the corresponding point z_2 on the z-axis – this is the position on the transmission line where the voltage picture exhibits an abrupt (step-like) change. For example, if the instant t_2 is chosen as in Fig. 12.45, so that P_2 belongs to the segment labeled with the component voltage magnitude v_{r1}, we read from the diagram that $v(z, t_2) = v_{i1}$ for $0 \leq z < z_2$ and $v(z, t_2) = v_{i1} + v_{r1}$ for $z_2 \leq z \leq l$. For the second case (fixed location), we look at the intersections of the vertical dashed line $z = z_3$ with the segments of the zigzag line, and the corresponding points on the t-axis, marked as $t^{(1)}, t^{(2)}, \ldots$ in Fig. 12.45. We then read that $v(z_3, t) = 0$ for $0 < t < t^{(1)}$, $v(z_3, t) = v_{i1}$ for $t^{(1)} < t < t^{(2)}$, $v(z_3, t) = v_{i1} + v_{r1}$ for $t^{(2)} < t < t^{(3)}$, and so on. Note that in this way all voltage snapshots and waveforms in Figs. 12.32–12.37 can now be directly reproduced – from the bounce diagrams. Finally, the current bounce diagram of a transmission line is constructed and used in the same way as the corresponding voltage diagram, the only difference being in the values of the reflection coefficients – the load and generator reflection coefficients for currents equal $-\Gamma_L$ and $-\Gamma_g$, respectively [see Eq. (12.28) and Fig. 12.39].

Example 12.36 Voltage and Current Waveforms from Bounce Diagrams

Construct voltage and current bounce diagrams for the coaxial cable and its excitation and load described in Example 12.27. Based on these diagrams, obtain the total voltage and current scans along the cable at a time instant $t = 5$ ns, as well as voltage and current waveforms for $0 \leq t \leq 9$ ns at the cable cross section whose distance from the generator is $z = 7.5$ cm.

Solution Using the data computed in Example 12.27 and having in mind Fig. 12.45 and the accompanying explanations, the voltage and current bounce diagrams of the cable, as well as v and i snapshots and waveforms at the given time and location on the cable, respectively, are sketched in Fig. 12.46.

Problems: 12.43–12.45; *MATLAB Exercises* (on Companion Website).

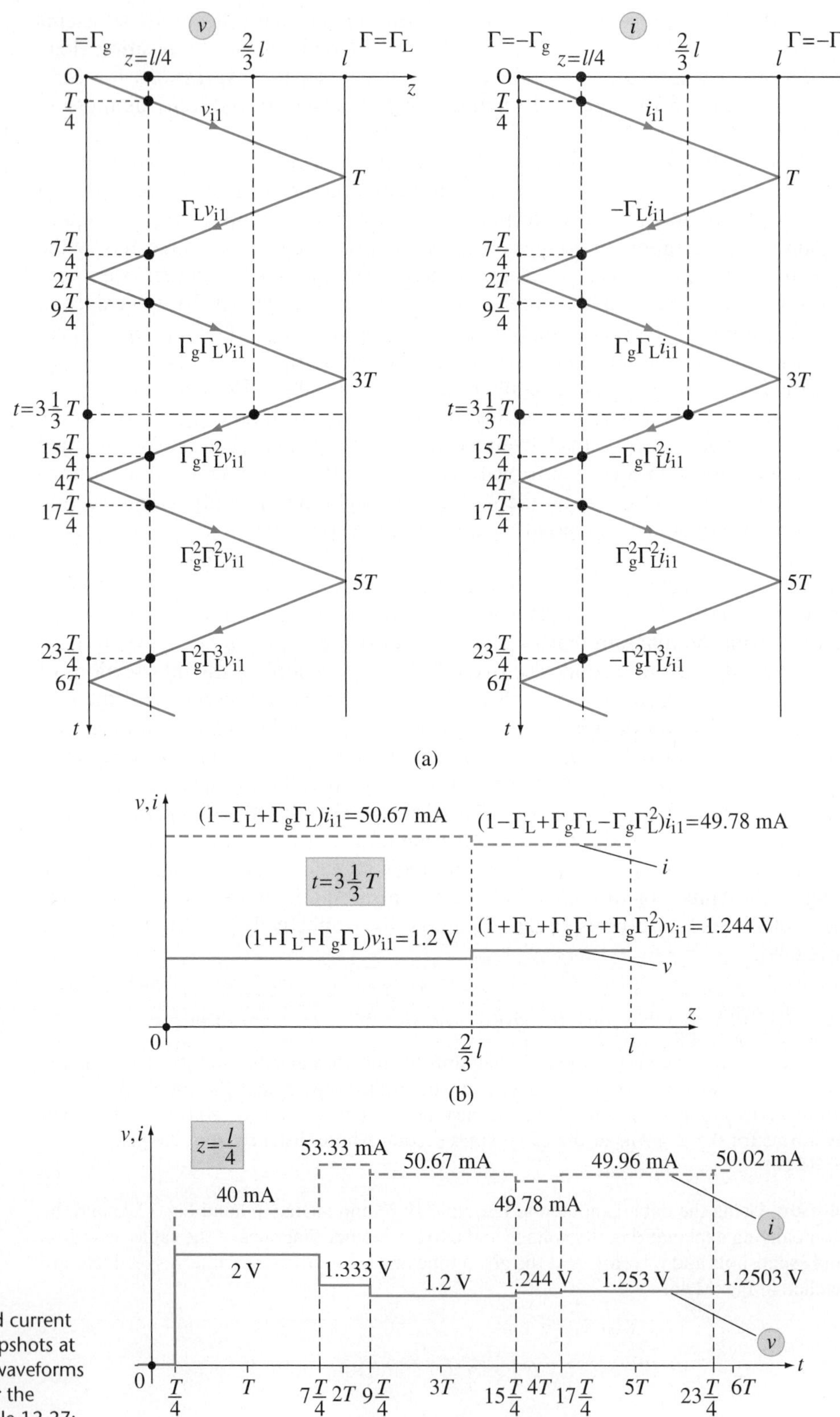

Figure 12.46 Voltage and current bounce diagrams (a), snapshots at $t = 5$ ns $= 3\frac{1}{3}\,T$ (b), and waveforms at $z = 7.5$ cm $= l/4$ (c) for the coaxial cable from Example 12.27; for Example 12.36.

12.18 TRANSIENT RESPONSE FOR REACTIVE OR NONLINEAR TERMINATIONS

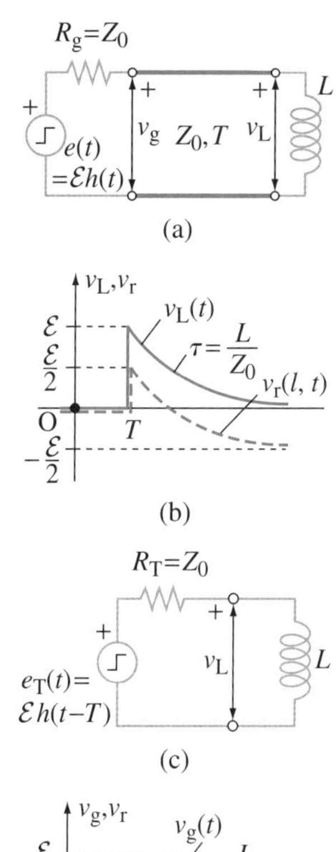

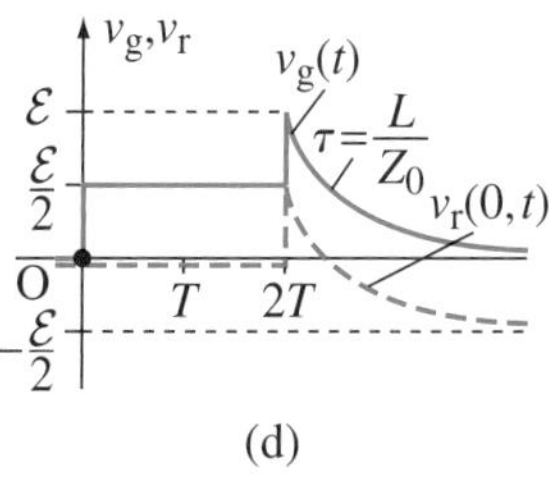

Figure 12.47 Step transient analysis of a lossless transmission line with a purely inductive load and matched generator: (a) circuit schematic diagram, (b) total [$v_L(t)$] and reflected [$v_r(l,t)$] voltage waveforms at the load, (c) Thévenin equivalent generator for the line as seen from the load, and (d) total [$v_g(t)$] and reflected [$v_r(0,t)$] voltage waveforms at the generator.

Often, transmission-line terminations involve reactive lumped elements, inductors and capacitors. In the analysis of high-speed digital circuits, these elements are used to model various lumped inductive and capacitive effects at line terminals that are not already included in the transmission-line model (with distributed inductors and capacitors). Some of examples are the inductance of bonding wires connecting two lines and that of resistor (load) leads, and capacitances of various packaging inclusions near line terminals, in the form of conducting wires (e.g., vias in a multilayer printed circuit board, Fig. 11.14, and packaging pins) and plates (e.g., nearby ground planes and plug-in cards in a computer). In addition, they can model, along with resistors, a complex input impedance of a device (e.g., a transmitting antenna) connected at the load end of the line. As the most notable difference when compared to the lines with purely resistive terminations, the reflected and total voltages and current intensities on lines with reactive loads do not have the same temporal shape as the incident ones. This is essentially due to integrating effects related to charging the reactive elements, that is, establishing the current through an inductor or voltage across a capacitor. In this section, we extend our transient analysis of transmission lines with resistive loads from previous sections to some simple characteristic cases of reactively loaded lines. We also discuss transient responses of lines with nonlinear loads.

Consider first a lossless transmission line terminated in an ideal inductor, of inductance L, as in Fig. 12.47(a). We assume, for simplicity, that the generator, with a step emf of magnitude $\mathcal{E}$, Eq. (12.205), is matched to the line, $R_g = Z_0$ ($\Gamma_g = 0$), so that there are no multiple reflections on the line, and the only forward propagating wave at all times (after the switch of the generator) is the initial one. The incident voltage is thus $v_i = v_{i1} = \mathcal{E}/2$, Eq. (12.214), much like in previous cases with a matched generator, in Figs. 12.34 and 12.37. Prior to the instant $t = T$, the load voltage, $v_L(t)$, is zero. At $t = T$, with T being the one-way time delay of the line, Eq. (12.193), the wavefront of the signal v_{i1} arrives to the load terminals, and the voltage of the inductor is abruptly changed. In general, for rapid variations of an applied voltage, an inductor behaves as an open circuit, and its current is zero. In other words, the current of an inductor cannot change instantaneously [note that a step-like jump of i in Eq. (7.3) would result in an infinite v]. So, at this time the incident voltage is reflected from the load in Fig. 12.47(a) as in the case of an open-circuited transmission line, and the total voltage of the load jumps to

$$v_L(T) = 2v_{i1} = \mathcal{E} \quad (\text{inductor} \to \text{open circuit}, \Gamma_L = 1), \tag{12.222}$$

as in Fig. 12.37(a). After this initial sudden change of the state of the inductor, its current starts to build up, from zero to a steady-state value. In the steady state, as in a dc regime, the inductor can be considered as a short circuit [Eq. (7.3) with $\mathrm{d}/\mathrm{d}t \equiv 0$, no time variations], and hence

$$v_L(t \to \infty) = v_{\text{steady}} = 0 \quad (\text{inductor} \to \text{short circuit}, \Gamma_L = -1), \tag{12.223}$$

as in Fig. 12.37(b). Between the time $t = T$ (abrupt variation, inductor open circuit) and $t \to \infty$ (steady state, inductor short circuit), the change (decrease) of the inductor voltage is an exponential one, as sketched in Fig. 12.47(b), and we can write

$$v_L(t) = \mathcal{E}h(t-T)\,\mathrm{e}^{-(t-T)/\tau} \qquad (0 < t < \infty), \tag{12.224}$$

where the time-shifted Heaviside function $h(t-T)$ makes the expression for $v_L(t)$ zero for $t < T$. Since the input dynamic impedance that the load sees looking into the transmission line equals the line characteristic impedance, Z_0, the time constant of this exponential change is

time constant of a line with a purely inductive load

$$\boxed{\tau = \frac{L}{Z_0}.} \quad (12.225)$$

Of course, the load current, $i_L(t)$, increases by the same exponential law, which is a typical charging curve for an inductor (as well as a capacitor, but for voltage). From Fig. 12.31(c), with $R_L = 0$ and $R_g = Z_0$, the current at $t \to \infty$ is

$$i_{\text{steady}} = \frac{\mathcal{E}}{R_g} = \frac{\mathcal{E}}{Z_0}. \quad (12.226)$$

Note that this same analytical result in Eqs. (12.224) and (12.225) and the plot of v_L in Fig. 12.47(b) can be obtained also using Thévenin's theorem, namely, replacing the circuit to the left of the load (inductor) in Fig. 12.47(a) by the Thévenin equivalent generator, as shown in Fig. 12.47(c). This is simply the Thévenin circuit on the right-hand side of Fig. 12.30, in the general Thévenin equivalent representation (Thévenin generator pair) for transients on transmission lines. Hence, the internal resistance of the equivalent generator in Fig. 12.47(c), R_T, equals Z_0, and its emf, e_T, twice the incident voltage at the load location in Fig. 12.47(a), $v_i(l, t)$, which, in turn, comes out to be the step-function incident voltage of magnitude $\mathcal{E}/2$ in Eq. (12.214) delayed by T (for the travel time over the distance l, to the load). However, even without taking the general result from Fig. 12.30, we can find e_T directly for this case as (by Thévenin's theorem) the output voltage of the line in Fig. 12.47(a) when open-circuited, and that is exactly the output voltage in Fig. 12.37(a) and Eq. (12.216). So, the parameters of the Thévenin equivalent generator are

Thévenin emf and resistance as seen from the load

$$\boxed{e_T(t) = 2v_i(l, t) = \mathcal{E}h(t-T), \quad R_T = Z_0,} \quad (12.227)$$

and we next show that the voltage $v_L(t)$ in Eq. (12.224) represents the solution of the differential equation for the simple RL circuit in Fig. 12.47(c), whose time constant τ is the one in Eq. (12.225). Furthermore, we have in mind that this equivalent representation of the line holds true for any load.

Kirchhoff's voltage law, Eq. (1.92) or (3.119), for the circuit in Fig. 12.47(c) gives

load line of the circuit

$$\boxed{v_L + Z_0 i_L = e_T.} \quad (12.228)$$

Taking the derivative with respect to time of both sides of this equation and using the current-voltage characteristic (differential equation) for an inductor in Eq. (7.3), we then obtain the following differential equation in time for v_L as unknown, valid for the time after the Thévenin emf is switched on:

$$v_L = L\frac{di_L}{dt} \quad \longrightarrow \quad \frac{dv_L}{dt} + \frac{Z_0}{L}v_L = 0 \quad (T < t < \infty) \quad (12.229)$$

[$de_T/dt = 0$ since $e_T = \text{const}$ for $t > T$, Eq. (12.227)]. The initial condition for v_L, right after the switch instant, $t = T$, is obtained from Eqs. (12.228) and (12.227), and the fact that there cannot be an abrupt change of current intensity through an inductor, so i_L initially stays the same (zero) as before the generator switch,

$$i_L(T^+) = i_L(T^-) = 0 \quad \longrightarrow \quad v_L(T^+) = e_T(T^+) = \mathcal{E}. \quad (12.230)$$

The solution of the first-order differential equation in Eq. (12.229) that satisfies this initial condition is exactly the expression for $v_L(t)$ in Eqs. (12.224) and (12.225). Eq. (12.228) then gives the solution for the current $i_L(t)$.

The input voltage of the line, $v_g(t)$, is given by Eq. (12.214) until the arrival of the reflected wave, at $t = 2T$. In order to compute and sketch its temporal form after that time, we first determine the voltage of the reflected wave, v_{r1}, on the line. Since this is the only one reflected voltage (for $T < t < \infty$), $v_r = v_{r1}$, it equals the difference of the total and incident voltages at a given location (z) along the line. At the load terminals ($z = l$), using the first equation in Eqs. (12.194),

$$v_r(l,t) = v(l,t) - v_i(l,t) = v_L(t) - \frac{\mathcal{E}}{2} h(t-T) = \mathcal{E} h(t-T)\left[\mathrm{e}^{-(t-T)/\tau} - \frac{1}{2}\right], \tag{12.231}$$

and this waveform is also shown in Fig. 12.47(b). For $t > 2T$, the reflected voltage at the generator terminals ($z = 0$), $v_r(0,t)$, has the same form as that at the load end, in Fig. 12.47(b), just with an additional delay T relative to the signal $v_r(l,t)$, so $v_r(0,t) = v_r(l, t-T)$, as shown in Fig. 12.47(d). Finally, the voltage of the generator is obtained as

$$v_g(t) = v_i(0,t) + v_r(0,t) = \frac{\mathcal{E}}{2} h(t) + \mathcal{E} h(t-2T)\left[\mathrm{e}^{-(t-2T)/\tau} - \frac{1}{2}\right], \tag{12.232}$$

and sketched in Fig. 12.47(d).

Next, let the reactive element at the farther end of the line be an ideal capacitor, of capacitance C, Fig. 12.48(a). Just opposite to an inductor, a capacitor acts as a short circuit for rapid variations of an applied signal (the voltage across a capacitor cannot change instantaneously), whereas as an open circuit (current is zero) in the steady (dc) state [see Eq. (3.45)]. Therefore, in place of Eqs. (12.222) and (12.223), we have

$$v_L(T) = 0 \quad (\Gamma_L = -1), \quad v_L(t \to \infty) = 2v_{i1} = \mathcal{E} \quad (\Gamma_L = 1). \tag{12.233}$$

During the time $T < t < \infty$, as the capacitor is being charged the signal $v_L(t)$ increases exponentially between these two values (characteristic charging curve for a capacitor), Fig. 12.48(b). It is zero for $t < T$, so that overall

$$v_L(t) = \mathcal{E} h(t-T)\left[1 - \mathrm{e}^{-(t-T)/\tau}\right] \quad (0 < t < \infty), \tag{12.234}$$

with the following time constant:

time constant for a purely capacitive load

$$\boxed{\tau = Z_0 C.} \tag{12.235}$$

This can as well be obtained solving the RC circuit in Fig. 12.48(c), where the structure to the left of the capacitor in Fig. 12.48(a) is replaced by the Thévenin equivalent generator, of parameters in Eqs. (12.227). Finally, shown in Figs. 12.48(b) and (d) are the corresponding reflected voltage waveforms at the load and generator terminals, $v_r(l,t)$ and $v_r(0,t)$, as well as the voltage of the generator, $v_g(t)$, in Fig. 12.48(a), obtained in the same way, using Eqs. (12.231) and (12.232), as for the inductive load in Fig. 12.47.

We see that, indeed, the total and reflected voltages at the load (inductor or capacitor) and generator in Figs. 12.47 and 12.48 do not exhibit the same temporal shape as the incident voltage (which is a step function). In specific, let us compare the waveforms for $v_L(t)$ and $v_g(t)$ in Fig. 12.47 for the inductor as load with the corresponding ones in Fig. 12.37(b) for the short-circuited line, and then the same diagrams in Fig. 12.48 for the capacitor with those for an open circuit as load, Fig. 12.37(a). We realize that essentially the only change in waveforms due to the presence of a reactive element in the circuit is that establishing the steady state is

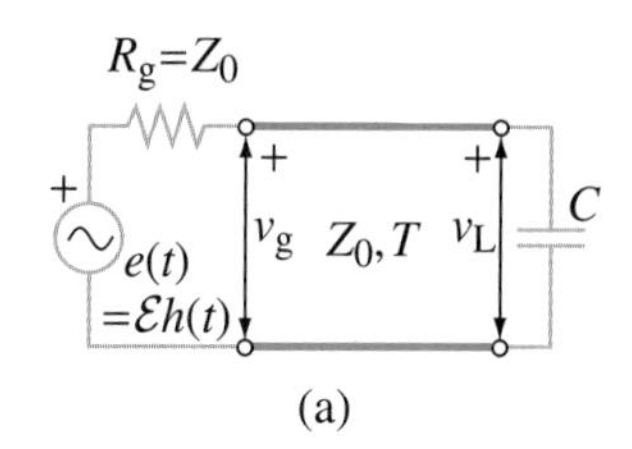

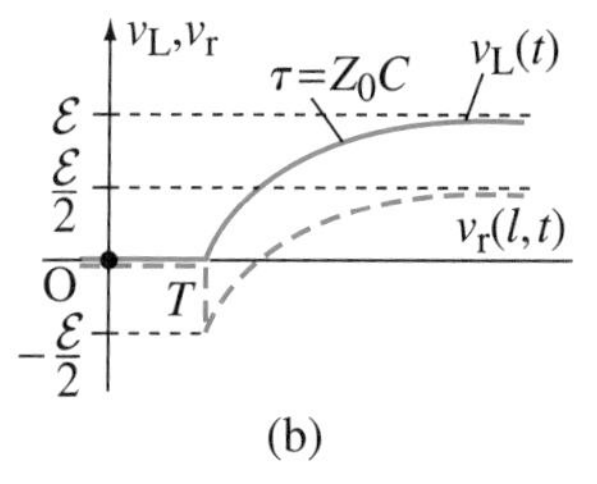

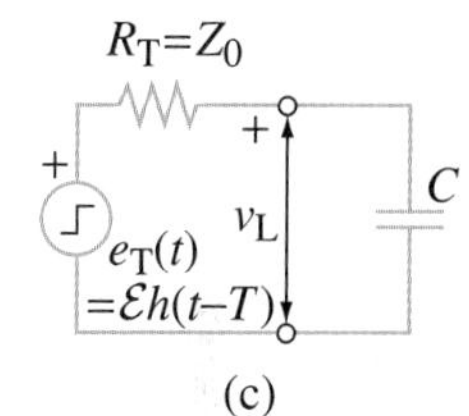

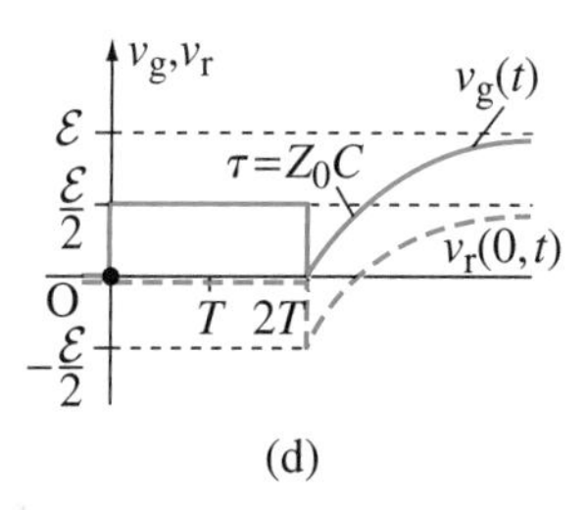

Figure 12.48 The same as in Fig. 12.47 but for a purely capacitive load.

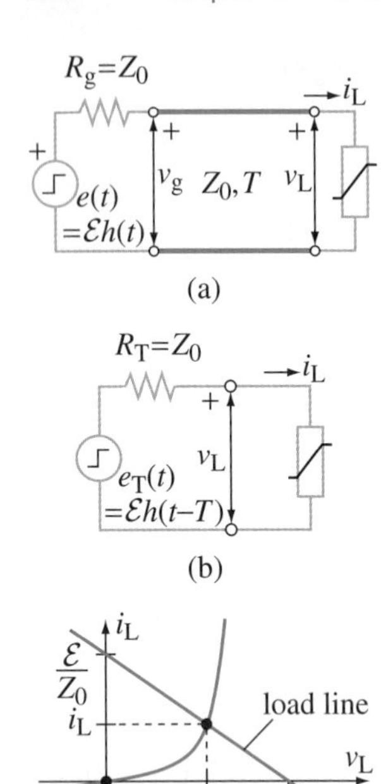

Figure 12.49 Transients on the line in Fig. 12.47(a) when terminated in a nonlinear load: (a) circuit schematic diagram, (b) Thévenin equivalent circuit at the load end, and (c) solution for the load voltage and current – at the intersection of the load line of the circuit in (b), Eq. (12.228), and (nonlinear) current-voltage characteristic of the load.

prolonged, theoretically indefinitely, to accommodate the time (theoretically infinite) for that element to charge. In other words, the integrating (charging) effects of the inductor and capacitor turn the sharp (rising or falling) edges of (positive or negative) step signals, i.e., the instantaneous voltage jumps (with infinite slopes), into smooth (exponential) changes over time. The lower the L and C the shorter the time constants τ in Eqs. (12.225) and (12.235) and faster the transitional period, which, in practice, may be said to be completed after several time constants.

Of course, similar computations of transients can be performed on transmission lines with various combined resistive and reactive terminations (i.e., lossy reactive loads), such as series or parallel combinations of a resistor and an inductor (capacitor). However, this simple technique, based on reflection tracking in conjunction with analysis of transient behavior of reactive loads from their current-voltage characteristics, can become exceedingly complicated for loads that involve more than one reactive element, as well as for lines with unmatched generators and/or emf time functions different from step and pulse ones. Moreover, reactive elements can be present also on the source end of the line. In general, the transient responses of a line can be found by solving the two (input and output) coupled Thévenin equivalent circuits in Fig. 12.30, that is, by simultaneously solving Eqs. (12.195), (12.200), (12.189), and (12.190) for the line together with differential equations describing boundary conditions imposed by the terminal networks. Such computations often use Fourier or Laplace transform methods to avoid the direct time-domain analysis.

Finally, as an illustration of transient analysis of transmission lines with nonlinear terminations, which are often encountered in high-speed digital circuits (nonlinear logic gates), let us assume that the load of a lossless transmission line with a matched step-voltage generator has a nonlinear current-voltage characteristic, as indicated in Fig. 12.49(a). Here, we can use the same Thévenin generator with parameters in Eqs. (12.227) to obtain the equivalent circuit in Fig. 12.49(b). The solution for the load voltage v_L (and current i_L) corresponds to the intersection of the load line of the circuit, dictated by the Thévenin generator and given in Eq. (12.228), and the nonlinear current-voltage characteristic of the load (e.g., a diode or transistor), as illustrated in Fig. 12.49(c).[10] This intersection can be determined analytically in some situations, and graphically or numerically for arbitrary nonlinear loads. The input voltage $v_g(t)$ can then be found via the reflected voltage, as in Eqs. (12.231) and (12.232). Note that, in general, the analysis can be extended to cases where the generator in Fig. 12.49(a) is unmatched, or when terminal networks at both ends of the line, Fig. 12.30, exhibit nonlinearities (e.g., driver and receiving logic gates in a digital circuit).

Example 12.37 TDR Measurement of Unknown Load Inductance

Consider the time-domain reflectometer and transmission line from Example 12.28 and assume that the load is an ideal inductor of unknown inductance. If the oscilloscope showing, $v_g(t)$, is now that in Fig. 12.47(d), where the area under the decaying exponential curve in the voltage display is measured to be $A = 12$ nVs [note that this "area," measured in a voltage (expressed in volts) over time (seconds) diagram, is expressed in Vs, rather than m^2], determine the value of the load inductance.

[10]Note the similarity with solutions for magnetic field intensities (H) and flux densities (B) in nonlinear magnetic circuits determined by intersections of circuit load lines and magnetization curves of core materials, Figs. 5.30 and 5.34.

Solution The analytical expression for the voltage v_g in the circuit in Fig. 12.47(a) is given by Eq. (12.232). Therefore, the "area" under the exponential curve in the TDR diagram in Fig. 12.47(d), and from it the inductance (L) we seek, can be computed as follows:

$$A = \int_{t=2T}^{\infty} v_g(t)\,dt = \mathcal{E}\int_{2T}^{\infty} e^{-(t-2T)/\tau}\,dt = \mathcal{E}\tau = \frac{L\mathcal{E}}{Z_0} \quad\longrightarrow\quad L = \frac{Z_0 A}{\mathcal{E}} = 150\ \text{nH}, \tag{12.236}$$

where the use is made of the fact that both Heaviside functions in Eq. (12.232) are unity for $t > 2T$, as well as of the expression for the time constant of the circuit in Eq. (12.225).

Note that a TDR response like the one in Fig. 12.47(d) tells us, in the first place, that the unknown impedance load at the farther end of a transmission line is purely inductive (ideal inductor). Other examples of TDR signatures that indicate the nature of a load on a transmission line are those in Figs. 12.34(a)–(b) and 12.37(a)–(b) for purely resistive loads with $R_L > Z_0$, $R_L < Z_0$, $R_L \to \infty$, and $R_L = 0$, respectively, whereas Fig. 12.48(d) shows the TDR signature of a purely capacitive load.

Example 12.38 Line Terminated in a Series Connection of a Resistor and an Inductor

A lossless transmission line, for which the one-way wave travel time is $T = 2$ ns, is fed by a voltage generator of rectangular pulse emf with magnitude $\mathcal{E} = 5$ V and width $t_0 = 1$ ns, applied at $t = 0$. The characteristic impedance of the line and internal resistance of the generator are $Z_0 = R_g = 50\ \Omega$. The other end of the line is terminated in a load consisting of a resistor of resistance $R = 30\ \Omega$ and an inductor of inductance $L = 80$ nH connected in series. Sketch the waveform for $0 \le t < \infty$ of the voltage across (a) the load and (b) the generator, respectively.

Solution

(a) Let us first find the step response of the line. At $t = T$, the inductor behaves like an open circuit, and so does the entire load (an open in series with a resistor is open as well), and the total voltage of the load is thus that in Eq. (12.222). At $t \to \infty$, the inductor can be replaced by a short circuit, Eq. (12.223), and the entire load reduces to the resistor (of resistance R) only, meaning that the expression for the steady-state voltage across the load is the same as in Fig. 12.34(b) ($R < Z_0$), given by Eqs. (12.215) and (12.199), so it amounts to

$$v_{\text{steady1}} = \left(1 + \frac{R - Z_0}{R + Z_0}\right)\frac{\mathcal{E}}{2} = 1.875\ \text{V}. \tag{12.237}$$

Between the initial abrupt variation of the load voltage and the steady state, this voltage decreases exponentially with the following time constant:

$$\tau = \frac{L}{R + Z_0} = 1\ \text{ns}, \tag{12.238}$$

in place of the one in Eq. (12.225), which can as well be obtained from the analysis of an equivalent circuit with the transmission line and generator replaced by the Thévenin equivalent generator in Fig. 12.47(c). Based on these data, the waveform of the step response of the line at the load terminals, $v_{L1}(t)$, is sketched in Fig. 12.50(a). Using Eq. (12.221), this waveform is then transformed to the corresponding pulse response, $v_L(t)$ – shown in the figure. Finally, since the step response $v_{L1}(t)$ can be written as [see also Eq. (12.224)]

$$v_{L1}(t) = \left[\left(\mathcal{E} - v_{\text{steady1}}\right) e^{-(t-T)/\tau} + v_{\text{steady1}}\right] h(t - T) \quad (0 < t < \infty), \tag{12.239}$$

we compute the voltage V_0 at the tip of the fall edge of the distorted pulse at the output, in the diagram of $v_L(t)$ in Fig. 12.50(a), as

$$V_0 = v_{L1}(T + t_0) = v_{L1}(3\ \text{ns}) = 3.025\ \text{V}. \tag{12.240}$$

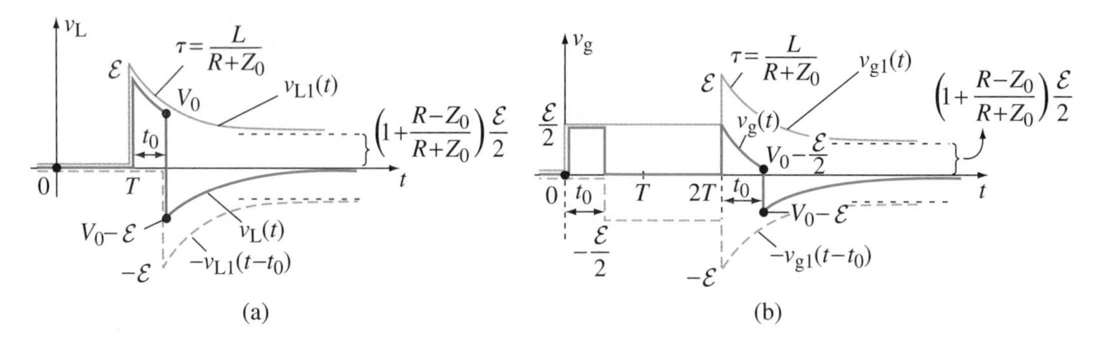

Figure 12.50 Step and pulse voltage responses of a transmission line terminated in a series connection of a resistor and an inductor observed at (a) the load, v_{L1} and v_L, and (b) the generator, v_{g1} and v_g, respectively; for Example 12.38.

(b) Fig. 12.50(b) shows the voltage step response of the line at its generator end, $v_{g1}(t)$, which is obtained, from $v_L(t)$ in Fig. 12.50(a), as in Fig. 12.47(b) and Fig. 12.47(d) using Eqs. (12.231) and (12.232); $v_{g1}(t)$ ends up being equal to $\mathcal{E}/2$ for $0 < t < 2T$ and to $v_L(t-T)$, so $v_L(t)$ delayed by T, for $2T < t < \infty$. The figure also shows the total input waveform $v_g(t)$ for the pulse excitation, determined from $v_{g1}(t)$ by means of Eq. (12.221).

Example 12.39 Line Terminated in a Resistor and a Capacitor in Series

Repeat the previous example but for a capacitor of capacitance $C = 20$ pF in place of the inductor in the line load.

Solution

(a) In the step analysis, at $t = T$, with the capacitor acting like a short circuit, we are left with the resistor (of resistance R) as load, and hence the reflection coefficient at the load is given by Eq. (12.199), namely, $\Gamma_L = (R - Z_0)/(R + Z_0)$, and the total output voltage actually equals the steady-state voltage in Eq. (12.237), $v_{L1}(T) = 1.875$ V. At $t \to \infty$, on the other hand, the capacitor, and thus the entire load, can be considered as an open circuit, so the output voltage in the steady state equals the initial voltage in Eq. (12.222), $v_{\mathrm{steady1}} = \mathcal{E} = 5$ V. Between the two voltage values, we have an exponential increase of $v_{L1}(t)$, similar to that in Fig. 12.48(b) but with a time constant $\tau = (R + Z_0)C = 1.6$ ns, which is shown in Fig. 12.51(a). Combining this waveform in accordance to Eq. (12.221), we then obtain and sketch the associated pulse response, $v_L(t)$, where the maximum value of the total load voltage, $V_0 = 3.33$ V in Fig. 12.51(a), is computed in a similar way to that in Eqs. (12.239) and (12.240).

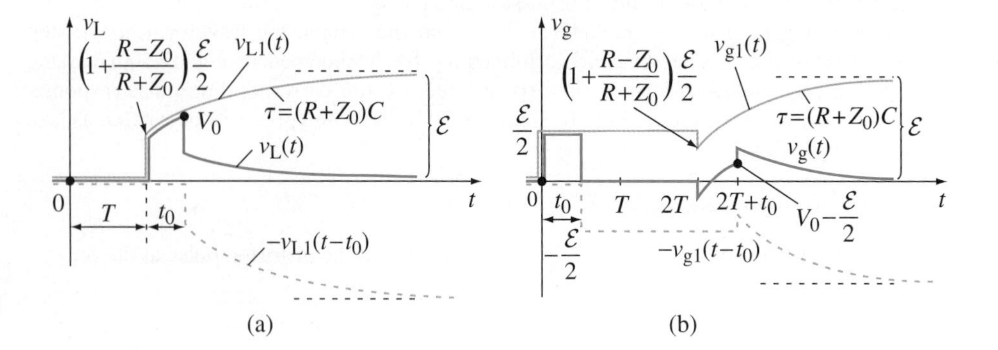

Figure 12.51 The same as in Fig. 12.50 but for the line terminated in a resistor and a capacitor connected in series; for Example 12.39.

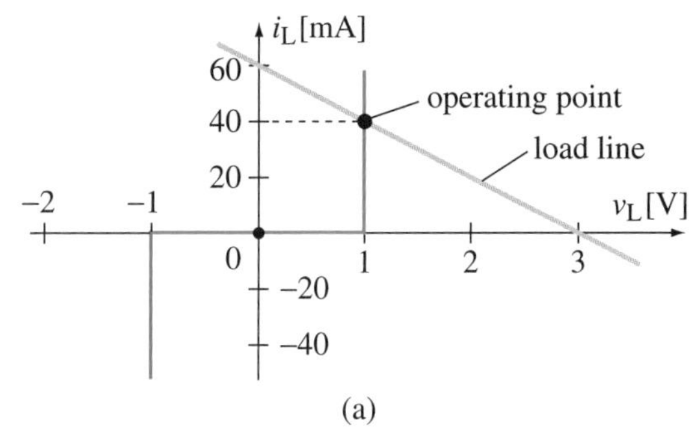

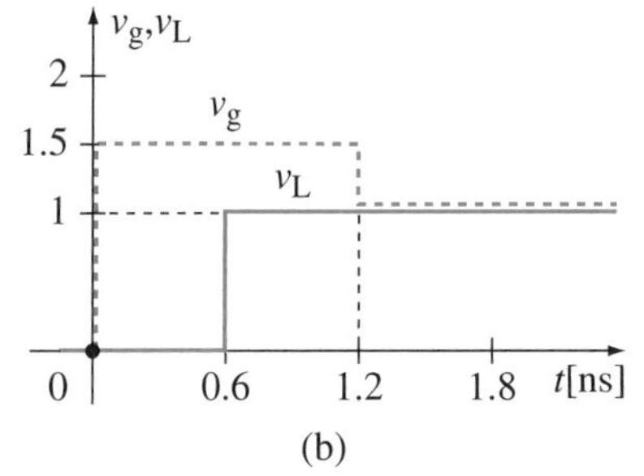

Figure 12.52 Transient analysis of a transmission line with a nonlinear load: (a) current-voltage characteristic of the load, and load line and operating point for the Thévenin equivalent circuit in Fig. 12.49(b), and (b) input and output step responses of the line; for Example 12.40.

(b) The step and pulse voltage responses of the line at the generator end are found as in Figs. 12.50(b) and 12.48(b) and (d), and plotted in Fig. 12.51(b).

Example 12.40 Step Response of a Transmission Line with a Nonlinear Load

A lossless transmission line of length $l = 9$ cm, dielectric relative permittivity $\varepsilon_r = 4$ ($\mu_r = 1$), and characteristic impedance $Z_0 = 50\ \Omega$ is driven by a matched voltage generator of step emf $\mathcal{E} = 3$ V applied at $t = 0$. The other end of the line is terminated in a nonlinear load whose current-voltage characteristic is shown in Fig. 12.52(a). Sketch the voltage waveforms at both ends of the line.

Solution With the given numerical data substituted, the equation of the load line for the Thévenin equivalent circuit in Fig. 12.49(b), Eq. (12.228), becomes

$$v_L + 50 i_L = 3 \quad (v_L \text{ in V; } i_L \text{ in A}). \tag{12.241}$$

The intersection of this line and the current-voltage characteristic of the load (operating point of the circuit) is as indicated in Fig. 12.52(a). We realize that for $T < t < \infty$, where $T = 0.6$ ns, from Eqs. (12.193) and (11.17), the load voltage equals $v_L = 1$ V, as shown in Fig. 12.52(b). By means of Eq. (12.241), the load current is $i_L = 40$ mA.

From $t = 0$ to $t = 2T = 1.2$ ns, the voltage at the generator end of the line equals the incident voltage in Eq. (12.214), namely, $v_g = v_i = \mathcal{E}/2 = 1.5$ V. As in Eq. (12.231), the voltage reflected from the load (for $T < t < \infty$) comes out to be

$$v_r = v_L - v_i = -0.5 \text{ V}. \tag{12.242}$$

It arrives to the generator at $t = 2T$, and hence $v_g = v_i + v_r = 1$ V for $2T < t < \infty$. The voltage waveform $v_g(t)$ is sketched in Fig. 12.52(b) as well.

Problems: 12.46–12.50; *Conceptual Questions* (on Companion Website): 12.44–12.47; *MATLAB Exercises* (on Companion Website).

Problems

12.1. Voltage and current standing wave patterns for resistive loads. Consider a lossless transmission line of characteristic impedance $Z_0 = 50\ \Omega$ and a time-harmonic traveling wave of rms voltage $V_{i0} = 10$ V on it. The wavelength along the line is $\lambda_z = 1$ m. Find the magnitude and phase angle of the load reflection coefficient, the maxima and minima of the total voltage and of the total current of the line and their locations, and the standing wave ratio of the line, and sketch the voltage and current standing wave patterns along the line – if it is terminated in a purely resistive load of resistance (a) $R_L = 250\ \Omega$ and (b) $R_L = 10\ \Omega$, respectively.

12.2. Standing wave patterns for reactive loads. Repeat the previous problem but for a purely reactive load of reactance (a) $X_L = 50\ \Omega$ and (b) $X_L = -50\ \Omega$, respectively.

12.3. Standing wave patterns for complex loads. Repeat Problem 12.1 but for a complex load of impedance (a) $\underline{Z}_L = (100 + j50)\ \Omega$ and (b) $\underline{Z}_L = (50 - j100)\ \Omega$, respectively.

12.4. Finding load impedance from voltage standing wave pattern. Using an air-filled slotted line of characteristic impedance $Z_0 = 100\ \Omega$ terminated in an unknown load, the measured standing wave ratio of the line turns out to be $s = 5$, the distance of the first voltage maximum from the load $l_{max} = 12.5$ cm, and that of the first voltage minimum $l_{min} = 37.5$ cm. The line can be considered to be lossless. (a) What is the operating frequency of the line and what the complex impedance of the load? (b) Repeat (a) but for another load that results in the first voltage maximum and minimum on the line switching places, namely, in $l_{min} = 12.5$ cm and $l_{max} = 37.5$ cm (and the same s).

12.5. More slotted-line measurements. (a) Find the operating frequency of an air-filled slotted line, whose characteristic impedance is $Z_0 = 60\ \Omega$ and losses in conductors are negligible, and the complex impedance of an unknown load terminating the line if measurements on it give a standing wave ratio of $s = 2$ and the location of the first voltage minimum at $l_{min} = 40$ cm from the load, which is also $\Delta l = 80$ cm to the next minimum. (b) Repeat (a) but for a load that results in the same for the first two voltage maxima, instead of minima, so in $l_{max} = 40$ cm and $\Delta l = 80$ cm.

12.6. Power flow on a low-loss microstrip line with a complex load. Take the low-loss microstrip line with a two-layer dielectric shown in Fig. 11.13 and described in Example 11.12, and assume that a time-harmonic wave of frequency $f = 1$ GHz is launched to propagate along the line, as well as that the line is terminated on its other end in a load of impedance $\underline{Z}_L = (75 + j30)\ \Omega$. The rms voltage of the incident wave at the load is $V_{i0} = 1$ V. Under these circumstances, compute the time-average power of the incident and reflected waves, and the net real power flow along the line at distances of 5 m, 50 cm, and 5 cm, respectively, from the load, and at the load.

12.7. Quarter-wave transformer for a resistive load. (a) Design a quarter-wave transformer to impedance-match a purely resistive load of resistance $R_L = 40\ \Omega$ to a transmission line with characteristic impedance $Z_0 = 160\ \Omega$ at a frequency of $f = 200$ MHz. (b) Then compute the percentage of the incident power that is reflected from the input terminals of the transformer, that is, from the combination of the transformer and the load, and the associated standing wave ratio at frequencies 10% above and below the design frequency, respectively.

12.8. Purely resistive line impedance for complex loads. For the transmission line from Problem 12.1 and each of the two complex loads specified in Problem 12.3, cases (a) and (b), determine the shortest distance from the load to a location where the transmission-line impedance is purely real (resistive), and compute that impedance (resistance). In both cases, also find the next shortest distance and the corresponding line resistance. What are the maximum and minimum line resistances, viewed along the entire line, for each of the loads?

12.9. Purely resistive line impedance for resistive loads. Repeat the previous problem but for the two purely resistive loads specified in Problem 12.1.

12.10. Maximum and minimum line impedances for reactive loads. Consider the two cases of purely reactive load terminations specified in Problem 12.2 (for the transmission line in Problem 12.1), and find the maximum and minimum values of the magnitude of the line impedance, and locations on the line (relative to the load) where these values are encountered.

12.11. Quarter-wave matching circuit for a capacitive complex load. If the load in Problem 12.7 is complex with impedance $Z_L = (100 - j60)\ \Omega$, design a matching circuit for it that includes a quarter-wave transformer.

12.12. Complete circuit and power analysis of a lossless line. In the transmission-line circuit in Fig. 12.3, let $l = 58$ cm, $Z_0 = 50\ \Omega$, $\underline{\mathcal{E}} = 12\, e^{j\pi/6}$ V, $\underline{Z}_g = (10 + j10)\ \Omega$, and $\underline{Z}_L = (70 - j100)\ \Omega$. In addition, let the wavelength along the line be $\lambda_z = 10$ cm, and the attenuation coefficient of the line be zero. Compute: (a) the total voltage and current along the line in both complex and time domains, (b) the net complex power flow (toward the load) along the line, (c) the time-average power of the reflected wave, (d) the time-average power that the generator delivers to the line, and (e) the instantaneous loss power in the load.

12.13. Complete circuit/power analysis of a low-loss line. Repeat the previous problem but for a nonzero attenuation coefficient along the line amounting to $\alpha = 0.5$ Np/m, which classifies this line as a low-loss one. Additionally, evaluate the total time-average loss power in the line conductors and dielectric.

12.14. Cascaded/series configuration of three transmission lines. Fig. 12.53 shows a combination of three lossless transmission lines and three complex impedance loads driven by a time-harmonic voltage generator. For the circuit parameters given in the figure, determine the time-average power delivered to each of the loads.

12.15. Cascaded/parallel configuration of three lines. Repeat the previous problem but for the circuit configuration with three transmission lines and three loads depicted in Fig. 12.54.

12.16. Combining impedances of a coaxial cable when shorted/open. By measurement at a frequency of $f = 313$ MHz on a coaxial cable of length $l = 1.3$ m, it is determined that the input impedance of the cable when short-circuited amounts to $\underline{Z}_{sc} = j130\ \Omega$, while $\underline{Z}_{oc} = -j43.4\ \Omega$ when open, as well as that the electrical length of the cable, $l_e = l/\lambda_z$ (λ_z being the wavelength along the cable) falls between 2 and 2.25. The losses along the cable can be neglected. What are (a) the characteristic impedance of the cable and (b) the relative permittivity of its dielectric ($\mu_r = 1$)?

12.17. Coaxial-cable inductor. What is the shortest section of the coaxial cable from the previous problem that would be equivalent to a lumped inductor of inductance $L_{eq} = 100$ nH at a frequency of $f = 400$ MHz when (a) short-circuited and (b) open-circuited, respectively?

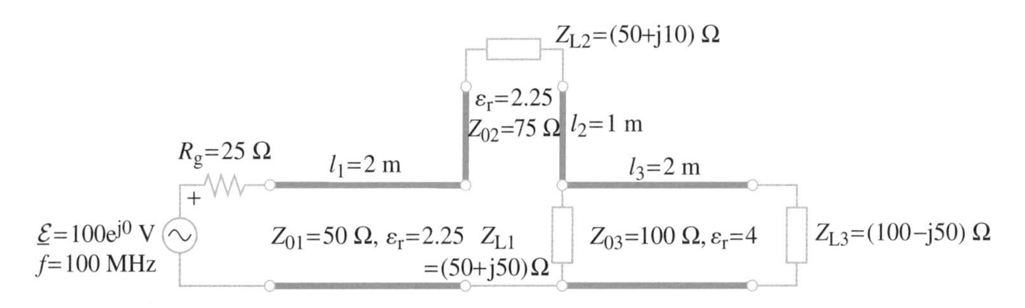

Figure 12.53 Circuit with three lossless transmission lines of different characteristic impedances and different electrical lengths in a cascaded/series configuration in a time-harmonic regime [ε_r designates the relative permittivity of line dielectrics ($\mu_r = 1$)]; for Problem 12.14.

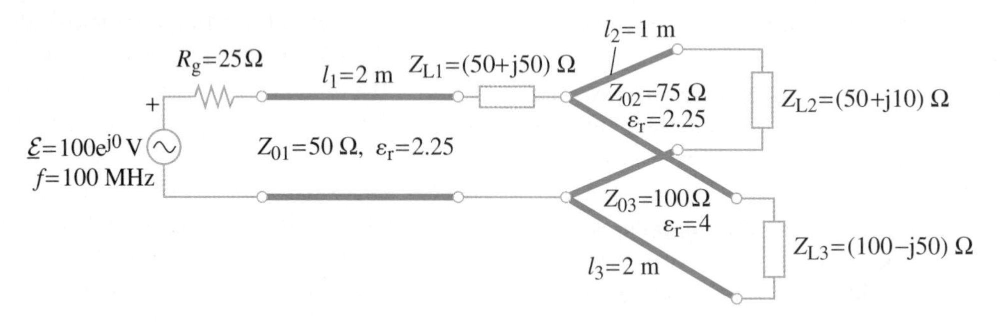

Figure 12.54 Transmission lines from Fig. 12.53 but in a cascaded/parallel configuration; for Problem 12.15.

(c) What lumped element is the section in (a) equivalent to if it is open (at the same frequency)? (d) What is obtained if the section in (b) is shorted?

12.18. Coaxial-cable capacitor. Repeat the previous problem but to obtain a lumped capacitor of capacitance $C_{eq} = 100$ pF (at the same frequency).

12.19. Admittance-matching by a shunt open-circuited stub. An antenna whose input impedance at a frequency of $f = 8$ GHz amounts to $\underline{Z}_L = (85 + j30)\ \Omega$ needs to be admittance-matched to a lossless microstrip transmission line of characteristic impedance $Z_0 = 50\ \Omega$ and phase velocity $v_p = 1.7 \times 10^8$ m/s – using a shunt open-circuited stub, in the same configuration as the one in Fig. 12.17. If the location on the main line at which the stub is attached is $l = 8.2$ mm away from the load and the stub is also a microstrip line, printed on the same substrate and having the same parameters as the main line, find the minimal required length of the stub.

12.20. Impedance-matching using a series two-wire stub. A series short-circuited stub is used in an impedance-matching circuit for an antenna with input impedance $\underline{Z}_L = (73 + j42)\ \Omega$ at a frequency of $f = 800$ MHz, as in Fig. 12.18. The stub is inserted at a location at $l = 112.05$ mm from the load. Both the main line and the stub are air two-wire lines of characteristic impedance $Z_0 = 300\ \Omega$. Determine the matching length of the stub.

12.21. Energy storage and damping in a coaxial-cable resonator. Consider the coaxial-cable resonator from Example 12.22, and assume that the rms voltage of the incident time-harmonic wave, as well as that of the reflected wave, in the structure amount to $V_{i0} = V_{r0} = 100$ V at a time instant $t = 0$. (a) What is the stored electromagnetic energy of the resonator at $t = 0$? (b) After what time is the energy reduced by 1% with respect to its value in (a)? (c) At what time the energy equals 1% of the value in (a)?

12.22. Quality factor of a microstrip-line resonator. A half-wavelength long section of the microstrip transmission line (with considerable fringing effects) described in Example 11.15 is short-circuited at both ends to form an electromagnetic resonator (at the resonant frequency of $f = 3$ GHz). The losses in the short circuits can be neglected. Find the quality factors of this resonator for its conductors and dielectric, respectively, as well as the total Q factor, using (a) the primary circuit parameters of the line, computed in Example 11.16, and (b) the secondary circuit parameters of the line, determined in Example 11.15. (c) What are the time constant (relaxation time) and damping factor of the resonator?

12.23. Transmission-line analysis using the Smith chart. A lossless transmission line of characteristic impedance $Z_0 = 100\ \Omega$, phase velocity $v_p = 2 \times 10^8$ m/s, and length $l = 34$ cm is terminated in a load of impedance $\underline{Z}_L = (30 - j40)\ \Omega$ at a frequency of $f = 1$ GHz. Using the Smith chart, determine: (a) the magnitude and phase angle of the load reflection coefficient, (b) the standing wave ratio of the line, (c) the line impedance and admittance halfway along the line, (d) the input impedance and admittance of the line, and (e) locations of all voltage maxima and minima on the line.

12.24. **Slotted-line measurements and the Smith chart.** Repeat Problem 12.4 but now using the Smith chart.

12.25. **One more load determination in the Smith chart.** Repeat Problem 12.5 with the help of the Smith chart.

12.26. **Purely resistive line impedances in the Smith chart.** Redo (a) Problem 12.8 and (b) Problem 12.9 graphically, in the Smith chart.

12.27. **Graphical analysis of lines with reactive loads.** Repeat Problem 12.10 but with the use of the Smith chart.

12.28. **Quarter-wave matching in the Smith chart.** Repeat Problem 12.11 using the Smith chart.

12.29. **Shunt stub design using the Smith chart.** With reference to Fig. 12.17, design a shunt short-circuited stub to admittance-match a load of impedance $\underline{Z}_L = (80 - j140)\ \Omega$ to a feeding line of characteristic impedance $Z_0 = 100\ \Omega$ at a frequency at which the wavelength along the line is $\lambda_z = 10$ cm. Losses along the main line and the stub can be neglected. Choose a stub with the same Z_0 and λ_z as the main line, and find the distance of the stub junction from the load and stub length to achieve matching.

12.30. **Series stub design using the Smith chart.** For the same parameters of the main line, load, and stub as in the previous problem, design an impedance-matching series short-circuited stub (Fig. 12.18).

12.31. **Step response of a line with unmatched load and generator.** A lossless transmission line of characteristic impedance $Z_0 = 100\ \Omega$, phase velocity $v_p = 1.75 \times 10^8$ m/s, and length $l =$ 35 cm, terminated in a purely resistive load, is driven by a voltage generator of step emf $\mathcal{E} = 4$ V applied at $t = 0$. The internal resistance of the generator is $R_g = 60\ \Omega$, while the load resistance is $R_L = 200\ \Omega$. Sketch (a) the voltage snapshots (scans) along the line at times $t = 3$ ns, $t = 5$ ns, and $t = 7$ ns, respectively, and (b) the voltage waveforms at both the generator and load within a time interval $0 \le t \le 12$ ns.

12.32. **Another combination of unmatched conditions.** Repeat the previous problem but for $R_g = 20\ \Omega$ and $R_L = 50\ \Omega$.

12.33. **Time-domain reflectometry on a buried coaxial cable.** A time-domain reflectometer is used to locate and evaluate damage (discontinuity) on a buried coaxial cable. The discontinuity behaves as a purely resistive load to the portion of the cable preceding the damage location. The cable dielectric is polyethylene, the relative permittivity of which is $\varepsilon_r = 2.25$, the characteristic impedance of the cable is $Z_0 = 50\ \Omega$, and the losses along it can be neglected. A step emf is launched to propagate along the cable by a matched voltage generator at $t = 0$, and, as a result, the oscilloscope displays a TDR voltage of the same form as that in Fig. 12.36 back, at the generator end of the cable. This voltage amounts to $v_g(t) = 25$ V for $0 < t < 100$ ns and $v_g(t) =$ 1 V for $t > 100$ ns. How far from the generator down the cable is the damage location? What is the resistance of the equivalent load at this location?

12.34. **Open- or short-circuited line with an unmatched generator.** A lossless air two-wire line with characteristic impedance $Z_0 = 300\ \Omega$ and length $l = 1$ m is fed by a step (at $t = 0$) voltage generator with emf magnitude $\mathcal{E} =$ 20 V and internal resistance $R_g = 500\ \Omega$.[11] Sketch the voltage waveforms at both ends of the line if it is (a) open and (b) shorted, respectively, for $0 \le t \le 20$ ns.

12.35. **Ideal generator and arbitrary resistive load.** Repeat Example 12.27 but for an ideal step voltage generator driving the coaxial cable.

12.36. **Transient diagrams for transmission-line current.** For the transmission line and its excitation and load described in Problem 12.31, sketch the current-intensity snapshots along the line at specified instants of time and current waveforms at both ends of the line within the given time interval – in two ways, as follows. (a) First, obtain the transient current by

[11]Note that such a high internal resistance (R_g) may be that of the Thévenin equivalent generator representing a more complex terminal network at the generator end of the transmission line.

tracing directly the line response for the current: starting with the current intensity of the initial incident wave launched by the generator at $t = 0$, and tracking multiple reflections at the load and generator with the use of the reflection coefficients for currents. (b) Then, obtain the same transient current diagrams directly from the corresponding voltage scans and waveforms, from Problem 12.31, using the voltage to current ratio for traveling waves on the line.

12.37. Transient current for an open/shorted line. Sketch the input/output current-intensity waveforms for the two-wire line from Problem 12.34, in both termination cases (open and short).

12.38. Pulse response of an open-circuited line, unmatched generator. Consider a lossless transmission line for which the one-way wave travel time is $T = 1$ ns, and assume that it is fed at one end by a voltage generator of rectangular pulse emf with magnitude $\mathcal{E} = 6$ V and width $t_0 = 1$ ns, applied at $t = 0$. The characteristic impedance of the line is $Z_0 = 50\ \Omega$, the internal resistance of the generator is $R_g = 250\ \Omega$, and the line is open-circuited at the other end. Under these circumstances, sketch the voltage waveforms at both ends of the line during time $0 \le t \le 10$ ns.

12.39. Pulse response for a matched generator or matched load. Repeat the previous problem but with (a) a matched generator ($R_g = 50\ \Omega$) and (b) matched load ($R_L = 50\ \Omega$), respectively.

12.40. Overlapping pulses at line ends. Repeat Problem 12.38 but for $t_0 = 3$ ns.

12.41. Trapezoidal pulse response of a transmission line. A lossless transmission line of length $l = 15$ cm and characteristic impedance $Z_0 = 75\ \Omega$ is driven by an ideal voltage generator with emf in the form of a trapezoidal pulse, like the one in Fig. 6.2(b), of magnitude $\mathcal{E} = 3$ V and total duration $t_0 = 2$ ns, applied at $t = 0$. The rise and fall times of the signal $e(t)$ are 0.5 ns each, and the duration of the constant part in between is 1 ns. The parameters of the line dielectric are $\varepsilon_r = 4$ and $\mu_r = 1$, and the resistance of a purely resistive load terminating the line at its other end amounts to $R_L = 150\ \Omega$. Sketch the voltage waveform at the load within a time interval $0 \le t \le 8$ ns.

12.42. Effects of a finite rise time for open and short load terminations. Consider the transmission-line circuit shown in Fig. 12.44(a) and described in Example 12.35, and assume that the line is open-circuited at its load end. Sketch the voltage waveform at the generator end of the line, for each of the two specified rise times of the applied emf. (b) Repeat (a) but for a short circuit as the load termination.

12.43. Bounce diagrams for a line with unmatched load and generator. (a) Construct voltage and current bounce diagrams for the transmission line with an unmatched load and generator described in Problem 12.31 (see also Problem 12.36). Use these diagrams to obtain (b) the total voltage and current scans along the line at a time instant $t = 9.5$ ns and (c) voltage and current waveforms for $0 \le t \le 12$ ns at a location on the line $z = 21$ cm away from the generator.

12.44. Bounce diagrams for an open/shorted line. Redo Problems 12.34 and 12.37 using bounce diagrams.

12.45. Pulse response using bounce diagram. Redo Example 12.32 but with the use of the voltage bounce diagram of the line.

12.46. TDR measurement of unknown load capacitance. Assume that the load in the TDR system from Example 12.28 is an ideal capacitor of unknown capacitance, and that the oscilloscope showing, $v_g(t)$, is that in Fig. 12.48(d). If the "area" (expressed in Vs) between the increasing exponential curve and its horizontal asymptote in the voltage display (starting with the time instant when $v_g = 0$) is measured to be $A = 8$ nVs, determine the load capacitance.

12.47. Line terminated in a resistor and a capacitor in parallel. A lossless transmission line of characteristic impedance $Z_0 = 50\ \Omega$, dielectric parameters $\varepsilon_r = 2.55$ and $\mu_r = 1$, and length $l = 22$ cm is fed by a matched voltage generator of rectangular pulse emf with magnitude $\mathcal{E} = 1$ V and duration $t_0 = 2$ ns, applied at $t = 0$. At the other end, the line is terminated in a parallel connection of a resistor of resistance $R = 50\ \Omega$ and a capacitor of capacitance $C = 70$ pF. Sketch the voltage waveforms at (a) the load and (b) the generator for $0 \leq t < \infty$.

12.48. **Parallel connection of a resistor and an inductor as load.** Repeat the previous problem but for a load consisting of a resistor of resistance $R = 150\ \Omega$ and an inductor of inductance $L = 90$ nH connected in parallel.

12.49. **Four different series-parallel resistive-reactive load terminations.** A lossless transmission line for which the one-way wave travel time is $T = 1$ ns and whose characteristic impedance is $Z_0 = 100\ \Omega$ is driven by a matched voltage generator of step emf $\mathcal{E} = 2$ V applied at $t = 0$. Consider four different series-parallel combinations of resistors of resistance $R = 100\ \Omega$, inductor of inductance $L = 50$ nH, and capacitor of capacitance $C = 50$ pF shown in Fig. 12.55(a)–(d) as a load termination for this line. For each of the combinations, sketch the voltage waveform at the load end of the line for $0 \leq t < \infty$.

12.50. **Nonlinear load with a square-law current-voltage characteristic.** Consider the transmission line and its step voltage excitation from the previous problem, and assume that the other end of the line is terminated in a nonlinear load whose current-voltage characteristic can be expressed as $i_L = av_L^2$, where a is a positive constant, for $v_L \geq 0$ and $i_L = 0$ for $v_L < 0$. Sketch the voltage waveforms at both ends of the line for $0 \leq t < \infty$ if (a) $a = 0.01\ \text{A/V}^2$ and (b) $a = 10\ \text{A/V}^2$, respectively.

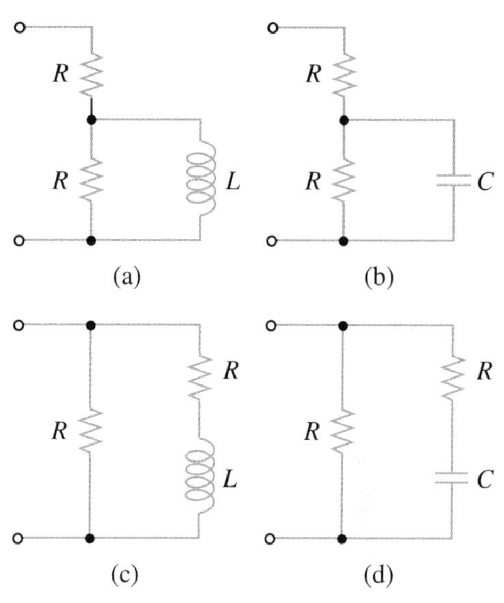

Figure 12.55 Four different lossy reactive load terminations for a lossless transmission line excited by a matched step voltage generator; for Problem 12.49.

13 Waveguides and Cavity Resonators

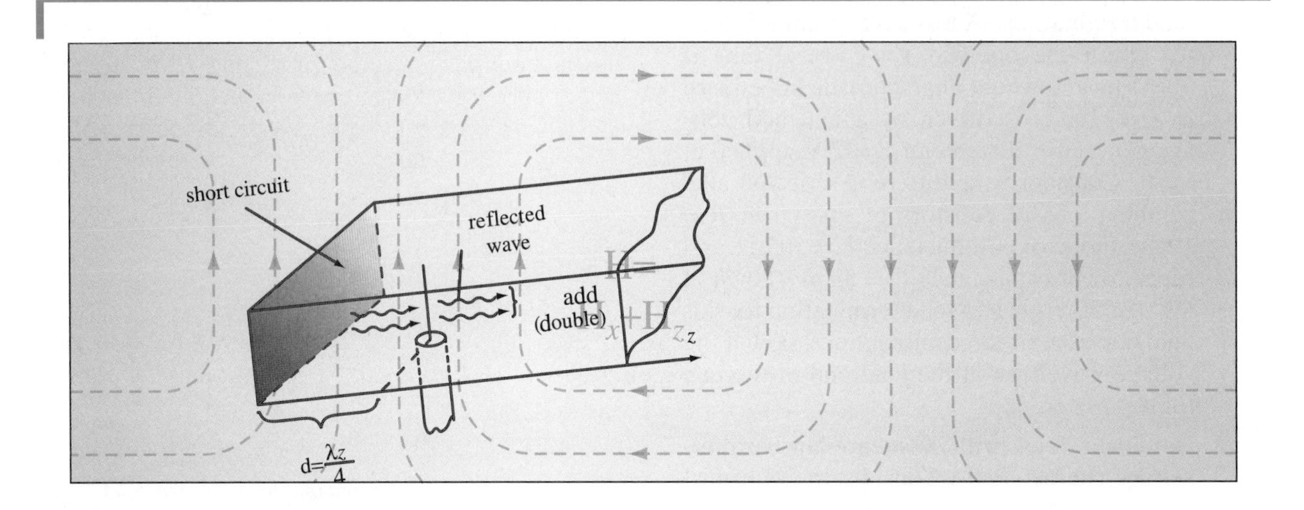

Introduction:

At frequencies in the microwave region (see Fig. 9.8), waveguides in the form of metallic tubes, as the one in Fig. 10.15, are used for energy and information transfer in electromagnetic devices and systems. Essentially, electromagnetic waves travel along such tubes by means of multiple reflections from the metallic walls, through the dielectric filling the tube (most frequently, air), so the waves are guided by the tube conductor. In general, the principal advantage of metallic waveguides, which have one conductor, over transmission lines (e.g., coaxial cables, Fig. 11.7), with two (or more) conductors, at frequencies above several GHz is considerably smaller attenuation along the structure[1] and its larger power transmission capacity. However, losses and wave attenuation are a limiting factor for practical usability of metallic waveguides too; for larger distances we thus normally use waves in free space (wireless links), transmitted and received by antennas,[2] or purely dielectric waveguides (e.g., optical fibers, Fig. 10.22). In addition to metallic waveguides for energy/information transmission, waveguide sections closed at both ends, thus forming rectangular metallic cavities, represent microwave resonators – also with widespread applications.

[1] We recall that losses in the inner conductor of a coaxial cable are considerably larger than losses in its outer conductor (see Examples 11.2 and 11.3), and that at frequencies above several GHz losses in the cable dielectric become significant as well [see the discussion of Eqs. (11.78) and (11.82)]. These two facts and the fact that a metallic waveguide has no inner conductor nor dielectric (if air-filled) provide an overly simplified but indicative explanation of why metallic waveguides exhibit smaller losses than coaxial cables.

[2] There is always a distance between the transmit and receive ends of an electromagnetic system below which the waveguide attenuation is smaller than the attenuation in free space (between antennas), and above which the power transfer in a wireless link becomes more efficient. This distance typically amounts to several tens of meters at a frequency of 1 GHz, and it decreases with an increase in frequency.

Although arbitrary cross sections of metallic tubes and cavities are theoretically possible, practical microwave devices and systems involve only those of rectangular and circular shapes, and – of the two – rectangular structures are used much more frequently; hence, our focus here will be on rectangular metallic waveguides and cavity resonators.

We shall first analyze wave propagation along rectangular metallic waveguides by basically "reformatting" and reinterpreting the solution to a wave reflection problem of a normally polarized uniform plane wave obliquely incident on a perfectly conducting boundary already obtained in Sections 10.5 and 10.6. Then, we shall generalize this solution to the theory of arbitrary modes (there are a double infinite number of such modes) within both TE (transverse electric) and TM (transverse magnetic) types of guided electromagnetic waves based on general Maxwell's and wave equations for the waveguide. We shall see that for each mode there is a cutoff frequency, below which the waves are evanescent (vanishing) – they cannot propagate, analogously to the plasma frequency discussed in Section 9.12. Using the general modal theory of waveguides, we shall study the TE and TM wave impedances (for the propagating waves), power flow along the waveguide, losses in both its walls and dielectric, waveguide dispersion and phase and group velocities, and waveguide couplers for excitation or reception of different wave modes. Studies of rectangular cavity resonators will include computation of cavity fields, resonant frequencies, stored electromagnetic energy, and the quality factor of rectangular cavities with small losses.

Metallic waveguides and cavity resonators are important parts of many technologies and practical applications, with radar antenna feeds and circuitry, waveguide slot antenna arrays, horn antennas, waveguide microwave filters and other circuit components, and microwave ovens being just some of the examples. In addition, many wireless applications and problems actually represent waveguide problems, such as radio propagation and coverage in tunnels. Also, understanding of resonant cavities is essential for some EMC and EMI studies and applications. For example, a metallic box used to shield an RF or digital device to reduce the electromagnetic coupling between the device and its environment, i.e., to fulfill the external EMC requirements,[3] behaves, however, as a resonant cavity. At the box resonant frequencies, even a weak coupling of some part of the device and the cavity can excite a strong resonant field, which, in turn, can generate a strong undesired signal at any other, possibly faraway, part of the circuit so that the internal EMC of the device is easily deteriorated. Finally, the importance of comprehensive understanding of wave guidance, wave modes, frequency cutoff, dispersion, and attenuation along metallic waveguides, as well as of resonant frequencies, fields, and the Q factor of microwave cavities, is well beyond the particular structures that will be studied in this chapter. Many of these concepts and associated analysis techniques and design approaches may be effectively used, directly or indirectly, in other applications of electromagnetic fields and waves, as well as in other areas of science and engineering.

13.1 ANALYSIS OF RECTANGULAR WAVEGUIDES BASED ON MULTIPLE REFLECTIONS OF PLANE WAVES

Consider an infinitely long uniform rectangular metallic waveguide with cross-sectional interior dimensions a and b, filled by a homogeneous dielectric of permittivity ε and permeability μ, as shown in Fig. 13.1. We assume that the waveguide is lossless, i.e., that its walls are made of a perfect electric conductor (PEC), and that the dielectric is also perfect. We would like to find a solution for a time-harmonic

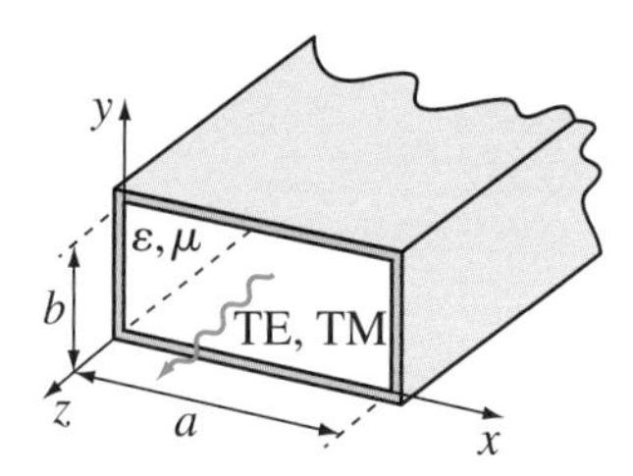

Figure 13.1 Rectangular waveguide with a TE or TM wave.

[3]External and internal EMC (electromagnetic compatibility) pertain to electromagnetic coupling between a device and other objects around it and coupling between various parts within the device, respectively. EMC studies are important not only because of stringent national and international EMC regulations but also for proper operation of devices.

electromagnetic wave, of frequency f, that propagates inside the waveguide, along the z-axis. However, we already have one such solution – see the waveguide in Fig. 10.15, obtained by placing three additional PEC planes in the field of a normally polarized uniform plane wave obliquely incident on a PEC boundary (fourth plane) in Fig. 10.13(a). The planes are positioned such that the tangential component of the electric field vector and normal component of the magnetic field vector of the resultant wave are zero at all waveguide walls,

boundary conditions at waveguide walls

$$\boxed{\underline{\mathbf{E}}_{\text{tang}} = 0, \quad \underline{\mathbf{H}}_{\text{norm}} = 0.} \tag{13.1}$$

The wave propagates in the positive x direction by bouncing back and forth, at an incident angle θ_{i}, between the walls at $z = 0$ and at one of the planes defined by Eq. (10.120). The electric and magnetic fields in the waveguide, $\underline{\mathbf{E}}$ and $\underline{\mathbf{H}}$, are given in Eqs. (10.99), (10.103), and (10.104). We note that the coordinate axes in Fig. 13.1 are set up differently from Fig. 10.15 [as in Fig. 11.1 for a transmission line, it is customary to have the z-axis of the adopted Cartesian coordinate system be along the waveguide], so let us rewrite the field expressions for the new coordinate system. Comparing Figs. 10.15 and 13.1, we see that the following exchange of variables between the two figures takes place:

$$x \to z, \quad z \to -x, \tag{13.2}$$

namely, x and z in Eqs. (10.99), (10.103), and (10.104) become z and $-x$, respectively, for the situation in Fig. 13.1, and the same transformation applies for the corresponding field components (for instance, $\underline{H}_z$ in Eq. (10.104) becomes $-\underline{H}_x$ in Fig. 13.1), whereas the y-coordinate is the same in both systems. Hence, the y-component (and the only existing one) of the electric field vector in Fig. 13.1, from Eqs. (10.99) and (13.2), reads

$$\underline{E}_y = 2\text{j}\underline{E}_{\text{i}0} \sin \beta_x x \, \text{e}^{-\text{j}\beta_z z} \tag{13.3}$$

($\underline{E}_x$, $\underline{E}_z = 0$). Here, β_x and β_z are the equivalent phase coefficients along the x- and z-axis, respectively, given by

$$\beta_x = \frac{\omega}{c} \cos \theta_{\text{i}}, \quad \beta_z = \frac{\omega}{c} \sin \theta_{\text{i}}, \quad c = \frac{1}{\sqrt{\varepsilon\mu}} \tag{13.4}$$

[see also Eqs. (10.119), (10.118), and (13.2)], $\omega = 2\pi f$ being the angular frequency of the wave, and c the intrinsic phase velocity of the dielectric in the waveguide, i.e., the velocity of uniform plane waves in an unbounded medium of the same parameters (ε and μ). Similarly, Eqs. (10.104), (10.103), and (13.2) give the following expressions for the x- and z-components of the magnetic field vector in Fig. 13.1:

$$\underline{H}_x = -\frac{2\text{j}\beta_z \underline{E}_{\text{i}0}}{\omega\mu} \sin \beta_x x \, \text{e}^{-\text{j}\beta_z z}, \quad \underline{H}_z = -\frac{2\beta_x \underline{E}_{\text{i}0}}{\omega\mu} \cos \beta_x x \, \text{e}^{-\text{j}\beta_z z} \tag{13.5}$$

($\underline{H}_y = 0$), where the use is made of the fact that $c/\eta = 1/\mu$, η being the intrinsic impedance of the waveguide dielectric, Eq. (9.21). Of course, this is a transverse electric (TE) wave, since $\underline{E}_z = 0$ and $\underline{H}_z \neq 0$.

From Eq. (10.120), i.e., the boundary condition for $\underline{E}_{\text{tang}} = \underline{E}_y$ or $\underline{H}_{\text{norm}} = \underline{H}_x$ at the waveguide wall defined by $x = a$ in Fig. 13.1, the transverse phase coefficient, β_x, is determined by

eigenvalues of a waveguide

$$\boxed{\beta_x a = m\pi \quad \longrightarrow \quad \beta_x = \frac{m\pi}{a} \quad (m = 1, 2, \ldots).} \tag{13.6}$$

We see that these boundary conditions can be satisfied only for a discrete set of values of β_x. These values are referred to as the eigenvalues (characteristic values) of the waveguide. Combining Eqs. (13.3) and (13.6), the electric field in the waveguide can now be written as

$$\boxed{\underline{E}_y = 2\mathrm{j}\underline{E}_{\mathrm{i}0} \sin\left(\frac{m\pi}{a}x\right) \mathrm{e}^{-\mathrm{j}\beta_z z} \quad (m = 1, 2, \ldots),} \tag{13.7}$$

TE_{m0} *modes*

with analogous expressions for the nonzero magnetic field components, $\underline{H}_x$ and $\underline{H}_z$. Each integer value of m determines a possible field solution in the waveguide. Hence, there are an infinite number of different field configurations satisfying the boundary conditions at the waveguide walls, Eqs. (13.1), along with the governing Maxwell's equations, Eqs. (11.1)–(11.4), or wave (Helmholtz) equations, Eqs. (9.8) and (9.9), in the dielectric. Helmholtz equations are satisfied for all time-harmonic uniform plane waves represented in the complex domain (see Section 9.4), and thus for the incident and reflected waves in Fig. 10.13(a), as well as for their superposition (resultant wave in the waveguide). These distinct waves that can exist in a waveguide are referred to as modes, and the corresponding electromagnetic field distributions as modal fields. Having another look at Eq. (10.120), rewritten with reference to Fig. 13.1 as

$$m\,\frac{\lambda_x}{2} = a, \tag{13.8}$$

where λ_x is the equivalent wavelength of the standing wave in the x direction, we realize that m actually equals the number of half-waves (half-wavelengths) along the x-axis that fit into the dimension a of the waveguide. Since the field does not vary at all along the y-axis (along the side b of the waveguide), we can say that the number of half-waves along the y-axis, marked as n, is zero. Accordingly, the wave corresponding to m (arbitrary positive integer) and $n = 0$ is called a TE_{m0} mode.[4] The lowest mode is TE_{10}, for $m = 1$.

We now emphasize an important difference in interpreting the situations in Figs. 10.15 and 13.1. In Fig. 10.15 and Eqs. (10.119) and (10.120), the incident angle, θ_i, is fixed, whereas the separation between the two parallel PEC planes at which the multiple reflections occur, z, varies for different integers m. In Fig. 13.1 and Eqs. (13.4) and (13.6), on the other hand, the separation between the planes, a, is fixed (a dimension of the waveguide), whereas the angle θ_i, since

$$\boxed{\cos\theta_\mathrm{i} = \frac{\beta_x c}{\omega} = m\,\frac{c}{2af},} \tag{13.9}$$

zigzag ray paths for different modes

is a function of m. So, each discrete value (for a given m) of θ_i corresponds to a different mode, TE_{m0}. The larger the m (higher mode) the smaller the θ_i (for $0 \le \theta_\mathrm{i} \le 90°$, $\cos\theta_\mathrm{i}$ is a decaying function) and longer the zigzag ray path in Fig. 10.15 between two locations along the waveguide.

Example 13.1 **Ray Paths of Several TE Modes in a Rectangular Waveguide**

For a rectangular metallic waveguide in Fig. 13.1, $a = 6$ cm and the dielectric is air. If the operating frequency of the guide is $f = 10$ GHz, sketch the ray paths, like the one in Fig. 10.15, corresponding to the first four TE_{m0} modes ($m = 1, 2, 3, 4$) in the structure.

[4] As we shall see in a later section, waveguide modes with both m and n being arbitrary nonnegative integers, i.e., with $m, n = 0, 1, 2, \ldots$, where the case $m = n = 0$ is excluded, are also possible (if properly excited) in the waveguide in Fig. 13.1.

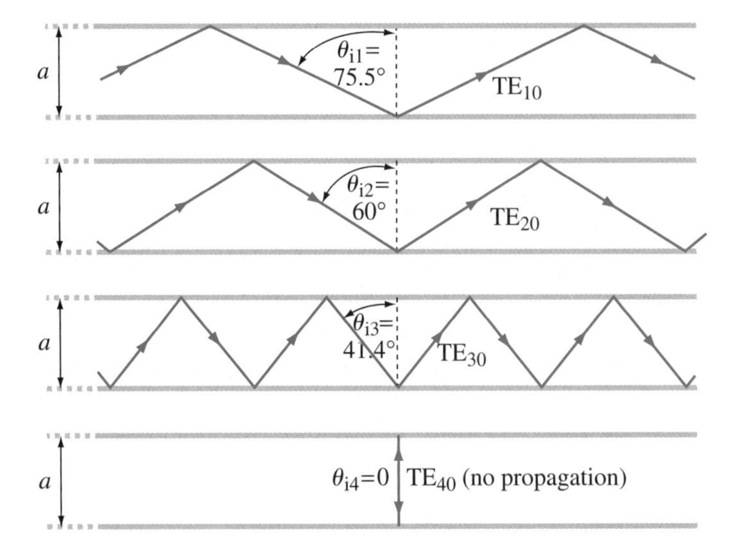

Figure 13.2 Ray paths of the first four TE_{m0} modes ($m = 1, 2, 3, 4$) in an air-filled rectangular waveguide (Fig. 13.1) for $a = 6$ cm and $f = 10$ GHz; for Example 13.1.

Solution Solving Eq. (13.9) for the incident angle, θ_i, on waveguide walls (Fig. 10.15), we have

$$\theta_\mathrm{i} = \arccos\left(m\,\frac{c_0}{2af}\right) \quad (m = 1, 2, 3, 4) \tag{13.10}$$

($\arccos \equiv \cos^{-1}$), where $c_0 = 3 \times 10^8$ m/s, Eq. (9.19), is the free-space (the structure is air-filled) wave velocity. This gives $\theta_{\mathrm{i}1} = 75.5°$ (for TE_{10} mode), $\theta_{\mathrm{i}2} = 60°$ (TE_{20}), $\theta_{\mathrm{i}3} = 41.4°$ (TE_{30}), and $\theta_{\mathrm{i}4} = 0$ (TE_{40}), and the corresponding ray paths traced by a uniform plane (TEM) wave bouncing back and forth, at angles θ_i, between the walls are illustrated in Fig. 13.2. We see that these zigzag ray paths are "denser" (break more frequently along the waveguide) for higher order modes, as expected. Ultimately, for $m = 4$, which results in $\cos\theta_\mathrm{i} = 1$ and $\theta_\mathrm{i} = 0$, they degenerate into a vertical line, and the wave does not progress at all along the guide (this phenomenon will be discussed in the next section). Note also that $\cos\theta_\mathrm{i} > 1$ for $m \geq 5$, implying that such waves cannot propagate either. So, only the first three ($m = 1, 2, 3$) are possible propagating TE_{m0} wave modes in this waveguide, at the given operating frequency.

13.2 PROPAGATING AND EVANESCENT WAVES

Using Eqs. (13.4) and (13.9), the longitudinal phase coefficient (in the z direction), β_z, of the waveguide in Fig. 13.1 is

$$\beta_z = \frac{\omega}{c}\sqrt{1 - \cos^2\theta_\mathrm{i}} = \frac{\omega}{c}\sqrt{1 - \left(\frac{mc}{2af}\right)^2}. \tag{13.11}$$

This coefficient is the principal phase coefficient for the waveguide, as it determines the propagation along the structure. Therefore, let us denote it simply by β, and call it the phase coefficient of the waveguide. Of course, we always have in mind that β depends on the mode index m. It can be written as

waveguide phase coefficient

$$\boxed{\beta = \beta_z = \frac{\omega}{c}\sqrt{1 - \frac{f_\mathrm{c}^2}{f^2}},} \tag{13.12}$$

where the frequency f_c, given by

$$f_c = (f_c)_{m0} = m\,\frac{c}{2a} \qquad (n = 0), \tag{13.13}$$

cutoff frequency, TE_{m0} mode

is called the cutoff or critical frequency of the waveguide – for a particular, TE_{m0}, mode. The corresponding wavelength, namely, the cutoff wavelength, λ_c, is defined as the intrinsic wavelength in the waveguide dielectric, Eq. (8.112), at the cutoff frequency,

$$\lambda_c = \frac{c}{f_c}. \tag{13.14}$$

cutoff wavelength

The frequency f_c has the same role as the plasma frequency (f_p), in Eq. (9.160). Analogously to a plasma medium (see Section 9.12), the waveguide in Fig. 13.1 behaves like a high-pass filter, letting only waves in Fig. 13.2 whose frequency is higher than the cutoff frequency,

$$f > f_c, \tag{13.15}$$

propagating waves

propagate through it (for $f > f_c$, β is purely real). Otherwise, if $f < f_c$, β is purely imaginary, and we have

$$\beta = \pm \mathrm{j}|\beta|, \quad |\beta| = \frac{\omega}{c}\sqrt{\frac{f_c^2}{f^2} - 1} \quad \longrightarrow \quad \mathrm{e}^{-\mathrm{j}\beta z} = \mathrm{e}^{\pm|\beta|z} = \mathrm{e}^{-|\beta|z}, \tag{13.16}$$

where the plus sign in the exponent is eliminated as it would imply an exponentially increasing (with z) wave amplitude. We see that $|\beta|$, for $f < f_c$, effectively acts like an attenuation coefficient (α). Therefore, the waves at these frequencies cannot propagate – they are so-called evanescent (vanishing) waves (as in plasmas, below f_p). The evanescent attenuation is not due to Joule's losses in the propagation medium, but is a consequence of the particular configuration of the guiding structure (namely, the particular separation a between waveguide walls in Fig. 13.1). In the border-line case, $f = f_c$, β is zero (no propagation), so that the complete evanescent (nonpropagating) frequency range is given by

$$f \le f_c. \tag{13.17}$$

evanescent waves

Note that the expression for the cutoff frequency in Eq. (13.13) is identical to that in Eq. (10.12) for the resonant frequency, f_{res}, of a Fabry-Perot resonator, in Fig. 10.3. Note also that $\beta = \beta_z = 0$ in Eq. (13.4) gives $\theta_i = 0$, which means that a wave in a rectangular waveguide with a frequency

$$f = f_c = f_{res} \qquad (\beta_z = 0\,,\ \theta_i = 0) \tag{13.18}$$

resonator

bounces up and down (normal incidence) between the waveguide walls, as in Fig. 10.3, and does not propagate along the z-axis, and this exactly is the case in Fig. 13.2 for $m = 4$. So, the waveguide in Fig. 13.1 at cutoff behaves like a Fabry-Perot resonator (at resonance), which is another explanation of nonpropagating waves at $f = f_c$ in Eq. (13.17).

Finally, note that the cutoff frequency of a TEM wave in a transmission line, Fig. 11.1, whose phase coefficient (β) is given in Eq. (11.16), is

$$f_c = 0. \tag{13.19}$$

TEM waves in transmission lines

Namely, unlike for TE (or TM) waves in waveguides, there is no theoretical lower frequency limit for the existence and propagation of TEM waves along transmission lines. Moreover, as we know from Section 3.12, even dc voltages and currents, at $f = 0$, can exist along lines, so their theoretical operating frequency range is actually $f \geq 0$.

Conceptual Questions (on Companion Website): 13.1–13.6

13.3 DOMINANT WAVEGUIDE MODE

From Eqs. (13.13) and (13.14), the cutoff frequency of the TE_{10} mode in a rectangular waveguide (Fig. 13.1) is

cutoff frequency, TE_{10} mode

$$(f_\mathrm{c})_{10} = \frac{c}{2a}, \tag{13.20}$$

and the cutoff wavelength

cutoff wavelength, TE_{10} mode

$$(\lambda_\mathrm{c})_{10} = \frac{c}{(f_\mathrm{c})_{10}} = 2a. \tag{13.21}$$

Note that the transverse dimension a of the waveguide equals a cutoff half-wave ($\lambda_\mathrm{c}/2$) for the TE_{10} mode. So, for the given operating frequency, f, of the wave, a must be greater than the intrinsic half-wave of the guide dielectric, $a > \lambda/2$, for propagation of this mode to be possible [see Eqs. (13.15) and (8.112)]. The next higher order mode (assuming $n = 0$), TE_{20} ($m = 2$), can propagate if the frequency f is above $(f_\mathrm{c})_{20} = c/a = 2(f_\mathrm{c})_{10}$. In the frequency range between $(f_\mathrm{c})_{20}$ and the next cutoff frequency in Eq. (13.13), for $m = 3$, both TE_{10} and TE_{20} are possible. As the frequency increases, more and more modes would propagate if excited. However, there is an exclusive frequency range, that between $(f_\mathrm{c})_{10}$ and $(f_\mathrm{c})_{20}$, in which only one mode, the TE_{10} mode, can propagate, and hence its name – the dominant mode. In practice, as a wave propagates along the waveguide, every discontinuity on its way [e.g., a waveguide bend (corner), a junction with another waveguide, a slot in the waveguide wall, or a metallic wire inserted in the guide] may cause a multitude of different modes to be excited (in order for the boundary conditions to be satisfied at the discontinuity). If f belongs to this exclusive range, referred to as the dominant range,

dominant frequency range

$$\frac{c}{2a} < f \leq \frac{c}{a}, \tag{13.22}$$

out of all excited modes only the dominant mode will propagate, while all other modes will be evanescent.

Multimode propagation along a waveguide (with several propagating modes existing simultaneously in the structure – if the frequency f is above the cutoffs of all of them) is, in general, undesirable, because each such propagating mode has a different phase coefficient, β, and thus different phase velocity, v_p [see Eqs. (13.12), (13.13), and (9.35)]. Note that v_p of a rectangular waveguide for arbitrary modes will be discussed in a later section. This means that different modal components of the total (multimode) field would propagate at different phase velocities, and thus arrive with different phase delays to the receiving end of the waveguide. Hence, the relative phases of the modal components in the receiving cross section of the waveguide would be changed, and the received field (signal) distorted (signal shape, i.e., waveform, in time would change). In addition, as the field configurations of individual modes differ from each other [see Eq. (13.7)], it is practically impossible to devise a receiving mechanism to extract the complete energy of the received multimode field. Namely, as we shall see in a later section, each mode requires a different coupling structure (e.g., a distinct set of electric or magnetic probes) in a waveguide for its excitation (generation) or reception. This is why most often only frequencies in Eq. (13.22), and only the dominant mode are used in waveguide applications. On the other hand, even in the so-called overmoded operation of a waveguide – at frequencies above the dominant range, mode filters (to also be discussed later) are used to remove undesired propagating modes, and ensure that, again, only one mode is present in the structure.

The multiplicative constant $\underline{E}_{i0}$, as the complex rms electric field intensity of the incident wave at the coordinate origin in Fig. 10.13(a), is present in the expressions for all field components in Eqs. (13.3) and (13.5). Its magnitude, $|\underline{E}_{i0}|$, determines the strength of the field at every point in the waveguide, and the overall power level carried by the wave, as determined, in turn, by the sources of the field, i.e., by the waveguide excitation (to be discussed in a later section). Simply, by multiplying $\underline{E}_{i0}$ by 2, for instance, the field will become twice stronger everywhere in the waveguide, and the transmitted power will quadruple. Of course, $\underline{E}_{i0}$, being complex in general, can also introduce a constant phase factor, due to the excitation, to all field components at every point in the waveguide. However, in the analysis of waveguides with TE waves it is customary to denote the "peak-value" in the complex expression for the axial component of the magnetic field vector, $\underline{H}_z$, in the cross section of the waveguide defined by $z = 0$ (Fig. 13.1) by $\underline{H}_0$, and to use this new constant to set the level of field strengths and transmitted power in the structure. Obviously, $\underline{H}_0$ is proportional to $\underline{E}_{i0}$; for $m = 1$, Eqs. (13.5) and (13.6) yield

$$\underline{H}_0 = \underline{H}_z\big|_{x=0,\ z=0} = -\frac{2\beta_x \underline{E}_{i0}}{\omega\mu} = -\frac{2\pi\,\underline{E}_{i0}}{\omega\mu a} \quad (m = 1). \tag{13.23}$$

With this, the field components of the dominant (TE_{10}) mode become, from Eqs. (13.7), (13.5), and (13.6),

$$\underline{E}_y = -\mathrm{j}\omega\mu\,\frac{a}{\pi}\,\underline{H}_0 \sin\left(\frac{\pi}{a}x\right) \mathrm{e}^{-\mathrm{j}\beta z} \tag{13.24}$$

$\underline{E}_y$ – *dominant mode* (TE_{10})

$$\underline{H}_x = \mathrm{j}\beta\,\frac{a}{\pi}\,\underline{H}_0 \sin\left(\frac{\pi}{a}x\right) \mathrm{e}^{-\mathrm{j}\beta z} \tag{13.25}$$

$\underline{H}_x$ – *dominant mode* (TE_{10})

$$\underline{H}_z = \underline{H}_0 \cos\left(\frac{\pi}{a}x\right) \mathrm{e}^{-\mathrm{j}\beta z} \tag{13.26}$$

$\underline{H}_z$ – *dominant mode* (TE_{10})

($\underline{E}_x$, $\underline{E}_z$, $\underline{H}_y = 0$). Fig. 13.3 shows the field distributions of the TE_{10} mode.

As we shall see in the following sections, TE_{m0} ($m = 1, 2, \ldots$) modes in Eqs. (13.7), (13.12), and (13.13) are only a subset of all possible modes in a rectangular waveguide (Fig. 13.1). The full set includes a double infinite series of TE_{mn} modes (with $m, n = 0, 1, 2, \ldots$, except the case $m = n = 0$), which are a combination of multiple reflections (see Fig. 13.2) from both pairs of parallel waveguide walls,

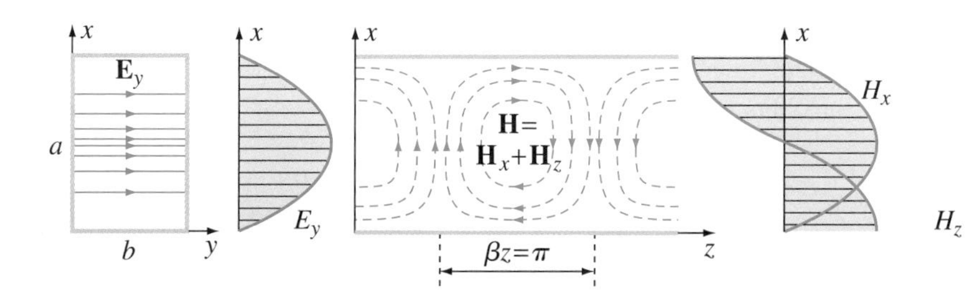

Figure 13.3 Field configurations of the dominant mode (TE_{10}) in a rectangular waveguide (Fig. 13.1).

as well as TM_{mn} (transverse magnetic) modes ($m, n = 1, 2, \ldots$), with $\underline{H}_z = 0$ and $\underline{E}_z \neq 0$. However, even within this extended list of possible propagating modes, the TE_{10} mode, as will be explained, remains to be the lowest and dominant mode. So, the generalization of the modal theory, that follows, to include all possible higher order modes does not affect the relevance and practical importance of the TE_{10} mode and its theory presented in this section. Note that, while the derivation of TE_{m0} (and TE_{10}) fields in this section is done simply taking and "reformatting" (and reinterpreting) the solution to a wave reflection problem in Sections 10.5 and 10.6, the theory of arbitrary modes will start with general Maxwell's and wave equations in the waveguide, and will involve mathematically more formal and complicated field derivations.

Example 13.2 Surface Current and Charge Distributions on Waveguide Walls

A TE_{10} wave of frequency f propagates along a rectangular metallic waveguide of transverse dimensions a and b, and dielectric parameters ε and μ (Fig. 13.1). The field components of the wave are given in Eqs. (13.24)–(13.26). Using boundary conditions, find the distributions of (a) surface currents and (b) surface charges on inner surfaces of waveguide walls. (c) Show that the currents and charges in (a) and (b) are interrelated by the continuity equation for high-frequency surface currents.

Solution

(a) The surface current density vector, $\underline{\mathbf{J}}_s$, on each of the interior walls of the waveguide is computed from the boundary condition for the magnetic field intensity vector, $\underline{\mathbf{H}}$, on the wall (PEC) surfaces as in Eq. (10.13), with the field components $\underline{H}_x$ and $\underline{H}_z$ in the structure being those in Eqs. (13.25) and (13.26). In specific, $\underline{\mathbf{J}}_s$ on the bottom wall in Fig. 13.1, where $y = 0$, amounts to

$$\begin{aligned}(\underline{\mathbf{J}}_s)_{\text{bottom}} &= \hat{\mathbf{n}} \times \underline{\mathbf{H}} = \hat{\mathbf{y}} \times \left(\underline{H}_x \hat{\mathbf{x}} + \underline{H}_z \hat{\mathbf{z}}\right)\big|_{y=0} = -\underline{H}_x\big|_{y=0}\,\hat{\mathbf{z}} + \underline{H}_z\big|_{y=0}\,\hat{\mathbf{x}} \\ &= \underbrace{-\mathrm{j}\beta \frac{a}{\pi}\underline{H}_0 \sin\left(\frac{\pi}{a}x\right)\mathrm{e}^{-\mathrm{j}\beta z}}_{\underline{J}_{sz}}\,\hat{\mathbf{z}} + \underbrace{\underline{H}_0 \cos\left(\frac{\pi}{a}x\right)\mathrm{e}^{-\mathrm{j}\beta z}}_{\underline{J}_{sx}}\,\hat{\mathbf{x}}. \end{aligned} \tag{13.27}$$

Since $\underline{\mathbf{H}}$ on the top wall ($y = b$) is the same as on the bottom of the guide, the associated current density vectors turn out to be opposite to each other,

$$(\underline{\mathbf{J}}_s)_{\text{top}} = (-\hat{\mathbf{y}}) \times \underline{\mathbf{H}}\big|_{y=b} = -\hat{\mathbf{y}} \times \underline{\mathbf{H}}\big|_{y=0} = -(\underline{\mathbf{J}}_s)_{\text{bottom}}. \tag{13.28}$$

The x-component of $\underline{\mathbf{H}}$ is normal to the left wall, and hence, for $x = 0$,

$$(\underline{\mathbf{J}}_s)_{\text{left}} = \hat{\mathbf{x}} \times \left(\underline{H}_z \hat{\mathbf{z}}\right)\big|_{x=0} = \underbrace{-\underline{H}_0\,\mathrm{e}^{-\mathrm{j}\beta z}}_{\underline{J}_{sy}}\,\hat{\mathbf{y}}. \tag{13.29}$$

Finally, on the right wall ($x = a$), $\underline{H}_z$ has the opposite value to that for $x = 0$, which yields the same result for $\underline{\mathbf{J}}_s$ on the two walls,

$$(\underline{\mathbf{J}}_s)_{\text{right}} = (-\hat{\mathbf{x}}) \times \left(\underline{H}_z \hat{\mathbf{z}}\right)\big|_{x=a} = -\underline{H}_0\,\mathrm{e}^{-\mathrm{j}\beta z}\,\hat{\mathbf{y}} = (\underline{\mathbf{J}}_s)_{\text{left}}. \tag{13.30}$$

(b) Similarly, applying the boundary condition for the vector $\underline{\mathbf{D}} = \varepsilon\underline{\mathbf{E}}$ as in Eq. (10.15), while having in mind that the only existing electric field component in Fig. 13.1 is $\underline{E}_y$, Eq. (13.24), we obtain the surface charge density, $\underline{\rho}_s$, on the inner wall surfaces, as follows:

$$\begin{aligned}(\underline{\rho}_s)_{\text{bottom}} &= \hat{\mathbf{n}} \cdot \underline{\mathbf{D}} = \varepsilon\,\hat{\mathbf{y}} \cdot \left(\underline{E}_y \hat{\mathbf{y}}\right)\Big|_{y=0} = -\mathrm{j}\omega\varepsilon\mu \frac{a}{\pi}\underline{H}_0 \sin\left(\frac{\pi}{a}x\right)\mathrm{e}^{-\mathrm{j}\beta z}, \\ (\underline{\rho}_s)_{\text{top}} &= \varepsilon(-\hat{\mathbf{y}}) \cdot \left(\underline{E}_y \hat{\mathbf{y}}\right)\Big|_{y=b} = -(\underline{\rho}_s)_{\text{bottom}}, \quad (\underline{\rho}_s)_{\text{left}} = (\underline{\rho}_s)_{\text{right}} = 0. \end{aligned} \tag{13.31}$$

(c) To verify that the current and charge densities in Eqs. (13.27)–(13.31) satisfy the corresponding continuity equation, Eq. (10.14), we compute the surface divergence of $\underline{\mathbf{J}}_s$ on each of the walls in Fig. 13.1 and show that it equals $-\mathrm{j}\omega$ times $\underline{\rho}_s$ at the same point. For the bottom wall,

$$\nabla_s \cdot (\underline{\mathbf{J}}_s)_{\text{bottom}} = \frac{\partial \underline{J}_{sx}}{\partial x} + \frac{\partial \underline{J}_{sz}}{\partial z} = -\underline{H}_0 \frac{\pi}{a} \sin\left(\frac{\pi}{a}x\right) \mathrm{e}^{-\mathrm{j}\beta z} + (-\mathrm{j}\beta)^2 \frac{a}{\pi} \underline{H}_0 \sin\left(\frac{\pi}{a}x\right) \mathrm{e}^{-\mathrm{j}\beta z}$$

$$= -\frac{a}{\pi} \underline{H}_0 \sin\left(\frac{\pi}{a}x\right) \mathrm{e}^{-\mathrm{j}\beta z} \underbrace{\left(\frac{\pi^2}{a^2} + \beta^2\right)}_{\omega^2 \varepsilon \mu}$$

$$= -\mathrm{j}\omega \left[-\mathrm{j}\omega\varepsilon\mu \frac{a}{\pi} \underline{H}_0 \sin\left(\frac{\pi}{a}x\right) \mathrm{e}^{-\mathrm{j}\beta z}\right] = -\mathrm{j}\omega (\underline{\rho}_s)_{\text{bottom}}, \tag{13.32}$$

since, using Eqs. (13.12), (13.20), (9.18), and (8.48),

$$\beta^2 = \frac{\omega^2}{c^2}\left(1 - \frac{c^2}{4a^2 f^2}\right) = \omega^2 \varepsilon\mu - \frac{\pi^2}{a^2} \quad \left(c = \frac{1}{\sqrt{\varepsilon\mu}}\right). \tag{13.33}$$

The relationship (and its derivation) for the top wall is just that in Eq. (13.32) multiplied by -1. Because $\underline{\mathbf{J}}_s$ on the remaining two walls is a y-directed vector depending only on the coordinate z, its divergence is zero,

$$\nabla_s \cdot (\underline{\mathbf{J}}_s)_{\text{left}} = \frac{\partial \underline{J}_{sy}(z)}{\partial y} = 0 = -\mathrm{j}\omega (\underline{\rho}_s)_{\text{left}}, \tag{13.34}$$

and the same for $(\underline{\mathbf{J}}_s)_{\text{right}}$, i.e., these currents are not associated with any excess surface charge, as found in (b).

Problems: 13.1 and 13.2; *Conceptual Questions* (on Companion Website): 13.7 and 13.8; *MATLAB Exercises* (on Companion Website).

13.4 GENERAL TE MODAL ANALYSIS OF RECTANGULAR WAVEGUIDES

We now redevelop the general field theory of a transmission line with a TEM wave in Fig. 11.1 from Section 11.1 to analyze arbitrary TE waves, namely, a complete set of TE_{mn} wave modes, in a rectangular waveguide, Fig. 13.1. Modal analysis of TM waves follows in the next section. In general, a solution for the fields in the waveguide must satisfy source-free Maxwell's equations for a (perfect) dielectric of parameters ε and μ (waveguide dielectric), Eqs. (11.1)–(11.4), as well as the corresponding Helmholtz equations, Eqs. (9.8) and (9.9), which, of course, are combinations of Maxwell's equations. The solution is also subject to the boundary conditions at the waveguide (PEC) walls, Eqs. (13.1). For a TE wave, we add a longitudinal (axial) component of the magnetic field, $\underline{\mathbf{H}}_z$, to Eqs. (11.8),

$$\boxed{\underline{\mathbf{E}} = \underline{\mathbf{E}}_\mathrm{t} \quad (\underline{\mathbf{E}}_z = 0) \quad \text{and} \quad \underline{\mathbf{H}} = \underline{\mathbf{H}}_\mathrm{t} + \underline{\mathbf{H}}_z.} \tag{13.35}$$

TE *wave*

This is the only existing axial field component in the system, and it is customary and convenient to use it as the pivotal unknown quantity in the solution procedure. In our analysis, we thus first express all other (nonzero) electric and magnetic field components in terms of $\underline{\mathbf{H}}_z$, then solve for it, and finally obtain the complete field picture from the solution for $\underline{\mathbf{H}}_z$.

For this first step, we need the two curl Maxwell's equations, Eqs. (11.1) and (11.2), and actually only the transverse components (projections) of these equations after their decomposition onto transverse and longitudinal components as in Eqs. (11.10) and (11.11). In other words, we need the first relationship in Eqs. (11.10), which remains the same as for the TEM wave, and the first one in Eqs. (11.11), which, in the TE case, acquires an additional term on the left-hand side of the equation, due to a nonzero $\underline{\mathbf{H}}_z$. The TE version of this latter relationship is obtained by a similar transverse vs. longitudinal decomposition as in Eq. (11.9), and we have

$$\underline{\gamma}\,\hat{\mathbf{z}} \times \underline{\mathbf{E}}_{\mathrm{t}} = \mathrm{j}\omega\mu\underline{\mathbf{H}}_{\mathrm{t}}, \quad \underline{\gamma}\,\hat{\mathbf{z}} \times \underline{\mathbf{H}}_{\mathrm{t}} - \nabla_{\mathrm{t}} \times \underline{\mathbf{H}}_z = -\mathrm{j}\omega\varepsilon\underline{\mathbf{E}}_{\mathrm{t}}. \tag{13.36}$$

Note that these relationships are also used in the TEM case (with $\underline{\mathbf{H}}_z = 0$) to compute, in Eq. (11.15), the propagation coefficient ($\underline{\gamma}$) of a transmission line, Eq. (11.16).

Obviously, the solution for waveguide fields we seek will be described in the Cartesian coordinate system in Fig. 13.1, as in Eqs. (13.24)–(13.26), and hence we further decompose the transverse fields, $\underline{\mathbf{E}}_{\mathrm{t}}$ and $\underline{\mathbf{H}}_{\mathrm{t}}$, as well as the transverse del operator, ∇_{t}, onto x- and y-components. The complete Cartesian representation of field vectors and ∇_{t} [see Eq. (11.6)] reads

$$\underline{\mathbf{E}}_{\mathrm{t}} = \underline{E}_x\,\hat{\mathbf{x}} + \underline{E}_y\,\hat{\mathbf{y}}, \quad \underline{\mathbf{H}}_{\mathrm{t}} = \underline{H}_x\,\hat{\mathbf{x}} + \underline{H}_y\,\hat{\mathbf{y}}, \quad \underline{\mathbf{H}}_z = \underline{H}_z\,\hat{\mathbf{z}}, \quad \nabla_{\mathrm{t}} = \frac{\partial}{\partial x}\,\hat{\mathbf{x}} + \frac{\partial}{\partial y}\,\hat{\mathbf{y}}. \tag{13.37}$$

Substituting these expressions in Eqs. (13.36), we take the indicated vector cross products as

$$\hat{\mathbf{z}} \times \underline{\mathbf{E}}_{\mathrm{t}} = \underline{E}_x\,\hat{\mathbf{y}} - \underline{E}_y\,\hat{\mathbf{x}}, \quad \nabla_{\mathrm{t}} \times \underline{\mathbf{H}}_z = -\frac{\partial \underline{H}_z}{\partial x}\,\hat{\mathbf{y}} + \frac{\partial \underline{H}_z}{\partial y}\,\hat{\mathbf{x}}, \tag{13.38}$$

and similarly for $\hat{\mathbf{z}} \times \underline{\mathbf{H}}_{\mathrm{t}}$. Equating then the x- and y-components, respectively, on the two sides of the equations gives

$$-\underline{\gamma}\underline{E}_y = \mathrm{j}\omega\mu\underline{H}_x, \quad \underline{\gamma}\underline{E}_x = \mathrm{j}\omega\mu\underline{H}_y, \quad \frac{\partial \underline{H}_z}{\partial y} + \underline{\gamma}\underline{H}_y = \mathrm{j}\omega\varepsilon\underline{E}_x,$$

$$\frac{\partial \underline{H}_z}{\partial x} + \underline{\gamma}\underline{H}_x = -\mathrm{j}\omega\varepsilon\underline{E}_y. \tag{13.39}$$

These four equations can easily be solved for $\underline{E}_x$, $\underline{E}_y$, $\underline{H}_x$, and $\underline{H}_y$ – in terms of $\underline{H}_z$, i.e., its derivatives with respect to x and y. Namely, combining the second and third equations on one side, and the first and fourth one on the other, we can eliminate $\underline{H}_y$ and $\underline{H}_x$, respectively, and solve for $\underline{E}_x$ and $\underline{E}_y$. Substituting these solutions back in the last two equations, we then solve for $\underline{H}_y$ and $\underline{H}_x$. Introducing a new constant k as

$$k^2 = \underline{\gamma}^2 + \omega^2\varepsilon\mu = -\beta^2 + \omega^2\varepsilon\mu \quad (\underline{\gamma} = \mathrm{j}\beta\,,\ \ \beta = \beta_z), \tag{13.40}$$

where $\beta = \beta_z$ is the phase coefficient of the waveguide ($\underline{\gamma}$ is purely imaginary, as the waveguide is assumed to exhibit no losses), we have

$$\underline{E}_x = -\frac{\mathrm{j}\omega\mu}{k^2}\frac{\partial \underline{H}_z}{\partial y}, \quad \underline{E}_y = \frac{\mathrm{j}\omega\mu}{k^2}\frac{\partial \underline{H}_z}{\partial x}, \quad \underline{H}_x = -\frac{\underline{\gamma}}{k^2}\frac{\partial \underline{H}_z}{\partial x}, \quad \underline{H}_y = -\frac{\underline{\gamma}}{k^2}\frac{\partial \underline{H}_z}{\partial y}. \tag{13.41}$$

In the next step, we seek a solution for the pivotal unknown, $\underline{H}_z$. To this end, we invoke the magnetic-field Helmholtz equation, Eq. (9.9),

$$\nabla^2\underline{\mathbf{H}} + \omega^2\varepsilon\mu\underline{\mathbf{H}} = 0, \tag{13.42}$$

which can be decoupled onto three scalar Helmholtz equations, for each of the components of $\underline{\mathbf{H}}$, as in Eqs. (9.10). For the z-component (here, we are not interested in the other two scalar equations),

$$\nabla^2 \underline{H}_z + \omega^2 \varepsilon \mu \underline{H}_z = 0. \tag{13.43}$$

From Eq. (11.6), the Laplacian for waves propagating along the z-axis is decomposed onto its transverse and axial components as

$$\nabla^2 = \nabla \cdot \nabla = (\nabla_{\mathrm{t}} - \underline{\gamma}\, \hat{\mathbf{z}}) \cdot (\nabla_{\mathrm{t}} - \underline{\gamma}\, \hat{\mathbf{z}}) = \nabla_{\mathrm{t}}^2 + \underline{\gamma}^2. \tag{13.44}$$

With this, and having in mind Eq. (13.40), Eq. (13.43) becomes

$$\nabla_{\mathrm{t}}^2 \underline{H}_z + k^2 \underline{H}_z = 0. \tag{13.45}$$

Writing k^2 as a sum of two other positive constants,

$$k^2 = \beta_x^2 + \beta_y^2, \tag{13.46}$$

and using Eq. (2.94), we are finally left with the following version of the scalar Helmholtz equation for the axial field $\underline{H}_z$, suitable for solution:

$$\frac{\partial^2 \underline{H}_z}{\partial x^2} + \frac{\partial^2 \underline{H}_z}{\partial y^2} + (\beta_x^2 + \beta_y^2)\underline{H}_z = 0. \tag{13.47}$$

This is a homogeneous (source-free) second-order differential equation in two variables (coordinates x and y), with $\underline{H}_z$ as unknown. However, it can be considered as a sum of two corresponding homogeneous second-order differential equations in a single variable (x or y), each one of the form

$$\frac{\mathrm{d}^2 f}{\mathrm{d}x^2} + A^2 f = 0 \quad \longrightarrow \quad f \sim \cos Ax, \ \sin Ax, \tag{13.48}$$

with cosine and sine of Ax, times a constant, as indicated, being their general solutions.[5] Since, moreover, Eq. (13.47) does not contain any derivative with respect to z, the propagation factor $\mathrm{e}^{-\gamma z}$ can readily be included, as a multiplicative constant (with respect to x and y), in the solutions [note that this dependence of the fields on the z-coordinate is already stipulated in the solution procedure through Eqs. (11.6) and (13.44)].[6] Consequently, the general solution for $\underline{H}_z$ in our problem is

$$\underline{H}_z = (\underline{A} \cos \beta_x x + \underline{B} \sin \beta_x x)(\underline{C} \cos \beta_y y + \underline{D} \sin \beta_y y)\, \mathrm{e}^{-\underline{\gamma} z}, \tag{13.49}$$

which can be easily verified by direct substitution in Eq. (13.47).[7]

[5] Of course, the exponential functions $\mathrm{e}^{\mathrm{j}Ax}$ and $\mathrm{e}^{-\mathrm{j}Ax}$ are also general solutions of Eq. (13.48), like the plane-wave expressions in Eqs. (9.36) satisfy the one-dimensional Helmholtz equations in Eqs. (9.37). However, from Eqs. (8.61) and (10.7), each of the two sets of solutions is a linear combination of the other. In other words, both traveling and standing waves are general solutions of the governing Helmholtz equations, and the final form of the field (wave) expressions (traveling vs. standing) depends on the relevant boundary conditions for the problem.

[6] In fact, we are solving a two-dimensional Helmholtz equation, in variables x and y only, with an already prescribed dependence on z in the solution.

[7] Note that the portion of the solution for $\underline{H}_z(x, y, z)$ in Eq. (13.49) that depends on transverse coordinates (x and y) only can be written as

$$\underline{H}_z(x, y, 0) = f(x) g(y),$$

i.e., as a product of functions that each depend on a single variable (x or y). This expression, in which the variables are formally separated between two independent functions (f and g), can actually be used as a

The constants $\underline{B}$, $\underline{D}$, β_x, and β_y in Eq. (13.49) are determined from the boundary conditions at waveguide walls, Eqs. (13.1). From Eqs. (13.41) and (13.49), the components of the electric field vector in the dielectric are

$$\underline{E}_x = -\frac{j\omega\mu\beta_y}{k^2}(\underline{A}\cos\beta_x x + \underline{B}\sin\beta_x x)(-\underline{C}\sin\beta_y y + \underline{D}\cos\beta_y y)\,e^{-\underline{\gamma}z}, \quad (13.50)$$

$$\underline{E}_y = \frac{j\omega\mu\beta_x}{k^2}(-\underline{A}\sin\beta_x x + \underline{B}\cos\beta_x x)(\underline{C}\cos\beta_y y + \underline{D}\sin\beta_y y)\,e^{-\underline{\gamma}z} \quad (13.51)$$

($\underline{E}_z = 0$). The condition $\mathbf{E}_{\text{tang}} = 0$ at the left wall in Fig. 13.1 (where $x = 0$, $0 \le y \le b$, and $-\infty < z < \infty$) and bottom wall ($y = 0$, $0 \le x \le a$, $-\infty < z < \infty$) gives

$$\underline{E}_y\Big|_{x=0} = 0 \quad\longrightarrow\quad \underline{B} = 0, \quad \underline{E}_x\Big|_{y=0} = 0 \quad\longrightarrow\quad \underline{D} = 0. \quad (13.52)$$

With $\underline{B} = \underline{D} = 0$ in Eqs. (13.50) and (13.51), the same condition at the other two walls results, as in Eq. (13.6), in

$$\underline{E}_y\Big|_{x=a} = 0 \quad\longrightarrow\quad \beta_x = \frac{m\pi}{a}, \quad \underline{E}_x\Big|_{y=b} = 0 \quad\longrightarrow\quad \beta_y = \frac{n\pi}{b} \quad (13.53)$$

($m, n = 0, 1, 2, \ldots$). Note that, from Eqs. (13.39), $\underline{E}_y$ is proportional to $\underline{H}_x$ and $\underline{E}_x$ to $\underline{H}_y$, which means that the boundary condition for $\underline{E}_y$ at $x = 0$ automatically sets the condition for $\underline{H}_x$, so for the normal component of $\underline{\mathbf{H}}$, at the same wall, and analogously for the other three conditions in Eqs. (13.52) and (13.53). In other words, by Eqs. (13.52) and (13.53) the condition $\underline{\mathbf{H}}_{\text{norm}} = 0$ is imposed as well at all waveguide walls, yielding the same values of the four determined transverse constants ($\underline{B}$, $\underline{D}$, β_x, and β_y) in Eq. (13.49). Merging the remaining two transverse constants into one, $\underline{H}_0 = \underline{A}\underline{C}$, the solution for $\underline{H}_z$ is

$\underline{H}_z$ – TE$_{mn}$ mode

$$\underline{H}_z = \underline{H}_0 \cos\left(\frac{m\pi}{a}x\right)\cos\left(\frac{n\pi}{b}y\right)e^{-j\beta z}, \quad (13.54)$$

where $\underline{H}_0$, as in Eq. (13.26), is the spatial complex "peak-value" (for $x = y = 0$) of $\underline{H}_z$ in the waveguide cross section defined by $z = 0$, used to set the level of field strengths at every point in the waveguide, and consequently of the transmitted power along the waveguide, as determined by the waveguide excitation (field sources). Using Eqs. (13.41), the other field components are

$\underline{E}_x$ – TE$_{mn}$ mode

$$\underline{E}_x = \frac{j\omega\mu}{k^2}\frac{n\pi}{b}\underline{H}_0\cos\left(\frac{m\pi}{a}x\right)\sin\left(\frac{n\pi}{b}y\right)e^{-j\beta z}, \quad (13.55)$$

$\underline{E}_y$ – TE$_{mn}$ mode

$$\underline{E}_y = -\frac{j\omega\mu}{k^2}\frac{m\pi}{a}\underline{H}_0\sin\left(\frac{m\pi}{a}x\right)\cos\left(\frac{n\pi}{b}y\right)e^{-j\beta z}, \quad (13.56)$$

$\underline{H}_x$ – TE$_{mn}$ mode

$$\underline{H}_x = \frac{j\beta}{k^2}\frac{m\pi}{a}\underline{H}_0\sin\left(\frac{m\pi}{a}x\right)\cos\left(\frac{n\pi}{b}y\right)e^{-j\beta z}, \quad (13.57)$$

$\underline{H}_y$ – TE$_{mn}$ mode

$$\underline{H}_y = \frac{j\beta}{k^2}\frac{n\pi}{b}\underline{H}_0\cos\left(\frac{m\pi}{a}x\right)\sin\left(\frac{n\pi}{b}y\right)e^{-j\beta z}, \quad (13.58)$$

---

starting point of a more general and formal procedure to solve the Helmholtz equation in our problem. Namely, its substitution in Eq. (13.47), assuming $z = 0$, and then division (of both sides) of thus obtained equation by fg, lead to the explicit decomposition of the Helmholtz equation onto two single-variable equations of the type in Eq. (13.48), which solved (separately) for f and g, give, of course, the same result as in Eq. (13.49). This constitutes the so-called method of separation of variables, which is a general analytical technique for solving wave and Helmholtz equations, and similar partial differential equations in multiple variables.

and $\underline{E}_z = 0$, where $m, n = 0, 1, 2, \ldots$, with the restriction that only one of the mode indices can be zero [the possibility of $m = n = 0$ is eliminated, because it would imply that all transverse field components, in Eqs. (13.55)–(13.58), are zero]. Eqs. (13.54)–(13.58) represent the field of a TE_{mn} mode in the waveguide (obviously, there are a double infinite number of such modes). The waveguide phase coefficient, β, and parameter k^2 in these expressions are also functions of m and n. Namely, Eqs. (13.46) and (13.53) give

$$k^2 = \beta_x^2 + \beta_y^2 = \left(\frac{m\pi}{a}\right)^2 + \left(\frac{n\pi}{b}\right)^2, \tag{13.59}$$

and β is then computed from Eq. (13.40). Its dependence on m and n will be discussed in a separate section. Field configurations for selected modes are illustrated in Fig. 13.4. Of course, the corresponding field expressions for the dominant mode, TE_{10}, in Eqs. (13.24)–(13.26) are a special case of those in Eqs. (13.54)–(13.58), with

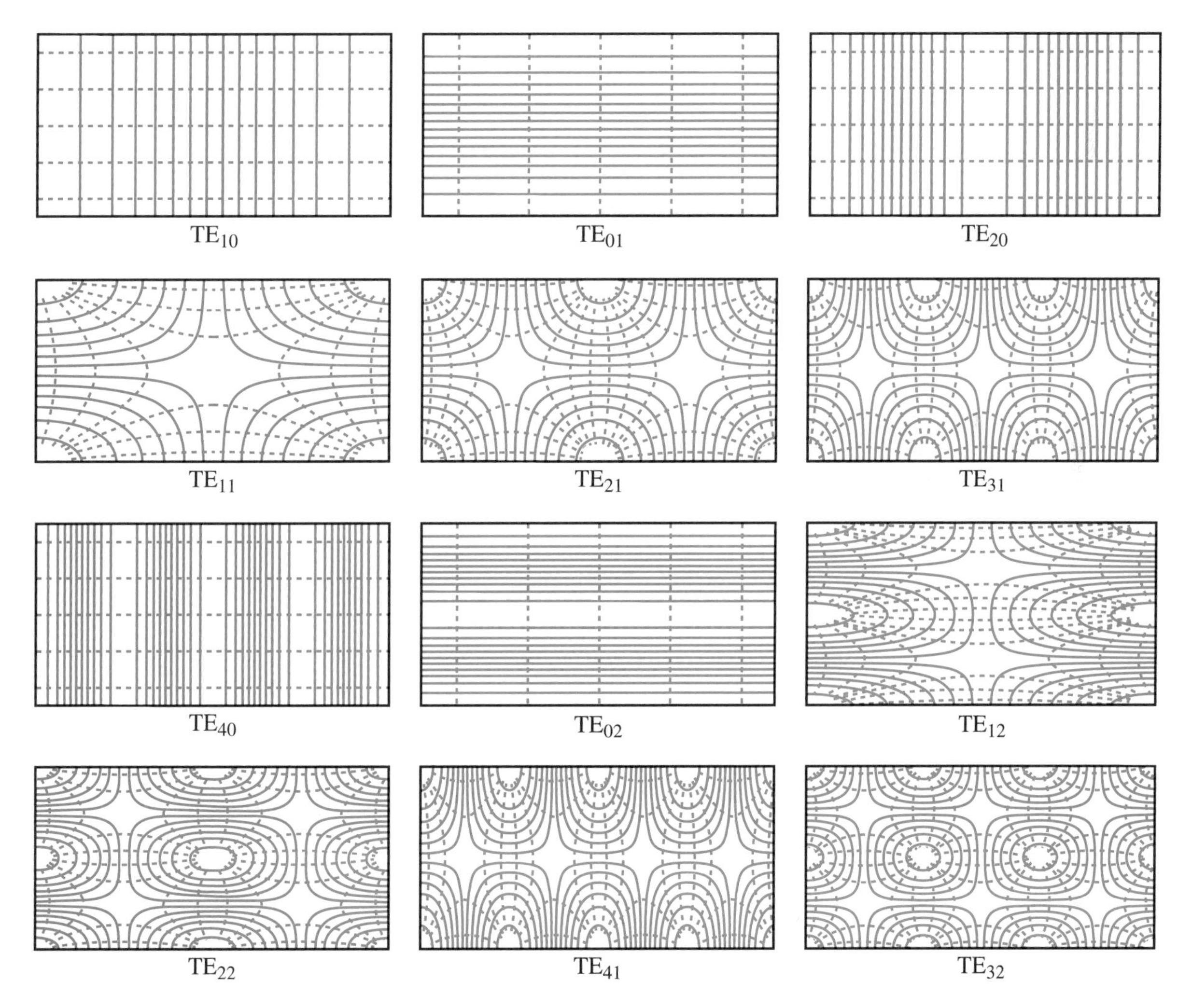

Figure 13.4 Selected TE_{mn} modal field distributions in a cross section of a rectangular waveguide (Fig. 13.1) with $a = 2b$ (electric field – solid line, magnetic field – dashed line).

$m = 1$ and $n = 0$; TE_{m0} $(m = 1, 2, \ldots)$ modes in Eq. (13.7) are a subset of TE_{mn} modes as well, for $n = 0$.

Problems: 13.3 and 13.4; *Conceptual Questions* (on Companion Website): 13.9; *MATLAB Exercises* (on Companion Website).

13.5 TM MODES IN A RECTANGULAR WAVEGUIDE

For a TM wave in a rectangular waveguide (Fig. 13.1), Eqs. (13.35) become

TM wave

$$\boxed{\underline{\mathbf{E}} = \underline{\mathbf{E}}_{\mathrm{t}} + \underline{\mathbf{E}}_z \quad \text{and} \quad \underline{\mathbf{H}} = \underline{\mathbf{H}}_{\mathrm{t}} \quad (\underline{\mathbf{H}}_z = 0),} \tag{13.60}$$

and the pivotal unknown quantity is now $\underline{\mathbf{E}}_z$. In place of Eqs. (13.36), we have

$$\underline{\gamma}\, \hat{\mathbf{z}} \times \underline{\mathbf{E}}_{\mathrm{t}} - \nabla_{\mathrm{t}} \times \underline{\mathbf{E}}_z = \mathrm{j}\omega\mu \underline{\mathbf{H}}_{\mathrm{t}}, \quad \underline{\gamma}\, \hat{\mathbf{z}} \times \underline{\mathbf{H}}_{\mathrm{t}} = -\mathrm{j}\omega\varepsilon \underline{\mathbf{E}}_{\mathrm{t}}, \tag{13.61}$$

and with analogous derivations as in Eqs. (13.37)–(13.41), the expressions for the other nonzero field components in terms of $\underline{E}_z$ are found to be

$$\underline{E}_x = -\frac{\underline{\gamma}}{k^2} \frac{\partial \underline{E}_z}{\partial x}, \quad \underline{E}_y = -\frac{\underline{\gamma}}{k^2} \frac{\partial \underline{E}_z}{\partial y}, \quad \underline{H}_x = \frac{\mathrm{j}\omega\varepsilon}{k^2} \frac{\partial \underline{E}_z}{\partial y},$$

$$\underline{H}_y = -\frac{\mathrm{j}\omega\varepsilon}{k^2} \frac{\partial \underline{E}_z}{\partial x}, \quad \underline{\gamma} = \mathrm{j}\beta, \quad k^2 = -\beta^2 + \omega^2 \varepsilon\mu. \tag{13.62}$$

Following then the procedure in Eqs. (13.42)–(13.49), accommodated to the TM case, the general solution of the transverse scalar Helmholtz equation for $\underline{E}_z$,

$$\nabla_{\mathrm{t}}^2 \underline{E}_z + k^2 \underline{E}_z = 0, \quad k^2 = \beta_x^2 + \beta_y^2, \tag{13.63}$$

is given by

$$\underline{E}_z = (\underline{A}' \cos \beta_x x + \underline{B}' \sin \beta_x x)(\underline{C}' \cos \beta_y y + \underline{D}' \sin \beta_y y)\, \mathrm{e}^{-\underline{\gamma} z}. \tag{13.64}$$

Similarly to Eqs. (13.52), from the boundary condition $\underline{E}_z = 0$ for $x = 0$ and $y = 0$, we obtain $\underline{A}' = 0$ and $\underline{C}' = 0$, respectively. The same condition ($\underline{E}_z = 0$) at $x = a$ and $y = b$ then gives the same solutions for β_x and β_y as in Eqs. (13.53). With Eqs. (13.62) and (13.63), this means that the waveguide phase coefficient, β, is the same for TM and TE waves, for the same mode number (m, n). In the next section, we shall discuss β and cutoff frequency of arbitrary TE and TM modes. Introducing $\underline{E}_0 = \underline{B}'\underline{D}'$, we finally have

$\underline{E}_z$ – TM_{mn} mode

$$\boxed{\underline{E}_z = \underline{E}_0 \sin\left(\frac{m\pi}{a} x\right) \sin\left(\frac{n\pi}{b} y\right) \mathrm{e}^{-\mathrm{j}\beta z},} \tag{13.65}$$

so that Eqs. (13.62) result in

$\underline{E}_x$ – TM_{mn} mode

$$\underline{E}_x = -\frac{\mathrm{j}\beta}{k^2} \frac{m\pi}{a} \underline{E}_0 \cos\left(\frac{m\pi}{a} x\right) \sin\left(\frac{n\pi}{b} y\right) \mathrm{e}^{-\mathrm{j}\beta z}, \tag{13.66}$$

$\underline{E}_y$ – TM_{mn} mode

$$\underline{E}_y = -\frac{\mathrm{j}\beta}{k^2} \frac{n\pi}{b} \underline{E}_0 \sin\left(\frac{m\pi}{a} x\right) \cos\left(\frac{n\pi}{b} y\right) \mathrm{e}^{-\mathrm{j}\beta z}, \tag{13.67}$$

$\underline{H}_x$ – TM_{mn} mode

$$\underline{H}_x = \frac{\mathrm{j}\omega\varepsilon}{k^2} \frac{n\pi}{b} \underline{E}_0 \sin\left(\frac{m\pi}{a} x\right) \cos\left(\frac{n\pi}{b} y\right) \mathrm{e}^{-\mathrm{j}\beta z}, \tag{13.68}$$

$\underline{H}_y$ – TM_{mn} mode

$$\underline{H}_y = -\frac{\mathrm{j}\omega\varepsilon}{k^2} \frac{m\pi}{a} \underline{E}_0 \cos\left(\frac{m\pi}{a} x\right) \sin\left(\frac{n\pi}{b} y\right) \mathrm{e}^{-\mathrm{j}\beta z} \tag{13.69}$$

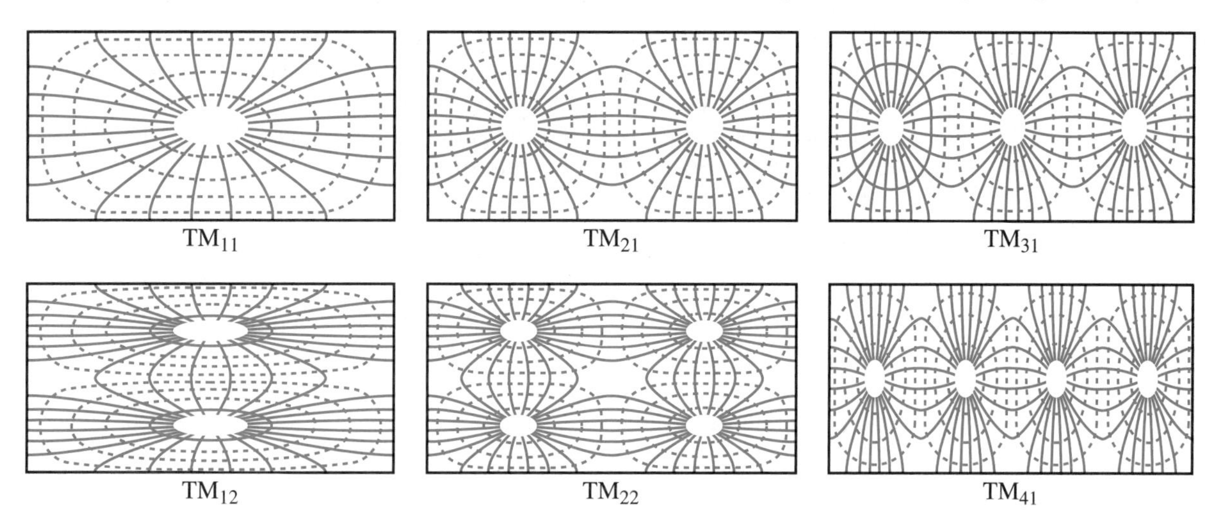

Figure 13.5 Field patterns in the xy-plane of selected rectangular ($a = 2b$) waveguide TM_{mn} modes (electric field – solid line, magnetic field – dashed line).

($\underline{H}_z = 0$). This is a TM_{mn} mode in the waveguide ($m, n = 1, 2, \ldots$). The field expressions are analogous to those in Eqs. (13.54)–(13.58), for the TE case. The lowest TM mode is TM_{11}, since the possibilities of either $m = 0$ or $n = 0$ would make all field components in Eqs. (13.65)–(13.69) zero, and are thus eliminated. Shown in Fig. 13.5 are selected TM_{mn} field configurations.

Problems: 13.5; *MATLAB Exercises* (on Companion Website).

13.6 CUTOFF FREQUENCIES OF ARBITRARY WAVEGUIDE MODES

As shown in the previous two sections, the phase coefficient β of a rectangular waveguide, Fig. 13.1, is a function of mode indices m and n, but does not depend on the type of the propagating wave, i.e., whether it is a TE or TM wave. From Eq. (13.40) or (13.62), $\beta^2 = \omega^2/c^2 - k^2$, and hence

$$\beta = \frac{\omega}{c}\sqrt{1 - \frac{1}{f^2}\frac{c^2k^2}{4\pi^2}}, \qquad c = \frac{1}{\sqrt{\varepsilon\mu}}, \tag{13.70}$$

where the modal expression for k^2 is given by Eq. (13.59) in both TE and TM cases. We see that β can be expressed in the same way as in Eq. (13.12), with the cutoff frequency, of the TE_{mn} or TM_{mn} mode, computed as

$$f_c = (f_c)_{mn} = \frac{c}{2}\sqrt{\left(\frac{m}{a}\right)^2 + \left(\frac{n}{b}\right)^2} \qquad (m, n = 0, 1, 2, \ldots). \tag{13.71}$$

cutoff frequency, TE_{mn} or TM_{mn} mode

Here, as explained in the preceding two sections, either m or n, but not both, may be zero for TE waves, whereas none can be zero for TM waves. As for the field expressions of a TE_{mn} mode in Eqs. (13.54)–(13.58), this expression for $(f_c)_{mn}$ is the general form, for a rectangular waveguide, of the corresponding expressions for $(f_c)_{m0}$ ($n = 0$) in Eq. (13.13) and $(f_c)_{10}$ ($m = 1$ and $n = 0$) in Eq. (13.20), obtained in the analysis of TE_{m0} and TE_{10} modes.

Having now the general expression for the cutoff frequency of an arbitrary mode, numbered (m, n), let us revisit the concept and relative size of the dominant frequency range defined as the range of frequencies f of a propagating wave in a waveguide in which only one propagating mode is possible. Let the relative size of the range be specified as the ratio of its upper and lower limits, f_2/f_1. The dominant range for TE_{m0} modes (for $n = 0$) is given in Eq. (13.22), where $f_2/f_1 = 2$ [often, we write $f_2 : f_1 = 2 : 1$, and call this a 2 : 1 (two-to-one) frequency range[8]]. We note here that it is customary to always denote the transverse dimensions of the waveguide in Fig. 13.1 such that $a \geq b$, where a and b are, respectively, the extents of the waveguide cross section in the x and y directions (as in the figure). Simply, in setting up the Cartesian coordinate system for the analysis, we adopt the x-axis along the larger dimension of the guide. The case $a = b$ (square waveguide) has very little practical significance, since it yields $(f_c)_{01} = (f_c)_{10}$, from Eq. (13.71), so that the dominant range essentially does not exist ($f_2 = f_1$). Therefore, we assume that $a > b$. With this, we have, in Eq. (13.71), that the TE_{10} mode has the lowest cutoff frequency among all values of m and n, and can, in general, be considered as the dominant mode in a rectangular waveguide. So, $f_1 = (f_c)_{10}$, as for TE_{m0} modes, also in the general (m, n) case. Eq. (13.71) then tells us that the next higher cutoff frequency, with respect to f_1, is one of the following two frequencies:

$$(f_c)_{20} = \frac{c}{a} \quad (\mathrm{TE}_{20} \text{ mode}), \qquad (f_c)_{01} = \frac{c}{2b} \quad (\mathrm{TE}_{01} \text{ mode}), \tag{13.72}$$

and which one defines the upper limit of the dominant range, f_2, depends on the ratio of the waveguide dimensions a/b (so-called waveguide aspect ratio). Normally, we would like to have an aspect ratio that maximizes f_2/f_1 (why multimode propagation along a waveguide is not desirable, in general, and the single-mode operation preferred is explained in Section 13.3). Since $(f_c)_{20}/f_1 = 2$, as in Eq. (13.22), regardless of b, the maximum possible f_2/f_1 is 2. On the other hand, the ratio $(f_c)_{01}/f_1$ does depend on b, and, to keep $f_2/f_1 = 2$, we seek an a/b that places $(f_c)_{01}$ at or above $(f_c)_{20}$. Using the values from Eqs. (13.72), we find that

$$f_2/f_1 = 2 \quad \longrightarrow \quad (f_c)_{01} \geq (f_c)_{20} \quad \longrightarrow \quad a/b \geq 2. \tag{13.73}$$

Thus, for $a \geq 2b$, the relative size of the dominant range is maximal (two), and the range is given by Eq. (13.22). In practice, the cross-sectional dimensions of rectangular waveguides are usually chosen to (approximately) be $a = 2b$ [the smallest aspect ratio in Eq. (13.73)],[9] which gives $(f_c)_{01} = (f_c)_{20}$. A waveguide with a 2 : 1 aspect ratio, or nearly so,

standard waveguide

$$\boxed{a = 2b \quad \text{or} \quad a \approx 2b,} \tag{13.74}$$

is often referred to as a standard waveguide (see Fig. 13.6 in the following example for a graphical representation along the frequency axis of the first several cutoff frequencies of TE and TM modes in a standard waveguide).

[8]Note that a frequency range from f_1 to $f_2 = 2f_1$ is also referred to as an octave; in music, eight ("octo") notes starting at an acoustic (sound) frequency f_1 and ending at $2f_1$ form an octave.

[9]As we shall see in later sections, too large aspect ratios a/b (above $a/b = 2$) for a rectangular waveguide, although in agreement with Eq. (13.73), are impractical due to an increased attenuation along the waveguide and reduced power transmission capacity of the structure.

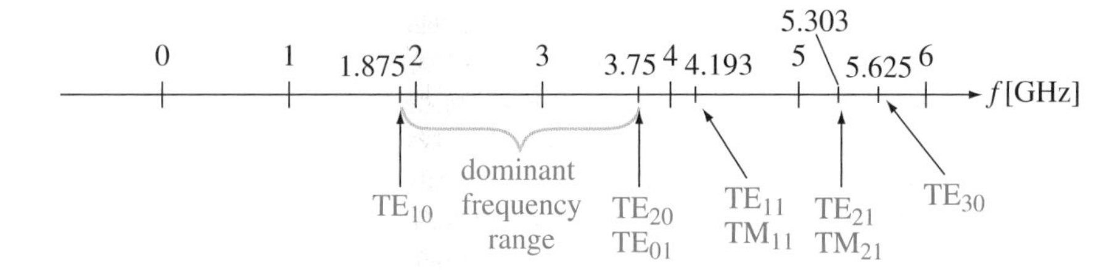

Figure 13.6 First several cutoff frequencies of TE_{mn} and TM_{mn} modes, Eq. (13.71), in an air-filled standard waveguide with $a = 8$ cm and $b = 4$ cm; for Example 13.3.

Example 13.3 First Several Modal Cutoff Frequencies in a Standard Waveguide

Consider a standard waveguide with transverse dimensions $a = 8$ cm and $b = 4$ cm, and air dielectric. (a) What wave modes can propagate along this waveguide at a frequency of $f = 4.5$ GHz? (b) What is the dominant frequency range of the waveguide? (c) Repeat (a) and (b) for $a = 4$ cm, $b = 2$ cm, and $f = 9$ GHz.

Solution

(a) Using Eqs. (13.20), (13.72), and (13.71), first several modal cutoff frequencies for the waveguide appear to be

$$(f_c)_{10} = \frac{c_0}{2a} = 1.875 \text{ GHz}, \quad (f_c)_{20} = (f_c)_{01} = \frac{c_0}{a} = \frac{c_0}{2b} = 3.75 \text{ GHz},$$

$$(f_c)_{11} = \frac{c_0}{2}\sqrt{\frac{1}{a^2} + \frac{1}{b^2}} = 4.193 \text{ GHz}, \quad (f_c)_{21} = \frac{c_0}{2}\sqrt{\frac{4}{a^2} + \frac{1}{b^2}} = 5.303 \text{ GHz}, \quad (13.75)$$

where c_0 is the free-space (the guide dielectric is air) wave velocity, Eq. (9.19), and, at the given operating frequency (f) of the structure, the propagating condition in Eq. (13.15) is satisfied for the first four of them, as shown in Fig. 13.6. Consequently, the list of possible propagating modes in the waveguide is as follows: TE_{10}, TE_{20}, TE_{01}, TE_{11}, and TM_{11}.

(b) From Fig. 13.6, the dominant frequency range, in which only one wave mode (the dominant mode, TE_{10}) can propagate, is given by $f_1 = 1.875 \text{ GHz} < f \leq f_2 = 3.75$ GHz. Of course, $f_2/f_1 = 2$ (for standard waveguides), as in Eq. (13.73), and this also complies with the dominant range specification in Eq. (13.22).

(c) If we proportionally decrease the waveguide dimensions (a and b) and increase the operating frequency (f) by the same factor, k ($k = 2$ in this case), the relative distribution of the cutoff frequencies in Eqs. (13.75) with respect to the new frequency f will remain the same as in Fig. 13.6, just with different (twice as large) absolute values of individual frequencies. This is as well obvious from the general expression for $(f_c)_{mn}$ in Eq. (13.71), where the substitution of a and b by a/k and b/k leads to a change of $(f_c)_{mn}$ to $k(f_c)_{mn}$. Finally, the same conclusion can be reached by having in mind Eqs. (12.73), from which we realize that the waveguides in (a) and (c) are electrically of the same size, i.e., have equal electrical dimensions a/λ_0 and b/λ_0, $\lambda_0 = c_0/f$ being the wavelength, Eq. (8.112) or (9.67), at the frequency f in the guide dielectric (air in our case), and hence the fields they support must be of equal forms (the same modes). So, the list of modes that would propagate if excited is the same as in (a).

The lower and upper frequency limits of the dominant frequency range are both doubled as compared to the case (b), namely, $f_1 = 3.75$ GHz and $f_2 = 7.5$ GHz. We note that, although the width of this range (bandwidth), $\Delta f = f_2 - f_1 = 3.75$ GHz, is twice that in (b), the bound frequency ratio f_2/f_1 remains the same (equal to 2), and, in this respect, both waveguides, in (b) and (c), can be said to have a 2 : 1 bandwidth.

Problems: 13.6–13.9; *Conceptual Questions* (on Companion Website): 13.10–13.13; *MATLAB Exercises* (on Companion Website).

13.7 WAVE IMPEDANCES OF TE AND TM WAVES

The first relationship in Eqs. (13.36) and the second one in Eqs. (13.61) tell us that the transverse electric and magnetic field vectors, $\underline{\mathbf{E}}_t$ and $\underline{\mathbf{H}}_t$, of a TE and TM wave, respectively, in a rectangular waveguide (Fig. 13.1) are perpendicular to each other, as in Eq. (11.14). Moreover, these relationships indicate that the ratio of the corresponding transverse complex field intensities, $\underline{E}_t$ and $\underline{H}_t$, for each of the wave types comes out to be independent of the coordinates in Fig. 13.1, and purely real, so a real constant. The ratio $\underline{E}_t/\underline{H}_t$, in turn, defines the wave impedance of a TE or TM wave, analogously to the TEM case (for a transmission line), in Eq. (11.19). Specifically, using Eqs. (13.36), (13.40), and (13.12), the TE wave impedance, Z_{TE}, equals

TE *wave impedance, arbitrary mode*

$$Z_{TE} = \left(\frac{\underline{E}_t}{\underline{H}_t}\right)_{TE} = \frac{j\omega\mu}{\underline{\gamma}} = \frac{\omega\mu}{\beta} = \frac{\eta}{\sqrt{1 - f_c^2/f^2}}, \tag{13.76}$$

since $\mu c = \eta$, where $c = 1/\sqrt{\varepsilon\mu}$ and $\eta = \sqrt{\mu/\varepsilon}$ are the intrinsic phase velocity and impedance, respectively, of the waveguide dielectric. Of course, the cutoff frequency of the waveguide, f_c, for a mode (m, n) is given in Eq. (13.71), and Z_{TE}, for a TE_{mn} mode, depends on mode indices m and n. Having in mind the expression for $\beta = \beta_z$ in Eqs. (13.4), we realize that the expression for the wave impedance $\underline{\eta}_w$ of a TE wave in Eq. (10.121), obtained from the analysis of multiple reflections in Fig. 10.15, reduces to $\omega\mu/\beta$, exactly as in Eq. (13.76). Similarly, Eqs. (13.61), (13.40), and (13.12) give the following expressions for the wave impedance of a TM_{mn} wave:

TM *wave impedance*

$$Z_{TM} = \left(\frac{\underline{E}_t}{\underline{H}_t}\right)_{TM} = \frac{\underline{\gamma}}{j\omega\varepsilon} = \frac{\beta}{\omega\varepsilon} = \eta\sqrt{1 - \frac{f_c^2}{f^2}}. \tag{13.77}$$

Unlike the TEM case, both TE and TM impedances[10] are functions of the frequency (f) of the propagating wave, as illustrated in Fig. 13.7.

In the entire propagating frequency region (for $f > f_c$), $Z_{TE} > Z_{TEM}$ and $Z_{TM} < Z_{TEM}$, and their product (for the same mode),

$$Z_{TE}Z_{TM} = Z_{TEM}^2, \tag{13.78}$$

is a constant, equal to Z_{TEM} or η squared, Z_{TEM} being the wave impedance of a TEM wave in a transmission line (with the same dielectric), Eq. (11.19). Note that Z_{TE} exhibits a frequency dependence of the same form as the intrinsic impedance of the plasma medium, Eq. (9.161). Actually, at propagating frequencies much farther away from the cutoff (for $f \gg f_c$), both Z_{TE} and Z_{TM} approach η asymptotically, and the TE or TM wave in the waveguide propagates much like a TEM wave (in free space). For TE_{m0} modes, this is also evident from Eq. (13.9), which tells us that the higher the frequency (of a TE_{m0} wave) the larger θ_i in Fig. 10.15, yielding $\theta_i = 90°$ for $f \gg f_c$ [$f_c = mc/(2a)$ for the TE_{m0} mode, Eq. (13.13)]. This in turn means practically no reflections in the waveguide, with an incident wave only, in Fig. 10.13(a),

[10]Note that, as for the TEM wave impedance in Eq. (11.19), and analogously to Eq. (10.11) for unbounded uniform plane waves, the expressions for the TE and TM wave impedances in Eqs. (13.76) and (13.77) are given only for a single traveling TE or TM wave in the waveguide. Namely, as we shall see in a later section, if there is also a backward (reflected) wave, propagating in the negative z direction in Fig. 13.1, the ratio $\underline{E}_t/\underline{H}_t$ for the resultant (forward plus backward) TE or TM wave is a function of z.

propagating parallel to the walls in the axial direction. Equivalently, $\theta_i = 90°$ yields $\underline{\eta}_w = \eta$ in Eq. (10.121).

On the other side, at frequencies below the cutoff (for $f < f_c$), both Z_{TE} and Z_{TM}, in Eqs. (13.76) and (13.77), become [see also Eq. (13.16)] purely imaginary (reactive). This is in agreement with the fact that there is no propagation of electromagnetic waves, and no net real power flow (by the waves) along the waveguide at evanescent frequencies, Eq. (13.17), which will be elaborated in the next section.

From Eqs. (13.36), (13.76), (13.61), (13.77), (11.18), and (11.19), the following vector relations between the transverse electric and magnetic field vectors of all three types of guided electromagnetic waves (TE, TM, and TEM) can be written:

$$\boxed{\underline{\mathbf{H}}_t = \frac{1}{Z}\,\hat{\mathbf{z}} \times \underline{\mathbf{E}}_t, \quad \underline{\mathbf{E}}_t = Z\underline{\mathbf{H}}_t \times \hat{\mathbf{z}} \quad (Z = Z_{TE}\,,\ Z_{TM}\,,\ \text{or}\ Z_{TEM}),} \tag{13.79}$$

general vector relations for transverse fields

which are the same in form as those in Eqs. (9.22) for free (unbounded) uniform plane waves as well. Alternatively, using the x- and y-components of the field vectors, we can write [see Eq. (9.191)]

$$\frac{\underline{E}_x}{\underline{H}_y} = -\frac{\underline{E}_y}{\underline{H}_x} = Z. \tag{13.80}$$

Note that, for TE and TM waves, these component relations also follow from Eqs. (13.41) and (13.62), respectively.

Conceptual Questions (on Companion Website): 13.14–13.18

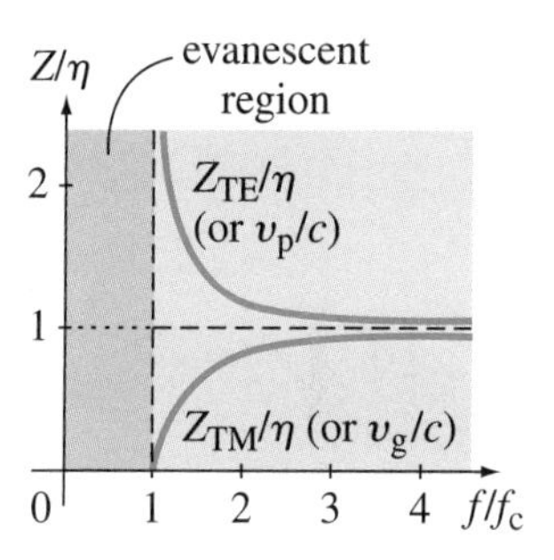

Figure 13.7 Sketch of wave impedances of TE and TM waves (Z_{TE} and Z_{TM}) in a rectangular waveguide (Fig. 13.1), Eqs. (13.76) and (13.77), normalized to the intrinsic impedance of the waveguide dielectric (η) as a function of the wave frequency (f) normalized to the waveguide cutoff frequency (f_c). These same graphs represent the waveguide phase and group velocities (v_p and v_g), given by Eqs. (13.109) and (13.111) in Section 13.10, normalized to the intrinsic phase velocity of the waveguide dielectric (c) versus f/f_c.

13.8 POWER FLOW ALONG A WAVEGUIDE

We now evaluate the power flow associated with a traveling TE or TM wave along a rectangular waveguide, in Fig. 13.1. Because the waveguide walls are considered to be perfectly conducting, the wave does not penetrate at all into them. Therefore, as in Eq. (11.21), the complex power carried by the wave along the z-axis equals the flux of the complex Poynting vector of the wave through a cross section of the guide dielectric, S_d. From Eq. (8.192), the real part of this complex power equals, in turn, the time-average transmitted power along the guide,

$$P = P_{ave} = \mathrm{Re}\left\{\int_{S_d} \underline{\boldsymbol{\mathcal{P}}} \cdot \mathrm{d}\mathbf{S}\right\} = \mathrm{Re}\left\{\int_{S_d} (\underline{\mathbf{E}} \times \underline{\mathbf{H}}^*) \cdot \mathrm{d}\mathbf{S}\right\} \tag{13.81}$$

$(\mathrm{d}\mathbf{S} = \mathrm{d}S\,\hat{\mathbf{z}})$.

Obviously, only the z-component of $\underline{\boldsymbol{\mathcal{P}}}$ contributes to the power in Eq. (13.81). Having in mind Eqs. (11.7) and (13.79), and adopting the reference directions for the vectors $\underline{\mathbf{E}}_t$ and $\underline{\mathbf{H}}_t$ such that their cross product is in the positive z direction, this component is given by

$$\underline{\boldsymbol{\mathcal{P}}}_z = \underline{\mathbf{E}}_t \times \underline{\mathbf{H}}_t^* = \frac{1}{Z}\underline{\mathbf{E}}_t \times (\hat{\mathbf{z}} \times \underline{\mathbf{E}}_t^*) = \frac{1}{Z}\underline{E}_t\underline{E}_t^*\,\hat{\mathbf{z}} = \frac{1}{Z}|\underline{E}_t|^2\,\hat{\mathbf{z}} = Z|\underline{H}_t|^2\,\hat{\mathbf{z}}$$

$$(Z = Z_{TE}\,,\ Z_{TM}\,,\ \text{or}\ Z_{TEM}), \tag{13.82}$$

since both $\underline{\mathbf{E}}_t \times \underline{\mathbf{H}}_z^*$ and $\underline{\mathbf{E}}_z \times \underline{\mathbf{H}}_t^*$, if nonzero, lie in a transversal plane, and $\hat{\mathbf{t}} \times (\hat{\mathbf{z}} \times \hat{\mathbf{t}}) = \hat{\mathbf{z}}$, where $\hat{\mathbf{t}}$ is the unit vector along $\underline{\mathbf{E}}_t$ [also see Fig. 11.2(a)]. Here,

$$|\underline{E}_t|^2 = |\underline{E}_x|^2 + |\underline{E}_y|^2, \quad |\underline{H}_t|^2 = |\underline{H}_x|^2 + |\underline{H}_y|^2, \tag{13.83}$$

with the x and y field components of a TE or TM wave, that is, an arbitrary TE$_{mn}$ or TM$_{mn}$ mode, being given in Eqs. (13.54)–(13.58) and (13.65)–(13.69), respectively. Note that the same result can as well be obtained from

$$\underline{\mathcal{P}}_z = \underbrace{(\underline{E}_x\,\hat{\mathbf{x}} + \underline{E}_y\,\hat{\mathbf{y}})}_{\underline{\mathbf{E}}_\mathrm{t}} \times \underbrace{(\underline{H}_x\,\hat{\mathbf{x}} + \underline{H}_y\,\hat{\mathbf{y}})}_{\underline{\mathbf{H}}_\mathrm{t}}{}^* = (\underline{E}_x\underline{H}_y^* - \underline{E}_y\underline{H}_x^*)\,\hat{\mathbf{z}}, \tag{13.84}$$

as in Eq. (9.200). Moreover, based on Eqs. (13.54)–(13.56), (13.65), (13.68), and (13.69), it is obvious that the transverse component of the complex Poynting vector, $\underline{\mathcal{P}}_\mathrm{t}$, of both TE and TM waves is purely imaginary. For instance, both $\underline{E}_x\underline{H}_z^*$ and $\underline{E}_y\underline{H}_z^*$ for a TE wave, Eqs. (13.54)–(13.56), are purely imaginary. This is in agreement with the facts that the TE and TM waves in the waveguide exhibit standing-wave behavior [see Eq. (10.20)] in x and y directions and that no power is delivered to the PEC walls of the guide. Expressions in Eq. (13.82) hold for all three types of guided electromagnetic waves (TE and TM waves along waveguides, and TEM waves along transmission lines), with the wave impedances Z_TE, Z_TM, and Z_TEM given in Eqs. (13.76), (13.77), and (11.19), respectively.

We see that, for a lossless guiding structure at propagating frequencies, $f > f_\mathrm{c}$, f_c being the cutoff frequency of the structure, $\underline{\mathcal{P}}_z$ in Eq. (13.82) is a purely real vector, and write $\underline{\mathcal{P}}_z = \mathcal{P}_z\,\hat{\mathbf{z}}$, so that the transmitted power along a waveguide (or a transmission line), Eq. (13.81), becomes

transmitted power

$$\boxed{P = \int_{S_\mathrm{d}} \mathcal{P}_z\,\mathrm{d}S = \frac{1}{Z}\int_{S_\mathrm{d}} |\underline{E}_\mathrm{t}|^2\,\mathrm{d}S = Z\int_{S_\mathrm{d}} |\underline{H}_\mathrm{t}|^2\,\mathrm{d}S} \tag{13.85}$$

[see Eq. (11.21) for the power along a transmission line]. Of course, P for TE or TM waves is a function of mode indices (m and n), $P = P_{mn}$. As for transmission lines, Eqs. (13.83), (13.55)–(13.58), and (13.66)–(13.69) tell us that $|\underline{E}_\mathrm{t}|$ and $|\underline{H}_\mathrm{t}|$ in waveguides do not depend on z ($|\mathrm{e}^{-\mathrm{j}\beta z}| = 1$), meaning that $P = \text{const}$ along the entire guide, which is in agreement with the no-loss assumption. However, at frequencies below the waveguide cutoff, $f < f_\mathrm{c}$, both Z_TE and Z_TM in Eqs. (13.76) and (13.77) are purely imaginary, and so is $\underline{\mathcal{P}}_z$ in Eq. (13.82). This means a zero real part of the complex transmitted power in Eq. (13.81), that is, $P = 0$ along the waveguide (no real power flow at evanescent frequencies). In addition, at $f = f_\mathrm{c}$, $Z_\mathrm{TE} \to \infty$ and $Z_\mathrm{TM} = 0$ (Fig. 13.7), so that Eq. (13.82) gives $\underline{\mathcal{P}}_z = 0$, i.e., no propagation at the cutoff as well.

Let us compute the integral in Eq. (13.85) for the dominant mode, TE$_{10}$. Using Eqs. (13.76), (13.83), (13.24), (13.20), and (8.48),

P – TE$_{10}$ mode

$$\boxed{\begin{aligned} P &= \frac{\beta}{\omega\mu}\int_{x=0}^{a} |\underline{E}_y|^2\,\underbrace{b\,\mathrm{d}x}_{\mathrm{d}S} = \frac{\omega\mu\beta a^2 b|\underline{H}_0|^2}{\pi^2}\underbrace{\int_0^a \sin^2\left(\frac{\pi}{a}x\right)\mathrm{d}x}_{a/2} \\ &= \frac{\omega\mu\beta a^3 b|\underline{H}_0|^2}{2\pi^2} = \eta\,\frac{ab}{2}\,\frac{f^2}{f_\mathrm{c}^2}\sqrt{1-\frac{f_\mathrm{c}^2}{f^2}}\,|\underline{H}_0|^2 \quad [f_\mathrm{c} = (f_\mathrm{c})_{10}], \end{aligned}} \tag{13.86}$$

where the elemental surface for integration, $\mathrm{d}S$, in the waveguide cross section is adopted in the form of a thin strip of length b and width $\mathrm{d}x$, as shown in Fig. 13.8 [this is similar to the flux evaluation in Eq. (6.64) and Fig. 6.12]. The integral in x is evaluated as in Eq. (12.138), and the result is $a/2$. Finally, the use is made of the fact that $\omega^2\mu a^2/\pi^2 c = \mu c f^2/[c/(2a)]^2 = \eta f^2/f_\mathrm{c}^2$. Similar integrations lead to the expressions for P of arbitrary TE$_{mn}$ and TM$_{mn}$ modes, respectively, in the waveguide.

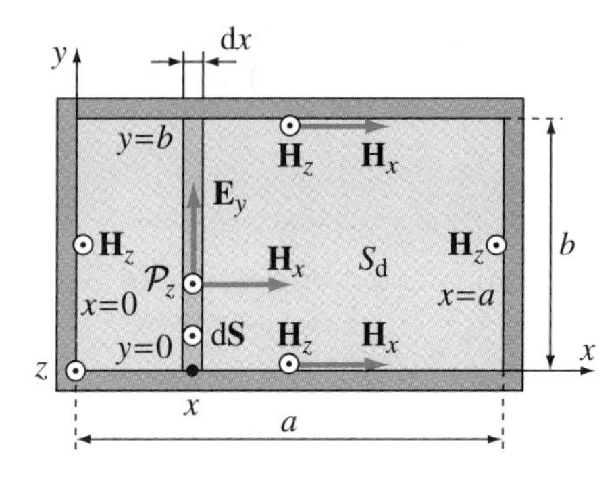

Figure 13.8 Evaluation of the time-average power carried by a TE_{10} traveling wave along a rectangular waveguide, as a flux of the axial component of $\underline{\mathcal{P}}$ ($\underline{\mathcal{P}}_z$ is a purely real vector) through the dielectric cross section; the figure also shows the existing tangential components of $\underline{\mathbf{H}}$ at waveguide walls for this wave, used in loss computation in the next section.

Example 13.4 Dielectric Breakdown and Power-Handling Capacity of a Waveguide

A lossless rectangular metallic waveguide of transverse dimensions a and b is filled with a homogeneous dielectric of permittivity ε, permeability μ, and dielectric strength E_{cr}. Find the maximum permissible time-average power, limited by an eventual dielectric breakdown in the structure, that can be carried along the waveguide by a TE_{10} wave of frequency f.

Solution The y-component (the only existing one) of the electric field intensity vector of the dominant waveguide mode (TE_{10}) is given in Eq. (13.24) and shown in Fig. 13.3. Obviously, this field is the strongest along the centerline of the larger transverse dimension of the waveguide, so for $x = a/2$, where its rms intensity amounts to

$$|\underline{E}_y|_{max} = |\underline{E}_y|_{x=a/2} = \omega\mu \frac{a}{\pi} |\underline{H}_0|. \tag{13.87}$$

Dielectric breakdown occurs when the corresponding amplitude (peak-value) of the field reaches the critical field value (dielectric strength), E_{cr}, for the dielectric filling the waveguide [most frequently, it is air, so $E_{cr} = E_{cr0} = 3$ MV/m, Eq. (2.53)]. Since $\mathbf{E}$ is a linearly polarized vector, its peak to rms value ratio equals $\sqrt{2}$, and the breakdown condition, such as the one in Eq. (2.225) for instance, becomes

$$\boxed{\underbrace{|\underline{E}_y|_{max}\sqrt{2}}_{\text{peak-value}} = E_{cr}.} \tag{13.88}$$

dielectric breakdown in a waveguide

Solving it for the magnitude of the complex constant $\underline{H}_0$ in Eq. (13.87), we have

$$|\underline{H}_0|_{cr} = \frac{\pi E_{cr}}{\sqrt{2}\omega\mu a}, \tag{13.89}$$

which, substituted in Eq. (13.86), gives the associated time-average power carried by the TE_{10} wave along the structure:

$$\boxed{P_{max} = P_{cr} = \eta \frac{ab}{2} \frac{f^2}{f_c^2} \sqrt{1 - \frac{f_c^2}{f^2}} |\underline{H}_0|_{cr}^2 = \frac{ab}{4} \frac{E_{cr0}^2}{\eta} \sqrt{1 - \frac{c^2}{4a^2 f^2}},} \tag{13.90}$$

waveguide power capacity

where the relationships $f_c = c/(2a)$, $\omega = 2\pi f$, and $c\mu = \eta$ [see Eqs. (13.20), (8.48), (9.18), and (9.21)] are used to transform the power expression into its final form. This is the maximum permissible power (P_{max}) that can be put through the system (before it breaks down), and it hence determines the power-handling capacity of the waveguide, in the dominant mode of operation.

As a numerical example, the power capacity at $f = 2$ GHz of an air-filled waveguide with $a = 10.922$ cm and $b = 5.461$ cm (WR-430 commercial rectangular waveguide) is as high as $P_{max} = 25.9$ MW (note that many high-power communication and radar systems require the transmission of very large microwave powers along waveguides). Of course, the power capacity values used in practice in designing and operating waveguides are always defined with a certain safety factor included in Eq. (13.90), which reduces the computed "ideal" value, to allow for any nonidealities in the system and its operation.

Example 13.5 Transmitted Power of an Arbitrary TM Wave Mode

Derive the expression for the time-average power carried through a cross section of a rectangular metallic waveguide, in Fig. 13.1, by an arbitrary TM_{mn} wave mode.

Solution Combining Eqs. (13.85), (13.77), (13.83), (13.66), and (13.67), the transmitted power of a TM_{mn} mode, for arbitrary $m, n = 1, 2, \ldots$, along the waveguide is given by

$$P = \frac{1}{Z_{\mathrm{TM}}} \int_{S_\mathrm{d}} |\underline{E}_\mathrm{t}|^2 \,\mathrm{d}S = \frac{\omega\varepsilon}{\beta} \int_{x=0}^{a} \int_{y=0}^{b} \left(|\underline{E}_x|^2 + |\underline{E}_y|^2\right) \underbrace{\mathrm{d}x\,\mathrm{d}y}_{\mathrm{d}S}$$

$$= \frac{\omega\varepsilon\beta|\underline{E}_0|^2}{k^4} \left[\left(\frac{m\pi}{a}\right)^2 \underbrace{\int_0^a \cos^2\left(\frac{m\pi}{a}x\right)\mathrm{d}x}_{a/2} \underbrace{\int_0^b \sin^2\left(\frac{n\pi}{b}y\right)\mathrm{d}y}_{b/2} \right.$$

$$\left. + \left(\frac{n\pi}{b}\right)^2 \underbrace{\int_0^a \sin^2\left(\frac{m\pi}{a}x\right)\mathrm{d}x}_{a/2} \underbrace{\int_0^b \cos^2\left(\frac{n\pi}{b}y\right)\mathrm{d}y}_{b/2} \right] \quad (m, n = 1, 2, \ldots), \quad (13.91)$$

where the integrals in x and y are computed in essentially the same way as the integral in time in Eq. (6.95), and the results are $a/2$ and $b/2$, respectively. Moreover, the two products of two integrals (in x and y) come out to be the same, $ab/4$, and can be taken as a common term out of the sum of the products. Finally, having in mind Eqs. (13.59), (13.12), and (8.48), we can write

$$P = \frac{\omega\varepsilon\beta|\underline{E}_0|^2}{k^4} \frac{ab}{4} \underbrace{\left[\left(\frac{m\pi}{a}\right)^2 + \left(\frac{n\pi}{b}\right)^2\right]}_{k^2} = \frac{ab}{4\eta} \frac{f^2}{f_\mathrm{c}^2} \sqrt{1 - \frac{f_\mathrm{c}^2}{f^2}}\, |\underline{E}_0|^2 \quad [f_\mathrm{c} = (f_\mathrm{c})_{mn}], \quad (13.92)$$

where the cutoff frequency $(f_\mathrm{c})_{mn}$ is given in Eq. (13.71), and the use is made (in the term $k^2/k^4 = 1/k^2$) also of the relationship $k^2 = 4\pi^2 f_\mathrm{c}^2/c^2$, which, in turn, is obtained comparing Eqs. (13.70) and (13.12), as well as of the fact that $c\varepsilon = 1/\eta$ [see Eqs. (13.70) and (9.21)].

Problems: 13.10–13.15; *Conceptual Questions* (on Companion Website): 13.19 and 13.20; *MATLAB Exercises* (on Companion Website).

13.9 WAVEGUIDES WITH SMALL LOSSES

To take into account conductor and dielectric losses in a rectangular waveguide (Fig. 13.8), with a TE or TM wave, we assume that these losses are small, and apply the perturbation method for evaluation of the time-average loss power and the associated signal attenuation in a transmission line with small losses described in Sections 11.5 and 11.6. As with transmission lines, the low-loss assumption in waveguide theory is based on the fact that, by design, waveguides are fabricated from good conductors and good dielectrics, at the relevant frequencies for a given application, so that the conditions in Eqs. (11.61) are satisfied. The main premise of the perturbation method is that the field distributions in every cross section of the waveguide are practically the same as if there were no losses, while the difference is in the axial (z)

direction, in which the fields attenuate as $e^{-\alpha z}$, α being the attenuation coefficient of the structure. For instance, the y-component (the only existing one) of the electric field intensity vector of the dominant waveguide mode (TE_{10}), Eqs. (13.24)–(13.26), is now

$$\underline{E}_y(x, y, z) = \underline{E}_y(x, y, 0)\,e^{-\alpha z}\,e^{-j\beta z} \tag{13.93}$$

[see also Eqs. (11.5) and (11.63)], and analogously for the nonzero magnetic field components of the mode. The distribution of $\underline{E}_y(x, y, 0)$ is the same as in Eq. (13.24), and the phase coefficient β is that in Eqs. (13.12) and (13.20) – as in the lossless case. Most importantly, α is computed using the no-loss field distributions.

From Eq. (11.66), the time-average power of Joule's losses in the waveguide conductor (i.e., in four waveguide walls) in Fig. 13.8 per unit length of the structure, P'_c, is given by

$$\boxed{P'_c = \oint_C R_s|\underline{\mathbf{H}}_{\rm tang}|^2\,dl \quad \left(R_s = \sqrt{\frac{\pi\mu_c f}{\sigma_c}}\right),} \tag{13.94}$$

p.u.l. conductor loss power

where C stands for the interior contour, with dimensions a and b, of the conductor in a cross section of the guide. $\underline{\mathbf{H}}_{\rm tang}$ is the tangential component of the complex rms magnetic field intensity vector of the propagating TE or TM wave on the conductor interior surface, along C, computed as if the conductor were perfect. At the given frequency, f, of the wave, R_s is the surface resistance of the conductor, Eq. (11.65), with σ_c and μ_c being its conductivity and permeability (most frequently, $\mu_c = \mu_0$), respectively.

As in Eq. (11.68), the time-average loss power in the dielectric in Fig. 13.8 per unit length of the waveguide, P'_d, is computed as

$$\boxed{P'_d = \int_{S_d} \sigma_d|\underline{\mathbf{E}}|^2\,dS.} \tag{13.95}$$

p.u.l. dielectric loss power

Here, S_d denotes a cross section of the dielectric, as in Eq. (13.81), σ_d is dielectric conductivity, and $\underline{\mathbf{E}}$ is the no-loss complex rms electric field intensity vector of the wave over S_d.

Once P'_c and P'_d are found, the respective attenuation coefficients, for the waveguide conductor and dielectric, α_c and α_d, are obtained using Eqs. (11.76) and the result for the time-average power transmitted along the guide, P, from Eq. (13.85). In what follows, we complete this procedure for the dominant mode.

To compute P'_c, we note that at the walls defined by $y = 0$ and $y = b$ in Fig. 13.8, both x- and z-components of the magnetic field vector of a TE_{10} wave are tangential to the wall surface, whereas only $\underline{H}_z$ is tangential to the conductor surfaces at $x = 0$ and $x = a$, as indicated in the figure. In addition, having in mind Eqs. (13.25) and (13.26), we realize that these specified tangential components of $\underline{\mathbf{H}}$ have the same magnitudes at each pair of parallel walls (field symmetry). With this, Eq. (13.94) yields

$$\begin{aligned} P'_c &= 2\int_{x=0}^{a} R_s\left(|\underline{H}_x|^2 + |\underline{H}_z|^2\right)\Big|_{y=0 \text{ or } b} dx + 2\int_{y=0}^{b} R_s\,|\underline{H}_z|^2_{x=0 \text{ or } a}\,dy \\ &= 2R_s|\underline{H}_0|^2\left\{\int_0^a\left[\left(\frac{\beta a}{\pi}\right)^2\sin^2\left(\frac{\pi}{a}x\right) + \cos^2\left(\frac{\pi}{a}x\right)\right]dx + \int_0^b dy\right\} \\ &= 2R_s|\underline{H}_0|^2\left(\frac{\beta^2 a^3}{2\pi^2} + \frac{a}{2} + b\right) = 2R_s|\underline{H}_0|^2\left(\frac{a}{2}\frac{f^2}{f_c^2} + b\right) \end{aligned} \tag{13.96}$$

[also see the integral in Eq. (13.86)], since, using Eqs. (13.12), (8.48), and (13.20),

$$\frac{\beta^2 a^3}{2\pi^2} = \frac{\omega^2 a^3}{2\pi^2 c^2}\left(1 - \frac{f_c^2}{f^2}\right) = \frac{a}{2}\frac{f^2}{\underbrace{(c/2a)^2}_{f_c}}\left(1 - \frac{f_c^2}{f^2}\right) = \frac{a}{2}\left(\frac{f^2}{f_c^2} - 1\right). \qquad (13.97)$$

The expression for α_c in Eqs. (11.76) and that for P in Eq. (13.86) then result in

α_c – TE_{10} *mode*

$$\boxed{\alpha_c = \frac{P'_c}{2P} = \frac{R_s}{\eta a}\frac{a/b + 2f_c^2/f^2}{\sqrt{1 - f_c^2/f^2}}.} \qquad (13.98)$$

Of course, α_c is given in Np/m; to convert it to dB/m, if so desired, we use the relationship in Eq. (9.89).

For a TE_{10} mode, $\underline{\mathbf{E}} = \underline{\mathbf{E}}_t$ [Eqs. (13.35)], so that comparison of Eqs. (13.95) and (13.85) leads to

$$P'_d = \sigma_d Z_{TE} P, \qquad (13.99)$$

where Z_{TE} is the corresponding wave impedance. We note that this relationship has the same form as the one in Eq. (11.81) for a TEM wave along a transmission line. Combining it with Eqs. (11.76), (13.76), (9.125), and (13.12), α_d can be written as

α_d – *any* TE *or* TM *mode*

$$\boxed{\alpha_d = \frac{P'_d}{2P} = \frac{\sigma_d Z_{TE}}{2} = \frac{\omega\mu\sigma_d}{2\beta} = \frac{\omega^2\varepsilon\mu\tan\delta_d}{2\beta} = \frac{\omega\tan\delta_d}{2c\sqrt{1 - f_c^2/f^2}},} \qquad (13.100)$$

$\tan\delta_d$ being the loss tangent of the dielectric, and this result is the same for any TE or TM mode in the waveguide (with cutoff frequency f_c).

Of course, the overall attenuation coefficient in Eq. (13.93) is $\alpha = \alpha_c + \alpha_d$, Eq. (11.75). However, in most applications of rectangular metallic waveguides, the dielectric is air, so that σ_d, $\tan\delta_d$, $\alpha_d = 0$ and $\alpha = \alpha_c$. In these cases, therefore, conductor losses and α_c, determined based on the waveguide parameters a, b, and σ_c ($\mu_c = \mu_0$), at a given operating frequency, f, in the propagating region, Eq. (13.15), represent a limiting factor for practical usability of the structure. Simply, if the waveguide is too long, the attenuation in the conductors becomes so large, i.e., the factor $e^{-\alpha z}$ in Eq. (13.93) so small, that the signal at the receiving end of the guide is unusable (or too weak to be efficiently used). In applications including high-power transfers along a waveguide, this may mean a prohibitively inefficient overall power balance of the system. In addition, we note a rather complex frequency dependence of α_c, for the dominant mode, in Eq. (13.98), where we have in mind that R_s [see Eq. (13.94)] is also a function of frequency, $R_s \propto \sqrt{f}$. This dependence, for a standard waveguide, Eq. (13.74), made of copper, is shown in Fig. 13.9. It is now obvious that the lower part of the dominant frequency range (ensuring a single-mode operation of the waveguide) in Eq. (13.22) is unusable, due to a very high signal attenuation [for frequencies just above the cutoff ($f = f_c^+$), $\alpha_c \to \infty$]. On the other side, while considering frequencies close to the upper limit of the dominant range, it is always preferable to have some safety margin (frequency separation) with respect to the next higher order mode. Consequently, a good rule of thumb is to design a waveguide such that the operating frequency (or frequencies) of the dominant mode be within the following range:

usable frequency range

$$\boxed{1.25\frac{c}{2a} = 0.625\frac{c}{a} < f < 0.95\frac{c}{a},} \qquad (13.101)$$

which is often called the usable frequency range of the waveguide.

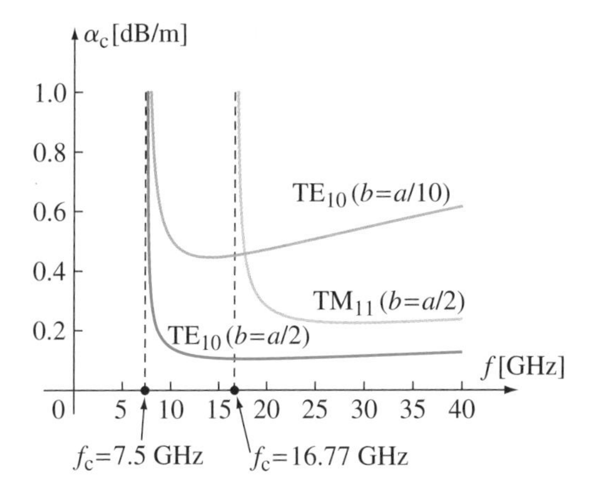

Figure 13.9 Attenuation coefficient $\alpha = \alpha_c$ versus frequency for the dominant mode, Eq. (13.98), in a standard ($a = 2$ cm, $b = 1$ cm) air-filled copper ($\sigma_c = 58$ MS/m, $\mu_c = \mu_0$) rectangular waveguide (Fig. 13.8); shown also is α_c for b decreased to $b = a/10$ and for the TM_{11} wave mode (and $a = 2b$).

Finally, we note, observing Eq. (13.98), that for a fixed larger transverse dimension of the waveguide, $a = \text{const}$ (and given frequency and material parameters of the structure), the larger the aspect ratio a/b (i.e., the smaller b) the larger the attenuation coefficient α_c (for the dominant mode), which is also illustrated in Fig. 13.9. In addition, even without considering losses, the expression for the maximum possible transmitted power along the waveguide (P_{max}) in Eq. (13.90), obtained based on considerations of an eventual dielectric breakdown in the guide, indicates that for a given a, a decrease in b means a reduction in the power transfer capacity of the structure. This is why too large aspect ratios (e.g., $a/b = 10$), although ensuring the maximum dominant frequency range of the waveguide, Eq. (13.73), are impractical, as already mentioned in relation with Eq. (13.74). On the other side, it certainly seems attractive to maximally increase b, i.e., reduce a/b (possibly to the limit of $a/b = 1$), in order to both increase P_{max} (regardless of losses) and reduce α_c. However, this is offset by the design requirement for as large as possible dominant range, with Eq. (13.73) defining the upper bound for b, for $a = \text{const}$. So, as a compromise between the two contradictory requirements, the typical choice for a/b is that in Eq. (13.74) – a standard waveguide.

Example 13.6 K_u-Band Waveguide Design

Design an air-filled K_u-band standard waveguide, such that the entire K_u-band (12–18 GHz) falls into the usable frequency range of the guide, as defined by Eq. (13.101).

Solution For a standard waveguide, the larger to smaller transverse dimension ratio (a/b) is given in Eq. (13.74). To cover all frequencies between $f_1 = 12$ GHz and $f_2 = 18$ GHz by the usable frequency range in Eq. (13.101), that is, to nest the K_u-band within the usable range of the waveguide, we require that

$$0.625\,\frac{c_0}{a} < f_1 \quad \text{and} \quad f_2 < 0.95\,\frac{c_0}{a} \tag{13.102}$$

($c_0 = 3 \times 10^8$ m/s). This translates into the following lower and upper bounds for the guide dimension a:

$$\frac{0.625c_0}{f_1} < a < \frac{0.95c_0}{f_2} \quad \longrightarrow \quad 1.562 \text{ cm} < a < 1.583 \text{ cm}. \tag{13.103}$$

Hence, any value for a within these bounds gives a standard waveguide whose usable frequency range includes the entire K_u-band. For instance, the choice of $a = 1.58$ cm ($b = 0.79$ cm) corresponds to a commercial K_u-band WR-62 rectangular waveguide.[11]

Example 13.7 Complex Propagation Coefficient of an X-Band Waveguide

An X-band (8–12 GHz) standard rectangular waveguide, with transverse dimensions $a = 15.63$ mm and $b = 7.81$ mm, is made of copper, whose conductivity is $\sigma_c = 58$ MS/m and permeability $\mu_c = \mu_0$, and filled with polyethylene, the relative permittivity of which is $\varepsilon_r = 2.25$ and loss tangent $\tan\delta_d = 10^{-4}$ ($\mu = \mu_0$). At an operating frequency of $f = 10$ GHz, compute the complex propagation coefficient of the waveguide for the dominant wave mode.

Solution The intrinsic phase velocity of the waveguide dielectric, Eq. (9.18), and cutoff frequency of the dominant mode (TE_{10}), Eq. (13.20), come out to be

$$c = \frac{c_0}{\sqrt{\varepsilon_r}} = 2 \times 10^8 \text{ m/s} \quad \longrightarrow \quad f_c = (f_c)_{10} = \frac{c}{2a} = 6.398 \text{ GHz} \tag{13.104}$$

($c_0 = 3 \times 10^8$ m/s). The fact that $(f_c)_{10} < f < 2(f_c)_{10}$, like in Eqs. (13.22) and (13.74), verifies that indeed only a TE_{10} wave can propagate along the waveguide. The phase coefficient of the waveguide, Eq. (13.12), amounts to

$$\beta = \frac{2\pi f}{c}\sqrt{1 - \frac{(f_c)_{10}^2}{f^2}} = 241.44 \text{ rad/m}. \tag{13.105}$$

Eq. (13.98) then tells us that the attenuation coefficient for the waveguide conductor is

$$\alpha_c = \frac{(R_s)_{Cu}\sqrt{\varepsilon_r}}{\eta_0 a} \frac{a/b + 2(f_c)_{10}^2/f^2}{\sqrt{1 - (f_c)_{10}^2/f^2}} = 0.0244 \text{ Np/m} = 0.212 \text{ dB/m} \tag{13.106}$$

($\eta_0 = 377\ \Omega$), where the surface resistance of copper, $R_s = (R_s)_{Cu} = 0.026\ \Omega$/square, is found from Eq. (10.80), and the use is made of the relationship in Eq. (9.89) to convert Np/m to dB/m. For the guide dielectric, Eq. (13.100) gives the coefficient

$$\alpha_d = \frac{\pi f \tan\delta_d}{c\sqrt{1 - (f_c)_{10}^2/f^2}} = 0.0204 \text{ Np/m} = 0.177 \text{ dB/m}. \tag{13.107}$$

Finally, using Eq. (11.75), the total attenuation coefficient of the waveguide is $\alpha = \alpha_c + \alpha_d = 0.0448$ Np/m $= 0.389$ dB/m, which, combined with Eq. (13.105), results in the following complex propagation coefficient:

$$\underline{\gamma} = \alpha + \mathrm{j}\beta = (0.0448 + \mathrm{j}241.44) \text{ m}^{-1}. \tag{13.108}$$

Problems: 13.16–13.20; *Conceptual Questions* (on Companion Website): 13.21–13.25; *MATLAB Exercises* (on Companion Website).

13.10 WAVEGUIDE DISPERSION AND WAVE VELOCITIES

Since the phase coefficient β in Eq. (13.12), of the rectangular waveguide in Fig. 13.1, is a nonlinear function of the angular frequency, ω, of a propagating TE or TM wave, the phase velocity of the wave, v_p, is frequency dependent, and

[11]Note that commercial rectangular waveguides are specified in WR-*xyz* numbers, with WR standing for "waveguide, rectangular" and *xyz* denoting that the larger transverse dimension (a) of the guide equals $xyz/100$ inches. Thus, a WR-62 waveguide has $a = 0.62$ inch $= 0.62 \times 2.54$ cm ≈ 1.58 cm.

the waveguide represents a dispersive propagation medium, as in Eq. (9.166). As explained in Section 9.13, a direct consequence of such frequency behavior of the medium, and normally the most troublesome one, is signal distortion in the time domain. Comparing Eqs. (13.12) and (9.160), we realize that the β-ω relationship of a waveguide is identical in form to that of a uniform plane wave in a plasma medium. Therefore, a typical dispersion diagram for a metallic waveguide is practically that in Fig. 9.17. Of course, each waveguide mode, (m, n), has a different diagram, according to Eq. (13.71). Moreover, in a waveguide, wave dispersion associated with Eq. (13.12) is, unlike in a plasma, not due to electromagnetic properties of the material through which the wave propagates, but is a result of the specific configuration of metallic boundaries (waveguide walls) confining the wave propagation. Therefore, to distinguish it from material dispersion, dispersion due to the waveguide structure (geometry), which is inherent for all metallic waveguides and present for all propagating frequencies, Eq. (13.15), and modes, is called waveguide dispersion. However, waveguides can also exhibit material dispersion, which occurs when the parameters of the dielectric inside the structure are frequency dependent.

So, in parallel to Eq. (9.178) for a plasma medium, or from Eqs. (9.35) and (13.12), the phase velocity of a wave mode with cutoff frequency f_c propagating, at frequency f $(f > f_c)$, through a rectangular metallic waveguide (Fig. 13.1) is given by

$$\boxed{v_p = \frac{\omega}{\beta} = \frac{c}{\sqrt{1 - f_c^2/f^2}}.} \qquad (13.109)$$

waveguide phase velocity

The wavelength along the structure (measured along the z-axis), λ_z, is then [see Eqs. (11.17)]

$$\boxed{\lambda_z = \lambda_{guide} = \frac{2\pi}{\beta} = \frac{v_p}{f} = \frac{\lambda}{\sqrt{1 - f_c^2/f^2}}.} \qquad (13.110)$$

guide wavelength

This wavelength is often referred to as the guide wavelength (λ_{guide}). In the above equations, $c = 1/\sqrt{\varepsilon\mu}$ and $\lambda = c/f$ are the intrinsic phase velocity and wavelength of the waveguide dielectric, i.e., the velocity of a uniform plane wave in an unbounded medium of parameters ε and μ, and the corresponding wavelength at the operating frequency f $(\omega = 2\pi f)$.[12] Fig. 13.10 illustrates the variation of λ_z with frequency for several modes (m, n) in a standard waveguide, Eq. (13.74), along with the frequency variation of λ. We see that, like the coefficient β in Fig. 9.17 and wave impedances Z_{TE} and Z_{TM} in Fig. 13.7, all guide wavelengths in Fig. 13.10 vary very rapidly near their respective cutoff frequencies. On the other side, for $f \gg f_c$, $\lambda_{guide} \to \lambda$ for all modes, which is another confirmation that TE_{mn} and TM_{mn} waves have characteristics of TEM waves in an unbounded medium when they are operated far above cutoff (for example, see also Fig. 13.7). Combining Eqs. (9.172) and (13.12), the group velocity (or energy velocity) along the waveguide, v_g, is given by the following

[12] If the polarization (and conduction) properties of the waveguide dielectric are specified using a complex permittivity ($\underline{\varepsilon}$), Eq. (9.127), whose real part is a function of frequency, then $c = 1/\sqrt{\varepsilon'(\omega)\mu}$, and we have a combination of waveguide and material dispersion in the structure. However, as the metallic waveguides in practice are mostly air-filled, this latter sort of dispersion in waveguide theory has much less practical importance than the former one.

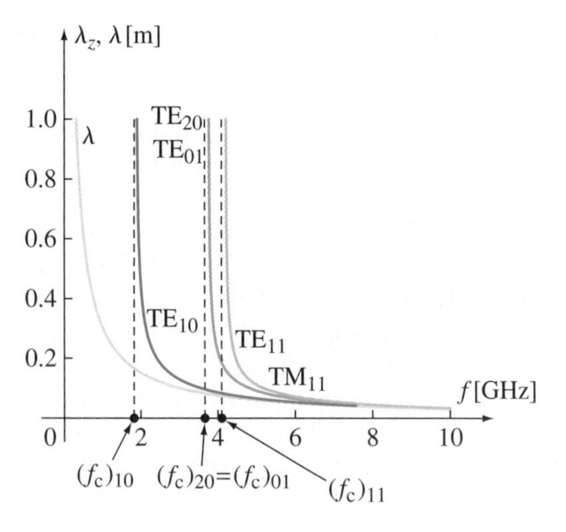

Figure 13.10 Frequency variation of the guide wavelength, Eq. (13.110), for several modes in a standard rectangular air-filled metallic waveguide ($a = 2b = 8$ cm); the intrinsic wavelength of the waveguide dielectric (air), $\lambda = c/f = c_0/f$, is also shown.

expression:

waveguide group velocity

$$v_g = \frac{1}{d\beta/d\omega} = c\sqrt{1 - \frac{f_c^2}{f^2}}, \tag{13.111}$$

which, again, has the same form as that in Eqs. (9.178) for plasmas.

Frequency dependences of v_p and v_g, Eqs. (13.109) and (13.111), are shown in Fig. 13.7 (in Section 13.7). The rapid variations in both velocities near cutoff are practically always undesirable. With such variations, problems with signal distortion due to waveguide dispersion, particularly in digital communication systems, are even more pronounced and difficult to overcome. This is one more reason to restrict the operating frequencies of waveguides to the usable frequency range in Eq. (13.101), rather than the entire dominant (single-mode) range in Eq. (13.22). At all propagating frequencies, $f > f_c$, the product of v_p and v_g is frequency independent,

$$v_p v_g = c^2. \tag{13.112}$$

In addition, each of the velocities approaches c asymptotically (and the wave propagates as if in an unbounded medium) for $f \to \infty$. Note that the fact that $v_p > v_g$ in Fig. 13.7 means that the waveguide is a normally dispersive propagation medium, like a plasma (Fig. 9.17). If the guide is air-filled (and this is most often the case), we have that $v_p > c_0$ in the entire propagating region, c_0 being the speed of light in free space, Eq. (9.19). However, as explained in Example 9.23, this does not violate the theory of special relativity, which only requires that $v_g \leq c_0$, and this holds true in Fig. 13.7.

Example 13.8 Relationship between Three Wavelengths for a Waveguide

For an arbitrary TE_{mn} or TM_{mn} wave mode in a rectangular metallic waveguide (Fig. 13.1), λ_z is the guide wavelength (along the z-axis), λ is the intrinsic wavelength of the waveguide dielectric, and λ_c is the cutoff wavelength of the mode. Derive the relationship between these three wavelengths.

Solution With the use of Eqs. (13.14) and (8.112), we have that the ratio of the modal cutoff frequency (f_c) and operating frequency (f) of the waveguide is given by $f_c/f = \lambda/\lambda_c$, which, substituted in Eq. (13.110), gives the following simple relationship in terms of the reciprocals of the three wavelengths:

$$\lambda_z = \frac{\lambda}{\sqrt{1-\lambda^2/\lambda_c^2}} \quad \longrightarrow \quad \lambda^2 = \lambda_z^2\left(1-\frac{\lambda^2}{\lambda_c^2}\right) \quad \longrightarrow \quad \frac{1}{\lambda_z^2} = \frac{1}{\lambda^2} - \frac{1}{\lambda_c^2}. \qquad (13.113)$$

Of course, both λ_z and λ_c are functions of mode indices m and n, $\lambda_z = (\lambda_z)_{mn}$ and $\lambda_c = (\lambda_c)_{mn}$.

Example 13.9 Measuring Dielectric Permittivity Using a Waveguide

If the larger transverse dimension and operating frequency of the waveguide in Fig. 13.1 are $a = 7.2$ cm and $f = 3$ GHz, respectively, find the relative permittivity of an unknown nonmagnetic lossless dielectric filling the structure – from the measured dominant-mode guide wavelength $\lambda_z = 7.62$ cm.

Solution Having in mind Eqs. (8.112) and (9.18), the intrinsic wavelength of the guide dielectric with unknown relative permittivity ε_r can be expressed as $\lambda = c_0/(\sqrt{\varepsilon_r}f)$, while the cutoff wavelength of the dominant (TE_{10}) mode is $\lambda_c = (\lambda_c)_{10} = 2a$, Eq. (13.21). Hence, solving Eq. (13.113) for ε_r, we obtain

$$\varepsilon_r = \frac{c_0^2}{f^2}\left(\frac{1}{\lambda_z^2} + \frac{1}{4a^2}\right) = 2.205, \qquad (13.114)$$

where $c_0 = 3 \times 10^8$ m/s, Eq. (9.19). Note that metallic waveguides are often used in measurement techniques for the characterization of unknown electromagnetic materials.

Example 13.10 Travel of Signals with Different Carrier Frequencies along a Waveguide

Two signals whose frequency spectra are confined to narrow bands around carrier frequencies $f_1 = 7$ GHz and $f_2 = 8$ GHz, respectively, are launched at the same instant of time at one end of an air-filled rectangular waveguide with transverse dimensions $a = 3$ cm and $b = 1.5$ cm, and length $l = 6$ m, to propagate along it. Find the time lag between the two signals as they are received on the other end of the waveguide.

Solution As the cutoff frequency of the dominant mode (TE_{10}), Eq. (13.20), amounts to $(f_c)_{10} = c_0/(2a) = 5$ GHz ($c_0 = 3 \times 10^8$ m/s), and this is a standard waveguide, Eq. (13.74), we realize that both carrier frequencies, f_1 and f_2, fall into the dominant frequency range of the waveguide, Eq. (13.22). Hence, we can assume that both signals travel along the guide as dominant wave modes. In addition, they travel with the corresponding group velocities (v_g). Using Eq. (13.111), these velocities come out to be

$$v_{g1} = c_0\sqrt{1-\frac{(f_c)_{10}^2}{f_1^2}} = 2.1 \times 10^8 \text{ m/s}, \quad v_{g2} = c_0\sqrt{1-\frac{(f_c)_{10}^2}{f_2^2}} = 2.34 \times 10^8 \text{ m/s}. \qquad (13.115)$$

For each signal, it takes the time $t = l/v_g$ to arrive to the receive end of the waveguide, so we obtain that signal 1 lags behind signal 2 for the following time interval:

$$\Delta t = t_1 - t_2 = \frac{l}{v_{g1}} - \frac{l}{v_{g2}} = 3 \text{ ns}. \qquad (13.116)$$

Example 13.11 TE Wave Group Velocity as Axial Projection of Ray Velocity

Interpreting a TE wave in the waveguide as a uniform plane (TEM) wave tracing the zigzag ray path in Fig. 10.15, show that the group velocity of the TE wave equals the axial component of the phase velocity of the TEM wave.

Solution Combining Eqs. (13.11), (13.12), and (13.111), we realize that the axial (z) projection of the phase velocity of the TEM wave (ray velocity) in Fig. 10.15 amounts to

$$v_{\text{axial}} = c \sin \theta_{\text{i}} = c\sqrt{1 - \cos^2 \theta_{\text{i}}} = c\sqrt{1 - \frac{f_{\text{c}}^2}{f^2}} = v_{\text{g}} \tag{13.117}$$

(ray velocity is c), i.e., it indeed is identical to the group velocity of the TE wave. This again verifies the fact that the energy (or information) travels along the guide at the velocity v_{g}.

Problems: 13.21; *Conceptual Questions* (on Companion Website): 13.26–13.29; *MATLAB Exercises* (on Companion Website).

13.11 WAVEGUIDE COUPLERS

In order to generate a particular TE_{mn} or TM_{mn} mode in a rectangular waveguide (Fig. 13.1), we need an electromagnetic coupling mechanism that feeds external energy into the guide, and excites that particular modal field. This field then travels along the structure carrying the input signal away from the feed. Conversely, the same mechanism can be used, in the reversed process, to extract the energy (signal) carried by the wave (in the same mode), and deliver it to an external device or system. Such signal transmitters or receivers based on electromagnetic coupling to waveguide fields are generally referred to as waveguide couplers. Most frequently, couplers convert input power from a coaxial cable, attached, externally, to a guide wall, into waveguide modes, and vice versa (coax-to-waveguide couplers). An extension of the inner conductor of the cable, called the probe, is inserted into the guide dielectric (usually air), with the outer conductor being connected to the wall, as in Fig. 13.11(a), showing a coupler with an electric probe and one with a magnetic probe. The electric probe, suitable for coupling to the electric field in the structure, is in the form of a short wire segment (straight extension of the cable conductor). The magnetic probe consists of a small wire loop (the conductor is folded and its tip connected back to the wall), and is better suited for magnetic field coupling. These probes are actually a short monopole wire antenna and small loop antenna, respectively, and both antenna types are to be studied in the next chapter. We recall that similar probes (antennas) are used in Fig. 10.5 to receive the signal carried by a uniform plane wave.

By the same reasoning as for the plane-wave receiving dipole in Fig. 10.5, electric probes aimed to launch or receive a TE or TM wave in the waveguide should be placed at the locations of the maxima of the guide electric field intensity, **E**, and

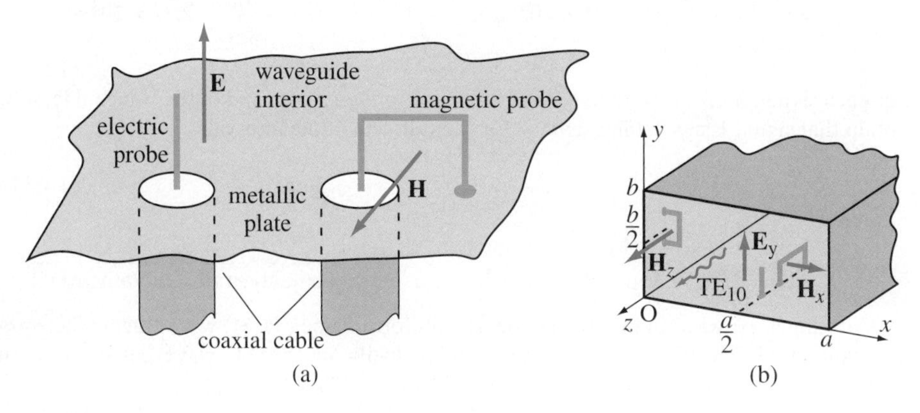

Figure 13.11 Coax-to-waveguide couplers in the form of an electric probe (short monopole antenna) and magnetic probe (small loop antenna) used to excite or receive wave modes in a rectangular metallic waveguide: (a) general geometry of couplers and (b) probes for the dominant (TE_{10}) mode.

directed in parallel to the electric field lines. Here, we have in mind that, while **E** is the existing electric field in the structure in the receiving mode of operation, in the transmitting mode it represents the anticipated field (of the desired mode) that we are about to generate using the probe. On the other side, Eqs. (10.24) and (10.25) tell us that magnetic probes should be positioned at the magnetic field maxima in the waveguide, and oriented such that **H** is perpendicular to the loop plane. Given the diversity of modal field configurations in Figs. 13.4 and 13.5, we realize that waveguide couplers are mode specific, i.e., each mode requires a different set of electric or magnetic probes for its excitation/reception. Of course, the operating frequency of the wave must be in the propagating region, Eq. (13.15), for the particular (desired or existing) mode in the structure.

For the dominant waveguide mode, whose field components are given in Eqs. (13.24)–(13.26), the magnitude of the only nonzero electric field component, $|\underline{E}_y|$, is maximum in the plane $x = a/2$ (see Fig. 13.3), so that a y-directed electric probe in that plane, shown in Fig. 13.11(b), is used to excite or receive a TE_{10} wave. Similarly, the maxima of the magnitudes of the nonzero magnetic field components, $|\underline{H}_x|$ and $|\underline{H}_z|$, occur for $x = a/2$ and $x = 0$ (or $x = a$), respectively, and hence the choice of magnetic probes (small loops) also shown in Fig. 13.11(b). Since $\underline{H}_z = 0$ for $x = a/2$ and $\underline{H}_x = 0$ for $x = 0$, each of the loops (operated one at a time) enables coupling with only one magnetic field component (at its maximum); this comes as well from the orientations of loops and mutual orthogonality of the two field components.

Each of the waveguide feeds in Fig. 13.11(b) actually launches two propagating TE_{10} waves, one in the positive (forward) z direction, as in Fig. 13.1 and Eqs. (13.24)–(13.26), and another in the negative (backward) z direction. However, Fig. 13.12 depicts a coupling configuration designed to transmit the TE_{10} mode in only one direction along the waveguide. To prevent the propagation in the other direction, the guide is closed (short-circuited), by inserting a metallic plate at its one end. The distance of the probe from the plate is $d = \lambda_z/4$, $\lambda_z = \lambda_{\text{guide}}$ being the guide wavelength, Eq. (13.110). With this, the backward propagating wave launched by the probe, after its reflection from the plate (PEC surface), adds constructively with the forward propagating wave. Namely, as $\beta d = \pi/2$, the phase difference resulting from the round trip of the backward wave, which travels a distance equal to $2d$ (from the probe to the plate and back), and the reflection phase shift of 180° at the plate [see Eq. (10.47)] cancel each other,

$$\beta d = \frac{2\pi}{\lambda_z}\frac{\lambda_z}{4} = \frac{\pi}{2} \quad \longrightarrow \quad e^{-j\beta d}\, e^{j\pi}\, e^{-j\beta d} = e^{j(\pi - 2\beta d)} = e^{j0} = 1, \qquad (13.118)$$

so that the two waves are in phase as they propagate to the right, away from the probe. The same constructive addition of the direct (incident) and reflected (from the plate) waves occur in the receiving mode of operation of the waveguide, when

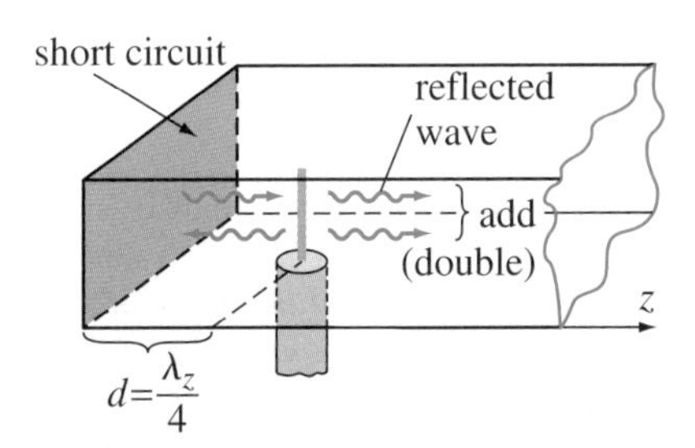

Figure 13.12 Electric-probe coupler transmitting the TE_{10} mode in only one direction along the waveguide.

the probe in Fig. 13.12 is used to receive the signal (and deliver it to the coaxial cable) from a TE_{10} wave propagating in the negative z direction (toward the probe).

Another general way to couple out the energy from a waveguide mode is to cut narrow slots in the guide walls so that they break the lines of the surface current density vector, $\mathbf{J}_s$, of that mode [see the surface current distribution on the walls for the dominant mode given by Eqs. (13.27)–(13.30)]. Fig. 13.13 shows one such slot, cut perpendicularly to the current lines. By the high-frequency continuity equation, more precisely, the surface version (for $\mathbf{J}_s$) of the boundary condition for normal components of the current density vector in Eq. (8.86) [see also Eq. (8.150)], line charges of equal magnitudes and opposite polarities, Q' and $-Q'$, are induced at the two long edges of the slot, as indicated in the figure. These charges, in turn, generate an electric field, $\mathbf{E}_{\rm slot}$, across the slot. The field $\mathbf{E}_{\rm slot}$ extends also to the external space near the slot region, and radiates the power from the mode under consideration (the mode with current $\mathbf{J}_s$ in Fig. 13.13) outside the waveguide. The structure in Fig. 13.13 turns out to be a slot antenna, used extensively in antenna practice either as a single radiator or in an antenna array.

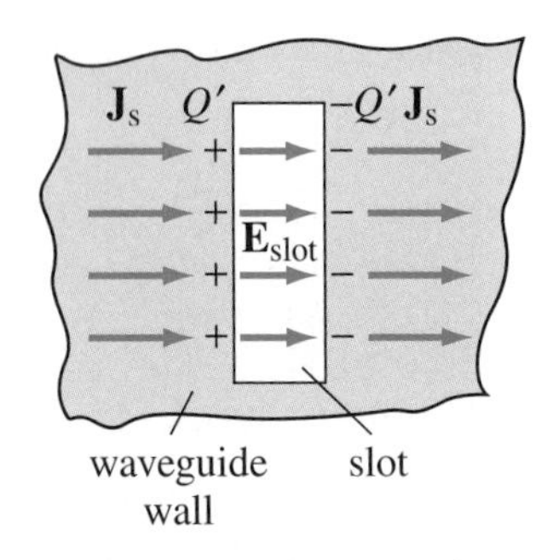

Figure 13.13 A narrow slot in a waveguide wall that breaks the wall surface-current lines, and radiates, via the slot electric field, the power of a propagating wave mode outside the structure (slot antenna).

We note that the slot, being narrow, would have practically no effect on $\mathbf{J}_s$ and would not act as a waveguide coupler if cut parallel to the current lines. This fact, in conjunction with the situation in Fig. 13.13, is a basis for the use of wall slots to remove (or substantially attenuate) undesired modes in the overmoded operation of a waveguide, at a frequency above the dominant range in Eq. (13.22), so as to ensure that only one mode is present in the structure. Namely, as the individual waveguide modes have different surface current configurations, the slots can, in some overmoded applications, be placed in a way that affects only the currents of the undesired propagating modes, while leaving the desired mode practically unchanged. Slots in waveguide walls thus act as mode filters.

Finally, we note that an ultimate way to deliver the energy from a guided propagating wave in a waveguide to the outside space (and to other devices in a communication or power-transfer system), and vice versa, is to leave one of its ends open, and radiate (or receive) a free-space propagating wave through this opening (aperture). Such open-ended waveguide antennas, Fig. 13.14(a), as well as horn antennas, Fig. 13.14(b), where the waveguide aperture is gradually increased to, essentially, better match the waveguide impedance $Z_{\rm TE}$ or $Z_{\rm TM}$, Fig. 13.1, to the free-space impedance η_0, Eq. (9.23), are also commonly used transmitting or receiving antenna elements.

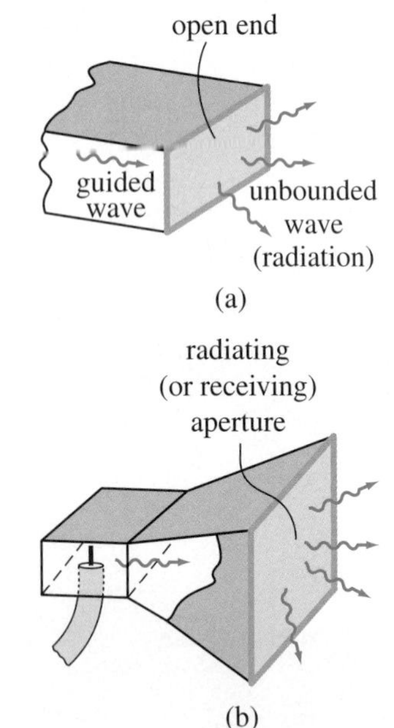

Figure 13.14 Radiation of a rectangular waveguide through its transversal aperture: (a) open-ended waveguide antenna and (b) horn antenna.

Example 13.12 Evaluation of the Emf Induced in Magnetic Probes in a Waveguide

A TE_{10} wave of frequency $f = 900$ MHz propagates along a lossless air-filled rectangular metallic waveguide of transverse dimensions $a = 24.765$ cm and $b = 12.383$ cm (WR-975 waveguide). The wave is received by a small wire loop (magnetic probe), of the loop surface area $S = 1\ \text{cm}^2$, attached to one of the guide walls. If the rms electromotive force (emf) induced in the loop is $\mathcal{E}_{\rm ind} = 2$ V, and the loop is attached to the conductor surface at either (a) $y = 0$ (bottom wall) or (b) $x = 0$ (left wall) as shown in Fig. 13.11(b), find the time-average power transported by the wave.

Solution Having in mind Eqs. (13.74), (13.22), and (13.20), we realize, since $(f_c)_{10} = 605.7$ MHz, that this propagating wave in the waveguide is a TE_{10} (dominant mode) one. The complex rms emf induced in any one of the loops, because they are small, is computed using the version of Faraday's law of electromagnetic induction in Eq. (10.24),

$$\underline{\mathcal{E}}_{\rm ind} = -j\omega\mu_0\underline{\mathbf{H}}\cdot\mathbf{S} \tag{13.119}$$

($\omega = 2\pi f$), where $\mathbf{S}$ is the loop surface area vector ($|\mathbf{S}| = S$), and $\underline{\mathbf{H}}$ the complex rms magnetic field intensity vector at the loop location.

(a) For the loop attached to the bottom wall ($y = 0$) at its center ($x = a/2$) and oriented in parallel to the left and right walls of the waveguide, the only magnetic field component generating the flux $\underline{\Phi}$ in Eq. (13.119) is $\underline{H}_x$, as indicated in Fig. 13.11(b). For the dominant mode, it is given in Eq. (13.25), and is maximum for $x = a/2$, so that the rms emf in the loop comes out to be

$$\mathcal{E}_{\text{ind}} = |\underline{\mathcal{E}}_{\text{ind}}| = \omega\mu_0 \left|\underline{H}_x\right|_{x=a/2} S = \omega\mu_0\beta \frac{a}{\pi} |\underline{H}_0| S. \qquad (13.120)$$

This equation can be solved for the magnitude of the complex constant $\underline{H}_0$ (present in all field expressions of the wave),

$$|\underline{H}_0| = \frac{\pi \mathcal{E}_{\text{ind}}}{\omega\mu_0\beta a S}, \qquad (13.121)$$

which is then substituted in Eq. (13.86) to obtain the time-average power of the wave,

$$P = \frac{\omega\mu_0\beta a^3 b|\underline{H}_0|^2}{2\pi^2} = \frac{ab\mathcal{E}_{\text{ind}}^2}{2\omega\mu_0\beta S^2} = 61.9 \text{ W} \quad (\beta = \beta_{10}), \qquad (13.122)$$

where the phase coefficient β_{10} is found from Eqs. (13.12) and (13.20). Of course, only a fraction of this power is actually received (collected) by the probe, but the probe, nevertheless, can receive all the relevant information (signal) carried by the TE_{10} wave along the waveguide.

(b) For the other loop in Fig. 13.11(b), we similarly have, using Eqs. (13.119) and (13.26),

$$\mathcal{E}_{\text{ind}} = \omega\mu_0 \left|\underline{H}_z\right|_{x=0} S = \omega\mu_0|\underline{H}_0|S, \qquad (13.123)$$

from which $P = (\beta a^3 b\mathcal{E}_{\text{ind}}^2)/(2\pi^2\omega\mu_0 S^2) = 74.8$ W.

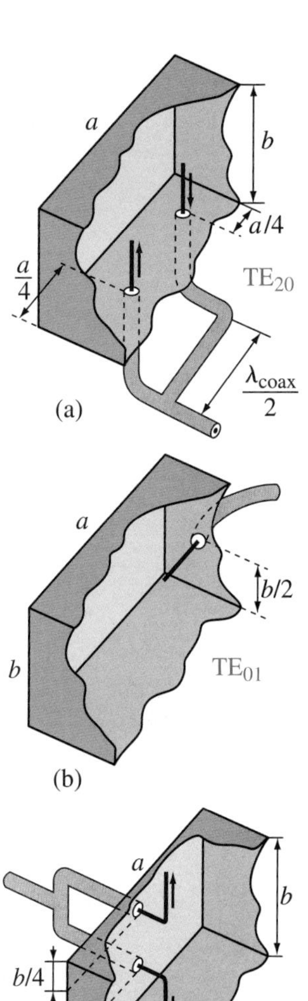

Example 13.13 Electric-Probe Waveguide Couplers for Higher Order Modes

Propose coupling configurations using electric probes for each of the higher order modes (modes higher than the dominant) that can propagate along the waveguide from Example 13.3.

Solution In analogy to the electric-probe waveguide coupler for the TE_{10} mode in Fig. 13.11(b), and given the electric-field distributions of TE_{mn} and TM_{mn} modes in Eqs. (13.55)–(13.56) and (13.65)–(13.67) and Figs. 13.4 and 13.5, Fig. 13.15 shows possible coupling configurations using electric probes for excitation/reception of TE_{20}, TE_{01}, TE_{11}, and TM_{11} modes (which can propagate under the specified conditions). It is a simple matter to verify that the probes are laid along the individual electric field components, at the locations of their maxima, to ensure the maximum coupling with the modal fields in the waveguide. For couplers consisting of two probes, the necessary relative reference directions of currents along the wires are indicated in the figure. In particular, note that the opposite relative directions of probe currents in the TE_{20} case are needed to couple with the peaks of the positive and negative half-waves in the function $\sin(2\pi x/a)$ in the expression for $\underline{E}_y$ for this mode at $x = a/4$ and $x = 3a/4$, respectively. Similarly, the oppositely directed currents in the TE_{11} case correspond to the positive and negative quarter-waves of $\cos(\pi y/b)$ along the y-axis in the $\underline{E}_y$ expression with $n = 1$. Note also that the phase shift of 180° between the two probes for the former case is easily achieved by inserting an extra half-wavelength long segment of coaxial cable between the probe feed points [$\beta_{\text{coax}}\lambda_{\text{coax}}/2 = (2\pi/\lambda_{\text{coax}})\,\lambda_{\text{coax}}/2 = \pi$].

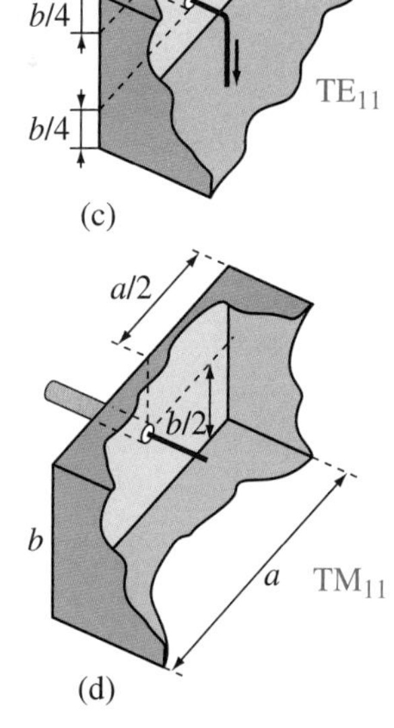

Figure 13.15 Coupling configurations using electric probes for several higher order modes in a rectangular waveguide; for Example 13.13.

Problems: 13.22–13.24; *MATLAB Exercises* (on Companion Website).

13.12 RECTANGULAR CAVITY RESONATORS

Next, we study electromagnetic resonators made from rectangular metallic waveguides with TE or TM waves, which are a TE/TM version of the plane-wave Fabry-Perot resonator, in Fig. 10.3, and TEM transmission-line resonators, Fig. 12.19. We recall that the Fabry-Perot resonator is obtained, essentially, by placing two parallel metallic (PEC) planes, that are a multiple of half-wavelengths apart, in the field of a uniform plane electromagnetic wave, perpendicularly to the direction of wave propagation, so that the resultant standing plane wave exists trapped between the two planes. Similarly, a section of a waveguide closed at both ends with new transversal conducting walls, thus forming a rectangular metallic box (cavity), represents a three-dimensional resonant wave structure (at certain resonant frequencies), called a rectangular cavity resonator. The resonant frequencies of the cavity depend, for an arbitrary wave mode, on all three dimensions of the parallelepiped, in addition to the parameters of the dielectric inside the structure. In general, any closed metallic structure (of an arbitrary shape), filled with an arbitrary dielectric medium, constitutes a cavity resonator, capable of storing high-frequency electromagnetic energy – at a discrete set of resonant frequencies. Note that the coaxial-cable resonator in Example 12.22 is a cavity resonator as well (a two-wire-line resonator is not, since it is open). Waveguide cavity resonators are typically used at frequencies higher than 1 GHz, and are important elements in a wide range of microwave applications, including oscillator circuits, filters, tuned amplifiers, frequency-meters (wavemeters), high-field generators, and microwave ovens. Laser cavities are also cavity resonators. In this section (and the following one), we assume that the resonator under consideration is lossless, i.e., that the cavity walls are perfectly conducting (PEC walls), and that the dielectric is perfect as well. Resonant cavities with small losses will be analyzed later in this chapter.

To determine the fields in a rectangular cavity resonator, let us first consider a rectangular waveguide short-circuited at only one end. In particular, let us assume that a PEC wall is placed in the plane $z = 0$ in the waveguide in Fig. 13.1, and that a TE_{10} wave (dominant mode) travels along the semi-infinite waveguide defined by $-\infty < z < 0$ in the positive z direction. The nonzero field components of this wave, which we refer to as the incident (or forward) wave, $\underline{E}_{\mathrm{i}y}$, $\underline{H}_{\mathrm{i}x}$, and $\underline{H}_{\mathrm{i}z}$, are given in Eqs. (13.24)–(13.26). The incident wave reflects back at the short circuit, so that, as in Fig. 10.1 for uniform plane waves, a reflected (or backward) wave, propagating in the negative z direction, also exists in the waveguide (it is radiated by the surface currents induced in the transversal PEC wall). The reflected wave is also a TE_{10} wave. To directly use Eqs. (13.24)–(13.26) for its field, we adopt a new rectangular coordinate system, $x'y'z'$ (attached to the waveguide), such that the z'-axis is in the backward direction ($z' = -z$). With reference to Fig. 13.16, the reflected field components are

$$\underline{E}'_{\mathrm{r}y} = -\mathrm{j}\omega\mu\,\frac{a}{\pi}\,\underline{H}_{\mathrm{r}0}\sin\left(\frac{\pi}{a}x'\right)\mathrm{e}^{-\mathrm{j}\beta z'},\quad \underline{H}'_{\mathrm{r}x} = \mathrm{j}\beta\,\frac{a}{\pi}\,\underline{H}_{\mathrm{r}0}\sin\left(\frac{\pi}{a}x'\right)\mathrm{e}^{-\mathrm{j}\beta z'},$$
$$\underline{H}'_{\mathrm{r}z} = \underline{H}_{\mathrm{r}0}\cos\left(\frac{\pi}{a}x'\right)\mathrm{e}^{-\mathrm{j}\beta z'}\quad (\underline{E}'_{\mathrm{r}x},\ \underline{E}'_{\mathrm{r}z},\ \underline{H}'_{\mathrm{r}y} = 0), \tag{13.124}$$

or, rewritten in the xyz coordinate system (used for the incident wave),

$$\underline{E}_{\mathrm{r}y} = -\underline{E}'_{\mathrm{r}y} = \mathrm{j}\omega\mu\,\frac{a}{\pi}\,\underline{H}_{\mathrm{r}0}\sin\left(\frac{\pi}{a}x\right)\mathrm{e}^{\mathrm{j}\beta z},\quad \underline{H}_{\mathrm{r}x} = \underline{H}'_{\mathrm{r}x} = \mathrm{j}\beta\,\frac{a}{\pi}\,\underline{H}_{\mathrm{r}0}\sin\left(\frac{\pi}{a}x\right)\mathrm{e}^{\mathrm{j}\beta z},$$
$$\underline{H}_{\mathrm{r}z} = -\underline{H}'_{\mathrm{r}z} = -\underline{H}_{\mathrm{r}0}\cos\left(\frac{\pi}{a}x\right)\mathrm{e}^{\mathrm{j}\beta z}. \tag{13.125}$$

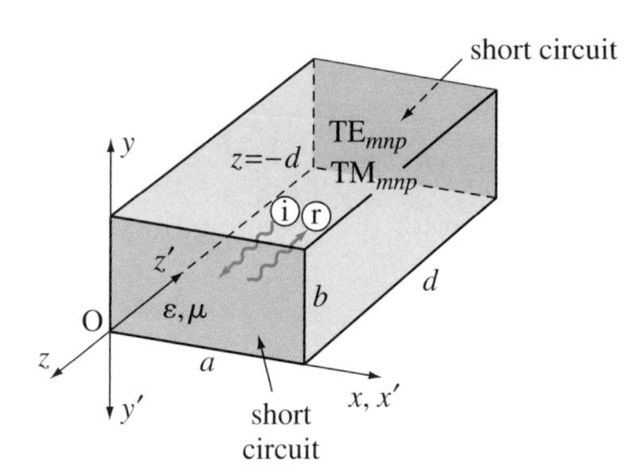

Figure 13.16 Rectangular cavity resonator, obtained by short-circuiting the rectangular metallic waveguide in Fig. 13.1 in two transversal planes.

The phase coefficient β is given by Eqs. (13.12) and (13.20) – for the dominant mode. The constant $\underline{H}_{r0}$, representing the complex "peak-value" of $\underline{H}_{rz}$ in the plane $z = 0$, is determined from the boundary condition for the tangential component of the total (incident plus reflected) electric field in this plane, as in Eq. (10.3) in the plane-wave case,

$$\left(\underline{E}_{iy} + \underline{E}_{ry}\right)\Big|_{z=0} = 0 \quad \longrightarrow \quad \underline{H}_{r0} = \underline{H}_0. \tag{13.126}$$

Using Eqs. (10.7), the expressions for the total field thus read

$$\underline{E}_y = \underline{E}_{\text{tot}\,y} = \underline{E}_{iy} + \underline{E}_{ry} = -2\omega\mu\,\frac{a}{\pi}\,\underline{H}_0 \sin\left(\frac{\pi}{a}x\right)\sin\beta z, \tag{13.127}$$

$$\underline{H}_x = \underline{H}_{\text{tot}\,x} = \underline{H}_{ix} + \underline{H}_{rx} = 2\text{j}\beta\,\frac{a}{\pi}\,\underline{H}_0 \sin\left(\frac{\pi}{a}x\right)\cos\beta z, \tag{13.128}$$

$$\underline{H}_z = \underline{H}_{\text{tot}\,z} = \underline{H}_{iz} + \underline{H}_{rz} = -2\text{j}\underline{H}_0 \cos\left(\frac{\pi}{a}x\right)\sin\beta z. \tag{13.129}$$

Of course, the boundary condition for the normal component of the total magnetic field at the transversal plate, $\underline{H}_{\text{tot}\,z} = 0$ for $z = 0$, is satisfied as well. We see that the resultant TE_{10} wave in a short-circuited waveguide exhibits standing-wave behavior in both x and z directions (the field is uniform in the y direction).

Let us now close the waveguide at its other end, introducing a transversal PEC wall also in a plane $z = -d$ in Fig. 13.16, with d designating the length of thus obtained rectangular PEC cavity (parallelepiped). However, in order to not disturb the field in Eqs. (13.127)–(13.129), this must be one of the planes where $\underline{E}_{\text{tot}\,y} = 0$ (and $\underline{H}_{\text{tot}\,z} = 0$), as in Eq. (10.10) and Fig. 10.3 for the plane-wave resonator, and similarly in Eq. (12.131) and Fig. 12.19 for the transmission-line resonator, and hence we have

$$\boxed{\beta d = p\pi \quad \longrightarrow \quad d = p\,\frac{\lambda_z}{2} \quad (p = 1, 2, \ldots),} \tag{13.130}$$

resonant length of a cavity

where λ_z is the guide wavelength, Eq. (13.110). With this, both $\underline{\mathbf{E}}_{\text{tang}} = 0$ and $\underline{\mathbf{H}}_{\text{norm}} = 0$, Eqs. (13.1), at all cavity walls. The wave described by Eqs. (13.127)–(13.129) and (13.130), corresponding to $m = 1$, $n = 0$, and an arbitrary p (positive integer), is denoted as a TE_{10p} cavity mode.

Although β in Eqs. (13.127)–(13.129) represents the phase coefficient of the dominant mode, let us generalize the condition in Eq. (13.130) to an arbitrary TE_{mn} or TM_{mn} mode in the waveguide, by assuming that the cutoff frequency in the expression for β in (13.12) is $f_c = (f_c)_{mn}$, Eq. (13.71). With this and Eq. (8.48),

Eq. (13.130), rewritten as $\beta/\pi = p/d$ and squared, results in

$$\frac{\beta^2}{\pi^2} = \frac{\omega^2}{\pi^2 c^2}\left(1 - \frac{f_c^2}{f^2}\right) = \left(\frac{2f}{c}\right)^2 - \left(\frac{m}{a}\right)^2 - \left(\frac{n}{b}\right)^2 = \left(\frac{p}{d}\right)^2 \tag{13.131}$$

($c = 1/\sqrt{\varepsilon\mu}$), which gives the following solution for the operating frequency, f, of the wave:

cavity resonance, TE_{mnp} or TM_{mnp} mode

$$\boxed{f = f_{res} = (f_{res})_{mnp} = \frac{c}{2}\sqrt{\left(\frac{m}{a}\right)^2 + \left(\frac{n}{b}\right)^2 + \left(\frac{p}{d}\right)^2} \quad (m, n, p = 0, 1, 2, \ldots).} \tag{13.132}$$

The same qualification regarding zero values of transverse mode indices m and n applies here as with the expression for the cutoff frequency $(f_c)_{mn}$ in Eq. (13.71), namely, that only one index can be zero for TE waves, and none for TM waves. Moreover, if we consider a cavity with no imposition of any given order between its dimensions a, b, and d, then the only restriction for integers m, n, and p in Eq. (13.132) is that not more than one of them can be zero at a time (for instance, a combination $m = n = 1$ and $p = 0$ is also possible). The frequency in Eq. (13.132) is the resonant frequency (f_{res}) of a mode (m, n, p), i.e., TE_{mnp} or TM_{mnp}, in the cavity. In other words, for given cavity dimensions (a, b, and d) and dielectric parameters (ε and μ), each mode can exist (oscillate) only at a single frequency, $f = (f_{res})_{mnp}$. We note that, in general, the oscillating condition for a cavity resonator ($f = f_{res}$) is much more stringent than the propagating condition for a waveguide ($f > f_c$). There are a triple infinite number of resonant frequencies, $(f_{res})_{mnp}$, and the corresponding resonance modal field configurations in the cavity, in both TE and TM versions. As in Eqs. (13.54)–(13.58) and (13.65)–(13.69), the integers m, n, and p (if nonzero) equal the number of half-wavelengths along the x-, y-, and z-axes that fit into a, b, and d, respectively. On the other hand, a zero index means that the field is uniform in that direction. Note that specifying $n \equiv 0$ and $p \equiv 0$ in the expression for $(f_{res})_{mnp}$ in Eq. (13.132), it becomes independent of b and d, and identical to that in Eq. (10.12), for the resonant frequency of the two-plane structure in Fig. 10.3. The Fabry-Perot resonator can therefore be considered as a one-dimensional (planar) version of the cavity resonator in Fig. 13.16 (with $b \to \infty$ and $d \to \infty$).

If the cavity dimensions are not all the same and the coordinate axes (x, y, and z) in Fig. 13.16 are chosen such that $a > b$ and $d > b$, out of all solutions the TE_{101} mode ($m = p = 1, n = 0$) has the lowest frequency, given by

resonant frequency, TE_{101} mode

$$\boxed{(f_{res})_{101} = \frac{c}{2}\sqrt{\frac{1}{a^2} + \frac{1}{d^2}},} \tag{13.133}$$

and is hence termed the dominant cavity mode.[13] Using Eqs. (13.127)–(13.129) and (13.130) with $p = 1$, the field components of this mode are

$\underline{E}_y - TE_{101}$ mode

$$\underline{E}_y = -2\omega\mu\,\frac{a}{\pi}\,\underline{H}_0 \sin\left(\frac{\pi}{a}x\right)\sin\left(\frac{\pi}{d}z\right), \tag{13.134}$$

$\underline{H}_x - TE_{101}$ mode

$$\underline{H}_x = 2\mathrm{j}\,\frac{a}{d}\,\underline{H}_0 \sin\left(\frac{\pi}{a}x\right)\cos\left(\frac{\pi}{d}z\right), \tag{13.135}$$

$\underline{H}_z - TE_{101}$ mode

$$\underline{H}_z = -2\mathrm{j}\underline{H}_0 \cos\left(\frac{\pi}{a}x\right)\sin\left(\frac{\pi}{d}z\right) \tag{13.136}$$

[13] For $a = b = d$ (cubical cavity), $(f_{res})_{101} = (f_{res})_{110} = (f_{res})_{011} = c\sqrt{2}/(2a)$.

($\underline{E}_x$, $\underline{E}_z$, $\underline{H}_y = 0$). They exhibit a single half-wavelength variation in both x and z directions ($m = p = 1$), and no variation in the y direction ($n = 0$). Field expressions of arbitrary higher order TE and TM modes in a rectangular cavity will be derived in examples.

Example 13.14 Wave Impedance in a Short-Circuited Waveguide

A TE_{10} wave propagates, at a frequency f, in the positive z direction along a lossless rectangular metallic waveguide of transverse dimensions a and b and dielectric parameters ε and μ, and is incident on a short-circuiting PEC plate placed in the plane $z = 0$. Find the wave impedance of the resultant wave in the waveguide.

Solution The total (incident plus reflected) wave in the short-circuited waveguide is a standing TE_{10} wave, whose transverse electric and magnetic field components, $\underline{E}_y$ and $\underline{H}_x$, are given in Eqs. (13.127) and (13.128), respectively. Using Eq. (13.80), the wave impedance of this wave is

$$\underline{Z}_{\text{standing}} = -\frac{\underline{E}_y}{\underline{H}_x} = -\mathrm{j}\,\frac{\omega\mu}{\beta}\tan\beta z = -\mathrm{j}Z_{\mathrm{TE}}\tan\beta z, \tag{13.137}$$

where $\omega = 2\pi f$ (angular frequency) and the phase coefficient β is computed by Eqs. (13.12) and (13.20). Note that $\underline{Z}_{\text{standing}}$, unlike the wave impedance (Z_{TE}) of each of the two traveling TE_{10} waves (incident and reflected waves) in the structure, in Eq. (13.76), is a function of the coordinate z. Note also that it is purely imaginary (reactive), and so is the corresponding component of the complex Poynting vector, $\underline{\mathcal{P}}_z = -\underline{E}_y\underline{H}_x^*\,\hat{\mathbf{z}}$ [Eq. (13.84)]. This is characteristic for all pure standing electromagnetic waves [e.g., see Eq. (10.20) for a pure standing uniform plane wave], and means no net real power flow by the wave (in the axial direction).

Example 13.15 Field Expressions for an Arbitrary TE Mode in a Resonant Cavity

Find the expressions for the electric and magnetic fields of an arbitrary TE_{mnp} wave mode in a lossless rectangular metallic cavity of dimensions a, b, and d (Fig. 13.16), filled with a homogeneous dielectric of permittivity ε and permeability μ.

Solution Considering first a short-circuited waveguide, with a PEC plate inserted at the guide end defined by $z = 0$ in Fig. 13.16, the incident TE_{mn} wave is that in Eqs. (13.54)–(13.58), and the field components of the reflected wave, which is also of the TE_{mn} form, are written in analogy to Eqs. (13.124) and (13.125) for the TE_{10} case. In doing so, we perform the conversion from the $x'y'z'$ coordinate system in Fig. 13.16, in which the reflected wave is originally represented, to the xyz coordinate system, used for the incident wave, as well as for the resultant (incident plus reflected) wave. For instance, the x-component of the reflected electric field vector is given by

$$\underline{E}_{\mathrm{r}x} = -\frac{\mathrm{j}\omega\mu}{k^2}\frac{n\pi}{b}\underline{H}_{\mathrm{r}0}\cos\left(\frac{m\pi}{a}x\right)\sin\left(\frac{n\pi}{b}y\right)\mathrm{e}^{\mathrm{j}\beta z} \quad (\underline{H}_{\mathrm{r}0} = \underline{H}_0), \tag{13.138}$$

where the constant $\underline{H}_{\mathrm{r}0}$ is determined from the boundary condition for $\underline{E}_{\mathrm{i}x} + \underline{E}_{\mathrm{r}x}$ for $z = 0$, as in Eq. (13.126). Therefore, the corresponding total field component, similarly to Eq. (13.127), comes out to be

$$\underline{E}_x = \underline{E}_{\mathrm{tot}\,x} = \underline{E}_{\mathrm{i}x} + \underline{E}_{\mathrm{r}x} = \frac{2\omega\mu}{k^2}\frac{n\pi}{b}\underline{H}_0\cos\left(\frac{m\pi}{a}x\right)\sin\left(\frac{n\pi}{b}y\right)\sin\beta z, \tag{13.139}$$

and the same can be done for other field components of the wave. Closing then the waveguide at the other end ($z = -d$), in Fig. 13.16, we obtain the condition in Eq. (13.130), from which the phase coefficient of the wave is $\beta = p\pi/d$. Hence, the final expressions for the TE_{mnp} field in the cavity read

$$\boxed{\underline{E}_x = \frac{2\omega\mu}{k^2}\frac{n\pi}{b}\underline{H}_0\cos\left(\frac{m\pi}{a}x\right)\sin\left(\frac{n\pi}{b}y\right)\sin\left(\frac{p\pi}{d}z\right),} \tag{13.140}$$

$\underline{E}_x$ – TE_{mnp} *mode*

$\underline{E}_y$ – TE_{mnp} mode

$$\underline{E}_y = -\frac{2\omega\mu}{k^2}\frac{m\pi}{a}\underline{H}_0 \sin\left(\frac{m\pi}{a}x\right)\cos\left(\frac{n\pi}{b}y\right)\sin\left(\frac{p\pi}{d}z\right), \tag{13.141}$$

$\underline{H}_x$ – TE_{mnp} mode

$$\underline{H}_x = \frac{2\mathrm{j}}{k^2}\frac{m\pi}{a}\frac{p\pi}{d}\underline{H}_0 \sin\left(\frac{m\pi}{a}x\right)\cos\left(\frac{n\pi}{b}y\right)\cos\left(\frac{p\pi}{d}z\right), \tag{13.142}$$

$\underline{H}_y$ – TE_{mnp} mode

$$\underline{H}_y = \frac{2\mathrm{j}}{k^2}\frac{n\pi}{b}\frac{p\pi}{d}\underline{H}_0 \cos\left(\frac{m\pi}{a}x\right)\sin\left(\frac{n\pi}{b}y\right)\cos\left(\frac{p\pi}{d}z\right), \tag{13.143}$$

$\underline{H}_z$ – TE_{mnp} mode

$$\underline{H}_z = -2\mathrm{j}\underline{H}_0 \cos\left(\frac{m\pi}{a}x\right)\cos\left(\frac{n\pi}{b}y\right)\sin\left(\frac{p\pi}{d}z\right) \tag{13.144}$$

($\underline{E}_z = 0$), where $m, n, p = 0, 1, 2, \ldots$, the parameter k^2 is computed from Eq. (13.59), and $\omega = 2\pi f$, with $f = (f_{\mathrm{res}})_{mnp}$ being the resonant frequency of the TE_{mnp} mode in the cavity, Eq. (13.132). Of course, Eqs. (13.140)–(13.144) for $m = p = 1$ and $n = 0$ reduce to Eqs. (13.134)–(13.136), describing the existing field components of the dominant cavity mode (TE_{101}).

Example 13.16 Field of an Arbitrary TM Cavity Mode

Repeat the previous example but for an arbitrary TM_{mnp} resonance mode in the cavity.

Solution We start with the field components of the incident TM_{mn} wave in the cavity (short-circuited waveguide), in Eqs. (13.65)–(13.69). Having in mind the expression for the x- or y-component of the electric field vector of the reflected wave, as in Eq. (13.138), it is obvious that the corresponding constant $\underline{E}_{\mathrm{r}0}$ (in place of $\underline{H}_{\mathrm{r}0}$) must be set to $\underline{E}_{\mathrm{r}0} = \underline{E}_0$ in order for the boundary condition for the tangential resultant electric field at the PEC surface at $z = 0$ in Fig. 13.16 to be satisfied. We then use the same procedure as in obtaining Eqs. (13.140)–(13.144), which yields the following for an arbitrary TM_{mnp} cavity mode:

$\underline{E}_x$ – TM_{mnp} mode

$$\underline{E}_x = -\frac{2}{k^2}\frac{m\pi}{a}\frac{p\pi}{d}\underline{E}_0 \cos\left(\frac{m\pi}{a}x\right)\sin\left(\frac{n\pi}{b}y\right)\sin\left(\frac{p\pi}{d}z\right), \tag{13.145}$$

$\underline{E}_y$ – TM_{mnp} mode

$$\underline{E}_y = -\frac{2}{k^2}\frac{n\pi}{b}\frac{p\pi}{d}\underline{E}_0 \sin\left(\frac{m\pi}{a}x\right)\cos\left(\frac{n\pi}{b}y\right)\sin\left(\frac{p\pi}{d}z\right), \tag{13.146}$$

$\underline{E}_z$ – TM_{mnp} mode

$$\underline{E}_z = 2\underline{E}_0 \sin\left(\frac{m\pi}{a}x\right)\sin\left(\frac{n\pi}{b}y\right)\cos\left(\frac{p\pi}{d}z\right), \tag{13.147}$$

$\underline{H}_x$ – TM_{mnp} mode

$$\underline{H}_x = \frac{2\mathrm{j}\omega\varepsilon}{k^2}\frac{n\pi}{b}\underline{E}_0 \sin\left(\frac{m\pi}{a}x\right)\cos\left(\frac{n\pi}{b}y\right)\cos\left(\frac{p\pi}{d}z\right), \tag{13.148}$$

$\underline{H}_y$ – TM_{mnp} mode

$$\underline{H}_y = -\frac{2\mathrm{j}\omega\varepsilon}{k^2}\frac{m\pi}{a}\underline{E}_0 \cos\left(\frac{m\pi}{a}x\right)\sin\left(\frac{n\pi}{b}y\right)\cos\left(\frac{p\pi}{d}z\right) \tag{13.149}$$

($\underline{H}_z = 0$). Here $m, n, p = 1, 2, \ldots$, and the case $p = 0$ can also be included, while the expressions for k^2 and $\omega = 2\pi(f_{\mathrm{res}})_{mnp}$ are the same as in the TE case, given by Eqs. (13.59) and (13.132), respectively.

Problems: 13.25–13.30; *Conceptual Questions* (on Companion Website): 13.30 and 13.31; *MATLAB Exercises* (on Companion Website).

13.13 ELECTROMAGNETIC ENERGY STORED IN A CAVITY RESONATOR

In order to start energy computations of waveguide cavity resonators, we recall that, during the course of time, the stored electromagnetic energy in any resonant electromagnetic structure periodically oscillates between the electric and magnetic fields, as they alternate between maximum and zero values. This energy fluctuation in the case of the Fabry-Perot resonator (Fig. 10.3) is illustrated in Fig. 10.4. To show that a complete periodic exchange of the stored energy between the electric and magnetic fields of a standing electromagnetic wave occurs also in a rectangular cavity resonator, Fig. 13.16, let us have a look at the field expressions in Eqs. (13.134)–(13.136), (13.140)–(13.144), and (13.145)–(13.149). We note that the complex electric and magnetic field intensity vectors of the dominant cavity mode (TE_{101}), as well as of an arbitrary higher order (TE_{mnp} or TM_{mnp}) mode, can be written as

$$\underline{\mathbf{E}} = \underline{H}_0 \mathbf{A}_1, \quad \underline{\mathbf{H}} = \mathrm{j}\underline{H}_0 \mathbf{A}_2, \tag{13.150}$$

where $\mathbf{A}_1$ and $\mathbf{A}_2$ are purely real vectors (that, of course, depend on coordinates x, y, and z), and the complex constant $\underline{H}_0$ is substituted by $\underline{E}_0$ for TM waves. Denoting the argument (phase angle) of $\underline{H}_0$ (or $\underline{E}_0$) by ξ ($\underline{H}_0 = |\underline{H}_0|\,\mathrm{e}^{\mathrm{j}\xi}$) and using Eq. (8.66), the instantaneous field intensity vectors in the cavity are then [similarly to the expressions in Eqs. (10.9)]

$$\boxed{\mathbf{E}(t) = |\underline{H}_0|\mathbf{A}_1\sqrt{2}\cos(\omega t + \xi), \quad \mathbf{H}(t) = -|\underline{H}_0|\mathbf{A}_2\sqrt{2}\sin(\omega t + \xi).} \tag{13.151}$$

time-phase quadrature of cavity field vectors

We see that $\mathbf{E}(t)$ and $\mathbf{H}(t)$ are in time-phase quadrature, i.e., are by 90° out of phase with respect to each other, at every point of the cavity, which results from the difference in "j" in complex expressions in Eqs. (13.150).[14] We also realize that both $\mathbf{E}(t)$ and $\mathbf{H}(t)$ are linearly polarized vectors – the tip of $\mathbf{E} = \mathbf{E}(x, y, z, t)$ at a point (x, y, z) oscillates, in the course of time, along the line defined by the time-constant vector $\mathbf{A}_1 = \mathbf{A}_1(x, y, z)$, and analogously for $\mathbf{H} = \mathbf{H}(x, y, z, t)$. We finally conclude, from Eqs. (13.151), that, like in Fig. 10.4, there are instants of time (t_1) at which the electric field is maximum [$\cos(\omega t + \xi) = \pm 1$], while the magnetic field is zero [$\sin(\omega t + \xi) = 0$],

$$\boxed{\mathbf{E}(x, y, z, t_1) = \pm|\underline{H}_0|\mathbf{A}_1(x, y, z)\sqrt{2} \quad \text{(maximum)}, \quad \mathbf{H}(x, y, z, t_1) = 0,} \tag{13.152}$$

energy all electric

at every point (x, y, z) in the cavity. At some other times ($t_2 = t_1 \pm T/4$), the situation is just opposite:

$$\boxed{\mathbf{E}(x, y, z, t_2) = 0, \quad \mathbf{H}(x, y, z, t_2) = \mp|\underline{H}_0|\mathbf{A}_2(x, y, z)\sqrt{2} \quad \text{(maximum)},} \tag{13.153}$$

energy all magnetic

everywhere in the resonator [T is the time period of the time-harmonic variation of the wave, Eq. (8.49)]. So, the instantaneous electromagnetic energy of the cavity, $W_{\mathrm{em}}(t)$, which is constant in time (assuming no Joule's losses in the resonator), is all electric at instants t_1, Eqs. (13.152), and all magnetic for $t = t_2$, Eqs. (13.153). At intermediate times, the energy is partly electric and partly magnetic, as it moves from the electric to the magnetic field, and vice versa.

We now restrict our attention to the dominant resonance mode, and compute its total energy, W_{em}, at one of the instants t_1, namely, as the maximum electric energy, $(W_{\mathrm{e}})_{\max}$, of the cavity. Since $\mathbf{E}(t)$ is linearly polarized, the peak-value ($E_{\max}$) equals $\sqrt{2}$ times rms (E_{rms}) of its instantaneous magnitude (this holds true for an

[14]Of course, the 90° phase shift between instantaneous electric and magnetic fields (difference in "j" in complex field expressions) at every point of the structure is characteristic for all electromagnetic resonators. This is apparent, for example, in Eqs. (10.8) and (10.9) for the Fabry-Perot resonator, Eqs. (12.111), (12.112), and (12.114) for transmission-line resonators, and Eqs. (8.69) and (3.45) for a simple resonant LC circuit.

arbitrary TE or TM resonance mode in the cavity, where **E** has two or three nonzero Cartesian components). The rms field intensity, in turn, equals the magnitude of the complex field expression ($\underline{E}_y$) in Eq. (13.134), and hence

peak-value of E-field, TE_{101} mode

$$\boxed{E_{\max} = E_{\mathrm{rms}}\sqrt{2} = |\underline{E}_y|\sqrt{2}.} \tag{13.154}$$

From Eq. (8.160), the electromagnetic energy (at any time) in the cavity is then obtained by integrating the peak electric energy density throughout the volume of the cavity dielectric, v_d, that is, the entire cavity interior (Fig. 13.16),

stored energy, TE_{101} mode

$$\begin{aligned} W_{\mathrm{em}} &= W_{\mathrm{em}}(t) = (W_\mathrm{e})_{\max} = \int_{v_\mathrm{d}} (w_\mathrm{e})_{\max}\,\mathrm{d}v = \int_{v_\mathrm{d}} \frac{1}{2}\varepsilon E_{\max}^2\,\mathrm{d}v \\ &= \varepsilon \int_{v_\mathrm{d}} |\underline{E}_y|^2\,\mathrm{d}v = \frac{4\omega^2\varepsilon\mu^2 a^2|\underline{H}_0|^2}{\pi^2}\int_{x=0}^{a}\int_{z=-d}^{0} \sin^2\left(\frac{\pi}{a}x\right)\sin^2\left(\frac{\pi}{d}z\right)\underbrace{b\,\mathrm{d}x\,\mathrm{d}z}_{\mathrm{d}v} \\ &= \frac{\omega^2\varepsilon\mu^2 a^3 bd|\underline{H}_0|^2}{\pi^2} = \mu abd\left(\frac{a^2}{d^2}+1\right)|\underline{H}_0|^2, \end{aligned} \tag{13.155}$$

where $\omega = 2\pi f$ and $f = (f_{\mathrm{res}})_{101}$ is the resonant frequency of the dominant cavity mode, given in Eq. (13.133). The elemental volume for integration, $\mathrm{d}v$, is adopted in the form of a thin rectangular prism of height b and basis dimensions $\mathrm{d}x$ and $\mathrm{d}y$. Finally, the integrals in x and z equal $a/2$ and $d/2$, respectively, from Eq. (13.86).

Example 13.17 General Proof of Peak Magnetic and Electric Energy Equality

Prove that the maximum (peak) magnetic and electric energies in a metallic cavity resonator are equal without knowing and using Eqs. (13.152) and (13.153), i.e., the fact that there are instants of time when either electric or magnetic field is zero in the entire cavity domain.

Solution The equality of $(W_\mathrm{m})_{\max}$ and $(W_\mathrm{e})_{\max}$ comes directly from Poynting's theorem in complex form, Eq. (8.196), applied to the cavity domain (v_d). Namely, the second integral on the right-hand side of Eq. (8.196) is exactly the difference of these energies, $\Delta W = (W_\mathrm{m})_{\max} - (W_\mathrm{e})_{\max}$. Since all other integrals in the equation are zero, assuming that there are no impressed sources in v_d, that the cavity dielectric is perfect, and that the walls are impenetrable (PEC), ΔW must be zero as well, which gives $(W_\mathrm{m})_{\max} = (W_\mathrm{e})_{\max}$.

Note that this equality and its proof are valid for an arbitrary lossless cavity resonator, of any shape and any material composition inside the cavity, and not necessarily for rectangular cavities with a homogeneous dielectric, in Fig. 13.16.

Example 13.18 Energy Probing by a Small Loop Attached to a Cavity Wall

A TE_{101} wave oscillates in an air-filled rectangular metallic cavity resonator of dimensions $a = 16$ cm, $b = 8$ cm, and $d = 18$ cm. A small wire loop, whose surface area is $S = 0.65\ \mathrm{cm}^2$, is attached at the center of one of the cavity walls of dimensions b and d, such that the loop plane is parallel to the walls of dimensions a and b, as shown in Fig. 13.17, and the measured rms value of the emf induced in the loop amounts to $\mathcal{E}_{\mathrm{ind}} = 12$ V. Under these circumstances, find the electromagnetic energy stored in the cavity.

Solution We note the similarity with computing the transported power in a waveguide from the loop emf in Example 13.12. Here, for the position and orientation of the loop in Fig. 13.17, the relevant component of the magnetic field intensity vector resulting in the flux

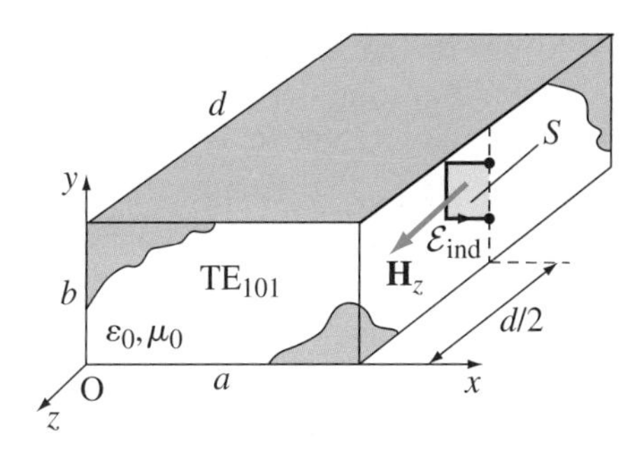

Figure 13.17 Computation of the emf induced in a small loop attached to a cavity wall; for Example 13.18.

$\underline{\Phi}$ in Eq. (13.119) is $\underline{H}_z$, given by Eq. (13.136) for the dominant cavity mode, and hence the following expression for the rms emf in the loop:

$$\mathcal{E}_{\text{ind}} = |\underline{\mathcal{E}}_{\text{ind}}| = \omega\mu_0 \left|\underline{H}_z\right|_{x=a,\, z=-d/2} S = \omega\mu_0 2|\underline{H}_0|S, \tag{13.156}$$

which, in turn, yields $|\underline{H}_0| = \mathcal{E}_{\text{ind}}/(2\omega\mu_0 S)$. Employing then Eq. (13.155), the electromagnetic energy stored in the cavity equals

$$W_{\text{em}} = \frac{\varepsilon_0 a^3 bd\mathcal{E}_{\text{ind}}^2}{4\pi^2 S^2} = 451 \text{ nJ}. \tag{13.157}$$

Problems: 13.31–13.37; *Conceptual Questions* (on Companion Website): 13.32–13.35; *MATLAB Exercises* (on Companion Website).

13.14 QUALITY FACTOR OF RECTANGULAR CAVITIES WITH SMALL LOSSES

Neglecting the losses in a cavity resonator, its electromagnetic energy, $W_{\text{em}}(t)$, established using some coupling configuration (e.g., Figs. 13.11–13.13), remains the same indefinitely (to $t \to \infty$), even though the excitation (source) might be removed (turned off). Here, of course, we also assume that the cavity is perfectly sealed (any opening used for excitation closed), to prevent energy leakage (radiation) outside. In a real (lossy) resonator, on the other hand, $W_{\text{em}}(t)$ decreases exponentially with time, as given by Eq. (12.141), with δ_{r} and τ being the damping factor and time constant, respectively, of the resonator (the fields in the cavity decay as $\text{e}^{-t/\tau}$). If Joule's losses in the structure are small, we can use the no-loss expressions for the energies of different resonance modes, e.g., Eq. (13.155), and just multiply them by $\text{e}^{-2t/\tau}$. From Eq. (12.148), τ is proportional to the quality factor, Q, of the structure, so the higher the Q the slower the damping of the resonator ($Q \to \infty$ for an ideal resonator). We recall that the Q factor also defines the bandwidth of a resonator, so the higher the Q of a resonant structure the sharper the resonance and higher the frequency selectivity of the device. In our analysis, moreover, Q (under the low-loss assumption) is determined using the field distributions for the lossless case (perturbation method).

In specific, the Q factor of a rectangular cavity resonator (Fig. 13.16) is found using Eq. (12.146), for which we need to compute the time-average loss power in the structure. This computation largely parallels the evaluation of conductor and dielectric losses in a real rectangular waveguide, with small losses, presented in Section 13.9. We consider the dominant resonance mode (TE_{101}) in the cavity.

The time-average power of Joule's losses in the cavity conductor (metallic walls), P_c, is determined from Eq. (10.90), by integrating the surface power density $R_s|\underline{\mathbf{H}}_{tang}|^2$ over the interior surface S of the cavity. Here, R_s is the surface resistance of the conductor, given in Eq. (13.94), and $\underline{\mathbf{H}}_{tang}$ is the tangential component of the complex rms magnetic field intensity vector on S, found from Eqs. (13.135) and (13.136) – for the TE_{101} mode. In analogy to the integration in Eq. (13.96), P_c equals a sum of the following integrals over the three pairs of parallel walls (plates) constituting the cavity:

$$P_c = \oint_S R_s|\underline{\mathbf{H}}_{tang}|^2 \, dS = \underbrace{2\int_{x=0}^{a}\int_{z=-d}^{0} R_s \left(|\underline{H}_x|^2 + |\underline{H}_z|^2\right)\Big|_{y=0 \text{ or } b} dx\, dz}_{\text{bottom and top plates}}$$

$$+ \underbrace{2\int_{y=0}^{b}\int_{z=-d}^{0} R_s \left|\underline{H}_z\right|^2_{x=0 \text{ or } a} dy\, dz}_{\text{left and right plates}} + \underbrace{2\int_{x=0}^{a}\int_{y=0}^{b} R_s \left|\underline{H}_x\right|^2_{z=0 \text{ or } -d} dx\, dy}_{\text{front and back plates}}, \quad (13.158)$$

where the use is made of the field symmetry, i.e., of the fact that the relevant field components in the surface integrals have the same magnitudes at each pair of parallel walls. Substituting the expressions for these components, we have

$$P_c = 8R_s|\underline{H}_0|^2 \left\{ \int_0^a \int_{-d}^0 \left[\frac{a^2}{d^2} \sin^2\left(\frac{\pi}{a}x\right)\cos^2\left(\frac{\pi}{d}z\right) \right.\right.$$

$$\left. + \cos^2\left(\frac{\pi}{a}x\right)\sin^2\left(\frac{\pi}{d}z\right) \right] dx\, dz + \int_{-d}^0 \sin^2\left(\frac{\pi}{d}z\right) b\, dz$$

$$\left. + \frac{a^2}{d^2}\int_0^a \sin^2\left(\frac{\pi}{a}x\right) b\, dx \right\} = 2R_s|\underline{H}_0|^2\left(\frac{a^3}{d} + ad + 2bd + \frac{2a^3 b}{d^2}\right) \quad (13.159)$$

[for the solution of individual integrals in x and z, see also Eq. (13.86)]. Combination of Eqs. (12.146), (8.48), (13.133), (13.155), and (13.159) then gives the quality factor Q_c of the cavity associated with conductor losses,

quality factor for cavity walls, TE_{101} mode

$$\boxed{Q_c = \omega_{res}\frac{W_{em}}{P_c} = \frac{\pi\eta}{2R_s}\frac{b(a^2+d^2)^{3/2}}{ad(a^2+d^2)+2b(a^3+d^3)},} \quad (13.160)$$

with $\omega_{res} = (\omega_{res})_{101} = \pi\sqrt{a^2+d^2}/(\sqrt{\varepsilon\mu}\, ad)$ being the resonant angular frequency of the dominant cavity mode, and $\eta = \sqrt{\mu/\varepsilon}$ the intrinsic impedance of the cavity dielectric. For a PEC cavity ($\sigma_c \to \infty$), $P_c = 0$ and $R_s = 0$, and hence $Q_c \to \infty$.

The time-average loss power in the imperfect dielectric in Fig. 13.16, P_d, is given by the volume version (with integration over the dielectric volume, v_d) of Eq. (13.95). We see that thus expressed P_d is proportional to the energy W_{em} in Eq. (13.155), provided that the dielectric is homogeneous. Accordingly, Eq. (12.146) tells us that the factor Q_d representing the losses in the dielectric is

quality factor for cavity dielectric, TE_{101} mode

$$\boxed{Q_d = \omega_{res}\frac{W_{em}}{P_d} = \omega_{res}\frac{\varepsilon\int_{v_d}|\underline{E}_y|^2\, dv}{\sigma_d\int_{v_d}|\underline{E}_y|^2\, dv} = \omega_{res}\frac{\varepsilon}{\sigma_d} = \frac{1}{\tan\delta_d} = \frac{\varepsilon'}{\varepsilon''},} \quad (13.161)$$

where σ_d and $\tan\delta_d$ are, respectively, the conductivity and loss tangent, and ε' and ε'' the real and imaginary part of the high-frequency complex permittivity of the dielectric [see Eqs. (9.125) and (9.130)]. These expressions for Q_d are valid for any TE or TM mode in the cavity. Note that they are the same as for a transmission-line resonator filled with the same dielectric, Eq. (12.154). If the dielectric is considered to be perfect (σ_d, $\varepsilon'' = 0$), $Q_d \to \infty$.

Finally, the total quality factor of the cavity, Q, is obtained from Q_c and Q_d using Eq. (12.156). In most applications, the cavity is air-filled, resulting in $Q_d \to \infty$ and $Q = Q_c$. As we shall see in an example, with waveguide cavities it is possible to achieve extremely high Q values, on the order of 30,000. This is by one and two orders of magnitude higher than the maximum Q values of transmission-line resonators and lumped resonant circuits (e.g., series and parallel resonant RLC circuits), respectively.

Example 13.19 Quality Factor of a Cubical Cavity

Compute the Q factor for the dominant resonance mode (TE_{101}) of an air-filled cubical cavity with edge length $a = 21.2$ cm and copper walls.

Solution The electromagnetic energy stored in the cavity and time-average power of conductor losses are evaluated using Eqs. (13.155) and (13.159), respectively, with $a = b = d$, and Eq. (12.146) then gives the following expression for the cavity quality factor:

$$W_{em} = 2\mu_0 a^3 |\underline{H}_0|^2 \quad \text{and} \quad P_c = 12 R_s a^2 |\underline{H}_0|^2 \quad \longrightarrow \quad Q = \omega_{res}\frac{W_{em}}{P_c} = \frac{\pi\mu_0 f_{res} a}{3R_s} \tag{13.162}$$

($\omega_{res} = 2\pi f_{res}$). By means of Eq. (13.133), the resonant frequency of the dominant cavity mode for the cube ($a = d$) equals $f_{res} = (f_{res})_{101} = c_0/(\sqrt{2}a) = 1$ GHz ($c_0 = 3 \times 10^8$ m/s), and the surface resistance of copper, Eq. (10.80), at this frequency amounts to $R_s = 8.25$ mΩ/square. Hence, the cube quality factor comes out to be as high as $Q = 33{,}816$.

Example 13.20 Quality Factor of a Teflon-Filled Cavity

What is the Q factor of the cubical cavity from the previous example if it is filled with teflon, whose relative permittivity is $\varepsilon_r = 2.1$ and loss tangent $\tan\delta_d = 10^{-4}$.

Solution Using Eq. (13.161), the quality factor associated with the losses in the cavity dielectric (teflon) is

$$Q_d = \frac{1}{\tan\delta_d} = 10{,}000 \qquad [(Q_d)_{air} \to \infty]. \tag{13.163}$$

Of course, this is a big change with respect to the case (infinite Q_d) with air dielectric. On the other side, although the conductor of the resonator is the same (copper) as for the empty cavity, and only dielectric is changed, the factor Q_c for the conductor losses changes as well. The reason is a different electrical size of the teflon-filled cavity, and thus different TE_{101} resonant frequency, as compared to its air-filled counterpart,

$$(f_{res})_{teflon} = \frac{c}{\sqrt{2}a} = \frac{c_0/\sqrt{\varepsilon_r}}{\sqrt{2}a} = \frac{(f_{res})_{air}}{\sqrt{\varepsilon_r}} = 690 \text{ MHz}, \tag{13.164}$$

with which, Eqs. (10.80) and (13.162) now result in the surface resistance $R_s = 6.86$ mΩ/square and $Q_c = 28{,}126$, respectively [note that $(Q_c)_{air} = 33{,}816$, from the previous example]. The overall quality factor of the cavity, Eq. (12.156), amounts to

$$Q = \frac{Q_c Q_d}{Q_c + Q_d} = 7{,}377. \tag{13.165}$$

So, Q of the teflon-filled cavity is considerably lower than that of the empty cavity, which is somewhat due to the lower resonant frequency and associated reduction in Q_c, but primarily because of the low Q_d (as compared to Q_c).

Example 13.21 Cooking Food in a Microwave Oven

A microwave oven in the form of a rectangular metallic cavity resonator of dimensions a, b, and d operates at the TE_{101} resonance of the cavity. A piece of food of rectangular shape and dimensions $a/2$, $b/2$, and $d/2$ is placed in the oven, centrally at the bottom plate, as shown in Fig. 13.18. The piece is a lossy homogeneous dielectric of conductivity σ_d. (a) Assuming that the electric field distribution in the cavity is the same as if it were empty, find the time-average power of Joule's losses in the food piece. (b) Repeat (a) if the dielectric completely fills up the cavity.

Solution

(a) The time-average loss power in the food piece, $(P_J)_{\text{in food}}$, is obtained by integrating the corresponding power density, $p_J = \sigma_d |\underline{E}_y|^2$, with $\underline{E}_y$ being the only existing electric field component of the dominant cavity mode (TE_{101}), given in Eq. (13.134), throughout the volume (v_{food}) of the piece [see Eq. (13.161)], so we have (Fig. 13.18)

$$(P_J)_{\text{in food}} = \int_{v_{\text{food}}} \sigma_d |\underline{E}_y|^2 \, dv = K \int_{x=a/4}^{3a/4} \int_{z=-3d/4}^{-d/4} \sin^2\left(\frac{\pi}{a}x\right) \sin^2\left(\frac{\pi}{d}z\right) \underbrace{\frac{b}{2}\, dx\, dz}_{dv}$$

$$= \frac{Kb}{8} \underbrace{\int_{a/4}^{3a/4} \left[1 - \cos\left(\frac{2\pi}{a}x\right)\right] dx}_{a/2 + a/\pi} \underbrace{\int_{-3d/4}^{-d/4} \left[1 - \cos\left(\frac{2\pi}{d}z\right)\right] dz}_{d/2 + d/\pi}$$

$$= \frac{1}{2\pi^2} (2+\pi)^2 f_{\text{res}}^2 \sigma_d \mu_0^2 a^3 b d |\underline{H}_0|^2 \quad \left(K = \frac{4\omega_{\text{res}}^2 \sigma_d \mu_0^2 a^2 |\underline{H}_0|^2}{\pi^2}\right), \tag{13.166}$$

where $f_{\text{res}} = (f_{\text{res}})_{101}$ is the resonant frequency of the dominant mode, Eq. (13.133), $\omega_{\text{res}} = 2\pi f_{\text{res}}$ is the associated angular frequency, and the use is made of the trigonometric identity $\sin^2\alpha = (1 - \cos 2\alpha)/2$. The power $(P_J)_{\text{in food}}$ is dissipated to heat throughout the dielectric object (food), and this is the power that, locally, cooks or heats up the food in the oven. Of course, we have completely neglected here the influence of the lossy dielectric object on the field distribution in the cavity.

(b) If the dielectric in Fig. 13.18 is extended to completely fill up the cavity, the volume element in Eq. (13.166) extends to $dv = b\, dx\, dz$, and integrals in x and z become equal

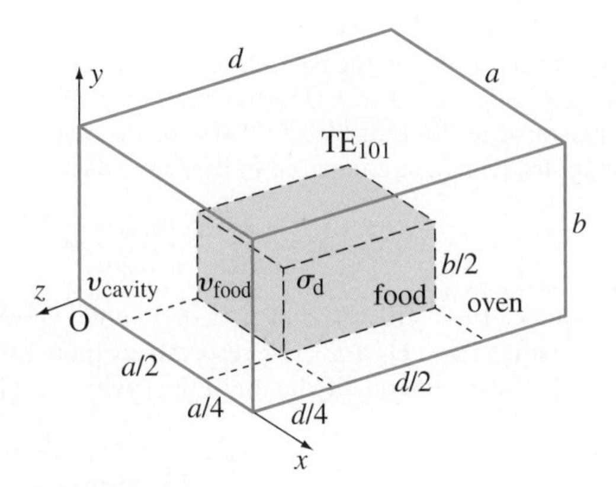

Figure 13.18 Microwave oven operating at the TE_{101} cavity resonance to cook a piece of food of rectangular shape and conductivity σ_d; for Example 13.21.

to $a/2$ and $d/2$, respectively, as in Eq. (13.86), so that the result for the power turns out to be $(P_\mathrm{J})_\mathrm{full\ oven} = 4f_\mathrm{res}^2\sigma_\mathrm{d}\mu_0^2 a^3 bd|\underline{H}_0|^2$. Note that this same result can be obtained by merely replacing ε by σ_d (and μ by μ_0) in the expression for the stored electromagnetic energy in the cavity (with a homogeneous dielectric in the entire cavity), in Eq. (13.155). Note also that $(P_\mathrm{J})_\mathrm{in\ food}/(P_\mathrm{J})_\mathrm{full\ oven} = (2+\pi)^2/(8\pi^2) \approx 1/3$, implying that in the case in Fig. 13.18 we practically use a third of the full power (heating) capacity of the oven, although the volume filling factor of the structure is only $v_\mathrm{food}/v_\mathrm{cavity} = 1/8$.

Example 13.22 Quality Factor of a Fabry-Perot Resonator

A plane-wave Fabry-Perot resonator, in Fig. 10.3, has low-loss conducting planes of conductivity σ_c and permeability $\mu_\mathrm{c} = \mu_0$, and a homogeneous lossless dielectric of permittivity ε and permeability μ. At a resonant frequency f_res, determine the quality factor of the resonator for an arbitrary number m of half-wavelengths between the planes.

Solution The time dependence of electric and magnetic energy densities in the resonator is illustrated in Fig. 10.4. Since the structure is (theoretically) infinite in both transverse directions (x and y directions) in Fig. 10.3, we consider only a part of it for a given (finite) area, S, of the conducting plane pair. Hence, we determine the total electromagnetic energy (W_em) stored in a finite volume (v_d) of a cylinder with base area S and length equal to the separation between the planes, a, positioned perpendicularly to the planes, as shown in Fig. 13.19. At instants when it is all electric, we compute W_em as in Eq. (13.155), using the total complex rms electric field intensity vector ($\underline{\mathbf{E}}_\mathrm{tot}$) in the structure, given in Eqs. (10.8). Having in mind that $a = m\lambda/2$ ($m = 1, 2, \ldots$), where λ is the intrinsic wavelength of the dielectric between the planes ($\beta = 2\pi/\lambda$), we obtain essentially the same integral in Eq. (13.91), as follows (Fig. 13.19):

$$W_\mathrm{em} = (W_\mathrm{e})_\mathrm{max} = \varepsilon \int_{v_\mathrm{d}} |\underline{\mathbf{E}}_\mathrm{tot}|^2 \,\mathrm{d}v = 4\varepsilon|\underline{E}_\mathrm{i0}|^2 \int_{z=-m\lambda/2}^{0} \sin^2 \beta z \underbrace{S\,\mathrm{d}z}_{\mathrm{d}v} = m\varepsilon\lambda|\underline{E}_\mathrm{i0}|^2 S \quad (13.167)$$

[see also Eq. (10.22)].

The time-average power of Joule's losses in the conductors (P_c) for the same part of the resonator, that is, P_c in the parts of the two conducting planes in Fig. 13.19 that are each S in area, is then evaluated similarly to Eq. (13.158), employing the total complex rms magnetic field intensity vector ($\underline{\mathbf{H}}_\mathrm{tot}$), Eqs. (10.8), on S. This vector is entirely tangential to the surfaces. Because $\beta a = m\pi$ and $\cos \beta a = \pm 1$, we have that the losses in the two planes are the same, so that P_c equals twice the power on one side of the cylinder in Fig. 13.19,

$$P_\mathrm{c} = 2R_\mathrm{s} \left|\underline{\mathbf{H}}_\mathrm{tot}\right|^2_{z=0 \text{ or } -a} S = 8R_\mathrm{s}\frac{|\underline{E}_\mathrm{i0}|^2}{\eta^2} S, \quad (13.168)$$

with the surface resistance of the conductors, R_s, and intrinsic impedance of the dielectric between them, η, being given in Eqs. (13.94) and (9.21), respectively.

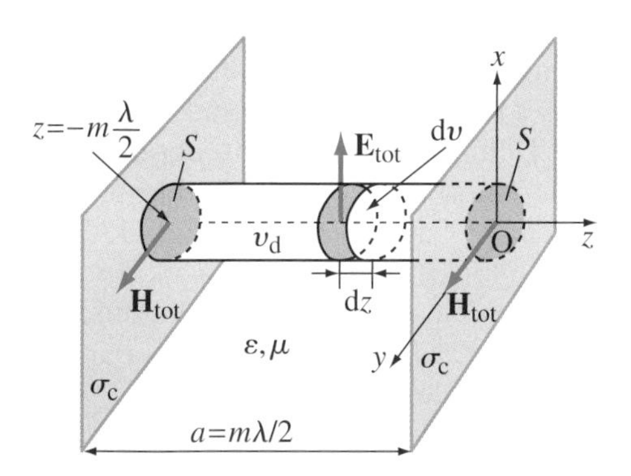

Figure 13.19 Evaluation of the Q factor of a plane-wave Fabry-Perot resonator (Fig. 10.3) from the stored electromagnetic energy and conductor losses in an imaginary cylinder with base area S positioned across the resonator; for Example 13.22.

Finally, from Eq. (12.146), the quality factor of the resonator associated with conductor losses (Q_c) comes out to be

Q_c – Fabry-Perot resonator

$$Q_c = \omega_{res} \frac{W_{em}}{P_c} = \frac{m\pi\varepsilon f_{res}\lambda\eta^2}{4R_s} = \frac{m\pi\eta}{4R_s} \quad (m = 1, 2, \ldots), \tag{13.169}$$

where the use is made of Eqs. (8.48) and (8.112), and of the relationship $\varepsilon f_{res}\lambda = \varepsilon c = 1/\eta$. Note that the factor Q_d for the dielectric losses (if nonzero and nonnegligible) of the structure is that in Eq. (13.161). As a numerical example, for $f = f_{res} = 3$ THz, $a = 1$ cm, copper conductors, Eq. (10.80), and air dielectric, Eqs. (9.19) and (9.23), we have $\lambda = 0.1$ mm, $m = 2a/\lambda = 200$, and $R_s = 0.452\ \Omega$/square, so that the quality factor is as high as $Q = Q_c =$ 131,015. In fact, since W_{em} depends on the number of half-wavelengths (m) fitting into the plane separation (a) in Fig. 13.19, and P_c does not, we can achieve practically arbitrarily high Q by increasing m (or a).

Problems: 13.38–13.42; *Conceptual Questions* (on Companion Website): 13.36; *MATLAB Exercises* (on Companion Website).

Problems

13.1. Waveguide phase coefficient from a Helmholtz equation. Consider a rectangular metallic waveguide of transverse dimensions a and b, and dielectric parameters ε and μ (Fig. 13.1) in a dominant mode of operation, at a frequency f. Requiring that the general electric-field Helmholtz equation be satisfied in the waveguide for the known TE_{10} electric field distribution, Eq. (13.24), obtain the waveguide phase coefficient (β) for the dominant mode, given by Eqs. (13.12) and (13.20), or by Eq. (13.33).

13.2. Maxwell's equations for dominant-mode fields. For the waveguide from the previous problem, assume that the electric field vector (of the dominant wave mode) is known, given in Eq. (13.24). (a) Use Maxwell's equations to derive the expression for the accompanying magnetic field vector in the waveguide. (b) Then start with the magnetic field in (a) and find the resulting electric-field expression, from Maxwell's equations; comparing this result with Eq. (13.24), obtain the waveguide phase coefficient.

13.3. Satisfaction of Maxwell's equations by TE_{mn} field vectors. Show that the electric and magnetic field vectors of an arbitrary TE_{mn} mode in a rectangular metallic waveguide, given by Eqs. (13.54)–(13.58), satisfy the corresponding set of general Maxwell's equations, Eqs. (11.1)–(11.4).

13.4. Currents and charges on guide walls, arbitrary TE mode. Repeat Example 13.2 but for an arbitrary TE_{mn} wave mode propagating through a rectangular metallic waveguide, in Fig. 13.1. Verify that the obtained solutions for the surface current and charge densities on waveguide walls reduce for $m = 1$ and $n = 0$ to those for the dominant mode in Eqs. (13.27)–(13.31).

13.5. Currents/charges for an arbitrary TM mode. Repeat Example 13.2 but for an arbitrary TM_{mn} mode.

13.6. Waveguide phase coefficient for an arbitrary TE or TM mode. (a) Repeat Problem 13.1 but for an arbitrary TE_{mn} wave mode – namely, use the electric-field expressions in Eqs. (13.55) and (13.56) and the corresponding Helmholtz equation to obtain the waveguide phase coefficient for arbitrary m and n, given by Eqs. (13.12) and Eq. (13.71), or by Eqs. (13.40) and (13.59). (b) Do the same using the magnetic field of the mode. (c) Do the same as in (a) and (b) but for an arbitrary TM_{mn} mode.

13.7. Modal cutoff frequencies in WR-975 and WR-340 waveguides. For an air-filled WR-975 commercial rectangular waveguide, with transverse dimensions $a = 24.766$ cm and $b = 12.383$ cm, determine (a) the dominant frequency range, (b) cutoff frequencies of the first three TE modes and first three TM modes, and (c) all possible propagating modes at a frequency of $f = 2$ GHz. (d) Repeat (a)–(c) for a WR-340 waveguide, with $a = 8.636$ cm and $b = 4.318$ cm (and air dielectric).

13.8. Modal cutoff frequencies in a square waveguide. Repeat the previous problem, parts (a)–(c), but for a square waveguide of cross-sectional interior dimension $a = 10$ cm if it is filled with (a) air and (b) dielectric of relative permittivity $\varepsilon_r = 2.5$, respectively.

13.9. FM and AM radio waves in a railway tunnel. A railway tunnel can be approximated by a rectangular waveguide with transverse dimensions $a = 7$ m and $b = 4$ m and nonpenetrable walls. What wave modes can propagate inside the tunnel (a) at an FM radio frequency of 100 MHz and (b) at an AM radio frequency of 1 MHz, respectively?

13.10. Poynting vector of the dominant mode. Consider a rectangular metallic waveguide with the dominant (TE_{10}) mode only, in Fig. 13.8. (a) Find the complex Poynting vector, $\underline{\mathcal{P}}$, in the structure. (b) Compute the flux of $\underline{\mathcal{P}}$ through the guide cross section. (c) What is the flux of $\underline{\mathcal{P}}$ into each of the waveguide walls, per unit length of the guide?

13.11. Poynting vector, based on multiple reflections. Using Eqs. (13.2), (13.4), (13.6), (13.12), and (13.23), show that the result for $\underline{\mathcal{P}}$ in the previous problem is equivalent to that in Eq. (10.122), for a TE wave generated by multiple reflections of a uniform plane wave in Fig. 10.15.

13.12. Power transfer by TE_{02} and TE_{11} wave modes. (a) Find the complex Poynting vector ($\underline{\mathcal{P}}$) of an arbitrary TE_{mn} wave mode traveling along a rectangular metallic waveguide (Fig. 13.8). (b) Compute the flux of $\underline{\mathcal{P}}$ through an arbitrary cross section of the structure for TE_{02} and TE_{11} waves, respectively. (c) Using the result in (a), determine the flux of $\underline{\mathcal{P}}$ into each of the waveguide walls (for an arbitrary mode).

13.13. Poynting vector for TM waves. Repeat the previous problem, parts (a) and (c), for an arbitrary TM_{mn} mode, and part (b) for a TM_{21} wave.

13.14. Dielectric breakdown and power capacity for a TE_{02} wave. Consider a TE_{02} wave propagating through an air-filled rectangular metallic waveguide of transverse dimensions $a = 38.1$ cm and $b = 19.05$ cm (WR-1500 waveguide). Find the power-handling capacity of the waveguide for this mode, i.e., the maximum time-average power that can be carried by the TE_{02} wave for the safe operation of the structure – prior to an eventual dielectric breakdown (dielectric strength of air is $E_{cr0} = 3$ MV/m), at a frequency of $f = 1.8$ GHz.

13.15. Power capacity for a TE_{11} wave. Repeat the previous problem but for the TE_{11} wave mode.

13.16. Attenuation coefficients for TE_{02} and TE_{11} modes. Compute the attenuation coefficient for (a) the TE_{02} wave mode and (b) the TE_{11} wave mode, respectively, of the WR-1500 waveguide from Problem 13.14 at the given frequency, assuming that the guide walls are made out of copper.

13.17. TE_{02} and TE_{11} attenuation including dielectric losses. Repeat the previous problem but for the WR-1500 waveguide filled with polyethylene, of relative permittivity $\varepsilon_r = 2.25$ and loss tangent $\tan\delta_d = 10^{-4}$.

13.18. Attenuation coefficient for the lowest TM waveguide mode. Find the attenuation coefficient for the lowest TM wave mode (TM_{11}) propagating at a frequency f along an air-filled rectangular metallic waveguide of transverse dimensions a and b, and skin-effect conductor surface resistance R_s. Check the result against data in Fig. 13.9.

13.19. X- and C-band waveguide designs. (a) Design an air-filled X-band (8–12 GHz) standard rectangular waveguide such that (if possible) this entire band is covered by the usable frequency range in Eq. (13.101). (b) Repeat (a) but for the C-band (4–8 GHz).

13.20. Analysis of a K-band waveguide. Consider a K-band standard rectangular waveguide with larger transverse dimension $a = 7$ mm, aluminum ($\sigma_c = 35$ MS/m and $\mu_c = \mu_0$) walls, and polyethylene ($\varepsilon_r = 2.25$, $\tan\delta_d = 10^{-4}$, and $E_{cr} = 47$ MV/m) dielectric. (a) What part of the K-band (18–27 GHz) falls into the usable frequency range of the waveguide [Eq. (13.101)]? At the central frequency of the K-band, find (b) the complex propagation coefficient of the traveling wave along the guide, (c) expressions for electric and magnetic field vectors in the structure, and (d) the power-handling capacity (maximum permissible time-average power, limited by an eventual dielectric breakdown in the structure) of the waveguide, including a safety factor of 2 in the computation.

13.21. Phase and group velocities in a K-band waveguide. For the waveguide from the previous problem, calculate the phase and group velocities and guide wavelength of the propagating wave, as well as the intrinsic wavelength of the waveguide dielectric, at the central frequency and at each of the end frequencies of the K-band.

13.22. Emf in a small loop in a waveguide close to dielectric breakdown. If the fields in the waveguide described in Example 13.12 are at a half of their intensities at dielectric breakdown, find the corresponding rms emf induced in a small loop attached to a wall of the guide – for each of the two loop positions/orientations, (a) and (b), considered in the example.

13.23. Magnetic-probe coupling above the dominant range. Consider an air-filled WR-650 waveguide, with transverse dimensions $a = 16.51$ cm and $b = 8.255$ cm, at a frequency of $f = 2$ GHz, and assume that all possible propagating modes are established in the structure. With reference to the coordinate system in Fig. 13.1, let a small wire loop of surface area $S = 0.25$ cm^2 be attached to the left wall of the waveguide (wall at $x = 0$) such that it lies in the plane $y = b/2$. Show that the loop couples to the magnetic field of only one of the established modes. If the time-average power carried by that mode amounts to $P = 1$ kW, determine the rms emf induced in the loop.

13.24. Electric-probe measurement on a slotted waveguide. A rectangular metallic waveguide (Fig. 13.1), with transverse dimensions $a = 10$ cm and $b = 5$ cm, has a narrow longitudinal slot in the upper wall, along the line defined by $x = a/2$ and $y = b$, through which a sliding electric probe (short wire antenna) is inserted to sample (measure) the electric field. In addition, the guide is short-circuited in its cross section defined by $z = 0$, as in Fig. 13.12. Finally, the structure is completely filled with a liquid dielectric whose losses can be neglected and is nonmagnetic, and whose permittivity is unknown, and measurements are carried out in the TE_{10} mode of operation of the waveguide, at a frequency of $f = 500$ MHz. By sliding the probe from the short-circuiting plate toward the interior of the structure, it is determined that the signal reception by the probe is maximum at a distance of $d = 4.5$ cm from the plate. (a) Explain why this longitudinal slot practically does not affect the field distribution in the waveguide. (b) Find the relative permittivity of the dielectric. (c) What is the cutoff frequency of the waveguide with and without the liquid, respectively?

13.25. Resonant frequency of a cavity from a Helmholtz equation. Consider a lossless rectangular metallic cavity of dimensions a, b, and d, filled with a homogeneous dielectric of permittivity ε and permeability μ, Fig. 13.16, and obtain the expression for the resonant frequency of the dominant cavity mode, given in Eq. (13.133) – by requiring that the general electric-field Helmholtz equation for the same dielectric be satisfied for the known TE_{101} electric field distribution, Eq. (13.134).

13.26. Resonant frequency from Maxwell's equations. (a) For the resonant cavity from the previous problem, start with the electric field in Eq. (13.134) and derive the expression for the accompanying magnetic field vector using Maxwell's equations. (b) Then derive back the electric field from the magnetic field in (a) and Maxwell's equations, and comparing the result with Eq. (13.134) obtain the expression for the resonant frequency of the TE_{101} mode.

13.27. Finding cavity dimensions from resonant frequencies. (a) If in the resonant cavity in Fig. 13.16, $b = a/2$ and the dielectric is air, find a and d such that the structure resonates in the dominant (TE_{101}) mode at a frequency of 8 GHz, whereas its TM_{111} resonance is at a frequency of 10 GHz. (b) For dimensions from (a), which TE_{10p} cavity modes resonate within the K_u-band (12–18 GHz)?

13.28. Satisfaction of Maxwell's equations by TM_{mnp} field. Show that the field vectors of an arbitrary TM_{mnp} resonance mode in a rectangular waveguide cavity (Fig. 13.16), given by Eqs. (13.145)–(13.149), satisfy all four pertinent Maxwell's equations, Eqs. (11.1)–(11.4).

13.29. Surface currents and charges on cavity walls. Consider an air-filled rectangular PEC cavity resonator of dimensions a, b, and d, in Fig. 13.16, and assume that a dominant (TE_{101}) standing wave is established in the cavity. The field components of this wave are given in Eqs. (13.134)–(13.136). Find the distributions of (a) surface currents and (b) surface charges on interior surfaces of all six sides of the cavity. (c) Show that the currents and charges in (a) and (b) satisfy the continuity equation for high-frequency surface currents (continuity equation for plates).

13.30. Poynting vector inside a cavity resonator. (a) Compute the complex Poynting vector, $\underline{\mathcal{P}}$, at an arbitrary point inside the cavity from the previous problem. (b) Show that the flux of $\underline{\mathcal{P}}$ into each of the cavity walls is zero, using the result in (a) and Poynting's theorem, respectively.

13.31. Computing the maximum magnetic energy of the cavity. Consider the dominant resonance wave mode (TE_{101}) in a lossless rectangular metallic cavity resonator, in Fig. 13.16, and obtain the expression for the electromagnetic energy stored in the structure at an instant when it is all magnetic, so as $W_{em} = (W_m)_{max}$.

13.32. Computing the energy of a resonator at arbitrary time. For the cavity from the previous problem, find the stored energy of a TE_{101} wave by evaluating the total (electric plus magnetic) energy at an arbitrary instant of time.

13.33. Instantaneous field vectors and energy in a cubical cavity. A TE_{101} wave is established in an air-filled cubical cavity with edge length $a = 50$ cm. The complex electric and magnetic field intensity vectors in the cavity are given by Eqs. (13.134)–(13.136), with $\underline{H}_0 = 5\,e^{j\pi/3}$ A/m. Determine the instantaneous field vectors in the cavity (a) at an arbitrary instant of time and (b) at an instant when the total stored energy in the resonator is exactly half electric and half magnetic.

13.34. Dielectric breakdown and maximum energy of a cavity. Find the maximum permissible electromagnetic energy, limited by an eventual dielectric breakdown in the structure, that can be stored in an air-filled rectangular metallic cavity resonator of dimensions $a = 20$ cm, $b = 10$ cm, and $d = 15$ cm with a TE_{101} wave (dielectric strength of air is $E_{cr0} = 3$ MV/m).

13.35. Emf in a small loop in a cavity close to dielectric breakdown. Consider an air-filled rectangular cavity with edge lengths $a = 35$ cm, $b = 14$ cm, and $d = 19$ cm in a dominant resonance mode of operation. With reference to the coordinate system in Fig. 13.16, a small wire loop of surface area $S = 1.8$ cm^2 is attached to the front wall of the cavity (wall at $z = 0$) such that its plane coincides with the plane $x = a/2$. If the TE_{101} fields in the cavity are at a half of their intensities at dielectric breakdown, determine the rms emf induced in the loop. What is the energy stored in the cavity?

13.36. Energy of a TE_{123} wave in a rectangular cavity. Derive the expression for the stored electromagnetic energy of a TE_{123} wave in a rectangular metallic cavity resonator, Fig. 13.16.

13.37. Stored energy of a TM_{111} wave. Repeat the previous problem but for a TM_{111} wave in the cavity.

13.38. Losses in copper walls for the TM_{111} cavity mode. Take the air-filled cubical cavity from Example 13.19, and assume its overmoded operation. (a) Compute the time-average power of Joule's losses in the copper walls of the cavity that are associated with the TM_{111} resonance mode, if $|\underline{E}_0| = 1$ kV/m in the modal field expressions. (b) Repeat (a) but for the teflon-filled cavity from Example 13.20.

13.39. Losses in the cavity dielectric for the TM_{111} mode. For the cavity and resonance mode from the previous problem, part (b), find the time-average power of Joule's losses in the cavity dielectric.

13.40. **Quality factor of a brass resonator in the dominant mode.** Calculate the Q factor of an air-filled cubical cavity with brass ($\sigma_c = 15$ MS/m and $\mu_c = \mu_0$) walls that resonates in the TE_{101} mode at a frequency of 500 MHz, 1 GHz, 10 GHz, and 50 GHz, respectively.

13.41. **Quality factor for the cavity dielectric, any TE/TM mode.** Show that the quality factor associated with the (small) losses in the dielectric of a rectangular cavity resonator equals $Q_d = 1/\tan\delta_d$ for an arbitrary TE_{mnp} or TM_{mnp} wave mode in the cavity, $\tan\delta_d$ being the loss tangent of the dielectric.

13.42. **Total quality factor for the TE_{123} cavity mode.** A rectangular waveguide cavity resonator of edge lengths $a = 1$ cm, $b = 2$ cm, and $d = 3$ cm has aluminum ($\sigma_c = 35$ MS/m and $\mu_c = \mu_0$) walls and polyethylene ($\varepsilon_r = 2.25$ and $\tan\delta_d = 10^{-4}$) dielectric. Find the Q factor for the TE_{123} resonance mode of this cavity.

14 Antennas and Wireless Communication Systems

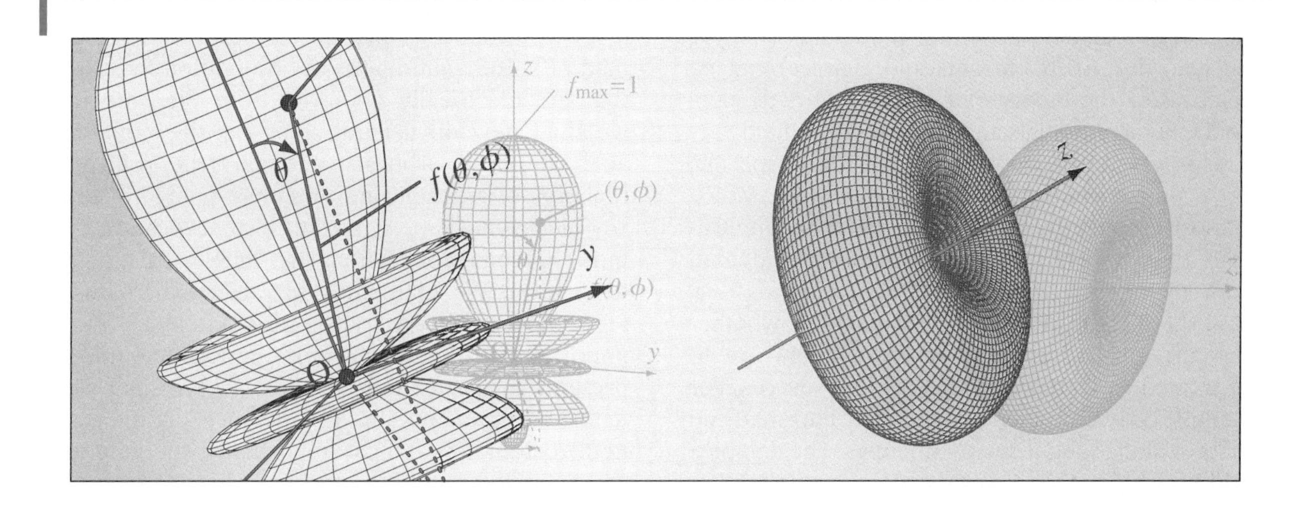

Introduction:

Although any conductor with a time-varying (e.g., time-harmonic) current radiates electromagnetic energy into the surrounding space, some conductor configurations are specially designed to maximize electromagnetic radiation, in desired directions at given frequencies. Such systems of conductors, which sometimes also include dielectric parts, are called antennas. In other words, antennas are electromagnetic devices designed and built to provide a means of efficient transmitting or receiving of radio waves. More precisely, they provide transition from a guided electromagnetic wave (in a transmission line or waveguide feeding the antenna) to a radiated unbounded electromagnetic wave (in free space or other ambient medium) in the transmitting (radiating) mode of operation, and vice versa for an antenna operating in the receiving mode. As has already been discussed in the previous three chapters, with an increase of both the distance and operating frequency in communication and power-transfer systems, the wave attenuation on transmission lines and metallic waveguides becomes, at some point, prohibitively large and/or their realization too costly, and wireless links using antennas are favored. Of course, even at large distances and high frequencies, wireline systems have their own advantages; for instance, coaxial cables and metallic waveguides are not susceptible to interference with external signals and other systems, which is often encountered in wireless systems, and dielectric waveguides, such as optical fibers, are also an alternative solution, due to their very low losses. However, in many applications antennas are practically the only choice for the type of transmission system, with examples spanning mobile communications involving aircraft, spacecraft, satellites, ships, and land vehicles, radio and TV broadcasting to unlimited numbers of receivers, mobile personal communication devices (e.g., cellular telephones), radar systems, etc.

In the general analysis of electromagnetic radiation, an arbitrary transmitting antenna is simply a distribution of rapidly time-varying (or high-frequency) currents and charges, as in Fig. 8.7. If the antenna is situated in free space (or some other linear, homogeneous, and lossless medium), the electromagnetic field that it radiates (that is, the field due to its currents and charges) can be found using the theory of Lorenz (retarded) electromagnetic potentials and the associated field vectors, provided in Sections 8.9 and 8.10. In specific, for a time-harmonic variation of sources, which is most frequently the case in antenna applications, the complex Lorenz potentials are evaluated from the respective expressions in Eqs. (8.113), (8.114), (8.116), and (8.117), depending whether the currents and charges of the antenna are to be treated as volume, surface, or line sources. For example, we normally assume line currents and charges along metallic wire antennas. The complex field vectors, on the other side, are computed either from potentials, using Eqs. (8.118) and (8.119), or directly through field integrals of sources, as in Eqs. (8.125) and (8.128). Of course, currents and charges are interrelated by means of different versions of the continuity equation, Eqs. (8.82), (10.14), and (8.133), and all potential and field expressions can thus be recast in terms of currents only. In fact, it is customary in antenna theory to explicitly consider solely antenna currents as sources of radiation, but always bearing in mind that there are accompanying charges as well on the structure. So, as it turns out, we already have in hand the general theory and analytical principles and procedures for evaluation of radiation by an arbitrary high-frequency volume, surface, or line current distribution, i.e., by an arbitrary antenna. In this chapter, we shall apply this knowledge and tools to concrete antenna structures, to describe and study various practical properties of antennas, and establish understanding of their operation and basis for their design, both as independent devices and as parts of communication (or other high-frequency, e.g., radar) systems.[1]

Our antenna theory will start with the analysis of a Hertzian dipole, namely, a capacitively loaded short wire dipole antenna, whose importance can hardly be overstated, given that an arbitrary transmitting antenna can be represented as a superposition of Hertzian dipoles. From the far field of a Hertzian dipole, at observation locations that are electrically far away from the antenna, we shall develop general steps for the radiation analysis of an arbitrary antenna (the principal function of transmitting antennas being to convey electromagnetic signals to distant locations, in most antenna applications we deal with the far field only). We shall study several circuit parameters of an arbitrary transmitting antenna – most importantly, the antenna input impedance, that it presents to the feed electric circuit. We shall also define and use a number of antenna radiation (far-field) parameters – most importantly, the antenna characteristic radiation function. This vector function represents the part of the general antenna electric and magnetic far-field expressions that is characteristic for individual antennas, i.e., that differs from antenna to antenna, while the remaining terms in the expressions are the same for all antennas. Out of many basic types of antennas for wireless communications, we shall focus on various wire antennas, including electrically short (loaded and nonloaded) antennas, arbitrary wire dipole antennas, wire monopole antennas attached to a metallic ground plane (analyzed by image theory), and electrically small loop (magnetic dipole) antennas. General theory of receiving antennas will be presented, where we shall show that both

[1]In addition to computing the radiation field of an antenna for its given current distribution, as well as some associated properties of antennas in both transmitting and receiving modes of operation, the major problem in antenna analysis and design is obtaining the current distribution of an antenna for its given geometry, material composition, and excitation. Although we shall discuss this general problem in several simple cases, its solution in more complex, real-world antenna designs and applications requires advanced numerical techniques that are beyond the scope of this text.

circuit (impedance) and directional properties of an antenna in the receiving mode of operation are directly related to its properties when transmitting. As the last topic, we shall investigate antenna arrays, that is, spatial arrangements of identical antennas (array elements), equally oriented in space but excited independently, and demonstrate that by varying the phases of feed currents of array elements, for example, we can change (steer) the direction of maximum radiation of an array antenna, without moving (slewing) the antenna. Note, finally, that in many discussions in this chapter we shall study not only antennas but entire wireless communication systems with antennas at the two ends. In this, we shall use many concepts and equations describing the propagation of uniform plane electromagnetic waves in unbounded media (e.g., free space) and in the presence of material interfaces (e.g., a perfectly conducting boundary), from Chapters 9 and 10.

14.1 ELECTROMAGNETIC POTENTIALS AND FIELD VECTORS OF A HERTZIAN DIPOLE

Consider the simplest antenna, a so-called Hertzian dipole, which is an electrically short straight metallic wire segment with a rapidly time-varying current that does not change along the wire. The dipole is named in honor of Heinrich Hertz, who invented the first source of radio waves (antenna), similar in concept and operation to what we now call a Hertzian dipole, and demonstrated the first radio system (link) consisting of a transmitting and receiving antenna. Assuming a time-harmonic regime of the dipole, let it be fed at its center by a lumped generator,[2] of frequency f, and let its instantaneous and complex rms current intensities be $i(t)$ and $\underline{I}$, respectively. In specific, as in Eq. (8.66),

$$i(t) = I\sqrt{2}\cos(\omega t + \psi) \quad \longleftrightarrow \quad \underline{I} = I\,\mathrm{e}^{\mathrm{j}\psi}, \tag{14.1}$$

where $I = |\underline{I}|$ and ψ are the rms value and initial phase (both uniform along the wire) of $i(t)$, and $\omega = 2\pi f$ is the angular (radian) frequency of the generator. We further assume that the ambient medium, in which the dipole resides, is linear, homogeneous, and lossless, of permittivity ε and permeability μ ($\sigma = 0$). Most frequently this is air (free space), so $\varepsilon = \varepsilon_0$ and $\mu = \mu_0$. Introducing a spherical coordinate system with the z-axis along the wire axis and origin (O) at the generator, as shown in Fig. 14.1, we would like to find the expressions for Lorenz electromagnetic potentials and field vectors of the antenna at a point P defined by (r, θ, ϕ). In addition to being electrically small, that is, small in comparison to the wavelength of the surrounding medium, λ, defined in Eq. (8.112), the length of the wire, l, is also much smaller than the distance r of the point P from the origin, so we have for the antenna

$$\boxed{\underline{I} = \text{const}, \quad l \ll \lambda, \quad l \ll r.} \tag{14.2}$$

Hertzian dipole

In fact, the current can be considered to be uniform along the wire because the dipole is electrically short. Apart from its physical manifestation and practical operation as an electrically short wire antenna, for theoretical purposes a Hertzian dipole can be identified to an infinitesimal (differentially short) line current element $\underline{I}\,\mathrm{d}\mathbf{l}$, as in Eqs. (4.10), and hence it is commonly known also as an infinitesimal dipole (of length $\mathrm{d}l$ rather than l).

[2] This generator may represent, for instance, an insertion of the output terminals of a two-wire (or some other two-conductor) transmission line between two halves of the dipole, so that the antenna current equals the output (load) current of the line.

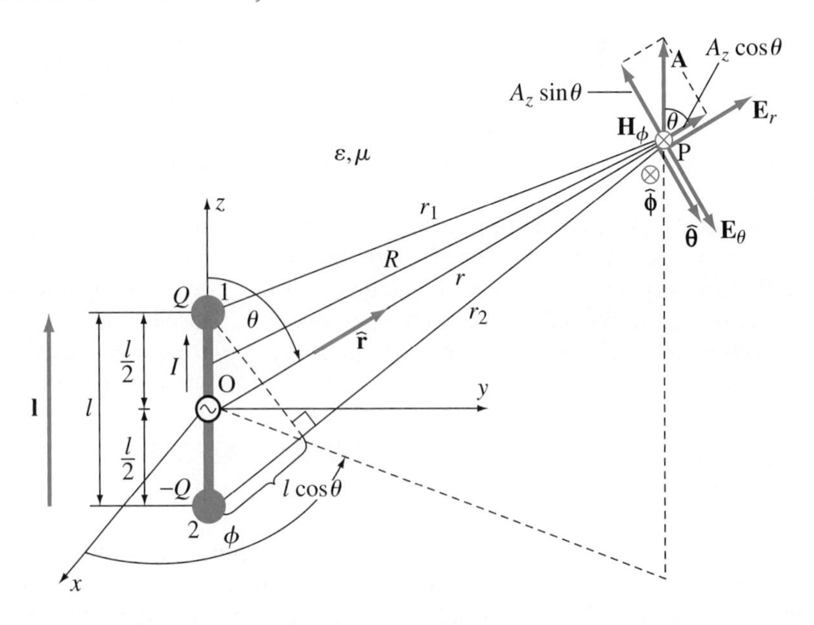

Figure 14.1 Hertzian dipole.

As $i(t)$ is nonzero at the wire ends, it must be terminated by time-harmonic charges $Q(t)$ and $-Q(t)$ that accumulate on a pair of small metallic spheres (Fig. 14.1), or conductors of other shapes, attached to these ends.[3] On the other side, a uniform antenna current, Eq. (14.2), implies that there is no charge distributed along the wire (i.e., on its surface), so the line charge density (charge per unit length) of the dipole is zero, $Q' = 0$, by virtue of Eq. (8.133). Applying the continuity equation for time-varying currents in integral form, Eq. (3.36), to a surface completely enclosing the upper sphere, much like in Figs. 8.1 and 3.5, we realize that i and Q are related by Eq. (8.2) or (3.44), and hence

dipole charge

$$\boxed{i = \frac{\mathrm{d}Q}{\mathrm{d}t} \quad \longrightarrow \quad \underline{I} = \mathrm{j}\omega \underline{Q} \quad \longrightarrow \quad \underline{Q} = -\frac{\mathrm{j}}{\omega}\underline{I},} \tag{14.3}$$

with $\underline{Q}$ standing for the complex rms charge of the sphere. Note that the system in Fig. 14.1 can be considered as a dynamic (high-frequency) generalization of an electric (electrostatic) dipole, in Fig. 1.28. In the dynamic case, the current is needed to drain, via the generator, the charges Q and $-Q$ in an oscillatory fashion, with the current intensity varying synchronously with the time derivative of the dipole charge. For $i(t)$ in Eqs. (14.1), $Q(t) = (I\sqrt{2}/\omega)\sin(\omega t + \psi)$. Note also that, since the conducting terminations (extensions) of the wire in Fig. 14.1 serve as charge containers (accumulators), like electrodes of a capacitor, and represent capacitive loads to the antenna, a Hertzian dipole is also referred to as a capacitively loaded short wire dipole.

The complex magnetic vector potential, $\underline{\mathbf{A}}$, at the point P in Fig. 14.1 is determined solving the integral in the second expression in Eqs. (8.117), where we can readily move $\underline{I}$, as a constant, outside the integral sign. Moreover, having in mind that the wire is short both electrically and relative to the location of the field

[3]In practical realization of Hertzian dipoles, conducting objects terminating the (vertical) wire antenna may, in addition to spheres, be in the form of (horizontal) circular plates, umbrella-like systems of radial wires, conductors of a two-wire transmission line, etc.

HISTORICAL ASIDE

Heinrich Rudolf Hertz (1857–1894), a German physicist and the first radio engineer, was a professor of physics at Universities of Kiel, Karlsruhe, and Bonn. Hertz was a student of Helmholtz (1821–1894) at the University of Berlin, from which he received his doctoral degree in 1880. He was the first to demonstrate experimentally the existence of electromagnetic waves, and of electromagnetic radiation, and he built the first antennas. In Hertz's famous 1887 experiment at Karlsruhe, the source of electrical disturbances was a capacitively loaded wire dipole radiator, a Hertzian dipole, and the receiver (detector) of disturbances was a circular loop of wire. The dipole was connected to an induction coil that produced sparks across the air gap between the dipole terminals, which, by radiation (generation and propagation) of electromagnetic waves, resulted in sparks at the air gap in the loop at a distance of several meters, in Hertz's laboratory. His wire dipole and loop were hence the first transmitting and receiving antennas, respectively, and the whole system was the first radio (wireless) link. Hertz's experiment was quickly confirmed by others and it laid a firm foundation for further discoveries and developments in what would become radio science and engineering in the years and decades to come. Hertz also demonstrated the first coaxial cable, and first observed the photoelectric effect (in 1887), which was later explained in different ways by several researchers and most notably by Einstein (1879–1955) in 1905 (Einstein's explanation of the photoelectric effect earned him the Nobel Prize for Physics in 1921). Hertz's name is further immortalized by the use of hertz (Hz) as the unit for frequency. *(Portrait: © Deutsches Museum)*

point (P), Eqs. (14.2), we can approximate the variable source-to-field distance R for an arbitrary point at the wire axis by the fixed distance r (from O to P), and take $R \approx r$ out of the integral as well. Consequently, we are left with

$$\underline{\mathbf{A}} = \frac{\mu}{4\pi}\int_l \frac{\underline{I}\,\mathrm{d}\mathbf{l}\,\mathrm{e}^{-\mathrm{j}\beta R}}{R} \approx \frac{\mu \underline{I}\,\mathrm{e}^{-\mathrm{j}\beta r}}{4\pi r}\int_l \mathrm{d}\mathbf{l} = \frac{\mu \underline{I}\,\mathbf{l}\,\mathrm{e}^{-\mathrm{j}\beta r}}{4\pi r} = \frac{\mu \underline{I}\,l\,\mathrm{e}^{-\mathrm{j}\beta r}}{4\pi r}\,\hat{\mathbf{z}} = \underline{A}_z\,\hat{\mathbf{z}} \quad (\beta = \omega\sqrt{\varepsilon\mu}), \tag{14.4}$$

magnetic potential of a Hertzian dipole

where β is the phase coefficient (wavenumber) for the ambient medium and given operating frequency, Eq. (8.111), and $\mathbf{l} = l\,\hat{\mathbf{z}}$ is the position vector of the charge Q with respect to $-Q$, i.e., the length vector of the dipole whose orientation coincides with the reference direction of the current i. On the other hand, if the antenna in Fig. 14.1 is treated as an infinitesimal dipole, characterized by $\underline{I}\,\mathrm{d}\mathbf{l}$, finding $\underline{\mathbf{A}}$ is even simpler, since then there is no integration in Eq. (14.4) in the first place and $R = r$. The vector $\underline{\mathbf{A}}$ is parallel to the dipole, so it has only a z-component, $\underline{A}_z$. From Fig. 14.1, $\underline{\mathbf{A}}$ can be decomposed onto an r- (radial) and θ-components in the spherical coordinate system, $\underline{A}_r$ and $\underline{A}_\theta$, as follows

$$\underline{\mathbf{A}} = \underline{A}_z \cos\theta\,\hat{\mathbf{r}} - \underline{A}_z \sin\theta\,\hat{\boldsymbol{\theta}} = \underline{A}_r\,\hat{\mathbf{r}} + \underline{A}_\theta\,\hat{\boldsymbol{\theta}}, \tag{14.5}$$

so $\underline{A}_r = \underline{A}_z \cos\theta$ and $\underline{A}_\theta = -\underline{A}_z \sin\theta$, with $\underline{A}_z$ given in Eq. (14.4).

Since $r \gg l$, in order to compute the complex electric scalar potential, $\underline{V}$, at the point P, using the first expression in Eqs. (8.116), the two charged spheres of the Hertzian dipole (charged over their surfaces) can be treated as point charges. In analogy to Eq. (1.114) for the electrostatic dipole, we can write

electric potential

$$\underline{V} = \frac{\underline{Q}}{4\pi\varepsilon}\underbrace{\left(\frac{\mathrm{e}^{-\mathrm{j}\beta r_1}}{r_1} - \frac{\mathrm{e}^{-\mathrm{j}\beta r_2}}{r_2}\right)}_{-\Delta(\mathrm{e}^{-\mathrm{j}\beta R}/R)} \approx -\frac{\underline{Q}}{4\pi\varepsilon}\,\frac{\mathrm{d}(\mathrm{e}^{-\mathrm{j}\beta R}/R)}{\mathrm{d}R}\bigg|_{R=r}\underbrace{l\cos\theta}_{\Delta R} = \frac{\underline{Q}\,l\cos\theta(1+\mathrm{j}\beta r)\,\mathrm{e}^{-\mathrm{j}\beta r}}{4\pi\varepsilon r^2}. \quad (14.6)$$

Here, the increment in the function $\mathrm{e}^{-\mathrm{j}\beta R}/R$ from the location of the charge $\underline{Q}$ to that of $-\underline{Q}$, with the distances of the point P from the two charges being r_1 and r_2, is approximated by the derivative of this function with respect to R [see Eq. (8.121)], at the dipole center ($R = r$), multiplied by the corresponding increment in R, $\Delta R = r_2 - r_1$, which is small. In addition, ΔR is approximately computed as $\Delta R \approx l\cos\theta$, like in Fig. 1.28. Alternatively, $\underline{V}$ can be found from the already known $\underline{\mathbf{A}}$, using the Lorenz condition for complex electromagnetic potentials, Eq. (8.115), and the formula for the divergence in spherical coordinates, Eq. (1.171),

$$\underline{V} = \frac{\mathrm{j}}{\omega\varepsilon\mu}\nabla\cdot\underline{\mathbf{A}} = \frac{\mathrm{j}}{\omega\varepsilon\mu r}\left[\frac{1}{r}\frac{\partial}{\partial r}\left(r^2\underline{A}_r\right) + \frac{1}{\sin\theta}\frac{\partial}{\partial\theta}\left(\sin\theta\,\underline{A}_\theta\right)\right], \quad (14.7)$$

which, substituting the expressions for $\underline{A}_r$ and $\underline{A}_\theta$ from Eqs. (14.5) and (14.4), and expressing $\underline{I}$ in terms of $\underline{Q}$ from Eq. (14.3), give the same result as in Eq. (14.6).

The complex electric field intensity vector, $\underline{\mathbf{E}}$, of the antenna can now be evaluated from both potentials, Eqs. (14.4)–(14.6), using Eq. (8.118) and the formula for the gradient in spherical coordinates, given by Eq. (1.108),

electric field of a Hertzian dipole

$$\underline{\mathbf{E}} = -\mathrm{j}\omega\underline{\mathbf{A}} - \nabla\underline{V} = -\mathrm{j}\omega\underline{\mathbf{A}} - \frac{\partial \underline{V}}{\partial r}\,\hat{\mathbf{r}} - \frac{1}{r}\frac{\partial \underline{V}}{\partial\theta}\,\hat{\boldsymbol{\theta}} = -\frac{\eta\beta^2\underline{I}\,l\,\mathrm{e}^{-\mathrm{j}\beta r}}{4\pi}\left\{\left[\frac{1}{(\mathrm{j}\beta r)^2} + \frac{1}{(\mathrm{j}\beta r)^3}\right]2\cos\theta\,\hat{\mathbf{r}} + \left[\frac{1}{\mathrm{j}\beta r} + \frac{1}{(\mathrm{j}\beta r)^2} + \frac{1}{(\mathrm{j}\beta r)^3}\right]\sin\theta\,\hat{\boldsymbol{\theta}}\right\} = \underline{E}_r\,\hat{\mathbf{r}} + \underline{E}_\theta\,\hat{\boldsymbol{\theta}}, \quad (14.8)$$

where $\eta = \sqrt{\mu/\varepsilon}$ is the intrinsic impedance of the medium, Eq. (9.21), and the use is made also of Eq. (14.3) and the facts that

$$\omega\mu = \eta\beta \quad \text{and} \quad \frac{1}{\omega\varepsilon} = \frac{\eta}{\beta}. \quad (14.9)$$

The expression for $\underline{\mathbf{E}}$ in terms of the magnetic vector potential only in Eq. (8.118) leads to this same result as well.

Similarly, we combine Eqs. (8.119), (5.60), (14.4), and (14.5), and apply the formula for the curl in spherical coordinates, Eq. (4.85), to obtain the following expression for the magnetic field intensity vector, $\underline{\mathbf{H}}$:

magnetic field

$$\underline{\mathbf{H}} = \frac{1}{\mu}\nabla\times\underline{\mathbf{A}} = \frac{1}{\mu r}\left[\frac{\partial}{\partial r}\left(r\underline{A}_\theta\right) - \frac{\partial \underline{A}_r}{\partial\theta}\right]\hat{\boldsymbol{\phi}} = -\frac{\beta^2\underline{I}\,l\,\mathrm{e}^{-\mathrm{j}\beta r}\sin\theta}{4\pi}\left[\frac{1}{\mathrm{j}\beta r} + \frac{1}{(\mathrm{j}\beta r)^2}\right]\hat{\boldsymbol{\phi}} = \underline{H}_\phi\,\hat{\boldsymbol{\phi}}. \quad (14.10)$$

The same result can also be obtained by directly computing $\underline{\mathbf{H}}$ due to the dipole current, by means of the high-frequency generalization of the Biot-Savart law,

Eq. (8.128) or (8.135),

$$\underline{\mathbf{H}} = \frac{1}{4\pi} \int_l \frac{\underline{I}\,\mathrm{d}\mathbf{l} \times \hat{\mathbf{R}}(1 + \mathrm{j}\beta R)\,\mathrm{e}^{-\mathrm{j}\beta R}}{R^2} \approx \frac{\underline{I}\mathbf{l} \times \hat{\mathbf{r}}(1 + \mathrm{j}\beta r)\,\mathrm{e}^{-\mathrm{j}\beta r}}{4\pi r^2}, \tag{14.11}$$

where [see Fig. 14.1 and Eq. (4.114)]

$$\mathbf{l} \times \hat{\mathbf{r}} = l\hat{\mathbf{z}} \times \hat{\mathbf{r}} = l \sin\theta\, \hat{\boldsymbol{\phi}}, \tag{14.12}$$

and $(1 + \mathrm{j}\beta r)/r^2$ multiplied by $-1/\beta^2 = 1/(\mathrm{j}\beta)^2$ gives the two terms in square brackets in Eq. (14.10).

Note that once we know one of the field vectors, the remaining one can be found from the appropriate curl Maxwell's equation. For instance, given that $\underline{\mathbf{H}}$ has only a ϕ-component, Eqs. (11.2) and (4.85) tell us that

$$\underline{\mathbf{E}} = -\frac{\mathrm{j}}{\omega\varepsilon} \nabla \times \underline{\mathbf{H}} = -\frac{\mathrm{j}}{\omega\varepsilon r} \left[\frac{1}{\sin\theta} \frac{\partial}{\partial\theta} (\sin\theta\, \underline{H}_\phi)\, \hat{\mathbf{r}} - \frac{\partial}{\partial r} (r \underline{H}_\phi)\, \hat{\boldsymbol{\theta}} \right], \tag{14.13}$$

and it is a simple matter to verify that the substitution of the result for $\underline{\mathbf{H}}$ from Eq. (14.10) into this expression leads to the result for $\underline{\mathbf{E}}$ in Eq. (14.8). So with Eqs. (14.11) and (14.13), we do not need potentials at all to get the field expressions for the dipole.

All of the different ways of evaluation of potentials and field vectors of a Hertzian dipole presented in this section are extremely important for understanding the operation of this fundamental antenna and relations between various quantities in its analysis. However, perhaps the simplest, conceptually, order of steps to get the expressions for $\underline{\mathbf{E}}$ and $\underline{\mathbf{H}}$ is the following: find $\underline{\mathbf{A}}$ in Eqs. (14.4) and (14.5), then $\underline{\mathbf{H}}$ from $\underline{\mathbf{A}}$ in Eq. (14.10), and finally $\underline{\mathbf{E}}$ from $\underline{\mathbf{H}}$ in Eq. (14.13). In fact, as we shall see in a later section, computing the potential $\underline{\mathbf{A}}$ due to currents of an arbitrary transmitting antenna is the focal point in finding its radiation characteristics.

Since the electric field in Eq. (14.8) has components $\underline{E}_r$ and $\underline{E}_\theta$ ($\underline{E}_\phi = 0$), and magnetic field in Eq. (14.10) only $\underline{H}_\phi$ ($\underline{H}_r = \underline{H}_\theta = 0$), as shown in Fig. 14.1, $\underline{\mathbf{E}}$ and $\underline{\mathbf{H}}$ are mutually orthogonal at every point of space (for every location of the field point, P, $\underline{\mathbf{E}}$ lies in a vertical plane defined by $\phi =$ const, and $\underline{\mathbf{H}}$ is normal to this plane). The field expressions do not depend on ϕ, $\underline{\mathbf{E}} = \underline{\mathbf{E}}(r, \theta)$ and $\underline{\mathbf{H}} = \underline{\mathbf{H}}(r, \theta)$, and the same is true for the potential expressions, Eqs. (14.4)–(14.6), as expected from the rotational symmetry (with respect to the z-axis) of the dipole (note that independence of ϕ, the azimuthal angle, is also referred to as the azimuthal symmetry). Therefore, on a sphere of radius r centered at the generator in Fig. 14.1, the field distribution (pattern) is determined only by the zenith angle, θ. On the other side, considering the field and potential dependences on r, we see that wherever we have r, it actually is βr. In addition, dipole fields and potentials depend on $\underline{I}$ and βl, and, having in mind Eq. (8.111), they can be expressed in terms of the electrical length [see Eqs. (12.73)] of the dipole, l/λ, as well as electrical distance from the dipole center, r/λ. As a result, the relative spatial field distributions do not change if we scale (increase or decrease) l, r, and λ by the same factor. This conclusion is quite important, given that an arbitrary transmitting antenna can be represented as a superposition of Hertzian dipoles (to be shown in a later section). It implies that antennas having (drastically) different physical dimensions that are equal electrically at the corresponding frequencies produce the same spatial field picture (pattern), which as well has to be expressed in electrical units. Thus, the physical length of an electrically short dipole may actually be very large, at lower frequencies (e.g., at $f = 100$ kHz, $\lambda = 3$ km (in air), and a dipole that is only one hundredth of a wavelength, is physically as long as $l = \lambda/100 = 30$ m).

Finally, combining Eqs. (8.194), (14.8), and (14.10), the complex Poynting vector at the point P in Fig. 14.1 is

complex Poynting vector of a Hertzian dipole

$$\underline{\mathcal{P}} = \underline{\mathbf{E}} \times \underline{\mathbf{H}}^* = (\underline{E}_r \,\hat{\mathbf{r}} + \underline{E}_\theta \,\hat{\boldsymbol{\theta}}) \times \underline{H}_\phi^* \,\hat{\boldsymbol{\phi}} = -\underline{E}_r \underline{H}_\phi^* \,\hat{\boldsymbol{\theta}} + \underline{E}_\theta \underline{H}_\phi^* \,\hat{\mathbf{r}}$$
$$= \eta \left(\frac{\beta^2 \underline{I} l}{4\pi}\right)^2 \left\{\left[\frac{\mathrm{j}}{(\beta r)^3} + \frac{\mathrm{j}}{(\beta r)^5}\right] \sin 2\theta \,\hat{\boldsymbol{\theta}} + \left[\frac{1}{(\beta r)^2} - \frac{\mathrm{j}}{(\beta r)^5}\right] \sin^2\theta \,\hat{\mathbf{r}}\right\}, \tag{14.14}$$

where $\underline{I}\underline{I}^* = |\underline{I}|^2 = I^2$. We see that the only real term in this expression is the one proportional to $1/r^2$, in the radial component of $\underline{\mathcal{P}}$. This term comes from $1/r$ terms in the electric and magnetic field expressions in Eqs. (14.8) and (14.10), and it equals the time average of the instantaneous Poynting vector due to the antenna, Eq. (8.195),

time-average Poynting vector

$$\mathcal{P}_{\text{ave}} = \text{Re}\{\underline{\mathcal{P}}\} = \frac{\eta \beta^2 I^2 l^2 \sin^2\theta}{16\pi^2 r^2} \,\hat{\mathbf{r}}. \tag{14.15}$$

Integrating $\mathcal{P}_{\text{ave}}$ over a closed surface placed about the dipole, which will be done in a later section, we can find the outward real (time-average) power flow emanating from the antenna. All other terms in Eq. (14.14) are imaginary, and thus represent the reactive power associated with the energy fluctuation back and forth between the electric and magnetic fields around the antenna.

Problems: 14.1–14.4; *Conceptual Questions* (on Companion Website): 14.1–14.8; *MATLAB Exercises* (on Companion Website).

14.2 FAR FIELD AND NEAR FIELD

This section introduces an important special case of the electromagnetic field due to a Hertzian dipole (Fig. 14.1): the far field, for observation locations that are electrically far away from the antenna. In this case, we do not need to use the exact expressions for dipole field vectors, as well as potentials and Poynting vector, derived in the previous section, which, of course, are valid only under qualifications in Eqs. (14.2), but much simpler approximate ones, specialized for large distances r of the field point P in Fig. 14.1 from the origin.

In specific, in the far zone r is much larger than the operating wavelength λ of the dipole (for the ambient medium), given in Eqs. (8.112) and (8.111), βr is much larger than unity, and we can write

far zone

$$r \gg \lambda \quad \longrightarrow \quad \beta r \gg 1 \quad \longrightarrow \quad \frac{1}{\beta r} \gg \frac{1}{(\beta r)^2} \gg \frac{1}{(\beta r)^3}. \tag{14.16}$$

In practice, a useful rule of thumb quantifying the far-field condition[4] is: $r > 10\lambda$. Therefore, the dominant terms in both field expressions in Eqs. (14.8) and (14.10) are those with the smallest inverse powers of r (or βr), that is, the $1/r$ terms. These expressions thus become

far electric and magnetic fields of a Hertzian dipole

$$\underline{\mathbf{E}} \approx \frac{\mathrm{j}\eta\beta \underline{I} l \,\mathrm{e}^{-\mathrm{j}\beta r} \sin\theta}{4\pi r} \,\hat{\boldsymbol{\theta}}, \quad \underline{\mathbf{H}} \approx \frac{\mathrm{j}\beta \underline{I} l \,\mathrm{e}^{-\mathrm{j}\beta r} \sin\theta}{4\pi r} \,\hat{\boldsymbol{\phi}}. \tag{14.17}$$

[4]For electrically large antennas, whose maximum dimension, D, is much larger than the wavelength, $D \gg \lambda$, the far-field condition is given by $r \gg D$, which then surpasses the condition in Eq. (14.16).

Fields decrease slowly with r (as $1/r$) and are actually proportional to $\mathrm{d}i/\mathrm{d}t$ in the time domain, enabling electromagnetic radiation by a rapidly time-varying dipole current to faraway distances – they constitute the so-called radiation field of a Hertzian dipole.[5] In addition, the far electric field vector has only a θ-component, while the far magnetic field vector is ϕ-directed, as shown in Fig. 14.2. Combining Eqs. (14.17), (14.5), (14.4), and (14.9), we can write for these components

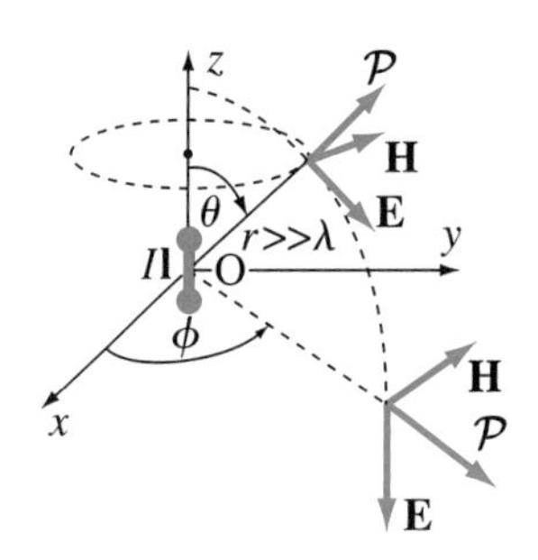

Figure 14.2 Far field of a Hertzian dipole (depicted in Fig. 14.1).

$$\underline{E}_\theta = -\mathrm{j}\omega \underline{A}_\theta, \quad \underline{H}_\phi = \frac{\underline{E}_\theta}{\eta}. \tag{14.18}$$

Hence, $\mathbf{E}$ can be expressed solely in terms of the magnetic vector potential, $\underline{\mathbf{A}}$, of the dipole, more precisely, in terms of its θ-component (the radial component of $\underline{\mathbf{A}}$ does not contribute to $\underline{\mathbf{E}}$ – in the far zone). The electric and magnetic complex field intensities are interrelated as in Eq. (9.20) for uniform plane waves. Once $\underline{E}$ is known, the corresponding $\underline{H}$ is found from it using this relationship. The radiated wave is a TEM (transverse electromagnetic) wave, as both $\mathbf{E}$ and $\mathbf{H}$ are transverse to the direction of propagation (radiation), that is, the radial direction in Fig. 14.2. In addition, $\mathbf{E}$ and $\mathbf{H}$ are perpendicular to each other. Both fields are proportional to $\mathrm{e}^{-\mathrm{j}\beta r}/r$, and the wavefronts are spherical, as in Fig. 8.7, so this is a spherical TEM wave.

Since $\mathbf{E}$ and $\mathbf{H}$ in the far zone are in phase, we expect that the associated complex Poynting vector is purely real. Indeed, as in the plane wave case, Eq. (9.40), we have

$$\underline{\mathcal{P}} = \underline{\mathbf{E}} \times \underline{\mathbf{H}}^* = \underline{E}_\theta \underline{H}_\phi^* \,\hat{\mathbf{r}} = \underline{E}_\theta \frac{\underline{E}_\theta^*}{\eta} \hat{\mathbf{r}} = \frac{|\underline{E}_\theta|^2}{\eta} \hat{\mathbf{r}} = \frac{\eta \beta^2 I^2 l^2 \sin^2\theta}{16\pi^2 r^2} \hat{\mathbf{r}} = \mathcal{P}_{\mathrm{ave}}, \tag{14.19}$$

far-zone Poynting vector of a dipole

and $\underline{\mathcal{P}}$ equals the time-average Poynting vector, $\mathcal{P}_{\mathrm{ave}}$, in Eq. (14.15). Of course, this can as well be obtained from Eq. (14.14), by eliminating all but the lowest term, proportional to $1/r^2$. The (radial) direction of $\underline{\mathcal{P}}$ (Fig. 14.2) coincides with the direction of the spherical TEM wave propagation (direction of radiation of the antenna) and its magnitude, $\underline{\mathcal{P}} = \mathcal{P}_{\mathrm{ave}}$, equals the time-average surface power density transported by the wave, i.e., the radiated power per unit area of the spherical wavefront. Note that the absence of any imaginary (reactive) component of $\underline{\mathcal{P}}$ in Eq. (14.19) means that, in the far-field region, all power is radiated power.

Just opposite to Eqs. (14.16) for the far field, the dipole-to-field distance r in the near zone in Fig. 14.1 is, by definition, small relative to the operating wavelength λ of the dipole (in practice, $r < 0.1\lambda$), but still $r \gg l$, and βr satisfies the condition in Eq. (8.130), which leads to

$$r \ll \lambda \quad \longrightarrow \quad \beta r \ll 1 \quad \longrightarrow \quad \frac{1}{(\beta r)^3} \gg \frac{1}{(\beta r)^2} \gg \frac{1}{\beta r}, \tag{14.20}$$

near zone

so only the dominant terms with the largest inverse powers of r in the expressions for each of the field components in Eqs. (14.8) and (14.10) need to be retained.

[5] In the general expressions for instantaneous electric and magnetic field intensities of a Hertzian dipole, obtained by converting the complex field expressions in Eqs. (14.8) and (14.10) by means of Eq. (8.66), the terms that are proportional to $1/r$ are also proportional to $\mathrm{d}i/\mathrm{d}t$ (of course, $\mathrm{d}i/\mathrm{d}t$ corresponds to $\mathrm{j}\omega\underline{I}$ in the complex domain). Being proportional to and due to the time rate of change of the dipole current, they are significant only if this current is rapidly varying in time, which is the case here, in Fig. 14.1, and for antennas in general. So, the $1/r$ terms determine the actual rapidly time-varying electromagnetic field of the Hertzian dipole, and distinguish it from its quasistatic version. This field, in turn, enables electromagnetic radiation by the antenna, and is termed the radiation field.

In addition, the retardation effect in this zone can be neglected, $e^{-j\beta r} \approx 1$, as in Eqs. (8.131) and (8.132). Consequently, the near electric field of a Hertzian dipole reduces to simply the electric field intensity vector of a quasistatic electric dipole, the same as an electrostatic dipole, Eq. (1.117), except that the dipole charge is slowly oscillating in time (and not time-constant), whose complex moment [see Eq. (1.116)] is $\underline{\mathbf{p}} = \underline{Q}\mathbf{l}$. On the other side, keeping only the dominant term for the near zone of the magnetic field, it is simplified to that of a quasistatic (slowly oscillating) current element $\underline{I}\,\mathbf{l}$ (producing the field with the same spatial distribution as a steady current one); placing this current element at the point P′ in Fig. 4.5 (with α replaced by θ), its field would be given by Eqs. (4.11) and (4.12). So, in evaluating the electric and magnetic fields in the near zone, a Hertzian dipole is equivalent to a quasistatic electric dipole and current element, respectively. Overall, by superposition, the near electromagnetic field of an arbitrary antenna is a quasistatic field. Since the charge and current intensity of the dipole, Q and i, are in time-phase quadrature (90° out of phase with respect to each other), which corresponds to the difference in "j" in their complex magnitudes, Eqs. (14.3), the same is true for the electric and magnetic field intensities of the dipole in the near zone. This, in turn, indicates reactive power associated with the fields, like in electromagnetic resonators [see, for example, Eqs. (8.69) and (3.45) for a simple resonant LC circuit or Eqs. (13.150) and (13.151) for waveguide cavity resonators]. We conclude that a Hertzian dipole (or any other antenna) in its near zone acts as a reactive, energy-storage, device. The energy oscillates between the near electric and magnetic field of the antenna, like in a resonant LC circuit.

As the principal function of transmitting antennas is to convey electromagnetic signals (energy or information) to distant locations (through free space or other material media), in most applications of antennas we deal with the far field only. Later in this chapter, we shall define and study a number of parameters that characterize and quantify the radiated field of an arbitrary antenna in different directions of radiation (i.e., for given angles θ and ϕ in Fig. 14.2). However, in addition to far-field parameters, the antenna input impedance, which determines the impedance matching properties of an antenna to its feed electric circuit, is of equal practical importance for antenna analysis and design, and the reactive part of this impedance implicitly includes (in an integral fashion) the near-field characteristics of the antenna.

Problems: 14.5 and 14.6; *Conceptual Questions* (on Companion Website): 14.9–14.16; *MATLAB Exercises* (on Companion Website).

14.3 STEPS IN FAR-FIELD EVALUATION OF AN ARBITRARY ANTENNA

An arbitrary transmitting wire metallic antenna can be represented as a chain of Hertzian dipoles, as illustrated in Fig. 14.3. The currents of individual dipoles, described by infinitesimal current elements $\underline{I}\,d\mathbf{l}$, constitute the current intensity of the wire, $\underline{I}$, which, in general, is a function of the location along the antenna axis. In other words, the elementary currents gradually change from dipole to dipole, following the current distribution of the antenna. As the current is not uniform along the wire, the charges of adjacent dipoles in the model do not entirely compensate each other, and hence a continuous charge distribution along the antenna. This distribution is described by the charge per unit length of the antenna, $\underline{Q}'$, which is related to the current $\underline{I}$ through the continuity equation for wires in Eq. (8.133). Only in the special case of a uniform antenna current ($\underline{I} = \text{const}$), all current elements in Fig. 14.3 are the same, and charge $\underline{Q}$ at the end of one dipole in the chain completely compensates charge $-\underline{Q}$ of the next dipole, etc., so that $\underline{Q}' = 0$ along the antenna, which is also apparent from Eq. (8.133). A similar representation by superposition of Hertzian dipoles applies to surface and volume antennas as well, with equivalent dipoles defined by surface and volume current elements $\underline{\mathbf{J}}_s \, \mathrm{d}S$ and $\underline{\mathbf{J}} \, \mathrm{d}v$, respectively. The dipoles are distributed and interconnected to each other in a two- or three-dimensional fashion, following the current distribution all over the surface S or volume v of the antenna.

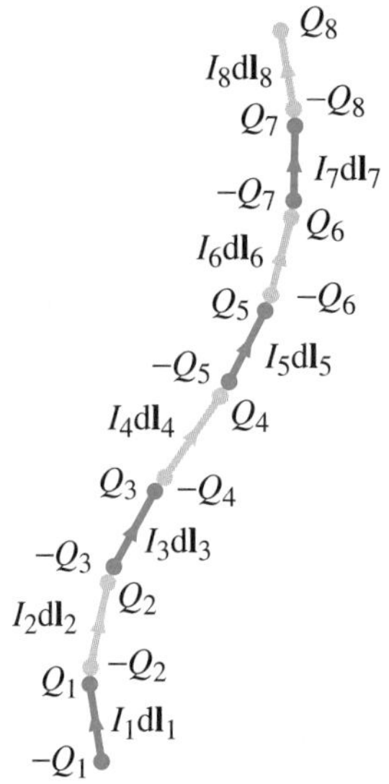

Figure 14.3 Representation of an arbitrary transmitting wire metallic antenna by a chain of Hertzian dipoles.

Representation in Fig. 14.3 means that all conclusions and expressions derived for a Hertzian dipole (Fig. 14.1) so far in this chapter can readily be generalized to arbitrary antennas. In the far zone, for instance, Eqs. (14.18) tell us that the electric field intensity vector, $\underline{\mathbf{E}}$, of an arbitrary current distribution can be expressed solely in terms of the corresponding magnetic vector potential, $\underline{\mathbf{A}}$. For a single dipole in Fig. 14.1, and for a straight wire antenna (of arbitrary length) along the z-axis, $\underline{\mathbf{E}}$ has only a θ-component, equal to $-\mathrm{j}\omega\underline{A}_\theta$. So, $\underline{\mathbf{E}}$ is proportional to a nonradial or transverse component of $\underline{\mathbf{A}}$. For an arbitrarily oriented Hertzian dipole, and for an arbitrary antenna, the transverse components of $\underline{\mathbf{A}}$ in the spherical coordinate system are both $\underline{A}_\theta$ and $\underline{A}_\phi$. In all cases, we can decompose $\underline{\mathbf{A}}$ onto a radial ($\underline{\mathbf{A}}_r$) and transverse ($\underline{\mathbf{A}}_\mathrm{t}$) vector components, analogously to the decomposition of the field vectors of guided electromagnetic waves onto longitudinal and transverse components, in Eqs. (11.7), and write for the far electric field

$$\boxed{\underline{\mathbf{A}} = \underline{\mathbf{A}}_r + \underline{\mathbf{A}}_\mathrm{t} \quad \longrightarrow \quad \underline{\mathbf{E}} = -\mathrm{j}\omega\underline{\mathbf{A}}_\mathrm{t},} \tag{14.21}$$

far E-field, arbitrary antenna

where $\underline{\mathbf{A}}_r = \underline{A}_r \, \hat{\mathbf{r}}$ and $\underline{\mathbf{A}}_\mathrm{t} = \underline{A}_\theta \, \hat{\boldsymbol{\theta}} + \underline{A}_\phi \, \hat{\boldsymbol{\phi}}$. Expressing $\underline{\mathbf{A}}$ in terms of all three components in the spherical coordinate system, it is a simple matter to show that the transverse vector $\underline{\mathbf{A}}_\mathrm{t}$ can be obtained from the total $\underline{\mathbf{A}}$ by means of the following transformation:

$$\underline{\mathbf{A}}_\mathrm{t} = \hat{\mathbf{r}} \times (\underline{\mathbf{A}} \times \hat{\mathbf{r}}). \tag{14.22}$$

The far magnetic field intensity vector, $\underline{\mathbf{H}}$, is also a purely transverse vector, perpendicular to $\underline{\mathbf{E}}$, Eqs. (14.18), and can be computed from $\underline{\mathbf{E}}$ using the vector relation

$$\boxed{\underline{\mathbf{H}} = \frac{1}{\eta} \hat{\mathbf{r}} \times \underline{\mathbf{E}} \quad \left(\eta = \sqrt{\frac{\mu}{\varepsilon}}\right),} \tag{14.23}$$

far H-field, arbitrary antenna

the same as for uniform plane waves, Eqs. (9.22). Alternatively, with

$$\underline{\mathbf{E}} = \underline{E}_\theta \, \hat{\boldsymbol{\theta}} + \underline{E}_\phi \, \hat{\boldsymbol{\phi}} \quad \text{and} \quad \underline{\mathbf{H}} = \underline{H}_\theta \, \hat{\boldsymbol{\theta}} + \underline{H}_\phi \, \hat{\boldsymbol{\phi}}, \tag{14.24}$$

Eqs. (14.21) and (14.23) can be written in terms of θ- and ϕ-components of the two field vectors,

$$\underline{E}_\theta = -\mathrm{j}\omega\underline{A}_\theta, \quad \underline{E}_\phi = -\mathrm{j}\omega\underline{A}_\phi, \quad \underline{H}_\phi = \frac{\underline{E}_\theta}{\eta}, \quad \underline{H}_\theta = -\frac{\underline{E}_\phi}{\eta}. \tag{14.25}$$

The associated complex Poynting vector in the far zone is purely real, equal to

$$\boxed{\begin{aligned} \underline{\mathcal{P}} &= \underline{\mathbf{E}} \times \underline{\mathbf{H}}^* = \left(\underline{E}_\theta \underline{H}_\phi^* - \underline{E}_\phi \underline{H}_\theta^*\right) \hat{\mathbf{r}} = \frac{\underline{E}_\theta \underline{E}_\theta^* + \underline{E}_\phi \underline{E}_\phi^*}{\eta} \, \hat{\mathbf{r}} \\ &= \frac{|\underline{E}_\theta|^2 + |\underline{E}_\phi|^2}{\eta} \, \hat{\mathbf{r}} = \eta \left(|\underline{H}_\phi|^2 + |\underline{H}_\theta|^2\right) \hat{\mathbf{r}} = \frac{|\underline{\mathbf{E}}|^2}{\eta} \, \hat{\mathbf{r}} = \eta |\underline{\mathbf{H}}|^2 \, \hat{\mathbf{r}}. \end{aligned}} \tag{14.26}$$

far-zone Poynting vector, arbitrary antenna

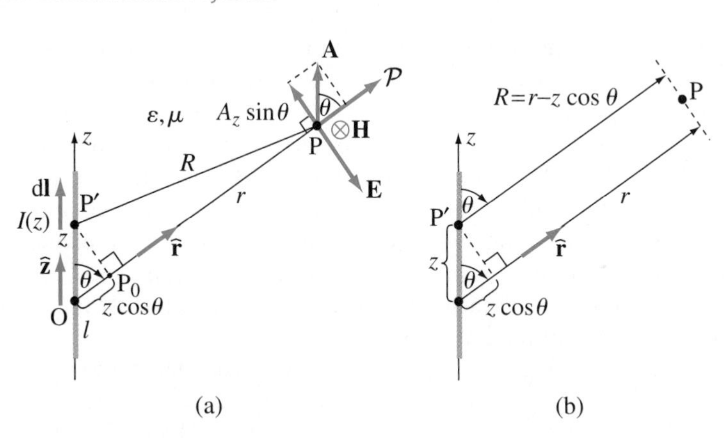

Figure 14.4 Straight wire antenna with an arbitrary current distribution: (a) evaluation of the magnetic vector potential in the far zone and (b) parallel-ray approximation for the phase calculation.

We see that the first and most important step in far-field evaluation of an arbitrary antenna is computing its magnetic vector potential, $\underline{\mathbf{A}}$. So, let us now focus on the integrals involved in this part of the analysis. Consider first a straight wire antenna along the z-axis, shown in Fig. 14.4(a). At an observation point P (field point) defined by (r, θ, ϕ), $\underline{\mathbf{A}}$ is given by the first integral in Eq. (14.4). Under the far-field assumption, Eq. (14.16), we apply different approximations for the magnitude and phase of the spherical-wave factor $\mathrm{e}^{-\mathrm{j}\beta R}/R$ in the integral, with R being the variable source-to-field distance for an arbitrary point P′ at the wire axis (source point). For the magnitude, $1/R$, we approximate R by the fixed radial coordinate r of the point P, in Fig. 14.4(a),

for magnitude approximation

$$\boxed{R \approx r,} \tag{14.27}$$

and bring $1/R \approx 1/r$ outside the integral sign. For the phase factor, $\mathrm{e}^{-\mathrm{j}\beta R}$, we approximate R by the distance from the projection point of P′ on the r-direction, point P_0 [Fig. 14.4(a)], to P,

for phase approximation, straight wire antenna

$$\boxed{R \approx r - z\cos\theta,} \tag{14.28}$$

where z is the coordinate along the wire axis defining the position of the source point ($z = 0$ at the coordinate origin, O). Note that a similar approximation is applied in Figs. 1.28 and 14.1 to express the difference of the distances of the field point from the two dipole charges, $\Delta R = r_2 - r_1$, in terms of the separation between the charges and the angle θ. With this, we are able to take the fixed part of the phase factor, $\mathrm{e}^{-\mathrm{j}\beta r}$, outside the integral sign, while the remaining, z-dependent, part, representing a phase correction for fields due to individual elementary dipoles in Fig. 14.3, must be integrated, and we have

radiation integral, straight wire antenna

$$\boxed{\underline{\mathbf{A}} = \frac{\mu}{4\pi}\int_l \frac{\underline{I}(z)\,\mathrm{d}\mathbf{l}\,\mathrm{e}^{-\mathrm{j}\beta R}}{R} \approx \frac{\mu\,\mathrm{e}^{-\mathrm{j}\beta r}}{4\pi r}\,\hat{\mathbf{z}}\int_l \underline{I}(z)\,\mathrm{e}^{\mathrm{j}\beta z\cos\theta}\,\mathrm{d}z,} \tag{14.29}$$

with l being the length of the antenna, and $\mathrm{d}\mathbf{l} = \mathrm{d}z\,\hat{\mathbf{z}}$. The resulting integral in terms of the coordinate z along the wire antenna is called the radiation integral. Its solutions, for given current distributions $\underline{I}(z)$, will be the basis for analysis of different types of wire antennas throughout the rest of this chapter. Obviously, the geometrical relation between R and r in Eq. (14.28) would be exact if the R and r lines (rays) ran parallel to each other, as in Fig. 14.4(b), and this would be the case if the field point P were at infinity. Hence this relation is known as the parallel-ray approximation for the far-field phase calculation.

It is very important to realize that, because of the significant phase differences between the far fields of individual elementary Hertzian dipoles constituting the antenna, the magnitude approximation in Eq. (14.27) is not accurate enough for the phase computation in Eq. (14.29), where Eq. (14.28) is used instead. Namely, although the source-to-field distances R for dipoles in the chain are almost the same in the far field, and can be taken as constant for the magnitude computation in Eq. (14.29), even small relative differences in R may have a large influence on the phase factor, $e^{-j\beta R}$. For example, if the distances R for two elementary dipoles are $R_1 = 100\lambda$ and $R_2 = 99.5\lambda$, i.e., they differ by $\Delta R = 0.5\lambda$, the corresponding magnitude factors are

$$\frac{1}{R_1} = \frac{1}{100\lambda} = \frac{0.01}{\lambda}, \quad \frac{1}{R_2} = \frac{1}{99.5\lambda} = \frac{0.01005}{\lambda}, \tag{14.30}$$

i.e., we make a 0.5% magnitude error by assuming that the two factors are the same. On the other side, the phase difference associated with this assumed difference in R is as large as 180°,

$$e^{-j\beta(R_1-R_2)} = e^{-j\frac{2\pi}{\lambda}0.5\lambda} = e^{-j\pi} = -1, \tag{14.31}$$

making the two fields be in counter-phase and actually cancel each other at the point P in Fig. 14.4(a). However, the approximation in Eq. (14.27), $R_1 \approx R_2$, if applied to the phases would make the fields add in phase [+1 instead of −1 in Eq. (14.31)], causing a very large error in the field integral. Only for electrically short antennas, $l \ll \lambda$, since then $\Delta R \ll \lambda$ for any two points along the wire, we can use the approximation in Eq. (14.27) even for the phase factor, $e^{-j\beta R} \approx e^{-j\beta r}$, as in Eq. (14.4), so that the radiation integral reduces to the integral of only the current intensity along the wire [$\beta z \approx 0$ in Eq. (14.29)].

Similar phase correction to that in Eq. (14.28) is used for arbitrary (wire, surface, and volume) antennas. For an arbitrary current distribution, in Fig. 14.5, let the source point P′ be defined by the position vector $\mathbf{r}'$ with respect to the coordinate origin. Thus, the difference between r and R is approximately equal to the projection of $\mathbf{r}'$ on the r-ray, i.e., to $r'\cos\alpha = \mathbf{r}'\cdot\hat{\mathbf{r}}$, with α denoting the angle between vectors $\mathbf{r}'$ and $\hat{\mathbf{r}}$. In other words, R is approximated by

$$\boxed{R \approx r - \mathbf{r}'\cdot\hat{\mathbf{r}},} \tag{14.32}$$

for phase approximation, arbitrary antenna

with which Eq. (14.29) becomes

$$\boxed{\underline{\mathbf{A}} = \frac{\mu\, e^{-j\beta r}}{4\pi r}\int_v \underline{\mathbf{J}}(\mathbf{r}')\, e^{j\beta\mathbf{r}'\cdot\hat{\mathbf{r}}}\, dv.} \tag{14.33}$$

radiation integral, arbitrary antenna

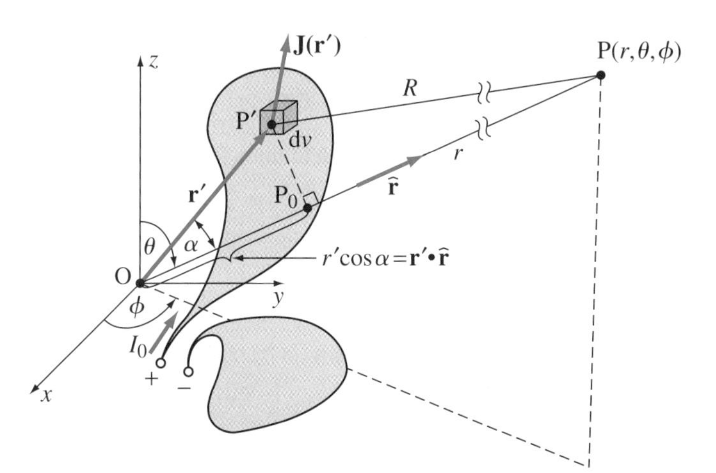

Figure 14.5 Far-field computation for an arbitrary transmitting antenna.

Analogous expressions hold for surface antennas and for arbitrary wire antennas. Note that Eqs. (14.28) and (14.29) are special cases of the parallel-ray approximation in Eq. (14.32) and radiation integral in Eq. (14.33). The equations for a straight wire antenna along the z-axis are obtained from general equations by specifying $\mathbf{r}' = z\,\hat{\mathbf{z}}$ (Fig. 14.4), so that $\mathbf{r}' \cdot \hat{\mathbf{r}} = z\cos\theta$.

Once $\underline{\mathbf{A}}$ is known, the remaining steps in the radiation analysis of the antenna are straightforward, and the same for any antenna type and geometry. Namely, for a general antenna in Fig. 14.5, $\underline{\mathbf{E}}$ and $\underline{\mathbf{H}}$ in the far zone are easily found from Eqs. (14.21) and (14.23), and Poynting vector from Eq. (14.26). Note that these expressions become even simpler for a wire antenna in Fig. 14.4, where the electric field has only a θ-component, which, using Eqs. (14.25) and (14.5), is given by

far E-field, straight wire antenna

$$\boxed{\underline{E}_\theta = -\mathrm{j}\omega\underline{A}_\theta = \mathrm{j}\omega\underline{A}_z\sin\theta,} \tag{14.34}$$

with $\underline{A}_z$ standing for the z-component of the magnetic vector potential in Eq. (14.29).

Finally, we see from Eqs. (14.33), (14.21), and (14.23) that the far electric and magnetic field intensity vectors due to an arbitrary antenna are proportional to the characteristic spherical-wave propagation function, $\mathrm{e}^{-\mathrm{j}\beta r}/r$. In each of the field expressions, this function of r is multiplied by a corresponding vector function of angles θ and ϕ (which define the direction of radiation) in Fig. 14.5. Globally, the radiated fields constitute a spherical TEM wave, centered at the coordinate origin, O. However, locally, in the far zone, the spherical wavefront appears to be approximately planar, as if it were a part of a plane wave. Moreover, if viewed only over a finite receiving aperture, i.e., over a small range of angles θ and ϕ (for a fixed r), the fields can be considered to be uniform (the same at every point), and the wave can be treated as if it were uniform as well, so a uniform plane wave, as illustrated in Fig. 9.1. The uniform-plane-wave approximation of nonuniform spherical waves radiated by antennas is of great theoretical and practical importance in the analysis and synthesis of antenna systems. It enables us to use the concepts and equations governing the propagation of uniform plane electromagnetic waves in unbounded media and in the presence of material interfaces, from Chapters 9 and 10, to describe and study the properties of far antenna fields and the associated waves as they propagate away from their sources. For instance, the theory of receiving antennas, to be presented in a later section, assumes that the electromagnetic field received by an antenna, which most frequently is originated by another (transmitting) antenna in a wireless link, is in the form of a uniform plane wave, arriving from a given direction.

Example 14.1 **Far Field of a Nonloaded Short Wire Dipole Antenna**

Consider a nonloaded electrically short symmetrical straight PEC wire dipole antenna of length l. If the antenna is (centrally) fed by a time-harmonic current of complex rms intensity $\underline{I}_0$ and frequency f, find the electric and magnetic field intensity vectors and Poynting vector in the far zone of the antenna.

Solution Since the antenna ends are free (with no capacitive loads), the current at them is zero. In addition, the antenna being electrically short ($l \ll \lambda$), we can readily assume that its current linearly varies between $\underline{I}_0$, at the antenna input terminals, and zero along each of the arms of the dipole, $h = l/2$ long. In other words, we consider a triangular current distribution of the antenna, as shown in Fig. 14.6(a). Analytically, placing the coordinate origin, O, at the center of the antenna, $\underline{I}(z)$ is given by

nonloaded short wire dipole antenna

$$\boxed{\underline{I}(z) = \underline{I}_0\left(1 - \frac{|z|}{h}\right) \quad (-h \le z \le h).} \tag{14.35}$$

As already discussed in this section, in the case of (arbitrary) electrically short straight wire antennas, the radiation integral in Eq. (14.29) reduces to

$$\underline{\mathbf{A}} = \frac{\mu_0 \, \mathrm{e}^{-\mathrm{j}\beta r}}{4\pi r} \, \hat{\mathbf{z}} \underbrace{\int_l \underline{I}(z) \, \mathrm{d}z}_{\underline{I}_{\text{mean}} \, l} \quad \left(\mathrm{e}^{\mathrm{j}\beta z \cos\theta} \approx 1\right), \tag{14.36}$$

arbitrary short wire antenna

with the phase coefficient being $\beta = 2\pi f \sqrt{\varepsilon_0 \mu_0}$, for free space. Substituting the current distribution from Eq. (14.35), the z-component (the only existing one) of the far-zone magnetic vector potential comes out to be

$$\underline{A}_z = \frac{\mu_0 \underline{I}_0 \, \mathrm{e}^{-\mathrm{j}\beta r}}{4\pi r} \, 2 \int_{z=0}^{h} \left(1 - \frac{z}{h}\right) \mathrm{d}z = \frac{\mu_0 (\underline{I}_0 \, l/2) \, \mathrm{e}^{-\mathrm{j}\beta r}}{4\pi r} \, \hat{\mathbf{z}} \quad (l = 2h), \tag{14.37}$$

where, given that $\underline{I}(z)$ in Fig. 14.6(a) is an even function in z (due to its actual dependence on the absolute value of z), the initial integral with symmetric integration limits, $-h$ and h, is solved as twice the integral from 0 to h, in which, in turn, the absolute value sign is removed from $|z|$, because $z \geq 0$. We see that $\underline{A}_z$ is the same as that, in Eq. (14.4), of a Hertzian dipole, Fig. 14.1, with a current-length product equal to $\underline{I}_0 \, l/2$, which can be interpreted as characterizing a dipole with the same current (uniform along the wire) as the feed current of the antenna in Fig. 14.6(a), but a half of it in length, as indicated in Fig. 14.6(b). Of course, an interpretation defining the equivalent Hertzian dipole as being of the same length (l) as the original antenna, but with a current that is a half of its feed current, is also possible. In fact, we note that this equivalent current ($\underline{I}_0/2$) is exactly the mean (computed along the antenna) of the current distribution in Fig. 14.6(a),

$$\underline{I}_{\text{mean}} = \frac{1}{l} \int_l \underline{I}(z) \, \mathrm{d}z. \tag{14.38}$$

This holds, as indicated in Eq. (14.36), for an arbitrary short straight wire antenna, which, as the magnetic potential in the far zone is concerned, can be replaced by a Hertzian dipole of the same length with a uniform current equaling the mean of the current of the original antenna.

The fact that in the far zone, the electric and magnetic field intensity vectors, as well as the Poynting vector, of an arbitrary transmitting antenna can all be computed from the potential $\underline{\mathbf{A}}$, using Eqs. (14.21), (14.23), and (14.26), obviously means that if the two antennas have the same $\underline{\mathbf{A}}$, like antennas in Fig. 14.6(a) and Fig. 14.6(b) – in our case, they also have the same far-zone $\underline{\mathbf{E}}$, $\underline{\mathbf{H}}$, and $\underline{\mathcal{P}}$. Therefore, these quantities for the antenna in Fig. 14.6(a) are given by Eqs. (14.17) and (14.19) with $\underline{I} \, l$ merely replaced by $\underline{I}_0 \, l/2$. On the other hand, note that both antennas in Fig. 14.6 have $\underline{\mathbf{E}}$, $\underline{\mathbf{H}}$, and $\underline{\mathcal{P}}$ amounting to a half of the respective field vectors and a quarter of the Poynting vector of the Hertzian dipole in Fig. 14.1, whose current is $\underline{I}$ and length l.

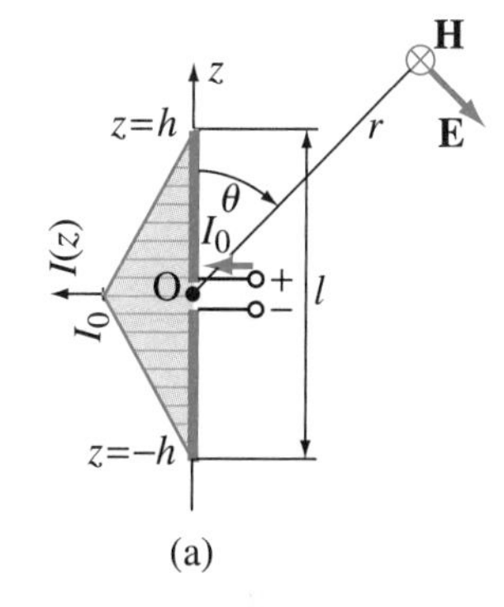

(a)

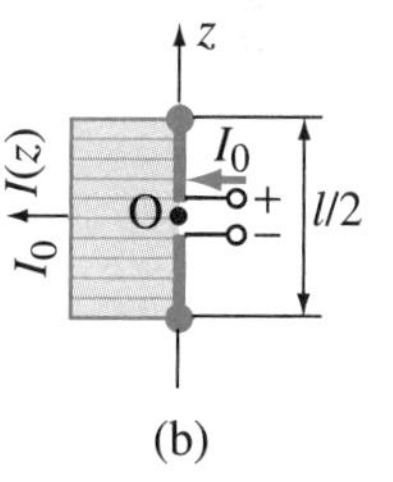

(b)

Figure 14.6 Analysis of radiation by a nonloaded electrically short symmetrical wire dipole antenna: (a) triangular current distribution along the antenna and (b) equivalent Hertzian dipole; for Example 14.1.

Example 14.2 Charge Distribution of a Nonloaded Short Dipole Antenna

For the nonloaded short dipole from the previous example, (a) evaluate the charge distribution along the dipole arms and (b) discuss the far-field equivalency of the two antennas in Fig. 14.6(a) and Fig. 14.6(b) from the standpoint of their charges.

Solution

(a) The charge per unit length of the antenna, $\underline{Q}'(z)$, is found from the antenna current, $\underline{I}(z)$, using the continuity equation for wires in Eq. (8.133). Since the current distribution, Eq. (14.35), is a linear function of z along each of the dipole arms, descending on the upper arm and ascending on the lower, the accompanying line charge densities are uniform (constant), and opposite to each other, as follows:

$$\underline{Q}'_1 = \frac{\mathrm{j}}{\omega} \frac{\mathrm{d}\underline{I}}{\mathrm{d}z} = -\frac{\mathrm{j}\underline{I}_0}{\omega h} \quad \text{for } 0 \leq z \leq h, \quad \underline{Q}'_2 = \frac{\mathrm{j}\underline{I}_0}{\omega h} = -\underline{Q}'_1 \quad \text{for } -h \leq z < 0, \tag{14.39}$$

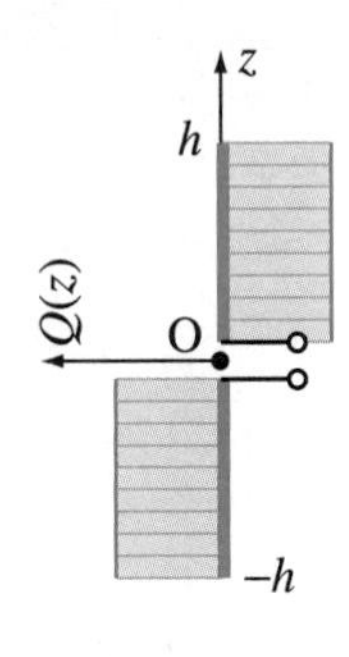

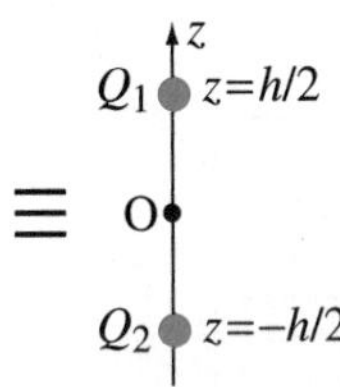

Figure 14.7 Evaluation of accompanying charges of the nonloaded short dipole antenna in Fig. 14.6(a); for Example 14.2.

where $\omega = 2\pi f$ is the operating angular frequency of the antenna. The charge distribution is shown in Fig. 14.7.

(b) The total charges of the two arms, $\underline{Q}_1$ and $\underline{Q}_2$, are given by

$$\underline{Q}_1 = \underline{Q}'_1 h = -\frac{\mathrm{j}\underline{I}_0}{\omega} = -\underline{Q}_2, \tag{14.40}$$

and these charges are exactly the same as $\underline{Q}$ and $-\underline{Q}$ on capacitive terminations of a Hertzian dipole with a uniform current intensity equal to $\underline{I}_0$, $\underline{Q} = -\mathrm{j}\underline{I}_0/\omega$, as in Eq. (14.3), which is also an expression of the continuity equation for the structure. Furthermore, if we place $\underline{Q}_1$ and $\underline{Q}_2$ at the centers of the respective (electrically short) dipole arms in Fig. 14.6(a), as an equivalent replacement of their uniformly distributed line charges, which as well is illustrated in Fig. 14.7, what we obtain is exactly the Hertzian dipole in Fig. 14.6(b), whose arms are $h/2$ long. We thus see that, indeed, the far-field equivalency due to the currents of the two antennas in Fig. 14.6(a) and Fig. 14.6(b), established in the previous example, automatically, i.e., by way of the continuity equation, extends itself to the equivalency in terms of charges.

Example 14.3 Traveling-Wave Wire Antenna

Fig. 14.8 shows an end-fed traveling-wave wire antenna. It consists of a horizontal PEC wire that is l long, driven at $z = 0$ by a current of complex rms intensity $\underline{I}_0$ and angular frequency ω, and terminated at $z = l$ by a purely resistive load, of resistance R_L, adopted such that the current distribution along the antenna is a wave traveling in the positive z direction with a velocity equal to the speed of light. Neglecting the influence of the ground plane, and thus assuming that the antenna operates in free space, as well as the radiation of currents in vertical wire pieces at the two antenna ends, calculate the magnitude of the far electric field vector of the antenna.

Solution As the current wave in Fig. 14.8 travels along the antenna with velocity c_0, Eq. (9.19), the associated phase coefficient of the wave is $\beta = \omega/c_0$, Eq. (8.111). In addition, the wave is unattenuated (there are no losses on the antenna, assumed to be a PEC one, and the ground plane is not taken into account), so that the current distribution, $\underline{I}(z)$, along the antenna can be written as

traveling current wave along an antenna

$$\boxed{\underline{I}(z) = \underline{I}_0\,\mathrm{e}^{-\mathrm{j}\beta z} \quad (0 \le z \le l).} \tag{14.41}$$

From Eq. (14.29), with the usual setup of the spherical coordinate system (Fig. 14.8), the far-zone magnetic potential of the antenna is given by

$$\begin{aligned}\underline{A}_z &= \frac{\mu_0\,\mathrm{e}^{-\mathrm{j}\beta r}}{4\pi r}\int_{z=0}^{l} \underline{I}_0\,\mathrm{e}^{-\mathrm{j}\beta z}\,\mathrm{e}^{\mathrm{j}\beta z\cos\theta}\,\mathrm{d}z = \frac{\mu_0 \underline{I}_0\,\mathrm{e}^{-\mathrm{j}\beta r}}{4\pi r}\int_0^l \mathrm{e}^{-\mathrm{j}\beta' z}\,\mathrm{d}z \\ &= \frac{\mathrm{j}\mu_0 \underline{I}_0\,\mathrm{e}^{-\mathrm{j}\beta r}}{4\pi\beta' r}\left(\mathrm{e}^{-\mathrm{j}\beta' l} - 1\right) = \frac{\mathrm{j}\mu_0 \underline{I}_0\,\mathrm{e}^{-\mathrm{j}\beta r}\,\mathrm{e}^{-\mathrm{j}\beta' l/2}}{4\pi\beta' r}\left(\mathrm{e}^{-\mathrm{j}\beta' l/2} - \mathrm{e}^{\mathrm{j}\beta' l/2}\right) \\ &= \frac{\mu_0 \underline{I}_0\,\mathrm{e}^{-\mathrm{j}\beta r}\,\mathrm{e}^{-\mathrm{j}\beta' l/2}}{2\pi\beta' r}\sin\frac{\beta' l}{2}, \quad \text{where} \quad \beta' = \beta\,(1 - \cos\theta).\end{aligned} \tag{14.42}$$

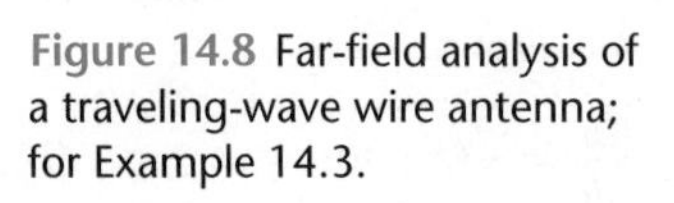
Figure 14.8 Far-field analysis of a traveling-wave wire antenna; for Example 14.3.

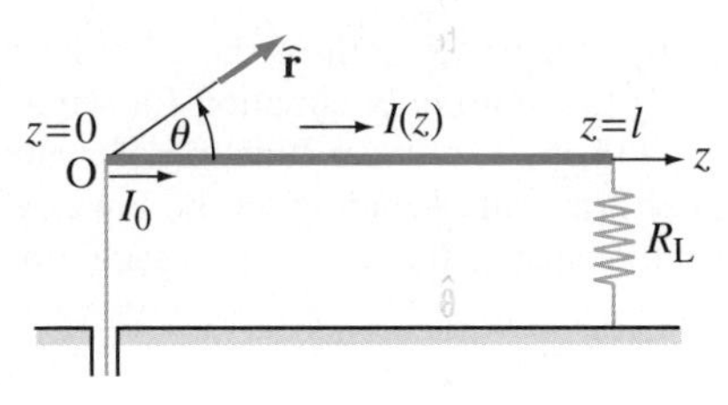

In the above transformations of the result of integration, the first identity in Eqs. (10.7) is used to obtain the final complex expression for $\underline{A}_z$ with magnitude conveniently standing out. Hence, combining Eqs. (14.34), (14.42), and (14.9), the magnitude of the far electric field vector comes out to be

$$|\mathbf{E}| = |\underline{E}_\theta| = \omega|\underline{A}_z|\sin\theta = \frac{\eta_0|\underline{I}_0|\sin\theta}{2\pi(1-\cos\theta)r}\left|\sin\left[\frac{\beta l}{2}(1-\cos\theta)\right]\right|. \quad (14.43)$$

far field of a traveling-wave antenna

As we shall see in a later section, for wire antennas that are open-ended (or eventually terminated in a reactive load), and are not electrically short, the current distribution along the antenna is a standing wave – simply, a superposition of a traveling current wave, in Eq. (14.41), and a reflected wave, progressing in the opposite direction. In the case in Fig. 14.8, on the other hand, a matched load of resistance R_L prevents reflection from the wire end at $z = l$ (the power incident on that end is entirely dissipated to heat in the load).

Example 14.4 Hertzian Dipole along the x-Axis

Find the expressions for the far electric and magnetic field intensity vectors and Poynting vector of a Hertzian dipole, with length l and current of intensity $\underline{I}$ and angular frequency ω, placed at the coordinate origin along the x-axis of a Cartesian coordinate system, in a medium of parameters ε and μ.

Solution The magnetic vector potential is parallel to the dipole, so it is an x-directed vector. Therefore, $\underline{\mathbf{A}}$ is the same as in Eq. (14.4) but with the x-axis taking the role of the z-axis,

$$\underline{\mathbf{A}} = \frac{\mu\underline{I}\,l\,\mathrm{e}^{-\mathrm{j}\beta r}}{4\pi r}\,\hat{\mathbf{x}} = \underline{A}_x\,\hat{\mathbf{x}} \quad (14.44)$$

$(\beta = \omega\sqrt{\varepsilon\mu})$. To find its components in the spherical coordinate system (adopted in the standard fashion), which are needed for the far-field computation, we employ the following decompositions as indicated in Fig. 14.9:

$$\hat{\mathbf{x}} = \cos\phi\,\hat{\boldsymbol{\rho}} + \sin\phi(-\hat{\boldsymbol{\phi}}), \qquad \hat{\boldsymbol{\rho}} = \sin\theta\,\hat{\mathbf{r}} + \cos\theta\,\hat{\boldsymbol{\theta}}, \quad (14.45)$$

which yield

$$\hat{\mathbf{x}} = \sin\theta\cos\phi\,\hat{\mathbf{r}} + \cos\theta\cos\phi\,\hat{\boldsymbol{\theta}} - \sin\phi\,\hat{\boldsymbol{\phi}}. \quad (14.46)$$

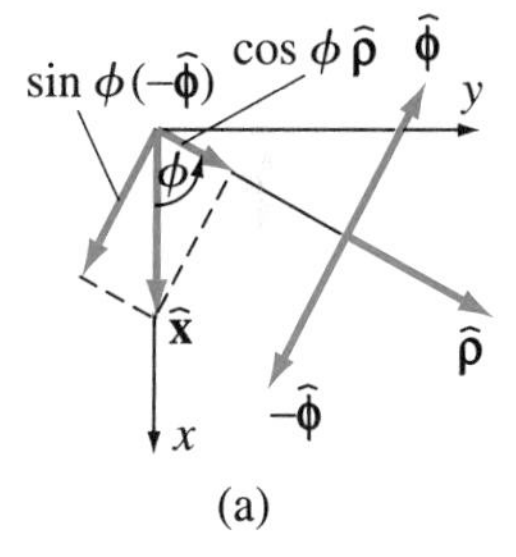

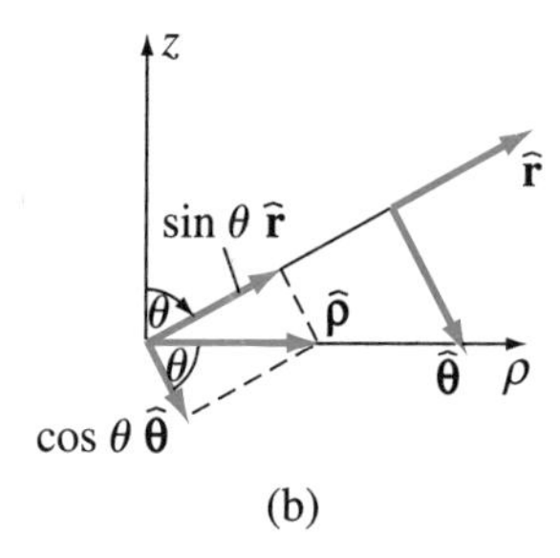

Figure 14.9 Resolving the unit vector $\hat{\mathbf{x}}$ into components in spherical coordinates, by first decomposing $\hat{\mathbf{x}}$ into cylindrical (polar) components (a), and then expressing the radial cylindrical unit vector, $\hat{\boldsymbol{\rho}}$, in terms of spherical unit vectors (b), Eqs. (14.45); for Example 14.4.

Using Eqs. (14.25), (14.44), and (14.46), the electric far-field components of the dipole are

$$\underline{E}_\theta = -\mathrm{j}\omega\underline{A}_\theta = -\mathrm{j}\omega\underline{A}_x\cos\theta\cos\phi, \quad \underline{E}_\phi = -\mathrm{j}\omega\underline{A}_\phi = -\mathrm{j}\omega\underline{A}_x(-\sin\phi), \quad (14.47)$$

so we can write

$$\underline{\mathbf{E}} = \frac{\mathrm{j}\omega\mu\underline{I}\,l\,\mathrm{e}^{-\mathrm{j}\beta r}}{4\pi r}(-\cos\theta\cos\phi\,\hat{\boldsymbol{\theta}} + \sin\phi\,\hat{\boldsymbol{\phi}}). \quad (14.48)$$

Note that $\underline{\mathbf{E}}$ can alternatively be computed from Eqs. (14.21), (14.22), and (14.44), that is, from the vector relationship

$$\underline{\mathbf{E}} = -\mathrm{j}\omega\underline{\mathbf{A}}_\mathrm{t} = -\mathrm{j}\omega\,\hat{\mathbf{r}}\times(\underline{\mathbf{A}}\times\hat{\mathbf{r}}) = -\mathrm{j}\omega\underline{A}_x\,\hat{\mathbf{r}}\times(\hat{\mathbf{x}}\times\hat{\mathbf{r}}), \quad (14.49)$$

which, upon substitution of $\hat{\mathbf{x}}$ from Eq. (14.46) and evaluation of the resulting cross products between unit vectors in the spherical coordinate system, gives the same expression in Eq. (14.48). Note also that the field for a fixed r is a function of both angles θ and ϕ, as expected, since the x-directed dipole is not azimuthally symmetrical (about the z-axis).

The magnetic far-field components are given by Eqs. (14.25), and hence

$$\underline{\mathbf{H}} = \frac{1}{\eta}\left(-\underline{E}_\phi\,\hat{\boldsymbol{\theta}} + \underline{E}_\theta\,\hat{\boldsymbol{\phi}}\right) = -\frac{\mathrm{j}\beta\underline{I}\,l\,\mathrm{e}^{-\mathrm{j}\beta r}}{4\pi r}(\sin\phi\,\hat{\boldsymbol{\theta}} + \cos\theta\cos\phi\,\hat{\boldsymbol{\phi}}), \quad (14.50)$$

where $\eta = \sqrt{\mu/\varepsilon}$ and the use is made of the first relationship in Eqs. (14.9). Finally, combining Eqs. (14.26) and Eq. (14.48), the Poynting vector due to the dipole in its far zone comes out to be

$$\underline{\mathcal{P}} = \mathcal{P}_{\text{ave}} = \frac{|\underline{\mathbf{E}}|^2}{\eta}\,\hat{\mathbf{r}} = \frac{\eta\beta^2 I^2 l^2}{16\pi^2 r^2}\left(\cos^2\theta\cos^2\phi + \sin^2\phi\right)\hat{\mathbf{r}} \tag{14.51}$$

($I = |\underline{I}|$).

Problems: 14.7–14.18; *Conceptual Questions* (on Companion Website): 14.17–14.22; *MATLAB Exercises* (on Companion Website).

14.4 RADIATED POWER, RADIATION RESISTANCE, ANTENNA LOSSES, AND INPUT IMPEDANCE

To find the time-average power radiated by a Hertzian dipole, in Fig. 14.1, we invoke Poynting's theorem in complex form, Eq. (8.196), and compute the flux of the real part of the complex Poynting vector, $\underline{\mathcal{P}}$, of the dipole, Eq. (14.14), through an arbitrary closed surface, S, placed about the dipole. The real part of $\underline{\mathcal{P}}$, that is, the time-average Poynting vector, $\mathcal{P}_{\text{ave}}$, of the antenna, is given in Eq. (14.15). Since the vector $\mathcal{P}_{\text{ave}}$ has only a radial component, the simplest flux evaluation will be for a spherical surface, of radius r, centered at the coordinate origin [see Eqs. (1.137)]. In addition, as the magnitude of $\mathcal{P}_{\text{ave}}$ depends on the angle θ only, for a fixed r (over the sphere S), we can use for integration an elementary surface dS in the form of a thin ring of radius $r\sin\theta$ and width $r\,d\theta$, as in Fig. 1.16 and Eq. (1.65). Thus, the radiated real power (or time-average power flow) through S is

radiated power, Hertzian dipole

$$P_{\text{rad}} = (P_{\text{flow}})_{\text{ave}} = \oint_S \mathcal{P}_{\text{ave}}\cdot d\mathbf{S} = \frac{\eta\beta^2 I^2 l^2}{16\pi^2 r^2}\int_{\theta=0}^{\pi}\sin^2\theta\,\underbrace{2\pi r\sin\theta\,r\,d\theta}_{dS}$$
$$= \frac{\eta\beta^2 I^2 l^2}{8\pi}\int_0^{\pi}\sin^3\theta\,d\theta = \frac{\eta\beta^2 I^2 l^2}{6\pi}, \tag{14.52}$$

where the integral in θ is solved in Eq. (5.46), and it equals 4/3. We see that the factor $1/r^2$ in the expression for $\mathcal{P}_{\text{ave}}$ cancels the factor r^2 in the expression for dS in Eq. (14.52), which gives rise to a constant radiated power of the dipole [the dependence $\mathcal{P}_{\text{ave}} \propto 1/r^2$ holds true for an arbitrary transmitting antenna, Eq. (14.26)]. That P_{rad} cannot depend on r (for any antenna) is also obvious from the conservation of power principle, in Eq. (8.196). Namely, as there are no impressed sources (generators) past the antenna itself in Fig. 14.1, and the surrounding medium is assumed to be lossless, the real power cannot be accumulated or depleted in the domain between two spherical surfaces with different radii, and outward power flow, P_{rad}, associated with both surfaces must be the same.

The radiated power in Eq. (14.52) is proportional to the dipole current intensity, $I = |\underline{I}|$, squared. For an arbitrary antenna (wire, surface, or volume), where the current distribution is not uniform over the antenna body, P_{rad} is proportional to the magnitude of the feed current at the antenna input terminals, $\underline{I}_0$, shown in Fig. 14.5, squared. The constant of proportionality is a resistance, called the radiation resistance of the antenna and denoted as R_{rad},

definition of radiation resistance (unit: Ω)

$$P_{\text{rad}} = R_{\text{rad}} I_0^2 \quad\longrightarrow\quad R_{\text{rad}} = \frac{P_{\text{rad}}}{I_0^2}, \tag{14.53}$$

where $I_0 = |\underline{I}_0|$. This resistance does not, of course, represent any ohmic losses in the ambient medium, i.e., it does not characterize the transformation of electromagnetic energy to heat. It is the resistance of an equivalent resistor that the transmitting antenna presents to its input terminals, so that the time-average radiated power of the antenna is absorbed in the load. In other words, R_{rad} characterizes the transformation of one form of electromagnetic energy to another, namely, energy of a guided electromagnetic wave (in a transmission line or waveguide feeding the antenna) to energy of a radiated unbounded electromagnetic wave (in free space or other ambient medium), and thus has the nature of a mutual resistance. However, viewed in a wider sense, the radiated power, which leaves the antenna and never returns, is a form of dissipation, but useful and desired dissipation, and we may say that R_{rad} represents antenna power dissipated by radiation. Sometimes, therefore, P_{rad} is referred to as the power of radiation losses. From Eqs. (14.53) and (14.52), the radiation resistance of a Hertzian dipole, with $\underline{I} = \underline{I}_0$, is

$$R_{\text{rad}} = \frac{\eta\,(\beta l)^2}{6\pi} = \frac{2\pi\eta}{3}\left(\frac{l}{\lambda}\right)^2. \tag{14.54}$$

radiation resistance, Hertzian dipole

If the ambient medium is free space, Eq. (9.23) gives

$$\eta = \eta_0 = 120\pi\ \Omega \quad\longrightarrow\quad R_{\text{rad}} = 20\,(\beta l)^2\ \Omega = 790\left(\frac{l}{\lambda}\right)^2\ \Omega. \tag{14.55}$$

Since the dipole is electrically short ($l \ll \lambda$), Eqs. (14.2), its radiation resistance is very small.

In addition to dissipation by radiation, all real antennas exhibit some real dissipation, i.e., some Joule's or ohmic losses in the lossy materials constituting the antenna body, because of the conduction current flow through the materials. These losses are associated with heating in the antenna structure. Although, in general, losses in dielectric parts of an antenna are present as well, losses in antenna conductors are of much greater practical importance. Assuming that the skin effect is pronounced, we use Eq. (10.90) to compute the time-average power of Joule's losses in metallic parts of the antenna, whose surface we mark as S_{metallic}. Moreover, we assume that the losses are small, so that a perturbation method for evaluation of losses in good conductors is applied and the tangential component of the complex rms magnetic field intensity vector near S_{metallic}, $\underline{\mathbf{H}}_{\text{tang}}$, in Eq. (10.90), is found as if the antenna were made of a perfect electric conductor (PEC). Finally, Eq. (10.91) tells us that the magnitude of $\underline{\mathbf{H}}_{\text{tang}}$ equals the magnitude of the complex rms surface current density vector, $\underline{\mathbf{J}}_{\text{s}}$, over S_{metallic},

$$|\underline{\mathbf{H}}_{\text{tang}}| = |\underline{\mathbf{J}}_{\text{s}}|. \tag{14.56}$$

Hence, the ohmic power, P_{ohmic}, in the antenna metallic parts can be expressed as

$$P_{\text{ohmic}} = (P_{\text{J}})_{\text{ave}} = \int_{S_{\text{metallic}}} R_{\text{s}}|\underline{\mathbf{H}}_{\text{tang}}|^2\,\mathrm{d}S = \int_{S_{\text{metallic}}} R_{\text{s}}|\underline{\mathbf{J}}_{\text{s}}|^2\,\mathrm{d}S, \tag{14.57}$$

ohmic power, arbitrary metallic antenna

with R_{s} being the surface resistance of the antenna conductors, Eq. (10.78), and $\underline{\mathbf{J}}_{\text{s}}$ the no-loss current distribution of the antenna.

For a straight wire antenna extending along the z-axis, in Fig. 14.4, the magnetic field close to the antenna cylindrical surface, S_{cylinder}, being quasistatic (near field), equals in form the magnetostatic field around a wire current conductor. Its lines are concentric circles centered on the wire axis, and its magnitude, from the generalized Ampère's law in integral form, Eq. (5.49), is $\underline{I}(z)/2\pi r$, where, again, $\underline{I}(z)$ is the current intensity along the antenna found in the lossless (PEC) case. On the surface, $r = a$, with a denoting the wire radius, and we have

$$\underline{H} = \underline{J}_{\mathrm{s}} = \frac{\underline{I}(z)}{2\pi a}, \tag{14.58}$$

which can as well be obtained through the surface current density, $\underline{J}_{\mathrm{s}}$, of the wire, from the definition of this density in Eq. (3.13), dividing the current intensity of the antenna by the wire circumference ($2\pi a$). With this, and Eq. (14.57), the ohmic power dissipated in the wire is

ohmic power, straight wire antenna

$$P_{\mathrm{ohmic}} = \int_{S_{\mathrm{cylinder}}} R_{\mathrm{s}} \left[\frac{|\underline{I}(z)|}{2\pi a}\right]^2 \underbrace{2\pi a\,\mathrm{d}z}_{\mathrm{d}S} = \int_l \frac{R_{\mathrm{s}}}{2\pi a} |\underline{I}(z)|^2\,\mathrm{d}z = \int_l R'|\underline{I}(z)|^2\,\mathrm{d}z, \tag{14.59}$$

where $\mathrm{d}S$ is the surface area of an elemental ring of width $\mathrm{d}z$ for integration over S_{cylinder}, and R' stands for

$$R' = \frac{R_{\mathrm{s}}}{2\pi a}. \tag{14.60}$$

Note that this expression turns out to be the high-frequency R' of a single conductor of a thin symmetrical two-wire transmission line or of the inner conductor of a coaxial cable, equaling the corresponding portions of the expressions in Eqs. (11.103) and (11.70), respectively.

The current intensity $\underline{I}(z)$ along the wire is proportional to the antenna feed current, $\underline{I}_0$. Therefore, $P_{\mathrm{ohmic}} \propto I_0^2$, and we can write

antenna ohmic resistance (unit: Ω)

$$P_{\mathrm{ohmic}} = R_{\mathrm{ohmic}} I_0^2 \quad \longrightarrow \quad R_{\mathrm{ohmic}} = \frac{P_{\mathrm{ohmic}}}{I_0^2}, \tag{14.61}$$

where R_{ohmic} is the total high-frequency ohmic resistance of the antenna. For a Hertzian dipole, the current is uniform along the wire, Eqs. (14.2), which leads to R_{ohmic} being simply R' times the length of the wire,

ohmic resistance, Hertzian dipole

$$\underline{I}(z) = \underline{I}_0 \quad \longrightarrow \quad P_{\mathrm{ohmic}} = R' I_0^2 \int_l \mathrm{d}z = R' l I_0^2 \quad \longrightarrow \quad R_{\mathrm{ohmic}} = R' l. \tag{14.62}$$

The complex input impedance of an antenna, $\underline{Z}_{\mathrm{A}}$, is, by definition, the complex voltage to current ratio at its input terminals, as illustrated in Fig. 14.10(a). Namely, if we feed the antenna by a voltage generator of complex rms electromotive force $\underline{\mathcal{E}}$ and complex internal (series) impedance $\underline{Z}_{\mathrm{g}}$ (see also Fig. 12.3), which, in general, represents the Thévenin equivalent generator of an arbitrary input network feeding into the antenna, and the complex rms voltage across the generator is $\underline{V}_{\mathrm{g}}$ and current intensity through it $\underline{I}_0$,

antenna input impedance

$$\underline{Z}_{\mathrm{A}} = \underline{Z}_{\mathrm{in}} = \frac{\underline{V}_{\mathrm{g}}}{\underline{I}_0} = R_{\mathrm{A}} + \mathrm{j}X_{\mathrm{A}}. \tag{14.63}$$

Of course, a feed current generator of current intensity $\underline{I}_0$ and internal (shunt) admittance $\underline{Y}_{\mathrm{g}}$ (Norton equivalent generator) is also possible. Analogously to the equivalent circuit in Fig. 12.10 in analysis of transmission lines, the transmitting antenna can thus be replaced, with respect to its input terminals (and to the generator), by a load of complex impedance $\underline{Z}_{\mathrm{A}}$, shown in Fig. 14.10(b).

The time-average (real) input power, $P_{\rm in}$, delivered by the generator to the antenna and input resistance of the antenna, $R_{\rm A}$, i.e., real part of $\underline{Z}_{\rm A}$, are [see Eq. (8.210)]

$$P_{\rm in} = R_{\rm A} I_0^2 = P_{\rm rad} + P_{\rm ohmic} \quad \longrightarrow \quad R_{\rm A} = \frac{P_{\rm in}}{I_0^2} = R_{\rm rad} + R_{\rm ohmic}, \tag{14.64}$$

with $R_{\rm A}$ coming out to be the sum of the antenna radiation and ohmic resistances. Given that, in general, $P_{\rm rad}$ and $P_{\rm ohmic}$ represent, respectively, the desired and undesired parts of $P_{\rm in}$, we can define the antenna power efficiency through the following ratio:

$$\boxed{\eta_{\rm rad} = \frac{P_{\rm rad}}{P_{\rm in}} = \frac{R_{\rm rad}}{R_{\rm rad} + R_{\rm ohmic}} \quad (0 \le \eta_{\rm rad} \le 1).} \tag{14.65}$$

radiation efficiency (dimensionless)

This efficiency is termed the radiation efficiency of the antenna ($\eta_{\rm rad}$ or $e_{\rm rad}$), and is customarily measured in percent. For an ideal (lossless) antenna, $\eta_{\rm rad} = 100\%$. Many antennas have $R_{\rm rad} \gg R_{\rm ohmic}$, and the radiation efficiency nearly 100%. However, electrically small antennas, like a Hertzian dipole, normally exhibit low efficiencies, since their ohmic losses are quite large relative to the low radiated power. As we shall see in examples throughout this chapter, efficient radiators are either of medium or large electrical sizes, that is, their physical dimensions are either comparable to or much larger than the wavelength (λ), Eq. (8.112).

The imaginary part of $\underline{Z}_{\rm A}$, the antenna input reactance, $X_{\rm A}$, represents reactive power stored in the near field, $Q_{\rm near\ field}$. Of course, this power equals the input reactive power at the antenna terminals, $Q_{\rm in}$, and we can write [see also Eqs. (8.211)]

$$Q_{\rm in} = Q_{\rm near\ field} = X_{\rm A} I_0^2 \quad \longrightarrow \quad X_{\rm A} = \frac{Q_{\rm near\ field}}{I_0^2}. \tag{14.66}$$

As will be illustrated in an example, electrically small antennas have a disproportionally large input reactance, along with a small input resistance, i.e., $|X_{\rm A}| \gg R_{\rm A}$.

In practice, we usually feed antennas by a two-conductor transmission line, and to maximize the transfer of power from the line to the antenna it is necessary to, as much as possible, match the antenna input impedance, Eq. (14.63), to the characteristic impedance of the line, Z_0. This latter impedance is most frequently purely real, Eq. (12.18), or nearly so, Eq. (12.23), and ensuring that the impedance match condition, $\underline{Z}_{\rm A} = Z_0$, is met (to a desired extent) is, in general, an extremely difficult task, especially over a broad band of frequencies. In addition to designing antennas with desirable $\underline{Z}_{\rm A}$, we also use impedance-matching networks to enhance the power transfer in antenna systems.

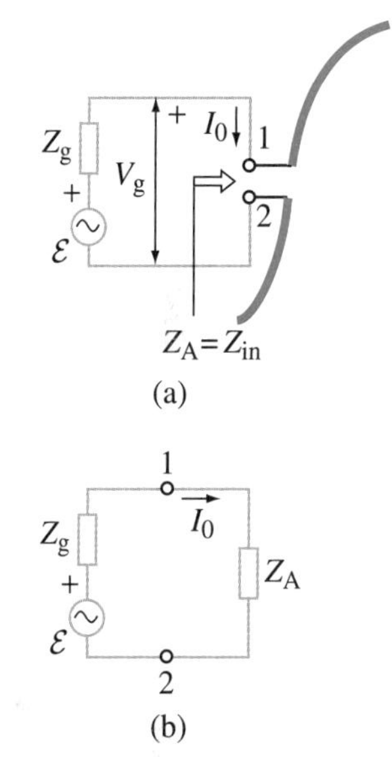

Figure 14.10 Illustration of the definition of the complex input impedance of a transmitting antenna, Eq. (14.63), as the impedance seen by a generator at the antenna input terminals (a), so that the antenna can be replaced by an equivalent load having the same impedance (b).

Example 14.5 Radiation and Ohmic Resistances of a Short Wire Dipole

Find (a) the radiation resistance and (b) the high-frequency ohmic resistance of the nonloaded short wire dipole antenna from Example 14.1, if the skin-effect surface resistance of the wire is $R_{\rm s}$ and wire radius is a.

Solution

(a) As concluded in Example 14.1, the Poynting vector of the nonloaded dipole in Fig. 14.6(a) equals a quarter of that due to the Hertzian dipole in Fig. 14.1. Hence, having in mind Eqs. (14.52) and (14.53), the same relationship holds for the radiation resistances of these two antennas, and using the expression for the resistance in Eq. (14.55), we have

$$(R_{\rm rad})_{\rm nonloaded\ dipole} = \frac{1}{4}(R_{\rm rad})_{\rm Hertzian\ dipole} = 5\,(\beta l)^2\ \Omega. \tag{14.67}$$

Of course, this same result can be obtained from the far-field equivalency of the two antennas in Fig. 14.6(a) and Fig. 14.6(b), so as R_{rad} of a Hertzian dipole with current $\underline{I}_0$ and length $l/2$, which substituted in Eq. (14.55) gives

$$R_{rad} = 20\left(\beta\frac{l}{2}\right)^2 \Omega. \tag{14.68}$$

(b) Combining Eqs. (14.61), (14.59), and (14.35), the ohmic resistance of the antenna in Fig. 14.6(a) amounts to

$$R_{ohmic} = \frac{P_{ohmic}}{I_0^2} = \frac{1}{I_0^2}\int_l R'|\underline{I}(z)|^2\,dz = R'\,2\int_{z=0}^{h}\left(1-\frac{z}{h}\right)^2 dz$$
$$= 2R'\left(z - \frac{z^2}{h} + \frac{z^3}{3h^2}\right)\Bigg|_0^h = \frac{R'l}{3}, \tag{14.69}$$

where $h = l/2$ and R' is computed from Eq. (14.60). This, compared with the expression for R_{ohmic} in Eq. (14.62), for the Hertzian dipole with the same R' and l, tells us that

$$(R_{ohmic})_{nonloaded\ dipole} = \frac{1}{3}(R_{ohmic})_{Hertzian\ dipole}. \tag{14.70}$$

Example 14.6 Efficiency Comparison of Nonloaded and Loaded Dipoles

Show that the short dipole in Fig. 14.6(a) is always less efficient than the Hertzian dipole in Fig. 14.1, assuming the same length, wire radius, operating frequency, and material properties of the two antennas.

Solution Denoting the nonloaded dipole as antenna 1 and Hertzian dipole as antenna 2, Eqs. (14.65), (14.67), and (14.70) result in

$$\eta_{rad1} = \frac{R_{rad1}}{R_{rad1} + R_{ohmic1}} = \frac{\frac{1}{4}R_{rad2}}{\frac{1}{4}R_{rad2} + \frac{1}{3}R_{ohmic2}} < \frac{\frac{1}{4}R_{rad2}}{\frac{1}{4}R_{rad2} + \frac{1}{4}R_{ohmic2}}$$
$$= \frac{R_{rad2}}{R_{rad2} + R_{ohmic2}} = \eta_{rad2}. \tag{14.71}$$

So, indeed, the radiation efficiency (η_{rad}) is always lower for the antenna in Fig. 14.6(a) than it is for the antenna in Fig. 14.1, for any set of antenna parameters and any frequency, provided that they are the same for the two antennas, as well as that the antennas are electrically short.

Example 14.7 Radiation Efficiencies of Steel and Copper AM Radio Antennas

An AM radio wire dipole antenna, operating at a frequency of $f = 1$ MHz, is $l = 1.5$ m in length and $a = 1.5$ mm in radius. Determine the radiation efficiency of the antenna, if it is made out either of steel, whose conductivity amounts to $\sigma = 2$ MS/m and relative permeability to $\mu_r = 2000$, or copper.

Solution Since the free-space wavelength, Eq. (9.67), at the operating frequency of the antenna is $\lambda_0 = c_0/f = 300$ m, c_0 being the wave velocity (speed of light) in free space, Eq. (9.19), we have that

$$\frac{l}{\lambda_0} = 0.005 \ll 1, \tag{14.72}$$

indicating an electrically short dipole, Fig. 14.6(a). This, in turn, means that the radiation resistance of the antenna is given by Eq. (14.67), from which

$$R_{rad} = 197.5\left(\frac{l}{\lambda_0}\right)^2 \Omega = 5\ \text{m}\Omega. \tag{14.73}$$

Using Eq. (10.78), the skin-effect surface resistances of steel and copper, respectively, at the operating frequency, are

$$(R_s)_{steel} = \sqrt{\frac{\pi \mu_r \mu_0 f}{\sigma}} = 63\ \text{m}\Omega/\text{square}, \quad (R_s)_{Cu} = 0.26\ \text{m}\Omega/\text{square}, \tag{14.74}$$

where the latter result can also be found directly from the specialized formula for copper in Eq. (10.80). Eqs. (14.69) and (14.60) then yield the following high-frequency ohmic resistances of the dipole:

$$(R_{ohmic})_{steel} = \frac{(R_s)_{steel}\, l}{6\pi a} = 3.33\ \Omega, \quad (R_{ohmic})_{Cu} = \frac{(R_s)_{Cu}\, l}{6\pi a} = 13.8\ \text{m}\Omega. \tag{14.75}$$

Finally, the corresponding radiation efficiencies of the antenna, Eq. (14.65), for the two materials come out to be

$$(\eta_{rad})_{steel} = \frac{R_{rad}}{R_{rad} + (R_{ohmic})_{steel}} = 0.15\%, \quad (\eta_{rad})_{Cu} = \frac{R_{rad}}{R_{rad} + (R_{ohmic})_{Cu}} = 26\%. \tag{14.76}$$

Obviously, these are quite low efficiencies, and especially if steel is used. Note that this particular antenna (for the given dimensions), in the steel version, is the dipole equivalent of a typical fender-mount AM car radio antenna (steel is used for mechanical sturdiness of the antenna) – with the upper half (called monopole) of the dipole in Fig. 14.6(a) attached to the car body (monopole antennas will be studied in a later section). Such an inefficiency of a receiving antenna in broadcast (e.g., AM radio) applications is usually compensated by the use of high-power transmitters.[6]

Example 14.8 Input Reactance of a Nonloaded Short Dipole

The input reactance of a nonloaded short wire dipole antenna, of length l, in free space can be approximated by

$$X_A = -\frac{120}{\pi l/\lambda_0}\left(\ln\frac{l}{2a} - 1\right)\ \Omega, \tag{14.77}$$

with λ_0 standing for the free-space wavelength at the operating frequency of the antenna. Using this expression, find the input impedance of the two antennas from the previous example.

Solution The dipole electrical length is computed in Eq. (14.72), and the length to diameter ratio is $l/(2a) = 500$, with which Eq. (14.77) gives $X_A = -39.837\ \text{k}\Omega$. From Eqs. (14.63), (14.64), (14.73), and (14.75), the complex input impedances of the steel and copper versions of the antenna come out to be $(\underline{Z}_A)_{steel} = (3.335 - \text{j}39{,}837)\ \Omega$ and $(\underline{Z}_A)_{Cu} = (0.0188 - \text{j}39{,}837)\ \Omega$, respectively. Evidently, the dipoles present to their terminals an extremely small resistance and extremely large capacitive (negative) reactance ($|X_A| \gg R_A$). Note that such a disproportionality between the imaginary and real parts of the antenna impedance, as well as between the reactive power stored in the near field around the antenna, Eq. (14.66), and the active power, i.e., the time-average radiated and loss powers combined, of the antenna, Eq. (14.64), are typical for electrically small antennas in general.

Problems: 14.19–14.21; *Conceptual Questions* (on Companion Website): 14.23 and 14.24; *MATLAB Exercises* (on Companion Website).

[6]Note that this is a quite common general approach to achieving overall efficiency and accessibility in broadcasting systems – concentrate (and invest) into complexity, efficiency, and power capacity of a few transmitting antenna stations in the system, allowing, on the other (consumers') side, for the simple (and low-cost) receiving antennas.

14.5 ANTENNA CHARACTERISTIC RADIATION FUNCTION AND RADIATION PATTERNS

The far electric field intensity vector, $\underline{\mathbf{E}}$, of an arbitrary antenna, Fig. 14.5, is proportional to the feed current at the antenna input terminals, $\underline{I}_0$. Having also in mind the spherical-wave dependence on r in Eq. (14.33), it is customary to write $\underline{\mathbf{E}}$, given by Eqs. (14.21), in the following form:

definition of the characteristic radiation function, $\underline{\mathbf{F}}$ (dimensionless)

$$\underline{\mathbf{E}}(r, \theta, \phi) = \underline{C}\,\underline{I}_0 \frac{\mathrm{e}^{-\mathrm{j}\beta r}}{r} \underline{\mathbf{F}}(\theta, \phi), \quad \text{where} \quad \underline{C} = \frac{\mathrm{j}\eta}{2\pi} \quad \left(\underline{C}_{\text{free space}} = \mathrm{j}60\ \Omega\right), \tag{14.78}$$

where $\underline{\mathbf{F}}(\theta, \phi)$ is termed the characteristic radiation function of the antenna. It is a simple matter to verify, based on a dimensional analysis of Eq. (14.78), that, since the constant $\underline{C}$ has the dimension of impedance (its unit is Ω), $\underline{\mathbf{F}}$ is a dimensionless quantity. The particular choice for the numerical value of $\underline{C}$ (e.g., j60 Ω for free space as the ambient medium) will become obvious in a later section. Most importantly, the characteristic radiation function represents the part of the field expression in Eq. (14.78) that is characteristic for individual antennas, i.e., that differs from antenna to antenna, while the remaining terms in the expression are the same for all antennas. Independent of r, and thus only a function of the direction of antenna radiation, defined by angles θ and ϕ, or, equivalently, by the radial unit vector $\hat{\mathbf{r}}$, in Fig. 14.5, $\underline{\mathbf{F}}(\theta, \phi)$ determines the directional properties of the antenna. In specific, being a complex vector function, it provides a complete picture of the dependence of the antenna radiated field, namely, the magnitude, phase, and polarization (see Section 9.14) of $\underline{\mathbf{E}}$, on the radiation direction. As an example, comparing Eqs. (14.17) and (14.78) we realize that $\underline{\mathbf{F}}$ of a Hertzian dipole (Fig. 14.1) is

radiation function of a Hertzian dipole

$$\underline{\mathbf{F}}(\theta) = \frac{\beta l}{2} \sin\theta\, \hat{\boldsymbol{\theta}}. \tag{14.79}$$

Of course, the radiation of the dipole is azimuthally symmetrical (does not depend on ϕ).

Different aspects of the characteristic radiation function, in Eq. (14.78), presented graphically, give different radiation patterns of the antenna under consideration. Namely, by definition, antenna radiation patterns are graphical representations of the angular (θ, ϕ) variation of radiation (far-field) properties around the antenna, when it is transmitting, over a far-zone sphere of radius r, Eq. (14.16), centered at the global coordinate origin, on or close to the antenna (Figs. 14.4 and 14.5). Although phase and polarization radiation patterns, representing the phase and polarization, respectively, of $\underline{\mathbf{F}}(\theta, \phi)$, are used as well, most frequently we plot only the magnitude of $\underline{\mathbf{F}}(\theta, \phi)$. Therefore, except when we explicitly specify otherwise, a radiation pattern (or, simply, pattern) means a field magnitude (for a fixed r) pattern of the antenna. Moreover, it is convenient and customary to normalize the function magnitude, $|\underline{\mathbf{F}}(\theta, \phi)|$, to its maximum value, $|\underline{\mathbf{F}}(\theta, \phi)|_{\max}$. We thus obtain the so-called normalized field pattern of the antenna, $f(\theta, \phi)$, whose maximum value is unity,

normalized field pattern (dimensionless)

$$f(\theta, \phi) = \frac{|\underline{\mathbf{F}}(\theta, \phi)|}{|\underline{\mathbf{F}}(\theta, \phi)|_{\max}} \quad \{[f(\theta, \phi)]_{\max} = 1\}. \tag{14.80}$$

From Eq. (14.79), f of a Hertzian dipole is

f – Hertzian dipole

$$f(\theta) = \sin\theta, \tag{14.81}$$

where the maximum radiation occurs in the equatorial plane (xy-plane) in Fig. 14.2,

$$f = f_{\text{max}} = 1 \quad \longrightarrow \quad \theta = \theta_{\text{max}} = 90°. \tag{14.82}$$

The normalized field pattern of an antenna can also be expressed in decibels, as in Eq. (9.88),

$$\boxed{f_{\text{dB}}(\theta, \phi) = 20 \log f(\theta, \phi),} \tag{14.83}$$

field pattern in dB

with the maximum normalized pattern level now being 0 dB.

Alternatively, we sometimes plot the normalized power pattern of an antenna, $p(\theta, \phi)$, defined as the normalized magnitude of the associated far-zone Poynting vector for $r = r_0 = \text{const}$, which, combining Eqs. (14.26), (14.78), and (14.80), comes out to be the square of $f(\theta, \phi)$,

$$\boxed{p(\theta, \phi) = \frac{|\underline{\mathcal{P}}(r_0, \theta, \phi)|}{|\underline{\mathcal{P}}(r_0, \theta, \phi)|_{\text{max}}} = \frac{|\underline{\mathbf{E}}(r_0, \theta, \phi)|^2}{|\underline{\mathbf{E}}(r_0, \theta, \phi)|^2_{\text{max}}} = \frac{|\underline{\mathbf{F}}(\theta, \phi)|^2}{|\underline{\mathbf{F}}(\theta, \phi)|^2_{\text{max}}} = f^2(\theta, \phi).} \tag{14.84}$$

normalized power pattern

Of course, the normalized power and field patterns are the same in decibels [see also Eq. (9.98)],

$$p_{\text{dB}}(\theta, \phi) = 10 \log p(\theta, \phi) = 10 \log f^2(\theta, \phi) = 20 \log f(\theta, \phi) = f_{\text{dB}}(\theta, \phi). \tag{14.85}$$

Fig. 14.11(a) shows, in an isometric view, a typical normalized field pattern, Eq. (14.80), of a directional antenna, as a three-dimensional (3-D) plot, in linear (absolute) units (rather than decibels). Such plots are referred to as solid polar patterns, since the pattern is a solid object, whose surface is defined by a function, f, of angles θ and ϕ, with the distance from the coordinate origin (O) to a point on the pattern surface along the direction (θ, ϕ) representing the value $f(\theta, \phi)$. We observe several radiation lobes in the pattern, bounded by pattern nulls, which are either zeros (no radiation) or deep minima of the function f, between adjacent lobes. The lobe containing the direction of maximum radiation (positive z direction or $\theta = 0$ in our case) is called the main (or major) lobe (or beam). Other lobes are side (or minor) lobes. Note that, due to symmetry of an antenna and its current distribution, the pattern may have two or more identical main lobes. In practice, 3-D radiation patterns are measured, computed, and used as a series of 2-D patterns, representing characteristic cuts (containing the coordinate origin) through the 3-D diagram. For virtually all practical applications, a few 2-D polar plots as functions of θ for some fixed values of ϕ, and/or vice versa, suffice. One such plot is shown in Fig. 14.11(b).

As a concrete simple example, let us plot, using Eq. (14.81), the f patterns of a Hertzian dipole, shown in Fig. 14.12(a) along with the polarization of its radiated electric and magnetic fields in two characteristic planes. Fig. 14.12(b) shows the polar pattern in a plane $\phi = \text{const}$, as a function of θ. Since such planes contain the vector $\mathbf{E}$ of the dipole [Fig. 14.12(a)], the pattern is called an E-plane pattern. The two curves in the plot, on the two sides of the z-axis, sharing the coordinate origin, are circles, and this will be proved analytically in an example. Most of the energy of the dipole is radiated in and near the equatorial plane ($\theta = 90°$), Eq. (14.82), whereas the radiation nulls are in directions $\theta = 0$ and $180°$, along the z-axis. Obviously, the pattern does not have any side lobes. The radiation pattern in the equatorial plane, Fig. 14.12(c), is uniform, all the way around the antenna, as there is no ϕ variation of the dipole fields, and $f = 1$ for $0 \leq \phi < 360°$ and $\theta = 90°$. Such patterns, the same in all directions in a given polar plot (plane), are said to be

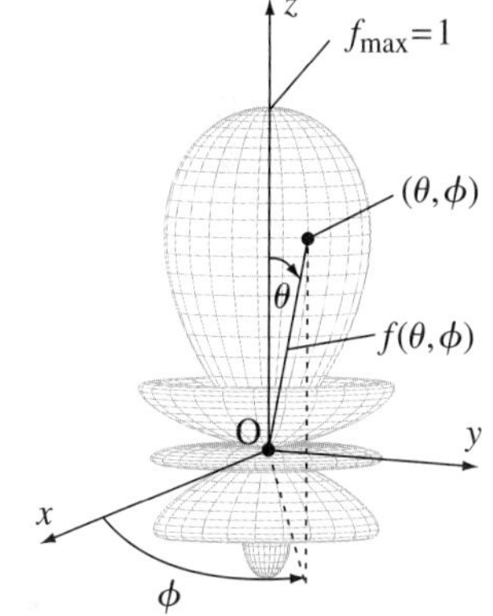

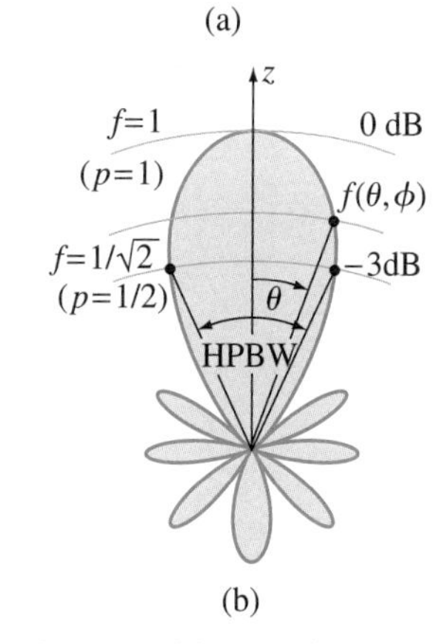

Figure 14.11 Typical normalized field pattern, Eq. (14.80), of a directional antenna: (a) three-dimensional isometric plot and (b) two-dimensional cut in a plane $\phi = \text{const}$.

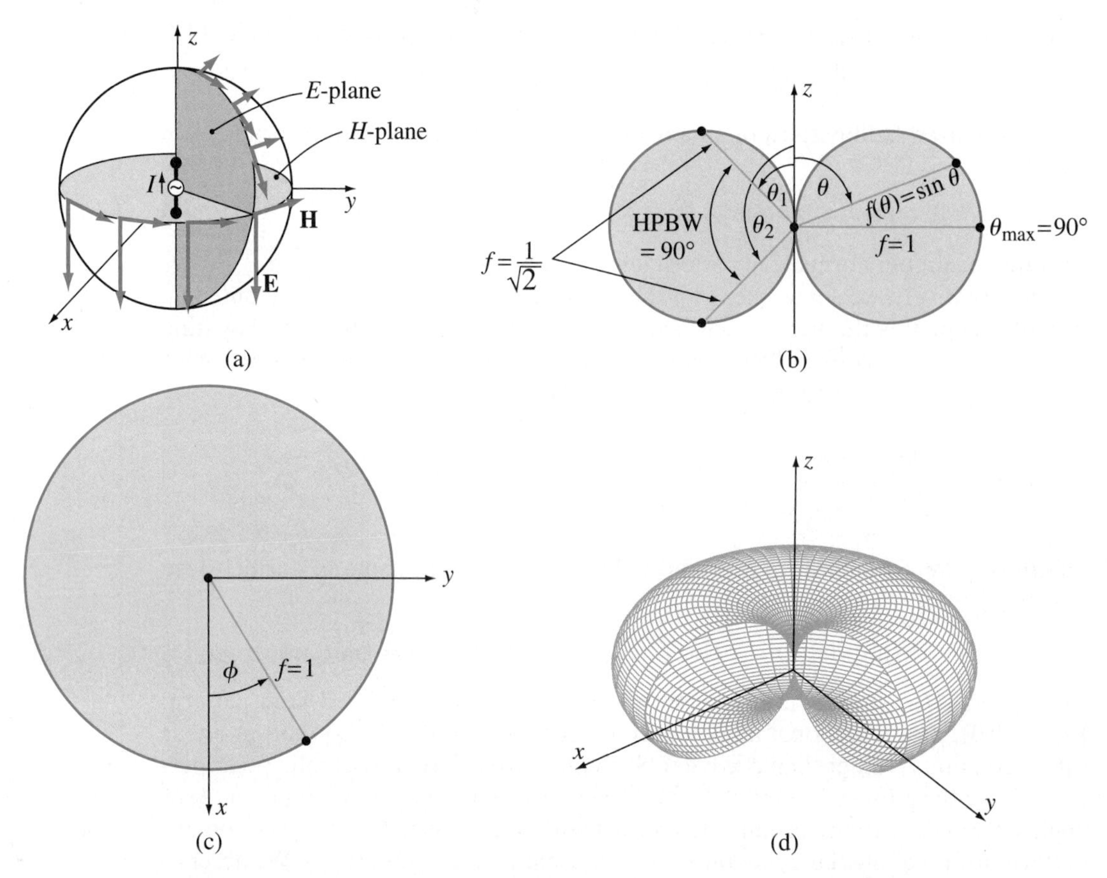

Figure 14.12 Radiation properties of a Hertzian dipole (Fig. 14.1): polarization of radiated electric and magnetic fields (a), and normalized field polar radiation patterns, using Eq. (14.81), in an E-plane (ϕ = const) (b), H-plane ($\theta = 90°$) (c), and in three dimensions (d).

omnidirectional.[7] On the other side, this is an H-plane pattern, because the vector **H** of the dipole lies in the xy-plane in Fig. 14.12(a). Finally, Fig. 14.12(d) depicts the 3-D polar radiation pattern of the dipole. Note that this characteristic solid pattern, having the shape of a "doughnut" with no hole, can be obtained by rotating, for $0 < \phi < 360°$, the 2-D plot in Fig. 14.12(b) about the z-axis, and this applies to all azimuthally symmetrical radiation patterns (the pattern represents a so-called body of revolution with respect to the z-axis).

Highly directional antennas have very narrow main lobes. This is quantified by the antenna half-power (or −3-dB) beamwidth (HPBW), which is the angular width in a 2-D radiation pattern of the central part of the main lobe that is at or above the half-power level. In other words, the HPBW is determined by the points on the main lobe curve where the antenna normalized power, Eq. (14.84), drops to a half of its maximum value, $p = 1/2$, which corresponds to a change of −3 dB [Eq. (14.85)]. On

[7] In general, antennas with omnidirectional radiation in the horizontal plane are extensively used in numerous ground-based radio and wireless applications. One of the examples is their use in broadcasting systems, as transmitting antennas, since they provide equally good coverage in all directions around them. Another example are receiving antennas in mobile telephony, where the incoming wave direction from the cellular tower to the device is generally unknown or randomly changing, and omnidirectional antennas ensure equally good reception from all directions.

the normalized field pattern, the half-power bounds are given by $f = 1/\sqrt{2} = 0.707$ [see Eq. (14.84)], as illustrated in Fig. 14.11(b). Of course, the HPBW can be computed and used also for antennas that do not exhibit highly directional properties, although it is of less practical relevance in such cases. For example, from the pattern f of a Hertzian dipole, Eq. (14.81), we have

$$\boxed{\begin{aligned} f = \sin\theta = \frac{1}{\sqrt{2}} \quad &\longrightarrow \quad \theta_1 = 45^\circ \quad \text{and} \quad \theta_2 = 135^\circ \\ &\longrightarrow \quad \text{HPBW} = \theta_2 - \theta_1 = 90^\circ, \end{aligned}} \tag{14.86}$$

half-power beamwidth (degrees)

and this is shown in Fig. 14.12(b).

Example 14.9 Radiation Functions of Three Different Antennas

Find the characteristic radiation functions of all antennas studied in Examples 14.1–14.4.

Solution With the use of the far-field equivalency of the two antennas in Fig. 14.6(a) and Fig. 14.6(b), the characteristic radiation function of a nonloaded short dipole of length l (Examples 14.1 and 14.2) is obtained by simply substituting l by $l/2$ in the corresponding radiation function for a Hertzian dipole, Eq. (14.79), which yields

$$\underline{\mathbf{F}}_1(\theta) = \frac{\beta l}{4} \sin\theta\,\hat{\boldsymbol{\theta}}. \tag{14.87}$$

Next, Eqs. (14.43), (14.42), and (14.78) tell us that the characteristic radiation function of a traveling-wave wire antenna (Example 14.3) is given by

$$\underline{\mathbf{F}}_2(\theta) = \frac{\sin\theta}{1-\cos\theta} \sin\left[\frac{\beta l}{2}(1-\cos\theta)\right] e^{-j\beta(1-\cos\theta)l/2}\,\hat{\boldsymbol{\theta}}. \tag{14.88}$$

Finally, for an x-directed Hertzian dipole (Example 14.4), whose far electric field vector is that in Eq. (14.48), we have

$$\underline{\mathbf{F}}_3(\theta,\phi) = \frac{\beta l}{2}(-\cos\theta\cos\phi\,\hat{\boldsymbol{\theta}} + \sin\phi\,\hat{\boldsymbol{\phi}}), \tag{14.89}$$

where the vector expression in terms of angles θ and ϕ in parentheses replaces $\sin\theta\,\hat{\boldsymbol{\theta}}$ in the radiation function of a z-directed dipole, in Eq. (14.79).

Example 14.10 Proof that *E*-plane Hertzian Dipole Pattern Cuts Are Circles

Prove that the two closed curves constituting an E-plane normalized field polar radiation pattern of a Hertzian dipole, in Fig. 14.12(b), are circles.

Solution With reference to Fig. 14.13, showing one of the two curves of the plot in Fig. 14.12(b), an application of the cosine rule to the triangle $\triangle$OCM gives

$$\xi^2 = p^2 + f^2 - 2pf\cos\gamma. \tag{14.90}$$

Now, taking into account that $f = \sin\theta$, Eq. (14.81), $p = f_{\max}/2 = 1/2$ (for $\theta = 90^\circ$), Eq. (14.82), and $\cos\gamma = \sin\theta$ ($\gamma = 90^\circ - \theta$), we have

$$\xi^2(\theta) = \left(\frac{1}{2}\right)^2 + \sin^2\theta - 2\times\frac{1}{2}\sin\theta\sin\theta = \left(\frac{1}{2}\right)^2 \quad \longrightarrow \quad \xi = \frac{1}{2} = \text{const} \quad \text{(circle)}. \tag{14.91}$$

So, indeed, as the distance of an arbitrary point M on the radiation pattern curve (defined by an arbitrary polar angle θ) from the point C in Fig. 14.13 turns out to be independent of θ, the curve is a circle, 1/2 in radius and centered at C.

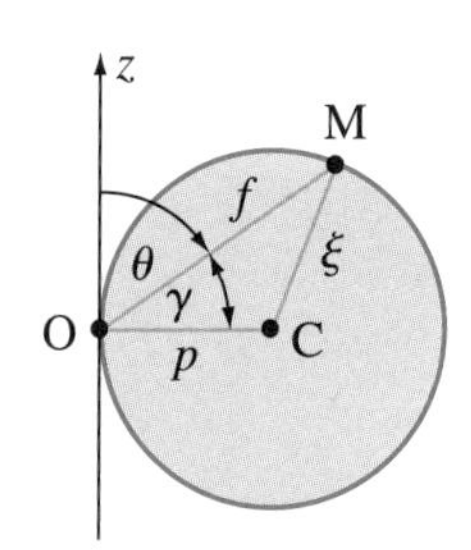

Figure 14.13 With the proof that the curves in an E-plane field pattern plot of a Hertzian dipole, Fig. 14.12(b), are circles (ξ = const); for Example 14.10.

Problems: 14.22; *Conceptual Questions* (on Companion Website): 14.25–14.27; *MATLAB Exercises* (on Companion Website).

14.6 ANTENNA DIRECTIVITY AND GAIN

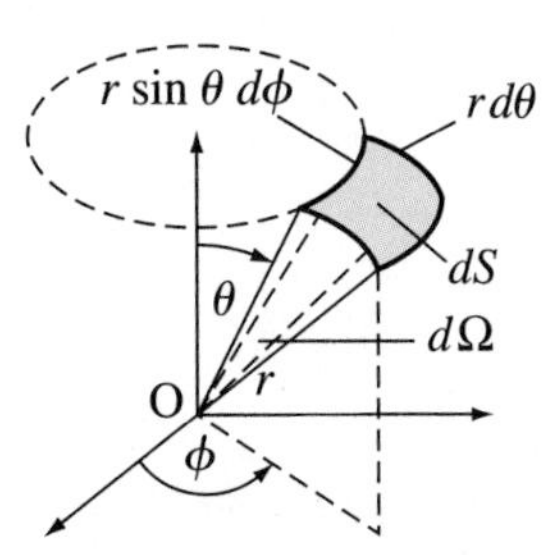

Figure 14.14 Element $\mathrm{d}S$, in a direction (θ, ϕ), of a far-field spherical surface of radius r, Eq. (14.92), around an arbitrary transmitting antenna, and the associated elementary solid angle $\mathrm{d}\Omega$, Eq. (14.93).

This section continues discussions of directional properties of transmitting antennas from the previous one, and defines several new concepts and quantities to describe and quantify these properties. To outline the geometrical relationships needed for the definitions, let us first consider an element $\mathrm{d}S$, in a direction (θ, ϕ), of a far-field spherical surface, S, of radius r [Eq. (14.16)] around an arbitrary transmitting antenna [as in Fig. 14.12(a)]. The patch $\mathrm{d}S$ is conveniently adopted in the form of a differentially small curvilinear quadrilateral whose sides are elementary arcs of lengths $r\,\mathrm{d}\theta$ in the θ direction (latitude) and $r \sin\theta\,\mathrm{d}\phi$ in the ϕ direction (longitude),

$$\mathrm{d}S = (r\,\mathrm{d}\theta)\,(r \sin\theta\,\mathrm{d}\phi), \tag{14.92}$$

as shown in Fig. 14.14 [see also Fig. 1.10 and Eq. (1.35)]. The elementary solid angle, $\mathrm{d}\Omega$, subtended by this surface is [see Eq. (1.124) and Fig. 1.30]

$$\mathrm{d}\Omega = \frac{\mathrm{d}S}{r^2} = \sin\theta\,\mathrm{d}\theta\,\mathrm{d}\phi. \tag{14.93}$$

We recall that the unit for a solid angle, Ω, is steradian or square radian (sr). The full solid angle, subtended by the whole sphere, is computed in Eq. (1.127) or (1.36), and it amounts to

$$\Omega_{\text{full}} = \oint_S \mathrm{d}\Omega = 4\pi \tag{14.94}$$

(as with radians, we usually just assume it and do not write "sr" next to the numerical value of a solid angle, so 4π here means 4π sr).

The far-zone complex Poynting vector of an arbitrary antenna, Eq. (14.26), is purely real, and thus the same as the corresponding time-average Poynting vector [Eq. (8.195)], and radial, $\underline{\mathcal{P}} = \mathcal{P}_{\text{ave}} = \mathcal{P}_{\text{ave}}\,\hat{\mathbf{r}}$ (note that $\mathcal{P}_{\text{ave}} \geq 0$, since the antenna radiates the power outward, in the positive radial direction). Having in mind Eqs. (14.52) and (14.93), the portion $\mathrm{d}P_{\text{rad}}$ of the total time-average radiated power of the antenna flowing through the surface $\mathrm{d}S$ in Fig. 14.14 is

$$\mathrm{d}P_{\text{rad}} = \mathcal{P}_{\text{ave}}\,\mathrm{d}S = \mathcal{P}_{\text{ave}} r^2\,\mathrm{d}\Omega. \tag{14.95}$$

The power $\mathrm{d}P_{\text{rad}}$ is independent of r, as it represents the power propagating within the solid angle $\mathrm{d}\Omega$ (in the far field), like through an imaginary waveguide defined by the bounds of $\mathrm{d}\Omega$, and is the same in every cross section $\mathrm{d}S$ of the "waveguide." This is also evident from the dependence $\mathcal{P}_{\text{ave}} \propto 1/r^2$ [see Eqs. (14.26) and (14.78)], because of which we find it convenient to multiply $\mathcal{P}_{\text{ave}}$ by r^2, to make it independent of r, and so form a new power density, denoted as $U(\theta, \phi)$, that is only a function of the direction of antenna radiation,

antenna radiation intensity (unit: W/sr)

$$\boxed{U(\theta, \phi) = r^2 \mathcal{P}_{\text{ave}}(r, \theta, \phi) = \frac{\mathrm{d}P_{\text{rad}}(\theta, \phi)}{\mathrm{d}\Omega}.} \tag{14.96}$$

The new quantity, which has the dimension of the power per unit solid angle, and is thus measured in W/sr, is termed the radiation intensity of the antenna, in a given direction. The total radiated power is now written as

$$P_{\text{rad}} = \oint_S \mathrm{d}P_{\text{rad}} = \int_{\theta=0}^{\pi} \int_{\phi=0}^{2\pi} U(\theta, \phi)\,\mathrm{d}\Omega. \tag{14.97}$$

Note that for an isotropic radiator, which is an important theoretical concept of a hypothetical antenna with uniform radiation in all directions, the radiation intensity, U_{iso}, as a constant, can be brought outside the integral sign in Eq. (14.97).

With the help of Eq. (14.94),

$$P_{\text{rad}} = U_{\text{iso}} \int_{\theta=0}^{\pi} \int_{\phi=0}^{2\pi} \text{d}\Omega = U_{\text{iso}} \Omega_{\text{full}} = U_{\text{iso}} 4\pi \qquad (U_{\text{iso}} = \text{const}), \tag{14.98}$$

and hence

$$\boxed{U_{\text{iso}} = \frac{P_{\text{rad}}}{4\pi}.} \tag{14.99}$$

isotropic radiator

From Eqs. (14.96), (14.26), and (14.78), the radiation intensity can be expressed in terms of the magnitude squared of the feed current, $\underline{I}_0$, and that of the characteristic radiation function, $\underline{\mathbf{F}}$, of the antenna,

$$U(\theta, \phi) = r^2 \frac{|\underline{\mathbf{E}}(r, \theta, \phi)|^2}{\eta} = \frac{\eta}{4\pi^2} I_0^2 \left|\underline{\mathbf{F}}(\theta, \phi)\right|^2 \tag{14.100}$$

($I_0 = |\underline{I}_0|$). This provides a very useful way to find U, as $|\underline{\mathbf{F}}|$ and I_0 are often at hand for a given antenna and its excitation. Combined with Eqs. (14.97) and (14.93), it leads to the corresponding integral expression for P_{rad},

$$\boxed{P_{\text{rad}} = \frac{\eta I_0^2}{4\pi^2} \int_{\theta=0}^{\pi} \int_{\phi=0}^{2\pi} \left|\underline{\mathbf{F}}(\theta, \phi)\right|^2 \sin\theta \, \text{d}\theta \, \text{d}\phi.} \tag{14.101}$$

total radiated power

In the case of a Hertzian dipole, for example, Eq. (14.79) can be used to verify that this integral gives the same result for the radiated power of the dipole as in Eq. (14.52).

The directivity of an antenna in a given direction, $D(\theta, \phi)$, is defined as the ratio of the antenna radiation intensity in that direction to the radiation intensity of an isotropic radiator, Eq. (14.99), for the same radiated power,

$$\boxed{D(\theta, \phi) = \frac{U(\theta, \phi)}{U_{\text{iso}}} = \frac{4\pi\, U(\theta, \phi)}{P_{\text{rad}}}.} \tag{14.102}$$

antenna directivity (dimensionless)

In this context, U_{iso} is equal to the average radiation intensity of the antenna under consideration – the total power P_{rad}, which is an integral of $U(\theta, \phi)$ in Eq. (14.97), divided by the full solid angle (the domain of integration), 4π. The directivity of an isotropic radiator is unity in all directions. Using Eqs. (14.100) and (14.53), the directivity in Eq. (14.102) can also be written as

$$\boxed{D(\theta, \phi) = \frac{\eta \left|\underline{\mathbf{F}}(\theta, \phi)\right|^2}{\pi R_{\text{rad}}} = \frac{120\ \Omega}{R_{\text{rad}}} \left|\underline{\mathbf{F}}(\theta, \phi)\right|^2,} \tag{14.103}$$

D in terms of $\underline{\mathbf{F}}$ and R_{rad}

where R_{rad} is the radiation resistance of the antenna, and the second expression holds true for antennas in free space [Eq. (9.23)]. Combining Eqs. (14.103), (14.53), and (14.101), $D(\theta, \phi)$ is then expressed solely in terms of the antenna characteristic radiation function,

$$D(\theta, \phi) = \frac{\left|\underline{\mathbf{F}}(\theta, \phi)\right|^2}{\frac{1}{4\pi} \int_{\theta=0}^{\pi} \int_{\phi=0}^{2\pi} \left|\underline{\mathbf{F}}(\theta, \phi)\right|^2 \sin\theta \, \text{d}\theta \, \text{d}\phi}, \tag{14.104}$$

with the denominator being equal to the average of $\left|\underline{\mathbf{F}}(\theta, \phi)\right|^2$ over all directions. For a Hertzian dipole, substituting the expressions for $\underline{\mathbf{F}}$ and R_{rad} from Eqs. (14.79) and (14.55) into Eq. (14.103) yields

$$\boxed{D(\theta, \phi) = \frac{120\ \Omega}{20\,(\beta l)^2\ \Omega} \frac{(\beta l)^2}{4} \sin^2\theta = 1.5 \sin^2\theta.} \tag{14.105}$$

D – Hertzian dipole

The maximum directivity, that is, the directivity in the direction of maximum radiation (Fig. 14.11), of an antenna is certainly of the most practical importance. It can be obtained either from the directivity function, as $D_{max} = [D(\theta, \phi)]_{max}$, or using in Eqs. (14.102) and (14.103) the corresponding maximum values of functions U and $|\underline{\mathbf{F}}|$,

$$D_{max} = \frac{U_{max}}{U_{iso}} = \frac{4\pi U_{max}}{P_{rad}} = \frac{\eta\,|\underline{\mathbf{F}}|^2_{max}}{\pi R_{rad}}. \tag{14.106}$$

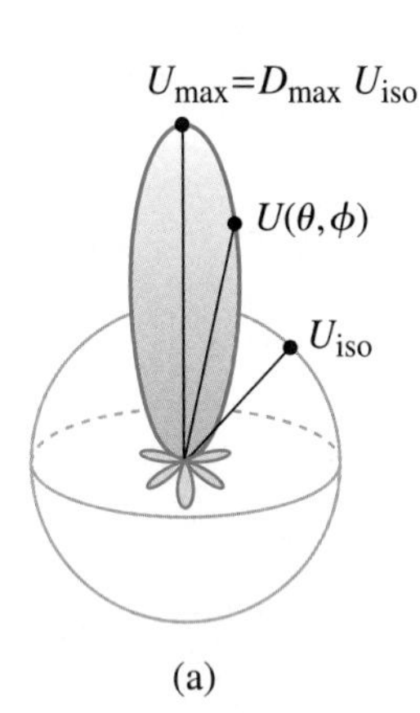

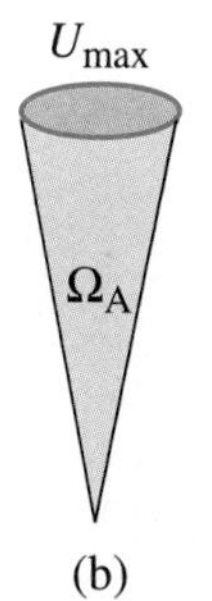

Figure 14.15 Illustrations of concepts of (a) antenna directivity, Eq. (14.106), with 3-D plots of the radiation intensity (U) of an actual directional antenna and of an isotropic radiator (U_{iso}), radiating the same power (P_{rad}), which means that the integral in Eq. (14.97) for the two patterns is the same, and (b) antenna beam solid angle (Ω_A), Eq. (14.108), with the U-plot of a fictitious antenna having a constant $U = U_{max}$ only within a cone of solid angle Ω_A, but radiating the same power as the actual antenna, Eq. (14.109).

To illustrate the concept of D_{max}, consider Fig. 14.15(a) with 3-D plots of the radiation intensity, U, of a given antenna and of an isotropic radiator, having the same radiated power, so that the integral over the full solid angle in Eq. (14.97) of the two patterns is the same. We see that D_{max} is a factor showing how much stronger the radiation in a preferred direction is if using a (highly) directional antenna, through the main beam of its radiation pattern, than what it would be if the same total power had been radiated uniformly in all directions. From Eq. (14.105), the maximum directivity of a Hertzian dipole is $D_{max} = 1.5$, for $\theta_{max} = 90°$, which is a 50% increase over an isotropic radiator. Frequently, D is used without specifying the direction of radiation, in which case the maximum directivity is implied, so $D \equiv D_{max}$.

Taking Eq. (14.104) for the direction of maximum radiation (with $|\underline{\mathbf{F}}|_{max}$ in the numerator), and dividing both the numerator and denominator of this expression by $|\underline{\mathbf{F}}|_{max}$, we are left, in the denominator, with the integral of the normalized field pattern of the antenna, $f(\theta, \phi)$, in Eq. (14.80), squared,

$$D_{max} = \frac{4\pi\,|\underline{\mathbf{F}}|^2_{max}}{\int_{\theta=0}^{\pi}\int_{\phi=0}^{2\pi}|\underline{\mathbf{F}}(\theta,\phi)|^2\,d\Omega} = \frac{4\pi}{\int_{\theta=0}^{\pi}\int_{\phi=0}^{2\pi}f^2(\theta,\phi)\,d\Omega} = \frac{4\pi}{\Omega_A}. \tag{14.107}$$

This integral, namely,

$$\Omega_A = \int_{\theta=0}^{\pi}\int_{\phi=0}^{2\pi} f^2(\theta,\phi)\,d\Omega \qquad \text{(beam solid angle)}, \tag{14.108}$$

is another basic parameter of transmitting antennas, denoted by Ω_A and called the beam solid angle of the antenna. Using Eq. (14.81), the beam solid angle for a Hertzian dipole comes out to be $\Omega_A = 8\pi/3$, that is, the integral in θ in Eqs. (14.52) and (5.46) times 2π (the integral in ϕ). Combining Eqs. (14.106) and (14.107),

$$P_{rad} = \frac{4\pi U_{max}}{D_{max}} = U_{max}\Omega_A. \tag{14.109}$$

Fig. 14.15(b) illustrates the concept of the beam solid angle by relating the radiation intensity plots of an actual antenna [Fig. 14.15(a)] and of a fictitious antenna that would uniformly radiate the same total power (P_{rad}) only into a cone of solid angle Ω_A with the radiation intensity equal to U_{max} of the actual pattern [Fig. 14.15(b)]. Since the radiation of the fictitious antenna is uniform within the cone in Fig. 14.15(b), its radiated power can be found as the radiation intensity (U_{max}) times the cone solid angle (Ω_A), which is exactly what Eq. (14.109) states.

The gain of an antenna in a certain direction, $G(\theta, \phi)$, is the ratio of the radiation intensity $U(\theta, \phi)$ to the constant radiation intensity that would be obtained if the time-average input power, P_{in}, delivered to the antenna at its terminals (Fig. 14.10) were radiated isotropically. In other words, antenna gain quantifies directional capabilities of a real antenna, that has ohmic losses, in relation to a lossless isotropic radiator. The isotropic radiation intensity equals $P_{in}/(4\pi)$ [see Eq. (14.99)], and thus

$$G(\theta,\phi) = \frac{4\pi U(\theta,\phi)}{P_{in}} \quad \longrightarrow \quad G_{max} = \frac{4\pi U_{max}}{P_{in}} = \frac{\eta\,|\underline{\mathbf{F}}|^2_{max}}{\pi(R_{rad}+R_{ohmic})}, \tag{14.110}$$

antenna gain (dimensionless)

where G_{max} is the maximum gain in the radiation pattern, R_{ohmic} is the ohmic resistance of the antenna, and the use is made of Eqs. (14.100) and (14.64). Basically, P_{rad} in expressions for the directivity should be replaced by P_{in} to obtain the corresponding expressions for the gain, and analogously for expressions involving antenna resistances. Again, if the direction of radiation is not stated, we assume that G denotes the maximum value ($G \equiv G_{max}$). From Eqs. (14.110), (14.106), and (14.65), we see that

$$G = \eta_{rad} D, \tag{14.111}$$

gain vs. directivity

i.e., the gain is reduced with respect to the directivity by a factor of η_{rad}, the radiation efficiency of the antenna. Of course, $G = D$ for lossless antennas, and $G \approx D$ for antennas whose losses can be neglected.

Both directivity and gain are frequently expressed in decibels,

$$D_{dB} = 10\log D, \quad G_{dB} = 10\log G \quad \text{(in dB)}, \tag{14.112}$$

decibel directivity and gain

where, since they are power ratios [see Eqs. (14.85) and (9.98)], we employ the scale factor of 10 (and not 20) in the definitions. For instance, the dB directivity of a Hertzian dipole appears to be

$$D = 1.5 \quad \longrightarrow \quad D_{dB} = 10\log 1.5 = 1.76 \text{ dB}, \tag{14.113}$$

whereas that of an isotropic radiator is 0 dB.

Example 14.11 Gains of Steel and Copper AM Radio Antennas

Evaluate the gains of the steel and copper nonloaded short dipole antennas from Example 14.7.

Solution Comparing the characteristic radiation functions in Eqs. (14.87) and (14.79), we conclude that the angular variation (with θ) of radiation properties of a nonloaded short wire dipole is the same as that of a Hertzian dipole. In other words, the normalized field pattern (f) given in Eq. (14.81) and plotted in Fig. 14.12 holds the same for the antenna in Fig. 14.6(a). Consequently, bearing also in mind Eq. (14.107), the two antennas have equal directivities, Eq. (14.113), $D = D_{max} = 1.5$ (for $\theta_{max} = 90°$).

With this, the antenna gain, Eq. (14.111), amounts to $G_{steel} = (\eta_{rad})_{steel}\,D = 0.00225$ for the steel dipole and $G_{Cu} = (\eta_{rad})_{Cu}\,D = 0.39$ for the copper one. In decibels, Eq. (14.112), the values are as low as $(G_{dB})_{steel} = 10\log G_{steel} = -26.5$ dB and $(G_{dB})_{Cu} = -4.1$ dB, respectively. Such extremely low negative gains obviously come from a low directivity of the antenna in the first place, but more from large ohmic losses relative to the radiated power of the electrically small antennas, that is, from low radiation efficiencies (especially for the steel antenna).

Example 14.12 Directivity of a Unidirectional Double-Sine Radiation Pattern

The normalized field pattern of an antenna is given by $f(\theta,\phi) = \sin\theta\sin\phi$ for $0 \le \theta \le 180°$ and $0 \le \phi \le 180°$, and is zero elsewhere. Find the directivity of the antenna.

Solution The antenna pattern is illustrated in Fig. 14.16. Note that such a plot, with radiation only into one half-space (defined by $y > 0$ in our case), is called a unidirectional pattern.

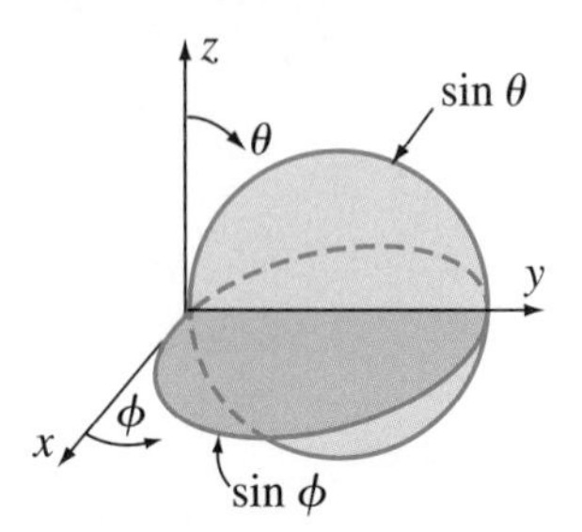

Figure 14.16 Unidirectional double-sine antenna radiation pattern; for Example 14.12.

Note also that the pattern cuts in both the xy- and yz-planes, i.e., for $\theta = 90°$ ($f = \sin\phi$) and $\phi = 90°$ ($f = \sin\theta$), respectively, are circles (see Fig. 14.13). Using Eqs. (14.108) and (14.93), the antenna beam solid angle is

$$\Omega_A = \int_{\theta=0}^{\pi}\int_{\phi=0}^{\pi} \underbrace{\sin^2\theta\sin^2\phi}_{f^2}\,\underbrace{\sin\theta\,d\theta\,d\phi}_{d\Omega} + \int_{\theta=0}^{\pi}\int_{\phi=\pi}^{2\pi} 0^2\,d\Omega = \frac{2\pi}{3}, \tag{14.114}$$

where the first (nonzero) integral in θ is exactly that in Eqs. (14.52) and (5.46), so 4/3, and the first integral in ϕ equals $\pi/2$, which is obtained similarly to the integration in Eq. (6.95), while the integral over the other half-space is, of course, zero. Eqs. (14.107) and (14.112) then yield the following for the directivity of the antenna:

$$D = \frac{4\pi}{\Omega_A} = 6 \quad (D_{\rm dB} = 10\log 6 = 7.78\ {\rm dB}). \tag{14.115}$$

Example 14.13 Isotropic, Cosine, and Cosine Squared Unidirectional Patterns

Compute (a) the directivity and (b) the half-power beamwidth for the normalized field antenna pattern given by $f_n(\theta) = \cos^n\theta$ for $0 \le \theta \le 90°$ and $f_n = 0$ for $90° < \theta \le 180°$ in three cases of $n = 0$, 1, and 2.

Solution The three radiation patterns are azimuthally symmetrical, and Fig. 14.17 shows their respective 2-D cuts in any one of the planes $\phi = $ const. We note that this is another example of unidirectional radiation, as in Fig. 14.16, but now into the upper half-space, defined by $z \ge 0$ ($0 \le \theta \le 90°$).

(a) For $n = 0$, $f_0(\theta) = 1$, which is an isotropic pattern – over a half-space. Therefore, the resulting beam solid angle, Eq. (14.108), of the pattern equals a half of the full solid angle, Eq. (14.94),

$$\Omega_{A0} = \int_{\text{half-space}} 1^2\,d\Omega = \frac{1}{2}\Omega_{\rm full} = 2\pi. \tag{14.116}$$

The associated antenna directivity, from Eqs. (14.107) and (14.112), is $D_0 = 4\pi/\Omega_{A0} = 2$ or $(D_0)_{\rm dB} = 10\log 2 = 3$ dB, which is twice (or 3 dB above) that of the fully isotropic radiator ($D_{\rm iso} = 1$ or 0 dB), as expected (no radiation in the lower half-space in Fig. 14.17).

For $n = 1$ and 2, the beam solid angles amount to

$$\Omega_{A1} = \int_{\theta=0}^{\pi/2}\int_{\phi=0}^{2\pi} \underbrace{\cos^2\theta}_{f_1^2}\sin\theta\,d\theta\,d\phi = \frac{2\pi}{3}, \quad \Omega_{A2} = 2\pi\int_0^{\pi/2} \underbrace{\cos^4\theta}_{f_2^2}\sin\theta\,d\theta = \frac{2\pi}{5}, \tag{14.117}$$

where the first integral in θ is a half of that in Eqs. (2.32), and the second one is evaluated in a similar fashion; the directivities are $D_1 = 4\pi/\Omega_{A1} = 6$ [$(D_1)_{\rm dB} = 7.78$ dB] and $D_2 = 4\pi/\Omega_{A2} = 10$ [$(D_2)_{\rm dB} = 10$ dB], respectively. Of course, the narrower the beam in Fig. 14.17 the higher the directivity ($D_2 > D_1 > D_0$).

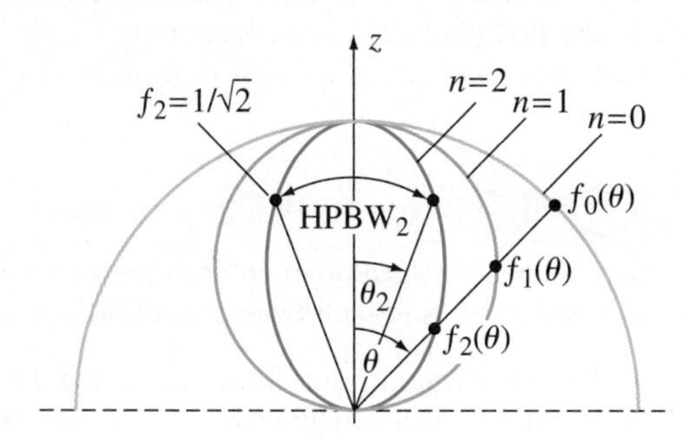

Figure 14.17 Three unidirectional normalized field patterns given by $f_n(\theta) = \cos^n\theta$ for $0 \le \theta \le 90°$ ($n = 0, 1, 2$); for Example 14.13.

(b) In the first case, $n = 0$, the HPBW cannot be defined, because the pattern is isotropic. Given that the middle curve in Fig. 14.17, representing the function $f_1(\theta) = \cos\theta$, is a circle (as in Fig. 14.13), the HPBW for $n = 1$ is the same as for a Hertzian dipole,[8] Eqs. (14.86), so $\text{HPBW}_1 = 90°$. Finally, the condition $f_2(\theta_2) = \cos^2\theta_2 = 1/\sqrt{2}$ (Fig. 14.17), for $n = 2$, yields $\theta_2 = 32.76°$, and $\text{HPBW}_2 = 2\theta_2 = 65.52°$.

Problems: 14.23–14.26; *Conceptual Questions* (on Companion Website): 14.28 and 14.29; *MATLAB Exercises* (on Companion Website).

14.7 ANTENNA POLARIZATION

The polarization of an antenna in a given direction, determined by angles θ and ϕ in Fig. 14.14, is defined as the polarization of the electromagnetic wave it radiates, so in the far zone, in that direction. Locally, this wave can be considered as a uniform plane wave (see Sections 14.3 and 9.2), which means that we can directly apply here the theory and discussions of linear, circular, and elliptical polarizations of uniform plane waves from Section 9.14. In particular, given that the polarization of a time-harmonic wave is determined by the polarization of its electric field intensity vector, as well as that the antenna polarization relates to the far field, and hence is independent of r, it is best represented by the polarization of the characteristic radiation function, $\underline{\mathbf{F}}(\theta, \phi)$, of an antenna, in Eq. (14.78).

Therefore, in place of studies of the antenna polarization, we simply invoke here what we already know about the (linear, circular, and elliptical) polarization of time-harmonic electromagnetic waves. In other words, we may refer to the material of Section 9.14 as an integral part of this chapter (on antennas and wireless systems).

14.8 WIRE DIPOLE ANTENNAS

We now consider a symmetrical (centrally fed) straight PEC wire dipole antenna with an arbitrary length, l, and free ends (no capacitive loads at wire ends), shown in Fig. 14.18(a). The current distribution, $\underline{I}(z)$, along the antenna can approximately be determined identifying it to the standing current wave on an open-circuited lossless two-wire transmission line [see Fig. 12.14(b) and Eqs. (12.114)], whose length equals the length of each of the arms of the dipole, $h = l/2$, and the dielectric is that surrounding the antenna. Namely, if we progressively bend out the conductors of such a line, as illustrated in Fig. 14.18(b), to ultimately make them coaxial (collinear) with respect to each other and form a dipole antenna, the current along the wires remains essentially unchanged throughout the progression. Of course, the current is zero at free wire ends. Hence, using the transmission-line current, Eqs. (12.114), as an approximation of the antenna current, we can write, for the coordinate z defined as in Fig. 14.18(a),

$$\boxed{\underline{I}(z) = \underline{I}_{\text{m}} \sin\beta(h - |z|) \quad (-h \le z \le h),} \tag{14.118}$$

sine-wave dipole current

[8]Note that the function $\cos\theta$, viewed (for simplicity) for all values of θ ($0 \le \theta \le 180°$), appears to have the same shape (two circles) as $\sin\theta$ (Hertzian dipole), just one extending along the z-axis (Fig. 14.17 for $n = 1$) and the other on the side of it [Fig. 14.12(b)] – but their shapes are equal only in a 2-D cut (for a specified angle ϕ). Their 3-D patterns, obtained by rotating (for $0 < \phi < 360°$) the corresponding 2-D plots about the z-axis, are, however, very different: $\sin\theta$ gives a "doughnut"-type pattern, whereas $\cos\theta$ sweeps out a "dumbbell" shape (two spheres touching at the coordinate origin). The latter 3-D pattern turns out to be twice as directive as the former one.

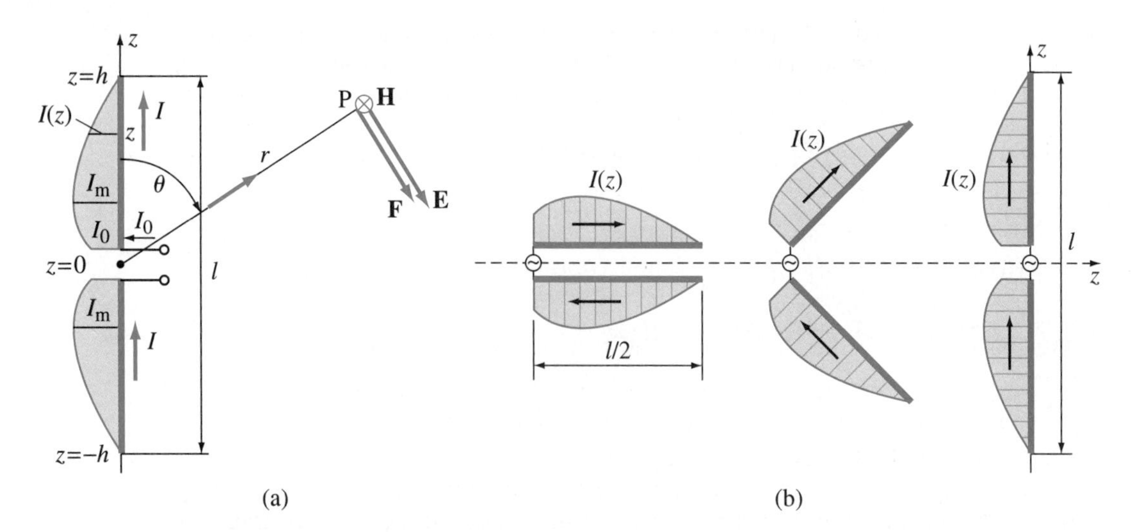

Figure 14.18 Symmetrical wire dipole antenna of arbitrary length with sinusoidal current approximation (a), which can be considered as if obtained by progressively bending out the conductors of an open-circuited two-wire transmission line with a standing current wave (b).

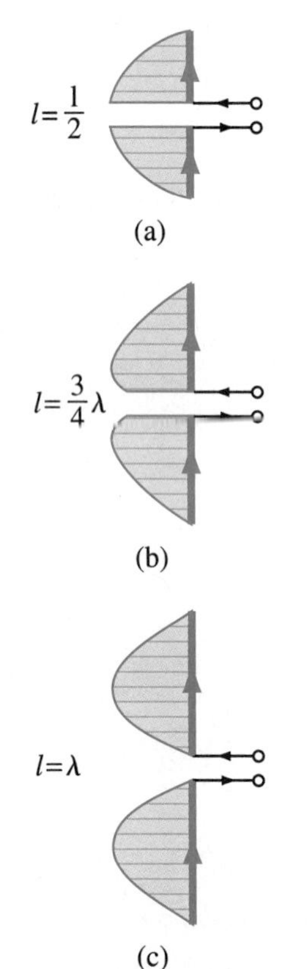

Figure 14.19 Normalized current intensity $\underline{I}(z)/\underline{I}_{\mathrm{m}}$, Eq. (14.118), along a wire dipole in Fig. 14.18(a) for dipole electrical lengths (l/λ) of (a) 1/2, (b) 3/4, (c) 1, (d) 5/4, and (e) 3/2.

where $\underline{I}_{\mathrm{m}}$ is the complex rms current intensity at the standing-wave maxima, and β is the phase coefficient of the antenna radiation. This is the so-called sinusoidal (sine-wave) current approximation for wire dipole antennas. It proves to be quite accurate (in comparison with measurements and results obtained by numerical techniques) for thin antennas, whose radius, a, is small relative to l ($a \ll l$). In fact, it can be shown analytically that the exact current distribution along the dipole approaches that in Eq. (14.118) as $a \to 0$ (infinitely thin dipole). Note that the current at the antenna input terminals, $z = 0$, is

$$\underline{I}_0 = \underline{I}(0) = \underline{I}_{\mathrm{m}} \sin \beta h. \tag{14.119}$$

Fig. 14.19 shows current distributions, Eq. (14.118), normalized as $\underline{I}(z)/\underline{I}_{\mathrm{m}}$, for wire dipoles of several characteristic electrical lengths, l/λ, λ being the operating wavelength of the antenna (for the surrounding medium), Eqs. (8.112) and (8.111). We see that for dipoles that are up to one wavelength long ($l \leq \lambda$), the current is in phase along the whole wire, that is, $\underline{I}(z)$ has the same reference direction, and thus the same phase, equal to the phase of $\underline{I}_{\mathrm{m}}$, for $-h \leq z \leq h$. On the other hand, $\underline{I}(z)$ is not all in phase on dipoles longer than a wavelength ($l > \lambda$), with the currents on adjacent half-wave (or shorter) sections (between the current zeros) flowing in opposite reference directions, i.e., being in counter-phase (phase shift of 180°), with respect to each other, which comes from opposite signs of the sine function on the two segments.

The dipole in Fig. 14.18(a) is a special case of an arbitrary straight wire antenna along the z-axis, Fig. 14.4, so its magnetic vector potential, $\underline{\mathbf{A}}$, in the far zone is given by Eq. (14.29). The vector $\underline{\mathbf{A}}$ has only a z-component, $\underline{A}_z$, which, substituting the current distribution from Eq. (14.118), comes out to be

$$\underline{A}_z = \frac{\mu \underline{I}_{\mathrm{m}} \, \mathrm{e}^{-\mathrm{j}\beta r}}{4\pi r} \int_{z=-h}^{h} \sin \beta (h - |z|) \, \mathrm{e}^{\mathrm{j}\beta z \cos\theta} \, \mathrm{d}z. \tag{14.120}$$

To solve this integral in z, we first express the complex exponential function in the integrand via its real and imaginary parts, using Euler's identity, Eq. (8.61), and realize that they are an even (cosine) and odd (sine) function, respectively, in z.

Since the sine function originally present in the integrand is an even function in z (note that it actually depends on the absolute value of z), the integrand can be represented as a sum of a product of two even functions, which results in an even function, and a product of an even and odd function, which gives an odd function. With symmetric integration limits, $-h$ and h, in Eq. (14.120), the integral of the first product is twice the integral from 0 to h, and the integral of the second product is zero, so we are left with

$$\underline{A}_z = \frac{\mu \underline{I}_{\rm m}\, {\rm e}^{-{\rm j}\beta r}}{4\pi r} 2 \int_0^h \sin\beta(h-z)\cos(\beta z\cos\theta)\, {\rm d}z, \quad (14.121)$$

where the absolute value sign is removed from $|z|$, as $z \geq 0$. We then apply the trigonometric identity $2\sin\alpha_1\cos\alpha_2 = \sin(\alpha_1+\alpha_2)+\sin(\alpha_1-\alpha_2)$ to transform the product of sine and cosine functions to easily integrable (sine) functions,

$$\underline{A}_z = \frac{\mu \underline{I}_{\rm m}\, {\rm e}^{-{\rm j}\beta r}}{4\pi r} \left\{ \int_0^h \sin\left[\beta h - \beta z(1-\cos\theta)\right] {\rm d}z + \int_0^h \sin\left[\beta h - \beta z(1+\cos\theta)\right] {\rm d}z \right\}. \quad (14.122)$$

The result of their integration is

$$\underline{A}_z = \frac{\mu \underline{I}_{\rm m}\, {\rm e}^{-{\rm j}\beta r}}{4\pi \beta r} \left[\cos(\beta h\cos\theta) - \cos\beta h\right] \left(\frac{1}{1-\cos\theta} + \frac{1}{1+\cos\theta} \right), \quad (14.123)$$

which can further be simplified by

$$\frac{1}{1-\cos\theta} + \frac{1}{1+\cos\theta} = \frac{2}{1-\cos^2\theta} = \frac{2}{\sin^2\theta}. \quad (14.124)$$

Combining Eqs. (14.34), (14.123), (14.124), (14.119), and (14.9), the far electric field intensity vector, $\underline{\mathbf{E}}$, of the dipole antenna, with only a θ-component, is

$$\underline{\mathbf{E}} = \underline{E}_\theta\, \hat{\boldsymbol{\theta}} = {\rm j}\omega \underline{A}_z \sin\theta\, \hat{\boldsymbol{\theta}} = \frac{{\rm j}\eta}{2\pi} \underline{I}_0 \frac{{\rm e}^{-{\rm j}\beta r}}{r} \frac{\cos(\beta h\cos\theta) - \cos\beta h}{\sin\beta h \sin\theta}\, \hat{\boldsymbol{\theta}}, \quad (14.125)$$

and comparing this expression with Eq. (14.78), we identify the characteristic radiation function, $\underline{\mathbf{F}}$, of the dipole,

$$\boxed{\underline{\mathbf{F}}(\theta) = \frac{\cos(\beta h\cos\theta) - \cos\beta h}{\sin\beta h\sin\theta}\, \hat{\boldsymbol{\theta}}.} \quad (14.126)$$

radiation function, arbitrary dipole

As expected, $\underline{\mathbf{F}}$ effectively depends on βh (and not just on h isolated from the operating frequency of the antenna), and thus on the electrical length of the dipole arm, h/λ, rather than on its physical length alone [see Eqs. (12.73)].

The most important dipole in Fig. 14.18(a) is by far that for $l = \lambda/2$. This simple wire antenna, known as a half-wave dipole, is, in fact, one of the most widely used of all antenna types. The dipole arm length being $h = \lambda/4$, we have

$$l = 2h = \frac{\lambda}{2} \quad \longrightarrow \quad \beta h = 2\pi\frac{h}{\lambda} = \frac{\pi}{2}, \quad (14.127)$$

with which the characteristic radiation function in Eq. (14.126) becomes

$$\boxed{\underline{\mathbf{F}}(\theta) = \frac{\cos\left(\frac{\pi}{2}\cos\theta\right)}{\sin\theta}\, \hat{\boldsymbol{\theta}}.} \quad (14.128)$$

radiation function, half-wave dipole

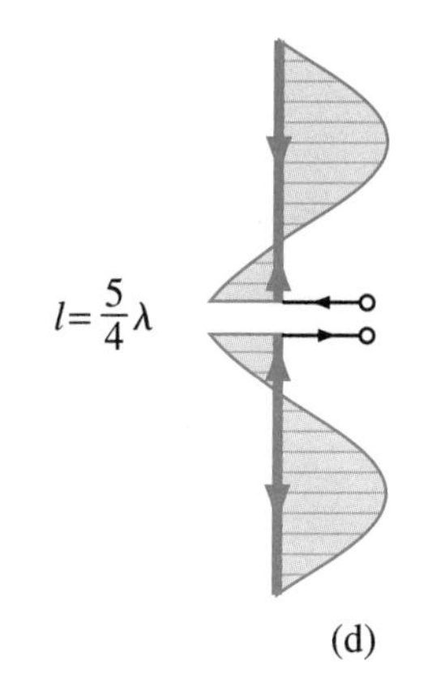

(d)

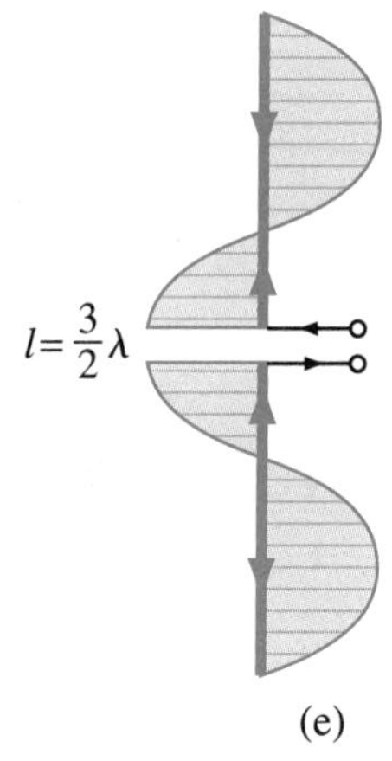

(e)

Figure 14.19 *(continued)*

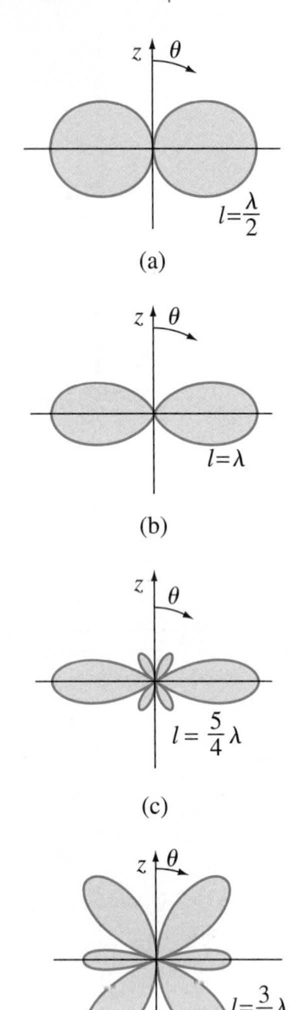

Figure 14.20 Normalized wire-dipole field patterns, Eqs. (14.80) and (14.126), in an E-plane corresponding to several current distributions in Fig. 14.19, with l/λ equal to (a) 1/2, (b) 1, (c) 5/4, and (d) 3/2.

The radiation is maximum in the equatorial plane ($\theta = 90°$), in which the antenna exhibits an omnidirectional radiation pattern, like in Fig. 14.12(c). The maximum magnitude of $\underline{\mathbf{F}}(\theta)$ is unity,

$$F_{\max} = 1 \quad \text{for} \quad \theta = 90° \quad \left(l = \frac{\lambda}{2}\right), \tag{14.129}$$

and this is why the constant $\underline{C}$ in the general expression for $\underline{\mathbf{F}}$ is adopted as in Eq. (14.78). Namely, that choice for the numerical value of $\underline{C}$ conveniently makes the characteristic radiation function of one of the most common antennas in engineering practice already normalized to its maximum value, so that the normalized field pattern in Eq. (14.80) of a half-wave dipole is $f(\theta) = |\underline{\mathbf{F}}(\theta)|$. Fig. 14.20(a) shows the radiation pattern of the dipole in a plane $\phi = \text{const}$ (E-plane), which apparently is very similar to that in Fig. 14.12(b) for a Hertzian dipole, although it is somewhat more directional, with a smaller half-power beamwidth of HPBW = 78° in comparison to that in Eq. (14.86).

Shown in Fig. 14.20 also are normalized field patterns, computed from Eqs. (14.80) and (14.126), for several other characteristic dipole lengths in Fig. 14.19. We observe a clear correspondence between the current distributions along the wire and far-field distributions in an E-plane. For dipoles of lengths $l \leq \lambda$, because of the phase uniformity of the antenna current along the whole wire, the radiation is strongest in a direction normal to the antenna ($\theta = 90°$). Namely, given the phase approximation in Eq. (14.28), all parallel rays originating from current elements that constitute the current distribution of an antenna (see Fig. 14.3) travel the same paths, and thus arrive in phase, to the far field at $\theta = 90°$ [$\cos\theta = 0$ in the phase correction term in the radiation integral, Eq. (14.120)]. However, when $l > \lambda$, multiple radiation lobes are formed in the pattern, due to cancelation effects of oppositely directed (counter-phase) currents on adjacent antenna sections, separated by the current zeros. Strong "lobing" effects, i.e., the formation of radiation lobes, are the principal reason for a limited practical use of wire antennas longer than a wavelength. On the other side, we note that all patterns, so irrespective of the current distribution of the antenna, have nulls along the axis of the dipole ($\theta = 0$ or $\theta = 180°$), which comes from the factor $\sin\theta$ in the expression for the transverse projection of the vector potential, $\underline{\mathbf{A}}$, in Eq. (14.125).

From Eqs. (14.53), (14.101), (14.128), and (9.23), the radiation resistance of a half-wave dipole situated in free space can be found via the radiated power, P_{rad}, as

$$R_{\text{rad}} = \frac{P_{\text{rad}}}{I_0^2} = \frac{\eta}{2\pi}\int_{\theta=0}^{\pi} |\underline{\mathbf{F}}(\theta)|^2 \sin\theta \, d\theta = 60\ \Omega \underbrace{\int_0^{\pi} \frac{\cos^2\left(\frac{\pi}{2}\cos\theta\right)}{\sin\theta}\, d\theta}_{1.22} \approx 73\ \Omega, \tag{14.130}$$

where the integral in ϕ in Eq. (14.101) is 2π. The integral in θ (which turns out to be about 1.22) cannot be evaluated analytically in a closed form, but can be transformed (through extensive mathematical manipulations) to a form that contains a special function [so-called cosine integral, $\text{Ci}(x)$], with available tabulated values (based on its series expansion), or can be computed numerically. We adopt here the latter approach, and evaluate this integral using a simple numerical-integration formula. Combining Eqs. (14.106), (14.129), (14.130), and (14.112), the directivity of the dipole amounts to

directivity, λ/2 dipole

$$\boxed{D = \frac{120\ \Omega}{73\ \Omega}\, 1^2 = 1.64 \quad \longrightarrow \quad D_{\text{dB}} = 2.15\ \text{dB},} \tag{14.131}$$

i.e., it is only slightly higher than that of a Hertzian dipole, Eq. (14.113). With further increasing the dipole length, Figs. 14.19 and 14.20, up to about $l = 5\lambda/4$, D also increases, and then drops sharply, due to large side lobes (broken pattern).

As we shall see in an example, the high-frequency ohmic resistance ($R_{\rm ohmic}$), Eq. (14.61), of a half-wave metallic (non-PEC) dipole is typically very small relative to $R_{\rm rad}$ in Eq. (14.130), which leads to a practically ideal antenna radiation efficiency, Eq. (14.65), of $\eta_{\rm rad} \approx 100\%$ (the same is true for dipoles longer than $\lambda/2$). Neglecting $R_{\rm ohmic}$ in Eq. (14.64), the real part of the antenna input impedance ($\underline{Z}_{\rm A}$), Eq. (14.63) and Fig. 14.10, of a half-wave dipole reduces to $R_{\rm A} = R_{\rm rad}$, and the antenna gain equals directivity, $G = D$, Eq. (14.111).[9] The imaginary (reactive) part ($X_{\rm A}$) of the impedance, which takes into account the reactive power stored in the near field around the dipole, Eq. (14.66), is nonzero for $l = \lambda/2$. Employing a quite complicated analytical technique based on Poynting's theorem in complex form, Eq. (8.196), the calculated impedance of an infinitely thin half-wave dipole is

$$\boxed{\underline{Z}_{\rm A} = (73 + {\rm j}42.5)\ \Omega.} \qquad (14.132)$$

input impedance, $\lambda/2$ dipole

So, $X_{\rm A}$ is inductive, and not very large in magnitude. Overall, this input impedance is very attractive for many applications.

Example 14.14 Superposition of Traveling Current Waves on a Dipole Antenna

Show that the current distribution along a symmetrical wire dipole antenna with an arbitrary length (l), in Fig. 14.18(a), can be represented as a superposition of two oppositely directed traveling waves.

Solution Let us first consider the upper half of the dipole [Fig. 14.18(a)], where $z \geq 0$ and $|z| = z$, so that, using the first identity in Eqs. (10.7) and a new coordinate z' to shift the coordinate origin to the wire end, as shown in Fig. 14.21, the expression for the complex rms current intensity in Eq. (14.118) along this dipole arm can be transformed to

$$\underline{I}(z') = \underline{I}_{\rm m} \sin\beta(h - z) = -\underline{I}_{\rm m} \sin\beta z' = \frac{\underline{I}_{\rm m}}{2{\rm j}}\, {\rm e}^{-{\rm j}\beta z'} - \frac{\underline{I}_{\rm m}}{2{\rm j}}\, {\rm e}^{{\rm j}\beta z'} \quad (z' = z - h). \qquad (14.133)$$

We see that the first term in the resulting current expression is a current wave traveling from the feed point outward along the wire to the location $z' = 0$, where it bounces back off the open wire end, creating a reflected current wave, the second term, which propagates toward the antenna feed. We also note the analogy with expressions for forward and backward traveling current waves on a transmission line, in Eq. (12.11). Given the minus sign in the second term in Eq. (14.133), we realize that the pertinent reflection coefficient is $\underline{\Gamma} = -1$, and this ensures a zero current at the end (for $z' = 0$). Note that such $\underline{\Gamma}$, being the current reflection coefficient ($\underline{\Gamma}_{\text{for currents}}$), Eq. (12.28), at an open circuit, is opposite to the customary value of positive unity for the voltage coefficient on an open-circuited transmission line, Eq. (12.113).

Similarly, with a coordinate transformation defined by $z'' = -z - h$, the current $\underline{I}(z'')$ on the other half of the dipole, where $z < 0$ and $|z| = -z$, can be written, from Eq. (14.118), in the same way as in Eq. (14.133) with z' now replaced by z''. As the z''-axis in Fig. 14.21 is directed in the negative z direction, this again is a sum of an incident current wave (term proportional to ${\rm e}^{-{\rm j}\beta z''}$) progressing from the generator outward to the end of the antenna at $z = -h$ or $z'' = 0$ and reflected wave (characterized by ${\rm e}^{{\rm j}\beta z''}$), traveling back.

Finally, we recall the case of a traveling-wave wire antenna, in Fig. 14.8, where only an incident current wave exists on the structure, given by Eq. (14.41). Reflection at the wire end is prevented by a matched load, Eq. (12.115).

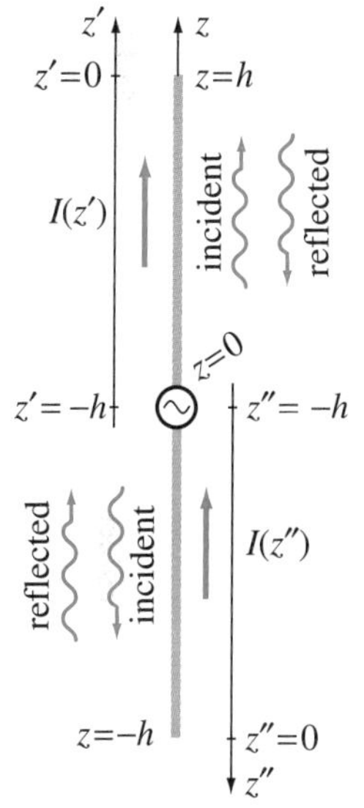

z' z $z'=0$ $z=h$ $I(z')$ incident reflected $z=0$ $z'=-h$ $z''=-h$ reflected incident $I(z'')$ $z=-h$ $z''=0$ z''

Figure 14.21 Representation of the current distribution along each of the arms of a wire dipole antenna in Fig. 14.18(a) as a superposition of an incident current wave traveling outward from the feed point to the respective antenna end and reflected wave bouncing back off the end; for Example 14.14.

[9] In what follows, we shall always assume that $R_{\rm ohmic} \approx 0$ in the analysis of antennas, except for electrically small antennas or when we explicitly indicate otherwise.

Example 14.15 **Short-Dipole Current/Field from Expressions for Arbitrary Dipole**

Show that for $l \ll \lambda$ both the current distribution and far electromagnetic field of the (arbitrary) dipole antenna in Fig. 14.18(a) reduce to those of the (short) dipole in Fig. 14.6(a).

Solution We start with the expression for the sinusoidal current distribution, $\underline{I}(z)$, in Eq. (14.118), along an arbitrary wire dipole, Fig. 14.18(a), with the antenna feed current, $\underline{I}_0$, incorporated from Eq. (14.119). If $l \ll \lambda$, then $h = l/2 \ll \lambda$ as well, and, having in mind Eq. (8.111), both $\beta h = 2\pi h/\lambda$ and $\beta(h - |z|)$ $(-h \le z \le h)$ are very small. Hence, given that $\sin x \approx x$ when $x \ll 1$, the current distribution approximately becomes

$$\underline{I}(z) = \underbrace{\underline{I}_0 \frac{\sin\beta(h-|z|)}{\sin\beta h}}_{\text{arbitrary dipole}} \approx \underline{I}_0 \frac{\beta(h-|z|)}{\beta h} = \underbrace{\underline{I}_0\left(1-\frac{|z|}{h}\right)}_{\text{short dipole}}, \tag{14.134}$$

and this exactly is $\underline{I}(z)$ in Eq. (14.35), namely, the triangular current along an electrically short dipole, in Fig. 14.6(a).

Similarly, since $\cos x \approx 1 - x^2/2$ when $x \ll 1$, the characteristic radiation function, $\underline{\mathbf{F}}(\theta)$, of an arbitrary dipole antenna, Eq. (14.126), reduces, for $l \ll \lambda$, to

$$\underline{\mathbf{F}}(\theta) = \underbrace{\frac{\cos(\beta h\cos\theta) - \cos\beta h}{\sin\beta h\sin\theta}\,\hat{\boldsymbol{\theta}}}_{\text{arbitrary dipole}} \approx \frac{1 - \frac{1}{2}(\beta h\cos\theta)^2 - \left[1 - \frac{1}{2}(\beta h)^2\right]}{\beta h\sin\theta}\,\hat{\boldsymbol{\theta}}$$

$$= \frac{\beta h}{2}\frac{1-\cos^2\theta}{\sin\theta}\,\hat{\boldsymbol{\theta}} = \underbrace{\frac{\beta l}{4}\sin\theta\,\hat{\boldsymbol{\theta}}}_{\text{short dipole}}, \tag{14.135}$$

i.e., to the expression in Eq. (14.87), for a (nonloaded) short dipole. Finally, the reduction to a short-dipole expression automatically translates from Eq. (14.135) to the far electric and magnetic field vectors, $\underline{\mathbf{E}}$ and $\underline{\mathbf{H}}$, of a dipole, since they can be directly expressed in terms of $\underline{\mathbf{F}}$, using Eqs. (14.78) and (14.23).

Example 14.16 **Ohmic Resistance and Radiation Efficiency of a Half-Wave Dipole**

A radio wire dipole antenna of length $l = 1.5$ m and radius $a = 4$ mm operates at a frequency of $f = 100$ MHz in free space. The antenna is made out of aluminum, with conductivity $\sigma_{\text{Al}} = 35$ MS/m. Find (a) the ohmic resistance and (b) the radiation efficiency of the antenna.

Solution Since, by means of Eq. (9.67), the wavelength at the operating frequency of the antenna, for free space, is $\lambda = \lambda_0 = c_0/f = 3$ m, we have $l = \lambda/2$, so this is a half-wave dipole. In addition, as the dipole length to radius ratio is rather large, $l/a = 375$, we can employ the sinusoidal current approximation for the dipole, in Eq. (14.118).

(a) For the dipole arm of $h = l/2 = \lambda/4$, a combination of Eqs. (14.118), (14.119), and (14.127) gives the following expression for the antenna current, for $-h \le z \le h$:

$$\underline{I}(z) = \underline{I}_0 \frac{\sin\beta(h-|z|)}{\sin\beta h} = \underline{I}_0 \sin\left(\frac{\pi}{2} - \beta|z|\right) = \underline{I}_0\cos\beta z = \underline{I}_0\cos\left(\frac{\pi}{l}z\right). \tag{14.136}$$

Of course, this agrees with the cosine current diagram in Fig. 14.19(a). From Eqs. (14.61), (14.59), and (14.136), the high-frequency ohmic resistances of the dipole is

ohmic resistance, $\lambda/2$ dipole

$$\boxed{R_{\text{ohmic}} = \frac{1}{I_0^2}\int_{z=-h}^{h} R'|\underline{I}(z)|^2\,\mathrm{d}z = R'2\int_0^h \cos^2\left(\frac{\pi}{l}z\right)\mathrm{d}z = R'h = R'\frac{l}{2} = R'\frac{\lambda}{4},} \tag{14.137}$$

where the integral in z, equaling $h/2$, is computed as in Eq. (6.95) or (13.86). Using Eqs. (10.78) and (14.60), we obtain, for the aluminum wire,

$$(R_s)_{Al} = \sqrt{\frac{\pi \mu_0 f}{\sigma_{Al}}} = 3.36\ \mathrm{m\Omega/square} \quad \longrightarrow \quad R' = \frac{(R_s)_{Al}}{2\pi a} = 0.134\ \Omega/\mathrm{m} \quad (14.138)$$

($\mu_{Al} = \mu_0$), which then results in $R_{ohmic} = 0.1\ \Omega$.

(b) The radiation resistance of a $\lambda/2$ dipole in free space being $R_{rad} = 73\ \Omega$, Eq. (14.130), its radiation efficiency, Eq. (14.65), comes out to be as high as $\eta_{rad} = R_{rad}/(R_{rad} + R_{ohmic}) = 99.86\%$. Hence, we can readily assume $\eta_{rad} = 100\%$ for the dipole, and treat it as if it were a PEC one. This, in turn, means that its gain, by way of Eqs. (14.111) and (14.131), practically equals $G = D = 1.64$ or 2.15 dB.

Problems: 14.27–14.30; *Conceptual Questions* (on Companion Website): 14.30–14.35; *MATLAB Exercises* (on Companion Website).

14.9 IMAGE THEORY FOR ANTENNAS ABOVE A PERFECTLY CONDUCTING GROUND PLANE

Often, we need to analyze antennas in the presence of conducting ground planes. Such a plane can be an approximation of the earth's surface, of a metallic plate (that may be isolated or a part of a larger structure, like an aircraft or an automobile) to which the antenna is attached (ground plate) or placed in parallel (reflector plate), or of a large conducting object (e.g., device housing) in the vicinity of the antenna. In many situations, the ground conductor can be assumed to be perfectly flat, of infinite extent, and perfectly conducting (PEC), so that the analysis model with a PEC ground plane suffices. This section presents image theory, analogous to techniques in Sections 1.21, 3.13, and 5.7, for the analysis of antennas above a PEC plane. Since an arbitrary antenna can be represented as a superposition of Hertzian dipoles, Fig. 14.3, image theory for a single Hertzian dipole with an arbitrary orientation with respect to the plane, as in Fig. 14.22(a), can be readily generalized to radiators with arbitrary current distributions.

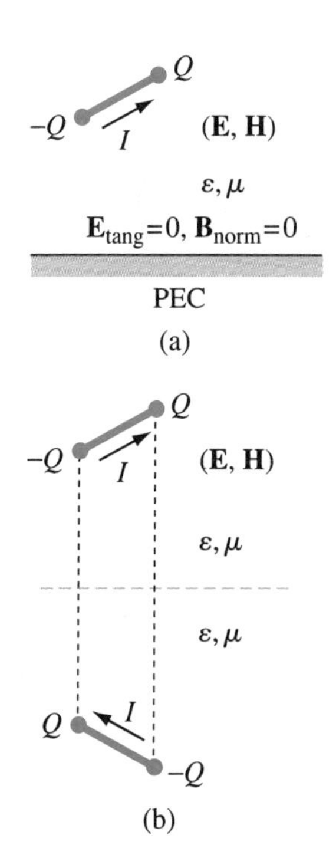

Figure 14.22 Image theory for antennas (or high-frequency current and charge) above a perfectly conducting ground plane: (a) an oblique Hertzian dipole as a constituting element of an arbitrary antenna above the ground and (b) equivalent system with the PEC plane replaced by a negative image of the primary current and charge configuration.

As a consequence of the electromagnetic field of the dipole, surface currents and charges are induced in the PEC plane, so that the total electromagnetic field in the upper half-space, which we would like to find, is the sum of the dipole field and the electromagnetic field due to the induced sources. This latter field can be indirectly computed in a simple manner – by image theory, as the field due to an image of the antenna in Fig. 14.22(a). In the equivalent system, the PEC plane is removed, and the lower half-space is filled with the ambient medium of the original system (most frequently, air), so that the two antennas, the actual one in the upper and virtual in the lower half-space, radiate in an unbounded, homogeneous medium. Applying the image theory for charges in Fig. 1.47 to dipole charges $\underline{Q}$ and $-\underline{Q}$ in Fig. 14.22(a), we obtain an equivalent system composed of charges $-\underline{Q}$ and $\underline{Q}$, respectively, shown in Fig. 14.22(b). The current $\underline{I}$ and charge $\underline{Q}$ (and $-\underline{Q}$) of the original antenna (dipole) in Fig. 14.22(a) are related by the continuity equation in Eq. (14.3), and the same relationship must hold for the current and charge of the image antenna. This means that the image antenna must be a Hertzian dipole as well, which fully and uniquely determines the magnitude and orientation of the image current in Fig. 14.22(b). Geometrically, the virtual dipole is the mirror image of the original in the (once PEC) symmetry plane. Electrically, it is a negative image of the charge and current of the actual antenna. In specific, the image of the charge $\underline{Q}$ is $-\underline{Q}$ and vice versa, whereas the image of the current $\underline{I}$ with reference direction away from the symmetry plane is the current of the same complex intensity flowing toward the plane, that is, the direction of the current image is opposite to the reflection in the symmetry plane of the direction of the primary current. Fig. 14.23 illustrates the images of three characteristic current elements above a PEC plane.

Figure 14.23 Image theory for three characteristic current elements above a PEC ground: (a) original system with the PEC interface and (b) equivalent system with actual and image elements radiating in an unbounded, homogeneous medium (e.g., free space).

Example 14.17 Proof of Image Theory for a Vertical Hertzian Dipole

Using the field expressions for a Hertzian dipole in Eqs. (14.8) and (14.10), show that a vertical dipole and its image obtained according to the image theory in Fig. 14.23 together yield a zero tangential electric field intensity vector and zero normal magnetic field intensity vector in the symmetry plane between the two dipoles.

Solution Inspecting the electric field expression in Eq. (14.8), we realize that the radial component of the vector $\underline{\mathbf{E}}$ can be written as $\underline{E}_r = \underline{f}_1(r)\cos\theta$, and θ-component as $\underline{E}_\theta = \underline{f}_2(r)\sin\theta$. Then we refer to Fig. 14.24, where, for the original dipole (antenna 1) and its image (antenna 2), $r_1 = r_2$ and $\theta_1 = 180° - \theta_2$, so it turns out that $\underline{E}_{r1} = -\underline{E}_{r2}$ and $\underline{E}_{\theta 1} = \underline{E}_{\theta 2}$, as $\cos\theta_1 = -\cos\theta_2$ and $\sin\theta_1 = \sin\theta_2$, respectively. Consequently, the vector sums $\underline{\mathbf{E}}_{r1} + \underline{\mathbf{E}}_{r2}$ and $\underline{\mathbf{E}}_{\theta 1} + \underline{\mathbf{E}}_{\theta 2}$ (Fig. 14.24) are both vertical (perpendicular to the symmetry plane), which proves that the tangential component of the total electric field vector, due to both dipoles acting together, is zero in this plane.

On the other side, given that the dipole magnetic field vector, in Eq. (14.10), has only a ϕ-component, vectors $\underline{\mathbf{H}}_1$ and $\underline{\mathbf{H}}_2$ due to the two dipoles in Fig. 14.24 are entirely horizontal, i.e., tangential to the symmetry plane. In other words, both $\underline{\mathbf{H}}_1$ and $\underline{\mathbf{H}}_2$, as well as the total field $\underline{\mathbf{H}}$, have no normal component in the plane.

We note that this elaboration can be considered as a proof of the image theory (Fig. 14.22) for a vertical Hertzian dipole above a PEC plane. Namely, the fact that $\underline{\mathbf{E}}_{\text{tang}} = 0$ and $\underline{\mathbf{B}}_{\text{norm}} = 0$ in the symmetry plane in Fig. 14.22(b) means that we can metalize that plane, i.e., insert a PEC foil in it, and nothing will change, because the boundary conditions in Eqs. (8.33) will be automatically satisfied at the foil. We thus obtain the original system in Fig. 14.22(a), which proves (for a vertical dipole) that, as far as the electromagnetic field in the upper half-space is concerned, systems in Fig. 14.22(a) and Fig. 14.22(b) are equivalent. Since the full field expressions, in Eqs. (14.8) and (14.10), are employed, the dipole can be at any distance from the plane. A proof for a horizontal dipole can be performed in a similar fashion.

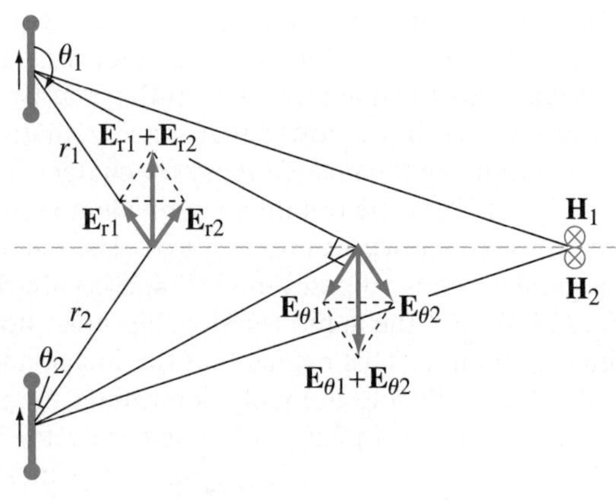

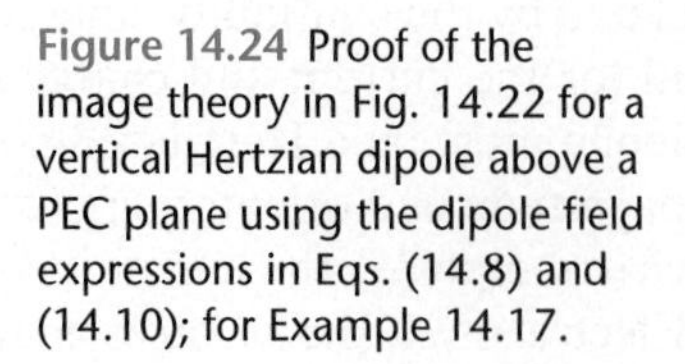

Figure 14.24 Proof of the image theory in Fig. 14.22 for a vertical Hertzian dipole above a PEC plane using the dipole field expressions in Eqs. (14.8) and (14.10); for Example 14.17.

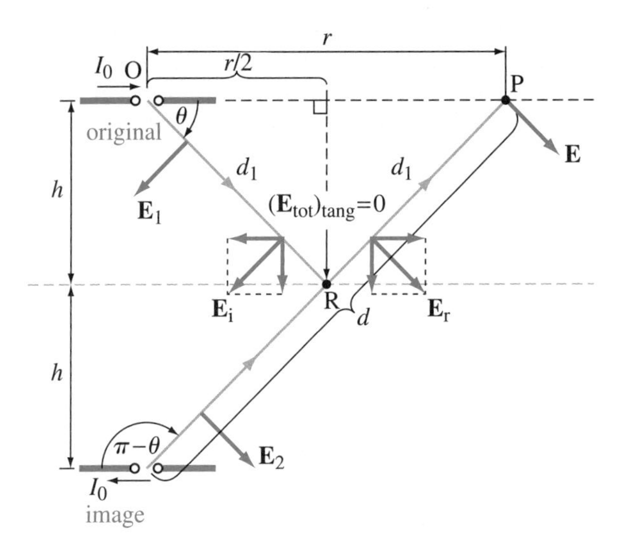

Figure 14.25 Horizontal half-wave dipole antenna above a PEC plane: field computation at a far-field point (P) along the dipole axis – by either tracking the reflected wave from the ground or employing the image theory; for Example 14.18.

Example 14.18 Reflected Wave of a Horizontal Dipole above a Ground Plane

A horizontal half-wave wire dipole antenna is placed at a height $h = 500$ m above a PEC ground plane, in free space, the rms intensity of its feed current is $I_0 = 1$ A, and the operating wavelength is $\lambda = 10$ m. Find the rms intensity of the electric field at a point (P) that is $r = 1$ km distant from the dipole center along its axis, Fig. 14.25.

Solution We first note that $r \gg \lambda$, so the point P in Fig. 14.25 is in the far zone of the antenna. If there were no ground plane, the field at this point would be zero, as the dipole exhibits no radiation in the directions of the wire axis (see Fig. 14.20). Consequently, our task reduces to computing the field of the reflected electromagnetic wave, caused by the presence of the plane. As shown in Fig. 14.25, this wave (viewed as a ray) is launched by the dipole antenna at an angle θ with respect to the dipole axis, and is reflected from the ground (at a point R) such that the total electric field intensity vector, which equals the sum of the incident and reflected vectors, $\underline{\mathbf{E}}_{\text{tot}} = \underline{\mathbf{E}}_{\text{i}} + \underline{\mathbf{E}}_{\text{r}}$, has no tangential component on the PEC surface,

$$\boxed{(\underline{\mathbf{E}}_{\text{i}} + \underline{\mathbf{E}}_{\text{r}})_{\text{tang}} = 0.} \tag{14.139}$$

boundary condition on the PEC plane

In other words, the tangential component of the field vector $\underline{\mathbf{E}}_{\text{i}}$ reverses its phase (acquires a 180° phase shift) at reflection, which results in the change of direction (polarization) of the vector indicated in Fig. 14.25.[10] Denoting by d the total distance traveled by the wave (ray) from the point O to P, which equals twice the distance (d_1) from O to R, and using Eqs. (14.78) and (14.128), the magnitude of the electric field vector $\underline{\mathbf{E}}$ at the receive point is given by

$$\boxed{|\underline{\mathbf{E}}| = \frac{\eta I_0}{2\pi d} F(\theta), \quad \text{where} \quad F(\theta) = \frac{\cos\left(\frac{\pi}{2}\cos\theta\right)}{\sin\theta}} \tag{14.140}$$

E-field magnitude due to a $\lambda/2$ dipole

[$I_0 = |\underline{I}_0|$; $\eta = \eta_0 = 377\ \Omega$, Eq. (9.23)]. From Fig. 14.25,

$$d = 2d_1 = 2\sqrt{\left(\frac{r}{2}\right)^2 + h^2} = \sqrt{r^2 + 4h^2} = 1.414\ \text{km}, \quad \theta = \arctan\frac{h}{r/2} = 45°, \tag{14.141}$$

and hence $|\underline{\mathbf{E}}| = 26.65$ mV/m.

[10]Note that this is the case of reflection at a PEC plane of an obliquely incident uniform plane wave with parallel polarization, in Fig. 10.13(b).

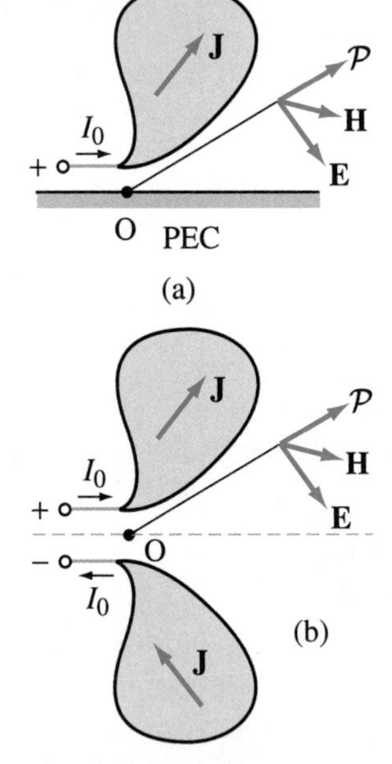

Figure 14.26 Field equivalency in the upper half-space of an arbitrary monopole antenna fed against a PEC plane (a) and the symmetrical dipole in an unbounded medium obtained by image theory (b).

Alternatively, $\underline{\mathbf{E}}$ can be obtained by means of the image theory, Fig. 14.23, as the field due to the negative image of the horizontal dipole in Fig. 14.25. The image antenna launches a wave toward the point P at an angle $180° - \theta$ measured from the local z-axis along the dipole directed as the reference current flow. Since, for the characteristic radiation function of a half-wave dipole, in Eqs. (14.140),

$$F(180° - \theta) = F(\theta), \tag{14.142}$$

and the straight distance traveled by the wave from the image to the receive point is exactly that, d, in Eq. (14.141), we obtain the same result for $|\underline{\mathbf{E}}|$ as in Eqs. (14.140).

Problems: 14.31–14.33; *Conceptual Questions* (on Companion Website): 14.36 and 14.37; *MATLAB Exercises* (on Companion Website).

14.10 MONOPOLE ANTENNAS

Perhaps the most important application of the image theory in Figs. 14.22 and 14.23 is to analyze so-called monopole antennas, i.e., antennas attached to a PEC ground plane and fed against it, as illustrated in Fig. 14.26(a). By image theory, the monopole is transformed to the equivalent dipole, in Fig. 14.26(b), with the current on the lower half of the dipole being the negative image of the current on its upper half, or of the current on the monopole, and the same electromagnetic field, including near and far field, in the upper half-space (e.g., air) in the two systems. Of course, the analysis of the dipole antenna is normally a much simpler task than the analysis of the antenna composed of the monopole and the ground plane.

From the equality of the far electric field vectors and input currents ($\underline{I}_0$) of the two antennas, monopole and equivalent dipole, we obtain that they have the same characteristic radiation function ($\underline{\mathbf{F}}$), Eq. (14.78),

radiation field of a monopole antenna by image theory

$$\boxed{\underline{\mathbf{E}}_{\text{monopole}} = \underline{\mathbf{E}}_{\text{dipole}} \quad \longrightarrow \quad \underline{\mathbf{F}}_{\text{monopole}} = \underline{\mathbf{F}}_{\text{dipole}} \quad (\underline{I}_0 = \text{const}),} \tag{14.143}$$

and this is also true for the time-average Poynting vector ($\mathcal{P}_{\text{ave}}$), Eq. (14.26), and radiation intensity (U), Eq. (14.96), of the antennas. The time-average radiated power (P_{rad}) of the monopole antenna, with the integration in Eq. (14.97) or (14.101) only over the upper hemisphere in Fig. 14.26(a), that is, only down to $\theta = \pi/2$ in the latitudinal direction, is a half that of the corresponding dipole,

radiated power of a monopole vs. dipole

$$\boxed{(P_{\text{rad}})_{\text{monopole}} = \int_{S_{\text{hemisphere}}} U(\theta, \phi)\, \mathrm{d}\Omega = \frac{1}{2} \oint_S U(\theta, \phi)\, \mathrm{d}\Omega = \frac{1}{2} (P_{\text{rad}})_{\text{dipole}}.} \tag{14.144}$$

This directly translates, using Eq. (14.53), to radiation resistances,

$$(R_{\text{rad}})_{\text{monopole}} = \frac{1}{2} (R_{\text{rad}})_{\text{dipole}}. \tag{14.145}$$

In addition, since the ohmic losses in the antenna body in Fig. 14.26(a) amount to a half the losses in Fig. 14.26(b),[11] we have the same relationship also for the ohmic resistances (R_{ohmic}), Eq. (14.61), in the two cases. Finally, the near-field

[11]Here, we neglect losses in the ground plane, in Fig. 14.26(a), which is assumed to be perfectly conducting.

zone, with the antenna reactive power, is only in the upper half-space for the monopole, so that its input reactance (X_A), Eq. (14.66), is a half the value for the dipole, which, combined with the results for the antenna resistances, gives that the complex input impedances ($\underline{Z}_A$), Eqs. (14.63) and (14.64), of the two antennas are related in the same way. This can also be obtained directly from the definition of the input impedance in Eq. (14.63) and Fig. 14.10, as the complex voltage ($\underline{V}_g$) to current ratio at antenna input terminals. Namely, observing the excitation of each of the antennas in Fig. 14.26, we realize that $(\underline{V}_g)_{\text{monopole}} = \underline{V}_0 - 0 = \underline{V}_0$ and $(\underline{V}_g)_{\text{dipole}} = \underline{V}_0 - (-\underline{V}_0) = 2\underline{V}_0$, where $\underline{V}_0$ is the electric potential with respect to the ground (or symmetry) plane, which is at potential zero, of the upper terminal of the antennas [note the analogy with capacitance computation in Eq. (2.146)], and hence

$$(\underline{Z}_A)_{\text{monopole}} = \frac{(\underline{V}_g)_{\text{monopole}}}{\underline{I}_0} = \frac{\frac{1}{2}(\underline{V}_g)_{\text{dipole}}}{\underline{I}_0} = \frac{1}{2}(\underline{Z}_A)_{\text{dipole}}. \tag{14.146}$$

Combining Eqs. (14.106), (14.143), and (14.145), the directivity (D) of the monopole is twice that of the equivalent dipole,

$$D_{\text{monopole}} = \frac{\eta\,|\underline{\mathbf{F}}|^2_{\max}}{\pi(R_{\text{rad}})_{\text{monopole}}} = \frac{\eta\,|\underline{\mathbf{F}}|^2_{\max}}{\pi\frac{1}{2}(R_{\text{rad}})_{\text{dipole}}} = 2D_{\text{dipole}}, \tag{14.147}$$

which is also obvious from Eqs. (14.106) and (14.144), and the same relationship holds for the antenna gain (G), Eq. (14.110).

The most frequently used monopole antenna is a quarter-wave vertical wire monopole, which is an upper half of a half-wave dipole, Fig. 14.18(a) for $l = \lambda/2$, positioned perpendicularly against a horizontal ground plane (the length of the monopole is $h = \lambda/4$), as shown in Fig. 14.27(a). The current intensity along the wire, $\underline{I}(z)$, is a quarter of a sine wave, i.e., the upper half (for $0 \le z \le h$) of the current distribution in Fig. 14.19(a). Using Eqs. (14.145), (14.130), (14.147), (14.112), and (14.131), we obtain the following values for the radiation resistance and dB directivity of the $\lambda/4$ monopole:

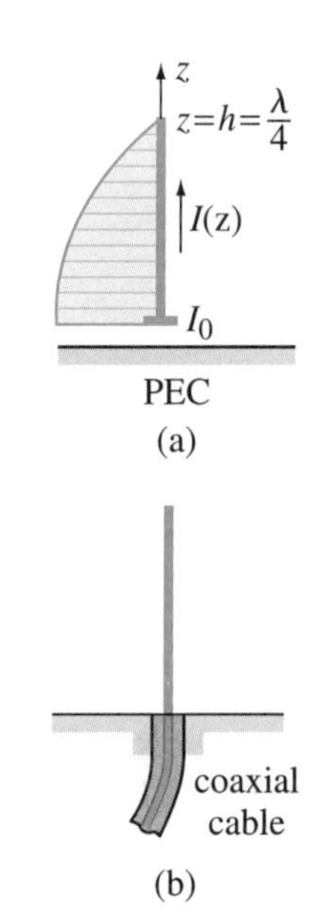

Figure 14.27 Vertical wire monopole antennas: (a) current distribution on a quarter-wave monopole and (b) excitation of a monopole antenna by a coaxial cable.

quarter-wave wire monopole

$$R_{\text{rad}} = \frac{1}{2} \times 73\ \Omega = 36.5\ \Omega, \quad D_{\text{dB}} = 2.15\ \text{dB} + 3\ \text{dB} = 5.15\ \text{dB}. \tag{14.148}$$

Note that the most common excitation of a vertical wire monopole [or an arbitrary monopole in Fig. 14.26(a)] in cases when the ground is a metallic plate (modeled here as a PEC plane) is by a coaxial cable. In such realizations, the monopole is an extension of the inner conductor of the cable, whose outer conductor is connected to the plate from the other side, as indicated in Fig. 14.27(b) [see also Fig. 13.11(a)].

Example 14.19 Short Monopole Antenna

An electrically short (nonloaded) aluminum ($\sigma = 35$ MS/m and $\mu = \mu_0$) wire monopole antenna of length $h = 10$ cm and radius $a = 3$ mm is attached to a PEC ground plane and fed against it by a time-harmonic current of rms intensity $I_0 = 0.5$ A and frequency $f = 10$ MHz. The monopole is perpendicular to the plane and the medium is air. Find the current distribution, characteristic radiation function, E- and H-plane radiation patterns, input impedance, radiation efficiency, directivity, and gain of the antenna.

Solution The current distribution of the short ($h/\lambda_0 = 0.0033 \ll 1$) monopole is a linear function of the coordinate z representing the upper half of the triangular function $\underline{I}(z)$ in

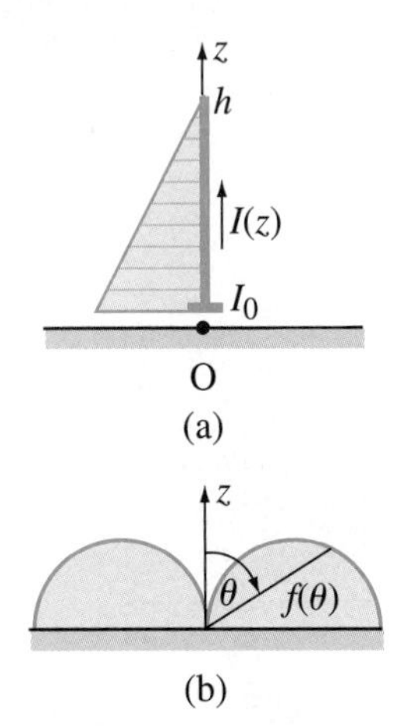

Figure 14.28 Analysis of a short wire monopole antenna attached to a PEC plane: (a) antenna current distribution and (b) normalized E-plane radiation pattern; for Example 14.19.

Fig. 14.6(a), so it is given by Eq. (14.35) for $0 \le z \le h$ and shown in Fig. 14.28(a), where a zero initial phase for the feed current is assumed. From Eqs. (14.143) and (14.87), the characteristic radiation function of the antenna is

$$\underline{\mathbf{F}}(\theta) = \frac{\beta h}{2} \sin\theta\,\hat{\boldsymbol{\theta}} = \pi \frac{h}{\lambda_0} \sin\theta\,\hat{\boldsymbol{\theta}} = 0.0105 \sin\theta\,\hat{\boldsymbol{\theta}} \quad (0 \le \theta \le 90°). \tag{14.149}$$

The resulting normalized field polar radiation pattern in a plane $\phi = \text{const}$ (E-plane) constitutes the upper half of the pattern in Fig. 14.12(b), and hence the two semicircles (see also Example 14.10) – in Fig. 14.28(b), while the pattern in the plane $\theta = 90°$ (H-plane) is omnidirectional, as in Fig. 14.12(c).

Combining Eqs. (14.145), (14.146), (14.68), (14.69), and (14.77), the radiation resistance, ohmic resistance, and input reactance of the antenna in Fig. 14.28 are as follows (see also computations in Example 14.7):

$$R_{\text{rad}} = 10\,(\beta h)^2\ \Omega = 4.4\ \text{m}\Omega, \qquad R_{\text{ohmic}} = \frac{1}{2}\frac{R'(2h)}{3} = \frac{R'h}{3} = \frac{R_s h}{6\pi a} = 2\ \text{m}\Omega,$$

$$X_A = \frac{1}{2}\left[-\frac{120}{\pi(2h)/\lambda_0}\left(\ln\frac{2h}{2a} - 1\right)\Omega\right] = -\frac{30}{\pi h/\lambda_0}\left(\ln\frac{h}{a} - 1\right)\Omega = -7.184\ \text{k}\Omega, \tag{14.150}$$

and its complex input impedance, Eqs. (14.63) and (14.64), and radiation efficiency, Eqs. (14.65), amount to $\underline{Z}_A = (0.0064 - \text{j}7{,}184)\ \Omega$ and $\eta_{\text{rad}} = 68.75\%$, respectively. Note that, having in mind Eqs. (14.145) and (14.146), η_{rad} of a monopole, in general, equals that of the equivalent dipole. The directivity of the short dipole being $D = 1.5$ (see Example 14.11), $D = 3$ or 4.76 dB, from Eq. (14.147), for the monopole. Finally, the antenna gain, Eq. (14.111), comes out to be $G = 2.063$ (3.14 dB).

Example 14.20 Radiation of a Quarter-Wave Monopole in the Presence of Ionosphere

A quarter-wave vertical wire monopole antenna is fed at its base against the earth's surface by a time-harmonic current of frequency $f = 8$ MHz. The input power of the antenna is $P_{\text{in}} = 1$ kW. Both the earth and ionosphere can be considered as PEC planes, and the perpendicular distance between them, the so-called virtual height of the ionosphere (symbolized by h_v), is $h_v = 100$ km, as shown in Fig. 14.29(a). Find the rms electric field intensity of (a) the surface wave and (b) the resultant ionospheric wave at a receive point on the ground at a distance $r = 500$ km from the antenna.

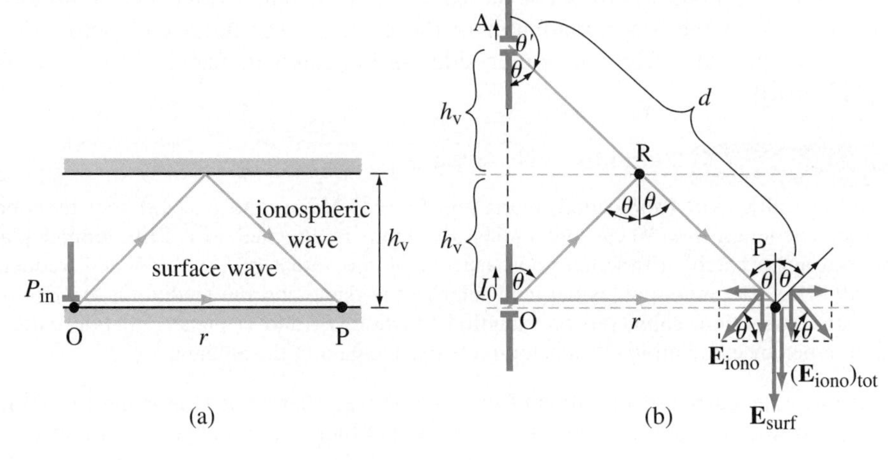

Figure 14.29 Radiation of a quarter-wave vertical monopole antenna at the earth's surface in the presence of ionosphere: (a) approximation of both the earth and ionosphere by PEC planes, separated by a distance equal to the virtual height of the ionosphere (h_v), and (b) evaluation of far fields due to the surface wave and the resultant ionospheric wave using image theory; for Example 14.20.

Solution

(a) Eq. (9.67) tells us that the operating wavelength of the transmitting antenna (for free space) equals $\lambda = \lambda_0 = c_0/f = 37.5$ m, so $r/\lambda = 13{,}333.3 \gg 1$, that is, the field point (P) in Fig. 14.29(a) is in the far zone of the antenna (note that the length of the monopole is $h = \lambda/4 = 9.375$ m). From Eqs. (14.64), (14.53), and (14.148), assuming that the losses in the monopole are negligible, the rms intensity of its feed current is

$$I_0 = |\underline{I}_0| = \sqrt{\frac{P_{\text{rad}}}{R_{\text{rad}}}} = \sqrt{\frac{P_{\text{in}}}{R_{\text{rad}}}} = 5.23 \text{ A} \qquad (R_{\text{rad}} = 36.5\ \Omega). \tag{14.151}$$

feed current of a lossless antenna

By image theory, the quarter-wave monopole is transformed to a half-wave dipole (of length $l = 2h = \lambda/2$), leading to the equivalent system in Fig. 14.29(b). The surface wave travels a straight distance r directly from the dipole to the receive point, in the direction perpendicular to the dipole. Therefore, using Eq. (14.78), the rms intensity of the electric field vector of this wave at the point P is

$$|\underline{\mathbf{E}}_{\text{surf}}| = \frac{\eta I_0}{2\pi r} F(90^\circ) = \frac{\eta I_0}{2\pi r} = 0.63 \text{ mV/m}, \tag{14.152}$$

surface wave

where $\eta = \eta_0 = 377\ \Omega$, Eq. (9.23), and the magnitude of the characteristic radiation function of a half-wave dipole, $F(\theta)$, is given in Eqs. (14.140).

(b) The wave that, launched by the dipole at an oblique angle, θ, with respect to the dipole axis, is reflected from the ionosphere at this same angle, Fig. 14.29(b), in the same way as the reflection from the ground plane in Fig. 14.25 occurs, travels a substantially longer path than the surface wave. It can as well be analyzed by another application of the image theory, now to replace the ionospheric PEC plane by an image of the radiating dipole antenna (at the ground level), which is introduced at the height $2h_{\text{v}}$ with respect to the ground. The total path traveled, d, then equals the hypotenuse of the right-angled triangle $\triangle$AOP, and, having also in mind Eq. (14.142), the rms electric field intensity of the ionospheric wave at the point P comes out to be

$$|\underline{\mathbf{E}}_{\text{iono}}| = \frac{\eta I_0}{2\pi d} F(\theta) = \frac{\eta I_0}{2\pi d} F(\theta') = 0.52 \text{ mV/m}, \quad d = \sqrt{r^2 + 4h_{\text{v}}^2} = 538.5 \text{ km},$$
$$\theta = \arctan\frac{r}{2h_{\text{v}}} = 68.2^\circ, \quad \theta' = 180^\circ - \theta. \tag{14.153}$$

However, upon arrival to the ground, this wave bounces off it, such that the electric field vector of the resultant (incident plus reflected) ionospheric wave is entirely normal to the PEC surface [see Eq. (14.139)], and its magnitude is [Fig. 14.29(b)]

$$|\underline{\mathbf{E}}_{\text{iono}}|_{\text{tot}} = 2|\underline{\mathbf{E}}_{\text{iono}}| \sin\theta = 0.966 \text{ mV/m}. \tag{14.154}$$

resultant ionospheric wave

Note that the two PEC planes in Fig. 14.29(a) constitute a parallel-plate waveguide, with plate separation equal to the virtual height of the ionosphere, through which the ionospheric wave can propagate to extremely long distances – by bouncing back and forth between the plates, with incidence at the angle θ, in Fig. 14.29(b), for both ionospheric and ground reflections. Note also that, in addition to the single-reflection ray path in Fig. 14.29, with the wave bouncing once off the ionosphere, multiple-reflection zigzag paths between points O and P may be possible as well, like in Fig. 13.2, the total lengths of these paths being larger and larger and the received field intensities lower and lower as compared to the field in Eq. (14.154). Note finally that the analysis of ionospheric waves that bounce more than once off each of the PEC planes in Fig. 14.29(a), within the same horizontal distance r, can be performed by additional applications of the image theory for both planes.

Problems: 14.34 and 14.35; *Conceptual Questions* (on Companion Website): 14.38–14.41; *MATLAB Exercises* (on Companion Website).

14.11 MAGNETIC DIPOLE (SMALL LOOP) ANTENNA

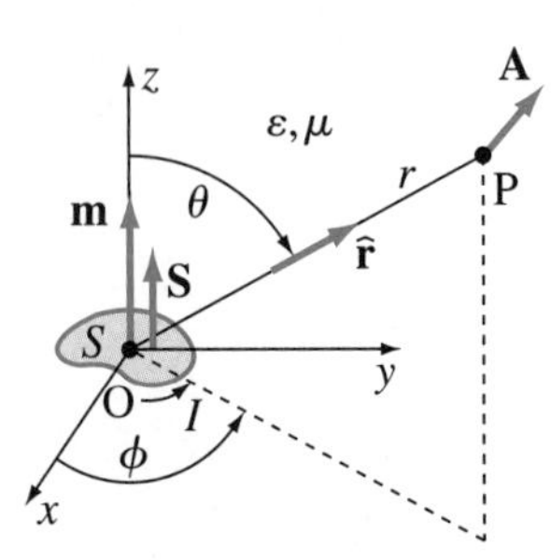

Figure 14.30 Magnetic dipole (small loop) antenna.

As another basic antenna type, of both theoretical and practical importance, consider a small wire loop antenna of arbitrary shape and surface area S, Fig. 14.30. The total length of the wire, l, is electrically small, i.e., small relative to the operating wavelength, λ, of the antenna (for the ambient medium), Eqs. (8.112) and (8.111). In practice, this qualification is usually expressed as $l < 0.1\lambda$. Because of the small electrical size, the complex rms current intensity of the antenna, $\underline{I}$, is constant (uniform) along the loop, and given in Eq. (14.1). In fact, the loop represents a high-frequency generalization of the magnetic dipole with a steady current in Fig. 4.30, and hence we refer to it as a magnetic dipole antenna, and base its analysis on the derivation and results from Section 4.11.

The magnetic moment of the loop, Eq. (4.134), is now complex,

magnetic moment of a small loop antenna

$$\boxed{\underline{\mathbf{m}} = \underline{I}\mathbf{S} = \underline{I}S\,\hat{\mathbf{z}} = \underline{m}\,\hat{\mathbf{z}},} \tag{14.155}$$

with $\mathbf{S}$ standing for the loop surface area vector, which is along the z-axis of a spherical coordinate system, in Fig. 14.30. Let the observation (field) point (P) be defined by (r, θ, ϕ) in this system. With no loss of generality, we assume a rectangular shape of the loop, as in Fig. 4.30 (as we shall see, the shape is irrelevant, as long as the loop is small). The loop side lengths, a and b, being electrically small, the current along each side is constituted by a single current element (Hertzian dipole, Fig. 14.1, without charge terminations), and the associated magnetic vector potential can be computed using Eq. (14.4). Of course, the restriction $r \gg a, b$ applies, as in both Figs. 14.1 and 4.30. Therefore, combining Eqs. (4.135) and (14.4), and then using practically identical approximations and transformations as in Eq. (14.6), the dynamic magnetic potential due to the pair of parallel current elements of length a in Fig. 4.30 is

$$\underline{\mathbf{A}}_{aa} = \frac{\mu\underline{I}\mathbf{a}}{4\pi}\left(\frac{\mathrm{e}^{-\mathrm{j}\beta r_1}}{r_1} - \frac{\mathrm{e}^{-\mathrm{j}\beta r_2}}{r_2}\right) \approx -\frac{\mu\underline{I}\mathbf{a}}{4\pi}\left.\frac{\mathrm{d}(\mathrm{e}^{-\mathrm{j}\beta R}/R)}{\mathrm{d}R}\right|_{R=r} \underbrace{(-\mathbf{b}\cdot\hat{\mathbf{r}})}_{\Delta R}$$
$$= -\frac{\mu\underline{I}\mathbf{a}(\mathbf{b}\cdot\hat{\mathbf{r}})(1+\mathrm{j}\beta r)\,\mathrm{e}^{-\mathrm{j}\beta r}}{4\pi r^2}, \tag{14.156}$$

with $\Delta R = r_2 - r_1 \approx \mathbf{d}\cdot\hat{\mathbf{r}}\ (\mathbf{d} = -\mathbf{b})$, as in Eq. (4.136). We see that this potential differs from that in Eq. (4.136) for the high-frequency factor $(1+\mathrm{j}\beta r)\,\mathrm{e}^{-\mathrm{j}\beta r}/r^2$, and the same is true for the contribution by the other pair of current elements, as well as for the total potential due to the loop antenna in Fig. 14.30, which, from Eq. (4.141), thus amounts to

magnetic potential of a magnetic dipole antenna

$$\boxed{\underline{\mathbf{A}} = \frac{\mu\underline{m}\sin\theta(1+\mathrm{j}\beta r)\,\mathrm{e}^{-\mathrm{j}\beta r}}{4\pi r^2}\,\hat{\boldsymbol{\phi}} \quad\longrightarrow\quad \underline{\mathbf{A}} \approx \frac{\mathrm{j}\mu\beta\underline{m}\sin\theta\,\mathrm{e}^{-\mathrm{j}\beta r}}{4\pi r}\,\hat{\boldsymbol{\phi}} = \underline{A}_\phi\,\hat{\boldsymbol{\phi}} \quad \text{for} \quad r \gg \lambda,} \tag{14.157}$$

where the second expression pertains to the far zone of the antenna. Indeed, $\underline{\mathbf{A}}$ does not depend on the shape of the loop, and not even on its surface area or current intensity separately, but on their product, the magnetic moment of the dipole, in Eq. (14.155).

The far electromagnetic field in Fig. 14.30 is evaluated from the far-zone $\underline{\mathbf{A}}$ of the loop carrying out the general steps in the radiation analysis of an arbitrary antenna, Eqs. (14.25), which give

$$\boxed{\begin{aligned}\underline{\mathbf{E}} &= -\mathrm{j}\omega\underline{A}_\phi\,\hat{\boldsymbol{\phi}} = \frac{\eta\beta^2\underline{m}\,\mathrm{e}^{-\mathrm{j}\beta r}\sin\theta}{4\pi r}\,\hat{\boldsymbol{\phi}} = \underline{E}_\phi\,\hat{\boldsymbol{\phi}}, \quad \underline{\mathbf{H}} = -\frac{\underline{E}_\phi}{\eta}\,\hat{\boldsymbol{\theta}} \\ &= -\frac{\beta^2\underline{m}\,\mathrm{e}^{-\mathrm{j}\beta r}\sin\theta}{4\pi r}\,\hat{\boldsymbol{\theta}} = \underline{H}_\theta\,\hat{\boldsymbol{\theta}},\end{aligned}} \tag{14.158}$$

far field of a small loop antenna

where the use is also made of the first relationship in Eqs. (14.9). Comparing these field expressions with the far-field expressions in Eqs. (14.17) for a Hertzian dipole, Fig. 14.1, we realize that they are almost the same, the difference being only in the multiplicative constants, but with the roles of $\underline{\mathbf{E}}$ and $\underline{\mathbf{H}}$ reversed. Based on such a reversed equivalency, where $\underline{\mathbf{E}}$ and $\underline{\mathbf{H}}$ of an electric source become $\underline{\mathbf{H}}$ and $-\underline{\mathbf{E}}$, respectively, of a magnetic source, we say that a Hertzian dipole and small loop antenna, i.e., dynamic electric and magnetic dipoles, are dual electromagnetic sources, that is, duals of each other. In electromagnetic theory, this is referred to as the general duality principle for electromagnetic fields due to electric and magnetic sources. Note that a similar duality applies for static dipoles as well [see Eqs. (1.117) and (4.142) and Fig. 4.31(a) and Fig. 4.31(b)].

Electrically small loops (magnetic dipoles) are extensively used as receiving antennas[12] in a multitude of applications, including AM, short-wave, and FM (see Table 9.1) broadcast receivers, pagers, direction-finding receivers, and radio navigation systems. They also find wide application as magnetic probes for field measurements and for excitation or reception of wave modes in waveguides and cavity resonators (see Fig. 13.11). In addition, the concept of a dynamic magnetic dipole can effectively be employed in electromagnetic interference (EMI) evaluations. Namely, in terms of electromagnetic disturbances it produces, practically every electrical or electronic device whose dimensions are smaller than the wavelength of the highest relevant component in the frequency spectrum of the disturbance can be approximated by a combination of a single equivalent Hertzian dipole and single equivalent magnetic dipole antenna. Finally, a special theoretical arrangement of the two dipoles, dynamic electric and magnetic dipoles, termed a Huygens radiator (or source) is the basis for the analysis of aperture antennas, such as slot antennas, Fig. 13.13, open-ended waveguide antennas, Fig. 13.14(a), and horn antennas, Fig. 13.14(b).

Example 14.21 Radiation Function and Resistance of a Small Loop Antenna

(a) Find the expression for the characteristic radiation function of a magnetic dipole (small loop) antenna, in Fig. 14.30. (b) What is the directivity of the antenna? (c) Discuss the radiation pattern of the antenna and its far-field duality with a Hertzian dipole. (d) Determine the radiation resistance of the magnetic dipole.

Solution

(a)–(c) From Eqs. (14.158) and (14.78), the characteristic radiation function of the magnetic dipole comes out to be

$$\boxed{\underline{\mathbf{F}}(\theta) = -\frac{\mathrm{j}\beta^2 S}{2}\sin\theta\,\hat{\boldsymbol{\phi}} \qquad [f(\theta) = \sin\theta].} \tag{14.159}$$

small loop, radiation function

[12]The first small loop antenna is constructed and used (in experiments) as a receiver by Hertz in 1887.

Of course, the normalized field pattern (f), Eq. (14.80), is the same as for a Hertzian dipole, Eq. (14.81), and thus, by virtue of Eq. (14.107), it has the same directivity, $D = 1.5$, as in Eq. (14.113). However, the two dipoles have mutually orthogonal polarizations: the far electric field vector of the magnetic dipole has only a ϕ-component, while the far magnetic field vector is θ-directed, which is just opposite to the situation in Fig. 14.2 for the electric dipole. The far-field duality of the two antennas is illustrated in Fig. 14.31, where we present in parallel their polar normalized patterns, along with the field polarizations, in a plane $\phi = \text{const}$, which is an E-plane for the electric dipole [see also Fig. 14.12(b)], while an H-plane for the magnetic dipole.

(d) On the other hand, the radiation resistance of a magnetic dipole antenna is quite different from its electric dual, because of different multiplicative constants in expressions for the characteristic radiation function for the two antennas. As the antenna radiated power in Eq. (14.101) is proportional to an integral over the full solid angle of the magnitude of the radiation function squared, $|\underline{\mathbf{F}}|^2$, and given that $|\underline{\mathbf{F}}|$ of the magnetic dipole can be obtained from $|\underline{\mathbf{F}}|$ of the electric dipole by replacing βl with $\beta^2 S$, we accordingly replace $(\beta l)^2$ in Eq. (14.55) by $\left(\beta^2 S\right)^2$ to get R_{rad} of the loop in free space,

small loop, radiation resistance

$$R_{\text{rad}} = 20\left(\beta^2 S\right)^2\ \Omega = 31{,}171\left(\frac{S}{\lambda^2}\right)^2\ \Omega. \tag{14.160}$$

Since $S \ll \lambda^2$, this is a very small resistance.

Problems: 14.36–14.38; *Conceptual Questions* (on Companion Website): 14.42–14.44; *MATLAB Exercises* (on Companion Website).

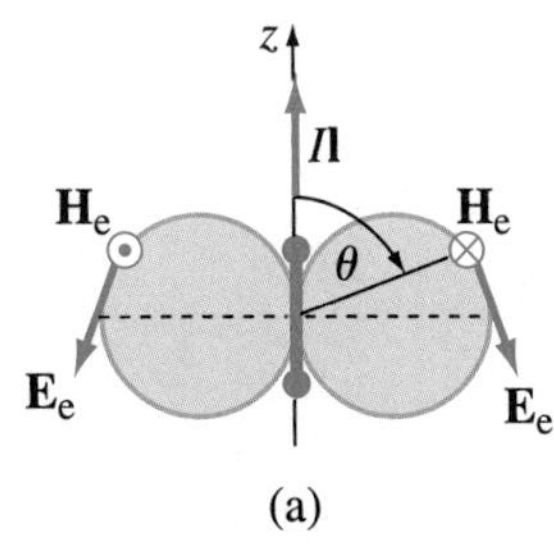

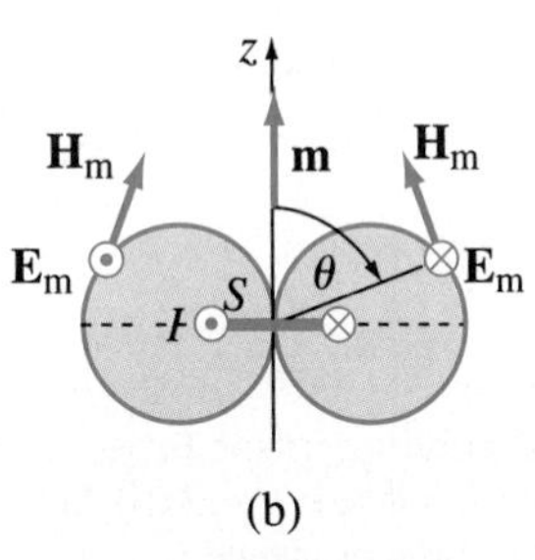

Figure 14.31 Duality in the far field (normalized field pattern and polarization in a plane $\phi = \text{const}$) of an electric (Hertzian) dipole (a) and magnetic (small loop) dipole (b), in Figs. 14.1 and 14.30, respectively; for Example 14.21.

14.12 THEORY OF RECEIVING ANTENNAS

So far, we have studied antennas in the transmitting mode of operation, i.e., when they radiate electromagnetic waves. Although the antenna operation in the receiving mode is just opposite, namely, an antenna is used to capture the power from an incident electromagnetic wave, and deliver it to a terminating device (load), both circuit (impedance) and directional properties of a receiving antenna are directly related to its properties when transmitting. In this section, we derive these relationships, and build a general theory of receiving antennas, which quantifies how good a receiver a given antenna is, that is, how capable it is to receive the power from an incoming wave for a certain direction of incidence and wave polarization, and to pass it to the load.

Consider an arbitrary receiving antenna illuminated by a uniform plane time-harmonic wave, of angular frequency ω, in a homogeneous and lossless medium, of permittivity ε and permeability μ ($\sigma = 0$). With reference to Fig. 14.32(a), the propagation unit vector of the wave, $\hat{\mathbf{n}}$, is directed toward the global coordinate origin, O, which is on or close to the antenna. As in Eq. (9.69), the complex rms electric field intensity vector of the wave at an arbitrary point defined by the position vector $\mathbf{r}$ with respect to the origin, is

$$\underline{\mathbf{E}}_{\text{i}}(\mathbf{r}) = \underline{\mathbf{E}}_0\, \text{e}^{-\text{j}\beta \mathbf{r}\cdot\hat{\mathbf{n}}}, \tag{14.161}$$

where $\underline{\mathbf{E}}_0 = \underline{\mathbf{E}}_{\text{i}}(0)$ (the field vector at the origin), and β is the phase coefficient of the wave, Eq. (8.111). This wave is, most frequently, originated by another (transmitting) antenna in a wireless link, which is far away from the receiving antenna, Eq. (14.16), so that the uniform-plane-wave approximation of the actual nonuniform spherical wave radiated by the other antenna, Fig. 9.1, applies. In addition, let the receiving antenna be terminated in a load of complex impedance $\underline{Z}_{\rm L}$. With respect to its output terminals (note that this same pair of terminals is used as input terminals in the transmitting mode), and to the load, the antenna can be replaced by the Thévenin equivalent generator, shown in Fig. 14.32(b). Our goal is to determine the parameters of this generator.

By definition, i.e., by Thévenin's theorem, the complex internal impedance of the generator, $\underline{Z}_{\rm T}$, equals the complex input impedance of the antenna in Fig. 14.32(a) (with the incident plane wave "turned off"). This is simply the impedance $\underline{Z}_{\rm A}$ in Eq. (14.63) and Fig. 14.10. Hence,

$$\boxed{\underline{Z}_{\rm T} = \underline{Z}_{\rm A}.} \qquad (14.162)$$

Thévenin impedance of a receiving antenna

So, the antenna presents to its input/output terminals the same complex impedance in the transmitting and receiving modes. However, while the transmitting antenna can be replaced by just a (passive) load, of impedance $\underline{Z}_{\rm A}$, the equivalent of the receiving antenna includes an electromotive force as well (active circuit).

This emf, and its complex rms value, $\underline{\mathcal{E}}_{\rm T}$, can, by definition, be found as the open-circuit voltage of the antenna in Fig. 14.32(a), that is, by computing the voltage across the open terminals (with the load removed) of the antenna (excited by the incoming wave). To simplify the analysis, let us find $\underline{\mathcal{E}}_{\rm T}$ for a small loop of surface area vector $\mathbf{S} = S\hat{\mathbf{z}}$ as the receiving antenna, as depicted in Fig. 14.33(a). The complex rms emf induced in the loop ($\underline{\mathcal{E}}_{\rm ind}$) is determined by Faraday's law of electromagnetic induction, and in particular its expression for an electrically small loop in Eq. (13.119), in terms of the complex rms magnetic field intensity vector at the loop location, that is, at the coordinate origin. This vector, $\underline{\mathbf{H}}_0$, is related to $\underline{\mathbf{E}}_0$ through Eq. (9.70), since the excitation is by a TEM wave. Using Eqs. (14.9) and (14.12), the relationship $\hat{\mathbf{n}} = -\hat{\mathbf{r}}$ between the unit vectors in Fig. 14.33(a), and cyclic-permutation equality for the scalar triple product of three vectors, we can write

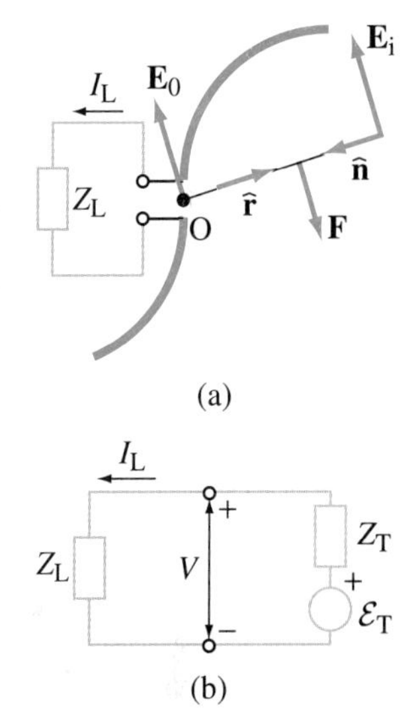

Figure 14.32 (a) Receiving antenna in the field of a uniform plane time-harmonic wave and (b) its Thévenin equivalent representation.

$$\underline{\mathcal{E}}_{\rm ind} = -{\rm j}\omega\mu\underline{\mathbf{H}}_0\cdot\mathbf{S} = -\frac{{\rm j}\omega\mu}{\eta}(\hat{\mathbf{n}}\times\underline{\mathbf{E}}_0)\cdot\mathbf{S} = {\rm j}\beta S(\hat{\mathbf{r}}\times\underline{\mathbf{E}}_0)\cdot\hat{\mathbf{z}} = {\rm j}\beta S(\hat{\mathbf{z}}\times\hat{\mathbf{r}})\cdot\underline{\mathbf{E}}_0$$

$$= {\rm j}\beta S\sin\theta\,\hat{\boldsymbol{\phi}}\cdot\underline{\mathbf{E}}_0 = -\frac{2}{\beta}\underline{\mathbf{E}}_0\cdot\underbrace{\left(-\frac{{\rm j}\beta^2S}{2}\sin\theta\,\hat{\boldsymbol{\phi}}\right)}_{\underline{\mathbf{F}}_{\rm loop}} \quad (\hat{\mathbf{r}} = -\hat{\mathbf{n}}), \qquad (14.163)$$

where $\underline{\mathbf{F}}_{\rm loop}$ is the characteristic radiation function of the loop antenna, Eq. (14.159), for the direction of the incident-wave arrival. So, having in mind Eq. (6.62) and the equivalent circuit in Fig. 14.32(b), and expressing β in terms of the corresponding wavelength, λ, by means of Eq. (8.111), the open-circuit voltage of the antenna equals

$$\boxed{\underline{V}_{\rm oc} = \underline{\mathcal{E}}_{\rm T} = -\underline{\mathcal{E}}_{\rm ind} = \frac{2}{\beta}\underline{\mathbf{E}}_0\cdot\underline{\mathbf{F}}_{\rm loop} = \frac{\lambda}{\pi}\underline{\mathbf{E}}_0\cdot\underline{\mathbf{F}}.} \qquad (14.164)$$

open-circuit voltage of a receiving antenna

Alternatively, let us obtain the same result assuming that the incident plane wave is being received by a Hertzian dipole of length vector $\mathbf{l} = l\hat{\mathbf{z}}$, shown in Fig. 14.33(b). Recalling Fig. 6.6 and Eq. (6.32), we realize that the incident field in Eq. (14.161) induces in an element $d\mathbf{l}$ of the wire of the dipole an emf equal to $d\underline{\mathcal{E}}_{\rm ind} = \underline{\mathbf{E}}_{\rm i}(\mathbf{r})\cdot d\mathbf{l}$, and the element can be replaced by an equivalent elementary voltage generator

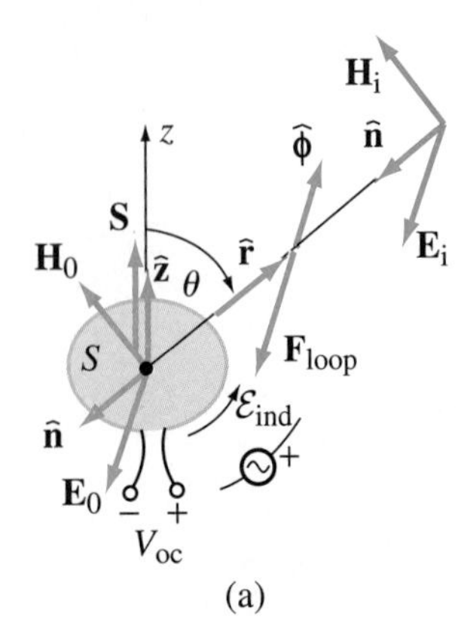

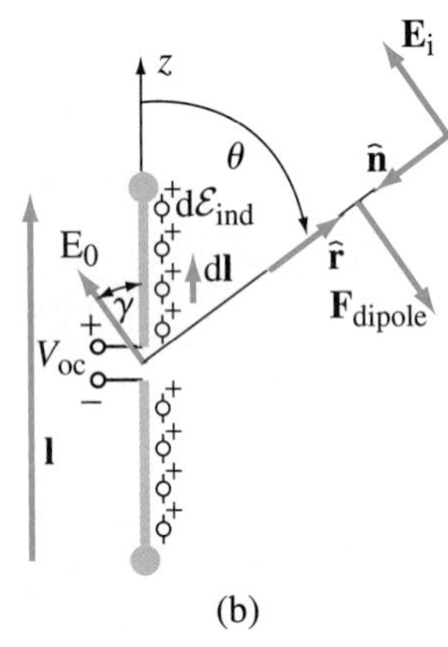

Figure 14.33 Computing the induced emf and open-circuit voltage of the receiving antenna in Fig. 14.32(a), if it is (a) a small loop and (b) a Hertzian dipole.

whose emf is $\mathrm{d}\underline{\mathcal{E}}_{\mathrm{ind}}$, as indicated in Fig. 14.33(b). As the dipole is electrically short, $\underline{\mathbf{E}}_{\mathrm{i}}(\mathbf{r}) \approx \underline{\mathbf{E}}_0$ along the entire antenna, so we have for the total induced emf $\underline{\mathcal{E}}_{\mathrm{ind}}$ of the antenna

$$\mathrm{d}\underline{\mathcal{E}}_{\mathrm{ind}} = \underline{\mathbf{E}}_0 \cdot \mathrm{d}\mathbf{l} \quad \longrightarrow \quad \underline{\mathcal{E}}_{\mathrm{ind}} = \int_l \mathrm{d}\underline{\mathcal{E}}_{\mathrm{ind}} = \underline{\mathbf{E}}_0 \cdot \mathbf{l} = \underline{E}_0 l \cos\gamma = \underline{E}_0 l \sin\theta$$
$$= \frac{2}{\beta}\underline{E}_0 \underbrace{\frac{\beta l}{2}\sin\theta}_{\underline{F}_{\mathrm{dipole}}} = -\frac{2}{\beta}(-\underline{E}_0\underline{F}_{\mathrm{dipole}}) = -\frac{2}{\beta}\underline{\mathbf{E}}_0 \cdot \underline{\mathbf{F}}_{\mathrm{dipole}} \qquad (\gamma = 90^\circ - \theta), \tag{14.165}$$

with $\underline{\mathbf{F}}_{\mathrm{dipole}}$ being the characteristic radiation function of the dipole, Eq. (14.79), again for the direction of the wave incidence [note that $\underline{\mathbf{E}}_0$ and $\underline{\mathbf{F}}_{\mathrm{dipole}}$ are oppositely directed vectors, in Fig. 14.33(b)]. This, obviously, leads to the same expression for $\underline{V}_{\mathrm{oc}}$ in Eq. (14.164), and this expression holds for an arbitrary receiving antenna (wire, surface, or volume).

Let us emphasize that $\underline{\mathbf{E}}_0$ in Eq. (14.164) is the electric field vector of the incident plane wave at the coordinate origin (reference point), which is usually adopted at the antenna terminals, and $\underline{\mathbf{F}}$ is the characteristic radiation function that the antenna would have if transmitting in the direction of the wave incidence, i.e., direction defined by $\hat{\mathbf{r}} = -\hat{\mathbf{n}}$, as indicated in Fig. 14.32(a). This direction is usually given by incident angles θ_{i} and ϕ_{i} in a spherical coordinate system (attached to the antenna), and hence $\underline{\mathbf{F}} = \underline{\mathbf{F}}(\theta_{\mathrm{i}}, \phi_{\mathrm{i}})$. So, the field pattern of a receiving antenna, showing how well it captures the incident signal in different directions in 3-D space, is identical to the radiation pattern of the antenna when in the transmitting mode, Fig. 14.11, and the same is true for power patterns, Eq. (14.84). In short, the transmit and receive patterns of an arbitrary antenna are identical, which essentially is a consequence of the linearity of the system in Fig. 14.32(a) and its reciprocity.[13] Since, in addition, the Thévenin impedance in the receive equivalent circuit of the antenna, in Fig. 14.32(b), is identical, by means of Eq. (14.162), to that in the transmit circuit, in Fig. 14.10(b), we conclude that the basic properties of a receiving antenna are fully determined by its properties as a transmitting antenna.

In Eq. (14.164), the polarization of both the receiving antenna and the incoming wave can be arbitrary (of linear, circular, or elliptical type, and with arbitrary parameters of the polarization ellipse), so we can have an arbitrary combination of polarizations of vectors $\underline{\mathbf{E}}_0$ and $\underline{\mathbf{F}}$. As already mentioned, most often the polarization of $\underline{\mathbf{E}}_0$ comes from a transmitting antenna launching the wave in a communication link. Hence, the dot product in Eq. (14.164) determines the polarization match (or mismatch) between the two antennas (transmitting and receiving) at the two ends of the link. If both antennas are linearly polarized, we can write

linear polarizations of the antenna and wave

$$\boxed{\underline{V}_{\mathrm{oc}} = \frac{\lambda}{\pi}\underline{E}_0\underline{F}\cos\alpha,} \tag{14.166}$$

where $\underline{E}_0$ and $\underline{F}$ are complex magnitudes, including the phase terms, of the two vectors in Eq. (14.164), and α is the angle between them. We see that the rms voltage

[13] Reciprocity of electromagnetic fields states that a response to a source remains the same if the source and observation locations are interchanged, as long as the electromagnetic media in the system are linear and isotropic. This can be proved using Maxwell's equations, and is one of the versions of what is known as the electromagnetic reciprocity theorem (first derived by H. A. Lorentz).

$|\underline{V}_{\rm oc}|$ is maximum for $\alpha = 0$ or $180°$, that is, for a (perfect) match of the two linear polarizations,

$$\alpha = 0 \ \text{or} \ 180° \quad \longrightarrow \quad |\underline{V}_{\rm oc}|_{\rm max} = \frac{\lambda}{\pi}\,|\underline{E}_0||\underline{F}| = \frac{\lambda}{\pi}\,|\underline{\mathbf{E}}_0||\underline{\mathbf{F}}|. \qquad (14.167)$$

maximum voltage – co-polarized case

An example is a system of two parallel distant wire dipole antennas communicating to each other. On the other hand, if the two distant dipoles are mutually orthogonal (i.e., cross-polarized), $\alpha = 90°$, so that the received voltage is zero.

Of course, we are usually interested in the antenna performance when it is loaded (and not open-circuited) in the receiving mode of operation. However, the analysis of the loaded antenna is now straightforward. Once the open-circuit voltage of a receiving antenna, $\underline{V}_{\rm oc} = \underline{\mathcal{E}}_{\rm T}$, is computed using Eq. (14.164) or (14.166), the current of the antenna load, $\underline{I}_{\rm L}$, in Fig. 14.32(a) is determined from the equivalent circuit in Fig. 14.32(b) and Eq. (14.162),

$$\underline{I}_{\rm L} = \frac{\underline{\mathcal{E}}_{\rm T}}{\underline{Z}_{\rm L} + \underline{Z}_{\rm T}} = \frac{\underline{V}_{\rm oc}}{\underline{Z}_{\rm L} + \underline{Z}_{\rm A}}. \qquad (14.168)$$

load current

Knowing the load (received) current, we can also find the complex and time-average powers delivered to the load, and other quantities of interest for the particular analysis or design.

Example 14.22 Wireless Link with Two Nonaligned Half-Wave Dipoles

Consider the wireless link in free space in Fig. 14.34(a). Both antennas are half-wave wire dipoles, and the distance between the transmit and receive ends is $r = 200$ m. The transmitting antenna, operating at a frequency of $f = 300$ MHz, is placed in the xz-plane at an angle of $\gamma_1 = 45°$ with respect to the z-axis, while the receiving antenna lies in the $x'y'$-plane, where it makes an angle of $\gamma_2 = 60°$ with the x'-axis. If the input power of the transmitting antenna is $P_{\rm in} = 10$ W, find the magnitude of the open-circuit voltage of the receiving antenna.

Solution Using Eq. (9.67), the operating free-space wavelength of the wireless link amounts to $\lambda = \lambda_0 = c_0/f = 1$ m, and the electrical distance between the antennas to $r/\lambda = 200 \gg 1$. This means that the receive end of the link is in the far zone of the transmitting dipole, as well as that the radiated electromagnetic wave can locally be considered as a uniform plane wave when evaluating the open-circuit voltage of the receiving dipole, $\underline{V}_{\rm oc}$. Therefore, $\underline{V}_{\rm oc}$ is given by Eq. (14.164) or (14.166), with $\underline{\mathbf{E}}_0 = \underline{\mathbf{E}}_{\rm t}$ being the far electric field intensity vector of the transmitting antenna computed at point O′ in Fig. 14.34(a), and $\underline{\mathbf{F}} = \underline{\mathbf{F}}_{\rm r}$ the characteristic radiation function of the receiving antenna that it would have if transmitting in the direction toward point O. With this notation,

$$\underline{V}_{\rm oc} = \frac{\lambda}{\pi}\,\underline{\mathbf{E}}_{\rm t} \cdot \underline{\mathbf{F}}_{\rm r} = \frac{\lambda}{\pi}\,\underline{E}_{\rm t}\underline{F}_{\rm r}\cos\alpha, \qquad (14.169)$$

where α is the angle between vectors $\underline{\mathbf{E}}_{\rm t}$ and $\underline{\mathbf{F}}_{\rm r}$.

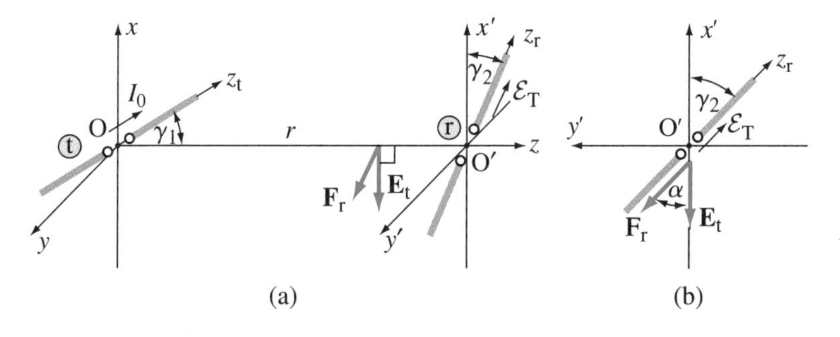

Figure 14.34 Wireless link with two nonaligned half-wave dipole antennas: (a) geometry of the system and (b) detail at the receive end, showing the far electric field vector of the transmitting antenna ($\underline{\mathbf{E}}_{\rm t}$) and characteristic radiation function of the receiving antenna ($\underline{\mathbf{F}}_{\rm r}$), in Eq. (14.169); for Example 14.22.

In a local spherical coordinate system with the z_t-axis along the transmitting dipole, $\underline{\mathbf{E}}_t$ has only a θ_t-component, where θ_t coincides with γ_1, so it turns out to be pointing in the negative x' direction, $\underline{\mathbf{E}}_t = \underline{E}_t(-\hat{\mathbf{x}}')$, in Fig. 14.34(a). By the same token, the dipole characteristic radiation function, Eq. (14.128), is determined by the local θ_t angle, and, having in mind Eqs. (14.78), (14.151), and (14.130), we can write

$$\underline{E}_t = \frac{j\eta}{2\pi} \underline{I}_0 \frac{e^{-j\beta r}}{r} F(\theta_t) \quad (\theta_t = \gamma_1), \quad I_0 = \sqrt{\frac{P_{in}}{R_{rad}}} = 0.37 \text{ A} \quad (R_{rad} = 73 \ \Omega), \quad (14.170)$$

where $F(\theta)$ is that in Eqs. (14.140).

On the other side, if we attach a local spherical coordinate system whose z_r-axis is along the receiving dipole, the local θ_r angle between the dipole and the direction toward the transmit end is a right angle, and $\underline{\mathbf{F}}_r$ is θ_r-directed, in parallel to the dipole, as shown in Fig. 14.34(b). So, from the fact that $\theta_r = 90°$ and the direction (polarization) of $\underline{\mathbf{F}}_r$ in Fig. 14.34(b), we have

$$\underline{F}_r = F(\theta_r) = F(90°) = 1 \quad \text{and} \quad \alpha = \gamma_2, \quad (14.171)$$

respectively.

Finally, combining Eqs. (14.169)–(14.171), the magnitude of the received voltage comes out to be

link with nonaligned λ/2 dipoles

$$|\underline{V}_{oc}| = \frac{\eta\lambda I_0}{2\pi^2 r} \frac{\cos\left(\frac{\pi}{2}\cos\gamma_1\right)}{\sin\gamma_1} \cos\gamma_2 = 11 \text{ mV}, \quad (14.172)$$

where $\eta = \eta_0 = 377 \ \Omega$, Eq. (9.23), $F(\gamma_1) = 0.628$, and $\cos\gamma_2 = 0.5$. Of course, this voltage can be maximized, for the two antennas used and the given length of the link, input power, and frequency, by orienting the dipoles such that $\gamma_1 = 90°$ and $\gamma_2 = 0$. The former condition implies pointing the transmitting antenna for its maximum directivity, i.e., orienting it such that the maximum of its radiation pattern is in the direction toward the receiving antenna, whereas the latter one indicates polarization match between the two dipoles. The maximum voltage would then be $|\underline{V}_{oc}|_{max} = \eta\lambda I_0/(2\pi^2 r) = 35.3$ mV. Note that the receiving dipole is already pointed in Fig. 14.34 for its maximum directivity [$\theta_r = 90°$ in Eq. (14.171)]. Note also that if the initial phase of the received voltage is needed, the complete complex expression for the field $\underline{E}_t$ in Eq. (14.170) is used to compute the complex voltage $\underline{V}_{oc}$ in Eq. (14.169).

Example 14.23 Reception of a Circularly Polarized Wave by a Wire Dipole

At the transmit end of a wireless link, two half-wave wire dipole antennas positioned along the x- and y-axis, as shown in Fig. 14.35(a), are fed with currents of complex rms intensities $\underline{I}_{01} = 3$ A and $\underline{I}_{02} = j3$ A, respectively, and the same frequency $f = 500$ MHz. At the receive end, $r = 30$ m away from the crossed dipoles, another half-wave dipole lies in the plane $x'y'$ and makes an angle γ with the x'-axis. Compute the rms voltage across the open terminals of the receiving dipole.

Solution The operating wavelength is $\lambda = c_0/f = 60$ cm (note that $r \gg \lambda$). The electric field intensity vectors radiated by dipoles 1 and 2, $\underline{\mathbf{E}}_{t1}$ and $\underline{\mathbf{E}}_{t2}$, have only θ-components in respective local spherical coordinate systems attached to dipoles (see Fig. 14.34), and at the receive location (point O′) in Fig. 14.35(a), each of them comes out to be parallel to the source dipole,

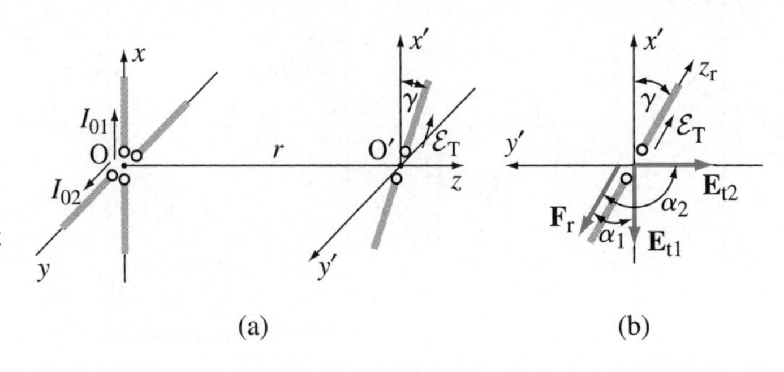

Figure 14.35 Wireless communication system with three half-wave dipole antennas: (a) two crossed dipoles radiating a circularly polarized electromagnetic wave ($\underline{I}_{02} = j\underline{I}_{01}$) and a receiving dipole in a transversal plane of the wave, and (b) detail at the receive end, showing the two incident electric field vectors and characteristic radiation function of the receiving antenna, in Eqs. (14.173); for Example 14.23.

as shown in Fig. 14.35(b). As both incident waves are approaching from the same direction, the characteristic radiation function of the receiving antenna (dipole 3), $\underline{\mathbf{F}}_\mathrm{r}$, defined for the supposed transmitting mode of operation in that direction (toward point O) is the same for the two waves, parallel to the dipole, Fig. 14.35(b). The angles that $\underline{\mathbf{E}}_\mathrm{t1}$ and $\underline{\mathbf{E}}_\mathrm{t2}$ make with $\underline{\mathbf{F}}_\mathrm{r}$ being $\alpha_1 = \gamma$ and $\alpha_2 = 90° + \gamma$, respectively, Eq. (14.166) tells us that the voltages received by dipole 3 due to the radiation of dipoles 1 and 2 are

$$\underline{V}_\mathrm{oc1} = \frac{\lambda}{\pi}\underline{E}_\mathrm{t1}\underline{F}_\mathrm{r}\cos\gamma, \quad \underline{V}_\mathrm{oc2} = \frac{\lambda}{\pi}\underline{E}_\mathrm{t2}\underline{F}_\mathrm{r}\cos(90° + \gamma) = -\frac{\lambda}{\pi}\underline{E}_\mathrm{t2}\underline{F}_\mathrm{r}\sin\gamma. \quad (14.173)$$

Given that the feed currents of the transmitting dipoles have the same magnitudes and are in time-phase quadrature ($\underline{I}_{02}$ is j times $\underline{I}_{01}$), from Eq. (14.78) the same relationship holds for fields $\underline{E}_\mathrm{t1}$ and $\underline{E}_\mathrm{t2}$,

$$\underline{I}_{02} = \mathrm{j}\underline{I}_{01} \quad \longrightarrow \quad \underline{E}_\mathrm{t2} = \mathrm{j}\underline{E}_\mathrm{t1}, \quad \text{where} \quad \underline{E}_\mathrm{t1} = \frac{\mathrm{j}\eta}{2\pi}\underline{I}_{01}\frac{\mathrm{e}^{-\mathrm{j}\beta r}}{r}\underline{F}_\mathrm{t}. \quad (14.174)$$

In addition, we realize that all three antennas are pointed, in Fig. 14.35, for their maximum directivities, namely, the direction OO′ or O′O falls in the equatorial plane ($\theta = 90°$) of each of the dipoles. In other words, the dipole characteristic radiation functions on both sides of the link are maximum, Eq. (14.129),

$$\underline{F}_\mathrm{t} = \underline{F}_\mathrm{r} = F(90°) = 1. \quad (14.175)$$

Hence, using Eqs. (14.173)–(14.175), the total received rms voltage is found as

$$\underline{V}_\mathrm{oc} = \underline{V}_\mathrm{oc1} + \underline{V}_\mathrm{oc2} = \frac{\lambda}{\pi}\underline{E}_\mathrm{t1}\left(\cos\gamma - \mathrm{j}\sin\gamma\right) \quad \longrightarrow \quad |\underline{V}_\mathrm{oc}| = \frac{\lambda}{\pi}|\underline{E}_\mathrm{t1}|\sqrt{\cos^2\gamma + \sin^2\gamma}$$
$$= \frac{\lambda}{\pi}|\underline{E}_\mathrm{t1}| = \frac{\eta\lambda|\underline{I}_{01}|}{2\pi^2 r} = 1.15\ \mathrm{V}. \quad (14.176)$$

reception of a CP wave by a wire diplole

We see that this voltage is the same for all angles γ in Fig. 14.35, which is expected, since the two crossed dipoles constitute a circularly polarized (CP) antenna system, and the resultant radiated field $\underline{\mathbf{E}}_\mathrm{t} = \underline{\mathbf{E}}_\mathrm{t1} + \underline{\mathbf{E}}_\mathrm{t2}$ represents a CP wave – the two mutually orthogonal linearly polarized field vectors are of equal amplitudes and 90° out of phase with respect to each other [see Eqs. (9.185) and Fig. 9.19]. Therefore, the reception at the point O′ of the vector $\mathbf{E}_\mathrm{t}(t)$ as it rotates with its tip describing a circle during the course of time (t) in the $x'y'$-plane does not depend on the orientation of a dipole in this plane.

Example 14.24 Reception of an Elliptically Polarized Wave by a Wire Dipole

Repeat the previous example but for $\underline{I}_{02} = \mathrm{j}9$ A ($\underline{I}_{01} = 3$ A). What γ makes the received voltage maximum, and what is the maximum voltage?

Solution With this change, Eqs. (14.174) and (14.176) now become

$$\underline{I}_{02} = 3\mathrm{j}\underline{I}_{01} \quad \longrightarrow \quad \underline{E}_\mathrm{t2} = 3\mathrm{j}\underline{E}_\mathrm{t1} \quad \longrightarrow \quad \underline{V}_\mathrm{oc} = \frac{\lambda}{\pi}\underline{E}_\mathrm{t1}\left(\cos\gamma - \mathrm{j}3\sin\gamma\right)$$
$$\longrightarrow \quad |\underline{V}_\mathrm{oc}| = \frac{\lambda}{\pi}|\underline{E}_\mathrm{t1}|\sqrt{\cos^2\gamma + 9\sin^2\gamma} = \frac{\lambda}{\pi}|\underline{E}_\mathrm{t1}|\sqrt{1 + 8\sin^2\gamma}, \quad (14.177)$$

reception of an EP wave

and the maximum of the received voltage magnitude is

$$|\underline{V}_\mathrm{oc}|_\mathrm{max} = \frac{\lambda}{\pi}|\underline{E}_\mathrm{t1}|\sqrt{9} = 3.44\ \mathrm{V} \quad \text{for} \quad \gamma = 90° \ \text{or} \ -90°. \quad (14.178)$$

Note that the only difference between the cases $\gamma = 90°$ and $\gamma = -90°$ (or $\gamma = 270°$) is a reversed reference orientation of the induced emf and received voltage along the y'-axis in Fig. 14.35, which, of course, does not affect the result for the voltage magnitude. So, the reception is maximum when the receiving dipole is aligned with dipole 2 in Fig. 14.35(a). This, again, is expected, given that the two crossed dipoles now radiate an elliptically polarized

(EP) resultant electromagnetic wave, as in Eqs. (9.187) and Fig. 9.20 [the two orthogonal component vectors have different amplitudes in Eqs. (14.177)]. Namely, a linearly polarized wire dipole at point O′ receives the maximum out of the incoming EP wave when placed along the major axis of its polarization ellipse (for the electric field vector) in the $x'y'$-plane.

Problems: 14.39–14.49; *Conceptual Questions* (on Companion Website): 14.45–14.47; *MATLAB Exercises* (on Companion Website).

14.13 ANTENNA EFFECTIVE APERTURE

One of the basic parameters of receiving antennas is the so-called effective aperture of an antenna in a given direction. With reference to Fig. 14.32(a), it is defined as the ratio of the power received by the load at the antenna output terminals and the surface power density of the incoming electromagnetic wave. In other words, it represents a portion of the incident wavefront from which the antenna, effectively, extracts power and delivers it to the load, and can thus be thought of as the effective (equivalent) collecting area of the antenna. In this section, we derive, as it is customary, the expressions for the evaluation of this important quantity for an ideal case when the load is impedance-matched (for the maximum power transfer) to the antenna and the antenna is polarization-matched to the wave.

From the equivalent circuit of the antenna in Fig. 14.32(b), and Eqs. (12.58), (14.168), and (14.63), the time-average power delivered to the load, whose impedance is expressed as $\underline{Z}_{\rm L} = R_{\rm L} + {\rm j}X_{\rm L}$, is

$$P_{\rm L} = R_{\rm L}|\underline{I}_{\rm L}|^2 = \frac{R_{\rm L}|\underline{\mathcal{E}}_{\rm T}|^2}{|\underline{Z}_{\rm L} + \underline{Z}_{\rm A}|^2} = \frac{R_{\rm L}|\underline{\mathcal{E}}_{\rm T}|^2}{(R_{\rm L} + R_{\rm A})^2 + (X_{\rm L} + X_{\rm A})^2}. \qquad (14.179)$$

For given antenna parameters and incident wave, this power is maximum under the conjugate matching condition, namely, when the load impedance is the complex conjugate of the antenna impedance, which will be proved in the example. With the conjugate-matched load, the expression for $(P_{\rm L})_{\rm max}$ acquires the following form:

conjugate-matched load – maximum power transfer

$$R_{\rm L} = R_{\rm A} \quad \text{and} \quad X_{\rm L} = -X_{\rm A} \quad (\underline{Z}_{\rm L} = \underline{Z}_{\rm A}^*) \quad \longrightarrow \quad (P_{\rm L})_{\rm max} = \frac{|\underline{\mathcal{E}}_{\rm T}|^2}{4R_{\rm A}}. \qquad (14.180)$$

Furthermore, we assume that the antenna is polarization-matched (co-polarized) to the wave, so that its open-circuit voltage, $\underline{V}_{\rm oc} = \underline{\mathcal{E}}_{\rm T}$, is that in Eq. (14.167). To shorten the writing, let the resulting power, under both impedance- and polarization-match conditions, be denoted simply by $P_{\rm r}$, so that we have

power received by a matched load in co-pol case

$$P_{\rm r} = P_{\rm received} = (P_{\rm L})_{\rm max}^{\text{co-pol}} = \frac{|\underline{V}_{\rm oc}|_{\rm max}^2}{4R_{\rm A}} = \frac{\lambda^2|\underline{\mathbf{E}}_0|^2|\underline{\mathbf{F}}|^2}{4\pi^2 R_{\rm A}}, \qquad (14.181)$$

where $\underline{\mathbf{E}}_0$ and $\underline{\mathbf{F}}$ are, respectively, the electric field vector of the incident wave at the coordinate origin and characteristic radiation function of the antenna in the direction of the wave arrival, Fig. 14.32(a).

Using Eqs. (9.200) and (14.161), the time-average Poynting vector of the incident wave is given by

$$(\mathcal{P}_{\rm i})_{\rm ave} = \frac{|\underline{\mathbf{E}}_0|^2}{\eta} \qquad [(\boldsymbol{\mathcal{P}}_{\rm i})_{\rm ave} = (\mathcal{P}_{\rm i})_{\rm ave}\,\hat{\mathbf{n}}], \qquad (14.182)$$

with which the time-average power received by the load, Eq. (14.181), can be written as a product of $(\mathcal{P}_\mathrm{i})_\mathrm{ave}$ and a surface area (measured in m^2). This exactly is the antenna effective aperture, A_eff, which is computed as

$$P_\mathrm{r} = (\mathcal{P}_\mathrm{i})_\mathrm{ave} A_\mathrm{eff} \quad \longrightarrow \quad A_\mathrm{eff} = \frac{\lambda^2 \eta |\underline{\mathbf{F}}|^2}{4\pi^2 R_\mathrm{A}}. \tag{14.183}$$

antenna effective aperture (in m^2)

Comparing this expression for A_eff with that for the antenna gain, G, in Eq. (14.110), we identify the following simple relationship between the two parameters:

$$A_\mathrm{eff}(\theta_\mathrm{i}, \phi_\mathrm{i}) = \frac{\lambda^2}{4\pi} G(\theta_\mathrm{i}, \phi_\mathrm{i}), \tag{14.184}$$

effective aperture vs. gain

where the angular dependence is inserted to emphasize that A_eff, just as G, is a function of the direction of the wave approach, defined by the incident angles θ_i and ϕ_i, i.e., of the unit vector $\hat{\mathbf{n}}$ or $\hat{\mathbf{r}}$ in Fig. 14.32(a). Most frequently, however, the antenna is pointed for the maximum gain (or directivity), namely, it is oriented so that the maximum of its radiation (or reception) pattern is in the direction $(\theta_\mathrm{i}, \phi_\mathrm{i})$. Of course, the maximum gain automatically means the maximum aperture, by Eq. (14.184). Overall, although G is defined, in Section 14.6, for the transmitting mode of operation of an antenna, and A_eff, here, for the receiving mode, they both can be used at both transmit and receive ends of antenna communication systems.

As an example, let us find A_eff of a half-wave wire dipole that is impedance-matched to the load, as well as pointed and polarized for the maximum voltage response to the incoming wave [the dipole is perpendicular to the direction of the wave incidence, $\theta_\mathrm{i} = 90^\circ$ in Eq. (14.184), and parallel to the incident electric field vector, $\alpha = 0$ in Eq. (14.166)]. A combination of Eqs. (14.184) and (14.131) gives

$$A_\mathrm{eff} = \frac{\lambda^2}{4\pi} 1.64 = 0.13\lambda^2, \tag{14.185}$$

A_eff – half-wave dipole

where we have also assumed no ohmic losses on the dipole (see Example 14.16), so the antenna gain equals directivity ($G = D$), Eq. (14.111). We see that the dipole effectively extracts power from a part of the incident wavefront that is approximately $\lambda/2 \times \lambda/4 = \lambda^2/8$ large, as illustrated in Fig. 14.36(a).

For antennas that have obvious physical apertures (openings), such as horn antennas, Fig. 13.14(b), and parabolic reflector antennas, the effective aperture of an antenna can be written as

$$A_\mathrm{eff} = \eta_\mathrm{aperture} A_\mathrm{physical}, \tag{14.186}$$

definition of aperture efficiency

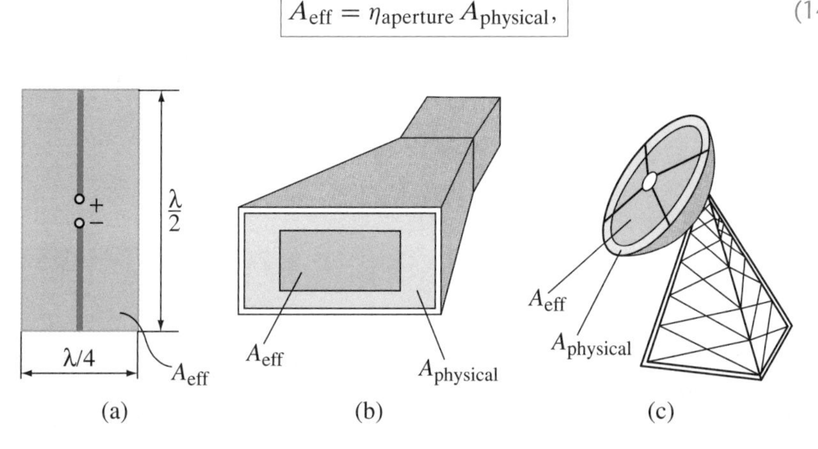

Figure 14.36 Illustration of the concept of antenna effective aperture (A_eff), Eq. (14.183), as an equivalent area from which an antenna extracts incident power and delivers it to a load – for a half-wave wire dipole [Eq. (14.185)] (a), pyramidal horn antenna (b), and circular parabolic reflector antenna (c); for cases (b) and (c), the antenna physical aperture area (A_physical) is also shown [see Eq. (14.186)].

where $A_{\rm physical}$ is the physical aperture area of the antenna, shown in Fig. 14.36(b) and Fig. 14.36(c), and $\eta_{\rm aperture}$ is a dimensionless coefficient termed the aperture efficiency, with $A_{\rm eff}$ being always smaller than or eventually equal to $A_{\rm physical}$ ($0 \le \eta_{\rm aperture} \le 1$). The aperture efficiency therefore represents a measure of how efficiently a receiving antenna utilizes its physical area to collect incident power and transfer it to a load. For instance, pyramidal horn antennas normally have $\eta_{\rm aperture} \approx 50\%$, whereas $\eta_{\rm aperture} = 60 - 80\%$ for typical parabolic reflector antennas. Combining Eqs. (14.186) and (14.184), we realize that an antenna has to be physically large, and thus have a large effective aperture, to attain a high gain. However, large $A_{\rm eff}$ is irrelevant if considered alone, isolated from the operating frequency of the wave, since

electrical effective aperture

$$\boxed{G \propto \frac{A_{\rm eff}}{\lambda^2} = (A_{\rm eff})_{\rm elec},} \tag{14.187}$$

and it is actually the electrical effective aperture, defined as the ratio of $A_{\rm eff}$ to the wavelength squared, what determines the gain, and vice versa. So, as a rule, for high gains (and narrow beams) antennas are electrically large. This is one of the most important general rules in antenna theory, which guides and motivates many different types of antenna and antenna-array designs when highly directional properties are needed.

Example 14.25 Maximum Power Transfer for Conjugate-Matched Load

Prove that the maximum power transfer, $P_{\rm L} = (P_{\rm L})_{\rm max}$, in Fig. 14.32(b) is achieved for the conjugate-matched load, as given in Eqs. (14.180).

Solution Since reactances can also be negative, it is obvious that the value for $X_{\rm L} + X_{\rm A}$ that makes the denominator of the expression for the load power, $P_{\rm L}$, in Eq. (14.179) minimum (and $P_{\rm L}$ maximum) for any fixed value for $R_{\rm L}$, and independently of it, is zero ($R_{\rm A}$, $X_{\rm A}$, and $\mathcal{E}_{\rm T}$ are assumed to be fixed, in the first place). Hence, $X_{\rm L} = -X_{\rm A}$, and we are left to maximize the following expression for $P_{\rm L}$:

$$X_{\rm L} = -X_{\rm A} \quad \longrightarrow \quad P_{\rm L} = \frac{R_{\rm L}|\mathcal{E}_{\rm T}|^2}{R_{\rm L}^2 + 2R_{\rm L}R_{\rm A} + R_{\rm A}^2} = \frac{|\mathcal{E}_{\rm T}|^2}{f(R_{\rm L})}, \tag{14.188}$$

where $f(R_{\rm L})$ stands for the denominator of this new expression ($R_{\rm A} = {\rm const}$). Its minimization gives

$$f(R_{\rm L}) = R_{\rm L} + 2R_{\rm A} + \frac{R_{\rm A}^2}{R_{\rm L}} \quad \longrightarrow \quad \frac{{\rm d}f}{{\rm d}R_{\rm L}} = 1 - \frac{R_{\rm A}^2}{R_{\rm L}^2} = 0 \quad \longrightarrow \quad R_{\rm L} = R_{\rm A}. \tag{14.189}$$

It is a simple matter to verify that this result for $R_{\rm L}$ is indeed a minimum (and not a maximum) of the function f (and maximum of $P_{\rm L}$), which concludes our proof.

Problems: 14.50 and 14.51; *Conceptual Questions* (on Companion Website): 14.48–14.53; *MATLAB Exercises* (on Companion Website).

14.14 FRIIS TRANSMISSION FORMULA FOR A WIRELESS LINK

In practice, we normally aim at ensuring antenna matching and orientation conditions that would maximize the power transfer in a wireless system, and in this section we restrict our attention to such an ideal case. Let us consider a general

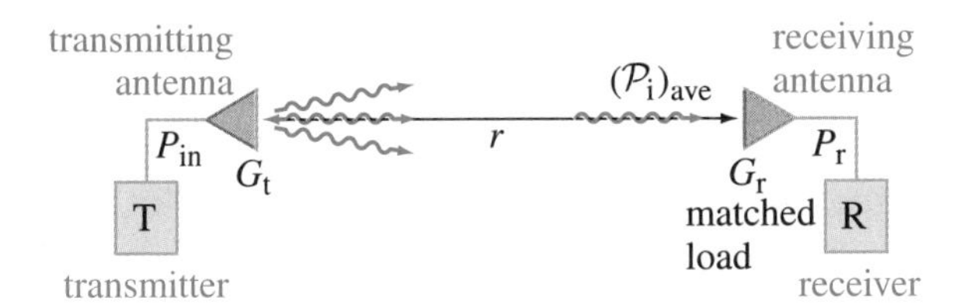

Figure 14.37 Wireless communication link with ideal antenna matching and orientation conditions (load impedance match to the receiving antenna, polarization match of antennas, and orientation of both antennas for maximum gains) – for deriving the Friis transmission formula, Eq. (14.190).

wireless (radio) communication link consisting of two antennas at a far distance r, Eq. (14.16), in free space (or any other homogeneous and lossless electromagnetic medium), as shown in Fig. 14.37. The transmitter and receiver attached to the antennas can be represented using the transmit and receive antenna equivalent circuits, Fig. 14.10(b) and 14.32(b), respectively. For an ideal case, we then assume an impedance match of the load (receiver) to the receiving antenna, polarization match between the antennas, and that both antennas are aligned and pointed toward each other for maximum gains, which equal G_t and G_r for the transmitting and receiving antennas, respectively. To completely determine the power budget in this link, let us find the ratio of the time-average power received by the load, P_r, to the time-average input power that the transmitting antenna accepts at its terminals, P_{in}.

Using Eqs. (14.183), (14.184), (14.110), and (14.96), both powers can be expressed in terms of the time-average Poynting vector magnitude $(\mathcal{P}_i)_{ave}$ of the wave radiated by the transmitting antenna computed at the receiving antenna terminals, $(\mathcal{P}_i)_{ave}$ thus being the connecting point between the two antennas, which is eliminated in the power ratio,

$$P_r = \frac{\lambda^2 G_r}{4\pi} (\mathcal{P}_i)_{ave} \quad \text{and} \quad P_{in} = \frac{4\pi r^2}{G_t} (\mathcal{P}_i)_{ave} \quad \longrightarrow \quad \frac{P_r}{P_{in}} = \left(\frac{\lambda}{4\pi r}\right)^2 G_t G_r, \tag{14.190}$$

Friis transmission formula

where λ is the operating wavelength in the system. This expression for P_r/P_{in} is known as the Friis transmission formula, and it is extremely useful in the evaluations of signal power levels in antenna communication systems, as part of the analysis or design of such systems. An equivalent formula with effective apertures of the two antennas instead of the gains can be written in a straightforward manner, employing Eq. (14.184).

HISTORICAL ASIDE

Harald T. Friis (1893–1976), an American electrical engineer, was born in Naestved, Denmark, and graduated from the Royal Technical College in Copenhagen in 1916. After a fellowship to study radio engineering at Columbia University, Friis joined Bell Laboratories in 1925, where he made pioneering contributions to radio propagation, antennas, radar, and radio astronomy. In his 1946 paper "A Note on a Simple Transmission Formula" in *Proceedings of IRE*, he presented a formula for the evaluation of the power budget in a radio link – the famous Friis transmission formula. Friis also performed highly influential work on the signal-to-noise ratio (SNR) in radio receivers, which culminated in the 1944 IRE paper "Noise Figures of Radio Receivers," where the relationship between the noise figure of a receiver as a whole and noise figures of its components was established.

In decibels, we rewrite Eq. (14.190) as a positive attenuation, A_{dB}, between the transmit and receive ends in the link,

dB attenuation in a wireless link

$$A_{\text{dB}} = 10 \log \frac{P_{\text{in}}}{P_{\text{r}}} = 20 \log \frac{4\pi r}{\lambda} - 10 \log G_{\text{t}} - 10 \log G_{\text{r}}$$
$$= (A_{\text{dB}})_{\text{free space}} - (G_{\text{t}})_{\text{dB}} - (G_{\text{r}})_{\text{dB}} \quad \text{(in dB)}. \tag{14.191}$$

The last two terms in the final expression are dB gains of the two antennas, Eq. (14.112), and the first one, $(A_{\text{dB}})_{\text{free space}}$, we refer to as the attenuation in free space. It can be written as

attenuation in free space

$$(A_{\text{dB}})_{\text{free space}} = 20 \log \frac{4\pi r}{\lambda} = 22 + 20 \log \frac{r}{\lambda} \quad \text{(in dB)} \tag{14.192}$$

$[20 \log(4\pi) = 22]$, and it is completely independent of the particular antennas in the system. This attenuation is an inherent manifestation of the nature of radiation (spherical wave) and reception of antennas, in general. It can be thought of as the inherent attenuation of free space (or another medium) if used for power transmission by antennas. If both antennas were isotropic radiators, Eq. (14.99), $(A_{\text{dB}})_{\text{free space}}$ would stand for the total attenuation in the link $[(G_{\text{t}})_{\text{dB}} = (G_{\text{r}})_{\text{dB}} = 0 \text{ dB}]$. Note that a very considerable fixed attenuation of 22 dB is present in the link always. The other component of the attenuation in Eq. (14.192) depends on the electrical separation between the antennas, r/λ, and it is even larger than the fixed part, given that the receive location is in the far field of the transmitting antenna (for instance, for $r = 1000\lambda$, it is as large as 60 dB). Gains $(G_{\text{t}})_{\text{dB}}$ and $(G_{\text{r}})_{\text{dB}}$ then serve to reduce, as much as possible, the attenuation in Eq. (14.191), relative to that with isotropic radiators. The more directional (the narrower main beam) the transmitting antenna the less power is lost by radiation in undesired directions (note that in the system in Fig. 14.37, all directions in 3-D space emanating from the transmitting antenna except one only, that toward the receiving antenna, are undesired). A highly directional receiving antenna "amplifies" the desired radiation and passes it to the load.

It is also common to express the input and received powers in units of decibels over a milliwatt (dBm), that is, to express them with respect to a reference power level of 1 mW, in the following way:

power in dB over a milliwatt

$$P_{\text{dBm}} = 10 \log \frac{P}{P_{\text{ref}}}, \quad \text{where} \quad P_{\text{ref}} = 1 \text{ mW} \tag{14.193}$$

(note that 1 W translates to 30 dBm), whereas the representation in dBW (decibels over a watt) is used as well. With this, the first equality in Eq. (14.191) becomes

$$(P_{\text{r}})_{\text{dBm}} = (P_{\text{in}})_{\text{dBm}} - A_{\text{dB}} \quad \text{(in dBm)}, \tag{14.194}$$

telling us that the dBm power level at the output of the link in Fig. 14.37, $(P_{\text{r}})_{\text{dBm}}$, is lower than that at the input, $(P_{\text{in}})_{\text{dBm}}$, by the amount of dB attenuation of the link, A_{dB}, given by the final expression in Eq. (14.191).

Example 14.26 Power Transmission in a Cellular Telephone System

The antenna of a cellular telephone base station transmitter operating at a frequency of $f = 869$ MHz has a gain of $(G_{\text{t}})_{\text{dB}} = 10$ dB. Assuming ideal antenna matching and orientation conditions (as in Fig. 14.37), find the required input power of the transmitter in

W in order for the power received by a mobile phone antenna with a gain of $(G_r)_{dB} = 2$ dB at a distance of $r = 20$ km to be at least $(P_r)_{dBm} = -60$ dBm.

Solution The operating free-space wavelength, Eq. (9.67), of the transmitter amounts to $\lambda = c_0/f = 34.5$ cm, and the electrical separation between the antennas to $r/\lambda = 57{,}933$. Employing Eqs. (14.191) and (14.192), the dB attenuation of the cellular system is

$$A_{dB} = 22 + 20 \log \frac{r}{\lambda} - (G_t)_{dB} - (G_r)_{dB} = 105.26 \text{ dB}. \tag{14.195}$$

Eq. (14.194) then tells us that the required input power in decibels over a milliwatt is

$$(P_{in})_{dBm} = (P_r)_{dBm} + A_{dB} = 45.26 \text{ dBm}, \tag{14.196}$$

and, from Eq. (14.193), this power expressed in watts is

$$P_{in} = 10^{(P_{in})_{dBm}/10} \times 1 \text{ mW} = 33.6 \text{ W}. \tag{14.197}$$

Example 14.27 Power Transmission in a Direct TV Satellite System

In a direct television satellite system, the transmitting and receiving parabolic reflector antennas have circular apertures of diameters $d_t = 0.5$ m and $d_r = 1.2$ m, and efficiencies $(\eta_{aperture})_t = 70\%$ and $(\eta_{aperture})_r = 60\%$, respectively. If the input power at the transmit end is $P_{in} = 150$ W and the distance between antennas $r = 35{,}800$ km, compute the received power in the system for ideal link conditions (Fig. 14.37) at a frequency of $f = 12.5$ GHz.

Solution From Eq. (14.186) and Fig. 14.36(c), effective apertures of the transmitting and receiving antennas in the system are

$$(A_{eff})_t = (\eta_{aperture})_t \, \pi \left(\frac{d_t}{2}\right)^2 = 0.137 \text{ m}^2, \quad (A_{eff})_r = (\eta_{aperture})_r \, \pi \left(\frac{d_r}{2}\right)^2 = 0.678 \text{ m}^2. \tag{14.198}$$

Using Eq. (14.184), we obtain the version of the Friis transmission formula with effective apertures in place of antenna gains, and hence the power of the receiving antenna

$$P_r = \frac{1}{\lambda^2 r^2} (A_{eff})_t (A_{eff})_r P_{in} = 18.9 \text{ pW}, \tag{14.199}$$

where the operating wavelength of the system is $\lambda = c_0/f = 2.4$ cm.

Example 14.28 Comparison of Wireless and Wireline Systems

In a point-to-point free-space communication system in Fig. 14.37, the frequency is $f = 1$ GHz, the antennas are the same, with gains $G_{dB} = 20$ dB, and the distance between them is $r = 100$ km. Compute the attenuation of this system and compare it to that of a wireline system using the coaxial cable described in Example 11.3, for the same distance and frequency.

Solution By means of Eq. (14.195), the decibel attenuation of the wireless system amounts to $(A_{dB})_{wireless} = 92.46$ dB.

On the other side, with the computed total attenuation coefficient of the coaxial cable in Example 11.3, $\alpha_{cable} = \alpha_c + \alpha_d = 0.16$ dB/m, the dB attenuation of the wireline system comes out to be as large as

$$(A_{dB})_{wireline} = \alpha_{cable} \, r = 16{,}000 \text{ dB} \gg (A_{dB})_{wireless}, \tag{14.200}$$

so much larger than in Fig. 14.37. We realize that repeater amplifiers would be necessary in the cable system to bring the received signal up to usable levels.

However, if a wireline system is realized using an optical fiber (Fig. 10.22), with a typical attenuation coefficient of $\alpha_{optical\ fiber} = 0.5$ dB/m, the total dB attenuation in the link (50 dB) would be comparable to, and actually lesser than, the wireless attenuation.

Problems: 14.52 and 14.53; *Conceptual Questions* (on Companion Website): 14.54–14.57; *MATLAB Exercises* (on Companion Website).

14.15 ANTENNA ARRAYS

Antenna arrays are spatial arrangements of identical antennas (array elements), equally oriented in space (e.g., wire array elements are parallel to each other or collinear), and excited independently, with feed currents of generally different magnitudes and phases, but of the same frequency. An array of a large number of electrically relatively small or medium-sized element antennas can be used to obtain a similar performance to that of a single electrically large antenna. For instance, such an array may have a large electrical effective aperture, Eq. (14.187), and high gain (narrow beam). Electrically large antenna arrays are usually much simpler to fabricate, maneuver, and maintain than similar single radiators (e.g., a large parabolic reflector antenna). In addition, arrays provide great flexibility and new degrees of freedom in synthesizing radiation patterns of desired shapes. For example, by varying the phases of feed currents of array elements, we can change the direction of maximum radiation, i.e., steer the main beam, of an array antenna as desired throughout space, without moving the antenna. Most importantly, the control of element input phases can be performed electronically, using an appropriate array feed network (beam-forming network), which gives rise to so-called electronic beam steering or scanning. Phase-scanned arrays, referred to simply as phased arrays, find many applications in radar and communication systems. Their advantage over realizations with mechanical slewing of the entire antenna structure toward the desired radiation direction using an appropriate positioning system (mechanical beam scanning) is quite obvious, especially if fast scanning is needed. An additional advantage of phased arrays is the capability of forming multiple main beams pointing in different directions simultaneously (which offers, for instance, a possibility to track multiple targets in radar systems). Finally, antenna arrays can be conformed to surfaces (platforms), i.e., formed in the shape dictated by the supporting structure, often with pronounced curvature. These antennas, termed conformal arrays, are mounted (or embedded) on the surfaces of aircraft, spacecraft, ships, automobiles, and other vehicles, or on the side of buildings or indoor structures. In what follows, we develop the basic theory of antenna arrays, and discuss their properties and applications.

Consider an array of N Hertzian dipoles (Fig. 14.1), with complex rms current intensities $\underline{I}_k$ and length vectors $\mathbf{l}_k$ ($k = 1, 2, \ldots, N$), whose (arbitrary) locations in the array are defined by position vectors $\mathbf{r}'_k$ of dipole centers with respect to a global coordinate origin, as shown in Fig. 14.38. This three-dimensional array is simply an antenna with discrete spatial current distribution, and we can evaluate its radiation (far-zone) magnetic vector potential, $\underline{\mathbf{A}}$, using the discrete form of the radiation integral in Eq. (14.33), originally aimed for continuous spatial current distributions over a volume v of the antenna. Namely, the integral over v now becomes a sum over array constituents, and continuously distributed current elements $\underline{\mathbf{J}}\,\mathrm{d}v$ are replaced by discrete vectors $\underline{I}_k\,\mathbf{l}_k$ [see Eq. (4.10)], so we have for the potential in the radiation direction defined by the radial unit vector $\hat{\mathbf{r}}$ at a (far) radial distance r:

magnetic potential due to a 3-D array of N Hertzian dipoles

$$\boxed{\underline{\mathbf{A}} = \frac{\mu\,\mathrm{e}^{-\mathrm{j}\beta r}}{4\pi r}\sum_{k=1}^{N}\underline{I}_k\,\mathbf{l}_k\,\mathrm{e}^{\mathrm{j}\beta\mathbf{r}'_k\cdot\hat{\mathbf{r}}}.} \tag{14.201}$$

It is customary to express element current intensities, $\underline{I}_k$, through relative magnitudes a_k and initial phases α_k with respect to a reference current intensity, $\underline{I}_{\mathrm{ref}}$, which may be one of the element currents. In other words, we use normalized complex currents of elements, $\underline{a}_k = a_k\,\mathrm{e}^{\mathrm{j}\alpha_k}$, relative to the reference current (coefficients

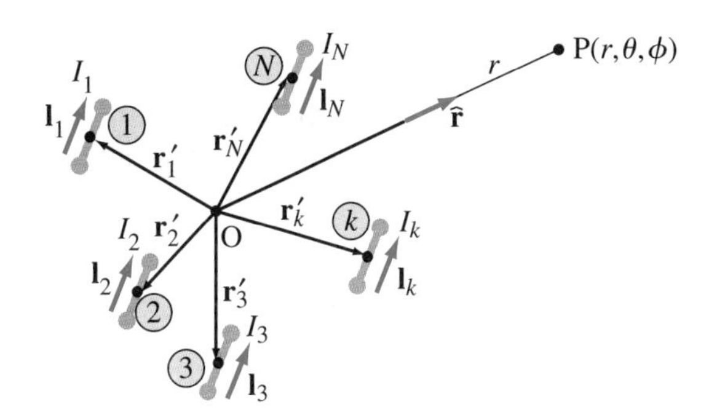

Figure 14.38 Antenna array with Hertzian dipoles as antenna elements.

$\underline{a}_k$ are dimensionless),

$$\boxed{\underline{I}_k = \underline{I}_{\rm ref}\,\underline{a}_k = \underline{I}_{\rm ref}\,a_k\,{\rm e}^{{\rm j}\alpha_k} \quad (k = 1, 2, \ldots, N).} \tag{14.202}$$

antenna array excitation

With this, and an assumption that all elements are equally oriented in space, $\mathbf{l}_1 = \mathbf{l}_2 = \ldots = \mathbf{l}_N = \mathbf{l}$ (length vectors of Hertzian dipoles are equal), the expression in Eq. (14.201) can be written as

$$\underline{\mathbf{A}} = \underbrace{\frac{\mu \underline{I}_{\rm ref}\,\mathbf{l}\,{\rm e}^{-{\rm j}\beta r}}{4\pi r}}_{\underline{\mathbf{A}}_{\rm element}} \underbrace{\sum_{k=1}^{N} \underline{a}_k\,{\rm e}^{{\rm j}\beta \mathbf{r}'_k\cdot\hat{\mathbf{r}}}}_{\text{array factor}} = \underline{\mathbf{A}}_{\rm element}\,\underline{F}_{\rm a}. \tag{14.203}$$

Here, having in mind Eq. (14.4), the total $\underline{\mathbf{A}}$ due to an antenna array is represented as a product of the vector potential $\underline{\mathbf{A}}_{\rm element}$ that a single array element antenna (Hertzian dipole in this case) would radiate if placed at the coordinate origin (reference point) and a complex scalar function $\underline{F}_{\rm a}$. This function provides a complete far-field characterization of the array itself (regardless of the characteristics of its elements), and is called accordingly the array factor (AF). Obviously, it is a dimensionless quantity. From Eq. (14.202),

$$\boxed{\underline{F}_{\rm a} = {\rm AF} = \sum_{k=1}^{N} a_k\,{\rm e}^{{\rm j}(\beta \mathbf{r}'_k\cdot\hat{\mathbf{r}}+\alpha_k)},} \tag{14.204}$$

array factor (dimensionless)

so the array factor is a sum of terms specifying the relative magnitudes and phases of radiated potentials (and fields) of individual antenna elements in the array due to, respectively, their feeds (for magnitudes) and both locations and feeds (for phases) with respect to the reference element. The reference element can be either virtual (nonexistent) or actual – one of the elements in the array. In the latter case, one of the terms in the sum in Eq. (14.204) equals unity (for the reference element, $a_k = 1$, $\alpha_k = 0$, and $\mathbf{r}'_k = 0$).

Since an arbitrary radiating current distribution can be represented as a superposition of Hertzian dipoles, Fig. 14.3, the derived array factor applies to an array of arbitrary antennas (in place of Hertzian dipoles in Fig. 14.38). Namely, given that all antennas in the array are identical, their current distributions normalized to the respective feed currents are all the same, although the feed currents, in general, are different. In computing the far-zone magnetic vector potential of the array, this, in turn, means that radiation integrals over antennas, as in Eq. (14.29) for wire antennas (and analogously for surface and volume antennas), pertaining to the normalized current distributions are also all the same. They thus constitute, as a common term, the magnetic potential of the reference antenna (array element), $\underline{\mathbf{A}}_{\text{element}}$, while the rest is the array factor, $\underline{F}_{\text{a}}$, in Eq. (14.204).

The far electric field vector, $\underline{\mathbf{E}}$, due to the antenna array is obtained from $\underline{\mathbf{A}}$, like for any transmitting antenna, using Eq. (14.21). Since $\underline{F}_{\text{a}}$ is a scalar, taking the transverse projection (component) of $\underline{\mathbf{A}}$ in Eq. (14.203) applies only to the vector potential of the reference element, $\underline{\mathbf{A}}_{\text{element}}$,

$$\underline{\mathbf{E}} = -\mathrm{j}\omega\underline{\mathbf{A}}_{\text{t}} = \underbrace{-\mathrm{j}\omega\left(\underline{\mathbf{A}}_{\text{element}}\right)_{\text{t}}}_{\underline{\mathbf{E}}_{\text{element}}}\underline{F}_{\text{a}} = \underline{\mathbf{E}}_{\text{element}}\,\underline{F}_{\text{a}}, \qquad (14.205)$$

and hence the total field of the array is the field of the reference element, $\underline{\mathbf{E}}_{\text{element}}$, times the array factor. Combining Eqs. (14.205) and (14.203), $\underline{\mathbf{E}}$, in turn, can be written as in Eq. (14.78),

$$\underline{\mathbf{E}} = \underline{C}\,\underline{I}_{\text{ref}}\,\frac{\mathrm{e}^{-\mathrm{j}\beta r}}{r}\,\underline{\mathbf{F}} \quad (\underline{C} = \mathrm{j}60\ \Omega \text{ for free space}), \qquad (14.206)$$

where $\underline{\mathbf{F}}$ is the characteristic radiation function of the antenna array in the radiation direction (defined by the vector $\hat{\mathbf{r}}$) in Fig. 14.38. Writing $\underline{\mathbf{E}}_{\text{element}}$ in the same way, in terms of the corresponding element radiation function, $\underline{\mathbf{F}}_{\text{element}}$, we have

pattern multiplication for antenna arrays

$$\boxed{\underline{\mathbf{F}} = \underline{\mathbf{F}}_{\text{element}}\,\underline{F}_{\text{a}}.} \qquad (14.207)$$

So, as the magnetic vector potential and electric field vector in the far zone, the radiation function (and radiation pattern) of an array can be factored into an element radiation function (pattern) and an array factor, which is known as the pattern multiplication theorem or principle for antenna arrays. Simply, multiplying the element pattern and array factor we obtain the overall array pattern, and this can be done analytically, graphically, or numerically. Noting once more that $\underline{F}_{\text{a}}$ is a complex scalar, we realize that by arraying multiple antenna elements we cannot change the polarization of the element antenna, but only the magnitude (field and power radiation patterns) and phase of its radiation field. We also note that the array factor can be understood as the characteristic (scalar) radiation function of an array of fictitious isotropic radiators or point sources (which have no polarization nor directional properties), Eq. (14.99), with the same locations (at the centers of dipoles in Fig. 14.38) and same relative magnitudes and phases of feed currents as the actual array. This is exactly how the AF is computed (as we shall see in examples) – as the radiation pattern of point sources (replacing array elements) for given locations of points and excitations (retained from the actual elements). Note finally that the directivity of an antenna array in a given direction, $D(\theta, \phi)$, can be found from the total characteristic radiation function of the array, in Eq. (14.207), using Eq. (14.104).

Example 14.29 Broadside Two-Element Array of Point Sources

Consider a two-element array of point sources (isotropic radiators) with feed currents of equal magnitudes and initial phases, and a half-wave interelement spacing. (a) Obtain the normalized array factor of this array and present it graphically in pertinent polar diagrams. (b) Show that the plots in (a) can be sketched also considering additions and cancelations of waves radiated by individual elements, without using the array factor.

Solution

(a) Let us start with an arbitrary distance, d, between element points (centers of array elements or point sources in array factor computation) of the array. For an array axis (straight line passing through the element points) coinciding with the z-axis of a spherical coordinate system, as shown in Fig. 14.39(a), position vectors of element points (see Fig. 14.38) with respect to the coordinate origin (O) placed midway between the elements, which is the array reference point, are

$$\mathbf{r}_1' = -\frac{d}{2}\,\hat{\mathbf{z}}, \quad \mathbf{r}_2' = \frac{d}{2}\,\hat{\mathbf{z}}. \tag{14.208}$$

The normalized feed currents $\underline{a}_k$ of point sources, in Eq. (14.202), being the same (their magnitudes can be adopted to be unity), we can write

$$a_1 = a_2 = 1, \quad \alpha_1 = \alpha_2 = 0. \tag{14.209}$$

By means of Eqs. (14.208) and (14.209), and the fact that $\hat{\mathbf{z}} \cdot \hat{\mathbf{r}} = \cos\theta$ in Fig. 14.39(a), the general array factor in Eq. (14.204) is simplified to the following function of the observation angle θ:

$$\underline{F}_\mathrm{a}(\theta) = \mathrm{e}^{-\mathrm{j}\beta(d/2)\cos\theta} + \mathrm{e}^{\mathrm{j}\beta(d/2)\cos\theta} = 2\cos\left(\beta\frac{d}{2}\cos\theta\right), \tag{14.210}$$

where the use is also made of the second identity in Eqs. (10.7). The array factor does not depend on ϕ, as expected from the rotational symmetry of the array about the z-axis. Normalizing it for the maximum value of unity, as in Eq. (14.80), gives

$$f_\mathrm{a}(\theta) = \frac{|\underline{F}_\mathrm{a}(\theta)|}{|\underline{F}_\mathrm{a}(\theta)|_\mathrm{max}} = \left|\cos\left(\beta\frac{d}{2}\cos\theta\right)\right|. \tag{14.211}$$

Finally, for the half-wave spacing, much like in Eqs. (14.127), and emphasizing that the phase shift between element feeds, $\delta = \alpha_2 - \alpha_1$, is zero in Eqs. (14.209), we have

$$d = \frac{\lambda}{2} \quad \text{and} \quad \delta = 0 \quad \longrightarrow \quad f_\mathrm{a}(\theta) = \cos\left(\frac{\pi}{2}\cos\theta\right) \quad \text{(broadside array)}. \tag{14.212}$$

The array pattern has nulls along the array axis, i.e., $f_\mathrm{a} = 0$ for $\theta = 0$ and $\theta = 180°$, respectively, whereas the peak radiation is in directions normal to the array axis, $f_\mathrm{a} = (f_\mathrm{a})_\mathrm{max} = 1$ for $\theta = 90°$, so on the broad (long) side of the array, and hence this array is called a broadside antenna array. A 2-D polar plot of f_a as a function of θ is presented in Fig. 14.39(b), and the corresponding 3-D pattern, obtained by rotating (for $0 < \phi < 360°$) the 2-D plot about the z-axis, is depicted in Fig. 14.39(c). Note a similar pattern shape ("doughnut" with no hole) to that of a Hertzian dipole, in Fig. 14.12(d).

(b) The broadside radiation pattern in Fig. 14.39(b) and Fig. 14.39(c) can be predicted also by tracking the individual electromagnetic waves (rays) launched by the two array elements (point sources), as indicated in Fig. 14.39(d).[14] Namely, looking into the positive z direction, wave 1 (emanating from source 1) travels a path equal to $d = \lambda/2$, and thus acquires an extra phase factor of $\mathrm{e}^{-\mathrm{j}\beta d} = \mathrm{e}^{-\mathrm{j}\pi} = -1$, before it reaches source 2 and joins wave 2. Consequently, the two waves propagating from there on together to the right in Fig. 14.39(d) arrive to the far zone in counter-phase (due to the half-wave interelement spacing) and with equal magnitudes (coming from the equal magnitudes of feed currents), causing a perfect cancelation of their fields and pattern null for $\theta = 0$. The same occurs in the negative z direction ($\theta = 180°$). On the other hand, for any broadside direction (e.g., in the positive or negative x direction) in Fig. 14.39(d), waves 1 and 2 travel exactly the same paths from the respective sources to a far-field point. In addition, given the equal initial phases of the feed currents, the rays arrive to the far field in phase, so that their fields add up together constructively (the total field is double that of one source). In other words, we have a perfect addition of far fields and pattern maximum (main lobe) for $\theta = 90°$. Finally, since the phase difference between waves 1 and 2 varies

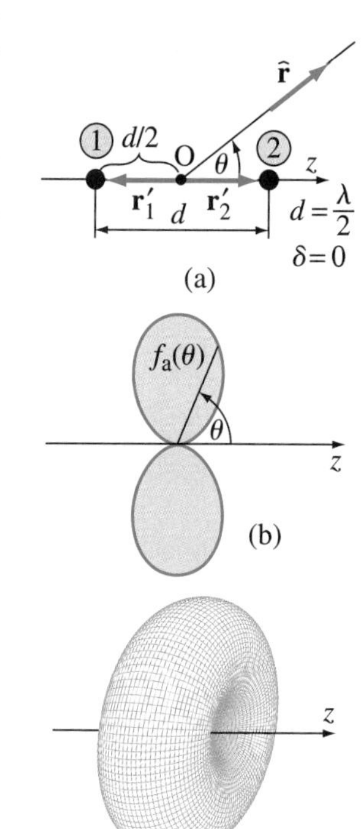

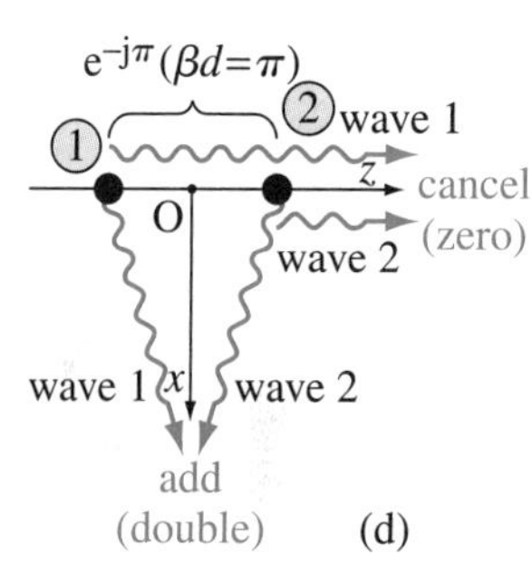

Figure 14.39 Radiation analysis of a broadside two-element antenna array: (a) array geometry and excitation (point sources with equal input powers, in phase, and half-wave apart), (b) 2-D polar plot of the normalized array factor, Eq. (14.212), (c) 3-D radiation pattern, and (d) determination of nulls and maxima of the array radiation in directions of perfect cancelations and additions, respectively, of individual waves launched by array elements; for Example 14.29.

[14] See the similar application of wave tracking and computation of field phases based on traveled paths in Fig. 13.12 and Eq. (13.118).

smoothly from 0 to 180° as the observation point moves from a broadside direction to an axial direction (along the array axis) over a far-zone sphere centered at the coordinate origin, there is a smooth pattern variation between the maximum and null points, making up the pattern curve in Fig. 14.39(b).

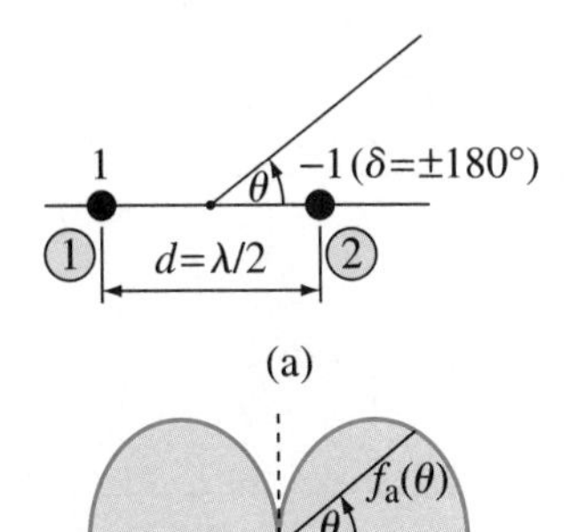

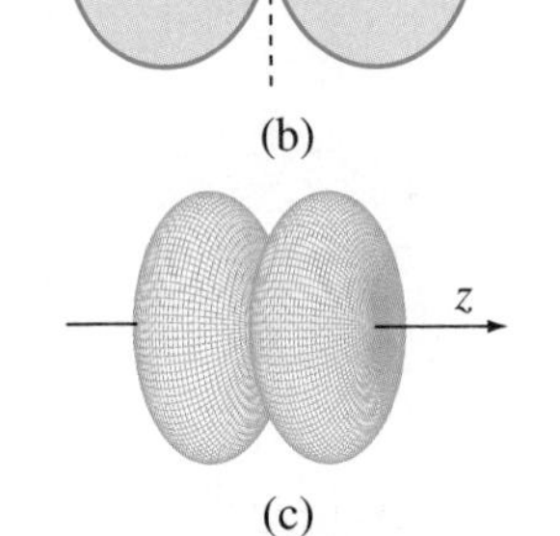

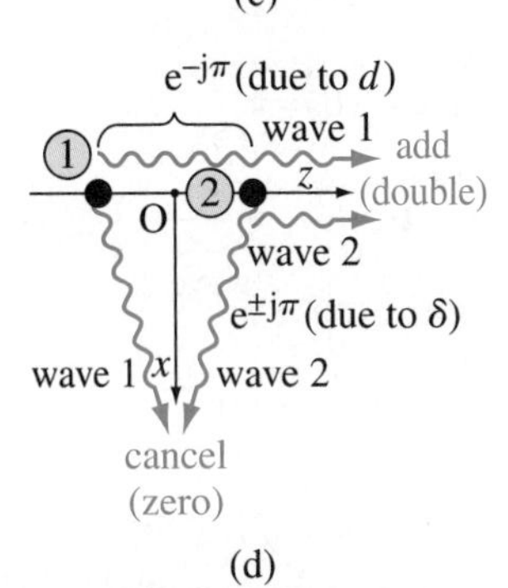

Figure 14.40 The same as in Fig. 14.39 but for an endfire two-element antenna array (point sources with equal input powers, in counter-phase, and half-wave apart); for Example 14.30.

Example 14.30 Endfire Two-Element Array of Point Sources

Repeat the previous example but for a two-element array of point sources in counter-phase (elements are still spaced a half-wavelength apart and fed with same current magnitudes), as indicated in Fig. 14.40(a).

Solution

(a) Adopting $\alpha_1 = \mp 180°$ and $\alpha_2 = 0$, so that the phase shift between feed currents of array elements amounts exactly to $\delta = \alpha_2 - \alpha_1 = \pm 180°$, we have $e^{j\alpha_1} = e^{j\mp\pi} = -1$ in Eq. (14.204), and the array factor in Eq. (14.210) becomes

$$\underline{F}_a(\theta) = -e^{-j\beta(d/2)\cos\theta} + e^{j\beta(d/2)\cos\theta} = 2j\sin\left(\beta\frac{d}{2}\cos\theta\right), \tag{14.213}$$

where the first identity in Eqs. (10.7) is invoked this time. The normalized array factor is now given by

$$d = \frac{\lambda}{2} \quad \text{and} \quad \delta = \pm 180° \quad \longrightarrow \quad f_a(\theta) = \left|\sin\left(\frac{\pi}{2}\cos\theta\right)\right| \quad \text{(endfire array).} \tag{14.214}$$

Just oppositely to Eq. (14.212), this array pattern exhibits zero radiation in broadside directions (for $\theta = 90°$), while maximum in both axial directions (for $\theta = 0$ and 180°). Such arrays whose main lobe maxima are along the array axis, so in directions toward array ends, are referred to as endfire arrays (note that it is possible to design endfire arrays with radiation only along one of the ends of the array). Fig. 14.40(b) and Fig. 14.40(c) show 2-D and 3-D polar plots, respectively, of the factor in Eq. (14.214), with the 3-D pattern having a characteristic shape of a "dumbbell" with no handle.

(b) As far as the technique of tracking individual waves launched by array elements to identify directions of their perfect cancelations and additions (nulls and maxima of the array radiation pattern) is concerned [see Fig. 14.39(d)], waves 1 and 2 now cancel each other in broadside directions, as illustrated in Fig. 14.40(d). Namely, the waves are launched with a 180° phase difference in the first place (phase shift δ between feed currents), and, traveling equal paths broadside (e.g., along the x-axis), they maintain this same phase relationship on out to the far field, yielding a pattern null. In axial directions (along the z-axis), on the other side, the phase factor reflecting the counter-phase feeds is compensated for by that due to the difference in paths the two waves travel, $e^{j\delta}\,e^{-j\beta d} = e^{\pm j\pi}\,e^{-j\pi} = 1$ [see the similar computation in Eq. (13.118)]. Hence, the two waves are in phase as they propagate away from the array, which results in a constructive addition of their fields (pattern maximum). Based on this discussion, we can sketch the endfire radiation pattern in Fig. 14.40(b) and Fig. 14.40(c), without actually knowing the array factor of the antenna array.

Example 14.31 Full-Wave Interelement Spacing and Grating Lobes

Repeat Example 14.29 but for a two-element array of point sources with a full-wave interelement spacing (elements are still fed with same current magnitudes and in phase), shown in Fig. 14.41(a).

Solution

(a) From $d = \lambda$ and $\beta d/2 = (2\pi/\lambda)\,d/2 = \pi$, the normalized array factor in Eq. (14.211) now takes the following form:

$$d = \lambda \quad \text{and} \quad \delta = 0 \quad \longrightarrow \quad f_a(\theta) = |\cos(\pi\cos\theta)|. \tag{14.215}$$

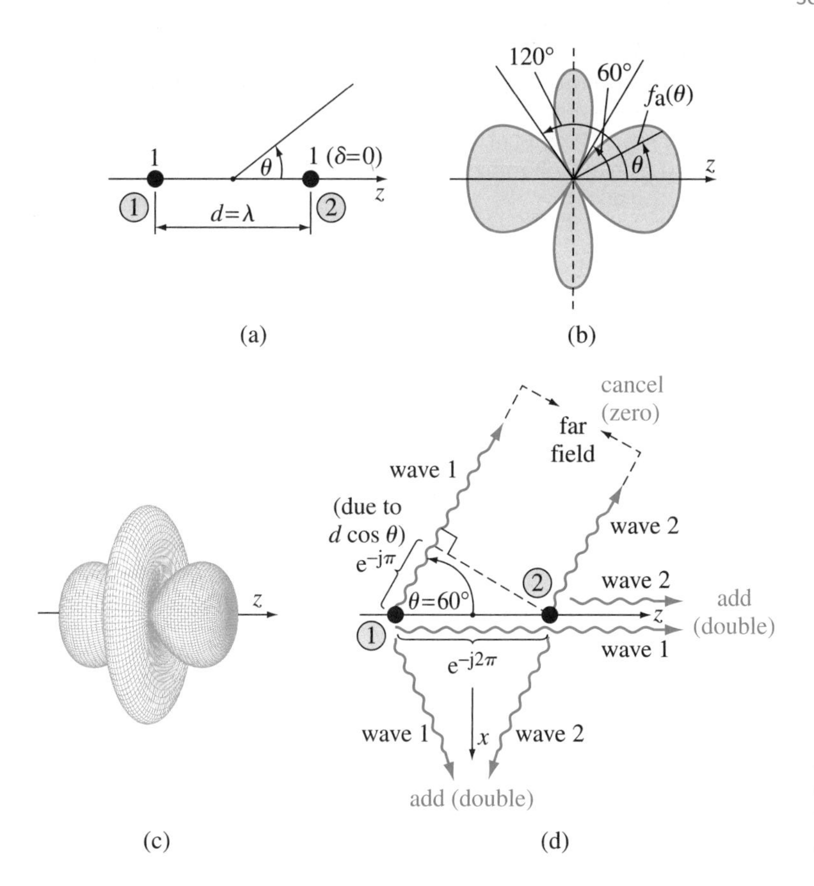

Figure 14.41 The same as in Fig. 14.39 but for a full-wave separation between in-phase point sources; for Example 14.31.

This array pattern has maxima (equal to unity) for $\theta = 0$, 90°, and 180°, and nulls for $\theta = 60^\circ$ and 120°, respectively, which is illustrated by radiation pattern plots in Fig. 14.41(b) and Fig. 14.41(c).

(b) Since $e^{-j\beta d} = e^{-j2\pi} = 1$, waves 1 and 2 launched by the array elements now add up in phase also in axial directions, Fig. 14.41(d), along with broadside perfect additions as in Fig. 14.39(d). To identify pattern nulls, we see in Fig. 14.41(d) that for $\theta = 60^\circ$ wave 1 travels, to the far zone, $d\cos\theta = \lambda/2$ more than wave 2, which results in an extra phase factor of $e^{-j\beta\lambda/2} = e^{-j\pi} = -1$ and a perfect cancelation in the far field. For $\theta = 120^\circ$, wave 2 travels that more, and the result is the same (pattern null). Combining this information on the maxima and zeros of the array radiation and filling in the smooth pattern variations between them, we can sketch the plots in Fig. 14.41(b) and Fig. 14.41(c).

This example is illustrative of the fact that multiple radiation lobes are formed in an array pattern for interelement spacings greater than $\lambda/2$. In addition, assuming that the array in Fig. 14.41 is principally meant as a broadside array, with a broadside 3-D main lobe, its endfire radiation lobes are considered as side lobes. However, they are of the same intensity at their peaks as the main lobe. Such additional "main" lobes are called grating lobes, and in the majority of array applications they are undesirable.

Example 14.32 Array of Two Collinear Hertzian Dipoles – Pattern Multiplication

An array of two collinear (coaxial) Hertzian dipole antennas, of length l, whose centers are spaced a half-wavelength apart, radiates in free space. The dipoles are fed with time-harmonic currents of frequency f and equal complex intensities. Determine the total characteristic radiation function of the antenna array and sketch the associated normalized radiation pattern.

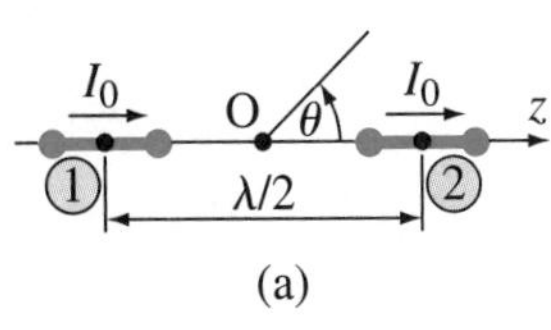

(a)

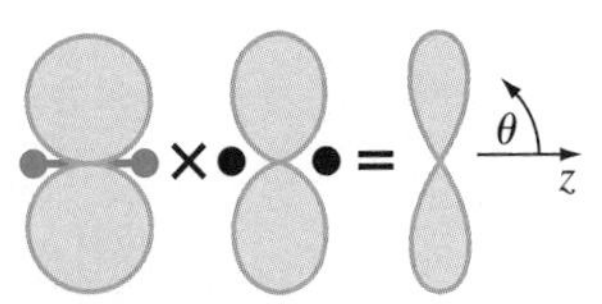

element pattern array factor total pattern

(b)

Figure 14.42 Radiation analysis of an array of two collinear Hertzian dipole antennas: (a) array spacing and excitation (for broadside radiation) and (b) graphical multiplication of the element pattern and array factor to obtain the overall array pattern; for Example 14.32.

Solution We introduce a spherical coordinate system with the z-axis along the axes of dipoles and origin (O) at the array center, as shown in Fig. 14.42(a). Hence, the element characteristic radiation function is that in Eq. (14.79), and the array factor (this is a broadside array) is given in Eqs. (14.210) and (14.212). Applying the pattern multiplication theorem, Eq. (14.207), we then obtain the following expression for the resultant characteristic radiation function of the antenna array:

$$\underline{\mathbf{F}}(\theta) = \underline{\mathbf{F}}_{\text{element}}(\theta)\, \underline{F}_{\text{a}}(\theta) = \frac{\beta l}{2} \sin\theta\, \hat{\boldsymbol{\theta}}\, 2\cos\left(\frac{\pi}{2}\cos\theta\right) = \beta l \sin\theta \cos\left(\frac{\pi}{2}\cos\theta\right)\hat{\boldsymbol{\theta}} \quad (14.216)$$

($\beta = 2\pi f\sqrt{\varepsilon_0\mu_0}$), whose normalized form, Eq. (14.80), reads

$$f(\theta) = \sin\theta\cos\left(\frac{\pi}{2}\cos\theta\right). \quad (14.217)$$

Although it is certainly possible and straightforward to plot the function $f(\theta)$ directly using Eq. (14.217), in Fig. 14.42(b) we instead graphically multiply the element pattern and array factor (both previously normalized), to obtain the overall array pattern. All three patterns are, of course, azimuthally symmetrical, and a 2-D polar plot in any one plane $\phi = \text{const}$, as a function of θ, suffices to fully describe the radiation. The total pattern exhibits broadside radiation as well, and is very similar to the array factor, $f_{\text{a}}(\theta)$. However, it is somewhat more directive, i.e., the resulting curve is narrower (sharper), than the $f_{\text{a}}(\theta)$ curve [analytically, the product of $f_{\text{a}}(\theta)$ and $\sin\theta$ is smaller than $f_{\text{a}}(\theta)$, except in the direction of maximum radiation].[15]

Example 14.33 Array of Two Parallel Dipoles – Three Pattern Cuts

Repeat the previous example but for an array of two parallel Hertzian dipole antennas (the dipole axes are perpendicular to the array axis); all other input parameters of the system are the same as in Fig. 14.42(a).

Solution Since the dipoles are not along the array axis, if we keep the array axis coincident with the z-axis of a global spherical coordinate system in which the analysis is performed, then the dipoles will not be z-directed, and vice versa. So, let us adopt the system as in Fig. 14.43(a), with dipole antennas parallel to the x-axis. Obviously, we cannot use now the expression in Eq. (14.79) for the element characteristic radiation function, and what we need rather is the one for an x-directed Hertzian dipole placed at the coordinate origin or reference point (O) of the array, in Fig. 14.43(a). The far electric field intensity vector of such a dipole is given in Eq. (14.48), which, compared with Eq. (14.78) while having in mind the first relationship in Eqs. (14.9), yields

$$\underline{\mathbf{F}}_{\text{element}}(\theta,\phi) = \frac{\beta l}{2}(-\cos\theta\cos\phi\,\hat{\boldsymbol{\theta}} + \sin\phi\,\hat{\boldsymbol{\phi}}). \quad (14.218)$$

The overall array pattern in Eq. (14.216) then becomes

$$\underline{\mathbf{F}}(\theta,\phi) = \underline{\mathbf{F}}_{\text{element}}(\theta,\phi)\,\underline{F}_{\text{a}}(\theta) = \beta l(-\cos\theta\cos\phi\,\hat{\boldsymbol{\theta}} + \sin\phi\,\hat{\boldsymbol{\phi}})\cos\left(\frac{\pi}{2}\cos\theta\right), \quad (14.219)$$

and its normalized magnitude (with a maximum value of unity) is

$$f(\theta,\phi) = \sqrt{\cos^2\theta\cos^2\phi + \sin^2\phi}\,\cos\left(\frac{\pi}{2}\cos\theta\right). \quad (14.220)$$

[15]Note that broadside collinear wire antenna arrays, but with half-wave dipoles rather than Hertzian dipoles as elements (element centers are spaced more than $\lambda/2$ apart, so that the antennas do not touch each other), are extensively used in base stations for land mobile communications. In such arrays, the array axis is oriented vertically, to produce an omnidirectional radiation pattern in the horizontal plane ($\theta = 90^\circ$), enabling communications between a base station and many scattered mobile units (so-called point-to-multipoint communications). In addition, the number of elements (N) in a base-station antenna array is increased (and the array lengthened), to narrow the beam of the array pattern, i.e., reduce the half-power beamwidth [see Fig. 14.11(b)] of the array, in vertical planes (containing the array axis). This, in turn, increases the array directivity, and extends the power "coverage" and the corresponding usable range of distances [see Eq. (14.190)] in the horizontal plane to mobile units in the system.

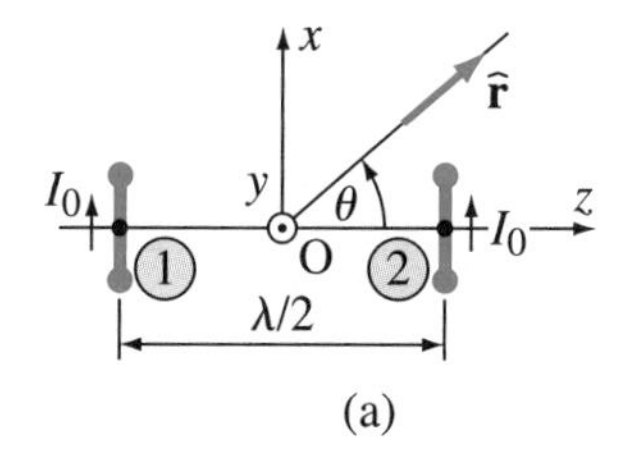

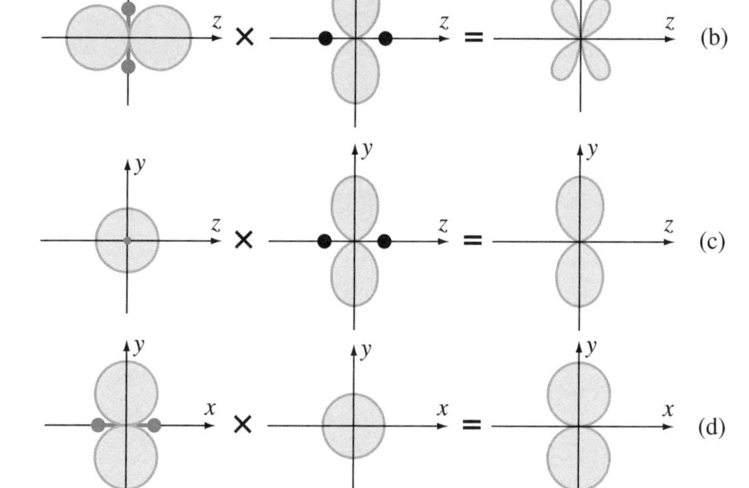

Figure 14.43 Radiation by an array of two parallel Hertzian dipole antennas: (a) array configuration and adopted coordinate system, and radiation pattern cuts, obtained by graphical pattern multiplication, in planes (b) xz, (c) yz, and (d) xy; for Example 14.33.

The total radiation pattern now being a function of both angles θ and ϕ, its plots in the plane in which dipoles lie (xz-plane, also defined by $\phi = 0$), the plane perpendicular to dipoles (yz-plane, where $\phi = 90°$), and plane perpendicular to the array axis (xy-plane, with $\theta = 90°$) are all different. Figs. 14.43(b)–(d) show these plots as obtained by graphical multiplication of the corresponding plots of the element pattern and array factor. We see that different distributions of radiation nulls in the element pattern and array factor in Fig. 14.43(b) cause formation of multiple lobes in the resulting array pattern cut. On the other hand, the total pattern in Fig. 14.43(c) is identical to the array factor, since the element pattern in the equatorial plane of a dipole is omnidirectional. Similarly, the cut of the array factor in Fig. 14.43(d) is a circle, so that the pattern of the antenna system equals that of an element in this cut. Of course, all these plots can as well be generated substituting the proper values for θ or ϕ in the function $f(\theta, \phi)$ in Eq. (14.220).

Example 14.34 Three-Element Array of Parallel Half-Wave Dipoles, Cardioid Pattern

Shown in Fig. 14.44(a) is an array of three parallel half-wave wire dipole antennas. The dipoles all lie in one plane, with a quarter-wave separation between adjacent element points. The feed currents of adjacent dipoles are in time-phase quadrature (90° out of phase with respect to each other), with dipole 1 lagging and dipole 3 advancing in phase with respect to dipole 2. The current magnitudes are in the ratio $1:2:1$ along the array, so this is a nonuniform array – more precisely, a nonuniformly excited, equally spaced array. Find the expression for the overall array pattern of this system and sketch its cuts in planes xy, xz, and yz, respectively.

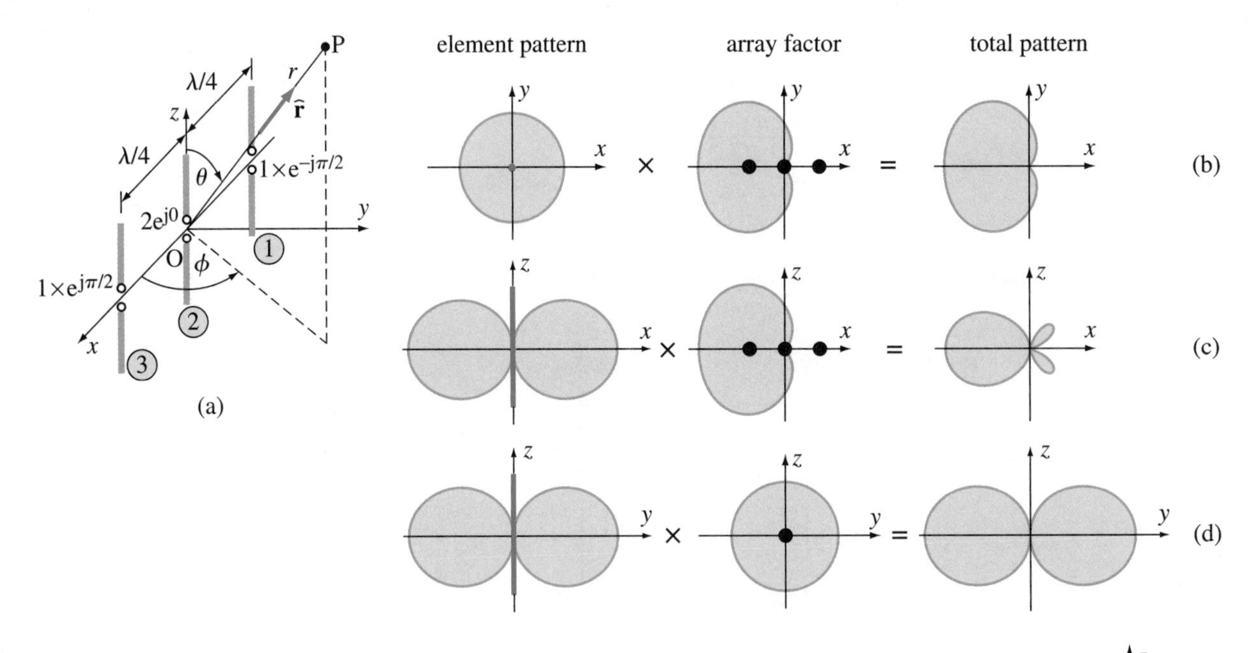

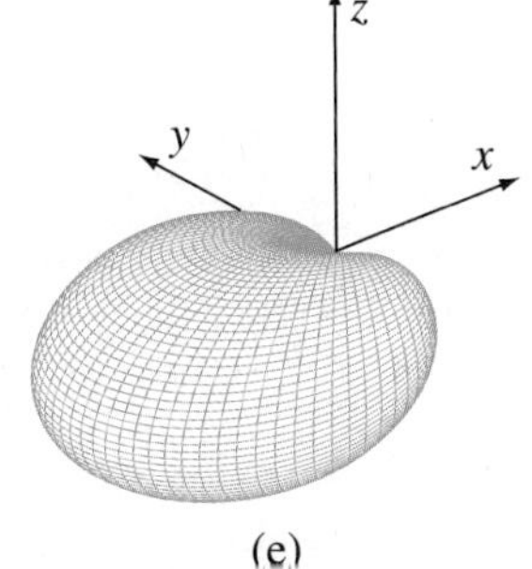

Figure 14.44 Analysis of a nonuniform array of three parallel half-wave wire dipole antennas: (a) quarter-wave spacing and excitation in time-phase quadrature of adjacent dipoles, element, array, and total pattern cuts in planes (b) xy, (c) xz, and (d) yz, and (e) 3-D polar plot of the radiation function; for Example 14.34.

Solution We now have z-directed dipoles and the array element points lying along the x-axis, so this situation is, in terms of description of the polarization of array elements and lineup of their centers in the adopted global coordinate system, just opposite to that in Fig. 14.43(a). The position vectors of element points, in Fig. 14.44(a), with respect to the array center, which falls into the center of the middle element, are

$$\mathbf{r}_1' = -\frac{\lambda}{4}\,\hat{\mathbf{x}}, \quad \mathbf{r}_2' = 0, \quad \mathbf{r}_3' = \frac{\lambda}{4}\,\hat{\mathbf{x}}. \tag{14.221}$$

To compute the array factor ($\underline{F}_{\rm a}$) in Eq. (14.204), we realize that

$$\beta\frac{\lambda}{4} = \frac{2\pi}{\lambda}\frac{\lambda}{4} = \frac{\pi}{2}, \quad \hat{\mathbf{x}}\cdot\hat{\mathbf{r}} = \sin\theta\cos\phi, \quad a_1 = 1, \quad a_2 = 2, \quad a_3 = 1,$$
$$\alpha_1 = -\frac{\pi}{2}, \quad \alpha_2 = 0, \quad \alpha_3 = \frac{\pi}{2}, \tag{14.222}$$

where the use is made of the decomposition of the unit vector $\hat{\mathbf{x}}$ into components in spherical coordinates, Eq. (14.46). With these relationships, the second identity in Eqs. (10.7), and the trigonometric formula $\cos(A+\pi/2) = -\sin A$, we have

$$\underline{F}_{\rm a}(\theta,\phi) = {\rm e}^{-{\rm j}(\pi/2)(\sin\theta\cos\phi+1)} + 2 + {\rm e}^{{\rm j}(\pi/2)(\sin\theta\cos\phi+1)}$$
$$= 2 + 2\cos\left(\frac{\pi}{2}\sin\theta\cos\phi + \frac{\pi}{2}\right) = 2\left[1 - \sin\left(\frac{\pi}{2}\sin\theta\cos\phi\right)\right]. \tag{14.223}$$

On the other side, the element characteristic radiation function ($\underline{\mathbf{F}}_{\rm element}$), for a half-wave dipole, is given in Eq. (14.128). The pattern multiplication theorem, Eq. (14.207), then yields the radiation function of the antenna array as follows:

$$\underline{\mathbf{F}}(\theta,\phi) = \underline{\mathbf{F}}_{\text{element}}(\theta)\,\underline{F}_{\text{a}}(\theta,\phi) = 2\,\frac{\cos\left(\frac{\pi}{2}\cos\theta\right)}{\sin\theta}\left[1-\sin\left(\frac{\pi}{2}\sin\theta\cos\phi\right)\right]\hat{\boldsymbol{\theta}}. \qquad (14.224)$$

Its magnitude in planes xy, xz, and yz, respectively, comes out to be

$$\theta = 90^\circ \quad\longrightarrow\quad |\underline{\mathbf{F}}(90^\circ,\phi)| = 2\left[1-\sin\left(\frac{\pi}{2}\cos\phi\right)\right] \quad (xy\text{-plane}), \qquad (14.225)$$

$$\phi = 0 \quad\longrightarrow\quad |\underline{\mathbf{F}}(\theta,0)| = 2\,\frac{\cos\left(\frac{\pi}{2}\cos\theta\right)}{\sin\theta}\left[1-\sin\left(\frac{\pi}{2}\sin\theta\right)\right] \quad (xz\text{-plane}), \qquad (14.226)$$

$$\phi = 90^\circ \quad\longrightarrow\quad |\underline{\mathbf{F}}(\theta,90^\circ)| = 2\,\frac{\cos\left(\frac{\pi}{2}\cos\theta\right)}{\sin\theta} \quad (yz\text{-plane}). \qquad (14.227)$$

The corresponding normalized overall array pattern cuts are sketched in Figs. 14.44(b)–(d), where the process of graphical multiplication of the constituting element pattern and array factor plots is also presented. We see that this is a predominantly endfire radiation pattern, with the maximum radiation only along one end of the array – in the negative x direction. In particular, the curve in the equatorial plane of the dipoles, in Fig. 14.44(b), is the so-called cardioid pattern (it resembles the heart shape). Fig. 14.44(e) shows a 3-D polar plot of the radiation function in Eq. (14.224).

Example 14.35 Uniform Linear Array with Arbitrary Number of Elements

Consider a linear array (array whose all element points lie along a straight line) with an arbitrary number, N, of point sources (isotropic radiators), radiating in an ambient medium for which the wavelength is λ. The points are equally spaced with respect to each other, and the interelement spacing is d, as shown in Fig. 14.45(a). Feed currents of sources in the array have equal magnitudes and equal phase shifts between adjacent points. Obtain (a) the normalized array factor of this array and (b) the interelement phase shift (α) that results in the direction of maximum radiation of the array at an angle $\theta = \theta_{\text{max}}$ with respect to the array axis (z-axis).

Solution

(a) Referring to Fig. 14.45(a), position vectors of element points with respect to the coordinate origin at the first point are $\mathbf{r}'_k = (k-1)d\,\hat{\mathbf{z}}$ $(k = 1, 2, \ldots, N)$. The phase of the kth source relative to the first one amounts to $\alpha_k = (k-1)\alpha$, i.e., it varies linearly with the source location on the array axis measured with respect to the first element, $(k-1)d$. Linear arrays with a uniform interelement spacing, uniform feed current magnitude, and uniform interelement phase difference along the array are referred to as uniform linear antenna arrays. Having in mind that $\hat{\mathbf{z}}\cdot\hat{\mathbf{r}} = \cos\theta$, as in Fig. 14.39(a), and adopting that

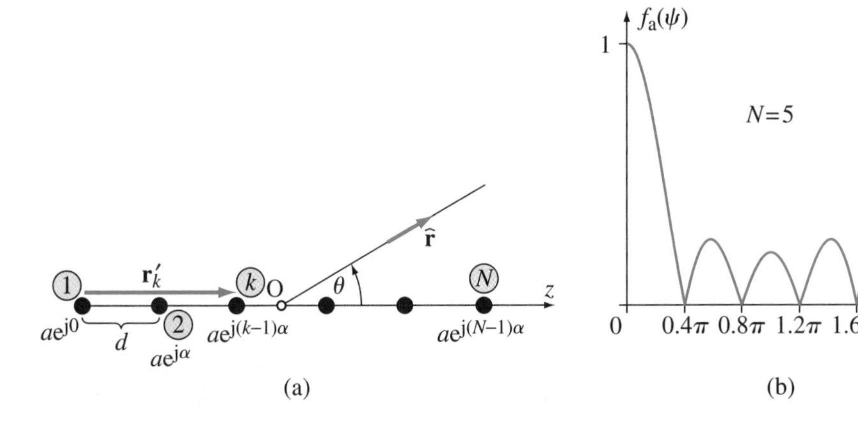

Figure 14.45 (a) Uniform linear antenna array with N point source elements and (b) plot of the normalized array factor, Eq. (14.230), against the variable ψ, given in Eq. (14.228), for $N = 5$; for Example 14.35.

the constant normalized current magnitudes of elements are $a_k = 1$ in Eq. (14.202), the array factor in Eq. (14.204) now becomes

$$\underline{F}_{\mathrm{a}} = \sum_{k=1}^{N} \mathrm{e}^{\mathrm{j}(k-1)\psi}, \quad \text{where} \quad \psi = \beta d \cos\theta + \alpha \quad (0 \le \theta \le 180°) \tag{14.228}$$

(ψ can be interpreted as the phase difference between far fields due to adjacent point sources in the array and $\beta = 2\pi/\lambda$). The factor can be further transformed as follows:

$$\underline{F}_{\mathrm{a}} = \frac{1 - \mathrm{e}^{\mathrm{j}N\psi}}{1 - \mathrm{e}^{\mathrm{j}\psi}} = \frac{\mathrm{e}^{\mathrm{j}N\psi/2}}{\mathrm{e}^{\mathrm{j}\psi/2}} \frac{\mathrm{e}^{-\mathrm{j}N\psi/2} - \mathrm{e}^{\mathrm{j}N\psi/2}}{\mathrm{e}^{-\mathrm{j}\psi/2} - \mathrm{e}^{\mathrm{j}\psi/2}} = \mathrm{e}^{\mathrm{j}(N-1)\psi/2} \frac{\sin(N\psi/2)}{\sin(\psi/2)}, \tag{14.229}$$

where the use is made of the formula for the sum of a geometric series, $1 + x + x^2 + \ldots + x^{N-1} = (1 - x^N)/(1 - x)$, with $x = \mathrm{e}^{\mathrm{j}\psi}$, and of the first identity in Eqs. (10.7). We then note that $(N-1)\psi/2$ is the far-field phase shift of an actual or virtual (depending whether N is an odd or even number) point source at the physical center of the array, with respect to the field of the first source. Therefore, by dropping the exponential phase factor $\mathrm{e}^{\mathrm{j}(N-1)\psi/2}$ from the result in Eq. (14.229), we actually shift the array reference point to the new coordinate origin (O) at the array center, as indicated in Fig. 14.45(a). With this, the final expressions for the array factor of a uniform linear antenna array and for its normalized version are

uniform linear array

$$\underline{F}_{\mathrm{a}} = \frac{\sin(N\psi/2)}{\sin(\psi/2)} \longrightarrow f_{\mathrm{a}} = \frac{|\underline{F}_{\mathrm{a}}|}{|\underline{F}_{\mathrm{a}}|_{\max}} = \frac{|\sin(N\psi/2)|}{N|\sin(\psi/2)|}, \tag{14.230}$$

as $|\underline{F}_{\mathrm{a}}|_{\max} = N$ for $\psi = 0$ ($\sin x \approx x$ when $x \to 0$). Fig. 14.45(b) shows a plot of the function $f_{\mathrm{a}}(\psi)$ in one period, for $0 \le \psi \le 2\pi$, for $N = 5$.

(b) Eqs. (14.228) and (14.230) provide an explicit expression of the property of main-beam scanning of a uniform linear antenna array by the phase control in its excitation network. Namely, with $\theta_{\max}$ denoting the polar angle at which a maximum of an array factor occurs, we have that the interelement phase shift, α, in the array excitation needed for steering the main beam peak to the (desired) angle $\theta = \theta_{\max}$ (so-called scan angle) is given by

main-beam steering phase

$$\psi = 0 \longrightarrow \alpha = -\beta d \cos\theta_{\max}. \tag{14.231}$$

Problems: 14.54–14.66; *Conceptual Questions* (on Companion Website): 14.58 and 14.59; *MATLAB Exercises* (on Companion Website).

Problems

14.1. **Hertzian dipole electric field from magnetic potential only.** Use the expression for the electric field vector in terms of the magnetic vector potential only in Eq. (8.118) to obtain $\mathbf{E}$ due to a Hertzian dipole from the result for $\underline{\mathbf{A}}$ in Eqs. (14.4) and (14.5).

14.2. **Satisfaction of complex Maxwell's equations by dipole fields.** Show that complex electric and magnetic field expressions for a Hertzian dipole in Eqs. (14.8) and (14.10) satisfy all four complex Maxwell's equations in differential form.

14.3. **Satisfaction of the wave equation for Lorenz scalar potential.** Show that the expression for the potential $\underline{V}$ of a Hertzian dipole in Eq. (14.6) satisfies the complex form of the source-free wave equation for the Lorenz scalar potential, in Eq. (8.92) with $\underline{\rho} = 0$.

14.4. **Potentials and field vectors of an x-directed Hertzian dipole.** Consider the x-directed Hertzian dipole described in Example 14.4, and use the expression for its magnetic vector potential, in Eqs. (14.44) and (14.46), to evaluate, as in Eq. (14.7), the accompanying electric scalar potential. Then find the electric and magnetic field vectors of the dipole in several ways, paralleling the computation for the z-directed dipole in Eqs. (14.8)–(14.13).

14.5. **Far-field approximation for an x-directed dipole.** Starting with the general field expressions for an x-directed Hertzian dipole found in the previous problem, obtain the corresponding far-field expressions for both $\underline{\mathbf{E}}$ and $\underline{\mathbf{H}}$ (check the results with those in Example 14.4).

14.6. **More complex definition of the far-field region.** The far-zone condition in Eq. (14.16), $r \gg \lambda$, combined with the condition that $D \gg \lambda$, D being the maximum dimension of the antenna, can be replaced in practical situations by the following combination of requirements for the distance r (from the center of the antenna) in the far zone: (i) $r > 5\lambda$ and (ii) $r > 5D$ and (iii) $r > 2D^2/\lambda$. Sketch a graph with r/λ on the ordinate and D/λ on the abscissa, and identify the far-field region where all three conditions (i)–(iii) are satisfied.

14.7. **Short wire antenna with a cosine current distribution.** An electrically short wire antenna of length l, placed in free space at the coordinate origin along the z-axis of a spherical coordinate system, has a current of intensity $\underline{I}(z) = \underline{I}_0 \cos(\pi z/l)$, for $-l/2 \le z \le l/2$, and frequency f. Compute (a) the electric and magnetic field intensity vectors of the antenna in its far zone and (b) the charge distribution along the antenna. From (a), find (c) the current of an equivalent (having the same far field) Hertzian dipole of the same length (l) as the original antenna and (d) the length of an equivalent Hertzian dipole with current equal to $\underline{I}_0$, respectively. (e) From (b), find the charge ($\underline{Q}$ and $-\underline{Q}$) of the equivalent dipole in (d).

14.8. **Short wire antenna with a quadratic current distribution.** A dipole copper wire antenna of length $l = 2$ m and radius $a = 5$ mm is fed at its center by a current of frequency $f = 10$ MHz and rms intensity $I_0 = 10$ A. The current along the antenna can be assumed to be given by the following function: $\underline{I}(z) = I_0[1 - (2z/l)^2]$, $-l/2 \le z \le l/2$. Calculate the far-zone Poynting vector of the antenna.

14.9. **Two traveling current waves on a wire antenna.** If the load of resistance $R_{\rm L}$ in Fig. 14.8 is unmatched, so that the current distribution along the antenna is given by $\underline{I}(z) = \underline{I}_{01}\,\mathrm{e}^{-\mathrm{j}\beta z} + \underline{I}_{02}\,\mathrm{e}^{\mathrm{j}\beta z}$ $(0 \le z \le l)$, where $\underline{I}_{01}$ and $\underline{I}_{02}$ are the magnitudes (rms values) of two current waves traveling in opposite directions along the antenna and $\beta = \omega/c_0$ is the free-space phase coefficient, find the far electric and magnetic field intensity vectors and Poynting vector of the antenna.

14.10. **Attenuated traveling current wave.** If the traveling current wave along the antenna in Fig. 14.8 is attenuated (due to losses on the antenna), and hence the current distribution in Eq. (14.41) becomes $\underline{I}(z) = \underline{I}_0\,\mathrm{e}^{-\alpha z}\,\mathrm{e}^{-\mathrm{j}\beta z}$ $(0 \le z \le l)$, where α is the associated attenuation coefficient, determine the Poynting vector in the far zone of the antenna.

14.11. **Uniform line source.** A uniform line source, defined as a uniform current of complex rms intensity $\underline{I}_{\rm u} =$ const and frequency f along a straight line of (arbitrary) length l, radiates in free space. Considering this source (line current) as a straight metallic wire antenna along the z-axis, Fig. 14.4, with a constant current ($\underline{I}_{\rm u}$), and adopting the coordinate origin at the center of the source, so that $-l/2 \le z \le l/2$ along the wire, find the far electric and magnetic field intensity vectors, and the associated Poynting vector, due to the source.

14.12. **Short uniform line source.** Show that for $l \ll \lambda_0$ (λ_0 being the free-space wavelength at the frequency f) the field expressions obtained in the previous problem reduce to those in Eqs. (14.17), for a Hertzian dipole (electrically short uniform line source).

14.13. **Long piece-wise uniform line source.** Consider a two-step, piece-wise uniform line source of (arbitrary) length l and frequency f, defined by the following current distribution along the z-axis of a spherical coordinate system: $\underline{I}(z) = \underline{I}_{\rm u}$ for $0 \le |z| \le l/4$ and $\underline{I}(z) = \underline{I}_{\rm u}/2$ for $l/4 < |z| \le l/2$ ($\underline{I}_{\rm u} =$ const). The permittivity and permeability of the ambient medium are ε and μ, respectively. (a) Find the far-field vectors due to this source. (b) What is the accompanying charge distribution of the source?

14.14. Hertzian dipole along the *y*-axis. Repeat Example 14.4 but for a Hertzian dipole along the *y*-axis of the Cartesian coordinate system.

14.15. Hertzian dipole in the *xy*-plane. Repeat Example 14.4 but for a Hertzian dipole positioned at the coordinate origin along the line defined by the vector $\hat{\mathbf{x}} + \hat{\mathbf{y}}$.

14.16. Straight wire antenna along the *x*-axis. Assume that the straight wire antenna of length l in Fig. 14.4 runs along the x-axis, instead of the z-axis, and that it carries an arbitrary current distribution given by a function $\underline{I}(x)$. The operating frequency of the antenna is f and the material parameters of the ambient medium are ε and μ. (a) Use Eqs. (14.32) and (14.46) to determine the expressions for the radiation integral and far-zone magnetic potential of the antenna, analogous to those in Eq. (14.29). Then, as in Example 14.4, find the expression for computing the far electric field intensity vector of the antenna, in place of that in Eq. (14.34).

14.17. Uniform line source along the *x*-axis. Find the far electric field due to an x-directed uniform line source of (arbitrary) length l, with a current of complex rms intensity $\underline{I}(x) = \underline{I}_{\mathrm{u}} = \text{const}$ for $-l/2 \leq x \leq l/2$ and frequency f, radiating in free space.

14.18. Wire antenna and uniform line source along the *y*-axis. (a) Repeat Problem 14.16 but for a straight wire antenna with an arbitrary current distribution $\underline{I}(y)$ along the y-axis. (b) Repeat Problem 14.17 but for an y-directed uniform line source of current intensity $\underline{I}(y) = \underline{I}_{\mathrm{u}} = \text{const}$ for $-l/2 \leq y \leq l/2$.

14.19. Radiation efficiency of a short dipole with cosine current. If the short wire antenna with a cosine current distribution described in Problem 14.7 is made out of steel ($\sigma = 2$ MS/m and $\mu_{\mathrm{r}} = 2000$), is $l = 1$ m long and $a = 3$ mm in radius, and operates at a frequency of $f = 15$ MHz, find (a) the radiation resistance, (b) the high-frequency ohmic resistance, and (c) the radiation efficiency of the antenna.

14.20. Losses in a short dipole with quadratic current. For the short copper wire dipole antenna with quadratic current from Problem 14.8, compute (a) the time-average radiated power of the antenna, (b) the time-average power of Joule's (ohmic) losses in the wire, and (c) the radiation efficiency of the antenna.

14.21. Computing the radiated power of *x*- and *y*-directed dipoles. (a) Starting with the far-zone complex Poynting vector of an x-directed Hertzian dipole in Eq. (14.51), and integrating the corresponding time-average surface power density through a spherical surface of radius r ($r \gg \lambda$) centered at the coordinate origin, as in Eq. (14.52), compute the radiated power and radiation resistance of the dipole. (b) Repeat (a) but for a y-directed Hertzian dipole (from Problem 14.14).

14.22. Radiation functions of 12 different antennas. Find the characteristic radiation functions of all antennas studied in Problems 14.7–14.18.

14.23. Gains of steel and copper short antennas. Compute the gain of steel and copper nonloaded short dipoles from Problems 14.19 and 14.20, respectively.

14.24. Sine squared, cosine squared unidirectional pattern. The normalized field pattern of an antenna is given by $f(\theta, \phi) = \sin^2\theta \cos^2\phi$ for $0 \leq \theta \leq 180°$ and $-90° \leq \phi \leq 90°$, and is zero elsewhere. Find (a) the directivity of the antenna and (b) the half-power beamwidths in planes $\phi = 0$ (elevation plane) and $\theta = 90°$ (azimuthal plane), respectively.

14.25. Sectoral radiation pattern. Compute the directivity of an antenna with a sectoral radiation pattern (having uniform radiation intensity over a specified angular region, while being zero elsewhere) given by the following normalized radiation function: $f(\theta) = 1$ for $\theta_1 \leq \theta \leq \theta_2$ and $f(\theta) = 0$ elsewhere – if (a) $\theta_1 = 0$ and $\theta_2 = \alpha$ (conical pattern) and (b) $\theta_1 = 90° - \alpha$ and $\theta_2 = 90° + \alpha$, respectively, where α is an arbitrary angle within the range $0 < \alpha < 90°$, and specifically for $\alpha = 30°$.

14.26. Aircraft radar antenna with a cosecant pattern. The antenna of a target-search ground-mapping radar placed on an aircraft flying parallel to the ground has a cosecant radiation pattern in the elevation plane, which, for the angle θ defined as in Fig. 14.46, is given by the normalized field function $f_\theta(\theta) = \csc\theta = 1/\sin\theta$ for $\theta_1 \leq \theta \leq \theta_2$, while $f_\theta(\theta) = 0$ elsewhere. (a) Use Eq. (14.78) and a trigonometric relationship between the antenna-to-ground distance, r, and θ in Fig. 14.46 to show that the magnitude of the radiated electric field at the ground surface is constant (not a function of r or θ) in the specified range of θ (from θ_1 to θ_2), namely, that the radar uniformly illuminates the area that is searched on the ground. (b) If, in addition, the radiation pattern is uniform in the azimuthal direction within a small range of the angle ϕ (sectoral pattern), so that the total normalized field pattern of the radar antenna is given by $f(\theta, \phi) = \csc\theta$ for $\theta_1 \leq \theta \leq \theta_2$ and $-\Delta\phi/2 \leq \phi \leq \Delta\phi/2$, and is zero elsewhere, express the antenna directivity in terms of θ_1, θ_2, and $\Delta\phi$, and compute its decibel value for $\theta_1 = 30°$, $\theta_2 = 90°$, and $\Delta\phi = 10°$.

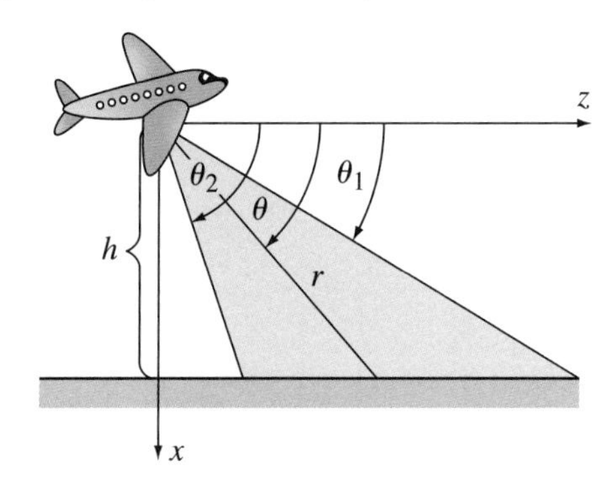

Figure 14.46 Aircraft radar antenna with a cosecant radiation pattern, which enables uniform illumination of the ground surface; for Problem 14.26.

14.27. **Current and radiation pattern of a wire dipole at six frequencies.** A symmetrical wire dipole antenna of length $l = 3$ m is fed by a current of rms intensity $I_0 = 1$ A, and radiates in free space. Sketch (a) the current distribution along the dipole arms and (b) E-plane radiation field pattern of the antenna if its operating frequency is $f_1 = 10$ MHz, $f_2 = 50$ MHz, $f_3 = 100$ MHz, $f_4 = 150$ MHz, $f_5 = 200$ MHz, and $f_6 = 250$ MHz, respectively. (c) Discuss the relationship between the respective diagrams in (a) and (b).

14.28. **Charge distribution of an arbitrary dipole antenna.** Consider a symmetrical wire dipole antenna with an arbitrary length, l, and sinusoidal current distribution, in Fig. 14.18(a). The antenna feed current is $\underline{I}_0$, its operating frequency is f, and the ambient medium is free space. (a) Evaluate the charge distribution along the antenna arms, and (b) show that for $l \ll \lambda$ it reduces to that of the short dipole – in Fig. 14.7.

14.29. **Ohmic resistance of an arbitrary dipole antenna.** Find the expression for the high-frequency ohmic resistance of the (arbitrary) dipole antenna in Fig. 14.18(a), if it radiates in free space at a frequency f, is l long, with radius a, and is made of a metal of conductivity σ and permeability μ.

14.30. **Shunt short-circuited stub connected to a half-wave dipole.** (a) Design a shunt short-circuited stub (see Example 12.19) to compensate (annul) the imaginary part of the complex input impedance, $\underline{Z}_A$, of a half-wave dipole in Eq. (14.132), at a frequency of $f = 500$ MHz. (b) What is the SWR with respect to $Z_0 = 75\ \Omega$ of thus obtained purely real impedance, R_A? (c) Design a quarter-wave transformer (see Example 12.8) to match R_A to $Z_0 = 50\Omega$.

14.31. **Proof of image theory for a horizontal Hertzian dipole.** Repeat Example 14.17 but for a horizontal Hertzian dipole and its image obtained by image theory in Fig. 14.23.

14.32. **Proof of image theory for an oblique Hertzian dipole.** (a) Repeat Example 14.17 but for an oblique Hertzian dipole, making an angle α with the normal to the ground plane, and its image (Fig. 14.23). (b) Show that the same conclusion about the satisfaction of boundary conditions on the ground surface is obtained also decomposing the dipole length vector, together with the associated current and charge, onto a vertical and horizontal components, and using the corresponding results for a vertical and horizontal Hertzian dipoles.

14.33. **Reflected wave of a vertical dipole above the ground plane.** If the half-wave wire dipole antenna in Fig. 14.25 is rotated about its center by 90° so that it becomes vertical, find (for the numerical data given in Example 14.18) the magnitude of the electric field vector at the point P due to (a) direct wave (propagating parallel to the ground) and (b) reflected wave (bouncing off the ground).

14.34. **Magnetic-field surface to ionospheric wave ratio.** Assume that the monopole antenna radiating at the earth's surface in the presence of ionosphere as described in Example 14.20 is $h = 1$ m long, and express the ratio of rms magnetic field intensities of the surface wave and resultant ionospheric wave, $|\underline{\mathbf{H}}_{\text{surf}}|/|\underline{\mathbf{H}}_{\text{iono}}|_{\text{tot}}$, on the ground in terms of the distance from the antenna, r, and virtual height of the ionosphere, h_{v}. Then compute this ratio for $r = n \times 50$ km and $n = 1, 2, \ldots, 10$, respectively.

14.35. Multiple reflections of an ionospheric wave. Consider the radiation of a quarter-wave monopole antenna in the presence of ionosphere in Fig. 14.29(a), and compute (for the numerical data from Example 14.20) at the point P the rms electric field intensity of the resultant ionospheric wave bouncing twice off the ionosphere, as illustrated by the multiple-reflection zigzag path between points O and P in Fig. 14.47(a). Perform the analysis both (a) by direct computations on the zigzag path and (b) by multiple applications of the image theory for the two PEC planes in Fig. 14.47(a) (find the image in one plane of the antenna representing the previously found image in the other plane, and so on) to establish the equivalent straight path from an image antenna to the receive point [analogous to the path in Fig. 14.29(b)], which is indicated in Fig. 14.47(b).

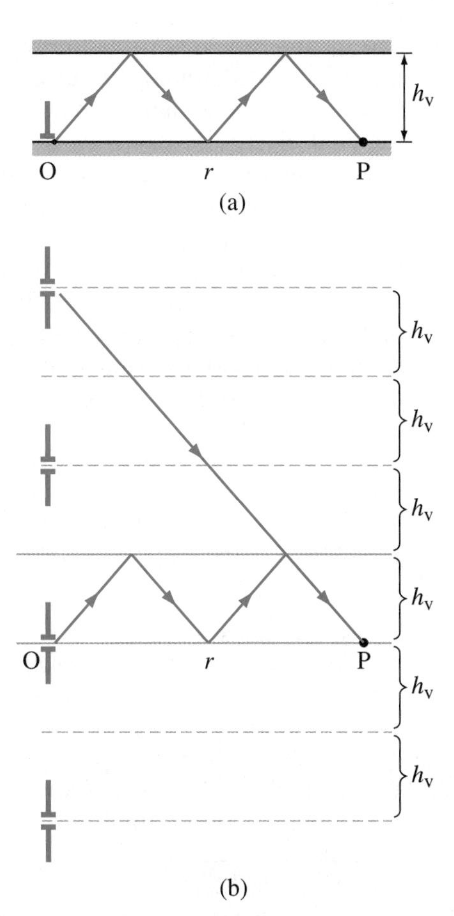

Figure 14.47 Reception of the ionospheric wave (launched by a quarter-wave monopole antenna) bouncing twice off the ionosphere: (a) multiple-reflection zigzag path of the wave and (b) multiple applications of the image theory for both the earth and ionospheric PEC planes to establish the equivalent straight path from an image antenna to the receive point; for Problem 14.35.

14.36. Poynting vector and radiated power of a small loop antenna. Find the expression for the far-zone complex Poynting vector of a magnetic dipole (small loop) antenna in Fig. 14.30. Starting with this expression, compute the time-average radiated power of the antenna. Compare the result to that obtained using the radiation resistance in Eq. (14.160).

14.37. Lorenz condition for potentials of a small loop antenna. (a) Find the line charge density, $\underline{Q}'$, along the wire of a magnetic dipole antenna (Fig. 14.30). (b) From the result in (a), compute the electric scalar potential, $\underline{V}$, of the antenna, at an arbitrary distance from it. (c) Show that $\underline{V}$ and the magnetic vector potential, $\underline{\mathbf{A}}$, in any field zone of the antenna satisfy the Lorenz condition for complex electromagnetic potentials.

14.38. General field expressions for a small loop antenna. (a) Using Eqs. (8.118) and (8.119), find – from potentials $\underline{V}$ and $\underline{\mathbf{A}}$ – the expressions for electric and magnetic field vectors, $\underline{\mathbf{E}}$ and $\underline{\mathbf{H}}$, of a magnetic dipole antenna valid for all field zones (not necessarily the far zone). (b) Compare the obtained field expressions with those in Eqs. (14.8) and (14.10) for a Hertzian dipole (Fig. 14.1) in light of the general duality principle for electromagnetic fields due to electric and magnetic sources.

14.39. Switching places of transmitting and receiving antennas. Assume that in the wireless link with two nonaligned half-wave dipole antennas shown in Fig. 14.34(a) and described in Example 14.22, the transmitting and receiving antennas switch places. Namely, the dipole on the right-hand side (lying in the $x'y'$-plane) is now fed by the input power $P_{\text{in}} = 10$ W and is transmitting at the frequency $f = 300$ MHz, while the terminals of the dipole on the left-hand side (in the xz-plane) are left open. Under these circumstances, find the magnitude of the open-circuit voltage received by the latter antenna – in the following two ways: (a) invoking the electromagnetic reciprocity theorem and (b) carrying out the analysis based on Eq. (14.166), like in Example 14.22, respectively.

14.40. Nonaligned half-wave dipole and small loop. In the wireless system in Fig. 14.34(a), the dipole antenna at the receive end is replaced by a small square loop antenna with edge length a, which is positioned such that its plane and the x'-axis make an angle γ_2, as shown in Fig. 14.48. (a) In this new system, where γ_1 and r ($r \gg \lambda$) are also given parameters, assume that the half-wave dipole antenna is fed by a time-harmonic current of complex rms intensity $\underline{I}_0$ and frequency f, and find the open-circuit complex rms voltage received by the loop antenna – using Faraday's law of electromagnetic induction and radiated magnetic field of the dipole. (b) Repeat (a) but using Eq. (14.166) and radiated electric field of the dipole. (c) Then assume that the loop antenna is fed by the same current as in (a), and compute the open-circuit complex rms voltage received by the half-wave dipole antenna – by means of Eq. (14.166) and radiated electric field of the loop. (d) Finally, repeat (c) but employing the electromagnetic reciprocity theorem.

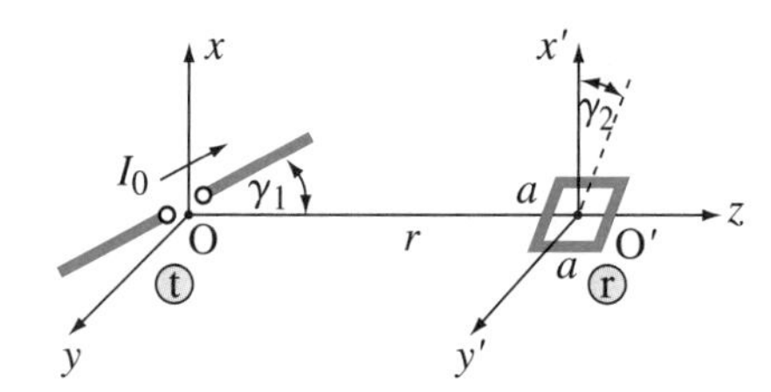

Figure 14.48 Wireless link with nonaligned half-wave dipole and small loop antennas; for Problem 14.40.

14.41. Reception of a CP wave by a magnetic dipole antenna. The feed currents of two crossed transmitting dipoles, both with length $l = 6$ cm, in the wireless link in free space in Fig. 14.49 have the same frequency, $f = 300$ MHz, and the same magnitude, $I_0 = 1$ A, but are in time-phase quadrature, with dipole 1 advancing by 90° in phase with respect to dipole 2. Find the magnitude of the open-circuit voltage of the receiving square loop antenna with edge length $a = 4$ cm, which is $r = 30$ m away from the transmit end and whose plane makes an angle of $\gamma = 30°$ with the x'-axis (note: first show that all three antennas are electrically small).

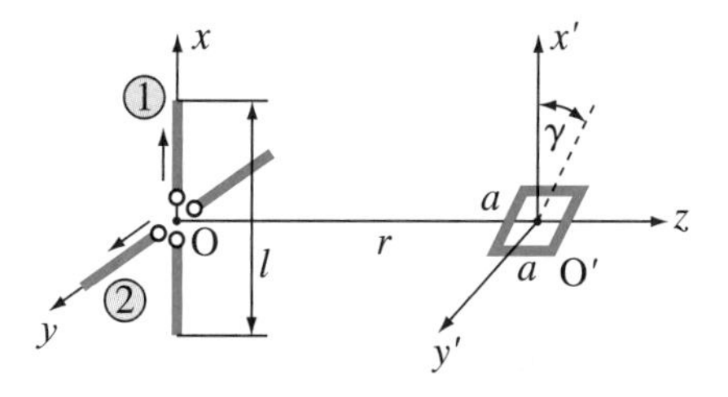

Figure 14.49 Wireless communication system with two crossed dipoles at the transmit and a square loop at the receive end; for Problem 14.41.

14.42. Large square plate scatterer. If in the wireless system in Fig. 14.49 the receiving loop antenna is replaced by an electromagnetic scatterer in the form of a square PEC plate, with edge length $a = 10$ m, which is centered at the point O′ and whose one pair of edges is parallel to the x'-axis and the other pair to the y'-axis, compute the complex rms surface induced current density vector ($\underline{\mathbf{J}}_s$) at the center of the plate. Note that, as the plate is electrically large, the induced current at its center can be evaluated as if the plate were infinitely large (PEC plane). Compare the result to that of Example 10.7, part (d).

14.43. Reception of an EP wave by a magnetic dipole antenna. Repeat Problem 14.41 but for the feed current of dipole 1 lagging in phase the feed current of dipole 2 by 45° (instead of advancing it by 90°). Then find the angle γ for which the received voltage is maximum, as well as that maximum voltage.

14.44. Receiving loop in the plane of transmitting crossed dipoles. Repeat Problem 14.41 but for the receiving loop antenna placed in the plane of crossed transmitting dipole antennas, as shown in Fig. 14.50, where the location of the loop center (point O′) in the plane is defined by an arbitrary angle ϕ ($r = 30$ m).

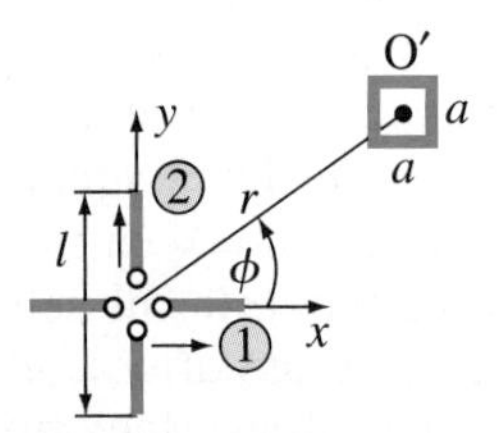

Figure 14.50 The same as in Fig. 14.49 but with all antennas in the same plane; for Problem 14.44.

14.45. Link with vertical and oblique dipoles above ground. In the wireless link in Fig. 14.51, the two half-wave dipole antennas lie in the same vertical plane, above a horizontal PEC ground. The transmitting dipole is vertical, while the receiving one makes an angle of $\gamma = 30°$ with the vertical direction. The ambient medium is air, the operating frequency is $f = 1$ GHz, and the input power of the transmitting antenna is $P_{\text{in}} = 730$ mW. The horizontal distance between the transmit and receive ends of the link is $r = 70$ m, and the height of centers of both antennas with respect to the ground is $h = 24.5$ m. Calculate the magnitude of the open-circuit voltage of the receiving dipole due to (a) direct wave and (b) reflected wave.

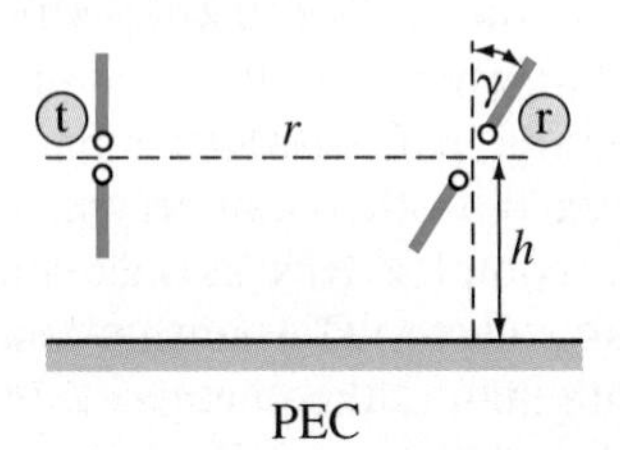

Figure 14.51 Communication between a vertical and an oblique half-wave dipole above a ground plane; for Problem 14.45.

14.46. Two collinear or parallel half-wave dipoles above ground. Both the transmitting and receiving antennas in a free-space wireless link at a frequency of $f = 500$ GHz are half-wave dipoles parallel to a PEC ground plane, at the same height, $h = 20$ m, with respect to it. The distance between dipole centers is $r = 40$ m, and the input power of the transmitting antenna is $P_{\text{in}} = 1$ W. Find the open-circuit rms voltage of the receiving antenna due to the direct and reflected waves, respectively, if the two dipoles are oriented to be (a) collinear (coaxial) with respect to each other, as in Fig. 14.52(a), or (b) parallel to each other, as in Fig. 14.52(b).

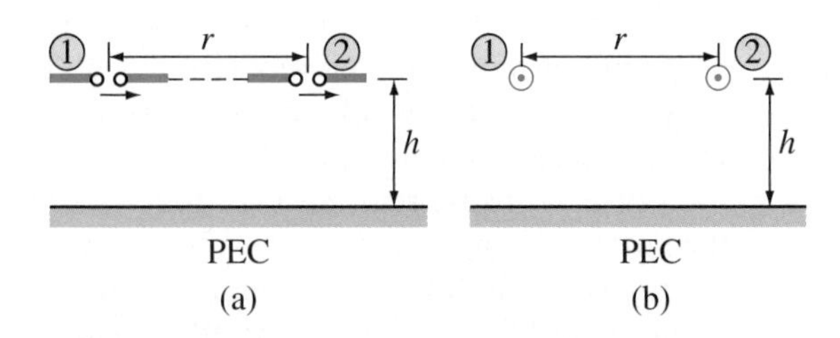

Figure 14.52 Communication between two half-wave dipole antennas above a ground plane: (a) collinear and (b) parallel dipoles; for Problem 14.46.

14.47. Communication between a monopole and loop via ionosphere. In the communication system in Fig. 14.53, the transmitting antenna is the quarter-wave monopole antenna described in Example 14.20, whereas the receiving antenna is a small square loop antenna with edge length $a = 6$ cm and $N = 20$ turns of wire. There is no direct line of sight between the antennas. All other parameters of the system are as in Fig. 14.29(a). Under these circumstances, find the magnitude of the received open-circuit voltage of the loop.

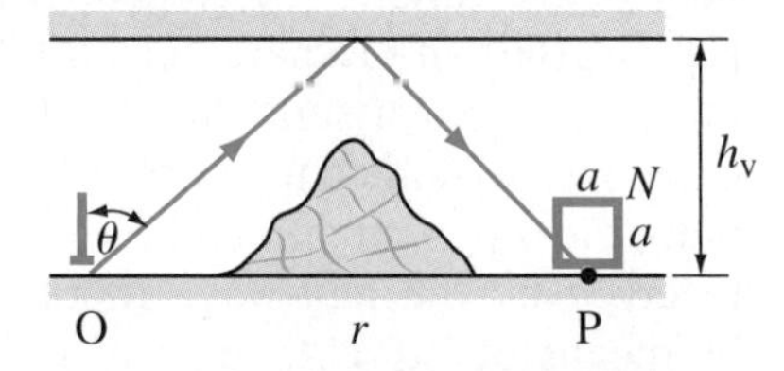

Figure 14.53 Communication system with a quarter-wave monopole and small loop antenna in the presence of ionosphere and no direct line of sight: ray path of the principal ionospheric wave, bouncing once off the ionosphere; for Problem 14.47.

14.48. Horizontal short dipole and small loop near ground. A time-harmonic signal emitted by a horizontal short wire dipole, of length $l = 3$ cm, positioned in air at height $h = 5$ m with respect to a PEC ground plane is being received by a small square loop antenna with edge length $a = 2.5$ cm and $N = 10$ turns of wire, placed immediately above the ground in the same vertical plane with the dipole, as depicted in Fig. 14.54 (see also Fig. 14.53). The rms intensity and frequency of the feed current of the transmitting dipole are $I_0 = 1$ A and $f = 1$ GHz, respectively, and the horizontal distance between the transmit and receive ends of the communication link is $d = 8$ m. Compute the rms voltage across open terminals of the receiving loop.

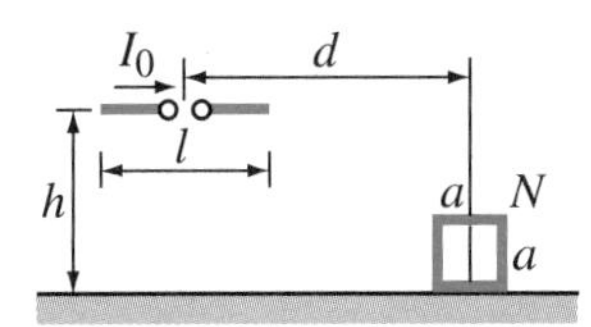

Figure 14.54 Communication between a horizontal short wire dipole and small square loop antenna placed immediately above a ground plane; for Problem 14.48.

14.49. **Reciprocity for a dipole-loop-ground system.** Assume that the small loop in Fig. 14.54 is fed by a time-harmonic current of rms intensity $I_0 = 2$ A (note a different current intensity with respect to the previous problem) to serve as a transmitting antenna, while the horizontal short dipole operates as a receiving antenna. Find the open-circuit rms voltage of the dipole – using the solution to the previous problem and electromagnetic reciprocity theorem, if the operating frequency of the loop is (a) $f_1 = 1$ GHz and (b) $f_2 = 750$ MHz, respectively.

14.50. **Effective aperture of a quarter-wave monopole.** Find the effective aperture (A_{eff}) of a quarter-wave wire monopole receiving antenna, Fig. 14.27(a), illuminated in free space by a uniform plane time-harmonic electromagnetic wave of frequency $f = 1$ GHz, for directions of the wave incidence given by $\theta_{i1} = 0$, $\theta_{i2} = 30°$, $\theta_{i3} = 60°$, and $\theta_{i4} = 90°$, respectively, assuming that the antenna is impedance-matched to its load and polarization-matched to the incident wave.

14.51. **Aperture efficiency and gain of a parabolic reflector antenna.** A parabolic reflector antenna [Fig. 14.36(c)] with aperture diameter $d = 3.7$ m operates at a frequency of $f = 11$ GHz in free space. The aperture efficiency of the reflector is $\eta_{\text{aperture}} = 70\%$. What is the dB gain of the antenna?

14.52. **Long-range microwave communication system.** A long-range microwave communication system operating at a frequency of $f = 20$ GHz in free space uses two identical antennas at the transmit and receive ends, each having gain of $G_{\text{dB}} = 20$ dB. For acceptable signal-to-noise ratio (SNR) of the receiver, the received power must be at least $P_{\text{r}} = 1\ \mu$W. For the input power of the transmitting antenna of $P_{\text{in}} = 100$ W and ideal antenna matching and orientation conditions (Fig. 14.37), find the largest separation (r) between antennas such that the receiver SNR is still acceptable.

14.53. **Two X-band horn antennas in an anechoic chamber.** Two identical X-band (8 – 12 GHz) pyramidal horn antennas [Fig. 13.14(b) or 14.36(b)] with aperture edge lengths $a_{\text{horn}} = 8$ cm and $b_{\text{horn}} = 6$ cm are placed as transmitting and receiving antennas in an anechoic chamber, at a distance of $r = 3$ m between them. When the input power of the transmitting antenna is $(P_{\text{in}})_{\text{dBm}} = 20$ dBm at a frequency of $f = 10$ GHz, the power received at the other end of the system is $P_{\text{r}} = 63\ \mu$W under ideal matching and orientation conditions. What is the aperture efficiency (η_{aperture}) of the two horns?

14.54. **Two-element array of point sources with cardioid pattern.** Repeat Example 14.29 but for a two-element array of point sources with a quarter-wave interelement spacing ($d = \lambda/4$) and in time-phase quadrature, where element 2 lags by 90° in phase with respect to element 1 (elements are still fed with same current magnitudes). Note that the obtained 2-D polar plot of the normalized array factor should be a cardioid pattern.

14.55. **Full-wave interelement spacing and counter-phase excitation.** Repeat Example 14.31 but for point sources in counter-phase ($d = \lambda$ and $\delta = \pm 180°$). In addition, find the polar angles $\theta_{\max}$ at which maxima of the array factor occur.

14.56. **Three-quarter-wave separation between in-phase sources.** Repeat Example 14.31 but for a three-quarter-wave separation between point sources ($d = 3\lambda/4$ and $\delta = 0$), including finding directions of radiation pattern nulls.

14.57. **Array of two collinear Hertzian dipoles fed in counter-phase.** Repeat Example 14.32 but for an array of two collinear Hertzian dipole antennas with currents in counter-phase, i.e., shifted in phase with respect to each other by 180° (element centers are still half-wave apart, with equal current magnitudes).

14.58. **Array of two parallel dipoles with counter-phase excitation.** Repeat Example 14.33 but for an array of two parallel Hertzian dipole antennas fed in counter-phase [all other input parameters of the system are the same as in Fig. 14.43(a)].

14.59. **Array of two collinear half-wave dipoles with full-wave spacing.** Consider an array of two collinear half-wave dipole antennas, with a full-wave separation between dipole centers, radiating at a frequency f in free space. The dipoles are fed in phase and with equal input powers. Find the total characteristic radiation function of this antenna array and sketch it as a normalized pattern.

14.60. **Two vertical quarter-wave monopoles, full-wave separation.** Two identical vertical wire monopole antennas of length $h = 25$ cm are attached to a horizontal PEC ground plane and fed against it with time-harmonic currents of frequency $f = 300$ MHz and equal complex intensities. The separation between the monopole axes is $d = 1$ m, and the medium above the ground is air. Determine the expression for the total radiation pattern of this system and sketch its cuts in three characteristic planes for the system, namely, the plane containing the two monopoles, the plane perpendicular to the array axis and containing the center of the array, and the plane perpendicular to the antennas and containing the array axis.

14.61. **Wire dipole parallel to a PEC plane.** A wire dipole antenna of length $l = 3$ m, radiating at a frequency of $f = 50$ MHz, is positioned in air parallel to a PEC plane, at a distance $d = 1.5$ m from it. Find the resultant characteristic radiation function of this antenna system and sketch the associated pattern cuts in three characteristic planes (all containing the projection of the dipole center on the PEC surface).

14.62. **Two small loops, quarter-wave spacing, quadrature excitation.** Two identical small wire loop antennas of surface area S operating at a frequency f are placed in the same plane in free space, at a quarter-wave separation between their centers, and the ratio of their complex current intensities is $\underline{I}_1/\underline{I}_2 = \mathrm{j}$. Compute and sketch the overall radiation pattern cuts of this antenna array in three characteristic planes.

14.63. **Nonuniform three-element array of collinear dipoles.** Consider an array of three collinear (coaxial) half-wave wire dipole antennas, with centers spaced a half-wavelength apart (however, the dipoles are not quite touching). The feed currents of dipoles are all in phase, but their magnitudes are in the ratio $1:2:1$ along the array. Under these circumstances, compute and sketch the total characteristic radiation function of the array.

14.64. **Nonuniform three-element array of parallel dipoles.** A nonuniform array of three in-phase parallel half-wave wire dipole antennas has a half-wave separation between adjacent element points, which lie on the x-axis of a Cartesian coordinate system, with the central point coinciding with the coordinate origin and the dipoles being z-directed. The ratio of current magnitudes along the array is $1:3:1$. Determine the expression for the total radiation pattern of this antenna array and sketch its cuts in planes xy, xz, and yz, respectively.

14.65. **Nonuniform, counter-phase array of parallel dipoles.** Repeat the previous problem but for a 180° phase shift between feed currents of adjacent elements in the array ($\alpha_1 = -\pi, \alpha_2 = 0$, and $\alpha_3 = \pi$); all other parameters of the array are the same.

14.66. **Broadside and endfire arrays with arbitrary numbers of elements.** For a free-space uniform linear antenna array with an arbitrary number (N) of point source elements, Fig. 14.45(a), and a given interelement spacing d and operating frequency f, find the excitation interelement phase shift, α, such that the array radiation is (a) broadside and (b) endfire.

Quantities, Symbols, Units, and Constants

APPENDIX 1

Symbol	Quantity or Parameter	SI Unit (and Value)	Defined In:
$\mathbf{A}$	Magnetic vector potential	T · m (tesla-meter)	Section 4.9
$A_{\rm dB}$	Decibel attenuation (attenuation in dB)	dB (decibel)	Section 9.7
$A_{\rm eff}$	Antenna effective aperture	m^2 (meter squared)	Section 14.13
$\mathbf{a}$	Acceleration (vector)	m/s^2 (s – second)	Section 6.1
$\mathbf{B}$	Magnetic flux density vector	T (tesla)	Section 4.1
B	Susceptance	S (siemens)	Section 12.6
b	Normalized transmission-line susceptance	dimensionless	Section 12.6
C	Capacitance	F (farad)	Section 2.12
C'	Capacitance per unit length (p.u.l.)	F/m	Section 2.12
c	Velocity of electromagnetic (EM) waves	m/s	Section 8.9
c_0	Velocity of EM waves in free space	299,792,458 m/s	Section 9.3
$\mathbf{D}$	Electric flux density (displacement) vector	C/m^2 (C – coulomb)	Section 2.5
D	Antenna directivity	dimensionless	Section 14.6
e	Charge of electron, magnitude	1.602×10^{-19} C	Section 1.1
$\mathbf{E}$	Electric field intensity vector	V/m (volt per meter)	Section 1.2
$E_{\rm cr}$	Dielectric strength of a material	V/m	Section 2.6
$E_{\rm cr0}$	Dielectric strength of air	3 MV/m $(M \equiv 10^6)$	Section 2.6
$\mathbf{E}_{\rm i}$	Impressed electric field intensity vector	V/m	Section 3.10
$\mathbf{E}_{\rm ind}$	Induced electric field intensity vector	V/m	Section 6.1
$\mathbf{E}_q$	Field due to excess charge	V/m	Section 6.2
$\mathcal{E}, e$	Electromotive force (emf) of a generator	V (volt)	Section 3.10
$e_{\rm ind}$	Induced electromotive force (emf)	V	Section 6.3
$\mathbf{F}_{\rm e}$	Electric (Coulomb) force	N (newton)	Section 1.1
$\mathbf{F}_{\rm m}$	Magnetic force	N	Section 4.1
f	Frequency	Hz (hertz)	Ch. 6, Intro
$f_{\rm p}$	Plasma frequency	Hz	Section 9.12
$f_{\rm res}$	Resonant frequency of an EM resonator	Hz	Section 10.1
$f_{\rm c}$	Cutoff frequency of a waveguide mode	Hz	Section 13.2
$\underline{\mathbf{F}}$	Antenna characteristic radiation function	dimensionless	Section 14.5
f	Antenna normalized field pattern	dimensionless	Section 14.5
$\underline{F}_{\rm a}$	Antenna array factor (AF)	dimensionless	Section 14.15
G	Conductance	S	Section 3.8

(*Continued*)

Symbol	Quantity or Parameter	SI Unit (and Value)	Defined In:
G'	Conductance per unit length	S/m	Section 3.12
g	Normalized transmission-line conductance	dimensionless	Section 12.6
G	Antenna gain	dimensionless	Section 14.6
g	Standard acceleration of free fall	9.81 $\mathrm{m/s^2}$	Example 7.23
$\mathbf{H}$	Magnetic field intensity vector	A/m (amp per meter)	Section 5.4
HPBW	Half-power beamwidth of an antenna	degree (°)	Section 14.5
I, i	Current intensity (or current)	A (ampere or amp)	Section 3.1
I_m	Magnetization current	A	Section 5.3
$\mathbf{J}$	Current density vector (conduction current)	$\mathrm{A/m^2}$	Section 3.1
$\mathbf{J}_\mathrm{i}$	Impressed current density vector	$\mathrm{A/m^2}$	Section 3.10
$\mathbf{J}_\mathrm{m}$	Magnetization volume current density vector	$\mathrm{A/m^2}$	Section 5.3
$\mathbf{J}_\mathrm{eddy}$	Density of eddy currents	$\mathrm{A/m^2}$	Section 6.8
$\mathbf{J}_\mathrm{d}$	Displacement current density vector	$\mathrm{A/m^2}$	Section 8.1
$\mathbf{J}_\mathrm{s}$	Conduction surface current density vector	A/m	Section 3.1
$\mathbf{J}_\mathrm{ms}$	Magnetization surface current density vector	A/m	Section 5.3
k	Coefficient of magnetic coupling	dimensionless	Section 7.3
l	Length	m (meter)	Section 1.3
l_e	Electrical length	dimensionless	Section 12.6
L	Self-inductance	H (henry)	Section 7.1
L_{21}	Mutual inductance	H	Section 7.2
L_i	Internal inductance	H	Section 7.6
L'	Inductance per unit length	H/m	Example 7.3
$\mathbf{m}$	Magnetic dipole moment	$\mathrm{A \cdot m^2}$	Section 4.11
$\mathbf{M}$	Magnetization vector	A/m	Section 5.1
m	Mass	kg (kilogram)	Section 5.2
N_v, N	Concentration (number/$\mathrm{m^3}$) of particles	$\mathrm{m^{-3}}$	Section 2.2
n	Index of refraction	dimensionless	Example 9.18
$\mathbf{p}$	Electric dipole moment	$\mathrm{C \cdot m}$	Section 1.11
$\mathbf{P}$	Polarization vector	$\mathrm{C/m^2}$	Section 2.2
p_e	Electric pressure	Pa = $\mathrm{N/m^2}$ (pascal)	Example 2.12
p_m	Magnetic pressure	Pa	Example 7.24
P	Power (instantaneous)	W (watt)	Section 8.12
P_J	Power of Joule's losses or ohmic losses	W	Section 3.3
P_mech	Mechanical power	W	Example 6.11
P_ave	Time-average power	W	Example 6.12
P_f	Power flow (transfer)	W	Section 8.11
p_J	Ohmic power density	$\mathrm{W/m^3}$	Section 3.3
P'_J	Power of Joule's losses per unit length	W/m	Section 3.12
$\mathcal{P}$	Poynting vector	$\mathrm{W/m^2}$	Section 8.11
$\underline{\mathcal{P}}$	Complex Poynting vector	$\mathrm{W/m^2}$	Section 8.12

Symbol	Quantity or Parameter	SI Unit (and Value)	Defined In:
$\mathcal{P}_{\text{ave}}$	Time-average Poynting vector	W/m^2	Section 8.12
Q, q	Charge	C (coulomb)	Section 1.1
Q_p	Bound (polarization) charge	C	Section 2.3
Q'	Line charge density (charge per unit length)	C/m	Section 1.3
Q	Reactive power	W	Section 8.12
Q	Quality factor of a resonator	dimensionless	Section 12.10
Q_c	Q factor for structure conductors	dimensionless	Section 12.10
Q_d	Q factor for structure dielectric	dimensionless	Section 12.10
R	Source-to-field distance	m	Section 1.2
R	Resistance	Ω (ohm)	Section 3.8
R_{gr}	Grounding resistance	Ω	Section 3.13
R_{rad}	Radiation resistance of an antenna	Ω	Section 14.4
R_{ohmic}	Ohmic resistance of an antenna	Ω	Section 14.4
R_s	Surface resistance of a good conductor	Ω/square (or Ω)	Section 10.3
R'	Resistance per unit length	Ω/m	Section 3.12
r	Normalized transmission-line resistance	dimensionless	Section 12.6
$\mathcal{R}$	Reluctance	H^{-1}	Section 5.10
S	Surface area	m^2	Section 1.3
$\underline{S}$	Complex power	W	Section 8.12
s	Standing wave ratio (SWR)	dimensionless	Example 10.9
t	Time	s (second)	Section 3.1
T	Period of time-harmonic oscillation	s	Ch. 6, Intro
T	One-way time delay period of a tr. line	s	Section 12.14
T	Temperature	K (kelvin) or °C	Section 3.2
$\mathbf{T}$	Torque	N · m	Section 2.1
U	Antenna radiation intensity	W/sr (W per steradian)	Section 14.6
v	Volume	m^3	Section 1.3
V	Electric scalar potential (at a point)	V	Section 1.6
V, v	Voltage (between two points)	V	Section 1.8
V_{cr}	Breakdown voltage	V	Section 2.17
V_{step}	Voltage of a step (for a grounding electrode)	V	Section 3.13
$\underline{V}$	Complex rms (root-mean-square) voltage	V	Section 8.7
$\underline{V}_{oc}$	Open-circuit voltage of a receiving antenna	V	Section 14.12
$\mathbf{v}$	Velocity (vector)	m/s	Section 3.1
v_p	Phase velocity	m/s	Section 9.4
v_g	Group velocity	m/s	Section 9.13
W	Work or energy	J (joule)	Section 1.6
W_e	Electric energy	J	Section 2.15
W_m	Magnetic energy	J	Section 7.4
W_{em}	Electromagnetic energy	J	Section 8.11

(*Continued*)

Symbol	Quantity or Parameter	SI Unit (and Value)	Defined In:
w_{e}	Electric energy density	$\mathrm{J/m^3}$	Section 2.16
w_{m}	Magnetic energy density	$\mathrm{J/m^3}$	Section 7.5
$\mathbf{w}$	Angular velocity (vector)	rad/s (radian per second)	Section 5.2
X	Reactance	Ω	Example 8.20
x	Normalized transmission-line reactance	dimensionless	Section 12.6
$\underline{Y}$	Complex admittance	S	Section 12.6
Y_0	Characteristic admittance of a tr. line	S	Section 11.4
$\underline{y}_{\mathrm{n}}$	Normalized transmission-line admittance	dimensionless	Section 12.6
$\underline{Z}$	Complex impedance	Ω	Example 8.20
Z_0	Characteristic impedance of a tr. line	Ω	Section 11.4
Z_{TEM}	TEM wave impedance	Ω	Section 11.1
Z_{TE}	TE wave impedance	Ω	Section 13.7
Z_{TM}	TM wave impedance	Ω	Section 13.7
$\underline{Z}_{\mathrm{A}}$	Antenna input impedance	Ω	Section 14.4
$\underline{z}_{\mathrm{n}}$	Normalized transmission-line impedance	dimensionless	Section 12.6
α	Attenuation coefficient	Np/m (neper per meter)	Section 9.7
α_{c}	Attenuation coeff. due to conductor losses	Np/m	Section 11.6
α_{d}	Attenuation coeff. due to dielectric losses	Np/m	Section 11.6
β	Phase coefficient or wavenumber	rad/m	Section 8.10
$\underline{\gamma}$	Complex propagation coefficient	$\mathrm{m^{-1}}$	Section 9.7
$\underline{\Gamma}$	Reflection coefficient	dimensionless	Section 10.2
δ	Skin depth	m	Section 9.11
$\tan\delta_{\mathrm{d}}$	Loss tangent	dimensionless	Section 9.9
ε	Permittivity of a dielectric material	F/m	Section 2.6
ε_0	Permittivity of a vacuum (free space)	$8.8542\ \mathrm{pF/m}\ \ (\mathrm{p} \equiv 10^{-12})$	Section 1.1
$\underline{\varepsilon}_{\mathrm{e}}$	Equivalent complex permittivity	F/m	Section 9.7
$\underline{\varepsilon}$	High-frequency complex permittivity	F/m	Section 9.9
ε_{p}	Plasma effective permittivity	F/m	Section 9.12
ε_{r}	Relative permittivity	dimensionless	Section 2.6
$\varepsilon_{\mathrm{reff}}$	Effective relative permittivity of a tr. line	dimensionless	Section 11.9
η	Intrinsic impedance of a medium	Ω	Section 9.3
η_0	Intrinsic impedance of free space	$\approx 120\pi\ \Omega \approx 377\ \Omega$	Section 9.3
$\underline{\eta}$	Complex intrinsic impedance	Ω	Section 9.7
$\underline{\eta}_{\mathrm{w}}$	Wave impedance	Ω	Example 10.11
η_{rad}	Radiation efficiency of an antenna	dimensionless	Section 14.4
η_{aperture}	Antenna aperture efficiency	dimensionless	Section 14.13
λ	Wavelength	m	Section 8.10
λ_0	Free-space wavelength	m	Section 9.5
λ_z	Wavelength along a tr. line or waveguide	m	Section 12.1
μ	Permeability of a magnetic material	H/m	Section 5.5

Symbol	Quantity or Parameter	SI Unit (and Value)	Defined In:
μ_0	Permeability of a vacuum (free space)	$4\pi \times 10^{-7}$ H/m	Section 4.1
μ_r	Relative permeability	dimensionless	Section 5.5
ρ	Volume charge density (free charge)	C/m^3	Section 1.3
ρ_p	Bound volume charge density	C/m^3	Section 2.3
ρ_s	Surface charge density	C/m^2	Section 1.3
ρ_{ps}	Bound surface charge density	C/m^2	Section 2.3
ρ	Resistivity of a medium	$\Omega \cdot$ m	Section 3.2
σ	Conductivity of a medium	S/m	Section 3.2
τ	Relaxation time or time constant	s	Section 3.7
$\underline{\tau}$	Transmission coefficient	dimensionless	Section 10.2
Φ	Magnetic flux	Wb (weber)	Section 4.8
χ_e	Electric susceptibility of a medium	dimensionless	Section 2.2
χ_m	Magnetic susceptibility of a medium	dimensionless	Section 5.2
Ψ	Electric flux	C	Section 2.5
ψ	Phase angle of a reflection coefficient	rad (radian)	Section 10.2
ω	Angular or radian frequency	rad/s	Example 6.4
Ω	Solid angle	sr (steradian)	Section 1.12
Ω_A	Beam solid angle of an antenna	sr	Section 14.6

Powers of Ten as Multipliers of Fundamental Units

Multiple	Prefix	Symbol	Multiple	Prefix	Symbol
10^{18}	Exa	E	10^{-2}	centi	c
10^{15}	Peta	P	10^{-3}	milli	m
10^{12}	Tera	T	10^{-6}	micro	μ
10^{9}	Giga	G	10^{-9}	nano	n
10^{6}	Mega	M	10^{-12}	pico	p
10^{3}	kilo	k	10^{-15}	femto	f
10^{2}	hecto	h	10^{-18}	atto	a

APPENDIX 2 Mathematical Facts and Identities

A2.1 TRIGONOMETRIC IDENTITIES

$$\sin(\alpha \pm \beta) = \sin\alpha\cos\beta \pm \cos\alpha\sin\beta$$

$$\cos(\alpha \pm \beta) = \cos\alpha\cos\beta \mp \sin\alpha\sin\beta$$

$$2\sin\alpha\sin\beta = \cos(\alpha - \beta) - \cos(\alpha + \beta)$$

$$2\sin\alpha\cos\beta = \sin(\alpha + \beta) + \sin(\alpha - \beta)$$

$$2\cos\alpha\cos\beta = \cos(\alpha + \beta) + \cos(\alpha - \beta)$$

$$\sin\alpha \pm \sin\beta = 2\sin\frac{\alpha \pm \beta}{2}\cos\frac{\alpha \mp \beta}{2}$$

$$\cos\alpha + \cos\beta = 2\cos\frac{\alpha + \beta}{2}\cos\frac{\alpha - \beta}{2}$$

$$\cos\alpha - \cos\beta = -2\sin\frac{\alpha + \beta}{2}\sin\frac{\alpha - \beta}{2}$$

$\sin 0 = 0$	$\cos 0 = 1$
$\sin 30^\circ = \frac{1}{2}$	$\cos 30^\circ = \frac{\sqrt{3}}{2}$
$\sin 45^\circ = \frac{\sqrt{2}}{2}$	$\cos 45^\circ = \frac{\sqrt{2}}{2}$
$\sin 60^\circ = \frac{\sqrt{3}}{2}$	$\cos 60^\circ = \frac{1}{2}$
$\sin 90^\circ = 1$	$\cos 90^\circ = 0$
$\sin 180^\circ = 0$	$\cos 180^\circ = -1$

$$\sin^2\alpha = \frac{1 - \cos 2\alpha}{2}$$

$$\cos^2\alpha = \frac{1 + \cos 2\alpha}{2}$$

$$\sin^2\alpha + \cos^2\alpha = 1$$

$$\sin 2\alpha = 2\sin\alpha\cos\alpha$$

$$\cos 2\alpha = \cos^2\alpha - \sin^2\alpha$$

$$\sin(-\alpha) = -\sin\alpha$$

$$\cos(-\alpha) = \cos\alpha$$

$$\sin(\alpha \pm 90^\circ) = \pm\cos\alpha$$

$$\cos(\alpha \pm 90^\circ) = \mp\sin\alpha$$

$$\tan\alpha = \frac{\sin\alpha}{\cos\alpha}$$

$$\cot\alpha = \frac{1}{\tan\alpha}$$

$$\sec\alpha = \frac{1}{\cos\alpha}$$

$$\csc\alpha = \frac{1}{\sin\alpha}$$

$c^2 = a^2 + b^2 - 2ab\cos\gamma$ (cosine formula, arbitrary triangle; angle γ is opposite to side c)

$c^2 = a^2 + b^2$ (Pythagorean theorem, right triangle; c is hypotenuse)

A2.2 EXPONENTIAL, LOGARITHMIC, AND HYPERBOLIC IDENTITIES

$e^x e^y = e^{x+y}$ $(e^x)^a = e^{ax}$	$e = 2.71828$ (base of natural logarithm) $j = \sqrt{-1}$ (imaginary unit)
$e^{jx} = \cos x + j\sin x$ (Euler's identity) $e^{jx} + e^{-jx} = 2\cos x$ $e^{jx} - e^{-jx} = 2j\sin x$	
$\log x = \log_{10} x$ (common logarithm) $\ln x = \log_e x$ (natural logarithm) $\ln e^x = x$	True for a logarithm of any base: $\log(xy) = \log x + \log y$ $\log \frac{x}{y} = \log x - \log y$ $\log x^a = a \log x$
Hyperbolic sine and cosine: $\sinh x = \frac{e^x - e^{-x}}{2}$ $\cosh x = \frac{e^x + e^{-x}}{2}$ $\sinh jx = j\sin x$ $\cosh jx = \cos x$	Hyperbolic tangent and cotangent: $\tanh x = \frac{\sinh x}{\cosh x}$ $\coth x = \frac{1}{\tanh x}$

A2.3 SOLUTION OF QUADRATIC EQUATION

$$ax^2 + bx + c = 0 \quad \longrightarrow \quad x = \frac{-b \pm \sqrt{b^2 - 4ac}}{2a}$$

A2.4 APPROXIMATIONS FOR SMALL QUANTITIES

For $|x| \ll 1$,

$(1+x)^a \approx 1 + ax$	$e^x \approx 1 + x$
$\sin x \approx x$	$\ln(1+x) \approx x$
$\cos x \approx 1 - \frac{x^2}{2}$	

A2.5 DERIVATIVES

$$\frac{\mathrm{d}}{\mathrm{d}x} x^c = cx^{c-1}$$

$$\frac{\mathrm{d}c}{\mathrm{d}x} = 0 \quad (c = \text{const})$$

$$\frac{\mathrm{d}}{\mathrm{d}x} \mathrm{e}^x = \mathrm{e}^x$$

$$\frac{\mathrm{d}}{\mathrm{d}x} \ln x = \frac{1}{x}$$

$$\frac{\mathrm{d}}{\mathrm{d}x} \sin x = \cos x$$

$$\frac{\mathrm{d}}{\mathrm{d}x} \cos x = -\sin x$$

$$\frac{\mathrm{d}}{\mathrm{d}x} \tan x = \sec^2 x$$

For $f = f(x)$ and $g = g(x)$,

$$\frac{\mathrm{d}}{\mathrm{d}x}(f + g) = \frac{\mathrm{d}f}{\mathrm{d}x} + \frac{\mathrm{d}g}{\mathrm{d}x}$$

$$\frac{\mathrm{d}}{\mathrm{d}x}(fg) = \frac{\mathrm{d}f}{\mathrm{d}x} g + f \frac{\mathrm{d}g}{\mathrm{d}x}$$

$$\frac{\mathrm{d}}{\mathrm{d}x}(cf) = c \frac{\mathrm{d}f}{\mathrm{d}x}$$

$$\frac{\mathrm{d}}{\mathrm{d}x}\left(\frac{f}{g}\right) = \frac{\frac{\mathrm{d}f}{\mathrm{d}x} g - f \frac{\mathrm{d}g}{\mathrm{d}x}}{g^2}$$

Chain rule for taking derivatives:

$$\frac{\mathrm{d}}{\mathrm{d}x} f[g(x)] = \frac{\mathrm{d}f}{\mathrm{d}g} \frac{\mathrm{d}g}{\mathrm{d}x}$$

A2.6 INTEGRALS

$$\int x^c \,\mathrm{d}x = \frac{x^{c+1}}{c+1} + C \quad (c \neq -1)$$

$$\int \frac{\mathrm{d}x}{x} = \ln |x| + C$$

$$\int \mathrm{e}^x \,\mathrm{d}x = \mathrm{e}^x + C$$

$$\int \sin x \,\mathrm{d}x = -\cos x + C$$

$$\int \cos x \,\mathrm{d}x = \sin x + C$$

$$\int \frac{\mathrm{d}x}{\sqrt{x^2 + c^2}} = \ln\left(x + \sqrt{x^2 + c^2}\right) + C$$

$$\int x\,\mathrm{e}^{-x} \,\mathrm{d}x = -(1 + x)\,\mathrm{e}^{-x} + C$$

$$\int cf \,\mathrm{d}x = c \int f \,\mathrm{d}x$$

$$\int (f + g)\,\mathrm{d}x = \int f \,\mathrm{d}x + \int g \,\mathrm{d}x$$

$$\int f \,\mathrm{d}g = fg - \int g \,\mathrm{d}f \quad \text{(integration by parts)}$$

A2.7 VECTOR ALGEBRAIC IDENTITIES

For vectors **a** and **b**, and angle α between them,

$\mathbf{a}\cdot\mathbf{b} = \lvert\mathbf{a}\rvert\lvert\mathbf{b}\rvert\cos\alpha$ (dot product of vectors)	
$\mathbf{a}\times\mathbf{b} = \lvert\mathbf{a}\rvert\lvert\mathbf{b}\rvert\sin\alpha\,\hat{\mathbf{n}}$ (cross product of vectors; $\hat{\mathbf{n}}$ is the unit vector normal to the plane of **a** and **b**, and its direction is determined by the right-hand rule when **a** is rotated by the shortest route toward **b**)	
$\mathbf{a}\cdot\mathbf{a} = \lvert\mathbf{a}\rvert^2 = a^2$	$\mathbf{a}\cdot\mathbf{b} = \mathbf{b}\cdot\mathbf{a}$
$\hat{\mathbf{a}} = \frac{\mathbf{a}}{a}$ (unit vector of **a**; $\lvert\hat{\mathbf{a}}\rvert = 1$)	$\mathbf{a}\times\mathbf{b} = -\mathbf{b}\times\mathbf{a}$
$(\mathbf{a}\times\mathbf{b})\cdot\mathbf{c} = (\mathbf{b}\times\mathbf{c})\cdot\mathbf{a} = (\mathbf{c}\times\mathbf{a})\cdot\mathbf{b}$ (scalar triple product)	
$\mathbf{a}\times(\mathbf{b}\times\mathbf{c}) = \mathbf{b}(\mathbf{a}\cdot\mathbf{c}) - \mathbf{c}(\mathbf{a}\cdot\mathbf{b})$ (vector triple product)	

For vectors in the Cartesian coordinate system,

$$\mathbf{a} = a_x\,\hat{\mathbf{x}} + a_y\,\hat{\mathbf{y}} + a_z\,\hat{\mathbf{z}} \quad (\hat{\mathbf{x}},\ \hat{\mathbf{y}},\text{ and }\hat{\mathbf{z}}\text{ are coordinate unit vectors})$$

$$a = \lvert\mathbf{a}\rvert = \sqrt{a_x^2 + a_y^2 + a_z^2}$$

$$\mathbf{a}+\mathbf{b} = (a_x+b_x)\,\hat{\mathbf{x}} + \left(a_y+b_y\right)\hat{\mathbf{y}} + (a_z+b_z)\,\hat{\mathbf{z}}$$

$$\mathbf{a}\cdot\mathbf{b} = a_xb_x + a_yb_y + a_zb_z$$

$$\mathbf{a}\times\mathbf{b} = \left(a_yb_z - a_zb_y\right)\hat{\mathbf{x}} + (a_zb_x - a_xb_z)\,\hat{\mathbf{y}} + \left(a_xb_y - a_yb_x\right)\hat{\mathbf{z}}$$

A2.8 VECTOR CALCULUS IDENTITIES

For a scalar function f (and g) and a vector function **a** (and **b**),

$\nabla f \equiv \text{grad}\, f$ (gradient of f) $\nabla\cdot\mathbf{a} \equiv \text{div}\,\mathbf{a}$ (divergence of **a**) $\nabla\times\mathbf{a} \equiv \text{curl}\,\mathbf{a}$ (curl of **a**) $\nabla\cdot(\nabla f) \equiv \nabla^2 f$ (Laplacian of f)	
$\nabla\times(\nabla f) = 0$	$\nabla\cdot(f\mathbf{a}) = (\nabla f)\cdot\mathbf{a} + f\nabla\cdot\mathbf{a}$
$\nabla\cdot(\nabla\times\mathbf{a}) = 0$	$\nabla\times(f\mathbf{a}) = (\nabla f)\times\mathbf{a} + f\nabla\times\mathbf{a}$
$\nabla(fg) = (\nabla f)\,g + f\nabla g$	$\nabla\cdot(\mathbf{a}\times\mathbf{b}) = \mathbf{b}\cdot(\nabla\times\mathbf{a}) - \mathbf{a}\cdot(\nabla\times\mathbf{b})$
$\nabla f(g) = \frac{\mathrm{d}f}{\mathrm{d}g}\nabla g$ (chain rule)	$\nabla\times(\nabla\times\mathbf{a}) = \nabla(\nabla\cdot\mathbf{a}) - \nabla^2\mathbf{a}$ ($\nabla^2\mathbf{a}$ – Laplacian of **a**)
$\int_v \nabla\cdot\mathbf{a}\,\mathrm{d}v = \oint_S \mathbf{a}\cdot\mathrm{d}\mathbf{S}$ (divergence theorem; S is the boundary surface of v) $\int_S(\nabla\times\mathbf{a})\cdot\mathrm{d}\mathbf{S} = \oint_C \mathbf{a}\cdot\mathrm{d}\mathbf{l}$ (Stokes' theorem; C is the boundary contour of S)	

A2.9 GRADIENT, DIVERGENCE, CURL, AND LAPLACIAN IN ORTHOGONAL COORDINATE SYSTEMS

Cartesian coordinate system $[f(x,y,z), \quad \mathbf{a} = a_x(x,y,z)\,\hat{\mathbf{x}} + a_y(x,y,z)\,\hat{\mathbf{y}} + a_z(x,y,z)\,\hat{\mathbf{z}}]$

$$\nabla = \frac{\partial}{\partial x}\,\hat{\mathbf{x}} + \frac{\partial}{\partial y}\,\hat{\mathbf{y}} + \frac{\partial}{\partial z}\,\hat{\mathbf{z}} \quad \text{(del operator)}$$

$$\nabla f = \frac{\partial f}{\partial x}\,\hat{\mathbf{x}} + \frac{\partial f}{\partial y}\,\hat{\mathbf{y}} + \frac{\partial f}{\partial z}\,\hat{\mathbf{z}}$$

$$\nabla \cdot \mathbf{a} = \frac{\partial a_x}{\partial x} + \frac{\partial a_y}{\partial y} + \frac{\partial a_z}{\partial z}$$

$$\nabla \times \mathbf{a} = \left(\frac{\partial a_z}{\partial y} - \frac{\partial a_y}{\partial z}\right)\hat{\mathbf{x}} + \left(\frac{\partial a_x}{\partial z} - \frac{\partial a_z}{\partial x}\right)\hat{\mathbf{y}} + \left(\frac{\partial a_y}{\partial x} - \frac{\partial a_x}{\partial y}\right)\hat{\mathbf{z}}$$

$$\nabla^2 f = \frac{\partial^2 f}{\partial x^2} + \frac{\partial^2 f}{\partial y^2} + \frac{\partial^2 f}{\partial z^2}$$

$$\nabla^2 \mathbf{a} = \nabla^2 a_x\,\hat{\mathbf{x}} + \nabla^2 a_y\,\hat{\mathbf{y}} + \nabla^2 a_z\,\hat{\mathbf{z}}$$

Cylindrical coordinate system $[f(r,\phi,z), \quad \mathbf{a} = a_r(r,\phi,z)\,\hat{\mathbf{r}} + a_\phi(r,\phi,z)\,\hat{\boldsymbol{\phi}} + a_z(r,\phi,z)\,\hat{\mathbf{z}}]$

$$\nabla f = \frac{\partial f}{\partial r}\,\hat{\mathbf{r}} + \frac{1}{r}\frac{\partial f}{\partial \phi}\,\hat{\boldsymbol{\phi}} + \frac{\partial f}{\partial z}\,\hat{\mathbf{z}}$$

$$\nabla \cdot \mathbf{a} = \frac{1}{r}\frac{\partial}{\partial r}(r a_r) + \frac{1}{r}\frac{\partial a_\phi}{\partial \phi} + \frac{\partial a_z}{\partial z}$$

$$\nabla \times \mathbf{a} = \left(\frac{1}{r}\frac{\partial a_z}{\partial \phi} - \frac{\partial a_\phi}{\partial z}\right)\hat{\mathbf{r}} + \left(\frac{\partial a_r}{\partial z} - \frac{\partial a_z}{\partial r}\right)\hat{\boldsymbol{\phi}} + \frac{1}{r}\left[\frac{\partial}{\partial r}(r a_\phi) - \frac{\partial a_r}{\partial \phi}\right]\hat{\mathbf{z}}$$

$$\nabla^2 f = \frac{1}{r}\frac{\partial}{\partial r}\left(r\frac{\partial f}{\partial r}\right) + \frac{1}{r^2}\frac{\partial^2 f}{\partial \phi^2} + \frac{\partial^2 f}{\partial z^2}$$

$$\nabla^2 \mathbf{a} = \nabla(\nabla \cdot \mathbf{a}) - \nabla \times (\nabla \times \mathbf{a})$$

Spherical coordinate system $[f(r,\theta,\phi), \quad \mathbf{a} = a_r(r,\theta,\phi)\,\hat{\mathbf{r}} + a_\theta(r,\theta,\phi)\,\hat{\boldsymbol{\theta}} + a_\phi(r,\theta,\phi)\,\hat{\boldsymbol{\phi}}]$

$$\nabla f = \frac{\partial f}{\partial r}\,\hat{\mathbf{r}} + \frac{1}{r}\frac{\partial f}{\partial \theta}\,\hat{\boldsymbol{\theta}} + \frac{1}{r\sin\theta}\frac{\partial f}{\partial \phi}\,\hat{\boldsymbol{\phi}}$$

$$\nabla \cdot \mathbf{a} = \frac{1}{r^2}\frac{\partial}{\partial r}\left(r^2 a_r\right) + \frac{1}{r\sin\theta}\frac{\partial}{\partial \theta}(\sin\theta\, a_\theta) + \frac{1}{r\sin\theta}\frac{\partial a_\phi}{\partial \phi}$$

$$\nabla \times \mathbf{a} = \frac{1}{r\sin\theta}\left[\frac{\partial}{\partial \theta}(\sin\theta\, a_\phi) - \frac{\partial a_\theta}{\partial \phi}\right]\hat{\mathbf{r}} + \frac{1}{r}\left[\frac{1}{\sin\theta}\frac{\partial a_r}{\partial \phi} - \frac{\partial}{\partial r}(r a_\phi)\right]\hat{\boldsymbol{\theta}} + \frac{1}{r}\left[\frac{\partial}{\partial r}(r a_\theta) - \frac{\partial a_r}{\partial \theta}\right]\hat{\boldsymbol{\phi}}$$

$$\nabla^2 f = \frac{1}{r^2}\frac{\partial}{\partial r}\left(r^2\frac{\partial f}{\partial r}\right) + \frac{1}{r^2\sin\theta}\frac{\partial}{\partial \theta}\left(\sin\theta\frac{\partial f}{\partial \theta}\right) + \frac{1}{r^2\sin^2\theta}\frac{\partial^2 f}{\partial \phi^2}$$

for $\nabla^2\mathbf{a}$, see the formula above

APPENDIX 3

Vector Algebra and Calculus Index

Concept	Page Numbers	Concept	Page Numbers
Unit vector	22	Line, surface, volume integrals	26–28
Magnitude of a vector	22	Line integral of a vector	34–35
Position vector of a point	22	Circulation of a vector	35
Decomposition of a vector	22–24	Flux (surface integral) of a vector	46–48
Graphical addition of vectors	24	Gradient	41–43
Dot product of vectors	34, 54	Directional derivative	43
Cross product of vectors	192, 213	Divergence	54–56
Scalar triple product	276–277	Divergence theorem	56
Vector triple product	222	Curl	213–215
Cartesian coordinate system	21–22	Stokes' theorem	215–216
Cylindrical coordinate system	42	Laplacian of a scalar	100–101
Spherical coordinate system	42–43	Laplacian of a vector	223

APPENDIX 4 Answers to Selected Problems

CHAPTER 1

1.2. $Q_3 = -4$ pC and $d = 2$ cm. **1.7.** $F_e = Q^2(6\sqrt{3} + 3\sqrt{6} + 2)/(24\pi\varepsilon_0 a^2)$ along the cube diagonal containing the charge. **1.11.** $Q = \pi\rho_{s0}a^2/2$. **1.15.** $E = Q(b - a)/\{2\pi\varepsilon_0 ab[\pi(a+b) + 2(b-a)]\}$, direction toward the midpoint of the larger semicircle. **1.19.** $\mathbf{E} = Q'[\hat{\mathbf{x}}/\sqrt{x^2+y^2} + (1 - x/\sqrt{x^2+y^2})\,\hat{\mathbf{y}}/y]/(4\pi\varepsilon_0)$. **1.23.** $\mathbf{E} = \rho_{s0}z[\sqrt{z^2+a^2} - |z| + z^2(1/\sqrt{z^2+a^2} - 1/|z|)]\,\hat{\mathbf{z}}/(2\varepsilon_0 a^2)$ (z-axis as in Fig. 1.14). **1.27.** $\mathbf{E} = Q'[(\theta_2 - \theta_1)\,\hat{\mathbf{x}} + \hat{\mathbf{y}}\ln(R_1/R_2)]/(2\pi\varepsilon_0 a)$ (notation in Fig. 4.11). **1.30.** $W_e = 526.5$ nJ. **1.35.** $V = \rho_s a/(2\varepsilon_0)$. **1.37.** $V = -99.27$ kV. **1.44.** (a) $0.793\hat{\mathbf{x}} + 0.637\hat{\mathbf{y}}$ and (b) $6.27°$. **1.46.** (a) $V = 0$ and $\mathbf{E} = -17.984\,\hat{\mathbf{z}}$ V/m, (b) $V = 0$ and $\mathbf{E} = -6.355\,\hat{\mathbf{z}}$ V/m, and (c) $V = 346\ \mu$V and $\mathbf{E} = 3.459(\hat{\mathbf{x}} + \hat{\mathbf{y}})\ \mu$V/m. **1.49.** $\mathbf{E} = p'(\cos\phi\,\hat{\mathbf{r}} + \sin\phi\,\hat{\boldsymbol{\phi}})/(2\pi\varepsilon_0 r^2)$. **1.55.** $V = \rho_0(4a^3 - r^3)/(12\varepsilon_0 a)$ $(r \le a)$ and $V = \rho_0 a^3/(4\varepsilon_0 r)$ $(r > a)$. **1.58.** $V = -\rho a^2/(4\varepsilon_0)$. **1.63.** (a) $\mathbf{E} = -\rho_0 a[1 + \cos(\pi x/a)]\,\hat{\mathbf{x}}/(\pi\varepsilon_0)$ $(|x| \le a)$ and $\mathbf{E} = 0$ $(|x| > a)$, and (b) $V = -2\rho_0 a^2/(\pi\varepsilon_0)$. **1.64.** $\mathbf{E} = \rho_0 a\,\mathrm{e}^{-|x|/a}\,\hat{\mathbf{x}}/\varepsilon_0$ $(-\infty < x < \infty)$. **1.72.** (a) For the notation as in Fig. 1.35, $\mathbf{E} = \rho x\,\hat{\mathbf{x}}/\varepsilon_0$ $(|x| \le d/2)$ and (b) $\mathbf{E} = \rho_s(x/|x|)\,\hat{\mathbf{x}}/(2\varepsilon_0)$, with $\rho_s = \rho d$ $(|x| > d/2)$. **1.76.** (a) $Q_a = -2Q$ and $Q_b = 3Q$, and (b) $V = 3Q/(4\pi\varepsilon_0 b)$. **1.79.** $Q_1' = -88.5$ nC/m and $Q_3' = 450$ nC/m. **1.80.** $Q_1 = 1.85$ pC and $Q_2 = 24.85$ pC. **1.82.** Adopting an x-axis such that the plane $x = 0$ coincides with the first surface of the first slab, (a) $\rho_s = 1\ \mu$C/m^2 at $x = 0$, $\rho_s = 1\ \mu$C/m^2 at $x = 1$ cm, $\rho_s = -1\ \mu$C/m^2 at $x = 4$ cm, and $\rho_s = 1\ \mu$C/m^2 at $x = 5$ cm, (b) $\mathbf{E} = -112.94\,\hat{\mathbf{x}}$ kV/m for $x < 0$, $\mathbf{E} = 0$ for $0 < x < 1$ cm, $\mathbf{E} = 112.94\,\hat{\mathbf{x}}$ kV/m for 1 cm $< x <$ 4 cm, $\mathbf{E} = 0$ for 4 cm $< x <$ 5 cm, and $\mathbf{E} = 112.94\,\hat{\mathbf{x}}$ kV/m for $x >$ 5 cm, and (c) $V = 3.39$ kV. **1.89.** $V = Q'\ln(h\sqrt{2}/a)/(2\pi\varepsilon_0)$.

CHAPTER 2

2.3. $\mathbf{E} = Ph\,\hat{\mathbf{z}}(1/\sqrt{b^2+h^2} - 1/\sqrt{a^2+h^2})/\varepsilon_0$. **2.6.** (a) $\rho_p = -2P_0x/a^2$, $\rho_{ps1} = -P_0$ $(x = -a)$, and $\rho_{ps2} = P_0$ $(x = a)$, (b) $\mathbf{E} = -P_0x^2\,\hat{\mathbf{x}}/(\varepsilon_0 a^2)$ $(|x| \le a)$ and $\mathbf{E} = 0$ $(|x| > a)$, and (c) $V = -2P_0a/(3\varepsilon_0)$. **2.9.** $\rho_p = -\rho$. **2.16.** (a) $\mathbf{E}_2 = (4\,\hat{\mathbf{x}} - 2\,\hat{\mathbf{y}} + 10\,\hat{\mathbf{z}})$ V/m and (b) $\mathbf{E}_2 = (4\,\hat{\mathbf{x}} - 2\,\hat{\mathbf{y}} + 7\,\hat{\mathbf{z}})$ V/m. **2.20.** (a) $\rho = -4\varepsilon_0V_0x^{-2/3}/(9d^{4/3})$, (b) $\rho_{s1} = 0$, (c) $\rho_{s2} = 4\varepsilon_0V_0/(3d)$, and (d) $Q = 0$. **2.22.** (a) $V = (-2.5 + 0.125/r)$ V and (b) $\mathbf{E} = (0.125/r^2)$ V/m (r in m). **2.30.** (a) $C = 132.8$ nF, (b) $V = 2.26$ GV, and (c) $E = 2.26$ MV/m. **2.32.** $C' = 2\pi\varepsilon_0/\ln[d^2/(ab)]$. **2.35.** $C' = 9.86$ pF/m. **2.42.** $V_{\text{new}} = 136.5$ V. **2.44.** (a) $C = 2.224$ pF, (b) $\rho_s = 1.77\ \mu$C/m^2, (c) $\rho_p = 0$, and (d) $\rho_{ps1} = -1.327\ \mu$C/m^2 $(r = a)$ and $\rho_{ps2} = 147.6$ nC/m^2 $(r = b)$. **2.46.** $V_{\text{new}} = 8.571$ kV. **2.50.** (a) $C' = 211$ pF/m and (b) $\rho_{s1} = 531.3$ nC/m^2, $\rho_{s2} = 177$ nC/m^2, $\rho_{s3} = 88.54$ nC/m^2, and $\rho_{s4} = 885.4$ nC/m^2 for the parts of the conductor surface interfacing dielectric sectors with ε_{r1}, ε_{r2}, ε_{r3}, and ε_{r4}, respectively. **2.55.** $C = 2ab\varepsilon_0(1 + 6/\pi)/d$. **2.59.** $b = 5$ cm. **2.63.** $W_{e1}/W_e = 66.1\%$. **2.68.** $W_e = \rho_0^2a^3S/(2\varepsilon)$. **2.71.** (a) $V_{cr} = 158.8$ kV, (b) $W_e' = 130$ mJ/m, and (c) $F_e' = 25$ mN/m. **2.73.** $V_{cr} = 80.67$ kV. **2.79.** $\varepsilon_{r1}E_{cr1}a = \varepsilon_{r2}E_{cr2}b$. **2.81.** (a) $V_{cr} = 440$ kV and (b) $W_e' = 8.15$ J/m.

CHAPTER 3

3.4. $G = 2\pi\sigma ab/(b - a)$. **3.8.** $R = a/[\sigma_0bc(1 + 18/\pi)]$. **3.12.** (a) $\mathbf{J} = 2\sigma_0V\,\hat{\mathbf{z}}/(5d)$, (b) $G = 2\pi\sigma_0a^2/(5d)$, (c) $P_J = 2\pi\sigma_0a^2V^2/(5d)$, (d) $\rho = 24\varepsilon_0V(1 + 3z/d)/(5d^2)$, $\rho_{s1} = 4\varepsilon_0V/(5d)$ $(z = 0)$, and $\rho_{s2} = -64\varepsilon_0V/(5d)$ $(z = d)$, and (e) $\rho_p = -18\varepsilon_0V(1 + 4z/d)/(5d^2)$, $\rho_{ps1} = -2\varepsilon_0V/(5d)$ $(z = 0)$, and $\rho_{ps2} = 56\varepsilon_0V/(5d)$ $(z = d)$. **3.16.** In the spherical coordinate system adopted as in Fig. 2.7, $\mathbf{J}_s = \sigma bV(1 + \cos\theta)\,\hat{\boldsymbol{\theta}}/[(b - a)\sin\theta]$. **3.18.** (a) $G' = 4\pi\sigma_0/[2\ln(b/a) + b^2/a^2 - 1]$ and (b) $\rho = 8\varepsilon_0V(1 + r^2/a^2)/\{a^2[2\ln(b/a) + b^2/a^2 - 1]\}$. **3.24.** (a) $R_{gr} = 31.85\ \Omega$, and (b) $P_{J1} = 15.92$ kJ and $P_{J2} = 63.69$ kJ. **3.27.** $R = 1/(2\pi\sigma a)$. **3.28.** $(V_{\text{step}})_{\max} = 119.2$ V.

CHAPTER 4

4.4. $\mathbf{B} = \mu_0J_{s0}a(1/\sqrt{z^2+a^2} - 1/\sqrt{z^2+b^2})\,\hat{\mathbf{z}}/2$. **4.8.** $B = \mu_0NI/(4a)$, the direction of $\mathbf{B}$ with respect to wire turns as in Fig. 4.10. **4.15.** With the rectangular coordinate system as in Fig. 6.25(b), $\mathbf{B} = \mu_0Ix\,\hat{\mathbf{y}}/(ad)$. **4.17.** With the spherical coordinate system as in Fig. 2.7,

$\mathbf{B} = -\mu_0 \sigma abV(1+\cos\theta)\,\hat{\boldsymbol{\phi}}/[(b-a)\,r\sin\theta]$. **4.18.** With the cylindrical coordinate system as in Fig. 3.20 [also see Fig. 8.12(a)], $\mathbf{B} = \mu_0 \mathcal{E}[G'(l-z)+1/R_\mathrm{L}]\,\hat{\boldsymbol{\phi}}/(2\pi r)$ ($G' = 6.84$ pS/m). **4.24.** (a) $\mathbf{J} = 795.8\,[x\,\hat{\mathbf{x}} + (-y+8z-8)\,\hat{\mathbf{y}} + 6x^2\,\hat{\mathbf{z}}]$ A/m^2 (x, y, z in m), (b) $I_C = 6.366$ kA, and (c) $\oint_C \mathbf{B}\cdot d\mathbf{l} = 8$ mT · m. **4.32.** With the Cartesian coordinate system as in Fig. 4.34, $\mathbf{F} = Q\rho_\mathrm{s}(\sqrt{2}-1)[\varepsilon_0\mu_0 vwa(\sqrt{2}-1)\,\hat{\mathbf{x}} + \hat{\mathbf{z}}]/(2\sqrt{2}\varepsilon_0)$. **4.34.** $F'_\mathrm{m} = 31.83$ mN/m (repulsive, symmetrically with respect to the semicylinder). **4.36.** $F'_\mathrm{m} = \mu_0 I^2\sqrt{2}(\pi/4 + \ln\sqrt{2})/(4\pi a)$ (repulsive, symmetrically with respect to the corner conductor). **4.38.** Adopting a Cartesian coordinate system with origin at the orthocenter of the triangle, x-axis along $\mathbf{B}$ ($\mathbf{B} = B\,\hat{\mathbf{x}}$), and z-axis out of the page, and denoting the loop side perpendicular to $\mathbf{B}$ as side 1, and the other two sides as 2 and 3 in the counterclockwise order, (a) $\mathbf{F}_{\mathrm{m}1} = IaB\,\hat{\mathbf{z}}$, $\mathbf{F}_{\mathrm{m}2} = -IaB\,\hat{\mathbf{z}}/2$, and $\mathbf{F}_{\mathrm{m}3} = -IaB\,\hat{\mathbf{z}}/2$, (b) $\mathbf{T}_1 = Ia^2B\sqrt{3}\,\hat{\mathbf{y}}/6$, $\mathbf{T}_2 = Ia^2B\sqrt{3}(\sqrt{3}\,\hat{\mathbf{x}} + \hat{\mathbf{y}})/24$, and $\mathbf{T}_3 = Ia^2B\sqrt{3}(-\sqrt{3}\,\hat{\mathbf{x}} + \hat{\mathbf{y}})/24$, (c) $\mathbf{F}_\mathrm{m} = 0$, and (d) $\mathbf{T} = Ia^2B\sqrt{3}\,\hat{\mathbf{y}}/4$.

CHAPTER 5

5.3. $\mathbf{B} = 2\sqrt{2}\mu_0 M_0 a^2 d\,\hat{\mathbf{z}}/[\pi(4z^2+a^2)\sqrt{2z^2+a^2}]$. **5.5.** (a) $\mathbf{J}_\mathrm{m} = -2M_0 r\,\hat{\boldsymbol{\phi}}/a^2$ and $\mathbf{J}_\mathrm{ms} = M_0\hat{\boldsymbol{\phi}}$, and (b) $\mathbf{B} = \mu_0 M_0 d\,\hat{\mathbf{z}}\{[|z| - \sqrt{z^2+a^2} + z^2(1/|z| - 1/\sqrt{z^2+a^2})]/a^2 + a^2/[2(z^2+a^2)^{3/2}]\}$. **5.9.** $\mathbf{J}_\mathrm{m} = -\mathbf{J}$. **5.16.** (a) $\mathbf{H}_2 = (5\,\hat{\mathbf{x}} - 3\,\hat{\mathbf{y}} + 4.8\,\hat{\mathbf{z}})$ A/m and (b) $\mathbf{H}_2 = (2\,\hat{\mathbf{x}} - 3\,\hat{\mathbf{y}} + 4.8\,\hat{\mathbf{z}})$ A/m. **5.19.** $\mathbf{B} = \mu_0 M_0\,\hat{\mathbf{z}}$ ($r < a$) and $\mathbf{B} = 0$ ($r > a$). **5.22.** For reference directions of vectors $\mathbf{B}$ and $\mathbf{H}$ as in Fig. 5.32(a), $B_1 = -0.375$ T, $H_1 = -250$ A/m, $B_2 = -1.5$ T, $H_2 = -1500$ A/m, $B_3 = 1.125$ T, and $H_3 = 750$ A/m. **5.24.** $N_2 I_2 = -107.3$ A turns.

CHAPTER 6

6.2. With the x- and y-axes oriented as in Fig. 6.2, $\mathbf{E}_\mathrm{ind} = 3.651\times10^{-8}(\mathrm{d}i/\mathrm{d}t)(-\,\hat{\mathbf{x}} + \hat{\mathbf{y}})$ V/m ($\mathrm{d}i/\mathrm{d}t$ in A/s). **6.7.** With the Cartesian coordinate system adopted as in Fig. 6.4, (a) $\tau = b/c_0 = 0.3$ ns $\ll T = 2\pi/\omega = 62.8$ ns, (b) $\mathbf{E}_\mathrm{ind} = 11\cos 10^8 t\,(-\,\hat{\mathbf{x}} + \hat{\mathbf{y}})$ V/m, and (c) $\mathbf{H} = 8.33\sin 10^8 t\,\hat{\mathbf{z}}$ A/m (t in s). **6.9.** $\mathbf{E}_\mathrm{ind} = \mu_0(\mathrm{d}i/\mathrm{d}t)\{\ln[(\sqrt{a^2+z^2}+a)/(\sqrt{a^2+z^2}-a)] - 2a/\sqrt{a^2+z^2}\}\,\hat{\mathbf{y}}/(4\pi)$. **6.13.** $v_\mathrm{MN} = \mu N' a^2\alpha(\mathrm{d}i/\mathrm{d}t)\{(S_1-S_2)/[S_1 + S_2\alpha/(2\pi-\alpha)]\}/2$. **6.19.** (a) $e_\mathrm{ind} = wBa^2/2$, (b) $E_q = -wBr$, with respect to the (outward) radial reference direction (r is the radial distance from the disk center), and (c) $V_{12} = wBa^2/2$. **6.21.** $B = 0.5$ T or $B = -1$ T, and $P_\mathrm{gen} = -25$ W (in both cases). **6.26.** In the cylindrical coordinate system adopted as in Fig. 3.20 [or Fig. 8.12(a)], (a) $\mathbf{B} = \mu_0 I_\mathrm{g}\,\hat{\boldsymbol{\phi}}/(2\pi r)$, (b) $\mathbf{E}_\mathrm{ind} = -v\mu_0 I_\mathrm{g}\,\hat{\mathbf{r}}/(2\pi r)$, (c) $\mathbf{J} = \sigma(\mathbf{E}_q + \mathbf{E}_\mathrm{ind})$, (d) $\mathbf{E}_q = \sigma\mu_0 R_\mathrm{V} I_\mathrm{g} vl\,\hat{\mathbf{r}}/\{r[2\pi\sigma R_\mathrm{V} l + \ln(b/a)]\}$, and (e) $V = \sigma\mu_0 R_\mathrm{V} I_\mathrm{g} vl\ln(b/a)/[2\pi\sigma R_\mathrm{V} l + \ln(b/a)]$. **6.30.** $e_\mathrm{ind} = \mu_0\omega bI_0\sin 2\omega t\,\{ac(c^2 + a^2/4)\cos\omega t\,/\,[(c^2 + a^2/4)^2 - a^2c^2\cos^2\omega t] + \ln[(c^2 + a^2/4 + ac\cos\omega t)/(c^2 + a^2/4 - ac\cos\omega t)]\}/(4\pi)$. **6.34.** $l_0 = l/\mu_\mathrm{r}$ and $[(P_\mathrm{J})_\mathrm{ave}]_\mathrm{max} = \pi\sigma\mu_\mathrm{r}\mu_0^2\omega^2N^2a^4I_0^2/(64l)$. **6.38.** (a) With the same reference direction as in Fig. 6.24, $J_\mathrm{eddy} = \sigma\omega\mu_0 Na^2I_0\sin\omega t/(2lr)$ and (b) $(P_\mathrm{J})_\mathrm{ave} = \pi\sigma\mu_0^2\omega^2N^2a^4hI_0^2\ln(c/b)/(4l^2)$. **6.42.** (a) $(P_\mathrm{J})_\mathrm{ave} = \sigma_0\omega^2a^3l\delta B_0^2/40$ and (b) $(\mathbf{T}_\mathrm{m})_\mathrm{ave} = -\sigma_0\omega a^3l\delta B_0^2\,\hat{\mathbf{y}}/40$. **6.45.** (a) $\omega = \mathbf{T}'_\mathrm{mech}/(\pi\sigma a^3\delta B^2)$ and (b) $P'_\mathrm{mech} = \mathbf{T}'^2_\mathrm{mech}/(\pi\sigma a^3\delta B^2)$.

CHAPTER 7

7.5. $L' = [\mu\ln 2 + \mu_0\ln(d/2a)]/\pi$. **7.7.** $L' = \mu_1 d_1/w + \mu_2 d_2/w$. **7.11.** $L_{21} = 332.7$ mH. **7.13.** $v = -\pi\mu\omega N'a^2I_0\sin\omega t$. **7.16.** (a) $L = 110$ μH, (b) $L = 22.5$ μH, and (c) $L = 49.5$ μH. **7.23.** (a) $I_0 = 183.3$ mA and (b) $I_0 = 85.3$ mA. **7.25.** Switch K open: $W_\mathrm{m}(t) = 706\cos^2 377t$ mJ (t in s) and $(W_\mathrm{m})_\mathrm{ave} = 353$ mJ, switch K closed: $W_\mathrm{m}(t) = 0$ and $(W_\mathrm{m})_\mathrm{ave} = 0$. **7.32.** (a) $w_\mathrm{m} = 250$ J/m^3 for $a \le r \le c$ and $w_\mathrm{m} = (0.2536/r^2)$ J/m^3 (r in m) for $c \le r \le b$ ($c = 3.2$ cm), and (b) $W_\mathrm{m} = 8.45$ mJ. **7.34.** (a) $w_\mathrm{h} = (2.026/r^2)$ J/m^3 (r in m) and (b) $(P_\mathrm{h})_\mathrm{ave} = 882.5$ W. **7.36.** $W'_\mathrm{mi} = 157.9$ nJ/m, $W'_\mathrm{me} = 346.6$ nJ/m, and $W'_\mathrm{m} = 504.5$ nJ/m.

CHAPTER 8

8.1. (a) $J_{\mathrm{d}01} = J_{\mathrm{d}02} = I_0/S$, (b) $E_{01} = I_0/(2\pi\varepsilon_1 fS)$ and $E_{02} = I_0/(2\pi\varepsilon_2 fS)$, and (c) $V_0 = (\varepsilon_1 d_2 + \varepsilon_2 d_1)I_0/(2\pi\varepsilon_1\varepsilon_2 fS)$. **8.5.** In the cylindrical coordinate system whose z-axis coincides with the cable axis, (a) $\mathbf{J}_\mathrm{d} = \varepsilon(\mathrm{d}v/\mathrm{d}t)\,\hat{\mathbf{r}}/[r\ln(b/a)]$ and $\mathbf{J}_\mathrm{tot} = \mathbf{J}_\mathrm{d}$, and (b) $\mathbf{J}_\mathrm{d}$ the same as in (a) and $\mathbf{J}_\mathrm{tot} = \sigma v\,\hat{\mathbf{r}}/[r\ln(b/a)] + \mathbf{J}_\mathrm{d}$. **8.9.** $\mathbf{H} = \sqrt{\varepsilon_0/\mu_0}\,E_0\sin\theta\cos(\omega t - \beta r)\,\hat{\boldsymbol{\phi}}/r$. **8.20.** (a) $\underline{Q}' = -\mathrm{j}I_0\sin\phi\,/(\omega a)$, (b) $\underline{V} = -\mathrm{j}I_0\,\mathrm{e}^{-\mathrm{j}\beta R}/(2\pi\varepsilon_0\omega R)$, (c) $\underline{\mathbf{A}} = \mu_0 aI_0\,\mathrm{e}^{-\mathrm{j}\beta R}\,\hat{\mathbf{x}}/(8R)$, (d) $\underline{\mathbf{E}} = -\mathrm{j}I_0\,\mathrm{e}^{-\mathrm{j}\beta R}[\omega\mu_0 a\,\hat{\mathbf{x}}/8 + z(1+\mathrm{j}\beta R)\,\hat{\mathbf{z}}/(2\pi\varepsilon_0\omega R^2)]/R$, (e) $\underline{\mathbf{H}} = -azI_0(1+\mathrm{j}\beta R)\,\mathrm{e}^{-\mathrm{j}\beta R}\,\hat{\mathbf{y}}/(8R^3)$, and (f) $\mathbf{A}(t) = \mu_0 aI_0\sqrt{2}\cos(\omega t - \beta R)\,\hat{\mathbf{x}}/(8R)$ and $\mathbf{E}(t) = \omega\mu_0 aI_0\sqrt{2}\cos(\omega t - \beta R - \pi/2)\,\hat{\mathbf{x}}/(8R) + zI_0\sqrt{2}\sqrt{1+\beta^2R^2}\cos(\omega t - \beta R + \arctan\beta R - \pi/2)\,\hat{\mathbf{z}}/(2\pi\varepsilon_0\omega R^3)$, where $R = \sqrt{z^2+a^2}$. **8.22.** (a) $\beta a = \omega/c_0 = 3.33$, so the condition $\beta a \ll 1$ is not met, (b) $\underline{Q}' = -1.061$

$\sin(3\phi/2)$ nC/m $(0 < \phi < 2\pi)$ and $\underline{Q}_0 = 141$ pC, (c) $\underline{V} = 0$ and $\underline{\mathbf{A}} = -\mathrm{j}11.31\, \mathrm{e}^{-\mathrm{j}33.33\sqrt{z^2+0.01}}\, \hat{\mathbf{x}}/\sqrt{z^2+0.01}$ nT·m (z in m), and (d) since $\underline{\mathbf{A}}$ has only an x-component and $\underline{V}$ is constant (zero) along the z-axis, $\underline{E}_z = -\mathrm{j}\omega\underline{A}_z - \partial\underline{V}/\partial z = 0$. **8.23.** $\underline{\mathbf{B}} = \mu_0 J_{s0} a(\mathrm{e}^{-\mathrm{j}\beta\sqrt{a^2+z^2}}/\sqrt{a^2+z^2} - \mathrm{e}^{-\mathrm{j}\beta\sqrt{b^2+z^2}}/\sqrt{b^2+z^2})\,\hat{\mathbf{z}}/2$. **8.25.** $\underline{\mathbf{A}} = \mathrm{j}2\mu_0 a J_{s0}\, \mathrm{e}^{-\mathrm{j}\beta a}\,\hat{\mathbf{z}}/3$. **8.32.** $\Psi_{\mathcal{P}} = 848.27$ kW, into the earth. **8.36.** (a) $\mathbf{H}(t) = \sqrt{\varepsilon/\mu}\, E_0 \cos(\omega t - \beta z)\,\hat{\boldsymbol{\phi}}/r$ and $\underline{\mathbf{H}} = \sqrt{\varepsilon/\mu}\, E_0\sqrt{2}\, \mathrm{e}^{-\mathrm{j}\beta z}\,\hat{\boldsymbol{\phi}}/(2r)$, (b) $\underline{\mathcal{P}} = \sqrt{\varepsilon/\mu}\, E_0^2\,\hat{\mathbf{z}}/(2r^2)$, (c) $\underline{S} = \pi\sqrt{\varepsilon/\mu}\, E_0^2 \ln(b/a)$, and (d) $P_{\mathrm{ave}} = \pi\sqrt{\varepsilon/\mu}\, E_0^2 \ln(b/a)$. **8.39.** First term (integral) in Eq. (8.196) equals zero, second term $\sigma\pi a^2 V_0^2/(2d)$, third one $-\mathrm{j}\omega\varepsilon\pi a^2 V_0^2/(2d)$ (approximately), and the last (complex power flow) $(-\sigma + \mathrm{j}\omega\varepsilon)\pi a^2 V_0^2/(2d)$. **8.41.** $\mathcal{P}(t) = \sqrt{\varepsilon_0/\mu_0}\, E_0^2 \sin^2\theta \cos^2(\omega t - \beta r)\,\hat{\mathbf{r}}/r^2$, $\underline{\mathcal{P}} = \sqrt{\varepsilon_0/\mu_0}\, E_0^2 \sin^2\theta\,\hat{\mathbf{r}}/(2r^2)$, and $\mathcal{P}_{\mathrm{ave}} = \mathrm{Re}\{\underline{\mathcal{P}}\} = \sqrt{\varepsilon_0/\mu_0}\, E_0^2 \sin^2\theta\,\hat{\mathbf{r}}/(2r^2)$ [see Eq. (6.95)].

CHAPTER 9

9.5. (a) $f = 1.5$ GHz, and $\theta_0 = 150°$ or $\theta_0 = 30°$, (b) $\underline{\mathbf{H}} = 28.2\, \mathrm{e}^{\mathrm{j}\theta_0}\, \mathrm{e}^{\mathrm{j}10\pi z}\,\hat{\mathbf{x}}$ mA/m (z in m), and (c) $\mathcal{P}_{\mathrm{ave}} = -0.3\,\hat{\mathbf{z}}$ W/m^2. **9.6.** (a) $\varepsilon_{\mathrm{r}} = 3.6$ and $\mu_{\mathrm{r}} = 2.5$. **9.11.** $I_{\mathrm{ind}} = 3.39$ mA. **9.15.** $W = 95.69$ mJ and (b) $W = 47.84$ mJ. **9.17.** $r = 18.2$ m. **9.24.** (a) $\varepsilon_{\mathrm{r}} = 81$ and $\sigma = 4$ S/m, (b) $\underline{\gamma} = (78.72 + \mathrm{j}204)$ m^{-1}, (c) $\mathbf{H} = 27.8\, \mathrm{e}^{78.72y} \cos(6.28 \times 10^9 t + 204y - 21°)\,\hat{\mathbf{z}}$ mA/m (t in s; y in m), and (d) $\mathcal{P}_{\mathrm{ave}} = -12.93\, \mathrm{e}^{157.44y}\,\hat{\mathbf{y}}$ mW/m^2. **9.26.** (a) $\varepsilon_{\mathrm{r}} = 3$ and $\tan\delta_{\mathrm{d}} = 10^{-3}$, (b) $\underline{\gamma} = (0.1814 + \mathrm{j}363)$ m^{-1}, and (c) $\mathcal{P}_{\mathrm{ave}} = 1.035\, \mathrm{e}^{-0.363z}\,\hat{\mathbf{z}}$ W/m^2. **9.31.** (a) $\delta_{1\%} = 1.91$ μm and (b) $A_{\mathrm{dB}} = 20{,}910$ dB (perfect shielding). **9.33.** $h_{\mathrm{b}} = 161.7$ km. **9.38.** (a) $\Delta t = 34.5$ ns and (b) $\Delta t = 0.0152$ ps. **9.42.** (a) RHEP, (b) LP, (c) LHEP, (d) LP, (e) LHEP, (f) LP, (g) LP, (h) RHEP, and (i) LHEP.

CHAPTER 10

10.3. $W_{\mathrm{em}}(t) = \varepsilon E_{\mathrm{i}0}^2 S(2\beta l + \sin 2\beta l \cos 2\omega t)/\beta$ $(\beta = 2\pi/\lambda$ and $\omega = \beta/\sqrt{\varepsilon\mu_0})$. **10.6.** $f = 375$ MHz and $E_{\mathrm{i}0} = 0.51$ V/m. **10.9.** (a) $f = 1.5$ GHz, (b) $\lambda_{\mathrm{glass}} = 10$ cm, (c) $\mathbf{E}_{\mathrm{air}}(t) = 1.777[\cos(9.425 \times 10^9 t + 31.42x + 90°) - \sin 31.42x \cos(9.425 \times 10^9 t)]\,\hat{\mathbf{y}}$ kV/m (t in s; x in m) and $\underline{\mathbf{E}}_{\mathrm{air}} = 1.257(\mathrm{j}\, \mathrm{e}^{\mathrm{j}31.42x} - \sin 31.42x)\,\hat{\mathbf{y}}$ kV/m, (d) $|\underline{\mathbf{H}}_{\mathrm{air}}|_{\mathrm{max}} = 6.67$ A/m, $|\underline{\mathbf{H}}_{\mathrm{air}}|_{\mathrm{min}} = 3.33$ A/m, and $|\underline{\mathbf{H}}_{\mathrm{glass}}|_{\mathrm{max}} = |\underline{\mathbf{H}}_{\mathrm{glass}}|_{\mathrm{min}} = 6.67$ A/m, (e) $\mathcal{P}_{\mathrm{ave}} = -8.378\,\hat{\mathbf{x}}$ kW/m^2 everywhere, and (f) $\underline{\eta}_{\mathrm{w}} = 188.5$ Ω. **10.14.** $d_{\mathrm{max}} = 38$ m. **10.19.** $(W_{\mathrm{em}})_{\mathrm{ave}} = 8.854 \times 10^{-14}$ J. **10.23.** $\delta_{\mathrm{violet}} = 53.64°$, $\delta_{\mathrm{blue}} = 52.19°$, $\delta_{\mathrm{green}} = 51.29°$, $\delta_{\mathrm{yellow}} = 50.83°$, $\delta_{\mathrm{orange}} = 50.68°$, and $\delta_{\mathrm{red}} = 50.16°$; $\gamma = 3.48°$. **10.28.** $\underline{\mathcal{P}} = (83.3\,\hat{\mathbf{x}} - 249.9\,\hat{\mathbf{z}})$ W/m^2. **10.31.** $\theta_{\mathrm{accept}} = \arcsin[(n_{\mathrm{core}}^2 - n_{\mathrm{cladding}}^2)^{1/2}/n_0]$; from air, $\theta_{\mathrm{accept}} = 17.31°$; from water, $\theta_{\mathrm{accept}} = 12.92°$. **10.34.** $h_{\mathrm{r}} = 4.59$ km. **10.38.** $(\Gamma_{\mathrm{e}})_{\mathrm{dB}} = -9.42$ dB.

CHAPTER 11

11.4. $(P_{\mathrm{cr}})_{\mathrm{max}} = 317$ MW and $P_l = 295$ MW. **11.7.** (a) $C' = 9.476$ pF/m, $L' = 1.174$ μH/m, $R' = 1.079$ Ω/m, and $G' = 0$, and (b) $Z_0 = 352.1$ Ω and $\alpha = 1.532 \times 10^{-3}$ Np/m. **11.10.** (a) $C' = 13.24$ pF/m, $L' = 840$ nH/m, $R' = 0.657$ Ω/m, and $G' = 0$, and (b) $|\underline{\rho}_{\mathrm{s}}| = 67.9$ nC/m^2 and $|\underline{\mathbf{J}}_{\mathrm{s}}| = 20.37$ A/m. **11.14.** (a) $(P_{\mathrm{cr}})_{\mathrm{max}} = 15.31$ MW and (b) $P_l = 5.426$ MW. **11.19.** $C' = 203.9$ pF/m, $L' = 218$ nH/m, $R' = 7.15$ Ω/m, and $G' = 384.5$ μS/m.

CHAPTER 12

12.4. (a) $f = 300$ MHz and $\underline{Z}_{\mathrm{L}} = (38.46 + \mathrm{j}92.31)$ Ω, and (b) $f = 300$ MHz and $\underline{Z}_{\mathrm{L}} = (38.46 - \mathrm{j}92.31)$ Ω. **12.8.** (a) $l_{\mathrm{max}} = 3.69$ cm and $R_{\mathrm{max}} = 130.9$ Ω, $l_{\mathrm{min}} = 28.69$ cm and $R_{\mathrm{min}} = 19.1$ Ω; (b) $l_{\mathrm{min}} = 18.75$ cm and $R_{\mathrm{min}} = 8.58$ Ω, $l_{\mathrm{max}} = 43.75$ cm and $R_{\mathrm{max}} = 291.4$ Ω. **12.11.** (a) With reference to Fig. 12.8, $l_{\mathrm{c}} = 12.08$ cm, $Z_{0\mathrm{t}} = 115.1$ Ω, and $l = 37.5$ cm, assuming that all transmission-line sections have air dielectric. **12.14.** $P_{\mathrm{L}1} = 22.41$ W, $P_{\mathrm{L}2} = 31.73$ W, and $P_{\mathrm{L}3} = 10.97$ W. **12.19.** $l_{\mathrm{stub}} = 2.09$ mm. **12.21.** (a) $W_{\mathrm{em}}(0) = 0.2$ μJ, (b) $t = 1.36$ ns, and (c) $t = 624.7$ ns. **12.29.** First solution: $l_1 = 9.3$ mm and $l_{\mathrm{stub}1} = 8.9$ mm; second solution: $l_2 = 24$ mm and $l_{\mathrm{stub}2} = 41.1$ mm. **12.33.** $l_{\mathrm{damage}} = 10$ m and $R_{\mathrm{eq}} = 1.02$ Ω. **12.35.** $v_{\mathrm{g}} = 5$ V $(0 \le t \le 9$ ns$)$; $v_{\mathrm{L}} = 0$ $(0 \le t < 1.5$ ns$)$, $v_{\mathrm{L}} = 3.33$ V $(1.5$ ns $< t < 4.5$ ns$)$, $v_{\mathrm{L}} = 4.44$ V $(4.5$ ns $< t < 7.5$ ns$)$, and $v_{\mathrm{L}} = 4.81$ V $(7.5$ ns $< t \le 9$ ns$)$. **12.38.** During time intervals $0 \le t < 1$ ns, 1 ns $< t < 2$ ns, ..., 9 ns $< t \le 10$ ns, v_{g} assumes values 6 V, 0, 10 V, 0, 6.66 V, 0, 4.45 V, 0, 2.96 V, and 0, respectively; the corresponding values of v_{L} are 0, 12 V, 0, 7.98 V, 0, 5.33 V, 0, 3.56 V, 0, and 2.37 V. **12.41.** Starting at $t = 1$ ns, v_{L} is a sequence of trapezoidal pulses of durations 2 ns and magnitudes 4 V, −1.33 V, 0.444 V, and −0.148 V, respectively (each new pulse starts exactly at the end of the preceding one). **12.46.** $C = 40$ pF. **12.48.** $v_{\mathrm{L}}(t) = 0.75[\mathrm{e}^{-(t-T)/\tau} h(t-T) - \mathrm{e}^{-(t-t_1)/\tau} h(t-t_1)]$ V and $v_{\mathrm{g}}(t) = 0.5[h(t) - h(t-t_0)] - 0.5[h(t-t_2) - h(t-t_3)] + 0.75[\mathrm{e}^{-(t-t_2)/\tau} h(t-t_2) - \mathrm{e}^{-(t-t_3)/\tau} h(t-t_3)]$ V, where $T = 1.17$ ns, $\tau = 2.4$ ns, $t_0 = 2$ ns, $t_1 = 3.17$ ns, $t_2 = 2.34$ ns, and $t_3 = 4.34$ ns.

CHAPTER 13

13.8. (a) For $\varepsilon_{\mathrm{r}} = 1$, dominant frequency range $1.5\ \mathrm{GHz} < f \leq 3\ \mathrm{GHz}$, first three TE modes TE_{10}, TE_{01}, and TE_{11}, first three TM modes TM_{11}, TM_{21}, and TM_{12}, $(f_{\mathrm{c}})_{10} = (f_{\mathrm{c}})_{01} = 1.5$ GHz, $(f_{\mathrm{c}})_{11} = 2.12$ GHz, and $(f_{\mathrm{c}})_{21} = (f_{\mathrm{c}})_{12} = 3.35$ GHz; at $f = 2$ GHz, possible TE_{10} and TE_{01}; (b) for $\varepsilon_{\mathrm{r}} = 2.5$, dominant frequency range $0.949\ \mathrm{GHz} < f \leq 1.897\ \mathrm{GHz}$, first three TE and first three TM modes as in (a), $(f_{\mathrm{c}})_{10} = (f_{\mathrm{c}})_{01} = 0.949$ GHz, $(f_{\mathrm{c}})_{11} = 1.34$ GHz, and $(f_{\mathrm{c}})_{21} = (f_{\mathrm{c}})_{12} = 2.12$ GHz; at $f = 2$ GHz, possible TE_{10}, TE_{01}, TE_{11}, TE_{20}, TE_{02}, and TM_{11}. **13.12.** (a) $\underline{\mathcal{P}} = -\mathrm{j}\omega\mu|\underline{H}_0|^2[(m\pi/a)\sin(2m\pi x/a)\cos^2(n\pi y/b)\,\hat{\mathbf{x}} + (n\pi/b)\cos^2(m\pi x/a)\sin(2n\pi y/b)\,\hat{\mathbf{y}}]/(2k^2) + \omega\mu\beta|\underline{H}_0|^2[(n\pi/b)^2\cos^2(m\pi x/a)\sin^2(n\pi y/b) + (m\pi/a)^2\sin^2(m\pi x/a)\cos^2(n\pi y/b)]\,\hat{\mathbf{z}}/k^4$, (b) $P = q\eta ab(f^2/f_{\mathrm{c}}^2)\sqrt{1 - f_{\mathrm{c}}^2/f^2}\,|\underline{H}_0|^2$, where $f_{\mathrm{c}} = (f_{\mathrm{c}})_{02} = c/b$ and $q = 1/2$ for TE_{02} and $f_{\mathrm{c}} = (f_{\mathrm{c}})_{11} = c\sqrt{1/a^2 + 1/b^2}/2$ and $q = 1/4$ for TE_{11}, and (c) the flux of $\underline{\mathcal{P}}$ into each wall is zero. **13.14.** $P_{\max} = P_{\mathrm{cr}} = 210.3$ MW. **13.16.** (a) $\alpha = 1.3 \times 10^{-3}$ Np/m and (b) $\alpha = 2.88 \times 10^{-4}$ Np/m. **13.17.** (a) $\alpha = 4.15 \times 10^{-3}$ Np/m and (b) $\alpha = 3.33 \times 10^{-3}$ Np/m. **13.21.** At $f = 22.5$ GHz: $v_{\mathrm{p}} = 2.59 \times 10^8$ m/s, $v_{\mathrm{g}} = 1.54 \times 10^8$ m/s, $\lambda_z = 11.5$ mm, and $\lambda = 8.9$ mm; at $f = 18$ GHz: $v_{\mathrm{p}} = 3.29 \times 10^8$ m/s, $v_{\mathrm{g}} = 1.22 \times 10^8$ m/s, $\lambda_z = 18.3$ mm, and $\lambda = 11.1$ mm; at $f = 27$ GHz: $v_{\mathrm{p}} = 2.36 \times 10^8$ m/s, $v_{\mathrm{g}} = 1.7 \times 10^8$ m/s, $\lambda_z = 8.7$ mm, and $\lambda = 7.4$ mm. **13.27.** (a) $a = 50$ mm and $d = 20.23$ mm, and (b) TE_{102}. **13.34.** $(W_{\mathrm{em}})_{\mathrm{cr}} = 29.9$ mJ. **13.35.** $\mathcal{E}_{\mathrm{ind}} = 3.16$ kV and $W_{\mathrm{em}} = 23.2$ mJ. **13.38.** (a) $P_{\mathrm{c}} = 17.37$ mW and (b) $P_{\mathrm{c}} = 30.19$ mW. **13.39.** $P_{\mathrm{d}} = 70.6$ mW.

CHAPTER 14

14.8. $\mathcal{P}_{\mathrm{ave}} = 18.62\sin^2\theta\,\hat{\mathbf{r}}/r^2\ \mathrm{W/m^2}$ (r in m). **14.10.** $\underline{\mathcal{P}} = \sqrt{\mu_0/\varepsilon_0}\,\beta^2|\underline{I}_0|^2\sin^2\theta\{1 + \mathrm{e}^{-2\alpha l} - 2\,\mathrm{e}^{-\alpha l}\cos[\beta l(1 - \cos\theta)]\}\,\hat{\mathbf{r}}/\{(4\pi r)^2[\alpha^2 + \beta^2(1 - \cos\theta)^2]\}$. **14.11.** $\underline{\mathbf{E}} = \underline{E}_\theta\,\hat{\boldsymbol{\theta}} = \mathrm{j}\omega\mu_0\underline{I}_{\mathrm{u}}\,l[\mathrm{e}^{-\mathrm{j}\beta r}/(4\pi r)]\sin\theta\,\hat{\boldsymbol{\theta}}\sin u/u$, where $u = \beta h\cos\theta$ ($h = l/2$), $\underline{\mathbf{H}} = H_\phi\,\hat{\boldsymbol{\phi}} = (\underline{E}_\theta/\eta_0)\,\hat{\boldsymbol{\phi}}$, and $\underline{\mathcal{P}} = \mathcal{P}_{\mathrm{ave}} = (|\underline{E}_\theta|^2/\eta_0)\,\hat{\mathbf{r}}$ ($\eta_0 = 377\ \Omega$). **14.15.** $\underline{\mathcal{P}} = \eta\beta^2 I^2 l^2(1 + \cos^2\theta - \sin^2\theta\sin 2\phi)\,\hat{\mathbf{r}}/(32\pi^2 r^2)$. **14.19.** $R_{\mathrm{rad}} = 0.8\ \Omega$, $R_{\mathrm{ohmic}} = 6.46\ \Omega$, and $\eta_{\mathrm{rad}} = 11\%$. **14.24.** (a) $D = 10$, and (b) $\mathrm{HPBW}_{\mathrm{elevation}} = 114.5^\circ$ and $\mathrm{HPBW}_{\mathrm{azimuth}} = 65.5^\circ$. **14.29.** $R_{\mathrm{ohmic}} = R'[l/2 - (\sin\beta l)/(2\beta)]$, where $R' = \sqrt{\pi\mu f/\sigma}/(2\pi a)$ and $\beta = 2\pi f/c_0$ ($c_0 = 3 \times 10^8$ m/s). **14.33.** (a) $|\underline{\mathbf{E}}_{\mathrm{direct}}| = 60$ mV/m and (b) $|\underline{\mathbf{E}}_{\mathrm{reflected}}| = 26.6$ mV/m. **14.35.** $|\underline{\mathbf{E}}_{\mathrm{iono\ twice}}|_{\mathrm{tot}} = 243.2\ \mu$V/m. **14.41.** $|\underline{V}_{\mathrm{oc}}| = 1.9$ mV. **14.44.** $|\underline{V}_{\mathrm{oc}}| = 1.9$ mV. **14.46.** (a) $|(\underline{V}_{\mathrm{oc}})_{\mathrm{direct}}| = 0$ and $|(\underline{V}_{\mathrm{oc}})_{\mathrm{reflected}}| = 9.35\ \mu$V; (b) $|(\underline{V}_{\mathrm{oc}})_{\mathrm{direct}}| = 33.5\ \mu$V and $|(\underline{V}_{\mathrm{oc}})_{\mathrm{reflected}}| = 23.7\ \mu$V. **14.52.** $r_{\max} = 1.191$ km. **14.57.** $f(\theta) = \sin\theta\sin[(\pi/2)\cos\theta]$. **14.60.** $f(\theta) = \{\cos[(\pi/2)\cos\theta]\cos(\pi\sin\theta\cos\phi)\}/\sin\theta$ for $0 \leq \theta \leq 90^\circ$, and $f(\theta) = 0$ for $90^\circ < \theta \leq 180^\circ$. **14.63.** $f(\theta) = \{\cos^3[(\pi/2)\cos\theta]\}/\sin\theta$. **14.66.** (a) $\theta_{\max} = 90^\circ \to \alpha = 0$ (broadside N-element array), and (b) $\theta_{\max} = 0 \to \alpha = -\beta d$ (endfire radiation in $+z$ direction) and $\theta_{\max} = 180^\circ \to \alpha = \beta d$ (endfire radiation in $-z$ direction).

Bibliography

Notaroš, B. M., *Electromagnetic Theory I*, University of Massachusetts, Dartmouth, MA, 2002.

Notaroš, B. M., *Electromagnetic Theory II*, University of Massachusetts, Dartmouth, MA, 2003.

Notaroš, B. M., V. V. Petrović, M. M. Ilić, and A. R. Djordjević, B. M. Kolundžija, and M. B. Dragović, *Collection of Examination Questions and Problems in Electromagnetics*, Department of Electrical Engineering, University of Belgrade, Belgrade, 1998 (in Serbian).

Djordjević, A. R., G. N. Božilović, and B. M. Notaroš, *Collection of Examination Problems in Fundamentals of Electrical Engineering with Solutions, Part I*, Department of Electrical Engineering, University of Belgrade, Belgrade, 1997 (in Serbian).

Djordjević, A. R., G. N. Božilović, and B. M. Notaroš, *Collection of Examination Problems in Fundamentals of Electrical Engineering with Solutions, Part II*, Department of Electrical Engineering, University of Belgrade, Belgrade, 1997 (in Serbian).

Popović, B. D., *Electromagnetics*, Gradjevinska Knjiga, Belgrade, 1986, 2nd edition (in Serbian).

Surutka, J., *Electromagnetics*, Gradjevinska Knjiga, Belgrade, 1978, 5th edition (in Serbian).

Djordjević, A. R., *Electromagnetics for Computer Engineering*, University of Belgrade, Belgrade, 1996 (in Serbian).

Popović, Z., and B. D. Popović, *Introductory Electromagnetics*, Prentice Hall, Upper Saddle River, NJ, 2000.

Inan U. S., and A. S. Inan, *Engineering Electromagnetics*, Addison Wesley Longman, Menlo Park, CA, 1999.

Inan U. S., and A. S. Inan, *Electromagnetic Waves*, Prentice Hall, Upper Saddle River, NJ, 2000.

Sadiku, M. N. O., *Elements of Electromagnetics*, Oxford University Press, New York, 2001, 3rd edition.

Hayt, W. H., Jr. and J. A. Buck, *Engineering Electromagnetics*, McGraw-Hill, New York, 2001, 6th edition.

Ulaby, F. T., *Fundamentals of Applied Electromagnetics*, Prentice Hall, Upper Saddle River, NJ, 1999 edition.

Asimov, I., *Asimov's Biographical Encyclopedia of Science and Technology*, Doubleday, New York, 1982, 2nd revised edition.

Popović, B. D., *Collection of Problems in Electromagnetics*, Gradjevinska Knjiga, Belgrade, 1985, 6th edition (in Serbian).

Dragović, M. B., *Antennas and Propagation of Radio Waves*, Kontekst, Belgrade, 1994 (in Serbian).

Kraus, J. D., and D. A. Fleisch, *Electromagnetics with Applications*, McGraw-Hill, New York, 1999, 5th edition.

Rao, N. N., *Elements of Engineering Electromagnetics*, Pearson Prentice Hall, Upper Saddle River, NJ, 2004, 6th edition.

Cheng, D. K., *Field and Wave Electromagnetics*, Addison-Wesley, Reading, MA, 1989, 2nd edition.

Božilović, H. A., Ž. A. Spasojević, and G. N. Božilović, *Collection of Problems in Fundamentals of Electrical Engineering – Electrostatics, Steady Currents*, Naučna Knjiga, Belgrade, 1983 (in Serbian).

Božilović, H., Ž. Spasojević, and G. Božilović, *Collection of Problems in Fundamentals of Electrical Engineering – Electromagnetism, ac Currents*, Department of Electrical Engineering, University of Belgrade, Belgrade, 1998, 7th edition (in Serbian).

Iskander, M. F., *Electromagnetic Fields and Waves*, Waveland Press, Prospect Hills, IL, 2000.

Haus, H. A., and J. R. Melcher, *Electromagnetic Fields and Energy*, Prentice Hall, Upper Saddle River, NJ, 1989.

Stutzman, W. L., and G. A. Thiele, *Antenna Theory and Design*, John Wiley & Sons, New York, 1998, 2nd edition.

Balanis, C. A., *Antenna Theory: Analysis and Design*, John Wiley & Sons, New York, 1997, 2nd edition.

Kraus, J. D., and R. J. Marhefka, *Antennas for All Applications*, McGraw-Hill, New York, 2002, 3rd edition.

Pozar, D. M., *Microwave Engineering*, John Wiley & Sons, New York, 2005, 3rd edition.

Demarest, K. R., *Engineering Electromagnetics*, Prentice Hall, Upper Saddle River, NJ, 1998.

Wentworth, S. M., *Fundamentals of Electromagnetics with Engineering Applications*, John Wiley & Sons, New York, 2005.

Paul, C. R., *Electromagnetics for Engineers with Applications*, John Wiley & Sons, New York, 2004.

Lonngren, K. E., S. V. Savov, and R. J. Jost, *Fundamentals of Electromagnetics with MATLAB*, SciTech Publishing, Raleigh, NC, 2007, 2nd edition.

Johnk, C. T. A., *Engineering Electromagnetic Fields and Waves*, John Wiley & Sons, New York, 1988, 2nd edition.

Ida, N., *Engineering Electromagnetics*, Springer, New York, 2004, 2nd edition.

Ramo, S., J. R. Whinnery, and T. Van Duzer, *Fields and Waves in Communication Electronics*, John Wiley & Sons, New York, 1994, 3rd edition.

Griffiths, D. J., *Introduction to Electrodynamics*, Pearson Addison Wesley, Upper Saddle River, NJ, 1999, 3rd edition.

Index

D

E

U

V

W

Thales of Miletus
(624 B.C.–546 B.C.)
Page 20

Benjamin Franklin[1]
(1706–1790)
Page 129

Charles Augustin de Coulomb
(1736–1806)
Page 20

Alessandro Volta[2]
(1745–1827)
Page 40

André-Marie Ampère[2]
(1775–1836)
Page 204

Johann Karl Friedrich Gauss
(1777–1855)
Page 47

Hans Christian Oersted[3]
(1777–1851)
Page 193

Georg Simon Ohm[2]
(1789–1854)
Page 159

Michael Faraday[2]
(1791–1867)
Page 290

Joseph Henry[4]
(1797–1878)
Page 331

Heinrich Friedrich Emil Lenz
(1804–1865)
Page 293

[1] *Library of Congress*
[2] *Edgar Fahs Smith Collection, University of Pennsylvania Libraries*
[3] *AIP Emilio Segrè Visual Archives*
[4] *Library of Congress, Brady-Handy Photograph Collection*

Wilhelm Eduard Weber[3]
(1804–1891)
Page 217

James Prescott Joule[5]
(1818–1889)
Page 162

Gustav Robert Kirchhoff[2]
(1824–1887)
Page 40

James Clerk Maxwell[3]
(1831–1879)
Page 376

Oliver Heaviside[3]
(1850–1925)
Page 554

John Henry Poynting[1]
(1852–1914)
Page 409

Hendrik Antoon Lorentz[3]
(1853–1928)
Page 228

Edwin Herbert Hall[6]
(1855–1938)
Page 228

Nikola Tesla[7]
(1856–1943)
Page 194

Heinrich Rudolf Hertz[8]
(1857–1894)
Page 735

[5] *National Bureau of Standards Archives, courtesy AIP Emilió Segrè Visual Archives, E. Scott Barr Collection*
[6] *"Voltiana," Como, Italy - Sept. 10, 1927 issue, courtesy AIP Emilio Segrè Visual Archives*